JANEWAY's
IMMUNO BIOLOGY

9TH EDITION

JANEWAY'S

IMMUNO BIOLOGY

9TH EDITION

Kenneth Murphy
Washington University School of Medicine, St. Louis

Casey Weaver
University of Alabama at Birmingham, School of Medicine

With contributions by:
Allan Mowat
University of Glasgow

Leslie Berg
University of Massachusetts Medical School

David Chaplin
University of Alabama at Birmingham, School of Medicine

With acknowledgment to:
Charles A. Janeway Jr.

Paul Travers
MRC Centre for Regenerative Medicine, Edinburgh

Mark Walport

 Garland Science
Taylor & Francis Group
NEW YORK AND LONDON

Vice President: Denise Schanck
Development Editor: Monica Toledo
Associate Editor: Allie Bochicchio
Assistant Editor: Claudia Acevedo-Quiñones
Text Editor: Elizabeth Zayetz
Production Editor: Deepa Divakaran
Typesetter: Deepa Divakaran and EJ Publishing Services
Illustrator and Design: Matthew McClements, Blink Studio, Ltd.
Copyeditor: Richard K. Mickey
Proofreader: Sally Livitt
Permission Coordinator: Sheri Gilbert
Indexer: Medical Indexing Ltd.

ISBN 978-0-8153-4505-3   978-0-8153-4551-0 (International Paperback)
     978-0-8153-4445-2 (Hardback)   978-0-8153-4550-3 (Looseleaf)

Library of Congress Cataloging-in-Publication Data
Names: Murphy, Kenneth (Kenneth M.), author. | Weaver, Casey, author.
Title: Janeway's immunobiology / Kenneth Murphy, Casey Weaver ; with contributions by Allan Mowat, Leslie Berg, David Chaplin ; with acknowledgment to Charles A. Janeway Jr., Paul Travers, Mark Walport.
Other titles: Immunobiology
Description: 9th edition. | New York, NY : Garland Science/Taylor & Francis Group, LLC, [2016] | Includes bibliographical references and index.
Identifiers: LCCN 2015050960| ISBN 9780815345053 (pbk.) | ISBN 9780815345510 (pbk.-ROW) | ISBN 9780815345503 (looseleaf)
Subjects: | MESH: Immune System--physiology | Immune System--physiopathology | Immunity | Immunotherapy
Classification: LCC QR181 | NLM QW 504 | DDC 616.07/9--dc23
LC record available at http://lccn.loc.gov/2015050960

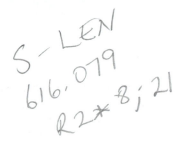

Published by Garland Science, Taylor & Francis Group, LLC, an informa business, 711 Third Avenue, New York, NY, 10017, USA, and 3 Park Square, Milton Park, Abingdon, OX14 4RN, UK.

Printed in the United States of America

15 14 13 12 11 10 9 8 7 6 5 4 3 2

SUSTAINABLE FORESTRY INITIATIVE

Certified Chain of Custody
Promoting Sustainable Forestry

www.sfiprogram.org
SFI-01681

This SFI Logo applies to the text stock only.

GS Garland Science
Taylor & Francis Group
NEW YORK AND LONDON

Visit our web site at http://www.garlandscience.com

Preface

Janeway's Immunobiology is intended for undergraduate and graduate courses and for medical students, but its depth and scope also make it a useful resource for trainees and practicing immunologists. Its narrative takes the host's perspective in the struggle with the microbial world—a viewpoint distinguishing 'immunology' from 'microbiology'. Other facets of immunology, such as autoimmunity, immunodeficiencies, allergy, transplant rejection, and new aspects of cancer immunotherapy are also covered in depth, and a companion book, *Case Studies in Immunology*, provides clinical examples of immune-related disease. In *Immunobiology*, symbols in the margin indicate where the basic immunological concepts related to *Case Studies* are discussed.

The ninth edition retains the previous organization of five major sections and sixteen chapters, but reorganizes content to clarify presentation and eliminate redundancies, updating each chapter and adding over 100 new figures. The first section (Chapters 1–3) includes the latest developments in innate sensing mechanisms and covers new findings in innate lymphoid cells and the concept of 'immune effector modules' that is used throughout the rest of the book. Coverage of chemokine networks has been updated throughout (Chapters 3 and 11). The second section (Chapters 4–6) adds new findings for $\gamma{:}\delta$ T cell recognition and for the targeting of activation-induced cytidine deaminase (AID) class switch recombination. The third section (Chapters 7 and 8) is extensively updated and covers new material on integrin activation, cytoskeletal reorganization, and Akt and mTOR signaling. The fourth section enhances coverage of CD4 T cell subsets (Chapter 9), including follicular helper T cells that regulate switching and affinity maturation (Chapter 10). Chapter 11 now organizes innate and adaptive responses to pathogens around the effector module concept, and features new findings for tissue-resident memory T cells. Chapter 12 has been thoroughly updated to keep pace with the quickly advancing field of mucosal immunity. In the last section, coverage of primary and secondary immunodeficiencies has been reorganized and updated with an expanded treatment of immune evasion by pathogens and HIV/AIDS (Chapter 13). Updated and more detailed consideration of allergy and allergic diseases are presented in Chapter 14, and for autoimmunity and transplantation in Chapter 15. Finally, Chapter 16 has expanded coverage of new breakthroughs in cancer immunotherapy, including 'checkpoint blockade' and chimeric antigen receptor (CAR) T-cell therapies.

End-of-chapter review questions have been completely updated in the ninth edition, posed in a variety of formats, with answers available online. Appendix I: The Immunologist's Toolbox has undergone a comprehensive revitalization with the addition of many new techniques, including the CRISPR/Cas9 system and mass spectrometry/proteomics. Finally, a new Question Bank has been created to aid instructors in the development of exams that require the student to reflect upon and synthesize concepts in each chapter.

Once again, we benefited from the expert revision of Chapter 12 by Allan Mowat, and from contributions of two new contributors, David Chaplin and Leslie Berg. David's combined clinical and basic immunologic strengths greatly improved Chapter 14, and Leslie applied her signaling expertise to Chapters 7 and 8, and Appendix I, and her strength as an educator in creating the new Question Bank for instructors. Many people deserve special thanks. Gary Grajales wrote all end-of-chapter questions. New for this edition, we enlisted input from our most important audience and perhaps best critics—students of immunology-in-training who provided feedback on drafts of individual chapters, and Appendices II–IV. We benefitted from our thoughtful colleagues who reviewed the eighth edition. They are credited in the Acknowledgments section; we are indebted to them all.

We have the good fortune to work with an outstanding group at Garland Science. We thank Monica Toledo, our development editor, who coordinated the entire project, guiding us gently but firmly back on track throughout the process, with efficient assistance from Allie Bochicchio and Claudia Acevedo-Quiñones. We thank Denise Schanck, our publisher, who, as always, contributed her guidance, support, and wisdom. We thank Adam Sendroff, who is instrumental in relaying information about the book to immunologists around the world. As in all previous editions, Matt McClements has contributed his genius—and patience—re-interpreting authors' sketches into elegant illustrations. We warmly welcome our new text editor Elizabeth Zayetz, who stepped in for Eleanor Lawrence, our previous editor, and guiding light. The authors wish to thank their most important partners—Theresa and Cindy Lou—colleagues in life who have supported this effort with their generosity of time, their own editorial insights, and their infinite patience.

As temporary stewards of Charlie's legacy, *Janeway's Immunobiology*, we hope this ninth edition will continue to inspire—as he did—students to appreciate immunology's beautiful subtlety. We encourage all readers to share with us their views on where we have come up short, so the next edition will further approach the asymptote. Happy reading!

Kenneth Murphy
Casey Weaver

Resources for Instructors and Students

The teaching and learning resources for instructors and students are available online. The homework platform is available to interested instructors and their students. Instructors will need to set up student access in order to use the dashboard to track student progress on assignments. The instructor's resources on the Garland Science website are password-protected and available only to adopting instructors. The student resources on the Garland Science website are available to everyone. We hope these resources will enhance student learning and make it easier for instructors to prepare dynamic lectures and activities for the classroom.

Online Homework Platform with Instructor Dashboard

Instructors can obtain access to the online homework platform from their sales representative or by emailing science@garland.com. Students who wish to use the platform must purchase access and, if required for class, obtain a course link from their instructor.

The online homework platform is designed to improve and track student performance. It allows instructors to select homework assignments on specific topics and review the performance of the entire class, as well as individual students, via the instructor dashboard. The user-friendly system provides a convenient way to gauge student progress, and tailor classroom discussion, activities, and lectures to areas that require specific remediation. The features and assignments include:

- *Instructor Dashboard* displays data on student performance: such as responses to individual questions and length of time spent to complete assignments.
- *Tutorials* explain essential or difficult concepts and are integrated with a variety of questions that assess student engagement and mastery of the material.

The tutorials were created by Stacey A. Gorski, University of the Sciences in Philadelphia.

Instructor Resources

Instructor Resources are available on the Garland Science Instructor's Resource Site, located at www.garlandscience.com/instructors. The website provides access not only to the teaching resources for this book but also to all other Garland Science textbooks. Adopting instructors can obtain access to the site from their sales representative or by emailing science@garland.com.

Art of Janeway's Immunobiology, Ninth Edition

The images from the book are available in two convenient formats: PowerPoint® and JPEG. They have been optimized for display on a computer. Figures are searchable by figure number, by figure name, or by keywords used in the figure legend from the book.

Figure-Integrated Lecture Outlines

The section headings, concept headings, and figures from the text have been integrated into PowerPoint® presentations. These will be useful for instructors who would like a head start creating lectures for their course. Like all of our PowerPoint® presentations, the lecture outlines can be customized. For example, the content of these presentations can be combined with videos and questions from the book or Question Bank, in order to create unique lectures that facilitate interactive learning.

Animations and Videos

The animations and videos that are available to students are also available on the Instructor's Website in two formats. The WMV-formatted movies are created for instructors who wish to use the movies in PowerPoint® presentations on Windows® computers; the QuickTime-formatted movies are for use in PowerPoint® for Apple computers or Keynote® presentations. The movies can easily be downloaded using the 'download' button on the movie preview page. The movies are related to specific chapters and callouts to the movies are highlighted in color throughout the textbook.

Question Bank

Written by Leslie Berg, University of Massachusetts Medical School, the Question Bank includes a variety of question formats: multiple choice, fill-in-the-blank, true-false, matching, essay, and challenging synthesis questions. There are approximately 30–40 questions per chapter, and a large number of the multiple-choice questions will be suitable for use with personal response systems (that is, clickers). The Question Bank provides a comprehensive sampling of questions that require the student to reflect upon and integrate information, and can be used either directly or as inspiration for instructors to write their own test questions.

Student Resources

The resources for students are available on the *Janeway's Immunobiology* Student Website, located at students.garlandscience.com.

Answers to End-of-Chapter Questions

Answers to the end-of-chapter questions are available to students for self-testing.

Animations and Videos

There are over 40 narrated movies, covering a range of immunology topics, which review key concepts and illuminate the experimental process.

Flashcards

Each chapter contains flashcards, built into the student website, that allow students to review key terms from the text.

Glossary

The comprehensive glossary of key terms from the book is online and can be searched or browsed.

Acknowledgments

We would like to thank the following experts who read parts or the whole of the eighth edition chapters and provided us with invaluable advice in developing this new edition.

Chapter 2: Teizo Fujita, Fukushima Prefectural General Hygiene Institute; Thad Stappenbeck, Washington University; Andrea J. Tenner, University of California, Irvine.

Chapter 3: Shizuo Akira, Osaka University; Mary Dinauer, Washington University in St. Louis; Lewis Lanier, University of California, San Francisco; Gabriel Nuñez, University of Michigan Medical School; David Raulet, University of California, Berkeley; Caetano Reis e Sousa, Cancer Research UK; Tadatsugu Taniguchi, University of Tokyo; Eric Vivier, Université de la Méditerranée Campus de Luminy; Wayne Yokoyama, Washington University.

Chapter 4: Chris Garcia, Stanford University; Ellis Reinherz, Harvard Medical School; Robyn Stanfield, The Scripps Research Institute; Ian Wilson, The Scripps Research Institute.

Chapter 5: Michael Lieber, University of Southern California Norris Cancer Center; Michel Neuberger, University of Cambridge; David Schatz, Yale University School of Medicine; Barry Sleckman, Washington University School of Medicine, St. Louis; Philip Tucker, University of Texas, Austin.

Chapter 6: Sebastian Amigorena, Institut Curie; Siamak Bahram, Centre de Recherche d'Immunologie et d'Hematologie; Peter Cresswell, Yale University School of Medicine; Mitchell Kronenberg, La Jolla Institute for Allergy & Immunology; Philippa Marrack, National Jewish Health; Hans-Georg Rammensee, University of Tuebingen, Germany; Jose Villadangos, University of Melbourne; Ian Wilson, The Scripps Research Institute.

Chapter 7: Oreste Acuto, University of Oxford; Francis Chan, University of Massachusetts Medical School; Vigo Heissmeyer, Helmholtz Center Munich; Steve Jameson, University of Minnesota; Pamela L. Schwartzberg, NIH; Art Weiss, University of California, San Francisco.

Chapter 8: Michael Cancro, University of Pennsylvania School of Medicine; Robert Carter, University of Alabama; Ian Crispe, University of Washington; Kris Hogquist, University of Minnesota; Eric Huseby, University of Massachusetts Medical School; Joonsoo Kang, University of Massachusetts Medical School; Ellen Robey, University of California, Berkeley; Nancy Ruddle, Yale University School of Medicine; Juan Carlos Zúñiga-Pflücker, University of Toronto.

Chapter 9: Francis Carbone, University of Melbourne; Shane Crotty, La Jolla Institute of Allergy and Immunology; Bill Heath, University of Melbourne, Victoria; Marc Jenkins, University of Minnesota; Alexander Rudensky, Memorial Sloan Kettering Cancer Center; Shimon Sakaguchi, Osaka University.

Chapter 10: Michael Cancro, University of Pennsylvania School of Medicine; Ann Haberman, Yale University School of Medicine; John Kearney, University of Alabama at Birmingham; Troy Randall, University of Alabama at Birmingham; Jeffrey Ravetch, Rockefeller University; Haley Tucker, University of Texas at Austin.

Chapter 11: Susan Kaech, Yale University School of Medicine; Stephen McSorley, University of California, Davis.

Chapter 12: Nadine Cerf-Bensussan, Université Paris Descartes-Sorbonne, Paris; Thomas MacDonald, Barts and London School of Medicine and Dentistry; Maria Rescigno, European Institute of Oncology; Michael Russell, University at Buffalo; Thad Stappenbeck, Washington University.

Chapter 13: Mary Collins, University College London; Paul Goepfert, University of Alabama at Birmingham; Paul Klenerman, University of Oxford; Warren Leonard, National Heart, Lung, and Blood Institute, NIH; Luigi Notarangelo, Boston Children's Hospital; Sarah Rowland-Jones, Oxford University; Harry Schroeder, University of Alabama at Birmingham.

Chapter 14: Cezmi A. Akdis, Swiss Institute of Allergy and Asthma Research; Larry Borish, University of Virginia Health System; Barry Kay, National Heart and Lung Institute; Harald Renz, Philipps University Marburg; Robert Schleimer, Northwestern University; Dale Umetsu, Genentech.

Chapter 15: Anne Davidson, The Feinstein Institute for Medical Research; Robert Fairchild, Cleveland Clinic; Rikard Holmdahl, Karolinska Institute; Fadi Lakkis, University of Pittsburgh; Ann Marshak-Rothstein, University of Massachusetts Medical School; Carson Moseley, University of Alabama at Birmingham; Luigi Notarangelo, Boston Children's Hospital; Noel Rose, Johns Hopkins Bloomberg School of Public Health; Warren Shlomchik, University of Pittsburgh School of Medicine; Laurence Turka, Harvard Medical School.

Chapter 16: James Crowe, Vanderbilt University; Glenn Dranoff, Dana–Farber Cancer Institute; Thomas Gajewski, University of Chicago; Carson Moseley, University of Alabama at Birmingham; Caetano Reis e Sousa, Cancer Research UK.

Appendix I: Lawrence Stern, University of Massachusetts Medical School.

We would also like to specially acknowledge and thank the students: Alina Petris, University of Manchester; Carlos Briseno, Washington University in St. Louis; Daniel DiToro, University of Alabama at Birmingham; Vivek Durai, Washington University in St. Louis; Wilfredo Garcia, Harvard University; Nichole Escalante, University of Toronto; Kate Jackson, University of Manchester; Isil Mirzanli, University of Manchester; Carson Moseley, University of Alabama at Birmingham; Daniel Silberger, University of Alabama at Birmingham; Jeffrey Singer, University of Alabama at Birmingham; Deepica Stephen, University of Manchester; Mayra Cruz Tleugabulova, University of Toronto.

Contents

Detailed Contents

PART III   THE DEVELOPMENT OF MATURE LYMPHOCYTE RECEPTOR REPERTOIRES

APPENDICES

Basic Concepts in Immunology

1

IN THIS CHAPTER

The origins of vertebrate immune cells.

Principles of innate immunity.

Principles of adaptive immunity.

The effector mechanisms of immunity.

Immunology is the study of the body's defense against infection. We are continually exposed to microorganisms, many of which cause disease, and yet become ill only rarely. How does the body defend itself? When infection does occur, how does the body eliminate the invader and cure itself? And why do we develop long-lasting immunity to many infectious diseases encountered once and overcome? These are the questions addressed by immunology, which we study to understand our body's defenses against infection at the cellular and molecular levels.

The beginning of immunology as a science is usually attributed to **Edward Jenner** for his work in the late 18th century (**Fig. 1.1**). The notion of immunity—that surviving a disease confers greater protection against it later—was known since ancient Greece. **Variolation**—the inhalation or transfer into superficial skin wounds of material from smallpox pustules—had been practiced since at least the 1400s in the Middle East and China as a form of protection against that disease and was known to Jenner. Jenner had observed that the relatively mild disease of cowpox, or vaccinia, seemed to confer protection against the often fatal disease of smallpox, and in 1796, he demonstrated that inoculation with cowpox protected the recipient against smallpox. His scientific proof relied on the deliberate exposure of the inoculated individual to infectious smallpox material two months after inoculation. This scientific test was his original contribution.

Jenner called the procedure **vaccination**. This term is still used to describe the inoculation of healthy individuals with weakened or attenuated strains of disease-causing agents in order to provide protection from disease. Although Jenner's bold experiment was successful, it took almost two centuries for smallpox vaccination to become universal. This advance enabled the World Health Organization to announce in 1979 that smallpox had been eradicated (**Fig. 1.2**), arguably the greatest triumph of modern medicine.

Jenner's strategy of vaccination was extended in the late 19th century by the discoveries of many great microbiologists. **Robert Koch** proved that infectious diseases are caused by specific microorganisms. In the 1880s, **Louis Pasteur**

Fig. 1.1 Edward Jenner. Portrait by John Raphael Smith. Reproduced courtesy of Yale University, Harvey Cushing/John Hay Whitney Medical Library.

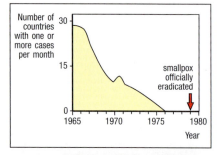

Fig. 1.2 The eradication of smallpox by vaccination. After a period of 3 years in which no cases of smallpox were recorded, the World Health Organization was able to announce in 1979 that smallpox had been eradicated, and vaccination stopped (upper panel). A few laboratory stocks have been retained, however, and some fear that these are a source from which the virus might reemerge. Ali Maow Maalin (lower panel) contracted and survived the last case of smallpox in Somalia in 1977. Photograph courtesy of Dr. Jason Weisfeld.

devised a vaccine against cholera in chickens, and developed a rabies vaccine that proved to be a spectacular success upon its first trial in a boy bitten by a rabid dog.

These practical triumphs led to a search for vaccination's mechanism of protection and to the development of the science of immunology. In the early 1890s, **Emil von Behring** and **Shibasaburo Kitasato** discovered that the serum of animals immune to diphtheria or tetanus contained a specific 'antitoxic activity' that could confer short-lived protection against the effects of diphtheria or tetanus toxins in people. This activity was later determined to be due to the proteins we now call **antibodies**, which bind specifically to the toxins and neutralize their activity. That these antibodies might have a crucial role in immunity was reinforced by **Jules Bordet's** discovery in 1899 of **complement**, a component of serum that acts in conjunction with antibodies to destroy pathogenic bacteria.

A specific response against infection by potential pathogens, such as the production of antibodies against a particular pathogen, is known as **adaptive immunity**, because it develops during the lifetime of an individual as an adaptation to infection with that pathogen. Adaptive immunity is distinguished from **innate immunity**, which was already known at the time von Behring was developing serum therapy for diphtheria chiefly through the work of the great Russian immunologist **Elie Metchnikoff,** who discovered that many microorganisms could be engulfed and digested by phagocytic cells, which thus provide defenses against infection that are nonspecific. Whereas these cells—which Metchnikoff called 'macrophages'—are always present and ready to act, adaptive immunity requires time to develop but is highly specific.

It was soon clear that specific antibodies could be induced against a vast range of substances, called **antigens** because they could stimulate *anti*body *gen*eration. **Paul Ehrlich** advanced the development of an **antiserum** as a treatment for diphtheria and developed methods to standardize therapeutic serums. Today the term antigen refers to any substance recognized by the adaptive immune system. Typically antigens are common proteins, glycoproteins, and polysaccharides of pathogens, but they can include a much wider range of chemical structures, for example, metals such as nickel, drugs such as penicillin, and organic chemicals such as the urushiol (a mix of pentadecylcatechols) in the leaves of poison ivy. Metchnikoff and Ehrlich shared the 1908 Nobel Prize for their respective work on immunity.

This chapter introduces the principles of innate and adaptive immunity, the cells of the immune system, the tissues in which they develop, and the tissues through which they circulate. We then outline the specialized functions of the different types of cells by which they eliminate infection.

The origins of vertebrate immune cells.

The body is protected from infectious agents, their toxins, and the damage they cause by a variety of effector cells and molecules that together make up the **immune system**. Both innate and adaptive immune responses depend upon the activities of white blood cells or **leukocytes**. Most cells of the immune system arise from the **bone marrow**, where many of them develop and mature. But some, particularly certain tissue-resident macrophage populations (for example, the microglia of the central nervous system), originate from the yolk sac or fetal liver during embryonic development. They seed tissues before birth and are maintained throughout life as independent, self-renewing populations. Once mature, immune cells reside within peripheral tissues, circulate in the bloodstream, or circulate in a specialized system of vessels called the **lymphatic system**. The lymphatic system drains extracellular fluid and immune cells from tissues and transports them as **lymph** that is eventually emptied back into the blood system.

All the cellular elements of blood, including the red blood cells that transport oxygen, the platelets that trigger blood clotting in damaged tissues, and the white blood cells of the immune system, ultimately derive from the **hematopoietic stem cells (HSCs)** of the bone marrow. Because these can give rise to all the different types of blood cells, they are often known as **pluripotent** hematopoietic stem cells. The hematopoietic stem cells give rise to cells of more limited developmental potential, which are the immediate progenitors of red blood cells, platelets, and the two main categories of white blood cells, the **lymphoid** and **myeloid** lineages. The different types of blood cells and their lineage relationships are summarized in **Fig. 1.3**.

Principles of innate immunity.

In this part of the chapter we will outline the principles of innate immunity and describe the molecules and cells that provide continuous defense against invasion by pathogens. Although a subset of white blood cells, known as **lymphocytes**, possess the most powerful ability to recognize and target pathogenic microorganisms, they need the participation of the innate immune system to initiate and mount their offensive. Indeed, the adaptive immune response and innate immunity use many of the same destructive mechanisms to eliminate invading microorganisms.

1-1   Commensal organisms cause little host damage while pathogens damage host tissues by a variety of mechanisms.

We recognize four broad categories of disease-causing microorganisms, or **pathogens**: **viruses**, **bacteria** and archaea, **fungi**, and the unicellular and multicellular eukaryotic organisms collectively termed **parasites** (**Fig. 1.4**). These microorganisms vary tremendously in size and in how they damage host tissues. The smallest are viruses, which range from five to a few hundred nanometers in size and are obligate intracellular pathogens. Viruses can directly kills cells by inducing lysis during their replication. Somewhat larger are intracellular bacteria and mycobacteria. These can kill cells directly or damage cells by producing toxins. Many single-celled intracellular parasites, such as members of the *Plasmodium* genus that cause malaria, also directly kill infected cells. Pathogenic bacteria and fungi growing in extracellular spaces can induce shock and sepsis by releasing toxins into the blood or tissues. The largest pathogens—parasitic worms, or helminths—are too large to infect host cells but can injure tissues by forming cysts that induce damaging cellular responses in the tissues into which the worms migrate.

Not all microbes are pathogens. Many tissues, especially the skin, oral mucosa, conjunctiva, and gastrointestinal tract, are constantly colonized by microbial communities—called the **microbiome**—that consist of archaea, bacteria, and fungi but cause no damage to the host. These are also called **commensal microorganisms**, since they can have a symbiotic relationship with the host. Indeed, some commensal organisms perform important functions, as in the case of the bacteria that aid in cellulose digestion in the stomachs of ruminants. The difference between commensal organisms and pathogens lies in whether they induce damage. Even enormous numbers of microbes in the intestinal microbiome normally cause no damage and are confined within the intestinal lumen by a protective layer of mucus, whereas pathogenic bacteria can penetrate this barrier, injure intestinal epithelial cells, and spread into the underlying tissues.

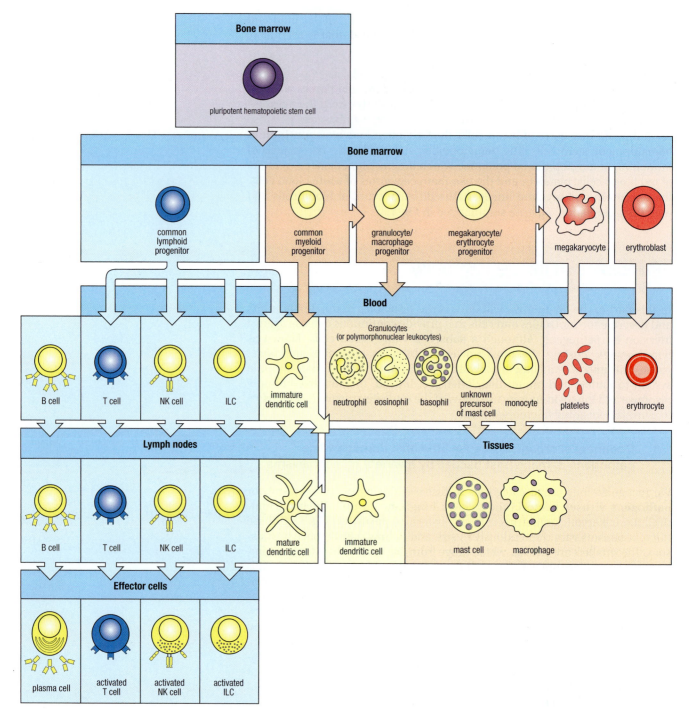

Fig. 1.3 All the cellular elements of the blood, including the cells of the immune system, arise from pluripotent hematopoietic stem cells in the bone marrow. These pluripotent cells divide to produce two types of stem cells. A common lymphoid progenitor gives rise to the lymphoid lineage (blue background) of white blood cells or leukocytes—the innate lymphoid cells (ILCs) and natural killer (NK) cells and the T and B lymphocytes. A common myeloid progenitor gives rise to the myeloid lineage (pink and yellow backgrounds), which comprises the rest of the leukocytes, the erythrocytes (red blood cells), and the megakaryocytes that produce platelets important in blood clotting. T and B lymphocytes are distinguished from the other leukocytes by having antigen receptors and from each other by their sites of differentiation—the thymus and bone marrow, respectively. After encounter with antigen, B cells differentiate into antibody-secreting plasma cells, while

T cells differentiate into effector T cells with a variety of functions. Unlike T and B cells, ILCs and NK cells lack antigen specificity. The remaining leukocytes are the monocytes, the dendritic cells, and the neutrophils, eosinophils, and basophils. The last three of these circulate in the blood and are termed granulocytes, because of the cytoplasmic granules whose staining gives these cells a distinctive appearance in blood smears, or polymorphonuclear leukocytes, because of their irregularly shaped nuclei. Immature dendritic cells (yellow background) are phagocytic cells that enter the tissues; they mature after they have encountered a potential pathogen. The majority of dendritic cells are derived from the common myeloid progenitor cells, but some may also arise from the common lymphoid progenitor. Monocytes enter tissues, where they differentiate into phagocytic macrophages or dendritic cells. Mast cells also enter tissues and complete their maturation there.

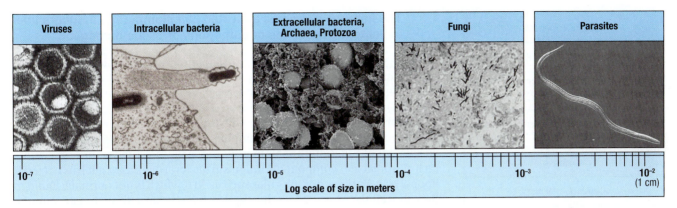

| Viruses | Intracellular bacteria | Extracellular bacteria, Archaea, Protozoa | Fungi | Parasites |

| 10^{-7} | 10^{-6} | 10^{-5} | 10^{-4} | 10^{-3} | 10^{-2} (1 cm) |

Log scale of size in meters

Fig. 1.4 Pathogens vary greatly in size and lifestyle.
Intracellular pathogens include viruses, such as herpes simplex (first panel), and various bacteria, such as *Listeria monocytogenes* (second panel). Many bacteria, such as *Staphylococcus aureus* (third panel), or fungi, such as *Aspergillus fumigates* (fourth panel), can grow in the extracellular spaces and directly invade through tissues, as do some archaea and protozoa (third panel). Many parasites, such as the nematode *Strongyloides stercoralis* (fifth panel), are large multicellular organisms that can move throughout the body in a complex life cycle. Second panel courtesy of Dan Portnoy. Fifth panel courtesy of James Lok.

1-2 Anatomic and chemical barriers are the first defense against pathogens.

The host can adopt three strategies to deal with the threat posed by microbes: **avoidance**, **resistance**, and **tolerance**. Avoidance mechanisms prevent exposure to microbes, and include both anatomic barriers and behavior modifications. If an infection is established, resistance is aimed at reducing or eliminating pathogens. To defend against the great variety of microbes, the immune system has numerous molecular and cellular functions, collectively called mediators, or **effector mechanisms**, suited to resist different categories of pathogens. Their description is a major aspect of this book. Finally, tolerance involves responses that enhance a tissue's capacity to resist damage induced by microbes. This meaning of the term 'tolerance' has been used extensively in the context of disease susceptibility in plants rather than animal immunity. For example, increasing growth by activating dormant meristems, the undifferentiated cells that generate new parts of the plant, is a common tolerance mechanism in response to damage. This should be distinguished from the term **immunological tolerance**, which refers to mechanisms that prevent an immune response from being mounted against the host's own tissues.

Anatomic and chemical barriers are the initial defenses against infection (**Fig. 1.5**). The skin and mucosal surfaces represent a kind of avoidance strategy that prevents exposure of internal tissues to microbes. At most anatomic barriers, additional resistance mechanisms further strengthen host defenses. For example, mucosal surfaces produce a variety of **antimicrobial proteins** that act as natural antibiotics to prevent microbes from entering the body.

If these barriers are breached or evaded, other components of the innate immune system can immediately come into play. We mentioned earlier the discovery by Jules Bordet of **complement**, which acts with antibodies to lyse bacteria. Complement is a group of around 30 different plasma proteins that act together and are one of the most important effector mechanisms in serum and interstitial tissues. Complement not only acts in conjunction with antibodies, but can also target foreign organisms in the absence of a specific antibody; thus it contributes to both innate and adaptive responses. We will examine anatomic barriers, the antimicrobial proteins, and complement in greater detail in Chapter 2.

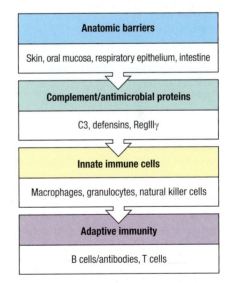

| **Anatomic barriers** |
| Skin, oral mucosa, respiratory epithelium, intestine |

| **Complement/antimicrobial proteins** |
| C3, defensins, RegIIIγ |

| **Innate immune cells** |
| Macrophages, granulocytes, natural killer cells |

| **Adaptive immunity** |
| B cells/antibodies, T cells |

Fig. 1.5 Protection against pathogens relies on several levels of defense.
The first is the anatomic barrier provided by the body's epithelial surfaces. Second, various chemical and enzymatic systems, including complement, act as an immediate antimicrobial barrier near these epithelia. If epithelia are breached, nearby various innate lymphoid cells can coordinate a rapid cell-mediated defense. If the pathogen overcomes these barriers, the slower-acting defenses of the adaptive immune system are brought to bear.

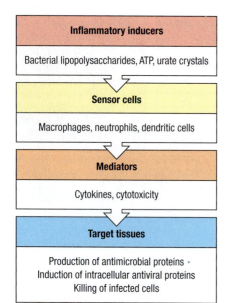

Fig. 1.6 Cell-mediated immunity proceeds in a series of steps. Inflammatory inducers are chemical structures that indicate the presence of invading microbes or the cellular damage produced by them. Sensor cells detect these inducers by expressing various innate recognition receptors, and in response produce a variety of mediators that act directly in defense or that further propagate the immune response. Mediators include many cytokines, and they act on various target tissues, such as epithelial cells, to induce antimicrobial proteins and resist intracellular viral growth; or on other immune cells, such as ILCs that produce other cytokines that amplify the immune response.

1-3   The immune system is activated by inflammatory inducers that indicate the presence of pathogens or tissue damage.

A pathogen that breaches the host's anatomic and chemical barriers will encounter the cellular defenses of innate immunity. Cellular immune responses are initiated when **sensor cells** detect **inflammatory inducers** (Fig. 1.6). Sensor cells include many cell types that detect **inflammatory mediators** through expression of many **innate recognition receptors**, which are encoded by a relatively small number of genes that remain constant over an individual's lifetime. Inflammatory inducers that trigger these receptors include molecular components unique to bacteria or viruses, such as bacterial lipopolysaccharides, or molecules such as ATP, which is not normally found in the extracellular space. Triggering these receptors can activate innate immune cells to produce various mediators that either act directly to destroy invading microbes, or act on other cells to propagate the immune response. For example, macrophages can ingest microbes and produce toxic chemical mediators, such as degradative enzymes or reactive oxygen intermediates, to kill them. Dendritic cells may produce cytokine mediators, including many cytokines that activate target tissues, such as epithelial or other immune cells, to resist or kill invading microbes more efficiently. We will discuss these receptors and mediators briefly below and in much greater detail in Chapter 3.

Innate immune responses occur rapidly on exposure to an infectious organism (Fig. 1.7). In contrast, responses by the adaptive immune system take days rather than hours to develop. However, the adaptive immune system is capable of eliminating infections more efficiently because of exquisite specificity

Phases of the immune response			
Response		**Typical time after infection to start of response**	**Duration of response**
Innate immune response	Inflammation, complement activation, phagocytosis, and destruction of pathogen	Minutes	Days
Adaptive immune response	Interaction between antigen-presenting dendritic cells and antigen-specific T cells: recognition of antigen, adhesion, co-stimulation, T-cell proliferation and differentiation	Hours	Days
	Activation of antigen-specific B cells	Hours	Days
	Formation of effector and memory T cells	Days	Weeks
	Interaction of T cells with B cells, formation of germinal centers. Formation of effector B cells (plasma cells) and memory B cells. Production of antibody	Days	Weeks
	Emigration of effector lymphocytes from peripheral lymphoid organs	A few days	Weeks
	Elimination of pathogen by effector cells and antibody	A few days	Weeks
Immunological memory	Maintenance of memory B cells and T cells and high serum or mucosal antibody levels. Protection against reinfection	Days to weeks	Can be lifelong

Fig. 1.7 Phases of the immune response.

of antigen recognition by its lymphocytes. In contrast to a limited repertoire of receptors expressed by innate immune cells, lymphocytes express highly specialized **antigen receptors** that collectively possess a vast repertoire of specificity. This enables the adaptive immune system to respond to virtually any pathogen and effectively focus resources to eliminate pathogens that have evaded or overwhelmed innate immunity. But the adaptive immune system interacts with, and relies on, cells of the innate immune system for many of its functions. The next several sections will introduce the major components of the innate immune system and prepare us to consider adaptive immunity later in the chapter.

1-4 The myeloid lineage comprises most of the cells of the innate immune system.

The **common myeloid progenitor (CMP)** is the precursor of the macrophages, granulocytes (the collective term for the white blood cells called neutrophils, eosinophils, and basophils), mast cells, and dendritic cells of the innate immune system. Macrophages, granulocytes, and dendritic cells make up the three types of phagocytes in the immune system. The CMP also generates megakaryocytes and red blood cells, which we will not be concerned with here. The cells of the myeloid lineage are shown in Fig. 1.8.

Macrophages are resident in almost all tissues. Many tissue-resident macrophages arise during embryonic development, but some macrophages that arise in the adult animal from the bone marrow are the mature form of **monocytes**, which circulate in the blood and continually migrate into tissues, where they differentiate. Macrophages are relatively long-lived cells and perform several different functions throughout the innate immune response and the subsequent adaptive immune response. One is to engulf and kill invading microorganisms. This phagocytic function provides a first defense in innate immunity. Macrophages also dispose of pathogens and infected cells targeted by an adaptive immune response. Both monocytes and macrophages are phagocytic, but most infections occur in the tissues, and so it is primarily macrophages that perform this important protective function. An additional and crucial role of macrophages is to orchestrate immune responses: they help induce inflammation, which, as we shall see, is a prerequisite to a successful immune response, and they produce many inflammatory mediators that activate other immune-system cells and recruit them into an immune response.

Local inflammation and the phagocytosis of invading bacteria can also be triggered by the activation of complement. Bacterial surfaces can activate the complement system, inducing a cascade of proteolytic reactions that coat the microbes with fragments of specific proteins of the complement system.

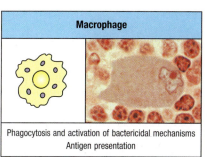

Macrophage

Phagocytosis and activation of bactericidal mechanisms
Antigen presentation

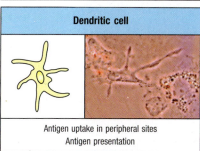

Dendritic cell

Antigen uptake in peripheral sites
Antigen presentation

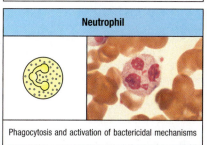

Neutrophil

Phagocytosis and activation of bactericidal mechanisms

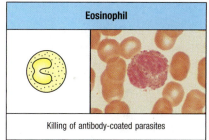

Eosinophil

Killing of antibody-coated parasites

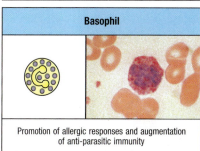

Basophil

Promotion of allergic responses and augmentation of anti-parasitic immunity

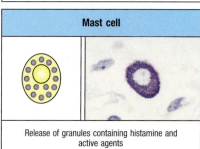

Mast cell

Release of granules containing histamine and active agents

Fig. 1.8 Myeloid cells in innate and adaptive immunity. In the rest of the book, these cells will be represented in the schematic form shown on the left. A photomicrograph of each cell type is shown on the right. Macrophages and neutrophils are primarily phagocytic cells that engulf pathogens and destroy them in intracellular vesicles, a function they perform in both innate and adaptive immune responses. Dendritic cells are phagocytic when they are immature and can take up pathogens; after maturing, they function as specialized cells that present pathogen antigens to T lymphocytes in a form they can recognize, thus activating T lymphocytes and initiating adaptive immune responses. Macrophages can also present antigens to T lymphocytes and can activate them. The other myeloid cells are primarily secretory cells that release the contents of their prominent granules upon activation via antibody during an adaptive immune response. Eosinophils are thought to be involved in attacking large antibody-coated parasites such as worms; basophils are also thought to be involved in anti-parasite immunity. Mast cells are tissue cells that trigger a local inflammatory response to antigen by releasing substances that act on local blood vessels. Mast cells, eosinophils, and basophils are also important in allergic responses. Photographs courtesy of N. Rooney, R. Steinman, and D. Friend.

Microbes coated in this way are recognized by specific **complement receptors** on macrophages and neutrophils, taken up by phagocytosis, and destroyed. In addition to their specialized role in the immune system, macrophages act as general scavenger cells in the body, clearing it of dead cells and cell debris.

The **granulocytes** are named for the densely staining granules in their cytoplasm; they are also called **polymorphonuclear leukocytes** because of their oddly shaped nuclei. The three types of granulocytes—neutrophils, eosinophils, and basophils— are distinguished by the different staining properties of their granules, which serve distinct functions. Granulocytes are all relatively short-lived, surviving for only a few days. They mature in the bone marrow, and their production increases during immune responses, when they migrate to sites of infection or inflammation. The phagocytic **neutrophils** are the most numerous and important cells in innate immune responses: they take up a variety of microorganisms by phagocytosis and efficiently destroy them in intracellular vesicles by using degradative enzymes and other antimicrobial substances stored in their cytoplasmic granules. Hereditary deficiencies in neutrophil function open the way to overwhelming bacterial infection, which is fatal if untreated. Their role is discussed further in Chapter 3.

Eosinophils and **basophils** are less abundant than neutrophils, but like neutrophils, they have granules containing a variety of enzymes and toxic proteins, which are released when these cells are activated. Eosinophils and basophils are thought to be important chiefly in defense against parasites, which are too large to be ingested by macrophages or neutrophils. They can also contribute to allergic inflammatory reactions, in which their effects are damaging rather than protective.

Mast cells begin development in the bone marrow, but migrate as immature precursors that mature in peripheral tissues, especially skin, intestines, and airway mucosa. Their granules contain many inflammatory mediators, such as histamine and various proteases, which play a role in protecting the internal surfaces from pathogens, including parasitic worms. We cover eosinophils, basophils, and mast cells and their role in allergic inflammation further in Chapters 10 and 14.

Dendritic cells were discovered in the 1970s by **Ralph Steinman**, for which he received half the 2011 Nobel Prize. These cells form the third class of phagocytic cells of the immune system and include several related lineages whose distinct functions are still being clarified. Most dendritic cells have elaborate membranous processes, like the dendrites of nerve cells. Immature dendritic cells migrate through the bloodstream from the bone marrow to enter tissues. They take up particulate matter by phagocytosis and also continually ingest large amounts of the extracellular fluid and its contents by a process known as **macropinocytosis**. They degrade the pathogens that they take up, but their main role in the immune system is not the clearance of microorganisms. Instead, dendritic cells are a major class of sensor cells whose encounter with pathogens triggers them to produce mediators that activate other immune cells. Dendritic cells were discovered because of their role in activating a particular class of lymphocytes—T lymphocytes—of the adaptive immune system, and we will return to this activity when we discuss T-cell activation in Section 1-15. But dendritic cells and the mediators they produce also play a critical role in controlling responses of cells of the innate immune system.

1-5   Sensor cells express pattern recognition receptors that provide an initial discrimination between self and nonself.

Long before the mechanisms of innate recognition were discovered, it was recognized that purified antigens such as proteins often did not evoke an immune response in an experimental immunization—that is, they were not

immunogenic. Rather, the induction of strong immune responses against purified proteins required the inclusion of microbial constituents, such as killed bacteria or bacterial extracts, famously called the immunologist's 'dirty little secret' by **Charles Janeway** (see Appendix I, Sections A-1–A-4). This additional material was termed an **adjuvant**, because it helped intensify the response to the immunizing antigen (*adjuvare* is Latin for 'to help'). We know now that adjuvants are needed, at least in part, to activate innate receptors on various types of sensor cells to help activate T cells in the absence of an infection.

Macrophages, neutrophils, and dendritic cells are important classes of sensor cells that detect infection and initiate immune responses by producing inflammatory mediators, although other cells, even cells of the adaptive immune system, can serve in this function. As mentioned in Section 1-3, these cells express a limited number of invariant innate recognition receptors as a means of detecting pathogens or the damage induced by them. Also called **pattern recognition receptors (PRRs)**, they recognize simple molecules and regular patterns of molecular structure known as **pathogen-associated molecular patterns (PAMPs)** that are part of many microorganisms but not of the host body's own cells. Such structures include mannose-rich oligosaccharides, peptidoglycans, and lipopolysaccharides of the bacterial cell wall, as well as unmethylated CpG DNA common to many pathogens. All of these microbial elements have been conserved during evolution, making them excellent targets for recognition because they do not change (**Fig. 1.9**). Some PRRs are transmembrane proteins, such as the **Toll-like receptors (TLRs)** that detect PAMPs derived from extracellular bacteria or bacteria taken into vesicular pathways by phagocytosis. The role of the Toll receptor in immunity was discovered first in *Drosophila melanogaster* by **Jules Hoffman**, and later extended to homologous TLRs in mice by Janeway and **Bruce Beutler**. Hoffman and Beutler shared the remaining half of the 2011 Nobel Prize (see Section 1-4) for their work in the activation of innate immunity. Other PRRs are cytoplasmic proteins, such as the **NOD-like receptors (NLRs)** that sense intracellular bacterial invasion. Yet other cytoplasmic receptors detect viral infection based on differences in the structures and locations of the host mRNA and virally derived RNA species, and between host and microbial DNA. Some receptors expressed by sensor cells detect cellular damage induced by pathogens, rather than the pathogens themselves. Much of our knowledge of innate recognition has emerged only within the past 15 years and is still an active area of discovery. We describe these innate recognition systems further in Chapter 3, and how adjuvants are used as a component of vaccines in Chapter 16.

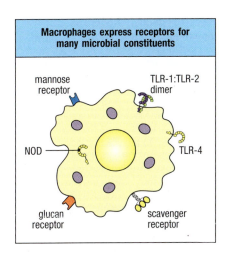

Fig. 1.9 Macrophages express a number of receptors that allow them to recognize different pathogens. Macrophages express a variety of receptors, each of which is able to recognize specific components of microbes. Some, like the mannose and glucan receptors and the scavenger receptor, bind cell-wall carbohydrates of bacteria, yeast, and fungi. The Toll-like receptors (TLRs) are an important family of pattern recognition receptors present on macrophages, dendritic cells, and other immune cells. TLRs recognize different microbial components; for example, a heterodimer of TLR-1 and TLR-2 binds certain lipopeptides from pathogens such as Gram-positive bacteria, while TLR-4 binds both lipopolysaccharides from Gram-negative and lipoteichoic acids from Gram-positive bacteria.

1-6 Sensor cells induce an inflammatory response by producing mediators such as chemokines and cytokines.

Activation of PRRs on sensor cells such as macrophages and neutrophils can directly induce effector functions in these cells, such as the phagocytosis and degradation of bacteria they encounter. But sensor cells serve to amplify the immune response by the production of inflammatory mediators. Two important categories of inflammatory mediators are the secreted proteins called **cytokines** and **chemokines**, which act in a manner similar to hormones to convey important signals to other immune cells.

'Cytokine' is a term for any protein secreted by immune cells that affects the behavior of nearby cells bearing appropriate receptors. There are more than 60 different cytokines; some are produced by many different cell types; others, by only a few specific cell types. Some cytokines influence many types of cells, while others influence only a few, through the expression pattern of each cytokine's specific receptor. The response that a cytokine induces in a target cell is typically related to amplifying an effector mechanism of the target cell, as illustrated in the next section.

Fig. 1.10 Infection triggers an inflammatory response. Macrophages encountering bacteria or other types of microorganisms in tissues are triggered to release cytokines (left panel) that increase the permeability of blood vessels, allowing fluid and proteins to pass into the tissues (center panel). Macrophages also produce chemokines, which direct the migration of neutrophils to the site of infection. The stickiness of the endothelial cells of the blood vessel wall is also changed, so that circulating cells of the immune system adhere to the wall and are able to crawl through it; first neutrophils and then monocytes are shown entering the tissue from a blood vessel (right panel). The accumulation of fluid and cells at the site of infection causes the redness, swelling, heat, and pain known collectively as inflammation. Neutrophils and macrophages are the principal inflammatory cells. Later in an immune response, activated lymphocytes can also contribute to inflammation.

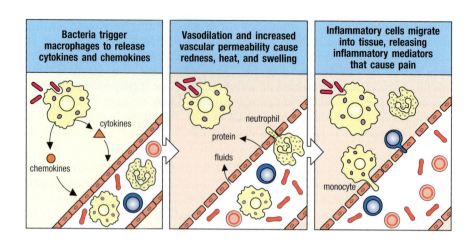

Instead of presenting all the cytokines together all at once, we introduce each cytokine as it arises during our description of cellular and functional responses. We list the cytokines, their producer and target cells, and their general functions in Appendix III.

Chemokines are a specialized subgroup of secreted proteins that act as chemoattractants, attracting cells bearing chemokine receptors, such as neutrophils and monocytes, out of the bloodstream and into infected tissue (Fig. 1.10). Beyond this role, chemokines also help organize the various cells in lymphoid tissues into discrete regions where specialized responses can take place. There are on the order of 50 different chemokines, which are all related structurally but fall into two major classes. Appendix IV lists the chemokines, their target cells, and their general functions. We will discuss chemokines as the need arises during our descriptions of particular cellular immune processes.

The cytokines and chemokines released by activated macrophages act to recruit cells from the blood into infected tissues, a process, known as **inflammation**, that helps to destroy the pathogen. Inflammation increases the flow of lymph, which carries microbes or cells bearing their antigens from the infected tissue to nearby lymphoid tissues, where the adaptive immune response is initiated. Once adaptive immunity has been generated, inflammation also recruits these effector components to the site of infection.

Inflammation is described clinically by the Latin words *calor, dolor, rubor,* and *tumor,* meaning heat, pain, redness, and swelling. Each of these features reflects an effect of cytokines or other inflammatory mediators on the local blood vessels. Heat, redness, and swelling result from the dilation and increased permeability of blood vessels during inflammation, leading to increased local blood flow and leakage of fluid and blood proteins into the tissues. Cytokines and complement fragments have important effects on the **endothelium** that lines blood vessels; the **endothelial cells** themselves also produce cytokines in response to infection. These alter the adhesive properties of the endothelial cells and cause circulating leukocytes to stick to the endothelial cells and migrate between them into the site of infection, to which they are attracted by chemokines. The migration of cells into the tissue and their local actions account for the pain.

The main cell types seen in the initial phase of an inflammatory response are macrophages and neutrophils, the latter being recruited into the inflamed, infected tissue in large numbers. Macrophages and neutrophils are thus also known as **inflammatory cells**. The influx of neutrophils is followed a short time later by the increased entry of monocytes, which rapidly differentiate into macrophages, thus reinforcing and sustaining the innate immune response. Later, if the inflammation continues, eosinophils also migrate into inflamed tissues and contribute to the destruction of the invading microorganisms.

1-7   Innate lymphocytes and natural killer cells are effector cells that share similarities with lymphoid lineages of the adaptive immune system.

The **common lymphoid progenitor (CLP)** in the bone marrow gives rise both to antigen-specific lymphocytes of the adaptive immune system and to several innate lineages that lack antigen-specific receptors. Although the B and T lymphocytes of the adaptive immune system were recognized in the 1960s, the **natural killer (NK) cells** (Fig. 1.11) of the innate immune system were not discovered until the 1970s. NK cells are large lymphocyte-like cells with a distinctive granular cytoplasm that were identified because of their ability to recognize and kill certain tumor cells and cells infected with herpesviruses. Initially, the distinction between these cells and T lymphocytes was unclear, but we now recognize that NK cells are a distinct lineage of cells that arise from the CLP in the bone marrow. They lack the antigen-specific receptors of the adaptive immune system cells, but express members of various families of innate receptors that can respond to cellular stress and to infections by very specific viruses. NK cells play an important role in the early innate response to viral infections, before the adaptive immune response has developed.

More recently, additional lineages of cells related to NK cells have been identified. Collectively, these cells are called **innate lymphoid cells (ILCs)**. Arising from the CLP, ILCs reside in peripheral tissues, such as the intestine, where they function as the sources of mediators of inflammatory responses. The functions of NK cells and ILC cells are discussed in Chapter 3.

Summary.

Strategies of avoidance, resistance, and tolerance represent different ways to deal with pathogens. Anatomic barriers and various chemical barriers such as complement and antimicrobial proteins may be considered a primitive form of avoidance, and they are the first line of defense against entry of both commensal organisms and pathogens into host tissues. If these barriers are breached, the vertebrate immune response becomes largely focused on resistance. Inflammatory inducers, which may be either chemical structures unique to microbes (PAMPs) or the chemical signals of tissue damage, act on receptors expressed by sensor cells to inform the immune system of infection. Sensor cells are typically innate immune cells such as macrophages or dendritic cells. Sensor cells can either directly respond with effector activity or produce inflammatory mediators, typically cytokines and chemokines that act on other immune cells, such as the innate NK cells and ILCs. These cells then are recruited into target tissues to provide specific types of immune-response effector activities, such as cell killing or production of cytokines that have direct antiviral activity, all aimed to reduce or eliminate infection by pathogens. Responses by mediators in target tissues can induce several types of inflammatory cells that are specially suited for eliminating viruses, intracellular bacteria, extracellular pathogens, or parasites.

Principles of adaptive immunity.

We come now to the components of adaptive immunity, the antigen-specific lymphocytes. Unless indicated otherwise, we shall use the term lymphocyte to refer only to the antigen-specific lymphocytes. Lymphocytes allow responses against a vast array of antigens from various pathogens encountered during a person's lifetime and confer the important feature of immunological memory. Lymphocytes make this possible through the highly variable antigen receptors

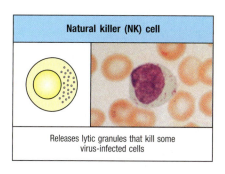

Natural killer (NK) cell

Releases lytic granules that kill some virus-infected cells

Fig. 1.11 Natural killer (NK) cells. These are large, granular, lymphoid-like cells with important functions in innate immunity, especially against intracellular infections, being able to kill other cells. Unlike lymphocytes, they lack antigen-specific receptors. Photograph courtesy of B. Smith.

on their surface, by which they recognize and bind antigens. Each lymphocyte matures bearing a unique variant of a prototype antigen receptor, so that the population of lymphocytes expresses a huge repertoire of receptors that are highly diverse in their antigen-binding sites. Among the billion or so lymphocytes circulating in the body at any one time there will always be some that can recognize a given foreign antigen.

A unique feature of the adaptive immune system is that it is capable of generating **immunological memory**, so that having been exposed once to an infectious agent, a person will make an immediate and stronger response against any subsequent exposure to it; that is, the individual will have protective immunity against it. Finding ways of generating long-lasting immunity to pathogens that do not naturally provoke it is one of the greatest challenges facing immunologists today.

1-8   The interaction of antigens with antigen receptors induces lymphocytes to acquire effector and memory activity.

There are two major types of lymphocytes in the vertebrate immune system, the **B lymphocytes (B cells)** and **T lymphocytes (T cells)**. These express distinct types of antigen receptors and have quite different roles in the immune system, as was discovered in the 1960s. Most lymphocytes circulating in the body appear as rather unimpressive small cells with few cytoplasmic organelles and a condensed, inactive-appearing nuclear chromatin (Fig. 1.12). Lymphocytes manifest little functional activity until they encounter a specific antigen that interacts with an antigen receptor on their cell surface. Lymphocytes that have not yet been activated by antigen are known as **naive lymphocytes**; those that have met their antigen, become activated, and have differentiated further into fully functional lymphocytes are known as **effector lymphocytes**.

B cells and T cells are distinguished by the structure of the antigen receptor that they express. The **B-cell antigen receptor**, or **B-cell receptor (BCR)**, is formed by the same genes that encode antibodies, a class of proteins also known as **immunoglobulins (Ig)** (Fig. 1.13). Thus, the antigen receptor of B lymphocytes is also known as **membrane immunoglobulin (mIg)** or **surface immunoglobulin (sIg)**. The **T-cell antigen receptor**, or **T-cell receptor (TCR)**, is related to the immunoglobulins but is quite distinct in its structure and recognition properties.

After antigen binds to a B-cell antigen receptor, or B-cell receptor (BCR), the B cell will proliferate and differentiate into **plasma cells**. These are the effector form of B lymphocytes, and they secrete antibodies that have the same antigen specificity as the plasma cell's B-cell receptor. Thus the antigen that activates a given B cell becomes the target of the antibodies produced by that B cell's progeny.

Fig. 1.12 Lymphocytes are mostly small and inactive cells. The left panel shows a light micrograph of a small lymphocyte in which the nucleus has been stained purple by hematoxylin and eosin dye, surrounded by red blood cells (which have no nuclei). Note the darker purple patches of condensed chromatin of the lymphocyte nucleus, indicating little transcriptional activity and the relative absence of cytoplasm. The right panel shows a transmission electron micrograph of a small lymphocyte. Again, note the evidence of functional inactivity: the condensed chromatin, the scanty cytoplasm, and the absence of rough endoplasmic reticulum. Photographs courtesy of N. Rooney.

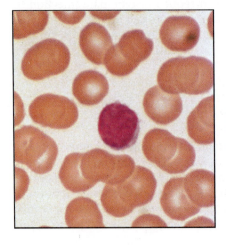
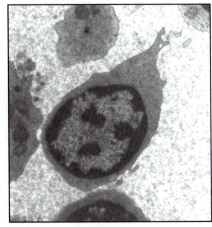

Fig. 1.13 Schematic structure of antigen receptors. Upper panel: an antibody molecule, which is secreted by activated B cells as an antigen-binding effector molecule. A membrane-bound version of this molecule acts as the B-cell antigen receptor (not shown). An antibody is composed of two identical heavy chains (green) and two identical light chains (yellow). Each chain has a constant part (shaded blue) and a variable part (shaded red). Each arm of the antibody molecule is formed by a light chain and a heavy chain, with the variable parts of the two chains coming together to create a variable region that contains the antigen-binding site. The stem is formed from the constant parts of the heavy chains and takes a limited number of forms. This constant region is involved in the elimination of the bound antigen. Lower panel: a T-cell antigen receptor. This is also composed of two chains, an α chain (yellow) and a β chain (green), each of which has a variable and a constant part. As with the antibody molecule, the variable parts of the two chains create a variable region, which forms the antigen-binding site. The T-cell receptor is not produced in a secreted form.

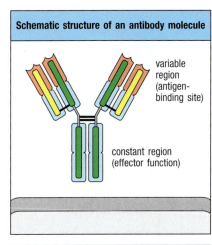

Schematic structure of an antibody molecule

variable region (antigen-binding site)

constant region (effector function)

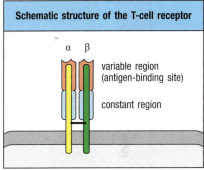

Schematic structure of the T-cell receptor

α   β

variable region (antigen-binding site)

constant region

When a T cell first encounters an antigen that its receptor can bind, it proliferates and differentiates into one of several different functional types of **effector T lymphocytes**. When effector T cells subsequently detect antigen, they can manifest three broad classes of activity. **Cytotoxic T cells** kill other cells that are infected with viruses or other intracellular pathogens bearing the antigen. **Helper T cells** provide signals, often in the form of specific cytokines that activate the functions of other cells, such as B-cell production of antibody and macrophage killing of engulfed pathogens. **Regulatory T cells** suppress the activity of other lymphocytes and help to limit the possible damage of immune responses. We discuss the detailed functions of cytotoxic, helper, and regulatory T cells in Chapters 9, 11, 12, and 15.

Some of the B cells and T cells activated by antigen will differentiate into **memory cells**, the lymphocytes that are responsible for the long-lasting immunity that can follow exposure to disease or vaccination. Memory cells will readily differentiate into effector cells on a second exposure to their specific antigen. Immunological memory is described in Chapter 11.

1-9   Antibodies and T-cell receptors are composed of constant and variable regions that provide distinct functions.

Antibodies were studied by traditional biochemical techniques long before recombinant DNA technology allowed the study of the membrane-bound forms of the antigen receptors on B and T cells. These early studies found that antibody molecules are composed of two distinct regions. One is a **constant region**, also called the **fragment crystallizable region**, or **Fc region**, which takes one of only four or five biochemically distinguishable forms (see Fig. 1.13). The **variable region**, by contrast, can be composed of a vast number of different amino acid sequences that allow antibodies to recognize an equally vast variety of antigens. It was the uniformity of the Fc region relative to the variable region that allowed its early analysis by X-ray crystallography by **Gerald Edelman** and **Rodney Porter**, who shared the 1972 Nobel Prize for their work on the structure of antibodies.

The antibody molecule is composed of two identical **heavy chains** and two identical **light chains**. Heavy and light chains each have variable and constant regions. The variable regions of a heavy chain and a light chain combine to form an **antigen-binding site** that determines the antigen-binding specificity of the antibody. Thus, both heavy and light chains contribute to the antigen-binding specificity of the antibody molecule. Also, each antibody has two identical variable regions, and so has two identical antigen-binding sites. The constant region determines the effector function of the antibody, that is, how the antibody will interact with various immune cells to dispose of antigen once it is bound.

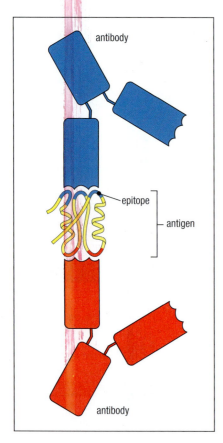

Fig. 1.14 Antigens are the molecules recognized by the immune response, while epitopes are sites within antigens to which antigen receptors bind. Antigens can be complex macromolecules such as proteins, as shown in yellow. Most antigens are larger than the sites on the antibody or antigen receptor to which they bind, and the actual portion of the antigen that is bound is known as the antigenic determinant, or epitope, for that receptor. Large antigens such as proteins can contain more than one epitope (indicated in red and blue) and thus may bind different antibodies (shown here in the same color as the epitopes they bind). Antibodies generally recognize epitopes on the surface of the antigen.

The T-cell receptor shows many similarities to the B-cell receptor and anti-body (see Fig. 1.13). It is composed of two chains, the TCR α and β chains, that are roughly equal in size and which span the T-cell membrane. Like antibody, each TCR chain has a variable region and a constant region, and the combination of the α- and β-chain variable regions creates a single site for binding antigen. The structures of both antibodies and T-cell receptors are described in detail in Chapter 4, and functional properties of antibody constant regions are discussed in Chapters 5 and 10.

1-10   Antibodies and T-cell receptors recognize antigens by fundamentally different mechanisms.

In principle, almost any chemical structure can be recognized as an antigen by the adaptive immune system, but the usual antigens encountered in an infection are the proteins, glycoproteins, and polysaccharides of pathogens. An individual antigen receptor or antibody recognizes a small portion of the antigen's molecular structure, and the part recognized is known as an **antigenic determinant** or **epitope** (Fig. 1.14). Typically, proteins and glycoproteins have many different epitopes that can be recognized by different antigen receptors.

Antibodies and B-cell receptors directly recognize the epitopes of native antigen in the serum or the extracellular spaces. It is possible for different antibodies to simultaneously recognize an antigen by its different epitopes; such simultaneous recognition increases the efficiency of clearing or neutralizing the antigen.

Whereas antibodies can recognize nearly any type of chemical structure, T-cell receptors usually recognize protein antigens and do so very differently from antibodies. The T-cell receptor recognizes a peptide epitope derived from a partially degraded protein, but only if the peptide is bound to specialized cell-surface glycoproteins called **MHC molecules** (Fig. 1.15). The members of this large family of cell-surface glycoproteins are encoded in a cluster of genes called the **major histocompatibility complex (MHC)**. The antigens recognized by T cells can be derived from proteins arising from intracellular pathogens, such as a virus, or from extracellular pathogens. A further difference from the antibody molecule is that there is no secreted form of the T-cell receptor; the T-cell receptor functions solely to signal to the T cell that it has bound its antigen, and the subsequent immunological effects depend on the actions of the T cells themselves. We will further describe how epitopes from antigens are placed on MHC proteins in Chapter 6 and how T cells carry out their subsequent functions in Chapter 9.

Fig. 1.15 T-cell receptors bind a complex of an antigen fragment and a self molecule. Unlike most antibodies, T-cell receptors can recognize epitopes that are buried within antigens (first panel). These antigens must first be degraded by proteases (second panel) and the peptide epitope delivered to a self molecule, called an MHC molecule (third panel). It is in this form, as a complex of peptide and MHC molecule, that antigens are recognized by T-cell receptors (TCRs; fourth panel).

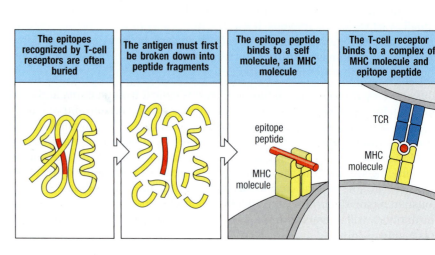

| The epitopes recognized by T-cell receptors are often buried | The antigen must first be broken down into peptide fragments | The epitope peptide binds to a self molecule, an MHC molecule | The T-cell receptor binds to a complex of MHC molecule and epitope peptide |

1-11 Antigen-receptor genes are assembled by somatic gene rearrangements of incomplete receptor gene segments.

The innate immune system detects inflammatory stimuli by means of a relatively limited number of sensors, such as the TLR and NOD proteins, numbering fewer than 100 different types of proteins. Antigen-specific receptors of adaptive immunity provide a seemingly infinite range of specificities, and yet are encoded by a finite number of genes. The basis for this extraordinary range of specificity was discovered in 1976 by **Susumu Tonegawa**, for which he was awarded the 1987 Nobel Prize. Immunoglobulin variable regions are inherited as sets of **gene segments**, each encoding a part of the variable region of one of the immunoglobulin polypeptide chains. During B-cell development in the bone marrow, these gene segments are irreversibly joined by a process of DNA recombination to form a stretch of DNA encoding a complete variable region. A similar process of antigen-receptor gene rearrangement takes place for the T-cell receptor genes during development of T cells in the thymus.

Just a few hundred different gene segments can combine in different ways to generate thousands of different receptor chains. This **combinatorial diversity** allows a small amount of genetic material to encode a truly staggering diversity of receptors. During this recombination process, the random addition or subtraction of nucleotides at the junctions of the gene segments creates additional diversity known as **junctional diversity**. Diversity is amplified further by the fact that each antigen receptor has two different variable chains, each encoded by distinct sets of gene segments. We will describe the gene rearrangement process that assembles complete antigen receptors from gene segments in Chapter 5.

1-12 Lymphocytes activated by antigen give rise to clones of antigen-specific effector cells that mediate adaptive immunity.

There are two critical features of lymphocyte development that distinguish adaptive immunity from innate immunity. First, the process described above that assembles antigen receptors from incomplete gene segments is carried out in a manner that ensures that each developing lymphocyte expresses only one receptor specificity. Whereas the cells of the innate immune system express many different pattern recognition receptors and recognize features shared by many pathogens, the antigen-receptor expression of lymphocytes is 'clonal,' so that each mature lymphocyte differs from others in the specificity of its antigen receptor. Second, because the gene rearrangement process irreversibly changes the lymphocyte's DNA, all its progeny inherit the same receptor specificity. Because this specificity is inherited by a cell's progeny, the proliferation of an individual lymphocyte forms a **clone** of cells with identical antigen receptors.

There are lymphocytes of at least 10^8 different specificities in an individual human at any one time, comprising the **lymphocyte receptor repertoire** of the individual. These lymphocytes are continually undergoing a process similar to natural selection: only those lymphocytes that encounter an antigen to which their receptor binds will be activated to proliferate and differentiate into effector cells. This selective mechanism was first proposed in the 1950s by **Macfarlane Burnet**, who postulated the preexistence in the body of many different potential antibody-producing cells, each displaying on its surface a membrane-bound version of the antibody that served as a receptor for the antigen. On binding antigen, the cell is activated to divide and to produce many identical progeny, a process known as **clonal expansion**; this clone of identical cells can now secrete **clonotypic** antibodies with a specificity identical to that of the surface receptor that first triggered activation and clonal expansion (**Fig. 1.16**). Burnet called this the **clonal selection theory**

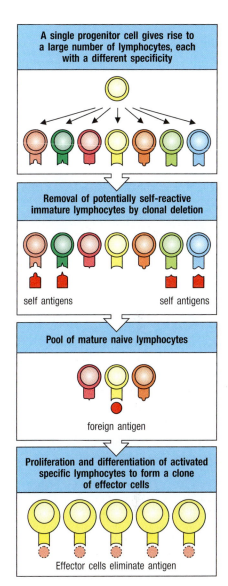

Fig. 1.16 Clonal selection. Each lymphoid progenitor gives rise to a large number of lymphocytes, each bearing a distinct antigen receptor. Lymphocytes with receptors that bind ubiquitous self antigens are eliminated before they become fully mature, ensuring tolerance to such self antigens. When a foreign antigen (red dot) interacts with the receptor on a mature naive lymphocyte, that cell is activated and starts to divide. It gives rise to a clone of identical progeny, all of whose receptors bind the same antigen. Antigen specificity is thus maintained as the progeny proliferate and differentiate into effector cells. Once antigen has been eliminated by these effector cells, the immune response ceases, although some lymphocytes are retained to mediate immunological memory.

Postulates of the clonal selection hypothesis
Each lymphocyte bears a single type of receptor with a unique specificity
Interaction between a foreign molecule and a lymphocyte receptor capable of binding that molecule with high affinity leads to lymphocyte activation
The differentiated effector cells derived from an activated lymphocyte will bear receptors of identical specificity to those of the parental cell from which that lymphocyte was derived
Lymphocytes bearing receptors specific for ubiquitous self molecules are deleted at an early stage in lymphoid cell development and are therefore absent from the repertoire of mature lymphocytes

Fig. 1.17 The four basic principles of clonal selection.

of antibody production; its four basic postulates are listed in Fig. 1.17. Clonal selection of lymphocytes is the single most important principle in adaptive immunity.

1-13 Lymphocytes with self-reactive receptors are normally eliminated during development or are functionally inactivated.

When Burnet formulated his theory, nothing was known of the antigen receptors or indeed the function of lymphocytes themselves. In the early 1960s, **James Gowans** discovered that removal of the small lymphocytes from rats resulted in the loss of all known adaptive immune responses, which were restored when the small lymphocytes were replaced. This led to the realization that lymphocytes must be the units of clonal selection, and their biology became the focus of the new field of **cellular immunology**.

Clonal selection of lymphocytes with diverse receptors elegantly explained adaptive immunity, but it raised one significant conceptual problem. With so many different antigen receptors being generated randomly during the lifetime of an individual, there is a possibility that some receptors might react against an individual's own **self antigens**. How are lymphocytes prevented from recognizing native antigens on the tissues of the body and attacking them? **Ray Owen** had shown in the late 1940s that genetically different twin calves with a common placenta, and thus a shared placental blood circulation, were immunologically unresponsive, or **tolerant**, to one another's tissues. **Peter Medawar** then showed in 1953 that exposure to foreign tissues during embryonic development caused mice to become immunologically tolerant to these tissues. Burnet proposed that developing lymphocytes that are potentially self-reactive are removed before they can mature, a process known as **clonal deletion**. Medawar and Burnet shared the 1960 Nobel Prize for their work on tolerance. This process was demonstrated to occur experimentally in the late 1980s. Some lymphocytes that receive either too much or too little signal through their antigen receptor during development are eliminated by a form of cell suicide called **apoptosis**—derived from a Greek word meaning the falling of leaves from trees—or **programmed cell death**. Other types of mechanisms of **immunological tolerance** have been identified since then that rely on the induction of an inactive state, called **anergy**, as well as mechanisms of active suppression of self-reactive lymphocytes. Chapter 8 will describe lymphocyte development and tolerance mechanisms that shape the lymphocyte receptor repertoire. Chapters 14 and 15 will discuss how immune tolerance mechanisms can sometimes fail.

1-14   Lymphocytes mature in the bone marrow or the thymus and then congregate in lymphoid tissues throughout the body.

Lymphocytes circulate in the blood and the lymph and are also found in large numbers in **lymphoid tissues** or **lymphoid organs**, which are organized aggregates of lymphocytes in a framework of nonlymphoid cells. Lymphoid organs can be divided broadly into the **central** or **primary lymphoid organs**, where lymphocytes are generated, and the **peripheral** or **secondary lymphoid organs**, where mature naive lymphocytes are maintained and adaptive immune responses are initiated. The central lymphoid organs are the bone marrow and the **thymus**, an organ in the upper chest. The peripheral lymphoid organs comprise the **lymph nodes**, the **spleen**, and the mucosal lymphoid tissues of the gut, the nasal and respiratory tract, the urogenital tract, and other mucosa. The locations of the main lymphoid tissues are shown schematically in Fig. 1.18; we describe the individual peripheral lymphoid organs in more detail later in the chapter. Lymph nodes are interconnected by a system of lymphatic vessels, which drain extracellular fluid from tissues, carry it through the lymph nodes, and deposit it back into the blood.

The progenitors that give rise to B and T lymphocytes originate in the bone marrow. B cells complete their development within the bone marrow. Although the 'B' in B lymphocytes originally stood for the **bursa of Fabricius**, a lymphoid organ in young chicks in which lymphocytes mature, it is a useful mnemonic for bone marrow. The immature precursors of T lymphocytes migrate to the thymus, from which they get their name, and complete their development there. Once they have completed maturation, both types of lymphocytes enter the bloodstream as mature naive lymphocytes and continuously circulate through the peripheral lymphoid tissues.

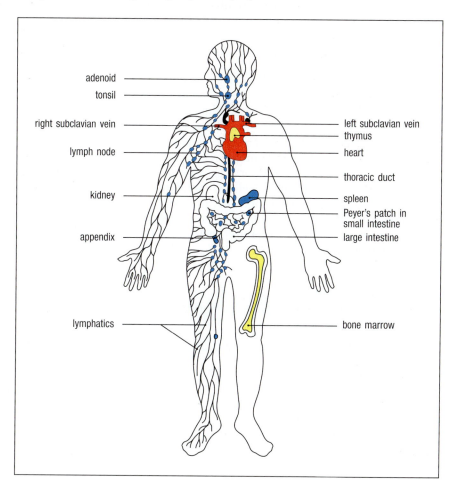

Fig. 1.18 The distribution of lymphoid tissues in the body. Lymphocytes arise from stem cells in bone marrow and differentiate in the central lymphoid organs (yellow)—B cells in the bone marrow and T cells in the thymus. They migrate from these tissues and are carried in the bloodstream to the peripheral lymphoid organs (blue). These include lymph nodes, spleen, and lymphoid tissues associated with mucosa, such as the gut-associated tonsils, Peyer's patches, and appendix. The peripheral lymphoid organs are the sites of lymphocyte activation by antigen, and lymphocytes recirculate between the blood and these organs until they encounter their specific antigen. Lymphatics drain extracellular fluid from the peripheral tissues, through the lymph nodes, and into the thoracic duct, which empties into the left subclavian vein. This fluid, known as lymph, carries antigen taken up by dendritic cells and macrophages to the lymph nodes, as well as recirculating lymphocytes from the lymph nodes back into the blood. Lymphoid tissue is also associated with other mucosa such as the bronchial linings (not shown).

Fig. 1.19 Dendritic cells initiate adaptive immune responses. Immature dendritic cells residing in a tissue take up pathogens and their antigens by macropinocytosis and by receptor-mediated endocytosis. They are stimulated by recognition of the presence of pathogens to migrate through the lymphatics to regional lymph nodes, where they arrive as fully mature nonphagocytic dendritic cells that express both antigen and the co-stimulatory molecules necessary to activate a naive T cell that recognizes the antigen. Thus the dendritic cells stimulate lymphocyte proliferation and differentiation.

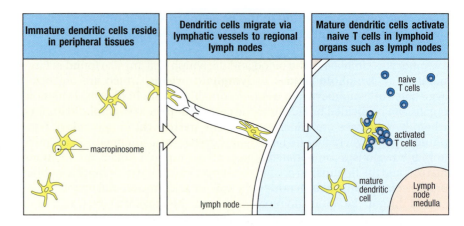

| Immature dendritic cells reside in peripheral tissues | Dendritic cells migrate via lymphatic vessels to regional lymph nodes | Mature dendritic cells activate naive T cells in lymphoid organs such as lymph nodes |

1-15 Adaptive immune responses are initiated by antigen and antigen-presenting cells in secondary lymphoid tissues.

Adaptive immune responses are initiated when B or T lymphocytes encounter antigens for which their receptors have specific reactivity, provided that there are appropriate inflammatory signals to support activation. For T cells, this activation occurs via encounters with dendritic cells that have picked up antigens at sites of infection and migrated to secondary lymphoid organs. Activation of the dendritic cells' PRRs by PAMPs at the site of infection stimulates the dendritic cells in the tissues to engulf the pathogen and degrade it intracellularly. They also take up extracellular material, including virus particles and bacteria, by receptor-independent macropinocytosis. These processes lead to the display of peptide antigens on the MHC molecules of the dendritic cells, a display that activates the antigen receptors of lymphocytes. Activation of PRRs also triggers the dendritic cells to express cell-surface proteins called **co-stimulatory molecules**, which support the ability of the T lymphocyte to proliferate and differentiate into its final, fully functional form (**Fig. 1.19**). For these reasons dendritic cells are also called **antigen-presenting cells (APCs)**, and as such, they form a crucial link between the innate immune response and the adaptive immune response (**Fig. 1.20**). In certain situations, macrophages and B cells can also act as antigen-presenting cells, but dendritic cells are the cells that are specialized in initiating the adaptive immune response. Free antigens can also stimulate the antigen receptors of B cells, but most B cells require 'help' from activated helper T cells for optimal antibody responses. The activation of naive T lymphocytes is therefore an essential first stage in virtually all adaptive immune responses. Chapter 6 returns to dendritic cells to discuss how antigens are processed for presentation to T cells. Chapters 7 and 9 discuss co-stimulation and lymphocyte activation. Chapter 10 describes how T cells help in activating B cells.

MOVIE 1.1

Fig. 1.20 Dendritic cells form a key link between the innate immune system and the adaptive immune system. Like the other cells of innate immunity, dendritic cells recognize pathogens via invariant cell-surface receptors for pathogen molecules and are activated by these stimuli early in an infection. Dendritic cells in tissues are phagocytic; they are specialized to ingest a wide range of pathogens and to display their antigens at the dendritic cell surface in a form that can be recognized by T cells.

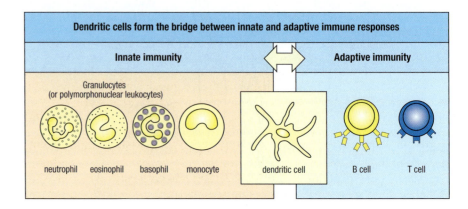

Dendritic cells form the bridge between innate and adaptive immune responses

Fig. 1.21 Circulating lymphocytes encounter antigen in peripheral lymphoid organs. Naive lymphocytes recirculate constantly through peripheral lymphoid tissue, here illustrated as a popliteal lymph node—a lymph node situated behind the knee. In the case of an infection in the foot, this will be the draining lymph node, where lymphocytes may encounter their specific antigens and become activated. Both activated and nonactivated lymphocytes are returned to the bloodstream via the lymphatic system.

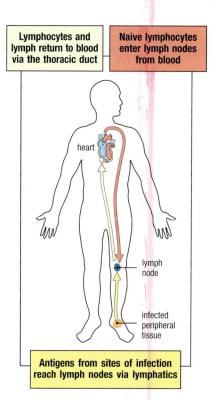

1-16   Lymphocytes encounter and respond to antigen in the peripheral lymphoid organs.

Antigen and lymphocytes eventually encounter each other in the peripheral lymphoid organs—the lymph nodes, spleen, and mucosal lymphoid tissues (see Fig. 1.18). Mature naive lymphocytes are continually recirculating through these tissues, to which pathogen antigens are carried from sites of infection, primarily by dendritic cells. The peripheral lymphoid organs are specialized to trap antigen-bearing dendritic cells and to facilitate the initiation of adaptive immune responses. Peripheral lymphoid tissues are composed of aggregations of lymphocytes in a framework of nonleukocyte stromal cells, which provide both the basic structural organization of the tissue and survival signals to help sustain the life of the lymphocytes. Besides lymphocytes, peripheral lymphoid organs also contain resident macrophages and dendritic cells.

When an infection occurs in a tissue such as the skin, free antigen and antigen-bearing dendritic cells travel from the site of infection through the afferent lymphatic vessels into the **draining lymph nodes** (Fig. 1.21)—peripheral lymphoid tissues where they activate antigen-specific lymphocytes. The activated lymphocytes then undergo a period of proliferation and differentiation, after which most leave the lymph nodes as effector cells via the efferent lymphatic vessel. This eventually returns them to the bloodstream (see Fig. 1.18), which then carries them to the tissues where they will act. This whole process takes about 4–6 days from the time that the antigen is recognized, which means that an adaptive immune response to an antigen that has not been encountered before does not become effective until about a week after infection (see Fig. 1.7). Naive lymphocytes that do not recognize their antigen also leave through the efferent lymphatic vessel and are returned to the blood, from which they continue to recirculate through lymphoid tissues until they recognize antigen or die.

The lymph nodes are highly organized lymphoid organs located at the points of convergence of vessels of the lymphatic system, which is the extensive system that collects extracellular fluid from the tissues and returns it to the blood (see Fig. 1.18). This extracellular fluid is produced continuously by filtration from the blood and is called **lymph**. Lymph flows away from the peripheral tissues under the pressure exerted by its continuous production, and is carried by **lymphatic vessels**, or **lymphatics**. One-way valves in the lymphatic vessels prevent a reverse flow, and the movements of one part of the body in relation to another are important in driving the lymph along.

As noted above, **afferent lymphatic vessels** drain fluid from the tissues and carry pathogens and antigen-bearing cells from infected tissues to the lymph nodes (Fig. 1.22). Free antigens simply diffuse through the extracellular fluid to the lymph node, while the dendritic cells actively migrate into the lymph node, attracted by chemokines. The same chemokines also attract lymphocytes from the blood, and these enter lymph nodes by squeezing through the walls of specialized blood vessels called **high endothelial venules (HEV)**, named for their thicker, more rounded appearance relative to flatter endothelial cells in other locations. In the lymph nodes, B lymphocytes are localized in **follicles**, which make up the outer **cortex** of the lymph node, with T cells more

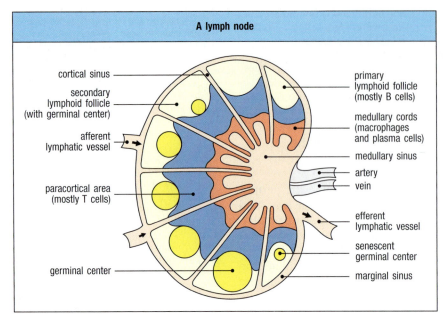

| **A lymph node** |

cortical sinus

secondary
lymphoid follicle
(with germinal center)

afferent
lymphatic vessel

paracortical area
(mostly T cells)

germinal center

primary
lymphoid follicle
(mostly B cells)

medullary cords
(macrophages
and plasma cells)

medullary sinus

artery

vein

efferent
lymphatic vessel

senescent
germinal center

marginal sinus

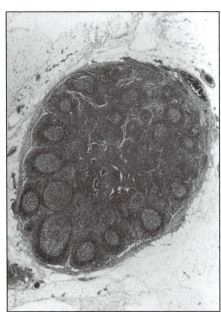

Fig. 1.22 Organization of a lymph node. As shown at left in the diagram of a lymph node in longitudinal section, a lymph node consists of an outermost cortex and an inner medulla. The cortex is composed of an outer cortex of B cells organized into lymphoid follicles and of adjacent, or paracortical, areas made up mainly of T cells and dendritic cells. When an immune response is under way, some of the follicles—known as secondary lymphoid follicles—contain central areas of intense B-cell proliferation called germinal centers. These reactions are very dramatic, but eventually die out as germinal centers become senescent. Lymph draining from the extracellular spaces of the body carries antigens in phagocytic dendritic cells and phagocytic macrophages from the tissues to the lymph node via the afferent lymphatics. These migrate directly from the sinuses into the cellular parts of the node. Lymph leaves via the efferent lymphatics in the medulla. The medulla consists of strings of macrophages and antibody-secreting plasma cells known as the medullary cords. Naive lymphocytes enter the node from the bloodstream through specialized postcapillary venules (not shown) and leave with the lymph through the efferent lymphatic. The light micrograph (right) shows a transverse section through a lymph node, with prominent follicles containing germinal centers. Magnification ×7. Photograph courtesy of N. Rooney.

diffusely distributed in the surrounding **paracortical areas**, also referred to as the deep cortex or **T-cell zones** (see Fig. 1.22). Lymphocytes migrating from the blood into lymph nodes enter the paracortical areas first, and because they are attracted by the same chemokines, antigen-presenting dendritic cells and macrophages also become localized there. Free antigen diffusing through the lymph node can become trapped on these dendritic cells and macrophages. This juxtaposition of antigen, antigen-presenting cells, and naive T cells in the T-cell zone creates an ideal environment in which naive T cells can bind their specific antigen and thus become activated.

As noted earlier, activation of B cells usually requires not only antigen, which binds to the B-cell receptor, but also the cooperation of activated helper T cells, a type of effector T cell. The location of B cells and T cells within the lymph node is dynamically regulated by their state of activation. When they become activated, T cells and B cells both move to the border of the follicle and T-cell zone, where T cells can first provide their helper function to B cells. Some of the B-cell follicles include **germinal centers**, where activated B cells are undergoing intense proliferation and differentiation into plasma cells. These mechanisms are described in detail in Chapter 10.

In humans, the spleen is a fist-sized organ situated just behind the stomach (see Fig. 1.18). It has no direct connection with the lymphatic system; instead, it collects antigen from the blood and is involved in immune responses to blood-borne pathogens. Lymphocytes enter and leave the spleen via blood vessels. The spleen also collects and disposes of senescent red blood cells. Its organization is shown schematically in **Fig. 1.23**. The bulk of the spleen is composed of **red pulp**, which is the site of red blood cell disposal. The lymphocytes surround the arterioles running through the spleen, forming isolated

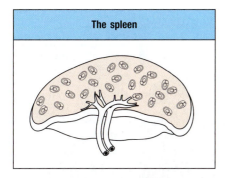

The spleen

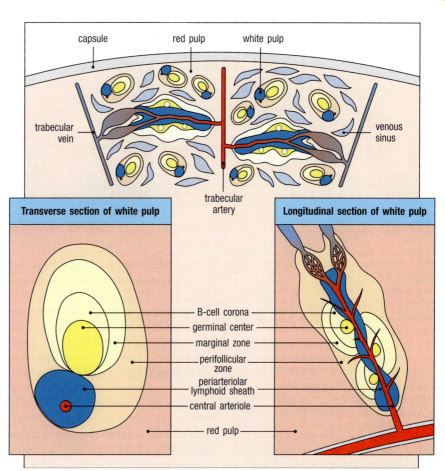

capsule   red pulp   white pulp

trabecular vein

venous sinus

trabecular artery

Transverse section of white pulp

Longitudinal section of white pulp

B-cell corona
germinal center
marginal zone
perifollicular zone
periarteriolar lymphoid sheath
central arteriole

red pulp

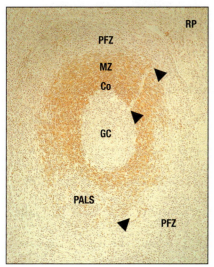

RP
PFZ
MZ
Co
GC
PALS
PFZ

Fig. 1.23 Organization of the lymphoid tissues of the spleen.
The schematic at top left shows that the spleen consists of red pulp (pink areas), which is a site of red blood cell destruction, interspersed with the lymphoid white pulp. An enlargement of a small section of a human spleen (top right) shows the arrangement of discrete areas of white pulp (yellow and blue) around central arterioles. Most of the white pulp is shown in transverse section, with two portions in longitudinal section. The two schematics below this diagram show enlargements of a transverse section (bottom center) and longitudinal section (bottom right) of white pulp. Surrounding the central arteriole is the periarteriolar lymphoid sheath (PALS), made up of T cells. Lymphocytes and antigen-loaded dendritic cells come together here. The follicles consist mainly of B cells; in secondary follicles, a germinal center is surrounded by a B-cell corona. The follicles are surrounded by a so-called marginal zone of lymphocytes. In each area of white pulp, blood carrying both lymphocytes and antigen flows from a trabecular artery into a central arteriole. From this arteriole smaller blood vessels fan out, eventually terminating in a specialized zone in the human spleen called the perifollicular zone (PFZ), which surrounds each marginal zone. Cells and antigen then pass into the white pulp through open blood-filled spaces in the perifollicular zone. The light micrograph at bottom left shows a transverse section of white pulp of human spleen immunostained for mature B cells. Both follicle and PALS are surrounded by the perifollicular zone. The follicular arteriole emerges in the PALS (arrowhead at bottom), traverses the follicle, goes through the marginal zone, and opens into the perifollicular zone (upper arrowheads). Co, follicular B-cell corona; GC, germinal center; MZ, marginal zone; RP, red pulp; arrowheads, central arteriole. Photograph courtesy of N. M. Milicevic.

areas of **white pulp.** The sheath of lymphocytes around an arteriole is called the **periarteriolar lymphoid sheath (PALS)** and contains mainly T cells. Lymphoid follicles occur at intervals along it, and these contain mainly B cells. An area called the **marginal zone** surrounds the follicle; it has few T cells, is rich in macrophages, and has a resident, noncirculating population of B cells known as **marginal zone B cells**. These B cells are poised to rapidly produce antibodies that have low affinity to bacterial capsular polysaccharides. These antibodies, which are discussed in Chapter 8, provide some degree of protection before the adaptive immune response is fully activated. Blood-borne microbes, soluble antigens, and antigen:antibody complexes are filtered from the blood by macrophages and immature dendritic cells within the marginal zone. Like the migration of immature dendritic cells from peripheral tissues to the T-cell areas of lymph nodes, dendritic cells in the marginal zones in the spleen migrate to the T-cell areas after taking up antigen and becoming activated; here they are able to present the antigens they carry to T cells.

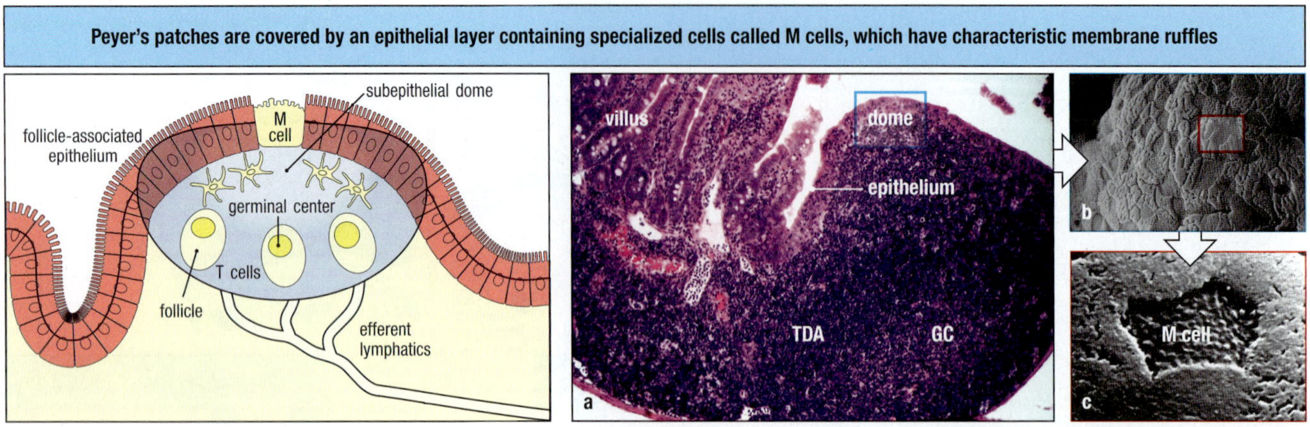

Peyer's patches are covered by an epithelial layer containing specialized cells called M cells, which have characteristic membrane ruffles

Fig. 1.24 Organization of a Peyer's patch in the gut mucosa. As the diagram on the left shows, a Peyer's patch contains numerous B-cell follicles with germinal centers. The areas between follicles are occupied by T cells and are therefore called the T-cell dependent areas. The layer between the surface epithelium and the follicles is known as the subepithelial dome, and is rich in dendritic cells, T cells, and B cells. Peyer's patches have no afferent lymphatics, and the antigen enters directly from the gut across a specialized epithelium made up of so-called microfold (M) cells. Although this tissue looks very different from other lymphoid organs, the basic divisions are maintained. As in the lymph nodes, lymphocytes enter Peyer's patches from the blood across the walls of high endothelial venules (not shown), and leave via the efferent lymphatic. The light micrograph in panel a shows a section through a Peyer's patch in the gut wall of a mouse. The Peyer's patch can be seen lying beneath the epithelial tissues. GC, germinal center; TDA, T-cell dependent area. Panel b, a scanning electron micrograph of the follicle-associated epithelium boxed in panel a, shows the M cells, which lack the microvilli and the mucus layer present on normal epithelial cells. Each M cell appears as a sunken area on the epithelial surface. Panel c, a higher-magnification view of the boxed area in panel b, shows the characteristic ruffled surface of an M cell. M cells are the portal of entry for many pathogens and other particles. Panel a, hematoxylin and eosin stain, magnification ×100; panel b, ×5000; panel c, ×23,000.

1-17   Mucosal surfaces have specialized immune structures that orchestrate responses to environmental microbial encounters.

Most pathogens enter the body through mucosal surfaces, and these are also exposed to a vast load of other potential antigens from the air, food, and the natural microbial flora of the body. Mucosal surfaces are protected by an extensive system of lymphoid tissues known generally as the **mucosal immune system** or **mucosa-associated lymphoid tissues (MALT)**. Collectively, the mucosal immune system is estimated to contain as many lymphocytes as all the rest of the body, and they form a specialized set of cells obeying somewhat different rules of recirculation from those in the other peripheral lymphoid organs. The **gut-associated lymphoid tissues (GALT)** include the **tonsils**, **adenoids**, **appendix**, and specialized structures in the small intestine called **Peyer's patches**, and they collect antigen from the epithelial surfaces of the gastrointestinal tract. In Peyer's patches, which are the most important and highly organized of these tissues, the antigen is collected by specialized epithelial cells called **microfold** or **M cells** (Fig. 1.24). The lymphocytes form a follicle consisting of a large central dome of B lymphocytes surrounded by smaller numbers of T lymphocytes. Dendritic cells resident within the Peyer's patch present the antigen to T lymphocytes. Lymphocytes enter Peyer's patches from the blood and leave through efferent lymphatics. Effector lymphocytes generated in Peyer's patches travel through the lymphatic system and into the bloodstream, from where they are disseminated back into mucosal tissues to carry out their effector actions.

Similar but more diffuse aggregates of lymphocytes are present in the respiratory tract and other mucosa: **nasal-associated lymphoid tissue (NALT)** and **bronchus-associated lymphoid tissue (BALT)** are present in the respiratory tract. Like the Peyer's patches, these mucosal lymphoid tissues are also overlaid by M cells, through which inhaled microbes and antigens that become trapped in the mucous covering of the respiratory tract can pass. The mucosal immune system is discussed in Chapter 12.

Although very different in appearance, the lymph nodes, spleen, and mucosa-associated lymphoid tissues all share the same basic architecture. They all operate on the same principle, trapping antigens and antigen-presenting cells from sites of infection in order to present antigen to migratory small lymphocytes, thus inducing adaptive immune responses. The peripheral lymphoid tissues also provide sustaining signals to lymphocytes that do not encounter their specific antigen immediately, so that they survive and continue to recirculate.

Because they are involved in initiating adaptive immune responses, the peripheral lymphoid tissues are not static structures but vary quite markedly, depending on whether or not infection is present. The diffuse mucosal lymphoid tissues may appear in response to infection and then disappear, whereas the architecture of the organized tissues changes in a more defined way during an infection. For example, the B-cell follicles of the lymph nodes expand as B lymphocytes proliferate to form germinal centers (see Fig. 1.22), and the entire lymph node enlarges, a phenomenon familiarly known as swollen glands.

Finally, specialized populations of lymphocytes and innate lymphoid cells can be found distributed throughout particular sites in the body rather than being found in organized lymphoid tissues. Such sites include the liver and the lamina propria of the gut, as well as the base of the epithelial lining of the gut, reproductive epithelia, and, in mice but not in humans, the epidermis. These lymphocyte populations seem to have an important role in protecting these tissues from infection, and are described further in Chapters 8 and 12.

1-18   Lymphocytes activated by antigen proliferate in the peripheral lymphoid organs, generating effector cells and immunological memory.

The great diversity of lymphocyte receptor repertoire means that there will usually be some lymphocytes bearing a receptor for any given foreign antigen. Recent experiments suggest this number to be perhaps a few hundred per mouse, certainly not enough to mount a response against a pathogen. To generate sufficient antigen-specific effector lymphocytes to fight an infection, a lymphocyte with an appropriate receptor specificity is activated first to proliferate. Only when a large clone of identical cells has been produced do these finally differentiate into effector cells, a process that requires 4 to 5 days. This means that the adaptive immune response to a pathogen occurs several days after the initial infection has occurred and been detected by the innate immune system.

On recognizing its specific antigen on an activated antigen-presenting cell, a naive lymphocyte stops migrating, the volume of the nucleus and cytoplasm increases, and new mRNAs and new proteins are synthesized. Within a few hours, the cell looks completely different and is known as a **lymphoblast**. Dividing lymphoblasts are able to duplicate themselves two to four times every 24 hours for 3–5 days, so that a single naive lymphocyte can produce a clone of around 1000 daughter cells of identical specificity. These then differentiate into effector cells. In the case of B cells, the differentiated effector cells are the plasma cells, which secrete antibody. In the case of T cells, the effector cells are either cytotoxic T cells, which are able to destroy infected cells, or helper T cells, which activate other cells of the immune system (see Section 1-8).

Effector lymphocytes do not recirculate like naive lymphocytes. Some effector T cells detect sites of infection and migrate into them from the blood; others stay in the lymphoid tissues to activate B cells. Some antibody-secreting plasma cells remain in the peripheral lymphoid organs, but most plasma cells generated in the lymph nodes and spleen will migrate to the bone marrow and take up residence there, secreting large amounts of antibodies into the blood system. Effector cells generated in the mucosal immune system generally stay within the mucosal tissues. Most lymphocytes generated by clonal expansion in an immune response will eventually die. However, a significant number of activated antigen-specific B cells and T cells persist after antigen has been eliminated. These cells are known as **memory cells** and form the basis of immunological memory. They can be reactivated much more quickly than naive lymphocytes, which ensures a more rapid and effective response on a second encounter with a pathogen and thereby usually provides lasting protective immunity.

Fig. 1.25 The course of a typical antibody response. The first encounter with an antigen produces a primary response. Antigen A introduced at time zero encounters little specific antibody in the serum. After a lag phase (light blue), antibody against antigen A (dark blue) appears; its concentration rises to a plateau and then gradually declines, typical of a primary response. When the serum is tested for antibody against another antigen, B (yellow), there is little preset. When the animal is later challenged with a mixture of antigens A and B, a very rapid and intense antibody secondary response to A occurs, illustrating immunological memory. This is the main reason for giving booster injections after an initial vaccination. Note that the response to B resembles the primary response to A, as this is the first encounter with antigen B.

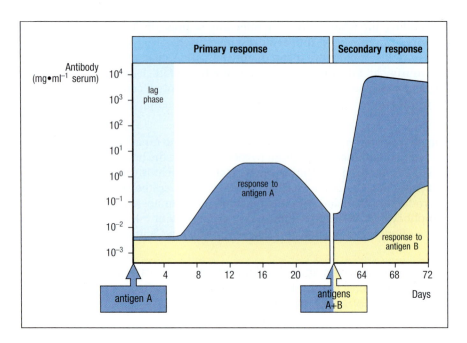

The characteristics of immunological memory are readily observed by comparing an individual's antibody response to a first or **primary immunization** with the response to a **secondary** or **booster immunization** with the same antigen. As shown in **Fig. 1.25**, the secondary antibody response occurs after a shorter lag phase and achieves a markedly higher level than in the primary response. During the secondary responses, antibodies can also acquire higher affinity, or strength of binding, for the antigen due to a process called **affinity maturation**, which takes place in the specialized germinal centers within B-cell follicles (see Section 1-16). Importantly, helper T cells are required for the process of affinity maturation, but T-cell receptors do not undergo affinity maturation. Compared with naive T cells, memory T cells show a lower threshold for activation, but as a result of changes in the responsiveness of the cell and not because of changes in the receptor. We describe the mechanisms of affinity maturation in Chapters 5 and 10.

The cellular basis of immunological memory is the clonal expansion and clonal differentiation of cells that have a specific attraction for the eliciting antigen, and the memory is therefore entirely antigen-specific. It is immunological memory that enables successful vaccination and prevents reinfection with pathogens that have been repelled successfully by an adaptive immune response. In Chapter 11, we will return to immunological memory, which is perhaps the most important biological consequence of adaptive immunity.

Summary.

While the innate immune system relies on invariant pattern recognition receptors to detect common microbial structures or the damage they cause, the adaptive immune system relies on a repertoire of antigen receptors to recognize structures that are specific to individual pathogens. This feature provides adaptive immunity with greater sensitivity and specificity. The clonal expansion of antigen-reactive lymphocytes also confers the property of immunological memory, which enhances protection against reinfection by the same pathogen.

Adaptive immunity relies on two major types of lymphocytes. B cells mature in the bone marrow and are the source of circulating antibodies. T cells mature in the thymus and recognize peptides from pathogens presented by MHC molecules on infected cells or antigen-presenting cells. An adaptive response involves the selection and amplification of clones of lymphocytes bearing receptors that recognize the foreign antigen. This clonal selection provides the theoretical framework for understanding all the key features of an adaptive immune response.

Each lymphocyte carries cell-surface receptors of a single antigen specificity. These receptors are generated by the random recombination of variable receptor gene segments and the pairing of distinct variable protein chains: heavy and light chains in immunoglobulins, or the two chains of T-cell receptors. The large antigen-receptor repertoire of lymphocytes can recognize virtually any antigen. Adaptive immunity is initiated when an innate immune response fails to eliminate a new infection and activated antigen-presenting cells—typically dendritic cells that bear antigens from pathogens and co-stimulatory receptors—migrate to the draining lymphoid tissues.

Immune responses are initiated in several peripheral lymphoid tissues. The spleen serves as a filter for blood-borne infections. Lymph nodes draining various tissues and the mucosal and gut-associated lymphoid tissues (MALT and GALT) are organized into specific zones where T and B cells can be activated efficiently by antigen-presenting cells or helper T cells. When a recirculating lymphocyte encounters its corresponding antigen in these peripheral lymphoid tissues, it proliferates, and its clonal progeny differentiate into effector T and B lymphocytes that can eliminate the infectious agent. A subset of these proliferating lymphocytes differentiates into memory cells, ready to respond rapidly to the same pathogen if it is encountered again. The details of these processes of recognition, development, and differentiation form the main material of the central three parts of this book.

The effector mechanisms of immunity.

For activated innate and adaptive immune cells to destroy pathogens, they must employ an appropriate effector mechanism suited to each infecting agent. The different types of pathogens noted in Fig. 1.26 have different lifestyles and require different responses for both their recognition and their destruction. Perhaps it is not surprising, then, that defenses against different pathogen types are organized into **effector modules** suited for these different lifestyles. In this sense, an effector module is a collection of cell-mediated and humoral mechanisms, both innate and adaptive, that act together to achieve

The immune system protects against four classes of pathogens		
Type of pathogen	**Examples**	**Diseases**
Viruses (intracellular)	Variola Influenza Varicella	Smallpox Flu Chickenpox
Intracellular bacteria, protozoa, parasites	*Mycobacterium leprae* *Leishmania donovani* *Plasmodium falciparum* *Toxoplasma gondii*	Leprosy Leishmaniasis Malaria Toxoplasmosis
Extracellular bacteria, parasites, fungi	*Streptococcus pneumoniae* *Clostridium tetani* *Trypanosoma brucei* *Pneumocystis jirovecii*	Pneumonia Tetanus Sleeping sickness *Pneumocystis* pneumonia
Parasitic worms (extracellular)	*Ascaris* *Schistosoma*	Ascariasis Schistosomiasis

Fig. 1.26 The major types of pathogens confronting the immune system, and some of the diseases they cause.

elimination of a particular category of pathogen. For example, defense against extracellular pathogens can involve both phagocytic cells and B cells, which recognize extracellular antigens and become plasma cells that secrete antibody into the extracellular environment. Defense against intracellular pathogens involves T cells that can detect peptides generated inside the infected cell. Some effector T cells directly kill cells infected with intracellular pathogens such as viruses. Moreover, activated T cells differentiate into three major subsets of **helper T cells**, which produce different patterns of cytokines. These three subsets, discussed below, generally specialize in promoting defenses against pathogens having three major lifestyles: they can defend against intracellular infection, destroy extracellular bacteria and fungi, or provide barrier immunity directed at parasites. T cells also promote defense against extracellular pathogens by helping B cells make antibody.

Most of the other effector mechanisms used by an adaptive immune response to dispose of pathogens are the same as those of innate immunity and involve cells such as macrophages and neutrophils, and proteins such as complement. Indeed, it seems likely that the vertebrate adaptive immune response evolved by the addition of specific recognition properties to innate defense mechanisms already existing in invertebrates. This is supported by recent findings that the innate lymphoid cells—the ILCs—show similar patterns of differentiation into different cytokine-producing subsets to those of T cells.

We begin this section by outlining the effector actions of antibodies, which depend almost entirely on recruiting cells and molecules of the innate immune system.

1-19   Innate immune responses can select from several effector modules to protect against different types of pathogens.

As we mentioned in Section 1-7, the innate immune system contains several types of cells—NK cells and ILCs—that have similarities to lymphocytes, particularly T cells. NK cells lack the antigen-specific receptors of T cells, but can exhibit the cytotoxic capacity of T cells and produce some of the cytokines that effector T cells produce. ILCs develop from the same progenitor cells in the bone marrow as NK cells, and they also lack antigen-specific receptors. Very recent discoveries indicate that ILCs actually comprise several closely related lineages that differ in the specific cytokines that they will produce when activated. Remarkably, there is a striking similarity between the patterns of cytokines produced by ILC subsets and helper T-cell subsets, as mentioned above. It appears that subsets of ILCs are the innate homologs of their helper T-cell counterparts, and NK cells are the innate homolog of cytotoxic T cells.

As mentioned in Section 1-6, there are a large number of cytokines with different functions (see Appendix III). A convenient way to organize the effects of cytokines is by the effector module that each cytokine promotes. Some cytokines tend to promote immunity to intracellular pathogens. One such cytokine is **interferon-γ**, which acts both by activating phagocytes to more efficiently kill intracellular pathogens and by inducing target tissues to resist intracellular pathogens. This is called **type 1 immunity**. IFN-γ is produced by some but not all subsets of innate and adaptive lymphocytes, and the subset of ILC making IFN-γ is called **ILC1**. Other ILC subsets produce cytokines favoring effector modules called **type 2** and **type 3**, which coordinate defense against parasitic and extracellular pathogens, respectively. The modular nature of immune effector functions will be encountered frequently throughout this book. One principle seems to be that activated sensor cells from either the innate or the adaptive immune system can activate different subsets of innate or adaptive lymphocytes that are specialized for amplifying particular effector modules that are directed against different categories of pathogens (Fig. 1.27).

Effector module	Cell types, functions, and mechanisms
Cytotoxicity	NK cells, CD8 T cells
	Elimination of virally infected and metabolically stressed cells
Intracellular immunity (Type 1)	ILC1, T_H1 cells
	Elimination of intracellular pathogens; activation of macrophages
Mucosal and barrier immunity (Type 2)	ILC2, T_H2 cells
	Elimination and expulsion of parasites; recruitment of eosinophils, basophils, and mast cells
Extracellular immunity (Type 3)	ILC3, T_H17 cells
	Elimination of extracellular bacteria and fungi; recruitment and activation of neutrophils

Fig. 1.27 Innate and adaptive lymphocyte cells share a variety of functions. The different effector modules are served by both innate and adaptive immune mechanisms. For each of the four major types of innate lymphocytes, there is a corresponding type of T cell with generally similar functional characteristics. Each set of innate lymphocyte and T cell exert an effector activity that is broadly directed at a distinct category of pathogen.

1-20 Antibodies protect against extracellular pathogens and their toxic products.

Antibodies are found in plasma—the fluid component of blood—and in extracellular fluids. Because body fluids were once known as humors, immunity mediated by antibodies is known as **humoral immunity**.

Antibodies are Y-shaped molecules with two identical antigen-binding sites and one constant, or Fc, region. As mentioned in Section 1-9, there are five forms of the constant region of an antibody, known as the antibody **classes** or **isotypes**. The constant region determines an antibody's functional properties—how it will engage with the effector mechanisms that dispose of antigen once it is recognized. Each class carries out its particular function by engaging a distinct set of effector mechanisms. We describe the antibody classes and their actions in Chapters 5 and 10.

The first and most direct way in which antibodies can protect against pathogens or their products is by binding to them and thereby blocking their access to cells that they might infect or destroy (Fig. 1.28, left panels). This is known as **neutralization** and is important for protection against viruses, which become prevented from entering cells and replicating, and against bacterial toxin and is the form of immunity elicited by most vaccines.

For bacteria, however, binding by antibodies is not sufficient to stop their replication. In this case, the function of the antibody is to enable a phagocytic cell such as a macrophage or a neutrophil to ingest and destroy the bacterium. Many bacteria evade the innate immune system because they have an outer coat that is not recognized by the pattern recognition receptors of phagocytes. However, antigens in the coat can be recognized by antibodies, and phagocytes have receptors, called **Fc receptors**, that bind the constant region and facilitate phagocytosis of the bacterium (see Fig. 1.28, center panels). The coating of pathogens and foreign particles in this way is known as **opsonization**.

The third function of antibodies is **complement activation**. In Section 1-2 we briefly mentioned Bordet's discovery of complement as a serum factor that 'complements' the activities of antibodies. Complement can be activated by microbial surfaces even without the help of antibodies, which leads to the covalent deposition of certain complement proteins onto the bacterial surface. But when an antibody binds first to the bacterial surface, its constant region provides a platform that is much more efficient in complement activation than

Fig. 1.28 Antibodies can participate in host defense in three main ways.
The left panels show antibodies binding to and neutralizing a bacterial toxin, thus preventing it from interacting with host cells and causing pathology. Unbound toxin can react with receptors on the host cell, whereas the toxin:antibody complex cannot. Antibodies also neutralize complete virus particles and bacterial cells by binding and inactivating them. The antigen:antibody complex is eventually scavenged and degraded by macrophages. Antibodies coating an antigen render it recognizable as foreign by phagocytes (macrophages and neutrophils), which then ingest and destroy it; this is called opsonization. The center panels show opsonization and phagocytosis of a bacterial cell. Antibody first binds to antigens (red) on the bacterial cell through the variable regions. Then the antibody's Fc region binds to Fc receptors (yellow) expressed by macrophages and other phagocytes, facilitating phagocytosis. The right panels show activation of the complement system by antibodies coating a bacterial cell. Bound antibodies form a platform that activates the first protein in the complement system, which deposits complement proteins (blue) on the surface of the bacterium. This can lead in some cases to formation of a pore that lyses the bacterium directly. More generally, complement proteins on the bacterium can be recognized by complement receptors on phagocytes; this stimulates the phagocytes to ingest and destroy the bacterium. Thus, antibodies target pathogens and their toxic products for disposal by phagocytes.

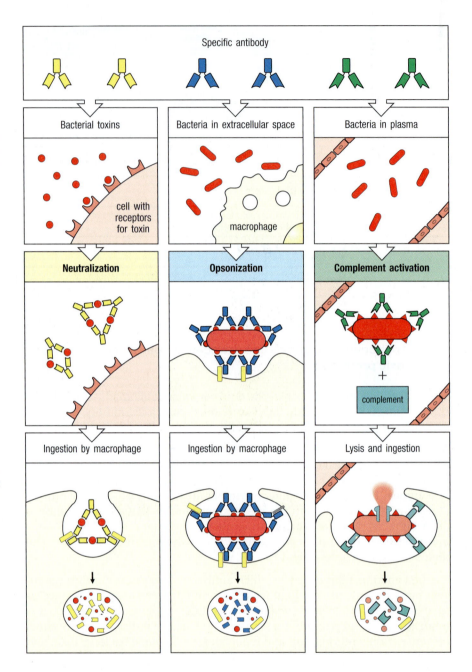

microbial activation alone. Thus, once antibodies are produced, complement activation against a pathogen can be substantially increased.

Certain complement components that are deposited on the bacterial surface can directly lyse the membranes of some bacteria, and this is important in a few bacterial infections (see Fig. 1.28, right panels). The major function of complement, however, is to enable phagocytes to engulf and destroy bacteria that the phagocytes would not otherwise recognize. Most phagocytes express receptors that bind certain complement proteins; called **complement receptors**, these receptors bind to the complement proteins deposited onto the bacterial surface and thus facilitate bacterial phagocytosis. Certain other complement proteins also enhance the phagocytes' bactericidal capacity. The end result is that all pathogens and free molecules bound by antibody are eventually delivered to phagocytes for ingestion, degradation, and removal from the body (see Fig. 1.28, bottom panels). The complement system and the phagocytes that antibodies recruit are not themselves antigen-specific; they depend upon antibody molecules to mark the particles as foreign.

1-21   T cells orchestrate cell-mediated immunity and regulate B-cell responses to most antigens.

Some bacteria and parasites, and all viruses, replicate inside cells, where they cannot be detected by antibodies, which access only the blood and extra-cellular space. The destruction of intracellular invaders is the function of the T lymphocytes, which are responsible for the **cell-mediated immune responses** of adaptive immunity. But T lymphocytes participate in responses to a wide variety of pathogens, including extracellular organisms, and so must exert a wide variety of effector activities.

T lymphocytes, of which there are several types, develop in the thymus. They are characterized by the type of T-cell receptors they express and by the expression of certain markers. The two main classes of T cells express either a cell-surface protein called **CD8** or another called **CD4**. These are not just random markers, but are important for a T cell's function, because they help to determine the interactions between the T cell and other cells. Recall from Section 1-10 that T cells detect peptides derived from foreign antigens that are displayed by MHC molecules on a cell's surface. CD8 and CD4 function in antigen recognition by recognizing different regions of MHC molecules and by being involved in the signaling of the T-cell receptor that is engaged with its antigen. Thus, CD4 and CD8 are known as **co-receptors** and they provide a functional difference between CD8 and CD4 T cells,

Importantly, there are two main types of MHC molecules, called **MHC class I** and **MHC class II**. These have slightly different structures, but both have an elongated groove on the outer surface that can bind a peptide (**Fig. 1.29**). The peptide becomes trapped in this groove during the synthesis and assembly of the MHC molecule inside the cell, and the peptide:MHC complex is then transported to the cell surface and displayed to T cells (**Fig. 1.30**). Because CD8 recognizes a region of the MHC class I protein while CD4 recognizes a region of MHC class II protein, the two co-receptors functionally distinguish T cells. Therefore, CD8 T cells selectively recognize peptides that are bound to MHC class I molecules, while CD4 T cells recognize peptides presented by MHC class II.

The most direct action of T cells is cytotoxicity. Cytotoxic T cells are effector T cells that act against cells infected with viruses. Antigens derived from the virus multiplying inside the infected cell are displayed on the cell's surface, where they are recognized by the antigen receptors of cytotoxic T cells. These T cells can then control the infection by directly killing the infected cell before viral replication is complete and new viruses are released (**Fig. 1.31**). Cytotoxic T cells carry CD8, and so recognize antigen presented by MHC class I molecules. Because MHC class I molecules are expressed on most cells of the body, they serve as an important mechanism to defend against viral infections.

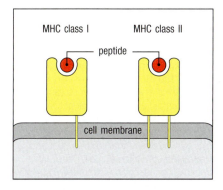

Fig. 1.29 MHC molecules on the cell surface display peptide fragments of antigens. MHC molecules are membrane proteins whose outer extracellular domains form a cleft in which a peptide fragment is bound. These fragments are derived from proteins degraded inside the cell and include both self and foreign protein antigens. The peptides are bound by the newly synthesized MHC molecule before it reaches the cell surface. There are two kinds of MHC molecules, MHC class I and MHC class II; they have related but distinct structures and functions. Although not shown here for simplicity, both MHC class I and MHC class II molecules are trimers of two protein chains and the bound self or nonself peptide.

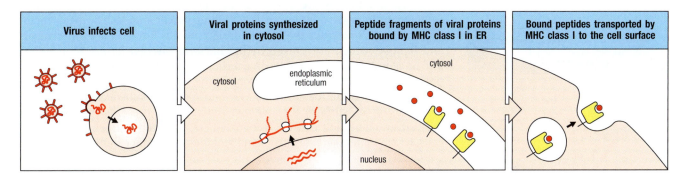

Fig. 1.30 MHC class I molecules present antigen derived from proteins in the cytosol. In cells infected with viruses, viral proteins are synthesized in the cytosol. Peptide fragments of viral proteins are transported into the endoplasmic reticulum (ER), where they are bound by MHC class I molecules, which then deliver the peptides to the cell surface.

Fig. 1.31 Mechanism of host defense against intracellular infection by viruses. Cells infected by viruses are recognized by specialized T cells called cytotoxic T cells, which kill the infected cells directly. The killing mechanism involves the activation of enzymes known as caspases, which contain cysteine in their active site and cleave target proteins at aspartic acid. The caspases in turn activate a cytosolic nuclease that cleaves host and viral DNA in the infected cell. Panel a is a transmission electron micrograph showing the plasma membrane of a cultured CHO cell (the Chinese hamster ovary cell line) infected with influenza virus. Many virus particles can be seen budding from the cell surface. Some of these have been labeled with a monoclonal antibody that is specific for a viral protein and is coupled to gold particles, which appear as the solid black dots in the micrograph. Panel b is a transmission electron micrograph of a virus-infected cell (V) surrounded by cytotoxic T lymphocytes. Note the close apposition of the membranes of the virus-infected cell and the T cell (T) in the upper left corner of the micrograph, and the clustering of the cytoplasmic organelles in the T cell between its nucleus and the point of contact with the infected cell. Panel a courtesy of M. Bui and A. Helenius; panel b courtesy of N. Rooney.

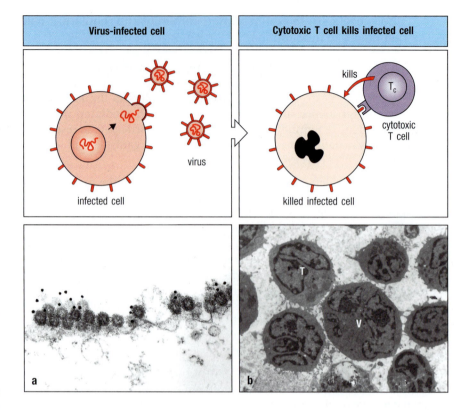

MHC class I molecules bearing viral peptides are recognized by CD8-bearing cytotoxic T cells, which then kill the infected cell (**Fig. 1.32**).

CD4 T cells recognize antigen presented by MHC class II proteins, which are expressed by the predominant antigen-presenting cells of the immune system: dendritic cells, macrophages, and B cells (**Fig. 1.33**). Thus CD4 T cells tend to recognize antigens taken up by phagocytosis from the extracellular environment. CD4 T cells are the helper T cells mentioned earlier in the chapter. They develop into a variety of different effector subsets, called T_H1 (for T helper type 1), T_H2, T_H17, and so on, and they produce cytokines in patterns similar to the subsets of ILCs mentioned earlier that activate effector modules protective against different pathogens. These subsets act primarily at sites of infection or injury in peripheral tissues. In the lymphoid tissues, a subset of CD4 T cells, called the **T follicular helper (T_{FH})** cell, interacts with B cells to regulate antibody production during the immune response. The various T helper subsets are described later, in Chapter 9.

For example, the T_H1 subset of CD4 T cells helps to control certain bacteria that take up residence in membrane-enclosed vesicles inside macrophages. They produce the same cytokine as ILC1 cells, IFN-γ, which activates macrophages to increase their intracellular killing power and destroy these bacteria. Important infections that are controlled by this function are tuberculosis and leprosy, which are caused by the bacteria *Mycobacterium tuberculosis* and *M. leprae*, respectively. Mycobacteria survive intracellularly because they prevent the vesicles they occupy from fusing with lysosomes, which contain a variety of degradative enzymes and antimicrobial substances (**Fig. 1.34**). However, on its surface, the infected macrophage presents mycobacteria-derived antigens that can be recognized by activated antigen-specific T_H1 cells, which in turn secrete particular cytokines that induce the macrophage to

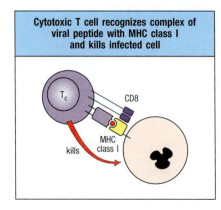

Fig. 1.32 Cytotoxic CD8 T cells recognize antigen presented by MHC class I molecules and kill the cell. The peptide:MHC class I complex on virus-infected cells is detected by antigen-specific cytotoxic T cells. Cytotoxic T cells are preprogrammed to kill the cells they recognize.

Fig. 1.33 CD4 T cells recognize antigen presented by MHC class II molecules. On recognition of their specific antigen on infected macrophages, T$_H$1 cells activate the macrophage, leading to the destruction of the intracellular bacteria (top panel). When T follicular helper (T$_{FH}$) cells recognize antigen on B cells (bottom panel), they activate these cells to proliferate and differentiate into antibody-producing plasma cells (not shown).

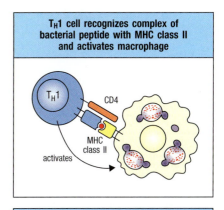

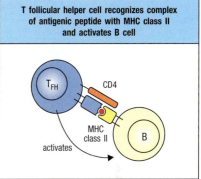

overcome the block on vesicle fusion. T$_H$2 and T$_H$17 subsets produce cytokines that are specialized for promoting responses against parasites or extracellular bacteria and fungi, respectively. CD4 T cells, and their specialized subsets, play a pervasive role in adaptive immunity, and we will be returning to them many times in this book, including in Chapters 8, 9, 11 and 12.

1-22   Inherited and acquired defects in the immune system result in increased susceptibility to infection.

We tend to take for granted the ability of our immune systems to free our bodies of infection and prevent its recurrence. In some people, however, parts of the immune system fail. In the most severe of these **immunodeficiency diseases**, adaptive immunity is completely absent, and death occurs in infancy from overwhelming infection unless heroic measures are taken. Other less catastrophic failures lead to recurrent infections with particular types of pathogens, depending on the particular deficiency. Much has been learned about the functions of the different components of the human immune system through the study of these immunodeficiencies, many of which are caused by inherited genetic defects. Because understanding the features of immunodeficiencies requires a detailed knowledge of normal immune mechanisms, we have postponed discussion of most of these diseases until Chapter 13, where they can be considered together.

More than 30 years ago, a devastating form of immunodeficiency appeared, the **acquired immune deficiency syndrome**, or **AIDS**, which is caused by an infectious agent, the human immunodeficiency viruses HIV-1 and HIV-2. This disease destroys T cells, dendritic cells, and macrophages bearing CD4, leading to infections caused by intracellular bacteria and other pathogens

Acquired Immune Deficiency Syndrome (AIDS)

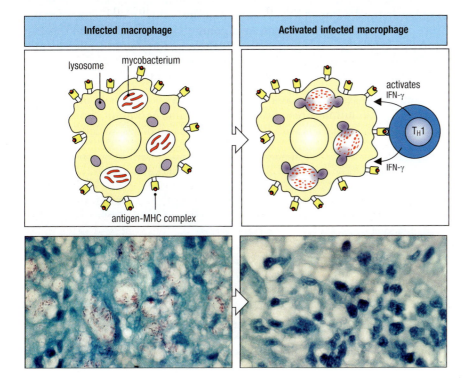

Fig. 1.34 Mechanism of host defense against intracellular infection by mycobacteria. Mycobacteria are engulfed by macrophages but resist being destroyed by preventing the intracellular vesicles in which they reside from fusing with lysosomes containing bactericidal agents. Thus the bacteria are protected from being killed. In resting macrophages, mycobacteria persist and replicate in these vesicles. When the phagocyte is recognized and activated by a T$_H$1 cell, however, the phagocytic vesicles fuse with lysosomes, and the bacteria can be killed. Macrophage activation is controlled by T$_H$1 cells, both to avoid tissue damage and to save energy. The light micrographs (bottom row) show resting (left) and activated (right) macrophages infected with mycobacteria. The cells have been stained with an acid-fast red dye to reveal mycobacteria. These are prominent as red-staining rods in the resting macrophages but have been eliminated from the activated macrophages. Photographs courtesy of G. Kaplan.

normally controlled by such cells. These infections are the major cause of death from this increasingly prevalent immunodeficiency disease, which is discussed fully in Chapter 13 together with the inherited immunodeficiencies.

1-23   Understanding adaptive immune responses is important for the control of allergies, autoimmune disease, and the rejection of transplanted organs.

The main function of our immune system is to protect the human host from infectious agents. However, many medically important diseases are associated with a normal immune response directed against an inappropriate antigen, often in the absence of infectious disease. Immune responses directed at noninfectious antigens occur in **allergy**, in which the antigen is an innocuous foreign substance; in **autoimmune disease**, in which the response is to a self antigen; and in **graft rejection**, in which the antigen is borne by a transplanted foreign cell (discussed in Chapter 15). The major antigens provoking graft rejection are, in fact, the MHC molecules, as each of these is present in many different versions in the human population—that is, they are highly **polymorphic**—and most unrelated people differ in the set of MHC molecules they express, a property commonly known as their 'tissue type.' The MHC was originally recognized by the work of Peter Goren in the 1930s as a gene locus in mice, the **H-2 locus**, that controlled the acceptance or rejection of transplanted tumors, and later by **George Snell**, who examined their role in tissue transplantation by developing mouse strains differing only at these histocompatibility loci. The human MHC molecules were first discovered during the Second World War, when attempts were made to use skin grafts from donors to repair badly burned pilots and bomb victims. The patients rejected the grafts, which were recognized by their immune systems as being 'foreign.' What we call a successful immune response or a failure, and whether the response is considered harmful or beneficial to the host, depends not on the response itself but rather on the nature of the antigen and the circumstances in which the response occurs (Fig. 1.35). Snell was awarded the 1980 Nobel Prize for his work on MHC, together with **Baruj Benacerraf** and **Jean Dausset**.

Allergic diseases, which include asthma, are an increasingly common cause of disability in the developed world. Autoimmunity is also now recognized as the cause of many important diseases. An autoimmune response directed against pancreatic β cells is the leading cause of diabetes in the young. In allergies and autoimmune diseases, the powerful protective mechanisms of the adaptive immune response cause serious damage to the patient.

Immune responses to harmless antigens, to body tissues, or to organ grafts are, like all other immune responses, highly specific. At present, the usual way to treat these responses is with immunosuppressive drugs, which inhibit all

Fig. 1.35 Immune responses can be beneficial or harmful, depending on the nature of the antigen. Beneficial responses are shown in white, harmful responses in red shaded boxes. Where the response is beneficial, its absence is harmful.

Antigen	Effect of response to antigen	
	Normal response	Deficient response
Infectious agent	Protective immunity	Recurrent infection
Innocuous substance	Allergy	No response
Grafted organ	Rejection	Acceptance
Self organ	Autoimmunity	Self tolerance
Tumor	Tumor immunity	Cancer

immune responses, desirable and undesirable alike. If it were possible to suppress only those lymphocyte clones responsible for the unwanted response, the disease could be cured or the grafted organ protected without impeding protective immune responses. At present, antigen-specific immunoregulation is outside the reach of clinical treatment. But as we shall see in Chapter 16, many new drugs have been developed recently that offer more selective immune suppression to control autoimmune and other unwanted immune responses. Among these, therapies using highly specific **monoclonal antibodies** were made possible by **Georges Köhler** and **César Milstein**, who shared the 1984 Nobel Prize for the discovery of their production. We shall discuss the present state of understanding of allergies, autoimmune disease, graft rejection, and immunosuppressive drugs and monoclonal antibodies in Chapters 14–16, and we shall see in Chapter 15 how the mechanisms of immune regulation are beginning to emerge from a better understanding of the functional subsets of lymphocytes and the cytokines that control them.

1-24 Vaccination is the most effective means of controlling infectious diseases.

The deliberate stimulation of an immune response by immunization, or vaccination, has achieved many successes in the two centuries since Jenner's pioneering experiment. Mass immunization programs have led to the virtual eradication of several diseases that used to be associated with significant morbidity (illness) and mortality (Fig. 1.36). Immunization is considered so safe

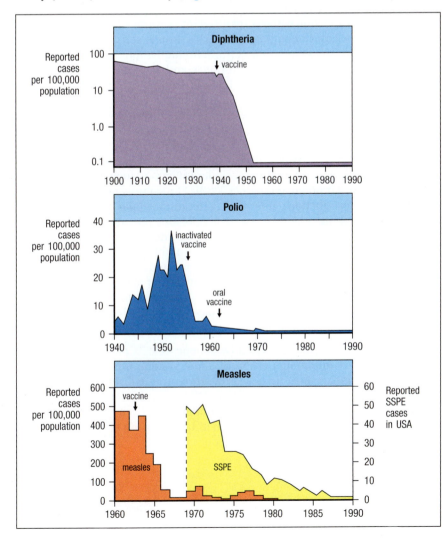

Fig. 1.36 Successful vaccination campaigns. Diphtheria, polio, and measles and their consequences have been virtually eliminated in the United States, as shown in these three graphs. SSPE stands for subacute sclerosing panencephalitis, a brain disease that is a late consequence of measles infection in a few patients. When measles was prevented, SSPE disappeared 15–20 years later. However, because these diseases have not been eradicated worldwide, immunization must be maintained in a very high percentage of the population to prevent their reappearance.

and so important that most states in the United States require children to be immunized against up to seven common childhood diseases. Impressive as these accomplishments are, there are still many diseases for which we lack effective vaccines. And even where vaccines for diseases such as measles can be used effectively in developed countries, technical and economic problems can prevent their widespread use in developing countries, where mortality from these diseases is still high.

The tools of modern immunology and molecular biology are being applied to develop new vaccines and improve old ones, and we discuss these advances in Chapter 16. The prospect of controlling these important diseases is tremendously exciting. The guarantee of good health is a critical step toward population control and economic development. At a cost of pennies per person, great hardship and suffering can be alleviated.

Many serious pathogens have resisted efforts to develop vaccines against them, often because they can evade or subvert the protective mechanisms of an adaptive immune response. We examine some of the evasive strategies used by successful pathogens in Chapter 13. The conquest of many of the world's leading diseases, including malaria and diarrheal diseases (the leading killers of children) as well as the more recent threat from AIDS, depends on a better understanding of the pathogens that cause them and their interactions with the cells of the immune system.

Summary.

The responses to infection can be organized into several effector modules that target the various types of pathogen lifestyles. Innate sensor cells that detect infection generate mediators that activate innate lymphoid cells (ILCs) and T cells, which amplify the immune response and also activate various effector modules. Innate lymphoid cells include subsets that produce different cytokines and activate distinct effector modules. T cells fall into two major classes that are based on the expression of the co-receptors CD8 and CD4; these T cells recognize antigen presented by MHC class I or MHC class II proteins, respectively. These subsets of T cells, like their ILC counterparts, also promote the actions of distinct effector modules. NK cells and CD8 T cells can exert cytotoxic activity to target intracellular infections such as viruses. Other subsets of innate lymphoid and helper T cells can secrete mediators that activate other effector functions, ones that target intracellular bacteria, extracellular bacteria and fungi, and parasites. T cells also provide signals that help regulate B cells and stimulate them to produce antibodies. Specific antibodies mediate the clearance and elimination of soluble toxins and extracellular pathogens. They interact not only with the toxins or the antigens on microbes, but also with the Fc region of specific receptors that are expressed by many types of phagocytes. Phagocytes also express receptors for complement proteins that are deposited on microbial surfaces, particularly in the presence of antibody.

Failures of immunity can be caused by genetic defects or by infections that target important components of the immune system. Misdirected immune responses can damage host tissues, as in autoimmune diseases or allergy, or lead to the failure of transplanted organs. While vaccination is still the greatest tool of immunology to fight diseases, modern approaches have added new tools, such as monoclonal antibodies, that have become progressively more important in the clinic over the past two decades.

Summary to Chapter 1.

The immune system defends the host against infection. Innate immunity serves as a first line of defense but lacks the ability to recognize certain pathogens and to provide the specific protective immunity that prevents reinfection. Adaptive immunity is based on clonal selection from a repertoire of lymphocytes bearing highly diverse antigen-specific receptors that enable the immune system to recognize any foreign antigen. In the adaptive immune response, antigen-specific lymphocytes proliferate and differentiate into clones of effector lymphocytes that eliminate the pathogen. Figure 1.7 summarizes the phases of the immune response and their approximate timings. Host defense requires different recognition systems and a wide variety of effector mechanisms to seek out and destroy the wide variety of pathogens in their various habitats within the body and at its external and internal surfaces. Not only can the adaptive immune response eliminate a pathogen, but, in the process, it also generates increased numbers of differentiated memory lymphocytes through clonal selection, and this allows a more rapid and effective response upon reinfection. The regulation of immune responses, whether to suppress them when unwanted or to stimulate them in the prevention of infectious disease, is the major medical goal of research in immunology.

Questions.

1.1 **Multiple Choice:** Which of the following examples can be considered an illustration of vaccination?

A. Inoculating an individual with cowpox in order to protect that individual against smallpox

B. Administering the serum of animals immune to diphtheria to protect against the effect of diphtheria toxin in an exposed individual

C. A bacterial infection that results in complement activation and destruction of the pathogen

D. An individual that becomes ill with chickenpox, but does not develop it again due to the development of immunologic memory

1.2 **Multiple Choice:** Which of the following is an appropriate definition for immunological memory?

A. The mechanism by which an organism prevents the development of an immune response against the host's own tissues

B. The mechanism by which an organism prevents exposure to microbes

C. The persistence of pathogen-specific antibodies and lymphocytes after the original infection has been eliminated so that reinfection can be prevented

D. The process of reducing or eliminating a pathogen

1.3 **True or False:** Toll-like receptors (TLRs) recognize intracellular bacteria, while NOD-like receptors (NLRs) recognize extracellular bacteria.

1.4 **Matching:** Classify the following as lymphoid or myeloid in origin:

A. Eosinophils

B. B cells

C. Neutrophils

D. NK cells

E. Mast cells

F. Macrophages

G. Red blood cells

1.5 **Multiple Choice:** The immunologist's 'dirty little secret' involves the addition of microbial constituents in order to stimulate a strong immune response against the desired protein antigen of interest. Which of the following is not a receptor or receptor family that can recognize microbial products in order to achieve a potent immune response?

A. Toll-like receptors (TLRs)

B. T-cell antigen receptor (TCR)

C. NOD-like receptors (NLRs)

D. Pattern recognition receptors (PRRs)

1.6 **True or False:** Hematopoietic stem cells can develop into any cell type in the body.

1.7 **Matching:** Match each of the following terms to the numbered phrase that describes it best:

A. Allergy __   1. Immunological response to an antigen present on a transplanted foreign cell

B. Immunological tolerance __   2. Immunological response to an antigen that is an innocuous foreign substance

C. Autoimmune disease __   3. Immunological process that prevents an immune response to self antigens

D. Graft rejection __   4. Immunological response to a self antigen

1.8 **Multiple Choice:** Which of the following processes is not a mechanism of maintaining immunologic tolerance?

A. Clonal deletion

B. Anergy

C. Clonal expansion

D. Suppression of self-reactive lymphocytes

1.9   Matching: Classify each of the following as a central/primary or peripheral/secondary lymphoid organ:

A. Bone marrow

B. Lymph node

C. Spleen

D. Thymus

E. Appendix

1.10   Matching: Match the following region, structure, or subcompartments with the numbered organ they are present in:

A. Lymph node __

B. Spleen __

C. Mucosa of the small intestine __

1. Periarteriolar lymphatic sheath (PALS)

2. Peyer's patches

3. High endothelial venules

1.11   Multiple Choice: Which of the following events do not occur during inflammation?

A. Cytokine secretion

B. Chemokine secretion

C. Recruitment of innate immune cells

D. Constriction of blood vessels

1.12   Fill-in-the-Blanks: _____ T cells are able to kill infected cells, while _____ T cells activate other cells of the immune system.

1.13   True or False: Both T-cell and B-cell receptors undergo the process of affinity maturation in order to acquire progressively higher affinity for an antigen during an immune response.

1.14   True or False: Each lymphocyte carries cell-surface receptors with multiple antigen specificity.

1.15   Multiple Choice: Which cell type forms an important link between the innate immune response and the adaptive immune response?

A. Dendritic cell

B. Neutrophil

C. B cell

D. Innate lymphoid cell (ILC)

1.16   Multiple Choice: Which of the following options is not a mechanism by which an antibody can protect against a pathogen?

A. Neutralization

B. Co-stimulation of T cells

C. Opsonization

D. Complement activation/deposition

1.17   True or False: T_H2 cells do not possess MHC class I molecules.

General references.

Historical background

Burnet, F.M.: *The Clonal Selection Theory of Acquired Immunity.* London: Cambridge University Press, 1959.

Gowans, J.L.: **The lymphocyte—a disgraceful gap in medical knowledge.** *Immunol. Today* 1996, **17**:288–291.

Landsteiner, K.: *The Specificity of Serological Reactions*, 3rd ed. Boston: Harvard University Press, 1964.

Metchnikoff, É.: *Immunity in the Infectious Diseases*, 1st ed. New York: Macmillan Press, 1905.

Silverstein, A.M.: *History of Immunology*, 1st ed. London: Academic Press, 1989.

Biological background

Alberts, B., Johnson, A., Lewis, J., Morgan, D., Raff, M., Roberts, K., and Walter, P.: *Molecular Biology of the Cell*, 6th ed. New York: Garland Science, 2015.

Berg, J.M., Stryer, L., and Tymoczko, J.L.: *Biochemistry*, 5th ed. New York: W.H. Freeman, 2002.

Geha, R.S., and Notarangelo, L.D.: *Case Studies in Immunology: A Clinical Companion*, 7th ed. New York: Garland Science, 2016.

Harper, D.R.: *Viruses: Biology, Applications, Control.* New York: Garland Science, 2012.

Kaufmann, S.E., Sher, A., and Ahmed, R. (eds): *Immunology of Infectious Diseases.* Washington, DC: ASM Press, 2001.

Lodish, H., Berk, A., Kaiser, C.A., Krieger, M., Scott, M.P., Bretscher, A., Ploegh, H., and Matsudaira, P.: *Molecular Cell Biology*, 6th ed. New York: W.H. Freeman, 2008.

Lydyard, P., Cole, M., Holton, J., Irving, W., Porakishvili, N., Venkatesan, P., and Ward, K.: *Case Studies in Infectious Disease.* New York: Garland Science, 2009.

Mims, C., Nash, A., and Stephen, J.: *Mims' Pathogenesis of Infectious Disease*, 5th ed. London: Academic Press, 2001.

Ryan, K.J. (ed): *Medical Microbiology*, 3rd ed. East Norwalk, CT: Appleton-Lange, 1994.

Advanced textbooks in immunology, compendia, etc.

Lachmann, P.J., Peters, D.K., Rosen, F.S., and Walport, M.J. (eds): *Clinical Aspects of Immunology*, 5th ed. Oxford: Blackwell Scientific Publications, 1993.

Mak, T.W., and Saunders, M.E.: *The Immune Response: Basic and Clinical Principles.* Burlington: Elsevier/Academic Press, 2006.

Mak, T.W., and Simard, J.J.L.: *Handbook of Immune Response Genes.* New York: Plenum Press, 1998.

Paul, W.E. (ed): *Fundamental Immunology*, 7th ed. New York: Lippincott Williams & Wilkins, 2012.

Roitt, I.M., and Delves, P.J. (eds): *Encyclopedia of Immunology*, 2nd ed. (4 vols). London and San Diego: Academic Press, 1998.

Innate Immunity: The First Lines of Defense

2

As introduced in Chapter 1, most microbial invaders can be detected and destroyed within minutes or hours by the body's defense mechanisms of **innate immunity**, which do not rely on expansion of antigen-specific lymphocytes. The innate immune system uses a limited number of secreted proteins and cell-associated receptors to detect infection and to distinguish between pathogens and host tissues. These are called innate receptors because they are inborn; they are encoded by genes directly inherited from an individual's parents, and do not need to be generated by the gene rearrangements used to assemble antigen receptors of lymphocytes described in Section 1-11. The importance of innate immunity is illustrated by several immunodeficiencies that result when it is impaired, discussed in Chapter 13, which increase susceptibility to infection even in the presence of an intact adaptive immune system.

As we saw in Fig. 1.5, an infection starts when a pathogen breaches one of the host's anatomic barriers. Some innate immune mechanisms start acting immediately (**Fig. 2.1**). These immediate defenses include several classes of preformed soluble molecules that are present in extracellular fluid, blood, and epithelial secretions and that can either kill the pathogen or weaken its effect. **Antimicrobial enzymes** such as lysozyme begin to digest bacterial cell walls; **antimicrobial peptides** such as the defensins lyse bacterial cell membranes directly; and a system of plasma proteins known as the **complement system** targets pathogens both for lysis and for phagocytosis by cells of the innate immune system such as macrophages. If these fail, innate immune cells become activated by pattern recognition receptors (PRRs) that detect molecules called pathogen-associated molecular patterns (PAMPs) (see Section 1-5) that are typical of microbes. The activated innate cells can engage various effector mechanisms to eliminate the infection. By themselves, neither the soluble nor the cellular components of innate immunity generate long-term protective immunological memory. Only if an infectious organism breaches these first two lines of defense will mechanisms be engaged to induce an adaptive immune response—the third phase of the response to a pathogen. This leads to the expansion of antigen-specific lymphocytes that target the pathogen specifically and to the formation of memory cells that provide long-lasting specific immunity.

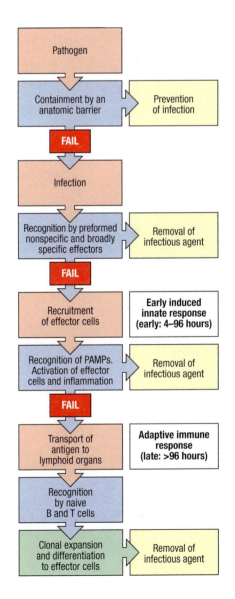

Fig. 2.1 The response to an initial infection occurs in three phases. These are the innate phase, the early induced innate response, and the adaptive immune response. The first two phases rely on the recognition of pathogens by germline-encoded receptors of the innate immune system, whereas adaptive immunity uses variable antigen-specific receptors that are produced as a result of gene segment rearrangements. Adaptive immunity occurs late, because the rare B cells and T cells specific for the invading pathogen must first undergo clonal expansion before they differentiate into effector cells that migrate to the site of infection and clear the infection. The effector mechanisms that remove the infectious agent are similar or identical in each phase.

This chapter considers the first phase of the innate immune response. We first describe the anatomic barriers that protect the host against infection and examine the immediate innate defenses provided by various secreted soluble proteins. The anatomic barriers are fixed defenses against infection and consist of the epithelia that line the internal and external surfaces of the body along with the phagocytes residing beneath all epithelial surfaces. These phagocytes act directly by engulfing and digesting invading microorganisms. Epithelia are also protected by many kinds of chemical defenses, including antimicrobial enzymes and peptides. Next we describe the complement system, which directly kills some microorganisms and interacts with others to promote their removal by phagocytic cells. The complement system together with other soluble circulating defensive proteins is sometimes referred to as **humoral** innate immunity, from the old word 'humor' for body fluids. If these early defenses fail, the phagocytes at the site of infection help recruit new cells and circulating effector molecules, a process called inflammation, which we will discuss in Chapter 3.

Anatomic barriers and initial chemical defenses.

Microorganisms that cause disease in humans and animals enter the body at different sites and produce disease symptoms by a variety of mechanisms. Microorganisms that cause disease and produce damage, or pathology, to tissues are referred to as **pathogenic microorganisms**, or simply **pathogens**. As innate immunity eliminates most microorganisms that may occasionally cross an anatomic barrier, pathogens are microorganisms that have evolved ways of overcoming the body's innate defenses more effectively than other microorganisms. Once infection is established, both innate and adaptive immune responses are typically required to eliminate pathogens from the body. Even in these cases, the innate immune system performs a valuable function by reducing pathogen numbers during the time needed for the adaptive immune system to gear up for action. In the first part of this chapter we briefly describe the different types of pathogens and their invasive strategies, and then examine the immediate innate defenses that, in most cases, prevent microorganisms from establishing an infection.

2-1   Infectious diseases are caused by diverse living agents that replicate in their hosts.

The agents that cause disease fall into five groups: viruses, bacteria, fungi, protozoa, and helminths (worms). Protozoa and worms are usually grouped together as parasites, and are the subject of the discipline of parasitology, whereas viruses, bacteria, and fungi are the subject of microbiology. Fig. 2.2 lists some examples of the different classes of microorganisms and parasites, and the diseases they cause. The characteristic features of each pathogen are its mode of transmission, its mechanism of replication, its mechanism of **pathogenesis**—the means by which it causes disease—and the response it elicits from the host. The distinct pathogen habitats and life cycles mean that a range of different innate and adaptive immune mechanisms have to be deployed for pathogen destruction.

Infectious agents can grow in all body compartments, as shown schematically in Fig. 2.3. We saw in Chapter 1 that two major compartments can be defined—extracellular and intracellular. Both innate and adaptive immune responses have different ways of dealing with pathogens found in these two

Routes of infection for pathogens				
Route of entry	**Mode of transmission**	**Pathogen**	**Disease**	**Type of pathogen**
Mucosal surfaces				
Mouth and respiratory tract	Inhalation or ingestion of infective material (e.g., saliva droplets)	Measles virus	Measles	Paramyxovirus
		Influenza virus	Influenza	Orthomyxovirus
		Varicella-zoster	Chickenpox	Herpesvirus
		Epstein–Barr virus	Mononucleosis	Herpesvirus
		Streptococcus pyogenes	Tonsillitis	Gram-positive bacterium
		Haemophilus influenzae	Pneumonia, meningitis	Gram-negative bacterium
		Neisseria meningitidis	Meningococcal meningitis	Gram-negative bacterium
	Spores	*Bacillus anthracis*	Inhalation anthrax	Gram-positive bacterium
Gastrointestinal tract	Contaminated water or food	Rotavirus	Diarrhea	Rotavirus
		Hepatitis A	Jaundice	Picornavirus
		Salmonella enteritidis, S. typhimurium	Food poisoning	Gram-negative bacterium
		Vibrio cholerae	Cholera	Gram-negative bacterium
		Salmonella typhi	Typhoid fever	Gram-negative bacterium
		Trichuris trichiura	Trichuriasis	Helminth
Reproductive tract and other routes	Sexual transmission/ infected blood	Hepatitis B virus	Hepatitis B	Hepadnavirus
		Human immunodeficiency virus (HIV)	Acquired immunodeficiency syndrome (AIDS)	Retrovirus
	Sexual transmission	*Neisseria gonorrhoeae*	Gonorrhea	Gram-negative bacterium
		Treponema pallidum	Syphilis	Bacterium (spirochete)
Opportunistic infections	Resident microbiota	*Candida albicans*	Candidiasis, thrush	Fungus
	Resident lung microbiota	*Pneumocystis jirovecii*	Pneumonia	Fungus
External epithelia				
External surface	Physical contact	*Trichophyton*	Athlete's foot	Fungus
Wounds and abrasions	Minor skin abrasions	*Bacillus anthracis*	Cutaneous anthrax	Gram-positive bacterium
	Puncture wounds	*Clostridium tetani*	Tetanus	Gram-positive bacterium
	Handling infected animals	*Francisella tularensis*	Tularemia	Gram-negative bacterium
Insect bites	Mosquito bites (*Aedes aegypti*)	Flavivirus	Yellow fever	Virus
	Deer tick bites	*Borrelia burgdorferi*	Lyme disease	Bacterium (spirochete)
	Mosquito bites (*Anopheles*)	*Plasmodium* spp.	Malaria	Protozoan

Fig. 2.2 A variety of microorganisms can cause disease. Pathogenic organisms are of five main types: viruses, bacteria, fungi, protozoa, and worms. Some well-known pathogens are listed.

Fig. 2.3 Pathogens can be found in various compartments of the body, where they must be combated by different host defense mechanisms. Virtually all pathogens have an extracellular phase in which they are vulnerable to the circulating molecules and cells of innate immunity and to the antibodies of the adaptive immune response. All these clear the microorganism mainly by promoting its uptake and destruction by the phagocytes of the immune system. Intracellular phases of pathogens such as viruses are not accessible to these mechanisms; instead, the infected cell is attacked by the NK cells of innate immunity or by the cytotoxic T cells of adaptive immunity. Activation of macrophages as a result of NK-cell or T-cell activity can induce the macrophage to kill pathogens that are living inside macrophage vesicles.

	Extracellular		Intracellular	
	Interstitial spaces, blood, lymph	Epithelial surfaces	Cytoplasmic	Vesicular
Site of infection				
Organisms	Viruses Bacteria Protozoa Fungi Worms	*Neisseria gonorrhoeae* *Streptococcus pneumoniae* *Vibrio cholerae* *Helicobacter pylori* *Candida albicans* Worms	Viruses *Chlamydia* spp. *Rickettsia* spp. Protozoa	*Mycobacterium* spp. *Yersinia pestis* *Legionella pneumophila* *Cryptococcus neoformans* *Leishmania* spp.
Protective immunity	Complement Phagocytosis Antibodies	Antimicrobial peptides Antibodies, especially IgA	NK cells Cytotoxic T cells	T-cell and NK-cell dependent macrophage activation

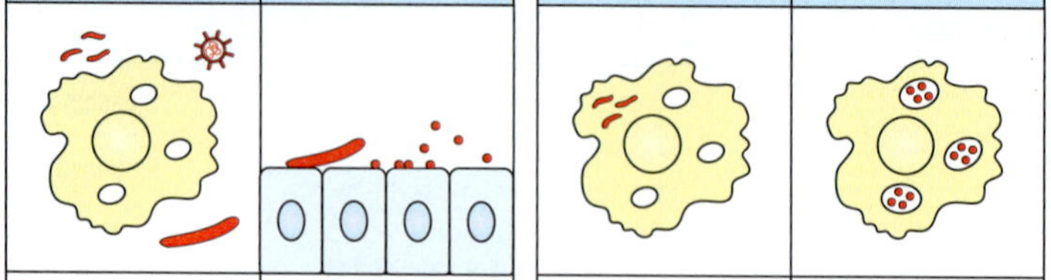

compartments. Many bacterial pathogens live and replicate in extracellular spaces, either within tissues or on the surface of the epithelia that line body cavities. Extracellular bacteria are usually susceptible to killing by phagocytes, an important arm of the innate immune system, but some pathogens, such as *Staphylococcus* and *Streptococcus* species, are protected by a polysaccharide capsule that resists engulfment. This can be overcome to some extent by the help of another component of innate immunity—complement—which renders the bacteria more susceptible to phagocytosis. In the adaptive immune response, bacteria are rendered more susceptible to phagocytosis by a combination of antibodies and complement.

Infectious diseases differ in their symptoms and outcome depending on where the causal pathogen replicates within the body—the intracellular or the extracellular compartment—and what damage it does to the tissues (**Fig. 2.4**). Pathogens that live intracellularly frequently cause disease by damaging or killing the cells they infect. Obligate intracellular pathogens, such as viruses, must invade host cells to replicate. Facultative intracellular pathogens, such as mycobacteria, can replicate either intracellularly or outside the cell. Two strategies of innate immunity defend against intracellular pathogens. One is to destroy pathogens before they infect cells. To this end, innate immunity includes soluble defenses such as antimicrobial peptides, as well as phagocytic cells that can engulf and destroy pathogens before they become intracellular. Alternatively, the innate immune system can recognize and kill cells infected by some pathogens. This is the role of the natural killer cells (NK cells), which are instrumental in keeping certain viral infections in check before cytotoxic T cells of the adaptive immune system become functional. Intracellular pathogens can be subdivided further into those that replicate freely in the cell, such as viruses and certain bacteria (for example, *Chlamydia, Rickettsia,* and *Listeria*), and those that replicate inside intracellular vesicles, such as mycobacteria. Pathogens that live inside macrophage vesicles may become more susceptible to being killed after activation of the macrophage as a result of NK-cell or T-cell actions (see Fig. 2.3).

Many of the most dangerous extracellular bacterial pathogens cause disease by releasing protein toxins; these secreted toxins are called **exotoxins** (see Fig. 2.4). The innate immune system has little defense against such toxins,

	Direct mechanisms of tissue damage by pathogens			Indirect mechanisms of tissue damage by pathogens		
	Exotoxin production	Endotoxin	Direct cytopathic effect	Immune complexes	Anti-host antibody	Cell-mediated immunity
Pathogenic mechanism						
Infectious agent	*Streptococcus pyogenes* *Staphylococcus aureus* *Corynebacterium diphtheriae* *Clostridium tetani* *Vibrio cholerae*	*Escherichia coli* *Haemophilus influenzae* *Salmonella typhi* *Shigella* *Pseudomonas aeruginosa* *Yersinia pestis*	Variola Varicella-zoster Hepatitis B virus Polio virus Measles virus Influenza virus Herpes simplex virus Human herpes virus 8 (HHV8)	Hepatitis B virus Malaria *Streptococcus pyogenes* *Treponema pallidum* Most acute infections	*Streptococcus pyogenes* *Mycoplasma pneumoniae*	Lymphocytic choriomeningitis virus Herpes simplex virus *Mycobacterium tuberculosis* *Mycobacterium leprae* *Borrelia burgdorferi* *Schistosoma mansoni*
Disease	Tonsillitis, scarlet fever Boils, toxic shock syndrome, food poisoning Diphtheria Tetanus Cholera	Gram-negative sepsis Meningitis, pneumonia Typhoid fever Bacillary dysentery Wound infection Plague	Smallpox Chickenpox, shingles Hepatitis Poliomyelitis Measles, subacute sclerosing panencephalitis Influenza Cold sores Kaposi's sarcoma	Kidney disease Vascular deposits Glomerulonephritis Kidney damage in secondary syphilis Transient renal deposits	Rheumatic fever Hemolytic anemia	Aseptic meningitis Herpes stromal keratitis Tuberculosis Tuberculoid leprosy Lyme arthritis Schistosomiasis

Fig. 2.4 Pathogens can damage tissues in a variety of different ways. The mechanisms of damage, representative infectious agents, and the common names of the diseases associated with each are shown. Exotoxins are released by microorganisms and act at the surface of host cells, for example, by binding to receptors. Endotoxins, which are intrinsic components of microbial structure, trigger phagocytes to release cytokines that produce local or systemic symptoms. Many pathogens are cytopathic, directly damaging the cells they infect. Finally, an adaptive immune response to the pathogen can generate antigen:antibody complexes that activate neutrophils and macrophages, antibodies that can cross-react with host tissues, or T cells that kill infected cells. All of these have some potential to damage the host's tissues. In addition, neutrophils, the most abundant cells early in infection, release many proteins and small-molecule inflammatory mediators that both control infection and cause tissue damage.

and highly specific antibodies produced by the adaptive immune system are required to neutralize their action (see Fig. 1.28). The damage caused by a particular infectious agent also depends on where it grows; *Streptococcus pneumoniae* in the lung causes pneumonia, for example, whereas in the blood it causes a potentially fatal systemic illness, pneumococcal sepsis. In contrast, nonsecreted constituents of bacterial structure that trigger phagocytes to release cytokines with local and systemic effects are called **endotoxins**. An endotoxin of major medical importance is the **lipopolysaccharide (LPS)** of the outer cell membrane of **Gram-negative bacteria**, such as *Salmonella*. Many of the clinical symptoms of infection by such bacteria—including fever, pain, rash, hemorrhage, septic shock—are due largely to LPS.

Most pathogenic microorganisms can overcome innate immune responses and continue to grow, making us ill. An adaptive immune response is required to eliminate them and to prevent subsequent reinfection. Certain pathogens are never entirely eliminated by the immune system, and persist in the body for years. But most pathogens are not universally lethal. Those that have lived for thousands of years in the human population are highly evolved to exploit their human hosts; they cannot alter their pathogenicity without upsetting the compromise they have achieved with the human immune system. Rapidly killing every host it infects is no better for the long-term survival of a pathogen than being wiped out by the immune response before the microbe has had time to infect someone else. In short, we have adapted to live with many microbes, and they with us. Nevertheless, the recent concern about highly pathogenic strains of avian influenza and the episode in 2002–2003 of SARS (severe acute respiratory syndrome), a severe pneumonia in humans that is caused by a coronavirus from bats, remind us that new and deadly infections can transfer from animal reservoirs to humans. Such transmission appears responsible for the Ebola virus epidemic in West Africa in 2014–2015. These are known as **zoonotic** infections—and we must be on the alert at all times for the emergence of new pathogens and new threats to health. The human immunodeficiency virus that causes AIDS (discussed in Chapter 13) serves as a warning that we remain constantly vulnerable.

Acquired Immune Deficiency Syndrome (AIDS)

2-2   Epithelial surfaces of the body provide the first barrier against infection.

Our body surfaces are defended by epithelia, which impose a physical barrier between the internal milieu and the external world that contains pathogens. Epithelia comprise the skin and the linings of the body's tubular structures— the respiratory, urogenital, and gastrointestinal tracts. Epithelia in these locations are specialized for their particular functions and possess unique innate defense strategies against the microbes they typically encounter (**Fig. 2.5** and **Fig. 2.6**).

Epithelial cells are held together by tight junctions, which effectively form a seal against the external environment. The internal epithelia are known as **mucosal epithelia** because they secrete a viscous fluid called **mucus**, which contains many glycoproteins called **mucins**. Mucus has a number of protective functions. Microorganisms coated in mucus may be prevented from adhering to the epithelium, and in the respiratory tract, microorganisms can be expelled in the outward flow of mucus driven by the beating of cilia on the mucosal epithelium (**Fig. 2.7**). The importance of mucus flow in clearing infection is illustrated by people with the inherited disease **cystic fibrosis**, in which the mucus becomes abnormally thick and dehydrated due to defects in a gene, *CFTR*, encoding a chloride channel in the epithelium. Such individuals frequently develop lung infections caused by bacteria that colonize the epithelial surface but do not cross it (see Fig. 2.7). In the gut, peristalsis is an important mechanism for keeping both food and infectious agents moving through the body. Failure of peristalsis is typically accompanied by the overgrowth of pathogenic bacteria within the lumen of the gut.

Fig. 2.5 Many barriers prevent pathogens from crossing epithelia and colonizing tissues. Surface epithelia provide mechanical, chemical, and microbiological barriers to infection.

	Skin	Gut	Lungs	Eyes/nose/oral cavity
Mechanical	Epithelial cells joined by tight junctions			
	Longitudinal flow of air or fluid	Longitudinal flow of air or fluid	Movement of mucus by cilia	Tears Nasal cilia
Chemical	Fatty acids	Low pH	Pulmonary surfactant	Enzymes in tears and saliva (lysozyme)
		Enzymes (pepsin)		
	β-defensins Lamellar bodies Cathelicidin	α-defensins (cryptdins) RegIII (lecticidins) Cathelicidin	α-defensins Cathelicidin	Histatins β-defensins
Microbiological	Normal microbiota			

Epidermis of skin

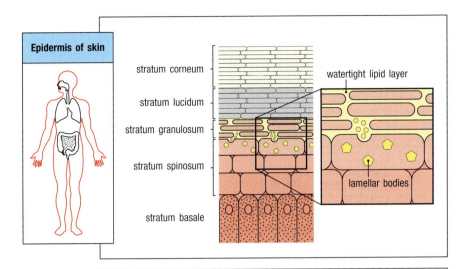

Bronchial ciliated epithelium

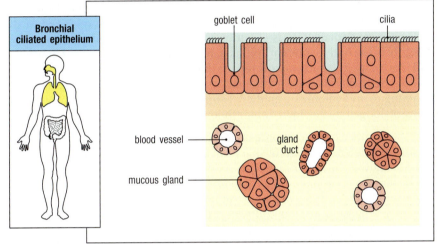

Gut epithelium

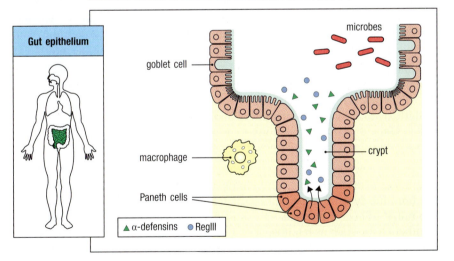

Fig. 2.6 Epithelia form specialized physical and chemical barriers that provide innate defenses in different locations. Top panel: the epidermis has multiple layers of keratinocytes in different stages of differentiation arising from the basal layer of stem cells. Differentiated keratinocytes in the stratum spinosum produce β-defensins and cathelicidins, which are incorporated into secretory organelles called lamellar bodies (yellow) and secreted into the intercellular space to form a waterproof lipid layer (the stratum corneum) containing antimicrobial activity. Center panel: in the lung, the airways are lined by ciliated epithelium. Beating of the cilia moves a continuous stream of mucus (green) secreted by goblet cells outward, trapping and ejecting potential pathogens. Type II pneumocytes in the lung alveoli (not shown) also produce and secrete antimicrobial defensins. Bottom panel: in the intestine, Paneth cells—specialized cells deep in the epithelial crypts—produce several kinds of antimicrobial proteins: α-defensins (cryptdins) and the antimicrobial lectin RegIII.

Most healthy epithelial surfaces are also associated with a large population of normally nonpathogenic bacteria, known as **commensal bacteria** or the **microbiota**, that help keep pathogens at bay. The microbiota can also make antimicrobial substances, such as the lactic acid produced by vaginal lactobacilli, some strains of which also produce antimicrobial peptides (bacteriocins). Commensal microorganisms also induce responses that help to strengthen the barrier functions of epithelia by stimulating the epithelial cells to produce

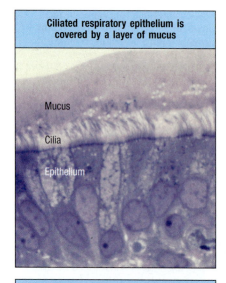

Ciliated respiratory epithelium is covered by a layer of mucus

Mucus

Cilia

Epithelium

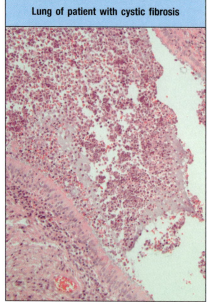

Lung of patient with cystic fibrosis

Fig. 2.7 Ciliated respiratory epithelium propels the overlying mucus layer for clearance of environmental microbes. Top panel: The ciliated respiratory epithelium in the airways of the lung is covered by a layer of mucus. The cilia propel the mucus outward and help prevent colonization of the airways by bacteria. Bottom panel: Section of a lung from a patient with cystic fibrosis. The dehydrated mucus layer impairs the ability of cilia to propel it, leading to frequent bacterial colonization and resulting inflammation of the airway. Courtesy of J. Ritter.

antimicrobial peptides. When commensal microorganisms are killed by antibiotic treatment, pathogens frequently replace them and cause disease (see Fig. 12.20). Under some circumstances commensal microbes themselves can cause disease if their growth is not kept in check or if the immune system is compromised. In Chapter 12, we will further discuss how commensal microorganisms play an important role in the setting of normal immunity, particularly in the intestine; and in Chapter 15, we will see how these normally nonpathogenic organisms can cause disease in the context of inherited immunodeficiencies.

2-3 Infectious agents must overcome innate host defenses to establish a focus of infection.

Our bodies are constantly exposed to microorganisms present in our environment, including infectious agents that have been shed by other individuals. Contact with these microorganisms may occur through external or internal epithelial surfaces. In order to establish an infection, a microorganism must first invade the body by binding to or crossing an epithelium (Fig. 2.8). With the epithelial damage that is common due to wounds, burns, or loss of the integrity of the body's internal epithelia, infection is a major cause of mortality and morbidity. The body rapidly repairs damaged epithelial surfaces, but even without epithelial damage, pathogens may establish infection by specifically adhering to and colonizing epithelial surfaces, using the attachment to avoid being dislodged by the flow of air or fluid across the surface.

Disease occurs when a microorganism succeeds in evading or overwhelming innate host defenses to establish a local site of infection, and then replicates there to allow its further transmission within our bodies. The epithelium lining the respiratory tract provides a route of entry into tissues for airborne microorganisms, and the lining of the gastrointestinal tract does the same for microorganisms ingested in food and water. The intestinal pathogens *Salmonella typhi*, which causes typhoid fever, and *Vibrio cholerae*, which causes cholera, are spread through fecally contaminated food and water, respectively. Insect bites and wounds allow microorganisms to penetrate the skin, and direct contact between individuals offers opportunities for infection through the skin, the gut, and the reproductive tract (see Fig. 2.2).

In spite of this exposure, infectious disease is fortunately quite infrequent. Most of the microorganisms that succeed in crossing an epithelial surface are efficiently removed by innate immune mechanisms that function in the underlying tissues, preventing infection from becoming established. It is difficult to know how many infections are repelled in this way, because they cause no symptoms and pass undetected.

In general, pathogenic microorganisms are distinguished from the mass of microorganisms in the environment by having special adaptations that evade the immune system. In some cases, such as the fungal disease athlete's foot, the initial infection remains local and does not cause significant pathology. In other cases, such as tetanus, the bacterium (*Clostridium tetani* in this case) secretes a powerful neurotoxin, and the infection causes serious illness as it spreads through the lymphatics or the bloodstream, invades and destroys tissues, and disrupts the body's workings.

The spread of a pathogen is often initially countered by an **inflammatory response** that recruits more effector cells and molecules of the innate immune system out of the blood and into the tissues, while inducing clotting in small blood vessels further downstream so that the microbe cannot spread through the circulation (see Fig. 2.8). The cellular responses of innate immunity act over several days. During this time, the adaptive immune response may also begin if antigens derived from the pathogen are delivered to local lymphoid tissues

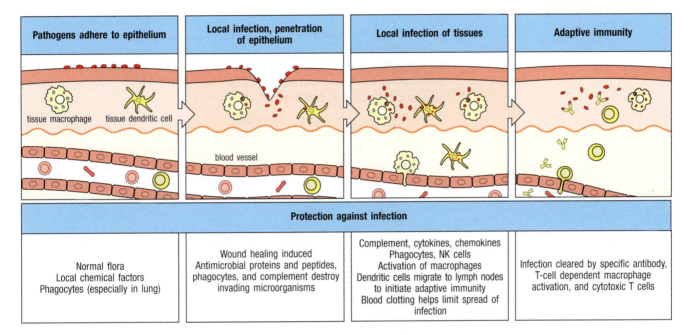

Pathogens adhere to epithelium	Local infection, penetration of epithelium	Local infection of tissues	Adaptive immunity

Protection against infection

| Normal flora Local chemical factors Phagocytes (especially in lung) | Wound healing induced Antimicrobial proteins and peptides, phagocytes, and complement destroy invading microorganisms | Complement, cytokines, chemokines Phagocytes, NK cells Activation of macrophages Dendritic cells migrate to lymph nodes to initiate adaptive immunity Blood clotting helps limit spread of infection | Infection cleared by specific antibody, T-cell dependent macrophage activation, and cytotoxic T cells |

Fig. 2.8 An infection and the response to it can be divided into a series of stages. These are illustrated here for an infectious microorganism entering through a wound in the skin. The infectious agent must first adhere to the epithelial cells and then cross the epithelium. A local immune response may prevent the infection from becoming established. If not, it helps to contain the infection and also delivers the infectious agent, carried in lymph and inside dendritic cells, to local lymph nodes. This initiates the adaptive immune response and eventual clearance of the infection.

by dendritic cells (see Section 1-15). While an innate immune response may eliminate some infections, an adaptive immune response can target particular strains and variants of pathogens and protect the host against reinfection by using either effector T cells or antibodies to generate immunological memory.

2-4   Epithelial cells and phagocytes produce several kinds of antimicrobial proteins.

Our surface epithelia are more than mere physical barriers to infection; they also produce a wide variety of chemical substances that are microbicidal or that inhibit microbial growth. For example, the acid pH of the stomach and the digestive enzymes, bile salts, fatty acids, and lysolipids present in the upper gastrointestinal tract create a substantial chemical barrier to infection (see Fig. 2.5). One important group of antimicrobial proteins comprises enzymes that attack chemical features specific to bacterial cell walls. Such antibacterial enzymes include **lysozyme** and **secretory phospholipase A$_2$**, which are secreted in tears and saliva and by phagocytes. Lysozyme is a glycosidase that breaks a specific chemical bond in the **peptidoglycan** component of the bacterial cell wall. Peptidoglycan is an alternating polymer of *N*-acetylglucosamine (GlcNAc) and *N*-acetylmuramic acid (MurNAc), strengthened by cross-linking peptide bridges (**Fig. 2.9**). Lysozyme selectively cleaves the β-(1,4) linkage between these two sugars and is more effective in acting against Gram-positive bacteria, in which the peptidoglycan cell wall is exposed, than against Gram-negative bacteria, which have an outer layer of LPS covering the peptidoglycan layer. Lysozyme is also produced by **Paneth cells**, specialized epithelial cells in the base of the crypts in the small intestine that secrete many antimicrobial proteins into the gut (see Fig. 2.6). Paneth cells also produce secretory phospholipase A$_2$, a highly basic enzyme that can enter the bacterial cell wall to access and hydrolyze phospholipids in the cell membrane, killing the bacteria.

The second group of antimicrobial agents secreted by epithelial cells and phagocytes is the **antimicrobial peptides**. These represent one of the most ancient forms of defense against infection. Epithelial cells secrete these peptides into the fluids bathing the mucosal surface, whereas phagocytes secrete them in tissues. Three important classes of antimicrobial peptides in mammals are **defensins**, **cathelicidins**, and **histatins**.

Defensins are an ancient, evolutionarily conserved class of antimicrobial peptides made by many eukaryotic organisms, including mammals, insects, and plants (Fig. 2.10). They are short cationic peptides of around 30–40 amino acids that usually have three disulfide bonds stabilizing a common **amphipathic** structure—a positively charged region separated from a hydrophobic region. Defensins act within minutes to disrupt the cell membranes of bacteria and fungi, as well as the membrane envelopes of some viruses. The mechanism is thought to involve insertion of the hydrophobic region into the membrane bilayer and the formation of a pore that makes the membrane leaky (see Fig. 2.10). Most multicellular organisms make many different defensins

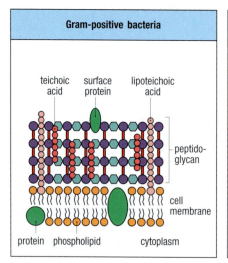
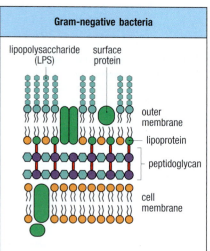
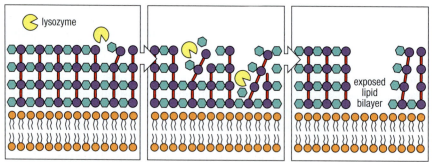

Fig. 2.9 Lysozyme digests the cell walls of Gram-positive and Gram-negative bacteria. Upper panels: the peptidoglycan of bacterial cell walls is a polymer of alternating residues of β-(1,4)-linked N-acetylglucosamine (GlcNAc) (large turquoise hexagons) and N-acetylmuramic acid (MurNAc) (purple circles) that are cross-linked by peptide bridges (red bars) into a dense three-dimensional network. In Gram-positive bacteria (upper left panel), peptidoglycan forms the outer layer in which other molecules are embedded such as teichoic acid and the lipoteichoic acids that link the peptidoglycan layer to the bacterial cell membrane itself. In Gram-negative bacteria (upper right panel), a thin inner wall of peptidoglycan is covered by an outer lipid membrane that contains proteins and lipopolysaccharide (LPS). Lipopolysaccharide is composed of a lipid, lipid A (turquoise circles), to which is attached a polysaccharide core (small turquoise hexagons). Lysozyme (lower panels) cleaves β-(1,4) linkages between GlcNAc and MurNAc, creating a defect in the peptidoglycan layer and exposing the underlying cell membrane to other antimicrobial agents. Lysozyme is more effective against Gram-positive bacteria because of the relatively greater accessibility of the peptidoglycan.

Fig. 2.10 Defensins are amphipathic peptides that disrupt the cell membranes of microbes. The structure of human β_1-defensin is shown in the top panel. It is composed of a short segment of α helix (yellow) resting against three strands of antiparallel β sheet (green), generating an amphipathic peptide with charged and hydrophobic residues residing in separate regions. This general feature is shared by defensins from plants and insects and allows the defensins to interact with the charged surface of the cell membrane and become inserted in the lipid bilayer (center panel). Although the details are still unclear, a transition in the arrangement of the defensins in the membrane leads to the formation of pores and a loss of membrane integrity (bottom panel).

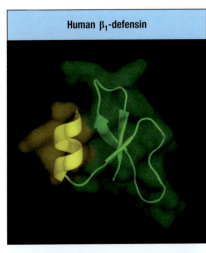

Human β_1-defensin

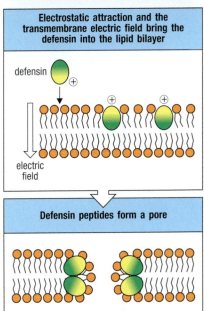

Electrostatic attraction and the transmembrane electric field bring the defensin into the lipid bilayer

defensin

electric field

Defensin peptides form a pore

—the plant *Arabidopsis thaliana* produces 13 and the fruitfly *Drosophila melanogaster* at least 15. Human Paneth cells make as many as 21 different defensins, many of which are encoded by a cluster of genes on chromosome 8.

Three subfamilies of defensins—α-, β-, and θ-defensins—are distinguished on the basis of amino acid sequence, and each family has members with distinct activities, some being active against Gram-positive bacteria and some against Gram-negative bacteria, while others are specific for fungal pathogens. All the antimicrobial peptides, including the defensins, are generated by proteolytic processing from inactive **propeptides** (Fig. 2.11). In humans, developing neutrophils produce **α-defensins** by the processing of an initial propeptide of about 90 amino acids by cellular proteases to remove an anionic propiece, generating a mature cationic defensin that is stored in so-called **primary granules**. The primary granules of neutrophils are specialized membrane-enclosed vesicles, rather similar to lysosomes, that contain a number of other antimicrobial agents as well as defensins. We will explain in Chapter 3 how these primary granules within neutrophils are induced to fuse with phagocytic vesicles (phagosomes) after the cell has engulfed a pathogen, helping to kill the microbe. The Paneth cells of the gut constitutively produce α-defensins, called **cryptdins**, which are processed by proteases such as the metalloprotease matrilysin in mice, or trypsin in humans, before being secreted into the gut lumen. The **β-defensins** lack the long propiece of α-defensins and are generally produced specifically in response to the presence of microbial products. β-Defensins (and some α-defensins) are made by epithelia outside the gut, primarily in the respiratory and urogenital tracts, skin, and tongue. β-Defensins made by keratinocytes in the epidermis and by type II pneumocytes in the lungs are packaged into **lamellar bodies** (see Fig. 2.6), lipid-rich secretory organelles that release their contents into the extracellular space to form a watertight lipid sheet in the epidermis and the pulmonary surfactant layer in the lung. The θ-defensins arose in the primates, but the single human θ-defensin gene has been inactivated by a mutation.

The antimicrobial peptides belonging to the cathelicidin family lack the disulfide bonds that stabilize the defensins. Humans and mice have one cathelicidin gene, but some other mammals, including cattle and sheep, have several. Cathelicidins are made constitutively by neutrophils and macrophages, and are made in response to infection by keratinocytes in the skin and epithelial cells in the lungs and intestine. They are made as inactive propeptides composed of two linked domains and are processed before secretion (see Fig. 2.11). In neutrophils, the inactive cathelicidin propeptides are stored in another type of specialized cytoplasmic granule called **secondary granules**. Cathelicidin is activated by proteolytic cleavage only when primary and secondary granules are induced to fuse with phagosomes, where it is cleaved by **neutrophil elastase** that has been stored in primary granules. Cleavage separates the two domains, and the cleavage products either remain in the phagosome or are released from the neutrophil by exocytosis. The carboxy-terminal peptide is a cationic amphipathic peptide that disrupts membranes and is toxic to a wide range of microorganisms. The amino-terminal peptide is similar in structure to a protein called **cathelin**, an inhibitor of cathepsin L (a lysosomal enzyme involved in antigen processing and protein degradation), but its role in immune defense is unclear. In keratinocytes, cathelicidins, like β-defensins, are stored and processed in the lamellar bodies.

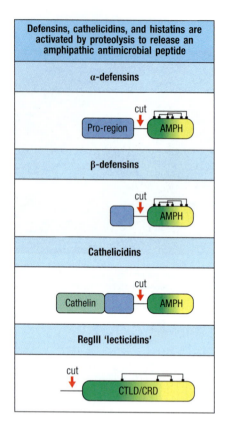

Defensins, cathelicidins, and histatins are activated by proteolysis to release an amphipathic antimicrobial peptide

α-defensins

cut

Pro-region | AMPH

β-defensins

cut

AMPH

Cathelicidins

cut

Cathelin | AMPH

RegIII 'lecticidins'

cut

CTLD/CRD

Fig. 2.11 Defensins, cathelicidins, and RegIII proteins are activated by proteolysis. When α- and β-defensins are first synthesized, they contain a signal peptide (not shown); a pro-region (blue), which is shorter in the β-defensins; and an amphipathic domain (AMPH, green–yellow); the pro-region represses the membrane-inserting properties of the amphipathic domain. After defensins are released from the cell, or into phagosomes, they undergo cleavage by proteases, which releases the amphipathic domain in active form. Newly synthesized cathelicidins contain a signal peptide, a cathelin domain, a short pro-region, and an amphipathic domain; they, too, are activated by proteolytic cleavage. RegIII contains a C-type lectin domain (CTLD), also known as a carbohydrate-recognition domain (CRD). After release of the signal peptide, further proteolytic cleavage of RegIII also regulates its antimicrobial activity.

A class of antimicrobial peptides called histatins are constitutively produced in the oral cavity by the parotid, sublingual, and submandibular glands. These short, histidine-rich, cationic peptides are active against pathogenic fungi such as *Cryptococcus neoformans* and *Candida albicans*. More recently histatins were found to promote the rapid wound healing that is typical in the oral cavity, but the mechanism of this effect is unclear.

Another type of bactericidal proteins made by epithelia is carbohydrate-binding proteins, or **lectins**. **C-type lectins** require calcium for the binding activity of their carbohydrate-recognition domain (CRD), which provides a variable interface for binding carbohydrate structures. C-type lectins of the RegIII family include several bactericidal proteins expressed by intestinal epithelium in humans and mice, comprising a family of 'lecticidins.' In mice, **RegIIIγ** is produced by Paneth cells and secreted into the gut, where it binds to peptidoglycans in bacterial cell walls and exerts direct bactericidal activity. Like other bactericidal peptides, RegIIIγ is produced in inactive form but is cleaved by the protease trypsin, which removes a short amino-terminal fragment to activate the bactericidal potential of RegIIIγ within the intestinal lumen (see Fig. 2.11). Human RegIIIα (also called **HIP/PAP** for hepatocarcinoma-intestine-pancreas/pancreatitis-associated protein) kills bacteria directly by forming a hexameric pore in the bacterial membrane (**Fig. 2.12**). RegIII family proteins preferentially kill Gram-positive bacteria, in which the peptidoglycan is exposed on the outer surface (see Fig. 2.9). In fact, the LPS of Gram-negative bacteria inhibits the pore-forming ability of RegIIIα, further enforcing the selectivity of RegIII proteins for Gram-positive bacteria.

Summary.

The mammalian immune response to invading organisms proceeds in three phases, beginning with immediate innate defenses, then the induced innate defenses, and finally adaptive immunity. The first phase of host defense consists of those mechanisms that are present and ready to resist an invader at any time. Epithelial surfaces provide a physical barrier against pathogen entry, but they also have other more specialized strategies. Mucosal surfaces have a protective barrier of mucus. Through particular cell-surface interactions, highly differentiated epithelia protect against both microbial colonization and invasion. Defense mechanisms of epithelia include the prevention of pathogen adherence, secretion of antimicrobial enzymes and bactericidal peptides, and the flow caused by the actions of cilia. Antimicrobial peptides and the bactericidal lectins of the RegIII family are made as inactive proproteins that require

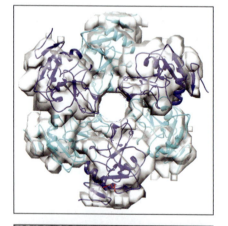

Fig. 2.12 Pore formation by human RegIIIα. Top: a model of the RegIIIα pore was generated by docking the human pro-RegIIIα structure (PDB ID: 1UV0), shown as individual purple and turquoise ribbon diagrams, into the cryo-electron microscopic map of the RegIIIα filament. LPS blocks the pore-forming activity of RegIIIα, explaining its selective bactericidal activity against Gram-positive but not Gram-negative bacteria. Bottom: electron microscopic images of RegIIIα pores assembled in the presence of lipid bilayers. Top structure courtesy of L. Hooper.

a proteolytic step to complete their activation, whereupon they become capable of killing microbes by forming pores in the microbial cell membranes. The actions of antimicrobial enzymes and peptides described in this section often involve binding to unique glycan/carbohydrate structures on the microbe. Thus, these soluble molecular defenses are both pattern recognition receptors and effector molecules at the same time, representing the simplest form of innate immunity.

The complement system and innate immunity.

When a pathogen breaches the host's epithelial barriers and initial antimicrobial defenses, it next encounters a major component of innate immunity known as the complement system, or **complement**. Complement is a collection of soluble proteins present in blood and other body fluids. It was discovered in the 1890s by **Jules Bordet** as a heat-labile substance in normal plasma whose activity could 'complement' the bactericidal activity of immune sera. Part of the process is **opsonization**, which refers to coating a pathogen with antibodies and/or complement proteins so that it can be more readily taken up and destroyed by phagocytic cells. Although complement was first discovered as an effector arm of the antibody response, we now understand that it originally evolved as part of the innate immune system and that it still provides protection early in infection, in the absence of antibodies, through more ancient pathways of complement activation.

The complement system is composed of more than 30 different plasma proteins, which are produced mainly by the liver. In the absence of infection, these proteins circulate in an inactive form. In the presence of pathogens or of antibody bound to pathogens, the complement system becomes 'activated.' Particular **complement proteins** interact with each other to form several different pathways of complement activation, all of which have the final outcome of killing the pathogen, either directly or by facilitating its phagocytosis, and inducing inflammatory responses that help to fight infection. There are three pathways of **complement activation**. As the antibody-triggered pathway of complement activation was discovered first, this became known as the classical pathway of complement activation. The next to be discovered was called the alternative pathway, which can be activated by the presence of the pathogen alone; and the most recently discovered is the lectin pathway, which is activated by lectin-type proteins that recognize and bind to carbohydrates on pathogen surfaces.

We learned in Section 2-4 that proteolysis can be used as a means of activating antimicrobial proteins. In the complement system, activation by proteolysis is inherent, with many of the complement proteins being proteases that successively cleave and activate one another. The proteases of the complement system are synthesized as inactive pro-enzymes, or **zymogens**, which become enzymatically active only after proteolytic cleavage, usually by another complement protein. The complement pathways are triggered by proteins that act as pattern recognition receptors to detect the presence of pathogens. This detection activates an initial zymogen, triggering a cascade of proteolysis in which complement zymogens are activated sequentially, each becoming an active protease that cleaves and activates many molecules of the next zymogen in the pathway, amplifying the signal as the cascade proceeds. This results in activation of three distinct effector pathways—**inflammation**, **phagocytosis**, and **membrane attack**—that help eliminate the pathogen. In this way, the detection of even a small number of pathogens produces a rapid response that is greatly amplified at each step. This overall scheme for complement is shown in **Fig. 2.13**.

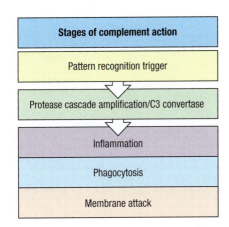

Fig. 2.13 The complement system proceeds in distinct phases in the elimination of microbes. Proteins that can distinguish self from microbial surfaces (yellow box) activate a proteolytic amplification cascade that ends in the formation of the critical enzymatic activity (green box) of C3 convertase, a family of proteases. This activity is the gateway to three effector arms of complement that produce inflammation (purple), enhance phagocytosis of the microbe (blue), and lyse microbial membranes (pink). We will use this color scheme in the figures throughout this chapter to illustrate which activity each complement protein serves.

Functional protein classes in the complement system	
Binding to antigen:antibody complexes and pathogen surfaces	C1q
Binding to carbohydrate structures such as mannose or GlcNAc on microbial surfaces	MBL Ficolins Properdin (factor P)
Activating enzymes*	C1r C1s C2a Bb D MASP-1 MASP-2 MASP-3
Surface-binding proteins and opsonins	C4b C3b
Peptide mediators of inflammation	C5a C3a C4a
Membrane-attack proteins	C5b C6 C7 C8 C9
Complement receptors	CR1 CR2 CR3 CR4 CRIg
Complement-regulatory proteins	C1INH C4BP CR1/CD35 MCP/CD46 DAF/CD55 H I P CD59

Fig. 2.14 Functional protein classes in the complement system. *In this book, C2a is used to denote the larger, active fragment of C2.

MOVIE 2.1

Nomenclature for complement proteins can seem confusing, so we will start by explaining their names. The first proteins discovered belong to the classical pathway, and they are designated by the letter C followed by a number. The native complement proteins—such as the inactive zymogens—have a simple number designation, for example, C1 and C2. Unfortunately, they were named in the order of their discovery rather than the sequence of reactions. The reaction sequence in the classical pathway, for example, is C1, C4, C2, C3, C5, C6, C7, C8, and C9 (note that not all of these are proteases). Products of cleavage reactions are designated by adding a lowercase letter as a suffix. For example, cleavage of C3 produces a small protein fragment called C3a and the remaining larger fragment, C3b. By convention, the larger fragment for other factors is designated by the suffix b, with one exception. For C2, the larger fragment was named **C2a** by its discoverers, and this system has been maintained in the literature, so we preserve it here. Another exception is the naming of C1q, C1r, and C1s: these are not cleavage products of C1 but are distinct proteins that together comprise C1. The proteins of the alternative pathway were discovered later and are designated by different capital letters, for example, factor B and factor D. Their cleavage products are also designated by the addition of lowercase a and b: thus, the large fragment of B is called Bb and the small fragment Ba. Activated complement components are sometimes designated by a horizontal line, for example, $\overline{C2a}$; however, we will not use this convention. All the components of the complement system are listed in **Fig. 2.14**.

Besides acting in innate immunity, complement also influences adaptive immunity. Opsonization of pathogens by complement facilitates their uptake by phagocytic antigen-presenting cells that express complement receptors; this enhances the presentation of pathogen antigens to T cells, which we discuss in more detail in Chapter 6. B cells express receptors for complement proteins that enhance their responses to complement-coated antigens, as we describe later in Chapter 10. In addition, several of the complement fragments can act to influence cytokine production by antigen-presenting cells, thereby influencing the direction and extent of the subsequent adaptive immune response, as we describe in Chapter 11.

2-5 The complement system recognizes features of microbial surfaces and marks them for destruction by coating them with C3b.

Fig. 2.15 gives a highly simplified preview of the initiation mechanisms and outcomes of complement activation. The three pathways of complement activation are initiated in different ways. The **lectin pathway** is initiated by soluble carbohydrate-binding proteins—mannose-binding lectin (MBL) and the ficolins—that bind to particular carbohydrate structures on microbial surfaces. Specific proteases, called MBL-associated serine proteases (MASPs), that associate with these recognition proteins then trigger the cleavage of complement proteins and activation of the pathway. The **classical pathway** is initiated when the complement component C1, which comprises a recognition protein (C1q) associated with proteases (C1r and C1s), either recognizes a microbial surface directly or binds to antibodies already bound to a pathogen. Finally, the **alternative pathway** can be initiated by spontaneous hydrolysis and activation of the complement component C3, which can then bind directly to microbial surfaces.

These three pathways converge at the central and most important step in complement activation. When any of the pathways interacts with a pathogen surface, the enzymatic activity of a **C3 convertase** is generated. There are various types of C3 convertase, depending on the complement pathway activated, but each is a multisubunit protein with protease activity that cleaves complement component 3 (C3). The C3 convertase is bound covalently to the pathogen surface, where it cleaves C3 to generate large amounts of **C3b**, the main effector molecule of the complement system; and **C3a**, a small peptide that binds to

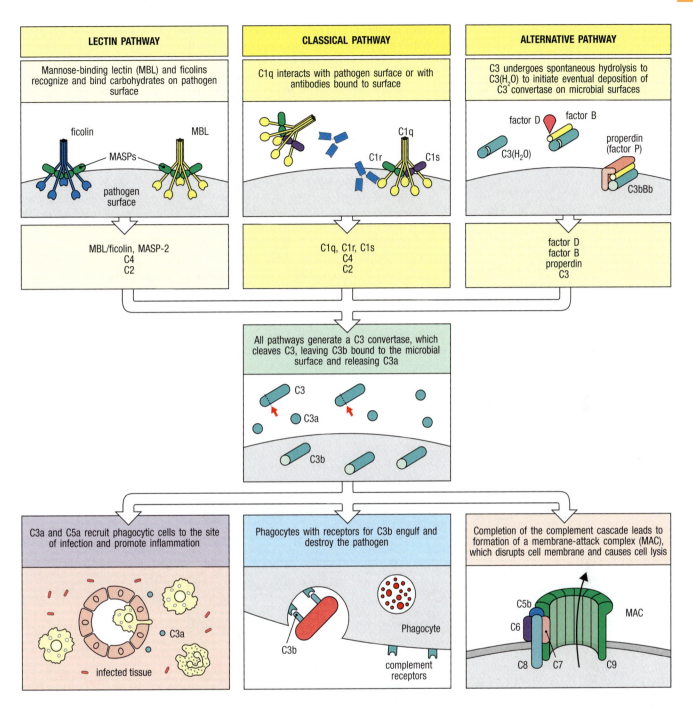

Fig. 2.15 Complement is a system of soluble pattern recognition receptors and effector molecules that detect and destroy microorganisms. The pathogen-recognition mechanisms of the three complement-activation pathways are shown in the top row, along with the complement components used in the proteolytic cascades leading to formation of a C3 convertase. This enzyme activity cleaves complement component C3 into the small soluble protein C3a and the larger component C3b, which becomes covalently bound to the pathogen surface (middle row). The components are listed by biochemical function in Fig. 2.14 and are described in detail in later figures. The lectin pathway of complement activation (top left) is triggered by the binding of mannose-binding lectin (MBL) or ficolins to carbohydrate residues in microbial cell walls and capsules. The classical pathway (top center) is triggered by binding of C1 either to the pathogen surface or to antibody bound to the pathogen. In the alternative pathway (top right), soluble C3 undergoes spontaneous hydrolysis in the fluid phase, generating C3(H₂O), which is augmented by the action of factors B, D, and P (properdin). All pathways thus converge on the formation of C3b bound to a pathogen and lead to all of the effector activities of complement, which are shown in the bottom row. C3b bound to a pathogen acts as an opsonin, enabling phagocytes that express receptors for C3b to ingest the complement-coated microbe more easily (bottom center). C3b can also bind to C3 convertases to produce another activity, a C5 convertase (detail not shown here), which cleaves C5 to C5a and C5b. C5b triggers the late events of the complement pathway in which the terminal components of complement—C6 to C9—assemble into a membrane-attack complex (MAC) that can damage the membrane of certain pathogens (bottom right). C3a and C5a act as chemoattractants that recruit immune-system cells to the site of infection and cause inflammation (bottom left).

specific receptors and helps induce inflammation. Cleavage of C3 is the critical step in complement activation and leads directly or indirectly to all the effector activities of the complement system (see Fig. 2.15). C3b binds covalently to the microbial surface and acts as an opsonin, enabling phagocytes that carry receptors for complement to take up and destroy the C3b-coated microbe. Later in the chapter, we will describe the different complement receptors that bind C3b that are involved in this function of complement and how C3b is degraded by a serum protease into inactive smaller fragments called **C3f** and **C3dg**. C3b can also bind to the C3 convertases produced by the classical and lectin pathways and form another multisubunit enzyme, the **C5 convertase**. This cleaves C5, liberating the highly inflammatory peptide **C5a** and generating **C5b**. C5b initiates the 'late' events of complement activation, in which additional complement proteins interact with C5b to form a **membrane-attack complex (MAC)** on the pathogen surface, creating a pore in the cell membrane that leads to cell lysis (see Fig. 2.15, bottom right).

The key feature of C3b is its ability to form a covalent bond with microbial surfaces, which allows the innate recognition of microbes to be translated into effector responses. Covalent bond formation is due to a highly reactive thioester bond that is hidden inside the folded C3 protein and cannot react until C3 is cleaved. When C3 convertase cleaves C3 and releases the C3a fragment, large conformational changes occur in C3b that allow the thioester bond to react with a hydroxyl or amino group on the nearby microbial surface (**Fig. 2.16**). If no bond is made, the thioester is rapidly hydrolyzed, inactivating C3b, which is one way the alternative pathway is inhibited in healthy individuals. As we will see below, some of the individual components of C3 and C5 convertases differ between the various complement pathways; the components that are different are listed in **Fig. 2.17**.

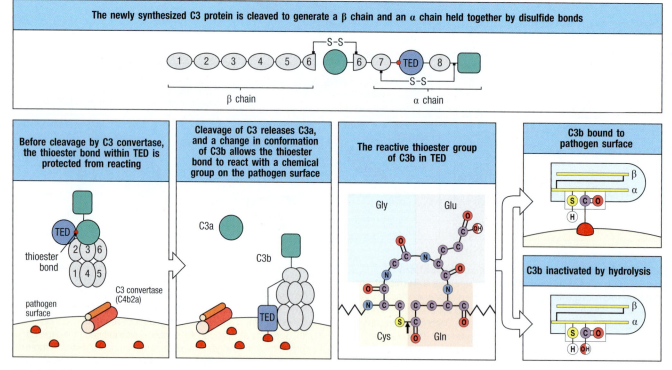

Fig. 2.16 C3 convertase activates C3 for covalent bonding to microbial surfaces by cleaving it into C3a and C3b and exposing a highly reactive thioester bond in C3b. Top panel: C3 in blood plasma consists of an α chain and a β chain (formed by proteolytic processing from the native C3 polypeptide) held together by a disulfide bond. The thioester-containing domain (TED) of the α chain contains a potentially highly reactive thioester bond (red spot). Bottom left panels: cleavage by C3 convertase (the lectin pathway convertase C4b2a is shown here) and release of C3a from the amino terminus of the α chain causes a conformational change in C3b that exposes the thioester bond. This can now react with hydroxyl or amino groups on molecules on microbial surfaces, covalently bonding C3b to the surface. Bottom right panels: schematic view of the thioester reaction. If a bond is not made with a microbial surface, the thioester is rapidly hydrolyzed (that is, cleaved by water), rendering C3b inactive.

Pathways leading to such potent inflammatory and destructive effects—and which have a series of built-in amplification steps—are potentially dangerous and must be tightly regulated. One important safeguard is that the key activated complement components are rapidly inactivated unless they bind to the pathogen surface on which their activation was initiated. There are also several points in the pathway at which regulatory proteins act to prevent the activation of complement on the surfaces of healthy host cells, thereby protecting them from accidental damage, as we shall see later in the chapter. Complement can, however, be activated by dying cells, such as those at sites of ischemic injury, and by cells undergoing apoptosis, or programmed cell death. In these cases, the complement coating helps phagocytes dispose of the dead and dying cells neatly, thus limiting the risk of cell contents being released and triggering an autoimmune response (discussed in Chapter 15).

Having introduced some of the main complement components, we are ready for a more detailed account of the three pathways. To help indicate the types of functions carried out by each of the complement components in the tables throughout the rest of the chapter, we will use the color code introduced in Fig. 2.13 and Fig. 2.14: yellow for recognition and activation, green for amplification, purple for inflammation, blue for phagocytosis, and pink for membrane attack.

2-6 The lectin pathway uses soluble receptors that recognize microbial surfaces to activate the complement cascade.

Microorganisms typically bear on their surface repeating patterns of molecular structures, known generally as pathogen-associated molecular patterns (PAMPs). The cell walls of Gram-positive and Gram-negative bacteria, for example, are composed of a matrix of proteins, carbohydrates, and lipids in a repetitive array (see Fig. 2.9). The lipoteichoic acids of Gram-positive bacterial cell walls and the lipopolysaccharide of the outer membrane of Gram-negative bacteria are not present on animal cells and are important in the recognition of bacteria by the innate immune system. Similarly, the glycans of yeast surface proteins commonly terminate in mannose residues rather than the sialic acid residues (N-acetylneuraminic acid) that terminate the glycans of vertebrate cells (Fig. 2.18). The lectin pathway uses these features of microbial surfaces to detect and respond to pathogens.

C3 convertase	
Lectin pathway	C4b2a
Classical pathway	C4b2a
Alternative pathway	C3bBb
Fluid phase	$C3(H_2O)Bb$

C5 convertase	
Lectin pathway	C4b2a3b
Classical pathway	C4b2a3b
Alternative pathway	$C3b_2Bb$

Fig. 2.17 C3 and C5 convertases of the complement pathways. Note the C5 convertase of the alternative pathway consists of two C3b subunits and one Bb subunit.

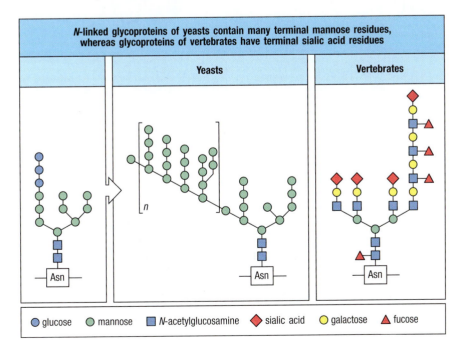

	N-linked glycoproteins of yeasts contain many terminal mannose residues, whereas glycoproteins of vertebrates have terminal sialic acid residues	
	Yeasts	Vertebrates

Asn — Asn — Asn

● glucose ● mannose ■ *N*-acetylglucosamine ◆ sialic acid ○ galactose ▲ fucose

Fig. 2.18 The carbohydrate side chains on yeast and vertebrate glycoproteins are terminated with different patterns of sugars. *N*-linked glycosylation in fungi and animals is initiated by the addition of the same precursor oligosaccharide, Glc_3-Man_9-$GlcNAc_2$ (left panel), to an asparagine residue. In many yeasts this is processed to high-mannose glycans (middle panel). In contrast, in vertebrates, the initial glycan is trimmed and processed, and the *N*-linked glycoproteins of vertebrates have terminal sialic acid residues (right panel).

The lectin pathway can be triggered by any of four different pattern recognition receptors that circulate in blood and extracellular fluids and recognize carbohydrates on microbial surfaces. The first such receptor to be discovered was **mannose-binding lectin (MBL)**, which is shown in **Fig. 2.19**, and which is synthesized in the liver. MBL is an oligomeric protein built up from a monomer that contains an amino-terminal collagen-like domain and a carboxy-terminal C-type lectin domain (see Section 2-4). Proteins of this type are called **collectins**. MBL monomers assemble into trimers through the formation of a triple helix by their collagen-like domains. Trimers then assemble into oligomers by disulfide bonding between the cysteine-rich collagen domains. The MBL present in the blood is composed of two to six trimers, with the major forms of human MBL being trimers and tetramers. A single carbohydrate-recognition domain of MBL has a low affinity for mannose, fucose, and N-acetylglucosamine (GlcNAc) residues, which are common on microbial glycans, but does not bind sialic acid residues, which terminate vertebrate glycans. Thus, multimeric MBL has high total binding strength, or **avidity**, for repetitive carbohydrate structures on a wide variety of microbial surfaces, including Gram-positive and Gram-negative bacteria, mycobacteria, yeasts, and some viruses and parasites, while not interacting with host cells. MBL is present at low concentrations in the plasma of most individuals, but in the presence of infection, its production is increased during the **acute-phase response**. This is part of the induced phase of the innate immune response and is discussed in Chapter 3.

The other three pathogen-recognition molecules used by the lectin pathway are known as **ficolins**. Although related in overall shape and function to MBL, they have a fibrinogen-like domain, rather than a lectin domain, attached to

Fig. 2.19 Mannose-binding lectin and ficolins form complexes with serine proteases and recognize particular carbohydrates on microbial surfaces. Mannose-binding lectin (MBL) (left panels) is an oligomeric protein in which two to six clusters of carbohydrate-binding heads arise from a central stalk formed from the collagen-like tails of the MBL monomers. An MBL monomer is composed of a collagen region (red), an α-helical neck region (blue), and a carbohydrate-recognition domain (yellow). Three MBL monomers associate to form a trimer, and between two and six trimers assemble to form a mature MBL molecule (bottom left panel). An MBL molecule associates with MBL-associated serine proteases (MASPs). MBL binds to bacterial surfaces that display a particular spatial arrangement of mannose or fucose residues. The ficolins (right panels) resemble MBL in their overall structure, are associated with MASP-1 and MASP-2, and can activate C4 and C2 after binding to carbohydrate molecules present on microbial surfaces. The carbohydrate-binding domain of ficolins is a fibrinogen-like domain, rather than the lectin domain present in MBL.

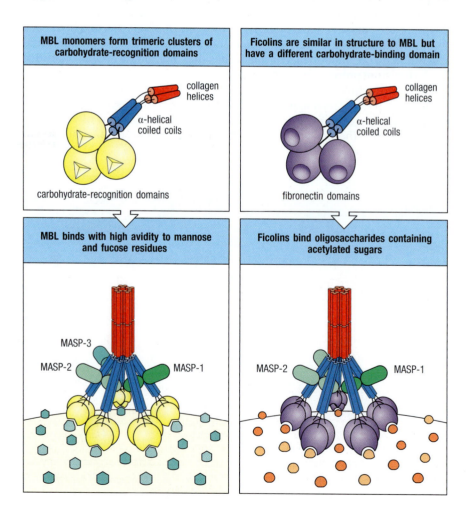

MBL monomers form trimeric clusters of carbohydrate-recognition domains

collagen helices

α-helical coiled coils

carbohydrate-recognition domains

Ficolins are similar in structure to MBL but have a different carbohydrate-binding domain

collagen helices

α-helical coiled coils

fibronectin domains

MBL binds with high avidity to mannose and fucose residues

MASP-3

MASP-2

MASP-1

Ficolins bind oligosaccharides containing acetylated sugars

MASP-2

MASP-1

the collagen-like stalk (see Fig. 2.19). The fibrinogen-like domain gives ficolins a general specificity for oligosaccharides containing acetylated sugars, but it does not bind mannose-containing carbohydrates. Humans have three ficolins: L-ficolin (ficolin-2), M-ficolin (ficolin-1), and H-ficolin (ficolin-3). L- and H-ficolin are synthesized by the liver and circulate in the blood; M-ficolin is synthesized and secreted by lung and blood cells.

MBL in plasma forms complexes with the **MBL-associated serine proteases MASP-1**, **MASP-2**, and **MASP-3**, which bind MBL as inactive zymogens. When MBL binds to a pathogen surface, a conformational change occurs in MASP-1 that enables it to cleave and activate a MASP-2 molecule in the same MBL complex. Activated MASP-2 can then cleave complement components C4 and C2 (**Fig. 2.20**). Like MBL, ficolins form oligomers that make a complex with MASP-1 and MASP-2, which similarly activate complement upon recognition of a microbial surface by the ficolin. **C4**, like C3, contains a buried thioester bond. When MASP-2 cleaves C4, it releases C4a, allowing a conformational change in C4b that exposes the reactive thioester as described for C3b (see Fig. 2.16). C4b bonds covalently via this thioester to the microbial surface nearby, where it then binds one molecule of C2 (see Fig. 2.20). C2 is cleaved by MASP-2, producing C2a, an active serine protease that remains bound to C4b to form **C4b2a**, which is the C3 convertase of the lectin pathway. (Remember, C2a is the exception in complement nomenclature.) C4b2a now cleaves many molecules of C3 into C3a and C3b. The C3b fragments bond covalently to the nearby pathogen surface, and the released C3a initiates a local inflammatory response. The complement-activation pathway initiated by ficolins proceeds like the MBL lectin pathway (see Fig. 2.20).

Individuals deficient in MBL or MASP-2 experience substantially more respiratory infections by common extracellular bacteria during early childhood, indicating the importance of the lectin pathway for host defense. This susceptibility illustrates the particular importance of innate defense mechanisms in early childhood, when adaptive immune responses are not yet fully developed

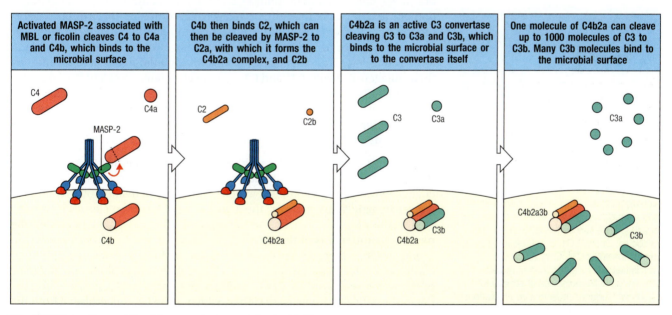

Fig. 2.20 The actions of the C3 convertase result in the binding of large numbers of C3b molecules to the pathogen surface. Binding of mannose-binding lectin or ficolins to their carbohydrate ligands on microbial surfaces induces MASP-1 to cleave and activate the serine protease MASP-2. MASP-2 then cleaves C4, exposing the thioester bond in C4b that allows it to react covalently with the pathogen surface. C4b then binds C2, making C2 susceptible to cleavage by MASP-2 and thus generating the C3 convertase C4b2a. C2a is the active protease component of the C3 convertase, and cleaves many molecules of C3 to produce C3b, which binds to the pathogen surface, and C3a, an inflammatory mediator. The covalent attachment of C3b and C4b to the pathogen surface is important in confining subsequent complement activity to pathogen surfaces.

Panel labels within figure:

Activated MASP-2 associated with MBL or ficolin cleaves C4 to C4a and C4b, which binds to the microbial surface

C4b then binds C2, which can then be cleaved by MASP-2 to C2a, with which it forms the C4b2a complex, and C2b

C4b2a is an active C3 convertase cleaving C3 to C3a and C3b, which binds to the microbial surface or to the convertase itself

One molecule of C4b2a can cleave up to 1000 molecules of C3 to C3b. Many C3b molecules bind to the microbial surface

but the maternal antibodies transferred across the placenta and present in the mother's milk are gone. Other members of the collectin family are the **surfactant proteins A** and **D** (**SP-A** and **SP-D**), which are present in the fluid that bathes the epithelial surfaces of the lung. There they coat the surfaces of pathogens, making them more susceptible to phagocytosis by macrophages that have left the subepithelial tissues to enter the alveoli. Because SP-A and SP-D do not associate with MASPs, they do not activate complement.

We have used MBL here as our prototype activator of the lectin pathway, but the ficolins are more abundant than MBL in plasma and so may be more important in practice. L-ficolin recognizes acetylated sugars such as GlcNAc and *N*-acetylgalactosamine (GalNAc), and particularly recognizes lipoteichoic acid, a component of the cell walls of Gram-positive bacteria that contains GalNAc. It can also activate complement after binding to a variety of capsulated bacteria. M-ficolin also recognizes acetylated sugar residues; H-ficolin shows a more restricted binding specificity, for D-fucose and galactose, and has only been linked to activity against the Gram-positive bacterium *Aerococcus viridans*, a cause of bacterial endocarditis.

2-7   The classical pathway is initiated by activation of the C1 complex and is homologous to the lectin pathway.

In its overall scheme, the classical pathway is similar to the lectin pathway, except that it uses a pathogen sensor known as the **C1 complex**, or **C1**. Because C1 interacts directly with some pathogens but can also interact with antibodies, C1 allows the classical pathway to function both in innate immunity, which we describe now, and in adaptive immunity, which we examine in more detail in Chapter 10.

Like the MBL–MASP complex, the C1 complex is composed of a large subunit (**C1q**), which acts as the pathogen sensor, and two serine proteases (**C1r** and **C1s**), which are initially in their inactive form (**Fig. 2.21**). C1q is a hexamer of trimers, composed of monomers that contain an amino-terminal globular domain and a carboxy-terminal collagen-like domain. The trimers assemble through interactions of the collagen-like domains, bringing the globular domains together to form a globular head. Six of these trimers assemble to form a complete C1q molecule, which has six globular heads held together by their collagen-like tails. C1r and C1s are closely related to MASP-2, and somewhat more distantly related to MASP-1 and MASP-3; all five enzymes are likely to have evolved from the duplication of a gene for a common precursor. C1r and C1s interact noncovalently and form tetramers that fold into the arms of C1q, with at least part of the C1r:C1s complex being external to C1q.

The recognition function of C1 resides in the six globular heads of C1q. When two or more of these heads interact with a ligand, this causes a conformational change in the C1r:C1s complex, which leads to the activation of an autocatalytic enzymatic activity in C1r; the active form of C1r then cleaves its associated C1s to generate an active serine protease. The activated C1s acts on the next two components of the classical pathway, C4 and C2. C1s cleaves C4 to produce C4b, which binds covalently to the pathogen surface as described earlier for the lectin pathway (see Fig. 2.20). C4b then also binds one molecule of C2, which is cleaved by C1s to produce the serine protease C2a. This produces the active C3 convertase C4b2a, which is the C3 convertase of both the lectin and the classical pathways. However, because it was first discovered as part of the classical pathway, C4b2a is often known as the **classical C3 convertase** (see Fig. 2.17). The proteins involved in the classical pathway, and their active forms, are listed in **Fig. 2.22**.

C1q can attach itself to the surface of pathogens in several different ways. One is by binding directly to surface components on some bacteria, including

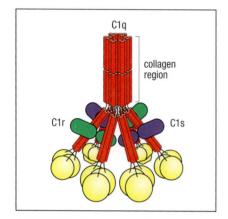

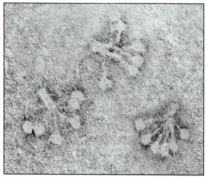

Fig. 2.21 The first protein in the classical pathway of complement activation is C1, which is a complex of C1q, C1r, and C1s. As shown in the micrograph and drawing, C1q is composed of six identical subunits with globular heads (yellow) and long collagen-like tails (red); it has been described as looking like "a bunch of tulips." The tails combine to bind to two molecules each of C1r and C1s, forming the C1 complex C1q:C1r$_2$:C1s$_2$. The heads can bind to the constant regions of immunoglobulin molecules or directly to the pathogen surface, causing a conformational change in C1r, which then cleaves and activates the C1s zymogen (proenzyme). The C1 complex is similar in overall structure to the MBL–MASP complex, and it has an identical function, cleaving C4 and C2 to form the C3 convertase C4b2a (see Fig. 2.20). Photograph (×500,000) courtesy of K.B.M. Reid.

Proteins of the classical pathway of complement activation		
Native component	**Active form**	**Function of the active form**
C1 (C1q: C1r$_2$:C1s$_2$)	C1q	Binds directly to pathogen surfaces or indirectly to antibody bound to pathogens, thus allowing autoactivation of C1r
	C1r	Cleaves C1s to active protease
	C1s	Cleaves C4 and C2
C4	C4b	Covalently binds to pathogen and opsonizes it. Binds C2 for cleavage by C1s
	C4a	Peptide mediator of inflammation (weak activity)
C2	C2a	Active enzyme of classical pathway C3/C5 convertase: cleaves C3 and C5
	C2b	Precursor of vasoactive C2 kinin
C3	C3b	Binds to pathogen surface and acts as opsonin. Initiates amplification via the alternative pathway. Binds C5 for cleavage by C2a
	C3a	Peptide mediator of inflammation (intermediate activity)

Fig. 2.22 The proteins of the classical pathway of complement activation.

certain proteins of bacterial cell walls and polyanionic structures such as the lipoteichoic acid on Gram-positive bacteria. A second is through binding to C-reactive protein, an acute-phase protein in human plasma that binds to phosphocholine residues in bacterial surface molecules such as pneumococcal C polysaccharide—hence the name C-reactive protein. We discuss the acute-phase proteins in detail in Chapter 3. However, a main function of C1q in an immune response is to bind to the constant, or Fc, regions of antibodies (see Section 1-9) that have bound pathogens via their antigen-binding sites. C1q thus links the effector functions of complement to recognition provided by adaptive immunity. This might seem to limit the usefulness of C1q in fighting the first stages of an infection, before the adaptive immune response has generated pathogen-specific antibodies. However, some antibodies, called **natural antibodies**, are produced by the immune system in the apparent absence of infection. These antibodies have a low affinity for many microbial pathogens and are highly cross-reactive, recognizing common membrane constituents such as phosphocholine and even recognizing some antigens of the body's own cells (that is, self antigens). Natural antibodies may be produced in response to commensal microbiota or to self antigens, but do not seem to be the consequence of an adaptive immune response to infection by pathogens. Most natural antibody is of the isotype, or class, known as IgM (see Sections 1-9 and 1-20) and represents a considerable amount of the total IgM circulating in humans. IgM is the class of antibody most efficient at binding C1q, making natural antibodies an effective means of activating complement on microbial surfaces immediately after infection and leading to the clearance of bacteria such as *Streptococcus pneumoniae* (the pneumococcus) before they become dangerous.

2-8 Complement activation is largely confined to the surface on which it is initiated.

We have seen that both the lectin and the classical pathways of complement activation are initiated by proteins that bind to pathogen surfaces. During the triggered enzyme cascade that follows, it is important that activating events are confined to this same site, so that C3 activation also occurs on the surface of the pathogen and not in the plasma or on host-cell surfaces. This is achieved principally by the covalent binding of C4b to the pathogen surface. In innate immunity, C4 cleavage is catalyzed by a ficolin or MBL complex that is bound to the pathogen surface, and so the C4b cleavage product can bind adjacent proteins or carbohydrates on the pathogen surface. If C4b does not rapidly form this bond, the thioester bond is cleaved by reaction with water and C4b is irreversibly inactivated. This helps to prevent C4b from diffusing from its site of activation on the microbial surface and becoming attached to healthy host cells.

C2 becomes susceptible to cleavage by C1s only when it is bound by C4b, and the active C2a serine protease is thereby also confined to the pathogen surface, where it remains associated with C4b, forming the C3 convertase C4b2a. Cleavage of C3 to C3a and C3b is thus also confined to the surface of the pathogen. Like C4b, C3b is inactivated by hydrolysis unless its exposed thioester rapidly makes a covalent bond (see Fig. 2.16), and it therefore opsonizes only the surface on which complement activation has taken place. Opsonization by C3b is more effective when antibodies are also bound to the pathogen surface, as phagocytes have receptors for both complement and Fc receptors that bind the Fc region of antibody (see Sections 1-20 and 10-20). Because the reactive forms of C3b and C4b are able to form a covalent bond with any adjacent protein or carbohydrate, when complement is activated by bound antibody, a proportion of the reactive C3b or C4b will become linked to the antibody molecules themselves. Antibody that is chemically cross-linked to complement is likely the most efficient trigger for phagocytosis.

2-9   The alternative pathway is an amplification loop for C3b formation that is accelerated by properdin in the presence of pathogens.

Although probably the most ancient of the complement pathways, the alternative pathway is so named because it was discovered as a second, or 'alternative,' pathway for complement activation after the classical pathway had been defined. Its key feature is its ability to be spontaneously activated. It has a unique C3 convertase, the **alternative pathway C3 convertase**, that differs from the C4b2a convertase of the lectin or classical pathways (see Fig. 2.17). The alternative pathway C3 convertase is composed of C3b itself bound to Bb, which is a cleavage fragment of the plasma protein **factor B**. This C3 convertase, designated **C3bBb**, has a special place in complement activation because, by producing C3b, it can generate more of itself. This means that once some C3b has been formed, by whichever pathway, the alternative pathway can act as an amplification loop to increase C3b production rapidly.

The alternative pathway can be activated in two different ways. The first is by the action of the lectin or classical pathway. C3b generated by either of these pathways and covalently linked to a microbial surface can bind factor B (Fig. 2.23). This alters the conformation of factor B, enabling a plasma protease

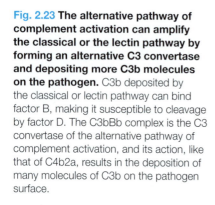

Fig. 2.23 The alternative pathway of complement activation can amplify the classical or the lectin pathway by forming an alternative C3 convertase and depositing more C3b molecules on the pathogen. C3b deposited by the classical or lectin pathway can bind factor B, making it susceptible to cleavage by factor D. The C3bBb complex is the C3 convertase of the alternative pathway of complement activation, and its action, like that of C4b2a, results in the deposition of many molecules of C3b on the pathogen surface.

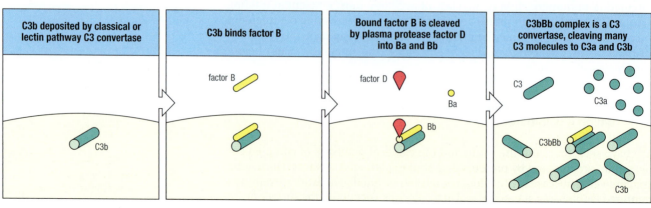

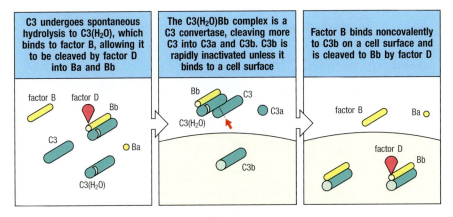

| C3 undergoes spontaneous hydrolysis to C3(H$_2$O), which binds to factor B, allowing it to be cleaved by factor D into Ba and Bb | The C3(H$_2$O)Bb complex is a C3 convertase, cleaving more C3 into C3a and C3b. C3b is rapidly inactivated unless it binds to a cell surface | Factor B binds noncovalently to C3b on a cell surface and is cleaved to Bb by factor D |

Fig. 2.24 The alternative pathway can be activated by spontaneous activation of C3. Complement component C3 hydrolyzes spontaneously in plasma to give C3(H$_2$O), which binds factor B and enables the bound factor B to be cleaved by factor D (first panel). The resulting 'soluble C3 convertase' cleaves C3 to give C3a and C3b, which can attach to host cells or pathogen surfaces (second panel). Covalently bound to the cell surface, C3b binds factor B; in turn, factor B is rapidly cleaved by factor D to Bb, which remains bound to C3b to form a C3 convertase (C3bBb), and Ba, which is released (third panel). This convertase functions in the alternative pathway as the C4b2a C3 convertase does in the lectin and classical pathways (see Fig. 2.17).

called **factor D** to cleave it into Ba and Bb. Bb remains stably associated with C3b, forming the C3bBb C3 convertase. The second way of activating the alternative pathway involves the spontaneous hydrolysis (known as '**tickover**') of the thioester bond in C3 to form C3(H$_2$O), as shown in **Fig. 2.24**. C3 is abundant in plasma, and tickover causes a steady, low-level production of C3(H$_2$O). This C3(H$_2$O) can bind factor B, which is then cleaved by factor D, producing a short-lived **fluid-phase C3 convertase, C3(H$_2$O)Bb**. Although formed in only small amounts by C3 tickover, fluid-phase C3(H$_2$O)Bb can cleave many molecules of C3 to C3a and C3b. Much of this C3b is inactivated by hydrolysis, but some attaches covalently via its thioester bond to the surfaces of any microbes present. C3b formed in this way is no different from C3b produced by the lectin or classical pathways and binds factor B, leading to the formation of C3 convertase and a stepping up of C3b production (see Fig. 2.23).

On their own, the alternative pathway C3 convertases C3bBb and C3(H$_2$O)Bb are very short-lived. They are, however, stabilized by binding the plasma protein **properdin (factor P)** (**Fig. 2.25**). Properdin is made by neutrophils and stored in secondary granules. It is released when neutrophils are activated by the presence of pathogens. Properdin may have some properties of a pattern recognition receptor, since it can bind to some microbial surfaces. Properdin-deficient patients are particularly susceptible to infections with *Neisseria meningitidis*, the main agent of bacterial meningitis. Properdin's ability to bind to bacterial surfaces may direct the activity of the alternative complement pathway to these pathogens, thus aiding their removal by phagocytosis. Properdin can also bind to mammalian cells that are undergoing apoptosis or have been damaged or modified by ischemia, viral infection, or antibody binding, leading to the deposition of C3b on these cells and facilitating their removal by phagocytosis. The distinctive components of the alternative pathway are listed in **Fig. 2.26**.

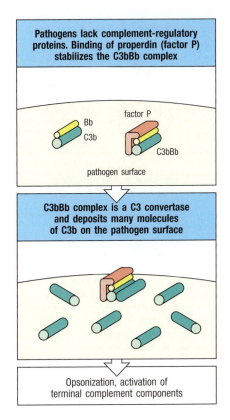

| Pathogens lack complement-regulatory proteins. Binding of properdin (factor P) stabilizes the C3bBb complex |

| C3bBb complex is a C3 convertase and deposits many molecules of C3b on the pathogen surface |

Opsonization, activation of terminal complement components

Fig. 2.25 Properdin stabilizes the alternative pathway C3 convertase on pathogen surfaces. Bacterial surfaces do not express complement-regulatory proteins and favor the binding of properdin (factor P), which stabilizes the C3bBb convertase. This convertase activity is the equivalent of C4b2a of the classical pathway. C3bBb then cleaves many more molecules of C3, coating the pathogen surface with bound C3b.

Fig. 2.26 The proteins of the alternative pathway of complement activation.

Proteins of the alternative pathway of complement activation		
Native component	**Active fragments**	**Function**
C3	C3b	Binds to pathogen surface; binds B for cleavage by D; C3bBb is a C3 convertase and C3b$_2$Bb is a C5 convertase
Factor B (B)	Ba	Small fragment of B, unknown function
	Bb	Bb is the active enzyme of the C3 convertase C3bBb and the C5 convertase C3b$_2$Bb
Factor D (D)	D	Plasma serine protease, cleaves B when it is bound to C3b to Ba and Bb
Properdin (P)	P	Plasma protein that binds to bacterial surfaces and stabilizes the C3bBb convertase

2-10   Membrane and plasma proteins that regulate the formation and stability of C3 convertases determine the extent of complement activation.

Several mechanisms ensure that complement activation will proceed only on the surface of a pathogen or on damaged host cells, and not on normal host cells and tissues. After initial complement activation by any pathway, the extent of amplification via the alternative pathway is critically dependent on the stability of the C3 convertase C3bBb. This stability is controlled by both positive and negative regulatory proteins. We have already described how properdin acts as a positive regulatory protein on foreign surfaces, such as those of bacteria or damaged host cells, by stabilizing C3bBb.

Several negative regulatory proteins, present in plasma and in host-cell membranes, protect healthy host cells from the injurious effects of inappropriate complement activation on their surfaces. These **complement-regulatory proteins** interact with C3b and either prevent the convertase from forming or promote its rapid dissociation (**Fig. 2.27**). For example, a membrane-attached protein known as **decay-accelerating factor** (**DAF** or **CD55**) competes with factor B for binding to C3b on the cell surface and can displace Bb from a convertase that has already formed. Convertase formation can also be prevented by cleaving C3b to an inactive derivative, **iC3b**. This is achieved by a plasma protease, **factor I**, in conjunction with C3b-binding proteins that act as cofactors, such as **membrane cofactor of proteolysis** (**MCP** or **CD46**), another host-cell membrane protein (see Fig. 2.27). Cell-surface complement receptor type 1 (**CR1**, also known as **CD35**) behaves similarly to DAF and MCP in that it inhibits C3 convertase formation and promotes the catabolism of C3b to inactive products, but it has a more limited tissue distribution. **Factor H** is another complement-regulatory protein in plasma that binds C3b, and like CR1, it is able to compete with factor B to displace Bb from the convertase; in addition, it acts as a cofactor for factor I. Factor H binds preferentially to C3b bound to vertebrate cells because it has an affinity for the sialic acid residues present on their cell surfaces (see Fig. 2.18). Thus, the amplification loop of the alternative pathway is allowed to proceed on the surface of a pathogen or on damaged host cells, but not on normal host cells or on tissues that express these negative regulatory proteins.

The C3 convertase of the classical and lectin pathways (C4b2a) is molecularly distinct from that of the alternative pathway (C3bBb). However, understanding of the complement system is simplified somewhat by recognition of the close evolutionary relationships between the different complement proteins (**Fig. 2.28**). Thus the complement zymogens, factor B and C2, are closely related proteins encoded by homologous genes located in tandem within the major

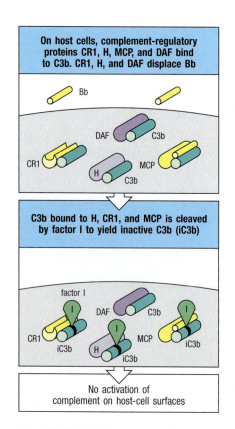

On host cells, complement-regulatory proteins CR1, H, MCP, and DAF bind to C3b. CR1, H, and DAF displace Bb

Bb

DAF   C3b

CR1   MCP
H
C3b

C3b bound to H, CR1, and MCP is cleaved by factor I to yield inactive C3b (iC3b)

factor I

DAF   C3b
I   I
CR1   MCP   iC3b
iC3b   H
iC3b

No activation of complement on host-cell surfaces

Fig. 2.27 Complement activation spares host cells, which are protected by complement-regulatory proteins. If C3bBb forms on the surface of host cells, it is rapidly inactivated by complement-regulatory proteins expressed by the host cell: complement receptor 1 (CR1), decay-accelerating factor (DAF), and membrane cofactor of proteolysis (MCP). Host-cell surfaces also favor the binding of factor H from plasma. CR1, DAF, and factor H displace Bb from C3b, and CR1, MCP, and factor H catalyze the cleavage of bound C3b by the plasma protease factor I to produce inactive C3b (known as iC3b).

Step in pathway	Protein serving function in pathway			Relationship
	Alternative	Lectin	Classical	
Initiating serine protease	D	MASP	C1s	Homologous (C1s and MASP)
Covalent binding to cell surface	C3b	C4b		Homologous
C3/C5 convertase	Bb	C2a		Homologous
Control of activation	CR1 H	CR1 C4BP		Identical Homologous
Opsonization	C3b			Identical
Initiation of effector pathway	C5b			Identical
Local inflammation	C5a, C3a			Identical
Stabilization	P	None		Unique

Fig. 2.28 **There is a close evolutionary relationship among the factors of the alternative, lectin, and classical pathways of complement activation.** Most of the factors are either identical or the homologous products of genes that have duplicated and then diverged in sequence. The proteins C4 and C3 are homologous and contain the unstable thioester bond by which their large fragments, C4b and C3b, bind covalently to membranes. The genes encoding proteins C2 and factor B are adjacent in the MHC region of the genome and arose by gene duplication. The regulatory proteins factor H, CR1, and C4BP share a repeat sequence common to many complement-regulatory proteins. The greatest divergence between the pathways is in their initiation: in the classical pathway the C1 complex binds either to certain pathogens or to bound antibody, and in the latter circumstance it serves to convert antibody binding into enzyme activity on a specific surface; in the lectin pathway, mannose-binding lectin (MBL) associates with a serine protease, activating MBL-associated serine protease (MASP), to serve the same function as C1r:C1s; in the alternative pathway this enzyme activity is provided by factor D.

histocompatibility complex (MHC) on human chromosome 6. Furthermore, their respective binding partners, C3 and C4, both contain thioester bonds that provide the means of covalently attaching the C3 convertases to a pathogen surface.

Only one component of the alternative pathway seems entirely unrelated to its functional equivalents in the classical and lectin pathways: the initiating serine protease, factor D. Factor D can also be singled out as the only activating protease of the complement system to circulate as an active enzyme rather than a zymogen. This is both necessary for the initiation of the alternative pathway (through the cleavage of factor B bound to spontaneously activated C3) and safe for the host, because factor D has no other substrate than factor B bound to C3b. This means that factor D finds its substrate only at pathogen surfaces and at a very low level in plasma, where the alternative pathway of complement activation can be allowed to proceed.

2-11   Complement developed early in the evolution of multicellular organisms.

The complement system was originally known only from vertebrates, but homologs of C3 and factor B and a prototypical 'alternative pathway' have been discovered in nonchordate invertebrates. This is not altogether surprising since C3, which is cleaved and activated by serine proteases, is evolutionarily related to the serine protease inhibitor α_2-macroglobulin, whose first appearance likely was in an ancestor to all modern vertebrates. The amplification loop of the alternative pathway also has an ancestral origin, as it is present in echinoderms (which include sea urchins and sea stars) and is based on a C3 convertase formed by the echinoderm homologs of C3 and factor B. These factors are expressed by phagocytic cells called **amoeboid coelomocytes** present in the coelomic fluid. Expression of C3 by these cells increases when bacteria are present. This simple system seems to function to opsonize bacterial cells and other foreign particles and facilitate their uptake by coelomocytes. C3 homologs in invertebrates are clearly related to each other. They all contain the distinctive thioester linkage and form a family of proteins, the **thioester proteins**, or **TEPs**. In the mosquito *Anopheles*, the production of protein TEP1 is induced in response to infection, and the protein may directly bind to bacterial surfaces to mediate phagocytosis of Gram-negative bacteria. Some form of C3 activity may even predate the evolution of the Bilateria—animals with bilateral symmetry, flatworms being the most primitive modern representatives—because genomic evidence of C3, factor B, and some later-acting complement components has been found in the Anthozoa (corals and sea anemones).

After its initial appearance, the complement system seems to have evolved by the acquisition of new activation pathways that allow specific targeting of microbial surfaces. The first to evolve was likely the ficolin pathway, which is present both in vertebrates and in some closely related invertebrates, such as the urochordates. Evolutionarily, the ficolins may predate the collectins, which are also first seen in the urochordates. Homologs of MBL and of the classical pathway complement component C1q, both collectins, have been identified in the genome of the ascidian urochordate *Ciona* (sea squirt). Two invertebrate homologs of mammalian MASPs also have been identified in *Ciona*, and it seems likely that they may be able to cleave and activate C3. Thus, the minimal complement system of the echinoderms appears to have been expanded in the urochordates by the recruitment of a specific activation system that may target C3 deposition onto microbial surfaces. This also suggests that when adaptive immunity evolved, much later, the ancestral antibody molecule used an already diversified C1q-like collectin member to activate the complement pathway, and that the complement activation system evolved further by use of this collectin and its associated MASPs to become the initiating components of the classical complement pathway, namely, C1q, C1r, and C1s.

2-12   Surface-bound C3 convertase deposits large numbers of C3b fragments on pathogen surfaces and generates C5 convertase activity.

We now return to the present-day complement system. The formation of C3 convertases is the point at which the three pathways of complement activation converge. The convertase of the lectin and classical pathways, C4b2a, and the convertase of the alternative pathway, C3bBb, initiate the same subsequent events—they cleave C3 to C3b and C3a. C3b binds covalently through its thioester bond to adjacent molecules on the pathogen surface; otherwise it is inactivated by hydrolysis. C3 is the most abundant complement protein in plasma, occurring at a concentration of 1.2 mg/ml, and up to 1000 molecules of C3b can bind in the vicinity of a single active C3 convertase (see Fig. 2.23). Thus, the main effect of complement activation is to deposit large quantities of C3b on the surface of the infecting pathogen, where the C3b forms a covalently bonded coat that can signal the ultimate destruction of the pathogen by phagocytes.

The next step in the complement cascade is the generation of the C5 convertases. C5 is a member of the same family of proteins as C3, C4, α_2-macroglobulin, and the thioester-containing proteins (TEPs) of invertebrates. C5 does not form an active thioester bond during its synthesis but, like C3 and C4, it is cleaved by a specific protease into C5a and C5b fragments, each of which exerts specific downstream actions that are important in propagating the complement cascade. In the classical and the lectin pathways, a C5 convertase is formed by the binding of C3b to C4b2a to yield **C4b2a3b**. The C5 convertase of the alternative pathway is formed by the binding of C3b to the C3bBb convertase to form **C3b$_2$Bb**. A C5 is captured by these C5 convertase complexes through binding to an acceptor site on C3b, and is thus rendered susceptible to cleavage by the serine protease activity of C2a or Bb. This reaction, which generates C5b and C5a, is much more limited than cleavage of C3, because C5 can be cleaved only when it binds to C3b that is in turn bound to C4b2a or C3bBb to form the active C5 convertase complex. Thus, complement activated by all three pathways leads to the binding of large numbers of C3b molecules on the surface of the pathogen, the generation of a more limited number of C5b molecules, and the release of C3a and a smaller amount of C5a (**Fig. 2.29**).

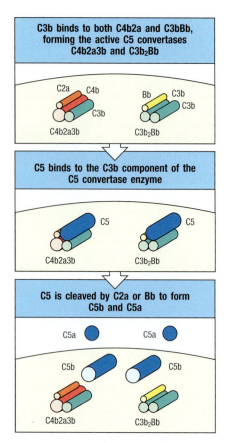

Fig. 2.29 Complement component C5 is cleaved when captured by a C3b molecule that is part of a C5 convertase complex. As shown in the top panel, C5 convertases are formed when C3b binds either the classical or lectin pathway C3 convertase C4b2a to form C4b2a3b, or the alternative pathway C3 convertase C3bBb to form C3b₂Bb. C5 binds to C3b in these complexes (center panel). The bottom panel shows that C5 is cleaved by the active enzyme C2a or Bb to form C5b and the inflammatory mediator C5a. Unlike C3b and C4b, C5b is not covalently bound to the cell surface. The production of C5b initiates the assembly of the terminal complement components.

In the figure (top panel): "C3b binds to both C4b2a and C3bBb, forming the active C5 convertases C4b2a3b and C3b₂Bb" with labels C2a, C4b, C3b, Bb, C3b, C4b2a3b, C3b₂Bb.

(center panel): "C5 binds to the C3b component of the C5 convertase enzyme" with labels C5, C5, C4b2a3b, C3b₂Bb.

(bottom panel): "C5 is cleaved by C2a or Bb to form C5b and C5a" with labels C5a, C5a, C5b, C5b, C4b2a3b, C3b₂Bb.

2-13   Ingestion of complement-tagged pathogens by phagocytes is mediated by receptors for the bound complement proteins.

The most important action of complement is to facilitate the uptake and destruction of pathogens by phagocytic cells. This occurs by the specific recognition of bound complement components by **complement receptors (CRs)** on phagocytes. These complement receptors bind pathogens opsonized with complement components: opsonization of pathogens is a major function of C3b and its proteolytic derivatives. C4b also acts as an opsonin but has a relatively minor role, largely because so much more C3b than C4b is generated.

The known receptors for bound complement components C5a and C3a are listed, with their functions and distributions, in Fig. 2.30. The C3b receptor CR1, described in Section 2-10, is a negative regulator of complement activation (see Fig. 2.27). CR1 is expressed on many types of immune cells, including macrophages and neutrophils. Binding of C3b to CR1 cannot by itself stimulate phagocytosis, but it can lead to phagocytosis in the presence of other immune mediators that activate macrophages. For example, the small complement fragment C5a can activate macrophages to ingest bacteria bound to their CR1 receptors (Fig. 2.31). C5a binds to another receptor expressed by macrophages, the **C5a receptor**, which has seven membrane-spanning domains. Receptors of this type transduce their signals via intracellular guanine-nucleotide-binding proteins called G proteins and are known generally as **G-protein-coupled receptors (GPCRs)**; they are discussed in Section 3-2. **C5L2 (GPR77)**, expressed by neutrophils and macrophages, is a

Receptor	Specificity	Functions	Cell types
CR1 (CD35)	C3b, C4bi	Promotes C3b and C4b decay Stimulates phagocytosis (requires C5a) Erythrocyte transport of immune complexes	Erythrocytes, macrophages, monocytes, polymorphonuclear leukocytes, B cells, FDC
CR2 (CD21)	C3d, iC3b, C3dg	Part of B-cell co-receptor Enhances B-cell response to antigens bearing C3d, iC3b, or C3dg Epstein–Barr virus receptor	B cells, FDC
CR3 (Mac-1) (CD11b: CD18)	iC3b	Stimulates phagocytosis	Macrophages, monocytes, polymorphonuclear leukocytes, FDC
CR4 (gp150, 95) (CD11c: CD18)	iC3b	Stimulates phagocytosis	Macrophages, monocytes, polymorphonuclear leukocytes, dendritic cells
CRIg	C3b, iC3b	Phagocytosis of circulating pathogens	Tissue-resident macrophages, hepatic sinusoid macrophages
C5a receptor (CD88)	C5a	Binding of C5a activates G protein	Neutrophils, macrophages, endothelial cells, mast cells
C5L2 (GPR77)	C5a	Decoy receptor, regulates C5a receptor	Neutrophils, macrophages
C3a receptor	C3a	Binding of C3a, activates G protein	Macrophages, endothelial cells, mast cells

Fig. 2.30 Distribution and function of cell-surface receptors for complement proteins. A variety of complement receptors are specific for bound C3b and its cleavage products (iC3b and C3dg). CR1 and CR3 are important in inducing the phagocytosis of bacteria with complement components bound to their surface. CR2 is found mainly on B cells, where it is part of the B-cell co-receptor complex CR1 and CR2 share structural features with the complement-regulatory proteins that bind C3b and C4b. CR3 and CR4 are integrins composed of integrin β2 paired with either integrin αM (CD11b) or integrin αX (CD11c), respectively (see Appendix II); CR3, also called Mac-1, is also important for leukocyte adhesion and migration, as we shall see in Chapter 3, whereas CR4 is only known to function in phagocytosis. The receptors for C5a and C3a are seven-span G-protein-coupled receptors. FDC, follicular dendritic cells; these are not involved in innate immunity and are discussed in later chapters.

Fig. 2.31 The anaphylatoxin C5a can enhance the phagocytosis of microorganisms opsonized in an innate immune response. Activation of complement leads to the deposition of C3b on the surface of microorganisms (left panel). C3b can be bound by the complement receptor CR1 on the surface of phagocytes, but this on its own is insufficient to induce phagocytosis (center panel). Phagocytes also express receptors for the anaphylatoxin C5a, and binding of C5a will now activate the cell to phagocytose microorganisms bound through CR1 (right panel).

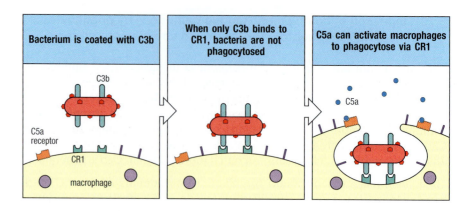

nonsignaling receptor that acts as a decoy receptor for C5a and may regulate activity of the C5a receptor. Proteins associated with the extracellular matrix, such as fibronectin, can also contribute to phagocyte activation; these are encountered when phagocytes are recruited to connective tissue and activated there.

Four other complement receptors—**CR2** (also known as **CD21**), **CR3** (**CD11b:CD18**), **CR4** (**CD11c:CD18**), and **CRIg** (complement receptor of the immunoglobulin family)—bind to forms of C3b that have been cleaved by factor I but that remain attached to the pathogen surface. Like several other key components of complement, C3b is subject to regulatory mechanisms that cleave it into derivatives, such as iC3b, that cannot form an active convertase. C3b bound to the microbial surface can be cleaved by factor I and MCP to remove the small fragment C3f, leaving the inactive iC3b form bound to the surface (Fig. 2.32). iC3b is recognized by several complement receptors—CR2, CR3, CR4, and CRIg. Unlike the binding of iC3b to CR1, the binding of iC3b to the receptor CR3 is sufficient on its own to stimulate phagocytosis. Factor I and CR1 cleave iC3b to release C3c, leaving C3dg bound to the pathogen. C3dg is recognized only by CR2. CR2 is found on B cells as part of a co-receptor complex that can augment the signal received through the antigen-specific immunoglobulin receptor. Thus, a B cell whose antigen receptor is specific for an antigen of a pathogen will receive a strong signal on binding this antigen if it or the pathogen is also coated with C3dg. The activation of complement can therefore contribute to producing a strong antibody response.

The importance of opsonization by C3b and its inactive fragments in destroying extracellular pathogens can be seen in the effects of various complement deficiencies. For example, individuals deficient in C3 or in molecules that catalyze C3b deposition show an increased susceptibility to infection by a wide range of extracellular bacteria, including *Streptococcus pneumoniae*. We describe the effects of various defects in complement and the diseases they cause in Chapter 13.

Fig. 2.32 The cleavage products of C3b are recognized by different complement receptors. After C3b is deposited on the surface of pathogens, it can undergo several conformational changes that alter its interaction with complement receptors. Factor I and MCP can cleave the C3f fragment from C3b, producing iC3b, which is a ligand for the complement receptors CR2, CR3, and CR4, but not CR1. Factor I and CR1 cleave iC3b to release C3c, leaving C3dg bound. C3dg is then recognized by CR2.

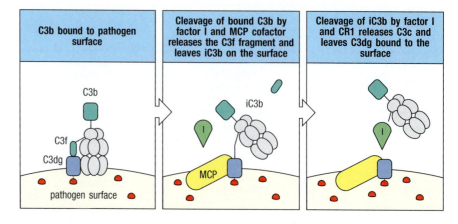

2-14   The small fragments of some complement proteins initiate a local inflammatory response.

The small complement fragments C3a and C5a act on specific receptors on endothelial cells and mast cells (see Fig. 2.30) to produce local inflammatory responses. Like C5a, C3a also signals through a G-protein-coupled receptor, discussed in more detail in Chapter 3. C4a, although generated during C4 cleavage, is not potent at inducing inflammation, is inactive at C3a and C5a receptors, and seems to lack a receptor of its own. When produced in large amounts or injected systemically, C3a and C5a induce a generalized circulatory collapse, producing a shocklike syndrome similar to that seen in a systemic allergic reaction involving antibodies of the IgE class, discussed in Chapter 14. Such a reaction is termed **anaphylactic shock**, and these small fragments of complement are therefore often referred to as **anaphylatoxins**. C5a has the highest specific biological activity, but both C3a and C5a induce the contraction of smooth muscle and increase vascular permeability and act on the endothelial cells lining blood vessels to induce the synthesis of adhesion molecules. In addition, C3a and C5a can activate the mast cells that populate submucosal tissues to release inflammatory molecules such as histamine and the cytokine tumor necrosis factor-α (TNF-α), which cause similar effects. The changes induced by C5a and C3a recruit antibody, complement, and phagocytic cells to the site of an infection (Fig. 2.33), and the increased fluid in the tissues hastens the movement of pathogen-bearing antigen-presenting cells to the local lymph nodes, contributing to the prompt initiation of the adaptive immune response.

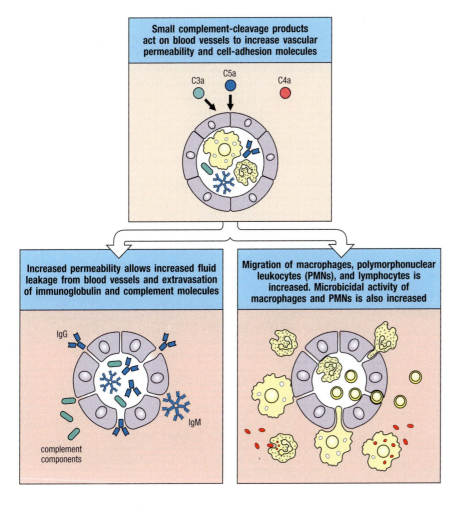

Fig. 2.33 Local inflammatory responses can be induced by small complement fragments, especially C5a. The small complement fragments are differentially active: C5a is more active than C3a; C4a is weak or inactive. C5a and C3a cause local inflammatory responses by acting directly on local blood vessels, stimulating an increase in blood flow, increased vascular permeability, and increased binding of phagocytes to endothelial cells. C3a and C5a also activate mast cells (not shown) to release mediators, such as histamine and TNF-α, that contribute to the inflammatory response. The increase in vessel diameter and permeability leads to the accumulation of fluid and protein in the surrounding tissue. Fluid accumulation increases lymphatic drainage, bringing pathogens and their antigenic components to nearby lymph nodes. The antibodies, complement, and cells thus recruited participate in pathogen clearance by enhancing phagocytosis. The small complement fragments can also directly increase the activity of the phagocytes.

C5a also acts directly on neutrophils and monocytes to increase their adherence to vessel walls, their migration toward sites of antigen deposition, and their ability to ingest particles; it also increases the expression of CR1 and CR3 on the surfaces of these cells. In this way, C5a, and to a smaller extent C3a and C4a, act in concert with other complement components to hasten the destruction of pathogens by phagocytes.

2-15   The terminal complement proteins polymerize to form pores in membranes that can kill certain pathogens.

One of the important effects of complement activation is the assembly of the terminal components of complement (Fig. 2.34) to form a membrane-attack complex. The reactions leading to the formation of this complex are shown schematically in Fig. 2.35. The end result is a pore in the lipid bilayer membrane that destroys membrane integrity. This is thought to kill the pathogen by destroying the proton gradient across the pathogen's cell membrane.

**Deficiency of the
C8 Complement
Component**

The first step in the formation of the membrane-attack complex is the cleavage of C5 by a C5 convertase to release C5b (see Fig. 2.29). In the next stages (see Fig. 2.35), C5b initiates the assembly of the later complement components and their insertion into the cell membrane. The process begins when one molecule of C5b binds one molecule of **C6**, and the C5b6 complex then binds one molecule of **C7**. This reaction leads to a conformational change in the constituent molecules, with the exposure of a hydrophobic site on C7, which inserts into the lipid bilayer. Similar hydrophobic sites are exposed on the later components **C8** and **C9** when they are bound to the complex, allowing these proteins also to insert into the lipid bilayer. C8 is a complex of two proteins, C8β and C8α-γ. The C8β protein binds to C5b, and the binding of C8β to the membrane-associated C5b67 complex allows the hydrophobic domain of C8α-γ to insert into the lipid bilayer. Finally, C8α-γ induces the polymerization of 10–16 molecules of C9 into a pore-forming structure called the membrane-attack complex. The membrane-attack complex has a hydrophobic external face, allowing it to associate with the lipid bilayer, but a hydrophilic internal channel. The diameter of this channel is about 10 nm, allowing the free passage of solutes and water across the lipid bilayer. The pore damage to the lipid bilayer leads to the loss of cellular homeostasis, the disruption of the proton gradient across the membrane, the penetration of enzymes such as lysozyme into the cell, and the eventual destruction of the pathogen.

Although the effect of the membrane-attack complex is very dramatic, particularly in experimental demonstrations in which antibodies against red blood cell membranes are used to trigger the complement cascade, the significance

The terminal complement components that form the membrane-attack complex		
Native protein	**Active component**	**Function**
C5	C5a	Small peptide mediator of inflammation (high activity)
	C5b	Initiates assembly of the membrane-attack system
C6	C6	Binds C5b; forms acceptor for C7
C7	C7	Binds C5b6; amphiphilic complex inserts into lipid bilayer
C8	C8	Binds C5b67; initiates C9 polymerization
C9	$C9_n$	Polymerizes to C5b678 to form a membrane-spanning channel, lysing the cell

Fig. 2.34 The terminal complement components.

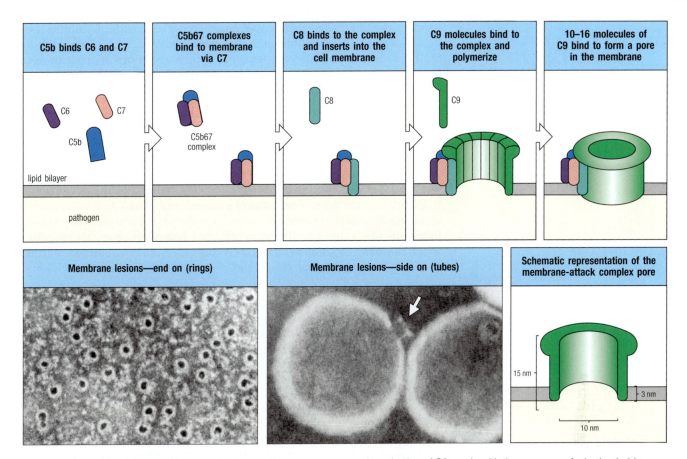

Fig. 2.35 Assembly of the membrane-attack complex generates a pore in the lipid bilayer membrane. The sequence of steps and their approximate appearance are shown here in schematic form. C5b triggers the assembly of a complex of one molecule each of C6, C7, and C8, in that order. C7 and C8 undergo conformational changes, exposing hydrophobic domains that insert into the membrane. This complex causes moderate membrane damage in its own right, and also serves to induce the polymerization of C9, again with the exposure of a hydrophobic site. Up to 16 molecules of C9 are then added to the assembly to generate a channel 10 nm in diameter in the membrane. This channel disrupts the bacterial cell membrane, killing the bacterium. The electron micrographs show erythrocyte membranes with membrane-attack complexes in two orientations, end on and side on. Photographs courtesy of S. Bhakdi and J. Tranum-Jensen.

of these components in host defense seems to be quite limited. So far, deficiencies in complement components C5–C9 have been associated with susceptibility only to *Neisseria* species, the bacteria that cause the sexually transmitted disease gonorrhea and a common form of bacterial meningitis. Thus, the opsonizing and inflammatory actions of the earlier components of the complement cascade are clearly more important for host defense against infection. Formation of the membrane-attack complex seems to be important only for the killing of a few pathogens, although, as we will see in Chapter 15, this complex might well have a major role in immunopathology.

2-16   Complement control proteins regulate all three pathways of complement activation and protect the host from their destructive effects.

Complement activation usually is initiated on a pathogen surface, and the activated complement fragments that are produced usually bind nearby on the pathogen surface or are rapidly inactivated by hydrolysis. Even so, all complement components are activated spontaneously at a low rate in plasma, and these activated complement components will sometimes bind proteins on host cells. Section 2-10 introduced the soluble host proteins factor I and factor H and the membrane-bound proteins MCP and DAF that regulate the alternative pathway of complement activation. In addition to these, several other soluble and membrane-bound complement-control proteins can regulate the complement cascade at various steps to protect normal host cells while allowing complement activation to proceed on pathogen surfaces (Fig. 2.36).

Hereditary Angioedema

The activation of C1 is controlled by the **C1 inhibitor (C1INH)**, which is a plasma **serine protease inhibitor**, or **serpin**. C1INH binds to the active enzymes C1r:C1s and causes them to dissociate from C1q, which remains bound to the pathogen (Fig. 2.37). In this way, C1INH limits the time during which active C1s is able to cleave C4 and C2. By the same means, C1INH limits the spontaneous activation of C1 in plasma. Its importance can be seen in the C1INH deficiency disease **hereditary angioedema (HAE)**, in which chronic spontaneous complement activation leads to the production of excess cleaved fragments of C4 and C2. The large activated fragments from this cleavage, which normally combine to form the C3 convertase, do not damage host cells in such patients because C4b is rapidly inactivated by hydrolysis in plasma,

Regulatory proteins of the classical and alternative pathways			
Soluble factors regulating complement			
Name	**Ligand/ binding factor**	**Action**	**Pathology if defective**
C1 inhibitor (C1INH)	C1r, C1s (C1q); MASP-2 (MBL)	Displaces C1r/s and MASP-2, inhibiting activation of C1q and MBL	Hereditary angiodema
C4-binding protein (C4BP)	C4b	Displaces C2a; cofactor for C4b cleavage by factor I	
CPN1 (Carboxypeptidase N)	C3a, C5a	Inactivates C3a and C5a	
Factor H	C3b	Displaces Bb, cofactor for factor I	Age-related macular degeneration, atypical hemolytic uremic syndrome
Factor I	C3b, C4b	Serine protease, cleaves C3b and C4b	Low C3 levels, hemolytic uremic syndrome
Protein S	C5b67 complex	Inhibits MAC formation	
Membrane-bound factors regulating complement			
Name	**Ligand/ binding factor**	**Action**	**Pathology if defective**
CRIg	C3b, iC3b, C3c	Inhibits activation of alternative pathway	Increased susceptibiity to blood-borne infections
Complement receptor 1 (CR1, CD35)	C3b, C4b	Cofactor for factor I; displaces Bb from C3b, and C2a from C4b	
Decay-accelerating factor (DAF, CD55)	C3 convertase	Displaces Bb and C2a from C3b and C4b, respectively	Paroxysmal nocturnal hemoglobinuria
Membrane-cofactor protein (MCP, CD46)	C3b, C4b	Cofactor for factor I	Atypical hemolytic anemia
Protectin (CD59)	C8	Inhibits MAC formation	Paroxysmal nocturnal hemoglobinuria

Fig. 2.36 The soluble and membrane-bound proteins that regulate the activity of complement.

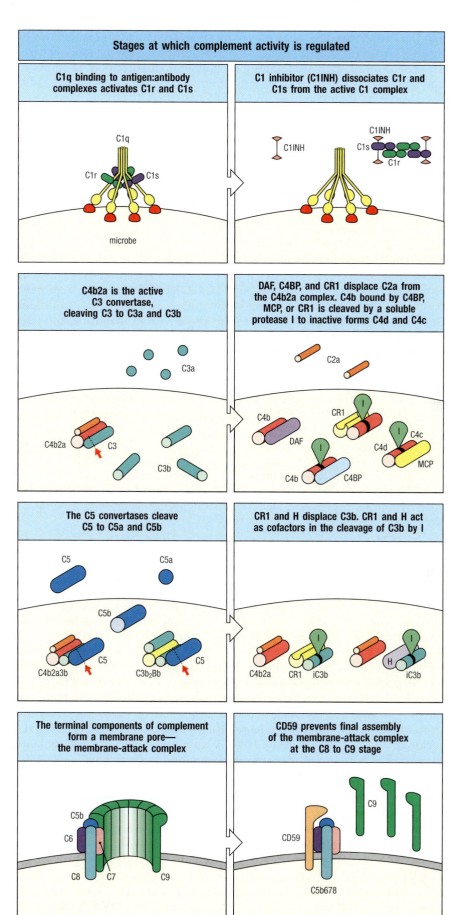

Fig. 2.37 Complement activation is regulated by a series of proteins that serve to protect host cells from accidental damage. These act on different stages of the complement cascade, dissociating complexes or catalyzing the enzymatic degradation of covalently bound complement proteins. Stages in the complement cascade are shown schematically down the left side of the figure, with the regulatory reactions on the right. The alternative-pathway C3 convertase is similarly regulated by DAF, CR1, MCP, and factor H.

and the convertase does not form. Bradykinin, which has similar actions to those of C2 kinin, is also produced in an uncontrolled fashion in this disease, as a result of the lack of inhibition of kallikrein, another plasma protease and component of the kinin system (discussed in Section 3-3). Kallikrein is activated by tissue damage and is also regulated by C1INH. Hereditary angioedema is fully corrected by replacing C1INH. A similar, extremely rare human disease stems from a partial deficiency of **carboxypeptidase N (CPN)**, a metalloproteinase that inactivates the anaphylatoxins C3a and C5a as well as bradykinin and kallikrein. Humans with partial CPN deficiency exhibit recurrent angioedema due to delayed inactivation of serum C3a and bradykinin.

Since the highly reactive thioester bond of activated C3 and C4 cannot distinguish between acceptor groups on a host cell from those on the surface of a pathogen, mechanisms have evolved that prevent the small amounts of C3 or C4 molecules deposited on host cells from fully triggering complement activation. We introduced these mechanisms in the context of control of the alternative pathway (see Fig. 2.27), but they are also important regulators of the classical pathway convertase (see Fig. 2.37, second and third rows). Section 2-10 described the proteins that inactivate any C3b or C4b that has bound to host cells. These are the plasma factor I and its cofactors MCP and CR1, which are membrane proteins. Circulating factor I is an active serine protease, but it can cleave C3b and C4b only when they are bound to MCP and CR1. In these circumstances, factor I cleaves C3b, first into iC3b and then further to C3dg, thus permanently inactivating it. C4b is similarly inactivated by cleavage into C4c and C4d. Microbial cell walls lack MCP and CR1 and thus cannot promote the breakdown of C3b and C4b, which instead act as binding sites for factor B and C2, promoting complement activation. The importance of factor I can be seen in people with genetically determined **factor I deficiency**. Because of uncontrolled complement activation, complement proteins rapidly become depleted and such people suffer repeated bacterial infections, especially with ubiquitous pyogenic bacteria.

There are also plasma proteins with cofactor activity for factor I, most notably **C4b-binding protein (C4BP)** (see Fig. 2.36). It binds C4b and acts mainly as a regulator of the classical pathway in the fluid phase. Another is factor H, which binds C3b in the fluid phase as well as at cell membranes and helps to distinguish the C3b that is bound to host cells from that bound to microbial surfaces. The higher affinity of factor H for sialic acid residues on host membrane glycoproteins allows it to displace factor B in binding to C3b on host cells. Also, C3b at cell membranes is bound by cofactor proteins DAF and MCP. Factor H, DAF, and MCP effectively compete with factor B for binding to C3b bound to host cells, so that the bound C3b is catabolized by factor I into iC3b and C3dg and complement activation is inhibited. In contrast, factor B is favored for binding C3b on microbial membranes, which do not express DAF or MCP and which lack the sialic acid modifications that attract factor H. The greater amount of factor B on a microbial surface stimulates formation of more C3bBb C3 convertase, and thus complement activation is amplified.

The critical balance between the inhibition and the activation of complement on cell surfaces is illustrated in individuals heterozygous for mutations in any of the regulatory proteins MCP, factor I, or factor H. In such individuals, the concentration of functional regulatory proteins is reduced, and the tipping of the balance toward complement activation leads to a predisposition to **atypical hemolytic uremic syndrome**, a condition characterized by damage to platelets and red blood cells and by kidney inflammation. Another serious health problem related to complement malfunction is a significantly increased risk of **age-related macular degeneration**, the leading cause of blindness in the elderly in developed countries, which has been predominantly linked to single-nucleotide polymorphisms in the factor H gene. Polymorphisms in other complement genes have also been found to be either detrimental or protective for this disease. Thus, even small alterations in the efficiency of either the activation or the regulation of this powerful effector system can contribute to the progression of degenerative or inflammatory disorders.

The competition between DAF or MCP and factor B for binding to surface-bound C3b is an example of the second mechanism for inhibiting complement activation on host cells. By binding C3b and C4b on the cell surface, these proteins competitively inhibit the binding of C2 to cell-bound C4b and of factor B to cell-bound C3b, thereby inhibiting convertase formation. DAF and MCP also mediate protection against complement through a third mechanism, which is to augment the dissociation of C4b2a and C3bBb convertases that have already formed. Like DAF, CR1 is among the host-cell membrane molecules that regulate complement through both these mechanisms—that is, by promoting the dissociation of convertase and exhibiting cofactor activity. All the proteins that bind the homologous C4b and C3b molecules share one or more copies of a structural element called the short consensus repeat (SCR), the complement control protein (CCP) repeat, or (especially in Japan) the sushi domain.

In addition to the mechanisms for preventing C3 convertase formation and C4 and C3 deposition on cell membranes, there are also inhibitory mechanisms that prevent the inappropriate insertion of the membrane-attack complex (MAC) into membranes. We saw in Section 2-15 that the membrane-attack complex polymerizes onto C5b molecules created by the action of C5 convertase. The MAC complex mainly inserts into cell membranes adjacent to the site of the C5 convertase, that is, close to the site of complement activation on a pathogen. However, some newly formed membrane-attack complexes may diffuse from the site of complement activation and insert into adjacent host-cell membranes. Several plasma proteins, including, notably, **vitronectin** (also known as **S-protein**), bind to the C5b67, C5b678, and C5b6789 complexes and thereby inhibit their random insertion into cell membranes. Host-cell membranes also contain an intrinsic protein, **CD59**, or **protectin**, that inhibits the binding of C9 to the C5b678 complex (see Fig. 2.37, bottom row). CD59 and DAF are both linked to the cell surface by a **glycosylphosphatidyl-inositol (GPI) tail**, like many other peripheral membrane proteins. One of the enzymes involved in the synthesis of GPI tails is encoded by a gene, *PIGA*, on the X chromosome. In people with a somatic mutation in this gene in a clone of hematopoietic cells, both CD59 and DAF fail to function. This causes the disease **paroxysmal nocturnal hemoglobinuria**, which is characterized by episodes of intravascular red blood cell lysis by complement. Red blood cells that lack only CD59 are also susceptible to destruction as a result of spontaneous activation of the complement cascade.

2-17 Pathogens produce several types of proteins that can inhibit complement activation.

Bacterial pathogens have evolved various strategies to avoid activation of complement, and thereby to avoid elimination by this first line of innate defense (Fig. 2.38). One strategy that many pathogens employ is to mimic host surfaces by attracting host complement regulators to their own surfaces. A mechanism to achieve this is for the pathogen to express surface proteins that bind to soluble complement-regulatory proteins such as C4BP and factor H. For example, the Gram-negative pathogen *Neisseria meningitidis* produces **factor H binding protein (fHbp)**, which recruits factor H (see Section 2-10), and the outer

Fig. 2.38 Complement evasion proteins produced by various pathogens.

Pathogen	Evasion molecule	Host target	Mechanism of action
Membrane proteins			
Neisseria meningitidis	Factor H binding protein (fHbp)	Factor H	Inactivates bound C3b
Borrelia burgdorferi	Outer surface protein E (OspE)	Factor H	Inactivates bound C3b
Streptococcus pneumoniae	Pneumococcal surface protein C (PspC)	Factor H	Inactivates bound C3b
Secreted proteins			
Neisseria meningitidis	PorA	C4BP	Inactivates bound C3b
Staphylococcus aureus	Clumping factor A (ClfA)	Factor I	Inactivates bound C3b
Staphylococcus aureus	Staphylococcus protein A (Spa)	Immunoglobulin	Binds to Fc regions and interferes with C1 activation
Staphylococcus aureus	Staphylokinase (SAK)	Immunoglobulin	Cleaves immunoglobulins
Staphylococcus aureus	Complement inhibitor (SCIN)	C3 convertase (C3b2a, C3bBb)	Inhibition of convertase activity

membrane protein **PorA**, which binds to C4BP. By recruiting factor H and C4BP to the pathogen membrane, the pathogen is able to inactivate C3b that is deposited on its surface and thereby avoid the consequences of complement activation. Complement is important in defense against *Neisseria* species, and several complement deficiencies are associated with increased susceptibility to this pathogen.

Another strategy employed by pathogens is to secrete proteins that directly inhibit components of complement. The Gram-positive pathogen *Staphylococcus aureus* provides several examples of this type of strategy. **Staphylococcal protein A (Spa)** binds to the Fc regions of immunoglobulins and interferes with the recruitment and activation of C1. This binding specificity was used as an early biochemical technique in the purification of antibodies. The staphylococcal protein **staphylokinase** (SAK) acts by cleaving immunoglobulins bound to the pathogen membrane, preventing complement activation and avoiding phagocytosis. The **staphylococcal complement inhibitor (SCIN)** protein binds to the classical C3 convertase, C4b2a, and the alternative pathway C3 convertase, C3bBb, and inhibits their activity. Other stages of complement activation, including formation of the C5 convertase, are targets of inhibition by proteins produced by these and other pathogens. We will return to this topic of complement regulation in Chapter 13 when we discuss how the immune system sometimes fails or is evaded by pathogens.

Summary.

The complement system is one of the major mechanisms by which pathogen recognition is converted into an effective host defense against initial infection. Complement is a system of plasma proteins that can be activated directly by pathogens or indirectly by pathogen-bound antibody, leading to a cascade of reactions that occurs on the surface of pathogens and generates active components with various effector functions. There are three pathways of complement activation: the lectin pathway, triggered by the pattern recognition receptors MBL and the ficolins; the classical pathway, triggered directly by antibody binding to the pathogen surface; and the alternative pathway, which utilizes spontaneous C3 deposition onto microbial surfaces, is augmented by properdin, and provides an amplification loop for the other two pathways. The early events in all pathways consist of a sequence of cleavage reactions in which the larger cleavage product binds covalently to the pathogen surface and contributes to the activation of the next component. The pathways converge with the formation of a C3 convertase enzyme, which cleaves C3 to produce the active complement component C3b. The binding of large numbers of C3b molecules to the pathogen is the central event in complement activation. Bound complement components, especially bound C3b and its inactive fragments, are recognized by specific complement receptors on phagocytic cells, which engulf pathogens opsonized by C3b and its inactive fragments. The small cleavage fragments of C3 and C5 act on specific trimeric G-protein-coupled receptors to recruit phagocytes, such as neutrophils, to sites of infection. Together, these activities promote the uptake and destruction of pathogens by phagocytes. The molecules of C3b that bind the C3 convertase itself initiate the late events of complement, binding C5 to make it susceptible to cleavage by C2a or Bb. The larger C5b fragment triggers the assembly of a membrane-attack complex, which can result in the lysis of certain pathogens. A system of soluble and membrane-bound complement-regulatory proteins act to limit complement activation on host tissues, in order to prevent tissue damage from the inadvertent binding of activated complement components to host cells or from the spontaneous activation of complement components in plasma. Many pathogens produce a variety of soluble and membrane-associated proteins that can counteract complement activation and contribute to infection of the host by the microbe.

Summary to Chapter 2.

This chapter has described the preexisting, constitutive components of innate immunity. The body's epithelial surfaces are a constant barrier to pathogen entry and have specialized adaptations, such as cilia, various antimicrobial molecules and mucus, that provide the simplest form of innate immunity. The complement system is a more specialized system that combines direct recognition of microbes with a complex effector system. Of the three pathways that can activate complement, two are devoted to innate immunity. The lectin pathway relies on pattern recognition receptors that detect microbial membranes, while the alternative pathway relies on spontaneous complement activation that is down-regulated by host molecules expressed on self membranes. The main event in complement activation is accumulation of C3b on microbial membranes, which is recognized by complement receptors on phagocytic cells to promote microbial clearance by cells recruited to sites of infection by C3a and C5a. In addition, C5b initiates the membrane-attack complex that is able to lyse some microbes directly. The complement cascade is under regulation to prevent attack on host tissues, and genetic variation in regulatory pathways can result in autoimmune syndromes and age-related tissue damage.

Questions.

2.1 **Multiple Choice:** The widely used β-lactam antibiotics are mainly active against Gram-positive bacteria. These inhibit the transpeptidation step in synthesis of peptidoglycan, a major component of the bacterial cell wall that is critical for the survival of the microorganism. Which of the following is an antimicrobial enzyme that functions to disrupt the same bacterial structure that β-lactams ultimately target?

    A. Phospholipase A

    B. Lysozyme

    C. Defensins

    D. Histatins

2.2 **Short Answer:** Why is the capacity of mannose-binding lectin (MBL) trimers to oligomerize important for their function?

2.3 **Multiple Choice:** Choose the option that correctly describes ficolins.

    A. C-type lectin domain, affinity for carbohydrates such as fucose and N-acetylglucosamine (GlcNAc), synthesized in the liver

    B. Fibrinogen-like domain, affinity for oligosaccharides containing acetylated sugars, synthesized in the liver

    C. C-type lectin domain, affinity for oligosaccharides containing acetylated sugars, synthesized in the liver

    D. Fibrinogen-like domain, affinity for carbohydrates such as fucose and N-acetylglucosamine (GlcNAc), synthetized in the liver and lungs

2.4 **Fill-in-the-Blanks:** For each of the following sentences, fill in the blanks with the best word selected from the list below. Not all words will be used; each word should be used only once.

Like MBLs, ficolins form oligomers with _____ and _____. Such interaction allows the oligomer to cleave the complement components _____ and _____. Once these are cleaved, they form _____, a C3 convertase, which cleaves _____ and permits the formation of the membrane-attack complex.

MASP-1	C2
MASP-2	C4a
C4	C4b2a
C4b2b	C3
C2a	C3b

2.5 **Short Answer:** One way in which the alternative pathway activates is by spontaneous hydrolysis of the C3 thioester bond that is normally used to covalently attach to the pathogen's surface. How can the alternative pathway proceed to form a membrane-attack complex if the C3 convertase that initiates this process is soluble?

2.6 **Fill-in-the-Blanks:** Paroxysmal nocturnal hemoglobinuria, a disease characterized by episodes of intravascular red blood cell lysis, is the result of red blood cells losing the expression of _____ and _____, which renders them susceptible to lysis by the _____ pathway of the complement system.

CD59	C3b
classical	DAF
lectin	alternative
factor I	C1 inhibitor (C1INH)

2.7 **Matching:** Match each of the following complement-regulatory proteins with the pathological manifestation that would develop if this factor were defective:

    A. C1INH   **1.** Atypical hemolytic uremic syndrome

    B. Factor H & factor I   **2.** Hereditary angioedema

    C. DAF   **3.** Paroxysmal nocturnal hemoglobinuria

2.8 **Multiple Choice:** Diseases such as cryoglobulinemia and systemic lupus erythematosus usually present with low C3 and C4 levels in blood due to the activation of the classical complement pathway. In contrast, diseases such as dense deposit disease or C3 glomerulonephritis generally have low C3 due to activation of the alternative complement pathway. What would be the expected levels of C2 and C4 in a patient suffering from dense deposit disease or C3 glomerulonephritis?

    A. Normal

    B. High

    C. Low

    D. High C4 and low C3

2.9 **True or False:** Mucins secreted at a mucosal surface exhibit direct microbicidal activities.

2.10 **Short Answer:** *Neisseria meningitidis* and *Staphylococcus aureus* each prevent complement activation in different ways. Explain how each does so.

2.11 **True or False:** Both neutrophils and Paneth cells of the gut secrete antimicrobial peptides, such as defensins, only upon stimulation.

2.12 **Short Answer:** What are two products of the C3 convertase? Name three downstream events that can result from the formation of these products and lead to clearance of the microbe.

2.13 **True or False:** CD21 (CR1) is a complement receptor expressed on B cells that binds C3dg (a C3b breakdown product) and serves as a co-receptor to augment signaling and trigger a stronger antibody response.

Section references.

2-1   Infectious diseases are caused by diverse living agents that replicate in their hosts.

Kauffmann, S.H.E., Sher, A., and Ahmed, R.: *Immunology of Infectious Diseases*. Washington, DC: ASM Press, 2002.

Mandell, G.L., Bennett, J.E., and Dolin, R. (eds): *Principles and Practice of Infectious Diseases*, 4th ed. New York: Churchill Livingstone, 1995.

2-2   Epithelial surfaces of the body provide the first line of defense against infection.

Gallo, R.L., and Hooper, L.V.: **Epithelial antimicrobial defense of the skin and intestine.** *Nat. Rev. Immunol.* 2012, **12**:503–516.

2-3   Infectious agents must overcome innate host defenses to establish a focus of infection.

Gorbach, S.L., Bartlett, J.G., and Blacklow, N.R. (eds): *Infectious Diseases*, 3rd ed. Philadelphia: Lippincott Williams & Wilkins, 2003.

Hornef, M.W., Wick, M.J., Rhen, M., and Normark, S.: **Bacterial strategies for overcoming host innate and adaptive immune responses.** *Nat. Immunol.* 2002, **3**:1033–1040.

2-4   Epithelial cells and phagocytes produce several kinds of antimicrobial proteins.

Cash, H.L., Whitham, C.V., Behrendt, C.L., and Hooper, L.H.: **Symbiotic bacteria direct expression of an intestinal bactericidal lectin.** *Science* 2006, **313**:1126–1130.

De Smet, K., and Contreras, R.: **Human antimicrobial peptides: defensins, cathelicidins and histatins.** *Biotechnol. Lett.* 2005, **27**:1337–1347.

Ganz, T.: **Defensins: antimicrobial peptides of innate immunity.** *Nat. Rev. Immunol.* 2003, **3**:710–720.

Mukherjee, S., Zheng, H., Derebe, M.G., Callenberg, K.M., Partch, C.L., Rollins, D., Propheter, D.C., Rizo, J., Grabe, M., Jiang, Q.X., and Hooper, L.V.: **Antibacterial membrane attack by a pore-forming intestinal C-type lectin.** *Nature* 2014, **505**:103–107.

Zanetti, M.: **The role of cathelicidins in the innate host defense of mammals.** *Curr. Issues Mol. Biol.* 2005, **7**:179–196.

2-5   The complement system recognizes features of microbial surfaces and marks them for destruction by coating them with C3b.

Gros, P., Milder, F.J., and Janssen, B.J.: **Complement driven by conformational changes.** *Nat. Rev. Immunol.* 2008, **8**:48–58.

Janssen, B.J., Christodoulidou, A., McCarthy, A., Lambris, J.D., and Gros, P.: **Structure of C3b reveals conformational changes that underlie complement activity.** *Nature* 2006, **444**:213–216.

Janssen, B.J., Huizinga, E.G., Raaijmakers, H.C., Roos, A., Daha, M.R., Nilsson-Ekdahl, K., Nilsson, B., and Gros, P.: **Structures of complement component C3 provide insights into the function and evolution of immunity.** *Nature* 2005, **437**:505–511.

2-6   The lectin pathway uses soluble receptors that recognize microbial surfaces to activate the complement cascade.

Bohlson, S.S., Fraser, D.A., and Tenner, A.J.: **Complement proteins C1q and MBL are pattern recognition molecules that signal immediate and long-term protective immune functions.** *Mol. Immunol.* 2007, **44**:33–43.

Fujita, T.: **Evolution of the lectin-complement pathway and its role in innate immunity.** *Nat. Rev. Immunol.* 2002, **2**:346–353.

Gál, P., Harmat, V., Kocsis, A., Bián, T., Barna, L., Ambrus, G., Végh, B., Balczer, J., Sim, R.B., Náray-Szabó, G., *et al*: **A true autoactivating enzyme. Structural insight into mannose-binding lectin-associated serine protease-2 activations.** *J. Biol. Chem.* 2005, **280**:33435–33444.

Héja, D., Kocsis, A., Dobó, J., Szilágyi, K., Szász, R., Závodszky, P., Pál, G., Gál, P.: **Revised mechanism of complement lectin-pathway activation revealing the role of serine protease MASP-1 as the exclusive activator of MASP-2.** *Proc. Natl Acad. Sci. USA* 2012, **109**:10498–10503.

Wright, J.R.: **Immunoregulatory functions of surfactant proteins.** *Nat. Rev. Immunol.* 2005, **5**:58–68.

2-7   The classical pathway is initiated by activation of the C1 complex and is homologous to the lectin pathway.

McGrath, F.D., Brouwer, M.C., Arlaud, G.J., Daha, M.R., Hack, C.E., and Roos, A.: **Evidence that complement protein C1q interacts with C-reactive protein through its globular head region.** *J. Immunol.* 2006, **176**:2950–2957.

2-8   Complement activation is largely confined to the surface on which it is initiated.

Cicardi, M., Bergamaschini, L., Cugno, M., Beretta, A., Zingale, L.C., Colombo, M., and Agostoni, A.: **Pathogenetic and clinical aspects of C1 inhibitor deficiency.** *Immunobiology* 1998, **199**:366–376.

2-9   The alternative pathway is an amplification loop for C3b formation that is accelerated by properdin in the presence of pathogens.

Fijen, C.A., van den Bogaard, R., Schipper, M., Mannens, M., Schlesinger, M., Nordin, F.G., Dankert, J., Daha, M.R., Sjoholm, A.G., Truedsson, L., *et al*: **Properdin deficiency: molecular basis and disease association.** *Mol. Immunol.* 1999, **36**:863–867.

Kemper, C., and Hourcade, D.E.: **Properdin: new roles in pattern recognition and target clearance.** *Mol. Immunol.* 2008, **45**:4048–4056.

Spitzer, D., Mitchell, L.M., Atkinson, J.P., and Hourcade, D.E.: **Properdin can initiate complement activation by binding specific target surfaces and providing a platform for de novo convertase assembly.** *J. Immunol.* 2007, **179**:2600–2608.

Xu, Y., Narayana, S.V., and Volanakis, J.E.: **Structural biology of the alternative pathway convertase.** *Immunol. Rev.* 2001, **180**:123–135.

2-10   Membrane and plasma proteins that regulate the formation and stability of C3 convertases determine the extent of complement activation.

Golay, J., Zaffaroni, L., Vaccari, T., Lazzari, M., Borleri, G.M., Bernasconi, S., Tedesco, F., Rambaldi, A., and Introna, M.: **Biologic response of B lymphoma cells to anti-CD20 monoclonal antibody rituximab** *in vitro*: **CD55 and CD59 regulate complement-mediated cell lysis.** *Blood* 2000, **95**:3900–3908.

Spiller, O.B., Criado-Garcia, O., Rodriguez De Cordoba, S., and Morgan, B.P.: **Cytokine-mediated up-regulation of CD55 and CD59 protects human hepatoma cells from complement attack.** *Clin. Exp. Immunol.* 2000, **121**:234–241.

Varsano, S., Frolkis, I., Rashkovsky, L., Ophir, D., and Fishelson, Z.: **Protection of human nasal respiratory epithelium from complement-mediated lysis by cell-membrane regulators of complement activation.** *Am. J. Respir. Cell Mol. Biol.* 1996, **15**:731–737.

2-11   Complement developed early in the evolution of multicellular organisms.

Fujita, T.: **Evolution of the lectin-complement pathway and its role in innate immunity.** *Nat. Rev. Immunol.* 2002, **2**:346–353.

Zhang, H., Song, L., Li, C., Zhao, J., Wang, H., Gao, Q., and Xu, W.: **Molecular cloning and characterization of a thioester-containing protein from Zhikong scallop** Chlamys farreri. *Mol. Immunol.* 2007, **44**:3492–3500.

2-12 Surface-bound C3 convertase deposits large numbers of C3b fragments on pathogen surfaces and generates C5 convertase activity.

Rawal, N., and Pangburn, M.K.: **Structure/function of C5 convertases of complement.** *Int. Immunopharmacol.* 2001, **1**:415–422.

2-13 Ingestion of complement-tagged pathogens by phagocytes is mediated by receptors for the bound complement proteins.

Gasque, P.: **Complement: a unique innate immune sensor for danger signals.** *Mol. Immunol.* 2004, **41**:1089–1098.

Helmy, K.Y., Katschke, K.J., Jr., Gorgani, N.N., Kljavin, N.M., Elliott, J.M., Diehl, L., Scales, S.J., Ghilardi, N., and van Lookeren Campagne, M.: **CRIg: a macrophage complement receptor required for phagocytosis of circulating pathogens.** *Cell* 2006, **124**:915–927.

2-14 The small fragments of some complement proteins initiate a local inflammatory response.

Barnum, S.R.: **C4a: an anaphylatoxin in name only**. *J. Innate Immun.* 2015, **7**:333-9.

Kohl, J.: **Anaphylatoxins and infectious and noninfectious inflammatory diseases.** *Mol. Immunol.* 2001, **38**:175–187.

Schraufstatter, I.U., Trieu, K., Sikora, L., Sriramarao, P., and DiScipio, R.: **Complement C3a and C5a induce different signal transduction cascades in endothelial cells.** *J. Immunol.* 2002, **169**:2102–2110.

2-15 The terminal complement proteins polymerize to form pores in membranes that can kill certain pathogens.

Hadders, M.A., Beringer, D.X., and Gros, P.: **Structure of C8α-MACPF reveals mechanism of membrane attack in complement immune defense.** *Science* 2007, **317**:1552–1554.

Parker, C.L., and Sodetz, J.M.: **Role of the human C8 subunits in complement-mediated bacterial killing: evidence that C8γ is not essential.** *Mol. Immunol.* 2002, **39**:453–458.

Scibek, J.J., Plumb, M.E., and Sodetz, J.M.: **Binding of human complement C8 to C9: role of the N-terminal modules in the C8α subunit.** *Biochemistry* 2002, **41**:14546–14551.

2-16 Complement control proteins regulate all three pathways of complement activation and protect the host from their destructive effects.

Ambati, J., Atkinson, J.P., and Gelfand, B.D.: **Immunology of age-related macular degeneration.** *Nat. Rev. Immunol.* 2013, **13**:438–451.

Atkinson, J.P., and Goodship, T.H.: **Complement factor H and the hemolytic uremic syndrome.** *J. Exp. Med.* 2007, **204**:1245–1248.

Jiang, H., Wagner, E., Zhang, H., and Frank, M.M.: **Complement 1 inhibitor is a regulator of the alternative complement pathway.** *J. Exp. Med.* 2001, **194**:1609–1616.

Miwa, T., Zhou, L., Hilliard, B., Molina, H., and Song, W.C.: **Crry, but not CD59 and DAF, is indispensable for murine erythrocyte protection in vivo from spontaneous complement attack.** *Blood* 2002, **99**:3707–3716.

Singhrao, S.K., Neal, J.W., Rushmere, N.K., Morgan, B.P., and Gasque, P.: **Spontaneous classical pathway activation and deficiency of membrane regulators render human neurons susceptible to complement lysis.** *Am. J. Pathol.* 2000, **157**:905–918.

Smith, G.P., and Smith, R.A.: **Membrane-targeted complement inhibitors.** *Mol. Immunol.* 2001, **38**:249–255.

Spencer, K.L., Hauser, M.A., Olson, L.M., Schmidt, S., Scott, W.K., Gallins, P., Agarwal, A., Postel, E.A., Pericak-Vance, M.A., and Haines, J.L.: **Protective effect of complement factor B and complement component 2 variants in age-related macular degeneration.** *Hum. Mol. Genet.* 2007, **16**:1986–1992.

Spencer, K.L., Olson, L.M., Anderson, B.M., Schnetz-Boutaud, N., Scott, W.K., Gallins, P., Agarwal, A., Postel, E.A., Pericak-Vance, M.A., and Haines, J.L.: **C3 R102G polymorphism increases risk of age-related macular degeneration.** *Hum. Mol. Genet.* 2008, **17**:1821–1824.

2-17 Pathogens produce several types of proteins that can inhibit complement activation.

Blom, A.M., Rytkonen, A., Vasquez, P., Lindahl, G., Dahlback, B., and Jonsson, A.B.: **A novel interaction between type IV pili of** Neisseria gonorrhoeae **and the human complement regulator C4B-binding protein.** *J. Immunol.* 2001, **166**:6764–6770.

Serruto, D., Rappuoli, R., Scarselli, M., Gros, P., and van Strijp, J.A.: **Molecular mechanisms of complement evasion: learning from staphylococci and meningococci.** *Nat. Rev. Microbiol.* 2010, **8**:393–399.

The Induced Responses of Innate Immunity

3

In Chapter 2 we introduced the innate defenses—such as epithelial barriers, secreted antimicrobial proteins, and the complement system—that act immediately upon encounter with microbes to protect the body from infection. We also introduced the phagocytic cells that lie beneath the epithelial barriers and stand ready to engulf and digest invading microorganisms that have been flagged for destruction by complement. These phagocytes also initiate the next phase of the innate immune response, inducing an inflammatory response that recruits new phagocytic cells and circulating effector molecules to the site of infection. In this chapter we describe how phagocytic cells of the innate immune system detect microbes or the cellular damage they cause, how they destroy these pathogens, and how they orchestrate downstream inflammatory responses through production of cytokines and chemokines (chemoattractant cytokines). We also introduce other cells of the innate immune system—a diverse array of specialized innate lymphoid cells (ILCs) including the natural killer cells (NK cells)—that contribute to innate host defenses against viruses and other intracellular pathogens. Also in this stage of infection dendritic cells initiate adaptive immune responses, so that if the infection is not cleared by innate immunity, a full immune response will ensue.

Pattern recognition by cells of the innate immune system.

The basis of the adaptive immune system's enormous capacity for antigen recognition has long been appreciated. In contrast, the basis of recognition of microbial products by innate immune sensors was discovered only in the late 1990s. Initially, innate recognition was considered to be restricted to relatively few **pathogen-associated molecular patterns**, or **PAMPs**, and we have already seen examples of such recognition of microbial surfaces by complement (see Chapter 2). In the last several years, with the discovery of an increasing number of innate receptors that are capable of discriminating among a number of closely related molecules, we have come to realize that a much greater flexibility in innate recognition exists than had been previously thought.

The first part of this chapter examines the cellular receptors that recognize pathogens and signal for a cellular innate immune response. Regular patterns of molecular structure are present on many microorganisms but do not occur on the host body's own cells. Receptors that recognize such features are expressed on macrophages, neutrophils, and dendritic cells, and they are similar to the secreted molecules, such as ficolins and histatins, described in Chapter 2. The general characteristics of these **pattern recognition receptors (PRRs)** are contrasted with those of the antigen-specific receptors of adaptive immunity in **Fig. 3.1**. A new insight is that self-derived host molecules can be induced that indicate cellular infection, damage, stress, or transformation, and that some innate receptors recognize such proteins to mediate responses by innate immune cells. Such indicator molecules have been termed '**damage-associated molecular patterns**,' or **DAMPs**, and some of the molecules in this class can be recognized by receptors also involved in pathogen recognition, such as the Toll-like receptors (TLRs).

Fig. 3.1 Comparison of the characteristics of recognition molecules of the innate and adaptive immune systems. The innate immune system uses germline-encoded receptors while the adaptive immune system uses antigen receptors of unique specificity assembled from incomplete gene segments during lymphocyte development. Antigen receptors of the adaptive immune system are clonally distributed on individual lymphocytes and their progeny. Typically, receptors of the innate immune system are expressed non-clonally, that is, they are expressed on all the cells of a given cell type. However, NK cells express various combinations of NK receptors from several families, making individual NK cells different from one another. A particular NK receptor may not be expressed on all NK cells.

Receptor characteristic	Innate immunity	Adaptive immunity
Specificity inherited in the genome	Yes	No
Expressed by all cells of a particular type (e.g., macrophages)	Variable	No
Triggers immediate response	Yes	No
Recognizes broad classes of pathogens	Yes	No
Interacts with a range of molecular structures of a given type	Yes	No
Encoded in multiple gene segments	No	Yes
Requires gene rearrangement	No	Yes
Clonal distribution	No	Yes
Able to discriminate between even closely related molecular structures	Yes	Yes

Coordination of the innate immune response relies on the information provided by many types of receptors. Pattern recognition receptors can be classified into four main groups on the basis of their cellular localization and their function: free receptors in the serum, such as ficolins and histatins (discussed in Chapter 2); membrane-bound phagocytic receptors; membrane-bound signaling receptors; and cytoplasmic signaling receptors. Phagocytic receptors primarily signal for phagocytosis of the microbes they recognize. A diverse group of receptors, including chemotactic receptors, help to guide cells to sites of infection, and other receptors, including PRRs and cytokine receptors, can control the activity of effector molecules at those sites.

In this part of the chapter we first look at the recognition properties of phagocytic receptors and of signaling receptors that activate phagocytic microbial killing mechanisms. Next we describe an evolutionarily ancient pathogen system of recognition and signaling, the Toll-like receptors (TLRs), the first of the innate sensor systems to be discovered, and several recently discovered systems that detect intracellular infections by sensing cytoplasmic microbial cell-wall components, foreign RNA, or foreign DNA.

3-1 After entering tissues, many microbes are recognized, ingested, and killed by phagocytes.

If a microorganism crosses an epithelial barrier and begins to replicate in the tissues of the host, in most cases, it is immediately recognized by resident phagocytic cells. The main classes of phagocytic cells in the innate immune system are macrophages and monocytes, granulocytes, and dendritic cells. **Macrophages** are the major phagocyte population resident in most normal tissues at homeostasis. They can arise either from progenitor cells that enter the tissues during embryonic development, and then self-renew at steady state during life, or from circulating **monocytes**. Studies suggest that the embryonic progenitors arise from either the fetal liver, the yolk sac, or an embryonic region near the dorsal aorta called the **aorta–gonad–mesonephros (AGM)**, although the relative contribution of these origins is still debated. Macrophages are found in especially large numbers in connective tissue: for example, in the submucosal layer of the gastrointestinal tract; in the submucosal layer of the bronchi, and in the lung interstitium—the tissue and intercellular spaces around the air sacs (alveoli)—and in the alveoli themselves; along some blood vessels in the liver; and throughout the spleen, where they remove senescent blood cells. Macrophages in different tissues were historically given different names, for example, **microglial cells** in neural tissue and **Kupffer cells** in the liver. The self-renewal of these two types of cells is dependent on a cytokine called interleukin 34 (IL-34) that is produced in these tissues and acts on the same receptor as macrophage-colony stimulating factor (M-CSF).

During infection or inflammation, macrophages can also arise from monocytes that leave from the circulation to enter into tissues. Monocytes in both mouse and human develop in the bone marrow and circulate in the blood as two main populations. In humans, 90% of circulating monocytes are the **'classical' monocyte** that expresses CD14, a co-receptor for a PRR described later, and function during infection by entering tissues and differentiating into activated inflammatory monocytes or macrophages. In mice, this monocyte population expresses high levels of the surface marker Ly6C. A smaller population are the '**patrolling monocytes**' that roll along the endothelium rather than circulating freely in the blood. In humans, they express CD14 and CD16, a type of Fc receptor (FcγRIII; see Section 10-21), and are thought to survey for injury to the endothelium but do not differentiate into tissue macrophages. In mice, they express low levels of Ly6C.

The second major family of phagocytes comprises the granulocytes, which include **neutrophils**, **eosinophils**, and **basophils**. Of these, neutrophils have the greatest phagocytic activity and are the cells most immediately involved in innate immunity against infectious agents. Also called polymorphonuclear neutrophilic leukocytes (PMNs, or polys), they are short-lived cells that are abundant in the blood but are not present in healthy tissues. Macrophages and granulocytes have an important role in innate immunity because they can recognize, ingest, and destroy many pathogens without the aid of an adaptive immune response. Phagocytic cells that scavenge incoming pathogens represent an ancient mechanism of innate immunity, as they are found in both invertebrates and vertebrates.

The third class of phagocytes in the immune system is the immature **dendritic cells** that reside in lymphoid organs and in peripheral tissues. There are two main functional types of dendritic cells: **conventional** (or **classical**) **dendritic cells (cDCs)** and **plasmacytoid dendritic cells (pDCs)**. Both types of cells arise from progenitors within the bone marrow that primarily branch from cells of myeloid potential, and they migrate via the blood to tissues throughout the body and to peripheral lymphoid organs. Dendritic cells ingest and break down microbes, but, unlike macrophages and neutrophils, their primary role in immune defense is not the front-line, large-scale direct killing of microbes. A major role of cDCs is to process ingested microbes in order to generate peptide antigens that can activate T cells and induce an adaptive immune response. They also produce cytokines in response to microbial recognition that activate other types of cells against infection. cDCs are thus considered to act as a bridge between innate and adaptive immune responses. pDCs are major producers of a class of cytokines known as type I interferons, or antiviral interferons, and are considered to be part of innate immunity; they are discussed in detail later in the chapter.

Because most microorganisms enter the body through the mucosa of the gut and respiratory system, skin, or urogenital tract, macrophages in the submucosal tissues are the first cells to encounter most pathogens, but they are soon reinforced by the recruitment of large numbers of neutrophils to sites of infection. Macrophages and neutrophils recognize pathogens by means of cell-surface receptors that can discriminate between the surface molecules of pathogens and those of the host. Although they are both phagocytic, macrophages and neutrophils have distinct properties and functions in innate immunity.

MOVIE 3.1

MOVIE 3.2

The process of **phagocytosis** is initiated when certain receptors on the surface of the cell—typically a macrophage, neutrophil, or dendritic cell—interacts with the microbial surface. The bound pathogen is first surrounded by the phagocyte plasma membrane and then internalized in a large membrane-enclosed endocytic vesicle known as a **phagosome**. The phagosome fuses with one or more lysosomes to generate a **phagolysosome**, in which the lysosomal contents are released. The phagolysosome also becomes acidified, acquires antimicrobial peptides and enzymes, and undergoes enzymatic processes that produce highly reactive superoxide and nitric oxide radicals, which together kill the microbe (**Fig. 3.2**). Neutrophils are highly specialized for the intracellular killing of microbes, and contain different types of cytoplasmic granules—the **primary granules** and **secondary granules** described in Section 2-4. These granules fuse with phagosomes, releasing additional enzymes and antimicrobial peptides that attack the microbe. Another pathway by which extracellular material, including microbial material, can be taken up into the endosomal compartment of cells and degraded is **receptor-mediated endocytosis**, which is not restricted to phagocytes. Dendritic cells and other phagocytes can also take up pathogens by a nonspecific process called **macro-pinocytosis**, in which large amounts of extracellular fluid and its contents are ingested.

Macrophages and neutrophils constitutively express a number of cell-surface receptors that stimulate the phagocytosis and intracellular killing of microbes bound to them, although some also signal through other pathways to trigger responses such as cytokine production. These phagocytic receptors include several members of the C-type lectin-like family (see Fig. 3.2). For example, **Dectin-1** is strongly expressed by macrophages and neutrophils and recognizes β-1,3-linked glucans (polymers of glucose), which are common components of fungal cell walls in particular. Dendritic cells also express Dectin-1, as well as several other C-type lectin-like phagocytic receptors, which will be discussed in relation to pathogen uptake for antigen processing and presentation in Chapter 9. Another C-type lectin, the **mannose receptor (MR)** expressed by macrophages and dendritic cells, recognizes various mannosylated ligands, including some present on fungi, bacteria, and viruses; it was once suspected to have an important role in resistance to microbes. However, experiments with mice that lack this receptor do not support this idea. The macrophage mannose receptor is now thought to function mainly as a clearance receptor for host glycoproteins such as β-glucuronidase and lysosomal hydrolases, which have mannose-containing carbohydrate side chains and whose extracellular concentrations are raised during inflammation.

A second set of phagocytic receptors on macrophages, called **scavenger receptors**, recognize various anionic polymers and acetylated low-density lipoproteins. These receptors are structurally heterogeneous, consisting of at least six different molecular families. Class A scavenger receptors are membrane proteins composed of trimers of collagen domains (see Fig. 3.2).

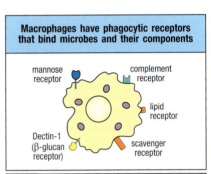

Macrophages have phagocytic receptors that bind microbes and their components

mannose receptor
complement receptor
lipid receptor
Dectin-1 (β-glucan receptor)
scavenger receptor

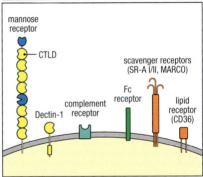

mannose receptor

CTLD

scavenger receptors (SR-A I/II, MARCO)

Fc receptor

complement receptor

Dectin-1

lipid receptor (CD36)

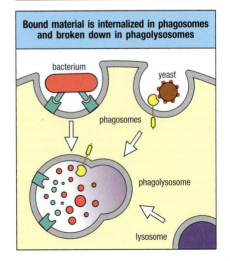

Bound material is internalized in phagosomes and broken down in phagolysosomes

bacterium
yeast
phagosomes
phagolysosome
lysosome

Fig. 3.2 Macrophages express receptors that enable them to take up microbes by phagocytosis. First panel: macrophages residing in tissues throughout the body are among the first cells to encounter and respond to pathogens. They carry cell-surface receptors that bind to various molecules on microbes, in particular carbohydrates and lipids, and induce phagocytosis of the bound material. Second panel: Dectin-1 is a member of the C-type lectin-like family built around a single C-type lectin-like domain (CTLD). Lectins in general are based on a carbohydrate-recognition domain (CRD). The macrophage mannose receptor contains many CTLDs, with a fibronectin-like domain and a cysteine-rich region at its amino terminus. Class A scavenger receptors such as MARCO are built from collagen-like domains and form trimers. The receptor protein CD36 is a class B scavenger receptor that recognizes and internalizes lipids. Various complement receptors bind and internalize complement-coated bacteria. Third panel: phagocytosis of receptor-bound material is taken into intracellular phagosomes, which fuse with lysosomes to form an acidified phagolysosome in which the ingested material is broken down by lysosomal hydrolases.

They include **SR-A I**, **SR-A II**, and **MARCO** (<u>m</u>acrophage <u>r</u>eceptor with a <u>c</u>ollagenous <u>s</u>tructure), which all bind various bacterial cell-wall components and help to internalize bacteria, although the basis of their specificity is poorly understood. Class B scavenger receptors bind high-density lipoproteins, and they internalize lipids. One of these receptors is CD36, which binds many ligands, including long-chain fatty acids.

A third set of receptors of crucial importance in macrophage and neutrophil phagocytosis is the complement receptors and Fc receptors introduced in Chapters 1 and 2. These receptors bind to complement-coated microbes or to antibodies that have bound to the surface of microbes and facilitate the phagocytosis of a wide range of microorganisms.

3-2 G-protein-coupled receptors on phagocytes link microbe recognition with increased efficiency of intracellular killing.

Phagocytosis of microbes by macrophages and neutrophils is generally followed by the death of the microbe inside the phagocyte. As well as the phagocytic receptors, macrophages and neutrophils have other receptors that signal to stimulate antimicrobial killing. These receptors belong to the evolutionarily ancient family of **G-protein-coupled receptors (GPCRs)**, which are characterized by seven membrane-spanning segments. Members of this family are crucial to immune system function because they also direct responses to anaphylatoxins such as the complement fragment C5a (see Section 2-14) and to many chemokines, recruiting phagocytes to sites of infection and promoting inflammation.

The **fMet-Leu-Phe (fMLF) receptor** is a G-protein-coupled receptor that senses the presence of bacteria by recognizing a unique feature of bacterial polypeptides. Protein synthesis in bacteria is typically initiated with an N-formylmethionine (fMet) residue, an amino acid present in prokaryotes but not in eukaryotes. The fMLF receptor is named after a tripeptide, formylmethionyl-leucyl phenylalanine, for which it has a high affinity, although it also binds other peptide motifs. Bacterial polypeptides binding to this receptor activate intracellular signaling pathways that direct the cell to move toward the most concentrated source of the ligand. Signaling through the fMLF receptor also induces the production of microbicidal **reactive oxygen species (ROS)** in the phagolysosome. The C5a receptor recognizes the small fragment of C5 generated when the classical or lectin pathways of complement are activated, usually by the presence of microbes (see Section 2-14), and signals by a similar pathway as the fMLF receptor. Thus, stimulation of these receptors both guides monocytes and neutrophils toward a site of infection and leads to increased antimicrobial activity; these cell responses can be activated by directly sensing unique bacterial products or by messengers such as C5a that indicate previous recognition of a microbe.

 MOVIE 3.3

The G-protein-coupled receptors are so named because ligand binding activates a member of a class of intracellular GTP-binding proteins called **G proteins**, sometimes referred to as **heterotrimeric G proteins** to distinguish them from the family of 'small' GTPases typified by Ras. Heterotrimeric G proteins are composed of three subunits: Gα, Gβ, and Gγ, of which the α subunit is similar to the small GTPases (**Fig. 3.3**). In the resting state, the G protein is inactive, not associated with the receptor, and a molecule of GDP is bound to the α subunit. Ligand binding induces conformational changes in the receptor that allow it to bind the G protein, which results in the displacement of the GDP from the G protein and its replacement with GTP. The active G protein dissociates into two components, the Gα subunit and a complex consisting of a Gβ and a Gγ subunit. Each of these components can interact with other intracellular signaling molecules to transmit and amplify the signal. G proteins can activate a wide variety of downstream enzymatic

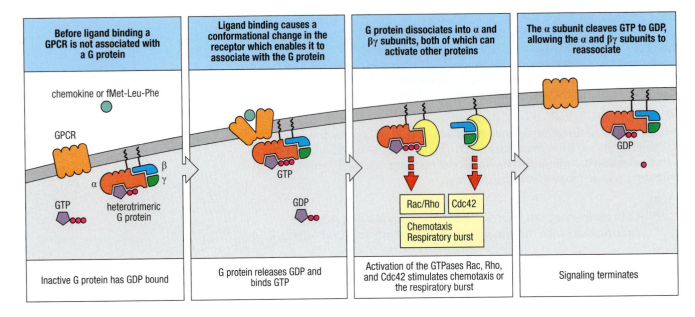

Before ligand binding a GPCR is not associated with a G protein	Ligand binding causes a conformational change in the receptor which enables it to associate with the G protein	G protein dissociates into α and βγ subunits, both of which can activate other proteins	The α subunit cleaves GTP to GDP, allowing the α and βγ subunits to reassociate
Inactive G protein has GDP bound	G protein releases GDP and binds GTP	Activation of the GTPases Rac, Rho, and Cdc42 stimulates chemotaxis or the respiratory burst	Signaling terminates

Fig. 3.3 G-protein-coupled receptors signal by coupling with intracellular heterotrimeric G proteins. First panel: G-protein-coupled receptors (GPCRs) such as the fMet-Leu-Phe (fMLF) and chemokine receptors signal through GTP-binding proteins known as heterotrimeric G proteins. In the inactive state, the α subunit of the G protein binds GDP and is associated with the β and γ subunits. Second panel: the binding of a ligand to the receptor induces a conformational change that allows the receptor to interact with the G protein, which results in the displacement of GDP and binding of GTP by the α subunit. Third panel: GTP binding triggers the dissociation of the G protein into the α subunit and the βγ subunit, each of which can activate other proteins at the inner face of the cell membrane. In the case of fMLF signaling in macrophages and neutrophils, the α subunit of the activated G protein indirectly activates the GTPases Rac and Rho, whereas the βγ subunit indirectly activates the GTPase Cdc42. The actions of these proteins result in the assembly of the NADPH oxidase, resulting in a respiratory burst. Chemokine signaling acts by a similar pathway and activates chemotaxis. Fourth panel: The activated response ceases when the intrinsic GTPase activity of the α subunit hydrolyzes GTP to GDP, and the α and βγ subunits reassociate. The intrinsic rate of GTP hydrolysis by α subunits is relatively slow, and signaling is regulated by additional GTPase-activating proteins (not shown), which accelerate the rate of GTP hydrolysis.

targets, such as adenylate cyclase, which produces the second messenger cyclic AMP; and phospholipase C, whose activation gives rise to the second messenger inositol 1,3,5-trisphosphate (IP_3) and the release of free Ca^{2+}.

Signaling by fMLF and C5a receptors influences cell motility, metabolism, gene expression, and cell division through activation of several **Rho family small GTPase proteins**. The α subunit of the activated G protein indirectly activates **Rac** and **Rho**, while the βγ subunit indirectly activates the small GTPase Cdc42 (see Fig. 3.3). Activation of these GTPases is controlled by **guanine nucleotide exchange factors (GEFs)** (see Fig. 7.4, which exchange GTP for GDP bound to the GTPase. The G proteins activated by fMLF activate the GEF protein **PREX1** (phosphatidylinositol 3,4,5-trisphosphate-dependent Rac exchanger 1 protein), which can directly activate Rac. Other GEFs, including members of the Vav family that are controlled by other types of receptors (see Section 7-19), can also activate Rac activity, and their activity synergizes with the actions of fMLF and C5a.

The activation of Rac and Rho helps to increase the microbicidal capacity of macrophages and neutrophils that have ingested pathogens. Upon phagocytosing microbes, macrophages and neutrophils produce a variety of toxic products that help to kill the engulfed microorganism (**Fig. 3.4**). The most important of these are the antimicrobial peptides described in Section 2-4, reactive nitrogen species such as nitric oxide (NO), and ROS, such as the superoxide anion (O_2^-) and hydrogen peroxide (H_2O_2). Nitric oxide is produced by a high-output form of nitric oxide synthase, inducible NOS2 (iNOS2), whose expression is induced by a variety of stimuli, including fMLF.

Activation of the fMLF and C5a receptors is directly involved in generating ROS. Superoxide is generated by a multicomponent, membrane-associated **NADPH oxidase**, also called **phagocyte oxidase**. In unstimulated phagocytes, this enzyme is inactive because it is not fully assembled. One set of subunits, the cytochrome b_{558} complex (composed of p22 and gp91), is localized in the plasma membranes of resting macrophages and neutrophils, and it appears in lysosomes after the maturation of phagolysosomes. The other components, p40, p47, and p67, are in the cytosol. Activation of phagocytes induces the cytosolic subunits to join with the membrane-associated cytochrome b_{558} to form a complete, functional NADPH oxidase in the phagolysosome membrane (**Fig. 3.5**). The fMLF and C5a receptors participate in the process by activating Rac, which functions to promote the movement of the cytosolic components to the membrane to assemble the active NADPH oxidase.

Antimicrobial mechanisms of phagocytes		
Class of mechanism	**Macrophage products**	**Neutrophil products**
Acidification	pH≈3.5–4.0, bacteriostatic or bactericidal	
Toxic oxygen-derived products	Superoxide O_2^-, hydrogen peroxide H_2O_2, singlet oxygen $^1O_2^{\bullet}$, hydroxyl radical $^{\bullet}OH$, hypohalite OCl^-	
Toxic nitrogen oxides	Nitric oxide NO	
Antimicrobial peptides	Cathelicidin, macrophage elastase-derived peptide	α-Defensins (HNP1–4), β-defensin HBD4, cathelicidin, azurocidin, bacterial permeability inducing protein (BPI), lactoferricin
Enzymes	Lysozyme: digests cell walls of some Gram-positive bacteria Acid hydrolases (e.g., elastase and other proteases): break down ingested microbes	
Competitors		Lactoferrin (sequesters Fe^{2+}), vitamin B_{12}-binding protein

Fig. 3.4 Bactericidal agents produced or released by phagocytes after uptake of microorganisms. Most of the agents listed are directly toxic to microbes and can act directly in the phagolysosome. They can also be secreted into the extracellular environment, and many of these substances are toxic to host cells. Other phagocyte products sequester essential nutrients in the extracellular environment, rendering them inaccessible to microbes and hindering microbial growth. Besides being directly bacteriostatic or bactericidal, acidification of lysosomes also activates the many acid hydrolases that degrade the contents of the vacuole.

The NADPH oxidase reaction results in a transient increase in oxygen consumption by the cell, which is known as the **respiratory burst**. It generates superoxide anion within the lumen of the phagolysosome, and this is converted by the enzyme **superoxide dismutase (SOD)** into H_2O_2. Further chemical and enzymatic reactions produce a range of toxic ROS from H_2O_2, including the hydroxyl radical ($^{\bullet}OH$), hypochlorite (OCl^-), and hypobromite (OBr^-). In this way, the direct recognition of bacterially derived polypeptides or previous pathogen recognition by the complement system activates a potent killing mechanism within macrophages and neutrophils that have ingested microbes via their phagocytic receptors. However, phagocyte activation can also cause extensive tissue damage because hydrolytic enzymes, membrane-disrupting peptides, and reactive oxygen species can be released into the extracellular environment and are toxic to host cells.

Neutrophils use the respiratory burst described above in their role as an early responder to infection. Neutrophils are not tissue-resident cells, and they need to be recruited to a site of infection from the bloodstream. Their sole function is to ingest and kill microorganisms. Although neutrophils are eventually present in much larger numbers than macrophages in some types of acute infection, they are short-lived, dying soon after they have accomplished a round of phagocytosis and used up their primary and secondary granules. Dead and dying neutrophils are a major component of the **pus** that forms in abscesses and in wounds infected by certain extracellular capsulated bacteria such as streptococci and staphylococci, which are thus known as **pus-forming**, or **pyogenic**, **bacteria**. Macrophages, in contrast, are long-lived cells and continue to generate new lysosomes.

Patients with a disease called **chronic granulomatous disease (CGD)** have a genetic deficiency of the NADPH oxidase, which means that their phagocytes do not produce the toxic oxygen derivatives characteristic of the respiratory burst and so are less able to kill ingested microorganisms and clear an infection. The most common form of CGD is an X-linked heritable disease that arises from inactivating mutations in the gene encoding the gp91 subunit of cytochrome b_{558}. People with this defect are unusually susceptible to bacterial and fungal infections, especially in infancy, though they remain susceptible for life. One autosomal recessive form of NADPH oxidase deficiency, p47phox deficiency, has very low but detectable activity and causes a milder form of CGD.

Chronic Granulomatous Disease

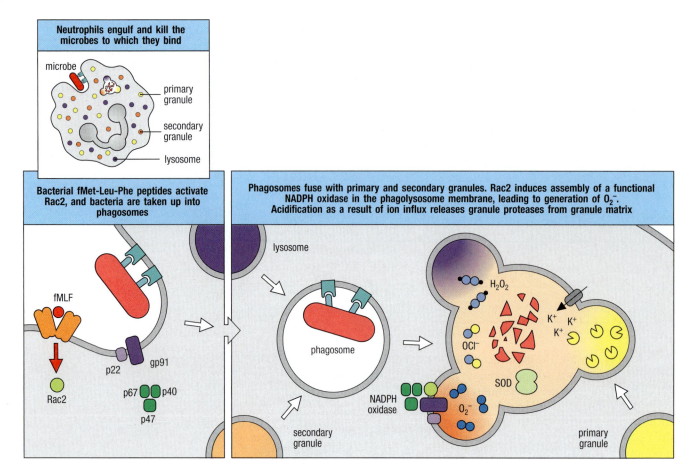

Fig. 3.5 The microbicidal respiratory burst in phagocytes is initiated by activation-induced assembly of the phagocyte NADPH oxidase. First panel: neutrophils are highly specialized for the uptake and killing of pathogens, and contain several different kinds of cytoplasmic granules, such as the primary and secondary granules shown in the first panel. These granules contain antimicrobial peptides and enzymes. Second panel: in resting neutrophils, the cytochrome b_{558} subunits (gp91 and p22) of the NADPH oxidase are localized in the plasma membrane; the other oxidase components (p40, p47, and p67) are located in the cytosol. Signaling by phagocytic receptors and by fMLF or C5a receptors synergizes to activate Rac2 and induce the assembly of the complete, active NADPH oxidase in the membrane of the phagolysosome, which has formed by the fusion of the phagosome with lysosomes and primary and secondary granules. Third panel: active NADPH oxidase transfers an electron from its FAD cofactor to molecular oxygen, forming the superoxide ion O_2^- (blue) and other free oxygen radicals in the lumen of the phagolysosome. Potassium and hydrogen ions are then drawn into the phagolysosome to neutralize the charged superoxide ion, increasing acidification of the vesicle. Acidification dissociates granule enzymes such as cathepsin G and elastase (yellow) from their proteoglycan matrix, leading to their cleavage and activation by lysosomal proteases. O_2^- is converted by superoxide dismutase (SOD) to hydrogen peroxide (H_2O_2), which can kill microorganisms, and can be converted by myeloperoxidase, a heme-containing enzyme, to microbicidal hypochlorite (OCl^-) and by chemical reaction with ferrous (Fe^{2+}) ions to the hydroxyl ($^\bullet OH$) radical.

MOVIE 3.4

In addition to killing microbes engulfed by phagocytosis, neutrophils use another rather novel mechanism of destruction that is directed at extracellular pathogens. During infection, some activated neutrophils undergo a unique form of cell death in which the nuclear chromatin, rather than being degraded as occurs during apoptosis, is released into the extracellular space and forms a fibril matrix known as **neutrophil extracellular traps**, or **NETs** (Fig. 3.6). NETs act to capture microorganisms, which may then be more efficiently phagocytosed by other neutrophils or macrophages. NET formation requires the generation of ROS, and patients with CGD have reduced NET formation, which may contribute to their susceptibility to microorganisms.

Macrophages can phagocytose pathogens and produce the respiratory burst immediately upon encountering an infecting microorganism, and this can be sufficient to prevent an infection from becoming established. In the nineteenth century, the immunologist **Élie Metchnikoff** believed that the innate response of macrophages encompassed all host defenses; indeed, invertebrates such as the sea star that he was studying rely entirely on innate immunity to overcome infection. Although this is not the case in humans and other vertebrates, the innate response of macrophages still provides an important front line of defense that must be overcome if a microorganism is to establish an infection that can be passed on to a new host.

Pathogens have, however, developed a variety of strategies to avoid immediate destruction by macrophages and neutrophils. Many extracellular pathogenic bacteria coat themselves with a thick polysaccharide capsule that is not recognized by any phagocytic receptor. In such cases, however, the complement system can recognize microbial surfaces and coat them with C3b, thereby flagging them for phagocytosis via complement receptors, as described in Chapter 2. Other pathogens, for example, mycobacteria, have evolved ways to grow inside macrophage phagosomes by inhibiting their acidification and fusion with lysosomes. Without such devices, a microorganism must enter the body in sufficient numbers to overwhelm the immediate innate host defenses and to establish a focus of infection.

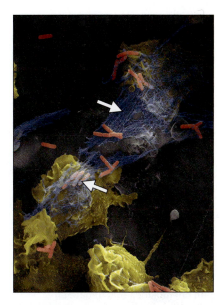

Fig. 3.6 Neutrophil extracellular traps (NETs) can trap bacteria and fungi. This scanning electron micrograph of activated human neutrophils infected with a virulent strain of *Shigella flexneri* (pink rods) shows the stimulated neutrophils forming NETs (blue, indicated by arrows). Bacteria trapped within NETs are visible (lower arrow). Photo courtesy of Arturo Zychlinsky.

3-3 Microbial recognition and tissue damage initiate an inflammatory response.

An important effect of the interaction between microbes and tissue macrophages is the activation of macrophages and other immune cells to release small proteins called **cytokines** and **chemokines**, and other chemical mediators. Collectively, these proteins induce a state of **inflammation** in the tissue, attract monocytes and neutrophils to the infection, and allow plasma proteins to enter the tissue from the blood. An inflammatory response is usually initiated within hours of infection or wounding. Macrophages are stimulated to secrete **pro-inflammatory** cytokines, such as TNF-α, and chemokines by interactions between microbes and microbial products and specific receptors expressed by the macrophage. We will examine how the cytokines interact with pathogens later in the chapter, but first we describe some general aspects of inflammation and how it contributes to host defense.

Inflammation has three essential roles in combating infection. The first is to deliver additional effector molecules and cells from the blood into sites of infection, and so increase the destruction of invading microorganisms. The second is to induce local blood clotting, which provides a physical barrier to the spread of the infection in the bloodstream. The third is to promote the repair of injured tissue.

Inflammatory responses are characterized by pain, redness, heat, and swelling at the site of an infection, reflecting four types of change in the local blood vessels, as shown in **Fig. 3.7**. The first is an increase in vascular diameter, leading to increased local blood flow—hence the heat and redness—and a reduction in the velocity of blood flow, especially along the inner walls of small blood vessels. The second change is the activation of endothelial cells lining the blood vessel to express **cell-adhesion molecules** that promote the binding of circulating leukocytes. The combination of slowed blood flow and adhesion molecules allows leukocytes to attach to the endothelium and migrate into the tissues, a process known as **extravasation**. All these changes are initiated by the pro-inflammatory cytokines and chemokines produced by activated macrophages and parenchymal cells.

Once inflammation has begun, the first white blood cells attracted to the site are neutrophils. These are followed by monocytes (**Fig. 3.8**), which upon activation are called **inflammatory monocytes** and can produce various pro-inflammatory cytokines, but are distinguishable from macrophages by their lack of expression of the adhesion G-protein-coupled receptor E1, commonly

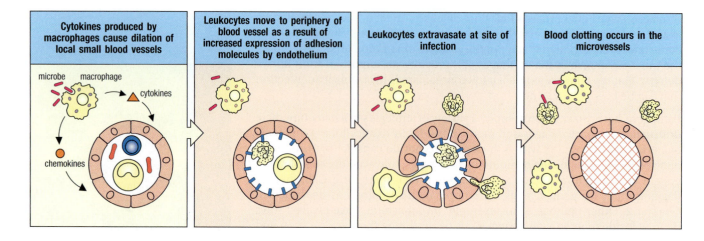

| Cytokines produced by macrophages cause dilation of local small blood vessels | Leukocytes move to periphery of blood vessel as a result of increased expression of adhesion molecules by endothelium | Leukocytes extravasate at site of infection | Blood clotting occurs in the microvessels |

Fig. 3.7 Infection stimulates macrophages to release cytokines and chemokines that initiate an inflammatory response. Cytokines produced by tissue macrophages at the site of infection cause the dilation of local small blood vessels and changes in the endothelial cells of their walls. These changes lead to the movement of leukocytes, such as neutrophils and monocytes, out of the blood vessel (extravasation) and into the infected tissue; this movement is guided by chemokines produced by the activated macrophages. The blood vessels also become more permeable, allowing plasma proteins and fluid to leak into the tissues. Together, these changes cause the characteristic inflammatory signs of heat, pain, redness, and swelling at the site of infection.

called F4/80. Monocytes are also able to give rise to dendritic cells in the tissues, depending on signals that they receive from their environment. In the later stages of inflammation, other leukocytes such as eosinophils and lymphocytes also enter the infected site.

The third major change in local blood vessels is an increase in vascular permeability. Thus, instead of being tightly joined together, the endothelial cells lining the blood vessel walls become separated, leading to an exit of fluid and proteins from the blood and their local accumulation in the tissue. This accounts for the swelling, or **edema**, and pain—as well as the accumulation in tissues of plasma proteins such as complement and MBL that aid in host defense. The changes that occur in endothelium as a result of inflammation are known generally as **endothelial activation**. The fourth change, clotting in microvessels in the site of infection, prevents the spread of the pathogen via the blood.

These changes are induced by a variety of inflammatory mediators released as a consequence of the recognition of pathogens by macrophages, and later by neutrophils and other white blood cells. Macrophages and neutrophils secrete lipid mediators of inflammation—**prostaglandins**, **leukotrienes**, and **platelet-activating factor (PAF)**—which are rapidly produced by enzymatic pathways that degrade membrane phospholipids. Their actions are followed by those of the chemokines and cytokines that are synthesized and secreted

Fig. 3.8 Monocytes circulating in the blood migrate into infected and inflamed tissues. Adhesion molecules on the endothelial cells of the blood vessel wall capture the monocyte and cause it to adhere to the vascular endothelium. Chemokines bound to the vascular endothelium then signal the monocyte to migrate across the endothelium into the underlying tissue. The monocyte, now differentiating into an inflammatory monocyte, continues to migrate, under the influence of chemokines released during inflammatory responses, toward the site of infection. Monocytes leaving the blood are also able to differentiate into dendritic cells (not shown), depending on the signals that they receive from their environment.

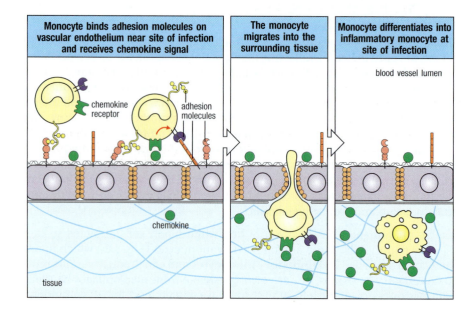

| Monocyte binds adhesion molecules on vascular endothelium near site of infection and receives chemokine signal | The monocyte migrates into the surrounding tissue | Monocyte differentiates into inflammatory monocyte at site of infection |

by macrophages and inflammatory monocytes in response to pathogens. The cytokine **tumor necrosis factor-α** (**TNF-α**, also known simply as **TNF**), for example, is a potent activator of endothelial cells. We describe TNF-α and related cytokines in more detail in Section 3-15.

Besides stimulating the respiratory burst in phagocytes and acting as a chemo-attractant for neutrophils and monocytes, C5a also promotes inflammation by increasing vascular permeability and inducing the expression of certain adhesion molecules on endothelium. C5a also activates local **mast cells** (see Section 1-4), which are stimulated to release their granules containing the small inflammatory molecule histamine as well as TNF-α and cathelicidins.

If wounding has occurred, the injury to blood vessels immediately triggers two protective enzyme cascades. One is the **kinin system** of plasma proteases that is triggered by tissue damage to generate several polypeptides that regulate blood pressure, coagulation, and pain. Although we will not fully describe its components here, one inflammatory mediator produced is the vaso-active peptide **bradykinin**, which increases vascular permeability to promote the influx of plasma proteins to the site of tissue injury. It also causes pain. Although unpleasant to the victim, pain draws attention to the problem and leads to immobilization of the affected part of the body, which helps to limit the spread of the infection.

The **coagulation system** is another protease cascade that is triggered in the blood after damage to blood vessels, although its full description is also outside our present scope. Its activation leads to the formation of a fibrin clot, whose normal role is to prevent blood loss. With regard to innate immunity, however, the clot physically encases the infectious microorganisms and prevents their entry into the bloodstream. The kinin and the coagulation cascades are also triggered by activated endothelial cells, and so they can have important roles in the inflammatory response to pathogens even if wounding or gross tissue injury has not occurred. Thus, within minutes of the penetration of tissues by a pathogen, the inflammatory response causes an influx of proteins and cells that may control the infection. Coagulation also forms a physical barrier in the form of blood clots to limit the spread of infection. Damage to tissues can occur in the absence of infection by microbes, such as physical trauma, ischemia, and metabolic or autoimmune disorders. In such **sterile injury**, many of the changes associated with infection, such as neutrophil recruitment, can also occur, in addition to activation of the kinin system and clot formation.

3-4   Toll-like receptors represent an ancient pathogen-recognition system.

Section 1-5 introduced pattern recognition receptors (PRRs), which function as sensors for pathogen-associated molecular patterns (PAMPs). Cytokine and chemokine production by macrophages is the result of signaling by these PRRs that is induced by a wide variety of pathogen components. The existence of these receptors was predicted by **Charles Janeway, Jr.,** before mechanisms of innate recognition were known, based on the requirement for adjuvants in driving immune responses to purified antigens. **Jules Hoffmann** discovered the first example of such a receptor, for which he was awarded part of the 2011 Nobel Prize in Physiology or Medicine. The receptor protein **Toll** was identified earlier as a gene controlling the correct dorso-ventral patterning embryo of the fruitfly *Drosophila melanogaster*. But in 1996, Hoffmann discovered that in the adult fly, Toll signaling induces the expression of several host-defense mechanisms, including antimicrobial peptides such as drosomycin, and is critical for defense against Gram-positive bacteria and fungal pathogens.

Mutations in *Drosophila* Toll or in signaling proteins activated by Toll decreased the production of antimicrobial peptides and led to susceptibility of the adult fly

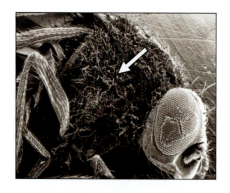

Fig. 3.9 Toll is required for antifungal responses in *Drosophila melanogaster*. Flies that are deficient in the Toll receptor are dramatically more susceptible than wild-type flies to fungal infection. This is illustrated here by the uncontrolled hyphal growth (arrow) of the normally weak pathogen *Aspergillus fumigatus* in a Toll-deficient fly. Photo courtesy of J.A. Hoffmann.

MOVIE 3.5

to fungal infections (Fig. 3.9). Subsequently, homologs of Toll, called **Toll-like receptors (TLRs)**, were found in other animals, including mammals, in which they are associated with resistance to viral, bacterial, and fungal infection. In plants, proteins with domains resembling the ligand-binding regions of TLR proteins are involved in the production of antimicrobial peptides, indicating the ancient association of these domains with this means of host defense.

3-5   Mammalian Toll-like receptors are activated by many different pathogen-associated molecular patterns.

There are 10 expressed *TLR* genes in humans and 12 in mice. Each TLR is devoted to recognizing a distinct set of molecular patterns that are essentially not found in healthy vertebrate cells. Initially called **pathogen-associated molecular patterns (PAMPs)**, these molecules are general components of both pathogenic and nonpathogenic microorganisms, and so are sometimes called microbial-associated molecular patterns, or MAMPs. Between them, the mammalian TLRs recognize molecules characteristic of Gram-negative and Gram-positive bacteria, fungi, and viruses. Among these, the **lipoteichoic acids** of Gram-positive bacterial cell walls and the **lipopolysaccharide (LPS)** of the outer membrane of Gram-negative bacteria (see Fig. 2.9) are particularly important in the recognition of bacteria by the innate immune system, and are recognized by TLRs. Other microbial components also have a repetitive structure. Bacterial flagella are made of a repeated **flagellin** subunit, and bacterial DNA has abundant repeats of **unmethylated CpG dinucleotides** (which are often methylated in mammalian DNA). In many viral infections, a double-stranded RNA intermediate is part of the viral life cycle, and frequently the viral RNA contains modifications that can be used to distinguish it from normal host RNA species.

The mammalian TLRs and their known microbial ligands are listed in Fig. 3.10. Because there are relatively few *TLR* genes, the TLRs have limited specificity compared with the antigen receptors of the adaptive immune system. However, they can recognize elements of most pathogenic microbes and are expressed by many types of cells, including macrophages, dendritic cells, B cells, stromal cells, and certain epithelial cells, enabling the initiation of antimicrobial responses in many tissues.

TLRs are sensors for microbes present in extracellular spaces. Some mammalian TLRs are cell-surface receptors similar to *Drosophila* Toll, but others are located intracellularly in the membranes of endosomes, where they detect pathogens or their components that have been taken into cells by phagocytosis, receptor-mediated endocytosis, or macropinocytosis (Fig. 3.11). TLRs are single-pass transmembrane proteins with an extracellular region composed of 18–25 copies of a **leucine-rich repeat (LRR)**. Each LLR of a TLR protein is composed of around 20–25 amino acids, and multiple LRRs create a horseshoe-shaped protein scaffold that is adaptable for ligand binding and recognition on both the outer (convex) and inner (concave) surfaces. Signaling by mammalian TLRs is activated when binding of a ligand induces formation of a dimer, or induces conformational changes in a preformed TLR dimer. All mammalian TLR proteins have in their cytoplasmic tail a **TIR** (for **Toll–IL-1 receptor**) **domain**, which interacts with other TIR-type domains, usually in other signaling molecules, and is also found in the cytoplasmic tail of the receptor for the cytokine **interleukin-1β (IL-1β)**. For years after the discovery of the mammalian TLRs it was not known whether they made direct contact with microbial products or whether they sensed the presence of microbes by some indirect means. *Drosophila* Toll, for example, does not recognize pathogen products directly, but instead it is activated when it binds a cleaved version of a self protein, Spätzle. *Drosophila* has other direct pathogen-recognition molecules, and these trigger the proteolytic cascade that ends in the cleavage

Innate immune recognition by mammalian Toll-like receptors		
Toll-like receptor	**Ligand**	**Hematopoietic cellular distribution**
TLR-1:TLR-2 heterodimer	Lipomannans (mycobacteria) Lipoproteins (diacyl lipopeptides; triacyl lipopeptides) Lipoteichoic acids (Gram-positive bacteria)	Monocytes, dendritic cells, mast cells, eosinophils, basophils
TLR-2:TLR-6 heterodimer	Cell-wall β-glucans (bacteria and fungi) Zymosan (fungi)	
TLR-3	Double-stranded RNA (viruses), poly I:C	Macrophages, dendritic cells, intestinal epithelium
TLR-4 (plus MD-2 and CD14)	LPS (Gram-negative bacteria) Lipoteichoic acids (Gram-positive bacteria)	Macrophages, dendritic cells, mast cells, eosinophils
TLR-5	Flagellin (bacteria)	Intestinal epithelium, macrophages, dendritic cells
TLR-7	Single-stranded RNA (viruses)	Plasmacytoid dendritic cells, macrophages, eosinophils, B cells
TLR-8	Single-stranded RNA (viruses)	Macrophages, neutrophils
TLR-9	DNA with unmethylated CpG (bacteria and herpesviruses)	Plasmacytoid dendritic cells, eosinophils, B cells, basophils
TLR-10 (human only)	Unknown	Plasmacytoid dendritic cells, eosinophils, B cells, basophils
TLR-11 (mouse only)	Profilin and profilin-like proteins (*Toxoplasma gondii*, uropathogenic bacteria)	Macrophages, dendritic cells (also liver, kidney, and bladder)
TLR-12 (mouse only)	Profilin (*Toxoplasma gondii*)	Macrophages, dendritic cells (also liver, kidney, bladder)
TLR-13 (mouse only)	Single-stranded RNA (bacterial ribosomal RNA)	Macrophages, dendritic cells

Fig. 3.10 Innate immune recognition by Toll-like receptors. Each of the human or mouse TLRs whose specificity is known recognizes one or more microbial molecular patterns, generally by direct interaction with molecules on the pathogen surface. Some Toll-like receptor proteins form heterodimers (e.g., TLR-1:TLR-2 and TLR-6:TLR-2). LPS, lipopolysaccharide.

of Spätzle. In this sense Toll is not a classical pattern recognition receptor. However, X-ray crystal structures of several mammalian dimeric TLRs bound to their ligands show that at least some mammalian TLRs make direct contact with microbial ligands.

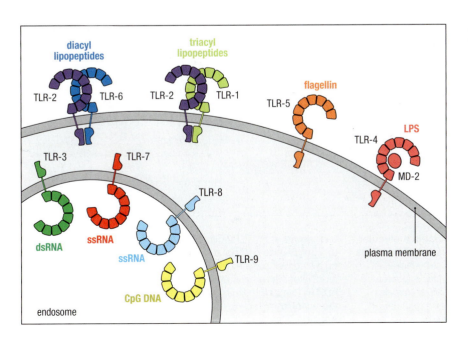

Fig. 3.11 The cellular locations of the mammalian Toll-like receptors. TLRs are transmembrane proteins whose extracellular region contains 18–25 copies of the leucine-rich repeat (LRR), but these cartoons depict only 9 LRRs for simplicity. Some TLRs are located on the cell surface of dendritic cells, macrophages, and other cells, where they are able to detect extracellular pathogen molecules. TLRs are thought to act as dimers. Only those that form heterodimers are shown in dimeric form here; the rest act as homodimers. TLRs located intracellularly, in the walls of endosomes, can recognize microbial components, such as DNA, that are accessible only after the microbe has been broken down. The diacyl and triacyl lipopeptides recognized by the heterodimeric receptors TLR-6:TLR-2 and TLR-1:TLR-2, respectively, are derived from the lipoteichoic acid of Gram-positive bacterial cell walls and the lipoproteins of Gram-negative bacterial surfaces.

Mammalian **TLR-1**, **TLR-2**, and **TLR-6** are cell-surface receptors that are activated by various ligands, including lipoteichoic acid and the **diacyl** and **triacyl lipoproteins** of Gram-negative bacteria. They are found on macrophages, dendritic cells, eosinophils, basophils, and mast cells. Ligand binding induces the formation of heterodimers of TLR-2 and TLR-1, or of TLR-2 and TLR-6. The X-ray crystal structure of a synthetic triacyl lipopeptide ligand bound to TLR-1 and TLR-2 shows exactly how it induces dimerization (Fig. 3.12). Two of the three lipid chains bind to the convex surface of TLR-2, while the third binds to the convex surface of TLR-1. Dimerization brings the cytoplasmic TIR domains of the TLR chains into close proximity with each other to initiate signaling. Similar interactions are presumed to occur with the diacyl lipopeptide ligands that induce the dimerization of TLR-2 and TLR-6. The scavenger receptor CD36, which binds long-chain fatty acids, and Dectin-1, which binds β-glucans (see Section 3-1), both cooperate with TLR-2 in ligand recognition.

TLR-5 is expressed on the cell surface of macrophages, dendritic cells, and intestinal epithelial cells; it recognizes flagellin, a protein subunit of bacterial flagella. TLR-5 recognizes a highly conserved site on flagellin that is buried and inaccessible in the assembled flagellar filament. This means that the receptor is activated only by monomeric flagellin, which is produced by the breakdown of flagellated bacteria in the extracellular space. Mice, but not humans, express **TLR-11** and **TLR-12**, which share with TLR-5 the ability to recognize an intact protein. TLR-11 is expressed by macrophages and dendritic cells, and also by liver, kidney, and bladder epithelial cells.

TLR-12 is also expressed in macrophages and dendritic cells, and is more broadly expressed in hematopoietic cells than TLR-11, but is not expressed by the epithelial tissues where TLR-11 is expressed. TLR-11-deficient mice develop urinary infections caused by uropathogenic strains of *Escherichia coli*, although the bacterial ligand for TLR-11 has not yet been identified. TLR-11 and TLR-12 have an overlapping function in that both recognize protozoan parasites such as *Toxoplasma gondii* and *Plasmodium falciparum*. They bind to protein motifs that are present in the protozoan actin-binding protein **profilin** but absent in mammalian profilins. TLR-11 and TLR-12 are both required in macrophages and conventional dendritic cells for activation by *T. gondii* profilin, but TLR-12 plays a more dominant role. Mice lacking TLR-11 develop more severe tissue injury than normal mice on infection with

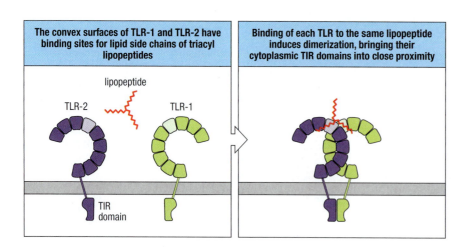

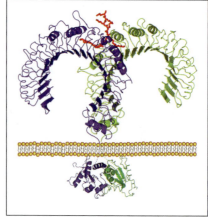

Fig. 3.12 Direct recognition of pathogen-associated molecular patterns by TLR-1 and TLR-2 induces dimerization of the TLRs and signaling. TLR-1 and TLR-2 are located on cell surfaces (left panel), where they can directly recognize bacterial triacyl lipoproteins (middle panel). The convex surfaces of their extracellular domains have binding sites for the lipid side chains of triacyl lipopeptides. In the crystal structure (right panel), the ligand is a synthetic lipid that can activate TLR1:TLR2 dimers; it has three fatty-acid chains bound to a polypeptide backbone. Two fatty-acid chains bind to a pocket on the convex surface of the TLR-2 ectodomain, and the third chain associates with a hydrophobic channel in the convex binding surface of TLR-1, inducing dimerization of the two TLR subunits and bringing their cytoplasmic Toll–IL-1 receptor (TIR) domains together to initiate signaling. Structure courtesy of Jie-Oh Lee.

Toxoplasma, whereas mice lacking TLR-12 die rapidly after infection. TLR-10 is expressed in humans, but *TLR-10* is a pseudogene in mice. Its ligand and function are currently not known.

Not all mammalian TLRs are cell-surface receptors. The TLRs that recognize nucleic acids are located in the membranes of endosomes, to which they are transported via the endoplasmic reticulum. **TLR-3** is expressed by macrophages, conventional dendritic cells, and intestinal epithelial cells; it recognizes **double-stranded RNA (dsRNA)**, which is a replicative intermediate of many types of viruses, not only those with RNA genomes. dsRNA is internalized either by the direct endocytosis of viruses with double-stranded RNA genomes, such as rotavirus, or by the phagocytosis of dying cells in which viruses are replicating, and it encounters the TLRs when the incoming endocytic vesicle or phagosome fuses with the TLR-containing endosome. Crystallographic analysis shows that TLR-3 binds directly to dsRNA. The TLR-3 ectodomain (the ligand-binding domain) has two contact sites for dsRNA: one on the amino terminus and a second near the membrane-proximal carboxy terminus. The twofold symmetry of dsRNA allows it to bind simultaneously to two TLR-3 ectodomains, inducing a dimerization that brings the TIR domains of TLR-3 together and activates intracellular signaling. This can be verified by using poly I:C to artificially induce signaling. A synthetic polymer composed of inosinic and cytidylic acid, poly I:C binds to TLR-3 and functions as an analog of dsRNA; poly I:C is often used experimentally to activate this pathway. Mutations in the ectodomain of human TLR-3, which produce a dominantly acting loss-of-function mutant receptor, have been associated with encephalitis that is caused by a failure to control the herpes simplex virus.

Recurrent Herpes Simplex Encephalitis

TLR-7, **TLR-8**, and **TLR-9**, like TLR-3, are endosomal nucleotide sensors involved in the recognition of viruses. TLR-7 and TLR-9 are expressed by plasmacytoid dendritic cells, B cells, and eosinophils; TLR-8 is expressed primarily by monocytes and macrophages. TLR-7 and TLR-8 are activated by **single-stranded RNA (ssRNA)**, which is a component of healthy mammalian cells, but it is normally confined to the nucleus and cytoplasm and is not present in endosomes. Many virus genomes, for example, those of orthomyxoviruses (such as influenza) and flaviviruses (such as West Nile virus), consist of ssRNA. When extracellular particles of these viruses are endocytosed by macrophages or dendritic cells, they are uncoated in the acidic environment of endosomes and lysosomes, exposing the ssRNA genome for recognition by TLR-7. Mice lacking TLR-7 have impaired immune responses to viruses such as influenza. In abnormal settings, TLR-7 may be activated by self-derived ssRNA. Normally, extracellular RNases degrade the ssRNA released from apoptotic cells during tissue injury. But in a mouse model of lupus nephritis, an inflammatory condition of the kidney, TLR-7 recognition of self ssRNA was observed to contribute to disease. Several studies have identified polymorphisms in the human TLR-7 gene that are associated with increased risk of the autoimmune disease systemic lupus erythematosus, suggesting a potential role in this disease. The role for TLR-8 has not been established as clearly from mouse model systems as for TLR-7. TLR-9 recognizes **unmethylated CpG dinucleotides**. In mammalian genomes, CpG dinucleotides in genomic DNA are heavily methylated on the cytosine by DNA methyltransferases. But in the genomes of bacteria and many viruses, CpG dinucleotides remain unmethylated and represent another pathogen-associated molecular pattern.

The delivery of TLR-3, TLR-7, and TLR-9 from the endoplasmic reticulum to the endosome relies on their interaction with a specific protein, **UNC93B1**, which is composed of 12 transmembrane domains. Mice lacking this protein have defects in signaling by these endosomal TLRs. Rare human mutations in UNC93B1 have been identified as causing susceptibility to herpes simplex encephalitis, similarly to TLR-3 deficiency, but do not impair immunity to many other viral pathogens, presumably because of the existence of other viral sensors, which are discussed later in this chapter.

3-6 **TLR-4 recognizes bacterial lipopolysaccharide in association with the host accessory proteins MD-2 and CD14.**

Not all mammalian TLRs bind their ligands so directly. **TLR-4** is expressed by several types of immune-system cells, including dendritic cells and macrophages, and is important in sensing and responding to numerous bacterial infections. TLR-4 recognizes the LPS of Gram-negative bacteria by a mechanism that is partly direct and partly indirect. The systemic injection of LPS causes a collapse of the circulatory and respiratory systems, a condition known as shock. These dramatic effects of LPS are seen in humans as **septic shock**, which results from an uncontrolled systemic bacterial infection, or **sepsis**. In this case, LPS induces an overwhelming secretion of cytokines, particularly TNF-α (see Section 3-15), causing systemic vascular permeability, an undesirable effect of its normal role in containing local infections. Mutant mice lacking TLR-4 function are resistant to LPS-induced septic shock but are highly sensitive to LPS-bearing pathogens such as *Salmonella typhimurium*, a natural pathogen of mice. In fact, TLR-4 was identified as the receptor for LPS by positional cloning of its gene from the LPS-resistant C3H/HeJ mouse strain, which harbors a naturally occurring mutation in the cytoplasmic tail of TLR-4 that interferes with the receptor's ability to signal. For this discovery, the 2011 Nobel Prize in Physiology or Medicine was partly awarded to **Bruce Buetler**.

LPS varies in composition among different bacteria but essentially consists of a polysaccharide core attached to an amphipathic lipid, lipid A, with a variable number of fatty-acid chains per molecule. To recognize LPS, the ectodomain of TLR-4 uses an accessory protein, **MD-2**. MD-2 initially binds to TLR-4 within the cell and is necessary both for the correct trafficking of TLR-4 to the cell surface and for the recognition of LPS. MD-2 associates with the central section of the curved ectodomain of TLR-4, binding off to one side as shown in **Fig. 3.13**. When the TLR4–MD-2 complex encounters LPS, five lipid chains of LPS bind to a deep hydrophobic pocket of MD-2, but not directly to TLR-4, while a sixth lipid chain remains exposed on the surface of MD-2. This last lipid chain and parts of the LPS polysaccharide backbone can then bind to the convex side of a second TLR-4 ectodomain, inducing TLR-4 dimerization that activates intracellular signaling pathways.

TLR-4 activation by LPS involves two other accessory proteins besides MD-2. While LPS is normally an integral component of the outer membrane of Gram-negative bacteria, during infections it can become detached from the membrane and be picked up by the host **LPS-binding protein** present in the blood and in extracellular fluid in tissues. LPS is transferred from LPS-binding protein to a second protein, CD14, which is present on the surface of macrophages, neutrophils, and dendritic cells. On its own, CD14 can act as a phagocytic receptor, but on macrophages and dendritic cells it also acts as an accessory protein for TLR-4.

3-7 **TLRs activate NFκB, AP-1, and IRF transcription factors to induce the expression of inflammatory cytokines and type I interferons.**

Signaling by mammalian TLRs in various cell types induces a diverse range of intracellular responses that together result in the production of inflammatory cytokines, chemotactic factors, antimicrobial peptides, and the antiviral cytokines **interferon-α** and **-β** (**IFN-α** and **IFN-β**), the **type I interferons**. TLR signaling achieves this by activating several different signaling pathways that each activate different transcription factors. As mentioned earlier, ligand-induced dimerization of two TLR ectodomains brings the cytoplasmic TIR domains together, allowing them to interact with the TIR domains of cytoplasmic adaptor molecules that initiate intracellular signaling. There are four such adaptors used by mammalian TLRs: **MyD88**, **MAL** (also known as TIRAP), **TRIF**, and **TRAM**. It is significant that the TIR domains of the different TLRs interact with different combinations of these adaptors (Fig. 3.14). Most TLRs interact only with MyD88, which is required for their signaling.

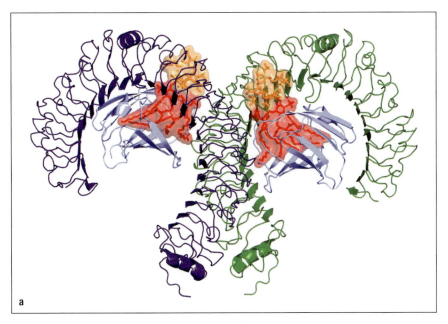

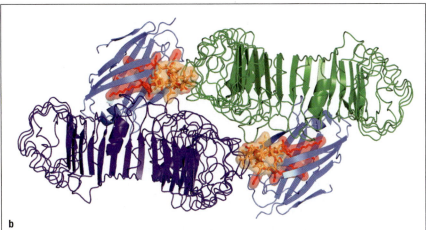

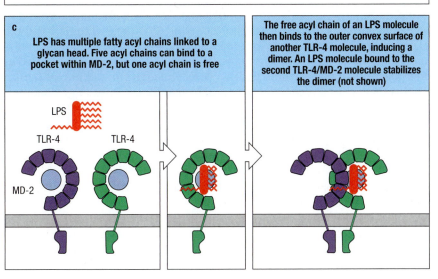

c

LPS has multiple fatty acyl chains linked to a glycan head. Five acyl chains can bind to a pocket within MD-2, but one acyl chain is free

The free acyl chain of an LPS molecule then binds to the outer convex surface of another TLR-4 molecule, inducing a dimer. An LPS molecule bound to the second TLR-4/MD-2 molecule stabilizes the dimer (not shown)

Fig. 3.13 TLR-4 recognizes LPS in association with the accessory protein MD-2. Panel a: a side view of the symmetrical complex of TLR-4, MD-2, and LPS. TLR-4 polypeptide backbones are shown in green and dark blue. The structure shows the entire extracellular region of TLR-4, composed of the LRR region (shown in green and dark blue), but lacks the intracellular signaling domain. The MD-2 protein is shown in light blue. Five of the LPS acyl chains (shown in red) are inserted into a hydrophobic pocket within MD-2. The remainder of the LPS glycan and one lipid chain (orange) make contact with the convex surface of a TLR-4 monomer. Panel b: the top view of the structure shows that an LPS molecule makes contact with one TLR-4 subunit on its convex (outer) surface, while binding to an MD-2 molecule that is attached to the other TLR-4 subunit. The MD-2 protein binds off to one side of the TLR-4 LRR region. Panel c: schematic illustration of relative orientation of LPS binding to MD-2 and TLR-4. Structures courtesy of Jie-Oh Lee.

TLR	Adaptor
TLR-2/1	MyD88/MAL
TLR-3	TRIF
TLR-4	MyD88/MAL TRIF/TRAM
TLR-5	MyD88
TLR-2/6	MyD88/MAL
TLR-7	MyD88
TLR-8	MyD88
TLR-9	MyD88
TLR11/12	MyD88
TLR-13	MyD88

Fig. 3.14 Mammalian TLRs interact with different TIR-domain adaptor molecules to activate downstream signaling pathways. The four signaling adaptor molecules used by mammalian TLRs are MyD88 (myeloid differentiation factor 88), MAL (MyD88 adaptor-like, also known as TIRAP, for TIR-containing adaptor protein), TRIF (TIR domain-containing adaptor-inducing IFN-β), and TRAM (TRIF-related adaptor molecule). All TLRs interact with MyD88, except TLR-3, which interacts only with TRIF. The table indicates the known pattern of adaptor interactions for the known TLRs.

**Interleukin 1
Receptor-Associated
Kinase 4 Deficiency**

TLR-3 interacts only with TRIF. Other TLRs use either MyD88 paired with MAL, or TRIF paired with TRAM. Signaling by the TLR-2 heterodimers (TLR-2/1 and TLR-2/6) requires MyD88/MAL. TLR-4 signaling uses both of these adaptor pairs, MyD88/MAL and TRIF/TRAM, which is used during endosomal signaling by TLR-4. Importantly, the choice of adaptor influences which of the several downstream signals will be activated by the TLR.

Signaling by most TLRs activates the transcription factor **NFκB** (Fig. 3.15), which is related to DIF, the factor activated by *Drosophila* Toll. Mammalian TLRs also activate several members of the **interferon regulatory factor (IRF)** transcription factor family through a second pathway, and they activate members of the **activator protein 1 (AP-1)** family, such as c-Jun, through yet another signaling pathway involving **mitogen-activated protein kinases (MAPKs)**. NFκB and AP-1 act primarily to induce the expression of pro-inflammatory cytokines and chemotactic factors. The IRF factors IRF3 and IRF7 are particularly important for inducing antiviral type I interferons, whereas a related factor, IRF5, is involved in the production of pro-inflammatory cytokines. Here we will describe how TLR signaling induces the transcription of various cytokine genes; later in the chapter, we will explain how those cytokines exert their various actions.

We consider first the signaling pathway triggered by TLRs that use MyD88. Two protein domains of MyD88 are responsible for its function as an adaptor. MyD88 has a TIR domain at its carboxy terminus that associates with the TIR domains in the TLR cytoplasmic tails. At its amino terminus, MyD88 has a **death domain**, so named because it was first identified in signaling proteins involved in apoptosis, a type of programmed cell death. The MyD88 death domain associates with a similar death domain present in other intracellular signaling proteins. Both MyD88 domains are required for signaling, since rare mutations in either domain are associated with immunodeficiency characterized by recurrent bacterial infections in humans. The MyD88 death domain recruits and activates two serine–threonine protein kinases—**IRAK4 (IL-1-receptor associated kinase 4)** and **IRAK1**—via their death domains. This IRAK complex performs two functions: it recruits enzymes that produce a **signaling scaffold**, and uses this scaffold to recruit other molecules that are then phosphorylated by the IRAKs.

To form a signaling scaffold, the IRAK complex recruits the enzyme **TRAF6 (tumor necrosis factor receptor-associated factor 6)**, which is an E3 **ubiquitin ligase** that acts in cooperation with **UBC13**, an E2 ubiquitin ligase, and its cofactor **Uve1A** (together called **TRIKA1**) (see Fig. 3.15). The combined activity of TRAF-6 and UBC13 is to ligate (unite with a chemical bond) one ubiquitin molecule to another protein, which can be another ubiquitin molecule, and thereby generate protein polymers. The polyubiquitin involved in signaling contains linkages between the lysine 63 on one ubiquitin and the carboxy terminus of the next, forming **K63 linkages**. This polyubiquitin polymer can be initiated on other proteins, including TRAF-6 itself, or produced as free linear ubiquitin polymers, and can be extended to produce **polyubiquitin chains** that act as a platform—or scaffold—that bind to other signaling molecules. Next, the scaffold recruits a signaling complex consisting of the poly-ubiquitin-binding adaptor proteins **TAB1**, **TAB2**, and the serine–threonine kinase **TAK1** (see Fig. 3.15). By being brought onto the scaffold, TAK1 is phosphorylated by the IRAK complex, and activated TAK1 propagates signaling by activating certain MAPKs, such as c-Jun terminal kinase (JNK) and MAPK14 (p38 MAPK). These then activate AP-1-family transcription factors that transcribe cytokine genes.

TAK1 also phosphorylates and activates the **IκB kinase (IKK)** complex, which is composed of three proteins: **IKKα**, **IKKβ**, and **IKKγ** (also known as **NEMO**, for NFκB essential modifier). NEMO functions by binding to polyubiquitin chains, which brings the IKK complex into proximity with TAK1. TAK1

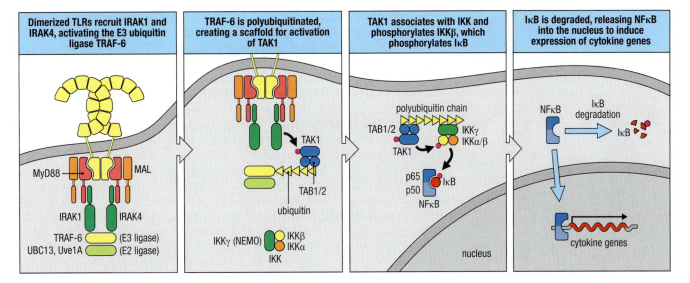

| Dimerized TLRs recruit IRAK1 and IRAK4, activating the E3 ubiquitin ligase TRAF-6 | TRAF-6 is polyubiquitinated, creating a scaffold for activation of TAK1 | TAK1 associates with IKK and phosphorylates IKKβ, which phosphorylates IκB | IκB is degraded, releasing NFκB into the nucleus to induce expression of cytokine genes |

Fig. 3.15 TLR signaling can activate the transcription factor NFκB, which induces the expression of pro-inflammatory cytokines. First panel: TLRs signal via their cytoplasmic TIR domains, which are brought into proximity to each other by ligand-induced dimerization of their ectodomains. Some TLRs use the adaptor protein MyD88, and others use the MyD88/MAL pair to initiate signaling. The MyD88 death domain recruits the serine–threonine kinases IRAK1 and IRAK4, in association with the ubiquitin E3 ligase TRAF-6. IRAK undergoes autoactivation and phosphorylates TRAF-6, activating its E3 ligase activity. Second panel: TRAF-6 cooperates with an E2 ligase (UBC13) and a cofactor (Uve1A) to generate polyubiquitin scaffolds (yellow triangles) by attachment of ubiquitin through its lysine 63 (K63). This scaffold recruits a complex of proteins composed of the kinase TAK1 (transforming growth factor-β-activated kinase 1) and two adaptor proteins, TAB1 (TAK1-binding protein 1) and TAB2. TAB1 and TAB2 function to bind to polyubiquitin, bringing TAK1 into proximity with IRAK to become phosphorylated (red dot). Third panel: activated TAK1 activates IKK, the IκB kinase complex. First, the IKKγ subunit (NEMO) binds to the polyubiquitin scaffold and brings the IKK complex into proximity to TAK1. TAK1 then phosphorylates and activates IKKβ. IKKβ then phosphorylates IκB, the cytoplasmic inhibitor of NFκB. Fourth panel: phosphorylated IκB is targeted by a process of ubiquitination (not shown) that leads to its degradation. This releases NFκB, which is composed of two subunits, p50 and p65, into the nucleus, driving the transcription of many genes including those encoding inflammatory cytokines. TAK1 also stimulates activation of the mitogen-activated protein kinases (MAPKs) JNK and p38, which phosphorylate and activate AP-1 transcription factors (not shown).

phosphorylates and activates IKKβ. IKKβ then phosphorylates **IκB** (inhibitor of κB), which is a distinct molecule whose name should not be confused with IKKβ. IκB is a cytoplasmic protein that constitutively binds to the transcription factor NFκB, which is composed of two subunits, **p50** and **p65**. The binding of IκB traps the NFκB proteins in the cytoplasm. Phosphorylation by IKK induces the degradation of IκB, and this releases NFκB into the nucleus, where it can drive transcription of genes for pro-inflammatory cytokines such as TNF-α, IL-1β, and **IL-6**. The actions of these cytokines in the innate immune response are described in the second half of this chapter. The outcome of TLR activation can also vary depending on the cell type in which it occurs. For example, activation of TLR-4 via MyD88 in specialized epithelial cells such as the Paneth cells of the intestine (see Section 2-4) results in the production of antimicrobial peptides, a mammalian example of the ancient function of Toll-like proteins.

The ability of TLRs to activate NFκB is crucial to their role of alerting the immune system to the presence of bacterial pathogens. Rare instances of inactivating mutations in IRAK4 in humans cause an immunodeficiency, **IRAK4 deficiency**, which, like MyD88 deficiency, is characterized by recurrent bacterial infections. Mutations in human NEMO produce a syndrome known as **X-linked hypohidrotic ectodermal dysplasia and immunodeficiency** or **NEMO deficiency**, which is characterized by both immunodeficiency and developmental defects.

The nucleic-acid-sensing TLRs—TLR-3, TLR-7, TLR-8, and TLR-9—activate members of the IRF family. IRF proteins reside in the cytoplasm and are inactive until they become phosphorylated on serine and threonine residues in

X-linked Hypohidrotic Ectodermal Dysplasia and Immunodeficiency

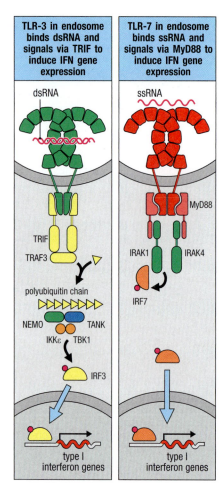

| TLR-3 in endosome binds dsRNA and signals via TRIF to induce IFN gene expression | TLR-7 in endosome binds ssRNA and signals via MyD88 to induce IFN gene expression |

Fig. 3.16 Expression of antiviral interferons in response to viral nucleic acids can be stimulated by two different pathways from different TLRs. Left panel: TLR-3, expressed by dendritic cells and macrophages, senses double-stranded viral RNA (dsRNA). TLR-3 signaling uses the adaptor protein TRIF, which recruits the E3 ligase TRAF3 to generate K63-linked polyubiquitin chains. This scaffold recruits NEMO and TANK (TRAF family member-associated NFκB activator), which associate with the serine–threonine kinases IKKε (IκB kinase ε) and TBK1 (TANK-binding kinase 1). TBK1 phosphorylates (red dot) the transcription factor IRF3, and IRF3 then enters the nucleus and induces expression of type I interferon genes. Right panel: TLR-7, expressed by plasmacytoid dendritic cells, detects single-stranded RNA (ssRNA) and signals through MyD88. Here, IRAK1 directly recruits and phosphorylates IRF7, which is also highly expressed in plasmacytoid dendritic cells. IRF7 then enters the nucleus to induce expression of type I interferons.

their carboxy termini. They then move to the nucleus as active transcription factors. Of the nine IRF family members, IRF3 and IRF7 are particularly important for TLR signaling and expression of antiviral type I interferons. For TLR-3, expressed by macrophages and conventional dendritic cells, the cytoplasmic TIR domain interacts with the adaptor protein TRIF. TRIF interacts with the E3 ubiquitin ligase **TRAF3**, which, like TRAF6, generates a polyubiquitin scaffold. In TLR-3 signaling, this scaffold recruits a multiprotein complex containing the kinases **IKKε** and **TBK1**, which phosphorylate IRF3 (**Fig. 3.16**). TLR-4 also triggers this pathway by binding TRIF, but the IRF3 response induced by TLR-4 is relatively weak compared with that induced by TLR-3, and its functional role *in vivo* remains elusive. In contrast to TLR-3, TLR-7, TLR-8, and TLR-9 signal uniquely through MyD88. For TLR-7 and TLR-9 signaling in plasmacytoid dendritic cells, the MyD88 TIR domain recruits the IRAK1/IRAK4 complex as described above. Here, the IRAK complex carries out a distinct function beyond recruiting TRAFs that generate a signaling scaffold. In these cells, IRAK1 can also physically associate with IRF7, which is highly expressed by plasmacytoid dendritic cells. This allows IRF7 to become phosphorylated by IRAK1, leading to induction of type I interferons (see Fig. 3.16). Not all IRF factors regulate type I interferon genes; IRF5, for example, plays a role in the induction of pro-inflammatory cytokines.

The collective ability of TLRs to activate both IRFs and NFκB means that they can stimulate either antiviral or antibacterial responses as needed. In human IRAK4 deficiency, for example, no extra susceptibility to viral infections has been noted. This would suggest that IRF activation is not impaired and the production of antiviral interferons is not affected. TLRs are expressed by different types of cells involved in innate immunity and by some stromal and epithelial cells, and the responses generated will differ in some respects depending on what type of cell is being activated.

3-8   The NOD-like receptors are intracellular sensors of bacterial infection and cellular damage.

The TLRs, being expressed on the cell's plasma membrane or endocytic vesicles, are primarily sensors of extracellular microbial products. Since the discovery of Toll and the mammalian TLRs, additional families of innate sensors have been identified that detect microbial products in the cytoplasm. One large group of cytoplasmic innate sensors has a centrally located **nucleotide-binding oligomerization domain (NOD)**, and other variable domains that detect microbial products or cellular damage or that activate signaling pathways; collectively, these are the **NOD-like receptors (NLRs)**. Some NLRs activate NFκB to initiate the same inflammatory responses as the TLRs, while other NLRs trigger a distinct pathway that induces cell death and the production of pro-inflammatory cytokines. The NLRs are considered a very ancient family of innate immunity receptors because the resistance (R) proteins that are part of plant defenses against pathogens are NLR homologs.

Subfamilies of NLRs can be distinguished on the basis of the other protein domains they contain. The **NOD** subfamily has an amino-terminal **caspase recruitment domain (CARD)** (**Fig. 3.17**). CARD was initially recognized in a family of proteases called **caspases** (for cysteine-aspartic acid proteases), which are important in many intracellular pathways, including those leading to cell death by apoptosis. CARD is structurally related to the TIR death domain in MyD88 and can dimerize with CARD domains on other proteins to induce signaling (**Fig. 3.18**). NOD proteins recognize fragments of bacterial cell-wall peptidoglycans, although it is not known whether this occurs through direct binding or via accessory proteins. **NOD1** senses **γ-glutamyl diaminopimelic acid (iE-DAP)**, a breakdown product of peptidoglycans of Gram-negative bacteria such as *Salmonella* and some Gram-positive bacteria such as *Listeria*,

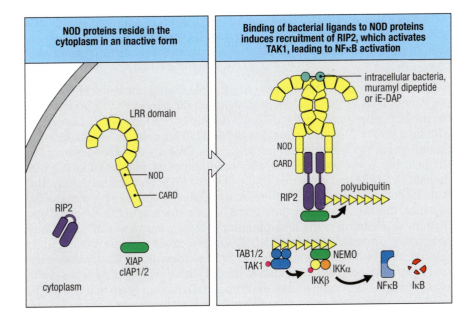

Fig. 3.17 Intracellular NOD proteins sense the presence of bacteria by recognizing bacterial peptidoglycans and activate NFκB to induce the expression of pro-inflammatory genes. First panel: NOD proteins reside in an inactive state in the cytoplasm, where they serve as sensors for various bacterial components. Second panel: degradation of bacterial cell-wall peptidoglycans produces muramyl dipeptide, which is recognized by NOD2. NOD1 recognizes γ-glutamyl diaminopimelic acid (iE-DAP), a breakdown product of Gram-negative bacterial cell walls. The binding of these ligands to NOD1 or NOD2 induces aggregation, allowing CARD-dependent recruitment of the serine–threonine kinase RIP2, which associates with E3 ligases, including XIAP (X-linked inhibitor of apoptosis protein), cIAP1 (cellular inhibitor of apoptosis 1), and cIAP2. This recruited E3 ligase activity produces a polyubiquitin scaffold, as in TLR signaling, and the association of TAK1 and the IKK complex with this scaffold leads to the activation of NFκB, as shown in Fig. 3.15. In this pathway, RIP2 acts as a scaffold to recruit XIAP, and RIP2 kinase activity is not required for signaling.

whereas **NOD2** recognizes **muramyl dipeptide (MDP)**, which is present in the peptidoglycans of most bacteria. NOD ligands may enter the cytoplasm as a result of intracellular infection, but may also be transported from materials captured by endocytosis, since mice lacking an oligopeptide transporter (SLC15A4) that is present in lysosomes have greatly reduced responses to NOD1 ligands.

When NOD1 or NOD2 recognizes its ligand, it recruits the CARD-containing serine–threonine kinase **RIP2** (also known as RICK and RIPK2) (see Fig. 3.17). RIP2 associates with the E3 ligases cIAP1, cIAP2, and XIAP, whose activity generates a polyubiquitin scaffold as in TLR signaling. This scaffold recruits TAK1 and IKK and results in activation of NFκB as shown in Fig. 3.15. NFκB then induces the expression of genes for inflammatory cytokines and for enzymes involved in the production of **nitric oxide (NO)**, which is toxic to bacteria and intracellular parasites. In keeping with their role as sensors of bacterial components, NOD proteins are expressed in cells that are routinely exposed to bacteria. These include epithelial cells forming the barrier that bacteria must cross to establish an infection in the body, and the macrophages and dendritic cells that ingest bacteria that have succeeded in entering the body. Macrophages and dendritic cells express TLRs as well as NOD1 and NOD2, and are activated by both pathways. In epithelial cells, NOD1 is an important activator of responses against bacterial infections, and NOD1 may also function as a systemic activator of innate immunity. It seems that peptidoglycans from intestinal microbiota are transported via blood in amounts sufficient to

Domain	Proteins
TIR	MyD88, MAL, TRIF, TRAM, all TLRs
CARD	Caspase 1, RIP2, RIG-I, MDA-5, MAVS, NODs, NLRC4, ASC, NLRP1
Pyrin	AIM2, IFI16, ASC, NLRP1-14
DD (death domain)	MyD88, IRAK1, IRAK4, DR4, DR5, FADD, FAS,
DED (death effector domain)	Caspase 8, caspase 10, FADD

Fig. 3.18 Protein-interaction domains contained in various immune signaling molecules. Signaling proteins contain protein-interaction domains that mediate the assembly of larger complexes. The table shows examples of proteins discussed in this chapter that contain the indicated domain. Proteins may have more than one domain, such as the adaptor protein MyD88, which can interact with TLRs via its TIR domain and with IRAK1/4 via its death domain (DD).

increase basal activation of neutrophils. A reduction in neutrophils activated in this way may explain why mice lacking NOD1 show increased susceptibility even to pathogens that lack NOD ligands, such as the parasite *Trypanosoma cruzi*.

NOD2 seems to have a more specialized role, being strongly expressed in the Paneth cells of the gut, where it regulates the expression of potent anti-microbial peptides such as the α- and β-defensins (see Chapter 2). Consistent with this, loss-of-function mutations in NOD2 in humans are associated with the inflammatory bowel condition known as **Crohn's disease** (discussed in Chapter 15). Some patients with this condition carry mutations in the LRR domain of NOD2 that impair its ability to sense MDP and activate NFκB. This is thought to diminish the production of defensins and other antimicrobial peptides, thereby weakening the natural barrier function of the intestinal epithelium and leading to the inflammation characteristic of this disease. Gain-of-function mutations in human NOD2 are associated with the inflammatory disorders **early-onset sarcoidosis** and **Blau syndrome**, which are characterized by spontaneous inflammation in tissues such as the liver, or in the joints, eyes, and skin. Activating mutations in the NOD domain seem to promote the signaling cascade in the absence of ligand, leading to an inappropriate inflammatory response in the absence of pathogens. Besides NOD1 and NOD2, there are other members of the NOD family, such as the proteins NLRX1 and NLRC5, but their function is currently less well understood.

3-9   NLRP proteins react to infection or cellular damage through an inflammasome to induce cell death and inflammation.

Another subfamily of NLR proteins has a **pyrin** domain in place of the CARD domain at their amino termini, and is known as the **NLRP** family. Pyrin domains are structurally related to the CARD and TIR domains, and interact with other pyrin domains (Fig. 3.19). Humans have 14 NLR proteins containing pyrin domains. The best characterized is **NLRP3** (also known as **NALP3** or cryopyrin), although the molecular details of its activation are still under active investigation. NLRP3 resides in an inactive form in the cytoplasm,

Crohn's Disease

Hereditary Periodic Fever Syndromes

MOVIE 3.6

Fig. 3.19 Cellular damage activates the NLRP3 inflammasome to produce pro-inflammatory cytokines. The LRR domain of NLRP3 associates with chaperones (HSP90 and SGT1) that prevent NLRP3 activation. Damage to cells caused by bacterial pore-forming toxins or activation of the P2X7 receptor by extracellular ATP allows efflux of K⁺ ions from the cell; this may dissociate these chaperones from NLRP3 and induce multiple NLRP3 molecules to aggregate through interactions of their NOD domains (also called the NACHT domain). Reactive oxygen intermediates (ROS) and disruption of lysosomes also can activate NLRP3 (see text). The aggregated NLRP3 conformation brings multiple NLRP3 pyrin domains into close proximity, which then interact with the pyrin domains of the adaptor protein ASC (PYCARD). This conformation aggregates the ASC CARD domains, which in turn aggregate the CARD domains of pro-caspase 1. This aggregation of pro-caspase 1 induces proteolytic cleavage of itself to form the active caspase 1, which cleaves the immature forms of pro-inflammatory cytokines, releasing the mature cytokines that are then secreted.

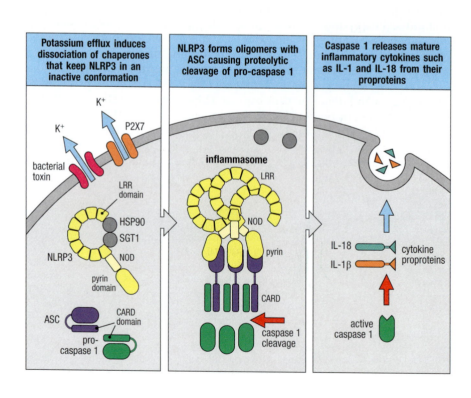

where its LRR domains are thought to bind the heat-shock chaperone protein HSP90 and the co-chaperone SGT1, which may hold NLRP3 in an inactive state (see Fig. 3.19). Several events seem to induce NLRP3 signaling: reduced intracellular potassium, the generation of reactive oxygen species (ROS), or the disruption of lysosomes by particulate or crystalline matter. The loss of intracellular potassium through efflux can occur during infection with, for example, intracellular bacteria such as *Staphylococcus aureus* that produce pore-forming toxins. Also, death of nearby cells can release ATP into the extracellular space; this ATP would activate the **purinergic receptor P2X7**, which itself is a potassium channel, and allow K$^+$ ion efflux. In one model, it is the reduction of intracellular K$^+$ concentration that triggers NLRP3 signaling by causing the dissociation of HSP90 and SGT1. A model proposed for ROS-induced NLRP3 activation involves the intermediate oxidization of sensor proteins collectively called **thioredoxin (TRX)**. Normally TRX proteins are bound to **thioredoxin-interacting protein (TXNIP)**, but oxidation of TRX by ROS causes the dissociation of TXNIP from TRX. The free TXNIP may then displace HSP90 and SGT1 from NLRP3, again causing its activation. In both of these cases, NLRP3 activation involves aggregation of multiple monomers via their LLR and NOD domains to induce signaling. Finally, phagocytosis of particulate matter, such as the adjuvant **alum**, a crystalline salt of aluminum potassium sulfate, may lead to the rupture of lysosomes and release of the active protease cathepsin B, which can activate NLRP3 by an unknown mechanism.

Rather than activating NFκB as in NOD1 and NOD2 signaling, NLRP3 signaling leads to the generation of pro-inflammatory cytokines and to cell death through formation of a multiprotein complex known as the **inflammasome** (see Fig. 3.19). Activation of the inflammasome proceeds in several stages. The first is the aggregation of LRR domains of several NLRP3 molecules, or other NLRP molecules, by a specific trigger or recognition event. This aggregation induces the pyrin domains of NLRP3 to interact with pyrin domains of another protein named **ASC** (also called PYCARD). ASC is an adaptor protein composed of an amino-terminal pyrin domain and a carboxy-terminal CARD domain. Pyrin and CARD domains are each able to form polymeric filamentous structures (Fig. 3.20). The interaction of NLRP3 with ASC further drives the formation of a polymeric ASC filament, with the pyrin domains in the center and CARD domains facing outward. These CARD domains then interact with CARD domains of the inactive protease **pro-caspase 1**, initiating

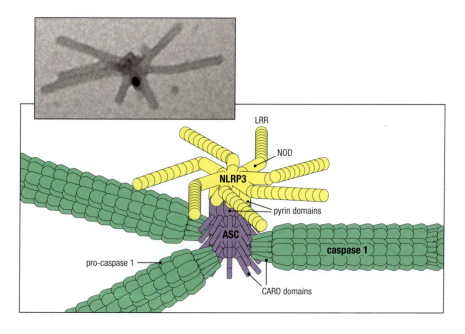

Fig. 3.20 The inflammasome is composed of several filamentous protein polymers created by aggregated CARD and pyrin domains. Top panel: an electron micrograph of structures formed by full-length ASC, the pyrin domain of AIM2, and the CARD domain of caspase 1. The central dark region represents anti-ASC staining with a gold-labeled (15 nm) antibody. The long outward filaments represent the polymer composed of the caspase 1 CARD domain. Bottom panel: Schematic interpretation of NLRP3 inflammasome assembly. In this model, CARD regions of ASC and caspase 1 aggregate into a filamentous structure. The adaptor ASC translates aggregation of NLRP3 into aggregation of pro-caspase 1. Electron micrograph courtesy of Hao Wu.

its CARD-dependent polymerization into discrete caspase 1 filaments. This aggregation seems to trigger the autocleavage of pro-caspase 1, which releases the active caspase 1 fragment from its autoinhibitory domains. Active caspase 1 then carries out the ATP-dependent proteolytic processing of pro-inflammatory cytokines, particularly IL-1β and IL-18, into their active forms (see Fig. 3.19). Caspase 1 activation also induces a form of cell death called **pyroptosis** ('fiery death') through an unknown mechanism that is associated with inflammation because of the release of these pro-inflammatory cytokines upon cell rupture.

For inflammasome activation to produce inflammatory cytokines, a priming step must first occur in which cells induce and translate the mRNAs that encode the pro-forms of IL-1β, IL-18, or other cytokines. This priming step can result from TLR signaling, which may help ensure that inflammasome activation proceeds primarily during infections. For example, the TLR-3 agonist poly I:C (see Section 3-5) can be used experimentally to prime cells for subsequent triggering of the inflammasome.

Several other NLR family members form inflammasomes with ASC and caspase 1 that activate these pro-inflammatory cytokines. **NLRP1** is highly expressed in monocytes and dendritic cells and is activated directly by MDP, similar to NOD2, but can also be activated by other factors. For example, *Bacillus anthracis* expresses an endopeptidase, called anthrax **lethal factor**, which allows the pathogen to evade the immune system by killing macrophages. Lethal factor does this by cleaving NLRP1, activating an NLRP1 inflammasome and inducing pyroptosis in the infected macrophages. **NLRC4** acts as an adaptor with two other NLR proteins, **NAIP2** and **NAIP5**, that serve to detect various bacterial proteins that enter cells through specialized secretion systems used by pathogens to transport materials into or access nutrients from host cells. One such protein, **PrgJ**, from the pathogen *Salmonella typhimurium*, is a component of the **type III secretion system** (T3SS), a needle-like macromolecular complex. Upon infection of host cells by *Salmonella*, PrgJ enters the cytoplasm and is recognized by NLRC4 functioning together with NAIP2. Extracellular bacterial flagellin is recognized by TLR5, but flagellin may also enter host cells with PrgJ via the T3SS, and in this case can be recognized by NLRC4 in conjunction with NAIP5. Some NLR proteins may negatively regulate innate immunity, such as **NLRP6**, since mice lacking this protein exhibit increased resistance to certain pathogens. However, NLRP6 is highly expressed in intestinal epithelium, where it appears to play a positive role in promoting normal mucosal barrier function and is required for the normal secretion of mucus granules into the intestine by goblet cells. **NLRP7**, which is present in humans but not mice, recognizes microbial acylated lipopeptides and forms an inflammasome with ASC and caspase 1 to produce IL-1β and IL-18. Less is known about **NLRP12**, but like NLRP6, it initially was proposed to have an inhibitory function. Subsequent studies of mice lacking NLRP12 suggest it has a possible role in the detection of and response to certain bacterial species, including *Yersinia pestis*, the bacterium that causes bubonic plague, although the basis of this recognition is still unclear.

Inflammasome activation can also involve proteins of the **PYHIN** family, which contain an N-terminal pyrin domain but lack the LRR domains present in the NLR family. In place of an LRR domain, PYHIN proteins have a HIN (H inversion) domain, so named for the HIN DNA recombinase of *Salmonella* that mediates DNA inversion between flagellar H antigens. There are four PYHIN proteins in humans, and 13 in mice. In one of these, **AIM2 (absent in melanoma 2)**, the HIN domain recognizes double-stranded DNA genomes and triggers caspase 1 activation through pyrin domain interactions with ASC. AIM2 is located in the cytoplasm and is important for responses *in vitro* to vaccinia virus, and its *in vivo* role has been demonstrated by the increased susceptibility of AIM2-deficient mice to infection by *Francisella tularensis*, the causative agent of tularemia. The related protein **IFI16 (interferon inducible protein 16)** contains two HIN domains; it is primarily located in the cell nucleus and recognizes viral double-stranded DNA, and will be described below in Section 3-11.

A **'non-canonical' inflammasome** (caspase 1-independent) pathway uses the protease **caspase 11** to detect intracellular LPS. The discovery of this pathway was initially confused as being dependent on caspase 1 because of a specific genetic difference between experimental mouse strains. Caspase 11 is encoded by the murine *Casp4* gene and is homologous to human caspases 4 and 5. The mice in which the caspase 1 gene (*Casp1*) was initially disrupted and studied were originally found to be resistant to lethal shock (see Section 3-20) induced by administration of LPS. This led researchers to conclude that caspase 1 acted in the inflammatory response to LPS. But researchers later discovered that this mouse strain also carried a natural mutation that inactivated the related *Casp4* gene. Because the *Casp1* and *Casp4* genes reside within 2 kilobases of each other on mouse chromosome 9, they failed to segregate independently during subsequent experimental genetic backcrosses to other mouse strains. Thus, mice initially thought to lack only caspase 1 protein in fact lacked both caspase 1 and caspase 11. Later, mice lacking only caspase 1 were generated by expressing functional *Casp4* as a transgene; these mice became susceptible to LPS-induced shock. Mice were also generated that lacked only caspase 11, and these were found to be resistant to LPS-induced shock. These results indicated that caspase 11, and not caspase 1 as originally thought, is responsible for LPS-induced shock. Caspase 11 is responsible for inducing pyroptosis, but not for processing of IL-1β or IL-18. It was suspected that TLR-4 was not the sensor for LPS that activated the non-canonical imflammasome, since mice lacking TLR-4 remain susceptible to LPS-induced shock. Recent evidence has suggested that caspase 11 itself is the intracellular LPS sensor, making it an example of a protein that is both a sensor and an effector molecule.

Inappropriate inflammasome activation has been associated with various diseases. **Gout** has been known for many years to cause inflammation in the cartilaginous tissues by the deposition of monosodium urate crystals, but how urate crystals caused inflammation was a mystery. Although the precise mechanism is still unclear, urate crystals are known to activate the NLRP3 inflammasome, which induces the inflammatory cytokines associated with the symptoms of gout. Mutations in the NOD domain of NLRP2 and NLRP3 can activate inflammasomes inappropriately, and they are the cause of some inherited **autoinflammatory diseases**, in which inflammation occurs in the absence of infection. Mutations in NLRP3 in humans are associated with hereditary periodic fever syndromes, such as **familial cold inflammatory syndrome** and **Muckle–Wells syndrome** (discussed in more detail in Chapter 13). Macrophages from patients with these conditions show spontaneous production of inflammatory cytokines such as IL-1β. We will also discuss how pathogens can interfere with formation of the inflammasome in Chapter 13.

Hereditary Periodic Fever Syndromes

3-10 The RIG-I-like receptors detect cytoplasmic viral RNAs and activate MAVS to induce type I interferon production and pro-inflammatory cytokines.

TLR-3, TLR-7, and TLR-9 detect extracellular viral RNAs and DNAs that enter the cell from the endocytic pathway. By contrast, viral RNAs produced within a cell are sensed by a separate family of proteins called the **RIG-I-like receptors (RLRs)**. These proteins serve as viral sensors by binding to viral RNAs using an RNA helicase-like domain in their carboxy terminal. The RLR helicase-like domain has a 'DExH' tetrapeptide amino acid motif and is a subgroup of DEAD-box family proteins. The RLR proteins also contain two amino-terminal CARD domains that interact with adaptor proteins and activate signaling to produce type I interferons when viral RNAs are bound. The first of these sensors to be discovered was **RIG-I (retinoic acid-inducible gene I)**. RIG-I is widely expressed across tissues and cell types and serves as an intracellular sensor for several kinds of infections. Mice deficient in RIG-I are highly susceptible to infection by several kinds of single-stranded RNA viruses, including paramyxoviruses, rhabdoviruses, orthomyxoviruses, and flaviviruses, but not picornaviruses.

RIG-I discriminates between host and viral RNA by sensing differences at the 5′ end of single-stranded RNA transcripts. Eukaryotic RNA is transcribed in the nucleus and contains a 5′-triphosphate group on its initial nucleotide that undergoes subsequent enzymatic modification called **capping** by the addition of a 7-methylguanosine to the 5′-triphosphate. Most RNA viruses, however, do not replicate in the nucleus, where capping normally occurs, and their RNA genomes do not undergo this modification. Biochemical studies have determined that RIG-I senses the unmodified 5′-triphosphate end of the ssRNA viral genome. Flavivirus RNA transcripts have the unmodified 5′-triphosphate, as do the transcripts of many other ssRNA viruses, and they are detected by RIG-I. In contrast, the picornaviruses, which include poliovirus and hepatitis A, replicate by a mechanism that involves the covalent attachment of a viral protein to the 5′ end of the viral RNA, so that the 5′-triphosphate is absent, which explains why RIG-I is not involved in sensing them.

MDA-5 (melanoma differentiation-associated 5), also called **helicard**, is similar in structure to RIG-I, but it senses dsRNA. In contrast to RIG-I-deficient mice, mice deficient in MDA-5 are susceptible to picornaviruses, indicating that these two sensors of viral RNAs have crucial but distinct roles in host defense. Inactivating mutations in alleles of human RIG-I or MDA-5 have been reported, but these mutations were not associated with immunodeficiency. The RLR family member **LGP2** (encoded by *DHX58*) retains a helicase domain but lacks CARD domains. LGP2 appears to cooperate with RIG-I and MDA-5 in the recognition of viral RNA, since mice lacking LGP2 have impaired antiviral responses normally mediated by RIG-I or MDA-5. This cooperative viral recognition by LGP2 appears to depend on its helicase domain, since in mice, mutations that disrupt its ATPase activity result in impaired IFN-β production in response to various RNA viruses.

Sensing of viral RNAs activates signaling by RIG-I and MDA-5 that leads to type I interferon production appropriate for defense against viral infection (Fig. 3.21). Before infection by viruses, RIG-I and MDA-5 are in the cytoplasm

Fig. 3.21 RIG-I and other RLRs are cytoplasmic sensors of viral RNA. First panel: before detecting viral RNA, RIG-I and MDA-5 are cytoplasmic and in inactive, auto-inhibited conformations. The adaptor protein MAVS is attached to the mitochondrial outer membrane. Second panel: detection of uncapped 5′-triphosphate RNA by RIG-I, or viral dsRNA by MDA-5, changes the conformation of their CARD domains to become free to interact with the amino-terminal CARD domain of MAVS. This interaction involves the generation of K63-linked polyubiquitin from the E3 ligases TRIM25 or Riplet, although structural details are still unclear. Third panel: the aggregation induces a proline-rich region of MAVS to interact with TRAFs (see text) and leads to the generation of additional K63-linked polyubiquitin scaffold. As in TLR signaling, this scaffold recruits TBK1 and IKK complexes (see Figs. 3.15 and 3.16) to activate IRF and NFκB, producing type I interferons and pro-inflammatory cytokines, respectively.

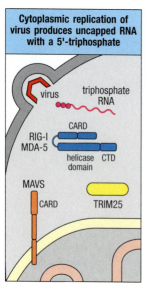

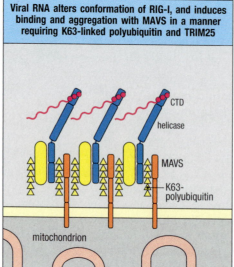

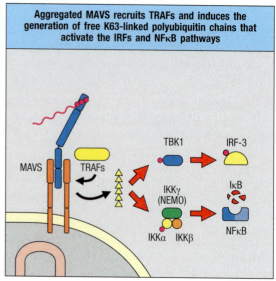

in an autoinhibited configuration that is stabilized by interactions between the CARD and helicase domains. These interactions are disrupted upon infection when viral RNA associates with the helicase domains of RIG-I or MDA-5, freeing the two CARD domains for other interactions. The more amino-proximal portion of the two CARD domains can then recruit E3 ligases, including **TRIM25** and **Riplet** (encoded by *RNF153*), which initiate K63-linked polyubiquitin scaffolds (see Section 3-7), either as free polyubiquitin chains or on linkages within the second CARD domain. Precise details are unclear, but this scaffold appears to help RIG-I and MDA-5 interact with a downstream adaptor protein called **MAVS (mitochondrial antiviral signaling protein)**. MAVS is attached to the outer mitochondrial membrane and contains a CARD domain that may bind RIG-I and MDA-5. This aggregation of CARD domains, as in the inflammasome, may initiate aggregation of MAVS. In this state, MAVS propagates signals by recruiting various TRAF family E3 ubiquitin ligases, including TRAF2, TRAF3, TRAF5, and TRAF6. The relative importance of each E3 ligase may differ between cell types, but their further production of K63-linked polyubiquitin leads to activation of TBK1 and IRF3 and production of type I interferons, as described for TLR-3 signaling (see Fig. 3.16), and also to activation of NFκB. Some viruses have evolved countermeasures to thwart the protection conferred by RLRs. For example, even though the negative-sense RNA genome of influenza virus replicates in the nucleus, some viral RNA transcripts produced during influenza infection are not capped but must be translated in the cytoplasm. The influenza A **nonstructural protein 1(NS1)** inhibits the activity of TRIM25, and thereby interrupts the antiviral actions that RIG-I might exert against infection.

3-11   Cytosolic DNA sensors signal through STING to induce production of type I interferons.

Innate sensors that recognize cytoplasmic RNA use specific modifications, such as the 5′ cap, to discriminate between host and viral origin. Host DNA is generally restricted to the nucleus, but viral, microbial, or protozoan DNA may become located in the cytoplasm during various stages of infection. Several innate sensors of cytoplasmic DNA have been identified that can lead to the production of type I interferon in response to infections. One component of the DNA-sensing pathway, **STING (stimulator of interferon genes)**, was identified in a functional screen for proteins that can induce expression of type I interferons. STING (encoded by *TMEM173*) is anchored to the endoplasmic reticulum membrane by an amino-terminal tetraspan transmembrane domain; its carboxy-terminal domain extends into the cytoplasm and interacts to form an inactive STING homodimer.

STING is known to serve as a sensor of intracellular infection, based on its recognition of bacterial **cyclic dinucleotides (CDNs)**, including cyclic diguanylate monophosphate (c-di-GMP) and cyclic diadenylate monophosphate (c-di-AMP). These molecules are bacterial second messengers and are produced by enzymes present in most bacterial genomes. CDNs activate STING signaling by changing the conformation of the STING homodimer. This homodimer recruits and activates TBK1, which in turn phosphorylates and activates IRF3, leading to type I interferon production (Fig. 3.22), similar to signaling by TLR-3 and MAVS (see Figs. 3.16 and 3.21). TRIF (downstream of TLR3), MAVS, and STING each contain a similar amino acid sequence motif at their carboxy termini that becomes serine-phosphorylated when these molecules are activated. It appears that this motif, when phosphorylated, recruits both TBK1 and IRF3, allowing IRF3 to be efficiently phosphorylated and activated by TBK1.

STING also plays a role in viral infections, since mice lacking STING are susceptible to infection by herpesvirus. But until recently, it was unclear whether

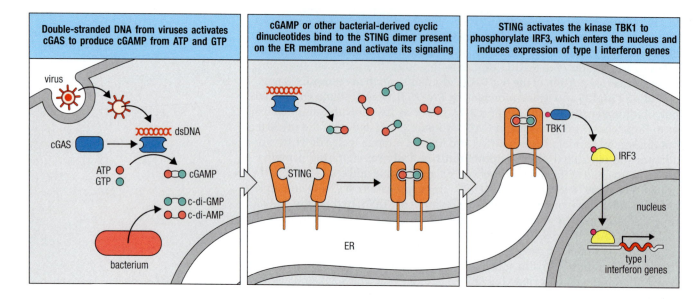

Double-stranded DNA from viruses activates cGAS to produce cGAMP from ATP and GTP	cGAMP or other bacterial-derived cyclic dinucleotides bind to the STING dimer present on the ER membrane and activate its signaling	STING activates the kinase TBK1 to phosphorylate IRF3, which enters the nucleus and induces expression of type I interferon genes

Fig. 3.22 cGAS is a cytosolic sensor of DNA that signals through STING to activate type I interferon production. First panel: cGAS resides in the cytoplasm and serves as a sensor of double-stranded DNA (dsDNA) from viruses. When cGAS binds dsDNA, its enzymatic activity is stimulated, leading to production of cyclic-GMP-AMP (cGAMP). Bacteria that infect cells produce second messengers such as cyclic dinucleotides, including cyclic diguanylate monophosphate (c-di-GMP) and cyclic diadenylate monophosphate (c-di-AMP). Second panel: cGAMP and other bacterial dinucleotides can bind and activate the STING dimer present on the ER membrane. Third panel: in this state STING activates TBK1, although the details of this interaction are still unclear. Active TBK1 activates IRF3, as described in Fig. 3.16.

STING recognized viral DNA directly or acted only downstream of an unknown viral DNA sensor. It was found that the introduction of DNA into cells, even without live infection, generated another second messenger molecule that activated STING. This second messenger was identified as **cyclic guanosine monophosphate-adenosine monophosphate** (cyclic GMP-AMP), or **cGAMP**. cGAMP, like bacterial CDNs, binds both subunits of the STING dimer and activates STING signaling. This result also suggested the presence of a DNA sensor acting upstream of STING. Purification of the enzyme that produces cGAMP in response to cytosolic DNA identified a previously unknown enzyme, which was named **cGAS**, for **cyclic GAMP synthase**. cGAS contains a protein motif present in the nucleotidyltransferase (NTase) family of enzymes, which includes adenylate cyclase (the enzyme that generates the second messenger molecule cyclic AMP) and various DNA polymerases. cGAS can bind directly to cytosolic DNA, and this stimulates its enzymatic activity to produce cGAMP from GTP and ATP in the cytoplasm, activating STING. Mice harboring an inactivated cGAS gene show increased susceptibility to herpesvirus infection, demonstrating its importance in immunity.

There are several other candidate DNA sensors, but less is known about the mechanism of their recognition and signaling, or their *in vivo* activity. IFI16 (IFN-γ-inducible protein 16) is a PYHIN family member related to AIM2, but appears to function in DNA sensing and acts through STING, TBK1, and IRF3 rather than activating an inflammasome pathway. **DDX41 (DEAD box polypeptide 41)** is an RLR related to RIG-I and is a member of the DEAD-box family, but appears to signal through STING rather than MAVS. **MRE11A (meitotic recombination 11 homolog a)** can sense cytosolic double-stranded DNA to activate the STING pathway, but its role in innate immunity is currently unknown.

3-12 Activation of innate sensors in macrophages and dendritic cells triggers changes in gene expression that have far-reaching effects on the immune response.

Besides activating effector functions and cytokine production, another outcome of the activation of innate sensing pathways is the induction of **co-stimulatory molecules** on tissue dendritic cells and macrophages (see Section 1-15). We will describe these in more detail later in the book, but mention them now because they provide an important link between innate and adaptive immune responses. Two important co-stimulatory molecules are the cell-surface proteins **B7.1 (CD80)** and **B7.2 (CD86)**, which are induced on macrophages and tissue dendritic cells by innate sensors such as TLRs in response to pathogen recognition (Fig. 3.23). B7.1 and B7.2 are recognized by specific **co-stimulatory receptors** expressed by cells of the adaptive immune response, particularly CD4 T cells, and their activation by B7 is an important step in activating adaptive immune responses.

Substances such as LPS that induce co-stimulatory activity have been used for years in mixtures that are co-injected with protein antigens to enhance their immunogenicity. These substances are known as **adjuvants** (see Appendix I, Section A-1), and it was found empirically that the best adjuvants contain microbial components that induce macrophages and tissue dendritic cells to express co-stimulatory molecules and cytokines. As we shall see in Chapters 9 and 11, the cytokines produced in response to infections influence the functional character of the adaptive immune response that develops. In this way the ability of the innate immune system to discriminate among different types of pathogens is used by the organism to ensure an appropriate module of adaptive immune response.

3-13   Toll signaling in *Drosophila* is downstream of a distinct set of pathogen-recognition molecules.

Before leaving innate sensing, we shall look briefly at how Toll, TLRs, and NODs are used in invertebrate innate immunity. Although Toll is central to defense against both bacterial and fungal pathogens in *Drosophila*, Toll itself is not a pattern recognition receptor, but is downstream of other proteins that detect pathogens (Fig. 3.24). In *Drosophila*, there are 13 genes encoding **peptidoglycan-recognition proteins (PGRPs)** that bind the peptidoglycan components of bacterial cell walls. Another family, the **Gram-negative binding proteins (GNBPs)**, recognizes LPS and β-1,3-linked glucans. GNBPs recognize Gram-negative bacteria and, unexpectedly, fungi, rather than Gram-positive bacteria. The family members GNBP1 and PGRP-SA cooperate in the recognition of peptidoglycan from Gram-positive bacteria. They interact with a serine protease called **Grass**, which initiates a proteolytic cascade that terminates in the cleavage of the protein Spätzle. One of the cleaved fragments forms a homodimer that binds to Toll and induces its dimerization, which in turn stimulates the antimicrobial response. A fungus-specific recognition protein, GNBP3, also activates the proteolytic cascade, causing cleavage of Spätzle and activation of Toll.

In *Drosophila*, fat-body cells and hemocytes are phagocytic cells that act as part of the fly's immune system. When the Spätzle dimer binds to Toll, hemocytes synthesize and secrete antimicrobial peptides. The Toll signaling pathway in *Drosophila* activates a transcription factor called DIF, which is related to mammalian NFκB. DIF enters the nucleus and induces the transcription of genes for antimicrobial peptides such as drosomycin. Another *Drosophila* factor in the NFκB family, **Relish**, induces the production of antimicrobial peptides in response to the **Imd (immunodeficiency) signaling pathway**, which is triggered in *Drosophila* by particular PGRPs that recognize Gram-negative bacteria. Relish induces expression of the antimicrobial peptides diptericin, attacin, and cecropin, which are distinct from the peptides induced by Toll signaling. Thus, the Toll and Imd pathways activate effector mechanisms to eliminate infection by different kinds of pathogens. Four mammalian PGRP homologs have been identified, but act differently than in *Drosophila*. One, PGLYRP-2, is secreted and functions as an amidase to hydrolyze bacterial peptidoglycans. The others are present in neutrophil granules and exert a bacteriostatic action through interactions with bacterial cell-wall peptidoglycan.

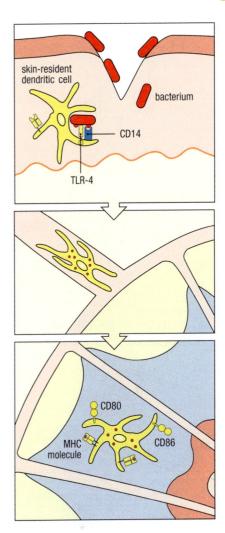

Fig. 3.23 Bacterial LPS induces changes in dendritic cells, stimulating them to migrate and to initiate adaptive immunity by activating T cells. Top panel: immature dendritic cells in the skin are highly phagocytic and macropinocytic, but lack the ability to activate T lymphocytes. Dendritic cells residing in the skin ingest microbes and their products and degrade them. During a bacterial infection, the dendritic cells are activated by various innate sensors and the activation induces two types of changes. Second panel: the dendritic cells migrate out of the tissues and enter the lymphatic system and begin to mature. They lose the ability to ingest antigen but gain the ability to stimulate T cells. Third panel: in the regional lymph nodes, they become mature dendritic cells. They change the character of their cell-surface molecules, increasing the number of MHC molecules on their surface and the expression of the co-stimulatory molecules CD80 (B7.1) and CD86 (B7.2).

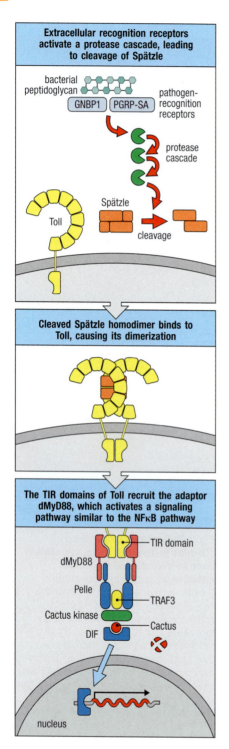

Extracellular recognition receptors activate a protease cascade, leading to cleavage of Spätzle

bacterial peptidoglycan

GNBP1 PGRP-SA — pathogen-recognition receptors

protease cascade

Toll

Spätzle

cleavage

Cleaved Spätzle homodimer binds to Toll, causing its dimerization

The TIR domains of Toll recruit the adaptor dMyD88, which activates a signaling pathway similar to the NFκB pathway

dMyD88 — TIR domain

Pelle

Cactus kinase — TRAF3

DIF — Cactus

nucleus

Fig. 3.24 *Drosophila* Toll is activated as a result of a proteolytic cascade initiated by pathogen recognition. The peptidoglycan-recognition protein PGRP-SA and the Gram-negative binding protein GNBP1 cooperate in the recognition of bacterial pathogens and activate the first protease in a protease cascade that leads to cleavage of the *Drosophila* protein Spätzle (first panel). Cleavage alters the conformation of Spätzle, enabling it to bind Toll and induce Toll dimerization (second panel). Toll's cytoplasmic TIR domains recruit the adaptor protein dMyD88 (third panel), which initiates a signaling pathway very similar to that leading to the release of NFκB from its cytoplasmic inhibitor in mammals. The Drosophila version of NFκB is the transcription factor DIF, which then enters the nucleus and activates the transcription of genes encoding antimicrobial peptides. Fungal recognition also leads to cleavage of Spätzle and the production of antimicrobial peptides by this pathway, although the recognition proteins for fungi are as yet unidentified.

3-14 TLR and NOD genes have undergone extensive diversification in both invertebrates and some primitive chordates.

There are only about a dozen mammalian TLR genes, but some organisms have diversified their repertoire of innate recognition receptors, especially those containing LRR domains, to a much greater degree. The sea urchin *Strongylocentrotus purpuratus* has an unprecedented 222 different *TLR* genes, more than 200 NOD-like receptor genes, and more than 200 scavenger receptor genes in its genome. The sea urchin also has an increased number of proteins that are likely to be involved in signaling from these receptors, there being, for example, four genes that are similar to the single mammalian *MyD88* gene. However, there is no apparent increase in the number of downstream targets, such as the family of NFκB transcription factors, suggesting that the ultimate outcome of TLR signaling in the sea urchin may be very similar to that in other organisms.

Sea urchin TLR genes fall into two broad categories. One is a small set of 11 divergent genes. The other is a large family of 211 genes, which show a high degree of sequence variation within particular LRR regions; this, together with the large number of pseudogenes in this family, indicates rapid evolutionary turnover, suggesting rapidly changing receptor specificities, in contrast with the few stable mammalian TLRs. Although the pathogen specificity of sea urchin TLRs is unknown, the hypervariability in the LRR domains could be used to generate a highly diversified pathogen-recognition system based on Toll-like receptors. A similar expansion of innate receptors has occurred in some chordates, the phylum to which vertebrates belong. Amphioxus (the lancelet) is a nonvertebrate chordate lacking an adaptive immune system. The amphioxus genome contains 71 TLRs, more than 100 NOD-like receptors, and more than 200 scavenger receptors. As we will see in Chapter 5, a primitive vertebrate lineage—the jawless fish, which lack immunoglobulin- and T-cell-based adaptive immunity—uses somatic gene rearrangement of LRR-containing proteins to provide a version of adaptive immunity (see Section 5-18).

Summary.

Innate immune cells express several receptor systems that recognize microbes and induce rapid defenses as well as delayed cellular responses. Several scavenger and lectin-like receptors on neutrophils, macrophages, and dendritic cells help rapidly eliminate microbes through phagocytosis. G-protein-coupled receptors for C5a (which can be produced by activation of the complement system's innate pathogen-recognition ability) and for the bacterial peptide fMLF synergize with phagocytic receptors in activating the NADPH oxidase in phagosomes to generate antimicrobial reactive oxygen intermediates. Toll-like receptors (TLRs) on the cell surface and in the membranes of endosomes detect microbes outside the cell and activate several host defense signaling pathways. The NFκB and IRF pathways downstream of these receptors induce pro-inflammatory cytokines, including TNF-α, IL-1β, and IL-6, and antiviral cytokines including type I interferons. Other receptor families detect microbial infection in the cytosol. NOD proteins detect bacterial products within the cytosol and activate NFκB and the production of pro-inflammatory cytokines. The related NLR family of proteins detects signs of cellular stress or damage, as well as certain microbial components. NLRs signal through the inflammasome, which generates pro-inflammatory cytokines and induces pyroptosis, a form of cell death. RIG-I and MDA-5 detect viral infection by sensing the presence of viral RNAs and activate the MAVS pathway, while sensors of cytosolic DNA, such as cGAS, activate the STING pathway; both of these pathways induce type I interferons. The signaling pathways activated by all of these primary sensors of pathogens induce a variety of genes, including those for cytokines, chemokines, and co-stimulatory molecules that have essential roles in immediate defense and in directing the course of the adaptive immune response later in infection.

Induced innate responses to infection.

We will now examine the responses of innate immunity induced as an immediate consequence of pathogen recognition by the sensors described in the last section. We will focus on the major phagocytes—neutrophils, macrophages, and dendritic cells—and the cytokines they produce that induce and maintain inflammation. First, we will introduce the families of cytokines and chemokines that coordinate many cellular responses, such as the recruitment of neutrophils and other immune cells to sites of infection. We will discuss the various adhesion molecules that are induced on immune cells circulating in the blood and on endothelial cells of blood vessels to coordinate movement of cells out of the blood and into infected tissues. We will consider in some detail how macrophage-derived chemokines and cytokines promote the continued destruction of infecting microbes. This is achieved both by stimulating the production and recruitment of fresh phagocytes and by inducing another phase of the innate immune response—the acute-phase response—in which the liver produces proteins that act as opsonizing molecules, helping to augment the actions of complement. We will also look at the mechanism of action of antiviral interferons, the type I interferons, and finally examine the growing class of innate lymphoid cells, or ILCs, which include the NK cells long known to contribute to innate immune defense against viruses and other intracellular pathogens. ILCs exert a diverse array of effector function that contribute to a rapid innate immune response to infection. They respond to early cytokine signals provided by innate sensor cells, and amplify the response by producing various types of effector cytokines. If an infection is not cleared by the induced innate response, an adaptive response will ensue that uses many of the same effector mechanisms used by the innate immune system but targets them with much greater precision. The effector mechanisms described here therefore serve as a primer for the focus on adaptive immunity in the later parts of this book.

3-15 Cytokines and their receptors fall into distinct families of structurally related proteins.

Cytokines are small proteins (about 25 kDa) that are released by various cells in the body, usually in response to an activating stimulus, and that induce responses through binding to specific receptors. Cytokines can act in an **autocrine** manner, affecting the behavior of the cell that releases the cytokine, or in a **paracrine** manner, affecting adjacent cells. Some cytokines are even stable enough to act in an **endocrine** manner, affecting distant cells, although this depends on their ability to enter the circulation and on their half-life in the blood. In an attempt to develop a standardized nomenclature for molecules secreted by, and acting on, leukocytes, many cytokines are called by the name **interleukin (IL)** followed by a number (for example, IL-1 or IL-2). However, not all cytokines are included in this system; thus students of immunology are still faced with a somewhat confusing and difficult task. The cytokines are listed alphabetically, together with their receptors, in Appendix III.

Cytokines can be grouped by structure into families—the IL-1 family, the hematopoietin superfamily, the interferons (described in Section 3-7), and the TNF family—and their receptors can likewise be grouped (Fig. 3.25). The **IL-1 family** contains 11 members, notably IL-1α, IL-1β, and IL-18. Most members of this family are produced as inactive proproteins that are cleaved (removing an amino-terminal peptide) to produce the mature cytokine. The exception to this rule is IL-1α, for which both the proprotein and its cleaved forms are biologically active. As discussed earlier, mature IL-1β and IL-18 are produced by macrophages through the action of caspase 1 in response to TLR signaling and inflammasome activation. The IL-1-family receptors have TIR domains in their cytoplasmic tails and signal by the NFκB pathway described earlier for TLRs. The IL-1 receptor functions in concert with a second transmembrane protein, the IL-1 receptor accessory protein (IL1RAP), that is required for IL-1 signal transduction.

The **hematopoietin superfamily** of cytokines is quite large and includes non-immune-system growth and differentiation factors such as erythropoietin (which stimulates red blood cell development) and growth hormone, as well as interleukins with roles in innate and adaptive immunity. IL-6 is a member of this superfamily, as is the cytokine GM-CSF, which stimulates the production of new monocytes and granulocytes in the bone marrow. Many of the soluble cytokines made by activated T cells are members of the hematopoietin family. The receptors for the hematopoietin cytokines are tyrosine kinase-associated receptors that form dimers when their cytokine ligand binds. Dimerization initiates intracellular signaling from the tyrosine kinases associated with the

Fig. 3.25 Cytokine receptors belong to families of receptor proteins, each with a distinctive structure. Many cytokines signal through receptors of the hematopoietin receptor superfamily, named after its first member, the erythropoietin receptor. The hematopoietin receptor superfamily includes homodimeric and heterodimeric receptors, which are subdivided into families on the basis of protein sequence and structure. Examples of these are given in the first three rows. Heterodimeric class I cytokine receptors have an α chain that often defines the ligand specificity of the receptor; they may share with other receptors a common β or γ chain that confers the intracellular signaling function. Heterodimeric class II cytokine receptors have no common chain and include receptors for interferons or interferon-like cytokines. All the cytokine receptors signal through the JAK-STAT pathway. The IL-1 receptor family have extracellular immunoglobulin domains and signal as dimers through TIR domains in their cytoplasmic tails and through MyD88. Other superfamilies of cytokine receptors are the tumor necrosis factor receptor (TNFR) family and the chemokine receptor family, the latter belonging to the very large family of G-protein-coupled receptors. The ligands of the TNFR family act as trimers and may be associated with the cell membrane rather than being secreted.

Homodimeric receptors		Receptors for erythropoietin and growth hormone
Heterodimeric receptors with a common chain	β_c	Receptors for IL-3, IL-5, GM-CSF share a common chain, CD131 or β_c (common β chain)
	γ_c	Receptors for IL-2, IL-4, IL-7, IL-9, and IL-15 share a common chain, CD132 or γ_c (common γ chain). IL-2 receptor also has a third chain, a high-affinity subunit IL-2Rα (CD25)
Heterodimeric receptors (no common chain)		IL-1 family receptors
		Receptors for IL-13, IFN-α, IFN-β, IFN-γ, IL-10
TNF receptor family		Tumor necrosis factor (TNF) receptors I and II, CD40, Fas (Apo1, CD95), CD30, CD27, nerve growth factor receptor
Chemokine receptor family		CCR1–10, CXCR1–5, XCR1, CX3CR1

cytoplasmic domains of the receptor. Some types of cytokine receptors are composed of two identical subunits, but others have two different subunits. An important feature of cytokine signaling is the large variety of different receptor subunit combinations that occur.

These cytokines and their receptors can also be further divided into subfamilies characterized by functional similarities and genetic linkage. For instance, IL-3, IL-4, IL-5, IL-13, and GM-CSF are related structurally, their genes are closely linked in the genome, and they are often produced together by the same kinds of cells. In addition, they bind to closely related receptors, which belong to the family of **class I cytokine receptors**. The IL-3, IL-5, and GM-CSF receptors form a subgroup that shares a **common β chain**. Another subgroup of class I cytokine receptors is defined by the use of the **common γ chain ($γ_c$)** of the IL-2 receptor. This chain is shared by receptors for the cytokines IL-2, IL-4, IL-7, IL-9, IL-15, and IL-21, and is encoded by a gene located on the X chromosome. Mutations that inactivate $γ_c$ cause an **X-linked severe combined immuno-deficiency (X-linked SCID)** due to inactivation of the signaling pathways for several cytokines—IL-7, IL-15, and IL-2—that are required for normal lymphocyte development (see Section 13-3). More distantly related, the receptor for IFN-γ is a member of a small family of heterodimeric cytokine receptors with some similarities to the hematopoietin receptor family. These so-called **class II cytokine receptors** (also known as interferon receptors) include the receptors for IFN-α and IFN-β, and the IL-10 receptor. The hematopoietin and interferon receptors all signal through the JAK–STAT pathway described below and activate different combinations of STATs with different effects.

X-linked Severe Combined Immunodeficiency

The **TNF family**, of which TNF-α is the prototype, contains more than 17 cytokines with important functions in adaptive and innate immunity. Unlike most of the other immunologically important cytokines, many members of the TNF family are transmembrane proteins, a characteristic that gives them distinct properties and limits their range of action. Some, however, can also be released from the membrane in some circumstances. They are usually found as homotrimers of a membrane-bound subunit, although some heterotrimers consisting of different subunits also occur. TNF-α (sometimes called simply TNF) is initially expressed as a trimeric membrane-bound cytokine but can be released from the membrane. The effects of TNF-α are mediated by either of two **TNF receptors**. TNF receptor I (TNFR-I) is expressed on a wide range of cells, including endothelial cells and macrophages, whereas TNFR-II is expressed largely by lymphocytes. The receptors for cytokines of the TNF family are structurally unrelated to the receptors described above and also have to cluster to become activated. Since TNF-family cytokines are produced as trimers, the binding of these cytokines induces the clustering of three identical receptor subunits. The signaling pathway activated by these receptors is described in Chapter 7, where we see that signaling uses members of the TRAF family to activate the so-called non-canonical NFκB pathway.

Members of the chemokine receptor family are listed in Appendix IV, along with the chemokines they recognize. These receptors have a 7-transmembrane structure and signal by interacting with G-proteins as described in Section 3-2.

3-16   Cytokine receptors of the hematopoietin family are associated with the JAK family of tyrosine kinases, which activate STAT transcription factors.

The signaling chains of the hematopoietin family of cytokine receptors are noncovalently associated with protein tyrosine kinases of the **Janus kinase (JAK) family**—so called because they have two tandem kinase-like domains and thus resemble the two-headed mythical Roman god Janus. There are four members of the JAK family: Jak1, Jak2, Jak3, and Tyk2. As mice deficient for individual JAK family members show different phenotypes, each kinase must

MOVIE 3.7

have a distinct function. For example, Jak3 is used by γ_c for signaling by several of the cytokines described above. Mutations that inactivate Jak3 cause a form of **SCID** that is not X-linked.

The dimerization or clustering of receptor signaling chains brings the JAKs into close proximity, causing phosphorylation of each JAK on a tyrosine residue that stimulates its kinase activity. The activated JAKs then phosphorylate their associated receptors on specific tyrosine residues. This phosphotyrosine, and the specific amino acid sequence surrounding it, creates a binding site that is recognized by **SH2 domains** found in other proteins, in particular members of a family of transcription factors known as **signal transducers and activators of transcription (STATs)** (Fig. 3.26).

There are seven STATs (1–4, 5a, 5b, and 6), which reside in the cytoplasm in an inactive form until activated by cytokine receptors. Before activation, most STATs form homodimers, due to a specific homotypic interaction between domains present at the amino termini of the individual STAT proteins. The receptor specificity of each STAT is determined by the recognition of the distinctive phosphotyrosine sequence on each activated receptor by the different SH2 domains within the various STAT proteins. Recruitment of a STAT to the activated receptor brings the STAT close to an activated JAK, which can then phosphorylate a conserved tyrosine residue in the carboxy terminus of the particular STAT. This leads to a rearrangement, in which the phosphotyrosine of each STAT protein binds to the SH2 domain of the other STAT, forming a configuration that can bind DNA with high affinity. Activated STATs predominantly form homodimers, with a cytokine typically activating one type of STAT. For example, IFN-γ activates STAT1 and generates STAT1 homodimers, whereas IL-4 activates STAT6, generating STAT6 homodimers. Other cytokine receptors can activate several STATs, and some STAT heterodimers can be formed. The phosphorylated STAT dimer enters the nucleus, where it acts as a transcription factor to initiate the expression of selected genes that can regulate growth and differentiation of particular subsets of lymphocytes.

Since signaling by these receptors depends on tyrosine phosphorylation, dephosphorylation of the receptor complex by **tyrosine phosphatases** is one way that cells can terminate signaling. A variety of tyrosine phosphatases have been implicated in the dephosphorylation of cytokine receptors, JAKs, and STATs. These include the nonreceptor tyrosine phosphatases SHP-1 and SHP-2 (encoded by *PTPN6* and *PTPN11*), and the transmembrane receptor

Fig. 3.26 Many cytokine receptors signal using a rapid pathway called the JAK–STAT pathway.
First panel: many cytokines act via receptors that are associated with cytoplasmic Janus kinases (JAKs). The receptor consists of at least two chains, each associated with a specific JAK. Second panel: binding of ligand brings the two chains together, allowing the JAKs to phosphorylate and activate each other, and then to phosphorylate (red dots) specific tyrosines in the receptor tails. The STAT (signal transducer and activator of transcription) family of proteins have an N-terminal domain that homodimerizes STATs in the cytosol before activation, and an SH2 domain that binds to the tyrosine-phosphorylated receptor tails. Third panel: upon binding, the STAT homodimers are phosphorylated by JAKs. Fourth panel: after phosphorylation, STAT proteins reconfigure into a dimer that is stabilized by SH2 domain binding to phosphotyrosine residues on the other STAT. They then translocate to the nucleus, where they bind to and activate the transcription of a variety of genes important for adaptive immunity.

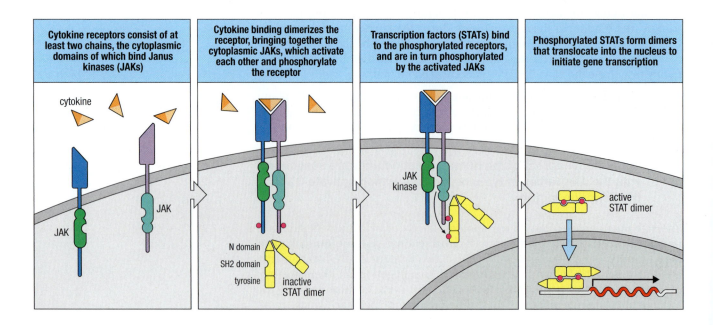

tyrosine phosphatase **CD45**, which is expressed as multiple isoforms on many hematopoietic cells. Cytokine signaling can also be terminated by negative feedback involving specific inhibitors that are induced by cytokine activation. The **suppressor of cytokine signaling (SOCS)** proteins are a class of inhibitors that terminate the signaling of many cytokine and hormone receptors. SOCS proteins contain an SH2 domain that can recruit them to the phosphorylated JAK kinase or receptor, and they can inhibit JAK kinases directly, compete for the receptor, and direct the ubiquitination and subsequent degradation of JAKs and STATs. SOCS proteins are induced by STAT activation, and thus inhibit receptor signaling after the cytokine has had its effect. Their importance can be seen in SOCS1-deficient mice, which develop a multiorgan inflammatory infiltrate caused by increased signaling from interferon receptors, γ_c-containing receptors, and TLRs. Another class of inhibitory proteins consists of the **protein inhibitors of activated STAT (PIAS)** proteins, which also seem to be involved in promoting the degradation of receptors and pathway components.

3-17   Chemokines released by macrophages and dendritic cells recruit effector cells to sites of infection.

All the cytokines produced by macrophages in innate immune responses have important local and systemic effects that contribute to both innate and adaptive immunity, and these are summarized in Fig. 3.27. The recognition of different classes of pathogens by phagocytes and dendritic cells may involve signaling through different receptors, such as the various TLRs, and can result in some variation in the cytokines expressed by stimulated macrophages and dendritic cells. This is one way in which appropriate immune responses can be selectively activated, as the released cytokines orchestrate the next phase of host defense. In response to activation by PRRs, macrophages and dendritic

Fig. 3.27 Important cytokines and chemokines secreted by dendritic cells and macrophages in response to bacterial products include IL-1β, IL-6, CXCL8, IL-12, and TNF-α. TNF-α is an inducer of a local inflammatory response that helps to contain infections. It also has systemic effects, many of which are harmful (discussed in Section 3-20). The chemokine CXCL8 is also involved in the local inflammatory response, helping to attract neutrophils to the site of infection. IL-1β, IL-6, and TNF-α have a crucial role in inducing the acute-phase response in the liver and inducing fever, which favors effective host defense in various ways. IL-12 activates natural killer (NK) cells and favors the differentiation of CD4 T cells into the T$_H$1 subset in adaptive immunity.

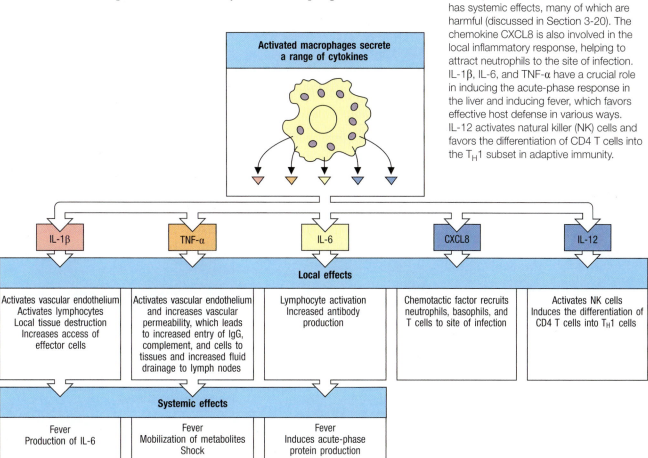

cells secrete a diverse group of cytokines that includes IL-1β, IL-6, IL-12, TNF-α, and the chemokine CXCL8 (formerly known as IL-8).

MOVIE 3.8

Among the cytokines released by tissues in the earliest phases of infection are members of a family of chemoattractant cytokines known as chemokines. These small proteins induce directed **chemotaxis** in nearby responsive cells, resulting in the movement of the cells toward the source of the chemokine. Because chemokines were first detected in functional assays, they were initially given a variety of names, which are listed along with their standardized nomenclature in Appendix IV. All the chemokines are related in amino acid sequence, and their receptors are G-protein-coupled receptors (see Section 3-2). The signaling pathway stimulated by chemokines causes changes in cell adhesiveness and changes in the cell's cytoskeleton that lead to directed migration. Chemokines can be produced and released by many different types of cells, not only those of the immune system. In the immune system they function mainly as chemoattractants for leukocytes, recruiting monocytes, neutrophils, and other effector cells of innate immunity from the blood into sites of infection. They also guide lymphocytes in adaptive immunity, as we will learn in Chapters 9–11. Some chemokines also function in lymphocyte development and migration and in angiogenesis (the growth of new blood vessels). There are more than 50 known chemokines, and this striking multiplicity may reflect their importance in delivering cells to their correct locations, which seems to be their main function in the case of lymphocytes. Some of the chemokines that are produced by or that affect human innate immune cells are listed in Fig. 3.28 along with their properties.

Fig. 3.28 Properties of selected human chemokines. Chemokines fall mainly into two related but distinct groups: the CC chemokines, which have two adjacent cysteine residues near the amino terminus; and the CXC chemokines, in which the equivalent cysteine residues are separated by a single amino acid. In humans, the genes for CC chemokines are mostly clustered in one region of chromosome 4. Genes for CXC chemokine genes are found mainly in a cluster on chromosome 17. The two groups of chemokines act on different sets of receptors, all of which are G-protein-coupled receptors. CC chemokines bind to receptors designated CCR1–10. CXC chemokines bind to receptors designated CXCR1–7. Different receptors are expressed on different cell types, and so a particular chemokine can be used to attract a particular cell type. In general, CXC chemokines with a Glu-Leu-Arg tripeptide motif immediately before the first cysteine promote the migration of neutrophils. CXCL8 is an example of this type. Most of the other CXC chemokines, including those that interact with receptors CXCR3, 4, and 5, lack this motif. Fractalkine is unusual in several respects: it has three amino acid residues between the two cysteines, and it exists in two forms, one that is tethered to the membrane of the endothelial and epithelial cells that express it, where it serves as an adhesion protein, and a soluble form that is released from the cell surface and acts as a chemoattractant for a wide range of cell types. A more comprehensive list of chemokines and their receptors is given in Appendix IV.

Class	Chemokine	Produced by	Receptors	Cells attracted	Major effects
CXC	CXCL8 (IL-8)	Monocytes Macrophages Fibroblasts Epithelial cells Endothelial cells	CXCR1 CXCR2	Neutrophils Naive T cells	Mobilizes, activates and degranulates neutrophils Angiogenesis
	CXCL7 (PBP, β-TG, NAP-2)	Platelets	CXCR2	Neutrophils	Activates neutrophils Clot resorption Angiogenesis
	CXCL1 (GROα) CXCL2 (GROβ) CXCL3 (GROγ)	Monocytes Fibroblasts Endothelium	CXCR2	Neutrophils Naive T cells Fibroblasts	Activates neutrophils Fibroplasia Angiogenesis
CC	CCL3 (MIP-1α)	Monocytes T cells Mast cells Fibroblasts	CCR1, 3, 5	Monocytes NK and T cells Basophils Dendritic cells	Competes with HIV-1 Antiviral defense Promotes T$_H$1 immunity
	CCL4 (MIP-1β)	Monocytes Macrophages Neutrophils Endothelium	CCR1, 3, 5	Monocytes NK and T cells Dendritic cells	Competes with HIV-1
	CCL2 (MCP-1)	Monocytes Macrophages Fibroblasts Keratinocytes	CCR2B	Monocytes NK and T cells Basophils Dendritic cells	Activates macrophages Basophil histamine release Promotes T$_H$2 immunity
	CCL5 (RANTES)	T cells Endothelium Platelets	CCR1, 3, 5	Monocytes NK and T cells Basophils Eosinophils Dendritic cells	Degranulates basophils Activates T cells Chronic inflammation
CXXXC (CX$_3$C)	CX3CL1 (Fractalkine)	Monocytes Endothelium Microglial cells	CX$_3$CR1	Monocytes T cells	Leukocyte–endothelial adhesion Brain inflammation

Chemokines fall mainly into two related but distinct groups. **CC chemokines** have two adjacent cysteine residues near the amino terminus, whereas in **CXC chemokines** the corresponding two cysteine residues are separated by a single amino acid. The CC chemokines promote the migration of monocytes, lymphocytes, and other cell types. One example relevant to innate immunity is CCL2, which attracts monocytes through the receptor CCR2B, inducing their migration from the bloodstream to become tissue macrophages. In contrast, neutrophil migration is promoted by CXC chemokines. CXCL8, acting through CXCR2, mobilizes neutrophils from bone marrow and induces them to leave the blood and migrate into the surrounding tissues. CCL2 and CXCL8 therefore have similar but complementary functions in the innate immune response, attracting monocytes and neutrophils respectively.

The role of chemokines in cell recruitment is twofold. First, they act on the leukocyte as it rolls along endothelial cells at sites of inflammation, converting this rolling into stable binding by triggering a change of conformation in the adhesion molecules known as leukocyte integrins. These conformational changes enable integrins to bind strongly to their ligands on the endothelial cells, which allows the leukocyte to cross the blood vessel walls by squeezing between the endothelial cells. Second, the chemokine directs the migration of the leukocyte along a gradient of chemokine molecules bound to the extracellular matrix and the surfaces of endothelial cells. This gradient increases in concentration toward the site of infection.

 MOVIE 3.9

Chemokines are produced by a wide variety of cell types in response to bacterial products, viruses, and agents that cause physical damage, such as silica, alum, or the urate crystals that occur in gout. Complement fragments such as C3a and C5a, and fMLF bacterial peptides, also act as chemoattractants for neutrophils. Thus, infection or physical damage to tissues induces the production of chemokine gradients that can direct phagocytes to the sites where they are needed. Neutrophils arrive rapidly in large numbers at a site of infection. The recruitment of monocytes occurs simultaneously, but they accumulate more slowly at the site of infection, perhaps because they are less abundant in the circulation. The complement fragment C5a and the chemokines CXCL8 and CCL2 activate their respective target cells, so that not only are neutrophils and monocytes brought to potential sites of infection but, in the process, they are armed to deal with the pathogens they encounter there. In particular, the signaling induced by C5a or CXCL8 in neutrophils serves to augment the respiratory burst that generates oxygen radicals and nitric oxide and to induce the neutrophils to release their stored antimicrobial granule contents (see Section 3-2).

Chemokines do not act alone in cell recruitment. They require the action of vasoactive mediators that bring leukocytes close to the blood vessel wall (see Section 3-3) and cytokines such as TNF-α to induce the necessary adhesion molecules on endothelial cells. We will return to the chemokines in later chapters, where they are discussed in the context of the adaptive immune response. Now, however, we turn to the molecules that enable leukocytes to adhere to the endothelium, and we shall then describe step by step the extravasation process by which monocytes and neutrophils enter infected sites.

3-18 Cell-adhesion molecules control interactions between leukocytes and endothelial cells during an inflammatory response.

The recruitment of activated phagocytes to sites of infection is one of the most important functions of innate immunity. Recruitment occurs as part of the inflammatory response and is mediated by cell-adhesion molecules that are induced on the surface of the endothelial cells of local blood vessels. Here we

MOVIE 3.10

consider those functions that participate in the recruitment of inflammatory cells in the hours to days after the establishment of an infection.

As with the complement components, a significant barrier to understanding the functions of cell-adhesion molecules is their nomenclature. Most adhesion molecules, especially those on leukocytes, which are relatively easy to analyze functionally, were originally named after the effects of specific monoclonal antibodies directed against them. Their names therefore bear no relation to their structural class. For instance, the **leukocyte functional antigens** LFA-1, LFA-2, and LFA-3 are actually members of two different protein families. In **Fig. 3.29**, the adhesion molecules relevant to innate immunity are grouped according to their molecular structure, which is shown in schematic form alongside their different names, sites of expression, and ligands. Three structural families of adhesion molecules are important for leukocyte recruitment. The **selectins** are membrane glycoproteins with a distal lectin-like domain that binds specific carbohydrate groups. Members of this family are induced on activated endothelium and initiate endothelium–leukocyte interactions by binding to fucosylated oligosaccharide ligands on passing leukocytes (see Fig. 3.29).

The next step in leukocyte recruitment depends on tighter adhesion, which is due to the binding of **intercellular adhesion molecules (ICAMs)** on the endothelium to heterodimeric proteins of the **integrin** family on leukocytes. ICAMs are single-pass membrane proteins that belong to the large superfamily of **immunoglobulin-like proteins**, which contain protein domains similar to those of immunoglobulins. The extracellular regions of ICAMs are composed of several immunoglobulin-like domains. An integrin molecule is composed

Fig. 3.29 Adhesion molecules involved in leukocyte interactions. Several structural families of adhesion molecules have a role in leukocyte migration, homing, and cell–cell interactions: the selectins, the integrins, and proteins of the immunoglobulin superfamily. The figure shows schematic representations of an example from each family, a list of other family members that participate in leukocyte interactions, their cellular distribution, and their ligand in adhesive interactions. The family members shown here are limited to those that participate in inflammation and other innate immune mechanisms. The same molecules and others participate in adaptive immunity and will be considered in Chapters 9 and 11. The nomenclature of the different molecules in these families is confusing because it often reflects the way in which the molecules were first identified rather than their related structural characteristics. Alternative names for each of the adhesion molecules are given in parentheses. Sulfated sialyl-Lewis^X, which is recognized by P- and E-selectin, is an oligosaccharide present on the cell-surface glycoproteins of circulating leukocytes.

		Name	Tissue distribution	Ligand
Selectins Bind carbohydrates. Initiate leukocyte–endothelial interaction	P-selectin	P-selectin (PADGEM, CD62P)	Activated endothelium and platelets	PSGL-1, sialyl-Lewis^x
		E-selectin (ELAM-1, CD62E)	Activated endothelium	Sialyl-Lewis^x
Integrins Bind to cell-adhesion molecules and extracellular matrix. Strong adhesion	LFA-1	$\alpha_L{:}\beta_2$ (LFA-1, CD11a:CD18)	Monocytes, T cells, macrophages, neutrophils, dendritic cells, NK cells	ICAM-1, ICAM-2
		$\alpha_M{:}\beta_2$ (CR3, Mac-1, CD11b:CD18)	Neutrophils, monocytes, macrophages, NK cells	ICAM-1, iC3b, fibrinogen
		$\alpha_X{:}\beta_2$ (CR4, p150.95, CD11c:CD18)	Dendritic cells, macrophages, neutrophils, NK cells	iC3b
		$\alpha_5{:}\beta_1$ (VLA-5, CD49d:CD29)	Monocytes, macrophages	Fibronectin
Immunoglobulin superfamily Various roles in cell adhesion. Ligand for integrins	ICAM-1	ICAM-1 (CD54)	Activated endothelium, activated leukocytes	LFA-1, Mac1
		ICAM-2 (CD102)	Resting endothelium, dendritic cells	LFA-1
		VCAM-1 (CD106)	Activated endothelium	VLA-4
		PECAM (CD31)	Activated leukocytes, endothelial cell–cell junctions	CD31

of two transmembrane protein chains, α and β, of which there are numerous different types. Subsets of integrins have a common β chain partnered with different α chains. The leukocyte integrins important for extravasation are **LFA-1** ($\alpha_L:\beta_2$, also known as **CD11a:CD18**) and **CR3** ($\alpha_M:\beta_2$, complement receptor type 3, also known as **CD11b:CD18** or Mac-1). We described CR3 in Section 2-13 as a receptor for iC3b, but it also binds other ligands. Both LFA-1 and CR3 bind to **ICAM-1** and to **ICAM-2** (Fig. 3.30). Even in the absence of infection, circulating monocytes are continuously leaving the blood and entering certain tissues, such as the intestine, where they become resident macrophages. To navigate out of the blood vessel, they may adhere to ICAM-2, which is expressed at low levels by unactivated endothelium. CR3 also binds to fibrinogen and factor X, both substrates of the coagulation cascade.

Strong adhesion between leukocytes and endothelial cells is promoted by the induction of ICAM-1 on inflamed endothelium together with a conformational change in LFA-1 and CR3 that occurs on the leukocyte. Integrins can switch between an 'active' state, in which they bind strongly to their ligands, and an 'inactive' state, in which binding is easily broken. This enables cells to make and break integrin-mediated adhesions in response to signals received by the cell either through the integrin itself or through other receptors. In the activated state, an integrin molecule is linked via the intracellular protein **talin** to the actin cytoskeleton. In the case of migrating leukocytes, chemokines binding to their receptors on the leukocyte generate intracellular signals that cause talin to bind to the cytoplasmic tails of the β chains of LFA-1 and CD3, forcing the integrin extracellular regions to assume an active binding conformation. The importance of leukocyte integrin function in inflammatory cell recruitment is illustrated by **leukocyte adhesion deficiencies**, which can be caused by defects in the integrins themselves or in the proteins required for modulating adhesion. People with these diseases suffer from recurrent bacterial infections and impaired healing of wounds.

Endothelial activation is driven by macrophage-produced cytokines, particularly TNF-α, which induce the rapid externalization of granules called **Weibel–Palade bodies** in the endothelial cells. These granules contain preformed **P-selectin**, which appears on the surfaces of local endothelial cells just minutes after macrophages have responded to the presence of microbes by producing TNF-α. Shortly after P-selectin gets to the cell surface, mRNA encoding **E-selectin** is synthesized, and within 2 hours the endothelial cells are expressing mainly E-selectin. Both P-selectin and E-selectin interact with **sulfated sialyl-Lewis^X**, a sulfated form of a carbohydrate structure that is also an important blood group antigen. Sulfated sialyl-Lewis^X is present on the surface of neutrophils, and its interactions with P-selectin and E-selectin are important for neutrophil rolling on the endothelium. Mutations in enzymes involved in its synthesis, such as fucosyltransferase, cause defective sialyl-Lewis^X expression that results in an immunodeficiency, **leukocyte adhesion deficiency type 2**.

Integrins are also convenient cell-surface markers for distinguishing different cell types. Dendritic cells, macrophages, and monocytes express different integrin α chains and thus display distinct β_2 integrins on their surface. The predominant leukocyte integrin on conventional dendritic cells is $\alpha_X:\beta_2$, also known as **CD11c:CD18** or complement receptor 4 (**CR4**) (see Fig. 3.29). This integrin is a receptor for the complement C3 cleavage product iC3b, fibrinogen, and ICAM-1. In contrast to conventional dendritic cells, most monocytes and macrophages express low levels of CD11c, and predominantly express the integrin $\alpha_M:\beta_2$ (CD11b:CD18; CR3). However, patterns of integrin expression can vary, with some tissue macrophages, such as those in the lung, expressing high levels of CD11c:CD18. In the mouse, the two major branches of conventional dendritic cells can be distinguished by expression of CD11b:CD18: one branch characterized by high expression of CD11b:CD18,

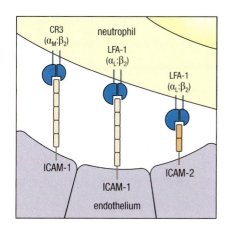

Fig. 3.30 Phagocyte adhesion to vascular endothelium is mediated by integrins. When vascular endothelium is activated by inflammatory mediators it expresses two adhesion molecules, namely ICAM-1 and ICAM-2. These are ligands for integrins expressed by phagocytes— $\alpha_M:\beta_2$ (also called CR3, Mac-1, or CD11b:CD18) and $\alpha_L:\beta_2$ (also called LFA-1 or CD11a:CD18).

Leukocyte Adhesion Deficiency

and a second branch that lacks CD11b:CD18. Plasmacytoid dendritic cells (pDCs) express lower levels of CD11c, but can be distinguished from conventional dendritic cells using other markers; human pDCs express the C-type lectin BDCA-2 **(blood dendritic cell antigen 2)**, and mouse pDCs express **BST2** (bone marrow stromal antigen), neither of which is expressed by conventional dendritic cells.

3-19   Neutrophils make up the first wave of cells that cross the blood vessel wall to enter an inflamed tissue.

The migration of leukocytes out of blood vessels, the process known as extravasation, occurs in response to signals generated at sites of infection. Under normal conditions, leukocytes travel in the center of small blood vessels, where blood flow is fastest. Within sites of inflammation, the vessels are dilated and the consequent slower blood flow allows leukocytes to interact in large numbers with the vascular endothelium. During an inflammatory response, the induction of adhesion molecules on the endothelial cells of blood vessels within the infected tissue, as well as induced changes in the adhesion molecules expressed on leukocytes, recruits large numbers of circulating leukocytes to the site of infection. We will describe this process with regard to monocytes and neutrophils (Fig. 3.31).

MOVIE 3.11

Extravasation proceeds in four stages. In the first, induction of selectins induces leukocyte rolling along the endothelium. P-selectin appears on endothelial cell surfaces within a few minutes of exposure to leukotriene B4, C5a, or histamine, which is released from mast cells in response to C5a. P-selectin can also be induced by TNF-α or LPS, and both of these induce synthesis of E-selectin, which appears on the endothelial cell surface a few hours later. When the sulfated sialyl-Lewis^X on monocytes and neutrophils contacts these exposed P-and E-selectins, these cells adhere reversibly to the vessel wall and begin to 'roll' along endothelium (see Fig. 3.31, top panel), permitting stronger interactions of the next step in leukocyte migration. Neutrophils are particularly efficient at rolling along endothelium even under flow rates that prevent rolling by other cells. Such '**shear-resistant rolling**' by neutrophils uses long extensions of plasma membrane, termed **slings**, that bind the endothelium and wrap around the cell as it rolls, serving to tether the cell firmly to the endothelium and to promote rapid entry to sites of infection.

MOVIE 3.12

The second step depends on interactions between the leukocyte integrins LFA-1 and CR3 with adhesion molecules such as ICAM-1 (which can be induced on endothelial cells by TNF-α) and ICAM-2 on endothelium (see Fig. 3.31, bottom panel). LFA-1 and CR3 normally bind their ligands only weakly, but CXCL8 (or other chemokines), bound to proteoglycans on the surface of endothelial cells, binds to specific chemokine receptors on the leukocyte and signals the cell to trigger a conformational change in LFA-1 and CR3 on the rolling leukocyte; this greatly increases the adhesive properties of the leukocyte, as discussed in Section 3-18. The cell then attaches firmly to the endothelium, and its rolling is arrested.

MOVIE 3.13

In the third step the leukocyte extravasates, or crosses the endothelial wall. This step also involves LFA-1 and CR3, as well as a further adhesive interaction involving an immunoglobulin-related molecule called **PECAM** or **CD31**, which is expressed both on the leukocyte and at the intercellular junctions of endothelial cells. These interactions enable the phagocyte to squeeze between the endothelial cells. It then penetrates the basement membrane with the aid of enzymes that break down the extracellular matrix proteins of the basement membrane. The movement through the basement membrane is known as **diapedesis**, and it enables phagocytes to enter the subendothelial tissues.

The fourth and final step in extravasation is the migration of leukocytes through the tissues under the influence of chemokines. Chemokines such as

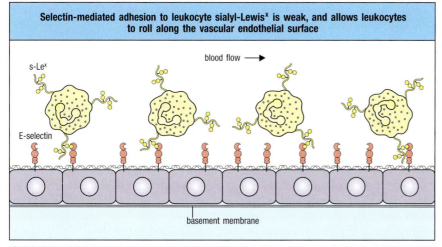

Selectin-mediated adhesion to leukocyte sialyl-Lewis^x is weak, and allows leukocytes to roll along the vascular endothelial surface

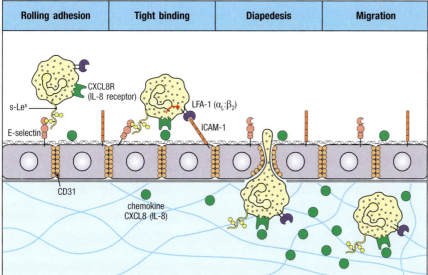

| Rolling adhesion | Tight binding | Diapedesis | Migration |

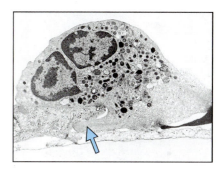

Fig. 3.31 Neutrophils leave the blood and migrate to sites of infection in a multi-step process involving adhesive interactions that are regulated by macrophage-derived cytokines and chemokines. Top panel: the first step involves the reversible binding of a neutrophil to vascular endothelium through interactions between selectins induced on the endothelium and their carbohydrate ligands on the neutrophil, shown here for E-selectin and its ligand, the sialyl-Lewis^X moiety (s-Le^x). This interaction cannot anchor the cells against the shearing force of the flow of blood, and thus they roll along the endothelium, continually making and breaking contact. Bottom panel: the binding does, however, eventually trigger stronger interactions, which result only when binding of a chemokine such as CXCL8 to its specific receptor on the neutrophil triggers the activation of the integrins LFA-1 and CR3 (Mac-1; not shown). Inflammatory cytokines such as TNF-α are also necessary to induce the expression of adhesion molecules such as ICAM-1 and ICAM-2, the ligands for these integrins, on the vascular endothelium. Tight binding between ICAM-1 and the integrins arrests the rolling and allows the neutrophil to squeeze between the endothelial cells forming the wall of the blood vessel (i.e., to extravasate). The leukocyte integrins LFA-1 and CR3 are required for extravasation and for migration toward chemoattractants. Adhesion between molecules of CD31, expressed on both the neutrophil and the junction of the endothelial cells, is also thought to contribute to extravasation. The neutrophil also needs to traverse the basement membrane; it penetrates this with the aid of a matrix metalloproteinase enzyme, MMP-9, that it expresses at the cell surface. Finally, the neutrophil migrates along a concentration gradient of chemokines (shown here as CXCL8) secreted by cells at the site of infection. The electron micrograph shows a neutrophil extravasating between endothelial cells. The blue arrow indicates the pseudopod that the neutrophil is inserting between the endothelial cells. Photograph (×5500) courtesy of I. Bird and J. Spragg.

CXCL8 and CCL2 (see Section 3-17) are produced at the site of infection and bind to proteoglycans in the extracellular matrix and on endothelial cell surfaces. In this way, a matrix-associated concentration gradient of chemokines is formed on a solid surface along which the leukocyte can migrate to the focus of infection (see Fig. 3.31). CXCL8 is released by the macrophages that first encounter pathogens; it recruits neutrophils, which enter the infected tissue in large numbers in the early part of the induced response. Their influx usually peaks within the first 6 hours of an inflammatory response. Monocytes are recruited through the action of CCL2, and accumulate more slowly than neutrophils. Once in the inflamed tissue, neutrophils are able to eliminate many pathogens by phagocytosis. In an innate immune response, neutrophils use their complement receptors and the direct pattern recognition receptors discussed earlier in this chapter (see Section 3-1) to recognize and phagocytose pathogens or pathogen components directly or after opsonization with complement (see Section 2-13). In addition, as we will see in Chapter 10, neutrophils act as phagocytic effectors in humoral adaptive immunity, taking up antibody-coated microbes by means of specific receptors.

The importance of neutrophils in immune defense is dramatically illustrated by diseases or medical treatments that severely reduce neutrophil numbers. Patients suffering this affliction are said to have **neutropenia**, and they are highly susceptible to deadly infection with a wide range of pathogens and commensal organisms. Restoring neutrophil levels in such patients by transfusion of neutrophil-rich blood fractions or by stimulating their production with specific growth factors largely corrects this susceptibility.

3-20 TNF-α is an important cytokine that triggers local containment of infection but induces shock when released systemically.

TNF-α acting on endothelial cells stimulates the expression of adhesion molecules and aids the extravasation of cells such as monocytes and neutrophils. Another important action of TNF-α is to stimulate endothelial cells to express proteins that trigger blood clotting in the local small vessels, occluding them and cutting off blood flow. This can be important in preventing the pathogen from entering the bloodstream and spreading through the blood to organs all over the body. The importance of TNF-α in the containment of local infection is illustrated by experiments in which rabbits were infected locally with a bacterium. Normally, the infection would be contained at the site of the inoculation; if, however, an injection of anti-TNF-α antibody was also given to block the action of TNF-α, the infection spread via the blood to other organs. In parallel, the fluid that has leaked into the tissue in the early phases of an infection carries the pathogen, usually enclosed in dendritic cells, via the lymph to the regional lymph nodes, where an adaptive immune response can be initiated.

Once an infection has spread to the bloodstream, however, the same mechanisms by which TNF-α so effectively contains local infection instead becomes catastrophic (Fig. 3.32). Although produced as a membrane-associated cytokine, TNF-α can be cleaved by a specific protease, **TACE** (TNF-α-converting enzyme, which is encoded by the *ADAM17* gene), and released from the membrane as a soluble cytokine. The presence of infection in the bloodstream, or sepsis, is accompanied by a massive release of soluble TNF-α from macrophages in the liver, spleen, and other sites throughout the body. The systemic release of TNF-α into the bloodstream causes vasodilation, which leads to a loss of blood pressure and increased vascular permeability; this in turn leads to a loss of plasma volume and eventually to shock, known in this case as septic shock because the underlying cause is a bacterial infection. The TNF-α released in septic shock also triggers blood clotting in small vessels throughout the body—known as **disseminated intravascular coagulation**—which leads to the massive consumption of clotting proteins, so that the patient's blood cannot clot appropriately. Disseminated intravascular coagulation frequently leads to the failure of vital organs such as the kidneys, liver, heart, and lungs, which are quickly compromised by the failure of normal blood perfusion; consequently, septic shock has a very high mortality rate.

Mice with defective or no TNF-α receptors are resistant to septic shock but are also unable to control local infection. Mice in which the *ADAM17* gene has been selectively inactivated in myeloid cells are also resistant to septic shock, confirming that the release of soluble TNF-α into the circulation both depends on TACE and is the main factor responsible for septic shock. Blockade of TNF-α activity, either with specific antibodies or with soluble proteins that mimic the receptor, is a successful treatment for several inflammatory disorders, including rheumatoid arthritis. However, these treatments have been found to reactivate tuberculosis in some apparently well patients with evidence of previous infection (as demonstrated by skin test), which is a direct demonstration of the importance of TNF-α in keeping infection local and in check.

3-21 Cytokines made by macrophages and dendritic cells induce a systemic reaction known as the acute-phase response.

As well as their important local effects, the cytokines produced by macrophages and dendritic cells have long-range effects that contribute to host defense. One of these is the elevation of body temperature, which is caused

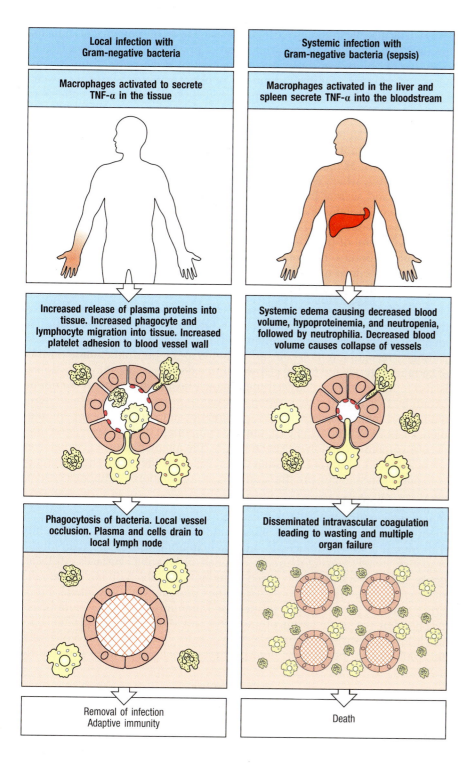

Fig. 3.32 The release of TNF-α by macrophages induces local protective effects, but TNF-α can be damaging when released systemically. The panels on the left show the causes and consequences of local release of TNF-α, and the panels on the right show the causes and consequences of systemic release. In both cases TNF-α acts on blood vessels, especially venules, to increase blood flow and vascular permeability to fluid, proteins, and cells, and to increase endothelial adhesiveness for leukocytes and platelets (center row). Local release thus allows an influx of fluid, cells, and proteins into the infected tissue, where they participate in host defense. Later, blood clots form in the small vessels (bottom left panel), preventing spread of infection via the blood, and the accumulated fluid and cells drain to regional lymph nodes, where an adaptive immune response is initiated. When there is a systemic infection, or sepsis, with bacteria that elicit TNF-α production, TNF-α is released into the blood by macrophages in the liver and spleen and acts in a similar way on all small blood vessels in the body (bottom right panel). The result is shock, disseminated intravascular coagulation with depletion of clotting factors, and consequent bleeding, multiple organ failure, and frequently death.

mainly by TNF-α, IL-1β, and IL-6. These cytokines are termed **endogenous pyrogens** because they cause fever and derive from an endogenous source rather than from bacterial components such as LPS, which also induces fever and is an **exogenous pyrogen**. Endogenous pyrogens cause fever by inducing the synthesis of prostaglandin E2 by the enzyme cyclooxygenase-2, the expression of which is induced by these cytokines. Prostaglandin E2 then acts on the hypothalamus, resulting in an increase in both heat production from the catabolism of brown fat and heat retention from vasoconstriction, which decreases the loss of excess heat through the skin. Exogenous pyrogens are able to induce fever by promoting the production of the endogenous pyrogens and also by directly inducing cyclooxygenase-2 as a consequence of signaling through TLR-4, leading to the production of prostaglandin E2. Fever is generally beneficial to host defense; most pathogens grow better at lower temperatures, whereas adaptive immune responses are more intense at elevated temperatures. Host cells are also protected from the deleterious effects of TNF-α at raised temperatures.

The effects of TNF-α, IL-1β, and IL-6 are summarized in Fig. 3.33. One of the most important of these occurs in the liver and is the initiation of a response known as the **acute-phase response** (Fig. 3.34). The cytokines act on hepatocytes, which respond by changing the profile of proteins that they synthesize and secrete into the blood. In the acute-phase response, blood levels of some proteins go down, whereas levels of others increase markedly. The proteins induced by TNF-α, IL-1β, and IL-6 are called the **acute-phase proteins**. Several of these are of particular interest because they mimic the action of antibodies, but unlike antibodies they have broad specificity for pathogen-associated molecular patterns and depend only on the presence of cytokines for their production.

One acute-phase protein, the **C-reactive protein**, is a member of the **pentraxin** protein family, so called because the proteins are formed from five identical subunits. C-reactive protein is yet another example of a multipronged pathogen-recognition molecule, and it binds to the phosphocholine portion of certain bacterial and fungal cell-wall lipopolysaccharides. Phosphocholine is also found in mammalian cell-membrane phospholipids, but it cannot be bound by C-reactive protein. When C-reactive protein binds to a bacterium, it not only is able to opsonize the bacterium but can also activate the complement cascade by binding to C1q, the first component of the classical pathway of complement activation (see Section 2-7). The interaction with C1q involves the collagen-like parts of C1q rather than the globular heads that make contact with pathogen surfaces, but the same cascade of reactions is initiated.

Mannose-binding lectin (MBL) is another acute-phase protein; it serves as an innate recognition molecule that can activate the lectin pathway of complement (see Section 2-6). MBL is present at low levels in the blood of healthy individuals, but it is produced in increased amounts during the acute-phase response. By recognizing mannose residues on microbial surfaces, MBL can act as an opsonin that is recognized by monocytes, which do not express the macrophage mannose receptor. Two other proteins with opsonizing properties that are also produced in increased amounts during an acute-phase response are the **surfactant proteins**, **SP-A** and **SP-D**. These are produced by the liver and a variety of epithelia. They are, for example, found along with macrophages in the alveolar fluid of the lung, where they are secreted by

Fig. 3.33 The cytokines TNF-α, IL-1β, and IL-6 have a wide spectrum of biological activities that help to coordinate the body's responses to infection. IL-1β, IL-6, and TNF-α activate hepatocytes to synthesize acute-phase proteins, and bone marrow endothelium to release neutrophils. The acute-phase proteins act as opsonins, whereas the disposal of opsonized pathogens is augmented by the enhanced recruitment of neutrophils from the bone marrow. IL-1β, IL-6, and TNF-α are also endogenous pyrogens, raising body temperature, which is believed to help in eliminating infections. A major effect of these cytokines is to act on the hypothalamus, altering the body's temperature regulation, and on muscle and fat cells, altering energy mobilization to increase the body temperature. At higher temperatures, bacterial and viral replication is less efficient, whereas the adaptive immune response operates more efficiently.

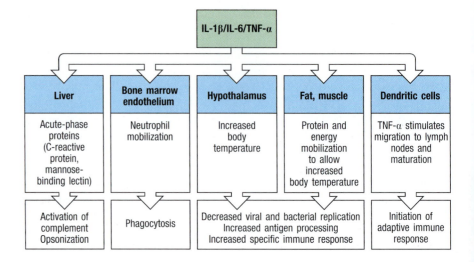

Fig. 3.34 The acute-phase response produces molecules that bind pathogens but not host cells. Acute-phase proteins are produced by liver cells in response to cytokines released by macrophages in the presence of bacteria (top panel). They include serum amyloid protein (SAP) (in mice but not humans), C-reactive protein (CRP), fibrinogen, and mannose-binding lectin (MBL). CRP binds phosphocholine on certain bacterial and fungal surfaces but does not recognize it in the form in which it is found in host cell membranes (middle panel). SAP and CRP are homologous in structure; both are pentraxins, forming five-membered discs, as shown for SAP (lower panel). SAP both acts as an opsonin in its own right and activates the classical complement pathway by binding C1q to augment opsonization. MBL is a member of the collectin family, which also includes the pulmonary surfactant proteins SP-A and SP-D. Like CRP, MBL can act as an opsonin in its own right, as can SP-A and SP-D. Model structure courtesy of J. Emsley.

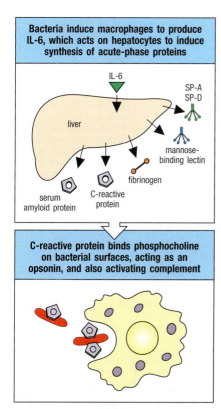

Bacteria induce macrophages to produce IL-6, which acts on hepatocytes to induce synthesis of acute-phase proteins

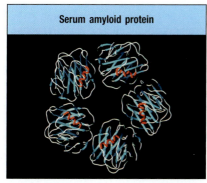

C-reactive protein binds phosphocholine on bacterial surfaces, acting as an opsonin, and also activating complement

Serum amyloid protein

pneumocytes, and are important in promoting the phagocytosis of opportunistic respiratory pathogens such as *Pneumocystis jirovecii* (formerly known as *P. carinii*), one of the main causes of pneumonia in patients with AIDS.

Thus, within a day or two, the acute-phase response provides the host with several proteins with the functional properties of antibodies but able to bind a broad range of pathogens. However, unlike antibodies, which we describe in Chapters 4 and 10, acute-phase proteins have no structural diversity and are made in response to any stimulus that triggers the release of TNF-α, IL-1, and IL-6. Therefore, unlike antibodies, their synthesis is not specifically induced and targeted.

A final, distant effect of the cytokines produced by macrophages is to induce **leukocytosis**, an increase in the numbers of circulating neutrophils. The neutrophils come from two sources: the bone marrow, from which mature leukocytes are released in increased numbers; and sites in blood vessels, where they are attached loosely to endothelial cells. Thus the effects of these cytokines contribute to the control of infection while the adaptive immune response is being developed. As shown in Fig. 3.33, TNF-α also has a role in stimulating the migration of dendritic cells from their sites in peripheral tissues to the lymph nodes and in their maturation into nonphagocytic but highly co-stimulatory antigen-presenting cells.

3-22 Interferons induced by viral infection make several contributions to host defense.

Viral infection induces the production of interferons, originally named because of their ability to interfere with viral replication in previously uninfected tissue culture cells. Interferons have a similar role *in vivo*, blocking the spread of viruses to uninfected cells. There are numerous genes encoding antiviral, or type I, interferons. Best understood are the IFN-α family of 12 closely related human genes and **IFN-β**, the product of a single gene; less well studied are IFN-κ, IFN-ε, and IFN-ω. **IFN-γ** is the sole **type II interferon**.

Type III interferons are a newly classified IFN family composed of the products of three **IFN-λ** genes, also known as IL-28A, IL-28B, and IL-29, which bind a heterodimeric **IFN-λ receptor** composed of a unique IL-28Rα subunit and the β subunit of the IL-10 receptor. While receptors for type I interferons and IFN-γ are widespread in their tissue distribution, type III receptors are more restricted, not expressed by fibroblasts or epithelial cells, but expressed on epithelial cells.

Type I interferons are inducible and are synthesized by many cell types after infection by diverse viruses. Almost all types of cells can produce IFN-α and IFN-β in response to activation of several innate sensors. For example, type I interferons are induced by RIG-I and MDA-5 (the sensors of cytoplasmic viral RNA) downstream of MAVS, and by signaling from cGAS (the sensor of cytoplasmic DNA) downstream of STING (see Sections 3-10 and 3-11).

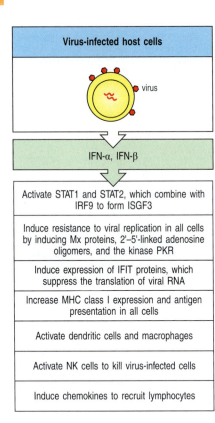

Fig. 3.35 Interferons are antiviral proteins produced by cells in response to viral infection. The interferons IFN-α and IFN-β have three major functions. First, they induce resistance to viral replication in uninfected cells by activating genes that cause the destruction of mRNA and inhibit the translation of viral proteins and some host proteins. These include the Mx proteins, oligoadenylate synthetase, PKR, and IFIT proteins. Second, they can induce MHC class I expression in most cell types in the body, thus enhancing their resistance to NK cells; they may also induce increased synthesis of MHC class I molecules in cells that are newly infected by virus, thus making them more susceptible to being killed by CD8 cytotoxic T cells (see Chapter 9). Third, they activate NK cells, which then selectively kill virus-infected cells.

Interferon-γ Receptor Deficiency

However, some immune cells seem to be specialized for this task. In Section 3-1 we introduced the plasmacytoid dendritic cell (pDC). Also called **interferon-producing cells (IPCs)** or **natural interferon-producing cells**, human plasmacytoid dendritic cells were initially recognized as rare peripheral blood cells that accumulate in peripheral lymphoid tissues during a viral infection and make abundant type I interferons (IFN-α and IFN-β)—up to 1000 times more than other cell types. This abundant production of type I interferon may result from the efficient coupling of viral recognition by TLRs to the pathways of interferon production (see Section 3-7). Plasmacytoid dendritic cells express a subset of TLRs that includes TLR-7 and TLR-9, which are endosomal sensors of viral RNA and of the nonmethylated CpG residues present in the genomes of many DNA viruses (see Fig. 3.11). The requirement for TLR-9 in sensing infections caused by DNA viruses has been demonstrated, for example, by the inability of TLR-9-deficient plasmacytoid dendritic cells to generate type I interferons in response to herpes simplex virus. Plasmacytoid dendritic cells express CXCR3, a receptor for the chemokines CXCL9, CXCL10, and CXCL11, which are produced by T cells. This allows pDCs to migrate from the blood into lymph nodes in which there is an ongoing inflammatory response to a pathogen.

Interferons help defend against viral infection in several ways (**Fig. 3.35**). IFN-β is particularly important because it induces cells to make IFN-α, thus amplifying the interferon response. Interferons act to induce a state of resistance to viral replication in all cells. IFN-α and IFN-β bind to a common cell-surface receptor, known as the **interferon-α receptor (IFNAR)**, which uses the **JAK** and **STAT** pathways described in Section 3-16. IFNAR uses the kinases Tyk2 and Jak1 to activate the factors **STAT1** and **STAT2**, which can interact with **IRF9** and form a complex called **ISGF3** that binds to the promoters of many **interferon stimulated genes (ISGs)**.

One ISG encodes the enzyme **oligoadenylate synthetase**, which polymerizes ATP into 2'–5'-linked oligomers (whereas nucleotides in nucleic acids are normally linked 3'–5'). These 2'–5'-linked oligomers activate an endoribonuclease that then degrades viral RNA. A second protein induced by IFN-α and IFN-β is a dsRNA-dependent protein kinase called **PKR**. This serine–threonine kinase phosphorylates the α subunit of **eukaryotic initiation factor 2 (eIF2α)**, thus suppressing protein translation and contributing to the inhibition of viral replication. **Mx** (myxoma resistant) **proteins** are also induced by type I interferons. Humans and wild mice have two highly similar proteins, **Mx1** and **Mx2**, which are GTPases belonging to the dynamin protein family, but how they interfere with viral replication is not understood. Oddly, most common laboratory strains of mice have inactivated both Mx genes, and in these mice, IFN-β cannot act to protect against influenza infection.

In the last few years, several novel ISGs have been identified and linked to antiviral functions. The **IFIT (IFN-induced protein with tetratricoid repeats)** family contains four human and three mouse proteins that function in restraining the translation of viral RNA into proteins. **IFIT1** and **IFIT2** can both suppress the translation of normal capped mRNAs by binding to subunits of the **eukaryotic initiation factor 3 (eIF3)** complex, which prevents eIF3 from interacting with eIF2 to form the 43S pre-initiation complex (**Fig. 3.36**). This action may be responsible in part for the reduction in cellular proliferation induced by type I interferons. Mice lacking IFIT1 or IFIT2 show increased susceptibility to infection by certain viruses, such as vesicular stomatitis virus.

Another function of IFIT1 is to suppress translation of viral RNA that lacks a normal host modification of the 5' cap. Recall that the normal mammalian 5' cap is initiated by linking a 7-methylguanosine nucleotide to the first ribose sugar of the mRNA by a 5'–5' triphosphate bridge, to produce a structure called cap-0. This structure is further modified by cytoplasmic methylation of the 2' hydroxyl groups on the first and second ribose sugars of the RNA. Methylation

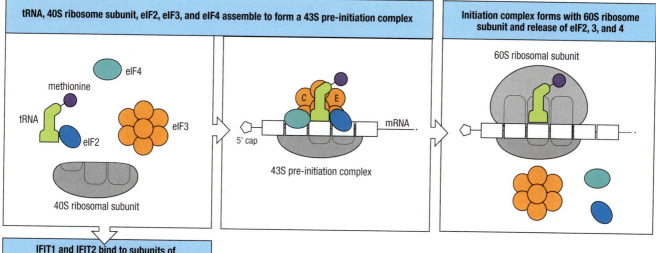

| tRNA, 40S ribosome subunit, eIF2, eIF3, and eIF4 assemble to form a 43S pre-initiation complex | Initiation complex forms with 60S ribosome subunit and release of eIF2, 3, and 4 |

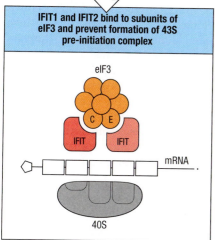

IFIT1 and IFIT2 bind to subunits of eIF3 and prevent formation of 43S pre-initiation complex

Fig. 3.36 IFIT proteins act as antiviral effector molecules by inhibiting steps in the translation of RNA. Top left panel: formation of a 43S pre-initiation complex is an early step in the translation of RNA into protein by the 80S ribosome that involves a charged methionine tRNA, the 40S ribosome subunit, and eukaryotic initiation factors (eIFs) eIF4, eIF2, and eIF3. Middle panel: eIFs and a charged methionine tRNA assemble into a 43S pre-initiation complex. Right panel: the pre-initiation complex mRNA recognizes the 5′ cap structure and joins with the 60S ribosomal subunit, releasing eIF2, eIF3, and eIF4 and forming a functional 80S ribosome. Lower panel: eIF3 has 13 subunits, a–m. IFIT proteins can inhibit several steps in protein translation. Mouse IFIT1 and IFIT2 interact with eIF3C, and human IFIT1 and IFIT2 interact with eIF3E, preventing formation of the 43S pre-initiation complex. IFITs can also interfere with other steps in translation, and can bind and sequester uncapped viral mRNAs to prevent their translation (not shown). Expression of IFIT proteins is induced in viral infection by signaling downstream of type I interferons.

of the first ribose sugar produces a structure called cap-1; methylation of the second generates cap-2. IFIT1 has a high affinity for cap-0, but much lower affinity for cap-1 and cap-2. Some viruses, such as Sindbis virus (family Togaviridae), lack 2′-*O*-methylation, and therefore are restricted by this action of IFIT1. Many viruses, such as West Nile virus and SARS coronavirus, have acquired a **2′e-*O*-methyltransferase (MTase)** that produce cap-1 or cap-2 on their viral transcripts. These viruses can thus evade restriction by IFIT1.

Members of the **interferon-induced transmembrane protein (IFITM)** family are expressed at a basal level on many types of tissues but are strongly induced by type I interferons. There are four functional IFITM genes in humans and in mice, and these encode proteins that have two transmembrane domains and are localized to various vesicular compartments of the cell. IFITM proteins act to inhibit, or restrict, viruses at early steps of infection. Although the molecular details are unclear, **IFITM1** appears to interfere with the fusion of viral membranes with the membrane of the lysosome, which is required for introducing some viral genomes into the cytoplasm. Viruses that must undergo this fusion event in lysosomes, such as the Ebola virus, are restricted by IFITM1. Similarly, **IFITM3** interferes with membrane fusion in late endosomes, and so restricts the influenza A virus, which undergoes fusion there. The importance of this mechanism is demonstrated by the increased viral load and higher mortality in mice lacking IFITM3 that are infected with the influenza A virus.

Interferons also stimulate production of the chemokines CXCL9, CXCL10, and CXCL11, which recruit lymphocytes to sites of infection. They also increase the expression of MHC class I molecules on all types of cells, which facilitates recognition of virally infected cells by cytotoxic T lymphocytes via the display of viral peptides complexed to MHC class I molecules on the infected cell surface (see Fig. 1.30). Through these effects, interferons indirectly help promote the killing of virus-infected cells by CD8 cytotoxic T cells. Another way in which interferons act is to activate populations of innate immune cells, such as NK cells, that can kill virus-infected cells, as described below.

3-23   Several types of innate lymphoid cells provide protection in early infection.

A defining feature of adaptive immunity is the clonal expression of antigen receptors, produced by somatic gene rearrangements, that provide the extraordinarily diverse specificities of T and B lymphocytes (see Section 1-11). However, for several decades, immunologists have recognized cells that have lymphoid characteristics but which lack specific antigen receptors. **Natural killer (NK) cells** have been known the longest, but in the past several years other distinct groups of such cells have been identified. Collectively, these are now called **innate lymphoid cells (ILCs)** and include NK cells (Fig. 3.37). ILCs develop in the bone marrow from the same **common lymphocyte progenitor (CLP)** that gives rise to B and T cells. Expression of the transcription factor Id2 (inhibitor of DNA binding 2) in the CLP represses B- and T-cell fates, and is required for the development of all ILCs. ILCs are identified by the absence of T- and B-cell antigen receptor and co-receptor complexes, but they express the receptor for IL-7. They migrate from the bone marrow and populate lymphoid tissues and peripheral organs, notably the dermis, liver, small intestine, and lung.

ILCs function in innate immunity as effector cells that amplify the signals delivered by innate recognition. They are stimulated by cytokines produced by other innate cells, such as macrophages or dendritic cells, that have been activated by innate sensors of microbial infection or cellular damage. Three major subgroups of ILCs are defined, largely on the basis of the types of cytokines that each produces. **Group 1 ILCs (ILC1s)** generate IFN-γ in response to activation by certain cytokines, in particular IL-12 and IL-18, made by dendritic cells and macrophages, and they function in protection against infection by viruses or intracellular pathogens. NK cells are now considered to be a type of ILC. ILC1s and NK cells are closely related, but have distinct functional properties and differ in the factors required for their development. NK cells are more similar to CD8 T cells in function, while ILC1s resemble more closely the T_H1 subset of CD4 T cells (see Section 3-24). NK cells can be distinguished from recently identified ILC1 cells in several ways. NK cells can be found within tissues, but they also circulate through the blood, while ILC1 cells appear to be largely non-circulating tissue-resident cells. In the mouse, conventional NK cells express the integrin α_2 (CD49b), while ILC1 cells, for example in the liver, lack

Fig. 3.37 The major categories of innate lymphoid cells (ILCs) and their properties.

The major categories of innate lymphoid cells (ILCs) and their properties			
Innate lymphoid subgroup	Inducing cytokine	Effector molecules produced	Function
NK cells	IL-12	IFN-γ, perforin, granzyme	Immunity against viruses, intracellular pathogens
ILC1	IL-12	IFN-γ	Defense against viruses, intracellular pathogens
ILC2	IL-25, IL-33, TSLP	IL-5, IL-13	Expulsion of extracellular parasites
ILC3, LTi cells	IL-23	IL-22, IL-17	Immunity to extracellular bacteria and fungi

CD49b but express the surface protein **Ly49a**. Both NK and ILC1 cells require the transcription factor Id2 for their development, but NK cells require the cytokine IL-15 and the transcription factors **Nfil3** and **eomesodermin**, while liver ILC1 cells require the cytokine IL-7 and the transcription factor **Tbet**.

ILC2s produce the cytokines IL-4, IL-5, and IL-13, in response to various cytokines, particularly **thymic stromal lymphopoietin (TSLP)** and IL-33. ILC2 cytokines function in promoting mucosal and barrier immunity and aid in protection against parasites. **ILC3s** respond to the cytokines IL-1β and IL-23 and produce several cytokines, including IL-17 and IL-22, which increase defenses against extracellular bacteria and fungi. IL-17 functions by stimulating the production of chemokines that recruit neutrophils, while IL-22 acts directly on epithelial cells to stimulate the production of antimicrobial peptides such as RegIIIγ (see Section 2-4).

The classification of ILC subtypes and the analysis of their development and function is still an active area, and studies to define the relative importance of these cells in immune responses are ongoing. The ILC subgroups identified so far appear to be highly parallel in structure to the subsets of effector CD8 and CD4 T cells that were defined over the last three decades. The transcription factors that control the development of different ILC subsets seem, for now at least, to be the same as those that control the corresponding T-cell subsets. Because of these similarities, we will postpone a detailed description of ILC development until Chapter 9, where we will cover this topic along with the development of T-cell subsets.

3-24 NK cells are activated by type I interferon and macrophage-derived cytokines.

NK cells are larger than T and B cells, have distinctive cytoplasmic granules containing cytotoxic proteins, and are functionally identified by their ability to kill certain tumor cell lines *in vitro* without the need for specific immunization. NK cells kill cells by releasing their cytotoxic granules, which are similar to those of cytotoxic T cells and have the same effects (discussed in Chapter 9). In brief, the contents of cytotoxic granules, which contain granzymes and the pore-forming protein perforin, are released onto the surface of the target cell, and penetrate the cell membrane and induce programmed cell death. However, unlike T cells, killing by NK cells is triggered by germline-encoded receptors that recognize molecules on the surface of infected or malignantly transformed cells. A second pathway used by NK cells to kill target cells involves the TNF family member known as **TRAIL** (tumor necrosis factor-related apoptosis-inducing ligand). NK cells express TRAIL on their cell surface. TRAIL interacts with two TNFR superfamily 'death' receptors, **DR4** and **DR5** (encoded by *TNFSF10A* and *B*), that are expressed by many types of cells. When NK cells recognize a target cell, TRAIL stimulates DR4 and DR5 to activate the pro-enzyme **caspase 8**, which leads to **apoptosis**. In contrast to pyroptosis, induced by caspase 1 following inflammasome activation (see Section 3-9), apoptosis is not associated with production of inflammatory cytokines. We will return to discuss more details of the mechanisms of caspase-induced apoptosis when we discuss killing by cytotoxic T cells in Chapter 9. Finally, NK cells express Fc receptors (see Section 1-20); binding of antibodies to these receptors activates NK cells to release their cytotoxic granules, a process known as **antibody-dependent cellular cytotoxicity**, or **ADCC**, to which we will return in Chapter 10.

The ability of NK cells to kill target cells can be enhanced by interferons or certain cytokines. NK cells that can kill sensitive targets can be isolated from uninfected individuals, but this activity is increased 20- to 100-fold when NK cells are exposed to IFN-α and IFN-β, or to IL-12, a cytokine produced by dendritic cells and macrophages during infection by many types of pathogens. Activated

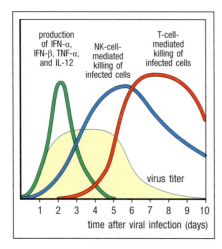

Fig. 3.38 Natural killer cells (NK cells) are an early component of the host response to virus infection. Experiments in mice have shown that IFN-α, IFN-β, and the cytokines TNF-α and IL-12 are produced first, followed by a wave of NK cells, which together control virus replication but do not eliminate the virus. Virus elimination is accomplished when virus-specific CD8 T cells and neutralizing antibodies are produced. Without NK cells, the levels of some viruses are much higher in the early days of the infection, and the infection can be lethal unless treated vigorously with antiviral compounds.

NK cells serve to contain virus infections while the adaptive immune response is generating antigen-specific cytotoxic T cells and neutralizing antibodies that can clear the infection (Fig. 3.38). A clue to the physiological function of NK cells in humans comes from rare patients deficient in these cells, who are frequently susceptible to herpesvirus infection. For example, a selective NK-cell deficiency results from mutations in the human MCM4 (minichromosome maintenance-deficient 4) protein, which is associated with predisposition to viral infections.

IL-12, acting in synergy with the cytokine IL-18 produced by activated macrophages, can also stimulate NK cells to secrete large amounts of interferon (IFN)-γ, and this is crucial in controlling some infections before the IFN-γ produced by activated CD8 cytotoxic T cells becomes available. IFN-γ, whose receptor activates only the STAT1 transcription factor, is quite distinct functionally from the antiviral type I interferons IFN-α and IFN-β, and is not directly induced by viral infection. The production of IFN-γ by NK cells early in an immune response can directly activate macrophages to enhance their capacity to kill pathogens, augmenting innate immunity, but also influences adaptive immunity through actions on dendritic cells and in regulating the differentiation of CD4 T cells into the pro-inflammatory T_H1 subset, which produces IFN-γ. NK cells also produce TNF-α, **granulocyte-macrophage stimulating factor (GM-CSF)**, and the chemokines CCL3 (MIF 1-α), CCL4, and CCL5 (RANTES), which act to recruit and activate macrophages.

3-25 NK cells express activating and inhibitory receptors to distinguish between healthy and infected cells.

For NK cells to defend against viruses or other pathogens, they should be able to distinguish infected cells from uninfected healthy cells. However, the mechanism used by NK cells is slightly more complicated than pathogen recognition by T or B cells. In general, it is thought that an individual NK cell expresses various combinations of germline-encoded **activating receptors** and **inhibitory receptors**. While the exact details are not clear in every case, it is thought that the overall balance of signaling by these receptors determines whether an NK cell engages and kills a target cell. The receptors on an NK cell are tuned to detect changes in expression of various surface proteins on a target cell, referred to as '**dysregulated self**.' The activating receptors generally recognize cell-surface proteins that are induced on target cells by metabolic stress, such as malignant transformation or microbial infection. These changes are referred to as '**stress-induced self**.' Specific cellular events, including DNA damage, signals related to proliferation, heat-shock related stress, and signaling by innate sensors including TLRs can lead to expression by host cells of surface proteins that bind to the activating receptors on NK cells. Stimulation of activating receptors will add to the chance that the NK cells will release cytokines such as IFN-γ and activate the killing of the stimulating cell through the release of cytotoxic granules.

By contrast, inhibitory receptors on NK cells recognize surface molecules that are constitutively expressed at high levels by most cells, and the loss of these molecules is referred to as '**missing self**.' Inhibitory receptors can recognize other molecules, but those recognizing **MHC class I molecules** have been studied the most so far. MHC molecules are glycoproteins expressed on nearly all cells of the body. We will discuss the role of MHC proteins in antigen presentation to T cells in Chapter 6, but for now we need only to introduce the two main classes of MHC molecules. MHC class I molecules are expressed on most of the cells of the body (except, notably, red blood cells), whereas the expression of **MHC class II molecules** is far more restricted, largely to immune cells.

Inhibitory receptors that recognize MHC class I molecules function to prevent NK cells from killing normal host cells. The greater the number of MHC class

I molecules on a cell surface, the better protected that cell is against attack by NK cells. Interferons induce expression of MHC class I molecules, and protect uninfected host cells from being killed by NK cells, while also activating NK cells to kill virus-infected cells. Viruses and some other intracellular pathogens can cause downregulation of MHC class I molecules as a strategy to prevent the display of antigens as peptides to T cells, also discussed in Chapter 6. NK cells are able to sense this reduction in expression of MHC class I molecules through reduced signaling from their inhibitory receptors. Reduction in MHC class I expression is an example of 'missing self,' and increases the chance that an NK cell will kill the target cell. It is thought that the balance of signals from 'stress-induced self' and 'missing self' determines whether an individual NK cell will be triggered to kill a particular target cell (Fig. 3.39). Thus receptors expressed on NK cells integrate the signals from two types of

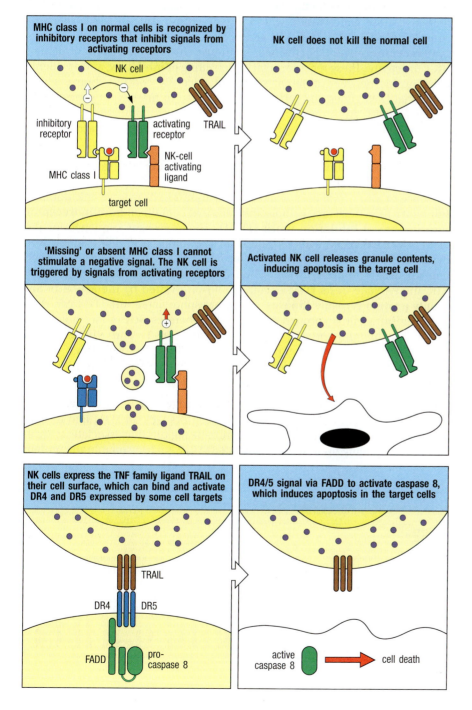

Fig. 3.39 Killing by NK cells depends on the balance between activating and inhibitory signals. NK cells have several different activating receptors that signal the NK to kill the bound cell. However, NK cells are prevented from a wholesale attack by another set of inhibitory receptors that recognize MHC class I molecules (which are present on almost all cell types) and that inhibit killing by overruling the actions of the activating receptors. This inhibitory signal is lost when the target cells do not express MHC class I, such as in cells infected with viruses, many of which specifically inhibit MHC class I expression or alter its conformation so as to avoid recognition by CD8 T cells. NK cells may also kill target cells through their expression of the TNF family member TRAIL, which binds to TNFR members DR4 and DR5 expressed by some types of cells. DR4 and DR5 signal through FADD, an adaptor that activates pro-caspase 8, leading to induction of apoptosis of the target cell.

Figure panels:

MHC class I on normal cells is recognized by inhibitory receptors that inhibit signals from activating receptors
NK cell
inhibitory receptor
activating receptor
TRAIL
NK-cell activating ligand
MHC class I
target cell

NK cell does not kill the normal cell

'Missing' or absent MHC class I cannot stimulate a negative signal. The NK cell is triggered by signals from activating receptors

Activated NK cell releases granule contents, inducing apoptosis in the target cell

NK cells express the TNF family ligand TRAIL on their cell surface, which can bind and activate DR4 and DR5 expressed by some cell targets
TRAIL
DR4   DR5
FADD   pro-caspase 8

DR4/5 signal via FADD to activate caspase 8, which induces apoptosis in the target cells
active caspase 8 → cell death

surface receptors, which together control the NK cell's cytotoxic activity and cytokine production.

3-26 NK-cell receptors belong to several structural families, the KIRs, KLRs, and NCRs.

The receptors that regulate the activity of NK cells fall into two large families that contain a number of other cell-surface receptors in addition to NK receptors (Fig. 3.40). Members of the **killer cell immunoglobulin-like receptor** family, or **KIRs**, have differing numbers of immunoglobulin domains. Some, such as KIR-2D, have two immunoglobulin domains, whereas others, such as KIR-3D, have three. The KIR genes form part of a larger cluster of immunoglobulin-like receptor genes known as the **leukocyte receptor complex (LRC)**. Another family, the **killer cell lectin-like receptors**, or **KLRs**, are C-type lectin-like proteins whose genes reside within a gene cluster called the **NK receptor complex (NKC)**. Mice lack KIR genes, and instead predominantly express **Ly49 receptors** encoded in the NKC on mouse chromosome 6 to control their NK-cell activity. These receptors can be activating or inhibitory, and are highly polymorphic between different strains of mice. By contrast, humans lack functional Ly49 genes and rely on KIRs encoded in the LRC to control their NK-cell activity. An important feature of the NK-cell population is that any given NK cell expresses only a subset of the receptors in its potential repertoire, and so not all NK cells in the individual are identical.

Activating and inhibitory receptors are present within the same structural family. Whether a KIR protein is activating or inhibitory depends on the presence or absence of particular signaling motifs in its cytoplasmic domain. Inhibitory KIRs have long cytoplasmic tails that contain an **immunoreceptor tyrosine-based inhibition motif (ITIM)**. The consensus sequence for the ITIM is V/I/LxYxxL/V, where x stands for any amino acid. For example, the cytoplasmic tails of the inhibitory receptors KIR-2DL and KIR-3DL each contain two ITIMs (Fig. 3.41). When ligands associate with an inhibitory KIR, the tyrosine in its ITIM becomes phosphorylated by the action of **Src family protein tyrosine kinases**. When phosphorylated, the ITIM can then bind the intracellular protein tyrosine phosphatases **SHP-1** (src homology region 2-containing protein tyrosine phosphatase-1) and **SHP-2**, which become localized near the cell membrane. These phosphatases inhibit signaling induced by other receptors by removing phosphates from tyrosine residues on other intracellular signaling molecules.

Activating KIRs have short cytoplasmic tails, designated, for example, as KIR-2DS and KIR-3DS (see Fig. 3.41). These receptors lack an ITIM and instead have a charged residue in their transmembrane regions that associates with an accessory signaling protein called **DAP12**. DAP12 is a transmembrane protein

Fig. 3.40 The genes that encode NK receptors fall into two large families. The first, the leukocyte receptor complex (LRC), comprises a large cluster of genes encoding a family of proteins composed of immunoglobulin-like domains. These include the killer cell immunoglobulin-like receptors (KIRs) expressed by NK cells, the ILT (immunoglobulin-like transcript) class, and the leukocyte-associated immunoglobulin-like receptor (LAIR) gene families. The sialic acid-binding Ig-like lectins (SIGLECs) and members of the CD66 family are located nearby. In humans, this complex is located on chromosome 19. The second gene cluster is called the NK receptor complex (NKC) and encodes killer cell lectin-like receptors (KLRs), a receptor family that includes the NKG2 proteins and CD94, with which some NKG2 molecules pair to form a functional receptor. This complex is located on human chromosome 12. Some NK receptor genes are found outside these two major gene clusters; for example, the genes for the natural cytotoxicity receptors NKp30 and NKp44 are located within the major histocompatibility complex on chromosome 6. Figure based on data courtesy of J. Trowsdale University of Cambridge.

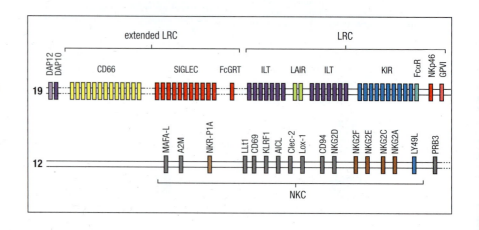

that contains an **immunoreceptor tyrosine-based activation motif** (**ITAM**, with consensus sequence YXX[L/I]X$_{6-9}$YXX[L/I]) in its cytoplasmic tail and forms a disulfide-linked homodimer in the membrane. When a ligand binds to an activating KIR, the tyrosine residues in the ITAM of DAP12 become phosphorylated, turning on intracellular signaling pathways that activate the NK cell and lead to release of the cytotoxic granules. The phosphorylated ITAMs bind and activate intracellular tyrosine kinases such as Syk or ZAP-70, leading to further signaling events similar to those described for T cells in Chapter 7.

The KLR family also has both activating and inhibitory members. In mice, inhibitory Ly49 receptors have an ITIM in their cytoplasmic tail that recruits SHP-1. The latter's importance is shown by the failure of Ly49 to inhibit NK activation upon binding to MHC class I in mice carrying the **motheaten** mutation, which inactivates SHP-1 protein. In humans and mice, NK cells express a heterodimer of two different C-type lectin-like receptors, **CD94** and **NKG2**. This heterodimer interacts with nonpolymorphic MHC class I-like molecules, including HLA-E in humans and Qa1 in mice. HLA-E and Qa1 are unusual in that instead of binding peptides derived from pathogens, they bind fragments of the **signal peptide** derived from other MHC class I molecules during processing in the endoplasmic reticulum. This enables CD94:NKG2 to detect the presence of several different MHC class I variants, whose expression may be targeted by viruses, and kill cells in which overall MHC molecule expression is diminished. In humans there are four closely related NKG2 family proteins, NKG2A, C, E, and F (encoded by *KLRC1–4*), and a more distantly related protein, NKG2D (encoded by *KLRK1*). Of these, for example, NKG2A contains an ITIM and is inhibitory, whereas NKG2C has a charged transmembrane residue, associates with DAP12, and is activating (see Fig. 3.41). NKG2D is also activating but quite distinct from the other NKG2 receptors, and we will discuss it separately below.

The overall response of NK cells to differences in MHC expression is further complicated by the extensive polymorphism of KIR genes, with different numbers of activating and inhibitory KIR genes being found in different people. This may explain why NK cells are a barrier to bone marrow transplantation, since the NK cells of the recipient may react more strongly to donor MHC molecules than to the self MHC with which they developed. A similar phenomenon may occur during pregnancy, because of differences between fetal and maternal MHC molecules (see Section 15-38). The advantage of such extensive KIR polymorphism is not yet clear, and some genetic epidemiologic studies even suggest an association between certain alleles of KIR genes and earlier onset (although not absolute frequency) of rheumatoid arthritis. The KIR gene cluster is not present in mice, but some species, including some primates, contain genes of both the KIR and KLR families. This might suggest that both gene clusters are relatively ancient and that for some reason, one or the other gene cluster was lost by mice and humans.

Signaling by the inhibitory NK receptors suppresses the killing activity and cytokine production of NK cells. This means that NK cells will not kill healthy, genetically identical cells with normal expression of MHC class I molecules, such as the other cells of the body. Virus-infected cells, however, can become susceptible to being killed by NK cells by a variety of mechanisms. First, some viruses inhibit all protein synthesis in their host cells, so that synthesis of MHC class I proteins would be blocked in infected cells, even while their production in uninfected cells is being stimulated by the actions of type I interferons. The reduced level of MHC class I expression in infected cells would make them correspondingly less able to inhibit NK cells through their MHC-specific receptors, and they would become more susceptible to being killed. Second, many viruses can selectively prevent the export of MHC class I molecules to the cell surface, or induce their degradation once there. This might allow the infected cell to evade recognition by cytotoxic T cells but would make it

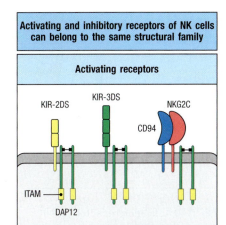

Activating and inhibitory receptors of NK cells can belong to the same structural family

Activating receptors

KIR-2DS   KIR-3DS   NKG2C   CD94   ITAM   DAP12

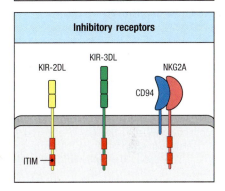

Inhibitory receptors

KIR-2DL   KIR-3DL   NKG2A   CD94   ITIM

Fig. 3.41 The structural families of NK receptors encode both activating and inhibitory receptors. The families of killer cell immunoglobulin-like receptors (KIRs) and killer cell lectin-like receptors (KLRs) have members that send activating signals to the NK cell (upper panel) and those that send inhibitory signals (lower panel). KIR family members are designated according to the number of immunoglobulin-like domains they possess and by the length of their cytoplasmic tails. Activating KIR receptors have short cytoplasmic tails and bear the designation 'S.' These associate with the signaling protein DAP12 via a charged amino acid residue in the transmembrane region. The cytoplasmic tails of DAP12 contain amino acid motifs called ITAMs, which are involved in signaling. NKG2 receptors belong to the KLR family, and, whether activating or inhibitory, form heterodimers with another C-type lectin-like family member, CD94. The inhibitory KIR receptors have longer cytoplasmic tails and are designated 'L'; these do not associate constitutively with adaptor proteins but contain a signaling motif called an ITIM, which when phosphorylated is recognized by inhibitory phosphatases.

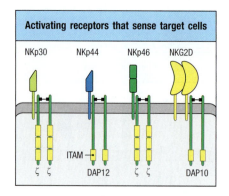

Activating receptors that sense target cells

NKp30   NKp44   NKp46   NKG2D

ITAM

ζ   ζ   DAP12   ζ   ζ   DAP10

Fig. 3.42 Activating receptors of NK cells include the natural cytotoxicity receptors and NKG2D. The natural cytotoxicity receptors are immunoglobulin-like proteins. NKp30 and NKp44, for example, have an extracellular domain that resembles a single variable domain of an immunoglobulin molecule. NKp30 and NKp46 activate the NK cell through their association with homodimers of the CD3ζ chain, or the Fc receptor γ chain (not shown). These signaling proteins also associate with other types of receptors that are described in Chapter 7. NKp44 activates the NK cell through their association with homodimers of DAP12. NKp46 resembles the KIR-2D molecules in having two domains that resemble the constant domains of an immunoglobulin molecule. NKG2D is a member of the C-type lectin-like family and forms a homodimer, and it associates with DAP10. In mice, an alternatively spliced form of NKG2D also associates with DAP12 (not shown).

susceptible to being killed by NK cells. Virally infected cells can still be killed by NK cells even if the cells do not downregulate MHC, provided that ligands for activating receptors are induced. However, some viruses target ligands for the activating receptors on NK cells, thwarting NK-cell recognition and killing of virus-infected cells

3-27 NK cells express activating receptors that recognize ligands induced on infected cells or tumor cells.

In addition to the KIRs and KLRs, which have a role in sensing the level of MHC class I proteins present on other cells, NK cells also express receptors that more directly sense the presence of infection or other perturbations in a cell. Activating receptors for the recognition of infected cells, tumor cells, and cells injured by physical or chemical damage include the **natural cytotoxicity receptors (NCRs)** NKp30, NKp44, and NKp46, which are immunoglobulin-like receptors, and the C-type lectin-like family members **Ly49H** and **NKG2D** (Fig. 3.42). Among NCRs, only NKp46 is conserved in humans and in mice, and it is the most selective marker of NK cells across mammalian species. The ligands recognized by the NCRs are still being defined, but some evidence suggests that they recognize viral proteins, including the **hemagglutinin** (HA) glycoprotein of the influenza virus. Ly49H is an activating receptor that recognizes the viral protein m157, an MHC class I-like structure encoded by the murine cytomegalovirus. The ligand for NKp30 is a protein named B7-H6, a member of the family of co-stimulatory proteins mentioned in Section 1-15, and is further described in Chapters 7 and 9.

NKG2D has a specialized role in activating NK cells. NKG2 family members form heterodimers with CD94 and bind the MHC class I molecule HLA-E. In contrast, two NKG2D molecules form a homodimer that binds to several MHC class I-like molecules that are induced by various types of cellular stress. These include the MIC molecules **MIC-A** and **MIC-B**, and the **RAET1** family of proteins, which are similar to the α_1 and α_2 domains of MHC class I molecules (Fig. 3.43). The RAET1 family has 10 members, 3 of which were initially characterized as ligands for the **cytomegalovirus UL16 protein** and are also called **UL16-binding proteins**, or **ULBPs**. Mice do not have equivalents of the MIC molecules; the ligands for mouse NKG2D have a very similar structure to that of the RAET1 proteins, and are probably orthologs of them. In fact, these ligands were first identified in mice as the **RAE1** (retinoic acid early inducible 1) protein family, and also include related proteins H60 and MULT1 (see Fig. 6.26). We will return to these MHC-like molecules when we discuss the structure of the MHC molecule in Section 6-18.

The ligands for NKG2D are expressed in response to cellular or metabolic stress, and so are upregulated on cells infected with intracellular bacteria and most viruses, as well as on incipient tumor cells that have become malignantly transformed. Thus, recognition by NKG2D acts as a generalized 'danger' signal to the immune system. In addition to expression by a subset of NK cells, NKG2D is expressed by various T cells, including all human CD8 T cells, γ:δ T cells, activated murine CD8 T cells, and invariant NKT cells (described in Chapter 8). In these cells, recognition of NKG2D ligands provides a potent co-stimulatory signal that enhances their effector functions.

The ligands for NKG2D are MHC-like molecules, MIC-A, MIC-B, or RAET1 family members, whose expression is induced by cellular stress

MIC-A or MIC-B   RAET1 family (includes MULT1, ULBPs)

Fig. 3.43 The ligands for the activating NK receptor NKG2D are proteins that are expressed in conditions of cellular stress. The MIC proteins MIC-A and MIC-B are MHC-like molecules induced on epithelial and other cells by stress, such as DNA damage, cellular transformation, or infection. RAET1 family members, including the subset designated as UL16-binding proteins (ULBPs), also resemble a portion of an MHC class I molecule—the α_1 and α_2 domains—and most (but not all) are attached to the cell via a glycophosphatidylinositol linkage. Unlike MHC class I molecules, the NKG2D ligands do not bind processed peptides.

NKG2D also differs from other activating receptors on NK cells in the signaling pathway it engages within the cell. The other activating receptors are associated intracellularly with signaling proteins such as the CD3ζ chain, the Fc receptor γ chain, and DAP12, which all contain ITAMs. In contrast, NKG2D binds a different adaptor protein, **DAP10**, which does not contain an ITAM sequence and instead activates the intracellular lipid kinase **phosphatidylinositol 3-kinase** (PI 3-kinase), initiating a different series of intracellular signaling events in the NK cell (see Section 7-4). Generally, PI 3-kinase is considered to enhance the survival of cells in which it is activated, thereby augmenting the cell's overall effector activity. In NK cells, activation of PI 3-kinase is directly linked to the induction of cytotoxic activity. In mice, the workings of NKG2D are even more complicated, because mouse NKG2D is produced in two alternatively spliced forms, one of which binds DAP12 and DAP10, whereas the other binds only DAP10. Mouse NKG2D can thus activate both signaling pathways, whereas human NKG2D seems to signal only through DAP10 to activate the PI 3-kinase pathway. Finally, NK cells express several receptors from the **SLAM** (signaling lymphocyte activation molecule) family, including **2B4**, which recognizes the cell-surface molecule **CD48** expressed by many cells including NK cells. Interactions between 2B4 and CD48 on nearby NK cells can release signals that promote survival and proliferation through **SAP** (SLAM-associated protein) and the Src kinase Fyn.

Summary.

Triggering of innate sensors on various cells—neutrophils, macrophages, and dendritic cells in particular—not only activates these cells' individual effector functions, but also stimulates the release of pro-inflammatory chemokines and cytokines that act together to recruit more phagocytic cells to the site of infection. Especially prominent is the early recruitment of neutrophils and monocytes. Furthermore, cytokines released by tissue phagocytic cells can induce more systemic effects, including fever and the production of acute-phase response proteins, including mannose-binding lectin, C-reactive protein, fibrinogen, and pulmonary surfactant proteins, which add to a general state of augmented innate immunity. These cytokines also mobilize antigen-presenting cells that induce the adaptive immune response. The innate immune system has at its service several recently recognized subtypes of innate lymphoid cells which join the ranks of the long-recognized NK cells. ILCs exhibit specialized effector activity in response to different signals, and act to amplify the strength of the innate response. The production of interferons in response to viral infections serves to inhibit viral replication and to activate NK cells. These in turn can distinguish healthy cells from those that are infected by virus or that are transformed or stressed in some way, based on the expression of class I MHC molecules and MHC-related molecules that are ligands for some NK receptors. As we will see later in the book, cytokines, chemokines, phagocytic cells, and NK cells are all effector mechanisms that are also employed in the adaptive immune response, which uses variable receptors to target specific pathogen antigens.

Summary to Chapter 3.

Innate immunity uses a variety of effector mechanisms to detect infection and eliminate pathogens, or hold them in check until an adaptive immune response develops. These effector mechanisms are all regulated by germline-encoded receptors on many types of cells that can detect molecules of microbial origin or that sense signs of host cellular damage. The induced responses of the innate immune system are based on several distinct components. After the initial barriers—the body's epithelia and the soluble antimicrobial molecules described in Chapter 2—have been breached, the most important innate defenses rely on tissue macrophages and other tissue-resident sensor cells, such as dendritic cells. Macrophages provide a double service: they mediate rapid cellular defense at the borders of infection through phagocytosis and antimicrobial actions, and they also use their various innate sensors to activate the process of inflammation, which involves recruiting additional cells to sites of infection. Innate sensors activate signaling pathways that lead to the production of pro-inflammatory and antiviral cytokines, which in turn stimulate innate effector responses while also helping to initiate an adaptive immune response. The uncovering of the pathogen-sensing mechanisms described in this chapter is still extremely active. It is providing new insights into human autoinflammatory conditions such as lupus, Crohn's disease, and gout. Indeed, the induction of powerful effector mechanisms by innate immune recognition based on germline-encoded receptors clearly has some dangers. It is a double-edged sword, as is illustrated by the effects of the cytokine TNF-α—beneficial when released locally, but disastrous when produced systemically. This illustrates the evolutionary knife edge along which all innate mechanisms of host defense travel. The innate immune system can be viewed as a defense system that mainly frustrates the establishment of a focus of infection; however, even when it proves inadequate in fulfilling this function, it has already set in motion—by recruiting and activating dendritic cells—the initiation of the adaptive immune response, which forms an essential part of humans' defenses against infection.

Having introduced immunology with a consideration of innate immune function, we next turn our attention to the adaptive immune response, beginning with an explanation of the structure and function of the antigen receptors expressed by lymphocytes.

Questions.

3.1 Matching: Match the Toll-like receptor (TLR) to its ligand:

A. TLR-2:TLR-1 or TLR-2:TLR-6	**i.** ssRNA
B. TLR-3	**ii.** Lipopolysaccharide
C. TLR-4	**iii.** Lipoteichoic acid and di-/triacyl lipoproteins
D. TLR-5	**iv.** dsRNA
E. TLR-7	**v.** Flagellin
F. TLR-9	**vi.** Unmethylated CpG DNA

3.2 Matching: Match the hereditary disorder to the gene affected:

A. Chronic granulomatous disease	**i.** NOD2
B. X-linked hypohidrotic ectodermal dysplasia and immunodeficiency	**ii.** IKKγ (NEMO)
C. Crohn's disease	**iii.** Jak3
D. X-linked SCID	**iv.** NAPDH oxidase
E. SCID (not X-linked)	**v.** NLRP3
F. Familial cold inflammatory syndrome	**vi.** γc

3.3 Multiple Choice: Which of the following does not occur during an inflammatory response?

A. Local blood clotting

B. Tissue injury repair

C. Endothelial cell activation

D. Decreased vascular permeability

E. Extravasation of leukocytes into inflamed tissue

3.4 Short Answer: What is the difference between conventional dendritic cells (cDCs) and plasmacytoid dendritic cells (pDCs)?

3.5 Multiple Choice: Which of the following is a G-protein-coupled receptor?

A. fMLF receptor

B. TLR-4

C. IL-1R

D. CD14

E. STING

F. B7.1 (CD80)

3.6 True or False: All forms of ubiquitination lead to proteasomal degradation.

3.7 Fill-in-the-Blanks:

A. Toll-like receptors (TLRs) have a cytoplasmic signaling domain called TIR that is also shared with _____.

B. Cytokine receptors of the hematopoietin family activate tyrosine kinases of the _____ family, in order to signal these recruit SH2-domain-containing transcriptions factors called _____.

C. Out of all the different TLRs, the only one that uses both MyD88/MAL and TRIF/TRAM adaptor pairs is _____.

3.8 True or False: Cytosolic DNA is sensed by cGAS, which signals through STING, while cytosolic ssRNA and dsRNA are sensed by RIG-I and MDA-5, respectively, which interact with the downstream adaptor protein MAVS.

3.9 Multiple Choice: Which of the following is not true?

A. CCL2 attracts macrophages through CCR2.

B. IL-3, IL-5, and GM-CSF are a subgroup of class I cytokine receptors that share a common β chain.

C. IL-2, IL-4, IL-7, IL-9, IL-15, and IL-21 share a common γ_c.

D. The inflammasome is a large oligomer composed of the sensor NLRP3, the adaptor ASC, and caspase 8.

E. CXCL8 attracts neutrophils through CXCR2.

F. ILC1s secrete IFN-γ, ILC2s secrete IL-4, IL-5, and IL-13, and ILC3s secrete IL-17 and IL-22.

3.10 True or False: Natural killer (NK) cells have killer-cell immunoglobulin-like receptors (KIRs), which detect pathogen peptides on self MHC molecules.

3.11 Matching: Match the step in neutrophil recruitment into inflamed tissues with the key effectors involved:

A. Endothelial cell activation

B. Rolling

C. Neutrophil integrin assuming 'active' state

D. Strong adhesion

E. Diapedesis

i. Neutrophil LFA-1 with endothelial ICAM-1

ii. Local secretion of TNF-α and other cytokines

iii. CXCL8 signaling through CXCR2 leading to talin activation

iv. Endothelial and neutrophil CD31

v. Interaction of endothelial P- and E-selectin with neutrophil sulfated sialyl-Lewis^X

3.12 Short Answer: What co-stimulatory molecules are induced on macrophages and dendritic cells upon pathogen recognition, and what is their function?

Section references.

3-1 After entering tissues, many microbes are recognized, ingested, and killed by phagocytes.

Aderem, A., and Underhill, D.M.: **Mechanisms of phagocytosis in macrophages.** *Annu. Rev. Immunol.* 1999, **17**:593–623.

Auffray, C., Fogg, D., Garfa, M., Elain, G., Join-Lambert, O., Kayal, S., Sarnacki, S., Cumano, A., Lauvau, G., and Geissmann, F.: **Monitoring of blood vessels and tissues by a population of monocytes with patrolling behavior.** *Science* 2007, **317**:666–670.

Cervantes-Barragan, L., Lewis, K.L., Firner, S., Thiel, V., Hugues, S., Reith, W., Ludewig, B., and Reizis, B.: **Plasmacytoid dendritic cells control T-cell response to chronic viral infection.** *Proc. Natl Acad. Sci. USA* 2012, **109**:3012–3017.

Goodridge, H.S., Wolf, A.J., and Underhill, D.M.: **Beta-glucan recognition by the innate immune system.** *Immunol. Rev.* 2009, **230**:38–50.

Greaves, D.R., and Gordon, S.: **The macrophage scavenger receptor at 30 years of age: current knowledge and future challenges.** *J. Lipid Res.* 2009, **50**:S282–S286.

Greter, M., Lelios, I., Pelczar, P., Hoeffel, G., Price, J., Leboeuf, M., Kündig, T.M., Frei, K., Ginhoux, F., Merad, M., and Becher, B.: **Stroma-derived interleukin-34 controls the development and maintenance of Langerhans cells and the maintenance of microglia.** *Immunity* 2012, **37**:1050–1060.

Harrison, R.E., and Grinstein, S.: **Phagocytosis and the microtubule cytoskeleton.** *Biochem. Cell Biol.* 2002, **80**:509–515.

Lawson, C.D., Donald, S., Anderson, K.E., Patton, D.T., and Welch, H.C.: **P-Rex1 and Vav1 cooperate in the regulation of formyl-methionyl-leucyl-phenylalanine-dependent neutrophil responses.** *J. Immunol.* 2011, **186**:1467–1476.

Lee, S.J., Evers, S., Roeder, D., Parlow, A.F., Risteli, J., Risteli, L., Lee, Y.C., Feizi, T., Langen, H., and Nussenzweig, M.C.: **Mannose receptor-mediated regulation of serum glycoprotein homeostasis.** *Science* 2002, **295**:1898–1901.

Linehan, S.A., Martinez-Pomares, L., and Gordon, S.: **Macrophage lectins in host defence.** *Microbes Infect.* 2000, **2**:279–288.

McGreal, E.P., Miller, J.L., and Gordon, S.: **Ligand recognition by antigen-presenting cell C-type lectin receptors.** *Curr. Opin. Immunol.* 2005, **17**:18–24.

Peiser, L., De Winther, M.P., Makepeace, K., Hollinshead, M., Coull, P., Plested, J., Kodama, T., Moxon, E.R., and Gordon, S.: **The class A macrophage scavenger receptor is a major pattern recognition receptor for *Neisseria meningitidis* which is independent of lipopolysaccharide and not required for secretory responses.** *Infect. Immun.* 2002, **70**:5346–5354.

Podrez, E.A., Poliakov, E., Shen, Z., Zhang, R., Deng, Y., Sun, M., Finton, P.J., Shan, L., Gugiu, B., Fox, P.L., *et al.*: **Identification of a novel family of oxidized phospholipids that serve as ligands for the macrophage scavenger receptor CD36.** *J. Biol. Chem.* 2002, **277**:38503–38516.

3-2 G-protein-coupled receptors on phagocytes link microbe recognition with increased efficiency of intracellular killing.

Bogdan, C., Rollinghoff, M., and Diefenbach, A.: **Reactive oxygen and reactive nitrogen intermediates in innate and specific immunity.** *Curr. Opin. Immunol.* 2000, **12**:64–76.

Brinkmann, V., and Zychlinsky, A.: **Beneficial suicide: why neutrophils die to make NETs.** *Nat. Rev. Microbiol.* 2007, **5**:577–582.

Dahlgren, C., and Karlsson, A.: **Respiratory burst in human neutrophils.** *J. Immunol. Methods* 1999, **232**:3–14.

Gerber, B.O., Meng, E.C., Dotsch, V., Baranski, T.J., and Bourne, H.R.: **An activation switch in the ligand binding pocket of the C5a receptor.** *J. Biol. Chem.* 2001, **276**:3394–3400.

Reeves, E.P., Lu, H., Jacobs, H.L., Messina, C.G., Bolsover, S., Gabella, G., Potma, E.O., Warley, A., Roes, J., and Segal, A.W.: **Killing activity of neutrophils is mediated through activation of proteases by K$^+$ flux.** *Nature* 2002, **416**:291–297.

Ward, P.A.: **The dark side of C5a in sepsis.** *Nat. Rev. Immunol.* 2004, **4**:133–142.

3-3   Microbial recognition and tissue damage initiate an inflammatory response.

Chertov, O., Yang, D., Howard, O.M., and Oppenheim, J.J.: **Leukocyte granule proteins mobilize innate host defenses and adaptive immune responses.** *Immunol. Rev.* 2000, **177**:68–78.

Kohl, J.: **Anaphylatoxins and infectious and noninfectious inflammatory diseases.** *Mol. Immunol.* 2001, **38**:175–187.

Mekori, Y.A., and Metcalfe, D.D.: **Mast cells in innate immunity.** *Immunol. Rev.* 2000, **173**:131–140.

Svanborg, C., Godaly, G., and Hedlund, M.: **Cytokine responses during mucosal infections: role in disease pathogenesis and host defence.** *Curr. Opin. Microbiol.* 1999, **2**:99–105.

Van der Poll, T.: **Coagulation and inflammation.** *J. Endotoxin Res.* 2001, **7**:301–304.

3-4   Toll-like receptors represent an ancient pathogen-recognition system.

Lemaitre, B., Nicolas, E., Michaut, L., Reichhart, J.M., and Hoffmann, J.A.: **The dorsoventral regulatory gene cassette spätzle/Toll/cactus controls the potent antifungal response in *Drosophila* adults.** *Cell* 1996, **86**:973–983.

Lemaitre, B., Reichhart, J.M., and Hoffmann, J.A.: ***Drosophila* host defense: differential induction of antimicrobial peptide genes after infection by various classes of microorganisms.** *Proc. Natl Acad. Sci. USA* 1997, **94**:14614–14619.

3-5   Mammalian Toll-like receptors are activated by many different pathogen-associated molecular patterns.

Beutler, B., and Rietschel, E.T.: **Innate immune sensing and its roots: the story of endotoxin.** *Nat. Rev. Immunol.* 2003, **3**:169–176.

Diebold, S.S., Kaisho, T., Hemmi, H., Akira, S., and Reis e Sousa, C..: **Innate antiviral responses by means of TLR7-mediated recognition of single-stranded RNA.** *Science* 2004, **303**:1529–1531.

Heil, F., Hemmi, H., Hochrein, H., Ampenberger, F., Kirschning, C., Akira, S., Lipford, G., Wagner, H., and Bauer, S.: **Species-specific recognition of single-stranded RNA via Toll-like receptor 7 and 8.** *Science* 2004, **303**:1526–1529.

Hoebe, K., Georgel, P., Rutschmann, S., Du, X., Mudd, S., Crozat, K., Sovath, S., Shamel, L., Hartung, T., Zähringer, U., *et al.*: **CD36 is a sensor of diacylglycerides.** *Nature* 2005, **433**:523–527.

Jin, M.S., Kim, S.E., Heo, J.Y., Lee, M.E., Kim, H.M., Paik, S.G., Lee, H., and Lee, J.O.: **Crystal structure of the TLR1-TLR2 heterodimer induced by binding of a tri-acylated lipopeptide.** *Cell* 2007, **130**:1071–1082.

Lee, Y.H., Lee, H.S., Choi, S.J., Ji, J.D., and Song, G.G.: **Associations between TLR polymorphisms and systemic lupus erythematosus: a systematic review and meta-analysis.** *Clin. Exp. Rheumatol.* 2012, **30**:262–265.

Liu, L., Botos, I., Wang, Y., Leonard, J.N., Shiloach, J., Segal, D.M., and Davies, D.R.: **Structural basis of Toll-like receptor 3 signaling with double-stranded RNA.** *Science* 2008, **320**:379–381.

Lund, J.M., Alexopoulou, L., Sato, A., Karow, M., Adams, N.C., Gale, N.W., Iwasaki, A., and Flavell, R.A.: **Recognition of single-stranded RNA viruses by Toll-like receptor 7.** *Proc. Natl Acad. Sci. USA* 2004, **101**:5598–5603.

Lund, J., Sato, A., Akira, S., Medzhitov, R., and Iwasaki, A.: **Toll-like receptor 9-mediated recognition of herpes simplex virus-2 by plasmacytoid dendritic cells.** *J. Exp. Med.* 2003, **198**:513–520.

Schulz, O., Diebold, S.S., Chen, M., Näslund, T.I., Nolte, M.A., Alexopoulou, L., Azuma, Y.T., Flavell, R.A., Liljeström, P., and Reis e Sousa, C.: **Toll-like receptor 3 promotes cross-priming to virus-infected cells.** *Nature* 2005, **433**:887–892.

Tanji, H., Ohto, U., Shibata, T., Miyake, K., and Shimizu, T.: **Structural reorganization of the Toll-like receptor 8 dimer induced by agonistic ligands.** *Science* 2013, **339**:1426–1429.

Yarovinsky, F., Zhang, D., Andersen, J.F., Bannenberg, G.L., Serhan, C.N., Hayden, M.S., Hieny, S., Sutterwala, F.S., Flavell, R.A., Ghosh, S., *et al.*: **TLR11 activation of dendritic cells by a protozoan profilin-like protein.** *Science* 2005, **308**:1626–1629.

3-6   TLR-4 recognizes bacterial lipopolysaccharide in association with the host accessory proteins MD-2 and CD14.

Beutler, B.: **Endotoxin, Toll-like receptor 4, and the afferent limb of innate immunity.** *Curr. Opin. Microbiol.* 2000, **3**:23–28.

Beutler, B., and Rietschel, E.T.: **Innate immune sensing and its roots: the story of endotoxin.** *Nat. Rev. Immunol.* 2003, **3**:169–176.

Kim, H.M., Park, B.S., Kim, J.I., Kim, S.E., Lee, J., Oh, S.C., Enkhbayar, P., Matsushima, N., Lee, H., Yoo, O.J., *et al.*: **Crystal structure of the TLR4–MD-2 complex with bound endotoxin antagonist Eritoran.** *Cell* 2007, **130**:906–917.

Park, B.S., Song, D.H., Kim, H.M., Choi, B.S., Lee, H., and Lee, J.O.: **The structural basis of lipopolysaccharide recognition by the TLR4–MD-2 complex.** *Nature* 2009, **458**:1191–1195.

3-7   TLRs activate NFκB, AP-1, and IRF transcription factors to induce the expression of inflammatory cytokines and type I interferons.

Fitzgerald, K.A., McWhirter, S.M., Faia, K.L., Rowe, D.C., Latz, E., Golenbock, D.T., Coyle, A.J., Liao, S.M., and Maniatis, T.: **IKKε and TBK1 are essential components of the IRF3 signaling pathway.** *Nat. Immunol.* 2003, **4**:491–496.

Häcker, H., Redecke, V., Blagoev, B., Kratchmarova, I., Hsu, L.C., Wang, G.G., Kamps, M.P., Raz, E., Wagner, H., Häcker, G., *et al.*: **Specificity in Toll-like receptor signalling through distinct effector functions of TRAF3 and TRAF6.** *Nature* 2006, **439**:204–207.

Hiscott, J., Nguyen, T.L., Arguello, M., Nakhaei, P., and Paz, S.: **Manipulation of the nuclear factor-κB pathway and the innate immune response by viruses.** *Oncogene* 2006, **25**:6844–6867.

Honda, K., and Taniguchi, T.: **IRFs: master regulators of signalling by Toll-like receptors and cytosolic pattern-recognition receptors.** *Nat. Rev. Immunol.* 2006, **6**:644–658.

Kawai, T., Sato, S., Ishii, K.J., Coban, C., Hemmi, H., Yamamoto, M., Terai, K., Matsuda, M., Inoue, J., Uematsu, S., *et al.*: **Interferon-alpha induction through Toll-like receptors involves a direct interaction of IRF7 with MyD88 and TRAF6.** *Nat. Immunol.* 2004, **5**:1061–1068.

Puel, A., Yang, K., Ku, C.L., von Bernuth, H., Bustamante, J., Santos, O.F., Lawrence, T., Chang, H.H., Al-Mousa, H., Picard, C., *et al.*: **Heritable defects of the human TLR signalling pathways.** *J. Endotoxin Res.* 2005, **11**:220–224.

Von Bernuth, H., Picard, C., Jin, Z., Pankla, R., Xiao, H., Ku, C.L., Chrabieh, M., Mustapha, I.B., Ghandil, P., Camcioglu, Y., *et al.*: **Pyogenic bacterial infections in humans with MyD88 deficiency.** *Science* 2008, **321**:691–696.

Werts, C., Girardinm, S.E., and Philpott, D.J.: **TIR, CARD and PYRIN: three domains for an antimicrobial triad.** *Cell Death Differ.* 2006, **13**:798–815.

3-8   The NOD-like receptors are intracellular sensors of bacterial infection and cellular damage.

Inohara, N., Chamaillard, M., McDonald, C., and Nunez, G.: **NOD-LRR proteins: role in host–microbial interactions and inflammatory disease.** *Annu. Rev. Biochem.* 2005, **74**:355–383.

Shaw, M.H., Reimer, T., Kim, Y.G., and Nuñez, G.: **NOD-like receptors (NLRs): bona fide intracellular microbial sensors.** *Curr. Opin. Immunol.* 2008, **20**:377–382.

Strober, W., Murray, P.J., Kitani, A., and Watanabe, T.: **Signalling pathways and molecular interactions of NOD1 and NOD2.** *Nat. Rev. Immunol.* 2006, **6**:9–20.

Ting, J.P., Kastner, D.L., and Hoffman, H.M.: **CATERPILLERs, pyrin and hereditary immunological disorders.** *Nat. Rev. Immunol.* 2006, **6**:183–195.

3-9   NLRP proteins react to infection or cellular damage through an inflammasome to induce cell death and inflammation.

Fernandes-Alnemri, T., Yu, J.W., Juliana, C., Solorzano, L., Kang, S., Wu, J., Datta, P., McCormick, M., Huang, L., McDermott, E., *et al.*: **The AIM2 inflammasome is critical for innate immunity to *Francisella tularensis*.** *Nat. Immunol.* 2010, **11**:385–393.

Hornung, V., Ablasser, A., Charrel-Dennis, M., Bauernfeind, F., Horvath, G., Caffrey, D.R., Latz, E., and Fitzgerald, K.A.: **AIM2 recognizes cytosolic dsDNA and forms a caspase-1-activating inflammasome with ASC.** *Nature* 2009, **458**:514–518.

Kofoed, E.M., and Vance, R.E.: **Innate immune recognition of bacterial ligands by NAIPs determines inflammasome specificity.** *Nature* 2011, **477**:592–595.

Martinon, F., Pétrilli, V., Mayor, A., Tardivel, A., and Tschopp, J.: **Gout-associated uric acid crystals activate the NALP3 inflammasome.** *Nature* 2006, **440**:237–241.

Mayor, A., Martinon, F., De Smedt, T., Pétrilli, V., and Tschopp, J.: **A crucial function of SGT1 and HSP90 in inflammasome activity links mammalian and plant innate immune responses.** *Nat. Immunol.* 2007, **8**:497–503.

Muñoz-Planillo, R., Kuffa, P., Martínez-Colón, G., Smith, B.L., Rajendiran, T.M., and Núñez, G.: **K+ efflux is the common trigger of NLRP3 inflammasome activation by bacterial toxins and particulate matter.** *Immunity* 2013, **38**:1142–1153.

3-10 The RIG-I-like receptors detect cytoplasmic viral RNAs and activate MAVS to induce type I interferon production and pro-inflammatory cytokines.

Brightbill, H.D., Libraty, D.H., Krutzik, S.R., Yang, R.B., Belisle, J.T., Bleharski, J.R., Maitland, M., Norgard, M.V., Plevy, S.E., Smale, S.T., *et al.*: **Host defense mechanisms triggered by microbial lipoproteins through Toll-like receptors.** *Science* 1999, **285**:732–736.

Hornung, V., Ellegast, J., Kim, S., Brzózka, K., Jung, A., Kato, H., Poeck, H., Akira, S., Conzelmann, K.K., Schlee, M., *et al.*: **5'-Triphosphate RNA is the ligand for RIG-I.** *Science* 2006, **314**:994–997.

Kato, H., Takeuchi, O., Sato, S., Yoneyama, M., Yamamoto, M., Matsui, K., Uematsu, S., Jung, A., Kawai, T., Ishii, K.J., *et al.*: **Differential roles of MDA5 and RIG-I helicases in the recognition of RNA viruses.** *Nature* 2006, **441**:101–105.

Konno, H., Yamamoto, T., Yamazaki, K., Gohda, J., Akiyama, T., Semba, K., Goto, H., Kato, A., Yujiri, T., Imai, T., *et al.*: **TRAF6 establishes innate immune responses by activating NF-κB and IRF7 upon sensing cytosolic viral RNA and DNA.** *PLoS ONE* 2009, **4**:e5674.

Martinon, F., Mayor, A., and Tschopp, J.: **The inflammasomes: guardians of the body.** *Annu. Rev. Immunol.* 2009, **27**:229–265.

Meylan, E., Curran, J., Hofmann, K., Moradpour, D., Binder, M., Bartenschlager, R., and Tschopp, J.: **Cardif is an adaptor protein in the RIG-I antiviral pathway and is targeted by hepatitis C virus.** *Nature* 2005, **437**:1167–1172.

Pichlmair, A., Schulz, O., Tan, C.P., Näslund, T.I., Liljeström, P., Weber, F., and Reis e Sousa, C.: **RIG-I-mediated antiviral responses to single-stranded RNA bearing 5'-phosphates.** *Science* 2006, **314**:935–936.

Takeda, K., Kaisho, T., and Akira, S.: **Toll-like receptors.** *Annu. Rev. Immunol.* 2003, **21**:335–376.

3-11 Cytosolic DNA sensors signal through STING to induce production of type I interferons.

Cai, X., Chiu, Y.H., and Chen, Z.J.: **The cGAS-cGAMP-STING pathway of cytosolic DNA sensing and signaling.** *Mol. Cell* 2014, **54**:289–296.

Ishikawa, H., and Barber, G.N.: **STING is an endoplasmic reticulum adaptor that facilitates innate immune signalling.** *Nature* 2008, **455**:674–678.

Li, X.D., Wu, J., Gao, D., Wang, H., Sun, L., and Chen, Z.J.: **Pivotal roles of cGAS-cGAMP signaling in antiviral defense and immune adjuvant effects.** *Science* 2013, **341**:1390–1394.

Liu, S., Cai, X., Wu, J., Cong, Q., Chen, X., Li, T., Du, F., Ren, J., Wu, Y.T., Grishin, N.V., and Chen, Z.J.: **Phosphorylation of innate immune adaptor proteins MAVS, STING, and TRIF induces IRF3 activation.** *Science* 2015, **347**:aaa2630.

Sun, L., Wu, J., Du, F., Chen, X., and Chen, Z.J.: **Cyclic GMP-AMP synthase is a cytosolic DNA sensor that activates the type I interferon pathway.** *Science* 2013, **339**:786–791.

3-12 Activation of innate sensors in macrophages and dendritic cells triggers changes in gene expression that have far-reaching effects on the immune response.
&
3-13 Toll signaling in *Drosophila* is downstream of a distinct set of pathogen-recognition molecules.

Dziarski, R., and Gupta, D.: **Mammalian PGRPs: novel antibacterial proteins.** *Cell Microbiol.* 2006, **8**:1059–1069.

Ferrandon, D., Imler, J.L., Hetru, C., and Hoffmann, J.A.: **The *Drosophila* systemic immune response: sensing and signalling during bacterial and fungal infections.** *Nat. Rev. Immunol.* 2007, **7**:862–874.

Gottar, M., Gobert, V., Matskevich, A.A., Reichhart, J.M., Wang, C., Butt, T.M., Belvin, M., Hoffmann, J.A., and Ferrandon, D.: **Dual detection of fungal infections in *Drosophila* via recognition of glucans and sensing of virulence factors.** *Cell* 2006, **127**:1425–1437.

Kambris, Z., Brun, S., Jang, I.H., Nam, H.J., Romeo, Y., Takahashi, K., Lee, W.J., Ueda, R., and Lemaitre, B.: ***Drosophila* immunity: a large-scale *in vivo* RNAi screen identifies five serine proteases required for Toll activation.** *Curr. Biol.* 2006, **16**:808–813.

Pili-Floury, S., Leulier, F., Takahashi, K., Saigo, K., Samain, E., Ueda, R., and Lemaitre, B.: ***In vivo* RNA interference analysis reveals an unexpected role for GNBP1 in the defense against Gram-positive bacterial infection in *Drosophila* adults.** *J. Biol. Chem.* 2004, **279**:12848–12853.

Royet, J., and Dziarski, R.: **Peptidoglycan recognition proteins: pleiotropic sensors and effectors of antimicrobial defences.** *Nat. Rev. Microbiol.* 2007, **5**:264–277.

3-14 TLR and NOD genes have undergone extensive diversification in both invertebrates and some primitive chordates.

Rast, J.P., Smith, L.C., Loza-Coll, M., Hibino, T., and Litman, G.W.: **Genomic insights into the immune system of the sea urchin.** *Science* 2006, **314**:952–956.

Samanta, M.P., Tongprasit, W., Istrail, S., Cameron, R.A., Tu, Q., Davidson, E.H., and Stolc, V.: **The transcriptome of the sea urchin embryo.** *Science* 2006, **314**:960–962.

3-15 Cytokines and their receptors fall into distinct families of structurally related proteins.

Basler, C.F., and Garcia-Sastre, A.: **Viruses and the type I interferon antiviral system: induction and evasion.** *Int. Rev. Immunol.* 2002, **21**:305–337.

Boulay, J.L., O'Shea, J.J., and Paul, W.E.: **Molecular phylogeny within type I cytokines and their cognate receptors.** *Immunity* 2003, **19**:159–163.

Collette, Y., Gilles, A., Pontarotti, P., and Olive, D.: **A co-evolution perspective of the TNFSF and TNFRSF families in the immune system.** *Trends Immunol.* 2003, **24**:387–394.

Ihle, J.N.: **Cytokine receptor signalling.** *Nature* 1995, **377**:591–594.

Proudfoot, A.E.: **Chemokine receptors: multifaceted therapeutic targets.** *Nat. Rev. Immunol.* 2002, **2**:106–115.

Taniguchi, T., and Takaoka, A.: **The interferon-α/β system in antiviral responses: a multimodal machinery of gene regulation by the IRF family of transcription factors.** *Curr. Opin. Immunol.* 2002, **14**:111–116.

3-16 Cytokine receptors of the hematopoietin family are associated with the JAK family of tyrosine kinases, which activate STAT transcription factors.

Fu, X.Y.: **A transcription factor with SH2 and SH3 domains is directly activated by an interferon α-induced cytoplasmic protein tyrosine kinase(s).** *Cell* 1992, **70**:323–335.

Krebs, D.L., and Hilton, D.J.: **SOCS proteins: Negative regulators of cytokine signaling.** *Stem Cells* 2001, **19**:378–387.

Leonard, W.J., and O'Shea, J.J.: **Jaks and STATs: biological implications.** *Annu. Rev. Immunol.* 1998, **16**:293–322.

Levy, D.E., and Darnell, J.E., Jr.: **Stats: transcriptional control and biological impact.** *Nat. Rev. Mol. Cell Biol.* 2002, **3**:651–662.

Ota, N., Brett, T.J., Murphy, T.L., Fremont, D.H., and Murphy, K.M.: **N-domain-dependent nonphosphorylated STAT4 dimers required for cytokine-driven activation.** *Nat. Immunol.* 2004, **5**:208–215.

Pesu, M., Candotti, F., Husa, M., Hofmann, S.R., Notarangelo, L.D., and O'Shea, J.J.: **Jak3, severe combined immunodeficiency, and a new class of immunosuppressive drugs.** *Immunol. Rev.* 2005, **203**:127–142.

Rytinki, M.M., Kaikkonen, S., Pehkonen, P., Jääskeläinen, T., and Palvimo, J.J.:

PIAS proteins: pleiotropic interactors associated with SUMO. *Cell. Mol. Life Sci.* 2009, **66**:3029–3041.

Schindler, C., Shuai, K., Prezioso, V.R., and Darnell, J.E., Jr.: **Interferon-dependent tyrosine phosphorylation of a latent cytoplasmic transcription factor.** *Science* 1992, **257**:809–813.

Shuai, K., and Liu, B.: **Regulation of JAK-STAT signalling in the immune system.** *Nat. Rev. Immunol.* 2003, **3**:900–911.

Yasukawa, H., Sasaki, A., and Yoshimura, A.: **Negative regulation of cytokine signaling pathways.** *Annu. Rev. Immunol.* 2000, **18**:143–164.

3-17 Chemokines released by macrophages and dendritic cells recruit effector cells to sites of infection.

Larsson, B.M., Larsson, K., Malmberg, P., and Palmberg, L.: **Gram-positive bacteria induce IL-6 and IL-8 production in human alveolar macrophages and epithelial cells.** *Inflammation* 1999, **23**:217–230.

Luster, A.D.: **The role of chemokines in linking innate and adaptive immunity.** *Curr. Opin. Immunol.* 2002, **14**:129–135.

Matsukawa, A., Hogaboam, C.M., Lukacs, N.W., and Kunkel, S.L.: **Chemokines and innate immunity.** *Rev. Immunogenet.* 2000, **2**:339–358.

Scapini, P., Lapinet-Vera, J.A., Gasperini, S., Calzetti, F., Bazzoni, F., and Cassatella, M.A.: **The neutrophil as a cellular source of chemokines.** *Immunol. Rev.* 2000, **177**:195–203.

Shortman, K., and Liu, Y.J.: **Mouse and human dendritic cell subtypes.** *Nat. Rev. Immunol.* 2002, **2**:151–161.

Svanborg, C., Godaly, G., and Hedlund, M.: **Cytokine responses during mucosal infections: role in disease pathogenesis and host defence.** *Curr. Opin. Microbiol.* 1999, **2**:99–105.

Yoshie, O.: **Role of chemokines in trafficking of lymphocytes and dendritic cells.** *Int. J. Hematol.* 2000, **72**:399–407.

3-18 Cell-adhesion molecules control interactions between leukocytes and endothelial cells during an inflammatory response.

Alon, R., and Feigelson, S.: **From rolling to arrest on blood vessels: leukocyte tap dancing on endothelial integrin ligands and chemokines at sub-second contacts.** *Semin. Immunol.* 2002, **14**:93–104.

Bunting, M., Harris, E.S., McIntyre, T.M., Prescott, S.M., and Zimmerman, G.A.: **Leukocyte adhesion deficiency syndromes: adhesion and tethering defects involving β2 integrins and selectin ligands.** *Curr. Opin. Hematol.* 2002, **9**:30–35.

D'Ambrosio, D., Albanesi, C., Lang, R., Girolomoni, G., Sinigaglia, F., and Laudanna, C.: **Quantitative differences in chemokine receptor engagement generate diversity in integrin-dependent lymphocyte adhesion.** *J. Immunol.* 2002, **169**:2303–2312.

Johnston, B., and Butcher, E.C.: **Chemokines in rapid leukocyte adhesion triggering and migration.** *Semin. Immunol.* 2002, **14**:83–92.

Ley, K.: **Integration of inflammatory signals by rolling neutrophils.** *Immunol. Rev.* 2002, **186**:8–18.

Vestweber, D.: **Lymphocyte trafficking through blood and lymphatic vessels: more than just selectins, chemokines and integrins.** *Eur. J. Immunol.* 2003, **33**:1361–1364.

3-19 Neutrophils make up the first wave of cells that cross the blood vessel wall to enter an inflamed tissue.

Bochenska-Marciniak, M., Kupczyk, M., Gorski, P., and Kuna, P.: **The effect of recombinant interleukin-8 on eosinophils' and neutrophils' migration *in vivo* and *in vitro*.** *Allergy* 2003, **58**:795–801.

Godaly, G., Bergsten, G., Hang, L., Fischer, H., Frendeus, B., Lundstedt, A.C., Samuelsson, M., Samuelsson, P., and Svanborg, C.: **Neutrophil recruitment, chemokine receptors, and resistance to mucosal infection.** *J. Leukoc. Biol.* 2001, **69**:899–906.

Gompertz, S., and Stockley, R.A.: **Inflammation—role of the neutrophil and the eosinophil.** *Semin. Respir. Infect.* 2000, **15**:14–23.

Lee, S.C., Brummet, M.E., Shahabuddin, S., Woodworth, T.G., Georas, S.N.,

Leiferman, K.M., Gilman, S.C., Stellato, C., Gladue, R.P., Schleimer, R.P., *et al.*: **Cutaneous injection of human subjects with macrophage inflammatory protein-1α induces significant recruitment of neutrophils and monocytes.** *J. Immunol.* 2000, **164**:3392–3401.

Sundd, P., Gutierrez, E., Koltsova, E.K., Kuwano, Y., Fukuda, S., Pospieszalska, M.K., Groisman, A., and Ley, K.: **'Slings' enable neutrophil rolling at high shear.** *Nature* 2012, **488**:399–403.

Worthylake, R.A., and Burridge, K.: **Leukocyte transendothelial migration: orchestrating the underlying molecular machinery.** *Curr. Opin. Cell Biol.* 2001, **13**:569–577.

3-20 TNF-α is an important cytokine that triggers local containment of infection but induces shock when released systemically.

Croft, M.: **The role of TNF superfamily members in T-cell function and diseases.** *Nat. Rev. Immunol.* 2009, **9**:271–285.

Dellinger, R.P.: **Inflammation and coagulation: implications for the septic patient.** *Clin. Infect. Dis.* 2003, **36**:1259–1265.

Georgel, P., Naitza, S., Kappler, C., Ferrandon, D., Zachary, D., Swimmer, C., Kopczynski, C., Duyk, G., Reichhart, J.M., and Hoffmann, J.A.: *Drosophila* **immune deficiency (IMD) is a death domain protein that activates antibacterial defense and can promote apoptosis.** *Dev. Cell* 2001, **1**:503–514.

Pfeffer, K.: **Biological functions of tumor necrosis factor cytokines and their receptors.** *Cytokine Growth Factor Rev.* 2003, **14**:185–191.

Rutschmann, S., Jung, A.C., Zhou, R., Silverman, N., Hoffmann, J.A., and Ferrandon, D.: **Role of *Drosophila* IKKγ in a *toll*-independent antibacterial immune response.** *Nat. Immunol.* 2000, **1**:342–347.

3-21 Cytokines made by macrophages and dendritic cells induce a systemic reaction known as the acute-phase response.

Bopst, M., Haas, C., Car, B., and Eugster, H.P.: **The combined inactivation of tumor necrosis factor and interleukin-6 prevents induction of the major acute phase proteins by endotoxin.** *Eur. J. Immunol.* 1998, **28**:4130–4137.

Ceciliani, F., Giordano, A., and Spagnolo, V.: **The systemic reaction during inflammation: the acute-phase proteins.** *Protein Pept. Lett.* 2002, **9**:211–223.

He, R., Sang, H., and Ye, R.D.: **Serum amyloid A induces IL-8 secretion through a G protein-coupled receptor, FPRL1/LXA4R.** *Blood* 2003, **101**:1572–1581.

Horn, F., Henze, C., and Heidrich, K.: **Interleukin-6 signal transduction and lymphocyte function.** *Immunobiology* 2000, **202**:151–167.

Manfredi, A.A., Rovere-Querini, P., Bottazzi, B., Garlanda, C., and Mantovani, A.: **Pentraxins, humoral innate immunity and tissue injury.** *Curr. Opin. Immunol.* 2008, **20**:538–544.

Mold, C., Rodriguez, W., Rodic-Polic, B., and Du Clos, T.W.: **C-reactive protein mediates protection from lipopolysaccharide through interactions with FcγR.** *J. Immunol.* 2002, **169**:7019–7025.

3-22 Interferons induced by viral infection make several contributions to host defense.

Baldridge, M.T., Nice, T.J., McCune, B.T., Yokoyama, C.C., Kambal, A., Wheadon, M., Diamond, M.S., Ivanova, Y., Artyomov, M., and Virgin, H.W.: **Commensal microbes and interferon-λ determine persistence of enteric murine norovirus infection.** *Science* 2015, **347**:266–269.

Carrero, J.A., Calderon, B., and Unanue, E.R.: **Type I interferon sensitizes lymphocytes to apoptosis and reduces resistance to *Listeria* infection.** *J. Exp. Med.* 2004, **200**:535–540.

Honda, K., Takaoka, A., and Taniguchi, T.: **Type I interferon gene induction by the interferon regulatory factor family of transcription factors.** *Immunity* 2006, **25**:349–360.

Kawai, T., and Akira, S.: **Innate immune recognition of viral infection.** *Nat. Immunol.* 2006, **7**:131–137.

Liu, Y.J.: **IPC: professional type 1 interferon-producing cells and plasmacytoid dendritic cell precursors.** *Annu. Rev. Immunol.* 2005, **23**:275–306.

Meylan, E., and Tschopp, J.: **Toll-like receptors and RNA helicases: two parallel ways to trigger antiviral responses.** *Mol. Cell* 2006, **22**:561–569.

Pietras, E.M., Saha, S.K., and Cheng, G.: **The interferon response to bacterial and viral infections.** *J. Endotoxin Res.* 2006, **12**:246–250.

Pott, J., Mahlakõiv, T., Mordstein, M., Duerr, C.U., Michiels, T., Stockinger, S., Staeheli, P., and Hornef, M.W.: **IFN-lambda determines the intestinal epithelial antiviral host defense.** *Proc. Natl Acad. Sci. USA* 2011, **108**:7944–7949.

3-23 Several types of innate lymphoid cells provide protection in early infection.

Cortez, V.S., Robinette, M.L., and Colonna, M.: **Innate lymphoid cells: new insights into function and development.** *Curr. Opin. Immunol.* 2015, **32**:71–77.

Spits, H., Artis, D., Colonna, M., Diefenbach, A., Di Santo, J.P., Eberl, G., Koyasu, S., Locksley, R.M., McKenzie, A.N., Mebius, R.E., *et al.*: **Innate lymphoid cells—a proposal for uniform nomenclature.** *Nat. Rev. Immunol.* 2013, **13**:145–149.

3-24 NK cells are activated by type I interferon and macrophage-derived cytokines.

Barral, D.C., and Brenner, M.B.: **CD1 antigen presentation: how it works.** *Nat. Rev. Immunol.* 2007, **7**:929–941.

Gineau, L., Cognet, C., Kara, N., Lach, F.P., Dunne, J., Veturi, U., Picard, C., Trouillet, C., Eidenschenk, C., Aoufouchi S., *et al.*: **Partial MCM4 deficiency in patients with growth retardation, adrenal insufficiency, and natural killer cell deficiency.** *J. Clin. Invest.* 2012, **122**:821–832.

Godshall, C.J., Scott, M.J., Burch, P.T., Peyton, J.C., and Cheadle, W.G.: **Natural killer cells participate in bacterial clearance during septic peritonitis through interactions with macrophages.** *Shock* 2003, **19**:144–149.

Lanier, L.L.: **Evolutionary struggles between NK cells and viruses.** *Nat. Rev. Immunol.* 2008, **8**:259–268.

Salazar-Mather, T.P., Hamilton, T.A., and Biron, C.A.: **A chemokine-to-cytokine-to-chemokine cascade critical in antiviral defense.** *J. Clin. Invest.* 2000, **105**:985–993.

Seki, S., Habu, Y., Kawamura, T., Takeda, K., Dobashi, H., Ohkawa, T., and Hiraide, H.: **The liver as a crucial organ in the first line of host defense: the roles of Kupffer cells, natural killer (NK) cells and NK1.1 Ag+ T cells in T helper 1 immune responses.** *Immunol. Rev.* 2000, **174**:35–46.

Yokoyama, W.M., and Plougastel, B.F.: **Immune functions encoded by the natural killer gene complex.** *Nat. Rev. Immunol.* 2003, **3**:304–316.

3-25 NK cells express activating and inhibitory receptors to distinguish between healthy and infected cells.

&

3-26 NK-cell receptors belong to several structural families, the KIRs, KLRs, and NCRs.

Borrego, F., Kabat, J., Kim, D.K., Lieto, L., Maasho, K., Pena, J., Solana, R., and Coligan, J.E.: **Structure and function of major histocompatibility complex (MHC) class I specific receptors expressed on human natural killer (NK) cells.** *Mol. Immunol.* 2002, **38**:637–660.

Boyington, J.C., and Sun, P.D.: **A structural perspective on MHC class I recognition by killer cell immunoglobulin-like receptors.** *Mol. Immunol.* 2002, **38**:1007–1021.

Brown, M.G., Dokun, A.O., Heusel, J.W., Smith, H.R., Beckman, D.L., Blattenberger, E.A., Dubbelde, C.E., Stone, L.R., Scalzo, A.A., and Yokoyama, W.M.: **Vital involvement of a natural killer cell activation receptor in resistance to viral infection.** *Science* 2001, **292**:934–937.

Long, E.O.: **Negative signalling by inhibitory receptors: the NK cell paradigm.** *Immunol. Rev.* 2008, **224**:70–84.

Robbins, S.H., and Brossay, L.: **NK cell receptors: emerging roles in host defense against infectious agents.** *Microbes Infect.* 2002, **4**:1523–1530.

Trowsdale, J.: **Genetic and functional relationships between MHC and NK receptor genes.** *Immunity* 2001, **15**:363–374.

Vilches, C., and Parham, P.: **KIR: diverse, rapidly evolving receptors of innate and adaptive immunity.** *Annu. Rev. Immunol.* 2002, **20**:217–251.

Vivier, E., Raulet, D.H., Moretta, A., Caligiuri, M.A., Zitvogel, L., Lanier, L.L., Yokoyama, W.M., and Ugolini, S.: **Innate or adaptive immunity? The example of natural killer cells.** *Science* 2011, **331**:44–49.

Vivier, E., Tomasello, E., Baratin, M., Walzer, T., and Ugolini, S.: **Functions of natural killer cells.** *Nat. Immunol.* 2008, **9**:503–510.

3-27 NK cells express activating receptors that recognize ligands induced on infected cells or tumor cells.

Brandt, C.S., Baratin, M., Yi, E.C., Kennedy, J., Gao, Z., Fox, B., Haldeman, B., Ostrander, C.D., Kaifu, T., Chabannon, C., *et al.*: **The B7 family member B7-H6 is a tumor cell ligand for the activating natural killer cell receptor NKp30 in humans.** *J. Exp. Med.* 2009, **206**:1495–1503.

Gasser, S., Orsulic, S., Brown, E.J., and Raulet, D.H.: **The DNA damage pathway regulates innate immune system ligands of the NKG2D receptor.** *Nature* 2005, **436**:1186–1190.

Gonzalez, S., Groh, V., and Spies, T.: **Immunobiology of human NKG2D and its ligands.** *Curr. Top. Microbiol. Immunol.* 2006, **298**:121–138.

Lanier, L.L.: **Up on the tightrope: natural killer cell activation and inhibition.** *Nat. Immunol.* 2008, **9**:495–502.

Moretta, L., Bottino, C., Pende, D., Castriconi, R., Mingari, M.C., and Moretta, A.: **Surface NK receptors and their ligands on tumor cells.** *Semin. Immunol.* 2006, **18**:151–158.

Parham, P.: **MHC class I molecules and KIRs in human history, health and survival.** *Nat. Rev. Immunol.* 2005, **5**:201–214.

Raulet, D.H., and Guerra, N.: **Oncogenic stress sensed by the immune system: role of natural killer cell receptors.** *Nat. Rev. Immunol.* 2009, **9**:568–580.

Upshaw, J.L., and Leibson, P.J.: **NKG2D-mediated activation of cytotoxic lymphocytes: unique signaling pathways and distinct functional outcomes.** *Semin. Immunol.* 2006, **18**:167–175.

Vivier, E., Nunes, J.A., and Vely, F.: **Natural killer cell signaling pathways.** *Science* 2004, **306**:1517–1519.

Antigen Recognition by B-cell and T-cell Receptors

4

Innate immune responses are the body's initial defense against infection, but these work only to control pathogens that have certain molecular patterns or that induce interferons and other nonspecific defenses. To effectively fight the wide range of pathogens an individual will encounter, the lymphocytes of the adaptive immune system have evolved to recognize a great variety of different **antigens** from bacteria, viruses, and other disease-causing organisms. An antigen is any molecule or part of a molecule that is specifically recognized by the highly specialized recognition proteins of lymphocytes. On B cells these proteins are the **immunoglobulins (Igs)**, which these cells produce in a vast range of antigen specificities, each B cell producing immunoglobulins of a single specificity (see Section 1-12). A membrane-bound form of immunoglobulin on the B-cell surface serves as the cell's receptor for antigen, and is known as the **B-cell receptor (BCR)**. A secreted form of immunoglobulin of the same antigen specificity is the **antibody** produced by terminally differentiated B cells—plasmablasts and plasma cells. The secretion of antibodies, which bind pathogens or their toxic products in the extracellular spaces of the body (see Fig. 1.25), is the main effector function of B cells in adaptive immunity.

Antibodies were the first proteins involved in specific immune recognition to be characterized, and are understood in great detail. The antibody molecule has two separate functions: one is to bind specifically to the pathogen or its products that have elicited the immune response; the other is to recruit other cells and molecules to destroy the pathogen once antibody has bound. For example, binding by antibodies can neutralize viruses and mark pathogens for destruction by phagocytes and complement, as described in Chapters 2 and 3. Recognition and effector functions are structurally separated in the antibody molecule, one part of which specifically binds to the antigen whereas the other engages the elimination mechanisms. The antigen-binding region varies extensively between antibody molecules and is known as the **variable region** or **V region**. The variability of antibody molecules allows each antibody to bind a different specific antigen, and the total repertoire of antibodies made by a single individual is large enough to ensure that virtually any structure can be recognized. The region of the antibody molecule that engages the effector functions of the immune system does not vary in the same way and is known as the **constant region** or **C region**. It comes in five main forms, called **isotypes**, each of which is specialized for activating different effector mechanisms. The membrane-bound B-cell receptor does not have these effector functions, because the C region remains inserted in the membrane of the B cell. The function of the B-cell receptor is to recognize and bind antigen via the V regions exposed on the surface of the cell, thus transmitting a signal that activates the B cell, leading to clonal expansion and antibody production. To this end, the B-cell receptor is associated with a set of intracellular signaling proteins, which will be described in Chapter 7. Antibodies have become an important class of drug due to their highly specific activities, and we return to discuss their therapeutic uses in Chapter 16.

The antigen-recognition molecules of T cells are made solely as membrane-bound proteins, which are associated with an intracellular signaling complex and function only to signal T cells for activation. These **T-cell receptors (TCRs)** are related to immunoglobulins both in their protein structure—having both V and C regions—and in the genetic mechanism that produces their great variability, which is discussed in Chapter 5. The T-cell receptor differs from the B-cell receptor in an important way, however: it does not recognize and bind antigen by itself, but instead recognizes short peptide fragments of protein antigens that are presented to them by proteins known as **MHC molecules** on the surface of host cells.

The MHC molecules are transmembrane glycoproteins encoded in the large cluster of genes known as the **major histocompatibility complex (MHC)**. The most striking structural feature of MHC molecules is a cleft in the extracellular face of the molecule in which peptides can be bound. MHC molecules are highly **polymorphic**—each type of MHC molecule occurs in many different versions—within the population. These are encoded by slightly different versions of individual genes called **alleles**. Most people are therefore heterozygous for the MHC molecules: that is, they express two different alleles for each type of MHC molecule, thus increasing the range of pathogen-derived peptides and self-peptides that can be bound. T-cell receptors recognize features of both the peptide antigen and the MHC molecule to which it is bound. This introduces an extra dimension to antigen recognition by T cells, known as **MHC restriction** because any given T-cell receptor is specific for a particular peptide bound to a particular MHC molecule.

In this chapter we focus on the structure and antigen-binding properties of immunoglobulins and T-cell receptors. Although B cells and T cells recognize foreign molecules in separate distinct fashions, the receptor molecules they use for this task are very similar in structure. We will see how this basic structure can accommodate great variability in antigen specificity, and how it enables immunoglobulins and T-cell receptors to perform their functions as the antigen-recognition molecules of the adaptive immune response. With this foundation, we will return to discuss the impact of MHC polymorphism on T-cell antigen recognition and T-cell development in Chapters 6 and 8, respectively.

The structure of a typical antibody molecule.

Antibodies are the secreted form of the B-cell receptor. Because they are soluble and secreted into the blood in large quantities, antibodies are easily obtained and easily studied. For this reason, most of what we know about the B-cell receptor comes from the study of antibodies.

Antibody molecules are roughly Y-shaped, as represented in **Fig. 4.1** using three different schematic styles. This part of the chapter will explain how this structure is formed and allows the antibody molecule to perform its dual tasks of binding to a wide variety of antigens while also binding to effector molecules

Fig. 4.1 Structure of an antibody molecule. In panel a, the X-ray crystallographic structure of an IgG antibody is illustrated as a ribbon diagram of the backbones of the polypeptide chains. The two heavy chains are shown in yellow and purple. The two light chains are both shown in red. Three globular regions form an approximate Y shape. The two antigen-binding sites are at the tips of the arms, which are tethered at their other end to the trunk of the Y by a flexible hinge region. The light-chain variable (V_L) and constant region (C_L) are indicated. The heavy-chain variable region (V_H) and V_L together form the antigen-binding site of the antibody. In panel b, a schematic representation of the same structure denotes each immunoglobulin domain as a separate rectangle. The hinge that tethers each heavy chain's first constant domain (C_H1) to its second (C_H2) is illustrated by a thin purple or yellow line, respectively. The antibody-binding sites are indicated by concave regions in V_L and V_H. Positions of carbohydrate modifications and disulfide linkages are indicated. In panel c, a more simplified schematic is shown that will be used throughout this book with the variable region in red and the constant region in blue. C terminus, carboxy terminus; N terminus, amino terminus. Structure courtesy of R.L. Stanfield and I.A. Wilson.

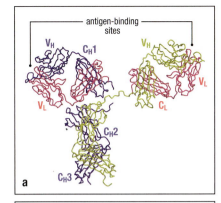

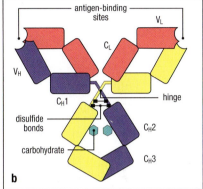

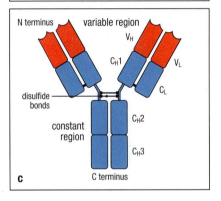

and to cells that destroy the antigen. Each of these tasks is performed by different parts of the molecule. The ends of the two arms of the Y—the V regions—are involved in antigen binding, and they vary in their detailed structure between different antibody molecules. The stem of the Y—the C region—is far less variable and is the part that interacts with effector molecules and cells. There are five different **classes** of immunoglobulins, distinguished in their being constructed from C regions that have different structures and properties. These are known as **immunoglobulin M (IgM)**, **immunoglobulin D (IgD)**, **immunoglobulin G (IgG)**, **immunoglobulin A (IgA)**, and **immunoglobulin E (IgE)**.

All antibodies are constructed in the same way from paired heavy and light polypeptide chains, and the generic term immunoglobulin is used for all such proteins. More subtle differences confined to the V region account for the specificity of antigen binding. We will use the IgG antibody molecule as an example to describe the general structural features of immunoglobulins.

4-1 IgG antibodies consist of four polypeptide chains.

IgG antibodies are large molecules with a molecular weight of approximately 150 kDa and are composed of two different kinds of polypeptide chains. One, of approximately 50 kDa, is called the **heavy** or **H chain**, and the other, of 25 kDa, is the **light** or **L chain** (Fig. 4.2). Each IgG molecule consists of two heavy chains and two light chains. The two heavy chains are linked to each other by disulfide bonds, and each heavy chain is linked to a light chain by a disulfide bond. In any given immunoglobulin molecule, the two heavy chains and the two light chains are identical, giving an antibody molecule two identical antigen-binding sites. This gives the antibody the ability to bind simultaneously to two identical antigens on a surface, thereby increasing the total strength of the interaction, which is called its **avidity**. The strength of the interaction between a single antigen-binding site and its antigen is called its **affinity**.

Two types of light chains, **lambda (λ)** and **kappa (κ)**, are found in antibodies. A given immunoglobulin has either κ chains or λ chains, never one of each. No functional difference has been found between antibodies having λ or κ light chains, and either type of light chain can be found in antibodies of any of the five major classes. The ratio of the two types of light chains varies from species to species. In mice, the average κ to λ ratio is 20:1, whereas in humans it is 2:1 and in cattle it is 1:20. The reason for this variation is unknown. Distortions of this ratio can sometimes be used to detect the abnormal proliferation of a

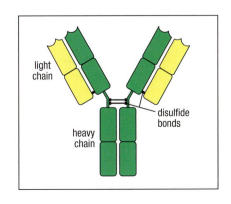

Fig. 4.2 Immunoglobulin molecules are composed of two types of protein chains: heavy chains and light chains. Each immunoglobulin molecule is made up of two hinged heavy chains (green) and two light chains (yellow) joined by disulfide bonds so that each heavy chain is linked to a light chain and the two heavy chains are linked together.

B-cell clone, since all progeny of a particular B cell will express an identical light chain. For example, an abnormally high level of λ light chains in a person might indicate the presence of a B-cell tumor that is producing λ chains.

The class, and thus the effector function, of an antibody is defined by the structure of its heavy chain. There are five main heavy-chain classes, or **isotypes**, some of which have several subtypes, and these determine the functional activity of an antibody molecule. The five major classes of immunoglobulin are IgM, IgD, IgG, IgA, and IgE, and their heavy chains are denoted by the lower-case Greek letters μ, δ, γ, α, and ε, respectively. For example, the constant region of IgM is denoted by Cμ. IgG is by far the most abundant immunoglobulin in serum and has several subclasses (IgG1, 2, 3, and 4 in humans). The distinctive functional properties of the different classes and subclasses of antibodies are conferred by the carboxy-terminal part of the heavy chain, where this chain is not associated with the light chain. The general structural features of all the isotypes are similar, particularly with respect to antigen binding. Here we will consider IgG as a typical antibody molecule, and we will return to discuss the structural and functional properties of the different heavy-chain isotypes in Chapter 5.

The structure of a B-cell receptor is identical to that of its corresponding antibody except for a small portion of the carboxy terminus of the heavy-chain C region. In the B-cell receptor, the carboxy terminus is a hydrophobic amino acid sequence that anchors the molecule in the membrane, and in the antibody it is a hydrophilic sequence that allows secretion.

4-2   Immunoglobulin heavy and light chains are composed of constant and variable regions.

The amino acid sequences of many immunoglobulin heavy and light chains have been determined and reveal two important features of antibody molecules. First, each chain consists of a series of similar, although not identical, sequences, each about 110 amino acids long. Each of these repeats corresponds to a discrete, compactly folded region of protein known as an **immunoglobulin domain**, or **Ig domain**. The light chain consists of two Ig domains, whereas the heavy chain of the IgG antibody contains four Ig domains (see Fig. 4.2). This suggests that the immunoglobulin chains have evolved by repeated duplication of ancestral gene segments corresponding to a single Ig domain.

The second important feature is that the amino-terminal amino acid sequences of the heavy and light chains vary greatly between different antibodies. The variability is limited to approximately the first 110 amino acids, corresponding to the first Ig domain, whereas the remaining domains are constant between immunoglobulin chains of the same isotype. The amino-terminal **variable Ig domains (V domains)** of the heavy and light chains (V_H and V_L, respectively) together make up the V region of the antibody and determine its antigen-binding specificity, whereas the **constant Ig domains (C domains)** of the heavy and light chains (C_H and C_L, respectively) make up the C region (see Fig. 4.1). The multiple heavy-chain C domains are numbered from the amino-terminal end to the carboxy terminus, for example, C_H1, C_H2, and so on.

4-3   The domains of an immunoglobulin molecule have similar structures.

Immunoglobulin heavy and light chains are composed of a series of Ig domains that have a similar overall structure. Within this basic structure, there are distinct differences between V and C domains that are illustrated for the light chain in **Fig. 4.3**. Each V or C domain is constructed from two **β sheets**. A β sheet is

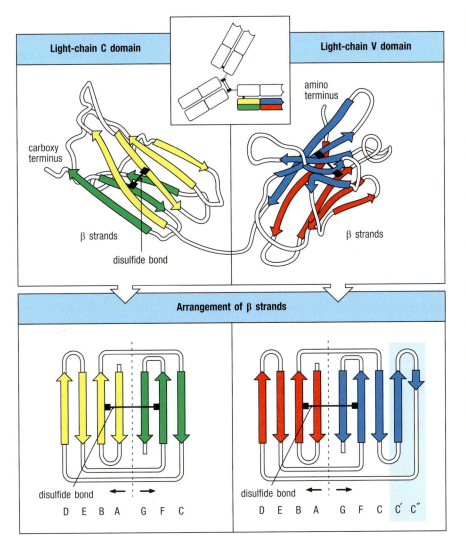

Light-chain C domain

carboxy
terminus

β strands

disulfide bond

amino
terminus

Light-chain V domain

β strands

Arrangement of β strands

disulfide bond   ◄─ : ─►

D E B A   G F C

disulfide bond   ◄─ : ─►

D E B A   G F C C′ C″

Fig. 4.3 The structure of immunoglobulin constant and variable domains. The upper panels show schematically the folding pattern of the constant (C) and variable (V) domains of an immunoglobulin light chain. Each domain is a barrel-shaped structure in which strands of polypeptide chain (β strands) running in opposite directions (antiparallel) pack together to form two β sheets (shown in yellow and green for the C domain and red and blue for the V domain), which are held together by a disulfide bond. The way in which the polypeptide chain folds to give the final structure can be seen more clearly when the sheets are opened out, as shown in the lower panels. The β strands are lettered sequentially with respect to the order of their occurrence in the amino acid sequence of the domains; the order in each β sheet is characteristic of immunoglobulin domains. The β strands C′ and C″ that are found in the V domains but not in the C domains are indicated by a blue-shaded background. The characteristic four-strand plus three-strand (C-region type domain) or four-strand plus five-strand (V-region type domain) arrangements are typical immunoglobulin superfamily domain building blocks, found in a whole range of other proteins as well as antibodies and T-cell receptors.

built from several **β strands**, which are regions of protein where several consecutive polypeptides have their peptide backbone bonds arranged in an extended, or flat, conformation. β strands in proteins are sometimes shown as 'ribbons with an arrow' to indicate the direction of the polypeptide backbone (see Fig. 4.3). β strands can pack together in a side-by-side manner, being stabilized laterally by two or three backbone hydrogen bonds between adjacent strands. This arrangement is called a β sheet. The Ig domain has two β sheets that are folded onto each other, like two pieces of bread, into a structure called a **β sandwich**, and are covalently linked by a disulfide bond between cysteine residues from each β sheet. This distinctive structure is known as the **immunoglobulin fold**.

The similarities and differences between V and C domains can be seen in the bottom panels of Fig. 4.3. Here, the Ig domains have been opened out to show how their respective polypeptide chains fold to create each of the β sheets and how each polypeptide chain forms flexible loops between adjacent β strands as it turns to change direction. The main difference between the V and C domains is that the V domain is larger and contains extra β strands, called C′ and C″. In the V domain, the flexible loops formed between some of the β strands contribute to the antigen-binding site of the immunoglobulin molecule.

Many of the amino acids that are common to the C and V domains are present in the core of the immunoglobulin fold and are essential for its stability. Other proteins with sequences similar to those of immunoglobulins have been found to have domains with a similar structure, called **immunoglobulin-like domains (Ig-like domains)**. These domains are present in many proteins of the immune system, such as the KIRs expressed by NK cells described in Chapter 3. They are also frequently involved in cell–cell recognition and adhesion, and together with the immunoglobulins and the T-cell receptors, these proteins make up the extensive **immunoglobulin superfamily**.

4-4   The antibody molecule can readily be cleaved into functionally distinct fragments.

When fully assembled, an antibody molecule comprises three equal-sized globular portions, with its two arms joined to its trunk by a flexible stretch of polypeptide chain known as the **hinge region** (see Fig. 4.1b). Each arm of this Y-shaped structure is formed by the association of a light chain with the amino-terminal half of a heavy chain; the V_H domain is paired with the V_L domain, and the C_H1 domain is paired with the C_L domain. The two antigen-binding sites are formed by the paired V_H and V_L domains at the ends of the two arms of the Y (see Fig. 4.1b). The trunk of the Y is formed by the pairing of the carboxy-terminal halves of the two heavy chains. The C_H3 domains pair with each other but the C_H2 domains do not interact. Carbohydrate side chains attached to the C_H2 domains lie between the two heavy chains.

Proteolytic enzymes (proteases) were an important tool in early studies of antibody structure, and it is valuable to review the terminology they generated. Limited digestion with the protease papain cleaves antibody molecules into three fragments (**Fig. 4.4**). **Papain** cuts the antibody molecule on the amino-terminal side of the disulfide bonds that link the two heavy chains, releasing the two arms of the antibody molecule as two identical fragments that contain the antigen-binding activity. These are called the **Fab fragments**, for fragment <u>a</u>ntigen <u>b</u>inding. The other fragment contains no antigen-binding activity, but because it crystallized readily, it was named the **Fc fragment** (fragment <u>c</u>rystallizable). It corresponds to the paired C_H2 and C_H3 domains. The Fc fragment is the part of the antibody molecule that does not interact with antigen, but rather interacts with effector molecules and cells, and it differs between heavy-chain isotypes. Another protease, **pepsin**, cuts on the carboxy-terminal side of the disulfide bonds (see Fig. 4.4). This produces a fragment, the **F(ab′)₂ fragment**, in which the two antigen-binding arms of the antibody molecule remain linked. Pepsin cuts the remaining part of the heavy chain into several small fragments. The F(ab′)₂ fragment has exactly the same antigen-binding characteristics as the original antibody but is unable to interact with any effector molecule, such as C1q or Fc receptors, and can be used experimentally to separate the antigen-binding functions from the antibody's other effector functions.

Many antibody-related molecules can be constructed using genetic engineering techniques, and many antibodies and antibody-related molecules are being used therapeutically to treat a variety of diseases. We will return to this topic in Chapter 16, where we discuss the various therapeutic uses of antibodies that have been developed over the last two decades.

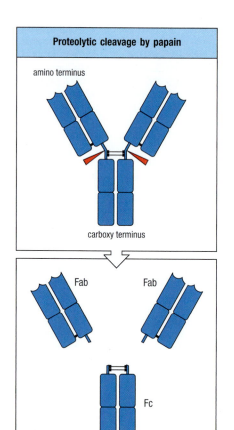

Proteolytic cleavage by papain

amino terminus

carboxy terminus

Fab   Fab

Fc

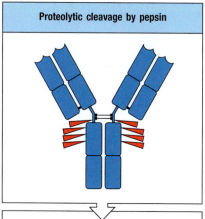

Proteolytic cleavage by pepsin

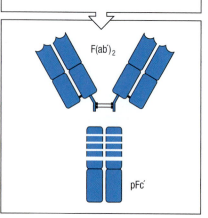

F(ab′)₂

pFc′

Fig. 4.4 The Y-shaped immunoglobulin molecule can be dissected by partial digestion with proteases. Upper panels: papain cleaves the immunoglobulin molecule into three pieces, two Fab fragments and one Fc fragment. The Fab fragment contains the V regions and binds antigen. The Fc fragment is crystallizable and contains C regions. Lower panels: pepsin cleaves immunoglobulin to yield one F(ab′)₂ fragment and many small pieces of the Fc fragment, the largest of which is called the pFc′ fragment. F(ab′)₂ is written with a prime because it contains a few more amino acids than Fab, including the cysteines that form the disulfide bonds.

(Micrograph ×300,000)	Angle between arms is 60°	Angle between arms is 90°
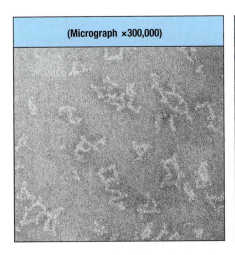	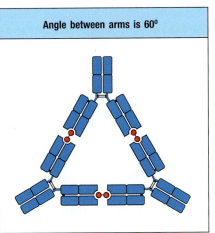	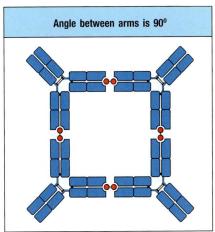

4-5   The hinge region of the immunoglobulin molecule allows flexibility in binding to multiple antigens.

The hinge region between the Fc and Fab portions of the IgG molecule allows for some degree of independent movement of the two Fab arms. For example, in the antibody molecule shown in Fig. 4.1a, not only are the two hinge regions clearly bent differently, but the angle between the V and C domains in each of the two Fab arms is also different. This range of motion has led to the junction between the V and C domains being referred to as a 'molecular ball-and-socket joint.' This flexibility can be revealed by studies of antibodies bound to small antigens known as **haptens**. These are molecules of various types that are typically about the size of a tyrosine side chain. Although haptens are specifically recognized by antibody, they can stimulate the production of anti-hapten antibodies only when linked to a protein (see Appendix I, Section A-1). Two identical hapten molecules joined by a short flexible region can link two or more anti-hapten antibodies, forming dimers, trimers, tetramers, and so on, which can be seen by electron microscopy (Fig. 4.5). The shapes formed by these complexes show that antibody molecules are flexible at the hinge region. Some flexibility is also found at the junction between the V and C domains, allowing bending and rotation of the V domain relative to the C domain. Flexibility at both the hinge and the V–C junction enables the two arms of an antibody molecule to bind to sites some distance apart, such as the repeating sites on bacterial cell-wall polysaccharides. Flexibility at the hinge also enables antibodies to interact with the antibody-binding proteins that mediate immune effector mechanisms.

Fig. 4.5 Antibody arms are joined by a flexible hinge. An antigen consisting of two hapten molecules (red balls in diagrams) that can cross-link two antigen-binding sites is used to create antigen:antibody complexes, which can be seen in the electron micrograph. Linear, triangular, and square forms are seen, with short projections or spikes. Limited pepsin digestion removes these spikes (not shown in the figure), which therefore correspond to the Fc portion of the antibody; the F(ab′)₂ pieces remain cross-linked by antigen. The interpretation of some of the complexes is shown in the diagrams. The angle between the arms of the antibody molecules varies. In the triangular forms, this angle is 60°, whereas it is 90° in the square forms, showing that the connections between the arms are flexible. Photograph courtesy of N.M. Green.

Summary.

The IgG antibody molecule is made up of four polypeptide chains, comprising two identical light chains and two identical heavy chains, and can be thought of as forming a flexible Y-shaped structure. Each of the four chains has a variable (V) region at its amino terminus, which contributes to the antigen-binding site, and a constant (C) region. The light chains are bound to the heavy chains by many noncovalent interactions and by disulfide bonds, and the V regions of the heavy and light chains pair in each arm of the Y to generate two identical antigen-binding sites, which lie at the tips of the arms of the Y. The possession of two antigen-binding sites allows antibody molecules to cross-link antigens and to bind them much more stably and with higher avidity. The trunk of the Y, also called the Fc fragment, is composed of the carboxy-terminal domains of the heavy chains, and it is these domains that determine the antibody's isotype. Joining the arms of the Y to the trunk are the flexible hinge regions. The Fc fragment and hinge regions differ in antibodies of different isotypes.

Different isotypes have different properties and therefore differ in their interactions with effector molecules and cell types. However, the overall organization of the domains is similar in all isotypes.

The interaction of the antibody molecule with specific antigen.

In this part of the chapter we look at the antigen-binding site of an immunoglobulin molecule in more detail. We discuss the different ways in which antigens can bind to antibody, and address the question of how variation in the sequences of the antibody V domains determines the specificity for antigen.

4-6   Localized regions of hypervariable sequence form the antigen-binding site.

The V regions of any given antibody molecule differ from those of every other. Sequence variability is not, however, distributed evenly throughout the V region but is concentrated in certain segments, as is clearly seen in a **variability plot** (Fig. 4.6), in which the amino acid sequences of many different antibody V regions are compared. Three particularly variable segments can be identified in both the V_H and V_L domains. They are designated **hypervariable regions** and are denoted HV1, HV2, and HV3. In the heavy chains they are located at residues 30 to 36, 49 to 65, and 95 to 103, respectively, while in the light chains they are located at residues 28 to 35, 49 to 59, and 92 to 103, respectively. The most variable part of the domain is in the HV3 region. The regions between the hypervariable regions comprise the rest of the V domain; they show less variability and are termed the **framework regions**. There are four such regions in each V domain, designated FR1, FR2, FR3, and FR4.

The framework regions form the β sheets that provide the structural framework of the immunoglobulin domain. The hypervariable sequences correspond to three loops and are positioned near one another in the folded domain at the outer edge of the β sandwich (Fig. 4.7). Thus, not only is diversity concentrated in particular parts of the V domain sequence, but it is also localized to a particular

Fig. 4.6 There are discrete regions of hypervariability in V domains. The hypervariability regions of both the heavy and the light chain contribute to antigen binding of an antibody molecule. A variability plot derived from comparison of the amino acid sequences of several dozen heavy-chain and light-chain V domains is shown. At each amino acid position, the degree of variability is the ratio of the number of different amino acids seen in all of the sequences together to the frequency of the most common amino acid. Three hypervariable regions (HV1, HV2, and HV3) are indicated in red. They are flanked by less variable framework regions (FR1, FR2, FR3, and FR4, shown in blue or yellow).

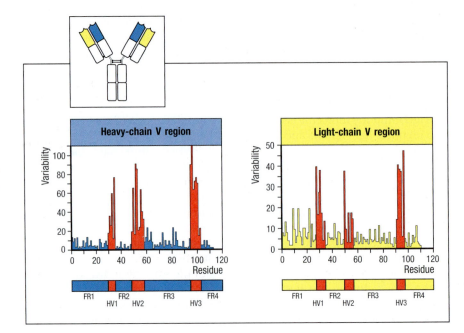

Fig. 4.7 The hypervariable regions lie in discrete loops of the folded structure. First panel: the hypervariable regions (red) are positioned on the structure of a map of the coding region of the V domain. Second panel: when shown as a flattened ribbon diagram, hypervariable regions are seen to occur in loops (red) that join particular β strands. Third panel: in the folded structure of the V domain, these loops (red) are brought together to form antigen-binding regions. Fourth panel: in a complete antibody molecule, the pairing of a heavy chain and a light chain brings together the hypervariable loops from each chain to create a single hypervariable surface, which forms the antigen-binding site at the tip of each arm. Because they are complementary to the antigen surface, the hypervariable regions are also commonly known as the complementarity-determining regions (CDRs). C, carboxy terminus; N, amino terminus.

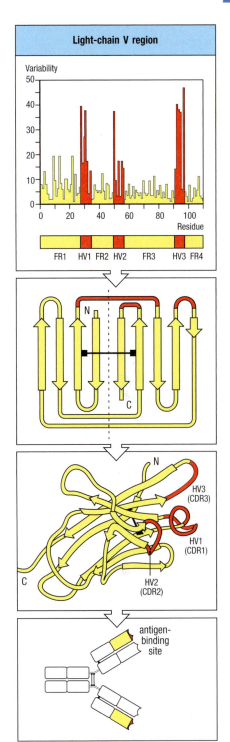

region on the surface of the molecule. When the V_H and V_L immunoglobulin domains are paired in the antibody molecule, the three hypervariable loops from each domain are brought together, creating a single hypervariable site at the tip of each arm of the molecule. This is the **antigen-binding site**, or **antibody-combining site**, which determines the antigen specificity of the antibody. These six hypervariable loops are more commonly termed the **complementarity-determining regions**, or **CDRs**, because the surface they form is complementary to that of the antigen they bind. There are three CDRs from each of the heavy and light chains, namely, CDR1, CDR2, and CDR3. In most cases, CDRs from both VH and VL domains contribute to the antigen-binding site; thus it is the combination of the heavy and the light chain that usually determines the final antigen specificity (see Fig. 4.6). However, there are some Fab crystal structures that show antigen interaction with just the heavy chain; for example, in one influenza Fab, antigen interaction involves binding mostly to the VH CDR3, and only minor contacts with other CDRs. Thus, one way in which the immune system is able to generate antibodies of different specificities is by generating different combinations of heavy-chain and light-chain V regions. This is known as combinatorial diversity; we will encounter a second form of **combinatorial diversity** in Chapter 5 when we consider how the genes encoding the heavy-chain and light-chain V regions are created from smaller segments of DNA during the development of B cells in the bone marrow.

4-7   Antibodies bind antigens via contacts in CDRs that are complementary to the size and shape of the antigen.

In early investigations of antigen binding to antibodies, the only available sources of large quantities of a single type of antibody molecule were tumors of antibody-secreting cells. The antigen specificities of these antibodies were unknown, and therefore many compounds had to be screened to identify ligands that could be used to study antigen binding. In general, the substances found to bind to these antibodies were haptens (see Section 4-5) such as phosphocholine or vitamin K_1. Structural analysis of complexes of antibodies with their hapten ligands provided the first direct evidence that the hypervariable regions form the antigen-binding site, and demonstrated the structural basis of specificity for the hapten. Subsequently, with the discovery of methods of generating **monoclonal antibodies** (see Appendix I, Section A-7), it became possible to make large amounts of pure antibody specific for a given antigen. This has provided a more general picture of how antibodies interact with their antigens, confirming and extending the view of antibody–antigen interactions derived from the study of haptens.

The surface of the antibody molecule formed by the juxtaposition of the CDRs of the heavy and light chains is the site to which an antigen binds. The amino acid sequences of the CDRs are different in different antibodies, and so too are the shapes and properties of the surfaces created by these CDRs. As a general principle, antibodies bind ligands whose surfaces are complementary to that of the antigen-binding site. A small antigen, such as a hapten or a short

Fig. 4.8 Antigens can bind in pockets, or grooves, or on extended surfaces in the binding sites of antibodies. The panels in the top row show schematic representations of the different types of binding sites in a Fab fragment of an antibody: first panel, pocket; second panel, groove; third panel, extended surface; and fourth panel, protruding surface. Below are examples of each type. Panel a: the top image shows the molecular surface of the interaction of a small hapten with the complementarity-determining regions (CDRs) of a Fab fragment as viewed looking into the antigen-binding site. The ferrocene hapten, shown in red, is bound into the antigen-binding pocket (yellow). In the bottom image (and in those of panels b, c, and d), the molecule has been rotated by about 90° to give a side-on view of the binding site. Panel b: in a complex of an antibody with a peptide from the human immunodeficiency virus (HIV), the peptide (red) binds along a groove (yellow) formed between the heavy-chain and light-chain V domains. Panel c: shown is a complex between hen egg-white lysozyme and the Fab fragment of its corresponding antibody (HyHel5). The surface on the antibody that comes into contact with the lysozyme is colored yellow. All six CDRs of the antigen-binding site are involved in the binding. Panel d: an antibody molecule against the HIV gp120 antigen has an elongated CDR3 loop (arrow) that protrudes into a recess on the side of the antigen. The structure of the complex between this antibody and gp120 has been solved, and in this case only the heavy chain interacts with gp120. Structures courtesy of R.L. Stanfield and I.A. Wilson.

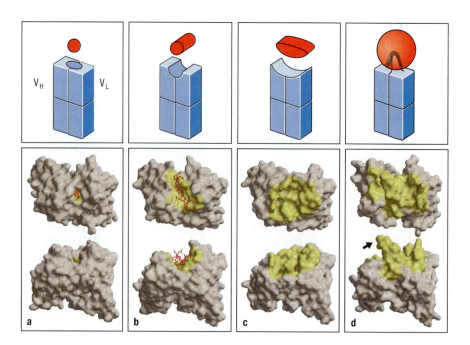

peptide, generally binds in a pocket or groove lying between the heavy-chain and light-chain V domains (**Fig. 4.8a** and b). Some antigens, such as proteins, can be the same size as, or larger than, the antibody itself. In these cases, the interface between antigen and antibody is often an extended surface that involves all the CDRs and, in some cases, part of the framework region as well (see Fig. 4.8c). This surface need not be concave but can be flat, undulating, or even convex. In some cases, antibody molecules with elongated CDR3 loops can protrude a 'finger' into recesses in the surface of the antigen, as shown in Fig. 4.8d, where an antibody binding to the HIV gp120 antigen projects a long loop into its target.

4-8   Antibodies bind to conformational shapes on the surfaces of antigens using a variety of noncovalent forces.

The biological function of antibodies is to bind to pathogens and their products, and to facilitate their removal from the body. An antibody generally recognizes only a small region on the surface of a large molecule such as a polysaccharide or protein. The structure recognized by an antibody is called an **antigenic determinant** or **epitope**. Some of the most important pathogens have polysaccharide coats, and antibodies that recognize epitopes formed by the sugar subunits of these molecules are essential in providing immune protection against such pathogens. In many cases, however, the antigens that provoke an immune response are proteins. For example, many protective antibodies against viruses recognize viral coat proteins. In all such cases, the structures recognized by the antibody are located on the surface of the protein. Such sites are likely to be composed of amino acids from different parts of the polypeptide chain that have been brought together by protein folding. Antigenic determinants of this kind are known as **conformational** or **discontinuous epitopes** because the structure recognized is composed of segments of the protein that are discontinuous in the amino acid sequence of the antigen but are brought together in the three-dimensional structure. In contrast, an epitope composed of a single segment of polypeptide chain is termed a **continuous** or **linear epitope**. Although most antibodies raised against intact, fully folded proteins recognize discontinuous epitopes, some will bind to peptide fragments of the protein. Conversely, antibodies raised against peptides of a protein or against synthetic peptides corresponding to part of its sequence are occasionally found to bind to the natural folded protein. This makes it possible, in some cases, to use synthetic peptides in vaccines that aim to raise antibodies against a pathogen's protein.

The interaction between an antibody and its antigen can be disrupted by high salt concentrations, by extremes of pH, by detergents, and sometimes by competition with high concentrations of the pure epitope itself. The binding is therefore a reversible noncovalent interaction. The forces, or bonds, involved in these noncovalent interactions are outlined in **Fig. 4.9**. **Electrostatic interactions** occur between charged amino acid side chains, as in salt bridges. Most antibody–antigen interactions involve at least one electrostatic interaction. Interactions also occur between electric dipoles, as in hydrogen bonds, or can involve short-range van der Waals forces. High salt concentrations and extremes of pH disrupt antigen–antibody binding by weakening electrostatic interactions and/or hydrogen bonds. This principle is employed in the purification of antigens by using affinity columns of immobilized antibodies (or in the purification of antibody by using antigens in a like manner) (see Appendix I, Section A-3). **Hydrophobic interactions** occur when two hydrophobic surfaces come together to exclude water. The strength of a hydrophobic interaction is proportional to the surface area that is hidden from water, and for some antigens, hydrophobic interactions probably account for most of the binding energy. In some cases, water molecules are trapped in pockets in the interface between antigen and antibody. These trapped water molecules, especially those between polar amino acid residues, may also contribute to binding and hence to the specificity of the antibody.

The contribution of each of these forces to the overall interaction depends on the particular antibody and antigen involved. A striking difference between antibody interactions with protein antigens and most other natural protein–protein interactions is that antibodies often have many aromatic amino acids

Noncovalent forces	Origin	
Electrostatic forces	Attraction between opposite charges	
Hydrogen bonds	Hydrogen shared between electronegative atoms (N, O)	
Van der Waals forces	Fluctuations in electron clouds around molecules polarize neighboring atoms oppositely	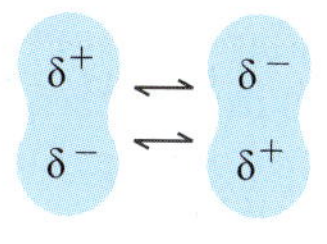
Hydrophobic forces	Hydrophobic groups interact unfavorably with water and tend to pack together to exclude water molecules. The attraction also involves van der Waals forces	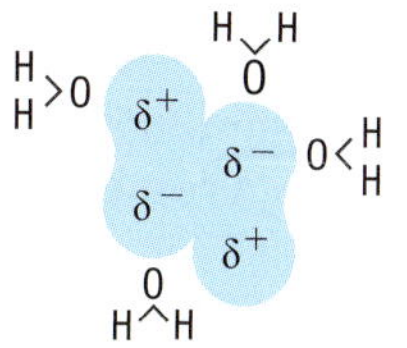
Cation-pi interaction	Non-covalent interaction between a cation and an electron cloud of a nearby aromatic group	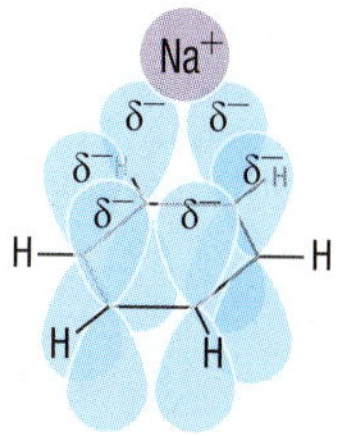

Fig. 4.9 The noncovalent forces that hold together the antigen:antibody complex. Partial charges found in electric dipoles are shown as δ^+ or δ^-. Electrostatic forces diminish as the inverse square of the distance separating the charges, whereas van der Waals forces, which are more numerous in most antigen–antibody contacts, fall off as the sixth power of the separation and therefore operate only over very short ranges. Covalent bonds never occur between antigens and naturally produced antibodies.

in their antigen-binding sites. These amino acids participate mainly in van der Waals and hydrophobic interactions, and sometimes in hydrogen bonds and **pi-cation interactions**. Tyrosine, for example, can take part in both hydrogen bonding and hydrophobic interactions; it is therefore particularly suitable for providing diversity in antigen recognition and is overrepresented in antigen-binding sites. In general, the hydrophobic and van der Waals forces operate over very short ranges and serve to pull together two surfaces that are complementary in shape: hills on one surface must fit into valleys on the other for good binding to occur. In contrast, electrostatic interactions between charged side chains, and hydrogen bonds bridging oxygen and/or nitrogen atoms, accommodate more specific chemical interactions while strengthening the interaction overall. The side chains of aromatic amino acids such as tyrosine can interact noncovalently through their pi-electron system with nearby cations, including nitrogen-containing side chains that may be in a protonated cationic state.

4-9 Antibody interaction with intact antigens is influenced by steric constraints.

An example of an antibody–antigen interaction involving a specific amino acid in the antigen can be seen in the complex of hen egg-white lysozyme with the antibody D1.3 (**Fig. 4.10**). In this structure, strong hydrogen bonds are formed between the antibody and a particular glutamine in the lysozyme molecule that protrudes between the V_H and V_L domains. Lysozymes from partridge and turkey have another amino acid in place of the glutamine and do not bind to this antibody. In the high-affinity complex of hen egg-white lysozyme with another antibody, HyHel5 (see Fig. 4.8c), two salt bridges between two basic arginines on the surface of the lysozyme interact with two glutamic acids, one each from the V_H CDR1 and CDR2 loops. Lysozymes that lack one of the two arginine residues show a 1000-fold decrease in affinity for HyHel5. Overall surface complementarity must have an important role in antigen–antibody interactions, but in most antibodies that have been studied at this level of detail, only a few residues make a major contribution to the binding energy and hence to the final specificity of the antibody. Although many antibodies naturally bind their ligands with high affinity, in the nanomolar range, genetic engineering by site-directed mutagenesis can tailor an antibody to bind even more strongly to its epitope.

Even when antibodies have high affinity for antigens on a larger structure, such as an intact viral particle, antibody binding may be prevented by their particular arrangement. For example, the intact West Nile virion is built from an icosahedral scaffold that has 90 homodimers of a membrane-anchored envelope glycoprotein, E, which has three domains, DI, DII, and DIII. The DIII domain has four polypeptide loops that protrude outward from the viral particle. A neutralizing antibody against West Nile virus, E16, recognizes these loops of DIII, as shown in **Fig. 4.11**. In theory, there should be 180 possible antigen-binding sites for the E16 antibody on the West Nile viral particle. However, a combination of crystallographic and electron micrographic studies show that even with an excess of the E16 Fab fragment, only about 120 of the total 180 DIII domains of E are able to bind E16 Fab fragment (see Fig. 4.11).

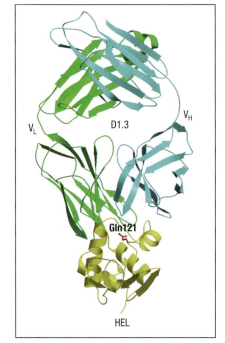

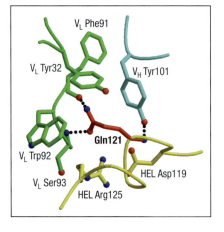

Fig. 4.10 The complex of lysozyme with the antibody D1.3. Top panel: The interaction of the Fab fragment of D1.3 with hen egg-white lysozyme (HEL) is shown. HEL is depicted in yellow, the heavy chain (V_H) in turquoise, and the light chain (V_L) in green. Bottom panel: A glutamine residue (Gln121) that protrudes from HEL (yellow) extends its side chain (shown in red) between the V_L (green) and V_H (turquoise) domains of the antigen-binding site and makes hydrogen bonds with the hydroxyl group (red dots) of the indicated amino acids of both domains. These hydrogen bonds are important to the antigen–antibody binding. Courtesy of R. Mariuzza and R.J. Poljak.

Fig. 4.11 Steric hindrance occludes the binding of antibody to native antigen in the intact West Nile viral particle. Top panel: the monoclonal antibody E16 recognizes DIII, one of the three structural domains in the West Nile virus glycoprotein E. Shown is a crystal structure of the E16 Fab bound to the DIII epitope. Bottom left panel: a computer model was used to dock E16 Fab to the mature West Nile virion. E16 Fabs were able to bind 120 of the 180 DIII epitopes. Sixty of the five-fold clustered DIII epitopes are sterically hindered by the binding of Fab to four nearby DIII epitopes. An example of an occluded epitope is the blue area indicated by the arrow. Bottom right panel: cryogenic electron microscopic reconstruction of saturating E16 Fab bound to West Nile virion confirmed the predicted steric hindrance. The vertices of the triangle shown in the figure indicate the icosahedral symmetry axes.

This appears to result from steric hindrance, with the presence of one Fab blocking the ability of another Fab to bind to some nearby E protein sites. Presumably, such steric hindrance would become more severe with intact antibody than is evident with the smaller Fab fragment. This study also showed that the Fab bound to the DIII region using only one of its antigen-binding arms, indicating that antibodies may not always bind to antigens with both antigen-binding sites, depending on the orientation of the antigens being recognized. These constraints will impact the ability of antibodies to neutralize their targets.

4-10 Some species generate antibodies with alternative structures.

Our focus in this chapter has been on the structure of antibodies produced by humans, which is generally similar in most mammalian species, including mice, an important model system for immunology research. However, some mammals have the ability to produce an alternative form of antibody that is based on the ability of a single V_H domain to interact with antigen in the absence of a V_L domain (Fig. 4.12). It has been known for some time that the serum of camels contained abundant immunoglobulin-like material composed of heavy-chain dimers that lack associated light chains but retain the capacity to bind antigens. These antibodies are called **heavy-chain-only IgGs (hcIgGs)**. This property is shared by other camelids, including llamas and alpacas. These species have retained the genes for the immunoglobulin light chains, and some IgG-like material in their sera remains associated with light chains, and so it is unclear what led to this particular adaptation during their evolution. In camelids, the ability to produce hcIgGs arises from mutations that allow the alternative splicing of the heavy-chain mRNA, with loss of the C_H1 exon and thus the joining of the V_H directly to the C_H2 domain in the protein. Other mutations stabilize this structure in the absence of V_L domains.

Cartilaginous fish, in particular the shark, also have an antibody molecule that differs substantially from human or murine antibodies (see Fig. 4.12). Like camelids, the shark also has genes encoding both immunoglobulin heavy and light chains, and does produce immunoglobulins containing both

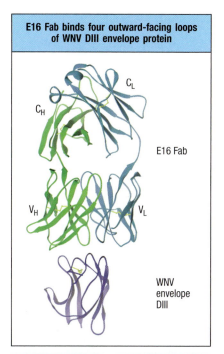

E16 Fab binds four outward-facing loops of WNV DIII envelope protein

C_L
C_H
E16 Fab
V_H
V_L
WNV envelope DIII

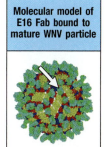

Molecular model of E16 Fab bound to mature WNV particle

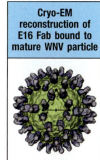

Cryo-EM reconstruction of E16 Fab bound to mature WNV particle

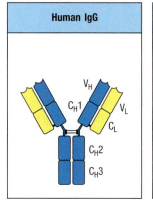

Human IgG

V_H
C_H1
V_L
C_L
C_H2
C_H3

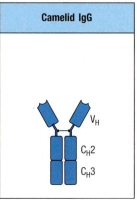

Camelid IgG

V_H
C_H2
C_H3

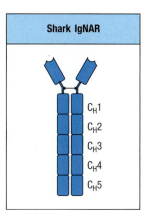

Shark IgNAR

C_H1
C_H2
C_H3
C_H4
C_H5

Fig. 4.12 Camelid and shark antibody can consist of heavy chain only. In the camelid heavy-chain-only antibody, a splicing event of the mature heavy chain can delete the exon encoding the C_H1 region and thereby create an in-frame hinge region linking the V_H1 to the C_H2 region. In the shark, the heavy-chain-only Ig molecule retains the C_H1 region, suggesting that this form of antibody may predate the evolution of light chains. For both, the repertoire of antigen-binding sites involves extensive variations in long CDR3 regions of the V_H domain relative to other types of antibody.

heavy and light chains. But sharks also produce an **immunoglobulin new antigen receptor (IgNAR)**, with heavy-chain-only antibody in which the V_H is spliced to the C_H1 exon, rather than the C_H1 exon being spliced out as in camelids. These differences suggest that hcIgG production by camelids and sharks represents an event of convergent evolution. The ability of camelid V_H domains to interact efficiently with antigens is the basis for producing so-called **single-chain antibody**. The simplification of using only a single domain for antigen recognition has prompted recent interest in single-chain monoclonal antibodies as an alternative to standard monoclonal antibodies, which we will discuss more in Chapter 16.

Summary.

X-ray crystallographic analyses of antigen:antibody complexes have shown that the hypervariable loops (complementarity-determining regions, CDRs) of immunoglobulin V regions determine the binding specificity of an antibody. Contact between an antibody molecule and a protein antigen usually occurs over a broad area of the antibody surface that is complementary to the surface recognized on the antigen. Electrostatic interactions, hydrogen bonds, van der Waals forces, and hydrophobic and pi-cation interactions can all contribute to binding. Depending on the size of the antigen, amino acid side chains in most or all of the CDRs make contact with antigen and determine both the specificity and the affinity of the interaction. Other parts of the V region normally play little part in the direct contact with the antigen, but they provide a stable structural framework for the CDRs and help to determine their position and conformation. Antibodies raised against intact proteins usually bind to the surface of the protein and make contact with residues that are discontinuous in the primary structure of the molecule; they may, however, occasionally bind peptide fragments of the protein, and antibodies raised against peptides derived from a protein can sometimes be used to detect the native protein molecule. Peptides binding to antibodies usually bind in a cleft or pocket between the V regions of the heavy and light chains, where they make specific contact with some, but not necessarily all, of the CDRs. This is also the usual mode of binding for carbohydrate antigens and small molecules such as haptens.

Antigen recognition by T cells.

In contrast to the immunoglobulins, which interact with pathogens and their toxic products in the extracellular spaces of the body, T cells recognize foreign antigens only when they are displayed on the surface of the body's own cells. These antigens can derive from pathogens such as viruses or intracellular bacteria, which replicate within cells, or from pathogens or their products that have been internalized by endocytosis from the extracellular fluid.

T cells detect the presence of an intracellular pathogen because the infected cells display peptide fragments of the pathogen's proteins on their surface. These foreign peptides are delivered to the cell surface by specialized host-cell glycoproteins—the MHC molecules. These are encoded in a large cluster of genes that were first identified by their powerful effects on the immune response to transplanted tissues. For that reason, the gene complex was called the major histocompatibility complex (MHC), and the peptide-binding glycoproteins are known as MHC molecules. The recognition of antigen as a small peptide fragment bound to an MHC molecule and displayed at the cell surface is one of the most distinctive features of T cells, and will be the focus of this part of the chapter. How the peptide fragments of antigen are generated and become associated with MHC molecules will be described in Chapter 6.

We describe here the structure and properties of the T-cell receptor (TCR). As might be expected from the T-cell receptors' function as highly variable antigen-recognition structures, the genes for TCRs are closely related to those for immunoglobulins. There are, however, important differences between T-cell receptors and immunoglobulins, and these differences reflect the special features of antigen recognition by T cells.

4-11 The TCRα:β heterodimer is very similar to a Fab fragment of immunoglobulin.

T-cell receptors were first identified by using monoclonal antibodies that bound to a single cloned T-cell line: such antibodies either specifically inhibit antigen recognition by the clone or specifically activate it by mimicking the antigen (see Appendix I, Section A-20). These **clonotypic** antibodies were then used to show that each T cell bears about 30,000 identical antigen receptors on its surface, each receptor consisting of two different polypeptide chains, termed the **T-cell receptor α (TCRα)** and **β (TCRβ)** chains. Each chain of the **α:β heterodimer** is composed of two Ig domains, and the two chains are linked by a disulfide bond, similar to the structure of the Fab fragment of an immunoglobulin molecule (Fig. 4.13). α:β heterodimers account for antigen recognition by most T cells. A minority of T cells bear an alternative, but structurally similar, receptor made up of a different pair of polypeptide chains designated γ and δ. The **γ:δ T-cell receptors** seem to have different antigen-recognition properties from the **α:β T-cell receptors**, and the functions of γ:δ T cells in immune responses are still being clarified as the various ligands they recognize are identified (see Section 6-20). In the rest of this chapter and elsewhere in the book we use the term T-cell receptor to mean the α:β receptor, except where specified otherwise. Both types of T-cell receptors differ from the membrane-bound immunoglobulin that serves as the B-cell receptor in two main ways. A T-cell receptor has only one antigen-binding site, whereas a B-cell receptor has two, and T-cell receptors are never secreted, whereas immunoglobulins can be secreted as antibodies.

Further insights into the structure and function of the α:β T-cell receptor came from studies of cloned cDNA encoding the receptor chains. The amino acid sequences predicted from the cDNA showed that both chains of the T-cell receptor have an amino-terminal variable (V) region with sequence homology to an immunoglobulin V domain, a constant (C) region with homology to an immunoglobulin C domain, and a short stalk segment containing a cysteine residue that forms the interchain disulfide bond (Fig. 4.14). Each chain spans the lipid bilayer by a hydrophobic transmembrane domain, and ends in a short cytoplasmic tail. These close similarities of T-cell receptor chains to the heavy and light immunoglobulin chains first enabled prediction of the structural resemblance of the T-cell receptor heterodimer to a Fab fragment of immunoglobulin.

The three-dimensional structure of the T-cell receptor determined by X-ray crystallography in Fig. 4.15a shows that T-cell receptor chains fold in much the same way as the regions comprising the Fab fragment in Fig. 4.1a.

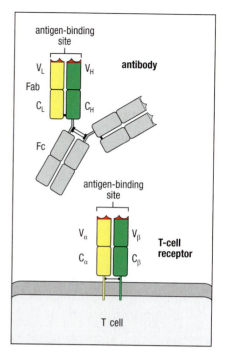

Fig. 4.13 The T-cell receptor resembles a membrane-bound Fab fragment. The Fab fragment of an antibody molecule is a disulfide-linked heterodimer, each chain of which contains one immunoglobulin C domain and one V domain; the juxtaposition of the V domains forms the antigen-binding site (see Section 4-6). The T-cell receptor is also a disulfide-linked heterodimer, with each chain containing an immunoglobulin C-like domain and an immunoglobulin V-like domain. As in the Fab fragment, the juxtaposition of the V domains forms the site for antigen recognition.

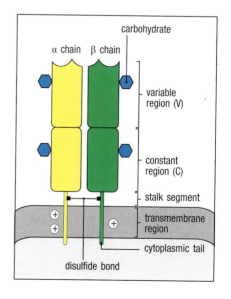

Fig. 4.14 Structure of the T-cell receptor. The T-cell receptor heterodimer is composed of two transmembrane glycoprotein chains, α and β. The extracellular portion of each chain consists of two domains, resembling immunoglobulin V and C domains, respectively. Both chains have carbohydrate side chains attached to each domain. A short stalk segment, analogous to an immunoglobulin hinge region, connects the Ig-like domains to the membrane and contains the cysteine residue that forms the interchain disulfide bond. The transmembrane helices of both chains are unusual in containing positively charged (basic) residues within the hydrophobic transmembrane segment. The α chain carries two such residues; the β chain has one.

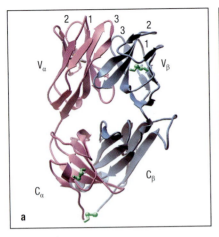

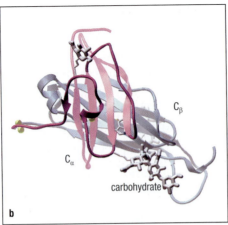

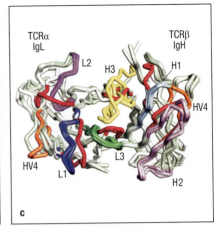

Fig. 4.15 The crystal structure of an α:β T-cell receptor resolved at 0.25 nm. In panels a and b, the α chain is shown in pink and the β chain in blue. Disulfide bonds are shown in green. In panel a, the T-cell receptor is viewed from the side as it would sit on a cell surface, with the CDR loops that form the antigen-binding site (labeled 1, 2, and 3) arrayed across its relatively flat top. In panel b, the C_α and C_β domains are shown. The C_α domain does not fold into a typical Ig-like domain; the face of the domain away from the C_β domain is mainly composed of irregular strands of polypeptide rather than β sheet. The intramolecular disulfide bond (far left) joins a β strand to this segment of α helix. The interaction between the C_α and C_β domains is assisted by carbohydrate (colored gray and labeled), with a sugar group from the C_α domain making hydrogen bonds to the C_β domain. In panel c, the T-cell receptor is shown aligned with the antigen-binding sites from three different antibodies. This view is looking down into the binding site. The V_α domain of the T-cell receptor is aligned with the V_L domains of the antigen-binding sites of the antibodies, and the V_β domain is aligned with the V_H domains. The CDRs of the T-cell receptor and immunoglobulin molecules are colored, with CDRs 1, 2, and 3 of the TCR shown in red and the HV4 loop in orange. For the immunoglobulin V domains, the CDR1 loops of the heavy chain (H1) and light chain (L1) are shown in light and dark blue, respectively, and the CDR2 loops (H2, L2) in light and dark purple, respectively. The heavy-chain CDR3 loops (H3) are in yellow; the light-chain CDR3s (L3) are in bright green. The HV4 loops of the TCR (orange) have no hypervariable counterparts in immunoglobulins. Model structures courtesy of I.A. Wilson.

There are, however, some distinct structural differences between T-cell receptors and Fab fragments. The most striking is in the C_α domain, where the fold is unlike that of any other Ig-like domain. The half of the C_α domain that is juxtaposed with the C_β domain forms a β sheet similar to that found in other Ig-like domains, but the other half of the domain is formed of loosely packed strands and a short segment of α helix (see Fig. 4.15b). In a C_α domain the intramolecular disulfide bond, which in Ig-like domains normally joins two β strands, joins a β strand to this segment of α helix.

There are also differences in the way in which the domains interact. The interface between the V and C domains of both T-cell receptor chains is more extensive than in most antibodies. The interaction between the C_α and C_β domains is distinctive, as it might be assisted by carbohydrates, with a sugar group from the C_α domain making a number of hydrogen bonds to the C_β domain (see Fig. 4.15b). Finally, a comparison of the variable binding sites shows that, although the CDR loops align fairly closely with those of antibody molecules, there is some relative displacement (see Fig. 4.15c). This is particularly marked in the V_α CDR2 loop, which is oriented at roughly right angles to the equivalent loop in antibody V domains, as a result of a shift in the β strand that anchors one end of the loop from one face of the domain to the other. A strand displacement also causes a change in the orientation of the V_β CDR2 loop in some V_β domains. These differences with antibodies influence the ability of the T-cell receptor to recognize their specific ligands, as we will discuss in the next section. In addition to the three hypervariable regions shared with immunoglobulins, the T-cell receptor has a fourth hypervariability region, HV4, in both of its chains (see Fig. 4.15c). These regions occur away from the antigen-binding face of the receptor, and have been implicated in other functions of the TCR, such as superantigen binding, which we will describe in Section 6-14.

4-12   A T-cell receptor recognizes antigen in the form of a complex of a foreign peptide bound to an MHC molecule.

Antigen recognition by T-cell receptors clearly differs from recognition by B-cell receptors and antibodies. The immunoglobulin on B cells binds directly to the intact antigen, and, as discussed in Section 4-8, antibodies typically bind to the surface of protein antigens, contacting amino acids that are discontinuous in the primary structure but are brought together in the folded protein. In contrast, αβ T cells respond to short, continuous amino acid sequences. As we described in Section 1-10, these peptide sequences are often buried within the native structure of the protein. Thus, antigens cannot be recognized directly by T-cell receptors unless the protein is unfolded and processed into peptide fragments (Fig. 4.16), and then presented by an MHC molecule (see Fig. 1.15). We will return to the issue of how this process occurs in Chapter 6.

The nature of the antigen recognized by T cells became clear with the realization that the peptides that stimulate T cells are recognized only when bound to an MHC molecule. The ligand recognized by the T cell is thus a complex of peptide and MHC molecule. The evidence for involvement of the MHC in T-cell recognition of antigen was at first indirect, but it has been proven conclusively by stimulating T cells with purified peptide:MHC complexes. The T-cell receptor interacts with this ligand by making contacts with both the MHC molecule and the antigen peptide.

4-13   There are two classes of MHC molecules with distinct subunit compositions but similar three-dimensional structures.

There are two classes of MHC molecules—**MHC class I** and **MHC class II**—and they differ in both their structure and their expression pattern in the tissues of the body. As shown in Figs. 4.17 and 4.18, MHC class I and MHC class II molecules are closely related in overall structure but differ in their subunit composition. In both classes, the two paired protein domains nearest to the membrane resemble immunoglobulin domains, whereas the two domains furthest away from the membrane fold together to create a long cleft, or groove, which is the site at which a peptide binds. Purified peptide:MHC class I and peptide:MHC class II complexes have been characterized structurally, allowing us to describe in detail both the MHC molecules themselves and the way in which they bind peptides.

MHC class I molecules (see Fig. 4.17) consist of two polypeptide chains. One chain, the α chain, is encoded in the MHC (on chromosome 6 in humans) and is noncovalently associated with a smaller chain, **β₂-microglobulin**, which is encoded on a different chromosome—chromosome 15 in humans. Only the class I α chain spans the membrane. The complete MHC class I molecule has four domains, three formed from the MHC-encoded α chain, and one contributed by β₂-microglobulin. The $α_3$ domain and β₂-microglobulin closely resemble Ig-like domains in their folded structure. The folded $α_1$ and $α_2$

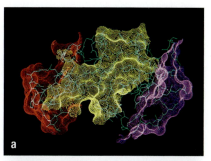

a

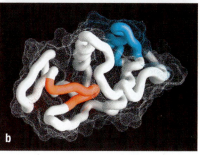

b

Fig. 4.16 Differences in the recognition of hen egg-white lysozyme by immunoglobulins and T-cell receptors. Antibodies can be shown by X-ray crystallography to bind epitopes on the surface of proteins, as shown in panel a, where the epitopes for three antibodies are shown in different colors on the surface of hen egg-white lysozyme (see also Fig. 4.10). In contrast, the epitopes recognized by T-cell receptors need not lie on the surface of the molecule, because the T-cell receptor recognizes not the antigenic protein itself but a peptide fragment of the protein. The peptides corresponding to two T-cell epitopes of lysozyme are shown in panel b. One epitope, shown in blue, lies on the surface of the protein, but a second, shown in red, lies mostly within the core and is inaccessible in the folded protein. This implies that T-cell receptors do not recognize their epitopes in the context of the native protein. Panel a courtesy of S. Sheriff.

domains form the walls of a cleft on the surface of the molecule; because this is where the peptide binds, this part of the MHC molecule is known as the **peptide-binding cleft** or **peptide-binding groove**. The MHC molecules are highly polymorphic, and the major differences between the different allelic

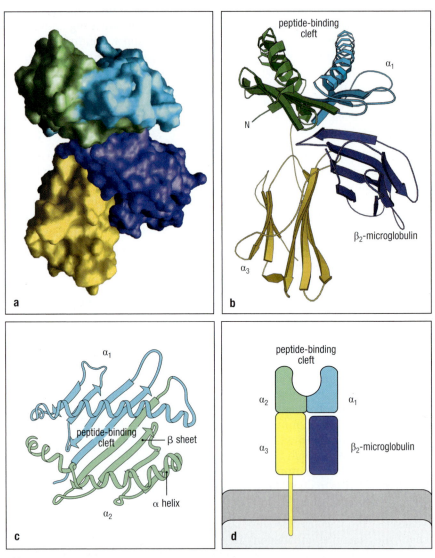

Fig. 4.17 The structure of an MHC class I molecule determined by X-ray crystallography. Panel a shows a computer graphic representation of a human MHC class I molecule, HLA-A2, which has been cleaved from the cell surface by the enzyme papain. The surface of the molecule is shown, colored according to the domains shown in panels b–d and described below. Panels b and c show a ribbon diagram of that structure. Shown schematically in panel d, the MHC class I molecule is a heterodimer of a membrane-spanning α chain (molecular weight 43 kDa) bound noncovalently to β_2-microglobulin (12 kDa), which does not span the membrane. The α chain folds into three domains: α_1, α_2, and α_3. The α_3 domain and β_2-microglobulin show similarities in amino acid sequence to immunoglobulin C domains and have similar folded structures, whereas the α_1 and α_2 domains are part of the same polypeptide and fold together into a single structure consisting of two separated α helices lying on a sheet of eight antiparallel β strands. The folding of the α_1 and α_2 domains creates a long cleft or groove, which is the site at which peptide antigens bind to the MHC molecules. For class I molecules, this groove is open at only one end. The transmembrane region and the short stretch of peptide that connects the external domains to the cell surface are not seen in panels a and b because they have been removed by the digestion with papain. As can be seen in panel c, looking down on the molecule from above, the sides of the cleft are formed from the inner faces of the two α helices; the β pleated sheet formed by the pairing of the α_1 and α_2 domains creates the floor of the cleft.

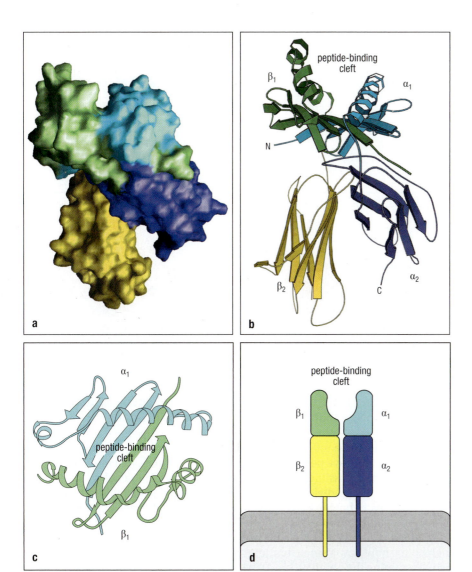

Fig. 4.18 MHC class II molecules resemble MHC class I molecules in overall structure. The MHC class II molecule is composed of two transmembrane glycoprotein chains, α (34 kDa) and β (29 kDa), as shown schematically in panel d. Each chain has two domains, and the two chains together form a compact four-domain structure similar to that of the MHC class I molecule (compare with panel d of Fig. 4.17). Panel a shows a computer graphic representation of the surface of the MHC class II molecule, in this case the human protein HLA-DR1, and panel b shows the equivalent ribbon diagram. N, amino terminus; C, carboxy terminus. The α_2 and β_2 domains, like the α_3 and β_2-microglobulin domains of the MHC class I molecule, have amino acid sequence and structural similarities to immunoglobulin C domains; in the MHC class II molecule the two domains forming the peptide-binding cleft are contributed by different chains and are therefore not joined by a covalent bond (see panels c and d). Another important difference, not apparent in this diagram, is that the peptide-binding groove of the MHC class II molecule is open at both ends.

forms are located in the peptide-binding cleft, influencing which peptides will bind and thus the specificity of the dual antigen presented to T cells. By contrast, β_2-microglobulin, which does not contribute directly to peptide binding, is not polymorphic.

An MHC class II molecule consists of a noncovalent complex of two chains, α and β, both of which span the membrane (see Fig. 4.18). The MHC class II α chain is a different protein from the class I α chain. The MHC class II α and β chains are both encoded within the MHC. The crystallographic structure of the MHC class II molecule shows that it is folded very much like the MHC class I molecule, but the peptide-binding cleft is formed by two domains from different chains, the α_1 and β_1 domains. The major differences lie at the ends of the peptide-binding cleft, which are more open in MHC class II than in MHC class I molecules. Consequently, the ends of a peptide bound to an MHC class I molecule are substantially buried within the molecule, whereas the ends of peptides bound to MHC class II molecules are not. In both MHC class I and class II molecules, bound peptides are sandwiched between the two α-helical segments of the MHC molecule (Fig. 4.19). The T-cell receptor interacts with this compound ligand, making contacts with both the MHC molecule and the peptide antigen. As in the case of MHC class I molecules, the sites of major polymorphism in MHC class II molecules are located in the peptide-binding cleft.

Fig. 4.19 MHC molecules bind peptides tightly within the cleft. When MHC molecules are crystallized with a single synthetic peptide antigen, the details of peptide binding are revealed. In MHC class I molecules (panels a and c), the peptide is bound in an elongated conformation with both ends tightly bound at either end of the cleft. In MHC class II molecules (panels b and d), the peptide is also bound in an elongated conformation but the ends of the peptide are not tightly bound and the peptide extends beyond the cleft. The upper surface of the peptide:MHC complex is recognized by T cells, and is composed of residues of the MHC molecule and the peptide. The amino acid side chains of the peptide insert into pockets in the peptide-binding groove of the MHC molecule; these pockets are lined with residues that are polymorphic within the MHC. In representations c and d, the surfaces of the different pockets for the different amino acids are depicted as areas of different colors. Structures courtesy of R.L. Stanfield and I.A. Wilson.

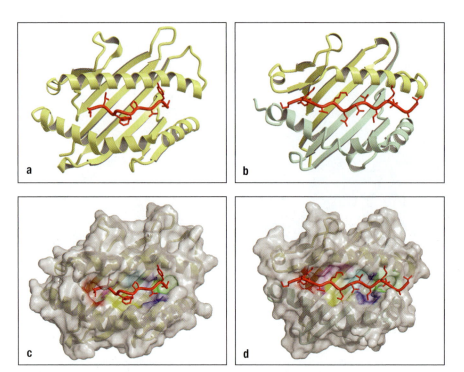

4-14 Peptides are stably bound to MHC molecules, and also serve to stabilize the MHC molecule on the cell surface.

An individual can be infected by a wide variety of pathogens, whose proteins will not generally have peptide sequences in common. For T cells to be able to detect the widest possible array of infections, the MHC molecules (both class I and class II) of an individual should be able to bind stably to many different peptides. This behavior is quite distinct from that of other peptide-binding receptors, such as those for peptide hormones, which usually bind only a single type of peptide. The crystal structures of peptide:MHC complexes have helped to show how a single binding site can bind a peptide with high affinity while retaining the ability to bind a wide variety of different peptides.

An important feature of the binding of peptides to MHC molecules is that the peptide is bound as an integral part of the MHC molecule's structure, and MHC molecules are unstable when peptides are not bound. This dependence on bound peptide applies to both MHC class I and MHC class II molecules. Stable peptide binding is important, because otherwise peptide exchanges occurring at the cell surface would prevent peptide:MHC complexes from being reliable indicators of infection or of uptake of a specific antigen. When MHC molecules are purified from cells, their stably bound peptides co-purify with them, and this fact has enabled the peptides bound by particular MHC molecules to be analyzed. Peptides are released from the MHC molecules by denaturing the complex in acid, and they are then purified and sequenced. Pure synthetic peptides can also be incorporated into empty MHC molecules and the structure of the complex determined, revealing details of the contacts between the MHC molecule and the peptide. From such studies a detailed picture of the binding interactions has been built up. We first discuss the peptide-binding properties of MHC class I molecules.

4-15 MHC class I molecules bind short peptides of 8–10 amino acids by both ends.

Binding of a peptide to an MHC class I molecule is stabilized at both ends of the peptide-binding cleft by contacts between atoms in the free amino and carboxy termini of the peptide and invariant sites that are found at each end

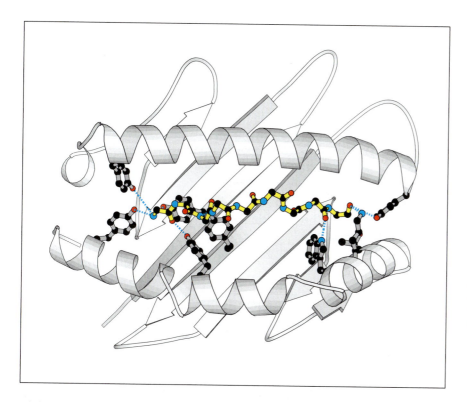

Fig. 4.20 Peptides are bound to MHC class I molecules by their ends.
MHC class I molecules interact with the backbone of a bound peptide (shown in yellow) through a series of hydrogen bonds and ionic interactions (shown as dotted blue lines) at each end of the peptide. The amino terminus of the peptide is to the left, the carboxy terminus to the right. Black circles are carbon atoms; red are oxygen; blue are nitrogen. The amino acid residues in the MHC molecule that form these bonds are common to all MHC class I molecules, and their side chains are shown in full (in gray) on a ribbon diagram of the MHC class I groove. A cluster of tyrosine residues common to all MHC class I molecules forms hydrogen bonds to the amino terminus of the bound peptide, while a second cluster of residues forms hydrogen bonds and ionic interactions with the peptide backbone at the carboxy terminus and with the carboxy terminus itself.

of the cleft in all MHC class I molecules (Fig. 4.20). These are thought to be the main stabilizing contacts for peptide:MHC class I complexes, because synthetic peptide analogs lacking terminal amino and carboxyl groups fail to bind stably to MHC class I molecules. Other residues in the peptide serve as additional anchors. Peptides that bind to MHC class I molecules are usually 8–10 amino acids long. Longer peptides are thought to bind, however, particularly if they can bind at their carboxy terminus, but they are subsequently shortened to 8–10 amino acids through cleavage by exopeptidases present in the endoplasmic reticulum, which is where MHC class I molecules bind peptides. The peptide lies in an elongated conformation along the cleft; variations in peptide length seem to be accommodated, in most cases, by a kinking in the peptide backbone. However, in some cases, length variation can also be accommodated in MHC class I molecules by allowing the peptide to extend out of the cleft at the carboxy terminus.

These interactions give MHC class I molecules a broad peptide-binding specificity. In addition, MHC molecules are highly polymorphic. As mentioned earlier, MHC genes are highly polymorphic and there are hundreds of different allelic variations of the MHC class I genes in the human population. Each individual carries only a small selection of these variants. The main differences between allelic MHC variants are found at certain sites in the peptide-binding cleft, resulting in different amino acids in key peptide-interaction sites. Because of this, different MHC variants preferentially bind different peptides. The peptides that can bind to a given MHC variant have the same or very similar amino acid residues at two or three particular positions along the peptide sequence. The amino acid side chains at these positions insert into pockets in the MHC molecule that are lined by the polymorphic amino acids. Because this binding anchors the peptide to the MHC molecule, the peptide residues involved are called the **anchor residues**, as illustrated in Fig. 4.21. Both the position and identity of these anchor residues can vary, depending on the particular MHC class I variant that is binding the peptide. However, most peptides that bind to MHC class I molecules have a hydrophobic (or sometimes basic) residue at the carboxy terminus that also serves to anchor the peptide in the groove. Whereas changing an anchor residue will in most cases prevent the peptide from binding, not every synthetic peptide of suitable length that

Fig. 4.21 Peptides bind to MHC molecules through structurally related anchor residues. Peptides eluted from two different MHC class I molecules are shown in the upper and lower panels, respectively. The anchor residues (green) differ for peptides that bind different allelic variants of MHC class I molecules but are similar for all peptides that bind to the same MHC molecule. The anchor residues that bind a particular MHC molecule need not be identical, but are always related: for example, phenylalanine (F) and tyrosine (Y) are both aromatic amino acids, whereas valine (V), leucine (L), and isoleucine (I) are all large hydrophobic amino acids. Peptides also bind to MHC class I molecules through their amino (blue) and carboxy (red) termini.

contains these anchor residues will bind the appropriate MHC class I molecule, and so the overall binding must also depend on the nature of the amino acids at other positions in the peptide. In some cases, particular amino acids are preferred in certain positions, whereas in others the presence of particular amino acids prevents binding. These additional amino acid positions are called 'secondary anchors.' These features of peptide binding enable an individual MHC class I molecule to bind a wide variety of different peptides, yet allow different MHC class I allelic variants to bind different sets of peptides. As we will see in Chapter 15, MHC polymorphisms also impact the binding of peptides derived from self-proteins and can influence the susceptibility of an individual to various autoimmune diseases.

4-16 The length of the peptides bound by MHC class II molecules is not constrained.

Like MHC class I molecules, MHC class II molecules that lack bound peptide are unstable. Peptide binding to MHC class II molecules has also been analyzed by elution of bound peptides and by X-ray crystallography, and differs in several ways from peptide binding to MHC class I molecules. Natural peptides that bind to MHC class II molecules are at least 13 amino acids long and can be much longer. The clusters of conserved residues that bind the two ends of a peptide in MHC class I molecules are not found in MHC class II molecules, and the ends of the peptide are not bound. Instead, the peptide lies in an extended conformation along the peptide-binding cleft. It is held there both by peptide side chains that protrude into shallow and deep pockets lined by polymorphic residues, and by interactions between the peptide backbone and the side chains of conserved amino acids that line the peptide-binding cleft in all MHC class II molecules (Fig. 4.22). Structural data show that amino acid side chains at residues 1, 4, 6, and 9 of an MHC class II-bound peptide can be held in these binding pockets.

The binding pockets of MHC class II molecules accommodate a greater variety of side chains than those of MHC class I molecules, making it more difficult to define anchor residues and to predict which peptides will be able to bind

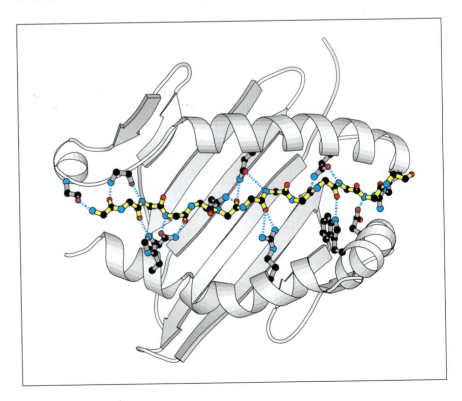

Fig. 4.22 Peptides bind to MHC class II molecules by interactions along the length of the binding groove. A peptide (yellow; shown as the peptide backbone only, with the amino terminus to the left and the carboxy terminus to the right) is bound by an MHC class II molecule through a series of hydrogen bonds (dotted blue lines) that are distributed along the length of the peptide. The hydrogen bonds toward the amino terminus of the peptide are made with the backbone of the MHC class II polypeptide chain, whereas throughout the peptide's length bonds are made with residues that are highly conserved in MHC class II molecules. The side chains of these residues are shown in gray on the ribbon diagram of the MHC class II groove.

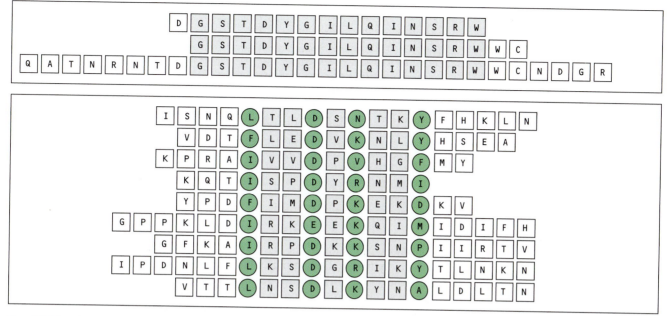

Fig. 4.23 Peptides that bind MHC class II molecules are variable in length and their anchor residues lie at various distances from the ends of the peptide. The sequences of a set of peptides that bind to the mouse MHC class II A^k allele are shown in the upper panel. All contain the same core sequence (shaded) but differ in length. In the lower panel, different peptides binding to the human MHC class II allele HLA-DR3 are shown. Anchor residues are shown as green circles. The lengths of these peptides can vary, and so by convention the first anchor residue is denoted as residue 1. Note that all of the peptides share a hydrophobic residue in position 1, a negatively charged residue [aspartic acid (D) or glutamic acid (E)] in position 4, and a tendency to have a basic residue [lysine (K), arginine (R), histidine (H), glutamine (Q), or asparagine (N)] in position 6 and a hydrophobic residue [for example, tyrosine (Y), leucine (L), phenylalanine (F)] in position 9.

a particular MHC class II variant (**Fig. 4.23**). Nevertheless, by comparing the sequences of known binding peptides it is usually possible to detect patterns of amino acids that permit binding to different MHC class II variants, and to model how the amino acids of this peptide sequence motif will interact with the amino acids of the peptide-binding cleft. Because the peptide is bound by its backbone and allowed to emerge from both ends of the binding groove, there is, in principle, no upper limit to the length of peptides that could bind to MHC class II molecules. An example of this is the protein known as **invariant chain**, part of which lies entirely across the peptide-binding groove of nascent MHC class II molecules during their synthesis in the endoplasmic reticulum. We will return in Chapter 6 to the role of the invariant chain in the loading of peptides onto MHC class II molecules. In most cases, long peptides bound to MHC class II molecules are trimmed by peptidases to a length of around 13–17 amino acids.

4-17 The crystal structures of several peptide:MHC:T-cell receptor complexes show a similar orientation of the T-cell receptor over the peptide:MHC complex.

At the time that the first X-ray crystallographic structure of a T-cell receptor was published, a structure of the same T-cell receptor bound to a peptide:MHC class I ligand was also produced. The orientation revealed by these structures showed that the T-cell receptor is aligned diagonally over the peptide and the peptide-binding cleft (**Fig. 4.24**). The TCRα chain lies over the α2 domain of the MHC molecule at the amino-terminal end of the bound peptide, as seen from the side view shown in Fig. 4.24a. The TCRβ chain lies over the MHC molecule's α1 domain, closer to the carboxy-terminal end of the peptide. Figure 4.24b shows a view of this structure as if looking down through a transparent T-cell receptor to indicate where it contacts the MHC molecule. The CDR3 loops of both TCRα and TCRβ chains come together and lie over

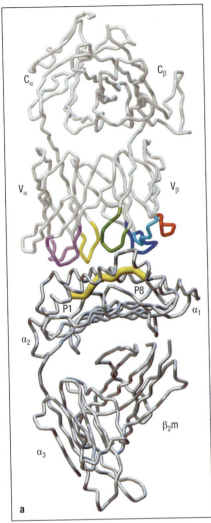

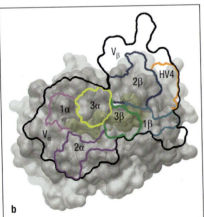

Fig. 4.24 The T-cell receptor binds to the peptide:MHC complex. Panel a: the T-cell receptor binds to the top of a peptide:MHC class I molecule, touching both the α_1 and α_2 domain helices. The CDRs of the T-cell receptor are shown in color: the CDR1 and CDR2 loops of the β chain are light and dark blue, respectively; and the CDR1 and CDR2 loops of the α chain are light and dark purple, respectively. The α-chain CDR3 loop is yellow, and the β-chain CDR3 loop is green. The β-chain HV4 loop is in red. The thick yellow line P1–P8 is the bound peptide. Panel b: the outline of the T-cell receptor's antigen-binding site (thick black line) is superimposed over the top surface of the peptide:MHC complex (the peptide is shaded dull yellow). The T-cell receptor lies at a somewhat diagonal angle across the peptide:MHC complex, with the α and β CDR3 loops of the T-cell receptor (3α, 3β, yellow and green, respectively) contacting the center of the peptide. The α-chain CDR1 and CDR2 loops (1α, 2α, light and dark purple, respectively) contact the MHC helices at the amino terminus of the bound peptide, whereas the β-chain CDR1 and CDR2 loops (1β, 2β, light and dark blue, respectively) make contact with the helices at the carboxy terminus of the bound peptide. Courtesy of I.A. Wilson.

central amino acids of the peptide. The T-cell receptor is threaded through a valley between the two high points on the two surrounding α helices that form the walls of the peptide-binding cleft. This can be seen in **Fig. 4.25**, which shows a view from the end of the peptide-binding groove of a peptide:MHC class II:T-cell receptor complex. Comparison of various peptide: MHC:T-cell receptor complexes shows that the axis of the TCR as it binds the surface of the MHC molecule is rotated somewhat relative to the peptide-binding groove of the MHC molecule (see Fig. 4.24b). In this orientation, the V_α domain makes contact primarily with the amino-terminal half of the bound peptide, whereas the V_β domain contacts primarily the carboxy-terminal half. Both chains also interact with the α helices of the MHC class I molecule (see Fig. 4.24). The T-cell receptor contacts are not symmetrically distributed over the MHC molecule: whereas the V_α CDR1 and CDR2 loops are in close contact with the helices of the peptide:MHC complex around the amino terminus of the bound peptide, the β-chain CDR1 and CDR2 loops, which interact with the complex at the carboxy terminus of the bound peptide, have variable contributions to the binding.

Comparison of the three-dimensional structure of an unliganded T-cell receptor and the same T-cell receptor complexed to its peptide:MHC ligand shows that the binding results in some degree of conformational change, or 'induced fit,' particularly within the V_α CDR3 loop. Subtle variations at amino acids that contact the T-cell receptor can have strikingly different effects on the recognition of an otherwise identical peptide:MHC ligand by the same T cell. The flexibility in the CDR3 loop demonstrated by these two structures helps to explain how the T-cell receptor can adopt conformations that recognize related, but different, peptide ligands.

The specificity of T-cell recognition involves both the peptide and its presenting MHC molecule. Kinetic analysis of T-cell receptor binding to peptide:MHC ligands suggests that the interactions with MHC molecules might predominate at the start of the contact, but that subsequent interactions with the peptide as well as the MHC molecule dictate the final outcome—binding or dissociation. As with antibody–antigen interactions, only a few amino acids at the interface might provide the essential contacts that determine the specificity and strength of binding. Simply changing a leucine to isoleucine in the peptide, for example, is sufficient to alter a T-cell response from strong killing to no response at all. Mutations of single residues in the presenting MHC molecules can have the same effect. This dual specificity for T-cell recognition of antigen underlies the MHC restriction of T-cell responses, a phenomenon that was observed long before the peptide-binding properties of MHC molecules were known. Another consequence of this dual specificity is a need for T-cell receptors to exhibit some inherent specificity for MHC molecules in order to be able to interact appropriately with the antigen-presenting surface of MHC

Fig. 4.25 The T-cell receptor interacts with MHC class I and MHC class II molecules in a similar fashion. Shown is the structure of a T-cell receptor, specific for a peptide derived from chicken cytochrome *c*, bound to an MHC class II molecule. This T-cell receptor's binding is at a site and orientation similar to that of the T-cell receptor bound to the MHC class I molecule shown in Fig. 4.24. The α and β chains of the T-cell receptor are colored in light and dark blue, respectively. The cytochrome *c* peptide is light orange. The T-cell receptor sits in a shallow saddle formed between the α-helical regions of the MHC class II α (brown) and β (yellow) chains at roughly 90° to the long axis of the MHC class II molecule and the bound peptide. Structure derived from PDB 3QIB. Courtesy of K.C. Garcia.

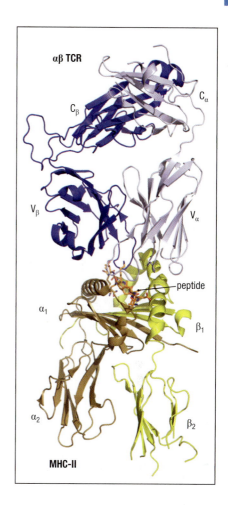

molecules. We will return to these issues in Chapter 6, where we recount the discovery of MHC restriction in the context of T-cell recognition and MHC polymorphisms, and in Chapter 8, where we discuss the impact of these phenomena on T-cell development in the thymus.

4-18 The CD4 and CD8 cell-surface proteins of T cells directly contact MHC molecules and are required to make an effective response to antigen.

As we introduced in Section 1-21, T cells fall into two major classes distinguished by the expression of the cell-surface proteins **CD4** and **CD8**. CD8 is expressed by cytotoxic T cells, while CD4 is expressed by T cells whose function is to activate other cells. CD4 and CD8 were known as markers for these functional sets for some time before it became clear that the distinction was based on the ability of T cells to recognize different classes of MHC molecules. We now know that CD8 recognizes MHC class I molecules and CD4 recognizes MHC class II. During antigen recognition, CD4 or CD8 (depending on the type of T cell) associates with the T-cell receptor on the T-cell surface and binds to invariant sites on the MHC portion of the composite peptide:MHC ligand, away from the peptide-binding site. This binding contributes to the overall effectiveness of the T-cell response, and so CD4 and CD8 are called **co-receptors**.

CD4 is a single-chain protein composed of four Ig-like domains (**Fig. 4.26**). The first two domains (D1 and D2) are packed tightly together to form a rigid

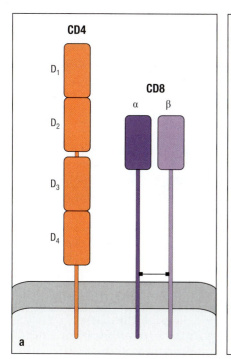
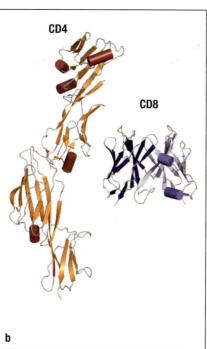

Fig. 4.26 The structures of the CD4 and CD8 co-receptor molecules. The CD4 molecule contains four Ig-like domains, shown in schematic form in panel a and as a ribbon diagram from the crystal structure in panel b. The amino-terminal domain, D_1, is similar in structure to an immunoglobulin V domain. The second domain, D_2, although related to an immunoglobulin domain, is different from both V and C domains and has been termed a C2 domain. The first two domains of CD4 form a rigid rodlike structure that is linked to the two carboxy-terminal domains by a flexible link. The binding site for MHC class II molecules involves mainly the D_1 domain. The CD8 molecule is a heterodimer of an α and a β chain covalently linked by a disulfide bond. An alternative form of CD8 exists as a homodimer of α chains. The heterodimer is depicted in panel a, whereas the ribbon diagram in panel b is of the homodimer. CD8α and CD8β chains have very similar structures, each having a single domain resembling an immunoglobulin V domain and a stretch of polypeptide chain, believed to be in a relatively extended conformation, that anchors the V-like domain to the cell membrane.

rod about 6 nm long, which is joined by a flexible hinge to a similar rod formed by the third and fourth domains (D3 and D4). The MHC-binding region on CD4 is located mainly on a lateral face of the D1 domain, and CD4 binds to a hydrophobic crevice formed at the junction of the α_2 and β_2 domains of the MHC class II molecule (**Fig. 4.27a**). This site is well away from the site where the T-cell receptor binds, as shown by the complete crystal structure of a T-cell receptor bound to peptide:MHC class II with bound CD4 (**Fig. 4.28**). This structure demonstrates that the CD4 molecule and the T-cell receptor can bind simultaneously to the same peptide:MHC class II complex. CD4 enhances sensitivity to antigen, as the T cell is about 100-fold more sensitive to the antigen when CD4 is present. The enhancement process results from the ability of the intracellular portion of CD4 to bind to a cytoplasmic tyrosine kinase called **Lck**. As we will discuss in detail Chapter 7, bringing Lck into proximity with the T-cell receptor complex helps activate the signaling cascade induced by antigen recognition.

The structure of CD8 is quite different. It is a disulfide-linked dimer of two different chains, called α and β, each containing a single Ig-like domain linked to the membrane by a segment of extended polypeptide (see Fig. 4.26). This segment

Fig. 4.27 The binding sites for CD4 and CD8 on MHC class II and class I molecules lie in the Ig-like domains. The binding sites for CD4 and CD8 on the MHC class II and class I molecules, respectively, lie in the Ig-like domains nearest to the membrane and distant from the peptide-binding cleft. The binding of CD4 to an MHC class II molecule is shown as a ribbon structure in panel a and schematically in panel c. The α chain of the MHC class II molecule is shown in pink, and the β chain in white, while CD4 is in gold. Only the D_1 and D_2 domains of the CD4 molecule are shown in panel a. The binding site for CD4 lies at the base of the β_2 domain of an MHC class II molecule, in the hydrophobic crevice between the β_2 and α_2 domains. The binding of CD8$\alpha\beta$ to an MHC class I molecule is shown in panel b and schematically in panel d. The class I heavy chain and β_2-microglobulin are shown in white and pink, respectively, and the two chains of the CD8 dimer are shown in light (CD8β) and dark (CD8α) purple. The binding site for CD8 on the MHC class I molecule lies in a similar position to that of CD4 in the MHC class II molecule, but CD8 binding also involves the base of the α_1 and α_2 domains, and thus the binding of CD8 to MHC class I is not completely equivalent to the binding of CD4 to MHC class II. Structures derived from PDB 3S4S (CD4/MHC class II) and PDB 3DMM (CD8$\alpha\beta$/MHC class I). Courtesy of K.C. Garcia.

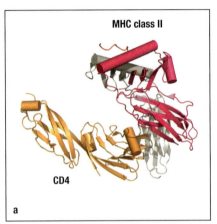

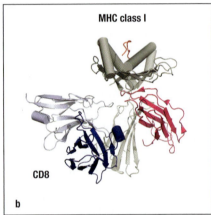

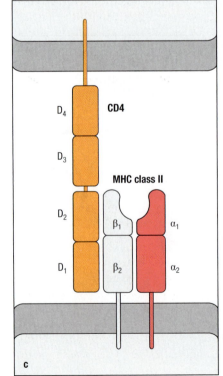

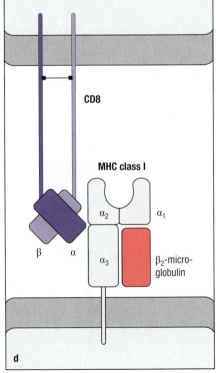

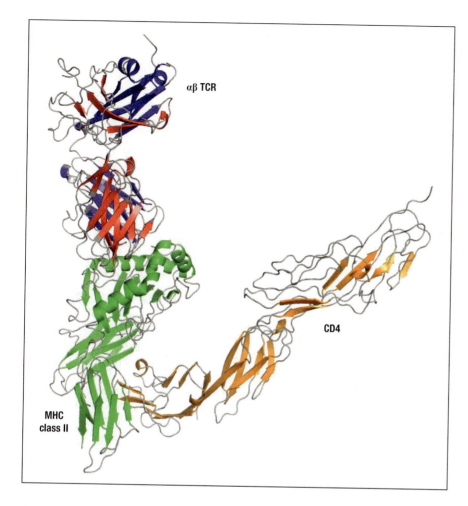

Fig. 4.28 CD4 and the T-cell receptor bind to distinct regions of the MHC class II molecule. A ribbon diagram is shown from a crystal structure of a complete α:β TCR:peptide-MHC:CD4 ternary complex. The α and β chains of the T-cell receptor (TCR) are blue and red, respectively. The MHC class II molecule is green, with the bound peptide shown in gray. CD4 is shown in orange. Structure derived from PDB 3T0E. Courtesy of K.C. Garcia.

is extensively glycosylated, which is thought to maintain it in an extended conformation and protect it from cleavage by proteases. CD8α chains can form homodimers, although these are usually not found when CD8β is expressed. Naive T cells express CD8αβ, but the CD8αα homodimer can be expressed by highly activated effector and memory T cells. CD8αα is also expressed by a population of intraepithelial lymphocytes known as **mucosal associated invariant T cells (MAIT cells)**; these cells recognize metabolites of folic acid that are produced by bacteria in association with the nonclassical MHC class I molecule **MR1**, which we will describe in Chapter 6.

CD8αβ binds weakly to an invariant site in the α3 domain of an MHC class I molecule (see Fig. 4.27b). The CD8β chain interacts with residues in the base of the α2 domain of the MHC class I molecule, while the α chain is in a lower position interacting with the α3 domain of the MHC class I molecule. The strength of binding of CD8 to the MHC class I molecule is influenced by the glycosylation state of the CD8 molecule; increasing the number of sialic acid residues on CD8 carbohydrate structures decreases the strength of the interaction. The pattern of sialylation of CD8 changes during the maturation of T cells and also on activation, and this likely has a role in modulating antigen recognition.

Like the interactions with MHC class II molecules, the T-cell receptor and CD8 can interact simultaneously with one MHC class I molecule (**Fig. 4.29**). Like CD4, CD8 binds Lck through the cytoplasmic tail of the α chain, and CD8αβ increases the sensitivity of T cells to antigen presented by MHC class I molecules about 100-fold. Although the molecular details are unclear, the CD8αα homodimer appears to function less efficiently than CD8αβ as a co-receptor, and may negatively regulate activation. In contrast to CD8, CD4 is not thought to dimerize.

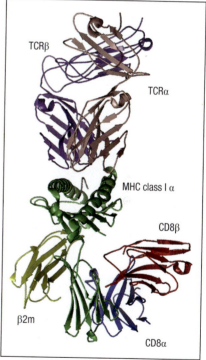

Fig. 4.29 CD8 binds to a site on MHC class I molecules that is distant from where the T-cell receptor binds. The relative binding positions of both the T-cell receptor and CD8 molecules can be seen in this hypothetical reconstruction of their interaction with MHC class I (α chain in dark green and β2-microglobulin in light green). The α and β chains of the T-cell receptor are shown in brown and purple, respectively. The CD8αβ heterodimer is shown bound to the MHC class I α3 domain. The CD8 α chain is in blue, and the CD8β chain is in red. Courtesy of Chris Nelson and David Fremont.

Tissue	MHC class I	MHC class II
Lymphoid tissues		
T cells	+++	+*
B cells	+++	+++
Macrophages	+++	++
Dendritic cells	+++	+++
Epithelial cells of thymus	+	+++

Other nucleated cells		
Neutrophils	+++	–
Hepatocytes	+	–
Kidney	+	–
Brain	+	–†

Nonnucleated cells		
Red blood cells	–	–

Fig. 4.30 The expression of MHC molecules differs among tissues. MHC class I molecules are expressed on all nucleated cells, although they are most highly expressed in hematopoietic cells. MHC class II molecules are normally expressed only by a subset of hematopoietic cells and by thymic stromal cells, although they may be expressed by other cell types on exposure to the inflammatory cytokine IFN-γ. *In humans, activated T cells express MHC class II molecules, whereas in mice all T cells are MHC class II-negative. †In the brain, most cell types are MHC class II-negative, but microglia, which are related to macrophages, are MHC class II-positive.

4-19   The two classes of MHC molecules are expressed differentially on cells.

MHC class I and MHC class II molecules have distinct distributions among cells, and these reflect the different effector functions of the T cells that recognize them (Fig. 4.30). MHC class I molecules present peptides from pathogens, commonly viruses, to CD8 cytotoxic T cells, which are specialized to kill any cell that they specifically recognize. Because viruses can infect any nucleated cell, almost all such cells express MHC class I molecules, although the level of constitutive expression varies from one cell type to the next. For example, cells of the immune system express abundant MHC class I on their surface, whereas liver cells (hepatocytes) express relatively low levels (see Fig. 4.30). Nonnucleated cells, such as mammalian red blood cells, express little or no MHC class I, and thus the interior of red blood cells is a site in which an infection can go undetected by cytotoxic T cells. Because red blood cells cannot support viral replication, this is of no great consequence for viral infection, but it might be the absence of MHC class I that allows the *Plasmodium* parasites that cause malaria to live in this privileged site.

In contrast, a major function of the CD4 T cells that recognize MHC class II molecules is to activate other effector cells of the immune system. Thus, MHC class II molecules are normally found on dendritic cells, B lymphocytes, and macrophages—antigen-presenting cells that participate in immune responses—but not on other tissue cells (see Fig. 4.30). The peptides presented by MHC class II molecules expressed by dendritic cells can function to activate naive CD4 T cells. When previously activated CD4 T cells recognize peptides bound to MHC class II molecules on B cells, the T cells secrete cytokines that can influence the isotype of antibody that those B cells will choose to produce. Upon recognizing peptides bound to MHC class II molecules on macrophages, CD4 T cells activate these cells, again in part through cytokines, to destroy the pathogens in their vesicles.

The expression of both MHC class I and MHC class II molecules is regulated by cytokines, in particular, interferons, released in the course of an immune response. Interferon-α (IFN-α) and IFN-β increase the expression of MHC class I molecules on all types of cells, whereas IFN-γ increases the expression of both MHC class I and MHC class II molecules, and can induce the expression of MHC class II molecules on certain cell types that do not normally express them. Interferons also enhance the antigen-presenting function of MHC class I molecules by inducing the expression of key components of the intracellular machinery that enables peptides to be loaded onto the MHC molecules.

4-20   A distinct subset of T cells bears an alternative receptor made up of γ and δ chains.

During the search for the gene for the TCRα chain, another T-cell receptor-like gene was unexpectedly discovered. This gene was named TCRγ, and its discovery led to a search for further T-cell receptor genes. Another receptor chain was identified by using antibody against the predicted sequence of the γ chain and was called the δ chain. It was soon discovered that a minority population of T cells bore a distinct type of T-cell receptor made up of γ:δ heterodimers rather than α:β heterodimers. The development of these cells is described in Sections 8-11 and 8-12.

Like α:β T cells, γ:δ T cells can be found in the lymphoid tissues of all vertebrates, but they are also prominent as populations of intraepithelial lymphocytes, particularly in the skin and female reproductive tracts, where their receptors display very limited diversity. Unlike α:β T cells, γ:δ T cells do not generally recognize antigen as peptides presented by MHC molecules, and γ:δ

T-cell receptors are not restricted by the 'classical' MHC class I and class II molecules that function in binding and presenting peptides to T cells. Instead, γ:δ T-cell receptors seem to recognize their target antigens directly and thus probably are able to recognize and respond rapidly to molecules expressed by many different cell types. Their ligands have been difficult to identify, but several have now been described and seem to indicate that γ:δ T cells play an intermediate, or transitional, role between wholly innate and fully adaptive immune responses.

Like NK-cell receptor ligands, such as the proteins MIC and RAET1 (see Section 3-27), many of the ligands seen by γ:δ T cells are induced by cellular stress or damage. γ:δ T cells may also bind antigens presented by 'nonclassical' **MHC class Ib molecules**, which we will discuss in Chapter 6. These proteins are related structurally to the MHC proteins we have already discussed, but have functions besides binding peptides for presentation to T cells. Additional ligands may include heat-shock proteins and nonpeptide ligands such as phosphorylated ligands or mycobacterial lipid antigens. γ:δ T cells can also respond to unorthodox nucleotides and phospholipids. Recognition of molecules expressed as a consequence of infection, rather than recognition of pathogen-specific antigens themselves, distinguishes intraepithelial γ:δ T cells from other lymphocytes, and this would place them in the innate-like class. For these reasons, the term '**transitional immunity**' has been proposed to clarify the role of γ:δ T cells, since the function of these cells seems to be someplace between innate and adaptive responses.

The crystallographic structure of a γ:δ T-cell receptor reveals that, as expected, it is similar in shape to α:β T-cell receptors. **Figure 4.31** shows a crystal structure of a γ:δ T-cell receptor complex bound to one of the nonclassical MHC class I molecules mentioned above, called **T22**. This structure shows that the overall orientation of the γ:δ T-cell receptor with the MHC molecule is strikingly different from that of an α:β T-cell receptor, in that it interacts primarily with one end of the T22 molecule. However, the CDR3 regions of the γ:δ T-cell receptor still play a critical role in recognition, similar to that of antibodies and of α:β T-cell receptors. Further, the CDR3 of the γ:δ T-cell receptor is longer than either of these other two antigen receptors, and this could have implications toward the type of ligand that the γ:δ T-cell receptor recognizes, as there is enormous combinatorial diversity from the CDR3 within the γ:δ T-cell receptor repertoire. We will return to discuss further the ligands and the development of γ:δ T cells in Chapters 6 and 8.

Summary.

The receptor for antigen on most T cells, the α:β T-cell receptor, is composed of two protein chains, TCRα and TCRβ, and resembles in many respects a single Fab fragment of immunoglobulin. α:β T-cell receptors are always membrane-bound and recognize a composite ligand of a peptide antigen bound to an MHC molecule. Each MHC molecule binds a wide variety of different peptides, but the different variants of MHC each preferentially recognize sets of peptides with particular sequences and physical features. The peptide antigen is generated intracellularly, and is bound stably in a peptide-binding cleft on the surface of the MHC molecule. There are two classes of MHC molecules, and these are bound in their nonpolymorphic domains by CD8 and CD4 molecules that distinguish two different functional classes of α:β T cells. CD8 binds MHC class I molecules and can bind simultaneously to the same peptide:MHC class I complex being recognized by a T-cell receptor, thus acting as a co-receptor and enhancing the T-cell response; CD4 binds MHC class II molecules and acts as a co-receptor for T-cell receptors that recognize peptide:MHC class II ligands. A T-cell receptor interacts directly both with the antigenic peptide

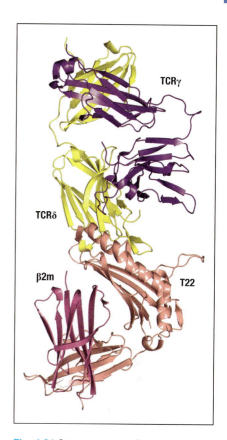

Fig. 4.31 Structures of γ:δ T-cell receptor bound to the nonclassical MHC class I molecule T22. The γ:δ T-cell receptor has a similar overall structure to the α:β T-cell receptor and the Fab fragment of an immunoglobulin. The C$_δ$ domain is more like an immunoglobulin domain than is the corresponding C$_α$ domain of the α:β T-cell receptor. In this structure, the overall orientation of the γ:δ T-cell receptor with respect to the nonclassical MHC molecule T22 is very different from the orientation of an α:β T-cell receptor with either MHC class I or class II molecules. Rather than lying directly over the peptide-binding groove, the γ:δ T-cell receptor is engaged with one end much more than the other; this is consistent with a lack of peptide contact and absence of MHC-restricted recognition.

and with polymorphic features of the MHC molecule that displays it, and this dual specificity underlies the MHC restriction of T-cell responses. A second type of T-cell receptor, composed of a γ and a δ chain, is structurally similar to the α:β T-cell receptor, but it binds to different ligands, including nonpeptide ligands, nonpolymorphic nonclassical MHC molecules, and certain lipids. The receptor is thought not to be MHC-restricted and is found on a minority population of lymphoid and intraepithelial T cells, the γ:δ T cells.

Summary to Chapter 4.

B cells and T cells use different, but structurally similar, molecules to recognize antigen. The antigen-recognition molecules of B cells are immunoglobulins, and are made both as a membrane-bound receptor for antigen, the B-cell receptor, and as secreted antibodies that bind antigens and elicit humoral effector functions. The antigen-recognition molecules of T cells, in contrast, are made only as cell-surface receptors and so elicit only cellular effector functions. Immunoglobulins and T-cell receptors are highly variable molecules, with the variability concentrated in that part of the molecule—the variable (V) region—that binds to antigen. Immunoglobulins bind a wide variety of chemically different antigens, whereas the major type of T-cell receptor, the α:β T-cell receptor, predominantly recognizes peptide fragments of foreign proteins bound to MHC molecules, which are ubiquitous on cell surfaces.

Binding of antigen by immunoglobulins has chiefly been studied with antibodies. The binding of antibody to its antigen is highly specific, and is determined by the shape and physicochemical properties of the antigen-binding site. Located at the other end of the antibody from the antigen-binding site is the constant, or Fc, region, which influences the types of effector function the antibody can elicit. There are five main functional classes of antibodies, each encoded by a different type of constant region. As we will see in Chapter 10, these interact with different components of the immune system to incite an inflammatory response and eliminate the antigen.

T-cell receptors differ in several respects from the B-cell immunoglobulins. One is the absence of a secreted form of T-cell receptor, reflecting the functional differences between T cells and B cells. B cells deal with pathogens and their protein products circulating within the body; secretion of a soluble antigen-recognition molecule enables the B cell to act in the clearance of antigen effectively throughout the extracellular spaces of the body. T cells, in contrast, are specialized for active surveillance of pathogens, and T-cell recognition does not involve a soluble, secreted receptor. Some, such as CD8 T cells, are able to detect intracellular infections and are able to kill infected cells that bear foreign antigenic peptides on their surface. Others, such as CD4 T cells, interact with cells of the immune system that have taken up foreign antigen and are displaying it on the cell surface.

T-cell receptors also recognize a composite ligand made up of the foreign peptide bound to a self MHC molecule, and not intact antigen. This means that T cells can interact only with a body cell displaying the antigen, not with the intact pathogen or protein. Each T-cell receptor is specific for a particular combination of peptide and a self MHC molecule. MHC molecules are encoded by a family of highly polymorphic genes. Expression of multiple variant MHC molecules, each with a different peptide-binding repertoire, helps to ensure that T cells from an individual will be able to recognize at least some peptides generated from nearly every pathogen.

Questions.

4.1 True or False: An antibody proteolytically cleaved by papain yields a fragment with higher avidity to the cognate antigen than an antibody cleaved by pepsin.

4.2 Short Answer: How is CD4 and CD8 co-receptor binding to MHC important for T-cell receptor signaling?

4.3 Short Answer: Why and how is it advantageous to have heterozygosity in the MHC locus?

4.4 Matching: Match the term to the best description:

A. Antigenic determinant

B. Conformational/ discontinuous epitope

C. Continuous/linear epitope

D. Hypervariable region

i. The structure recognized by an antibody (that is, the epitope)

ii. Regions of the V region that have signiicant sequence variation

iii. An epitope composed of a single segment of a polypeptide chain

iv. An epitope composed of amino acids from different parts of a polypeptide chain brought together by protein folding

4.5 Fill-in-the-Blanks: Most vertebrates, including humans and mice, produce antibodies composed of _____ and _____ chains. These bear _____ regions that recognize the antigen and _____ regions that dictate the antibody class and isotype. Camelids and cartilaginous fish, however, produce _____ and _____, respectively, which are forming the basis for single-chain antibody production for clinical applications.

4.6 Multiple Choice: Which of the following statements is *not* true?

A. T-cell receptor α and β chains pair together, but the α chain can be switched out for a γ or a δ chain.

B. Electrostatic interactions (for example, a salt bridge) occur between charged amino acids.

C. Hydrophobic interactions occur between two hydrophobic surfaces and exclude water.

D. Antibodies often have many aromatic amino acids such as tyrosine in their antigen-binding sites.

E. MHC restriction is the phenomenon where T cells will recognize a unique set of peptides bound to a particular MHC molecule.

4.7 Multiple Choice: Which of the following is the most abundant immunoglobulin class in healthy adult humans and mice?

A. IgA

B. IgD

C. IgE

D. IgG

E. IgM

4.8 Multiple Choice: Which of the following describes the structure of an immunoglobulin fold?

A. Two antiparallel β sheets with an α-helical linker and a disulfide bond link

B. Two β strands linked by a disulfide bond

C. Four α helices linked by two disulfide bonds

D. Seven antiparallel α helices in series

E. One β sandwich of two β sheets folded together and linked by a disulfide bond

4.9 Multiple Choice: Antibodies have flexibility at various points in the molecule, particularly the hinge region between the Fc and Fab portion and, to some extent, the junction between the V and C regions. Which of the following properties of an antibody are not affected by its flexibility?

A. Binding to small antigens (haptens)

B. Avidity to antigen

C. Affinity to antigen

D. Interaction with antibody-binding proteins

E. Binding to distantly spaced antigens

4.10 Multiple Choice: Which region of the antigen receptor of B cells and T cells is most critical in antigen recognition and specificity?

A. FR1

B. CDR1

C. FR2

D. CDR2

E. FR3

F. CDR3

G. FR4

General references.

Garcia, K.C., Degano, M., Speir, J.A., and Wilson, I.A.: **Emerging principles for T cell receptor recognition of antigen in cellular immunity.** *Rev. Immunogenet.* 1999, **1**:75–90.

Garcia, K.C., Teyton, L., and Wilson, I.A.: **Structural basis of T cell recognition.** *Annu. Rev. Immunol.* 1999, **17**:369–397.

Moller, G. (ed): **Origin of major histocompatibility complex diversity.** *Immunol. Rev.* 1995, **143**:5–292.

Poljak, R.J.: **Structure of antibodies and their complexes with antigens.** *Mol. Immunol.* 1991, **28**:1341–1345.

Rudolph, M.G., Stanfield, R.L., and Wilson, I.A: **How TCRs bind MHCs, peptides, and coreceptors.** *Annu. Rev. Immunol.* 2006, **24**:419–466.

Sundberg, E.J., and Mariuzza, R.A.: **Luxury accommodations: the expanding role of structural plasticity in protein-protein interactions.** *Structure* 2000, **8**:R137–R142.

Section references.

4-1   IgG antibodies consist of four polypeptide chains.

Edelman, G.M.: **Antibody structure and molecular immunology.** *Scand. J. Immunol.* 1991, **34**:4–22.

Faber, C., Shan, L., Fan, Z., Guddat, L.W., Furebring, C., Ohlin, M., Borrebaeck, C.A.K., and Edmundson, A.B.: **Three-dimensional structure of a human Fab with high affinity for tetanus toxoid.** *Immunotechnology* 1998, **3**:253–270.

Harris, L.J., Larson, S.B., Hasel, K.W., Day, J., Greenwood, A., and McPherson, A.: **The three-dimensional structure of an intact monoclonal antibody for canine lymphoma.** *Nature* 1992, **360**:369–372.

4-2   Immunoglobulin heavy and light chains are composed of constant and variable regions.

&

4-3   The domains of an immunoglobulin molecule have similar structures.

Barclay, A.N., Brown, M.H., Law, S.K., McKnight, A.J., Tomlinson, M.G., and van der Merwe, P.A. (eds): *The Leukocyte Antigen Factsbook*, 2nd ed. London: Academic Press, 1997.

Brummendorf, T., and Lemmon, V.: **Immunoglobulin superfamily receptors: *cis*-interactions, intracellular adapters and alternative splicing regulate adhesion.** *Curr. Opin. Cell Biol.* 2001, **13**:611–618.

Marchalonis, J.J., Jensen, I., and Schluter, S.F.: **Structural, antigenic and evolutionary analyses of immunoglobulins and T cell receptors.** *J. Mol. Recog.* 2002, **15**:260–271.

Ramsland, P.A., and Farrugia, W.: **Crystal structures of human antibodies: a detailed and unfinished tapestry of immunoglobulin gene products.** *J. Mol. Recog.* 2002, **15**:248–259.

4-4   The antibody molecule can readily be cleaved into functionally distinct fragments.

Porter, R.R.: **Structural studies of immunoglobulins.** *Scand. J. Immunol.* 1991, **34**:382–389.

Yamaguchi, Y., Kim, H., Kato, K., Masuda, K., Shimada, I., and Arata, Y.: **Proteolytic fragmentation with high specificity of mouse IgG—mapping of proteolytic cleavage sites in the hinge region.** *J. Immunol. Methods.* 1995, **181**:259–267.

4-5   The hinge region of the immunoglobulin molecule allows flexibility in binding to multiple antigens.

Gerstein, M., Lesk, A.M., and Chothia, C.: **Structural mechanisms for domain movements in proteins.** *Biochemistry* 1994, **33**:6739–6749.

Jimenez, R., Salazar, G., Baldridge, K.K., and Romesberg, F.E.: **Flexibility and molecular recognition in the immune system.** *Proc. Natl Acad. Sci. USA* 2003, **100**:92–97.

Saphire, E.O., Stanfield, R.L., Crispin, M.D., Parren, P.W., Rudd, P.M., Dwek, R.A., Burton, D.R., and Wilson, I.A.: **Contrasting IgG structures reveal extreme asymmetry and flexibility.** *J. Mol. Biol.* 2002, **319**:9–18.

4-6   Localized regions of hypervariable sequence form the antigen-binding site.

Chitarra, V., Alzari, P.M., Bentley, G.A., Bhat, T.N., Eiselé, J.-L., Houdusse, A., Lescar, J., Souchon, H., and Poljak, R.J.: **Three-dimensional structure of a heteroclitic antigen-antibody cross-reaction complex.** *Proc. Natl Acad. Sci. USA* 1993, **90**:7711–7715.

Decanniere, K., Muyldermans, S., and Wyns, L.: **Canonical antigen-binding loop structures in immunoglobulins: more structures, more canonical classes?** *J. Mol. Biol.* 2000, **300**:83–91.

Gilliland, L.K., Norris, N.A., Marquardt, H., Tsu, T.T., Hayden, M.S., Neubauer, M.G., Yelton, D.E., Mittler, R.S., and Ledbetter, J.A.: **Rapid and reliable cloning of antibody variable regions and generation of recombinant single-chain antibody fragments.** *Tissue Antigens* 1996, **47**:1–20.

Johnson, G., and Wu, T.T.: **Kabat Database and its applications: 30 years after the first variability plot.** *Nucleic Acids Res.* 2000, **28**:214–218.

Wu, T.T., and Kabat, E.A.: **An analysis of the sequences of the variable regions of Bence Jones proteins and myeloma light chains and their implications for antibody complementarity.** *J. Exp. Med.* 1970, **132**:211–250.

Xu, J., Deng, Q., Chen, J., Houk, K.N., Bartek, J., Hilvert, D., and Wilson, I.A.: **Evolution of shape complementarity and catalytic efficiency from a primordial antibody template.** *Science* 1999, **286**:2345–2348.

4-7   Antibodies bind antigens via contacts in CDRs that are complementary to the size and shape of the antigen.

&

4-8   Antibodies bind to conformational shapes on the surfaces of antigens using a variety of noncovalent forces.

Ban, N., Day, J., Wang, X., Ferrone, S., and McPherson, A.: **Crystal structure of an anti-anti-idiotype shows it to be self-complementary.** *J. Mol. Biol.* 1996, **255**:617–627.

Davies, D.R., and Cohen, G.H.: **Interactions of protein antigens with antibodies.** *Proc. Natl Acad. Sci. USA* 1996, **93**:7–12.

Decanniere, K., Desmyter, A., Lauwereys, M., Ghahroudi, M.A., Muyldermans, S., and Wyns, L.: **A single-domain antibody fragment in complex with RNase A: non-canonical loop structures and nanomolar affinity using two CDR loops.** *Structure Fold. Des.* 1999, **7**:361–370.

Padlan, E.A.: **Anatomy of the antibody molecule.** *Mol. Immunol.* 1994, **31**:169–217.

Saphire, E.O., Parren, P.W., Pantophlet, R., Zwick, M.B., Morris, G.M., Rudd, P.M., Dwek, R.A., Stanfield, R.L., Burton, D.R., and Wilson, I.A.: **Crystal structure of a neutralizing human IgG against HIV-1: a template for vaccine design.** *Science* 2001, **293**:1155–1159.

Stanfield, R.L., and Wilson, I.A.: **Protein–peptide interactions.** *Curr. Opin. Struct. Biol.* 1995, **5**:103–113.

Tanner, J.J., Komissarov, A.A., and Deutscher, S.L.: **Crystal structure of an antigen-binding fragment bound to single-stranded DNA.** *J. Mol. Biol.* 2001, **314**:807–822.

Wilson, I.A., and Stanfield, R.L.: **Antibody–antigen interactions: new structures and new conformational changes.** *Curr. Opin. Struct. Biol.* 1994, **4**:857–867.

4-9   Antibody interaction with intact antigens is influenced by steric constraints.

Braden, B.C., Goldman, E.R., Mariuzza, R.A., and Poljak, R.J.: **Anatomy of an antibody molecule: structure, kinetics, thermodynamics and mutational studies of the antilysozyme antibody D1.3.** *Immunol. Rev.* 1998, **163**:45–57.

Braden, B.C., and Poljak, R.J.: **Structural features of the reactions between antibodies and protein antigens.** *FASEB J.* 1995, **9**:9–16.

Diamond, M.S., Pierson, T.C., and Fremont, D.H.: **The structural immunology of antibody protection against West Nile virus.** *Immunol Rev.* 2008, **225**:212–225.

Lok, S.M., Kostyuchenko, V., Nybakken, G.E., Holdaway, H.A., Battisti, A.J., Sukupolvi-Petty, S., Sedlak, D., Fremont, D.H., Chipman, P.R., Roehrig, J.T., *et al.*: **Binding of a neutralizing antibody to dengue virus alters the arrangement of surface glycoproteins.** *Nat. Struct. Mol. Biol.* 2008, **15**:312–317.

Ros, R., Schwesinger, F., Anselmetti, D., Kubon, M., Schäfer, R., Plückthun, A., and Tiefenauer, L.: **Antigen binding forces of individually addressed single-chain Fv antibody molecules.** *Proc. Natl Acad. Sci. USA* 1998, **95**:7402–7405.

4-10 Some species generate antibodies with alternative structures.

Hamers-Casterman, C., Atarhouch, T., Muyldermans, S., Robinson, G., Hamers, C., Songa, E.B., Bendahman, N., and Hamers, R.: **Naturally occurring antibodies devoid of light chains.** *Nature* 1993, **363**:446–448.

Muyldermans, S.: **Nanobodies: natural single-domain antibodies.** *Annu. Rev. Biochem.* 2013, **82**:775–797.

Nguyen, V.K., Desmyter, A., and Muyldermans, S.: **Functional heavy-chain antibodies in Camelidae.** *Adv. Immunol.* 2001, **79**:261–296.

4-11 The TCRα:β heterodimer is very similar to a Fab fragment of immunoglobulin.

Al-Lazikani, B., Lesk, A.M., and Chothia, C.: **Canonical structures for the hypervariable regions of T cell αβ receptors.** *J. Mol. Biol.* 2000, **295**:979–995.

Kjer-Nielsen, L., Clements, C.S., Brooks, A.G., Purcell, A.W., McCluskey, J., and Rossjohn, J.: **The 1.5 Å crystal structure of a highly selected antiviral T cell receptor provides evidence for a structural basis of immunodominance.** *Structure (Camb.)* 2002, **10**:1521–1532.

Machius, M., Cianga, P., Deisenhofer, J., and Ward, E.S.: **Crystal structure of a T cell receptor Vα11 (AV11S5) domain: new canonical forms for the first and second complementarity determining regions.** *J. Mol. Biol.* 2001, **310**:689–698.

4-12 A T-cell receptor recognizes antigen in the form of a complex of a foreign peptide bound to an MHC molecule.

Garcia, K.C., and Adams, E.J.: **How the T cell receptor sees antigen—a structural view.** *Cell* 2005, **122**:333–336.

Hennecke, J., and Wiley, D.C.: **Structure of a complex of the human αβ T cell receptor (TCR) HA1.7, influenza hemagglutinin peptide, and major histocompatibility complex class II molecule, HLA-DR4 (DRA*0101 and DRB1*0401): insight into TCR cross-restriction and alloreactivity.** *J. Exp. Med.* 2002, **195**:571–581.

Luz, J.G., Huang, M., Garcia, K.C., Rudolph, M.G., Apostolopoulos, V., Teyton, L., and Wilson, I.A.: **Structural comparison of allogeneic and syngeneic T cell receptor–peptide–major histocompatibility complex complexes: a buried alloreactive mutation subtly alters peptide presentation substantially increasing Vβ interactions.** *J. Exp. Med.* 2002, **195**:1175–1186.

Reinherz, E.L., Tan, K., Tang, L., Kern, P., Liu, J., Xiong, Y., Hussey, R.E., Smolyar, A., Hare, B., Zhang, R., *et al.*: **The crystal structure of a T cell receptor in complex with peptide and MHC class II.** *Science* 1999, **286**:1913–1921.

Rudolph, M.G., Stanfield, R.L., and Wilson, I.A.: **How TCRs bind MHCs, peptides, and coreceptors.** *Annu. Rev. Immunol.* 2006, **24**:419–466.

4-13 There are two classes of MHC molecules with distinct subunit compositions but similar three-dimensional structures.

&

4-14 Peptides are stably bound to MHC molecules, and also serve to stabilize the MHC molecule on the cell surface.

Bouvier, M.: **Accessory proteins and the assembly of human class I MHC molecules: a molecular and structural perspective.** *Mol. Immunol.* 2003, **39**:697–706.

Dessen, A., Lawrence, C.M., Cupo, S., Zaller, D.M., and Wiley, D.C.: **X-ray crystal structure of HLA-DR4 (DRA*0101, DRB1*0401) complexed with a peptide from human collagen II.** *Immunity* 1997, **7**:473–481.

Fremont, D.H., Hendrickson, W.A., Marrack, P., and Kappler, J.: **Structures of an MHC class II molecule with covalently bound single peptides.** *Science* 1996, **272**:1001–1004.

Fremont, D.H., Matsumura, M., Stura, E.A., Peterson, P.A., and Wilson, I.A.: **Crystal structures of two viral peptides in complex with murine MHC class 1 H-2Kᵇ.** *Science* 1992, **257**:919–927.

Fremont, D.H., Monnaie, D., Nelson, C.A., Hendrickson, W.A., and Unanue, E.R.: **Crystal structure of I-Ak in complex with a dominant epitope of lysozyme.** *Immunity* 1998, **8**:305–317.

Macdonald, W.A., Purcell, A.W., Mifsud, N.A., Ely, L.K., Williams, D.S., Chang, L., Gorman, J.J., Clements, C.S., Kjer-Nielsen, L., Koelle, D.M., *et al.*: **A naturally selected dimorphism within the HLA-B44 supertype alters class I structure, peptide repertoire, and T cell recognition.** *J. Exp. Med.* 2003, **198**:679–691.

Zhu, Y., Rudensky, A.Y., Corper, A.L., Teyton, L., and Wilson, I.A.: **Crystal structure of MHC class II I-Ab in complex with a human CLIP peptide: prediction of an I-Ab peptide-binding motif.** *J. Mol. Biol.* 2003, **326**:1157–1174.

4-15 MHC class I molecules bind short peptides of 8–10 amino acids by both ends.

Bouvier, M., and Wiley, D.C.: **Importance of peptide amino and carboxyl termini to the stability of MHC class I molecules.** *Science* 1994, **265**:398–402.

Govindarajan, K.R., Kangueane, P., Tan, T.W., and Ranganathan, S.: **MPID: MHC-Peptide Interaction Database for sequence–structure–function information on peptides binding to MHC molecules.** *Bioinformatics* 2003, **19**:309–310.

Saveanu, L., Fruci, D., and van Endert, P.: **Beyond the proteasome: trimming, degradation and generation of MHC class I ligands by auxiliary proteases.** *Mol. Immunol.* 2002, **39**:203–215.

Weiss, G.A., Collins, E.J., Garboczi, D.N., Wiley, D.C., and Schreiber, S.L.: **A tricyclic ring system replaces the variable regions of peptides presented by three alleles of human MHC class I molecules.** *Chem. Biol.* 1995, **2**:401–407.

4-16 The length of the peptides bound by MHC class II molecules is not constrained.

Conant, S.B., and Swanborg, R.H.: **MHC class II peptide flanking residues of exogenous antigens influence recognition by autoreactive T cells.** *Autoimmun. Rev.* 2003, **2**:8–12.

Guan, P., Doytchinova, I.A., Zygouri, C., and Flower, D.R.: **MHCPred: a server for quantitative prediction of peptide–MHC binding.** *Nucleic Acids Res.* 2003, **31**:3621–3624.

Lippolis, J.D., White, F.M., Marto, J.A., Luckey, C.J., Bullock, T.N., Shabanowitz, J., Hunt, D.F., and Engelhard, V.H.: **Analysis of MHC class II antigen processing by quantitation of peptides that constitute nested sets.** *J. Immunol.* 2002, **169**:5089–5097.

Park, J.H., Lee, Y.J., Kim, K.L., and Cho, E.W.: **Selective isolation and identification of HLA-DR-associated naturally processed and presented epitope peptides.** *Immunol. Invest.* 2003, **32**:155–169.

Rammensee, H.G.: **Chemistry of peptides associated with MHC class I and class II molecules.** *Curr. Opin. Immunol.* 1995, **7**:85–96.

Rudensky, A.Y., Preston-Hurlburt, P., Hong, S.C., Barlow, A., and Janeway, C.A., Jr.: **Sequence analysis of peptides bound to MHC class II molecules.** *Nature* 1991, **353**:622–627.

Sercarz, E.E., and Maverakis, E.: **MHC-guided processing: binding of large antigen fragments.** *Nat. Rev. Immunol.* 2003, **3**:621–629.

Sinnathamby, G., and Eisenlohr, L.C.: **Presentation by recycling MHC class II molecules of an influenza hemagglutinin-derived epitope that is revealed in the early endosome by acidification.** *J. Immunol.* 2003, **170**:3504–3513.

4-17 The crystal structures of several peptide:MHC:T-cell receptor complexes show a similar orientation of the T-cell receptor over the peptide:MHC complex.

Buslepp, J., Wang, H., Biddison, W.E., Appella, E., and Collins, E.J.: **A correlation between TCR Vα docking on MHC and CD8 dependence: implications for T cell selection.** *Immunity* 2003, **19**:595–606.

Ding, Y.H., Smith, K.J., Garboczi, D.N., Utz, U., Biddison, W.E., and Wiley, D.C.: **Two human T cell receptors bind in a similar diagonal mode to the HLA-A2/Tax peptide complex using different TCR amino acids.** *Immunity* 1998, **8**:403–411.

Garcia, K.C., Degano, M., Pease, L.R., Huang, M., Peterson, P.A., Leyton, L., and Wilson, I.A.: **Structural basis of plasticity in T cell receptor recognition of a self peptide-MHC antigen.** *Science* 1998, **279**:1166–1172.

Kjer-Nielsen, L., Clements, C.S., Purcell, A.W., Brooks, A.G., Whisstock, J.C., Burrows, S.R., McCluskey, J., and Rossjohn, J.: **A structural basis for the selection of dominant αβ T cell receptors in antiviral immunity.** *Immunity* 2003, **18**:53–64.

Newell, E.W., Ely, L.K., Kruse, A.C., Reay, P.A., Rodriguez, S.N., Lin, A.E., Kuhns, M.S., Garcia, K.C., and Davis, M.M.: **Structural basis of specificity and cross-reactivity in T cell receptors specific for cytochrome c-I-E(k).** *J. Immunol.* 2011, **186**:5823–5832.

Reiser, J.B., Darnault, C., Gregoire, C., Mosser, T., Mazza, G., Kearney, A., van der Merwe, P.A., Fontecilla-Camps, J.C., Housset, D., and Malissen, B.: **CDR3 loop flexibility contributes to the degeneracy of TCR recognition.** *Nat. Immunol.* 2003, **4**:241–247.

Sant'Angelo, D.B., Waterbury, G., Preston-Hurlburt, P., Yoon, S.T., Medzhitov, R., Hong, S.C., and Janeway, C.A., Jr.: **The specificity and orientation of a TCR to its peptide-MHC class II ligands.** *Immunity* 1996, **4**:367–376.

Teng, M.K., Smolyar, A., Tse, A.G.D., Liu, J.H., Liu, J., Hussey, R.E., Nathenson, S.G., Chang, H.C., Reinherz, E.L., and Wang, J.H.: **Identification of a common docking topology with substantial variation among different TCR–MHC–peptide complexes.** *Curr. Biol.* 1998, **8**:409–412.

4-18 The CD4 and CD8 cell-surface proteins of T cells directly contact MHC molecules and are required to make an effective response to antigen.

Chang, H.C., Tan, K., Ouyang, J., Parisini, E., Liu, J.H., Le, Y., Wang, X., Reinherz, E.L., and Wang, J.H.: **Structural and mutational analyses of CD8αβ heterodimer and comparison with the CD8αα homodimer.** *Immunity* 2005, **6**:661–671.

Cheroutre, H., and Lambolez, F.: **Doubting the TCR coreceptor function of CD8αα.** *Immunity* 2008, **28**:149–159.

Gao, G.F., Tormo, J., Gerth, U.C., Wyer, J.R., McMichael, A.J., Stuart, D.I., Bell, J.I., Jones, E.Y., and Jakobsen, B.Y.: **Crystal structure of the complex between human CD8αα and HLA-A2.** *Nature* 1997, **387**:630–634.

Gaspar, R., Jr., Bagossi, P., Bene, L., Matko, J., Szollosi, J., Tozser, J., Fesus, L., Waldmann, T.A., and Damjanovich, S.: **Clustering of class I HLA oligomers with CD8 and TCR: three-dimensional models based on fluorescence resonance energy transfer and crystallographic data.** *J. Immunol.* 2001, **166**:5078–5086.

Kim, P.W., Sun, Z.Y., Blacklow, S.C., Wagner, G., and Eck, M.J.: **A zinc clasp structure tethers Lck to T cell coreceptors CD4 and CD8.** *Science* 2003, **301**:1725–1728.

Moody, A.M., North, S.J., Reinhold, B., Van Dyken, S.J., Rogers, M.E., Panico, M., Dell, A., Morris, H.R., Marth, J.D., and Reinherz, E.L.: **Sialic acid capping of CD8β core 1-O-glycans controls thymocyte-major histocompatibility complex class I interaction.** *J. Biol. Chem.* 2003, **278**:7240–7260.

Walker, L.J., Marrinan, E., Muenchhoff, M., Ferguson, J., Kloverpris, H., Cheroutre, H., Barnes, E., Goulder, P., and Klenerman, P.: **CD8αα expression marks terminally differentiated human CD8+ T cells expanded in chronic viral infection.** *Front Immunol.* 2013, **4**:223.

Wang, J.H., and Reinherz, E.L.: **Structural basis of T cell recognition of peptides bound to MHC molecules.** *Mol. Immunol.* 2002, **38**:1039–1049.

Wang, R., Natarajan, K., and Margulies, D.H.: **Structural basis of the CD8αβ/ MHC class I interaction: focused recognition orients CD8β to a T cell proximal position.** *J. Immunol.* 2009, **183**:2554–2564.

Wang, X.X., Li, Y., Yin, Y., Mo, M., Wang, Q., Gao, W., Wang, L., and Mariuzza, R.A.: **Affinity maturation of human CD4 by yeast surface display and crystal structure of a CD4-HLA-DR1 complex.** *Proc. Natl Acad. Sci. USA* 2011, **108**:15960–15965.

Wu, H., Kwong, P.D., and Hendrickson, W.A.: **Dimeric association and segmental variability in the structure of human CD4.** *Nature* 1997, **387**:527–530.

Yin, Y., Wang, X.X., and Mariuzza, R.A.: **Crystal structure of a complete ternary complex of T-cell receptor, peptide-MHC, and CD4.** *Proc. Natl Acad. Sci. USA* 2012, **109**:5405–5410.

Zamoyska, R.: **CD4 and CD8: modulators of T cell receptor recognition of antigen and of immune responses?** *Curr. Opin. Immunol.* 1998, **10**:82–86.

4-19 The two classes of MHC molecules are expressed differentially on cells.

Steimle, V., Siegrist, C.A., Mottet, A., Lisowska-Grospierre, B., and Mach, B.: **Regulation of MHC class II expression by interferon-γ mediated by the transactivator gene CIITA.** *Science* 1994, **265**:106–109.

4-20 A distinct subset of T cells bears an alternative receptor made up of γ and δ chains.

Adams, E.J., Chien, Y.H., and Garcia, K.C.: **Structure of a γδ T cell receptor in complex with the nonclassical MHC T22.** *Science* 2005, **308**:227–231.

Allison, T.J., and Garboczi, D.N.: **Structure of γδ T cell receptors and their recognition of non-peptide antigens.** *Mol. Immunol.* 2002, **38**:1051–1061.

Allison, T.J., Winter, C.C., Fournie, J.J., Bonneville, M., and Garboczi, D.N.: **Structure of a human γδ T-cell antigen receptor.** *Nature* 2001, **411**:820–824.

Das, H., Wang, L., Kamath, A., and Bukowski, J.F.: **Vγ2Vδ2 T-cell receptor-mediated recognition of aminobisphosphonates.** *Blood* 2001, **98**:1616–1618.

Luoma, A.M., Castro, C.D., Mayassi, T., Bembinster, L.A., Bai, L., Picard, D., Anderson, B., Scharf, L., Kung, J.E., Sibener, L.V., *et al.*: **Crystal structure of Vδ1 T cell receptor in complex with CD1d-sulfatide shows MHC-like recognition of a self-lipid by human γδ T cells.** *Immunity* 2013, **39**:1032–1042.

Vantourout, P., and Hayday, A.: **Six-of-the-best: unique contributions of γδ T cells to immunology.** *Nat. Rev. Immunol.* 2013, **13**:88–100.

Wilson, I.A., and Stanfield, R.L.: **Unraveling the mysteries of γδ T cell recognition.** *Nat. Immunol.* 2001, **2**:579–581.

Wingren, C., Crowley, M.P., Degano, M., Chien, Y., and Wilson, I.A.: **Crystal structure of a γδ T cell receptor ligand T22: a truncated MHC-like fold.** *Science* 2000, **287**:310–314.

Wu, J., Groh, V., and Spies, T.: **T cell antigen receptor engagement and specificity in the recognition of stress-inducible MHC class I-related chains by human epithelial γδ T cells.** *J. Immunol.* 2002, **169**:1236–1240.

The Generation of Lymphocyte Antigen Receptors

5

A lymphocyte expresses many exact copies of a single antigen receptor that has a unique antigen-binding site (see Section 1-12). The clonal expression of antigen receptors means that each lymphocyte is unique among the billions of lymphocytes that each person possesses. Chapter 4 described the structural features of immunoglobulins and T-cell receptors, the antigen receptors on B cells and T cells, respectively. We saw that the vast repertoire of antigen receptors results from variations in the amino acid sequence at the antigen-binding site, which is composed of the two **variable regions** from the two chains of the receptor. In immunoglobulins, these are the **heavy-chain variable region** (V_H) and the **light-chain variable region** (V_L), and in T-cell receptors, the V_α and V_β regions. The immunoglobulin domains of these regions contain three loops that comprise three **hypervariable regions**, or **complementarity-determining regions (CDRs)** (see Section 4-6) that determine the receptor's antigen binding site and allow for seemingly limitless diversity in specificity.

In the 1960s and 1970s, immunologists recognized that the limited size of the genome (at roughly 3 billion nucleotides) meant that the genome could not directly encode a sufficient number of genes to account for the observed diversity of antigen receptors. For example, encoding each distinct antibody by its own gene could easily fill the genome with nothing but antibody genes. As we will see, variable regions of the receptor chains are not directly encoded as a complete immunoglobulin domain by a single DNA segment. Instead, the variable regions are initially specified by so-called **gene segments** that encode only a part of the immunoglobulin domain. During the development of each lymphocyte, these gene segments are rearranged by a process of **somatic DNA recombination** to form a complete and unique variable-region coding sequence. This process is known generally as **gene rearrangement**. A fully assembled variable region sequence is produced by combining two or three types of gene segments, each of which is present in multiple copies in the germline genome. The final diversity of the receptor repertoire is the result of assembling complete antigen receptors from the many different gene segments of each type during the development of each individual lymphocyte. This process gives each new lymphocyte only one of many possible combinations of antigen receptors, providing the repertoire of diverse antigen specificities of naive B cells and T cells.

The first and second parts of this chapter describe the gene rearrangements that generate the primary repertoire of immunoglobulins and T-cell receptors. The mechanism of gene rearrangement is common to both B cells and T cells, and its evolution was probably critical to the evolution of the vertebrate adaptive immune system. The third part of the chapter explains how the transition from production of transmembrane immunoglobulins by activated B cells results in the production of secreted antibodies by plasma cells. Immunoglobulins can be synthesized as either transmembrane receptors or secreted antibodies, unlike T-cell receptors, which exist only as transmembrane receptors. Antibodies can also be produced with different types of constant regions, or isotypes (see Section 4-1). Here, we describe how the

IN THIS CHAPTER

Primary immunoglobulin gene rearrangement.

T-cell receptor gene rearrangement.

Structural variation in immunoglobulin constant regions.

Evolution of the adaptive immune response.

expression of the isotypes IgM and IgD is regulated, but we postpone describing how isotype switching occurs until Chapter 10, since that process and the affinity maturation of antibodies occurs normally in the context of an immune response. The last part of this chapter briefly examines alternative evolutionary forms of gene rearrangements that give rise to different forms of adaptive immunity in other species.

Primary immunoglobulin gene rearrangement.

Virtually any substance can be the target of an antibody response, and the response to even a single epitope comprises many different antibody molecules, each with a subtly different specificity for the epitope and a unique **affinity**, or binding strength. The total number of antibody specificities available to an individual is known as the **antibody repertoire** or **immunoglobulin repertoire**, and in humans is at least 10^{11} and probably several orders of magnitude greater. The number of antibody specificities present at any one time is, however, limited by the total number of B cells in an individual, as well as by each individual's previous encounters with antigens.

Before it was possible to examine the immunoglobulin genes directly, there were two main hypotheses for the origin of this diversity. The **germline theory** held that there is a separate gene for each different immunoglobulin chain and that the antibody repertoire is largely inherited. In contrast, **somatic diversification theories** proposed that the observed repertoire is generated from a limited number of inherited V-region sequences that undergo alteration within B cells during the individual's lifetime. Cloning of the immunoglobulin genes revealed that elements of both theories were correct and that the DNA sequence encoding each variable region is generated by rearrangements of a relatively small group of inherited gene segments. Diversity is further enhanced by the process of **somatic hypermutation** in mature activated B cells. Thus, the somatic diversification theory was essentially correct, although the germline theory concept of the existence of multiple germline genes also proved true.

5-1 Immunoglobulin genes are rearranged in the progenitors of antibody-producing cells.

Figure 5.1 shows the relationships between a light-chain variable region's antigen-binding site, its domain structure, and the gene that encodes it. The variable regions of immunoglobulin heavy and light chains are based on the **immunoglobulin fold**, which is composed of nine β sheets. The antibody-binding site is formed by three loops of amino acids known as hypervariable regions HV1, HV2, and HV3, or also CDR1, CDR2, and CDR3 (see Fig. 5.1a). These loops are located between the pairs of β sheets B and C, C' and C'', and F and G (see Fig. 5.1b). In a mature B cell, the variable regions for heavy and light chains are encoded by a single exon, but are separated from one another within this coding sequence (see Fig. 5.1c). This exon is the gene's second exon (exon 2). The first exon of the variable regions encodes the antibody's leader sequence, which directs the antibody into the endoplasmic reticulum for surface expression or secretion.

Unlike most genes, the complete DNA sequence of the variable-region exon is not present in the germline of the individual, but is originally encoded by two separate DNA segments, as illustrated in **Fig. 5.2**. These two DNA segments are spliced together to form the complete exon 2 as the B cell develops in the bone marrow. The first 95–101 amino acids of the variable region, encoding β sheets A–F and the first two complete hypervariable regions, originate from a

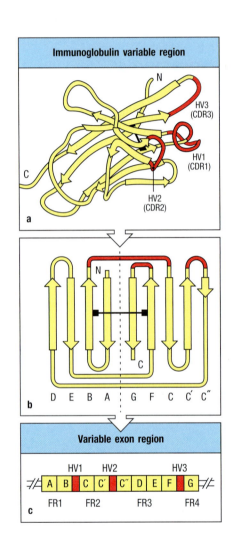

Fig. 5.1 Three hypervariable regions are encoded within a single V-region exon. Panel a: the variable region is based on the immunoglobulin (Ig) fold that is supported by framework regions (yellow) composed of nine β sheets and contains three hypervariable (HV) regions (red) that determine its antigen specificity. Panel b: the three HV regions exist as loops of amino acids between the β sheets of B and C, between C' and C'', and between F and G. Panel c: a complete variable region in a lymphocyte is encoded within a single exon of the full antigen-receptor gene. The three HV regions are interspersed between four framework regions (FRs) made up of the β sheets of the Ig domain.

Fig. 5.2 The CDR3 region originates from two or more individual gene segments that are joined during lymphocyte development. Panel a: a complete light-chain variable region encoding the CDR1, CDR2, and CDR3 loops resides in a single exon. Panel b: the complete variable region is derived from distinct germline DNA sequences. A V gene segment encodes the CDR1 and CDR2 loops, and the CDR3 loop is formed by sequences from the end of the V gene segment and the beginning of the J gene segment, and by nucleotides added or lost when these gene segments are joined during lymphocyte development. The exon for the CDR3 loop of the heavy chain is formed by the joining of sequences from V, D, and J gene segments (not shown).

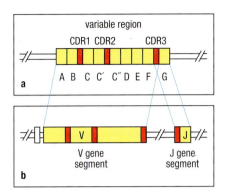

variable or **V gene segment** (see Fig. 5.2). This segment also contributes part of the third hypervariable region. Other parts of the third hypervariable region, and the remainder of the variable region including β sheet G (up to 13 amino acids), originate from a **joining** or **J gene segment**. By convention, we will refer to the exon encoding the complete variable region formed by the splicing together of these gene segments as the **V-region** gene.

In nonlymphoid cells, the V-region gene segments remain in their original germline configuration, and are a considerable distance away from the sequence encoding the C region. In mature B lymphocytes, however, the assembled V-region sequence lies much closer to the C region, as a consequence of a splicing event of the gene's DNA. Rearrangement within the immunoglobulin genes was originally discovered almost 40 years ago, when the techniques of restriction enzyme analysis first made it possible to study the organization of the immunoglobulin genes in both B cells and nonlymphoid cells. Such experiments showed that segments of genomic DNA within the immunoglobulin genes are rearranged in cells of the B-lymphocyte lineage, but not in other cells. This process of rearrangement is known as 'somatic' DNA recombination to distinguish it from the meiotic recombination that takes place during the production of gametes.

5-2 Complete genes that encode a variable region are generated by the somatic recombination of separate gene segments.

The rearrangements that produce the complete immunoglobulin light-chain and heavy-chain genes are shown in Fig. 5.3. For the light chain, the joining of a V_L and a J_L gene segment creates an exon that encodes the whole light-chain V_L region. In the unrearranged DNA, the V_L gene segments are located relatively far away from the exons encoding the constant region of the light chain (C_L region). The J_L gene segments are located close to the C_L region, however, and the joining of a V_L gene segment to a J_L gene segment also brings the V_L gene segment close to a C_L-region sequence. The J_L gene segment of the rearranged V_L region is separated from a C_L-region sequence only by a short intron. To make a complete immunoglobulin light-chain messenger RNA, the V-region exon is joined to the C-region sequence by RNA splicing after transcription.

For the heavy-chain, there is one additional complication. The heavy-chain V region (V_H) is encoded in three gene segments, rather than two. In addition to the V and J gene segments (denoted V_H and J_H to distinguish them from the light-chain V_L and J_L), the heavy chain uses a third gene segment called the **diversity** or **D_H gene segment**, which lies between the V_H and J_H gene segments. The recombination process that generates a complete heavy-chain V region is shown in Fig. 5.3 (right panel), and occurs in two separate stages. First, a D_H gene segment is joined to a J_H gene segment; then a V_H gene segment rearranges to DJ_H to make a complete V_H-region exon. As with the light-chain genes, RNA splicing joins the assembled V-region sequence to the neighboring C-region gene.

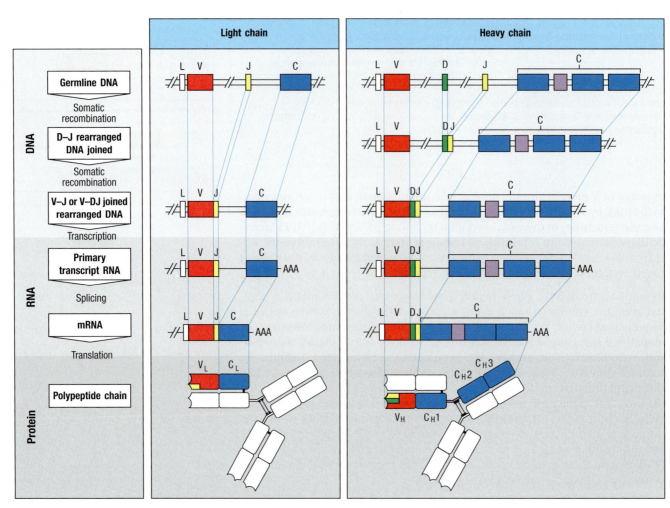

Fig. 5.3 V-region genes are constructed from gene segments. Light-chain V-region genes are constructed from two segments (center panel). A variable (V) and a joining (J) gene segment in the genomic DNA are joined to form a complete light-chain V-region exon. Immunoglobulin chains are extracellular proteins, and the V gene segment is preceded by an exon encoding a leader peptide (L), which directs the protein into the cell's secretory pathways and is then cleaved. The light-chain C region is encoded in a separate exon and is joined to the V-region exon by splicing of the light-chain RNA to remove the L-to-V and the J-to-C introns. Heavy-chain V regions are constructed from three gene segments (right panel). First, the diversity (D) and J gene segments join, and then the V gene segment joins to the combined DJ sequence, forming a complete V$_H$ exon. A heavy-chain C-region gene is encoded by several exons. The C-region exons, together with the leader sequence, are spliced to the V-domain sequence during processing of the heavy-chain RNA transcript. The leader sequence is removed after translation, and the disulfide bonds that link the polypeptide chains are formed. The hinge region is shown in purple.

5-3 Multiple contiguous V gene segments are present at each immunoglobulin locus.

For simplicity we have discussed the formation of a complete V-region sequence as though there were only a single copy of each gene segment. In fact, there are multiple copies of the V, D, and J gene segments in germline DNA. It is the random selection of just one gene segment of each type that produces the great diversity of V regions among immunoglobulins. The numbers of functional gene segments of each type in the human genome, as determined by gene cloning and sequencing, are shown in Fig. 5.4. Not all the gene segments discovered are functional, as some have accumulated mutations that prevent them from encoding a functional protein. Such genes are termed '**pseudogenes**.' Because there are many V, D, and J gene segments in germline DNA, no single gene segment is essential, resulting in a relatively large number of pseudogenes. Since some of these can undergo rearrangement just like a functional gene segment, a significant proportion of rearrangements incorporate a pseudogene and will thus be nonfunctional.

We saw in Section 4-1 that there are three sets of immunoglobulin chains—the heavy chain, and two equivalent types of light chains, the κ and λ **chains**. The immunoglobulin gene segments that encode these chains are organized into three clusters or **genetic loci**—the κ, λ, and heavy-chain loci—each of which can assemble a complete V-region sequence. Each locus is on a different chromosome and is organized slightly differently, as shown for the human loci in Fig. 5.5. At the λ light-chain locus, located on human chromosome 22, a cluster of $V_λ$ gene segments is followed by four (or in some individuals five) sets of $J_λ$ gene segments each linked to a single $C_λ$ gene. In the κ light-chain locus, on chromosome 2, the cluster of $V_κ$ gene segments is followed by a cluster of $J_κ$ gene segments, and then by a single $C_κ$ gene.

The organization of the heavy-chain locus, on chromosome 14, contains separate clusters of V_H, D_H, and J_H gene segments and of C_H genes. The heavy-chain locus differs in one important way: instead of a single C region, it contains a series of C regions arrayed one after the other, each of which corresponds to a different immunoglobulin isotype (see Fig. 5.19). While the $C_λ$ locus contains several distinct C regions, these encode similar proteins, which function similarly, whereas the different heavy-chain isotypes are structurally quite distinct and have different functions.

B cells initially express the heavy-chain isotypes μ and δ (see Section 4-1), which is accomplished by alternative mRNA splicing and which leads to the expression of immunoglobulins IgM and IgD, as we shall see in Section 5-14. The expression of other isotypes, such as γ (giving IgG), occurs through DNA rearrangements referred to as **class switching**, and takes place at a later stage, after a B cell is activated by antigen in an immune response. We describe class switching in Chapter 10.

The human V gene segments can be grouped into families in which each member shares at least 80% DNA sequence identity with all others in the

Number of functional gene segments in human immunoglobulin loci			
Segment	**Light chains**		**Heavy chain**
	κ	λ	H
Variable (V)	34–38	29–33	38–46
Diversity (D)	0	0	23
Joining (J)	5	4–5	6
Constant (C)	1	4–5	9

Fig. 5.4 The number of functional gene segments for the V regions of human heavy and light chains. The numbers shown are derived from exhaustive cloning and sequencing of DNA from one individual and exclude all pseudogenes (mutated and nonfunctional versions of a gene sequence). As a result of genetic polymorphism, the numbers will not be the same for all people.

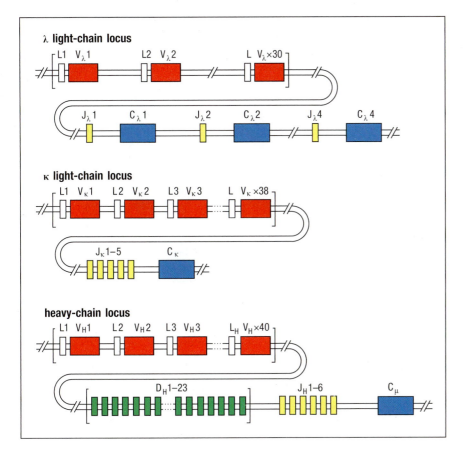

λ light-chain locus

κ light-chain locus

heavy-chain locus

Fig. 5.5 The germline organization of the immunoglobulin heavy- and light-chain loci in the human genome. Depending on the individual, the genetic locus for the λ light chain (chromosome 22) has between 29 and 33 functional $V_λ$ gene segments and four or five pairs of functional $J_λ$ gene segments and $C_λ$ genes. The κ locus (chromosome 2) is organized in a similar way, with about 38 functional $V_κ$ gene segments accompanied by a cluster of five $J_κ$ gene segments but with a single $C_κ$ gene. In approximately 50% of individuals, the entire cluster of $V_κ$ gene segments has undergone an increase by duplication (not shown, for simplicity). The heavy-chain locus (chromosome 14) has about 40 functional V_H gene segments and a cluster of around 23 D_H segments lying between these V_H gene segments and 6 J_H gene segments. The heavy-chain locus also contains a large cluster of C_H genes (see Fig. 5.19). For simplicity, all V gene segments have been shown in the same chromosomal orientation; only the first C_H gene (for $C_μ$) is shown, without illustrating its separate exons; and all pseudogenes have been omitted. This diagram is not to scale: the total length of the heavy-chain locus is more than 2 megabases (2 million bases), whereas some of the D gene segments are only 6 bases long.

family. Both the heavy-chain and the κ-chain V gene segments can be subdivided into seven families, and there are eight families of V_λ gene segments. The families can be grouped into clans, made up of families that are more similar to each other than to families in other clans. Human V_H gene segments fall into three clans. All the V_H gene segments identified from amphibians, reptiles, and mammals also fall into the same three clans, suggesting that these clans existed in a common ancestor of these modern animal groups. Thus, the V gene segments that we see today have arisen by a series of gene duplications and diversification through evolutionary time.

5-4 Rearrangement of V, D, and J gene segments is guided by flanking DNA sequences.

For a complete immunoglobulin or T-cell receptor chain to be expressed, DNA rearrangements must take place at the correct locations relative to the V, D, or J gene segment coding regions. In addition, these DNA rearrangements must be regulated such that a V gene segment is joined to a D or a J and not joined to another V gene segment. DNA rearrangements are guided by conserved noncoding DNA sequences, called **recombination signal sequences (RSSs)**, that are found adjacent to the points at which recombination takes place. The structure and arrangements of the RSSs are shown in **Fig. 5.6** for the λ and κ light-chain loci and the heavy-chain loci. An RSS consists of a conserved block of seven nucleotides—the **heptamer** 5′CACAGTG3′, which is always contiguous with the coding sequence; followed by a nonconserved region known as the **spacer**, which is either 12 or 23 base pairs (bp) long; followed by a second conserved block of nine nucleotides, the **nonamer** 5′ACAAAAACC3′.

The sequences given here are the consensus sequences, but they can vary substantially from one gene segment to another, even in the same individual, as there is some flexibility in the recognition of these sequences by the enzymes that carry out the recombination. The spacers vary in sequence, but their conserved lengths correspond to one turn (12 bp) or two turns (23 bp) of the DNA double helix. This is thought to bring the heptamer and nonamer sequences to the same side of the DNA helix to allow interactions with proteins that catalyze recombination, but this concept still lacks structural proof. The heptamer–spacer–nonamer sequence motif—the RSS—is always found directly adjacent to the coding sequence of V, D, or J gene segments. Recombination normally occurs between gene segments located on the same chromosome. A gene segment flanked by an RSS with a 12-bp spacer typically can be joined only to one flanked by a 23-bp spacer RSS. This is known as the **12/23 rule**.

Fig. 5.6 Recombination signal sequences are conserved heptamer and nonamer sequences that flank the gene segments encoding the V, D, and J regions of immunoglobulins. Recombination signal sequences (RSSs) are composed of heptamer (CACAGTG) and nonamer (ACAAAAACC) sequences that are separated by either 12 bp or approximately 23 bp of nucleotides. The heptamer–12-bp spacer–nonamer motif is depicted here as an orange arrowhead; the motif that includes the 23-bp spacer is depicted as a purple arrowhead. Joining of gene segments almost always involves a 12-bp and a 23-bp RSS—the 12/23 rule. The arrangement of RSSs in the V (red), D (green), and J (yellow) gene segments of heavy (H) and light (λ and κ) chains of immunoglobulin is shown here. The RAG-1 recombinase (see Section 5-5) cuts the DNA precisely between the last nucleotide of the V gene segment and the first C of the heptamer; or between the last G of the heptamer and the first nucleotide of the D or J gene segment. Note that according to the 12/23 rule, the arrangement of RSSs in the immunoglobulin heavy-chain gene segments precludes direct V-to-J joining.

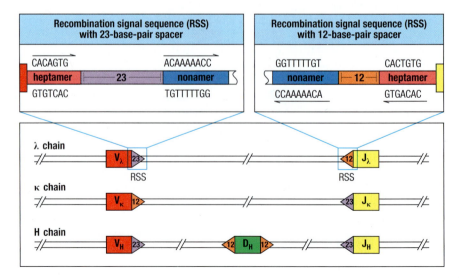

It is important to recognize that the pattern of 12- and 23-bp spacers used by the various gene segments is different between the λ, κ, and heavy-chain loci (see Fig. 5.6). Thus, for the heavy chain, a D_H gene segment can be joined to a J_H gene segment and a V_H gene segment to a D_H gene segment, but V_H gene segments cannot be joined to J_H gene segments directly, as both V_H and J_H gene segments are flanked by 23-bp spacers. However, they can be joined with a D_H gene segment between them, as D_H segments have 12-bp spacers on both sides (see Fig. 5.6).

In the antigen-binding region of an immunoglobulin, CDR1 and CDR2 are encoded directly in the V gene segment (see Fig. 5.2). CDR3 is encoded by the additional DNA sequence that is created by the joining of the V and J gene segments for the light chain, and the V, D, and J gene segments for the heavy chain. Further diversity in the antibody repertoire can be supplied by CDR3 regions that result from the joining of one D gene segment to another D gene segment, before being joined by a J gene segment. Such D–D joining is infrequent and seems to violate the 12/23 rule, suggesting that such violations of the 12/23 rule can occur at low frequency. In humans, D–D joining is found in approximately 5% of antibodies and is the major mechanism accounting for the unusually long CDR3 loops found in some heavy chains.

The mechanism of DNA rearrangement is similar for the heavy- and light-chain loci, although only one joining event is needed to generate a light-chain gene but two are required for a heavy-chain gene. When two gene segments are in the same transcriptional orientation in the germline DNA, their rearrangement involves the looping out and deletion of the DNA between them (Fig. 5.7, left panels). By contrast, when the gene segments have opposite transcriptional orientations, the rearrangement retains the intervening DNA in the chromosome but with an inverted orientation (see Fig. 5.7, right panels). This mode of recombination is less common, but it accounts for about half of all V_κ to J_κ joins in humans because the orientation of half the V_κ gene segments is opposite to that of the J_κ gene segments.

MOVIE 5.1

5-5 The reaction that recombines V, D, and J gene segments involves both lymphocyte-specific and ubiquitous DNA-modifying enzymes.

The overall enzymatic mechanisms involved in V-region rearrangement, or **V(D)J recombination**, are illustrated in Fig. 5.8. Two RSSs are brought together by interactions between proteins that specifically recognize the length of the spacers and thus enforce the 12/23 rule for recombination. The DNA molecule is then precisely cleaved by endonuclease activity at two locations and is then rejoined in a different configuration. The ends of the heptamer sequences are joined in a head-to-head fashion to form a **signal joint**. In the majority of cases, no nucleotides are lost or added between the two heptamer sequences, creating a double-heptamer sequence 5′CACAGTGCACAGTG3′ within the DNA molecule. When the joining segments are in the same orientation, the signal joint is contained in a circular piece of **extrachromosomal DNA** (see Fig. 5.7, left panels), which is lost from the genome when the cell divides. The V and J gene segments, which remain on the chromosome, join to form what is called the **coding joint**. When the joining segments are in the opposite relative orientation to each other within the chromosome (see Fig. 5.7, right panels), the signal joint is also retained within the chromosome, and the region of DNA between the V gene segment and the RSS of the J gene segment is inverted to form the coding joint. This situation leads to **rearrangement by inversion**. As we shall see later, the coding joint junction is imprecise, meaning that nucleotides can be added or lost between joined segments during the rearrangement process. This imprecise nature of coding joint formation adds to the variability in the V-region sequence, called **junctional diversity**.

Fig. 5.7 V-region gene segments are joined by recombination. Top panel: in every V-region recombination event, the recombination signal sequences (RSSs) flanking the gene segments are brought together to allow recombination to take place. The 12-bp-spaced RSSs are shown in orange, the 23-bp-spaced RSSs in purple. For simplicity, the recombination of a light-chain gene is illustrated; for a heavy-chain gene, two separate recombination events are required to generate a functional V region. Left panels: in most cases, the two segments undergoing rearrangement (the V and J gene segments in this example) are arranged in the same transcriptional orientation in the chromosome, and juxtaposition of the RSSs results in the looping out of the intervening DNA. Recombination occurs at the ends of the heptamer sequences in the RSSs, creating the so-called signal joint and releasing the intervening DNA in the form of a closed circle. Subsequently, the joining of the V and J gene segments creates the coding joint in the chromosomal DNA. Right panels: in other cases, the V and J gene segments are initially oriented in opposite transcriptional directions. In this case, alignment of the RSSs requires the coiled configuration shown, rather than a simple loop, so that joining the ends of the two heptamer sequences now results in the inversion and integration of the intervening DNA into a new position on the chromosome. Again, the joining of the V and J segments creates a functional V-region exon.

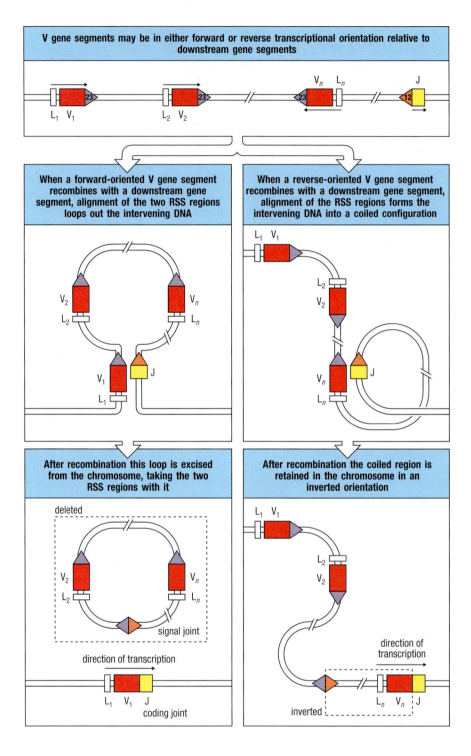

The complex of enzymes that act in concert to carry out somatic V(D)J recombination is termed the **V(D)J recombinase**. The lymphoid-specific components of the recombinase are called **RAG-1** and **RAG-2**, and they are encoded by two recombination-activating genes, *RAG1* and *RAG2*. This pair of genes is essential for V(D)J recombination, and they are expressed in developing lymphocytes only while the lymphocytes are engaged in assembling their antigen receptors, as described in more detail in Chapter 8. Indeed, the *RAG* genes expressed together can confer on nonlymphoid cells such as fibroblasts the capacity to rearrange exogenous segments of DNA containing the appropriate RSSs; this is how RAG-1 and RAG-2 were initially discovered.

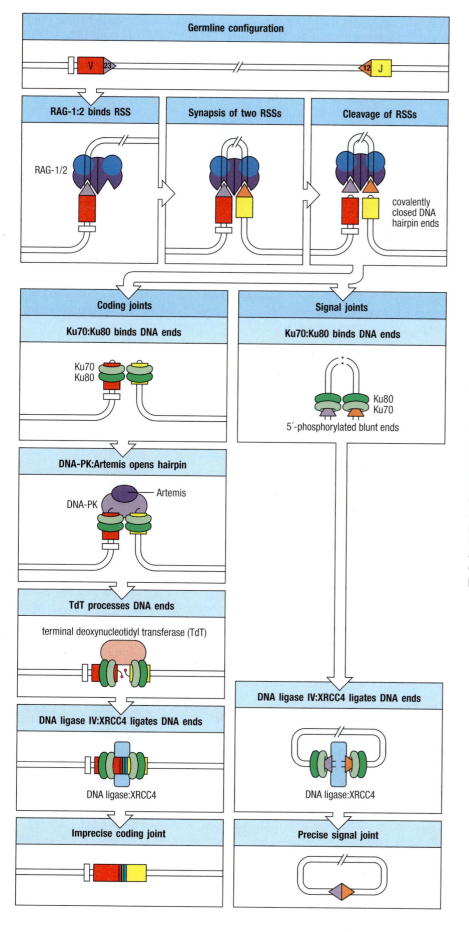

Fig. 5.8 Enzymatic steps in RAG-dependent V(D)J rearrangement. Recombination of gene segments containing recombination signal sequences (RSSs, triangles) begins with the binding of a complex of RAG-1 (purple), RAG-2 (blue), and high-mobility-group (HMG) proteins (not shown) to one of the RSSs flanking the coding sequences to be joined (second row). The RAG complex then recruits the other RSS. In the cleavage step, the endonuclease activity of RAG makes single-stranded cuts in the DNA backbone precisely between each coding segment and its RSS. At each cutting point this creates a 3′-OH group, which then reacts with a phosphodiester bond on the opposite DNA strand to generate a hairpin, leaving a blunt double-stranded break at the end of the RSS. These two types of DNA ends are resolved in different ways. At the coding ends (left panels), essential repair proteins such as Ku70:Ku80 (green) bind to the hairpin. Ku70:80 forms a ringlike structure as a heterodimer, but the monomers do not encircle the DNA. The DNA-PK:Artemis complex (purple) then joins the complex, and its endonuclease activity opens the DNA hairpin at a random site, yielding either two flush-ended DNA strands or a single-strand extension. The cut end is then modified by terminal deoxynucleotidyl transferase (TdT, pink) and exonuclease, which randomly add and remove nucleotides, respectively (this step is shown in more detail in Fig. 5.11). The two coding ends are finally ligated by DNA ligase IV in association with XRCC4 (turquoise). At the signal ends (right panels), Ku70:Ku80 binds to the RSS but the ends are not further modified. Instead, a complex of DNA ligase IV:XRCC4 joins the two ends precisely to form the signal joint.

The other proteins in the recombinase complex are members of the ubiquitously expressed **nonhomologous end joining (NHEJ)** pathway of DNA repair known as **double-strand break repair (DSBR)**. In all cells, this process is responsible for rejoining the two ends at the site of a double-strand break in DNA. The DSBR joining process is imprecise, meaning that nucleotides are frequently gained or lost at the site of joining. This has evolutionary relevance as in most cells it would not be advantageous to gain or lose nucleotides when repairing DSBs. However, in lymphocytes, the imprecise nature of DSBR is critical for junctional diversity and adaptive immunity. Thus, this may be the driving pressure for NHEJ to mediate imprecise joining. One ubiquitous protein contributing to DSBR is **Ku**, which is a heterodimer (Ku70:Ku80); this forms a ring around the DNA and associates tightly with a protein kinase catalytic subunit, DNA-PKcs, to form the **DNA-dependent protein kinase (DNA-PK)**. Another protein that associates with DNA-PKcs is **Artemis**, which has nuclease activity. The DNA ends are finally joined together by the enzyme **DNA ligase IV**, which forms a complex with the DNA repair protein **XRCC4**. DNA polymerases μ and λ participate in DNA-end fill-in synthesis. In addition, polymerase μ can add nucleotides in a template-independent manner. In summary, lymphocytes have adapted several enzymes used in common DNA repair pathways to help complete the process of somatic V(D)J recombination that is initiated by the RAG-1 and RAG-2 V(D)J recombinases.

The first reaction is an endonucleolytic cleavage that requires the coordinated activity of both RAG proteins. Initially, a complex of RAG-1 and RAG-2 proteins, together with high-mobility group chromatin protein HMGB1 or HMGB2, recognizes and aligns the two RSSs that are the target of the cleavage reaction. RAG-1 operates as a dimer, with RAG-2 acting as a cofactor (Fig. 5.9). RAG-1 specifically recognizes and binds the heptamer and the nonamer of the RSS and contains the Zn^{2+}-dependent endonuclease activity of the RAG protein complex. As a dimer, RAG-1 seems to align the two RSSs that will undergo rearrangement. Recent models suggest that the 12/23 rule may be established because an essential asymmetric orientation of the RAG-1:RAG-2 complex induces a preference for binding to RSS elements of different types (Fig. 5.10). The bound RAG complex makes a single-strand DNA break at the nucleotide just 5' of the heptamer of the RSS, thus creating a free 3'-OH group at the end of the coding segment. This nucleophilic 3'-OH group immediately attacks the phosphodiester bond on the opposite DNA strand, making a double-strand break and creating a DNA 'hairpin' at the coding region and a flush double-strand break at the end of the heptamer sequence. This cutting process occurs twice, once for the each gene segment being joined, producing four ends: two hairpin ends at the coding regions and two flush ends at both heptamer sequences (see Fig. 5.8). These DNA ends do not float apart, however, but are held tightly in the complex until a joining step has been completed. The blunt ends of the heptamer sequence are precisely joined by a complex of DNA ligase IV and XRCC4 to form the signal joint.

Formation of the coding joint is more complex. The two coding hairpin ends are each bound by Ku, which recruits the DNA-PKcs subunit. Artemis is recruited into this complex and is phosphorylated by DNA-PK. Artemis then opens the DNA hairpins by making a single-strand nick in the DNA. This nicking can happen at various points along the hairpin, which leads to sequence variability in the final joint. The DNA repair enzymes in the complex modify the opened

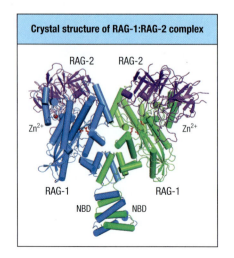

Crystal structure of RAG-1:RAG-2 complex

RAG-2   RAG-2

Zn^{2+} Zn^{2+}

RAG-1   RAG-1

NBD   NBD

Fig. 5.9 RAG-1 and RAG-2 form a heterotetramer capable of binding to two RSSs. Shown as ribbon diagrams, the RAG-1:RAG-2 complex contains two RAG-1 (green and blue) and two RAG-2 proteins (purple). The first 383 amino acids of RAG-1 were truncated before crystallization. The N-terminal nonamer binding domain (NBD) of the two RAG-1 proteins undergoes domain swapping and mediates dimerization of the two proteins. The remainder of the RAG-1 protein contains the endonuclease activity that is dependent on the binding of a Zn^{2+} ion. Each RAG-1 protein binds a separate RAG-2 protein. Courtesy of Martin Gellert.

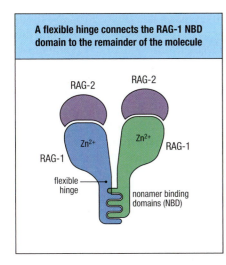

A flexible hinge connects the RAG-1 NBD domain to the remainder of the molecule

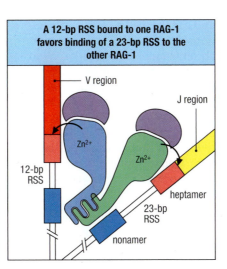

A 12-bp RSS bound to one RAG-1 favors binding of a 23-bp RSS to the other RAG-1

Fig. 5.10 The 12/23 base pair rule may result from asymmetric binding of RSSs to the RAG-1:RAG-2 dimer. Left panel: This cartoon of the structure shown in Fig 5.9 illustrates the flexibility of the hinge connecting the NBD to the catalytic domain of RAG-1. Right panel: the NBD domain of RAG-1 interacts with the RSS nonamer sequence (blue), while the RSS heptamer sequence (red) is bound to the portion of RAG-1 that contains the Zn^{2+} endonuclease activity. In this cartoon model, the interaction of a 12-bp RSS with one of the RAG-1 subunits induces the NBD domain to rotate toward the catalytic domain of RAG-1, to accommodate the length of the RSS. Since the two NBD domains are coupled by domain swaps, this induced conformation pulls the other NBD away from its RAG-1 subunit, which then prefers binding of the 23-bp RSS. The endonucleolytic cleavage (arrows) of the DNA by RAG-1 occurs precisely at the junction between the heptamer and the respective V, D, or J gene segment.

hairpins by removing nucleotides, while at the same time the lymphoid-specific enzyme **terminal deoxynucleotidyl transferase** (**TdT**), which is also part of the recombinase complex, adds nucleotides randomly to the single-strand ends. Addition and deletion of nucleotides can occur in any order, and one does not necessarily precede the other. Finally, DNA ligase IV joins the processed ends together, thus reconstituting a chromosome that includes the rearranged gene. This repair process creates diversity in the joint between gene segments while ensuring that the RSS ends are ligated without modification and that unintended genetic damage such as a chromosome break is avoided. Despite the use of some ubiquitous mechanisms of DNA repair, adaptive immunity based on the RAG-mediated generation of antigen receptors by somatic recombination seems to be unique to the jawed vertebrates, and its evolution is discussed in the last part of this chapter.

The *in vivo* roles of the enzymes involved in V(D)J recombination have been established through both natural and artificially induced mutations. Mice lacking TdT have about 10% of the normal level of non-templated nucleotides added to the joints between gene segments. This small remainder may result from the template-independent activity of DNA polymerase μ.

Mice in which either of the *RAG* genes has been inactivated, or which lack DNA-PKcs, Ku, or Artemis, suffer a complete block in lymphocyte development at the gene-rearrangement stage or make only trivial numbers of B and T cells. They are said to suffer from **severe combined immune deficiency** (**SCID**). The original *scid* mutation was discovered some time before the components of the recombination pathway were identified and was subsequently identified as a mutation in DNA-PKcs. In humans, mutations in *RAG1* or *RAG2* that result in partial V(D)J recombinase activity are responsible for an inherited disorder called **Omenn syndrome**, which is characterized by an absence of circulating B cells and an infiltration of skin by activated oligoclonal T lymphocytes. Mice deficient in components of ubiquitous DNA repair pathways, such as DNA-PKcs, Ku, or Artemis, are defective in double-strand break repair in general and are therefore also hypersensitive to ionizing radiation (which produces double-strand breaks). Defects in Artemis in humans produce a combined immunodeficiency of B and T cells that is associated with increased radiosensitivity. SCID caused by mutations in DNA repair pathways is called **irradiation-sensitive SCID** (**IR-SCID**) to distinguish it from SCID due to lymphocyte-specific defects.

Another genetic condition in which radiosensitivity is associated with some degree of immunodeficiency is **ataxia telangiectasia**, which is due to mutations in the protein kinase **ATM** (ataxia telangiectasia mutated), which are also associated with cerebellar degeneration and increased radiation sensitivity

X-linked Severe Combined Immunodeficiency

Ataxia Telangiectasia

and cancer risk. ATM is a serine/threonine kinase, like DNA-PKcs, and functions during V(D)J recombination by activating pathways that prevent the chromosomal translocations and large DNA deletions that can sometimes occur during resolution of DNA double-strand breaks. Some V(D)J recombination can occur in the absence of ATM, since the immune deficiencies seen in ataxia telangiectasia, which include low numbers of B and T cells and/or a deficiency in antibody class switching, are variable in their severity and are less severe than in SCID. Evidence that ATM and DNA-PKcs are partially redundant in their functions comes from the observation that B cells lacking both kinases show much more severely abnormal signal joining sequences compared with B cells lacking either enzyme alone.

5-6   The diversity of the immunoglobulin repertoire is generated by four main processes.

The gene rearrangements that combine gene segments to form a complete V-region exon generate diversity in two ways. First, there are multiple different copies of each type of gene segment, and different combinations of gene segments can be used in different rearrangement events. This **combinatorial diversity** is responsible for a substantial part of the diversity of V regions. Second, **junctional diversity** is introduced at the joints between the different gene segments as a result of the addition and subtraction of nucleotides by the recombination process. A third source of diversity is also combinatorial, arising from the many possible different combinations of heavy- and light-chain V regions that pair to form the antigen-binding site in the immunoglobulin molecule. The two means of generating combinatorial diversity alone could give rise, in theory, to approximately 1.9×10^6 different antibody molecules, as we will see below. Coupled with junctional diversity, it is estimated that at least 10^{11} different receptors could make up the repertoire of receptors expressed by naive B cells, and diversity could even be several orders of magnitude greater, depending on how one calculates junctional diversity. Finally, somatic hypermutation, which we describe in Chapter 10, occurs only in B cells after the initiation of an immune response and introduces point mutations into the rearranged V-region genes. This process generates further diversity in the antibody repertoire that can be selected for enhanced binding to antigen.

5-7   The multiple inherited gene segments are used in different combinations.

There are multiple copies of the V, D, and J gene segments, each of which can contribute to an immunoglobulin V region. Many different V regions can therefore be made by selecting different combinations of these segments. For human κ light chains, there are approximately 40 functional V_κ gene segments and 5 J_κ gene segments, and thus potentially 200 different combinations of complete V_κ regions. For λ light chains there are approximately 30 functional V_λ gene segments and 4 to 5 J_λ gene segments, yielding at least 120 possible V_λ regions (see Fig. 5.4). So, in all, 320 different light chains can be made as a result of combining different light-chain gene segments. For the heavy chains of humans, there are 40 functional V_H gene segments, approximately 25 D_H gene segments, and 6 J_H gene segments, and thus around 6000 different possible V_H regions ($40 \times 25 \times 6 = 6000$). During B-cell development, rearrangement at the heavy-chain gene locus to produce a heavy chain is followed by several rounds of cell division before light-chain gene rearrangement takes place, resulting in the same heavy chain being paired with different light chains in different cells. Because both the heavy- and the light-chain V regions contribute to antibody specificity, each of the 320 different light chains could be combined with each of the approximately 6000 heavy chains to give around 1.9×10^6 different antibody specificities.

This theoretical estimate of combinatorial diversity is based on the number of germline V gene segments contributing to functional antibodies (see Fig. 5.4); the total number of V gene segments is larger, but the additional gene segments are pseudogenes and do not appear in expressed immunoglobulin molecules. In practice, combinatorial diversity is likely to be less than one might expect from the calculations above. One reason is that not all V gene segments are used at the same frequency; some are common in antibodies, while others are found only rarely. This bias for or against certain V gene segments relates to their proximity with **intergenic control regions** within the heavy-chain locus that activate V(D)J recombination in developing B cells. Also, not every heavy chain can pair with every light chain: certain combinations of V_H and V_L regions will not form a stable molecule. Cells in which heavy and light chains fail to pair may undergo further light-chain gene rearrangement until a suitable chain is produced or they will be eliminated. Nevertheless, it is thought that most heavy and light chains can pair with each other, and that this type of combinatorial diversity has a major role in forming an immunoglobulin repertoire with a wide range of specificities.

5-8   Variable addition and subtraction of nucleotides at the junctions between gene segments contributes to the diversity of the third hypervariable region.

As noted earlier, of the three hypervariable loops in an immunoglobulin chain, CDR1 and CDR2 are encoded within the V gene segment. CDR3, however, falls at the joint between the V gene segment and the J gene segment, and in the heavy chain it is partly encoded by the D gene segment. In both heavy and light chains, the diversity of CDR3 is significantly increased by the addition and deletion of nucleotides at two steps in the formation of the junctions between gene segments. The added nucleotides are known as P-nucleotides and N-nucleotides, and their addition is illustrated in **Fig. 5.11**.

P-nucleotides are so called because they make up palindromic sequences added to the ends of the gene segments. As described in Section 5-5, the RAG proteins generate DNA hairpins at the coding ends of the V, D, or J segments, after which Artemis catalyzes a single-stranded cleavage at a random point within the coding sequence but near where the hairpin was first formed. When this cleavage occurs at a different point from the initial break induced by the RAG1/2 complex, a single-stranded tail is formed from a few nucleotides of the coding sequence plus the complementary nucleotides from the other DNA

Fig. 5.11 The introduction of P- and N-nucleotides diversifies the joints between gene segments during immunoglobulin gene rearrangement. The process is illustrated for a D_H to J_H rearrangement (first panel); however, the same steps occur in V_H to D_H and in V_L to J_L rearrangements. After formation of the DNA hairpins (second panel), the two heptamer sequences are ligated to form the signal joint (not shown here), while the Artemis:DNA-PK complex cleaves the DNA hairpin at a random site (indicated by the arrows) to yield a single-stranded DNA end (third panel). Depending on the site of cleavage, this single-stranded DNA may contain nucleotides that were originally complementary in the double-stranded DNA and which therefore form short DNA palindromes, such as TCGA and ATAT, as indicated by the light blue-shaded box. For example, the sequence GA at the end of the D segment shown is complementary to the preceding sequence TC. Such stretches of nucleotides that originate from the complementary strand are known as P-nucleotides. Where the enzyme terminal deoxynucleotidyl transferase (TdT) is present, nucleotides are added at random to the ends of the single-stranded segments (fourth panel); these nontemplated, or N, nucleotides are indicated by the shaded box. The two single-stranded ends then pair (fifth panel). Exonuclease trimming of unpaired nucleotides (sixth panel) and repair of the coding joint by DNA synthesis and ligation (bottom panel) leaves both P- and N-nucleotides (indicated by light blue shading) in the final coding joint. The randomness of insertion of P- and N-nucleotides makes an individual P–N region virtually unique and a valuable marker for following an individual B-cell clone as it develops, for instance in studies of somatic hypermutation.

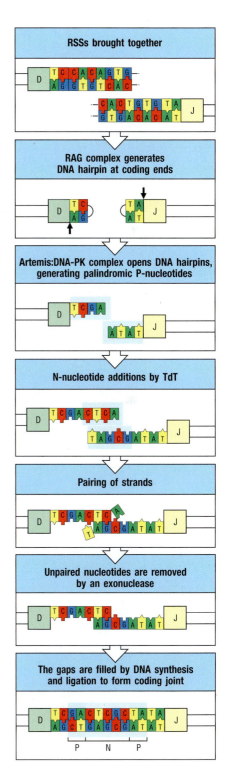

strand (see Fig. 5.11). In many light-chain gene rearrangements, DNA repair enzymes then fill in complementary nucleotides on the single-stranded tails, which would leave short palindromic sequences (the P-nucleotides) at the joint if the ends were rejoined without any further exonuclease activity.

In heavy-chain gene rearrangements and in a proportion of human light-chain gene rearrangements, however, **N-nucleotides** are added by a quite different mechanism before the ends are rejoined. N-nucleotides are so called because they are non-template-encoded. They are added by the enzyme TdT to the single-stranded ends of the coding DNA after hairpin cleavage. After the addition of up to 20 nucleotides, single-stranded stretches may have some complementary base pairs. Repair enzymes then trim off nonmatching nucleotides, synthesize complementary DNA to fill in the remaining single-stranded gaps, and ligate the new DNA to the palindromic region (see Fig. 5.11). TdT is maximally expressed during the period in B-cell development when the heavy-chain gene is being assembled, and so N-nucleotides are common in heavy-chain V–D and D–J junctions. N-nucleotides are less common in light-chain genes, which undergo rearrangement after heavy-chain genes, when TdT expression has been shut off, as we will explain further in Chapter 8 when discussing the specific developmental stages of B and T cells.

Nucleotides can also be deleted at gene segment junctions. This is accomplished by exonucleases, and although these have not yet been identified, Artemis has dual endonuclease and exonuclease activity and so could well be involved in this step. Thus, a heavy-chain CDR3 can be shorter than even the smallest D segment. In some instances it is difficult, if not impossible, to recognize the D segment that contributed to CDR3 formation because of the excision of most of its nucleotides. Deletions may also erase the traces of P-nucleotide palindromes introduced at the time of hairpin opening. For this reason, many completed VDJ joins do not show obvious evidence of P-nucleotides. As the total number of nucleotides added by these processes is random, the added nucleotides often disrupt the reading frame of the coding sequence beyond the joint. Such frameshifts will lead to a nonfunctional protein, and DNA rearrangements leading to such disruptions are known as **nonproductive rearrangements**. As roughly two in every three rearrangements will be nonproductive, many B-cell progenitors never succeed in producing functional immunoglobulin and therefore never become mature B cells. Thus, junctional diversity is achieved only at the expense of considerable loss of cells during B-cell development. In Chapter 8, we return to this topic when we discuss the cellular stages of B-cell development and how they relate to the temporal sequence of rearrangement of the V, D, and J gene segments of the antigen receptor chains.

Summary.

The extraordinary diversity of the immunoglobulin repertoire is achieved in several ways. Perhaps the most important factor enabling this diversity is that V regions are encoded by separate gene segments (V, D, and J gene segments), which are brought together by a somatic recombination process—V(D)J recombination—to produce a complete V-region exon. Many different gene segments are present in the genome of an individual, thus providing a heritable source of diversity that this combinatorial mechanism can use. Unique lymphocyte-specific recombinases, the RAG proteins, are absolutely required to catalyze this rearrangement, and the evolution of RAG proteins coincided with the appearance of the modern vertebrate adaptive immune system. Another substantial fraction of the functional diversity of immunoglobulins comes from the imprecise nature of the joining process itself. Variability at the coding joints between gene segments is generated by the insertion of random numbers of P- and N-nucleotides and by the variable deletion of nucleotides at the ends of some segments. These are brought about by the random opening of the hairpin by Artemis and by the actions of TdT. The association of different light- and heavy-chain V regions to form the antigen-binding site of an immunoglobulin molecule contributes further diversity. The combination of all of these sources of diversity generates a vast primary repertoire of antibody specificities.

T-cell receptor gene rearrangement.

The mechanism by which B-cell antigen receptors are generated is such a powerful means of creating diversity that it is not surprising that the antigen receptors of T cells bear structural resemblances to immunoglobulins and are generated by the same mechanism. In this part of the chapter we describe the organization of the T-cell receptor loci and the generation of the genes for the individual T-cell receptor chains.

5-9 The T-cell receptor gene segments are arranged in a similar pattern to immunoglobulin gene segments and are rearranged by the same enzymes.

Like immunoglobulin light and heavy chains, T-cell receptor (TCR) α and β chains each consist of a variable (V) amino-terminal region and a constant (C) region (see Section 4-10). The organization of the TCRα and TCRβ loci is shown in **Fig. 5.12**. The organization of the gene segments is broadly homologous to that of the immunoglobulin gene segments (see Sections 5-2 and 5-3). The TCRα locus, like the loci of the immunoglobulin light chains, contains V and J gene segments (V_α and J_α). The TCRβ locus, like the locus of the immunoglobulin heavy chain, contains D gene segments in addition to V_β and J_β gene segments. The T-cell receptor gene segments rearrange during T-cell development to form complete V-domain exons (**Fig. 5.13**). T-cell receptor gene rearrangement takes place in the thymus; the order and regulation of the rearrangements are dealt with in detail in Chapter 8. Essentially, however, the mechanics of gene rearrangement are similar for B and T cells. The T-cell receptor gene segments are flanked by 12-bp and 23-bp spacer recombination signal sequences (RSSs) that are homologous to those flanking immunoglobulin

Fig. 5.12 The germline organization of the human T-cell receptor α and β loci. The arrangement of the gene segments for the T-cell receptor resembles that at the immunoglobulin loci, with separate variable (V), diversity (D), and joining (J) gene segments, and constant (C) genes. The TCRα locus (chromosome 14) consists of 70–80 V_α gene segments, each preceded by an exon encoding the leader sequence (L). How many of these V_α gene segments are functional is not known exactly. A cluster of 61 J_α gene segments is located a considerable distance from the V_α gene segments. The J_α gene segments are followed by a single C gene, which contains separate exons for the constant and hinge domains and a single exon encoding the transmembrane and cytoplasmic regions (not shown). The TCRβ locus (chromosome 7) has a different organization, with a cluster of 52 functional V_β gene segments located distant from two separate clusters that each contain a single D gene segment together with six or seven J gene segments and a single C gene. Each TCRβ C gene has separate exons encoding the constant domain, the hinge, the transmembrane region, and the cytoplasmic region (not shown). The TCRα locus is interrupted between the J and V gene segments by another T-cell receptor locus—the TCRδ locus (not shown here; see Fig. 5.17).

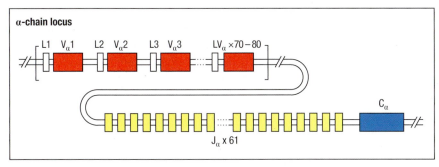

α-chain locus

L1 V_α1   L2 V_α2   L3 V_α3 LV_α ×70–80

C_α

J_α × 61

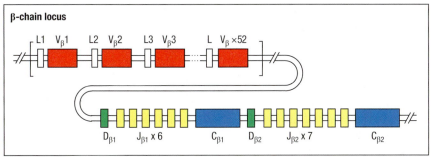

β-chain locus

L1 V_β1   L2 V_β2   L3 V_β3   L V_β ×52

$D_{\beta1}$ $J_{\beta1}$ × 6 $C_{\beta1}$ $D_{\beta2}$ $J_{\beta2}$ × 7 $C_{\beta2}$

Fig. 5.13 T-cell receptor α- and β-chain gene rearrangement and expression. The TCRα- and β-chain genes are composed of discrete segments that are joined by somatic recombination during development of the T cell. Functional α- and β-chain genes are generated in the same way that complete immunoglobulin genes are created. For the α chain (upper part of figure), a V_α gene segment rearranges to a J_α gene segment to create a functional V-region exon. Transcription and splicing of the VJ_α exon to C_α generates the mRNA that is translated to yield the T-cell receptor α-chain protein. For the β chain (lower part of figure), like the immunoglobulin heavy chain, the variable domain is encoded in three gene segments, V_β, D_β, and J_β. Rearrangement of these gene segments generates a functional VDJ_β V-region exon that is transcribed and spliced to join to C_β; the resulting mRNA is translated to yield the T-cell receptor β chain. The α and β chains pair soon after their synthesis to yield the α:β T-cell receptor heterodimer. Not all J gene segments are shown, and the leader sequences preceding each V gene segment are omitted for simplicity.

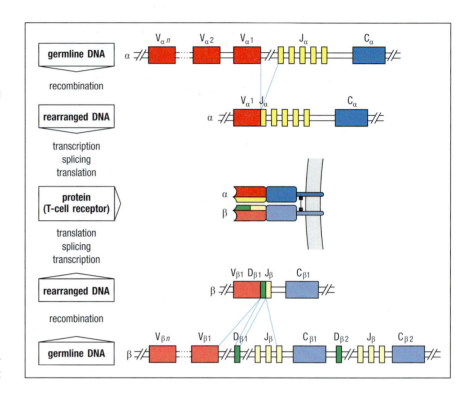

Fig. 5.14 Recombination signal sequences flank T-cell receptor gene segments. As in the immunoglobulin gene loci (see Fig. 5.6), the individual gene segments at the TCRα and TCRβ loci are flanked by heptamer–spacer–nonamer recombination signal sequences (RSSs). RSS motifs containing 12-bp spacers are depicted here as orange arrowheads, and those containing 23-bp spacers are shown in purple. Joining of gene segments almost always follows the 12/23 rule. Because of the disposition of heptamer and nonamer RSSs in the TCRβ and TCRδ loci, direct V_β to J_β joining is in principle allowed by the 12/23 rule (unlike in the immunoglobulin heavy-chain gene), although this occurs very rarely owing to other types of regulation.

gene segments (Fig. 5.14; see Section 5-4) and are recognized by the same enzymes. The DNA circles resulting from gene rearrangement (see Fig. 5.7) are known as **T-cell receptor excision circles** (**TRECs**) and are used as markers for T cells that have recently emigrated from the thymus. All known defects in genes that control V(D)J recombination affect T cells and B cells equally, and animals with these genetic defects lack functional B and T lymphocytes altogether (see Section 5-5). A further shared feature of immunoglobulin and T-cell receptor gene rearrangement is the presence of P- and N-nucleotides in the junctions between the V, D, and J gene segments of the rearranged TCRβ gene. In T cells, P- and N-nucleotides are also added between the V and J gene segments of all rearranged TCRα genes, whereas only about half of the V–J joints in immunoglobulin light-chain genes are modified by N-nucleotide addition, and these are often left without any P-nucleotides as well (Fig. 5.15; see Section 5-8).

The main differences between the immunoglobulin genes and those encoding T-cell receptors reflect the differences in how B cells and T cells function. All the effector functions of B cells depend upon secreted antibodies whose different heavy-chain C-region isotypes trigger distinct effector mechanisms. The effector functions of T cells, in contrast, depend upon cell–cell contact and are not mediated directly by the T-cell receptor, which serves only for antigen

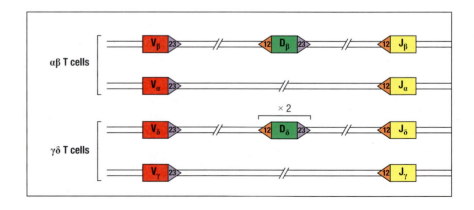

Element	Immunoglobulin		α:β T-cell receptors	
	H	**κ+λ**	**β**	**α**
Number of variable segments (V)	~40	~70	52	~70
Number of diversity segments (D)	23	0	2	0
Number of D segments read in three frames	rarely	–	often	–
Number of joining segments (J)	6	5(κ) 4(λ)	13	61
Number of joints with N- and P-nucleotides	2 (VD and DJ)	50% of joints	2 (VD and DJ)	1 (VJ)
Number of V gene pairs	1.9×10^6		5.8×10^6	
Number of junctional diversity	$\sim 3 \times 10^7$		$\sim 2 \times 10^{11}$	
Number of total diversity	$\sim 5 \times 10^{13}$		$\sim 10^{18}$	

Fig. 5.15 The number of human T-cell receptor gene segments and the sources of T-cell receptor diversity compared with those of immunoglobulins. Note that only about half of human κ chains contain N-nucleotides. Somatic hypermutation as a source of diversity is not included in this figure because it does not occur in T cells.

recognition. Thus, the C regions of the TCRα and TCRβ loci are much simpler than those of the immunoglobulin heavy-chain locus. There is only one Cα gene, and although there are two Cβ genes, they are very closely homologous and there is no known functional distinction between their products. The T-cell receptor C-region genes encode only transmembrane polypeptides.

Another difference between the rearrangement of immunoglobulin genes and T-cell receptor genes is in the nature of the RSSs surrounding the D gene segments. For the immunoglobulin heavy chain, the D segment is surrounded by two RSSs, both with a 12-bp spacing (see Fig. 5.6), whereas the D segments in the TCRβ and TCRγ loci have a 5′ 12-bp RSS and a 3′ 23-bp RSS (see Fig. 5.14). The arrangement in the immunoglobulin locus naturally enforces the inclusion of D segments in the heavy-chain V region, since direct V to J joining would violate the 12/23 rule. However, in the T-cell receptor loci, direct V to J joining would not violate this rule, since the 23-bp RSS of the $V_β$ or $V_γ$ segment is compatible with the 12-bp RSS of the J gene segment, and yet normally, little to no such direct joining is observed. Instead, regulation of gene rearrangements appears to be controlled by mechanisms beyond the 12/23 rule, and these mechanisms are still being investigated.

5-10 T-cell receptors concentrate diversity in the third hypervariable region.

The three-dimensional structure of the antigen-recognition site of a T-cell receptor looks much like that of the antigen-recognition site of an antibody molecule (see Sections 4-10 and 4-7, respectively). In an antibody, the center of the antigen-binding site is formed by the CDR3 loops of the heavy and light chains. The structurally equivalent third hypervariable loops of the T-cell receptor α and β chains, to which the D and J gene segments contribute, also form the center of the antigen-binding site of a T-cell receptor; the periphery of the site consists of the CDR1 and CDR2 loops, which are encoded within the germline V gene segments for the α and β chains. The extent and pattern of variability in T-cell receptors and immunoglobulins reflect the distinct nature of their ligands. Whereas the antigen-binding sites of immunoglobulins must conform to the surfaces of an almost infinite variety of different antigens, and thus come in a wide variety of shapes and chemical properties, the ligand for the major class of human T-cell receptors (α:β) is always a peptide bound to an MHC molecule. As a group, the antigen-recognition sites of T-cell receptors

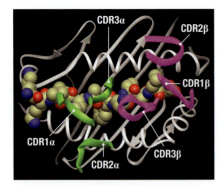

Fig. 5.16 The most variable parts of the T-cell receptor interact with the peptide of a peptide:MHC complex. The CDR loops of a T-cell receptor are shown as colored tubes, which in this figure are superimposed on the peptide:MHC complex (MHC, gray; peptide, yellow-green with O atoms in red and N atoms in blue). The CDR loops of the α chain are in green, while those of the β chain are in magenta. The CDR3 loops lie in the center of the interface between the T-cell receptor and the peptide:MHC complex, and make direct contact with the antigenic peptide.

would therefore be predicted to have a less variable shape, with most of the variability focused on the bound antigenic peptide occupying the center of the surface in contact with the receptor. Indeed, the less variable CDR1 and CDR2 loops of a T-cell receptor mainly contact the relatively less variable MHC component of the ligand, whereas the highly variable CDR3 regions mainly contact the unique peptide component (Fig. 5.16).

The structural diversity of T-cell receptors is attributable mainly to combinatorial and junctional diversity generated during the process of gene rearrangement. It can be seen from Fig. 5.15 that most of the variability in T-cell receptor chains is in the junctional regions, which are encoded by V, D, and J gene segments and modified by P- and N-nucleotides. The TCRα locus contains many more J gene segments than either of the immunoglobulin light-chain loci: in humans, 61 J_α gene segments are distributed over about 80 kb of DNA, whereas immunoglobulin light-chain loci have only 5 J gene segments at most (see Fig. 5.15). Because the TCRα locus has so many J gene segments, the variability generated in this region is even greater for T-cell receptors than for immunoglobulins. Thus, most of the diversity resides in the CDR3 loops that contain the junctional region and form the center of the antigen-binding site.

5-11 γ:δ T-cell receptors are also generated by gene rearrangement.

A minority of T cells bear T-cell receptors composed of γ and δ chains (see Section 4-20). The organization of the TCRγ and TCRδ loci (Fig. 5.17) resembles that of the TCRα and TCRβ loci, although there are important differences. The cluster of gene segments encoding the δ chain is found entirely within the TCRα locus, between the V_α and the J_α gene segments. V_δ genes are

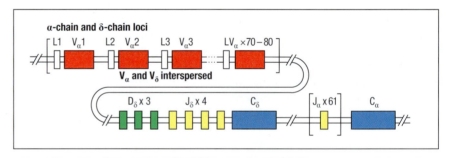

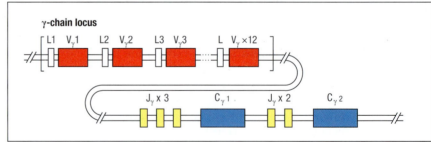

Fig. 5.17 The organization of the T-cell receptor γ- and δ-chain loci in humans. The TCRγ and TCRδ loci, like the TCRα and TCRβ loci, have discrete V, D, and J gene segments, and C genes. Uniquely, the locus encoding the δ chain is located entirely within the α-chain locus. The three D_δ gene segments, four J_δ gene segments, and the single δ C gene lie between the cluster of V_α gene segments and the cluster of J_α gene segments. There are two V_δ gene segments (not shown) located near the δ C gene, one just upstream of the D regions and one in inverted orientation just downstream of the C gene. In addition, there are six V_δ gene segments interspersed among the V_α gene segments. Five are shared with V_α and can be used by either locus, and one is unique to the δ locus. The human TCRγ locus resembles the TCRβ locus in having two C genes, each with its own set of J gene segments. The mouse γ locus (not shown) has a more complex organization and there are three functional clusters of γ gene segments, each containing V and J gene segments and a C gene. Rearrangement at the γ and δ loci proceeds as for the other T-cell receptor loci, with the exception that during TCRδ rearrangement two D segments can be used in the same gene. The use of two D segments greatly increases the variability of the δ chain, mainly because extra N-region nucleotides can be added at the junction between the two D gene segments as well as at the V–D and D–J junctions.

interspersed with the V_α genes but are located primarily in the 3′ region of the locus. Because all V_α gene segments are oriented such that rearrangement will delete the intervening DNA, any rearrangement at the α locus results in the loss of the δ locus (**Fig. 5.18**). There are substantially fewer V gene segments at the TCRγ and TCRδ loci than at either the TCRα or TCRβ loci or any of the immunoglobulin loci. Increased junctional variability in the δ chains may compensate for the small number of V gene segments and has the effect of focusing almost all the variability in the γ:δ receptor in the junctional region. As we have seen for the α:β T-cell receptors, the amino acids encoded by the junctional regions lie at the center of the T-cell receptor binding site. T cells bearing γ:δ receptors are a distinct lineage of T cells, and as discussed in Chapter 4, some γ:δ T cells recognize nonclassical MHC class I molecules and other molecules whose expression may be an indication of cellular damage or infection. As we saw in Section 4-20, the CDR3 of a γ:δ T cell is frequently longer than the CDR3 in an α:β T-cell receptor; this permits the CDR of γ:δ T cell receptors to interact directly with ligand and also contributes to the great diversity of these receptors. We will discuss the regulation of the fate choice between the α:β and γ:δ T-cell lineages in Chapter 8.

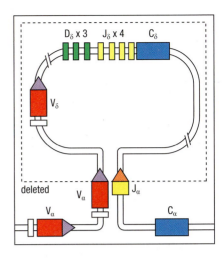

Fig. 5.18 Deletion of the TCRδ locus is induced by rearrangement of a V_α to J_α gene segment. The TCRδ locus is entirely contained within the chromosomal region containing the TCRα locus. When any V region in the V_α/V_δ region rearranges to any one of the J_α segments, the intervening region, and the entire V_δ locus, is deleted. Thus, V_α rearrangement prevents any continued expression of a V_δ gene and precludes lineage development down the γ:δ pathway.

Summary.

T-cell receptors are structurally similar to immunoglobulins and are encoded by homologous genes. T-cell receptor genes are assembled by somatic recombination from sets of gene segments in the same way that the immunoglobulin genes are. Diversity is, however, distributed differently in immunoglobulins and T-cell receptors: the T-cell receptor loci have roughly the same number of V gene segments as the immunoglobulin loci but more J gene segments, and there is greater diversification of the junctions between gene segments during the process of gene rearrangement. Thus, the greatest diversity of the T-cell receptor is in the central part of the receptor, within the CDR3, which in the case of α:β T-cell receptors contacts the bound peptide fragment of the ligand. Most of the diversity among γ:δ T-cell receptors is also within the CDR3, which is frequently longer than the CDR3 of α:β T-cell receptors and can also directly interact with ligands recognized by the γ:δ T cells.

Structural variation in immunoglobulin constant regions.

This chapter so far has focused on the mechanisms of assembly of the V regions for immunoglobulins and T-cell receptors. We now turn to the C regions. The C regions of T-cell receptors act only to support the V regions and anchor the receptor into the membrane, and they do not vary after assembly of a complete receptor gene. Immunoglobulins, in contrast, can be made as both a transmembrane receptor and a secreted antibody, and they can be made in several different classes, depending on the different C regions used by the heavy chain. The light-chain C regions (C_L) provide only structural attachment for V regions, and there seem to be no functional differences between λ and κ light chains. The heavy-chain locus encodes different C regions (C_H) that are present as separate genes located downstream of the V-region segments. Initially, naive B cells use only the first two of these, the C_μ and C_δ genes, which are expressed along with the associated assembled V-region sequence to produce transmembrane IgM and IgD on the surface of the naive B cell.

In this section, we introduce the different heavy-chain isotypes and discuss some of their special properties as well as the structural features that distinguish the C_H regions of antibodies of the five major classes. We explain how naive B cells express both C_μ and C_δ isotypes at the same time and how the same antibody gene can generate both membrane-bound immunoglobulin and secreted immunoglobulin through alternative mRNA splicing. During an antibody response, activated B cells can switch to the expression of C_H genes other than C_μ and C_δ by a type of somatic recombination known as class switching (discussed in Chapter 10) that links different heavy-chain C regions (C_H) to the rearranged VDJ_H gene segment.

5-12 Different classes of immunoglobulins are distinguished by the structure of their heavy-chain constant regions.

The five main classes of immunoglobulins are IgM, IgD, IgG, IgE, and IgA, all of which can occur as transmembrane antigen receptors or secreted antibodies (Fig. 5.19). In humans, IgG is found as four subclasses (IgG1, IgG2, IgG3, and IgG4), named by decreasing order of their abundance in serum, and IgA antibodies are found as two subclasses (IgA1 and IgA2). The different heavy chains that define these classes are known as isotypes and are designated by the lowercase Greek letters μ, δ, γ, ε, and α. The different heavy chains are encoded by different immunoglobulin C_H genes located in a gene cluster that is 3′ of the J_H segments as illustrated in Fig. 5.19. Figure 5.20 lists the major physical and functional properties of the different human antibody classes.

The functions of the immunoglobulin classes are discussed in detail in Chapter 10, in the context of the humoral immune response; here, we just touch on them briefly. IgM is the first class of immunoglobulin produced after activation of a B cell, and the IgM antibody is secreted as a pentamer (see Section 5-14 and Fig. 5.21). This accounts for the high molecular weight of IgM and the fact that it is normally present in the bloodstream but not in tissues. Being a pentamer

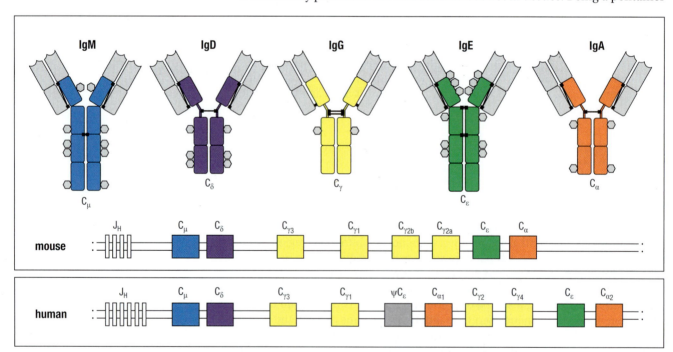

Fig. 5.19 The immunoglobulin isotypes are encoded by a cluster of immunoglobulin heavy-chain C-region genes.
The general structure of the main immunoglobulin isotypes (above in upper panel) is indicated, with each rectangle denoting an immunoglobulin domain. These isotypes are encoded by separate heavy-chain C-region genes arranged in a cluster in both mouse and human (lower panel). The constant region of the heavy chain for each isotype is indicated by the same color as the C-region gene segment that encodes it. IgM and IgE lack a hinge region but each contains an extra heavy-chain domain. Note the differences

in the number and location of the disulfide bonds (black lines) linking the chains. The isotypes also differ in the distribution of N-linked carbohydrate groups, shown as hexagons. In humans, the gene cluster shows evidence of evolutionary duplication of a unit consisting of two γ genes, an ε gene, and an α gene. One of the ε genes is a pseudogene (ψ); hence only one subtype of IgE is expressed. For simplicity, other pseudogenes are not illustrated, and the exon details within each C gene are not shown. The classes of immunoglobulins found in mice are called IgM, IgD, IgG1, IgG2a, IgG2b, IgG3, IgA, and IgE.

	Immunoglobulin								
	IgG1	IgG2	IgG3	IgG4	IgM	IgA1	IgA2	IgD	IgE
Heavy chain	γ_1	γ_2	γ_3	γ_4	μ	α_1	α_2	δ	ε
Molecular weight (kDa)	146	146	165	146	970	160	160	184	188
Serum level (mean adult mg/ml)	9	3	1	0.5	1.5	3.0	0.5	0.03	5×10^{-5}
Half-life in serum (days)	21	20	7	21	10	6	6	3	2
Classical pathway of complement activation	++	+	+++	−	++++	−	−	−	−
Alternative pathway of complement activation	−	−	−	−	−	+	−	−	−
Placental transfer	+++	+	++	−/+	−	−	−	−	−
Binding to macrophage and phagocyte Fc receptors	+	−	+	−/+	−	+	+	−	+
High-affinity binding to mast cells and basophils	−	−	−	−	−	−	−	−	+++
Reactivity with staphylococcal Protein A	+	+	−/+	+	−	−	−	−	−

Fig. 5.20 The physical and functional properties of the human immunoglobulin isotypes. IgM is so called because of its size: although monomeric IgM is only 190 kDa, it normally forms pentamers, known as macroglobulin (hence the M), of very large molecular weight (see Fig. 5.23). IgA dimerizes to give an approximate molecular weight of around 390 kDa in secretions. IgE antibody is associated with immediate-type hypersensitivity. When fixed to tissue mast cells, IgE has a much longer half-life than its half-life in plasma shown here. The relative activities of the various isotypes are compared for several functions, ranging from inactive (−) to most active (++++).

also increases the avidity of IgM for antigens before its affinity is increased through the process of affinity maturation.

IgG isotypes produced during an immune response are found in the bloodstream and in the extracellular spaces in tissues. IgM and most IgG isotypes can interact with the complement component C1 to activate the classical complement pathway (described in Section 2-7). IgA and IgE do not activate complement. IgA can be found in the bloodstream, but it also acts in the defense of mucosal surfaces; it is secreted into the gut and respiratory tract, and also into mother's milk. IgE is particularly involved in defense against multicellular parasites (for example, schistosomes), but it is also the antibody involved in common allergic diseases such as allergic asthma. IgG and IgE are always monomers, but IgA can be secreted as either a monomer or a dimer.

Sequence differences in the constant regions of the immunoglobulin heavy chains produce the distinct characteristics of each antibody isotype. These characteristics include the number and location of interchain disulfide bonds, the number of attached carbohydrate groups, the number of C domains, and the length of the hinge region (see Fig. 5.19). IgM and IgE heavy chains contain an extra C domain that replaces the hinge region found in γ, δ, and α chains. The absence of the hinge region does not imply that IgM and IgE molecules lack flexibility; electron micrographs of IgM molecules binding to ligands show that the Fab arms can bend relative to the Fc portion. However, such a difference in structure may have functional consequences that are not yet characterized. Different isotypes and subtypes also differ in their ability to engage various effector functions, as described below.

5-13   The constant region confers functional specialization on the antibody.

Antibodies can protect the body in a variety of ways. In some cases it is enough for the antibody simply to bind antigen. For instance, by binding tightly to a toxin or virus, an antibody can prevent it from recognizing its receptor on a host cell (see Fig. 1.25). The antibody V regions are sufficient for this activity. The C region is essential, however, for recruiting the help of other cells and molecules to destroy and dispose of pathogens to which the antibody has bound.

The Fc region contains all C regions of an antibody and has three main effector functions: Fc-receptor binding, complement activation, and regulation of secretion. First, the Fc region of certain isotypes binds to specialized Fc receptors expressed by immune effector cells. Fcγ receptors expressed on the surface of macrophages and neutrophils bind the Fc portions of IgG1 and IgG3 antibodies, facilitating the phagocytosis of pathogens coated with these antibodies. The Fc region of IgE binds to a high-affinity Fcε receptor on mast cells, basophils, and activated eosinophils, triggering the release of inflammatory mediators in response to antigens. We will return to this topic in Section 10-19.

Second, the Fc regions in antigen:antibody complexes can bind to the C1q complement protein (see Section 2-7) and initiate the classical complement cascade, which recruits and activates phagocytes to engulf and destroy pathogens. Third, the Fc portion can deliver antibodies to places they would not reach without active transport. These include transport of IgA into mucous secretions, tears, and milk, and the transfer of IgG from the pregnant mother into the fetal blood circulation. In both cases, the Fc portion of IgA or IgG engages a specific receptor, the neonatal Fc receptor (FcRn), that actively transports the immunoglobulin through cells to reach different body compartments. Podocytes in the kidney glomerulus express FcRn to help remove IgG that has been filtered from the blood and accumulated at the glomerular basement membrane.

The role of the Fc portion in these effector functions has been demonstrated by studying immunoglobulins that have had one or more Fc domains cleaved off enzymatically or modified genetically. Many microorganisms have responded to the destructive potential of the Fc portion by evolving proteins that either bind it or cleave it, and so prevent the Fc region from working; examples are Protein A and Protein G of *Staphylococcus* and Protein D of *Haemophilus*. Researchers have exploited these proteins to help map the Fc region and also as immunological reagents. Not all immunoglobulin classes have the same capacity to engage each of the effector functions (see Fig. 5.20). For example, IgG1 and IgG3 have a higher affinity than IgG2 for the most common type of Fc receptor.

5-14 IgM and IgD are derived from the same pre-mRNA transcript and are both expressed on the surface of mature B cells.

The immunoglobulin C_H genes form a large cluster spanning about 200 kb to the 3′ side of the J_H gene segments (see Fig. 5.19). Each C_H gene is split into several exons (not shown in the figure), with each exon corresponding to an individual immunoglobulin domain in the folded C region. The gene encoding the μ C region lies closest to the J_H gene segments, and therefore closest to the assembled V_H-region exon (VDJ exon) after DNA rearrangement. Once rearrangement is completed, transcription from a promoter just 5′ to the rearranged VDJ exon produces a complete μ heavy-chain transcript. Any J_H gene segments remaining between the assembled V gene and the C_μ gene are removed during RNA processing to generate the mature mRNA. The μ heavy chains are therefore the first to be expressed, and IgM is the first immunoglobulin to be produced during B-cell development.

Immediately 3′ to the μ gene lies the δ gene, which encodes the C region of the IgD heavy chain (see Fig. 5.19). IgD is coexpressed with IgM on the surface of almost all mature B cells, but is secreted in only small amounts by plasma cells. The unique function of IgD is still unclear and a matter of active research. Because IgD has hinge regions that are more flexible than those in IgM, IgD

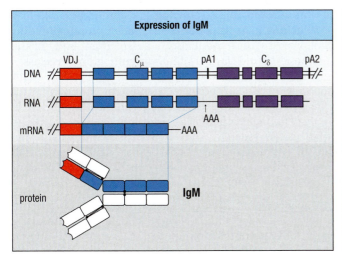

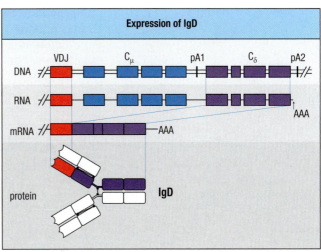

Fig. 5.21 Coexpression of IgD and IgM is regulated by RNA processing. In mature B cells, transcription initiated at the V_H promoter extends through both C_μ and C_δ exons. This long primary transcript is then processed by cleavage and polyadenylation (AAA), and by splicing. Cleavage and polyadenylation at the μ site (pA1) and splicing between C_μ exons yields an mRNA encoding the μ heavy chain (left panel). Cleavage and polyadenylation at the δ site (pA2) and a different pattern of splicing that joins the V region exon to the C_δ exons and removes the C_μ exons yields mRNA encoding the complete δ heavy chain (right panel). For simplicity we have not shown all the individual C-region exons.

has been suggested to be an auxiliary receptor that may facilitate the binding of antigens by naive B cells. Mice lacking the C_δ exons show normal B-cell development and can generate largely normal antibody responses, but show a delay in the process of affinity maturation of antibody for antigens. We return to this topic in Chapter 10, when we discuss somatic hypermutation.

B cells expressing IgM and IgD have not undergone class switching, which requires irreversible changes to the DNA. Instead, these B cells produce a long primary mRNA transcript that is differentially spliced to yield either of two distinct mRNA molecules (Fig. 5.21). In one transcript, the VDJ exon is spliced to the C_μ exons and undergoes polyadenylation from a nearby site (pA1), to encode a complete IgM molecule. The second RNA transcript extends well beyond this site and includes the downstream C_δ exons. In this transcript, the VDJ exon is spliced to these C_δ exons and polyadenylation occurs at a separate site downstream (pA2). This transcript encodes an IgD molecule.

It has been known since the 1980s that the processing of the long mRNA transcript is developmentally regulated, with immature B cells making mostly the μ transcript and mature B cells making mostly the δ along with some μ, although until recently there was little to no molecular explanation. A recent forward genetic screen of *N*-ethyl-*N*-nitrosourea (ENU)-induced mutagenesis in mice identified a gene involved in IgD expression that regulates the alternative splicing process. The gene encodes **ZFP318**, a protein structurally related to the U1 small nuclear ribonucleoprotein of the spliceosome, the RNA–protein complex that is required for mRNA splicing. ZFP318 is not expressed in immature B cells, where the IgD transcript is not produced, but becomes expressed in mature and activated B cells that coexpress IgD with IgM. ZFP318 is required for alternative splicing of the long pre-mRNA from the VDJ exon to the C_δ exons, as mice with a fully inactivated ZFP318 gene fail to express IgD and express increased levels of IgM. While the precise mechanism is unclear, it seems likely that ZFP318 may act directly on the pre-mRNA transcript during elongation, by suppressing splicing of the VDJ exon to the C_μ exons, allowing transcript elongation and promoting splicing to the C_δ exons. In short, expression of ZFP318 promotes IgD expression, although how ZFP318 expression itself is regulated in immature and mature B cells is still unknown.

5-15 Transmembrane and secreted forms of immunoglobulin are generated from alternative heavy-chain mRNA transcripts.

Each of the immunoglobulin isotypes can be produced either as a membrane-bound receptor or as secreted antibodies. B cells initially express the transmembrane form of IgM; after stimulation by antigen, some of their progeny differentiate into plasma cells producing IgM antibodies, whereas others undergo class switching to express transmembrane immunoglobulins of a different class, followed by the production of secreted antibody of the new class. The membrane-bound forms of all immunoglobulin classes are monomers comprising two heavy and two light chains. IgM and IgA polymerize only when they have been secreted. The membrane-bound form of immunoglobulin heavy chain has at the carboxy terminus a hydrophobic transmembrane domain of about 25 amino acid residues that anchors it to the surface of the B lymphocyte. The secreted form replaces this transmembrane domain with a carboxy terminus composed of a hydrophilic secretory tail. These two forms of carboxy termini are encoded by different exons found at the end of each C_H gene as these exons undergo alternative RNA processing.

For example, the IgM heavy-chain gene contains four exons—$C_\mu 1$ to $C_\mu 4$—that encode its four heavy-chain Ig domains (Fig. 5.22). The end of the $C_\mu 4$ exon also encodes the carboxy terminus for the secreted form. Two additional downstream exons, M1 and M2, encode the transmembrane forms. If the primary transcript is cleaved at the polyadenylation site (pA_s) located just downstream of the $C_\mu 4$ exon but before the last two exons, then only the secreted molecule can be produced. If the polymerase transcribes through this first polyadenylation site, then splicing can occur from a non-consensus splice-donor site within the $C_\mu 4$ exon to the M1 exon. In this case, polyadenylation occurs at a downstream site (pA_m) and the cell-surface form of immunoglobulin can be produced. This alternative splicing is incompletely understood, but

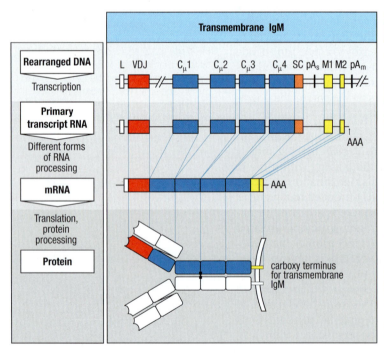
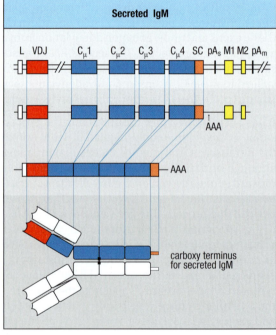

Fig. 5.22 Transmembrane and secreted forms of immunoglobulins are derived from the same heavy-chain sequence by alternative RNA processing. At the end of the heavy-chain C gene, there are two exons (M1 and M2, yellow) that together encode the transmembrane region and cytoplasmic tail of the transmembrane form. Within the last C-domain exon, a secretion-coding (SC) sequence (orange) encodes the carboxy terminus of the secreted form. In the case of IgD, the SC sequence is in a separate exon (not shown), but for the other isotypes, including IgM as shown here, the SC sequence is contiguous with the last C-domain exon. The events that dictate whether a heavy-chain RNA will result in a secreted or a transmembrane immunoglobulin occur during processing of the pre-mRNA transcript. Each heavy-chain C gene has two potential polyadenylation sites (shown as pA_s and pA_m). Left panel: the transcript is cleaved and polyadenylated (AAA) at the second site (pA_m). Splicing occurs from a site located within the last $C_\mu 4$ exon just upstream of the SC sequence (orange), to a second site at the 5' end of the M1 exons (yellow). This results in removal of the SC sequence and joining of the $C_\mu 4$ exon to the exons M1 and M2 and generates the transmembrane form of the heavy chain. Right panel: polyadenylation occurs at the first poly(A) addition site (pA_s), and transcription terminates before the exons M1 and M2, preventing the generation of the transmembrane form of the heavy chain, and producing the secreted form.

may involve the regulation of RNA polymerase activity as the polymerase transcribes through the IgM locus. One factor that regulates the polyadenylation of RNA transcripts is a **cleavage stimulation factor** subunit, **CstF-64**, which favors production of the transcript for secreted IgM. The transcription elongation factor **ELL2**, which is induced in plasma cells, also promotes polyadenylation at the pA_s site and favors the secreted form. CstF-64 and ELL2 co-associate with RNA polymerase within the immunoglobulin locus. This differential RNA processing is illustrated for C_μ in Fig. 5.22, but it occurs in the same way for all isotypes. In activated B cells that differentiate to become antibody-secreting plasma cells, most of the transcripts are spliced to yield the secreted rather than the transmembrane form of whichever heavy-chain isotype the B cell is expressing.

5-16 IgM and IgA can form polymers by interacting with the J chain.

Although all immunoglobulin molecules are constructed from a basic unit of two heavy and two light chains, both IgM and IgA can form multimers of these basic units (Fig. 5.23). C regions of IgM and IgA can include a 'tailpiece' of 18 amino acids that contains a cysteine residue essential for polymerization. A separate 15-kDa polypeptide chain called the **J chain** promotes polymerization by linking to the cysteine of this tailpiece, which is found only in the secreted forms of the μ and α chains. (This J chain should not be confused with the immunoglobulin J region encoded by a J gene segment; see Section 5-2.) In the case of IgA, dimerization is required for transport through epithelia, as we will discuss in Chapter 10. IgM molecules are found as pentamers, and occasionally hexamers (without J chain), in plasma, whereas IgA is found mainly as a dimer in mucous secretions but as a monomer in plasma.

Immunoglobulin polymerization is also thought to be important in the binding of antibody to repetitive epitopes. An antibody molecule has at least two identical antigen-binding sites, each of which has a given affinity, or binding strength, for antigen. If the antibody attaches to multiple identical epitopes on a target antigen, it will dissociate only when all binding sites dissociate. The dissociation rate of the whole antibody will therefore be much slower than the dissociation rate for a single binding site; multiple binding sites thus give the antibody a greater total binding strength, or avidity. This consideration is particularly relevant for pentameric IgM, which has 10 antigen-binding sites.

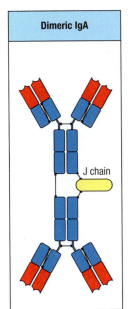

Dimeric IgA

J chain

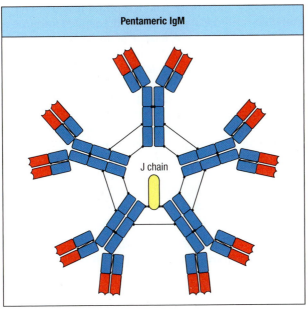

Pentameric IgM

J chain

Fig. 5.23 The IgM and IgA molecules can form multimers. IgM and IgA are usually synthesized as multimers in association with an additional polypeptide chain, the J chain. In dimeric IgA (left panel), the monomers have disulfide bonds to the J chain as well as to each other. In pentameric IgM (right panel), the monomers are cross-linked by disulfide bonds to each other and to the J chain. IgM can also form hexamers that lack a J chain (not shown).

IgM antibodies frequently recognize repetitive epitopes such as those on bacterial cell-wall polysaccharides, but individual binding sites are often of low affinity because IgM is made early in immune responses, before somatic hypermutation and affinity maturation. Multisite binding makes up for this, markedly improving the overall functional binding strength. This implies that binding of a single IgM pentamer to a target could be sufficient to mediate biological effector activity, whereas in the case of IgGs, two independent target molecules may need to be located in close proximity.

Summary.

The classes of immunoglobulins are defined by their heavy-chain C regions, with the different heavy-chain isotypes being encoded by different C-region genes. The heavy-chain C-region genes are present in a cluster 3' to the V, D, and J gene segments. A productively rearranged V-region exon is initially expressed in association with μ and δ C_H genes, which are coexpressed in naive B cells by alternative splicing of an mRNA transcript that contains both the μ and δ C_H exons. In addition, B cells can express any class of immunoglobulin as a membrane-bound antigen receptor or as a secreted antibody. This is achieved by differential splicing of mRNA to include exons that encode either a hydrophobic membrane anchor or a secretable tailpiece. The antibody that a B cell secretes upon activation thus recognizes the antigen that initially activated the B cell via its antigen receptor. The same V-region exon can subsequently be associated with any one of the other isotypes to direct the production of antibodies of different classes by the process of class switching, which is described in Chapter 10.

Evolution of the adaptive immune response.

The form of adaptive immunity that we have discussed so far in this book depends on the action of the RAG-1/RAG-2 recombinase to generate an enormously diverse clonally distributed repertoire of immunoglobulins and T-cell receptors. This system is found only in the jawed vertebrates, the **gnathostomes**, which split off from the other vertebrates around 500 million years ago. Adaptive immunity seems to have arisen abruptly in evolution. Even the cartilaginous fishes, the earliest group of jawed fishes to survive to the present day, have organized lymphoid tissue, T-cell receptors and immunoglobulins, and the ability to mount adaptive immune responses. The diversity generated within the vertebrate adaptive immune system was once viewed as unique among animal immune systems. But we now know that organisms as different as insects, echinoderms, and mollusks use a variety of genetic mechanisms to increase their repertoires of pathogen-detecting molecules, although they do not achieve true adaptive immunity. Nearer to home, it has been found that the surviving species of jawless vertebrates, the **agnathans**—the lampreys and hagfish—have a form of adaptive or 'anticipatory' immunity that is based on non-immunoglobulin 'antibody'-like proteins and involves a system of somatic gene rearrangement that is quite distinct from RAG-dependent V(D)J rearrangement. So we should now view our adaptive immune system as only one solution, albeit the most powerful, to the problem of generating highly diverse systems for pathogen recognition.

5-17   Some invertebrates generate extensive diversity in a repertoire of immunoglobulin-like genes.

Until very recently, it was thought that invertebrate immunity was limited to an innate system that had a very restricted diversity in recognizing pathogens.

This idea was based on the knowledge that innate immunity in vertebrates relied on around 10 distinct Toll-like receptors and a similar number of other receptors that also recognize PAMPs, and also on the assumption that the number of receptors in invertebrates was no greater. Recent studies have, however, uncovered at least two invertebrate examples of extensive diversification of an immunoglobulin superfamily member, which could potentially provide an extended range of recognition of pathogens.

In *Drosophila*, fat-body cells and hemocytes act as part of the immune system. Fat-body cells secrete proteins, such as the antimicrobial defensins (see Chapters 2 and 3), into the hemolymph. Another protein found in hemolymph is the **Down syndrome cell adhesion molecule** (**Dscam**), a member of the immunoglobulin superfamily. Dscam was originally discovered in the fly as a protein involved in specifying neuronal wiring. It is also made in fat-body cells and hemocytes, which can secrete it into the hemolymph, where it is thought to recognize invading bacteria and aid in their engulfment by phagocytes.

The Dscam protein contains multiple, usually 10, immunoglobulin-like domains. The gene that encodes Dscam has, however, evolved to contain a large number of alternative exons for several of these domains (Fig. 5.24). Exon 4 of the gene encoding the Dscam protein can be any 1 of 12 different exons, each specifying an immunoglobulin domain of differing sequence. Exon cluster 6 has 48 alternative exons, cluster 9 another 33, and cluster 17 a further 2: it is estimated that the Dscam gene could encode around 38,000 protein isoforms. A role for Dscam in immunity was proposed when it was found that *in vitro* phagocytosis of *Escherichia coli* by isolated hemocytes lacking Dscam was less efficient than by normal hemocytes. These observations suggest that at least some of this extensive repertoire of alternative exons may have evolved to diversify insects' ability to recognize pathogens. This role for Dscam has been confirmed in the mosquito *Anopheles gambiae*, in which silencing of the *Dscam* homolog *AgDscam* has been shown to weaken the mosquito's normal resistance to bacteria and to the malaria parasite *Plasmodium*. There is also evidence from the mosquito that some *Dscam* exons have specificity for particular pathogens. It is not clear whether Dscam isoforms are expressed in a clonal manner.

Another invertebrate, this time a mollusk, uses a different strategy to diversify an immunoglobulin superfamily protein for use in immunity. The freshwater snail *Biomphalaria glabrata* expresses a small family of **fibrinogen-related proteins** (**FREPs**) thought to have a role in innate immunity. FREPs have one or two immunoglobulin domains at their amino-terminal end and a fibrinogen domain at their carboxy terminus. The immunoglobulin domains may interact with pathogens, while the fibrinogen domain may confer on the FREP lectin-like properties that help precipitate the complex. FREPs are produced by

Fig. 5.24 The Dscam protein of Drosophila innate immunity contains multiple immunoglobulin domains and is highly diversified through alternative splicing. The gene encoding Dscam in *Drosophila* contains several large clusters of alternative exons. The clusters encoding exon 4 (green), exon 6 (light blue), exon 9 (red), and exon 17 (orange) contain 12, 48, 33, and 2 alternative exons, respectively. For each of these clusters, only one alternative exon is used in the complete *Dscam* mRNA. There is some differential usage of exons in neurons, fat-body cells, and hemocytes. All three cell types use the entire range of alternative exons for exons 4 and 6. For exon 9, there is a restricted use of alternative exons in hemocytes and fat-body cells. The combinatorial use of alternative exons in the *Dscam* gene makes it possible to generate more than 38,000 protein isoforms. Adapted from Anastassiou, D.: *Genome Biol.* 2006, **7**:R2.

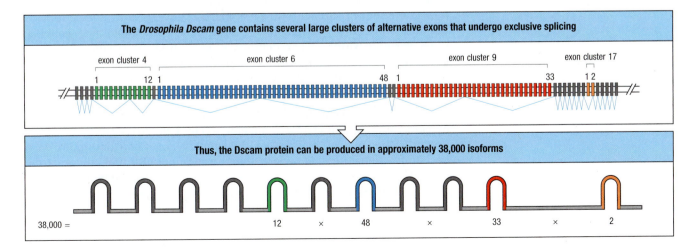

hemocytes and secreted into the hemolymph. Their concentration increases when the snail is infected by parasites—it is, for example, the intermediate host for the parasitic schistosomes that cause human schistosomiasis.

The *B. glabrata* genome contains many copies of FREP genes that can be divided into approximately 13 subfamilies. A study of the sequences of expressed FREP3 subfamily members has revealed that the FREPs expressed in an individual organism are extensively diversified compared with the germline genes. There are fewer than five genes in the FREP3 subfamily, but an individual snail was found to generate more than 45 distinct FREP3 proteins, all with slightly different sequences. An analysis of the protein sequences suggested that this diversification was due to the accumulation of point mutations in one of the germline FREP3 genes. Although the precise mechanism of this diversification, and the cell type in which it occurs, are not yet known, it does suggest some similarity to somatic hypermutation that occurs in the immunoglobulins. Both the insect and *Biomphalaria* examples seem to represent a way of diversifying molecules involved in immune defense, but although they resemble in some ways the strategy of an adaptive immune response, there is no evidence of clonal selection—the cornerstone of true adaptive immunity.

5-18 Agnathans possess an adaptive immune system that uses somatic gene rearrangement to diversify receptors built from LRR domains.

Since the early 1960s it has been known that certain jawless fishes, the hagfish and the lamprey, could mount a form of accelerated rejection of transplanted skin grafts and exhibit a kind of immunological delayed-type hypersensitivity. Their serum also seemed to contain an activity that behaved as a specific agglutinin, increasing in titer after secondary immunizations, in a similar way to an antibody response in higher vertebrates. Although these phenomena seemed reminiscent of adaptive immunity, there was no evidence of a thymus or of immunoglobulins, but these animals did have cells that could be considered to be genuine lymphocytes on the basis of morphological and molecular analysis. Analysis of the genes expressed by lymphocytes of the sea lamprey *Petromyzon marinus* revealed none related to T-cell receptor or immunoglobulin genes. However, these cells expressed large amounts of mRNAs from genes encoding proteins with multiple LRR domains, the same protein domain from which the pathogen-recognizing Toll-like receptors (TLRs) are built (see Section 3-5).

This might simply have meant that these cells are specialized for recognizing and reacting to pathogens, but the LRR proteins expressed had some surprises in store. Instead of being present in a relatively few forms (like the invariant TLRs), they were found to have highly variable amino acid sequences, with a large number of variable LRR units placed between less variable amino-terminal and carboxy-terminal LRR units. These LRR-containing proteins, called **variable lymphocyte receptors** (**VLRs**), have an invariant stalk region connecting them to the plasma membrane by a glycosylphosphatidylinositol linkage, and they can either be tethered to the cell or, at other times, like antibodies, be secreted into the blood.

Analysis of the expressed lamprey VLR genes indicates that they are assembled by a process of somatic gene rearrangement (**Fig. 5.25**). In the germline configuration, there are three incomplete VLR genes, *VLRA*, *VLRB*, and *VLRC*, each encoding a signal peptide, a partial amino-terminal LRR unit, and a partial carboxy-terminal LRR unit, but these three blocks of coding sequence are separated by noncoding DNA that contains neither typical signals for RNA splicing nor the RSSs present in immunoglobulin genes (see Section 5-4). Instead, the regions flanking the incomplete VLR genes include a large number of DNA 'cassettes' that contain LRR units—one, two, or three LRR domains at a time.

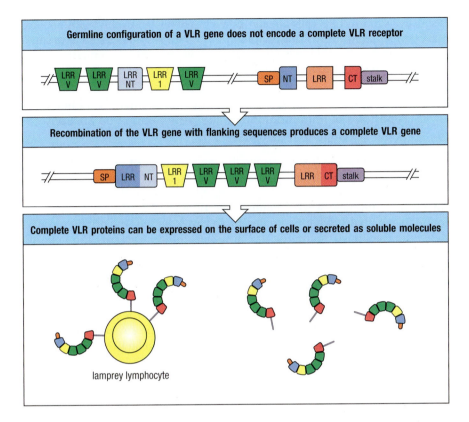

Germline configuration of a VLR gene does not encode a complete VLR receptor

Recombination of the VLR gene with flanking sequences produces a complete VLR gene

Complete VLR proteins can be expressed on the surface of cells or secreted as soluble molecules

lamprey lymphocyte

Fig. 5.25 Somatic recombination of an incomplete germline variable lymphocyte receptor (VLR) gene generates a diverse repertoire of complete VLR genes in the lamprey. Top panel: an incomplete germline copy of a lamprey VLR gene contains a framework (right) for the complete gene: the portion encoding the signal peptide (SP), part of an amino-terminal LRR unit (NT, dark blue), and a carboxy-terminal LRR unit that is split into two parts (LRR, light red; and CT, red) by intervening noncoding DNA sequences. Nearby flanking regions (left) contain multiple copies of VLR gene-'cassettes' with single or double copies of variable LRR domains (green) and cassettes that encode part of the amino-terminal LRR domains (light blue, yellow). Middle panel: somatic recombination causes various LRR units to be copied into the original VLR gene. This creates a complete VLR gene that contains the assembled amino-terminal LRR cassette (LRR NT) and first LRR (yellow), followed by several variable LRR units (green) and the completed carboxy-terminal LRR unit, and ends with the portion that encodes the stalk region of the VLR receptor. The cytidine deaminases PmCDA1 and PmCDA2 from the lamprey *P. marinus* are candidates for enzymes that may initiate this gene rearrangement. Expression of the rearranged gene results in a complete receptor that can be attached to the cell membrane by glycosylphosphatidylinositol (GPI) linkage of its stalk. Bottom panel: an individual lymphocyte undergoes somatic gene rearrangement to produce a unique VLR receptor. These receptors can be tethered to the surface of the lymphocyte via the GPI linkage or can be secreted into the blood. Unique somatic rearrangement events in each developing lymphocyte generate a repertoire of VLR receptors of differing specificities. Adapted from Pancer, Z., and Cooper, M.D.: *Annu. Rev. Immunol.* 2006, **24**:497–518.

Each mature lamprey lymphocyte expresses a complete and unique VLR gene, either *VLRA*, *VLRB*, or *VLRC*, which has undergone recombination of these flanking regions with the germline VLR gene.

The creation of a complete VLR gene is currently thought to occur during replication of lamprey lymphocyte DNA by a 'copy-choice' mechanism that is similar, but not identical, to gene conversion (described in Section 5-20). During DNA replication, LRR units flanking the VLR gene are copied into the VLR gene—presumably when a DNA strand being synthesized switches templates and copies sequences from one of these LRR units. Although final proof is still lacking, this template-switching mechanism may be triggered by enzymes of the AID-APOBEC family that are expressed by lamprey lymphocytes, and whose **cytidine deaminase activity (CDA)** could cause the single-strand DNA breaks that can start the copy-choice process. Lampreys possess two such enzymes: CDA1, which is expressed in *VLRA*-lineage lymphocytes, and CDA2, which is expressed in *VLRB*-lineage lymphocytes. It is not yet known if CDA1 or CDA2 is expressed in *VLRC*-expressing lymphocytes. The final VLR gene contains a complete amino-terminal capping LRR subunit, followed by the addition of up to seven internal LRR domains, each 24 amino acids long, and the removal of the internal noncoding regions to complete the formation of the carboxy-terminal LRR domain (see Fig. 5.25).

It is estimated that this somatic rearrangement mechanism can generate as much diversity in the VLR proteins as is possible for immunoglobulins. Indeed, the crystal structure of a VLR protein shows that the concave surface formed by the series of LRR repeats interacts with a variable insert in the carboxy-terminal LRR to form a surface capable of interacting with a great diversity of antigens. Thus, the diversity of the anticipatory repertoire of agnathans may be limited not by the numbers of possible receptors they can generate but by the number of lymphocytes present in any individual, as in the adaptive immune system of their evolutionary cousins, the gnathostomes. As noted above, each lamprey lymphocyte rearranges only one of the two germline VLR genes, expressing either a complete VLRA or VLRB or VLRC protein. The first two cell populations seem to have some characteristics of mammalian T and B lymphocytes, respectively, and VLRC cells appear more closely related to the VLRA lineage. For example, VLRA-expressing lymphocytes also express genes similar to some mammalian T-cell cytokine genes, suggesting an even closer similarity to our own RAG-dependent adaptive immune system than was previously appreciated.

5-19 RAG-dependent adaptive immunity based on a diversified repertoire of immunoglobulin-like genes appeared abruptly in the cartilaginous fishes.

Within the vertebrates, we can trace the development of immune functions from the agnathans through the cartilaginous fishes (sharks, skates, and rays) to the bony fishes, then to the amphibians, to reptiles and birds, and finally to mammals. RAG-dependent V(D)J recombination has not been found in agnathans, other chordates, or any invertebrate. The origins of RAG-dependent adaptive immunity are now becoming clearer as the genome sequences of many more animals become available. The first clue was that RAG-dependent recombination shares many features with the transposition mechanism of **DNA transposons**—mobile genetic elements that encode their own **transposase**, an enzymatic activity that allows them to excise from one site in the genome and reinsert themselves elsewhere. The mammalian RAG complex can act as a transposase *in vitro*, and even the structure of the *RAG* genes, which lie close together in the chromosome and lack the usual introns of mammalian genes, is reminiscent of a transposon.

All this provoked speculation that the origin of RAG-dependent adaptive immunity was the invasion of a DNA transposon into a gene similar to an immunoglobulin or a T-cell receptor V-region gene, an event that would have occurred in some ancestor of the jawed vertebrates (**Fig. 5.26**). DNA transposons carry inverted repeated sequences at either end, which are bound by the transposase for transposition to occur. These terminal repeats are considered to be the ancestors of the RSSs in present-day antigen-receptor genes (see Section 5-4), while the RAG-1 protein is believed to have evolved from a transposase. Subsequent duplication, reduplication, and recombination of the immune-receptor gene and its inserted RSSs eventually led to the separation of the RAG genes from the rest of the relic transposon and to the multisegmented immunoglobulin and T-cell receptor loci of present-day vertebrates.

The ultimate origins of the RSSs and the RAG-1 catalytic core are now thought to lie in the **Transib** superfamily of DNA transposons, and genome sequencing has led to the discovery of sequences related to *RAG1* in animals as distantly related to vertebrates as the sea anemone *Nematostella*. The origin of *RAG2* is

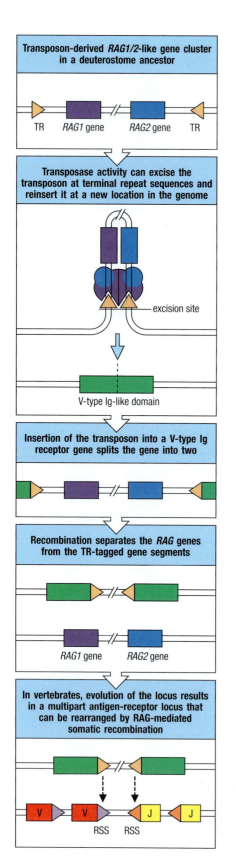

Transposon-derived *RAG1/2*-like gene cluster in a deuterostome ancestor

TR   *RAG1* gene   *RAG2* gene   TR

Transposase activity can excise the transposon at terminal repeat sequences and reinsert it at a new location in the genome

— excision site

V-type Ig-like domain

Insertion of the transposon into a V-type Ig receptor gene splits the gene into two

Recombination separates the *RAG* genes from the TR-tagged gene segments

RAG1 gene   *RAG2* gene

In vertebrates, evolution of the locus results in a multipart antigen-receptor locus that can be rearranged by RAG-mediated somatic recombination

V   V   J   J
RSS   RSS

Fig. 5.26 Integration of a transposon into a V-type immunoglobulin receptor gene is thought to have given rise to the T-cell receptor and immunoglobulin genes. Top panel: a DNA transposon in an ancestor of the deuterostomes (the large group of phyla to which the chordates belong) is thought to have had genes related to *RAG1* and *RAG2*—prototype *RAG1* (purple) and *RAG2* (blue), which acted as its transposase. DNA transposons are bounded by terminal inverted repeat (TR) sequences. Second panel: to excise a transposon from DNA, the transposase proteins (purple and blue) bind the TRs, bringing them together, and the transposase enzymatic activity cuts the transposon out of the DNA, leaving a footprint in the host DNA that resembles the TRs. After excision from one site, the transposon reinserts elsewhere in the genome, in this case into a V-type immunoglobulin receptor (green). The enzymatic activity of the transposase enables the transposon to insert into DNA in a reaction that is the reverse of the excision reaction. Third panel: the integration of the *RAG1/2*-like transposon into the middle of the gene for a V-type immunoglobulin receptor splits the V exon into two parts. Fourth and fifth panels: in the evolution of the immunoglobulin and T-cell receptor genes, the initial integration event has been followed by DNA rearrangements that separate the transposase genes (now known as the *RAG1* and *RAG2* genes) from the transposon TRs, which we now term the recombination signal sequences (RSSs). The purple sea urchin (an invertebrate deuterostome) has a *RAG1/2*-like gene cluster (not shown) and expresses proteins similar to RAG-1 and RAG-2 proteins, but does not have immunoglobulins, T-cell receptors, or adaptive immunity. The RAG-like proteins presumably retain some other cellular function (so far unknown) in this animal.

more obscure, but a *RAG1–RAG2*-related gene cluster was recently discovered in sea urchins, invertebrate relatives of the chordates. Sea urchins themselves show no evidence of immunoglobulins, T-cell receptors, or adaptive immunity, but the proteins expressed by the sea-urchin *RAG* genes form a complex with each other and with RAG proteins from the bull shark (*Carcharhinus leucas*), a primitive jawed vertebrate, but not with those from mammals. This suggests that these proteins could indeed be related to the vertebrate RAGs, and that RAG-1 and RAG-2 were already present in a common ancestor of chordates and echinoderms (the group to which sea urchins belong), presumably fulfilling some other cellular function.

The origin of somatic gene rearrangement in the excision of a transposable element makes sense of an apparent paradox in the rearrangement of immune-system genes. This is that the RSSs are joined precisely in the excised DNA (see Section 5-5), which has no further function and whose fate is irrelevant to the cell, whereas the cut ends in the genomic DNA, which form part of the immunoglobulin or T-cell receptor gene, are joined by an error-prone process, which could be viewed as a disadvantage. However, when looked at from the transposon's point of view, this makes sense, because the transposon preserves its integrity by this excision mechanism, whereas the fate of the DNA it leaves behind is of no significance to it. As it turned out, the error-prone joining in the primitive immunoglobulin gene generated useful diversity in antigen-recognition molecules and was strongly selected for. The RAG-based rearrangement system also provided something else that mutations could not—a means of rapidly modifying the size of the coding region, not just its diversity.

The next question is what sort of gene the transposon inserted into. Proteins containing Ig-like domains are ubiquitous throughout the plant, animal, and bacterial kingdoms, making this one of the most abundant protein superfamilies; in species whose genomes have been fully sequenced, the immunoglobulin superfamily is one of the largest families of protein domains in the genome. The functions of the members of this superfamily are very disparate, and they are a striking example of natural selection taking a useful structure—the basic Ig-domain fold—and adapting it to different purposes.

The immunoglobulin superfamily domains can be divided into four families on the basis of differences in structure and sequence of the immunoglobulin domain. These are V (resembling an immunoglobulin variable domain), C1 and C2 (resembling constant-region domains), and a type of immunoglobulin domain called an I domain (for intermediate). The target of the RSS-containing element is likely to have been a gene encoding a cell-surface receptor containing an Ig-like V domain, most probably a type similar to present-day VJ domains. These domains are found in some invariant receptor proteins and are so called because of the resemblance of one of the strands to a J segment. It is possible to imagine how transposon movement into such a gene could produce separate V and J gene segments (see Fig. 5.26). On the basis of phylogenetic analysis, **agnathan paired receptors resembling Ag receptors**, or **APARs**, which are encoded by a multigene family found in hagfish and lamprey, are currently the best candidates for being relatives of the ancestor of the antigen receptor. Their DNA sequences predict single-pass transmembrane proteins with a single extracellular VJ domain and a cytoplasmic region containing signaling modules. APARs are expressed in leukocytes.

5-20 Different species generate immunoglobulin diversity in different ways.

Most of the vertebrates we are familiar with generate a large part of their antigen receptor diversity in the same way as mice and humans, by putting together gene segments in different combinations. There are exceptions, however, even within the mammals. Some animals use gene rearrangement to always join together the same V and J gene segment initially, and then diversify this recombined V region. In birds, rabbits, cows, pigs, sheep, and horses, there is little or no germline diversity in the V, D, and J gene segments that are rearranged to form the genes for the initial B-cell receptors, and the rearranged V-region sequences are identical or similar in most immature B cells. These immature B cells migrate to specialized microenvironments—the **bursa of Fabricius** in the gut of chickens, and another intestinal lymphoid organ in rabbits. Here, B cells proliferate rapidly, and their rearranged immunoglobulin genes undergo further diversification.

In birds and rabbits this occurs mainly by gene conversion, a process by which short sequences in the expressed rearranged V-region gene are replaced with sequences from an upstream V gene segment pseudogene. The germline arrangement of the chicken heavy-chain locus is a single set of rearranging V, J, D, and C gene segments and multiple copies of V-segment pseudogenes. Diversity in this system is created by gene conversion in which sequences from the V_H pseudogenes are copied into the single rearranged V_H gene (**Fig. 5.27**). It seems that gene conversion is related to somatic hypermutation

Fig. 5.27 The diversification of chicken immunoglobulins occurs through gene conversion. In chickens, the immunoglobulin diversity that can be created by V(D)J recombination is extremely limited. Initially, there are only one active V, one J, and 15 D gene segments at the chicken heavy-chain locus and one active V and one J gene segment at the single light-chain locus (top left panel). Primary gene rearrangement can thus produce only a very limited number of receptor specificities (second panels). Immature B cells expressing this receptor migrate to the bursa of Fabricius, where the cross-linking of surface immunoglobulin (sIg) induces cell proliferation (second panels). The chicken genome contains numerous pseudogenes with a prerearranged VH–D structure. Gene conversion events introduce sequences from these adjacent V gene segment pseudogenes into the expressed gene, creating diversity in the receptors (third panels). Some of these gene conversions will inactivate the previously expressed gene (not shown). If a B cell can no longer express sIg after such a gene conversion, it is eliminated. Repeated gene conversion events can continue to diversify the repertoire (bottom panels).

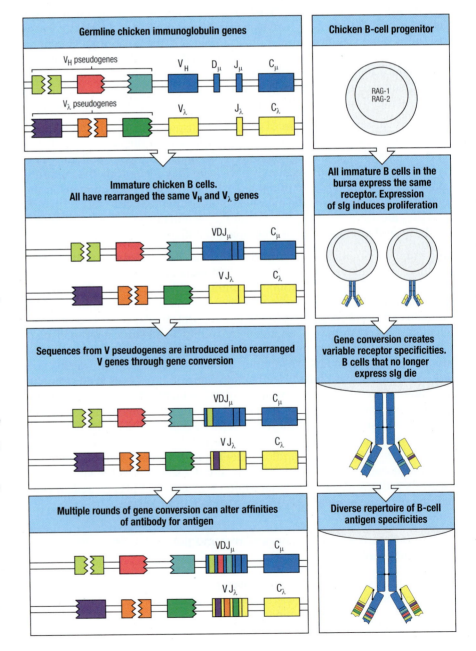

in its mechanism, because gene conversion in a chicken B-cell line has been shown to require the enzyme **activation-induced cytidine deaminase** (**AID**). In Chapter 10, we will see that this same enzyme is involved in class switching and affinity maturation of the antibody response. For gene conversion, it is thought that single-strand cuts in DNA generated by the endonuclease **apurinic/apyrimidinic endonuclease-1** (**APE1**) after the actions of AID are the signal that initiates a homology-directed repair process in which a homologous V pseudogene segment is used as the template for the DNA replication that repairs the V-region gene.

In sheep and cows, immunoglobulin diversification is the result of somatic hypermutation, which occurs in an organ known as the ileal Peyer's patch. Somatic hypermutation, independent of T cells and a particular driving antigen, also contributes to immunoglobulin diversification in birds, sheep, and rabbits.

A more fundamentally different organization of immunoglobulin genes is found in the cartilaginous fish, the most primitive jawed vertebrates. Sharks have multiple copies of discrete V_L-J_L-C_L and V_H-D_H-J_H-C_H cassettes, and activate rearrangement within individual cassettes (**Fig. 5.28**). Although this is somewhat different from the kind of combinatorial gene rearrangement of higher vertebrates, in most cases there is still a requirement for a RAG-mediated somatic rearrangement event. As well as rearranging genes, cartilaginous fish have multiple 'rearranged' V_L regions (and sometimes rearranged V_H regions) in the germline genome (see Fig. 5.28) and apparently generate diversity by activating the transcription of different copies. Even here, some diversity is also contributed by combinatorial means by the subsequent pairing of heavy and light chains.

This 'germline-joined' organization of the light-chain loci is unlikely to represent an intermediate evolutionary stage, because in that case the heavy-chain and light-chain genes would have had to independently acquire the capacity for rearrangement by convergent evolution. It is much more likely that, after the divergence of the cartilaginous fishes, some immunoglobulin loci became rearranged in the germline of various ancestors through activation of the *RAG* genes in germ cells, with the consequent inheritance of the rearranged loci by the offspring. In these species, the rearranged germline loci might confer some advantages, such as ensuring rapid responses to common pathogens by producing a preformed set of immunoglobulin chains.

The IgM antibody isotype is thought to go back to the origins of adaptive immunity. It is the predominant form of immunoglobulin in cartilaginous

Activation-Induced Cytidine Deaminase Deficiency

Fig. 5.28 The organization of immunoglobulin genes is different in different species, but all can generate a diverse repertoire of receptors. The organization of the immunoglobulin heavy-chain genes in mammals, in which there are separated clusters of repeated V, D, and J gene segments, is not the only solution to the problem of generating a diverse repertoire of receptors. Other vertebrates have found alternative solutions. In 'primitive' groups, such as the sharks, the locus consists of multiple repeats of a basic unit composed of a V gene segment, one or two D gene segments, a J gene segment, and a C gene segment. A more extreme version of this organization is found in the κ-like light-chain locus of some cartilaginous fishes such as the rays and the carcharhine sharks, in which the repeated unit consists of already rearranged VJ–C genes, from which a random choice is made for expression. In chickens, there is a single rearranging set of gene segments at the heavy-chain locus but there are multiple copies of pre-integrated V_H–D segment pseudogenes. Diversity in this system is created by gene conversion, in which sequences from the V_H-–D pseudogenes are copied onto the single rearranged V_H gene.

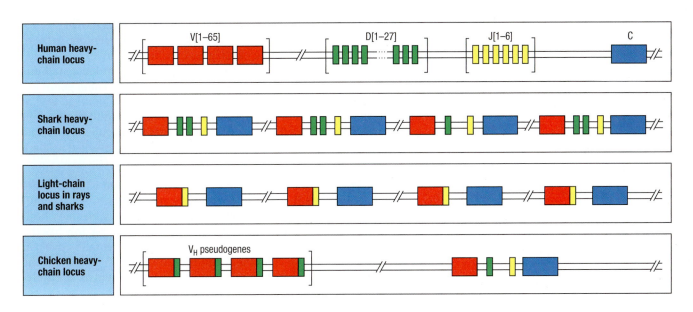

fishes and bony fishes. The cartilaginous fishes also have at least two other heavy-chain isotypes not found in more recently evolved species. One, **IgW**, has a constant region composed of six immunoglobulin domains, whereas the second, **IgNAR**, which we described in Section 4-10, seems to be related to IgW but has lost the first constant-region domain and does not pair with light chains. Instead, it forms a homodimer in which each heavy-chain V domain forms a separate antigen-binding site. IgW seems to be related to IgD (which is first found in bony fish) and, like IgM, seems to go back to the origin of adaptive immunity.

5-21 Both α:β and γ:δ T-cell receptors are present in cartilaginous fishes.

Neither the T-cell receptors nor the immunoglobulins have been found in any species evolutionarily earlier than the cartilaginous fishes, in which they have essentially the same form that we see in mammals. The identification of TCRβ-chain and δ-chain homologs from sharks, and of distinct TCRα, β, γ, and δ chains from a skate, show that even at the earliest time that these adaptive immune system receptors can be identified, they had already diversified into at least two recognition systems. Moreover, each lineage shows diversity resulting from combinatorial somatic rearrangement. The identification of many ligands recognized by γ:δ T cells has helped clarify their role in the immune response. Although a complete list is still lacking, the trend appears to be more similar to a kind of innate sensing rather than the fine peptide specificity of the α:β T cells. Ligands of γ:δ T cells include various lipids that may derive from microbes and nonclassical MHC class Ib molecules whose expression may be an indication of infection or cellular stress (see Section 6-17). Even certain α:β T cells appear to participate in a form of innate recognition, such as the mucosa-associated invariant T cells described in Section 4-18. This could indicate that early in the evolution of RAG-dependent adaptive immunity, the receptors generated by excision of the primordial retrotransposon were useful in innate sensing of infections, and this role has persisted in certain minor T-cell populations to this day. In any case, the very early divergence of these two classes of T-cell receptors and their conservation through subsequent evolution suggests an important early separation of functions.

5-22 MHC class I and class II molecules are also first found in the cartilaginous fishes.

One would expect to see the specific ligands of T-cell receptors, the MHC molecules, emerge at around the same time in evolution as the receptors. Indeed, MHC molecules are present in the cartilaginous fishes and in all higher vertebrates, but, like the T-cell receptors, they have not been found in agnathans or invertebrates. Both MHC class I and class II α-chain and β-chain genes are present in sharks, and their products seem to function in an identical way to mammalian MHC molecules. The key residues of the peptide-binding cleft that interact with the ends of the peptide in MHC class I molecules or with the central region of the peptide in MHC class II molecules are conserved in shark MHC molecules.

Moreover, the MHC genes are also polymorphic in sharks, with multiple alleles of class I and class II loci. In some species, more than 20 MHC class I alleles have been identified so far. For the shark MHC class II molecules, both the class II α and the class II β chains are polymorphic. Thus, not only has the function of the MHC molecules in selecting peptides for presentation evolved during the divergence of the agnathans and the cartilaginous fishes, but the continuous selection imposed by pathogens has also resulted in the polymorphism that is a characteristic feature of the MHC.

Section 4-20 introduced the division between **classical MHC class I** genes (sometimes called class Ia) and the **nonclassical MHC class Ib genes**, which will be discussed in Chapter 6. This division is also present in cartilaginous fishes, because the class I genes of sharks include some that resemble mammalian class Ib molecules. However, it is thought that the shark class Ib genes are not the direct ancestors of the mammalian class Ib genes. For the class I genes, it seems that within each of the five major vertebrate lineages studied (cartilaginous fishes, lobe-finned fishes, ray-finned fishes, amphibians, and mammals), these genes have independently separated into classical and nonclassical loci.

Thus, the characteristic features of the MHC molecules are all present when these molecules are first encountered, and there are no intermediate forms to guide our understanding of their evolution. Although we can trace the evolution of the components of the innate immune system, the mystery of the origin of the adaptive immune system still largely persists. But although we may not have a sure answer to the question of what selective forces led to RAG-dependent elaboration of adaptive immunity, it has never been clearer that, as Charles Darwin remarked about evolution in general, "from so simple a beginning endless forms most beautiful and most wonderful have been, and are being, evolved."

Summary.

Evolution of RAG-dependent adaptive immunity in jawed vertebrates was once considered a wholly unique and inexplicable 'immunological Big Bang.' However, we now understand that adaptive immunity has also evolved independently at least one other time during evolution. Our close vertebrate cousins, the jawless fishes, have evolved an adaptive immune system built on a completely different basis—the diversification of LRR domains rather than immunoglobulin domains—but which otherwise seems to have the essential features of clonal expression of receptors produced through a somatic rearrangement and with a form of immunological memory, all features of an adaptive immune system. We now appreciate that evolution of the RAG-dependent adaptive immune system is probably related to the insertion of a transposon into a member of a primordial immunoglobulin superfamily gene, which must have occurred in a germline cell in an ancestor of the vertebrates. By chance, the transposon terminal sequences, the forerunners of the RSSs, were placed in an appropriate location within this primordial antigen-receptor gene to enable intramolecular somatic recombination, thus paving the way for the full-blown somatic gene rearrangement seen in present-day immunoglobulin and T-cell receptor genes. The MHC molecules that are the ligands for T-cell receptors first appear in the cartilaginous fishes, suggesting coevolution with RAG-dependent adaptive immunity. The transposase genes (the *RAG* genes) could have already been present and active in some other function in the genome of this ancestor. *RAG1* seems to be of very ancient origin, as *RAG1*-related sequences have been found in a wide variety of animal genomes.

Summary to Chapter 5.

The antigen receptors of lymphocytes are remarkably diverse, and developing B cells and T cells use the same basic mechanism to achieve this diversity. In each cell, functional genes for the immunoglobulin and T-cell receptor chains are assembled by somatic recombination from sets of separate gene segments that together encode the V region. The substrates for the joining process are arrays of V, D, and J gene segments, which are similar in all the antigen-receptor gene loci. The lymphoid-specific proteins RAG-1 and RAG-2 direct the site-specific cleavage of DNA at RSSs flanking the V, D, and J segments to form double-strand breaks that initiate the recombination process in both T and B cells. These proteins function in concert with ubiquitous DNA-modifying enzymes acting in the double-strand break repair pathway, and with at least one other lymphoid-specific enzyme, TdT, to complete the gene rearrangements. As each type of gene segment is present in multiple, slightly different, versions, the random selection of one gene segment from each set is a source of substantial potential diversity. During assembly, the imprecise joining mechanisms at the coding junctions create a high degree of diversity concentrated in the CDR3 loops of the receptor, which lie at the center of the antigen-binding sites. The independent association of the two chains of immunoglobulins or T-cell receptors to form a complete antigen receptor multiplies the overall diversity available. An important difference between immunoglobulins and T-cell receptors is that immunoglobulins exist in both membrane-bound forms (B-cell receptors) and secreted forms (antibodies). The ability to express both a secreted and a membrane-bound form of the same molecule is due to alternative splicing of the heavy-chain mRNA to include exons that encode different forms of the carboxy terminus. Heavy-chain C regions of immunoglobulins contain three or four domains, whereas the T-cell receptor chains have only one. Other species have developed strategies to diversify receptors involved in immunity, and the agnathans use a system of VLRs that undergo somatic rearrangement that has some specific similarities to our own adaptive immune system. Adaptive immunity in jawed vertebrates—gnathostomes—appears to have arisen by the integration of a retrotransposon that encoded prototype *RAG1/2* genes into a preexisting V-type immunoglobulin-like gene that subsequently diversified to generate T- and B-cell receptor genes.

Questions.

5.1 **True or False:** A developing T cell may by chance express both an αβ heterodimer and a γδ heterodimer if all the loci recombine successfully.

5.2 **Multiple Choice:** Which of the following factors involved in antigen-receptor recombination could be deleted without completing ablating antigen receptor formation?

   A. Artemis
   B. TdT
   C. RAG-2
   D. Ku
   E. XRCC4

5.3 **True or False:** Both B and T cells can undergo somatic hypermutation of their antigen receptor in the context of an immune response in order to enhance antigen affinity.

5.4 **Short Answer:** What four processes contribute to the vast diversity of antibodies and B-cell receptors?

5.5 **Matching:** Match the protein(s) to its (their) function:

   A. RAG-1 and RAG-2   **i.** Nontemplate addition of N-nucleotides

   B. Artemis   **ii.** Nuclease activity to open the DNA hairpin and generate P-nucleotides

   C. TdT   **iii.** Recognize(s) RSS and create(s) single-stranded break

   D. DNA ligase IV and XRCC4   **iv.** Join(s) DNA ends

   E. DNA-PKcs   **v.** Form(s) a complex with Ku to hold DNA together and phosphorylate Artemis

5.6 **Short Answer:** What is the 12/23 rule and how does it ensure proper V(D)J segment joining?

5.7 **Matching:** Match the clinical disorder to the gene defects:

   A. Ataxia telangiectasia   **i.** RAG-1 or RAG-2 mutations resulting in decreased recombinase activity

   B. Irradiation-sensitive SCID (IR-SCID)   **ii.** ATM mutations

   C. Omenn syndrome   **iii.** Artemis mutations

5.8 Matching: Match the immunoglobulin class to its main function:

A. IgA   **i.** Most abundant in serum and strongly induced during an immune response

B. IgD   **ii.** First one produced after B-cell activation

C. IgE   **iii.** Defense at mucosal sites

D. IgG   **iv.** Defense against parasites but also involved in allergic diseases

E. IgM   **v.** Function not well known; may serve as auxiliary BCR

5.9 Fill-in-the-Blanks: Out of the five different antibody classes, two are secreted as multimers. _____ is secreted as a dimer and _____ is secreted as a pentamer, both of which have a(n) _____ as part of the multimeric complex. IgM and _____ are both expressed at the surface of mature B cells and are derived from the same pre-mRNA transcript. The balance of expression between these two is determined by alternative _____ and is regulated by the snRNP _____. The process that regulates membrane-bound versus secreted forms of antibodies is determined by two factors: _____ and _____. Fcγ receptors on macrophages and neutrophils bind to the Fc portions of _____ and _____ isotype antibodies of the IgG class. Mast cells, basophils, and activated eosinophils, however, will bear Fcε receptors that bind to _____ class antibodies. IgA and IgG class antibodies are able to bind to _____, which actively transports them to different body tissues and recycles them at the kidney glomerulus to prevent their loss and prolong their half-lives.

5.10 Multiple Choice: Which of the following is not true concerning the evolutionary history of the adaptive immune system?

A. Adaptive immunity arose abruptly in evolution.

B. Fruitflies and mosquitoes exhibit diversity in the secreted Dscam protein by alternative splicing of a vast array of different exons, while freshwater snails exhibit diversity in FREP genes by differential accumulation of genomic mutations in these genes.

C. Jawless fish recombine VLR genes during DNA replication to engender diversity in these genes, which are expressed on lymphocytes and have GPI-anchored and secreted forms.

D. RAG-1 arose from transposases while the RSSs it recognizes arose from terminal repeats from DNA transposons.

E. MHC class I and class II genes arose before T cells and immunoglobulins in cartilaginous fish.

General references.

Fugmann, S.D., Lee, A.I., Shockett, P.E., Villey, I.J., and Schatz, D.G.: **The RAG proteins and V(D)J recombination: complexes, ends, and transposition.** *Annu. Rev. Immunol.* 2000, **18**:495–527.

Jung, D., Giallourakis, C., Mostoslavsky, R., and Alt, F.W.: **Mechanism and control of V(D)J recombination at the immunoglobulin heavy chain locus.** *Annu. Rev. Immunol.* 2006, **24**:541–570.

Schatz, D.G.: **V(D)J recombination.** *Immunol. Rev.* 2004, **200**:5–11.

Schatz, D.G., and Swanson, P.C.: **V(D)J recombination: mechanisms of initiation.** *Annu. Rev. Genet.* 2011, **45**:167–202.

Section references.

5-1 Immunoglobulin genes are rearranged in the progenitors of antibody-producing cells.

Hozumi, N., and Tonegawa, S.: **Evidence for somatic rearrangement of immunoglobulin genes coding for variable and constant regions.** *Proc. Natl Acad. Sci. USA* 1976, **73**:3628–3632.

Seidman, J.G., and Leder, P.: **The arrangement and rearrangement of antibody genes.** *Nature* 1978, **276**:790–795.

Tonegawa, S., Brack, C., Hozumi, N., and Pirrotta, V.: **Organization of immunoglobulin genes.** *Cold Spring Harbor Symp. Quant. Biol.* 1978, **42**:921–931.

5-2 Complete genes that encode a variable region are generated by the somatic recombination of separate gene segments.

Early, P., Huang, H., Davis, M., Calame, K., and Hood, L.: **An immunoglobulin heavy chain variable region gene is generated from three segments of DNA: V_H, D and J_H.** *Cell* 1980, **19**:981–992.

Tonegawa, S., Maxam, A.M., Tizard, R., Bernard, O., and Gilbert, W.: **Sequence of a mouse germ-line gene for a variable region of an immunoglobulin light chain.** *Proc. Natl Acad. Sci. USA* 1978, **75**:1485–1489.

5-3 Multiple contiguous V gene segments are present at each immunoglobulin locus.

Maki, R., Traunecker, A., Sakano, H., Roeder, W., and Tonegawa, S.: **Exon shuffling generates an immunoglobulin heavy chain gene.** *Proc. Natl Acad. Sci. USA* 1980, **77**:2138–2142.

Matsuda, F., and Honjo, T.: **Organization of the human immunoglobulin heavy-chain locus.** *Adv. Immunol.* 1996, **62**:1–29.

Thiebe, R., Schable, K.F., Bensch, A., Brensing-Kuppers, J., Heim, V., Kirschbaum, T., Mitlohner, H., Ohnrich, M., Pourrajabi, S., Roschenthaler, F., *et al.*: **The variable genes and gene families of the mouse immunoglobulin kappa locus.** *Eur. J. Immunol.* 1999, **29**:2072–2081.

5-4 Rearrangement of V, D, and J gene segments is guided by flanking DNA sequences.

Grawunder, U., West, R.B., and Lieber, M.R.: **Antigen receptor gene rearrangement.** *Curr. Opin. Immunol.* 1998, **10**:172–180.

Lieber, M. R.: **The mechanism of human nonhomologous DNA end joining.** *J. Biol. Chem.* 2008, **283**:1–5.

Sakano, H., Huppi, K., Heinrich, G., and Tonegawa, S.: **Sequences at the somatic recombination sites of immunoglobulin light-chain genes.** *Nature* 1979, **280**:288–294.

5-5   The reaction that recombines V, D, and J gene segments involves both lymphocyte-specific and ubiquitous DNA-modifying enzymes.

Agrawal, A., and Schatz, D.G.: **RAG1 and RAG2 form a stable postcleavage synaptic complex with DNA containing signal ends in V(D)J recombination.** *Cell* 1997, **89**:43–53.

Ahnesorg, P., Smith, P., and Jackson, S.P.: **XLF interacts with the XRCC4-DNA ligase IV complex to promote nonhomologous end-joining.** *Cell* 2006, **124**:301–313.

Blunt, T., Finnie, N.J., Taccioli, G.E., Smith, G.C.M., Demengeot, J., Gottlieb, T.M., Ma, Y., Pannicke, U., Schwarz, K., and Lieber, M.R.: **Hairpin opening and overhang processing by an Artemis:DNA-PKcs complex in V(D)J recombination and in nonhomologous end joining.** *Cell* 2002, **108**:781–794.

Buck, D., Malivert, L., deChasseval, R., Barraud, A., Fondaneche, M.-C., Xanal, O., Plebani, A., Stephan, J.-L., Hufnagel, M., le Diest, F., *et al.*: **Cernunnos, a novel nonhomologous end-joining factor, is mutated in human immunodeficiency with microcephaly.** *Cell* 2006, **124**:287–299.

Jung, D., Giallourakis, C., Mostoslavsky, R., and Alt, F.W.: **Mechanism and control of V(D)J recombination at the immunoglobulin heavy chain locus.** *Annu. Rev. Immunol.* 2006, **24**:541–570.

Kim, M.S., Lapkouski, M., Yang, W., and Gellert, M.: **Crystal structure of the V(D)J recombinase RAG1-RAG2.** *Nature* 2015, **518**:507–511.

Li, Z.Y., Otevrel, T., Gao, Y.J., Cheng, H.L., Seed, B., Stamato, T.D., Taccioli, G.E., and Alt, F.W.: **The XRCC4 gene encodes a novel protein involved in DNA double-strand break repair and V(D)J recombination.** *Cell* 1995, **83**:1079–1089.

Mizuta, R., Varghese, A.J., Alt, F.W., Jeggo, P.A., and Jackson, S.P.: **Defective DNA-dependent protein kinase activity is linked to V(D)J recombination and DNA-repair defects associated with the murine *scid* mutation.** *Cell* 1995, **80**:813–823.

Moshous, D., Callebaut, I., de Chasseval, R., Corneo, B., Cavazzana-Calvo, M., le Deist, F., Tezcan, I., Sanal, O., Bertrand, Y., Philippe, N., *et al.*: **Artemis, a novel DNA double-strand break repair/V(D)J recombination protein, is mutated in human severe combined immune deficiency.** *Cell* 2001, **105**:177–186.

Oettinger, M.A., Schatz, D.G., Gorka, C., and Baltimore, D.: **RAG-1 and RAG-2, adjacent genes that synergistically activate V(D)J recombination.** *Science* 1990, **248**:1517–1523.

Villa, A., Santagata, S., Bozzi, F., Giliani, S., Frattini, A., Imberti, L., Gatta, L.B., Ochs, H.D., Schwarz, K., Notarangelo, L.D., *et al.*: **Partial V(D)J recombination activity leads to Omenn syndrome.** *Cell* 1998, **93**:885–896.

Yin, F.F., Bailey, S., Innis, C.A., Ciubotaru, M., Kamtekar, S., Steitz, T.A., and Schatz, D.G.: **Structure of the RAG1 nonamer binding domain with DNA reveals a dimer that mediates DNA synapsis.** *Nat. Struct. Mol. Biol.* 2009, **16**:499–508.

5-6   The diversity of the immunoglobulin repertoire is generated by four main processes.

Weigert, M., Perry, R., Kelley, D., Hunkapiller, T., Schilling, J., and Hood, L.: **The joining of V and J gene segments creates antibody diversity.** *Nature* 1980, **283**:497–499.

5-7   The multiple inherited gene segments are used in different combinations.

Lee, A., Desravines, S., and Hsu, E.: **IgH diversity in an individual with only one million B lymphocytes.** *Dev. Immunol.* 1993, **3**:211–222.

5-8   Variable addition and subtraction of nucleotides at the junctions between gene segments contributes to the diversity of the third hypervariable region.

Gauss, G.H., and Lieber, M.R.: **Mechanistic constraints on diversity in human V(D)J recombination.** *Mol. Cell. Biol.* 1996, **16**:258–269.

Gilfillan, S., Dierich, A., Lemeur, M., Benoist, C., and Mathis, D.: **Mice lacking TdT: mature animals with an immature lymphocyte repertoire.** *Science* 1993, **261**:1755–1759.

Komori, T., Okada, A., Stewart, V., and Alt, F.W.: **Lack of N regions in antigen receptor variable region genes of TdT-deficient lymphocytes.** *Science* 1993, **261**:1171–1175.

Weigert, M., Gatmaitan, L., Loh, E., Schilling, J., and Hood, L.: **Rearrangement of genetic information may produce immunoglobulin diversity.** *Nature* 1978, **276**:785–790.

5-9   The T-cell receptor gene segments are arranged in a similar pattern to immunoglobulin gene segments and are rearranged by the same enzymes.

Bassing, C.H., Alt, F.W., Hughes, M.M., D'Auteuil, M., Wehrly, T.D., Woodman, B.B., Gärtner, F., White, J.M., Davidson, L., and Sleckman, B.P.: **Recombination signal sequences restrict chromosomal V(D)J recombination beyond the 12/23 rule.** *Nature* 2000, **405**:583–586.

Bertocci, B., DeSmet, A., Weill, J.-C., and Reynaud, C.A. **Non-overlapping functions of polX family DNA polymerases, pol μ, pol λ, and TdT, during immunoglobulin V(D)J recombination *in vivo*.** *Immunity* 2006, **25**:31–41.

Lieber, M.R.: **The polymerases for V(D)J recombination.** *Immunity* 2006, **25**:7–9.

Rowen, L., Koop, B.F., and Hood, L.: **The complete 685-kilobase DNA sequence of the human β T cell receptor locus.** *Science* 1996, **272**:1755–1762.

Shinkai, Y., Rathbun, G., Lam, K.P., Oltz, E.M., Stewart, V., Mendelsohn, M., Charron, J., Datta, M., Young, F., Stall, A.M., *et al.*: **RAG-2 deficient mice lack mature lymphocytes owing to inability to initiate V(D)J rearrangement.** *Cell* 1992, **68**:855–867.

5-10   T-cell receptors concentrate diversity in the third hypervariable region.

Davis, M.M., and Bjorkman, P.J.: **T-cell antigen receptor genes and T-cell recognition.** *Nature* 1988, **334**:395–402.

Garboczi, D.N., Ghosh, P., Utz, U., Fan, Q.R., Biddison, W.E., and Wiley, D.C.: **Structure of the complex between human T-cell receptor, viral peptide and HLA-A2.** *Nature* 1996, **384**:134–141.

Hennecke, J., Carfi, A., and Wiley, D.C.: **Structure of a covalently stabilized complex of a human αβ T-cell receptor, influenza HA peptide and MHC class II molecule, HLA-DR1.** *EMBO J.* 2000, **19**:5611–5624.

Hennecke, J., and Wiley, D.C.: **T cell receptor–MHC interactions up close.** *Cell* 2001, **104**:1–4.

Jorgensen, J.L., Esser, U., Fazekas de St. Groth, B., Reay, P.A., and Davis, M.M.: **Mapping T-cell receptor–peptide contacts by variant peptide immunization of single-chain transgenics.** *Nature* 1992, **355**:224–230.

5-11   γ:δ T-cell receptors are also generated by gene rearrangement.

Chien, Y.H., Iwashima, M., Kaplan, K.B., Elliott, J.F., and Davis, M.M.: **A new T-cell receptor gene located within the alpha locus and expressed early in T-cell differentiation.** *Nature* 1987, **327**:677–682.

Lafaille, J.J., DeCloux, A., Bonneville, M., Takagaki, Y., and Tonegawa, S.: **Junctional sequences of T cell receptor gamma delta genes: implications for gamma delta T cell lineages and for a novel intermediate of V-(D)-J joining.** *Cell* 1989, **59**:859–870.

Tonegawa, S., Berns, A., Bonneville, M., Farr, A.G., Ishida, I., Ito, K., Itohara, S., Janeway, C.A., Jr., Kanagawa, O., Kubo, R., *et al.*: **Diversity, development, ligands, and probable functions of gamma delta T cells.** *Adv. Exp. Med. Biol.* 1991, **292**:53–61.

5-12   Different classes of immunoglobulins are distinguished by the structure of their heavy-chain constant regions.

Davies, D.R., and Metzger, H.: **Structural basis of antibody function.** *Annu. Rev. Immunol.* 1983, **1**:87–117.

5-13   The constant region confers functional specialization on the antibody.

Helm, B.A., Sayers, I., Higginbottom, A., Machado, D.C., Ling, Y., Ahmad, K., Padlan, E.A., and Wilson, A.P.M.: **Identification of the high affinity receptor** binding region in human IgE. *J. Biol. Chem.* 1996, **271**:7494–7500.

Nimmerjahn, F., and Ravetch, J.V.: **Fc-receptors as regulators of immunity.** *Adv. Immunol.* 2007, **96**:179–204.

Sensel, M.G., Kane, L.M., and Morrison, S.L.: **Amino acid differences in the N-terminus of C_H2 influence the relative abilities of IgG2 and IgG3 to activate complement.** *Mol. Immunol.* **34**:1019–1029.

5-14   **IgM and IgD are derived from the same pre-mRNA transcript and are both expressed on the surface of mature B cells.**

Abney, E.R., Cooper, M.D., Kearney, J.F., Lawton, A.R., and Parkhouse, R.M.: **Sequential expression of immunoglobulin on developing mouse B lymphocytes: a systematic survey that suggests a model for the generation of immunoglobulin isotype diversity.** *J. Immunol.* 1978, **120**:2041–2049.

Blattner, F.R., and Tucker, P.W.: **The molecular biology of immunoglobulin D.** *Nature* 1984, **307**:417–422.

Enders, A., Short, A., Miosge, L.A., Bergmann, H., Sontani, Y., Bertram, E.M., Whittle, B., Balakishnan, B., Yoshida, K., Sjollema, G., *et al.*: **Zinc-finger protein ZFP318 is essential for expression of IgD, the alternatively spliced Igh product made by mature B lymphocytes.** *Proc. Natl Acad. Sci. USA* 2014, **111**:4513–4518.

Goding, J.W., Scott, D.W., and Layton, J.E.: **Genetics, cellular expression and function of IgD and IgM receptors.** *Immunol. Rev.* 1977, **37**:152–186.

5-15   **Transmembrane and secreted forms of immunoglobulin are generated from alternative heavy-chain mRNA transcripts.**

Early, P., Rogers, J., Davis, M., Calame, K., Bond, M., Wall, R., and Hood, L.: **Two mRNAs can be produced from a single immunoglobulin μ gene by alternative RNA processing pathways.** *Cell* 1980, **20**:313–319.

Martincic, K., Alkan, S.A., Cheatle, A., Borghesi, L., and Milcarek, C.: **Transcription elongation factor ELL2 directs immunoglobulin secretion in plasma cells by stimulating altered RNA processing.** *Nat. Immunol.* 2009, **10**:1102–1109.

Peterson, M.L., Gimmi, E.R., and Perry, R.P.: **The developmentally regulated shift from membrane to secreted μ mRNA production is accompanied by an increase in cleavage-polyadenylation efficiency but no measurable change in splicing efficiency.** *Mol. Cell. Biol.* 1991, **11**:2324–2327.

Rogers, J., Early, P., Carter, C., Calame, K., Bond, M., Hood, L., and Wall, R.: **Two mRNAs with different 3′ ends encode membrane-bound and secreted forms of immunoglobulin μ chain.** *Cell* 1980, **20**:303–312.

Takagaki, Y., and Manley, J.L.: **Levels of polyadenylation factor CstF-64 control IgM heavy chain mRNA accumulation and other events associated with B cell differentiation.** *Mol. Cell.* 1998, **2**:761–771.

Takagaki, Y., Seipelt, R.L., Peterson, M.L., and Manley, J.L.: **The polyadenylation factor CstF-64 regulates alternative processing of IgM heavy chain pre-mRNA during B cell differentiation.** *Cell* 1996, **87**:941–952.

5-16   **IgM and IgA can form polymers by interacting with the J chain.**

Hendrickson, B.A., Conner, D.A., Ladd, D.J., Kendall, D., Casanova, J.E., Corthesy, B., Max, E.E., Neutra, M.R., Seidman, C.E., and Seidman, J.G.: **Altered hepatic transport of IgA in mice lacking the J chain.** *J. Exp. Med.* 1995, **182**:1905–1911.

Niles, M.J., Matsuuchi, L., and Koshland, M.E.: **Polymer IgM assembly and secretion in lymphoid and nonlymphoid cell-lines—evidence that J chain is required for pentamer IgM synthesis.** *Proc. Natl Acad. Sci. USA* 1995, **92**:2884–2888.

5-17   **Some invertebrates generate extensive diversity in a repertoire of immunoglobulin-like genes.**

Dong, Y., Taylor, H.E., and Dimopoulos, G.: **AgDscam, a hypervariable immunoglobulin domain-containing receptor of the *Anopheles gambiae* innate immune system.** *PLoS Biol.* 2006, **4**:e229.

Loker, E.S., Adema, C.M., Zhang, S.M., and Kepler, T.B.: **Invertebrate immune systems—not homogeneous, not simple, not well understood.** *Immunol. Rev.* 2004, **198**:10–24.

Watson, F.L., Puttmann-Holgado, R., Thomas, F., Lamar, D.L., Hughes, M., Kondo, M., Rebel, V.I., and Schmucker, D.: **Extensive diversity of Ig-superfamily proteins in the immune system of insects.** *Science* 2005, **309**:1826–1827.

Zhang, S.M., Adema, C.M., Kepler, T.B., and Loker, E.S.: **Diversification of Ig superfamily genes in an invertebrate.** *Science* 2004, **305**:251–254.

5-18   **Agnathans possess an adaptive immune system that uses somatic gene rearrangement to diversify receptors built from LRR domains.**

Boehm, T., McCurley, N., Sutoh, Y., Schorpp, M., Kasahara, M., and Cooper, M.D.: **VLR-based adaptive immunity.** *Annu. Rev. Immunol.* 2012, **30**:203–220.

Finstad, J., and Good, R.A.: **The evolution of the immune response. 3. Immunologic responses in the lamprey.** *J. Exp. Med.* 1964, **120**:1151–1168.

Guo, P., Hirano, M., Herrin, B.R., Li, J., Yu, C., Sadlonova, A., and Cooper, M.D.: **Dual nature of the adaptive immune system in lampreys.** *Nature* 2009, **459**:796–801. [Erratum: *Nature* 2009, **460**:1044.]

Han, B.W., Herrin, B.R., Cooper, M.D., and Wilson, I.A.: **Antigen recognition by variable lymphocyte receptors.** *Science* 2008, **321**:1834–1837.

Hirano, M., Guo, P., McCurley, N., Schorpp, M., Das, S., Boehm, T., and Cooper, M.D.: **Evolutionary implications of a third lymphocyte lineage in lampreys.** *Nature* 2013, **501**:435–438.

Litman, G.W., Finstad, F.J., Howell, J., Pollara, B.W., and Good, R.A.: **The evolution of the immune response. 3. Structural studies of the lamprey immunoglobulin.** *J. Immunol.* 1970, **105**:1278–1285.

5-19   **RAG-dependent adaptive immunity based on a diversified repertoire of immunoglobulin-like genes appeared abruptly in the cartilaginous fishes.**

Fugmann, S.D., Messier, C., Novack, L.A., Cameron, R.A., and Rast, J.P.: **An ancient evolutionary origin of the *Rag1/2* gene locus.** *Proc. Natl Acad. Sci. USA* 2006, **103**:3728–3733.

Kapitonov, V.V., and Jurka, J.: **RAG1 core and V(D)J recombination signal sequences were derived from Transib transposons.** *PLoS Biol.* 2005, **3**:e181.

Litman, G.W., Rast, J.P., and Fugmann, S.D.: **The origins of vertebrate adaptive immunity.** *Nat. Rev. Immunol.* 2010, **10**:543–553.

Suzuki, T., Shin-I, T., Fujiyama, A., Kohara, Y., and Kasahara, M.: **Hagfish leukocytes express a paired receptor family with a variable domain resembling those of antigen receptors.** *J. Immunol.* 2005, **174**:2885–2891.

5-20   **Different species generate immunoglobulin diversity in different ways.**

Becker, R.S., and Knight, K.L.: **Somatic diversification of immunoglobulin heavy chain VDJ genes: evidence for somatic gene conversion in rabbits.** *Cell* 1990, **63**:987–997.

Knight, K.L., and Crane, M.A.: **Generating the antibody repertoire in rabbit.** *Adv. Immunol.* 1994, **56**:179–218.

Kurosawa, K., and Ohta, K.: **Genetic diversification by somatic gene conversion.** *Genes* (Basel) 2011, **2**:48–58.

Reynaud, C.A., Bertocci, B., Dahan, A., and Weill, J.C.: **Formation of the chicken B-cell repertoire—ontogeny, regulation of Ig gene rearrangement, and diversification by gene conversion.** *Adv. Immunol.* 1994, **57**:353–378.

Reynaud, C.A., Garcia, C., Hein, W.R., and Weill, J.C.: **Hypermutation generating the sheep immunoglobulin repertoire is an antigen independent process.** *Cell* 1995, **80**:115–125.

Vajdy, M., Sethupathi, P., and Knight, K.L.: **Dependence of antibody somatic diversification on gut-associated lymphoid tissue in rabbits.** *J. Immunol.* 1998, **160**:2725–2729.

Winstead, C.R., Zhai, S.K., Sethupathi, P., and Knight, K.L.: **Antigen-induced somatic diversification of rabbit IgH genes: gene conversion and point mutation.** *J. Immunol.* 1999, **162**:6602–6612.

5-21   **Both α:β and γ:δ T-cell receptors are present in cartilaginous fishes.**

Rast, J.P., Anderson, M.K., Strong, S.J., Luer, C., Litman, R.T., and Litman, G.W.: **α, β, γ, and δ T-cell antigen receptor genes arose early in vertebrate phylo-**

geny. *Immunity* 1997, **6**:1–11.

Rast, J.P., and Litman, G.W.: **T-cell receptor gene homologs are present in the most primitive jawed vertebrates.** *Proc. Natl Acad. Sci. USA* 1994, **91**:9248–9252.

5-22   MHC class I and class II molecules are also first found in the cartilaginous fishes.

Hashimoto, K., Okamura, K., Yamaguchi, H., Ototake, M., Nakanishi, T., and Kurosawa, Y.: **Conservation and diversification of MHC class I and its related molecules in vertebrates.** *Immunol. Rev.* 1999, **167**:81–100.

Kurosawa, Y., and Hashimoto, K.: **How did the primordial T cell receptor and MHC molecules function initially?** *Immunol. Cell Biol.* 1997, **75**:193–196.

Ohta, Y., Okamura, K., McKinney, E.C., Bartl, S., Hashimoto, K., and Flajnik, M.F.: **Primitive synteny of vertebrate major histocompatibility complex class I and class II genes.** *Proc. Natl Acad. Sci. USA* 2000, **97**:4712–4717.

Okamura, K., Ototake, M., Nakanishi, T., Kurosawa, Y., and Hashimoto, K.: **The most primitive vertebrates with jaws possess highly polymorphic MHC class I genes comparable to those of humans.** *Immunity* 1997, **7**:777–790.

Antigen Presentation to T Lymphocytes

6

Vertebrate adaptive immune cells possess two types of antigen receptors: the immunoglobulins that serve as antigen receptors on B cells, and the T-cell receptors. While immunoglobulins can recognize native antigens, T cells recognize only antigens that are displayed by MHC complexes on cell surfaces. The conventional **α:β T cells** recognize antigens as peptide:MHC complexes (see Section 4-13). The peptides recognized by α:β T cells can be derived from the normal turnover of self proteins, from intracellular pathogens, such as viruses, or from products of pathogens taken up from the extracellular fluid. Various tolerance mechanisms normally prevent self peptides from initiating an immune response; when these mechanisms fail, self peptides can become the target of autoimmune responses, as discussed in Chapter 15. Other classes of T cells, such as **MAIT cells** and **γ:δ T cells** (see Sections 4-18 and 4-20), recognize different types of surface molecules whose expression may indicate infection or cellular stress.

The first part of this chapter describes the cellular pathways used by various types of cells to generate peptide:MHC complexes recognized by α:β T cells. This process participates in adaptive immunity in at least two different ways. In somatic cells, peptide:MHC complexes can signal the presence of an intracellular pathogen for elimination by armed effector T cells. In dendritic cells, which may not themselves be infected, peptide:MHC complexes serve to activate antigen-specific effector T cells. We will also introduce mechanisms by which certain pathogens defeat adaptive immunity by blocking the production of peptide:MHC complexes.

The second part of this chapter focuses on the MHC class I and II genes and their tremendous variability. The MHC molecules are encoded within a large cluster of genes that were first identified by their powerful effects on the immune response to transplanted tissues and were therefore called the **major histocompatibility complex (MHC)**. There are several different MHC molecules in each class, and each of their genes is highly polymorphic, with many variants present in the population. MHC polymorphism has a profound effect on antigen recognition by T cells, and the combination of multiple genes and polymorphism greatly extends the range of peptides that can be presented to T cells in each individual and in populations as a whole, thus enabling individuals to respond to the wide range of potential pathogens they will encounter. The MHC also contains genes other than those for the MHC molecules; some of these genes are involved in the processing of antigens to produce peptide:MHC complexes.

The last part of the chapter discusses the ligands for unconventional classes of T cells. We will examine a group of proteins similar to MHC class I molecules that have limited polymorphism, some encoded within the MHC and others encoded outside the MHC. These so-called **nonclassical MHC class I proteins** serve various functions, some acting as ligands for γ:δ T-cell receptors and MAIT cells, or as ligands for NKG2D expressed by T cells and NK cells. In addition, we will introduce a special subset of α:β T cells known as invariant NKT cells that recognize microbial lipid antigens presented by these proteins.

The generation of α:β T-cell receptor ligands.

The protective function of T cells depends on their recognition of cells harboring intracellular pathogens or that have internalized their products. As we saw in Chapter 4, the ligand recognized by an α:β T-cell receptor is a peptide bound to an MHC molecule and displayed on a cell surface. The generation of peptides from native proteins is commonly referred to as **antigen processing**, while peptide display at the cell surface by the MHC molecule is referred to as **antigen presentation**. We have already described the structure of MHC molecules and seen how they bind peptide antigens in a cleft, or groove, on their outer surface (see Sections 4-13 to 4-16). We will now look at how peptides are generated from the proteins derived from pathogens and how they are loaded onto MHC class I or MHC class II molecules.

6-1   Antigen presentation functions both in arming effector T cells and in triggering their effector functions to attack pathogen-infected cells.

The processing and presentation of pathogen-derived antigens has two distinct purposes: inducing the development of armed effector T cells, and triggering the effector functions of these armed cells at sites of infection. MHC class I molecules bind peptides that are recognized by CD8 T cells, and MHC class II molecules bind peptides that are recognized by CD4 T cells, a pattern of recognition determined by specific binding of the CD8 or CD4 molecules to the respective MHC molecules (see Section 4-18). The importance of this specificity of recognition lies in the different distributions of MHC class I and class II molecules on cells throughout the body. Nearly all somatic cells (except red blood cells) express MHC class I molecules. Consequently, the CD8 T cell is primarily responsible for pathogen surveillance and cytolysis of somatic cells. Also called **cytotoxic T cells**, their function is to kill the cells they recognize. CD8 T cells are therefore an important mechanism in eliminating sources of new viral particles and bacteria that live only in the cytosol, and thus freeing the host from infection.

By contrast, MHC class II molecules are expressed primarily only on cells of the immune system, and particularly by dendritic cells, macrophages, and B cells. Thymic cortical epithelial cells and activated, but not naive, T cells can express MHC class II molecules, which can also be induced on many cells in response to the cytokine IFN-γ. Thus, CD4 T cells can recognize their cognate antigens during their development in the thymus, on a limited set of 'professional' antigen-presenting cells, and on other somatic cells under specific inflammatory conditions. **Effector CD4 T** cells comprise several subsets with different activities that help eliminate the pathogens. Importantly, naive CD8 and CD4 T cells can become armed effector cells only after encountering their cognate antigen once it has been processed and presented by activated dendritic cells.

In considering antigen processing, it is important to distinguish between the various cellular compartments from which antigens can be derived (**Fig. 6.1**). These compartments, which are separated by membranes, include the **cytosol** and the various **vesicular compartments** involved in endocytosis and secretion. Peptides derived from the cytosol are transported into the endoplasmic reticulum and directly loaded onto newly synthesized MHC class I molecules on the same cell for recognition by T cells, as we will discuss below in greater detail. Because viruses and some bacteria replicate in the cytosol or in the contiguous nuclear compartment, peptides from their components can be loaded onto MHC class I molecules by this process (**Fig. 6.2**, first upper panel).

Fig. 6.1 There are two categories of major intracellular compartments, separated by membranes. One compartment is the cytosol, which communicates with the nucleus via pores in the nuclear membrane. The other is the vesicular system, which comprises the endoplasmic reticulum, Golgi apparatus, endosomes, lysosomes, and other intracellular vesicles. The vesicular system can be thought of as being continuous with the extracellular fluid. Secretory vesicles bud off from the endoplasmic reticulum and are transported via fusion with Golgi membranes to move vesicular contents out of the cell. Extracellular material is taken up by endocytosis or phagocytosis into endosomes or phagosomes, respectively. The fusion of incoming and outgoing vesicles is important both for pathogen destruction in cells such as neutrophils and for antigen presentation. Autophagosomes surround components in the cytosol and deliver them to lysosomes in a process known as autophagy.

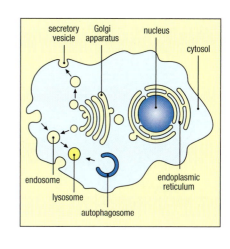

This pathway of recognition is sometimes referred to as **direct presentation**, and can identify both somatic and immune cells that are infected by a pathogen.

Certain pathogenic bacteria and protozoan parasites survive ingestion by macrophages and are able to replicate inside the intracellular vesicles of the endosomal–lysosomal system (Fig. 6.2, second panel). Other pathogenic bacteria proliferate outside cells, and can be internalized, along with their toxic products, by phagocytosis, receptor-mediated endocytosis, or macropinocytosis into endosomes and lysosomes, where they are broken down by digestive enzymes. For example, receptor-mediated endocytosis by B cells can efficiently internalize extracellular antigens through B-cell receptors (Fig. 6.2, third panel). Virus particles and parasite antigens in extracellular fluids can also be taken up by these routes and degraded, and their peptides presented to T cells.

Some pathogens may infect somatic cells but not directly infect phagocytes such as dendritic cells. In this case, dendritic cells must acquire antigens from exogenous sources in order to process and present antigens to T cells. For example, to eliminate a virus that infects only epithelial cells, activation of CD8 T cells will require that dendritic cells load MHC class I molecules with peptides derived from viral proteins taken up from virally infected cells. This exogenous pathway of loading MHC class I molecules is called **cross-presentation**, and is carried out very efficiently by some specialized types of dendritic cells (Fig. 6.3). The activation of naive T cells by this pathway is called **cross-priming**.

	Cytosolic pathogens	Intravesicular pathogens	Extracellular pathogens and toxins
	any cell	macrophage	B cell
Degraded in	Cytosol	Endocytic vesicles (low pH)	Endocytic vesicles (low pH)
Peptides bind to	MHC class I	MHC class II	MHC class II
Presented to	Effector CD8 T cells	Effector CD4 T cells	Effector CD4 T cells
Effect on presenting cell	Cell death	Activation to kill intravesicular bacteria and parasites	Activation of B cells to secrete Ig to eliminate extracellular bacteria/toxins

Fig. 6.2 Cells become targets of T-cell recognition by acquiring antigens from either the cytosolic or the vesicular compartments. Top, first panel: viruses and some bacteria replicate in the cytosolic compartment. Their antigens are presented by MHC class I molecules to activate killing by cytotoxic CD8 T cells. Second panel: other bacteria and some parasites are taken up into endosomes, usually by specialized phagocytic cells such as macrophages. Here they are killed and degraded, or in some cases are able to survive and proliferate within the vesicle. Their antigens are presented by MHC class II molecules to activate cytokine production by CD4 T cells. Third panel: proteins derived from extracellular pathogens may bind to cell-surface receptors and enter the vesicular system by endocytosis, illustrated here for antigens bound by the surface immunoglobulin of B cells. These antigens are presented by MHC class II molecules to CD4 helper T cells, which can then stimulate the B cells to produce antibody.

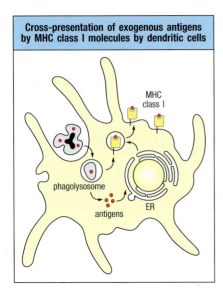

Cross-presentation of exogenous antigens by MHC class I molecules by dendritic cells

Fig. 6.3 Cross-presentation of extracellular antigens on MHC class I molecules by dendritic cells. Certain subsets of dendritic cells are efficient in capturing exogenous proteins and loading peptides derived from them onto MHC class I molecules. There is evidence that several cellular pathways may be involved. One route may involve the translocation of ingested proteins from the phagolysosome into the cytosol for degradation by the proteasome, with the resultant peptides then passing through TAP (see Section 6-3) into the endoplasmic reticulum, where they load onto MHC class I molecules in the usual way. Another route may involve direct transport of antigens from the phagolysosome into a vesicular loading compartment—without passage through the cytosol—where peptides are allowed to be bound to mature MHC class I molecules.

For loading peptides onto MHC class II molecules, dendritic cells, macrophages, and B cells are able to capture exogenous proteins via endocytic vesicles and through specific cell-surface receptors. For B cells, this process of antigen capture can include the B-cell receptor. The peptides that are derived from these proteins are loaded onto MHC class II molecules in specially modified endocytic compartments in these antigen-presenting cells, which we will discuss in more detail later. In dendritic cells, this pathway operates to activate naive CD4 T cells to become effector T cells. Macrophages take up particulate material by phagocytosis and so mainly present pathogen-derived peptides on MHC class II molecules. In macrophages, such antigen presentation may be used to indicate the presence of a pathogen within its vesicular compartment. Effector CD4 T cells, on recognizing antigen, produce cytokines that can activate the macrophage to destroy the pathogen. Some intravesicular pathogens have adapted to resist intracellular killing, and the macrophages in which they live require these cytokines to kill the pathogen: this is one of the roles of the T_H1 subset of CD4 T cells. Other CD4 T cell subsets have roles in regulating other aspects of the immune response, and some CD4 T cells even have cytotoxic activity. In B cells, antigen presentation may serve to recruit help from CD4 T cells that recognize the same protein antigen as the B cell. By efficiently endocytosing a specific antigen via their surface immunoglobulin and presenting the antigen-derived peptides on MHC class II molecules, B cells can activate CD4 T cells that will in turn serve as helper T cells for the production of antibodies against that antigen.

Beyond the presentation of exogenous proteins, MHC class II molecules can also be loaded with peptides derived from cytosolic proteins by a ubiquitous pathway of **autophagy**, in which cytoplasmic proteins are delivered into the endocytic system for degradation in lysosomes (**Fig. 6.4**). This pathway can serve in the presentation of self-cytosolic proteins for the induction of tolerance to self antigens, and also as a means for presenting antigens from pathogens, such as herpes simplex virus, that have accessed the cell's cytosol.

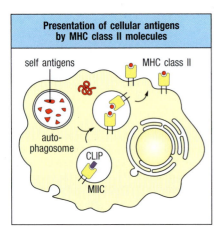

Presentation of cellular antigens by MHC class II molecules

Fig. 6.4 Autophagy pathways can deliver cytosolic antigens for presentation by MHC class II molecules. In the process of autophagy, portions of the cytoplasm are taken into autophagosomes, specialized vesicles that are fused with endocytic vesicles and eventually with lysosomes, where the contents are catabolized. Some of the resulting peptides of this process can be bound to MHC class II molecules and presented on the cell surface. In dendritic cells and macrophages, this can occur in the absence of activation, so that immature dendritic cells may express self peptides in a tolerogenic context, rather than inducing T-cell responses to self antigens.

6-2 Peptides are generated from ubiquitinated proteins in the cytosol by the proteasome.

Proteins in cells are continually being degraded and replaced with newly synthesized proteins. Much cytosolic protein degradation is carried out by a large, multicatalytic protease complex called the **proteasome** (**Fig. 6.5**). A typical proteasome is composed of one **20S catalytic core** and two **19S regulatory caps**, one at each end; both the core and the caps are multisubunit complexes of proteins. The 20S core is a large cylindrical complex of some 28 subunits, arranged in four stacked rings of seven subunits each around a hollow core. The two outer rings are composed of seven distinct α subunits and are noncatalytic. The two inner rings of the 20S proteasome core are composed of seven distinct β subunits. The constitutively expressed **proteolytic subunits** are **β1, β2,** and **β5,** which form the catalytic chamber. The 19S regulator is composed of a base containing nine subunits that binds directly to the α ring of the 20S core particle and a lid that has up to 10 different subunits. The association of the 20S core with a 19S cap requires ATP as well as the ATPase activity of many of the caps' subunits. One of the 19S caps binds and delivers proteins into the proteasome, while the other keeps them from exiting prematurely.

Proteins in the cytosol are tagged for degradation via the **ubiquitin–proteasome system (UPS)**. This begins with the attachment of a chain of several ubiquitin molecules to the target protein, a process called **ubiquitination**. First, a lysine residue on the targeted protein is chemically linked to the glycine at the carboxy terminus of one ubiquitin molecule. Ubiquitin chains are then formed by linking the lysine at residue 48 (K48) of the first ubiquitin to the carboxy-terminal glycine of a second ubiquitin, and so on until at least 4 ubiquitin molecules are bound. This K48-linked type of ubiquitin chain is recognized by the 19S cap of the proteasome, which then unfolds the tagged protein so that it can be introduced into the proteasome's catalytic core. There the protein chain is degraded with a general lack of sequence specificity into short peptides, which are subsequently released into the cytosol. The general degradative functions of the proteasome have been co-opted for antigen presentation, so that MHC molecules have evolved to work with the peptides that the proteasome can produce.

Various lines of evidence implicate the proteasome in the production of peptide ligands for MHC class I molecules. Experimentally tagging proteins with ubiquitin results in more efficient presentation of their peptides by MHC class I molecules, and inhibitors of the proteolytic activity of the proteasome inhibit antigen presentation by MHC class I molecules. Whether the proteasome is the only cytosolic protease capable of generating peptides for transport into the endoplasmic reticulum is not known.

The constitutive β1, β2, and β5 subunits of the catalytic chamber are sometimes replaced by three alternative catalytic subunits that are induced by interferons. These induced subunits are called **β1i (or LMP2)**, **β2i (or MECL-1)**, and **β5i (or LMP7)**. Both β1i and β5i are encoded by the *PSMB9* and *PSMB8* genes, which are located in the MHC locus, whereas β2i is encoded by *PSMB10* outside the MHC locus. Thus, the proteasome can exist both as both a constitutive proteasome present in all cells and as the **immunoproteasome**, which is present in cells stimulated with interferons. MHC class I proteins are also induced by interferons. The replacement of the β subunits by their interferon-inducible counterparts alters the enzymatic specificity of the proteasome such that there is increased cleavage of polypeptides after hydrophobic residues, and decreased cleavage after acidic residues. This produces peptides with carboxy-terminal residues that are preferred anchor residues for binding to most MHC class I molecules (see Chapter 4) and are also the preferred structures for transport by TAP.

Another substitution for a β subunit in the catalytic chamber has been found to occur in cells in the thymus. Epithelial cells of the thymic cortex (**cTECs**) express a unique β subunit, called **β5t**, that is encoded by *PSMB11*. In cTECs, β5t becomes a component of the proteasome in association with β1i and β2i, and this specialized type of proteasome is called the **thymoproteasome**. Mice lacking expression of β5t have reduced numbers of CD8 T cells, indicating that the peptide:MHC complexes produced by the thymoproteasome are important in CD8 T-cell development in the thymus.

Interferon-γ (IFN-γ) can further increase the production of antigenic peptides by inducing expression of the **PA28 proteasome-activator complex** that binds to the proteasome. PA28 is a six- or seven-membered ring composed of two proteins, PA28α and PA28β, both of which are induced by IFN-γ. A PA28 ring, which can bind to either end of the 20S proteasome core in place of the 19S regulatory cap, acts to increase the rate at which peptides are released (**Fig. 6.6**). In addition to simply providing more peptides, the increased rate of

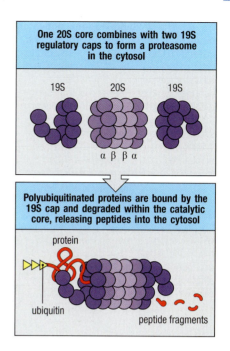

Fig. 6.5 Cytosolic proteins are degraded by the ubiquitin–proteasome system into short peptides. The proteasome is composed of a 20S catalytic core, which consists of four multisubunit rings (see text), and two 19S regulatory caps on either end. Proteins (orange) that are targeted become covalently tagged with K48-linked polyubiquitin chains (yellow) through the actions of various E3 ligases. The 19S regulatory cap recognizes polyubiquitin and draws the tagged protein inside the catalytic chamber; there, the protein is degraded, giving rise to small peptide fragments that are released back into the cytoplasm.

Fig. 6.6 The PA28 proteasome activator binds to either end of the proteasome. Panel a: in this side view cross-section, the heptamer rings of the PA28 proteasome activator (yellow) interact with the α subunits (pink) at either end of the core proteasome (the β subunits that make up the catalytic cavity of the core are in blue). Within this region is the α-annulus (green), a narrow ringlike opening that is normally blocked by other parts of the α subunits (shown in red). Panel b: a close-up view from the top, looking into the α-annulus without PA28 bound. Panel c: with the same perspective, the binding of PA28 to the proteasome changes the conformation of the α subunits, moving those parts of the molecule that block the α-annulus, and opening the end of the cylinder. For simplicity, PA28 is not shown. Structures courtesy of F. Whitby.

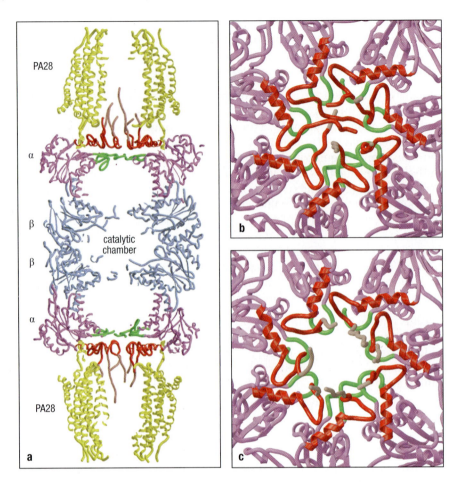

flow allows potentially antigenic peptides to escape additional processing that might destroy their antigenicity.

Translation of self or pathogen-derived mRNAs in the cytoplasm generates not only properly folded proteins but also a significant quantity—possibly up to 30%—of peptides and proteins that are known as **defective ribosomal products (DRiPs)**. These include peptides translated from introns in improperly spliced mRNAs, translations of frameshifts, and improperly folded proteins, which are tagged by ubiquitin for rapid degradation by the proteasome. This seemingly wasteful process provides another source of peptides and ensures that both self proteins and proteins derived from pathogens generate abundant peptide substrates for eventual presentation by MHC class I proteins.

6-3 **Peptides from the cytosol are transported by TAP into the endoplasmic reticulum and further processed before binding to MHC class I molecules.**

The polypeptide chains of proteins destined for the cell surface, such as the two chains of MHC molecules, are translocated during synthesis into the lumen of the endoplasmic reticulum, where two chains fold correctly and assemble with each other. This means that the peptide-binding site of the MHC class I molecule is formed in the lumen of the endoplasmic reticulum and is never exposed to the cytosol. The antigen fragments that bind to MHC class I molecules, however, are typically derived from proteins made in the cytosol. This raises the question, How are these peptides able to bind to MHC class I molecules and be delivered to the cell surface?

The answer was aided by analysis of mutant cells that had a defect in antigen presentation by MHC class I molecules. These cells expressed far fewer MHC class I proteins than normal on their surface despite normal synthesis of these molecules in the cytoplasm. This defect could be corrected by adding synthetic peptides to the culture medium, suggesting that the supply of peptides to the MHC class I molecules in the endoplasmic reticulum might be the limiting factor. Analysis of the DNA of the mutant cells identified the problem responsible for this phenotype to be in genes for members of the **ATP-binding cassette (ABC)** family of proteins; the ABC proteins mediate the ATP-dependent transport of ions, sugars, amino acids, and peptides across membranes.

Missing from the mutant cells were two ABC proteins, called **transporters associated with antigen processing-1 and -2 (TAP1 and TAP2)**, that are normally associated with the endoplasmic reticulum membrane. Transfection of the mutant cells with the missing genes restored the presentation of peptides by the cell's MHC class I molecules. The two TAP proteins form a heterodimer in the membrane (Fig. 6.7), and mutations in either TAP gene can prevent antigen presentation by MHC class I molecules. The genes *TAP1* and *TAP2* are located in the MHC locus (see Section 6-10), near the *PSMB9* and *PSMB8* genes, and their basal level of expression is further enhanced by interferons produced in response to viral infection, similar to MHC class I and β1, β2, and β5 subunits of the proteasome. This induction results in increased delivery of cytosolic peptides into the endoplasmic reticulum.

Microsomal vesicles from non-mutant cells can mimic the endoplasmic reticulum in assays *in vitro*, by internalizing peptides that then bind to MHC class I molecules present in the microsome lumen. In contrast, vesicles from TAP1- or TAP2-deficient cells do not take up peptides. Peptide transport into normal microsomes requires ATP hydrolysis, confirming that the TAP1:TAP2 complex is an ATP-dependent peptide transporter. The TAP complex has limited specificity for the peptides it will transport, transporting peptides of between 8 and 16 amino acids in length and preferring peptides that have hydrophobic or basic residues at the carboxy terminus—the precise features of peptides that bind MHC class I molecules (see Section 4-15). The TAP complex has a bias against proline in the first three amino-terminal residues, but lacks in any true peptide-sequence specificity. The discovery of TAP explained how viral peptides from proteins synthesized in the cytosol gain access to the lumen of the endoplasmic reticulum and are bound by MHC class I molecules.

Peptides produced in the cytosol are protected from complete degradation by cellular chaperones such as the TCP-1 ring complex (TRiC), but many of these peptides are longer than can be bound by MHC class I molecules. Evidence indicates that the carboxy terminus of peptide antigens is produced by cleavage in the proteasome. However, the amino terminus of peptides that are too long to bind MHC class I molecules can be trimmed by an enzyme called the **endoplasmic reticulum aminopeptidase associated with antigen processing (ERAAP)**. Like other components of the antigen-processing pathway, expression of ERAAP is increased by IFN-γ stimulation. Mice lacking the enzyme ERAAP have an altered repertoire of peptides loaded onto MHC class I molecules. Although the loading of some peptides is not affected by the absence of ERAAP, other peptides fail to load normally, and many unstable and immunogenic peptides not normally present are found bound to MHC molecules on the cell surface. This causes cells from ERAAP-deficient mice to be immunogenic for T cells from wild-type mice, demonstrating that ERAAP is an important editor of the normal peptide:MHC repertoire.

6-4 Newly synthesized MHC class I molecules are retained in the endoplasmic reticulum until they bind a peptide.

Binding a peptide is an important step in the assembly of a stable MHC class I molecule. When the supply of peptides into the endoplasmic reticulum is disrupted, as in *TAP*-mutant cells, newly synthesized MHC class I molecules

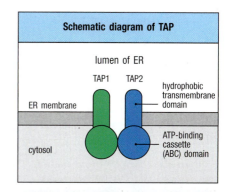

Schematic diagram of TAP

lumen of ER

TAP1   TAP2

hydrophobic transmembrane domain

ER membrane

ATP-binding cassette (ABC) domain

cytosol

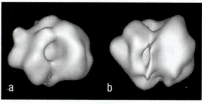

a   b

Fig. 6.7 TAP1 and TAP2 form a peptide transporter in the endoplasmic reticulum membrane. Upper panel: TAP1 and TAP2 are individual polypeptide chains, each with one hydrophobic and one ATP-binding domain. The two chains assemble into a heterodimer to form a four-domain transporter typical of the ATP-binding cassette (ABC) family. The hydrophobic transmembrane domains have multiple transmembrane regions (not shown here). The ATP-binding domains lie within the cytosol, whereas the hydrophobic domains project through the membrane into the lumen of the endoplasmic reticulum (ER) to form a channel through which peptides can pass. Lower panel: electron microscopic reconstruction of the structure of the TAP1:TAP2 heterodimer. Panel a shows the surface of the TAP transporter as seen from the lumen of the ER, looking down onto the top of the transmembrane domains, while panel b shows a lateral view of the TAP heterodimer in the plane of the membrane. The ATP-binding domains form two lobes beneath the transmembrane domains; the bottom edges of these lobes are just visible at the back of the lateral view. TAP structures courtesy of G. Velarde.

MHC Class I Deficiency

are held in the endoplasmic reticulum in a partly folded state. This explains why the rare human patients who have been identified with immunodeficiency due to defects in *TAP1* and *TAP2* have few MHC class I molecules on their cell surfaces, a condition known as **MHC class I deficiency**. The folding and assembly of a complete MHC class I molecule (see Fig. 4.19) depends on the association of the MHC class I α chain first with β_2-microglobulin and then with peptide, and this process involves a number of accessory proteins with chaperone-like functions. Only after peptide has bound is the MHC class I molecule released from the endoplasmic reticulum and transported to the cell surface.

Newly synthesized MHC class I α chains that enter the endoplasmic reticulum membranes bind to **calnexin**, a general-purpose chaperone protein that retains the MHC class I molecule in a partly folded state (**Fig. 6.8**). Calnexin also associates with partly folded T-cell receptors, immunoglobulins, and MHC class II molecules, and so has a central role in the assembly of many immunological as well as non-immunological proteins. When β_2-microglobulin binds to the α chain, the partly folded MHC class I α:β_2-microglobulin heterodimer dissociates from calnexin and binds to an assembly of proteins called the MHC class I **peptide-loading complex (PLC)**. One component of

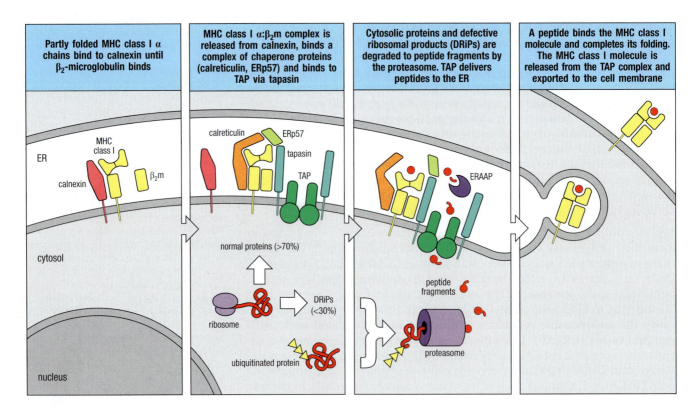

Fig. 6.8 MHC class I molecules do not leave the endoplasmic reticulum unless they bind peptides. Newly synthesized MHC class I α chains assemble in the endoplasmic reticulum (ER) with the membrane-bound protein calnexin. When this complex binds β_2-microglobulin (β_2m), the MHC class I α:β_2m dimer dissociates from calnexin, and the partly folded MHC class I molecule then binds to the TAP-associated protein tapasin. Two MHC:tapasin complexes may bind with the TAP dimer at the same time. The chaperone molecules ERp57, which forms a heterodimer with tapasin, and calreticulin also bind to form the MHC class I peptide-loading complex. The MHC class I molecule is retained within the ER until released by the binding of a peptide, which completes the folding of the MHC molecule. Even in the absence of infection, there is a continual flow of peptides from the cytosol into the ER. Defective ribosomal products (DRiPs) and proteins marked for destruction by K48-linked polyubiquitin (yellow triangles) are degraded in the cytoplasm by the proteasome to generate peptides that are transported into the lumen of the endoplasmic reticulum by TAP. Some of these peptides will bind to MHC class I molecules. The aminopeptidase ERAAP trims the peptides at their amino termini, allowing peptides that are too long to bind to MHC class I molecules and thereby increasing the repertoire of potential peptides for presentation. Once a peptide has bound to the MHC molecule, the peptide:MHC complex leaves the endoplasmic reticulum and is transported through the Golgi apparatus and finally to the cell surface.

the PLC—**calreticulin**—is similar to calnexin and probably also has a general chaperone function, like calnexin. A second component of the complex is the TAP-associated protein **tapasin**, encoded by a gene within the MHC. Tapasin forms a bridge between MHC class I molecules and TAP, allowing the partly folded α:β$_2$-microglobulin heterodimer to await the transport of a suitable peptide from the cytosol. A third component of this complex is the chaperone **ERp57**, a thiol oxidoreductase that may have a role in breaking and re-forming the disulfide bond in the MHC class I α$_2$ domain during peptide loading (Fig. 6.9). ERp57 forms a stable disulfide-linked heterodimer with tapasin. Tapasin seems to be a component of the PLC that is specific to antigen processing, while calnexin, ERp57, and calreticulin bind various other glycoproteins assembling in the endoplasmic reticulum and seem to be part of the cell's general quality control machinery. TAP itself is the final component of the PLC, and it delivers peptides to the partially folded MHC class I molecule.

The PLC maintains the MHC class I molecule in a state that is receptive to peptide binding and mediates the exchange of low-affinity peptides bound to the MHC molecule for peptides of higher affinity, a process called **peptide editing**. The ERp57:tapasin heterodimer functions in editing peptides binding to MHC class I. Cells lacking calreticulin or tapasin show defects in the assembly of MHC class I molecules, and those molecules that reach the cell surface are bound to suboptimal, low-affinity peptides. The binding of a peptide to the partly folded MHC class I molecule releases it from the PLC, and the peptide:MHC complex leaves the endoplasmic reticulum and is transported to the cell surface. Most of the peptides transported by TAP will not bind to the MHC molecules and are rapidly cleared out of the endoplasmic reticulum; these appear to be transported back into the cytosol by **Sec61**, an ATP-dependent transport complex distinct from TAP.

As mentioned above, the MHC class I molecule must bind a peptide in order to be released from the PLC. In cells lacking functional *TAP* genes, the MHC class I molecules fail to exit the endoplasmic reticulum, and so must be degraded instead. Since the ubiquitin–proteasome system is located in the cytosol, these terminally misfolded MHC molecules must somehow be transported back into the cytoplasm for degradation. This is achieved by a system of quality control pathways called **endoplasmic reticulum-associated protein degradation (ERAD)**. ERAD comprises several general cellular pathways that involve the recognition and delivery of misfolded proteins to a **retrotranslocation complex** that unfolds and translocates the proteins across the membrane of the endoplasmic reticulum and into the cytosol. The proteins are ubiquitinated during this process and so are targeted to the ubiquitin–proteasome system (UPS) for eventual degradation. We shall not delve deeply into the details of ERAD here, since these pathways are not unique to MHC

MOVIE 6.1

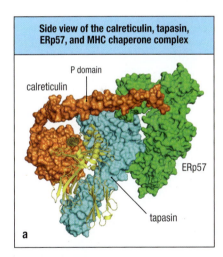

Side view of the calreticulin, tapasin, ERp57, and MHC chaperone complex

P domain
calreticulin
ERp57
tapasin

a

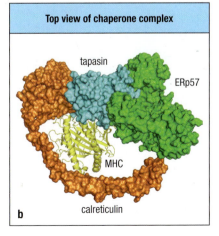

Top view of chaperone complex

tapasin
ERp57
MHC
calreticulin

b

Fig. 6.9 The MHC class I peptide-loading complex includes the chaperones calreticulin, ERp57, and tapasin. This model shows a side (a) and top view (b) of the peptide-loading complex (PLC) oriented as it extends from the luminal surface of the endoplasmic reticulum. The newly synthesized MHC class I and β$_2$-microglobulin are shown as yellow ribbons, with the α helices of the MHC peptide-binding groove clearly identifiable. The MHC and tapasin (cyan) would be tethered to the membrane of the endoplasmic reticulum by carboxy-terminal extensions not shown here. Tapasin and ERp57 (green) form a heterodimer linked by a disulfide bond, and tapasin makes contacts with the MHC molecule that stabilize the empty conformation of the peptide-binding groove; they function in editing peptides binding to the MHC class I molecule. Calreticulin (orange), like the calnexin it replaces (see Fig. 6.8), binds to the monoglucosylated *N*-linked glycan at asparagine 86 of the immature MHC molecule. The long, flexible P domain of calreticulin extends around the top of the peptide-binding groove of the MHC molecule to make contact with ERp57. The transmembrane region of tapasin (not shown) associates the PLC with TAP (see Fig. 6.8), bringing the empty MHC molecules into proximity with peptides arriving into the endoplasmic reticulum from the cytosol. Structure based on PDB file provided by Karin Reinisch and Peter Cresswell.

class I assembly or antigen processing. However, we will see in Chapter 13 how many viral pathogens co-opt the ERAD pathways to block assembly of MHC class I molecules as a way to evade recognition by CD8 T cells.

In uninfected cells, peptides derived from self proteins fill the peptide-binding groove of the mature MHC class I molecules and are carried to the cell surface. In normal cells, MHC class I molecules are retained in the endoplasmic reticulum for some time, which suggests that they are present in excess of peptide. This is important for the immunological function of MHC class I molecules, which must be immediately available to transport viral peptides to the cell surface if the cell becomes infected.

6-5 Dendritic cells use cross-presentation to present exogenous proteins on MHC class I molecules to prime CD8 T cells.

The pathway described above explains how proteins synthesized in the cytosol can generate peptides that become displayed as complexes with MHC class I molecules on the cell surface. This pathway is sufficient to ensure detection and destruction of pathogen-infected cells by cytotoxic T cells. But how do these cytotoxic T cells first become activated? Our explanation so far would require that dendritic cells become infected as well, so that they express the peptide:MHC class I complex needed to activate naive CD8 T cells. But many viruses exhibit a restricted **tropism** for different cells types, and not all viruses will infect dendritic cells. This creates the chance that antigens from such pathogens might never be displayed by dendritic cells, and that cytotoxic T cells that recognize them might not be activated. As it turns out, certain dendritic cells are able to generate peptide:MHC class I complexes from peptides that were not generated within their own cytosol. Peptides from extracellular sources—such as viruses, bacteria, and phagocytosed dying cells infected with cytosolic pathogens—can be presented on MHC class I molecules on the surface of these dendritic cells by the process of **cross-presentation**.

Long before its role in priming T-cell responses to viruses was appreciated, cross-presentation was observed in studies of minor histocompatibility antigens. These are non-MHC gene products that can elicit strong responses between mice of different genetic backgrounds. When spleen cells from B10 mice of MHC type H-2^b were injected into BALB mice of MHC type H-2$^{b×d}$ (which express both b and d MHC types), BALB mice generated cytotoxic T cells reactive against minor antigens of the B10 background. Some of these cytotoxic T cells recognized minor antigens presented by the H-2^b B10 cells used for immunization, as one might expect from direct priming of T cells by the B10 antigen-presenting cells. But other cytotoxic T cells recognized minor B10 antigens only when presented by cells of the H-2^d MHC type. This meant that these CD8 T cells had been activated *in vivo* by recognizing the minor B10 antigens presented by the BALB host's own H-2^d molecules. In other words, the minor histocompatibility antigens must have become transferred from the original immunizing B10 cells to the BALB host's dendritic cells and processed for MHC class I presentation. We now know that cross-presentation by MHC class I molecules occurs not only for antigens on tissue or cell grafts, as in the original experiment described above, but also for viral and bacterial antigens.

It appears that the capacity for cross-presentation is not equally distributed across all antigen-presenting cells. While still an area of active study, it seems that cross-presentation is most efficiently performed by certain subsets of dendritic cells that are present in both humans and mice. Dendritic cell subsets are not identified by the same markers in humans and mice, but in both species, one strongly cross-presenting dendritic cell subset requires the transcription factor **BATF3** for its development, and these cells uniquely express the chemokine receptor **XCR1**. In lymphoid tissues such as the spleen, this lineage of dendritic cells expresses the CD8α molecule on the cell surface, and migratory dendritic cells in lymph nodes capable of cross-presentation are identified by their expression of the **α$_E$ integrin (CD103)**. Mice lacking a functional BATF3 gene lack these types of dendritic cells and are also unable to generate normal CD8 T-cell responses to many viruses, including herpes simplex virus.

The biochemical mechanisms enabling cross-presentation are still unclear, and there may be several different pathways at work. It is not clear whether all proteins captured by phagocytic receptors and taken into endosomes need to be transported into the cytosol and degraded by the proteasome in order to be cross-presented. Some evidence supports a direct pathway in which the PLC is transported from the endoplasmic reticulum to the endosomal compartments, allowing exogenous antigens to be loaded onto newly synthesized MHC class I molecules in phagosomes (see Fig. 6.3). Another pathway of cross-presentation by dendritic cells may involve an interferon-γ-induced GTPase known as **IRGM3** (short for immune-related GTPase family M protein 3). IRGM3 interacts with **adipose differentiation related protein (ADRP)** in the endoplasmic reticulum and regulates the generation of neutral lipid storage organelles called **lipid bodies**, which are thought to originate from ER membranes. Dendritic cells from mice lacking IRGM3 are selectively deficient in cross-presentation of antigens to CD8 T cells, but have a normal process for presenting antigens on MHC class II molecules. The relationship between this and other pathways remains an area of active research.

6-6 Peptide:MHC class II complexes are generated in acidified endocytic vesicles from proteins obtained through endocytosis, phagocytosis, and autophagy.

The immunological function of MHC class II molecules is to bind peptides generated in the intracellular vesicles of dendritic cells, macrophages, and B cells, and to present these peptides to CD4 T cells. The purpose for this pathway is different for each cell type. Dendritic cells primarily are concerned with activating CD4 T cells, while macrophages and B cells are concerned with receiving various forms of help from these CD4 T cells. For example, the intracellular vesicles of macrophages are the sites of replication for several types of pathogens, including the protozoan parasite *Leishmania* and the mycobacteria that cause leprosy and tuberculosis. Because these pathogens reside in membrane-enclosed vesicles, the proteins of these pathogens are not usually accessible to proteasomes in the cytosol. Instead, after activation of the macrophage, the pathogens are degraded by activated intravesicular proteases into peptide fragments that can bind to MHC class II molecules, which pass through this compartment on their way from the endoplasmic reticulum to the cell surface. Like all membrane proteins, MHC class II molecules are first delivered into the endoplasmic reticulum membrane, and are then transported onward as part of membrane-enclosed vesicles that bud off the endoplasmic reticulum and are directed to intracellular vesicles containing internalized antigens. Complexes of peptides and MHC class II molecules are formed there and are then delivered to the cell surface, where they can be recognized by CD4 T cells.

Antigen processing for MHC class II molecules begins when extracellular pathogens and proteins are internalized into endocytic vesicles (Fig. 6.10). Proteins that bind to surface immunoglobulin on B cells and are internalized by receptor-mediated endocytosis are processed by this pathway. Larger particulate materials, such as fragments of dead cells, are internalized by phagocytosis, particularly by macrophages and dendritic cells. Soluble proteins, such as secreted toxins, are taken up by macropinocytosis. Proteins that enter cells through endocytosis are delivered to endosomes, which become

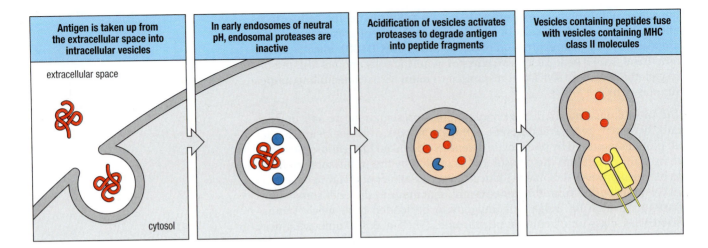

| Antigen is taken up from the extracellular space into intracellular vesicles | In early endosomes of neutral pH, endosomal proteases are inactive | Acidification of vesicles activates proteases to degrade antigen into peptide fragments | Vesicles containing peptides fuse with vesicles containing MHC class II molecules |

Fig. 6.10 Peptides that bind to MHC class II molecules are generated in acidified endocytic vesicles. In the case illustrated here, extracellular foreign antigens, such as bacteria or bacterial antigens, have been taken up by an antigen-presenting cell such as a macrophage or an immature dendritic cell. In other cases, the source of the peptide antigen may be bacteria or parasites that have invaded the cell to replicate in intracellular vesicles. In both cases the antigen-processing pathway is the same. The pH of the endosomes containing the engulfed pathogens decreases progressively, activating proteases within the vesicles to degrade the engulfed material. At some point on their pathway to the cell surface, newly synthesized MHC class II molecules pass through such acidified vesicles and bind peptide fragments of the antigen, transporting the peptides to the cell surface.

increasingly acidic as they progress into the interior of the cell, eventually fusing with lysosomes. The endosomes and lysosomes contain proteases, known as acid proteases, that are activated at low pH and eventually degrade the protein antigens contained in the vesicles.

Drugs such as chloroquine that raise the pH of endosomes, making them less acidic, inhibit the presentation of intravesicular antigens, suggesting that acid proteases are responsible for processing internalized antigen. These proteases include the cysteine proteases—so called because they use a cysteine in their catalytic site—known as **cathepsins** B, D, S, and L, of which L is the most active. Antigen processing can be mimicked to some extent by the digestion of proteins with these enzymes *in vitro* at acid pH. Cathepsins S and L may be the predominant proteases in the processing of vesicular antigens; mice that lack cathepsin B or cathepsin D process antigens normally, whereas mice with no cathepsin S show some deficiencies, including in cross-presentation. Asparagine endopeptidase (AEP), a cysteine protease cleaving after asparagines, is important for processing some antigens, such as the tetanus toxin antigen for MHC class II presentation, but is not required in all cases where antigens contain asparagine residues near their relevant epitopes. It is likely that the overall repertoire of peptides produced within the vesicular pathway reflects the activities of the many proteases present in endosomes and lysosomes. Disulfide bonds, particularly intramolecular disulfide bonds, help in the denaturation process and facilitate proteolysis in endosomes. The enzyme **IFN-γ-induced lysosomal thiol reductase (GILT)** is present in endosomes and functions by breaking and re-forming disulfide bonds in the antigen-processing pathway. The various endosomal proteases act in a largely redundant and nonspecific manner to digest regions of the polypeptide that have become accessible to proteolysis by denaturation and previous steps of degradation. The peptides generated vary in sequence and abundance throughout the endocytic pathway, so that MHC class II molecules can bind and present many different peptides from these compartments.

A significant number of the self-peptides bound to MHC class II molecules arise from common proteins that are cytosolic in location, such as actin and ubiquitin. The most likely way in which cytosolic proteins are processed for MHC class II presentation is by the natural process of protein turnover known as **autophagy**, in which damaged organelles and cytosolic proteins are delivered to lysosomes for degradation. Here their peptides could encounter MHC class II molecules present in the lysosome membranes, and the resulting peptide:MHC class II complex could be transported to the cell surface via endolysosomal tubules (see Fig. 6.4). Autophagy is constitutive, but it is increased by cellular stresses such as starvation, when the cell catabolizes intracellular proteins to obtain energy. In **microautophagy**, cytosol is continuously internalized into the vesicular system by lysosomal invaginations, whereas in **macroautophagy**, which is induced by starvation, a double-membraned autophagosome engulfs cytosol and fuses with lysosomes. A third autophagic pathway uses the heat-shock cognate protein 70 (Hsc70) and the lysosome-associated membrane protein-2 (LAMP-2) to transport cytosolic proteins to lysosomes. Autophagy has been shown to be involved in the processing of the Epstein–Barr virus nuclear antigen 1 (EBNA-1) for presentation on MHC class II molecules. Such presentation enables cytotoxic CD4 T cells to recognize and kill B cells infected with Epstein–Barr virus.

6-7 The invariant chain directs newly synthesized MHC class II molecules to acidified intracellular vesicles.

The biosynthetic pathway for MHC class II molecules begins with their translocation into the endoplasmic reticulum. Here, it is important to prevent them from prematurely binding to peptides transported into the endoplasmic reticulum lumen or to the cell's own newly synthesized polypeptides. The endoplasmic reticulum is full of unfolded and partly folded polypeptide chains, and so a general mechanism is needed to prevent these from binding in the open-ended peptide-binding groove of the MHC class II molecule. Premature peptide binding is prevented by the assembly of newly synthesized MHC class II molecules with a membrane protein known as the MHC class II-associated **invariant chain (Ii, CD74)**. Ii is a type II membrane glycoprotein; its amino terminus resides in the cytosol and its transmembrane region spans the membrane of the endoplasmic reticulum (**Fig. 6.11**). The remainder of Ii and its carboxy terminus reside within the endoplasmic reticulum. Ii has a unique cylindrical domain that mediates formation of stable Ii trimers. Near this domain, Ii contains a peptide sequence, the **class II-associated invariant chain peptide (CLIP)**, with which each Ii subunit of the trimer binds noncovalently to an MHC class II α:β heterodimer. Each Ii subunit binds to an MHC class II molecule with CLIP lying within the peptide-binding groove, thus blocking the groove and preventing the binding of either peptides or partly folded proteins. The binding site of an MHC class II molecule is open relative to the binding site of an MHC class I molecule.

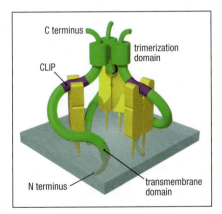
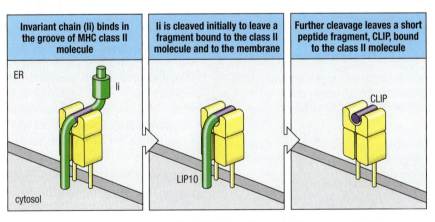

Fig. 6.11 The invariant chain is cleaved to leave a peptide fragment, CLIP, bound to the MHC class II molecule.
A model of the trimeric invariant chain bound to MHC class II α:β heterodimers is shown on the left. The CLIP portion is shown in purple, the rest of the invariant chain is shown in green, and the MHC class II molecules are shown in yellow (model, and leftmost of the three panels). In the endoplasmic reticulum, the invariant chain (Ii) binds to MHC class II molecules with the CLIP section of its polypeptide chain lying along the peptide-binding groove. After transport into an acidified vesicle, Ii is cleaved, initially just at one side of the MHC class II molecule (center panel), first by non-cysteine proteases to give a remaining portion of Ii known as the leupeptin-induced peptide LIP22 (not shown), and then by cysteine protease to the LIP10 fragment shown. LIP10 retains the transmembrane and cytoplasmic segments that contain the signals that target Ii:MHC class II complexes to the endosomal pathway. Subsequent cleavage (right panel) of LIP10 leaves only a short peptide still bound by the class II molecule; this peptide is the CLIP fragment. Model structure courtesy of P. Cresswell.

This allows MHC class II molecules to more easily allow the CLIP region of Ii to pass through their binding sites. While this complex is being assembled in the endoplasmic reticulum, its component parts are associated with calnexin. Only when a nine-chain complex—three Ii chains, three α chains, and three β chains—has been assembled is the complex released from calnexin for transport out of the endoplasmic reticulum. As part of the nine-chain complex, the MHC class II molecules cannot bind peptides or unfolded proteins, so that peptides present in the endoplasmic reticulum are not usually presented by MHC class II molecules. There is evidence that in the absence of Ii many MHC class II molecules are retained in the endoplasmic reticulum as complexes with misfolded proteins.

Trafficking of membrane proteins is controlled by cytosolic sorting tags. In this regard, Ii has a second function, which is to target delivery of the MHC class II molecules to a low-pH endosomal compartment where peptide loading can occur. The complex of MHC class II α:β heterodimers with Ii trimers is retained for 2–4 hours in this compartment (see Fig. 6.11). During this time, the Ii molecule undergoes an initial cleavage by acid proteases to remove the trimerization domain, generating a truncated 22-kDa fragment of Ii called **LIP22**. This is further cleaved by cysteine proteases into a 10-kDa fragment called **LIP10**, which remains bound to the MHC class II molecule and retains it within the proteolytic compartment. A subsequent cleavage of LIP10 releases the MHC class II molecule from the membrane-associated Ii, leaving the CLIP fragment bound to the MHC molecule. This cleavage is carried out by cathepsin S in most MHC class II-positive cells but by cathepsin L in thymic epithelial cells. Being associated with CLIP, the MHC class II molecules cannot yet bind other peptides. However, since CLIP does not carry the Ii-encoded signals that retain the complex in the endocytic compartment, the MHC–CLIP complex is now free to escape to the cell surface.

To allow another peptide to bind to the MHC class II molecule, CLIP must either dissociate or be displaced. Newly synthesized MHC class II molecules are brought toward the cell surface in vesicles, most of which at some point fuse with incoming endosomes. However, some MHC class II:Ii complexes may first be transported to the cell surface and reinternalized into endosomes. In either case, MHC class II:Ii complexes enter the endosomal pathway, where they encounter and bind peptides derived from either internalized pathogen proteins or self proteins. Initially, specialized endosomal compartments were thought to exist for antigen-presenting cells. One was an early endosomal compartment in dendritic cells that was called the **CIIV (MHC class II vesicle)**. Another, a late endosomal compartment containing Ii and MHC class II molecules, was the **MHC class II compartment**, or **MIIC** (Fig. 6.12). The current view is that MHC class II molecules use many common endocytic compartments, including lysosomes, to allow for the exchange of CLIP for as many peptides as possible. MHC class II molecules that do not bind peptide after dissociation from CLIP are unstable in the acidic pH after fusion with lysosomes, and they are rapidly degraded.

6-8 The MHC class II-like molecules HLA-DM and HLA-DO regulate exchange of CLIP for other peptides.

Because an MHC class II:CLIP complex cannot be released to the cell surface unless another peptide replaces it, antigen-presenting cells possess a mechanism that facilitates the efficient exchange of CLIP for other peptides. This process was uncovered by analysis of mutant human B-cell lines with a defect in antigen presentation. MHC class II molecules in these mutant cells assemble correctly with Ii and seem to follow the normal vesicular route, but fail to bind peptides derived from internalized proteins and often arrive at the cell surface with the CLIP peptide still bound. The defect in these cells lies in an MHC class

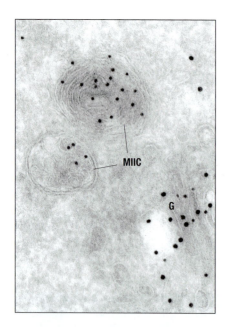

Fig. 6.12 MHC class II molecules are loaded with peptide in a late endosomal compartment called the MIIC. MHC class II molecules are transported from the Golgi apparatus (labeled G in this electron micrograph of an ultrathin section of a B cell) to the cell surface via intracellular vesicles called the MHC class II compartment (MIIC). These have a complex morphology, showing internal vesicles and sheets of membrane. Antibodies labeled with different-sized gold particles identify the presence of both MHC class II molecules (visible as small dark spots) and the invariant chain (large dark spots) in the Golgi, whereas only MHC class II molecules are detectable in the MIIC. This compartment is thought to be a late endosome, an acidified compartment of the endocytic system (pH 4.5–5) in which the invariant chain is cleaved and peptide loading occurs. Photograph (×135,000) courtesy of H.J. Geuze.

II-like molecule called **HLA-DM** in humans (**H-2DM** in mice). The HLA-DM genes (see Section 6-10) are found near the *TAP* and *PSMB8/9* genes in the MHC class II region (see Fig. 6.16); they encode an α chain and a β chain that closely resemble those of other MHC class II molecules. The HLA-DM molecule is not present at the cell surface, however, but is found predominantly in the endosomal compartment that contains Ii and MHC class II molecules. HLA-DM binds to and stabilizes empty MHC class II molecules and catalyzes the release of CLIP, thus allowing the binding of other peptides to the empty MHC class II molecule (Fig. 6.13). The HLA-DM molecule does not contain the open groove found in other MHC class II molecules, and it does not bind peptides. Instead, HLA-DM binds to the α chain of the MHC class II molecule near the region of the floor of the peptide-binding site (Fig. 6.14). This binding induces changes in the structure of the MHC class II molecule, and holds this part of the peptide binding groove in a partially 'open' configuration (see Fig. 6.14, right panel). In this way, HLA-DM catalyzes the release of CLIP and of other unstably bound peptides from MHC class II molecules.

In the presence of a mixture of peptides capable of binding to MHC class II molecules, HLA-DM continuously binds and rebinds to newly formed peptide:MHC class II complexes, allowing for the dissociation of weakly bound peptides and for other peptides to replace them. Antigens presented by MHC class II molecules may have to persist on the surface of antigen-presenting cells for some days before encountering T cells able to recognize them. The ability of HLA-DM to remove unstably bound peptides, sometimes called peptide editing (see Section 6-4), ensures that the peptide:MHC class II complexes displayed on the surface of the antigen-presenting cell survive long enough to stimulate the appropriate CD4 T cells. During this process, it is likely that some peptides are captured first as longer polypeptides that undergo amino-terminal trimming by exopeptidases, further increasing the number of possible peptides that can be bound.

A second atypical MHC class II molecule, called **HLA-DO** in humans (**H-2O** in mice), is produced in thymic epithelial cells, B cells, and dendritic cells.

▶ **MOVIE 6.2**

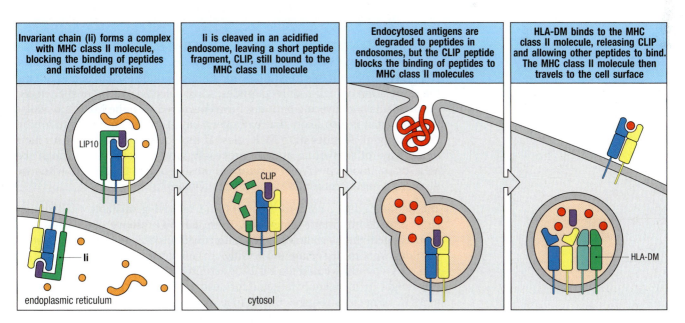

| Invariant chain (Ii) forms a complex with MHC class II molecule, blocking the binding of peptides and misfolded proteins | Ii is cleaved in an acidified endosome, leaving a short peptide fragment, CLIP, still bound to the MHC class II molecule | Endocytosed antigens are degraded to peptides in endosomes, but the CLIP peptide blocks the binding of peptides to MHC class II molecules | HLA-DM binds to the MHC class II molecule, releasing CLIP and allowing other peptides to bind. The MHC class II molecule then travels to the cell surface |

Fig. 6.13 HLA-DM facilitates the loading of antigenic peptides onto MHC class II molecules. First panel: the invariant chain (Ii) binds to newly synthesized MHC class II molecules and blocks peptides from binding class II molecules in the endoplasmic reticulum and during their transport to acidified endosomes. Second panel: in late endosomes, proteases cleave the invariant chain, leaving CLIP bound to the MHC class II molecules. Third panel: pathogens and their proteins are broken down into peptides within acidified endosomes, but these peptides cannot bind to MHC class II molecules that are occupied by CLIP. Fourth panel: the class II-like molecule HLA-DM binds to MHC class II:CLIP complexes, catalyzing the release of CLIP and the binding of antigenic peptides.

HLA-DM bound to HLA-DR	HLA-DM bound to HLA-DO	HLA-DR binding a peptide	HLA-DM bound to HLA-DR
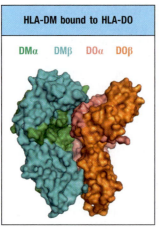	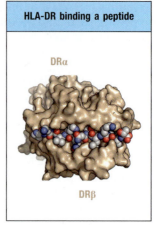	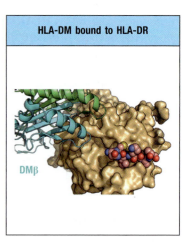	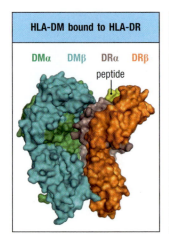

Fig. 6.14 HLA-DM and HLA-DO regulate loading of peptides into MHC class II molecules. First panel: the HLA-DM dimer, composed of α (green) and β (turquoise) chains, binds to the HLA-DR MHC class II molecule (side view). HLA-DM contacts the MHC molecule near the peptide-binding groove where the peptide amino terminus would reside. Second panel: HLA-DO binds to HLA-DM in a similar configuration as HLA-DR, thus blocking DM's peptide-editing activity. Third panel: top view of HLA-DR with bound peptide in the absence of HLA-DM. Fourth panel: top view of HLA-DR with HLA-DM bound. The amino-terminal end of the MHC peptide-binding groove is open and devoid of bound peptide, enabling peptide exchange.

This molecule is a heterodimer of the HLA-DOα chain and the HLA-DOβ chain. HLA-DO is not present at the cell surface, being found only in intracellular vesicles, and it does not seem to bind peptides. HLA-DO acts as a negative regulator of HLA-DM. HLA-DO binds to HLA-DM in the same manner as MHC class II molecules (see Fig. 6.14), and it must be bound to HLA-DM in order to leave the endoplasmic reticulum. When the DM–DO dimer reaches an acidified endocytic compartment, HLA-DO appears to dissociate slowly from HLA-DM, which is then free to catalyze peptide editing for MHC class II molecules. Moreover, IFN-γ increases the expression of HLA-DM, but not of the HLA-DOβ chain. Thus, during inflammatory responses, IFN-γ produced by T cells and NK cells can increase the expression of HLA-DM, and so overcome the inhibitory effects of HLA-DO. Why HLA-DO is expressed in this way remains obscure. The loss of HLA-DO in mice does not dramatically alter adaptive immunity, but does cause a spontaneous production of autoantibodies with age. As thymic epithelial cells function in the selection of developing CD4 T cells, perhaps HLA-DO influences the repertoire of self peptides that these T cells encounter at different stages, as discussed further in Chapter 8.

The role of HLA-DM in peptide editing for MHC class II molecules parallels the role of tapasin in facilitating peptide binding to MHC class I molecules. HLA-DM carries out this function by mediating peptide exchange and driving the association of high-affinity peptides. Thus, it seems likely that specialized mechanisms of delivering peptides have coevolved with the MHC molecules themselves. It is also likely that pathogens have evolved strategies to inhibit the loading of peptides onto MHC class II molecules, much as viruses have found ways of subverting antigen processing and presentation through the MHC class I molecules. We will return to these topics in Chapter 13 when we discuss pathogen immunoevasion mechanisms.

The peptide editing conferred by DM and removing unstable MHC molecules provide important safeguards. To reveal the presence of an intracellular pathogen, the peptide:MHC complex must be stable at the cell surface. If peptides were to dissociate too readily, an infected cell could escape detection, and if peptides could too easily be acquired from other cells, then healthy cells might be mistakenly targeted for destruction. Tight binding of peptides to MHC molecules reduces the chance of these unwanted outcomes. MHC class I molecules display peptides derived largely from cytosolic proteins, so it is important that dissociation of a peptide from a cell-surface MHC molecule does not allow extracellular peptides to bind in the empty peptide-binding site. Fortunately, when an MHC class I molecule at the surface of a living cell loses its peptide, its conformation changes, the β₂-microglobulin dissociates, and the α chain is internalized and rapidly degraded. Thus, most empty MHC class I molecules are quickly lost from the cell surface, largely preventing them from acquiring peptides directly from the surrounding extracellular fluid. This helps ensure that primed T cells target infected cells while sparing surrounding healthy cells.

Empty MHC class II molecules are also removed from the cell surface. Although at neutral pH, empty MHC class II molecules are more stable than empty MHC class I molecules, they aggregate readily, and internalization of such aggregates may account for their removal. Moreover, peptide loss from MHC class II molecules is most likely when the molecules are transiting through acidified endosomes as part of the normal process of cell-membrane recycling. At acidic pH, MHC class II molecules are able to bind peptides that are present in the vesicles, but those that fail to do so are rapidly degraded.

Some binding of extracellular peptides to MHC molecules at the cell surface can occur, however, as the addition of peptides to living or even chemically fixed cells *in vitro* can generate peptide:MHC complexes that are recognized by T cells specific for those peptides. This has been readily demonstrated for many peptides that bind MHC class II and class I molecules. Whether this phenomenon is due to the presence of empty MHC proteins on the cells or to peptide exchange is not clear. Nevertheless it can happen and is a widely used technique to load synthetic peptides for analyzing the specificity of T cells.

6-9 Cessation of antigen processing occurs in dendritic cells after their activation through reduced expression of the MARCH-1 E3 ligase.

Dendritic cells that have not yet been activated by infection carry out active surveillance of the antigens in their location, for example, through macropinocytosis of soluble proteins. Peptides derived from proteins are continuously processed and loaded onto MHC class II molecules for expression on the cell surface. In addition, peptide:MHC complexes are also continuously being recycled from the surface and degraded in cells by ubiquitination and proteasomal degradation. MHC class II molecules contain a conserved lysine residue in the cytoplasmic tails of the β chain; this lysine residue is a target of an **E3 ligase** (see Section 3-7) called **membrane associated ring finger (C3HC4) 1**, or **MARCH-1**, expressed in B cells, dendritic cells, and macrophages. MARCH-1 is expressed constitutively in B cells and induced by the cytokine IL-10 in other cells. It resides in the membrane of a recycling endosomal compartment, where it ubiquitinates the cytoplasmic tail of MHC class II molecules, leading to their eventual degradation in lysosomes, thereby regulating their steady-state level of expression (**Fig. 6.15**).

The MARCH-1 pathway is shut down during infection to increase the stability of peptide:MHC complexes. Dendritic cells that capture antigens at sites of infection must first migrate to local lymph nodes in order to activate naive T cells; this may take many hours. Since continuous recycling limits the lifetime of peptide:MHC complexes on the cell surface, pathogen-derived peptide:MHC complexes could be lost during this migration, preventing T-cell activation. To prevent this situation, when dendritic cells are activated by pathogens, expression of MARCH-1 is shut off. This may be mediated directly by innate pathogen sensors, since TLR signaling in dendritic cells rapidly reduces the level of mRNA for MARCH-1. The MARCH-1 protein half-life is only around 30 minutes, so that activated dendritic cells soon accumulate peptide:MHC complexes on their cell surface produced at the time of encounter with pathogen.

In addition to regulating MHC class II expression in dendritic cells, MARCH-1 similarly regulates expression in dendritic cells of the co-stimulatory molecule (see Section 1-15) **CD86** (or **B7-2**), which like MHC class II molecules, is also

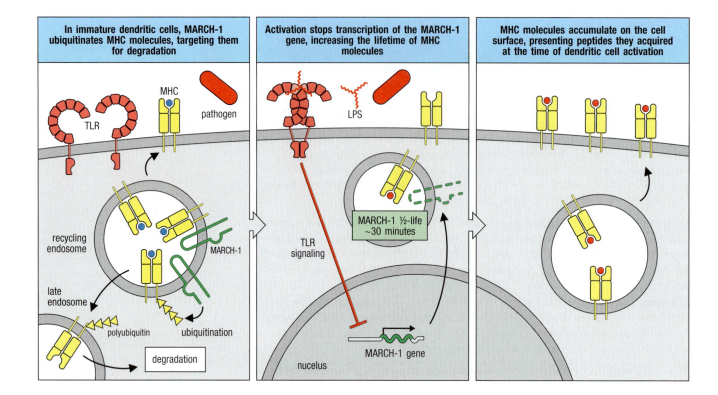

Fig. 6.15 Activation of dendritic cells reduces MARCH-1 expression, thus increasing the lifetime of MHC molecules. Before activation by innate recognition of pathogens, dendritic cells express the membrane-associated E3 ligase MARCH-1, which resides in the recycling endosomes compartment, where it attaches K-48-linked ubiquitin chains to the β chain of MHC class II molecules. This causes MHC molecules to move from the recycling endosomes and eventually to be degraded, leading to a reduced overall half-life and level of MHC expression on the cell surface. Signals emanating from innate sensors, such as TLR-4, reduce the level of MARCH-1 mRNA, and with the half-life of MARCH-1 now being short, MHC molecules are free to accumulate on the cell surface. Because innate signaling also triggers acidification of endocytic compartments and activates caspases associated with antigen processing, the MHC molecules that accumulate on the cell surface will bear peptides from the pathogens captured around the time of innate activation of the dendritic cells.

regulated by ubiquitination. This means that by the time dendritic cells arrive at lymph nodes, they express peptides derived from the pathogens that activated them and have higher CD86 levels that provide signals for greater CD4 T-cell activation. However, we will see in Chapter 13 that viral pathogens have taken advantage of this pathway by producing MARCH-1-like proteins to downregulate MHC class II molecules as a means of evading adaptive immunity.

Summary.

The ligand recognized by the T-cell receptor is a peptide bound to an MHC molecule. MHC class I and MHC class II molecules acquire peptides at different intracellular sites and activate either CD8 or CD4 T cells, respectively. Infected cells presenting peptides derived from virus replication in the cytosol are thus recognized by CD8 cytotoxic T cells, which are specialized to kill any cells displaying foreign antigens. MHC class I molecules are synthesized in the endoplasmic reticulum and typically acquire their peptides at this location. The peptides loaded onto MHC class I molecules are derived from proteins degraded in the cytosol by the proteasome, transported into the endoplasmic reticulum by the heterodimeric ATP-binding protein TAP, and further processed by the aminopeptidase ERAAP before being loaded onto the MHC molecules. Peptide binding to MHC class I molecules is required for them to be released from chaperones in the endoplasmic reticulum and to travel to the cell surface. Certain subsets of dendritic cells are able to produce peptides from exogenous proteins and load them onto MHC class I molecules. Such cross-presentation of antigens ensures that CD8 T cells can be activated by pathogens that may not directly infect antigen-presenting cells.

MHC class II molecules do not acquire their peptide ligands in the endoplasmic reticulum, because the invariant chain (Ii) first inserts CLIP into their peptide-binding groove. Association with Ii targets these MHC molecules to an acidic endosomal compartment where active proteases cleave Ii, and HLA-DM helps to catalyze dissociation of CLIP. The MHC molecules can then associate with peptides derived from proteins that have entered the vesicular compartments of macrophages, dendritic cells, or B cells. The process of autophagy can deliver cytosolic proteins to the vesicular system for presentation by MHC class II molecules. The CD4 T cells that recognize peptide:MHC class II complexes have a variety of specialized effector activities. Subsets of CD4 T cells activate macrophages to kill the intravesicular pathogens they harbor, help B cells to secrete immunoglobulins against foreign molecules, and regulate immune responses.

The major histocompatibility complex and its function.

The function of MHC molecules is to bind peptide fragments derived from pathogens and display them on the cell surface for recognition by the appropriate T cells. The consequences are almost always deleterious to the pathogen—virus-infected cells are killed, macrophages are activated to kill bacteria living in their intracellular vesicles, and B cells are activated to produce antibodies that eliminate or neutralize extracellular pathogens. Thus, there is strong selective pressure in favor of any pathogen that has mutated in such a way that it escapes presentation by an MHC molecule.

Two separate properties of the major histocompatibility complex (MHC) make it difficult for pathogens to evade immune responses in this way. First, the MHC is **polygenic**: it contains several different MHC class I and MHC class II genes, so that every individual possesses a set of MHC molecules with different ranges of peptide-binding specificities. Second, the MHC is highly **polymorphic**; that is, there are multiple variants, or alleles, of each gene within the population as a whole. The MHC genes are, in fact, the most polymorphic genes known. In this section we describe the organization of the genes in the MHC and discuss how the variation in MHC molecules arises. We also consider how the effect of polygeny and polymorphism on the range of peptides that can be bound contributes to the ability of the immune system to respond to the multitude of different and rapidly evolving pathogens.

6-10 Many proteins involved in antigen processing and presentation are encoded by genes within the MHC.

The MHC is located on chromosome 6 in humans and chromosome 17 in the mouse and extends over at least 4 million base pairs. In humans it contains more than 200 genes. As work continues to define the genes within and around the MHC, it becomes difficult to establish precise boundaries for this genetic region, which is now thought to span as many as 7 million base pairs. The genes encoding the α chains of MHC class I molecules and the α and β chains of MHC class II molecules are linked within the complex; the genes for β_2-microglobulin and the invariant chain are on different chromosomes (chromosomes 15 and 5, respectively, in humans, and chromosomes 2 and 18 in the mouse). **Figure 6.16** shows the general organization of the MHC class I and II genes in human and mouse. In humans these genes are called **human leukocyte antigen** or **HLA** genes, because they were first discovered through antigenic differences between white blood cells from different individuals; in the mouse they are known as the **H-2 genes**. The mouse MHC class II genes were in fact first identified as genes that controlled whether an immune response was made to a given antigen and were originally called **Ir (immune response) genes**. Because of this, the mouse MHC class II *A* and *E* genes were in the past referred to as *I-A* and *I-E*, but this terminology could be confused with MHC class I genes and it is no longer used.

Fig. 6.16 The genetic organization of the major histocompatibility complex (MHC) in humans and mice. The organization of the MHC genes is shown. In humans, the cluster is called HLA (short for human leukocyte antigen) and is on chromosome 6, and in mice, it is called H-2 (for histocompatibility) and is on chromosome 17. The organization is similar in both species, with separate clusters of MHC class I genes (red) and MHC class II genes (yellow). In mice, the MHC class I gene *H-2K* has been translocated relative to the human MHC, splitting the class I region in two. Both species have three main class I genes, which are called *HLA-A*, *HLA-B*, and *HLA-C* in humans, and *H2-K*, *H2-D*, and *H2-L* in the mouse. These encode the α chain of the respective MHC class I proteins, HLA-A, HLA-B, and so on. The other subunit of an MHC class I molecule, β₂-microglobulin, is encoded by a gene located on a different chromosome—chromosome 15 in humans and chromosome 2 in the mouse. The class II region includes the genes for the α and β chains (designated *A* and *B*) of the MHC class II molecules HLA-DR, -DP, and -DQ (H-2A and -E in the mouse). Also in the MHC class II region are the genes for the TAP1:TAP2 peptide transporter, the *PSMB* (or *LMP*) genes that encode proteasome subunits, the genes encoding the DMα and DMβ chains (*DMA* and *DMB*), the genes encoding the α and β chains of the DO molecule (DOA and DOB, respectively), and the gene encoding tapasin (*TAPBP*). The so-called class III genes encode various other proteins with functions in immunity (see Fig. 6.17).

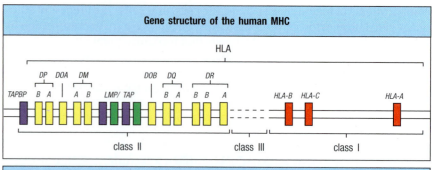

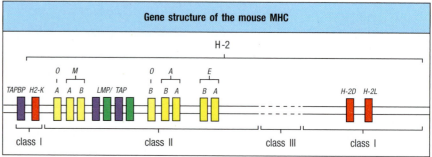

There are three class I α-chain genes in humans, called *HLA-A*, *-B*, and *-C*. There are also three pairs of MHC class II α- and β-chain genes, called *HLA-DR*, *-DP*, and *-DQ*. In many people, however, the HLA-DR cluster contains an extra β-chain gene whose product can pair with the DRα chain. This means that the three sets of genes can give rise to four types of MHC class II molecules. All the MHC class I and class II molecules can present peptides to T cells, but each protein binds a different range of peptides (see Sections 4-14 and 4-15). Thus, the presence of several different genes for each MHC class means that any one individual is equipped to present a much broader range of peptides than if only one MHC molecule of each class were expressed at the cell surface.

Figure 6.17 shows a more detailed map of the human MHC region. Many genes within this locus participate in antigen processing or antigen presentation, or have other functions related to the innate or adaptive immune response. The two *TAP* genes lie in the MHC class II region near the *PSMB8* and *PSMB9* genes, whereas the gene encoding tapasin (*TAPBP*) lies at the edge of the MHC nearest the centromere. The genetic linkage of the MHC class I genes (whose products deliver cytosolic peptides to the cell surface) with the *TAP*, tapasin, and proteasome (*PSMB* or *LMP*) genes (whose products deliver cytosolic peptides into the endoplasmic reticulum) suggests that the entire MHC has been selected during evolution for antigen processing and presentation.

When cells are treated with the interferons IFN-α, -β, or -γ, there is a marked increase in the transcription of MHC class I α-chain and β₂-microglobulin genes and of the proteasome, tapasin, and *TAP* genes. Interferons are produced early in viral infections as part of the innate immune response, as described in Chapter 3. The increase in MHC expression they produce helps all cells to process viral proteins and present the resulting virus-derived peptides on their surface (except for red blood cells). On dendritic cells, this helps to activate the appropriate T cells and initiate the adaptive immune response to the virus. The coordinated regulation of the genes encoding these components may be facilitated by the linkage of many of them in the MHC.

The *DMA* and *DMB* genes encoding the subunits of the HLA-DM molecule that catalyzes peptide binding to MHC class II molecules are clearly related to the MHC class II genes, as are the *DOA* and *DOB* genes that encode the subunits of the regulatory HLA-DO molecule. Gene expression of the classical

MHC class II proteins, along with the invariant-chain, DMα, DMβ, and DOα, but not DOβ, is coordinately increased by **IFN-γ**, which is produced by activated T_H1 cells, CD8 T cells, and NK cells. This form of regulation allows dendritic cells and macrophages to upregulate molecules involved in processing of intravesicular antigens when presenting antigens to T cells and NK cells. Expression of all these molecules is induced by IFN-γ (but not by IFN-α or -β), via the production of a protein known as **MHC class II transactivator (CIITA)**, which acts as a positive transcriptional co-activator of MHC class II genes. An absence of CIITA causes severe immunodeficiency due to the nonproduction of MHC class II molecules—**MHC class II deficiency**. Finally, the MHC contains many 'non-classical' MHC genes, so-called because while they resemble MHC genes in structure, their products do not function in presenting peptides to conventional α:β T cells. Many of these genes are now referred to as MHC class Ib genes, and their protein products have a variety of different functions, which we will describe in Section 6-16 , following our discussion of the conventional MHC genes.

MHC Class II Deficiency

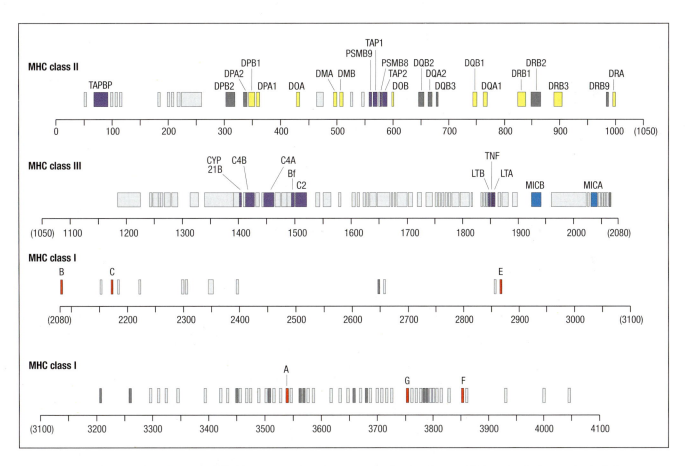

Fig. 6.17 Detailed map of the human MHC. The organization of the class I, class II, and class III regions of the human MHC is shown, with approximate genetic distances given in thousands of base pairs. Most of the genes in the class I and class II regions are mentioned in the text. The additional genes indicated in the class I region (for example, E, F, and G) are class I-like genes encoding class Ib molecules; the additional class II genes are pseudogenes. The genes shown in the class III region encode the complement proteins C4 (two genes, shown as C4A and C4B), C2, and factor B (shown as Bf), as well as genes that encode the cytokines tumor necrosis factor-α (TNF) and lymphotoxin (LTA, LTB). Closely linked to the C4 genes is the gene encoding 21-hydroxylase (shown as CYP 21B), an enzyme involved in steroid biosynthesis. Immunologically important functional protein-coding genes mentioned in the text are color coded, with the MHC class I genes being shown in red, except for the MIC genes, which are shown in blue; these are distinct from the other class I-like genes and are under different transcriptional control. The immunologically important MHC class II genes are shown in yellow. Genes in the MHC region that have immune functions but are not related to the MHC class I and class II genes are shown in purple. Genes in dark gray are pseudogenes related to immune-function genes. Unnamed genes unrelated to immune function are shown in light gray.

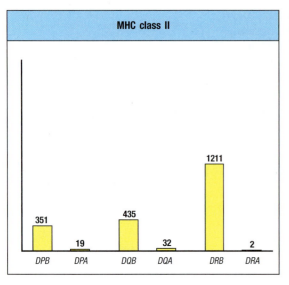

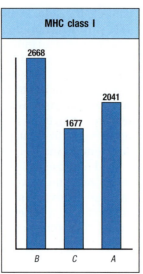

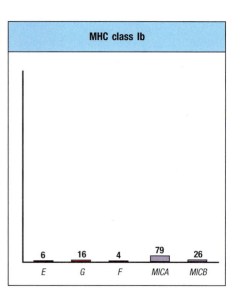

Fig. 6.18 Human MHC genes are highly polymorphic. With the notable exception of the DRα locus, which is functionally monomorphic, each gene locus has many alleles. The number of functional proteins encoded is less than the total number of alleles. Shown in this figure as the heights of the bars are the number of different HLA proteins assigned by the WHO Nomenclature Committee for Factors of the HLA System as of January 2010.

6-11   The protein products of MHC class I and class II genes are highly polymorphic.

Because of the polygeny of the MHC, every person expresses at least three different MHC class I molecules and three (or sometimes four) MHC class II molecules on his or her cells. In fact, the number of different MHC molecules expressed by most people is greater because of the extreme polymorphism of the MHC (Fig. 6.18).

The term **polymorphism** comes from the Greek *poly*, meaning many, and *morphe*, meaning shape or structure. As used here, it means within-species variation at a gene locus, and thus in the gene's protein product; the variant genes that can occupy the locus are termed **alleles**. For several MHC class I and class II genes, there are more than 1000 alleles in the human population, far more than the number of alleles for other genes found within the MHC region. Each MHC class I and class II allele is relatively frequent in the population, so there is only a small chance that the corresponding gene loci on both homologous chromosomes of an individual will have the same allele; most individuals will be **heterozygous** for the genes encoding MHC class I and class II molecules. The particular combination of MHC alleles found on a single chromosome is known as an **MHC haplotype**. Expression of MHC alleles is **codominant**, meaning that the protein products of both of the alleles at a locus are expressed equally in the cell, and both gene products can present antigens to T cells. The number of MHC alleles discovered that do not code for a functional protein is remarkably small. The extensive polymorphism at each locus thus has the potential to double the number of different MHC molecules expressed in an individual and thereby increase the diversity already available through polygeny (Fig. 6.19).

Because most individuals are heterozygous, most matings will produce offspring that receive one of four possible combinations of the parental MHC haplotypes. Thus siblings are also likely to differ in the MHC alleles they express, there being one chance in four that an individual will share both haplotypes with a sibling. One consequence of this is the difficulty of finding suitable donors for tissue transplantation, even among siblings.

All MHC class I and II proteins are polymorphic to a greater or lesser extent, with the exception of the DRα chain and its mouse homolog Eα. These chains do not vary in sequence between different individuals and are said to be **monomorphic**. This might indicate a functional constraint that prevents variation in the DRα and Eα proteins, but no such special function has been found.

Polymorphism	Polygeny	Polymorphism and polygeny
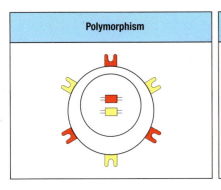	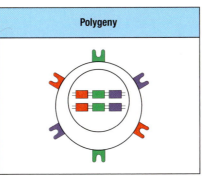	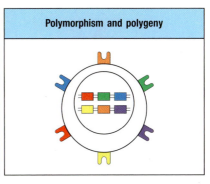

Many mice, both domestic and wild, have a mutation in the Eα gene that prevents synthesis of the Eα protein. They thus lack cell-surface H-2E molecules, so if H2-E does have a special function it is unlikely to be essential.

MHC polymorphisms at individual MHC genes seem to have been strongly selected by evolutionary pressures. Several genetic mechanisms contribute to the generation of new alleles. Some new alleles arise by point mutations and others by gene conversion, a process in which a sequence in one gene is replaced, in part, by sequences from a different gene (Fig. 6.20). The effects of selective pressure in favor of polymorphism can be seen clearly in the pattern of point mutations in the MHC genes. Point mutations can be classified as replacement substitutions, which change an amino acid, or silent substitutions, which simply change the codon but leave the amino acid the same. Replacement substitutions occur within the MHC at a higher frequency relative to silent substitutions than would be expected, providing evidence that polymorphism has been actively selected for in the evolution of the MHC.

Fig. 6.19 Polymorphism and polygeny both contribute to the diversity of MHC molecules expressed by an individual. The high polymorphism of the classical MHC genes ensures diversity in MHC gene expression in the population as a whole. However, no matter how polymorphic a gene is, no individual can express more than two alleles at a single gene locus. Polygeny, the presence of several different related genes with similar functions, ensures that each individual produces a number of different MHC molecules. The combination of polymorphism and polygeny produces the diversity of MHC molecules seen both within an individual and in the population at large.

6-12 MHC polymorphism affects antigen recognition by T cells by influencing both peptide binding and the contacts between T-cell receptor and MHC molecule.

The next few sections describe how MHC polymorphisms benefit the immune response and how pathogen-driven selection can account for the large number of MHC alleles. The products of individual MHC alleles, often known as protein **isoforms**, can differ from one another by up to 20 amino acids, making each variant protein quite distinct. Most of the differences are localized to exposed surfaces of the extracellular domain furthest from the membrane, and to the peptide-binding groove in particular (Fig. 6.21). We have seen that peptides bind to MHC class I and class II molecules through the interaction of specific anchor residues with peptide-binding pockets in the peptide-binding groove (see Sections 4-15 and 4-16). Many of the polymorphisms in MHC molecules alter the amino acids that line these pockets and thus change

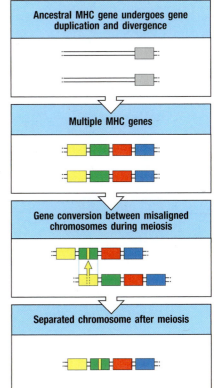

Fig. 6.20 Gene conversion can create new alleles by copying sequences from one MHC gene to another. Multiple MHC genes of generally similar structure were derived over evolutionary time by duplication of an unknown ancestral MHC gene (gray) followed by genetic divergence. Further interchange between these genes can occur by a process known as gene conversion, in which sequences can be transferred from part of one gene to a similar gene. For this to happen, the two genes must become apposed during meiosis. This can occur as a consequence of the misalignment of the two paired homologous chromosomes when there are many copies of similar genes arrayed in tandem—somewhat like buttoning in the wrong buttonhole. During the process of crossing-over and DNA recombination, a DNA sequence from one chromosome is sometimes copied to the other, replacing the original sequence. In this way, several nucleotide changes can be inserted all at once into a gene and can cause several simultaneous amino acid changes in the gene product. Because of the similarity of the MHC genes to each other and their close linkage, gene conversion has occurred many times in the evolution of MHC alleles.

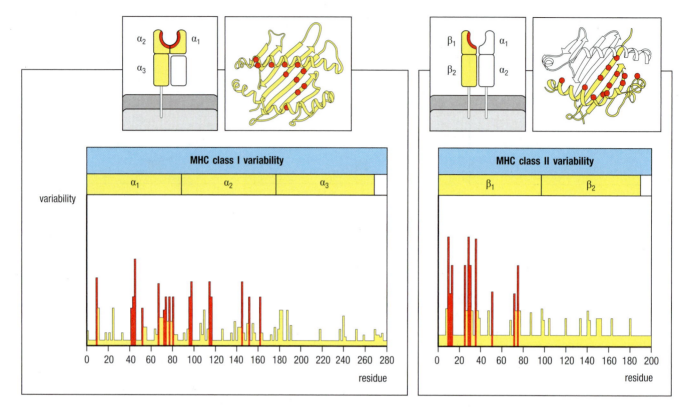

Fig. 6.21 Allelic variation in MHC molecules occurs predominantly within the peptide-binding region. Variability plots of the amino acid sequences of MHC molecules show that the variation arising from genetic polymorphism is restricted to the amino-terminal domains (α_1 and α_2 domains of MHC class I molecules, and α_1 and β_1 domains of MHC class II molecules). These are the domains that form the peptide-binding groove.

Moreover, allelic variability is clustered in specific sites within the amino-terminal domains, lying in positions that line the peptide-binding groove, either on the floor of the groove or inward from the walls. For the MHC class II molecule, the variability of the HLA-DR alleles is shown. For HLA-DR, and its homologs in other species, the α chain is essentially invariant and only the β chain shows significant polymorphism.

the pockets' binding specificities. This in turn changes the anchor residues of peptides that can bind to each MHC isoform. The set of anchor residues that allows binding to a given isoform of an MHC class I or class II molecule is called a **sequence motif**, and this can be used to predict peptides within a protein that might bind that variant (Fig. 6.22). Such predictions may be very important in designing peptide vaccines, as we will see in Chapter 16, where we discuss recent progress in cancer immunotherapy.

In rare cases, processing of a protein does not generate any peptides with a suitable sequence motif for binding to any of the MHC molecules expressed by an individual. This individual fails to respond to the antigen. Such failures in responsiveness to simple antigens were first reported in inbred animals and were called immune response (Ir) gene defects. These defects were mapped to genes within the MHC long before the structure or function of MHC molecules was understood, and they were the first clue to the antigen-presenting function of MHC molecules. We now understand that Ir gene defects are common in inbred strains of mice because the mice are homozygous at all their MHC gene loci, which limits the range of peptides they can present to T cells. Ordinarily, MHC polymorphism guarantees a sufficient number of different MHC molecules in a single individual to make this type of nonresponsiveness unlikely, even to relatively simple antigens such as small toxins.

Initially, the only evidence linking Ir gene defects to the MHC was genetic—mice of one MHC genotype could make antibody in response to a particular antigen, whereas mice of a different MHC genotype, but otherwise genetically identical, could not. The MHC genotype was somehow controlling the ability

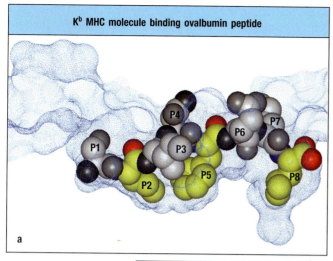

K^b MHC molecule binding ovalbumin peptide

a

	P1	P2	P3	P4	—	P5	P6	P7	P8
Ovalbumin (257–264)	S	I	I	N		F	E	K	L
HBV surface antigen (208–215)	I	L	S	P		F	L	P	L
Influenza NS2 (114–121)	R	T	F	S		F	Q	L	I
LCMV NP (205–212)	Y	T	V	K		Y	P	N	L
VSV NP (52–59)	R	G	Y	V		Y	Q	G	L
Sendai virus NP (324–332)	F	A	P	G	N	Y	P	A	L

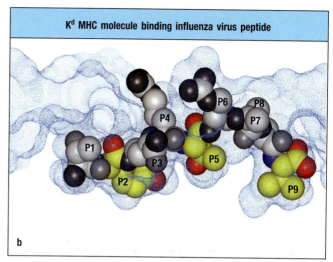

K^d MHC molecule binding influenza virus peptide

b

	P1	P2	P3	P4	P5	P6	P7	P8	P9
Influenza NP (147–155)	T	Y	Q	R	T	R	A	L	V
ERK4 (136–144)	Q	Y	I	H	S	A	N	V	L
P198 (14–22)	K	Y	Q	A	V	T	T	T	L
P. yoelii CSP (280–288)	S	Y	V	P	S	A	E	Q	I
P. berghei CSP (25)	G	Y	I	P	S	A	E	K	I
JAK1 (367–375)	S	Y	F	P	E	I	T	H	I

Fig. 6.22 Different allelic variants of an MHC class I molecule bind different peptides. Shown are cutaway views of (a) ovalbumin peptide bound to the mouse H2-K^b MHC class I molecule and (b) influenza nucleoprotein (NP) peptide bound to the H2-K^d MHC class I molecule. The solvent-accessible surface of the MHC molecules is shown as a blue dotted surface. Class I MHC molecules typically have six pockets in the peptide-binding groove, which are conventionally called A–F. The bound peptides, shown as space-filling models, fit into the peptide-binding groove, with side chains from the anchor residues extending to fill the pockets. H2-K^b is binding SIINFEKL (single-letter amino acid code), a peptide of eight residues (P1–8) from ovalbumin, and H2-K^d is binding TYQRTRALV, a peptide of nine residues (P1–9) from the influenza nucleoprotein (NP). Anchor residues (shown in yellow) may be primary or secondary in their influence on peptide binding. For H2-K^b, the sequence motif is determined by two primary anchors,

P5 and P8; the C pocket binds the P5 side chain of the peptide [a tyrosine (Y) or a phenylalanine (F)], and the F pocket binds the P8 residue [a non-aromatic hydrophobic side chain from leucine (L), isoleucine (I), methionine (M), or valine (V)]. The B pocket binds P2, a secondary anchor residue in H-2K^b. For H2-K^d, the sequence motif is primarily determined by the two primary anchors, P2 and P9. The B pocket accommodates a tyrosine side chain. The F pocket binds leucine, isoleucine, or valine. Beneath the structures are shown sequence motifs from peptides that are known to bind to the MHC molecule. CSP, circumsporozoite antigen; ERK4, extracellular signal-related kinase 4; HBV, hepatitis B virus; JAK1, Janus-associated kinase 1; LCMV, lymphocytic choriomeningitis virus; NS2, NS2 protein; P198, modified tumor-cell antigen; *P. berghei, Plasmodium berghei; P. yoelii, Plasmodium yoelii*; VSV, vesicular stomatitis virus. An extensive collection of motifs can be found at http://www.syfpeithi.de. Structures courtesy of V.E. Mitaksov and D. Fremont.

of the immune system to detect or respond to specific antigens, but it was not clear at the time that direct recognition of MHC molecules was involved.

Later experiments showed that the antigen specificity of T-cell recognition was controlled by MHC molecules. The immune responses affected by the Ir genes were known to depend on T cells, and this led to a series of experiments in mice to ascertain how MHC polymorphism might control T-cell responses. The earliest of these experiments showed that T cells could be activated only by macrophages or B cells that shared MHC alleles with the mouse in which the T cells originated. This was the first evidence that antigen recognition by T cells depends on the presence of specific MHC molecules in the antigen-presenting cell—the phenomenon we now know as **MHC restriction**.

We first mentioned MHC restriction in Section 4-17 in the context of the crystal structure of the T-cell receptor bound to peptide:MHC complexes. But the phenomenon of MHC restriction was discovered much earlier and is illustrated by the studies of virus-specific cytotoxic T cells carried out by **Peter Doherty** and **Rolf Zinkernagel**, for which they received the Nobel Prize in 1996. When mice are infected with a virus, they generate cytotoxic T cells that kill the virus-infected cells while sparing both uninfected cells and cells infected with unrelated viruses. The cytotoxic T cells are thus virus-specific. The additional and striking outcome of these experiments was the demonstration that the ability of cytotoxic T cells to kill virus-infected cells was also affected by the polymorphism of MHC molecules. Cytotoxic T cells induced by viral infection in mice of MHC genotype a (MHC^a) would kill any MHC^a cell infected with that virus. But these same T cells would not kill cells of MHC genotype b, or c, and so on, even if they were infected with the same virus. In other words, cytotoxic T cells kill cells infected by virus only if those cells express the same MHC by which the T cells were primed. Because the MHC genotype 'restricts' the antigen specificity of the T cells, this effect was called MHC restriction. Together with the earlier studies on both B cells and macrophages, this work showed that MHC restriction is a critical feature of antigen recognition by all functional classes of T cells.

We now know that MHC restriction is due to the fact that the binding specificity of an individual T-cell receptor is not for its peptide antigen alone but for the complex of peptide and MHC molecule, as discussed in Section 4-17. MHC restriction is explained in part by the fact that different MHC molecules bind different peptides. In addition, some of the polymorphic amino acids in MHC molecules are located in the α helices that flank the peptide-binding groove but have side chains oriented toward the exposed surface of the peptide:MHC complex that can directly contact the T-cell receptor (see Figs. 6.21 and 4.24). In retrospect, it is therefore not surprising that T cells can distinguish between a peptide bound to MHC^a and the same peptide bound to MHC^b. This restricted recognition may sometimes be caused both by differences in the conformation of the bound peptide imposed by the different MHC molecules and by direct recognition of polymorphic amino acids in the MHC molecule itself. Thus, the specificity of a T-cell receptor is defined both by the peptide it recognizes and by the MHC molecule bound to it (**Fig. 6.23**).

Fig. 6.23 T-cell recognition of antigens is MHC-restricted. The antigen-specific T-cell receptor (TCR) recognizes a complex consisting of an antigenic peptide and a self MHC molecule. One consequence of this is that a T cell specific for peptide x and an MHC molecule that is the product of a particular MHC allele, MHC^a (left panel), will usually not recognize the complex of peptide x bound to a different MHC allele product, MHC^b (center panel), or the complex of a different peptide, peptide y, bound to MHC^a (right panel). The co-recognition of a foreign peptide and an MHC molecule is known as MHC restriction because the particular MHC allele product is said to restrict the ability of the T cell to recognize antigen. This restriction may either result from direct contact between the MHC molecule and T-cell receptor or be an indirect effect of MHC polymorphism on the peptides that bind or on their bound conformation.

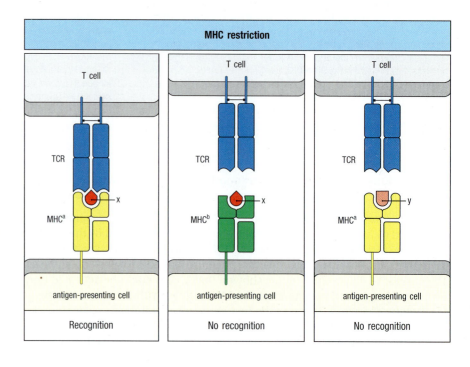

MHC restriction

6-13   Alloreactive T cells recognizing nonself MHC molecules are very abundant.

The discovery of MHC restriction also helped to explain the otherwise puzzling phenomenon of recognition of nonself MHC in the rejection of organs and tissues transplanted between members of the same species. Transplanted organs from donors bearing MHC molecules that differ from those of the recipient—even by as little as one amino acid—are rapidly rejected owing to the presence in any individual of large numbers of T cells that react to nonself, or **allogeneic**, MHC molecules. Early studies on T-cell responses to allogeneic MHC molecules used the **mixed lymphocyte reaction**, in which T cells from one individual are mixed with lymphocytes from a second individual. If the T cells of this individual recognize the other individual's MHC molecules as 'foreign,' the T cells will divide and proliferate. The lymphocytes from the second individual are usually prevented from dividing by irradiation or treatment with the cytostatic drug mitomycin C. Such studies have shown that roughly 1–10% of all T cells in an individual will respond to stimulation by cells from another, unrelated, member of the same species. This type of T-cell response is called an **alloreaction** or **alloreactivity**, because it represents the recognition of allelic polymorphisms in MHC molecules.

Before the role of the MHC molecules in antigen presentation was understood, it was a mystery why so many T cells should recognize nonself MHC molecules, as there is no reason that the immune system should have evolved a defense against tissue transplants. Once it was realized that T-cell receptors have evolved to recognize foreign peptides in combination with polymorphic MHC molecules, alloreactivity became easier to explain. We now know of at least two processes that can contribute to the high frequency of alloreactive T cells. The first is the process of **positive selection**. When developing in the thymus, T cells whose T-cell receptors interact weakly with the self MHC molecules receive survival signals, and so are favored for representation in the peripheral repertoire. It is thought that T-cell receptors that interact with one type of MHC molecule are more likely to cross-react with other (nonself) MHC variants. We discuss positive selection in greater detail in Chapter 8.

But positive selection is not the only basis for alloreactivity. This conclusion was implied by observations that T cells artificially driven to mature in animals lacking MHC class I and class II, in which positive selection in the thymus cannot occur, still display frequent alloreactivity. It appears that T-cell receptor genes encode the inherent ability to recognize MHC molecules. X-ray crystallographic studies of T-cell receptors bound to MHC molecules provide a structural basis for an inherent binding interaction (Fig. 6.24). Specific amino acid residues within the germline-encoded region of certain TCRβ genes interact with conserved features of the MHC molecule, implying a type of germline-encoded affinity. Given the large number of variable-region sequences in T-cell receptors, each T-cell receptor may bind MHC molecules in its own idiosyncratic way using both germline-encoded regions and variable regions.

In principle, alloreactive T cells might depend on recognizing either a foreign peptide antigen or the nonself MHC molecule to which it is bound for their reactivity against nonself MHC; these options have been called peptide-dependent and peptide-independent allorecognition. But as the number of individual alloreactive T-cell clones studied has increased, it seems that most alloreactive T cells actually recognize both; that is, most individual alloreactive T-cell clones respond to a foreign MHC molecule only when a particular peptide is bound to it. In this sense, the structural basis of allorecognition may be quite similar to normal MHC-restricted peptide recognition and be dependent on contacts with both peptide and MHC molecule (see Fig. 6.23, left panel), but in this case a foreign MHC molecule. In practice, alloreactive responses against a transplanted organ are likely to represent the combined

Fig. 6.24 Germline-encoded residues in CDR1 and CDR2 of V-region genes confer on T-cell receptors an inherent affinity for MHC molecules. Shown is the structure for several T-cell receptors bound to a class II MHC molecule. Conserved residues (Lys39, Gln57, and Gln61) within the α1 helix of the MHC (green) make an extended hydrogen-bonded network with germline-encoded and nonpolymorphic residues located in the CDR1 (Asn31) and CDR2 (Glu56, Tyr50) regions of the Vβ 8.2 gene, respectively. The configuration of these contacts is very similar between different structures, implying that the germline sequence of the CDR1 and CDR2 confers an inherent bias for T-cell receptor affinity for MHC. Courtesy of K.C. Garcia.

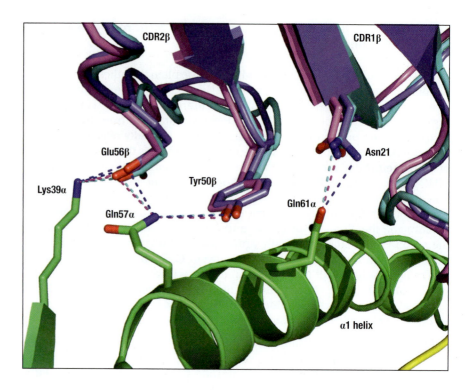

activity of many alloreactive T cells, and it is not possible to determine what peptides from the donor might be involved in recognition by the alloreactive T cells. We will return to alloreactivity when we discuss organ transplantation in more detail in Chapter 15.

6-14 Many T cells respond to superantigens.

Superantigens are a distinct class of antigens that stimulate a primary T-cell response similar in magnitude to a response to allogeneic MHC molecules. Such responses were first observed in mixed lymphocyte reactions using lymphocytes from strains of mice that were MHC-identical but otherwise genetically distinct. The antigens provoking this reaction were originally designated as **minor lymphocyte stimulating (Mls) antigens**, and it seemed reasonable to suppose that they might be functionally similar to the MHC molecules themselves. We now know that this is not true. The Mls antigens in these mouse strains are encoded by retroviruses, such as the mouse mammary tumor virus, that have become stably integrated at various sites in the mouse chromosomes. Superantigens are produced by many different pathogens, including bacteria, mycoplasmas, and viruses, and the responses they provoke are helpful to the pathogen rather than the host.

Mls proteins act as superantigens because they have a distinctive mode of binding to both MHC and T-cell receptor molecules that enables them to stimulate very large numbers of T cells. Superantigens are unlike other protein antigens, in that they are recognized by T cells without being processed into peptides that are captured by MHC molecules. Indeed, fragmentation of a superantigen destroys its biological activity, which depends on binding as an intact protein to the outside surface of an MHC class II molecule that has already bound peptide. In addition to binding MHC class II molecules, superantigens are able to bind the V_β region of many T-cell receptors (**Fig. 6.25**). Bacterial superantigens bind mainly to the V_β CDR2 loop, and, to a smaller extent, to the V_β CDR1 loop and an additional loop called the hypervariable 4 or HV4 loop. The HV4 loop is the predominant binding site for viral superantigens, at least for the Mls antigens encoded by the endogenous mouse mammary tumor viruses.

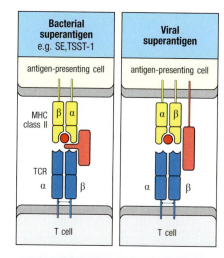

Fig. 6.25 Superantigens bind directly to T-cell receptors and to MHC molecules. Superantigens can bind independently to MHC class II molecules and to T-cell receptors. As shown in the top panels, the superantigens (red bars) can bind to the V_β domain of the T-cell receptor (TCR), away from the complementarity-determining regions, and to the outer faces of the MHC class II molecule, outside the peptide-binding site. In the bottom panel, a reconstruction of the interaction between a T-cell receptor, an MHC class II molecule, and a staphylococcal enterotoxin (SE) superantigen is shown by superimposing separate structures of an enterotoxin:MHC class II complex onto an enterotoxin:T-cell receptor complex. The two enterotoxin molecules (actually SEC3 and SEB) are shown in turquoise and blue, binding to the α chain of the MHC class II molecule (yellow) and to the β chain of the T-cell receptor (colored gray for the V_β domain and pink for the C_β domain). Molecular model courtesy of H.M. Li, B.A. Fields, and R.A. Mariuzza.

Thus, the α-chain V region and the CDR3 of the β chain of the T-cell receptor have little effect on superantigen recognition, which is determined largely by the germline-encoded V gene segments that encode the expressed V_β chain. Each superantigen is specific for one or a few of the different V_β gene products, of which there are 20–50 in mice and humans; a superantigen can thus stimulate 2–20% of all T cells.

This mode of stimulation does not prime an adaptive immune response specific for the pathogen. Instead, it causes a massive production of cytokines by CD4 T cells, the predominant responding population of T cells. These cytokines have two effects on the host: systemic toxicity and suppression of the adaptive immune response. Both these effects contribute to microbial pathogenicity. Among the bacterial superantigens are the **staphylococcal enterotoxins (SEs)**, which cause food poisoning, and the **toxic shock syndrome toxin-1 (TSST-1)** of *Staphylococcus aureus*, the etiologic principle in **toxic shock syndrome**, which can be caused by a localized infection with toxin-producing strains of the bacterium. The role of viral superantigens in human disease is less clear.

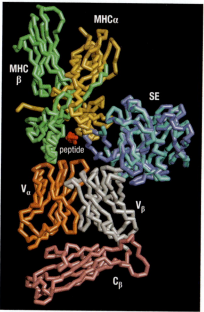

6-15 MHC polymorphism extends the range of antigens to which the immune system can respond.

Most polymorphic genes encode proteins that vary by only one or a few amino acids, whereas the allelic variants of MHC proteins differ from each other by up to 20 amino acids. The extensive polymorphism of the MHC proteins has almost certainly evolved to outflank the evasive strategies of pathogens. The requirement that pathogen antigens must be presented by an MHC molecule provides two possible ways in which pathogens could evolve to evade detection. One is through mutations that eliminate from the pathogen's proteins all peptides able to bind MHC molecules. The Epstein–Barr virus provides an example. There are small isolated populations in southeast China and Papua New Guinea in which about 60% of the people carry the HLA-A11 allele. Many isolates of the Epstein–Barr virus obtained from these populations have mutations in a dominant peptide epitope normally presented by HLA-A11; the mutant peptides no longer bind to HLA-A11 and cannot be recognized by HLA-A11-restricted T cells. This strategy is clearly much less successful if there are many different MHC molecules, and the polygeny of the MHC may have evolved in response.

Toxic Shock Syndrome

In addition, in large outbred populations, polymorphism at each locus can potentially double the number of different MHC molecules expressed by an individual, as most individuals will be heterozygotes. Polymorphism has the additional advantage that individuals in a population will differ in the combinations of MHC molecules that they express and will therefore present different sets of peptides from each pathogen. This makes it unlikely that all individuals in a population will be equally susceptible to a given pathogen, and its spread will therefore be limited. The fact that exposure to pathogens over an evolutionary timescale can select for particular MHC alleles is indicated by the strong association of the HLA-B53 allele with recovery from a potentially lethal form of malaria. This allele is very common in people from West Africa, where malaria is endemic, and rare elsewhere, where lethal malaria is uncommon.

Similar arguments apply to a second strategy by which pathogens could evade recognition. Pathogens that can block the presentation of their peptides by MHC molecules can avoid the adaptive immune response. Adenoviruses encode a protein that binds to MHC class I molecules in the endoplasmic reticulum and prevents their transport to the cell surface, thus preventing the recognition of viral peptides by CD8 cytotoxic T cells. This viral MHC-binding protein interacts with a polymorphic region of the MHC class I molecule, as some allelic variants are retained in the endoplasmic reticulum by the adeno-viral protein, whereas others are not. Increasing the variety of MHC molecules expressed reduces the likelihood that a pathogen will be able to block presentation by all of them and completely evade an immune response.

These arguments raise a question: if having three MHC class I loci is better than having one, why are there not far more? The probable explanation is that each time a distinct MHC molecule is added to the repertoire, all T cells that can respond to self peptides bound to that MHC molecule must be removed to maintain self tolerance. It seems that the number of MHC genes present in humans and mice is about optimal to balance the advantages of presenting an increased range of foreign peptides with the disadvantages of losing T cells from the repertoire.

Summary.

The major histocompatibility complex (MHC) of genes consists of a linked set of genetic loci encoding many of the proteins involved in antigen presentation to T cells, most notably the MHC class I and class II glycoproteins (the MHC molecules) that present peptides to the T-cell receptor. The outstanding feature of the MHC molecules is their extensive polymorphism. This polymorphism is of critical importance in antigen recognition by T cells. A T cell recognizes antigen as a peptide bound by a particular allelic variant of an MHC mole-cule, and will not recognize the same peptide bound to other MHC molecules. This behavior of T cells is called MHC restriction. Most MHC alleles differ from one another by multiple amino acid substitutions, and these differences are focused on the peptide-binding site and the surface-exposed regions that make direct contact with the T-cell receptor. At least three properties of MHC molecules are affected by MHC polymorphism: the range of peptides bound; the conformation of the bound peptide; and the direct interaction of the MHC molecule with the T-cell receptor. Thus, the highly polymorphic nature of the MHC has functional consequences, and the evolutionary selection for this polymorphism suggests that it is critical to the role of the MHC molecules in the immune response. Powerful genetic mechanisms generate the variation that is seen among MHC alleles, and a compelling argument can be made that selective pressure to maintain a wide variety of MHC molecules in the popu-lation comes from infectious agents. As a consequence, the immune system is highly individualized—each individual responds differently to a given antigen.

Generation of ligands for unconventional T-cell subsets.

So far we have focused on how peptide:MHC complexes—the ligands for α:β T cells—are generated. We now turn to the question of how other types of T cells recognize their ligands and how these ligands are generated. Our current knowledge in this area is still incomplete, and is perhaps most apparent in the area of γ:δ T cells, where a growing list of ligands for individual γ:δ T cells suggests an innate-like pattern of recognition. The recent discovery that the mucosal associated invariant T (MAIT) cells (see Section 4-18) recognize a microbial metabolite when it is presented by a nonpolymorphic MHC class I-like molecule solved a long-standing mystery regarding the function of this particular T-cell subset. Another invariant subset, the invariant NKT cells, provides a system for detecting and responding to lipid rather than peptide antigens. These findings suggest that these invariant and unconventional T cells operate somewhere between innate and adaptive immunity. In this part of the chapter, we will discuss the ligands they recognize and what is known about how they are generated or expressed.

6-16   A variety of genes with specialized functions in immunity are also encoded in the MHC.

In addition to the highly polymorphic 'classical' MHC class I and class II genes, there are many 'nonclassical' MHC genes, many encoded in the MHC but others encoded outside this region. The MHC class I-type molecules show comparatively little polymorphism; many of these have yet to be assigned a function. They are linked to the class I region of the MHC, and their exact number varies greatly among species and even among members of the same species. These genes have been termed **MHC class Ib** genes; like MHC class I genes, many, but not all, associate with β_2-microglobulin when expressed on the cell surface. Their expression on cells is variable, both in the amount present at the cell surface and in tissue distribution. The characteristics of several MHC class Ib gene products are shown in **Fig. 6.26**.

One mouse MHC class Ib molecule, **H2-M3**, can present peptides with N-formylated amino termini, which is of interest because all bacteria initiate protein synthesis with N-formylmethionine. Cells infected with cytosolic bacteria can be killed by CD8 T cells that recognize N-formylated bacterial peptides bound to H2-M3. Whether an equivalent MHC class Ib molecule exists in humans is not known.

Two other closely related mouse MHC class Ib genes, *T22* and *T10*, are expressed by activated lymphocytes and are recognized by a subset of γ:δ T cells. Although the precise purpose remains unclear, it has been proposed that this interaction allows γ:δ T cells to regulate these activated lymphocytes expressing T22 and T10 proteins.

The other genes that map within the MHC include some that encode complement components (for example, C2, C4, and factor B) and some that encode cytokines—for example, tumor necrosis factor-α (TNF-α) and lymphotoxin—all of which have important functions in immunity. These genes lie in the so-called 'MHC class III' region (see Fig. 6.17), a somewhat misleading name, since genes in this region are not MHC molecules at all.

Many studies have established associations between susceptibility to certain diseases and particular alleles of MHC genes (see Chapter 15), and we now have considerable insight into how polymorphism in the classical MHC class I and class II genes can affect disease resistance or susceptibility. Most MHC-influenced traits or diseases are known or suspected to have an immunological cause, but this is not so for all of them: some genes residing within the MHC have no known or suspected immunological function. For example, the class Ib gene *M10* encodes a protein that functions in the vomeronasal organ as a chaperone to escort certain types of pheromone receptors to the cell surface. *M10* could potentially influence mating preference, a trait that has been linked to the MHC region in rodents.

The **HFE** gene, encoding the **hemochromatosis protein**, lies some 4 million base pairs from HLA-A. Its protein product is expressed in cells of the intestinal

	Class 1b molecule					Receptors or interacting proteins			
Human	Mouse	Expression pattern	Associates with β_2m	Poly-morphism	Ligand	T-cell receptor	NK receptor	Other	Biological function
HLA-C (class 1a)		Ubiquitous	Yes	High	Peptide	TCR	KIRs		Activate T cells Inhibit NK cells
	H2-M3	Limited	Yes	Low	fMet peptide	TCR			Activate CTLs with bacterial peptides
	T22 T10	Splenocytes	Yes	Low	None	γ:δ TCR			Regulation of activated splenocytes
HLA-E	Qa-1	Ubiquitous	Yes	Low	MHC leader peptides (Qdm)		NKG2A NKG2C		NK cell inhibition
HLA-F		Widely expressed	Yes	Low	Peptide?		LILRB1 LILRB2		Unknown
HLA-G		Maternal/fetal interface	Yes	Low	Peptide	TCR	LILRB1		Modulate maternal/fetal interaction
MIC-A MIC-B		Widely expressed	No	Moderate	None		NKG2D		Stress-induced activation of NK cells, γ:δ and CD8 T cells
	TL	Small intestine epithelium	Yes	Low	None	CD8α:α			Potential modulation of T-cell activation
	M10	Vomeronasal neurons	Yes	Low	Unknown			Vomeronasal receptor V2R	Pheromone detection
ULBPs	MULT1 H60, Rae1	Limited	No	Low	None		NKG2D		Induced NK-cell-activating ligand
MR1	MR1	Ubiquitous	Yes	None	Vitamin B9 metabolite	α:β TCR			Control of inflammatory response
CD1a–CD1e	CD1d	Limited	Yes	None	Lipids glycolipids	α:β TCR			Activate T cells against bacterial lipids
	Mill1 Mill2	Ubiquitous	Yes?	Low	Unknown	Unknown			Unknown
HFE	HFE	Liver and gut	Yes	Low	None			Transferrin receptor	Iron homeostasis
FcRn	FcRn	Maternal/fetal interface	Yes	Low	None			Fc (IgG)	Shuttle maternal IgG to fetus (passive immunity)
ZAG	ZAG	Bodily fluid	No	None	Fatty acid				Lipid homeostasis
EPCR	EPCR	Endothelial cells	No	Low		γ:δ TCR		Protein C	Blood coagulation

Left-margin group labels: **MHC encoded** (rows HLA-C through M10); **Non-MHC encoded** (rows ULBPs through EPCR).

Fig. 6.26 Mouse and human MHC class Ib proteins and their functions. MHC class Ib proteins are encoded both within the MHC region and on other chromosomes. The functions of some MHC class Ib proteins are unrelated to the adaptive immune response, but many have a role in innate immunity by interacting with receptors on NK cells (see the text and Section 3-24). HLA-C, which is a classical MHC molecule (class Ia), is included here because, in addition to presenting peptides to T-cell receptors, all HLA-C isoforms interact with the KIR class of NK-cell receptors to regulate NK-cell function in the innate immune response. CTL, cytotoxic T lymphocyte.

tract and acts in iron metabolism by regulating the uptake of dietary iron into the body. It seems to interact with the transferrin receptor and decrease the receptor's affinity for iron-loaded transferrin. Individuals defective for this gene have an iron-storage disease, **hereditary hemochromatosis**, in which an abnormally high level of iron is retained in the liver and other organs. Mice lacking β_2-microglobulin have defective expression of all class I molecules and thus also show a similar iron overload. Another MHC gene with a nonimmune function encodes the enzyme **21-hydroxylase**, which, when deficient, causes congenital adrenal hyperplasia and, in severe cases, salt-wasting syndrome. Even where a disease-related gene is clearly homologous to immune-system genes, as is the case with *HFE*, the disease mechanism may not be immune-related. Disease associations mapping to the MHC must therefore be interpreted with caution, in the light of a detailed understanding of its genetic structure and the functions of its individual genes. Much remains to be learned about the significance of all the genetic variation localized within the MHC. For instance, the human complement component C4 comes in two versions, C4A and C4B, not to be confused with the C4 convertase cleavage products C4a and C4b, and different individuals have variable numbers of the gene for each type in their genomes, but the adaptive significance of this genetic variability is not well understood.

6-17   Specialized MHC class I molecules act as ligands for the activation and inhibition of NK cells and unconventional T-cell subsets.

In Sections 3-24 to 3-27, we introduced NK cells and briefly discussed their activation by members of the **MIC** gene family. These are MHC class Ib genes that are under a different regulatory control than the classical MHC class I genes and are induced in response to cellular stress (such as heat shock). There are seven MIC genes, but only two—*MICA* and *MICB*—are expressed and produce protein products (see Fig. 6.26). They are expressed in fibroblasts and epithelial cells, particularly in intestinal epithelial cells, and have a role in innate immunity or in the induction of immune responses in circumstances in which interferons are not produced. The MICA and MICB proteins are recognized by the NKG2D receptor expressed by NK cells. But in addition, NKG2D is also expressed by $\gamma{:}\delta$ T cells and some CD8 T cells, and it can activate these cells to kill MIC-expressing targets. NKG2D is an 'activating' member of the NKG2 family of NK-cell receptors (see Fig. 3.42); its cytoplasmic domain lacks the inhibitory sequence motif found in other members of this family, which act as inhibitory receptors (see Section 3-26). NKG2D is coupled to the adaptor protein DAP10, which transmits the signal into the interior of the cell by interacting with and activating intracellular phosphatidylinositol 3-kinase.

Even more distantly related to MHC class I genes is a small family of proteins known in humans as the **UL16-binding proteins** (ULBPs) or the **RAET1** proteins (see Fig. 6.26); the homologous proteins in mice are known as Rae1 (retinoic acid early inducible 1) and H60. These proteins also bind NKG2D (see Section 3-27). They seem to be expressed under conditions of cellular stress, such as when cells are infected with pathogens (UL16 is a human cytomegalovirus protein) or have undergone transformation to tumor cells. By expressing ULBPs, stressed or infected cells can bind and activate NKG2D molecules expressed on NK cells, $\gamma{:}\delta$ T cells, and CD8 cytotoxic $\alpha{:}\beta$ T cells, and so be recognized and eliminated.

The human MHC class Ib molecule HLA-E and its mouse counterpart Qa-1 (see Fig. 6.26) have an unusual and somewhat puzzling role in cell recognition by NK cells and CD8 T cells. HLA-E and Qa-1 bind a very restricted subset of nonpolymorphic peptides, called **Qa-1 determinant modifiers (Qdm)**, that are derived from the leader peptides of other HLA class I molecules. These peptide:HLA-E complexes can bind to the inhibitory receptor NKG2A:CD94 expressed on NK cells, and so should inhibit the cytotoxic activity of NK cells. This function might seem redundant, since expression of other MHC class I molecules by cells should prevent NK-cell activation (see Section 3-25). Nonetheless, it has been shown that Qa-1 expression by activated CD4 T cells protects them from lysis by NK cells and so Qa-1 expression by other host cells may provide them with additional protection from being killed by NK cells. HLA-E and Qa-1 can also bind leader peptides from the heat shock protein Hsp60sp, and CD8 T cells that are specific for these complexes have been identified in mice and humans. Some recent evidence suggests that CD8 T cells restricted by HLA-E/Qa-1 may help maintain self-tolerance by killing or suppressing potentially autoreactive T cells.

In Section 3-26, we introduced the **killer cell immunoglobulin-like receptors (KIRs)** expressed by NK cells. Members of the KIRs recognize the classical class Ia MHC molecules HLA-A, -B, and -C, which present a diverse repertoire of peptides to CD8 T cells. Although KIRs interact with the same face of the MHC class I molecule as do T-cell receptors, the KIRs bind only at one end, and not over the whole area recognized by the T-cell receptor. Like MHC molecules, KIRs themselves are highly polymorphic, and they have undergone rapid evolution in humans. Only a few *HLA-A* and *HLA-B* alleles code for proteins that bind KIRs, but all *HLA-C* alleles express proteins that bind KIRs, indicating a specialization of HLA-C for regulating NK cells in humans.

Two other MHC class Ib molecules, HLA-F and HLA-G (see Fig. 6.26), can also inhibit cell killing by NK cells. HLA-G is expressed on fetus-derived placental cells that migrate into the uterine wall. These cells express no classical MHC class I molecules and cannot be recognized by CD8 T cells but, unlike other cells lacking such proteins, they are not killed by NK cells. This seems to be because HLA-G is recognized by another inhibitory receptor on NK cells, the leukocyte immunoglobulin-like receptor subfamily B member 1 (LILRB1), also called ILT-2 or LIR-1, which prevents the NK cell from killing the placental cells. HLA-F is expressed in a variety of tissues, although it is usually not detected at the cell surface except, for example, on some monocyte cell lines or on virus-transformed lymphoid cells. HLA-F is also thought to interact with LILRB1.

6-18 Members of the CD1 family of MHC class I-like molecules present microbial lipids to invariant NKT cells.

Some MHC class I-like genes map outside the MHC region. One small family of such genes is called **CD1** and is expressed on dendritic cells, monocytes, and some thymocytes. Humans have five CD1 genes, CD1a through e, whereas mice express only two highly homologous versions of CD1d, namely, CD1d1 and CD1d2. CD1 proteins can present antigens to T cells, but they have two features that distinguish them from classical MHC class I molecules. The first is that CD1, although similar to an MHC class I molecule in its subunit organization and association with β_2-microglobulin, behaves like an MHC class II molecule. It is not retained within the endoplasmic reticulum by association with the TAP complex but is targeted to vesicles, where it binds its ligand. The second unusual feature is that, unlike MHC class I, CD1 molecules have a hydrophobic channel that is specialized for binding hydrocarbon alkyl chains. This confers on CD1 molecules an ability to bind and present a variety of glycolipids.

CD1 molecules are classified into group 1, comprising CD1a, CD1b, and CD1c, and group 2, containing only CD1d; CD1e is considered intermediate. Group 1 molecules bind various **microbial glycolipids**, phospholipids, and **lipopeptide antigens**, such as the mycobacterial membrane components mycolic acid, glucose monomycolate, phosphoinositol mannosides, and lipoarabinomannan (Fig. 6.27). Group 2 CD1 molecules are thought to bind mainly self

Fig. 6.27 CD1c binds microbial lipids for presentation to iNKT cells. Top panel: the structure of mannosyl-β1-phosphomycoketides (MPMs) from the cell walls of *Mycobacterium tuberculosis* (Mtb) (R = C_7H_{15}) and *M. avium* (R = C_5H_{11}). Middle panel: MPM (stick figure) bound to CD1c (purple) as viewed from the top, facing the surface of the cell bearing CD1c. Bottom panel: side view of MPM bound to CD1c. The general resemblance with peptide:MHC complexes is apparent. Note: the long acyl chain of MPM extends deep into the binding groove of CD1c, to beneath the α1 helical domain. Courtesy of E. Adams.

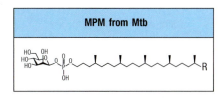

MPM from Mtb

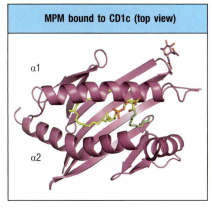

MPM bound to CD1c (top view)

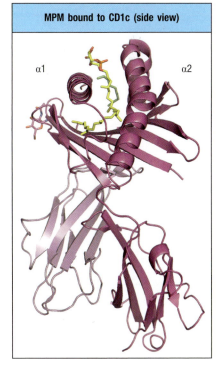

MPM bound to CD1c (side view)

lipid antigens such as **sphingolipids** and **diacylglycerols**. Structural studies show that the CD1 molecule has a deep binding groove into which the gly-colipid antigens bind (Fig. 6.28). Unlike the binding of peptide to MHC, in which the peptide takes on a linear, extended conformation, CD1 molecules bind their antigens by anchoring the alkyl chains in the hydrophobic groove, which orients the variable carbohydrate headgroups (or other hydrophilic parts of these molecules) so that they protrude from the end of the binding groove, allowing recognition by the T-cell receptors on CD1-restricted T cells.

The T cells that recognize lipids presented by CD1 molecules are largely negative for CD4 and CD8 expression, although some express CD4. Most of the T cells recognizing lipids presented by group 1 CD1 molecules have a diverse repertoire of α:β receptors, and respond to these lipids presented by CD1a, CD1b, and CD1c. In contrast, CD1d-restricted T cells are less diverse, many using the same TCRα chain ($V_\alpha24$-$J_\alpha18$ in humans), but they also express NK-cell receptors. These CD1-restricted T cells are called **invariant NKT (iNKT) cells**.

One recognized ligand for CD1d molecules is **α-galactoceramide (α-GalCer)**, which was isolated from an extract of marine sponge. Related glycosphingo-lipids are produced by various bacteria, including *Bacteroides fragilis*, which is present in the normal human microbiota. When α-galactoceramide is bound to CD1d, it forms a structure that is recognized by many iNKT cells. The ability of iNKT cells to recognize different glycolipid constituents from microorganisms presented by CD1d molecules places them in an 'innate' category, while their possession of a fully rearranged T-cell receptor, despite its relatively limited repertoire, makes them 'adaptive.'

CD1 proteins have evolved as a separate lineage of antigen-presenting molecules able to present microbial lipids and glycolipids to T cells. Just as peptides are loaded onto classical MHC proteins at various cellular locations, the various CD1 proteins are transported differently through the endoplasmic reticulum and endocytic compartments; this provides access to different lipid antigens. Transport is regulated by an amino acid sequence motif at the terminus of the cytoplasmic domain of the CD1 protein through interaction with adaptor-protein (AP) complexes. CD1a lacks this binding motif and moves to the cell surface, where it is transported only through the early endocytic compartment. CD1c and CD1d have motifs that interact with the adaptor AP-2 and are transported through early and late endosomes; CD1d is also targeted to lysosomes. CD1b and mouse CD1d bind AP-2 and AP-3 and can be transported through late endosomes, lysosomes, and the MIIC. CD1 proteins can thus bind lipids delivered into and processed within the endocytic pathway, such as by the internalization of mycobacteria or the ingestion of mycobacterial lipoarabinomannans mediated by mannose receptors.

From an evolutionary perspective it is interesting that some class Ib genes seem to have evolved early, before the divergence of the cartilaginous fishes from the vertebrate line, and are likely to have homologs in all vertebrates. Other class I genes have independently evolved into classical and nonclassical loci within the vertebrate lineages that have been studied (for example, cartilaginous fishes, lobe-finned fishes, ray-finned fishes, amphibians, and mammals). Sequence data have also revealed homologs of the mammalian MHC-I and MHC-II gene families in virtually all jawed vertebrates including sharks, bony fishes, reptiles, and birds. In contrast, CD1 genes may not be as old as

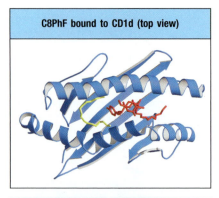

C8PhF bound to CD1d (top view)

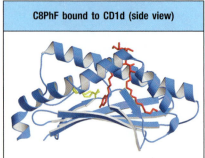

C8PhF bound to CD1d (side view)

Fig. 6.28 Structure of CD1 binding to a lipid antigen. Shown are top and side views of the structure of mouse CD1d bound to C8PhF, a synthetic lipid that is an analog of α-GalCer. The helical side chains of CD1d (blue) form a binding pocket that is generally similar in shape to the binding pockets of MHC class I and II molecules. However, the C8PhF (red) ligand binds to CD1 molecules in a distinctly different conformation from that of peptides. The two long alkyl side chains extend deep inside the binding groove (see side view), where they make contacts with hydrophobic residues. This orientation of the alkyl side chains places the carbohydrate component of α-GalCer to the outer surface of CD1, where it can be recognized by the T-cell receptor. In addition, the CD1 molecule contains an endogenous lipid molecule (yellow) derived from cellular sources that binds to a distinct region within the groove and prevents a large pocket adjacent to the α-GalCer-binding region from collapsing. The ability to incorporate additional ligands into the binding groove may provide flexibility to CD1d in accommodating a variety of exogenous glycosphingolipids from microorganisms. Courtesy of I.A. Wilson.

other MHC class Ib genes. They have been found only in a subset of these animal groups and appear to be missing in fish. This pattern of CD1 occurrence in the genomes of living species suggests the emergence of CD1 in an early terrestrial vertebrate.

6-19 The nonclassical MHC class I molecule MR1 presents microbial folate metabolites to MAIT cells.

Another nonclassical MHC class Ib molecule is **MR1 (MHC-related protein 1)**. MR1 associates with β_2-microglobulin and is encoded outside the MHC, but its function was originally known only in relation to a conserved population of α:β T cells known as **mucosal associated invariant T cells (MAIT cells)**. In Section 4-18, we introduced MAIT cells as one population of T cells expressing the CD8α homodimer, but they are uniquely characterized by expressing an invariant α chain of the T-cell receptor, specifically human $V_\alpha 7.2J2$–$J_\alpha 33$ (or in mouse, $V_\alpha 19$). This α chain pairs with a limited number of V_β chains, typically $V_\beta 2$ or $V_\beta 13$. MAIT cells are very abundant in humans and can comprise up to 10% of the lymphocytes in the peripheral blood and tissues such as the liver. They are also present in mesenteric lymph nodes and the mucosa of the intestine. Studies of MAIT cells revealed that their development requires the expression of MR1, and further, that a wide spectrum of microbes, including diverse bacteria and yeast, can activate MAIT cells. However, when they were originally identified around a decade ago, it was unclear what, if any, ligand is being recognized by these cells.

Structural studies of MR1 uncovered an important clue. The MR1 protein was unstable when produced *in vitro* by cell lines grown in typical tissue culture conditions. It was discovered that the protein was stabilized when it was refolded in media containing B vitamins or **folic acid** (vitamin B_9). Chemical analysis revealed that a small molecule—identified as a folate derivative, 6-formyl pterin (6-FP)—was bound to the stabilized MR1. X-ray crystallographic studies showed that 6-FP was bound in the central groove of the MR1 molecule; this helped explain how folate derivatives might stabilize MR1. However, MAIT cells were not activated by cells expressing the 6-FP:MR1 complex, suggesting other molecules might be the physiological ligands to activate MAIT cells. Analysis of MR1 proteins that were refolded in the presence of supernatants from cultures of *Salmonella typhimurium* eventually led to the identification of several riboflavin metabolites that are formed by biosynthetic pathways in most bacteria and yeast. These metabolites not only bind to MR1, but also activate MAIT cells. Thus, MAIT cells are activated in response to infection by these organisms by detecting products specific to their folate metabolism. As such, MAIT cells appear to hold an intermediate place in the spectrum of innate and adaptive immunity, similar to iNKT cells, in that they use an antigen receptor assembled by somatic gene rearrangement, but recognize a molecular structure that falls within the definition of a PAMP.

6-20   γ:δ T cells can recognize a variety of diverse ligands.

γ:δ T cells and α:β T cells have been known to be distinct developmental lineages almost since the T-cell receptor genes were identified. But unlike α:β T cells, the function of γ:δ T cells has remained somewhat obscure, due primarily to difficulty in identifying the ligands they recognize. Yet the abundance of γ:δ T cells across vertebrate species, their rapid expansion to form more than 50% of the blood lymphocytes during infections, and their abundant cytokine production all argue for an important role in immunity. Over time, many different ligands recognized by γ:δ T-cell clones have been identified (Fig. 6.29), and their diversity suggests that, like iNKT and MAIT cells, they hold an intermediate, or transitional, position in the spectrum of innate versus adaptive immunity.

In Section 4-20, we discussed how one γ:δ T-cell receptor binds to the non-classical MHC class I molecule **T22**. Instead of binding centrally over the MHC binding groove, similar to an α:β T-cell receptor, the γ:δ T-cell receptor interacts obliquely from one side of the T22 molecule. However, fewer than 1% of γ:δ T cells recognize this ligand. Other antigens recognized by murine γ:δ T cells are the protein **phycoerythrin (PE)** from algae, the inner mitochondrial membrane lipid **cardiolipin**, glycoprotein I of herpes simplex virus, and a peptide derived from the hormone insulin. Among antigens that can activate human γ:δ T cells are the nonclassical MHC class I proteins **MICA** and **ULBP4** and the **endothelial protein C receptor (EPCR)**, which is expressed by endothelial cells. Like MICA and ULBPs, EPCR appears to be induced upon stress, such as during infection of cells by cytomegalovirus, suggesting that reactive γ:δ T cells could serve in an innate capacity similar to that of NK cells activated by stress-induced nonclassical MHC class Ib molecules. Several other antigens can activate human γ:δ T cells (see Fig. 6.29), although there is still limited structural information about their interaction with the T-cell receptor, and even reservations about whether such an interaction is always

Ligands that activate γ:δ T cell		
Ligands	**Species**	**γ:δ subset**
T22, T10	Mouse	Various
I-E (MHC class II)	Mouse	Clones
Phycoerythrin (PE)	Mouse	Various
Cardiolipin	Mouse	Various
Keratinocytes	Mouse	DETC Vγ5Vδ1
HSV-gl	Mouse	Clone
Skint-1	Mouse	Vγ5Vδ1
MICA/MICB	Human	Clones
ULBP4	Human	Vγ9Vδ2
CD1-sulfatide	Human	Vδ1
EPCR (endothelial protein C receptor)	Human	Clones
Phosphoantigens, amino-bisphosphonates	Human	Vγ9Vδ2
Alkylamines	Human	Vγ9Vδ2

Fig. 6.29 Ligands that activate γ:δ T cells.

the basis for activation. Among these activating antigens is **Skint-1** (selection and upkeep of intraepithelial T cells 1), an immunoglobulin superfamily member that is expressed by thymic epithelial cells and by keratinocytes. Skint-1 seems to be required for the generation of a subset of $V_\gamma 5:V_\delta 1$ T cells that develop in the thymus and home to the skin to become '**dendritic epidermal T cells' (DETCs)**. Some evidence suggests a direct interaction between Skint-1 and the γ:δ T-cell receptor, although structural studies are not yet available. Conceivably, DETCs localize to the skin due to recognition by their T-cell receptor of Skint-1 expressed by keratinocytes. There, they might provide a 'transitional' mode of immune defense, becoming activated through innate receptors that are triggered locally during infections.

Summary.

Antigen presentation to various nonconventional T-cell subsets and γ:δ T cells generally does not involve the generation of peptide:MHC complexes. Instead, these cells recognize surface proteins, such as ULBPs and RAET-1 proteins, that may indicate cellular stress, transformation, or intracellular infection, or nonpeptide antigens, such as microbial glycolipids or folate metabolites presented by CD1 molecules. The MHC region contains many genes whose structure is closely related to the MHC class I molecules—the so-called nonclassical, or class Ib, MHC. Some of these genes serve purposes that are unrelated to the immune system, but many are involved in recognition by activating and inhibitory receptors expressed by NK cells, γ:δ T cells, and α:β T cells. MHC class Ib proteins called CD1 molecules are encoded outside the MHC region. CD1c and CD1d can bind lipids and glycolipid antigens for presentation to iNKT cells expressing invariant T-cell receptors. The T-cell population called MAIT cells, which are abundant in humans, recognize vitamin B_9 metabolites presented by the MR1 MHC class Ib molecule, suggesting that the MAIT cells have a 'transitional' role between innate and adaptive immunity. Likewise, many antigens that activate γ:δ T cells may be indicators of stress or infection, and these cells are able to generate cytokines that amplify immune defense pathways.

Summary to Chapter 6.

T-cell receptors on conventional α:β T cells recognize peptides bound to MHC molecules. In the absence of infection, MHC molecules are occupied by self peptides, which do not normally provoke a T-cell response, because of various tolerance mechanisms. But during infections, pathogen-derived peptides become bound to MHC molecules and are displayed on the cell surface, where they can be recognized by T cells that have been previously activated and armed for the specific peptide:MHC complex. Naive T cells become activated when they encounter their specific antigen presented on activated dendritic cells. MHC class I molecules in most cells bind to peptides derived from proteins that have been synthesized and then degraded in the cytosol. Some dendritic cells can obtain and process exogenous antigens and present them on MHC class I molecules. This process of cross-presentation is important for priming CD8 T cells to many viral infections.

Through assembly with the invariant chain (Ii), MHC class II molecules bind peptides derived from proteins degraded in endocytic vesicles, but they can also acquire self antigens through autophagy. Stable peptides are bound after a process of peptide editing in the endocytic compartment involving HLA-DM and HLA-DO. CD8 T cells recognize peptide:MHC class I complexes and are activated to kill cells displaying foreign peptides derived from cytosolic pathogens, such as viruses. CD4 T cells recognize peptide:MHC class II complexes and are specialized to activate other immune effector cells, for example, B cells or macrophages, to act against the foreign antigens or pathogens that they have taken up.

For each class of MHC molecule, there are several genes arranged in clusters within a larger region known as the major histocompatibility complex (MHC). Within the MHC, the genes for the MHC molecules are closely linked to genes involved in the degradation of proteins into peptides, the formation of the complex of peptide and MHC molecule, and the transport of these complexes to the cell surface. Because the several different genes for the MHC class I and class II molecules are highly polymorphic and are expressed in a codominant fashion, each individual expresses a number of different MHC class I and class II molecules. Each different MHC molecule can bind stably to a range of different peptides, and thus the MHC repertoire of each individual can recognize and bind many different peptide antigens. Because the T-cell receptor binds a combined peptide:MHC ligand, T cells show MHC-restricted antigen recognition, such that a given T cell is specific for a particular peptide bound to a particular MHC molecule.

Unconventional T-cell subsets include iNKT cells, MAIT cells, and γ:δ T cells, which recognize nonpeptide ligands of various types. Some CD1 molecules bind self lipids and pathogen-derived lipid molecules and present them to iNKT cells. MAIT cells recognize vitamin metabolites that are specific to bacteria and yeast and that are presented by MR1. γ:δ T cells are activated by a diverse array of ligands, including MHC class Ib molecules and EPCR, that are induced by infection or cellular stress. These T-cell subsets function in the transitional area between innate and adaptive immunity, relying on a repertoire of receptors produced by somatic gene rearrangement but recognizing ligands in a manner somewhat similar to the way PAMPs are recognized by TLRs and other fully innate receptors.

Questions.

6.1 Short Answer: Dendritic cells are capable of efficiently acquiring antigens from exogenous sources and presenting these them to T cells on MHC class I molecules. How is this different from every other cell in the body and why is it important?

6.2 Matching: Match the following terms with the appropriate description:

A. Proteasome

B. 20S core

C. LMP2, LMP7, MECL-1

D. PA28

E. Lysine 48 ubiquitin

i. Displace the constitutive β subunits of the catalytic chamber as a response to interferons

ii. Composed of one catalytic core and two 19S regulatory caps

iii. Large cylindrical complex of 28 subunits arranged in four stacked rings

iv. Targets protein for degradation

v. Binds the proteasome and increases the rate of protein release from the proteasome

6.3 True or False: MHC class I surface expression is not affected by the cell's capacity to transport peptides into the endoplasmic reticulum.

6.4 Fill-in-the-Blanks: Cell membrane-destined polypeptides are translocated to the lumen of the endoplasmic reticulum, which is intriguing because the MHC class I presented peptides are found in the _____. Further research revealed that presentation of cytosolic peptides is possible due to a family of ABC transporters, _____, that mediate the ATP-dependent transport of peptides into the lumen of the _____. This transporter complex has limited specificities for the transported peptides; for example, peptides are generally _____ amino acids in length and transport is biased in favor of _____ residues in the carboxy terminus and against _____ residues within the first _____ amino-terminal residues.

6.5   **Multiple Choice:** CD8 dendritic cells are uniquely capable of strongly cross-presenting antigens. Which of the following options correctly matches a transcription factor essential for CD8 dendritic cell development and a surface marker uniquely expressed by these cells?

A. CIITA, CD74

B. BATF3, CD4

C. CIITA, CD94

D. BATF3, XCR1

6.6   **Matching:** Match the following terms with the appropriate description:

A. TRiC	i.	Retains the MHC class I molecule α chain in a partly folded state
B. ERAAP	ii.	Protects peptides produced in the cytosol from complete degradation
C. Calnexin	iii.	Forms a bridge between the MHC class I molecule and the TAP complex
D. ERp57	iv.	Trims the amino terminus of peptides that are too long for MHC binding
E. Tapasin	v.	Breaks and re-forms disulfide bonds in the MHC class I α domain during peptide loading

6.7   **True or False:** Cytosolic antigens are not presented through MHC class II molecules.

6.8   **Matching:** Order the following events in the sequence in which MHC class II processing happens in an antigen-presenting cell:

____The CD74 trimerization domain is cleaved.

____MHC class II is translocated into the endoplasmic reticulum.

____Cathepsin S cleaves LIP22 and leaves the CLIP fragment on the MHC molecule.

____CD74 trimers bind non-covalently to MHC class II α:β heterodimers.

____HLA-DM catalyzes the release of CLIP and promotes peptide editing.

____MHC class II heterodimers are released from calnexin for transport to a low-pH endosomal compartment.

6.9   **Multiple Choice:** Defective function of which of the following proteins will result in failed CD8 T-cell priming?

A. HLA-DM

B. Cathepsin S

C. TAP1/2

D. CD74

6.10   **Multiple Choice:** Defective function of which of the following proteins will result in decreased cytosolic peptide presentation on MHC class II?

A. IRGM3

B. BATF3

C. MARCH-1

D. TAP1/2

6.11   **True or False:** Superantigens do not induce an adaptive immune response and are independent of peptide-specific MHC–TCR interactions.

6.12   **Multiple Choice:** Which of the following statements is *false*?

A. Polymorphisms at each locus can potentially double the number of different MHC molecules expressed by an individual.

B. Pathogens can evade the immune system by mutating the immunodominant epitope, which results in loss of affinity for the specific MHC allele.

C. Pathogens do not cause evolutionary pressure to select MHC alleles that confer protection against them.

D. The DRα chain and its mouse homolog, Eα, are monomorphic.

6.13   **True or False:** Classical MHC class I molecules are highly polymorphic, as opposed to MHC class Ib, which are oligomorphic.

6.14   **Matching:** Match the following MHC class Ib genes with their appropriate description:

A. H2-M3	i.	Presents microbial folate metabolites
B. MICA	ii.	Binds α-GalCer
C. CD1d	iii.	Presents *N*-formylated peptides
D. MR1	iv.	Binds NKG2D

General references.

Germain, R.N.: **MHC-dependent antigen processing and peptide presentation: providing ligands for T lymphocyte activation.** *Cell* 1994, **76**:287–299.

Klein, J.: **Natural History of the Major Histocompatibility Complex.** New York: Wiley, 1986.

Moller, G. (ed.): **Origin of major histocompatibility complex diversity.** *Immunol. Rev.* 1995, **143**:5–292.

Trombetta, E.S., and Mellman, I.: **Cell biology of antigen processing *in vitro* and *in vivo*.** *Annu. Rev. Immunol.* 2005, **23**:975–1028.

Section references.

6-1   **Antigen presentation functions both in arming effector T cells and in triggering their effector functions to attack pathogen-infected cells.**

Guermonprez, P., Valladeau, J., Zitvogel, L., Théry, C., and Amigorena, S.: **Antigen presentation and T cell stimulation by dendritic cells.** *Annu. Rev. Immunol.* 2002, **20**:621–667.

Lee, H.K., Mattei, L.M., Steinberg, B.E., Alberts, P., Lee, Y.H., Chervonsky, A., Mizushima, N., Grinstein, S., and Iwasaki, A.: ***In vivo* requirement for Atg5 in antigen presentation by dendritic cells. *Immunity* 2010, **32**:227–239.

Segura, E., and Villadangos, J.A.: **Antigen presentation by dendritic cells *in vivo*.** *Curr. Opin. Immunol.* 2009, **21**:105–110.

Vyas, J.M., Van der Veen, A.G., and Ploegh, H.L.: **The known unknowns of antigen processing and presentation.** *Nat. Rev. Immunol.* 2008, **8**:607–618.

6-2   Peptides are generated from ubiquitinated proteins in the cytosol by the proteasome.

Basler, M., Kirk. C.J., and Groettrup, M.: **The immunoproteasome in antigen processing and other immunological functions.** *Curr. Opin. Immunol.* 2013, **25**:74–80.

Brocke, P., Garbi, N., Momburg, F., and Hammerling, G.J.: **HLA-DM, HLA-DO and tapasin: functional similarities and differences.** *Curr. Opin. Immunol.* 2002, **14**:22–29.

Cascio, P., Call, M., Petre, B.M., Walz, T., and Goldberg, A.L.: **Properties of the hybrid form of the 26S proteasome containing both 19S and PA28 complexes.** *EMBO J.* 2002, **21**:2636–2645.

Gromme, M., and Neefjes, J.: **Antigen degradation or presentation by MHC class I molecules via classical and non-classical pathways.** *Mol. Immunol.* 2002, **39**:181–202.

Goldberg, A.L., Cascio, P., Saric, T., and Rock, K.L.: **The importance of the proteasome and subsequent proteolytic steps in the generation of antigenic peptides.** *Mol. Immunol.* 2002, **39**:147–164.

Hammer, G.E., Gonzalez, F., Champsaur, M., Cado, D., and Shastri, N.: **The aminopeptidase ERAAP shapes the peptide repertoire displayed by major histocompatibility complex class I molecules.** *Nat. Immunol.* 2006, **7**:103–112.

Hammer, G.E., Gonzalez, F., James, E., Nolla, H., and Shastri, N.: **In the absence of aminopeptidase ERAAP, MHC class I molecules present many unstable and highly immunogenic peptides.** *Nat. Immunol.* 2007, **8**:101–108.

Murata, S., Sasaki, K., Kishimoto, T., Niwa, S., Hayashi, H., Takahama, Y., and Tanaka, K.: **Regulation of CD8+ T cell development by thymus-specific proteasomes.** *Science* 2007, 316:1349–1353.

Schubert, U., Anton, L.C., Gibbs, J., Norbury, C.C., Yewdell, J.W., and Bennink, J.R.: **Rapid degradation of a large fraction of newly synthesized proteins by proteasomes.** *Nature* 2000, **404**:770–774.

Serwold, T., Gonzalez, F., Kim, J., Jacob, R., and Shastri, N.: **ERAAP customizes peptides for MHC class I molecules in the endoplasmic reticulum.** *Nature* 2002, **419**:480–483.

Shastri, N., Schwab, S., and Serwold, T.: **Producing nature's gene-chips: the generation of peptides for display by MHC class I molecules.** *Annu. Rev. Immunol.* 2002, **20**:463–493.

Sijts, A., Sun, Y., Janek, K., Kral, S., Paschen, A., Schadendorf, D., and Kloetzel, P.M.: **The role of the proteasome activator PA28 in MHC class I antigen processing.** *Mol. Immunol.* 2002, **39**:165–169.

Vigneron, N., Stroobant, V., Chapiro, J., Ooms, A., Degiovanni, G., Morel, S., van der Bruggen, P., Boon, T., and Van den Eynde, B.J.: **An antigenic peptide produced by peptide splicing in the proteasome.** *Science* 2004, **304**:587–590.

Villadangos, J.A.: **Presentation of antigens by MHC class II molecules: getting the most out of them.** *Mol. Immunol.* 2001, **38**:329–346.

Williams, A., Peh, C.A., and Elliott, T.: **The cell biology of MHC class I antigen presentation.** *Tissue Antigens* 2002, **59**:3–17.

6-3   Peptides from the cytosol are transported by TAP into the endoplasmic reticulum and further processed before binding to MHC class I molecules.

Gorbulev, S., Abele, R., and Tampe, R.: **Allosteric crosstalk between peptide-binding, transport, and ATP hydrolysis of the ABC transporter TAP.** *Proc. Natl Acad. Sci. USA* 2001, **98**:3732–3737.

Kelly, A., Powis, S.H., Kerr, L.A., Mockridge, I., Elliott, T., Bastin, J., Uchanska-Ziegler, B., Ziegler, A., Trowsdale, J., and Townsend, A.: **Assembly and function of the two ABC transporter proteins encoded in the human major histocompatibility complex.** *Nature* 1992, **355**:641–644.

Lankat-Buttgereit, B., and Tampe, R.: **The transporter associated with antigen processing: function and implications in human diseases.** *Physiol. Rev.* 2002, **82**:187–204.

Powis, S.J., Townsend, A.R., Deverson, E.V., Bastin, J., Butcher, G.W., and Howard, J.C.: **Restoration of antigen presentation to the mutant cell line RMA-S by an MHC-linked transporter.** *Nature* 1991, **354**:528–531.

Townsend, A., Ohlen, C., Foster, L., Bastin, J., Lunggren, H.G., and Karre, K.: **A mutant cell in which association of class I heavy and light chains is induced by viral peptides.** *Cold Spring Harbor Symp. Quant. Biol.* 1989, **54**:299–308.

6-4   Newly synthesized MHC class I molecules are retained in the endoplasmic reticulum until they bind a peptide.

Bouvier, M.: **Accessory proteins and the assembly of human class I MHC molecules: a molecular and structural perspective.** *Mol. Immunol.* 2003, **39**:697–706.

Gao, B., Adhikari, R., Howarth, M., Nakamura, K., Gold, M.C., Hill, A.B., Knee, R., Michalak, M., and Elliott, T.: **Assembly and antigen-presenting function of MHC class I molecules in cells lacking the ER chaperone calreticulin.** *Immunity* 2002, **16**:99–109.

Grandea, A.G. III, and Van Kaer, L.: **Tapasin: an ER chaperone that controls MHC class I assembly with peptide.** *Trends Immunol.* 2001, **22**:194–199.

Van Kaer, L.: **Accessory proteins that control the assembly of MHC molecules with peptides.** *Immunol. Res.* 2001, **23**:205–214.

Williams, A., Peh, C.A., and Elliott, T.: **The cell biology of MHC class I antigen presentation.** *Tissue Antigens* 2002, **59**:3–17.

Williams, A.P., Peh, C.A., Purcell, A.W., McCluskey, J., and Elliott, T.: **Optimization of the MHC class I peptide cargo is dependent on tapasin.** *Immunity* 2002, **16**:509–520.

Zhang, W., Wearsch, P.A., Zhu, Y., Leonhardt, R.M., and Cresswell P.: **A role for UDP-glucose glycoprotein glucosyltransferase in expression and quality control of MHC class I molecules.** *Proc. Natl Acad. Sci. USA* 2011, **108**:4956–4961.

6-5   Dendritic cells use cross-presentation to present exogenous proteins on MHC class I molecules to prime CD8 T cells.

Ackerman, A.L., and Cresswell, P.: **Cellular mechanisms governing cross-presentation of exogenous antigens.** *Nat. Immunol.* 2004, **5**:678–684.

Bevan, M.J.: **Minor H antigens introduced on H-2 different stimulating cells cross-react at the cytotoxic T cell level during *in vivo* priming.** *J. Immunol.* 1976, **117**:2233–2238.

Bevan, M.J.: **Helping the CD8+ T cell response.** *Nat. Rev. Immunol.* 2004, **4**:595–602.

Hildner, K., Edelson, B.T., Purtha, W.E., Diamond, M., Matsushita, H., Kohyama, M., Calderon, B., Schraml, B.U., Unanue, E.R., Diamond, M.S., *et al.*: **Batf3 deficiency reveals a critical role for CD8α+ dendritic cells in cytotoxic T cell immunity.** *Science* 2008, **322**:1097–1100.

Segura, E., and Villadangos, J.A.: **A modular and combinatorial view of the antigen cross-presentation pathway in dendritic cells.** *Traffic* 2011, **12**:1677–1685.

6-6   Peptide:MHC class II complexes are generated in acidified endocytic vesicles from proteins o.btained through endocytosis, phagocytosis, and autophagy.

Dengjel, J., Schoor, O., Fischer, R., Reich, M., Kraus, M., Müller, M., Kreymborg, K., Altenberend, F., Brandenburg, J., Kalbacher, H., *et al.*: **Autophagy promotes MHC class II presentation of peptides from intracellular source proteins.** *Proc. Natl Acad. Sci. USA* 2005, **102**:7922–7927.

Deretic, V., Saitoh, T., and Akira, S.: **Autophagy in infection, inflammation and immunity.** *Nat. Rev. Immunol.* 2013, **13**:722–737.

Godkin, A.J., Smith, K.J., Willis, A., Tejada-Simon, M.V., Zhang, J., Elliott, T., and Hill, A.V.: **Naturally processed HLA class II peptides reveal highly conserved immunogenic flanking region sequence preferences that reflect** antigen processing rather than peptide–MHC interactions. *J. Immunol.* 2001, **166**:6720–6727.

Hiltbold, E.M., and Roche, P.A.: **Trafficking of MHC class II molecules in the late secretory pathway.** *Curr. Opin. Immunol.* 2002, **14**:30–35.

Hsieh, C.S., deRoos, P., Honey, K., Beers, C., and Rudensky, A.Y.: **A role for cathepsin L and cathepsin S in peptide generation for MHC class II presentation.** *J. Immunol.* 2002, **168**:2618–2625.

Lennon-Duménil, A.M., Bakker, A.H., Wolf-Bryant, P., Ploegh, H.L., and Lagaudrière-Gesbert, C.: **A closer look at proteolysis and MHC-class-II–restricted antigen presentation.** *Curr. Opin. Immunol.* 2002, **14**:15–21.

Li, P., Gregg, J.L., Wang, N., Zhou, D., O'Donnell, P., Blum, J.S., and Crotzer, V.L.: **Compartmentalization of class II antigen presentation: contribution of cytoplasmic and endosomal processing.** *Immunol. Rev.* 2005, **207**:206–217.

Maric, M., Arunachalam, B., Phan, U.T., Dong, C., Garrett, W.S., Cannon, K.S., Alfonso, C., Karlsson, L., Flavell, R.A., and Cresswell, P.: **Defective antigen processing in GILT-free mice.** *Science* 2001, **294**:1361–1365.

Münz, C.: **Enhancing immunity through autophagy.** *Annu. Rev. Immunol.* 2009, **27**:423–449.

Pluger, E.B., Boes, M., Alfonso, C., Schroter, C.J., Kalbacher, H., Ploegh, H.L., and Driessen, C.: **Specific role for cathepsin S in the generation of antigenic peptides *in vivo*.** *Eur. J. Immunol.* 2002, **32**:467–476.

6-7   The invariant chain directs newly synthesized MHC class II molecules to acidified intracellular vesicles.

Gregers, T.F., Nordeng, T.W., Birkeland, H.C., Sandlie, I., and Bakke, O.: **The cytoplasmic tail of invariant chain modulates antigen processing and presentation.** *Eur. J. Immunol.* 2003, **33**:277–286.

Hiltbold, E.M., and Roche, P.A.: **Trafficking of MHC class II molecules in the late secretory pathway.** *Curr. Opin. Immunol.* 2002, **14**:30–35.

Kleijmeer, M., Ramm, G., Schuurhuis, D., Griffith, J., Rescigno, M., Ricciardi-Castagnoli, P., Rudensky, A.Y., Ossendorp, F., Melief, C.J., Stoorvogel, W., *et al.*: **Reorganization of multivesicular bodies regulates MHC class II antigen presentation by dendritic cells.** *J. Cell Biol.* 2001, **155**:53–63.

van Lith, M., van Ham, M., Griekspoor, A., Tjin, E., Verwoerd, D., Calafat, J., Janssen, H., Reits, E., Pastoors, L., and Neefjes, J.: **Regulation of MHC class II antigen presentation by sorting of recycling HLA-DM/DO and class II within the multivesicular body.** *J. Immunol.* 2001, **167**:884–892.

6-8   The MHC class II-like molecules HLA-DM and HLA-DO regulate exchange of CLIP for other peptides.

Alfonso, C., and Karlsson, L.: **Nonclassical MHC class II molecules.** *Annu. Rev. Immunol.* 2000, **18**:113–142.

Apostolopoulos, V., McKenzie, I.F., and Wilson, I.A.: **Getting into the groove: unusual features of peptide binding to MHC class I molecules and implications in vaccine design.** *Front. Biosci.* 2001, **6**:D1311–D1320.

Buslepp, J., Zhao, R., Donnini, D., Loftus, D., Saad, M., Appella, E., and Collins, E.J.: **T cell activity correlates with oligomeric peptide-major histocompatibility complex binding on T cell surface.** *J. Biol. Chem.* 2001, **276**:47320–47328.

Gu, Y., Jensen, P.E., and Chen, X.: **Immunodeficiency and autoimmunity in H2-O-deficient mice.** *J. Immunol.* 2013, **190**:126–137.

Hill, J.A., Wang, D., Jevnikar, A.M., Cairns, E., and Bell, D.A.: **The relationship between predicted peptide-MHC class II affinity and T-cell activation in a HLA-DRβ1*0401 transgenic mouse model.** *Arthritis Res. Ther.* 2003, **5**:R40–R48.

Mellins, E.D., and Stern, L.J.: **HLA-DM and HLA-DO, key regulators of MHC-II processing and presentation.** *Curr. Opin. Immunol.* 2014, **26**:115–122.

Nelson, C.A., Vidavsky, I., Viner, N.J., Gross, M.L., and Unanue, E.R.: **Amino-terminal trimming of peptides for presentation on major histocompatibility complex class II molecules.** *Proc. Natl Acad. Sci. USA* 1997, **94**:628–633.

Pathak, S.S., Lich, J.D., and Blum, J.S.: **Cutting edge: editing of recycling class II:peptide complexes by HLA-DM.** *J. Immunol.* 2001, **167**:632–635.

Pos, W., Sethi, D.K., Call, M.J., Schulze, M.S., Anders, A.K., Pyrdol, J., and Wucherpfennig, K.W.: **Crystal structure of the HLA-DM–HLA-DR1 complex defines mechanisms for rapid peptide selection.** *Cell* 2012, **151**:1557–1568.

Qi, L., and Ostrand-Rosenberg, S.: **H2-O inhibits presentation of bacterial superantigens, but not endogenous self antigens.** *J. Immunol.* 2001, **167**:1371–1378.

Su, R.C., and Miller, R.G.: **Stability of surface H-2Kb, H-2Db, and peptide-receptive H-2Kᵇ on splenocytes.** *J. Immunol.* 2001, **167**:4869–4877.

Zarutskie, J.A., Busch, R., Zavala-Ruiz, Z., Rushe, M., Mellins, E.D., and Stern, L.J.: **The kinetic basis of peptide exchange catalysis by HLA-DM.** *Proc. Natl Acad. Sci. USA* 2001, **98**:12450–12455.

6-9   Cessation of antigen processing occurs in dendritic cells after their activation through reduced expression of the MARCH-1 E3 ligase.

Baravalle, G., Park, H., McSweeney, M., Ohmura-Hoshino, M., Matsuki, Y., Ishido, S., and Shin, J.S.: **Ubiquitination of CD86 is a key mechanism in regulating antigen presentation by dendritic cells.** *J. Immunol.* 2011, **187**:2966–2973.

De Gassart, A., Camosseto, V., Thibodeau, J., Ceppi, M., Catalan, N., Pierre, P., and Gatti, E.: **MHC class II stabilization at the surface of human dendritic cells is the result of maturation-dependent MARCH I down-regulation.** *Proc. Natl Acad. Sci. USA* 2008, **105**:3491–3496.

Jiang, X., and Chen, Z.J.: **The role of ubiquitylation in immune defence and pathogen evasion.** *Nat. Rev. Immunol.* 2012, **12**:35–48.

Ma, J.K., Platt, M.Y., Eastham-Anderson, J., Shin, J.S., and Mellman, I.: **MHC class II distribution in dendritic cells and B cells is determined by ubiquitin chain length.** *Proc. Natl Acad. Sci. USA* 2012, **109**:8820–8827.

Ohmura-Hoshino, M., Matsuki, Y., Mito-Yoshida, M., Goto, E., Aoki-Kawasumi, M., Nakayama, M., Ohara, O., and Ishido, S.: **Cutting edge: requirement of MARCH-I-mediated MHC II ubiquitination for the maintenance of conventional dendritic cells.** *J. Immunol.* 2009, **183**:6893–6897.

Walseng, E., Furuta, K., Bosch, B., Weih, K.A., Matsuki, Y., Bakke, O., Ishido, S., and Roche, P.A.: **Ubiquitination regulates MHC class II-peptide complex retention and degradation in dendritic cells.** *Proc. Natl Acad. Sci. USA* 2010, **107**:20465–20470.

6-10   Many proteins involved in antigen processing and presentation are encoded by genes within the MHC.

Aguado, B., Bahram, S., Beck, S., Campbell, R.D., Forbes, S.A., Geraghty, D., Guillaudeux, T., Hood, L., Horton, R., Inoko, H., *et al.* (the MHC Sequencing Consortium): **Complete sequence and gene map of a human major histocompatibility complex.** *Nature* 1999, **401**:921–923.

Chang, C.H., Gourley, T.S., and Sisk, T.J.: **Function and regulation of class II transactivator in the immune system.** *Immunol. Res.* 2002, **25**:131–142.

Kumnovics, A., Takada, T., and Lindahl, K.F.: **Genomic organization of the mammalian MHC.** *Annu. Rev. Immunol.* 2003, **21**:629–657.

Lefranc, M.P.: **IMGT, the international ImMunoGeneTics database.** *Nucleic Acids Res.* 2003, **31**:307–310.

6-11   The protein products of MHC class I and class II genes are highly polymorphic.

Gaur, L.K., and Nepom, G.T.: **Ancestral major histocompatibility complex DRB genes beget conserved patterns of localized polymorphisms.** *Proc. Natl Acad. Sci. USA* 1996, **93**:5380–5383.

Marsh, S.G.: **Nomenclature for factors of the HLA system, update December 2002.** *Eur. J. Immunogenet.* 2003, **30**:167–169.

Robinson, J., and Marsh, S.G.: **HLA informatics. Accessing HLA sequences from sequence databases.** *Methods Mol. Biol.* 2003, **210**:3–21.

6-12   MHC polymorphism affects antigen recognition by T cells by influencing both peptide binding and the contacts between T-cell receptor and MHC molecule.

Falk, K., Rotzschke, O., Stevanovic, S., Jung, G., and Rammensee, H.G.: **Allele-specific motifs revealed by sequencing of self-peptides eluted from MHC molecules.** *Nature* 1991, **351**:290–296.

Garcia, K.C., Degano, M., Speir, J.A., and Wilson, I.A.: **Emerging principles for T cell receptor recognition of antigen in cellular immunity.** *Rev. Immunogenet.* 1999, **1**:75–90.

Katz, D.H., Hamaoka, T., Dorf, M.E., Maurer, P.H., and Benacerraf, B.: **Cell interactions between histoincompatible T and B lymphocytes. IV. Involvement of immune response (Ir) gene control of lymphocyte interaction controlled by the gene.** *J. Exp. Med.* 1973, **138**:734–739.

Kjer-Nielsen, L., Clements, C.S., Brooks, A.G., Purcell, A.W., Fontes, M.R., McCluskey, J., and Rossjohn, J.: **The structure of HLA-B8 complexed to an immunodominant viral determinant: peptide-induced conformational changes and a mode of MHC class I dimerization.** *J. Immunol.* 2002, **169**:5153–5160.

Wang, J.H., and Reinherz, E.L.: **Structural basis of T cell recognition of peptides bound to MHC molecules.** *Mol. Immunol.* 2002, **38**:1039–1049.

Zinkernagel, R.M., and Doherty, P.C.: **Restriction of *in vivo* T-cell mediated cytotoxicity in lymphocytic choriomeningitis within a syngeneic or semiallogeneic system.** *Nature* 1974, **248**:701–702.

6-13 Alloreactive T cells recognizing nonself MHC molecules are very abundant.

Felix, N.J., and Allen, P.M.: **Specificity of T-cell alloreactivity.** *Nat. Rev. Immunol.* 2007, **7**:942–953.

Feng, D., Bond, C.J., Ely, L.K., Maynard, J., and Garcia, K.C.: **Structural evidence for a germline-encoded T cell receptor–major histocompatibility complex interaction 'codon.'** *Nat. Immunol.* 2007, **8**:975–993.

Hennecke, J., and Wiley, D.C.: **Structure of a complex of the human α/β T cell receptor (TCR) HA1.7, influenza hemagglutinin peptide, and major histocompatibility complex class II molecule, HLA-DR4 (DRA*0101 and DRB1*0401): insight into TCR cross-restriction and alloreactivity.** *J. Exp. Med.* 2002, **195**:571–581.

Jankovic, V., Remus, K., Molano, A., and Nikolich-Zugich, J.: **T cell recognition of an engineered MHC class I molecule: implications for peptide-independent alloreactivity.** *J. Immunol.* 2002, **169**:1887–1892.

Nesic, D., Maric, M., Santori, F.R., and Vukmanovic, S.: **Factors influencing the patterns of T lymphocyte allorecognition.** *Transplantation* 2002, **73**:797–803.

Reiser, J.B., Darnault, C., Guimezanes, A., Gregoire, C., Mosser, T., Schmitt-Verhulst, A.M., Fontecilla-Camps, J.C., Malissen, B., Housset, D., and Mazza, G.: **Crystal structure of a T cell receptor bound to an allogeneic MHC molecule.** *Nat. Immunol.* 2000, **1**:291–297.

Rötzschke, O., Falk, K., Faath, S., Rammensee, H.G.: **On the nature of peptides involved in T cell alloreactivity.** *J. Exp. Med.* 1991, **174**:1059–1071.

Speir, J.A., Garcia, K.C., Brunmark, A., Degano, M., Peterson, P.A., Teyton, L., and Wilson, I.A.: **Structural basis of 2C TCR allorecognition of H-2Ld peptide complexes.** *Immunity* 1998, **8**:553–562.

6-14 Many T cells respond to superantigens.

Acha-Orbea, H., Finke, D., Attinger, A., Schmid, S., Wehrli, N., Vacheron, S., Xenarios, I., Scarpellino, L., Toellner, K.M., MacLennan, I.C., *et al.*: **Interplays between mouse mammary tumor virus and the cellular and humoral immune response.** *Immunol. Rev.* 1999, **168**:287–303.

Kappler, J.W., Staerz, U., White, J., and Marrack, P.: **T cell receptor Vb elements which recognize Mls-modified products of the major histocompatibility complex.** *Nature* 1988, **332**:35–40.

Rammensee, H.G., Kroschewski, R., and Frangoulis, B.: **Clonal anergy induced in mature Vβ6+ T lymphocytes on immunizing Mls-1b mice with Mls-1a expressing cells.** *Nature* 1989, **339**:541–544.

Spaulding, A.R., Salgado-Pabón, W., Kohler, P.L., Horswill, A.R., Leung, D.Y., and Schlievert, P.M.: **Staphylococcal and streptococcal superantigen exotoxins.** *Clin. Microbiol. Rev.* 2013, **26**:422–447.

Sundberg, E.J., Li, H., Llera, A.S., McCormick, J.K., Tormo, J., Schlievert, P.M., Karjalainen, K., and Mariuzza, R.A.: **Structures of two streptococcal superantigens bound to TCR β chains reveal diversity in the architecture of T cell signaling complexes.** *Structure* 2002, **10**:687–699.

Torres, B.A., Perrin, G.Q., Mujtaba, M.G., Subramaniam, P.S., Anderson, A.K., and

Johnson, H.M.: **Superantigen enhancement of specific immunity: antibody production and signaling pathways.** *J. Immunol.* 2002, **169**:2907–2914.

White, J., Herman, A., Pullen, A.M., Kubo, R., Kappler, J.W., and Marrack, P.: **The V$_\beta$-specific super antigen staphylococcal enterotoxin B: stimulation of mature T cells and clonal deletion in neonatal mice.** *Cell* 1989, **56**:27–35.

6-15 MHC polymorphism extends the range of antigens to which the immune system can respond.

Hill, A.V., Elvin, J., Willis, A.C., Aidoo, M., Allsopp, C.E.M., Gotch, F.M., Gao, X.M., Takiguchi, M., Greenwood, B.M., Townsend, A.R.M., *et al.*: **Molecular analysis of the association of B53 and resistance to severe malaria.** *Nature* 1992, **360**:435–440.

Martin, M.P., and Carrington, M.: **Immunogenetics of viral infections.** *Curr. Opin. Immunol.* 2005, **17**:510–516.

Messaoudi, I., Guevara Patino, J.A., Dyall, R., LeMaoult, J., and Nikolich-Zugich, J.: **Direct link between *mhc* polymorphism, T cell avidity, and diversity in immune defense.** *Science* 2002, **298**:1797–1800.

Potts, W.K., and Slev, P.R.: **Pathogen-based models favouring MHC genetic diversity.** *Immunol. Rev.* 1995, **143**:181–197.

6-16 A variety of genes with specialized functions in immunity are also encoded in the MHC.

Alfonso, C., and Karlsson, L.: **Nonclassical MHC class II molecules.** *Annu. Rev. Immunol.* 2000, **18**:113–142.

Hofstetter, A.R., Sullivan, L.C., Lukacher, A.E., and Brooks, A.G.: **Diverse roles of non-diverse molecules: MHC class Ib molecules in host defense and control of autoimmunity.** *Curr. Opin. Immunol.* 2011, **23**:104–110.

Loconto, J., Papes, F., Chang, E., Stowers, L., Jones, E.P., Takada, T., Kumánovics, A., Fischer Lindahl, K., and Dulac, C.: **Functional expression of murine V2R pheromone receptors involves selective association with the M10 and M1 families of MHC class Ib molecules.** *Cell* 2003, **112**:607–118.

Powell, L.W., Subramaniam, V.N., and Yapp, T.R.: **Haemochromatosis in the new millennium.** *J. Hepatol.* 2000, **32**:48–62.

6-17 Specialized MHC class I molecules act as ligands for the activation and inhibition of NK cells and unconventional T-cell subsets.

Borrego, F., Kabat, J., Kim, D.K., Lieto, L., Maasho, K., Pena, J., Solana, R., and Coligan, J.E.: **Structure and function of major histocompatibility complex (MHC) class I specific receptors expressed on human natural killer (NK) cells.** *Mol. Immunol.* 2002, **38**:637–660.

Boyington, J.C., Riaz, A.N., Patamawenu, A., Coligan, J.E., Brooks, A.G., and Sun, P.D.: **Structure of CD94 reveals a novel C-type lectin fold: implications for the NK cell-associated CD94/NKG2 receptors.** *Immunity* 1999, **10**:75–82.

Braud, V.M., Allan, D.S., O'Callaghan, C.A., Söderström, K., D'Andrea, A., Ogg, G.S., Lazetic, S., Young, N.T., Bell, J.I., Phillips, J.H., *et al.*: **HLA-E binds to natural killer cell receptors CD94/NKG2A, B and C.** *Nature* 1998, **391**:795–799.

Braud, V.M., and McMichael, A.J.: **Regulation of NK cell functions through interaction of the CD94/NKG2 receptors with the nonclassical class I molecule HLA-E.** *Curr. Top. Microbiol. Immunol.* 1999, **244**:85–95.

Jiang, H., Canfield, S.M., Gallagher, M.P., Jiang, H.H., Jiang, Y., Zheng, Z., and Chess, L.: **HLA-E-restricted regulatory CD8(+) T cells are involved in development and control of human autoimmune type 1 diabetes.** *J. Clin. Invest.* 2010, **120**:3641–3650.

Lanier, L.L.: **NK cell recognition.** *Annu. Rev. Immunol.* 2005, **23**:225–274.

Lopez-Botet, M., and Bellon, T.: **Natural killer cell activation and inhibition by receptors for MHC class I.** *Curr. Opin. Immunol.* 1999, **11**:301–307.

Lopez-Botet, M., Bellon, T., Llano, M., Navarro, F., Garcia, P., and de Miguel, M.: **Paired inhibitory and triggering NK cell receptors for HLA class I molecules.** *Hum. Immunol.* 2000, **61**:7–17.

Lopez-Botet, M., Llano, M., Navarro, F., and Bellon, T.: **NK cell recognition of non-classical HLA class I molecules.** *Semin. Immunol.* 2000, **12**:109–119.

Lu, L., Ikizawa, K., Hu, D., Werneck, M.B., Wucherpfennig, K.W., and Cantor, H.:

Regulation of activated CD4+ T cells by NK cells via the Qa-1-NKG2A inhibitory pathway. *Immunity* 2007, **26**:593–604.

Pietra, G., Romagnani, C., Moretta, L., and Mingari, M.C.: **HLA-E and HLA-E-bound peptides: recognition by subsets of NK and T cells.** *Curr. Pharm. Des.* 2009, **15**:3336–3344.

Rodgers, J.R., and Cook, R.G.: **MHC class Ib molecules bridge innate and acquired immunity.** *Nat. Rev. Immunol.* 2005, **5**:459–471.

6-18 Members of the CD1 family of MHC class I-like molecules present microbial lipids to invariant NKT cells.

Gendzekhadze, K., Norman, P.J., Abi-Rached, L., Graef, T., Moesta, A.K., Layrisse, Z., and Parham, P.: **Co-evolution of KIR2DL3 with HLA-C in a human population retaining minimal essential diversity of KIR and HLA class I ligands.** *Proc. Natl Acad. Sci. USA* 2009, **106**:18692–18697.

Godfrey, D.I., Stankovic, S., and Baxter, A.G.: **Raising the NKT cell family.** *Nat. Immunol.* 2010, **11**:197–206.

Hava, D.L., Brigl, M., van den Elzen, P., Zajonc, D.M., Wilson, I.A., and Brenner, M.B.: **CD1 assembly and the formation of CD1-antigen complexes.** *Curr. Opin. Immunol.* 2005, **17**:88–94.

Moody, D.B., and Besra, G.S.: **Glycolipid targets of CD1-mediated T-cell responses.** *Immunology* 2001, **104**:243–251.

Moody, D.B., and Porcelli, S.A.: **CD1 trafficking: invariant chain gives a new twist to the tale.** *Immunity* 2001, **15**:861–865.

Moody, D.B., and Porcelli, S.A.: **Intracellular pathways of CD1 antigen presentation.** *Nat. Rev. Immunol.* 2003, **3**:11–22.

Scharf, L., Li, N.S., Hawk, A.J., Garzón, D., Zhang, T., Fox, L.M., Kazen, A.R., Shah, S., Haddadian, E.J., Gumperz, J.E., *et al.*: **The 2.5Å structure of CD1c in complex with a mycobacterial lipid reveals an open groove ideally suited for diverse antigen presentation.** *Immunity* 2010, **33**:853–862.

Schiefner, A., Fujio, M., Wu, D., Wong, C.H., and Wilson, I.A.: **Structural evaluation of potent NKT cell agonists: implications for design of novel stimulatory ligands.** *J. Mol. Biol.* 2009, **394**:71–82.

6-19 The nonclassical MHC class I molecule MR1 presents microbial folate metabolites to MAIT cells.

Birkinshaw, R.W., Kjer-Nielsen, L., Eckle, S.B., McCluskey, J., and Rossjohn, J.: **MAITs, MR1 and vitamin B metabolites.** *Curr. Opin. Immunol.* 2014, **26**:7–13.

Kjer-Nielsen, L., Patel, O., Corbett, A.J., Le Nours, J., Meehan, B., Liu, L., Bhati, M., Chen, Z., Kostenko, L., Reantragoon, R., *et al.*: **MR1 presents microbial vitamin B metabolites to MAIT cells.** *Nature* 2012, **491**:717–723.

López-Sagaseta, J., Dulberger, C.L., Crooks, J.E., Parks, C.D., Luoma, A.M., McFedries, A., Van Rhijn, I., Saghatelian, A., and Adams, E.J.: **The molecular basis for Mucosal-Associated Invariant T cell recognition of MR1 proteins.** *Proc. Natl Acad. Sci. USA* 2013, **110**:E1771–1778.

6-20 γ:δ T cells can recognize a variety of diverse ligands.

Chien, Y.H., Meyer, C., and Bonneville, M.: **γδ T cells: first line of defense and beyond.** *Annu. Rev. Immunol.* 2014, **32**:121–155.

Turchinovich, G., and Hayday, A.C.: **Skint-1 identifies a common molecular mechanism for the development of interferon-γ-secreting versus interleukin-17-secreting γδ T cells.** *Immunity* 2011, **35**:59–68.

Uldrich, A.P., Le Nours, J., Pellicci, D.G., Gherardin, N.A., McPherson, K.G., Lim, R.T., Patel, O., Beddoe, T., Gras, S., Rossjohn, J., *et al.*: **CD1d-lipid antigen recognition by the γδ TCR.** *Nat. Immunol.* 2013, **14**:1137–1145.

Willcox, C.R., Pitard, V., Netzer, S., Couzi, L., Salim, M., Silberzahn, T., Moreau, J.F., Hayday, A.C., Willcox, B.E., and Déchanet-Merville, J.: **Cytomegalovirus and tumor stress surveillance by binding of a human γδ T cell antigen receptor to endothelial protein C receptor.** *Nat. Immunol.* 2012, **13**:872–879.

Lymphocyte Receptor Signaling

7

T and B lymphocytes are cells of the adaptive immune system, each of which expresses a unique antigen receptor. These cells circulate between the blood, the lymph, and, most importantly, the secondary lymphoid organs, where they survey antigen-presenting cells for their specific antigen. Once that antigen is encountered, signals from the antigen receptor activate several downstream pathways that convert quiescent naive lymphocytes into metabolically active cells that are reorganizing their actin cytoskeleton, activating transcription factors, and synthesizing a wide range of new proteins. As a result of these events, naive T and B cells undergo rapid cell division and differentiate into armed effector cells, thus expanding lymphocyte populations during an immune response and equipping these cells with the machinery to combat infections.

We begin by discussing some general principles of intracellular signaling, and then outline the pathways activated when a naive lymphocyte encounters its specific antigen. Next, we briefly discuss the co-stimulatory signaling that is necessary to activate naive T cells and, in most cases, naive B cells. In the last part of the chapter we focus on inhibitory receptors, and their roles in down-regulating signaling pathways in T and B cells.

IN THIS CHAPTER

General principles of signal transduction and propagation.

Antigen receptor signaling and lymphocyte activation.

Co-stimulatory and inhibitory receptors modulate antigen receptor signaling in T and B lymphocytes.

General principles of signal transduction and propagation.

In this part of the chapter we review briefly some general principles of receptor action and signal transduction that are common to many of the pathways discussed here. All cell-surface receptors that have a signaling function either are transmembrane proteins themselves or form parts of protein complexes that link the exterior and interior of the cell. Different classes of receptors transduce extracellular signals in a variety of ways. A common theme among the receptors covered in this chapter is that ligand binding results in the activation of an intracellular enzymatic activity.

7-1 Transmembrane receptors convert extracellular signals into intracellular biochemical events.

The enzymes most commonly associated with receptor activation are the **protein kinases**. This large group of enzymes catalyzes the covalent attachment of a phosphate group to a protein, a reversible process called **phosphorylation**. For receptors that use protein kinases, the binding of ligand to the extracellular part of the receptor allows the receptor-associated protein kinase to become 'active'—that is, to phosphorylate its intracellular substrate—and thus to propagate the signal. As we shall see, receptor-associated kinases can become activated in various ways, such as by undergoing modifications to the kinase itself that alter its intrinsic catalytic efficiency or by changes in subcellular localization that increase access to its biochemical substrates.

In animals, protein kinases phosphorylate proteins on three amino acids— tyrosine, serine, or threonine. Most of the enzyme-linked receptors we discuss in detail in this chapter activate **tyrosine protein kinases**. Tyrosine kinases are specific for tyrosine residues, whereas serine/threonine kinases phosphorylate serine and threonine residues; less common are dual-specificity kinases that phosphorylate both tyrosine and serine/threonine residues in their substrates. Protein tyrosine phosphorylation is much less common than serine/threonine phosphorylation in general, and is employed mainly in signaling pathways. One large group of receptors—the so-called **receptor tyrosine kinases**— carry a kinase activity within the cytoplasmic region of the receptor itself (**Fig. 7.1**, top panel). This group contains receptors for many growth factors; lymphocyte receptors of this type include Kit and FLT3, which are expressed on developing lymphocytes in addition to other hematopoietic progenitor cells and are discussed in Chapter 8. The receptor for transforming growth factor-β (TGF-β), an important regulatory cytokine produced by many cells, is a **receptor serine/threonine kinase**.

Even more important to the function of mature lymphocytes are receptors that have no intrinsic enzymatic activity themselves but associate with intracellular tyrosine kinases. The antigen receptors on B lymphocytes and T lymphocytes are of this type, as are the receptors for some types of cytokines. Ligand binding to the extracellular domain of such receptors causes particular amino acid residues in their cytoplasmic domains to become phosphorylated by specific cytoplasmic tyrosine kinases (see Fig. 7.1, bottom panel). These **nonreceptor kinases** either can be constitutively associated with the cytoplasmic domains of the receptors, as with many cytokine receptors, or may become associated with the receptors when they bind their ligands, as is the case for the antigen receptors.

For many cytokine receptors, ligand binding causes dimerization or clustering of individual receptor molecules, bringing the associated kinases together and enabling them to phosphorylate the cytoplasmic tail of adjacent receptors— thus initiating an intracellular signal. In the case of the lymphocyte antigen receptors, association with cytoplasmic tyrosine kinases occurs after ligand binding but is unlikely to be due to a simple clustering mechanism. Instead, the actions of co-receptors are required: these bring cytoplasmic tyrosine kinases into proximity with the cytoplasmic regions of the antigen receptor, a complex process that we will describe later.

Signaling is usually not a simple 'on or off' switch. Depending on the affinity of the receptor for the ligand, the abundance of the ligand, the concentrations of intracellular signaling components, and a complex network of positive- and negative-feedback pathways, receptor activation and downstream signaling occur when a minimum threshold determined by all of these factors is exceeded. These features are often merged into the simple term 'signal

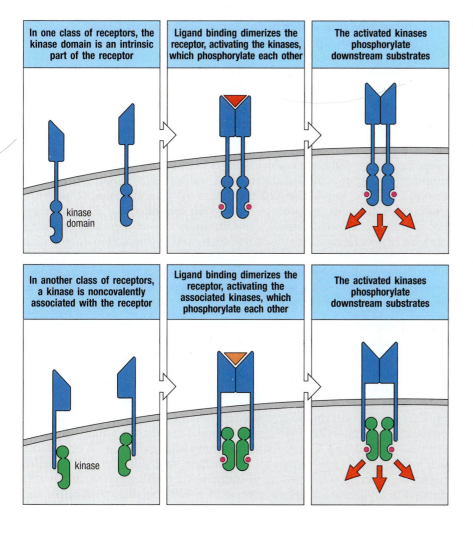

In one class of receptors, the kinase domain is an intrinsic part of the receptor	Ligand binding dimerizes the receptor, activating the kinases, which phosphorylate each other	The activated kinases phosphorylate downstream substrates
kinase domain		

In another class of receptors, a kinase is noncovalently associated with the receptor	Ligand binding dimerizes the receptor, activating the associated kinases, which phosphorylate each other	The activated kinases phosphorylate downstream substrates
kinase		

Fig. 7.1 Enzyme-associated receptors of the immune system can use intrinsic or associated protein kinases to signal. These receptors activate a protein kinase on the cytoplasmic side of the membrane to convey the information that a ligand has bound to their extracellular portion. Receptor tyrosine kinases (top panels) contain the kinase activity as part of the receptor itself. Ligand binding results in clustering of the receptor, activation of catalytic activity, and the consequent phosphorylation (denoted by red dots) of the receptor tails and other substrates, transmitting the signal onward. Receptors that lack intrinsic kinase activity associate with nonreceptor kinases (bottom panels). Receptor dimerization or clustering after ligand binding activates the associated enzyme. In all receptors of these types encountered in this chapter, the enzyme is a tyrosine kinase.

strength.' It is important to keep in mind that variations in signal strength will determine the magnitude of cellular responses—some will be all-or-nothing, whereas others will increase as the strength of signaling increases.

The role of protein kinases in cell signaling is not confined to receptor activation, as they act at many different stages in intracellular signaling pathways. Protein kinases figure largely in cell signaling because phosphorylation and **dephosphorylation**—the removal of a phosphate group—are the means of regulating the activity of many enzymes, transcription factors, and other proteins. Equally important to the workings of signaling pathways is the fact that phosphorylation generates sites on proteins to which other signaling proteins can bind.

Phosphate groups are removed from proteins by a large class of enzymes called **protein phosphatases**. Different classes of protein phosphatases remove phosphate groups from phosphotyrosine or from phosphoserine/ phosphothreonine, or both (as in dual-specificity phosphatases). Specific dephosphorylation by phosphatases is one important means of regulating signaling pathways by resetting a protein to its original state and thus switching signaling off. Dephosphorylation does not always inhibit a protein's activity. In many instances the removal of a particular phosphate group by a specific phosphatase is needed to activate an enzyme. In other cases, the extent of phosphorylation of an enzyme determines its activity, and represents a balance between the activity of kinases and phosphatases.

7-2   Intracellular signal propagation is mediated by large multiprotein signaling complexes.

As we learned in Chapter 3, binding of a ligand to its receptor can initiate a cascade of events involving intracellular proteins that sequentially convey signaling information onward. The unique enzymes and other components assembled into a particular multiprotein receptor complex will determine the character of the signal it generates. These components may be shared by several receptor pathways, or they may be exclusive to one receptor pathway, thus allowing distinct signaling pathways to be built up from a relatively limited number of components. The assembly of multisubunit signaling complexes involves specific interactions of a number of distinct types of **protein-interaction domains**, or **protein-interaction modules**, carried by the signaling proteins. **Figure 7.2** gives a few examples of such domains. Signaling proteins in general contain at least one such protein-interaction domain, but many contain multiple domains. These protein modules cooperate with each other, for example, to organize signaling proteins into the correct subcellular localizations, to enable specific binding between protein partners, and to modify enzymatic activity.

For the pathways considered in this chapter, the most important mechanism underlying the formation of signaling complexes is the phosphorylation of protein tyrosine residues. Phosphotyrosines are binding sites for a number of protein-interaction domains, including the **SH2 (Src homology 2) domain** (see Fig. 7.2). SH2 domains, built from approximately 100 amino acids, are present in many intracellular signaling proteins, where they are frequently linked to other types of enzymatic or other functional domains. SH2 domains recognize the phosphorylated tyrosine (pY) and, typically, the amino acid three positions away (pYXXZ, where X is any amino acid and Z is a specific amino acid); they bind in a sequence-specific fashion, with different SH2 domains preferring different combinations of amino acids. In this way, the unique SH2 domain of a signaling molecule can act as a 'key' that allows inducible and specific association with a protein containing the appropriate pY-containing amino acid sequence.

Tyrosine kinase-associated receptors can assemble multiprotein signaling complexes by using proteins called **scaffolds** and **adaptors**. Scaffolds and adaptors lack enzymatic activity, and they function by recruiting other proteins into a signaling complex so that interactions among these proteins can take place.

Fig. 7.2 Signaling proteins interact with each other and with lipid signaling molecules via modular protein domains. A few of the most common protein domains used by immune-system signaling proteins are listed, together with some proteins that contain them and the general class of ligand bound by the interaction domain. The right-hand column lists specific examples of a protein motif (in single-letter amino acid code) or, for the phosphoinositide-binding domains, the particular phosphoinositide that they bind. All these domains are used in many other nonimmune signaling pathways as well.

Protein domain	Found in	Ligand class	Example of ligand
SH2	Lck, ZAP-70, Fyn, Src, Grb2, PLC-γ, STAT, Cbl, Btk, Itk, SHIP, Vav, SAP, PI3K	phosphotyrosine	pYXXZ
SH3	Lck, Fyn, Src, Grb2, Btk, Itk, Tec, Fyb, Nck, Gads	proline	PXXP
PH	Tec, PLC-γ, Akt, Btk, Itk, Sos	phosphoinositides	PIP_3
PX	P40^{phox}, P47^{phox}, PLD	phosphoinositides	PI(3)P
PDZ	CARMA1	C termini of proteins	IESDV, VETDV
C1	RasGRP, PKC-θ	membrane lipid	diacylglycerol (DAG) phorbol ester
NZF	TAB2	polyubiquitin (K63-linked)	polyubiquitinated RIP, TRAF-6, or NEMO

Scaffolds are relatively large proteins that can, for example, become tyrosine phosphorylated on multiple sites in order to recruit many different proteins (Fig. 7.3, top panel). By specifying which proteins are recruited, scaffolds can define the character of a particular signaling response. This is accomplished by several mechanisms. For example, scaffolds can regulate the specificity of a recruited enzyme by recruiting one of the enzyme's substrates. Binding to a scaffold can also change the conformation of a recruited protein, thereby revealing sites for protein modifications, such as phosphorylation or ubiquitination, or for protein–protein interactions. Finally, scaffolds can function to promote membrane localization of the signaling complex.

Adaptors are membrane-anchored or cytoplasmic proteins containing several signaling modules that serve to link two or more proteins together. The adaptor proteins Grb2 and Gads, for example, each contain an SH2 domain and two copies of another module called the SH3 domain (see Fig. 7.2). This arrangement of modules can be used to link tyrosine phosphorylation of a receptor to molecules acting in the next stage of signaling. For example, the SH2 domain of Grb2 binds to a phosphotyrosine residue on a receptor or a scaffold protein, while its two SH3 domains bind to proline-rich motifs on other signaling proteins (see Fig. 7.3, bottom panel), such as Sos, which we discuss in the next section.

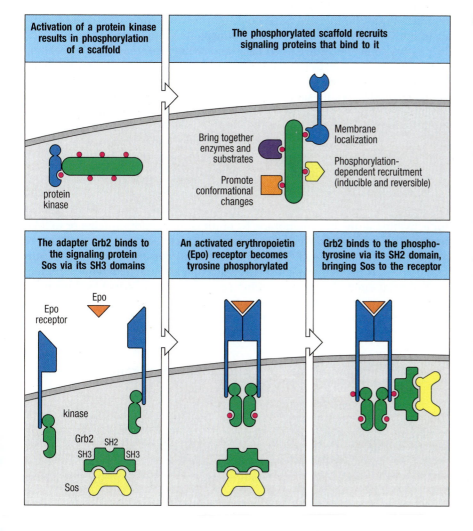

Fig. 7.3 Assembly of signaling complexes is mediated by scaffold and adaptor proteins. Assembly of signaling complexes is an important aspect of signal transduction. This is often achieved through scaffold and adaptor proteins. In general, scaffolds have numerous sites of phosphorylation that function to bring many different signaling proteins together (top panel). Scaffolds may also function to promote membrane localization, to bring enzymes into close proximity with their substrates, and to induce conformational changes in proteins that regulate their functions. An adaptor protein functions to bring two different proteins together (bottom panel). When erythropoietin (Epo) binds to its receptor, associated tyrosine kinases phosphorylate (red dots) sites on the Epo receptor cytoplasmic domain, generating binding sites for the SH2 domain of an adaptor protein. The adaptor protein (green) shown here contains two SH3 domains in addition to an SH2 domain. With the SH3 domains it can, for example, bind proline-rich sites on an intracellular signaling molecule (yellow).

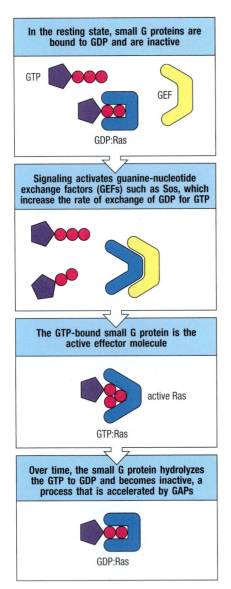

In the resting state, small G proteins are bound to GDP and are inactive

GTP

GEF

GDP:Ras

Signaling activates guanine-nucleotide exchange factors (GEFs) such as Sos, which increase the rate of exchange of GDP for GTP

The GTP-bound small G protein is the active effector molecule

active Ras

GTP:Ras

Over time, the small G protein hydrolyzes the GTP to GDP and becomes inactive, a process that is accelerated by GAPs

GDP:Ras

Fig. 7.4 Small G proteins are switched from inactive to active states by guanine-nucleotide exchange factors (GEFs) and the binding of GTP. Ras is a small GTP-binding protein with intrinsic GTPase activity. In its resting state, Ras is bound to GDP. Receptor signaling activates guanine-nucleotide exchange factors (GEFs), such as Sos, which can bind to small G proteins such as Ras and increase the rate of exchange of GDP for GTP (center panels). The GTP-bound form of Ras can then bind to a large number of effectors, recruiting them to the membrane. Over time, the intrinsic GTPase activity of Ras will result in the hydrolysis of GTP to GDP. GTPase-activating proteins (GAPs) can accelerate the hydrolysis of GTP to GDP, thus shutting off the signal more rapidly.

7-3   Small G proteins act as molecular switches in many different signaling pathways.

Monomeric GTP-binding proteins known as **small G proteins** or **small GTPases** are important in the signaling pathways leading from many tyrosine kinase-associated receptors. The small GTPases are distinct from the larger heterotrimeric G proteins associated with G-protein-coupled receptors such as the chemokine receptors discussed in Chapter 3. The superfamily of small GTPases comprises more than 100 different proteins, and many are important in lymphocyte signaling. One of these, **Ras**, is involved in many pathways leading to cell proliferation. Other small GTPases include Rac, Rho, and Cdc42, which control changes in the actin cytoskeleton caused by signals received through the T-cell receptor or B-cell receptor. We will describe their actions in more detail in Section 7-19.

Small GTPases exist in two states, depending on whether they are bound to GTP or to GDP. The GDP-bound form is inactive but is converted into the active form by exchange of the GDP for GTP. This reaction is mediated by proteins known as **guanine-nucleotide exchange factors**, or **GEFs**, which cause the GTPase to release GDP and to bind the more abundant GTP (Fig. 7.4). Sos, which is recruited to signaling pathways by the adaptor Grb2 (see Section 7-2), is one of the GEFs for Ras. The binding of GTP induces a conformational change in the small GTPase that enables it to bind to and induce effector activity in a variety of target proteins. Thus, GTP binding functions as an on/off switch for small GTPases.

This GTP-bound form does not remain permanently active but is eventually converted into the inactive GDP-bound form by the intrinsic GTPase activity in the G protein, which removes a phosphate group from the bound GTP. Regulatory cofactors known as **GTPase-activating proteins** (**GAPs**) accelerate the conversion of GTP to GDP, thus rapidly downregulating the activity of the small GTPase. Because of GAP activity, small GTPases are usually present in the inactive GDP-bound state and are activated only transiently in response to a signal from an activated receptor. *RAS* is frequently mutated in cancer cells, and the mutated Ras protein is thought to be a significant contributor to the cancerous state. The importance of GAPs in signaling regulation is indicated by the fact that some mutations in Ras found in cancer act by preventing GAP from enhancing the intrinsic GTPase activity of Ras, thus prolonging the duration for which Ras exists in the active GTP-bound state.

GEFs are the key to G-protein activation and are recruited to the site of receptor activation at the cell membrane by binding to adaptor proteins or to lipid metabolites produced by receptor activation. Once recruited, they are able to activate Ras or other small G proteins, which are localized to the inner surface of the plasma membrane via fatty acids that are attached to the G protein post-translationally. Thus, G proteins act as molecular switches, becoming switched on when a cell-surface receptor is activated and then being switched off. Each G protein has its own specific GEFs and GAPs, which help to confer specificity on the pathway.

7-4   Signaling proteins are recruited to the membrane by a variety of mechanisms.

We have seen how receptors can recruit intracellular signaling proteins to the plasma membrane through tyrosine phosphorylation of the receptor itself or of an associated scaffold, followed by recruitment of SH2-domain-containing signaling proteins or adaptors, such as Grb2 (Fig. 7.5). A second mechanism for membrane recruitment of signaling proteins is via binding to small GTPases, such as Ras, following their activation. As described in Section 7-3, small GTPases are constitutively bound to the cytoplasmic surface of the plasma

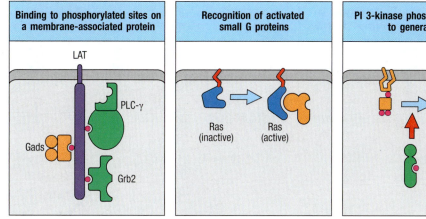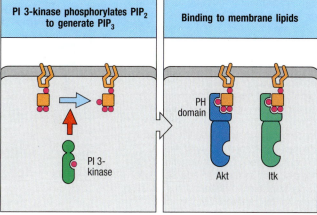

| Binding to phosphorylated sites on a membrane-associated protein | Recognition of activated small G proteins | PI 3-kinase phosphorylates PIP$_2$ to generate PIP$_3$ | Binding to membrane lipids |

Fig. 7.5 **Signaling proteins can be recruited to the membrane in a variety of ways.** Recruitment of signaling proteins to the plasma membrane is important in signal propagation because this is where receptors are usually located. Left panel: tyrosine phosphorylation of membrane-associated proteins, such as the scaffold LAT, recruits phosphotyrosine-binding proteins. This can also protect the scaffold from dephosphorylation by tyrosine phosphatases, which inhibit signaling. Second panel: small G proteins such as Ras can associate with the membrane by having lipid modifications (shown in red). When activated, they can bind a variety of signaling proteins. Right two panels: modifications to the membrane itself that result from receptor activation can recruit signaling proteins. In this example the membrane lipid PIP$_3$ has been produced in the inner leaflet of the plasma membrane by the phosphorylation of PIP$_2$ by PI 3-kinase. PIP$_3$ is recognized by the PH domains of signaling proteins such as the protein kinases Akt and Itk.

membrane due to their fatty acid modifications. Once activated by exchange of GDP for GTP, the activated GTPases bind to signaling proteins such as Sos, relocalizing the bound proteins to the plasma membrane (see Fig. 7.5).

Another way in which receptors can recruit signaling molecules to the plasma membrane is by the local production of modified membrane lipids. These lipids are produced by phosphorylation of the membrane phospholipid phosphatidylinositol by enzymes known as **phosphatidylinositol kinases**, which are activated as a result of receptor signaling. The inositol headgroup of phosphatidylinositol is a carbohydrate ring that can be phosphorylated at one or more positions to generate a wide variety of derivatives. The ones that we will focus on in this chapter are phosphatidylinositol 4,5-bisphosphate (**PIP$_2$**) and phosphatidylinositol 3,4,5-trisphosphate (**PIP$_3$**), the latter of which is generated from PIP$_2$ by the enzyme **phosphatidylinositol 3-kinase** (**PI 3-kinase**) (see Fig. 7.5). PI 3-kinase is often recruited by binding of the SH2 domain of its regulatory subunit to phosphotyrosines in a receptor tail, bringing its catalytic subunit into proximity with inositol phospholipid substrates in the membrane. In this way, membrane phosphoinositides such as PIP$_3$ are rapidly produced after receptor activation. This, combined with their short life span, makes them ideal signaling molecules. PIP$_3$ is recognized specifically by proteins containing a pleckstrin homology (PH) domain or, less commonly, a PX domain (see Fig. 7.2), and one of its functions is to recruit such proteins to the membrane and in some cases contribute to enzyme activation.

7-5 Post-translational modifications of proteins can both activate and inhibit signaling responses.

Protein phosphorylation is a common mechanism for transducing signals from cellular receptors to downstream pathways. These signals are terminated by the action of protein phosphatases, which dephosphorylate signaling intermediates (Fig. 7.6). The importance of protein phosphatases in terminating signaling is underscored by the existence of diseases, such as autoimmunity and cancer, which may result from absent or deficient protein phosphatase activity. However, protein dephosphorylation can also function as a mechanism of activation. Dephosphorylation can regulate protein–protein interactions, protein subcellular localization, or nucleic acid binding, thereby promoting downstream signaling events.

Another general mechanism of protein regulation by post-translational modification is the covalent attachment of one or more molecules of the small protein **ubiquitin**. Ubiquitination is a potent means of signal termination, as it often leads to protein degradation. Ubiquitin is attached by its carboxy-terminal glycine to lysine residues of target proteins in a multi-step process.

Fig. 7.6 Signaling must be turned off as well as turned on. The inability to terminate a signaling pathway can result in serious diseases such as autoimmunity or cancer. As a significant proportion of signaling events depend on protein phosphorylation, protein phosphatases, such as SHP, have an important part in shutting down signaling pathways (left panel). Another common mechanism for terminating signaling is regulated protein degradation (center and right panels). Phosphorylated proteins recruit ubiquitin ligases, such as Cbl, that add the small protein ubiquitin to proteins, thus targeting them for degradation. Cytoplasmic proteins are targeted for destruction in the proteasome by the addition of polyubiquitin chains, linked through lysine 48 (K48) of ubiquitin (center panel). Membrane receptors that become ubiquitinated by individual ubiquitin molecules or di-ubiquitin are internalized and transported to the lysosome for destruction (right panel).

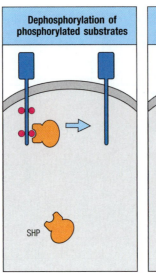
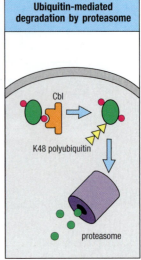
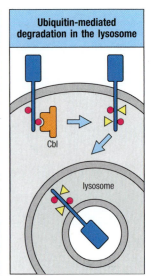

First, an E1 ubiquitin-activating enzyme promotes attachment of ubiquitin to an E2 ubiquitin-conjugating enzyme. The ubiquitin is then transferred to the protein substrate by an enzyme known as an E3 **ubiquitin ligase**. Ubiquitin ligases can continue to add ubiquitin molecules to form polyubiquitin. Importantly, different ubiquitin ligases add the carboxy terminus of one ubiquitin molecule to different lysine residues of the conjugated ubiquitin, typically either lysine 48 (K48) or lysine 63 (K63). These different forms of polyubiquitin produce divergent outcomes for signaling pathways.

When polyubiquitin chains are formed using K48 linkages, the outcome is to target the protein for degradation by the proteasome (see Fig 7.6). An important ubiquitin ligase of this kind in lymphocytes is Cbl, which selects its targets via its SH2 domain. Cbl can thus bind to specific tyrosine-phosphorylated targets, causing them to become ubiquitinated via K48 linkages. Proteins that recognize this form of polyubiquitin then target the ubiquitinated proteins to degradative pathways via the proteasome. Membrane proteins such as receptors can be tagged by single ubiquitin molecules or by di-ubiquitin. These are not recognized by the proteasome, but instead, are recognized by specific ubiquitin-binding proteins that target proteins for degradation in lysosomes (see Fig. 7.6). Thus, ubiquitination of proteins can inhibit signaling. Unlike phosphatases, where the mechanism of inhibition is reversible, inhibition by ubiquitin-mediated protein degradation is a more permanent means of terminating signaling.

Ubiquitination can also be used to activate signaling pathways. We have already discussed this aspect in Section 3-7 in connection with the NFκB signaling pathway from TLRs. There, the ubiquitin ligase TRAF-6 produces K63-linked polyubiquitin chains on TRAF-6 and NEMO. In lymphocytes, K63-linked polyubiquitination is a key step in signaling through tumor necrosis factor (TNF) receptor family members, as will be discussed in Section 7-23 (and Fig. 7.31). This form of polyubiquitin is recognized by specific domains in signaling proteins that recruit additional signaling molecules to the pathway (see Fig. 3.15).

7-6   The activation of some receptors generates small-molecule second messengers.

In many cases, intracellular signaling involves the activation of enzymes that produce small-molecule biochemical mediators known as **second messengers** (Fig. 7.7). These mediators can diffuse throughout the cell, enabling the

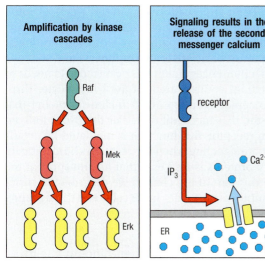

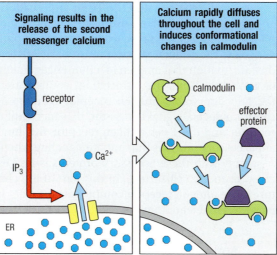

| Amplification by kinase cascades | Signaling results in the release of the second messenger calcium | Calcium rapidly diffuses throughout the cell and induces conformational changes in calmodulin |

Fig. 7.7 Signaling pathways amplify the initial signal. Amplification of the initial signal is an important element of most signal transduction pathways. One means of amplification is a kinase cascade (left panel), in which protein kinases successively phosphorylate and activate each other. In this example, taken from a commonly used kinase cascade (see Fig. 7.19), activation of the kinase Raf results in the phosphorylation and activation of a second kinase, Mek, that phosphorylates yet another kinase, Erk. As each kinase can phosphorylate many different substrate molecules, the signal is amplified at each step, resulting in a huge amplification of the initial signal. Another method of signal amplification is the generation of second messengers (center and right panels). In this example, signaling results in the release of the second messenger calcium (Ca^{2+}) from intracellular stores into the cytosol or its influx from the extracellular environment. Ca^{2+} release from the endoplasmic reticulum (ER) is shown here. The sharp increase in free Ca^{2+} in the cytoplasm can potentially activate many downstream signaling molecules, such as the calcium-binding protein calmodulin. Calcium binding induces a conformational change in calmodulin, which allows it to bind to and regulate a variety of effector proteins.

signal to activate a variety of target proteins. The enzymatic production of second messengers also serves the dual purpose of achieving concentrations of them sufficient to activate the next stage of the pathway and of amplifying the signaling cascade. The second messengers generated by receptors that signal via tyrosine kinases include calcium ions (Ca^{2+}) and a variety of membrane lipids and their soluble derivatives. Although some of these lipid messengers are confined to membranes, they can move within them. A second messenger binding to its target protein typically induces a conformational change that allows the protein to be activated.

Summary.

Cell-surface receptors serve as the front line of a cell's interaction with its environment; they sense extracellular events and convert them into biochemical signals for the cell. As most receptors sit in the plasma membrane, a critical step in the transduction of extracellular signals to the interior of the cell is recruitment of intracellular proteins to the membrane and changes in the composition of the membrane surrounding the receptor. Many immune receptors operate by activating tyrosine kinases to transmit their signals onward, often using scaffolds and adaptors to form large multiprotein signaling complexes. Both the qualitative and quantitative changes that take place in the composition of these signaling complexes determine the character of the response and biological outcomes. Formation of signaling complexes is mediated by a wide variety of protein-interaction domains, or modules, including the SH2, SH3, and PH domains found in proteins. In many cases, the increase in enzymatically produced small-molecule signaling intermediates called second messengers regulates and amplifies the signaling cascade. Termination of signaling involves protein dephosphorylation as well as ubiquitin-mediated protein degradation.

Antigen receptor signaling and lymphocyte activation.

The ability of T cells and B cells to recognize and respond to their specific antigen is central to adaptive immunity. As described in Chapters 4 and 5, the B-cell antigen receptor (BCR) and the T-cell antigen receptor (TCR) are made up of antigen-binding chains—the heavy and light immunoglobulin chains

T-cell receptor complex

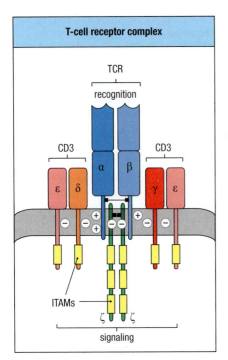

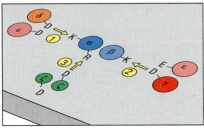

Fig. 7.8 The T-cell receptor complex is made up of variable antigen-recognition proteins and invariant signaling proteins. Upper panel: the functional T-cell receptor (TCR) complex is composed of the antigen-binding TCRα:β heterodimer associated with six signaling chains: two ε, one δ, and one γ collectively called CD3, and a homodimer of ζ. Cell-surface expression of the antigen-binding chains requires assembly of TCRα:β with the signaling subunits. Each CD3 chain has one immunoreceptor tyrosine-based activation motif (ITAM), shown as a yellow segment, whereas each ζ chain has three. The transmembrane regions of each chain have unusual acidic or basic residues as shown. Lower panel: the transmembrane regions of the various TCR subunits are represented in cross-section. It is thought that one of the positive charges, from a lysine (K) of the α chain, interacts with the two negative charges of aspartic acid (D) of the CD3δ:ε dimer, while the other positive charge, of arginine (R), interacts with aspartic acids of the ζ homodimer. The positive arginine (K) charge of the β chain interacts with the negative charges of aspartic acid and glutamic acid (E) in the CD3γ:ε dimer.

in the B-cell receptor, and the TCRα and TCRβ chains in the T-cell receptor. These variable chains have exquisite specificity for antigen, allowing each lymphocyte to detect the presence of one type of pathogen. However, binding of antigen to the antigen receptor is not sufficient for a lymphocyte to respond—the information that antigen receptor engagement has occurred also needs to be transduced into the intracellular compartment of the lymphocyte. Thus, the fully functional antigen receptor complex must include proteins that can transduce a signal across the plasma membrane. For the B-cell antigen receptor and the T-cell antigen receptor, this function is mediated by invariant accessory proteins that initiate signaling when the receptors bind antigen. Assembly with these accessory proteins is also essential for transport of the receptor to the cell surface. In this part of the chapter we describe the structure of the antigen receptor complexes on T cells and B cells, and the signaling pathways that lead from them.

7-7   Antigen receptors consist of variable antigen-binding chains associated with invariant chains that carry out the signaling function of the receptor.

In T cells, the highly variable TCRα:β heterodimer (see Chapter 5) is not sufficient on its own to make up a complete cell-surface antigen receptor. When cells were transfected with cDNAs encoding the TCRα and TCRβ chains, the heterodimers formed were degraded and did not appear on the cell surface. This implied that other molecules are required for the T-cell receptor to be expressed on the cell surface. In the T-cell receptor, the other required molecules are the **CD3γ**, **CD3δ**, and **CD3ε** chains, which together form the **CD3 complex**; and the **ζ chain**, which is present as a disulfide-linked homodimer (**Fig. 7.8**). The CD3 proteins have an extracellular immunoglobulin-like domain, whereas the ζ chain contains only a short extracellular domain. Throughout the remainder of the chapter, we will use the term T-cell receptor to refer to the entire T-cell receptor complex, including these associated signaling subunits.

Although the exact stoichiometry of the complete T-cell receptor is not definitively established, it is thought that the receptor α chain interacts with one CD3δ:CD3ε dimer and one ζ dimer, while the receptor β chain interacts with one CD3γ:CD3ε dimer (see Fig. 7.8). These interactions are mediated by reciprocal charge interactions between basic and acidic intramembrane amino acids of the receptor subunits. There are two positive charges in the TCRα transmembrane region and one in the TCRβ transmembrane domain. Negative charges in the CD3 and ζ transmembrane domains interact with the positive charges in α and β. Assembly of CD3 and a ζ dimer with the α:β heterodimer stabilizes the α and β dimer during its production in the endoplasmic reticulum and allows the complex to be transported to the plasma membrane. These associations ensure that all T-cell receptors present on the plasma membrane are properly assembled.

Signaling from the T-cell receptor is initiated by tyrosine phosphorylation within cytoplasmic regions called **immunoreceptor tyrosine-based activation motifs (ITAMs)** in the CD3ε, γ, δ, and ζ chains. CD3γ, δ, and ε each contain one ITAM, and each ζ chain contains three, giving the T-cell receptor a total of 10 ITAMs. This motif is also present in the signaling chains of the B-cell receptor and in the NK-cell receptors described in Chapter 3, as well as in the receptors for the immunoglobulin constant region (Fc receptors) that are present on mast cells, macrophages, monocytes, neutrophils, and NK cells (see Section 7-11 below).

Each ITAM contains two tyrosine residues that become phosphorylated by specific protein tyrosine kinases when the receptor binds its ligand, providing

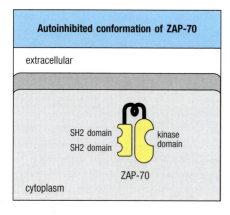

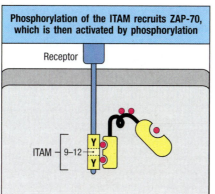

Fig. 7.9 ITAMs recruit signaling proteins that have tandem SH2 domains. The ITAMs of the T-cell receptor (TCR) and B-cell receptor (BCR) contain tyrosine residues within the motif …YXX[L/I] X_{6-9}YXX[L/I]…. The spacing between the tyrosines is important in binding to tandem SH2-containing proteins such as Syk and ZAP-70. Left panel: prior to TCR or BCR stimulation, these kinases are in an inactive conformation, known as the autoinhibited conformation. The autoinhibited conformation is stabilized by interactions between the tandem SH2 domain-kinase domain linker region and the kinase domain that hold the enzyme in a catalytically inactive state. Right panel: after phosphorylation of both tyrosines within one ITAM (here depicted as two tyrosines separated by 9-12 amino acids), the tandem SH2 domains of Syk or ZAP-70 can dock cooperatively to both phosphotyrosines, as shown here for ZAP-70. By being recruited into the active signaling complex, ZAP-70 can itself be phosphorylated, so that it becomes an active kinase that can then phosphorylate its substrates. This final step of ZAP-70 activation requires phosphorylation on two tyrosines in the linker region between the tandem SH2 domains and the kinase domain, together with phosphorylation of a tyrosine in the catalytic site of the kinase domain.

sites for the recruitment of the SH2 domains of signaling proteins as described earlier in the chapter. Two YXXL/I motifs are separated by about six to nine amino acids within each ITAM, so the canonical ITAM sequence is …YXX[L/I] X_{6-9}YXX[L/I]…, where Y is tyrosine, L is leucine, I is isoleucine, and X represents any amino acid. The two tyrosines of the ITAM make it particularly efficient in recruiting signaling proteins that contain two tandem SH2 domains (Fig. 7.9). When both tyrosines of the ITAM are phosphorylated, tandem SH2 domain-containing proteins such as Syk or ZAP-70 are recruited. This leads to the phosphorylation of Syk or ZAP-70, an essential step in the activation of these kinases, as will be discussed further below (see Section 7-10).

The antigen-binding immunoglobulin on the B-cell surface is also associated with invariant protein chains that mediate signaling. These two polypeptides, called **Igα** and **Igβ**, are required for transport of the receptor to the surface and for B-cell receptor signaling (Fig. 7.10). Igα and Igβ are single-chain proteins composed of an extracellular immunoglobulin-like domain connected by a transmembrane domain to a cytoplasmic tail. They form a disulfide-linked heterodimer that becomes associated with immunoglobulin heavy chains and enables their transport to the cell surface. The Igα:Igβ dimer associates with the B-cell receptor through hydrophilic rather than charge interactions between their transmembrane regions. The complete B-cell receptor is thought to be a complex of six chains—two identical light chains, two identical heavy chains, and one associated heterodimer of Igα and Igβ. Like CD3 and the ζ chains of the T-cell receptor, Igα and Igβ have ITAMs, but just one each, and these are essential for the ability of the B-cell receptor to signal.

7-8 Antigen recognition by the T-cell receptor and its co-receptors transduces a signal across the plasma membrane to initiate signaling.

To make an effective immune response, T cells and B cells must be able to respond to their specific antigen even when it is present at extremely low levels. This is especially important for T cells, as an antigen-presenting cell will display on its surface many different peptides from both self and foreign proteins, and the number of peptide:MHC complexes specific for a particular T-cell receptor is likely to be very low. Some estimates suggest that a naive CD4 T

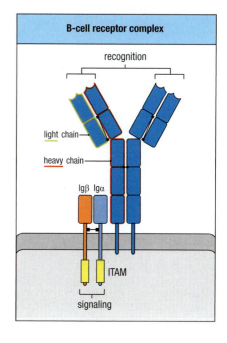

Fig. 7.10 The B-cell receptor complex is made up of cell-surface immunoglobulin with one each of the invariant signaling proteins Igα and Igβ. The immunoglobulin recognizes and binds antigen but cannot itself generate a signal. It is associated with antigen-nonspecific signaling molecules—Igα and Igβ. These each have a single ITAM (yellow segment) in their cytosolic tails that enables them to signal when the B-cell receptor is ligated with antigen. Igα and Igβ form a disulfide-linked heterodimer that is non-covalently associated with the heavy chains.

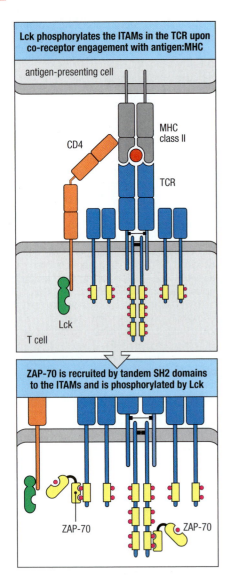

Lck phosphorylates the ITAMs in the TCR upon co-receptor engagement with antigen:MHC

antigen-presenting cell

CD4

MHC class II

TCR

Lck

T cell

ZAP-70 is recruited by tandem SH2 domains to the ITAMs and is phosphorylated by Lck

ZAP-70

ZAP-70

Fig. 7.11 Engagement of co-receptors with the T-cell receptor enhances phosphorylation of the ITAMs. Upper panel: for simplicity, we show the CD4 co-receptor engaging the same MHC molecule as the T-cell receptor (TCR), although signaling within receptor microclusters may differ from this arrangement. When T-cell receptors and co-receptors are brought together by binding to peptide:MHC complexes on the surface of an antigen-presenting cell, recruitment of the co-receptor-associated kinase Lck leads to phosphorylation of ITAMs in CD3γ, δ, and ε, and in the ζ chains. Lower panel: the tyrosine kinase ZAP-70 binds to phosphorylated ITAMs through its SH2 domains, enabling ZAP-70 to be phosphorylated and activated by Lck. ZAP-70 then phosphorylates other intracellular signaling molecules.

cell can become activated by fewer than 50 antigenic peptide:MHC complexes displayed by an antigen-presenting cell, and effector CD8 cytotoxic T cells may be even more sensitive. B cells become activated when about 20 B-cell receptors are engaged. These estimates based on *in vitro* studies may not be precise for cells *in vivo*, but it is clear that the antigen receptors on T cells and B cells confer remarkable sensitivity to antigen.

For a peptide:MHC complex to activate T cells, it must bind directly to the T-cell receptor (Fig. 7.11, upper panel, and see Fig. 4.22). However, it remains unclear precisely how this extracellular recognition event is transmitted across the T-cell membrane to initiate signaling. Questions remain as to the stoichiometry and physical arrangement of T-cell receptors and peptide:MHC complexes required to initiate the signaling cascade. We will discuss this area of active research only briefly before moving on to explain well-understood intracellular events that occur after antigen recognition.

One suggestion is that signaling is initiated by T-cell receptor dimerization through formation of **'pseudo-dimeric' peptide:MHC complexes** containing one antigen peptide:MHC molecule and one self peptide:MHC molecule on the surface of the antigen-presenting cell. This model relies on a weak interaction occurring between the T-cell receptor and self peptide:MHC complexes, stabilized by the CD4 or CD8 co-receptor interactions with the self peptide:MHC complexes, but it would explain signaling induced by low densities of antigenic peptides. An additional possibility is that the antigenic peptide:MHC complex induces a conformational change in the T-cell receptor, or its associated CD3 and ζ chains, that promotes receptor phosphorylation; however, direct structural evidence supporting this model is still lacking.

It has also been suggested that signaling might involve receptor oligomerization, or clustering, as antibodies that bind to and cross-link T-cell receptors can activate T cells. Since antigenic peptides are vastly outnumbered by other peptides displayed on the surface of the antigen-presenting cell, it is unlikely that physiologic amounts of antigen induce conventional oligomerization as observed with antibodies. However, assemblies of small numbers of T-cell receptors called **microclusters** have been observed in the zone of contact between the T cell and the antigen-presenting cell. These microclusters form rapidly following TCR stimulation, and quickly merge with microclusters containing downstream signaling components, such as scaffolds and adaptors. Current evidence indicates that signaling is initiated in these microclusters. One popular model proposes that signal initiation takes place when inhibitory signaling proteins are excluded from these complexes. A key component of this model is that prior to TCR signaling, activating and inhibitory enzymes are in a balanced equilibrium; the initiation of signaling occurs when this equilibrium is perturbed in favor of the activating modifications.

7-9 Antigen recognition by the T-cell receptor and its co-receptors leads to phosphorylation of ITAMs by Src-family kinases, generating the first intracellular signal in a signaling cascade.

The first intracellular signal generated after the T cell has detected its specific antigen is the phosphorylation of both tyrosines in the ITAMs of the T-cell receptor. This signal is initiated with the help of the CD4 and CD8 co-receptors, which bind to MHC class II molecules and class I molecules, respectively, via their extracellular domains (see Section 4-18), and associate with nonreceptor kinases via their intracellular domains. The Src-family kinase **Lck** is constitutively associated with the cytoplasmic domains of CD4 and CD8 and is thought to be the kinase primarily responsible for phosphorylation of the ITAMs of the T-cell receptor (see Fig. 7.11). Evidence suggests that binding of the co-receptor to the peptide:MHC complex that binds the T-cell receptor enhances the recruitment of Lck to the engaged T-cell receptor, leading to more efficient phosphorylation of the T-cell receptor ITAMs. The importance of this event is demonstrated by the profound reduction in T-cell development in Lck-deficient mice. This indicates the essential role of Lck in T-cell receptor signaling during the selection of developing T cells in the thymus (discussed in Chapter 8). Lck is important for T-cell receptor signaling in naive T cells and effector T cells, but is less important for the activation or maintenance of memory CD8 T cells by their specific antigen. A related tyrosine kinase, **Fyn**, is weakly associated with the ITAMs of the T-cell receptor and may have some role in signaling. Whereas mice lacking Fyn develop normal CD4 and CD8 T cells that respond in essentially normal fashion to antigen, mice lacking both Lck and Fyn show a more complete loss of T-cell development than mice lacking Lck alone.

Another role of the co-receptors in T-cell receptor signaling may be to stabilize interactions between the receptor and the peptide:MHC complex. Affinities of individual receptors for their specific peptide:MHC complexes are in the micromolar range, which means that the T-cell receptor:peptide:MHC complexes have half-lives of less than 1 second and dissociate rapidly. The additional binding of a co-receptor to the MHC molecule is thought to stabilize the interaction by increasing its duration, thereby providing time for an intracellular signal to be generated.

The Lck bound to the cytoplasmic tails of CD4 or CD8 is brought near its substrate ITAMs in the T-cell receptor when the co-receptor binds the receptor:peptide:MHC complex (see Fig. 7.11). Lck's activity is also regulated allosterically by phosphorylation of a tyrosine in its carboxy terminus by the **C-terminal Src kinase** (**Csk**). The resulting phosphotyrosine interacts with Lck's SH2 domain and helps maintain Lck in a closed conformation, resulting in a catalytically inactive state (**Fig. 7.12**). The absence of Csk during T-cell development causes T cells to mature autonomously in the thymus without needing to bind peptide:MHC, presumably as a result of abnormal activation of TCR signaling by hyperactive Lck in Csk-deficient thymocytes. This suggests that Csk normally acts to reduce Lck activity and to attenuate TCR signaling. Dephosphorylation of the tyrosine or engagement of the SH2 or SH3 domains by their ligands releases Lck from its inactive conformation, resulting in a primed, but not fully active, Lck kinase (see Fig. 7.12). Full activation of catalytic activity also requires Lck autophosphorylation of a tyrosine in its kinase domain. In nonstimulated lymphocytes, phosphorylation of Lck is counteracted by the **tyrosine phosphatase CD45**, which can dephosphorylate both of the Lck tyrosine phosphorylation sites. Prior to TCR stimulation, multiple phosphorylated species of Lck are found in T cells, but antigen receptor stimulation is required to stabilize the activated form of Lck and lead to ITAM phosphorylation.

Fig. 7.12 Lck activity is regulated by tyrosine phosphorylation and dephosphorylation. Src kinases such as Lck contain SH3 (blue) and SH2 (orange) domains preceding the kinase domain (green). Lck also contains a unique amino-terminal motif (yellow) with two cysteine residues that bind a Zn ion that is also bound to a similar motif in the cytoplasmic domain of CD4 or CD8. Upper panel: in inactive Lck, the two lobes of the kinase domain are constrained by interactions with both the SH2 and SH3 domains. The SH2 domain interacts with a phosphorylated tyrosine (red) at the carboxy-terminal end of the kinase domain. The SH3 domain interacts with a proline-rich sequence in the linker between the SH2 domain and the kinase domain. Middle panel: dephosphorylation of the carboxy-terminal tyrosine by the phosphatase CD45 (not shown) releases the SH2 domain. Binding of other ligands to the SH3 region can facilitate release of the linker region (not shown). In this state, Lck is considered primed, but not fully activated. Lower panel: full activation of Lck catalytic activity requires autophosphorylation on the activation loop in the kinase domain. Active Lck can then phosphorylate ITAMs in the signaling chains of the nearby T-cell receptor. Rephosphorylation of the carboxy-terminal tyrosine by the C-terminal Src kinase (Csk) or loss of the SH3 ligand returns Lck to the inactive state.

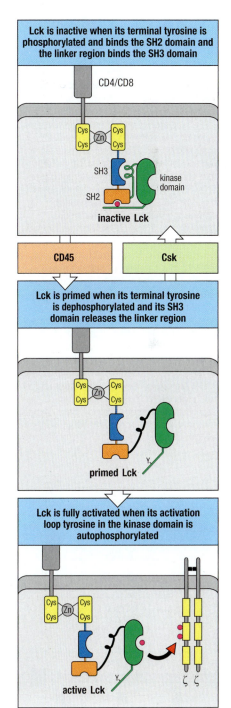

Lck is inactive when its terminal tyrosine is phosphorylated and binds the SH2 domain and the linker region binds the SH3 domain

CD4/CD8

Cys Cys
Zn
Cys Cys

SH3
SH2
kinase domain

inactive Lck

CD45   **Csk**

Lck is primed when its terminal tyrosine is dephosphorylated and its SH3 domain releases the linker region

Cys Cys
Zn
Cys Cys

primed Lck

Lck is fully activated when its activation loop tyrosine in the kinase domain is autophosphorylated

Cys Cys
Zn
Cys Cys

active Lck

ζ ζ

Structure of autoinhibited ZAP-70

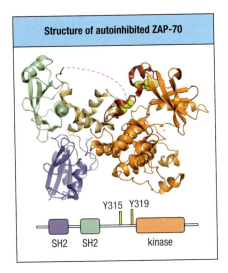

Fig. 7.13 Structure of the autoinhibited ZAP-70 kinase. The structure of the inactive autoinhibited conformation of ZAP-70 is shown with the protein domains color-coded according to the domain map shown at the bottom; the dashed red line indicates a region of the protein that was not detected in the structural analysis. Prior to T-cell receptor stimulation, the ZAP-70 kinase is in this inactive conformation, based on interactions between the tandem SH2 domain-kinase domain linker region (thick red line) and the kinase domain. This interaction stabilizes ZAP-70 in a catalytically inactive state, referred to as the autoinhibited form of ZAP-70, by locking the kinase domain in an inactive conformation. Following T-cell receptor stimulation, Lck phosphorylates two tyrosine residues in this linker region, Y315 and Y319, shown in yellow. Lck also phosphorylates a tyrosine residue in the catalytic (kinase) domain. When Y315 and Y319 are phosphorylated, the linker region no longer binds to the kinase domain, allowing the phosphorylated kinase domain to adopt an active conformation. Courtesy of Arthur Weiss.

7-10   Phosphorylated ITAMs recruit and activate the tyrosine kinase ZAP-70.

The precise spacing of the two YXXL/I motifs in an ITAM suggests that the ITAM is a binding site for a signaling protein with two SH2 domains. In the case of the T-cell receptor, this protein is the tyrosine kinase **ZAP-70** (ζ**-chain-associated protein**), which carries the activation signal onward. ZAP-70 has two tandem SH2 domains that can be simultaneously engaged by the two phosphorylated tyrosines in an ITAM (see Fig. 7.9). The affinity of the phosphorylated YXXL sequence for a single SH2 domain is low; binding of both SH2 domains to the ITAM is significantly stronger and confers specificity on ZAP-70 binding. Thus, when Lck has sufficiently phosphorylated an ITAM in the T-cell receptor, ZAP-70 binds to it. Once bound, ZAP-70 is phosphorylated by Lck at three tyrosine residues, two in the linker region between the tandem SH2 domains and the kinase domain, and a third residue in the catalytic domain. Together these phosphorylations activate ZAP-70 by disrupting the autoinhibited form of inactive ZAP-70, allowing the ZAP-70 kinase domain to adopt the active conformation (Fig. 7.13). ZAP-70 can also be activated by autophosphorylation.

7-11   ITAMs are also found in other receptors on leukocytes that signal for cell activation.

The signaling subunits of the T-cell receptor and the B-cell receptor each contain ITAMs, which are essential for T-cell receptor and B-cell receptor signaling. Phosphorylation of both tyrosines in the ITAM functions to recruit a tyrosine kinase with tandem SH2 domains—ZAP-70 in the case of T cells, and a closely related kinase, Syk, in B cells. Other immune-system receptors also use ITAM-containing accessory chains to transduce activating signals (Fig. 7.14). One example is **FcγRIII** (CD16); this is a receptor for IgG that triggers antibody-dependent cell-mediated cytotoxicity (ADCC) by NK cells, which we consider in Chapter 11; CD16 is also found on macrophages and neutrophils, where it facilitates the uptake and destruction of antibody-bound pathogens. To signal, FcγRIII must associate either with the ζ chain found also in the T-cell receptor or with another member of the same protein family known as the Fcγ chain. The Fcγ chain is also the signaling component of another Fc receptor—the Fcε receptor I (FcεRI) on mast cells. As we discuss in Chapter 14, this receptor binds IgE antibodies, and on cross-linking by allergens, it triggers the degranulation of mast cells. Last, many activating receptors on NK cells are associated with DAP12, another ITAM-containing protein (see Section 3-26). Each of these additional ITAM-containing signaling chains becomes tyrosine phosphorylated following stimulation of its associated receptor, leading to the recruitment of a tyrosine kinase, either Syk or ZAP-70. With the exception of

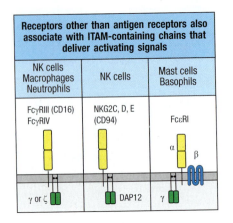

Receptors other than antigen receptors also associate with ITAM-containing chains that deliver activating signals		
NK cells Macrophages Neutrophils	NK cells	Mast cells Basophils
FcγRIII (CD16) FcγRIV	NKG2C, D, E (CD94)	FcεRI
γ or ζ	DAP12	γ

Fig. 7.14 Other receptors that pair with ITAM-containing chains can deliver activating signals. Cells other than B and T cells have receptors that pair with accessory chains containing ITAMs, which are phosphorylated when the receptor is cross-linked. These receptors deliver activating signals. The Fcγ receptor III (FcγRIII, or CD16) is found on NK cells, macrophages, and neutrophils. Binding of IgG to this receptor activates the killing function of the NK cell, leading to the process known as antibody-dependent cell-mediated cytotoxicity (ADCC). Activating receptors on NK cells, such as NKG2C, NKG2D, and NKG2E, also associate with ITAM-containing signaling chains. The Fcε receptor (FcεRI) is found on mast cells and basophils. The α subunit binds to IgE antibodies with very high affinity. The β subunit is a four-spanning transmembrane protein. When antigen subsequently binds to the IgE, the mast cell is triggered to release granules containing inflammatory mediators. The γ chain associated with the Fc receptors, and the DAP12 chain that associates with the NK killer-activating receptors, also contain one ITAM per chain and are present as homodimers.

T cells, Syk is broadly expressed in all leukocyte subsets; in contrast, ZAP-70 has been found only in T cells and NK cells.

Several viral pathogens seem to have acquired ITAM-containing receptors from their hosts. These include the Epstein–Barr virus (EBV), whose *LMP2A* gene encodes a membrane protein with a cytoplasmic tail containing an ITAM. This enables EBV to trigger B-cell proliferation by the signaling pathways discussed in Section 7-20, an important step in the development of EBV-induced malignancies. Another virus that expresses an ITAM-containing protein is the Kaposi sarcoma herpesvirus (KSHV or HHV8), which causes malignant transformation and proliferation of the cells it infects.

▶ **MOVIE 7.1**

7-12 Activated ZAP-70 phosphorylates scaffold proteins and promotes PI 3-kinase activation.

As described in Section 7-10, phosphorylation of the tyrosines in the T-cell receptor ITAMs leads to the recruitment and activation of ZAP-70. This provides ZAP-70 proximity to the cell membrane, where it phosphorylates the scaffold protein **LAT** (**linker for activated T cells**), a transmembrane protein with a large cytoplasmic domain (**Fig. 7.15**). ZAP-70 also phosphorylates another adaptor protein, **SLP-76**. LAT and SLP-76 can be linked by the adaptor protein Gads; this three-protein complex, referred to as the LAT:Gads:SLP-76 complex, plays a central role in T-cell activation. This is illustrated by the profound TCR signaling and T-cell development defects seen in mice lacking any

Fig. 7.15 ZAP-70 phosphorylates LAT and SLP-76, initiating four downstream signaling modules. Activated ZAP-70 phosphorylates the scaffold proteins LAT and SLP-76 and recruits them to the activated T-cell receptor (TCR) complex. An adaptor protein, Gads, holds the tyrosine-phosphorylated LAT and SLP-76 together. The multiple binding sites on these scaffold proteins recruit several additional adaptors and enzymes that initiate four essential downstream signaling modules. One key component required for several of these modules is the activation of PI 3-kinase, which phosphorylates PIP_2 in the plasma membrane to generate PIP_3. These four modules include the activation of the serine/threonine kinase Akt, which promotes enhanced cellular metabolic activity; the activation of PLC-γ, which leads to transcription factor activation; the activation of Vav, which induces actin polymerization and cytoskeletal reorganization; and the recruitment of ADAP, an adaptor protein that promotes enhanced integrin adhesiveness and clustering.

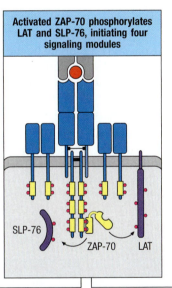

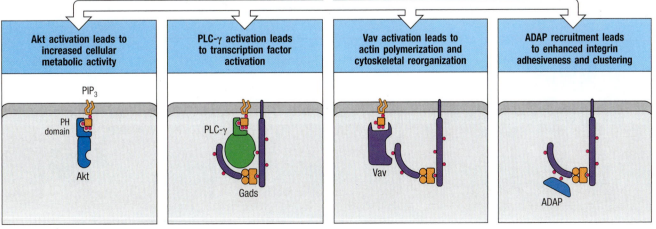

one of these components, and in humans lacking ZAP-70. A second essential event that occurs rapidly following ZAP-70 activation is the recruitment and activation of the enzyme PI 3-kinase (see Section 7.4); while the detailed mechanism linking PI 3-kinase activation to T-cell receptor stimulation is not well understood, current evidence suggests a role for the small GTPase Ras. In this case, Ras may be activated by recruitment of the Ras-GEF Sos to LAT via Sos binding to the small adaptor protein Grb2, forming a second three-protein complex containing LAT and Sos, bridged by Grb2.

Following formation of the LAT:Gads:SLP-76 complex and the activation of PI 3-kinase, the T-cell receptor signaling pathway branches into several downstream modules, each of which induces cellular changes that contribute to optimal T-cell activation (see Figure 7.15). Each module is initiated by the recruitment of a key intermediate to the active signaling complexes, either via binding to the LAT:Gads:SLP-76 complex, to the PIP$_3$ generated by PI 3-kinase, or to both. In brief, these modules lead to the activation of phospholipase C-γ (PLC-γ), which affects transcription; the activation of the serine/threonine kinase Akt, which affects metabolism, among other things; the recruitment of the adaptor protein ADAP, which upregulates cell adhesion; and the activation of the protein Vav, which initiates actin polymerization. Each of these modules will be described in detail in the sections below.

7-13 Activated PLC-γ generates the second messengers diacylglycerol and inositol trisphosphate that lead to transcription factor activation.

One important module of T-cell receptor signaling is the activation of the enzyme **phospholipase C-γ (PLC-γ)**. First, PLC-γ is brought to the inner face of the plasma membrane by the binding of its PH domain to the PIP$_3$ that has been formed by the phosphorylation of PIP$_2$ by PI 3-kinase; it then binds to phosphorylated LAT and SLP-76. The actions of PLC-γ produce two second messengers that activate three distinct terminal branches of the T-cell receptor pathway leading to transcription factor activation.

Due to this crucial role in T-cell activation, PLC-γ activation is controlled at several different levels. Recruitment to the membrane, while necessary, is not sufficient to activate PLC-γ. PLC-γ activation requires phosphorylation by Itk, a member of the Tec family of cytoplasmic tyrosine kinases. Like PLC-γ, Tec kinases contain PH, SH2, and SH3 domains and are recruited to the plasma membrane by interactions via these domains; specifically, the PH domain interacts with PIP$_3$ on the inner face of the membrane (**Fig. 7.16**), and the SH2

Fig. 7.16 The recruitment of phospholipase C-γ by LAT and SLP-76, and its phosphorylation and activation by the protein kinase Itk, are crucial steps in T-cell activation. ZAP-70 phosphorylates the scaffold proteins LAT and SLP-76, which are brought together to form a complex at the activated T-cell receptor by the adaptor protein Gads. This complex also promotes the activation of PI 3-kinase, leading to the production of PIP$_3$ (formed by the phosphorylation of PIP$_2$ by PI 3-kinase). Phospholipase C-γ (PLC-γ) is recruited to the membrane by its PH domain binding to PIP$_3$ and then binds to phosphorylated sites in LAT and to the proline-rich domain of SLP-76. To be activated, PLC-γ must be phosphorylated by the Tec-family kinase Itk. Itk is recruited to the membrane by interaction of its PH domain with PIP$_3$, and by interactions with phosphorylated SLP-76. Once phosphorylated by Itk, phospholipase C-γ is active.

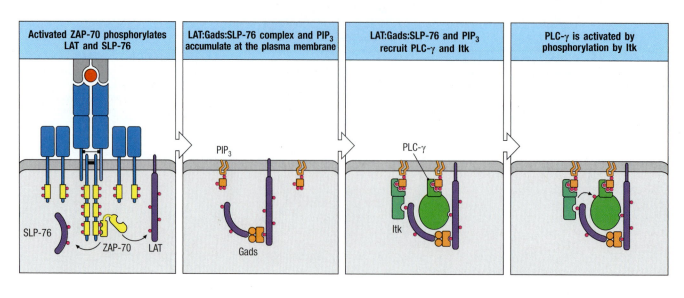

| Activated ZAP-70 phosphorylates LAT and SLP-76 | LAT:Gads:SLP-76 complex and PIP$_3$ accumulate at the plasma membrane | LAT:Gads:SLP-76 and PIP$_3$ recruit PLC-γ and Itk | PLC-γ is activated by phosphorylation by Itk |

and SH3 domains interact with SLP-76. These interactions serve to localize Itk in close proximity to its substrate, PLC-γ.

Once PLC-γ has been recruited to the inner face of the plasma membrane and has been activated, it can catalyze the breakdown of the membrane lipid PIP$_2$ (see Section 7-4 and Fig. 7.5) to generate two products, the membrane lipid **diacylglycerol** (**DAG**) and the diffusible second messenger **inositol 1,4,5-trisphosphate** (**IP$_3$**) (not to be confused with the membrane lipid PIP$_3$) (Fig. 7.17). DAG is confined to the membrane, but diffuses in the plane of the membrane and serves as a molecular target that recruits other signaling molecules to the membrane. IP$_3$ diffuses into the cytosol and binds to IP$_3$ receptors on the endoplasmic reticulum (ER) membrane. These receptors are Ca^{2+} channels, which open and release calcium stored in the ER into the cytosol. The consequent low levels of calcium in the ER then cause a conformational change in the transmembrane protein **STIM1**, promoting its clustering within the ER membrane. The STIM1 oligomers bind to the plasma membrane, where they interact directly with ORAI1, the plasma membrane calcium channel (also known as the **CRAC channel**: calcium release-activated calcium channel). Binding of STIM1 to ORAI1 triggers calcium channel opening, allowing extracellular calcium to flow into the cell to activate further signaling pathways and to replenish ER calcium stores.

The activation of PLC-γ marks an important step in T-cell activation, because after this point the PLC-γ signaling module splits into three distinct branches—the stimulation of Ca^{2+} entry, the activation of Ras, and the activation of **protein kinase C-θ** (**PKC-θ**)—each of which ends in the activation of a different transcription factor. These signaling pathways are utilized by many cell types, in addition to lymphocytes. Their importance in T-cell activation is shown by the observation that treatment of T cells with phorbol myristate acetate (an analog of DAG) and ionomycin (a pore-forming drug that allows extracellular calcium to flow into the cell) can reconstitute many of the effects of T-cell receptor stimulation. Additionally, deficiencies in several of these signaling components, including Lck, Zap-70, Itk, CD45, Carma1, and ORAI1, have been found to be mutated in cases of **severe combined immunodeficiency** (**SCID**).

7-14   Ca^{2+} entry activates the transcription factor NFAT.

One of the three signaling pathways leading from PLC-γ results in an influx of calcium ions into the cytosol. An important outcome of the increased cytosolic Ca^{2+} resulting from T-cell receptor signaling via PLC-γ is the activation of a family of transcription factors called **NFAT** (**nuclear factor of activated T cells**). NFAT is something of a misnomer, because the five members of this family are expressed in many different tissues. In the absence of activating

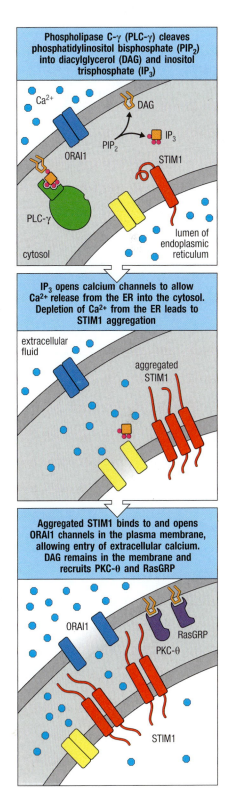

Phospholipase C-γ (PLC-γ) cleaves phosphatidylinositol bisphosphate (PIP$_2$) into diacylglycerol (DAG) and inositol trisphosphate (IP$_3$)

Ca^{2+} DAG IP$_3$ PIP$_2$ ORAI1 STIM1 PLC-γ lumen of endoplasmic reticulum cytosol

IP$_3$ opens calcium channels to allow Ca^{2+} release from the ER into the cytosol. Depletion of Ca^{2+} from the ER leads to STIM1 aggregation

extracellular fluid aggregated STIM1

Aggregated STIM1 binds to and opens ORAI1 channels in the plasma membrane, allowing entry of extracellular calcium. DAG remains in the membrane and recruits PKC-θ and RasGRP

ORAI1 RasGRP PKC-θ STIM1

Fig. 7.17 Phospholipase C-γ cleaves inositol phospholipids to generate two important signaling molecules. Top panel: phosphatidylinositol bisphosphate (PIP$_2$) is a component of the inner leaflet of the plasma membrane. When PLC-γ is activated by phosphorylation, it cleaves PIP$_2$ into two parts: inositol trisphosphate (IP$_3$), which diffuses away from the membrane into the cytosol, and diacylglycerol (DAG), which stays in the membrane. Both of these molecules are important in signaling. Middle panel: there are two phases of calcium release. IP$_3$ binds to a receptor in the endoplasmic reticulum (ER) membrane, opening calcium channels (yellow) and allowing the early phase of calcium ions (Ca^{2+}) to enter the cytosol from the ER. The depletion of Ca^{2+} stores in the ER causes the aggregation of an ER calcium sensor, STIM1. Bottom panel: aggregated STIM1 stimulates the second phase of calcium entry by binding to and opening calcium channels called ORAI1 in the plasma membrane. This further increases cytosolic calcium and restores ER Ca^{2+} stores. DAG binds and recruits signaling proteins to the membrane, most importantly the Ras-GEF called RasGRP and a serine/threonine kinase called protein kinase C-θ (PKC-θ). Recruitment of RasGRP to the plasma membrane activates Ras, and PKC-θ activation results in the activation of the transcription factor NFκB.

Fig. 7.18 The transcription factor NFAT is regulated by calcium signaling. Left panel: NFAT is maintained in the cytoplasm by phosphorylation on serine and threonine residues. Center panel: after antigen receptor stimulation, calcium enters the cytosol, first from the endoplasmic reticulum (not shown; see Fig. 7.17) and later from the extracellular space (shown). After entering the cytosol, calcium binds to calmodulin, and the Ca^{2+}:calmodulin complex binds to the serine/threonine phosphatase calcineurin, activating it to dephosphorylate NFAT. Right panel: once dephosphorylated, NFAT moves into the nucleus, where it binds to promoter elements and activates the transcription of various genes.

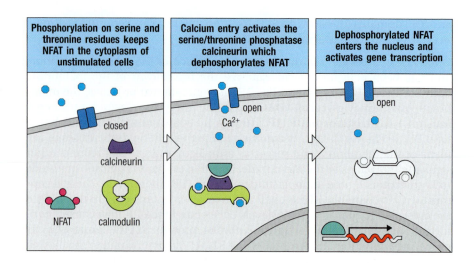

signals, NFAT is kept in the cytosol of resting T cells by phosphorylation on serine/threonine residues. This phosphorylation is mediated by serine/threonine kinases, including glycogen synthase kinase 3 (GSK3) and casein kinase 2 (CK2). When phosphorylated, the nuclear localization sequence of NFAT is not recognized by nuclear transporters, and NFAT is unable to enter into the nucleus (Fig. 7.18).

The cytoplasmic Ca^{2+} resulting from T-cell receptor signaling binds to and induces a conformational change in a protein called **calmodulin**, which is then able to bind to and activate a wide range of different target enzymes. In T cells, an important target of calmodulin is **calcineurin**, a protein phosphatase that acts on NFAT. Dephosphorylation of NFAT by calcineurin allows the nuclear localization sequence to be recognized by nuclear transporters, and NFAT enters the nucleus (see Fig. 7.18). There it participates in turning on many of the genes crucial for T-cell activation, such as the gene for the cytokine interleukin-2 (IL-2).

The importance of NFAT in T-cell activation is illustrated by the effects of selective inhibitors of calcineurin called **cyclosporin A (CsA)** and **tacrolimus** (also known as **FK506**). CsA forms a complex with the protein cyclophilin A, and this complex inhibits calcineurin. Tacrolimus binds a different protein, FK-binding protein (FKBP), making a complex that similarly inhibits calcineurin. By inhibiting calcineurin, these drugs prevent the formation of active NFAT. T cells express low levels of calcineurin, so they are more sensitive to inhibition of this pathway than are many other cell types. Both cyclosporin A and tacrolimus thus act as effective immunosuppressants, and are widely used to prevent the rejection of organ transplants (see Chapter 16, Section 16-3).

7-15 Ras activation stimulates the mitogen-activated protein kinase (MAPK) relay and induces expression of the transcription factor AP-1.

MOVIE 7.2

A second branch of the PLC-γ signaling module is the activation of the small GTPase Ras. This can occur by various means. The most efficient mechanism for Ras activation in T cells is via the DAG generated by PLC-γ, which diffuses in the plasma membrane and activates a variety of proteins. One of these is the protein RasGRP, which is a guanine-nucleotide exchange factor that specifically activates Ras. RasGRP contains a protein-interaction module called a C1 domain that binds to DAG. This interaction recruits RasGRP to the membrane near active signaling complexes (Fig. 7.19), where it activates Ras by promoting the exchange of GDP for GTP. Ras is also activated in the T-cell receptor

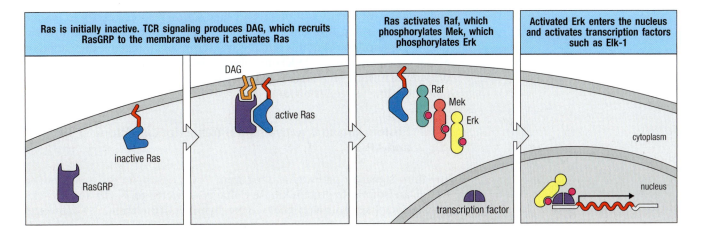

| Ras is initially inactive. TCR signaling produces DAG, which recruits RasGRP to the membrane where it activates Ras | Ras activates Raf, which phosphorylates Mek, which phosphorylates Erk | Activated Erk enters the nucleus and activates transcription factors such as Elk-1 |

signaling pathway by the guanine-exchange factor Sos, which is recruited by the adaptor protein Grb2 (see Sections 7-2 and 7-3), which has itself been recruited by binding to phosphorylated LAT and SLP-76.

Activated Ras then triggers a three-kinase relay that ends in the activation of a serine/threonine kinase known as a **mitogen-activated protein kinase** or **MAP kinase** (**MAPK**) (see Fig. 7.19). In the case of antigen receptor signaling, the first member of the relay is a MAPK kinase kinase (MAP3K) called **Raf**. Raf is a serine/threonine kinase that phosphorylates the next member of the series, a MAPK kinase (MAP2K) called **MEK1**. MEK1 is a dual-specificity protein kinase that phosphorylates a tyrosine and a threonine residue on the last member of the relay, a MAPK which in T cells and B cells is **Erk** (**extracellular signal-related kinase**).

Signaling by MAPK cascades is facilitated by specialized scaffold proteins that bind to all three kinases in a particular MAPK relay, thereby accelerating their interactions. The scaffold protein **kinase suppressor of Ras** (**KSR**) functions in the Raf/MEK1/Erk pathway. During T-cell receptor signaling, KSR associates with Raf, MEK1, and Erk and localizes itself and its cargo to the membrane. In that location, activated Ras can engage with the Raf bound to KSR and trigger the kinase relay (see Fig. 7.19).

An important function of MAPKs is to phosphorylate and activate transcription factors that can then induce new gene expression. Erk acts indirectly to generate the transcription factor **AP-1**, which is a heterodimer composed of one monomer each from the Fos and Jun families of transcription factors (**Fig. 7.20**). Active Erk phosphorylates the transcription factor Elk-1, which

Fig. 7.19 DAG activates MAPK cascades, leading to transcription factor activation. All MAPK cascades are initiated by the activation of small G proteins, such as Ras in this example. Ras is switched from an inactive state (first panel) to an active state (second panel) by a guanine-nucleotide exchange factor (GEF), RasGRP, which is recruited to the membrane by DAG. Ras activates the first enzyme of the cascade, a protein kinase called Raf, a MAPK kinase kinase (MAP3K) (third panel). Raf phosphorylates Mek, a MAP2K, which in turn phosphorylates and activates Erk, a MAPK. The scaffold protein KSR associates with Raf, Mek, and Erk to facilitate their efficient interactions (not shown). Phosphorylation and activation of Erk releases it from the complex so that it can diffuse within the cell and enter the nucleus (fourth panel). Phosphorylation of transcription factors by Erk results in new gene transcription.

Fig. 7.20 The transcription factor AP-1 is formed as a result of the Ras/MAPK signaling pathway. Left panel: phosphorylation of the MAPK Erk activated as a result of the Ras–MAPK cascade allows Erk to enter the nucleus, where it phosphorylates the transcription factor Elk-1. Elk-1, along with serum response factor (SRF), binds to the serum response element (SRE) in the promoter of the gene (*FOS*) for the transcription factor c-Fos, stimulating its transcription. Right panel: the protein kinase PKC-θ can induce the phosphorylation of another MAPK called Jun kinase (JNK). This enables JNK to enter the nucleus and phosphorylate the transcription factor c-Jun, which forms a dimer with c-Fos. The phosphorylated c-Jun/Fos dimer is an active AP-1 transcription factor that binds to AP-1 sites and promotes transcription of many target genes.

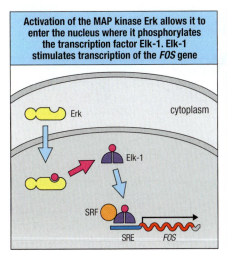

Activation of the MAP kinase Erk allows it to enter the nucleus where it phosphorylates the transcription factor Elk-1. Elk-1 stimulates transcription of the *FOS* gene

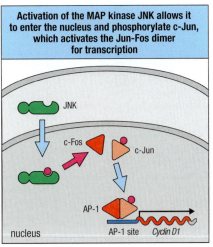

Activation of the MAP kinase JNK allows it to enter the nucleus and phosphorylate c-Jun, which activates the Jun-Fos dimer for transcription

cooperates with a transcription factor called serum response factor to initiate transcription of the *FOS* gene. Fos protein then associates with Jun to form the AP-1 heterodimer, but this remains transcriptionally inactive until another MAPK called **Jun kinase** (**JNK**) phosphorylates Jun. Similar to NFAT, AP-1 participates in turning on transcription of many genes important for T-cell activation, including the gene encoding the cytokine IL-2.

7-16   Protein kinase C activates the transcription factors NFκB and AP-1.

The third signaling pathway leading from PLC-γ results in the activation of PKC-θ, an isoform of protein kinase C that is restricted to T cells and muscle. Mice lacking PKC-θ develop T cells in the thymus, but their mature T cells have a defect in the activation of two crucial transcription factors, NFκB and AP-1, in response to signaling by the T-cell receptor and CD28. These transcription factors also participate in turning on genes required for T-cell activation. For example, transcription of the gene for IL-2 requires NFκB in addition to NFAT and AP-1, making PKC-θ activation an important component of T-cell activation.

PKC-θ has a C1 domain and is recruited to the membrane when DAG is generated by activated PLC-γ (see Fig. 7.17). In this location, the kinase activity of PKC-θ initiates a series of steps that results in the activation of NFκB (Fig. 7.21). PKC-θ phosphorylates the large membrane-localized scaffold protein CARMA1, causing it to oligomerize and form a multisubunit complex with other proteins. This complex recruits and activates TRAF-6, the same protein that we encountered in Chapter 3 in its role in activating NFκB in the TLR signaling pathway (see Fig. 3.13).

NFκB is the general name for a member of a family of homo- and heterodimeric transcription factors made up of the Rel family of proteins. The most common NFκB activated in lymphocytes is a heterodimer of p50:p65Rel.

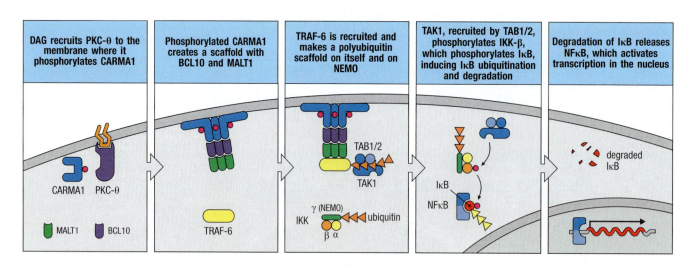

Fig. 7.21 Activation of the transcription factor NFκB by antigen receptors is mediated by protein kinase C. Diacylglycerol (DAG), produced as a result of T-cell receptor signaling activating PLC-γ, recruits a protein kinase C (PKC-θ) to the membrane, where it phosphorylates a scaffold protein called CARMA1. This forms a complex with BCL10 and MALT1 that recruits the E3 ubiquitin ligase TRAF-6. As described in Fig. 3.13, the kinase TAK1 is recruited by the polyubiquitin scaffold produced by TRAF-6 and phosphorylates

the IκB kinase (IKK) complex [IKKα:IKKβ:IKKγ (NEMO)]. IKK phosphorylates IκB, stimulating its ubiquitination, which targets IκB for degradation by the proteasome. This releases NFκB to enter the nucleus and stimulate transcription of its target genes. A defect in NEMO that prevents NFκB activation causes immunodeficiency that results in susceptibility to extracellular bacterial infections, along with a skin disease known as ectodermal dysplasia.

The dimer is held in an inactive state in the cytoplasm by binding to an inhibitory protein called inhibitor of κB (IκB). As described for TLR signaling (see Fig. 3.13), TRAF-6 stimulates the degradation of IκB by first activating the kinase TAK1, which activates a complex of serine kinases, IκB kinase (IKK). IKK phosphorylates IκB, causing its ubiquitination and subsequent degradation, and the consequent release and entry into the nucleus of active NFκB. Inherited deficiency of the IKKγ subunit (also called **NEMO**) leads to a syndrome known as **X-linked hypohidrotic ectodermal dysplasia and immunodeficiency**, which is characterized by developmental defects in ectodermal structures such as skin and teeth, as well as immunodeficiency.

PKC-θ can also activate JNK, and might be able to activate the transcription factor AP-1 by this route. However, T cells lacking PKC-θ have a defect in AP-1 activation in addition to their defect in NFκB activation, but no defect in JNK activation, indicating that our understanding of this pathway is still incomplete.

7-17 PI 3-kinase activation upregulates cellular metabolic pathways via the serine/threonine kinase Akt.

While the activation of transcription factors is an important outcome of antigen receptor signaling, a productive T-cell response also requires substantial changes in cellular metabolism, needed to accommodate the energetic and macromolecular demands of rapidly dividing cells. The PI 3-kinase pathway plays a central role in this response via the recruitment and activation of the second important signaling module, initiated by the serine/threonine kinase Akt (also known as protein kinase B). Akt, via its PH domain, binds to PIP$_3$ in the membrane, which is generated by PI 3-kinase (**Fig. 7.22**; see also Fig. 7.5). In that location, Akt is phosphorylated by PDK1, and once activated, phosphorylates a variety of downstream proteins. One of its effects is to promote cell survival by inhibiting cell death via multiple mechanisms. A major mechanism is the phosphorylation of the pro-apoptotic protein Bad. When phosphorylated, Bad can no longer bind to and inhibit the anti-apoptotic (pro-survival) protein Bcl-2 (see Fig. 7.22). Another effect of activated Akt is to regulate the expression of homing and adhesion receptors that orchestrate the migratory properties of activated T cells (discussed in detail in Chapters 9 and 11). Activated Akt also functions to stimulate the cell's metabolism by increasing the utilization of glucose; this is mediated by increasing the activity of glycolytic enzymes and by inducing the upregulation of nutrient transporters on the T-cell membrane.

X-linked Hypohidrotic Ectodermal Dysplasia and Immunodeficiency

Fig. 7.22 The serine/threonine kinase Akt is activated by TCR signaling and promotes cell survival and enhanced metabolic activity via mTOR. Panel one: T-cell receptor (TCR) signaling activates PI 3-kinase (not shown), generating PIP$_3$ in the plasma membrane; PIP$_3$ recruits and activates the kinase PDK1. Akt, a second serine/threonine kinase, binds to PIP$_3$ via its PH domain, and is phosphorylated and activated by PDK1. Panel two: active Akt phosphorylates the pro-apoptotic protein Bad, which is binding to and inhibiting the anti-apoptotic protein Bcl-2 at the mitochondrial membrane. Panel three: phosphorylated Bad binds to 14-3-3, releasing Bcl-2 to promote cell survival. Panel four: a second function of active Akt is to phosphorylate the TSC1/2 complex, a GAP for the small GTPase Rheb. Panel five: When TSC1/2 is phosphorylated, it releases the inactive Rheb protein, leading to Rheb activation. Active Rheb binds to and activates mTOR (mammalian target of rapamycin). Once activated, mTOR acts on multiple pathways that lead to increased lipid production, ribosome biosynthesis, mRNA synthesis, and protein translation.

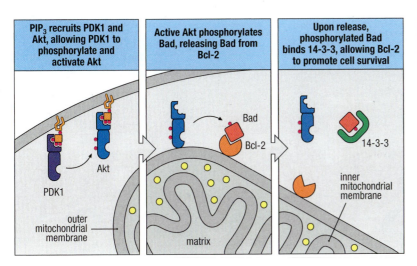
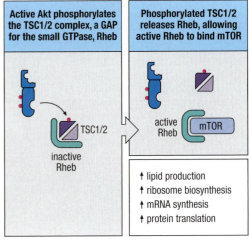

Yet another important function of activated Akt is to stimulate the **mTOR** (mammalian target of rapamycin) pathway, a key regulator of macromolecular biosynthesis (see Fig. 7.22). In this case, Akt phosphorylates and inactivates the TSC complex, a GAP for the small GTPase Rheb. This leads to Rheb activation, and in turn, to the activation of mTOR. The mTOR pathway has multiple effects on cellular metabolism; collectively, these changes are essential to provide the raw materials needed to carry out the increased gene expression, protein production, and cell division that accompany T-cell activation. Specifically, mTOR activation leads to increased lipid production, ribosome biosynthesis, mRNA synthesis, and protein translation.

7-18 T-cell receptor signaling leads to enhanced integrin-mediated cell adhesion.

The third signaling module induced by TCR stimulation leads to enhanced integrin adhesion. Together with cytoskeletal changes (discussed in the next section), this process promotes stability of the T cell-APC interaction and localizes active signaling complexes into a structure known as the 'immune synapse' described below in Section 7-19 (and see Fig. 7.25). The immune synapse, the region of the T-cell membrane that is in direct and stable contact with the APC or target cell, is formed within minutes of T-cell receptor recognition of MHC/peptide ligands. One important component of this is increased adhesiveness of the T-cell integrin LFA-1. On nonstimulated T cells, LFA-1 resides in a low-affinity state and is well dispersed on the T-cell membrane, resulting in weak binding to its ligand, ICAM-1. Following T-cell receptor stimulation, LFA-1 molecules aggregate at the synapse, and also undergo a conformational change that converts each LFA-1 molecule into a high-affinity binding partner for ICAM-1. Together, these changes lead to enhanced adhesion between the T cell and the APC, and to stabilization of this cell–cell interaction. These effects on LFA-1 are induced by the recruitment of the adaptor protein ADAP to the LAT:Gads:SLP-76 scaffold complex (Fig. 7.23). In turn, ADAP recruits two additional proteins, SKAP55 and RIAM. The ADAP:SKAP55:RIAM complex binds to the small GTPase Rap1, activating Rap1 at the site of T-cell receptor signaling. GTP-bound Rap1 then promotes LFA-1 aggregation and the conformational change that converts LFA-1 into a high-affinity binding partner for ICAM-1. The importance of this pathway is underscored by the finding that ADAP-deficient T cells show impaired proliferation and cytokine production following T-cell receptor stimulation.

Fig. 7.23 Recruitment of ADAP to the LAT:Gads:SLP-76 complex activates integrin adhesion and aggregation. Left panel: prior to T-cell receptor (TCR) signaling, the integrin LFA-1 is present on the T-cell membrane in a low-affinity conformation that binds weakly to ICAM-1 on antigen-presenting-cells. Middle panel: following TCR signaling, the adaptor protein ADAP is recruited to the LAT:Gads:SLP-76 complex by an interaction between tyrosine-phosphorylated ADAP and the SH2 domain of SLP-76. ADAP then recruits a complex of SKAP and RIAM (Rap1-GTP-interacting adaptor molecule), activating the small GTPase Rap1. Right panel: active Rap1 induces aggregation of LFA-1 and a conformational change in LFA-1 that leads to high-affinity binding to ICAM-1.

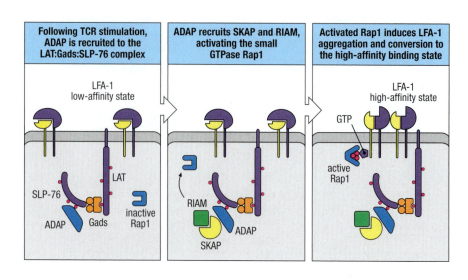

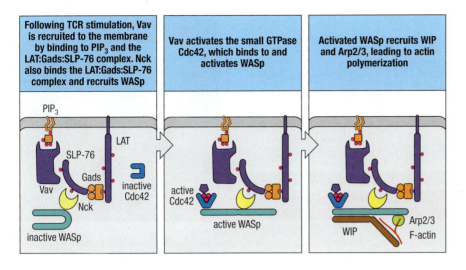

Following TCR stimulation, Vav is recruited to the membrane by binding to PIP$_3$ and the LAT:Gads:SLP-76 complex. Nck also binds the LAT:Gads:SLP-76 complex and recruits WASp

Vav activates the small GTPase Cdc42, which binds to and activates WASp

Activated WASp recruits WIP and Arp2/3, leading to actin polymerization

Fig. 7.24 Recruitment of Vav to the LAT:Gads:SLP-76 complex induces activation of Cdc42, leading to actin polymerization. Left panel: Vav, a guanine-nucleotide exchange factor (GEF) for the small GTPase Cdc42, is recruited to the activated T-cell receptor (TCR) complex by binding via its PH domain to PIP$_3$ in the membrane and by binding to phosphorylated SLP-76. The small adaptor protein Nck binds to an adjacent phosphorylated tyrosine on SLP-76 and recruits the inactive form of the protein WASp. Middle panel: Vav activates Cdc42, which binds to and activates WASp. Right panel: active WASp binds to WIP, recruiting Arp2/3 and inducing actin polymerization. The importance of this pathway is illustrated by the discovery of WASp as the protein encoded by the gene responsible for the human immunodeficiency disease Wiskott–Aldrich syndrome.

7-19 T-cell receptor signaling induces cytoskeletal reorganization by activating the small GTPase Cdc42.

The fourth TCR signaling module, also involved in the formation of a stable immune synapse, leads to reorganization of the actin cytoskeleton. Without this process, integrin aggregation would not occur, interactions between the T cell and the APC would not stabilize, and in fact, T-cell activation would completely fail. A major component of this T-cell receptor signal is transduced by Vav, a GEF that activates Rho-family GTPases, including Cdc42. Like PLC-γ and Itk, Vav is recruited to the site of receptor activation by interactions of the Vav PH domain with PIP$_3$ and of the Vav SH2 domain with the LAT:Gads:SLP-76 scaffold complex (Fig. 7.24). When Cdc42 is activated by Vav, the GTP-bound Cdc42 induces a conformational change in the protein **WASp** (**Wiskott–Aldrich syndrome** protein), which is also recruited to the LAT:Gads:SLP-76 scaffold complex by binding to the small adaptor protein Nck. This active form of WASp binds to WIP, and together, these proteins recruit Arp2/3, leading to actin polymerization. The importance of this pathway is underscored by the fact that defects in WASp are the basis for the immunodeficiency disease Wiskott–Aldrich syndrome. Due to the widespread expression of WASp, individuals suffering from this disease have impairments in multiple immune cell types, all of which depend on WASp-dependent actin polymerization for their functions. One major defect in this disease is in T cell-dependent antibody responses, due to the requirement for actin polymerization to ensure effective interactions between CD4 T cells and B cells. Thus, the failure of WASp-deficient T cells to provide adequate 'help' to B cells most likely results from a defect in the formation of the immune synapse, which is normally required to ensure directed secretion of cytokines from the T cell onto the B-cell membrane (Fig. 7.25).

 **Wiskott–Aldrich Syndrome**

7-20 The logic of B-cell receptor signaling is similar to that of T-cell receptor signaling, but some of the signaling components are specific to B cells.

There are many similarities between signaling from T-cell receptors and signaling from B-cell receptors. As with the T-cell receptor, the B-cell receptor is composed of antigen-specific chains associated with ITAM-containing signaling chains, in this case Igα and Igβ (see Fig. 7.10). In B cells, three protein tyrosine kinases of the Src family—Fyn, Blk, and Lyn—are thought to be responsible for phosphorylation of the ITAMs (Fig. 7.26). These kinases associate with

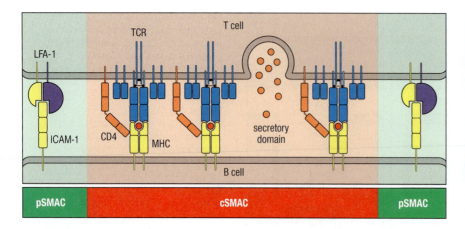

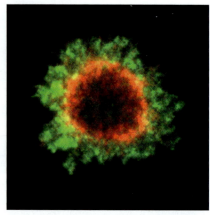

Fig. 7.25 The immune synapse provides a structure for directed secretion of T-cell cytokines. When the T-cell receptors (TCRs) on a T cell recognize peptide:MHC on an antigen-presenting cell, a process of receptor reorganization takes place on the plasma membranes of the two interacting cells. Left panel: when a CD4+ T cell recognizes its peptide:MHC ligand on a B cell, the immune synapse functions to direct T-cell-secreted cytokines onto the B-cell surface at the site of closest contact between the plasma membranes of the two cells. Right panel: confocal microscopy images of the TCR/peptide:MHC (red) and the LFA-1/ICAM-1 (green) proteins 30 minutes after the initiation of signaling show a central accumulation of the TCR/peptide:MHC complexes and a peripheral ring of the LFA-1/ICAM-1 complexes. These structures have been called the central supermolecular activation complex (cSMAC, red) and the peripheral supermolecular activation complex (pSMAC, green). The combined structure is known as the immune synapse. Photograph courtesy of Y. Kaizuka.

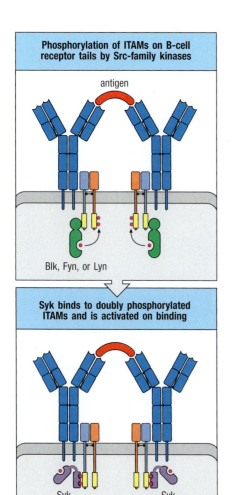

Fig. 7.26 Src-family kinases are associated with B-cell receptors and phosphorylate the tyrosines in ITAMs to create binding sites for Syk and Syk activation via transphosphorylation. The membrane-bound Src-family kinases Fyn, Blk, and Lyn associate with the B-cell antigen receptor by binding to ITAMs, either (as shown in the figure) through their amino-terminal domains or by binding a single phosphorylated tyrosine through their SH2 domains. After ligand binding and receptor clustering, the associated kinases phosphorylate tyrosines in the ITAMs on the cytoplasmic tails of Igα and Igβ. Subsequently, Syk binds to the phosphorylated ITAMs of the Igβ chain. Because there are at least two receptor complexes in each cluster, Syk molecules become bound in close proximity and can activate each other by transphosphorylation, thus initiating further signaling.

resting receptors via a low-affinity interaction with unphosphorylated ITAMs in Igα and Igβ. After the receptors have bound a multivalent antigen, which cross-links them, the receptor-associated kinases are activated and phosphorylate the tyrosine residues in the ITAMs. B cells do not express ZAP-70; instead, a closely related tyrosine kinase, **Syk**, containing two SH2 domains, is recruited to the phosphorylated ITAM. In contrast to ZAP-70, which requires additional Lck phosphorylation for activation, Syk is activated simply by its binding to the phosphorylated site.

For B cells, the co-receptor and co-stimulatory receptor functions are combined into one accessory receptor that is a complex of cell-surface proteins—**CD19**, **CD21**, and **CD81**—often referred to as the **B-cell co-receptor** (Fig. 7.27). As with T cells, antigen-dependent signaling from the B-cell receptor is enhanced if the B-cell co-receptor is simultaneously bound by its ligand and clusters with the antigen receptor. CD21 (also known as complement receptor 2, CR2) is a receptor for the C3dg fragment of complement. This means that antigens such as bacterial pathogens on which C3dg is bound (see Fig. 7.27) can cross-link the B-cell receptor with the CD21:CD19:CD81 complex. This induces phosphorylation of the cytoplasmic tail of CD19 by B-cell receptor-associated tyrosine kinases, which in turn leads to the binding of additional Src-family kinases, the augmentation of signaling through the B-cell receptor itself, and the recruitment of PI 3-kinase (see Section 7-4). PI 3-kinase

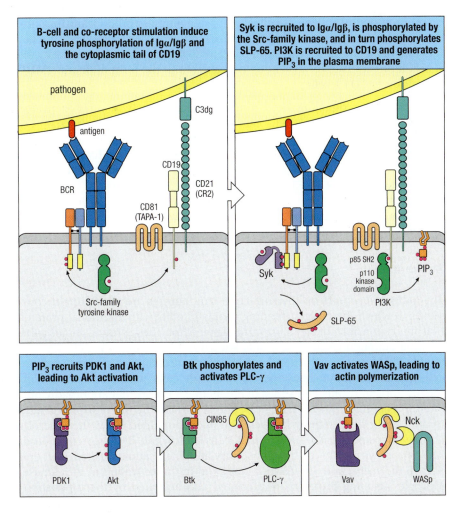

B-cell and co-receptor stimulation induce tyrosine phosphorylation of Igα/Igβ and the cytoplasmic tail of CD19

Syk is recruited to Igα/Igβ, is phosphorylated by the Src-family kinase, and in turn phosphorylates SLP-65. PI3K is recruited to CD19 and generates PIP₃ in the plasma membrane

PIP₃ recruits PDK1 and Akt, leading to Akt activation

Btk phosphorylates and activates PLC-γ

Vav activates WASp, leading to actin polymerization

Fig. 7.27 B-cell antigen receptor plus co-receptor engagement activate downstream signaling modules leading to activation of Akt, PLC-γ, and WASp. B-cell receptor (BCR) signaling is greatly enhanced when the antigen is tagged by complement fragments, engaging the B-cell co-receptor together with the B-cell antigen receptor. Cleavage of the antigen-bound complement component C3 to C3dg (see Fig. 2.30) allows the tagged antigen to bind to the cell-surface protein CD21 (complement receptor 2, CR2), a component of the B-cell co-receptor complex, which also includes CD19 and CD81 (TAPA-1). Cross-linking and clustering of the co-receptor with the antigen receptor result in the phosphorylation of tyrosine residues in the ITAM sequences of the cytoplasmic domains of the BCR signaling subunits, Igα and Igβ. The Src-family kinase also phosphorylates tyrosine residues in the cytoplasmic domain of CD19. The phosphorylated ITAMs in Igα and Igβ recruit and activate the tyrosine kinase Syk, which functions similarly to ZAP-70 in T cells. The phosphorylated tail of CD19 recruits PI 3-kinase, leading to PIP₃ generation in the plasma membrane. Activated Syk phosphorylates the membrane-associated scaffold protein SLP-65, which associates with the plasma membrane by binding CIN85. PIP₃ recruits PDK1 and Akt, leading to Akt activation. Phosphorylated SLP-65 and PIP₃ recruit the Tec-family tyrosine kinase Btk and PLC-γ, leading to Btk phosphorylation and activation of PLC-γ. Phosphorylated SLP-65 and PIP₃ also recruit Vav, Nck, and inactive WASp. Vav activates small GTPases that activate WASp, leading to actin polymerization; the activated GTPases also induce integrin aggregation and conversion of LFA-1 to the high-affinity binding state.

initiates an additional signaling pathway leading from the B-cell receptor (see Fig. 7.27). Thus, the B-cell co-receptor serves to strengthen the signal resulting from antigen recognition. The role of the third component of the B-cell receptor complex, CD81 (TAPA-1), is as yet unknown.

Once activated, Syk phosphorylates the scaffold protein **SLP-65** (also known as **BLNK**). Like LAT and SLP-76 in T cells, SLP-65 functions as a composite of these two proteins, providing multiple sites for tyrosine phosphorylation and recruiting a variety of SH2-containing proteins, including enzymes and adaptor proteins, to form several distinct multiprotein signaling complexes that can act in concert. As in T cells, a key signaling protein is the phospholipase PLC-γ, which is activated with the aid of the B cell-specific Tec kinase **Bruton's tyrosine kinase** (**Btk**) and hydrolyzes PIP₂ to form DAG and IP₃ (see Fig. 7.27). As discussed for the T-cell receptor, signaling by calcium and DAG leads to the activation of downstream transcription factors. A deficiency in Btk (which is encoded by a gene on the X chromosome) prevents the development and functioning of B cells, resulting in the disease **X-linked agammaglobulinemia**, which is characterized by a lack of antibodies. Besides Btk, mutations in other signaling molecules in B cells, including receptor chains and SLP-65, have been linked to B-cell immunodeficiencies (see Chapter 8).

Several other downstream pathways described for TCR signaling are also shared with BCR signaling, and are dependent on the adaptor protein SLP-65. These include the Vav-dependent induction of actin polymerization by Cdc42 and WASp, and the recruitment and activation of small GTPases that promote integrin adhesion (see Fig. 7.27). In the case of B-cell recognition of

X-linked Agammaglobulinemia

membrane-bound antigen, B-cell receptor signaling also produces an immune synapse that localizes signaling complexes to the cell–cell interface. One key function of the immune synapse in B cells is to promote antigen uptake by the B cell, a prerequisite to presenting that antigen in the form of peptide:MHC complexes to CD4 T cells.

B-cell receptor signaling also induces metabolic changes in activated B cells. As is the case for T cells, this response is dependent on the action of PI 3-kinase, whose activation leads to formation of the membrane phospho-inositide PI(3,4,5)P$_3$ at the site of the activated B-cell receptor. This response is augmented by combined signaling through the B-cell receptor plus the B-cell co-receptor complex of CD19/CD21/CD81. PI(3,4,5)P$_3$ recruits Akt, which is then phosphorylated and activated, leading to downstream activation of mTOR, as well as additional Akt-dependent pathways promoting cell survival and proliferation (see Fig. 7.27).

Summary.

The antigen receptors on the surface of lymphocytes are multiprotein complexes in which the antigen-binding chains interact with additional proteins that are responsible for signaling from the receptor. These protein chains carry tyrosine-containing signaling motifs known as ITAMs. Signaling chains containing ITAM motifs are widely used by activating receptors in many immune cell types in addition to lymphocytes. In lymphocytes, activation of the receptors by antigen results in a series of biochemical events that are broadly outlined in **Fig. 7.28**. This signaling cascade is initiated by the phosphorylation of the ITAMs by Src-family kinases. The phosphorylated ITAM then recruits another tyrosine kinase, ZAP-70 in T cells and Syk in B cells. Activation of ZAP-70 or Syk results in the phosphorylation of scaffolds called LAT and SLP-76 in T cells, and SLP-65 in B cells, and in the activation of PI 3-kinase. Multiple signaling proteins are recruited and activated by these phosphorylated scaffolds, including phospholipase C-γ, ADAP, and Vav, whereas Akt is recruited and activated by the action of PI 3-kinase generating PIP$_3$ at the plasma membrane. PLC-γ generates inositol trisphosphate (IP$_3$) and diacylglycerol (DAG). IP$_3$ has an important role in inducing changes in intracellular calcium concentrations, and DAG is involved in activating protein kinase C-θ and the small G protein Ras. In T cells, these pathways ultimately result in the activation of three transcription factors, namely, AP-1, NFAT, and NFκB; together, these transcription factors induce transcription of the cytokine IL-2, which is essential for the proliferation and further differentiation of the activated lymphocyte. In addition to transcription factor activation, antigen receptor signaling in both T cells and B cells leads to enhanced cell survival, metabolic activity, adhesiveness, and cytoskeletal reorganization. Signaling by antigen receptors is facilitated by co-receptors that become engaged as a result of receptor–antigen binding. These co-receptors are the MHC-binding CD4 and CD8 transmembrane proteins in T cells and the complement-binding B-cell co-receptor complex containing CD19 in B cells.

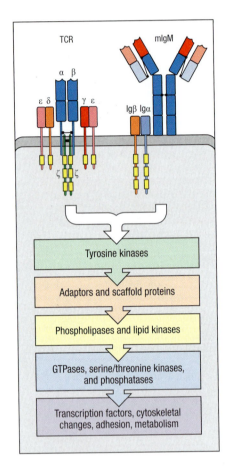

Fig. 7.28 Summary of antigen receptor signaling pathways. As outlined in this section, the signal transduction pathways downstream of the T-cell and B-cell receptors occur in an orchestrated series of stages involving many categories of proteins that produce widespread changes in the cells. The first detectable events following antigen receptor stimulation are the activation of tyrosine kinases. Following this, adaptor proteins and scaffolds are modified, recruiting phospholipases and lipid kinases to the activated receptor complexes. The next level of signaling amplifies these earlier stages by activating multiple small GTPases, serine/threonine kinases, and protein phosphatases. Together, these lead to transcription factor activation, cytoskeletal changes, and increases in cellular adhesion and metabolism, all of which contribute to T- and B-cell activation.

Co-stimulatory and inhibitory receptors modulate antigen receptor signaling in T and B lymphocytes.

Signals initiated by the T-cell and B-cell antigen receptors are essential for lymphocyte activation, and determine the specificity of the adaptive immune response that is initiated. However, signaling from the antigen receptor is not on its own sufficient to activate a naive T cell or B cell. These naive lymphocytes require additional signals to achieve full activation. Receptors on T cells and B cells that can provide this necessary second signal are called **co-stimulatory receptors**, and are members of either the CD28 family of proteins or of the TNF receptor superfamily. While naive T cells primarily utilize CD28 as the co-stimulatory receptor, naive B cells use the TNF receptor family member CD40. The overall function of signaling through these co-stimulatory receptors is to enhance the antigen receptor signals that induce transcription factor activation and PI 3-kinase activation, thereby ensuring activation of the T cell or B cell. In contrast to these activating co-stimulatory receptor signals, other cell-surface receptors on T cells and B cells function to downregulate activation signals. These inhibitory receptors are important in preventing excessive immune responses that can lead to destructive inflammatory or autoimmune conditions, particularly in the case of chronic infections that are inefficiently controlled by the immune system.

7-21 The cell-surface protein CD28 is a required co-stimulatory signaling receptor for naive T-cell activation.

The signaling through the T-cell receptor complex described in the previous sections is not by itself sufficient to activate a naive T cell. As noted in Chapter 1, antigen-presenting cells that can activate naive T cells bear cell-surface proteins known as **co-stimulatory molecules** or co-stimulatory ligands. These interact with cell-surface receptors, known as co-stimulatory receptors, on the naive T cell to transmit a signal that is required, along with antigen stimulation, for T-cell activation—this signal is sometimes called 'signal 2.' We discuss the immunological consequences of this requirement for co-stimulation in detail in Chapter 9.

The best understood of these co-stimulatory receptors is the cell-surface protein **CD28**. CD28 is present on the surface of all naive T cells and binds the co-stimulatory ligands **B7.1** (**CD80**) and **B7.2** (**CD86**), which are expressed mainly on specialized antigen-presenting cells such as dendritic cells. To become activated, the naive lymphocyte must engage both antigen and a co-stimulatory ligand on the same antigen-presenting cell. CD28 signaling aids antigen-dependent T-cell activation mainly by promoting T-cell proliferation, cytokine production, and cell survival. All these effects are mediated by signaling motifs present in the cytoplasmic domain of CD28.

After engagement by B7 molecules, CD28 becomes tyrosine phosphorylated by Lck in its cytoplasmic domain on tyrosine residues in a YXN motif that can recruit the adaptor protein Grb2, and in a non-ITAM motif YMNM. The cytoplasmic tail of CD28 also carries a proline-rich motif (PXXP) that binds the SH3 domains of Lck and Itk. Although the details are uncertain, a major effect of CD28 phosphorylation is to activate PI 3-kinase to generate PIP_3 (**Fig. 7.29**). By this mechanism, the co-stimulatory signal induced by CD28 cooperates with the T-cell receptor signal to ensure maximal activation of three of the four T-cell receptor signaling modules described above. Specifically, a high concentration of PIP_3 recruits Itk to the membrane, where Lck can phosphorylate it, thereby enhancing PLC-γ activation. PIP_3 also functions to recruit and activate Akt, which promotes cell survival and increased cellular metabolism (see Section 7.17). An additional function of Akt is to phosphorylate the RNA-binding protein NF-90; when phosphorylated, NF-90 translocates from the nucleus to the cytoplasm and binds to and stabilizes the IL-2 mRNA, leading to increased IL-2 synthesis. Finally, PIP_3 recruits Vav, leading to cytoskeletal reorganization (see Section 7.19). Thus, co-stimulatory signaling through CD28 functions to amplify most of the downstream responses to T-cell receptor stimulation (see Fig. 7.29).

 MOVIE 7.3

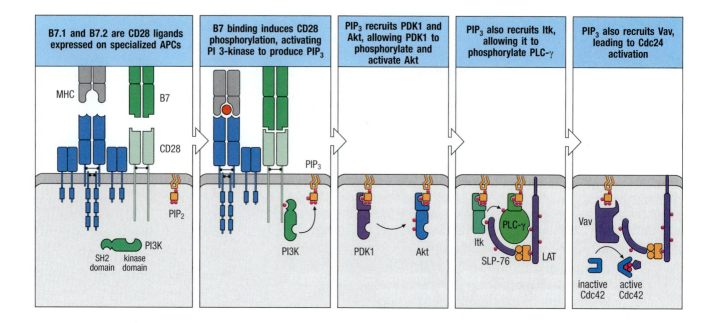

| B7.1 and B7.2 are CD28 ligands expressed on specialized APCs | B7 binding induces CD28 phosphorylation, activating PI 3-kinase to produce PIP₃ | PIP₃ recruits PDK1 and Akt, allowing PDK1 to phosphorylate and activate Akt | PIP₃ also recruits Itk, allowing it to phosphorylate PLC-γ | PIP₃ also recruits Vav, leading to Cdc24 activation |

Fig. 7.29 The T-cell co-stimulatory protein CD28 transduces signals that enhance antigen receptor signaling pathways. The ligands for CD28, namely B7.1 and B7.2, are expressed only on specialized antigen-presenting cells (APCs) such as dendritic cells (first panel). Engagement of CD28 induces its tyrosine phosphorylation, which activates PI 3-kinase (PI3K), with subsequent production of PIP₃ that recruits several enzymes via their PH domains, thus bringing them together with their substrates in the membrane. The protein kinase Akt, which becomes phosphorylated by phosphoinositide-dependent protein kinase-1 (PDK1), is activated and enhances cell survival and upregulates cell metabolism (see Fig. 7.22). Recruitment of the kinase Itk to the membrane is critical for the full activation of PLC-γ (see Fig. 7.16). PIP₃ also recruits Vav, leading to Cdc42 activation and inducing actin polymerization (see Fig. 7.24).

7-22   Maximal activation of PLC-γ, which is important for transcription factor activation, requires a co-stimulatory signal induced by CD28.

One important function of co-stimulatory signaling through CD28 is to promote the maximal activation of PLC-γ via the local production of PIP₃. This recruits Itk by its PH domain, enhancing Itk phosphorylation by Lck. Activated Itk is then recruited to the phosphorylated LAT:Gads:SLP-76 complex by its SH2 and SH3 domains binding to SLP-76, where it phosphorylates and activates PLC-γ (see Fig. 7.16). Activated PLC-γ cleaves PIP₂ to produce the two second messengers DAG and IP₃, ultimately leading to the activation of transcription factors NFAT, AP-1, and NFκB. Thus, the full activation of PLC-γ leading to transcription factor activation requires signals emanating from both the T-cell receptor and CD28.

In T cells, one of the major functions of NFAT, AP-1, and NFκB is to act together to stimulate expression of the gene for the cytokine IL-2, which is essential for promoting T-cell proliferation and differentiation into effector cells. The promoter for the *IL-2* gene contains multiple regulatory elements that must be bound by transcription factors to initiate *IL2* expression. Some control sites are already bound by transcription factors, such as Oct1, that are produced constitutively in lymphocytes, but this is not sufficient to switch on *IL-2*. Only when AP-1, NFAT, and NFκB are all activated and are bound to their control sites in the *IL-2* promoter is the gene expressed. NFAT and AP-1 bind to the promoter cooperatively and with higher affinity by forming a heterotrimer of NFAT, Jun, and Fos. In addition, CD28 co-stimulation further enhances *IL-2* transcription by increasing NFκB activation. Thus, the *IL-2* promoter integrates signals from both the T-cell receptor and CD28 signaling pathways to ensure that IL-2 is produced only in appropriate circumstances (**Fig. 7.30**). Together with the CD28-induced phosphorylation of NF-90 leading to increased IL-2 mRNA stability, CD28 co-stimulation leads to substantially increased production of IL-2 by activated T cells.

7-23   TNF receptor superfamily members augment T-cell and B-cell activation.

While naive T- and B-cell activation requires signaling through the antigen receptors on these cells, T-cell receptor or B-cell receptor signaling,

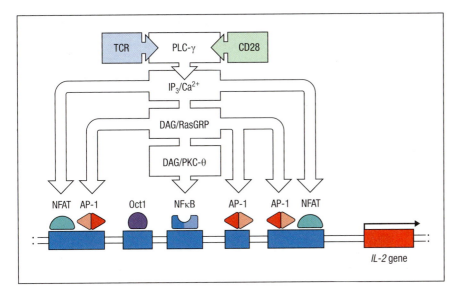

Fig. 7.30 Simplified scheme depicting multiple signaling pathways that converge on the *IL-2* promoter. AP-1, NFAT, and NFκB binding to the promoter of the *IL-2* gene integrate multiple signaling pathways emanating from the T-cell receptor (TCR) and CD28 into a single output, the production of the cytokine IL-2. The MAPK pathway activates AP-1; calcium activates NFAT; and protein kinase C activates NFκB. All three pathways are required to stimulate *IL-2* transcription. Activation of the gene requires both the binding of NFAT and AP-1 to a specific promoter element, and the additional binding of AP-1 on its own to another site. Oct1 is a transcription factor that is required for *IL-2* transcription. Unlike the other transcription factors, it is constitutively bound to the promoter and is therefore not regulated by T-cell receptor or CD28 signaling.

respectively, is not sufficient to activate the cells. For naive T cells, an additional co-stimulatory signal is required and, as discussed above (see Sections 7-21 and 7-22), is frequently provided by the CD28 receptor. For naive B cells, the additional activation signal can be contributed by direct interactions between the pathogen and an innate pattern recognition receptor (PRR), such as a TLR, on the B cell. However, for more effective B-cell activation leading to the production of all classes of antibodies and to the formation of memory B cells, additional B-cell activation signals are contributed by CD4 T cells. One component of this is the production of T-cell cytokines, which bind to and stimulate their receptors on the B-cell surface (see Chapter 10). The second and more essential component provided by the CD4 T cells is the stimulation of CD40 on the B cell by the CD40 ligand expressed on the T cell. The importance of the CD40–CD40-ligand interaction for B-cell responses to protein antigens is highlighted by the discovery that a severe immunodeficiency disease resulting from impaired antibody responses is caused by the absence of CD40 ligand expression on a patient's CD4 T cells.

CD40 Ligand Deficiency

CD40 is a member of the large TNF receptor superfamily, which consists of more than 20 members. While some members of this family, such as Fas (see Chapter 11), are specialized to induce cell death, the majority of TNF receptor superfamily members, including CD40, activate both the NFκB and the PI 3-kinase pathways following receptor stimulation (Fig. 7.31). While NFκB activation leads to enhanced cell survival, the PI 3-kinase pathway has widespread and pleiotropic effects on B-cell physiology, and is a central feature of CD40 signaling. The major mediator of the PI 3-kinase signal is the serine/threonine kinase Akt, which is recruited and activated following the generation of $PI(3,4,5)P_3$ at the B-cell membrane. Akt then stimulates multiple downstream pathways that induce cell survival, cell cycle progression, glucose uptake and metabolism, and mTOR activation, all of which are essential for the productive response of the activated B cell. In general terms, CD40 on the B cell functions in a manner analogous to CD28 on the T cell, as both receptors serve to enhance the levels of Akt activation induced by the B-cell receptor or T-cell receptor signaling pathways, respectively.

TNF receptor superfamily members, including CD40, signal by a mechanism distinct from that of antigen receptors, as it does not involve the activation of tyrosine kinases. Instead, stimulation of TNF receptors recruits adaptor proteins known as TRAFs (TNF receptor-associated factors). In addition to serving as simple adaptors that promote the assembly of multiprotein complexes, five

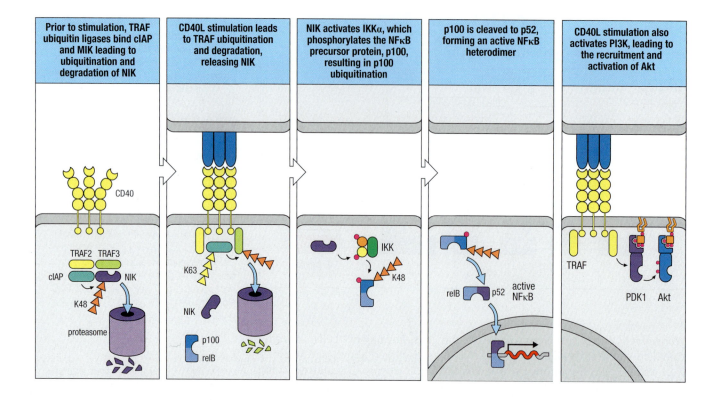

Prior to stimulation, TRAF ubiquitin ligases bind cIAP and MIK leading to ubiquitination and degradation of NIK	CD40L stimulation leads to TRAF ubiquitination and degradation, releasing NIK	NIK activates IKKα, which phosphorylates the NFκB precursor protein, p100, resulting in p100 ubiquitination	p100 is cleaved to p52, forming an active NFκB heterodimer	CD40L stimulation also activates PI3K, leading to the recruitment and activation of Akt

Fig. 7.31 The TNF receptor superfamily member CD40 is an important co-stimulatory molecule on B cells. Several members of the TNF receptor superfamily are expressed on T cells and B cells. A key function of these receptors is the activation of NFκB, which occurs by a pathway distinct from the one initiated by antigen receptor stimulation and is often referred to as the non-canonical NFκB pathway. TNF receptor superfamily members also activate PI 3-kinase signaling pathways. One important TNF receptor superfamily member on B cells is CD40. Prior to stimulation, TRAF molecules, which are ubiquitin ligases, are associated with cIAP, another ubiquitin ligase, and the NFκB-inducing kinase NIK. Under these steady-state conditions, TRAF binding promotes ubiquitination and degradation of NIK. When CD40 is stimulated by binding to CD40L, this complex is recruited to the intracellular domain of CD40. TRAF2 catalyzes K63-linked ubiquitination of cIAP, leading to cIAP-mediated K48-linked ubiquitination of TRAF3. This leads to TRAF3 degradation, releasing NIK, and allowing NIK to phosphorylate and activate IκB kinase α (IKKα). IKKα phosphorylates the NFκB precursor protein p100, inducing its cleavage to form the active p52 subunit, which binds to relB to form the active NFκB transcription factor. CD40L stimulation of CD40 also activates PI 3-kinase, which leads to the activation of Akt by PDK1.

of the six known TRAFs also function as E3 ubiquitin ligases. This activity contributes to the ability of most TNF receptor superfamily members to activate the NFκB pathway, using a pathway distinct from the one initiated by antigen receptor stimulation and often referred to as the **non-canonical NFκB pathway** (see Fig. 7.31). In contrast, the precise biochemical mechanism by which TNF receptors and TRAFs induce PI 3-kinase activation is not yet known.

CD40 is constitutively expressed on B cells and functions during B-cell activation in response to antigen recognition by the B-cell receptor. Additional TNF receptor superfamily members are also expressed on B cells, and each of them is important in B-cell survival at a particular stage of B-cell maturation, including B cells that have differentiated into antibody-secreting cells or memory cells. Similarly, TNF receptor superfamily members are expressed on T cells, many of which are upregulated following T-cell activation. These molecules, such as OX40, 4-1BB, CD30, and CD27, contribute important survival signals and function to enhance cellular metabolism at later stages of the T-cell response to infection (see Chapter 11).

7-24   Inhibitory receptors on lymphocytes downregulate immune responses by interfering with co-stimulatory signaling pathways.

CD28 belongs to a family of structurally related receptors that are expressed by lymphocytes and bind B7-family ligands. Some, such as the receptor ICOS, which is discussed in Chapter 9, act as activating receptors, but others inhibit signaling by the antigen receptors, can stimulate apoptosis, and are important in regulating the immune response. Inhibitory receptors related to CD28 and expressed by T cells include **CTLA-4** (CD152) and **PD-1 (programmed death-1)**, while the **B and T lymphocyte attenuator (BTLA)** is expressed by both T cells and B cells. Of these, CTLA-4 seems to be the most important: mice lacking CTLA-4 die at a young age from an uncontrolled proliferation of T cells in multiple organs, whereas loss of PD-1 or BTLA causes less marked changes that alter the magnitude of responses following lymphocyte activation, rather than causing widespread spontaneous lymphoproliferation. Both CTLA-4 and PD-1 have been targeted for the development of protein-based therapeutics that function to block the activities of these receptors. The goal of these therapeutics is to enhance T-cell responses by inhibiting these inhibitory receptors, a therapeutic strategy referred to as **checkpoint blockade** (see Chapter 16). Recent clinical trials demonstrate that both CTLA-4 and PD-1 blockade have remarkable efficacy in the treatment of cancer by enhancing the patient's own antitumor T-cell responses.

CTLA-4 is induced on activated T cells and binds to the same co-stimulatory ligands (B7.1 and B7.2) as CD28, but CTLA-4 engagement is inhibitory for T-cell activation, rather than enhancing (**Fig. 7.32**). The function of CTLA-4 is controlled largely by regulation of its surface expression. Initially, CTLA-4 resides on intracellular membranes but moves to the cell surface after T-cell receptor signaling. The surface expression of CTLA-4 is controlled by phosphorylation of the tyrosine-based motif GVYVKM in its cytoplasmic tail. When this motif is not phosphorylated, it is able to bind to the clathrin adaptor molecule AP-2, which removes CTLA-4 from the surface. When it is phosphorylated, this motif cannot bind AP-2, and CTLA-4 remains in the membrane, where it can bind B7 molecules on antigen-presenting cells.

CTLA-4 has a higher affinity for its B7 ligands than does CD28, and, apparently of importance for its inhibitory function, it engages B7 molecules in a different orientation. CD28, CTLA-4, and B7.1 are all expressed as homodimers. A CD28 dimer engages one B7.1 dimer in a direct one-to-one correspondence, but a CTLA-4 dimer engages two different B7 dimers in a configuration that allows for extended cross-linkages that confer high avidity to the interaction (see Fig. 7.32). CTLA-4 was once presumed to act by recruiting inhibitory phosphatases, like some of the other inhibitory receptors described later, but this is no longer thought to be so. It is still not clear whether CTLA-4 directly activates inhibitory signaling pathways. Instead, its actions may result in part from blocking the binding of CD28 to B7, thereby reducing CD28-dependent co-stimulation.

CTLA-4-expressing T cells can also exert an inhibitory effect on the activation of other T cells. How they do this is not yet clear, but it might result from CTLA-4 binding to B7 molecules on antigen-presenting cells, in effect stealing the ligand for CD28 required by the other T cells. Direct actions of CTLA-4 on T cells have not been excluded, however. Notably, the regulatory T cells that are needed to suppress autoimmunity express high levels of CTLA-4 on their surface, and they require CTLA-4 to function normally. Regulatory cells are described in detail in Chapter 9.

7-25 Inhibitory receptors on lymphocytes downregulate immune responses by recruiting protein or lipid phosphatases.

Some other receptors that can inhibit lymphocyte activation possess motifs in their cytoplasmic regions that are known as the **immunoreceptor tyrosine-based inhibitory motif** (**ITIM**, consensus sequence [I/V]XYXX[L/I], where X is any amino acid) (**Fig. 7.33**) or the related **immunoreceptor tyrosine-based switch motif** (**ITSM**, consensus sequence TXYXX[V/I]). When the tyrosine in an ITIM or ITSM is phosphorylated, it can recruit either of two inhibitory phosphatases, called **SHP** (SH2-containing phosphatase) and **SHIP** (SH2-containing inositol phosphatase), via their SH2 domains. SHP is a protein tyrosine phosphatase that removes phosphate groups added by tyrosine kinases to a variety of proteins. SHIP is an inositol phosphatase and removes the phosphate from PIP_3 to generate PIP_2, thus reversing the recruitment of proteins such as Tec kinases and Akt to the cell membrane and thereby inhibiting signaling.

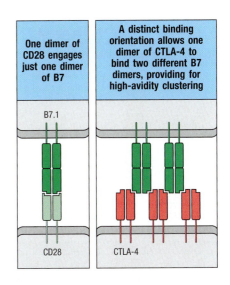

One dimer of CD28 engages just one dimer of B7

A distinct binding orientation allows one dimer of CTLA-4 to bind two different B7 dimers, providing for high-avidity clustering

B7.1

CD28   CTLA-4

Fig. 7.32 CTLA-4 has a higher affinity than CD28 for B7 and engages it in a multivalent orientation. CD28 and CTLA-4 are both expressed as dimers on the cell surface and both bind to two ligands of B7.1, which is a dimer, and B7.2, which is not. However, the orientations of the B7 binding of CD28 and of CTLA-4 differ in a way that contributes to the inhibitory action of CTLA-4. One dimer of CD28 engages just one dimer of B7.1. But one dimer of CTLA-4 binds in such a way that two different dimers of B7.1 are engaged at once, allowing these molecules to cluster into complexes of high avidity. This, and the higher affinity of CTLA-4 for B7 molecules, may give it an advantage in competing for available B7 molecules on an antigen-presenting cell, providing one mechanism by which it could block the co-stimulation of T cells.

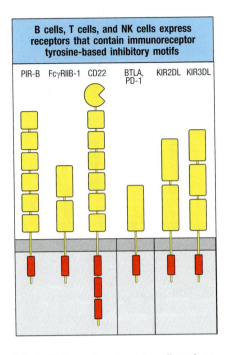

Fig. 7.33 Some lymphocyte cell-surface receptors contain motifs involved in downregulating activation. Several receptors that transduce signals that inhibit lymphocyte or NK-cell activation contain one or more ITIMs (immunoreceptor tyrosine-based inhibitory motifs) in their cytoplasmic tails (red rectangles). ITIMs bind to various phosphatases that, when activated, inhibit signals derived from ITAM-containing receptors.

One ITIM-containing receptor is PD-1 (see Fig. 7.33), which is induced transiently on activated T cells, B cells, and myeloid cells. It can bind to the B7-family ligands **PD-L1** (programmed death ligand-1, B7-H1) and **PD-L2** (programmed death ligand-2, B7-DC). Despite their names, we now understand that these proteins function as ligands for the inhibitory receptor PD-1, rather than acting directly in cell death. PD-L1 is constitutively expressed by a wide variety of cells, whereas PD-L2 expression is induced on antigen-presenting cells during inflammation. Because PD-L1 is expressed constitutively, regulation of PD-1 expression could have a critical role in controlling T-cell responses. For example, signaling by pro-inflammatory cytokines can repress PD-1, thus enhancing the T-cell response. Mice lacking PD-1 gradually develop autoimmunity, presumably because of an inability to regulate T-cell activation. In chronic infections, the widespread expression of PD-1 reduces the effector activity of T cells; this helps to limit potential damage to bystander cells, but at the expense of pathogen clearance.

BTLA contains an ITIM and an ITSM and is expressed on activated T cells and B cells, as well as on some cells of the innate immune system. Unlike other CD28-family members, however, BTLA does not interact with B7 ligands but binds a member of the TNF receptor family; called the **herpesvirus entry molecule** (**HVEM**), this receptor is highly expressed on resting T cells and immature dendritic cells. When BTLA and HVEM are co-expressed on the same cell, BTLA utilizes a second mechanism that further inhibits lymphocyte activation. In this configuration, BLTA binds to HVEM and prevents HVEM from binding to alternative partners that would stimulate NFκB-dependent pro-survival signaling pathways downstream of HVEM. Alternatively, when BTLA and HVEM are expressed on different cells, the interaction of these two receptors functions to stimulate the positive pro-survival signal in the HVEM-expressing cell.

Other receptors on B cells and T cells also contain ITIMs and can inhibit cell activation when ligated along with the antigen receptors. One example is the receptor **FcγRIIB-1** on B cells, which binds the Fc region of IgG antibodies. As a result, antigens present as immune complexes containing IgG antibodies are poor at activating naive B cells, due to the co-engagement of the B-cell receptor with this inhibitory Fc receptor. The ITIM in FcγRIIB-1 recruits SHIP into a complex with the B-cell receptor to block the actions of PI 3-kinase (**Fig. 7.34**). Another inhibitory receptor on B cells is **CD22**, a transmembrane protein that recognizes sialic acid-modified glycoproteins commonly found on mammalian cells but rarely on microbial pathogen surfaces. CD22 contains an ITIM that interacts with SHP, a phosphatase that can dephosphorylate adaptors such as SLP-65 that associate with CD22, thereby inhibiting signaling from the B-cell receptor.

The ITIM motif is also an important motif in signaling by receptors on NK cells that inhibit the killer activity of these cells (see Section 3-26). These inhibitory receptors recognize MHC class I molecules and transmit signals that inhibit the release of the NK cell's cytotoxic granules when the NK cell recognizes a healthy uninfected cell. In NK cells, ITIM-containing receptors play an important role in setting the threshold for NK cell activation by balancing positive signals from ITAM-containing receptors.

Summary.

Signaling through the antigen receptors on T cells and B cells is essential for the activation of these cells. However, for naive T and B cells, the signal through the T-cell receptor or B-cell receptor, respectively, is not sufficient to initiate a response. In addition to the antigen receptor signals, these cells require signals through accessory receptors that serve to monitor the environment of the cell to ensure the presence of an infection. An important secondary signaling

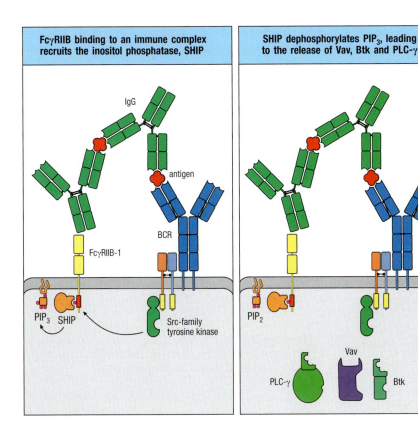

FcγRIIB binding to an immune complex recruits the inositol phosphatase, SHIP

SHIP dephosphorylates PIP₃, leading to the release of Vav, Btk and PLC-γ

Fig. 7.34 The ITIM-containing Fc receptor inhibits B-cell receptor signaling by recruiting the inositol phosphatase SHIP. When the B-cell receptor binds an antigen that is already present in immune complexes with IgG, the ITIM-containing Fc receptor FcγRIIB is engaged at the same time as the B-cell receptor. The Src-family kinase present at the B-cell receptor (BCR) phosphorylates the ITIM motif of FcγRIIB, which then recruits the SH2 domain-containing inositol phosphatase SHIP. SHIP dephosphorylates PIP_3 in the plasma membrane, generating PIP_2. PH domain-containing enzymes, such as Vav, Btk, and PLC-γ, depend on their PH domain binding to PIP_3 for their stable recruitment to the activated B-cell receptor complex. The loss of PIP_3 terminates the recruitment of these enzymes and inhibits B-cell receptor signaling.

system in naive T cells is provided by the CD28 family of co-stimulatory proteins, which bind members of the B7 family of proteins. Activating members of the CD28 family provide co-stimulatory signals that amplify the signal from the T-cell receptor and are important in ensuring the activation of naive T cells by the appropriate target cell. In B cells, these secondary signals are provided by members of the TNF receptor superfamily, such as CD40. Inhibitory members of the CD28 and other receptor families function to attenuate or completely block signaling by activating receptors. The regulated expression of activating and inhibitory receptors and their ligands generates a sophisticated level of control of immune responses that is only beginning to be understood.

Summary to Chapter 7.

Signaling by cell-surface receptors of many different sorts is crucial to the ability of the immune system to respond appropriately to foreign pathogens. The importance of these signaling pathways is demonstrated by the many diseases that are due to aberrant signaling, which include both immunodeficiency diseases and autoimmune diseases. Common features of many signaling pathways are the generation of second messengers such as calcium and phosphoinositides and the activation of both serine/threonine and tyrosine kinases. An important concept in the initiation of signaling pathways by receptor proteins is the recruitment of signaling proteins to the plasma membrane and the assembly of multiprotein signaling complexes. In many cases, signal transduction leads to the activation of transcription factors that lead directly or indirectly to the proliferation, differentiation, and effector function of activated lymphocytes. Other roles of signal transduction are to mediate changes in the cytoskeleton that are important for cell functions such as migration and shape changes. These steps of the T-cell and B-cell receptor signaling pathways are summarized in Fig. 7.28.

While we are beginning to understand the basic circuitry of signal transduction pathways, it is important to keep in mind that we do not yet understand why these pathways are so complex. One reason might be that the signaling pathways have roles in properties such as amplification, robustness, diversity, and efficiency of signaling responses. An important goal for the future will be to understand how the design of each signaling pathway contributes to the particular quality and sensitivity needed for specific signaling responses.

Questions.

7.1 **True or False:** Antigen receptors bear intrinsic kinase activity that allows for phosphorylation of cytoplasmic proteins and subsequent downstream signaling events.

7.2 **Matching:** Indicate whether the following receptors are receptor tyrosine kinases (**RTKs**), are receptor serine/threonine kinases (**RSTKs**), or have no intrinsic enzymatic activity (null).

   A. ___ Kit

   B. ___ B-cell receptor

   C. ___ FLT3

   D. ___ TGF-β receptor

7.3 **Short Answer:** How can scaffolds and adaptors modulate signaling responses if they have no intrinsic enzymatic activity?

7.4 **Multiple Choice:** Which of the following alterations would result in increased activity of Ras (one or more may apply):

   A. A mutation in Ras that enhances its GTPase activity

   B. Overexpression of GEFs

   C. Depletion of GTP in the cytoplasm

   D. Overexpression of GAPs

   E. A mutation in Ras that renders it unsusceptible to the activities of GAPs

7.5 **Matching:** Order (by numbering 1–5) the downstream signaling events that occur immediately after T-cell receptor engagement:

   ____ LAT and SLP-76, scaffold proteins linked by Gads, are phosphorylated

   ____ ZAP-70, a tandem SH2 domain containing kinase, binds to ITAMs

   ____ Recruitment and activation of SH2, PH, and PX domain-containing proteins

   ____ PI 3-kinase is activated and produces PIP_3

   ____ Phosphorylation of ITAMs by Lck, a Src-family kinase

7.6 **Fill-in-the-Blanks:** For each of the following sentences, fill in the blanks with the best word selected from the list below. Each word should be used only once.

PH/PX	PLC-γ
SH2	ADAP
Vav	PI 3-kinase
LAT:Gads:SLP-76	Akt

Antigen receptor signaling leads to many downstream events that branch out into many signaling pathways or modules. These can be activated by the scaffold complex _____, the generation of PIP_3 from PIP_2 by the enzyme _____, or both. Phosphorylated tyrosine residues on the scaffold recruit proteins containing _____ domains, while PIP_3 recruits proteins containing _____ domains. These four modules are the activation of (1) _____, which cleaves PIP_2 to produce DAG and IP_3, (2) _____, which binds to PIP_3 and activates the mTOR pathway by phosphorylating and inactivating the TSC complex, (3) _____, an adaptor that recruits SKAP55 and RIAM, and (4) _____, a GEF that leads to activation of WASp. These pathways ultimately lead to increased transcription of key genes, increased cellular metabolism, increased cellular adhesion, and actin polymerization, respectively.

7.7 **Matching:** Match the small G protein (GTPase) to its function.

   A. _____ Ras

   B. _____ Cdc42 (Rho family)

   C. _____ Rap1

   D. _____ Rheb

   i. WASp; actin polymerization

   ii. mTOR; cellular metabolism

   iii. LFA-1 aggregation; cellular adhesion

   iv. MAPK pathway; cellular proliferation

7.8 Multiple Choice: Which of the follow statements is false?

A. K63 polyubiquitination leads to downstream cellular signaling.

B. K48 polyubiquitination leads to degradation by the proteasome.

C. Out of the three families of enzymes involved in ubiquitination—E1 (ubiquitin-activating) enzymes, E2 (ubiquitin-conjugating) enzymes, and E3 enzymes (ubiquitin ligases)—Cbl is an E3 enzyme that selects its target via its SH2 domain.

D. Mono- or di-ubiquitination of surface receptors leads to degradation by the proteasome.

7.9 Matching: Match the human disease with the gene that is defective:

A. ____ X-linked agammaglobulinemia	**i.** ORAI1	
B. ____ Wiscott–Aldrich syndrome	**ii.** NEMO	
C. ____ Severe combined immunodeficiency	**iii.** Btk	
D. ____ X-linked hypohidrotic ectodermal dysplasia and immunodeficiency	**iv.** WASp	

7.10 Fill-in-the-Blanks: Name the corresponding receptor or signaling component in its respective T/B cell counterpart:

T cell	B cell
CD3ε:CD3δ:(CD3γ)2:(CD3ζ)$_2$	**A.** _____
B. _____	CD21:CD19:CD81
CD28	**C.** _____
D. _____	Fyn, Blk, Lyn
E. _____	Syk
LAT:Gads:SLP-76	**F.** _____

7.11 True or False: CTLA-4 and PD-1 are both ITIM-containing inhibitory receptors that interfere with co-stimulatory signaling pathways by activating intracellular protein and/or lipid phosphatases.

7.12 Multiple Choice: Intravenous administration of exogenous immunoglobulin is a widely used therapy for autoimmune disorders that involve the production of autoantibodies (antibodies against self antigens). Researchers have discovered that the presence of sialic acids on the infused immunoglobulins is critical for the inhibition of autoantibody production in the patient's own B cells. Which of the following receptors could potentially be responsible for the inhibition of antibody production by B cells, given this finding?

A. FcγRIIB

B. CD22

C. PD-1

D. CD40

E. BTLA

General references.

Alberts, B., Johnson, A., Lewis, J., Raff, M., Roberts, K., and Walter, P.: *Molecular Biology of the Cell*, 6th ed. New York: Garland Science, 2015.

Gomperts, B., Kramer, I., and Tatham, P.: *Signal Transduction*. San Diego: Elsevier, 2002.

Marks, F., Klingmüller, U., and Müller-Decker, K.: *Cellular Signal Processing*. New York: Garland Science, 2009.

Samelson, L., and Shaw, A. (eds.): *Immunoreceptor Signaling*. Cold Spring Harbor, NY: Cold Spring Harbor Laboratory Press, 2010.

Section references.

7-1 Transmembrane receptors convert extracellular signals into intracellular biochemical events.

Lin, J., and Weiss, A.: **T cell receptor signalling**. *J. Cell Sci.* 2001, **114**:243–244.

Smith-Garvin, J.E., Koretzky, G.A., and Jordan, M.S.: **T cell activation**. *Annu. Rev. Immunol.* 2009, **27**:591–619.

7-2 Intracellular signal propagation is mediated by large multiprotein signaling complexes.

Balagopalan, L., Coussens, N.P., Sherman, E., Samelson, L.E., and Sommers, C.L.: **The LAT story: a tale of cooperativity, coordination, and choreography.** *Cold Spring Harbor Perspect. Biol.* 2010, **2**:a005512.

Jordan, M.S., and Koretzky, G.A.: **Coordination of receptor signaling in multiple hematopoietic cell lineages by the adaptor protein SLP-76.** *Cold Spring Harbor Perspect. Biol.* 2010, **2**:a002501.

Lim, W.A., and Pawson, T.: **Phosphotyrosine signaling: evolving a new cellular communication system.** *Cell* 2010, **142**:661–667.

Scott, J.D., and Pawson, T.: **Cell signaling in space and time: where proteins come together and when they're apart.** *Science* 2009, **326**:1220–1224.

7-3 Small G proteins act as molecular switches in many different signaling pathways.

Etienne-Manneville, S., and Hall, A.: **Rho GTPases in cell biology.** *Nature* 2002, **420**:629–635.

Mitin, N., Rossman, K.L., and Der, C.J.: **Signaling interplay in Ras superfamily function.** *Curr. Biol.* 2005, **15**:R563–R574.

7-4   Signaling proteins are recruited to the membrane by a variety of mechanisms.

Buday, L.: **Membrane-targeting of signalling molecules by SH2/SH3 domain-containing adaptor proteins.** *Biochim. Biophys. Acta* 1999, **1422**:187–204.

Kanai, F., Liu, H., Field, S.J., Akbary, H., Matsuo, T., Brown, G.E., Cantley, L.C., and Yaffe, M.B.: **The PX domains of p47phox and p40phox bind to lipid products of PI(3)K.** *Nat. Cell Biol.* 2001, **3**:675–678.

Lemmon, M.A.: **Membrane recognition by phospholipid-binding domains.** *Nat. Rev. Mol. Cell Biol.* 2008, **9**:99–111.

7-5   Post-translational modifications of proteins can both activate and inhibit signaling responses.

Ciechanover, A.: **Proteolysis: from the lysosome to ubiquitin and the proteasome.** *Nat. Rev. Mol. Cell Biol.* 2005, **6**:79–87.

Hurley, J.H., Lee, S., and Prag, G.: **Ubiquitin-binding domains.** *Biochem. J.* 2006, **399**:361–372.

Liu, Y.C., Penninger, J., and Karin, M.: **Immunity by ubiquitylation: a reversible process of modification.** *Nat. Rev. Immunol.* 2005, **5**:941–952.

Wertz, I.E., and Dixit, V.M.: **Signaling to NFκB: regulation by ubiquitination.** *Cold Spring Harbor Perspect. Biol.* 2010, **2**:a003350.

7-6   The activation of some receptors generates small-molecule second messengers.

Kresge, N., Simoni, R.D., and Hill, R.L.: **Earl W. Sutherland's discovery of cyclic adenine monophosphate and the second messenger system.** *J. Biol. Chem.* 2005, **280**:39–40.

Rall, T.W., and Sutherland, E.W.: **Formation of a cyclic adenine ribonucleotide by tissue particles.** *J. Biol. Chem.* 1958, **232**:1065–1076.

7-7   Antigen receptors consist of variable antigen-binding chains associated with invariant chains that carry out the signaling function of the receptor.

Brenner, M.B., Trowbridge, I.S., and Strominger, J.L.: **Cross-linking of human T cell receptor proteins: association between the T cell idiotype beta subunit and the T3 glycoprotein heavy subunit.** *Cell* 1985, **40**:183–190.

Call, M.E., Pyrdol, J., Wiedmann, M., and Wucherpfennig, K.W.: **The organizing principle in the formation of the T cell receptor-CD3 complex.** *Cell* 2002, **111**:967–979.

Kuhns, M.S., and Davis, M.M.: **TCR signaling emerges from the sum of many parts.** *Front. Immunol.* 2012, **3**:159.

Samelson, L.E., Harford, J.B., and Klausner, R.D.: **Identification of the components of the murine T cell antigen receptor complex.** *Cell* 1985, **43**:223–231.

Tolar, P., Sohn, H.W., Liu, W., and Pierce, S.K.: **The molecular assembly and organization of signaling active B-cell receptor oligomers.** *Immunol. Rev.* 2009, **232**:34–41.

7-8   Antigen recognition by the T-cell receptor and its co-receptors transduces a signal across the plasma membrane to initate signaling.

Klausner, R.D., and Samelson, L.E.: **T cell antigen receptor activation pathways: the tyrosine kinase connection.** *Cell* 1991, **64**:875–878.

Weiss, A., and Littman, D.R.: **Signal transduction by lymphocyte antigen receptors.** *Cell* 1994, **76**:263–274.

7-9   Antigen recognition by the T-cell receptor and its co-receptors leads to phosphorylation of ITAMs by Src-family kinases, generating the first intracellular signal in a signaling cascade.

Au-Yeung, B.B., Deindl, S., Hsu, L.-Y., Palacios, E.H., Levin, S.E., Kuriyan, J., and Weiss, A.: **The structure, regulation, and function of ZAP-70.** *Immunol. Rev.* 2009, **228**:41–57.

Bartelt, R.R., and Houtman, J.C.D.: **The adaptor protein LAT serves as an integration node for signaling pathways that drive T cell activation.** *Wiley Interdiscip. Rev. Syst. Biol. Med.* 2013, **5**:101–110.

Chan, A.C., Iwashima, M., Turck, C.W., and Weiss, A.: **ZAP-70: a 70 Kd protein-tyrosine kinase that associates with the TCR zeta chain.** *Cell* 1992, **71**:649–662.

Iwashima, M., Irving, B.A., van Oers, N.S., Chan, A.C., and Weiss, A.: **Sequential interactions of the TCR with two distinct cytoplasmic tyrosine kinases.** *Science* 1994, **263**:1136–1139.

Okkenhaug, K., and Vanhaesebroeck, B.: **PI3K in lymphocyte development, differentiation and activation.** *Nat. Rev. Immunol.* 2003, **3**:317–330.

Zhang, W., Sloan-Lancaster, J., Kitchen, J., Trible, R.P., and Samelson, L.E.: **LAT: the ZAP-70 tyrosine kinase substrate that links T cell receptor to cellular activation.** *Cell* 1998, **92**:83–92.

7-10   Phosphorylated ITAMs recruit and activate the tyrosine kinase ZAP-70.

Yan, Q., Barros, T., Visperas, P.R., Deindl, S., Kadlecek, T.A., Weiss, A., and Kuriyan, J.: **Structural basis for activation of ZAP-70 by phosphorylation of the SH2-kinase linker.** *Mol. Cell. Biol.* 2013, **33**:2188–2201.

7-11   ITAMs are also found in other receptors on leukocytes that signal for cell activation.

Lanier, L.L.: **Up on the tightrope: natural killer cell activation and inhibition.** *Nat. Immunol.* 2008, **9**:495–502.

7-12   Activated ZAP-70 phosphorylates scaffold proteins and promotes PI 3-kinase activation.

Zhang, W., Sloan-Lancaster, J., Kitchen, J., Trible, R.P., and Samelson, L.E.: **LAT: the ZAP-70 tyrosine kinase substrate that links T cell receptor to cellular activation.** *Cell* 1998, **92**:83–92.

7-13   Activated PLC-γ generates the second messengers diacylglycerol and inositol trisphosphate that lead to transcription factor activation.

Berg, L.J., Finkelstein, L.D., Lucas, J.A., and Schwartzberg, P.L.: **Tec family kinases in T lymphocyte development and function.** *Annu. Rev. Immunol.* 2005, **23**:549–600.

Yang, Y.R., Choi, J.H., Chang, J.-S., Kwon, H.M., Jang, H.-J., Ryu, S.H., and Suh, P.-G.: **Diverse cellular and physiological roles of phospholipase C-γ1.** *Adv. Biol. Regul.* 2012, **52**:138–151.

7-14   Ca²⁺ entry activates the transcription factor NFAT.

Hogan, P.G., Chen, L., Nardone, J., and Rao, A.: **Transcriptional regulation by calcium, calcineurin, and NFAT.** *Genes Dev.* 2003, **17**:2205–2232.

Hogan, P.G., Lewis, R.S., and Rao, A.: **Molecular basis of calcium signaling in lymphocytes: STIM and ORAI.** *Annu. Rev. Immunol.* 2010, **28**:491–533.

Picard, C., McCarl, C.A., Papolos, A., Khalil, S., Lüthy, K., Hivroz, C., LeDeist, F., Rieux-Laucat, F., Rechavi, G., Rao, A., *et al.* **STIM1 mutation associated with a syndrome of immunodeficiency and autoimmunity.** *N. Engl. J. Med.* 2009, **360**:1971–1980.

Prakriya, M., Feske, S., Gwack, Y., Srikanth, S., Rao, A., and Hogan, P.G.: **Orai1 is an essential pore subunit of the CRAC channel.** *Nature* 2006, **443**:230–233.

7-15   Ras activation stimulates the mitogen-activated protein kinase (MAPK) relay and induces expression of the transcription factor AP-1.

Downward, J., Graves, J.D., Warne, P.H., Rayter, S., and Cantrell, D.A.: **Stimulation of P21ras upon T-cell activation.** *Nature* 1990, **346**:719–723.

Leevers, S.J., and Marshall, C.J.: **Activation of extracellular signal-regulated kinase, ERK2, by P21ras oncoprotein.** *EMBO J.* 1992, **11**:569–574.

Roskoski, R.: **ERK1/2 MAP kinases: structure, function, and regulation.** *Pharmacol. Res.* 2012, **66**:105–143.

Shaw, A.S., and Filbert, E.L.: **Scaffold proteins and immune-cell signalling.** *Nat. Rev. Immunol.* 2009, **9**:47–56.

7-16 Protein kinase C activates the transcription factors NFκB and AP-1.

Blonska, M., and Lin, X.: **CARMA1-mediated NFκB and JNK activation in lymphocytes.** *Immunol. Rev.* 2009, **228**:199–211.

Matsumoto, R., Wang, D., Blonska, M., Li, H., Kobayashi, M., Pappu, B., Chen, Y., Wang, D., and Lin, X.: **Phosphorylation of CARMA1 plays a critical role in T cell receptor-mediated NFκB activation.** *Immunity* 2005, **23**:575–585.

Rueda, D., and Thome, M.: **Phosphorylation of CARMA1: the link(Er) to NFκB activation.** *Immunity* 2005, **23**:551–553.

Sommer, K., Guo, B., Pomerantz, J.L., Bandaranayake, A.D., Moreno-García, M.E., Ovechkina, Y.L., and Rawlings, D.J.: **Phosphorylation of the CARMA1 linker controls NFκB activation.** *Immunity* 2005, **23**:561–574.

Thome, M., and Weil, R.: **Post-translational modifications regulate distinct functions of CARMA1 and BCL10.** *Trends Immunol.* 2007, **28**:281–288.

7-17 PI 3-kinase activation upregulates cellular metabolic pathways via the serine/threonine kinase Akt.

Gamper, C.J., and Powell, J.D.: **All PI3Kinase signaling is not mTOR: dissecting mTOR-dependent and independent signaling pathways in T cells.** *Front. Immunol.* 2012, **3**:312.

Kane, L.P., and Weiss, A.: **The PI-3 kinase/Akt pathway and T cell activation: pleiotropic pathways downstream of PIP3.** *Immunol. Rev.* 2003, **192**:7–20.

Pearce, E.L.: **Metabolism in T cell activation and differentiation.** *Curr. Opin. Immunol.* 2010, **22**:314–320.

7-18 T-cell receptor signaling leads to enhanced integrin-mediated cell adhesion.

Bezman, N., and Koretzky, G.A.: **Compartmentalization of ITAM and integrin signaling by adapter molecules.** *Immunol. Rev.* 2007, **218**:9–28.

Mor, A., Dustin, M.L., and Philips, M.R.: **Small GTPases and LFA-1 reciprocally modulate adhesion and signaling.** *Immunol. Rev.* 2007, **218**:114–125.

7-19 T-cell receptor signaling induces cytoskeletal reorganization by activating the small GTPase Cdc42.

Burkhardt, J.K., Carrizosa, E., and Shaffer, M.H.: **The actin cytoskeleton in T cell activation.** *Annu. Rev. Immunol.* 2008, **26**:233–259.

Tybulewicz, V.L.J., and Henderson, R.B.: **Rho family GTPases and their regulators in lymphocytes.** *Nat. Rev. Immunol.* 2009, **9**:630–644.

7-20 The logic of B-cell receptor signaling is similar to that of T-cell receptor signaling, but some of the signaling components are specific to B cells.

Cambier, J.C., Pleiman, C.M., and Clark, M.R.: **Signal transduction by the B cell antigen receptor and its coreceptors.** *Annu. Rev. Immunol.* 1994, **12**:457–486.

DeFranco, A.L., Richards, J.D., Blum, J.H., Stevens, T.L., Law, D.A., Chan, V.W., Datta, S.K., Foy, S.P., Hourihane, S.L., and Gold, M.R.: **Signal transduction by the B-cell antigen receptor.** *Ann. N.Y. Acad. Sci.* 1995, **766**:195–201.

Harwood, N.E., and Batista, F.D.: **Early events in B cell activation.** *Annu. Rev. Immunol.* 2010, **28**:185–210.

Koretzky, G.A., Abtahian, F., and Silverman, M.A.: **SLP76 and SLP65: complex regulation of signalling in lymphocytes and beyond.** *Nat. Rev. Immunol.* 2006, **6**:67–78.

Kurosaki, T., and Hikida, M.: **Tyrosine kinases and their substrates in B lymphocytes.** *Immunol. Rev.* 2009, **228**:132–148.

7-21 The cell-surface protein CD28 is a required co-stimulatory signaling receptor for naive T cell activation.

Acuto, O., and Michel, F.: **CD28-mediated co-stimulation: a quantitative support for TCR signalling.** *Nat. Rev. Immunol.* 2003, **3**:939–951.

Frauwirth, K.A., Riley, J.L., Harris, M.H., Parry, R.V., Rathmell, J.C., Plas, D.R., Elstrom, R.L., June, C.H., and Thompson, C.B.: **The CD28 signaling pathway regulates glucose metabolism.** *Immunity* 2002, **16**: 769–777.

Sharpe, A.H.: **Mechanisms of costimulation.** *Immunol. Rev.* 2009, **229**:5–11.

7-22 Maximal activation of PLC-γ, which is important for transcription factor activation, requires a co-stimulatory signal induced by CD28.

Chen, L., and Flies, D.B.: **Molecular mechanisms of T cell co-stimulation and co-inhibition.** *Nat. Rev. Immunol.* 2013, **13**:227–242.

7-23 TNF receptor superfamily members augment T-cell and B-cell activation.

Chen, L., and Flies, D.B.: **Molecular mechanisms of T cell co-stimulation and co-inhibition.** *Nat. Rev. Immunol.* 2013, **13**:227–242.

Rickert, R.C., Jellusova, J., and Miletic, A.V.: **Signaling by the tumor necrosis factor receptor superfamily in B-cell biology and disease.** *Immunol. Rev.* 2011, **244**:115–133.

7-24 Inhibitory receptors on lymphocytes downregulate immune responses by interfering with co-stimulatory signaling pathways.

Qureshi, O.S., Zheng, Y., Nakamura, K., Attridge, K., Manzotti, C., Schmidt, E.M., Baker, J., Jeffery, L.E., Kaur, S., Briggs, Z., *et al.*: **Trans-endocytosis of CD80 and CD86: a molecular basis for the cell-extrinsic function of CTLA-4.** *Science* 2011, **332**:600–603.

Rudd, C.E., Taylor, A., and Schneider, H.: **CD28 and CTLA-4 coreceptor expression and signal transduction.** *Immunol. Rev.* 2009, **229**:12–26.

7-25 Inhibitory receptors on lymphocytes downregulate immune responses by recruiting protein or lipid phosphatases.

Acuto, O., Di Bartolo, V., and Michel, F.: **Tailoring T-cell receptor signals by proximal negative feedback mechanisms.** *Nat. Rev. Immunol.* 2008, **8**:699–712.

Chen, L., and Flies, D.B.: **Molecular mechanisms of T cell co-stimulation and co-inhibition.** *Nat. Rev. Immunol.* 2013, **13**:227–242.

The Development of B and T Lymphocytes

8

The production of new lymphocytes, or **lymphopoiesis**, takes place in specialized lymphoid tissues—the **central** (or **primary**) **lymphoid tissues**—which are the bone marrow for most B cells and the thymus for most T cells. Precursors for both populations originate in the bone marrow, but whereas B cells complete most of their development there, the precursors of most T cells migrate to the thymus, where they develop into mature T cells. A major goal of lymphopoiesis is to generate a diverse repertoire of B-cell receptors and T-cell receptors on circulating B and T cells, respectively, thereby enabling an individual to make adaptive immune responses against the wide range of pathogens encountered during a lifetime. In the fetus and the juvenile, the central lymphoid tissues are the sources of large numbers of new lymphocytes, which migrate to populate the **peripheral lymphoid tissues** (also called **secondary lymphoid tissues**) such as lymph nodes, spleen, and mucosal lymphoid tissue. In mature individuals, the development of new T cells in the thymus slows down, and peripheral T-cell numbers are maintained by the division of mature T cells outside the central lymphoid organs. New B cells, in contrast, are continually produced from the bone marrow, even in adults. This chapter will focus on the development of T cells and B cells from their uncommitted progenitors, with an emphasis on the major populations of CD4$^+$ and CD8$^+$ T cells and B cells. The development of additional subsets of T cells and B cells, such as invariant NKT (iNKT) cells, T_{reg} cells, $\gamma{:}\delta$ TCR$^+$ T cells, B-1 B cells, and marginal zone B cells will be briefly discussed.

The structure of the antigen-receptor genes expressed by B cells and T cells, and the mechanisms by which a complete antigen receptor is assembled, were described in Chapters 4 and 5. Once an antigen receptor has been formed, rigorous testing is required to select lymphocytes that carry useful antigen receptors—that is, antigen receptors that can recognize a wide spectrum of pathogens and yet will not react against an individual's own cells. Given the incredible diversity of receptors that the rearrangement process can generate, it is important that those lymphocytes that mature are likely to be useful in recognizing and responding to foreign antigens, especially as an individual can express only a small fraction of the total possible receptor repertoire in his or her lifetime. We describe how the specificity and affinity of the receptor for self ligands are tested to determine whether the immature lymphocyte will either survive and join the mature repertoire, or die. In general, it seems that developing lymphocytes whose receptors interact weakly with self antigens, or that bind self antigens in a particular way, receive a signal that enables them to survive. This process, known as **positive selection**, is particularly critical in the development of $\alpha{:}\beta$ T cells, which recognize composite antigens consisting of peptides bound to MHC molecules, because it ensures that an individual's T cells will be able to respond to peptides bound to one's own MHC molecules.

In contrast, lymphocytes with strongly self-reactive receptors must be eliminated to prevent autoimmune reactions; this process of **negative selection** is one of the ways in which the immune system is made self-tolerant. The default fate of developing lymphocytes, in the absence of any signal being received from the receptor, is death by apoptosis, and as we will see, the vast majority of developing lymphocytes die before emerging from the central lymphoid organs or before completing maturation in the peripheral lymphoid organs.

In this chapter we describe the different stages of the development of B cells and T cells in mice and humans from the uncommitted stem cell in the bone marrow up to the mature, functionally specialized lymphocyte with its unique antigen receptor ready to respond to a foreign antigen. The final stages in the life history of a mature lymphocyte, in which an encounter with its antigen activates it to become an effector or memory lymphocyte, are discussed in Chapters 9–11. We now know that the B- and T-cell development that predominates during late fetal life and after birth is distinct from waves of lymphocyte development that take place earlier in fetal ontogeny. These earlier waves originate from stem cells found in the fetal liver and in even more primitive hematopoietic tissues in the developing embryo. Unlike the lymphocytes that develop from bone marrow stem cells, B and T cells that develop from these early fetal progenitors generally populate mucosal and epithelial tissues and function in innate immune responses. In the adult, these subsets of lymphocytes are minority populations in secondary lymphoid tissues. This chapter will focus on B and T cells that develop from bone marrow stem cells and that comprise the cells of the adaptive immune response (see Figs 1.7 and 1.20). The chapter is divided into three parts. The first two describe B-cell and T-cell development, respectively. In the third section, we discuss the positive and negative selection of T cells in the thymus.

Fig. 8.1 B cells develop in the bone marrow and migrate to peripheral lymphoid organs, where they can be activated by antigens. In the first phase of development, progenitor B cells in the bone marrow rearrange their immunoglobulin genes. This phase is independent of antigen but is dependent on interactions with bone marrow stromal cells (first panels). It ends in an immature B cell that carries an antigen receptor in the form of cell-surface IgM (second panels), and in the second phase it can now interact with antigens in its environment. Immature B cells that are strongly stimulated by antigen at this stage either die or are inactivated in a process of negative selection, thus removing many self-reactive B cells from the repertoire. In the third phase of development, the surviving immature B cells emerge into the periphery and mature to express IgD as well as IgM. They can now be activated by encounter with their specific foreign antigen in a peripheral lymphoid organ (third panels). Activated B cells proliferate, and differentiate into antibody-secreting plasma cells and long-lived memory cells (fourth panels).

Development of B lymphocytes.

The main phases of a B lymphocyte's life history are shown in **Fig. 8.1**. The stages in both B-cell and T-cell development are defined mainly by the successive steps in the assembly and expression of functional antigen-receptor genes.

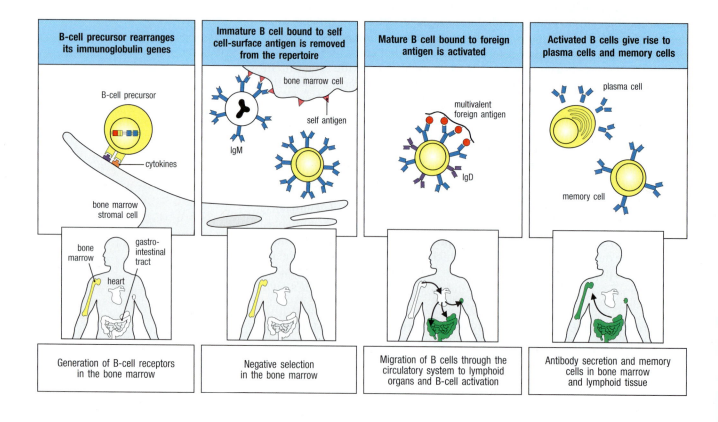

At each step of lymphocyte development, the progress of gene rearrangement is monitored; the major recurring theme is that successful gene rearrangement leads to the production of a protein chain that serves as a signal for the cell to progress to the next stage. We will see that a developing B cell is presented with opportunities for multiple rearrangements that increase the likelihood of expressing a functional antigen receptor, but that there are also checkpoints that reinforce the requirement that each B cell express just one receptor specificity. We will start by looking at how the earliest recognizable cells of the B-cell lineage develop from the multipotent hematopoietic stem cells in the bone marrow, and at what point the B-cell and T-cell lineages diverge.

8-1   Lymphocytes derive from hematopoietic stem cells in the bone marrow.

The cells of the lymphoid lineage—B cells, T cells, and innate lymphoid cells (ILCs)—are all derived from common lymphoid progenitor cells, which themselves derive from the multipotent **hematopoietic stem cells (HSCs)** that give rise to all blood cells (see Fig. 1.3). Development from the precursor stem cell into cells that are committed to becoming B cells or T cells follows the basic principles of cell differentiation. Properties that are essential for the function of the mature cell are gradually acquired, along with the loss of properties that are more characteristic of the immature cell. In the case of lymphocyte development, cells become committed first to the lymphoid lineage, as opposed to the myeloid, and then to either the B-cell or the T-cell lineage (Fig. 8.2).

The specialized microenvironment of the bone marrow provides signals both for the development of lymphocyte progenitors from hematopoietic stem cells and for the subsequent differentiation of B cells. Such signals act on the developing lymphocytes to switch on key genes that direct the developmental program and are produced by the network of specialized nonlymphoid connective tissue **stromal cells** that are in intimate contact with the developing lymphocytes (Fig. 8.3). The contribution of the stromal cells is twofold. First, they form specific adhesive contacts with the developing lymphocytes by interactions between cell-adhesion molecules and their ligands. Second, they provide soluble and membrane-bound cytokines and chemokines that control lymphocyte differentiation and proliferation.

The hematopoietic stem cells first differentiate into **multipotent progenitor cells (MPPs)**, which can produce both lymphoid and myeloid cells but are no longer self-renewing stem cells. Multipotent progenitors express a cell-surface receptor tyrosine kinase known as FLT3 that binds the membrane-bound FLT3 ligand on stromal cells. Additionally, MPPs express transcription factors and

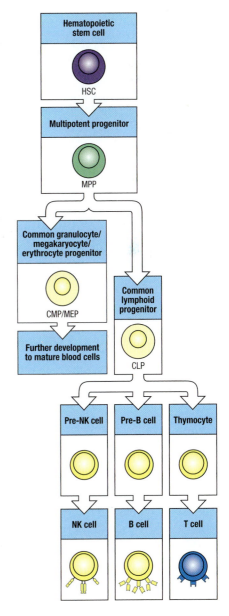

Fig. 8.2 A multipotent hematopoietic stem cell generates all the cells of the immune system. In the bone marrow or other hematopoietic sites, the multipotent stem cell gives rise to cells with progressively more limited potential. A simplified progression is shown here. The multipotent progenitor (MPP), for example, has lost its stem-cell properties. The first branch leads to cells with myeloid and erythroid potential, on the one hand (CMPs and MEPs), and, on the other, to the common lymphoid progenitors (CLPs), with lymphoid potential. The former give rise to all nonlymphoid cellular blood elements, including circulating monocytes and granulocytes, as well as the macrophages and dendritic cells that reside in tissues and peripheral lymphoid organs (not shown). The CLP population is heterogeneous and single cells can give rise to NK cells, T cells, or B cells through successive stages of differentiation in either the bone marrow or thymus. There may be considerable plasticity in these pathways, in that in certain circumstances progenitor cells may switch their commitment. For example, a progenitor cell may give rise to either B cells or macrophages; however, for simplicity these alternative pathways are not shown. Some dendritic cells are also thought to be derived from the lymphoid progenitor.

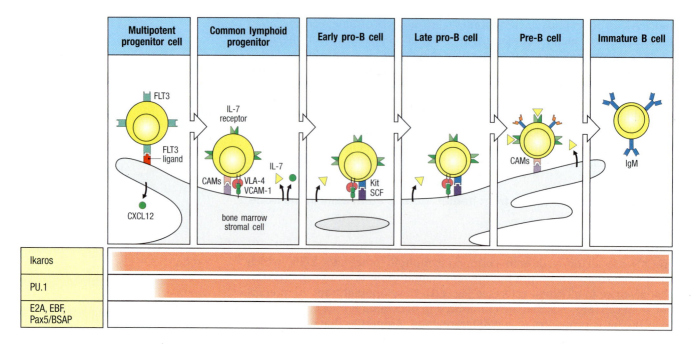

	Multipotent progenitor cell	Common lymphoid progenitor	Early pro-B cell	Late pro-B cell	Pre-B cell	Immature B cell

Ikaros						
PU.1						
E2A, EBF, Pax5/BSAP						

Fig. 8.3 The early stages of B-cell development are dependent on bone marrow stromal cells. Interaction of B-cell progenitors with bone marrow stromal cells is required for development to the immature B-cell stage. The designations pro-B cell and pre-B cell refer to defined phases of B-cell development, as described in Fig. 8.4. Multipotent progenitor cells express the receptor tyrosine kinase FLT3, which binds to its ligand on stromal cells. Signaling through FLT3 is required for differentiation to the next stage, the common lymphoid progenitor. The chemokine CXCL12 (SDF-1) acts to retain stem cells and lymphoid progenitors at appropriate stromal cells in the bone marrow. The receptor for interleukin-7 (IL-7) is present at this stage, and IL-7 produced by stromal cells is required for the development of B-lineage cells. Progenitor cells bind to the adhesion molecule VCAM-1 on stromal cells through the integrin VLA-4 and also interact through other cell-adhesion molecules (CAMs). The adhesive interactions promote the binding of the receptor tyrosine kinase Kit (CD117) on the surface of the pro-B cell to stem-cell factor (SCF) on the stromal cell, which activates the kinase and induces the proliferation of B-cell progenitors. The actions of the listed transcription factors in B-cell development are discussed in the text. The pink horizontal bands denote the expression of particular proteins at the indicated stages of development.

receptors that are required for the development of multiple hematopoietic lineages, such as the transcription factor PU.1 and the receptor c-kit. In the next stage, MPPs produce two subsets of progenitor cells that give rise to all the lymphocyte lineages. One progenitor cell, as yet unnamed, produces the ILC subsets, ILC1, ILC2, and ILC3 cells. A second progenitor cell arising from the MPP is known as the **common lymphoid progenitor (CLP)**. Differentiation of MPPs into CLPs requires signaling through the FLT3 receptor expressed on MPPs. Progenitor cell transfer and lineage repopulation experiments have shown that the CLP population is actually heterogeneous and represents a continuum of cells with decreasing multipotent potential. A subset of CLP cells with the broadest potential is able to generate B cells, T cells, and NK cells. A second subset of CLPs is able to generate only B cells and T cells, and a third subset of CLPs is committed exclusively to the B-cell lineage. B-cell-committed CLPs give rise to **pro-B cells** (see Fig. 8.3).

The production of lymphocyte progenitors from the multipotent progenitor cell is accompanied by expression of the receptor for interleukin-7 (IL-7), which is induced by FLT3 signaling together with the activity of PU.1. The cytokine IL-7, secreted by bone marrow stromal cells, is essential for the growth and survival of developing B cells in mice (but possibly not in humans). The IL-7 receptor is composed of two polypeptides, the IL-7 receptor α chain and the common cytokine receptor γ chain (γ-c), so called because it is also a subunit of five additional cytokine receptors. This family of cytokine receptors includes the receptors for IL-2, IL-4, IL-9, IL-15, and IL-21, in addition to IL-7. These receptors also share the tyrosine kinase Jak3, a signaling protein that binds exclusively to γ-c and is required for productive signaling by each of the receptors.

Due to the importance of IL-7 for murine B-cell development, mice with a genetic deficiency in IL-7, IL-7 receptor α, γ-c, or Jak3 all exhibit a severe block in B-cell development.

Another essential factor for B-cell development is stem-cell factor (SCF), a membrane-bound cytokine present on bone marrow stromal cells that stimulates the growth of hematopoietic stem cells and the earliest B-lineage progenitors. SCF interacts with the receptor tyrosine kinase Kit on the precursor cells (see Fig. 8.3). The chemokine CXCL12 (stromal cell-derived factor 1, SDF-1) is also essential for the early stages of B-cell development. It is produced constitutively by bone marrow stromal cells, and one of its roles may be to retain developing B-cell precursors in the marrow microenvironment. **Thymic stroma-derived lymphopoietin (TSLP)** resembles IL-7 and binds a receptor that includes the IL-7 receptor α chain, but not γ-c. Despite its name, TSLP may promote B-cell development in the embryonic liver and, in the perinatal period at least, in the mouse bone marrow.

A definitive B-cell stage, the pro-B cell, is specified by induction of the B-lineage-specific transcription factor E2A. It is not clear what initiates the expression of E2A in some progenitors, but it is known that the transcription factors PU.1 and Ikaros are required for E2A expression. E2A then induces the expression of the early B-cell factor (EBF). IL-7 signaling promotes the survival of these committed progenitors, while E2A and EBF act together to drive the expression of proteins that determine the pro-B-cell state.

As B-lineage cells mature, they migrate within the marrow, remaining in contact with the stromal cells. The earliest stem cells lie in a region called the **endosteum**, which lines the inner cavity of the long bones such as the femur and tibia. Developing B-lineage cells make contact with reticular stromal cells in the trabecular spaces, and as they mature they move toward the central sinus of the marrow cavity. The final stages of development of immature B cells into mature B cells occur in peripheral lymphoid organs such as the spleen, which we describe in Sections 8-7 and 8-8 of this chapter.

X-linked Severe Combined Immunodeficiency

8-2   B-cell development begins by rearrangement of the heavy-chain locus.

The stages of B-cell development are, in the order they occur, **early pro-B cell**, **late pro-B cell**, **large pre-B cell**, **small pre-B cell**, **immature B cell**, and **mature B cell** (Fig. 8.4). Rearrangement of the heavy-chain locus is initiated in the pro-B cell when E2A and EBF induce the expression of several key proteins that enable gene rearrangement to occur, including the RAG-1 and RAG-2 components of the V(D)J recombinase (see Chapter 5). Only one gene locus is rearranged at a time, in a fixed sequence. The first rearrangement to take place is the joining of a D gene segment to a J segment at the immunoglobulin heavy-chain (IgH) locus. D to J_H rearrangement takes place mostly in the early pro-B-cell stage, but can be seen as early as the common lymphoid progenitor. In the absence of E2A or EBF this initial rearrangement event fails to occur. Another key protein induced by E2A and EBF is the transcription factor Pax5, one isoform of which is known as the B-cell activator protein (BSAP) (see Fig. 8.3). Among the targets of Pax5 are the gene for the B-cell co-receptor component CD19 and the gene for Igα, a signaling component of both the pre-B-cell receptor and the B-cell receptor (see Section 7-7). In the absence of Pax5, pro-B cells fail to develop further down the B-cell pathway but can be induced to give rise to T cells and myeloid cell types, indicating that Pax5 is required for commitment of the pro-B cell to the B-cell lineage. Pax5 also induces the expression of the B-cell linker protein (BLNK), an SH2-containing scaffold protein that is required for further development of the pro-B cell and for signaling from the mature B-cell antigen receptor

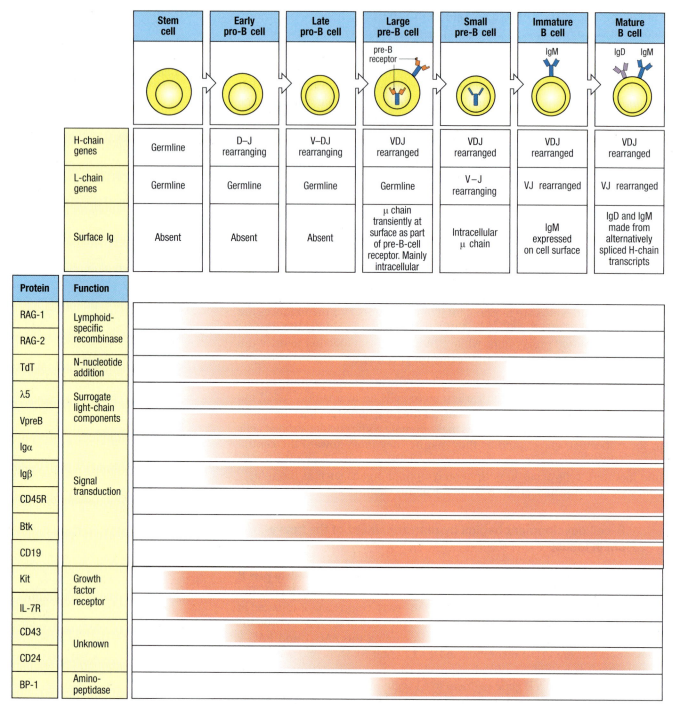

Protein	Function	Stem cell	Early pro-B cell	Late pro-B cell	Large pre-B cell	Small pre-B cell	Immature B cell	Mature B cell
					pre-B receptor		IgM	IgD IgM
H-chain genes		Germline	D–J rearranging	V–DJ rearranging	VDJ rearranged	VDJ rearranged	VDJ rearranged	VDJ rearranged
L-chain genes		Germline	Germline	Germline	Germline	V–J rearranging	VJ rearranged	VJ rearranged
Surface Ig		Absent	Absent	Absent	μ chain transiently at surface as part of pre-B-cell receptor. Mainly intracellular	Intracellular μ chain	IgM expressed on cell surface	IgD and IgM made from alternatively spliced H-chain transcripts

The development of a B-lineage cell proceeds through several stages marked by the rearrangement and expression of the immunoglobulin genes.

Fig. 8.4 **The development of a B-lineage cell proceeds through several stages marked by the rearrangement and expression of the immunoglobulin genes.** The stem cell has not yet begun to rearrange its immunoglobulin (Ig) gene segments; they are in the germline configuration found in all nonlymphoid cells. The heavy-chain (H-chain) locus rearranges first. Rearrangement of a D gene segment to a J$_H$ gene segment starts in the common lymphoid progenitor and occurs mostly in early pro-B cells, generating late pro-B cells in which V$_H$ to DJ$_H$ rearrangement occurs. A successful VDJ$_H$ rearrangement leads to the expression of a complete immunoglobulin heavy chain as part of the pre-B-cell receptor, which signals via Igα, Igβ, and Btk (see Fig 7.27). Once this occurs, the cell is stimulated to become a large pre-B cell, which proliferates to become small resting pre-B cells; at this point the cells cease expression of the surrogate light chains (λ5 and VpreB) and express the μ heavy chain alone in the cytoplasm. Small pre-B cells reexpress the RAG proteins and start to rearrange the light-chain (L-chain) genes. Upon successfully assembling a light-chain gene, a cell becomes an immature B cell that expresses a complete IgM molecule at the cell surface, which also signals via Igα and Igβ. Mature B cells produce a δ heavy chain as well as a μ heavy chain, by a mechanism of alternative mRNA splicing (see Fig. 5.17), and are marked by the additional appearance of IgD on the cell surface. All stages through the development of immature B cells takes place in the bone marrow; the final maturation to IgM⁺IgD⁺ mature B cells occurs in the spleen. The earliest B-lineage surface markers are CD19 and CD45R (B220 in the mouse), which are expressed throughout B-cell development. A pro-B cell is also distinguished by the expression of CD43 (a marker of unknown function), Kit (CD117), and the IL-7 receptor. A late pro-B cell starts to express CD24 (a marker of unknown function). A pre-B cell is phenotypically distinguished by the expression of the enzyme BP-1, whereas Kit is no longer expressed.

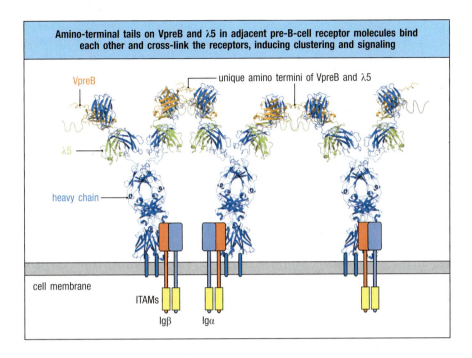

Amino-terminal tails on VpreB and λ5 in adjacent pre-B-cell receptor molecules bind each other and cross-link the receptors, inducing clustering and signaling

VpreB

unique amino termini of VpreB and λ5

λ5

heavy chain

cell membrane

ITAMs

Igβ   Igα

Fig. 8.6 The pre-B-cell receptor initiates signaling through spontaneous dimerization induced by the unique regions of VpreB and λ5. Two surrogate protein chains, VpreB (orange) and λ5 (green), substitute for a light chain and bind to a heavy chain, thus allowing its surface expression. VpreB substitutes for the light-chain V region in this surrogate interaction, while λ5 takes the part of the light-chain constant region. Both VpreB and λ5 contain 'unique' amino-terminal regions that are not present in other immunoglobulin-like domains, shown here as unstructured tails extending out from the globular domains. These amino-terminal regions associated with one pre-B-cell receptor can interact with the corresponding regions on the adjacent pre-B-cell receptor, promoting the spontaneous formation of pre-B-cell receptor dimers on the cell surface. Dimerization generates signaling from the pre-B-cell receptor that is dependent on the presence of the ITAM-containing signaling chains Igα and Igβ. The signals cause the inhibition of RAG-1 and RAG-2 expression and the proliferation of the large pre-B cell. Courtesy of Chris Garcia.

as they function to transduce signals through the antigen receptor on mature B cells (see Section 7-7).

Formation of the pre-B-cell receptor and signaling through this receptor provide an important checkpoint that mediates the transition between the pro-B cell and the pre-B cell. In mice that either lack λ5 or have mutant heavy-chain genes that cannot produce the transmembrane domain, the pre-B-cell receptor cannot be formed and B-cell development is blocked after heavy-chain gene rearrangement. In normal B-cell development, the pre-B-cell receptor complex is expressed transiently, perhaps because the production of λ5 mRNA stops as soon as pre-B-cell receptors begin to be formed. Although present at only low levels on the cell surface, the pre-B-cell receptor generates signals required for the transition from pro-B cell to pre-B cell. No antigen or other external ligand seems to be involved in signaling by the receptor. Instead, pre-B-cell receptors are thought to interact with each other, forming dimers or oligomers that generate signals as described in Section 7-16. Dimerization involves 'unique' regions in the amino termini of λ5 and VpreB proteins that are not present in other immunoglobulin-like domains and which mediate the cross-linking of adjacent pre-B-cell receptors on the cell surface (**Fig. 8.6**). Pre-B-cell receptor signaling requires the scaffold protein BLNK and Bruton's tyrosine kinase (Btk), an intracellular Tec-family tyrosine kinase (see Section 7-20). In humans and mice, deficiency of BLNK leads to a block in B-cell development at the pro-B-cell stage. In humans, mutations in the *BTK* gene cause a profound B-lineage-specific immune deficiency, **Bruton's X-linked agamma-globulinemia (XLA)**, in which no mature B cells are produced. The block in B-cell development caused by mutations in *BTK* is almost total, interrupting the transition from pre-B cell to immature B cell. A similar, but less severe, defect called **X-linked immunodeficiency**, or **xid**, arises from mutations in the *Btk* gene in mice.

X-linked Agammaglobulinemia

8-4   Pre-B-cell receptor signaling inhibits further heavy-chain locus rearrangement and enforces allelic exclusion.

The signaling generated by pre-B-cell receptor clustering halts further rearrangement of the heavy-chain locus and allows the pro-B cell to become sensitive to IL-7. This induces cell proliferation, initiating the transition to the large

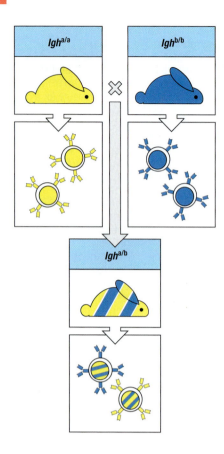

Fig. 8.7 Allelic exclusion in individual B cells. Most species have genetic polymorphisms of the constant regions of their immunoglobulin heavy-chain and light-chain genes; these polymorphisms lead to amino acid differences between the encoded proteins. These variants of heavy-chain or light-chain proteins expressed by different individuals in a species are known as allotypes. In rabbits, for example, all of the B cells in an individual homozygous for the *a* allele of the immunoglobulin heavy-chain locus (*Igh*$^{a/a}$) will express immunoglobulin of allotype a, whereas in an individual homozygous for the *b* allele (*Igh*$^{b/b}$) all the B cells make immunoglobulin of allotype b. In a heterozygous animal (*Igh*$^{a/b}$), which carries the *a* allele at one of the Igh loci and the *b* allele at the other, individual B cells can be shown to express surface immunoglobulin of either the a-allotype or the b-allotype, but not both (bottom panel). This allelic exclusion reflects the productive rearrangement of only one of the two Igh alleles in the B cell, because the production of a successfully rearranged immunoglobulin heavy chain forms a pre-B-cell receptor, which signals the cessation of further heavy-chain gene rearrangement.

pre-B cell. Successful rearrangements at both heavy-chain alleles could result in a B cell producing two receptors of different antigen specificities. To prevent this, signaling by the pre-B-cell receptor enforces **allelic exclusion**, the state in which only one of the two alleles of a gene is expressed in a diploid cell. Allelic exclusion, which occurs at both the heavy-chain locus and the light-chain loci, was discovered nearly 50 years ago and provided one of the original pieces of experimental support for the theory that one lymphocyte expresses one type of antigen receptor (Fig. 8.7).

Signaling from the pre-B-cell receptor promotes heavy-chain allelic exclusion in three ways. First, it reduces V(D)J recombinase activity by directly reducing the expression of the *RAG-1* and *RAG-2* genes. Second, it further reduces levels of *RAG-2* by indirectly causing this protein to be targeted for degradation, which occurs when RAG-2 is phosphorylated in response to the entry of the pro-B cell into S phase (the DNA synthesis phase) of the cell cycle. Finally, pre-B-cell receptor signaling reduces access of the heavy-chain locus to the recombinase machinery, although the precise details of this are not clear. At a later stage of B-cell development, RAG proteins will again be expressed in order to carry out light-chain locus rearrangement, but at that point the heavy-chain locus does not undergo further rearrangement. In the absence of pre-B-cell receptor signaling, allelic exclusion of the heavy-chain locus does not occur. Since a second important role of pre-B-cell receptor signaling is to stimulate proliferative expansion of B-cell precursors with a successful heavy-chain rearrangement, a deficiency in this signal causes a profound reduction in the numbers of pre-B cells and mature B cells that develop.

8-5 Pre-B cells rearrange the light-chain locus and express cell-surface immunoglobulin.

The transition from the pro-B-cell to the large pre-B-cell stage is accompanied by several rounds of cell division, expanding the population of cells with successful in-frame joins by about 30- to 60-fold before they become resting small pre-B cells. A large pre-B cell with a particular rearranged heavy-chain gene therefore gives rise to numerous small pre-B cells. RAG proteins are produced again in the small pre-B cells, and rearrangement of the light-chain locus begins. Each of these cells can make a different rearranged light-chain gene, and so cells with many different antigen specificities are generated from a single pre-B cell, which makes an important contribution to overall B-cell receptor diversity.

Light-chain rearrangement also exhibits allelic exclusion. Rearrangements at the light-chain locus generally take place at only one allele at a time, a process regulated by a mechanism not currently understood. The light-chain loci lack D segments, and rearrangement occurs by V to J joining; and if a particular VJ rearrangement fails to produce a functional light chain, repeated rearrangements of unused V and J gene segments at the same allele can occur (Fig. 8.8). Several attempts at productive rearrangement of a light-chain gene can therefore be made on one chromosome before initiating any rearrangements on the second chromosome. This greatly increases the chances of eventually generating an intact light chain, especially as there are two different light-chain loci. As a result, many cells that reach the pre-B-cell stage succeed in generating progeny that bear intact IgM molecules and can be classified as immature B cells. Figure 8.4 lists some of the proteins involved in V(D)J recombination and shows how their expression is regulated throughout B-cell development. Figure 8.5 summarizes the stages of B-cell development up to the point of assembly of a complete surface immunoglobulin. Developing B cells that fail to assemble a complete surface immunoglobulin undergo apoptosis in the bone marrow, and are eliminated from the B-cell pool.

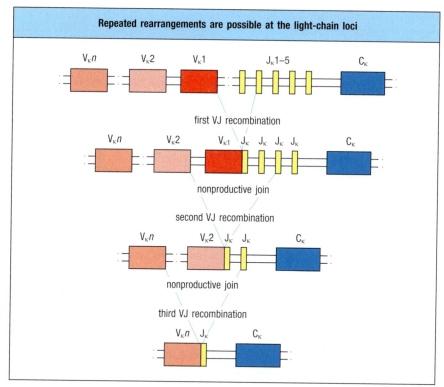

Repeated rearrangements are possible at the light-chain loci

first VJ recombination

nonproductive join

second VJ recombination

nonproductive join

third VJ recombination

Fig. 8.8 Nonproductive light-chain gene rearrangements can be rescued by further rearrangement. The organization of the light-chain loci in mice and humans offers many opportunities for the rescue of pre-B cells that initially make an out-of-frame rearrangement. Light-chain rescue is illustrated here at the human κ locus. If the first rearrangement is nonproductive, a 5′ V_κ gene segment can recombine with a 3′ J_κ gene segment to remove the out-of-frame join located between them and to replace it with a new rearrangement. In principle, this can happen up to five times on each chromosome, because there are five functional J_κ gene segments in humans. If all rearrangements of κ-chain genes fail to yield a productive light-chain join, λ-chain gene rearrangement may succeed (not shown).

As well as allelic exclusion, light chains also display **isotypic exclusion**, that is, the expression of only one type of light chain—κ or λ—by an individual B cell. Again, the mechanism regulating this process is not known. In mice and humans, the κ light-chain locus tends to rearrange before the λ locus. This was first deduced from the observation that myeloma cells secreting λ light chains generally have both their κ and λ light-chain genes rearranged, whereas in myelomas secreting κ light chains, generally only the κ genes are rearranged. This order is occasionally reversed, however, and λ gene rearrangement does not absolutely require the previous rearrangement of the κ genes. The ratios of κ-expressing versus λ-expressing mature B cells vary from one extreme to the other in different species. In mice and rats it is 95% κ to 5% λ, in humans it is typically 65%:35%, and in cats it is 5%:95%, the opposite of that in mice. These ratios correlate most strongly with the number of functional V_κ and V_λ gene segments in the genome of the species. They also reflect the kinetics and efficiency of gene segment rearrangements. The κ:λ ratio in the mature lymphocyte population is useful in clinical diagnostics, because an aberrant κ:λ ratio indicates the dominance of one clone and the presence of a lymphoproliferative disorder.

8-6   Immature B cells are tested for autoreactivity before they leave the bone marrow.

Once a rearranged light chain has paired with a μ chain, IgM can be expressed on the cell surface (as a surface IgM, or sIgM) and the pre-B cell becomes an immature B cell. At this stage, the antigen receptor is first tested for reactivity to self antigens, or autoreactivity. The elimination or inactivation of autoreactive B cells ensures that the B-cell population as a whole will be tolerant of self antigens. The **tolerance** produced at this stage of B-cell development is known as **central tolerance** because it arises in a central lymphoid organ, the bone marrow. However, B cells leaving the bone marrow are not fully mature and require additional maturation steps that take place in peripheral lymphoid organs (see Section 8-8). As we shall see later in the chapter and in Chapter 15, self-reactive B cells that escape central tolerance may still be removed from the repertoire after they have left the bone marrow, a process that takes place during the final peripheral stages of B-cell maturation, and is referred to as peripheral tolerance, described in Section 8-7.

sIgM associates with Igα and Igβ to form a functional B-cell receptor complex, and the fate of an immature B cell in the bone marrow depends on signals delivered from this receptor complex when it interacts with ligands in the environment. Igα signaling is particularly important in dictating the emigration of B cells from the bone marrow and/or their survival in the periphery: mice that express Igα with a truncated cytoplasmic domain that cannot signal show a fourfold reduction in the number of immature B cells in the marrow, and a hundredfold reduction in the number of peripheral B cells. The release of immature B cells from the bone marrow into the circulation is also dependent on their expression of **S1PR1**, a G-protein-coupled receptor that binds to the lipid ligand S1P and promotes cell migration towards the high concentrations of S1P that exist in the blood (see Section 8-27).

Immature B cells that have no strong reactivity to self antigens continue to mature (**Fig. 8.9**, first panel). They leave the marrow via sinusoids that enter the central sinus, enter the circulation, and are carried by the venous blood supply to the spleen. If, however, the newly expressed receptor encounters a strongly cross-linking antigen in the bone marrow—that is, if the B cell is strongly self-reactive—development is arrested at this stage.

Experiments using genetically modified mice that enforce the expression of self-reactive B-cell receptors have shown that there are four possible fates for self-reactive immature B cells (see Fig. 8.9, last three panels). These fates are the production of a new receptor by a process known as receptor editing; cell death by apoptosis, resulting in clonal deletion; the induction of a permanent state of unresponsiveness to antigen, or anergy; and a state of immunological ignorance in which antigen concentrations are too low to stimulate B-cell receptor signaling. The outcome for each self-reactive B cell is dependent on the interaction of the B-cell receptor with the self antigen.

Fig. 8.9 Binding to self molecules in the bone marrow can lead to the death or inactivation of immature B cells. First panels: immature B cells that do not encounter antigen mature normally; they migrate from the bone marrow to the peripheral lymphoid tissues, where they may become mature recirculating B cells bearing both IgM and IgD on their surface. Second panels: when developing B cells express receptors that recognize multivalent ligands, for example, ubiquitous cell-surface self molecules such as those of the MHC, these receptors are deleted from the repertoire. The B cells either undergo receptor editing (see Fig. 8.10), thereby eliminating the self-reactive receptor, or the cells themselves undergo programmed cell death (apoptosis), resulting in clonal deletion. Third panels: immature B cells that bind soluble self antigens able to cross-link the B-cell receptor are rendered unresponsive to the antigen (anergic) and bear little surface IgM. They migrate to the periphery, where they express IgD but remain anergic; if in competition with other B cells in the periphery, anergic B cells fail to receive survival signals and die. Fourth panels: immature B cells whose antigen is inaccessible to them, or which bind monovalent or soluble self antigens with low affinity, do not receive any signal and mature normally. Such cells are potentially self-reactive, however, and are said to be clonally ignorant because their ligand is present but is unable to activate them.

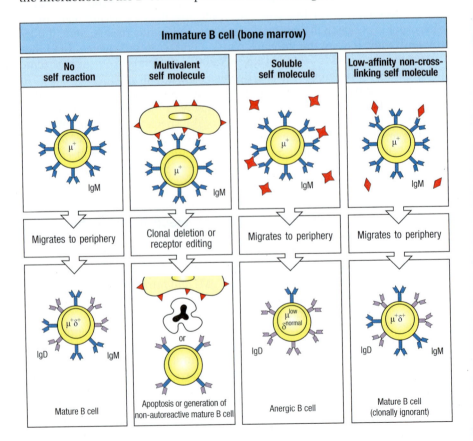

Fig. 8.10 Replacement of light chains by receptor editing can rescue some self-reactive B cells by changing their antigen specificity. When a developing B cell expresses antigen receptors that are strongly cross-linked by multivalent self antigens such as MHC molecules on cell surfaces (top panel), its development is arrested. The cell decreases surface expression of IgM and does not turn off the *RAG* genes (second panel). Continued synthesis of RAG proteins allows the cell to continue light-chain gene rearrangement. This usually leads to a new productive rearrangement and the expression of a new light chain, which combines with the previous heavy chain to form a new receptor; the process is called receptor editing (third panel). If this new receptor is not self-reactive, the cell is 'rescued' and continues normal development, much like a cell that had never reacted with self antigen (bottom right panel). If the cell remains self-reactive, it may be rescued by another cycle of rearrangement; however, if it continues to react strongly with self antigen, it will undergo apoptosis, resulting in clonal deletion from the repertoire of B cells (bottom left panel).

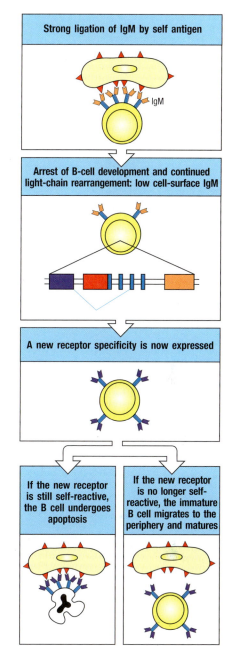

Immature B cells that express an autoreactive receptor recognizing a multivalent self antigen can be rescued by further gene rearrangements that replace the autoreactive receptor with a new receptor that is not self-reactive. This mechanism is termed **receptor editing** (Fig. 8.10). When an immature B cell first produces sIgM, RAG proteins are still being made. If the receptor is not self-reactive, the absence of sIgM cross-linking allows gene rearrangement to cease and B-cell development continues, with RAG proteins eventually disappearing. For an autoreactive receptor, however, an encounter with the self antigen results in strong cross-linking of sIgM; RAG expression continues, and light-chain gene rearrangement can continue, as described in Fig. 8.8. These secondary rearrangements can rescue immature self-reactive B cells by deleting the self-reactive light-chain gene and replacing it with another sequence. If the new light chain is not autoreactive, the B cell continues normal development. If the receptor remains autoreactive, rearrangement continues until a non-autoreactive receptor is produced or until no additional light-chain V and J gene segments are available for recombination. The importance of receptor editing as a mechanism of tolerance is well established, as defects in this process contribute to the human autoimmune diseases systemic lupus erythematosus and rheumatoid arthritis, two diseases characterized by high levels of autoreactive antibodies (see Chapter 15).

It was originally thought that the successful production of a heavy chain and a light chain caused the almost instantaneous shutdown of light-chain locus rearrangement and that this ensured both allelic and isotypic exclusion. The unexpected ability of self-reactive B cells to continue to rearrange their light-chain genes, even after having made a productive rearrangement, suggests an alternative mechanism of allelic exclusion, where the fall in the level of RAG proteins that follows a successful non-autoreactive rearrangement could be the principal means by which light-chain rearrangement is terminated. It is now apparent that allelic exclusion is not absolute, as there are rare B cells that express two different light chains.

Cells that remain autoreactive when receptor editing efforts fail to generate a non-autoreactive receptor undergo a process known as **clonal deletion**, in which they are subjected to cell death by apoptosis to eliminate their specific autoreactivity from the repertoire. Early experiments using transgenic mice expressing both chains of an immunoglobulin specific for H-2K^b MHC class I molecules, in which nearly all developing B cells expressed the anti-MHC immunoglobulin as sIgM, suggested that clonal deletion was a predominant mechanism of B-cell tolerance. These studies found that transgenic mice not expressing H-2K^b had normal numbers of B cells, all bearing the transgene-encoded anti-H-2K^b receptors. However, in mice expressing both H-2K^b and the immunoglobulin transgenes, B-cell development was blocked. Normal numbers of pre-B cells and immature B cells were found, but B cells expressing the anti-H-2K^b immunoglobulin as sIgM never matured to populate the spleen and lymph nodes; instead, most of these immature B cells died in the bone marrow by apoptosis. However, more recent studies, using

Rheumatoid Arthritis

Systemic Lupus Erythematosus

mice bearing transgenes for autoantibody heavy and light chains that have been placed within the immunoglobulin loci by homologous recombination (see Appendix I, Section A-35, for details of this method), indicate that receptor editing, rather than clonal deletion, is the more likely outcome for immature autoreactive B cells.

We have so far discussed the fate of newly formed B cells that undergo multivalent cross-linking of their sIgM. Immature B cells that encounter more weakly cross-linking self antigens of low valence, such as small soluble proteins, respond differently. In this situation, some self-reactive B cells are inactivated and enter a state of permanent unresponsiveness, or **anergy**, but do not immediately die (see Fig. 8.9). Anergic B cells cannot be activated by their specific antigen even with help from antigen-specific T cells. Again, this phenomenon was elucidated using transgenic mice. Hen egg-white lysozyme (HEL) was expressed in soluble form from a transgene in mice that were also transgenic for high-affinity anti-HEL immunoglobulin. The HEL-specific B cells matured and emigrated from the bone marrow, but could not respond to antigen. Furthermore, the migration of anergic B cells is impaired, as the cells are detained in the T-cell areas of peripheral lymphoid tissues and are excluded from lymphoid follicles, thereby reducing their life-span and their ability to compete with immunocompetent B cells (described further in Section 8-8). Under normal circumstances, where few self-reactive anergic B cells successfully mature, these cells die relatively quickly. This mechanism ensures that the long-lived pool of peripheral B cells is purged of potentially self-reactive cells.

The fourth potential fate of self-reactive immature B cells is that nothing happens to them; they remain in a state of **immunological ignorance** of their self antigen (see Fig. 8.9). Immunologically ignorant cells have affinity for a self antigen but for various reasons do not sense and respond to it. The antigen may not be accessible to developing B cells in the bone marrow or spleen, or may be in low concentration, or may bind so weakly to the B-cell receptor that it does not generate an activating signal. Because some ignorant cells can be (and in fact are) activated under certain conditions such as inflammation or when the self antigen becomes available or reaches an unusually high concentration, they should not be considered inert, and they are fundamentally different from cells with non-autoreactive receptors that could never be activated by self antigens.

The fact that central tolerance is not perfect and some self-reactive B cells are allowed to mature reflects the balance that the immune system strikes between purging all self-reactivity and maintaining the ability to respond to pathogens. If the elimination of self-reactive cells were too efficient, the receptor repertoire might become too limited and thus unable to recognize a wide variety of pathogens. Some autoimmune disease is the price of this balance: we shall see in Chapter 15 that ignorant self-reactive lymphocytes can be activated and cause disease under certain circumstances. Normally, however, ignorant B cells are held in check by a lack of T-cell help, the continued inaccessibility of the self antigen, or the tolerance that can be induced in mature B cells following their emigration from the bone marrow, which is described below.

8-7 Lymphocytes that encounter sufficient quantities of self antigens for the first time in the periphery are eliminated or inactivated.

While large numbers of autoreactive B cells are purged from the population of new lymphocytes in the bone marrow, only lymphocytes specific for autoantigens that are expressed in or can reach this organ are affected. Some antigens, like the thyroid product thyroglobulin, are highly tissue specific, or are compartmentalized so that little if any is available in the circulation. Therefore, newly emigrated self-reactive B cells that encounter their specific autoantigen

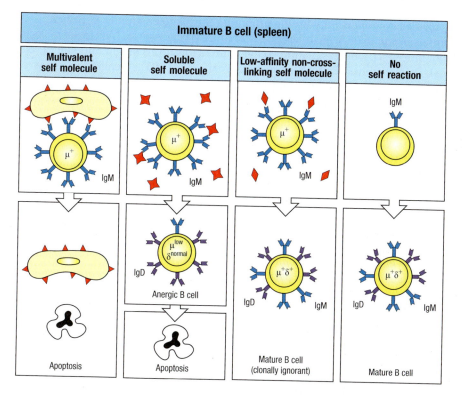

Fig. 8.11 Transitional B cells that recognize self antigens undergo peripheral tolerance. After emigrating from the bone marrow and entering the circulation, immature B cells are known as transitional B cells. Not yet fully mature, these cells are still subject to tolerance in the spleen following engagement of their sIgM receptor by a self antigen. Transitional B cells that encounter a multivalent self antigen receive a strong B-cell receptor signal and undergo cell death. Transitional B cells with sIgM that binds to a soluble self molecule are rendered anergic, and ultimately die within a few days due to being excluded from the B-cell follicles in the spleen (see Fig. 8.12). Transitional B cells that bind with low affinity to a soluble self molecule remain clonally ignorant of the self antigen and continue their maturation. Transitional B cells with no self reaction also continue their maturation into mature B cells. The final stages of B-cell maturation lead to upregulation of sIgD, and take place in the B-cell follicles in the spleen.

for the first time in the periphery must be eliminated or inactivated also. This tolerance mechanism, which acts on newly emigrated B cells that are still immature, is known as **peripheral tolerance**. Like self-reactive lymphocytes in the central lymphoid organs, lymphocytes that encounter self antigens *de novo* in the periphery can have several fates: deletion, anergy, or survival (**Fig. 8.11**).

In the absence of an infection, newly emigrated B cells that encounter a strongly cross-linking antigen in the periphery will undergo clonal deletion. This was elegantly shown in studies of B cells expressing B-cell receptors specific for H-2K^b MHC class I molecules. These B cells are deleted even when, in transgenic animals, the expression of the H-2K^b molecule is restricted to the liver by the use of a liver-specific gene promoter. There is no receptor editing: B cells that encounter strongly cross-linking antigens in the periphery undergo apoptosis directly, unlike their counterparts in the bone marrow, which attempt further receptor rearrangements. This difference may be due to the fact that the B cells in the periphery are somewhat more mature and can no longer rearrange their light-chain loci.

As with immature B cells in the bone marrow, newly developed peripheral B cells that encounter and bind an abundant soluble antigen become unresponsive. This was demonstrated in mice by placing the *HEL* transgene under the control of an inducible promoter that can be regulated by changes in the diet. It is thus possible to induce the production of lysozyme at any time and thereby study its effects on HEL-specific B cells at different stages of maturation. These experiments have shown that both peripheral and immature bone marrow B cells are inactivated when they are chronically exposed to soluble antigen.

8-8   Immature B cells arriving in the spleen turn over rapidly and require cytokines and positive signals through the B-cell receptor for maturation and long-term survival.

When B cells emerge from the bone marrow into the periphery, they are still functionally immature. As discussed above, their final maturation in the periphery provides an opportunity for the immature B cells to encounter peripheral self antigens and to undergo tolerance. Immature B cells express high levels of sIgM but little sIgD, whereas mature B cells express low levels of IgM and high levels of IgD; while the changes in expression of sIgM and sIgD as B cells mature is well documented, the function of sIgD on mature B cells is not known.

Most immature B cells leaving the bone marrow will not survive to become fully mature B cells. **Figure 8.12** shows the possible fates of newly produced B cells that enter the periphery. The daily output of new B cells from the bone marrow is roughly 5–10% of the total B-lymphocyte population in the steady-state peripheral pool. In unimmunized animals, the size of this pool seems to remain constant, due to homeostasis, which means that the stream of new B cells needs to be balanced by the removal of an equal number of peripheral B cells. However, the majority of peripheral mature B cells are long-lived, and only 1–2% of these die each day. Thus, most of the B cells that die are in the rapidly turning-over immature peripheral B-cell population, of which more than 50% die every 3 days. The failure of most newly formed B cells to survive for more than a few days in the periphery seems to be due to competition between peripheral B cells for access to the follicles in the spleen. If newly produced immature B cells do not enter a follicle, their passage through the periphery is halted and they eventually die. The limited number of lymphoid follicles cannot accommodate all of the B cells generated each day, and so there is continual competition for entry.

The follicle provides signals necessary for B-cell survival. In particular, the TNF-family member **BAFF** (for B-cell activating factor belonging to the TNF family) is made by several cell types, but is produced abundantly by the follicular dendritic cells (FDCs). FDCs are non-hematopoietic cells resident in the B-cell follicles that are specialized to capture antigens for recognition by B-cell antigen receptors (see Section 9-1). B cells express three different receptors for BAFF, namely BAFF-R, BCMA, and TACI. The BAFF-R is the most important for follicular B-cell survival, because mutant mice lacking BAFF-R have mainly immature B cells and few long-lived peripheral B cells. BCMA and TACI also bind the related TNF family cytokine APRIL, which is not required for the survival of immature B cells but is important for IgA antibody production, as we shall see in Chapter 10.

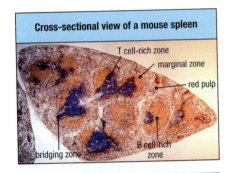

Cross-sectional view of a mouse spleen

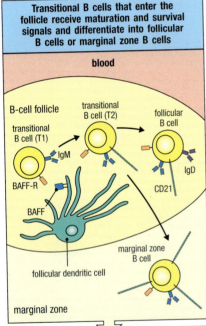

Transitional B cells that enter the follicle receive maturation and survival signals and differentiate into follicular B cells or marginal zone B cells

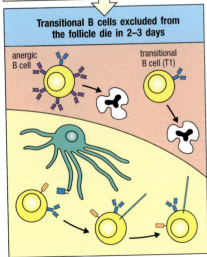

Transitional B cells excluded from the follicle die in 2–3 days

Fig. 8.12 Transitional B cells complete their maturation in B-cell follicles in the spleen. The micrograph in the top panel shows a cross-sectional view of a mouse spleen indicating the distribution of B cells (anti-B220, brown) and T cells (anti-CD3, blue), comprising the white pulp. Surrounding the B-cell-rich follicles (intense brown staining) are the marginal zones (also brown due to presence of B220+ B cells). The white pulp cords sit within the red pulp, which is rich in myeloid cells (mostly macrophages), plasma cells, and passing red blood cells. Transitional B cells that have left the bone marrow must complete their maturation in the B-cell follicles of the spleen, where they receive necessary maturation and survival signals (middle panel). One essential component is a low level of signaling through the B-cell receptor. A second essential factor is the expression of BAFF, a TNF-family member, on follicular dendritic cells (FDCs). BAFF stimulates the BAFF-R on transitional B cells, promoting B-cell survival. Newly emigrated transitional B cells (T1) exhibit high levels of surface IgM, little IgD, and the BAFF-R. In the B-cell follicles, these cells upregulate CD21 to become transitional stage 2 B cells (T2). Finally, the cells upregulate surface IgD, and become long-lived mature B cells. The majority of long-lived B cells are recirculating B cells, known as follicular B cells. A second, less numerous subset is the marginal zone B-cell population. Marginal zone B cells are thought to be weakly self-reactive and express very high levels of the complement receptor CD21. These cells migrate to the marginal zones of the splenic white pulp, an area at the white pulp/red pulp junctions. In this location, marginal zone B cells are poised to make rapid responses to blood-borne antigens or pathogens. Transitional T1 B cells that are excluded from the follicles fail to receive maturation and survival signals and will die within 2–3 days of leaving the bone marrow (bottom panel). Self-reactive anergic B cells are also excluded from the follicles and undergo cell death. Micrograph courtesy of Xiaoming Wang and Jason Cyster. Howard Hughes Medical Institute and Department of Microbiology and Immunology, UCSF.

Immature B cells in the spleen proceed through two distinct **transitional stages**, called T1 and T2, defined by the absence or presence of the B-cell co-receptor component CD21 (complement receptor 2) (see Section 2-13 and Section 7-20). In mice lacking BAFF, immature B cells progress to the T1 stage in the spleen but fail to express CD21, and the mice lack mature B cells. Signaling through the B-cell receptor is also required for immature B cells in the spleen to progress through the T1 and T2 stages and enter the long-lived peripheral B-cell pool. In this case, the B-cell receptor signals do not arise from high-affinity interactions between the sIgM of the B-cell receptor with an antigen, which would induce strong signals; instead, these B-cell receptor signals are thought to be weak, constitutive signals that are developmentally programmed into the maturing B cells, although the mechanism responsible for this constitutive signaling is not known. These weak B-cell receptor signals together with the BAFF-R signals are essential to promote the final stages of B-cell maturation in the periphery. Disregulation of the appropriate balance between B-cell receptor and BAFF-R signaling occurs in individuals who overexpress BAFF, and has been linked to the development of autoimmune diseases, such as Sjögren's syndrome, that result from a failure to purge auto-reactive B cells.

The majority of peripheral B cells that reside in the spleen and other secondary lymphoid organs are known as **follicular B cells**, often referred to as B-2 B cells. A second, minor population of B cells found in the spleen consists of **marginal zone B cells**, named for their predominance at the marginal zones that lie at the white pulp/red pulp junctions (see Fig. 8.12). Follicular and marginal zone B cells both derive from a common lineage that develops in the bone marrow and bifurcates during the final stage of B-cell maturation in the splenic follicles. Experiments reconstituting the signals promoting peripheral B-cell maturation in cell culture, starting from immature B-cell precursors, indicate that follicular versus marginal zone B-cell lineages diverge at the transitional T2 stage, as cells transition to the fully mature stage. Like follicular B cells, marginal zone B cells are dependent on BAFF signals, and are missing in mice lacking BAFF expression. Marginal zone B cells can be identified by their expression of very high levels of the complement receptor CD21. Studies using rearranged immunoglobulin gene knock-in mice that express a single B-cell receptor specificity on all developing B cells have demonstrated that some B-cell receptors predominantly generate follicular B cells whereas others generate marginal zone B cells. These findings indicate that the specificity of the B-cell receptor is a major factor in determining the final commitment of transitional B cells to the follicular versus the marginal zone lineage; however, the details of this process are still not fully understood. Due to their location, marginal zone B cells are poised to make rapid responses to antigens or pathogens filtered from the blood. Therefore, it is thought that marginal zone B cells represent an early line of defense for blood-borne pathogens.

Peripheral B cells also include memory B cells, which are generated in addition to antibody-producing plasma cells from mature B cells after their first encounter with antigen; we will return to B-cell memory in Chapter 11. Competition for follicular entry favors mature B cells that are already established in the relatively long-lived and stable peripheral B-cell pool. Mature B cells have undergone phenotypic changes that might make their access to the follicles easier; for example, they express CXCR5, the receptor for CXCL13, which is expressed by FDCs (see Section 10-3). They also have increased expression of CD21 compared with immature newly developed B cells, which enhances the signaling capacity of the B cell.

The B-cell receptor plays a positive role in the maturation and continued recirculation of peripheral B cells. Mice that lack the tyrosine kinase Syk, which is involved in signaling from the B-cell receptor (see Section 7-20), have immature B cells but fail to develop mature B cells. Thus, a Syk-transduced signal may be required for final B-cell maturation or for the survival of mature B cells. Furthermore, continuous expression of the B-cell receptor is required for B-cell survival, as is evident from the loss of all B cells in mice whose BCR is conditionally deleted specifically in mature B cells. Although each B-cell receptor has a unique specificity, antigen-specific interactions may not induce the signals used for final B-cell maturation and survival; the receptor could, for example, be responsible for 'tonic' signaling, in which a weak but important signal is generated by the assembly of the receptor complex and infrequently triggers some or all of the downstream signaling events.

8-9 B-1 B cells are an innate lymphocyte subset that arises early in development.

Thus far, this chapter has focused on the development of the majority populations of B cells that reside in secondary lymphoid organs, such as follicular (B-2) B cells and marginal zone B cells. These two populations comprise the B-cell arm of the adaptive immune response. A third important subset of B cells, called **B-1 B cells**, is part of the innate immune system. These cells are present only in low numbers in secondary lymphoid organs, and are found in large numbers in the peritoneal and pleural cavities instead. B-1 B cells are the major source of 'natural' antibodies, which are constitutively produced circulating antibodies that are secreted by these B cells prior to any infections. Most antibodies made by B-1 B cells recognize capsular polysaccharide antigens, and B-1 B cells are important in controlling infections of pathogenic viruses and bacteria.

One important feature of B-1 B cells is that they can produce antibodies of the IgM class without 'help' from T cells. Although this response can be enhanced by T-cell cooperation, the antibodies first appear within 48 hours of exposure to antigen, when T cells cannot be involved. The lack of an antigen-specific interaction with helper T cells might explain why immunological memory is not generated as a result of B-1 cell responses: repeated exposures to the same antigen elicit similar, or decreased, responses with each exposure. While the precise functions of B-1 B cells are still not clear, mice deficient in B-1 cells are more susceptible to infection with *Streptococcus pneumoniae* because they fail to produce an anti-phosphocholine antibody that provides protection against this bacterium. Since a significant fraction of the B-1 cells can make antibodies of this specificity, and because no antigen-specific T-cell help is required, a potent response can be produced early in infection with this pathogen. Whether human B-1 cells have the same role is not certain.

Unlike follicular and marginal zone B cells that develop from bone marrow stem cells, the majority of B-1 B cells are generated from progenitor cells found in the fetal liver (Fig. 8.13). During late fetal and early neonatal stages in mice, B-1 B cells are produced in large numbers. After birth, the development of follicular and marginal zone B cells predominates, and few B-1 B cells are made. Current evidence indicates that the progenitor cells giving rise to B-1 B cells are committed to this lineage, and are distinct from those producing B-2 B cells. Whereas B-2 B cells are absent in mice lacking BAFF or the BAFF-R, these deficiencies have no effect on the development or survival of B-1 B cells. Additionally, the weak B-cell receptor signals that promote the final stages of B-2 B-cell maturation in the spleen require the non-canonical NF-κB activation pathway (see Section 7-23), a signaling pathway that is dispensable for B-1 B-cell development. Cytokine requirements also differ between these developmental pathways. B-1 B cells develop normally in mice lacking IL-7 or IL-7R signaling components, defects that prevent the development of B-2 B cells. B-2 B-cell development also requires the transcription factor PU.1, which is not needed for the development of B-1 B cells.

Property	B-1 cells	B-2 cells	
		Follicular B cells	Marginal zone B cells
When first produced	Fetus	After birth	After birth
N regions in VDJ junctions	Few	Extensive	Yes
V-region repertoire	Restricted	Diverse	Partly restricted
Primary location	Body cavities (peritoneal, pleural)	Secondary lymphoid organs	Spleen
Dependence on BAFF	No	Yes	Yes
Dependence on IL-7	No	Yes	Yes
Mode of renewal	Self-renewing	Replaced from bone marrow	Long-lived
Spontaneous production of immunoglobulin	High	Low	Low
Isotypes secreted	IgM >> IgG	IgG > IgM	IgM > IgG
Response to carbohydrate antigen	Yes	Maybe	Yes
Response to protein antigen	Maybe	Yes	Yes
Requirement for T-cell help	No	Yes	Sometimes
Somatic hypermutation	Low to none	High	?
Memory development	Little or none	Yes	?

Fig. 8.13 A comparison of the properties of B-1 cells, follicular B cells (B-2 cells), and marginal zone B cells. In addition to developing in the liver, B-1 cells can develop in unusual sites in the fetus, such as the omentum. B-1 cells predominate in the young animal, although they probably can be produced throughout life. Being produced mainly during fetal and neonatal life, their rearranged variable-region sequences contain few N-nucleotides. In contrast, marginal zone B cells accumulate after birth and do not reach peak levels in the mouse until 8 weeks of age. Follicular B-2 cells and marginal zone B cells share a common precursor population, the transitional T2 B cells in the spleen; as a consequence, both subsets are dependent on IL-7 and BAFF signals for their development. In contrast, B-1 cell development does not require IL-7 or BAFF. B-1 cells are best thought of as a partly activated self-renewing pool of lymphocytes that are selected by ubiquitous self and foreign antigens. Because of this selection, and possibly because the cells are produced early in life, the B-1 cells have a restricted repertoire of variable regions and antigen-binding specificities. Marginal zone B cells also have a restricted repertoire of V-region specificities that may be selected by a set of antigens similar to those that select B-1 cells. B-1 cells seem to be the major population of B cells in certain body cavities, most probably because of exposure at these sites to antigens that drive B-1 cell proliferation. Marginal zone B cells remain in the marginal zone of the spleen and are not thought to recirculate. Partial activation of B-1 cells leads to the secretion of mainly IgM antibody; B-1 cells contribute much of the IgM that circulates in the blood. The limited diversity of both the B-1 and marginal zone B-cell repertoire and the propensity of these cells to react with common bacterial carbohydrate antigens suggest that they carry out a more primitive, less adaptive immune response than follicular B cells (B-2 cells). In this regard they are comparable to $\gamma:\delta$ T cells.

Summary.

In this section, we have followed B-cell development from the earliest progenitors in the bone marrow to the long-lived mature peripheral B-cell pool (Fig. 8.14). The heavy-chain locus is rearranged first and, if this is successful, a μ heavy chain is produced that combines with surrogate light chains to form the pre-B-cell receptor; this is the first checkpoint in B-cell development. Production of the pre-B-cell receptor signals successful heavy-chain gene rearrangement and causes cessation of this rearrangement, thus enforcing allelic exclusion. It also initiates pre-B-cell proliferation, generating numerous progeny in which subsequent light-chain rearrangement can be attempted. If the initial light-chain gene rearrangement is productive, a complete immunoglobulin B-cell receptor is formed, gene rearrangement again ceases, and the B cell continues its development. If the first light-chain gene rearrangement is unsuccessful, rearrangement continues until either a productive rearrangement is made or all available J regions are used up. If no productive rearrangement is made, the developing B cell dies. Once a complete immunoglobulin receptor is expressed on the surface of the cell, immature B cells undergo tolerance to self antigens. This process begins in the bone marrow and continues for a short time after immature B cells emigrate to the periphery. For the majority population of B cells, the final stages of their maturation occur in the B-cell follicles of the spleen, and require the TNF family member BAFF as well as signals through the B-cell receptor.

Fig. 8.14 A summary of the development of human conventional B-lineage cells. The state of the immunoglobulin genes, the expression of some essential intracellular proteins, and the expression of some cell-surface molecules are shown for successive stages of conventional B-2 B-cell development. During antigen-driven B-cell differentiation, the immunoglobulin genes undergo further changes, such as class switching and somatic hypermutation (see Chapter 5), which are evident in the immunoglobulins produced by memory cells and plasma cells. These antigen-dependent stages are described in more detail in Chapter 9.

B cells		Heavy-chain genes	Light-chain genes	Intra-cellular proteins	Surface marker proteins
Stem cell		Germline	Germline		CD34 CD45 AA4.1
Early pro-B cell		D–J rearranged	Germline	RAG-1 RAG-2 TdT λ5, VpreB	CD34 CD45R AA4.1, IL-7R MHC class II CD10, CD19 CD38
Late pro-B cell		V–DJ rearranged	Germline	TdT λ5, VpreB	CD45R AA4.1, IL-7R MHC class II CD10, CD19 CD38, CD20 CD40
Large pre-B cell	pre-B-cell receptor	VDJ rearranged	Germline	λ5, VpreB	CD45R AA4.1, IL-7R MHC class II pre-B-R CD19, CD38 CD20, CD40
Small pre-B cell	cytoplasmic μ	VDJ rearranged	V–J rearrangement	μ RAG-1 RAG-2	CD45R AA4.1 MHC class II CD19, CD38 CD20, CD40
Immature B cell	IgM	VDJ rearranged. μ heavy chain produced in membrane form	VJ rearranged		CD45R AA4.1 MHC class II IgM CD19, CD20 CD40
Mature naive B cell	IgD IgM	VDJ rearranged. μ chain produced in membrane form. Alternative splicing yields μ + δ mRNA			CD45R MHC class II IgM, IgD CD19, CD20 CD21, CD40
Lympho-blast	IgM	Alternative splicing yields secreted μ chains		Ig	CD45R MHC class II CD19, CD20 CD21, CD40
Memory B cell	IgG	Isotype switch to Cγ, Cα, or Cε. Somatic hypermutation	Somatic hypermutation		CD45R MHC class II IgG, IgA CD19, CD20 CD21, CD40
Plasma blast and plasma cell	IgG	Alternative splicing yields both membrane and secreted Ig	VJ rearranged	Ig	CD135 Plasma cell antigen-1 CD38

ANTIGEN INDEPENDENT

ANTIGEN DEPENDENT

TERMINAL DIFFERENTIATION

BONE MARROW

PERIPHERY

Development of T lymphocytes.

Like B cells, T lymphocytes derive from the multipotent hematopoietic stem cells in the bone marrow. However, their progenitor cells migrate from the bone marrow via the blood to the thymus, where they mature (Fig. 8.15); this is the reason for the name thymus-dependent (T) lymphocytes, or T cells. T-cell development parallels that of B cells in many ways, including the orderly and stepwise rearrangement of antigen-receptor genes, the sequential testing for successful gene rearrangement, and the eventual assembly of a heterodimeric antigen receptor. Nevertheless, T-cell development in the thymus has some features not seen for B cells, such as the generation of two distinct lineages of T cells expressing antigen receptors encoded by distinct genes, the γ:δ lineage and the α:β lineage. Developing T cells, which are known generally as **thymocytes**, also undergo rigorous selection that depends on interactions with thymic cells and that shapes the mature repertoire of T cells to ensure self MHC restriction as well as self-tolerance. We begin with a general overview of the stages of thymocyte development and its relationship to thymic anatomy before considering gene rearrangement and the mechanisms of selection.

8-10 T-cell progenitors originate in the bone marrow, but all the important events in their development occur in the thymus.

The thymus is situated in the upper anterior thorax, just above the heart. It consists of numerous lobules, each clearly differentiated into an outer cortical region—the **thymic cortex**—and an inner **medulla** (Fig. 8.16). In young individuals, the thymus contains large numbers of developing T-cell precursors embedded in a network of epithelia known as the **thymic stroma**. This provides a unique microenvironment for T-cell development analogous to that provided for B cells by the stromal cells of the bone marrow.

Fig. 8.15 T cells undergo development in the thymus and migrate to the peripheral lymphoid organs, where they are activated by foreign antigens. T-cell precursors migrate from the bone marrow to the thymus, where they commit to the T-cell lineage following Notch receptor signaling. In the thymus, T-cell receptor genes are rearranged (top first panel); α:β T-cell receptors that are compatible with self MHC molecules transmit a survival signal on interacting with thymic epithelium, leading to positive selection of the cells that bear them. Self-reactive receptors transmit a signal that leads to cell death, and cells bearing them are removed from the repertoire in a process of negative selection (top second panel). T cells that survive selection mature and leave the thymus to circulate in the periphery; they repeatedly leave the blood to migrate through the peripheral lymphoid organs, where they may encounter their specific foreign antigen and become activated (top third panel). Activation leads to clonal expansion and differentiation into effector T cells. Some of these are attracted to sites of infection, where they can kill infected cells or activate macrophages (top fourth panel); others are attracted into B-cell areas, where they help to activate an antibody response (not shown).

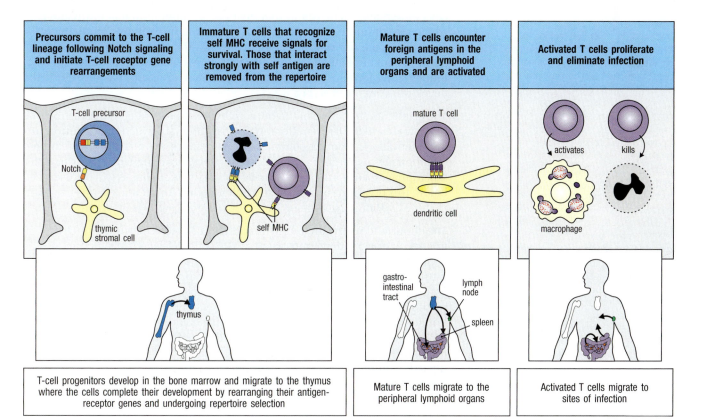

| Precursors commit to the T-cell lineage following Notch signaling and initiate T-cell receptor gene rearrangements | Immature T cells that recognize self MHC receive signals for survival. Those that interact strongly with self antigen are removed from the repertoire | Mature T cells encounter foreign antigens in the peripheral lymphoid organs and are activated | Activated T cells proliferate and eliminate infection |

T-cell precursor

Notch

thymic stromal cell

self MHC

mature T cell

dendritic cell

activates   kills

macrophage

thymus

gastro-intestinal tract   lymph node

spleen

| T-cell progenitors develop in the bone marrow and migrate to the thymus where the cells complete their development by rearranging their antigen-receptor genes and undergoing repertoire selection | Mature T cells migrate to the peripheral lymphoid organs | Activated T cells migrate to sites of infection |

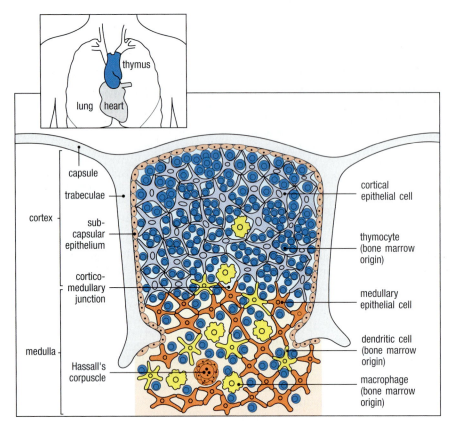

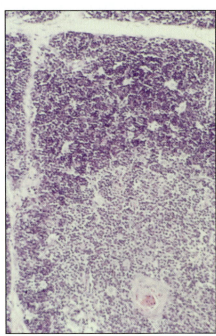

Fig. 8.16 The cellular organization of the human thymus. The thymus, which lies in the midline of the body, above the heart, is made up of several lobules, each of which contains discrete cortical (outer) and medullary (central) regions. As shown in the diagram on the left, the cortex consists of immature thymocytes (dark blue); branched cortical epithelial cells (pale blue), with which the immature cortical thymocytes are closely associated; and scattered macrophages (yellow), which are involved in clearing apoptotic thymocytes. The medulla consists of mature thymocytes (dark blue) and medullary epithelial cells (orange), along with macrophages (yellow) and dendritic cells (yellow) of bone marrow origin. Hassall's corpuscles are probably also sites of cell degradation. The thymocytes in the outer cortical cell layer are proliferating immature cells, whereas the deeper cortical thymocytes are mainly immature T cells undergoing thymic selection. The photograph shows the equivalent section of a human thymus, stained with hematoxylin and eosin. The cortex is darkly stained, whereas the medulla is lightly stained. The large body in the medulla is a Hassall's corpuscle. Photograph courtesy of C.J. Howe.

The thymic epithelium arises early in embryonic development from endoderm-derived structures known as the third pharyngeal pouches. These epithelial tissues form a rudimentary thymus, or **thymic anlage**. This is colonized by cells of hematopoietic origin that give rise to large numbers of thymocytes, which are committed to the T-cell lineage, and to **intrathymic dendritic cells**. Thymocytes are not simply passengers within the thymus: they influence the arrangement of the thymic epithelial cells on which they depend for survival, inducing the formation of a reticular epithelial structure that surrounds the developing thymocytes (**Fig. 8.17**).

The cellular architecture of the human thymus is illustrated in Fig. 8.16. Bone marrow-derived cells are differentially distributed between the thymic cortex and medulla. The cortex contains only immature thymocytes and scattered macrophages, whereas more mature thymocytes, along with dendritic cells, macrophages, and some B cells, are found in the medulla. As will be discussed below, this organization reflects the different developmental events that occur in these two compartments.

The importance of the thymus in immunity was first discovered through experiments on mice; indeed, most of our knowledge of T-cell development in the thymus comes from the mouse. It was found that surgical removal of the thymus (**thymectomy**) at birth resulted in immunodeficient mice, focusing interest on this organ at a time when the difference between T and B lymphocytes in mammals had not yet been defined. Much evidence, including observations in immunodeficient children, has since confirmed the importance of the thymus in T-cell development. In **DiGeorge syndrome** in humans and in mice with the ***nude*** mutation, the thymus does not form and the affected individual produces B lymphocytes but few T lymphocytes. DiGeorge syndrome is a complex combination of cardiac, facial, endocrine, and immune defects associated with deletions of chromosome 22q11. The *nude* mutation in mice is due to a defect in the gene for Foxn1, a transcription factor required for terminal epithelial cell differentiation; the name *nude* was given to this mutation because it also causes hairlessness. Rare cases of a defect in the human *FOXN1* gene (which is on chromosome 17) have been associated with T-cell immunodeficiency, absence of a thymus, congenital alopecia, and nail dystrophy.

In mice, the thymus continues to develop for 3–4 weeks after birth, whereas in humans it is fully developed at birth. The rate of T-cell production by the thymus is greatest before puberty. After puberty, the thymus begins to shrink, and the production of new T cells in adults is reduced, although it does continue throughout life. In both mice and humans, removal of the thymus after puberty is not accompanied by any notable loss of T-cell function or numbers. Thus, it seems that once the T-cell repertoire is established, immunity can be sustained without the production of significant numbers of new T cells; the pool of peripheral T cells is instead maintained by long-lived T cells and also by division of some mature T cells.

8-11   Commitment to the T-cell lineage occurs in the thymus following Notch signaling.

T lymphocytes develop from a lymphoid progenitor in the bone marrow that also gives rise to B lymphocytes. Some of these progenitors leave the bone marrow and migrate to the thymus. In the thymus, the progenitor cell receives a signal from thymic epithelial cells that is transduced through a receptor called Notch1 to switch on specific genes. Notch signaling is widely used in animal development to specify tissue differentiation; in lymphocyte development, the Notch signal instructs the precursor to commit to the T-cell lineage rather than the B-cell lineage. Notch signaling is required throughout T-cell development and is also thought to help regulate other T-cell lineage choices, including the α:β versus γ:δ choice.

Notch signaling in thymic progenitor cells is essential to initiate the T-cell-specific gene expression program and commitment to the T-cell lineage (Fig. 8.18). First, Notch signaling induces the expression of two transcription factors, T-cell factor-1 (TCF1) and GATA3, each of which is required for T-cell development. Together, TCF1 and GATA3 initiate expression of several T-lineage-specific genes, such as those encoding components of the CD3 complex, as well as *Rag1*, a gene required for T-cell receptor and B-cell receptor gene rearrangements (see Fig. 8.18). However, TCF1 and GATA3 are not sufficient to induce the entire program of T-cell-specific gene expression. A third transcription factor, Bcl11b, is required to induce T-lineage commitment by restricting progenitor cells from adopting alternative fates; this final phase of T-cell commitment is a necessary prerequisite for activating the complete T-cell gene expression program.

8-12   T-cell precursors proliferate extensively in the thymus, but most die there.

T-cell precursors arriving in the thymus from the bone marrow spend up to a week differentiating there before they enter a phase of intense proliferation. In a young adult mouse the thymus contains about 10^8 to 2×10^8 thymocytes. About 5×10^7 new cells are generated each day; however, only about 10^6 to 2×10^6 (roughly 2–4%) of these leave the thymus each day as mature T cells. Despite the disparity between the number of T cells generated in the thymus and the number leaving, the thymus does not continue to grow in size or cell number. This is because about 98% of the thymocytes that develop in the thymus also die in the thymus by apoptosis (see Section 1-14). Cells undergoing apoptosis are recognized and ingested by macrophages, and apoptotic bodies, which are the residual condensed chromatin of apoptotic cells, are seen inside macrophages throughout the thymic cortex (Fig. 8.19). This apparently

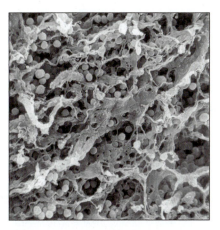

Fig. 8.17 The epithelial cells of the thymus form a network surrounding developing thymocytes. In this scanning electron micrograph of the thymus, the developing thymocytes (the spherical cells) occupy the interstices of an extensive network of epithelial cells. Photograph courtesy of W. van Ewijk.

DiGeorge Syndrome

Fig. 8.18 The stages of α:β T-cell development in the mouse thymus correlate with the program of gene rearrangement, and the expression of cell-surface proteins, signaling proteins, and transcription factors. Lymphoid precursors are triggered to proliferate and become thymocytes committed to the T-cell lineage through interactions with Notch ligands expressed on the thymic stroma. T-cell commitment requires Notch signaling to induce the expression of TCF1 and GATA3, which in turn induce the expression of Bcl11b. This gene expression program begins in the double-negative (DN1) cells that express CD44 and Kit. Cells become irreversibly committed to the T-cell lineage at the subsequent (DN2) stage, which is marked by expression of CD25, the α chain of the IL-2 receptor. After this, the DN2 (CD44$^+$CD25$^+$) cells begin to rearrange the β-chain locus, becoming CD44^{low} and Kit^{low} as this occurs, and they become DN3 cells. The DN3 cells are arrested in the CD44^{low}CD25$^+$ stage until they productively rearrange the β-chain locus; the in-frame β chain then pairs with a surrogate chain called pTα to form the pre-T-cell receptor (pre-TCR), which is expressed on the cell surface and triggers entry into the cell cycle. Expression of small amounts of pTα:β on the cell surface in association with CD3 signals the cessation of β-chain gene rearrangement and triggers rapid cell proliferation, which causes the loss of CD25. The cells are then known as DN4 cells. Eventually, the DN4 cells cease to proliferate and CD4 and CD8 are expressed. The small CD4$^+$CD8$^+$ double-positive cells begin efficient rearrangement at the α-chain locus. The cells then express low levels of an α:β T-cell receptor and the associated CD3 complex and are ready for selection. Most cells die by failing to be positively selected or as a consequence of negative selection, but some are selected to mature into CD4 or CD8 single-positive cells and eventually to leave the thymus. Maturation of CD4$^+$CD8$^+$ double-positive cells into CD4 or CD8 single-positive cells is regulated by transcription factors ThPOK and Runx3, respectively. KLF2 is first expressed at the single-positive stage; if it is absent, thymocytes exhibit a defect in emigrating to peripheral lymphoid tissues, due in part to their failure to express receptors involved in trafficking, such as the sphingosine 1-phosphate (S1P) receptor, S1PR1 (see Fig. 8.32). The individual contributions to T-cell development of the other proteins are discussed in the text.

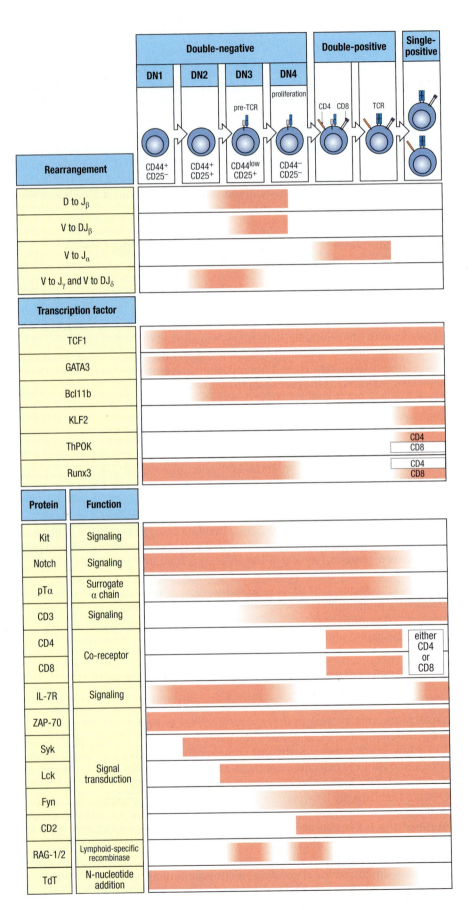

Fig. 8.19 Developing T cells that undergo apoptosis are ingested by macrophages in the thymic cortex. The left panel shows a section through the thymic cortex and part of the medulla in which cells have been stained for apoptosis with a red dye. The thymic cortex is to the right in the photograph. Apoptotic cells are scattered throughout the cortex but are rare in the medulla. The right panel shows at higher magnification a section of thymic cortex that has been stained red for apoptotic cells and blue for macrophages. The apoptotic cells can be seen within macrophages. Magnifications: left panel, ×45; right panel, ×164. Photographs courtesy of J. Sprent and C. Surh.

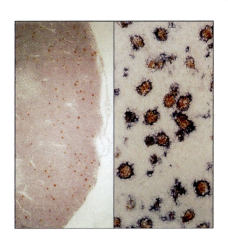

profligate waste of thymocytes is a crucial part of T-cell development because it reflects the intensive screening that each thymocyte undergoes for the ability to recognize self peptide:self MHC complexes and for self-tolerance.

8-13 Successive stages in the development of thymocytes are marked by changes in cell-surface molecules.

Like developing B cells, developing thymocytes pass through a series of distinct stages. These are marked by changes in the status of the T-cell receptor genes and in the expression of the T-cell receptor, and by changes in the expression of cell-surface proteins such as the CD3 complex (see Section 7-7) and the co-receptor proteins CD4 and CD8 (see Section 4-18). These surface changes reflect the state of functional maturation of the cell, and particular combinations of cell-surface proteins are used as markers for T cells at different stages of differentiation. The principal stages are summarized in **Fig. 8.20**. Two distinct lineages of T cells—α:β and γ:δ, which have different types of T-cell receptor chains—are produced early in T-cell development. Later, α:β T cells develop into two distinct functional subsets—CD4 T cells and CD8 T cells.

When progenitor cells first enter the thymus from the bone marrow, they lack most of the surface molecules characteristic of mature T cells, and their receptor genes are not rearranged. These cells give rise to the major population of α:β T cells and the minor population of γ:δ T cells. If injected into the peripheral circulation, these lymphoid progenitors can even give rise to B cells and NK cells, although it is uncertain whether individual thymic progenitor cells retain this multipotency, or whether the progenitor cell population consists of a mixture of cells, only some of which are fully committed to the αβ or γδ T-cell lineage.

Interactions with the thymic stroma trigger an initial phase of differentiation along the T-cell lineage pathway, followed by cell proliferation and the expression of the first cell-surface molecules specific for T cells, for example, CD2 and (in mice) Thy-1. At the end of this phase, which can last about a week, the thymocytes bear distinctive markers of the T-cell lineage but do not express any of the three cell-surface markers that define mature T cells. These are the CD3:T-cell receptor complex and the co-receptors CD4 or CD8. Because of the absence of CD4 and CD8, such cells are called **double-negative thymocytes** (see Fig. 8.20).

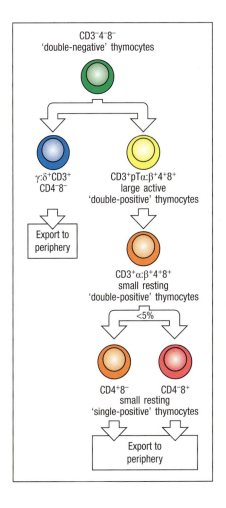

Fig. 8.20 Two distinct lineages of thymocytes are produced in the thymus. CD4, CD8, and T-cell receptor complex molecules (CD3, and the T-cell receptor α and β chains) are important cell-surface molecules for identifying thymocyte subpopulations. The earliest cell population in the thymus does not express any of these proteins, and because these cells do not express CD4 or CD8, they are called 'double-negative' thymocytes. These cells include precursors that give rise to two T-cell lineages: the minority population of γ:δ T cells (which lack CD4 or CD8 even when mature), and the majority α:β T-cell lineage. The development of prospective α:β T cells proceeds through stages in which both CD4 and CD8 are expressed by the same cell; these are known as 'double-positive' thymocytes. These cells enlarge and divide. Later, they become small resting double-positive cells that express low levels of the T-cell receptor. Most thymocytes die within the thymus after becoming small double-positive cells, but those cells whose receptors can interact with self peptide:self MHC molecular complexes lose expression of either CD4 or CD8 and increase the level of expression of the T-cell receptor. The outcome of this process is the 'single-positive' thymocytes, which, after maturation, are exported from the thymus as mature single-positive CD4 or CD8 T cells.

In the fully developed thymus, only ~60% of the double-negative thymocytes are immature T cells. The double-negative thymocyte pool (about 5% of all thymocytes) also includes two populations of more mature T cells that belong to minority lineages, including T cells expressing γ:δ T-cell receptors (see Section 8-16) and T cells bearing α:β T-cell receptors of very limited diversity (iNKT cells; see Section 6-19). In this and subsequent discussions, we reserve the term 'double-negative thymocytes' for the immature thymocytes that do not yet express a complete T-cell receptor molecule. These cells give rise to both γ:δ and α:β T cells (see Fig. 8.20), although most of them develop along the α:β pathway.

The α:β pathway is shown in more detail in Fig. 8.18. The double-negative stage can be further subdivided into four stages on the basis of expression of the adhesion molecule CD44, CD25 (the α chain of the IL-2 receptor), and Kit, the receptor for SCF (see Section 8-1). At first, double-negative thymocytes express Kit and CD44 but not CD25 and are called **DN1** cells; in these cells, the genes encoding both chains of the T-cell receptor are in the germline configuration. As thymocytes mature, they begin to express CD25 on their surface and are called **DN2** cells; later, expression of CD44 and Kit is reduced, and they are called **DN3** cells.

Rearrangement of the T-cell receptor β-chain locus begins in DN2 cells with some D_β to J_β rearrangements and continues in DN3 cells with V_β to DJ_β rearrangement. Cells that fail to make a successful rearrangement of the β-chain locus remain at the DN3 ($CD44^{low}CD25^+$) stage and soon die, whereas cells that make productive β-chain gene rearrangements and express the β-chain protein lose expression of CD25 once again and progress to the **DN4** stage, in which they proliferate. The functional significance of the transient expression of CD25 is unclear: T cells develop normally in mice in which the IL-2 gene has been deleted by gene knockout (see Appendix I, Section A-35). By contrast, Kit is quite important for the development of the earliest double-negative thymocytes, in that mice lacking Kit have a much smaller number of double-negative T cells. In addition, continuous Notch signaling is important for progression through each stage of T-cell development. A second essential factor is IL-7, which is produced by the thymic stroma. In the absence of IL-7, IL-7 receptor α, γ-c, or the IL-7 receptor signaling protein Jak3, a severe block in T-cell development occurs in both mice and humans. In fact, the human primary immunodeficiency disease characterized by defects in T cells and NK cells, X-linked SCID (severe combined immunodeficiency disease), is caused by a genetic deficiency leading to the absence of γ-c protein expression.

X-linked Severe Combined Immunodeficiency

In DN3 thymocytes (see Fig. 8.18), the expressed β chains pair with a surrogate pre-T-cell receptor α chain called **pTα** (pre-T-cell α), which allows the assembly of a complete **pre-T-cell receptor** (pre-TCR) that is analogous in structure and function to the pre-B-cell receptor. The pre-TCR is expressed on the cell surface in a complex with the CD3 molecules that provide the signaling components of T-cell receptors (see Section 7-7). As with the pre-B-cell receptor, the assembly of the CD3:pre-T-cell receptor complex causes constitutive signaling that does not require interaction with a ligand. Recent structural evidence shows that the pre-TCR forms dimers in a manner similar to pre-BCR dimerization. The pTα Ig domain makes two important contacts. It associates with the constant-region Ig domain of the V_β subunit to form the pre-TCR itself. A distinct surface of the pre-Tα then binds to a V_β domain from another pre-TCR molecule, forming a bridge between two different pre-TCRs. The region of contact with the V_β involves residues that are highly conserved across many V_β families. In this way, expression of the pre-TCR induces ligand-independent dimerization, which leads to cell proliferation, the arrest of further β-chain gene rearrangement, and the expression of both CD8 and CD4. These **double-positive thymocytes** make up the vast majority of thymocytes. Once the large double-positive thymocytes have ceased to proliferate and have become small double-positive cells, the α-chain locus begins to rearrange.

As we will see later in this chapter, the structure of the α locus (see Section 5-9) allows multiple successive attempts at rearrangement, so that it is successfully rearranged in most developing thymocytes. Thus, most double-positive cells produce an α:β T-cell receptor during their relatively short life-span.

Small double-positive thymocytes initially express low levels of the T-cell receptor. Most of these receptors cannot recognize self peptide:self MHC molecular complexes; they will fail positive selection and the cells will die. In contrast, those double-positive cells that recognize self peptide:self MHC complexes, and can therefore be positively selected, go on to mature, and express high levels of the T-cell receptor. At the same time they cease to express one or the other of the two co-receptor molecules, becoming either CD4 or CD8 **single-positive thymocytes** (see Fig. 8.18). Thymocytes also undergo negative selection during and after the double-positive stage, a mechanism that eliminates those cells capable of responding to self antigens. About 2% of the double-positive thymocytes survive this dual screening and mature as single-positive T cells that are gradually exported from the thymus to form the peripheral T-cell repertoire. The time between the entry of a T-cell progenitor into the thymus and the export of its mature progeny is estimated to be about 3 weeks in the mouse.

8-14   Thymocytes at different developmental stages are found in distinct parts of the thymus.

The thymus is divided into two main regions, a peripheral cortex and a central medulla (see Fig. 8.16). Most T-cell development takes place in the cortex; only mature single-positive thymocytes are seen in the medulla. Initially, progenitors from the bone marrow enter the thymus from the blood at the cortico-medullary junction and migrate to the outer cortex (Fig. 8.21). At the outer edge of the cortex, in the subcapsular region of the thymus, large immature double-negative thymocytes proliferate vigorously; these cells are thought to represent committed thymocyte progenitors and their immediate progeny and will give rise to all subsequent thymocyte populations. Deeper in the cortex, most of the thymocytes are small double-positive cells. The cortical stroma is composed of epithelial cells with long branching processes that express both MHC class II and MHC class I molecules on their surface. The thymic cortex is densely packed with thymocytes, and the branching processes of the thymic cortical epithelial cells make contact with almost all cortical thymocytes (see Fig. 8.17). Contact between the MHC molecules on

 MOVIE 8.1

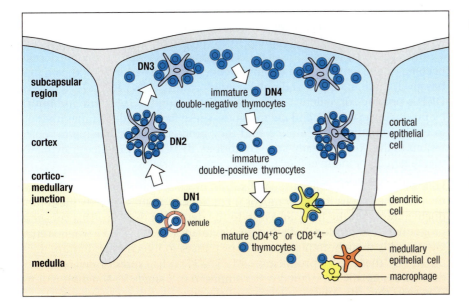

Fig. 8.21 Thymocytes at different developmental stages are found in distinct parts of the thymus. The earliest precursor thymocytes enter the thymus from the bloodstream via venules near the cortico-medullary junction. Ligands that interact with the receptor Notch1 are expressed in the thymus and act on the immigrant cells to commit them to the T-cell lineage. As these cells differentiate through the early CD4⁻CD8⁻ double-negative (DN) stages described in the text, they migrate through the cortico-medullary junction and to the outer cortex. DN3 cells reside near the subcapsular region, where they undergo proliferation. As the progenitor matures further to the CD4⁺CD8⁺ double positive stage, it migrates back through the cortex. Finally, the medulla contains only mature single-positive T cells, which eventually leave the thymus.

thymic cortical epithelial cells and the receptors of developing T cells has a crucial role in positive selection, as we will see later in this chapter.

After positive selection, developing T cells migrate from the cortex to the medulla. The medulla contains fewer lymphocytes, and those present are predominantly the newly matured single-positive T cells that will eventually leave the thymus. The medulla plays a role in negative selection. The antigen-presenting cells in this environment include dendritic cells that express co-stimulatory molecules, which are generally absent from the cortex. In addition, specialized medullary epithelial cells present peripheral antigens for the negative selection of T cells reactive for these self antigens.

8-15   T cells with α:β or γ:δ receptors arise from a common progenitor.

T cells bearing γ:δ receptors differ from α:β T cells in that they are found primarily in epithelial and mucosal sites and lack expression of the CD4 and CD8 co-receptors; in comparison with α:β T cells, relatively little is known about the ligands recognized by the γ:δ T-cell receptors, which are thought not to be MHC restricted (see Section 4-20). Recall from Section 5-11 that different genetic loci are used to make these two types of T-cell receptors. The γ and δ loci are the first to undergo rearrangement, followed shortly thereafter by the β locus. In addition, the δ locus is contained within the α locus, so rearrangements at the α locus eliminate the δ coding sequences on the chromosome. While the mechanism regulating commitment of individual precursor cells to the α:β versus the γ:δ lineage is still not understood, there is some plasticity in this process. This can be deduced from the pattern of gene rearrangements found in thymocytes and in mature γ:δ and α:β T cells. Mature γ:δ T cells can contain rearranged β-chain genes, although 80% of these are nonproductive, and mature α:β T cells often contain rearranged, but mostly out-of-frame, γ-chain genes.

8-16   T cells expressing γ:δ T-cell receptors arise in two distinct phases during development.

Although γ:δ T cells arise from the same progenitors as α:β T cells, most mature γ:δ T cells are components of the innate rather than the adaptive immune system. When their maturation in the thymus is complete, the cells have acquired a defined effector function that can be rapidly elicited following their activation. After emigration from the thymus, most γ:δ T cells home to mucosal and epithelial sites in the body, and take up stable residence in these locations.

In mice, the majority of γ:δ T cells in the body arise during embryonic development and the early neonatal period. In the fetal thymus, the first T cells to develop are γ:δ T cells that all express T-cell receptors assembled from the same V_γ and V_δ regions (**Fig 8.22**). These cells populate the epidermis; the T cells become wedged among the keratinocytes and adopt a dendritic-like form that has given them the name of **dendritic epidermal T cells (dETCs)** (**Fig. 8.23**). dETCs provide surveillance of the skin and respond to infection and injury by producing cytokines and chemokines. These factors induce inflammation to enhance pathogen clearance, and they promote wound healing to repair lesions in the skin. In steady-state conditions, dETCs also produce growth factors that help maintain epidermal growth and survival.

Following the dETC cells, a second subset of γ:δ T cells develops in the fetal thymus. These cells home to mucosal epithelia of tissues such as the reproductive tract and the lung, and also to the dermis of the skin. This subset is programmed to produce inflammatory cytokines such as IL-17 when stimulated, and is thought to play a role in responses to infection and injury. Like the dETCs, these IL-17-producing γ:δ T ($T_{\gamma:\delta}$-17) cells express T-cell receptors that are essentially invariant, being composed of a single V_γ–V_δ combination. However, the two subsets, dETCs and the fetal $T_{\gamma:\delta}$-17 cells, express T-cell receptors that use distinct V_γ gene segments—$V_\gamma 5$ in the dETCs and $V_\gamma 6$ in the $T_{\gamma:\delta}$-17 cells. As fetal thymocytes do not express the enzyme TdT, there are no N-nucleotides contributing additional diversity at the junctions between V, D, and J gene segments of the T-cell receptors in these two fetally derived

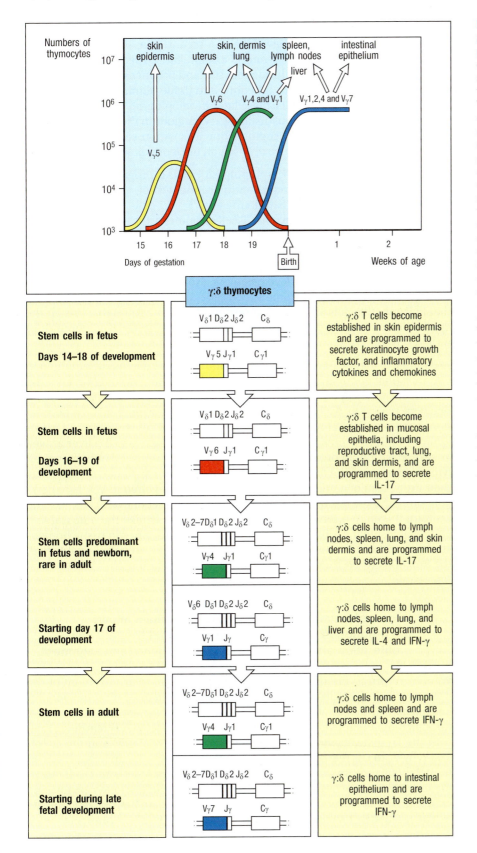

Fig. 8.22 The rearrangement of T-cell receptor γ and δ genes in the mouse proceeds in waves of cells expressing different V_γ and V_δ gene segments. At about 2 weeks of gestation in the mouse, the $C_\gamma 1$ locus is expressed with its closest V gene ($V_\gamma 5$). After a few days, $V_\gamma 5$-bearing cells decline in numbers in the thymus (first row of panels) and are replaced by cells expressing the next most proximal gene, $V_\gamma 6$. Both these rearranged γ chains are expressed with the same rearranged δ-chain gene, as shown in the lower panels, and there is little junctional diversity in either the V_γ or the V_δ chain. As a consequence, most of the γ:δ T cells produced in each of these early waves share the same specificity, although the antigen recognized in each case is not known. The $V_\gamma 5$-bearing cells become established selectively in the epidermis; they are programmed to secrete keratinocyte growth factor and inflammatory cytokines and chemokines. In contrast, $V_\gamma 6$-bearing cells become established in the lung, the dermis of the skin, and the epithelium of the reproductive tract, and are programmed to secrete IL-17. The next wave of γδ development begins on day 17 of gestation, and produces two different populations. One population rearranges and expresses the $V_\gamma 4$ chain, which pairs with heterogeneous delta chains. These $V_\gamma 4$-bearing cells are the second subset of $T_{\gamma:\delta}$-17 (IL-17-secreting) cells, and home to lymph nodes, spleen, lung, and the dermis of the skin. The second population in this wave expresses $V_\gamma 1$, and homes to lymph nodes, spleen, and liver. Some of these cells are paired with $V_\gamma 6$ chains and are programmed to secrete IL-4 and IFN-γ, and represent γ:δ NKT cells. Finally, the last wave of γ:δ T-cell development begins late during fetal development, and persists into adulthood. This last wave includes a heterogeneous population of cells bearing $V_\gamma 1$, $V_\gamma 2$, and $V_\gamma 4$ chains paired with many different delta chains. These cells home to lymphoid organs and are programmed to secrete IFN-γ. The other population in this last wave are cells bearing the $V_\gamma 7$ chain paired with heterogeneous delta chains. These γ:δ cells home to the intestinal epithelium and are programmed to secrete IFN-γ as well as antimicrobial compounds. Although γ:δ T cells continue to be produced after birth, at this stage the α:β T-cell lineage becomes the dominant population developing in the thymus.

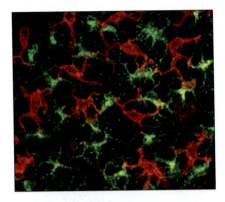

Fig. 8.23 Dendritic epidermal T cells reside within the epithelial layer, forming an interdigitating network with Langerhans cells. This face-on view of a murine epidermal sheet shows Langerhans cells (green) and dendritic epidermal T cells (dETCs; red) forming an interdigitating network within the layers of the epidermis. The epidermal epithelial cells are not visible in this fluorescence image. The branching dendritic-like form of these γ:δ T cells is the source of their name. Although the ligands for all γ:δ T-cell receptors are not known, some γ:δ T cells recognize nonclassical MHC molecules (see Sections 6-16 and 6-17), which can be induced in epithelia by stresses such as UV damage or pathogens. Thus, dETCs may serve as sentinels of such damage, producing cytokines that activate the innate immune response and, in turn, adaptive immunity. Courtesy of Adrian Hayday.

subsets of γ:δ T cells. Why certain V, D, and J gene segments are selected for rearrangement at particular times during embryonic development remains incompletely understood.

dETCs and the $V_\gamma6$-positive $T_{\gamma:\delta}$-17 cells develop exclusively from the early wave of hematopoietic stem cells derived from the fetal liver (see Fig. 8.22). Consequently, these two γ:δ T-cell subsets arise only for a brief period of time in the fetal thymus, and then never again. A second phase of γ:δ T-cell development is initiated in the fetal thymus just before birth. This phase persists at a low level in the adult thymus throughout life, and produces several subsets of cells, each with distinct effector functions and tissue homing properties. Like the dETCs and fetal $T_{\gamma:\delta}$-17 cells, these later-developing γ:δ T cells can be generally classified by their usage of distinct V_γ–V_δ regions in their T-cell receptors (see Fig. 8.22), although the receptor sequences within each population are more diverse due to the presence of N-region nucleotides added by TdT.

One population of these later-developing γ:δ T cells is programmed to secrete IL-17 when activated; these represent a second subset of $T_{\gamma:\delta}$-17 cells that express a different V_γ region than do the fetal $T_{\gamma:\delta}$-17 cells. Specifically, these later-developing $T_{\gamma:\delta}$-17 cells express T-cell receptors using the $V_\gamma4$ region. This $T_{\gamma:\delta}$-17 subset is found in all lymphoid organs, as well as in the skin dermis and the intestinal epithelium, where the cells provide rapid inflammatory signals in response to bacterial and parasitic infections. In addition, γ:δ T cells using the $V_\gamma7$ region in their T-cell receptors also develop in this second phase. The $V_\gamma7$-positive γ:δ T cells home specifically to the intestinal epithelium. In that location, the cells are poised to respond to gut microbes that breach the epithelial barrier, and are important producers of antibacterial compounds as well as IFN-γ.

In contrast to the γ:δ T-cell subsets that reside in barrier tissues such as the skin and intestinal epithelium, γ:δ T cells are also found in lymphoid organs. The majority of the lymphoid-resident γ:δ T cells arise during the late fetal–early neonatal period as well as thereafter, and represent a more diverse population expressing the $V_\gamma1$ region. $V_\gamma1$-positive T cells are composed of two major groups—an IFN-γ- plus IL-4-producing subset that homes to the liver as well as several lymphoid organs, and an IFN-γ-producing subset that homes to all lymphoid organs. The former population of cells, which can be identified by their expression of a unique TCRδ chain ($V_\delta6$) that is paired to $V_\gamma1$, is remarkably similar to the α:β T-cell receptor-expressing subset of iNKT cells, and the cells are therefore often referred to as γ:δ NKT cells. Unlike the mucosal and epithelial resident γ:δ T-cell populations, whose importance in tissue homeostasis, repair, and innate defense against infections has been well established, the functions of γ:δ T cells within secondary lymphoid organs is still not well understood.

8-17   Successful synthesis of a rearranged β chain allows the production of a pre-T-cell receptor that triggers cell proliferation and blocks further β-chain gene rearrangement.

We now return to the development of α:β T cells. The rearrangement of the β- and α-chain loci closely parallels the rearrangement of immunoglobulin heavy-chain and light-chain loci during B-cell development (see Sections 8-2 through 8-5). As shown in Fig. 8.24, the β-chain gene segments rearrange first, with the D_β gene segments rearranging to J_β gene segments, and this is followed by V_β to DJ_β rearrangement. If no functional β chain can be synthesized from this rearrangement, the cell will not be able to produce a pre-T-cell receptor and will die. However, unlike B cells with nonproductive heavy-chain gene rearrangements, thymocytes with nonproductive β-chain VDJ rearrangements can be rescued by further rearrangement, which is possible because of the two clusters of D_β and J_β gene segments upstream of two C_β genes (see Fig. 5.13).

The likelihood of a productive VDJ join at the β locus is therefore somewhat higher than the 55% chance for a productive immunoglobulin heavy-chain gene arrangement.

Once a productive β-chain gene rearrangement has occurred, the β chain is expressed together with the invariant pTα and the CD3 molecules (see Fig. 8.24) and is transported in this complex to the cell surface. The β:pTα complex is a functional pre-T-cell receptor analogous to the μ:VpreB:λ5 pre-B-cell receptor complex in B-cell development (see Section 8-3). Expression of

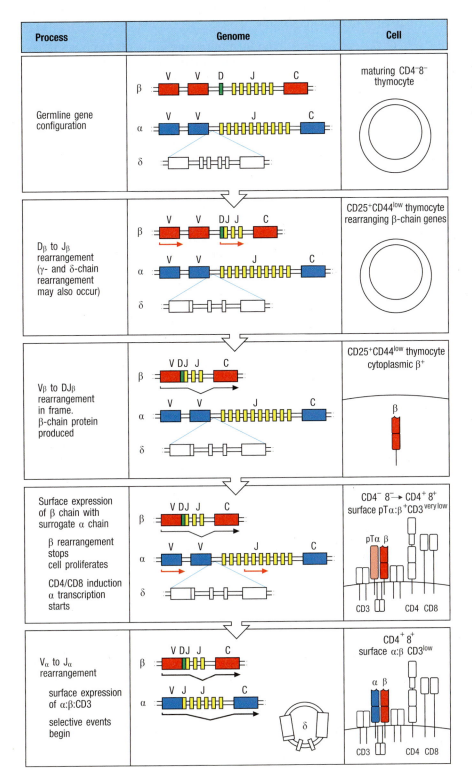

Fig. 8.24 The stages of gene rearrangement in α:β T cells. The sequence of gene rearrangements is shown, together with an indication of the stage at which the events take place and the nature of the cell-surface receptor molecules expressed at each stage. The β-chain locus rearranges first, in CD4⁻CD8⁻ double-negative thymocytes expressing CD25 and low levels of CD44. As with immunoglobulin heavy-chain genes, D to J gene segments rearrange before V gene segments rearrange to DJ (second and third panels). It is possible to make up to four attempts to generate a productive rearrangement at the β-chain locus, because there are four D gene segments with two sets of J gene segments associated with each TCRβ chain locus (not shown). The productively rearranged gene is expressed initially within the cell and then at low levels on the cell surface. It associates with pTα, a surrogate 33-kDa α chain that is equivalent to λ5 in B-cell development, and this pTα:β heterodimer forms a complex with the CD3 chains (fourth panel). The expression of the pre-T-cell receptor signals the developing thymocytes to halt β-chain gene rearrangement and to undergo multiple cycles of division. At the end of this proliferative burst, the CD4 and CD8 molecules are expressed, the cell ceases cycling, and the α chain is now able to undergo rearrangement. The first α-chain gene rearrangement deletes all δ D, J, and C gene segments on the chromosome, although these are retained as a circular DNA, indicating that these are nondividing cells (bottom panel). This permanently inactivates the δ-chain gene. Rearrangements at the α-chain locus can proceed through several cycles, because of the large number of V_α and J_α gene segments, so that productive rearrangements almost always occur. When a functional α chain is produced that pairs efficiently with the β chain, the CD3^low CD4⁺CD8⁺ thymocyte is ready to undergo selection for its ability to recognize self peptides in association with self-MHC molecules.

the pre-T-cell receptor at the DN3 stage of thymocyte development induces signals that cause the phosphorylation and degradation of RAG-2, thus halting β-chain gene rearrangement and ensuring allelic exclusion at the β locus. These signals also induce the DN4 stage, in which rapid cell proliferation occurs, and eventually the co-receptor proteins CD4 and CD8 are expressed. The pre-T-cell receptor signals constitutively via the cytoplasmic protein kinase Lck, an Src-family tyrosine kinase (see Fig. 7.12), but seems not to require a ligand on the thymic epithelium. Lck subsequently associates with the co-receptor proteins. In mice genetically deficient in Lck, T-cell development is arrested before the CD4+CD8+ double-positive stage, and no α-chain gene rearrangements can be made.

The role of the expressed β chain in suppressing further β-locus rearrangement can be demonstrated in mice containing a rearranged TCRβ transgene: these mice express the transgenic β chain on virtually 100% of their T cells, and rearrangement of their endogenous β-chain genes is strongly suppressed. The importance of pTα has been shown in mice deficient in pTα, in which there is a hundredfold decrease in α:β T cells and an absence of allelic exclusion at the β locus.

During the proliferation of DN4 cells triggered by expression of the pre-T-cell receptor, the *RAG-1* and *RAG-2* genes are repressed (see Fig. 8.18). Hence, no rearrangement of the α-chain locus occurs until the proliferative phase ends, at which time *RAG-1* and *RAG-2* are transcribed again, and the functional RAG-1:RAG-2 complex accumulates. This ensures that each cell in which a β-chain gene has been successfully rearranged gives rise to many CD4+CD8+ thymocytes. Once the cells stop dividing, each of them can independently rearrange its α-chain genes, so that a single functional β chain can be associated with many different α chains in the progeny cells. During the period of α-chain gene rearrangement, α:β T-cell receptors are first expressed and selection by self peptide:self MHC complexes on the thymus cells can begin.

The progression of thymocytes from the double-negative to the double-positive and finally to the single-positive stage is accompanied by a distinct pattern of expression of proteins involved in DNA rearrangement, signaling, and T-cell-specific gene expression (see Fig. 8.18). TdT, the enzyme responsible for the insertion of N-nucleotides, is expressed throughout T-cell receptor gene rearrangement; N-nucleotides are found at the junctions of all rearranged α and β genes. Lck and another tyrosine kinase, ZAP-70, are both expressed from an early stage in thymocyte development. As well as its key role in signaling from the pre-T-cell receptor, Lck is also important for γ:δ T-cell development. In contrast, gene knockout studies (see Appendix I, Section A-35) show that ZAP-70, although expressed from the double-negative stage onward, is not essential for pre-T-cell receptor signaling, as double-negative thymocytes also express the related Syk kinase, which is capable of fulfilling this role. Instead, ZAP-70 is required later, to promote the development of single-positive thymocytes from double-positive thymocytes; at this stage, Syk is no longer expressed. Fyn, an Src-family kinase similar to Lck, is expressed at increasing levels from the double-positive stage onward. It is not essential for the development of α:β thymocytes as long as Lck is present, but is required for the development of iNKT cells (see Section 8-26).

8-18   T-cell α-chain genes undergo successive rearrangements until positive selection or cell death intervenes.

The T-cell receptor α-chain genes are comparable to the immunoglobulin κ and λ light-chain genes in that they do not have D gene segments and are rearranged only after their partner receptor chain has been expressed. As with the light-chain genes, repeated attempts at α-chain gene rearrangement are possible, as illustrated in Fig. 8.25. The presence of multiple V_α gene segments,

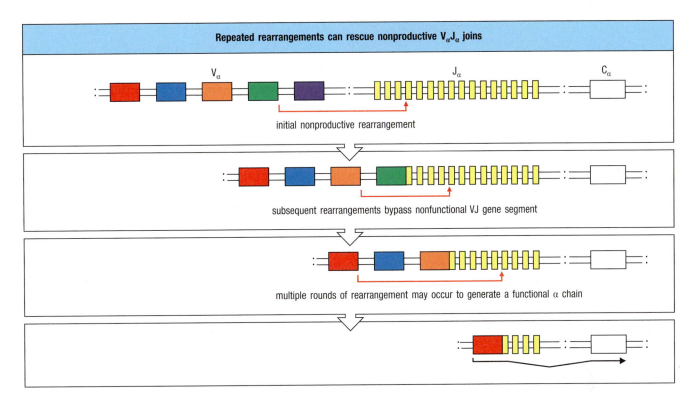

Repeated rearrangements can rescue nonproductive V$_\alpha$J$_\alpha$ joins

initial nonproductive rearrangement

subsequent rearrangements bypass nonfunctional VJ gene segment

multiple rounds of rearrangement may occur to generate a functional α chain

and about 60 J$_\alpha$ gene segments spread over some 80 kilobases of DNA, allows many successive V$_\alpha$ to J$_\alpha$ rearrangements to take place at both α-chain alleles. This means that T cells with an initial nonproductive α-gene rearrangement are much more likely to be rescued by a subsequent rearrangement than are B cells with a nonproductive light-chain gene rearrangement.

One key difference between B and T cells is that the final assembly of an immunoglobulin leads to the cessation of gene rearrangement and initiates the further differentiation of the B cell, whereas rearrangement of the V$_\alpha$ gene segments continues in T cells unless there is signaling by a self peptide:self MHC complex that positively selects the receptor (see Section 8-19 below). This means that many T cells have in-frame rearrangements on both chromosomes and so can produce two types of α chains. This is possible because expression of the T-cell receptor is not in itself sufficient to shut off gene rearrangement. Continued rearrangements on both chromosomes can allow several different α chains to be produced successively as well as simultaneously in each developing T cell and to be tested for self peptide:self MHC recognition in partnership with the same β chain. This phase of gene rearrangement lasts for 3 or 4 days in the mouse and ceases only when positive selection occurs as a consequence of receptor engagement, or when the cell dies. One can predict that if the frequency of positive selection is sufficiently low, roughly one in three mature T cells will express two productively rearranged α chains at the cell surface. This has been confirmed for both human and mouse T cells. Thus, in the strict sense, T-cell receptor α-chain genes are not subject to allelic exclusion.

T cells with dual specificity might be expected to give rise to inappropriate immune responses if the cell is activated through one receptor yet can act upon target cells recognized by the second receptor. However, only one of the two receptors is likely to be able to recognize peptide presented by a self MHC molecule, and so the T cell will have only a single functional specificity. This is because once a thymocyte has been positively selected by self peptide:self MHC recognition, α-chain gene rearrangement ceases. Thus, the existence of cells with two α-chain genes productively rearranged and two α chains expressed at the cell surface does not truly challenge the idea that a single functional specificity is expressed by each cell.

Fig. 8.25 Multiple successive rearrangement events can rescue nonproductive T-cell receptor α-chain gene rearrangements. The multiplicity of V and J gene segments at the α-chain locus allows successive rearrangement events to 'leapfrog' over previously rearranged VJ segments, deleting any intervening gene segments. The α-chain rescue pathway resembles that of the immunoglobulin κ light-chain genes (see Section 8-5), but the number of possible successive rearrangements is greater. α-chain gene rearrangement continues until either a productive rearrangement leads to positive selection or the cell dies.

Summary.

The thymus provides a specialized and architecturally organized microenvironment for the development of mature T cells. Precursors of T cells migrate from the bone marrow to the thymus, where they interact with environmental cues, such as ligands for the Notch receptor, that drive commitment to the T lineage. Developing thymocytes develop along one of several T-cell lineages: the most prominent subsets in the thymus are γ:δ T cells, conventional α:β T cells, and α:β T cells with receptors of very limited diversity, such as iNKT cells.

T-cell progenitors develop along the γ:δ or the α:β T-cell lineages. Early in ontogeny, the production of γ:δ T cells predominates over α:β T cells, and these cells populate several peripheral tissues, including the skin, the intestine, and other mucosal and epithelial surfaces. These subsets predominantly develop from fetal liver, rather than bone marrow, stem cells. Later, more than 90% of thymocytes express α:β T-cell receptors. In developing thymocytes, the γ, δ, and β genes are the first to rearrange. Cells of the α:β lineage that rearrange a functional beta chain form a pre-T-cell receptor that signals thymocyte proliferation, α-chain gene rearrangement, and CD4 and CD8 expression. Most steps in T-cell development take place in the thymic cortex, whereas the medulla contains mainly mature T cells.

Positive and negative selection of T cells.

Up to the stage at which an α:β receptor is produced, T-cell development is independent of MHC proteins or antigen. From this point onward, developmental decisions in the α:β T-cell lineage depend on the interaction of the receptor with peptide:MHC ligands it encounters in the thymus, and we now consider this phase of T-cell development.

T-cell precursors committed to the α:β lineage at the DN3 stage undergo vigorous proliferation in the subcapsular region and progress to the DN4 stage. These cells then rapidly transit through an immature CD8 single-positive stage and become double-positive cells that express low levels of the T-cell receptor and both the CD4 and CD8 co-receptors as they move deeper into the thymic cortex. These double-positive cells have a life-span of only about 3–4 days unless they are rescued by engagement of their T-cell receptor. The rescue of double-positive cells from programmed cell death and their maturation into CD4 or CD8 single-positive cells is the process known as positive selection. Only about 10–30% of the T-cell receptors generated by gene rearrangement will be able to recognize self peptide:self MHC complexes and thus function in self MHC-restricted responses to foreign antigens (see Chapter 4); those that have this capability are selected for survival in the thymus. Double-positive cells also undergo negative selection: T cells whose receptors recognize self peptide:self MHC complexes too strongly undergo apoptosis, thus eliminating potentially self-reactive cells. In this section, we examine the interactions between developing double-positive thymocytes and different thymic components and discuss the mechanisms by which these interactions shape the mature T-cell repertoire.

8-19 Only thymocytes whose receptors interact with self peptide:self MHC complexes can survive and mature.

Early experiments using bone marrow chimeras (see Appendix I, Section A-32) and thymic grafting provided evidence that MHC molecules in the thymus influence the MHC-restricted T-cell repertoire. However, mice transgenic for rearranged T-cell receptor genes provided the first conclusive evidence that the interaction of the T cell with self peptide:self MHC complexes is necessary for the survival of immature T cells and their maturation into naive CD4 or CD8 T cells. For these experiments, the rearranged α- and β-chain genes were cloned from a T-cell clone (see Appendix I, Section A-20) whose origin, antigen specificity, and MHC restriction were known. When such genes are introduced into the mouse genome, they are expressed early during thymocyte development. As a consequence of expressing functional transgene-encoded TCRα- and β-chain proteins in developing T cells, the rearrangement of endogenous T-cell receptor genes is inhibited, albeit to different degrees. In general, endogenous β-chain gene rearrangement is inhibited completely but that of α-chain genes is inhibited only incompletely. The result is that most of the developing thymocytes in TCR transgenic mouse lines express the T-cell receptor encoded by the transgenes.

By introducing T-cell receptor transgenes specific for a known peptide:MHC complex, the effect of allelic variations in MHC molecules on the maturation of thymocytes with receptors of known specificity can be studied directly, without the need for immunization and analysis of effector function. Such studies showed that thymocytes bearing a particular T-cell receptor could develop to the double-positive stage in thymuses that expressed different MHC molecules from those present in the mouse from which the original T-cell clone was isolated. However, these transgenic thymocytes developed into mature CD4 or CD8 single-positive thymocytes only if the thymus expressed the same self MHC molecule as that on which the original T-cell clone was selected (Fig. 8.26).

Such experiments also discovered the fate of T cells that fail positive selection. Rearranged receptor genes from a mature T cell specific for a peptide presented by a particular MHC molecule were introduced into a recipient mouse lacking that MHC molecule, and the fate of thymocytes was investigated by staining with antibodies specific for the transgenic receptor. Antibodies against other molecules, such as CD4 and CD8, were used at the same time to mark the stages of T-cell development. It was found that cells that fail to recognize the MHC molecules present on the thymic epithelium never progress further than the double-positive stage and die in the thymus within 3 or 4 days of their last division.

8-20   Positive selection acts on a repertoire of T-cell receptors with inherent specificity for MHC molecules.

Positive selection acts on a repertoire of T-cell receptors whose specificity is determined by randomly generated combinations of V, D, and J gene segments (see Section 5-7). Despite this, T-cell receptors exhibit a bias toward recognition of MHC molecules even before positive selection. If the specificity of the unselected repertoire were completely random, only a very small proportion of thymocytes would be expected to recognize any MHC molecule. However, an inherent specificity of T-cell receptors for MHC molecules has been detected by examining mature T cells that represent the unselected receptor repertoire. Such T cells can be produced *in vitro* from fetal thymuses that lack expression of MHC class I and MHC class II molecules by triggering generalized 'positive selection' using antibodies that bind to the V_β chain of T-cell receptors and to the CD4 co-receptor. When such antibody-selected CD4 T cells are examined, roughly 5% can respond to any one MHC class II genotype. Because these cells developed without selection by MHC molecules, this reactivity must reflect an inherent MHC-specificity encoded in the germline V gene segments. This specificity should significantly increase the proportion of receptors that can be positively selected by any individual's MHC molecules.

The germline-encoded reactivity seems to be due to specific amino acids in the CDR1 and CDR2 regions of T-cell receptor V_β and V_α regions. The CDR1 and CDR2 regions are encoded in the germline V gene segments and are highly

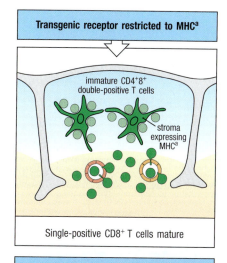

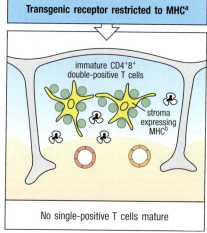

Fig. 8.26 Positive selection is demonstrated by the development of T cells expressing rearranged T-cell receptor transgenes. In mice transgenic for rearranged α:β T-cell receptor genes, the maturation of T cells depends on the MHC haplotype expressed in the thymus. If the transgenic mice express the same MHC haplotype in their thymic stromal cells as the mouse from which the rearranged TCRα-chain and TCRβ-chain genes originally developed (both MHCᵃ, top panel), then the T cells expressing the transgenic T-cell receptor will develop from the double-positive stage (pale green) into mature T cells (dark green), in this case mature CD8⁺ single-positive cells. If the MHCᵃ-restricted TCR transgenes are genetically crossed into a different MHC background (MHCᵇ, yellow, bottom panel), then developing T cells expressing the transgenic receptor will progress to the double-positive stage but will fail to mature further. This failure is due to the absence of an interaction between the transgenic T-cell receptor with MHC molecules on the thymic cortex, and thus no signal for positive selection is delivered, leading to apoptotic death by neglect.

variable (see Section 5-8). But among this variability, certain amino acids are conserved and common to many V segments. Analysis of numerous crystal structures has revealed that when the T-cell receptor binds a peptide:MHC complex, specific amino acids of the V_β region interact with a particular part of the MHC protein. For example, in many human and mouse V_β regions, the CDR2 has a tyrosine at position 48, and this interacts with a region in the middle of the $\alpha1$ helix of MHC class I and class II proteins. Two other amino acids commonly found in other V_β regions (tyrosine at 46 and glutamic acid at 54) interact with the same region of MHC. T cells expressing V_β genes with mutations at any of these positions showed reduced positive selection, demonstrating that the interaction of such V regions with MHC molecules contributes to T-cell development.

8-21   Positive selection coordinates the expression of CD4 or CD8 with the specificity of the T-cell receptor and the potential effector functions of the T cell.

At the time of positive selection, the thymocyte expresses both CD4 and CD8 co-receptor molecules. By the end of thymic selection, mature $\alpha{:}\beta$ T cells ready for export to the periphery have stopped expressing one of these co-receptors. The majority of these cells belong to the conventional CD4 or CD8 T-cell lineages. Less abundant subsets, such as iNKT cells and a subset of regulatory T cells expressing CD4 and high levels of CD25, also develop in the thymus from CD4+CD8+ cells. Moreover, almost all mature T cells that express CD4 have receptors that recognize peptides bound to self MHC class II molecules and are programmed to become cytokine-secreting helper T cells. In contrast, most of the cells that express CD8 have receptors that recognize peptides bound to self MHC class I molecules and are programmed to become cytotoxic effector cells. Thus, positive selection also determines the cell-surface phenotype and functional potential of the mature T cell, selecting the appropriate co-receptor for efficient antigen recognition and the appropriate program for the T cell's eventual functional differentiation in an immune response.

Experiments with mice transgenic for rearranged T-cell receptor genes show clearly that the specificity of the T-cell receptor for self peptide:self MHC complexes determines which co-receptor a mature T cell will express. If the transgenes encode a T-cell receptor specific for antigen presented by self MHC class I molecules, mature T cells that express the transgenic receptor are CD8 T cells. Similarly, in mice made transgenic for a receptor that recognizes antigen with self MHC class II molecules, mature T cells that express the transgenic receptor are CD4 T cells (**Fig. 8.27**).

The importance of MHC molecules in this selection is illustrated by the human immunodeficiency diseases caused by mutations that lead to an absence of MHC molecules on lymphocytes and thymic epithelial cells. People who lack MHC class II molecules have CD8 T cells but only a few, highly abnormal CD4 T cells; a similar result has been obtained in mice in which MHC class II expression has been eliminated by targeted gene disruption (see Appendix I, Section A-35). Likewise, mice and humans that lack MHC class I molecules lack CD8 T cells. Thus, MHC class II molecules are absolutely required for CD4 T-cell development, whereas MHC class I molecules are similarly required for CD8 T-cell development.

In mature T cells, the co-receptor functions of CD8 and CD4 depend on their respective abilities to bind invariant sites on MHC class I and MHC class II molecules (see Section 4-18). Co-receptor binding to an MHC molecule is also required for normal positive selection, as shown for CD4 in the experiment discussed in the next section. In thymocytes, nearly all of the Lck is associated with CD4 and CD8 co-receptors, providing a mechanism to ensure that signaling is initiated only in thymocytes bearing T-cell receptors that bind to MHC

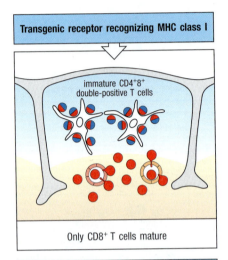

Transgenic receptor recognizing MHC class I

immature CD4+8+ double-positive T cells

Only CD8+ T cells mature

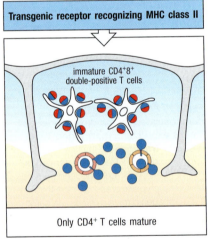

Transgenic receptor recognizing MHC class II

immature CD4+8+ double-positive T cells

Only CD4+ T cells mature

Fig. 8.27 The MHC molecules that induce positive selection determine co-receptor specificity. In mice transgenic for T-cell receptors restricted by an MHC class I molecule (top panel), the mature T cells that develop all have the CD8 (red) phenotype. In mice transgenic for receptors restricted by an MHC class II molecule (bottom panel), all mature T cells have the CD4 (blue) phenotype. In both cases, normal numbers of immature, double-positive thymocytes (half blue, half red) are found. The specificity of the T-cell receptor determines the outcome of the developmental pathway, ensuring that the only T cells that mature are those equipped with a co-receptor that is able to bind the same self MHC molecule as the T-cell receptor.

molecules. Thus, positive selection depends on engagement of both the antigen receptor and co-receptor by an MHC molecule, and this signal determines the survival of single-positive cells that express only the appropriate co-receptor. Commitment to either the CD4 or CD8 lineage is coordinated with receptor specificity, and it seems that the developing thymocyte integrates signals from both the antigen receptor and the co-receptor. Co-receptor-associated Lck signals are most effectively delivered when CD4 rather than CD8 is engaged as a co-receptor, and these Lck signals play a large part in the decision to become a mature CD4 cell.

T-cell receptor signaling regulates this choice of the CD4 versus the CD8 lineage by controlling the expression of two transcription factors, ThPOK and Runx3 (see Fig. 8.18). The role of ThPOK was identified through a naturally occurring loss-of-function mutation in mice that lacked CD4 T-cell development. In mice lacking ThPOK, MHC class II-restricted thymocytes are redirected toward the CD8 lineage. ThPOK is not expressed in pre-selection double-positive thymocytes, but strong T-cell receptor signaling at this stage of development induces its expression. ThPOK, in turn, reinforces its own expression and represses expression of Runx3; together, the expression of ThPOK and absence of Runx3 lead to CD4 commitment and the ability to express cytokine genes typical of CD4 cells. If T-cell signaling is of insufficient strength or duration, however, ThPOK is not induced, and Runx3 is allowed to be expressed. This leads to silencing of CD4 expression, maintenance of CD8 expression, and the expression of genes typical of CD8 T cells, namely, genes that encode proteins involved in target-cell killing.

While the majority of double-positive thymocytes that undergo positive selection develop into either CD4 or CD8 single-positive T cells, the thymus also generates less numerous populations of other T-cell subsets with specialized functions; these will be discussed further in Section 8-26.

8-22 Thymic cortical epithelial cells mediate positive selection of developing thymocytes.

Thymus transplantation studies indicate that stromal cells are important for positive selection. These cells form a web of cell processes that make close contacts with the double-positive T cells undergoing positive selection (see Fig. 8.17), and T-cell receptors can be seen clustering with MHC molecules at the sites of contact. Direct evidence that thymic cortical epithelial cells mediate positive selection comes from an ingenious manipulation of mice whose MHC class II genes have been eliminated by targeted gene disruption (Fig. 8.28).

MHC Class II Deficiency

MHC Class I Deficiency

Fig. 8.28 Thymic cortical epithelial cells mediate positive selection. In the thymus of normal mice (first panels), which express MHC class II molecules on epithelial cells in the thymic cortex (blue) as well as on medullary epithelial cells (orange) and bone marrow-derived cells (yellow), both CD4 (blue) and CD8 (red) T cells mature. Double-positive thymocytes are shown as half red and half blue. The second panels represent mutant mice in which MHC class II expression has been eliminated by targeted gene disruption; in these mice, few CD4 T cells develop, although CD8 T cells develop normally. In MHC class II-negative mice containing an MHC class II transgene engineered so that it is expressed only on the epithelial cells of the thymic cortex (third panels), normal numbers of CD4 T cells mature. In contrast, if a mutant MHC class II molecule with a defective CD4-binding site is expressed (fourth panel), positive selection of CD4 T cells does not take place. This indicates that the cortical epithelial cells are the critical cell type mediating positive selection and that the MHC class II molecule needs to be able to interact with the CD4 protein.

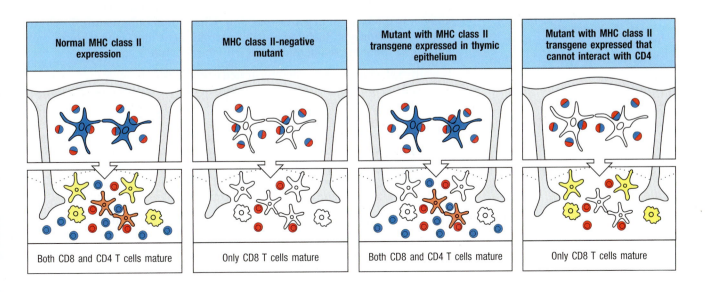

Normal MHC class II expression	MHC class II-negative mutant	Mutant with MHC class II transgene expressed in thymic epithelium	Mutant with MHC class II transgene expressed that cannot interact with CD4
Both CD8 and CD4 T cells mature	Only CD8 T cells mature	Both CD8 and CD4 T cells mature	Only CD8 T cells mature

Mutant mice that lack MHC class II molecules do not normally produce CD4 T cells. To test the role of the thymic epithelium in positive selection, an MHC class II gene was placed under the control of a promoter that restricted the gene's expression to thymic cortical epithelial cells. This was then introduced as a transgene into the MHC class II-mutant mice, and CD4 T-cell development was restored. A variant of this experiment showed that, to promote the development of CD4 T cells, the MHC class II molecule on the thymic cortical epithelium must be able to interact effectively with CD4. Thus, when the MHC class II transgene expressed in the thymus contains a mutation that prevents binding of the MHC to CD4, very few CD4 T cells develop. Equivalent studies of CD8 interaction with MHC class I molecules showed that co-receptor binding is also necessary for the positive selection of CD8 cells.

The critical role of the thymic cortical epithelium in positive selection raises the question whether there is anything distinctive about the antigen-presenting properties of these cells. The thymic stromal cells may simply be in closest proximity to the developing thymocytes, as there are very few macrophages and dendritic cells in the cortex to perform the antigen presentation. In addition, however, thymic epithelium differs from other tissues in the expression of key proteases that are involved in MHC class I and II antigen processing (see Section 6-8). Cortical epithelial cells express cathepsin L as opposed to the more widely expressed cathepsin S, and mice deficient in cathepsin L have severely impaired CD4 T-cell development. Thymic epithelial cells from mice lacking cathepsin L exhibit a relatively high proportion of MHC class II molecules on their surface that retain the class II invariant chain-associated peptide (CLIP) (see Fig. 6.11). Cortical epithelial cells also express a unique proteasome subunit, β5T, whereas other cells express β5 or β5i. Mice deficient in β5T have severely impaired CD8 T-cell development. Because mice that lack either cathepsin L or β5T still have normal levels of MHC on the surface of their thymic cortical cells, it would seem that it is the peptide repertoire displayed by the MHC molecules on cortical epithelial cells that is responsible for altering CD8 T-cell development, although the mechanism is still unclear.

8-23 T cells that react strongly with ubiquitous self antigens are deleted in the thymus.

When the T-cell receptor of a mature naive T cell is strongly ligated by a peptide:MHC complex displayed on a specialized antigen-presenting cell in a peripheral lymphoid organ, the T cell is activated to proliferate and produce effector T cells. In contrast, when the T-cell receptor of a developing thymocyte is similarly ligated by a self peptide:self MHC complex in the thymus, it dies by apoptosis (Fig. 8.29). The response of immature T cells to stimulation by antigen is the basis of negative selection. Elimination of immature T cells in the thymus prevents their potentially harmful activation later, should they encounter the same self peptides when they are mature T cells.

Negative selection has been demonstrated using TCR-transgenic mice expressing T-cell receptors specific for self peptides derived from proteins encoded on the Y chromosome, and thus expressed only in male mice. Thymocytes bearing these receptors disappear from the developing T-cell population in male mice at the double-positive stage of development, and no single-positive cells bearing the transgenic receptors mature. By contrast, in female mice, which lack the male-specific peptide, T cells bearing the transgenic receptors mature normally. Negative selection to male-specific peptides has also been demonstrated in nontransgenic mice and also occurs by deletion of T cells.

TCR transgenic mice were very useful for the classic experiments above, but they express a functional T-cell receptor earlier during development than normal mice and have a very high frequency of cells reactive to any particular peptide. A more realistic system for evaluating negative selection involves the

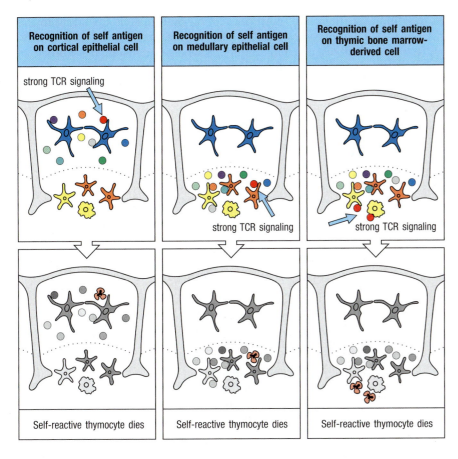

Fig. 8.29 Negative selection of thymocytes can occur in the cortex or the medulla. When the T-cell receptor (TCR) on a developing thymocyte is strongly stimulated by recognition of self peptide:self MHC complexes (red cell), the thymocyte is induced to die, a process known as negative selection. Negative selection can occur in the cortex when a CD4+CD8+ double-positive thymocyte has strong reactivity to peptide:MHC complexes found on cortical epithelial cells (left panel). Negative selection can also occur in the medulla when an immature CD4 or CD8 single-positive thymocyte receives strong T-cell receptor signaling following recognition of peptide:MHC complexes on medullary epithelial cells (middle panel) or on bone marrow-derived macrophages or dendritic cells (right panel).

transgenic expression of only the β chain of a T-cell receptor reactive to a given peptide antigen. In such mice, the β chain pairs with endogenous α chains, yet the frequency of peptide-reactive T cells is sufficient for detection using peptide:MHC tetramers (see Appendix I, Section A-24). These and other more physiologic approaches showed that clonal deletion can occur at either the double-positive or the single-positive stage, presumably depending on where the T cell encounters the antigen that causes deletion.

These experiments illustrate the principle that self peptide:self MHC complexes encountered in the thymus purge the mature T-cell repertoire of cells bearing self-reactive receptors. One obvious problem with this scheme is that many tissue-specific proteins, such as pancreatic insulin, would not be expected to be expressed in the thymus. However, it is now clear that many such 'tissue-specific' proteins are expressed by certain stromal cells in the thymic medulla; thus, intrathymic negative selection could apply even to proteins that are otherwise restricted to tissues outside the thymus. The expression of some, but not all, tissue-specific proteins in the thymic medulla is controlled by a gene called ***AIRE* (autoimmune regulator)**. *AIRE* is expressed in medullary stromal cells (Fig. 8.30), interacts with many proteins involved in transcription, and seems to lengthen transcripts that would otherwise terminate earlier. Mutations in *AIRE* give rise to the human autoimmune disease known as **autoimmune polyendocrinopathy–candidiasis–ectodermal dystrophy (APECED)** or **autoimmune polyglandular syndrome type I**, highlighting the important role of intrathymic expression of tissue-specific proteins in maintaining tolerance to self. Negative selection of developing T cells involves interactions with ubiquitous and tissue-restricted self antigens, and can take place in both the thymic cortex and the thymic medulla (see Fig. 8.29).

It is unlikely that all possible self proteins are expressed in the thymus. Thus, negative selection in the thymus may not remove all T cells reactive to self antigens that appear exclusively in other tissues or are expressed at different stages

Autoimmune Polyendocrinopathy–Candidiasis–Ectodermal Dystrophy (APECED)

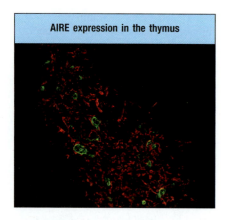

Fig. 8.30 AIRE is expressed in the medulla of the thymus and promotes the expression of proteins normally expressed in peripheral tissues. Expression of AIRE by thymic medullary cells is limited to the medullary region of the thymus, where it is expressed in a subset of epithelial-like cells. The expression of the thymic medullary epithelial marker MTS10 is shown in red. AIRE expression is shown in green by immunofluorescence, and is present in only a fraction of the medullary epithelial cells. Photograph courtesy of R.K. Chin and Y.-X. Fu.

in development. There are, however, several mechanisms operating in the periphery that can prevent mature T cells from responding to tissue-specific antigens; these are discussed in Chapter 15, when we consider the problem of autoimmune responses and their avoidance.

8-24 Negative selection is driven most efficiently by bone marrow-derived antigen-presenting cells.

As discussed above, negative selection occurs throughout thymocyte development, both in the thymic cortex and in the medulla, and so is likely to be mediated by antigen presentation by several different cell types (see Fig. 8.29). There does seem to be a hierarchy in the effectiveness of cells in mediating negative selection. At the top are bone marrow-derived dendritic cells and macrophages. These are antigen-presenting cells that also activate mature T cells in peripheral lymphoid tissues, as we shall see in Chapter 9. The self antigens presented by these cells are therefore the most important source of potential autoimmune responses, and T cells responding to such self peptides must be eliminated in the thymus.

In addition, both the thymocytes themselves and the thymic epithelial cells can cause the deletion of self-reactive cells. The medullary epithelial cells expressing AIRE, and thus presenting a wide range of self-antigens, are one population that has been shown to directly induce thymocyte negative selection. More generally, in patients undergoing bone marrow transplantation from an unrelated donor, where all the thymic macrophages and dendritic cells are of donor type, negative selection mediated by thymic epithelial cells is of critical importance in maintaining tolerance to the recipient's own antigens.

8-25 The specificity and/or the strength of signals for negative and positive selection must differ.

T cells undergo both positive selection for self MHC restriction and negative selection for self-tolerance by interacting with self peptide:self MHC complexes expressed on stromal cells in the thymus. An unresolved issue is how the interaction of the T-cell receptor with self peptide:self MHC complexes distinguishes between these opposite outcomes. First, more receptor specificities must be positively selected than are negatively selected. Otherwise, all the cells that were positively selected in the thymic cortex would be eliminated by negative selection, and no T cells would ever be produced. Second, the consequences of the interactions that lead to positive and negative selection must differ: cells that recognize self peptide:self MHC complexes on cortical epithelial cells are induced to mature, whereas those whose receptors might confer strong and potentially damaging autoreactivity are induced to die.

Currently, the choice between positive and negative selection is thought to hinge on the strength of self peptide:self MHC binding by the T-cell receptor, an idea known as the **affinity hypothesis** (Fig. 8.31). Low-affinity interactions rescue the cell from death by neglect, leading to positive selection; high-affinity interactions induce apoptosis and thus negative selection. Because more complexes are likely to bind weakly than strongly, this model explains the positive selection of a larger repertoire of cells than are negatively selected. Using T-cell receptor transgenic thymocytes, it was shown that variants of the antigenic peptide could induce positive selection in thymic organ cultures or *in vivo*. Peptide variants that induced positive selection had a lower affinity for the T-cell receptor than did antigenic peptide. How this quantitative difference in receptor affinity translates into a qualitatively distinct cell fate is still an area of active investigation. Many of the biochemical signals induced by low-affinity interactions are weaker or of shorter duration than those from high-affinity interactions. However, low-affinity interactions lead to sustained

Fig. 8.31 The affinity model of T-cell positive and negative selection. Random TCRα and β chain gene rearrangements generate a large pool of immature thymocytes expressing a varied repertoire of specificities. The T-cell receptors on many of these cells fail to have sufficient binding strength to the self peptide:self MHC complexes on thymic epithelium and so receive no signals. These cells die by neglect. Another fraction of immature thymocytes are positively selected because their T-cell receptors bind with sufficient strength to the self peptide:self MHC complexes on thymic epithelium to generate T-cell receptor-dependent survival signals. From this cohort of positively selected thymocytes, negative selection removes those thymocytes whose receptors have excessively strong reactivity to self peptides complexed with self MHC molecules (resulting in clonal deletion), thereby establishing self-tolerance of the mature T-cell population. A small subset of positively selected cells receiving signals slightly weaker than those inducing negative selection differentiate into regulatory T cells (T_{regs}), a process referred to as agonist selection.

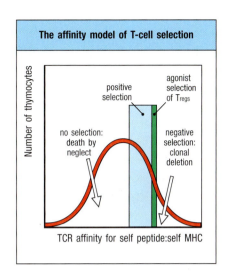

activation of the protein kinase Erk, whereas high-affinity interactions lead to only transient activation of Erk, suggesting that differential activation of this or other MAPKs might determine the outcome of thymic selection. Indeed, experiments showed that developing T cells need to engage low-affinity ligands for more than 24 hours for positive selection to occur.

8-26 Self-recognizing regulatory T cells and innate T cells develop in the thymus.

Additional populations besides the conventional CD4$^+$ and CD8$^+$ α:β T cells discussed above emerge from the thymus; they are numerically minor but functionally important. Two of these subsets, the T_{reg} cells (see Section 9-23) and the iNKT cells (see Section 6.18), have been well studied, and found to each have unique developmental requirements.

Thymically derived T_{reg} cells are a subset of CD4$^+$ T cells that function to maintain self-tolerance. These cells arise from CD4$^+$CD8$^+$ thymocytes, as do conventional T cells. During their maturation, they upregulate the transcription factor FoxP3. T_{reg} cell development also depends on IL-2 receptor signaling, a cytokine signal that is not required for the development of conventional T cells. The repertoire of T-cell receptors expressed on T_{reg} cells is thought to be composed of receptors with high affinity for self MHC:self peptide complexes. Evidence supporting this conclusion comes from studies showing that some lines of TCR transgenic mice generate large numbers of T_{reg} cells when the mice also express the antigen for this T-cell receptor. In addition, studies using mice expressing a fluorescent reporter that monitors T-cell receptor signal strength have shown that T_{reg} cells express high levels of the fluorescent reporter, both during their development and after their export from the thymus, indicating that they likely express T-cell receptors with high affinity for self. This process of positive selection following high-affinity T-cell receptor interactions with self peptide:self MHC complexes has been termed **agonist selection**—in other words, agonist selection refers to interactions of a T-cell receptor with a self peptide:self MHC that would normally activate a mature T cell expressing that T-cell receptor.

A second specialized subset of T cells that develops from CD4$^+$CD8$^+$ thymocyte precursors is a lineage known as **invariant NKT cells (iNKT cells)**, based on their expression of the NK1.1 receptor commonly found on NK cells. iNKT cells are activated as part of the early response to many infections; they differ from the major lineage of α:β T cells in recognizing CD1 molecules rather than MHC class I or MHC class II molecules (see Section 6-18). Unlike other T cells, iNKT cells require for their development a T-cell receptor interaction with CD1 molecules expressed on thymocytes and a signal through the adaptor protein SAP. iNKT cells, like γ:δ T cells, acquire a defined effector program during their development in the thymus. Therefore, these cells exhibit a memory-cell phenotype when they leave the thymus and

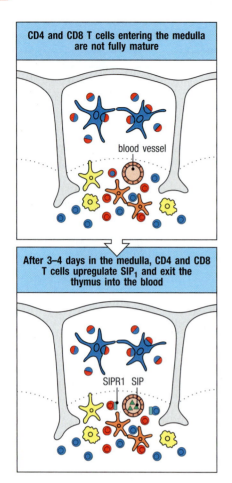

CD4 and CD8 T cells entering the medulla are not fully mature

blood vessel

After 3–4 days in the medulla, CD4 and CD8 T cells upregulate SIP₁ and exit the thymus into the blood

S1PR1 SIP

Fig. 8.32 Thymocyte emigration is induced by signaling through the sphingosine 1-phosphate receptor, S1PR1. CD4 and CD8 single-positive thymocytes that have successfully survived positive and negative selection are found in the medulla but are not yet fully mature. At the termination of the maturation process, which takes 3–4 days, the CD4 and CD8 single-positive thymocytes upregulate the sphingosine 1-phosphate (S1P) receptor, known as S1PR1. S1PR1 is a G-protein-coupled receptor that promotes chemotaxis of the cells toward the ligand S1P. Due to the high levels of S1P in the blood, single-positive thymocytes are induced to leave the thymus by entering the blood, where they become part of the recirculating naive T-cell population.

migrate to peripheral lymphoid tissues and mucosal surfaces. iNKT cells have been suggested to develop in response to 'agonist' signaling. Recent studies have revealed that CD1-binding lipid antigens produced by the commensal microbes in the gut are an important source of these agonist ligands, and that the composition of the gut microbiome regulates the development of iNKT cells early in life. Since agonist stimulation of immature T cells is also known to cause clonal deletion, it is not yet clear which activating interactions lead to clonal deletion in the thymus and which lead to selection of T_{reg} cells or the nonconventional iNKT cells.

8-27 The final stage of T-cell maturation occurs in the thymic medulla.

After surviving positive and negative selection, thymocytes complete their maturation in the thymic medulla and then emigrate to peripheral lymphoid organs. Their final maturation results in changes to the T-cell receptor signaling machinery. Whereas an immature double-positive or single-positive thymocyte stimulated through the T-cell receptor will undergo apoptosis, a mature single-positive thymocyte responds by proliferating. The final maturation stage takes less than 4 days, and functionally competent T cells then emigrate from the thymus into the bloodstream (**Fig. 8.32**). Emigration requires recognition of the lipid molecule sphingosine 1-phosphate (S1P) by the G-protein-coupled receptor S1PR1, which is expressed by thymocytes during their final maturation. S1P is present in high concentration in blood and lymph, and mature thymocytes seem to be drawn toward it. Mature thymocytes also express CD62L (L-selectin), a lymph-node homing receptor that facilitates the localization of mature naive T cells to peripheral lymphoid organs after their emigration from the thymus.

8-28 T cells that encounter sufficient quantities of self antigens for the first time in the periphery are eliminated or inactivated.

Many autoreactive T cells are purged during their development in the thymus. As discussed in Section 8-23, this negative selection process is facilitated by the AIRE protein, which promotes the expression of many tissue-specific antigens in thymic medullary epithelial cells. Nonetheless, not all self antigens are expressed in the thymus, and some autoreactive T cells complete their maturation and migrate to the periphery. Our understanding of the fates of autoreactive T cells in the periphery comes mainly from the study of mice transgenic for self-reactive T-cell receptors. In some cases, T cells reacting to self antigens in the periphery are eliminated. This usually follows a brief period of activation and cell division, and so is known as **activation-induced cell death**. In other cases, the self-reactive cells may be rendered anergic. When studied *in vitro*, these anergic T cells prove refractory to signals delivered through the T-cell receptor.

The question immediately arises: if the encounter of a mature naive lymphocyte with a self antigen leads to cell death or anergy, why does this not also happen to a mature lymphocyte that recognizes a pathogen-derived antigen? The answer is that infection sets up inflammation, which induces the expression of co-stimulatory molecules on the antigen-presenting dendritic cells and the production of cytokines promoting lymphocyte activation. The outcome of an encounter with antigen in these conditions is the activation, proliferation, and differentiation of the lymphocyte to effector-cell status. In the absence of infection or inflammation, dendritic cells still process and present self antigens, but in the absence of co-stimulatory and other signals, any interaction of a mature lymphocyte with its specific antigen seems to result in a tolerance-inducing (**tolerogenic**) signal from the antigen receptor.

Summary.

The stages of thymocyte development up to the expression of the pre-T-cell receptor—including the decision between commitment to either the α:β or the γ:δ lineage—are all independent of interactions with peptide:MHC antigens. With the successful rearrangement of α-chain genes and expression of the T-cell receptor, α:β thymocytes undergo further development that is determined by the interactions of their T-cell receptors with self peptides presented by the MHC molecules on the thymic stroma. CD4$^+$CD8$^+$ double-positive thymocytes whose receptors interact with self peptide:self MHC complexes on thymic cortical epithelial cells are positively selected, and will eventually mature into CD4 or CD8 single-positive cells. T cells that react too strongly with self antigens are deleted in the thymus, a process driven by bone marrow-derived antigen-presenting cells and AIRE-expressing epithelial cells in the medullary region of the thymus. The outcome of positive and negative selection is the generation of a mature conventional T-cell repertoire that is both MHC-restricted and self-tolerant. Some non-conventional T-cell lineages undergo 'agonist' selection following strong T-cell receptor signaling. Precisely how the recognition of self peptide:self MHC ligands by the T-cell receptor leads to either positive or negative selection remains an unsolved problem.

Summary to Chapter 8.

In this chapter we have learned about the formation of the B-cell and T-cell lineages from an uncommitted hematopoietic stem cell. The somatic gene rearrangements that generate the highly diverse repertoire of antigen receptors—immunoglobulin for B cells, and the T-cell receptor for T cells—occur in the early stages of the development of T cells and B cells from a common bone marrow-derived lymphoid progenitor. Mammalian B-cell development takes place in fetal liver and, after birth, in the bone marrow; T cells also originate from stem cells in the fetal liver or the bone marrow, but undergo most of their development in the thymus. Much of the somatic recombination machinery, including the RAG proteins that are an essential part of the V(D)J recombinase, is common to both B and T cells. In both B and T cells, gene rearrangements begin with the loci that contain D gene segments, and proceed successively at each locus. The first step in B-cell development is the rearrangement of the locus for the immunoglobulin heavy chain, and for T cells the β chain. In each case, the developing cell is allowed to proceed to the next stage of development only if the rearrangement has produced an in-frame sequence that can be translated into a protein expressed on the cell surface: either the pre-B-cell receptor or the pre-T-cell receptor. Cells that do not generate successful rearrangements for both receptor chains die by apoptosis. The course of conventional B-cell development is summarized in Fig. 8.14, and that of α:β T cells in Fig. 8.33.

Once a functional antigen receptor has appeared on the cell surface, the lymphocyte is tested in two ways. Positive selection tests for the potential usefulness of the antigen receptor, whereas negative selection removes self-reactive cells from the lymphocyte repertoire, rendering it tolerant to the antigens of the body. Positive selection is particularly important for T cells, because it ensures that only cells bearing T-cell receptors that can recognize antigen in combination with self MHC molecules will continue to mature. Positive selection also coordinates the choice of co-receptor expression. CD4 becomes expressed by T cells harboring MHC class II-restricted receptors, and CD8 by cells harboring MHC class I-restricted receptors. This ensures the optimal use of these receptors in responses to pathogens. For B cells, positive selection seems to occur at the final transition from immature to mature B cells, which occurs in peripheral lymphoid tissues. Tolerance to self antigens is enforced by negative selection at different stages throughout the development of both B and T cells, and positive selection likewise seems to represent a continuous process.

Fig. 8.33 A summary of the development of human α:β T cells.
The state of the T-cell receptor genes, the expression of some essential intracellular proteins, and the expression of some cell-surface molecules are shown for successive stages of α:β T-cell development. Note that because the T-cell receptor genes do not undergo further changes during antigen-driven development, only the phases during which they are actively undergoing rearrangement in the thymus are indicated. The antigen-dependent phases of CD4 and CD8 cells are depicted separately, and are detailed in Chapter 9.

	T cells	β-chain gene rearrangements	α-chain gene rearrangements	Intracellular proteins	Surface marker proteins	
ANTIGEN INDEPENDENT	Stem cell	Germline	Germline		CD34?	BONE MARROW
ANTIGEN INDEPENDENT	Early double-negative thymocyte	D–J rearranged	Germline	RAG-1 RAG-2 TdT Lck ZAP-70	CD2 HSA CD44^{hi}	THYMUS
ANTIGEN INDEPENDENT	Late double-negative thymocyte	V–DJ rearranged	Germline	RAG-1 RAG-2 TdT Lck ZAP-70	CD25 CD44^{lo} HSA	THYMUS
ANTIGEN INDEPENDENT	Early double-positive thymocyte (pre-T receptor)		V–J rearranged	RAG-1 RAG-2	PTα CD4 CD8 HSA	THYMUS
ANTIGEN INDEPENDENT	Late double-positive thymocyte (T-cell receptor)			Lck ZAP-70	CD69 CD4 CD8 HSA	THYMUS
ANTIGEN DEPENDENT	Naive CD4 T cell			Lck ZAP-70 LKLF	CD4 CD62L CD45RA CD5	PERIPHERY
ANTIGEN DEPENDENT	Memory CD4 T cell			Lck ZAP-70	CD4 CD45RO CD44	PERIPHERY
TERMINAL DIFFERENTIATION	Effector CD4 T cell			T$_H$17: IL-17 / T$_H$1: IFN-γ / T$_H$2: IL-4	CD4 CD45RO CD44^{hi} Fas FasL (type 1)	PERIPHERY
ANTIGEN DEPENDENT	Naive CD8 T cell				CD8 CD45RA	PERIPHERY
ANTIGEN DEPENDENT	Memory CD8 T cell				CD8 CD45RO CD44	PERIPHERY
TERMINAL DIFFERENTIATION	Effector CD8 T cell			IFN-γ granzyme perforin	FasL Fas CD8 CD44^{hi}	PERIPHERY

Questions.

8.1 **True or False:** B-cell development is not affected in mice that are lacking the cytokine receptor common γ chain (γ-c).

8.2 **Fill-in-the-Blanks:** B-cell development is regulated by the expression of various transcription factors that enable gene rearrangement and the successful progression into a new developmental stage. For example, during the _____ stage, Rag-1 and Rag-2 expression is induced by _____, which permits the successful D to J rearrangement and then V to DJ rearrangement of the heavy-chain locus. As a consequence, a functional _____ is expressed, and upon signaling, the cell is instructed to perform _____ and progress toward the next developmental step and rearrange the light-chain locus.

8.3 **True or False:** Self antigen recognition is needed in order to cross-link the pre-B-cell receptor, which in turn allows this complex to signal and permit the transition from pro-B cell to pre-B cell.

8.4 **Matching:** Match the B-cell stage with the proper description:

A. Early pro-B cell   i. V-DJ rearranging (heavy chain)

B. Small pre-B cell   ii. D-J rearranging (heavy chain)

C. Immature B cell   iii. Expressing pre-B-cell receptor

D. Late pro-B cell   iv. V-J rearranging (light chain)

E. Large pre-B cell   v. Surface IgM

8.5 **Short Answer:** How does the process of allelic exclusion prevent the rearrangement of the second heavy-chain locus, and why is this important?

8.6 **Short Answer:** How can one large pre-B cell give rise to multiple B cells with different antigen specificities?

8.7 **Matching:** Match the following terms to the appropriate definition:

A. Receptor editing   i. Result of persistent autoreactivity after failure of successful receptor editing

B. Isotypic exclusion   ii. Selection of either the κ or the λ light chain

C. Clonal deletion   iii. Result of a peripheral encounter of a weakly cross-linking or low-valence antigen

D. Anergy   iv. Process by which the light-chain locus is rearranged in order to produce a non-autoreactive receptor

E. Immunological ignorance   v. B cells that have affinity for a self-antigen but for various reasons do not respond to it

8.8 **True or False:** All CD4 and CD8 double-negative thymocytes are immature T cells.

8.9 **Matching:** Match the correct expression of CD44 and CD25 and the T-cell receptor locus rearrangement status with the appropriate DN T-cell stage:

A. DN1   i. CD44$^+$CD25$^+$, D to J TCRβ-chain locus rearrangement

B. DN2   ii. CD44$^+$CD25$^-$, germline T-cell receptor locus

C. DN3   iii. CD44^{low}CD25$^+$, V to DJ β-chain locus rearrangement

D. DN4   iv. CD44$^-$CD25$^-$, functional β-chain rearrangement

8.10 **Fill-in-the-Blanks:** Successful rearrangement of the _____ during the DN__ stage permits the formation of the pre-T-cell receptor, which is analogous in structure and function to the pre-B-cell receptor. The TCRβ chain associates itself with the _____, which allows ligand-independent cross-linking of the pre-T-cell receptor, causing _____, the arrest of further _____ gene rearrangement, and the expression of both _____. As with the B-cell light-chain locus, the _____ can undergo multiple rearrangements to produce a functional protein.

8.11 **Matching:** Match the following subsets of murine γ:δ T cells with the appropriate description:

A. Dendritic epidermal T cells   i. Can be divided in two groups: IFN-γ and IL-4 producing, and IFN-γ producing subsets

B. V$_γ$4$^+$ ii. Cells that home to the reproductive tract, lung, and dermis; upon stimulation these can produce inflammatory cytokines

C. V$_γ$6$^+$ T cells   iii. A population of later developing γ:δ T cells programmed to secrete IL-17 when activated, and can be found in all lymphoid organs, as well as the dermis

D. V$_γ$1$^+$ T cells   iv. Cells that, as a response to a pathogen or a wound, can induce inflammation, promote wound healing, and produce growth factors; also characterized by their T-cell receptors' use of the V$_γ$5 segment

E. V$_γ$7$^+$ T cells   v. Specifically home to the intestinal epithelium

8.12 **Multiple Choice:** Which of the following options correctly describes a difference between the B-cell receptor and the T-cell receptor?

A. VDJ rearrangement of the T-cell receptor β chain occurs first in T-cell development, as opposed to the B-cell receptor, which undergoes VDJ rearrangement after light-chain VJ rearrangement.

B. T cells do not require the formation of a pre-T-cell receptor in order to advance their development, as opposed to B cells, which require signaling through the pre-B-cell receptor in order to undergo allelic exclusion and continue development.

C. Expression of the B-cell receptor stops further light-chain rearrangement and enforces strict allelic exclusion, while expression of the T-cell receptor does not restrict further rearrangements of the alpha chain until there is signaling through peptide:MHC binding, resulting in many T cells that express two different TCRα chains.

D. TCRα chains cannot undergo successive rearrangements, as opposed to B-cell receptors, which go through the process of receptor editing.

8.13   **Multiple Choice:** Which of the following correctly describes T regulatory cells (T_{reg} cells)?

A. T_{reg} cells are a subset of CD8[+] T cells that express cytotoxic activity against cells infected by intracellular pathogens.

B. The T_{reg} T-cell receptor is characterized by its weak affinity for self MHC so that self-tolerance can be mediated.

C. T_{reg} cells express FoxP3.

D. In many cases, autoimmunity is the product of overactive T_{reg} cells.

8.14   **Multiple Choice:** Which of the following would *not* lead to a defect in CD8[+] T-cell development in the thymus?

A. Genetic deletion of cathepsin.

B. An inactivating mutation in the gene for the transcription factor Runx3.

C. Overexpression of the transcription factor ThPOK.

D. Genetic deletion of MHC class I genes.

E. Genetic deletion of the proteasomal subunit β5T.

8.15   **Multiple Choice:** Which of the following best explains MHC restriction of mature T cells?

A. TCRα and TCRβ CDR1 and CDR2 regions exhibit a germline-encoded bias to recognize MHC.

B. Apoptosis is induced in thymocytes when they receive a strong T-cell receptor signal.

C. CD4 and CD8 bind up all almost all of intracellular Lck.

D. Medullary thymic epithelial cells express AIRE, which promotes the expression of tissue-specific proteins.

E. Bone marrow-derived dendritic cells and macrophages are much more effective at mediating negative selection of thymocytes than thymic epithelial cells and thymocytes themselves.

8.16   **Short Answer:** What is the affinity hypothesis for thymocyte development?

General references.

Loffert, D., Schaal, S., Ehlich, A., Hardy, R.R., Zou, Y.R., Muller, W., and Rajewsky, K.: **Early B-cell development in the mouse—insights from mutations introduced by gene targeting.** *Immunol. Rev.* 1994, **137**:135–153.

Melchers, F., ten Boekel, E., Seidl, T., Kong, X.C., Yamagami, T., Onishi, K., Shimizu, T., Rolink, A.G., and Andersson, J.: **Repertoire selection by pre-B-cell receptors and B-cell receptors, and genetic control of B-cell development from immature to mature B cells.** *Immunol. Rev.* 2000, **175**:33–46.

Starr, T.K., Jameson, S.C., and Hogquist, K.A.: **Positive and negative selection of T cells.** *Annu. Rev. Immunol.* 2003, **21**:139–176.

von Boehmer, H.: **The developmental biology of T lymphocytes.** *Annu. Rev. Immunol.* 1993, **6**:309–326.

Weinberg, R.A.: **The Biology of Cancer**, 2nd ed. New York: Garland Science, 2014.

Section references.

8-1   Lymphocytes derive from hematopoietic stem cells in the bone marrow.

Busslinger, M.: **Transcriptional control of early B cell development.** *Annu. Rev. Immunol.* 2004, **22**:55–79.

Chao, M.P., Seita, J., and Weissman, I.L.: **Establishment of a normal hematopoietic and leukemia stem cell hierarchy.** *Cold Spring Harb. Symp. Quant. Biol.* 2008, **73**:439–449.

Funk, P.E., Kincade, P.W., and Witte, P.L.: **Native associations of early hematopoietic stem-cells and stromal cells isolated in bone-marrow cell aggregates.** *Blood* 1994, **83**:361–369.

Jacobsen, K., Kravitz, J., Kincade, P.W., and Osmond, D.G.: **Adhesion receptors on bone-marrow stromal cells—*in vivo* expression of vascular cell adhesion molecule-1 by reticular cells and sinusoidal endothelium in normal and γ-irradiated mice.** *Blood* 1996, **87**:73–82.

Kiel, M.J., and Morrison, S.J.: **Uncertainty in the niches that maintain haematopoietic stem cells.** *Nat. Rev. Immunol.* 2008, **8**:290–301.

8-2   B-cell development begins by rearrangement of the heavy-chain locus.

Allman, D., Li, J., and Hardy, R.R.: **Commitment to the B lymphoid lineage occurs before DH-JH recombination.** *J. Exp. Med.* 1999, **189**:735–740.

Allman, D., Lindsley, R.C., DeMuth, W., Rudd, K., Shinton, S.A., and Hardy, R.R.: **Resolution of three nonproliferative immature splenic B cell subsets reveals multiple selection points during peripheral B cell maturation.** *J. Immunol.* 2001, **167**:6834–6840.

Hardy, R.R., Carmack, C.E., Shinton, S.A., Kemp, J.D., and Hayakawa, K.: **Resolution and characterization of pro-B and pre-pro-B cell stages in normal mouse bone marrow.** *J. Exp. Med.* 1991, **173**:1213–1225.

Osmond, D.G., Rolink, A., and Melchers, F.: **Murine B lymphopoiesis: towards a unified model.** *Immunol. Today* 1998, **19**:65–68.

Welinder, E., Ahsberg, J., and Sigvardsson, M.: **B-lymphocyte commitment: identifying the point of no return.** *Semin. Immunol.* 2011, **23**:335–340.

8-3   The pre-B-cell receptor tests for successful production of a complete heavy chain and signals for the transition from the pro-B cell to the pre-B cell stage.

Bankovich, A.J., Raunser, S., Juo, Z.S., Walz, T., Davis, M.M., and Garcia, K.C.: **Structural insight into pre-B cell receptor function.** *Science* 2007, **316**:291–294.

Grawunder, U., Leu, T.M.J., Schatz, D.G., Werner, A., Rolink, A.G., Melchers, F., and Winkler, T.H.: **Down-regulation of Rag1 and Rag2 gene expression in pre-B cells after functional immunoglobulin heavy-chain rearrangement.** *Immunity* 1995, **3**:601–608.

Monroe, J.G.: **ITAM-mediated tonic signalling through pre-BCR and BCR complexes.** *Nat. Rev. Immunol.* 2006, **6**:283–294.

8-4 Pre-B-cell receptor signaling inhibits further heavy-chain locus rearrangement and enforces allelic exclusion.

Geier, J.K., and Schlissel, M.S.: **Pre-BCR signals and the control of Ig gene rearrangements.** *Semin. Immunol.* 2006, **18**:31–39.

Loffert, D., Ehlich, A., Muller, W., and Rajewsky, K.: **Surrogate light-chain expression is required to establish immunoglobulin heavy-chain allelic exclusion during early B-cell development.** *Immunity* 1996, **4**:133–144.

Melchers, F., ten Boekel, E., Yamagami, T., Andersson, J., and Rolink, A.: **The roles of preB and B cell receptors in the stepwise allelic exclusion of mouse IgH and L chain gene loci.** *Semin. Immunol.* 1999, **11**:307–317.

8-5 Pre-B cells rearrange the light-chain locus and express cell-surface immunoglobulin.

Arakawa, H., Shimizu, T., and Takeda, S.: **Reevaluation of the probabilities for productive rearrangements on the κ-loci and λ-loci.** *Int. Immunol.* 1996, **8**:91–99.

Gorman, J.R., van der Stoep, N., Monroe, R., Cogne, M., Davidson, L., and Alt, F.W.: **The Igk 3' enhancer influences the ratio of Igκ versus Igλ B lymphocytes.** *Immunity* 1996, 5:241–252.

Hesslein, D.G., and Schatz, D.G.: **Factors and forces controlling V(D)J recombination.** *Adv. Immunol.* 2001, **78**:169–232.

Kee, B.L., and Murre, C.: **Transcription factor regulation of B lineage commitment.** *Curr. Opin. Immunol.* 2001, **13**:180–185.

Sleckman, B.P., Gorman, J.R., and Alt, F.W.: **Accessibility control of antigen receptor variable region gene assembly—role of** *cis*-acting elements. *Annu. Rev. Immunol.* 1996, **14**:459–481.

Takeda, S., Sonoda, E., and Arakawa, H.: **The κ–λ ratio of immature B cells.** *Immunol. Today* 1996, **17**:200–201.

8-6 Immature B cells are tested for autoreactivity before they leave the bone marrow.

Casellas, R., Shih, T.A., Kleinewietfeld, M., Rakonjac, J., Nemazee, D., Rajewsky, K., and Nussenzweig, M.C.: **Contribution of receptor editing to the antibody repertoire.** *Science* 2001, **291**:1541–1544.

Chen, C., Nagy, Z., Radic, M.Z., Hardy, R.R., Huszar, D., Camper, S.A., and Weigert, M.: **The site and stage of anti-DNA B-cell deletion.** *Nature* 1995, **373**:252–255.

Cornall, R.J., Goodnow, C.C., and Cyster, J.G.: **The regulation of self-reactive B cells.** *Curr. Opin. Immunol.* 1995, **7**:804–811.

Melamed, D., Benschop, R.J., Cambier, J.C., and Nemazee, D.: **Developmental regulation of B lymphocyte immune tolerance compartmentalizes clonal selection from receptor selection.** *Cell* 1998, **92**:173–182.

Nemazee, D.: **Receptor editing in lymphocyte development and central tolerance.** *Nat. Rev. Immunol.* 2006, **6**:728–740.

Prak, E.L., and Weigert, M.: **Light-chain replacement—a new model for antibody gene rearrangement.** *J. Exp. Med.* 1995, **182**:541–548.

8-7 Lymphocytes that encounter sufficient quantities of self antigens for the first time in the periphery are eliminated or inactivated.

Cyster, J.G., Hartley, S.B., and Goodnow, C.C.: **Competition for follicular niches excludes self-reactive cells from the recirculating B-cell repertoire.** *Nature* 1994, **371**:389–395.

Goodnow, C.C., Crosbie, J., Jorgensen, H., Brink, R.A., and Basten, A.: **Induction of self-tolerance in mature peripheral B lymphocytes.** *Nature* 1989, **342**:385–391.

Lam, K.P., Kuhn, R., and Rajewsky, K.: *In vivo* **ablation of surface immunoglobulin on mature B cells by inducible gene targeting results in rapid cell death.** *Cell* 1997, **90**:1073–1083.

Russell, D.M., Dembic, Z., Morahan, G., Miller, J.F.A.P., Burki, K., and Nemazee, D.: **Peripheral deletion of self-reactive B cells.** *Nature* 1991, **354**:308–311.

Steinman, R.M., and Nussenzweig, M.C.: **Avoiding horror autotoxicus: the importance of dendritic cells in peripheral T cell tolerance.** *Proc. Natl Acad. Sci. USA* 2002, **99**:351–358.

8-8 Immature B cells arriving in the spleen turn over rapidly and require cytokines and positive signals through the B-cell receptor for maturation and survival.

Allman, D.M., Ferguson, S.E., Lentz, V.M., and Cancro, M.P.: **Peripheral B cell maturation. II. Heat-stable antigen**[hi] **splenic B cells are an immature developmental intermediate in the production of long-lived marrow-derived B cells.** *J. Immunol.* 1993, **151**:4431–4444.

Harless, S.M., Lentz, V.M., Sah, A.P., Hsu, B.L., Clise-Dwyer, K., Hilbert, D.M., Hayes, C.E., and Cancro, M.P.: **Competition for BLyS-mediated signaling through Bcmd/BR3 regulates peripheral B lymphocyte numbers.** *Curr. Biol.* 2001, **11**:1986–1989.

Levine, M.H., Haberman, A.M., Sant'Angelo, D.B., Hannum, L.G., Cancro, M.P., Janeway, C.A., Jr., and Shlomchik, M.J.: **A B-cell receptor-specific selection step governs immature to mature B cell differentiation.** *Proc. Natl Acad. Sci. USA* 2000, **97**:2743–2748.

Loder, F., Mutschler, B., Ray, R.J., Paige, C.J., Sideras, P., Torres, R., Lamers, M.C., and Carsetti, R.: **B cell development in the spleen takes place in discrete steps and is determined by the quality of B cell receptor-derived signals.** *J. Exp. Med.* 1999, **190**:75–89.

Rolink, A.G., Tschopp, J., Schneider, P., and Melchers, F.: **BAFF is a survival and maturation factor for mouse B cells.** *Eur. J. Immunol.* 2002, **32**:2004–2010.

Schiemann, B., Gommerman, J.L., Vora, K., Cachero, T.G., Shulga-Morskaya, S., Dobles, M., Frew, E., and Scott, M.L.: **An essential role for BAFF in the normal development of B cells through a BCMA-independent pathway.** *Science* 2001, **293**:2111–2114.

Stadanlick, J.E., and Cancro, M.P.: **BAFF and the plasticity of peripheral B cell tolerance.** *Curr. Opin. Immunol.* 2008, **20**:158–161.

Wen, L., Brill-Dashoff, J., Shinton, S.A., Asano, M., Hardy, R.R., and Hayakawa, K.: **Evidence of marginal-zone B cell-positive selection in spleen.** *Immunity* 2005, **23**:297–308.

8-9 B-1 B cells are an innate lymphocyte subset that arises early in development

Montecino-Rodriguez, E., and Dorshkind, K.: **B-1 B cell development in the fetus and adult.** *Immunity* 2012, **36**:13–21.

8-10 T-cell progenitors originate in the bone marrow, but all the important events in their development occur in the thymus.

Anderson, G., Moore, N.C., Owen, J.J.T., and Jenkinson, E.J.: **Cellular interactions in thymocyte development.** *Annu. Rev. Immunol.* 1996, **14**:73–99.

Carlyle, J.R., and Zúñiga-Pflücker, J.C.: **Requirement for the thymus in α:β T lymphocyte lineage commitment.** *Immunity* 1998, **9**:187–197.

Gordon, J., Wilson, V.A., Blair, N.F., Sheridan, J., Farley, A., Wilson, L., Manley, N.R., and Blackburn, C.C.: **Functional evidence for a single endodermal origin for the thymic epithelium.** *Nat. Immunol.* 2004, **5**:546–553.

Nehls, M., Kyewski, B., Messerle, M., Waldschütz, R., Schüddekopf, K., Smith, A.J.H., and Boehm, T.: **Two genetically separable steps in the differentiation of thymic epithelium.** *Science* 1996, **272**:886–889.

Rodewald, H.R.: **Thymus organogenesis.** *Annu. Rev. Immunol.* 2008, **26**:355–388.

van Ewijk, W., Hollander, G., Terhorst, C., and Wang, B.: **Stepwise development of thymic microenvironments** *in vivo* **is regulated by thymocyte subsets.** *Development* 2000, **127**:1583–1591.

8-11 Commitment to the T-cell lineage occurs in the thymus following Notch signaling.

Pui, J.C., Allman, D., Xu, L., DeRocco, S., Karnell, F.G., Bakkour, S., Lee, J.Y., Kadesch, T., Hardy, R.R., Aster, J.C., *et al.*: **Notch1 expression in early**

lymphopoiesis influences B versus T lineage determination. *Immunity* 1999, **11**:299–308.

Radtke, F., Fasnacht, N., and Macdonald, H.R.: **Notch signaling in the immune system.** *Immunity* 2010, **32**:14–27.

Radtke, F., Wilson, A., Stark, G., Bauer, M., van Meerwijk, J., MacDonald, H.R., and Aguet, M.: **Deficient T cell fate specification in mice with an induced inactivation of Notch1.** *Immunity* 1999, **10**:547–558.

Rothenberg, E.V.: **Transcriptional drivers of the T-cell lineage program.** *Curr. Opin. Immunol.* 2012, **24**:132–138.

8-12   T-cell precursors proliferate extensively in the thymus, but most die there.

Shortman, K., Egerton, M., Spangrude, G.J., and Scollay, R.: **The generation and fate of thymocytes.** *Semin. Immunol.* 1990, **2**:3–12.

Surh, C.D., and Sprent, J.: **T-cell apoptosis detected** *in situ* **during positive and negative selection in the thymus.** *Nature* 1994, **372**:100–103.

8-13   Successive stages in the development of thymocytes are marked by changes in cell-surface molecules.

Borowski, C., Martin, C., Gounari, F., Haughn, L., Aifantis, I., Grassi, F., and von Boehmer, H.: **On the brink of becoming a T cell.** *Curr. Opin. Immunol.* 2002, **14**:200–206.

Pang, S.S., Berry, R., Chen, Z., Kjer-Nielsen, L., Perugini, M.A., King, G.F., Wang, C., Chew, S.H., La Gruta, N.L., Williams, N.K., *et al.*: **The structural basis for autonomous dimerization of the pre-T-cell antigen receptor.** *Nature* 2010, **467**:844–848.

Saint-Ruf, C., Ungewiss, K., Groettrup, M., Bruno, L., Fehling, H.J., and von Boehmer, H.: **Analysis and expression of a cloned pre-T-cell receptor gene.** *Science* 1994, **266**:1208–1212.

Shortman, K., and Wu, L.: **Early T lymphocyte progenitors.** *Annu. Rev. Immunol.* 1996, **14**:29–47.

8-14   Thymocytes at different developmental stages are found in distinct parts of the thymus.

Benz, C., Heinzel, K., and Bleul, C.C.: **Homing of immature thymocytes to the subcapsular microenvironment within the thymus is not an absolute requirement for T cell development.** *Eur. J. Immunol.* 2004, **34**:3652–3663.

Bleul, C.C., and Boehm, T.: **Chemokines define distinct microenvironments in the developing thymus.** *Eur. J. Immunol.* 2000, **30**:3371–3379.

Nitta, T., Murata, S., Ueno, T., Tanaka, K., and Takahama, Y.: **Thymic microenvironments for T-cell repertoire formation.** *Adv. Immunol.* 2008, **99**:59–94.

Ueno, T., Saito F., Gray, D.H.D., Kuse, S., Hieshima, K., Nakano, H., Kakiuchi, T., Lipp, M., Boyd, R.L., and Takahama, Y.: **CCR7 signals are essential for cortex–medulla migration of developing thymocytes.** *J. Exp. Med.* 2004, **200**:493–505.

8-15   T cells with α:β or γ:δ receptors arise from a common progenitor.

Fehling, H.J., Gilfillan, S., and Ceredig, R.: **αβ/γδ lineage commitment in the thymus of normal and genetically manipulated mice.** *Adv. Immunol.* 1999, **71**:1–76.

Hayday, A.C., Barber, D.F., Douglas, N., and Hoffman, E.S.: **Signals involved in γδ T cell versus αβ T cell lineage commitment.** *Semin. Immunol.* 1999, **11**:239–249.

Hayes, S.M., and Love, P.E.: **Distinct structure and signaling potential of the γδ TCR complex.** *Immunity* 2002, **16**:827–838.

Kang, J., and Raulet, D.H.: **Events that regulate differentiation of αβ TCR⁺ and γδ TCR⁺ T cells from a common precursor.** *Semin. Immunol.* 1997, **9**:171–179.

Kreslavsky, T., Garbe, A.I., Krueger, A., and von Boehmer, H.: **T cell receptor-instructed αβ versus γδ lineage commitment revealed by single-cell analysis.** *J. Exp. Med.* 2008, **205**:1173–1186.

Lauritsen, J.P., Haks, M.C., Lefebvre, J.M., Kappes, D.J., and Wiest, D.L.: **Recent insights into the signals that control αβ/γδ-lineage fate.** *Immunol. Rev.* 2006, **209**:176–190.

Livak, F., Petrie, H.T., Crispe, I.N., and Schatz, D.G.: **In-frame TCRδ gene rearrangements play a critical role in the αβ/γδ T cell lineage decision.** *Immunity* 1995, **2**:617–627.

Xiong, N., and Raulet, D.H.: **Development and selection of γδ T cells.** *Immunol. Rev.* 2007, **215**:15–31.

8-16   T cells expressing γ:δ T-cell receptors arise in two distinct phases during development.

Carding, S.R., and Egan, P.J.: **γδ T cells: functional plasticity and heterogeneity.** *Nat. Rev. Immunol.* 2002, **2**:336–345.

Ciofani, M., Knowles, G.C., Wiest, D.L., von Boehmer, H., and Zúñiga-Pflücker, J.C.: **Stage-specific and differential notch dependency at the α:β and γ:δ T lineage bifurcation.** *Immunity* 2006, **25**:105–116.

Dunon, D., Courtois, D., Vainio, O., Six, A., Chen, C.H., Cooper, M.D., Dangy, J.P., and Imhof, B.A.: **Ontogeny of the immune system: γ:δ and α:β T cells migrate from thymus to the periphery in alternating waves.** *J. Exp. Med.* 1997, **186**:977–988.

Haas, W., Pereira, P., and Tonegawa, S.: **Gamma/delta cells.** *Annu. Rev. Immunol.* 1993, **11**:637–685.

Lewis, J.M., Girardi, M., Roberts, S.J., Barbee, S.D., Hayday, A.C., and Tigelaar, R.E.: **Selection of the cutaneous intraepithelial γδ⁺ T cell repertoire by a thymic stromal determinant.** *Nat. Immunol.* 2006, **7**:843–850.

Narayan, K., Sylvia, K.E., Malhotra, N., Yin, C.C., Martens, G., Vallerskog, T., Kornfeld, H., Xiong, N., Cohen, N.R., Brenner, M.B., *et al.*: **Intrathymic programming of effector fates in three molecularly distinct gamma:delta T cell subtypes.** *Nat. Immunol.* 2012, **13**:511–518.

Strid, J., Tigelaar, R.E., and Hayday, A.C.: **Skin immune surveillance by T cells—a new order?** *Semin. Immunol.* 2009, **21**:110–120.

8-17   Successful synthesis of a rearranged β chain allows the production of a pre-T-cell receptor that triggers cell proliferation and blocks further β-chain gene rearrangement.

Borowski, C., Li, X., Aifantis, I., Gounari, F., and von Boehmer, H.: **Pre-TCRα and TCRα are not interchangeable partners of TCRβ during T lymphocyte development.** *J. Exp. Med.* 2004, **199**:607–615.

Dudley, E.C., Petrie, H.T., Shah, L.M., Owen, M.J., and Hayday, A.C.: **T-cell receptor β chain gene rearrangement and selection during thymocyte development in adult mice.** *Immunity* 1994, **1**:83–93.

Philpott, K.I., Viney, J.L., Kay, G., Rastan, S., Gardiner, E.M., Chae, S., Hayday, A.C., and Owen, M.J.: **Lymphoid development in mice congenitally lacking T cell receptor αβ-expressing cells.** *Science* 1992, **256**:1448–1453.

von Boehmer, H., Aifantis, I., Azogui, O., Feinberg, J., Saint-Ruf, C., Zober, C., Garcia, C., and Buer, J.: **Crucial function of the pre-T-cell receptor (TCR) in TCRβ selection, TCRβ allelic exclusion and α:β versus γ:δ lineage commitment.** *Immunol. Rev.* 1998, **165**:111–119.

8-18   T-cell α-chain genes undergo successive rearrangements until positive selection or cell death intervenes.

Buch, T., Rieux-Laucat, F., Förster, I., and Rajewsky, K.: **Failure of HY-specific thymocytes to escape negative selection by receptor editing.** *Immunity* 2002, **16**:707–718.

Hardardottir, F., Baron, J.L., and Janeway, C.A., Jr.: **T cells with two functional antigen-specific receptors.** *Proc. Natl Acad. Sci. USA* 1995, **92**:354–358.

Huang, C.-Y., Sleckman, B.P., and Kanagawa, O.: **Revision of T cell receptor α chain genes is required for normal T lymphocyte development.** *Proc. Natl Acad. Sci. USA* 2005, **102**:14356–14361.

Marrack, P., and Kappler, J.: **Positive selection of thymocytes bearing α:β cell receptors.** *Curr. Opin. Immunol.* 1997, **9**:250–255.

Padovan, E., Casorati, G., Dellabona, P., Meyer, S., Brockhaus, M., and Lanzavecchia, A.: **Expression of two T-cell receptor α chains: dual receptor T cells.** *Science* 1993, **262**:422–424.

Petrie, H.T., Livak, F., Schatz, D.G., Strasser, A., Crispe, I.N., and Shortman, K.: **Multiple rearrangements in T-cell receptor α-chain genes maximize the production of useful thymocytes.** *J. Exp. Med.* 1993, **178**:615–622.

8-19 Only thymocytes whose receptors interact with self peptide:self MHC complexes can survive and mature.

Hogquist, K.A., Tomlinson, A.J., Kieper, W.C., McGargill, M.A., Hart, M.C., Naylor, S., and Jameson, S.C.: **Identification of a naturally occurring ligand for thymic positive selection.** *Immunity* 1997, **6**:389–399.

Kisielow, P., Teh, H.S., Blüthmann, H., and von Boehmer, H.: **Positive selection of antigen-specific T cells in thymus by restricting MHC molecules.** *Nature* 1988, **335**:730–733.

8-20 Positive selection acts on a repertoire of T-cell receptors with inherent specificity for MHC molecules.

Marrack, P., Scott-Browne, J.P., Dai, S., Gapin, L., and Kappler, J.W.: **Evolutionarily conserved amino acids that control TCR-MHC interaction.** *Annu. Rev. Immunol.* 2008, **26**:171–203.

Merkenschlager, M., Graf, D., Lovatt, M., Bommhardt, U., Zamoyska, R., and Fisher, A.G.: **How many thymocytes audition for selection?** *J. Exp. Med.* 1997, **186**:1149–1158.

Scott-Browne, J.P., White, J., Kappler, J.W., Gapin, L., and Marrack, P.: **Germline-encoded amino acids in the αβ T-cell receptor control thymic selection.** *Nature* 2009, **458**:1043–1046.

Zerrahn, J., Held, W., and Raulet, D.H.: **The MHC reactivity of the T cell repertoire prior to positive and negative selection.** *Cell* 1997, **88**:627–636.

8-21 Positive selection coordinates the expression of CD4 or CD8 with the specificity of the T-cell receptor and the potential effector functions of the T cell.

Egawa, T., and Littman, D.R.: **ThPOK acts late in specification of the helper T cell lineage and suppresses Runx-mediated commitment to the cytotoxic T cell lineage.** *Nat. Immunol.* 2008, **9**:1131–1139.

He, X., Xi, H., Dave, V.P., Zhang, Y., Hua, X., Nicolas, E., Xu, W., Roe, B.A., and Kappes, D.J.: **The zinc finger transcription factor Th-POK regulates CD4 versus CD8 T-cell lineage commitment.** *Nature* 2005, **433**:826–833.

Singer, A., Adoro, S., and Park, J.H.: **Lineage fate and intense debate: myths, models and mechanisms of CD4- versus CD8-lineage choice.** *Nat. Rev. Immunol.* 2008, **8**:788–801.

von Boehmer, H., Kisielow, P., Lishi, H., Scott, B., Borgulya, P., and Teh, H.S.: **The expression of CD4 and CD8 accessory molecules on mature T cells is not random but correlates with the specificity of the α:β receptor for antigen.** *Immunol. Rev.* 1989, **109**:143–151.

8-22 Thymic cortical epithelial cells mediate positive selection of developing thymocytes.

Cosgrove, D., Chan, S.H., Waltzinger, C., Benoist, C., and Mathis, D.: **The thymic compartment responsible for positive selection of CD4+ T cells.** *Int. Immunol.* 1992, **4**:707–710.

Ernst, B.B., Surh, C.D., and Sprent, J.: **Bone marrow-derived cells fail to induce positive selection in thymus reaggregation cultures.** *J. Exp. Med.* 1996, **183**:1235–1240.

Murata, S., Sasaki, K., Kishimoto, T., Niwa, S.-I., Hayashi, H., Takahama, Y., and Tanaka, K.: **Regulation of CD8+ T cell development by thymus-specific proteasomes.** *Science* 2007, **316**:1349–1353.

Nakagawa, T., Roth, W., Wong, P., Nelson, A., Farr, A., Deussing, J., Villadangos, J.A., Ploegh, H., Peters, C., and Rudensky, A.Y.: **Cathepsin L: critical role in Ii degradation and CD4 T cell selection in the thymus.** *Science* 1998, **280**:450–453.

8-23 T cells that react strongly with ubiquitous self antigens are deleted in the thymus.

Anderson, M.S., Venanzi, E.S., Klein, L., Chen, Z., Berzins, S.P., Turley, S.J., von Boehmer, H., Bronson, R., Dierich, A., Benoist, C., *et al.*: **Projection of an immunological self shadow within the thymus by the aire protein.** *Science* 2002, **298**:1395–1401.

Kishimoto, H., and Sprent, J.: **Negative selection in the thymus includes semimature T cells.** *J. Exp. Med.* 1997, **185**:263–271.

Zal, T., Volkmann, A., and Stockinger, B.: **Mechanisms of tolerance induction in major histocompatibility complex class II-restricted T cells specific for a blood-borne self antigen.** *J. Exp. Med.* 1994, **180**:2089–2099.

8-24 Negative selection is driven most efficiently by bone marrow-derived antigen-presenting cells.

Anderson, M.S., and Su, M.A.: **Aire and T cell development.** *Curr. Opin. Immunol.* 2011, **23**:198–206.

McCaughtry, T.M., Baldwin, T.A., Wilken, M.S., and Hogquist, K.A.: **Clonal deletion of thymocytes can occur in the cortex with no involvement of the medulla.** *J. Exp. Med.* 2008, **205**:2575–2584.

Sprent, J., and Webb, S.R.: **Intrathymic and extrathymic clonal deletion of T cells.** *Curr. Opin. Immunol.* 1995, **7**:196–205.

Webb, S.R., and Sprent, J.: **Tolerogenicity of thymic epithelium.** *Eur. J. Immunol.* 1990, **20**:2525–2528.

8-25 The specificity and/or the strength of signals for negative and positive selection must differ.

Alberola-Ila, J., Hogquist, K.A., Swan, K.A., Bevan, M.J., and Perlmutter, R.M.: **Positive and negative selection invoke distinct signaling pathways.** *J. Exp. Med.* 1996, **184**:9–18.

Ashton-Rickardt, P.G., Bandeira, A., Delaney, J.R., Van Kaer, L., Pircher, H.P., Zinkernagel, R.M., and Tonegawa, S.: **Evidence for a differential avidity model of T-cell selection in the thymus.** *Cell* 1994, **76**:651–663.

Bommhardt, U., Scheuring, Y., Bickel, C., Zamoyska, R., and Hunig, T.: **MEK activity regulates negative selection of immature CD4+CD8+ thymocytes.** *J. Immunol.* 2000, **164**:2326–2337.

Hogquist, K.A., Jameson, S.C., Heath, W.R., Howard, J.L., Bevan, M.J., and Carbone, F.R.: **T-cell receptor antagonist peptides induce positive selection.** *Cell* 1994, **76**:17–27.

8-26 Self-recognizing regulatory T cells and innate T cells develop in the thymus.

Jordan, M.S., Boesteanu, A., Reed, A.J., Petrone, A.L., Holenbeck, A.E., Lerman, M.A., Naji, A., and Caton, A.J.: **Thymic selection of CD4+CD25+ regulatory T cells induced by an agonist self-peptide.** *Nat. Immunol.* 2001, **2**:301–306.

Moran, A.E., and Hogquist, K.A.: **T-cell receptor affinity in thymic development.** *Immunology* 2012, **135**:261–267.

Zheng, Y., and Rudensky, A.Y.: **Foxp3 in control of the regulatory T cell lineage.** *Nat. Immunol.* 2007, **8**:457–462.

8-27 The final stage of T-cell maturation occurs in the thymic medulla.

Matloubian, M., Lo, C.G., Cinamon, G., Lesneski, M.J., Xu, Y., Brinkmann, V., Allende, M.L., Proia, R.L., and Cyster, J.G.: **Lymphocyte egress from thymus and peripheral lymphoid organs is dependent on S1P receptor 1.** *Nature* 2004, **427**:355–360.

Zachariah, M.A., and Cyster, J.G.: **Neural crest-derived pericytes promote egress of mature thymocytes at the corticomedullary junction.** *Science* 2010, **328**:1129–1135.

8-28 T cells that encounter sufficient quantities of self antigens for the first time in the periphery are eliminated or inactivated.

Fink, P.J., and Hendricks, D.W.: **Post-thymic maturation: young T cells assert their individuality.** *Nat. Rev. Immunol.* 2011, **11**:544–549.

Steinman, R.M., and Nussenzweig, M.C.: **Avoiding horror autotoxicus: the importance of dendritic cells in peripheral T cell tolerance.** *Proc. Natl Acad. Sci. USA* 2002, **99**:351–358.

Xing, Y., and Hogquist, K.A.: **T-cell tolerance: central and peripheral.** *Cold Spring Harb. Perspect. Biol.* 2012, **4**.pii:a006957

T-cell-Mediated Immunity

9

IN THIS CHAPTER

Development and function of secondary lymphoid organs—sites for the initiation of adaptive immune responses.

Priming of naive T cells by pathogen-activated dendritic cells.

General properties of effector T cells and their cytokines.

T-cell-mediated cytotoxicity.

An adaptive immune response is initiated when a pathogen overwhelms innate defense mechanisms. As the pathogen replicates and antigen accumulates, sensor cells of the innate immune system become activated to trigger the adaptive immune response. While some infections may be dealt with solely by innate immunity, as discussed in Chapters 2 and 3, host defense against most pathogens, almost by definition, requires recruitment of adaptive immunity. This is shown by the immunodeficiency syndromes that are associated with failure of particular parts of the adaptive immune response; these will be discussed in Chapter 13. In the next three chapters, we will learn how the adaptive immune response involving antigen-specific T cells and B cells is initiated and deployed. T-cell responses that lead to cellular immunity will be considered first, in this chapter; and B-cell responses that lead to antibody-mediated, or humoral, immunity will be considered in Chapter 10. In Chapter 11 we will consider the dynamics of T-cell and B-cell responses in the context of their integration with innate immunity and how this culminates in one of the most important features of adaptive immunity—immunological memory.

Once T cells have completed their primary development in the thymus, they enter the bloodstream. On reaching a secondary lymphoid organ, they leave the blood to migrate through the lymphoid tissue, returning via the lymphatics to the bloodstream to recirculate between blood and secondary lymphoid tissues. Mature recirculating T cells that have not yet encountered their specific antigens are known as **naive T cells**. To participate in an adaptive immune response, a naive T cell must meet its specific antigen, presented to it as a peptide:MHC complex on the surface of an antigen-presenting cell, and be induced to proliferate and differentiate into progeny with new activities that contribute to removal of antigen. These progeny cells are called **effector T cells** and, unlike naive T cells, perform their functions as soon as they encounter their specific antigen on other cells—generally without requirement for further differentiation. Because of their requirement to recognize peptide antigens presented by MHC molecules, all effector T cells act on other host cells, not on the pathogen itself. The cells on which effector T cells act will be referred to as their target cells.

On recognizing antigen, naive T cells differentiate into several functional classes of effector T cells that are specialized for different activities. CD8 T cells recognize pathogen peptides presented by MHC class I molecules, and naive CD8 T cells differentiate into cytotoxic effector T cells that recognize and kill infected cells. CD4 T cells have a more flexible repertoire of effector activities. After recognizing pathogen peptides presented by MHC class II molecules, naive CD4 T cells can differentiate down distinct pathways that generate effector subsets with different immunological functions. The main CD4 effector subsets are T_H1, T_H2, T_H17, and T_{FH}, which activate their target cells; and regulatory T cells, or T_{reg} cells, which inhibit the extent of immune activation.

Effector T cells differ from their naive precursors in ways that equip them to respond quickly and efficiently when they encounter specific antigen on target cells. Among the changes that occur are alterations in the expression of surface molecules that alter the patterns of migration of effector T cells, directing them to exit the secondary lymphoid tissues and move to sites of inflammation where pathogens have entered, or to B-cell zones within secondary lymphoid tissues, where they help generate pathogen-specific antibodies. The interactions with target cells in these sites are mediated both by direct T-cell–target cell contact and the release of cytokines, which can act locally on target cells; and at a distance to orchestrate the clearance of antigen. Some of the effector functions of T cells will be considered in this chapter; others will be discussed in Chapters 10 and 11 in the context of T-cell help for B cells and heightened activation of effector cells of the innate immune system.

The activation and clonal expansion of a naive T cell on its initial encounter with antigen is often called **priming**, to distinguish this process from the responses of effector T cells to antigen on their target cells and the responses of primed memory T cells. The initiation of adaptive immunity is one of the most compelling narratives in immunology. As we will learn, the activation of naive T cells is controlled by a variety of signals. The primary signal that a naive T cell must recognize is antigen in the form of a peptide:MHC complex on the surface of a specialized antigen-presenting cell, as discussed in Chapter 6. Activation of the naive T cell also requires that it recognize co-stimulatory molecules that are displayed by antigen-presenting cells. Finally, cytokines that control differentiation into different types of effector cells are delivered to the activated naive T cell. All these events are set in motion by earlier signals that arise from the initial detection of the pathogens by the innate immune system. Microbe-derived signals are delivered to cells of the innate immune system by receptors such as the Toll-like receptors (TLRs), which recognize microbe-associated molecular patterns, or MAMPs, that signify the presence of nonself (see Chapters 2 and 3). As we will see in this chapter, these signals are essential to activate antigen-presenting cells so that they are able, in turn, to activate naive T cells.

By far the most important antigen-presenting cells in the activation of naive T cells are dendritic cells, whose major function is to ingest and present antigen. Tissue dendritic cells take up antigen at sites of infection and are activated as part of the innate immune response. This induces their migration to local lymphoid tissue and their maturation into cells that are highly effective at presenting antigen to recirculating naive T cells. In the first part of this chapter we will consider the development and organization of secondary lymphoid tissues and discuss how naive T cells and dendritic cells meet in these sites to initiate adaptive immunity.

Development and function of secondary lymphoid organs—sites for the initiation of adaptive immune responses.

As discussed in Chapter 8, the primary lymphoid organs—the thymus and bone marrow—are the tissue sites where antigenic receptor repertoires of T and B cells, respectively, are selected. Adaptive immune responses are initiated in the secondary lymphoid organs—lymph nodes, spleen, and the mucosa-associated lymphoid tissues (MALTs) such as the Peyer's patches in the gut. The architecture of these tissues is similar throughout the body and is structured to provide a crossroads for the interaction of rare clonal precursors of recirculating T and B cells with their cognate antigens—whether delivered by dendritic cells in the case of T cells, or as free antigens in the case of B cells. In view of the rarity of naive T cells that recognize a specific peptide:MHC complex—roughly 50–500 cells in the entire immune repertoire of approximately 100 million T cells in the mouse—and the large area over which an infectious agent can invade, the antigens derived from the pathogen, or in some instances the pathogen itself, must be brought from sites of entry to secondary lymphoid organs to facilitate their recognition by lymphocytes. In this part of the chapter we shall first consider the development and structure of secondary lymphoid organs that enable these interactions. We shall then discuss how naive T cells are directed to exit the blood and enter the lymphoid organs. This will be followed by considering how dendritic cells pick up antigen and travel to local lymphoid organs, where they can both present antigen to T cells and activate them.

9-1   T and B lymphocytes are found in distinct locations in secondary lymphoid tissues.

The various secondary lymphoid organs are organized roughly along the same lines (see Chapter 1), with distinct areas in which B cells and T cells are concentrated—the B-cell and T-cell zones. They also contain macrophages, dendritic cells, and nonleukocyte stromal cells. In the case of the spleen, which is specialized for the capture of antigens that enter the bloodstream, the lymphoid tissue component is called the **white pulp** (Fig. 9.1).

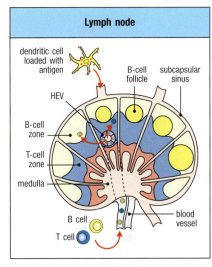

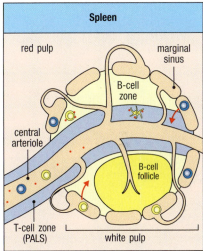

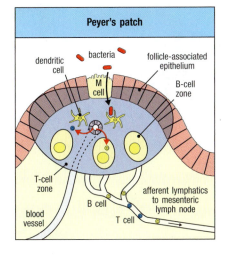

Fig. 9.1 Secondary lymphoid tissues serve as anatomical crossroads for interactions of antigens and lymphocytes. Secondary lymphoid tissues are specialized to serve as sites that facilitate interactions between lymphocytes and antigens. In lymph nodes (upper panel), antigen (denoted by red dots) is delivered in lymph either free or as cargo of dendritic cells that have taken up the antigen in the tissues drained by the lymph node. The antigen is conducted via afferent lymphatics to the subcapsular sinus, from which it is delivered to T-cell zones, where T cells can recognize it on the surface of dendritic cells; or, in the case of B cells, it is detected as a free antigen at the border of the T-cell zone and B-cell follicles. T and B cells enter the lymph node via high endothelial venules (HEVs) in T-cell zones, and then diverge into T-cell and B-cell zones. In the spleen (middle panel), antigen is delivered via arterioles that branch from the central arteriole to the marginal sinus, which is the boundary between the white pulp and red pulp, with which the marginal sinus communicates. In the marginal sinus, antigen can be taken up by marginal zone B cells, macrophages, or dendritic cells, which can transport antigen into either T-cell zones (periarteriolar lymphoid sheath, or PALS) or B-cell follicles. T and B cells enter the spleen via the same route as antigen, and leave the marginal sinus to travel to either the PALS or the B-cell follicles. In the intestines (lower panel), antigens are transported from the lumen via the microfold or M cells—specialized epithelium that overlays Peyer's patches—to dendritic cells that reside in the subepithelial dome. Antigen-loaded dendritic cells are then surveyed by T cells in T-cell zones, and if the antigens they bear are not recognized by T cells locally, the dendritic cells can migrate to mesenteric lymph nodes to be further surveyed. As for lymph nodes, T and B cells enter the Peyer's patches via HEVs in the T-cell zones.

Each area of white pulp is demarcated from the **red pulp** by a **marginal sinus**, a vascular network that is formed from branches of the central arteriole. Circulating T and B cells are initially delivered to the marginal sinus, which is a highly organized region of cells that is specialized for the capture of blood-borne antigens or intact microbes, such as viruses and bacteria. It is rich in macrophages and contains a unique population of B cells, the **marginal zone B cells**, which do not recirculate. Pathogens reaching the bloodstream are efficiently trapped in the marginal zone by macrophages, and it could be that marginal zone B cells are adapted to provide the first responses to such pathogens.

From the marginal sinus, T and B cells migrate centrally toward the central arteriole, where they bifurcate into T-cell zones that are clustered around the central arteriole—the so-called **periarteriolar lymphoid sheath (PALS)**—and B-cell zones, or follicles, that are located more peripherally. Some follicles may contain **germinal centers**, in which B cells involved in an adaptive immune response are proliferating and undergoing somatic hypermutation (see Section 1-16). The antigen-driven production of germinal centers will be described in detail when we consider B-cell responses in Chapter 10.

Other types of cells are found within the B-cell and T-cell areas. The B-cell zone contains a network of **follicular dendritic cells (FDCs)**, which are concentrated mainly in the area of the follicle most distant from the central arteriole. FDCs have long processes that are in contact with B cells. FDCs are a distinct type of cell from the dendritic cells we encountered previously (see Section 1-3), in that they are not leukocytes and are not derived from bone marrow precursors; in addition, they are not phagocytic and do not express MHC class II proteins. FDCs are specialized for the capture of antigen in the form of immune complexes—complexes of antigen, antibody, and complement. The immune complexes are not internalized but remain intact on the surface of the FDC for prolonged periods of time, where the antigen can be recognized by B cells. FDCs are also important in the development of B-cell follicles.

T-cell zones contain a network of bone marrow-derived dendritic cells, sometimes known as **interdigitating dendritic cells** from the way in which their processes interweave among T cells. There are two major subtypes of these dendritic cells, distinguished by characteristic cell-surface proteins: one expresses the α chain of CD8, whereas the other is CD8-negative but expresses CD11b:CD18, an integrin that is also expressed by macrophages.

As in the spleen, the T cells and B cells in lymph nodes are organized into discrete T-cell and B-cell areas (see Fig. 9.1). B-cell follicles have a similar structure and composition to those in the spleen and are located just under the outer capsule of the lymph node. T-cell zones surround the follicles in the paracortical areas. Unlike the spleen, lymph nodes have connections to both the blood system and the lymphatic system. Lymph conducted to lymph nodes by afferent lymphatic vessels enters into the subcapsular space, which is also known as the marginal sinus, and brings in antigen and antigen-bearing dendritic cells from the tissues. T and B cells enter the lymph node via specialized blood vessels called **high endothelial venules (HEVs)** that are found in T-cell zones, as will be discussed further in Section 9-3.

The **mucosa-associated lymphoid tissues (MALTs)** are associated with the body's epithelial surfaces, which provide physical barriers against infection. Peyer's patches are part of the MALT and are lymph node-like structures interspersed at intervals just beneath the gut epithelium. They have B-cell follicles and T-cell zones (see Fig. 9.1), and the epithelium overlying them contains specialized M cells that are adapted to channel antigens and pathogens directly from the gut lumen to the underlying lymphoid tissue (see Section 1-16 and Chapter 12). Peyer's patches and similar tissue present in the tonsils provide specialized sites where B cells can become committed to the synthesis of IgA. The mucosal immune system is discussed in more detail in Chapter 12.

9-2 The development of secondary lymphoid tissues is controlled by lymphoid tissue inducer cells and proteins of the tumor necrosis factor family.

Before discussing how T cells and B cells become partitioned into their respective zones in secondary lymphoid organs, we shall briefly look at how these organs develop in the first place. Lymphatic vessels are formed during embryonic development from endothelial cells that originate in blood vessels. Some endothelial cells in the early venous system begin to express the homeobox transcription factor Prox1. These cells bud from the vein, migrate away, and reassociate to form a parallel network of lymphatic vessels. Mice lacking Prox1 have normal arteries and veins, but fail to form a lymphatic system, showing this factor to be critical in establishing the identity of lymphatic endothelium. As the lymphatic vessels form, hematopoietic cells called **lymphoid tissue inducer (LTi) cells** arise in the fetal liver and are carried in the bloodstream to sites of prospective lymph nodes and Peyer's patches. LTi cells initiate the formation of lymph nodes and Peyer's patches by interacting with stromal cells and inducing the production of cytokines and chemokines, which recruit other lymphoid cells to these sites. Members of the tumor necrosis factor (TNF)/TNF receptor (TNFR) family of cytokines turn out to be critically involved in the interactions between LTi cells and stromal cells.

The role of this family of cytokines in the formation of secondary lymphoid organs has been demonstrated in a series of studies involving knockout mice in which either the TNF-family ligand or its receptor was inactivated (**Fig. 9.2**). These knockout mice have complicated phenotypes, which is partly due to the fact that individual TNF-family proteins can bind to multiple receptors and, conversely, many receptors can bind more than one ligand. In addition, it seems clear that there is some overlapping function or cooperation between TNF-family proteins. Nonetheless, some general conclusions can be drawn.

Lymph-node development depends on the expression of TNF-family proteins known as the **lymphotoxins (LTs)**, and different types of lymph nodes depend on signals from different LTs. LT-α_3, a soluble homotrimer of the LT-α chain, supports the development of cervical and mesenteric lymph nodes, and possibly lumbar and sacral lymph nodes. All these lymph nodes drain mucosal sites. LT-α_3 probably exerts its effects by binding to TNFR-I. The membrane-bound heterotrimer consisting of two molecules of LT-α and one molecule of the distinct transmembrane protein LT-β (that is, LT-α_2:β_1), often known as LT-β, binds only to the LT-β receptor and supports the development of all the other lymph nodes. Peyer's patches also do not form in the absence of LT-β. The effects of the LT knockouts are not reversible in adult animals; there are certain critical developmental periods during which the absence or inhibition of these LT-family proteins will permanently prevent the development of lymph nodes and Peyer's patches.

LTi cells express LT-β, which engages the LT-β receptors on stromal cells in the prospective lymphoid site, activating the non-canonical NFκB pathway

Fig. 9.2 The role of TNF family members in the development of peripheral lymphoid organs. The role of TNF family members in the development of peripheral lymphoid organs has been deduced mainly from the study of knockout mice deficient in one or more TNF-family ligands or receptors. Some receptors bind more than one ligand, and some ligands bind more than one receptor, complicating elucidation of the effects of their deletion. (Note that receptors are named for the first ligand known to bind them.) The defects are organized here with respect to the two main receptors, TNFR-I and the LT-β receptor, and their ligands, TNF-α and the lymphotoxins (LTs). Note that in some cases, the losses of individual ligands out of several that bind the same receptor lead to different respective phenotypes, as indicated in the figure. This is due to the ability of the different ligands to bind different sets of receptors. The LT-α protein chain contributes to two distinct ligands, the trimer LT-α_3 and the heterodimer LT-α_2:β_1, each of which acts through a distinct receptor. In general, signaling through the LT-β receptor is required for lymph-node and follicular dendritic cell (FDC) development and for normal splenic architecture, whereas signaling through TNFR-I is also required for FDCs and normal splenic architecture but not for lymph-node development.

Receptor	Ligands	Effects seen in knockout (KO) mice				
		Spleen	Peripheral lymph node	Mesenteric lymph node	Peyer's patch	Follicular dendritic cells
TNFR-I	TNF-α LT-α_3	Distorted architecture	Present in TNF-α KO Absent in LT-α KO owing to lack of LT-β signals	Present	Reduced	Absent
LT-β receptor	TNF-α LT-α_2:β_1	Distorted architecture No marginal zones	Absent	Present in LT-β KO Absent in LT-β receptor KO	Absent	Absent

MOVIE 9.1

(see Section 7-23). This induces the stromal cells to express adhesion molecules and chemokines such as CXCL13 (B-lymphocyte chemokine, BLC), which in turn recruits more LTi cells, which have receptors for these molecules, eventually generating large clusters of cells that will become lymph nodes or Peyer's patches. The chemokines also attract cells such as lymphocytes and other hematopoietic-lineage cells with appropriate receptors to populate the forming lymphoid organ. The principles, and even some of the molecules, underlying the development of secondary lymphoid organs in the fetus are very similar to those that maintain the organization of lymphoid organs in the adult, as we shall see in the next section.

Although the spleen will develop in mice deficient in any of the known TNF or TNFR family members, its architecture will be abnormal in many of these mutants (see Fig. 9.2). LT (most probably the membrane-bound LT-β) is required for the normal segregation of T-cell and B-cell zones in the spleen. TNF-α, binding to TNFR-I, also contributes to the organization of the white pulp: when TNF-α signals are disrupted, B cells surround T-cell zones in a ring rather than forming discrete follicles, and the marginal zones are not well defined.

Fig. 9.3 The development of secondary lymphoid organs is orchestrated by chemokines. The cellular organization of lymphoid organs is initiated by stromal cells and vascular endothelial cells, which secrete the chemokine CCL21 (first panel). Dendritic cells with CCR7—a receptor for CCL21—are attracted to the site of the developing lymph node by CCL21 (second panel); it is not known whether at the earliest stages of lymph-node development immature dendritic cells enter from the bloodstream or via the lymphatics, as they do later in life. Once in the lymph node, the dendritic cells express the chemokine CCL19, which is also bound by CCR7. Together, the chemokines secreted by stromal cells and dendritic cells attract T cells to the developing lymph node (third panel). The same combination of chemokines also attracts B cells into the developing lymph node (fourth panel). The B cells are able to either induce the differentiation of the nonleukocyte FDCs (which are a lineage distinct from the bone marrow-derived dendritic cells) or direct their recruitment into the lymph node. Once present, the FDCs secrete CXCL13, a chemokine that is a chemoattractant for B cells. The production of CXCL13 drives the organization of B cells into discrete B-cell areas (follicles) around the FDCs and contributes to the further recruitment of B cells from the circulation into the lymph node (fifth panel).

Perhaps the most important role of TNF-α and TNFR-I in lymphoid organ development is in the development of FDCs, as these cells are lacking in mice with knockouts of either TNF-α or TNFR-I (see Fig. 9.2). The knockout mice do have lymph nodes and Peyer's patches, because they express LTs, but these structures lack FDCs. LT-β is also required for FDC development: mice that cannot form LT-β or signal through its receptor lack normal FDCs in the spleen and any residual lymph nodes. Unlike the disruption of lymph-node development, the disorganized lymphoid architecture in the spleen is reversible if the missing TNF-family member is restored. B cells are the likely source of the LT-β, because normal B cells can restore FDCs and follicles when transferred to RAG-deficient recipients (which lack lymphocytes).

9-3 T and B cells are partitioned into distinct regions of secondary lymphoid tissues by the actions of chemokines.

Circulating T and B cells seed secondary lymphoid tissues from the blood by a common route, but are then directed into their respective compartments under the control of distinct chemokines that are produced by both stromal cells and bone marrow-derived cells resident in the T- and B-cell zones (Fig. 9.3). The localization of T cells into T-cell zones involves two chemokines, CCL19 (MIP-3β) and CCL21 (secondary lymphoid chemokine, SLC). Both of these bind the receptor CCR7, which is expressed by T cells; mice that lack

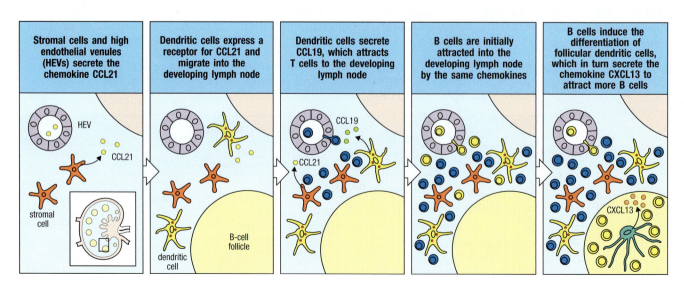

| Stromal cells and high endothelial venules (HEVs) secrete the chemokine CCL21 | Dendritic cells express a receptor for CCL21 and migrate into the developing lymph node | Dendritic cells secrete CCL19, which attracts T cells to the developing lymph node | B cells are initially attracted into the developing lymph node by the same chemokines | B cells induce the differentiation of follicular dendritic cells, which in turn secrete the chemokine CXCL13 to attract more B cells |

CCR7 do not form normal T-cell zones and have impaired primary immune responses. CCL21 is produced by stromal cells of T-cell zones in secondary lymphoid tissues, and is displayed on endothelial cells of high endothelial venules (HEVs). Another source of CCL21 is interdigitating dendritic cells, which also produce CCL19 and are prominent in T-cell zones. Indeed, dendritic cells themselves express CCR7 and will localize to secondary lymphoid tissues even in RAG-deficient mice, which lack lymphocytes and therefore defined T-cell zones. Thus, during normal lymph-node development, the T-cell zone might be organized first through the attraction of dendritic cells and T cells by CCL21 produced by stromal cells. This organization would then be reinforced by CCL21 and CCL19 secreted by resident dendritic cells, which in turn attract more T cells and migratory dendritic cells.

Like T cells, circulating B cells express CCR7, which initially directs them into the lymph node across HEVs. Because they also constitutively express the chemokine receptor CXCR5, they are then attracted to the follicles by the ligand for this receptor, CXCL13. The most likely source of CXCL13 is the FDC, possibly along with other follicular stromal cells. This is reminiscent of the expression of CXCL13 by stromal cells during the formation of the lymph node (Section 9-2). B cells are, in turn, the source of the LT that is required for the development of FDCs, which is reminiscent of LTi cells expressing the LT required to activate stromal cells. The reciprocal dependence of B cells and FDCs, and LTis and stromal cells, illustrates the complex web of interactions that organizes secondary lymphoid tissues. A subset of CD4 T cells called **T follicular helper**, or T_{FH}, **cells** can also express CXCR5 following their activation by antigen, allowing them to enter B-cell follicles to participate in the formation of germinal centers (see Chapter 10).

9-4 Naive T cells migrate through secondary lymphoid tissues, sampling peptide:MHC complexes on dendritic cells.

Naive T cells perpetually circulate from the bloodstream into lymph nodes, spleen, and mucosa-associated lymphoid tissues and back to the blood (see Fig. 1.21). This allows them to contact thousands of dendritic cells every day and sample the peptide:MHC complexes on the surfaces of these cells. Because of their high rates of recirculation and their concentration in T-cell zones where incoming dendritic cells dwell, each T cell has a high probability of encountering antigens derived from any pathogen that has set up an infection anywhere in the body (Fig. 9.4). Within hours of their arrival, naive T cells that do not encounter their specific antigen exit from the lymphoid tissue and reenter the bloodstream, where they continue to recirculate—via the efferent lymphatics in lymph nodes or MALTs, or directly back to the blood in the spleen, which has no connection with the lymphatic system.

When a naive T cell recognizes its specific antigen on the surface of an activated dendritic cell, however, it ceases to migrate. It remains in the T-cell zone,

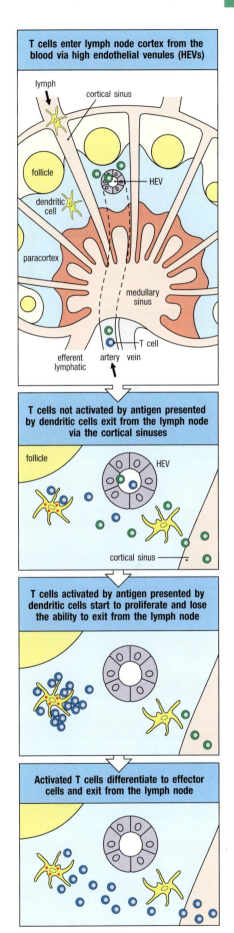

Fig. 9.4 Naive T cells encounter antigen during their recirculation through peripheral lymphoid organs. Naive T cells recirculate through peripheral lymphoid organs, such as a lymph node (shown here), entering from the arterial blood via the specialized vascular endothelium of high endothelial venules (HEVs). Entry into the lymph node is regulated by chemokines (not shown) that direct the T cells' migration through the HEV wall and into the paracortical areas, where the T cells encounter mature dendritic cells (top panel). Those T cells shown in green do not encounter their specific antigen; they receive a survival signal through their interaction with self peptide:self MHC complexes and IL-7, and leave the lymph node through the lymphatics to return to the circulation (second panel). T cells shown in blue encounter their specific antigen on the surface of mature dendritic cells; they lose their ability to exit from the node and become activated to proliferate and to differentiate into effector T cells (third panel). After several days, these antigen-specific effector T cells regain the expression of receptors needed to exit from the node, leave via the efferent lymphatics, and enter the circulation in greatly increased numbers (bottom panel).

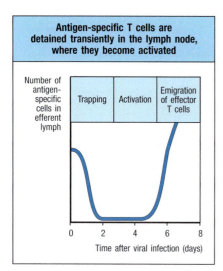

Antigen-specific T cells are detained transiently in the lymph node, where they become activated

Number of antigen-specific cells in efferent lymph

| Trapping | Activation | Emigration of effector T cells |

Time after viral infection (days)

Fig. 9.5 Trapping and activation of antigen-specific naive T cells in lymphoid tissue. Naive T cells entering the lymph node from the blood encounter antigen-presenting dendritic cells in T-cell zones. T cells that recognize their specific antigen bind stably to the dendritic cells and are activated through their T-cell receptors, resulting in their retention within the lymph node as they develop into effector T cells. By 5 days after the arrival of antigen, activated effector T cells are leaving the lymph node in large numbers via the efferent lymphatics. Lymphocyte recirculation and recognition are so effective that all the naive T cells in the peripheral circulation specific for a particular antigen can be trapped by that antigen in one node within 2 days.

where it proliferates for several days, undergoing clonal expansion and differentiation to give rise to effector T cells and memory cells of identical antigen specificity. At the end of this period, most effector T cells exit the lymphoid organ and reenter the bloodstream, through which they migrate to the sites of infection (see Chapter 11). Some effector T cells that are fated to interact with B cells migrate instead to B-cell zones, where they participate in the germinal center response (see Chapter 10).

The efficiency with which T cells screen antigen-presenting cells in lymph nodes is very high, as can be seen by the rapid trapping of antigen-specific T cells in a single lymph node containing antigen; within 48 hours, all antigen-specific T cells in the body can be trapped in the lymph node draining a site of antigen injection (**Fig. 9.5**). Such efficiency is crucial for the initiation of an adaptive immune response, as only one naive T cell in 10^5–10^6 is likely to be specific for a particular antigen, and adaptive immunity depends on the activation and expansion of these rare cells.

9-5   Lymphocyte entry into lymphoid tissues depends on chemokines and adhesion molecules.

Migration of naive T cells into secondary lymphoid tissues depends on their binding to high endothelial venules (HEVs) through cell–cell interactions that are not antigen-specific but are governed by cell-adhesion molecules. The main classes of adhesion molecules involved in lymphocyte interactions are the selectins, the integrins, members of the immunoglobulin superfamily, and some mucin-like molecules (see Fig. 3.30). Entry of lymphocytes into lymph nodes occurs in distinct stages that include initial rolling of lymphocytes along the endothelial surface, activation of integrins, firm adhesion, and transmigration or diapedesis across the endothelial layer into the paracortical areas, the T-cell zones (**Fig. 9.6**). These stages are regulated by a coordinated interplay of adhesion molecules and chemokines that resembles the recruitment of leukocytes to sites of inflammation (see Chapter 3). Adhesion molecules have fairly broad roles in immune responses, being involved not only in lymphocyte migration but also in interactions between naive T cells and antigen-presenting cells (see Section 9-14).

The selectins (**Fig. 9.7**) are important for specifically guiding leukocytes to particular tissues, a phenomenon known as leukocyte **homing**. L-selectin (CD62L) is expressed on leukocytes, whereas P-selectin (CD62P) and E-selectin (CD62E) are expressed on vascular endothelium (see Section 3-18). L-selectin on naive T cells guides their exit from the blood into secondary lymphoid tissues by initiating a light attachment to the wall of the HEV that results in the T cells' rolling along the endothelial surface (see Fig. 9.6). P-selectin and

Fig. 9.6 Lymphocyte entry into a lymph node from the blood occurs in distinct stages involving the activity of adhesion molecules, chemokines, and chemokine receptors. Naive T cells are induced to roll along the surface of a high endothelial venule (HEV) by the interactions of selectins expressed by the T cells with vascular addressins on the endothelial cell membranes. Chemokines present at the HEV surface activate receptors on the T cell, and chemokine receptor signaling leads to an increase in the affinity of integrins on the T cell for the adhesion molecules expressed on the HEV. This induces strong adhesion. After adhesion, the T cells follow gradients of chemokines to pass through the HEV wall into the paracortical region of the lymph node.

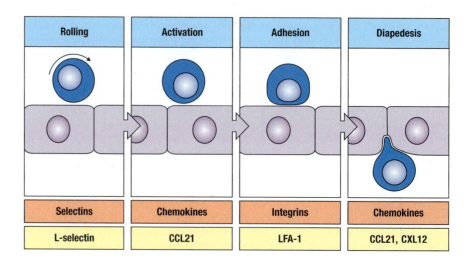

Rolling	Activation	Adhesion	Diapedesis
Selectins	Chemokines	Integrins	Chemokines
L-selectin	CCL21	LFA-1	CCL21, CXL12

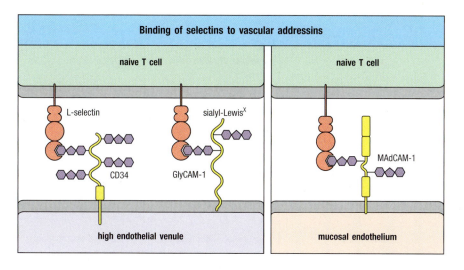

Fig. 9.7 L-selectin binds to mucin-like vascular addressins. L-selectin is expressed on naive T cells and recognizes carbohydrate motifs. Its binding to sulfated sialyl-Lewis^X moieties on the vascular addressins CD34 and GlyCAM-1 on HEVs binds the lymphocyte weakly to the endothelium. The relative importance of CD34 and GlyCAM-1 in this interaction is unclear. CD34 has a transmembrane anchor and is expressed in appropriately glycosylated form only on HEV cells, although it is found in other forms on other endothelial cells. GlyCAM-1 is expressed on HEVs but has no transmembrane region and may be secreted into the HEVs. The addressin MAdCAM-1 is expressed on mucosal endothelium and guides lymphocytes to mucosal lymphoid tissue. The configuration shown represents mouse MAdCAM-1, which contains an IgA-like domain closest to the cell membrane; human MAdCAM-1 has an elongated mucin-like domain and lacks the IgA-like domain.

E-selectin are expressed on the vascular endothelium at sites of infection, and serve to recruit effector cells into the infected tissue. Selectins are cell-surface molecules with a common core structure and are distinguished from each other by the presence of different lectin-like domains in their extracellular portion. The lectin domains bind to particular sugar groups, and each selectin binds to a cell-surface carbohydrate. L-selectin binds to the carbohydrate moiety—sulfated sialyl-Lewis^X—of mucin-like molecules called vascular addressins, which are expressed on the surface of vascular endothelial cells. Two of these addressins, CD34 and GlyCAM-1 (see Fig. 9.7), are expressed on high endothelial venules in lymph nodes. A third, MAdCAM-1, is expressed on endothelium in mucosae, and guides lymphocyte entry into mucosal lymphoid tissue such as the Peyer's patches in the gut.

The interaction between L-selectin and the vascular addressins is responsible for the specific homing of naive T cells to lymphoid organs. On its own, however, it does not enable the cell to cross the endothelial barrier into the lymphoid tissue. This requires the concerted action of chemokines and integrins.

9-6 Activation of integrins by chemokines is responsible for the entry of naive T cells into lymph nodes.

Naive T cells rolling on the endothelium of HEVs via selectins require two additional types of cell-adhesion molecules to enter secondary lymphoid organs—integrins, and members of the immunoglobulin superfamily. Integrins bind tightly to their ligands after receiving signals that induce a change in their conformation. Signaling by chemokines activates integrins on leukocytes to bind tightly to the vascular wall in preparation for the migration of the leukocytes into sites of inflammation (see Section 3-18). Similarly, chemokines present at the luminal surface of the HEV activate integrins expressed on naive T cells during migration into lymphoid organs (see Fig. 9.6).

An integrin molecule consists of a large α chain that pairs noncovalently with a smaller β chain. There are several integrin subfamilies, broadly defined by their common β chains. We will be concerned here chiefly with the leukocyte integrins, which have a common β_2 chain paired with distinct α chains (**Fig. 9.8**). All T cells express the integrin $\alpha_L{:}\beta_2$ (CD11a:CD18), better known as leukocyte functional antigen-1 (LFA-1). It enables migration of both naive and effector T cells out of the blood. This integrin is also present on macrophages and neutrophils, and is involved in their recruitment to sites of infection (see Section 3-18).

LFA-1 is also important in the adhesion of both naive and effector T cells to their target cells. Nevertheless, T-cell responses can be normal in individuals genetically lacking the β_2 integrin chain and hence all β_2 integrins,

MOVIE 9.2

Fig. 9.8 Integrins are important in T-lymphocyte adhesion. Integrins are heterodimeric proteins containing a β chain, which defines the class of integrin, and an α chain, which defines the different integrins within a class. The α chain is larger than the β chain and contains binding sites for divalent cations that may be important in signaling. LFA-1 (integrin $\alpha_L{:}\beta_2$) is expressed on all leukocytes. It binds ICAMs and is important in cell migration and in the interactions of T cells with antigen-presenting cells (APCs) or target cells; it is expressed at higher levels on effector T cells than on naive T cells. Lymphocyte Peyer's patch adhesion molecule (LPAM-1, or integrin $\alpha_4{:}\beta_7$) is expressed by a subset of naive T cells and contributes to lymphocyte entry into mucosal lymphoid tissues by supporting adhesive interactions with vascular addressin MAdCAM-1. VLA-4 (integrin $\alpha_4{:}\beta_1$) is expressed strongly after T-cell activation. It binds to VCAM-1 on activated endothelium and is important for recruiting effector T cells into sites of infection.

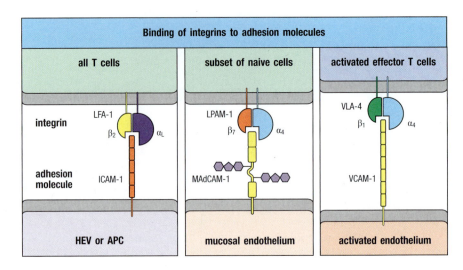

including LFA-1. This is probably because T cells also express other adhesion molecules, including the immunoglobulin superfamily member CD2 and β_1 integrins, which may compensate for the absence of LFA-1. Expression of the β_1 integrins increases significantly at a late stage in T-cell activation, and they are thus often called VLAs, for very late activation antigens; they are important in directing effector T cells to inflamed target tissues.

At least five members of the immunoglobulin superfamily are especially important in T-cell activation (Fig. 9.9). Three very similar intercellular adhesion molecules (ICAMs)—ICAM-1, ICAM-2, and ICAM-3—all bind to the T-cell integrin LFA-1. ICAM-1 and ICAM-2 are expressed on endothelium as well as on antigen-presenting cells, and binding to these molecules enables lymphocytes to migrate through blood vessel walls. ICAM-3 is expressed only on naive T cells and is thought to have an important role in the adhesion of T cells to antigen-presenting cells by binding to LFA-1 expressed on dendritic cells. The two remaining immunoglobulin superfamily adhesion molecules, CD58 (formerly known as LFA-3) on the antigen-presenting cell and CD2 on the T cell, bind to each other; this interaction synergizes with that of ICAM-1 or ICAM-2 with LFA-1.

As discussed above in the context of lymphoid tissue development (see Section 9-3), naive T cells are specifically attracted into the T-cell zones of secondary lymphoid tissues by chemokines. The chemokines bind to proteoglycans in the extracellular matrix and high endothelial venule wall, forming a chemical gradient, and are recognized by receptors on the naive T cell. The extravasation of naive T cells is prompted by the chemokine CCL21, which is expressed by vascular high endothelial cells and the stromal cells of lymphoid tissues, as well as by dendritic cells that reside in T-cell zones. It binds to the chemokine receptor CCR7 on naive T cells, stimulating activation of the intracellular receptor-associated G-protein subunit $G\alpha_i$. The resulting intracellular signaling rapidly increases the affinity of integrin binding (see Section 3-18).

Fig. 9.9 Immunoglobulin superfamily adhesion molecules involved in leukocyte interactions. Adhesion molecules of the immunoglobulin superfamily bind to adhesion molecules of various types, including integrins (LFA-1 and VLA-4) and other immunoglobulin superfamily members [the CD2–CD58 (LFA-3) interaction]. These interactions have a role in lymphocyte migration, homing, and cell–cell interactions; see Fig. 3.24 for the other molecules listed here.

Immunoglobulin superfamily	Name	Tissue distribution	Ligand
ICAM1/3, VCAM1 CD58 CD2	CD2 (LFA-2)	T cells	CD58 (LFA-3)
	ICAM-1 (CD54)	Activated vessels, lymphocytes, dendritic cells	LFA-1, Mac-1
	ICAM-2 (CD102)	Resting vessels	LFA-1
	ICAM-3 (CD50)	Naive T cells	LFA-1
	LFA-3 (CD58)	Lymphocytes, antigen-presenting cells	CD2
	VCAM-1 (CD106)	Activated endothelium	VLA-4

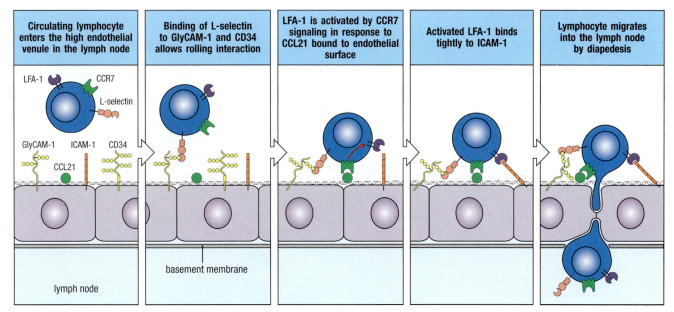

| Circulating lymphocyte enters the high endothelial venule in the lymph node | Binding of L-selectin to GlyCAM-1 and CD34 allows rolling interaction | LFA-1 is activated by CCR7 signaling in response to CCL21 bound to endothelial surface | Activated LFA-1 binds tightly to ICAM-1 | Lymphocyte migrates into the lymph node by diapedesis |

Fig. 9.10 Lymphocytes in the blood enter lymphoid tissue by crossing the walls of high endothelial venules. The first step is the binding of L-selectin on the lymphocyte to sulfated carbohydrates (sulfated sialyl-Lewis^X) of GlyCAM-1 and CD34 on the HEV. Local chemokines such as CCL21 bound to a proteoglycan matrix on the HEV surface stimulate chemokine receptors on the T cell, leading to the activation of LFA-1. This causes the T cell to bind tightly to ICAM-1 on the endothelial cell, allowing migration across the endothelium. As in the case of neutrophil migration (see Fig. 3.31), matrix metalloproteinases on the lymphocyte surface (not shown) enable the lymphocyte to penetrate the basement membrane.

The entry of a naive T cell into a lymph node is shown in detail in Fig. 9.10. Initial rolling of the T cell along the surface of HEVs is mediated by L-selectin. Recognition of CCL21 on the endothelial surface of the HEV by CCR7 on the T cell causes LFA-1 to become activated, increasing its affinity for ICAM-2 and ICAM-1. ICAM-2 is expressed constitutively on all endothelial cells, whereas in the absence of inflammation, ICAM-1 is expressed only on the high endothelial cells of secondary lymphoid tissues. The organization of LFA-1 molecules in the T-cell membrane is also altered by chemokine stimulation, such that they become concentrated in areas of cell–cell contact. This produces stronger binding, which arrests the T cell on the endothelial surface and thus enables it to enter the lymphoid tissue.

Once naive T cells have arrived in the T-cell zone via high endothelial venules, CCR7 directs their retention in this location, as they are attracted to dendritic cells that produce CCL21 and CCL19 in the T-cell zone. The naive T cells scan the surfaces of dendritic cells for specific peptide:MHC complexes, and if they find their antigen and bind to it, they are trapped in the lymph node. If they are not activated by antigen, naive T cells soon leave the lymph node (see Fig. 9.4).

9-7   The exit of T cells from lymph nodes is controlled by a chemotactic lipid.

T cells exit from a lymph node via the cortical sinuses, which lead into the medullary sinus and then the efferent lymphatic vessel. The egress of T cells from secondary lymphoid organs involves the lipid molecule **sphingosine 1-phosphate (S1P)** (Fig. 9.11). This lipid has chemotactic activity and signaling properties similar to those of chemokines, in that the receptors for S1P are G-protein-coupled receptors. A concentration gradient of S1P between the lymphoid tissues and lymph or blood acts to draw unactivated naive T cells expressing an S1P receptor away from the lymphoid tissues and back into circulation.

Fig. 9.11 The egress of lymphocytes from lymphoid tissue is mediated by a sphingosine 1-phosphate (S1P) gradient. The level of sphingosine 1-phosphate (S1P) within lymphoid tissue is low compared with efferent lymph, thereby forming an S1P gradient (indicated by shading). The S1P receptor 1 (S1PR1) expressed on naive T cells is responsive to the S1P gradient. In the absence of antigen recognition, S1PR1 signaling promotes T-cell egress from the T-cell zones into the efferent lymphatic vessel. T cells activated by an antigen-expressing dendritic cell upregulate CD69, which causes a decrease in S1PR1 expression and retention in the T-cell zone. Effector T cells eventually reexpress S1PR1 as CD69 expression decreases, and thereby egress from the lymph node. FTY720 inhibits T-cell egression by downmodulating expression of S1PR1 by ligand-induced internalization and by S1PR1-mediated closure of egress ports on the endothelium by enhancement of junctional contacts (not shown).

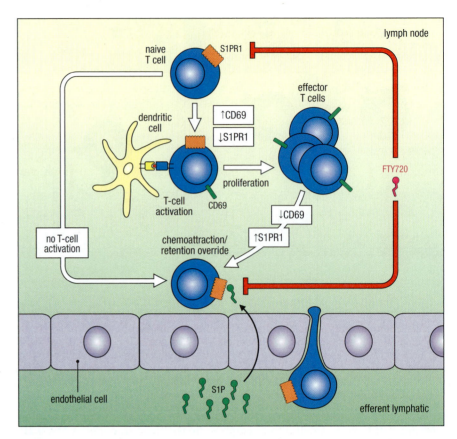

T cells activated by antigen in lymphoid organs downregulate the surface expression of the S1P receptor, S1PR1, for several days. This loss of S1PR1 surface expression is caused by CD69, a surface protein whose expression is induced by T-cell receptor signaling and which acts to internalize S1PR1. During this period, T cells cannot respond to the S1P gradient and do not exit the lymphoid organ. After several days of proliferation, as T-cell activation wanes, CD69 expression decreases and S1PR1 reappears on the surface of effector T cells, allowing them to migrate out of the lymphoid tissue in response to the S1P gradient.

The regulation of the exit of both naive and effector lymphocytes from secondary lymphoid organs by S1P is the basis for a new kind of potential immunosuppressive drug, FTY720 (fingolimod). FTY720 inhibits immune responses by preventing lymphocytes from returning to the circulation, thereby sequestering them in lymphoid tissues and causing rapid onset of lymphopenia (a lack of lymphocytes in the blood). *In vivo*, FTY720 becomes phosphorylated and mimics S1P as an agonist at S1P receptors. Phosphorylated FTY720 may inhibit lymphocyte exit by effects on endothelial cells that increase tight junction formation and close exit portals, or by chronic activation of S1P receptors, leading to inactivation and downregulation of the receptor.

9-8   T-cell responses are initiated in secondary lymphoid organs by activated dendritic cells.

Secondary lymphoid organs were first shown to be important in the initiation of adaptive immune responses by ingenious experiments in which a flap of skin was isolated from the body wall so that it had blood circulation but no lymphatic drainage. Antigen placed in the flap did not elicit a T-cell response, showing that T cells do not become sensitized in the infected tissue itself. Rather, pathogens and their products must be transported to lymphoid tissues. Antigens introduced directly into the bloodstream are picked up by antigen-presenting cells in the spleen. Pathogens infecting other sites, such as a skin wound, are transported in lymph via lymphatic vessels and trapped in the lymph nodes nearest to the site of infection (see Section 1-16). Pathogens infecting mucosal surfaces are transported directly across the mucosa into lymphoid tissues such as the tonsils or Peyer's patches, as well as draining lymph nodes.

In this chapter we will focus on T-cell activation by dendritic cells as it occurs in organs of the systemic immune system—lymph nodes and spleen. The activation of T cells by dendritic cells in the mucosal immune system follows the same principles, but differs in some details, described in Chapter 12, such as the route by which antigen is delivered and the subsequent circulation patterns of the effector cells.

The delivery of antigen from a site of infection to lymphoid tissue is actively aided by the innate immune response. One effect of innate immunity is an inflammatory reaction that increases the rate of entry of blood plasma into infected tissues and thus increases the drainage of extracellular fluid into the lymph, taking with it free antigen that is carried to lymphoid tissues. Even more important for initiation of the adaptive response is the activation of tissue dendritic cells that have taken up particulate and soluble antigens at the site of infection (**Fig. 9.12**).

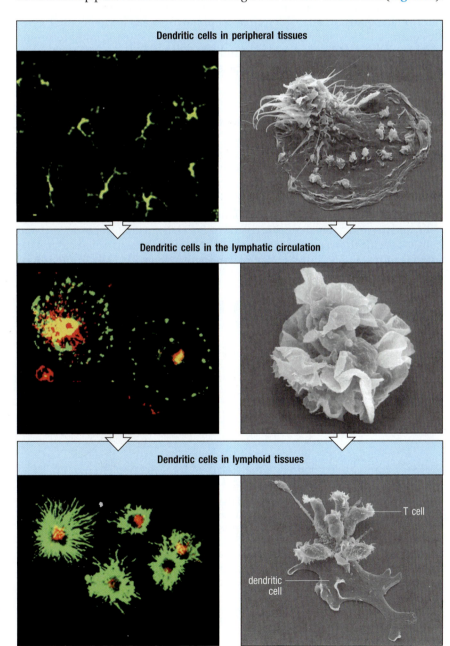

Fig. 9.12 Dendritic cells in different stages of activation and migration. The left panels show fluorescence micrographs of dendritic cells stained for MHC class II molecules in green and for a lysosomal protein in red. The right panels show scanning electron micrographs of single dendritic cells. Unactivated dendritic cells (top panels) have many long processes, or dendrites, from which the cells get their name. The cell bodies are difficult to distinguish in the left panel, but the cells contain many endocytic vesicles that stain both for MHC class II molecules and for the lysosomal protein; when these two colors overlap they give rise to a yellow fluorescence. Activated dendritic cells leave the tissues to migrate through the lymphatics to secondary lymphoid tissues. During this migration their morphology changes. The dendritic cells stop phagocytosing antigen, and staining for the lysosomal protein is beginning to be distinct from that for MHC class II molecules (center left panel). The dendritic cell now has many folds of membrane (center right panel), which gave these cells their original name of 'veil' cells. Finally, in the lymph nodes, dendritic cells express high levels of peptide:MHC complexes and co-stimulatory molecules, and are very good at stimulating naive CD4 and naive CD8 T cells. At this stage, the activated dendritic cells do not phagocytose, and the red staining of the lysosomal protein is quite distinct from the green-stained MHC class II molecules displayed at high density on many dendritic processes (bottom left panel). The typical morphology of a mature dendritic cell is shown in the bottom right panel, as it interacts with a T cell. Fluorescent micrographs courtesy of I. Mellman, P. Pierre, and S. Turley. Scanning electron micrographs courtesy of K. Dittmar.

Labels within figure:
Dendritic cells in peripheral tissues
Dendritic cells in the lymphatic circulation
Dendritic cells in lymphoid tissues
T cell
dendritic cell

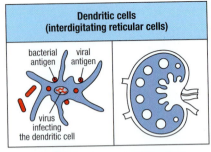

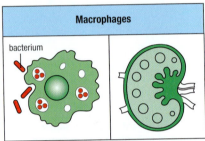

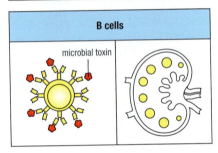

Fig. 9.13 Antigen-presenting cells are distributed by type in specific areas of the lymph node. Dendritic cells are found throughout the cortex of the lymph node in the T-cell areas. Mature dendritic cells are by far the strongest activators of naive T cells, and can present antigens from many types of pathogens, such as bacteria or viruses as shown here. Macrophages are distributed throughout the lymph node but are concentrated mainly in the marginal sinus, where the afferent lymph collects before percolating through the lymphoid tissue, and also in the medullary cords, where the efferent lymph collects before passing via the efferent lymphatics into the blood. B cells are found mainly in the follicles and can contribute to neutralizing soluble antigens such as toxins.

Dendritic cells can be activated via their TLRs and other pathogen-recognition receptors (see Chapter 3), by tissue damage, or by cytokines produced during the inflammatory response. Activated dendritic cells migrate to the lymph node and express the co-stimulatory molecules that are required, in addition to antigen, for the activation of naive T cells. In the lymphoid tissues, these dendritic cells present antigen to naive T lymphocytes and prime antigen-specific T cells to divide and mature into effector cells that reenter the circulation.

Macrophages, which are found in most tissues including lymphoid tissue, and B cells, which are located primarily in lymphoid tissue, can be similarly activated by pathogen-recognition receptors to express co-stimulatory molecules and act as antigen-presenting cells. The distribution of dendritic cells, macrophages, and B cells in a lymph node is shown schematically in Fig. 9.13. Only these three cell types express co-stimulatory molecules required to efficiently activate T cells, and they express these molecules only when activated in the context of infection. However, these cells activate T-cell responses in distinct ways. Dendritic cells take up, process, and present antigens from all types of sources, and are present mainly in the T-cell areas where they drive the initial clonal expansion and differentiation of naive T cells into effector T cells. By contrast, B cells and macrophages specialize in processing and presenting soluble antigens and antigens from intracellular pathogens, respectively; they interact mainly with effector CD4 T cells already primed by dendritic cells to recruit helper functions of those T cells.

9-9 Dendritic cells process antigens from a wide array of pathogens.

Dendritic cells primarily arise from myeloid progenitors within the bone marrow (see Fig. 1.3). They emerge from the bone marrow to migrate via the blood to tissues throughout the body, or directly to secondary lymphoid organs. There are two major classes of dendritic cells: conventional dendritic cells, and plasmacytoid dendritic cells (Fig. 9.14). The cell-surface markers and subset-specific transcription factors that distinguish these two classes, and the interferon-producing functions of plasmacytoid dendritic cells in the innate

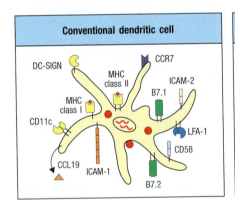

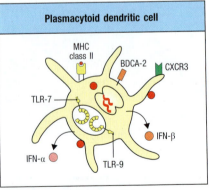

Fig. 9.14 Conventional and plasmacytoid dendritic cells have different roles in the immune response. Mature conventional dendritic cells (left panel) are primarily concerned with the activation of naive T cells. There are several subsets of conventional dendritic cells, but these all process antigen efficiently, and when they are mature they express MHC proteins and co-stimulatory molecules for priming naive T cells. The cell-surface proteins expressed by the mature dendritic cell are described in the text. Immature dendritic cells lack many of the cell-surface molecules shown here but have numerous surface receptors that recognize pathogen molecules, including most of the Toll-like receptors (TLRs). Plasmacytoid dendritic cells (right panel) are sentinels primarily for viral infections, and secrete large amounts of class I interferons. This category of dendritic cell is less efficient in priming naive T cells, but they express the intracellular receptors TLR-7 and TLR-9 for sensing viral infections.

immune response, are discussed in Chapter 3. In this chapter we shall focus on the role of conventional dendritic cells in the adaptive immune response—presenting antigens to and activating naive T cells.

Conventional dendritic cells are abundant at barrier tissue sites, such as the intestines, lung, and skin, where they are in close contact with surface epithelia. They are also present in most solid organs such as the heart and kidneys. In the absence of infection or tissue injury, dendritic cells have low levels of co-stimulatory molecules, and so are not yet equipped to stimulate naive T cells. Like macrophages, dendritic cells are very active in ingesting antigens by phagocytosis using complement receptors and Fc receptors (which recognize the constant regions of antibodies in antigen:antibody complexes), and C-type lectins, which recognize carbohydrates and on dendritic cells include the mannose receptor, DEC 205, langerin, and Dectin-1. Other extracellular antigens are taken up nonspecifically by the process of macropinocytosis, in which large volumes of surrounding fluid are engulfed. In this way microbes that have evolved strategies to escape recognition by phagocytic receptors, such as bacteria with thick polysaccharide capsules, can be ingested. The versatility in pathways for antigen uptake enables dendritic cells to present antigens from virtually any type of microbe, including fungi, parasites, viruses, and bacteria (Fig. 9.15). Uptake of extracellular antigens by these pathways directs them into the endocytic pathway, where they are processed and presented on MHC class II molecules (see Chapter 6) for recognition by CD4 T cells.

A second route of antigen handling by dendritic cells occurs when antigen directly enters the cytosol, for example, through viral infection. Dendritic cells are directly susceptible to infection by some viruses, which enter the cytoplasm by binding to cell-surface molecules that act as entry receptors. Viral proteins synthesized in the cytoplasm of dendritic cells are processed in the proteasome and presented on the cell surface as peptides loaded onto MHC class I molecules after transport into the endoplasmic reticulum, as in any other type of virus-infected cell (see Chapter 6). This enables dendritic cells to present antigen to and activate naive CD8 T cells, which then differentiate into cytotoxic effector CD8 T cells that recognize and kill any virus-infected cell.

Routes of antigen processing and presentation by dendritic cells				
Receptor-mediated phagocytosis	Macropinocytosis	Viral infection	Cross-presentation after phagocytic or macropinocytic uptake	Transfer from incoming dendritic cell to resident dendritic cell
Type of pathogen presented Extracellular bacteria	Extracellular bacteria, soluble antigens, virus particles	Viruses	Viruses	Viruses
MHC molecules loaded MHC class II	MHC class II	MHC class I	MHC class I	MHC class I
Type of naive T cell activated CD4 T cells	CD4 T cells	CD8 T cells	CD8 T cells	CD8 T cells

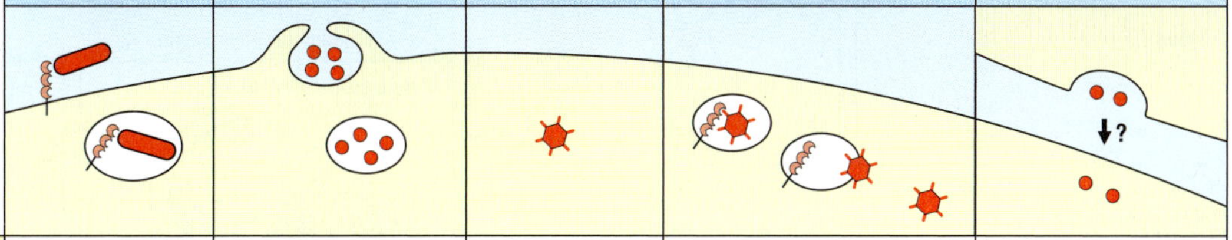

Fig. 9.15 The different routes by which dendritic cells can take up, process, and present protein antigens. Uptake of antigens into the endocytic system, either by receptor-mediated phagocytosis or by macropinocytosis, is considered to be the major route for delivering peptides to MHC class II molecules for presentation to CD4 T cells (first two panels). Production of antigens in the cytosol, for example, as a result of viral infection, is thought to be the major route for delivering peptides to MHC class I molecules for presentation to CD8 T cells (third panel). It is possible, however, for exogenous antigens taken into the endocytic pathway to be delivered into the cytosol for eventual delivery to MHC class I molecules for presentation to CD8 T cells, a process called cross-presentation (fourth panel). Finally, it seems that antigens can be transmitted from one dendritic cell to another, particularly for presentation to CD8 T cells, although the details of this route are still unclear (fifth panel).

MOVIE 9.3

Uptake of extracellular virus particles or virus-infected cells into the endocytic pathway by macropinocytosis or phagocytosis can also result in the presentation of viral peptides on MHC class I molecules. This phenomenon, known as cross-presentation, is an alternative to the usual cytosolic pathway for MHC class I antigen processing and is discussed in Section 6-5. Here, viral antigens that enter dendritic cells via endocytic or phagocytic vesicles may be diverted to the cytosol for proteasomal degradation and transferred to the endoplasmic reticulum for loading onto MHC class I molecules. The result is that viruses that do not directly infect dendritic cells can stimulate the activation of CD8 T cells. Cross-presentation is performed most efficiently by a subset of conventional dendritic cells that is specialized for stimulating T-cell responses to intracellular pathogens (see Section 6-5). Any viral infection can therefore lead to the generation of cytotoxic effector CD8 T cells, whether the virus can directly infect dendritic cells or not. In addition, viral peptides presented on the dendritic cell's MHC class II molecules activate naive CD4 T cells, which leads to the production of effector CD4 T cells that stimulate the production of antiviral antibodies by B cells and produce cytokines that enhance the immune response.

In some cases, such as infections with herpes simplex or influenza viruses, the dendritic cells that migrate to the lymph nodes from peripheral tissues may not be the same cells that finally present antigen to naive T cells. In herpes simplex infection, for example, dendritic cells residing in the skin capture antigen and transport it to the draining lymph nodes (**Fig. 9.16**). There, some antigen is transferred to resident CD8α-positive dendritic cells, which are the dominant dendritic cells responsible for priming naive CD8 T cells in this infection.

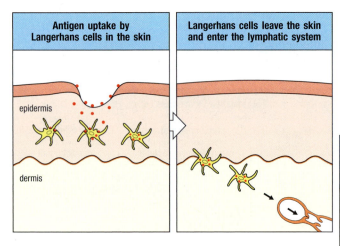

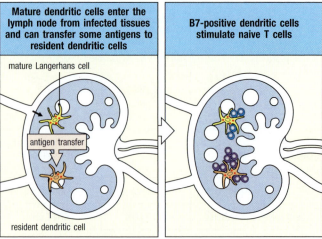

Fig. 9.16 Langerhans cells take up antigen in the skin, migrate to the peripheral lymphoid organs, and present foreign antigens to T cells. Langerhans cells (yellow) are one type of immature dendritic cell that resides in the epidermis. They ingest antigen in various ways but have no co-stimulatory activity (first panel). In the presence of infection, they take up antigen locally and then migrate to the lymph nodes (second panel). There they differentiate into mature dendritic cells that can no longer ingest antigen but have co-stimulatory activity. Now they can prime both naive CD8 and CD4 T cells. In the case of some viral infections, for example, with herpes simplex virus, some dendritic cells arriving from the site of infection seem able to transfer antigen to resident dendritic cells (orange) in the lymph nodes (third panel) for presentation of class I MHC-restricted antigens to naive CD8 T cells (fourth panel).

This type of transfer means that antigens from viruses that infect but rapidly kill dendritic cells can still be presented by uninfected dendritic cells that have been activated via their TLRs and can take up the dying dendritic cells and cross-present this material.

9-10   Microbe-induced TLR signaling in tissue-resident dendritic cells induces their migration to lymphoid organs and enhances antigen processing.

A critical step in the induction of adaptive immunity is the activation of dendritic cell maturation. When an infection occurs, dendritic cells capture pathogens by means of phagocytic receptors or macropinocytosis, and then activate responses to these pathogens through pattern recognition receptors such as TLRs (Fig. 9.17, top panel). Multiple members of the TLR family are expressed on tissue dendritic cells and are thought to be involved in detecting and signaling the presence of the various classes of pathogens (see Fig. 3.16). In humans, conventional dendritic cells express all known TLRs except for TLR-9, which is, however, expressed by plasmacytoid dendritic cells along with TLR-1 and TLR-7, and other TLRs to a lesser degree. In addition to the pattern recognition receptors described in Chapter 3, several of the phagocytic receptors used by dendritic cells to take up pathogens also provide maturation signals. Examples include the lectin **DC-SIGN**, which binds mannose and fucose residues present on a wide range of pathogens; and Dectin-1, which recognizes β-1,3-linked glucans found in fungal cell walls (see Fig. 3.2). Other receptors that can bind pathogens, such as receptors for complement, or phagocytic receptors such as the mannose receptor, may contribute to dendritic cell activation as well as to phagocytosis.

TLR signaling results in a significant alteration in the chemokine receptors expressed by dendritic cells, which facilitates their migration into secondary lymphoid tissues. This change in dendritic cell behavior is often called **licensing**, as the cells are now embarked on the program of differentiation that will enable them to activate T cells. TLR signaling induces expression of the receptor CCR7, which makes the activated dendritic cells sensitive to the chemokine CCL21 produced by lymphoid tissue and induces their migration through the lymphatics and into the local lymphoid tissues. Whereas T cells must cross the wall of high endothelial venules to leave the blood and reach the T-cell zones, dendritic cells entering via the afferent lymphatics migrate directly into the T-cell zones from the marginal sinus.

Fig. 9.17 Conventional dendritic cells are activated through at least two definable stages to become potent antigen-presenting cells in peripheral lymphoid tissue. Dendritic cells originate from bone marrow progenitors and migrate via the blood, from which they enter and populate most tissues, including peripheral lymphoid tissues, into which they can make direct entry. Entry to particular tissues is based on the particular chemokine receptors they express: CCR1, CCR2, CCR5, CCR6, CXCR1, and CXCR2 (not all shown here, for simplicity). Tissue-resident dendritic cells are highly phagocytic via receptors such as Dectin-1, DEC 205, DC-SIGN, and langerin, and are actively macropinocytic, but they do not express co-stimulatory molecules. They carry most of the different types of Toll-like receptors (TLRs; see the text). At sites of infection, dendritic cells are exposed to pathogens, leading to activation of their TLRs (top panel). TLR signaling causes the dendritic cells to become activated ('licensed'), which involves induction of the chemokine receptor CCR7 (second panel). TLR signaling also increases the processing of antigens taken up into phagosomes. Dendritic cells expressing CCR7 are sensitive to CCL19 and CCL21, which direct them to the draining lymphoid tissue (third panel). CCL19 and CCL21 provide further maturation signals, which result in higher levels of co-stimulatory B7 molecules and MHC molecules. By the time they arrive in the draining lymph node, conventional dendritic cells have become powerful activators of naive T cells but are no longer phagocytic. They express B7.1, B7.2, and high levels of MHC class I and class II molecules, as well as high levels of the adhesion molecules ICAM-1, ICAM-2, LFA-1, and CD58 (bottom panel).

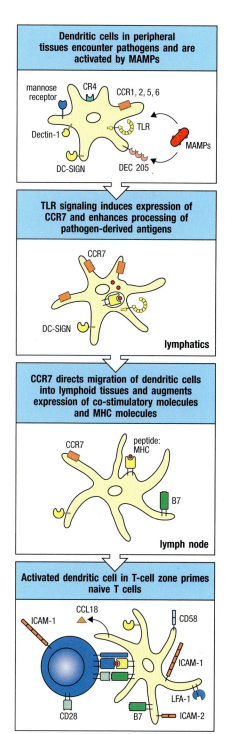

CCL21 signaling through CCR7 not only induces the migration of dendritic cells into lymphoid tissue, but it also contributes to their enhanced antigen-presenting function (see Fig. 9.17, third panel). By the time activated dendritic cells arrive within lymphoid tissues, they are no longer able to engulf antigens by phagocytosis or macropinocytosis. They instead express very high levels of long-lived MHC class I and MHC class II molecules, which enable them to stably present peptides from pathogens already taken up and processed. Of equal importance, they express high levels of co-stimulatory molecules on their surface. There are two main co-stimulatory molecules: the structurally related transmembrane glycoproteins B7.1 (CD80) and B7.2 (CD86), which deliver co-stimulatory signals by interacting with receptors on naive T cells (see Section 7-21). Activated dendritic cells also express very high levels of adhesion molecules, including DC-SIGN, and they secrete the chemokine CCL19, which specifically attracts naive T cells. Together, these properties enable the dendritic cell to stimulate strong responses in naive T cells (see Fig. 9.17, bottom panel).

Despite their enhanced presentation of pathogen-derived antigens, activated dendritic cells also present some self peptides, which could present a problem for the maintenance of self-tolerance. The T-cell receptor repertoire has, however, been purged of receptors that recognize self peptides presented in the thymus (see Chapter 8), so that T-cell responses against most ubiquitous self antigens are avoided. In addition, dendritic cells in the lymphoid tissues that have not been activated by infection will bear self-peptide:MHC complexes on their surface, derived from the breakdown of their own proteins and tissue proteins present in the extracellular fluid. Because these cells do not express the appropriate co-stimulatory molecules, however, they do not have the same capacity to activate naive T cells as do activated dendritic cells. Although the details are still unclear, the presentation of self peptides by lymph node-resident, or 'unlicensed,' dendritic cells may induce an alternative program of activation in naive T cells that favors immune regulation rather than immune activation.

Intracellular degradation of pathogens reveals pathogen components other than peptides that trigger dendritic cell activation. For example, bacterial or viral DNA containing unmethylated CpG dinucleotide motifs induces the rapid activation of plasmacytoid dendritic cells as a consequence of recognition of the DNA by TLR-9, which is present in intracellular vesicles (see Fig. 3.10). Exposure to unmethylated DNA activates NFκB and mitogen-activated protein kinase (MAPK) signaling pathways (see Figs. 7.19–7.21), leading to the production of pro-inflammatory cytokines such as IL-6, IL-12, IL-18, and interferon (IFN)-α and IFN-β by dendritic cells. In turn, these cytokines act on the dendritic cells themselves to augment the expression of co-stimulatory molecules. Heat-shock proteins are another internal bacterial constituent that can activate the antigen-presenting function of dendritic cells. Similarly, some viruses are recognized by TLRs inside the dendritic cell via double-stranded RNA produced during viral replication.

The induction of co-stimulatory activity in antigen-presenting cells by common microbial constituents is believed to allow the immune system to distinguish antigens borne by infectious agents from antigens associated with innocuous proteins, including self proteins. Indeed, many foreign proteins do not induce an immune response when injected on their own because they fail to induce co-stimulatory activity in antigen-presenting cells. When such protein antigens are mixed with bacteria, however, they become immunogenic, because the bacteria induce the essential co-stimulatory activity in cells that ingest the protein. Bacteria or bacterial components used in this way are known as **adjuvants** (see Appendix I, Section A-1). We will see in Chapter 15 how self proteins mixed with bacterial adjuvants can induce autoimmune disease, illustrating the crucial importance of the regulation of co-stimulatory activity in the discrimination of self from nonself.

9-11 Plasmacytoid dendritic cells produce abundant type I interferons and may act as helper cells for antigen presentation by conventional dendritic cells.

Plasmacytoid dendritic cells are thought to act as sentinels in early defense against viral infection on the basis of their expression of TLRs and the intracellular nucleic acid-sensing RIG-I-like helicases, and their high production of antiviral type I interferons (see Sections 3-10 and 3-22). For several reasons, they are not thought to be involved in a major way in the antigen-specific activation of naive T cells. Plasmacytoid dendritic cells express fewer MHC class II and co-stimulatory molecules on their surface, and they process antigens less efficiently than conventional dendritic cells. In addition, unlike conventional dendritic cells, plasmacytoid dendritic cells do not cease the synthesis and recycling of MHC class II molecules after being activated. This means that they rapidly recycle their surface MHC II molecules and so cannot present pathogen-derived peptide:MHC complexes to T cells for extended periods, as conventional dendritic cells do.

Plasmacytoid dendritic cells may, however, act as helper cells for antigen presentation by conventional dendritic cells. This activity was revealed by studies in mice infected with the intracellular bacterium *Listeria monocytogenes*. Normally, IL-12 made by conventional dendritic cells induces CD4 T cells to produce abundant IFN-γ, which helps macrophages kill the bacteria. When plasmacytoid dendritic cells were experimentally eliminated, IL-12 production by conventional dendritic cells decreased, and the mice became susceptible to *Listeria*. It appears that plasmacytoid dendritic cells interact with conventional dendritic cells to sustain IL-12 production. Activation of plasmacytoid dendritic cells through TLR-9 induces the expression of CD40 ligand (CD40L or CD154), a TNF-family transmembrane cytokine, which binds to CD40, a TNF-family receptor that is expressed by activated conventional dendritic cells. This interaction enables conventional dendritic cells to sustain production of the pro-inflammatory cytokine IL-12, strengthening the IL-12-induced production of IFN-γ by T cells. Plasmacytoid dendritic cells can also produce IL-12 themselves, although in smaller amounts than conventional dendritic cells do.

9-12 Macrophages are scavenger cells that can be induced by pathogens to present foreign antigens to naive T cells.

The two other cell types that can act as antigen-presenting cells to T cells are macrophages and B cells, although there is an important distinction between the function of antigen presentation by these cells and that of dendritic cells. It is unlikely that macrophages and B cells present antigen to activate naive T cells. Rather, these cells present antigen to T cells that have already been primed by conventional dendritic cells as a means to recruit the effector, or 'helper,' functions of T cells that, in turn, provide signals to enhance their own effector functions. In this way, naive B cells that are activated by antigen bound to their surface immunoglobulin receptor present peptides derived from that antigen to elicit help from effector T cells in order to differentiate into immunoglobulin-secreting cells. And, as we learned in Chapter 3, while many microorganisms that enter the body are engulfed and destroyed by phagocytes, which provide an innate, antigen-nonspecific first line of defense against infection, some pathogens have developed mechanisms to avoid elimination by innate immunity, such as resisting the killing properties of phagocytes. Macrophages that have ingested microorganisms but have failed to destroy them can use antigen presentation to recruit the adaptive immune response to enhance their microbicidal activities, as we will discuss further in Chapter 11.

Resting macrophages have few or no MHC class II molecules on their surface and do not express B7. The expression of both MHC class II molecules and B7 is induced by the ingestion of microorganisms and recognition of their microbe-associated molecular patterns (MAMPs). Macrophages, like dendritic cells, have a variety of pattern recognition receptors that recognize microbial surface components (see Chapter 3). Receptors such as Dectin-1, scavenger receptors, and complement receptors take up microorganisms into phagosomes, where they are degraded to produce peptides for presentation, while recognition of pathogen components via TLRs triggers intracellular signaling that contributes to the expression of MHC class II molecules and B7. However, unlike conventional dendritic cells, tissue-resident macrophages are generally nonmigratory; they do not traffic to T-cell zones of lymphoid tissues when activated by pathogens. It is thus likely that increased expression of MHC class II molecules and co-stimulatory molecules by activated macrophages is more important for locally amplifying T-cell responses already initiated by dendritic cells. This appears to be important for the maintenance and functioning of effector or memory T cells that enter a site of infection.

In addition to residing in tissues, macrophages are found in lymphoid organs (see Fig. 9.13). They are present in many areas of the lymph node, including the marginal sinus, where the afferent lymph enters the lymphoid tissue, and in the medullary cords, where the efferent lymph collects before flowing into the blood. However, they are largely sequestered from T-cell zones and are inefficient activators of naive T cells. Rather, their main function in lymphoid tissues appears to be the ingestion of microbes and particulate antigens to prevent them from entering the blood. They are also important scavengers of apoptotic lymphocytes.

Macrophages in other sites also continuously scavenge dead or dying cells, which are rich sources of self antigens, so it is particularly important that they should not activate naive T cells. The Kupffer cells of the liver sinusoids and the macrophages of the splenic red pulp, in particular, remove large numbers of dying cells from the blood daily. Kupffer cells express little MHC class II and no TLR-4, the receptor that signals the presence of bacterial LPS. Thus, although they generate large amounts of self peptides in their endosomes, these macrophages are not likely to elicit an immune response.

9-13 B cells are highly efficient at presenting antigens that bind to their surface immunoglobulin.

B cells are uniquely adapted to bind specific soluble molecules through their membrane-bound antigenic receptor, or B-cell receptor (BCR), the antigen-binding component of which is membrane-associated IgM, which is highly efficient at internalizing the bound molecules by receptor-mediated endocytosis. If the antigen contains a protein component, the B cell will process the internalized protein to peptide fragments and then display the fragments as peptide:MHC class II complexes. Through this mechanism B cells are able to take up and present even low concentrations of specific antigen to T cells. B cells also constitutively express high levels of MHC class II molecules, and so high levels of specific peptide:MHC class II complexes appear on the B-cell surface (Fig. 9.18). As we will see in Chapter 10, this pathway of antigen presentation allows the B cell to specifically interact with a CD4 T cell that has been previously activated by the same antigen as a mechanism to receive signals from the T cell to drive the B cell's differentiation into an antibody-producing cell.

B cells do not constitutively express co-stimulatory molecules, but, as with dendritic cells and macrophages, they can be induced by various microbial constituents to express B7 molecules. In fact, B7.1 was first identified as a protein on B cells activated by LPS, and B7.2 is predominantly expressed by B cells *in vivo*. Soluble protein antigens are not abundant during infections;

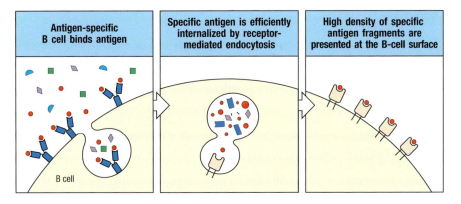

| Antigen-specific B cell binds antigen | Specific antigen is efficiently internalized by receptor-mediated endocytosis | High density of specific antigen fragments are presented at the B-cell surface |

Fig. 9.18 B cells can use their surface immunoglobulin to present specific antigen very efficiently to T cells. Surface immunoglobulin allows B cells to bind and internalize specific antigen very efficiently, especially if the antigen is present as a soluble protein, as most toxins are. The internalized antigen is processed in intracellular vesicles, where it binds to MHC class II molecules. The vesicles are transported to the cell surface, where the foreign-peptide:MHC class II complexes can be recognized by T cells. When the protein antigen is not specific for the B-cell receptor, its internalization is inefficient and only a few fragments of such proteins are subsequently presented at the B-cell surface (not shown).

most natural antigens, such as bacteria and viruses, are particulate, and many soluble bacterial toxins act by binding to cell surfaces and so are present only at low concentrations in solution. Some natural immunogens enter the body as soluble molecules, however; examples are bacterial toxins, anticoagulants injected by blood-sucking insects, snake venoms, and many allergens. Nevertheless, it is unlikely that B cells are important in priming naive T cells to soluble antigens in natural immune responses. Tissue dendritic cells can take up soluble antigens by macropinocytosis, and although they cannot concentrate these antigens as antigen-specific B cells do, dendritic cells are more likely to encounter a naive T cell with the appropriate antigen specificity than are the extremely limited number of antigen-specific B cells. The chances of a B cell encountering a T cell that can recognize the peptide antigens it displays are greatly increased once a naive T cell has been detained in lymphoid tissue by finding its antigen on the surface of a dendritic cell and has undergone clonal expansion.

The three types of antigen-presenting cells are compared in Fig. 9.19. In each of these cell types the expression of co-stimulatory activity is controlled so as to provoke responses against pathogens while avoiding immunization against self.

	Dendritic cells	Macrophages	B cells
Antigen uptake	+++ Macropinocytosis and phagocytosis by tissue dendritic cells	+++ Macropinocytosis +++ Phagocytosis	Antigen-specific receptor (Ig) ++++
MHC expression	Low on tissue-resident dendritic cells High on dendritic cells in lymphoid tissues	Inducible by bacteria and cytokines − to +++	Constitutive Increases on activation +++ to ++++
Co-stimulation delivery	Inducible High on dendritic cells in lymphoid tissues ++++	Inducible − to +++	Inducible − to +++
Location	Ubiquitous throughout the body	Lymphoid tissue Connective tissue Body cavities	Lymphoid tissue Peripheral blood
Effect	Results in activation of naive T cells	Results in activation of macrophages	Results in delivery of help to B cell

Fig. 9.19 The properties of the various antigen-presenting cells. Dendritic cells, macrophages, and B cells are the main cell types involved in the presentation of foreign antigens to T cells. These cells vary in their means of antigen uptake, MHC class II expression, co-stimulator expression, the type of antigen they present effectively, their locations in the body, and their surface adhesion molecules (not shown). Antigen presentation by dendritic cells is primarily involved in activating naive T cells for expansion and differentiation. Macrophages and B cells present antigen primarily to receive specific help from effector T cells in the form of cytokines or surface molecules.

Summary.

An adaptive immune response is generated when naive T cells contact activated antigen-presenting cells in the secondary lymphoid organs. These tissues have a specialized architecture that facilitates efficient interaction between circulating lymphocytes and their target antigens. The formation and organization of the peripheral lymphoid organs are controlled by proteins of the TNF family and their receptors (TNFRs). Lymphoid tissue inducer (LTi) cells expressing lymphotoxin-β (LT-β) interact with stromal cells expressing the LT-β receptor in the developing embryo to induce chemokine production, which in turn initiates formation of the lymph nodes and Peyer's patches. Similar interactions between lymphotoxin-expressing B cells and TNFR-I-expressing follicular dendritic cells (FDCs) establish the normal architecture of the spleen and lymph nodes. B and T cells are partitioned into distinct areas within lymphoid tissue by specific chemokines.

To ensure that rare antigen-specific T cells survey the body effectively for pathogen-bearing antigen-presenting cells, T cells continuously recirculate through the lymphoid organs and thus can sample antigens brought in by antigen-presenting cells from many different tissue sites. The migration of naive T cells into lymphoid organs is guided by the chemokine receptor CCR7, which binds CCL21 that is produced by stromal cells in the T-cell zones of secondary lymphoid tissues and is displayed on the specialized endothelium of HEVs. L-selectin expressed by naive T cells initiates their rolling along the specialized surfaces of high endothelial venules, where contact with CCL21 induces a switch in the integrin LFA-1 expressed by T cells to a configuration with affinity for the ICAM-1 expressed on the venule endothelium. This initiates strong adhesion, diapedesis, and migration of the T cells into the T-cell zone. There, naive T cells meet antigen-bearing dendritic cells, of which there are two main populations: conventional dendritic cells, and plasmacytoid dendritic cells. Conventional dendritic cells continuously survey secondary tissues for invading pathogens and are the dendritic cells responsible for activating naive T cells. Contact with pathogens delivers signals to dendritic cells via TLRs and other receptors that accelerate antigen processing and the production of foreign-peptide:self MHC complexes. TLR signaling also induces expression of CCR7 by dendritic cells, which directs their migration to T-cell zones of secondary lymphoid organs, where they encounter and activate naive T cells.

Macrophages and B cells can also process particulate or soluble antigens from pathogens to be presented as peptide:MHC complexes to T cells. However, whereas antigen presentation to naive T cells is uniquely mediated by dendritic cells, antigen presentation by macrophages and B cells enables the latter two cell types to recruit the effector activities of previously activated antigen-specific T cells. For example, as discussed in Chapter 11, by presenting antigens of ingested pathogens, macrophages recruit help from IFN-γ-producing CD4 T cells to augment their intracellular killing of these pathogens. Presentation of antigens by B cells recruits help from T cells to stimulate antibody production and class switching, a topic discussed further in Chapter 10. In all three types of antigen-presenting cells, the expression of co-stimulatory molecules is activated in response to signals from receptors that also function in innate immunity to signal the presence of infectious agents.

Priming of naive T cells by pathogen-activated dendritic cells.

T-cell responses are initiated when a mature naive CD4 or CD8 T cell encounters an activated antigen-presenting cell displaying the appropriate peptide:MHC ligand. We will now describe the generation of effector T cells from naive T cells. The activation and differentiation of naive T cells, often called **priming**, is distinct from the later responses of effector T cells to antigen on their target cells, and from the responses of primed memory T cells to subsequent encounters with the same antigen. Priming of naive CD8 T cells generates cytotoxic T cells capable of directly killing pathogen-infected cells. CD4 cells develop into a diverse array of effector cell types depending on the nature of the signals they receive during priming. CD4 effector activity can also include cytotoxicity, but more frequently it involves the secretion of a set of cytokines, which direct target cells toward a more pathogen-specific response.

9-14 Cell-adhesion molecules mediate the initial interaction of naive T cells with antigen-presenting cells.

As they migrate through the cortical region of the lymph node, naive T cells bind transiently to each antigen-presenting cell that they encounter. Activated dendritic cells bind naive T cells very efficiently through interactions between LFA-1 and CD2 on the T cell and ICAM-1, ICAM-2, and CD58 on the dendritic cell (**Fig. 9.20**). Perhaps because of this synergy, the precise role of each adhesion molecule has been difficult to distinguish. People lacking LFA-1 can have normal T-cell responses, and this also seems to be true for genetically engineered mice lacking CD2, suggesting substantial redundancy in the function of these molecules.

The transient binding of naive T cells to antigen-presenting cells is crucial in providing time for a T cell to sample large numbers of MHC molecules for the presence of its cognate antigenic peptide. In those rare cases in which a naive T cell recognizes its peptide:MHC ligand, signaling through the T-cell receptor induces a conformational change in LFA-1 that greatly increases its affinity for ICAM-1 and ICAM-2. This conformational change is the same as that induced by signaling through CCR7 during the migration of naive T cells into a secondary lymphoid organ (see Section 9-6). The change in LFA-1 stabilizes the association between the antigen-specific T cell and the antigen-presenting cell (**Fig. 9.21**). The association can persist for several days, during which time the naive T cell proliferates and its progeny, which can also adhere to the antigen-presenting cell, differentiate into effector T cells.

Most encounters of T cells with antigen-presenting cells do not, however, result in the recognition of an antigen. In this case, the T cell must be able to separate efficiently from the antigen-presenting cell so that it can continue to migrate through the lymphoid tissue, eventually exiting to reenter the blood and continue circulating. Dissociation, like stable binding, may also involve signaling between the T cell and the antigen-presenting cells, but little is known of its mechanism.

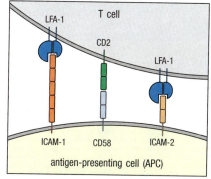

Fig. 9.20 Cell-surface molecules of the immunoglobulin superfamily are important in the interactions of lymphocytes with antigen-presenting cells. In the initial encounter of T cells with antigen-presenting cells, CD2 binding to CD58 on the antigen-presenting cell synergizes with LFA-1 binding to ICAM-1 and ICAM-2. LFA-1 is the $\alpha_L{:}\beta_2$ integrin heterodimer CD11a:CD18. ICAM-1 and ICAM-2 are also known as CD54 and CD102, respectively.

 MOVIE 9.4

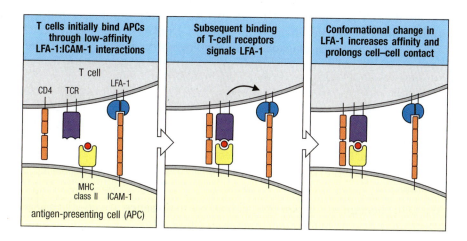

Fig. 9.21 Transient adhesive interactions between T cells and antigen-presenting cells are stabilized by specific antigen recognition. When a T cell binds to its specific ligand on an antigen-presenting cell, intracellular signaling through the T-cell receptor (TCR) induces a conformational change in LFA-1 that causes it to bind with higher affinity to ICAMs on the antigen-presenting cell. The T cell shown here is a CD4 T cell.

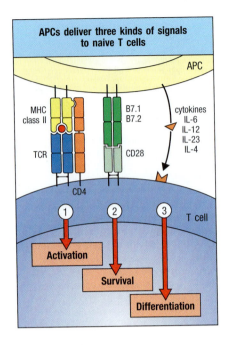

APCs deliver three kinds of signals to naive T cells

MOVIE 9.5

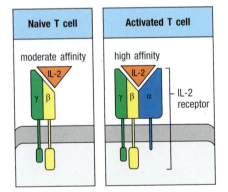

Fig. 9.23 High-affinity IL-2 receptors are three-chain structures that are present only on activated T cells. On resting T cells, the β and γ chains are expressed constitutively. They bind IL-2 with moderate affinity. Activation of T cells induces the synthesis of the α chain and the formation of the high-affinity heterotrimeric receptor. The β and γ chains show similarities in amino acid sequence to cell-surface receptors for growth hormone and prolactin, each of which also regulates cell growth and differentiation.

Fig. 9.22 Three kinds of signals are involved in activation of naive T cells by antigen-presenting cells. Binding of the foreign-peptide:self MHC complex by the T-cell receptor and, in this example, a CD4 co-receptor transmits a signal (arrow 1) to the T cell that antigen has been encountered. Effective activation of naive T cells requires a second signal (arrow 2), the co-stimulatory signal, to be delivered by the same antigen-presenting cell (APC). In this example, CD28 on the T cell encountering B7 molecules on the antigen-presenting cell delivers signal 2, whose net effect is the increased survival and proliferation of the T cell that has received signal 1. ICOS and various members of the TNF receptor family may also provide co-stimulatory signals. For CD4 T cells in particular, different pathways of differentiation produce subsets of effector T cells that carry out different effector responses, depending on the nature of a third signal (arrow 3) delivered by the antigen-presenting cell. Cytokines are commonly, but not exclusively, involved in directing this differentiation.

9-15 Antigen-presenting cells deliver multiple signals for the clonal expansion and differentiation of naive T cells.

When discussing the activation of naive T cells, it is useful to consider at least three different types of signals (Fig. 9.22). The first signal is generated from the interaction of a specific peptide:MHC complex with the T-cell receptor. Engagement of the T-cell receptor with its specific peptide antigen is essential for activating a naive T cell. However, even if the co-receptor—CD4 or CD8—is also ligated, this does not, on its own, stimulate the T cell to fully proliferate and differentiate into effector T cells. Expansion and differentiation of naive T cells involve at least two other kinds of signals: co-stimulatory signals that promote the survival and expansion of the T cells, and cytokines that direct T-cell differentiation into one of the different subsets of effector T cells. Additional signals, such as Notch ligands, can contribute to the effector differentiation of naive T cells, although these signals appear to be of lesser importance than those of lineage-specifying cytokines.

The best-characterized co-stimulatory molecules are the B7 molecules. These homodimeric members of the immunoglobulin superfamily are found exclusively on the surfaces of cells, such as dendritic cells, that stimulate naive T-cell proliferation (see Section 9-8). The receptor for B7 molecules on the T cell is CD28, a member of the immunoglobulin superfamily (see Section 7-21). Ligation of CD28 by B7 molecules is necessary for the optimal clonal expansion of naive T cells; targeted deficiency of B7 molecules or experimental blockade of the binding of B7 molecules to CD28 has been shown to inhibit T-cell responses.

9-16 CD28-dependent co-stimulation of activated T cells induces expression of interleukin-2 and the high-affinity IL-2 receptor.

Naive T cells are found as small resting cells with condensed chromatin and scanty cytoplasm, and they synthesize little RNA or protein. On activation, they reenter the cell cycle and divide rapidly to produce large numbers of progeny as they undergo antigen-driven differentiation. Unlike effector T cells, which can produce a diversity of cytokines depending on the mature effector phenotype, naive T cells primarily produce interleukin-2 (IL-2) when activated. Based on *in vitro* studies, IL-2 was long thought to be required for the proliferation of naive T cells. However, *in vivo* studies indicate that while IL-2 can augment T-cell proliferation and survival, in many cases it is dispensable and other functions of IL-2 might be more important. In particular, IL-2 is essential for the maintenance of regulatory T cells, which do not produce their own IL-2 when activated. IL-2 also appears to affect the balance of effector and memory T cells that develop in a primary response to antigen, as will be discussed in Chapter 11.

The initial encounter with specific antigen in the presence of a co-stimulatory signal triggers entry of the T cell into the G_1 phase of the cell cycle; at the same time, it also induces the synthesis of IL-2 along with the α chain of the IL-2 receptor (also known as CD25). The IL-2 receptor is composed of three chains: α, β, and γ (Fig. 9.23). Prior to activation, naive T cells express a form of the receptor composed only of the β and γ chains, which has only moderate affinity for IL-2 binding. Within hours of activation, naive T cells upregulate the expression of CD25. Association of CD25 with the β and γ heterodimer creates a receptor with a much higher affinity for IL-2, allowing the T cell to respond to very low concentrations of IL-2.

In contrast to naive T cells, regulatory T, or T_{reg}, cells constitutively express CD25, and thus the high-affinity, trimeric form of the IL-2 receptor (see Fig. 9.23). As is discussed later (see Section 9-23), it is thought that by expressing the high-affinity form of the IL-2 receptor, T_{reg} cells can outcompete T cells that express only the low-affinity form of the receptor for binding of the limited quantities of IL-2 that are available early in the response to antigen. In this way, T_{reg} cells act as a 'sink' for IL-2 to limit its availability to other cells. However, once activated naive T cells have upregulated CD25, they form the high-affinity receptor and compete with T_{reg} cells for binding of IL-2. The binding of IL-2 by these activated naive T cells triggers signaling that supports their activation and differentiation, and can enhance their proliferation (Fig. 9.24). T cells activated in this way can divide up to four times a day for several days, allowing one precursor cell to give rise to thousands of clonal progeny that all bear the same antigenic receptor.

Antigen recognition by the T-cell receptor induces the synthesis or activation of the transcription factors NFAT, AP-1, and NFκB, which bind to the promoter region of the IL-2 gene in naive T cells to activate its transcription (see Sections 7-14 and 7-16). Co-stimulation through CD28 contributes to the production of IL-2 in at least three ways. First, CD28 signaling activates PI 3-kinase, which increases production of the AP-1 and NFκB transcription factors, thereby increasing the transcription of IL-2 mRNA. However, the mRNAs of many cytokines, including IL-2, are very short-lived because of an 'instability' sequence (AUUUAUUUA) in the 3' untranslated region. CD28 signaling prolongs the lifetime of an IL-2 mRNA molecule by inducing the expression of proteins that block the activity of the instability sequence, resulting in increased translation and more IL-2 protein. Finally, PI 3-kinase helps activate the protein kinase Akt (see Section 7-17), which generally promotes cell growth and survival, increasing the total production of IL-2 by activated T cells.

9-17 Additional co-stimulatory pathways are involved in T-cell activation.

Once a naive T cell is activated, it expresses a number of proteins in addition to CD28 that contribute to sustaining or modifying the co-stimulatory signal. These other co-stimulatory receptors generally belong to either the CD28 or the TNF receptor family.

CD28-related proteins are expressed on activated T cells and modify the co-stimulatory signal as the T-cell response develops. One such protein is the inducible co-stimulator (ICOS), which binds a ligand known as ICOSL (ICOS ligand, or B7-H2), a structural relative of B7.1 and B7.2. ICOSL is produced on activated dendritic cells, monocytes, and B cells. Although ICOS resembles CD28 in driving T-cell proliferation, it does not induce IL-2 but seems to regulate the expression of other cytokines, such as IL-4 and IFN-γ, made by CD4 T-cell subsets. ICOS is particularly important for enabling CD4 T cells to function as helper cells for B-cell responses such as isotype switching. ICOS is expressed on T cells in germinal centers within lymphoid follicles, and mice lacking ICOS fail to develop germinal centers and have severely diminished antibody responses.

Another receptor for B7 molecules is CTLA-4 (CD152), which is related in sequence to CD28. CTLA-4 binds B7 molecules about 20 times more avidly

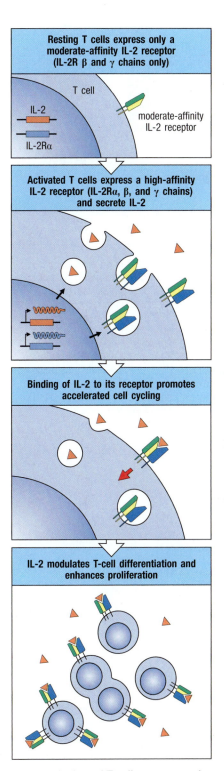

Fig. 9.24 Activated T cells secrete and respond to IL-2. Activation of naive T cells induces the expression and secretion of IL-2 and the expression of high-affinity IL-2 receptors. IL-2 binds to the high-affinity IL-2 receptors to enhance T-cell growth and differentiation.

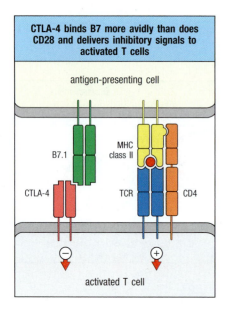

CTLA-4 binds B7 more avidly than does CD28 and delivers inhibitory signals to activated T cells

Fig. 9.25 CTLA-4 is an inhibitory receptor for B7 molecules. Naive T cells express CD28, which delivers a co-stimulatory signal on binding B7 molecules (see Fig. 9.22), thereby driving the survival and expansion of the T cells. Activated T cells express increased levels of CTLA-4 (CD152), which has a higher affinity than CD28 for B7 molecules and thus binds most or all of the B7 molecules. CTLA-4 thereby serves to regulate the proliferative phase of the T-cell response.

than CD28, but its effect is to inhibit, rather than activate, the T cell (Fig. 9.25). CTLA-4 does not contain an ITIM motif, and it is suggested to inhibit T-cell activation by competing with CD28 for interaction with B7 molecules expressed by antigen-presenting cells. Activation of naive T cells induces the surface expression of CTLA-4, making activated T cells less sensitive than naive T cells to stimulation by the antigen-presenting cell, thereby restricting IL-2 production. Thus, binding of CTLA-4 to B7 molecules is essential for limiting the proliferative response of activated T cells to antigen and B7. This was confirmed by producing mice with a disrupted CTLA-4 gene; such mice develop a fatal disorder characterized by a massive overgrowth of lymphocytes. Antibodies that block CTLA-4 from binding to B7 molecules markedly increase T cell-dependent immune responses.

Several TNF-family molecules also deliver co-stimulatory signals. These all seem to function by activating NFκB through a TRAF-dependent pathway (see Section 7-23). The binding of CD70 on dendritic cells to its constitutively expressed CD27 receptor on naive T cells delivers a potent co-stimulatory signal to T cells early in the activation process. The receptor CD40 on dendritic cells binds to CD40 ligand expressed on T cells, initiating two-way signaling that transmits activating signals to the T cell and also induces the dendritic cell to express increased B7, thus stimulating further T-cell proliferation. The role of the CD40–CD40 ligand pair in sustaining a T-cell response is demonstrated in mice lacking CD40 ligand; when these mice are immunized, the clonal expansion of responding T cells is curtailed at an early stage. The T-cell molecule 4-1BB (CD137) and its ligand 4-1BBL, which is expressed on activated dendritic cells, macrophages, and B cells, make up another pair of TNF-family co-stimulators. The effects of this interaction are also bidirectional, with both the T cell and the antigen-presenting cell receiving activating signals; this type of interaction is sometimes referred to as the T-cell:antigen-presenting cell dialog. Another co-stimulatory receptor and its ligand, OX40 and OX40L, are expressed on activated T cells and dendritic cells, respectively. Mice deficient in OX40 show reduced CD4 T-cell proliferation in response to viral infection, indicating that OX40 has a role in sustaining ongoing T-cell responses by enhancing T-cell survival and proliferation.

9-18 Proliferating T cells differentiate into effector T cells that do not require co-stimulation to act.

During the 4–5 days of rapid cell division that follow naive T-cell activation, T cells differentiate into effector T cells that acquire the ability to synthesize molecules required for their specialized helper or cytotoxic functions when they re-encounter their specific antigen. Effector T cells undergo additional changes that distinguish them from naive T cells. One of the most important is in their activation requirements: once a T cell has differentiated into an effector cell, encounter with its specific antigen results in immune attack without the need for co-stimulation (Fig. 9.26). This distinction is particularly easy to understand for CD8 cytotoxic T cells, which must be able to act on any cell infected with a virus, whether or not the infected cell can express co-stimulatory molecules. However, this feature is also important for the effector function of CD4 cells, as effector CD4 T cells must be able to activate B cells and macrophages that have taken up antigen even if these cells are not initially expressing co-stimulatory molecules.

Changes are also seen in the cell-adhesion molecules and receptors expressed by effector T cells. They lose cell-surface L-selectin and therefore cease to recirculate through lymph nodes. Instead, they express glycans that serve as ligands for P- and E-selectins (for example, P-selectin glycoprotein-1, or PSGL-1), which are upregulated on inflamed vascular endothelial cells and

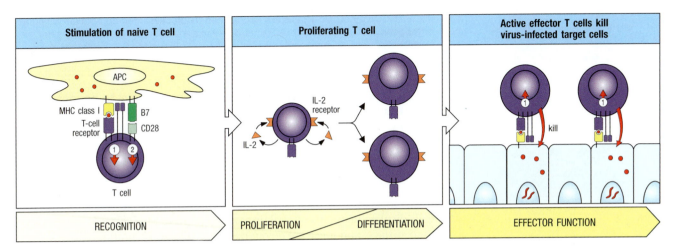

Fig. 9.26 Effector T cells can respond to their target cells without co-stimulation. A naive T cell that recognizes antigen on the surface of an antigen-presenting cell and receives the required two signals (arrows 1 and 2, left panel) becomes activated, and it both secretes and responds to IL-2. IL-2 signaling enhances clonal expansion and contributes to the differentiation of the T cells to effector cell status (central panel). Once the cells have differentiated into effector T cells, any encounter with specific antigen triggers their effector actions without the need for co-stimulation. Thus, as illustrated here, a cytotoxic T cell can kill any virus-infected target cells, including those that do not express co-stimulatory molecules.

allow effector T cells to roll on the blood vessels at sites of inflammation. They also express higher levels of LFA-1 and CD2 than do naive T cells, as well as the integrin VLA-4, which allows them to bind to vascular endothelium bearing the adhesion molecule VCAM-1, which is also expressed on the inflamed endothelium. This allows effector T cells to exit the bloodstream and enter sites of infection, where they orchestrate the local immune response. These changes in the T-cell surface are summarized in **Fig. 9.27**, and will be discussed further in Chapter 11.

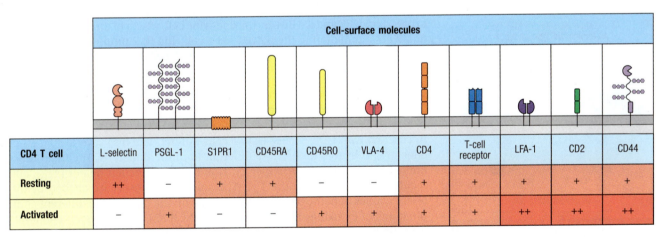

Cell-surface molecules											
CD4 T cell	L-selectin	PSGL-1	S1PR1	CD45RA	CD45RO	VLA-4	CD4	T-cell receptor	LFA-1	CD2	CD44
Resting	++	–	+	+	–	–	+	+	+	+	+
Activated	–	+	–	–	+	+	+	+	++	++	++

Fig. 9.27 Activation of T cells changes the expression of several cell-surface molecules. The example here is a CD4 T cell. Resting naive T cells express L-selectin, through which they home to lymph nodes, but express relatively low levels of other adhesion molecules such as CD2 and LFA-1. Upon activation, expression of L-selectin ceases and, instead, expression of ligands for P- and E-selectins are induced (e.g., PSGL-1), which allow the activated T cells to roll on P- and E-selectins expressed on endothelium at sites of inflammation. Increased amounts of the integrin LFA-1 are also produced, which is activated to bind its ligands, ICAM-1 and ICAM-2. A newly expressed integrin called VLA-4, which allows T cells to arrest on inflamed vascular endothelium, ensures that activated T cells enter peripheral tissues at sites where they are likely to encounter infection. Activated T cells also have on their surface a higher density of the adhesion molecule CD2, increasing the avidity of their interaction with potential target cells, as well as a higher density of the adhesion molecule CD44. By alternative splicing of the RNA transcript of the CD45 gene, a change occurs in the isoform of CD45 that is expressed, with activated T cells expressing the CD45RO isoform, which associates with the T-cell receptor and CD4. This change makes the T cell more sensitive to stimulation by low concentrations of peptide:MHC complexes. Finally, the sphingosine 1-phosphate receptor 1(S1PR1) is expressed by resting naive T cells, allowing the egress from lymphoid tissues of cells that do not become activated (see Fig. 9.11). Downregulation of S1PR1 for several days after activation prevents T-cell egress during the period of proliferation and differentiation. After several days, it is expressed again, allowing effector cells to exit from the lymphoid tissues.

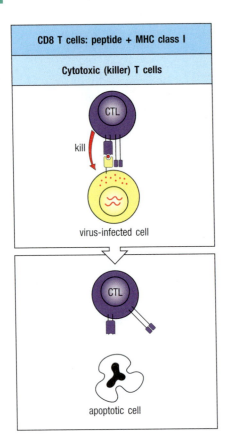

CD8 T cells: peptide + MHC class I
Cytotoxic (killer) T cells

Fig. 9.28 CD8 cytotoxic T cells are specialized to kill cells infected with intracellular pathogens. CD8 cytotoxic cells kill target cells that display at their cell surface peptide fragments of cytosolic pathogens, most notably viruses, bound to MHC class I molecules.

9-19 CD8 T cells can be activated in different ways to become cytotoxic effector cells.

Naive T cells fall into two large classes, of which one carries the co-receptor CD8 on its surface and the other bears the co-receptor CD4. CD8 T cells all differentiate into CD8 cytotoxic T cells (sometimes called cytotoxic lymphocytes, or CTLs), which kill their target cells (Fig. 9.28). They are important in the defense against intracellular pathogens, especially viruses. Virus-infected cells display fragments of viral proteins as peptide:MHC class I complexes on their surface, and these are recognized by cytotoxic T lymphocytes.

Perhaps because the effector actions of these cells are so destructive, naive CD8 T cells require more co-stimulatory activity to drive them to become activated effector cells than do naive CD4 T cells. This requirement can be met in two ways. The simplest is priming by activated dendritic cells, which have high intrinsic co-stimulatory activity. In some viral infections, dendritic cells become sufficiently activated to directly induce CD8 T cells to produce the IL-2 required for their differentiation into cytotoxic effector cells, without help from CD4 T cells. This property of dendritic cells has been exploited to generate cytotoxic T-cell responses against tumors, as we will see in Chapter 16.

In the majority of viral infections, however, CD8 T-cell activation requires additional help, which is provided by CD4 effector T cells. CD4 T cells that recognize related antigens presented by the antigen-presenting cell can amplify the activation of naive CD8 T cells by further activating the antigen-presenting cell (Fig. 9.29). B7 expressed by the dendritic cell first activates the CD4 T cells to express IL-2 and CD40 ligand (see Sections 9-16 and 9-17). CD40 ligand binds CD40 on the dendritic cell, delivering an additional signal that increases the expression of B7 and 4-1BBL by the dendritic cell, which in turn provides additional co-stimulation to the naive CD8 T cell. The IL-2 produced by activated CD4 T cells also acts to promote effector CD8 T-cell differentiation.

9-20 CD4 T cells differentiate into several subsets of functionally different effector cells.

In contrast with CD8 T cells, CD4 T cells differentiate into several subsets of effector T cells that orchestrate different immune functions. The main functional subsets are T_H1 (T helper 1), T_H2, T_H17, T follicular helper (T_{FH}), and regulatory T (T_{reg}) cells. The T_H1, T_H2, and T_H17 subsets are elicited by different classes of pathogens and are defined on the basis of the different combinations of cytokines that they secrete (Fig. 9.30). These subsets cooperate with different innate cells of the myelomonocytic series and with innate lymphoid cells (ILCs) to form integrated 'immune modules' specialized for the clearance of the different classes of pathogens (see Fig. 3.37). One or the other of these subsets will typically become predominant as an immune response progresses, especially in persistent infections, autoimmunity, or allergies. As

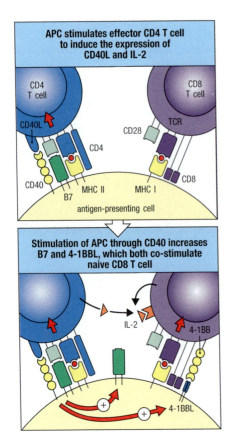

APC stimulates effector CD4 T cell to induce the expression of CD40L and IL-2

Stimulation of APC through CD40 increases B7 and 4-1BBL, which both co-stimulate naive CD8 T cell

Fig. 9.29 Most CD8 T-cell responses require CD4 T cells. CD8 T cells recognizing antigen on weakly co-stimulatory cells may become activated only in the presence of CD4 T cells interacting with the same antigen-presenting cell (APC). This happens mainly by an effector CD4 T cell recognizing antigen on the antigen-presenting cell and being triggered to induce increased levels of co-stimulatory activity on the antigen-presenting cell. The CD4 T cells also produce abundant IL-2 and thus help drive CD8 T-cell proliferation. This may in turn activate the CD8 T cell to make its own IL-2.

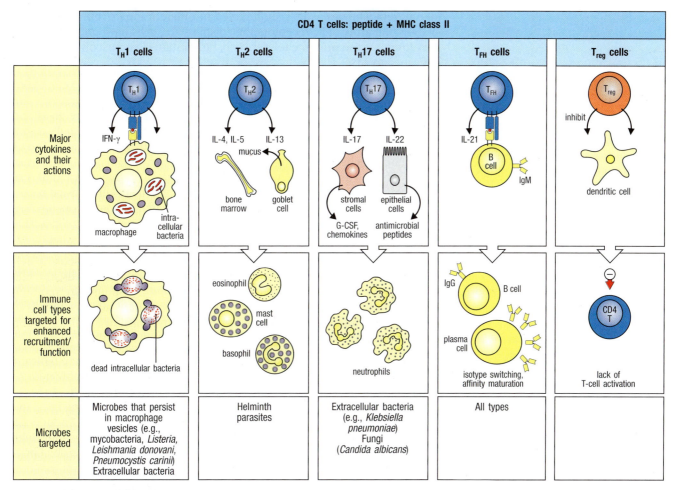

Fig. 9.30 Subsets of CD4 effector T cells are specialized to provide help to different target cells for the eradication of different classes of pathogens. Unlike CD8 T cells, which act directly on infected target cells to eliminate pathogens, CD4 T cells typically enhance the effector functions of other cells that eradicate pathogens—whether cells of the innate immune system, or, in the case of T_{FH} cells, antigen-specific B cells. T_H1 cells (first panels) produce cytokines, such as IFN-γ, which activate macrophages, enabling them to destroy intracellular microorganisms more efficiently. T_H2 cells (second panels) produce cytokines that recruit and activate eosinophils (IL-5) and mast cells and basophils (IL-4), and promote enhanced barrier immunity at mucosal surfaces (IL-13) to eradicate helminths. T_H17 cells (third panels) secrete IL-17-family cytokines that induce local epithelial and stromal cells to produce chemokines that recruit neutrophils to sites of infection. T_H17 cells also produce IL-22, which along with IL-17 can activate epithelial cells at the barrier site to produce antimicrobial peptides that kill bacteria. T_{FH} cells (fourth panels) form cognate interactions with naive B cells through linked recognition of antigen and traffic to B-cell follicles, where they promote the germinal center response. T_{FH} cells produce cytokines characteristic of other subsets and participate in type 1, 2, and 3 responses that are recruited against different types of pathogens. T_{FH} cells producing IFN-γ activate B cells to produce strongly opsonizing antibodies belonging to certain IgG subclasses (IgG1 and IgG3 in humans, and their homologs, IgG2a and IgG2b, in the mouse) in type 1 responses. Those T_{FH} cells producing IL-4 drive B cells to differentiate and produce immunoglobulin IgE, which arms mast cells and basophils for granule release in type 2 responses. T_{FH} cells that produce IL-17 appear to be important for generating opsonizing antibodies directed against extracellular pathogens in the context of type 3/T_H17 immunity. Regulatory T cells (right panels) generally suppress T-cell and innate immune cell activity and help prevent the development of autoimmunity during immune responses.

we shall discuss further in Chapter 11, the functional features of these T-cell subsets parallel in many ways those of innate lymphoid cells (ILCs), which, although lacking antigenic receptors, produce many of the same patterns of effector cytokines or cytotoxins.

The first CD4 T-cell subsets to be distinguished were the T_H1 and T_H2 subsets, hence their names. T_H1 cells are characterized by the production of IFN-γ, whereas T_H2 cells are characterized by the production of IL-4, IL-5, and IL-13. T_H17 cells are so named because they produce the cytokines IL-17A and IL-17F; they also produce IL-22. T_{FH} cells develop in concert with T_H1, T_H2, or T_H17 cells to help B cells generate class-switched immunoglobulins of different isotypes, which are targeted to different innate immune effector cells by the array of Fc receptors they display. T_{reg} cells have immunoregulatory function and promote tolerance to, rather than clearance of, the antigens they recognize.

T_H1 cells help to eradicate infections by microbes that can survive or replicate within macrophages. Examples include certain viruses, protozoans, and intracellular bacteria, such as the mycobacteria that cause tuberculosis and leprosy. These bacteria are phagocytosed by macrophages in the usual way but can evade the intracellular killing mechanisms described in Chapter 3. If a T_H1 cell recognizes bacterial antigens displayed on the surface of an infected macrophage, it will activate the macrophage further through the release of IFN-γ, which enhances the macrophage's microbicidal activity to kill ingested bacteria. Type 1 responses also promote B-cell class switching that favors production of opsonizing IgG antibodies, such as IgG2a in mouse. We shall describe the macrophage-activating functions of T_H1 cells in more detail in Chapter 11.

T_H2 cells help to control infections by extracellular parasites, particularly helminths, by promoting responses mediated by eosinophils, mast cells, and IgE. In particular, cytokines produced as part of a type 2 response are required for class switching of B cells to produce IgE, the primary role of which is to fight parasitic infections. IgE is also the antibody responsible for allergies and asthma, making T_H2 differentiation of additional medical interest.

The third major effector subset of CD4 T cells is T_H17. T_H17 cells are typically induced in response to extracellular bacteria and fungi, and amplify neutrophilic responses that help to clear such pathogens (see Fig. 9.30). T_H17, or type 3, responses also promote B-cell class switching to opsonizing IgG2 and IgG3 antibodies. Cytokines produced by T_H17 cells, including IL-17 and IL-22, are also important in activating barrier epithelial cells in the gastrointestinal, respiratory, and urogenital tracts and the skin, to produce antimicrobial peptides that resist microbial invasion.

In contrast to T_H1, T_H2, or T_H17 cells, T_{FH} cells contribute to the eradication of most classes of pathogens through their unique role in providing help to B cells to promote germinal center responses—irrespective of the pattern of immune response with which they are associated. Thus, T_{FH} cells are elicited in the context of either type 1, type 2, or type 3 responses, where they play a central role in the development of distinct patterns of class-switched antibodies. T_{FH} cells are identified mainly by their expression of certain markers, such as CXCR5 and PD-1, and their localization to lymphoid follicles.

Prior to the discovery of T_{FH} cells, a point of controversy had been the role of CD4 T effector subsets in providing B-cell help. Although it was originally implied that this was primarily the function of T_H2 cells, it is now thought that the T_{FH} cell, rather than T_H1, T_H2, or T_H17 cells, is the primary effector T cell that provides B-cell help for high-affinity antibody production in lymphoid follicles. Nevertheless, T_{FH} cells develop as a component of type 1, 2, or 3 responses, and share production of some of the same lineage-defining cytokines of T_H1, T_H2, and T_H17 cells to drive the differentiation of naive B cells to alternative patterns of isotype switching. This explains how, in the course of an infection, B cells can receive help to switch to IgE through the presence of 'T_H2' cytokines, or switch to other isotypes such as IgG2a through the presence of 'T_H1' cytokines. Thus, while the developmental relationship of T_{FH} to other CD4 subsets is still a matter of active research, T_{FH} cells appear to represent a distinct branch of effector T cells that remain within the lymphoid tissues and are specialized for providing B-cell help. We will return to the helper functions of T_{FH} cells in more detail in Chapters 10 and 11.

All the effector T cells described above are involved in activating their target cells to make responses that help clear pathogens from the body. Other CD4 T cells have a different function. These are called regulatory T cells, or T_{reg}

cells, because their function is to suppress T-cell responses rather than activate them. Thus, T_{reg} cells are involved in limiting the immune response and preventing autoimmunity. Two main subsets of regulatory T cells are currently recognized. One subset becomes committed to a regulatory fate while still in the thymus, and is known as natural, or thymically derived, T_{reg} cells (nT_{reg} and tT_{reg}, respectively; see Section 8-26). The other subset of T_{reg} cells differentiates from naive CD4 T cells in the periphery under the influence of particular environmental conditions. This group is known as induced, or peripherally derived, T_{reg} cells (iT_{reg} and pT_{reg}, respectively). These cells will be discussed further in Section 9-23.

9-21 Cytokines induce the differentiation of naive CD4 T cells down distinct effector pathways.

Having briefly noted the types and functions of CD4 T-cell subsets, we will now consider how they are derived from naive T cells. The fate of the progeny of a naive CD4 T cell is largely defined during the initial priming period and is regulated by signals provided by the local environment, whether by the priming antigen-presenting cell or other innate immune cells that have been activated by a pathogen. As noted previously, the principal determinants of the developmental fate of naive CD4 T cells are the combination and balance of lineage-specifying cytokines, which are integrated with TCR and co-stimulatory signaling during priming. The five main subsets into which naive CD4 T cells may develop—T_H1, T_H2, T_H17, T_{FH}, and induced regulatory T cells (iT_{reg} cells)—are associated with distinct signals that induce their formation, different transcription factors that drive their differentiation, and unique cytokines and surface markers that define their identity (**Figs. 9.31** and **9.32**).

T_H1 development is induced when there is a predominance of the cytokines IFN-γ and IL-12 during the early stages of naive T-cell activation. As described in Section 3-16, many key cytokines, including IFN-γ and IL-12, stimulate the JAK–STAT intracellular signaling pathway, resulting in the activation of specific gene networks. Different members of the JAK and STAT families are activated by different cytokines. Each of the effector pathways is dependent on a distinct pattern of STAT activation downstream of the lineage-specifying cytokines to program a unique transcription factor network that defines the gene-expression profile of mature effector T cells (see Fig. 9.32). For T_H1 development, STAT1 and STAT4 are critical and are sequentially activated by interferons (type 1—IFN-α and IFN-β; or type 2—IFN-γ) and IL-12, respectively, which are produced by innate immune cells early during infection. Activated group 1 ILCs, such as NK cells, may also be an important source of IFN-γ. Finally, T_H1 cells themselves may provide IFN-γ, thus reinforcing the signal for the differentiation of more T_H1 cells through a positive feedback loop.

Fig. 9.31 Cytokines are the principal determinants of alternative programs of CD4 T-cell effector differentiation. Antigen-presenting cells, principally dendritic cells, as well as other innate immune cells can provide various cytokines that induce the development of naive CD4 T cells into distinct subsets. The environmental conditions, such as the exposure to various pathogens, determine which cytokines innate sensor cells will produce. T_H1 cells differentiate in response to sequential IFN-γ and IL-12 signaling, whereas T_H2 cells differentiate in response to IL-4. IL-6 produced by dendritic cells acts with transforming growth factor-β (TGF-β) to induce differentiation of T_H17 cells, which upregulate expression of the IL-23 receptor and become responsive to IL-23. T_{FH} cells also require IL-6 for their development, although it is not currently understood what additional signals might induce their differentiation from naive precursors. When pathogens are absent, the presence of TGF-β and IL-2, and the lack of IL-6, favor the development of induced T_{reg} cells.

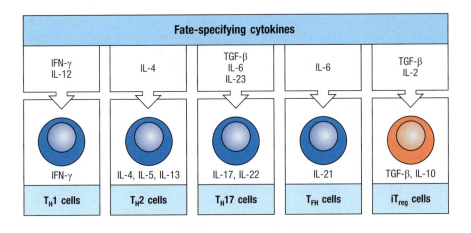

Fate-specifying cytokines				
IFN-γ IL-12	IL-4	TGF-β IL-6 IL-23	IL-6	TGF-β IL-2
IFN-γ	IL-4, IL-5, IL-13	IL-17, IL-22	IL-21	TGF-β, IL-10
T_H1 cells	T_H2 cells	T_H17 cells	T_{FH} cells	iT_{reg} cells

Fig. 9.32 Different members of the STAT family of transcription factors act immediately downstream of cytokines that determine CD4 T-cell subset development. With the exception of TGF-β, which participates in both T_H17 and iT_{reg} development, each of the cytokines that specify the development of distinct effector cells activates different members of the STAT family of transcription factors. T_H1 cell differentiation is dependent on sequential activation of STAT1 and STAT4 by binding of IFN-γ and IL-12 to their respective receptors on antigen-activated naive CD4 T cells. Both of these STAT factors participate in the induction of T-bet expression, which then cooperates with the STATs to program T_H1 differentiation. T_H2 cell differentiation is dependent on STAT6 activation downstream of IL-4 receptor signaling. STAT6 acts to increase GATA3 expression, which cooperates with STAT6 to program T_H2 differentiation. IL-6 activates STAT3, which, in concert with TGF-β, participates in the induction of RORγt expression and T_H17 differentiation. IL-23, which acts later in T_H17 differentiation, also activates STAT3 to sustain and amplify the T_H17 program. The programming of T_{FH} cell differentiation by STAT factors is not fully understood, although STAT3 actions upstream of Bcl-6 expression are essential. Activation of STAT5 by IL-2 is important in iT_{reg} differentiation and acts upstream of FoxP3 expression.

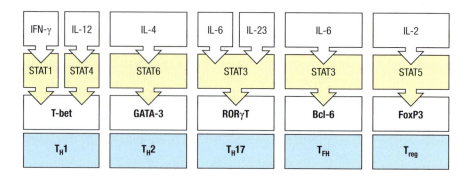

The activation of STAT1 by interferon induces in activated naive CD4 T cells induces the expression of another transcription factor, T-bet, which switches on the genes for IFN-γ and the inducible component of the IL-12 receptor, IL-12Rβ2 (the other component of the receptor, IL-12Rβ1, is already expressed on naive T cells). These T cells are now committed to becoming T_H1 cells, and can be further activated by IL-12 produced by dendritic cells and macrophages to induce STAT4 signaling. STAT4 further upregulates T-bet expression and completes T_H1 programming. Due to its central role in programming T_H1 development, T-bet is sometimes referred to as a 'master regulator' of T_H1 cell differentiation.

T_H2 development requires IL-4. When an antigen-activated naive T cell encounters IL-4, its receptor activates STAT6, which promotes expression of the transcription factor GATA3. GATA3 is a powerful activator of the genes encoding several cytokines produced by T_H2 cells, such as IL-4 and IL-13. GATA3 also induces its own expression, thereby stabilizing T_H2 differentiation via cell-intrinsic positive feedback. The initial source of IL-4 that triggers a T_H2 response has long been debated. Eosinophils, basophils, and mast cells are each attractive possibilities because they can produce abundant IL-4 when activated by chitin, a polysaccharide present in helminth parasites, as well as in insects and crustaceans, which induces T_H2 responses. In mice treated with chitin, eosinophils and basophils are recruited into tissues and are activated to produce IL-4. In humans, group 2 ILCs can also produce IL-4, suggesting that these cells might contribute to T_H2 differentiation, although this is unproven. Clearly, there are several innate immune cells that might contribute IL-4 for T_H2 development, and the cellular source might differ, contingent on the inciting antigen. Similar to the positive feedback for T_H1 cell development provided by IFN-γ produced by activated T_H1 cells, IL-4 produced by activated T_H2 cells may amplify T_H2 development from naive T-cell precursors.

T_H17 cells arise when the cytokines IL-6 and transforming growth factor (TGF)-β predominate during naive CD4 T-cell activation (see Figs. 9.31 and 9.32). Development of T_H17 cells requires the actions of STAT3, which is activated by IL-6 signaling. Developing T_H17 cells express the receptor for the cytokine IL-23, rather than the IL-12 receptor typical of T_H1 cells, and the expansion and further development of T_H17 effector activity seem to require IL-23, similar to the requirement for IL-12 in effective T_H1 responses (see Figs. 9.31 and 9.32). The signature transcription factor, or master regulator, of T_H17 cell differentiation is RORγt, a nuclear hormone receptor that is central to stabilizing the development of T_H17 cells. The source of IL-6 and TGF-β required for T_H17 cell differentiation is primarily derived from innate immune cells activated by microbial products. Unlike T_H1 or T_H2 cells, T_H17 cells do not appear to directly induce further T_H17 cell development from naive CD4 T cells via positive feedback, as they do not produce IL-6. However, IL-17 produced by T_H17 cells appears to enhance IL-6 production by innate immune cells and provide an indirect mechanism for reinforcing T_H17 differentiation from naive precursors.

Induced regulatory T cells (iT_{reg} cells) differ from nT_{reg} cells in that they develop upon antigen recognition in secondary lymphoid tissues, and not the thymus. They develop when naive T cells are activated in the presence of the cytokine transforming growth factor-β (TGF-β) and in the absence of IL-6 and other pro-inflammatory cytokines. Thus, it is the presence or absence of IL-6 that determines whether TGF-β co-signaling leads to the development of immunosuppressive T_{reg} cells or of T_H17 cells, which promote inflammation and the generation of immunity (Fig. 9.33). The generation of IL-6 by innate immune cells is regulated by the presence or absence of pathogens, with pathogen products tending to stimulate its production. In the absence of pathogens, IL-6 production is low, favoring differentiation of the immunosuppressive T_{reg} cells and so preventing unwanted immune responses. Like nT_{reg} cells, iT_{reg} cells are distinguished by expression of the transcription factor FoxP3 and cell-surface CD25, and appear to be functionally equivalent to nT_{reg} cells. Both iT_{reg} and nT_{reg} cells themselves can produce TGF-β, as well as IL-10, which act in an inhibitory manner to suppress immune responses and inflammation, and may act to support further iT_{reg} differentiation.

T_{FH} cells, unlike the subsets described above, have not been produced efficiently *in vitro*, and so the requirements for their differentiation are not yet clearly established. IL-6 seems to be important for T_{FH} development, but much remains to be learned about the control of this subset. One transcription factor important for T_{FH} development is Bcl-6, which is required for the expression of CXCR5, the receptor for the chemokine CXCL13, which is produced by the stromal cells of the B-cell follicle. CXCR5 is essential for T_{FH} localization in follicles, and is not expressed by other effector T-cell subsets. T_{FH} cells also express ICOS, the ligand for which is expressed abundantly by B cells. ICOS seems crucial for the helper activity of T_{FH} cells, because mice lacking ICOS show a severe defect in T-cell-dependent antibody responses. In addition to production of low amounts of cytokines characteristic of the effector T-cell subsets with which they develop in parallel (for example, IFN-γ, IL-4, or IL-17), and which promote different patterns of B-cell class switching, T_{FH} cells produce high amounts of IL-21, a cytokine that supports the proliferation and differentiation of B cells into antibody-producing plasma cells.

9-22 CD4 T-cell subsets can cross-regulate each other's differentiation through the cytokines they produce.

The various subsets of effector CD4 T cells each have very different functions. For the immune response to efficiently control different types of pathogens it must orchestrate a coordinated effector response that is dominated by one of these subsets. A principal means of achieving this is through the distinct ensemble of cytokines that are produced by the different subsets. Importantly, some of these same cytokines also participate in positive and negative feedback loops that control the differentiation of effector T cells from naive precursors, thereby providing a mechanism to promote one pattern of effector response while suppressing others. For example, both IFN-γ (produced by T_H1 cells) and IL-4 (produced by T_H2 cells) potently inhibit T_H17 development, promoting T_H1 or T_H2 development, respectively (Fig. 9.34). Similarly, there is cross-regulation between T_H1 and T_H2 cells. IL-4 produced by T_H2 cells potently inhibits T_H1 development. Conversely, IFN-γ, a product of T_H1 cells, can inhibit the proliferation of T_H2 cells (see Fig. 9.34). TGF-β produced by T_{reg} cells inhibits the development of both T_H1 and T_H2 cells. In this way, cytokines produced by effector T cells reinforce the differentiation of their own kind from naive precursor cells.

T_H1 cells generate copious amounts of IFN-γ when they recognize antigen on a target cell, thus reinforcing the signal for the differentiation of more T_H1 cells through a positive feedback loop. In this way, recognition of a particular type

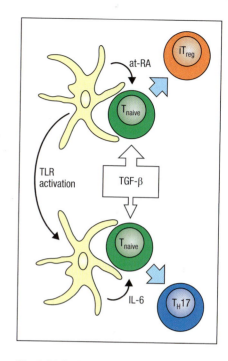

Fig. 9.33 A shared requirement for TGF-β in the differentiation of iT_{reg} and T_H17 cells provides a developmental link that reflects their complementary roles in promoting mutualism with the microbiota. A major site for the deployment of iT_{reg} and T_H17 cells is mucosal tissues, particularly the intestines, where the immune system must cope with an extraordinarily high density of microbial organisms that comprise the microbiota. While the microbiota provides its host with important metabolic functions, it also represents a potential threat as some of its constituents are opportunistic pathogens that can cause serious infections if they breach the mucosal barrier. As an adaptation to restrain untoward inflammation directed against the microbiota while retaining the capacity to mount a host-protective immune response should barrier breach occur, the developmental balance between iT_{reg} cells, which suppress inflammatory responses against the microbiota, and T_H17 cells, which promote host-protective inflammatory responses, is determined by the balance between production of the vitamin A metabolite all-*trans* retinoic acid (at-RA) and production of the pro-inflammatory cytokine IL-6 by mucosal dendritic cells. At homeostasis, antigens derived from the microbiota are presented by a specialized subset of resident dendritic cells that produce at-RA, but no IL-6. However, when antigens are recognized in the context of TLR-stimulating signals, at-RA production is suppressed in favor of IL-6, thereby favoring the development of T_H17 effector cells.

Fig. 9.34 The subsets of CD4 T cells each produce cytokines that can negatively regulate the development or effector activity of other subsets. Under homeostatic conditions, TGF-β produced by T_{reg} cells represses T_H1 and T_H2 responses in order to promote T_{reg} development. Under inflammatory conditions that favor IL-6 production, TGF-β production by T_{reg} cells similarly inhibits the activation of T_H1 or T_H2 responses (upper panels) in order to facilitate the development of T_H17 cells, which otherwise would be potently inhibited by IFN-γ or IL-4. Conversely, if signals are present to induce T_H1 or T_H2 cells, the cytokines IFN-γ or IL-4 produced by them can override the effect of IL-6 and inhibit T_H17 development (lower center panel). IFN-γ produced by T_H1 cells blocks the growth of T_H2 cells (right panels). On the other hand, IL-4 produced by T_H2 cells dominantly prevents T_H1 cell development in favor of T_H2 (left panels). Although not shown, all T-cell subsets can produce IL-10 under conditions of chronic antigen stimulation, which inhibits the production of IL-12, IL-4, and IL-23 by dendritic cells and macrophages, thereby suppressing the development and/or maintenance of T_H1, T_H2, and T_H17 cells.

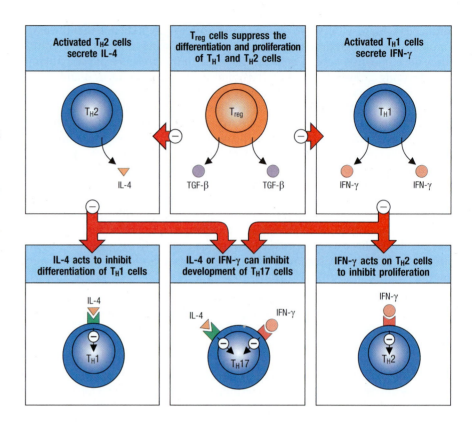

of pathogen by the innate immune system initiates a chain reaction that links the innate response to the adaptive immune response, which in turn amplifies the innate response. Thus, certain intracellular bacterial infections (for example, mycobacteria and *Listeria*) induce dendritic cells and macrophages to produce IL-12, favoring the emergence of T_H1 effector cells. T_H1 cells, in turn, promote enhanced macrophage activation that clears these intracellular pathogens.

The adverse consequences of inappropriate cross-regulation of effector T-cell responses by cytokines have been demonstrated in a number of infectious models in mice. Such studies reinforce the notion that induction of the appropriate effector CD4 T-cell subset is crucial for pathogen clearance, and show that subtle differences in CD4 T-cell responses can have a significant impact on the outcome of infection. One example of this is the murine model of infection by the protozoan parasite *Leishmania major*, which requires a T_H1 response and activation of macrophages for clearance. C57BL/6 mice produce T_H1 cells that protect the animal by activating infected macrophages to kill *L. major*. In BALB/c mice infected with *L. major*, however, CD4 T cells fail to differentiate into T_H1 cells; instead, they become T_H2 cells, which are unable to activate macrophages to inhibit *Leishmania* growth. This difference seems to result from a population of memory T cells that are specific for gut-derived antigens but cross-react with an antigen, LACK (*Leishmania* analog of the receptors of activated C kinase), expressed by the *Leishmania* parasite. These memory cells are present in both strains of mice, but for unknown reasons they produce IL-4 in BALB/c mice but not in C57BL/6 mice. In BALB/c mice, the small amount of IL-4 secreted by these memory cells during *Leishmania* infection drives new *Leishmania*-specific CD4 T cells to become T_H2 cells instead of T_H1 cells, leading to failure of pathogen elimination and death. The preferential development of T_H2 rather than T_H1 cells in BALB/c mice can be reversed if IL-4 is blocked early during infection by anti-IL-4 antibody, but this treatment is ineffective after a week or so of infection, demonstrating the crucial importance of cytokines early in developmental decisions made by naive T cells (Fig. 9.35).

9-23 Regulatory CD4 T cells are involved in controlling adaptive immune responses.

Regulatory T cells play a central role in preventing autoreactive immune responses and fall into different groups that are defined by their different developmental origins and functions. Natural T-regulatory (nT$_{reg}$) cells develop in the thymus (see Section 8-26) and are CD4-positive cells that constitutively express CD25 and high levels of the L-selectin receptor CD62L and of CTLA-4. Induced T$_{reg}$ (iT$_{reg}$) cells arise in the periphery from naive CD4 T cells and also express CD25 and CTLA-4 (see Section 9-20). Collectively, T$_{reg}$ cells represent about 5–10% of the CD4 T cells in circulation. A hallmark of both natural and induced T$_{reg}$ cells is expression of the transcription factor FoxP3, which, among other actions, interferes with the interaction between AP-1 and NFAT at the IL-2 gene promoter, preventing transcriptional activation of the gene and production of IL-2.

Natural T$_{reg}$ cells develop from potentially self-reactive T cells that express conventional α:β T-cell receptors and are selected in the thymus by high-affinity binding to MHC molecules containing self-peptides. It is not currently known whether they are activated to express their regulatory function in the periphery by the same self ligands that selected them in the thymus or by other self or non-self antigens. Multiple mechanisms appear to contribute to the ability of T$_{reg}$ cells to inhibit responses of other T cells, but principal among these are interactions with antigen-presenting cells that interfere with the capacity of antigen-presenting cells to provide activating signals. The expression of high levels of CTLA-4 on the surface of natural T$_{reg}$ cells is thought to permit them to compete for B7 expressed by antigen-presenting cells, thereby preventing adequate co-stimulation of naive T cells. Indeed, it has been proposed that CTLA4 expressed on T$_{reg}$ cells can physically remove B7 molecules from the surface of antigen-presenting cells, thereby depleting them of co-stimulatory activity. Similarly, by expressing CD25, and thus the high-affinity receptor for IL-2, and lacking the ability to produce IL-2, T$_{reg}$ cells appear to sequester IL-2 from naive T cells, which lack CD25 expression until fully activated.

Other functions of T$_{reg}$ cells are mediated by their production of immunosuppressive cytokines. TGF-β produced by T$_{reg}$ cells can inhibit T-cell proliferation (see Fig. 9.34). IL-10, which is produced by T$_{reg}$ cells late in an immune response, inhibits the expression of MHC molecules and co-stimulatory molecules by antigen-presenting cells. As a means of limiting the responses of effector T cells, IL-10 also inhibits the production of pro-inflammatory cytokines by antigen-presenting cells. For example, IL-10 potently inhibits the production of IL-12 and IL-23 by antigen-presenting cells and thus impairs their ability to promote the differentiation and maintenance of T$_H$1 and T$_H$17 cells, respectively. The critical role of T$_{reg}$ cells in immune regulation is highlighted by several autoimmune syndromes (described in Chapter 15) that are caused by deficiency in different aspects of T$_{reg}$ cell function.

Although they differentiate in secondary lymphoid tissues after export from the thymus, induced T$_{reg}$ cells also express FoxP3 and share most of the phenotypic and functional features of natural T$_{reg}$ cells. A major function of iT$_{reg}$ cells is the prevention of inflammatory immune responses to the commensal microbiota, particularly microbes resident in mucosal tissues such as the intestines. Here, iT$_{reg}$ cells appear to be the dominant source of IL-10, deficiency of which causes inflammatory bowel disease, an immune-mediated disease of the intestines characterized by chronic reactivity against antigens of the intestinal microbiota (see also Section 15-23). As will be discussed in more detail in Chapter 12, the differentiation of induced T$_{reg}$ cells in the intestines is favored by the presence of antigen-presenting cells that produce retinoic acid, which is derived from vitamin A. Retinoic acid produced by intestinal dendritic cells acts with TGF-β to induce T$_{reg}$ differentiation while suppressing the differentiation of T$_H$17 cells (see Fig. 9.33). The antagonistic balance of retinoic acid and IL-6 therefore

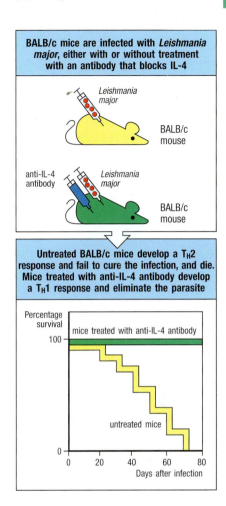

Fig. 9.35 The development of CD4 T-cell subsets can be manipulated by altering the cytokines acting during the early stages of infection. Elimination of infection with the intracellular protozoan parasite *Leishmania major* requires a T$_H$1 response, because IFN-γ is needed to activate the macrophages that provide protection. BALB/c mice are normally susceptible to *L. major* because they generate a T$_H$2 response to the pathogen. This is because they produce IL-4 early during infection and this induces naive T cells to develop into the T$_H$2 lineage (see the text). Treatment of BALB/c mice with neutralizing anti-IL-4 antibodies at the beginning of infection inhibits this IL-4 and prevents the diversion of naive T cells toward the T$_H$2 lineage; these mice develop a protective T$_H$1 response.

controls the differentiation of induced T_{reg} cells and T_H17 cells, respectively, in the intestinal mucosa-associated lymphoid tissues (MALT).

CD4 T cells that lack FoxP3 expression but produce immunosuppressive cytokines characteristic of T_{reg} cells have also been described. One such population, referred to as T_R1 cells, has been defined largely by their production of IL-10, but absence of expression of FoxP3. However, we now recognize that many different cells, including T_H1, T_H2, T_H17, and B cells, can produce IL-10 under certain circumstances, such as during chronic responses to persistent antigen. Therefore, it is uncertain whether T_R1 cells represent a distinct subset of T cells, and if so, whether they have unique functions in immune regulation.

Summary.

The crucial first step in adaptive immunity is the activation, or priming, of naive antigen-specific T cells by antigen-presenting cells within the lymphoid tissues through which they constantly circulate. The most distinctive feature of antigen-presenting cells is the expression of cell-surface co-stimulatory molecules, of which the B7 molecules are the most important. Naive T cells will respond to antigen only when the antigen-presenting cell presents both a specific antigen to the T-cell receptor and a B7 molecule to CD28 on the T cell. This dual requirement for both receptor ligation and co-stimulation by the same antigen-presenting cell helps to prevent naive T cells from responding to self antigens on tissue cells, which lack co-stimulatory activity.

Activation of naive T cells leads to their proliferation and differentiation into effector T cells, the critical event in most adaptive immune responses. Various combinations of cytokines regulate the type of effector T cell that develops in response to antigen. In turn, the cytokines present during primary T-cell activation are influenced by the innate immune system. Once an expanded clone of T cells acquires effector function, its progeny can act on any target cell that displays antigen on its surface. Effector T cells have a variety of functions. CD8 cytotoxic T cells recognize virus-infected cells and kill them. T_H1 effector cells promote the activation of macrophages to enhance their killing of intracellular pathogens. T_H2 cells promote mucosal barrier immunity against pathogens, such as helminths, requiring the effector activities of cells such as eosinophils and mast cells for their elimination. The elimination of certain types of bacteria and fungi is orchestrated by T_H17 cells, particularly at barrier sites, where they recruit neutrophils to sites of infection and promote the production of antimicrobial peptides by epithelial cells. T_{FH} cells are specialized for interactions with B cells and localization to the B-cell follicle and germinal centers, where they provide help for antibody production and isotype switching. Regulatory CD4 T-cell subsets restrain the immune response by preventing the activation of self-reactive naive T cells by antigen-presenting cells and producing inhibitory cytokines that limit the effector responses of other T-cell subsets.

General properties of effector T cells and their cytokines.

T-cell effector functions involve the interaction of an effector T cell with a target cell displaying specific antigen. Effector proteins expressed by the T cell, whether cell-associated (for example, CD40L) or secreted (for example, cytokines), are focused on the target by mechanisms that are activated by antigen recognition. The focusing mechanism is common to all types of effector T cells, whereas their effector actions depend on the type of effector T cell that is engaged.

9-24 Effector T-cell interactions with target cells are initiated by antigen-nonspecific cell-adhesion molecules.

Once an effector T cell has completed its differentiation in the lymphoid tissue, it must find target cells that are displaying the peptide:MHC complex that it recognizes. T_FH cells encounter their B-cell targets without leaving the lymphoid tissue. However, most other effector T cells emigrate from their site of activation in lymphoid tissues and enter the blood, either directly if primed by antigen in the spleen or via the efferent lymphatics and thoracic duct if primed in lymph nodes. Because of the cell-surface changes that have occurred during their differentiation, effector T cells can now migrate into tissues, particularly at sites of infection. They are guided to these sites by changes in the adhesion molecules expressed on the endothelium of the local blood vessels as a result of infection, and by local chemotactic factors, as will be discussed further in Chapter 11.

The initial binding of an effector T cell to its target, like that of a naive T cell to an antigen-presenting cell, is an antigen-nonspecific interaction mediated by LFA-1 and CD2. The levels of LFA-1 and of CD2 are two- to fourfold higher on effector T cells than on naive T cells, and so effector T cells can bind efficiently to target cells that have less ICAM and CD58 on their surface than do antigen-presenting cells. This interaction is transient unless recognition of antigen on the target cell by the T-cell receptor triggers an increase in the affinity of the T-cell's LFA-1 for its ligands. The T cell then binds more tightly to its target and remains bound long enough to release its effector molecules. Effector CD4 T cells, which activate macrophages or induce B cells to secrete antibody, have to switch on new genes and synthesize new proteins to carry out their effector actions and so must maintain contact with their targets for relatively long periods. Cytotoxic T cells, by contrast, can be observed under the microscope attaching to and dissociating from successive targets relatively rapidly as they kill them (Fig. 9.36). Killing of the target, or some local change in the T cell, allows the effector T cell to detach and address new targets. How CD4 effector T cells disengage from their antigen-negative targets is not known, although evidence suggests that CD4 binding to MHC class II molecules without engagement of the T-cell receptor provides a signal for the cell to detach.

9-25 An immunological synapse forms between effector T cells and their targets to regulate signaling and to direct the release of effector molecules.

When binding to their specific antigenic peptide:self MHC complexes or to self peptide:self MHC complexes, the T-cell receptors and their associated co-receptors cluster at the site of cell–cell contact, forming what is called the **supramolecular activation complex (SMAC)** or **the immunological synapse**. Other cell-surface molecules also cluster here. For example, the

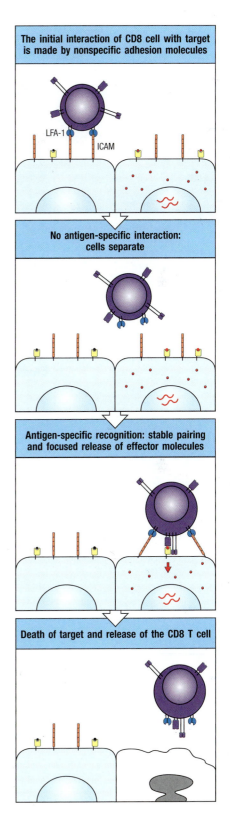

The initial interaction of CD8 cell with target is made by nonspecific adhesion molecules

LFA-1   ICAM

No antigen-specific interaction: cells separate

Antigen-specific recognition: stable pairing and focused release of effector molecules

Death of target and release of the CD8 T cell

Fig. 9.36 Interactions of T cells with their targets initially involve nonspecific adhesion molecules. The major initial interaction is between LFA-1 on the T cell, illustrated here as a cytotoxic CD8 T cell, and ICAM-1 or ICAM-2 on the target cell (top panel). This binding allows the T cell to remain in contact with the target cell and to scan its surface for the presence of specific peptide:MHC complexes. If the target cell does not carry the specific antigen, the T cell disengages (second panel) and can scan other potential targets until it finds the specific antigen (third panel). Signaling through the T-cell receptor increases the strength of the adhesive interactions, prolonging the contact between the two cells and stimulating the T cell to deliver its effector molecules. The T cell then disengages (bottom panel).

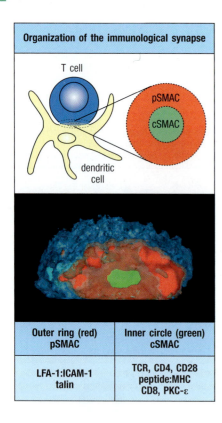

Organization of the immunological synapse

Outer ring (red) pSMAC	Inner circle (green) cSMAC
LFA-1:ICAM-1 talin	TCR, CD4, CD28 peptide:MHC CD8, PKC-ε

Fig. 9.37 The area of contact between an effector T cell and another cell forms an immunological synapse. A confocal fluorescence micrograph of the area of contact between a CD4 T cell and an antigen-presenting cell (APC) (as viewed through one of the cells) is shown. Proteins in the contact area between the T cell and the APC form a structure called the immunological synapse, also known as the supramolecular activation complex (SMAC), which is organized into two distinct regions: the outer, or peripheral SMAC (pSMAC), indicated by the red ring; and the inner, or central SMAC (cSMAC), indicated in bright green. The cSMAC is enriched in the T-cell receptor (TCR), CD4, CD8, CD28, CD2, and PKC-ε. The pSMAC is enriched for the integrin LFA-1 and the cytoskeletal protein talin. Photograph courtesy of A. Kupfer.

MOVIE 9.6

tight binding of LFA-1 to ICAM-1 induced by ligation of the T-cell receptor creates a molecular seal that surrounds the T-cell receptor and its co-receptor (Fig. 9.37). In some cases, the contact surface organizes into two zones: a central zone known as the central supramolecular activation complex (cSMAC) and an outer zone known as the peripheral supramolecular activation complex (pSMAC). The cSMAC contains most of the signaling proteins known to be important in T-cell activation. The pSMAC is notable mainly for the presence of the LFA-1 and the cytoskeletal protein talin, which connects LFA-1 to the actin cytoskeleton (see Section 3-18). The immunological synapse is not a static structure as implied by Fig. 9.37, but is quite dynamic. T-cell receptors move from the periphery into the cSMAC, where they undergo endocytosis through ubiquitin-mediated degradation involving the E3 ligase Cbl (see Section 7-5). Because T-cell receptors are being degraded in the cSMAC, signaling is actually weaker there than in the peripheral contact areas, where microclusters of T-cell receptors are being formed and are highly active (see Section 7-8).

Clustering of the T-cell receptors signals a reorientation of the cytoskeleton that polarizes the effector cell and focuses the release of effector molecules at the site of contact with the target cell. This is illustrated for a cytotoxic T cell in Fig. 9.38. An important intermediary in the effect of T-cell signaling on the cytoskeleton is the Wiskott–Aldrich syndrome protein (WASp), defects in which result in the inability of T cells to become polarized, among other effects, and cause an immune deficiency syndrome for which the protein is named (see Sections 7-19 and 13-6). Activation and recruitment of WASp by T-cell receptor signaling is mediated by the adaptor protein Vav (see Section 7-19). Polarization starts with the local reorganization of the cortical actin cytoskeleton at the site of contact; this in turn leads to the reorientation of the microtubule-organizing center (MTOC), the center from which the microtubule cytoskeleton is produced, and reorientation of the Golgi apparatus (GA), through which most proteins destined for secretion travel. In the cytotoxic T cell, the cytoskeletal reorientation focuses exocytosis of the preformed cytotoxic granules at the site of T-cell contact with its target cell. The polarization of a T cell also focuses the secretion of newly synthesized effector molecules induced by ligation of the T-cell receptor. For example, the secreted cytokine IL-4, which is the principal effector molecule of T_H2 cells, is confined and concentrated at the site of contact with the target cell.

Thus, the T-cell receptor controls the delivery of effector signals in three ways: it induces tight binding of effector cells to their target cells to create a narrow space in which effector molecules can be concentrated; it focuses delivery of effector molecules at the site of contact by inducing a reorientation of the secretory apparatus of the effector cell; and it triggers the synthesis and/or release of the effector molecules. All these mechanisms contribute to targeting actions of effector molecules onto the cell bearing specific antigen. Effector T-cell activity is thus highly selective for appropriate target cells, even though effector molecules themselves are not antigen-specific.

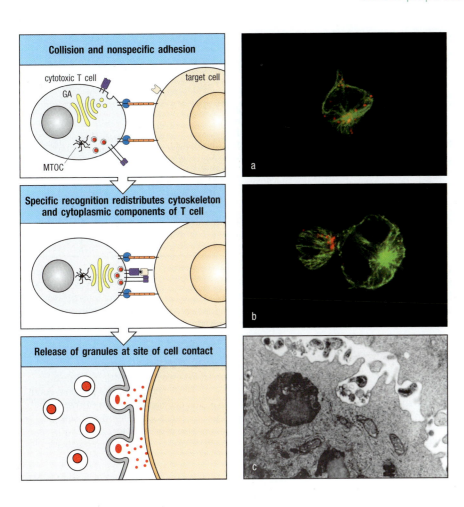

Fig. 9.38 The cellular polarization of T cells during specific antigen recognition allows effector molecules to be focused on the antigen-bearing target cell. The example illustrated here is a CD8 cytotoxic T cell. Cytotoxic T cells contain specialized lysosomes called cytotoxic granules (shown in red in the left panels), which contain cytotoxic proteins. Initial binding to a target cell through adhesion molecules does not have any effect on the location of the cytotoxic granules. Binding of the T-cell receptor causes the T cell to become polarized: reorganization within the cortical actin cytoskeleton at the site of contact aligns the microtubule-organizing center (MTOC), which in turn aligns the secretory apparatus, including the Golgi apparatus (GA), toward the target cell. Proteins stored in cytotoxic granules derived from the Golgi are then directed specifically onto the target cell. The photomicrograph in panel a shows an unbound, isolated cytotoxic T cell. The microtubule cytoskeleton is stained in green and the cytotoxic granules in red. Note how the granules are dispersed throughout the T cell. Panel b depicts a cytotoxic T cell bound to a (larger) target cell. The granules are now clustered at the site of cell–cell contact in the bound T cell. The electron micrograph in panel c shows the release of granules from a cytotoxic T cell. Panels a and b courtesy of G. Griffiths. Panel c courtesy of E. Podack.

9-26 The effector functions of T cells are determined by the array of effector molecules that they produce.

The effector molecules produced by effector T cells fall into two broad classes: cytotoxins, which are stored in specialized cytotoxic granules and released by CD8 cytotoxic T cells (see Fig. 9.38), and cytokines and related membrane-associated proteins, which are synthesized *de novo* by all effector T cells. Cytotoxins are the principal effector molecules of cytotoxic T cells and are discussed in Section 9-31. Their release in particular must be tightly regulated because they are not specific: they can penetrate the lipid bilayer and trigger apoptosis in any cell. By contrast, CD4 effector T cells act mainly through the production of cytokines and membrane-associated proteins, and their actions are largely restricted to cells bearing MHC class II molecules and expressing receptors for these proteins.

The main effector molecules of T cells are summarized in Fig. 9.39. The cytokines are a diverse group of proteins and we will review them briefly before discussing the T-cell cytokines and their actions. Secreted cytokines and membrane-associated molecules often act in concert to mediate these effects.

9-27 Cytokines can act locally or at a distance.

Cytokines are small soluble proteins secreted by cells that can alter the behavior or properties of the secreting cell itself (autocrine actions) or of another cell (paracrine actions). Cytokines are produced by many cell types in addition to those of the immune system. We have already introduced the families of

MOVIE 9.7

CD8 T cells: peptide + MHC class I		CD4 T cells: peptide + MHC class II							
Cytotoxic (killer) T cells		T$_H$1 cells		T$_H$2 cells		T$_H$17 cells		T$_{reg}$ cells	
Cytotoxic effector molecules	Others	Macrophage-activating effector molecules	Others	Barrier immunity activating effector molecules	Others	Barrier immunity activating effector molecules, neutrophil recruitment	Others	Suppressive cytokines	Others
Perforin Granzymes Granulysin Fas ligand	IFN-γ LT-α TNF-α	IFN-γ GM-CSF TNF-α CD40 ligand Fas ligand	IL-3 LT-α CXCL2 (GROβ)	IL-4 IL-5 IL-13 CD40 ligand	IL-3 GM-CSF IL-10 TGF-β CCL11 (eotaxin) CCL17 (TARC)	IL-17A IL-17F IL-22 CD40 ligand	IL-3 TNF-α CCL20	IL-10 TGF-β	IL-35

Fig. 9.39 The different types of effector T-cell subsets produce different effector molecules. CD8 T cells are predominantly killer T cells that recognize peptide:MHC class I complexes. They release perforin (which helps deliver granzymes into the target cell) and granzymes (which are pro-proteases that are activated intracellularly to trigger apoptosis in the target cell), and often also produce the cytokine IFN-γ. They also carry the membrane-bound effector molecule Fas ligand (CD178). When this binds to Fas (CD95) on a target cell it activates apoptosis in the Fas-bearing cell. The various functional subsets of CD4 T cells recognize peptide:MHC class II complexes. T$_H$1 cells are specialized to activate macrophages that are infected by or have ingested pathogens; they secrete IFN-γ to activate the infected cell, as well as other effector molecules. They can express membrane-bound CD40 ligand and/or Fas ligand.

CD40 ligand triggers activation of the target cell, whereas Fas ligand triggers the death of Fas-bearing targets, and so which molecule is expressed strongly influences T$_H$1 function. T$_H$2 cells are specialized for promoting immune responses to parasites and also promote allergic responses. They provide help in B-cell activation and secrete the B-cell growth factors IL-4, IL-5, IL-9, and IL-13. The principal membrane-bound effector molecule expressed by T$_H$2 cells is CD40 ligand, which binds to CD40 on B cells and induces B-cell proliferation and isotype switching (see Chapter 10). T$_H$17 cells produce members of the IL-17 family and IL-22, and promote acute inflammation by helping to recruit neutrophils to sites of infection. T$_{reg}$ cells produce inhibitory cytokines such as IL-10 and TGF-β that may act at a distance, but also exert inhibitory actions such as sequestration of B7 and IL-2, which act via cell–cell interactions.

cytokines and their receptors that are important in innate and adaptive immunity in Chapters 3 and 7 (see Sections 3-15 and 7-1). Here we are concerned with cytokines that mediate the effector functions of T cells. Many cytokines produced by T cells are given the name interleukin (IL) followed by a number. The cytokines produced by T cells are shown in Fig. 9.40, and a more comprehensive list of cytokines of immunological interest is in Appendix III. Although many cytokines can have diverse biological effects when tested *in vitro*, targeted disruption of the genes for cytokines and cytokine receptors in mice (see Appendix I, Section A-35) has helped to clarify their physiological roles.

Binding of the T-cell receptor orchestrates the polarized release of cytokines so that they are concentrated at the site of contact with the target cell (see Section 9-25). Furthermore, most of the soluble cytokines have local actions that synergize with those of the membrane-bound effector molecules. The effect of all these molecules is therefore combinatorial, and, because the membrane-bound effectors can bind only to receptors on an interacting cell, this is another mechanism by which selective effects of cytokines are focused on the target cell. The effects of some cytokines are further confined to target cells by tight regulation of their synthesis: the synthesis of IL-2, IL-4, and IFN-γ, for example, is controlled by mRNA instability (see Section 9-16), so that their secretion by T cells does not continue after the interaction with a target cell has ended.

Some cytokines have distant effects. IL-3 and GM-CSF (see Fig. 9.39) are released by T$_H$1, T$_H$2, and T$_H$17 cells and act on bone marrow cells to stimulate the production of macrophages and granulocytes, which are important innate effector cells in both antibody- and T-cell-mediated immunity. IL-3 and GM-CSF also stimulate the production of dendritic cells from bone marrow precursors. IL-17A and IL-17F produced by T$_H$17 cells act primarily on stromal cells, activating them to produce G-CSF, which enhances production

Cytokine	T-cell source	Effects on					Effect of gene knockout
		B cells	T cells	Macrophages	Hematopoietic cells	Other tissue cells	
Interleukin-2 (IL-2)	Naive, T_H1, some CD8	Stimulates growth and J-chain synthesis	Growth and differentiation	–	Stimulates NK cell growth	–	Impaired T_{reg} cell development and function
Interferon-γ (IFN-γ)	T_H1, T_{FH}, CTL	Differentiation IgG2a synthesis (mouse)	Inhibits T_H2 and T_H17 cell differentiation	Activation, ↑MHC class I and class II	Activates NK cells	Antiviral ↑MHC class I and class II	Susceptible to mycobacteria, some viruses
Lymphotoxin-α (LT-α, TNF-β)	T_H1, some CTL	Inhibits	Kills	Activates, induces NO production	Activates neutrophils	Kills fibroblasts and tumor cells	Absence of lymph nodes Disorganized spleen
Interleukin-4 (IL-4)	T_H2, T_{FH}	Activation, growth IgG1, IgE ↑MHC class II induction	Growth, survival	Promotes marginal zone macrophage activation	↑Growth of mast cells	–	No T_H2
Interleukin-5 (IL-5)	T_H2	Mouse: Differentiation IgA synthesis	–	–	↑Eosinophil growth and differentiation	–	Reduced eosinophilia
Interleukin-13 (IL-13)	T_H2	IgG1, IgE class switch	–	Promotes marginal zone macrophage	–	↑Production of mucus (goblet cell)	Impaired helminth expulsion
Interleukin-17 (IL-17)	T_H17	Promotes IgG2a, IgG2b, IgG3 (mouse)	–	–	Stimulates neutrophil recruitment (indirect)	Stimulates fibroblasts and epithelial cells to secrete chemokines	Impaired antibacterial defense
Interleukin-22 (IL-22)	T_H17	–	–	–	–	Stimulates mucosal epithelium and skin to produce antimicrobial peptides	Impaired antibacterial defense
Transforming growth factor-β (TGF-β)	T_{reg}	Inhibits growth IgA switch factor	T_H17 and iT_{reg} differentiation, inhibits T_H1 and T_H2	Inhibits activation	Activates neutrophils	Inhibits/ stimulates cell growth	Impaired T_{reg} cell development Multi-organ auto-immunity and death ~10 weeks
Interleukin-10 (IL-10)	T_{reg}, some T_H1, T_H2, T_H17, CTL	↑MHC class II	Inhibits T_H1	Inhibits inflammatory cytokine release	Co-stimulates mast cell growth	–	IBD
Interleukin-3 (IL-3)	T_H1, T_H2, T_H17, some CTL	–	–	–	Growth factor for progenitor hematopoietic cells (multi-CSF)	–	–
Tumor necrosis factor-α (TNF-α)	T_H1, T_H17, some T_H2, some CTL	–	–	Activates, induces NO production	–	Activates microvascular endothelium	Susceptibility to Gram –ve sepsis
Granulocyte-macrophage colony-stimulating factor (GM-CSF)	T_H1, T_H17, some T_H2, some CTL	Differentiation	Inhibits growth?	Activation Differentiation to dendritic cells	↑Production of granulocytes and macrophages (myelopoiesis) and dendritic cells	–	–

Fig. 9.40 The nomenclature and functions of well-defined T-cell cytokines. Each cytokine has multiple activities on different cell types. Major activities of effector cytokines are highlighted in red. The mixture of cytokines secreted by a given cell type produces many effects through what is called a 'cytokine network.' ↑ increase; ↓, decrease; CTL, cytotoxic lymphocyte; NK cells, natural killer cells; CSF, colony-stimulating factor; IBD, inflammatory bowel disease; NO, nitric oxide.

of neutrophils by the bone marrow. T_H2 cells produce IL-5, which stimulates bone marrow production of eosinophils. Whether a given cytokine effect is local or more distant is likely to reflect the amounts released, the degree to which this release is focused on the target cell, and the stability of the cytokine *in vivo*.

9-28 T cells express several TNF-family cytokines as trimeric proteins that are usually associated with the cell surface.

Most effector T cells express members of the TNF family as membrane-associated proteins on the cell surface. These include TNF-α, the lymphotoxins (LTs), Fas ligand (CD178), and CD40 ligand, the latter two always being cell-surface associated. TNF-α is made by T cells in soluble and membrane-associated forms and assembles into a homotrimer. Secreted LT-α is a homotrimer, but in its membrane-bound form, LT-α is linked to a third, transmembrane member of this family called LT-β to form heterotrimers, called simply LT-β (see Section 9-2). The receptors for TNF-α and LT-α, TNFR-I and TNFR-II, form homotrimers when bound to their ligands. The trimeric structure is characteristic of all members of the TNF family, and the ligand-induced trimerization of their receptors seems to be the critical event in initiating signaling.

Fas ligand and CD40 ligand bind respectively to the transmembrane proteins Fas (CD95) and CD40 on target cells. Fas contains a 'death' domain in its cytoplasmic tail, and binding of Fas by Fas ligand induces death by apoptosis in the Fas-bearing cell (see Fig. 11.22). Other TNFR-family members, including TNFR-I, are also associated with death domains and can also induce apoptosis. Thus, TNF-α and LT-α can induce apoptosis by binding to TNFR-I.

CD40 Ligand Deficiency

CD40 ligand is particularly important for CD4 T-cell effector function; its expression is induced on T_H1, T_H2, T_H17, and T_{FH} cells, and it delivers activating signals to B cells and innate immune cells through CD40. The cytoplasmic tail of CD40 lacks a death domain; instead, it is linked downstream to proteins called TRAFs (TNF-receptor-associated factors). CD40 is involved in the activation of B cells and macrophages; the ligation of CD40 on B cells promotes growth and isotype switching, whereas CD40 ligation on macrophages induces them to secrete higher amounts of pro-inflammatory cytokines (for example, TNF-α) and become receptive to much lower concentrations of IFN-γ. Deficiency in CD40 ligand expression is associated with immunodeficiency, as we will learn in Chapter 13.

Summary.

Interactions between effector T cells and their targets are initiated by transient antigen-nonspecific adhesion. T-cell effector functions are elicited only when peptide:MHC complexes on the surface of the target cell are recognized by the receptor on an effector T cell. This recognition event triggers the effector T cell to adhere more strongly to the antigen-bearing target cell and to release its effector molecules directly at the target cell, leading to the activation or death of the target. The immunological consequences of antigen recognition by an effector T cell are determined largely by the set of effector molecules that the T cell produces on binding a specific target cell. CD8 cytotoxic T cells store preformed cytotoxins in specialized cytotoxic granules whose release is tightly focused at the site of contact with the infected target cell, thus killing it without killing any uninfected cells nearby. Cytokines and members of the TNF family of membrane-associated effector proteins are synthesized *de novo* by most effector T cells. Membrane-associated effector molecules can deliver signals only to an interacting cell bearing the appropriate receptor, whereas soluble cytokines can act on cytokine receptors expressed locally on the target cell, or on other cells at a distance. The actions of cytokines and membrane-associated effector molecules through their specific receptors, together with the effects of the cytotoxins released by CD8 cells, account for most of the effector functions of T cells.

T-cell-mediated cytotoxicity.

All viruses, and some bacteria, multiply in the cytoplasm of infected cells; indeed, a virus is a highly sophisticated parasite that has no biosynthetic or metabolic apparatus of its own and, in consequence, can replicate only inside cells. Although susceptible to antibody-mediated clearance before they enter cells, once they enter cells these pathogens are not accessible to antibodies and can be eliminated only by the destruction or modification of the infected cells in which they replicate. This role in host defense is largely filled by CD8 cytotoxic T cells, although T_H1 cells may also acquire cytotoxic capacities. The crucial role of cytotoxic T cells in limiting such infections is seen in the increased susceptibility of animals artificially depleted of these T cells, or of mice or humans that lack the MHC class I molecules that present antigen to CD8 T cells. The elimination of infected cells without the destruction of healthy tissue requires the cytotoxic mechanisms of CD8 T cells to be both powerful and accurately targeted.

9-29   Cytotoxic T cells induce target cells to undergo programmed cell death via extrinsic and intrinsic pathways of apoptosis.

To deprive cytosolic pathogens of their cellular host, cytotoxic T cells target the infected host cells for death. Cells can die in various ways. Physical or chemical injury, such as the deprivation of oxygen that occurs in heart muscle during a heart attack or membrane damage with antibody and complement, leads to cell disintegration or **necrosis**. This form of cell death is often accompanied by local inflammation and stimulates a wound healing response. The other form of cell death is known as programmed cell death, which can occur by apoptosis or autophagy. **Apoptosis** is a regulated process that is induced either by specific extracellular signals or by the lack of signals required for survival, and proceeds by a series of cellular events that include plasma membrane blebbing, changes in the distribution of membrane lipids, and enzymatic fragmentation of chromosomal DNA. A hallmark of apoptosis is the fragmentation of nuclear DNA into pieces 200 base pairs long through the activation of nucleases that cleave the DNA between nucleosomes. As described in Chapter 6, **autophagy** is the process of degrading senescent or abnormal proteins and organelles. In autophagic cell death, large vacuoles degrade cellular organelles before the condensation and destruction of the nucleus that is characteristic of apoptosis.

Cytotoxic T cells kill by inducing their targets to undergo apoptosis (Fig. 9.41). Two general pathways are involved in signaling apoptotic cell death. One, called the **extrinsic pathway of apoptosis**, is mediated by the activation of so-called death receptors by extracellular ligands. Engagement of ligand stimulates apoptosis in receptor-bearing cells. The other pathway is known as the **intrinsic** or **mitochondrial pathway of apoptosis** and is induced in response to noxious stimuli (for example, ultraviolet irradiation or chemotherapeutic drugs), or lack of the growth factors required for survival. Common to both pathways is the activation of specialized proteases called aspartic acid-specific cysteine proteases, or caspases, which were introduced in Chapter 3 for their role in processing the cytokines IL-1 and IL-18 to their mature forms.

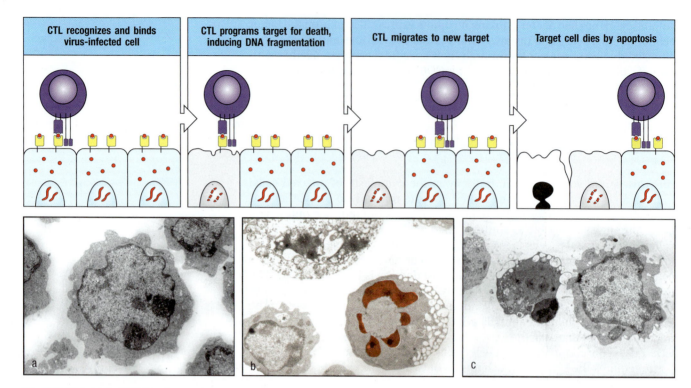

| CTL recognizes and binds virus-infected cell | CTL programs target for death, inducing DNA fragmentation | CTL migrates to new target | Target cell dies by apoptosis |

Fig. 9.41 Cytotoxic CD8 T cells can induce apoptosis in target cells. Specific recognition of peptide:MHC complexes on a target cell (top panels) by a cytotoxic CD8 T cell (CTL) leads to the death of the target cell by apoptosis. Cytotoxic T cells can recycle to kill multiple targets. Each killing requires the same series of steps, including receptor binding and the directed release of cytotoxic proteins stored in granules. The process of apoptosis is shown in the micrographs (bottom panels), where panel a shows a healthy cell with a normal nucleus. Early in apoptosis (panel b) the chromatin becomes condensed (red) and, although the cell sheds membrane vesicles, the integrity of the cell membrane is retained, in contrast to the necrotic cell in the upper part of the same field. In late stages of apoptosis (panel c), the cell nucleus (middle cell) is very condensed, no mitochondria are visible, and the cell has lost much of its cytoplasm and membrane through the shedding of vesicles. Photographs (×3500) courtesy of R. Windsor and E. Hirst.

Like many other proteases, caspases are synthesized as inactive pro-enzymes, in this case, pro-caspases, in which the catalytic domain is inhibited by an adjacent pro-domain. Pro-caspases are activated by other caspases that cleave the protein to release the inhibitory pro-domain. There are two classes of caspases involved in the apoptotic pathway: **initiator caspases** promote apoptosis by cleaving and activating other caspases; **effector caspases** initiate the cellular changes associated with apoptosis. The extrinsic pathway uses two related initiator caspases, caspase 8 and caspase 10, whereas the intrinsic pathway uses caspase 9. Both pathways use caspases 3, 6, and 7 as effector caspases. The effector caspases cleave a variety of proteins that are critical for cellular integrity and also activate enzymes that promote the death of the cell. For example, they cleave and degrade nuclear proteins that are required for the structural integrity of the nucleus, and activate the endonucleases that fragment the chromosomal DNA.

Cytotoxic T cells can induce target-cell death by either the extrinsic or the intrinsic apoptotic pathway. The extrinsic pathway is mediated by expression of FasL and TNF-α or LT-α, receptors for which (Fas, or CD95, and TNFR-I) are expressed by other cells of the immune system, as well as non-immune-system cells. Because the distribution of these receptors is somewhat restricted, cytotoxic T cells have acquired a more universal mechanism for inducing cell death in antigen-specific targets: the directional release of cytotoxic granules that activate the intrinsic pathway of apoptosis. When cytotoxic T cells are mixed with target cells and rapidly brought into contact by centrifugation, they can induce antigen-specific target cells to die within 5 minutes, although death can take hours to become fully evident. The rapidity of this response reflects the release of preformed effector molecules that are delivered to the target cell. In addition to killing the host cell, the apoptotic mechanism may also act directly on cytosolic pathogens. For example, the nucleases that are activated in apoptosis to destroy cellular DNA can also degrade viral DNA. This prevents the assembly of virions and the release of infectious virus, which could otherwise infect nearby cells. Other enzymes activated in the course of apoptosis may destroy nonviral cytosolic pathogens. Apoptosis is therefore preferable to necrosis as a means of killing infected cells; in cells dying by necrosis, intact pathogens are released from the dead cell, and these can continue to infect healthy cells or parasitize the macrophages that ingest them.

9-30 The intrinsic pathway of apoptosis is mediated by the release of cytochrome *c* from mitochondria.

Apoptosis by the intrinsic pathway is triggered by the release of cytochrome *c* from mitochondria, which triggers the activation of caspases. Once in the cytoplasm, cytochrome *c* binds to a protein called Apaf-1 (apoptotic protease activating factor-1), stimulating its oligomerization to form the **apoptosome**. The apoptosome then recruits an initiator caspase, pro-caspase 9, aggregation of which promotes its self-cleavage and frees its catalytic domain to activate effector caspases (Fig. 9.42).

 MOVIE 9.8

The release of cytochrome *c* is controlled by interactions between members of the Bcl-2 family of proteins. The Bcl-2 family of proteins is defined by the presence of one or more Bcl-2 homology (BH) domains and can be divided into two general groups: members that promote apoptosis, and members that

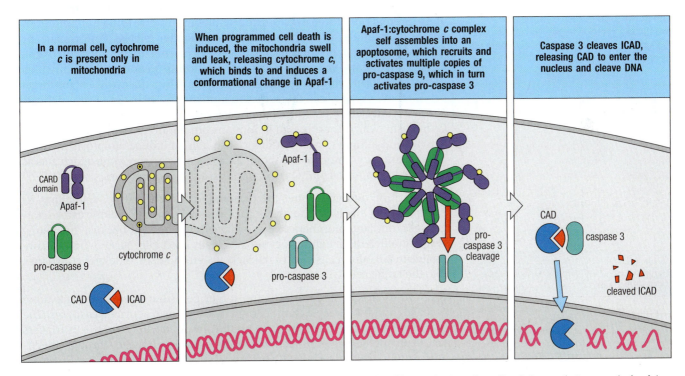

Fig. 9.42 In the intrinsic pathway, cytochrome *c* release from mitochondria induces formation of the apoptosome, which activates pro-caspase 9 to initiate programmed cell death. In normal cells, cytochrome *c* is confined to the mitochondria (first panel). However, during stimulation of the intrinsic pathway, the mitochondria swell, allowing the cytochrome *c* to leak out into the cytosol (second panel), where cytochrome *c* is bound by Apaf-1. The resultant conformational change that ensues in Apaf-1 induces self-assembly of the multimeric apoptosome, which recruits pro-caspase 9 (third panel). Clustering of pro-caspase 9 by the apoptosome activates it, allowing it to cleave downstream caspases, such as caspase 3; this results in the activation of enzymes such as ICAD, which can cleave DNA (fourth panel).

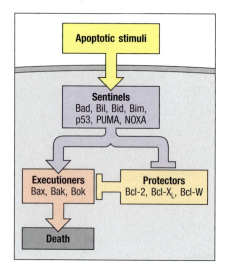

Fig. 9.43 General scheme of intrinsic pathway regulation by the Bcl-2 family of proteins. Extracellular apoptotic stimuli activate a group of pro-apoptotic (sentinel) proteins. Sentinel proteins can function either to block the protection provided by pro-survival, protector proteins or to directly activate pro-apoptotic, executioner proteins. In mammalian cells, apoptosis is mediated by the executioner proteins Bax, Bak, and Bok. In normal cells, these proteins are prevented from acting by the protector proteins (Bcl-2, Bcl-X$_L$, and Bcl-W). The release of activated executioner proteins causes the release of cytochrome *c* and subsequent cell death, as shown in Fig. 9.42.

Protein in granules of cytotoxic T cells	Actions on target cells
Perforin	Aids in delivering contents of granules into the cytoplasm of target cell
Granzymes	Serine proteases, which activate apoptosis once in the cytoplasm of the target cell
Granulysin	Has antimicrobial actions and can induce apoptosis

Fig. 9.44 Cytotoxic effector proteins released by cytotoxic T cells.

inhibit apoptosis (Fig. 9.43). Pro-apoptotic Bcl-2 family members, such as Bax, Bak, and Bok (referred to as executioners), bind to mitochondrial membranes and can directly cause cytochrome *c* release. How they do this is still not known, but they may form pores in the membranes.

The anti-apoptotic Bcl-2 family members are induced by stimuli that promote cell survival. The best known of the anti-apoptotic proteins is Bcl-2 itself. The *Bcl2* gene was first identified as an oncogene in a B-cell lymphoma, and its overexpression in tumors makes the cells more resistant to apoptotic stimuli and thus more likely to progress to an invasive cancer. Other members of the inhibitory family include Bcl-X$_L$ and Bcl-W. Anti-apoptotic proteins function by binding to the mitochondrial membrane to block the release of cytochrome *c*. The precise mechanism of inhibition is not clear, but they may function by directly blocking the function of the pro-apoptotic family members.

A second family of pro-apoptotic Bcl-2 family members are termed 'sentinels' and are activated by apoptotic stimuli. Once activated, these proteins, which include Bad, Bid, and PUMA, can either act to block the activity of the anti-apoptotic proteins or act directly to stimulate the activity of the executioner pro-apoptotic proteins.

9-31   Cytotoxic effector proteins that trigger apoptosis are contained in the granules of CD8 cytotoxic T cells.

The principal mechanism of cytotoxic T-cell action is the calcium-dependent release of specialized cytotoxic granules upon recognition of antigen on the surface of a target cell. Cytotoxic granules are modified lysosomes that contain at least three distinct classes of cytotoxic effector proteins that are expressed specifically in cytotoxic T cells: perforin, granzymes, and granulysin (Fig. 9.44). These proteins are stored in cytotoxic granules in an active form, but conditions within the granules prevent their actions until after release. Perforin acts by forming pores in, or perforating, the target-cell plasma membrane, which both causes direct damage to the target cell and forms a conduit through which other contents of cytotoxic granules are delivered into the cytosol of the target cell. Granzymes, of which there are 5 in humans and 10 in the mouse, activate apoptosis once delivered to the target-cell cytosol via pores formed by perforin. Granulysin, which is expressed in humans but not in mice, has antimicrobial activity and at high concentrations is also able to induce apoptosis in target cells. Cytotoxic granules also contain the proteoglycan serglycin, which acts as a scaffold, forming a complex with perforin and the granzymes.

Both perforin and granzymes are required for effective target-cell killing. In cytotoxic cells that lack granzymes, the presence of perforin alone can kill target cells, but large numbers of cytotoxic cells are needed because the killing is very inefficient. In contrast, cytotoxic T cells from mice lacking perforin are unable to kill other cells, due to the lack of a mechanism to deliver granzymes into the target cell.

Granzymes trigger apoptosis in the target cell both by directly activating caspases and by damaging mitochondria, which also activates caspases. The two most abundant granzymes are granzymes A and B. Granzyme A triggers cell death by caspase-independent mitochondrial damage, through mechanisms that are not completely understood. Granzyme B, like the caspases, cleaves proteins after aspartic acid residues and activates caspase 3, thereby activating a caspase proteolytic cascade, which eventually activates the caspase-activated deoxyribonuclease (CAD) by cleaving an inhibitory protein (ICAD) that binds to and inactivates CAD. This nuclease is believed to be the enzyme that degrades DNA in target cells (Fig. 9.45). Granzyme B also targets mitochondria to activate the intrinsic apoptotic pathway; it cleaves the protein BID (for BH3-interacting domain death agonist protein), either directly

Fig. 9.45 Perforin, granzymes, and serglycin are released from cytotoxic granules and deliver granzymes into the cytosol of target cells to induce apoptosis. Recognition of its antigen on a virus-infected cell by a cytotoxic CD8 T cell induces the T cell to release the contents of its cytotoxic granules in a directed fashion. Perforin and granzymes, in a complex with the proteoglycan serglycin, are delivered to the membrane of the target cell (top panel). By an unknown mechanism, perforin directs the entry of the granule contents into the cytosol of the target cell without apparent pore formation, and the introduced granzymes then act on specific intracellular targets such as the proteins BID and pro-caspase 3 (second panel). Either directly or indirectly, the granzymes cause the cleavage of BID into truncated BID (tBID) and the cleavage of pro-caspase 3 into an active caspase (third panel). tBID acts on mitochondria to release cytochrome *c* into the cytosol. This promotes apoptosis by inducing the formation of the apoptosome that activates pro-caspase 9, which in turn further amplifies caspase 3 activation. Activated caspase 3 targets ICAD to release caspase-activated DNase (CAD), which fragments the DNA (bottom panel).

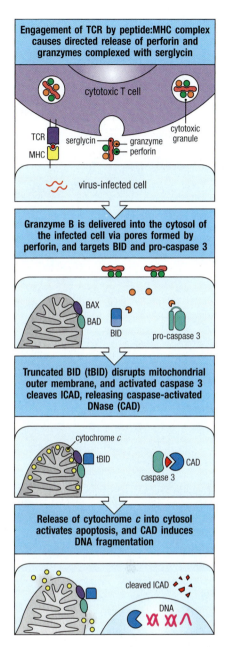

or indirectly by activated caspase 3, causing disruption of the mitochondrial outer membrane and the release from the mitochondrial intermembrane space of pro-apoptotic molecules such as cytochrome *c*. As discussed above (Section 9-30), cytochrome *c* is central to amplification of the intrinsic apoptotic cascade, as it initiates assembly of the apoptosome with Apaf-1, which in turn activates the initiator caspase 9. Thus, granzyme B acts directly to activate the effector caspase 3, and indirectly to activate the initiator caspase 9.

Cells undergoing programmed cell death are rapidly ingested by phagocytic cells, which recognize a change in the cell membrane: phosphatidylserine, which is normally found only in the inner leaflet of the membrane, replaces phosphatidylcholine as the predominant phospholipid in the outer leaflet. The ingested cell is broken down and completely digested by the phagocyte without the induction of co-stimulatory proteins. Thus, apoptosis is normally an immunologically 'quiet' process; that is, apoptotic cells do not normally contribute to or stimulate immune responses.

9-32   Cytotoxic T cells are selective serial killers of targets expressing a specific antigen.

When cytotoxic T cells are offered a mixture of equal amounts of two target cells, one bearing a specific antigen and the other not, they kill only the target cell bearing the specific antigen. The 'innocent bystander' cells and the cytotoxic T cells themselves are not killed. The cytotoxic T cells are probably not killed because release of the cytotoxic effector molecules is highly polarized. As we saw in Fig. 9.38, cytotoxic T cells orient their Golgi apparatus and microtubule-organizing center to focus secretion on the point of contact with a target cell. Granule movement toward the point of contact is shown in Fig. 9.46. Cytotoxic T cells attached to several different target cells reorient their secretory apparatus toward each cell in turn and kill them one by one, strongly suggesting that the mechanism whereby cytotoxic mediators are released allows attack at only one point of contact at any one time. The narrowly focused action of CD8 cytotoxic T cells allows them to kill single infected cells in a tissue without creating widespread tissue damage (Fig. 9.47) and is of crucial importance in tissues where cell regeneration does not occur, as with the neurons of the central nervous system, or is very limited, as in the pancreatic islets.

Cytotoxic T cells can kill their targets rapidly because they store preformed cytotoxic proteins in forms that are inactive in the environment of the cytotoxic granule. Cytotoxic proteins are synthesized and loaded into the granules soon after the first encounter of a naive cytotoxic precursor T cell with its specific antigen. Ligation of the T-cell receptor similarly induces *de novo* synthesis of perforin and granzymes in effector CD8 T cells, so that the supply of cytotoxic granules is replenished. This makes it possible for a single CD8 T cell to kill a series of targets in succession.

 MOVIE 9.9

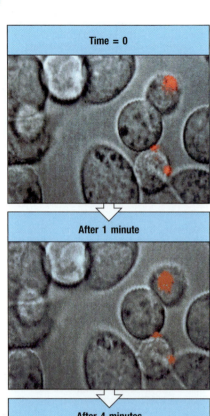

Time = 0

After 1 minute

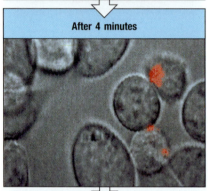

After 4 minutes

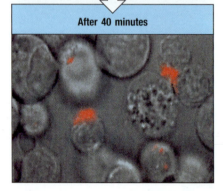

After 40 minutes

Fig. 9.46 Effector molecules are released from T-cell granules in a highly polar fashion. The granules of cytotoxic T cells can be labeled with fluorescent dyes, allowing the granules to be seen under the microscope and their movements to be followed by time-lapse photography. Here we show a series of pictures taken during the interaction of a cytotoxic T cell with a target cell, which is eventually killed. In the top panel, at time 0, the T cell (upper right) has just made contact with a target cell (diagonally below). At this time, the granules of the T cell, labeled with a red fluorescent dye, are distant from the point of contact. In the second panel, after 1 minute has elapsed, the granules have begun to move toward the target cell, a move that has essentially been completed in the third panel, after 4 minutes. After 40 minutes, in the last panel, the granule contents have been released into the space between the T cell and the target, which has begun to undergo apoptosis (note the fragmented nucleus). The T cell will now disengage from the target cell, whereupon it can go on to recognize and kill other targets. Photographs courtesy of G. Griffiths.

9-33   Cytotoxic T cells also act by releasing cytokines.

Inducing apoptosis in target cells is the main way in which CD8 cytotoxic T cells eliminate infection. However, most CD8 cytotoxic T cells also release the cytokines IFN-γ, TNF-α, and LT-α, which contribute to host defense in other ways. IFN-γ inhibits viral replication directly, and induces the increased expression of MHC class I molecules and of other proteins that are involved in peptide loading of these newly synthesized MHC class I molecules in infected cells. This increases the chance that infected cells will be recognized as target cells for cytotoxic attack. IFN-γ also activates macrophages, recruiting them to sites of infection, where they serve both as effector cells and as antigen-presenting cells. TNF-α and LT-α can synergize with IFN-γ in macrophage activation via TNFR-II, and can kill some target cells through their interaction with TNFR-I, which can induce apoptosis (see Sections 9-28 and 9-29). Thus, effector CD8 cytotoxic T cells act in a variety of ways to limit the spread of cytosolic pathogens.

Summary.

Effector CD8 cytotoxic T cells are essential in host defense against pathogens that reside in the cytosol: most commonly these will be viruses. These cytotoxic T cells can kill any cell harboring such pathogens by recognizing foreign peptides that are transported to the cell surface bound to MHC class I molecules. CD8 cytotoxic T cells perform their killing function by releasing three types of preformed cytotoxic proteins: granzymes, which use multiple mechanisms to induce apoptosis in any type of target cell; perforin, which acts in the delivery of granzymes into the target cell; and granulysin, which has antimicrobial activity and is pro-apoptotic. These properties allow the cytotoxic T cell to attack and destroy virtually any cell infected with a cytosolic pathogen. The membrane-bound Fas ligand, expressed by CD8 and some CD4 T cells, may also induce apoptosis by binding to Fas, which is expressed on some target cells. However, this pathway is less important in most infections than that mediated by cytotoxic granules. CD8 cytotoxic T cells also produce IFN-γ, which inhibits viral replication and is an important inducer of MHC class I molecule expression and macrophage activation. Cytotoxic T cells kill infected targets with great precision, sparing adjacent normal cells. This precision is crucial in minimizing tissue damage while allowing the eradication of infected cells.

Summary to Chapter 9.

An adaptive immune response is initiated when naive T cells encounter specific antigen on the surface of an antigen-presenting cell in T-cell zones of secondary lymphoid tissues. In most cases, the antigen-presenting cells responsible

Fig. 9.47 Cytotoxic T cells kill target cells bearing specific antigen while sparing neighboring uninfected cells. All the cells in a tissue are susceptible to killing by the cytotoxic proteins of armed effector CD8 T cells, but only infected cells are killed. Specific recognition by the T-cell receptor identifies which target cell to kill, and the polarized release of the cytotoxic granules (not shown) ensures that neighboring cells are spared.

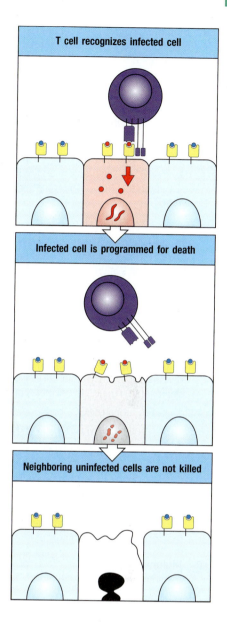

for activating naive T cells, and inducing their clonal expansion, are conventional dendritic cells that express the co-stimulatory molecules B7.1 and B7.2. Conventional dendritic cells not only reside in lymphoid tissues, but they also survey the periphery, where they encounter pathogens, take up antigen at sites of infection, become activated through innate recognition, and migrate to local lymphoid tissue. The dendritic cell may become a potent direct activator of naive T cells, or it may transfer antigen to dendritic cells resident in secondary lymphoid organs for cross-presentation to naive CD8 T cells. Plasmacytoid dendritic cells contribute to rapid responses against viruses by producing type I interferons. Activated T cells produce IL-2, which is important in modulating early proliferation and differentiation of T cells; various other signals drive the differentiation of several types of effector T cells, which primarily act by releasing mediators directly onto their target cells. This triggering of effector T cells by peptide:MHC complexes occurs independently of co-stimulation, so that any infected target cell can be activated or destroyed by an effector T cell. CD8 cytotoxic T cells kill target cells infected with cytosolic pathogens, thus removing sites of pathogen replication. CD4 T cells can become specialized effectors that in turn promote distinct arms of the immune response by targeting different innate and adaptive immune cells for enhanced effector function: macrophages (T_H1); eosinophils, basophils, and mast cells (T_H2); neutrophils (T_H17); or B cells (T_{FH}). Thus, effector T cells control virtually all known effector mechanisms of the adaptive and innate immune response. In addition, subsets of CD4 regulatory T cells are produced that help control and limit immune responses by suppressing T-cell activity.

Questions.

9.1   Multiple Choice: Which of the following statements is *true*?

A. Development of the arterial and venous system is regulated by the homeobox transcription factor Prox1.

B. Arteries deliver lymphotoxin to the non-hematopoietic stromal LTi cell to induce lymph-node development.

C. Lymphotoxin-α3 signaling represses NFκB to induce chemokines such as CXCL13.

D. Lymphotoxin-α3 binds TNFR-I and supports development of the cervical and mesenteric lymph nodes.

9.2   Fill-in-the-Blanks: T and B cells are distributed to the secondary lymphoid organs through the blood. These are then directed to their respective compartments as instructed by chemokines. For example, CCL21 is secreted by _____ of the T-cell zone in the spleen and displayed by the _____ in the lymph nodes. Signaling of this chemokine as well as _____ through CCR7 directs the T cells into the respective T-cell zone. In contrast, _____ is the ligand for CXCR5, which is secreted by _____ and attracts B cells to the _____. T cells can also respond to CXCL13 as a subset of T cells expresses _____, which allows them to enter the B-cell follicle and participate in the formation of the germinal center.

9.3 **Multiple Choice:** Which of the following correctly describes events necessary for naive T-cell entry into the lymph node?

A. CCR7 signaling induces Gα$_i$, which results in lowered affinity for integrin binding.

B. Upregulation of S1P receptor on naive T cells promotes migration into the lymph node.

C. Rolling in the HEV exposes the T cell to CCL21, which activates LFA-1 and promotes migration.

D. MAdCAM-1 expression on the HEV interacts with CD62L on the T cell and promotes migration into the lymph node.

9.4 **Short Answer:** In some cases, HSV or influenza viruses infect antigen-presenting cells from peripheral tissues that do not present the viral antigens to naive T cells. How is the immune system able to develop an adaptive immune response to such pathogens?

9.5 **True or False:** TLR stimulation induces CCR7 expression in the dendritic cells, which promotes migration to the lymph node through the bloodstream.

9.6 **Matching:** Classify each of the following activation signatures as a conventional dendritic cell (cDC) or plasmacytoid dendritic cell (pDC) pathogen response.

____1. Production of CCL18

____2. Continuous MHC recycling upon activation

____3. Expression of DC-SIGN

____4. Expression of CD80 and CD86

____5. CD40L expression upon TLR-9 stimulation

9.7 **Short Answer:** How does the process of antigen presentation differ among B cells, dendritic cells, and macrophages in the context of an immune response?

9.8 **Multiple Choice:** Which of the following is a common consequence of TCR and CCR7 signaling?

A. Integrin activation

B. Positive selection

C. T$_H$1 induction

D. T$_H$2 induction

9.9 **Multiple Choice:** Which of the following describes a mechanism by which CD28 signaling can increase IL-2 production?

A. CD28 signaling induces the expression of proteins that stabilize the IL-2 mRNA sequence.

B. PI 3-kinase inhibits Akt, supporting IL-2 production by cell cycle arrest.

C. PI 3-kinase suppresses the production of AP-1 and NFκB, thereby increasing IL-2 production.

9.10 **True or False:** In the majority of viral infections, CD8 T-cell activation requires CD4 T-cell help.

9.11 **Matching:** Match each CD4 T cell subset-specific secreted cytokine with its respective effector function.

A. IL-17	**i.** Eradication of intracellular infections
B. IL-4	**ii.** Response to extracellular bacteria
C. IFN-γ	**iii.** Control of extracellular parasites
D. IL-10	**iv.** Suppression of T-cell responses

9.12 **Matching:** The following cytokines drive CD4 T$_H$ subset effector differentiation. Match each with its respective subset-specific transcription factor.

A. IFN-γ	**i.** RORγt
B. IL-4	**ii.** FoxP3
C. IL-6 and TGF-β	**iii.** T-bet
D. TGF-β	**iv.** GATA3

9.13 **Multiple Choice:** Which of the following statements is *false*?

A. TCR signaling is strongest at the cSMAC.

B. Cb1, an E3 ligase, mediates degradation of TCRs in the cSMAC.

C. Cytoskeletal reorganization directs effector molecule release at the immunological synapse.

D. Integrins such as LFA-1 associate in the SMAC.

9.14 **Fill-in-the-Blanks:** For each of the following sentences, fill in the blanks with the best word selected from the list below. Not all words will be used; each word should be used only once.

CD8 T cells can specifically mediate the destruction of infected or malignant cells. In order to do this, CD8 T cells induce _____ cell death, which can be induced in two different ways. First, CD8 T cells possess ligands such as _____, _____, or _____ that can induce the _____ apoptosis pathway. In contrast, cell death can also be induced through an intrinsic pathway. To initiate this mechanism, _____ are released, which allow the entrance of granzymes into the cell. Once the granzymes have gained access to the cell's cytoplasm, these can cleave and activate _____, which in turn cleaves _____, allowing _____ to degrade DNA. Granzyme B also cleaves _____, and as a consequence disrupts the mitochondrial membrane, allowing for release of _____ and formation of the _____.

CAD	necrotic	caspase 3
intrinsic	LT-α	proton gradient
apoptotic	caspase 9	ICAD
apoptosome	extrinsic	TNF-α
FasL	perforins	BID
cytochrome *c*	hypoxia	

General references.

Coffman, R.L.: **Origins of the T$_H$1-T$_H$2 model: a personal perspective.** *Nat. Immunol.* 2006, **7**:539–541.

Griffith, J.W., Sokol, C.L., and Luster, A.D.: **Chemokines and chemokine receptors: positioning cells for host defense and immunity.** *Annu. Rev. Immunol.* 2014, **32**:659–702.

Heath, W.R., and Carbone, F.R.: **Dendritic cell subsets in primary and secondary T cell responses at body surfaces.** *Nat. Immunol.* 2009, **10**:1237–1244.

Jenkins, M.K., Chu, H.H., McLachlan, J.B., and Moon, J.J.: **On the composition of the preimmune repertoire of T cells specific for peptide-major histocompatibility complex ligands.** *Annu. Rev. Immunol.* 2010, **28**:275–294.

Springer, T.A.: **Traffic signals for lymphocyte recirculation and leukocyte emigration: the multistep paradigm.** *Cell* 1994, **76**:301–314.

Zhu, J., Yamane, H., and Paul, W.E.: **Differentiation of effector CD4 T cell populations.** *Annu. Rev. Immunol.* 2010 **28**:445–489.

Section references.

9-1 **T and B lymphocytes are found in distinct locations in secondary lymphoid tissues.**

Liu, Y.J.: **Sites of B lymphocyte selection, activation, and tolerance in spleen.** *J. Exp. Med.* 1997, **186**:625–629.

Loder, F., Mutschler, B., Ray, R.J., Paige, C.J., Sideras, P., Torres, R., Lamers, M.C., and Carsetti, R.: **B cell development in the spleen takes place in discrete steps and is determined by the quality of B cell receptor-derived signals.** *J. Exp. Med.* 1999, **190**:75–89.

Mebius, R.E.: **Organogenesis of lymphoid tissues.** *Nat. Rev. Immunol.* 2003, **3**:292–303.

9-2 **The development of secondary lymphoid tissues is controlled by lymphoid tissue inducer cells and proteins of the tumor necrosis factor family.**

Douni, E., Akassoglou, K., Alexopoulou, L., Georgopoulos, S., Haralambous, S., Hill, S., Kassiotis, G., Kontoyiannis, D., Pasparakis, M., Plows, D., *et al.*: **Transgenic and knockout analysis of the role of TNF in immune regulation and disease pathogenesis.** *J. Inflamm.* 1996, **47**:27–38.

Fu, Y.X., and Chaplin, D.D.: **Development and maturation of secondary lymphoid tissues.** *Annu. Rev. Immunol.* 1999, **17**:399–433.

Mariathasan, S., Matsumoto, M., Baranyay, F., Nahm, M.H., Kanagawa, O., and Chaplin, D.D.: **Absence of lymph nodes in lymphotoxin-α (LTα)-deficient mice is due to abnormal organ development, not defective lymphocyte migration.** *J. Inflamm.* 1995, **45**:72–78.

Mebius, R.E., and Kraal, G.: **Structure and function of the spleen.** *Nat. Rev. Immunol.* 2005, **5**: 606–616.

Mebius, R.E., Rennert, P., and Weissman, I.L.: **Developing lymph nodes collect CD4$^+$CD3$^-$ LTβ$^+$ cells that can differentiate to APC, NK cells, and follicular cells but not T or B cells.** *Immunity* 1997, **7**:493–504.

Roozendaal, R., and Mebius, R.E.: **Stromal-immune cell interactions.** *Annu. Rev. Immunol.* 2011, **29**:23–43.

Wigle, J.T., and Oliver, G.: **Prox1 function is required for the development of the murine lymphatic system.** *Cell* 1999, **98**:769–778.

9-3 **T and B cells are partitioned into distinct regions of secondary lymphoid tissues by the actions of chemokines.**

Ansel, K.M., and Cyster, J.G.: **Chemokines in lymphopoiesis and lymphoid organ development.** *Curr. Opin. Immunol.* 2001, **13**:172–179.

Cyster, J.G.: **Chemokines and cell migration in secondary lymphoid organs.** *Science* 1999, **286**:2098–2102.

Cyster, J.G.: **Leukocyte migration: scent of the T zone.** *Curr. Biol.* 2000, **10**:R30–R33.

Cyster, J.G., Ansel, K.M., Reif, K., Ekland, E.H., Hyman, P.L., Tang, H.L., Luther, S.A., and Ngo, V.N.: **Follicular stromal cells and lymphocyte homing to follicles.** *Immunol. Rev.* 2000, **176**:181–193.

9-4 **Naive T cells migrate through secondary lymphoid tissues, sampling peptide:MHC complexes on dendritic cells.**

Caux, C., Ait-Yahia, S., Chemin, K., de Bouteiller, O., Dieu-Nosjean, M.C., Homey, B., Massacrier, C., Vanbervliet, B., Zlotnik, A., and Vicari, A.: **Dendritic cell biology and regulation of dendritic cell trafficking by chemokines.** *Springer Semin. Immunopathol.* 2000, **22**:345–369.

Itano, A.A., and Jenkins, M.K.: **Antigen presentation to naïve CD4 T cells in the lymph node.** *Nat. Immunol.* 2003, **4**:733–739.

Mackay, C.R., Kimpton, W.G., Brandon, M.R., and Cahill, R.N.: **Lymphocyte subsets show marked differences in their distribution between blood and the afferent and efferent lymph of secondary lymph nodes.** *J. Exp. Med.* 1988, **167**:1755–1765.

Picker, L.J., and Butcher, E.C.: **Physiological and molecular mechanisms of lymphocyte homing.** *Annu. Rev. Immunol.* 1993, **10**:561–591.

Steptoe, R.J., Li, W., Fu, F., O'Connell, P.J., and Thomson, A.W.: **Trafficking of APC from liver allografts of Flt3L-treated donors: augmentation of potent allostimulatory cells in recipient lymphoid tissue is associated with a switch from tolerance to rejection.** *Transpl. Immunol.* 1999, **7**:51–57.

Yoshino, M., Yamazaki, H., Nakano, H., Kakiuchi, T., Ryoke, K., Kunisada, T., and Hayashi, S.: **Distinct antigen trafficking from skin in the steady and active states.** *Int. Immunol.* 2003, **15**:773–779.

9-5 **Lymphocyte entry into lymphoid tissues depends on chemokines and adhesion molecules.**

Hogg, N., Henderson, R., Leitinger, B., McDowall, A., Porter, J., and Stanley, P.: **Mechanisms contributing to the activity of integrins on leukocytes.** *Immunol. Rev.* 2002, **186**:164–171.

Kunkel, E.J., Campbell, D.J., and Butcher, E.C.: **Chemokines in lymphocyte trafficking and intestinal immunity.** *Microcirculation* 2003, **10**:313–323.

Madri, J.A., and Graesser, D.: **Cell migration in the immune system: the evolving interrelated roles of adhesion molecules and proteinases.** *Dev. Immunol.* 2000, **7**:103–116.

Rasmussen, L.K., Johnsen, L.B., Petersen, T.E., and Sørensen, E.S.: **Human GlyCAM-1 mRNA is expressed in the mammary gland as splicing variants and encodes various aberrant truncated proteins.** *Immunol. Lett.* 2002, **83**:73–75.

Rosen, S.D.: **Ligands for L-selectin: homing, inflammation, and beyond.** *Annu. Rev. Immunol.* 2004, **22**:129–156.

von Andrian, U.H., and Mempel, T.R.: **Homing and cellular traffic in lymph nodes.** *Nat. Rev. Immunol.* 2003, **3**:867–878.

9-6 **Activation of integrins by chemokines is responsible for the entry of naive T cells into lymph nodes.**

Cyster, J.G.: **Chemokines, sphingosine-1-phosphate, and cell migration in secondary lymphoid organs.** *Annu. Rev. Immunol.* 2005, **23**:127–159.

Laudanna, C., Kim, J.Y., Constantin, G., and Butcher, E.: **Rapid leukocyte integrin activation by chemokines.** *Immunol. Rev.* 2002, **186**:37–46.

Lo, C.G., Lu, T.T., and Cyster, J.G.: **Integrin-dependence of lymphocyte entry into the splenic white pulp.** *J. Exp. Med.* 2003, **197**:353–361.

Luo, B.H., Carman, C.V., and Springer, T.A.: **Structural basis of integrin regulation and signaling.** *Annu. Rev. Immunol.* 2007, **25**:619–647.

Rosen, H., and Goetzl, E.J.: **Sphingosine 1-phosphate and its receptors: an autocrine and paracrine network.** *Nat. Rev. Immunol.* 2005, **5**:560–570.

9-7 **The exit of T cells from lymph nodes is controlled by a chemotactic lipid.**

Cyster, J.G., and Schwab, S.R.: **Sphingosine-1-phosphate and lymphocyte egress from lymphoid organs.** *Annu. Rev. Immunol.* 2012, **30**:69-94.

9-8   T-cell responses are initiated in secondary lymphoid organs by activated dendritic cells.

Germain, R.N., Miller, M.J., Dustin, M.L., and Nussenzweig, M.C.: **Dynamic imaging of the immune system: progress, pitfalls and promise.** *Nat. Rev. Immunol.* 2006, **6**:497–507.

Miller, M.J., Wei, S.H., Cahalan, M.D., and Parker, I.: **Autonomous T cell trafficking examined *in vivo* with intravital two-photon microscopy.** *Proc. Natl Acad. Sci. USA* 2003, **100**:2604–2609.

Schlienger, K., Craighead, N., Lee, K.P., Levine, B.L., and June, C.H.: **Efficient priming of protein antigen-specific human CD4+ T cells by monocyte-derived dendritic cells.** *Blood* 2000, **96**:3490–3498.

Thery, C., and Amigorena, S.: **The cell biology of antigen presentation in dendritic cells.** *Curr. Opin. Immunol.* 2001, **13**:45–51.

9-9   Dendritic cells process antigens from a wide array of pathogens.

Belz, G.T., Carbone, F.R., and Heath, W.R.: **Cross-presentation of antigens by dendritic cells.** *Crit. Rev. Immunol.* 2002, **22**:439–448.

Guermonprez, P., Valladeau, J., Zitvogel, L., Thery, C., and Amigorena, S.: **Antigen presentation and T cell stimulation by dendritic cells.** *Annu. Rev. Immunol.* 2002, **20**:621–667.

Mildner, A., and Jung, S.: **Development and function of dendritic cells.** *Immunity* 2014, **40**:642–656.

Satpathy, A.T., Wu, X., Albring, J.C., and Murphy, K.M.: **Re(de)fining the dendritic cell lineage.** *Nat. Immunol.* 2012, **13**:1145–1154.

Shortman, K., and Heath, W.R.: **The CD8+ dendritic cell subset.** *Immunol. Rev.* 2010, **234**:18–31.

Shortman, K., and Naik, S.H.: **Steady-state and inflammatory dendritic-cell development.** *Nat. Rev. Immunol.* 2007, **7**:19–30.

9-10   Microbe-induced TLR signaling in tissue-resident dendritic cells induces their migration to lymphoid organs and enhances antigen processing.

Allan, R.S., Waithman, J., Bedoui, S., Jones, C.M., Villadangos, J.A., Zhan, Y., Lew, A.M., Shortman, K., Heath, W.R., and Carbone, F.R.: **Migratory dendritic cells transfer antigen to a lymph node-resident dendritic cell population for efficient CTL priming.** *Immunity* 2006, **25**:153–162.

Bachman, M.F., Kopf, M., and Marsland, B.J.: **Chemokines: more than just road signs.** *Nat. Rev. Immunol.* 2006, **6**:159–164.

Blander, J.M., and Medzhitov, R.: **Toll-dependent selection of microbial antigens for presentation by dendritic cells.** *Nature* 2006, **440**:808–812.

Iwasaki, A., and Medzhitov, R.: **Toll-like receptor control of adaptive immune responses.** *Nat. Immunol.* 2004, **10**:988–995.

Reis e Sousa, C.: **Toll-like receptors and dendritic cells: for whom the bug tolls.** *Semin. Immunol.* 2004, **16**:27–34.

9-11   Plasmacytoid dendritic cells produce abundant type I interferons and may act as helper cells for antigen presentation by conventional dendritic cells.

Asselin-Paturel, C., and Trinchieri, G.: **Production of type I interferons: plasmacytoid dendritic cells and beyond.** *J. Exp. Med.* 2005, **202**:461–465.

Krug, A., Veeraswamy, R., Pekosz, A., Kanagawa, O., Unanue, E.R., Colonna, M., and Cella, M.: **Interferon-producing cells fail to induce proliferation of naïve T cells but can promote expansion and T helper 1 differentiation of antigen-experienced unpolarized T cells.** *J. Exp. Med.* 2003, **197**:899–906.

Kuwajima, S., Sato, T., Ishida, K., Tada, H., Tezuka, H., and Ohteki, T.: **Interleukin 15-dependent crosstalk between conventional and plasmacytoid dendritic cells is essential for CpG-induced immune activation.** *Nat. Immunol.* 2006, **7**:740–746.

Swiecki, M., and Colonna, M.: **Unraveling the functions of plasmacytoid dendritic cells during viral infections, autoimmunity, and tolerance.** *Immunol. Rev.* 2010, **234**:142–162.

9-12   Macrophages are scavenger cells that can be induced by pathogens to present foreign antigens to naive T cells.

Barker, R.N., Erwig, L.P., Hill, K.S., Devine, A., Pearce, W.P., and Rees, A.J.: **Antigen presentation by macrophages is enhanced by the uptake of necrotic, but not apoptotic, cells.** *Clin. Exp. Immunol.* 2002, **127**:220–225.

Underhill, D.M., Bassetti, M., Rudensky, A., and Aderem, A.: **Dynamic interactions of macrophages with T cells during antigen presentation.** *J. Exp. Med.* 1999, **190**:1909–1914.

Zhu, F.G., Reich, C.F., and Pisetsky, D.S.: **The role of the macrophage scavenger receptor in immune stimulation by bacterial DNA and synthetic oligonucleotides.** *Immunology* 2001, **103**:226–234.

9-13   B cells are highly efficient at presenting antigens that bind to their surface immunoglobulin.

Guermonprez, P., England, P., Bedouelle, H., and Leclerc, C.: **The rate of dissociation between antibody and antigen determines the efficiency of antibody-mediated antigen presentation to T cells.** *J. Immunol.* 1998, **161**:4542–4548.

Shirota, H., Sano, K., Hirasawa, N., Terui, T., Ohuchi, K., Hattori, T., and Tamura, G.: **B cells capturing antigen conjugated with CpG oligodeoxynucleotides induce Th1 cells by elaborating IL-12.** *J. Immunol.* 2002, **169**:787–794.

Zaliauskiene, L., Kang, S., Sparks, K., Zinn, K.R., Schwiebert, L.M., Weaver, C.T., and Collawn, J.F.: **Enhancement of MHC class II-restricted responses by receptor-mediated uptake of peptide antigens.** *J. Immunol.* 2002, **169**:2337–2345.

9-14   Cell-adhesion molecules mediate the initial interaction of naive T cells with antigen-presenting cells.

Dustin, M.L.: **T-cell activation through immunological synapses and kinapses.** *Immunol. Rev.* 2008, **221**:77–89.

Friedl, P., and Brocker, E.B.: **TCR triggering on the move: diversity of T-cell interactions with antigen-presenting cells.** *Immunol. Rev.* 2002, **186**:83–89.

Gunzer, M., Schafer, A., Borgmann, S., Grabbe, S., Zanker, K.S., Brocker, E.B., Kampgen, E., and Friedl, P.: **Antigen presentation in extracellular matrix: interactions of T cells with dendritic cells are dynamic, short lived, and sequential.** *Immunity* 2000, **13**:323–332.

Montoya, M.C., Sancho, D., Vicente-Manzanares, M., and Sanchez-Madrid, F.: **Cell adhesion and polarity during immune interactions.** *Immunol. Rev.* 2002, **186**:68–82.

Wang, J., and Eck, M.J.: **Assembling atomic resolution views of the immunological synapse.** *Curr. Opin. Immunol.* 2003, **15**:286–293.

9-15   Antigen-presenting cells deliver multiple signals for the clonal expansion and differentiation of naive T cells.

Bour-Jordan, H., and Bluestone, J.A.: **CD28 function: a balance of costimulatory and regulatory signals.** *J. Clin. Immunol.* 2002, **22**:1–7.

Chen, L., and Flies, D.B.: **Molecular mechanisms of T cell co-stimulation and co-inhibition.** *Nat. Rev. Immunol.* 2013, **13**:227–242.

Gonzalo, J.A., Delaney, T., Corcoran, J., Goodearl, A., Gutierrez-Ramos, J.C., and Coyle, A.J.: **Cutting edge: the related molecules CD28 and inducible costimulator deliver both unique and complementary signals required for optimal T-cell activation.** *J. Immunol.* 2001, **166**:1–5.

Greenwald, R.J., Freeman, G.J., and Sharpe, A.H.: **The B7 family revisited.** *Annu. Rev. Immunol.* 2005, **23**:515–548.

Kapsenberg, M.L.: **Dendritic-cell control of pathogen-driven T-cell polarization.** *Nat. Rev. Immunol.* 2003, **3**:984–993.

Wang, S., Zhu, G., Chapoval, A.I., Dong, H., Tamada, K., Ni, J., and Chen, L.: **Costimulation of T cells by B7-H2, a B7-like molecule that binds ICOS.** *Blood* 2000, **96**:2808–2813.

9-16   CD28-dependent co-stimulation of activated T cells induces expression of interleukin-2 and the high-affinity IL-2 receptor.

Acuto, O., and Michel, F.: **CD28-mediated co-stimulation: a quantitative support for TCR signalling.** *Nat. Rev. Immunol.* 2003, **3**:939–951.

Gaffen, S.L.: **Signaling domains of the interleukin 2 receptor.** *Cytokine* 2001, **14**:63–77.

Seko, Y., Cole, S., Kasprzak, W., Shapiro, B.A., and Ragheb, J.A.: **The role of cytokine mRNA stability in the pathogenesis of autoimmune disease.** *Autoimmun. Rev.* 2006, **5**:299–305.

Zhou, X.Y., Yashiro-Ohtani, Y., Nakahira, M., Park, W.R., Abe, R., Hamaoka, T., Naramura, M., Gu, H., and Fujiwara, H.: **Molecular mechanisms underlying differential contribution of CD28 versus non-CD28 costimulatory molecules to IL-2 promoter activation.** *J. Immunol.* 2002, **168**:3847–3854.

9-17 Additional co-stimulatory pathways involved in T-cell activation.

Croft, M.: **The role of TNF superfamily members in T-cell function and diseases.** *Nat. Rev. Immunol.* 2009, **9**:271–285.

Greenwald, R.J., Freeman, G.J., and Sharpe, A.H.: **The B7 family revisited.** *Annu. Rev. Immunol.* 2005, **23**:515–548.

Watts, T.H.: **TNF/TNFR family members in costimulation of T cell responses.** *Annu. Rev. Immunol.* 2005, **23**:23–68.

9-18 Proliferating T cells differentiate into effector T cells that do not require co-stimulation to act.

Gudmundsdottir, H., Wells, A.D., and Turka, L.A.: **Dynamics and requirements of T cell clonal expansion** *in vivo* **at the single-cell level: effector function is linked to proliferative capacity.** *J. Immunol.* 1999, **162**:5212–5223.

London, C.A., Lodge, M.P., and Abbas, A.K.: **Functional responses and costimulator dependence of memory CD4+ T cells.** *J. Immunol.* 2000, **164**:265–272.

Schweitzer, A.N., and Sharpe, A.H.: **Studies using antigen-presenting cells lacking expression of both B7-1 (CD80) and B7-2 (CD86) show distinct requirements for B7 molecules during priming versus restimulation of Th2 but not Th1 cytokine production.** *J. Immunol.* 1998, **161**:2762–2771.

9-19 CD8 T cells can be activated in different ways to become cytotoxic effector cells.

Andreasen, S.O., Christensen, J.E., Marker, O., and Thomsen, A.R.: **Role of CD40 ligand and CD28 in induction and maintenance of antiviral CD8+ effector T cell responses.** *J. Immunol.* 2000, **164**:3689–3697.

Blazevic, V., Trubey, C.M., and Shearer, G.M.: **Analysis of the costimulatory requirements for generating human virus-specific** *in vitro* **T helper and effector responses.** *J. Clin. Immunol.* 2001, **21**:293–302.

Liang, L., and Sha, W.C.: **The right place at the right time: novel B7 family members regulate effector T cell responses.** *Curr. Opin. Immunol.* 2002, **14**:384–390.

Seder, R.A., and Ahmed, R.: **Similarities and differences in CD4+ and CD8+ effector and memory T cell generation.** *Nat. Immunol.* 2003, **4**:835–842.

Weninger, W., Manjunath, N., and von Andrian, U.H.: **Migration and differentiation of CD8+ T cells.** *Immunol. Rev.* 2002, **186**:221–233.

9-20 CD4 T cells differentiate into several subsets of functionally different effector cells.

Basu, R., Hatton, R.D., and Weaver, C.T.: **The Th17 family: flexibility follows function.** *Immunol. Rev.* 2013, **252**:89–103.

Bluestone, J.A., and Abbas, A.K.: **Natural versus adaptive regulatory T cells.** *Nat. Rev. Immunol.* 2003, **3**:253–257.

Crotty, S.: **Follicular helper T cells (T_FH).** *Annu. Rev. Immunol.* 2011, **29**:621–663.

King, C.: **New insights into the differentiation and function of T follicular helper cells.** *Nat. Rev. Immunol.* 2009, **9**:757–766.

Littman, D.R., and Rudensky, A.Y.: **Th17 and regulatory T cells in mediating and restraining inflammation.** *Cell* 2010, **140**:845–858.

Murphy, K.M., and Reiner, S.L.: **The lineage decisions of helper T cells.** *Nat. Rev. Immunol.* 2002, **2**:933–944.

Nurieva, R.I., and Chung, Y.: **Understanding the development and function of T follicular helper cells.** *Cell Mol. Immunol.* 2010, **7**:190–197.

9-21 Cytokines induce the differentiation of naive CD4 T cells down distinct effector pathways.

Nath, I., Vemuri, N., Reddi, A.L., Jain, S., Brooks, P., Colston, M.J., Misra, R.S., and Ramesh, V.: **The effect of antigen presenting cells on the cytokine profiles of stable and reactional lepromatous leprosy patients.** *Immunol. Lett.* 2000, **75**:69–76.

O'Shea, J.J., and Paul, W.E.: **Mechanisms underlying lineage commitment and plasticity of helper CD4+ T cells.** *Science* 2010, **327**:1098–1102.

Reese, T.A., Liang, H.E., Tager, A.M., Luster, A.D., Van Rooijen, N., Voehringer, D., and Locksley, R.M.: **Chitin induces the accumulation in tissue of innate immune cells associated with allergy.** *Nature* 2007, **447**:92–96.

Szabo, S.J., Sullivan, B.M., Peng, S.L., and Glimcher, L.H.: **Molecular mechanisms regulating Th1 immune responses.** *Annu. Rev. Immunol.* 2003, **21**:713–758.

Weaver, C.T., Harrington, L.E., Mangan, P.R., Gavrieli, M., and Murphy, K.M.: **Th17: an effector CD4 lineage with regulatory T cell ties.** *Immunity* 2006, **24**:677–688.

9-22 CD4 T-cell subsets can cross-regulate each other's differentiation through the cytokines they produce.

Croft, M., Carter, L., Swain, S.L., and Dutton, R.W.: **Generation of polarized antigen-specific CD8 effector populations: reciprocal action of interleukin-4 and IL-12 in promoting type 2 versus type 1 cytokine profiles.** *J. Exp. Med.* 1994, **180**:1715–1728.

Grakoui, A., Donermeyer, D.L., Kanagawa, O., Murphy, K.M., and Allen, P.M.: **TCR-independent pathways mediate the effects of antigen dose and altered peptide ligands on Th cell polarization.** *J. Immunol.* 1999, **162**:1923–1930.

Harrington, L.E., Hatton, R.D., Mangan, P.R., Turner, H., Murphy, T.L., Murphy, K.M., and Weaver, C.T.: **Interleukin 17-producing CD4+ effector T cells develop via a lineage distinct from the T helper type 1 and 2 lineages.** *Nat. Immunol.* 2005, **6**:1123–1132.

Julia, V., McSorley, S.S., Malherbe, L., Breittmayer, J.P., Girard-Pipau, F., Beck, A., and Glaichenhaus, N.: **Priming by microbial antigens from the intestinal flora determines the ability of CD4+ T cells to rapidly secrete IL-4 in BALB/c mice infected with** *Leishmania major.* *J. Immunol.* 2000, **165**:5637–5645.

Martin-Fontecha, A., Thomsen, L.L., Brett, S., Gerard, C., Lipp, M., Lanzavecchia, A., and Sallusto, F.: **Induced recruitment of NK cells to lymph nodes provides IFN-γ for T_H1 priming.** *Nat. Immunol.* 2004, **5**:1260–1265.

Nakamura, T., Kamogawa, Y., Bottomly, K., and Flavell, R.A.: **Polarization of IL-4- and IFN-γ-producing CD4+ T cells following activation of naïve CD4+ T cells.** *J. Immunol.* 1997, **158**:1085–1094.

Seder, R.A., and Paul, W.E.: **Acquisition of lymphokine producing phenotype by CD4+ T cells.** *Annu. Rev. Immunol.* 1994, **12**:635–673.

9-23 Regulatory CD4 T cells are involved in controlling adaptive immune responses.

Fontenot, J.D., and Rudensky, A.Y.: **A well adapted regulatory contrivance: regulatory T cell development and the forkhead family transcription factor Foxp3.** *Nat. Immunol.* 2005, **6**:331–337.

Roncarolo, M.G., Bacchetta, R., Bordignon, C., Narula, S., and Levings, M.K.: **Type 1 T regulatory cells.** *Immunol. Rev.* 2001, **182**:68–79.

Sakaguchi, S.: **Naturally arising Foxp3-expressing CD25+CD4+ regulatory T cells in immunological tolerance to self and non-self.** *Nat. Immunol.* 2005, **6**:345–352.

Sakaguchi, S., Yamaguchi, T., Nomura, T., and Ono, M.: **Regulatory T cells and immune tolerance.** *Cell* 2008, **133**:775–787.

Saraiva, M., and O'Garra, A.: **The regulation of IL-10 production by immune cells.** *Nat. Rev. Immunol.* 2010, **10**:170–181.

9-24 Effector T-cell interactions with target cells are initiated by antigen-nonspecific cell-adhesion molecules.

Dustin, M.L.: **T-cell activation through immunological synapses and kinases.** *Immunol. Rev.* 2008, **221**:77–89.

van der Merwe, P.A., and Davis, S.J.: **Molecular interactions mediating T cell antigen recognition.** *Annu. Rev. Immunol.* 2003, **21**:659–684.

9-25   An immunological synapse forms between effector T cells and their targets to regulate signaling and to direct the release of effector molecules.

Bossi, G., Trambas, C., Booth, S., Clark, R., Stinchcombe, J., and Griffiths, G.M.: **The secretory synapse: the secrets of a serial killer.** *Immunol. Rev.* 2002, **189**:152–160.

Montoya, M.C., Sancho, D., Vicente-Manzanares, M., and Sanchez-Madrid, F.: **Cell adhesion and polarity during immune interactions.** *Immunol. Rev.* 2002, **186**:68–82.

Trambas, C.M., and Griffiths, G.M.: **Delivering the kiss of death.** *Nat. Immunol.* 2003, **4**:399–403.

9-26   The effector functions of T cells are determined by the array of effector molecules that they produce.
&
9-27   Cytokines can act locally or at a distance.

Basler, C.F., and Garcia-Sastre, A.: **Viruses and the type I interferon antiviral system: induction and evasion.** *Int. Rev. Immunol.* 2002, **21**:305–337.

Boulay, J.L., O'Shea, J.J., and Paul, W.E.: **Molecular phylogeny within type I cytokines and their cognate receptors.** *Immunity* 2003, **19**:159–163.

Guidotti, L.G., and Chisari, F.V.: **Cytokine-mediated control of viral infections.** *Virology* 2000, **273**:221–227.

Harty, J.T., Tvinnereim, A.R., and White, D.W.: **CD8+ T cell effector mechanisms in resistance to infection.** *Annu. Rev. Immunol.* 2000, **18**:275–308.

Proudfoot, A.E.: **Chemokine receptors: multifaceted therapeutic targets.** *Nat. Rev. Immunol.* 2002, **2**:106–115.

9-28   T cells express several TNF-family cytokines as trimeric proteins that are usually associated with the cell surface.

Bekker, L.G., Freeman, S., Murray, P.J., Ryffel, B., and Kaplan, G.: **TNF-alpha controls intracellular mycobacterial growth by both inducible nitric oxide synthase-dependent and inducible nitric oxide synthase-independent pathways.** *J. Immunol.* 2001, **166**:6728–6734.

Hehlgans, T., and Mannel, D.N.: **The TNF–TNF receptor system.** *Biol. Chem.* 2002, **383**:1581–1585.

Ware, C.F.: **Network communications: lymphotoxins, LIGHT, and TNF.** *Annu. Rev. Immunol.* 2005, **23**:787–819.

9-29   Cytotoxic T cells induce target cells to undergo programmed cell death via extrinsic and intrinsic pathways of apoptosis.

Aggarwal, B.B.: **Signalling pathways of the TNF superfamily: a double-edged sword.** *Nat. Rev. Immunol.* 2003, **3**:745–756.

Ashton-Rickardt, P.G.: **The granule pathway of programmed cell death.** *Crit. Rev. Immunol.* 2005, **25**:161–182.

Bishop, G.A.: **The multifaceted roles of TRAFs in the regulation of B-cell function.** *Nat. Rev. Immunol.* 2004, **4**:775–786.

Green, D.R., Droin, N., and Pinkoski, M.: **Activation-induced cell death in T cells.** *Immunol. Rev.* 2003, **193**:70–81.

Russell, J.H., and Ley, T.J.: **Lymphocyte-mediated cytotoxicity.** *Annu. Rev. Immunol.* 2002, **20**:323–370.

Siegel, R.M.: **Caspases at the crossroads of immune-cell life and death.** *Nat. Rev. Immunol.* 2006, **6**:308–317.

Wallin, R.P., Screpanti, V., Michaelsson, J., Grandien, A., and Ljunggren, H.G.: **Regulation of perforin-independent NK cell-mediated cytotoxicity.** *Eur. J. Immunol.* 2003, **33**:2727–2735.

9-30   The intrinsic pathway of apoptosis is mediated by the release of cytochrome *c* from mitochondria.

Borner, C.: **The Bcl-2 protein family: sensors and checkpoints for life-or-death decisions.** *Mol. Immunol.* 2003, **39**:615–647.

Bratton, S.B., and Salvesen, G.S.: **Regulation of the Apaf-1-caspase-9 apoptosome.** *J. Cell Sci.* 2010, **123**:3209–3214.

Chowdhury, D., and Lieberman, J.: **Death by a thousand cuts: granzyme pathways of programmed cell death.** *Annu. Rev. Immunol.* 2008, **26**:389–420.

Hildeman, D.A., Zhu, Y., Mitchell, T.C., Kappler, J., and Marrack, P.: **Molecular mechanisms of activated T cell death *in vivo*.** *Curr. Opin. Immunol.* 2002, **14**:354–359.

Strasser, A.: **The role of BH3-only proteins in the immune system.** *Nat. Rev. Immunol.* 2005, **5**:189–200.

9-31   Cytotoxic effector proteins that trigger apoptosis are contained in the granules of CD8 cytotoxic T cells.

Barry, M., Heibein, J.A., Pinkoski, M.J., Lee, S.F., Moyer, R.W., Green, D.R., and Bleackley, R.C.: **Granzyme B short-circuits the need for caspase 8 activity during granule-mediated cytotoxic T-lymphocyte killing by directly cleaving Bid.** *Mol. Cell Biol.* 2000, **20**:3781–3794.

Grossman, W.J., Revell, P.A., Lu, Z.H., Johnson, H., Bredemeyer, A.J., and Ley, T.J.: **The orphan granzymes of humans and mice.** *Curr. Opin. Immunol.* 2003, **15**:544–552.

Lieberman, J.: **The ABCs of granule-mediated cytotoxicity: new weapons in the arsenal.** *Nat. Rev. Immunol.* 2003, **3**:361–370.

Pipkin, M.E., and Lieberman, J.: **Delivering the kiss of death: progress on understanding how perforin works.** *Curr. Opin. Immunol.* 2007, **19**:301–308.

Yasukawa, M., Ohminami, H., Arai, J., Kasahara, Y., Ishida, Y., and Fujita, S.: **Granule exocytosis, and not the Fas/Fas ligand system, is the main pathway of cytotoxicity mediated by alloantigen-specific CD4+ as well as CD8+ cytotoxic T lymphocytes in humans.** *Blood* 2000, **95**:2352–2355.

9-32   Cytotoxic T cells are selective and serial killers of targets expressing a specific antigen.

Stinchcombe, J.C., and Griffiths, G.M.: **Secretory mechanisms in cell-mediated cytotoxicity.** *Annu. Rev. Cell Dev. Biol.* 2007, **23**:495–517.

Veugelers, K., Motyka, B., Frantz, C., Shostak, I., Sawchuk, T., and Bleackley, R.C.: **The granzyme B-serglycin complex from cytotoxic granules requires dynamin for endocytosis.** *Blood* 2004, **103**:3845–3853.

9-33   Cytotoxic T cells also act by releasing cytokines.

Amel-Kashipaz, M.R., Huggins, M.L., Lanyon, P., Robins, A., Todd, I., and Powell, R.J.: **Quantitative and qualitative analysis of the balance between type 1 and type 2 cytokine-producing CD8− and CD8+ T cells in systemic lupus erythematosus.** *J. Autoimmun.* 2001, **17**:155–163.

Dobrzanski, M.J., Reome, J.B., Hollenbaugh, J.A., and Dutton, R.W.: **Tc1 and Tc2 effector cell therapy elicit long-term tumor immunity by contrasting mechanisms that result in complementary endogenous type 1 antitumor responses.** *J. Immunol.* 2004, **172**:1380–1390.

Prezzi, C., Casciaro, M.A., Francavilla, V., Schiaffella, E., Finocchi, L., Chircu, L.V., Bruno, G., Sette, A., Abrignani, S., and Barnaba, V.: **Virus-specific CD8+ T cells with type 1 or type 2 cytokine profile are related to different disease activity in chronic hepatitis C virus infection.** *Eur. J. Immunol.* 2001, **31**:894–906.

The Humoral Immune Response

10

Many pathogens multiply in the body's extracellular spaces, and even intracellular pathogens can spread by moving through the extracellular fluids. The extracellular spaces are protected by the **humoral immune response**, in which antibodies produced by B cells act to destroy extracellular microorganisms and their products, and prevent the spread of intracellular infections. As we introduced in Section 1-20, antibodies contribute to immunity in three main ways: **neutralization**, **opsonization**, and **complement activation** (Fig. 10.1). Antibodies can bind to pathogens and prevent their ability to enter and infect cells, and therefore are thus said to neutralize the pathogen; antibodies may also bind bacterial toxins, preventing their action or ability to enter cells. Antibodies also facilitate opsonization, the uptake of the pathogens by phagocytes, by binding to Fc receptors through their constant regions (C regions). Finally, antibodies bound to pathogens can activate proteins of the classical pathway of the complement system, as we described in Chapter 2. This can increase opsonization by placing other complement proteins onto the pathogen's surface, help recruit phagocytic cells to the site of infection, and activate the membrane-attack complex, which can directly lyse certain microorganisms by forming pores in their membranes. The choice of which effector mechanisms are used is influenced by the heavy-chain isotype of the antibodies produced, which determines their class (see Section 5-12).

In the first part of this chapter, we describe the interactions of naive B cells with antigen and with helper T cells that lead to the activation of B cells and antibody production. Some microbial antigens can provoke antibody production without T-cell help, but activation of naive B cells by antigens usually involves help from **T follicular helper** (T_{FH}) cells (see Section 9-20). Activated B cells then differentiate into antibody-secreting **plasma cells** and memory B cells. Most antibody responses undergo a process called affinity maturation, in which antibodies of greater affinity for their target antigen are produced by the somatic hypermutation of antibody variable-region (V-region) genes. We examine the molecular mechanism of somatic hypermutation and its immunological consequences, as well as class switching—a process that generates the different classes of antibodies that confer functional diversity on the antibody response. Both affinity maturation and class switching occur only in B cells and require T-cell help. In the second part of the chapter, we introduce the distributions and functions of various classes of antibody, in particular those secreted into mucosal sites. In the third part of the chapter, we discuss in detail how the Fc region of the antibody engages various effector mechanisms to contain and eliminate infections. Like the T-cell response, the humoral immune response produces immunological memory, and this is discussed in Chapter 11.

IN THIS CHAPTER

B-cell activation by antigen and helper T cells.

The distributions and functions of immunoglobulin classes.

The destruction of antibody-coated pathogens via Fc receptors.

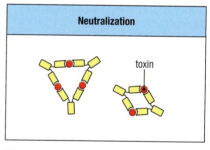

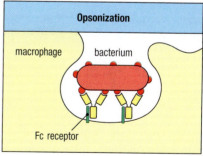

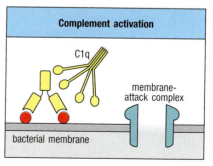

Fig. 10.1 Antibodies mediate the humoral immune response through neutralization, opsonization, and complement activation. After being secreted by plasma cells, antibodies protect the host from infection in three main ways. They can inhibit the toxic effects or infectivity of pathogens or their products by binding to them, a process called neutralization (top panel). When bound to pathogens, the antibody's Fc region can bind to Fc receptors on accessory cells, such as macrophages and neutrophils, helping these cells to ingest and kill the pathogen. This process is called opsonization (middle panel). Antibodies can trigger complement by activating C1, the first step in the classical complement pathway. Deposition of complement proteins enhances opsonization and can also directly kill certain bacterial cells by activating the membrane-attack complex (bottom panel).

B-cell activation by antigen and helper T cells.

The surface immunoglobulin that serves as the **B-cell receptor** (**BCR**) plays two roles in B-cell activation in response to pathogens. Like the antigen receptor on T cells, the BCR initiates a signaling cascade upon binding antigens derived from the microbe. In addition, the BCR can deliver the antigen to intracellular sites for antigen processing, so that antigenic peptides bound to MHC class II molecules can be returned to the B-cell surface. These peptide:MHC class II complexes are recognized by antigen-specific helper T cells that have already differentiated in response to the same pathogen. The effector T cells express surface molecules and cytokines that help the B cell to proliferate and to differentiate into antibody-secreting cells and into memory B cells, and a structure called the germinal center (see Section 10-6) is formed during an intermediate phase of the antibody response, before the emergence of long-term plasma cells that generate antibody or of memory B cells. Some microbial antigens can activate B cells directly in the absence of T-cell help, and the ability of B cells to respond directly to these antigens provides a rapid response to many important pathogens. However, the fine tuning of antibody responses to increase the affinity of the antibody for the antigen and the switching to most immunoglobulin classes other than IgM depend on the interaction of antigen-stimulated B cells with helper T cells and other cells in the peripheral lymphoid organs. Thus, antibodies induced by microbial antigens alone tend to have lower affinity and to be less functionally versatile than those induced with T-cell help.

10-1 Activation of B cells by antigen involves signals from the B-cell receptor and either T$_{FH}$ cells or microbial antigens.

As we learned in Chapter 8, activation of naive T cells requires signals derived from the T-cell receptor as well as co-stimulatory signals provided by professional antigen-presenting cells. Similarly, in addition to signals derived from the B-cell receptor, naive B cells also require accessory signals that can arise either from a helper T cell or, in some cases, directly from microbial constituents (Fig. 10.2).

Fig. 10.2 A second signal is required for B-cell activation by either thymus-dependent or thymus-independent antigens. The first signal (indicated as 1) required for B-cell activation is delivered through its antigen receptor (BCR) and activates several pathways as described in Chapter 7. Signaling by the BCR is enhanced by the co-receptors CD21 and CD19, which interact with C3b on opsonized microbial surfaces. For thymus-dependent antigens (first panel), a second signal (indicated as 2) is delivered by a helper T cell (T$_{FH}$) that recognizes degraded fragments of the antigen as peptides bound to MHC class II molecules on the B-cell surface. CD40L on the T$_{FH}$ cell binds to CD40 on the B cell, activating the non-canonical NFκB signaling pathway via NFκB-inducing kinase (NIK). This induces expression of pro-survival genes such as Bcl-2 (see Section 7-17). For thymus-independent antigens (second panel), a second signal can be delivered through Toll-like receptors that recognize antigen-associated TLR ligands, such as bacterial lipopolysaccharide (LPS) or bacterial DNA, as described in Chapter 3.

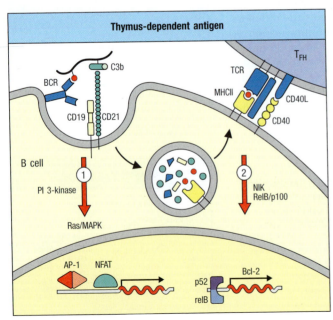
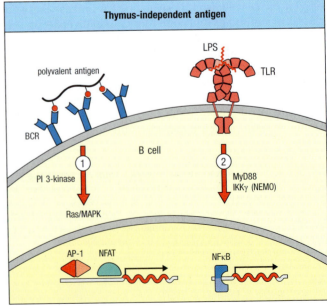

Protein antigens alone are unable to induce antibody responses in animals or humans who lack T cells, and they are therefore known as **thymus-dependent** or **TD antigens**, and typically involve antigen-specific T-cell help. The T cells involved are T_{FH} cells that reside in the lymphoid tissues and are not fully differentiated T_H1, T_H2, or T_H17 effector cells. To receive T-cell help, the B cell must display antigen on its surface in a form that a T cell can recognize. This occurs when antigen bound by surface immunoglobulin on the B cell is internalized and degraded within the B cell and peptides derived from it are returned to the cell surface in a complex with MHC class II molecules (see Fig. 10.2, first panel). When the T_{FH} cell recognizes these peptide:MHC complexes, it provides the B cell with signals that favor survival and induce proliferation. These signals include the activation of **CD40** on B cells by T_{FH} expression of its ligand, **CD40L** (CD154), and production of various cytokines by T_{FH} cells, including **IL-21** (Fig. 10.3). CD40 signaling activates the **non-canonical NFκB pathway** (see Section 7-23) and enhances B-cell survival by inducing the expression of anti-apoptotic molecules such as Bcl-2. IL-21 signaling activates STAT3 and enhances cellular proliferation and differentiation into plasma cells and memory B cells. Other cytokines produced by T_{FH} cells include IL-6, TGF-β, IFN-γ, and IL-4, which provide signals that can regulate the type of antibody produced, as we will see in Section 10-12. These cytokines are also made by other differentiated effector subsets (described in Chapter 9), but T_{FH} cells are distinct from these. For example, T_{FH} cells transcribe the IL-4 gene using regulatory elements that are independent of the transcription factors GATA-3 and STAT6, which are responsible for IL-4 production by T_H2 cells.

While B-cell responses to protein antigens rely on help from T cells, some microbial constituents can induce antibody production in the absence of helper T cells. These microbial antigens are known as **thymus-independent** or **TI antigens** because they can induce antibody responses in individuals who have no T lymphocytes. Such antigens are typically highly repetitive molecules, such as the polysaccharides of bacterial cell walls, and can cross-link the BCR on B cells. In such cases, a second signal can be derived from direct recognition of a common microbial constituent such as LPS that can activate TLR signaling in the B cell (see Fig. 10.2, second panel), activating the **NFκB pathway**, as described in Chapter 3. Thymus-independent antibody responses provide some protection against extracellular bacteria, and we will return to them later.

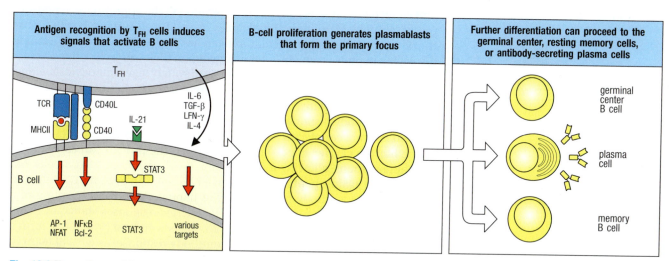

Fig. 10.3 T_{FH} cells provide several signals that activate B cells and control their subsequent differentiation. After antigen binding to the B-cell receptor delivers the first signal for B-cell activation (not shown), the T_{FH} cell delivers additional signals when it recognizes a peptide:MHC class II complex on the B-cell surface (first panel). Besides expression of CD40 ligand, the T_{FH} cell secretes several important cytokines. Included among them is IL-21, which activates the transcription factor STAT3 to enhance B-cell proliferation and survival. T_{FH} cells can also produce cytokines that will regulate isotype switching (see Section 10-12). After receiving these signals, activated B cells begin to proliferate (second panel), enter the germinal center, and eventually become plasma cells or memory B cells (third panel).

10-2 Linked recognition of antigen by T cells and B cells promotes robust antibody responses.

B-cell activation by antigens on microbial surfaces can be greatly stimulated by the concurrent deposition of complement on these pathogens. The **B-cell co-receptor complex** contains the cell-surface proteins CD19, CD21, and CD81 (see Fig. 7.27). When **CD21**, or complement receptor 2 (CR2), binds to the complement fragments C3d and C3dg that are deposited on microbial surfaces (see Section 2-13), it is brought near to the activated B-cell receptor bound to the same surface. CD21 and CD19 are associated with each other, and CD19 becomes phosphorylated by the activated B-cell receptor. This recruits PI 3-kinase, which then stimulates several downstream pathways, enhancing proliferation, differentiation, and antibody production (see Fig. 10.2, arrow 1). This effect is shown dramatically when mice are immunized with the experimental antigen hen egg-white lysozyme that is coupled to three linked molecules of C3dg. In this case the dose of modified lysozyme needed to induce antibody in the absence of added adjuvant is as little as 1/10,000 of that needed with the unmodified lysozyme.

For T-dependent antibody responses, the T cells involved are activated by the same antigen as is recognized by the B cells; this is called **linked recognition**. However, the peptide recognized by the T_{FH} cell is likely to differ from the protein epitope recognized by the B cell's antigen receptor. Natural antigens, such as viruses and bacteria, contain multiple proteins and carry both protein and carbohydrate epitopes. For linked recognition to occur, the peptide recognized by the T cell must be physically associated with the antigen recognized by the B cell's receptor, so that the B cell can take up and present the appropriate peptide to the T cell. For example, a B cell that recognizes an epitope on a viral coat protein will internalize the complete virus particle. The B cell can degrade multiple viral proteins into peptides for display on MHC class II molecules on the B-cell surface. CD4 T cells specific for such viral peptides may have been activated by dendritic cells earlier in the infection, and some will have differentiated into T_{FH} cells. When these T_{FH} cells are activated by B cells presenting their peptide, they are stimulated to provide specific signals that help B cells to generate antibodies against the viral coat protein (Fig. 10.4).

Linked recognition relies on the concentration of the appropriate peptide for presentation by MHC class II molecules on the B-cell surface. B cells whose B-cell receptor binds a particular antigen are 10,000 times more efficient at displaying peptide fragments of that antigen on their MHC class II molecules than are B cells that process the antigen through macropinocytosis alone. Linked recognition was originally discovered through studies of the production of antibodies against haptens, which are small chemical groups that cannot elicit antibody responses on their own (see Appendix I, Section A-1). But haptens that are coupled to a carrier protein become immunogenic—known as the **hapten carrier effect**—for two reasons. The protein can carry multiple hapten groups, allowing it to cross-link B-cell receptors. Also, T cells that are activated against peptides of the carrier protein can become T_{FH} cells and strengthen the antibody response to the hapten. Accidental coupling of a hapten to a protein is responsible for the allergic responses shown by many people to the antibiotic penicillin, which reacts with host proteins to form a coupled hapten that can stimulate an antibody response, as we will learn in Chapter 14.

Linked recognition works to preserve self-tolerance, since autoreactive antibodies will arise only if self-reactive T_{FH} and self-reactive B cells are present at the same time. This is discussed further in Chapter 15. Vaccine design can take advantage of linked recognition, as in the vaccine used to immunize infants against *Haemophilus influenzae* type b (see Section 16-26).

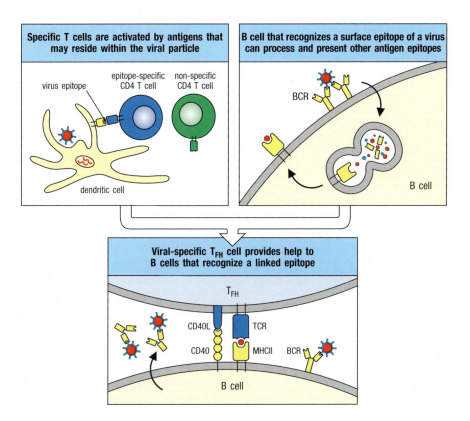

Specific T cells are activated by antigens that may reside within the viral particle

virus epitope
epitope-specific CD4 T cell
non-specific CD4 T cell
dendritic cell

B cell that recognizes a surface epitope of a virus can process and present other antigen epitopes

BCR
B cell

Viral-specific T$_{FH}$ cell provides help to B cells that recognize a linked epitope

T$_{FH}$
CD40L
TCR
CD40
MHCII
BCR
B cell

Fig. 10.4 T cells and B cells must recognize antigens contained within the same molecular complex in order to interact. In this example, an internal viral protein harbors a peptide epitope (shown as red) that is presented by MHC class II molecules and is recognized by a CD4 T cell. The virus also harbors a native epitope on an external viral coat protein (shown as blue) that is recognized by the surface immunoglobulin on a B cell. If the virus is captured and presented by a dendritic cell, a peptide-specific CD4 T cell (blue) becomes activated (top left panel), whereas nonspecific T cells (green) remain inactive. If the virus is recognized by a specific B cell (top right panel), peptides derived from internal viral proteins are processed and presented by MHC class II molecules. When the activated T cell recognizes its peptide on this B cell (bottom panel), the T cell will deliver various accessory signals to the B cell that promote antibody production against the coat protein. This process is known as linked recognition.

10-3   B cells that encounter their antigens migrate toward the boundaries between B-cell and T-cell areas in secondary lymphoid tissues.

The frequency of naive lymphocytes specific for a particular antigen is extremely low (less than 1 in 10,000). Thus, the chance of a random encounter between a T and a B cell with the same antigen specificity should be less than 1 in 10^8, making it remarkable that B cells ever interact with T$_{FH}$ cells with similar antigen specificity. For these reasons, linked recognition requires a precise regulation of the migration of activated B and T cells—orchestrated by several sets of ligands and receptors—into specific locations within the lymphoid tissues, which serves to increase the chances of a productive interaction (Fig. 10.5).

Naive T cells and B cells express the **sphingosine 1-phosphate receptor, S1PR1**, which they use to egress from the peripheral lymphoid tissues (see Section 9-7). However, before they exit, they are retained and initially occupy two distinct zones, the **T-cell areas** and the **primary lymphoid follicles** (or B-cell areas or B-cell zones), respectively (see Figs 1.18–1.20). These zones are established by different patterns of chemokine receptor expression and chemokine production. Naive T cells express the chemokine receptor **CCR7**, and localize to zones where its ligands, **CCL19** and **CCL21**, are highly expressed by stromal cells and dendritic cells (see Section 9-3). Circulating naive B cells express **CXCR5**, and when they migrate into lymphoid tissues, they enter the primary lymphoid follicles, where the chemokine **CXCL13** is abundant. Within the follicle, stromal cells and a specialized cell type, the **follicular dendritic cell** (**FDC**), secrete CXCL13. The FDC is a nonphagocytic cell of nonhematopoietic origin that bears numerous long processes; it functions by trapping antigen using complement receptors on its cell surface for access by B cells in the follicle.

Once in the follicle, naive B cells encounter the soluble TNF-family cytokine **BAFF** (see Section 8-8), which is secreted by FDCs, stromal cells, and dendritic cells and which acts as a survival factor for B cells. BAFF can act through three

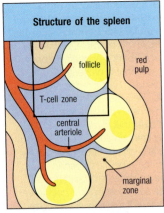

Structure of the spleen

follicle

red pulp

T-cell zone

central arteriole

marginal zone

Fig. 10.5 Antigen-binding B cells meet T cells at the border between the T-cell area and a B-cell follicle in secondary lymphoid tissues. Antigens enter the spleen from the blood and collect in T-cell zones and follicles (first panel). Naive CCR7-positive T cells and CXCR5-positive B cells migrate to distinct regions where the chemokines CCL19 and CCL21, or CXCL13 and 7α, 25-hydroxycholesterol (7α, 25-HC), respectively, are being produced (second panel). If a B cell encounters its antigen, either on a follicular dendritic cell (FDC) or a macrophage, it increases expression of CCR7 and migrates toward the border with the T-cell zone (third panel). T cells activated by antigen-presenting dendritic cells induce expression of CXCR5 and migrate to this same border, where linked recognition induces further B-cell proliferation. After 2 to 3 days, B cells reduce expression of CCR7, but retain EBI2 and migrate in response to 7α, 25-HC to the outer follicle and interfollicular regions (fourth panel). After another day or so, some B cells cluster in the interfollicular regions near the red pulp, proliferate, and differentiate into plasmablasts, forming a primary focus with terminal differentiation into antibody-secreting plasma cells. T cells that retain EBI2 expression may remain in the follicle and induce Bcl-6 expression to become T_{FH} cells that participate with B cells there to form a germinal center reaction.

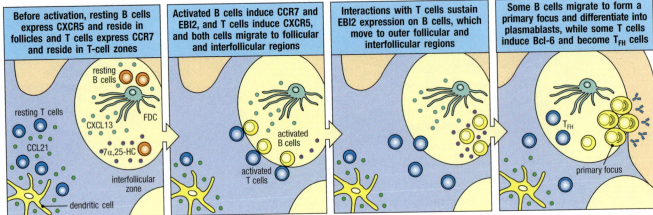

Before activation, resting B cells express CXCR5 and reside in follicles and T cells express CCR7 and reside in T-cell zones

resting B cells

resting T cells

CXCL13   FDC

CCL21

7α,25-HC

interfollicular zone

dendritic cell

Activated B cells induce CCR7 and EBI2, and T cells induce CXCR5, and both cells migrate to follicular and interfollicular regions

activated B cells

activated T cells

Interactions with T cells sustain EBI2 expression on B cells, which move to outer follicular and interfollicular regions

Some B cells migrate to form a primary focus and differentiate into plasmablasts, while some T cells induce Bcl-6 and become T_{FH} cells

T_{FH}

primary focus

receptors, but its major actions in promoting survival are through **BAFF-R** (Fig. 10.6). BAFF-R signals through **TRAF3** (see Section 3-7) to activate the non-canonical NFκB pathway, as described for CD40 (see Fig. 7.31), and, like CD40 signaling, induces expression of Bcl-2. Two other receptors for BAFF are **TACI** and **BCMA**, although BAFF has a relatively low affinity for BCMA. TACI and BCMA also bind the related cytokine **APRIL**, and they signal through TRAF2, 5, and 6 to induce signaling pathways involved in B-cell activation.

Antigens derived from microorganisms and viruses are transported into lymph nodes via the afferent lymph, and into the spleen via the blood. Opsonized antigens bearing C3b or C3dg accumulate in the B-cell follicles because they are trapped by complement receptors CR1 and CR2 expressed on the surface of FDCs. Opsonized particulate antigens can also be taken up by specialized macrophages residing in the **subcapsular sinus** (SCS) of lymph nodes and the **marginal sinus** of the spleen, regions that are both adjacent to the B-cell

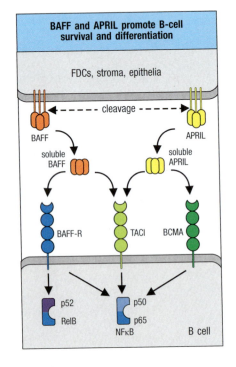

BAFF and APRIL promote B-cell survival and differentiation

FDCs, stroma, epithelia

cleavage

BAFF   APRIL

soluble BAFF   soluble APRIL

BAFF-R   TACI   BCMA

p52   p50
RelB   p65
         NFκB   B cell

Fig. 10.6 BAFF and APRIL promote B-cell survival and regulate differentiation. BAFF (B-cell activating factor, also called B-lymphocyte stimulator, or BLyS) and APRIL (a proliferation-inducing ligand) are both members of the TNF superfamily of cytokines. They are initially produced as membrane-bound trimers by several cell types. BAFF is produced by FDCs and other cells in the B-cell follicle, where it supports B-cell survival. Its main receptor, BAFF-R, signals in a manner similar to CD40 (see Fig. 7.31) through TRAF3 and NIK to activate both the non-canonical NFκB pathway, leading to the RelB:p52 transcription factor, and the canonical p50:p65 NFκB pathway. BAFF also binds to the receptors TACI (transmembrane activator and calcium modulator and cyclophilin ligand interactor) and BCMA (B-cell maturation antigen), although its affinity for the latter is relatively weak. These receptors activate the canonical NFκB pathway.

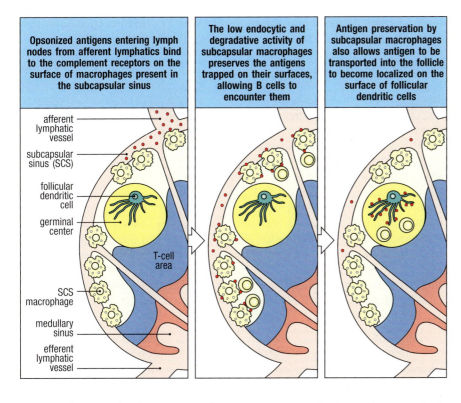

Fig. 10.7 Opsonized antigens are captured and preserved by subcapsular sinus macrophages. Macrophages residing in the lymph node subcapsular sinus (SCS) express complement receptors 1 and 2 (CR1 and CR2, respectively), are poorly endocytic, and have reduced levels of lysosomal enzymes compared with macrophages in the medulla. Opsonized antigen arriving from the afferent lymphatics binds to CR1 and CR2 on the surface of SCS macrophages. Instead of being completely degraded by these macrophages, some antigen is retained on the cell surface, where it can be presented and transferred to the surface of follicular B cells. B cells are then able to transport the antigen into the follicle, where it can be trapped on the surfaces of follicular dendritic cells.

follicles (Fig. 10.7). These macrophages seem to retain the antigen on their surface rather than ingesting and degrading it. These antigens can then be sampled and carried by antigen-specific follicular B cells. B cells of any antigen specificity could also acquire antigen from these macrophages via their complement receptors and transport it within the follicle. In the spleen, marginal zone B cells shuttle between that site and the follicle, carrying antigen trapped in the marginal zone for deposition on FDCs. SCS macrophages can also actively function to restrict the dissemination of infection. In mice, infection of these macrophages in lymph nodes by vesicular stomatitis virus (VSV), a relative of rabies virus, triggers the cells to produce interferon and to recruit plasmacytoid dendritic cells (pDCs). Type I interferon produced by pDCs restricts further viral spread, which would otherwise eventually pass on to the central nervous system.

After a naive follicular B cell first encounters specific antigen displayed by FDCs or macrophages, within a few hours it will become positioned in the outer follicles of lymphoid tissue close to the sites where antigen enters the lymph node or spleen. This positioning is orchestrated by the B cell's expression of a chemokine receptor, **EBI2** (GPR183), whose ligands are oxysterols such as 7α, 25-dihydroxycholesterol. The precise source of these ligands is still unclear, but they are abundant in the outer follicular and interfollicular regions. After sampling antigens there for 6 hours to 1 day, the B cell induces expression of CCR7, which functions together with EBI2 to distribute activated B cells along the interface between the B-cell follicle and the T-cell zone, where CCL21 is expressed.

During an immune response, T cells are activated within the T-cell zones by dendritic cells. When naive T cells are activated, some will proliferate, differentiate into effector cells, downregulate expression of S1P1, and exit the lymphoid tissue. However, others will induce expression of CXCR5 and migrate to the border with the B-cell follicle. There, T cells can encounter B cells activated during the same response, increasing the chance that they might recognize linked antigens presented by activated B cells that have recently moved to this location (see Fig. 10.5).

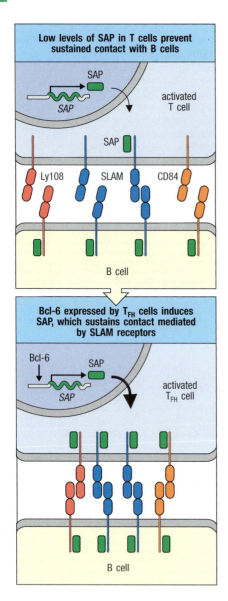

Fig. 10.8 Induction of SAP in T$_{FH}$ cells allows SLAM family receptors to mediate sustained contact with B cells. The SLAM receptor family members SLAM, Ly108, and CD84 are expressed on T cells and B cells and mediate homotypic interactions that lead to adhesion between cells. SLAM can also enhance signaling by the T-cell receptor to augment production of cytokines such as IL-21 that help B cells. The SLAM-associated molecule SAP is a signaling adapter that is required for one SLAM receptor to sustain binding with another. T cells initially express SAP at low levels that are insufficient for sustained adhesion between T and B cells. Fully differentiated T$_{FH}$ cells express high levels of the transcription factor Bcl-6, which induces higher levels of SAP expression. This level is sufficient to sustain cell–cell interactions and allow for the delivery of CD40L and cytokine signals to B cells.

10-4　T cells express surface molecules and cytokines that activate B cells, which in turn promote T$_{FH}$-cell development.

When T$_{FH}$ cells encounter an activating peptide presented by B cells, the T$_{FH}$ cells respond by expressing receptors and cytokines that in turn activate B cells. As mentioned above, the induced expression of CD40L on T$_{FH}$ cells activates CD40 on B cells to increase B-cell survival, and also induces B-cell expression of co-stimulatory molecules, especially those of the B7 family. Activated T cells also express **CD30 ligand** (**CD30L**), which binds to **CD30** expressed on B cells and promotes B-cell activation. Mice lacking CD30 show reduced proliferation of activated B cells in lymphoid follicles and weaker secondary humoral responses than normal. T$_{FH}$ cells also secrete several cytokines that regulate B-cell proliferation and antibody production. Primary among these is **IL-21**, which is produced early in immune responses by T$_{FH}$ cells and which activates the transcription factor **STAT3** in B cells to support proliferation and differentiation. IL-21 exerts similar autocrine effects on T$_{FH}$ cells. Later in the antibody response, T$_{FH}$ cells will produce other cytokines, such as IL-4 and IFN-γ, that are characteristic of the other T helper subsets (described in Chapter 9). These will impact B-cell differentiation, particularly class switching, as we discuss later.

The ability of T$_{FH}$ cells to successfully deliver these signals to B cells depends on intimate contact between these cells. Specific adhesion molecules, including several Ig superfamily receptors of the **SLAM** (**signaling lymphocyte activation molecule**) family, are involved that prolong and stabilize cell–cell contact. T$_{FH}$ cells and B cells both express SLAM (CD150), **CD84**, and **Ly108**, which promote cell adhesion through homotypic binding interactions (**Fig. 10.8**). The cytoplasmic regions of these SLAM family receptors all interact with an adaptor protein, **SAP** (SLAM-associated protein), which is expressed highly by T$_{FH}$ cells and which is necessary for prolonging cell–cell contact mediated by these receptors. The SAP gene is inactivated in **X-linked lymphoproliferative syndrome**, which is associated with a T-cell and NK-cell lymphoproliferative disorder and with a defect in antibody production due to failed interactions between T$_{FH}$ cells and B cells in the germinal center, discussed below. The regulated migration of activated B cells and T$_{FH}$ cells to the same location in the peripheral lymphoid organ increases the chance that linked recognition will occur and deliver appropriate help for B-cell differentiation. Antigen-stimulated B cells that fail to interact with T cells that recognize the same antigen die within 24 hours.

This first interaction between T and B cells not only provides important help to B cells, but also influences T-cell differentiation by signals provided by the B cell. Activated B cells express **ICOSL**, a member of the B7 family of co-stimulatory molecules and a ligand for **ICOS** (inducible co-stimulatory protein), which is expressed by T cells. This T- and B-cell interaction, provided by linked recognition, activates ICOS signaling in T cells and is important for the completion of T$_{FH}$ differentiation (see Section 7-21), leading to induction of the transcription factors **Bcl-6** and **c-Maf**. These transcription factors are required for SAP production and the consequent sustained contact between B and T$_{FH}$ cells.

10-5　Activated B cells differentiate into antibody-secreting plasmablasts and plasma cells.

After their initial encounter, B cells that have received T-cell help migrate from the follicle border to continue to proliferate and differentiate. Two to three days after activation, B cells begin to decrease expression of CCR7 and to increase expression of EBI2 (see Fig. 10.5). Decreased expression of CCR7 causes B cells to move away from the boundary with the T-cell zone: EBI2 directs their migration back to the interfollicular regions and the subcapsular sinus in the lymph nodes, or, in the spleen, to the splenic bridging channels, a region between the T-cell area and the red pulp. Here, some B cells will form an emerging aggregate of differentiating B cells called the **primary focus**, which in lymph nodes is located in the medullary cords, where lymph drains out of the node, and in the spleen can be seen as extrafollicular foci in the splenic red pulp. Primary foci are apparent by about 5 days after an infection or immunization with an antigen not previously encountered.

B cells proliferate in the primary focus for several days, and this constitutes the first phase of the primary humoral immune response. Some of these proliferating B cells differentiate into antibody-synthesizing **plasmablasts** in the primary focus. Not all B cells activated by the initial interaction with T_{FH} cells will move into the primary focus. Some will migrate into the lymphoid follicle, where they may eventually differentiate into plasma cells, as described below. Plasmablasts are cells that have begun to secrete antibody, yet are still dividing and express many of the characteristics of activated B cells that allow their interaction with T cells. After a few more days, the plasmablasts in the primary focus stop dividing and may eventually die. Subsequently, long-lived plasma cells will develop and migrate to the bone marrow, where they will continue antibody production. Since many long-lived plasma cells are generated long after the primary focus has dissipated, it is likely that they do not arise directly from plasmablasts in the primary focus, but rather from B cells that entered the germinal center reaction.

The properties of resting B cells, plasmablasts, and plasma cells are compared in Fig. 10.9. The differentiation of a B cell into a plasma cell is accompanied by many morphological changes that reflect a commitment to the production of large amounts of secreted antibody, which can constitute up to 20% of all the protein synthesized by a plasma cell. Plasmablasts and plasma cells have a prominent perinuclear Golgi apparatus and an extensive rough endoplasmic reticulum that is rich in immunoglobulin molecules that are being synthesized and exported into the lumen of the endoplasmic reticulum for secretion. Plasmablasts have relatively large numbers of B-cell receptors on the cell surface, whereas plasma cells have many fewer. This low level of surface immunoglobulin on plasma cells may still be physiologically important, since their survival seems to be determined in part by their ability to continue to bind antigen. Plasmablasts still express B7 co-stimulatory molecules and MHC class II molecules; by contrast, plasma cells turn down the expression of MHC class II molecules. Nevertheless, T cells still provide important signals for plasma-cell differentiation and survival, such as IL-6 and CD40 ligand.

X-linked Lymphoproliferative Syndrome

Fig. 10.9 Plasma cells secrete antibody at a high rate but can no longer respond to antigen. Resting naive B cells have membrane-bound immunoglobulin (usually IgM and IgD) and MHC class II molecules on their surface. Although their V genes do not carry somatic mutations, B cells can take up antigen and present it to helper T cells. The T cells in return induce the B cells to proliferate and to undergo isotype switching and somatic hypermutation, but B cells do not secrete significant amounts of antibody during this period. Plasmablasts have an intermediate phenotype. They secrete antibody but retain substantial surface immunoglobulin and MHC class II molecules and so can continue to take up and present antigen to T cells. Plasmablasts early in the immune response and those activated by T-independent antigens have usually not undergone somatic hypermutation and class switching, and therefore secrete IgM. Plasma cells are terminally differentiated cells that secrete antibodies. Plasma cells have very low levels of surface immunoglobulin but can express MHC class II molecules and may suppress T_{FH} activity in a negative feedback pathway while differentiating. Early in the immune response they differentiate from unswitched activated B cells and secrete IgM; later in the response they derive from activated B cells that entered the germinal center reaction and underwent class switching and somatic hypermutation. Plasma cells have lost the ability to change the class of their antibody or undergo further somatic hypermutation.

	Intrinsic properties			Inducible by antigen stimulation		
B-lineage cell	**Surface Ig**	**Surface MHC class II**	**High-rate Ig secretion**	**Growth**	**Somatic hyper-mutation**	**Class switch**
Resting B cell	High	Yes	No	Yes	Yes	Yes
Plasmablast	High	Yes	Yes	Yes	Unknown	Yes
Plasma cell	Low	Yes	Yes	No	No	No

Recent evidence indicates that even the low level of MHC class II molecules expressed on plasma cells functions to present cognate antigen to T_{FH} cells, and acts to suppress IL-21 production and Bcl-6 expression, thus acting as a feedback pathway to regulate ongoing B-cell responses. While some plasma cells survive for only days to a few weeks after their final differentiation, others are very long lived and account for the persistence of antibody responses.

10-6 The second phase of a primary B-cell immune response occurs when activated B cells migrate into follicles and proliferate to form germinal centers.

Not all B cells activated by T_{FH} cells will migrate to the outer follicle and eventually establish a primary focus. Instead, some move into a primary lymphoid follicle together with their associated T_{FH} cells (Fig. 10.10), where they continue to proliferate and ultimately form a **germinal center**; follicles with germinal centers are also called **secondary lymphoid follicles**. Downregulating EBI2 by B cells appears to favor germinal center formation. In mice lacking EBI2 expression in B cells, antigen-activated B cells remain near the border with the T-cell zone and are able to form germinal centers, but generate fewer plasmablasts.

Germinal centers are composed mainly of proliferating B cells, but antigen-specific T cells make up about 10% of germinal center lymphocytes and provide indispensable help to the B cells. The germinal center is an area of active cell division that forms within a surrounding region of resting B cells in the primary follicle. Proliferating germinal center B cells displace the resting B cells toward the periphery of the follicle, forming a **mantle zone** of resting cells around the two distinguishable areas of activated B cells, called the **light zone** and the **dark zone** (Fig. 10.11, left panel). The germinal center grows in size as the immune response proceeds, and then shrinks and finally disappears when the infection is cleared. Germinal centers are present for about 3–4 weeks after initial antigen exposure.

The primary focus and the germinal center reaction differ in the quality of antibody that they produce. Plasmablasts, germinal center B cells, and early memory B cells begin to emerge during the first 4–5 days of an immune response. Plasmablasts in primary foci primarily secrete antibodies of the IgM isotype that offer some immediate protection. In contrast, B cells in the germinal center reaction undergo several processes that produce antibodies that are more effective in eliminating infections. These processes include **somatic hypermutation**, which alters the V regions of immunoglobulin genes (see below), and which enables a process called **affinity maturation**, which selects for the survival of mutated B cells that have a high affinity for the antigen.

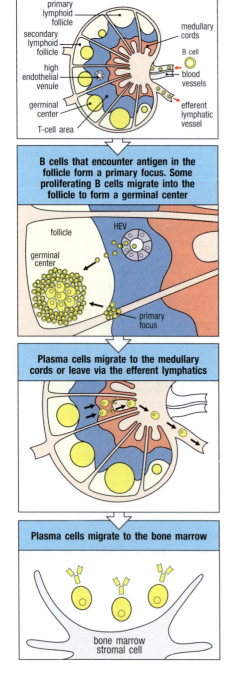

Fig. 10.10 Activated B cells form germinal centers in lymphoid follicles. Activation of B cells in a lymph node is shown here. Top panel: naive circulating B cells enter lymph nodes from the blood via high endothelial venules (HEV) and are attracted by chemokines into the primary lymphoid follicle; if these B cells do not encounter antigen in the follicle, they leave via the efferent lymphatic vessel. Second panel: B cells that have bound antigen move to the border with the T-cell area, where they may encounter activated helper T cells specific for the same antigen; these T cells interact with the B cells and activate them to start proliferation and differentiation into plasmablasts. Some B cells activated at the T-cell–B-cell border migrate to form a primary focus of antibody-secreting plasmablasts in the interfollicular regions (spleen) or medullary cords (lymph nodes), whereas others move back into the follicle, where they continue to proliferate and form a germinal center. Germinal centers are sites of sustained B-cell proliferation and differentiation. Follicles in which germinal centers have formed are known as secondary follicles. Within the germinal center, B cells begin their differentiation into either antibody-secreting plasma cells or memory B cells. Third and fourth panels: plasma cells leave the germinal center and migrate to the medullary cords, or leave the lymph node altogether via the efferent lymphatics and migrate to the bone marrow.

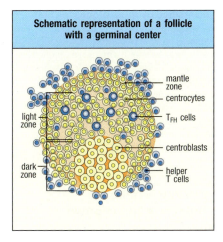

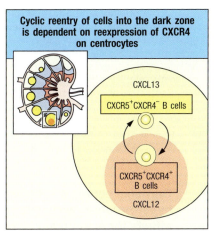

Fig. 10.11 The structure of a germinal center. The germinal center is a specialized microenvironment in which B-cell proliferation, somatic hypermutation, and selection for strength of antigen binding all occur. Closely packed centroblasts, which express CXCR4 and CXCR5, form the 'dark zone' of the germinal center; the less densely packed 'light zone' contains centrocytes, which express only CXCR5. Stromal cells in the dark zone produce CXCL12, which attracts the CXCR4-expressing centroblasts. Cyclic reentry describes the process by which B cells can lose and gain expression of CXCR4 and thus move from the light zone to the dark zone and back again.

In addition, **class switching** allows the selected B cells to produce antibodies with a variety of effector functions. These B cells will differentiate either into plasma cells that secrete higher-affinity and class-switched antibody in the latter part of the primary immune response, or into memory B cells as described in Chapter 11.

B cells in the germinal center divide rapidly, every 6–8 hours. Initially, these rapidly proliferating B cells, called **centroblasts**, express the chemokine receptors CXCR4 and CXCR5 but markedly reduce their expression of surface immunoglobulin, particularly of IgD. Centroblasts proliferate in the dark zone of the germinal center, named for its densely packed appearance (**Fig. 10.12**). Stromal cells in the dark zone produce **CXCL12** (SDF-1), a ligand for CXCR4 that acts to retain centroblasts in this region. As time goes on, some centroblasts reduce their rate of cell division, enter the growth phase, pausing at the G_2/M phase of the cell cycle, reduce CXCR4 expression, and begin to produce higher levels of surface immunoglobulin. These B cells are termed **centrocytes**. The loss of CXCR4 allows centrocytes to move into the light zone, a less densely packed area containing abundant FDCs that produce the chemokine CXCL13 (BLC), a ligand for CXCR5 (see Fig. 10.11, right panel). The B cells proliferate in the light zone, but to a lesser extent than in the dark zone.

Fig. 10.12 Germinal centers are sites of intense cell proliferation and cell death. The photomicrograph (first panel) shows a section through a human tonsillar germinal center. Closely packed centroblasts, seen in the lower part of this photomicrograph, form the so-called dark zone of the germinal center. Above this region is the less densely packed light zone. The second panel shows immunofluorescent staining of a germinal center. B cells are found in the dark zone, light zone, and mantle zone. Proliferating cells are stained green for Ki67, an antigen expressed in nuclei of dividing cells, revealing the rapidly proliferating centroblasts in the dark zone. The dense network of follicular dendritic cells, stained red, mainly occupies the light zone. Centrocytes in the light zone proliferate to a lesser degree than centroblasts. Small recirculating B cells occupy the mantle zone at the edge of the B-cell follicle. Large masses of CD4 T cells, stained blue, can be seen in the T-cell zones, which separate the follicles. There are also significant numbers of T cells in the light zone of the germinal center; CD4 staining in the dark zone is associated mainly with CD4-positive phagocytes, that digest B cells that die there. Photographs courtesy of I. MacLennan.

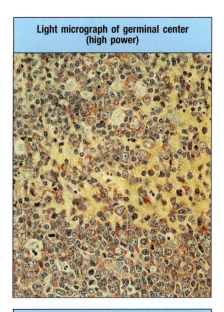

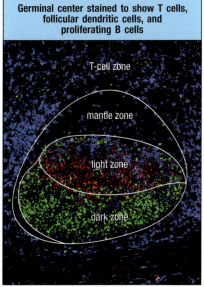

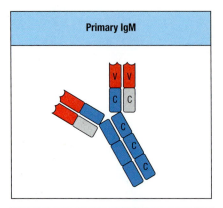

Primary IgM

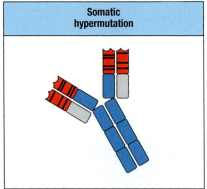

Somatic hypermutation

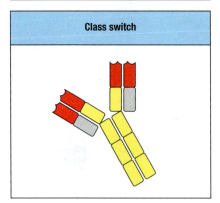

Class switch

MOVIE 10.1

Fig. 10.13 The primary antibody repertoire is diversified by three processes that modify the rearranged immunoglobulin gene. First panel: the primary antibody repertoire is initially composed of IgM-containing variable regions (red) produced by V(D)J recombination and constant regions (blue) from the μ gene segment. The range of reactivity of this primary repertoire can be further modified by somatic hypermutation, by class switch recombination at the immunoglobulin loci, and in some species by gene conversion (not shown). Second panel: somatic hypermutation results in mutations (shown as black lines) being introduced into the heavy-chain and light-chain V regions (red), altering the affinity of the antibody for its antigen. Third panel: in class switch recombination, the initial μ heavy-chain C regions (blue) are replaced by heavy-chain regions of another isotype (shown as yellow), modifying the effector activity of the antibody but not its antigen specificity.

10-7 Germinal center B cells undergo V-region somatic hypermutation, and cells with mutations that improve affinity for antigen are selected.

Somatic hypermutation introduces mutations that change anywhere from one to a few amino acids in the immunoglobulin, producing closely related B-cell clones that differ subtly in specificity and antigen affinity (Fig. 10.13). These mutations in the V genes are initiated by an enzyme called **activation-induced cytidine deaminase**, or **AID**, which is expressed only by germinal center B cells. Before describing the enzymatic mechanisms initiated by AID, we first present a general overview of this process in which random mutations can improve antibody affinity.

The immunoglobulin V-region genes accumulate mutations at a rate of about one base pair change per 10^3 base pairs per cell division, while the rate of mutations in the rest of the cell's DNA is much lower: around one base pair change per 10^{10} base pairs per cell division. Somatic hypermutation also affects some DNA flanking the rearranged V gene, but does not generally extend into the C-region exons. Since each V region is encoded by about 360 base pairs and about three out of every four base changes will alter the amino acid encoded, there is about a 50% chance during each B-cell division that a mutation will occur to the receptor.

The point mutations accumulate in a stepwise manner as the descendants of each B cell proliferate in the germinal center to form B-cell clones (Fig. 10.14). An altered receptor can affect the ability of a B cell to bind antigen and thus will affect the fate of the B cell in the germinal center. Most mutations have a negative impact on the ability of the B-cell receptor to bind the original antigen, by preventing the correct folding of the immunoglobulin molecule or by blocking the complementarity-determining regions from binding antigen. Detrimental mutations may alter conserved framework regions (see Fig. 4.7) and disrupt basic immunoglobulin structure. Cells that harbor such detrimental mutations are eliminated by apoptosis in a process of negative selection, either because they can no longer make a functional B-cell receptor or because they cannot take up antigen as well as sibling B cells (Fig. 10.15). Germinal centers are filled with apoptotic B cells that are quickly engulfed by macrophages, giving rise to the characteristic **tingible body macrophages**. These contain dark-staining nuclear debris in their cytoplasm. Negative selection is implied by the relative scarcity of amino acid replacements in the framework regions, reflecting the loss of cells that had mutated any one of the many residues that are critical for immunoglobulin V-region folding. This process prevents rapidly dividing B cells from expanding to numbers that would overwhelm the lymphoid tissues. Less frequently, mutations may improve the affinity of a B-cell receptor for antigen, and these mutations will be selectively expanded (see Fig. 10.15) because the cells expressing receptors with such mutations will have an increased survival rate compared with low-affinity cells. Positive selection is evident in the accumulation of numerous amino acid replacements in the complementarity-determining regions, which determine antibody specificity

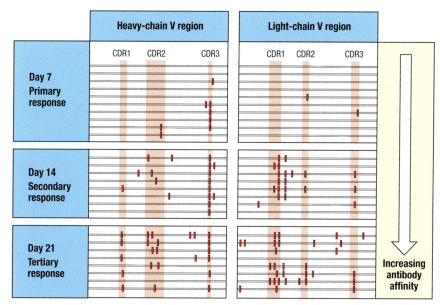

Fig. 10.14 Somatic hypermutation introduces mutations into the rearranged immunoglobulin variable (V) regions that improve antigen binding. The process of somatic hypermutation can be tracked by sequencing immunoglobulin V regions from hybridomas (clones of antibody-producing cells; see Appendix I, Section A-7) established at different time points after the experimental immunization of mice. The result of one experiment is depicted here. Each V region sequenced is represented by a horizontal line. The complementarity-determining regions CDR1, CDR2, and CDR3 are shown by pink shading. Mutations that change the amino acid sequence are represented by red bars. Within a few days of immunization, the V regions within a particular clone of responding B cells begin to acquire mutations, and over the course of the next week more mutations accumulate (top panels). B cells whose V regions have accumulated deleterious mutations and can no longer bind antigen die. B cells whose V regions have acquired mutations that improve the affinity of the B-cell receptor for antigen are able to compete more effectively for antigen, and receive signals that drive their proliferation and expansion. The antibodies they produce also have this improved affinity. This process of mutation and selection can continue in the lymph node germinal center through multiple cycles in response to secondary and tertiary immune responses elicited by further immunization with the same antigen (center and bottom panels). In this way, the antigen-binding efficiency of the antibody response is improved over time.

and affinity (see Fig. 10.14), a process we discuss in the next section. The result of selection for enhanced binding to antigen is that the nucleotide changes that alter amino acid sequences, and thus protein structure, tend to be clustered in the CDRs of the immunoglobulin V-region genes, whereas silent, or neutral, mutations that preserve amino acid sequence and do not alter protein structure are scattered throughout the V region.

Fig. 10.15 Selection for high-affinity mutants in the germinal center relies on help provided by T_FH cells. After activated B cells interact with T_FH cells at the follicle border, they migrate to germinal centers (GCs), where the following events depicted here occur. In the dark zone of the GC, somatic hypermutation alters the immunoglobulin V regions (first panel). In some B cells (yellow), the mutated B-cell receptor (BCR) will have low or no affinity for the antigen, while in other B cells (orange) the mutated BCR affinity may be higher. After exiting the dark zone, the B cells with higher-affinity BCRs will capture antigen (red) trapped on follicular dendritic cells (FDCs) and then process and present it on MHC class II molecules (second panel). B cells with low-affinity BCRs will fail to capture and present antigen. B cells that present linked antigen epitopes to T_FH cells will receive help through CD40L and IL-21, which promote survival and proliferation. B cells that lack antigen on MHC class II molecules receive no help and will eventually die (third panel). Some of the proliferating B cells undergo repeated cycles of entry to the dark zone, mutation, and selection (fourth panel), and other progeny B cells undergo differentiation to either memory B cells or plasma cells (not shown).

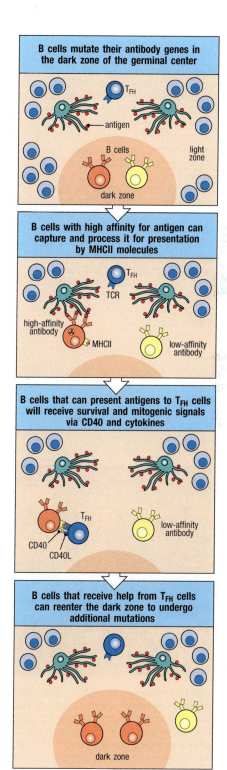

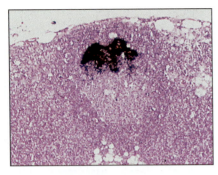

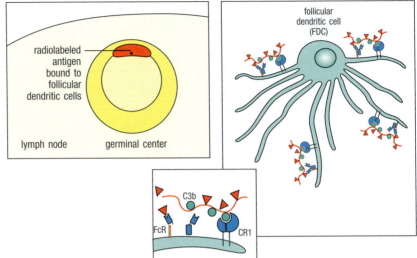

Fig. 10.16 Antigens are trapped in immune complexes that bind to the surface of follicular dendritic cells. Radiolabeled antigen localizes to, and persists in, lymphoid follicles of draining lymph nodes (see the light micrograph and the schematic representation (middle panel), showing a germinal center in a lymph node). The intense dark staining shows the localization in the germinal center of radiolabeled antigen that had been injected 3 days previously. The antigen is in the form of antigen:antibody:complement complexes bound to Fc receptors and to complement receptors CR1 or CR2 on the surface of the follicular dendritic cell (FDC), as depicted in the right-hand panel and inset. These complexes are not internalized, as such antigen can persist in this form for long periods. Photograph courtesy of J. Tew.

10-8 Positive selection of germinal center B cells involves contact with T_FH cells and CD40 signaling.

Selection of B cells with improved affinity for antigen occurs in increments. It was originally discovered *in vitro* that resting B cells could be kept alive by simultaneously cross-linking their B-cell receptors and ligating their cell-surface CD40. *In vivo* these signals are delivered by antigen and by T_{FH} cells, respectively. The details of selection in the germinal center have become more clear recently from *in vivo* two-photon microscopic studies (see Appendix I, A-10) that show that positive selection of a B cell depends on the B cell's ability to take up antigen, and to receive signals delivered by T_{FH} cells. It is thought that somatic hypermutation occurs in the centroblasts in the dark zone; when a centroblast reduces its rate of proliferation and becomes a centrocyte, it increases the number of B-cell receptors on its surface and moves to the light zone, where there are abundant FDCs. Antigen can be trapped and stored for long periods in the form of immune complexes on FDCs (Fig. 10.16 and Fig. 10.17). The centrocyte's ability to bind antigen determines its relative ability to acquire

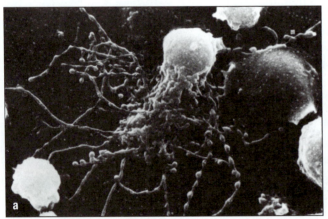

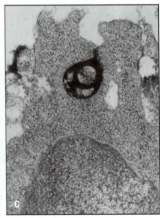

Fig. 10.17 Immune complexes bound to FDCs form iccosomes, which are released and can be taken up by B cells in the germinal center. FDCs have a prominent cell body and many dendritic processes. Immune complexes, bound to complement receptors and Fc receptors on the FDC surface, become clustered, forming prominent 'beads' along the dendrites (a). An intermediate form of FDC is shown, which has both straight filiform dendrites and others that are becoming beaded. These beads are shed from the cell as iccosomes (immune complex-coated bodies), which can bind to a B cell in the germinal center (b) and be taken up by it (c). In panels b and c, the iccosome has been formed with immune complexes containing horseradish peroxidase, which is electron-dense and therefore appears dark in the transmission electron micrographs. Photographs courtesy of A.K. Szakal.

antigen, in competition with the other clonally related centrocytes harboring different mutations. Centrocytes whose receptors bind antigen better will capture and present more peptides on their surface MHC class II molecules. T_{FH} cells in the germinal center recognize these peptides and, as before, are activated to deliver signals to the B cell that promote survival. Centrocytes whose mutations reduce antigen-binding affinity will take up less antigen, and so will receive weaker survival signals from T_{FH} cells. Successful B cells will reexpress CXCR4 and return to the dark zone, where they will undergo additional rounds of division, in effect becoming centroblasts again. Germinal center B cells that fail to acquire sufficient antigen from FDCs to engage T_{FH} cells will become apoptotic and be lost. This process of B-cell migration within the germinal center is known as the **cyclic reentry model** (see Fig. 10.11, right panel). In this way, the affinity and specificity of B cells are continually refined during the germinal center response, through affinity maturation (see Section 10-6). The selection process can be quite stringent: although 50–100 B cells may seed the germinal center, most of them leave no progeny, and by the time the germinal center reaches maximum size, it is typically composed of the descendants of only one or a few B cells.

In the germinal center, T_{FH} cells and B cells interact to deliver signals that are important for both cells (see Section 10-4). Mice that lack ICOS are deficient in the germinal center reaction and have severely reduced class-switched antibody responses due to defective T_{FH}-cell function. CD40 signaling in B cells is activated by CD40L on T_{FH} cells and increases expression of the survival molecule Bcl-X$_L$, a relative of Bcl-2. These interactions also include signaling by SLAM family receptors through the adapter protein SAP, as discussed above. Two-photon intravital microscopy has revealed that mice lacking the SLAM receptor CD84 have reduced numbers of conjugates between antigen-specific T cells and B cells in germinal centers, and these mice also have a reduced humoral response to antigen.

10-9 Activation-induced cytidine deaminase (AID) introduces mutations into genes transcribed in B cells.

Now that we have discussed the cellular processes involved in somatic hypermutation and affinity maturation, we will delve into the details of the mutation process itself. The enzyme AID is important for both somatic hypermutation and class switch recombination, as mice lacking AID have defects in both processes. People with mutations in the *AID* gene that inactivate the enzyme—that is, have **activation-induced cytidine deaminase deficiency**, or **AID deficiency**—also lack both somatic hypermutation and class switching. This condition leads to the production of predominantly IgM antibodies and the absence of affinity maturation, a syndrome known as **hyper IgM type 2 immunodeficiency** (discussed in Chapter 13).

AID is related to enzymes that deaminate cytosine to uracil in making nucleotide precursors for RNA and DNA synthesis. Its closest homolog, **APOBEC1** (apolipoprotein B mRNA editing catalytic polypeptide 1), is an RNA-editing enzyme that deaminates cytosine in the context of RNA. However, AID fulfills its activity in antibody gene diversification by acting on cytosine in the DNA of the immunoglobulin locus. When AID deaminates cytidine residues in the immunoglobulin V regions, somatic hypermutation is initiated; when cytidine residues in switch regions are deaminated, class switch recombination is initiated.

AID can deaminate cytidine residues in single-stranded DNA but not double-stranded DNA (**Fig. 10.18**). For AID to act, AID target genes are typically being transcribed, so that the DNA double helix is temporarily unwound. Since AID is expressed only in germinal center B cells, targeting of the immunoglobulin genes takes place only in these cells and in the actively transcribed rearranged

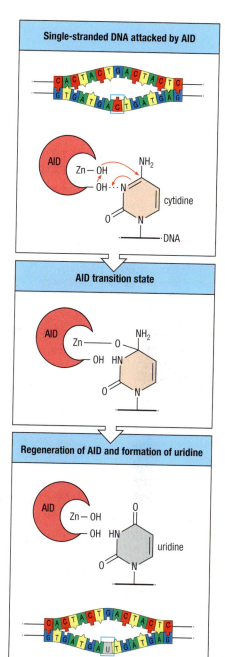

Fig. 10.18 Activation-induced cytidine deaminase (AID) is the initiator of mutations in somatic hypermutation, gene conversion, and class switching. The activity of AID, which is expressed only in B cells, requires access to the cytidine side chain of a single-stranded DNA molecule (first panel), which is normally prevented by the hydrogen bonding in double-stranded DNA. AID initiates a nucleophilic attack on the exposed cytosine ring (second panel), which is resolved by the deamination of the cytidine to form uridine (third panel).

Activation-Induced Cytidine Deaminase Deficiency

V regions where RNA polymerase generates transient single-stranded regions. Somatic hypermutation does not occur in loci that are not being actively transcribed. Rearranged V_H and V_L genes are mutated even if they are 'nonproductive' rearrangements and are not being expressed as protein, as long as they are being transcribed. Some actively transcribed genes in B cells besides those for immunoglobulins can also be affected by the somatic mutation process, but at a much lower rate.

10-10 Mismatch and base-excision repair pathways contribute to somatic hypermutation following initiation by AID.

The uridine produced by AID represents a dual lesion in DNA; not only is uridine foreign to normal DNA, but it is now a mismatch with the guanosine nucleoside on the opposite DNA strand. The presence of uridine in DNA can trigger several types of DNA repair—including the **mismatch repair** and the **base-excision repair** pathways—which further alter the DNA sequence. The various repair processes lead to different mutational outcomes (**Fig. 10.19**). In the mismatch repair pathway, the presence of uridine is detected by the mismatch repair proteins **MSH2** and **MSH6** (MSH2/6). They recruit nucleases that remove the complete uridine nucleotide along with several adjacent nucleotides from the damaged DNA strand. This is followed by a fill-in 'patch repair' by a DNA polymerase; unlike the process in all other cells, in B cells this DNA synthesis is error-prone and tends to introduce mutations at nearby A:T base pairs.

The initial steps in the base-excision repair pathway are shown in **Fig. 10.20**. In this pathway, the enzyme **uracil-DNA glycosylase** (**UNG**) removes the uracil base from the uridine to create an abasic site in the DNA. If no further modification is made, this will result at the next round of DNA replication in the random insertion of a nucleotide opposite the abasic site by DNA polymerase, leading to mutation. The action of UNG may, however, be followed by the action of another enzyme, **apurinic/apyrimidinic endonuclease 1** (**APE1**), which excises the abasic residue to create a single-strand discontinuity (known as a single-strand nick) in the DNA at the site of the original cytidine. Repair of the single-strand nick proceeding through double-strand breaks may result in gene conversion. Gene conversion is not used in the diversification of immunoglobulin genes in humans and mice, but is of importance in some other mammals and in birds.

Fig. 10.19 AID initiates DNA lesions whose repair leads to somatic hypermutation, class switch recombination, or gene conversion. When AID converts a cytidine (C) to uridine (U) in the DNA of an immunoglobulin gene, the final mutation produced depends on which repair pathways are used. Somatic hypermutation can result from either the mismatch repair (MSH2/6) pathway combined with error-prone polymerase activity of Polη, or the base-excision repair (UNG) pathway. Acting together, these can generate point mutations at and around the site of the original C:G pair. REV1 is a DNA repair enzyme that can synthesize DNA, or recruit other enzymes that can synthesize DNA, over the abasic sites in damaged DNA. REV1 itself will insert only C opposite the abasic site, but it can help recruit other polymerases that can also insert A, G, and T. The end result is insertion of a random nucleotide at the C:G residues where AID initially acted. Both class switch recombination and gene conversion require the formation of a single-strand break in the DNA. A single-strand break is formed when apurinic/apyrimidinic endonuclease 1 (APE1) removes a damaged residue from the DNA as part of the repair process (see Fig. 10.20, bottom two panels). In class switch recombination, single-strand breaks made in two of the so-called switch regions upstream of the C-region genes are converted to double-strand breaks. The cell's machinery for repairing double-strand breaks, which is very similar to the later stages of V(D)J recombination, then rejoins the DNA ends in a way that leads to a recombination event in which a different C-region gene is brought adjacent to the rearranged V region. Gene conversion results from the broken DNA strand using homologous sequences flanking the immunoglobulin gene as a template for repair DNA synthesis, thus replacing part of the gene with new sequences.

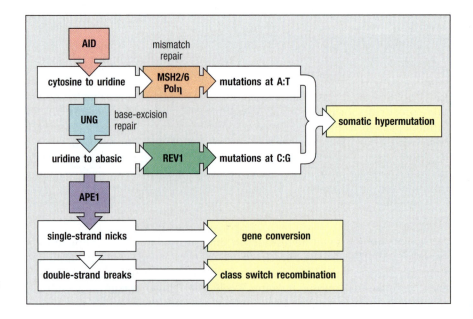

Fig. 10.20 The base-excision repair pathway produces single-strand nicks in DNA by the sequential actions of AID, uracil-DNA glycosylase (UNG), and apurinic/ apyrimidinic endonuclease 1 (APE1). Double-stranded DNA (first panel) can be made accessible to AID by transcription that unwinds the DNA helix locally (second panel). AID, which is specifically expressed in activated B cells, converts cytidine residues to uridines (third panel). The ubiquitous base-excision repair enzyme UNG can then remove the uracil ring from uridine, creating an abasic site (fourth panel). The repair endonuclease APE1 then cuts the sugar–phosphate DNA backbone next to the abasic residue (fifth panel), thereby forming a single-strand nick in the DNA (sixth panel). APE1 does not excise ribose to form a single-strand nick in DNA, but rather cuts the DNA backbone to yield a 5'-deoxyribosephosphate terminus that is then removed by, for example, DNA polymerase b.

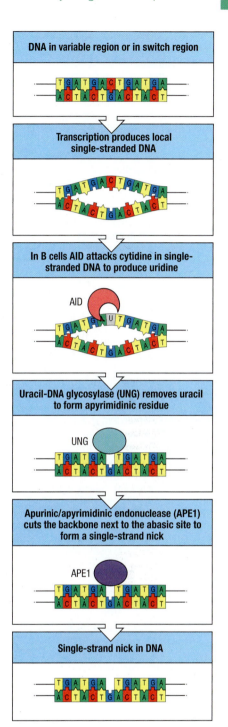

Somatic hypermutation involves both mutation at the original cytidines targeted by AID and mutation at nearby non-cytidine nucleotides. If the original U:G mismatch is recognized by UNG, then an abasic site will be generated in the DNA (see Fig. 10.19). If no further modification is made to this site, it can be replicated without instructive base pairing from the template strand by a class of **error-prone 'translesion' DNA polymerases** that normally repair gross damage to DNA, such as that caused by ultraviolet (UV) radiation. These polymerases can incorporate any nucleotide into the new DNA strand opposite the abasic site, and after a further round of DNA replication this can result in a stable mutation at the site of the original C:G base pair.

In the mismatch repair pathway in B cells, but not in other cell types, the DNA lesion is repaired by error-prone DNA polymerases rather than by more accurate polymerases that faithfully copy the undamaged template strand. Individuals with a defect in the translesion polymerase **Polη** have relatively fewer mutations than usual at A:T, but not at C:G, in their hypermutated immunoglobulin V regions. This fact suggests that Polη is the repair polymerase involved in this pathway of somatic hypermutation. These individuals also have a form of **xeroderma pigmentosum**, a condition resulting from the inability of their cells to repair DNA damage caused by UV radiation.

10-11 AID initiates class switching to allow the same assembled V_H exon to be associated with different C_H genes in the course of an immune response.

All the progeny of a particular B cell activated in an immune response will express the same V_H gene that was generated during its development in the bone marrow, although the gene may be modified by somatic hypermutation. In contrast, that B cell's progeny may express several different C-region isotypes as the cells mature and proliferate during the immune response. The first antigen receptors expressed by B cells are IgM and IgD, and the first antibody produced in an immune response is always IgM. Later in the immune response, the same assembled V region may be expressed in IgG, IgA, or IgE antibodies. This change is known as class switching (or **isotype switching**), and, unlike the expression of IgD, it involves irreversible DNA recombination. It is stimulated in the course of an immune response by external signals such as cytokines released by T_{FH} cells.

Switching from IgM to the other immunoglobulin classes occurs only after B cells have been stimulated by antigen. It is achieved through **class switch recombination**, which is a type of nonhomologous DNA recombination that is guided by stretches of repetitive DNA known as **switch regions**. Switch regions lie in the intron between the J_H gene segments and the C_μ gene, and at equivalent sites upstream of the genes for each of the other heavy-chain isotypes, with the exception of the δ gene, which does not require DNA rearrangement for its expression (**Fig. 10.21**, first panel). When a B cell switches from the coexpression of IgM and IgD to the expression of another subtype, DNA recombination occurs between S_μ and the S region immediately upstream of

MOVIE 10.2

Fig. 10.21 Class switching involves recombination between specific switch signals. The top panel shows the organization of a rearranged immunoglobulin heavy-chain locus before class switching. Second panel: this figure illustrates switching between the μ and ε isotypes in the mouse heavy-chain locus. Switch regions (S) are repetitive DNA sequences that guide class switching and are found upstream of each of the immunoglobulin C-region genes, with the exception of the δ gene. Switching is guided by the initiation of transcription by RNA polymerase (shaded circle) through these regions from promoters (shown as arrows) located upstream of each S. Due to the repetitive sequences, RNA polymerase can stall within the S regions, allowing these regions to serve as substrates for AID, and subsequently for UNG and APE1. Third panel: these enzymes introduce a high density of single-strand nicks into the non-template DNA strand and the template strand. Staggered nicks are converted to double-strand breaks by a mechanism that is not yet understood. Fourth panel: these breaks are then recognized by the cell's double-strand break repair machinery, which involves DNA-PKcs, Ku proteins, and other repair proteins. Bottom two panels: the two switch regions, in this case S_μ and S_ε, are brought together by the repair proteins, and class switching is completed by excision of the intervening region of DNA (including C_μ and C_δ) and ligation of the S_μ and S_ε regions.

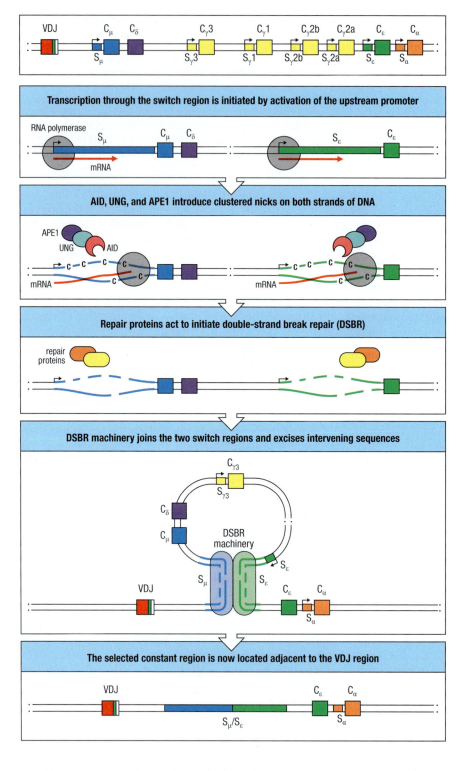

the new constant-region gene. In such a recombination event, the C_μ coding regions and the entire intervening DNA between C_μ and the S region undergoing rearrangement are deleted. Figure 10.21 illustrates switching from C_μ to C_ε in the mouse. All switch recombination events produce genes that can encode a functional protein, because the switch sequences lie in introns and therefore cannot cause frameshift mutations.

The enzyme AID initiates class switch recombination, and acts only on regions of DNA being transcribed. Certain properties of the switch region sequences

Fig. 10.22 RNA processed from switch region introns interacts with AID and guides its activity. Top panel: promoters upstream of each switch region initiate transcription by RNA polymerase upstream of a rearranged V_H gene, as in the case of C_μ, shown here, or a noncoding exon for all other constant regions. In all cases, the switch region itself lies within an intron upstream of the exons encoding the constant regions. This intronic switch region RNA is removed from the primary RNA transcript by splicing at specific splice acceptor and donor sites. Middle panel: after splicing, the switch region RNA is further processed and its repetitive elements allow the formation of putative G-quadruplex structures. Evidence indicates that these RNAs are able to bind AID, as implied in the cartoon. Bottom panel: the RNA acts as a guide to bring AID to the switch region by the ability of the G-quadruplex to hybridize with the original DNA template strand from which it was transcribed.

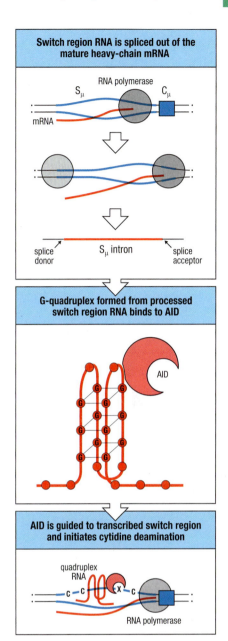

promote the accessibility to AID when they are being transcribed. Each switch region consists of many repeats of a G-rich sequence element on the non-template strand. For example, S_μ consists of about 150 repeats of the sequence (GAGCT)n(GGGGGT), where n is usually 3 but can be as many as 7. The sequences of the other switch regions (S_γ, S_α, and S_ε) differ in exact sequence but all contain repeats of the GAGCT and GGGGGT sequences. It appears that movement of RNA polymerase through this highly repetitive region is occasionally halted—called **polymerase stalling**. This may be caused by bubble-like structures, called **R-loops**, that form when the transcribed RNA displaces the non-template strand of the DNA double helix (see Fig. 10.21, third panel) due to having many G residues in tandem on one strand.

Polymerase stalling seems closely connected with the recruitment of AID to specific switch regions being transcribed. A multisubunit RNA processing/degradation complex, the **RNA exosome**, associates with AID and accumulates on transcribed switch regions, and the protein **Spt5** associates with the stalled polymerase; both are necessary for AID to generate double-stranded breaks. Recent evidence indicates that AID is selectively guided to the transcribed switch by an additional mechanism. After an RNA polymerase has completed transcription of one RNA template, the intron RNA harboring the switch region is spliced out. This RNA is processed to generate an RNA structure, called a **G-quadruplex**, that is based on the G-rich repetitive element of the switch region (Fig. 10.22). This G-quadruplex serves a dual purpose, both binding to AID and also associating with the switch region from which it was transcribed, based on its sequence complementarity. Thus the G-quadruplex guides AID to the appropriate switch region, where particular palindromic sequences, such as AGCT, act as good substrates to allow its cytidine deaminase activity to act on both strands concurrently. In this way, the G-quadruplex functions in a manner similar to the synthetic guide RNAs that deliver the Cas9 endonuclease to specific genomic regions, as described in Appendix I, Section A-35).

Following the generation of double-stranded breaks in switch regions, general cellular mechanisms for repairing these breaks lead to the nonhomologous recombination between switch regions that results in class switching (see Fig. 10.21, fourth and fifth panels). The ends to be joined are brought together by the alignment of repetitive sequences common to the different switch regions, and rejoining of the DNA ends then leads to excision of all DNA between the two switch regions and the formation of a chimeric region at the junction. Loss of AID completely blocks class switching, but deficiency of UNG in both mice and humans severely impairs class switching, suggesting sequential actions of AID and UNG in generating DNA breaks. Joining of DNA ends is probably mediated by classic nonhomologous end joining (as in V(D)J recombination) as well as by a poorly understood alternative end-joining pathway. Class switching is sometimes impaired in the disease **ataxia telangiectasia**, which is caused by mutations in the DNA-PKcs-family kinase **ATM**, a known DNA repair protein. The role of ATM in class switching is not yet entirely clear, however.

Ataxia Telangiectasia

10-12 Cytokines made by T$_{FH}$ cells direct the choice of isotype for class switching in T-dependent antibody responses.

Now that we understand the general mechanisms that control DNA rearrangements of class switching, we are ready to explain how a particular heavy-chain is selected during an immune response. It is the choice of antibody isotype that ultimately determines the effector function of antibodies, and we will see that this choice is largely controlled by the cytokines that are produced by T$_{FH}$ cells in the germinal center reaction.

As discussed above, interactions between germinal center B cells and T$_{FH}$ cells are essential for class switching to occur. The required interactions occur through the interplay of CD40 on B cells with CD40 ligand on activated helper T cells. Genetic deficiency of CD40 ligand greatly reduces class switching and causes abnormally high levels of plasma IgM, a condition known as **hyper IgM syndrome**. People with this defect lack antibodies of classes other than IgM and exhibit severe humoral immunodeficiency, manifested as repeated infections with common bacterial pathogens. Much of the IgM in hyper IgM syndromes may be induced by thymus-independent antigens on the pathogens that chronically infect these patients. Nevertheless, people with CD40 ligand deficiency can make IgM antibodies in response to thymus-dependent antigens, which indicates that in the B-cell response, CD40L–CD40 interactions are most important in enabling a sustained response that includes class switching and affinity maturation, rather than in the initial activation of B cells.

CD40 Ligand Deficiency

The selection of the particular C region for class switch recombination is not random but is regulated by the cytokines produced by T$_{FH}$ cells and other cells during the immune response. Different cytokines preferentially induce switching to different isotypes (Fig. 10.23). Cytokines induce class switching in part by inducing the production of RNA transcripts through the switch regions that lie 5′ to each heavy-chain C gene segment. When activated B cells are exposed to IL-4, for example, transcription from promoters that lie upstream of the switch regions of C$_\gamma$1 and C$_\varepsilon$ can be detected a day or two before switching occurs. This will make it possible for switch to occur to either of these two heavy-chain C genes, but in any particular germinal center B cell, recombination will occur in only one. In the example of class switching shown in Fig. 10.21, transcription through the S$_\varepsilon$ regions caused the rearrangement between the S$_\mu$ and S$_\varepsilon$ regions, making the IgE isotype antibody. This results because IL-4 signaling activates the transcription factor **STAT6**, which initiates transcription of the Iε promoter upstream of the S$_\varepsilon$ region. Other cytokines activate other promoters upstream of other switch regions to produce other antibody classes. T$_{FH}$ cells also produce IL-21, which promotes switching to IgG1 and IgG3. Transforming growth factor (TGF)-β induces switching to IgG2b (C$_\gamma$2b) and IgA (C$_\alpha$). IL-5 promotes switching to IgA, and interferon (IFN)-γ induces switching to IgG2a and IgG3.

Fig. 10.23 Different cytokines induce switching to different antibody classes. The individual cytokines induce (violet) or inhibit (red) the production of certain antibody classes. Much of the inhibitory effect is probably the result of directed switching to a different class. The actions of IL-21 on class switching are regulated by IL-4. These data are drawn from experiments with mouse cells.

Role of cytokines in regulating expression of antibody classes							
Cytokines	**IgM**	**IgG3**	**IgG1**	**IgG2b**	**IgG2a**	**IgE**	**IgA**
IL-4	Inhibits	Inhibits	Induces		Inhibits	Induces	
IL-5							Augments production
IFN-γ	Inhibits	Induces	Inhibits		Induces	Inhibits	
TGF-β	Inhibits	Inhibits		Induces			Induces
IL-21		Induces	Induces				Induces

10-13 B cells that survive the germinal center reaction eventually differentiate into either plasma cells or memory cells.

When B cells have undergone affinity maturation and class switching, some eventually exit from the light zone and start to differentiate into plasma cells that produce large amounts of antibody. In B cells, the transcription factors Pax5 and **Bcl-6** inhibit the expression of transcription factors required for plasma-cell differentiation, and both Pax5 and Bcl-6 are downregulated when the B cell starts differentiating. The transcription factor IRF4 then induces the expression of **BLIMP-1**, a transcriptional repressor that switches off genes required for B-cell proliferation, class switching, and affinity maturation. B cells in which BLIMP-1 is induced become plasma cells; they cease proliferating, increase the synthesis and secretion of immunoglobulins, and change their cell-surface properties. Plasma cells downregulate CXCR5 and upregulate CXCR4 and $\alpha_4{:}\beta_1$ integrins so that they can leave the germinal centers and home to peripheral tissues.

Some plasma cells deriving from germinal centers in lymph nodes or spleen migrate to the bone marrow, where a subset live for a long period, whereas others migrate to the medullary cords in lymph nodes or splenic red pulp. B cells that have been activated in germinal centers in mucosal tissues, and which are predominantly switched to IgA production, stay within the mucosal system. A splice variant of **XBP1** (X-box binding protein 1) is expressed in plasma cells and helps to regulate their secretory capacity. Plasma cells in bone marrow receive signals from stromal cells that are essential for their survival, and they can be very long lived, whereas plasma cells in the medullary cords or red pulp are not long lived. XBP1 is also required for plasma cells to colonize bone marrow successfully. Plasma cells in the bone marrow are the source of long-lasting high-affinity class-switched antibody.

Other germinal center B cells differentiate into **memory B cells**. Memory B cells are long-lived descendants of cells that were once stimulated by antigen and had proliferated in the germinal center. They divide very slowly if at all; they express surface immunoglobulin but secrete no antibody, or do so only at a low rate. Because the precursors of some memory B cells arise from the germinal center reaction, memory B cells can inherit the genetic changes that occur there, including somatic hypermutation and the gene rearrangements that result in a class switch. The signals that control which path of differentiation a B cell takes are still being investigated. We will briefly return to memory B cells in Chapter 11.

10-14 Some antigens do not require T-cell help to induce B-cell responses.

Humans and mice with T-cell deficiencies are able to produce antibodies against thymus-independent (TI) antigens, which we introduced in Section 10-1. These antigens include certain bacterial polysaccharides, polymeric proteins, and lipopolysaccharides, which are able to stimulate naive B cells in the absence of T-cell help. These nonprotein bacterial products cannot elicit classical T-cell responses, yet they induce antibody responses in normal individuals. In addition, there are TI antigens that are not derived from bacteria; these include plant-derived mitogens and lectins, viral antigens, and superantigens, and some parasite-derived antigens.

Thymus-independent antigens fall into two classes, TI-1 and TI-2, which activate B cells by two different mechanisms. **TI-1 antigens** rely on activity that can directly induce B-cell division without T-cell help. We now understand that TI-1 antigens contain molecules that cause the proliferation and differentiation of most B cells regardless of their antigen specificity; this is known as **polyclonal activation** (Fig. 10.24, top panels). TI-1 antigens are therefore

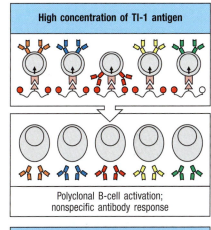

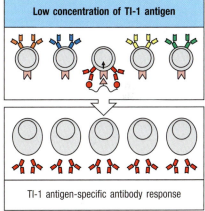

Fig. 10.24 TI-1 antigens induce polyclonal B-cell responses at high concentrations, and antigen-specific antibody responses at low concentrations. At high antigen concentration, the signal delivered by the B-cell-activating moiety of TI-1 antigens is sufficient to induce B-cell proliferation and antibody secretion in the absence of specific antigen binding to surface immunoglobulin. Thus, all B cells respond (top panels). At low concentration, only B cells specific for the TI-1 antigen bind enough of it to focus its B-cell activating properties onto the B cell; this gives a specific antibody response to epitopes on the TI-1 antigen (lower panels).

often called **B-cell mitogens**, a mitogen being a substance that induces cells to undergo mitosis. For example, LPS and bacterial DNA are both TI-1 antigens because they activate TLRs expressed by B cells (see Section 3-5) and can act as a mitogen. Naive murine B cells express most TLRs constitutively, but naive human B cells do not express high levels of most TLRs until they receive stimulation through the B-cell receptor. So by the time a B cell has been stimulated by antigen through its B-cell receptor, it is likely to express several TLRs and be responsive to stimulation by TLR ligands that accompany the antigens. Thus, when B cells are exposed to concentrations of TI-1 antigens that are 10^3–10^5 times lower than those used for polyclonal activation, only those B cells whose B-cell receptors specifically bind the TI-1 antigen become activated. At these low concentrations, amounts of TI-1 antigen sufficient for B-cell activation can only be concentrated on the B-cell surface with the aid of this specific binding (see Fig. 10.24, bottom panels). B-cell responses to TI-1 antigens in the early stages of an infection may be important in defense against several extracellular pathogens, but they do not lead to affinity maturation or memory B cells, both of which require antigen-specific T-cell help.

The second class of thymus-independent antigens—**TI-2 antigens**—consists of molecules that have highly repetitive structures, such as bacterial capsular polysaccharides. These contain no intrinsic B-cell-stimulating activity. Whereas TI-1 antigens can activate both immature and mature B cells, TI-2 antigens can activate only mature B cells; immature B cells, as we saw in Section 8-6, are inactivated by encounter with repetitive epitopes. Infants and young children up to about 5 years of age do not make fully effective antibody responses against polysaccharide antigens, and this might be because most of their B cells are immature.

Responses to several TI-2 antigens are prominently made by **marginal zone B cells**, a subset of nonrecirculating B cells that line the border of the splenic white pulp, and by **B-1 cells** (see Section 8-9). Marginal zone B cells are rare at birth and accumulate with age; they might therefore be responsible for most physiological TI-2 responses, which increase in efficiency with age. TI-2 antigens probably act by simultaneously cross-linking a critical number of B-cell receptors on the surface of antigen-specific mature B cells (**Fig. 10.25**, left

Fig. 10.25 B-cell activation by TI-2 antigens requires, or is greatly enhanced by, cytokines. Multiple cross-linking of the B-cell receptor by TI-2 antigens can lead to IgM antibody production (left panels), but there is evidence that in addition cytokines greatly augment these responses, and also lead to isotype switching (right panels). It is not clear where such cytokines originate, but one possibility is that dendritic cells, which may be able to bind the antigen through innate immune-system receptors on their surface and so present it to the B cells, secrete a soluble TNF-family cytokine called BAFF, which can activate class switching by the B cell.

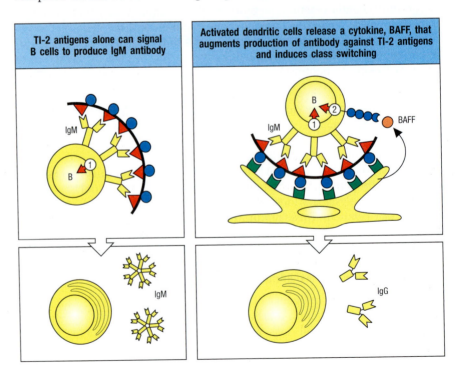

TI-2 antigens alone can signal B cells to produce IgM antibody

Activated dendritic cells release a cytokine, BAFF, that augments production of antibody against TI-2 antigens and induces class switching

panels). Dendritic cells and macrophages can provide co-stimulatory signals for activation of B cells by TI-2 antigens. One of these co-stimulatory signals is BAFF, which can be secreted by dendritic cells and interacts with the receptor TACI on the B cell (see Fig. 10.25, right panels). The density of TI-2 antigen epitopes is critical; excessive cross-linking of B-cell receptors renders mature B cells unresponsive or anergic, as in immature B cells, while too low a density may be insufficient for activation.

An important class of TI-2 antigens arises during infection by **capsulated bacteria**. Many common extracellular bacterial pathogens are surrounded by a polysaccharide capsule that enables them to resist ingestion by phagocytes. The bacteria not only escape direct destruction by phagocytes but also avoid stimulating T-cell responses against bacterial peptides presented by macrophages. IgM antibodies rapidly produced against the capsular polysaccharide independent of peptide-specific T-cell help will coat the bacteria, promoting their ingestion and destruction by phagocytes early in the infection.

Not all antibodies against bacterial polysaccharides are produced strictly through this TI-2 mechanism. We mentioned earlier the importance of antibodies against the capsular polysaccharide of *Haemophilus influenzae* type b in protective immunity to this bacterium. The immunodeficiency disease **Wiskott–Aldrich syndrome** is caused by defects in T cells that impair their interaction with B cells (see Chapter 13). Patients with Wiskott–Aldrich syndrome respond poorly to protein antigens, but, unexpectedly, also fail to make IgM and IgG antibody against polysaccharide antigens and are highly susceptible to infection with encapsulated bacteria such as *H. influenzae*. The failure to make IgM seems to be due in part to greatly reduced development of the marginal zone of the spleen, which contains B cells responsible for making much of the 'natural' IgM antibody against ubiquitous carbohydrate antigens. Thus, IgM and IgG antibodies induced by TI-2 antigens are likely to be an important part of the humoral immune response in many bacterial infections, and in humans at least, the production of class-switched antibodies to TI-2 antigens might normally rely on some degree of T-cell help.

Wiskott–Aldrich Syndrome

As well as producing IgM, TI responses can include switching to certain other antibody classes, such as IgG3 in the mouse. This is probably the result of help from dendritic cells (see Fig. 10.25, right panels), which provide secreted cytokines such as BAFF and membrane-bound signals to proliferating plasmablasts as they respond to TI antigens. The distinguishing features of thymus-dependent, TI-1, and TI-2 antibody responses are summarized in **Fig. 10.26**.

Summary.

B-cell activation by many antigens requires both binding of the antigen by the B-cell surface immunoglobulin—the B-cell receptor—and interaction of the B cell with antigen-specific helper T cells. Helper T cells recognize peptide fragments derived from the antigen internalized by the B cells and displayed by the B cells as peptide:MHC class II complexes. Follicular helper T cells stimulate B cells by conjugation in germinal centers, with binding of CD40 ligand on the T cell to CD40 on the B cell, and by their release of cytokines, such as IL-21. Activated B cells also express molecules, such as ICOSL, that can stimulate T cells. The initial interaction between B and T cells occurs at the border of the T-cell and B-cell areas of secondary lymphoid tissue, to which antigen-activated helper T cells and B cells migrate in response to chemokines. Further interactions between T cells and B cells continue after migration into the follicle and the formation of a germinal center.

Fig. 10.26 Properties of different classes of antigen that elicit antibody responses. Some data indicate a minor role for T cells in antibody responses to TI-2 antigens; robust responses to TI-2 antigens can be observed in T-cell-deficient mice.

	TD antigen	TI-1 antigen	TI-2 antigen
Antibody response in infants	Yes	Yes	No
Antibody production in congenitally athymic individual	No	Yes	Yes
Antibody response in absence of all T cells	No	Yes	Yes
Primes T cells	Yes	No	No
Polyclonal B-cell activation	No	Yes	No
Requires repeating epitopes	No	No	Yes
Examples of antigen	Diphtheria toxin Viral hemagglutinin Purified protein derivative (PPD) of *Mycobacterium tuberculosis*	Bacterial lipopoly-saccharide *Brucella abortus*	Pneumococcal polysaccharide *Salmonella* polymerized flagellin Dextran Hapten-conjugated Ficoll (polysucrose)

T cells induce a phase of vigorous B-cell proliferation in the germinal center reaction and direct the differentiation of clonally expanded B cells into either antibody-secreting plasma cells or memory B cells. Immunoglobulin genes expressed in B cells are diversified in the germinal center reaction by somatic hypermutation and class switching, initiated by activation-induced cytidine deaminase (AID). Unlike V(D)J recombination, these processes occur only in B cells. Somatic hypermutation diversifies the V region through the introduction of point mutations that are selected for providing greater affinity for the antigen as the immune response proceeds. Class switching does not affect the V region but increases the functional diversity of immunoglobulins by replacing the C_μ region in the immunoglobulin gene, which is first expressed with another heavy-chain C region to produce IgG, IgA, or IgE antibodies. Class switching provides antibodies with the same antigen specificity but distinct effector capacities. The switching to different antibody isotypes is regulated by cytokines released from helper T cells. Some nonprotein antigens stimulate B cells in the absence of linked recognition by peptide-specific helper T cells. Responses to these thymus-independent antigens are accompanied by only limited class switching and do not induce memory B cells. However, such responses have a crucial role in host defense against pathogens whose surface antigens cannot elicit peptide-specific T-cell responses.

The distributions and functions of immunoglobulin classes.

Extracellular pathogens can invade most sites within the body, and so antibodies must be equally widely distributed to combat them. Most classes of antibodies are distributed by diffusion from their site of synthesis, but specialized transport mechanisms are required to deliver antibodies across the epithelial surfaces lining the mucosa of organs such as the lungs and intestine. The particular heavy-chain isotype of the antibody can either limit antibody diffusion or engage specific transporters that deliver the antibody across epithelia. This part of the chapter describes these mechanisms and the antibody classes that use them to enter compartments of the body where their particular effector functions are appropriate. Here we restrict our discussion to the protective functions of antibodies that result solely from their binding to pathogens, and in the next part of the chapter, we discuss the effector cells and molecules that are specifically engaged by different antibody classes.

10-15 Antibodies of different classes operate in distinct places and have distinct effector functions.

Pathogens most commonly enter the body across the epithelial barriers of the mucosa lining the respiratory, digestive, and urogenital tracts, or through damaged skin. Less often, insects, wounds, or hypodermic needles introduce microorganisms directly into the blood. Antibodies protect all the body's mucosal surfaces, tissues, and blood from such infections; these antibodies serve to neutralize the pathogen or promote its elimination before it can establish a significant infection.

The different classes of antibodies (see Fig. 5.19) are adapted to function in different compartments of the body. Their functional activities and distributions are listed in Fig. 10.27. Because a given V region can become associated with any C region through class switching, the progeny of a single B cell can produce antibodies that share the same specificity yet provide all of the protective functions appropriate for each body compartment. All naive B cells express cell-surface IgM and IgD. IgM is the first antibody secreted by activated B cells but is less than 10% of the immunoglobulin found in plasma. Little IgD antibody is produced at any time, while IgE contributes a small but biologically important part of the immune response. IgG and IgA are the predominant antibody classes. IgE contributes a small but biologically important part of the response. The overall predominance of IgG is also due in part to its longer lifetime in the plasma (see Fig. 5.20).

Functional activity	IgM	IgD	IgG1	IgG2	IgG3	IgG4	IgA	IgE
Neutralization	+	–	++	++	++	++	++	–
Opsonization	+	–	++	*	++	+	+	–
Sensitization for killing by NK cells	–	–	++	–	++	–	–	–
Sensitization of mast cells	–	–	+	–	+	–	–	+++
Activates complement system	+++	–	++	+	+++	–	+	–

Distribution	IgM	IgD	IgG1	IgG2	IgG3	IgG4	IgA	IgE
Transport across epithelium	+	–	–	–	–	–	+++ (dimer)	–
Transport across placenta	–	–	+++	+	++	+/–	–	–
Diffusion into extravascular sites	+/–	–	+++	+++	+++	+++	++ (monomer)	+
Mean serum level (mg•ml^{-1})	1.5	0.04	9	3	1	0.5	2.1	3×10^{-5}

Fig. 10.27 Each human immunoglobulin class has specialized functions and a unique distribution. The major effector functions of each class (+++) are shaded in dark red, whereas lesser functions (++) are shown in dark pink, and very minor functions (+) in pale pink. The distributions are marked similarly, with actual average levels in serum being shown in the bottom row. IgA has two subclasses, IgA1 and IgA2. The IgA column refers to both. *IgG2 can act as an opsonin in the presence of an Fc receptor of the appropriate allotype, found in about 50% of people of Caucasian descent.

IgM antibodies are produced first in a humoral immune response and tend to be of low affinity. However, IgM molecules form pentamers that are stabilized by a single J-chain molecule (see Fig. 5.23) and have 10 antigen-binding sites, conferring higher overall avidity when binding to multivalent antigens such as bacterial capsular polysaccharides. This higher avidity of the pentamer compensates for the low affinity of the individual antigen-binding site within the IgM monomers. Because of the large size of the pentamers, IgM is found mainly in the bloodstream and, to a lesser extent, in the lymph, rather than in intercellular spaces within tissues. The pentameric structure of IgM makes it especially effective in activating the complement system, as we will see in the last part of this chapter. IgM hexamers can also form, and these fix complement much more efficiently than pentamers, possibly because C1q is also a hexamer. However, the *in vivo* role of IgM hexamers in protecting against infections has not been fully established.

Infection of the bloodstream has serious consequences unless it is controlled quickly, and the rapid production of IgM and its efficient activation of the complement system are important in controlling such infections. Some IgM is produced by conventional B cells that have not undergone class switching, but most is produced by B-1 cells residing in the peritoneal cavity and pleural spaces and by marginal zone B cells of the spleen. These cells secrete antibodies against commonly encountered carbohydrate antigens, including those of bacteria, and do not require T-cell help; they therefore provide a preformed repertoire of IgM antibodies in blood and body cavities that can recognize invading pathogens (see Section 8-9).

Antibodies of the other classes—IgG, IgA, and IgE—are smaller, and diffuse easily out of the blood into the tissues. IgA can form dimers (see Fig. 5.23), but IgG and IgE are always monomeric. The affinity of the individual antigen-binding sites for their antigen is therefore critical for the effectiveness of these antibodies, and most of the B cells expressing these classes have been selected in the germinal centers for their increased affinity for antigen after somatic hypermutation. IgG4 is the least abundant of the IgG subclasses, but has the unusual ability to form hybrid antibodies. One IgG4 heavy chain and attached light chain can split from the original heavy-chain dimer and reassociate with a different IgG4 heavy chain–light chain pair, forming a bivalent IgG4 antibody with two distinct antigen specificities.

IgG is the principal class of antibody in blood and extracellular fluid, whereas IgA is the principal class in secretions, the most important being those from the epithelia lining the intestinal and respiratory tracts. IgG efficiently opsonizes pathogens for engulfment by phagocytes and activates the complement system, but IgA is a less potent opsonin and a weak activator of complement. IgG operates mainly in the tissues, where accessory cells and molecules are available, whereas dimeric IgA operates mainly on epithelial surfaces, where complement and phagocytes are not normally present; therefore IgA functions chiefly as a neutralizing antibody. Monomeric IgA can be produced by plasma cells that differentiate from class-switched B cells in lymph nodes and spleen, and it acts as a neutralizing antibody in extracellular spaces and in the blood. This monomeric IgA is predominantly of the subclass IgA1; the ratio of IgA1 to IgA2 in the blood is 10:1. The IgA antibodies produced by plasma cells in the gut are dimeric and predominantly of subclass IgA2; the ratio of IgA2 to IgA1 in the gut is 3:2.

Finally, IgE antibody is present only at very low levels in blood or extracellular fluid, but is bound avidly by receptors on **mast cells** that are found just beneath the skin and mucosa and along blood vessels in connective tissue. Antigen binding to this cell-associated IgE triggers mast cells to release powerful chemical mediators that induce reactions such as coughing, sneezing, and vomiting, which in turn can expel infectious agents, as discussed later in this chapter.

10-16 Polymeric immunoglobulin receptor binds to the Fc regions of IgA and IgM and transports them across epithelial barriers.

In the mucosal immune system, IgA-secreting plasma cells are found predominantly in the lamina propria, which lies immediately below the basement membrane of many surface epithelia. From there the IgA antibodies can be transported across the epithelium to its external surface, for example to the lumen of the gut or of the bronchi (Fig. 10.28). IgA antibody synthesized in the lamina propria is secreted as a dimeric IgA molecule associated with a single J chain. This polymeric form of IgA binds specifically to a receptor called the **polymeric immunoglobulin receptor (pIgR)**, which is present on the basolateral surfaces of the overlying epithelial cells. When the pIgR has bound a molecule of dimeric IgA, the complex is internalized and carried in a transport vesicle through the cytoplasm of the epithelial cell to its luminal surface. This process is called transcytosis. IgM also binds to the pIgR and can be secreted into the gut by the same mechanism. Upon reaching the luminal surface of the enterocyte, the antibody is released into the mucous layer covering the gut lining by proteolytic cleavage of the extracellular domain of the pIgR. The cleaved extracellular domain of the pIgR is known as **secretory component** (frequently abbreviated to **SC**) and remains associated with the antibody. Secretory component is bound to the part of the Fc region of IgA that contains the binding site for the Fcα receptor I, which is why secretory IgA does not bind to this receptor. Secretory component serves several physiological roles. It binds to mucins in mucus, acting as 'glue' to bind secreted IgA to the mucous layer on the luminal surface of the gut epithelium, where the antibody binds and neutralizes gut pathogens and their toxins (see Fig. 10.28). Secretory component also protects the antibodies against cleavage by gut enzymes.

The principal sites of IgA synthesis and secretion are the gut, the respiratory epithelium, the lactating breast, and various other exocrine glands such as the salivary and tear glands. It is believed that the primary functional role of IgA antibodies is to protect epithelial surfaces from infectious agents, just as IgG antibodies protect the extracellular spaces inside tissues. By binding bacteria, virus particles, and toxins, IgA antibodies prevent the attachment of bacteria and viruses to epithelial cells and the uptake of toxins, and provide the first line of defense against a wide variety of pathogens. IgA is also thought to have an additional role in the gut, that of regulating the gut microbiota (see Chapter 12). The alveolar spaces in the lower respiratory tract lack the thicker mucosal layer characteristic of the upper respiratory tract, because efficient gas diffusion would be impeded by a mucous layer covering the alveolar epithelium. IgG can rapidly transudate into these spaces and is the major isotype responsible for protection there.

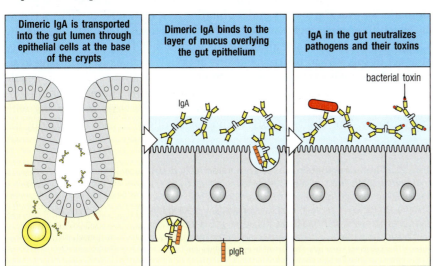

| Dimeric IgA is transported into the gut lumen through epithelial cells at the base of the crypts | Dimeric IgA binds to the layer of mucus overlying the gut epithelium | IgA in the gut neutralizes pathogens and their toxins |

bacterial toxin

IgA

pIgR

Fig. 10.28 Dimeric IgA is the major class of antibody present in the lumen of the gut. IgA is synthesized by plasma cells in the lamina propria and transported into the lumen of the gut through epithelial cells at the base of the crypts. Dimeric IgA binds to the layer of mucus overlying the gut epithelium and acts as an antigen-specific barrier to pathogens and toxins in the gut lumen

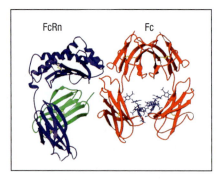

Fig. 10.29 The neonatal Fc receptor (FcRn) binds to the Fc portion of IgG. The structure of a molecule of FcRn (blue) is shown bound to one chain of the Fc portion of IgG (red), at the interface of the C$_\gamma$2 and C$_\gamma$3 domains, with the C$_\gamma$2 region at the top. The β$_2$-microglobulin component of the FcRn is green. The dark-blue structure attached to the Fc portion of IgG is a carbohydrate chain, reflecting glycosylation. FcRn transports IgG molecules across the placenta in humans and also across the gut in rats and mice. It also has a role in maintaining the levels of IgG in adults. Although only one molecule of FcRn is shown binding to the Fc portion, it is thought that it takes two molecules of FcRn to capture one molecule of IgG. Courtesy of P. Björkman.

10-17   The neonatal Fc receptor carries IgG across the placenta and prevents IgG excretion from the body.

Newborn infants are especially vulnerable to infection, having had no previous exposure to the microbes in the environment they enter at birth. IgA antibodies are secreted in breast milk and are transferred to the gut of the newborn infant, where they provide protection from newly encountered bacteria until the infant can synthesize its own protective antibody. IgA is not the only protective antibody that a mother passes on to her baby. Maternal IgG is transported across the placenta directly into the bloodstream of the fetus during intrauterine life; human babies at birth have as high levels of plasma IgG as their mothers, and with the same range of antigen specificities. The selective transport of IgG from mother to fetus is due to an IgG transport protein in the placenta, **FcRn (neonatal Fc receptor)**, which is closely related in structure to MHC class I molecules. Despite this similarity, FcRn binds IgG quite differently from the binding of peptide to MHC class I molecules, because its peptide-binding groove is occluded. It binds to the Fc portion of IgG molecules (Fig. 10.29). Two molecules of FcRn bind one molecule of IgG, bearing it across the placenta. Maternal IgG is ingested by the newborn animal from its mother's milk and colostrum, the protein-rich fluid secreted by the early postnatal mammary gland. In this case, FcRn transports the IgG from the lumen of the neonatal gut into the blood and tissues. Interestingly, FcRn is also found in adults in the gut and liver and on endothelial cells. Its function in adults is to maintain the levels of IgG in plasma, which it does by binding antibody, endocytosing it, and recycling it to the blood, thus preventing its excretion from the body.

By means of these specialized transport systems, mammals are supplied from birth with antibodies against pathogens common in their environments. As they mature and make their own antibodies of all isotypes, these are distributed selectively to different sites in the body (Fig. 10.30). Thus, throughout life, class switching and the distribution of antibody classes throughout the body provide effective protection against infection in extracellular spaces.

10-18   High-affinity IgG and IgA antibodies can neutralize toxins and block the infectivity of viruses and bacteria.

Pathogens can cause damage to a host by producing toxins or by infecting cells directly, and antibodies can protect by blocking both of these actions. Many bacteria cause disease by secreting toxins that damage or disrupt the function of the host's cells (Fig. 10.31). To affect cells, many toxins consist of separate domains for exerting toxicity and for binding to specific cell-surface receptors by which they enter cells. Antibodies that bind a toxin's receptor-binding site can prevent cell entry and protect cells from attack (Fig. 10.32). Antibodies that act in this way to neutralize toxins are referred to as **neutralizing antibodies**. Most toxins are active at nanomolar concentrations: a single molecule of diphtheria toxin can kill a cell. To neutralize toxins, therefore, antibodies must be able to diffuse into the tissues and bind the toxin rapidly and with high affinity. The ability of IgG antibodies to diffuse easily throughout the

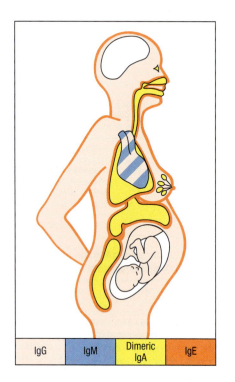

Fig. 10.30 Immunoglobulin classes are selectively distributed in the body. IgG and IgM predominate in blood (shown here for simplicity by IgM and IgG in the heart), whereas IgG and monomeric IgA are the major antibodies in extracellular fluid within the body. Dimeric IgA predominates in secretions across epithelia, including breast milk. The fetus receives IgG from the mother by transplacental transport. IgE is found mainly associated with mast cells just beneath epithelial surfaces (especially of the respiratory tract, gastrointestinal tract, and skin). The brain is normally devoid of immunoglobulin.

Disease	Organism	Toxin	Effects *in vivo*
Tetanus	*Clostridium tetani*	Tetanus toxin	Blocks inhibitory neuron action, leading to chronic muscle contraction
Diphtheria	*Corynebacterium diphtheriae*	Diphtheria toxin	Inhibits protein synthesis, leading to epithelial cell damage and myocarditis
Gas gangrene	*Clostridium perfringens*	Clostridial toxin	Phospholipase activation, leading to cell death
Cholera	*Vibrio cholerae*	Cholera toxin	Activates adenylate cyclase, elevates cAMP in cells, leading to changes in intestinal epithelial cells that result in loss of water and electrolytes
Anthrax	*Bacillus anthracis*	Anthrax toxic complex	Increases vascular permeability, leading to edema, hemorrhage, and circulatory collapse
Botulism	*Clostridium botulinum*	Botulinum toxin	Blocks release of acetylcholine, leading to paralysis
Whooping cough	*Bordetella pertussis*	Pertussis toxin	ADP-ribosylation of G proteins, leading to lymphoproliferation
		Tracheal cytotoxin	Inhibits cilia and causes epithelial cell loss
Scarlet fever	*Streptococcus pyogenes*	Erythrogenic toxin	Vasodilation, leading to scarlet fever rash
		Leukocidin Streptolysins	Kill phagocytes, allowing bacterial survival
Food poisoning	*Staphylococcus aureus*	Staphylococcal enterotoxin	Acts on intestinal neurons to induce vomiting. Also a potent T-cell mitogen (SE superantigen)
Toxic-shock syndrome	*Staphylococcus aureus*	Toxic-shock syndrome toxin	Causes hypotension and skin loss. Also a potent T-cell mitogen (TSST-1 superantigen)

Fig. 10.31 Many common diseases are caused by bacterial toxins. The toxins shown here are all exotoxins—proteins secreted by the bacteria. High-affinity IgG and IgA antibodies protect against these toxins. Bacteria also have nonsecreted endotoxins, such as lipopolysaccharide, which are released when the bacterium dies and may also mediate pathogenesis of disease. Host responses to exotoxins are more complex because the innate immune system has receptors for some endotoxins, such as TLR-4 (see Chapter 3).

extracellular fluid, and their high affinity for antigen once affinity maturation has taken place, make them the principal antibodies that neutralize toxins in tissues. High-affinity IgA antibodies similarly neutralize toxins at the mucosal surfaces of the body.

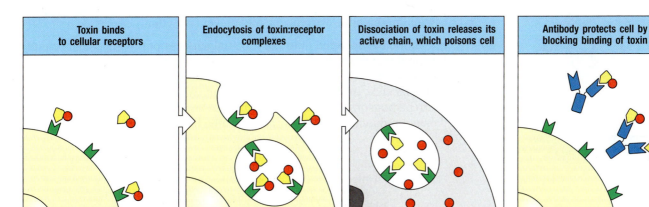

| Toxin binds to cellular receptors | Endocytosis of toxin:receptor complexes | Dissociation of toxin releases its active chain, which poisons cell | Antibody protects cell by blocking binding of toxin |

Fig. 10.32 Neutralization of toxins by IgG antibodies protects cells from damage. The damaging effects of many bacteria are due to the toxins they produce (see Fig. 10.31). These toxins are usually composed of several distinct moieties. One part of the toxin molecule binds a cell-surface receptor, which enables the molecule to be internalized. Another part of the toxin molecule then enters the cytoplasm and poisons the cell. Antibodies that inhibit toxin binding can prevent, or neutralize, these effects.

Diphtheria and tetanus toxins are two bacterial toxins in which the toxic and receptor-binding functions are on separate protein chains. It is therefore possible to immunize individuals, usually as infants, with modified toxin molecules in which the toxic chain has been denatured. These modified toxins, called **toxoids**, lack toxic activity but retain the receptor-binding site. Thus, immunization with the toxoid induces neutralizing antibodies that protect against the native toxin.

Some insect or animal venoms are so toxic that a single exposure can cause severe tissue damage or death. For these the adaptive immune response is too slow to be protective. Exposure to these venoms is a rare event, and protective vaccines have not been developed for use in humans. Instead, neutralizing antibodies are generated by immunizing other species, such as horses, with insect and snake venoms to produce anti-venom antibodies, or **antivenins**. The antivenins are injected into exposed individuals to protect them against the toxic effects of the venom. Transfer of antibodies in this way is known as **passive immunization** (see Appendix I, Section A-30).

Animal viruses infect cells by binding to a particular cell-surface receptor. These are often cell-type-specific proteins that determine which cells a virus can infect, or its **tropism**. Many antibodies that neutralize viruses do so by directly blocking the binding of virus to surface receptors (**Fig. 10.33**). The **hemagglutinin** of influenza virus, for example, binds to terminal sialic acid residues on the carbohydrates of glycoproteins present on epithelial cells of the respiratory tract. It is known as hemagglutinin because it recognizes and binds to similar sialic acid residues on chicken red blood cells and agglutinates these red blood cells. Antibodies against the hemagglutinin can prevent infection by the influenza virus. Such antibodies are called **virus-neutralizing antibodies**, and, as with the neutralization of toxins, high-affinity IgA and IgG antibodies are particularly important. However, antibodies can also neutralize viruses by interfering with the fusion mechanisms used to enter the cell's cytoplasm after binding to surface receptors.

Many bacteria have cell-surface molecules called **adhesins** that enable them to bind to the surface of host cells. This adherence is crucial to the ability of these bacteria to cause disease, whether they subsequently enter the cell, as do *Salmonella* species, or remain attached to the cell surface as extracellular pathogens (**Fig. 10.34**). *Neisseria gonorrhoeae*, the causative agent of the sexually transmitted disease gonorrhea, has a cell-surface protein known as **pilin** that enables the bacterium to adhere to the epithelial cells of the urinary and reproductive tracts and is essential to its infectivity. Antibodies against pilin can inhibit this adhesive reaction and prevent infection.

IgA antibodies secreted onto the mucosal surfaces of the intestinal, respiratory, and reproductive tracts are particularly important in inhibiting the colonization of these surfaces by pathogens and in preventing infection of the epithelial cells. Adhesion of bacteria to cells within tissues can also contribute to pathogenesis, and IgG antibodies against adhesins protect tissues from damage in much the same way as IgA antibodies protect mucosal surfaces.

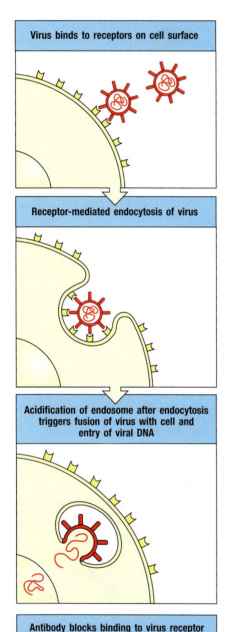

Virus binds to receptors on cell surface

Receptor-mediated endocytosis of virus

Acidification of endosome after endocytosis triggers fusion of virus with cell and entry of viral DNA

Antibody blocks binding to virus receptor and can also block fusion event

Fig. 10.33 Viral infection of cells can be blocked by neutralizing antibodies. For a virus to multiply within a cell, it must introduce its genes into the cell. The first step in entry is usually the binding of the virus to a receptor on the cell surface. For enveloped viruses, as shown in the figure, entry into the cytoplasm requires fusion of the viral envelope and the cell membrane. For some viruses this fusion event takes place on the cell surface (not shown); for others it can occur only within the more acidic environment of endosomes, as shown here. Non-enveloped viruses must also bind to receptors on cell surfaces, but they enter the cytoplasm by disrupting endosomes. Antibodies bound to viral surface proteins neutralize the virus, inhibiting either its initial binding to the cell or its subsequent entry.

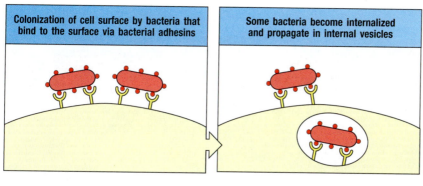

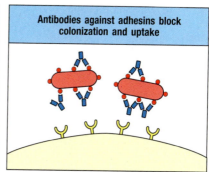

10-19 Antibody:antigen complexes activate the classical pathway of complement by binding to C1q.

Chapter 2 introduced the complement system as an essential component of innate immunity. Complement activation can proceed in the absence of antibody via the **lectin pathway** through the actions of mannose-binding lectin (MBL) and ficolins. But complement is also an important effector of antibody responses via the **classical pathway**. The different pathways of complement activation converge to coat pathogen surfaces or antigen:antibody complexes with covalently attached complement fragment C3b, which acts as an opsonin to promote uptake and removal by phagocytes. In addition, the terminal complement components can form a membrane-attack complex that damages some bacteria.

In the classical pathway, complement activation is triggered by C1, a complex of C1q and the serine proteases C1r and C1s (see Section 2-7). Complement activation is initiated when antibodies that are attached to the surface of a pathogen then bind to C1 via C1q (**Fig. 10.35**). C1q can be bound by either

Fig. 10.34 Antibodies can prevent the attachment of bacteria to cell surfaces. Many bacterial infections require an interaction between the bacterium and a cell-surface receptor. This is particularly true for infections of mucosal surfaces. The attachment process involves very specific molecular interactions between bacterial adhesins and their receptors on host cells; antibodies against bacterial adhesins can block such infections.

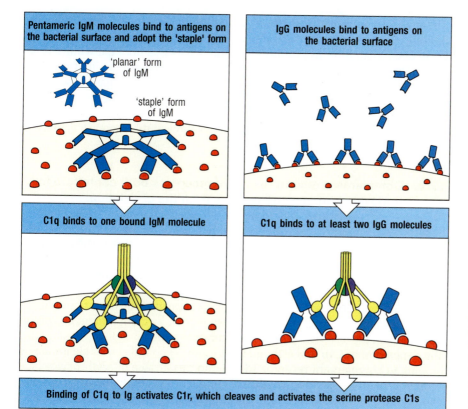

Fig. 10.35 The classical pathway of complement activation is initiated by the binding of C1q to antibody on a pathogen surface. When a molecule of IgM binds several identical epitopes on a pathogen surface, it is bent into the 'staple' conformation, which allows the globular heads of C1q to bind to the Fc regions of IgM (left panels). Multiple molecules of IgG bound on the surface of a pathogen allow the binding of a single molecule of C1q to two or more Fc regions (right panels). In both cases, the binding of C1q to the Fc regions induces a conformational change that activates the associated C1r, which becomes an active enzyme that cleaves the pro-enzyme C1s, generating a serine protease that initiates the classical complement cascade (see Chapter 2).

Fig. 10.36 The two conformations of IgM. The left panel shows the planar conformation of soluble IgM; the right panel shows the 'staple' conformation of IgM bound to a bacterial flagellum. Photographs (×760,000) courtesy of K.H. Roux.

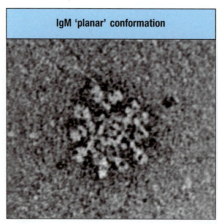

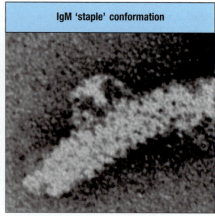

IgM or IgG antibodies, but, because of the structural requirements of binding to C1q, neither of these antibody classes can activate complement in solution; the complement reactions are initiated only when the antibodies are already bound to multiple sites on a cell surface, normally that of a pathogen.

Each globular head of a C1q molecule can bind to one Fc region, and binding of two or more heads activates the C1 complex. In plasma, the **pentameric IgM** molecule has a planar conformation that does not bind C1q (Fig. 10.36, left panel); however, binding to the surface of a pathogen deforms the IgM pentamer so that it looks like a staple (see Fig. 10.36, right panel), and this distortion exposes binding sites for the C1q heads. As mentioned in Section 10-15, IgM hexamers can also form but comprise less than 5% of total serum IgM. Hexameric IgM activates complement about 20 times more efficiently than its pentameric form, possibly because C1q is also a hexamer. The *in vivo* role of IgM hexamers in protecting against infections has not been fully established, and it has even been suggested that IgM hexamers are too reactive and may be harmful.

Although C1q binds with low affinity to some subclasses of IgG in solution, the binding energy required for C1q activation is achieved only when a single molecule of C1q can bind two or more IgG molecules that are held within 30–40 nm of each other as a result of binding antigen. This requires that multiple molecules of IgG be bound to a single pathogen or to an antigen in solution. For this reason, IgM is much more efficient than IgG in activating complement. The binding of C1q to a single bound IgM molecule, or to two or more bound IgG molecules (see Fig. 10.35), leads to activation of the protease activity of C1r, triggering the complement cascade.

10-20 Complement receptors and Fc receptors both contribute to removal of immune complexes from the circulation.

Fc receptors confer the distinct effector functions to the various antibody isotypes by interacting with their Fc regions. One such function is the clearance from the circulation of antigen:antibody complexes (immune complexes), which can include toxins, or debris from dead host cells and microorganisms, bound by neutralizing antibodies. Immune complexes can be cleared by the binding of the antibody's Fc region to Fc receptors expressed on various phagocytic cells in tissues. This clearance is also helped by complement activation (described in the last section), which occurs when the Fc region activates C1q. The deposition of C4b and C3b onto the immune complex aids clearance by binding to complement receptor 1 (CR1) on the surface of erythrocytes (see Section 2-13 for a description of the different types of complement receptors). The erythrocytes transport the bound complexes of antigen, antibody, and complement to the liver and spleen. Here, macrophages bearing CR1 and Fc

Fig. 10.37 Erythrocyte CR1 helps to clear immune complexes from the circulation. CR1 on the erythrocyte surface has an important role in the clearance of immune complexes from the circulation. Immune complexes bind to CR1 on erythrocytes, which transport them to the liver and spleen, where they are removed by macrophages expressing receptors for both Fc and bound complement components.

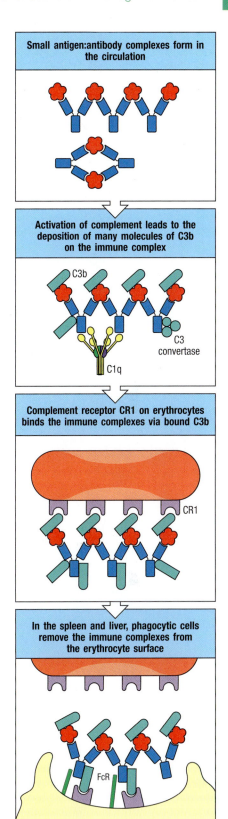

receptors remove the complexes from the erythrocyte surface without destroying the cell, and then degrade the complexes (Fig. 10.37). Even larger aggregates of particulate antigen, such as bacteria, viruses, and cell debris, can be coated with complement, picked up by erythrocytes, and transported to the spleen for destruction.

Complement-coated immune complexes that are not removed from the circulation tend to deposit in the basement membranes of small blood vessels, most notably those of the renal glomerulus, where the blood is filtered to form urine. Immune complexes that pass through the basement membrane of the glomerulus bind to CR1 present on the renal podocytes, cells that lie beneath the basement membrane. The functional significance of these receptors in the kidney is unknown; however, they have an important role in the pathology of some autoimmune diseases. In the autoimmune disease **systemic lupus erythematosus** (SLE) (see Section 15-16), excessive levels of circulating immune complexes lead to their deposition in large amounts on the podocytes, damaging the glomerulus; kidney failure is the principal danger in this disease. The strongest genetic risk factor for SLE is C1q deficiency, although this is very rare. Mutations in complement receptors 2 and 3 and the Fc receptor FcγRIIIa are also associated with increased susceptibility to develop lupus, implying the involvement of both complement receptors and FcR pathways in clearing immune complexes.

Antigen:antibody complexes can also be a cause of pathology in patients with deficiencies in the early components of complement (C1, C2, and C4). These deficiencies result in the classical complement pathway not being activated properly, and immune complexes not being cleared effectively because they do not become tagged with complement. These patients also suffer tissue damage as a result of immune-complex deposition, especially in the kidneys.

Summary.

The T-cell-dependent antibody response begins with IgM secretion but quickly progresses to the production of additional antibody classes. Each class is specialized both in its localization in the body and in the functions it can perform. IgM antibodies are found mainly in blood; they are pentameric in structure. IgM is specialized to activate complement efficiently upon binding antigen and to compensate for the low affinity of a typical IgM antigen-binding site. IgG antibodies are usually of higher affinity and are found in blood and in extracellular fluid, where they can neutralize toxins, viruses, and bacteria, opsonize them for phagocytosis, and activate the complement system. IgA antibodies are synthesized as monomers, which enter blood and extracellular fluids, or they are secreted as dimeric molecules by plasma cells in the lamina propria of various mucosal tissues. IgA dimers are selectively transported across the epithelial layer into sites such as the lumen of the gut, where they neutralize toxins and viruses and block the entry of bacteria across the intestinal epithelium. Most IgE antibody is bound to the surface of mast cells that reside mainly just below the body surface; antigen binding to this IgE triggers local defense reactions. Antibodies can defend the body against extracellular pathogens and their toxic products in several ways. The simplest is by direct interactions with pathogens or their products, for example, by binding to the active sites of toxins and neutralizing them or by blocking their ability to bind to host cells through specific receptors. When antibodies of the appropriate isotype bind to antigens, they can activate the classical pathway of complement, which leads

Systemic Lupus Erythematosus

to the elimination of the pathogen by the various mechanisms described in Chapter 2. Soluble immune complexes of antigen and antibody also fix complement and are cleared from the circulation via complement receptors on red blood cells.

The destruction of antibody-coated pathogens via Fc receptors.

The neutralization of toxins, viruses, or bacteria by high-affinity antibodies can protect against infection but does not, on its own, solve the problem of how to remove the pathogens and their products from the body. Moreover, many pathogens cannot be neutralized by antibody and must be destroyed by other means. Many pathogen-specific antibodies do not bind to neutralizing targets on pathogen surfaces and thus need to be linked to other effector mechanisms to play their part in host defense. We have already seen how the binding of antibody to antigen can activate complement. Another important defense mechanism is the activation of a variety of **accessory effector cells** bearing receptors called Fc receptors because they are specific for the Fc portion of antibodies. These receptors facilitate the phagocytosis of antibody-bound extracellular pathogens by macrophages, dendritic cells, and neutrophils. Other, non-phagocytic cells of the immune system—NK cells, eosinophils, basophils, and mast cells (see Fig. 1.8)—are triggered to secrete stored mediators when their Fc receptors are engaged by antibody-coated pathogens. These mechanisms maximize the effectiveness of all antibodies regardless of where they bind.

10-21 The Fc receptors of accessory cells are signaling receptors specific for immunoglobulins of different classes.

The **Fc receptors** are a family of cell-surface molecules that bind the Fc portion of immunoglobulins. Each member of the Fc family recognizes immunoglobulin of one or a few closely related heavy-chain isotypes through a recognition domain on the α chain of the Fc receptor. Most Fc receptors are themselves members of the immunoglobulin gene superfamily. Different cell types bear different sets of Fc receptors, and the isotype of the antibody thus determines which types of cells will be engaged in a given response. The different Fc receptors, the cells that express them, and their specificities for different antibody classes are shown in Fig. 10.38.

Most Fc receptors function as part of a multisubunit complex. Only the α chain is required for antibody recognition; the other chains are required for transport of the receptor to the cell surface and for signal transduction when an Fc region is bound. Some Fcγ receptors, the Fcα receptor I, and the high-affinity receptor for IgE (FcεRI) all use a γ chain for signaling. This chain, which is closely related to the ζ chain of the T-cell receptor complex (see Section 7-7), associates noncovalently with the Fc-binding α chain. The human FcγRII-A is a single-chain receptor in which the cytoplasmic domain of the α chain replaces the function of the γ chain. FcγRII-B1 and FcγRII-B2 are also single-chain receptors, but function as inhibitory receptors because they contain an ITIM that engages the inositol 5′-phosphatase SHIP (see Section 7-25). The most prominent function of Fc receptors is the activation of accessory cells to attack pathogens, but they also contribute in other ways to immune responses. For example, FcγRII-B receptors negatively regulate the activities of B cells, mast cells, macrophages, and neutrophils by adjusting the threshold at which immune complexes will activate these cells. Fc receptors expressed by dendritic cells enable them to ingest antigen:antibody complexes efficiently and thus process these antigens and present their peptides to T cells.

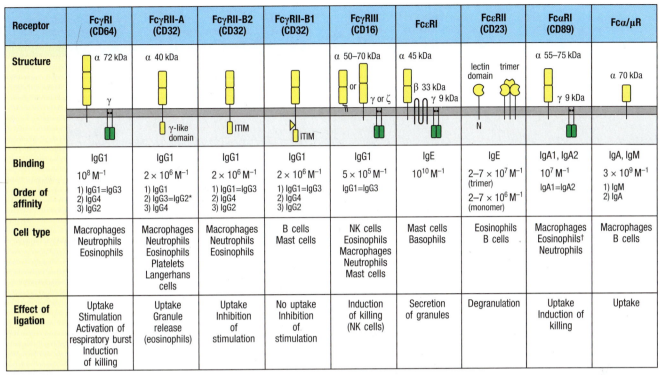

Receptor	FcγRI (CD64)	FcγRII-A (CD32)	FcγRII-B2 (CD32)	FcγRII-B1 (CD32)	FcγRIII (CD16)	FcεRI	FcεRII (CD23)	FcαRI (CD89)	Fcα/μR
Structure	α 72 kDa, γ	α 40 kDa, γ-like domain	ITIM	ITIM	α 50–70 kDa, or, γ or ζ	α 45 kDa, β 33 kDa, γ 9 kDa	lectin domain trimer, N	α 55–75 kDa, γ 9 kDa	α 70 kDa
Binding / Order of affinity	IgG1, $10^8\,M^{-1}$, 1) IgG1=IgG3, 2) IgG4, 3) IgG2	IgG1, $2 \times 10^6\,M^{-1}$, 1) IgG1, 2) IgG3=IgG2*, 3) IgG4	IgG1, $2 \times 10^6\,M^{-1}$, 1) IgG1=IgG3, 2) IgG4, 3) IgG2	IgG1, $2 \times 10^6\,M^{-1}$, 1) IgG1=IgG3, 2) IgG4, 3) IgG2	IgG1, $5 \times 10^5\,M^{-1}$, IgG1=IgG3	IgE, $10^{10}\,M^{-1}$	IgE, 2–$7 \times 10^7\,M^{-1}$ (trimer), 2–$7 \times 10^6\,M^{-1}$ (monomer)	IgA1, IgA2, $10^7\,M^{-1}$, IgA1=IgA2	IgA, IgM, $3 \times 10^9\,M^{-1}$, 1) IgM, 2) IgA
Cell type	Macrophages Neutrophils Eosinophils	Macrophages Neutrophils Eosinophils Platelets Langerhans cells	Macrophages Neutrophils Eosinophils	B cells Mast cells	NK cells Eosinophils Macrophages Neutrophils Mast cells	Mast cells Basophils	Eosinophils B cells	Macrophages Eosinophils† Neutrophils	Macrophages B cells
Effect of ligation	Uptake Stimulation Activation of respiratory burst Induction of killing	Uptake Granule release (eosinophils)	Uptake Inhibition of stimulation	No uptake Inhibition of stimulation	Induction of killing (NK cells)	Secretion of granules	Degranulation	Uptake Induction of killing	Uptake

Fig. 10.38 Distinct receptors for the Fc region of the different immunoglobulin classes are expressed on different accessory cells. The subunit structure and binding properties of these receptors and the cell types expressing them are shown. All are immunoglobulin superfamily members except FcεRII, which is a lectin and can form trimers. The exact chain composition of any receptor can vary from one cell type to another. For example, FcγRIII in neutrophils is expressed as a molecule with a glycosylphosphatidylinositol membrane anchor without γ chains, whereas in NK cells it is a transmembrane molecule associated with γ chains. The FcγRII-B1 differs from the FcγRII-B2 by the presence of an additional exon in the intracellular region (indicated by yellow triangle). This exon prevents the FcγRII-B1 from being internalized after cross-linking. The binding affinities are taken from data on human receptors. *Only some allotypes of FcγRII-A bind IgG2. †In eosinophils, the molecular weight of the CD89α chain is 70–100 kDa.

Antibody-coated viruses that enter the cytoplasm are cleared by a system that employs a novel class of Fc receptor called **TRIM21** (tripartite motif-containing 21) that is expressed by a variety of immune and nonimmune cell types. TRIM21 is a cytosolic IgG receptor that has a higher affinity for IgG than any other Fc receptor, and it also has **E3 ligase** activity. When a virus that has bound IgG enters the cytoplasm, TRIM21 attaches to the antibody and uses its E3 ligase activity to ubiquitinate viral proteins. This leads to proteasomal degradation of virions in the cytosol before translation of virally encoded genes can occur.

10-22 Fc receptors on phagocytes are activated by antibodies bound to the surface of pathogens and enable the phagocytes to ingest and destroy pathogens.

The most important Fc-bearing cells in humoral immune responses are the phagocytic cells of the monocytic and myelocytic lineages, particularly macrophages and neutrophils. Many bacteria are directly recognized, ingested, and destroyed by phagocytes, and these bacteria are not pathogenic in normal individuals. However, some bacterial pathogens have **polysaccharide capsules**, a large structure that lies outside the bacterial cell membrane and resists direct engulfment by phagocytes. Such pathogens become susceptible to phagocytosis only when they are coated with antibodies and complement that engage the Fcγ or Fcα receptors and the complement receptor CR1 on phagocytic cells, triggering bacterial uptake (**Fig. 10.39**). The stimulation

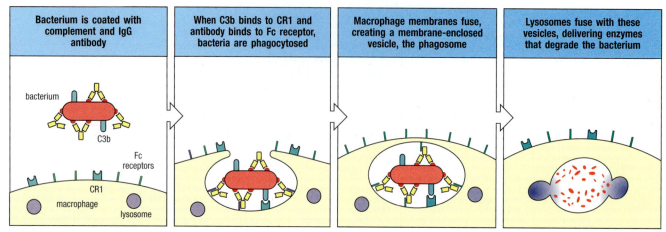

| Bacterium is coated with complement and IgG antibody | When C3b binds to CR1 and antibody binds to Fc receptor, bacteria are phagocytosed | Macrophage membranes fuse, creating a membrane-enclosed vesicle, the phagosome | Lysosomes fuse with these vesicles, delivering enzymes that degrade the bacterium |

Fig. 10.39 Fc and complement receptors on phagocytes trigger the uptake and degradation of antibody-coated bacteria. Many bacteria resist phagocytosis by macrophages and neutrophils. Antibodies bound to these bacteria, however, enable the bacteria to be ingested and degraded through the interaction of the multiple Fc domains arrayed on the bacterial surface with Fc receptors on the phagocyte surface. Antibody coating also induces activation of the complement system and the binding of complement components to the bacterial surface. These can interact with complement receptors (for example, CR1) on the phagocyte. Fc receptors and complement receptors synergize in inducing phagocytosis. Bacteria coated with IgG antibody and complement are therefore more readily ingested than those coated with IgG alone. Binding of Fc and complement receptors signals the phagocyte to increase the rate of phagocytosis, to fuse lysosomes with phagosomes, and to increase its bactericidal activity.

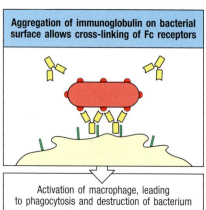

| Free immunoglobulin does not cross-link Fc receptors |

No activation of macrophage, no destruction of bacterium

| Aggregation of immunoglobulin on bacterial surface allows cross-linking of Fc receptors |

Activation of macrophage, leading to phagocytosis and destruction of bacterium

of phagocytosis by complement-coated antigens binding to complement receptors is particularly important early in the immune response, before isotype-switched antibodies have been made. Capsular polysaccharides belong to the TI-2 class of thymus-independent antigens, and can therefore stimulate the early production of IgM antibodies, which are very effective at activating the complement system. IgM binding to encapsulated bacteria thus triggers the opsonization of these bacteria by complement and their prompt ingestion and destruction by phagocytes bearing complement receptors. Recently, Fcα/μR was discovered as a receptor that binds both IgA and IgM. Fcα/μR is expressed primarily on macrophages and B cells in the lamina propria of the intestine and in germinal centers. It is thought to have a role in the endocytosis of IgM antibody complexed with bacteria such as *Staphylococcus aureus*.

Phagocyte activation can initiate an inflammatory response that causes tissue damage, and so Fc receptors on phagocytes must be able to distinguish antibody molecules bound to a pathogen from the much larger number of free antibody molecules that are not bound to anything. This distinction is made possible by the aggregation of antibodies that occurs when they bind to multimeric antigens or to multivalent particulate antigens such as viruses and bacteria. Individual Fc receptors on a cell surface bind monomers of free antibody with low affinity, but when presented with an antibody-coated particle, the simultaneous binding by multiple Fc receptors results in binding of high avidity, and this is the principal mechanism by which bound antibodies are distinguished from free immunoglobulin (**Fig. 10.40**). The result is that Fc receptors enable cells to detect pathogens via the antibody molecules bound to them. Fc receptors therefore give phagocytic cells that lack intrinsic specificity the ability to identify and remove specific pathogens and their products from the extracellular spaces.

Fig. 10.40 Bound antibody is distinguishable from free immunoglobulin by its state of aggregation. Free immunoglobulin molecules bind most Fc receptors with very low affinity and cannot cross-link Fc receptors. Antigen-bound immunoglobulin, however, binds to Fc receptors with high avidity because several antibody molecules that are bound to the same surface bind to multiple Fc receptors on the surface of the accessory cell. This Fc receptor cross-linking sends a signal to activate the cell bearing it. With Fc receptors that have ITIMs, the result is inhibition.

Phagocytosis is greatly enhanced by interactions between the molecules coating an opsonized microorganism and receptors on the phagocyte surface. When an antibody-coated pathogen binds to Fcγ receptors, for example, the cell surface of the phagocyte extends around the surface of the pathogen through successive binding of the Fcγ receptors to the antibody Fc regions bound to the pathogen. This is an active process that is triggered by the stimulation of the Fcγ receptors. Phagocytosis leads to enclosure of the pathogen (or particle) in an acidified cytoplasmic vesicle—the phagosome. This then fuses with one or more lysosomes to generate a phagolysosome; lysosomal enzymes are released into the vesicle interior, where they destroy the bacterium (see Fig. 10.39). The process of intracellular killing by phagocytes was described in more detail in Chapter 3.

Some particles are too large for a phagocyte to ingest; parasitic worms are one example. In this case the phagocyte attaches to the surface of the antibody-coated parasite via its Fcγ, Fcα, or Fcε receptors, and the contents of the secretory granules or lysosomes of the phagocyte are released by exocytosis. The contents are discharged directly onto the surface of the parasite and damage it. Thus, stimulation of Fcγ and Fcα receptors can trigger either the internalization of external particles by phagocytosis or the externalization of internal vesicles by exocytosis. The principal leukocytes involved in the destruction of bacteria are macrophages and neutrophils, whereas large parasites such as helminths are usually attacked by eosinophils (Fig. 10.41), nonphagocytic cells that can bind antibody-coated parasites via several different Fc receptors, including the low-affinity Fcε receptor for IgE, CD23 (see Fig. 10.38). Cross-linking of these receptors by antibody-coated surfaces activates the eosinophil to release its granule contents, which include proteins toxic to parasites (see Fig. 14.10). Cross-linking by antigen of IgE bound to the high-affinity FcεRI on mast cells and basophils also results in exocytosis of their granule contents, as we describe below.

Fig. 10.41 Eosinophils attacking a schistosome larva in the presence of serum from an infected patient. Large parasites, such as worms, cannot be ingested by phagocytes; however, when the worm is coated with antibody, eosinophils can attack it through binding via their Fc receptors for IgG and IgA. Similar attacks on large targets can be mounted by other Fc receptor-bearing cells. These cells release the toxic contents of their granules directly onto the target, a process known as exocytosis. Photograph courtesy of A. Butterworth.

10-23 Fc receptors activate NK cells to destroy antibody-coated targets.

Virus-infected cells are usually destroyed by T cells that recognize virus-derived peptides bound to cell-surface MHC molecules. Cells infected by some viruses also signal the presence of intracellular infection by expressing on their surface proteins, such as viral envelope proteins, that can be recognized by antibodies originally produced against the virus particle. Host cells with antibodies bound to them can be killed by a specialized non-T, non-B cell of the lymphoid lineage called a natural killer cell (NK cell), which we met in Chapter 3. NK cells are large cells with prominent intracellular granules and make up a small fraction of peripheral blood lymphocytes. Although belonging to the lymphoid lineage, NK cells express a limited repertoire of invariant receptors recognizing a range of ligands that are induced on abnormal cells, such as those infected with viruses; NK cells are considered to be part of innate immunity (see Section 3-25). On recognition of a ligand, the NK cell kills the target cell directly without the need for antibody. Although first discovered for their ability to kill some tumor cells, NK cells play an important role in innate immunity in the early stages of virus infection.

As well as this innate function, NK cells can recognize and destroy antibody-coated target cells in a process called **antibody-dependent cell-mediated cytotoxicity** (**ADCC**). This is triggered when antibody bound to the surface of a cell interacts with Fc receptors on the NK cell (Fig. 10.42). NK cells express the receptor FcγRIII (CD16), which recognizes the IgG1 and IgG3 subclasses. The killing mechanism is analogous to that of cytotoxic T cells, involving the release of cytoplasmic granules containing perforin and granzymes (see Section 9-31). ADCC has been shown to have a role in the defense against

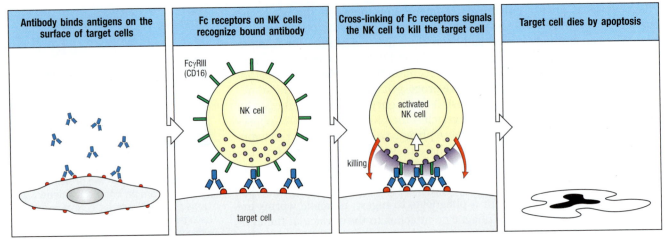

Antibody binds antigens on the surface of target cells	Fc receptors on NK cells recognize bound antibody	Cross-linking of Fc receptors signals the NK cell to kill the target cell	Target cell dies by apoptosis

Fig. 10.42 Antibody-coated target cells can be killed by NK cells in antibody-dependent cell-mediated cytotoxicity (ADCC). NK cells (see Chapter 3) are large granular non-T, non-B lymphoid cells that have FcγRIII (CD16) on their surface. When these cells encounter cells coated with IgG antibody, they rapidly kill the target cell. ADCC is only one way in which NK cells can contribute to host defense.

infection by viruses, and represents another mechanism by which antibodies can direct an antigen-specific attack by an effector cell that itself lacks specificity for antigen.

10-24 Mast cells and basophils bind IgE antibody via the high-affinity Fcε receptor.

When pathogens cross epithelial barriers and establish a local focus of infection, the host must mobilize its defenses and direct them to the site of pathogen growth. One way in which this is achieved is to activate the cells known as **mast cells**. Mast cells are large cells containing distinctive cytoplasmic granules that contain a mixture of chemical mediators, including histamine, that act rapidly to make local blood vessels more permeable. Mast cells have a distinctive appearance after staining with the dye toluidine blue that makes them readily identifiable in tissues (see Fig. 1.8). They are found in particularly high concentrations in vascularized connective tissues just beneath epithelial surfaces, including the submucosal tissues of the gastrointestinal and respiratory tracts and the dermis of the skin.

Mast cells have Fc receptors specific for IgE (FcεRI) and IgG (FcγRIII), and can be activated to release their granules and to secrete lipid inflammatory mediators and cytokines via antibody bound to these receptors. Most Fc receptors bind stably to the Fc regions of antibodies only when the antibodies have themselves bound antigen, and cross-linking of multiple Fc receptors is needed for strong binding. In contrast, FcεRI binds IgE antibody monomers with a very high affinity—approximately 10^{10} M^{-1}. Thus, even at the low levels of circulating IgE present in normal individuals, a substantial portion of the total IgE is bound to the FcεRI on mast cells in tissues and on circulating basophils.

Although mast cells are usually stably associated with bound IgE, this on its own does not activate them, nor will the binding of monomeric antigen to the IgE. Mast-cell activation occurs only when the bound IgE is cross-linked by multivalent antigens. This signal activates the mast cell to release the contents of its granules, which occurs in seconds (**Fig. 10.43**), to synthesize and release lipid mediators such as prostaglandin D_2 and leukotriene C4, and to secrete cytokines such as TNF-α, thereby initiating a local inflammatory response. Degranulation also releases stored histamine, which increases local blood flow and vascular permeability; this quickly leads to an accumulation

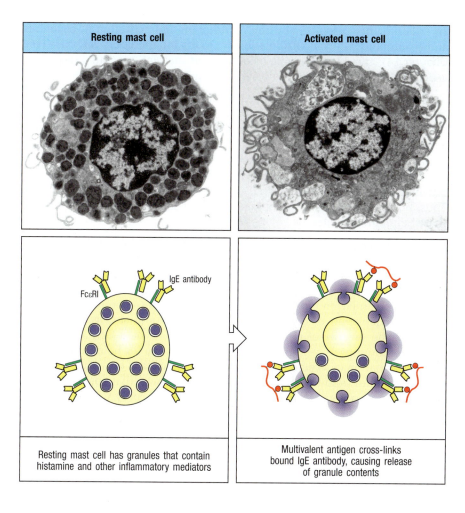

Resting mast cell	Activated mast cell

FcεRI

IgE antibody

Resting mast cell has granules that contain histamine and other inflammatory mediators	Multivalent antigen cross-links bound IgE antibody, causing release of granule contents

Fig. 10.43 IgE antibody cross-linking on mast-cell surfaces leads to a rapid release of inflammatory mediators. Mast cells are large cells found in connective tissue and can be distinguished by their secretory granules, which contain many inflammatory mediators. They bind stably to monomeric IgE antibodies through the very high-affinity receptor FcεRI. Antigen cross-linking of the bound IgE antibody molecules triggers rapid degranulation, releasing inflammatory mediators into the surrounding tissue. These mediators trigger local inflammation, which recruits cells and proteins required for host defense to sites of infection. These cells are also triggered during allergic reactions when allergens bind to IgE on mast cells. Photographs courtesy of A.M. Dvorak.

of fluid and blood proteins, including antibodies, in the surrounding tissue. Shortly afterward there is an influx of blood-borne cells such as neutrophils and, later, monocytes, eosinophils, and effector lymphocytes. This influx can last from a few minutes to a few hours and produces an inflammatory response at the site of infection. Thus, mast cells are part of the front-line host defenses against pathogens that enter the body across epithelial barriers. They are also of medical importance because of their involvement in IgE-mediated allergic responses, which are discussed in Chapter 14. In allergic responses, mast cells are activated in the way described above by exposure to normally innocuous antigens (allergens), such as pollen, to which the individual has previously mounted a sensitizing immune response that produces allergen-specific IgE.

10-25 IgE-mediated activation of accessory cells has an important role in resistance to parasite infection.

Mast cells are thought to serve at least three important functions in host defense. First, their location near body surfaces allows them to recruit both pathogen-specific elements, such as antigen-specific lymphocytes, and non-specific effector elements, such as neutrophils, macrophages, basophils, and eosinophils, to sites where infectious agents are most likely to enter the internal milieu. Second, the inflammation they cause increases the flow of lymph from sites of antigen deposition to the regional lymph nodes, where naive lymphocytes are first activated. Third, the ability of mast-cell products to trigger muscular contraction can contribute to the physical expulsion of pathogens from the lungs or the gut. Mast cells respond rapidly to the binding of antigen to surface-bound IgE antibodies, and their activation leads to the initiation of an inflammatory response and the recruitment and activation of basophils and eosinophils, which contribute further to the inflammatory response (see Chapter 14). There is increasing evidence that such IgE-mediated responses are crucial to defense against parasite infestation.

A role for mast cells in the clearance of parasites is suggested by the accumulation of mast cells in the intestine, known as **mastocytosis**, that accompanies helminth infection, and by observations in W/W^V mutant mice, which have a profound mast-cell deficiency caused by a mutation in the gene c-*kit*. These mutant mice show impaired clearance of the intestinal nematodes *Trichinella spiralis* and *Strongyloides* species. Clearance of *Strongyloides* is even more impaired in W/W^V mice that lack IL-3 and so also fail to produce basophils. Thus, both mast cells and basophils seem to contribute to defense against these helminth parasites.

Other evidence points to the importance of IgE antibodies and eosinophils in the defense against parasites. Infection with certain types of multicellular parasites, particularly helminths, is strongly associated with the production of IgE antibodies and the presence of abnormally large numbers of eosinophils (eosinophilia) in blood and tissues. Furthermore, experiments in mice show that depletion of eosinophils by polyclonal anti-eosinophil antisera increases the severity of infection with the parasitic helminth *Schistosoma mansoni*. Eosinophils seem to be directly responsible for helminth destruction; examination of infected tissues shows degranulated eosinophils adhering to helminths, and experiments *in vitro* have shown that eosinophils can kill *S. mansoni* in the presence of anti-schistosome IgG or IgA antibodies (see Fig. 10.41).

The role of IgE, mast cells, basophils, and eosinophils can also be seen in resistance to the feeding of blood-sucking ixodid ticks. Skin at the site of a tick bite has degranulated mast cells and an accumulation of degranulated basophils and eosinophils, an indicator of recent activation. Subsequent resistance to feeding by these ticks develops after the first exposure, suggesting a specific immunological mechanism. Mice deficient in mast cells show no such acquired resistance to ticks, and in guinea pigs the depletion of either basophils or eosinophils by specific polyclonal antibodies also reduces resistance to tick feeding. Finally, experiments in mice showed that resistance to ticks is mediated by specific IgE antibody. Thus, many clinical studies and experiments support a role for this system of IgE bound to the high-affinity FcεRI in host resistance to pathogens that enter across epithelia or exoparasites such as ticks that breach it.

Summary.

Antibody-coated pathogens are recognized by effector cells through Fc receptors that bind to an array of constant regions (Fc portions) provided by the pathogen-bound antibodies. Binding activates the cell and triggers destruction of the pathogen, through either phagocytosis, granule release, or both. Fc receptors comprise a family of proteins, each of which recognizes immunoglobulins of particular isotypes. Fc receptors on macrophages and neutrophils recognize the constant regions of IgG or IgA antibodies bound to a pathogen and trigger the engulfment and destruction of such bacteria. Binding to the Fc receptor also induces the production of microbicidal agents in the intracellular vesicles of the phagocyte. Eosinophils are important in the elimination of parasites too large to be engulfed; they bear Fc receptors specific for the constant region of IgG, as well as receptors for IgE; aggregation of these receptors triggers the release of toxic substances onto the surface of the parasite. NK cells, tissue mast cells, and blood basophils also release their granule contents when their Fc receptors are engaged. The high-affinity receptor for

IgE is expressed constitutively by mast cells and basophils. It differs from other Fc receptors in that it can bind free monomeric antibody, thus enabling an immediate response to pathogens at their site of first entry into the tissues. When IgE bound to the surface of a mast cell is aggregated by binding to antigen, it triggers the release of histamine and many other mediators that increase the blood flow to sites of infection; it thereby recruits antibodies and effector cells to these sites. Mast cells are found principally below epithelial surfaces of the skin and beneath the basement membrane of the digestive and respiratory tracts. Their activation by innocuous substances is responsible for many of the symptoms of acute allergic reactions, as will be described in Chapter 14.

Summary to Chapter 10.

The humoral immune response to infection involves the production of antibody by plasma cells derived from B lymphocytes, the binding of this antibody to the pathogen, and the elimination of the pathogen by phagocytic cells and molecules of the humoral immune system. The production of antibody usually requires the action of helper T cells specific for a peptide fragment of the antigen recognized by the B cell, a phenomenon called linked recognition. An activated B cell first moves to the T-zone–B-zone boundary in secondary lymphoid tissues, where it may encounter its cognate T cell and begin to proliferate. Some B cells become plasmablasts, while others move to the germinal center, where somatic hypermutation and class switch recombination take place. B cells that bind antigen with the highest affinity are selected for survival and further differentiation, leading to affinity maturation of the antibody response. Cytokines made by helper T cells direct class switching, leading to the production of antibody of various classes that can be distributed to various body compartments.

IgM antibodies are produced early in an infection by conventional B cells and are also made in the absence of infection by subsets of nonconventional B cells in particular locations (as natural antibodies). IgM has a major role in protecting against infection in the bloodstream, whereas isotypes secreted later in an adaptive immune response, such as IgG, diffuse into the tissues. Antigens that have highly repeating antigenic determinants and that contain mitogens—called TI antigens—can elicit IgM and some IgG independently of T-cell help, and this provides an early protective immune response. Multimeric IgA is produced in the lamina propria and is transported across epithelial surfaces, whereas IgE is made in small amounts and binds avidly to receptors on the surface of basophils and mast cells.

Antibodies that bind with high affinity to critical sites on toxins, viruses, and bacteria can neutralize them. However, pathogens and their products are destroyed and removed from the body largely through uptake into phagocytes and degradation inside these cells. Antibodies that coat pathogens bind to Fc receptors on phagocytes, which are thereby triggered to engulf and destroy the pathogen. Binding of antibody C regions to Fc receptors on other cells leads to the exocytosis of stored mediators; this is particularly important in parasite infections, in which Fcε-expressing mast cells are triggered by the binding of antigen to IgE antibody to release inflammatory mediators directly onto parasite surfaces. Antibodies can also initiate the destruction of pathogens by activating the complement system. Complement components can opsonize pathogens for uptake by phagocytes, and recruit phagocytes to sites of infection. Receptors for complement components and Fc receptors often synergize in activating the uptake and destruction of pathogens and immune complexes. Thus, the humoral immune response is targeted to the infecting pathogen through the production of specific antibody; however, the effector actions of that antibody are determined by its heavy-chain isotype.

Questions.

10.1 Multiple Choice: Which of the following is not an antibody effector function?

   A. Opsonization

   B. Neutralization

   C. Complement activation

   D. Linked recognition

   E. NK-cell cytotoxicity

   F. Mast-cell degranulation

10.2 Short Answer: The *Haemophilus influenzae* type b (Hib) vaccine was initially composed only of the polysaccharide capsule of the organism, but this failed to mount potent antibody responses. Directly conjugating the Hib polysaccharide to a tetanus or diphtheria toxoid, however, yielded very potent antibody responses to Hib, and is the current vaccine formulation. Indicate which immunological phenomenon is taken advantage of by conjugating the Hib capsule-derived polysaccharide to a toxoid, and how it works to elicit a potent antibody response.

10.3 Matching: During T-dependent antibody responses, numerous receptor/ligand interactions and cytokine signaling events occur between T_{FH} cells and activated B cells. For the following list of surface receptors/ligands and cytokines, indicate whether they are produced by T cells (T), B cells (B), both (TB), or neither (N) in this context.

   A. IL-21

   B. ICOSL

   C. CD40L

   D. CD30L

   E. Peptide:MHC II

   F. CCL21

   G. SLAM

10.4 Matching: Match the human disease to the associated genetic defect.

A. X-linked lymphoproliferative disorder	**i.** Translesion polymerase Polη
B. Hyper IgM type 2 immunodeficiency	**ii.** ATM (a DNA-PKcs-family kinase)
C. Xeroderma pigmentosum	**iii.** SLAM-associated protein (SAP)
D. Ataxia telangiectasia	**iv.** Activation-induced cytidine deaminase (AID)

10.5 Matching: Indicate whether the following properties apply to IgA, IgD, IgE, IgG, and/or IgM.

   A. First produced during humoral response

   B. Monomeric (predominantly)

   C. Dimeric (predominantly)

   D. Pentameric (predominantly)

   E. Contains a J chain

   F. Capable of eliciting complement deposition

   G. Most abundant in mucosal surfaces and secretions

   H. Low-affinity

   I. Bound onto mast cells

   J. Binds to polymeric immunoglobulin receptor (pIgR)

   K. Binds the neonatal Fc receptor (FcRn)

10.6 Short Answer: How is TRIM21, a novel class of Fc receptor, different from other Fc receptors?

10.7 Multiple Choice: Which of the following functions is not elicited by antibody binding to Fcγ receptors?

   A. Antibody-dependent cell-mediated cytotoxicity (ADCC) via NK cells

   B. Phagocytosis by neutrophils

   C. Mast-cell degranulation

   D. Downregulation of B-cell activity

   E. Ingestion of immune complexes by dendritic cells

10.8 Multiple Choice: Which of the following is a false statement?

   A. Naive B-cell survival in follicles is dependent on BAFF, which signals through BAFF-R, TACI, and BCMA to induce Bcl-2 expression.

   B. Subcapsular sinuses of lymph nodes and marginal sinuses of the spleen are functionally similar areas filled with specialized macrophages that retain but do not digest antigens.

   C. ICOS signaling in T cells is essential for their completion of T_{FH} differentiation and expression of the transcription factors Bcl-6 and c-Maf.

   D. Both plasmablasts and plasma cells express B7 co-stimulatory molecules, MHC class II molecules, and high levels of B-cell receptors.

   E. T_{FH} cells determine the choice of isotype for class switching in T-dependent antibody responses.

10.9 True or False: Germinal centers contain a light and a dark zone. In the light zone, B cells proliferate extensively and are called centroblasts. They are maintained there by CXCL12–CXCR4 chemokine signaling and undergo somatic hypermutation leading to affinity maturation and class switching. In the dark zone, B cells cease proliferation and are called centrocytes. Here, they are maintained by CXCL13–CXCR5 chemokine signaling, express higher levels of B-cell receptor, and interact extensively with T_{FH} cells.

10.10 Multiple Choice: Choose the correct statement:

   A. R-loops are structures formed during somatic hypermutation that promote accessibility of the immunoglobulin V regions to AID.

   B. APE1 removes deaminated cytosine to create an abasic residue that results in the random insertion of a base during the next round of DNA replication.

C. Frameshift mutations during class switch recombination do not occur because switch regions lie in introns.

D. The error-prone MSH2/6 polymerase repairs DNA lesions and causes mutations that promote somatic hypermutation.

10.11 Fill-in-the-Blanks: Fc receptors diversify the effector functions of the distinct antibody isotypes. Most Fc receptors can bind the Fc regions of antibodies with _____ affinity. In contrast, FcεRI binds with _____ affinity. Multivalent antigen-bound IgE can bind _____ in mast cells and cause release of lipid mediators such as _____ and _____. Mast cells also degranulate in response to cross-linking of the FC receptor-bound IgE, which causes release of _____, and as a consequence local blood flow and _____ are increased, initiating an inflammatory response.

General references.

Batista, F.D., and Harwood, N.E.: **The who, how and where of antigen presentation to B cells.** *Nat. Rev. Immunol.* 2009, **9**:15–27.

Nimmerjahn, F., and Ravetch, J.V.: **Fcγ receptors as regulators of immune responses.** *Nat. Rev. Immunol.* 2008, **8**:34–47.

Rajewsky, K.: **Clonal selection and learning in the antibody system.** *Nature* 1996, **381**:751–758.

Section references.

10-1 Activation of B cells by antigen involves signals from the B-cell receptor and either T$_{FH}$ cells or microbial antigens.

Crotty, S.: **T follicular helper cell differentiation, function, and roles in disease.** *Immunity* 2014, **41**:529–542.

Maglione, P.J., Simchoni, N., Black, S., Radigan, L., Overbey, J.R., Bagiella, E., Bussel, J.B., Bossuyt, X., Casanova, J.L., Meyts, I., *et al.*: **IRAK-4- and MyD88 deficiencies impair IgM responses against T-independent bacterial antigens.** *Blood* 2014, **124**:3561–3571.

Pasare, C., and Medzhitov, R.: **Control of B-cell responses by Toll-like receptors.** *Nature* 2005, **438**:364–368.

Vijayanand, P., Seumois, G., Simpson, L.J., Abdul-Wajid, S., Baumjohann, D., Panduro, M., Huang, X., Interlandi, J., Djuretic, I.M., Brown, D.R., *et al.*: **Interleukin-4 production by follicular helper T cells requires the conserved Il4 enhancer hypersensitivity site V.** *Immunity* 2012, **36**:175–187.

10-2 Linked recognition of antigen by T cells and B cells promotes robust antibody responses.

Barrington, R.A., Zhang, M., Zhong, X., Jonsson, H., Holodick, N., Cherukuri, A., Pierce, S.K., Rothstein, T.L., and Carroll, M.C.: **CD21/CD19 coreceptor signaling promotes B cell survival during primary immune responses.** *J. Immunol.* 2005, **175**:2859–2867.

Eskola, J., Peltola, H., Takala, A.K., Kayhty, H., Hakulinen, M., Karanko, V., Kela, E., Rekola, P., Ronnberg, P.R., Samuelson, J.S., *et al.*: **Efficacy of *Haemophilus influenzae* type b polysaccharide-diphtheria toxoid conjugate vaccine in infancy.** *N. Engl. J. Med.* 1987, **317**:717–722.

Kalled, S.L.: **Impact of the BAFF/BR3 axis on B cell survival, germinal center maintenance and antibody production.** *Semin. Immunol.* 2006, **18**:290–296.

Mackay, F., and Browning, J.L.: **BAFF: a fundamental survival factor for B cells.** *Nat. Rev. Immunol.* 2002, **2**:465–475.

Mackay, F., and Schneider, P.: **Cracking the BAFF code.** *Nat. Rev. Immunol.* 2009, **9**:491–502.

MacLennan, I.C.M., Gulbranson-Judge, A., Toellner, K.M., Casamayor-Palleja, M.,

Chan, E., Sze, D.M.Y., Luther, S.A., and Orbea, H.A.: **The changing preference of T and B cells for partners as T-dependent antibody responses develop.** *Immunol. Rev.* 1997, **156**:53–66.

McHeyzer-Williams, L.J., Malherbe, L.P., and McHeyzer-Williams, M.G.: **Helper T cell-regulated B cell immunity.** *Curr. Top. Microbiol. Immunol.* 2006, **311**:59–83.

Nitschke, L.: **The role of CD22 and other inhibitory co-receptors in B-cell activation.** *Curr. Opin. Immunol.* 2005, **17**:290–297.

Rickert, R.C.: **Regulation of B lymphocyte activation by complement C3 and the B cell coreceptor complex.** *Curr. Opin. Immunol.* 2005, **17**:237–243.

Teichmann, L.L., Kashgarian, M., Weaver, C.T., Roers, A., Müller, W., and Shlomchik, M.J.: **B cell-derived IL-10 does not regulate spontaneous systemic autoimmunity in MRL.Fas(lpr) mice.** *J. Immunol.* 2012, **188**:678–685.

10-3 B cells that encounter their antigens migrate toward the boundaries between B-cell and T-cell areas in secondary lymphoid tissues.

Cinamon, G., Zachariah, M.A., Lam, O.M., Foss, F.W., Jr., and Cyster, J.G.: **Follicular shuttling of marginal zone B cells facilitates antigen transport.** *Nat. Immunol.* 2008, **9**:54–62.

Fang, Y., Xu, C., Fu, Y.X., Holers, V.M., and Molina, H.: **Expression of complement receptors 1 and 2 on follicular dendritic cells is necessary for the generation of a strong antigen-specific IgG response.** *J. Immunol.* 1998, **160**:5273–5279.

Okada, T., and Cyster, J.G.: **B cell migration and interactions in the early phase of antibody responses.** *Curr. Opin. Immunol.* 2006, **18**:278–285.

Phan, T.G., Gray, E.E., and Cyster, J.G.: **The microanatomy of B cell activation.** *Curr. Opin. Immunol.* 2009, **21**:258–265.

10-4 T cells express surface molecules and cytokines that activate B cells, which in turn promote T$_{FH}$-cell development.

Choi, Y.S., Kageyama, R., Eto, D., Escobar, T.C., Johnston, R.J., Monticelli, L., Lao, C., and Crotty, S.: **ICOS receptor instructs T follicular helper cell versus effector cell differentiation via induction of the transcriptional repressor Bcl6.** *Immunity* 2011, **34**:932–946.

Gaspal, F.M., Kim, M.Y., McConnell, F.M., Raykundalia, C., Bekiaris, V., and Lane, P.J.: **Mice deficient in OX40 and CD30 signals lack memory antibody responses because of deficient CD4 T cell memory.** *J. Immunol.* 2005, **174**:3891–3896.

Iannacone, M., Moseman, E.A., Tonti, E., Bosurgi, L., Junt, T., Henrickson, S.E., Whelan, S.P., Guidotti, L.G., and von Andrian, U.H.: **Subcapsular sinus macrophages prevent CNS invasion on peripheral infection with a neurotropic virus.** *Nature* 2010, **465**:1079–1083.

Yoshinaga, S.K., Whoriskey, J.S., Khare, S.D., Sarmiento, U., Guo, J., Horan, T., Shih, G., Zhang, M., Coccia, M.A., Kohno, T., *et al.*: **T-cell co-stimulation through B7RP-1 and ICOS.** *Nature* 1999, **402**:827–832.

10-5   Activated B cells differentiate into antibody-secreting plasmablasts and plasma cells.

Moser, K., Tokoyoda, K., Radbruch, A., MacLennan, I., and Manz, R.A.: **Stromal niches, plasma cell differentiation and survival.** *Curr. Opin. Immunol.* 2006, **18**:265–270.

Pelletier, N., McHeyzer-Williams, L.J., Wong, K.A., Urich, E., Fazilleau, N., and McHeyzer-Williams, M.G.: **Plasma cells negatively regulate the follicular helper T cell program.** *Nat. Immunol.* 2010, **11**:1110–1118.

Radbruch, A., Muehlinghaus, G., Luger, E.O., Inamine, A., Smith, K.G., Dorner, T., and Hiepe, F.: **Competence and competition: the challenge of becoming a long-lived plasma cell.** *Nat. Rev. Immunol.* 2006, **6**:741–750.

Sciammas, R., and Davis, M.M.: **Blimp-1; immunoglobulin secretion and the switch to plasma cells.** *Curr. Top. Microbiol. Immunol.* 2005, **290**:201–224.

Shapiro-Shelef., M., and Calame, K.: **Regulation of plasma-cell development.** *Nat. Rev. Immunol.* 2005, **5**:230–242.

10-6   The second phase of a primary B-cell immune response occurs when activated B cells migrate into follicles and proliferate to form germinal centers.

Allen, C.D., Okada, T., and Cyster, J.G.: **Germinal-center organization and cellular dynamics.** *Immunity* 2007, **27**:190–202.

Cozine, C.L., Wolniak, K.L., and Waldschmidt, T.J.: **The primary germinal center response in mice.** *Curr. Opin. Immunol.* 2005, **17**:298–302.

Kunkel, E.J., and Butcher, E.C.: **Plasma-cell homing.** *Nat. Rev. Immunol.* 2003, **3**:822–829.

Victora, G.D., Schwickert, T.A., Fooksman, D.R., Kamphorst, A.O., Meyer-Hermann, M., Dustin, M.L., and Nussenzweig, M.C.: **Germinal center dynamics revealed by multiphoton microscopy with a photoactivatable fluorescent reporter.** *Cell* 2010, **143**:592–605.

10-7   Germinal center B cells undergo V-region somatic hypermutation, and cells with mutations that improve affinity for antigen are selected.

Anderson, S.M., Khalil, A., Uduman, M., Hershberg, U., Louzoun, Y., Haberman, A.M., Kleinstein, S.H., and Shlomchik, M.J.: **Taking advantage: high-affinity B cells in the germinal center have lower death rates, but similar rates of division, compared to low-affinity cells.** *J. Immunol.* 2009, **183**:7314–7325.

Gitlin, A.D., Shulman, Z., and Nussenzweig, M.C.: **Clonal selection in the germinal centre by regulated proliferation and hypermutation.** *Nature* 2014, **509**:637–640.

Jacob, J., Kelsoe, G., Rajewsky, K., and Weiss, U.: **Intraclonal generation of antibody mutants in germinal centres.** *Nature* 1991, **354**:389–392.

Li, Z., Woo, C.J., Iglesias-Ussel, M.D., Ronai, D., and Scharff, M.D.: **The generation of antibody diversity through somatic hypermutation and class switch recombination.** *Genes Dev.* 2004, **18**:1–11.

Wang, Z., Karras, J.G., Howard, R.G., and Rothstein, T.L.: **Induction of bcl-x by CD40 engagement rescues sIg-induced apoptosis in murine B cells.** *J. Immunol.* 1995, **155**:3722–3725.

10-8   Positive selection of germinal center B cells involves contact with T$_{FH}$ cells and CD40 signaling.

Bannard, O., Horton, R.M., Allen, C.D., An, J., Nagasawa, T., and Cyster, J.G.: **Germinal center centroblasts transition to a centrocyte phenotype according to a timed program and depend on the dark zone for effective selection.** *Immunity* 2013, **39**:912–924.

Cannons, J.L., Qi, H., Lu, K.T., Dutta, M., Gomez-Rodriguez, J., Cheng, J., Wakeland, E.K., Germain, R.N., and Schwartzberg, P.L.: **Optimal germinal center responses require a multistage T cell:B cell adhesion process involving integrins, SLAM-associated protein, and CD84.** *Immunity* 2010, **32**:253–265.

Hauser, A.E., Junt, T., Mempel, T.R., Sneddon, M.W., Kleinstein, S.H., Henrickson, S.E., von Andrian, U.H., Shlomchik, M.J., and Haberman, A.M.: **Definition of germinal-center B cell migration in vivo reveals predominant intrazonal circulation patterns.** *Immunity* 2007, **26**:655–667.

Jumper, M., Splawski, J., Lipsky, P., and Meek, K.: **Ligation of CD40 induces sterile transcripts of multiple Ig H chain isotypes in human B cells.** *J. Immunol.* 1994, **152**:438–445.

Litinskiy, M.B., Nardelli, B., Hilbert, D.M., He, B., Schaffer, A., Casali, P., and Cerutti, A.: **DCs induce CD40-independent immunoglobulin class switching through BLyS and APRIL.** *Nat. Immunol.* 2002, **3**:822–829.

Shulman, Z., Gitlin, A.D., Weinstein, J.S., Lainez, B., Esplugues, E., Flavell, R.A., Craft, J.E., and Nussenzweig, M.C.: **Dynamic signaling by T follicular helper cells during germinal center B cell selection.** *Science* 2014, **345**:1058–1062.

10-9   Activation-induced cytidine deaminase (AID) introduces mutations into genes transcribed in B cells.

Bransteitter, R., Pham, P., Scharff, M.D., and Goodman, M.F.: **Activation-induced cytidine deaminase deaminates deoxycytidine on single-stranded DNA but requires the action of RNase.** *Proc. Natl Acad. Sci. USA* 2003, **100**:4102–4107.

Muramatsu, M., Kinoshita, K., Fagarasan, S., Yamada, S., Shinkai, Y., and Honjo, T.: **Class switch recombination and hypermutation require activation-induced cytidine deaminase (AID), a potential RNA editing enzyme.** *Cell* 2000, **102**:553–563.

Petersen-Mahrt, S.K., Harris, R.S., and Neuberger, M.S.: **AID mutates *E. coli* suggesting a DNA deamination mechanism for antibody diversification.** *Nature* 2002, **418**:99–103.

Pham, P., Bransteitter, R., Petruska, J., and Goodman, M.F.: **Processive AID-catalyzed cytosine deamination on single-stranded DNA stimulates somatic hypermutation.** *Nature* 2003, **424**:103–107.

Yu, K., Huang, F.T., and Lieber, M.R.: **DNA substrate length and surrounding sequence affect the activation-induced deaminase activity at cytidine.** *J. Biol. Chem.* 2004, **279**:6496–6500.

10-10 Mismatch and base-excision repair pathways contribute to somatic hypermutation following initiation by AID.

Basu, U., Chaudhuri, J., Alpert, C., Dutt, S., Ranganath, S., Li, G., Schrum, J.P., Manis, J.P., and Alt, F.W.: **The AID antibody diversification enzyme is regulated by protein kinase A phosphorylation.** *Nature* 2005, **438**:508–511.

Chaudhuri, J., Khuong, C., and Alt, F.W.: **Replication protein A interacts with AID to promote deamination of somatic hypermutation targets.** *Nature* 2004, **430**:992–998.

Di Noia, J.M., and Neuberger, M.S.: **Molecular mechanisms of antibody somatic hypermutation.** *Annu. Rev. Biochem.* 2007, **76**:1–22.

Odegard, V.H., and Schatz, D.G.: **Targeting of somatic hypermutation.** *Nat. Rev. Immunol.* 2006, **6**:573–583.

Weigert, M.G., Cesari, I.M., Yonkovich, S.J., and Cohn, M.: **Variability in the lambda light chain sequences of mouse antibody.** *Nature* 1970, **228**:1045–1047.

10-11 AID initiates class switching to allow the same assembled V$_H$ exon to be associated with different C$_H$ genes in the course of an immune response.

Basu, U., Meng, F.L., Keim, C., Grinstein, V., Pefanis, E., Eccleston, J., Zhang, T., Myers, D., Wasserman, C.R., Wesemann, D.R., *et al.*: **The RNA exosome targets the AID cytidine deaminase to both strands of transcribed duplex DNA substrates.** *Cell* 2011, **144**:353–363.

Chaudhuri, J., and Alt, F.W.: **Class-switch recombination: interplay of transcription, DNA deamination and DNA repair.** *Nat. Rev. Immunol.* 2004, **4**:541–552.

Pavri, R., Gazumyan, A., Jankovic, M., Di Virgilio, M., Klein, I., Ansarah-Sobrinho, C., Resch, W., Yamane, A., Reina San-Martin, B., Barreto, V., *et al.*: **Activation-induced cytidine deaminase targets DNA at sites of RNA polymerase II stalling by interaction with Spt5.** *Cell* 2010, **143**:122–133.

Revy, P., Muto, T., Levy, Y., Geissmann, F., Plebani, A., Sanal, O., Catalan, N., Forveille, M., Dufourcq-Lagelouse, R., Gennery, A., *et al.*: **Activation-induced cytidine deaminase (AID) deficiency causes the autosomal recessive form of the hyper-IgM syndrome (HIGM2).** *Cell* 2000, **102**:565–575.

Shinkura, R., Tian, M., Smith, M., Chua, K., Fujiwara, Y., and Alt, F.W.: **The influence of transcriptional orientation on endogenous switch region function.** *Nat. Immunol.* 2003, **4**:435–441.

Yu, K., Chedin, F., Hsieh, C.-L., Wilson, T.E., and Lieber, M.R.: **R-loops at immunoglobulin class switch regions in the chromosomes of stimulated B cells.** *Nat. Immunol.* 2003, **4**:442–451.

10-12 Cytokines made by T_{FH} cells direct the choice of isotype for class switching in T-dependent antibody responses.

Avery, D.T., Bryant, V.L., Ma, C.S., de Waal Malefyt, R., and Tangye, S.G.: **IL-21-induced isotype switching to IgG and IgA by human naive B cells is differentially regulated by IL-4.** *J. Immunol.* 2008, **181**:1767–1779.

Francke, U., and Ochs, H.D.: **The CD40 ligand, gp39, is defective in activated T cells from patients with X-linked hyper-IgM syndrome.** *Cell* 1993, **72**:291–300.

Park, S.R., Seo, G.Y., Choi, A.J., Stavnezer, J., and Kim, P.H.: **Analysis of transforming growth factor-beta1-induced Ig germ-line gamma2b transcription and its implication for IgA isotype switching.** *Eur. J. Immunol.* 2005, **35**:946–956.

Ray, J.P., Marshall, H.D., Laidlaw, B.J., Staron, M.M., Kaech, S.M., and Craft, J.: **Transcription factor STAT3 and type I interferons are corepressive insulators for differentiation of follicular helper and T helper 1 cells.** *Immunity* 2014, **40**:367–377.

Seo, G.Y., Park, S.R., and Kim, P.H.: **Analyses of TGF-beta1-inducible Ig germline gamma2b promoter activity: involvement of Smads and NF-kappaB.** *Eur. J. Immunol.* 2009, **39**:1157–1166.

Stavnezer, J.: **Immunoglobulin class switching.** *Curr. Opin. Immunol.* 1996, **8**:199–205.

Vijayanand, P., Seumois, G., Simpson, L.J., Abdul-Wajid, S., Baumjohann, D., Panduro, M., Huang, X., Interlandi, J., Djuretic, I.M., Brown, D.R., *et al.*: **Interleukin-4 production by follicular helper T cells requires the conserved Il4 enhancer hypersensitivity site V.** *Immunity* 2012, **36**:175–187.

10-13 B cells that survive the germinal center reaction eventually differentiate into either plasma cells or memory cells.

Hu, C.C., Dougan, S.K., McGehee, A.M., Love, J.C., and Ploegh, H.L.: **XBP-1 regulates signal transduction, transcription factors and bone marrow colonization in B cells.** *EMBO J.* 2009, **28**:1624–1636.

Nera, K.P., and Lassila, O.: **Pax5—a critical inhibitor of plasma cell fate.** *Scand. J. Immunol.* 2006, **64**:190–199.

Omori, S.A., Cato, M.H., Anzelon-Mills, A., Puri, K.D., Shapiro-Shelef, M., Calame, K., and Rickert, R.C.: **Regulation of class-switch recombination and plasma cell differentiation by phosphatidylinositol 3-kinase signaling.** *Immunity* 2006, **25**:545–557.

Radbruch, A., Muehlinghaus, G., Luger, E.O., Inamine, A., Smith, K.G., Dorner, T., and Hiepe, F.: **Competence and competition: the challenge of becoming a long-lived plasma cell.** *Nat. Rev. Immunol.* 2006, **6**:741–750.

Schebesta, M., Heavey, B., and Busslinger, M.: **Transcriptional control of B-cell development.** *Curr. Opin. Immunol.* 2002, **14**:216–223.

10-14 Some antigens do not require T-cell help to induce B-cell responses.

Anderson, J., Coutinho, A., Lernhardt, W., and Melchers, F.: **Clonal growth and maturation to immunoglobulin secretion in vitro of every growth-inducible B lymphocyte.** *Cell* 1977, **10**:27–34.

Balazs, M., Martin, F., Zhou, T., and Kearney, J.: **Blood dendritic cells interact with splenic marginal zone B cells to initiate T-independent immune responses.** *Immunity* 2002, **17**:341–352.

Bekeredjian-Ding, I., and Jego, G.: **Toll-like receptors—sentries in the B-cell response.** *Immunology* 2009, **128**:311–323.

Craxton, A., Magaletti, D., Ryan, E.J., and Clark, E.A.: **Macrophage- and dendritic cell-dependent regulation of human B-cell proliferation requires the TNF family ligand BAFF.** *Blood* 2003, **101**:4464–4471.

Fagarasan, S., and Honjo, T.: **T-independent immune response: new aspects of B cell biology.** *Science* 2000, **290**:89–92.

Garcia De Vinuesa, C., Gulbranson-Judge, A., Khan, M., O'Leary, P., Cascalho, M., Wabl, M., Klaus, G.G., Owen, M.J., and MacLennan, I.C.: **Dendritic cells associated with plasmablast survival.** *Eur. J. Immunol.* 1999, **29**:3712–3721.

MacLennan, I., and Vinuesa, C.: **Dendritic cells, BAFF, and APRIL: innate players in adaptive antibody responses.** *Immunity* 2002, **17**:341–352.

Mond, J.J., Lees, A., and Snapper, C.M.: **T cell-independent antigens type 2.** *Annu. Rev. Immunol.* 1995, **13**:655–692.

Ruprecht, C.R., and Lanzavecchia, A.: **Toll-like receptor stimulation as a third signal required for activation of human naive B cells.** *Eur. J. Immunol.* 2006, **36**:810–816.

Snapper, C.M., Shen, Y., Khan, A.Q., Colino, J., Zelazowski, P., Mond, J.J., Gause, W.C., and Wu, Z.Q.: **Distinct types of T-cell help for the induction of a humoral immune response to** *Streptococcus pneumoniae*. *Trends Immunol.* 2001, **22**:308–311.

Yanaba, K., Bouaziz, J.D., Matsushita, T., Tsubata, T., and Tedder, T.F.: **The development and function of regulatory B cells expressing IL-10 (B10 cells) requires antigen receptor diversity and TLR signals.** *J. Immunol.* 2009, **182**:7459–7472.

Yoshizaki, A., Miyagaki, T., DiLillo, D.J., Matsushita, T., Horikawa, M., Kountikov, E.I., Spolski, R., Poe, J.C., Leonard, W.J., and Tedder, T.F.: **Regulatory B cells control T-cell autoimmunity through IL-21-dependent cognate interactions.** *Nature* 2012, **491**:264–268.

10-15 Antibodies of different classes operate in distinct places and have distinct effector functions.

Diebolder, C.A., Beurskens, F.J., de Jong, R.N., Koning, R.I., Strumane, K., Lindorfer, M.A., Voorhorst, M., Ugurlar, D., Rosati, S., Heck, A.J., *et al.*: **Complement is activated by IgG hexamers assembled at the cell surface.** *Science* 2014, **343**:1260–1263.

Hughey, C.T., Brewer, J.W., Colosia, A.D., Rosse, W.F., and Corley, R.B.: **Production of IgM hexamers by normal and autoimmune B cells: implications for the physiologic role of hexameric IgM.** *J. Immunol.* 1998, **161**:4091–4097.

Petrušić, V., Živkovic, I., Stojanovic, M., Stojicevic, I., Marinkovic, E., and Dimitrijevic, L.: **Hexameric immunoglobulin M in humans: desired or unwanted?** *Med. Hypotheses* 2011, **77**:959–961.

Rispens, T., den Bleker, T.H., and Aalberse, R.C.: **Hybrid IgG4/IgG4 Fc antibodies form upon 'Fab-arm' exchange as demonstrated by SDS-PAGE or size-exclusion chromatography.** *Mol. Immunol.* 2010, **47**:1592–1594.

Suzuki, K., Meek, B., Doi, Y., Muramatsu, M., Chiba, T., Honjo, T., and Fagarasan, S.: **Aberrant expansion of segmented filamentous bacteria in IgA-deficient gut.** *Proc. Natl Acad. Sci. USA* 2004, **101**:1981–1986.

Ward, E.S., and Ghetie, V.: **The effector functions of immunoglobulins: implications for therapy.** *Ther. Immunol.* 1995, **2**:77–94.

10-16 Polymeric immunoglobulin receptor binds to the Fc regions of IgA and IgM and transports them across epithelial barriers.

Ghetie, V., and Ward, E.S.: **Multiple roles for the major histocompatibility complex class I-related receptor FcRn.** *Annu. Rev. Immunol.* 2000, **18**:739–766.

Johansen, F.E., and Kaetzel, C.S.: **Regulation of the polymeric immunoglobulin receptor and IgA transport: new advances in environmental factors that stimulate pIgR expression and its role in mucosal immunity.** *Mucosal Immunol.* 2011, **4**:598–602.

Lamm, M.E.: **Current concepts in mucosal immunity. IV. How epithelial transport of IgA antibodies relates to host defense.** *Am. J. Physiol.* 1998, **274**:G614–G617.

Mostov, K.E.: **Transepithelial transport of immunoglobulins.** *Annu. Rev. Immunol.* 1994, **12**:63–84.

10-17 The neonatal Fc receptor carries IgG across the placenta and prevents IgG excretion from the body.

Akilesh, S., Huber, T.B., Wu, H., Wang, G., Hartleben, B., Kopp, J.B., Miner, J.H., Roopenian, D.C., Unanue, E.R., and Shaw, A.S.: **Podocytes use FcRn to clear IgG from the glomerular basement membrane.** *Proc. Natl Acad. Sci. USA* 2008, **105**:967–972.

Burmeister, W.P., Gastinel, L.N., Simister, N.E., Blum, M.L., and Bjorkman, P.J.: **Crystal structure at 2.2 Å resolution of the MHC-related neonatal Fc receptor.** *Nature* 1994, **372**:336–343.

Roopenian, D.C., and Akilesh, S.: **FcRn: the neonatal Fc receptor comes of age.** *Nat. Rev. Immunol.* 2007, **7**:715–725.

10-18 High-affinity IgG and IgA antibodies can neutralize toxins and block the infectivity of viruses and bacteria.

Brandtzaeg, P.: **Role of secretory antibodies in the defence against infections.** *Int. J. Med. Microbiol.* 2003, **293**:3–15.

Haghi, F., Peerayeh, S.N., Siadat, S.D., and Zeighami, H.: **Recombinant outer membrane secretin PilQ(406-770) as a vaccine candidate for serogroup B Neisseria meningitidis.** *Vaccine* 2012, **30**:1710–1714.

Kaufmann, B., Chipman, P.R., Holdaway, H.A., Johnson, S., Fremont, D.H., Kuhn, R.J., Diamond, M.S., and Rossmann, M.G.: **Capturing a flavivirus pre-fusion intermediate.** *PLoS Pathog.* 2009, **5**:e1000672.

Nybakken, G.E., Oliphant, T., Johnson, S., Burke, S., Diamond, M.S., and Fremont, D.H.: **Structural basis of West Nile virus neutralization by a therapeutic antibody.** *Nature* 2005, **437**:764–769.

Sougioultzis, S., Kyne, L., Drudy, D., Keates, S., Maroo, S., Pothoulakis, C., Giannasca, P.J., Lee, C.K., Warny, M., Monath, T.P., *et al.*: **Clostridium difficile toxoid vaccine in recurrent *C. difficile*-associated diarrhea.** *Gastroenterology* 2005, **128**:764–770.

10-19 Antibody:antigen complexes activate the classical pathway of complement by binding to C1q.

Cooper, N.R.: **The classical complement pathway. Activation and regulation of the first complement component.** *Adv. Immunol.* 1985, **37**:151–216.

Perkins, S.J., and Nealis, A.S.: **The quaternary structure in solution of human complement subcomponent C1r2C1s2.** *Biochem. J.* 1989, **263**:463–469.

Sörman, A., Zhang, L., Ding, Z., and Heyman, B.: **How antibodies use complement to regulate antibody responses.** *Mol. Immunol.* 2014, **61**:79–88

10-20 Complement receptors and Fc receptors both contribute to removal of immune complexes from the circulation.

Dong, C., Ptacek, T.S., Redden, D.T., Zhang, K., Brown, E.E., Edberg, J.C., McGwin, G., Jr., Alarcón, G.S., Ramsey-Goldman, R., Reveille, J.D., *et al.*: **Fcγ receptor IIIa single-nucleotide polymorphisms and haplotypes affect human IgG binding and are associated with lupus nephritis in African Americans.** *Arthritis Rheumatol.* 2014, **66**:1291–1299.

Leffler, J., Bengtsson, A.A., and Blom, A.M.: **The complement system in systemic lupus erythematosus: an update.** *Ann. Rheum. Dis.* 2014, **73**:1601–1606.

Nash, J.T., Taylor, P.R., Botto, M., Norsworthy, P.J., Davies, K.A., and Walport, M.J.: **Immune complex processing in C1q-deficient mice.** *Clin. Exp. Immunol.* 2001, **123**:196–202.

Walport, M.J., Davies, K.A., and Botto, M.: **C1q and systemic lupus erythematosus.** *Immunobiology* 1998, **199**:265–285.

10-21 The Fc receptors of accessory cells are signaling receptors specific for immunoglobulins of different classes.

Kinet, J.P., and Launay, P.: **Fcα/μR: single member or first born in the family?** *Nat. Immunol.* 2000, **1**:371–372.

Mallery, D.L., McEwan, W.A., Bidgood, S.R., Towers, G.J., Johnson, C.M., and James, L.C.: **Antibodies mediate intracellular immunity through tripartite motif-containing 21 (TRIM21).** *Proc. Natl Acad. Sci. USA* 2010, **107**:19985–19990.

Ravetch, J.V., and Bolland, S.: **IgG Fc receptors.** *Annu. Rev. Immunol.* 2001, **19**:275–290.

Ravetch, J.V., and Clynes, R.A.: **Divergent roles for Fc receptors and complement *in vivo*.** *Annu. Rev. Immunol.* 1998, **16**:421–432.

Shibuya, A., Sakamoto, N., Shimizu, Y., Shibuya, K., Osawa, M., Hiroyama, T., Eyre, H.J., Sutherland, G.R., Endo, Y., Fujita, T., *et al.*: **Fcα/μ receptor mediates endocytosis of IgM-coated microbes.** *Nat. Immunol.* 2000, **1**:441–446.

Stefanescu, R.N., Olferiev M., Liu, Y., and Pricop, L.: **Inhibitory Fc gamma receptors: from gene to disease.** *J. Clin. Immunol.* 2004, **24**:315–326.

10-22 Fc receptors on phagocytes are activated by antibodies bound to the surface of pathogens and enable the phagocytes to ingest and destroy pathogens.

Dierks, S.E., Bartlett, W.C., Edmeades, R.L., Gould, H.J., Rao, M., and Conrad, D.H.: **The oligomeric nature of the murine Fc epsilon RII/CD23. Implications for function.** *J. Immunol.* 1993, **150**:2372–2382.

Hogan, S.P., Rosenberg, H.F., Moqbel, R., Phipps, S., Foster, P.S., Lacy, P., Kay, A.B., and Rothenberg, M.E.: **Eosinophils: biological properties and role in health and disease.** *Clin. Exp. Allergy* 2008, **38**:709–750.

Karakawa, W.W., Sutton, A., Schneerson, R., Karpas, A., and Vann, W.F.: **Capsular antibodies induce type-specific phagocytosis of capsulated *Staphylococcus aureus* by human polymorphonuclear leukocytes.** *Infect. Immun.* 1986, **56**:1090–1095.

10-23 Fc receptors activate NK cells to destroy antibody-coated targets.

Chung, A.W., Rollman, E., Center, R.J., Kent, S.J., and Stratov, I.: **Rapid degranulation of NK cells following activation by HIV-specific antibodies.** *J. Immunol.* 2009, **182**:1202–1210.

Lanier, L.L., and Phillips, J.H.: **Evidence for three types of human cytotoxic lymphocyte.** *Immunol. Today* 1986, **7**:132.

Leibson, P.J.: **Signal transduction during natural killer cell activation: inside the mind of a killer.** *Immunity* 1997, **6**:655–661.

Sulica, A., Morel, P., Metes, D., and Herberman, R.B.: **Ig-binding receptors on human NK cells as effector and regulatory surface molecules.** *Int. Rev. Immunol.* 2001, **20**:371–414.

Takai, T.: **Multiple loss of effector cell functions in FcRγ-deficient mice.** *Int. Rev. Immunol.* 1996, **13**:369–381.

10-24 Mast cells and basophils bind IgE antibody via the high-affinity Fcε receptor.

Beaven, M.A., and Metzger, H.: **Signal transduction by Fc receptors: the FcεRI case.** *Immunol. Today* 1993, **14**:222–226.

Kalesnikoff, J., Huber, M., Lam, V., Damen, J.E., Zhang, J., Siraganian, R.P., and Krystal, G.: **Monomeric IgE stimulates signaling pathways in mast cells that lead to cytokine production and cell survival.** *Immunity* 2001, **14**:801–811.

Sutton, B.J., and Gould, H.J.: **The human IgE network.** *Nature* 1993, **366**:421–428.

10-25 IgE-mediated activation of accessory cells has an important role in resistance to parasite infection.

Capron, A., Riveau, G., Capron, M., and Trottein, F.: **Schistosomes: the road from host-parasite interactions to vaccines in clinical trials.** *Trends Parasitol.* 2005, **21**:143–149.

Grencis, R.K.: **Th2-mediated host protective immunity to intestinal nematode infections.** *Philos. Trans. R. Soc. Lond. B* 1997, **352**:1377–1384.

Grencis, R.K., Else, K.J., Huntley, J.F., and Nishikawa, S.I.: **The in vivo role of stem cell factor (c-kit ligand) on mastocytosis and host protective immunity to the intestinal nematode *Trichinella spiralis* in mice.** *Parasite Immunol.* 1993, **15**:55–59.

Kasugai, T., Tei, H., Okada, M., Hirota, S., Morimoto, M., Yamada, M., Nakama, A., Arizono, N., and Kitamura, Y.: **Infection with *Nippostrongylus brasiliensis* induces invasion of mast cell precursors from peripheral blood to small intestine.** *Blood* 1995, **85**:1334–1340.

Ushio, H., Watanabe, N., Kiso, Y., Higuchi, S., and Matsuda, H.: **Protective immunity and mast cell and eosinophil responses in mice infested with larval *Haemaphysalis longicornis* ticks.** *Parasite Immunol.* 1993, **15**:209–214.

Integrated Dynamics of Innate and Adaptive Immunity

11

Throughout this book we have examined the separate ways in which the innate and the adaptive immune responses protect against invading pathogens. In this chapter, we consider how the cells and molecules of the immune system work as an integrated defense system to eliminate or control different types of infectious agents, and also how the adaptive immune system provides long-lasting protective immunity. In Chapters 2 and 3, we saw how innate immunity is brought into play in the earliest phases of an infection and is probably sufficient to prevent colonization of the body by most of the microorganisms encountered in the environment. We also introduced **innate lymphoid cells** (**ILCs**), which, although lacking antigen-specific receptors, share overlapping developmental and functional features with effector CD4 T-cell subsets and cytotoxic CD8 T cells and act early during infection to generate distinct types of immune responses that target specific types of pathogens. Unlike naive T and B cells, ILCs reside in barrier tissues, such as the intestinal and respiratory mucosae, where they are poised to respond rapidly to pathogens to impair or eliminate their spread.

However, most pathogens have developed strategies to evade innate immune defenses and establish a focus of infection. In these circumstances, the innate immune response sets the scene for the induction of an adaptive immune response, which is orchestrated by signals that emanate from innate sensor cells and is coordinated with innate effector cells to bring about pathogen clearance. In the **primary immune response**, which occurs against a pathogen encountered for the first time, ILCs respond to innate sensor cells to mount a rapid response over the first few hours to days of pathogen invasion. Concurrent with this response, clonal expansion and differentiation of naive lymphocytes into effector T cells and antibody-secreting B cells is initiated and guided by innate sensor cells and ILCs. However, the adaptive response requires several days to weeks to fully mature, largely due to the rarity of antigen-specific precursor cells. Following expansion and differentiation in the secondary lymphoid tissues, effector T cells migrate to sites of infection and, along with pathogen-specific antibodies, enhance the effector functions of innate immune cells, and, in most cases, effectively target the pathogen for elimination (**Fig. 11.1**).

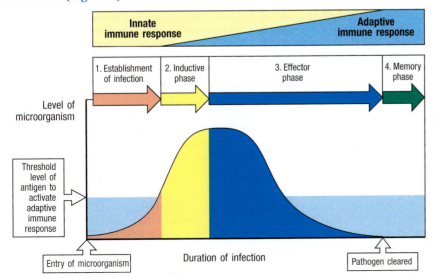

Fig. 11.1 The course of a typical acute infection that is cleared by an adaptive immune response. 1. The infectious agent colonizes and its numbers increase as it replicates. The innate response is initiated immediately following detection of the pathogen. 2. When numbers of the pathogen exceed the threshold dose of antigen required for an adaptive response, the response is initiated; the pathogen continues to grow, restrained by responses of the innate immune system. At this stage, immunological memory also starts to be induced. 3. After 4–7 days, effector cells and molecules of the adaptive response start to clear the infection. 4. When the infection has been cleared and the dose of antigen has fallen below the response threshold, the response ceases, but antibody, residual effector cells, and immunological memory provide lasting protection against reinfection in most cases.

During this period, specific **immunological memory** is also established by adaptive immune cells. This ensures a rapid reinduction of antigen-specific antibody and effector T cells on encounter with the same pathogen during a **secondary immune response**, thus providing long-lasting and often lifelong protection against the pathogen. Immunological memory is discussed in the last part of the chapter. Memory responses differ in several ways from primary responses. We discuss the reasons for this, and what is known about how immunological memory is maintained.

Integration of innate and adaptive immunity in response to specific types of pathogens.

The immune response is a dynamic process, and both its nature and its intensity change over time. It begins with the antigen-independent responses of innate immunity and becomes both more focused on the pathogen and more powerful as the antigen-specific adaptive immune response matures. The character of the response differs contingent on the type of pathogen. Different types of pathogens (for example, intracellular and extracellular bacteria, viruses, helminthic parasites, and fungi) elicit different types of immune response (for example, type 1, 2, or 3), so that the most effective immune response is induced for effective elimination of the pathogen. The innate immune system not only anticipates and initiates the adaptive T- and B-cell responses, but continues to provide effector cells and reinforcing pathways of the different types of immunity throughout infection. Early in the response, different subsets of innate lymphoid cells (ILCs) are activated by cytokines produced by innate sensor cells. This early response acts to constrain pathogen entry at the initial site of infection to prevent dissemination while the adaptive response develops. However, the more sensitive and specific actions of effector T cells and class-switched, affinity-matured antibodies are often required for the complete elimination of infection, or **sterilizing immunity**. In this part of the chapter, we provide an overview of how the different phases of an immune response are orchestrated in space and time and then discuss how distinct cytokines from innate sensor cells activate different innate lymphoid cell subsets to restrain pathogen invasion and direct pathogen-specific defenses while the adaptive response is developing.

11-1   The course of an infection can be divided into several distinct phases.

Although some of the microbe-associated molecular patterns (MAMPs) of different types of pathogens are shared, others are not, and these differences underlie the induction of distinct patterns of innate and adaptive immunity that can be broadly grouped into type 1, type 2, and type 3 responses, as will be discussed below. Irrespective of the inciting pathogen and the pattern of immune response it provokes, however, the tempo of the host response is similar and can be broken down into various stages (Fig. 11.1, and see Fig. 3.38).

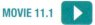
MOVIE 11.1

In the first stage of infection, a new host is exposed to infectious particles either shed by an infected individual or present in the environment. The numbers, route, mode of transmission, and stability of an infectious agent outside the host determine its infectivity. The first contact with a new host occurs through an epithelial surface, such as the skin or the mucosal surfaces of the respiratory, gastrointestinal, or urogenital tracts. After making contact, an infectious agent must establish a focus of infection. It must either adhere to the epithelial surface and colonize it, or penetrate it to replicate in the tissues (**Fig. 11.2**). Bites by arthropods (insects and ticks) and wounds breach the epidermal barrier and help some microorganisms gain entry through the skin.

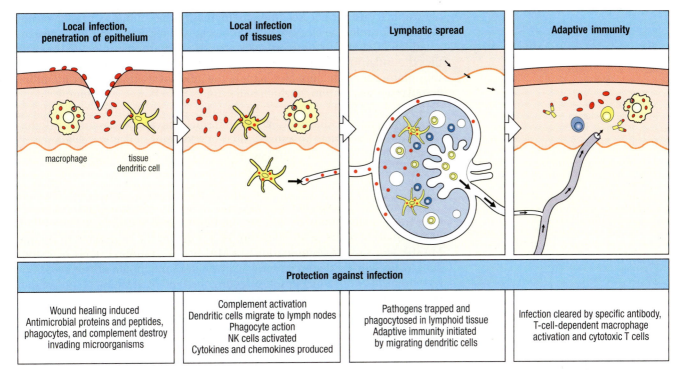

Local infection, penetration of epithelium	Local infection of tissues	Lymphatic spread	Adaptive immunity
macrophage tissue dendritic cell			

Protection against infection			
Wound healing induced Antimicrobial proteins and peptides, phagocytes, and complement destroy invading microorganisms	Complement activation Dendritic cells migrate to lymph nodes Phagocyte action NK cells activated Cytokines and chemokines produced	Pathogens trapped and phagocytosed in lymphoid tissue Adaptive immunity initiated by migrating dendritic cells	Infection cleared by specific antibody, T-cell-dependent macrophage activation and cytotoxic T cells

Fig. 11.2 Infections and the responses to them can be divided into a series of stages. These are illustrated here for a pathogenic microorganism (red) entering across a wound in an epithelium. The microorganism first adheres to epithelial cells and then invades beyond the epithelium into underlying tissues (first panel). A local innate immune response helps to contain the infection, and delivers antigen and antigen-loaded dendritic cells to lymphatics (second panel) and thence to local lymph nodes (third panel). This leads to an adaptive immune response in the lymph node that involves the activation and further differentiation of B cells and T cells with the eventual production of antibody and effector T cells, which clear the infection (fourth panel).

Only when a microorganism has successfully established a focus of infection in the host does disease occur (see Fig. 11.2). With few exceptions, little damage will be caused unless the pathogen spreads from the original focus or secretes toxins that spread to other parts of the body. Extracellular pathogens spread by direct extension of the infection through the lymphatics or the bloodstream. Spread into the bloodstream usually occurs only after the lymphatic system has been overwhelmed. Obligate intracellular pathogens spread from cell to cell; they do so either by direct transmission from one cell to the next or by release into the extracellular fluid and reinfection of both adjacent and distant cells. Facultative intracellular pathogens can do the same after a period of survival in the extracellular environment. In contrast, some of the bacteria that cause gastroenteritis exert their effects without spreading into the tissues. They establish a site of infection on the luminal surface of the epithelium lining the gut and cause pathology by damaging the epithelium or by secreting toxins that cause damage either *in situ* or after crossing the epithelial barrier and entering the circulation.

The establishment of a focus of infection in tissues and the response of the innate immune system produce changes in the immediate environment. Many microorganisms are repelled or kept in check at this stage by innate defenses, which are triggered by stimulation of the various germline-encoded pattern recognition receptors expressed by innate sensor cells—such as epithelial cells, tissue-resident mast cells, macrophages, and dendritic cells (see Chapters 2 and 3). Cytokines and chemokines produced by pathogen-activated innate sensor cells initiate local inflammation, and also activate ILCs. These responses are activated within minutes to hours, and are sustained for at least several days. The inflammatory response is induced through activation of the endothelium of local post-capillary venules (see Fig. 3.31). This leads to the recruitment of circulating innate effector cells, particularly neutrophils and monocytes, thereby increasing the numbers of phagocytes available for microbe clearance. As monocytes enter tissue and become activated, additional inflammatory cells are attracted into the infected tissue so that the inflammatory response is maintained and reinforced. Leakiness of the inflamed endothelium also leads to the influx of serum proteins, including complement, activation of which in a primary infection occurs mainly via the alternative and lectin pathways (see Fig. 2.15). This results in the production of the anaphylatoxins C3a and C5a, which further activate the vascular endothelium; and C3b, which opsonizes microbes for more effective clearance by recruited phagocytes. This early phase of the inflammatory response is largely nonspecific for the type of pathogen.

Concordant with the production of pro-inflammatory cytokines such as TNF-α, which activate nonspecific inflammation, innate sensor cells produce additional cytokines that differentially activate specific subsets of ILCs within the first hours of an infection. This is due to the expression of unique MAMPs—or unique combinations of MAMPs—by different types of pathogens, which elicit different patterns of cytokines from innate sensor cells. This has important consequences in directing the type of immune response that will be mounted against the pathogen, as subsets of ILCs are differentially activated to produce their own effector cytokines and chemokines contingent on the pattern of cytokines produced by innate sensor cells (Fig. 11.3). The products of activated ILCs amplify and coordinate local innate responses that are better tailored to resist specific types of pathogens, and also alter the recruitment and maturation of different **myelomonocytic** innate effector cells (that is, granulocytes like neutrophils, eosinophils, and basophils, or monocytes) at the site of infection. Cytokines produced by ILCs may also guide the development of naive T cells into distinct effector subsets (for example, T_H1, T_H2, or T_H17 cells)—either directly by acting on naive T cells themselves or indirectly by modulating the activation of dendritic cells that migrate to regional lymphoid tissues to prime naive T cells. In this way, ILCs perform an important bridging function during the first few days of an immune response, by both providing for innate defense and influencing the type of adaptive response that follows.

Adaptive immunity is triggered when an infection eludes or overwhelms the innate defense mechanisms and generates a threshold level of antigen (see Fig. 11.1). Adaptive immune responses are then initiated in the local lymphoid

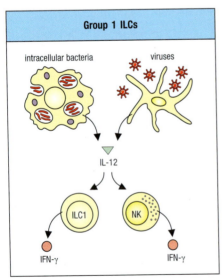
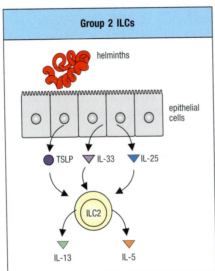
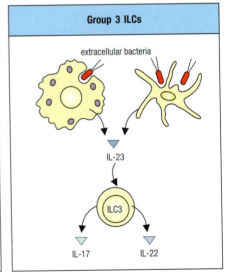

Fig. 11.3 Cytokines produced by innate sensor cells activate innate lymphoid cells (ILCs). The microbe-associated molecular patterns (MAMPs) expressed by different types of pathogens stimulate distinct cytokine responses from innate sensor cells. These, in turn, stimulate different subsets of ILCs to produce different effector cytokines that act to coordinate and amplify the innate response.

tissue, in response to antigens presented by dendritic cells activated during the course of the innate immune response (see Fig. 11.2, second and third panels). Antigen-specific effector T cells and antibody-secreting B cells are generated by clonal expansion and differentiation over several days, during which innate responses orchestrated by ILCs 'buy time' for the adaptive response to mature. Within several days of infection, antigen-specific T cells and then antibodies are released into the blood, and from there can enter the site of infection (see Fig. 11.2, fourth panel). Adaptive responses are more powerful because antigen-specific targeting of innate effector mechanisms can eliminate pathogens more precisely. For example, antibodies can activate complement to directly kill pathogens; they can opsonize pathogens for enhanced phagocytosis; and they can arm Fc-bearing innate effector cells for release of microbicidal factors or recruit the cytocidal actions of natural killer (NK) cells—these last capabilities known as **antibody-dependent cell-mediated cytotoxicity** (**ADCC**). Effector CD8 T cells can directly kill antigen-bearing target cells via similar cytocidal cytotoxic actions, and effector CD4 T cells can directly release cytokines onto macrophages to enhance their microbicidal actions.

Resolution of an infection typically involves complete clearance of the pathogen, and thus the source of antigens, over the course of days to weeks, following which most effector lymphocytes die—a stage known as clonal contraction (see Section 11-16). What remain are long-lived antibody-producing plasma cells that sustain circulating antibodies for months to years, and small numbers of memory B and T cells that may also persist for years, poised for an accelerated adaptive response in the event of future encounters with the same pathogen. Thus, in addition to clearing the infectious agent, an effective adaptive immune response prevents reinfection. For some infectious agents, this protection is essentially absolute, whereas for others infection is only reduced or attenuated on reexposure to the pathogen.

It is not known how many infections are dealt with solely by the nonadaptive mechanisms of innate immunity, because such infections are eliminated early and produce little in the way of symptoms or pathology. Innate immunity does, however, seem to be essential for effective host defense, as shown by the progression of infection in mice that lack components of innate immunity but have an intact adaptive immune system (Fig. 11.4). Conversely, many infections can be curbed—but not cleared—in the absence of adaptive immunity.

After many types of infection, little or no residual pathology follows an effective primary adaptive response. In some cases, however, the infection itself, or the response it induces, causes significant tissue damage. In yet other cases, such as infection with cytomegalovirus or *Mycobacterium tuberculosis*, the pathogen is contained but not eliminated, and can persist in a latent form. If the adaptive immune response is later weakened, as it is in acquired immune deficiency syndrome (AIDS), these pathogens may resurface to cause virulent systemic infections. We will focus on the strategies used by certain pathogens to evade or subvert adaptive immunity, and thereby establish a persistent, or chronic, infection, in Chapter 13.

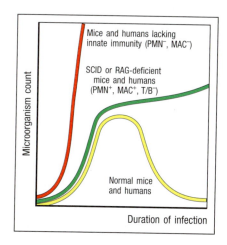

Fig. 11.4 The time course of infection in normal and immunodeficient mice and humans. The red curve shows the rapid growth of microorganisms in the absence of innate immunity, when macrophages (MAC) and polymorphonuclear leukocytes (PMN) are lacking. The green curve shows the course of infection in mice and humans that have innate immunity but have no T or B lymphocytes and so lack adaptive immunity. The yellow curve shows the normal course of an infection in immunocompetent mice or humans.

11-2   The effector mechanisms that are recruited to clear an infection depend on the infectious agent.

Most infections ultimately engage both T- and B-cell-mediated adaptive immunity, and in many cases both are helpful in clearing or containing the pathogen and setting up protective immunity. However, the relative importance of the different effector mechanisms, and the effective classes of antibody involved, vary with different pathogens. An emerging concept is that there are different types of immune responses that are focused on the activation of distinct **immune effector modules** (see Section 1-19). In each type of immune response, a specific collection of innate and adaptive mechanisms act together to eliminate a specific type of pathogen. Each effector module includes subsets of innate sensor cells, ILCs, effector T cells, and antibody isotypes, which coordinate with subsets of circulating or tissue-resident myelomonocytic cells whose microbicidal functions they recruit and enhance (Fig. 11.5). Circulating myelomonocytic cells are important innate effector cells that are targeted for heightened functions by ILCs, effector T cells, and antibodies following their recruitment into sites of infection. In their order of abundance in circulating blood, these include neutrophils, monocytes (which enter inflamed tissues and differentiate into activated macrophages), eosinophils, and basophils. Tissue-resident mast cells, which share many functions with basophils, are also targeted for heightened function.

It appears that each of the three major ILC and effector CD4 T-cell subsets (ILC1/ILC2/ILC3 and $T_H1/T_H2/T_H17$, respectively) evolved to enhance and coordinate the functions of, and integrate adaptive immunity with, different arms of the myelomonocytic pathway for optimal eradication of different classes of pathogens: monocyte and macrophages are enhanced by T_H1 cells; eosinophils, basophils, and mast cells by T_H2 cells; and neutrophils by T_H17 cells. The three major types of immune responses are controlled by cytokine and chemokine networks, as discussed below.

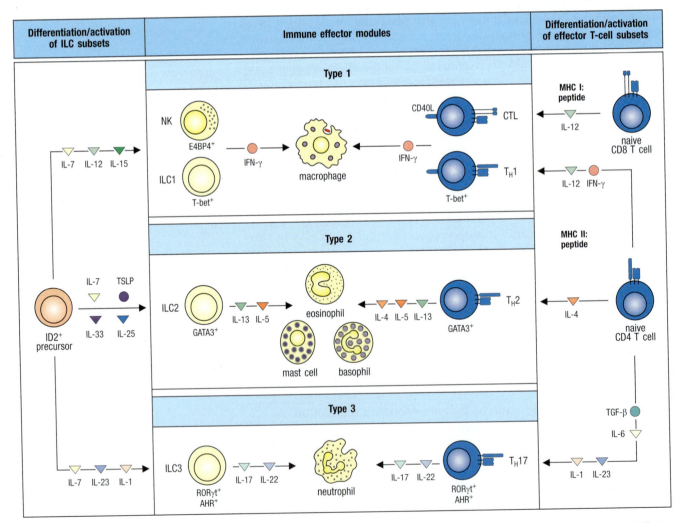

Fig. 11.5 Integration of ILCs, T-cell subsets, and innate effector cells into immune effector modules. The major inductive and effector cytokines, and transcription factors (e.g., ID2, T-bet, GATA3, RORγt, and AHR), that are associated with each effector module are shown. See text for details.

Type 1 responses are characterized by the actions of group 1 ILCs (ILC1), T_H1 cells, opsonizing IgG isotypes (for example, IgG1 and IgG2), and macrophages in response to intracellular pathogens, including intracellular bacteria, viruses, and parasites (see Fig. 11.5). **Type 2** responses are characterized by the actions of group 2 ILCs (ILC2), T_H2 cells, IgE, and the innate effector cells eosinophils, basophils, and tissue mast cells, the latter two of which are armed for function by IgE bound to surface Fcε receptors. Type 2 responses are induced by, and target, multicellular parasites, or helminths. **Type 3** responses are characterized by the actions of group 3 ILCs (ILC3), T_H17 cells, opsonizing IgG isotypes, and neutrophils in response to extracellular bacteria and fungi. It is the activation of different subsets of ILCs early in the innate response that sets the stage for polarized type 1, type 2, or type 3 responses. Unlike the effector CD4 T cells with which they share overlapping functional features, ILCs do not require priming and differentiation in order to acquire their effector functions; hence they are able to respond rapidly to amplify the activities of resident and recruited innate effector cells. Here we will consider in more detail the induction and actions of ILC subsets, as these responses precede and are integrated with adaptive T-cell responses.

As discussed in Chapter 3, ILC1s and related NK cells are characterized by their production of IFN-γ in response to IL-12 and IL-18 produced by pathogen-activated dendritic cells and macrophages. Functionally, ILC1s and NK cells most closely resemble T_H1 cells and CTLs, respectively. ILC1s lack the cytolytic granules that are characteristic of NK cells and CTLs, and appear to promote clearance of intracellular pathogens through their activation of infected macrophages by the release of IFN-γ. Thus, through their production of IL-12 and IL-18, macrophages can rapidly induce ILC1 production of IFN-γ, which acts back on the macrophage to induce its heightened killing of intracellular pathogens several days prior to the development and recruitment of T_H1 cells. Moreover, the production of IFN-γ by ILC1s may contribute to the early polarization of T_H1 cells, linking the effector function of these cells to the induction of the T_H1 cell response that follows. Similarly, the rapid induction of the cytolytic activity of NK cells enables the killing of a range of pathogen-infected cells, through the recognition of surface molecules that are expressed on the target cells (see Section 3-23), in advance of the antigen-driven development and deployment of cytolytic CD8 T cells. Also, similar to the effect on T_H1 cells of IFN-γ production by ILC1s, the production of IFN-γ by activated NK cells may contribute to the enhanced differentiation of cytolytic CD8 T cells.

ILC2s that reside in mucosal tissues are preferentially activated by the cytokines **thymic stromal lymphopoietin** (**TSLP**), a STAT5-activating cytokine, and IL-33 or IL-25, each of which are produced in response to helminths. These cytokines are primarily produced by epithelial cells that sense molecular patterns common to helminths, such as chitin, a polysaccharide polymer of β-1,4-*N*-acetylglucosamine that is a widespread constituent of helminths, the exoskeletons of insects, and some fungi. Activated ILC2s rapidly produce large amounts of IL-13 and IL-5; IL-13 stimulates mucus production by goblet cells in the epithelium and mucosal smooth muscle contractions that facilitate worm expulsion; and IL-5 stimulates the production and activation of eosinophils that can kill worms. Unlike T_H2 cells, with which they share functional features, ILC2s appear to produce little or no IL-4 *in vivo*, suggesting that they might not directly promote T_H2 differentiation. However, eosinophils and basophils that are recruited by chemokines produced by ILC2s are activated to produce IL-4 in response to the IL-5 and IL-13 produced by ILC2s, possibly providing an indirect mechanism by which T_H2 differentiation is directed by ILC2s. Moreover, IL-13 produced by ILC2s appears to regulate the activation and migration to regional lymphoid tissues of dendritic cells that promote T_H2 differentiation, although it is unclear whether these dendritic cells can also produce IL-4.

ILC3s play a critical early role in defense against extracellular bacteria and fungi at barrier tissues. Similarly toT_H17 cells, ILC3s are responsive to IL-23 and IL-1β; these cytokines elicit the production of IL-17 and IL-22, which promote early type 3 responses. IL-17 is a pro-inflammatory cytokine that acts on a variety of cells, including stromal cells, epithelial cells, and myeloid cells, to stimulate the production of other pro-inflammatory cytokines (for example, IL-6 and IL-1β), hematopoietic growth factors (G-CSF and GM-CSF), and chemokines that recruit neutrophils and monocytes. IL-22 acts on epithelial cells to induce their production of antimicrobial peptides (AMPs) and promote enhanced barrier integrity. As with other ILCs, the cytokines produced by ILC3s act indirectly via IL-6 and IL-1β in a positive feedback loop to enhance type 3 responses by increasing local production of IL-23 and IL-1β. Through induction of elevated IL-6, IL-1β, and IL-23, ILC3s may also promote the differentiation of T_H17 cells in mucosal lymphoid tissues, where they can be found in substantial numbers.

In a further parallel with effector CD4 T cells, an important feature of ILCs is that they 'license' other innate immune cells for killing or expulsion of microbes, but they do not do so themselves. Instead, myelomonocytic cells, and even cells of the mucosal epithelium, are the agents of ILCs and effector CD4 T cells; they are recruited and/or activated through the pro-inflammatory cytokines and chemokines that these lymphoid cells produce. An exception is NK cells, which, like effector CD8 T cells, directly kill target cells that harbor intracellular pathogens. As we will discuss below, because of their ability to focus effector cytokines on antigen-bearing target cells and to induce B-cell maturation and the production of class-switched antibodies, effector CD4 T cells provide an additional layer of licensing of innate effector cells that increases their lethality and ability to achieve microbial clearance.

Summary.

Integration of the innate and adaptive immune responses is required for effective protection of the host against pathogenic microorganisms. The responses of the innate immune system act early to restrain pathogens while simultaneously helping to initiate the adaptive immune response, which takes a longer time to fully develop. Different types of pathogens cause the activation of different patterns of cytokine production by innate sensor cells. This, in turn, promotes the activation of different patterns of innate lymphoid cells (ILCs), which recruit innate effector cells to sites of infection and contribute to the differentiation of parallel programs of CD4 T-cell differentiation. Coordination of the induction of different immune effector modules that are composed of related subsets of ILCs, innate effector cells, CD4 effector T cells, and class-switched antibodies underlies the different types of immunity directed against different types of pathogens.

Effector T cells augment the effector functions of innate immune cells.

In Chapter 9 we described how dendritic cells loaded with their antigen cargo migrate away from the infected tissue through the lymphatics to enter secondary lymphoid tissues, where they initiate the adaptive immune response. We discussed how CD8 T cells are primed to become cytotoxic effectors that are specialized for killing infected target cells that express MHC class I molecules. We also saw how transcription factor networks activated by specific cytokines direct the differentiation of naive CD4 T cells into distinct classes of CD4 effector T cells—T_H1, T_H2, and T_H17 (see Fig. 9.31). In Chapter 10 we discussed the specialized role of T_{FH} cells, which engage antigen-bearing B cells to control antibody class switching and B-cell maturation in germinal centers in the context of type 1, type 2, and type 3 responses. We now turn our attention to the specialized roles of subsets of effector CD4 T cells that emigrate from secondary lymphoid tissues after their differentiation to orchestrate the functions of innate immune cells at sites of infection.

As discussed in the preceding sections, the pattern of cytokines produced by the innate immune response during the early course of infection is determined by how the microorganism influences the behavior of innate sensor cells and the different subsets of ILCs they engage. The local inflammatory conditions produced by these interactions have a major impact on how T cells differentiate during their initial contact with dendritic cells, thus determining the subsets of effector T cells that are generated (see Chapter 9). In turn, the recruitment of effector T cells to sites of infection sustains and amplifies innate effector cell responses initiated by ILCs through effector mechanisms that require antigen-specific recognition, whether through cell–cell contact between CD4 or CD8 T cells and cell targets bearing their cognate antigens, or via pathogen-specific antibodies. In this part of the chapter, we will discuss how the differentiation of effector CD4 T cells generated during the adaptive immune response alters their expression of surface receptors and causes them to leave secondary lymphoid tissues and home to sites of infection. We will then consider how T_H1, T_H2, and T_H17 cells interact with innate immune cells at sites of infection to bring about the clearance of the specific pathogens that elicit their development and recruitment. Finally, we will consider how the primary effector response is terminated as the pathogen is eliminated.

11-3   Effector T cells are guided to specific tissues and sites of infection by changes in their expression of adhesion molecules and chemokine receptors.

When naive T cells differentiate into effector T cells, changes occur in the expression of specific surface molecules that redirect their trafficking from T-cell zones into B-cell zones, in the case of T_{FH} cells, or from lymphoid to nonlymphoid tissues, in the case of other effector T cells. During the 3–5 days required for differentiation of naive T cells into effector T cells within secondary lymphoid tissues, marked changes occur in the expression of these trafficking molecules, including alterations in the display of selectins and their ligands, integrins, and chemokine receptors. As we will see, some of these changes are generic, and are common to all CD4 and CD8 effector T cells. Others are tissue-specific, facilitating the recruitment of T cells back to the tissues in which they were primed. Yet others are T-cell subset-specific, particularly the patterns of chemokine receptor expression, which are important in directing T_{FH} cells to germinal centers, where they provide help to developing B cells, or in directing T_H1, T_H2, and T_H17 cells to the same tissue sites as the myelomonocytic cells whose effector functions they will recruit and enhance.

Naive CD4 T cells that are activated by antigen and become T_{FH} cells acquire expression of CXCR5 and lose expression of CCR7 and **S1PR1**, the receptor for the chemotactic lipid **sphingosine 1-phosphate** (see Section 9-7). The constitutive expression of CXCL13 by follicular dendritic cells establishes a gradient that attracts developing T_{FH} cells first to the border of the T-cell zone with a B-cell follicle, where they can interact with B cells that present their cognate antigen, and then into the B-cell follicle, where they provide help to germinal center B cells. Unlike T_{FH} cells, other effector CD4 and CD8 T cells must leave the lymphoid tissue in which they have developed to interact with myelomonocytic cells at sites of infection in nonlymphoid tissues. The exit of effector T cells is induced by their loss of CCR7 and their reexpression of S1PR1. S1PR1 is normally rapidly downregulated by CD69 following antigenic stimulation

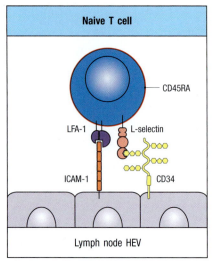

Lymph node HEV

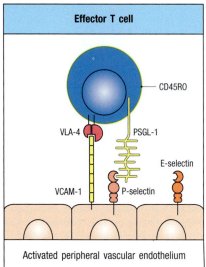

Activated peripheral vascular endothelium

Leukocyte Adhesion Deficiency

Fig. 11.6 Effector T cells change their surface molecules, allowing them to home to sites of infection. Naive T cells home to lymph nodes through the binding of L-selectin to sulfated carbohydrates displayed by various proteins, such as CD34 and GlyCAM-1 (not shown), on the high endothelial venule (HEV, upper panel). After encounter with antigen, many of the differentiated effector T cells lose expression of L-selectin, leave the lymph node about 4–5 days later, and now express the integrin VLA-4 and increased levels of LFA-1 (not shown). These bind to VCAM-1 and ICAM-1, respectively, on peripheral vascular endothelium at sites of inflammation (lower panel). On differentiating into effector cells, T cells also alter their splicing of the mRNA encoding the cell-surface protein CD45. The CD45RO isoform expressed by effector T cells lacks one or more exons that encode extracellular domains present in the CD45RA isoform expressed by naive T cells, and somehow makes effector T cells more sensitive to stimulation by specific antigen.

of naive T cells in order to retain the developing effector cells in the lymphoid tissue as they undergo differentiation and clonal expansion (see Section 9-6). Most effector T cells also shed L-selectin, which mediates rolling on the high endothelial venules of secondary lymphoid tissues such as lymph nodes, in favor of **P-selectin glycoprotein ligand-1** (**PSGL-1**), a homodimeric sialo-glycoprotein that is the major ligand for tethering and rolling on the P- and E-selectin expressed by activated endothelial cells at sites of inflammation (**Fig. 11.6**). In contrast to granulocytes and monocytes, which constitutively express glycosyltransferases necessary for biosynthesis of selectin ligands, T cells express these enzymes only after effector T-cell development. Effector differentiation induces expression of the glycosyltransferase α1,3-fucosyl-transferase VII (FucT-VII), a key enzyme required for both P- and E-selectin ligand generation. Thus, although PSGL-1 is expressed by both naive and effector T cells, it is appropriately glycosylated for selectin binding only by effector T cells.

The expression of other adhesion molecules such as integrins that are important for the recruitment of effector T cells to inflamed tissues is also increased (see Fig. 11.6). Naive T cells mainly express **LFA-1** ($\alpha_L\beta_2$), which is retained on effector T cells as they develop from naive T-cell precursors. However, LFA-1 is not the only integrin that these cells express. Effector T cells also synthesize the integrin $\alpha_4{:}\beta_1$, or VLA-4, which binds to the adhesion molecule **VCAM-1**, a member of the immunoglobulin superfamily related to ICAM-1. When T cells are activated by chemokine signaling, VLA-4 is altered so that it can bind to VCAM-1 with greater affinity, similar to chemokine-induced binding of acti-vated LFA-1 to ICAM-1 (see Section 3-18). Thus, chemokines activate VCAM-1 to bind VLA-4 on vascular endothelial cells near sites of inflammation, allow-ing extravasation of effector T cells. Although both VCAM-1 and ICAM-1 are expressed on activated endothelial cell surfaces, there appears to be preferen-tial utilization of one of the two adhesion pairs in some inflamed tissue vascu-lar beds: recruitment of effector T cells is more dependent on VLA-4 in some tissues, and more dependent on LFA-1 in others.

Induction of expression of some adhesion molecules is compartmentalized so that effector T cells primed within lymphoid compartments of those tissues home back to them, whether during an active immune response or at homeostasis. Thus, the site of priming appears to imprint effector T cells with the ability to traffic to particular tissues. This is achieved by the expression of adhesion molecules that bind selectively to tissue-specific addressins. In this context, the adhesion molecules are often known as **homing receptors** (**Fig. 11.7**). As we shall see in Chapter 12, dendritic cells that prime T cells in the gut-associated lymphoid tissues (GALT) induce expression of the $\alpha_4{:}\beta_7$ integrin, which binds to the mucosal vascular addressin **MAdCAM-1** that is constitutively expressed by endothelial cells of blood vessels within the gut mucosa (see Fig. 11.7, lower left panel).

T cells primed in the GALT also express specific chemokine receptors that bind chemokines produced constitutively—and specifically—by the gut epi-thelium. Thus, at homeostasis CCR9 expressed on T cells primed in lymphoid

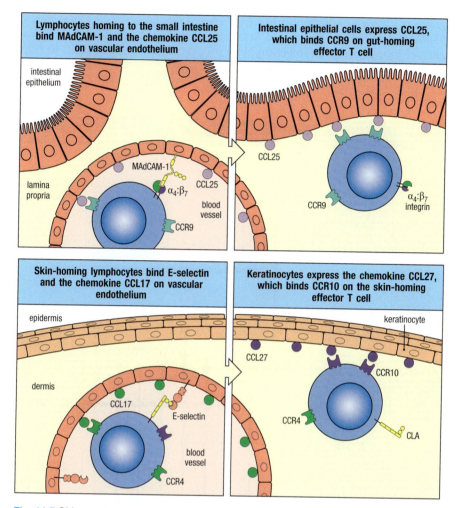

Fig. 11.7 Skin- and gut-homing T cells use particular combinations of integrins and chemokines to migrate specifically to their target tissues. $\alpha_4\beta_7$ expressed on circulating lymphocytes primed in gut-associated lymphoid tissues initally binds MAdCAM-1 (upper left panel), then uses CCR9 to move along a CCL25 chemokine gradient to traverse the endothelium and migrate to the intestinal epothelium (upper right panel). Similarly, circulating lymphocytes primed in lymph nodes draining the skin bind to the endothelium lining a cutaneous blood vessel by interactions between cutaneous lymphocyte antigen (CLA) and constitutively expressed E-selectin on the endothelial cells (lower left panel). The adhesion is strengthened by an interaction between lymphocyte chemokine receptor CCR4 and the endothelial chemokine CCL17. Once through the endothelium, effector T lymphocytes are attracted to keratinocytes of the epidermis by the chemokine CCL27, which binds to the receptor CCR10 on lymphocytes (lower right panel).

tissues of the small intestine recruits those T cells back to the lamina propria subjacent to the epithelium of the small intestine along a CCL25 gradient (see Fig. 11.7, upper right panel). In contrast, T cells primed in skin-draining lymph nodes preferentially home back to the skin. They are induced to express the adhesion molecule **cutaneous lymphocyte antigen (CLA)**, an isoform of PSGL-1 that differs in its pattern of glycosylation and binds to E-selectin on cutaneous vascular endothelium (see Fig. 11.7, lower panels). CLA-expressing T lymphocytes also express the chemokine receptors CCR4 and CCR10, which bind sequentially the chemokines CCL17 (TARC) and CCL27 (CTACK), respectively, which are present at highest levels in cutaneous blood vessels and the epidermis. Because these tissue-homing chemokines are produced at steady state, they are referred to as **homeostatic chemokines**. They are analogous to chemokines produced constitutively in lymphoid tissues at steady state, such as CCL19 and CCL21, that direct CCR7-bearing naive T cells along a gradient from the endothelium of HEVs to T-cell zones (Fig. 11.8). Homeostatic chemokines are to be contrasted with **inflammatory chemokines**, which are elicited in the context of infection to recruit circulating immune cells to sites of inflammation.

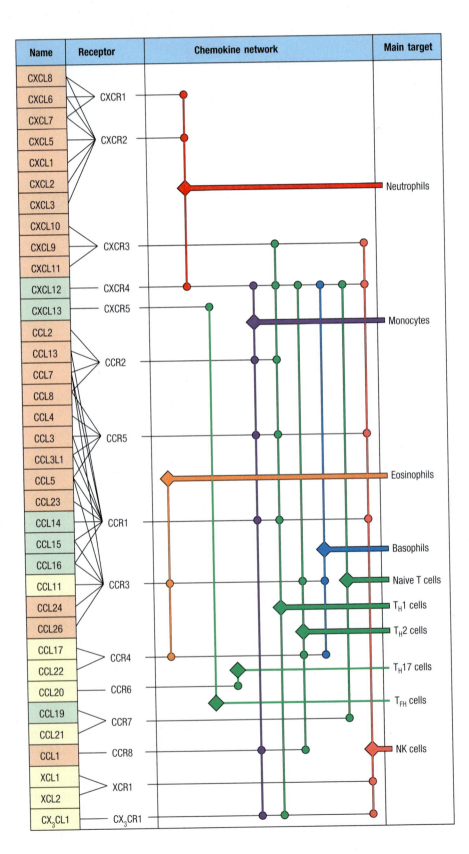

Fig. 11.8 Chemokine networks coordinate the interactions of innate and adaptive immune-cell populations. Chemokines are classified into four families on the basis of structural differences: CXCL, CCL, XCL, and CXC_3CL. Chemokines can also be classified as pro-inflammatory (red), homeostatic (green), and mixed function (yellow). Chemokines bind to a subfamily of seven-transmembrane G-protein-coupled receptors, which are classified as CXCR, CCR, XCR, and CX_3CR on the basis of the class of chemokines they bind. Many, but not all, of the chemokine-chemokine receptor networks that coordinate immune modules are represented here. The connection between receptors and cell types on which they are expressed is indicated by the 'circuit' representation of lines and connecting nodes. To connect chemokines and their receptors to target cells, follow a horizontal line and turn on a vertical one at each node; the rhomboids (diamond shapes) link vertical lines to the cell type. Note that most chemokine receptors can bind multiple chemokines. Modified from Mantovani et al., *Nat. Rev. Immunol.* 2006, **6**:907-918.

In addition to the general and tissue-specific changes in trafficking molecules induced during effector T-cell differentiation, there is subset-specific expression of chemokine receptors that accompanies the loss of CCR7 expression. This results in distinct patterns of chemokine receptor expression by T_H1, T_H2, and T_H17 cells that guide their differential recruitment to sites of inflammation contingent on the local patterns of inflammatory chemokines induced by the innate immune response to different types of pathogens (Fig. 11.9). For example, T_H1 cells express CCR5, which is also expressed on monocytes that mature into macrophages as they enter the inflammatory site. Thus, both T_H1 cells and the innate effector cells whose effector functions they enhance are recruited to the same tissue site by the same chemokines (see Fig. 11.8). As is the case for many other chemokine receptors, CCR5 has multiple ligands (CCL3, CCL4, CCL5, and CCL8), which may be induced by different cellular sources and by different pathogens targeted by type 1 immunity. Some of these are produced by activated macrophages themselves following their recruitment to the inflammatory site. This provides a feed-forward mechanism by which the emerging innate response is amplified and then contributes to the recruitment of T_H1 cells, which then provide antigen-dependent 'help' to further activate the macrophages, as we discuss in the next section. T_H1 cells also express CXCR3, which is shared by NK cells and cytotoxic CD8 T cells. In response to CXCR3 ligands— CXCL9 and CXCL10—these cells are recruited to the same inflammatory site to coordinate the cell-mediated killing of targets infected by intracellular pathogens, such as *Listeria monocytogenes*, or certain viruses.

T_H2 and T_H17 cells display different patterns of inflammatory chemokine receptors, some of which, like those expressed by T_H1 cells, are shared with the myelomonocytic cells with which they interact in inflamed tissues (see Figs. 11.8 and 11.9). The shared expression pattern of chemokine receptors by innate and adaptive effector cells represents an important mechanism for the spatiotemporal coordination and integration of immune effector modules in response to different types of pathogen (see Fig. 11.8). Thus, the local release of cytokines and chemokines at the site of infection has far-reaching consequences. In addition to recruiting granulocytes and monocytes, which constitutively express their specific complement of chemokine receptors while in circulation, changes induced in the blood vessel walls also enable newly generated effector T lymphocytes to enter infected tissues. Once in the tissue, recruited T cells produce T helper cell type–specific cytokines that further increase specific chemokine production by innate immune cells in an additional feed-forward mechanism that results in further effector T cell and innate effector cell trafficking into the tissue. Because the cytokines that differentially drive local production of effector module-specific chemokines are similarly produced by ILCs, this represents another major function of ILCs in coordinating the early polarization of pathogen-specific responses.

11-4   Pathogen-specific effector T cells are enriched at sites of infection as adaptive immunity progresses.

In the early stage of the adaptive immune response, only a minority of the effector T cells that enter infected tissues will be specific for pathogen. This is because activation of the endothelium of local blood vessels by inflammatory cytokines induces expression of selectins, integrin ligands, and chemokines that can recruit any circulating effector or memory T cell that expresses the appropriate trafficking receptors, irrespective of its antigenic specificity. However, specificity of the reaction is rapidly increased as the number of pathogen-specific T cells increases and recognition of antigen within the inflamed tissue retains them there. Although the precise mechanisms controlling retention of antigen-activated effector T cells in the inflamed tissue are not entirely understood, it is thought that the same mechanisms that retain antigen-activated naive T cells within secondary lymphoid tissues during effector

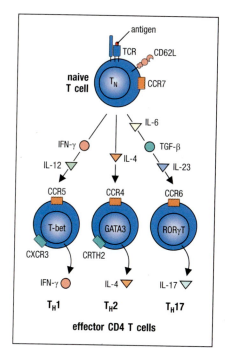

Fig. 11.9 The expression of adhesion and chemokine receptors is altered during effector T cell differentiation. During a primary immune response, specific cytokines derived from the innate immune system (indicated along the three diverging arrows) and unique master transcription factors (T-bet, GATA3, and RORγt) direct naive CD4 T cells to differentiate into T_H1, T_H2, or T_H17 effector cells. Effector T cells of each subset lose expression of L-selectin (CD62L) and CCR7, and express characteristic chemokine receptors.

T-cell development might be at play. This includes a role for the S1P pathway, although other chemokine signals may participate. By the peak of an adaptive immune response, after several days of clonal expansion and differentiation, a large fraction of the recruited T cells will be specific for the infecting pathogen.

Effector T cells that enter tissues but do not recognize their cognate antigen are not retained there. They either undergo apoptosis locally or enter the afferent lymphatics and migrate to the draining lymph nodes and eventually return to the bloodstream. Thus, T cells in the afferent lymph that drains tissues are memory or effector T cells, which characteristically express the CD45RO isoform of the cell-surface molecule CD45 and lack L-selectin (see Fig. 11.6). Effector T cells and some memory T cells have similar trafficking phenotypes, as we discuss later (see Section 11-22), and both seem to be committed to migration through, and in some cases, retention within, barrier tissues that are the primary sites of infection. In addition to allowing effector T cells to clear all sites of infection, this pattern of migration allows them to contribute, along with memory cells, to protecting the host against reinfection with the same pathogen.

11-5   T$_H$1 cells coordinate and amplify the host response to intracellular pathogens through classical activation of macrophages.

Type 1 responses (see Fig. 11.5) are important for the eradication of those pathogens that have evolved mechanisms to survive and replicate within macrophages—for example, viruses, and bacterial and protozoan pathogens that can survive inside macrophage intracellular vesicles. In the case of viruses, a T$_H$1 response is generally involved in helping to activate the CD8 cytotoxic T cells that will recognize virus-infected cells and destroy them (see Chapter 9). T$_{FH}$ cells that differentiate in type 1 responses induce the production of subclasses of IgG antibodies that neutralize virus particles in the blood and extracellular fluid. In the case of intracellular bacteria such as mycobacteria and *Salmonella*, and of protozoa such as *Leishmania* and *Toxoplasma*, which all take up residence inside macrophages, the role of T$_H$1 cells is to activate macrophages to heighten their microbicidal function (**Fig 11.10**).

Pathogens of all types are ingested by macrophages from the extracellular fluid, and are often destroyed without the need for additional macrophage activation. In several clinically important infections, such as those caused by mycobacteria, ingested pathogens are not killed, and can even set up a chronic infection in macrophages and incapacitate them. Such microorganisms are able to maintain themselves in the hostile environment of phagosomes—shielded from the effects of both antibodies and cytotoxic T cells—by inhibiting the fusion of phagosomes and lysosomes, or by preventing the acidification required to activate lysosomal proteases. Nevertheless, peptides derived from such microorganisms can be displayed by MHC class II molecules on the macrophage surface, where they are recognized by antigen-specific effector T$_H$1 cells. The T$_H$1 cell is stimulated to synthesize membrane-associated proteins and soluble cytokines that enhance the macrophage's antimicrobial defenses and enable it to either eliminate the pathogen or control its growth and spread. This boost to antimicrobial mechanisms is known as 'classical' macrophage activation, the result of which is the so-called **classically-activated, or M1, macrophage** (**Fig. 11.11**).

Macrophages require two main signals for classical activation, and effector T$_H$1 cells can deliver both. One signal is the cytokine IFN-γ; the other, CD40L, sensitizes the macrophage to respond to IFN-γ (see Fig. 11.10). T$_H$1 cells also secrete lymphotoxin, which can substitute for CD40 ligand in M1 macrophage activation. The M1 macrophage is a potent antimicrobial effector cell. Phagosomes fuse with lysosomes, and microbicidal reactive oxygen and

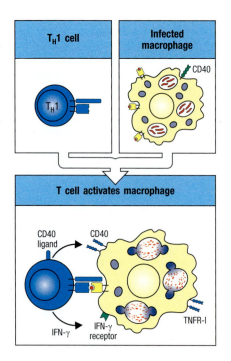

Fig. 11.10 T$_H$1 cells activate macrophages to become highly microbicidal. When an effector T$_H$1 cell specific for a bacterial peptide contacts an infected macrophage, the T cell is induced to secrete the macrophage-activating factor IFN-γ and to express CD40 ligand. Together these newly synthesized T$_H$1 proteins activate the macrophage.

Fig. 11.11 Macrophages activated by T_H1 cells undergo changes that greatly increase their antimicrobial effectiveness and amplify the immune response. Activated macrophages increase their expression of CD40 and of TNF receptors, and are stimulated to secrete TNF-α. This autocrine stimulus synergizes with IFN-γ secreted by T_H1 cells to induce classical, or M1, macrophage activation characterized by the production of nitric oxide (NO) and superoxide (O_2^-). The macrophage also upregulates its B7 molecules in response to binding to CD40 ligand on the T cell, and increases its expression of MHC class II molecules in response to IFN-γ, thus allowing further activation of resting CD4 T cells.

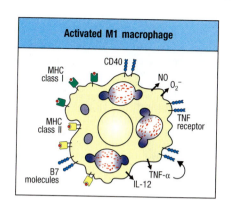

nitrogen species are generated, as described in Section 3-2. When T_H1 cells stimulate macrophages through these molecules, the M1 macrophage also secretes TNF-α, further stimulating macrophages through the TNFR-I, the same receptor activated by LT-α. TNF receptor signalling seems to be required to maintain the viability of the macrophage in this setting; in mice lacking TNFR-I (see Section 9-28), infection by *Mycobacterium avium*, an opportunistic intracellular pathogen that does not normally cause disease, leads to excessive apoptosis of macrophages that results in the release and dissemination of the pathogen before it can be killed within the infected macrophage. CD8 T cells also produce IFN-γ and can activate macrophages presenting antigens derived from cytosolic proteins on MHC class I molecules. Macrophages can also be made more sensitive to IFN-γ by very small amounts of bacterial LPS, and this latter pathway may be particularly important when CD8 T cells are the primary source of the IFN-γ.

In addition to increased intracellular killing, T_H1 cells induce other changes in macrophages that help to amplify the adaptive immune response against intracellular pathogens. These changes include an increase in the number of MHC class II molecules, B7 molecules, CD40, and TNF receptors on the M1 macrophage surface (see Figs. 11.10 and 11.11), making the cell more effective at presenting antigen to T cells, and more responsive to CD40 ligand and TNF-α. In addition, M1 macrophages secrete IL-12, which increases the amount of IFN-γ produced by ILC1s and T_H1 cells. This also promotes the differentiation of activated naive CD4 T cells into T_H1 effector cells, and naive CD8 T cells into cytotoxic effectors (see Sections 9-20 and 9-18).

Another important function of T_H1 cells is the recruitment of additional phagocytic cells to sites of infection. T_H1 cells recruit macrophages by two mechanisms (**Fig. 11.12**). First, they make the hematopoietic growth factors IL-3 and GM-CSF, which stimulate the production of new monocytes in the bone marrow. Second, the TNF-α and lymphotoxin secreted by T_H1 cells at sites of infection change the surface properties of endothelial cells so that monocytes adhere to them. Chemokines such as CCL2, which are induced by T_H1 cells at inflammatory sites, direct the migration of monocytes through the vascular endothelium and into the infected tissue, where they differentiate into macrophages (see Section 3-17). Cytokines and chemokines secreted by M1 macrophages themselves are also important in recruiting other monocytes to sites of infection. Collectively, these T_H1-mediated effects provide a positive feedback loop that amplifies and sustains type 1 responses until the pathogen is controlled or eliminated.

Certain intravesicular bacteria, including some mycobacteria and *Listeria monocytogenes*, escape from phagocytic vesicles and enter the cytoplasm, where they are no longer susceptible to the microbicidal actions of activated macrophages. Their presence can, however, be detected by cytotoxic CD8 T cells. The pathogens released when macrophages are killed by these CTLs can be killed in the extracellular environment by antibody-mediated mechanisms, or can be phagocytosed by freshly recruited macrophages. In this circumstance, the provision of T_H1-mediated 'help' for the development of CTLs, such as the provision of IL-2, may play an important role in coordinating the T_H1 and CTL responses.

 MOVIE 11.2

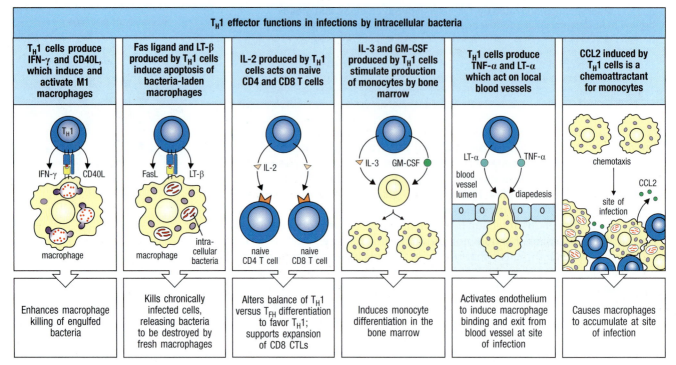

Fig. 11.12 The immune response to intracellular bacteria is coordinated by activated T$_H$1 cells. The activation of T$_H$1 cells by infected macrophages results in the synthesis of cytokines that both induce M1 macrophage and coordinate the immune response to intracellular pathogens. IFN-γ and CD40 ligand synergize in activating the macrophage, which allows it to kill engulfed pathogens. Chronically infected macrophages lose the ability to kill intracellular bacteria, and membrane-bound Fas ligand or LT-β produced by the T$_H$1 cell can kill these macrophages, releasing the engulfed bacteria, which are taken up and killed by fresh macrophages. In this way, IFN-γ and LT-β synergize in the removal of intracellular bacteria. IL-2 produced by T$_H$1 cells augments effector T-cell differentiation and potentiates the release of other cytokines. IL-3 and GM-CSF stimulate the production of new monocytes by acting on hematopoietic stem cells in the bone marrow. New macrophages are recruited to the site of infection by the actions of secreted TNF-α, LT-α, and other cytokines on vascular endothelium, which signal monocytes to leave the bloodstream and enter the tissues where they become macrophages. A chemokine with monocyte chemotactic activity (CCL2) signals monocytes to migrate into sites of infection and accumulate there. Thus, the T$_H$1 cell coordinates a macrophage response that is highly effective in destroying intracellular infectious agents.

11-6 Activation of macrophages by T$_H$1 cells must be tightly regulated to avoid tissue damage.

As discussed in Chapter 9, distinguishing features of effector T cells are their capacity for antigen-induced activation of effector functions without the requirement for co-stimulation, and also their efficient delivery of effector molecules through polarized secretion or expression of cytokines and cell-surface molecules—often through formation of an immunological synapse with an antigen-bearing cell (see Section 9-25). After a T$_H$1 cell recognizes its cognate antigen expressed by a macrophage, the secretion of effector molecules requires several hours. T$_H$1 cells must therefore adhere to their target cells far longer than do cytotoxic CD8 T cells. Similarly to cytotoxic T cells, the secretory machinery of the T$_H$1 cell becomes oriented toward the site of contact with the macrophage and newly synthesized cytokines are secreted there (see Fig. 9.38). CD40 ligand also seems to be delivered to the same contact site. So although all macrophages have receptors for IFN-γ, the infected macrophage that presents antigen to the T$_H$1 cell is far more likely to become activated than nearby uninfected macrophages.

In addition to more efficiently focusing activating signals on infected macrophages, the antigen-specific induction of macrophage activation may play an important role in limiting tissue injury. By targeting only infected macrophages through MHC:peptide recognition, T$_H$1 cells minimize 'collateral damage' that might otherwise result to normal components of the inflamed tissue: oxygen radicals, NO, and proteases that are toxic to host cells as well as to the pathogen that is targeted for destruction. Thus, antigen-specific macrophage activation by T_H1 cells is a means of deploying this powerful defensive mechanism to maximum effect while minimizing local tissue damage. In this regard, it is notable that although ILC1s are also producers of IFN-γ, they lack antigen receptors that can focus the cytokine on infected macrophages for more efficient activation. It is not yet known whether ILC1 cells have other mechanisms with which to direct IFN-γ onto macrophages, or whether they play a more limited role in macrophage activation, but the IFN-γ they produce is important in indirectly enhancing the local inflammatory response.

11-7 Chronic activation of macrophages by T_H1 cells mediates the formation of granulomas to contain intracellular pathogens that cannot be cleared.

Some intracellular pathogens, most notably *Mycobacterium tuberculosis*, are sufficiently resistant to the microbicidal effects of activated macrophages that they are incompletely eliminated by a type 1 response. This gives rise to chronic, low-level infection that requires an ongoing T_H1 response to prevent pathogen proliferation and spread. In this circumstance, chronic coordination between T_H1 cells and macrophages underlies the formation of the immunological reaction called the **granuloma**, in which microbes are held in check within a central area of macrophages surrounded by activated lymphocytes (**Fig. 11.13**). A characteristic feature of granulomas is the fusion of several macrophages to form multinucleated giant cells, which can be found at the border of the central focus of activated macrophages and the lymphocytes that surround them and which appear to have heightened antimicrobial activity. A granuloma serves to 'wall off' pathogens that resist destruction. In tuberculosis, the centers of large granulomas can become isolated and the cells there die, probably from a combination of lack of oxygen and the cytotoxic effects of activated macrophages. As the dead tissue in the center resembles cheese, this process is called 'caseous' necrosis. Thus, the chronic activation of T_H1 cells can cause significant pathology. The absence of the T_H1 response, however, leads to the more serious consequence of death from disseminated infection, which is now seen frequently in patients with AIDS and concomitant mycobacterial infection.

11-8 Defects in type 1 immunity reveal its important role in the elimination of intracellular pathogens.

In mice whose gene for IFN-γ or CD40 ligand has been deleted by gene targeting, classical macrophage activation is impaired; consequently, the animals succumb to sublethal doses of *Mycobacterium*, *Salmonella*, and *Leishmania* species. Classical (M1) macrophage activation is also crucial in controlling vaccinia virus. However, although IFN-γ and CD40 ligand are probably the most important effector molecules synthesized by T_H1 cells, the immune response to pathogens that proliferate in macrophage vesicles is complex, and other cytokines secreted by T_H1 cells may also be crucial (see Fig. 11.12).

The depletion of CD4 T cells in people with HIV/AIDS causes ineffective T_H1 responses that can lead to the dissemination of microbes that are normally cleared by macrophages. This is the case with the opportunist fungal pathogen *Pneumocystis jirovecii* (see also Chapter 13). The lungs of healthy people are kept clear of *P. jirovecii* by phagocytosis and intracellular killing by alveolar macrophages. Pneumonia caused by *P. jirovecii* is, however, a frequent cause of death in people with AIDS. In the absence of CD4 T cells, phagocytosis of *P. jirovecii* and intracellular killing by lung macrophages are impaired, and the

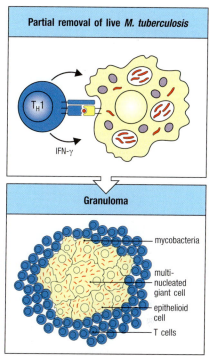

Partial removal of live *M. tuberculosis*

Granuloma

mycobacteria

multi-nucleated giant cell

epithelioid cell

T cells

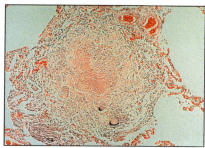

Fig. 11.13 Granulomas form when an intracellular pathogen or its constituents cannot be completely eliminated. When mycobacteria (red) resist the effects of macrophage activation, a characteristic localized inflammatory response called a granuloma develops. This consists of a central core of infected macrophages. The core may include multinucleate giant cells, which are fused macrophages, surrounded by large macrophages often called epithelioid cells, but in granulomas caused by mycobacteria the core usually becomes necrotic. Mycobacteria can persist in the cells of the granuloma. The central core is surrounded by T cells, many of which are CD4-positive. The exact mechanisms by which this balance is achieved, and how it breaks down, are unknown. Granulomas, as seen in the bottom panel, also form in the lungs and elsewhere in a disease known as sarcoidosis, which may be caused by inapparent mycobacterial infection. Photograph courtesy of J. Orrell.

pathogen colonizes the surface of the lung epithelium and invades lung tissue. The requirement for CD4 T cells seems to be due, at least in part, to a requirement for the macrophage-activating cytokines IFN-γ and TNF-α produced by T_H1 cells.

11-9 T_H2 cells coordinate type 2 responses to expel intestinal helminths and repair tissue injury.

Type 2 immunity is directed against parasitic helminths: roundworms (nematodes) and two types of flatworms—tapeworms (cestodes) and flukes (trematodes). Unlike microbial pathogens, or 'micropathogens' (bacteria, viruses, fungi, and protozoa), which replicate rapidly and can overwhelm host defenses by sheer numbers, most helminths do not replicate in their mammalian host. Moreover, helminths are multicellular; they are metazoan 'macropathogens' that are far too large—ranging in size from approximately 1 mm to over 1 meter—to be engulfed by host phagocytic cells, and therefore require very different strategies for host defense. In the developing world, the intestines of virtually all animals and humans are colonized by helminth parasites (Fig. 11.14). Many of these infections may be cleared rapidly by the generation of an effective type 2 response, although often the host response is successful in reducing worm burden, but not in completely clearing the parasite, resulting in chronic disease. In these circumstances, the parasite persists for long periods despite the host's attempts to expel it, and causes disease by competing with the host for nutrients, or by causing local tissue injury.

Irrespective of the type of helminth involved, or its site of host entry, the host adaptive response is orchestrated by T_H2 cells (Fig. 11.15; see also Fig. 9.30). The T_H2 response is induced by the actions of worm products on a variety of different innate cells: epithelial cells, ILC2 cells, mast cells, and dendritic cells. Dendritic cells required for the presentation of helminth antigens to naive CD4 T cells appear to be activated by IL-13 produced by ILC2 cells and innate cytokines, such as epithelium-derived TSLP, which repress the development of T_H1- and T_H17-inducing dendritic cells in favor of dendritic cells that promote T_H2 cell differentiation. The initial source of the IL-4 required for T_H2 cell differentiation appears to be context-specific and redundant. Thus, although several cell types have been proposed as the source, including iNKT cells, mast cells, and basophils, none of these has been proven to be essential.

The development of T_H2 cells in draining lymphoid tissues is followed by their exodus to sites of helminth invasion, where they enhance the recruitment and function of circulating type 2 innate effector cells—eosinophils,

Fig. 11.14 Intestinal helminth infection. Panel a: the whipworm *Trichuris trichiura* is a helminth parasite that lives partly embedded in intestinal epithelial cells. This scanning electron micrograph of mouse colon shows the head of the parasite buried in an epithelial cell and its posterior lying free in the lumen. Panel b: a cross-section of crypts from the colon of a mouse infected with *T. trichiura* shows the markedly increased production of mucus by goblet cells in the intestinal epithelium. The mucus is seen as large droplets in vesicles inside the goblet cells and stains dark blue with periodic acid–Schiff reagent. Magnification ×400.

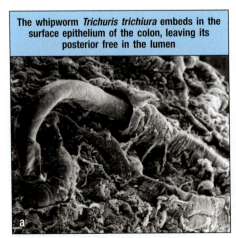

The whipworm *Trichuris trichiura* embeds in the surface epithelium of the colon, leaving its posterior free in the lumen

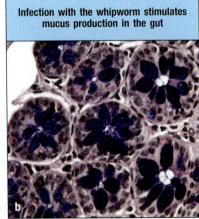

Infection with the whipworm stimulates mucus production in the gut

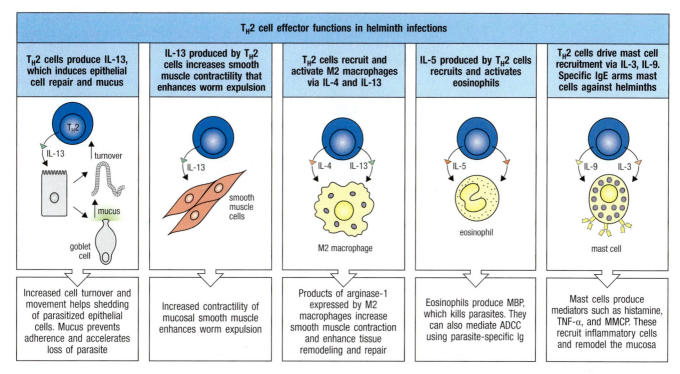

T_H2 cell effector functions in helminth infections

T_H2 cells produce IL-13, which induces epithelial cell repair and mucus	IL-13 produced by T_H2 cells increases smooth muscle contractility that enhances worm expulsion	T_H2 cells recruit and activate M2 macrophages via IL-4 and IL-13	IL-5 produced by T_H2 cells recruits and activates eosinophils	T_H2 cells drive mast cell recruitment via IL-3, IL-9. Specific IgE arms mast cells against helminths
Increased cell turnover and movement helps shedding of parasitized epithelial cells. Mucus prevents adherence and accelerates loss of parasite	Increased contractility of mucosal smooth muscle enhances worm expulsion	Products of arginase-1 expressed by M2 macrophages increase smooth muscle contraction and enhance tissue remodeling and repair	Eosinophils produce MBP, which kills parasites. They can also mediate ADCC using parasite-specific Ig	Mast cells produce mediators such as histamine, TNF-α, and MMCP. These recruit inflammatory cells and remodel the mucosa

Fig. 11.15 Protective responses to intestinal helminths are mediated by T_H2 cells. Most intestinal helminths induce both protective and pathological immune responses by CD4 T cells. T_H2 responses tend to be protective, creating an unfriendly environment for the parasite, and leading to its expulsion and the generation of protective immunity (see the text for details). M2 macrophage, alternatively activated macrophage.

basophils, tissue mast cells, and macrophages. Like T_H1 and T_H17 cells, T_H2 cells express a distinct complement of chemokine receptors that is shared with the circulating innate effector cells with which they interact, thus selectively guiding them to sites of ongoing type 2 responses (see Figs. 11.8 and 11.9). CCR3 and CCR4 are expressed both by T_H2 cells and by eosinophils and basophils, as is CRTH2, the ligand for which is prostaglandin D_2, a lipid mediator that is produced by activated tissue mast cells. Ligands for CCR3 (for example, the eotaxins CCL11, CCL24, and CCL26) are produced by multiple innate immune cells in tissue sites of helminth infection and are induced by IL-4 and IL-13 signaling. Hence, ILC2 cells, T_H2 cells, and eosinophils and basophils can each amplify the recruitment of other type 2 cells through this chemokine network.

Although the T_H2 effector response can coordinate the direct killing of some worms by enhancing innate effector cell functions, a major focus of the anti-helminth response is expelling the worms and limiting the tissue damage they cause when they invade the host—functions that are both mediated by type 2 cytokines. IL-13 directly enhances the production of mucus by goblet cells, activates smooth muscle cells in mucosal tissues for hypermotility, and increases the migration and turnover of epithelial cells in the mucosa (see Fig. 11.15, first panel). In the intestines, which are the most common site of worm infestation, each of these actions is a critical component of the host response, as it helps to eliminate parasites that have attached to the epithelium and decreases the surface area available for colonization.

The response to helminths generates high levels of IgE antibody, induced by IL-4-producing T_FH cells that develop in concert with T_H2 cells (see Section 9-20). IgE binds to Fcε receptors expressed by mast cells, eosinophils, and basophils, which arms them for antigen-specific recognition and activation. Type 2 adaptive responses also promote production of IgG1, which is recognized by macrophages and engages them in the type 2 response. IL-4 and IL-13 produced by T_H2 cells also result in the differentiation of **alternatively activated macrophages** (also called **M2 macrophages**). Unlike classically activated, M1 macrophages, which differentiate after interaction with T_H1 cells and are potent activators of inflammation (see Fig. 11.10), M2 macrophages participate in worm killing and expulsion, and also promote tissue remodeling and repair (see Fig. 11.15). A major difference between M1 and M2 macrophages is their different metabolism of arginine to produce antipathogen products. Whereas M1 macrophages express iNOS, which produces the potent intracellular microbicide nitric oxide (NO) (see Section 3-2), M2 macrophages express arginase-1, which produces ornithine and proline from arginine. Along with other factors, ornithine increases the contractility of mucosal smooth muscle and promotes tissue remodeling and repair (see Fig. 11.15). Through a mechanism that is unclear, ornithine has also been found to be directly toxic to IgG-coated larvae of certain helminths. Because helminths that have invaded tissues are too large to be ingested by macrophages, the targeted release of toxic mediators directly onto the worm by antibody-dependent cell-mediated cytotoxicity (ADCC) enables macrophages, as well as eosinophils (see below), to attack these large extracellular pathogens.

Macrophages activated by T_H2 cells also appear to be important in walling off invading worms, as well as repairing tissue damage caused as worms migrate through host tissues. These 'tissue repair' functions of M2 macrophages are dependent on secreted factors important in tissue remodeling and include the stimulation of collagen production, formation of which requires proline that is generated by arginase-1 activity. Moreover, T_H2-activated macrophages can also induce the formation of granulomas that entrap worm larvae in tissues. In this regard, antigen-specific macrophage activation by T_H2 cells has non-redundant function in type 2 responses. Although ILC2 cells, and innate effector cells, may promote M2 macrophage activation via IL-13, they are unable to sustain this response. Thus, in several models of worm infection, antihelminth responses are considerably impaired in RAG-deficient or T-cell-depleted mice, demonstrating that sustained alternative activation of macrophages requires T_H2 cells.

The IL-5 produced by T_H2 cells and ILC2 cells recruits and activates eosinophils (see Fig. 11.15), which have direct toxic effects on worms by releasing cytotoxic molecules stored in their secretory granules, such as major basic protein (MBP). In addition to Fcε receptors that arm them for degranulation with IgE, eosinophils bear Fc receptors for IgG and can mediate ADCC against IgG-coated parasites (see Fig. 10.38). They also express the Fcα receptor (CD89) and degranulate in response to stimulation by secretory IgA.

IL-3 and IL-9 produced by T_H2 cells in the mucosae recruit, expand, and activate a specialized population of mast cells known as **mucosal mast cells** (see Fig. 11.15). The innate cytokines IL-25 and IL-33 also activate mucosal mast cells early in a response to helminths. Mucosal mast cells differ from their counterparts in other tissues by having only small numbers of IgE receptors and producing very little histamine. When activated by cytokines, or by the binding of worm antigens to receptor-bound IgE, mucosal mast cells release large amounts of preformed inflammatory mediators that are stored in secretory granules: prostaglandins, leukotrienes, and several proteases, including the mucosal mast cell protease (MMCP-1), which can degrade the epithelial tight junctions to increase permeability and fluid flow into the mucosal lumen. Together, the mast-cell-derived mediators increase vascular and epithelial permeability, increase intestinal motility, stimulate the production of mucus by goblet cells, and induce leukocyte recruitment, all of which contribute to the 'weep and sweep' response that helps to expel parasites from the host.

11-10 T$_H$17 cells coordinate type 3 responses to enhance the clearance of extracellular bacteria and fungi.

The subset of effector T cells generated in response to infection by extracellular bacteria and fungi is T$_H$17. At homeostasis, T$_H$17 cells are deployed almost exclusively in the intestinal mucosa, where they contribute to the mutualistic relationship between the host and the intestinal microbiota—which is composed of extracellular bacteria and some fungi. However, they are also critical for defense against pathogenic extracellular bacteria and fungi that invade at barrier sites, as well as components of the normal microbiota that may enter the host when epithelial barrier function is compromised, whether as the result of trauma or pathogenic infection. In these settings, a principal function of T$_H$17 cells is the orchestration of type 3 responses, in which neutrophils are the principal type of innate effector cell.

As discussed in Chapter 9, the development of T$_H$17 cells is induced by the combined actions of TGF-β and the pro-inflammatory cytokines IL-6, IL-1, and IL-23 (see Fig. 9.31). The latter is preferentially produced by CD103$^+$CD11b$^+$ conventional dendritic cells that recognize MAMPs produced by extracellular bacteria, such as flagellin, which is recognized by TLR5; or MAMPs produced by fungi, such as β-glucan polymers of glucose expressed by yeast and fungi that are recognized by Dectin-1. As for T$_H$1 and T$_H$2 cells, the egress of T$_H$17 cells from secondary lymphoid tissues is associated with altered chemokine expression: primarily the induction of CCR6, the ligand for which (CCL20) is produced by activated epithelial cells in mucosal tissues and skin, as well as T$_H$17 cells themselves and ILC3 cells (see Figs. 11.8 and 11.9).

T$_H$17 cells are stimulated to release IL-17A and IL-17F when they encounter antigen at sites of infection (Fig. 11.16). A primary effect of these cytokines is the enhanced production and recruitment of neutrophils. The receptor for

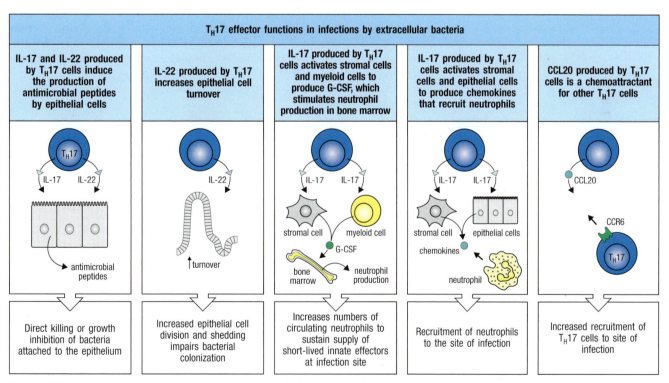

Fig. 11.16 The immune response to extracellular bacteria and some fungi is coordinated by activated T$_H$17 cells. T$_H$17 cells activated by antigen-bearing macrophages and dendritic cells in barrier tissues (e.g., intestinal or respiratory mucosa, and skin) produce cytokines that activate local epithelial and stromal cells to coordinate the immune response to extracellular bacteria and some types of fungi.

IL-17A and IL-17F is expressed widely on cells such as fibroblasts, epithelial cells, and keratinocytes. IL-17 induces these cells to secrete various cytokines, including IL-6, which amplifies the T$_H$17 response, and the hematopoietic factor granulocyte colony-stimulating factor (G-CSF), which increases the production of neutrophils by the bone marrow. IL-17 also stimulates production of the chemokines CXCL8 and CXCL2, the receptors for which (CXCR1 and CXCR2) are uniquely expressed by neutrophils (see Fig. 11.8). Thus, one important action of IL-17 at sites of infection is to induce local cells to secrete cytokines and chemokines that attract neutrophils.

T$_H$17 cells also produce IL-22, a member of the IL-10 family that acts cooperatively with IL-17 to induce the expression by epithelial cells of antimicrobial proteins (see Fig. 11.16). These include β-defensins and the C-type lectins RegIIIβ and RegIIIγ, all of which can directly kill bacteria (see Section 2-4). IL-22 and IL-17 can also induce epithelial cells to produce metal-binding proteins that are bacteriostatic and fungistatic. **Lipocalin-2** limits iron availability to bacterial pathogens; S100A8 and S100A9 are two antimicrobial peptides that heterodimerize to form the antimicrobial protein **calprotectin**, which sequesters zinc and manganese from microbes. Many of these antimicrobial agents are also produced by neutrophils recruited to the site of infection; calprotectin has been reported to comprise up to a third of the cytosolic protein of neutrophils. IL-22 also stimulates the proliferation and shedding of epithelial cells as a mechanism to deprive bacteria and fungi of a 'foothold' for colonization at epithelial surfaces. While ILC3 cells in barrier tissues respond rapidly to pathogens to produce IL-22, pathogen-specific T$_H$17 cells have been shown to amplify and sustain the production of IL-22 at sites of infection.

As in type 1 and type 2 responses, integration of innate and adaptive effector cells in the type 3 response is mediated in large part by the production of pathogen-specific antibodies that opsonize extracellular bacteria and fungi for destruction by neutrophils, macrophages, and complement. T$_{FH}$ cells that develop coordinately with T$_H$17 cells promote the production of high-affinity IgG and IgA antibodies by plasma cells that can express CCR6 and thereby localize to sites of type 3 responses in barrier tissues, where they can arm neutrophils and macrophages 'on-site.' Antibodies are the principal immune reactants that clear primary infections by common extracellular bacteria that elicit type 3 responses, such as *Staphylococcus aureus* and *Streptococcus pneumoniae*.

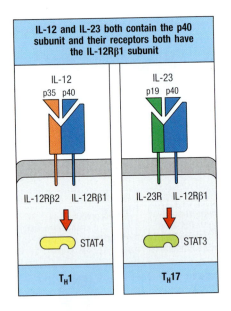

Fig. 11.17 The cytokines IL-12 and IL-23 have a component in common, as do their receptors. The heterodimeric cytokines IL-12 and IL-23 both contain the p40 subunit, and the receptors for IL-12 and IL-23 have the IL-12Rβ1 subunit in common, which binds the p40 subunit. IL-12 signaling primarily activates the transcriptional activator STAT4, which increases IFN-γ production. IL-23 primarily activates STAT3, but also activates STAT4 weakly (not shown). Both cytokines augment the activity and proliferation of the CD4 subsets that express receptors for them; T$_H$1 cells express IL-12R, and T$_H$17 cells express primarily IL-23R, but can also express low levels of IL-12R (not shown). Mice deficient in p40 lack expression of both of these cytokines, and manifest immune defects as a result of deficiencies in both T$_H$1 and T$_H$17 activities.

11-11 Differentiated effector T cells continue to respond to signals as they carry out their effector functions.

The commitment of CD4 T cells to distinct lineages of effector cells occurs in peripheral lymphoid tissues, such as lymph nodes. However, the effector activities of these cells are not defined simply by the signals received in the lymphoid tissues. Evidence suggests that there is continuous regulation of the expansion and the effector activities of differentiated CD4 cells, in particular those of T$_H$17 and T$_H$1 cells, once they enter sites of infection.

As noted in Chapter 9, commitment of naive T cells to become T$_H$17 cells is triggered by exposure to TGF-β and IL-6; commitment to become T$_H$1 cells is initially triggered by IFN-γ. These initial conditions are not, however, sufficient to generate complete or effective T$_H$17 or T$_H$1 responses. In addition, each T-cell subset also requires stimulation by another cytokine: IL-23 in the case of T$_H$17 cells, and IL-12 in the case of T$_H$1 cells. IL-23 and IL-12 are closely related in structure; each is a heterodimer and they have a subunit in common (Fig. 11.17). IL-23 is composed of one p40 (IL-12 p40) and one p19 (IL-23 p19) subunit, whereas IL-12 has the p40 (IL-12 p40) subunit and a unique p35 (IL-12 p35) subunit. Committed T$_H$17 cells express a receptor for IL-23, and, as will be discussed below, low levels of the receptor for IL-12. T$_H$1 cells express the receptor for IL-12. The receptors for IL-12 and IL-23 are also related. They have a common subunit, IL-12Rβ1, which is expressed by naive T cells. Upon receipt of differentiating cytokine signals, developing T$_H$17 cells synthesize IL-23R, the inducible component of the mature IL-23 receptor heterodimer; T$_H$1 cells synthesize IL-12Rβ2, the inducible component of the mature IL-12 receptor.

IL-23 and IL-12 amplify the activities of T$_H$17 and T$_H$1 cells, respectively. Like many other cytokines, they both act through the JAK–STAT intracellular signaling pathway (see Fig. 9.32). IL-23 signaling primarily activates the intracellular transcriptional activator STAT3, but also activates STAT4. In contrast, IL-12 strongly activates STAT4, with minimal activation of STAT3. IL-23 does not initiate the commitment of naive CD4 T cells to become T$_H$17 cells, but it does stimulate their expansion and contributes to their maintenance. Many *in vivo* responses that depend on IL-17 are diminished in the absence of IL-23. For example, mice lacking the IL-23-specific subunit p19 show decreased production of IL-17A and IL-17F in the lung after infection by *Klebsiella pneumoniae*.

IL-12 regulates the effector activity of committed T$_H$1 cells at sites of infection. Studies of two different pathogens have shown that the initial differentiation of T$_H$1 cells is not sufficient for protection, and that continuous signals are required. Mice deficient in IL-12 p40 can resist initial infection by *Toxoplasma gondii* as long as IL-12 is administered continuously. If IL-12 is administered during the first 2 weeks of infection, p40-deficient mice survive the initial infection and establish a latent chronic infection characterized by cysts containing the pathogen. When IL-12 administration is stopped, however, these mice gradually reactivate the latent cysts and the animals eventually die of toxoplasmic encephalitis. IFN-γ production by pathogen-specific T cells decreases in the absence of IL-12 but can be restored by IL-12 administration. Similarly, the adoptive transfer of differentiated T$_H$1 cells from mice cured of *Leishmania major* protects RAG-deficient mice infected by *L. major*, but cannot protect IL-12 p40-deficient mice (**Fig. 11.18**). Together, these experiments indicate that T$_H$1 cells continue to respond to signals during an infection, and that continuous IL-12 is needed to sustain the effectiveness of differentiated T$_H$1 cells against at least some pathogens.

11-12 Effector T cells can be activated to release cytokines independently of antigen recognition.

As we have seen, a central paradigm in adaptive immunity is the requirement for antigen recognition by the cognate receptors of naive lymphocytes to induce their differentiation into mature effector cells. However, effector T cells also acquire the ability to be activated by pairs of cytokines, independently of antigen recognition by their T-cell receptor. The cytokine pairs that mediate this 'noncognate' function of differentiated effector cells appear to be the same as those that activate the ILC subset that parallels each T-cell subset (**Fig. 11.19**). And in each case, the pair of stimulating cytokines includes one cytokine that activates a receptor that signals via a STAT factor, and one that activates a receptor that signals via NFκB—typically a member of the IL-1 receptor family. Thus, for both T$_H$1 cells and ILC1 cells, stimulation by IL-12 (STAT4) plus IL-18 induces production of IFN-γ. Similarly, stimulation of T$_H$2 and ILC2 cells by TSLP (STAT5) plus IL-33 produces IL-5 and IL-13, and both T$_H$17 and ILC3 cells stimulated by IL-23 (STAT3) plus IL-1 produce IL-17 and IL-22. In this way, mature effector CD4 T cells acquire innate-like functional properties that allow them to amplify different types of immune responses without the requirement for antigen recognition. Note that in the case of type 1 and type 3 cells, the IL-1 family member involved (IL-18 and IL-1, respectively) is produced by inflammasome activation in myeloid cells. Conversely, IL-33,

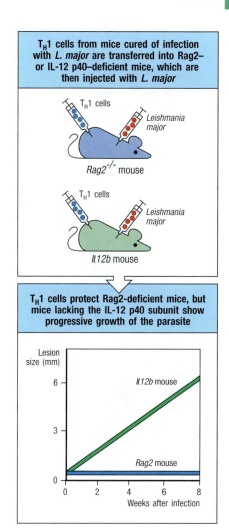

Fig. 11.18 Continuous IL-12 is necessary for resistance to pathogens requiring T$_H$1 responses. Mice that have eliminated an infection with *Leishmania major* and have generated T$_H$1 cells specific to the pathogen were used as a source of T cells that were adoptively transferred either into Rag2-deficient mice, which lack T cells and B cells and cannot control *L. major* infection but can produce IL-12, or into mice lacking IL-12 p40, which cannot produce IL-12. On subsequent infection of the Rag2-deficient mice, lesions did not enlarge because the transferred T$_H$1 cells conferred immunity. But despite the fact that the transferred cells were already differentiated T$_H$1 cells, they did not confer resistance to IL-12 p40-deficient mice, which lacked a source of IL-12 to support T$_H$1 function.

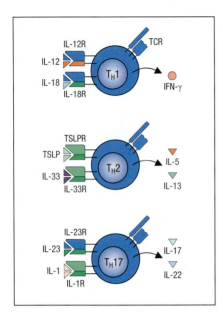

Fig. 11.19 Effector T cells can be activated to release cytokines independently of antigen recognition. Analogous to ILCs, effector T cells can be stimulated to produce effector cytokines independently of T-cell receptor signaling via the coordinate actions of pairs of cytokines.

which activates type 2 responses, is inactivated by the inflammasome, indicating another basis for counterregulation between type 2 and type 1 or type 3 immune responses. Although the precise role of noncognate activation of effector T cells by cytokines is not clearly defined, it may provide a mechanism by which tissue-resident memory T cells could be rapidly recruited in recall, or memory, responses (see Section 11-22).

11-13   Effector T cells demonstrate plasticity and cooperativity that enable adaptation during anti-pathogen responses.

Thus far we have discussed subsets of effector CD4 T cells as though they are intrinsically stable, such that their functional phenotype is unchanging after their development. Similarly, we have discussed the different types of immunity as though they are unimodal, that is, only one type of response is recruited to clear a given pathogen. Although this is often the case, it is not always so. Just as the pathogens can modify their tactics to evade destruction, so too can the elicited effector T cells adapt in order to clear the host of these pathogens. Adaptation can occur by flexibility in the programming of individual T cells, referred to as **T-cell plasticity**, wherein effector T cells can transition into different cytokine phenotypes contingent on changes in the local inflammatory environment. It can also occur as a result of cooperation between different subsets of T cells. Plasticity applies to cells of the same clonal origin and identical antigenic specificity, whereas cooperation applies to cells that develop from different clonal precursors and target different antigens, typically at different stages of an infection.

Although some degree of plasticity has been experimentally demonstrated for each of the major effector CD4 subsets, it appears to be most prevalent in type 3 responses. It is common for T_H17 cells to deviate, or be 'reprogrammed,' into T_H1-type cells (Fig. 11.20). This was originally discovered using cytokine reporter mice, in which T_H17 cells that express IL-17F could be identified and isolated based on their expression of a cell-associated reporter molecule controlled by the *Il17f* gene. When T_H17 cells isolated using the reporter were restimulated in the presence of the T_H1-polarizing cytokine IL-12, their progeny rapidly lost expression of IL-17 and acquired expression of IFN-γ. Moreover, repetitive restimulation of T_H17 cells with the T_H17 lineage cytokine IL-23 could lead to a subset of progeny that also acquired features of T_H1 cells. In both cases, reprogramming of T_H17 cells into T_H1 cells required expression of the T_H1-associated transcription factor T-bet and loss of the T_H17-associated transcription factor RORγt, both of which were contingent on activation of STAT4 by the IL-12 and IL-23 receptors. Thus, T_H17 cells deficient for either T-bet or STAT4 failed to demonstrate transition into T_H1 cells, or 'T_H17 cell plasticity.'

An example of the importance of effector T-cell plasticity and cooperativity between subsets is found in host protection against facultative intracellular bacterial pathogens, such as *Salmonella*, which, unlike obligate extracellular bacteria, have also evolved mechanisms to survive within macrophages that are not activated by IFN-γ. Early in infection, *Salmonella* can colonize the intestinal epithelium similarly to other enteric Gram-negative pathogens. During this period, a T_H17 response dominates, resulting in a robust IL-17-induced influx of neutrophils that engulf extracellular bacteria and IL-22-induced release of antimicrobial proteins that restrain bacterial growth in the intestinal lumen. During this intestinal phase of infection, much of the T-cell response appears to be directed against antigenic epitopes within bacterial flagellins, which are potent activators of TLR5. Activation of this innate sensor promotes IL-23 expression by CD11b$^+$ classical dendritic cells in the intestine, and thereby induces a type 3 immune response. Flagellin-specific T_H1 cells also emerge during the early intestinal phase of infection, and may arise from T_H17 cell precursors as a result of plasticity. To escape destruction by macrophages activated for intracellular killing by these 'ex-T_H17' T_H1 cells, *Salmonella* simultaneously downregulates the expression of flagellin and begins synthesizing new proteins, such as SseI and SseJ, that allow it to suppress intracellular killing within macrophages. This allows *Salmonella* to both evade detection by flagellin-specific T cells and use the host macrophage as a safe haven to shield it from extracellular killing—at least temporarily—as the infection spreads systemically.

During the systemic phase of the infection, the T-cell response shifts to become focused on those antigens that enable the intracellular lifestyle of the pathogen. Some of these newly expressed antigens appear to activate cytosolic sensors within $CD8\alpha^+$ classical dendritic cells, which produce IL-12 to activate pathogen-specific T_H1 cells and a type 1 response. The pathogen can now be cleared by T_H1-induced macrophage activation directed against these newly expressed antigens. Because the anti-pathogen response now includes both type 3 and type 1 immunity to different sets of antigens that the bacterium requires for its extracellular and intracellular lifestyles, *Salmonella* is deprived of a niche for its survival and is cleared from the host.

11-14 Integration of cell- and antibody-mediated immunity is critical for protection against many types of pathogens.

The type of effector T cell or antibody required for host protection depends on the infectious strategy and lifestyle of the pathogen. As we learned in Chapter 9, cytotoxic T cells are important in destroying virus-infected cells, and in some viral diseases they are the predominant class of lymphocytes present in the blood during a primary infection. Nevertheless, the role of antibodies in clearing viruses from the body and preventing them from establishing another infection can be essential. Ebola virus causes a hemorrhagic fever and is one of the most lethal viruses known, but patients who do survive are protected and asymptomatic if they become infected again. In both the initial and recurrent infection, a strong, rapid antiviral IgG response against the virus is essential for survival. The antibody response clears the virus from the bloodstream and gives the patient time to activate cytotoxic T cells. This antibody response does not occur in infections that prove fatal; the virus continues to replicate, and even though there is activation of T cells, the disease progresses.

Cytotoxic T cells are also required for the destruction of cells infected with some intracellular bacterial pathogens, such as *Rickettsia* (the causative agent of typhus) or *Listeria*, which can escape from phagocytic vesicles to avoid the killing mechanisms of activated macrophages. In contrast, mycobacteria, which resist phagolysosomal killing and can live inside macrophage vesicles, are mainly kept in check by T_H1 cells, which activate infected macrophages to kill the bacteria. Nevertheless, antibodies are induced in these infections and can contribute to pathogen killing when organisms are released from dying phagocytes, and are important in resistance to reinfection.

In many cases the most efficient protective immunity is mediated by neutralizing antibodies that can prevent pathogens from establishing an infection, and most of the established vaccines against acute childhood viral infections work primarily by inducing protective antibodies. Effective immunity against polio virus, for example, requires preexisting antibody, because the virus rapidly infects motor neurons and destroys them unless it is immediately neutralized by antibody and prevented from spreading within the body. In polio, specific IgA on mucosal epithelial surfaces also neutralizes the virus before it enters the tissues. Thus, protective immunity can involve effector mechanisms (IgA in this case) that do not operate in the elimination of the primary infection.

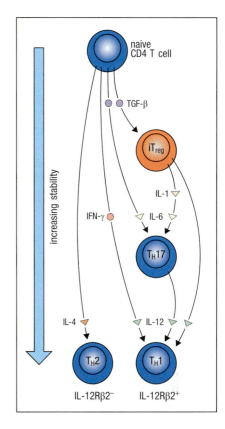

Fig. 11.20 Plasticity of CD4+ T-cell subsets. There is a hierarchy of stability of effector and regulatory CD4 T cells. Naive CD4+ T cells are multipotent, whereas T_H1 and T_H2 cells appear to be relatively stable, or at 'ground state'; that is, they are highly resistant to transitioning into other effector cell phenotypes. iT_{reg} cells and T_H17 cells are less stable, and can transition into other subsets depending on prevailing cytokines. When acted on by IL-6 and IL-1, iT_{reg} cells can transition into T_H17 cells, or when acted on by IL-12 can become T_H1 cells. T_H17 cells acted on by IL-12 can transition into T_H1 cells. Notably, the transitions of iT_{reg} cells into T_H17 cells, and T_H17 cells into T_H1 cells, appear to be unidirectional, or irreversible. Developing T_H2 cells (left) repress expression of the inducible component of the IL-12 receptor (IL-12Rβ) and are unresponsive to IL-12; iTreg, T_H17, and T_H1 subsets (right) remain responsive to IL-12.

11-15 Primary CD8 T-cell responses to pathogens can occur in the absence of CD4 T-cell help.

Many CD8 T-cell responses are deficient in the absence of help from CD4 T cells (see Section 9-19). In such circumstances, CD4 T-cell help is required to activate dendritic cells for them to become able to stimulate a complete CD8 T-cell response, an activity that has been described as licensing of the antigen-presenting cell (see Section 9-10). Licensing involves the induction of co-stimulatory molecules such as B7, CD40, and 4-1BBL on the dendritic cell, which can then deliver signals that fully activate naive CD8 T cells (see Fig. 9.29). Licensing enforces a requirement for dual recognition of an antigen by the immune system by both CD4 and CD8 T cells, which provides a useful safeguard against autoimmunity. Dual recognition is also seen in the cooperation between T cells and B cells for antibody generation (see Chapter 10). However, not all CD8 T-cell responses require such help.

Some infectious agents, such as the intracellular Gram-positive bacterium *Listeria monocytogenes* and the Gram-negative bacterium *Burkholderia pseudomallei*, appear to be able to directly license dendritic cells to induce primary CD8 T-cell responses without requirement for CD4 T-cell help (Fig. 11.21). Primary CD8 T-cell responses to *L. monocytogenes* were examined in mice that were genetically deficient in MHC class II molecules and thus lacked CD4 T cells (see Section 11-23). The numbers of CD8 T cells specific for a particular antigen expressed by the pathogen were measured by using **tetrameric peptide:MHC complexes**, or **peptide:MHC tetramers** (see Appendix I, Section A-24), which can identify CD4 or CD8 T cells on the basis of the antigenic specificity of their T-cell receptors. On day 7 of infection, wild-type mice and mice lacking CD4 T cells showed equivalent expansion, and

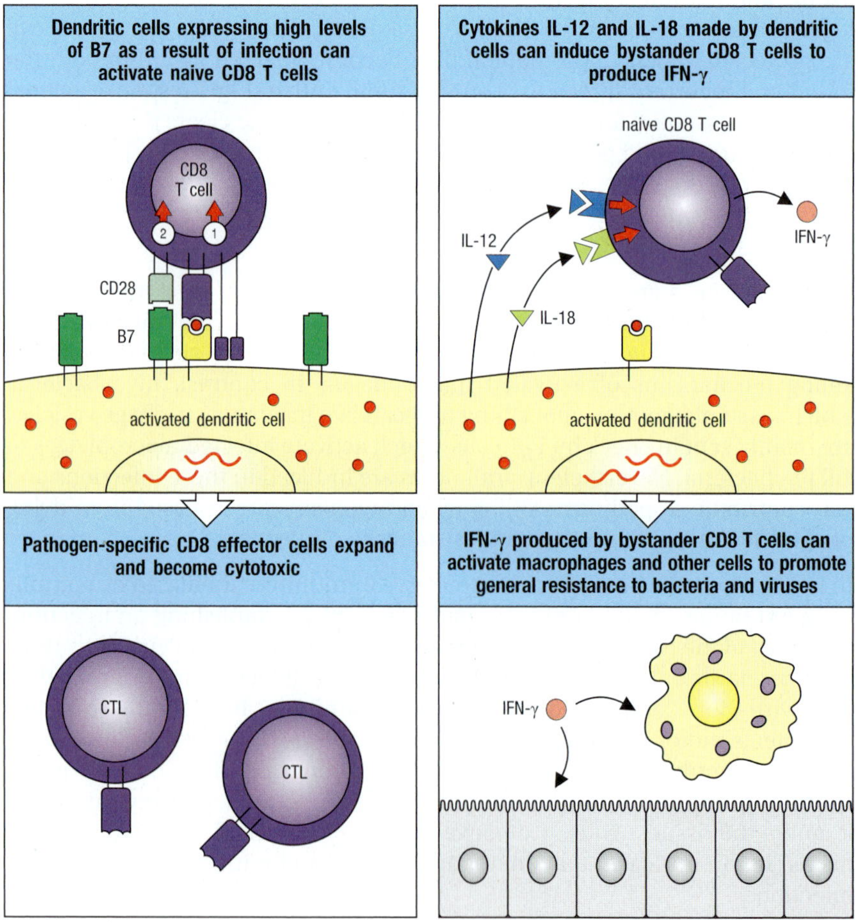

Fig. 11.21 Naive CD8 T cells can be activated directly by potent antigen-presenting cells through their T-cell receptor or through the action of cytokines. Left panels: naive CD8 T cells that encounter peptide:MHC class I complexes on the surface of dendritic cells expressing high levels of co-stimulatory molecules as a result of the inflammatory environment produced by some pathogens (upper left panel) are activated to proliferate in response, eventually differentiating into cytotoxic CD8 T cells (lower left panel). Right panels: activated dendritic cells also produce the cytokines IL-12 and IL-18, whose combined effect on CD8 T cells rapidly induces the production of IFN-γ (upper right panel). This activates macrophages for the destruction of intracellular bacteria and can promote antiviral responses in other cells (lower right panel).

equivalent cytotoxic capacity, of pathogen-specific CD8 T cells. Mice lacking CD4 T cells cleared the initial infection by *L. monocytogenes* as effectively as wild-type mice. These experiments clearly show that protective responses can be generated by pathogen-specific CD8 T cells without CD4 T-cell help. However, as we will see later, the nature of the CD8 memory response is different and is diminished in the absence of CD4 T-cell help.

Naive CD8 T cells can also undergo 'bystander' activation by IL-12 and IL-18 to produce IFN-γ very early during infection (see Fig. 11.21). Mice infected with *L. monocytogenes* or *B. pseudomallei* rapidly produce a strong IFN-γ response, which is essential for their survival. The source of this IFN-γ seems to be both NK cells and naive CD8 T cells, which begin to secrete it within the first few hours after infection. This is believed to be too soon for any significant expansion of pathogen-specific CD8 T cells, which would initially be too rare to contribute in an antigen-specific manner. The production of IFN-γ by both NK and CD8 T cells at this early time can be blocked experimentally by antibodies against IL-12 and IL-18, suggesting that these cytokines are responsible. These experiments indicate that naive CD8 T cells can contribute nonspecifically in a kind of innate defense, one not requiring CD4 T cells, in response to early signals of infection.

11-16 Resolution of an infection is accompanied by the death of most of the effector cells and the generation of memory cells.

When an infection is effectively repelled by the adaptive immune system, two things occur. First, the actions of effector cells remove the pathogen and, with them, the antigens that originally stimulated their differentiation. Second, in the absence of antigen, most effector T cells undergo 'death by neglect,' removing themselves by apoptosis. The resulting 'clonal contraction' of effector T cells appears to be due both to the loss of pro-survival cytokines that are produced by antigenic stimulation, such as IL-2, and to the loss of expression of receptors for these cytokines. CD25, the IL-2 receptor component that mediates high-affinity binding, is transiently upregulated on antigen-activated T cells, but then declines, thus limiting IL-2 signaling in the absence of antigenic restimulation. Also, as is discussed in Section 11-21, most effector T cells lose expression of the specific component of the IL-7 receptor **IL-7Rα** (**CD127**) soon after activation. Like IL-2 signaling, IL-7 signaling activates STAT5, which promotes the expression of anti-apoptotic survival factors such as Bcl-2. Effector cells that lose responsiveness to IL-2 and IL-7 lose Bcl-2 and express Bim, which is a pro-apoptotic factor that acts via the intrinsic, or mitochondrial, pathway of apoptosis that leads to assembly of the apoptosome (see Sections 9-29 and 9-30).

While many effector T cells die from the loss of survival signals and the activation of the Bim-mediated intrinsic pathway of apoptosis, effector T-cell death can also occur via the extrinsic pathway of apoptosis that is activated by signaling via members of the TNF receptor superfamily, particularly Fas (CD95) (Fig. 11.22). Activation of the extrinsic pathway (or death receptor pathway) leads to the formation of the **death-inducing signaling complex** (**DISC**). The first step in Fas-mediated formation of the DISC is binding of the trimeric FasL, which results in the trimerization of Fas. This causes the death domains of Fas to bind to the death domain of FADD (Fas-associated via death domain), an adaptor protein introduced in Section 3-25. FADD contains a death domain and an additional domain called a **death effector domain** (**DED**) that can bind to DEDs present in other proteins. When FADD is recruited to Fas, the DED of FADD then recruits the initiator caspases pro-caspase 8 and pro-caspase 10 via interaction with a DED in the pro-caspases. The high local concentration of these caspases in association with activated receptors allows the caspases to cleave themselves, resulting in their activation. Once activated, caspases 8 and 10 are released from the receptor complex and can activate the downstream

 MOVIE 11.3

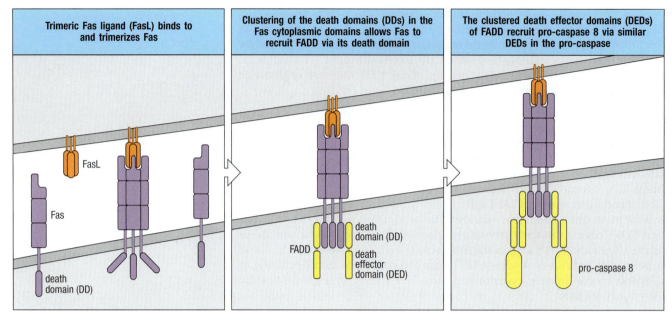

| Trimeric Fas ligand (FasL) binds to and trimerizes Fas | Clustering of the death domains (DDs) in the Fas cytoplasmic domains allows Fas to recruit FADD via its death domain | The clustered death effector domains (DEDs) of FADD recruit pro-caspase 8 via similar DEDs in the pro-caspase |

Fig. 11.22 Binding of Fas ligand to Fas initiates the extrinsic pathway of apoptosis. The cell-surface receptor Fas contains a so-called death domain (DD) in its cytoplasmic tail. When Fas ligand (FasL) binds Fas, this trimerizes the receptor (left panel). The adaptor protein FADD (also known as MORT-1) also contains a death domain and can bind to the clustered death domains of Fas (center panel). FADD also contains a domain called a death effector domain (DED) that allows it to recruit pro-caspase 8 or pro-caspase 10, (not shown), which also contains a DED domain (right panel). Clustered pro-caspase 8 activates itself to release an active caspase into the cytoplasm (not shown).

Autoimmune Lymphoproliferative Syndrome (ALPS)

effector caspases that induce apoptosis. Loss-of-function mutations in Fas lead to the increased survival of lymphocytes and are one cause of the disease **autoimmune lymphoproliferative syndrome** (**ALPS**). This disease can also be due to mutations in FasL and in caspase 10.

The relative contributions of the Bim- and Fas-mediated apoptotic pathways to effector T-cell loss depend on the infectious agent, but they appear to be complementary mechanisms, as mice with specific deficiencies of Bim or Fas have milder defects in T-cell clearance than mice with deficiencies of both. Thus, the two pathways appear to be nonredundant; what aspects of infection contribute to the dominance of one mechanism over the other in response to different pathogens is unclear. Irrespective of whether their death is induced by the intrinsic or extrinsic pathway, dying T cells are rapidly cleared by phagocytes that recognize on their surface the membrane lipid phosphatidylserine. This lipid is normally found only on the inner surface of the plasma membrane, but in apoptotic cells it rapidly redistributes to the outer surface, where it can be recognized by specific receptors on many cells. Thus, not only is the pathogen removed at the end of infection, but most of the pathogen-specific effector cells are also removed. Some of the pathogen-specific effector cells survive, however, and provide the basis for memory T-cell and B-cell responses, as will be discussed in the next section.

Summary.

CD4 T cells develop in response to, and subsequently amplify and sustain, innate immune responses that are induced by pathogens. Pathogen antigens are transported to local lymphoid organs by migrating dendritic cells and are presented to antigen-specific naive T cells that continuously recirculate through the lymphoid organs. T-cell priming and the differentiation of effector T cells occurs here, and the effector T cells either leave the lymphoid organ to provide cell-mediated immunity at sites of infection in the tissues or remain in the lymphoid organ to participate in humoral immunity by activating antigen-binding B cells. Distinct types of CD4 T cells develop in response to infection by different types of pathogens, and their development is influenced largely by the cytokines produced by innate sensor cells and ILCs that are activated early in the response.

Effector CD4 T cells serve to amplify and enhance early innate responses orchestrated by ILCs, while T_{FH} cells that develop in concert with each subset of effector T cells direct the production of high-affinity antibodies that arm innate effector cells for heightened pathogen elimination. T_H1 responses promote the development and activation of classical, M1 macrophages to protect against intracellular pathogens. T_H2 responses are directed against infections by parasites such as helminths, and promote the development and activation of alternative, M2 macrophages and the recruitment of eosinophils and basophils to sites of infection. T_H17 cells are integral to the clearance of extracellular bacteria and fungi, by orchestrating sustained neutrophil recruitment and production of antimicrobial peptides by epithelial cells of barrier tissues such as the intestines, lungs, and skin. CD8 T cells have an important role in protective immunity, especially in protecting the host against infection by viruses and intracellular infections by *Listeria* and other microbial pathogens that have special means for entering the host cell's cytoplasm. Primary CD8 T-cell responses to pathogens usually require CD4 T-cell help, but can occur in response to some pathogens without such help. The patterns of anti-pathogenic response are not fixed, and effector T cells retain plasticity for adapting their response as pathogens alter their survival strategy in response to pressure from the immune system. Ideally, the adaptive immune response eliminates the infectious agent, at which time the expanded clonal populations of effector T cells contract, retaining only small populations of long-lived memory cells that provide the host with a state of protective immunity against reinfection with the same pathogen.

Immunological memory.

In this part of the chapter, we will examine how long-lasting protective immunity is maintained after an infection is successfully eliminated. Immunological memory is perhaps the most important consequence of an adaptive immune response, because it enables the immune system to respond more rapidly and effectively to pathogens that have been encountered previously, and prevents them from causing disease. Memory responses, called **secondary immune responses**, **tertiary immune responses**, and so on, depending on the number of exposures to antigen, also differ qualitatively from primary responses. This is particularly clear for B-cell responses, in that antibodies made during secondary and subsequent responses exhibit distinct characteristics, such as higher affinity to antigen compared with antibodies made during the primary response. Memory T-cell responses can also be distinguished qualitatively from the responses of naive or effector T cells, in terms of location, trafficking patterns, and effector functions.

11-17 Immunological memory is long lived after infection or vaccination.

Most children in developed countries are now vaccinated against measles virus; before vaccination was widespread, most were naturally exposed to this virus and developed an acute, unpleasant, and potentially dangerous illness. Whether through vaccination or infection, children exposed to the virus acquire long-term protection from measles, lasting for most people for the whole of their life. The same is true of many other acute infectious diseases (see Chapter 16): this state of protection is a consequence of immunological memory.

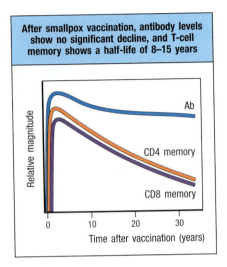

Fig. 11.23 Antiviral immunity after smallpox vaccination is long lived. Because smallpox has been eradicated, recall responses measured in people who were vaccinated for smallpox can be taken to represent true memory in the absence of reinfection. After smallpox vaccination, antibody levels show an early peak with a period of rapid decay, which is followed by long-term maintenance that shows no significant decay. CD4 and CD8 T-cell memory is long lived but gradually decays, with a half-life in the range of 8–15 years.

The basis of immunological memory has been difficult to explore experimentally. Although the phenomenon was first recorded by the ancient Greeks and has been exploited routinely in vaccination programs for more than 200 years, it was recognized only within the last 30 years that this memory phenomenon reflects a small population of specialized **memory cells** that are formed during the adaptive immune response and can persist in the absence of the antigen that originally induced them. This mechanism of maintaining memory is consistent with the finding that only individuals who were themselves previously exposed to a given infectious agent are immune. The idea that immunological memory does not depend on repeated exposure to infection as a result of contact with other infected individuals has been supported by observations made of populations of people living on remote islands. In that setting, a virus such as measles can cause an epidemic, infecting all people living on the island at that time, after which the virus disappears for many years. On reintroduction from outside the island, the virus does not affect the original population but causes disease in those people born since the first epidemic.

The duration of immunological memory has been estimated by examining responses in people who received vaccinia, the virus used to immunize against smallpox (Fig. 11.23). Because smallpox was eradicated in 1978, it is presumed that their responses represent true immunological memory and are not due to restimulation from time to time by the smallpox virus. One study found strong vaccinia-specific CD4 and CD8 T-cell memory responses as long as 75 years after the original immunization, and from the strength of these responses it was estimated that the memory response had an approximate half-life of between 8 and 15 years. Half-life represents the time required for a response to diminish to 50% of its original strength. By contrast with T-cell memory, titers of antivirus antibody remained stable, almost without measurable decline.

These findings show that immunological memory need not be maintained by repeated exposure to infectious virus. Instead, it is likely that memory is sustained by long-lived antigen-specific lymphocytes that were induced by the original exposure and that persist until a second encounter with the pathogen. Although most of the memory cells are in a resting state, a small percentage of memory cells are dividing at any one time. It appears that this turnover is maintained by cytokines, such as IL-7 and IL-15, produced either constitutively or during antigen-specific immune responses directed at other, non-cross-reactive antigens. The number of memory cells for a given antigen is highly regulated, persisting with a relatively long half-life balanced by cell proliferation and cell death.

Immunological memory can be measured experimentally in various ways. Adoptive transfer assays (see Appendix I, Section A-30) of lymphocytes from animals immunized with simple, nonliving antigens have been favored for such studies, because the antigen cannot proliferate. In these experiments, the existence of memory cells is measured purely in terms of the transfer of specific responsiveness from an immunized, or 'primed,' animal to a nonimmunized recipient, as tested by a subsequent immunization with the antigen. Animals that received memory cells have a faster and more robust response to antigen challenge than do controls that did not receive cells, or that received cells from nonimmune donors.

Experiments like these have shown that when an animal is first immunized with a protein antigen, functional helper T-cell memory against that antigen appears abruptly and reaches a maximum after 5 days or so. Functional antigen-specific B-cell memory appears some days later, then enters a phase of cell proliferation and selection in lymphoid tissues. By 1 month after immunization, memory B cells are present at their maximum level. These levels of memory cells are then maintained, with little alteration, for the lifetime of the animal. It is important to recognize that the functional memory elicited in these experiments can be due to the precursors of memory cells as well as the memory cells themselves. These precursors are probably activated T cells and B cells, some of whose progeny will later differentiate into memory cells. Thus, precursors to memory cells can appear very shortly after immunization, even though resting memory-type lymphocytes may not yet have developed.

In the following sections we look in more detail at the changes that occur in lymphocytes after antigen priming and lead to the development of resting memory lymphocytes, and discuss the mechanisms that might account for these changes.

11-18 Memory B-cell responses are more rapid and have higher affinity for antigen compared with responses of naive B cells.

Immunological memory in B cells can be examined *in vitro* by isolating B cells from immunized or unimmunized mice and restimulating them with antigen in the presence of helper T cells specific for the same antigen (Fig. 11.24). B cells from immunized mice produce responses that differ both quantitatively and qualitatively compared with naive B cells from unimmunized mice. B cells that respond to the antigen increase in frequency by up to 100-fold after their initial priming in the primary immune response. Further, due to the process of affinity maturation (described in Chapter 10), antibodies produced by B cells of immunized mice typically have higher affinity for antigen than antibodies produced by unprimed B lymphocytes. The response from immunized mice is due to **memory B cells** that arise in the primary response. Memory B cells can arise from the germinal center reaction during a primary response, and may have undergone isotype switching and somatic mutations there. But memory B cells can also arise independently of the germinal center reaction from short-lived plasma cells produced in the primary response. In either case, they circulate through the blood and take up residence in the spleen and lymph nodes. Memory B cells express some markers that distinguish them from naive B cells and plasma cells. One marker of memory B cells is simply

	Source of B cells	
	Unimmunized donor Primary response	**Immunized donor Secondary response**
Frequency of antigen-specific B cells	$1:10^4 - 1:10^5$	$1:10^2 - 1:10^3$
Isotype of antibody produced	IgM > IgG	IgG, IgA
Affinity of antibody	Low	High
Somatic hypermutation	Low	High

Fig. 11.24 The generation of secondary antibody responses from memory B cells is distinct from the generation of the primary antibody response. These responses can be studied and compared by isolating B cells from immunized and unimmunized donor mice, and stimulating them in culture in the presence of antigen-specific effector T cells. The primary response usually consists of antibody molecules made by plasma cells derived from a quite diverse population of precursor B cells specific for different epitopes of the antigen and with receptors that have a range of affinities for the antigen. The antibodies are of relatively low affinity overall, with few somatic mutations. The secondary response derives from a far more limited population of high-affinity B cells, which have, however, undergone significant clonal expansion. Their receptors and antibodies are of high affinity for the antigen and show extensive somatic mutation. The overall effect is that although there is usually only a 10- to 100-fold increase in the frequency of activatable B cells after priming, the quality of the antibody response is radically altered, in that these precursors induce a far more intense and effective response.

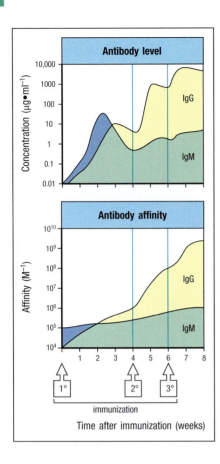

Fig. 11.25 Both the affinity and the amount of antibody increase with repeated immunization. The upper panel shows the increase in antibody concentration with time after a primary (1°), followed by a secondary (2°) and a tertiary (3°), immunization; the lower panel shows the increase in the affinity of the antibodies (affinity maturation). Affinity maturation is seen largely in IgG antibody (as well as in IgA and IgE, which are not shown) coming from mature B cells that have undergone isotype switching and somatic hypermutation to yield higher-affinity antibodies. The blue shading represents IgM on its own, the yellow shading IgG, and the green shading the presence of both IgG and IgM. Although some affinity maturation occurs in the primary antibody response, most arises in later responses to repeated antigen injections. Note that these graphs are on a logarithmic scale; it would otherwise be impossible to represent the overall increase of around 1 millionfold in the concentration of specific IgG antibody from its initial level.

the expression of a switched surface immunoglobulin isotype, compared with naive B cells that express surface IgM and IgD. In contrast, plasma cells have very low surface immunoglobulin altogether. In humans, a marker of memory B cells is **CD27**, a member of the TNF receptor family that is also expressed by naive T cells and binds the TNF-family ligand **CD70**, which is expressed by dendritic cells (see Section 9-17).

A primary antibody response is characterized by the initial rapid production of IgM and a slightly delayed IgG response due to time required for class switching (**Fig. 11.25**). The secondary antibody response is characterized in its first few days by the production of relatively small amounts of IgM antibody and much larger amounts of IgG antibody, with some IgA and IgE. At the beginning of the secondary response, antibodies are made by memory B cells that were generated in the primary response and have already switched from IgM to another isotype, and that express IgG, IgA, or IgE on their surface. Memory B cells express somewhat higher levels of MHC class II molecules and the co-stimulatory ligand **B7.1** than do naive B cells. This helps memory B cells acquire and present antigen more efficiently to T_{FH} cells compared with naive B cells, and to T_{FH} cells through the B7.1 receptor **CD28**, so that they can in turn help in antibody production so it begins earlier after antigen exposure compared with primary responses. The secondary response is characterized by a more vigorous and earlier generation of plasma cells than in the primary response, thus accounting for the almost immediate abundant production of IgG (see Fig. 11.25).

11-19 Memory B cells can reenter germinal centers and undergo additional somatic hypermutation and affinity maturation during secondary immune responses.

During a secondary exposure to infection, antibodies persisting from a primary immune response are available immediately to bind the pathogen and mark it for degradation by complement or by phagocytes. If the antibody can completely neutralize the pathogen, a secondary immune response may not occur. Otherwise, excess antigens will bind to receptors on B cells and initiate a secondary response in the peripheral lymphoid organs. Memory B cells recirculate through the same secondary lymphoid compartments as naive B cells, principally the follicles of the spleen, lymph nodes, and the Peyer's patches of the gut mucosa. B cells with the highest avidity for antigen are activated first. Thus memory B cells, which have been selected previously for their avidity to antigen, make up a substantial component of the secondary response.

Besides responding more rapidly, memory B cells can reenter **germinal centers** during secondary immune responses and undergo additional **somatic hypermutation** and **affinity maturation**, as described in Sections 10-6 through 10-8. As in primary responses, secondary B-cell responses begin at the interface between the T-cell and B-cell zones, where memory B cells that have acquired antigen can present peptide:MHC class II complexes to helper T cells. This interaction initiates proliferation of both the B cells and T cells.

Reactivated memory B cells that have not yet undergone differentiation into plasma cells migrate into the follicle and become germinal center B cells, undergoing additional rounds of proliferation and somatic hypermutation before differentiating into antibody-secreting plasma cells. Since B cells with the higher-affinity antigen receptors will more efficiently acquire and present antigen to antigen-specific T_{FH} cells in the germinal center, the affinity of the antibodies produced during secondary and tertiary responses rises progressively (see Fig. 10.14).

11-20 MHC tetramers identify memory T cells that persist at an increased frequency relative to their frequency as naive T cells.

Until relatively recently, analysis of T-cell memory relied on assays of T-cell function rather than a direct identification of antigen-specific memory T cells. Some assays of T-cell effector function, such as providing help to B cells or macrophages, can take several days to perform. Because of this, such assays are not optimal for distinguishing memory T cells from preexisting effector cells, since memory cells can be reactivated during the time frame of the assay. This is a problem particularly for effector CD4 T cells, but does not apply as much for effector CD8 T cells, which can program a target cell for lysis in 5 minutes. In contrast, memory CD8 T cells need more time than this to be reactivated to become cytotoxic, so that the actions of memory CD8 T cells will appear much later than those of preexisting effector cells.

Examining T-cell memory has been made easier by the development of MHC tetramers (see Appendix I, Section A-24). Before MHC tetramers, effector and memory responses were studied using naive T cells from mice carrying specific T-cell receptor (TCR) transgenes. Such TCR-transgenic T cells could be uniquely identified by antibodies to their rearranged T-cell receptors, but were not part of the host's natural T-cell repertoire. MHC tetramers measure the *in vivo* frequency of all clones with a given antigen specificity, but do not distinguish between different T-cell clones of the same specificity. MHC tetramers were generated for MHC class I molecules first, but are now also available for some MHC class II molecules, allowing the study of both CD8 and CD4 T cells both in normal mice and in humans.

MHC tetramers have allowed a direct analysis of the formation of memory T cells. In the example shown in **Fig. 11.26**, T-cell responses to infection by the intracellular bacteria *Listeria monocytogenes* are analyzed with MHC class II tetramers specific to the toxin listeriolysin O (LLO). In the naive T-cell repertoire of the mouse, there are approximately 100 LLO-specific CD4 T cells, which undergo 1000-fold expansion into effector T cells during the expansion phase of 6 days after infection. When the infection is eliminated,

Fig. 11.26 Generation of memory T cells after an infection. After an infection, in this case with an attenuated strain of *Listeria monocytogenes*, the number of T cells specific for the listeriolysin (LLO) toxin increases dramatically and then falls back to give a sustained low level of memory T cells. T-cell responses are detected by binding of an MHC tetramer consisting of an LLO peptide bound by I-A^b. The left panel shows the primary response of CD4 T cells that are LLO-specific, and the right panel shows the contraction and memory phase. Approximately 100 T cells in the naive T-cell repertoire expand to about 100,000 effector cells by day 7, and then contract to about 7000 memory cells by day 25. These memory cells then slowly decay to 500 cells by day 450. Data courtesy of Marc Jenkins.

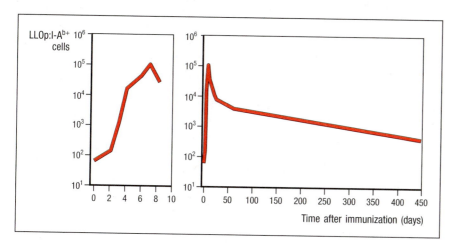

Time after immunization (days)

a slower contraction phase follows in which these T cells are reduced by about 100-fold within a few weeks. This leaves a population of memory T cells present at a frequency of about 10-fold higher than in the naive repertoire, and this population persists with a half-life of about 60 days.

11-21   Memory T cells arise from effector T cells that maintain sensitivity to IL-7 or IL-15.

Naive and memory T cells can be distinguished by differences in their expression of various cell-surface proteins, by their distinct responses to stimuli, and by their expression of certain genes. Overall, memory cells continue to express many markers of activated T cells, such as **phagocytic glycoprotein-1 (Pgp1, CD44)**, but they stop expressing other activation markers, such as **CD69**. Memory T cells express more **Bcl-2**, a protein that promotes cell survival and may be responsible for their long half-life. Figure 11.27 lists several molecules by which naive, effector, and memory T cells can be distinguished.

Protein	Naive	Effector	Memory	Comments
CD44	+	+++	+++	Cell-adhesion molecule
CD45RO	+	+++	+++	Modulates T-cell receptor signaling
CD45RA	+++	+	+++	Modulates T-cell receptor signaling
CD62L	+++	–	Some +++	Receptor for homing to lymph node
CCR7	+++	+/–	Some +++	Chemokine receptor for homing to lymph node
CD69	–	+++	–	Early activation antigen
Bcl-2	++	+/–	+++	Promotes cell survival
Interferon-γ	–	+++	+++	Effector cytokine; mRNA present and protein made on activation
Granzyme B	–	+++	+/–	Effector molecule in cell killing
FasL	–	+++	+	Effector molecule in cell killing
CD122	+/–	++	++	Part of receptor for IL-15 and IL-2
CD25	–	++	–	Part of receptor for IL-2
CD127	++	–	+++	Part of receptor for IL-7
Ly6C	+	+++	+++	GPI-linked protein
CXCR4	+	+	++	Receptor for chemokine CXCL12; controls tissue migration
CCR5	+/–	++	Some +++	Receptor for chemokines CCL3 and CCL4; tissue migration
KLRG1	–	+++	Some +++	Cell surface receptor

Fig. 11.27 Expression of many proteins alters when naive T cells become memory T cells. Proteins that are expressed differently in naive T cells, effector T cells, and memory T cells include adhesion molecules, which govern interactions with antigen-presenting cells and endothelial cells; chemokine receptors, which affect migration to lymphoid tissues and sites of inflammation; proteins and receptors that promote the survival of memory cells; and proteins that are involved in effector functions, such as granzyme B. Some changes also increase the sensitivity of the memory T cell to antigen stimulation. Many of the changes that occur in memory T cells are also seen in effector cells, but some, such as expression of the cell-surface proteins CD25 and CD69, are specific to effector T cells; others, such as expression of the survival factor Bcl-2, are limited to long-lived memory T cells. This list represents a general picture that applies to both CD4 and CD8 T cells in mice and humans, but some details that may differ between these sets of cells have been omitted for simplicity.

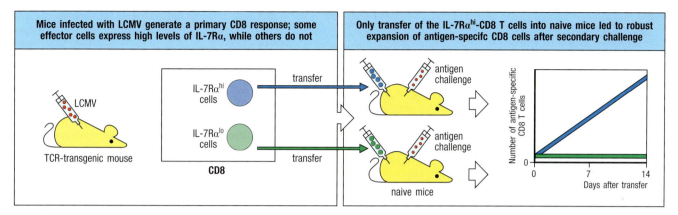

Fig. 11.28 Expression of the IL-7 receptor (IL- 7Rα) indicates which CD8 effector T cells can generate robust memory responses. Mice expressing a T-cell receptor (TCR) transgene specific for a viral antigen from lymphocytic choriomeningitis virus (LCMV) were infected with the virus, and effector cells were collected on day 11. Effector CD8 T cells expressing high levels of IL-7Rα (IL-7Rα^{hi}, blue) were separated and transferred into one group of naive mice, and effector CD8 T cells expressing low IL-7Rα (IL-7Rα^{lo}, green) were transferred into another group. Three weeks after transfer, the mice were challenged with a bacterium engineered to express the original viral antigen, and the numbers of responding transferred T cells (detected by their expression of the transgenic TCR) were measured at various times after challenge. Only the transferred IL-7Rα^{hi} effector cells could generate a robust expansion of CD8 T cells after the secondary challenge.

Among the important markers of memory T cells is the α subunit of the **IL-7 receptor** (**IL-7Rα** or **CD127**). Naive T cells express IL-7Rα, but it is rapidly lost upon activation and is not expressed by most effector T cells. For example, the experiment shown in Fig. 11.28 examines mice infected with lymphocytic choriomeningitis virus (LCMV). Around day 7 after infection, a small population of approximately 5% of CD8 effector T cells expressed high levels of IL-7Rα. Adoptive transfer of these IL-7Rα^{hi} cells, but not IL-7Rα^{lo} effector T cells, could provide functional CD8 T-cell memory to uninfected mice. This experiment suggests that memory T cells arise from effector T cells that maintain or reexpress IL-7Rα, perhaps because they compete more effectively for the survival signals delivered by IL-7.

The homeostatic mechanisms governing the survival of memory T cells also differ from those for naive T cells. Memory T cells divide more frequently than naive T cells, and their expansion is controlled by a shift in the balance between proliferation and cell death. As illustrated in Fig. 11.29, naive T cells require contact with self peptide:self MHC complexes in addition to cytokine stimulation for their long-term survival in the periphery (see Fig. 9.4). As with naive cells, the survival of memory T cells requires signaling by the receptors for the cytokines IL-7 and IL-15. IL-7 is required for the survival of both CD4 and CD8 memory T cells. In addition, IL-15 is critical for the long-term survival and proliferation of CD8 memory T cells under normal conditions. It also appears that memory T cells are less dependent on contact with self peptide:self MHC than naive T cells and are more sensitive to cytokines.

Memory T cells still require contact with peptide:MHC complexes to become reactivated during a secondary encounter with pathogen, but are also more sensitive to restimulation by antigen than are naive T cells. Furthermore, they more quickly and more vigorously produce several cytokines such as IFN-γ, TNF-α, and IL-2 in response to such stimulation. A similar progression occurs for T cells in humans after immunization with a vaccine against yellow fever virus.

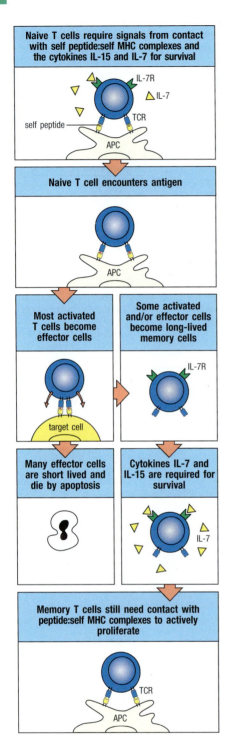

Naive T cells require signals from contact with self peptide:self MHC complexes and the cytokines IL-15 and IL-7 for survival

IL-7R

IL-7

self peptide

TCR

APC

Naive T cell encounters antigen

APC

Most activated T cells become effector cells

Some activated and/or effector cells become long-lived memory cells

IL-7R

target cell

Many effector cells are short lived and die by apoptosis

Cytokines IL-7 and IL-15 are required for survival

IL-7

Memory T cells still need contact with peptide:self MHC complexes to actively proliferate

TCR

APC

Fig. 11.29 Memory and naive T cells have different requirements for survival. For their survival in the periphery, naive T cells require periodic stimulation with the cytokines IL-7 and IL-15 and with self-antigens presented by self MHC molecules. On priming with its specific antigen, a naive T cell divides and differentiates. Most of the progeny differentiate into relatively short-lived effector cells that have lost expression of the IL-7 receptor (yellow), but some effector cells retain or reexpress the receptor and become long-lived memory T cells. These memory cells can be maintained by IL-7 and IL-15 and are less dependent on contact with self peptide:self MHC complexes for survival compared with naive T cells. However, contact with self antigens may be necessary for some memory T cells to keep up their numbers in the memory pool, but this may vary between different clones and is the subject of ongoing investigation.

11-22 Memory T cells are heterogeneous and include central memory, effector memory, and tissue-resident subsets.

Changes in other cell-surface proteins that occur on memory CD4 T cells after exposure to antigen are significant (see Fig. 11.27). **L-selectin** (**CD62L**) is the homing receptor that directs T cells into secondary lymphoid tissues, and it is lost by effector cells and most memory CD4 T cells. CD44 is a receptor for hyaluronic acid and other ligands expressed in peripheral tissues, and it is induced on effector and memory T cells. The change in expression of these two molecules helps memory T cells migrate from the blood into the peripheral tissues rather than migrating directly into lymphoid tissues, as would naive T cells. Different isoforms of **CD45**, a cell-surface protein tyrosine phosphatase expressed on all hematopoietic cells, are useful in distinguishing naive from effector and memory T cells. The **CD45RO** isoform is produced because of changes in the alternative splicing of exons that encode the CD45 extracellular domain and identifies effector and memory cells, although it is unclear what functional consequences this change may impose. Some surface receptors, such as **CD25**, the α subunit of the IL-2 receptor, are expressed on activated effector cells, but not memory cells; however, they can be reexpressed when memory cells are again reactivated by antigen and become effector T cells.

Memory T cells are heterogeneous, and both CD4 and CD8 T cells are classified into three major subsets. Each type exhibits a distinct pattern of receptors, for example, for different chemokines and adhesion molecules, and shows different activation characteristics (**Fig. 11.30**). **Central memory T cells** (T_{CM}) express the chemokine receptor **CCR7**, which allows their recirculation to be similar to that of naive T cells and allows them to traffic through the T-cell zones of peripheral lymphoid tissues. Central memory cells are very sensitive to cross-linking of their T-cell receptors and rapidly express **CD40** ligand in response; however, they are relatively slower compared with other memory subsets to acquire effector functions such as production of cytokines early after restimulation. Central memory cells primarily migrate from the blood, into the secondary lymphoid organs, then into the lymphatic system and back into the blood, a route very similar to the migration pattern of naive T cells. By contrast, **effector memory T cells** (T_{EM}) lack the chemokine receptor CCR7, but express high levels of β_1 and β_2 integrins, and so are specialized for rapidly entering inflamed tissues. They also express receptors for inflammatory chemokines and can rapidly mature into effector T cells and secrete large amounts of IFN-γ, IL-4, and IL-5 early after restimulation. Effector memory T cells migrate from the blood primarily into peripheral nonlymphoid tissues, then through the lymphatic system and finally into secondary lymphoid tissues. There they can reenter the lymphatic system and reach the blood again. In contrast to central and effector memory cells, **tissue-resident memory T cells** (T_{RM}) comprise a substantial fraction of memory T cells that do not migrate, but rather take up long-term residency in various epithelial sites (**Fig. 11.31**). Like T_{EM} cells, T_{RM} cells lack CCR7 but express other chemokine receptors (for example, CXCR3, CCR9) that allow migration into peripheral tissues such as the dermis or the lamina propria of the intestine. In these sites, T_{RM} cells induce CD69, which

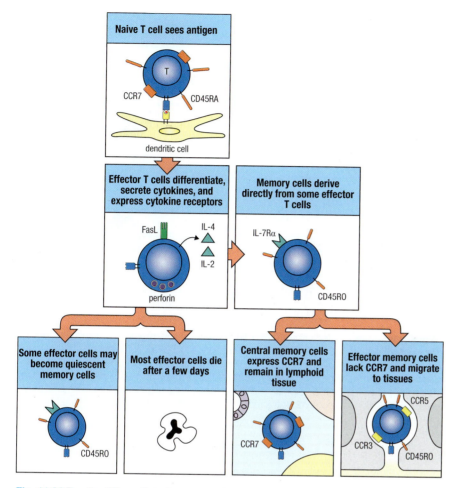

Fig. 11.30 T cells differentiate into central memory and effector memory subsets, which are distinguished by the expression of the chemokine receptor CCR7. Quiescent memory cells bearing the characteristic CD45RO surface protein can arise from activated effector cells (right half of diagram) or directly from activated naive T cells (left half of diagram). Two types of quiescent memory T cells can derive from the primary T-cell response: central memory cells and effector memory cells. Central memory cells express CCR7 and remain in peripheral lymphoid tissues after restimulation. Memory cells of the other type—effector memory cells—mature rapidly into effector T cells after restimulation, and secrete large amounts of IFN-γ, IL-4, and IL-5. They do not express the receptor CCR7, but express receptors (CCR3 and CCR5) for inflammatory chemokines.

reduces S1PR expression, thereby promoting retention in tissues. T_{RM} cells, particularly CD8 T_{RM} cells, enter and reside within the epithelium. TGF-β production by epithelial cells induces T_{RM} cells to express the integrin $α_E$:$β_7$, which binds E-cadherin expressed by epithelium and is required for T_{RM} retention.

The distinction between T_{CM}, T_{EM}, and T_{RM} memory populations has been made both in humans and in the mouse. However, each subset itself is not strictly a homogeneous population. For example, within the CCR7-expressing T_{CM} cells, there are cells with differing expression of other markers, particularly chemokine receptors. A subset of the CCR7-positive T_{CM} cells also express CXCR5, similarly to T_{FH} cells, although it is not yet clear whether these memory cells can provide help to B cells in the germinal center.

On stimulation by antigen, T_{CM} cells rapidly lose expression of CCR7 and differentiate into T_{EM} cells. T_{EM} cells are also heterogeneous in the chemokine receptors they express, and have been classified according to chemokine receptors typical of T_H1 cells (CCR5), T_H17 cells (CCR6), and T_H2 cells (CCR4). Central memory cells do not appear committed to particular effector lineages,

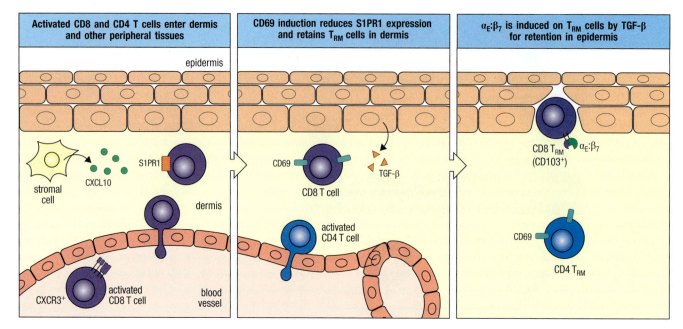

Fig. 11.31 Tissue-resident memory T cells are a major immune compartment that surveys peripheral tissues for reinfection by pathogens. After activation and priming in lymphoid tissues, activated CD8 and CD4 T cells enter the blood and enter tissues in response to various chemokines, as shown here for entry into the dermis, guided by expression of CXCR3. The reexpression of CD69 by T cells caused by antigen or other unknown signals leads to decreased S1PR1 surface expression, promoting retention of these cells in the dermis. In response to TGF-β, some cells induce integrin $\alpha_E:\beta_7$ (CD103), which binds E-cadherin expressed by epithelial cells, promoting entry and retention of T cells in the epidermis, where many CD8 T_{RM} cells reside. Recent estimates indicate that TRM cells may outnumber the recirculating T cells that migrate through the body.

and even effector memory cells are not fully committed to the T_H1, T_H17, or T_H2 lineage, although there is some correlation between their eventual output of T_H1, T_H17, or T_H2 cells and the chemokine receptors expressed. Further stimulation with antigen seems to drive the differentiation of effector memory cells gradually into the distinct effector T-cell lineages.

11-23   CD4 T-cell help is required for CD8 T-cell memory and involves CD40 and IL-2 signaling.

There is experimental evidence that CD4 T cells play an important role in programming optimal CD8 T-cell memory. In the experiment shown in **Fig. 11.32**, the primary and memory CD8 T-cell responses were compared between wild-type mice and mice that lack expression of MHC class II, and which therefore have a defect in CD4 T cells. In this experiment, the CD8 T-cell response was measured against a protein, ovalbumin, carried by an experimental strain of *Listeria monocytogenes*. After 7 days of infection, both types of mice showed equivalent expansion and activity of antigen-specific CD8 effector T cells. But mice with a defect in CD4 T cells generated much weaker secondary responses, characterized by the presence of far fewer expanding memory CD8 T cells after a secondary challenge. These results imply a role for CD4 T cells either in the initial programming of CD8 T cells or during the secondary memory response.

Further experiments suggest that this CD4 T-cell help is necessary for the initial programming of naive CD8 T cells. Memory CD8 T cells that developed in the absence of CD4 help were transferred into wild-type mice. After transfer, the recipient mice were challenged again, whereupon the CD8 T cells showed a reduced ability to proliferate even though the recipient mice expressed MHC class II. This result indicates that CD4 T-cell help is required during the priming of CD8 T cells and not simply at the time of secondary responses.

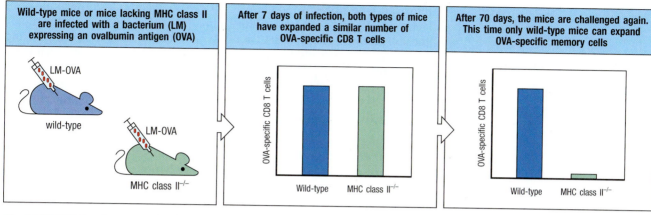

Fig. 11.32 CD4 T cells are required for the development of functional CD8 memory T cells. Mice that do not express MHC class II molecules (MHC II⁻/⁻) fail to develop CD4 T cells. Wild-type and MHC II⁻/⁻ mice were infected with *Listeria monocytogenes* expressing the model antigen ovalbumin (LM-OVA). After 7 days, the number of OVA-specific CD8 T cells can be measured by using specific MHC tetramers that contain an OVA peptide, and therefore bind to T-cell receptors that react with this antigen. After 7 days of infection, mice lacking CD4 T cells were found to have the same number of OVA-specific CD8 T cells as wild-type mice. However, when mice were allowed to recover for 60 days—a time during which memory T cells could develop—and were then re-challenged with LM-OVA, the mice lacking CD4 T cells failed to expand CD8 memory cells specific to OVA, whereas there was a strong CD8 memory response in the wild-type mice.

This requirement for CD4 help in CD8 memory generation has also been demonstrated by experiments in which CD4 T cells were depleted by treatment with antibody or in which mice were deficient in the CD4 gene.

The mechanism underlying this requirement for CD4 T cells is not completely understood. It may involve two types of signals received by the CD8 T cell—those received through CD40 and those received through the IL-2 receptor. CD8 T cells that do not express CD40 are unable to generate memory T cells. Although many cells could potentially express the CD40 ligand needed to stimulate CD40, it is most likely that CD4 T cells are the source of this signal.

The requirement for IL-2 signaling in programming CD8 memory was discovered by using CD8 T cells that were unable to respond to IL-2 because of a genetic deficiency in the IL-2Rα subunit. Because IL-2Rα signaling is required for the development of T_{reg} cells, mice lacking IL-2Rα develop a lymphoproliferative disorder. However, this disorder does not develop in mice that are mixed bone marrow chimeras harboring both wild-type and IL-2Rα-deficient cells, and these chimeras can be used to study the behavior of IL-2Rα-deficient cells. When these chimeric mice were infected with LCMV and their responses were tested, memory CD8 responses were found to be defective specifically in the T cells lacking IL-2Rα.

The experiment shown in **Fig. 11.33** indicates that, distinct from their effect in programming naive CD8 T cells, CD4 T cells also provide help in maintaining the number of CD8 memory T cells. In this case, CD8 memory T cells that had been programmed in normal mice were transferred into immunologically naive mice that either expressed or lacked MHC class II. Transfer of CD8 memory cells into mice lacking MHC class II resulted in a more rapid decrease in the number of memory CD8 T cells in comparison with a similar transfer

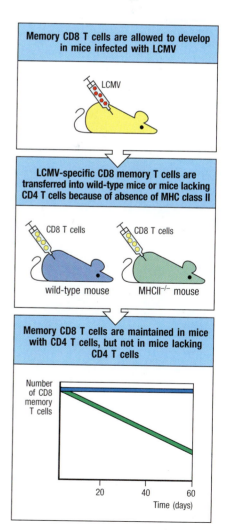

Fig. 11.33 CD4 T cells promote the maintenance of CD8 memory cells. The dependence of memory CD8 T cells on CD4 T cells is shown by the different lifetimes of the memory cells after their transfer into host mice that either have normal CD4 T cells (wild-type) or lack CD4 T cells (MHC II⁻/⁻). In the absence of MHC class II proteins, CD4 T cells fail to develop in the thymus. When CD8 memory T cells specific for LCMV were isolated from donor mice 35 days after infection with the virus and transferred into these hosts, memory cells were maintained only in mice that had CD4 T cells. The basis for this action of CD4 T cells is not yet clear, but has implications for conditions such as HIV/AIDS in which the number of CD4 T cells is diminished.

into wild-type mice. In addition, CD8 effector cells transferred into mice lacking MHC class II had a relative impairment of CD8 effector functions. These experiments imply that CD4 T cells activated by MHC class II–expressing antigen-presenting cells during an immune response have a significant impact on the quantity and quality of the CD8 T-cell response, even when they are not needed for the initial CD8 T-cell activation. CD4 T cells help to program naive CD8 T cells to be able to generate memory T cells, help to promote efficient effector activity, and help to maintain memory T-cell numbers.

11-24   In immune individuals, secondary and subsequent responses are mainly attributable to memory lymphocytes.

In the normal course of an infection, a pathogen proliferates to a level sufficient to elicit an adaptive immune response and then stimulates the production of antibodies and effector T cells that eliminate the pathogen from the body. Most of the effector T cells then die, and antibody levels gradually decline, because the antigens that elicited the response are no longer present at the level needed to sustain it. We can think of this as feedback inhibition of the response. Memory T and B cells remain, however, and maintain a heightened ability to mount a response to a recurrence of infection with the same pathogen.

Hemolytic Disease of the Newborn

Antibody and memory lymphocytes remaining in an immunized individual can have the effect of reducing the activation of naive B and T cells on a subsequent encounter with the same antigen. In fact, passively transferring antibody to a naive recipient can be used to inhibit naive B-cell responses to that same antigen. This phenomenon has been put to practical use to prevent Rh⁻ mothers from making an immune response to an Rh⁺ fetus, which can result in **hemolytic disease of the newborn** (see Appendix I, Section A-6). If anti-Rh antibody is given to the mother before she is first exposed to her child's Rh⁺ red blood cells, her response will be inhibited. The mechanism of this suppression is likely to involve the antibody-mediated clearance and destruction of fetal red blood cells that have entered the mother, thus preventing naive B cells and T cells from mounting an immune response. Presumably, the anti-Rh antibody is in excess over antigen, so that not only is antigen eliminated, but immune complexes are not formed to stimulate naive B cells through Fc receptors. Memory B-cell responses are, however, not inhibited by antibody, so the Rh⁻ mothers at risk must be identified and treated before a primary response has occurred. Because of their high affinity for antigen and alterations in their B-cell receptor signaling requirements, memory B cells are much more sensitive to the small amounts of antigen that cannot be efficiently cleared by the passive anti-Rh antibody. The ability of memory B cells to be activated to produce antibody, even when exposed to preexisting antibody, also allows secondary antibody responses to occur in individuals who are already immune.

These suppressive mechanisms might also explain a phenomenon called **original antigenic sin**. This term was coined to describe the tendency of people to make antibodies only against the epitopes expressed on the first influenza virus variant to which they are exposed, even in subsequent infections with viral variants that have additional, highly immunogenic epitopes (**Fig. 11.34**). Antibodies against the original virus will tend to suppress responses of naive B cells specific for the new epitopes. This might benefit the host by using only those B cells that can respond most rapidly and effectively to the virus. This pattern is broken only if the person is exposed to an influenza virus that lacks all epitopes seen in the original infection, because now no preexisting antibodies bind the virus, and naive B cells are able to respond.

A similar suppression of naive T-cell responses by antigen-specific memory T cells can occur in the setting of infection by lymphocytic choriomeningitis

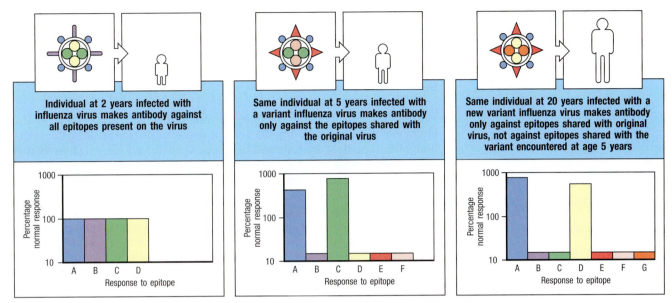

Fig. 11.34 When individuals who have been infected with one variant of influenza virus are infected with a second or third variant, they make antibodies only against epitopes that were present on the initial virus. A child infected for the first time with an influenza virus at 2 years of age makes a response to all epitopes (left panel). At age 5 years, the same child exposed to a different influenza virus responds preferentially to those epitopes shared with the original virus, and makes a smaller than normal response to new epitopes on the virus (center panel). Even at age 20 years, this commitment to respond to epitopes shared with the original virus, and the subnormal response to new epitopes, is retained (right panel). This phenomenon is called 'original antigenic sin.'

virus (LCMV) in the mouse or dengue virus in humans. Mice that were primed with one strain of LCMV responded to a subsequent infection with a second strain of LCMV by expanding CD8 T cells that reacted to antigens specific for the first strain. However, this type of effect was not observed when responses to variable ovalbumin antigenic epitopes were examined in the setting of recurrent infections using the bacterial pathogen *Listeria monocytogenes*, suggesting that the suppression caused by 'original antigenic sin' does not occur in all immune responses.

Summary.

Protective immunity against reinfection is one of the most important consequences of adaptive immunity, and results from the establishment of populations of long-lived memory B cells and memory T cells. These antigen-specific memory cells arise from the populations of lymphocytes that expand dramatically during the primary infection, and that survive at higher frequencies than in the naive lymphocyte repertoires. Both their increased frequency and their capacity to respond more rapidly to restimulation to the same antigen contribute to protective immunity, which can be transferred to naive recipients by memory B and T cells. Memory lymphocytes are maintained by their expression of receptors for cytokines, such as IL-7 and IL-15, that provide survival signals. Memory B cells can be distinguished by changes in their immunoglobulin genes because of isotype switching and somatic hypermutation, and secondary and subsequent immune responses are characterized by antibodies with increasing affinity for the antigen. The advent of receptor-specific reagents—MHC tetramers—has allowed for the direct analysis of the expansion and differentiation of effector and memory T cells. We now recognize that T-cell memory is complex, and memory T cells are quite heterogeneous, having central memory, effector memory, and tissue-resident memory subtypes. While CD8 T cells can generate effective primary responses in the absence of help from CD4 T cells, it is becoming clear that CD4 T cells have an integral role in regulating CD8 T-cell memory. These issues will be critical in understanding, for example, how to design effective vaccines for diseases such as HIV/AIDS.

Summary to Chapter 11.

Vertebrates resist infection by pathogenic microorganisms in several ways. The innate defenses can act immediately and may succeed in repelling the infection, but if not, they are followed by a series of induced early responses that help to contain the infection as adaptive immunity develops. These first two phases of the immune response rely on recognizing the presence of infection by using the nonclonotypic receptors of the innate immune system. They are summarized in Fig. 11.35 and covered in detail in Chapter 3. Several specialized subsets of immune cells, which can be viewed as intermediates between innate and adaptive immunity, act next and include the innate lymphoid cells, or ILCs, which are rapid responders to cytokines produced by innate sensor cells and can help to bias the CD4 T-cell response toward parallel subsets of effector T cells; and NK cells, which can be recruited to lymph nodes and secrete IFN-γ, and thus promote a T_H1 response. The third phase of an immune response is the adaptive immune response (see Fig. 11.35), which is mounted in the peripheral lymphoid tissue that serves the particular site of infection and takes several days to develop, because T and B lymphocytes must encounter their specific antigen, proliferate, and differentiate into effector cells. T-cell dependent B-cell responses cannot be initiated until antigen-specific T_{FH} cells have had a chance to proliferate and differentiate. Once an adaptive immune response has occurred, the antibodies and effector T cells are dispersed via the circulation and recruited into the infected tissues; the infection is usually controlled and the pathogen is contained or eliminated. The final effector mechanisms used to clear an infection depend on the type of infectious agent, and in most cases they are the same as those employed in the early phases of immune defense; only the recognition mechanism changes and is now more selective (see Fig. 11.35).

Fig. 11.35 The components of the three phases of the immune response against different classes of microorganisms. The mechanisms of innate immunity that operate in the first two phases of the immune response are described in Chapters 2 and 3, and thymus-independent (T-independent) B-cell responses are covered in Chapter 10. The early phases contribute to the initiation of adaptive immunity, and they influence the functional character of the antigen-specific effector T cells and antibodies that appear on the scene in the late phase of the response. There are striking similarities in the effector mechanisms at each phase of the response; the main change is in the recognition structure used.

	Phases of the immune response		
	Immediate (0–4 hours)	**Early (4–96 hours)**	**Late (96–100 hours)**
	Nonspecific Innate No memory No specific T cells	Nonspecific + specific Inducible No memory No specific T cells	Specific Inducible Memory Specific T cells
Barrier functions	Skin, epithelia, mucins, acid	Local inflammation (C5a) Local TNF-α	IgA antibody in luminal spaces IgE antibody on mast cells Local inflammation
Response to extracellular pathogens	Phagocytes Alternative and MBL complement pathway Lysozyme Lactoferrin Peroxidase Defensins	Mannan-binding lectin C-reactive protein T-independent B-cell antibody Complement	IgG antibody and Fc receptor-bearing cells IgG, IgM antibody + classical complement pathway
Response to intracellular bacteria	Macrophages	Activated NK-dependent macrophage activation IL-1, IL-6, TNF-α, IL-12	T-cell activation of macrophages by IFN-γ
Response to virus-infected cells	Natural killer (NK) cells	IFN-α and IFN-β IL-12-activated NK cells	Cytotoxic T cells IFN-γ

An effective adaptive immune response leads to a state of protective immunity. This state consists of the presence of effector cells and molecules produced in the initial response, and of immunological memory. Immunological memory is manifested as a heightened ability to respond to pathogens that have previously been encountered and successfully eliminated. Memory T and B lymphocytes have the property of being able to transfer immune memory to naive recipients. The mechanisms that maintain immunological memory include certain cytokines, such as IL-7 and IL-15, as well as homeostatic interactions between the T-cell receptors on memory cells with self peptide:self MHC complexes. The artificial induction of protective immunity, which includes immunological memory, by vaccination is the most outstanding accomplishment of immunology in the field of medicine. The understanding of how this is accomplished is now catching up with its practical success. However, as we will see in Chapter 13, many pathogens do not induce protective immunity that completely eliminates the pathogen, so we will need to learn what prevents this before we can prepare effective vaccines against these pathogens.

Questions.

11.1 True or False: The immune response is a dynamic process that initiates with an antigen-independent response, which becomes more focused and powerful as it develops antigen specificity. Once the adaptive immune system develops, a single type of response is capable of eliminating any type of pathogen.

11.2 Multiple Choice: Which statement is incorrect?

A. IL-12 and IL-18 production by macrophages and dendritic cells induces IFN-γ secretion by ILC1 to induce heightened killing of intracellular pathogens.

B. ILC3s are activated by thymic stromal lymphopoietin (TSLP), which activates STAT5 and induces IL-17 production.

C. Molecular patterns common to helminths activate IL-33 and IL-25 production, which in turn activates ILC2s to induce mucus production by goblet cells and mucosal smooth muscle contraction.

D. ILC3-derived IL-22 acts on epithelial cells to induce production of antimicrobial peptides and promotes an enhanced barrier integrity.

11.3 Matching: Match the following proteins with their effect on T-cell migration.

A. CXCR5 _____   **i.** Interacts with P- and E-selectin, expressed at activated endothelial cells

B. PSGL-1 _____   **ii.** CXCL13 binding attracts T$_{FH}$ cells to the B-cell follicle

C. FucT-VII _____   **iii.** Interacts with VCAM-1 to initiate extravasation of the effector T cells

D. VLA-4 _____   **iv.** Necessary for the production of P- and E-selectin

11.4 Fill-in-the-Blanks: Expression of selective adhesion molecules among effector T cells helps compartmentalize their distribution. For example, T cells that are primed in the GALT induce the expression of the _____ integrin, which binds to _____, which is constitutively expressed by the gut mucosal endothelial cells. These T cells also express the chemokine receptor _____, which attracts T cells to the lamina propria subjacent to the small intestine epithelium via a _____ gradient. This compartmentalization capacity is not unique to the gut and can be observed in other organs such as the skin. For example, expression of a glycosylated form of PSGL-1, _____, binds _____ on cutaneous vascular endothelium.

11.5 Multiple Choice: Which of the following statements incorrectly describes T$_H$1 macrophage activation?

A. CD40 ligand sensitizes the macrophage to respond to IFN-γ

B. LT-α can substitute for CD40 ligand in macrophage activation

C. TNFR-I activation is antagonized by activated T$_H$1 cells

D. Macrophages are sensitized to IFN-γ by small amounts of bacterial LPS

11.6 Short Answer: How do M2 macrophages stimulate collagen production in order to promote tissue repair?

11.7 Multiple Choice: Which of the following is not true concerning type 3 responses?

A. The primary innate effector cells are neutrophils, which are recruited by CXCL8 and CXCL2 and have an increased output due to G-CSF and GM-CSF

B. At homeostasis, T$_H$17 cells are present almost exclusively in the intestinal mucosa

C. IL-17 is the central cytokine

D. IL-22 is produced to induce antimicrobial peptide production and epithelial cell proliferation and shedding of natural killer cells

E. IL-23 initiates the commitment of naive CD4$^+$ T cells to the T$_H$17 fate

11.8 Multiple Choice: Which of the following pathogens is able to induce a robust CD8$^+$ T-cell response independent of CD4$^+$ T-cell help?

A. *Streptococcus pneumoniae*

B. Lymphocytic choriomeningitis virus (LCMV)

C. *Listeria monocytogenes*

D. *Staphylococcus aureus*

E. *Salmonella*

F. *Toxoplasma*

11.9 Fill-in-the-Blanks: During an immune response to a pathogen, activated T cells express _____, a high-affinity IL-2 receptor component, and lose expression of _____, an IL-7 receptor component. The activated cells also generate different isoforms of _____, a protein tyrosine phosphatase expressed by all hematopoietic cells. Effector and central memory cells develop, and these can be distinguished by high expression of _____ in the former and _____ in the latter. The survival of both CD4$^+$ and CD8$^+$ memory T cells is dependent on _____, although the survival of CD8$^+$ memory T cells is additionally dependent on _____.

11.10 True or False: CD27 is a marker of naive B cells as well as memory T cells.

11.11 Short Answer: How is inflammasome activation able to help induce type 1 and type 3 responses while blunting type 2 responses?

11.12 Matching: Match the cytokine to the downstream STAT.

A. ___ IL-4 and IL-13   **i.** STAT3

B. ___ IL-12   **ii.** STAT4

C. ___ IL-23   **iii.** STAT5

D. ___ TSLP, IL-2, and IL-7   **iv.** STAT6

Section references.

11-1   The course of an infection can be divided into several distinct phases.

Mandell, G., Bennett, J., and Dolin, R. (eds): **Principles and Practice of Infectious Diseases**, 5th ed. New York, Churchill Livingstone, 2000.

Zhang, S.Y., Jouanguy, E., Sancho-Shimizu, V., von Bernuth, H., Yang, K., Abel, L., Picard, C., Puel, A., and Casanova, J.L.: **Human Toll-like receptor-dependent induction of interferons in protective immunity to viruses.** *Immunol. Rev.* 2007, **220**:225–236.

11-2   The effector mechanisms that are recruited to clear an infection depend on the infectious agent.

Bernink, J., Mjösberg, J., and Spits, H.: **Th1- and Th2-like subsets of innate lymphoid cells.** *Immunol. Rev.* 2013, **252**:133–138.

Fearon, D.T., and Locksley, R.M.: **The instructive role of innate immunity in the acquired immune response.** *Science* 1996, **272**:50–53.

Gasteiger, G., and Rudensky, A.Y.: **Interactions between innate and adaptive lymphocytes.** *Nat. Rev. Immunol.* 2014, **14**:631–639.

Janeway, C.A., Jr: **The immune system evolved to discriminate infectious nonself from noninfectious self.** *Immunol. Today* 1992, **13**:11–16.

McKenzie, A.N.J., Spits, H., and Eberl, G.: **Innate lymphoid cells in inflammation and immunity.** *Immunity* 2014, **41**:366–374.

Neill, D.R., Wong, S.H., Bellosi, A., Flynn, R.J., Daly, M., Langford, T.K., Bucks, C., Kane, C.M., Fallon, P.G., Pannell, R., *et al.*: **Nuocytes represent a new innate effector leukocyte that mediates type-2 immunity.** *Nature* 2010, **464**:1367–1370.

Oliphant, C.J., Hwang, Y.Y., Walker, J.A., Salimi, M., Wong, S.H., Brewer, J.M., Englezakis, A., Barlow, J.L., Hams, E., Scanlon, S.T., *et al.*: **MHCII-mediated dialog between group 2 innate lymphoid cells and CD4$^+$ T cells potentiates type 2 immunity and promotes parasitic helminth expulsion.** Immunity 2014, **41**:283–295.

Walker, J.A., Barlow, J.L., and McKenzie, A.N.J.: **Innate lymphoid cells—how did we miss them?** *Nat. Rev. Immunol.* 2013, **13**:75–87.

11-3   Effector T cells are guided to specific tissues and sites of infection by changes in their expression of adhesion molecules and chemokine receptors.

Griffith, J.W., Sokol, C.L., and Luster, A.D.: **Chemokines and chemokine receptors: positioning cells for host defense and immunity.** *Annu. Rev. Immunol.* 2014, **32**:659–702.

Hidalgo, A., Peired, A.J., Wild, M.K., Vestweber, D., and Frenette, P.S.: **Complete identification of E-selectin ligands on neutrophils reveals distinct functions of PSGL-1, ESL-1, and CD44.** *Immunity* 2007, **26**:477–489.

MacKay, C.R., Marston, W., and Dudler, L.: **Altered patterns of T-cell migration through lymph nodes and skin following antigen challenge.** *Eur. J. Immunol.* 1992, **22**:2205–2210.

Mantovani, A., Bonecchi, R., and Locati, M.: **Tuning inflammation and immunity by chemokine sequestration: decoys and more.** *Nat. Rev. Immunol.* 2006, **6**:907–918.

Mueller, S.N., Gebhardt, T., Carbone, F.R., and Heath, W.R.: **Memory T cell subsets, migration patterns, and tissue residence.** *Annu. Rev. Immunol.* 2013, **31**:137–161.

Sallusto, F., Kremmer, E., Palermo, B., Hoy, A., Ponath, P., Qin, S., Forster, R., Lipp, M., and Lanzavecchia, A.: **Switch in chemokine receptor expression upon TCR stimulation reveals novel homing potential for recently activated T cells.** *Eur. J. Immunol.* 1999, **29**:2037–2045.

11-4   Pathogen-specific effector T cells are enriched at sites of infection as adaptive immunity progresses.

Jenkins, M.K., Khoruts, A., Ingulli, E., Mueller, D.L., McSorley, S.J., Reinhardt, R.L., Itano, A., and Pape, K.A.: **In vivo activation of antigen-specific CD4 T cells.** *Annu. Rev. Immunol.* 2001, **19**:23–45.

11-5 T_H1 cells coordinate and amplify the host response to intracellular pathogens through classical activation of macrophages.

Bekker, L.G., Freeman, S., Murray, P.J., Ryffel, B., and Kaplan, G.: **TNF-α controls intracellular mycobacterial growth by both inducible nitric oxide synthase-dependent and inducible nitric oxide synthase-independent pathways.** *J. Immunol.* 2001, **166**:6728–6734.

Ehlers, S., Kutsch, S., Ehlers, E.M., Benini, J., and Pfeffer, K.: **Lethal granuloma disintegration in mycobacteria-infected TNFRp55^{−/−} mice is dependent on T cells and IL-12.** *J. Immunol.* 2000, **165**:483–492.

Hsieh, C.S., Macatonia, S.E., Tripp, C.S., Wolf, S.F., O'Garra, A., and Murphy, K.M.: **Development of T_H1 CD4⁺ T cells through IL-12 produced by *Listeria*-induced macrophages.** *Science* 1993, **260**:547–549.

Muñoz-Fernández, M.A., Fernández, M.A., and Fresno, M.: **Synergism between tumor necrosis factor-α and interferon-γ on macrophage activation for the killing of intracellular *Trypanosoma cruzi* through a nitric oxide-dependent mechanism.** *Eur. J. Immunol.* 1992, **22**:301–307.

Murray, P.J., and Wynn, T.A.: **Protective and pathogenic functions of macrophage subsets.** *Nat. Rev. Immunol.* 2011, **11**:723–737.

Stout, R.D., Suttles, J., Xu, J., Grewal, I.S., and Flavell, R.A.: **Impaired T cell-mediated macrophage activation in CD40 ligand-deficient mice.** *J. Immunol.* 1996, **156**:8–11.

11-6 Activation of macrophages by T_H1 cells must be tightly regulated to avoid tissue damage.

Duffield, J.S.: **The inflammatory macrophage: a story of Jekyll and Hyde.** *Clin. Sci.* 2003, **104**:27–38.

Labow, R.S., Meek, E., and Santerre, J.P.: **Model systems to assess the destructive potential of human neutrophils and monocyte-derived macrophages during the acute and chronic phases of inflammation.** *J. Biomed. Mater. Res.* 2001, **54**:189–197.

Wigginton, J.E., and Kirschner, D.: **A model to predict cell-mediated immune regulatory mechanisms during human infection with Mycobacterium tuberculosis.** *J. Immunol.* 2001, **166**:1951–1967.

11-7 Chronic activation of macrophages by T_H1 cells mediates the formation of granulomas to contain intracellular pathogens that cannot be cleared.

James, D.G.: **A clinicopathological classification of granulomatous disorders.** *Postgrad. Med. J.* 2000, **76**:457–465.

11-8 Defects in type 1 immunity reveal its important role in the elimination of intracellular pathogens.

Berberich, C., Ramirez-Pineda, J.R., Hambrecht, C., Alber, G., Skeiky, Y.A., and Moll, H.: **Dendritic cell (DC)-based protection against an intracellular pathogen is dependent upon DC-derived IL-12 and can be induced by molecularly defined antigens.** *J. Immunol.* 2003, **170**:3171–3179.

Biedermann, T., Zimmermann, S., Himmelrich, H., Gumy, A., Egeter, O., Sakrauski, A.K., Seegmuller, I., Voigt, H., Launois, P., Levine, A.D., *et al.*: **IL-4 instructs T_H1 responses and resistance to *Leishmania major* in susceptible BALB/c mice.** *Nat. Immunol.* 2001, **2**:1054–1060.

Neighbors, M., Xu, X., Barrat, F.J., Ruuls, S.R., Churakova, T., Debets, R., Bazan, J.F., Kastelein, R.A., Abrams, J.S., and O'Garra, A.: **A critical role for interleukin 18 in primary and memory effector responses to *Listeria monocytogenes* that extends beyond its effects on interferon gamma production.** *J. Exp. Med.* 2001, **194**:343–354.

11-9 T_H2 cells coordinate type 2 responses to expel intestinal helminths and repair tissue injury.

Artis, D., and Grencis, R.K.: **The intestinal epithelium: sensors to effectors in nematode infection.** *Mucosal Immunol.* 2008, **1**:252–264.

Fallon, P.G., Ballantyne, S.J., Mangan, N.E., Barlow, J.L., Dasvarma, A., Hewett, D.R., McIlgorm, A., Jolin, H.E., and McKenzie, A.N.J.: **Identification of an interleukin (IL)-25-dependent cell population that provides IL-4, IL-5, and IL-13 at the onset of helminth expulsion.** *J. Exp. Med.* 2006, **203**:1105–1116.

Finkelman, F.D., Shea-Donohue, T., Goldhill, J., Sullivan, C.A., Morris, S.C., Madden, K.B., Gauser, W.C., and Urban, J.F., Jr: **Cytokine regulation of host defense against parasitic intestinal nematodes.** *Annu. Rev. Immunol.* 1997, **15**:505–533.

Humphreys, N.E., Xu, D., Hepworth, M.R., Liew, F.Y., and Grencis, R.K.: **IL-33, a potent inducer of adaptive immunity to intestinal nematodes.** *J. Immunol.* 2008, **180**:2443–2449.

Liang, H.-E., Reinhardt, R.L., Bando, J.K., Sullivan, B.M., Ho, I.-C., and Locksley, R.M.: **Divergent expression patterns of IL-4 and IL-13 define unique functions in allergic immunity.** *Nat. Immunol.* 2012, **13**:58–66.

Maizels, R.M., Pearce, E.J., Artis, D., Yazdanbakhsh, M., and Wynn, T.A.: **Regulation of pathogenesis and immunity in helminth infections.** *J. Exp. Med.* 2009, **206**:2059–2066.

Ohnmacht, C., Schwartz, C., Panzer, M., Schiedewitz, I., Naumann, R., and Voehringer, D.: **Basophils orchestrate chronic allergic dermatitis and protective immunity against helminths.** *Immunity* 2010, **33**:364–374.

Saenz, S.A., Noti, M., and Artis, D.: **Innate immune cell populations function as initiators and effectors in Th2 cytokine responses.** *Trends Immunol.* 2010, **31**:407–413.

Sullivan, B.M., and Locksley, R.M.: **Basophils: a nonredundant contributor to host immunity.** *Immunity* 2009, **30**:12–20.

Van Dyken, S.J., and Locksley, R.M.: **Interleukin-4- and interleukin-13-mediated alternatively activated macrophages: roles in homeostasis and disease.** *Annu. Rev. Immunol.* 2013, **31**:317–343.

11-10 T_H17 cells coordinate type 3 responses to enhance the clearance of extracellular bacteria and fungi.

Aujla, S.J., Chan, Y.R., Zheng, M., Fei, M., Askew, D.J., Pociask, D.A., Reinhart, T.A., Mcallister, F., Edeal, J., Gaus, K., *et al.*: **IL-22 mediates mucosal host defense against Gram-negative bacterial pneumonia.** *Nat. Med.* 2008, **14**:275–281.

Fossiez, F., Djossou, O., Chomarat, P., Flores-Romo, L., Ait-Yahia, S., Maat, C., Pin, J.J., Garrone, P., Garcia, E., Saeland, S., *et al.*: **T cell interleukin-17 induces stromal cells to produce proinflammatory and hematopoietic cytokines.** *J. Exp. Med.* 1996, **183**:2593–2603.

Happel, K.I., Zheng, M., Young, E., Quinton, L.J., Lockhart, E., Ramsay, A.J., Shellito, J.E., Schurr, J.R., Bagby, G.J., Nelson, S., *et al.*: **Cutting edge: roles of Toll-like receptor 4 and IL-23 in IL-17 expression in response to *Klebsiella pneumoniae* infection.** *J. Immunol.* 2003, **170**:4432–4436.

LeibundGut-Landmann, S., Gross, O., Robinson, M.J., Osorio, F., Slack, E.C., Tsoni, S.V., Schweighoffer, E., Tybulewicz, V., Brown, G.D., Ruland, J., *et al.*: **Syk- and CARD9-dependent coupling of innate immunity to the induction of T helper cells that produce interleukin 17.** *Nat. Immunol.* 2007, **8**:630–638.

Ouyang, W., Kolls, J.K., and Zheng, Y.: **The biological functions of T helper 17 cell effector cytokines in inflammation.** *Immunity* 2008, **28**:454–467.

Romani, L.: **Immunity to fungal infections.** *Nat. Rev. Immunol.* 2011, **11**:275–288.

Sonnenberg, G.F., Monticelli, L.A., Elloso, M.M., Fouser, L.A., and Artis, D.: **CD4⁺ lymphoid tissue-inducer cells promote innate immunity in the gut.** *Immunity* 2011, **34**:122–134.

Zheng, Y., Valdez, P.A., Danilenko, D.M., Hu, Y., Sa, S.M., Gong, Q., Abbas, A.R., Modrusan, Z., Ghilardi, N., De Sauvage, F.J., *et al.*: **Interleukin-22 mediates early host defense against attaching and effacing bacterial pathogens.** *Nat. Med.* 2008, **14**:282–289.

11-11 Differentiated effector T cells continue to respond to signals as they carry out their effector functions.

Cua, D.J., Sherlock, J., Chen, Y., Murphy, C.A., Joyce, B., Seymour, B., Lucian, L., To, W., Kwan, S., Churakova, T., *et al.*: **Interleukin-23 rather than interleukin-12 is** the critical cytokine for autoimmune inflammation of the brain. *Nature* 2003, **421**:744–748.

Ghilardi, N., Kljavin, N., Chen, Q., Lucas, S., Gurney, A.L., and De Sauvage, F.J.: **Compromised humoral and delayed-type hypersensitivity responses in IL-23-deficient mice.** *J. Immunol.* 2004, **172**:2827–2833.

Park, A.Y., Hondowics, B.D., and Scott, P.: **IL-12 is required to maintain a Th1 response during *Leishmania major* infection.** *J. Immunol.* 2000, **165**:896–902.

Yap, G., Pesin, M., and Sher, A.: **Cutting edge: IL-12 is required for the maintenance of IFN-γ production in T cells mediating chronic resistance to the intracellular pathogen *Toxoplasma gondii*.** *J. Immunol.* 2000, **165**:628–631.

11-12 Effector T cells can be activated to release cytokines independently of antigen recognition.

Guo, L., Junttila, I.S., and Paul, W.E.: **Cytokine-induced cytokine production by conventional and innate lymphoid cells.** *Trends Immunol.* 2012, **33**:598–606.

Kohno, K., Kataoka, J., Ohtsuki, T., Suemoto, Y., Okamoto, I., Usui, M., Ikeda, M., and Kurimoto, M.: **IFN-gamma-inducing factor (IGIF) is a costimulatory factor on the activation of Th1 but not Th2 cells and exerts its effect independently of IL-12.** *J. Immunol.* 1997, **158**:1541–1550.

11-13 Effector T cells demonstrate plasticity and cooperativity that enable adaptation during anti-pathogen responses.

Basu, R., Hatton, R.D., and Weaver, C.T.: **The Th17 family: flexibility follows function.** *Immunol. Rev.* 2013, **252**:89–103.

Lee, S.-J., McLachlan, J.B., Kurtz, J.R., Fan, D., Winter, S.E., Bäumler, A.J., Jenkins, M.K., and McSorley, S.J.: **Temporal expression of bacterial proteins instructs host CD4 T cell expansion and Th17 development.** *PLoS Pathog.* 2012, **8**:e1002499.

Lee, Y.K., Turner, H., Maynard, C.L., Oliver, J.R., Chen, D., Elson, C.O., and Weaver, C.T.: **Late developmental plasticity in the T helper 17 lineage.** *Immunity* 2009, **30**:92–107.

Murphy, K.M., and Stockinger, B.: **Effector T cell plasticity: flexibility in the face of changing circumstances.** *Nat. Immunol.* 2010, **11**:674–680.

O'Shea, J.J., and Paul, W.E.: **Mechanisms underlying lineage commitment and plasticity of helper CD4⁺ T cells.** *Science* 2010, **327**:1098–1102.

11-14 Integration of cell- and antibody-mediated immunity is critical for protection against many types of pathogens.

Baize, S., Leroy, E.M., Georges-Courbot, M.C., Capron, M., Lansoud-Soukate, J., Debre, P., Fisher-Hoch, S.P., McCormick, J.B., and Georges, A.J.: **Defective humoral responses and extensive intravascular apoptosis are associated with fatal outcome in Ebola virus-infected patients.** *Nat. Med.* 1999, **5**:423–426.

11-15 Primary CD8 T-cell responses to pathogens can occur in the absence of CD4 T-cell help.

Lertmemongkolchai, G., Cai, G., Hunter, C.A., and Bancroft, G.J.: **Bystander activation of CD8 T cells contributes to the rapid production of IFN-γ in response to bacterial pathogens.** *J. Immunol.* 2001, **166**:1097–1105.

Rahemtulla, A., Fung-Leung, W.P., Schilham, M.W., Kundig, T.M., Sambhara, S.R., Narendran, A., Arabian, A., Wakeham, A., Paige, C.J., Zinkernagel, R.M., *et al.*: **Normal development and function of CD8⁺ cells but markedly decreased helper cell activity in mice lacking CD4.** *Nature* 1991, **353**:180–184.

Schoenberger, S.P., Toes, R.E., van der Voort, E.I., Offringa, R., and Melief, C.J.: **T-cell help for cytotoxic T lymphocytes is mediated by CD40–CD40L interactions.** *Nature* 1998, **393**:480–483.

Sun, J.C., and Bevan, M.J.: **Defective CD8 T cell memory following acute infection without CD4 T-cell help.** *Science* 2003, **300**:339–349.

11-16 Resolution of an infection is accompanied by the death of most of the effector cells and the generation of memory cells.

Bouillet, P., and O'Reilly, L.A.: **CD95, BIM and T cell homeostasis.** *Nat. Rev. Immunol.* 2009, **9**:514-519.

Chowdhury, D., and Lieberman, J.: **Death by a thousand cuts: granzyme pathways of programmed cell death.** *Annu. Rev. Immunol.* 2008, **26**:389–420.

Siegel, R.M.: **Caspases at the crossroads of immune-cell life and death.** *Nat. Rev. Immunol.* 2006, **6**:308–317.

Strasser, A.: **The role of BH3-only proteins in the immune system.** *Nat. Rev. Immunol.* 2005, **5**:189–200.

11-17 Immunological memory is long lived after infection or vaccination.

Black, F.L., and Rosen, L.: **Patterns of measles antibodies in residents of Tahiti and their stability in the absence of re-exposure.** *J. Immunol.* 1962, **88**:725–731.

Hammarlund E., Lewis, M.W., Hanifin, J.M., Mori, M., Koudelka, C.W., and Slifka, M.K.: **Antiviral immunity following smallpox virus infection: a case-control study.** *J Virol.* 2010, **84**:12754–60.

Hammarlund, E., Lewis, M.W., Hansen, S.G., Strelow, L.I., Nelson, J.A., Sexton, G.J., Hanifin, J.M., and Slifka, M.K.: **Duration of antiviral immunity after smallpox vaccination.** *Nat. Med.* 2003, **9**:1131–1137.

MacDonald, H.R., Cerottini, J.C., Ryser, J.E., Maryanski, J.L., Taswell, C., Widmer, M.B., and Brunner, K.T.: **Quantitation and cloning of cytolytic T lymphocytes and their precursors.** *Immunol. Rev.* 1980, **51**:93–123.

Murali-Krishna, K., Lau, L.L., Sambhara, S., Lemonnier, F., Altman, J., and Ahmed, R.: **Persistence of memory CD8 T cells in MHC class I-deficient mice.** *Science* 1999, **286**:1377–1381.

Seddon, B., Tomlinson, P., and Zamoyska, R.: **Interleukin 7 and T cell receptor signals regulate homeostasis of CD4 memory cells.** *Nat. Immunol.* 2003, **4**:680–686.

11-18 Memory B-cell responses are more rapid and have higher affinity for antigen compared with responses of naive B cells.

Andersson, B.: **Studies on the regulation of avidity at the level of the single antibody-forming cell: The effect of antigen dose and time after immunization.** *J. Exp. Med.* 1970, **132**:77–88.

Berek, C., and Milstein, C.: **Mutation drift and repertoire shift in the maturation of the immune response.** *Immunol. Rev.* 1987, **96**:23–41.

Bergmann, B., Grimsholm, O., Thorarinsdottir, K., Ren, W., Jirholt, P., Gjertsson, I., and Mårtensson, I.L.: **Memory B cells in mouse models.** *Scand. J. Immunol.* 2013, **78**:149–156.

Davie, J.M., and Paul, W.E.: **Receptors on immunocompetent cells. V. Cellular correlates of the "maturation" of the immune response.** *J. Exp. Med.* 1972, **135**:660–674.

Eisen, H.N., and Siskind, G.W.: **Variations in affinities of antibodies during the immune response.** *Biochemistry* 1964, **3**:996–1008.

Good-Jacobson, K.L., and Tarlinton, D.M.: **Multiple routes to B-cell memory.** *Int. Immunol.* 2012, **24**:403–408.

Klein, U., Rajewsky, K., and Küppers, R.: **Human immunoglobulin (Ig)M+IgD+ peripheral blood B cells expressing the CD27 cell surface antigen carry somatically mutated variable region genes: CD27 as a general marker for somatically mutated (memory) B cells.** *J. Exp. Med.* 1998, **188**:1679–1689.

Takemori, T., Kaji, T., Takahashi, Y., Shimoda, M., and Rajewsky, K.: **Generation of memory B cells inside and outside germinal centers.** *Eur. J. Immunol.* 2014, **44**:1258–1264.

11-19 Memory B cells can reenter germinal centers and undergo additional somatic hypermutation and affinity maturation during secondary immune responses.

Bende, R.J., van Maldegem, F., Triesscheijn, M., Wormhoudt, T.A., Guijt, R., and van Noesel, C.J.: **Germinal centers in human lymph nodes contain reactivated memory B cells.** *J. Exp. Med.* 2007, **204**:2655–2665.

Dal Porto, J.M., Haberman, A.M., Kelsoe, G., and Shlomchik, M.J.: **Very low affinity B cells form germinal centers, become memory B cells, and participate in secondary immune responses when higher affinity competition is reduced.** *J. Exp. Med.* 2002, **195**:1215–1221.

Goins, C.L., Chappell, C.P., Shashidharamurthy, R., Selvaraj, P., and Jacob, J.: **Immune complex-mediated enhancement of secondary antibody responses.** *J. Immunol.* 2010, **184**:6293–6298.

Kaji, T., Furukawa, K., Ishige, A., Toyokura, I., Nomura, M., Okada, M., Takahashi, Y., Shimoda, M., and Takemori, T.: **Both mutated and unmutated memory B cells accumulate mutations in the course of the secondary response and develop a new antibody repertoire optimally adapted to the secondary stimulus.** *Int. Immunol.* 2013, **25**:683–695.

11-20 MHC tetramers identify memory T cells that persist at an increased in frequency relative to their frequency as naive T cells.

Hataye, J., Moon, J.J., Khoruts, A., Reilly, C., and Jenkins, M.K.: **Naive and memory CD4+ T cell survival controlled by clonal abundance.** *Science* 2006, **312**:114–116.

Pagán, A.J., Pepper, M., Chu, H.H., Green, J.M., and Jenkins, M.K.: **CD28 promotes CD4+ T cell clonal expansion during infection independently of its YMNM and PYAP motifs.** *J. Immunol.* 2012, **189**:2909–2917.

Pepper, M., Pagán, A.J., Igyártó, B.Z., Taylor, J.J., and Jenkins, M.K.: **Opposing signals from the Bcl6 transcription factor and the interleukin-2 receptor generate T helper 1 central and effector memory cells.** *Immunity* 2011, **35**:583–595.

11-21 Memory T cells arise from effector T cells that maintain sensitivity to IL-7 or IL-15.

Akondy, R.S., Monson, N.D., Miller, J.D., Edupuganti, S., Teuwen, D., Wu, H., Quyyumi, F., Garg, S., Altman, J.D., Del Rio, C., *et al.*: **The yellow fever virus vaccine induces a broad and polyfunctional human memory CD8+ T cell response.** *J. Immunol.* 2009, **183**:7919–7930.

Bradley, L.M., Atkins, G.G., and Swain, S.L.: **Long-term CD4+ memory T cells from the spleen lack MEL-14, the lymph node homing receptor.** *J. Immunol.* 1992, **148**:324–331.

Kaech, S.M., Hemby, S., Kersh, E., and Ahmed, R.: **Molecular and functional profiling of memory CD8 T cell differentiation.** *Cell* 2002, **111**:837–851.

Kassiotis, G., Garcia, S., Simpson, E., and Stockinger, B.: **Impairment of immunological memory in the absence of MHC despite survival of memory T cells.** *Nat. Immunol.* 2002, **3**:244–250.

Ku, C.C., Murakami, M., Sakamoto, A., Kappler, J., and Marrack, P.: **Control of homeostasis of CD8+ memory T cells by opposing cytokines.** *Science* 2000, **288**:675–678.

Rogers, P.R., Dubey, C., and Swain, S.L.: **Qualitative changes accompany memory T cell generation: faster, more effective responses at lower doses of antigen.** *J. Immunol.* 2000, **164**:2338–2346.

Wherry, E.J., Teichgraber, V., Becker, T.C., Masopust, D., Kaech, S.M., Antia, R., von Andrian, U.H., and Ahmed, R.: **Lineage relationship and protective immunity of memory CD8 T cell subsets.** *Nat. Immunol.* 2003, **4**:225–234.

11-22 Memory T cells are heterogeneous and include central memory, effector memory, and tissue-resident subsets.

Cerottini, J.C., Budd, R.C., and MacDonald, H.R.: **Phenotypic identification of memory cytolytic T lymphocytes in a subset of Lyt-2+ cells.** *Ann. N. Y. Acad. Sci.* 1988, **532**:68–75.

Kaech, S.M., Tan, J.T., Wherry, E.J., Konieczny, B.T., Surh, C.D., and Ahmed, R.: **Selective expression of the interleukin 7 receptor identifies effector CD8 T cells that give rise to long-lived memory cells.** *Nat. Immunol.* 2003, **4**:1191–1198.

Lanzavecchia, A., and Sallusto, F.: **Understanding the generation and function of memory T cell subsets.** *Curr. Opin. Immunol.* 2005, **17**:326–332.

Mueller, S.N., Gebhardt, T., Carbone, F.R., and Heath, W.R.: **Memory T cell subsets, migration patterns, and tissue residence.** *Annu. Rev. Immunol.* 2012, **31**:137–161.

Sallusto, F., Geginat, J., and Lanzavecchia, A.: **Central memory and effector memory T cell subsets: function, generation, and maintenance.** *Annu. Rev. Immunol.* 2004, **22**:745–763.

Sallusto, F., Lenig, D., Forster, R., Lipp, M., and Lanzavecchia, A.: **Two subsets of memory T lymphocytes with distinct homing potentials and effector functions.** *Nature* 1999, **401**:708–712.

Skon, C.N., Lee, J.Y., Anderson, K.G., Masopust, D., Hogquist, K.A., and Jameson, S.C.: **Transcriptional downregulation of S1pr1 is required for the establishment of resident memory CD8+ T cells.** *Nat. Immunol.* 2013, **14**:1285–1293.

11-23 CD4 T-cell help is required for CD8 T-cell memory and involves CD40 and IL-2 signaling.

Bourgeois, C., and Tanchot, C.: **CD4 T cells are required for CD8 T cell memory generation.** *Eur. J. Immunol.* 2003, **33**:3225–3231.

Bourgeois, C., Rocha, B., and Tanchot, C.: **A role for CD40 expression on CD8 T cells in the generation of CD8 T cell memory.** *Science* 2002, **297**:2060–2063.

Janssen, E.M., Lemmens, E.E., Wolfe, T., Christen, U., von Herrath, M.G., and Schoenberger, S.P.: **CD4 T cells are required for secondary expansion and memory in CD8 T lymphocytes.** *Nature* 2003, **421**:852–856.

Shedlock, D.J., and Shen, H.: **Requirement for CD4 T cell help in generating functional CD8 T cell memory.** *Science* 2003, **300**:337–339.

Sun, J.C., Williams, M.A., and Bevan, M.J.: **CD4 T cells are required for the maintenance, not programming, of memory CD8 T cells after acute infection.** *Nat. Immunol.* 2004, **5**:927–933.

Tanchot, C., and Rocha, B.: **CD8 and B cell memory: same strategy, same signals.** *Nat. Immunol.* 2003, **4**:431–432.

Williams, M.A., Tyznik, A.J., and Bevan, M.J.: **Interleukin-2 signals during priming are required for secondary expansion of CD8 memory T cells.** *Nature* 2006, **441**:890–893.

11-24 In immune individuals, secondary and subsequent responses are mainly attributable to memory lymphocytes.

Fazekas de St Groth, B., and Webster, R.G.: **Disquisitions on original antigenic sin. I. Evidence in man.** *J. Exp. Med.* 1966, **140**:2893–2898.

Fridman, W.H.: **Regulation of B cell activation and antigen presentation by Fc receptors.** *Curr. Opin. Immunol.* 1993, **5**:355–360.

Klenerman, P., and Zinkernagel, R.M.: **Original antigenic sin impairs cytotoxic T lymphocyte responses to viruses bearing variant epitopes.** *Nature* 1998, **394**:482–485.

Mongkolsapaya, J., Dejnirattisai, W., Xu, X.N., Vasanawathana, S., Tangthawornchaikul, N., Chairunsri, A., Sawasdivorn, S., Duangchinda, T., Dong, T., Rowland-Jones, S., *et al.*: **Original antigenic sin and apoptosis in the pathogenesis of dengue hemorrhagic fever.** *Nat. Med.* 2003, **9**:921–927.

Pollack, W., Gorman, J.G., Freda, V.J., Ascari, W.Q., Allen, A.E., and Baker, W.J.: **Results of clinical trials of RhoGAm in women.** *Transfusion* 1968, **8**:151–153.

Zehn, D., Turner, M.J., Lefrançois, L., and Bevan, M.J.: **Lack of original antigenic sin in recall CD8+ T cell responses.** *J. Immunol.* 2010, **184**:6320–6326.

The Mucosal Immune System

12

Adaptive immune responses are typically initiated in the peripheral lymph nodes that drain the infected tissues. While most internal tissues are free of active microbial growth, the skin and the various mucosae lining organs that directly contact the external world will have continuous encounters with environmental microbes. These surfaces are where most pathogens invade. In this chapter, we will discuss the specialized features of the immune system that serves these mucosal surfaces—the **mucosal immune system**.

The mucosal immune system, in particular, that of the gut, may well have been the first part of the vertebrate adaptive immune system to evolve, possibly linked to the need to deal with the vast populations of commensal bacteria that coevolved with the vertebrates. Organized lymphoid tissues and immunoglobulin antibodies are first found in vertebrates in the gut of primitive cartilaginous fishes, and two important central lymphoid organs—the **thymus** and the avian **bursa of Fabricius**—derive from the embryonic intestine. Fish also have a primitive form of secretory antibody that protects their body surface and may be the forerunner of IgA in mammals. It has therefore been suggested that the mucosal immune system represents the original vertebrate immune system, and that the spleen and lymph nodes are later specializations

IN THIS CHAPTER

The nature and structure of the mucosal immune system.

The mucosal response to infection and regulation of mucosal immune responses.

The nature and structure of the mucosal immune system.

The first line of defense against invasion by potential pathogens and commensal microorganisms is the thin layer of epithelium that covers all these surfaces. However, the epithelium can be breached relatively easily, and so its barrier function needs to be supplemented by defenses provided by the cells and molecules of the mucosal immune system. The innate defenses of mucosal tissues, such as antimicrobial peptides and cells bearing invariant pathogen-recognition receptors, are described in Chapters 2 and 3. In this chapter we concentrate on the adaptive mucosal immune system, highlighting only those innate responses that are of particular importance to our discussion. Many of the anatomical and immunological principles underlying the mucosal immune system apply to all its constituent tissues; here we will use the intestine as our example, and the reader is referred to the general references at the end of this chapter for further details of the other sites.

12-1 The mucosal immune system protects the internal surfaces of the body.

The mucosal immune system comprises the internal body surfaces that are lined by a mucus-secreting epithelium—the gastrointestinal tract, the upper and lower respiratory tract, the urogenital tract, and the middle ear. It also includes the exocrine glands associated with these organs, such as the conjunctivae and lacrymal glands of the eye, the salivary glands, and the lactating breast (Fig. 12.1). The mucosal surfaces represent an enormous area to

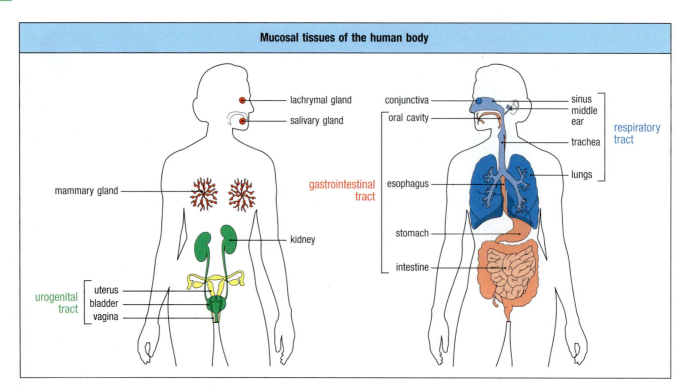

Fig. 12.1 The mucosal immune system. The tissues of the mucosal immune system are the lymphoid organs and cells associated with the intestine, respiratory tract, and urogenital tract, as well as the oral cavity, pharynx, middle ear, and the glands associated with these tissues, such as the salivary glands and lachrymal glands. The lactating breast is also part of the mucosal immune system.

be protected. The human small intestine, for instance, has a surface area of almost 400 m², which is 200 times that of the skin.

The mucosal immune system forms the largest part of the body's immune tissues, containing approximately three-quarters of all lymphocytes and producing the majority of immunoglobulin in healthy individuals. It is also exposed continuously to antigens and other materials entering from the environment. When compared with lymph nodes and spleen (which in this chapter we will call the **systemic immune system**), the mucosal immune system has many unique and unusual features (**Fig. 12.2**).

Fig. 12.2 Distinctive features of the mucosal immune system. The mucosal immune system is bigger, encounters a wider range of antigens, and encounters them much more frequently than the rest of the immune system—what we call in this chapter the systemic immune system. This is reflected in distinctive anatomical features, specialized mechanisms for the uptake of antigen, and unusual effector and regulatory responses that are designed to prevent unwanted immune responses to food and other innocuous antigens.

Distinctive features of the mucosal immune system	
Anatomical features	Intimate interactions between mucosal epithelia and lymphoid tissues
	Discrete compartments of diffuse lymphoid tissue and more organized structures such as Peyer's patches, isolated lymphoid follicles, and tonsils
	Specialized antigen-uptake mechanisms, e.g., M cells in Peyer's patches, adenoids, and tonsils
Effector mechanisms	Activated/memory T cells predominate even in the absence of infection
	Multiple activated 'natural' effector/regulatory T cells present
	Secretory IgA antibodies
	Presence of distinctive microbiota
Immunoregulatory environment	Active downregulation of immune responses (e.g., to food and other innocuous antigens) predominates
	Inhibitory macrophages and tolerance-inducing dendritic cells

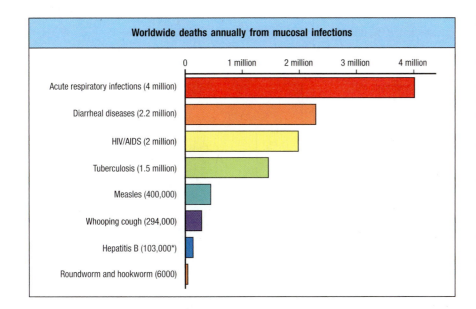

Worldwide deaths annually from mucosal infections

- Acute respiratory infections (4 million)
- Diarrheal diseases (2.2 million)
- HIV/AIDS (2 million)
- Tuberculosis (1.5 million)
- Measles (400,000)
- Whooping cough (294,000)
- Hepatitis B (103,000*)
- Roundworm and hookworm (6000)

Fig. 12.3 Mucosal infections comprise one of the biggest health problems worldwide. Most of the pathogens that cause death throughout the world either are those of mucosal surfaces or enter the body through these routes. Respiratory infections are caused by numerous bacteria (such as *Streptococcus pneumoniae* and *Haemophilus influenzae*, which cause pneumonia; and *Bordetella pertussis*, the cause of whooping cough) and viruses (such as influenza and respiratory syncytial virus). Diarrheal diseases are caused by both bacteria (such as the cholera bacterium *Cholera vibrio*) and viruses (such as rotaviruses). The human immunodeficiency virus (HIV) that causes AIDS enters through the mucosa of the urogenital tract or is secreted into breast milk and passed from mother to child in this way. The bacterium *Mycobacterium tuberculosis*, which causes tuberculosis, also enters through the respiratory tract. Measles manifests itself as a systemic disease, but it originally enters via the oral/respiratory route. Hepatitis B is also a sexually transmitted virus. Finally, parasitic worms inhabiting the intestine cause chronic debilitating disease and premature death. Most of these deaths, especially those from acute respiratory and diarrheal diseases, occur in children under 5 years old in the developing world, and there are still no effective vaccines against many of these pathogens. Numbers shown are the most recent estimated figures available (*The Global Burden of Disease*: 2004 Update. World Health Organization, 2008). *Does not include deaths from liver cancer or cirrhosis resulting from chronic infection.

Because of their physiological functions in gas exchange (the lungs), food absorption (the gut), sensory activities (eyes, nose, mouth, and throat), and reproduction (uterus and vagina), the mucosal surfaces are thin and permeable barriers to the interior of the body. The importance of these tissues to life means that effective defense mechanisms are essential to protect them from invasion. Equally significant is that their fragility and permeability create obvious vulnerability to infection, and it is not surprising that the vast majority of infectious agents invade the human body by these routes (**Fig. 12.3**). Diarrheal diseases, acute respiratory infections, pulmonary tuberculosis, measles, whooping cough, and worm infestations continue to be the major causes of death throughout the world, especially in infants in developing countries. To these must be added the human immunodeficiency virus (HIV), a pathogen whose natural route of entry via a mucosal surface is often overlooked, as well as other sexually transmitted infections such as syphilis.

The mucosal surfaces are also portals of entry for a vast array of foreign antigens that are not pathogenic. This is best seen in the gut, which is exposed to enormous quantities of food proteins—an estimated 30–35 kg per year per person. At the same time, the healthy large intestine is colonized by at least a thousand species of bacteria that live in symbiosis with their host and are known as **commensal microorganisms**, or the **microbiota**. These bacteria are present at levels of at least 10^{12} organisms per milliliter in the colon contents, making them the most numerous cells in the body by a factor of 10. Substantial populations of viruses and fungi are also found in the healthy intestine. In normal circumstances these organisms do no harm, and many are beneficial to their hosts, having important metabolic functions, as well as being essential for normal immune function. The other mucosal surfaces are also colonized by substantial populations of resident commensal organisms (**Fig. 12.4**).

As food proteins and the microbiota contain many foreign antigens, they are capable of being recognized by the adaptive immune system. Generating protective immune responses against these harmless agents would, however, be inappropriate and wasteful. Indeed, aberrant immune responses of this kind are now believed to be the cause of some relatively common diseases, including **celiac disease** (caused by a response to the wheat protein gluten; discussed in Chapter 14) and inflammatory bowel diseases such as **Crohn's disease** (a response to commensal bacteria). As we shall see, the intestinal mucosal immune system has evolved means of distinguishing harmful pathogens from antigens in food and the normal microbiota. Similar issues are faced at other mucosal surfaces, such as the respiratory tract and female genital

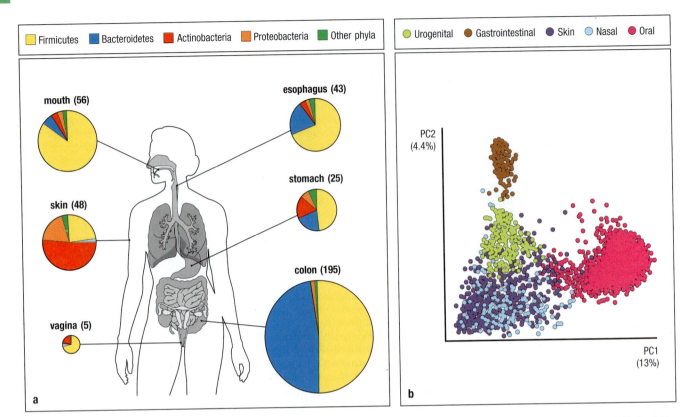

Fig. 12.4 Composition of the commensal microbiota at different mucosal surfaces in healthy humans. Panel a: the different sizes of the pie charts for different sites reflect the number of distinct bacterial species typically present at those sites. The colon contains the greatest number of different species (over 1000 as estimated from individual surveys). The color key indicates the four bacterial phyla that contain the majority of commensal species. Ubiquitous commensal bacteria include *Lactobacillus* and *Clostridium* spp. (Firmicutes), *Bifidobacterium* spp. (Actinobacteria), *Bacteroides fragilis* (Bacteroidetes), and *Escherichia coli* (Proteobacteria). Panel b: principal component analysis of the microbiomes isolated from the indicated human tissues, plotting the first and second principal components. The primary component of microbiome variation is due to body area and accounts for 13% of the variation in microbial identity between samples taken from these sites. a, adapted from Dethlefsen, L., *et al.*: *Nature* 2007, **449**:811-818. b, from Huttenhower, C., *et al.*: *Nature* 2012, **486**:207–214.

tract. Here, protective immunity against pathogens is essential, but many of the antigens entering these tissues are also harmless, being derived from commensal organisms, pollen, other innocuous environmental material, and, in the lower urogenital tract, seminal fluid. The fetus is a further important source of foreign antigen encountered by the normal mucosal immune system to which immune responses must be controlled.

12-2 Cells of the mucosal immune system are located both in anatomically defined compartments and scattered throughout mucosal tissues.

Lymphocytes and other immune-system cells such as macrophages and dendritic cells are found throughout the intestinal tract, both in organized tissues and scattered throughout the surface epithelium of the mucosa and in the underlying layer of connective tissue called the **lamina propria**. The organized secondary lymphoid tissues in the gut comprise a group of organs known as the **gut-associated lymphoid tissues** (**GALT**), together with the draining **mesenteric** and caudal lymph nodes (**Fig. 12.5**). The GALT and the mesenteric lymph nodes have the anatomically compartmentalized structure typical of peripheral lymphoid organs, and are sites at which immune responses are initiated. The cells scattered throughout the epithelium and the lamina propria comprise the effector cells of the local immune response.

Intestinal lymphocytes are found in organized tissues where immune responses are induced, and scattered throughout the intestine, where they carry out effector functions	
Scattered lymphoid cells	Organized lymphoid tissues

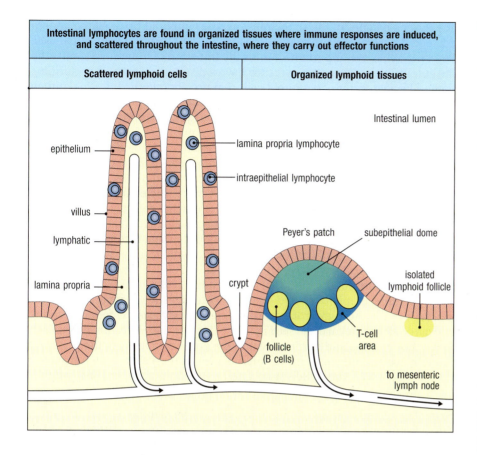

Fig. 12.5 Gut-associated lymphoid tissues and lymphocyte populations.
The intestinal mucosa of the small intestine is made up of fingerlike processes (villi) covered by a thin layer of epithelial cells (red) that are responsible for digestion of food and absorption of nutrients. These epithelial cells are replaced continually by new cells that derive from stem cells in the crypts. The tissue layer underlying the epithelium is called the lamina propria, and is colored pale yellow throughout this chapter. Lymphocytes are found in several discrete compartments in the intestine, with the organized lymphoid tissues such as Peyer's patches and isolated lymphoid follicles forming what is known as the gut-associated lymphoid tissues (GALT). These tissues lie in the wall of the intestine itself, separated from the contents of the intestinal lumen by the single layer of epithelium. The draining lymph nodes for the gut are the mesenteric lymph nodes, which are connected to Peyer's patches and the intestinal mucosa by afferent lymphatic vessels and are the largest lymph nodes in the body. Together, these organized tissues are the sites of antigen presentation to T cells and B cells and are responsible for the induction phase of immune responses. Peyer's patches and mesenteric lymph nodes contain discrete T-cell areas (blue) and B-cell follicles (yellow), while the isolated follicles comprise mainly B cells. Many lymphocytes are found scattered throughout the mucosa outside the organized lymphoid tissues: these are effector cells—effector T cells and antibody-secreting plasma cells, as well as innate lymphoid cells (ILCs). Effector lymphocytes are found both in the epithelium and in the lamina propria. Lymphatics also drain from the lamina propria to the mesenteric lymph nodes.

The GALT comprises the **Peyer's patches**, which are present only in the small intestine; **isolated lymphoid follicles** (**ILF**), which are found throughout the intestine; the appendix (in humans); and the tonsils and adenoids in the throat. The **palatine tonsils**, **adenoids**, and **lingual tonsils** are large aggregates of lymphoid tissue covered by a layer of squamous epithelium and form a ring, known as Waldeyer's ring, at the back of the mouth around the entrance of the gut and airways (**Fig. 12.6**). They often become extremely enlarged in childhood because of recurrent infections, and in the past were frequently removed as a result. A reduced IgA response to oral polio vaccination has been seen in individuals who have had their tonsils and adenoids removed.

The tonsils and adenoids form a ring of lymphoid tissues, Waldeyer's ring, around the entrance of the gut and airway

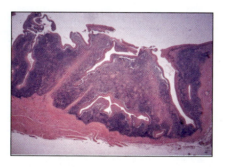

Fig. 12.6 A ring of lymphoid organs called Waldeyer's ring surrounds the entrance to the intestine and respiratory tract. The adenoids lie at either side of the base of the nose, while the palatine tonsils lie at either side of the back of the oral cavity. The lingual tonsils are discrete lymphoid organs on the base of the tongue. The micrograph shows a section through an inflamed human tonsil, where the areas of organized lymphoid tissue are covered by a layer of squamous epithelium (at top of photo). The surface contains deep crevices (crypts) that increase the surface area but can easily become sites of infection. Hematoxylin and eosin staining. Magnification ×100.

The Peyer's patches of the small intestine, the lymphoid tissue of the appendix, and the isolated lymphoid follicles are located within the intestinal wall. Peyer's patches are important sites for the initiation of immune responses in the gut. Visible to the naked eye, they have a distinctive appearance, forming dome-like aggregates of lymphoid cells that project into the intestinal lumen (see Fig. 1.24). There are 100–200 Peyer's patches in the human small intestine. They are much richer in B cells than the systemic peripheral lymphoid organs, each Peyer's patch consisting of a large number of B-cell follicles with germinal centers, with small T-cell areas between and immediately below the follicles (see Fig. 12.5). The subepithelial dome area lies immediately beneath the epithelium and is rich in dendritic cells, T cells, and B cells. Separating the lymphoid tissues from the gut lumen is a layer of **follicle-associated epithelium**. This contains conventional intestinal epithelial cells known as enterocytes and a smaller number of specialized epithelial cells called **microfold cells** (**M cells**), as shown in Fig. 1.24. M-cell development is controlled by local B cells and RANK ligand (RANKL), which is a member of the tumor necrosis factor (TNF) superfamily like CD40L (see Section 7-23). Unlike the enterocytes that make up most of the intestinal epithelium, M cells have a folded luminal surface instead of microvilli and do not secrete digestive enzymes or mucins, and so lack the thick layer of surface mucus (the glycocalyx) found covering conventional epithelial cells (see Fig. 1.24). They are therefore directly exposed to microorganisms and particles within the gut lumen and are the preferred route by which antigens such as microbes enter the Peyer's patch from the lumen (Fig. 12.7). The follicle-associated epithelium also contains lymphocytes and dendritic cells.

Several thousand isolated lymphoid follicles can be identified microscopically throughout the small and large intestines, but they are more frequent in the large intestine, correlating with the load of local microorganisms. Like Peyer's patches, these follicles have an epithelium containing M cells that lies over the organized lymphoid tissue. However, they contain mainly B cells and develop only after birth in response to antigen stimulation due to colonization of the gut by commensal microorganisms. Peyer's patches, in contrast, are already present in the fetal gut, although their full development is not completed until after birth. In the mouse gut, isolated lymphoid follicles seem to arise from small aggregates in the intestinal wall called **cryptopatches**, which contain dendritic cells and **lymphoid tissue inducer** (LTi) **cells** (see Section 9-2). Cryptopatches have not yet been identified in the human gut. Peyer's patches and isolated lymphoid follicles are connected by lymphatics to the draining lymph nodes.

Fig. 12.7 Transport of antigens by M cells facilitates antigen presentation. The first panel illustrates the passage of antigen through M cells in the follicle-associated epithelium of Peyer's patches. M cells have convoluted basal membranes that form 'pockets' within the epithelial layer, allowing close contact with lymphocytes and other cells. This favors the local transport of antigens that have been taken up from the intestine by the M cells and their delivery to dendritic cells for antigen presentation. The top micrograph of a Peyer's patch stained with fluorescently labeled antibodies shows epithelial cells (cytokeratin, dark blue), with M-cell pockets inferred from the presence of T cells (CD3, red) and B cells (CD20, green). At bottom right, Peyer's patch follicle epithelium shows CX3CR1-expressing myeloid cells (green), which include some dendritic cells, interacting with M cells identified by expression of peptidoglycan recognition protein-S (red) and apical staining for the UEA-1 lectin (cyan). Some CX3CR1-expressing cells extend processes into the M cells (arrows). Top right, micrograph from Espen, S., *et al.*: *Immunol. Today* 1999, **20**:141–151. Bottom right, micrograph from Wang *et al.*: *J. Immunol.* 2011, **187**:5277–5285.

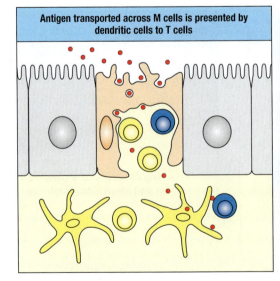

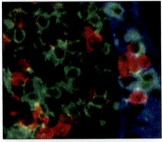

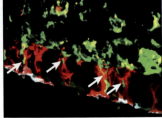

The tissues of the small intestine drain to the mesenteric lymph nodes, which are located in the connective tissue that tethers the intestine to the rear wall of the abdomen. These are the largest lymph nodes in the body and play a crucial role in initiating and shaping immune responses to intestinal antigens. The mucosal surface and lymphoid aggregates of the large intestine drain to part of the mesenteric lymph node and to a separate node known as the caudal lymph node, found close to the bifurcation of the aorta.

The mesenteric lymph nodes and Peyer's patches differentiate independently of the systemic immune system during fetal development, and their development involves distinct chemokines and receptors of the tumor necrosis factor (TNF) family (see Section 9-2). The differences between the GALT and the systemic lymphoid organs are thus imprinted early in life.

In some species such as mice, isolated lymphoid follicles are also found in the lining of the nose, and in the wall of the upper respiratory tract; those in the nose are called **nasal-associated lymphoid tissues** (**NALT**), while those in the upper respiratory tract are known as **bronchus-associated lymphoid tissues** (**BALT**). The term mucosa-associated lymphoid tissues (MALT) is sometimes used to refer collectively to all such tissues found in mucosal organs, although defined organized lymphoid tissues are not found in the nose or respiratory tract in adult humans unless infection is present.

12-3 The intestine has distinctive routes and mechanisms of antigen uptake.

Antigens present at mucosal surfaces must be transported across an epithelial barrier before they can stimulate the mucosal immune system. Peyer's patches and isolated lymphoid follicles are highly adapted for the uptake of antigen from the intestinal lumen. The M cells in the follicle-associated epithelium are continually taking up molecules and particles from the gut lumen by endocytosis or phagocytosis (see Fig. 12.7). For several bacteria this may involve specific recognition of the bacterial FimH protein found in type 1 pili by a glycoprotein (GP2) on the M cell. This material is transported through the interior of the cell in membrane-bound vesicles to the basal cell membrane, where it is released into the extracellular space—a process known as **transcytosis**. Because M cells lack a glycocalyx and so are much more accessible than enterocytes, a number of pathogens target M cells to gain access to the subepithelial space, even though they then find themselves in the heart of the intestinal adaptive immune system. These include *Salmonella enterica* serotype Typhi, the causative agent of typhoid fever; other *Salmonella enterica* serotypes, which are major causes of bacterial food poisoning; *Shigella* species that cause dysentery; and *Yersinia pestis*, which causes plague. Poliovirus, reoviruses, some retroviruses such as HIV, and prions such as the causal agent of scrapie follow the same entry route. After entry into the M cell, bacteria produce proteins that reorganize the M-cell cytoskeleton in a manner that encourages their transcytosis.

The basal cell membrane of an M cell is extensively folded, forming a pocket that encloses lymphocytes and which makes close contacts with local myeloid cells, including dendritic cells (see Fig. 12.7). Macrophages and dendritic cells take up the transported material released from the M cells and process it for presentation to T lymphocytes. Local dendritic cells are in a favorable position to acquire gut antigens, and they are recruited toward, or even into, the follicle-associated epithelium in response to chemokines that are released constitutively by the epithelial cells. The chemokines include **CCL20** (MIP-3α) and **CCL9** (MIP-1γ), which bind to the receptors **CCR6** and **CCR1**, respectively, on dendritic cells (see Appendix IV for a listing of chemokines and their receptors). The antigen-loaded dendritic cells then migrate from the dome region to the T-cell areas of the Peyer's patch, where they meet naive, antigen-specific T cells. Together, the dendritic cells and primed T cells then activate B cells and initiate class switching to IgA. All these processes—the uptake of antigen by M cells, the migration of dendritic cells into the epithelial layer, the production of chemokines, and the subsequent migration of dendritic cells into T-cell areas—are markedly increased in the presence of pathogenic organisms and their products due to the ligation of pattern recognition receptors on epithelial cells and immune cells (see Section 3-5). Similar processes also underlie the induction of immune responses in isolated lymphoid follicles of the gut and in the MALT of other mucosal surfaces.

12-4 The mucosal immune system contains large numbers of effector lymphocytes even in the absence of disease.

In addition to the organized lymphoid organs, mucosal surfaces such as the gut and lung contain enormous numbers of lymphocytes and other leukocytes scattered throughout the tissue. Most of the scattered lymphocytes have the appearance of cells that have been activated by antigen, and they comprise the effector T cells and plasma cells of the mucosal immune system. In the intestine, effector cells are found in two main compartments: the epithelium and the lamina propria (see Fig. 12.5).

These tissues are quite distinct in immunological terms, despite being separated by only a thin layer of basement membrane. The lymphoid component of the epithelium consists mainly of lymphocytes, which in the small intestine are virtually all CD8 T cells. The lamina propria contains many types of immune cells, including IgA-producing plasma cells, conventional CD4 and CD8 T cells with effector and memory phenotypes, innate lymphoid cells, dendritic cells, macrophages, and mast cells. T cells in the lamina propria of the small intestine express the **integrin $\alpha_4{:}\beta_7$** and the chemokine receptor **CCR9** (**Fig. 12.8**), which attracts them into the tissue from the bloodstream. **Intraepithelial lymphocytes** (**IELs**) are mostly CD8 T cells and express either the conventional

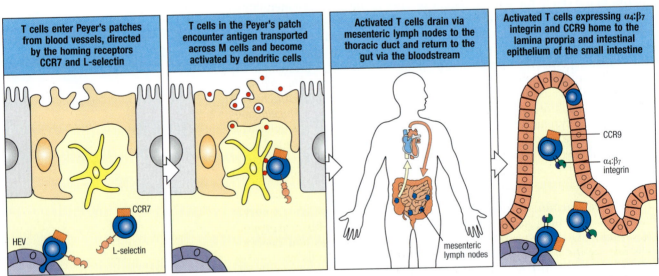

| T cells enter Peyer's patches from blood vessels, directed by the homing receptors CCR7 and L-selectin | T cells in the Peyer's patch encounter antigen transported across M cells and become activated by dendritic cells | Activated T cells drain via mesenteric lymph nodes to the thoracic duct and return to the gut via the bloodstream | Activated T cells expressing $\alpha_4{:}\beta_7$ integrin and CCR9 home to the lamina propria and intestinal epithelium of the small intestine |

Fig. 12.8 Priming of naive T cells and the redistribution of effector T cells in the intestinal immune system. Naive T cells carry the chemokine receptor CCR7 and L-selectin, which direct their entry into Peyer's patches via high endothelial venules (HEVs). In the T-cell area they encounter antigen that has been transported into the lymphoid tissue by M cells and is presented by local dendritic cells. During activation, and under the selective control of gut-derived dendritic cells, the T cells lose L-selectin and acquire the chemokine receptor CCR9 and the integrin $\alpha_4{:}\beta_7$. After activation, but before full differentiation, the primed T cells exit from the Peyer's patch via the draining lymphatics, passing through the mesenteric lymph node to enter the thoracic duct. The thoracic duct empties into the bloodstream, delivering the activated T cells back to the wall of the small intestine. Here T cells bearing CCR9 and $\alpha_4{:}\beta_7$ are attracted specifically to leave the bloodstream and enter the lamina propria of the villus.

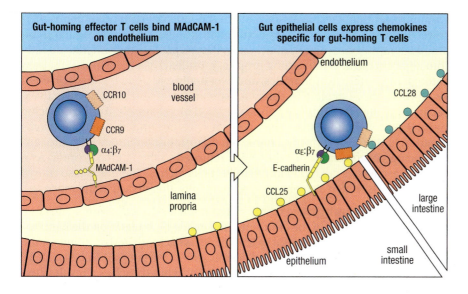

Gut-homing effector T cells bind MAdCAM-1 on endothelium

CCR10
CCR9
α₄:β₇
MAdCAM-1
blood vessel
lamina propria

Gut epithelial cells express chemokines specific for gut-homing T cells

endothelium
CCL28
αE:β₇
E-cadherin
CCL25
epithelium
large intestine
small intestine

Fig. 12.9 Molecular control of intestine-specific homing of lymphocytes.
Left panel: T and B lymphocytes primed by antigen in the Peyer's patches or mesenteric lymph nodes arrive as effector lymphocytes in the bloodstream supplying the intestinal wall (see Fig. 12.8). The lymphocytes express the integrin $\alpha_4{:}\beta_7$, which binds specifically to MAdCAM-1 expressed selectively on the endothelium of blood vessels in mucosal tissues. This provides the adhesion signal needed for the emigration of cells into the lamina propria. Right panel: if primed in the small intestine, the effector lymphocytes also express the chemokine receptor CCR9, which allows them to respond to CCL25 (yellow circles) produced by epithelial cells of the small intestine; this enhances selective recruitment. Effector lymphocytes that have been primed in the large intestine do not express CCR9 but instead express CCR10. CCR10 may respond to CCL28 (blue circles) produced by colon epithelial cells to fulfill a similar function. Lymphocytes destined to enter the epithelial layer stop expressing the $\alpha_4{:}\beta_7$ integrin and instead express the αE:β7 integrin. The receptor for this is E-cadherin on the epithelial cells. These interactions may help keep lymphocytes in the epithelium once they have entered it.

α:β form of CD8 or the CD8α:α homodimer, which may act to dampen T-cell activation. These IELs express CCR9 and the **integrin $\alpha_E{:}\beta_7$ (CD103)**, which binds to **E-cadherin** on epithelial cells (**Fig. 12.9**). By contrast, in the lamina propria it is CD4 T cells that predominate.

The healthy intestinal mucosa therefore displays many characteristics of a chronic inflammatory response—namely, the presence of numerous effector lymphocytes and other leukocytes in the tissues. The presence of such large numbers of effector cells would be unusual for a healthy nonlymphoid tissue, but in the gut does not necessarily signify infection. Rather, it is the local response to the myriad innocuous antigens normally present at mucosal surfaces, and is essential for maintaining the beneficial symbiosis between host and microbiota. It involves a balanced generation of effector and regulatory T cells, but, when required, can be refocused to produce a full adaptive immune response to invading pathogens.

12-5 The circulation of lymphocytes within the mucosal immune system is controlled by tissue-specific adhesion molecules and chemokine receptors.

The entry of effector lymphocytes into the mucosa results from changes in their homing characteristics as they become activated. Naive T cells and B cells circulating in the bloodstream are not predetermined as to which compartment of the immune system they will end up in, and they enter Peyer's patches and mesenteric lymph nodes through **high endothelial venules (HEVs)** (see Fig. 9.4). As in the systemic immune system, this process is controlled largely by the chemokines **CCL21** and **CCL19**, which are released from the lymphoid tissues and bind the receptor **CCR7** on naive lymphocytes. In the Peyer's patches, this is assisted by binding of the mucosal vascular addressin **MAdCAM-1** on HEVs to the **L-selectin** expressed on naive T cells. **CXCR5** responding to **CXCL13** produced in B-cell follicles is also important for recruitment of naive B cells to Peyer's patches and isolated lymphoid follicles of the intestine. As in other secondary lymphoid tissues, if the naive lymphocytes do not see their antigen, they exit via the lymphatics and return to the bloodstream. If they encounter antigen in the GALT, the lymphocytes become activated and lose expression of CCR7 and L-selectin. This means that they have lost their ability to home to secondary lymphoid organs, because they cannot enter them via the HEVs (see Section 9-5).

Lymphocytes that have been activated in the mucosal lymphoid organs then travel to the mucosa, where they accumulate as effector cells. Although some T and B lymphocytes initially activated in Peyer's patches may migrate directly into adjacent parts of the lamina propria, most leave via the lymphatics, pass through mesenteric lymph nodes, and eventually end up in the thoracic duct. From there they circulate in the bloodstream (see Fig. 12.8) and selectively reenter the intestinal lamina propria via small blood vessels. Antigen-specific naive B cells in the follicular areas of Peyer's patches undergo isotype switching from IgM to IgA production there, but they only differentiate fully into IgA-producing plasma cells once they have returned to the lamina propria. As a result, plasma cells are rarely found in Peyer's patches, and this is also true of effector T cells, which also only differentiate fully after arrival in the mucosa.

Gut-specific homing by antigen-stimulated T and B cells is determined in large part by the expression of the adhesion molecule $\alpha_4{:}\beta_7$ integrin on the lymphocytes. This binds to the mucosal vascular addressin MAdCAM-1, found on the endothelial cells that line the blood vessels within the gut wall (see Fig. 12.9). Lymphocytes originally primed in the gut are also lured back as a result of tissue-specific expression of chemokines by the gut epithelium. In the case of the small intestine, **CCL25** (TECK) produced constitutively by epithelial cells is a ligand for the receptor CCR9 expressed on gut-homing T cells and B cells. Within the intestine there seems to be regional specialization of chemokine expression, as CCL25 is not expressed outside the small intestine and CCR9 is not required for migration of lymphocytes to the colon. However, the colon, lactating mammary gland, and salivary glands express **CCL28** (MEC, mucosal epithelial chemokine), which is a ligand for the receptor **CCR10** on gut-primed lymphocytes and attracts IgA-producing B lymphoblasts to these tissues. The addressins and chemokine receptors involved in migration of activated lymphocytes to other mucosal surfaces are unknown.

Under most normal circumstances, only lymphocytes that first encounter antigen in a gut-associated secondary lymphoid organ are induced to express gut-specific homing receptors and integrins. As we shall see in the next sections, these molecules are induced or 'imprinted' on T lymphocytes by intestinal dendritic cells during antigen presentation. In contrast, dendritic cells from nonmucosal lymphoid tissues induce lymphocytes to express other adhesion molecules and chemokine receptors—for example, **α4:β1 integrin** (VLA-4), which binds VCAM-1; **cutaneous lymphocyte antigen** (CLA), which binds E-selectin; and the chemokine receptor CCR4—which direct them to tissues such as the skin (see Section 11-3). The tissue-specific consequences of lymphocyte priming in the GALT explain why effective vaccination against intestinal infections requires immunization by a mucosal route, because other routes, such as subcutaneous or intramuscular immunization, do not involve dendritic cells with the correct imprinting properties.

12-6 Priming of lymphocytes in one mucosal tissue may induce protective immunity at other mucosal surfaces.

Not all parts of the mucosal immune system exploit the same tissue-specific chemokines, allowing localized compartmentalization of lymphocyte recirculation within the system. Thus, effector T and B cells primed in lymphoid organs draining the small intestine (mesenteric lymph nodes and Peyer's patches) are most likely to return to the small intestine; similarly, those primed in the respiratory tract migrate most efficiently back to the respiratory mucosa. This homing is obviously useful in returning antigen-specific effector cells to the mucosal organ in which they will be most effective in fighting an infection or in controlling immune responses against foreign proteins and commensals. Nevertheless, some lymphocytes that have been primed in the GALT, for example, can also recirculate as effector cells to other mucosal tissues such as the respiratory tract, urogenital tract, and lactating breast. This over-lap between mucosal recirculation routes gave rise to the idea of a **common mucosal immune system**, which is distinct from other parts of the immune system. Although this is now understood to be an oversimplification, it does have important implications for vaccine development, because it may enable immunization by one mucosal route to be used to protect against infection at another mucosal surface. An important example of this is the induction of IgA antibody production in the lactating breast by natural infection or vaccination of mucosal surfaces such as the intestine. This is because the vasculature of the lactating breast expresses MAdCAM-1 and the phenomenon is a crucial means of generating protective immunity that can be transmitted to infants by passive transfer of the antibodies in milk. A further example has been shown in experimental animals, in which nasal immunization has a special ability to prime immune responses in the urogenital tract against HIV. The mechanisms behind this are unknown.

12-7   Distinct populations of dendritic cells control mucosal immune responses.

As elsewhere, dendritic cells are important for initiating and shaping immune responses in mucosal tissues, and are located both in secondary mucosal lymphoid organs and scattered throughout the mucosal surfaces. Within Peyer's patches, dendritic cells are found in two main areas. In the subepithelial dome region, dendritic cells can acquire antigen from M cells (**Fig. 12.10**). Both of the major subtypes of dendritic cells are present in the intestine (see Sections 6-5 and 9-1). In mice, the most abundant subset of dendritic cells in the Peyer's patch expresses **CD11b** (α_M integrin) and, when activated, tends to produce IL-23. This promotes development of T_H17 cells and stimulates ILC3 cells, both of which produce IL-17 and IL-22 (see Sections 3-23 and 11-2). These dendritic cells express CCR6, the receptor for CCL20 produced by follicle-associated epithelial cells. In resting conditions, they reside beneath the epithelium and produce IL-10 in response to antigen uptake, maintaining a noninflammatory environment. However, during infection by a pathogen such as *Salmonella*, dendritic cells are rapidly recruited into the epithelial layer of the Peyer's patch in response to the CCL20 that is released in increased quantities by epithelial cells in the presence of the bacteria. Bacterial products also activate the dendritic cells to express co-stimulatory molecules, allowing them to induce pathogen-specific naive T cells to differentiate into effector cells. Also in the T-cell area of Peyer's patches are the less abundant CD11b-negative subset of dendritic cells, whose development requires the factor **BATF3** and which produce the cytokine IL-12 (see Sections 6-5 and 9-9). CD11b-expressing dendritic cells are protective in many intestinal infections.

Dendritic cells are also abundant in the wall of the small intestine outside Peyer's patches, mainly in the lamina propria. These sample antigens from the lumen and surrounding tissue, and they spend a relatively short time in the intestine before migrating in afferent lymph to the draining mesenteric lymph node, where they present antigen to naive T cells. As elsewhere, migration of dendritic cells depends on the chemokine receptor CCR7 (see Fig. 9.17). It is estimated that 5–10% of the mucosal dendritic cell population emigrates to the mesenteric lymph nodes every day in the resting intestine, allowing constant delivery of antigens from the intestinal surface to T cells. In the absence of infection or inflammation, the encounter between migrating dendritic cells and naive T cells in mesenteric lymph nodes results in the generation of antigen-specific FoxP3$^+$ regulatory T cells that express the gut-homing molecules CCR9 and integrin α_4:β_7 described above (see Section 12-4). These 'primed' T_{reg} cells then leave the lymph node, return to the wall of the small intestine, and suppress the production of inflammatory responses to harmless antigens in food.

Fig. 12.10 Capture of antigens from the intestinal lumen by mononuclear cells in the lamina propria. Top row, first panel: soluble antigens such as food proteins can be transported directly across or between enterocytes, or taken up by M cells in the surface epithelium outside Peyer's patches. Second panel: enterocytes can capture and internalize antigen:antibody complexes by means of the neonatal Fc receptor (FcRn) on their surface and transport them across the epithelium by transcytosis. Lamina propria dendritic cells express FcRn and other Fc receptors and capture and internalize the complexes. Third panel: an enterocyte infected with an intracellular pathogen undergoes apoptosis and is phagocytosed by a dendritic cell. Lower rows, left panels: mononuclear cells can extend cellular processes between the cells of the epithelium without disturbing its integrity. These cells, now thought to be macrophages, may internalize antigen and pass it to neighboring dendritic cells for presentation to T cells. The micrograph shows mononuclear cells stained for CD11c (green) in the lamina propria of a villus of mouse small intestine. The epithelium appears black but its luminal (outer) surface is shown by the white line. A cell process stretches between two epithelial cells and extends its tip into the lumen of the intestine. Magnification ×200. Center panels: mucus-secreting goblet cells can transport soluble antigens to lamina propria dendritic cells. The micrograph shows the soluble marker dextran (purple) being transported across goblet cells (white in the bottom right panel) in the epithelium (nuclei stained blue) to underlying dendritic cells (stained for CD11c in green). Scale bar 10 μm. Right panels: dendritic cells (purple) may enter the epithelial layer and capture bacteria before returning to the lamina propria. Dendritic cells and macrophages remaining in the lamina propria are stained blue or green. Scale bar 10 μm. Bottom left, micrograph from Niess, J.H., et al.: Science 2005, **307**:254–258. Bottom center, micrograph from McDole, J.R., et al.: Nature 2012, **483**:345–349. Bottom right, micrograph from Farache et al.: Immunity 2013, **38**:581–595.

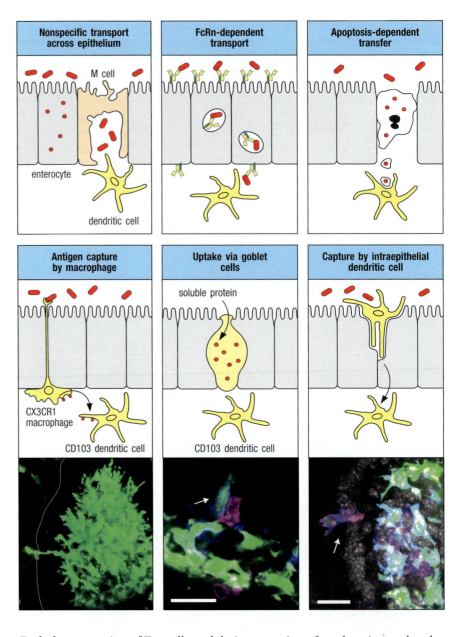

Both the generation of T_{reg} cells and their expression of gut-homing molecules require that the dendritic cells produce **retinoic acid**, which is derived from the metabolism of dietary vitamin A through the action of retinal dehydrogenases. Retinoic acid is also produced by stromal cells in the mesenteric lymph node, further enhancing the effects of the migratory dendritic cells. Retinoic acid-producing dendritic cells are also found in Peyer's patches, and may also be important for generating regulatory T cells either in the Peyer's patch itself, or after they migrate to the mesenteric lymph node. The induction of regulatory T cells in intestinal tissues is assisted by transforming growth factor-β (TGF-β), which is produced by dendritic cells. Migratory populations of dendritic cells that continuously take up local antigens in the tissue and transport them to the draining lymph nodes are also found in the large intestine and other mucosal surfaces such as the lung. Although it is believed dendritic cells from these tissues are also involved in maintaining tolerance to harmless materials such as commensal bacteria, they do not produce retinoic acid, and it is not clear how they influence T-cell differentiation and homing.

The dendritic cells in the intestinal lamina propria also include the two major subsets described above. Collectively, the properties of intestinal dendritic cells result in a dominantly tolerogenic environment that prevents unnecessary and damaging reactions to foods and commensal microorganisms. The anti-inflammatory behavior of mucosal dendritic cells in the healthy gut is promoted by factors that are constitutively produced in the mucosal environment. These include thymic stromal lymphopoietin (TSLP), TGF-β produced by dendritic and epithelial cells, prostaglandin PGE_2 produced by stromal cells, and IL-10 produced by intestinal macrophages and CD4 T cells. Retinol stored in the liver and delivered into the small intestine via the bile provides an additional source for the local generation of retinoic acid for conditioning dendritic cells in the wall of the small intestine.

12-8 Macrophages and dendritic cells have different roles in mucosal immune responses.

The lamina propria of the healthy intestine contains the largest population of macrophages in the body. Like dendritic cells, they express CD11c and class II MHC, but unlike dendritic cells in this site, they lack CD103 expression, but express **FcγR1** (CD64; see Fig. 10.38) and **CX3CR1**, the receptor for CX3CL1 (fractalkine). Macrophages are also unable to migrate from the intestine to the draining lymph nodes and cannot present antigen to naive T cells. Unlike many other tissue-resident macrophages, such as those in the brain or liver, which develop from embryonic precursors (see Section 3-1), those in the intestine require constant replenishment by blood monocytes.

Macrophages are important for maintaining a healthy intestine. They are positioned immediately under the epithelium and are highly phagocytic, and thus ideally suited to ingest and degrade any microbes that penetrate across the epithelial barrier. They can also clear away dying epithelial cells, which are found in large numbers in the intestine, an inevitable consequence of such a rapidly dividing tissue. However, unlike macrophages in other parts of the body, intestinal macrophages do not produce significant quantities of inflammatory cytokines or reactive oxygen or nitrogen species in response to phagocytosis or exposure to stimuli such as bacteria or TLR ligands. This is because they produce large amounts of IL-10 constitutively, allowing them to limit inflammation while acting as powerful scavengers. Macrophage-derived IL-10 also contributes to maintaining antigen-specific tolerance in the mucosa, as it is needed to sustain the survival and secondary expansion of $FoxP3^+$ T_{reg} cells that have migrated back to the intestine after being primed by tolerogenic dendritic cells in the lymph node. Indeed, they have features of both these populations, and their functions are specifically adapted to the conditions of their local environment. Thus macrophages and dendritic cells play distinct but complementary roles in the steady-state intestine. Migratory dendritic cells carry out the initial priming and shaping of T-cell responses in secondary lymphoid organs, and sessile macrophages scavenge cellular debris and microbes and may tune the activity of already primed T cells in the mucosa itself.

12-9 Antigen-presenting cells in the intestinal mucosa acquire antigen by a variety of routes.

The total surface area provided by M cells in the epithelium of Peyer's patches for antigen to the intestinal immune system is limited, and the lamina propria itself is covered by an intact epithelium. Various additional mechanisms have been proposed to explain how antigen crosses the epithelium to gain access to macrophages and dendritic cells (see Fig. 12.10). Soluble antigens such as food proteins might be transported across epithelial cells or between gaps formed in the epithelium where dying cells are being shed. Alternatively, M cells may reside in the surface epithelium of the mucosa outside Peyer's patches. Some intestinal bacteria, such as enteropathogenic and enterohe-molytic strains of *E. coli*, have specialized means of attaching to and invading epithelial cells, allowing them to enter the underlying lamina propria directly. Antigen from the lumen can be delivered to lamina propria dendritic cells by uptake of antibody-coated antigens by epithelial cells expressing the **neonatal Fc receptor** (FcRn). Antigen derived from apoptotic epithelial cells may be processed by cross-presenting dendritic cells (see Section 6-5) for induction of immune responses against enteric viruses, such as rotaviruses, which cause diarrheal disease because of their specialized ability to infect enterocytes (see Fig. 12.10).

Macrophages in the lamina propria may also participate in local antigen uptake by sending transepithelial dendrites that extend between epithelial cells and reach the lumen to sample bacteria (see Fig. 12.10). Lamina propria macrophages have also been reported to take up soluble antigen from the lumen, passing it on to dendritic cells for subsequent presentation to T cells. Some experiments also suggest that dendritic cells or macrophages might make their way into the lumen to acquire antigens such as bacteria, before returning with them to the lamina propria.

12-10 Secretory IgA is the class of antibody associated with the mucosal immune system.

The dominant class of antibody in the mucosal immune system is **IgA**, which is produced locally by plasma cells in the mucosal wall. The nature of IgA differs between the two main compartments in which it is found—the blood and mucosal secretions. IgA in the blood is mainly in the form of a monomer (mIgA) that is produced in the bone marrow by plasma cells derived from B cells activated in lymph nodes. In mucosal tissues, IgA is produced almost exclusively as a polymer, usually as a dimer, in which the two immunoglobulin monomers are linked by a **J chain** (see Section 5-16).

The naive B-cell precursors of the IgA-secreting mucosal plasma cells are activated in Peyer's patches and mesenteric lymph nodes. Class switching of activated B cells to IgA is controlled by the cytokine TGF-β. In the human gut, this class switching is entirely T-cell dependent and occurs only in organized lymphoid tissues, where **follicular helper T cells** (T_{FH}) instruct B cells by the same mechanisms as described in Chapter 10. The subsequent expansion and differentiation of IgA-switched B cells are driven by IL-5, IL-6, IL-10, and IL-21. Upward of 75,000 IgA-producing plasma cells are present in the normal human intestine, and 3–4 g of IgA is secreted by the mucosal tissues each day, and is the major immunoglobulin class produced there. This continuous production of large quantities of IgA occurs in the absence of pathogenic invasion and is driven almost entirely by recognition of the commensal microbiota.

In humans, monomeric and dimeric IgA are both found as two isotypes, IgA1 and IgA2. The ratio of IgA1 to IgA2 varies markedly depending on the tissue, being about 10:1 in the blood and upper respiratory tract, about 3:2 in the small intestine, and 2:3 in the colon. Some common pathogens of the respiratory mucosa (such as *Haemophilus influenzae*) and the genital mucosa (such as *Neisseria gonorrhoeae*) produce proteolytic enzymes that can cleave IgA1, whereas IgA2 is resistant to cleavage. The higher proportion of plasma cells secreting IgA2 in the large intestine might result because the high density of commensal microorganisms at this site drives the production of cytokines that cause selective class switching. In mice, only one IgA isotype is found, and it is most closely similar to IgA2 in humans.

After activation and differentiation, the resulting IgA-expressing B lymphoblasts express the mucosal homing integrin $\alpha_4{:}\beta_7$, as well as the chemokine receptors CCR9 and CCR10, and localize to mucosal tissues by the mechanisms discussed above. Once in the lamina propria, the B cells undergo final differentiation into plasma cells, which synthesize IgA dimers and secrete them into the subepithelial space (Fig. 12.11). To reach its target antigen in the gut lumen, the IgA has to be transported across the epithelium by the **polymeric immunoglobulin receptor** (**pIgR**), which we introduced in Section 10-16. pIgR is expressed constitutively on the basolateral surfaces of the immature epithelial cells located at the base of the intestinal crypts and binds covalently to the Fc portion of J-chain-linked polymeric immunoglobulins such as dimeric IgA and pentameric IgM, and it transports the antibody by transcytosis to the luminal surface of the epithelium, where it is released by proteolytic cleavage of the extracellular domain of the receptor. Part of the cleaved pIgR remains associated with the IgA and is known as **secretory component** (frequently abbreviated to **SC**). The resulting antibody is protected from proteolytic cleavage and is referred to as **secretory IgA** (**SIgA**).

In some animals there is a second route of IgA secretion into the intestine—the **hepatobiliary route**. Dimeric IgA that has not bound pIgR is taken up into venules in the lamina propria, which drain intestinal blood to the liver via the portal vein. In the liver these small veins (sinusoids) are lined by an endothelium that allows the antibodies access to underlying hepatocytes, which have pIgR on their surface. IgA is taken up into the hepatocytes and transported by transcytosis into an adjacent bile duct. In this way, secretory IgA can be delivered directly into the upper small intestine via the common bile duct. This hepatobiliary route allows dimeric IgA to eliminate antigens that have invaded the lamina propria and have been bound there by IgA. Although highly efficient in rats and other rodents, this route does not seem to be of great significance in humans and other primates, in whom hepatocytes do not express pIgR.

IgA secreted into the gut lumen binds to the layer of mucus coating the epithelial surface via carbohydrate determinants in secretory component. There it is involved in preventing invasion by pathogenic organisms and, just as important, it also has a crucial role in maintaining the homeostatic balance

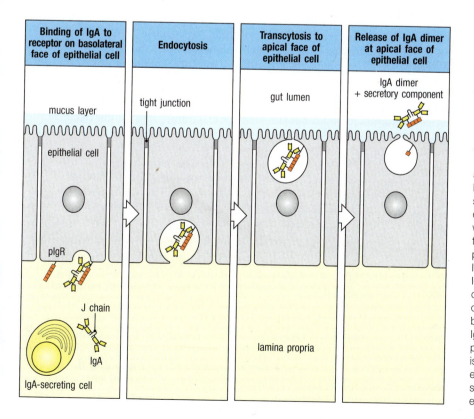

Fig. 12.11 Transcytosis of IgA antibody across epithelia is mediated by the polymeric Ig receptor (pIgR), a specialized transport protein. Most IgA antibody is synthesized in plasma cells lying just beneath epithelial basement membranes of the gut, the respiratory epithelia, the tear and salivary glands, and the lactating mammary gland. The IgA dimer linked by a J chain diffuses across the basement membrane and is bound by the pIgR on the basolateral surface of the epithelial cell. The bound complex undergoes transcytosis, by which it is transported in a vesicle across the cell to the apical surface. There the pIgR is cleaved, leaving the extracellular IgA-binding component bound to the IgA molecule as the so-called secretory component. Although not shown, carbohydrate on the secretory component binds to mucins in mucus and holds the IgA at the epithelial surface. The residual piece of the pIgR is nonfunctional and is degraded. IgA is transported across epithelia in this way into the lumina of several organs that are in contact with the external environment.

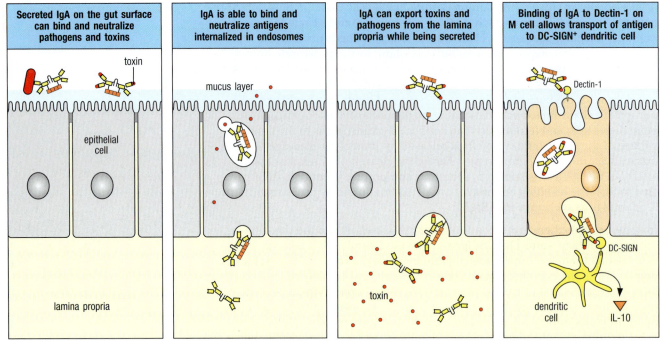

Fig. 12.12 Mucosal IgA has several functions in epithelial surfaces. First panel: IgA adsorbs on the layer of mucus covering the epithelium, where it can neutralize pathogens and their toxins, preventing their access to tissues and inhibiting their functions. Second panel: antigen internalized by the epithelial cell can meet and be neutralized by IgA in endosomes. Third panel: toxins or pathogens that have reached the lamina propria encounter pathogen-specific IgA there, and the resulting complexes are reexported into the lumen across the epithelial cell as the dimeric IgA is secreted. Fourth panel: antigen bound to secretory IgA in the lumen can bind via carbohydrate residues on the Fc portion of IgA to Dectin-1 on M cells in Peyer's patches and be transported to underlying dendritic cells. Binding of the IgA-containing complex to DC-SIGN on the dendritic cells induces them to produce anti-inflammatory IL-10.

between the host and the commensal microbiota. It does this in a number of ways (Fig. 12.12). First, it inhibits microbial adherence to the epithelium, its ability to bind bacteria being assisted by the unusually wide and flexible angle between the Fab pieces of the IgA molecule, particularly the IgA1 isotype (see Section 5-12), allowing very efficient bivalent binding to large antigens such as bacteria. Secretory IgA can also neutralize microbial toxins or enzymes.

In addition to its activities in the lumen, IgA can neutralize bacterial lipopolysaccharide and viruses it encounters within endosomes inside epithelial cells, as well as across the epithelial barrier in the lamina propria after bacteria and viruses have penetrated there. The resulting IgA:antigen complexes are then reexported into the gut lumen, from where they are excreted from the body (see Fig. 12.12). Complexes containing dimeric IgA formed in the lamina propria can also be excreted via the hepatobiliary route described above. In addition to enabling the elimination of antigens, the formation of IgA:antigen complexes can enhance the uptake of luminal antigen by M cells and local dendritic cells, via binding of carbohydrate residues on IgA to lectin receptors such as Dectin-1 and DC-SIGN. As well as these antigen-specific effects, secretory IgA can restrict the entry of bacteria in a nonspecific manner. This is because the high carbohydrate content of the Fc part of the IgA heavy chain allows it to act as a decoy for receptors that bacteria use to bind carbohydrates on the epithelial surface. Secretory IgA has little capacity to activate the classical pathway of complement or to act as an opsonin, and so does not induce inflammation. Uptake of IgA:antigen complexes by dendritic cells also induces these cells to produce anti-inflammatory IL-10. Together these properties mean that IgA can limit the penetration of microbes into the mucosa without risking inflammatory damage to these fragile tissues, something that would be potentially harmful in the intestine. For the same reasons, secretory

IgA is crucial to the beneficial symbiosis between an individual and gut commensal bacteria (see Section 12-20).

12-11 T-independent processes can contribute to IgA production in some species.

In mice, unlike humans, a significant proportion of intestinal IgA is derived from T-cell-independent B-cell activation and class switching. This depends on activation of the innate immune system by the products of commensal microbes and may result from the direct interaction of B cells with conventional dendritic cells and follicular dendritic cells in solitary lymphoid follicles. This antibody production seems to involve lymphocytes of the B-1 subset (see Section 8-9), which arise from precursor B cells in the peritoneal cavity and migrate to the intestinal wall in response to microbial constituents such as lipopolysaccharide. Once in the mucosa, TGF-β-dependent class switching to IgA occurs under the influence of local factors, including IL-6, retinoic acid, and BAFF and APRIL (see Fig. 10.6), which bind to TACI on B cells substituting for signals otherwise supplied by CD4 helper T cells (see Section 10-1). Intestinal epithelial cells can produce BAFF and APRIL, while local eosinophils may contribute by producing APRIL, IL-6, and TGF-β. Other myeloid cells may produce nitric oxide (NO) and TNF-α, both of which assist in the processing and activation of TGF-β.

The IgA antibodies produced in these T-cell-independent responses are of limited diversity and of generally low affinity, with little evidence of somatic hypermutation. They are nevertheless an important source of 'natural' antibodies directed at commensal bacteria. As yet, there is little evidence for this source of IgA in humans, in whom all secretory IgA responses involve somatic hypermutation and seem to be T-cell dependent. The enzyme activation-induced cytidine deaminase (AID), which is required for class switching (see Chapter 5), cannot be detected in human intestinal lamina propria, indicating that class switching is unlikely to occur there. Nevertheless, its occurrence in lamina propria B cells in mice may offer a glimpse into the evolutionary history of specific antibody responses in the mucosa, and might indicate pathways that could be activated when T-cell-dependent IgA production is compromised in humans, as it is in AIDS. Nonetheless, it is likely that secondary reactivation of IgA-committed B lymphoblasts occurs in the lamina propria for full differentiation of plasma cells, which likely involves production by myeloid and epithelial cells of APRIL, BAFF, and other mediators

12-12 IgA deficiency is relatively common in humans but may be compensated for by secretory IgM.

Selective deficiency of IgA production is the commonest primary immune deficiency in humans, occurring in about 1 in 500 to 700 individuals in populations of Caucasian origin, although it is somewhat rarer in other ethnic groups. The most frequent genetic mutation that has been identified in this condition is in the TACI receptor for BAFF. A slightly higher incidence of respiratory infections, atopy (a tendency for allergic reactions to harmless environmental antigens), and autoimmune disease has been reported in older people with IgA deficiency. However, most individuals with IgA deficiency are not overly susceptible to infections unless there is also a deficiency in IgG2 production. The dispensability of IgA probably reflects the ability of IgM to replace IgA as the predominant antibody in secretions, and increased numbers of IgM-producing plasma cells are indeed found in the intestinal mucosa of IgA-deficient people. Because IgM is a J-chain-linked polymer, IgM produced in the gut mucosa is bound efficiently by the pIgR and is transported across epithelial cells into the gut lumen as secretory IgM. The importance of this backup mechanism has been shown in knockout mice. Animals lacking IgA alone have a normal phenotype, but those lacking the pIgR are susceptible to mucosal infections. They also show increased penetration of commensal bacteria into tissues and a consequent systemic immune response to these bacteria. Genetic absence of the pIgR has never been reported in humans, suggesting that such a defect is lethal.

12-13 The intestinal lamina propria contains antigen-experienced T cells and populations of unusual innate lymphoid cells.

Most of the T cells in the healthy lamina propria have been activated by dendritic cells and express markers of effector or memory T cells, such as CD45RO in humans, and express gut-homing markers such as CCR9 and $\alpha_4{:}\beta_7$ integrin, as well as receptors for pro-inflammatory chemokines such as CCL5 (RANTES). The T-cell population of the lamina propria has a ratio of CD4 to CD8 T cells of 3:1 or more, similar to that in systemic lymphoid tissues.

Lamina propria CD4 T cells secrete large amounts of cytokines such as interferon (IFN)-γ, IL-17, and IL-22, even in the absence of overt inflammation. This likely reflects the constant state of immune recognition of the microbiota and other environmental antigens that takes place in the intestine, and their importance is underlined by the frequent opportunistic infections of the intestine that occur in individuals lacking CD4 T cells, such as those with HIV infection (see Section 13-24). Effector T_H17 cells are prominent in the intestinal mucosa, and their products are important components of local immune defense. IL-17 is needed for full expression of the poly-immunoglobulin receptor involved in secretion of IgA into the lumen, while IL-22 stimulates intestinal epithelial cells to produce antimicrobial peptides that help maintain epithelial barrier integrity. Effector CD8 T cells are also present in the normal lamina propria and are capable of both cytokine production and cytotoxic activity when a protective immune response to a pathogen is required.

In any other situation, the presence of such large numbers of differentiated effector T cells would suggest the presence of a pathogen and likely would lead to inflammation. The fact that it does not in the healthy lamina propria is because the generation of T_H1, T_H17, and cytotoxic T cells is balanced by the presence of substantial numbers of IL-10-producing regulatory T cells. In the small intestine, these are mostly FoxP3-negative, whereas in the colon, FoxP3-positive T_{reg} cells dominate. Many of the inducible T_{reg} cells recognize antigens derived from organisms within the microbiota.

The healthy lamina propria also contains many innate lymphoid cells (ILCs) (see Sections 1-19 and 9-20). The ILC3 subset is prominent in both human and mouse intestinal mucosa. Mature ILC3s produce IL-17 and IL-22, and some express the NK-cell receptors NKp44 and NKp46. Their development is controlled by the aryl hydrocarbon receptor and the transcription factor RORγT (see Section 9-21). ILC3s are present in secondary lymphoid organs in the intestine and are important for their lymphoid tissue development there. In response to IL-23 secreted by local dendritic cells, ILC3s produce IL-22, which stimulates the epithelium to generate antimicrobial peptides that promote local defense against bacterial and fungal pathogens in the intestine. During the course of inflammatory diseases, ILC3s can acquire the ability to produce IFN-γ in response to IL-12, and combined with their production of IL-17, this endows them with significant pathological properties. IL-5 and IL-13 produced by ILC2s form an important layer of T-cell-independent responses to helminth parasites in the intestine, and an equivalent population is involved in allergic reactions in the respiratory tract.

CD1-restricted **iNKT** cells (see Section 6-18) and **mucosal invariant T (MAIT) cells** (see Section 6-19) are also present in the lamina propria, and account for 2–3% of lamina propria T cells in human small intestine. MAIT cells express an invariant TCRα chain paired with a limited range of TCRβ chain and recognize metabolites of vitamin B derived mainly from the riboflavin metabolism pathway in microbes presented by MR1.

12-14   The intestinal epithelium is a unique compartment of the immune system.

We have already introduced the fact that there are abundant intraepithelial lymphocytes (IELs) present in intestine. In the healthy small intestine, there are 10–15 lymphocytes for every 100 epithelial cells, making the IELs one of the single largest populations of lymphocytes in the body (Fig. 12.13). More than 90% of the IELs in the small intestine are T cells, and around 80% of these carry CD8, in complete contrast to the lymphocytes in the lamina propria. IELs are also present in the large intestine, although there are fewer of them relative to the number of epithelial cells and the proportion of CD4 T cells is greater than in the small intestine.

Like the lymphocytes in the lamina propria, most IELs have an activated appearance even in the absence of infection by a pathogen, and they contain intracellular granules containing perforin and granzymes, like those in conventional effector CD8 cytotoxic T cells. However, the T-cell receptors of most CD8 IELs show evidence of oligoclonality, with restricted use of V(D)J gene segments, an indication that they may expand locally in response to a relatively small number of antigens. The IELs of the small intestine express the chemokine receptor CCR9, and the α_E:β_7 integrin (CD103), which interacts with E-cadherin expressed on epithelial cells and assists their retention in the epithelium (see Fig.12.9).

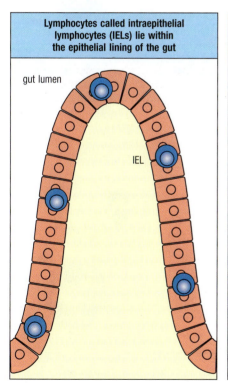
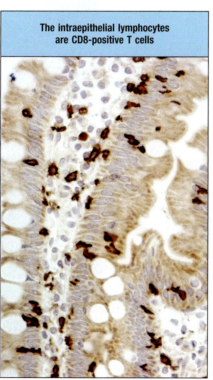
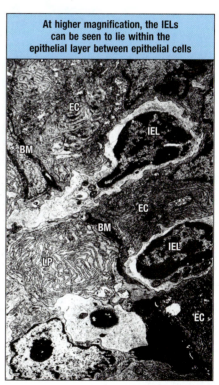

Fig.12.13 Intraepithelial lymphocytes. The epithelium of the small intestine contains a large population of lymphocytes known as intraepithelial lymphocytes (IELs; left panel). The micrograph in the center is of a section through human small intestine in which CD8 T cells have been stained brown with a peroxidase-labeled monoclonal antibody. Most of the lymphocytes in the epithelium are CD8 T cells. Magnification ×400. The electron micrograph on the right shows that the IELs lie between epithelial cells (EC) on the basement membrane (BM) separating the lamina propria (LP) from the epithelium. One IEL can be seen having crossed the basement membrane into the epithelium, leaving a trail of cytoplasm in its wake. Magnification ×8000.

There are two main subsets of CD8 intraepithelial T cells—type a ('inducible') and type b ('natural')—identified based on which form of CD8 is expressed. The relative proportions of the subsets vary with age, strain (in mice), and number of bacteria in the intestine. Type a (inducible) IELs express α:β T-cell receptors and the CD8α:β heterodimer. They are derived from naive CD8 T cells that were activated by antigen in the Peyer's patches or mesenteric lymph nodes, and they function as conventional class I MHC-restricted cytotoxic T cells, killing virus-infected cells, for example (Fig. 12.14, top panels). They also secrete effector cytokines such as IFN-γ.

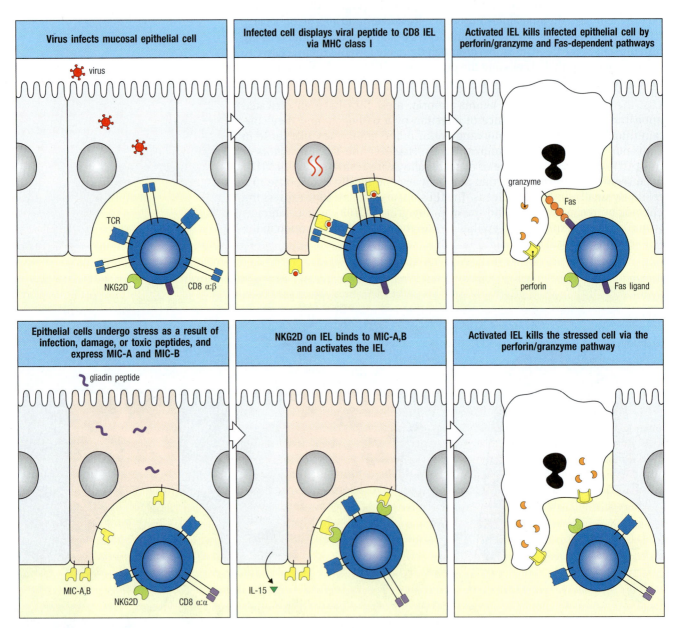

Fig. 12.14 Effector functions of intraepithelial lymphocytes. Type a IELs (top panels) are conventional CD8 cytotoxic T cells that recognize peptides derived from viruses or other intracellular pathogens bound to classical MHC class I molecules on infected epithelial cells. Type a IELs express an α:β T-cell receptor and the CD8α:β heterodimer co-receptor. Type b IELs carrying the CD8α:α homodimer (bottom panels) recognize MIC-A and MIC-B using the receptor NKG2D and are activated by IL-15. Human epithelial cells that have been stressed by infection or altered cell growth or by a toxic peptide from the protein α-gliadin (a component of gluten) upregulate expression of the nonclassical MHC class I molecules MIC-A and MIC-B and produce IL-15. Both types of IELs can kill by the release of perforin and granzyme. Apoptosis of epithelial cells can also be induced by the binding of Fas ligand on the T cell to Fas on the epithelial cell.

Type b (natural) CD8 IELs can express either an α:β or a γ:δ T-cell receptor, but are distinguished by their expression of the CD8α:α homodimer. The γ:δ T cells in the intestine are enriched for particular Vγ and Vδ genes and are distinct from those found in other tissues (see Fig. 8.23). Some of the α:β receptors expressed by IELs bind nonconventional ligands, including those presented by MHC class Ib molecules (see Section 6-17). Type b IELs also express molecules typical of natural killer cells, such as the activating C-type lectin NKG2D, which binds to the two MHC-like molecules MIC-A and MIC-B. These are induced on intestinal epithelial cells in response to cellular injury, stress, or ligation of TLRs (see Section 6-16). The injured cells can then be recognized and killed by the IELs, a process that is enhanced by the production of IL-15 by the damaged epithelial cells. Like innate immune cells, type b IELs constitutively express genes associated with inflammation, such as the production of high levels of cytotoxic molecules, NO, and pro-inflammatory cytokines and chemokines. Their role in the gut may be the rapid recognition and elimination of epithelial cells that express an abnormal phenotype as a result of stress or infection (see Fig. 12.14, bottom panels). Type b IELs are also thought to be important in aiding the repair of the mucosa after inflammatory damage: they stimulate the release of antimicrobial peptides, thus helping to remove the source of the inflammation; and they release cytokines such as keratinocyte growth factor, which promotes epithelial barrier function, and TGF-β, which assists tissue repair, as well as inhibiting local inflammatory reactions.

Type b IELs are kept in check by their co-expression of signaling inhibitors, including the immunomodulatory cytokine TGF-β and inhibitory receptors like those found on NK cells. The importance of these control processes is shown by the fact that inappropriate or excess activation of type b IELs may give rise to disease. For example, increased numbers of IELs expressing a γ:δ T-cell receptor are found in celiac disease, which is caused by an abnormal immune response to the wheat protein gluten (see Section 14-17). MIC-A-dependent cytotoxic activity of intraepithelial T cells contributes to the intestinal damage in this condition, as certain components of gluten can stimulate the production of IL-15 by epithelial cells and increase expression of MIC-A. These processes lead to killing of epithelial cells by the activated IELs, as described above (see Fig. 12.14, bottom panels).

Celiac Disease

The origin and development of type b IELs has been controversial and is unexplored in humans. Unlike type a IELs, many type b IELs expressing an α:β T-cell receptor seem not to have undergone conventional positive and negative selection (see Chapter 8), and express apparently autoreactive T-cell receptors. The absence of the CD8α:β heterodimer, however, means that these T cells have low affinity for conventional peptide:MHC complexes, because the CD8β chain binds more strongly than the CD8α chain to classical MHC molecules. Type b α:β T-cell receptor-expressing IELs therefore cannot act as self-reactive effector cells. This low affinity for self MHC molecules is also probably the reason that these cells escape negative selection in the thymus. Rather, they appear to develop via a process of so-called **agonist selection**, in which late double-negative/early double-positive T cells are positively selected in the thymus by unknown ligands and are released immediately to the intestine. Here they mature and are induced to express the CD8α:α homodimer under the influence of TGF-β produced by epithelial cells. Nonclassical MHC molecules expressed on the intestinal epithelium are also important for the maturation of these type b IELs. One example of this kind of selection molecule is the **thymus leukemia antigen (TL)**, another nonclassical MHC class I molecule (see Fig. 6.26) found in certain mouse strains which does not present peptides. TL is expressed by intestinal epithelial cells and directly binds CD8α:α with high affinity.

Type b IELs expressing a γ:δ T-cell receptor also develop via agonist selection in the thymus, as part of the programmed wave of γ:δ T-cell development (see

Fig. 8.23). The expression of this receptor is driven by specific ligands in the thymus and endows these cells with the specific ability to migrate to the intestinal epithelium, where they may be further programmed by the same agonist ligand.

The local differentiation events involved in the development of type b IELs require the presence of the cytokine IL-15, which is produced in response to the microbiota and 'trans-presented' to IELs in a complex with the IL-15 receptor present on epithelial cells. Type b IEL development is dependent on the **aryl hydrocarbon receptor** (**AhR**), a transcription factor activated by various environmental ligands that are derived from brassica and other dietary vegetables. Mice that lack the AhR have reduced numbers of ILC3 and type b IELs and show abnormalities in epithelial barrier repair, reinforcing the view that these unusual lymphocytes play important roles in the innate immune response to local materials in the intestine.

Summary.

The mucosal tissues of the body such as the intestine and respiratory tract are continuously exposed to enormous amounts of different antigens, which can be either pathogenic invaders or harmless materials such as foods and commensal organisms. Potential immune responses to this antigen load are controlled by a distinct compartment of the immune system, the mucosal immune system, which is the largest in the body. Its unique features include distinctive routes and processes for the uptake and presentation of antigens, exploitation of M cells to transport antigens across the epithelium of Peyer's patches, and retinoic acid-producing dendritic cells that imprint the T and B cells they activate with gut-homing properties. Dendritic cells also favor the generation of FoxP3-positive T_{reg} cells in the normal gut. Tissue-resident intestinal macrophages contribute to these regulatory processes by phagocytosing antigens without causing inflammation, due to their production of IL-10. Lymphocytes primed in the mucosa-associated lymphoid tissues acquire specific homing receptors, allowing them to redistribute preferentially back to mucosal surfaces as effector cells. The adaptive immune response in mucosal tissues is characterized by the production of secretory dimeric IgA, and by the presence of distinct populations of memory/effector T cells in the epithelium and lamina propria. CD4 T cells in the lamina propria produce pro-inflammatory cytokines such as IL-17 and IFN-γ even in the absence of overt infection, but this is normally balanced by the presence of IL-10-producing T_{reg} cells. IELs exhibit cytolytic activity and other innate functions that help maintain a healthy epithelial barrier.

The mucosal response to infection and regulation of mucosal immune responses.

The major role of the mucosal immune response is defense against infectious agents, which include all forms of microorganisms from viruses to multicellular parasites. This means that the host must be able to generate a wide spectrum of immune responses tailored to meet the challenge of individual pathogens; unsurprisingly, many microbes have evolved means of adapting to and subverting the host response. To ensure an adequate response to pathogens, the mucosal immune system needs to be able to recognize and respond to any foreign antigen, but it must not produce the same effector response to a harmless antigen (from food or commensals) as it would to a pathogen. A major role of the mucosal immune system is to balance these competing demands, and how it does this will be the focus of this part of the chapter.

12-15 Enteric pathogens cause a local inflammatory response and the development of protective immunity.

Despite the array of innate immune mechanisms in the gut, and stiff competition from the indigenous microbiota, the gut is a frequent site of infection by a wide variety of pathogenic organisms. These include many viruses; enteric bacteria such as *Vibrio*, *Salmonella*, and *Shigella* species; protozoans such as *Entamoeba histolytica*; and multicellular helminth parasites such as tapeworms and pinworms. These pathogens cause disease in many ways, and as elsewhere in the body, the key to generating protective immunity is the activation of appropriate aspects of the innate immune system.

The effector mechanisms of the innate immune system can themselves eliminate most intestinal infections rapidly and without significant spread of the infection beyond the intestine. The essential features of these responses in epithelial surfaces are discussed in Section 2-2 and here we highlight only aspects that are unique or unusual to the intestine. Of these, the most important involve the epithelial cells themselves (**Fig. 12.15**). The tight junctions between

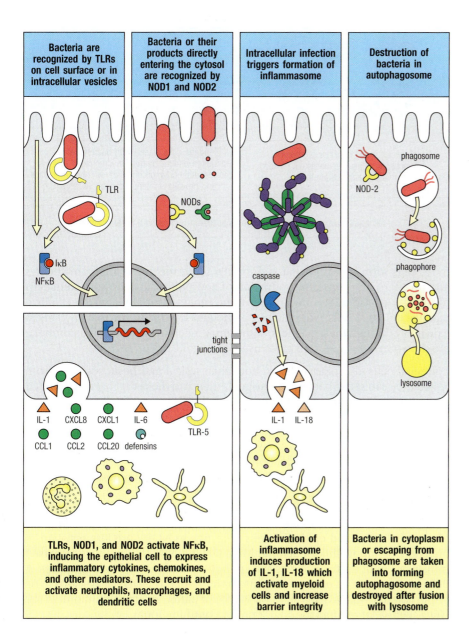

Fig. 12.15 Epithelial cells have a crucial role in innate defense against pathogens. TLRs are present in intracellular vesicles or on the basolateral or apical surfaces of epithelial cells, where they recognize different components of invading bacteria. Cytoplasmic NOD1 and NOD2 detect cell-wall peptides from bacteria. TLRs and NODs activate NFκB (see Fig. 3.15), inducing epithelial cells to produce CXCL8, CXCL1 (GROα), CCL1, and CCL2, which attract neutrophils and macrophages, CCL20, which attracts dendritic cells, and IL-1 and IL-6 that activate macrophages. Many types of cell damage can activate inflammasome (see Section 3-9) that activates pro-caspase 1 and produces IL-1 and IL-18. Bacteria that invade the epithelial-cell cytoplasm or escape into the cytosol from phagosomes can induce autophagy. The organisms become ubiquitinated, leading to the recruitment of adaptor proteins that attract the phagophore, forming an autophagosome. Fusion with lysosomes then leads to destruction of the cargo within the autophagosome. NOD2 can also trigger autophagosome formation by binding directly to adaptor proteins, including the Crohn's disease-associated molecule ATGL16L1.

Fig. 12.16 *Salmonella enterica* serovar Typhimurium is an important cause of food poisoning and penetrates the epithelial layer in three ways. *Salmonella* Typhimurium adheres to and enters M cells, which it then kills by causing apoptosis (top left). It then can infect macrophages and gut epithelial cells. TLR-5 expressed by the epithelial basal membrane can bind salmonella flagellin, activating the NFκB pathway. After uptake by macrophages in the lamina propria, invasive *Salmonella* induces caspase 1 activation, promoting production of IL-1 and IL-8. CXCL8 is also produced by the infected macrophages, and together these mediators recruit and activate neutrophils (lower left panel). Salmonellae can also invade gut epithelial cells directly by adherence of fine threadlike protrusions on the luminal epithelial surface called fimbriae (top middle panel). The cell processes extended between epithelial cells by mononuclear phagocytes may be infected by salmonellae in the lumen and thus effectively breach the epithelial layer (top right panel). Dendritic cells in lamina propria may become infected from infected macrophages and carry them to the draining mesenteric lymph node to prime the adaptive immune response (lower right panel). If containment processes in the lymph node fail, *Salmonella* can invade beyond the intestine and its lymphoid tissues and enter the bloodstream to cause systemic infection.

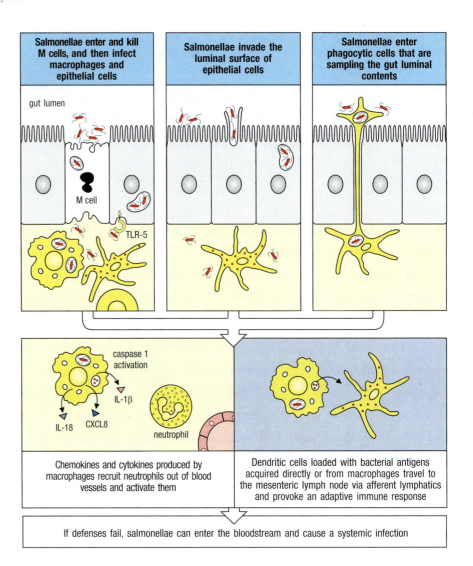

these cells form a barrier that is normally impermeable to macromolecules and invaders. The constant production of new epithelial cells from stem cells in the crypts also allows the barrier to be repaired rapidly after mechanical damage or loss of cells. Nonetheless, pathogens have acquired mechanisms to gain entry through these barriers; some entry mechanisms used by salmonella are shown in **Fig. 12.16**, and those used by shigella in **Fig. 12.17**.

Epithelial cells also bear TLRs on both their apical and basal surfaces, from which they can sense bacteria in the gut lumen and those that have penetrated across the epithelium. In addition, epithelial cells carry TLRs in intracellular vacuoles that can detect intracellular pathogens or extracellular pathogens and their products that have been internalized by endocytosis. Epithelial cells also have intracellular sensors, described in Chapter 3, and can react when microorganisms or their products enter the cytoplasm. These sensors include the **nucleotide-binding oligomerization domain** (NOD) proteins NOD1 and NOD2 (see Section 3-8 and Fig. 3.17). NOD1 recognizes a diaminopimelic acid-containing peptide that is found only in the cell walls of Gram-negative bacteria. NOD2 recognizes a muramyl dipeptide found in the peptidoglycans of most bacteria, and epithelial cells defective in NOD2 are less resistant to infection by intracellular bacteria. Mice lacking NOD2 also show increased translocation of bacteria across the epithelium and out of Peyer's patches. A defect in recognition of the commensal microbiota by NOD2 also seems to be important in Crohn's disease, as up to 25% of patients carry a mutation in the *NOD2* gene that renders the NOD2 protein nonfunctional.

Crohn's Disease

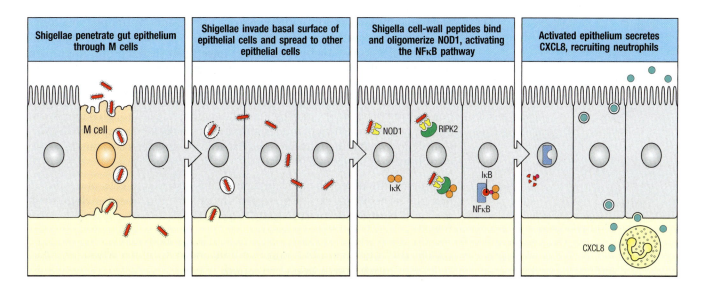

| Shigellae penetrate gut epithelium through M cells | Shigellae invade basal surface of epithelial cells and spread to other epithelial cells | Shigella cell-wall peptides bind and oligomerize NOD1, activating the NFκB pathway | Activated epithelium secretes CXCL8, recruiting neutrophils |

Fig. 12.17 *Shigella flexneri* **infects intestinal epithelial cells to cause bacterial dysentery.** *Shigella flexneri* binds to M cells and is translocated beneath the gut epithelium (first panel). The bacteria infect intestinal epithelial cells from their basal surface and are released into the cytoplasm (second panel). Muramyl tripeptides containing diaminopimelic acid in the cell walls of the shigellae bind to and oligomerize the protein NOD1. Oligomerized NOD1 binds the serine/threonine kinase RIPK2 and activates the NFκB pathway (see Fig. 3.17), leading to the transcription of genes for chemokines and cytokines (third panel). Activated epithelial cells release the chemokine CXCL8, which acts as a neutrophil chemoattractant (fourth panel). IκB, inhibitor of NFκB; IκK, IκB kinase.

Ligation of TLRs or NOD proteins in epithelial cells stimulates the production of cytokines, such as IL-1 and IL-6, and the production of chemokines. The chemokines include CXCL8, which is a potent neutrophil chemoattractant, and CCL2, CCL3, CCL4, and CCL5, which attract monocytes, eosinophils, and T cells out of the blood. Stimulated epithelial cells also increase their production of the chemokine CCL20, which attracts immature dendritic cells toward the epithelial surface (see Sections 12-4 and 12-7).

Epithelial cells also express members of the intracellular **NOD-like receptor (NLR)** family, including gNLRP3, NLRC4, and NLRP6, that can form **inflammasomes** (see Fig. 12.15). As described in Section 3-9, formation of an inflammasome leads to activation of caspase 1, which cleaves pro-IL-1 and pro-IL-18 to produce the active cytokines (see Fig. 3.19). Both these cytokines contribute to epithelial defense against bacterial invasion by promoting barrier integrity, but can cause tissue damage if present for long periods.

One mechanism recently recognized as important for epithelial defense against infection is **autophagy**, which we discussed in Section 6-6 for its relationship to antigen processing. In this process, a crescent-shaped double-membrane fragment in the cytoplasm, called the **isolation membrane**, or **phagophore**, engulfs various cytoplasmic contents to form a complete vesicle, the **autophagosome**, which fuses with lysosomes to degrade the contents (see Fig. 12.15). When autophagy is disrupted, bacteria cannot be contained effectively, and epithelial cells become stressed. This can lead to increased penetration of bacteria into the body and to NFκB-mediated inflammation. Autophagy is promoted by the NOD1 and NOD2 intracellular bacterial sensors. As with NOD2, mutations in the autophagy-related genes *ATG16L1* and *IRGM1* are associated with susceptibility to Crohn's disease in humans.

Certain specialized populations of epithelial cells have particularly important roles in innate immune defense of the intestine. **Paneth cells** are found only in the small intestine, where they produce antimicrobial peptides such as RegIIIγ and defensins when exposed to IL-22 released by CD4 T_H17 cells or ILC3s. They can also respond directly to microbes, as they express TLRs and NODs and they are highly autophagic. Defects in Paneth cell function lead to reduced bacterial defense and are believed to be important in susceptibility to inflammatory bowel disease in humans. **Goblet cells** are a further kind of specialized epithelial cell and produce mucus in response to cytokines derived from CD4 T_H2 cells or ILC2s. Mucus consists of a complex mixture of highly charged glycoproteins (mucins) and forms an essential component of immune defense in all mucosal surfaces. Its density, charge, and stickiness mean that it presents a formidable barrier to invasion, by trapping microbes and other particles. At the same time, it acts as a scaffolding to retain IgA antibodies and antimicrobial peptides that have been secreted into the lumen across the epithelium. Mucus is also slippery in nature, meaning that trapped materials can then be expelled easily by normal peristaltic movements. In the intestine, there are two layers of mucus, an outer loose layer and a much denser inner layer, found mostly in the large intestine. Although bacteria can penetrate the loose layer of mucus, they are normally kept away from the surface of the epithelial cells by the inner dense layer, and defects in this structure compromise antimicrobial defense.

As we have discussed, the intestinal mucosa is also rich in cells of the innate immune system that can respond rapidly to infection. These include macrophages, eosinophils, mast cells, ILCs, MAIT cells, NKT cells, and $\gamma{:}\delta$ T cells.

12-16 Pathogens induce adaptive immune responses when innate defenses have been breached.

If pathogenic bacteria and viruses gain access to the subepithelial space, they may interact with TLRs on inflammatory cells in the underlying tissue. Together with the cascade of inflammatory mediators released by epithelial cells, this dramatically alters the environment of the mucosa and changes the behavior of local antigen-presenting cells such as dendritic cells. As described in Section 9-8, activated dendritic cells will express high levels of co-stimulatory molecules and cytokines such as IL-1, IL-6, IL-12, and IL-23, and promote development of effector T cells. Dendritic cells activated in Peyer's patches migrate to the T-cell-dependent areas of the patch, whereas dendritic cells that encounter antigen in the lamina propria migrate to the mesenteric lymph node under control of CCR7. The effector T cells activated in these ways acquire gut-homing molecules such as $\alpha_4{:}\beta_7$ and CCR9 due to the presence of retinoic acid, ensuring that they return to the gut wall to encounter the invading organisms. Similarly, IgA-producing B lymphocytes are generated in Peyer's patches and mesenteric lymph nodes, generating plasma cells that accumulate in the lamina propria. IgA secretion into the lumen is enhanced in response to infection because pIgR expression is enhanced by TLR ligands and pro-inflammatory cytokines. In some infections, IgG antibodies can now be found in intestinal secretions, but these are derived from serum and require invading organisms to reach systemic immune tissues.

The activated myeloid cells found in the inflamed mucosa also contribute to sustaining the functions of effector T and B cells after their arrival in the mucosa. IL-1 and IL-6 produced by recently arrived monocytes are important for maintaining the survival and function of local T_H17 cells. Pro-inflammatory myeloid cells also produce mediators such as IL-6, TNF-α, and nitric oxide that help drive IgA switching and secondary expansion of mucosal B cells.

12-17 Effector T-cell responses in the intestine protect the function of the epithelium.

Once activated, the effector T cells that accumulate in the intestine behave much like their counterparts elsewhere in the body, producing cytokines and generating cytolytic activity as appropriate to the pathogen. What is different is that the aim of the protective immune response in the intestine is tailored to preserving the integrity and function of the epithelial barrier. This is achieved in a number of ways, depending on the nature of the pathogen. In virus infections, CD8 cytotoxic T cells among intraepithelial lymphocytes kill infected epithelial cells (see Fig 12.14), triggering their replacement by uninfected cells derived from the rapidly dividing stem cells in the crypts. A similar process can occur during other forms of protective immune responses, with cytokines from CD4 effector T cells directly stimulating epithelial cell division. This forces the replacement of infected cells and generates a moving target for organisms that are attempting to attach to the surface of the epithelium. An example of a cytokine of this kind is IL-13, produced by T_H2 cells (and ILC2s) during parasitic infections. In addition to its ability to stimulate antimicrobial peptide production by Paneth cells, IL-22 produced by T_H17 cells contributes to defense against extracellular bacteria and fungi by enhancing the tight junctions between epithelial cells that keep the barrier intact. Mucus is crucially important in protecting the epithelial barrier, and its production by goblet cells is enhanced by CD4 T-cell-derived cytokines such as IL-13 and IL-22, as well as by products of mast cells and other innate effector cells recruited by T cells. Finally, these mediators and others can enhance the peristaltic action of the intestine and its outward secretion of fluid, washing out pathogens within the lumen of the intestine. Together these processes aim to generate a hostile and unstable environment for the pathogen, reducing its ability to invade and damage the epithelial barrier.

12-18 The mucosal immune system must maintain tolerance to harmless foreign antigens.

Antigens within food and commensal bacteria normally do not induce an inflammatory immune response, despite the lack of central tolerance to them (Fig. 12.18). The mucosal immune system's environment is inherently tolerogenic, and this is a barrier to the development of nonliving vaccines, which need to overcome local regulatory mechanisms. Food proteins are not digested completely in the intestine; significant amounts are absorbed into the body in an immunologically relevant form. The default response to oral administration of a protein antigen is the development of a phenomenon known as **oral tolerance**. This is a form of **peripheral tolerance** that renders the systemic and mucosal immune systems relatively unresponsive to the same antigen. It can be demonstrated experimentally in mice by feeding them a foreign protein such as ovalbumin (Fig. 12.19). When the animals are then challenged with the antigen by a nonmucosal route, such as injection into the skin, the immune response one would expect is blunted. This suppression of systemic immune responses is long lasting and is antigen specific: responses to other antigens are not affected. A similar suppression of subsequent immune responses is observed after the administration of proteins into the respiratory tract, giving rise to the concept of **mucosal tolerance**, as the usual response to such antigens is delivered via a mucosal surface. Systemic T-cell responses can also be inhibited by feeding humans protein antigens that they have not encountered previously.

Oral tolerance can impact all aspects of the peripheral immune response, including T-cell-dependent effector responses and IgE production. Effector T-cell responses in the mucosa are also downregulated in oral tolerance, although low levels of secretory IgA antibodies directed at food proteins can be found in healthy humans, but do not lead to inflammation.

Various mechanisms are likely to account for oral tolerance to protein antigens, including anergy, deletion of antigen-specific T cells, and the generation of regulatory T cells induced in the mesenteric lymph node to become gut-homing, antigen-specific FoxP3-positive T_{reg} cells via the production of retinoic acid and TGF-β by migratory dendritic cells (see Section 12-7). Although it is known that all these events are also essential for the suppression of systemic immune responses, the mechanisms responsible for this link between the mucosal and peripheral immune system are not yet understood. At times, oral tolerance can fail, as is believed to occur in celiac disease (discussed in more detail in Section 14-17) or peanut allergies (discussed in Sections 14-10 and 14-12).

	Protective immunity	Mucosal tolerance
Antigen	Invasive bacteria, viruses, toxins	Food proteins; commensal bacteria
Primary Ig production	Intestinal IgA Specific Ab present in serum	Some local IgA Low or no Ab in serum
Primary T-cell response	Local and systemic effector and memory T cells	No local effector T-cell response
Response to antigen reexposure	Enhanced (memory) response	Low or no response or systemic response

Fig. 12.18 Immune priming and tolerance are different outcomes of intestinal exposure to antigen. The intestinal immune system generates protective immunity against antigens that are presented during infections by pathogenic organisms. IgA antibodies are produced locally, serum IgG and IgA are made, and the appropriate effector T cells are activated in the intestine and elsewhere. When the antigen is encountered again, there is effective memory, ensuring rapid protection. Antigens from food proteins induce tolerance locally and systemically, with little or no IgA antibody production. T cells are not activated, and subsequent responses to challenge are suppressed. In the case of commensal bacteria, there may be some local IgA production, but no primary systemic antibody responses, and effector T cells are not activated.

Celiac Disease

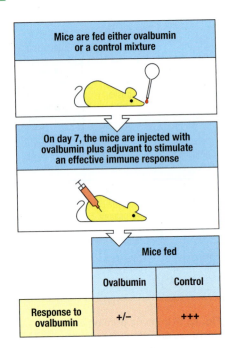

Fig. 12.19 Tolerance to antigens can be experimentally generated by oral administration. Mice are fed for 2 weeks with 25 mg of either ovalbumin, the experimental protein, or a second protein as a control. Seven days later, the mice are immunized subcutaneously with ovalbumin plus an adjuvant, and after 2 weeks, the serum antibodies and T-cell function are measured. Mice that were fed ovalbumin have a lower ovalbumin-specific systemic immune response than those fed the control protein.

Although mucosal tolerance can be used to avoid inflammatory disease in experimental animal models of type 1 diabetes mellitus, arthritis, and encephalomyelitis, clinical trials in humans have been less successful, and have been superseded by other therapies, such as monoclonal antibodies, that we will discuss in greater detail in Chapter 16.

12-19 The normal intestine contains large quantities of bacteria that are required for health.

The surfaces of the healthy body are colonized by large numbers of microorganisms, collectively referred to as the **microbiota**, or **microbiome**, composed mostly of bacteria, but also archaea, viruses, fungi, and protozoa . The intestine is the largest source of these organisms, although all the other mucosal tissues harbor their own, distinct populations of microbes. We each harbor more than 1000 species of commensal bacteria in our intestine, and they are present in greatest numbers in the colon and lower ileum. As many of the species cannot be grown in culture, their exact numbers and identities are only now being established by high-throughput sequencing techniques. In humans, there are several major phyla of bacteria, plus the archaea—in descending order, Firmicutes, Bacteroidetes, Proteobacteria, Actinobacteria, and Archaea. There are at least 10^{14} of these microorganisms that collectively weigh about 1 kg. The intestinal microbiota normally exists in a mutually beneficial, or **symbiotic**, relationship known as **mutualism** that has been established in humans over many millennia and has coevolved with vertebrates throughout their history. As a result, the populations of these microbes found in different groups of animals are distinctive and are highly adapted to their individual host species.

The microbiota has an essential role in maintaining health. Its members assist in the metabolism of dietary constituents such as cellulose, as well as degrade toxins and produce essential cofactors such as vitamin K_1. Short-chain fatty acids (SCFAs), such as acetate, propionate, and especially **butyrate**, produced by anaerobic metabolism of dietary carbohydrates by commensal bacteria are an essential source of energy for colonic enterocytes through their entry as substrates into the tricarboxylic acid (TCA) cycle. Surgical procedures, such as ileostomy, that remove the normal fecal flow to the colon can cause a syndrome called **diversion colitis**, in which enterocytes starved of SCFAs undergo inflammation and necrosis. Providing SCFAs to the affected colon segment can reverse this condition. Another important property of commensal organisms is that they interfere with the ability of pathogenic bacteria to colonize and invade the gut, partly by competing for space and nutrients. They can also directly inhibit the pro-inflammatory signaling pathways that pathogens stimulate in epithelial cells and that are needed for invasion. Perturbations in the balance between the various species of bacteria present in the microbiota (**dysbiosis**) have been found to increase susceptibility to a variety of diseases (see Sections 12-21 and 12-22).

The protective role of the commensal microbiota is dramatically illustrated by the adverse effects of broad-spectrum antibiotics. These antibiotics can kill large numbers of commensal gut bacteria, thereby creating an ecological niche for bacteria that would not otherwise be able to compete successfully. One example of a bacterium that grows in the antibiotic-treated gut and can cause a severe infection is ***Clostridium difficile*** (**Fig. 12.20**). This organism is an increasing problem in countries where broad-spectrum antibiotic use is prevalent, as it produces toxins that cause severe diarrhea and mucosal injury. Restoring the normal microbiota by a transplant of feces from healthy individuals can be used to treat *C. difficile* infection.

The importance of the local defense mechanisms against commensal bacteria for health is shown by experiments in animals that lack one or more of the factors involved. For instance, mice without secretory antibodies have increased

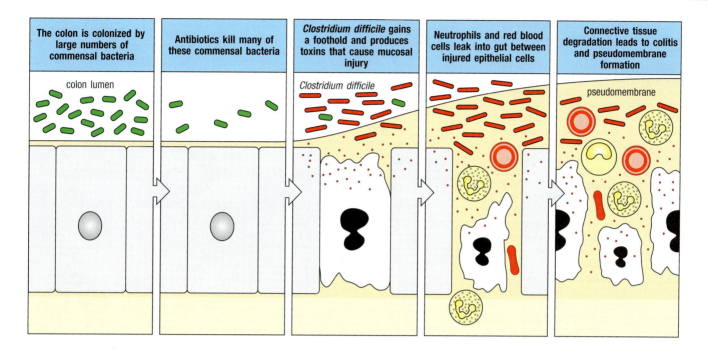

| The colon is colonized by large numbers of commensal bacteria | Antibiotics kill many of these commensal bacteria | *Clostridium difficile* gains a foothold and produces toxins that cause mucosal injury | Neutrophils and red blood cells leak into gut between injured epithelial cells | Connective tissue degradation leads to colitis and pseudomembrane formation |

colon lumen

Clostridium difficile

pseudomembrane

numbers of commensal bacteria that have penetrated the intestinal mucosa and have disseminated beyond its draining lymphoid tissues. The composition of the microbiota in these mice is also altered, with increased numbers of bacteria, but decreased species diversity. Similar dysbiosis has been described in mice lacking FoxP3+ regulatory T cells or eosinophils.

12-20 Innate and adaptive immune systems control microbiota while preventing inflammation without compromising the ability to react to invaders.

Despite their beneficial effects, commensal bacteria are a potential threat, as is shown when the integrity of the intestinal epithelium is damaged. In these circumstances, normally innocuous gut bacteria, such as nonpathogenic *E. coli*, can cross the mucosa, invade the bloodstream, and cause fatal systemic infection. Therefore, the immune system in the intestine has to mount some form of response to control commensal microbes (**Fig. 12.21**). Since inappropriate reactions may lead to chronic inflammation and damage to the intestine, the immune system must balance the recognition and response to commensal bacteria with the cost of damaging tissues from inflammation. Commensal bacteria elicit antigen-specific responses that maintain the local balance between host and microbiota and are largely confined to the intestine itself. Unlike soluble food antigens, commensal bacteria do not induce a state of systemic immune unresponsiveness, and when these organisms enter the bloodstream, they can stimulate a normal primary systemic immune response.

Recognition of the microbiota by the adaptive immune system is dependent on the uptake and intracellular transport of organisms by local dendritic cells that remain in Peyer's patches or migrate no further than the mesenteric lymph node (see Fig. 12.21), which acts to prevent wider dissemination of the microbiota. Because commensal microbes are noninvasive, dendritic cells are not fully activated and induce a finely balanced response, comprising secretory IgA antibodies that enter intestinal secretions and are directed at commensal bacteria. Up to 75% of commensal organisms living in the lumen appear to be coated by IgA (see Fig. 12.21), limiting their adherence to and penetration of the epithelium. In addition, coating of the microbiota with SIgA can alter their gene expression. Many of the large numbers of fully differentiated T_H1 and

Fig. 12.20 Infection by *Clostridium difficile*. Treatment with antibiotics causes massive death of the commensal bacteria that normally colonize the colon. This allows pathogenic bacteria to proliferate and to occupy an ecological niche that is normally occupied by harmless commensal bacteria. *Clostridium difficile* is an example of a pathogen producing toxins that can cause severe bloody diarrhea in patients treated with antibiotics.

Fig. 12.21 Several local processes ensure peaceful homeostasis between host and microbiota. Commensal bacteria in the lumen gain access to the immune system via M cells. Antigens are taken up by dendritic cells in Peyer's patches and isolated follicles under noninflammatory conditions (left panel). Presentation of these antigens generates IgA-switched B cells that localize in the lamina propria as IgA-producing plasma cells (right panel). IgA then binds commensal bacteria, altering their gene expression, limiting their access to the epithelium, and blocking their binding to the surface. Interference with penetration of the epithelium is assisted by the presence of thick layers of mucus, which also contain mucin glycoproteins that have antibacterial properties. In addition, stimulation of pattern recognition receptors on Paneth cells induces the production of antimicrobial peptides such as RegIIIγ and defensins (see Section 2-4), which are also stimulated by IL-22 derived from T_H17 CD4 T cells and ILC3s. IL-22 also tightens the epithelial barrier. Phagocytic macrophages found immediately under the epithelium can ingest and kill bacteria that penetrate the surface.

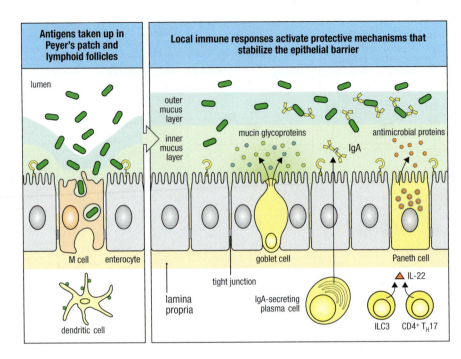

T_H17 cells found in healthy intestine are also directed at the microbiota. While these cells produce mediators that can assist bacterial clearance by macrophages and epithelial cells, they come with the risk of producing inflammation and collateral damage. This does not occur, because of IL-10 produced by T cells and FoxP3+ regulatory T cells present in the mucosa. T_H17 and FoxP3+ regulatory T cells in the intestine can enter germinal centers in Peyer's patches and acquire the functions of follicular helper T cells, leading to selective IgA switching.

The endotoxin present on commensal bacteria also seems unusually sensitive to neutralization by gut enzymes such as alkaline phosphatase, leading to weaker immune activation. If commensal bacteria do cross the epithelium in small numbers, their lack of virulence factors means they cannot resist uptake and killing by phagocytic cells, and they are rapidly destroyed. In contrast to what happens in other tissues, ingestion of commensal bacteria in the intestine does not lead to inflammation. If macrophages cannot respond to the inhibitory effects of IL-10, intestinal inflammation develops spontaneously. Eosinophils in the healthy intestine assist antigen-specific IgA switching by producing APRIL, IL-6, and TGF-β when exposed to commensal microbes (see Section 12-11). Thus, commensal organisms associate with the mucosal surface without invading or provoking inflammation. This symbiosis involves many innate and adaptive immune effector cells that are usually associated with chronic inflammation but in the intestine create a state sometimes referred to as **physiological inflammation**.

12-21 The intestinal microbiota plays a major role in shaping intestinal and systemic immune function.

Commensal bacteria and their products play an essential role in normal development of the immune system. This effect is illustrated in **germ-free**, or **gnotobiotic**, mice, in which there is no colonization of the gut by microorganisms. These animals have marked reductions in the size of all lymphoid organs, low serum immunoglobulin levels, fewer mature T cells, and markedly reduced immune responses, especially T_H1 and T_H17 responses. Such mice are prone to make T_H2-type responses such as IgE antibodies and are more susceptible

Fig 12.22 Effects of the microbiota on disease and systemic immune function. The presence and composition of the microbiota have many downstream consequences for the function of the immune system and other body tissues, some of which may be secondary to the events in the mucosa, while others may reflect the ability of products of intestinal microbes to enter the circulation. The microbiota is also known to have many effects on susceptibility to a wide range of diseases in humans and experimental animals.

Remote effects of microbiota
Decrease IgE, T_H2 responses
Increase T_{reg}
Increase bone remodeling
Carbohydrate lipid metabolism
Insulin sensitivity
Myelopoiesis
Hypothalamus-pituitary-adrenal axis

Modulation of disease
Arthritis
Experimental autoimmune encephalomyelitis
Inflammatory bowel disease
Atopy, asthma
Metabolic disease
Cardiovascular disease
Type I diabetes (reduced by microbiota)

to certain immunological diseases such as type 1 diabetes. In the intestine, Peyer's patches do not develop normally and isolated lymphoid follicles are absent. Germ-free mice also have severely reduced numbers of T lymphocytes and ILCs in the lamina propria and epithelium, nearly absent IgA-secreting plasma cells, and reduced mediators of local immunity, such as antimicrobial peptides, retinoic acid, IL-7, IL-22, IL-25, IL-33, and TSLP. In contrast, invariant NKT cells (iNKTs) are more abundant in the germ-free intestine, perhaps contributing to the T_H2 bias seen in germ-free animals.

The effects of the intestinal microbiota extend far beyond the intestine (Fig. 12.22). For example, several autoimmune diseases are more frequent in germ-free animals. The germ-free state greatly increases the severity of symptoms in a genetic model of type 1 diabetes. The composition of the microbiota influences susceptibility to many different immunological diseases, metabolic disorders such as obesity, cancer, cardiovascular disease, and even psychiatric disorders. The basis for these associations is unclear and few individual commensal species have been identified in disease susceptibility. However, some affected individuals have unusual compositions of the major bacterial species that normally make up the microbiota, a form of **dysbiosis**, as we saw in Section 12-19. In experimental models, disease susceptibility can be conferred by transferring intestinal bacteria between affected and unaffected animals, supporting the idea that the change in microbiota is a causal factor, rather than being secondary to preexisting disease. This observation underlies the use of probiotics, which are particular mixtures of live bacteria and yeast that are considered beneficial. Their use may manipulate the intestinal microbiota to prevent disease and promote health, although much remains to be understood about their potential benefits.

Many different mechanisms are probably involved in the effects of the microbiota (Fig. 12.23). Ligation of TLRs and NLRs is undoubtedly important for many of the local effects on epithelial cells and myeloid cells. Flagellin present on many intestinal bacterial species can stimulate TLR-5 on mucosal CD11b-expressing dendritic cells, inducing the production of IL-6 and IL-23 and favoring T_H17 and IgA responses. There are also examples of individual bacterial species that have specific effects on immune function. Colonization of mice with **segmented filamentous bacteria** (SFB) enhances IgA production, accumulation of IELs, and the number of intestinal effector T_H17 T cells (see Fig. 12.23). Conversion of dietary tryptophan by lactobacilli into **kynurenine metabolites** can activate the AhR (see Section 12-14) and enhance IL-22 production by ILC3. Polysaccharide A (PSA) from *Bacteroides fragilis* drives the differentiation of T_{reg} cells in a TLR-2-dependent manner. Also, several *Clostridium* species stimulate the preferential generation of FoxP3+ regulatory T cells in the colon, perhaps by promoting a TGF-β-rich environment and by producing SCFAs. The mechanism by which SCFAs directly alter immune cell function is currently unclear. As yet, few specific organisms have been identified to explain the effects of dysbiosis on human disease, although certain *E coli* species, collectively called **enteroadherent Escherichia coli**, have been found to be prevalent in patients with Crohn's disease. Recent studies have also shown increased abundance of *Prevotella copri* in a number of patients with newly diagnosed rheumatoid arthritis, but much more work needs to be done to confirm these associations and to find if there are similar effects in other diseases.

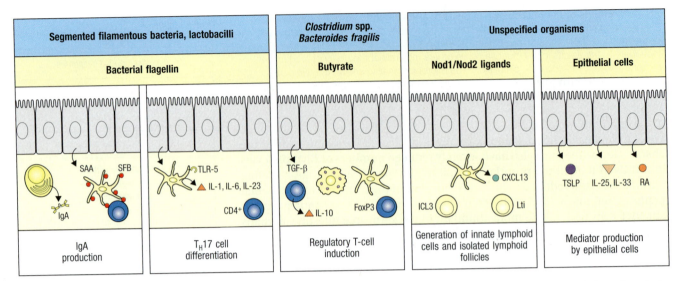

Fig 12.23 The microbiota tune local and systemic immune responses. The microbiota has local and distant effects on immune function, although only a few individual organisms and mechanisms have been identified. Segmented filamentous bacteria (SFB) potently induce SFB-specific T_H17 cells, perhaps by inducing epithelial cells to produce serum amyloid A (SAA) protein that may act on dendritic cells. Bacterial flagellin favors T_H17 and IgA responses by stimulating TLR-5 on mucosal CD11b-expressing dendritic cells. The microbiota is also needed for the presence of isolated lymphoid follicles and ILCs, especially ILC3 cells, but inhibits the accumulation of invariant NKT cells (iNKTs). Besides providing energy to colonic enterocytes, butyrate and other SCFAs may also act to drive the generation of FoxP3$^+$ T_{reg}s, although the molecular mechanism is still unclear. Clostridia also induce production of TGF-β by epithelial cells. The polysaccharide antigen (PSA) from *Bacteroides fragilis* stimulates the preferential generation of regulatory T cells, possibly by binding to TLR-2 on CD4$^+$ T cells. Unidentified members of the microbiota are needed to maintain the production of TSLP, IL-25, IL-33, and RA.

12-22   Full immune responses to commensal bacteria provoke intestinal disease.

Elegant experiments in the 1990s led to the now generally accepted idea that potentially aggressive T cells that can respond to commensal bacteria are present in normal animals but are usually kept in check by active regulation (Fig. 12.24). If these regulatory mechanisms fail, unrestricted immune responses to commensal bacteria can lead to **inflammatory bowel diseases** such as Crohn's disease. Many genes that are associated with susceptibility to Crohn's disease in humans encode proteins that regulate innate immunity. When these regulatory processes fail, systemic immune responses are generated against antigens from commensal bacterial, such as flagellin. T-cell responses are also generated in the mucosa, leading to severe intestinal damage. IL-23 is important in this process, promoting differentiation of T_H17 effector cells. IL-23 and IL-12 in concert can also induce inflammatory T_H1 responses in the intestine, with some CD4 effector T cells found to produce both IFN-γ and IL-17 under these circumstances. These experimental results are consistent with clinical evidence for a linkage between polymorphisms in the IL-23 receptor and Crohn's disease in humans. In all experimental models, the intestinal damage depends on the presence of commensal bacteria, can be prevented by treatment with antibiotics, and does not occur in germ-free animals.

Patients with Crohn's disease and the related disorder **ulcerative colitis** exhibit dysbiosis and harbor unusual populations of intestinal microbiota. However, with the exception of enteroadherent *Escherichia coli* mentioned above, no individual species of commensal bacteria have yet been proven to be responsible for causing the damage. There is also experimental evidence that local responses to certain pathogenic viruses or parasites such as *Toxoplasma gondii* may trigger bystander activation of effector T cells specific to commensal organisms and produce persistent inflammation.

Cells transferred	TGF-β neutralized	Microbiota	Disease
Unpurified CD4+ T cells	–	+	No
Purified CD4+CD45RB^hi T cells	–	+	Colitis
Purified CD4+CD45RB^lo T cells + (CD25+/FoxP3+) T_reg	–	+	No
CD4+CD45RB^hi T cells + CD4+CD45RB^lo T_reg	–	+	No
CD4+CD45RB^hi T cells + CD4+CD45RB^lo T_reg	+	+	Colitis
CD4+CD45RB^hi T cells	–	–	No

Fig. 12.24 T cells with the potential to produce inflammation in response to commensal bacteria are present in normal animals, but are controlled by regulatory T cells. Transfer of unseparated CD4+ T cells from a normal mouse into an immunodeficient mouse, such as one lacking the *rag* gene (*rag*-/-), will lead to reconstitution of the CD4+ T-cell compartment. However if 'naive' CD4+ T cells (CD4+CD45RB^hi) are purified and transferred, the host mice develop severe inflammation of the colon. This can be prevented by co-transferring the CD4+CD25+ FoxP3+ T cells that were removed during the purification of the naive CD4+ T-cell population. The effects of these regulatory T cells are blocked by neutralizing TGF-β *in vivo* and are also dependent on IL-10. The intestinal inflammation caused by naive CD4+ T cells requires the presence of the microbiota, as it is prevented in germ-free mice, or by treatment with antibiotics. These experiments demonstrate that some CD4+ T cells in normal animals are capable of provoking inflammatory responses against the intestinal microbiota, but that these are normally held in check by regulatory T cells. Micrographs from Powrie, F., *et al.*: *J. Exp Med.* 1996, **183**:2669–2674.

Summary.

The immune system in the mucosa has to distinguish between potential pathogens and harmless antigens, generating strong effector responses to pathogens but remaining unresponsive to foods and commensals. Food proteins induce an active form of immunological tolerance in the systemic and mucosal immune systems; this tolerance may be mediated by regulatory T cells producing IL-10 and/or TGF-β. Commensal bacteria are also recognized by the immune system, but this is limited to the mucosa and its draining lymphoid tissues because commensal antigens are presented to T cells by semi-mature dendritic cells that migrate from the intestinal wall to draining mesenteric lymph nodes. This results in active mucosal tolerance and the production of local IgA antibodies that restrict colonization by the microorganisms, but 'ignorance' of these antigens by the systemic immune system. Because commensal bacteria have many beneficial effects for the host, these immunoregulatory processes are important in allowing the bacteria to coexist with the immune system. When the normal regulatory processes break down, local dendritic cells become fully activated and induce differentiation of naive T cells into effector T cells in the mesenteric lymph node. This is important for protective immunity against pathogens, but when it occurs under the wrong circumstances, it can lead to inflammatory diseases such as Crohn's disease or celiac disease. As a consequence of these competing, but interacting, needs of the immune response, the intestine normally has the appearance of physiological inflammation, which helps maintain normal function of the gut and immune system. This process is driven mostly by the need to control the intestinal microbiota without eliminating it completely or causing damaging inflammation, and results in the coordinated production of IgA, activation of regulatory and effector T cells, and several innate immune responses. Abnormalities in the host response can alter the composition and behavior of the microbiota, while changes in the microbiota can also influence the development and outcome of many diseases outside the intestine.

Summary to Chapter 12.

The mucosal immune system is a large and complex apparatus that has a crucial role in health, not just by protecting physiologically vital organs but also by helping to regulate the tone of the entire immune system and prevent disease. The peripheral lymphoid organs focused on by most immunologists may be a recent specialization of an original template that evolved in mucosal tissues.

The mucosal surfaces of the body are highly vulnerable to infection and possess a complex array of innate and adaptive mechanisms of immunity. The adaptive immune system of the mucosa-associated lymphoid tissues differs from that of the rest of the peripheral lymphoid system in several respects: the immediate juxtaposition of mucosal epithelium and lymphoid tissue; diffuse lymphoid tissue as well as more organized lymphoid organs; specialized antigen uptake mechanisms and distinctive dendritic cells and macrophages; the predominance of activated/memory lymphocytes and distinctive innate lymphoid cells (ILCs) even in the absence of infection; the production of dimeric secretory IgA as the predominant antibody; and the downregulation of immune responses to innocuous antigens such as food antigens and commensal microorganisms. No systemic immune response can normally be detected to these antigens. In contrast, pathogenic microorganisms induce strong protective responses. The key factor in the decision between tolerance and the development of powerful adaptive immune responses is the context in which antigen is presented to T lymphocytes in the mucosal immune system. When there is no inflammation, presentation of antigen to T cells by dendritic cells induces the differentiation of regulatory T cells. By contrast, pathogenic microorganisms crossing the mucosa induce an inflammatory response in the tissues, which stimulates the maturation of antigen-presenting cells and their expression of co-stimulatory molecules, thus favoring a protective T-cell response. This decision-making process is controlled mostly by the way in which specialized dendritic cells react to their environment before migrating to present antigen to naive T cells. The mutualistic loop formed between the host immune response and the local microbiota plays a central role in maintaining health and in the development of disease.

Questions.

12.1 Multiple Choice: Which of the following is an incorrect statement?

A. Microfold cells have a folded luminal surface and possess a thick layer of mucus that allows the entry of microbes to Peyer's patches.

B. Microfold cells recognize several bacterial proteins by GP2 and release the material to the extracellular space by a process called transcytosis.

C. Gut-associated lymphoid tissues attract dendritic cells trough chemokines such as CCL20 and CCL9.

D. Pathogens such as *Yersinia pestis* and *Shigella* target microfold cells to gain access to the subepithelial space.

12.2 True or False: Intraepithelial lymphocytes are mostly CD4 T cells, in contrast to the lamina propria, where CD8 T cells predominate.

12.3 Matching: Match each chemokine or chemokine receptor to its tissue homing function.

A. CXCL13

i. Recruitment of lymphocytes to the colon, lactating mammary gland, and salivary glands

B. CCL25

ii. Recruitment of B and T cells to the small intestine

C. CCL28

iii. Directing lymphocytes to the skin

D. CCR4

iv. Recruitment of naive B cells to Peyer's patches

12.4 Multiple Choice: Which of the following is a correct statement?

A. CD11b⁺ dendritic cells stimulate ILC3s and are the main source for IL-12 in Peyer's patches.

B. CD11b⁻ dendritic cells require BATF3 for their development.

C. Retinoic acid production by naive T cells is required for dendritic cells for the generation of Treg cells.

D. CCL20 prevents entrance of dendritic cells into the epithelial layer of Peyer's patches.

12.5 Short Answer: IgA:antigen complexes can be reexported to the gut lumen in order to enhance pathogen excretion from the organism. In contrast, the formation of IgA:antigen complexes can also enhance the uptake of luminal antigen. How can uptake of antigen be beneficial to the organism?

12.6 Short Answer: Large amounts of IgA are produced by intestinal B cells and plasma cells and secreted into the lumen as a means to keep the microbiota in check and prevent invasion from pathogens, yet most individuals with IgA deficiency are not overly susceptible to infections. Explain why this is the case.

12.7 Multiple Choice: Which of the following best describes intraepithelial lymphocytes (IELs)?

A. Express CCR9 and $\alpha_4\beta_7$ integrin

B. Express CCR9 and $\alpha_E\beta_7$ integrin (CD103)

C. Have a CD4 to CD8 T cell ratio of 3:1

D. Contain CD4$^+$ T cells that produce IFN-γ, IL-17, and IL-22

E. Consist of 90% T cells, 80% of which express CD8 as an α:α homodimer or an α:β heterodimer

F. A and C

G. B and E

H. A, C, and D

12.8 Multiple Choice: Which of the following cell types depend on aryl hydrocarbon receptor expression for proper development?

A. Type b intraepithelial lymphocyte (EIL)

B. ILC1

C. B cell

D. Macrophage

E. ILC2

F. Neutrophil

12.9 Matching: Match the human disease with the pathophysiology:

A. Abnormal response to the wheat protein gluten, causing increased frequency of IELs with MIC-A-dependent cytotoxic activity against intestinal epithelial cells

i. *Clostridium difficile* infection

B. Interruption of normal fecal flow to colon, causing enterocytes to undergo inflammation and necrosis due to absence of short-chain fatty acids (SCFAs) produced by commensals

ii. Celiac disease

C. Antibiotic treatment that eliminates the bulk of commensal flora allowing a particular species to overgrow and produce toxins that lead to severe diarrhea and mucosal injury

iii. Inflammatory bowel disease (Crohn's disease and ulcerative colitis)

D. Hyperactive immune responses to commensal bacteria due to defects in innate immunity genes

iv. Diversion colitis

12.10 True or False: Lamina propria CD4$^+$ T cells secrete large amounts of cytokines such as IFN-γ, IL-17, and IL-22 only in response to pathogens and inflammatory insults.

12.11 True or False: Most T$_{reg}$s in the small intestine do not express FoxP3.

General references.

Hooper, L.V., Littman, D.R., and Macpherson, A.J.: **Interactions between the microbiota and the immune system.** *Science* 2012, **336**:1268–1273.

MacDonald, T.T., Monteleone, I., Fantini, M.C., and Monteleone, G.: **Regulation of homeostasis and inflammation in the intestine.** *Gastroenterology* 2011, **140**: 1768–1775.

Mowat, A.M.: **Anatomical basis of tolerance and immunity to intestinal antigens.** *Nat. Rev. Immunol.* 2003, **3**:331–341.

Society for Mucosal Immunology: *Principles of Mucosal Immunology*, 1st ed. New York, Garland Science, 2013.

Section references.

12-1 The mucosal immune system protects the internal surfaces of the body.

Brandtzaeg, P.: **Function of mucosa-associated lymphoid tissue in antibody formation.** *Immunol. Invest.* 2010, **39**:303–355.

Cerutti, A., Chen, K., and Chorny, A.: **Immunoglobulin responses at the mucosal interface.** *Annu. Rev. Immunol.* 2011, **29**:273–293.

Corthesy, B.: **Role of secretory IgA in infection and maintenance of homeostasis.** *Autoimmun. Rev.* 2013, **12**:661–665.

Fagarasan, S., Kawamoto, S., Kanagawa, O., and Suzuki, K.: **Adaptive immune regulation in the gut: T cell-dependent and T cell-independent IgA synthesis.** *Annu. Rev. Immunol.* 2010, **28**:243–273.

Matsunaga, T., and Rahman, A.: **In search of the origin of the thymus: the thymus and GALT may be evolutionarily related.** *Scand. J. Immunol.* 2001, **53**:1–6.

Naz, R.K.: **Female genital tract immunity: distinct immunological challenges for vaccine development.** *J .Reprod. Immunol.* 2012, **93**:1–8.

Randall, T.D.: **Bronchus-associated lymphoid tissue (BALT) structure and function.** *Adv. Immunol.* 2010, **107**:187–241.

Sato, S., and Kiyono, H. **The mucosal immune system of the respiratory tract.** *Curr. Opin. Virol.* 2012, **2**:225–232.

12-2 Cells of the mucosal immune system are located both in anatomically defined compartments and scattered throughout mucosal tissues.

Baptista, A.P., Olivier, B.J., Goverse, G., Greuter, M., Knippenberg, M., Kusser, K., Domingues, R.G., Veiga-Fernandes, H., Luster, A.D., Lugering, A., *et al.*: **Colonic patch and colonic SILT development are independent and differentially regulated events.** *Mucosal Immunol* 2013, **6**:511–521.

Brandtzaeg, P., Kiyono, H., Pabst, R., and Russell, M.W.: **Terminology: nomenclature of mucosa-associated lymphoid tissue.** *Mucosal Immunol.* 2008, **1**:31–37.

Eberl, G., and Sawa, S.: **Opening the crypt: current facts and hypotheses on the function of cryptopatches.** *Trends Immunol.* 2010, **31**:50–55.

Lee, J.S., Cella, M., McDonald, K.G., Garlanda, C., Kennedy, G.D., Nukaya, M., Mantovani, A., Kopan, R., Bradfield, C.A., Newberry, R.D., *et al.*: **AHR drives the development of gut ILC22 cells and postnatal lymphoid tissues via pathways dependent on and independent of Notch.** *Nat. Immunol.* 2012, **13**:144–151.

Macpherson, A.J., McCoy, K.D., Johansen, F.E., and Brandtzaeg, P.: **The immune geography of IgA induction and function.** *Mucosal. Immunol.* 2008, **1**:11–22.

Pabst, O., Herbrand, H., Worbs, T., Friedrichsen, M., Yan, S., Hoffmann, M.W., Korner, H., Bernhardt, G., Pabst, R., and Forster, R.: **Cryptopatches and isolated lymphoid follicles: dynamic lymphoid tissues dispensable for the generation of intraepithelial lymphocytes.** *Eur. J. Immunol.* 2005, **35**:98–107.

Randall, T.D.: **Bronchus-associated lymphoid tissue (BALT) structure and function.** *Adv. Immunol.* 2010, **107**:187–241.

Randall, T.D., and Mebius, R.E.: **The development and function of mucosal lymphoid tissues: a balancing act with micro-organisms.** *Mucosal Immunol.* 2014, **7**: 455–466.

Suzuki, K., Kawamoto, S., Maruya, M., and Fagarasan, S.: **GALT: organization and dynamics leading to IgA synthesis.** *Adv. Immunol.* 2010, **107**:153–185.

12-3 The intestine has distinctive routes and mechanisms of antigen uptake.

Anosova, N.G., Chabot, S., Shreedhar, V., Borawski, J.A., Dickinson, B.L., and Neutra, M.R.: **Cholera toxin, *E. coli* heat-labile toxin, and non-toxic derivatives induce dendritic cell migration into the follicle-associated epithelium of Peyer's patches.** *Mucosal Immunol.* 2008, **1**:59–67.

Hase, K., Kawano, K., Nochi, T., Pontes, G.S., Fukuda, S., Ebisawa, M., Kadokura, K., Tobe, T., Fujimura, Y., Kawano, S., *et al.*: **Uptake through glycoprotein 2 of FimH+ bacteria by M cells initiates mucosal immune response.** *Nature* 2009, **462**:226–230.

Jang, M.H., Kweon, M.N., Iwatani, K., Yamamoto, M., Terahara, K., Sasakawa, C., Suzuki, T., Nochi, T., Yokota, Y., Rennert, P.D., *et al.*: **Intestinal villous M cells: an antigen entry site in the mucosal epithelium.** *Proc. Natl Acad. Sci. USA* 2004, **101**:6110–6115.

Lelouard, H., Fallet, M., de Bovis, B., Meresse, S., and Gorvel, J.P.: **Peyer's patch dendritic cells sample antigens by extending dendrites through M cell-specific transcellular pores.** *Gastroenterology* 2012, **142**:592–601.

Mabbott, N.A., Donaldson, D.S., Ohno, H., Williams, I.R., and Mahajan, A.: **Microfold (M) cells: important immunosurveillance posts in the intestinal epithelium.** *Mucosal Immunol.* 2013, **6**:666–677.

Sato, S., Kaneto, S., Shibata, N., Takahashi, Y., Okura, H., Yuki, Y., Kunisawa, J., and Kiyono, H.: **Transcription factor Spi-B-dependent and -independent pathways for the development of Peyer's patch M cells.** *Mucosal Immunol.* 2013, **6**:838–846.

Salazar-Gonzalez, R.M., Niess, J.H., Zammit, D.J., Ravindran, R., Srinivasan, A., Maxwell, J.R., Stoklasek, T., Yadav, R., Williams, I.R., Gu, X., *et al.*: **CCR6-mediated dendritic cell activation of pathogen-specific T cells in Peyer's patches.** *Immunity* 2006, **24**:623–632.

Zhao, X., Sato, A., Dela Cruz, C.S., Linehan, M., Luegering, A., Kucharzik, T., Shirakawa, A.K., Marquez, G., Farber, J.M., Williams, I., *et al.*: **CCL9 is secreted by the follicle-associated epithelium and recruits dome region Peyer's patch CD11b+ dendritic cells.** *J. Immunol.* 2003, **171**:2797–2803.

12-4 The mucosal immune system contains large numbers of effector lymphocytes even in the absence of disease.

Belkaid, Y., Bouladoux, N., and Hand, T.W.: **Effector and memory T cell responses to commensal bacteria.** *Trends Immunol.* 2013, **34**:299–306.

Brandtzaeg, P.: **Mucosal immunity: induction, dissemination, and effector functions.** *Scand. J. Immunol.* 2009, **70**:505–515.

Cao A.T., Yao S., Gong B., Elson C.O., and Cong Y.: **Th17 cells upregulate polymeric Ig receptor and intestinal IgA and contribute to intestinal homeostasis.** *J. Immunol.* 2012, **189**:4666–4673.

Cheroutre, H., and Lambolez, F.: **Doubting the TCR coreceptor function of CD8$\alpha\alpha$.** *Immunity* 2008, **28**:149–159.

Cauley, L.S., and Lefrancois, L.: **Guarding the perimeter: protection of the mucosa by tissue-resident memory T cells.** *Mucosal. Immunol.* 2013, **6**:14–23.

Maynard, C.L., and Weaver, C.T.: **Intestinal effector T cells in health and disease.** *Immunity* 2009, **31**:389–400.

Sathaliyawala, T., Kubota, M., Yudanin, N., Turner, D., Camp, P., Thome, J.J., Bickham, K.L., Lerner, H., Goldstein, M., Sykes, *et al.*: **Distribution and compartmentalization of human circulating and tissue-resident memory T cell subsets.** *Immunity* 2013, **38**:187–197.

12-5 The circulation of lymphocytes within the mucosal immune system is controlled by tissue-specific adhesion molecules and chemokine receptors.

Agace, W.: **Generation of gut-homing T cells and their localization to the small intestinal mucosa.** *Immunol. Lett.* 2010, **128**:21–23.

Hu, S., Yang, K., Yang, J., Li, M., and Xiong, N.: **Critical roles of chemokine receptor CCR10 in regulating memory IgA responses in intestines.** *Proc. Natl Acad. Sci. USA* 2011, **108**: E1035–1044.

Kim, S.V., Xiang, W.V., Kwak, C., Yang, Y., Lin, X.W., Ota, M., Sarpel, U., Rifkin, D.B., Xu, R., and Littman, D.R.: **GPR15-mediated homing controls immune homeostasis in the large intestine mucosa.** *Science* 2013, **340**:1456–1459.

Macpherson, A.J., Geuking, M.B., Slack, E., Hapfelmeier, S., and McCoy, K.D.: **The habitat, double life, citizenship, and forgetfulness of IgA.** *Immunol. Rev.* 2012, **245**:132–146.

Mikhak, Z., Strassner, J.P., and Luster, A.D.: **Lung dendritic cells imprint T cell lung homing and promote lung immunity through the chemokine receptor CCR4.** *J. Exp. Med.* 2013, **210**:1855–1869.

Mora, J.R., and von Andrian, U.H.: **Differentiation and homing of IgA-secreting cells.** *Mucosal Immunol.* 2008, **1**:96–109.

Pabst, O., and Bernhardt, G.: **On the road to tolerance--generation and migration of gut regulatory T cells.** *Eur. J. Immunol.* 2013, **43**:1422–1425.

12-6 Priming of lymphocytes in one mucosal tissue may induce protective immunity at other mucosal surfaces.

Agnello, D., Denimal, D., Lavaux, A., Blondeau-Germe, L., Lu, B., Gerard, N.P., Gerard, C., and Pothier, P.: **Intrarectal immunization and IgA antibody-secreting cell homing to the small intestine.** *J. Immunol.* 2013, **190**:4836–4847.

Brandtzaeg, P.: **Induction of secretory immunity and memory at mucosal surfaces.** *Vaccine* 2007, **25**:5467–5484.

Czerkinsky, C., and Holmgren, J.: **Mucosal delivery routes for optimal immunization: targeting immunity to the right tissues.** *Curr. Top. Microbiol. Immunol.* 2012, **354**:1–18.

Ruane, D., Brane, L., Reis, B.S., Cheong, C., Poles, J., Do, Y., Zhu, H., Velinzon, K., Choi, J.H., Studt, N., *et al.*: **Lung dendritic cells induce migration of protective T cells to the gastrointestinal tract.** *J. Exp. Med.* 2013, **210**:1871–1888.

12-7 Distinct populations of dendritic cells control mucosal immune responses.

Cerovic, V., Bain, C.C., Mowat, A.M., and Milling, S.W.F.: **Intestinal macrophages and dendritic cells: what's the difference?** *Trends Immunol.* 2014, **35**:270–277.

Goto, Y., Panea, C., Nakato, G., Cebula, A., Lee, C., Diez, M.G., Laufer, T.M., Ignatowicz, L., and Ivanov, I.I.: **Segmented filamentous bacteria antigens presented by intestinal dendritic cells drive mucosal Th17 cell differentiation.** *Immunity* 2014, **40**:594-607.

Guilliams, M., Lambrecht, B.N., and Hammad, H.: **Division of labor between lung dendritic cells and macrophages in the defense against pulmonary infections.** *Mucosal Immunol.* 2013, **6**:464–473.

Jaensson-Gyllenback, E., Kotarsky, K., Zapata, F., Persson, E.K., Gundersen, T.E., Blomhoff, R., and Agace, W.W.: **Bile retinoids imprint intestinal CD103+ dendritic cells with the ability to generate gut-tropic T cells.** *Mucosal Immunol.* 2011, **4**:438–447.

Matteoli, G.: **Gut CD103+ dendritic cells express indoleamine 2,3-dioxygen-ase which influences T regulatory/T effector cell balance and oral tolerance induction.** *Gut* 2010, **59**:595–604.

Schlitzer, A., McGovern, N., Teo, P., Zelante, T., Atarashi, K., Low, D., Ho, A.W., See, P., Shin, A., Wasan, P.S., *et al.*: **IRF4 transcription factor-dependent CD11b+ dendritic cells in human and mouse control mucosal IL-17 cytokine responses.** *Immunity* 2013, **38**:970–983.

Scott, C.L., Bain, C.C., Wright, P.B., Schien, D., Kotarsky, K., Persson, E.K., Luda, K., Guilliams, M., Lambrecht, B.N., Agace, W.W., *et al.*: **CCR2+CD103- intestinal dendritic cells develop from DC-committed precursors and induce interleukin-17 production by T cells.** *Mucosal Immunol.* 2015, **8**:327–239.

Travis, M.A., Reizis, B., Melton, A.C., Masteller, E., Tang, Q., Proctor, J.M., Wang, Y., Bernstein, X., Huang, X., Reichardt, L.F., *et al.*: **Loss of integrin $\alpha_V\beta_8$ on dendritic cells causes autoimmunity and colitis in mice.** *Nature* 2007, **449**:361–365.

Vicente-Suarez, I., Larange, A., Reardon, C., Matho, M., Feau, S., Chodaczek, G., Park, Y., Obata, Y., Gold, R., Wang-Zhu, Y., *et al.*: **Unique lamina propria stromal cells imprint the functional phenotype of mucosal dendritic cells.** *Mucosal Immunol.* 2015, **8**:141–151.

Watchmaker, P.B., Lahl, K., Lee, M., Baumjohann, D., Morton, J., Kim, S.J., Zeng, R., Dent, A., Ansel, K.M., Diamond, B., *et al.*: **Comparative transcriptional and functional profiling defines conserved programs of intestinal DC differentiation in humans and mice.** *Nat. Immunol.* 2014, **15**:98–108.

12-8 Macrophages and dendritic cells have different roles in mucosal immune responses.

Bain, C.C., Bravo-Blas, A., Scott, C.L., Geissmann, F., Henri, S., Malissen, B., Osborne, L.C., Artis, D., and Mowat, A.M.: **Constant replenishment from circulating monocytes maintains the macrophage pool in adult intestine.** *Nat. Immunol.* 2014, **15**:929–937.

Guilliams, M., Lambrecht, B.N., and Hammad, H.: **Division of labor between lung dendritic cells and macrophages in the defense against pulmonary infections.** *Mucosal Immunol* 2013, **6**:464–473.

Hadis, U., Wahl, B., Schulz, O., Hardtke-Wolenski, M., Schippers, A., Wagner, N., Muller, W., Sparwasser, T., Forster, R., and Pabst, O.: **Intestinal tolerance requires gut homing and expansion of FoxP3+ regulatory T cells in the lamina propria.** *Immunity* 2011, **34**:237–246.

Mortha, A., Chudnovskiy, A., Hashimoto, D., Bogunovic, M., Spencer, S.P., Belkaid, Y., and Merad, M.: **Microbiota-dependent crosstalk between macrophages and ILC3 promotes intestinal homeostasis.** *Science* 2014, **343**:1249288.

12-9 Antigen-presenting cells in the intestinal mucosa acquire antigen by a variety of routes.

Farache, J., Zigmond, E., Shakhar, G., and Jung, S.: **Contributions of dendritic cells and macrophages to intestinal homeostasis and immune defense.** *Immunol. Cell Biol.* 2013, **91**:232–239.

Jang, M.H., Kweon, M.N., Iwatani, K., Yamamoto, M., Terahara, K., Sasakawa, C., Suzuki, T., Nochi, T., Yokota, Y., Rennert, P.D., *et al.*: **Intestinal villous M cells: an antigen entry site in the mucosal epithelium.** *Proc. Natl Acad. Sci. USA* 2004, **101**:6110–6115.

Mazzini, E., Massimiliano, L., Penna, G., and Rescigno, M.: **Oral tolerance can be established via gap junction transfer of fed antigens from CX3CR1+ macrophages to CD103+ dendritic cells.** *Immunity* 2014, **40**:248–261.

McDole, J.R., Wheeler, L.W., McDonald, K.G., Wang, B., Konjufca, V., Knoop, K.A., Newberry, R.D., and Miller, M.J.: **Goblet cells deliver luminal antigen to CD103+ dendritic cells in the small intestine.** *Nature* 2012, **483**:345–349.

Schulz, O., and Pabst, O.: **Antigen sampling in the small intestine.** *Trends Immunol.* 2013, **34**:155–161.

Yoshida, M., Claypool, S.M., Wagner, J.S., Mizoguchi, E., Mizoguchi, A., Roopenian, D.C., Lencer, W.I., and Blumberg, R.S.: **Human neonatal Fc receptor mediates transport of IgG into luminal secretions for delivery of antigens to mucosal dendritic cells.** *Immunity* 2004, **20**:769–783.

12-10 Secretory IgA is the class of antibody associated with the mucosal immune system.

Fritz, J.H., Rojas, O.L., Simard, N., McCarthy, D.D., Hapfelmeier, S., Rubino, S., Robertson, S.J., Larijani, M., Gosselin, J., Ivanov, II, *et al.*: **Acquisition of a multifunctional IgA+ plasma cell phenotype in the gut.** *Nature* 2012, **481**:199–203.

Kawamoto, S., Maruya, M., Kato, L.M., Suda, W., Atarashi, K., Doi, Y., Tsutsui, Y., Qin, H., Honda, K., Okada, T., *et al.*: **Foxp3 T cells regulate immunoglobulin A selection and facilitate diversification of bacterial species responsible for immune homeostasis.** *Immunity* 2014, **41**:152–165.

Lin, M., Du, L., Brandtzaeg, P., and Pan-Hammarstrom, Q.: **IgA subclass switch recombination in human mucosal and systemic immune compartments.** *Mucosal Immunol.* 2014, **7**:511–520.

Woof, J.M., and Russell, M.W.: **Structure and function relationships in IgA.** *Mucosal Immunol.* 2011, **4**:590–597.

12-11 T-independent processes can contribute to IgA production in some species

Barone, F., Vossenkamper, A., Boursier, L., Su, W., Watson, A., John, S., Dunn-Walters, D.K., Fields, P., Wijetilleka, S., Edgeworth, J.D., *et al.*: **IgA-producing plasma cells originate from germinal centers that are induced by B-cell receptor engagement in humans.** *Gastroenterology* 2011, **140**:947–956.

Fagarasan, S., Kawamoto, S., Kanagawa, O., and Suzuki, K.: **Adaptive immune regulation in the gut: T cell-dependent and T cell-independent IgA synthesis.** *Annu. Rev. Immunol.* 2010, **28**:243–273.

Lin, M., Du, L., Brandtzaeg, P., and Pan-Hammarstrom, Q.: **IgA subclass switch recombination in human mucosal and systemic immune compartments.** *Mucosal Immunol.* 2014, **7**:511–520.

Tezuka, H., Abe, Y., Asano, J., Sato, T., Liu, J., Iwata, M., and Ohteki, T.: **Prominent role for plasmacytoid dendritic cells in mucosal T cell-independent IgA induction.** *Immunity* 2011, **34**:247–257.

12-12 IgA deficiency is relatively common in humans but may be compensated for by secretory IgM.

Karlsson, M.R., Johansen, F.E., Kahu, H., Macpherson, A., and Brandtzaeg, P.: **Hypersensitivity and oral tolerance in the absence of a secretory immune system.** *Allergy* 2010, **65**:561–570.

Yel, L.: **Selective IgA deficiency.** *J. Clin. Immunol.* 2010, **30**:10–16.

12-13 The intestinal lamina propria contains antigen-experienced T cells and populations of unusual innate lymphoid cells.

Buonocore, S., Ahern, P.P., Uhlig, H.H., Ivanov, I.I., Littman, D.R., Maloy, K.J., and Powrie, F.: **Innate lymphoid cells drive interleukin-23-dependent innate intestinal pathology.** *Nature* 2010, **464**:1371–1375.

Satpathy, A.T., Briseño, C.G., Lee, J.S., Ng, D., Manieri, N.A., Kc, W., Wu, X., Thomas, S.R., Lee, W.L., Turkoz, M., *et al.*: **Notch2-dependent classical dendritic cells orchestrate intestinal immunity to attaching-and-effacing bacterial pathogens.** *Nat. Immunol.* 2013, **14**:937–948.

Klose, C.S., Kiss, E.A., Schwierzeck, V., Ebert, K., Hoyler, T., d'Hargues, Y., Goppert, N., Croxford, A.L., Waisman, A., Tanriver, Y., *et al.*: **A T-bet gradient controls the fate and function of CCR6–RORγt+ innate lymphoid cells.** *Nature* 2013, **494**:261–265.

Kruglov, A.A., Grivennikov, S.I., Kuprash, D.V., Winsauer, C., Prepens, S., Seleznik, G.M., Eberl, G., Littman, D.R., Heikenwalder, M., Tumanov, A.V., *et al.*: **Nonredundant function of soluble LTα3 produced by innate lymphoid cells in intestinal homeostasis.** *Science* 2013, **342**:1243–1246.

Le Bourhis, L., Dusseaux, M., Bohineust, A., Bessoles, S., Martin, E., Premel, V., Core, M., Sleurs, D., Serriari, N.E., and Treiner, E.: **MAIT cells detect and efficiently lyse bacterially-infected epithelial cells.** *PLoS Pathog.* 2013, **9**:e1003681.

Spits, H., and Cupedo, T.: **Innate lymphoid cells: emerging insights in development, lineage relationships, and function.** *Annu. Rev. Immunol.* 2012, **30**:647–675.

12-14   The intestinal epithelium is a unique compartment of the immune system.

Agace, W.W., Roberts, A.I., Wu, L., Greineder, C., Ebert, E.C., and Parker, C.M.: **Human intestinal lamina propria and intraepithelial lymphocytes express receptors specific for chemokines induced by inflammation.** *Eur. J. Immunol.* 2000, **30**:819–826.

Cheroutre, H., Lambolez, F., and Mucida, D.: **The light and dark sides of intestinal intraepithelial lymphocytes.** *Nat. Rev. Immunol.* 2011, **11**:445–456.

Eberl, G., and Sawa, S.: **Opening the crypt: current facts and hypotheses on the function of cryptopatches.** *Trends Immunol.* 2010, **31**:50–55.

Hayday, A., Theodoridis, E., Ramsburg, E., and Shires, J.: **Intraepithelial lymphocytes: exploring the Third Way in immunology.** *Nat. Immunol.* 2001, **2**:997–1003.

Jiang, W., Wang, X., Zeng, B., Liu, L., Tardivel, A., Wei, H., Han, J., MacDonald, H.R., Tschopp, J., Tian, Z., *et al:* **Recognition of gut microbiota by NOD2 is essential for the homeostasis of intestinal intraepithelial lymphocytes.** *J. Exp. Med.* 2013, **210**:2465–2476.

Li, Y., Innocentin, S., Withers, D.R., Roberts, N.A., Gallagher, A.R., Grigorieva, E.F., Wilhelm, C., and Veldhoen, M.: **Exogenous stimuli maintain intraepithelial lymphocytes via aryl hydrocarbon receptor activation.** *Cell* 2011, **147**:629–640.

Pobezinsky, L.A., Angelov, G.S., Tai, X., Jeurling, S., Van Laethem, F., Feigenbaum, L., Park, J.H., and Singer, A.: **Clonal deletion and the fate of autoreactive thymocytes that survive negative selection.** *Nat. Immunol.* 2012, **13**:569–578.

12-15   Enteric pathogens cause a local inflammatory response and the development of protective immunity.

Clevers, H.C., and Bevins, C.L.: **Paneth cells: maestros of the small intestinal crypts.** *Annu. Rev. Physio.* 2013, **75**:289–311.

Conway, K.L., Kuballa, P., Song, J.H., Patel, K.K., Castoreno, A.B., Yilmaz, O.H., Jijon, H.B., Zhang, M., Aldrich, L.N., Villablanca, E.J., *et al.:* **Atg16l1 is required for autophagy in intestinal epithelial cells and protection of mice from Salmonella infection.** *Gastroenterology* 2013, **145**:1347–1357.

Geddes, K., Rubino, S.J., Magalhaes, J.G., Streutker, C., Le Bourhis, L., Cho, J.H., Robertson, S.J., Kim, C.J., Kaul, R., Philpott, D.J., *et al:* **Identification of an innate T helper type 17 response to intestinal bacterial pathogens.** *Nat. Med.* 2011, **17**:837–844.

Lassen, K.G., Kuballa, P., Conway, K.L., Patel, K.K., Becker, C.E., Peloquin, J.M., Villablanca, E.J., Norman, J.M., Liu, T.C., Heath, R.J., *et al.:* **Atg16L1 T300A variant decreases selective autophagy resulting in altered cytokine signaling and decreased antibacterial defense.** *Proc. Natl Acad. Sci. USA* 2014, **111**:7741–7746.

Prescott, D., Lee, J., and Philpott, D.J.: **An epithelial armamentarium to sense the microbiota.** *Semin. Immunol.* 2013, **25**:323–333.

Song-Zhao, G.X., Srinivasan, N., Pott, J., Baban, D., Frankel, G., and Maloy, K.J.: **Nlrp3 activation in the intestinal epithelium protects against a mucosal pathogen.** *Mucosal Immunol.* 2014, **7**:763–774.

12-16   Pathogens induce adaptive immune responses when innate defenses have been breached.

Bain, C.C., and Mowat, A.M.: **Macrophages in intestinal homeostasis and inflammation.** *Immunol. Rev.* 2014: **260**:102–117.

Farache, J., Koren, I., Milo, I., Gurevich, I., Kim, K.W., Zigmond, E., Furtado, G.C., Lira, S.A., and Shakhar, G.: **Luminal bacteria recruit CD103⁺ dendritic cells into the intestinal epithelium to sample bacterial antigens for presentation.** *Immunity* 2013, **38**:581–595.

Persson, E.K., Scott, C.L., Mowat, A.M., and Agace, W.W.: **Dendritic cell subsets in the intestinal lamina propria: Ontogeny and function.** *Eu. J. Immunol.* 2013, **43**:3098–3107.

Salazar-Gonzalez, R.M., Niess, J.H., Zammit, D.J., Ravindran, R., Srinivasan, A., Maxwell, J.R., Stoklasek, T., Yadav, R., Williams, I.R., Gu, X., *et al.:* **CCR6-mediated dendritic cell activation of pathogen-specific T cells in Peyer's patches.** *Immunity* 2006, **24**:623–632.

Uematsu, S., Jang, M.H., Chevrier, N., Guo, Z., Kumagai, Y., Yamamoto, M., Kato, H., Sougawa, N., Matsui, H., Kuwata, H., *et al.:* **Detection of pathogenic intestinal bacteria by Toll-like receptor 5 on intestinal CD11c⁺ lamina propria cells.** *Nat. Immunol.* 2006, **7**:868–874.

12-17   Effector T-cell responses in the intestine protect the function of the epithelium.

Cliffe, L.J., Humphreys, N.E., Lane, T.E., Potten, C.S., Booth, C., and Grencis, R.K.: **Accelerated intestinal epithelial cell turnover: a new mechanism of parasite expulsion.** *Science* 2005, **308**:1463–1465.

Kinnebrew, M.A., Buffie, C.G., Diehl, G.E., Zenewicz, L.A., Leiner, I., Hohl, T.M., Flavell, R.A., Littman, D.R., and Pamer, E.G.: **Interleukin 23 production by intestinal CD103⁺CD11b⁺ dendritic cells in response to bacterial flagellin enhances mucosal innate immune defense.** *Immunity* 2012, **36**:276–287.

Sokol, H., Conway, K.L., Zhang, M., Choi, M., Morin, B., Cao, Z., Villablanca, E.J., Li, C., Wijmenga, C., Yun, S.H., *et al:* **Card9 mediates intestinal epithelial cell restitution, T-helper 17 responses, and control of bacterial infection in mice.** *Gastroenterology* 2013, **145**:591–601.

Sonnenberg, G.F., Fouser, L.A., and Artis, D.: **Functional biology of the IL-22-IL-22R pathway in regulating immunity and inflammation at barrier surfaces.** *Adv. Immunol.* 2010, **107**:1–29.

Turner, J.E., Stockinger, B., and Helmby, H.: **IL-22 mediates goblet cell hyperplasia and worm expulsion in intestinal helminth infection.** *PLoS Patho.* 2013, **9**:e1003698.

12-18   The mucosal immune system must maintain tolerance to harmless foreign antigens.

Cassani, B., Villablanca, E.J., Quintana, F.J., Love, P.E., Lacy-Hulbert, A., Blaner, W.S., Sparwasser, T., Snapper, S.B., Weiner, H.L., and Mora, J.R.: **Gut-tropic T cells that express integrin α4β7 and CCR9 are required for induction of oral immune tolerance in mice.** *Gastroenterology* 2011, **141**:2109–2118.

Coombes, J.L., Siddiqui, K.R., Arancibia-Carcamo, C.V., Hall, J., Sun, C.M., Belkaid, Y., and Powrie, F.: **A functionally specialized population of mucosal CD103⁺ DCs induces Foxp3⁺ regulatory T cells via a TGF-β and retinoic acid-dependent mechanism.** *J. Exp. Med.* 2007, **204**:1757–1764.

Du Toit, G., Roberts, G., Sayre, P.H., Bahnson, H.T., Radulovic, S., Santos, A.F., Brough, H.A., Phippard, D., Basting, M., Feeney, M., *et al:* **Randomized trial of peanut consumption in infants at risk for peanut allergy.** *N. Engl. J. Med.* 2015, **372**:803–813.

Huang, G., Wang, Y., and Chi, H.: **Control of T cell fates and immune tolerance by p38α signaling in mucosal CD103⁺ dendritic cells.** *J. Immunol.* 2013, **191**:650–659.

Mowat, A.M., Strobel, S., Drummond, H.E., and Ferguson, A.: **Immunological responses to fed protein antigens in mice. I. Reversal of oral tolerance to ovalbumin by cyclophosphamide.** *Immunology* 1982, **45**:105–113.

12-19   The normal intestine contains large quantities of bacteria that are required for health.

&

12-20   Innate and adaptive immune systems control microbiota while preventing inflammation without compromising the ability to react to invaders.

Arpaia, N., Campbell, C., Fan, X., Dikiy, S., van der Veeken, J., deRoos, P., Liu, H., Cross, J.R., Pfeffer, K., Coffer, P.J., *et al.:* **Metabolites produced by commensal bacteria promote peripheral regulatory T-cell generation.** *Nature* 2013, **504**:451–455.

Atarashi, K., Tanoue, T., Oshima, K., Suda, W., Nagano, Y., Nishikawa, H., Fukuda, S., Saito, T., Narushima, S., Hase, K., *et al.:* **T_reg induction by a rationally selected mixture of Clostridia strains from the human microbiota.** *Nature* 2013, **500**:232–236.

Belkaid, Y., and Hand, T.W.: **Role of the microbiota in immunity and inflammation.** *Cell* 2014, **157**:121–141.

Harig, J.M., Soergel, K.H., Komorowski, R.A., and Wood, C.M.: **Treatment of diversion colitis with short-chain-fatty acid irrigation.** *N. Engl. J. Med.* 1989, **320**:23–28.

Hirota, K., Turner, J.E., Villa, M., Duarte, J.H., Demengeot, J., Steinmetz, O.M., and Stockinger, B.: **Plasticity of Th17 cells in Peyer's patches is responsible for the induction of T cell-dependent IgA responses.** *Nat. Immunol.* 2013, **14**:372–379.

Kato, L.M., Kawamoto, S., Maruya, M., and Fagarasan, S.: **The role of the adaptive immune system in regulation of gut microbiota.** *Immunol. Rev.* 2014, **260**:67–75.

Macia, L., Thorburn, A.N., Binge, L.C., Marino, E., Rogers, K.E., Maslowski, K.M., Vieira, A.T., Kranich, J., and Mackay, C.R.: **Microbial influences on epithelial integrity and immune function as a basis for inflammatory diseases.** *Immunol. Rev.* 2012, **245**:164–176.

Maynard, C.L., Elson, C.O., Hatton, R.D., and Weaver, C.T.: **Reciprocal interactions of the intestinal microbiota and immune system.** *Nature* 2012, **489**:231–241.

Peterson, D.A., McNulty, N.P., Guruge, J.L., and Gordon, J.I.: **IgA response to symbiotic bacteria as a mediator of gut homeostasis.** *Cell Host Microbe* 2007, **2**:328–339.

Round, J.L., and Mazmanian, S.K.: **Inducible Foxp3+ regulatory T-cell development by a commensal bacterium of the intestinal microbiota.** *Proc. Natl Acad. Sci. USA* 2010, **107**:12204–12209.

Scher, J.U., Sczesnak, A., Longman, R.S., Segata, N., Ubeda, C., Bielski, C., Rostron, T., Cerundolo, V., Pamer, E.G., Abramson, S.B., *et al.*: **Expansion of intestinal *Prevotella copri* correlates with enhanced susceptibility to arthritis.** *eLife* 2013, **2**:e01202.

Zigmond, E., Bernshtein, B., Friedlander, G., Walker, C.R., Yona, S., Kim, K.W., Brenner, O., Krauthgamer, R., Varol, C., Müller, W., *et al.*: **Macrophage-restricted interleukin-10 receptor deficiency, but not IL-10 deficiency, causes severe spontaneous colitis.** *Immunity* 2014, **40**:720–733.

12-21 The intestinal microbiota plays a major role in shaping intestinal and systemic immune function.

&

12-22 Full immune responses to commensal bacteria provoke intestinal disease.

Adolph, T.E., Tomczak, M.F., Niederreiter, L., Ko, H.J., Bock, J., Martinez-Naves, E., Glickman, J.N., Tschurtschenthaler, M., Hartwig, J., Hosomi, S., *et al.*: **Paneth cells as a site of origin for intestinal inflammation.** *Nature* 2013, **503**:272–276.

Alexander, K.L., Targan, S.R., and Elson, C.O.: **Microbiota activation and regulation of innate and adaptive immunity.** *Immunol. Rev.* 2014, **260**:206–220.

Arenas-Hernández, M.M., Martínez-Laguna, Y., and Torres, A.G.: **Clinical implications of enteroadherent Escherichia coli.** *Curr. Gastroenterol. Rep.* 2012, **14**:386–394.

Chung, H., Pamp, S.J., Hill, J.A., Surana, N.K., Edelman, S.M., Troy, E.B., Reading, N.C., Villablanca, E.J., Wang, S., Mora, J.R., *et al.*: **Gut immune maturation depends on colonization with a host-specific microbiota.** *Cell* 2012, **149**:1578–1593.

Coccia, M., Harrison, O.J., Schiering, C., Asquith, M.J., Becher, B., Powrie, F., and Maloy, K.J. **IL-1β mediates chronic intestinal inflammation by promoting the accumulation of IL-17A secreting innate lymphoid cells and CD4+ Th17 cells.** *J. Exp. Med.* 2012, **209**:1595–1609.

Knights, D., Lassen, K.G., and Xavier, R.J.: **Advances in inflammatory bowel disease pathogenesis: linking host genetics and the microbiome.** *Gut* 2013, **62**:1505–1510.

Kullberg, M.C., Jankovic, D., Feng, C.G., Hue, S., Gorelick, P.L., McKenzie, B.S., Cua, D.J., Powrie, F., Cheever, A.W., Maloy, K.J., *et al.*: **Intestinal epithelial cells: regulators of barrier function and immune homeostasis.** *J. Exp. Med.* 2006, **203**:2485–2494.

Peterson, L.W., and Artis, D.: **Intestinal epithelial cells: regulators of barrier function and immune homeostasis.** *Nat. Rev. Immunol.* 2014, **14**:141–153.

Shale, M., Schiering, C., and Powrie, F.: **CD4+ T-cell subsets in intestinal inflammation.** *Immunol. Rev.* 2013, **252**:164–182.

Zelante, T., Iannitti, R.G., Cunha, C., De Luca, A., Giovannini, G., Pieraccini, G., Zecchi, R., D'Angelo, C., Massi-Benedetti, C., Fallarino, F., *et al.*: **Tryptophan catabolites from microbiota engage aryl hydrocarbon receptor and balance mucosal reactivity via interleukin-22.** *Immunity* 2013, **39**:372–385.

Failures of Host Defense Mechanisms

13

In the normal course of an infection, the infectious agent first triggers an innate immune response. The foreign antigens of the infectious agent, enhanced by signals from innate immune cells, then induce an adaptive immune response that ultimately clears the infection and establishes a state of protective immunity. This does not always happen, however. In this chapter we examine circumstances in which there are failures of host defense against infectious agents, whether due to immune defects in an abnormal host, as occurs in immunodeficiency, or to the evasion or subversion of immune defenses in normal hosts by pathogens. Finally, we will consider the special case in which the immune defenses of a normal host are impaired by one infectious agent that leads to more generalized susceptibility to infection, as occurs in the acquired immune deficiency syndrome (AIDS) caused by human immunodeficiency virus (HIV).

In the first part of the chapter, we examine **primary**, or **inherited**, **immunodeficiency diseases**, in which host defense fails due to an inherited defect in a gene that results in the elimination or impaired function of one or more components of the immune system, leading to heightened susceptibility to infection with particular classes of pathogens. Immunodeficiency diseases caused by defects in T- or B-lymphocyte development, phagocyte function, and complement components have all been discovered. In the second part of the chapter, we briefly consider mechanisms by which pathogens evade or subvert specific components of the immune response to avoid elimination, so-called **immune evasion**. In the last part of the chapter, we consider how persistent infection by HIV leads to AIDS, an example of **secondary**, or **acquired**, **immunodeficiency**. The study of circumstances and mechanisms by which the immune system can fail has already contributed greatly to our understanding of host defense mechanisms and, in the longer term, might help to provide new methods of controlling or preventing infectious diseases, including AIDS.

IN THIS CHAPTER

Immunodeficiency diseases.

Evasion and subversion of immune defenses.

Acquired immune deficiency syndrome.

Immunodeficiency diseases.

Immunodeficiencies occur when one or more components of the immune system are defective; immunodeficiencies are classified as primary (inherited, or congenital) or secondary (acquired). **Primary immunodeficiencies** are caused by inherited mutations in any of a large number of genes that are involved in or control immune responses. Well over 150 primary immuno-deficiencies have now been described that affect the development of immune cells, their function, or both. Clinical features of these disorders are therefore highly variable, although a common feature is recurrent and often overwhelming infections in very young children. In contrast, **secondary immuno-deficiencies** are acquired as a consequence of other diseases, or are secondary to environmental factors such as starvation, or are an adverse consequence of medical intervention. Some forms of immunodeficiency principally affect immune-regulatory pathways. Defects of this type can lead to allergy, abnormal proliferation of lymphocytes, autoimmunity, and certain types of cancer, and will be discussed in other chapters. Here, we will mainly focus on those immunodeficiencies that predispose to infection.

Primary immunodeficiencies can be classified on the basis of the components of the immune system involved. However, because of the integration of many aspects of immune defense, defects in one component of the immune system can impact the function of others. Therefore, primary defects in innate immunity can lead to defects in adaptive immunity, and vice versa. Nevertheless, it is instructive to consider immune defects in the context of the major types of immunity affected, as these can lead to distinct patterns of infection and clinical disease. By examining which infectious diseases accompany a particular immunodeficiency, we gain insights into components of the immune system that are important in the response to particular agents. The inherited immunodeficiencies also reveal how interactions between different immune cell types contribute to the immune response and to the development of T and B lymphocytes. Finally, these inherited diseases can lead us to the defective gene, often revealing new information about the molecular basis of immune processes and providing the necessary information for diagnosis, genetic counseling, and eventually the possibility of gene therapy for cure.

13-1 A history of repeated infections suggests a diagnosis of immunodeficiency.

Patients with immune deficiency are usually detected clinically by a history of recurrent infection, often by the same or similar pathogens. The type of infection is a guide to which part of the immune system is deficient. Recurrent infection by **pyogenic**, or **pus-forming**, **bacteria** suggests a defect in antibody, complement, or phagocyte function, reflecting the role of these parts of the immune system in defense against such infections. Alternatively, a history of persistent fungal skin infection, such as cutaneous candidiasis, or recurrent viral infections suggests a defect in host defense mediated by T lymphocytes.

13-2 Primary immunodeficiency diseases are caused by inherited gene defects.

Before the advent of antibiotics, most individuals with inherited immune defects died in infancy or early childhood because of their susceptibility to particular classes of pathogens. Such cases were not easily identified, because many normal infants also died of infection. Most of the gene defects that cause inherited immunodeficiencies are recessive, and many are caused by mutations in genes on the X chromosome. As males have only one X chromosome, all males who inherit an X chromosome carrying a defective gene will be affected by the disease. In contrast, female carriers with one defective X chromosome are usually healthy.

Gene knockout techniques in mice (see Appendix I, Section A-35) have created many immunodeficient states that are adding rapidly to our knowledge of the contribution of individual proteins to normal immune function. Nevertheless, human immunodeficiency diseases remain the best source of insight into the normal pathways of defense against infectious diseases. For example, deficiencies of antibody, of complement, or of phagocytic function each increase the risk of infection by certain types of bacteria. This reflects the fact that the normal pathway of host defense against such bacteria is the binding of antibody followed by the fixation of complement, which allows the opsonized bacteria to be taken up by phagocytic cells and killed. Breaking any of the links in this chain of events causes a similar immunodeficient state.

Immunodeficiencies also teach us about the redundancy of defense mechanisms against infectious disease. By chance, the first person to be reported with a hereditary deficiency of complement (C2 deficiency) was a healthy immunologist. This teaches us that there are multiple protective immune mechanisms against infection, such that a defect in one component of immunity might be compensated for by other components. Thus, although there is abundant evidence that complement deficiency increases susceptibility to pyogenic infection, not every human with complement deficiency suffers from recurrent infections.

Examples of immunodeficiency diseases are listed in Fig. 13.1. None is very common (a selective deficiency in IgA being the most frequently reported), and some are extremely rare. These diseases are described in subsequent sections, and we have grouped the diseases according to where the specific causal defect lies in the adaptive or innate immune systems.

13-3 Defects in T-cell development can result in severe combined immunodeficiencies.

The developmental pathways leading to circulating naive T cells and B cells are summarized in Fig. 13.2. Patients with defects in T-cell development are highly susceptible to a broad range of infectious agents. This demonstrates the central role of T-cell differentiation and maturation in adaptive immune responses to virtually all antigens. Because such patients exhibit neither T-cell-dependent antibody responses nor cell-mediated immune responses, and thus cannot develop immunological memory, they are said to suffer from **severe combined immunodeficiency** (**SCID**).

X-linked SCID (**XSCID**) is the most frequent form of SCID and is caused by mutations in the gene *IL2RG* on the human X chromosome, which encodes the interleukin-2 receptor (IL-2R) common gamma chain (γ_c). γ_c is required in all receptors of the IL-2 cytokine family (IL-2, IL-4, IL-7, IL-9, IL-15, and IL-21). Patients with XSCID thus have defects in signaling of all IL-2-family cytokines, and, owing to the defects in IL-7 and IL-15, their T cells and NK cells fail to develop normally (see Fig. 13.2). B-cell numbers, on the other hand, are normal, but due to absence of T-cell help, their function is not. XSCID patients are overwhelmingly male; in females who are carriers of the mutation, T-cell and NK-cell progenitors in which X-inactivation has preserved the wild-type *IL2RG* allele progress through development to establish a normal mature immune repertoire. XSCID is known as the 'bubble boy disease' after a boy with XSCID who lived in a protective bubble for more than a decade before he died from complications of a bone marrow transplant. A clinically and immunologically indistinguishable type of SCID is associated with an inactivating mutation in the kinase Jak3 (see Section 8-1), which physically associates with γ_c and transduces signaling through γ_c-chain cytokine receptors. This autosomal recessive mutation also impairs the development of T and NK cells, but the development of B cells is unaffected.

X-linked Severe Combined Immunodeficiency

Other immunodeficiencies in mice have pinpointed more precisely the roles of individual cytokines and their receptors in T-cell and NK-cell development. For example, mice with targeted mutations in the β_c gene (*IL2RB*) defined a key role for IL-15 as a growth factor for the development of NK cells, as well as

Fig. 13.1 Human immunodeficiency syndromes. The specific gene defect, the consequence for the immune system, and the resulting disease susceptibilities are listed for some common and some rare human immunodeficiency syndromes. Severe combined immunodeficiency (SCID) can be due to many different defects, as summarized in Fig. 13.2 and described in the text. AID, activation-induced cytidine deaminase; ATM, ataxia telangiectasia-mutated protein; EBV, Epstein–Barr virus; IKK, inhibitor of κ B kinase; STAT3, signal transducer and activator of transcription 3; TAP, transporters associated with antigen processing; UNG, uracil-DNA glycosylase.

Name of deficiency syndrome	Specific abnormality	Immune defect	Susceptibility
Severe combined immune deficiency	See text and Fig. 13.2		General
DiGeorge's syndrome	Thymic aplasia	Variable numbers of T cells	General
MHC class I deficiency	Mutations in TAP1, TAP2, and tapasin	No CD8 T cells	Chronic lung and skin inflammation
MHC class II deficiency	Lack of expression of MHC class II	No CD4 T cells	General
Wiskott–Aldrich syndrome	X-linked; defective WASp gene	Defective anti-polysaccharide antibody, impaired T-cell activation responses, and T$_{reg}$ dysfunction	Encapsulated extracellular bacteria Herpesvirus infections (e.g., HSV, EBV)
X-linked agamma-globulinemia	Loss of BTK tyrosine kinase	No B cells	Extracellular bacteria, enteroviruses
Hyper-IgM syndrome	CD40 ligand deficiency CD40 deficiency NEMO (IKK) deficiency	No isotype switching and/or no somatic hypermutation plus T-cell defects	Extracellular bacteria *Pneumocystis jirovecii* *Cryptosporidium parvum*
Hyper-IgM syndrome—B-cell intrinsic	AID deficiency UNG deficiency	No isotype switching +/− normal somatic hypermutation	Extracellular bacteria
Hyper-IgE syndrome (Job's syndrome)	Defective STAT3	Block in T$_H$17 cell differentiation Elevated IgE	Extracellular bacteria and fungi
Common variable immunodeficiency	Mutations in TACI, ICOS, CD19, etc.	Defective IgA and IgG production	Extracellular bacteria
Selective IgA	Unknown; MHC-linked	No IgA synthesis	Respiratory infections
Phagocyte deficiencies	Many different	Loss of phagocyte function	Extracellular bacteria and fungi
Complement deficiencies	Many different	Loss of specific complement components	Extracellular bacteria especially *Neisseria* spp.
X-linked lympho-proliferative syndrome	Mutations in SAP or XIAP	Inability to control B-cell growth	EBV-driven B-cell tumors Fatal infectious mononucleosis
Ataxia telangiectasia	Mutations in ATM	T cells reduced	Respiratory infections
Bloom's syndrome	Defective DNA helicase	T cells reduced Reduced antibody levels	Respiratory infections

a role for the cytokine in T-cell maturation and trafficking. Mice with targeted mutations in IL-15 itself or the α chain of its receptor also have no NK cells and relatively normal T-cell development, but they show a more specific T-cell defect, primarily limited to impaired maintenance of memory CD8 T cells.

Humans with a deficiency of the IL-7 receptor α chain have no T cells but normal levels of NK cells, illustrating that IL-7 signaling, while essential for T-cell development, is not essential for the development of NK cells (see Fig. 13.2). Interestingly, mice with a gene-targeted deficiency of the IL-7R

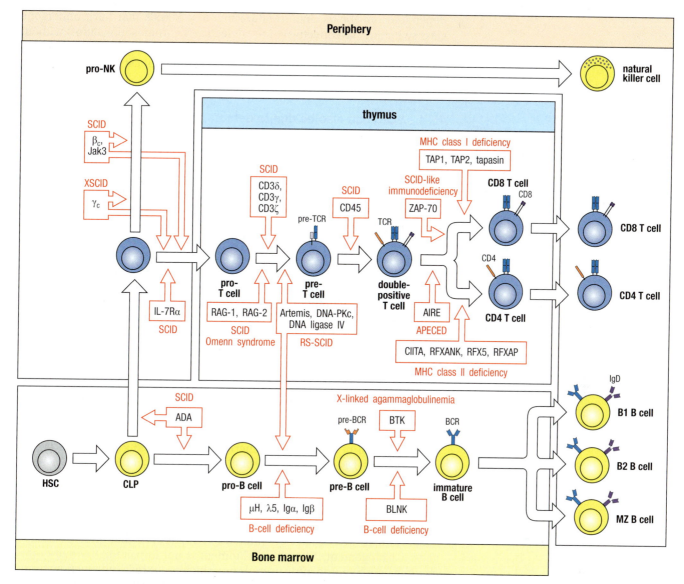

Fig. 13.2 Defects in T-cell and B-cell development that cause immunodeficiency. The pathways leading to circulating naive T cells and B cells are shown here. Mutations in genes that encode the proteins (indicated in red boxes) are known to cause human immunodeficiency diseases. BCR, B-cell receptor; CLP, common lymphoid progenitor; HSC, hematopoietic stem cell; MZ B cell, marginal zone B cell; pre-BCR, pre-B-cell receptor; pre-TCR, pre-T-cell receptor; RS-SCID, radiation-sensitive SCID; SCID, severe combined immunodeficiency; TCR, T-cell receptor; XSCID, X-linked SCID. Immunodeficiency can also be caused by mutations in genes in the thymic epithelium that impair thymic development, and thus T-cell development.

share with humans a deficiency of T cells, but also lack B cells, which is not the case in humans. This illustrates the species-specific role of certain cytokines, and provides a cautionary note against extrapolating findings from mice to humans. In humans and mice whose T cells show defective production of IL-2 after receptor stimulation, most T-cell development itself is normal, although there is impaired development of FoxP3+ T$_{reg}$ cells that predisposes to immune-regulatory abnormalities and autoimmunity (see Chapter 15). The more limited effects of individual cytokine signaling defects are in contrast to the global defects in T- and NK-cell development in patients with XSCID.

As in all serious T-cell deficiencies, patients with XSCID do not make effective antibody responses to most antigens, although their B cells seem normal. Most, but not all, naive IgM-positive B cells from female carriers of XSCID have inactivated the defective X chromosome rather than the normal one (see Section 13-3), showing that B-cell development is affected by, but is not wholly dependent on, the γ_c chain. Mature memory B cells that have undergone class switching have inactivated the defective X chromosome almost without exception. This might reflect the fact that the γ_c chain is also part of the receptor for IL-21, which is important for the maturation of class-switched B cells (see Section 10-4).

13-4 SCID can also be due to defects in the purine salvage pathway.

Adenosine Deaminase Deficiency

Variants of autosomal recessive SCID that arise from defects in enzymes of the salvage pathway of purine synthesis include **adenosine deaminase (ADA) deficiency** (see Fig. 13.2) and **purine nucleotide phosphorylase (PNP) deficiency**. ADA catalyzes the conversion of adenosine and deoxyadenosine to inosine and deoxyinosine, respectively, and its deficiency results in the accumulation of deoxyadenosine and its precursor, *S*-adenosylhomocysteine, which are toxic to developing T and B cells. PNP catalyzes the conversion of inosine and guanosine to hypoxanthine and guanine, respectively. PNP deficiency, which is a rarer form of SCID, also causes the accumulation of toxic precursors but affects developing T cells more severely than B cells. In both diseases, the development of lymphopenia, or decreased numbers of lymphocytes, is progressive after birth, resulting in profound lymphopenia within the first few years of life. Because both enzymes are housekeeping proteins expressed by many cell types, the immune deficiency associated with each of these inherited defects is part of a broader clinical syndrome.

13-5 Defects in antigen receptor gene rearrangement can result in SCID.

Another group of autosomally inherited defects leading to SCID is caused by failures of DNA rearrangement in developing lymphocytes. Mutations in either the *RAG1* or *RAG2* gene that result in nonfunctional proteins cause arrest of lymphocyte development at the pro- to pre-T-cell and B-cell transitions because of a failure of V(D)J recombination (see Fig. 13.2). Thus, there is a complete lack of both T cells and B cells in these patients. Because the effects of RAG deficiencies are limited to lymphocytes that undergo antigen gene rearrangement, NK-cell development is not impaired. There are other children with **hypomorphic mutations** (which cause reduced, but not absent, function) in either *RAG1* or *RAG2* who can make a small amount of functional RAG protein, allowing limited V(D)J recombination. This latter group includes patients with a distinctive and severe disease called **Omenn syndrome**, which,

Omenn Syndrome

in addition to increased susceptibility to multiple opportunistic infections, has clinical features very similar to graft-versus-host disease characterized by rashes, eosinophilia, diarrhea, and enlargement of the lymph nodes (see Section 15-36). Normal or increased numbers of activated T cells are found in these children. An explanation for this phenotype is that low levels of *RAG* activity allow some limited T-cell receptor gene recombination. No B cells are found, however, suggesting that B cells have more stringent requirements for *RAG* activity. Due to the limited number of T-cell receptors that are successfully rearranged, the repertoire of T cells is highly restricted in patients with Omenn syndrome, and there is activation and clonal expansion of the limited number of specificities present. The clinical features strongly suggest that these peripheral T cells are autoreactive and are responsible for the graft-versus-host phenotype. In addition to Omenn syndrome, which is manifested very early in life, other forms of immunodeficiency have also been associated with reduced but not absent RAG activity, and are often characterized by granulomatous disease that is not evident until late childhood or adolescence.

A subset of patients with autosomal recessive SCID are characterized by an abnormal sensitivity to ionizing radiation. They produce very few mature B and T cells because there is a failure of DNA rearrangement in their developing lymphocytes; only rare VJ or VDJ joints are seen, and most of these are abnormal. This type of SCID is due to defects in ubiquitous DNA repair proteins involved in repairing DNA double-strand breaks, which are generated not only during antigen receptor gene rearrangement (see Section 5-5) but also by ionizing radiation. Owing to the increased radiosensitivity in these patients, this class of SCID is called **radiation-sensitive SCID** (**RS-SCID**) to distinguish it from SCID due to lymphocyte-specific defects. Defects in the genes for Artemis, DNA protein-kinase catalytic subunit (DNA-PKcs), and DNA ligase IV cause RS-SCID (see Fig. 13.2). Because defects in repair of DNA breaks increase the risk of translocations during cell division that can lead to malignant transformation, patients with RS-SCID variants are also more likely to develop cancer.

13-6 Defects in signaling from T-cell antigen receptors can cause severe immunodeficiency.

Several gene defects have been described that interfere with signaling through the T-cell receptor (TCR) and thus block the activation of T cells early in thymic development. Patients with mutations in the CD3δ, CD3ε, or CD3ζ chains of the CD3 complex have defective pre-T-cell receptor signaling and fail to progress to the double-positive stage of thymic development (see Fig. 13.2), resulting in SCID. Another lymphocyte signaling defect that leads to severe immunodeficiency is caused by mutations in the tyrosine phosphatase CD45. Humans and mice with CD45 deficiency show a marked reduction in peripheral T-cell numbers and also abnormal B-cell maturation. Severe immunodeficiency also occurs in patients who make a defective form of the cytosolic protein tyrosine kinase ZAP-70, which transmits signals from the T-cell receptor (see Section 7-7). CD4 T cells emerge from the thymus in normal numbers, whereas CD8 T cells are absent. However, the CD4 T cells that mature fail to respond to stimuli that normally activate the cells through the T-cell receptor.

Wiskott–Aldrich syndrome (**WAS**), which is caused by a defect in the *WAS* gene on the X chromosome that encodes WAS protein (WASp), has shed new light on the molecular basis of signaling and immune synapse formation between various cells in the immune system. Although the disease also affects platelets and was first described as a blood-clotting disorder, it also causes immunodeficiency that is characterized by reduced T-cell numbers, defective NK-cell cytotoxicity, and a failure of antibody responses (see Section 7-19). WASp is expressed in all hematopoietic cell lineages and is a key regulator of lymphocyte and platelet development and function through its transduction of receptor-mediated signals that induce reorganization of the cytoskeleton (see Section 9-25). Several signaling pathways downstream of the T-cell receptor are known to activate WASp (see Section 7.19). Activation of WASp in turn activates the Arp2/3 complex, which is essential for initiating actin polymerization that is critical for immune synapse formation and the polarized release of effector molecules by T cells. In patients with WAS, and in mice whose *Was* gene has been knocked out, T cells fail to respond normally to T-cell receptor cross-linking. It has also recently been suggested that WASp is required for the suppressive function of natural T_{reg} cells, and this may help explain why patients with WASp are susceptible to autoimmune diseases.

 **Wiskott–Aldrich Syndrome**

13-7 Genetic defects in thymic function that block T-cell development result in severe immunodeficiencies.

A disorder of thymic development associated with SCID and a lack of body hair has been known for many years in mice; the mutant strain is descriptively named ***nude*** (see Section 8-10). A small number of children have been described with the same phenotype. In both mice and humans this syndrome is caused by mutations in the gene *FOXN1*, which encodes a transcription factor selectively expressed in skin and thymus. FOXN1 is necessary for the differentiation of thymic epithelium and the formation of a functional thymus. In patients with a mutation in *FOXN1*, the lack of thymic function prevents normal T-cell development. B-cell development is normal in individuals with the mutation, yet B-cell responses are deficient because of the lack of T cells, and the response to nearly all pathogens is profoundly impaired.

DiGeorge syndrome is another disorder in which the thymic epithelium fails to develop normally, resulting in SCID. The genetic abnormality underlying this complex developmental disorder is a deletion within one copy of chromosome 22. The deletion varies between 1.5 and 5 megabases in size, with the smallest deletion that causes the syndrome containing approximately 24 genes. The relevant gene within this interval is *TBX1*, which encodes the transcription factor T-box 1. DiGeorge syndrome is caused by the deletion of a single copy of this gene, such that patients with this disorder are **haploinsufficient** for *TBX1*. Without the proper inductive thymic environment, T cells cannot mature, and both cell-mediated immunity and T-cell-dependent antibody production are impaired. Patients with this syndrome have normal levels of serum immunoglobulin but an absence of, or incomplete development of, the thymus and parathyroid glands, with varying degrees of T-cell immunodeficiency.

Defects in the expression of MHC molecules can lead to severe immunodeficiency as a result of effects on the positive selection of T cells in the thymus (see Fig. 13.2). Individuals with **bare lymphocyte syndrome** lack expression of all MHC class II molecules; the disease is now called **MHC class II deficiency**. Because the thymus lacks MHC class II molecules, CD4 T cells cannot be positively selected and few develop. The antigen-presenting cells in these individuals also lack MHC class II molecules and so the few CD4 T cells that do develop cannot be stimulated by antigen. MHC class I expression is normal, and CD8 T cells develop normally. However, such people suffer from severe immunodeficiency, illustrating the central importance of CD4 T cells in adaptive immunity to most pathogens.

MHC class II deficiency is caused not by mutations in the MHC genes themselves but by mutations in one of several genes encoding gene-regulatory proteins that are required for the transcriptional activation of MHC class II genes. Four complementing gene defects (known as groups A, B, C, and D) have been defined in patients who fail to express MHC class II molecules, indicating that the products of at least four different genes are required for the normal expression of these proteins. Genes corresponding to each complementation group have been identified: the *MHC class II transactivator*, or *CIITA*, is mutated in group A, and the genes *RFXANK*, *RFX5*, and *RFXAP* are mutated in groups B, C, and D, respectively (see Fig. 13.2). These last three encode proteins that are components of a multimeric complex, RFX, which is involved in the control of gene transcription. RFX binds a DNA sequence named an X-box, which is present in the promoter region of all MHC class II genes.

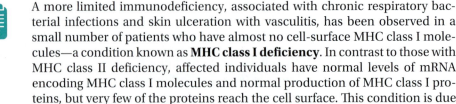

DiGeorge Syndrome

MHC Class II Deficiency

MHC Class I Deficiency

A more limited immunodeficiency, associated with chronic respiratory bacterial infections and skin ulceration with vasculitis, has been observed in a small number of patients who have almost no cell-surface MHC class I molecules—a condition known as **MHC class I deficiency**. In contrast to those with MHC class II deficiency, affected individuals have normal levels of mRNA encoding MHC class I molecules and normal production of MHC class I proteins, but very few of the proteins reach the cell surface. This condition is due either to mutations in *TAP1* or *TAP2*, which encode the subunits of the peptide transporter responsible for transporting peptides generated in the cytosol into the endoplasmic reticulum, where they are loaded into nascent MHC I molecules, or to mutations in *TAPBP*, which encodes tapasin, another component of the peptide transporter complex (see Section 6-4). Although the reduction in MHC class I molecules on the surface of thymic epithelial cells results in reduced numbers of CD8 T cells (see Fig. 13.2), people with MHC class I deficiency are not abnormally susceptible to viral infections. This is surprising given the key role of MHC class I presentation and of cytotoxic CD8 T cells in combating viral infections. There is, however, evidence for TAP-independent pathways for the presentation of certain peptides by MHC class I molecules, and the clinical phenotype of *TAP1*- and *TAP2*-deficient patients indicates that these pathways can compensate to allow sufficient development and function of CD8 T cells to control viruses.

Some defects in thymic cells lead to a phenotype with other effects besides those of immunodeficiency. The gene *AIRE* encodes a transcription factor that enables thymic epithelial cells to express many self proteins and so to mediate efficient negative selection. Defects in *AIRE* lead to a complex syndrome called **APECED** (autoimmune polyendocrinopathy-candidiasis-ectodermal dystrophy), which is characterized by autoimmunity, developmental defects, and immunodeficiency (see Section 8-23, and Chapter 15).

Autoimmune Polyendocrinopathy–Candidiasis–Ectodermal Dystrophy (APECED)

13-8 Defects in B-cell development result in deficiencies in antibody production that cause an inability to clear extracellular bacteria and some viruses.

In addition to inherited defects in proteins that are crucial to both T-cell and B-cell development, such as RAG-1 and RAG-2, defects that are specific to B-cell development have also been identified (see Fig. 13.2). Patients with these defects are characterized by an inability to cope with extracellular bacteria and some viruses whose efficient clearance requires specific antibodies. Pyogenic bacteria, such as staphylococci and streptococci, have polysaccharide capsules that are not directly recognized by the receptors on macrophages and neutrophils that stimulate phagocytosis. The bacteria escape elimination by the innate immune response and are successful extracellular pathogens, but can be cleared by an adaptive immune response. Opsonization by antibody and complement enables phagocytes to ingest and destroy the bacteria (see Section 10-22). The principal effect of deficiencies in antibody production is therefore a failure to control infections by pyogenic bacteria. Susceptibility to some viral infections, notably those caused by enteroviruses, is also increased because of the importance of antibodies in neutralizing viruses that enter the body through the gut.

The first description of an immunodeficiency disease was **Ogden C. Bruton's** account, in 1952, of the failure of a male child to produce antibody. Because inheritance of this condition is X-linked and is characterized by the absence of immunoglobulin in the serum (**agammaglobulinemia**), it was called **Bruton's X-linked agammaglobulinemia** (**XLA**) (see Fig. 13.2). Since then, autosomal recessive variants of agammaglobulinemia have been described. Infants with these diseases are usually identified as a result of recurrent infections with pyogenic bacteria, such as *Streptococcus pneumoniae*, and enteroviruses. In this regard, it should be noted that normal infants have a transient deficiency in immunoglobulin production in the first 3–12 months of life. The newborn infant has antibody levels comparable to those of the mother because of the transplacental transport of maternal IgG (see Section 10-17). As this IgG is catabolized, antibody levels gradually decrease until the infant begins to produce significant amounts of its own IgG at about 6 months (**Fig. 13.3**). Thus, IgG levels are quite low between the ages of 3 months and 1 year. This can lead to a period of heightened susceptibility to infection, especially in premature babies, who begin with lower levels of maternal IgG and also reach immune competence later after birth. Because of the transient protection afforded

X-linked Agammaglobulinemia

Fig. 13.3 Immunoglobulin levels in newborn infants fall to low levels at about 6 months of age. Babies are born with high levels of maternal IgG, which is actively transported across the placenta from the mother during gestation. After birth, the production of IgM starts almost immediately; the production of IgG, however, does not begin for about 6 months, during which time the total level of IgG falls as the maternally acquired IgG is catabolized. Thus, IgG levels are low from about the age of 3 months to 1 year, which can lead to susceptibility to disease.

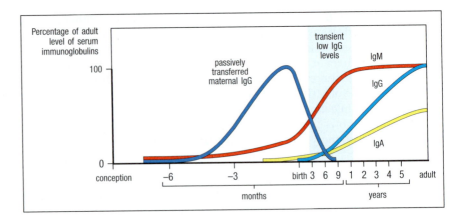

newborn infants by maternal antibodies, XLA is typically detected several months after birth, when maternal antibody levels in the infant have declined.

The defective gene in XLA encodes a protein tyrosine kinase called BTK (Bruton's tyrosine kinase), which is a member of the Tec family of kinases that transduce signals through the pre-B-cell receptor (pre-BCR; see Section 7-20). As discussed in Section 8-3, the pre-B-cell receptor is composed of successfully rearranged μ heavy chains complexed with the surrogate light chain composed of λ5 and VpreB, and with the signal-transducing subunits Igα and Igβ. Stimulation of the pre-B-cell-receptor recruits cytoplasmic proteins, including BTK, which convey signals required for the proliferation and differentiation of pre-B cells. In the absence of BTK function, B-cell maturation is largely arrested at the pre-B-cell stage (see Fig. 13.2; see also Section 8-3), resulting in profound B-cell deficiency and agammaglobulinemia. Some B cells do mature, however, perhaps as a result of compensation by other Tec kinases.

During embryonic development, females randomly inactivate one of their two X chromosomes. Because BTK is required for B-lymphocyte development, only cells in which the normal allele of *BTK* is active develop into mature B cells. Thus, in all B cells in female carriers of a mutant *BTK* gene, the active X chromosome is the normal one and the abnormal X chromosome is inactivated. This fact allowed female carriers of XLA to be identified even before the nature of the BTK protein was known. In contrast, the active X chromosomes in the T cells and macrophages of carriers are an equal mixture of normal and *BTK* mutant X chromosomes. Nonrandom X-inactivation only in B cells shows conclusively that the *BTK* gene is required for the development of B cells but not the other cell types, and that BTK must act in the B cells themselves rather than in stromal or other cells required for B-cell development (Fig. 13.4).

Autosomal recessive deficiencies in other components of the pre-BCR also block early B-cell development and cause severe B-cell deficiency and congenital agammaglobulinemia similar to that of XLA. These disorders are much rarer than XLA, and include mutations in the genes that encode the μ heavy chain (*IGHM*), which is the second most common cause of agammaglobulinemia; λ5 (*IGLL1*); and Igα (*CD79A*) and Igβ (*CD79B*) (see Fig. 13.2). Mutations that cripple the B-cell receptor signaling adaptor, B-cell linker protein (encoded by *BLNK*), also cause the arrest of early B-cell development that results in selective B-cell deficiency.

Patients with pure B-cell defects resist many pathogens other than pyogenic bacteria. Fortunately, the latter can be suppressed with antibiotics and with monthly infusions of human immunoglobulin collected from a large pool of donors. Because there are antibodies against many common pathogens in this pooled immunoglobulin, it serves as a fairly successful shield against infection.

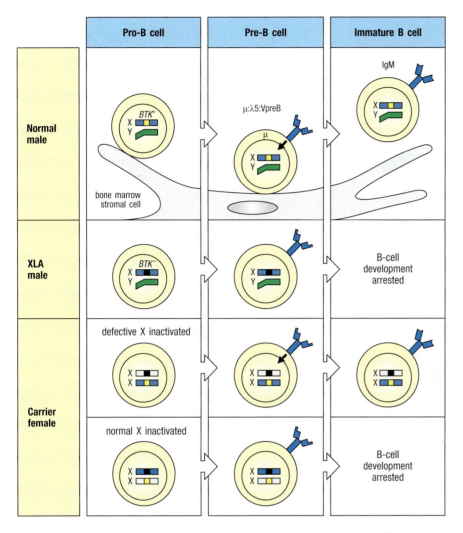

| Pro-B cell | Pre-B cell | Immature B cell |

Normal male

bone marrow stromal cell

μ:λ5:VpreB

IgM

XLA male

B-cell development arrested

Carrier female

defective X inactivated

normal X inactivated

B-cell development arrested

Fig. 13.4 The product of the *BTK* gene is important for B-cell development. In X-linked agammaglobulinemia (XLA), a protein tyrosine kinase of the Tec family called BTK, which is encoded on the X chromosome, is defective. In normal individuals, B-cell development proceeds through a stage in which the pre-B-cell receptor, consisting of μλ5:VpreB (see Section 8-3), transduces a signal via BTK, triggering further B-cell development. In males with XLA, no signal can be transduced and, although the pre-B-cell receptor is expressed, the B cells develop no further. In female mammals, including humans, one of the two X chromosomes in each cell is permanently inactivated early in development. Because the choice of which chromosome to inactivate is random, half of the pre-B cells in a carrier female will have inactivated the chromosome with the wild-type *BTK* gene, meaning that they can express only the defective *BTK* gene and cannot develop further. Therefore, in the carrier, mature B cells always have the nondefective X chromosome active. This is in sharp contrast to all other cell types, which have the nondefective X chromosome active in only half of their cells. Nonrandom X-chromosome inactivation in a particular cell lineage is a clear indication that the product of the X-linked gene is required for the development of cells of that lineage. It is also sometimes possible to identify the stage at which the gene product is required, by detecting the point in development at which X-chromosome inactivation develops bias. Using this kind of analysis, one can identify carriers of X-linked traits such as XLA without needing to know the nature of the mutant gene.

13-9   Immune deficiencies can be caused by defects in B-cell or T-cell activation and function that lead to abnormal antibody responses.

After their development in the bone marrow or thymus, B and T cells require antigen-driven activation and differentiation to mount effective immune responses. Analogous to defects in early T-cell development, defects in T-cell activation and differentiation that occur after thymic selection have an impact on both cell-mediated immunity and antibody responses (Fig. 13.5). Defects specific to the activation and differentiation of B cells can impair their ability to undergo class switching to IgG, IgA, and IgE while leaving cell-mediated immunity largely intact. Depending on where in the process of T- or B-cell differentiation these defects occur, the characteristics of the immune deficiency that results can be either profound or relatively circumscribed.

A common feature of patients with defects that affect B-cell class switching is **hyper-IgM syndrome** (see Fig. 13.5). These patients have normal B- and T-cell development and normal or high serum levels of IgM, but make very limited antibody responses against antigens that require T-cell help. Thus immunoglobulin isotypes other than IgM and IgD are produced only in trace amounts. This renders these patients highly susceptible to infection with extracellular pathogens. Several causes for hyper-IgM syndromes have been distinguished, and these have helped to elucidate the pathways that are essential for normal class-switch recombination and somatic hypermutation in B cells. Defects have been found in both T-cell helper function and in the B cells themselves.

Fig. 13.5 Defects in T-cell and B-cell activation and differentiation cause immunodeficiencies. The pathways leading to activation and differentiation of naive T cells and B cells are shown here. The protein products of genes known to be mutated in the relevant human immunodeficiency diseases are indicated in red boxes. BCR, B-cell receptor; CVIDs, common variable immunodeficiencies; TCR, T-cell receptor. Note that the defect in cytoskeletal function in Wiskott–Aldrich syndrome (WAS) affects immune-cell function at many steps in this schema, and is not included in the figure for the sake of clarity.

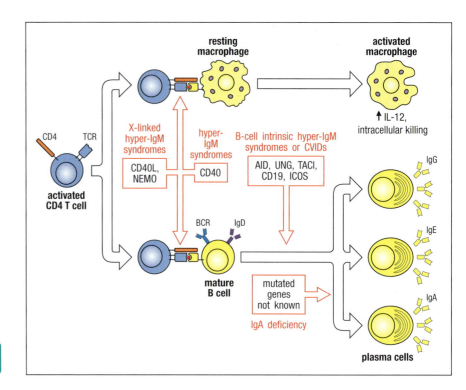

CD40 Ligand Deficiency

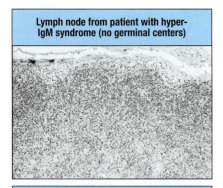

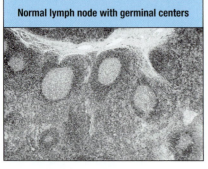

Fig. 13.6 Patients with CD40 ligand deficiency are unable to activate their B cells fully. Lymphoid tissues in patients with CD40 ligand deficiency, which manifests as a hyper-IgM syndrome, are devoid of germinal centers (top panel), unlike a normal lymph node (bottom panel). B-cell activation by T cells is required both for isotype switching and for the formation of germinal centers, where extensive B-cell proliferation takes place. Photographs courtesy of R. Geha and A. Perez-Atayde.

The most common form of hyper-IgM syndrome is **X-linked hyper-IgM syndrome**, or **CD40 ligand deficiency**, which is caused by mutations in the gene encoding CD40 ligand (CD154) (see Fig. 13.5). CD40 ligand is normally expressed on activated T cells, enabling them to engage the CD40 protein on antigen-presenting cells, including B cells, dendritic cells, and macrophages (see Section 10-4). In males with CD40 ligand deficiency, B cells are normal, but in the absence of engagement of CD40, their B cells do not undergo isotype switching or initiate the formation of germinal centers (Fig. 13.6). These patients therefore have severe reductions in circulating levels of all antibody isotypes except IgM and are highly susceptible to infections by pyogenic extracellular bacteria.

Because CD40 signaling is also required for the activation of dendritic cells and macrophages for optimal production of IL-12, which is important for the production of IFN-γ by T_H1 cells and NK cells, patients with CD40 ligand deficiency also have defects in type 1 immunity and thus manifest a form of combined immunodeficiency. Inadequate cross-talk between T cells and dendritic cells via CD40L–CD40 interaction can lead to lower levels of co-stimulatory molecules on dendritic cells, thus impairing their ability to stimulate naive T cells (see Section 9-17). These patients are therefore susceptible to infections by extracellular pathogens that require class-switched antibodies, such as pyogenic bacteria, but also have defects in the clearance of intracellular pathogens, such as mycobacteria, and are particularly prone to opportunistic infections by *Pneumocystis jirovecii*, which is normally killed by activated macrophages.

A similar syndrome has been identified in patients with mutations in two other genes. Not unexpectedly, one is the gene encoding CD40, mutations in which have been found in a few patients with an autosomal recessive variant of hyper-IgM syndrome (see Fig. 13.5). In another form of X-linked hyper-IgM syndrome, known as **NEMO deficiency**, mutations occur in the gene encoding the protein NEMO ('NFκB essential modulator'; also known as IKKγ, a subunit of the kinase IKK), which is an essential component of the intracellular signaling pathway downstream of CD40 that leads to activation of the transcription factor NFκB (see Fig. 3.15). This group of hyper-IgM syndromes shows that mutations at different points in the CD40L–CD40 signaling pathway result in a similar combined immunodeficiency syndrome. In view of the role of NFκB signaling in many other pathways, NEMO deficiency results in additional immune dysfunction beyond its impairment of B-cell class switching (see Section 13-15), as well as nonimmune manifestations, including abnormalities of the skin.

Other variants of hyper-IgM syndrome are due to intrinsic defects in the process of B-cell class-switch recombination. Patients having these defects are susceptible to severe extracellular bacterial infections, but because T-cell differentiation and function are spared, they do not show increased susceptibility to intracellular pathogens or opportunistic agents such as *P. jirovecii*. One class-switching defect is due to mutations in the gene for activation-induced cytidine deaminase (AID), which is required for both somatic hypermutation and class switching (see Section 10-7). Patients with autosomally inherited defects in the AID gene (*AICDA*) fail to switch antibody isotype and also have greatly reduced somatic hypermutation (see Fig. 13.5). Immature B cells accumulate in abnormal germinal centers, causing enlargement of the lymph nodes and spleen. Another variant of B-cell-intrinsic hyper-IgM syndrome was identified recently in a small number of patients with an autosomal recessive defect in the DNA repair enzyme uracil-DNA glycosylase (UNG; see Section 10-10), which is also involved in class switching; these patients have normal AID function and normal somatic hypermutation, but defective class switching.

Other examples of predominantly antibody deficiency include the most common forms of primary immunodeficiency, referred to as **common variable immunodeficiencies** (**CVIDs**). CVIDs are a clinically and genetically heterogeneous group of disorders that typically go undiagnosed until late childhood or adulthood, because the immune deficiency is relatively mild. Unlike other causes of immunoglobulin deficiency, patients with CVID can have defects in immunoglobulin production that are limited to one or more isotypes (see Fig. 13.5). **IgA deficiency**, the most common primary immunodeficiency, exists in both sporadic and familial forms, and both autosomal recessive and autosomal dominant inheritance have been described. The etiology of IgA deficiency in most patients is not understood, and these patients are asymptomatic. In IgA-deficient patients who do develop recurrent infections, an associated defect in one of the IgG subclasses is often found.

A small minority of CVID patients have mutations in the transmembrane protein TACI (TNF-like receptor transmembrane activator and CAML interactor), which is encoded by the gene *TNFRSF13B*. TACI is the receptor for the cytokines BAFF and APRIL, which are produced by T cells, dendritic cells, and macrophages, and which can provide co-stimulatory and survival signals for B-cell activation and class switching (see Section 10-3). Other patients with selective deficiencies in IgG subclasses have also been described. B-cell numbers are typically normal in these patients, but serum levels of the affected immunoglobulin isotype are depressed. Although some of these patients have recurrent bacterial infections, as in IgA deficiency, many are asymptomatic. CVID patients with other defects that affect immunoglobulin class switching have been identified. Included in this group are patients with inherited defects in CD19, which is a component of the B-cell co-receptor (see Fig. 13.5). A genetic defect that has been linked to a small percentage of people with CVID is deficiency of the co-stimulatory molecule ICOS. As described in Section 9-17, ICOS is upregulated on T cells when they are activated. The effects of ICOS deficiency have confirmed its essential role in T-cell help for the later stages of B-cell differentiation, including class switching and the formation of memory cells.

The final immunodeficiency to be considered in this section is **hyper-IgE syndrome** (**HIES**), also called **Job's syndrome**. This disease is characterized by

X-linked Hypohidrotic Ectodermal Dysplasia and Immunodeficiency

Activation-Induced Cytidine Deaminase Deficiency

Common Variable Immunodeficiency

Hyper IgE Syndrome

recurrent skin and pulmonary infections caused by pyogenic bacteria, chronic mucocutaneous candidiasis (noninvasive fungal infection of the skin and mucosal surfaces), very high serum concentrations of IgE, and chronic eczematous dermatitis or skin rash. HIES is inherited in an autosomal recessive or dominant pattern, with the latter manifesting skeletal and dental abnormalities not found in the recessive variant. The inherited defect in the autosomal dominant variant of HIES is in the transcription factor STAT3, which is activated downstream of several cytokine receptors, including those for IL-6, IL-22, and IL-23, and which is central to the differentiation of T_H17 cells and activation of ILC3 cells. STAT3 signaling activated by IL-6 and IL-22 is also important for enhanced antimicrobial resistance of epithelial cells of the skin and mucosal barriers. Because differentiation of T_H17 cells is deficient in these patients, the recruitment of neutrophils normally orchestrated by the T_H17 response is also defective, as is production of IL-22, an important cytokine in activating epithelial cell production of antimicrobial peptides. This is thought to underlie the impaired defense against extracellular bacteria and fungi at barrier epithelia, such as the skin and mucosae. The cause of the elevated IgE is not understood, but it might be due to an abnormal accentuation of skin and mucosal T_H2 responses as a result of T_H17 deficiency. In an autosomal recessive variant of HIES, the mutation is in the gene that encodes the protein DOCK8 (dedicator of cytokinesis 8), the function of which in immune cells is poorly characterized. However, because DOCK8 is thought to play a broader role in T-cell function, as well as NK-cell function, this variant of HIES is distinguished from that caused by STAT3 defects by the additional occurrence of opportunistic infections and recurrent cutaneous viral infections (for example, herpes simplex), as well as allergic and autoimmune manifestations.

Dedicator of Cytokinesis 8 Deficiency

13-10   Normal pathways for host defense against different infectious agents are pinpointed by genetic deficiencies of cytokine pathways central to type 1/T_H1 and type 3/T_H17 responses.

Inherited defects in cytokines that are involved in the development and function of different effector T-cell subsets have been defined, as have defects in the receptors or the signaling pathways through which they act. In contrast to the T-cell immunodeficiencies considered above, here we consider those deficiencies that do not have major defects in antibody production. A small number of families have been discovered with individuals who suffer from persistent and sometimes fatal attacks by intracellular pathogens normally restrained by type 1 immunity, especially *Mycobacterium*, *Salmonella*, and *Listeria* species. These microbes are specialized for survival within macrophages, and their eradication requires enhanced microbicidal activities induced by IFN-γ produced by type 1 cells: NK cells, ILC1 cells, and T_H1 cells (see Section 11-2). Accordingly, susceptibility to these infections is conferred by a variety of inherited mutations that impair or abolish the function of IL-12 or IFN-γ, key cytokines in the development and functions of type 1 cells (Fig. 13.7). Patients with mutations in the genes encoding the p40 subunit of IL-12 (*IL12B*), the IL-12 receptor β₁ chain (*IL12RB1*), and the two subunits (R1 and R2) of the IFN-γ receptor (*IFNGR1* and *IFNGR2*) have been identified. Although affected individuals have heightened susceptibility to the more virulent *M. tuberculosis*, they suffer more frequently from the nontuberculous, or atypical, strains of mycobacteria, such as *Mycobacterium avium*, likely due to the greater prevalence of atypical strains in the environment. They may also develop disseminated infection after vaccination with *Mycobacterium bovis* bacillus Calmette–Guérin (BCG), the strain of *M. bovis* that is used as a live vaccine against *M. tuberculosis*. Because the IL-12 p40 subunit is shared by IL-12 and IL-23, IL-12 p40 deficiency results in a broader infectious disease risk due to defective type 1 and type 3 (T_H17) functions (see Fig. 13.7).

Interferon-γ Receptor Deficiency

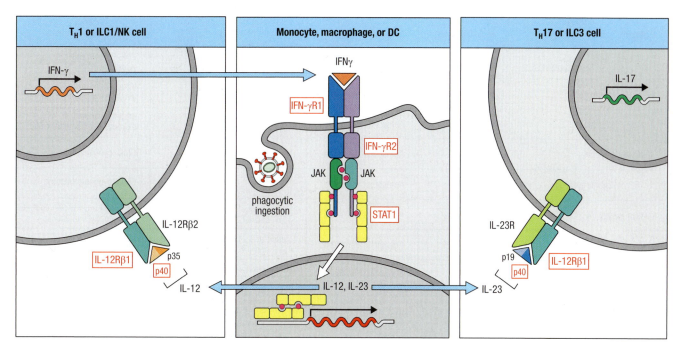

Fig. 13.7 Inherited defects in effector cytokine pathways that impair type 1/T$_H$1 and type 3/T$_H$17 immunity. Shown are pathways in IL-12, IL-23, and IFN-γ signaling for which inherited defects have been described. Note that defects in IL-12p40 (p40) and IL-12Rβ1 result in impaired function of both ILC1 cells/NK cells/T$_H$1 cells and ILC3/T$_H$17 cells because these subunits are shared, respectively, by IL-12 and IL-23 and their receptors. Also, because STAT1 is activated is by the receptors of type II interferon (IFN-γ) and type I interferon (IFN-α and IFN-β, not shown), defects in STAT1 result in impaired antibacterial and antiviral immune defense, whereas deficiency in either of the IFN-γ receptor subunits (IFN-γR1 or IFN-γR2) primarily results in impaired defense against intracellular bacteria.

Similarly, deficiency of the IL-12Rβ$_1$ chain, which is shared by the IL-12 and IL-23 receptors, results in broader susceptibilities than the deficiencies of IFN-γ or its receptor.

Autosomal recessive loss-of-function mutations in STAT1 impair IFN-γ receptor signaling and are also associated with increased susceptibility to infections by mycobacterial and other intracellular bacteria (see Fig. 13.7). However, due to its shared function in IFN-α receptor signaling in response to IFN-α and IFN-β (type I interferons), patients with STAT1 defects are also susceptible to viral infections. Interestingly, patients with partial loss of function of STAT1 have been identified who are susceptible to mycobacterial infections, but not viral infections, suggesting a more stringent requirement of STAT1 in protection against the former.

In addition to the T$_H$17-related defects described above for STAT3-deficient HIES (see Section 13-9), other defects in cytokine-mediated functions of this pathway have been identified that lack the hyper-IgE component (**Fig. 13.8**). Whereas heightened susceptibility to intracellular bacteria is a feature common to immunodeficiencies that impair type 1 immunity, heightened susceptibility to infections by *Candida* spp. and pyogenic bacteria, particularly *C. albicans* and *S. aureus*, respectively, is characteristic of these type 3 deficiencies. This reflects the specialized function of T$_H$17 and ILC3 cells in barrier defense against fungi and extracellular bacteria. Inherited deficiencies in IL-17F and IL-17RA, the shared receptor component for homo- and heterodimeric IL-17F–IL-17A ligands, confer susceptibility to these infectious agents and thus identify a key role for IL-17 cytokines in host defense against them. A similar susceptibility to chronic mucocutaneous candidiasis and pyogenic bacteria has been found in patients with autosomal dominant, gain-of-function mutations in STAT1. Because T$_H$17 cell development is impaired by STAT1 signaling downstream of several cytokine receptors (for example, type I and type II IFN receptors), individuals with these mutations have impaired

Increased Susceptibility to *Candida* Infections

Disease	Mutated gene	Inheritance	Immune phenotype	Associated infections
STAT3 deficiency; hyper-IgE syndrome (Job's disease)	STAT3	Autosomal dominant	Deficit of IL-17 producing T_H17 and ILC3 cells; hyper-IgE	CMC, *Staph. aureus*, *Aspergillus*
IL-12p40 deficiency	IL12B	Autosomal recessive	Deficit of IL-17 producing T_H17 and ILC3 cells*	Intracellular and extracellular bacteria, CMC
IL-12Rβ deficiency	IL12RB1	Autosomal recessive	Deficit of IL-17 producing T_H17 and ILC3 cells*	Intracellular and extracellular bacteria, CMC
IL-17RA deficiency	IL17RA	Autosomal recessive	No IL-17 response	CMC, pyogenic bacteria
IL-17F deficiency (partial)	IL17F	Autosomal recessive	Impaired IL-17F and IL-17A/F function	CMC, pyogenic bacteria
CARD9 deficiency	CARD9	Autosomal recessive	Deficit of IL-17 producing T_H17 and ILC3 cells	CMC and severe *Candida/* dermatophyte infections
STAT1 gain-of-function (GOF) mutation	STAT1	Autosomal dominant	Deficit of IL-17 producing T_H17 and ILC3 cells**	CMC, pyogenic bacteria
APECED syndrome	AIRE	Autosomal recessive	Neutralizing antibodies: IL-17A, IL-17F +/− IL-22	CMC

Fig. 13.8 Immune deficiencies with defects in T_H17/ILC3 function. Nearly all T_H17/ILC3 immune deficiencies result in chronic mucocutaneous candidiasis (CMC) and most also cause defects in defense against extracellular bacteria. *Deficiencies of IL-12p40 and IL-12Rβ1 also result in T_H1/ILC1/NK cell deficits. **It is currently unclear whether STAT1 gain-of-function mutations cause ILC3 deficits in addition to reduced numbers of T_H17 cells.

type 3 defenses. Note that this is in contrast to patients with loss-of-function STAT1 mutations, who are predisposed to intracellular bacterial infections due to defective type 1 immunity.

Autoimmune Polyendocrinopathy–Candidiasis–Ectodermal Dystrophy (APECED)

Increased Susceptibility to *Candida* Infections

In addition to inherited defects in effector cytokine genes, some forms of immunodeficiency result in the production of neutralizing autoantibodies against these cytokines. This results in infectious risks similar to those caused by primary cytokine deficiencies. Most patients with APECED syndrome (caused by defects in the *AIRE* gene; see Section 13-7) develop chronic mucocutaneous candidiasis that is due to the development of autoantibodies against IL-17A, IL-17F, and/or IL-22. Patients with neutralizing autoantibodies against IFN-γ who have impaired protection against atypical mycobacteria have also been reported, although the basis for this is unknown.

13-11 Inherited defects in the cytolytic pathway of lymphocytes can cause uncontrolled lymphoproliferation and inflammatory responses to viral infections.

Cytolytic granules are formed from components of late endosomes and lysosomes. Once formed, there are multiple steps in the exocytosis of cytolytic granules from cytotoxic cells and their delivery to target cells. The importance of the cytolytic pathway in immune regulation is highlighted by inherited defects that impair key steps in either the formation or the exocytosis of cytolytic granules (**Fig. 13.9**). These result in a severe and often fatal condition, known as **hemophagocytic lymphohistiocytic (HLH) syndrome**, which manifests as uncontrolled activation and expansion of CD8 T lymphocytes and macrophages that infiltrate multiple organs, causing tissue necrosis and organ failure. The hyperactive immune response is thought to reflect the inability of cytotoxic cells to destroy infected targets, and possibly themselves, following an initiating viral infection, particularly by a herpes family virus such as Epstein–Barr virus (EBV). In this regard it is notable that despite the impaired release of cytolytic granules, the release of IFN-γ by CTLs and NK cells of patients with this disorder is not typically impaired, which contributes to the heightened activity of macrophages and associated inflammatory disorder caused by an increased release of pro-inflammatory cytokines such as TNF, IL-6, and macrophage colony-stimulating factor (M-CSF). The activated

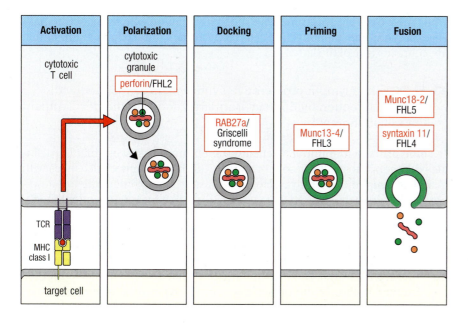

Activation	Polarization	Docking	Priming	Fusion

Fig. 13.9 Defects in components involved in the exocytosis of cytotoxic granules cause familial hemophagocytic lymphohistiocytosis (FHL) syndromes. Following antigen recognition, there is polarization of perforin-containing cytotoxic granules of the CTL toward the target cell at the site of immunological synapse formation. Cytotoxic granules are transported along microtubules to the plasma membrane, where they dock through a RAB27a-dependent interaction. Docked vesicles are primed by Munc13-4-mediated triggering of a conformational change in syntaxin 11, which is part of a large SNARE (soluble *N*-ethylmaleimide-sensitive factor accessory protein receptor) complex connecting the docked vesicle with the plasma membrane. Through the actions of Munc18-2, a fusion reaction is initiated via the syntaxin 11-containing SNARE complex that releases the contents of the cytotoxic granules into the synapse-bounded intercellular space, allowing perforin-mediated pore formation in the target-cell plasma membrane. The sites in the exocytic pathway that are affected by inherited defects in each of the proteins highlighted in red are indicated, as is the associated familial hemophagocytic lymphohistiocytosis syndrome.

macrophages phagocytose blood cells, including erythrocytes and leukocytes, which gives the syndrome its name.

There are multiple autosomal recessive variants of HLH, also referred to as **familial hemophagocytic lymphohistiocytosis** (**FHL**), that are distinguished by the protein in the cytolytic pathway that is affected (see Fig. 13.9). Examples include inherited deficiencies of the cytolytic granule protein perforin, which is required for pore formation in the target cell (and when defective is associated with the syndrome FHL2); of the priming protein, Munc13-4 (FHL3); of syntaxin 11 (FHL4), a member of the SNARE family of proteins that mediate membrane fusion; and of Munc18-2 (FHL5), a protein involved in the reorganization of SNARE proteins to activate the fusion process. Because components of the biogenesis and exocytosis of cytolytic granules are shared with other secretory vesicles, such as lysosomes, additional immune defects can occur in affected individuals, as can nonimmune defects. In particular, a subset of the immunodeficiencies that affect cytolytic granule function are also characterized by partial loss of skin pigmentation. This is due to defects in vesicular transport proteins that are also required for the formation or exocytosis of melanosomes, organelles that store the skin pigment melanin in melanocytes. Examples of these immunodeficiencies include **Chediak–Higashi syndrome**, caused by mutations in a protein, CHS1, which regulates lysosomal trafficking, and **Griscelli syndrome**, caused by mutations in a small GTPase, RAB27a (see Fig. 13.9), which is integral to the tethering of certain vesicles, including cytolytic granules, to cytoskeletal structures to enable their intracellular trafficking.

In patients with Chediak–Higashi syndrome, abnormal giant lysosomes and granules accumulate in T lymphocytes, myeloid cells, platelets, and melanocytes. The hair is typically a metallic silver color, vision is poor because of abnormalities in retinal pigment cells, and platelet dysfunction causes increased bleeding. Because the phagocytes of these patients have defective phagolysosomal fusion, the patients experience defective killing of intracellular and extracellular pathogens in addition to defective cytolytic activity of CTLs and NK cells. Affected children therefore suffer early in life from severe, recurrent infections by a range of bacteria and fungi. This is typically followed by development of hemophagocytic lymphohistiocytosis, often triggered by viral infection such as EBV, leading to an accelerated phase of the disease. Three variants of Griscelli syndrome have been identified, each caused by a distinct gene defect: in the type 2 variant (*RAB27A* mutation), the defect results in both immunodeficiency and pigment abnormalities; in types 1 and 3, only the pigment abnormalities are found. Although the immune defects in children

Hemophagocytic Lymphohistiocytosis

Chediak–Higashi Syndrome

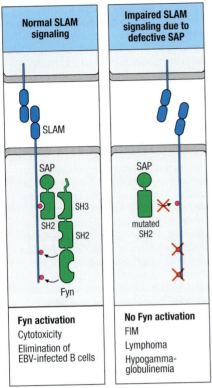

Fyn activation
Cytotoxicity
Elimination of
EBV-infected B cells

No Fyn activation
FIM
Lymphoma
Hypogamma-
globulinemia

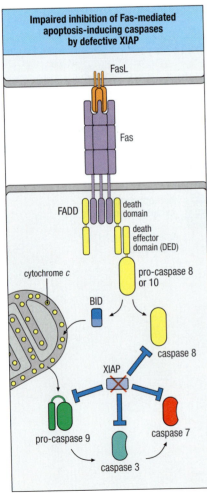

with type 2 Griscelli syndrome have much in common with those of Chediak–Higashi syndrome, their myeloid cells lack the giant intracellular granules that are characteristic of the latter.

13-12 X-linked lymphoproliferative syndrome is associated with fatal infection by Epstein–Barr virus and with the development of lymphomas.

In some primary immunodeficiencies, there is unique susceptibility to a single pathogen. Such is the case for two rare X-linked immunodeficiencies with a similar lymphoproliferative defect that results from vulnerability to a herpes family virus, the Epstein–Barr virus (EBV), albeit via independent mechanisms. EBV specifically infects B cells and typically causes a self-limiting infection in normal individuals due to active control of the virus by the actions of NK cells, NKT cells, and cytotoxic T cells with specificity for B cells expressing EBV antigens. Following the development of immunity to EBV, the virus is not completely cleared, but is maintained in a latent state in infected B cells (see Section 13-24). In the presence of certain types of immunodeficiency, however, this control can break down, resulting in overwhelming EBV infection (severe infectious mononucleosis) that is accompanied by unrestrained proliferation of EBV-infected B cells and cytotoxic T cells, hypogammaglobulinemia (low levels of circulating immunoglobulins), and the potential for the development of lethal, non-Hodgkin's B-cell lymphomas. These occur in the rare immunodeficiency **X-linked lymphoproliferative (XLP) syndrome**. XLP syndrome results from mutations in one of two X-linked genes: the SH2 domain-containing gene 1A (*SH2D1A*), which encodes SAP [signaling lymphocyte activation molecule (SLAM)-associated protein], or the X-linked inhibitor of apoptosis gene (*XIAP*).

In XLP1, which accounts for approximately 80% of patients with this syndrome, the defect in SAP results in defective coupling of immune-cell receptors of the SLAM family with the Src-family tyrosine kinase Fyn in T cells, NKT cells, and NK cells (**Fig. 13.10**). SLAM family members interact through homotypic or heterotypic binding to modulate the outcome of interactions between T cells and antigen-presenting cells and between NK cells and their target cells. In the absence of SAP, ineffective EBV-specific cytotoxic T-cell and NK-cell responses are made, and there is severe deficiency of NKT cells, indicating that SAP has a nonredundant role in the control of EBV infection and the development of NKT cells. There is unchecked proliferation of EBV-reactive cytotoxic T cells and NK cells that results in systemic macrophage activation, inflammation, and hemophagocytic features similar to those that occur in immunodeficiencies caused by defects in the cytolytic pathway (see

Fig.13.10 X-linked lymphoproliferative disease (XLP) is caused by inherited defects in SAP and XIAP, resulting in abnormal SLAM and TNF family receptor signaling, respectively. SLAM is an immune receptor family, members of which are expressed by T cells, B cells, natural killer (NK) cells, dendritic cells, and macrophages. Signaling is initiated via homotypic or heterotypic interactions between family members. SLAM signaling recruits the Src homology 2 (SH2) domain-containing factor SAP, which recognizes tyrosine-based motifs in the cytoplasmic domain of SLAM to activate the Src-related tyrosine kinase Fyn (left upper panel). Fyn then phosphorylates additional tyrosine residues of SLAM to recruit additional signaling components. Mutation in SAP in patients with XLP1 (right upper panel) disrupts activation of Fyn and SLAM signaling, thereby impairing T- and NK-cell cytotoxicity, which results in severe Epstein–Barr virus (EBV) infections and lymphoma. Defective SLAM signaling also impairs upregulation of inducible T-cell co-stimulator (ICOS) expression in T_FH cells, which results in impaired antibody responses. Activation of apoptosis-inducing caspases by TNF family receptors, such as Fas, is normally inhibited by XIAP (lower panel). XIAP interacts with both initiator caspases (8 and 9) and executioner caspases (3 and 7) through its baculoviral inhibitory repeat (BIR) domain to inhibit their actions. In XLP2 patients, XIAP is defective, resulting in abnormal regulation of caspase activation that leads to the complex clinical phenotype that includes lymphoproliferation and defective control of EBV.

Section 13-11). Moreover, defective SLAM signaling between T_{FH} cells and B cells of XLP1 patients causes impaired T-dependent antibody responses and hypogammaglobulinemia.

Defects in the XIAP protein, which normally binds the TNF-receptor-associated factors TRAF-1 and TRAF-2 and inhibits the activation of apoptosis-inducing caspases (see Section 7-23), lead to a similar X-linked syndrome called XLP2 (see Fig. 13.10). XIAP deficiency results in the enhanced apoptosis and turnover of activated T cells and NK cells. Paradoxically, this leads to a lymphoproliferative phenotype similar to XLP1, the basis of which is not well understood. As in XLP1, there is also severe depletion of NKT cells, indicating that, like SAP, XIAP is required for the normal maintenance of these cells. As in XLP1, there is also defective control of EBV infection in XLP2, although this is less prominent. The exact reason for the impaired suppression of EBV latency in these immunodeficiencies remains to be defined.

13-13 Immunodeficiency is caused by inherited defects in the development of dendritic cells.

Our understanding of the diversity and function of dendritic cells has been advanced from studies of mice with gene-targeted deletion of transcription factors that result in the loss of subsets of these cells, and from the defects in protection against specific pathogens that result as a consequence of the loss of these subsets. In humans, where study of the development and function of dendritic cells is more challenging, the identification of primary immune deficiencies that result from defects in genes that encode the transcription factors GATA2 and IRF8 has begun to provide insights into the relative roles of these cells in different species.

An autosomal dominant mutation of GATA2 has been identified in the largest group of patients with an inherited deficiency of dendritic cells. Affected individuals develop a progressive loss of all subsets of dendritic cells (conventional and plasmacytoid) and monocytes, as well as reduced numbers of B and NK lymphoid cells, a condition that has been termed DCML deficiency. Although T-cell numbers are normal in these patients, their function becomes impaired as dendritic cells are lost. The loss of products of several hematopoietic lineages, but not all, suggests redundancy of the function of GATA2 in unaffected lineages. The basis for the progressive decline in products of the affected lineages is unknown, but is thought to reflect a role for GATA2 in maintaining stem-cell progenitors that seed these populations. Given the deficiency of all dendritic and monocyte cells, the immune defects in these individuals are diverse, as are pathogen susceptibilities. These patients also have a substantial risk of hematologic malignancies.

Two inherited defects in interferon regulatory factor 8 (IRF8) are the first to be described that are associated with specific defects in the development of dendritic cells. In both variants, the mutation is in the DNA-binding domain of the transcription factor. In an autosomal recessive form, there is loss of monocytes and all types of circulating dendritic cells; all conventional and plasmacytoid dendritic cells are absent. Because dendritic cells are the primary antigen-presenting cells for naive T cells, their deficiency results in impaired development of effector T cells, and patients with these defects are susceptible to a range of severe opportunistic infections early in life, including those caused by intracellular bacteria, viruses, and fungi. They also develop a striking increase in the number of circulating immature granulocytes; this is thought to be due to diversion of myeloid precursors down the granulocytic pathway in the absence of the monocyte–dendritic cell developmental pathway. In contrast, in patients with autosomal dominant inheritance of a dominant-negative mutant allele of *IRF8*, there is a less severe phenotype, one that is characterized by a more selective deficiency of the CD1c-positive subset of dendritic cells (thought to be the equivalent of the CD11b-positive subset of mouse dendritic cells). This results in heightened susceptibility to intracellular bacteria, particularly atypical *Mycobacterium* spp., without the myeloproliferative syndrome seen in patients with the autosomal recessive variant.

13-14 Defects in complement components and complement-regulatory proteins cause defective humoral immune function and tissue damage.

The diseases discussed so far are mainly due to disturbances of the adaptive immune system. In the next few sections we look at some immunodeficiency diseases that affect cells and molecules of the innate immune system. We start with the complement system, which can be activated by any of three pathways that converge on the cleavage and activation of complement component C3, allowing it to bind covalently to pathogen surfaces, where it acts as an opsonin (discussed in Chapter 2). Not surprisingly, the spectrum of infections associated with complement deficiencies overlaps substantially with that seen in patients with deficiencies in antibody production. In particular, there is increased susceptibility to extracellular bacteria that require opsonization by antibody and/or complement for efficient clearance by phagocytes (Fig. 13.11). Defects in the activation of C3 by any of the three pathways, as well as defects in C3 itself, are associated with increased susceptibility to infection by a range of pyogenic bacteria, including *S. pneumoniae*, emphasizing the role of C3 as a central effector that promotes the phagocytosis and clearance of capsulated bacteria.

Deficiency of the C8 Complement Component

In contrast, defects in the membrane-attack components of complement (C5-C9) downstream of C3 activation have more limited effects, and result almost exclusively in susceptibility to *Neisseria* species. A similar susceptibility to *Neisseria* species is found in patients with defects in the alternative complement pathway components factor D and properdin, indicating that defense against these bacteria, which can survive intracellularly, is largely mediated via antibody-independent extracellular lysis by the membrane-attack complex. Data from large-population studies in Japan, where endemic *N. meningitidis* infection is rare, show that the risk each year of infection with this organism

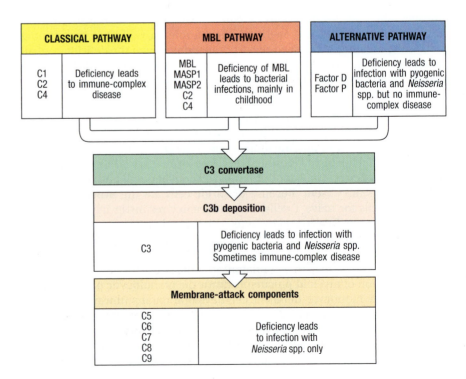

Fig. 13.11 Defects in complement components are associated with susceptibility to certain infections and the accumulation of immune complexes. Defects in the early components of the alternative pathway and in C3 lead to susceptibility to extracellular pathogens, particularly pyogenic bacteria. Defects in the early components of the classical pathway predominantly affect the processing of immune complexes (see Section 10-20) and the clearance of apoptotic cells, leading to immune-complex disease. Deficiency of mannose-binding lectin (MBL), the recognition molecule of the mannose-binding lectin pathway, is associated with bacterial infections, mainly in early childhood. Defects in the membrane-attack components are associated only with susceptibility to strains of *Neisseria* species, the causative agents of meningitis and gonorrhea, implying that the effector pathway is important chiefly in defense against these organisms.

is approximately 1 in 2,000,000 to a normal person. This compares with a risk of 1 in 200 to a person in the same population with an inherited deficiency of one of the membrane-attack complex proteins—a 10,000-fold increase in risk.

The early components of the classical complement pathway are particularly important for the elimination of immune complexes (discussed in Section 10-20) and apoptotic cells, which can cause significant pathology in auto-immune diseases such as systemic lupus erythematosus. This aspect of inherited complement deficiency is discussed in Chapter 15. Deficiencies in mannose-binding lectin (MBL), which initiates complement activation independently of antibody (see Section 2-6), are relatively common (found in 5% of the population). MBL deficiency may be associated with a mild immunodeficiency that results in an increased incidence of bacterial infection in early childhood. A similar phenotype is found in patients with defects in the gene that encodes the MBL-associated serine protease-2 (*MASP2*).

Another set of complement-related diseases is caused by defects in complement-control proteins (Fig. 13.12). Deficiencies in decay-accelerating factor (DAF) or protectin (CD59), membrane-associated control proteins that protect the surfaces of the body's cells from complement activation, lead to destruction of red blood cells, resulting in the disease **paroxysmal nocturnal hemoglobinuria**, as discussed in Section 2-16. Deficiencies in soluble complement-regulatory proteins such as factor I and factor H have various outcomes. Homozygous **factor I deficiency** is a rare defect that results in uncontrolled activity of the alternative pathway C3 convertase, leading to a *de facto* C3 deficiency (see Section 2-16). Deficiencies in MCP, factor I, or factor H can also cause a condition known as **atypical hemolytic–uremic syndrome**, so called because it leads to lysis of red blood cells (hemolysis) and impaired kidney function (uremia).

A striking consequence of the loss of a complement-regulatory protein is seen in patients with C1-inhibitor defects, which cause the syndrome known as **hereditary angioedema** (**HAE**) (see Section 2-16). Deficiency of C1 inhibitor results in a failure to regulate both the blood clotting and complement activation pathways, leading to excessive production of vasoactive mediators that cause fluid accumulation in the tissues (edema) and local laryngeal swelling that can result in suffocation.

Complement protein	Effects of deficiency
C1, C2, C4	Immune-complex disease
C3	Susceptibility to encapsulated bacteria
C5–C9	Susceptibility to *Neisseria*
Factor D, prosperdin (factor P)	Susceptibility to encapsulated bacteria and *Neisseria* but no immune-complex disease
Factor I	Similar effects to deficency of C3
MCP, factor I or factor H	Atypical hemolytic-uremic syndrome
Polymorphisms in factor H	Age-related macular degeneration
DAF, CD59	Autoimmune-like conditions, including paroxysmal nocturnal hemoglobinuria
C1INH	Herditary angiodema (HAE)

Fig. 13.12 Defects in complement-control proteins are associated with a range of diseases.

Hereditary Angiodema

13-15 Defects in phagocytic cells permit widespread bacterial infections.

Deficiencies in phagocyte numbers or function can be associated with severe immunodeficiency; indeed, a total absence of neutrophils is incompatible with survival in a normal environment. Phagocyte immunodeficiencies can be grouped into four general types: deficiencies in phagocyte production, phagocyte adhesion, phagocyte activation, and phagocyte killing of microorganisms (Fig. 13.13). We consider each in turn.

Inherited deficiencies of neutrophil production (**neutropenias**) are classified either as **severe congenital neutropenia** (**SCN**) or **cyclic neutropenia**. In severe congenital neutropenia, which can be inherited as a dominant or recessive trait, the neutrophil count is persistently less than 0.5×10^9 per liter of blood (normal numbers are 3×10^9 to 5.5×10^9 per liter). Cyclic neutropenia is characterized by fluctuation in neutrophil numbers from near normal to very low or undetectable with an approximate cycle time of 21 days, resulting in periodicity of infectious risk. The most common causes of SCN are sporadic or autosomal dominant mutations of the gene that encodes neutrophil elastase (*ELA2*), a component of the azurophilic, or primary, granules involved in the degradation of phagocytosed microbes. Altered targeting of defective elastase 2 to granules causes apoptosis of developing myelocytes and a developmental block at the promyelocyte–myelocyte stage. Some mutations of *ELA2* cause

Severe Congenital Neutropenia

Fig. 13.13 Defects in phagocytic cells are associated with persistence of bacterial infection. Defects in neutrophil development caused by congenital neutropenias result in profound defects in antibacterial defense. Impairment of the leukocyte integrins with a common β_2 subunit (CD18) or defects in the selectin ligand sialyl-Lewis^X, prevent phagocytic cell adhesion and migration to sites of infection (leukocyte adhesion deficiency). Inability to transmit signals through Toll-like receptors (TLRs), due to defects in MyD88 or IRAK4, for example, impairs the proximal sensing of many infectious agents by innate immune cells. The respiratory burst is defective in chronic granulomatous disease, glucose-6-phosphate dehydrogenase (G6PD) deficiency, and myeloperoxidase deficiency. In chronic granulomatous disease, infections persist because macrophage activation is defective, leading to chronic stimulation of CD4 T cells and hence to granulomas. Vesicle fusion in phagocytes is defective in Chediak–Higashi syndrome. These diseases illustrate the critical role of phagocytes in removing and killing pathogenic bacteria.

Type of defect/name of syndrome	Associated infections or other diseases
Congenital neutropenias (e.g., elastase 2 deficiency)	Widespread pyogenic bacterial infections
Leukocyte adhesion deficiency	Widespread pyogenic bacterial infections
TLR signaling defects (e.g., MyD88 or IRAK4)	Severe cold pyogenic bacterial infections
Chronic granulomatous disease	Intracellular and extracellular infection, granulomas
G6PD deficiency	Defective respiratory burst, chronic infection
Myeloperoxidase deficiency	Defective intracellular killing, chronic infection
Chediak–Higashi syndrome	Intracellular and extracellular infection, granulomas

cyclic neutropenia; how the mutant elastase causes a 21-day cycle in neutropenia is still a mystery. A rare autosomal dominant form of SCN is caused by mutations in the oncogene *GFI1*, which encodes a transcriptional repressor that acts on *ELA2*. This finding arose from the unexpected observation that mice lacking the protein Gfi1 are neutropenic due to overexpression of *Ela2*.

Autosomal recessive forms of SCN have also been identified. Deficiency of the mitochondrial protein HAX1 leads to increased apoptosis in developing myeloid cells, resulting in a severe neutropenia referred to as **Kostmann's disease**. The heightened sensitivity of developing neutrophils to apoptosis is highlighted by SCN associated with genetic defects in glucose metabolism. Patients with recessive mutations in the genes encoding the glucose-6-phosphatase catalytic subunit 3 (*G6PC3*) or the glucose-6-phosphate translocase 1 (*SLC37A4*) also demonstrate increased apoptosis during granulocyte development that results in neutropenia. Acquired neutropenia associated with chemotherapy, malignancy, or aplastic anemia is also associated with a similar spectrum of severe pyogenic bacterial infections. Finally, neutropenia can also be a feature of other primary immunodeficiency diseases, including CD40 ligand deficiency, CVID, XLA, Wiskott–Aldrich syndrome, and GATA2 deficiency. Some patients exhibit formation of autoantibodies that lead to accelerated destruction of neutrophils.

Defects in the migration of phagocytic cells to extravascular sites of infection can cause serious immunodeficiency. Leukocytes reach such sites by emigrating from blood vessels in a tightly regulated process (see Fig. 3.31). Deficiencies in the molecules involved in each stage of this process can prevent neutrophils and macrophages from penetrating infected tissues, and are referred to as **leukocyte adhesion deficiencies** (**LADs**). Deficiency in the leukocyte integrin common β_2 subunit CD18, which is a component of LFA-1, MAC-1, and p150:95, prevents the migration of leukocytes into an infected site by abolishing the cells' ability to adhere tightly to the endothelium. Because it was the first LAD to be characterized, it is now referred to as type 1 LAD, or LAD-1, and is the most common LAD variant. Reduced rolling of leukocytes on the endothelium has been described in rare patients who lack the sialyl-Lewis^X antigen owing to a deficiency in the GDP-fucose-specific transporter that is involved in the biosynthesis of sialyl-Lewis^X and other fucosylated ligands for the selectins. This is referred to as type 2 LAD or LAD-2. LAD-3 results from deficiency of Kindlin-3, a protein involved in the induction of the high-affinity binding state of β integrins required for firm adhesion. Each variant of LAD has an autosomal recessive pattern of inheritance and causes severe, life-threatening bacterial or fungal infections early in life that are characterized by impaired wound healing and, in pyogenic bacterial infections, the absence of

Leukocyte Adhesion Deficiency

pus formation. The infections that occur in these patients are resistant to antibiotic treatment. LAD-3 is also associated with defects in platelet aggregation that cause increased bleeding.

A key step in the activation of innate immune cells, including phagocytes, is their recognition of microbe-associated molecular patterns through Toll-like receptors (TLRs; see Section 3-5). Several primary immunodeficiencies have been identified that are caused by defects in intracellular signaling components of TLRs. Signaling through all TLRs except TLR-3 requires the adaptor protein MyD88, which recruits and activates the kinases IRAK4 and IRAK1, which are required for downstream activation of the NFκB and MAP kinase pathways (see Section 3-7). Autosomal recessive mutations in the genes encoding MyD88 or IRAK4 have a similar phenotype: recurrent, severe peripheral and invasive infections by pyogenic bacteria that elicit little inflammation, a situation known as a 'cold' infection. Note that many of the signaling functions of MyD88 and IRAK4 molecules are shared with IL-1 family receptors. Thus, at least part of the immune defect in patients with inherited abnormalities in these molecules may be attributed to aberrant signaling by IL-1 family members. Note also that NEMO deficiency, which impairs B-cell class switching (see Section 13-9), also impairs TLR and IL-1 receptor family signaling through its block of normal NFκB activation. Immunodeficiency associated with defects in NEMO therefore affects both adaptive and innate immune function. Interestingly, increased viral infections are not typical in patients with MyD88 mutations, despite this protein's role in signaling by each of the nucleic acid-sensing TLRs except TLR-3 (for example, TLR-7, TLR-8, and TLR-9). This indicates that activation of interferon regulating factors (IRFs) that induce interferon responses downstream of these TLRs remains intact despite the defects in MyD88.

Remarkably, of the ten TLRs found in humans, defects in only one—TLR-3— have been linked to immunodeficiency. While defects in other TLRs have been identified (for example, TLR-5), they have not been associated with an overt immunodeficiency phenotype, likely reflecting a substantial level of redundancy. On the other hand, patients with hemizygous (dominant) and homozygous (recessive) mutations in the gene encoding TLR-3, which senses double-stranded RNA, typically have recurrent herpes simplex virus-1 (HSV-1) infections of the central nervous system (herpes simplex encephalitis) due to impaired production of type I interferons by cells of the nervous system. Those with inherited deficiencies in molecules involved in TLR-3 signaling (for example, TRIF, TRAF3, or TBK1) are similarly susceptible to HSV-1 encephalitis, as are patients with mutations in the TLR-transport protein UNC93B1, which is required for the transport of TLR-3 from the endoplasmic reticulum to the endolysosome. Interestingly, leukocytes from these patients have no defect in their response to TLR-3 ligands or HSV-1, indicating redundancy of TLR-3 function in these cells, but not those of the central nervous system. Similarly, these patients show only a limited predisposition to other viral infections, implying the existence of TLR-3-independent protection against most other types of viral infection.

Interleukin 1 Receptor-Associated Kinase 4 Deficiency

Recurrent Herpes Simplex Encephalitis

Genetic defects that affect signaling by pattern recognition receptors (PRRs) other than TLRs have been described. CARD9 is an adaptor molecule involved in signaling downstream of C-type lectin receptors expressed on myeloid cells. Dectin-1, Dectin-2, and macrophage-inducible C-type lectin (MINCLE) each recognize fungal-associated molecular patterns that signal through CARD9 to induce secretion of pro-inflammatory cytokines, including IL-6 and IL-23 (see Section 3-1). Autosomal recessive CARD9 deficiency results in impaired T_H17 cell responses to fungi, with the result that patients with this deficiency, like patients with inborn errors of IL-17 immunity (for example, IL-17RA deficiency and IL-17F deficiency; see Section 13-10), suffer chronic mucocutaneous candidiasis. However, in addition, these patients can suffer infections by dermatophytes, which are ubiquitous filamentous fungi that normally cause common superficial skin and nail infections, such as tinea pedis (athlete's foot).

Chronic Granulomatous
Disease

Most of the other known defects in phagocytic cells affect their ability to ingest microbes and destroy them once ingested (see Fig. 13.13). Patients with **chronic granulomatous disease** (**CGD**) are highly susceptible to bacterial and fungal infections and form granulomas as a result of an inability to kill bacteria ingested by phagocytes (see Fig. 11.13). The defect in this case is in the production of reactive oxygen species (ROS) such as the superoxide anion (see Section 3-2). Discovery of the molecular defect in this disease gave weight to the idea that these agents killed bacteria directly; this notion has since been challenged by the finding that the generation of ROS is not itself sufficient to kill target microorganisms. It is now thought that ROS cause an influx of K^+ ions into the phagocytic vacuole, increasing the pH to the optimal level for the action of microbicidal peptides and proteins, which are the key agents in killing the invading microorganism.

Genetic defects affecting any of the constituent proteins of the NADPH oxidase expressed in neutrophils and monocytes (see Section 3-2) can cause chronic granulomatous disease. Patients with the disease have chronic bacterial infections, which in some cases lead to the formation of granulomas. Deficiencies in glucose-6-phosphate dehydrogenase (G6PD) and myeloperoxidase (MPO) also impair intracellular bacterial killing and lead to a similar, although less severe, phenotype.

13-16 Mutations in the molecular regulators of inflammation can cause uncontrolled inflammatory responses that result in 'autoinflammatory disease.'

Hereditary Periodic
Fever Syndromes

There are a small number of diseases in which mutations in genes that control the life, death, and activities of inflammatory cells are associated with severe inflammatory disease. Although they do not lead to immunodeficiency, we have included them in this chapter because they are single-gene defects affecting a crucial aspect of innate immunity—the inflammatory response. These conditions represent a failure of the normal mechanisms that limit inflammation, and are known as **autoinflammatory diseases**: they can lead to inflammation even in the absence of infection (Fig. 13.14). **Familial Mediterranean fever** (**FMF**) is characterized by episodic attacks of severe inflammation, which can occur at various sites throughout the body and are

Fig. 13.14 The autoinflammatory diseases.

Disease (common abbreviation)	Clinical features	Inheritance	Mutated gene	Protein (alternative name)
Familial Mediterranean fever (FMF)	Periodic fever, serositis (inflammation of the pleural and/or peritoneal cavity), arthritis, acute-phase response	Autosomal recessive	*MEFV*	Pyrin
TNF-receptor associated periodic syndrome (TRAPS) (also known as familial Hibernian fever)	Periodic fever, myalgia, rash, acute-phase response	Autosomal dominant	*TNFRSF1A*	TNF-α 55 kDa receptor (TNFR-I)
Pyogenic arthritis, pyoderma gangrenosum, and acne (PAPA)		Autosomal dominant	*PSTPIP1*	CD2-binding protein 1
Muckle–Wells syndrome	Periodic fever, urticarial rash, joint pains, conjunctivitis, progressive deafness	Autosomal dominant	*NLRP3*	Cryopyrin
Familial cold autoinflammatory syndrome (FCAS) (familial cold urticaria)	Cold-induced periodic fever, urticarial rash, joint pains, conjunctivitis			
Chronic infantile neurologic cutaneous and articular syndrome (CINCA)	Neonatal onset recurrent fever, urticarial rash, chronic arthropathy, facial dysmorphia, neurologic involvement			
Hyper-IgD syndrome (HIDS)	Periodic fever, elevated IgD levels, lymphadenopathy	Autosomal recessive	*MVK*	Mevalonate synthase
Blau syndrome	Granulomatous inflammation of skin, eye, and joints	Autosomal dominant	*NOD2*	NOD2

associated with fever, an acute-phase response (see Section 3-18), and severe malaise. The pathogenesis of FMF was a mystery until its cause was discovered to be mutations in the gene *MEFV*, which encodes a protein called pyrin, to reflect its association with fever. Pyrin and pyrin domain-containing proteins are involved in pathways that lead to the apoptosis of inflammatory cells, and in pathways that inhibit the secretion of pro-inflammatory cytokines such as IL-1β. It is proposed that an absence of functional pyrin leads to unregulated cytokine activity and defective apoptosis, resulting in a failure to control inflammation. In mice, an absence of pyrin causes increased sensitivity to lipopolysaccharide and a defect in macrophage apoptosis. A disease with similar clinical manifestations, known as **TNF-receptor associated periodic syndrome** (**TRAPS**), is caused by mutations in quite a different gene, that encoding the TNF-α receptor TNFR-I. Patients with TRAPS have reduced levels of TNFR-I, which lead to increased levels of pro-inflammatory TNF-α in the circulation because it is not regulated by binding to this receptor. The disease responds to therapeutic blockade with anti-TNF agents such as etanercept, a soluble TNF receptor developed primarily to treat patients with rheumatoid arthritis (see Section 16-8). Mutations in the gene encoding PSTPIP1 (proline-serine-threonine phosphatase-interacting protein 1), which interacts with pyrin, are associated with another dominantly inherited autoinflammatory syndrome—**pyogenic arthritis, pyoderma gangrenosum, and acne** (**PAPA**). The mutations increase binding of pyrin to PSTPIP1, and it has been proposed that the interaction sequesters pyrin and limits its normal regulatory function.

The episodic autoinflammatory diseases **Muckle–Wells syndrome** and **familial cold autoinflammatory syndrome** (**FCAS**) are clearly linked to the inappropriate stimulation of inflammation, because they are due to mutations in NLRP3, a component of the 'inflammasome' that normally senses cell damage and stress as a result of infection (see Section 3-9). The mutations lead to the activation of NLRP3 in the absence of such stimuli and the unregulated production of pro-inflammatory cytokines. Patients with these dominantly inherited syndromes present with episodes of fever—which is induced by exposure to cold in the case of FCAS—as well as urticarial rash, joint pains, and conjunctivitis. Mutations in *NLRP3* are also associated with the autoinflammatory disorder **chronic infantile neurologic cutaneous and articular syndrome** (**CINCA**), in which short, recurrent fever episodes are common, although severe arthropathic, neurologic, and dermatologic symptoms predominate. Both pyrin and NLRP3 are predominantly expressed in leukocytes and in cells that act as innate barriers to pathogens, such as intestinal epithelial cells. The stimuli that modulate pyrin and related molecules include inflammatory cytokines and stress-related changes in cells. Indeed, Muckle–Wells syndrome responds dramatically to the drug anakinra, an antagonist of the receptor for IL-1.

13-17 Hematopoietic stem cell transplantation or gene therapy can be useful to correct genetic defects.

It is frequently possible to correct the defects in lymphocyte development that lead to SCID and some other immunodeficiency phenotypes by replacing the defective component, generally by hematopoietic stem cell (HSC) transplantation (see Section 15-36). The main difficulties in these therapies result from human leukocyte antigen (HLA) polymorphism. To be useful, the graft must share some HLA alleles with the host. As we learned in Section 8-21, the HLA alleles expressed by the thymic epithelium determine which T cells can be positively selected. When HSCs are used to restore immune function to individuals with a normal thymic stroma, both the T cells and the antigen-presenting cells are derived from the graft. Therefore, unless the graft shares at least some HLA alleles with the recipient, the T cells that are selected on host thymic epithelium cannot be activated by graft-derived antigen-presenting cells (**Fig. 13.15**).

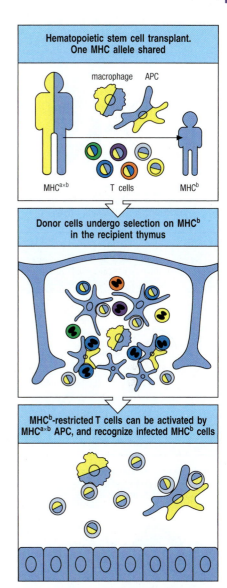

Fig. 13.15 The donor and the recipient of a hematopoietic stem cell (HSC) graft must share at least some MHC molecules to restore immune function. An HSC transplant from a genetically different donor is illustrated in which the donor marrow cells share some MHC molecules with the recipient. The shared MHC type is designated 'b' and illustrated in blue; the MHC type of the donor HSCs that is not shared is designated 'a' and shown in yellow. In the recipient, developing donor lymphocytes are positively selected on MHC^b on thymic epithelial cells and negatively selected by the recipient's stromal epithelial cells and at the corticomedullary junction by encounter with dendritic cells derived from both the donor HSCs and residual recipient dendritic cells. The negatively selected cells are shown as apoptotic cells. The donor-derived antigen-presenting cells (APCs) in the periphery can activate T cells that recognize MHC^b molecules; the activated T cells can then recognize the recipient's infected MHC^b-bearing cells.

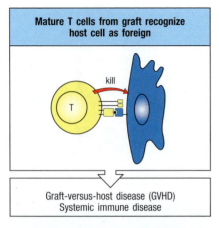

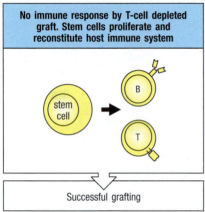

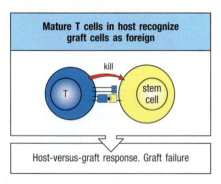

Fig. 13.16 Grafting of bone marrow can be used to correct immunodeficiencies caused by defects in lymphocyte maturation, but two problems can arise. First, if there are mature T cells in the bone marrow, they can attack cells of the host by recognizing their MHC antigens, causing graft-versus-host disease (top panel). This can be prevented by T-cell depletion of the donor bone marrow (center panel). Second, if the recipient has competent T cells, these can attack the bone marrow stem cells (bottom panel). This causes failure of the graft by the usual mechanism of transplant rejection (see Chapter 15).

There is also a danger that mature, post-thymic T cells that contaminate donor HSCs prepared from the peripheral blood or bone marrow might recognize the host as foreign and attack it, causing **graft-versus-host disease** (**GVHD**) (Fig. 13.16, top panel). This can be overcome by depleting the donor graft of mature T cells. For immunodeficiency diseases other than SCID where there are residual T cells and NK cells in the recipient, some form of myeloablative treatment (destruction of the bone marrow, typically using cytotoxic drugs) is usually carried out before transplantation, both to generate space for engraftment of the transplanted HSCs and to minimize the threat of **host-versus-graft disease** (**HVGD**) (see Fig. 13.16, third panel). The intensity of the myeloablative regimen is dependent on the nature of the immunodeficiency. For diseases where persistence of the patient's cells can be tolerated, engraftment of only a fraction of donor HSCs is sufficient for cure and nonmyeloablative chemotherapy may suffice prior to HSC transplantation. In other conditions such as XLP, which require complete elimination of the patient's blood cells and full engraftment of donor cells, more intensive (myeloablative) chemotherapy may be required.

Because many specific gene defects that cause inherited immunodeficiencies have been identified, an alternative therapeutic approach is **somatic gene therapy**. This strategy involves the isolation of HSCs from the patient's bone marrow or peripheral blood, introduction of a normal copy of the defective gene with the use of a viral vector, and reinfusion of the corrected stem cells into the patient. Initially, retroviral vectors were used for gene therapy trials, but were ceased due to severe complications in some patients. Although the genetic defect was corrected in patients with X-linked SCID, CGD, and WAS who received this treatment, some patients developed leukemia due to insertion of the retrovirus within a proto-oncogene. The inability to control the site in the genome in which retrovirally encoded genes insert and the use of viral vectors with strong promoters that can transactivate neighboring genes is therefore problematic. More recently, the use of self-inactivating retroviral and lentiviral vectors for gene correction has shown promise as a means of avoiding this complication. Also, a technique for the generation of **induced pluripotent stem cells** (**iPS cells**) from a patient's own somatic cells has been demonstrated. Forced expression of a set of transcription factors can reprogram somatic cells to become pluripotent progenitors that can give rise to HSCs. This approach offers the promise of 'repairing' specific defective genes in patient-derived stem cells by gene targeting *ex vivo* before reinfusion, but is not yet established. Until better methods for introduction of corrected genes into self-renewing stem cells are identified, allogeneic HSC transplantation will remain the mainstay of treatment for many primary immunodeficiencies.

13-18   Noninherited, secondary immunodeficiencies are major predisposing causes of infection and death.

The primary immunodeficiencies have taught us much about the biology of specific proteins of the immune system. Fortunately, these conditions are rare. In contrast, secondary immunodeficiency is relatively common. Malnutrition devastates many populations around the world, and a major feature of malnutrition is secondary immunodeficiency. This particularly affects cell-mediated immunity, and death in famines is frequently caused by infection. Measles, which itself causes immunosuppression, is an important cause of death in malnourished children. In the developed world, measles is an unpleasant illness but major complications are uncommon. In contrast, measles in the malnourished has a high mortality. Tuberculosis is another important infection in the malnourished. In mice, protein starvation causes immunodeficiency by affecting antigen-presenting cell function, but in humans it is not clear how malnourishment specifically affects immune responses. Links between the endocrine and immune systems may provide part of the answer.

Adipocytes (fat cells) produce the hormone leptin, levels of which are related directly to the amount of fat present in the body; leptin levels therefore fall during starvation when fat is consumed. Both mice and humans with genetic leptin deficiency have reduced T-cell responses, and in mice the thymus atrophies. In both starved mice and those with inherited leptin deficiency, these abnormalities can be reversed by the administration of leptin.

Secondary immunodeficiency states are also associated with hematopoietic tumors such as leukemia and lymphomas. Myeloproliferative diseases, such as leukemia, can be associated with deficiencies of neutrophils (neutropenia) or an excess of immature myeloid progenitors that lack functional properties of mature neutrophils, either of which increases susceptibility to bacterial and fungal infections. Destruction or invasion of peripheral lymphoid tissue by lymphomas or metastases from other cancers can promote opportunistic infections.

Congenital asplenia (a rare inherited absence of the spleen), surgical removal of the spleen, and destruction of spleen function by certain diseases are associated with a lifelong predisposition to overwhelming infection by *S. pneumoniae*, graphically illustrating the role of mononuclear phagocytic cells within the spleen in the clearance of this organism from blood. Patients who have lost spleen function should be vaccinated against pneumococcal infection and are often recommended to take prophylactic antibiotics throughout their life.

 **Congential Asplenia**

Secondary immunodeficiency is also a complication of certain medical therapies. A major complication of cytotoxic drugs used to treat cancer is immunosuppression and increased susceptibility to infection. Many of these drugs kill all dividing cells, including normal cells of the bone marrow and lymphoid systems. Infection is thus one of the major side effects of cytotoxic drug therapy. Immune suppression to induce host tolerance of solid organ allografts, such as kidney or heart transplants, also carries a substantial increased risk for infection and even for malignancy. The recent introduction of biologic therapies for some forms of autoimmunity has led to an increased risk of infection because of their immunosuppressive effects. For example, administration of antibodies that block TNF-α in patients with rheumatoid arthritis or other forms of autoimmunity has been associated with infrequent, but increased, instances of infectious complications.

Summary.

Genetic defects can occur in almost any molecule involved in the immune response. These defects give rise to characteristic deficiency diseases, which, although rare, provide much information about the development and functioning of the immune system in normal humans. Inherited immunodeficiencies illustrate the vital role of the adaptive immune response and T cells in particular, without which both cell-mediated and humoral immunity fail. They have provided information about the separate roles of B lymphocytes in humoral immunity and of T lymphocytes in cell-mediated immunity, the importance of phagocytes and complement in humoral and innate immunity, and the specific functions of a growing number of cell-surface or signaling molecules in the adaptive immune response. There are also some inherited immune disorders whose causes we still do not understand. The study of these diseases will undoubtedly teach us more about the normal immune response and its control. Acquired defects in the immune system, the secondary immunodeficiencies, are much more common than the primary, inherited immunodeficiencies. In the next sections we briefly consider general mechanisms by which successful pathogens evade or subvert immune defenses and then detail how extreme subversion of the immune system by one pathogen, the human immunodeficiency virus (HIV), has created a major pandemic characterized by the acquired immune deficiency syndrome (AIDS) in affected individuals.

Evasion and subversion of immune defenses.

In the previous section, we learned how specific defects in immune pathways lead to infection, often by microbes that would normally be defeated by a healthy immune system. These 'opportunistic' infections often dominate the clinical expression of inherited immunodeficiencies because the causative organisms are ubiquitous and abundant in the environment. A minority of microbes are true pathogens that can infect those with normal immune defenses. A defining feature of pathogens is their ability to avoid immune destruction by components of the innate and adaptive immune systems, at least long enough to replicate in the infected host and spread to new hosts. At one end of the spectrum are pathogens that establish an acute infection, replicate quickly, and find a new host before they are cleared by a successful immune response; at the other are pathogens that establish chronic infections, persisting long term in the host while evading elimination by immune defenses. Successful pathogens use different strategies to achieve these ends, and over millions of years of coevolution with their hosts have evolved a remarkable diversity of strategies for avoiding detection and destruction by the immune system, often employing several strategies that subvert immunity at multiple points. The anti-immune strategies employed by pathogens are as sophisticated as the immune system itself. They must be for any pathogens to achieve success against the diverse strategies that vertebrates have evolved to ensure host protection.

Viruses, bacteria, and protozoan (unicellular) or metazoan (multicellular) parasites can all act as pathogens. While fungi and helminthic (metazoan) parasites are major causes of common skin infections and intestinal worm infestation, respectively, they do not typically cause life-threatening infections in normal people and will not be considered further here. In contrast, a select number of viruses, bacteria, and protozoan parasites are the major causes of morbidity and mortality caused by infectious agents. AIDS, tuberculosis, and malaria, caused by the human immunodeficiency virus (HIV), *Mycobacterium tuberculosis*, and *Plasmodium falciparum*, respectively, are the three largest infectious disease threats to humans; each of these pathogens infects over 100 million people worldwide and kills 1 to 2 million people annually. Although the strategies for survival within—and propagation between—hosts differ for each type of pathogen, many of the innate and adaptive immune mechanisms employed to thwart the pathogen are the same. Here we briefly consider the lifestyles of, and the principal immune responses elicited by, different types of pathogens, and the strategies the pathogens employ to evade or subvert the immune system.

13-19 Extracellular bacterial pathogens have evolved different strategies to avoid detection by pattern recognition receptors and destruction by antibody, complement, and antimicrobial peptides.

Extracellular bacterial pathogens replicate outside of host cells, whether on the surfaces of barrier tissues they colonize (for example, gastrointestinal or respiratory tract), or in tissue spaces or blood following invasion across barrier epithelia. Both Gram-negative and -positive species are pathogenic, and, as discussed in Section 11-10, they typically elicit type 3 immunity, which orchestrates neutrophilic responses, the development of opsonizing and complement-fixing antibodies, and the production of antimicrobial peptides by barrier epithelial cells and immune cells that clear these microbes from barrier epithelia and prevent their invasion. Some of the MAMPS expressed by Gram-negative and -positive bacteria are distinct, but share similar immune-cell activating properties. Gram-negative pathogens contain LPS, a potent activator of TLR-4, in their outer cell membrane, whereas the cell wall of Gram-positive pathogens contains peptidoglycans, which activate TLR-2 and

NOD1 and NOD2. One strategy of immune evasion used by these pathogens is therefore shielding of surface MAMPs to avoid their detection by pattern recognition receptors on immune cells (Fig. 13.17). Several Gram-negative pathogens modify the lipid A core of LPS with carbohydrates and other moieties that interfere with TLR-4 binding. Indeed, some bacteria produce variants of lipid A that act as TLR-4 antagonists rather than agonists. Select Gram-positive pathogens have evolved mechanisms to modulate peptidoglycan recognition by NODs, or to produce hydrolases that degrade peptidoglycan.

Bacterial strategy	Mechanism	Result	Examples
Extracellular bacteria			
Shielding or inhibition of MAMPs	Capsular polysaccharide	Block detection of lipopolysaccharide (LPS)	S. pneumoniae
	Hypoacylation of lipid A	Antagonism of TLR-4	P. gingivalis
	Coating of bacterium by self proteins (e.g., fibrin)	Block detection of peptidoglycan	S. aureus
Antigenic variation	Modulation of expressed pili, fimbriae	Antibodies that block bacterial attachment become ineffective	N. gonorrhoeae, E. coli
Inhibition of opsonization	Secretion of complement-degrading factors	Cleavage of complement components	N. meningitidis, P. aeruginosa, S. aureus
	Capsular polysaccharide	Block fixation of complement	S. pneumoniae, H. influenzae, K. pneumoniae
	Expression of Fc-binding surface molecules (e.g., Protein A)	Prevents binding of antibody to Fc receptors of phagocytes	S. aureus
Inhibition/scavenging of reactive oxygen species (ROS)	Secretion of catalase and superoxide dismutase	Neutralize ROS produced by NADPH and myeloperoxidase (MPO)	S. aureus, B. abortus
Resistance to antimicrobial peptides (AMPs)	Secretion of AMP-degrading peptidases	Cleavage of AMPS	E. coli
	Modulation of cell membrane phospholipids	Prevents binding, functional insertion of AMPs in cell membrane	S. aureus
Intracellular bacteria			
Antigenic variation	Modulation of expressed pili, fimbriae	Antibodies that block bacterial attachment become ineffective	Salmonella spp.
Inhibition of MAMP recognition/ signaling	Production of peptidoglycan hydrolase	Block detection of peptidoglycan by NODs	L. monocytogenes
	Secretion of intracellular toxins	Block NFκB and MAP kinase signaling pathways	Y. pestis
Resistance to anti-microbial peptides	Secretion of AMP-degrading peptidases	Cleavage of AMPS	Y. pestis
	Modulation of cell membrane phospholipids	Prevents binding, functional insertion of AMPs in cell membrane	Salmonella spp.
Inhibition of fusion of phagosome with lysosome	Release of bacterial cell wall components	Inhibits phago-lysosomal fusion	M. tuberculosis, M. leprae, L. pneumophila
Survival within phagolysosome	Waxy, hydrophobic cell wall containing mycolic acids and other lipids	Resistance against lysosomal enzymes	M. tuberculosis, M. leprae
Escape from phagosome	Production of hemolysins (e.g., listeriolysin O)	Lysis of phagosome; escape into cytosol	L. monocytogenes, Shigella spp.

Fig. 13.17 Mechanisms used by bacteria to subvert the host immune system. Listed are examples of immune evasion/subversion mechanisms used by different strains of extracellular and intracellular bacterial pathogens. Examples of the strains of bacteria that employ each mechanism are listed in the far right column (e.g., *Streptococcus pneumoniae*, *Porphyromonus gingivalis*, *Pseudomonas aeruginosa*, *Brucella abortus*, *Yersinia pestis*).

A limited number of Gram-positive pathogens can also shield their outer cell membrane with a thick carbohydrate capsule. In addition to inhibiting recognition of peptidoglycan and activation of the alternative pathway of complement, the capsule prevents antibody and complement deposition on the bacterial surface, thereby avoiding direct damage by the membrane-attack complex of the complement cascade. The capsule also impairs clearance of the pathogen by phagocytes (see Fig. 13.17). In the case of *Streptococcus pneumoniae*, an important cause of bacterial pneumonia, the carbohydrate capsule also serves as a platform for **antigenic variation** to modulate the expression of surface antigenic epitopes recognized by antibody. Over 90 known types of *S. pneumoniae* are distinguished by differences in the structure of their polysaccharide capsules. The different types are distinguished by using specific antibodies as reagents in serological tests and so are often known as **serotypes**. Infection with one serotype can lead to type-specific immunity, which protects against reinfection with that type but not with a different serotype. Thus, from the point of view of the adaptive immune system, each serotype of *S. pneumoniae* represents a distinct organism, with the result that essentially the same pathogen can cause disease many times in the same individual (**Fig. 13.18**). Similarly, antigenic variation mediated by DNA rearrangement also occurs in bacteria and helps to account for the success of enteropathogenic *E. coli*, or of *Neisseria* species, which cause gonorrhea and meningitis. Fimbriae or pili are expressed on the bacterial surface and used for attachment to host-cell surfaces and are major antigenic targets for antibody-mediated blockade of bacterial attachment and colonization. The gene locus encoding the expressed *Neisseria* pilus (*pilE*) can undergo recombination with partial pilin genes stored in 'silent' (*pilS*) loci to generate a constantly shifting pilus for display on the bacterial surface. By constantly changing the expressed pilus, the bacterium evades antibody-mediated immune clearance.

Among other anti-immune strategies used by extracellular pathogens are mechanisms to inactivate the C3 convertase of the complement cascade; the expression of Fc-binding proteins that block functional antibody binding to the bacterium (for example, Protein A); and the decoration of the bacterial surface with host inhibitors of complement (for example, factor H). These bacteria

Fig. 13.18 Host defense against *Streptococcus pneumoniae* is type specific. The different strains of *S. pneumoniae* have antigenically distinct capsular polysaccharides. The capsule prevents effective phagocytosis until the bacterium is opsonized by specific antibody and complement, allowing phagocytes to destroy it. Antibody against one type of *S. pneumoniae* does not cross-react with the other types, so an individual immune to one type has no protective immunity to a subsequent infection with a different type. An individual must generate a new adaptive immune response each time he or she is infected with a different type of *S. pneumoniae*.

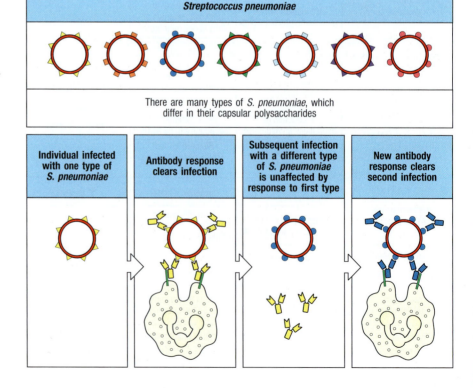

have also evolved mechanisms to defeat antimicrobial peptides (AMPs; for example, defensins and cathelicidins). These small cationic and amphipathic peptides have significant antimicrobial activity by inserting into negatively charged cell membranes to generate pores that lyse the bacterium. Pathogens can alter their membrane composition to minimize AMP binding, and can produce proteases that degrade the AMPs.

An unusual feature of Gram-negative pathogens, including both extracellular and intracellular bacteria, is their capacity to inject immune modulatory bacterial proteins directly into host cells via specialized structures: the type III and type IV secretion systems (T3SS and T4SS, respectively) (Fig. 13.19). These needle-like structures, or injectisomes, assemble on the bacterial surface and provide a conduit through which bacterial proteins are secreted directly into the cytosol of target cells. A range of bacterial virulence factors that aid in subverting the host immune response are delivered via this mechanism, including bacterial factors that block signaling cascades central to the inflammatory response: NFκB and MAP kinases. Among the most remarkable of these are the *Yersinia* outer proteins (Yops) produced by *Yersinia pestis*, the causative agent of bubonic plague. Secretion of several of these factors (for example, YopH, YopE, YopO, and YopT) into phagocytes disrupts the actin cytoskeleton, which is essential for phagocytosis. The essential roles played by T3SS or T4SS in immune subversion by a number of Gram-negative pathogens are demonstrated by the loss of pathogenicity of mutant bacteria lacking components of these structures.

13-20 Intracellular bacterial pathogens can evade the immune system by seeking shelter within phagocytes.

To avoid the major effectors directed against extracellular bacteria—complement and antibodies—some bacterial pathogens have evolved specialized mechanisms for surviving within macrophages, using these phagocytes as their primary host cell, as well as a vehicle for dissemination within the host. This 'Trojan horse' strategy is achieved by three general strategies: blockade of phagosome–lysosome fusion; escape from the phagosome into the cytosol; and resistance to killing mechanisms within the phagolysosome. *Mycobacterium tuberculosis*, for example, is taken up by macrophages but prevents the fusion of the phagosome with the lysosome, protecting itself from the bactericidal actions of the lysosomal contents. Other microorganisms, such as the bacterium *Listeria monocytogenes*, escape from the phagosome into the cytoplasm of the macrophage, where they multiply. They then spread to adjacent cells in the tissue without emerging into the extracellular environment. They do this by hijacking the cytoskeletal protein actin, which assembles into filaments at the rear of the bacterium. The actin filaments drive the bacteria forward into vacuolar projections to adjacent cells; the vacuoles are then lysed by the listeria, releasing the bacteria into the cytoplasm of the adjacent cell. Moreover, *Listeria* has been shown to induce the formation of bacteria-containing blebs on the surface of infected cells. These blebs which express phosphatidyl serine on the outer membrane leaflet. This membrane phospholipid is normally restricted to the inner membrane leaflet, and when exposed on the outer membrane leaflet is normally recognized by phagocytes as a signal for the uptake of apoptotic cell debris. In this way, *Listeria* is delivered directly to phagocytic cells, thereby avoiding attack by antibodies.

Following uptake, *Salmonella* species use a type III secretion system (see Fig. 13.19) to secrete effectors such as *SifA* into the host cytosol and membranes in order to alter the composition of the *Salmonella*-containing vacuole so as to avoid destruction. Remarkably, *Salmonella* can also inject factors that delay the apoptotic death of host macrophages, prolonging the phagocytes' lifespans until their bacterial cargo can be delivered to new cellular hosts. Other actions of intracellular bacteria counter reactive oxygen species or microbicidal peptides delivered into the phagolysosome by the ingesting phagocyte.

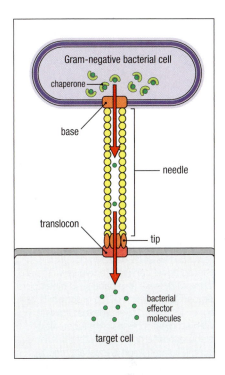

Fig. 13.19 Pathogenic bacteria use specialized secretion systems to inject effector molecules into host cells. A number of pathogenic Gram-negative bacteria use a complex, needle-like protein assembly—a type III or IV secretion system, or injectisome—to inject virulence proteins into target cells to compromise host defenses and establish infection. These 'nanoinjectors' are assembled from more than 20 proteins, and are composed of a base that spans the two bacterial membranes, a needle that is anchored in the base and is formed by the polymerization of repeating α-helical subunits, and a tip complex that serves as a docking structure for the translocon, which penetrates the host-cell membrane to allow bacterial effector proteins to pass into the host cell.

As a tradeoff for their intracellular lifestyle, intracellular bacteria risk activation of immune effectors that target these pathogens: NK cells and T cells. As discussed in Section 11-5, a major function of the type 1 immune response is activation of NK cells and T$_H$1 cells that activate phagocytes for enhanced intracellular killing by secretion of IFN-γ or expression of CD40L. Additionally, those intracellular pathogens that have evolved mechanisms for escape of the phagosome, such as *Listeria*, generate cytosolic peptides that feed the MHC class I antigen presentation pathway and thus induce cytotoxic T-cell responses that target their host cell for destruction. In leprosy, a disease caused by skin and peripheral nerve infection with *Mycobacterium leprae*, effective host defense requires macrophage activation by NK and T$_H$1 cells (**Fig. 13.20**).

Fig. 13.20 T-cell and macrophage responses to *Mycobacterium leprae* are sharply different in the two polar forms of leprosy. Infection with *M. leprae*, whose cells stain as small dark red dots in the photographs, can lead to two very different forms of disease (top panels). In tuberculoid leprosy (left), growth of the organism is well controlled by T$_H$1-like cells that activate infected macrophages. The tuberculoid lesion contains granulomas and is inflamed, but the inflammation is local and causes only local effects, such as peripheral nerve damage. In lepromatous leprosy (right), infection is widely disseminated and the bacilli grow uncontrolled in macrophages; in the late stages of disease there is major damage to connective tissues and to the peripheral nervous system. There are several intermediate stages between these two polar forms (not shown). The lower panel shows Northern blots demonstrating that the cytokine patterns in the two polar forms of the disease are sharply different, as shown by the analysis of RNA isolated from lesions of four patients with lepromatous leprosy and four patients with tuberculoid leprosy. Cytokines typically produced by T$_H$2 cells (IL-4, IL-5, and IL-10) dominate in the lepromatous form, whereas cytokines produced by T$_H$1 cells (IL-2, IFN-γ, and TNF-β) dominate in the tuberculoid form. It therefore seems that T$_H$1-like cells predominate in tuberculoid leprosy, and T$_H$2-like cells in lepromatous leprosy. IFN-γ would be expected to activate macrophages, enhancing the killing of *M. leprae*, whereas IL-4 can actually inhibit the induction of bactericidal activity in macrophages. Photographs courtesy of G. Kaplan; cytokine patterns courtesy of R.L. Modlin.

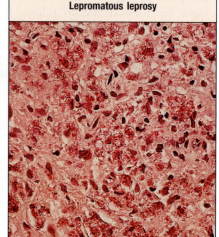

Infection with *Mycobacterium leprae* can result in different clinical forms of leprosy	
There are two polar forms, tuberculoid and lepromatous leprosy, but several intermediate forms also exist	
Tuberculoid leprosy	**Lepromatous leprosy**
Organisms present at low to undetectable levels	Organisms show florid growth in macrophages
Low infectivity	High infectivity
Granulomas and local inflammation. Peripheral nerve damage	Disseminated infection. Bone, cartilage, and diffuse nerve damage
Normal serum immunoglobulin levels	Hypergammaglobulinemia
Normal T-cell responsiveness. Specific response to *M. leprae* antigens	Low or absent T-cell responsiveness. No response to *M. leprae* antigens

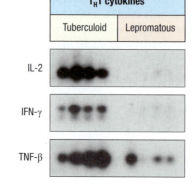

Cytokine patterns in leprosy lesions

T$_H$1 cytokines			T$_H$2 cytokines		
	Tuberculoid	Lepromatous		Tuberculoid	Lepromatous
IL-2			IL-4		
IFN-γ			IL-5		
TNF-β			IL-10		

Like *M. tuberculosis*, *M. leprae* is able to persist and grow in macrophage vesicles and is normally restrained, but not cleared, by a type 1 host response. In patients who mount normal type 1 immune responses, few live bacteria are found, little antibody is produced, and, although skin and peripheral nerves are damaged by the inflammatory responses associated with macrophage activation, the disease progresses slowly and the patients typically survive. Because of its similarities to tuberculosis, this variant is called tuberculoid leprosy. This is in contrast to lepromatous leprosy, in which type 1 responses to *M. leprae* are deficient and an ineffective type 2 response is mounted instead. This results in abundant growth of the bacterium in macrophages and gross tissue destruction that is eventually fatal, if untreated. Although high levels of antibacterial antibodies are produced in patients with lepromatous leprosy, probably due to the high bacterial load, the antibodies are ineffective at controlling infection because they do not reach the intracellular bacteria.

Lepromatous Leprosy

13-21   Immune evasion is also practiced by protozoan parasites.

Most common protozoan parasites, such as *Plasmodium* and *Trypanosoma* species, have complex life cycles, part of which occurs in humans and part of which occurs in an intermediate host, such as an arthropod vector (for example, mosquitoes, flies, or ticks). The route of transmission of these organisms by their intermediate host is unusual, in that the normal barriers to infection are bypassed when the infectious agent is directly delivered to blood by a bite or the taking of a blood meal. Thus, many of the normal innate immune defenses associated with barrier function are completely bypassed during infection. Further, the most successful of these organisms have developed complex and varied immune evasion strategies that often result in 'hide-and-seek' chronic infections characterized by episodic disease manifestations, despite their eliciting antibody- and cell-mediated adaptive immune responses.

As described above for some bacterial pathogens (see Section 13-19), *Trypanosoma brucei*, a causative agent of trypanosomiasis, or sleeping sickness, has evolved a remarkable capacity for antigenic variation to evade the antibody response elicited in infected humans. The trypanosome is coated with a single type of glycoprotein, the variant-specific glycoprotein (VSG), which elicits a potent protective antibody response that rapidly clears most of the parasites. The trypanosome genome, however, contains about 1000 VSG genes, each encoding a protein with distinct antigenic properties. A VSG gene is expressed by being placed into the active expression site in the parasite genome. Only one VSG gene is expressed at a time, and it can be changed by a gene rearrangement that places a new VSG gene into the expression site (**Fig. 13.21**). So, under the selective pressure of an effective host antibody response, the few trypanosomes within the population that express a different VSG escape elimination and multiply to cause a recurrence of disease (see Fig. 13.21, bottom panel). Antibodies are then made against the new VSG, and the whole cycle is repeated, resulting in periods of active and quiescent disease. The chronic cycles of antigen clearance lead to immune-complex damage and

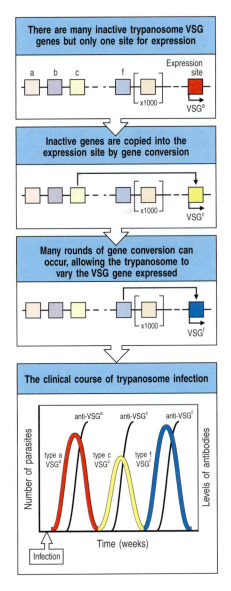

Fig. 13.21 Antigenic variation in trypanosomes allows them to escape immune surveillance. The surface of a trypanosome is covered with a variant-specific glycoprotein (VSG). Each trypanosome has about 1000 genes encoding different VSGs, but only the gene in a specific expression site within the telomere at one end of the chromosome is active. Although several genetic mechanisms have been observed for changing the VSG gene expressed, the usual mechanism is gene conversion. An inactive gene, which is not at the telomere, is copied and transposed into the telomeric expression site, where it becomes active. When an individual is first infected, antibodies are raised against the VSG initially expressed by the trypanosome population. A small number of trypanosomes spontaneously switch their VSG gene to a new type, and although the host antibody eliminates the initial variant, the new variant is unaffected. As the new variant grows, the whole sequence of events is repeated.

inflammation, and eventually to neurological damage, resulting finally in the coma that gives sleeping sickness its name. The cycles of evasive action make trypanosome infections very difficult for the immune system to defeat, and they are a major health problem in Africa.

Caused by *Plasmodium* species, malaria is another serious and widespread disease. Like *Trypanosoma brucei*, *Plasmodium* species vary their antigens to avoid elimination by the immune system. In addition, plasmodia undergo different parts of their life cycle in different cellular hosts within humans. Initial infection is by the sporozoite form of the organism, which is transmitted by the bite of an infected mosquito and targets hepatocytes in the liver. Here, the organism replicates rapidly to produce merozoites that burst from infected hepatocytes to infect circulating red blood cells. Thus, as the immune system focuses it efforts on the eradication of the parasite in the liver, the parasite morphs and escapes to its second cellular host, red blood cells. And because red blood cells are the only cells in the body that lack MHC class I molecules, antigens produced by the merozoites within infected red blood cells escape detection by CD8 T cells, preventing cytotoxic destruction of the infected cells. This represents one of the most elegant adaptations to evade cell-mediated immunity.

Immune subversion is also practiced by protozoan parasites. *Leishmania major*, which is transmitted to dermal tissues of humans by the bite of sandflies, is an obligate intracellular parasite that replicates within tissue macrophages. As is true of other intracellular pathogens that reside within phagocytic vesicles, eradication of *L. major* is dependent on a type 1 immune response. Through mechanisms that are incompletely understood, *L. major* specifically inhibits the production of IL-12 by its macrophage host, thereby inhibiting the production of IFN-γ by NK cells and inhibiting the differentiation and function of T_H1 cells. In addition, *L. major* has been shown to actively induce IL-10-producing T_{reg} cells that suppress clearance of the infection.

13-22 RNA viruses use different mechanisms of antigenic variation to keep a step ahead of the adaptive immune system.

Viruses are both the simplest and most diverse of pathogens. They can replicate only within living cells, relying on host cellular machinery to replicate and propagate themselves. As obligate intracellular pathogens, they activate intracellular PRRs that sense viral genetic material and provoke cytolytic immune responses by innate and adaptive immune cells—NK cells and CD8 T cells, respectively. They also induce type I interferon responses, which activate cell-intrinsic mechanisms to limit viral replication in both infected and uninfected cells. Although many cells produce type I interferons, plasmacytoid dendritic cells are innate sensor cells that are specialized for high levels of type I interferon production early in viral infections and, along with NK cells, play a central role in early antiviral host defense before the adaptive response matures. The latter involves all arms of adaptive immunity: induction of T_H1 cells that provide help for production of opsonizing and complement-fixing virus-specific antibodies that block viral entry into uninfected cells and activate complement to destroy enveloped viruses; and cytolytic CD8 T cells, which destroy virally infected cells and produce IFN-γ.

The strategies used by viruses to defeat immune defenses are as varied as the pathogens themselves. However, some general strategies relate to the type of viral genome. RNA viruses must replicate their genomes using RNA polymerases, which lack the proofreading ability of DNA polymerase. A consequence of this is that RNA viruses have a greater rate of mutation than DNA viruses, with the practical consequence that RNA viruses cannot support large genomes. However, this also affords them opportunities for rapidly altering antigenic epitopes targeted by the adaptive immune system as a mechanism for immune evasion. Further, some RNA viruses have segmented genomes, which lend themselves to reassortment during viral replication. Both of these mechanisms are used by influenza virus, a common seasonal viral pathogen that causes acute infections and has been responsible for several major pandemics. At any one time, a single virus type is responsible for most cases of influenza throughout the world. The human population gradually develops protective immunity to this virus type, chiefly through the production of neutralizing antibody directed against the viral hemagglutinin, the main surface protein of the influenza virus. Because the virus is rapidly cleared from immune individuals, it might be in danger of running out of potential hosts were it not able to use both mutation mechanisms to alter its antigenic type (Fig. 13.22).

MOVIE 13.1

The first of these, caused by point mutations in the genes encoding the two major viral surface glycoproteins—hemagglutinin and neuraminidase—is called **antigenic drift**. Every 2–3 years a variant flu virus arises with mutations that allow it to evade neutralization by the antibodies present in the population. Other mutations may affect epitopes in viral proteins that are recognized by T cells, particularly CD8 cytotoxic T cells, with the consequence that cells infected with the mutant virus also escape destruction. People immune to the old flu virus are thus susceptible to the new variant, but because the changes in the viral proteins are relatively minor, there is still some cross-reaction with antibodies and memory T cells produced against the previous variant, and most of the population still has some level of immunity. Thus, epidemics resulting from antigenic drift are typically mild.

MOVIE 13.2

Antigenic changes in influenza virus that result from reassortment of the segmented RNA genome are known as **antigenic shift**, and result in major changes in the hemagglutinin expressed by the virus. Antigenic shifts cause global pandemics of severe disease, often with substantial mortality, because the new hemagglutinin is recognized poorly, if at all, by antibodies and T cells

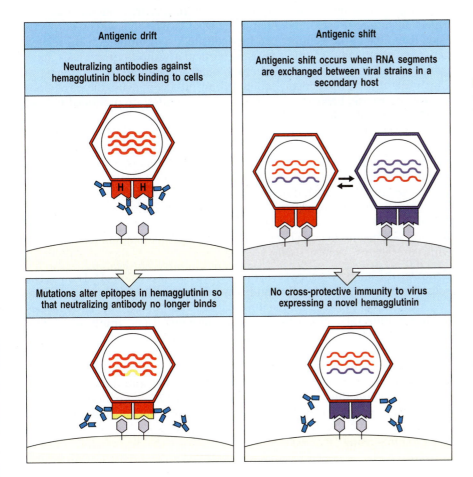

Fig. 13.22 Two types of variation allow repeated infection with type A influenza virus. Neutralizing antibody that mediates protective immunity is directed at the viral surface protein hemagglutinin (H), which is responsible for viral binding to and entry into cells. Antigenic drift (left panels) involves the emergence of point mutants with altered binding sites for protective antibodies on the hemagglutinin. The new virus can grow in a host that is immune to the previous strain of virus, but as T cells and some antibodies can still recognize epitopes that have not been altered, the new variants cause only mild disease in previously infected individuals. Antigenic shift (right panels) is a rare event involving the reassortment of the segmented RNA viral genomes of two different influenza viruses, probably in a bird or a pig. These antigen-shifted viruses have large changes in their hemagglutinin, and therefore T cells and antibodies produced in earlier infections are not protective. These shifted strains cause severe infection that spreads widely, causing the influenza pandemics that occur every 10–50 years. There are eight RNA molecules in each viral genome, but for simplicity only three are shown.

directed against the previous variant. Antigenic shift is due to reassortment of the segmented RNA genome of the human influenza virus and animal influenza viruses in an animal host, in which the hemagglutinin gene from the animal virus replaces the hemagglutinin gene in the human virus (see Fig. 13.22).

Hepatitis C virus (HCV) is an RNA virus that can cause both acute and chronic infections of the liver. It is the most common cause of blood-borne chronic infection in the United States, and the leading viral cause of liver cirrhosis. As with influenza virus, HCV has a high capacity for mutation of immune epitopes that allow it to evade elimination. However, unlike influenza, the viral surface glycoprotein responsible for binding of HCV to hepatocytes (E2, which binds CD81) presents a difficult target against which to produce effective neutralizing antibodies, due both to its heavy glycosylation in the region of CD81 binding and its high rate of mutation. Antibody responses against HCV are therefore of limited effectiveness. Similarly, high rates of mutation of T-cell epitopes select for escape variants of HCV that evade cytolytic T-cell responses. Finally, there is evidence that HCV also expresses factors that subvert the function of dendritic cells, thereby impairing the induction of T-cell immunity.

13-23 DNA viruses use multiple mechanisms to subvert NK-cell and CTL responses.

Of all the pathogens, DNA viruses that can establish chronic infections have evolved the greatest diversity of mechanisms for subverting or escaping immune defenses. Unlike RNA viruses, DNA viruses have relatively low mutation rates and are thus less able to employ antigenic variation to evade immune defenses. However, because their lower rate of mutation allows them to support much larger genomes, these viruses have been able to accommodate a remarkable number of viral genes encoding proteins that can subvert nearly every aspect of antiviral defense. In the case of poxvirus, adenovirus, and especially herpesviruses, all of them large DNA viruses that will be our focus here, over 50% of the genome can be dedicated to immune evasion-related genes. Further, some of these viruses, particularly the herpesviruses, have evolved mechanisms that allow them to enter a state known as **latency**, in which the virus is not actively replicated. In the latent state, the virus does not cause disease; however, because no viral peptides are produced to load MHC class I molecules that signal the virus's presence to cytolytic T cells, it cannot be eliminated, and can establish lifelong infections. As we will discuss in Section 13-24, latent infections can be reactivated, resulting in recurrent illness. Of the eight types of herpesvirus that infect humans, at least one of the five most common types—herpes simplex virus (HSV)-1 and -2 (both of which can cause labial and genital herpes), Epstein–Barr virus (EBV, which causes infectious mononucleosis), varicella-zoster (which causes chickenpox and shingles), and cytomegalovirus (CMV)—infects nine out of ten people, and typically establishes lifelong latency. Here we highlight major mechanisms by which these viruses succeed (Fig. 13.23).

Central to the long-term survival of the DNA viruses is evasion of CTLs and NK cells. The presentation of viral peptides by MHC class I molecules at a cell surface signals CD8 T cells to kill the infected cell. Many of the large DNA viruses evade immune recognition by producing proteins called **immunoevasins**, which prevent the appearance of viral peptide:MHC class I complexes on the infected cell (Fig. 13.24). Indeed, at least one viral inhibitor of every key step in the processing and presentation of peptide:MHC class I complexes has been described. Some immunoevasins block peptide entry into the endoplasmic reticulum by targeting the TAP transporter (Fig. 13.25, left panel). Viral proteins can also prevent peptide:MHC complexes from reaching the cell surface by retaining MHC class I molecules in the endoplasmic reticulum (see Fig. 13.25, middle panel). Several viral proteins catalyze the degradation of newly synthesized MHC class I molecules by a process known as **dislocation**, which

Viral strategy	Specific mechanism	Result	Virus examples
Inhibition of humoral immunity	Virally encoded Fc receptor	Blocks effector functions of antibodies bound to infected cells	Herpes simplex Cytomegalovirus
	Virally encoded complement receptor	Blocks complement-mediated effector pathways	Herpes simplex
	Virally encoded complement control protein	Inhibits complement activation by infected cell	Vaccinia
Inhibition of inflammatory response	Virally encoded chemokine receptor homolog, e.g., β-chemokine receptor	Sensitizes infected cells to effects of β-chemokine; advantage to virus unknown	Cytomegalovirus
	Virally encoded soluble cytokine receptor, e.g., IL-1 receptor homolog, TNF receptor homolog, interferon-γ receptor homolog	Blocks effects of cytokines by inhibiting their interaction with host receptors	Vaccinia Rabbit myxoma virus
	Viral inhibition of adhesion molecule expression, e.g., LFA-3 ICAM-1	Blocks adhesion of lymphocytes to infected cells	Epstein–Barr virus
	Protection from NFκB activation by short sequences that mimic TLRs	Blocks inflammatory responses elicited by IL-1 or bacterial pathogens	Vaccinia
Blocking of antigen processing and presentation	Inhibition of MHC class I expression	Impairs recognition of infected cells by cytotoxic T cells	Herpes simplex Cytomegalovirus
	Inhibition of peptide transport by TAP	Blocks peptide association with MHC class I	Herpes simplex
Immunosuppression of host	Virally encoded cytokine homolog of IL-10	Inhibits T_H1 lymphocytes Reduces interferon-γ production	Epstein–Barr virus

Fig. 13.23 Mechanisms used by viruses of the herpes and pox families to subvert the host immune system.

initiates the pathway normally used to degrade misfolded endoplasmic reticulum proteins by directing them back into the cytosol (Fig. 13.25, right panel). By preventing the formation of stably assembled/folded peptide:MHC class I complexes, these viral proteins divert the peptide:MHC class I complexes into the ER-associated degradation (ERAD) pathway for disposal. Through these multiple mechanisms, viral factors impair or completely block the presentation of viral peptides to CTLs. The actions of viral inhibitors are not limited to the MHC class I pathway, as viral inhibitors of the class II processing pathway have also been described; these inhibitors ultimately target CD4 T cells. Finally, as many viruses target cells other than dendritic cells, their antigens come to the attention of CD8 T cells via cross-presentation. Viral mechanisms that interfere with this pathway are not well described, although it is known that because the viruses are not required to persist in dendritic cells, the viruses can block recognition and destruction of their cellular hosts even after primed CTL effectors have been generated.

In addition to their role in the acute innate response to viral infection, a major function of NK cells is the recognition and cytolysis of cells that have downregulated MHC class I molecules as a foil to pathogen attempts to evade detection by CTLs. Accordingly, viruses that target the MHC class I pathway have also evolved mechanisms to repress the cytolytic activity of NK cells. Strategies here include, but are not limited to, expression of viral homologs of MHC class I that engage killer inhibitory receptors (KIRs) and leukocyte inhibitory receptors (LIRs). For example, human CMV produces a homolog of HLA class I called

Fig. 13.24 Immunoevasins produced by viruses interfere with the processing of antigens that bind to MHC class I molecules.

Virus	Protein	Category	Mechanism
Herpes simplex virus 1	ICP47	Blocks peptide entry to endoplasmic reticulum	Blocks peptide binding to TAP
Human cytomegalovirus (HCMV)	US6		Inhibits TAP ATPase activity and blocks peptide release into endoplasmic reticulum
Bovine herpes virus	UL49.5		Inhibits TAP peptide transport
Adenovirus	E19	Retention of MHC class I in endoplasmic reticulum	Competitive inhibitor of tapasin
HCMV	US3		Blocks tapasin function
Murine cytomegalovirus (CMV)	m152		Downregulation of host MHC class I
HCMV	US2	Degradation of MHC class I (dislocation)	Transports some newly synthesized MHC class I molecules into cytosol
Murine gamma herpes virus 68	mK3		E3-ubiquitin ligase activity
Murine CMV	m4	Binds MHC class I at cell surface	Interferes with recognition by cytotoxic lymphocytes by an unknown mechanism

MOVIE 13.3 ▶

UL18, which binds LIR-1 on NK cells and provides an inhibitory signal that blocks NK-cell cytolysis. Viral products have been defined that also antagonize activating receptors on NK cells, as well as inhibit NK-cell effector pathways.

DNA viruses have evolved mechanisms to subvert additional functions of the immune system. The mechanisms used include the expression of viral homologs of cytokines or chemokines and their receptors, or viral proteins

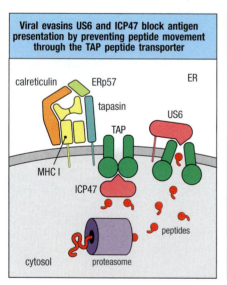

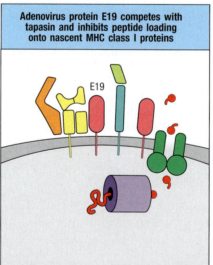

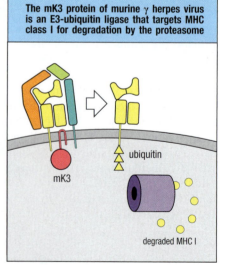

Fig. 13.25 The peptide-loading complex in the endoplasmic reticulum is targeted by viral immunoevasins. The left panel shows blockade of peptide entry to the endoplasmic reticulum (ER). The cytosolic ICP47 protein from herpes simplex virus (HSV)-1 prevents peptides from binding to TAP in the cytosol, whereas the US6 protein from human CMV interferes with the ATP-dependent transfer of peptides through TAP. The middle panel shows the retention of MHC class I molecules in the ER by the adenovirus E19 protein. This binds certain MHC molecules and retains them in the ER through an ER-retention motif, at the same time competing with tapasin to prevent association with TAP and peptide loading. The right panel shows how the murine herpes virus mK3 protein, an E3-ubiquitin ligase, targets newly synthesized MHC class I molecules. mK3 associates with tapasin:TAP complexes and directs the addition of ubiquitin subunits with K48 linkages (see Section 7-5) to the cytoplasmic tail of the MHC class I molecule. The polyubiquitination of the cytoplasmic tail of MHC initiates the process of degradation of the MHC molecule by the proteasomal pathway.

that bind cytokines or their receptors to inhibit their actions. As type I and II interferons are major effector cytokines in antiviral defense, diverse viral strategies are centered on the inhibition of this family of cytokines, whether by production of decoy receptors or inhibitory binding proteins, inhibition of JAK/STAT signaling by IFN receptors, inhibition of cytokine transcription, or interference with transcription factors induced by IFNs. Some DNA viruses also produce antagonists of the pro-inflammatory cytokines IL-1, IL-18, and TNF-α, among others. Viral homologs of immunosuppressive cytokines are also produced. CMV impairs antiviral responses by producing a homolog of the cytokine IL-10, called cmvIL-10, which downregulates the production of several pro-inflammatory cytokines by immune cells, including IFN-γ, IL-12, IL-1, and TNF-α, to promote tolerogenic rather than immunogenic adaptive responses to viral antigens.

Several viruses also interfere with chemokine responses by producing either decoy chemokine receptors or chemokine homologs that interfere with natural ligand-induced signaling through chemokine receptors. Collectively, herpesviruses and poxviruses produce over 40 viral homologs of receptors belonging to the seven transmembrane-spanning G-protein-coupled chemokine receptor (vGPCR) superfamily. Finally, CMV has been shown to promote chronic infection that is associated with 'exhaustion' of antiviral CD8 T cells. CD8 T cells induced in this setting are characterized by expression of an inhibitory receptor of the CD28 superfamily, the programmed death-1 (PD-1) receptor (see Section 7-24), activation of which by its ligand PD-L1 suppresses CD8 T-cell effector function. Blockade of the PD-L1–PD-1 interaction restores antiviral CD8 effector function and decreases the viral load, indicating that ongoing activation of this pathway is involved in impaired viral clearance. A similar mechanism has been implicated in RNA viruses that can establish chronic infections, such as hepatitis C virus (HCV). Suffice it to say that the range of strategies that viruses have evolved to subvert immune clearance mechanisms is quite remarkable, and the discovery of these mechanisms continues to have a major impact on our understanding of host–pathogen relationships.

13-24 Some latent viruses persist *in vivo* by ceasing to replicate until immunity wanes.

As mentioned in the previous section, a major class of viral agents that cause latent infections in humans are the herpesviruses, large, enveloped DNA viruses that are characterized by their ability to establish lifelong infections. While we have considered a number of strategies by which these viruses subvert immunity, they have also evolved mechanisms to maintain their genome within the nucleus of infected cells indefinitely without replicating. In contrast to an actively **lytic**, or **productive**, **phase** of the viral life cycle, wherein the virus replicates and lyses its cellular host, herpesviruses can establish latency, or a **lysogenic phase**, by expression of a small region of their genome called the latency associated transcript (LAT). In addition to suppressing the transcription of the remaining viral genome, the LAT produces factors that interfere with apoptotic death of the host cell, both interfering with immune mechanisms that might clear the cell and prolonging the cell's life-span—and that of the viral genome it harbors. An example is herpes simplex virus (HSV), the cause of cold sores, which infects epithelial cells and then spreads to sensory neurons serving the infected area. An effective immune response controls the epithelial infection, but the virus persists in a latent state in the sensory neurons. Factors such as sunlight, bacterial infection, or hormonal changes reactivate the virus, which then travels down the axons of the sensory neuron and reinfects the epithelial tissues (Fig. 13.26). At this point, the immune response again becomes active and controls the local infection by killing the epithelial cells, producing a new sore. This cycle can be repeated many times.

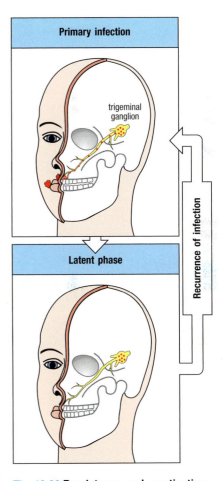

Fig. 13.26 Persistence and reactivation of herpes simplex virus infection. The initial infection in the skin is cleared by an effective immune response, but residual infection persists in sensory neurons such as those of the trigeminal ganglion, whose axons innervate the lips. When the virus is reactivated, usually by some environmental stress and/or alteration in immune status, the skin in the area served by the nerve is reinfected from virus in the ganglion and a new cold sore results. This process can be repeated many times.

There are two reasons why the sensory neuron remains infected: first, the virus is quiescent and generates few virus-derived peptides to present on MHC class I molecules; second, neurons carry very low levels of MHC class I molecules, which makes it harder for CD8 cytotoxic T cells to recognize infected neurons and attack them. The low level of MHC class I expression is beneficial because it reduces the risk that neurons, which have a limited capacity for regeneration, will be targeted inappropriately by cytotoxic T cells. It does, however, make neurons attractive as cellular reservoirs for persistent infections. Herpesviruses often enter latency. Herpes zoster (varicella-zoster), which causes chickenpox, remains latent in one or a few dorsal root ganglia after the acute illness is over, and can be reactivated by stress or immunosuppression. It then spreads down the nerve and reinfects the skin to cause the disease herpes zoster, or **shingles**, which is marked by the reappearance of the classic varicella rash in the area of skin served by the infected dorsal root. Unlike herpes simplex, which reactivates frequently, herpes zoster usually reactivates only once in a lifetime in an immunocompetent host.

Acute Infectious Mononucleosis

Yet another herpesvirus, the Epstein–Barr virus (EBV), establishes a persistent infection in most individuals. EBV enters latency in B cells after a primary infection that often passes without being diagnosed. In a minority of infected individuals, the initial acute infection of B cells is more severe, causing the disease known as **infectious mononucleosis** or glandular fever. EBV infects B cells by binding to CR2 (CD21), a component of the B-cell co-receptor complex, and to MHC class II molecules. In the primary infection, most of the infected cells proliferate and produce virus, leading in turn to the proliferation of antigen-specific T cells and the excess of mononuclear white cells in the blood that gives the disease its name. Virus is released from the B cells, destroying them in the process, and virus can be recovered from saliva. The infection is eventually controlled by virus-specific CD8 cytotoxic T cells, which kill the infected proliferating B cells. A fraction of memory B lymphocytes become latently infected, however, and EBV remains quiescent in these cells.

These two forms of infection are accompanied by quite different patterns of expression of viral genes. EBV has a large DNA genome encoding more than 70 proteins. Many of these are required for viral replication and are expressed by the replicating virus, providing a source of viral peptides by which infected cells can be recognized. In a latent infection, in contrast, the virus survives within the host B cells without replicating, and a very limited set of viral proteins is expressed. One of these is the Epstein–Barr nuclear antigen 1 (EBNA1), which is needed to maintain the viral genome. EBNA1 interacts with the proteasome (see Section 6-2) to prevent its own degradation into peptides that would otherwise elicit a T-cell response.

Latently infected B cells can be isolated by culturing B cells from individuals who have apparently cleared their EBV infection: in the absence of T cells, latently infected cells retaining the EBV genome become transformed into so-called immortal cell lines, the equivalent of tumorigenesis *in vitro*. Infected B cells occasionally undergo malignant transformation *in vivo*, giving rise to a B-cell lymphoma called Burkitt's lymphoma. In this lymphoma, expression of the peptide transporters TAP1 and TAP2 is downregulated (see Section 6-3), and so cells are unable to process endogenous antigens for presentation on HLA class I molecules (the human MHC class I). This deficiency provides one explanation for how these tumors escape attack by CD8 cytotoxic T cells. Patients with acquired and inherited immunodeficiencies of T-cell function have an increased risk of developing EBV-associated lymphomas, presumably as a result of a failure of immune surveillance.

The viruses hepatitis B (HBV, a DNA virus) and hepatitis C (HCV, an RNA virus) infect the liver and cause acute and chronic hepatitis, liver cirrhosis, and in some cases hepatocellular carcinoma. Immune responses probably have an important role in the clearance of both types of hepatitis infection, but in many cases HBV and HCV set up a chronic infection. Although HCV mainly infects the liver during the early stage of a primary infection, the virus subverts the adaptive immune response by interfering with dendritic-cell activation and maturation. This leads to inadequate activation of CD4 T cells and a consequent lack of T_H1 cell differentiation, which is thought to be responsible for the infection becoming chronic, most probably because of the lack of CD4 T-cell help to activate naive CD8 cytotoxic T cells. There is evidence that the decrease in levels of viral antigen seen after antiviral treatment improves CD4 T-cell help and allows the restoration of cytotoxic CD8 T-cell function and memory CD8 T-cell function. The delay in dendritic-cell maturation caused by HCV is thought to synergize with another property of the virus that helps it to evade an immune response: the RNA polymerase that the virus uses to replicate its genome lacks proofreading capacity. This contributes to a high viral mutation rate and thus a change in its antigenicity, which allows it to evade adaptive immunity.

Summary.

Infectious agents can cause recurrent or persistent disease by avoiding normal host defense mechanisms or by subverting them to promote their own replication. There are many different ways of evading or subverting the immune response. Antigenic variation, latency, resistance to immune effector mechanisms, and suppression of the immune response all contribute to persistent and medically important infections. In some cases the immune response is part of the problem: some pathogens use immune activation to spread infection, and others would not cause disease if it were not for the immune response. Each of these mechanisms teaches us something about the nature of the immune response and its weaknesses, and each requires a different medical approach to prevent or to treat infection.

Acquired immune deficiency syndrome.

The most extreme example of immune subversion by a pathogen is the **acquired immune deficiency syndrome (AIDS)** caused by the **human immunodeficiency virus (HIV)**. The disease is characterized by a progressive loss of CD4 T cells, which, when they have become sufficiently depleted, results in high susceptibility to opportunistic infections and certain malignancies. The earliest documented case of HIV infection in humans to date was reported in a sample of serum from Kinshasa (Democratic Republic of the Congo) that was stored in 1959. It was not until 1981, however, that the first cases of AIDS were officially reported. Because the disease seemed to be spread by contact with body fluids, the cause was suspected to be a new virus, and by 1983 the causative agent, HIV, was isolated and identified.

Acquired Immune Deficiency Syndrome (AIDS)

There are at least two types of HIV—HIV-1 and HIV-2—that are closely related. While both types are transmitted by sexual contact and blood-borne exposure (for example, blood transfusion, shared needles), HIV-1 replicates to higher viral loads in blood and is therefore more easily transmitted; HIV-1 has a high rate of transmission from mother to child, which is uncommon in HIV-2. Although disease is indistinguishable in patients who progress to AIDS, HIV-1 progresses to AIDS more rapidly and with greater incidence than HIV-2. HIV-1 is therefore by far the most prevalent cause of AIDS worldwide. And while both HIV-1 and HIV-2 are endemic to West Africa, HIV-2 is rarely found elsewhere.

Both viruses seem to have originally spread to humans from other primate species in Africa. Viral genome sequencing of isolates suggests that the primate precursor of HIV-1, simian immunodeficiency virus (SIV), has passed

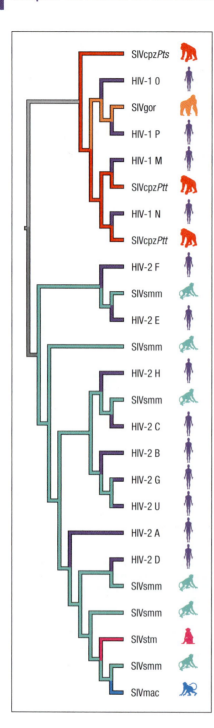

SIVcpz*Pts*

HIV-1 0

SIVgor

HIV-1 P

HIV-1 M

SIVcpz*Ptt*

HIV-1 N

SIVcpz*Ptt*

HIV-2 F

SIVsmm

HIV-2 E

SIVsmm

HIV-2 H

SIVsmm

HIV-2 C

HIV-2 B

HIV-2 G

HIV-2 U

HIV-2 A

HIV-2 D

SIVsmm

SIVsmm

SIVstm

SIVsmm

SIVmac

Fig. 13.27 Phylogenetic origins of HIV-1 and HIV-2. HIV-1 shows marked genetic variability and is classified on the basis of genomic sequence into four major groups: M (main), O (outlier), N (non-M, non-O), and P (non-M, non-N, non-O), which are further diversified into subtypes, or clades, that are designated by the letters A to K. In different parts of the world, different subtypes predominate. Phylogenetic analyses of chimpanzee simian immunodeficiency virus (SIVcpz), gorilla SIV (SIVgor), and HIV-1 sequences demonstrate that the four groups of HIV-1 (M, N, O, and P) originated from four independent cross-species transmission events: two transfers of SIVcpz*Ptt* from central chimpanzees (subspecies *Pan troglodytes troglodytes*, or *Ptt*) gave rise to HIV-1 groups M and N, while two transfers of SIVgor from western lowland gorillas (subspecies *Gorilla gorilla gorilla*) gave rise to HIV-1 groups O and P. Similarly, separate zoonotic transmissions of SIVsmm from sooty mangabey monkeys to humans are responsible for at least nine different lineages of HIV-2 (groups A–H and a newly described lineage, U). SIVstm and SIVmac resulted from experimental infections of stump-tailed macaques and rhesus macaques with SIVsmm, respectively. Abbreviations: cpz*Pts*, chimpanzee *Pan troglodytes schweinfurthii*; cpz*Ptt*, chimpanzee *Pan troglodytes troglodytes*; mac, macaque; SIV, simian immunodeficiency virus; smm, sooty mangabey monkey; stm, stump-tailed macaque. Figure courtesy of Drs. Beatrice Hahn and Gerald Learn.

to humans on at least four independent occasions from chimpanzees or western lowland gorillas, whereas HIV-2 originated in the sooty mangabey (**Fig. 13.27**). The best estimate is that the most prevalent of the four major variants of HIV-1, group M (responsible for ~99% of HIV-1 infections worldwide), was transmitted to humans from chimpanzees in the first half of the twentieth century; transmission of group O also dates to the early twentieth century, whereas the two other HIV-1 variants (groups N and P) appear to have been transmitted more recently. As is true for other zoonotic infections where there has been insufficient time for pathogen and host to coevolve to an equilibrium that attenuates virulence, SIV is generally less pathogenic in its nonhuman primate host than is HIV in its human host. Thus, whereas development of AIDS is nearly universal in HIV-1-infected humans that do not receive treatment, development of AIDS in SIV-infected nonhuman primates is considerably more variable, with some primates failing to develop disease at all.

HIV infection does not immediately cause AIDS. Without treatment, the average time to development of AIDS following infection of adults is several years. The long delay between infection and development of symptomatic immune deficiency reflects the unusual tropism of the virus for CD4 T cells of the immune system, as well as the nature of the immune response to the virus. HIV is now pandemic, and despite great strides in treatment and prevention that have followed from a greater understanding of the pathogenesis and epidemiology of the disease, 1.6 million people died of AIDS-related causes in 2012 and an estimated 35.3 million are infected by HIV worldwide, presaging the death of many from AIDS for years to come (**Fig. 13.28**). In sub-Saharan Africa, which accounts for over two-thirds of the global incidence, 1 in every 20 adults is infected. Indeed, HIV/AIDS has emerged as the most deadly single infectious agent in the short time since its identification as a new human pathogen. Nevertheless, there is cause for optimism: the incidence of new cases of HIV infections worldwide has declined annually since its peak in 1997, and the number of annual deaths from HIV/AIDS has steadily declined since its peak in the mid-2000s. Among the regions with the most rapid declines in incidence of new infections is sub-Saharan Africa. Still, there are focal areas of increasing incidence (for example, Eastern Europe and Central Asia).

13-25 HIV is a retrovirus that establishes a chronic infection that slowly progresses to AIDS.

HIV is an enveloped RNA virus whose structure is shown in **Fig. 13.29**. Each virus particle, or virion, is decorated with two viral envelope proteins that are used by the virus to infect target cells, and contains two copies of an RNA

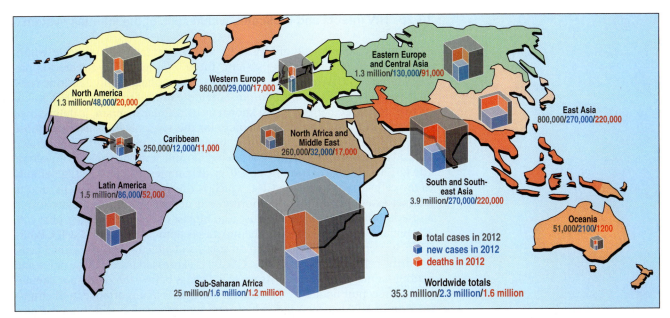

Fig. 13.28 The incidence of new HIV infection is increasing more slowly in many regions of the world, but AIDS is still a major disease burden. The number of individuals living with HIV/AIDS is large and continues to grow, but the number of new infections in 2012 decreased by over one-third since the peak of the epidemic. Worldwide, in 2012, it is estimated that there were around 35.3 million individuals infected with HIV, including some 2.4 million new cases, and approximately 1.6 million deaths from AIDS, a decrease of 30% since the peak in 2005. New infections in children have declined approximately 50% since 2001, with 260,000 new cases in 2012. (*AIDS Epidemic Update*, UNAIDS/World Health Organization, 2013.)

genome and numerous copies of viral enzymes that are required to establish infection in the cellular host. HIV is an example of a **retrovirus**, so named because the viral genome must be transcribed from RNA into DNA in the infected cell—the reverse (*retro*) of the usual pattern of transcription—by the viral **reverse transcriptase** enzyme. This generates a DNA intermediate that is integrated into the host-cell chromosomes to enable viral replication. RNA transcripts produced from the integrated viral DNA serve both as mRNAs to direct the synthesis of viral proteins and later as the RNA genomes of new viral particles. These escape from the cell by budding from the plasma membrane, each enclosed in a membrane envelope.

HIV belongs to a group of retroviruses called the **lentiviruses**, from the Latin *lentus*, meaning slow, because of the gradual course of the diseases that they cause. These viruses persist and continue to replicate for many years before causing overt signs of disease. In the case of HIV, the virus targets cells of the immune system itself, producing an initial acute infection that is controlled to the point that infection is not apparent, but rarely leading to an immune response that can prevent ongoing replication of the virus. Thus, although the initial acute infection does seem to be controlled by the immune system, HIV establishes latency within cells of the immune system and continues to replicate and infect new cells for many years. As will be discussed below, this ultimately exhausts the immune system, resulting in immune deficiency, or AIDS, that leads to opportunistic infections and/or malignancy that cause death.

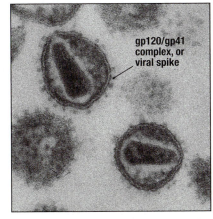

Fig. 13.29 The virion of human immunodeficiency virus (HIV). The virus illustrated is HIV-1, the leading cause of AIDS. The virion is roughly spherical and measures 120 nm in diameter, or about 60 times smaller than the T cells it infects. The three viral enzymes that are packaged in the virion—reverse transcriptase, integrase, and protease—are shown schematically in the viral capsid. In reality, many molecules of these enzymes are contained in each virion. Photograph courtesy of H. Gelderblom.

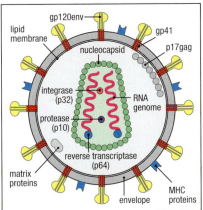

13-26 HIV infects and replicates within cells of the immune system.

MOVIE 13.4

A defining characteristic of HIV is its ability to infect and replicate within activated cells of the immune system. Three immune cell types are the primary targets of HIV infection: CD4 T cells, macrophages, and dendritic cells. Of these, CD4 T cells support the great majority of viral replication. HIV's ability to enter particular cell types, known as its cellular **tropism**, is determined by the expression of specific receptors for the virus on the surface of those cells. HIV enters cells by means of a complex of two noncovalently associated viral glycoproteins, gp120 and gp41, which form trimers in the viral envelope. The gp120 subunits of trimeric gp120/gp40 complexes bind with high affinity to the cell-surface molecule CD4, which is expressed on CD4 T cells, and to a lesser extent on subsets of dendritic cells and macrophages. Before fusion and entry of the virus, gp120 must also bind a co-receptor on the host cell. The major co-receptors are the chemokine receptors CCR5 and CXCR4. While CCR5 is predominantly expressed on subsets of effector memory CD4 T cells, dendritic cells, and macrophages, CXCR4 is expressed primarily by naive and central memory CD4 T cells. As we will discuss below, the particular chemokine co-receptor bound by a given viral particle is of importance in the transmission of HIV between individuals and its propagation within an infected person. After binding CD4, gp120 undergoes a conformational change that exposes a high-affinity site that is bound by the co-receptor. This, in turn, causes gp41 to unfold and insert a portion of its structure (fusion peptide) into the plasma membrane of the target cell, inducing fusion of the viral envelope with the cell's plasma membrane. This allows the viral nucleocapsid, composed of the viral genome and associated viral proteins, to enter the host-cell cytoplasm (Fig. 13.30).

Once the virus has entered cells, it replicates similarly to other retroviruses. Reverse transcriptase transcribes the viral RNA into a complementary DNA (cDNA) copy. The viral cDNA, which encodes nine genes (Fig. 13.31), is then integrated into the host-cell genome by viral integrase, which recognizes and partially cleaves repetitive DNA sequences, called long terminal repeats (LTRs), that reside at each end of the viral genome. LTRs are required for the integration of the provirus into the host-cell DNA and contain binding sites for gene-regulatory proteins that control expression of the viral genes. The integrated cDNA copy is known as the **provirus**.

Like other retroviruses, the HIV genome is small, with three major genes—*gag*, *pol*, and *env*. The *gag* gene encodes the structural proteins of the viral nucleocapsid core, *pol* encodes the enzymes involved in viral replication, and *env* encodes the viral envelope glycoproteins. The *gag* and *pol* mRNAs are translated to give polyproteins—long polypeptide chains that are then cleaved by the **viral protease** (encoded by *pol*) into individual functional proteins. Thus, *pol* alone encodes the virion's three major enzymes that are required for viral replication: reverse transcriptase, integrase, and viral protease. The product of the *env* gene, gp160, has to be cleaved by a host-cell protease into gp120 and gp41, which are then assembled as trimers in the viral envelope. HIV has six other, smaller, regulatory genes encoding proteins that affect viral replication and infectivity in various ways. Two of these, Tat and Rev, perform regulatory functions that are essential early in the viral replication cycle. The remaining four—Nef, Vif, Vpr, and Vpu—are essential for efficient virus production *in vivo*.

HIV can complete its replication cycle in the host cell to produce progeny virus, or, like other retroviruses and herpesviruses, establish a latent infection in which the provirus remains quiescent. What determines whether infection of a cell results in latency or a productive infection is unclear, but is thought to be related to the activation state of the cell infected. As we will discuss in the next section, transcription of the provirus following integration is initiated by host transcription factors, which are induced by immune-cell activation.

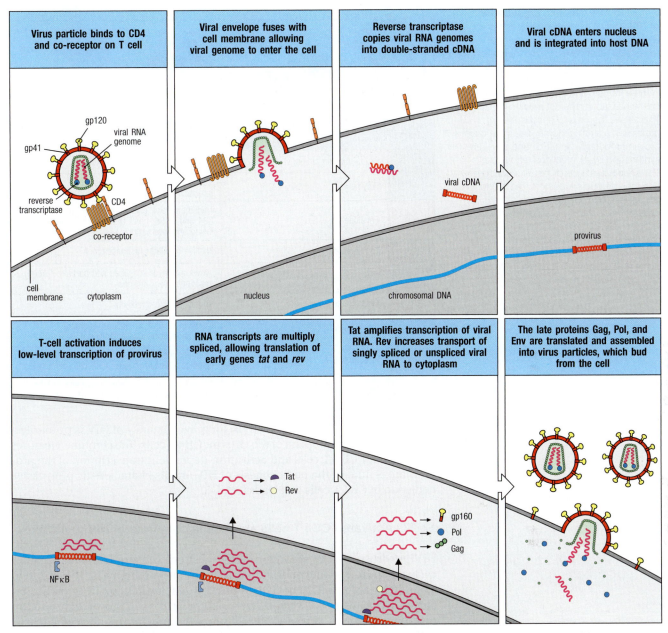

Fig. 13.30 The life cycle of HIV. Top row: the virus binds to CD4 using gp120, which is altered by CD4 binding so that it now also binds a chemokine receptor that acts as a co-receptor for viral entry. This binding releases gp41, which causes fusion of the viral envelope with the cell membrane and release of the viral core into the cytoplasm. Once in the cytoplasm, the viral core releases the RNA genome, which is reverse-transcribed into double-stranded cDNA using the viral reverse transcriptase. The double-stranded cDNA migrates to the nucleus in association with the viral integrase and the Vpr protein and is integrated into the cell genome, becoming a provirus. Bottom row: activation of CD4 T cells induces the expression of the transcription factors NFκB and NFAT, which bind to the proviral LTR and initiate transcription of the HIV genome. The first viral transcripts are extensively processed, producing spliced mRNAs encoding several regulatory proteins, including Tat and Rev. Tat both enhances transcription from the provirus and binds to the RNA transcripts, stabilizing them in a form that can be translated. Rev binds the RNA transcripts and transports them to the cytosol. As levels of Rev increase, less extensively spliced and unspliced viral transcripts are transported out of the nucleus. The singly spliced and unspliced transcripts encode the structural proteins of the virus, and unspliced transcripts, which are also the new viral genomes, are packaged with these proteins to form many new virus particles.

Thus, infection of a cell that becomes dormant soon after infection might favor viral latency, whereas infection of activated cells favors productive viral replication. This has important consequences in the case of CD4 T cells, which, unlike macrophages and dendritic cells, are very long-lived and can provide a reservoir of latent HIV provirus that can be activated when the T cells are reactivated, even years after initial infection. Because macrophages and dendritic cells in tissues are short-lived cells that do not divide, latency in these

Fig. 13.31 The genomic organization of HIV. Like all retroviruses, HIV-1 has an RNA genome flanked by long terminal repeats (LTRs) involved in viral integration and in regulation of transcription of the viral genome. The genome can be read in three frames, and several of the viral genes overlap in different reading frames. This allows the virus to encode many proteins in a small genome. The three main protein products—Gag, Pol, and Env—are synthesized by all infectious retroviruses. The known functions of the different genes and their products are listed. The products of *gag*, *pol*, and *env* are known to be present in the mature viral particle, together with the viral RNA. The mRNAs for Tat, Rev, and Nef proteins are produced by splicing of viral transcripts, so their genes are split in the viral genome. In the case of Nef, only one exon, shown in yellow, is translated.

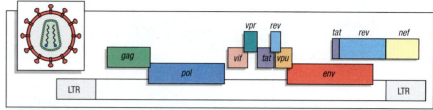

Gene		Gene product/function
gag	Group-specific antigen	Core proteins and matrix proteins
pol	Polymerase	Reverse transcriptase, protease, and integrase enzymes
env	Envelope	Transmembrane glycoproteins. gp120 binds CD4 and CCR5; gp41 is required for virus fusion and internalization
tat	Transactivator	Positive regulator of transcription
rev	Regulator of viral expression	Allows export of unspliced and partially spliced transcripts from nucleus
vif	Viral infectivity	Affects particle infectivity
vpr	Viral protein R	Transport of DNA to nucleus. Augments virion production. Cell-cycle arrest
vpu	Viral protein U	Promotes intracellular degradation of CD4 and enhances release of virus from cell membrane
nef	Negative-regulation factor	Augments viral replication *in vivo* and *in vitro*. Decreases CD4, MHC class I and II expression

host cells would be short-lived. Thus, long-lived latency of HIV is primarily a consequence of the tropism of the virus for CD4 T cells. The combined features of tropism for CD4 T cells and activation-dependent transcription of the provirus are central to the pathogenesis of HIV and its characteristic progressive depletion of CD4 T cells that leads to AIDS.

13-27 Activated CD4 T cells are the major source of HIV replication.

HIV provirus requires activation of the host cell to complete its replication cycle and produce infectious virions that can infect other cells. This is due to a requirement for transactivation of proviral gene expression by transcription factors of the host cell. Two host transcription factors can initiate transcription of the viral genome: NFκB and NFAT. Both of these factors require cellular activation for their translocation to the nucleus, where they bind DNA and induce gene transcription (see Sections 7-14 and 7-16). While NFκB is expressed in all of the immune cells infected by HIV, NFAT is primarily activated in CD4 T cells, enabling transactivation of the provirus by an additional factor in this cell host. This, coupled with the fact that CD4 T cells are long-lived and abundant in immune tissues, contributes to CD4 T cells being the major cellular source for HIV replication. Here we consider the mechanism by which transcription of the HIV provirus is regulated in CD4 T cells.

As discussed in Sections 7-14 and 7-16, activation of T cells by antigen induces activation and nuclear translocation of NFAT and NFκB; activation of effector-memory T cells by cytokines can also activate NFκB in the absence of antigen (Section 11-12). Thus, in addition to antigen-dependent activation of HIV provirus transcription by NFAT and NFκB, provirus might be activated independently of T-cell receptor stimulation in memory CD4 T cells via NFκB alone, as it is in infected macrophages and dendritic cells. Binding of NFAT and NFκB initiates transcription of viral RNA by binding to promoters in the proviral LTR. The viral transcript is spliced in various ways to produce mRNAs for translation of viral proteins (see Fig. 13.26).

At least two of the viral proteins—Tat and Rev—serve to enhance production of the viral genome (see Fig. 13.30). Tat binds to a transcriptional activation region (TAR) in the 5′ LTR. This recruits cellular cyclin T1 and its partner, cyclin-dependent kinase 9 (CDK9), to form a complex that phosphorylates RNA polymerase and enhances its ability to generate full-length transcripts of the viral genome. In this way, Tat provides a positive feedback circuit for amplification of productive viral replication. Rev is important for shuttling unspliced viral RNA transcripts out of the nucleus by binding to a specific viral RNA sequence, the Rev response element (RRE). Eukaryotic cells have mechanisms to prevent the export from the cell nucleus of incompletely spliced mRNA transcripts. This could pose a problem for the retrovirus, which is dependent on export of unspliced mRNA species that encode the full complement of viral proteins, as well as the viral RNA genome. While export of a fully spliced mRNA transcript that encodes Tat and Rev occurs early after viral infection by means of the normal host cellular mechanisms of mRNA export, the export of later, unspliced viral transcripts requires Rev to prevent their destruction by the host cell.

The success of viral replication also depends on the proteins Nef, Vif, Vpu, and Vpr. These viral products appear to have evolved to defeat host immune mechanisms of viral clearance, as well as antiviral **restriction factors**—host cellular proteins that function in a cell-autonomous manner to inhibit replication of retroviruses. Nef (negative regulation factor) performs multiple critical functions in the viral life cycle. It acts early in the viral life cycle to sustain T-cell activation and the establishment of a persistent state of HIV infection, in part by lowering the threshold for T-cell receptor signaling and downregulating expression of the inhibitory co-stimulatory receptor CTLA4. Combined, these actions result in greater and more sustained T-cell activation that promotes viral replication. Nef also contributes to immune evasion of infected cells by downregulating expression of MHC class I and class II molecules, making actively infected cells less likely to induce an antiviral immune response or be killed by cytotoxic T cells. Nef also promotes the clearance of surface CD4 molecules, which otherwise would bind to the virion during budding and interfere with virion release. Vif (viral infectivity factor) acts to overcome a cytidine deaminase called APOBEC, which catalyzes the conversion of deoxycytidine to deoxyuridine in reverse-transcribed viral cDNA, thereby destroying its ability to encode viral proteins. Vpu (viral protein U) is unique to HIV-1 and variants of SIV, and is required to overcome a cellular factor called tetherin, which inserts into both the plasma membrane of the host cell and the envelope of the mature virion to block its release. The function of Vpr (viral protein R) is not fully understood, but it appears to target the restriction factor SAMHD1, a cellular protein that inhibits HIV-1 infection in myeloid cells and quiescent CD4 T cells by limiting the intracellular pool of deoxynucleotides (dNTPs) available for viral cDNA synthesis by reverse transcriptase.

13-28 There are several routes by which HIV is transmitted and establishes infection.

Infection with HIV occurs after the transfer of body fluids from an infected person to an uninfected one. HIV infection is most commonly spread by sexual intercourse. Transmission by the exchange of contaminated needles used for intravenous drug delivery or by the therapeutic use of infected blood or blood products also occurs, although the latter route of transmission has largely been eliminated in countries where blood products are routinely screened for HIV. An important route of virus transmission is from an infected mother to child, which can occur *in utero*, during birth, or through breast milk. Rates of transmission from an untreated infected mother to her child vary (from about 15% to 45%), depending on the viral load in the mother and whether the mother breastfeeds the child, as breastfeeding increases the risk of transmission.

The use of antiretroviral drugs to decrease maternal viral load during pregnancy dramatically reduces the transmission rate to the child (see Section 13-35).

The virus can be transmitted as free infectious particles or via infected cells for which the virus has tropism (for example, CD4 T cells and macrophages). Infected cells are found in blood, but can also be present in semen or vaginal secretions, as well as breast milk; free virus is present in blood, semen, vaginal fluid, or mother's milk. As we discuss in the next section, HIV virions can differentially express gp120 variants that bind either CCR5 or CXCR4, thereby influencing the cell types infected. In the genital and gastrointestinal mucosae, which are the dominant sites of primary infection by sexual transmission, HIV virions establish infection initially in a small number of mucosal immune cells that express CCR5—effector memory CD4 T cells, dendritic cells, and macrophages. The virus replicates locally in these cells before spreading via T cells or dendritic cells (mucosal macrophages are nonmigratory) to lymph nodes draining the mucosa. The lymphoid compartment of mucosal tissues is enriched for T_H1 and T_H17 cells, which express CCR5 (naive T cells and T_H2 cells do not), so initial viral replication is favored in these subsets of CD4 T cells. After accelerated replication in regional lymph nodes, where there is a high concentration of CD4 T cells, the virus disseminates widely via the bloodstream, and gains broader access to the gut-associated lymphoid tissues (GALT), where the highest number of CD4 T cells in the body reside.

13-29 HIV variants with tropism for different co-receptors play different roles in transmission and progression of disease.

To establish infection in a new host, HIV must make contact with a CD4-expressing immune cell. The cell type targeted is determined by the affinity of viral gp120 for the different chemokine co-receptors: CCR5 or CXCR4. Accordingly, the two major tropism variants of HIV are referred to as 'R5' and 'X4,' respectively. Because CCR5 dominates on CD4-expressing immune cells resident at the major sites of viral transmission—sites that are constantly exposed to commensal microbes and thus harbor large numbers of activated immune cells (mucosal tissues of the female and male genital tracts or rectum for sexual transmission; upper gastrointestinal tract for mother-to-child transmission)—CCR5-tropic R5 strains of virus are typically required for transmission, and dominate early in infection.

Before HIV can contact CD4-expressing immune cells in the genital and intestinal mucosae, it must traverse the epithelium of these tissues. Here, the CCR5-tropic variants of the virus also have an advantage. Infection occurs across two types of epithelium: the stratified, or multilayered, squamous epithelium that lines the mucosae of the vagina, foreskin of the penis, ectocervix, rectum, oropharynx, and esophagus; or the single-layered columnar epithelium lining endocervix, rectum, and upper GI tract. Epithelial cells of the rectum and endocervix can express CCR5 and have been shown to selectively translocate R5, but not X4, HIV variants through the epithelial monolayer. Other molecules expressed by epithelial cells also participate; gp120-binding glycosphingolipids expressed by epithelial cells of the vaginal or ectocervical mucosae also foster transcytosis of virus across the epithelium. The rate at which virus can transit epithelial barriers to establish infection is fast. SIV virus has been shown to penetrate the cervicovaginal epithelium within 30 to 60 minutes of exposure.

In addition to direct transcytosis across epithelial cells, the interdigitating processes of dendritic cells that ramify between epithelial cells provide an avenue for HIV to traverse the epithelium. A complex ferrying mechanism seems to transfer HIV picked up by dendritic cells to CD4 T cells in lymphoid tissue. HIV can attach to dendritic cells by the binding of viral gp120 to C-type lectin receptors such as langerin (CD207), the mannose receptor (CD206), and

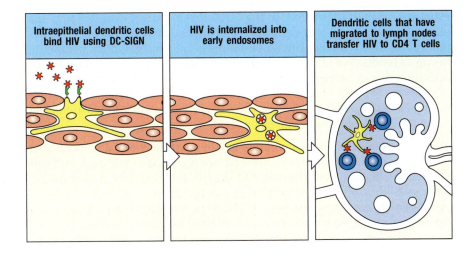

Fig. 13.32 Dendritic cells can initiate infection by transporting HIV from mucosal surfaces to lymphoid tissue.
HIV adheres to the surface of intraepithelial dendritic cells by the binding of viral gp120 to DC-SIGN (left panel). It gains access to dendritic cells at sites of mucosal injury or possibly to dendritic cells that have protruded between epithelial cells to sample the external world; HIV can also bind directly to some epithelial cells and is transported across them to subepithelial dendritic cells (not shown). Dendritic cells internalize HIV virions into mildly acidic early endosomes and migrate to lymphoid tissue (center panel). HIV virions are translocated back to the cell surface, and when the dendritic cell encounters CD4 T cells in a secondary lymphoid tissue, the HIV is transferred to the T cell (right panel).

DC-SIGN. A portion of the bound virus is rapidly taken up into vacuoles, where it can remain for days in an infectious state. In this way the virus is protected and remains stable until it encounters a susceptible CD4 T cell, whether in the local mucosal environment or after being carried to draining lymphoid tissue (Fig. 13.32). Finally, at some mucosal sites, CCR5-expressing CD4 T cells reside within the epithelium (intraepithelial T cells), and have been shown to be sites of early viral replication. Thus, HIV can infect CD4 T cells either directly or via dendritic cells that interact with CD4 T cells.

During the **acute phase** of infection, which typically lasts for several weeks and is characterized by an influenza-like illness, there is rapid replication of the virus, primarily in CCR5-expressing CD4 T cells (Fig. 13.33). This period is marked by an abundance of virus circulating in the blood (viremia) and the rapid decline of CCR5-expressing CD4 T cells, the latter due primarily to the

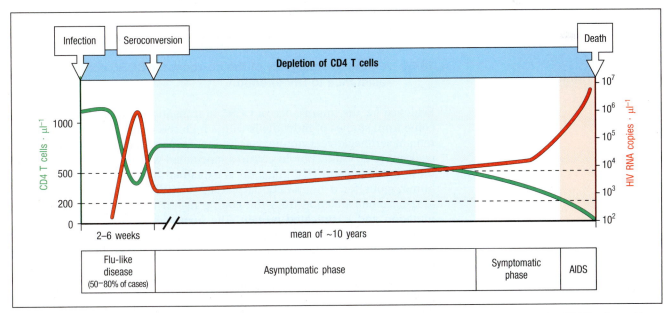

Fig. 13.33 The typical course of untreated infection with HIV.
The first few weeks are typified by an acute influenza-like viral illness, sometimes called seroconversion disease, with high titers of virus in the blood. An adaptive immune response follows, which controls the acute illness and largely restores levels of CD4 T cells but does not eradicate the virus. This is the asymptomatic phase, which typically lasts 5–10 years without treatment. Opportunistic infections and other symptoms become more frequent as the CD4 T-cell count in peripheral blood falls, starting at about 500 cells · μl^{-1}. The disease then enters the symptomatic phase. When CD4 T-cell counts fall below 200 cells · μl^{-1}, the patient is said to have AIDS. Note that CD4 T-cell counts are measured for clinical purposes in cells per microliter (cells · μl^{-1}), rather than cells per milliliter (cells · ml^{-1}), the unit used elsewhere in this book.

extensive death of CD4 T cells in the GALT that are killed by viral cytopathic effects (macrophages and dendritic cells appear more resistant to lysis by replicating virus). The depletion of immune cells in the gut might compound the rapid production of virus in the GALT by fostering increased immune-cell activation due to barrier breakdown and translocation of constituents of the microbiota. Because of the high viral titers and preponderance of R5 strains during the acute phase of infection, the risk of viral transmission to uninfected contacts during this time is especially high.

The acute phase and its high viremia reside in virtually all patients once an adaptive immune response is established (see Fig. 13.33). Cytolytic CD8 T cells specific for viral antigens develop and kill HIV-infected cells, and virus-specific antibodies become detectable in the serum of those infected (**seroconversion**). The development of a CTL response results in early control of the virus, resulting in a sharp drop in viral titers and a rebound of CD4 T-cell counts. The level of virus that persists in blood plasma at this stage of infection, referred to as the **viral set point**, is usually a good indicator of future disease progression. At this point, the disease transitions to a clinically latent, or **asymptomatic**, **phase** marked by low viremia and slowly declining CD4 T-cell numbers, typically over several years. During this time the virus continues to actively replicate, but it is held in check, principally by HIV-specific CD8 T cells and antibodies.

Under strong selective pressure brought by the antiviral immune response, there is selection for HIV **escape mutants** that are no longer detected by adaptive immune cells. This gives rise to many different viral variants in a single infected person and to even broader variation within the population as a whole. Late in infection, in approximately 50% of cases, the dominant viral type switches from R5 to X4 variants that infect T cells via CXCR4 co-receptors. This is followed by a rapid decline in CD4 T-cell count and progression to AIDS. The exact mechanism by which this shift in viral tropism leads to accelerated loss of CD4 T cells is unknown. On balance, then, R5 variants appear critical for transmission of the virus from infected to uninfected individuals, whereas X4 variants that emerge under selective pressure exerted by the antiviral immune response contribute to the progression of disease within an infected individual.

13-30 A genetic deficiency of the co-receptor CCR5 confers resistance to HIV infection.

Evidence for the importance of CCR5 in transmission of HIV infection has come from studies of individuals with a high risk of exposure to HIV-1 who remain seronegative. Lymphocytes and macrophages from these people are resistant to HIV infection in cultures inoculated with HIV. The resistance of these individuals to HIV infection is explained by discovery that they are homozygous for a nonfunctional variant of CCR5 called Δ32, caused by a 32-base-pair deletion from the coding region that leads to a frameshift mutation and a truncated protein. The frequency of this mutant allele in Caucasians is high at 0.09 (that is, about 10% of the population are heterozygous carriers of the allele and about 1% are homozygous). The mutant allele has not been found in Japanese or in black Africans from Western or Central Africa. Whether heterozygous deficiency of CCR5 provides some protection against infection by HIV is controversial, but it appears to contribute to a modest, if any, reduction in progression rates. In addition to the structural polymorphism of the gene, variations in the promoter region of the CCR5 gene have been associated with different rates of disease progression. The high incidence of the *CCR5Δ32* allele in Caucasians predating the HIV pandemic suggests selection for this variant in a past epidemic. Both smallpox and bubonic plague have been put forward as possible selective agents, but this is as yet unproven.

13-31   An immune response controls but does not eliminate HIV.

Infection with HIV generates an immune response that contains the virus but only very rarely, if ever, eliminates it. A time-course of the response of various adaptive immune elements to HIV in adults is shown, together with the levels of infectious virus in plasma, in **Fig. 13.34**. As was noted earlier, in the acute phase, virus-mediated cytopathicity results in a substantial depletion of CD4 T cells, particularly in mucosal tissues. There is a good initial recovery of T-cell numbers and transition to the asymptomatic phase of disease as the immune response develops and curbs viral replication (see Fig. 13.33). However, replication of the virus persists, and, after a variable period lasting from a few months to more than 20 years, the CD4 T-cell numbers fall too low to maintain effective immunity, and AIDS develops (defined as less than 200 CD4 T cells per microliter in peripheral blood). Several factors conspire to progressively deplete CD4 T cells until they can no longer maintain immunity: destruction by cytotoxic lymphocytes directed against HIV-infected cells, immune activation (direct and indirect) that induces activation of latent virus, ongoing viral cytopathic effects, and insufficient T-cell regeneration in the thymus. In this section we consider in turn the roles of CD8 cytotoxic T cells, CD4 T cells, antibodies, and soluble factors in mounting the immune response to HIV infection that initially contains the infection but ultimately fails.

Studies of peripheral blood cells from infected individuals reveal cytotoxic T cells specific for viral peptides that can kill infected cells *in vitro*. *In vivo*, cytotoxic T cells traffic to sites of HIV replication, where they are thought to kill many productively infected cells before any infectious virus is released, thereby containing viral load at the quasi-stable levels that are characteristic of the asymptomatic period. Evidence for the clinical importance of the control of HIV-infected cells by CD8 cytotoxic T cells comes from studies relating the numbers and activity of CD8 T cells to viral load. There is also direct evidence from experiments in macaques infected with SIV that CD8 cytotoxic T cells control retrovirus-infected cells; treatment of infected animals with monoclonal antibodies that remove CD8 T cells is rapidly followed by a large increase in viral load.

In addition to direct cytotoxicity mediated by recognition of cells infected with virus, a variety of factors produced by CD4, CD8, and NK cells are important in antiviral immunity. Chemokines that bind CCR5, such as CCL5, CCL3, and CCL4, are released at the site of infection by CD8 T cells and inhibit virus spread by competing with R5 strains of HIV-1 for the engagement of co-receptor CCR5, whereas factors still unknown compete with R4 strains for binding to CXCR4. Cytokines such as IFN-α and IFN-γ may also be involved in controlling virus spread.

Evidence shows that in addition to being a major target for HIV infection, CD4 T cells also have an important role in the host response to HIV-infected cells.

Fig. 13.34 The immune response to HIV. Infectious virus is present at relatively low levels in the peripheral blood of infected individuals during a prolonged asymptomatic phase, during which the virus is replicated persistently in lymphoid tissues. During this period, CD4 T-cell counts gradually decline (see Fig. 13.33), although antibodies and CD8 cytotoxic T cells directed against the virus remain at high levels. Two different antibody responses are shown in the figure, one to the envelope protein (Env) of HIV, and one to the core protein p24. Eventually, the levels of antibody and HIV-specific cytotoxic T lymphocytes (CTLs) also decline, and there is a progressive increase in infectious HIV in the peripheral blood.

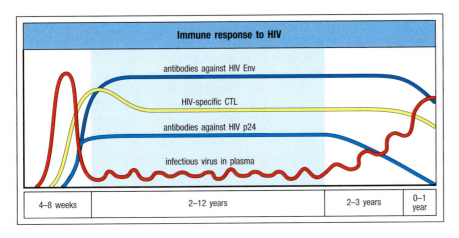

Immune response to HIV

antibodies against HIV Env

HIV-specific CTL

antibodies against HIV p24

infectious virus in plasma

| 4–8 weeks | 2–12 years | 2–3 years | 0–1 year |

An inverse correlation is found between the strength of CD4 T-cell proliferative responses to HIV antigen and viral load. In addition, the type of effector CD4 T-cell response mounted against the virus appears important. There is an inverse correlation with viral load and control of acute infection in patients whose CD4 T cells express greater T_H1 type activity, including production of IFN-γ and granzyme B. Moreover, CD4 T cells from patients who do not progress to AIDS long after infection by HIV show strong antiviral proliferative responses. Finally, early treatment of acutely infected individuals with anti-retroviral drugs is associated with a recovery in CD4 proliferative responses to HIV antigens. If antiretroviral therapy is stopped, the CD4 responses persist in some of these people and are associated with reduced levels of viremia. However, infection continues to persist in these patients and immunological control of the infection will ultimately fail. If CD4 T-cell responses are essential for the control of HIV infection, then the fact that HIV is tropic for these cells and kills them may explain the inability of the host immune response to control the infection in the long term.

Antibodies against HIV proteins are generated early in the course of infection, but, like T cells, are ultimately unable to clear the virus. As for viral T-cell epitopes, the virus shows a high capacity for generating escape mutants under the selective pressure of the antibody response. Two aspects of the antibody response appear to be important: (1) generating neutralizing antibodies against gp120 and gp41 envelope viral antigens in order to block viral attachment or entry into CD4-positive target cells, and (2) generating nonneutralizing antibodies that target infected cells for antibody-dependent cellular cytotoxicity (ADCC). Although neutralizing antibodies are eventually produced in nearly all who are HIV-infected, the relative inaccessibility of viral epitopes that bind CD4 and chemokine co-receptors hampers the development of such antibodies for a prolonged period (typically months); this buys the virus time to generate escape mutants before the neutralizing antibodies can be produced. Indeed, the generation of so-called **broadly neutralizing antibodies**, which can block infection by multiple viral strains, is typically found in those with high viral titers, emphasizing the fact that these antibodies cannot significantly modify established disease. Analyses of effective neutralizing antibodies to HIV indicate that they have undergone extensive somatic hypermutation that is rarely induced before the first year following infection. Nevertheless, the passive administration of some antibodies against HIV can protect experimental animals from mucosal infection by HIV, offering hope that an effective vaccine might be developed that could prevent new infections.

In contrast to neutralizing antibodies, which develop late in infection and appear to play a modest role in restraining viral replication, there is growing evidence that nonneutralizing antibodies that recruit ADCC by NK cells, macrophages, and neutrophils develop early in infections and are important in restraining viral replication in concert with the actions of cytolytic CD8 T cells. Again, however, the high rate of mutability of the virus allows it to stay a step ahead and persist. The mutations that occur as HIV replicates can allow resulting virus variants to escape recognition by CTLs or antibodies, and are important in contributing to the long-term failure of the immune system to contain the infection. An immune response is often dominated by T or B cells specific for particular epitopes—**immunodominant** epitopes—and mutations in immunodominant HIV peptides presented by MHC class I molecules have been found, as have mutations in the epitopes targeted by neutralizing and nonneutralizing antibodies. Mutant peptides have been found to inhibit T cells responsive to the wild-type epitope, thus allowing both the mutant and wild-type viruses to survive.

While the immune response to HIV is ultimately unsuccessful, its importance in restraining disease progression is clear. This is perhaps best exemplified in the tragic case of children infected with HIV perinatally, in whom the course of disease is much more fulminant than in adults. This reflects a poor immune response to the virus in the acute phase of infection due to immaturity of the neonatal immune system, as well as infection by a viral strain that has already evaded an immune system that is genetically close to that of the child. In essence, the poor immune response results in the lack of a latent phase, leading rapidly to AIDS.

13-32 Lymphoid tissue is the major reservoir of HIV infection.

In view of the active, ongoing immune response to HIV infections and the advent of antiretroviral therapies that efficiently blunt viral replication (see Section 13-35), it is important to identify the reservoirs that allow the virus to persist. Although HIV load and turnover are usually measured in terms of the RNA present in virions in the blood, the major reservoir of HIV infection appears to be lymphoid tissue. Here, in addition to infected CD4 T cells, macrophages, and dendritic cells, HIV is also trapped in the form of immune complexes on the surface of follicular dendritic cells in germinal centers. These cells are not themselves infected but may act as a reservoir of infective virions that can persist for months, if not longer. Although tissue macrophages and dendritic cells seem able to harbor replicating HIV without being killed by it, these cells are short-lived and are not thought to be major reservoirs of latent infection. However, they appear to be important in spreading virus to other tissues, such as the brain, where infected cells in the central nervous system may contribute to the virus's long-term persistence.

From studies of patients receiving antiretroviral treatment, it is estimated that more than 95% of the virus that can be detected in the plasma is derived from productively infected CD4 T cells that have a very short half-life—about 2 days. Virus-producing CD4 T cells are found in the T-cell areas of lymphoid tissues, and these T cells are thought to succumb to infection while being activated in an immune response. Latently infected CD4 memory T cells that become reactivated by antigen also produce virus that can spread to other activated CD4 T cells. In addition to cells that are productively or latently infected, a further large population of cells is infected by defective proviruses, which do not produce infectious virus. Unfortunately, latently infected CD4 memory T cells have an extremely long mean half-life of around 44 months. This means that drug therapy that effectively eliminates viral replication would have to be administered for over 70 years to completely clear the virus. Practically, then, infected patients will never be able to eliminate an HIV infection, and require treatment for life.

13-33 Genetic variation in the host can alter the rate of disease progression.

It became clear early in the HIV/AIDS pandemic that the course of the disease could vary widely. Indeed, while nearly all untreated HIV-infected individuals progress to AIDS and ultimately die from opportunistic infections or cancer, not all do. A small percentage of people exposed to the virus seroconvert, but do not seem to have progressive disease. Their CD4 T-cell counts and other measures of immune competence are maintained for decades without antiretroviral therapy. Among these **long-term nonprogressors**, one subgroup, called **elite controllers**, have unusually low levels of circulating virus (undetectable by standard clinical assays, despite ongoing low-level viral replication) and represent approximately 1 in 300 infected individuals. They are being studied intensively to discover how they are able to control their infection. A second group consists of individuals who engage in high-risk behaviors that repeatedly expose them to infection yet remain virus- and disease-free. Although evidence of prior HIV infection has been reported in such individuals, it is unclear whether these individuals were ever truly infected with infectious virus or were exposed to highly attenuated or defective strains unable to successfully establish infection. In any case, study of these individuals is of considerable interest as it could provide a better understanding of how the host immune response might better control the virus and define what genetic factors might predispose to a protective host response. It might also provide mechanistic insights that could guide development of better vaccines.

Although genetic variation in the virus itself can affect the outcome of infection, a growing number of host gene variants are being defined that impact the rate of progression of HIV infection toward AIDS. The implementation of genome-wide association studies (GWAS) and, more recently, better high-throughput tools to define individual genetic variation (for example, exome and whole-genome sequencing) are accelerating discovery of genetic variations that distinguish highly susceptible and resistant individuals (Fig. 13.35). As discussed in Section 13-30, one of the clearest cases of host genetic variation affecting

Fig. 13.35 Genes that influence progression to AIDS in humans. E, an effect that acts early in progression to AIDS; L, acts late in AIDS progression; ?, plausible mechanism of action with no direct support. From O'Brien, S.J., and Nelson, G.W.: *Nat. Genet.* 36:565–574. Reprinted with permission from Macmillan Publishers Ltd. © 2004.

Genes that influence progression to AIDS				
Gene	**Allele**	**Mode**	**Effect**	**Mechanism of action**
HIV entry				
CCR5	Δ32	Recessive	Prevents infection	Knockout of CCR5 expression
		Dominant	Prevents lymphoma (L)	Decreases available CCR5
			Delays AIDS	
	P1	Recessive	Accelerates AIDS (E)	Increases CCR5 expression
CCR2	I64	Dominant	Delays AIDS	Interacts with and reduces CXCR4
CCL5	In1.1c	Dominant	Accelerates AIDS	Decreases CCL5 expression
CXCL12	3′A	Recessive	Delays AIDS (L)	Impedes CCR5–CXCR4 transition (?)
CXCR6	E3K	Dominant	Accelerates *P. jirovecii* pneumonia (L)	Alters T-cell activations (?)
CCL2-CCL7-CCL11	H7	Dominant	Enhances infection	Stimulates immune response (?)
Cytokine anti-HIV				
IL10	5′A	Dominant	Limits infection	Decreases IL-10 expression
			Accelerates AIDS	
IFNG	−179T	Dominant	Accelerates AIDS (E)	
Acquired immunity, cell-mediated				
HLA	A, B, C	Homozygous	Accelerates AIDS	Decreases breadth of HLA class I epitope recognition
	B*27	Codominant	Delays AIDS	Delays HIV-1 escape
	B*57			
	B*35-Px		Accelerates AIDS	Deflects CD8-mediated T-cell clearance of HIV-1
Acquired immunity, innate				
KIR3DS1	3DS1	Epistatic with HLA-Bw4	Delays AIDS	Clears HIV+, HLA− cells (?)

HIV infection is a mutant allele of *CCR5*, *CCR5Δ32*, that when homozygous effectively blocks HIV-1 infection, and when heterozygous may slow AIDS progression. Genetic polymorphisms in the HLA class I locus, particularly in *HLA-B* and *HLA-C* alleles, are another major factor in determining disease progression and are currently the strongest predictors of HIV control. Evidence from GWAS has mapped polymorphisms to the peptide-binding groove of HLA class I molecules as key determinants defining disease progression. Polymorphisms outside of the peptide-binding groove, as well as those in noncoding regions that control the expression levels of HLA molecules, are also implicated. The HLA class I alleles *HLA-B57*, *HLA-B27*, and *HLA-B13*, among others, are associated with a better prognosis, whereas *HLA-B35* and *HLA-B07* are associated with more rapid disease progression. Homozygosity of HLA class I alleles (*HLA-A*, *HLA-B*, and *HLA-C*) is also associated with more rapid progression, presumably because the T-cell response to infection is less diverse. Remarkably, one of the strongest associations with viral control is a single-nucleotide polymorphism (SNP) 35 kb upstream of the *HLA-C* locus; this polymorphism confers greater immune control that correlates with increased expression levels of HLA-C—with the greater control presumably being due to enhanced presentation of viral peptides to CD8 T cells. Certain polymorphisms of the killer-cell immunoglobulin-like receptors (KIRs) present on NK cells (see Section 3-26), in particular the receptor KIR-3DS1 in combination with certain alleles of *HLA-B*, also delay the progression to AIDS. Mutations that affect the production of cytokines such as IFN-γ and IL-10 have also been implicated in the restriction of HIV progression.

13-34 The destruction of immune function as a result of HIV infection leads to increased susceptibility to opportunistic infection and eventually to death.

When CD4 T-cell numbers decline below a critical level, cell-mediated immunity is lost, and infections with a variety of opportunistic microbes appear (Fig. 13.36). Typically, resistance is lost early to oral *Candida* species and to *M. tuberculosis*, which cause thrush (oral candidiasis) and tuberculosis, respectively. Later, patients suffer from shingles (caused by the activation of latent herpes zoster), from aggressive EBV-induced B-cell lymphomas, and from Kaposi sarcoma, a tumor of endothelial cells that probably results both from a response to cytokines produced in the infection and a herpesvirus called Kaposi sarcoma-associated herpesvirus (KSHV, or HHV8). Since the earliest recognition of AIDS, pneumonia caused by *P. jirovecii* (previously called *P. carinii*) has been the most common opportunistic infection; it was typically fatal before effective antifungal therapy was introduced. Co-infection by hepatitis C virus is common and associated with more rapid progression of hepatitis. In the final stages of AIDS, infection with cytomegalovirus or a member of the *Mycobacterium avium* group of bacteria is more prominent. It is important to note that not all patients with AIDS get all of these infections or tumors, and there are other tumors and infections that are less prominent but still significant. Figure 13.36 lists the most common opportunistic infections and tumors, which are typically controlled until the CD4 T-cell count drops toward zero.

Increased Susceptibility to *Candida* Infections

Infections	
Parasites	*Toxoplasma* spp. *Cryptosporidium* spp. *Leishmania* spp. *Microsporidium* spp.
Intracellular bacteria	*Mycobacterium tuberculosis* *Mycobacterium avium intracellulare* *Salmonella* spp.
Fungi	*Pneumocystis jirovecii* *Cryptococcus neoformans* *Candida* spp. *Histoplasma capsulatum* *Coccidioides immitis*
Viruses	Herpes simplex Cytomegalovirus Herpes zoster

Malignancies
Kaposi sarcoma – (HHV8) Non-Hodgkin's lymphoma, including EBV-positive Burkitt's lymphoma Primary lymphoma of the brain

Fig. 13.36 A variety of opportunistic pathogens and cancers can kill AIDS patients. Infections are the major cause of death in AIDS, the most prominent being respiratory infection with *P. jirovecii* and mycobacteria. Host defense against most of these pathogens requires effective macrophage activation by CD4 T cells or effective cytotoxic T cells. Opportunistic pathogens are present in the normal environment, but cause severe disease primarily in immunocompromised hosts, such as AIDS patients and cancer patients. AIDS patients are also susceptible to several rare cancers, such as Kaposi sarcoma [associated with human herpesvirus 8 (HHV8)] and various lymphomas, suggesting that immune surveillance of the causative herpesviruses by T cells can normally prevent such tumors (see Chapter 16).

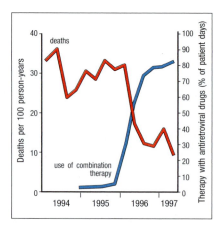

Fig. 13.37 The mortality of patients with advanced HIV infection fell in the United States in parallel with the introduction of combination antiretroviral drug therapy. The graph shows the number of deaths, expressed each calendar quarter as the deaths per 100 person-years. Figure based on data from F. Palella.

Fig. 13.38 The time-course of reduction of HIV circulating in the blood during drug treatment. The production of new HIV particles can be arrested for prolonged periods by treatment using combinations of protease inhibitors and viral reverse-transcriptase inhibitors. After the initiation of such treatment, virus production is curtailed as the infected cells die and no new cells are infected. The half-life of virus decay occurs in three phases. In the first phase the half-life ($t_{1/2}$) is about 2 days, reflecting the half-life of productively infected CD4 T cells; this phase lasts about 2 weeks, during which time viral production declines as the lymphocytes that were productively infected at the onset of treatment die. Released virus is rapidly cleared from the circulation, where it has a half-life of 6 hours, and there is a decrease in virus levels in plasma of more than 95% during this first phase. The second phase lasts for about 6 months; the virus now has a half-life of about 2 weeks. During this phase, virus is released from infected macrophages and from resting, latently infected CD4 T cells stimulated to divide and develop productive infection. It is thought that there is then a third phase of unknown length that results from the reactivation of integrated provirus in memory T cells and other long-lived reservoirs of infection. This reservoir of latently infected cells might remain present for many years. Measurement of this phase of viral decay is impossible at present because viral levels in plasma are below detectable levels (dotted line). Data courtesy of G.M. Shaw.

13-35 Drugs that block HIV replication lead to a rapid decrease in titer of infectious virus and an increase in CD4 T cells.

Studies with drugs that block HIV replication indicate that the virus is replicating rapidly at all phases of infection, including the asymptomatic phase. Three viral proteins in particular have been the target of drugs aimed at arresting viral replication. These are viral reverse transcriptase, which is required for synthesis of the provirus; viral integrase, which is required for insertion of the viral provirus into the host genome; and viral protease, which cleaves viral polyproteins to produce virion proteins and viral enzymes. Reverse transcriptase is inhibited by nucleoside analogs such as zidovudine (AZT), which was the first anti-HIV drug to be licensed in the United States. Inhibitors of reverse transcriptase, integrase, and protease prevent infection of uninfected cells; cells that are already infected can continue to produce virions because, once the provirus is established, reverse transcriptase and integrase are not needed to make new virus particles, and while the viral protease acts at a very late maturation step of the virus, inhibition of the protease does not prevent virus from being released. However, in all cases, further cycles of infection by released virions are blocked, and replication is therefore prevented.

The introduction of combination therapy with a cocktail of viral protease inhibitors and nucleoside analogs, also known as **highly active antiretroviral therapy (HAART)**, dramatically reduced mortality and morbidity in patients with advanced HIV infection in the United States between 1995 and 1997 (**Fig. 13.37**). Many patients treated with HAART show a rapid and dramatic reduction in viremia, eventually maintaining levels of HIV RNA close to the limit of detection (50 copies per ml of plasma) for a long period (**Fig. 13.38**). It is unclear how the virus particles are removed so rapidly from the circulation after the initiation of HAART. It seems most likely that they are opsonized by specific antibody and complement and removed by cells of the mononuclear phagocyte system. Opsonized HIV particles may also be trapped in lymphoid follicles on the surface of follicular dendritic cells.

HAART is also accompanied by a slow but steady increase in CD4 T cells, despite the fact that many other compartments of the immune system remain compromised. Three complementary mechanisms have been established for the recovery in CD4 T-cell numbers. The first is a redistribution of CD4 T memory cells from lymphoid tissues into the circulation as viral replication

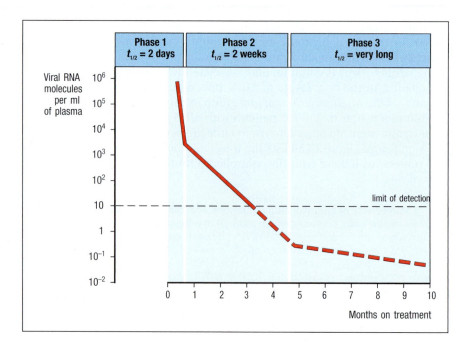

is controlled; this occurs within weeks of starting treatment. The second is a reduction in the abnormal levels of immune activation as HIV infection is controlled; this is associated with reduced killing of infected CD4 T cells by cytotoxic T lymphocytes. The third is much slower and is caused by the emergence of new naive T cells from the thymus, which is indicated by the presence of T-cell receptor excision circles (TRECs) in these later-arriving cells (see Section 5-9).

Although HAART is effective at inhibiting HIV replication, thereby preventing the progression to AIDS and greatly decreasing transmission by those infected, it is ineffective at eradicating all viral stores. Cessation of HAART therefore leads to a rapid rebound of virus multiplication, so that patients require treatment indefinitely. This, coupled with the side-effects and high cost of HAART, have stimulated investigation into other targets to block viral replication (**Fig. 13.39**) as well as of eliminating viral reservoirs to eradicate infection permanently. New classes of anti-HIV replication drugs include **viral entry inhibitors**, which block the binding of gp120 to CCR5 or block viral fusion by inhibiting gp41; and **viral integrase inhibitors**, which block the insertion of the reverse-transcribed viral genome into the host DNA. Another approach under development is to enhance the activity of HIV restriction factors, including APOBEC (see Section 13-27) and TRIM 5α. APOBEC causes extensive mutation of newly formed HIV cDNA to destroy its coding and replicative capacity, and TRIM 5α limits HIV-1 infections by targeting the viral nucleocapsid and preventing the uncoating and release of viral RNA after it enters cells.

Given the success of HAART in blocking active viral replication, the inability of existing therapies to purge reservoirs of latently infected cells has become the greatest barrier to a cure. To overcome this, strategies are being considered that would induce viral replication in latently infected cells in combination with measures to enhance immune clearance of virus and infected cells. Examples of ways to activate latent virus include the administration of cytokines that activate viral transcription and replication (for example, IL-2, IL-6, and TNF-α), or the use of agents that target epigenetic modifiers, such as histone deacetylase (HDAC) inhibitors, that can activate latent provirus. To date, however, no clinical trial using agents that target latent viral reservoirs has shown a significant reduction in viral load over that gained from HAART alone. Indeed, it was recently discovered that the activation of viral replication in latently infected cells is intrinsically stochastic, so that many immune cells

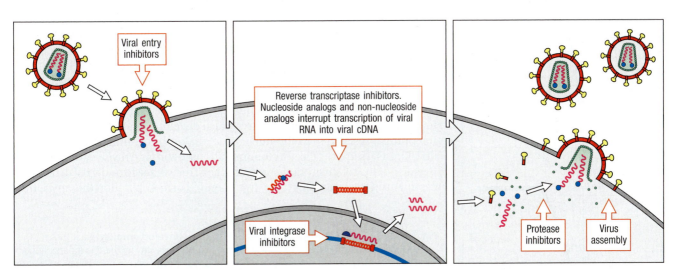

Fig. 3.39 Possible targets for interference with the HIV life cycle. In principle, HIV could be attacked by therapeutic drugs at multiple points in its life cycle: virus entry, reverse transcription of viral RNA, insertion of viral cDNA into cellular DNA by the viral integrase, cleavage of viral polyproteins by the viral protease, and assembly and budding of infectious virions. As yet, only drugs that inhibit reverse transcriptase and protease action have been developed. Combination therapy using different kinds of drugs is more effective than using a single drug.

harboring latent provirus will fail to activate viral replication in any given cycle of cellular activation. This adaptation of HIV to avoid elimination of latently infected cells could represent a formidable barrier to strategies aimed at 'flushing out' latent virus so that it can be eliminated.

An alternative strategy for cure has been highlighted in a single HIV patient who underwent hematopoietic stem-cell transplantation (HSCT) for treatment of leukemia in Berlin (hence referred to as the **Berlin patient**). By using a stem-cell donor who was homozygous for the *CCR5Δ32* co-receptor mutation, the patient was reconstituted with immune cells resistant to viral propagation. The patient's CD4 T-cell counts rebounded and he was found to be free of any evidence of HIV infection (or leukemia) following cessation of antiretroviral therapy post-transplant. He has remained so for more than 5 years, suggesting that he has been cured of infection. In view of the large number of infected individuals worldwide, the risk of complications with HSCT, and rarity of HLA-matched CCR5-deletant donors, this will never be a practical approach for cure at the population level. Moreover, there is a risk of progression or reinfection post-transplant by CXCR4-tropic viral variants. This outcome does, however, dramatically establish that eradication of a latency reservoir (in this case by inductive chemoirradiation therapy for leukemia) combined with blockade of viral replication—whether by genetic or therapy-mediated interventions—might achieve a permanent cure.

13-36 In the course of infection HIV accumulates many mutations, which can result in the outgrowth of drug-resistant variants.

The rapid replication of HIV, with the generation of 10^9 to 10^{10} virions every day, is coupled with a mutation rate of approximately 3×10^{-5} substitutions per nucleotide per cycle of replication, and thus leads to the generation of many variants of HIV in a single infected patient in the course of a day. This high mutation rate arises from the error-prone nature of retroviral replication and poses a formidable challenge to the immune system. Reverse transcriptase lacks the proofreading mechanisms of cellular DNA polymerases, and the RNA genomes of retroviruses are copied into DNA with relatively low fidelity. Thus, although primary infection is typically established by a single founder virus, numerous variants of HIV, called **quasi-species**, rapidly develop within an infected individual. This phenomenon was first recognized in HIV and has since proved to be common to all lentiviruses.

As a consequence of its high variability, HIV rapidly develops resistance to antiviral drugs, much as it develops escape mutants that evade T-cell recognition (see Section 13-31). When drug is administered, viral variants with mutations that confer resistance emerge and multiply until the previous levels of virus are regained. Resistance to some viral protease inhibitors requires only a single mutation and appears after only a few days (Fig. 13.40); resistance to some inhibitors of reverse transcriptase develops in a similarly short time. In contrast, resistance to the nucleoside analog zidovudine takes months to develop, as it requires three or four mutations in the viral reverse transcriptase. Because of the relatively rapid appearance of resistance to anti-HIV drugs, successful drug treatment has typically depended on combination therapy, where

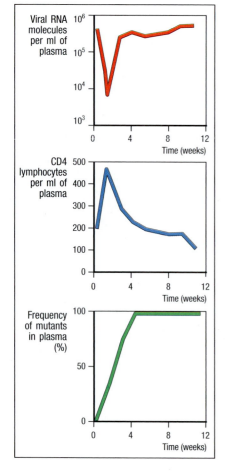

Fig. 13.40 Resistance of HIV to protease inhibitors develops rapidly. After the administration of a single protease inhibitor drug to a patient with HIV there is a precipitous fall in plasma levels of viral RNA, with a half-life of about 2 days (top panel). This is accompanied by an initial rise in the number of CD4 T cells in peripheral blood (center panel). Within days of starting the drug, mutant drug-resistant variants can be detected in plasma (bottom panel) and in peripheral blood lymphocytes. After only 4 weeks of treatment, viral RNA levels and CD4 lymphocyte levels have returned to their original pre-drug levels, and 100% of plasma HIV is present as the drug-resistant mutant.

the chance of simultaneous resistance mutations in multiple HIV proteins is virtually nil. Nevertheless, monotherapy with newer generation antiretroviral agents has proven effective in patients with low viral loads at the onset of treatment.

13-37 Vaccination against HIV is an attractive solution but poses many difficulties.

Although the effectiveness of HAART in restraining HIV replication has profoundly altered the natural history and transmission rates of HIV infection, a safe and effective vaccine for the prevention of HIV infection and AIDS remains the ultimate goal. Ideally, an effective vaccine would elicit both broadly neutralizing antibodies that block viral entry into target cells (that is, anti-gp120) and effective cytolytic T-cell responses, to both prevent and control HIV infection, respectively. However, no such vaccine has yet been developed, and its attainment is fraught with difficulties that have not been faced in the development of vaccines against other diseases.

The main problem is the nature of the infection itself, featuring a virus that directly undermines the central component of adaptive immunity—the CD4 T cell—and that proliferates and mutates extremely rapidly to cause sustained infection in the face of strong cytotoxic T-cell and antibody responses. The development of vaccines that could be administered to patients already infected, to boost immune responses and prevent progression to AIDS, has been considered, as have prophylactic vaccines that would prevent initial infection. The development of therapeutic vaccination in those already infected would be extremely difficult. As discussed in the previous section, HIV evolves in individual patients by the selective proliferation of mutant viruses that escape recognition by antibodies and cytotoxic T cells. The ability of the virus to persist in latent form as a transcriptionally silent provirus invisible to the immune system might also prevent even an immunized person from clearing an infection once it has been established.

There has been more hope for prophylactic vaccination to prevent new infection. But even here, the lack of protection of the normal immune response and the sheer scale of sequence diversity among HIV strains in the infected population—there are currently thousands of different HIV strains circulating in the human population—remain significant challenges. Patients infected with one strain of virus do not seem to be resistant to closely related strains, and cases of superinfection, where two strains simultaneously infect the same cell, have also been described. This is compounded by the intrinsic difficulty in generating broadly neutralizing antibodies against HIV envelope glycoproteins (see Section 13-31). Further, there remains uncertainty over what form protective immunity to HIV might take. It is now felt that induction of both effective antibody and T-cell responses will be required to achieve protective immunity, although which epitopes might provide the best targets and how best to induce them remain undefined. Finally, the time from conception to design to performance of complete clinical trials of vaccines against HIV takes years, slowing the rate of progress; to date, few major clinical vaccine trials have been completed, and those have failed.

However, against this pessimistic background, progress has been made and there remains hope that successful vaccines might yet be developed. Various strategies are being tried in an attempt to develop vaccines against HIV, varying among delivery of recombinant HIV proteins, plasmid DNA vaccination with HIV genes (see Section 16-30), delivery of HIV genes in viral vectors, or combinations thereof. Many successful vaccines against other viral diseases contain a live attenuated strain of the virus, which raises an immune response but does not cause disease (see Section 16-23). There are substantial difficulties in the development of live attenuated vaccines against HIV, not least the worry of recombination between vaccine strains and wild-type viruses that would lead to a reversion to virulence. An alternative approach is the use of other viruses, such as vaccinia or adenovirus, to deliver and express HIV genes that elicit B- and T-cell responses against HIV antigens. Because these viral vectors have already demonstrated safety in other human vaccination studies, they have been obvious choices for initial trials. Recently, there has been encouraging, albeit limited, success with this type of approach in combination with boosts using recombinant gp120. Delivery of HIV *gag*, *pol*, and *env* genes via a canarypox viral vector followed by boosts with HIV gp120 was shown to reduce the risk of infection in a modest but significant number of high-risk vaccine recipients. This represents the first demonstration of any degree of efficacy in a large HIV vaccine trial to date. Perhaps as important, data from this study provided insights into the type of immune response that correlated with protection, indicating that the induction of non neutralizing antibodies that elicit ADCC (for example, IgG3 isotype) might provide protection. Because neutralizing antibodies against HIV have proven so difficult to elicit, this provides encouragement that they might not be required. Further, a study that used a cytomegalovirus (CMV) vector to deliver SIV genes to rhesus monkeys showed that potent CTL responses were induced. Although these CTL responses did not prevent infection by a pathogenic SIV strain, they did result in clearance of virus in about half of the monkeys vaccinated after its systemic spread. This unprecedented result suggests that the viral vector used for delivery of HIV genes—in this case a vector that produces HIV antigens for prolonged periods after vaccination—might play an important role in the type and amplitude of the antiviral CD8 T-cell response elicited, and protection might be achieved by an effective T-cell response alone. Additional studies will be needed to determine whether combination vaccines that elicit the appropriate non neutralizing antibodies and robust CD8 T-cell responses might achieve protection even in the absence of neutralizing antibodies.

In addition to the biological obstacles to developing effective HIV vaccines, there are difficult ethical issues. It would be unethical to conduct a vaccine trial without trying at the same time to minimize the exposure of a vaccinated population to the virus itself. The effectiveness of a vaccine can, however, only be assessed in a population in which the exposure rate to the virus is high enough to assess whether vaccination protects against infection. This means that initial vaccine trials might have to be conducted in countries where the incidence of infection is very high and public health measures have not yet succeeded in reducing the spread of HIV.

13-38 Prevention and education are important in controlling the spread of HIV and AIDS.

The spread of HIV can be prevented if precautions are taken by those already infected and those who are uninfected but at risk for exposure. The advent of HAART represents a major advance in blocking the transmission of HIV from infected people due to its ability to greatly reduce viral titers in body fluids. However, most who are infected with HIV do not have access to HAART, as it is expensive and requires lifelong treatment, and many of those infected are unaware that they carry the virus. Even where HAART is unavailable, access to regular screening for those at risk is critical to inform those infected so they can take measures to avoid passing the virus to others. This, in turn, requires strict confidentiality and mutual trust. A barrier to the control of HIV is reluctance of individuals to find out whether they are infected, especially as one of the consequences of a positive HIV test is stigmatization by society. Here, education becomes an important component of the prevention strategy, both to remove the stigma and to provide guidance on how transmission of the virus can be prevented.

Precautions that can be taken by uninfected individuals are relatively inexpensive and involve measures to protect against contact with body fluids, such as semen, blood, blood products, or milk, from people who are infected. It has been repeatedly demonstrated that this is sufficient to prevent infection, as exemplified by healthcare workers who care for AIDS patients for long periods without seroconversion or signs of infection. The routine use of condoms greatly reduces the risk of HIV transmission, as does restraint from breastfeeding by infected mothers of newborns. Male circumcision also reduces transmission rates, as the foreskin is a major site of viral entry in uncircumcised males. Additional measures that have been considered include the use of microbicidal gels or suppositories, improvements in which have led to products that show promise in recent trials. Some of these agents can also reduce the transmission of other sexually transmitted diseases (for example, genital herpes) that increase the risk of HIV transmission. Finally, there is increasing interest in the prophylactic use of antiretroviral drugs (referred to as pre-exposure prophylaxis, or PrEP), which are administered either topically or orally to individuals at high risk for contracting HIV. To date, two reverse transcriptase inhibitors have shown efficacy in trials, and combined daily use of both drugs taken orally has demonstrated over 90% reduction in the risk of HIV infection. Moreover, use of antiretroviral therapy immediately post-exposure—for example, in hospital workers exposed to contaminated blood by accidental needlestick—substantially reduces the risk of acquiring HIV. One concern with this approach is the risk of developing drug resistance in those who do contract HIV while on PrEP, particularly in individuals with poor adherence to the dosing regimen. Although the significance of this risk has yet to be established, it remains an issue. Nevertheless, testing of new PrEP strategies based on additional antiretrovirals or of long-acting formulations that reduce the risk of poor compliance represents areas of considerable promise.

Summary.

Infection with the human immunodeficiency virus (HIV) is the cause of acquired immune deficiency syndrome (AIDS). Although substantial gains in curbing the rate of spread of HIV have been made, this worldwide epidemic continues to spread, especially through heterosexual contact in less-developed countries. HIV is an enveloped retrovirus that replicates in cells of the immune system. Viral entry requires the presence of CD4 and a particular chemokine receptor, and the viral cycle is dependent on transcription factors found in activated T cells. Infection with HIV causes a loss of CD4 T cells and an acute viremia that rapidly subsides as cytotoxic T-cell responses develop, but HIV infection is not eliminated by this immune response. HIV establishes a state of persistent infection in which the virus is continually replicating in newly infected cells. Current treatment consists of combinations of antiviral drugs that block viral replication and cause a rapid decrease in virus levels and a slow increase in CD4 T-cell counts. The main effect of HIV infection is the destruction of CD4 T cells, which occurs through the direct cytopathic effects of HIV infection and through killing by CD8 cytotoxic T cells. As CD4 T-cell counts wane, the body becomes progressively more susceptible to opportunistic infection. Eventually, most untreated HIV-infected individuals develop AIDS and die; however, a small minority, so-called long-term nonprogressors, remain healthy for many years with no apparent ill effects of infection. We hope to be able to learn from these people how infection with HIV can be controlled. The existence of these people, and of others who seem to have been naturally immunized against infection, gives hope that it will be possible to develop effective vaccines against HIV.

Summary to Chapter 13.

Whereas most infections elicit protective immunity, successful pathogens have developed some means of at least partly resisting the immune response and can cause serious, sometimes persistent, disease. Some individuals have inherited deficiencies in different components of the immune system, making them highly susceptible to certain classes of infectious agents. Persistent infections and inherited immunodeficiency diseases illustrate the importance of innate and adaptive immunity in effective host defense, and present ongoing challenges for immunological research. The human immunodeficiency virus (HIV), which leads to acquired immune deficiency syndrome (AIDS), combines the characteristics of a persistent infectious agent with the ability to create immunodeficiency in its human host, a combination that is usually slowly lethal to the patient. The key to fighting new pathogens such as HIV is to increase our understanding of the basic properties of the immune system and its role in combating infection.

Questions.

13.1 Matching: Match the following gene defects with the associated primary immunodeficiency.

___ **A.** Common γ chain mutations

    i. Omenn syndrome

___ **B.** RAG1 or RAG2 hypomorphic mutations

    ii. SCID associated with abnormal thymic development

___ **C.** Defects in DNA-PKcs or Artemis

    iii. X-linked SCID

___ **D.** FOXN1 mutations

    iv. Autoimmune polyendocrinopathy-candidiasis-ectodermal dystrophy

___ **E.** Mutations in TAP1 or TAP2

    v. MHC I deficiency

___ **F.** Defects in AIRE

    vi. Radiation-sensitive SCID

13.2 True or False: Individuals with mutations in the genes encoding the IL-12 p40 subunit are susceptible not only to pathogens such as *M. tuberculosis* that require a T_H1 response, but type 3 (T_H17) responses are also affected.

13.3 Short Answer: Name two genetic defects that lead to the absence of CD8+ T cells with CD4+ T cells preserved, and one genetic defect that leads to the absence of CD4+ T cells with CD8+ T cells preserved.

13.4 Short Answer: Both CD40L deficiency and AID deficiency cause hyper-IgM syndrome, but T-cell function is severely impaired in CD40L deficiency and preserved in AID deficiency. Why?

13.5 True or False: Common variable immunodeficiency (CVID) severely impairs both T-cell and antibody responses.

13.6 Multiple Choice: Which of the following hereditary immune disorders does not have an autoimmune or autoinflammatory phenotype?

A. Autoimmune polyendocrinopathy-candidiasis-ectodermal dystrophy (APECED), caused by defects in *AIRE*

B. Familial Mediterranean fever (FMF), caused by pyrin mutations

C. Ommen syndrome, caused by *RAG1* or *RAG2* hypomorphic mutations

D. Wiskott–Aldrich syndrome (WAS), caused by WAS deficiency

E. Hyper-IgE syndrome (also called Job's syndrome), caused by STAT3 or DOCK8 mutations

F. Chronic granulomatous disease (CGD), caused by production of reactive oxygen species in phagocytes

13.7 Multiple Choice: Pyogenic bacteria are protected by polysaccharide capsules against recognition by receptors on macrophages and neutrophils. Antibody-dependent opsonization is one of the mechanisms utilized by phagocytes to ingest and destroy these bacteria. Which of the following diseases or deficiencies directly affects a mechanism by which the immune system controls infection by these pathogens?

A. IL-12 p40 deficiency

B. Defects in *AIRE*

C. WASp deficiency

D. Defects in C3

13.8 Multiple Choice: Defects in which of the following genes have a phenotype similar to defects in *ELA2*, the gene that encodes neutrophil elastase?

A. *GFI1*

B. *CD55* (encodes DAF)

C. *CD59*

D. *XIAP*

13.9 Matching: Match each protein to the associated phagocytic cell function.

A. Kindlin-3   i. Production

B. Neutrophil elastase   ii. Adhesion

C. Myeloperoxidase   iii. Activation

D. MyD88   iv. Microbial killing

13.10 Multiple Choice: Which of the following pathogens primarily evade(s) the immune system by antigenic variation?

A. Influenza A virus

B. Herpes simplex virus-1

C. Cytomegalovirus

D. *Trypanosoma brucei*

E. *Plasmodium falciparum*

F. Hepatitis B virus

13.11 Human immunodeficiency virus (HIV) produces various immunoevasins. One of these, Nef, is exceptionally pleiotropic and a major target of CD8+ T-cell responses. Which of the following is not one of the functions of Nef?

A. Inhibition of the restriction factor SAMHD1

B. MHC class I downregulation

C. CD4 downregulation

D. MHC class II downregulation

E. Sustaining T-cell activation

13.12 Fill-in-the-Blanks: Human immunodeficiency virus (HIV) is a retrovirus that is classified as such because it contains a _____ enzyme. It infects host cells via its envelope binding to the _____ receptor and either the _____ or _____ co-receptor. When an individual becomes infected, they mount an immune response that results in the production of anti-HIV antibodies, a process called _____. CD8+ T-cell responses are also generated, but HIV can acquire _____ that allow it to evade recognition by these CTLs.

13.13 Multiple Choice: Which of the following is not a genetic variant that reduces susceptibility to HIV infection or slows protections to AIDS?

A. Mutant CCR5 allele

B. Mutant CXCR4 allele

C. Certain HLA class I alleles

D. Possessing KIR3DS1 with certain HLA-B alleles

General references.

Alcami, A., and Koszinowski, U.H.: **Viral mechanisms of immune evasion.** *Immunol. Today* 2000, **21**:447–455.

De Cock, K.M., Mbori-Ngacha, D., and Marum, E.: **Shadow on the continent: public health and HIV/AIDS in Africa in the 21st century.** *Lancet* 2002, **360**:67–72.

Finlay, B.B., and McFadden, G.: **Anti-immunology: evasion of the host immune system by bacterial and viral pathogens.** *Cell* 2006, **124**:767–782.

Hill, A.V.: **The immunogenetics of human infectious diseases.** *Annu. Rev. Immunol.* 1998, **16**:593–617.

Lederberg, J.: **Infectious history.** *Science* 2000, **288**:287–293.

Notarangelo, L.D.: **Primary immunodeficiencies.** *J. Allergy Clin. Immunol.* 2010, **125**:S182–S194.

Xu, X.N., Screaton, G.R., and McMichael, A.J.: **Virus infections: escape, resistance, and counterattack.** *Immunity* 2001, **15**:867–870.

Section references.

13-1 A history of repeated infections suggests a diagnosis of immunodeficiency.

Carneiro-Sampaio, M., and Coutinho, A.: **Immunity to microbes: lessons from primary immunodeficiencies.** *Infect. Immun.* 2007, **75**:1545–1555.

Cunningham-Rundles, C., and Ponda, P.P.: **Molecular defects in T- and B-cell primary immunodeficiency diseases.** *Nat. Rev. Immunol.* 2005, **5**:880–892.

13-2 Primary immunodeficiency diseases are caused by inherited gene defects.

Bolze, A., Mahlaoui, N., Byun, M., Turner, B., Trede, N., Ellis, S.R., Abhyankar, A., Itan, Y., Patin, E., Brebner, S., *et al.*: **Ribosomal protein SA haploinsufficiency in humans with isolated congenital asplenia.** *Science* 2013, **340**:976–978.

Cunningham-Rundles, C., and Ponda, P.P.: **Molecular defects in T- and B-cell primary immunodeficiency diseases.** *Nat. Rev. Immunol.* 2005, **5**:880–892.

Kokron, C.M., Bonilla, F.A., Oettgen, H.C., Ramesh, N., Geha, R.S., and Pandolfi, F.: **Searching for genes involved in the pathogenesis of primary immunodeficiency diseases: lessons from mouse knockouts.** *J. Clin. Immunol.* 1997, **17**:109–126.

Koss, M., Bolze, A., Brendolan, A., Saggese, M., Capellini, T.D., Bojilova, E., Boisson, B., Prall, O.W.J., Elliott, D.A., Solloway, M., *et al.*: **Congenital asplenia in mice and humans with mutations in a Pbx/Nkx2-5/p15 module.** *Dev. Cell* 2012, **22**:913–926.

Marodi, L., and Notarangelo, L.D.: **Immunological and genetic bases of new primary immunodeficiencies.** *Nat. Rev. Immunol.* 2007, **7**:851–861.

13-3 Defects in T-cell development can result in severe combined immunodeficiencies.

Buckley, R.H., Schiff, R.I., Schiff, S.E., Markert, M.L., Williams, L.W., Harville, T.O., Roberts, J.L., and Puck, J.M.: **Human severe combined immunodeficiency: genetic, phenotypic, and functional diversity in one hundred eight infants.** *J. Pediatr.* 1997, **130**:378–387.

Leonard, W.J.: **The molecular basis of X linked severe combined immuno-deficiency.** *Annu. Rev. Med.* 1996, **47**:229–239.

Leonard, W.J.: **Cytokines and immunodeficiency diseases.** *Nat. Rev. Immunol.* 2001, **1**:200–208.

Stephan, J.L., Vlekova, V., Le Deist, F., Blanche, S., Donadieu, J., De Saint-Basile, G., Durandy, A., Griscelli, C., and Fischer, A.: **Severe combined immuno-deficiency: a retrospective single-center study of clinical presentation and outcome in 117 patients.** *J. Pediatr.* 1993, **123**:564–572.

13-4 SCID can also be due to defects in the purine salvage pathway.

Hirschhorn, R.: **Adenosine deaminase deficiency: molecular basis and recent developments.** *Clin. Immunol. Immunopathol.* 1995, **76**:S219–S227.

13-5 Defects in antigen receptor gene rearrangement can result in SCID.

Bosma, M.J., and Carroll, A.M.: **The SCID mouse mutant: definition, charac-terization, and potential uses.** *Annu. Rev. Immunol.* 1991, **9**:323–350.

Fugmann, S.D.: **DNA repair: breaking the seal.** *Nature* 2002, **416**:691–694.

Gennery, A.R., Cant, A.J., and Jeggo, P.A.: **Immunodeficiency associated with DNA repair defects.** *Clin. Exp. Immunol.* 2000, **121**:1–7.

Moshous, D., Callebaut, I., de Chasseval, R., Corneo, B., Cavazzana-Calvo, M., Le Deist, F., Tezcan, I., Sanal, O., Bertrand, Y., Philippe, N., *et al.*: **Artemis, a novel DNA double-strand break repair/V(D)J recombination protein, is mutated in human severe combined immune deficiency.** *Cell* 2001, **105**:177–186.

13-6 Defects in signaling from T-cell antigen receptors can cause severe immunodeficiency.

Castigli, E., Pahwa, R., Good, R.A., Geha, R.S., and Chatila, T.A.: **Molecular basis of a multiple lymphokine deficiency in a patient with severe combined immu-nodeficiency.** *Proc. Natl Acad. Sci. USA* 1993, **90**:4728–4732.

Kung, C., Pingel, J.T., Heikinheimo, M., Klemola, T., Varkila, K., Yoo, L.I., Vuopala, K., Poyhonen, M., Uhari, M., Rogers, M., *et al.*: **Mutations in the tyrosine phos-phatase CD45 gene in a child with severe combined immunodeficiency dis-ease.** *Nat. Med.* 2000, **6**:343–345.

Roifman, C.M., Zhang, J., Chitayat, D., and Sharfe, N.: **A partial deficiency of interleukin-7R α is sufficient to abrogate T-cell development and cause severe combined immunodeficiency.** *Blood* 2000, **96**:2803–2807.

13-7 Genetic defects in thymic function that block T-cell development result in severe immunodeficiencies.

Coffer, P.J., and Burgering, B.M.: **Forkhead-box transcription factors and their role in the immune system.** *Nat. Rev. Immunol.* 2004, **4**:889–899.

DiSanto, J.P., Keever, C.A., Small, T.N., Nicols, G.L., O'Reilly, R.J., and Flomenberg, N.: **Absence of interleukin 2 production in a severe combined immunodefi-ciency disease syndrome with T cells.** *J. Exp. Med.* 1990, **171**:1697–1704.

DiSanto, J.P., Rieux Laucat, F., Dautry Varsat, A., Fischer, A., and de Saint Basile, G.: **Defective human interleukin 2 receptor γ chain in an atypical X chromo-some-linked severe combined immunodeficiency with peripheral T cells.** *Proc. Natl Acad. Sci. USA* 1994, **91**:9466–9470.

Gadola, S.D., Moins-Teisserenc, H.T., Trowsdale, J., Gross, W.L., and Cerundolo, V.: **TAP deficiency syndrome.** *Clin. Exp. Immunol.* 2000, **121**:173–178.

Gilmour, K.C., Fujii, H., Cranston, T., Davies, E.G., Kinnon, C., and Gaspar, H.B.: **Defective expression of the interleukin-2/interleukin-15 receptor β subunit leads to a natural killer cell-deficient form of severe combined immunodefi-ciency.** *Blood* 2001, **98**:877–879.

Grusby, M.J., and Glimcher, L.H.: **Immune responses in MHC class II-deficient mice.** *Annu. Rev. Immunol.* 1995, **13**:417–435.

Pignata, C., Gaetaniello, L., Masci, A.M., Frank, J., Christiano, A., Matrecano, E., and Racioppi, L.: **Human equivalent of the mouse Nude/SCID phenotype: long-term evaluation of immunologic reconstitution after bone marrow transplan-tation.** *Blood* 2001, **97**:880–885.

Steimle, V., Reith, W., and Mach, B.: **Major histocompatibility complex class II deficiency: a disease of gene regulation.** *Adv. Immunol.* 1996, **61**:327–340.

13-8 Defects in B-cell development result in deficiencies in antibody production that cause an inability to clear extracellular bacteria and some viruses.

Bruton, O.C.: **Agammaglobulinemia.** *Pediatrics* 1952, **9**:722–728.

Conley, M.E.: **Genetics of hypogammaglobulinemia: what do we really know?** *Curr. Opin. Immunol.* 2009, **21**:466–471.

Lee, M.L., Gale, R.P., and Yap, P.L.: **Use of intravenous immunoglobulin to prevent or treat infections in persons with immune deficiency.** *Annu. Rev. Med.* 1997, **48**:93–102.

Notarangelo, L.D.: **Immunodeficiencies caused by genetic defects in protein kinases.** *Curr. Opin. Immunol.* 1996, **8**:448–453.

Preud'homme, J.L., and Hanson, L.A.: **IgG subclass deficiency.** *Immunodefic. Rev.* 1990, **2**:129–149.

13-9 Immune deficiencies can be caused by defects in B-cell or T-cell activation and function that lead to abnormal antibody responses.

Burrows, P.D., and Cooper, M.D.: **IgA deficiency.** *Adv. Immunol.* 1997, **65**:245–276.

Doffinger, R., Smahi, A., Bessia, C., Geissmann, F., Feinberg, J., Durandy, A., Bodemer, C., Kenwrick, S., Dupuis-Girod, S., Blanche, S., *et al.*: **X-linked anhidrotic ectodermal dysplasia with immunodeficiency is caused by impaired NF-κB signaling.** *Nat. Genet.* 2001, **27**:277–285.

Durandy, A., and Honjo, T.: **Human genetic defects in class-switch recombi-nation (hyper-IgM syndromes).** *Curr. Opin. Immunol.* 2001, **13**:543–548.

Ferrari, S., Giliani, S., Insalaco, A., Al Ghonaium, A., Soresina, A.R., Loubser, M., Avanzini, M.A., Marconi, M., Badolato, R., Ugazio, A.G., *et al.*: **Mutations of CD40 gene cause an autosomal recessive form of immunodeficiency with hyper IgM.** *Proc. Natl Acad. Sci. USA* 2001, **98**:12614–12619.

Harris, R.S., Sheehy, A.M., Craig, H.M., Malim, M.H., and Neuberger, M.S.: **DNA deamination: not just a trigger for antibody diversification but also a mecha-nism for defense against retroviruses.** *Nat. Immunol.* 2003, **4**:641–643.

Minegishi, Y.: **Hyper-IgE syndrome.** *Curr. Opin. Immunol.* 2009, **21**:487–492.

Park, M.A., Li, J.T., Hagan, J.B., Maddox, D.E., and Abraham, R.S.: **Common variable immunodeficiency: a new look at an old disease.** *Lancet* 2008, **372**:489–503.

Thrasher, A.J., and Burns, S.O.: **WASP: a key immunological multitasker.** *Nat. Rev. Immunol.* 2010, **10**:182–192.

Yel, L.: **Selective IgA deficiency.** *J. Clin. Immunol.* 2010, **30**:10–16.

Yong, P.F., Salzer, U., and Grimbacher, B.: **The role of costimulation in anti-body deficiencies: ICOS and common variable immunodeficiency.** *Immunol. Rev.* 2009, **229**:101–113.

13-10 Normal pathways for host defense against different infectious agents are pinpointed by genetic deficiencies of cytokine pathways central to type 1/T$_H$1 and type 3/T$_H$17 responses.

Browne, S.K.: **Anticytokine autoantibody-associated immunodeficiency.** *Annu. Rev. Immunol.* 2014, **32**:635–657.

Casanova, J.L., and Abel, L.: **Genetic dissection of immunity to mycobacte-ria: the human model.** *Annu. Rev. Immunol.* 2002, **20**:581–620.

Dupuis, S., Dargemont, C., Fieschi, C., Thomassin, N., Rosenzweig, S., Harris, J., Holland, S.M., Schreiber, R.D., and Casanova, J.L.: **Impairment of mycobacterial but not viral immunity by a germline human STAT1 mutation.** *Science* 2001, **293**:300–303.

Lammas, D.A., Casanova, J.L., and Kumararatne, D.S.: **Clinical consequences of defects in the IL-12-dependent interferon-γ (IFN-γ) pathway.** *Clin. Exp. Immunol.* 2000, **121**:417–425.

Lanternier, F., Cypowyj, S., Picard, C., Bustamante, J., Lortholary, O., Casanova, J.-L., and Puel, A.: **Primary immunodeficiencies underlying fungal infections.** *Curr. Opin. Pediatr.* 2013, **25**:736–747.

Lanternier, F., Pathan, S., Vincent, Q.B., Liu, L., Cypowyj, S., Prando, C., Migaud, M., Taibi, L., Ammar-Khodja, A., Boudghene Stambouli, O., *et al.*: **Deep dermato-phytosis and inherited CARD9 deficiency.** *N. Engl. J. Med.* 2013, **369**:1704–1714.

Newport, M.J., Huxley, C.M., Huston, S., Hawrylowicz, C.M., Oostra, B.A., Williamson, R., and Levin, M.: **A mutation in the interferon-γ-receptor gene and susceptibility to mycobacterial infection.** *N. Engl. J. Med.* 1996, **335**:1941–1949.

Puel, A., Döffinger, R., Natividad, A., Chrabieh, M., Barcenas-Morales, G., Picard, C., Cobat, A., Ouachée-Chardin, M., Toulon, A., Bustamante, J., *et al.*: **Autoantibodies against IL-17A, IL-17F, and IL-22 in patients with chronic mucocutaneous candidiasis and autoimmune polyendocrine syndrome type I.** *J. Exp. Med.* 2010, **207**:291–297.

Van de Vosse, E., Hoeve, M.A., and Ottenhoff, T.H.: **Human genetics of intracellular infectious diseases: molecular and cellular immunity against mycobacteria and salmonellae.** *Lancet Infect. Dis.* 2004, **4**:739–749.

13-11 Inherited defects in the cytolytic pathway of lymphocytes can cause uncontrolled lymphoproliferation and inflammatory responses to viral infections.

de Saint Basile, G., Ménasché, G., and Fischer, A.: **Molecular mechanisms of biogenesis and exocytosis of cytotoxic granules.** *Nat. Rev. Immunol.* 2010, **10**:568–579.

de Saint Basille, G., and Fischer, A.: **The role of cytotoxicity in lymphocyte homeostasis.** *Curr. Opin. Immunol.* 2001, **13**:549–554.

13-12 X-linked lymphoproliferative syndrome is associated with fatal infection by Epstein–Barr virus and with the development of lymphomas.

Latour, S., Gish, G., Helgason, C.D., Humphries, R.K., Pawson, T., and Veillette, A.: **Regulation of SLAM-mediated signal transduction by SAP, the X-linked lymphoproliferative gene product.** *Nat. Immunol.* 2001, **2**:681–690.

Morra, M., Howie, D., Grande, M.S., Sayos, J., Wang, N., Wu, C., Engel, P., and Terhorst, C.: **X-linked lymphoproliferative disease: a progressive immunodeficiency.** *Annu. Rev. Immunol.* 2001, **19**:657–682.

Rigaud, S., Fondaneche, M.C., Lambert, N., Pasquier, B., Mateo, V., Soulas, P., Galicier, L., Le Deist, F., Rieux-Laucat, F., Revy, P., *et al.*: **XIAP deficiency in humans causes an X-linked lymphoproliferative syndrome.** *Nature* 2006, **444**:110–114.

13-13 Immunodeficiency is caused by inherited defects in the development of dendritic cells.

Collin, M., Bigley, V., Haniffa, M., and Hambleton, S.: **Human dendritic cell deficiency: the missing ID?** *Nat. Rev. Immunol.* 2011, **11**:575–583.

Hambleton, S., Salem, S., Bustamante, J., Bigley, V., Boisson-Dupuis, S., Azevedo, J., Fortin, A., Haniffa, M., Ceron-Gutierrez, L., Bacon, C.M., *et al.*: **IRF8 mutations and human dendritic-cell immunodeficiency.** *N. Engl. J. Med.* 2011, **365**:127–138.

13-14 Defects in complement components and complement-regulatory proteins cause defective humoral immune function and tissue damage.

Colten, H.R., and Rosen, F.S.: **Complement deficiencies.** *Annu. Rev. Immunol.* 1992, **10**:809–834.

Dahl, M., Tybjaerg-Hansen, A., Schnohr, P., and Nordestgaard, B.G.: **A population-based study of morbidity and mortality in mannose-binding lectin deficiency.** *J. Exp. Med.* 2004, **199**:1391–1399.

Walport, M.J.: **Complement. First of two parts.** *N. Engl. J. Med.* 2001, **344**:1058–1066.

Walport, M.J.: **Complement. Second of two parts.** *N. Engl. J. Med.* 2001, **344**:1140–1144.

13-15 Defects in phagocytic cells permit widespread bacterial infections.

Andrews, T., and Sullivan, K.E.: **Infections in patients with inherited defects in phagocytic function.** *Clin. Microbiol. Rev.* 2003, **16**:597–621.

Etzioni, A.: **Genetic etiologies of leukocyte adhesion defects.** *Curr. Opin. Immunol.* 2009, **21**:481–486.

Fischer, A., Lisowska Grospierre, B., Anderson, D.C., and Springer, T.A.: **Leukocyte adhesion deficiency: molecular basis and functional consequences.** *Immunodefic. Rev.* 1988, **1**:39–54.

Goldblatt, D., and Thrasher, A.J.: **Chronic granulomatous disease.** *Clin. Exp. Immunol.* 2000, **122**:1–9.

Klein, C., and Welte, K.: **Genetic insights into congenital neutropenia.** *Clin. Rev. Allergy Immunol.* 2010, **38**:68–74.

Netea, M.G., Wijmenga, C., and O'Neill, L.A.J.: **Genetic variation in Toll-like receptors and disease susceptibility.** *Nat. Immunol.* 2012, **13**:535–542.

Spritz, R.A.: **Genetic defects in Chediak–Higashi syndrome and the beige mouse.** *J. Clin. Immunol.* 1998, **18**:97–105.

Suhir, H., and Etzioni, A.: **The role of Toll-like receptor signaling in human immunodeficiencies.** *Clin. Rev. Allergy Immunol.* 2010, **38**:11–19.

13-16 Mutations in the molecular regulators of inflammation can cause uncontrolled inflammatory responses that result in 'autoinflammatory disease.'

Delpech, M., and Grateau, G.: **Genetically determined recurrent fevers.** *Curr. Opin. Immunol.* 2001, **13**:539–542.

Dinarello, C.A.: **Immunological and inflammatory functions of the interleukin-1 family.** *Annu. Rev. Immunol.* 2009, **27**:519–550.

Drenth, J.P., and van der Meer, J.W.: **Hereditary periodic fever.** *N. Engl. J. Med.* 2001, **345**:1748–1757.

Kastner, D.L., and O'Shea, J.J.: **A fever gene comes in from the cold.** *Nat. Genet.* 2001, **29**:241–242.

Stehlik, C., and Reed, J.C.: **The PYRIN connection: novel players in innate immunity and inflammation.** *J. Exp. Med.* 2004, **200**:551–558.

13-17 Hematopoietic stem cell transplantation or gene therapy can be useful to correct genetic defects.

Fischer, A., Hacein-Bey, S., and Cavazzana-Calvo, M.: **Gene therapy of severe combined immunodeficiencies.** *Nat. Rev. Immunol.* 2002, **2**:615–621.

Fischer, A., Le Deist, F., Hacein-Bey-Abina, S., Andre-Schmutz, I., de Saint, B.G., de Villartay, J.P., and Cavazzana-Calvo, M.: **Severe combined immunodeficiency. A model disease for molecular immunology and therapy.** *Immunol. Rev.* 2005, **203**:98–109.

Hacein-Bey-Abina, S., Le Deist, F., Carlier, F., Bouneaud, C., Hue, C., De Villartay, J.P., Thrasher, A.J., Wulffraat, N., Sorensen, R., Dupuis-Girod, S., *et al.*: **Sustained correction of X-linked severe combined immunodeficiency by *ex vivo* gene therapy.** *N. Engl. J. Med.* 2002, **346**:1185–1193.

Hacein-Bey-Abina, S., Von Kalle, C., Schmidt, M., McCormack, M.P., Wulffraat, N., Leboulch, P., Lim, A., Osborne, C.S., Pawliuk, R., Morillon, E., *et al.*: **LMO2-associated clonal T cell proliferation in two patients after gene therapy for SCID-X1.** *Science* 2003, **302**:415–419.

Rosen, F.S.: **Successful gene therapy for severe combined immunodeficiency.** *N. Engl. J. Med.* 2002, **346**:1241–1243.

13-18 Noninherited, secondary immunodeficiencies are major predisposing causes of infection and death.

Chandra, R.K.: **Nutrition, immunity and infection: from basic knowledge of dietary manipulation of immune responses to practical application of ameliorating suffering and improving survival.** *Proc. Natl Acad. Sci. USA* 1996, **93**:14304–14307.

Lord, G.M., Matarese, G., Howard, J.K., Baker, R.J., Bloom, S.R., and Lechler, R.I.: **Leptin modulates the T-cell immune response and reverses starvation-induced immunosuppression.** *Nature* 1998, **394**:897–901.

13-19 Extracellular bacterial pathogens have evolved different strategies to avoid detection by pattern recognition receptors and destruction by antibody, complement, and antimicrobial peptides.

Bhavsar, A.P., Guttman, J.A., and Finlay, B.B.: **Manipulation of host-cell pathways by bacterial pathogens.** *Nature* 2007, **449**:827–834.

Blander, J.M., and Sander, L.E.: **Beyond pattern recognition: five immune checkpoints for scaling the microbial threat.** *Nat. Rev. Immunol.* 2012, **12**:215–225.

Hajishengallis, G., and Lambris, J.D.: **Microbial manipulation of receptor crosstalk in innate immunity.** *Nat. Rev. Immunol.* 2011, **11**:187–200.

Hornef, M.W., Wick, M.J., Rhen, M., and Normark, S.: **Bacterial strategies for overcoming host innate and adaptive immune responses.** *Nat. Immunol.* 2002, **3**:1033–1040.

Lambris, J.D., Ricklin, D., and Geisbrecht, B.V.: **Complement evasion by human pathogens.** *Nat. Rev. Microbiol.* 2008, **6**:132–142.

Phillips, R.E.: **Immunology taught by Darwin.** *Nat. Immunol.* 2002, **3**:987–989.

Raymond, B., Young, J.C., Pallett, M., Endres, R.G., Clements, A., and Frankel, G.: **Subversion of trafficking, apoptosis, and innate immunity by type III secretion system effectors.** *Trends Microbiol.* 2013, **21**:430–441.

Vance, R.E., Isberg, R.R., and Portnoy, D.A.: **Patterns of pathogenesis: discrimination of pathogenic and nonpathogenic microbes by the innate immune system.** *Cell Host Microbe* 2009, **6**:10–21.

Yeaman, M.R., and Yount, N.Y.: **Mechanisms of antimicrobial peptide action and resistance.** *Pharmacol. Rev.* 2003, **55**:27–55.

13-20 Intracellular bacterial pathogens can evade the immune system by seeking shelter within phagocytes.

Cambier, C.J., Takaki, K.K., Larson, R.P., Hernandez, R.E., Tobin, D.M., Urdahl, K.B., Cosma, C.L., and Ramakrishnan, L.: **Mycobacteria manipulate macrophage recruitment through coordinated use of membrane lipids.** *Nature* 2014, **505**:218–222.

Clegg, S., Hancox, L.S., and Yeh, K.S.: *Salmonella typhimurium* **fimbrial phase variation and FimA expression.** *J. Bacteriol.* 1996, **178**:542–545.

Cossart, P.: **Host/pathogen interactions. Subversion of the mammalian cell cytoskeleton by invasive bacteria.** *J. Clin. Invest.* 1997, **99**:2307–2311.

Young, D., Hussell, T., and Dougan, G.: **Chronic bacterial infections: living with unwanted guests.** *Nat. Immunol.* 2002, **3**:1026–1032.

13-21 Immune evasion is also practiced by protozoan parasites.

Donelson, J.E., Hill, K.L., and El-Sayed, N.M.: **Multiple mechanisms of immune evasion by African trypanosomes.** *Mol. Biochem. Parasitol.* 1998, **91**:51–66.

Sacks, D., and Sher, A.: **Evasion of innate immunity by parasitic protozoa.** *Nat. Immunol.* 2002, **3**:1041–1047.

13-22 RNA viruses use different mechanisims of antigenic variation to keep a step ahead of the adaptive immune system.

Bowie, A.G., and Unterholzner, L.: **Viral evasion and subversion of pattern-recognition receptor signalling.** *Nat. Rev. Immunol.* 2008, **8**:911–922.

Brander, C., and Walker, B.D.: **Modulation of host immune responses by clinically relevant human DNA and RNA viruses.** *Curr. Opin. Microbiol.* 2000, **3**:379–386.

Gibbs, M.J., Armstrong, J.S., and Gibbs, A.J.: **Recombination in the hemagglutinin gene of the 1918 'Spanish flu.'** *Science* 2001, **293**:1842–1845.

Hatta, M., Gao, P., Halfmann, P., and Kawaoka, Y.: **Molecular basis for high virulence of Hong Kong H5N1 influenza A viruses.** *Science* 2001, **293**:1840–1842.

Hilleman, M.R.: **Strategies and mechanisms for host and pathogen survival in acute and persistent viral infections.** *Proc. Natl Acad. Sci. USA* 2004, **101**:14560–14566.

Laver, G., and Garman, E.: **Virology. The origin and control of pandemic influenza.** *Science* 2001, **293**:1776–1777.

13-23 DNA viruses use multiple mechanisms to subvert NK and CTL cell responses.

Alcami, A.: **Viral mimicry of cytokines, chemokines and their receptors.** *Nat. Rev. Immunol.* 2003, **3**:36–50.

Hansen, T.H., and Bouvier, M.: **MHC class I antigen presentation: learning from viral evasion strategies.** *Nat. Rev. Immunol.* 2009, **9**:503–513.

McFadden, G., and Murphy, P.M.: **Host-related immunomodulators encoded by poxviruses and herpesviruses.** *Curr. Opin. Microbiol.* 2000, **3**:371–378.

Paludan, S.R., Bowie, A.G., Horan, K.A., and Fitzgerald, K.A.: **Recognition of herpesviruses by the innate immune system.** *Nat. Rev. Immunol.* 2011, **11**:143–154.

Yewdell, J.W., and Hill, A.B.: **Viral interference with antigen presentation.** *Nat. Immunol.* 2002, **2**:1019–1025.

13-24 Some latent viruses persist *in vivo* by ceasing to replicate until immunity wanes.

Cohen, J.I.: **Epstein–Barr virus infection.** *N. Engl. J. Med.* 2000, **343**:481–492.

Hahn, G., Jores, R., and Mocarski, E.S.: **Cytomegalovirus remains latent in a common precursor of dendritic and myeloid cells.** *Proc. Natl Acad. Sci. USA* 1998, **95**:3937–3942.

Ho, D.Y.: **Herpes simplex virus latency: molecular aspects.** *Prog. Med. Virol.* 1992, **39**:76–115.

Kuppers, R.: **B cells under the influence: transformation of B cells by Epstein–Barr virus.** *Nat. Rev. Immunol.* 2003, **3**:801–812.

Lauer, G.M., and Walker, B.D.: **Hepatitis C virus infection.** *N. Engl. J. Med.* 2001, **345**:41–52.

Macsween, K.F., and Crawford, D.H.: **Epstein–Barr virus—recent advances.** *Lancet Infect. Dis.* 2003, **3**:131–140.

Nash, A.A.: **T cells and the regulation of herpes simplex virus latency and reactivation.** *J. Exp. Med.* 2000, **191**:1455–1458.

13-25 HIV is a retrovirus that establishes a chronic infection that slowly progresses to AIDS.

Baltimore, D.: **The enigma of HIV infection.** *Cell* 1995, **82**:175–176.

Barre-Sinoussi, F.: **HIV as the cause of AIDS.** *Lancet* 1996, **348**:31–35.

Campbell-Yesufu, O.T., and Gandhi, R.T.: **Update on human immunodeficiency virus (HIV)-2 infection.** *Clin. Infect .Dis.* 2011, **52**:780–787.

Heeney, J.L., Dalgleish, A.G., and Weiss, R.A.: **Origins of HIV and the evolution of resistance to AIDS.** *Science* 2006, **313**:462–466.

Sharp, P.M., and Hahn, B.H.: **Origins of HIV and the AIDS pandemic.** *Cold Spring Harb. Perspect. Med.* 2011, **1**: a006841.

13-26 HIV infects and replicates within cells of the immune system.

Grouard, G., and Clark, E.A.: **Role of dendritic and follicular dendritic cells in HIV infection and pathogenesis.** *Curr. Opin. Immunol.* 1997, **9**:563–567.

Moore, J.P., Trkola, A., and Dragic, T.: **Co-receptors for HIV-1 entry.** *Curr. Opin. Immunol.* 1997, **9**:551–562.

Pohlmann, S., Baribaud, F., and Doms, R.W.: **DC-SIGN and DC-SIGNR: helping hands for HIV.** *Trends Immunol.* 2001, **22**:643–646.

Root, M.J., Kay, M.S., and Kim, P.S.: **Protein design of an HIV-1 entry inhibitor.** *Science* 2001, **291**:884–888.

Sol-Foulon, N., Moris, A., Nobile, C., Boccaccio, C., Engering, A., Abastado, J.P., Heard, J.M., van Kooyk, Y., and Schwartz, O.: **HIV-1 Nef-induced upregulation of DC-SIGN in dendritic cells promotes lymphocyte clustering and viral spread.** *Immunity* 2002, **16**:145–155.

Unutmaz, D., and Littman, D.R.: **Expression pattern of HIV-1 coreceptors on T cells: implications for viral transmission and lymphocyte homing.** *Proc. Natl Acad. Sci. USA* 1997, **94**:1615–1618.

Wyatt, R., and Sodroski, J.: **The HIV-1 envelope glycoproteins: fusogens, antigens, and immunogens.** *Science* 1998, **280**:1884–1888.

13-27 Activated CD4 T cells are the major source of HIV replication.

Chiu, Y.L., Soros, V.B., Kreisberg, J.F., Stopak, K., Yonemoto, W., and Greene, W.C.: **Cellular APOBEC3G restricts HIV-1 infection in resting CD4+ T cells.** *Nature* 2005, **435**:108–114.

Cullen, B.R.: **HIV-1 auxiliary proteins: making connections in a dying cell.** *Cell* 1998, **93**:685–692.

Cullen, B.R.: **Connections between the processing and nuclear export of mRNA: evidence for an export license?** *Proc. Natl Acad. Sci. USA* 2000, **97**:4–6.

Emerman, M., and Malim, M.H.: **HIV-1 regulatory/accessory genes: keys to unraveling viral and host cell biology.** *Science* 1998, **280**:1880–1884.

Ho, Y.-C., Shan, L., Hosmane, N.N., Wang, J., Laskey, S.B., Rosenbloom, D.I.S., Lai, J., Blankson, J.N., Siliciano, J.D., and Siliciano, R.F.: **Replication-competent noninduced proviruses in the latent reservoir increase barrier to HIV-1 cure.** *Cell* 2013, **155**:540–551.

Kinoshita, S., Su, L., Amano, M., Timmerman, L.A., Kaneshima, H., and Nolan, G.P.: **The T-cell activation factor NF-ATc positively regulates HIV-1 replication and gene expression in T cells.** *Immunity* 1997, **6**:235–244.

Malim, M.H., and Bieniasz, P.D.: **HIV restriction factors and mechanisms of evasion.** *Cold Spring Harb Perspect Med* 2012, **2**:a006940.

Trono, D.: **HIV accessory proteins: leading roles for the supporting cast.** *Cell* 1995, **82**:189–192.

13-28 There are several routes by which HIV is transmitted and establishes infection.

Bomsel, M., and David, V.: **Mucosal gatekeepers: selecting HIV viruses for early infection.** *Nat. Med.* 2002, **8**:114–116.

Kwon, D.S., Gregorio, G., Bitton, N., Hendrickson, W.A., and Littman, D.R.: **DC-SIGN-mediated internalization of HIV is required for trans-enhancement of T cell infection.** *Immunity* 2002, **16**:135–144.

Pantaleo, G., Menzo, S., Vaccarezza, M., Graziosi, C., Cohen, O.J., Demarest, J.F., Montefiori, D., Orenstein, Peckham, C., and Gibb, D.: **Mother-to-child transmission of the human immunodeficiency virus.** *N. Engl. J. Med.* 1995, **333**:298–302.

Royce, R.A., Sena, A., Cates, W., Jr, and Cohen, M.S.: **Sexual transmission of HIV.** *N. Engl. J. Med.* 1997, **336**:1072–1078.

13-29 HIV variants with tropism for different co-receptors play different roles in transmission and progression of disease.

Berger, E.A., Murphy, P.M., and Farber, J.M.: **Chemokine receptors as HIV-1 coreceptors: roles in viral entry, tropism, and disease.** *Annu. Rev. Immunol.* 1999, **17**:657–700.

Connor, R.I., Sheridan, K.E., Ceradini, D., Choe, S., and Landau, N.R.: **Change in coreceptor use correlates with disease progression in HIV-1—infected individuals.** *J. Exp. Med.* 1997, **185**:621–628.

Littman, D.R.: **Chemokine receptors: keys to AIDS pathogenesis?** *Cell* 1998, **93**:677–680.

13-30 A genetic deficiency of the co-receptor CCR5 confers resistance to HIV infection.

Gonzalez, E., Kulkarni, H., Bolivar, H., Mangano, A., Sanchez, R., Catano, G., Nibbs, R.J., Freedman, B.I., Quinones, M.P., Bamshad, M.J., *et al.*: **The influence of CCL3L1 gene-containing segmental duplications on HIV-1/AIDS susceptibility.** *Science* 2005, **307**:1434–1440.

Liu, R., Paxton, W.A., Choe, S., Ceradini, D., Martin, S.R., Horuk, R., Macdonald, M.E., Stuhlmann, H., Koup, R.A., and Landau, N.R.: **Homozygous defect in HIV-1 coreceptor accounts for resistance of some multiply exposed individuals to HIV 1 infection.** *Cell* 1996, **86**:367–377.

Samson, M., Libert, F., Doranz, B.J., Rucker, J., Liesnard, C., Farber, C.M., Saragosti, S., Lapoumeroulie, C., Cognaux, J., Forceille, C., *et al.*: **Resistance to HIV-1 infection in Caucasian individuals bearing mutant alleles of the CCR 5 chemokine receptor gene.** *Nature* 1996, **382**:722–725.

13-31 An immune response controls but does not eliminate HIV.

Baltimore, D.: **Lessons from people with nonprogressive HIV infection.** *N. Engl. J. Med.* 1995, **332**:259–260.

Barouch, D.H., and Letvin, N.L.: **CD8+ cytotoxic T lymphocyte responses to lentiviruses and herpesviruses.** *Curr. Opin. Immunol.* 2001, **13**:479–482.

Haase, A.T.: **Targeting early infection to prevent HIV-1 mucosal transmission.** *Nature* 2010, **464**:217–223.

Ho, D.D., Neumann, A.U., Perelson, A.S., Chen, W., Leonard, J.M., and Markowitz, M.: **Rapid turnover of plasma virions and CD4 lymphocytes in HIV-1 infection.** *Nature* 1995, **373**:123–126.

Liao, H.-X., Lynch, R., Zhou, T., Gao, F., Alam, S.M., Boyd, S.D., Fire, A.Z., Roskin, K.M., Schramm, C.A., Zhang, Z., *et al.*: **Co-evolution of a broadly neutralizing HIV-1 antibody and founder virus.** *Nature* 2013, **496**:469–476.

Johnson, W.E., and Desrosiers, R.C.: **Viral persistence: HIV's strategies of immune system evasion.** *Annu. Rev. Med.* 2002, **53**:499–518.

McMichael, A.J., Borrow, P., Tomaras, G.D., Goonetilleke, N., and Haynes, B.F.: **The immune response during acute HIV-1 infection: clues for vaccine development.** *Nat. Rev. Immunol.* 2010, **10**:11–23.

Price, D.A., Goulder, P.J., Klenerman, P., Sewell, A.K., Easterbrook, P.J., Troop, M., Bangham, C.R., and Phillips, R.E.: **Positive selection of HIV-1 cytotoxic T lymphocyte escape variants during primary infection.** *Proc. Natl Acad. Sci. USA* 1997, **94**:1890–1895.

Schmitz, J.E., Kuroda, M.J., Santra, S., Sasseville, V.G., Simon, M.A., Lifton, M.A., Racz, P., Tenner-Racz, K., Dalesandro, M., Scallon, B.J., *et al.*: **Control of viremia in simian immunodeficiency virus infection by CD8+ lymphocytes.** *Science* 1999, **283**:857–860.

Siliciano, R.F., and Greene, W.C.: **HIV latency.** *Cold Spring Harb. Perspect. Med.* 2011, **1**:a007096.

Pantaleo, G., Menzo, S., Vaccarezza, M., Graziosi, C., Cohen, O.J., Demarest, JF, Montefiori, D, Orenstein, J.M., Fox, C., Schrager, L.K., *et al.*: **Studies in subjects with long-term nonprogressive human immunodeficiency virus infection.** *N. Engl. J. Med.* 1995, **332**:209–216.

13-32 Lymphoid tissue is the major reservoir of HIV infection.

Burton, G.F., Masuda, A., Heath, S.L., Smith, B.A., Tew, J.G., and Szakal, A.K.: **Follicular dendritic cells (FDC) in retroviral infection: host/pathogen perspectives.** *Immunol. Rev.* 1997, **156**:185–197.

Chun, T.W., Carruth, L., Finzi, D., Shen, X., DiGiuseppe, J.A., Taylor, H., Hermankova, M., Chadwick, K., Margolick, J., Quinn, T.C., *et al.*: **Quantification of latent tissue reservoirs and total body viral load in HIV-1 infection.** *Nature* 1997, **387**:183–188.

Haase, A.T.: **Population biology of HIV-1 infection: viral and CD4+ T cell demographics and dynamics in lymphatic tissues.** *Annu. Rev. Immunol.* 1999, **17**:625–656.

Pierson, T., McArthur, J., and Siliciano, R.F.: **Reservoirs for HIV-1: mechanisms for viral persistence in the presence of antiviral immune responses and antiretroviral therapy.** *Annu. Rev. Immunol.* 2000, **18**:665–708.

13-33 Genetic variation in the host can alter the rate of disease progression.

Bream, J.H., Ping, A., Zhang, X., Winkler, C., and Young, H.A.: **A single nucleotide polymorphism in the proximal IFN-gamma promoter alters control of gene transcription.** *Genes Immun.* 2002, **3**:165–169.

Martin, M.P., Gao, X., Lee, J.H., Nelson, G.W., Detels, R., Goedert, J.J., Buchbinder, S., Hoots, K., Vlahov, D., Trowsdale, J., *et al.*: **Epistatic interaction between KIR3DS1 and HLA-B delays the progression to AIDS.** *Nat. Genet.* 2002, **31**:429–434.

Shin, H.D., Winkler, C., Stephens, J.C., Bream, J., Young, H., Goedert, J.J., O'Brien, T.R., Vlahov, D., Buchbinder, S., Giorgi, J., *et al.*: **Genetic restriction of HIV-1 pathogenesis to AIDS by promoter alleles of IL10.** *Proc. Natl Acad. Sci. USA* 2000, **97**:14467–14472.

Walker, B.D., and Yu, X.G.: **Unravelling the mechanisms of durable control of HIV-1.** *Nat. Rev. Immunol.* 2013, **13**:487–498.

13-34 The destruction of immune function as a result of HIV infection leads to increased susceptibility to opportunistic infection and eventually to death.

Kedes, D.H., Operskalski, E., Busch, M., Kohn, R., Flood, J., and Ganem, D.R.: **The seroepidemiology of human herpesvirus 8 (Kaposi's sarcoma associated herpesvirus): distribution of infection in KS risk groups and evidence for sexual transmission.** *Nat. Med.* 1996, **2**:918–924.

Miller, R.: **HIV-associated respiratory diseases.** *Lancet* 1996, **348**:307–312.

Zhong, W.D., Wang, H., Herndier, B., and Ganem, D.R.: **Restricted expression of Kaposi sarcoma associated herpesvirus (human herpesvirus 8) genes in Kaposi sarcoma.** *Proc. Natl Acad. Sci. USA* 1996, **93**:6641–6646.

13-35 Drugs that block HIV replication lead to a rapid decrease in titer of infectious virus and an increase in CD4 T cells.

Allers, K., Hütter, G., Hofmann, J., Loddenkemper, C., Rieger, K., Thiel, E., and Schneider, T.: **Evidence for the cure of HIV infection by CCR5Δ32/Δ32 stem cell transplantation.** *Blood* 2011, **117**:2791–2799.

Barouch, D.H., and Deeks, S.G.: **Immunologic strategies for HIV-1 remission and eradication.** *Science* 2014, **345**:169–174.

Boyd, M., and Reiss, P.: **The long-term consequences of antiretroviral therapy: a review.** *J. HIV Ther.* 2006, **11**:26–35.

Cammack, N.: **The potential for HIV fusion inhibition.** *Curr. Opin. Infect. Dis.* 2001, **14**:13–16.

Carcelain, G., Debre, P., and Autran, B.: **Reconstitution of CD4+ lymphocytes in HIV-infected individuals following antiretroviral therapy.** *Curr. Opin. Immunol.* 2001, **13**:483–488.

Farber, J.M., and Berger, E.A.: **HIV's response to a CCR5 inhibitor: I'd rather tighten than switch!** *Proc. Natl Acad. Sci. USA* 2002, **99**:1749–1751.

Ho, D.D.: **Perspectives series: host/pathogen interactions. Dynamics of HIV-1 replication *in vivo*.** *J. Clin. Invest.* 1997, **99**:2565–2567.

Kordelas, L., Verheyen, J., and Esser, S.: **Shift of HIV tropism in stem-cell transplantation with CCR5 Delta32 mutation.** *N. Engl. J. Med.* 2014, **371**:880–882.

Lundgren, J.D., and Mocroft, A.: **The impact of antiretroviral therapy on AIDS and survival.** *J. HIV Ther.* 2006, **11**:36–38.

Perelson, A.S., Essunger, P., Cao, Y.Z., Vesanen, M., Hurley, A., Saksela, K., Markowitz, M., and Ho, D.D.: **Decay characteristics of HIV-1-infected compartments during combination therapy.** *Nature* 1997, **387**:188–191.

Wei, X., Ghosh, S.K., Taylor, M.E., Johnson, V.A., Emini, E.A., Deutsch, P., Lifson, J.D., Bonhoeffer, S., Nowak, M.A., Hahn, B.H., *et al.*: **Viral dynamics in human immunodeficiency virus type 1 infection.** *Nature* 1995, **373**:117–122.

13-36 In the course of infection HIV accumulates many mutations, which can result in the outgrowth of drug-resistant variants.

Condra, J.H., Schleif, W.A., Blahy, O.M., Gabryelski, L.J., Graham, D.J., Quintero, J.C., Rhodes, A., Robbins, H.L., Roth, E., Shivaprakash, M., *et al.*: ***In vivo* emergence of HIV-1 variants resistant to multiple protease inhibitors.** *Nature* 1995, **374**:569–571.

Finzi, D., and Siliciano, R.F.: **Viral dynamics in HIV-1 infection.** *Cell* 1998, **93**:665–671.

Katzenstein, D.: **Combination therapies for HIV infection and genomic drug resistance.** *Lancet* 1997, **350**:970–971.

Moutouh, L., Corbeil, J., and Richman, D.D.: **Recombination leads to the rapid emergence of HIV 1 dually resistant mutants under selective drug pressure.** *Proc. Natl Acad. Sci. USA* 1996, **93**:6106–6111.

13-37 Vaccination against HIV is an attractive solution but poses many difficulties.

Baba, T.W., Liska, V., Hofmann-Lehmann, R., Vlasak, J., Xu, W., Ayehunie, S., Cavacini, L.A., Posner, M.R., Katinger, H., Stiegler, G., *et al.*: **Human neutralizing monoclonal antibodies of the IgG1 subtype protect against mucosal simian–human immunodeficiency virus infection.** *Nat. Med.* 2000, **6**:200–206.

Barouch, D.H.: **The quest for an HIV-1 vaccine--moving forward.** *N. Engl. J. Med.* 2013, **369**:2073–2076.

Barouch, D.H., Kunstman, J., Kuroda, M.J., Schmitz, J.E., Santra, S., Peyerl, F.W., Krivulka, G.R., Beaudry, K., Lifton, M.A., Gorgone, D.A., *et al.*: **Eventual AIDS vaccine failure in a rhesus monkey by viral escape from cytotoxic T lymphocytes.** *Nature* 2002, **415**:335–339.

Isitman, G., Stratov, I., and Kent, S.J.: **Antibody-dependent cellular cytotoxicity and Nk cell-driven immune escape in HIV Infection: Implications for HIV vaccine development.** *Adv. Virol.* 2012, **212**:637208.

Letvin, N.L.: **Progress and obstacles in the development of an AIDS vaccine.** *Nat. Rev. Immunol.* 2006, **6**:930–939.

McMichael, A.J., and Koff, W.C.: **Vaccines that stimulate T cell immunity to HIV-1: the next step.** *Nat. Immunol.* 2014, **15**:319–322.

13-38 Prevention and education are important in controlling the spread of HIV and AIDS.

Coates, T.J., Aggleton, P., Gutzwiller, F., Des-Jarlais, D., Kihara, M., Kippax, S., Schechter, M., and van-den-Hoek, J.A.: **HIV prevention in developed countries.** *Lancet* 1996, **348**:1143–1148.

Decosas, J., Kane, F., Anarfi, J.K., Sodji, K.D., and Wagner, H.U.: **Migration and AIDS.** *Lancet* 1995, **346**:826–828.

Dowsett, G.W.: **Sustaining safe sex: sexual practices, HIV and social context.** *AIDS* 1993, **7** Suppl. 1:S257–S262.

Kirby, M.: **Human rights and the HIV paradox.** *Lancet* 1996, **348**:1217–1218.

Nelson, K.E., Celentano, D.D., Eiumtrakol, S., Hoover, D.R., Beyrer, C., Suprasert, S., Kuntolbutra, S., and Khamboonruang, C.: **Changes in sexual behavior and a decline in HIV infection among young men in Thailand.** *N. Engl. J. Med.* 1996, **335**:297–303.

Weniger, B.G., and Brown, T.: **The march of AIDS through Asia.** *N. Engl. J. Med.* 1996, **335**:343–345.

14

Allergy and Allergic Diseases

The adaptive immune response is a critical component of host defense against infection and is essential for normal health. Antigens not associated with infectious agents sometimes elicit adaptive immune responses, and this can cause disease. One circumstance in which this happens is when harmful immunologically mediated hypersensitivity reactions known generally as **allergic reactions** occur in response to inherently harmless 'environmental' antigens such as pollen, food, and drugs.

Historically, hypersensitivity reactions due to immunological responses were classified by Gell and Coombs into four broad types, of which type I hypersensitivity reactions represented immediate-type allergic reactions mediated by **IgE** antibodies, with mast-cell activation the major final effector mechanism. Type II and III hypersensitivity responses were defined as those that were driven by antigen-specific IgG antibodies, the final effector mechanism being complement (type II) or FcR-bearing cellular effectors (type III). Finally, type IV hypersensitivity responses were depicted as being driven by cellular effectors, including lymphocytes and a variety of myeloid cell types. While the Gell and Coombs classification system provides an effective framework for understanding the mechanisms underlying some prototypic immunologic reactions, it is now becoming clear that most normal and pathologic host immune responses involve both the humoral and cellular arms of the immune system, and that the definitions provided in Chapter 11 for types 1, 2, and 3 immune response modules provide a more thorough mechanistic context for understanding disease pathogenesis, including allergic responses (see Fig. 11.5). In most allergic reactions, such as those to food, pollen, or house dust, reactions occur because the individual has become **sensitized** to an innocuous antigen—the **allergen**—by producing IgE antibodies against it. This is usually a result of the formation of an unwanted type 2 immune response to the allergen. Subsequent exposure to the allergen triggers the activation of IgE-binding cells, chiefly mast cells and basophils, in the exposed tissue, leading to a series of responses that are characteristic of this type of allergic reaction. In hay fever (allergic rhinoconjunctivitis), for example, symptoms occur when allergenic proteins leached out of grass or weed pollen grains come into contact with the mucous membrane of the nose and eyes. In contrast, other hypersensitivity disorders, such as allergic contact dermatitis, serum sickness, or celiac disease, are not dependent on IgE antibodies and represent unwanted immune responses driven by IgG antibodies and/or cellular immune responses.

We are all exposed regularly to common environmental agents that can cause allergic reactions in some individuals. While most of the population does not develop clinically significant allergic reactions to the majority of potential allergens, in some surveys over half of the population shows an allergic response to at least one substance in the environment. Some individuals manifest allergic responses to multiple common antigens. A predisposition to become IgE-sensitized to environmental allergens is called **atopy**, and later in the chapter we discuss the various factors—both genetic and environmental—that may contribute to this predisposition. Genetic factors clearly play a role in predisposing an individual to IgE-mediated allergic disease. If both parents are atopic, a child has a 40–60% chance of developing an IgE-mediated allergy, whereas the risk is much lower, on the order of 10%, if neither parent is atopic.

IN THIS CHAPTER

IgE and IgE-mediated allergic diseases.

Effector mechanisms in IgE-mediated allergic reactions.

Non-IgE-mediated allergic diseases.

Fig. 14.1 IgE-mediated reactions to extrinsic antigens. All IgE-mediated responses involve mast-cell degranulation, but the symptoms experienced by the patient can be very different depending, for example, on whether the allergen is injected directly into the bloodstream, is eaten, or comes into contact with the mucosa of the ocular or respiratory tract.

IgE-mediated allergic reactions			
Reaction or disease	**Common stimuli**	**Route of entry**	**Response**
Systemic anaphylaxis	Drugs Venoms Food, e.g., peanuts Serum	Intravenous (either directly or following absorption into the blood after oral intake)	Edema Increased vascular permeability Laryngeal edema Circulatory collapse Death
Acute urticaria (wheal-and-flare)	Post-viral Animal hair Bee stings Allergy testing	Through skin Systemic	Local increase in blood flow and vascular permeability Edema
Seasonal rhinoconjunctivitis (hay fever)	Pollens (ragweed, trees, grasses) Dust-mite feces	Contact with conjunctiva of eye and nasal mucosa	Edema of conjunctiva and nasal mucosa Sneezing
Asthma	Dander (cat) Pollens Dust-mite feces	Inhalation leading to contact with mucosal lining of lower airways	Bronchial constriction Increased mucus production Airway inflammation Bronchial hyperreactivity
Food allergy	Peanuts Tree nuts Shellfish Fish Milk Eggs Soy Wheat	Oral	Vomiting Diarrhea Pruritus (itching) Urticaria (hives) Anaphylaxis (rarely)

IgE is prominent in the defense against extracellular parasitic organisms, especially helminths and protozoa (see Section 11-9). These parasitic organisms are prevalent in developing nations, but most serum IgE in developed nations is directed against innocuous antigens, sometimes causing allergic symptoms (**Fig. 14.1**). Almost half the population of North America and Europe is sensitized to one or more common environmental antigens, and, although rarely life-threatening, allergic diseases initiated by contact with a specific allergen can cause much distress and lost time from school and work. The burden of allergic diseases in the Western world is considerable, with their prevalence having more than doubled in the past 20 years. Consequently, most clinical and scientific attention paid to IgE has been directed toward its pathologic roles in allergic disease rather than its protective capacity. Until the last decade, developing countries in Africa and the Middle East reported a relatively low prevalence of allergy; however, this situation is rapidly changing, probably as a result of Western-style modernization.

In this chapter we first consider the mechanisms that favor the sensitization of an individual to an allergen, resulting in the production of antigen-specific IgE. We then describe the IgE-mediated allergic reaction itself—the pathological consequences of the interaction between allergen and the IgE bound to the high-affinity Fcε receptor on mast cells and basophils. Finally, we consider the causes and consequences of other types of immunological hypersensitivity reaction.

IgE and IgE-mediated allergic diseases.

Immediate hypersensitivity reactions are those allergic reactions caused by activation of mast cells and basophils by multivalent antigen bridging IgE bound to their cell surfaces. IgE differs from other antibody isotypes in being predominantly localized in the tissues, where it is tightly bound to the surfaces

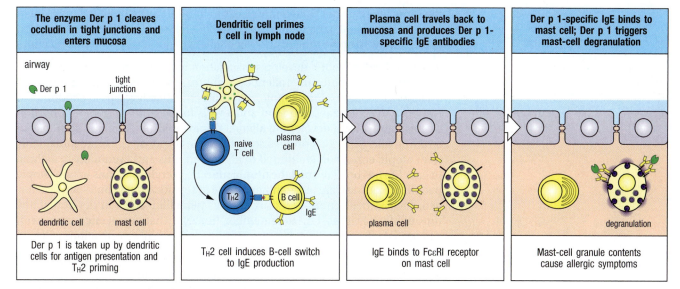

The enzyme Der p 1 cleaves occludin in tight junctions and enters mucosa	Dendritic cell primes T cell in lymph node	Plasma cell travels back to mucosa and produces Der p 1-specific IgE antibodies	Der p 1-specific IgE binds to mast cell; Der p 1 triggers mast-cell degranulation
Der p 1 is taken up by dendritic cells for antigen presentation and T$_H$2 priming	T$_H$2 cell induces B-cell switch to IgE production	IgE binds to FcεRI receptor on mast cell	Mast-cell granule contents cause allergic symptoms

Fig. 14.2 Sensitization to an inhaled allergen. Der p 1 is a common respiratory allergen that is found in fecal pellets of the house dust mite. When an atopic individual first encounters Der p 1, subepithelial dendritic cells ingest the allergenic protein and traffic to the draining lymph node, where T$_H$2 cells specific for Der p 1 are produced (first and second panels). Interaction of these T cells with Der p 1-specific B cells leads to the production of class-switched plasma cells producing Der p 1-specific IgE in the mucosal tissues (third panel), and this IgE becomes bound to Fc receptors on resident submucosal mast cells. On a subsequent encounter with Der p 1, the allergen binds to the mast cell-bound IgE, triggering mast-cell activation and the release of mast-cell granule contents, which cause the symptoms of the allergic reaction (last panel). Der p 1 is a protease that cleaves occludin, a protein that helps to maintain epithelial tight junctions; the enzymatic activity of Der p 1 is thought to help it pass through the epithelium.

of mast cells and some other cell types through the high-affinity IgE receptor **FcεRI** (see Section 10-24). Binding of antigen to IgE cross-links the high-affinity IgE receptors, causing the release of chemical mediators from the mast cells that can lead to allergic disease (Fig. 14.2). How an initial antibody response to environmental antigens comes to be dominated by IgE production in atopic individuals is still being worked out. In this part of the chapter we describe the current understanding of the factors that contribute to this process.

14-1 Sensitization involves class switching to IgE production on first contact with an allergen.

To produce an allergic reaction against a given antigen, an individual must first be exposed to the antigen under conditions that result in the production of IgE antibodies. Allergic symptoms occur when an individual who has been sensitized in this fashion has subsequent exposure to the antigen. Exposure can lead to different clusters of symptoms, characterized by the tissues that are most prominently affected. The most common forms of allergic response in developed countries are to airborne allergens, causing symptoms that affect predominantly the nasal passages (allergic rhinitis), the eyes (allergic conjunctivitis), or the lower airways and lungs (asthma). Ingested allergens can lead to food allergy, sometimes affecting only the gastrointestinal tract (for example, eosinophilic esophagitis), but not infrequently involving locations distant from the site of antigen entry. Reactions that occur at locations distant from the site of entry of the challenging antigen are considered to be systemic reactions, and are thought to occur because of spread of the antigen throughout the body via the blood circulation. Systemic reactions can be limited to a single distant organ, causing hives (also called urticaria) when they target the skin, wheezing (or bronchospasms) when they involve the lungs, and life-threatening lowering of the blood pressure when they target the vascular system. Serious systemic reactions are designated by the term anaphylaxis. It is not known why sensitization with a particular allergen in one individual leads to local reactions at the time of allergen challenge, whereas sensitization with the same allergen can yield anaphylaxis in another individual. In fact, even in a single individual, a challenge that usually yields only a mild local reaction can be followed at the time of another challenge by a severe systemic reaction.

Atopic individuals often develop sensitization to many different antigens, and can express multiple forms of allergic symptoms, which depend on the route and quantity of allergen—for example, atopic eczema that develops in childhood in response to sensitization to food antigens is followed in a sizable proportion of those individuals developing allergic rhinitis and/or asthma in response to airborne allergens. This progression of allergic responses in some individuals from atopic eczema in childhood to allergic rhinitis and eventually to asthma in later life has been termed the **atopic march**. Allergic reactions in non-atopic people, in contrast, are predominantly due to sensitization to one specific allergen, such as bee venom or a drug such as penicillin, and can develop at any time of life. It is important to remember, however, that not all encounters with a potential allergen will lead to sensitization, and not all sensitizations will lead to a symptomatic allergic response, even in atopic individuals.

The immune response leading to IgE production in response to antigen is driven by two main groups of signals that together are typical of type 2 immune reactions. The first consists of signals that favor the differentiation of naive T cells to a T_H2 phenotype. The second comprises T_H2 cytokines and co-stimulatory signals that stimulate B cells to switch to the production of IgE. As described in Section 9-21, the fate of a naive CD4 T cell responding to an antigenic peptide presented by a dendritic cell is determined by the cytokines it is exposed to before and during this response, and by the intrinsic properties of the antigen, the antigen dose, and the route of presentation. Exposure to IL-4, IL-5, IL-9, and IL-13 favors the development of T_H2 cells, whereas exposure to IFN-γ and IL-12 (and its relative IL-27) favors T_H1-cell development.

Immune defenses against multicellular parasites are found mainly at the sites of parasite entry, namely, under the skin and in the mucosal tissues of the airways and the gut. Cells of the innate and adaptive immune systems at these sites are specialized to secrete cytokines that promote a type 2 response to parasitic infection. In the presence of an invading parasite, dendritic cells taking up antigens in these tissues migrate to regional lymph nodes, where they tend to drive antigen-specific naive CD4 T cells to become effector T_H2 cells. T_H2 cells themselves secrete IL-4, IL-5, IL-9, and IL-13, thus maintaining an environment in which further differentiation of T_H2 cells is favored. The cytokine IL-33, which can be produced by activated mast cells and by damaged or injured epithelial cells, also contributes to amplification of the T_H2 response. IL-33 can act directly on T_H2 cells via the IL-33 receptors that these cells express. Allergic responses against common environmental antigens are normally avoided because mucosal dendritic cells that encounter antigen in the absence of danger signals such as those provoked by microbial infection generally induce naive CD4 T cells to differentiate into antigen-specific regulatory T cells (T_{reg} cells). The T_{reg} cells suppress T-cell responses and contribute to a state of tolerance to the antigen (see Section 12-8) rather than allowing the production of effector or helper cells that might support the production of an allergic response.

CD40 Ligand Deficiency

The cytokines and chemokines produced by T_H2 cells both amplify the T_H2 response and stimulate the class switching of activated B cells to IgE production. As we saw in Chapter 10, IL-4 or IL-13 provides the first signal that switches B cells to IgE production. IL-4 and IL-13 acting on T and B lymphocytes activate the Janus-family tyrosine kinases Jak1 and Jak3 (see Section 7-20), ultimately leading to phosphorylation (and thereby activation) of the transcriptional regulator STAT6. Mice lacking functional IL-4, IL-13, or STAT6 have impaired T_H2 responses and an impaired ability to switch to production of IgE, demonstrating the key importance of these cytokines and their signaling pathways in the IgE response. The second signal for IgE production is a co-stimulatory interaction between CD40 ligand on the T-cell surface and CD40 on the B-cell surface. This interaction is essential for all antibody class switching. Patients with a genetic deficiency of CD40 ligand produce no IgG, IgA, or IgE, and display a hyper-IgM syndrome phenotype (see Section 13-9).

Murine mast cells and basophils can also produce the signals that drive IgE production by B cells. Mast cells and basophils express FcεRI, and when they are activated by antigen cross-linking their FcεRI-bound IgE, they express cell-surface CD40 ligand and secrete IL-4. Similar data exist for human basophils that have also been primed by inflammatory stimuli (Fig. 14.3). Like T_H2 cells, they can induce class switching and IgE production by B cells. Generally, class switching to IgE occurs in lymph nodes (secondary lymphoid organs) that drain the site of antigen entry or in inducible lymphoid follicles (also called tertiary lymphoid tissues) that form in mucosal and other tissues at sites of persistent inflammation. The potential for formation in mucosal tissues of tertiary lymphoid follicles with germinal centers containing B cells that have switched to production of IgE means that mast cells or basophils can amplify the B-cell response close to the site of the allergic reaction. One goal of therapy for allergies is to block this amplification process and thus prevent allergic reactions from becoming self-sustaining.

In humans, the IgE response, once initiated, can also be amplified by the capture of IgE by Fcε receptors on dendritic cells. Some populations of human immature dendritic cells—for example, the Langerhans cells of the skin—express surface FcεRI in an inflammatory setting, and once anti-allergen IgE antibodies have been produced, they can bind to these receptors. The bound IgE forms a highly effective trap for allergen, which is then efficiently processed by the dendritic cell for presentation to naive T cells, thus maintaining and reinforcing the T_H2 response to the allergen. Eosinophils have also been reported to express IgE receptors, but this is still controversial. Eosinophils may act as antigen-presenting cells to T cells in a standard fashion after upregulation of eosinophil MHC class II and co-stimulatory molecules; however, this probably occurs in tissues where activated T cells have migrated rather than in lymph nodes where naive T cells are primed by dendritic cells.

14-2 Although many types of antigens can cause allergic sensitization, proteases are common sensitizing agents.

Most airborne allergens are relatively small, highly soluble proteins that are carried on dry particles such as pollen grains or mite feces (Fig. 14.4). On contact with the mucus-covered epithelia of the eyes, nose, or airways, the soluble allergen is eluted from the particle and diffuses into the mucosa, where it can be picked up by dendritic cells and provoke sensitization (see Fig. 14.2). At mucosal surfaces, allergens are typically presented to the immune system at low concentrations. It has been estimated that the maximum exposure of a person to the common pollen allergens in ragweed (*Ambrosia* species) does not exceed 1 μg per year. It is thought that low-dose sensitization favors formation of a strong T_H2 response. Consequently, these minute doses of allergen can provoke irritating and even life-threatening T_H2-driven IgE antibody responses in atopic individuals.

Antigen exposures that lead to allergic responses do not always involve such low doses of antigen, especially at other tissue sites. For example, bee venom is a frequent cause of allergic sensitization, and individual bee stings result in the injection into the skin of 20–75 μg of bee venom (1 to 2 orders of magnitude more than the total dose of ragweed antigen that is inhaled into the airways). In the case of food allergy, ingestion of many grams of an allergenic food into the gastrointestinal tract over prolonged periods of time can lead to sensitization. Sensitization can also occur in response to small or large doses of injected antigens. For example, before the introduction of recombinant

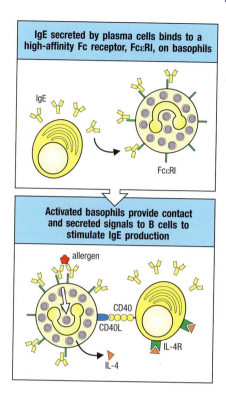

Fig. 14.3 Antigen binding to IgE on basophils or mast cells leads to amplification of IgE production. Top panel: IgE secreted by plasma cells binds to the high-affinity IgE receptor on basophils (illustrated here) and mast cells. Bottom panel: when the surface-bound IgE is cross-linked by antigen, these cells express CD40 ligand (CD40L) and secrete IL-4, which in turn binds to IL-4 receptors (IL-4R) on the activated B cell. Together with ligation of B-cell CD40 by basophil CD40L, this activates class switching by the B cell and the production of more IgE. These interactions can occur *in vivo* at the site of allergen-triggered inflammation, for example, in bronchus-associated lymphoid tissue.

Features of airborne allergens that may promote the priming of T_H2 cells that drive IgE responses	
Protein, often with carbohydrate side chains	Protein antigens induce T-cell responses
Low dose	Favors activation of IL-4-producing CD4 T cells
Low molecular weight	Allergen can diffuse out of particles into the mucosa
Highly soluble	Allergen can be readily eluted from particle
Stable	Allergen can survive in desiccated particle
Contains peptides that bind host MHC class II	Required for T-cell priming

Fig. 14.4 Properties of inhaled allergens. The typical characteristics of inhaled allergens are described in this table.

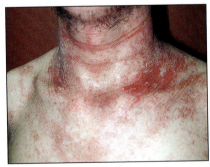

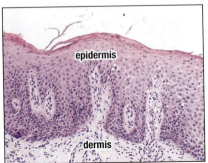

Fig. 14.5 Netherton's syndrome illustrates the association of proteases with the development of high levels of IgE and allergy. This 26-year-old man with Netherton's syndrome, caused by a deficiency in the protease inhibitor SPINK5, had persistent erythroderma (redness of the skin), recurrent infections of the skin and other tissues, and multiple food allergies associated with high serum IgE levels. In the top photograph, large erythematous plaques covered with scales and erosions are visible over the upper trunk. The lower panel shows a section through the skin of the same patient. Note the psoriasis-like hyperplasia of the epidermis. Neutrophils are also present in the epidermis. In the dermis, a perivascular infiltrate is evident. Although not discernible at this magnification, the infiltrate contains both mononuclear cells and neutrophils. Source: Sprecher, E., *et al.*: *Clin. Exp. Dermatol.* 2004, **29**:513–517.

human insulin, individuals with diabetes could develop allergy to porcine insulin, usually administered in doses of 1–2 milligrams per injection. In contrast, doses of penicillin-type drugs (including cephalosporins and other β-lactam-containing antibiotics) that can lead to sensitization when administered by intramuscular or intravenous injection are usually in the 1- to 2-gram per injection range.

Considerable effort has been directed toward identifying physical, chemical, or functional characteristics that might be common for all allergens, but no common characteristics of all allergens have emerged. Thus, it appears that in a susceptible host, essentially any antigenic molecule can elicit an allergic response.

While any type of molecule appears to be able to elicit an allergic response, the search for common features of allergenic molecules has demonstrated that some clinically significant allergens are proteases. One ubiquitous protease allergen is the cysteine protease Der p 1, which is present in the feces of the house dust mite *Dermatophagoides pteronyssinus*. Der p 1 provokes allergic reactions in about 20% of the North American population. This enzyme has been found to cleave occludin, a protein component of intercellular tight junctions in the airway mucosa. This reveals one possible reason for the allergenicity of certain enzymes. By destroying the integrity of the tight junctions between epithelial cells, Der p 1 may gain abnormal access to subepithelial antigen-presenting cells (see Fig. 14.2). The tendency of proteases to induce IgE production is highlighted by individuals with Netherton's syndrome (Fig. 14.5), which is characterized by high levels of IgE and multiple allergies. This disease is caused by a mutation in *SPINK5* (serine protease inhibitor Kazal-type 5), which encodes the serine protease inhibitor LEKTI (lymphoepithelial Kazal type-related inhibitor). LEKTI is expressed in the most differentiated viable layer of the skin (the granular cell layer), just internal to the cornified cell layer of the epidermis. Absence of LEKTI in Netherton's syndrome results in overly active epidermal kallikreins, proteases that can cleave desmosomes in the skin, leading to keratinocyte shedding and disturbed skin barrier function. Overly active kallikrein 5 leads to overexpression in the skin of TNF-α, ICAM-1, IL-8, and thymic stromal lymphopoietin (TSLP). TSLP is a major agonist of allergic manifestations in the skin, and is essential for the development of both the eczematous skin lesions and the allergic manifestations (including food allergy) seen in Netherton's syndrome. Additionally, LEKTI is thought to inhibit the proteases released by bacteria such as *Staphylococcus aureus*. This may be of special significance in the eczematous process, since a very large fraction of individuals with chronic eczema show persistent colonization with *S. aureus* and resolution of the eczema is facilitated by elimination of the *Staphylococcus*, in addition to suppression of the inflammatory response.

The observation that loss-of-function mutations in a protease inhibitor in Netherton's syndrome led to the development of multiple allergies provides additional support for the possibility that protease inhibitors might be novel therapeutic targets in some allergic disorders. Furthermore, the cysteine protease papain, derived from the papaya fruit, is used as a meat. Papain can cause allergic reactions in workers preparing the enzyme. Allergies caused by environmental allergens present in the workplace are called **occupational allergies.** Although Der p 1 and papain are potent allergens, not all allergens are enzymes. In fact, two allergens identified from filarial worms are enzyme inhibitors, and, in general, most allergenic pollen-derived proteins do not seem to possess enzymatic activity.

Knowledge of the identity of allergenic proteins can be important to public health and can have economic significance, as illustrated by the following cautionary tale. Some years ago, the gene for 2S albumin from Brazil nuts, a protein that is rich in methionine and cysteine, was transferred by genetic engineering into soybeans intended for animal feed. This was done to improve the nutritional value of soybeans, which are intrinsically poor in these sulfur-containing amino acids. This experiment led to the discovery that 2S albumin is the major Brazil nut allergen. Injection of extracts of the genetically modified soybeans into the epidermis triggered an allergic skin response in people allergic to Brazil nuts. As there could be no guarantee that the modified soybeans could be kept out of the human food chain if they were produced on a large scale, development of this genetically modified food was abandoned.

14-3 Genetic factors contribute to the development of IgE-mediated allergic disease.

Susceptibility to development of allergic disease has both genetic and environmental components. In studies performed in Western industrialized countries, up to 40% of the test population shows an exaggerated tendency to mount IgE responses to a wide variety of common environmental allergens. Atopic individuals often develop two or more allergic diseases such as allergic rhinoconjunctivitis, allergic asthma, or allergic eczema. Individuals who manifest all three of these disorders are said to express the atopic triad.

Genome-wide association studies (GWASs) have uncovered more than 40 susceptibility genes for the allergic skin condition atopic eczema (also known as atopic dermatitis) and for allergic asthma (**Fig. 14.6**). Some of the susceptibility genes are common to both atopic eczema and allergic asthma, suggesting that some aspects of the atopic diathesis (that is, predisposition) are governed by similar genetic factors regardless of the organs that are the targets of the allergic response. For example, specific alleles at the IL-33 receptor and IL-13 loci show strong association with both allergic asthma and atopic eczema. This sharing of genetic risk alleles by allergic asthma and atopic eczema is consistent with the finding that these two disorders are commonly found together in atopic families, with some family members manifesting both disorders, while other family members may manifest only atopic eczema or allergic asthma, but not both. There are, however, alleles of many genes (especially genes that regulate skin-barrier function) that show linkage to atopic eczema without enhancing the risk for allergic asthma or allergic rhinoconjunctivitis, indicating that other genetic factors contribute importantly to the phenotype of allergic responsiveness any individual may express. In addition, there are many ethnic differences in the susceptibility genes for a given allergic disease. Several of the chromosome regions associated with allergy or asthma are also associated with the inflammatory disease psoriasis and with autoimmune diseases, suggesting that these loci contain genes that are involved in exacerbating inflammation.

One candidate susceptibility gene for both allergic asthma and atopic eczema resides at chromosome 11q12–13 and encodes the β subunit of the high-affinity IgE receptor FcεRI. Another region of the genome associated with allergic disease, 5q31–33, contains at least four types of candidate genes that might be responsible for increased susceptibility. First, there is a cluster of tightly linked genes for cytokines that enhance IgE class switching, eosinophil

Asthma susceptibility loci
Genes expressed in airway epithelial cells
Chemokines: *CCL5, CCL11, CCL24, CCL26*
Antimicrobial peptides: *DEFB1*
Secretoglobin family: *SCGB1A1*
Epithelial barrier protein: *FLG*
Genes regulating CD4 T-cell and ILC2 differentiation and function
Transcription factors: *GATA3, TBX21, RORA, STAT3, PHF11, IKZF4*
Cytokines: *IL4, IL5, IL10, IL13, IL25, IL33, TGFβ1*
Cytokine receptors: *IL2RB, IL4RA, IL5RA, IL6R, IL18R, IL1RL1, FCER1B*
Pattern recognition receptors: *CD14, TLR2, TLR4, TLR6, TLR10, NOD1, NOD2*
Antigen presentation: *HLA-DRB1, HLA-DRB3, HLA-DQA, HLA-DQB, HLA-DPA, HLA-DPB, HLA-G*
Prostaglandin receptors: *PDFER2, PTGDR*
Genes with other functions
Proteinase or proteinase inhibitor: *ADAM33, USP38, SPINK5*
Signaling proteins: *IRAKM, SMAD3, PYHIN1, NOTCH4, GAB1, TNIP1*
Receptors: *ADRB2, P2X7*
Other: *DPP10, GPRA, COL29A1, ORMDL3, GSDMB, WDR36, DENND1B, RAD50, PBX2, LRRC32, AGER, CDK2*

Fig. 14.6 Susceptibility loci for asthma. Loci that have shown linkage on the basis of GWAS or targeted gene analysis are listed, separated into genes that are expressed in epithelial cells of the airways, genes that regulate the differentiation and/or function of CD4 T cells and ILC2s, and genes with other miscellaneous or unknown functions.

survival, and mast-cell proliferation, all of which help to produce and maintain an IgE-mediated allergic response. This cluster includes the genes for IL-3, IL-4, IL-5, IL-9, IL-13, and granulocyte–macrophage colony-stimulating factor (GM-CSF). In particular, genetic variation in the promoter region of the gene encoding IL-4 has been associated with raised IgE levels in atopic individuals. The variant promoter directs increased expression of a reporter gene in experimental systems and thus might produce increased IL-4 *in vivo*. Atopy has also been associated with a gain-of-function mutation of the α subunit of the IL-4 receptor, with the mutation causing increased signaling after ligation of the receptor.

A second set of genes in this region of chromosome 5 belongs to the TIM family (for $\underline{T}$ cell, $\underline{i}$mmunoglobulin domain, and $\underline{m}$ucin domain). The genes in this set encode three T-cell-surface proteins (Tim-1, -2, and -3) and one protein expressed primarily on antigen-presenting cells (Tim-4). In mice, Tim-3 protein is specifically expressed on T_H1 cells and negatively regulates T_H1 responses, whereas Tim-2 (and to a lesser extent Tim-1) is preferentially expressed in T_H2 cells and negatively regulates them. Mouse strains that carry different variants of the *Tim* genes differ both in their susceptibility to allergic inflammation of the airways and in the production of IL-4 and IL-13 by their T cells. Although no homolog of the mouse Tim-2 gene has been found in humans, inherited variation in the three human *TIM* genes has been correlated with **airway hyperreactivity** or **hyperresponsiveness**. In this condition, contact not only with allergen but also with nonspecific irritants causes airway narrowing (bronchoconstriction), with wheezy breathlessness similar to that seen in asthma. The third candidate susceptibility gene in this part of the genome encodes p40, one of the two subunits of IL-12 and IL-23. These cytokines promote T_H1 and T_H17 responses, and genetic variation in p40 expression that could cause reduced production of IL-12 and IL-23 was found to be associated with more severe asthma. A fourth candidate susceptibility gene which encodes the β-adrenergic receptor is also located in this region. Variation in this receptor might be associated with alteration in smooth muscle responsiveness to endogenous and pharmacological ligands.

The detection of multiple potential susceptibility genes illustrates a common challenge in identifying the genetic basis of complex disease traits. Relatively small regions of the genome, identified as containing genes for altered disease susceptibility, may contain many good candidates, judging by their known physiological activities. Identifying the gene, or genes, that truly lead to expression of the disease may require studies of several very large populations of patients and controls. For chromosome 5q31–33, for example, it is still too early to know how important each of the different polymorphisms is in the complex genetics of atopy.

A second type of inherited variation in IgE responses is linked to the HLA class II region (the human MHC class II region), and it affects responses to specific allergens, rather than a general susceptibility to atopy. IgE production in response to particular allergens is associated with certain HLA class II alleles, implying that particular peptide:MHC combinations might favor a strong T_H2 response; for example, IgE responses to several ragweed pollen allergens are associated with haplotypes containing the HLA class II allele *DRB1*1501*. Many people are therefore generally predisposed to make T_H2 responses and are specifically predisposed to respond to some allergens more than others. Allergic responses to drugs such as penicillin were originally thought to show no association with HLA class II or with the presence or absence of atopy. Recent studies, however, have shown evidence that some drugs can interact with specific HLA alleles in a way that changes the structure of peptide antigens bound in the groove of the HLA molecule so that these altered peptides elicit an autoimmune-type reaction. An example of this is the binding of the seizure medication carbamazepine with HLA-B15:02 and peptide bound in this HLA-B allele. The immune response to this carbamazepine:peptide:HLA-B complex can lead to development of toxic epidermal necrolysis, a severe immune-mediated skin reaction in which widespread skin loss occurs due to necrosis, leaving the skin looking scalded.

There are also likely to be genes that affect only particular aspects of allergic disease. In asthma, for example, there is evidence that different genes affect at least three aspects of the disease—IgE production, the inflammatory response, and clinical responses to particular treatments. On chromosome 20, polymorphism of the gene encoding ADAM33, a metalloproteinase expressed by bronchial smooth muscle cells and lung fibroblasts, has been associated with asthma and bronchial hyperreactivity. This is likely to be an example of genetic variation in the pulmonary inflammatory response and in the pathological anatomical changes that occur in the airways (airway remodeling). In the skin, filaggrin importantly contributes to normal skin barrier function by binding keratin molecules into the lipid envelope of cornifying keratinocytes. Loss-of-function mutations in the gene encoding filaggrin lead to the development of eczema. Through unknown mechanisms, filaggrin mutations also can contribute to the development of asthma. Almost half of people in the United States who suffer from severe eczema have at least one mutated filaggrin allele. Between 7 and 10% of Caucasians carry a loss-of-function mutation in filaggrin, and the frequency of this mutation is considerably higher in individuals with asthma.

14-4 Environmental factors may interact with genetic susceptibility to cause allergic disease.

Studies of susceptibility suggest that environmental factors and genetic variation each account for about 50% of the risk of developing atopy. The prevalence of atopic allergic diseases, and of asthma in particular, is increasing in economically advanced regions of the world, and this is likely due to the impact of changes in certain environmental factors on individuals with genetic backgrounds that predispose them toward atopy. Interestingly, although the incidence of asthma is lower in economically underdeveloped regions of Africa, Americans of African ancestry show increased asthma frequency and severity compared with Americans not of African ancestry. This shows a clear impact of environment on the expressivity of genetic influences.

The prevalence of atopy and especially allergic asthma has been steadily increasing in the developed world for the past 50–60 years. One hypothesis for this steady increase is changes in exposure to infectious diseases in early childhood as our population has moved increasingly from rural into urban environments. This shift has meant less early-life exposure to microorganisms associated with farm animals and microorganisms in the soil. This change in exposure is thought to lead to alterations in the intestinal microbiota, which performs an important immunomodulatory function (discussed in Chapter 12). Changes in exposure to ubiquitous microorganisms as a possible cause of increased atopy was first suggested in 1989, ultimately giving rise to the **hygiene hypothesis** (Fig. 14.7). The original proposition was that less hygienic environments, particularly those encountered in less developed rural settings, predispose to infections early in childhood, which help to protect against the development of atopy and allergic asthma. It was originally proposed that the protective effect might be due to mechanisms that skew immune responses away from the production of T_H2 cells (and their associated cytokines, which dispose toward IgE production) and toward the production of T_H1 cells. This would impede responses that favor the production of IgE and promote responses that suppress class switching to IgE.

Suggesting that this interpretation was overly simplistic was the strong negative correlation between infection by helminths (such as hookworms and schistosomes) and the development of allergic disease. A study in Venezuela showed that children treated for a prolonged period with antihelminthic agents had

Allergic Asthma

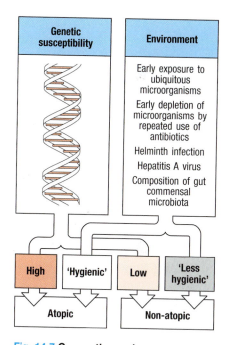

Fig. 14.7 Genes, the environment, and atopic allergic diseases. Both inherited and environmental factors are important determinants of the likelihood of developing allergic disease. Many genes are known to influence the development of asthma (see Fig. 14.6). The postulate of the 'hygiene hypothesis' is that exposure to some infections and to common environmental microorganisms in infancy and childhood drives the immune system toward a general state of non-atopy. In contrast, children who have a genetic susceptibility to atopy and who live in an environment with low exposure to infectious disease and environmental microorganisms, or who received multiple courses of antibiotics in infancy and early childhood, are thought not to develop efficient immunoregulatory mechanisms and to be most susceptible to the development of atopic allergic disease.

a higher prevalence of atopy than did untreated and heavily parasitized children. As helminths provoke a strong T_H2-mediated IgE response, this seemed inconsistent with the hygiene hypothesis.

One possible explanation for this apparent inconsistency suggests that all types of infection might protect against the development of atopy because the host responses they elicit include the production of cytokines such as IL-10 and TGF-β, perhaps as part of the homeostatic responses that occur as the infection is being controlled. IL-10 and TGF-β suppress the formation of both T_H1 and T_H2 responses, and IL-10 suppresses T_H17 responses (see Sections 9-21 and 9-23). A large proportion of allergic reactions are initiated by antigens that enter though mucosal surfaces such as the respiratory or intestinal epithelium. As described in Chapter 12, the human mucosal immune system has evolved mechanisms of regulating responses to commensal flora and environmental antigens (such as food antigens) that involve the generation of IL-10/TGF-β-producing T_{reg} cells. The idea underlying the current version of the hygiene hypothesis is that decreased early exposure to common microbial pathogens and commensals in some way makes the body less efficient at producing these T_{reg} cells, thus increasing the risk of making an allergic response to a common environmental antigen.

In support of a role for disturbed immunoregulatory pathways in susceptibility to asthma is evidence that exposure to certain types of childhood infection, with the important exception of some respiratory infections that we consider below, helps to protect against the development of allergic disease. Younger children from families with three or more older siblings and children aged less than 6 months who are exposed to other children in daycare facilities—situations linked to a greater exposure to infections—appear to be partially protected against atopy and asthma. Additionally, children with early-life exposure to a farm or ones having a dog are also somewhat protected against development of atopy and asthma, presumably because of their exposure to farm- or pet-associated microbes. Furthermore, early colonization of the gut by commensal bacteria such as lactobacilli and bifidobacteria, or infection by gut pathogens such as *Toxoplasma gondii* or *Helicobacter pylori*, is associated with a reduced prevalence of allergic disease. There is also emerging evidence that, conversely, repeated exposure to antibiotics in early life increases the risk of developing asthma.

A history of infection with hepatitis A virus also seems to have a negative association with atopy. A possible explanation for this association is that the human counterpart of the murine Tim-1 protein (see Section 14-3) is the cellular receptor for hepatitis A virus (designated HAVCR1). The infection of T cells by hepatitis A virus could thus directly influence their differentiation and cytokine production, limiting the development of an IgE-generating response.

In contrast to these negative associations between childhood infection and the development of atopy and asthma is evidence that children who have had attacks of bronchiolitis associated with respiratory syncytial virus (RSV) infection are more prone to developing asthma later on. Children hospitalized with RSV infection have a skewed ratio of cytokine production away from IFN-γ toward IL-4, presumably increasing their likelihood of developing T_H2 responses and increased production of IgE. This effect of RSV appears to depend on age at first infection. Experimental infection of neonatal mice with RSV was followed by lower increases in the production of IFN-γ compared with mice that received experimental RSV infection at 4 or 8 weeks of age. When the mice were rechallenged at 12 weeks of age with RSV infection, animals that had been primarily infected as neonates had more severe lung inflammation than animals first infected at 4 or 8 weeks of age.

Other environmental factors that might contribute to the increase in atopic disease are changes in diet, allergen exposure, atmospheric pollution, and tobacco smoke. Pollution has been blamed for an increase in the prevalence of nonallergic cardiopulmonary diseases such as chronic bronchitis, but an association with allergic disease has been more difficult to demonstrate. There is, however, increasing evidence for an interaction between allergens and pollution, particularly in genetically susceptible individuals. Diesel exhaust particles are the best-studied pollutant in this context; they increase IgE production 20- to 50-fold when combined with allergen, with an accompanying shift to T_H2 cytokine production. Reactive oxidant chemicals such as ozone are generated as a result of such pollution, and individuals less able to deal with this onslaught may be at increased risk of allergic disease.

Genes that might be governing this aspect of susceptibility are *GSTP1* and *GSTM1*, members of the glutathione-*S*-transferase superfamily that are important in preventing oxidant stress. Individuals who were allergic to ragweed pollen and who carried particular variant alleles of these genes showed an increased airway hyperreactivity when challenged with the allergen-plus-diesel exhaust particles, compared with the allergen alone. A study in Mexico City on the effects of atmospheric ozone levels on atopic children with allergic asthma also found that the children carrying the null allele of *GSTM1* were more susceptible than noncarriers to airway hyperreactivity when exposed to given levels of ozone. Underscoring the potential for reactive oxygen species such as ozone and superoxide to contribute to asthma exacerbation, studies using mice indicate that airway myeloid cells that produce high levels of superoxide worsen antigen-induced airway hyperreactivity. Inhibitors of NADPH oxidase, required for the production of superoxide, reduce antigen-induced airway hyperreactivity in sensitized and challenged animals, whereas adoptive transfer of superoxide-producing myeloid cells into the airways of sensitized and challenged mice causes marked exacerbation of hyperresponsiveness.

14-5 Regulatory T cells can control allergic responses.

The observation that treatment of peripheral blood mononuclear cells (consisting primarily of lymphocytes and monocytes) from atopic individuals with anti-CD3 and anti-CD28 stimulates production of substantial quantities of T_H2 cytokines, whereas similar treatment of cells from non-atopic individuals does not, suggests that circulating leukocytes in atopic individuals have been previously stimulated in a fashion that programs them to generate type 2 responses. An increasing number of studies suggest that regulatory mechanisms that normally serve to suppress overly aggressive type 2 responses are also abnormal in subjects with atopy. When peripheral blood $CD4^+CD25^+$ T_{reg} cells from atopic individuals are co-cultured with polyclonally activated $CD4^+$ T cells, they are less effective at suppressing T_H2 cytokine production compared with similar T_{reg} cells from non-atopic individuals, and this defect is even more pronounced during the pollen season. More evidence for a role for T_{reg} cells in atopy comes from mice deficient in the transcription factor FoxP3, the master switch for producing both natural (thymus-derived) and some types of induced T_{reg} cells. These mice develop several manifestations of atopy, including increased numbers of blood eosinophils and increased levels of circulating IgE, as well as spontaneous allergic airway inflammation. Manipulation of the T_{reg} pathway can ameliorate experimental asthmatic inflammation in mice. Increasing expression of the anti-inflammatory enzyme indoleamine 2,3-dioxygenase (IDO) by treatment with IFN-γ or by unmethylated CpG DNA can induce the generation or activation of T_{reg} cells. Induction of IDO activity in resident dendritic cells in the lung by stimulation with CpG DNA enhances T_{reg} activity and ameliorates experimental asthma in mice. These findings suggest that therapies aiming to enhance T_{reg} function could be beneficial in asthma and other atopic disorders. Other immunoregulatory molecules that might serve as immunotherapeutic agents for treatment of asthma include the cytokines IL-35 and IL-27, which, like IL-10, can inhibit T_H2 responses. Alternatively, blockade of the cytokine IL-31 is anticipated to be therapeutically beneficial since IL-31 promotes T_H2-driven inflammation.

Summary.

Allergens are generally innocuous antigens that commonly provoke an IgE antibody response in susceptible individuals. Such antigens normally enter the body at very low doses by diffusion across mucosal surfaces and trigger a type 2 immune response. The differentiation of naive allergen-specific T cells into T_H2 cells is favored by cytokines such as IL-4 and IL-13. Allergen-specific T_H2 cells producing IL-4 and IL-13 drive allergen-specific B cells to produce IgE. The specific IgE produced in response to the allergen binds to the high-affinity receptor for IgE on mast cells and basophils. IgE production can be amplified by these cells because, upon activation, they produce IL-4 and express CD40 ligand. The tendency to IgE overproduction is influenced by both genetic and environmental factors. Once IgE has been produced in response to an allergen, reexposure to the allergen triggers an allergic response. We describe the mechanism and pathology of the allergic responses themselves in the next part of the chapter.

Effector mechanisms in IgE-mediated allergic reactions.

Allergic reactions are triggered when allergens cross-link preformed IgE bound to the high-affinity receptor FcεRI on mast cells. Mast cells line external mucosal surfaces and serve to alert the immune system to local infection. Once activated, they induce inflammatory reactions by secreting pharmacological mediators such as **histamine** stored in preformed granules and by synthesizing prostaglandins, leukotrienes, and platelet-activating factor from the plasma membrane. They also release various cytokines and chemokines after activation. In the case of an allergic reaction, they provoke unpleasant reactions to innocuous antigens that are not associated with invading pathogens that need to be expelled. The consequences of IgE-mediated mast-cell activation depend on the dose of antigen and its route of entry; symptoms range from the swollen eyes and rhinitis associated with contact of pollen with the conjunctiva of the eye and the nasal epithelium, to the life-threatening circulatory collapse that occurs in anaphylaxis (**Fig. 14.8**). The immediate reaction

Fig. 14.8 Mast-cell activation has different effects on different tissues.

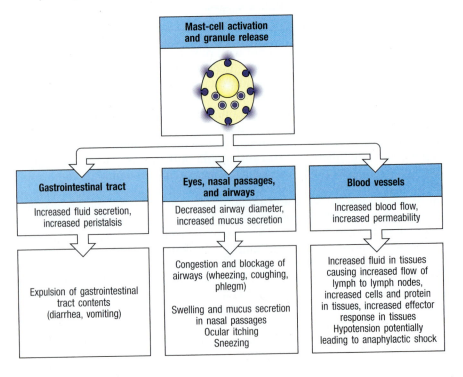

caused by mast-cell degranulation is followed, to a greater or lesser extent depending on the disease, by a more sustained inflammation, which is due to the recruitment of other effector leukocytes, notably T_H2 lymphocytes, eosinophils, and basophils.

14-6 Most IgE is cell-bound and engages effector mechanisms of the immune system by pathways different from those of other antibody isotypes.

Antibodies engage effector cells such as mast cells by binding to receptors specific for their Fc constant regions. Most antibodies engage Fc receptors only after their antigen-binding sites have bound specific antigen, forming an immune complex of antigen and antibody. IgE is an exception, because it is captured by the high-affinity Fcε receptor (FcεRI) in the absence of bound antigen. This means that, unlike other antibodies, which are found mainly in body fluids, IgE is mostly found fixed on cells that carry this receptor—mast cells in tissues, and basophils in the circulation and at sites of inflammation. The ligation of the cell-bound IgE antibody by specific multivalent antigen triggers the activation of these cells at the sites of antigen entry into the tissues. The release from these activated mast cells of inflammatory lipid mediators, cytokines, and chemokines at sites of IgE-triggered reactions recruits eosinophils and basophils to augment the allergic response. It also recruits T_H2 cells, which can then mount a local type 2 cellular response.

There are two types of IgE-binding Fc receptors. The first, FcεRI, expressed on mast cells and basophils, is a high-affinity receptor of the immunoglobulin superfamily (see Section 10-24). When IgE bound to this receptor is cross-linked by specific antigen, it transduces an activating signal through the receptor-bound Lyn tyrosine kinase, which phosphorylates ITAMs on the intracellular domain of the receptor. This recruits and activates the amplifying tyrosine kinase Syk, which phosphorylates and activates a broad range of downstream effector pathways. High levels of IgE, such as those that exist in people with allergic diseases or parasite infections, can result in a marked increase in FcεRI on the surface of mast cells, an enhanced sensitivity of such cells to activation by low concentrations of specific antigen, and a markedly increased, IgE-dependent release of chemical mediators and cytokines.

The second IgE receptor, FcεRII, usually known as **CD23**, is a C-type lectin and is structurally unrelated to FcεRI; it binds IgE with low affinity. CD23 is present on many cell types, including B cells, activated T cells, monocytes, eosinophils, platelets, follicular dendritic cells, and some thymic epithelial cells. This receptor was thought to be crucial for the regulation of IgE levels, but mouse strains in which the gene for CD23 has been inactivated still develop relatively normal polyclonal IgE responses. Nevertheless, CD23 does seem to be involved in enhancing IgE antibody levels in some situations. Responses against a specific antigen are known to be increased in the presence of the antigen complexed with IgE, but such enhancement fails to occur in mice that lack the gene for CD23. This has been interpreted to indicate that CD23 on antigen-presenting cells has a role in the capture of antigen that is complexed with IgE.

14-7 Mast cells reside in tissues and orchestrate allergic reactions.

When **Paul Ehrlich** described mast cells found in the mesentery of rabbits, he called them *Mastzellen* ('fattened cells'). Like basophils, mast cells contain granules rich in acidic proteoglycans that take up basic dyes. Mast cells are derived from hematopoietic stem cells but mature locally, often residing near surfaces exposed to pathogens and allergens, such as mucosal tissues and the connective tissues surrounding blood vessels. Mucosal mast cells differ in some of their properties from submucosal or connective tissue mast cells, but both can be involved in allergic reactions.

The major factors for mast-cell growth and development include stem-cell factor (the ligand for the receptor tyrosine kinase Kit), IL-3, and T_H2-associated cytokines such as IL-4 and IL-9. Mice with defective Kit lack differentiated mast cells, and although they produce IgE, they cannot make IgE-mediated inflammatory responses. This shows that such responses depend almost exclusively on mast cells. Mast-cell activation depends on the activation of phosphatidylinositol 3-kinase (PI 3-kinase) in mast cells by Kit, and pharmacological inactivation of the p110δ isoform of PI 3-kinase has been shown to protect mice against allergic responses. Inhibitors of the amplifying tyrosine kinase Syk are also showing promise as blockers of IgE-dependent mast-cell responses.

Mast cells express FcεRI constitutively on their surface and are activated when antigens cross-link IgE bound to these receptors (see Fig. 10.43). A relatively low level of allergen is sufficient to trigger degranulation. There are many mast-cell precursors in tissues, and they can rapidly differentiate into mature mast cells in conditions of allergic inflammation, thus aiding the continuation of the allergic response. Mast-cell degranulation begins within seconds of antigen binding, releasing an array of preformed and newly generated inflammatory mediators (Fig. 14.9). Granule contents include the short-lived vasoactive amine histamine, serine esterases, and proteases such as chymase and tryptase.

Histamine has four known receptors through which it acts—H_1 through H_4, each a G-protein-coupled receptor. Histamine acts via the H_1 receptor on local blood vessels to cause an immediate increase in local blood flow and vessel permeability. This leads to edema and local inflammation. Histamine is also a major stimulus for itching and sneezing, by virtue of its activation of neural receptors. Acting through the H_1 receptor on dendritic cells, histamine can increase antigen-presenting capacity and T_H1 cell priming; acting through the H_1 receptor on T cells, it can enhance T_H1 cell proliferation and IFN-γ production. By acting through H_2, H_3, and H_4 receptors on a variety of leukocytes and tissue cells, histamine participates in atopic dermatitis, chronic urticaria, and several autoimmune disorders.

Fig. 14.9 Molecules released by activated mast cells. Mast cells release a wide variety of biologically active proteins and other chemical mediators. The enzymes and toxic mediators listed in the first two rows are released from the preformed granules. The cytokines, chemokines, and lipid mediators are mostly synthesized after activation.

Class of product	Examples	Biological effects
Enzyme	Tryptase, chymase, cathepsin G, carboxypeptidase	Remodel connective tissue matrix
Toxic mediator	Histamine, heparin	Toxic to parasites Increase vascular permeability Cause smooth muscle contraction Anticoagulation
Cytokine	IL-4, IL-13, IL-33	Stimulate and amplify T_H2-cell response
Cytokine	IL-3, IL-5, GM-CSF	Promote eosinophil production and activation
Cytokine	TNF-α (some stored preformed in granules)	Promotes inflammation, stimulates cytokine production by many cell types, activates endothelium
Chemokine	CCL3	Attracts monocytes, macrophages, and neutrophils
Lipid mediator	Prostaglandins D_2, E_2 Leukotrienes C4, D4, E4	Smooth muscle contraction Chemotaxis of eosinophils, basophils, and T_H2 cells Increase vascular permeability Stimulate mucus secretion Bronchoconstriction
Lipid mediator	Platelet-activating factor	Attracts leukocytes Amplifies production of lipid mediators Activates neutrophils, eosinophils, and platelets

Human mast cells are classified into subtypes on the basis of their protease content and tissue location. Mast cells in mucosal epithelia express tryptase as their primary serine protease. These cells are designated MC_T. Mast cells in the submucosa and other connective tissues predominantly express chymase, tryptase, carboxypeptidase A, and cathepsin G and are designated MC_{CT}. The proteases released by the mast cells activate matrix metalloproteinases, which break down extracellular matrix proteins, causing tissue disintegration and damage. These proteases can exert beneficial effects such as degrading snake and bee venoms, thus helping to suppress allergic responses to these agents.

Following activation through FcεRI, in addition to releasing preformed mediators such as histamine and serine proteases that are stored in their intracellular granules, mast cells also synthesize *de novo* and release chemokines, cytokines, and lipid mediators—prostaglandins, leukotrienes, thromboxanes (collectively called eicosanoids), and platelet-activating factor. MC_T and MC_{CT} mast cells, for example, produce the cytokine IL-4, which helps perpetuate type 2 immune responses. These secreted products contribute to both acute and chronic inflammation. The lipid mediators, in particular, can act both rapidly and persistently to cause smooth muscle contraction, increased vascular permeability, and the secretion of mucus, as well as induce the influx and activation of leukocytes, which contribute to allergic inflammation.

Eicosanoids derive mainly from the membrane-associated fatty acid arachidonic acid. This is cleaved from membrane phospholipids by phospholipase A2, which is activated at the plasma membrane as a result of cell activation. Arachidonic acid can be modified by either of two pathways to give rise to lipid mediators. Modification via the cyclooxygenase pathway produces the prostaglandins and thromboxanes, whereas leukotrienes are produced via the lipoxygenase pathway. Prostaglandin D_2 is the major prostaglandin produced by mast cells and recruits T_H2 cells, eosinophils, and basophils, all of which express its receptor (PTGDR). Prostaglandin D_2 is critical to the development of allergic diseases such as asthma, and polymorphisms in the *PTGDR* gene have been linked to an increased risk of developing asthma. The leukotrienes, especially C4, D4, and E4, are also important in sustaining inflammatory responses in tissues. Nonsteroidal anti-inflammatory drugs such as aspirin and ibuprofen exert their effects by preventing prostaglandin production. They inhibit the cyclooxygenases that act on arachidonic acid to form the ring structure present in prostaglandins.

Large amounts of the cytokine tumor necrosis factor (TNF)-α are also released by mast cells after activation. Some comes from stores in the granules; some is newly synthesized by the activated mast cells. TNF-α activates endothelial cells, resulting in increased expression of adhesion molecules, which in turn promotes the influx of pro-inflammatory leukocytes and lymphocytes into the affected tissue (see Chapter 3). Additionally, mast-cell TNF-α contributes importantly to the influx of leukocytes into regional lymph nodes in response to microbial infection of peripheral tissues.

Through the action of all of these mediators, IgE-mediated mast-cell activation orchestrates a broad inflammatory cascade that is amplified by the recruitment of several types of leukocytes including eosinophils, basophils, T_H2 lymphocytes, and B cells. The biological role of this reaction in normal host immunity is as a defense against parasite infection (see Section 10-25). In an allergic reaction, however, the acute and chronic inflammatory reactions triggered by mast-cell activation have important pathophysiological consequences, as seen in the diseases associated with allergic responses to environmental antigens. The role of mast cells is not, however, limited to IgE-driven pro-inflammatory responses. Increasingly, mast cells are also considered to have a role in immunoregulation. They can be stimulated by neuropeptides such as substance P and by TLR ligands. In response to multiple stimuli, they can secrete the immunosuppressive cytokine IL-10, suppressing

T-cell responses. Conversely, interactions between mast cells and regulatory T cells can prevent mast-cell degranulation.

14-8 Eosinophils and basophils cause inflammation and tissue damage in allergic reactions.

Eosinophils are granulocytic leukocytes that originate in bone marrow. They are so called because their granules, which contain arginine-rich basic proteins, are colored bright orange by the acidic stain eosin. In healthy humans, these cells represent less than 6% of the leukocytes in the circulation; most eosinophils are found in tissues, especially in the connective tissue immediately underneath respiratory, gut, and urogenital epithelium, implying a likely role for these cells in defense against invading organisms at these sites. They possess numerous cell-surface receptors, including receptors for cytokines (such as IL-5), Fcγ and Fcα receptors, and the complement receptors CR1 and CR3, through which they can be activated and stimulated to degranulate. For example, parasites coated with IgG, C3b, or IgA can cause eosinophil degranulation. In allergic tissue reactions, the large concentrations of IL-5, IL-3, and GM-CSF that are typically present are likely to contribute to degranulation.

When activated, eosinophils express two kinds of effector function. First, they can release highly toxic granule proteins and free radicals, which can kill microorganisms and parasites but also can cause significant damage to host tissues in allergic reactions (Fig. 14.10). Second, they can synthesize chemical mediators, including prostaglandins, leukotrienes, and cytokines. These amplify the inflammatory response by activating epithelial cells and by recruiting and activating more eosinophils and leukocytes. In chronic inflammatory responses, eosinophils can contribute to airway tissue remodeling.

What were later to be defined as eosinophils were observed in the 19th century in the first pathological description of fatal status asthmaticus (an episode of severe asthma that does not respond to treatment and leads to respiratory

Fig. 14.10 Eosinophils secrete a range of highly toxic granule proteins and other inflammatory mediators. As for mast cells (see Fig. 14.9), enzymes and toxic mediators released by eosinophils are largely stored preformed in granules. In contrast, cytokines, chemokines, and lipid mediators are largely synthesized after eosinophil activation.

Class of product	Examples	Biological effects
Enzyme	Eosinophil peroxidase	Toxic to targets by catalyzing halogenation Triggers histamine release from mast cells
	Eosinophil collagenase	Remodels connective tissue matrix
	Matrix metalloproteinase-9	Matrix protein degradation
Toxic protein	Major basic protein	Toxic to parasites and mammalian cells Triggers histamine release from mast cells
	Eosinophil cationic protein	Ribonuclease Toxic to parasites Neurotoxin
	Eosinophil-derived neurotoxin	Neurotoxin
Cytokine	IL-3, IL-5, GM-CSF	Amplify eosinophil production by bone marrow Eosinophil activation
	TGF-α, TGF-β	Epithelial proliferation, myofibroblast formation
Chemokine	CXCL8 (IL-8)	Promotes influx of leukocytes
Lipid mediator	Leukotrienes C4, D4, E4	Smooth muscle contraction Increase vascular permeability Increase mucus secretion Bronchoconstriction
	Platelet-activating factor	Attracts leukocytes Amplifies production of lipid mediators Activates neutrophils, eosinophils, and platelets

failure and death), but the precise role of these cells in allergic disease generally is still unclear. In allergic tissue reactions, for example, those that lead to chronic asthma, mast-cell degranulation, and T_H2 activation cause eosinophils to accumulate in large numbers and to become activated. Among other things, eosinophils secrete T_H2-type cytokines and *in vitro* can promote the apoptosis of T_H1 cells by their expression of IDO and consequent production of kynurenine, which acts on the T_H1 cells. Their apparent promotion of T_H2-cell expansion may thus be partly due to a relative reduction in T_H1-cell numbers. The continued presence of eosinophils is characteristic of chronic allergic inflammation, and eosinophils are thought to be major contributors to tissue damage. However, the observation that eosinophils accumulate at sites where there are high levels of cell turnover and considerable local stem-cell activity bolsters a growing consensus that eosinophils play an important role in restoring tissue homeostasis after infection and other types of tissue damage.

The activation and degranulation of eosinophils is strictly regulated, because their inappropriate activation is harmful to the host. The first level of control acts on the production of eosinophils by the bone marrow. Few eosinophils are produced in the absence of infection or other immune stimulation. But when T_H2 cells are activated, cytokines they produce such as IL-5 and GM-CSF increase the production of eosinophils in the bone marrow and their release into the circulation. However, transgenic animals overexpressing IL-5 have increased numbers of eosinophils (**eosinophilia**) in the circulation but not in their tissues, indicating that the migration of eosinophils from the circulation into tissues is regulated separately, by a second set of controls. The key molecules in this case are the CC chemokines that have been named **eotaxins** because of their specificity for eosinophils: CCL11 (eotaxin 1), CCL24 (eotaxin 2), and CCL26 (eotaxin 3).

The eotaxin receptor on eosinophils, CCR3, is quite promiscuous and binds other CC chemokines, including CCL5, CCL7, and CCL13, which also induce eosinophil chemotaxis and activation. Identical or similar chemokines stimulate mast cells and basophils. For example, eotaxins attract basophils and cause their degranulation. T_H2 cells also carry the receptor CCR3 and migrate toward eotaxins.

Basophils are also present at the site of an inflammatory reaction, and growth factors for basophils are very similar to those for eosinophils; they include IL-3, IL-5, and GM-CSF. There is evidence for reciprocal control of the maturation of the stem-cell population into basophils or eosinophils. For example, TGF-β in the presence of IL-3 suppresses eosinophil differentiation and enhances that of basophils. Basophils are normally present in very low numbers in the circulation and seem to have a similar role to that of eosinophils in defense against pathogens. Like eosinophils, they are recruited to the sites of IgE-mediated allergic reactions. Basophils express high-affinity FcεRI on their cell surfaces and so have IgE bound. On activation by antigen binding to IgE or by cytokines, they release histamine from their granules and also produce IL-4 and IL-13.

Eosinophils, mast cells, and basophils can interact with each other. Eosinophil degranulation releases **major basic protein** (see Fig. 14.10), which in turn causes the degranulation of mast cells and basophils. This effect is augmented by any of the cytokines that affect eosinophil and basophil growth, differentiation, and activation, such as IL-3, IL-5, and GM-CSF.

14-9 IgE-mediated allergic reactions have a rapid onset but can also lead to chronic responses.

Under laboratory conditions, the clinical response of a sensitized individual to challenge by intradermal allergen or inhalation of allergen can be divided into an 'immediate reaction' and a 'late-phase reaction' (Fig. 14.11). The immediate

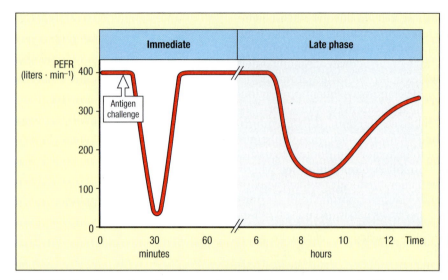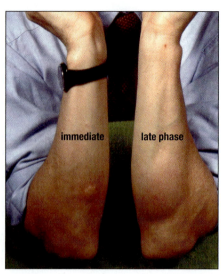

Fig. 14.11 Allergic reactions in response to test antigens can be divided into an immediate response and a late-phase response. Left panel: the response to an inhaled antigen can be divided into early and late responses. An asthmatic response in the lungs with narrowing of the airways caused by the constriction of bronchial smooth muscle and development of edema can be measured as a fall in the peak expiratory flow rate (PEFR). The immediate response peaks within minutes after antigen inhalation and then subsides, returning to near baseline PEFR. Six to eight hours after antigen challenge, there can be a late-phase response that also results in a fall in the PEFR. The immediate response is caused by the direct effects on blood vessels, nerves, and smooth muscle of rapidly metabolized mediators such as histamine and lipid mediators released by mast cells. The late-phase response is caused by the continued production of these mediators, by the production of vasoactive compounds that dilate blood vessels, and by recruitment of lymphocytes and myeloid cells, which together lead to the production of edema. Right panel: a wheal-and-flare allergic reaction develops within a minute or two of intradermal injection of antigen and lasts for up to 30–60 minutes. The more widespread edematous response characteristic of the late phase develops approximately 6 hours later and can persist for up to 2 or 3 days. The photograph shows an intradermal skin challenge with allergen resulting in a wheal-and-flare (early-phase) reaction observed 15 minutes after allergen challenge (left) and a late-phase reaction occurring 6 hours after challenge (right). The allergen was grass pollen extract. Photograph courtesy of S.R. Durham.

reaction is due to IgE-mediated mast-cell activation and starts within seconds of allergen exposure. It is the result of the actions of histamine, prostaglandins, and other preformed or rapidly synthesized mediators released by mast cells. These mediators cause a rapid increase in vascular permeability, resulting in visible edema and reddening of the skin (in a skin response) and airway narrowing as result of edema and the constriction of smooth muscle (in an airway response). In the skin, histamine acting on H_1 receptors on local blood vessels causes an immediate increase in vascular permeability, which leads to extravasation of fluid and edema. Histamine also acts on H_1 receptors on local nerve endings, leading to reflex vasodilation of cutaneous blood vessels and local reddening of the skin. The resulting skin lesion is called a **wheal-and-flare reaction** (see Fig. 14.11, right panel).

Whether a **late-phase reaction** occurs depends on allergen dose and on aspects of the cellular immune activation that are difficult to quantify. At doses of intradermally administered allergen that are deemed safe for skin testing of subjects with allergic asthma, for example, a late reaction occurs in about 50% of individuals who show an immediate response (see Fig. 14.11, right panel). The late reaction peaks between 3 and 9 hours after antigen challenge, and in skin tests becomes obvious as a much increased area and degree of edema (see Fig. 14.11, right panel) that can persist for 24 hours or longer. The late-phase reaction is caused by the continued synthesis and release of inflammatory mediators by mast cells, especially vasoactive mediators such as calcitonin gene-related peptide (CGRP) and vascular endothelial growth factor (VEGF), which cause vasodilation and vascular leakage that result in edema and the recruitment of eosinophils, basophils, monocytes, and lymphocytes. The importance of this cellular influx is shown by the ability of glucocorticoid medications to block the late-phase response through their inhibition of cell recruitment, whereas glucocorticoids do not block the immediate response. A late-phase reaction can also occur after aerosol exposure to allergen, and is characterized by a second phase of airway narrowing with sustained edema and cellular infiltration into the peribronchial spaces (see Fig. 14.11, left panel).

In patients with a clinical history of allergic disease, allergists use the immediate response to help assess and confirm sensitization, and to determine which allergens are responsible. Minute amounts of potential allergens are introduced into the skin by a skin prick—one site for each allergen—and if the individual is sensitive to any of the allergens tested, a wheal-and-flare reaction will occur at the site within a few minutes (see Fig. 14.11, right panel). Although the reaction after the administration of such small amounts of allergen is usually very localized, there is a small risk of inducing anaphylaxis. Another standard test for allergy is to measure the circulating concentration of IgE antibody specific for a particular allergen in a sandwich ELISA (see Appendix I, Section A-4).

The late-phase reaction described above occurs under controlled experimental conditions to a single, relatively high dose of allergen and so does not reflect all the effects of long-term natural exposure. In IgE-mediated allergic diseases, a long-term consequence of allergen exposure can be chronic allergic inflammation, which consists of a persistent type 2 immune response with a dominant cellular quality that is driven by T_H2 lymphocytes, basophils, eosinophils, and macrophages. These chronic reactions contribute importantly to serious long-term illnesses, such as chronic asthma. In long-standing asthma, for example, the cytokines released by T_H2 cells and vasoactive mediators such as calcitonin gene-related peptide and vascular endothelial growth factor result in persistent edema, which results in persistent narrowing of the airways. They can also lead to **airway tissue remodeling**, which changes the bronchial tissue via smooth muscle hypertrophy (an increase in the size of the muscle cells) and hyperplasia (an increase in the number of cells), subepithelial deposition of collagen, and often goblet cell hyperplasia. Although T_H2 cytokines appear to predominate in this chronic phase of asthma, T_H1 cytokines (such as IFN-γ) and T_H17 cytokines (IL-17, IL-21, and IL-22) can also participate.

In the natural situation, the clinical symptoms produced by an IgE-mediated allergic reaction depend critically on several variables: the amount of allergen-specific IgE present, the route by which the allergen is introduced, the dose of allergen, and most probably some underlying defect in barrier function in the particular tissue or organ affected. The outcomes produced by different combinations of allergen dose and route of entry are summarized in Fig. 14.12. When exposure to allergen in a sensitized individual triggers an allergic reaction, both the immediate and the chronic effects are focused on the site at which mast-cell degranulation occurs and they involve the recruitment of many soluble and cellular components of effector pathways.

14-10 Allergen introduced into the bloodstream can cause anaphylaxis.

If allergen is introduced directly into the bloodstream, for example, by a bee or wasp sting, or is rapidly absorbed into the bloodstream from the gut in a sensitized individual, connective-tissue mast cells associated with blood vessels throughout the body can become immediately activated, resulting in a widespread release of histamine and other mediators that causes the systemic reaction called **anaphylaxis**. The symptoms of anaphylaxis can range in severity from mild **urticaria** (hives) to fatal **anaphylactic shock** (see Fig. 14.12, first and last panels). Acute urticaria is a response to foreign allergens that are delivered to the skin via the systemic blood circulation. Activation of mast cells in the skin by allergen causes them to release histamine, which in turn causes itchy, red swellings all over the body—a disseminated version of the wheal-and-flare reaction. Although acute urticaria is commonly caused by an IgE-mediated reaction against an allergen, the causes of chronic urticaria, in which the urticarial rash persists or recurs over long periods, remain incompletely defined. Some cases of chronic urticaria are caused by autoantibodies

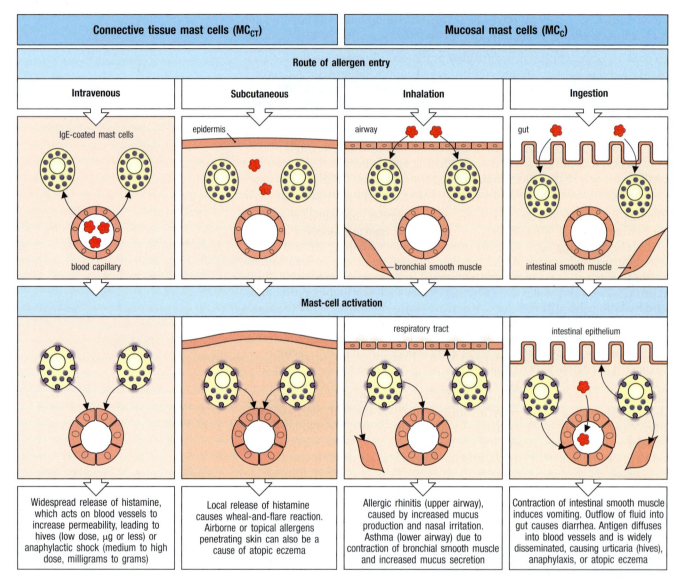

Connective tissue mast cells (MC_CT)		Mucosal mast cells (MC_C)	
Route of allergen entry			
Intravenous	Subcutaneous	Inhalation	Ingestion

Fig. 14.12 The route of administration of allergen determines the type of IgE-mediated allergic reaction that results. There are two main anatomical distributions of mast cells: those associated with vascularized connective tissues and called connective tissue mast cells (MC_CT), and those found in submucosal layers of the gut and respiratory tract and called mucosal mast cells (MC_C). In an allergic individual, all of these mast cells are coated through their cell-surface Fcε receptors with IgE directed against specific allergens. The response to an allergen then depends on which mast cells are activated. Allergen in the bloodstream (intravenous) activates connective tissue mast cells throughout the body, resulting in the systemic release of histamine and other mediators. Entry of allergen through the skin activates local connective tissue mast cells, leading to a local inflammatory reaction. After an experimental skin prick with allergen or bite from an insect to which the individual is sensitized, this manifests as a wheal-and-flare reaction. In atopic individuals, airborne or topically applied allergens that penetrate the skin may lead to atopic eczema. Inhaled allergen that penetrates respiratory mucosal epithelia activates mainly mucosal mast cells, causing increased secretion of mucus by the mucosal epithelium and irritation in the nasal mucosa, leading to allergic rhinitis—or to asthma if constriction of smooth muscle in the lower airways occurs. Ingested allergen penetrates the gut epithelium, causing vomiting due to intestinal smooth muscle contraction and diarrhea due to outflow of fluid across the gut epithelium. Food allergens can also be disseminated in the bloodstream, causing widespread urticaria (hives) when the allergen reaches the skin, or they may cause eczema.

against either the α chain of FcεRI or against IgE itself, and thus can be considered a form of autoimmunity. Interaction of the autoantibody with the receptor triggers mast-cell degranulation, with resulting urticaria. In some patients, treatment with omalizumab (a therapeutic monoclonal anti-IgE antibody) leads to resolution of hives, demonstrating a role for IgE in these individuals, even though the antigen that elicited the IgE often cannot be identified.

In anaphylactic shock, a widespread increase in vascular permeability and smooth muscle contraction results from a massive release of histamine and other mast cell- and basophil-derived mediators such as leukotrienes.

The consequences are a catastrophic reduction of blood pressure, culminating in hypotensive shock, (a condition in which low blood pressure leads to inadequate supply of blood to vital organs, often leading to death), and constriction of the airways, culminating in respiratory failure. The most common causes of anaphylaxis are allergic reactions to wasp and bee stings, ingested or injected medications, or allergic responses to foods in sensitized individuals. For example, anaphylaxis in individuals allergic to peanuts is relatively common. Severe anaphylactic shock can be rapidly fatal if untreated, but can usually be controlled by the immediate injection of epinephrine, which via stimulation of β-adrenergic receptors causes relaxation of airway smooth muscles, and via stimulation of α-adrenergic receptors reverses the life-threatening cardiovascular effects.

Systemic allergic reactions can occur following repeated treatment with many classes of drugs. A relatively common inducer of IgE-mediated allergic reaction is penicillin and other drugs that share aspects of its structure and immunological reactivity. In people who have developed IgE antibodies against penicillin, injection of the drug can cause anaphylaxis and even death. While administration of oral penicillin to an allergic individual can also cause anaphylaxis, the symptoms after oral ingestion are usually less severe and very rarely result in death. One of the reasons that penicillin is particularly prone to inducing allergic reactions is that it acts as a hapten (see Appendix I, Section A-1); it is a small molecule with a highly reactive β-lactam ring that is crucial for its antibacterial activity. This ring reacts with amino groups on host proteins to form covalent conjugates. When penicillin is ingested or injected, it forms conjugates with self proteins, and the penicillin-modified self peptides are recognized as foreign and elicit a host immune response. A large proportion of individuals who are treated with intravenous penicillin develop IgG antibodies against the drug, but these usually cause no symptoms. In some individuals, self proteins conjugated with penicillin provoke a T_H2 response that activates penicillin-binding B cells to produce IgE antibody against the penicillin hapten. Thus, penicillin acts both as the B-cell antigen and, by modifying self peptides, as the T-cell antigen. When penicillin is injected intravenously into an allergic individual, the penicillin-modified proteins can cross-link IgE molecules on tissue mast cells and circulating basophils and thus cause anaphylaxis. Great care should be taken to avoid giving a drug to patients with a past history of allergy to that drug or a close structural relative.

As is true for individuals with sensitivity to inhalant allergens, patients with a past history of anaphylactic-type reactions to penicillin or other β-lactam antibiotics can be evaluated by skin-prick testing. A positive skin test, manifested by the formation of a wheal-and-flare reaction at the site of the test, is associated with a substantial risk of developing an anaphylactic reaction when treated with therapeutic doses of the drug.

14-11 Allergen inhalation is associated with the development of rhinitis and asthma.

The respiratory tract is an important route of allergen entry (see Fig. 14.12, third panels). Many atopic people react to airborne allergens with an IgE-mediated allergic reaction known as **allergic rhinitis**. This results from the activation of mucosal mast cells beneath the nasal epithelium by allergens such as pollens that, when they contact the epithelium, release their soluble protein contents, which then diffuse across the mucous membranes of the nasal passages. Allergic rhinitis is characterized by intense itching and sneezing; local edema leading to blocked nasal passages; a nasal discharge, which is typically rich in eosinophils; and irritation of the nasal mucosa as a result of histamine release. A similar reaction to airborne allergens deposited on the conjunctiva of the eye is called **allergic conjunctivitis**. Allergic rhinitis and conjunctivitis are commonly caused by environmental allergens that are present only during certain seasons of the year. For example, hay fever (known

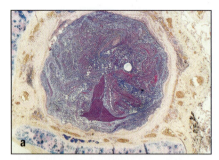

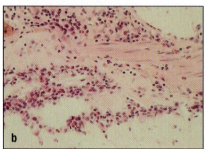

Fig. 14.13 Histologic evidence of chronic inflammation in the airways of an asthmatic patient. Panel a shows a section through a bronchus of a patient who died of asthma; there is almost total occlusion of the airway by a mucus plug. In panel b, a close-up view of the bronchial wall shows injury to the epithelium lining the bronchus, accompanied by a dense inflammatory infiltrate. Although not discernable at this magnification, the infiltrate includes eosinophils, neutrophils, and lymphocytes. Photographs courtesy of T. Krausz.

Allergic Asthma

clinically as **seasonal allergic rhinoconjunctivitis**) is caused by a variety of allergens, including certain grass and tree pollens. Symptoms in the late summer or autumn are commonly caused by weed pollen, such as that of ragweed, or the spores of fungi such as *Alternaria*. Ubiquitous allergens such as Fel d 1 in cat dander, Der p 1 in the feces of house dust mites, and Bla g 1 in cockroach can be a cause of year-round, or perennial, allergic rhinoconjunctivitis.

A more serious IgE-mediated respiratory disease is **allergic asthma**, which is triggered by allergen-induced activation of submucosal mast cells in the lower airways. This can lead within seconds to bronchial constriction and an increased secretion into the airways of fluid and mucus, making breathing more difficult by trapping inhaled air in the lungs. Patients with allergic asthma usually need treatment, and severe asthmatic attacks can be life threatening. The same allergens that cause allergic rhinitis and conjunctivitis commonly cause asthma attacks. For example, respiratory arrest caused by severe attacks of asthma in the summer or autumn has been associated with the inhalation of *Alternaria* spores.

Chronic allergen exposure leads to an important feature of asthma, namely, chronic inflammation of the airways, which is characterized by the continued presence of increased numbers of pathologic lymphocytes, eosinophils, neutrophils, basophils, and other leukocytes (Fig. 14.13). The concerted actions of these cells cause airway hyperreactivity and remodeling—a thickening of the airway walls due to hyperplasia and hypertrophy of the smooth muscle layer, with the eventual development of fibrosis. Fibrotic remodeling leads to a permanent narrowing of the airways, and is responsible for many of the clinical manifestations of chronic allergic asthma. In chronic asthmatics, a general hyperreactivity of the airways to nonimmunological stimuli such as perfumes or volatile irritants also often develops.

It has become apparent that there are many different phenotypic subtypes of asthma. These subtypes are being recognized because patients differ widely in responsiveness to different therapies, in the nature of the inflammatory cell infiltrates that are present in their airways, and in the molecular signature of inflammatory mediators that can be recovered from the airways. Many investigators refer to these subtypes as asthma 'endotypes.' The expectation is that classification of patients' asthmas by endotype will elucidate differences in the underlying pathophysiology of their disease and will improve therapeutic outcomes by permitting their therapy to be matched to the underlying molecular disorder that is leading to symptoms. Some of the most common endotypes include common allergic asthma, exercise-induced asthma, neutrophil-predominant (as opposed to eosinophil-predominant) asthma, and steroid-resistant severe asthma. The fundamental driver of the allergic response in allergic asthma is thought to be pathologically activated T_H2 cells, and eosinophils and basophils are prominent in the inflammatory infiltrates in the lungs. In severe, steroid-resistant asthma, T_H17 cells appear to play a larger role, and neutrophils are prominent in the inflammatory infiltrates. T_H17 cells also appear to be major inducers of the asthmatic syndrome allergic bronchopulmonary aspergillosis (ABPA). Other endotypes are characterized by the participation of additional leukocyte subsets and different effector-cell populations. The endotype of asthma of any individual patient is thought to be the result of the specific conditions under which the individual was sensitized to allergen and the specific predisposition of the individual based on inherited genetic factors and environmentally determined epigenetic factors.

For the following discussion of mechanisms of asthma, we will focus on the most common endotype, common allergic asthma. In patients with allergic asthma, allergen challenge can cause activation of mast cells in an antigen-specific, IgE-dependent fashion, leading to mast-cell mediator release. Allergens can also stimulate the airway epithelium directly, through TLRs and other damage receptors, to release IL-25 and IL-33. These cytokines can lead to the activation of submucosal type 2 innate lymphoid cells (ILC2s), inducing them to release IL-4, IL-5, IL-9, and IL-13. At the same time, bronchial epithelial cells can produce at least two of the chemokine ligands—CCL5 and CCL11—that bind to the receptor CCR3 expressed on T_H2 cells, macrophages, eosinophils, and basophils. Thus, these chemokines, together with the products of activated ILC2s, enhance the type 2 response by attracting more T_H2 cells and eosinophils to the damaged lungs. The direct effects of ILC2- and T_H2 cell-derived cytokines and chemokines on airway smooth muscle cells and fibroblasts lead to the apoptosis of epithelial cells and airway remodeling. The remodeling is induced in part by the production of TGF-β, which has numerous effects on the epithelium, ranging from inducing apoptosis to stimulating cell proliferation. The direct action of additional T_H2-type cytokines such as IL-9 and IL-13 on airway epithelial cells may also have a dominant role in another major feature of chronic allergic asthma, the induction of goblet-cell metaplasia, which is the increased differentiation of epithelial cells into goblet cells, and a consequent increase in mucus secretion. CD1d-restricted invariant NKT cells (iNKTs, a type of innate-like lymphocyte; see Sections 3-27, 6-18, and 8-26) also seem to have an important role in the development of airway hyperreactivity, whether allergen-induced or nonspecific, and this function can be enhanced by the cooperation of ILC2s. Animal models of asthma have shown that airway hyperreactivity is exacerbated by the presence of iNKT cells. Additionally, in mouse models, superoxide-producing myeloid lineage regulatory cells also appear to play pathological roles in the establishment of airway hyperreactivity.

Mice do not naturally develop asthma, but a disease resembling human asthma develops in mice that lack the transcription factor T-bet. This transcription factor is required for T_H1 differentiation (see Section 9-21). When T-bet is absent, T-cell responses are skewed toward the T_H2 phenotype. T-bet-deficient mice have increased levels of the T_H2 cytokines IL-4, IL-5, and IL-13, and develop airway inflammation involving lymphocytes and eosinophils (**Fig. 14.14**).

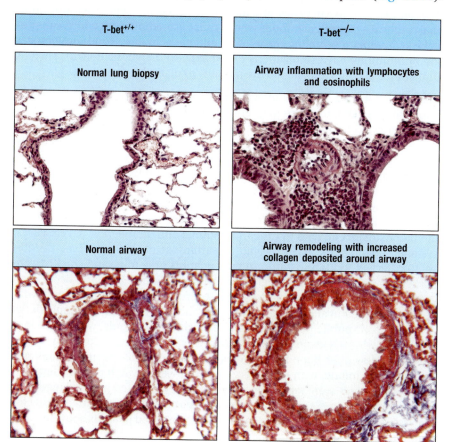

Fig. 14.14 Mice lacking the transcription factor T-bet develop allergic airway inflammation and T-cell responses polarized toward T_H2. T-bet binds to the promoter of the gene encoding IL-2 and is present in T_H1 but not T_H2 cells. Mice with a gene-targeted deletion of T-bet (T-bet$^{-/-}$) manifest impaired T_H1 responses, and show spontaneous differentiation of T_H2 cells and development of an asthma-like phenotype in the lungs. Left-hand panels: lung and airways in normal mice. Right-hand panels: T-bet-deficient mice show lung inflammation, with lymphocytes and eosinophils around the airway and blood vessels (top) and airway remodeling with increased collagen around the airway (bottom). Photographs courtesy of L. Glimcher.

They also develop nonspecific airway hyperreactivity to nonimmunological stimuli, similar to what is seen in human asthma. These changes occur in the absence of any exogenous inflammatory stimulus and show that, in extreme circumstances, a genetic imbalance toward T_H2 responses can cause allergic disease. The availability of a large number of genetically deficient mouse strains has permitted testing of the roles of many inflammatory effector cells and cytokines in this experimental model, providing hypotheses that are now being tested in human asthma.

Although allergic asthma is initially driven by a response to a specific allergen, the subsequent chronic inflammation seems to be perpetuated even in the apparent absence of ongoing exposure to allergen. The airways become characteristically hyperreactive, and factors other than antigen can trigger asthma attacks. Asthmatics characteristically show hyperresponsiveness to environmental chemical irritants such as cigarette smoke and sulfur dioxide. Viral agents, especially rhinovirus, or, to a smaller extent, bacterial respiratory tract infections can also exacerbate the disease. Both irritant agents and the infectious agents can induce IL-25 and IL-33 release from airway epithelial cells, leading to activation of ILC2s and exacerbation of the chronic asthmatic inflammation. The significance of viral augmentation of the asthmatic response is evident from the fact that rhinovirus infection is one of the main causes of hospitalizations for asthma and is associated with the majority of deaths from asthma.

14-12 Allergy to particular foods causes systemic reactions as well as symptoms limited to the gut.

Adverse reactions to particular foods are common, but only some are due to an immune reaction. 'Food allergy' can be classified into IgE-mediated allergic reactions, non-IgE-mediated food allergy (celiac disease, discussed in Section 14-17), idiosyncrasies, and food intolerance. Idiosyncrasies are abnormal responses to particular foods whose cause is unknown but which can provoke symptoms resembling those of an allergic response. Food intolerances are nonimmune adverse reactions due mainly to metabolic deficits, such as intolerance of cow's milk due to the inability to digest lactose.

IgE-mediated food allergies affect about 1–4% of American and European adults and are slightly more frequent in children (around 5%). About 25% of food allergy in children is due to peanuts, and peanut allergy is increasing in incidence. Figure 14.15 lists risk factors for developing IgE-mediated food allergy. IgE-mediated food allergy can manifest itself in a variety of ways, ranging from a swelling of the lips and oral tissue on contact with the allergen, to gastrointestinal cramping, diarrhea, or vomiting. Local gastrointestinal symptoms are due to activation of mucosal mast cells, leading to transepithelial fluid loss and smooth muscle contractions. Food allergens that subsequently reach the bloodstream can lead to urticaria, asthma, and, in the most severe cases, systemic anaphylaxis that can lead to cardiovascular collapse (see Section 14-10). Certain foods, most importantly peanuts, tree nuts, and shellfish, are particularly associated with severe anaphylaxis. Around 150 deaths occur each year in the United States as a result of a severe allergic reaction to food, with peanut and tree nut allergies accounting for most of the deaths. Peanut allergy is a significant public health problem, especially in schools, where children may be unwittingly exposed to peanuts, which are present in many foods. Recent studies offer hope for reducing the incidence of severe food allergy. In one study, infants with severe eczema who were at high risk for developing peanut allergy were randomly assigned either to be fed peanuts regularly, starting between ages 4 and 11 months, or to be on a peanut-avoidance diet for 5 years. At the age of 5, the children who had consumed peanuts showed more than a threefold reduction in the frequency of peanut allergy; the reduction

Risk factors for the development of food allergy
Immature mucosal immune system
Early introduction of solid food
Hereditary increase in mucosal permeability
IgA deficiency or delayed IgA production
Inadequate colonization of the intestinal immune system by commensal flora
Birth by cesarean section
Genetically determined bias toward a T_H2 environment
Polymorphisms of T_H2 cytokine or IgE receptor genes
Impaired enteric nervous system
Immune alterations (e.g., low levels of TGF-β)
Gastrointestinal infections

Fig. 14.15 Risk factors for the development of food allergy.

was associated with decreased production of peanut-specific IgE. This suggests that deliberate introduction into the diet of allergen at the appropriate time to at-risk individuals may suppress the development of food allergy.

Of interest is that one of the characteristic features of food allergens is a high degree of resistance to digestion by pepsin in the stomach. This allows them to reach the mucosal surface of the small intestine as intact allergens. Cases of IgE-mediated food allergies arising in small numbers of previously unaffected adults who were taking antacids or proton-pump inhibitors for ulcers or acid reflux have been proposed to be due to impaired digestion of potential allergens in the less acidic stomach conditions produced by these medications.

14-13 IgE-mediated allergic disease can be treated by inhibiting the effector pathways that lead to symptoms or by desensitization techniques that aim at restoring biological tolerance to the allergen.

Most of the current drugs that are used to treat allergic disease either treat only the symptoms—examples of such drugs are antihistamines and β-agonists—or are general anti-inflammatory or immunosuppressive drugs such as the corticosteroids (Fig. 14.16). Treatment is largely palliative, rather than curative, and the drugs often need to be taken throughout life. Anaphylactic reactions are treated with epinephrine, which stimulates the re-formation of endothelial

Treatments for allergic disease		
Target	**Mechanism of treatment**	**Specific approach**
In clinical use		
Mediator action	Inhibit effects of mediators on specific receptors Inhibit synthesis of specific mediators	Antihistamines, β-agonists Leukotriene receptor blockers Lipoxygenase inhibitors
Chronic inflammatory reactions	General anti-inflammatory effects	Corticosteroids
T_H2 response	Induction of regulatory T cells	Desensitization therapy by injections of specific antigen
IgE binding to mast cell	Bind to IgE Fc region and prevent IgE binding to Fc receptors on mast cells	Anti-IgE antibodies (omalizumab)
Proposed or under investigation		
T_H2 activation	Induction of regulatory T cells	Injection of specific antigen peptides Administration of cytokines, e.g., IFN-γ, IL-10, IL-12, TGF-β Use of adjuvants such as CpG oligodeoxynucleotides to stimulate T_H1 response
Activation of B cell to produce IgE	Block co-stimulation Inhibit T_H2 cytokines	Inhibit CD40L Inhibit IL-4 or IL-13
Mast-cell activation	Inhibit effects of IgE binding to mast cell	Blockade of IgE receptor
Eosinophil-dependent inflammation	Block cytokine and chemokine receptors that mediate eosinophil recruitment and activation	Inhibit IL-5 Block CCR3

Fig. 14.16 Approaches to the treatment of allergic disease. Examples of treatments in current clinical use for allergic reactions are listed in the top half of the table, with approaches under investigation listed below.

tight junctions, promotes the relaxation of constricted bronchial muscle, and stimulates the heart. Antihistamines that target the H_1 receptor reduce the symptoms that follow the release of histamine from mast cells in allergic rhinoconjunctivitis and IgE-triggered urticaria. In urticaria, for example, the relevant H_1 receptors include those on blood vessels and unmyelinated nerve fibers in the skin. Anticholinergic drugs bronchodilate constricted airways and reduce respiratory secretions. Antileukotriene drugs act as antagonists of leukotriene receptors on smooth muscle, endothelial cells, and mucous-gland cells, and are also used to relieve the symptoms of allergic rhinoconjunctivitis and asthma. Inhaled bronchodilators that act on β-adrenergic receptors to relax constricted muscle relieve acute asthma attacks. In chronic allergic disease it is extremely important to treat and prevent the chronic inflammatory injury to tissues, and regular use of inhaled corticosteroids is now recommended in persistent asthma to help suppress inflammation. Topical corticosteroids are used to suppress the chronic inflammatory changes seen in eczema.

A new type of allergy suppressive therapy that is beginning to gain significant use is blockade of IgE function by treatment with monoclonal anti-IgE antibodies, omalizumab being an example. This antibody binds the Fc portion of IgE at the same site that binds the FcεRI on basophils and mast cells. The portion of the Fc domain of IgE that binds to the low-affinity IgE receptor (FcεRII) that is expressed on a variety of leukocytes other than basophils and mast cells is different from the domain that binds the high-affinity FcεRI; omalizumab by steric hindrance blocks binding of IgE to the low-affinity receptor as well as to FcεRI. Prevention of binding of IgE to its receptors on basophils results in downregulation of these receptors on these cells, making them less easily activated by exposure to allergens. Omalizumab also appears to act in chronic allergic asthma to reduce IgE-mediated antigen trapping and presentation by dendritic cells, thus preventing the activation of new allergen-specific T_H2 cells. Altogether, these actions lead to suppression of the late-phase response to allergen challenge (see Section 14-9). The antibody is administered by subcutaneous injections once every 2 to 4 weeks. This treatment has been shown to be highly effective for patients with chronic urticaria and also appears to be effective in individuals with severe chronic asthma. Of special interest is that in studies of children with moderate to severe asthma who were treated for 4 years with omalizumab, most remained symptom free without any anti-asthma treatment, suggesting that the anti-IgE therapy modified the natural history of the disease.

Another, more routinely used approach that aims to permanently eliminate the allergic response is **allergen desensitization**. This form of immunotherapy aims to restore the patient's ability to tolerate exposure to the allergen. Patients are desensitized by injection with escalating doses of allergen, starting with tiny amounts. The mechanism by which desensitization occurs is not definitively established, but for most successfully desensitized patients, the procedure results in a change in the antibody response from one that is IgE predominant to one dominated by an IgG subclass. Successful desensitization appears to depend on the induction of T_{reg} cells secreting IL-10 and/or TGF-β, which skew the response away from IgE production (see Section 14-4). For example, beekeepers exposed to repeated stings (paralleling the therapeutic desensitization process) are often naturally protected from severe allergic reactions such as anaphylaxis through a mechanism that involves IL-10-secreting T cells. Similarly, specific allergen immunotherapy for sensitivity to insect venom and airborne allergens induces the increased production of IL-10 and in some cases TGF-β, as well as the production of IgG isotypes, particularly IgG4, an isotype selectively promoted by IL-10. Recent evidence shows that desensitization is also associated with a reduction in the numbers of inflammatory cells at the site of the allergic reaction. A potential complication of the desensitization approach is that in spite of starting with extremely small doses of allergen, some patients can experience an IgE-mediated allergic response, sometimes including bronchospasms. Thus, many physicians feel that allergen immunotherapy is contraindicated in patients with severe asthma. For patients who experience resolution of their allergy symptoms during allergen immunotherapy, weekly or every other week injections are continued for 3 years, and then the therapy is discontinued. In approximately half of patients treated in this fashion, symptoms do not recur following cessation of the injections. These patients experience durable ability to tolerate the allergen without symptoms. Recent studies suggest that administration of immunotherapy via the sublingual route is equally or more effective than administration by subcutaneous injection, offering the possibility of less expensive and perhaps more effective immunotherapy in the future.

When a patient is allergic to a drug that is essential for treatment of a disease (such as an antibiotic, insulin, or a chemotherapeutic agent), it is often possible to achieve a state of temporary **acute desensitization** by treating the individual with progressively increasing doses of the drug, starting at a very low dose that causes no allergic symptoms and increasing the dose every half hour until the therapeutic dose is reached. It is common for individuals undergoing drug desensitization to manifest mild to moderate allergic symptoms (itching, urticaria, mild wheezing) at some time during the procedure. If this occurs, the physician reduces the dose to the previous tolerated dose and then advances the dose again. This procedure is thought to lead to subclinical activation of mast cells and basophils that have been sensitized with IgE against the drug, inducing them to gradually release their intracellular mediators at a rate that does not cause severe symptoms; eventually all of the cell-bound IgE is consumed by this process, leaving insufficient IgE available to cause an allergic reaction when subsequent therapeutic doses are administered. In order to maintain the desensitized state, the patient must receive daily therapeutic doses of the drug. If treatment is interrupted, then newly formed mast cells and basophils can be charged with newly secreted drug-specific IgE and can accumulate at levels sufficient to yield a new anaphylactic reaction.

An alternative, and still experimental, immunotherapy approach is a vaccination strategy using allergen coupled to oligodeoxynucleotides rich in unmethylated CpG. The oligonucleotide mimics the CpG motifs in bacterial DNA and strongly promotes T_H1 responses while suppressing T_H2 responses. This appears to be useful for chronic treatment of an antigen-specific allergic response, but is not effective for acute desensitization.

A further approach to the treatment of allergic disease may be to block the recruitment of eosinophils to sites of allergic inflammation. The eotaxin receptor CCR3 is a potential target in this context. In experimental animals, the production of eosinophils in bone marrow and their exit into the circulation is reduced by blocking IL-5 action. Anti-IL-5 antibody (mepolizumab) is of benefit in treating human patients with the **hypereosinophilic syndrome**, in which chronic overproduction of eosinophils causes severe organ damage. Clinical trials of anti-IL-5 treatment of asthma, however, show that, in practice, any beneficial effect is likely to be limited to a small subset of asthma patients with prednisone-dependent eosinophilic asthma; in these patients, IL-5 blockade seems to reduce the number of asthma attacks when the corticosteroid dose is reduced.

Summary.

The allergic response to innocuous antigens reflects the pathophysiological aspects of a defensive immune response whose physiological role is to protect against helminth parasites. It is triggered by the binding of antigen to IgE antibodies bound to the high-affinity IgE receptor FcεRI on mast cells and basophils. Mast cells are strategically distributed beneath the mucosal surfaces of the body and in connective tissue. Antigen cross-linking the IgE on the surface of mast cells causes them to release large amounts of inflammatory mediators. The resulting inflammation can be divided into early events that are characterized by short-lived mediators such as histamine, and later events that involve leukotrienes, cytokines, and chemokines, which recruit and activate eosinophils, basophils, and other leukocytes. This response can evolve into chronic inflammation, which is characterized by the presence of effector T cells and eosinophils, and is most clearly seen in chronic allergic asthma.

Non-IgE-mediated allergic diseases.

In this part of the chapter we focus on immunological hypersensitivity responses involving IgG antibodies and type 1 or type 3 immune responses that involve antigen-specific T_H1 or T_H17 cells or CD8 T cells. These effector arms of the immune response occasionally react with noninfectious antigens to produce acute or chronic allergic reactions. Although the mechanisms initiating the various forms of hypersensitivity are different, much of the pathology is due to the same immunological effector mechanisms.

14-14 Non-IgE dependent drug-induced hypersensitivity reactions in susceptible individuals occur by binding of the drug to the surface of circulating blood cells.

Antibody-mediated destruction of red blood cells (hemolytic anemia) or platelets (thrombocytopenia) can be caused by some drugs, including the β-lactam antibiotics penicillin and cephalosporin. In these reactions, the drug binds covalently to the cell surface and is a target for anti-drug IgG antibodies that cause destruction of the cell. The anti-drug antibodies are made in only a minority of people, and it is not clear why these individuals make them. The cell-bound antibody triggers the clearance of the cell from the circulation, predominantly by tissue macrophages in the spleen, which bear Fcγ receptors.

14-15 Systemic disease caused by immune-complex formation can follow the administration of large quantities of poorly catabolized antigens.

Hypersensitivity reactions can arise following treatment with soluble antigens such as animal antisera. The pathology is caused by the deposition of antigen:antibody aggregates, or **immune complexes**, in particular tissues and sites. Immune complexes are generated in all antibody responses, but their pathogenic potential is determined, in part, by their size and by the amount, affinity, and isotype of the responding antibody. Larger aggregates fix complement and are readily cleared from the circulation by the mononuclear phagocyte system. However, the small complexes that form when antigen is in excess tend to be deposited in blood vessel walls. There they can ligate Fc receptors on leukocytes, leading to leukocyte activation and tissue injury.

A local hypersensitivity reaction called an **Arthus reaction** (Fig. 14.17) can be triggered in the skin of sensitized individuals who possess IgG antibodies against the sensitizing antigen. When antigen is injected into the skin, circulating IgG antibody that has diffused into the skin forms immune complexes locally. The immune complexes bind Fc receptors such as FcγRIII on mast cells and other leukocytes, generating a local inflammatory response and increased vascular permeability. Fluid and cells, especially polymorphonuclear leukocytes, then enter the site of inflammation from local blood vessels. The immune complexes also activate complement, leading to the production

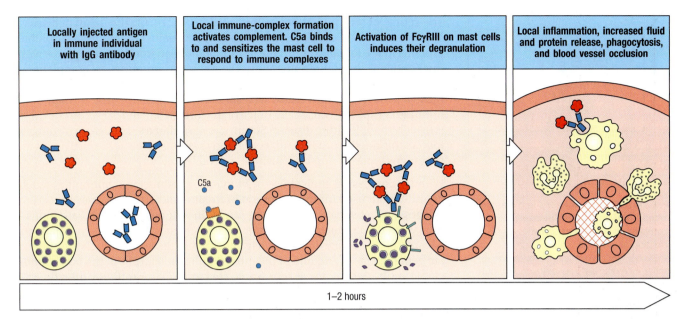

| Locally injected antigen in immune individual with IgG antibody | Local immune-complex formation activates complement. C5a binds to and sensitizes the mast cell to respond to immune complexes | Activation of FcγRIII on mast cells induces their degranulation | Local inflammation, increased fluid and protein release, phagocytosis, and blood vessel occlusion |

1–2 hours

Fig. 14.17 The deposition of immune complexes in tissues causes a local inflammatory response known as an Arthus reaction. In individuals who have already made IgG antibody against an antigen, the same antigen injected into the skin forms immune complexes with IgG antibody that has diffused out of the capillaries. Because the dose of antigen is low, the immune complexes are formed only close to the site of injection, where they activate mast cells bearing Fcγ receptors (FcγRIII). The immune complex induces activation of complement, and the complement component C5a contributes to sensitizing the mast cell to respond to immune complexes. As a result of mast-cell activation, inflammatory cells invade the site, and blood vessel permeability and blood flow are increased. Platelets also accumulate inside the vessel at the site, ultimately leading to vessel occlusion. If the reaction is severe, all these changes can lead to tissue necrosis.

of the complement fragment C5a. This is a key participant in the inflammatory reaction because it interacts with C5a receptors on leukocytes to activate these cells and attract them to the site of inflammation (see Section 2-5). Both C5a and FcγRIII have been shown to be required for the experimental induction of an Arthus reaction in the lung by macrophages in the walls of the alveoli, and they are probably required for the same reaction induced by mast cells in the skin and the synovial linings of joints. Recruitment and activation of C5a receptor-bearing leukocytes leads to tissue injury, sometimes resulting in frank necrosis.

A systemic reaction known as **serum sickness** can result from the injection of large quantities of a poorly catabolized foreign antigen. This illness was so named because it frequently followed the administration of therapeutic horse antiserum. In the pre-antibiotic era, antiserum made by immunizing horses with *Streptococcus pneumoniae* was often used to treat pneumococcal pneumonia; the specific anti-pneumococcal antibodies in the horse serum would help the patient to clear the infection. In much the same way, **antivenin** (serum from horses immunized with snake venoms) is still used today as a source of neutralizing antibodies to treat people suffering from the bites of poisonous snakes.

Serum sickness occurs 7–10 days after the injection of horse serum, an interval that corresponds to the time required to mount an IgG-switched primary immune response against the foreign horse serum antigens. The clinical features of serum sickness are chills, fever, rash, arthritis, and sometimes glomerulonephritis (inflammation of the glomeruli of the kidneys). Urticaria is a prominent feature of the rash, implying a role for histamine derived from mast-cell degranulation. In this case, the mast-cell degranulation is triggered by the ligation of cell-surface FcγRIII by IgG-containing immune complexes and by the anaphylatoxins C3a and C5a released due to complement activation by these complexes.

Drug-induced Serum Sickness

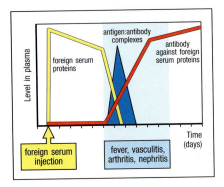

Fig. 14.18 Serum sickness is a classic example of a transient immune complex-mediated syndrome. An injection of a foreign protein, such as horse antitoxin, leads to an anti-horse serum antibody response. These antibodies form immune complexes with the circulating foreign proteins. The complexes are deposited in small blood vessels and activate complement and phagocytes, inducing fever and inflammatory lesions in blood vessels in the skin and connective tissues (vasculitis), in the kidney (nephritis), and in joints (arthritis). All these effects are transient and resolve when the foreign protein is cleared.

The course of serum sickness is illustrated in **Fig. 14.18**. The onset of disease coincides with the development of antibodies against the abundant soluble proteins in the foreign serum; these antibodies form immune complexes with their antigens throughout the body. The immune complexes fix complement and can bind to and activate leukocytes bearing Fc and complement receptors; these leukocytes in turn cause widespread tissue damage. The formation of immune complexes causes clearance of the foreign antigen, so serum sickness is usually a self-limiting disease. Serum sickness after a second dose of antigen follows the kinetics of a secondary antibody response (see Section 10-14), with symptoms typically appearing within a day or two.

With the increasing clinical use of humanized monoclonal antibodies (such as anti-TNF-α used for the treatment of rheumatoid arthritis), cases of serum sickness are observed, fortunately rarely, in settings where the attempt to humanize the monoclonal antibody was not successful for selected patients because they produce uncommon Ig allotypes. In these individuals, symptoms are generally mild, and one of the most significant features of the anti-monoclonal antibody response is more rapid clearance of the antibody from the circulation, leading to reduction of its therapeutic effects.

Pathological immune-complex deposition is seen in other situations in which antigen persists. One is when an adaptive antibody response fails to clear the infecting pathogen, as occurs in subacute bacterial endocarditis or chronic viral hepatitis. In these situations, the replicating pathogen is continuously generating new antigen in the presence of a persistent antibody response, with the consequent formation of abundant immune complexes. These are deposited within small blood vessels and result in injury in many tissues and organs, including the skin, kidneys, and nerves.

Immune-complex disease also occurs when inhaled allergens provoke IgG rather than IgE antibody responses, perhaps because they are present at relatively high levels in the air. When a person is reexposed to high doses of such allergens, immune complexes form in the walls of alveoli in the lungs. This leads to the accumulation of fluid, protein, and cells in the alveolar wall, slowing blood:air exchange of O_2 and CO_2, compromising lung function. This type of reaction is more likely to occur in occupations such as farming, in which there is repeated exposure to hay dust or mold spores, and the resulting disease is known as **farmer's lung**. If exposure to antigen is sustained, the lining of the lungs can be permanently damaged.

14-16 Hypersensitivity reactions can be mediated by T_H1 cells and CD8 cytotoxic T cells.

Unlike the immediate hypersensitivity reactions, which are mediated by antibodies, **cellular hypersensitivity reactions** such as **delayed-type hypersensitivity** reactions are mediated by antigen-specific effector T cells. We have already seen the involvement of T_H2 effector cells and the cytokines they produce in the chronic response of IgE-initiated allergic reactions. Here we consider the hypersensitivity diseases caused by T_H1 and CD8 cytotoxic T cells (**Fig. 14.19**). These cells function in hypersensitivity in essentially the same way as when they respond to a pathogen (described in Chapter 9), and the responses can be transferred between experimental animals by purified T cells or cloned T-cell lines. Much of the chronic inflammation seen in some of the allergic diseases described earlier is due to cellular hypersensitivity reactions mediated by antigen-specific T_H1 cells acting in concert with T_H2 cells.

The prototypic delayed-type hypersensitivity reaction is the **Mantoux test**—the standard tuberculin test that is used to determine whether an individual has previously been infected with *Mycobacterium tuberculosis*. In the Mantoux test, small amounts of tuberculin—a complex extract of peptides

Cellular hypersensitivity reactions are mediated by antigen-specific effector T cells		
Syndrome	**Antigen**	**Consequence**
Delayed-type hypersensitivity	Proteins: Insect venom Mycobacterial proteins (tuberculin, lepromin)	Local skin swelling: Erythema Induration Cellular infiltrate Dermatitis
Contact hypersensitivity	Haptens: Pentadecacatechol (poison ivy) DNFB Small metal ions: Nickel Chromate	Local epidermal reaction: Erythema Cellular infiltrate Vesicles Intraepidermal abscesses
Gluten-sensitive enteropathy (celiac disease)	Gliadin	Villous atrophy in small bowel Malabsorption

Fig. 14.19 Cellular hypersensitivity reactions. These responses are mediated by T cells and require 3–5 days or more to develop. They can be grouped into three syndromes, according to the route by which antigen passes into the body. In delayed-type hypersensitivity the antigen is injected into the skin; in contact hypersensitivity it is absorbed into the skin; and in gluten-sensitive enteropathy it is absorbed by the gut. In contact hypersensitivity, vesicles commonly form. They represent accumulations of fluid in small blister-like lesions at the level of the basement membrane between the dermis and epidermis. Their formation at this location is probably the result of antigen penetrating the epidermis, accumulating at the basement membrane, and inducing a local inflammatory response with edema fluid. DNFB (dinitrofluorobenzene) is a sensitizing agent that can cause contact hypersensitivity.

and carbohydrates derived from *M. tuberculosis*—are injected intradermally. In people who have been exposed to the bacterium, either by infection or by immunization with the BCG vaccine (an attenuated form of *M. tuberculosis*), a local T-cell-mediated inflammatory reaction evolves over 24–72 hours. The response is caused by T_H1 cells, which enter the site of antigen injection, recognize complexes of peptide:MHC class II molecules on antigen-presenting cells, and release inflammatory cytokines such as IFN-γ, TNF-α, and lymphotoxin. These stimulate the expression of adhesion molecules on endothelium and increase local blood vessel permeability, allowing plasma and accessory cells to enter the site, thus causing a visible swelling (**Fig. 14.20**). Each of these phases takes several hours and so the fully developed response only appears 24–48 hours after challenge. The cytokines produced by the activated T_H1 cells and their actions are shown in **Fig. 14.21**.

Very similar reactions are observed in **allergic contact dermatitis** (also called contact hypersensitivity), which is an immune-mediated local inflammatory reaction in the skin caused by direct skin contact with certain antigens. It is important to note that not all contact dermatitis is immune-mediated and allergic in nature; it can also be caused by direct damage to the skin by irritant or toxic chemicals.

 **MOVIE 14.1**

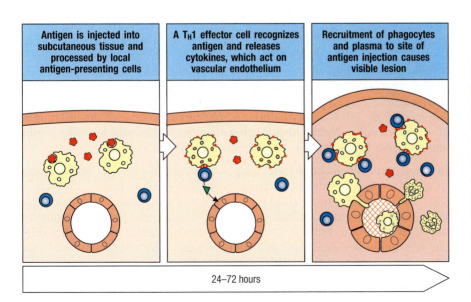

Antigen is injected into subcutaneous tissue and processed by local antigen-presenting cells	A T_H1 effector cell recognizes antigen and releases cytokines, which act on vascular endothelium	Recruitment of phagocytes and plasma to site of antigen injection causes visible lesion

24–72 hours

Fig. 14.20 The stages of a delayed-type hypersensitivity reaction. The first phase involves uptake, processing, and presentation of the antigen by local antigen-presenting cells. In the second phase, T_H1 cells that have been primed by a previous exposure to the antigen migrate into the site of injection and become activated. Because these specific cells are rare, and because there is little inflammation to attract cells into the site, it can take several hours for a T cell of the correct specificity to arrive. These cells release mediators that activate local endothelial cells, which recruit an inflammatory cell infiltrate dominated by macrophages and cause the accumulation of fluid, serum proteins, and more leukocytes, thus producing a visible lesion.

Fig. 14.21 The delayed-type hypersensitivity response is directed by chemokines and cytokines released by antigen-stimulated T$_H$1 cells. Antigen in the local tissues is internalized and processed by antigen-presenting cells and presented on MHC class II molecules. Antigen-specific T$_H$1 cells that recognize the antigen:MHC complexes locally at the site of antigen injection release chemokines and cytokines that recruit macrophages and other leukocytes to the site. Antigen presentation by the newly recruited macrophages then amplifies the response. T cells can also affect local blood vessels through the release of TNF-α and lymphotoxin (LT), and stimulate the production of macrophages through the release of IL-3 and GM-CSF. T$_H$1 cells activate macrophages through the release of IFN-γ and TNF-α, and kill macrophages and other sensitive cells through the cell-surface expression of the Fas ligand.

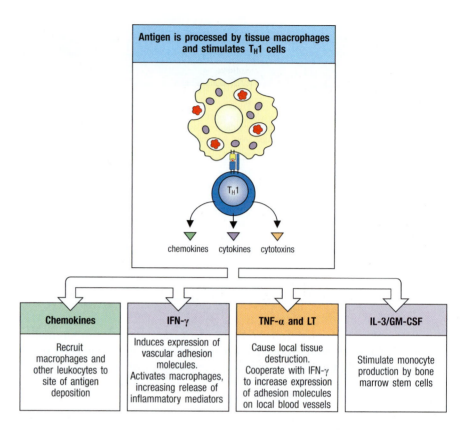

Antigen is processed by tissue macrophages and stimulates T$_H$1 cells

Chemokines	IFN-γ	TNF-α and LT	IL-3/GM-CSF
Recruit macrophages and other leukocytes to site of antigen deposition	Induces expression of vascular adhesion molecules. Activates macrophages, increasing release of inflammatory mediators	Cause local tissue destruction. Cooperate with IFN-γ to increase expression of adhesion molecules on local blood vessels	Stimulate monocyte production by bone marrow stem cells

Allergic contact dermatitis can be caused by the activation of CD4 or CD8 T cells, depending on the pathway by which the antigen is processed. Typical antigens that cause allergic contact dermatitis are highly reactive small molecules that can easily penetrate intact skin, especially if they cause itching that leads to scratching and its consequent injury to the skin barrier. These chemicals then react with self proteins, creating haptenated proteins that can be proteolytically processed in antigen-presenting cells to haptenated peptides capable of being presented by MHC molecules and recognized by T cells as foreign antigens. As with other allergic responses, there are two phases to a cutaneous allergic response: sensitization and elicitation. During the sensitization phase, Langerhans cells in the epidermis and dendritic cells in the dermis take up and process antigen, and migrate to regional lymph nodes, where they activate T cells (see Fig. 9.13) with the consequent production of memory T cells, which localize in the dermis. In the elicitation phase, a subsequent exposure to the sensitizing chemical leads to antigen presentation to memory T cells in the dermis, with the release of T-cell cytokines such as IFN-γ and IL-17. This stimulates the keratinocytes of the epidermis to release IL-1, IL-6, TNF-α, GM-CSF, the chemokine CXCL8, and the interferon-inducible chemokines CXCL11 (IP-9), CXCL10 (IP-10), and CXCL9 (Mig, a monokine induced by IFN-γ). These cytokines and chemokines enhance the inflammatory response by inducing the migration of monocytes into the lesion and their maturation into macrophages, and by attracting more T cells (**Fig. 14.22**).

Contact Sensitivity to Poison Ivy

The rash produced by contact with the poison ivy plant (**Fig. 14.23**) is a common example of allergic contact dermatitis in the United States and is caused by a CD8 T-cell response to urushiol oil (a mixture of pentadecacatechols) in the plant. These chemicals are lipid-soluble and so can cross the cell membrane and attach to intracellular proteins. The modified proteins are recognized by the immunoproteasome, and following cleavage, they are translocated into the endoplasmic reticulum and delivered to the cell surface bound to MHC class I molecules. CD8 T cells recognizing the peptides cause damage either by killing the eliciting cell or by secreting cytokines such as IFN-γ.

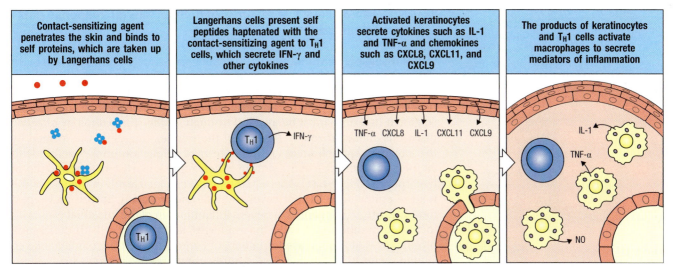

Contact-sensitizing agent penetrates the skin and binds to self proteins, which are taken up by Langerhans cells	Langerhans cells present self peptides haptenated with the contact-sensitizing agent to T_H1 cells, which secrete IFN-γ and other cytokines	Activated keratinocytes secrete cytokines such as IL-1 and TNF-α and chemokines such as CXCL8, CXCL11, and CXCL9	The products of keratinocytes and T_H1 cells activate macrophages to secrete mediators of inflammation

Fig. 14.22 Elicitation of a delayed-type hypersensitivity response to a contact-sensitizing agent. A contact-sensitizing agent is a small, highly reactive molecule that can penetrate intact skin. It binds covalently as a hapten to a variety of endogenous proteins, altering their structures so they become antigenic. These modified proteins are internalized, processed by Langerhans cells (the major antigen-presenting cells of skin), and presented to effector T_H1 cells (which were primed in lymph nodes as a result of prior antigen exposure). The activated T_H1 cells then secrete cytokines such as IFN-γ that stimulate keratinocytes to secrete additional cytokines and chemokines, which in turn attract monocytes and induce their maturation into activated tissue macrophages, further contributing to inflammatory lesions like those caused by poison ivy (see Fig. 14.23). NO, nitric oxide.

The ability of CD4 T cells to mediate contact hypersensitivity responses is established by experimental exposure to the strong sensitizing chemical picryl chloride. Picryl chloride modifies extracellular self proteins by haptenation. These haptenated proteins can then be proteolytically processed by APCs, yielding haptenated peptides that bind to self-MHC class II molecules and are recognized by T_H1 cells. When sensitized T_H1 cells recognize these complexes, they produce extensive inflammation by activating macrophages (see Fig. 14.22). Common clinical features of allergic contact hypersensitivity responses are erythema of the affected skin; development of a dermal and epidermal infiltrate consisting of monocytes, macrophages, lymphocytes, scant neutrophils, and mast cells; formation of intraepidermal abscesses; and vesicles (blisterlike collections of edema fluid between the dermis and epidermis).

Some insect proteins also elicit a delayed-type hypersensitivity response. One example of this in the skin is a severe reaction to mosquito bites. Instead of a small itchy bump, people allergic to proteins in mosquito saliva can develop an immediate hypersensitivity reaction such as urticaria and swelling or, much more rarely, anaphylactic shock (see Section 14-10). Some allergic individuals subsequently develop a delayed reaction (consisting of a late-phase response) that can include profound swelling that can involve an entire limb.

Contact hypersensitivity responses to divalent cations such as nickel have also been observed. These divalent cations can alter the conformation or the peptide binding of MHC class II molecules, and thus provoke a T-cell response. In humans, nickel can also bind to the receptor TLR-4 and produce a pro-inflammatory signal. Sensitization to nickel is widespread as a result of prolonged contact with nickel-containing items such as jewelry, buttons, and clothing fasteners, but some countries now have standards that specify that such products must have non-nickel coatings, and this is reducing the prevalence of nickel allergy in those countries.

Finally, although this section has focused on the role of T_H1 and cytotoxic T cells in inducing cellular hypersensitivity reactions, there is evidence that antibody and complement might also have a role. Mice deficient in B cells, antibody, or complement show impaired contact hypersensitivity reactions. In particular, IgM antibodies (produced in part by B1 cells), which activate the complement cascade, facilitate the initiation of these reactions.

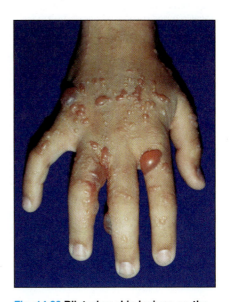

Fig. 14.23 Blistering skin lesions on the hand of a patient with allergic contact dermatitis caused by poison ivy. Photograph courtesy of R. Geha.

14-17 Celiac disease has features of both an allergic response and autoimmunity.

Celiac Disease

Celiac disease is a chronic condition of the upper small intestine caused by an immune response directed at gluten, a complex of proteins present in wheat, oats, and barley. Elimination of gluten from the diet restores normal gut function, but to date no approach for desensitization to gluten has been developed, so gluten ingestion must be avoided throughout life. The pathology of celiac disease is characterized by the loss of the slender, fingerlike villi formed by the intestinal epithelium (a condition termed villous atrophy), together with an increase in the size of the sites in which epithelial cells are renewed (crypt hyperplasia) (Fig. 14.24). These pathological changes result in the loss of the mature epithelial cells that cover the villi and normally absorb and digest food, and are accompanied by severe inflammation of the intestinal wall, with increased numbers of T cells, macrophages, and plasma cells in the lamina propria, as well as increased numbers of lymphocytes in the epithelial layer. Gluten seems to be the only food component that provokes intestinal inflammation in this way, a property that reflects gluten's ability to stimulate both innate and specific immune responses in genetically susceptible individuals. The incidence of celiac disease has increased fourfold in the past 60 years, correlating with changes in baking practice that include adding large amounts of extra gluten to dough to decrease the time required for the dough to rise and to improve the texture.

Celiac disease shows an extremely strong genetic predisposition, with more than 95% of patients expressing the HLA-DQ2 class II MHC allele. In monozygotic twins, if one twin develops it, there is an 80% probability that the other will, but only a 10% concordance is observed in dizygotic twins. Nevertheless, most individuals expressing HLA-DQ2 do not develop celiac disease despite the almost universal presence of gluten in the Western diet. Thus, other genetic or environmental factors must make important contributions to susceptibility.

Fig. 14.24 The pathological features of celiac disease. Left: the surface of the normal small intestine is folded into fingerlike villi, which provide an extensive surface for nutrient absorption. Right: the local immune response against the food protein α-gliadin, a prominent component of wheat, oat, and barley gluten, leads to massive infiltration of the lamina propria (in the deeper, inner portion of the villi) with CD4 T cells, plasma cells, macrophages, and smaller numbers of other leukocytes, ultimately leading to destruction of the villi. In parallel, there is lengthening and increased mitotic activity in the underlying crypts, where new epithelial cells are produced. Because the villi contain all the mature epithelial cells that digest and absorb foodstuffs, their loss can result in life-threatening malabsorption and diarrhea. Photographs courtesy of Allan Mowat.

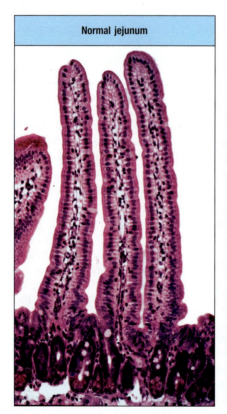
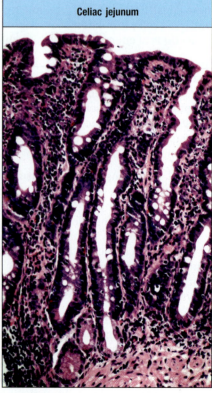

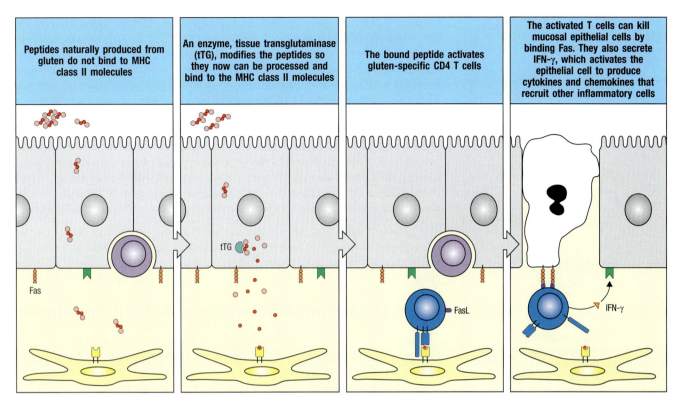

Fig. 14.25 Molecular basis of immune recognition of gluten in celiac disease. After the digestion of gluten by gut digestive enzymes, deamidation of epitopes by tissue transglutaminase renders the gluten more susceptible to being readily processed by local antigen-presenting cells, ultimately leading to its binding to HLA-DQ molecules and priming of the immune system.

Most evidence indicates that celiac disease requires the aberrant priming of IFN-γ-producing CD4 T cells by antigenic peptides present in α-gliadin, one of the major proteins in gluten. It is generally accepted that only a limited number of peptides can provoke an immune response leading to celiac disease. This is likely due to the unique structure of the peptide-binding groove of the HLA-DQ2 molecule. The key step in the immune recognition of α-gliadin is the deamidation of its peptides by the enzyme tissue transglutaminase (tTG), which converts selected glutamine residues to negatively charged glutamic acid. Only peptides containing negatively charged residues in certain positions bind strongly to HLA-DQ2, and thus the transamination reaction promotes the formation of peptide:HLA-DQ2 complexes, which can activate antigen-specific CD4 T cells (Fig. 14.25). Activated gliadin-specific CD4 T cells accumulate in the lamina propria, producing IFN-γ, a cytokine that when present in this location leads to intestinal inflammation.

Celiac disease is entirely dependent on the presence of the foreign antigen, gluten. It is not associated with a specific immune response against self antigens in the tissue—the intestinal epithelium—that is damaged during the immune response. Thus, celiac disease is not a classical autoimmune disease. But it does have some features of autoimmunity. Autoantibodies against tissue transglutaminase are found in all patients with celiac disease; indeed, the presence of serum IgA antibodies against this enzyme is used as a sensitive and specific test for the disease. Interestingly, no tTG-specific T cells have been found, and it has been proposed that gluten-reactive T cells provide help to B cells that are reactive to tissue transglutaminase. In support of this hypothesis, gluten can complex with the enzyme and therefore could be taken up and presented by tTG-reactive B cells (Fig. 14.26). There is no evidence, however, that these autoantibodies contribute directly to tissue damage.

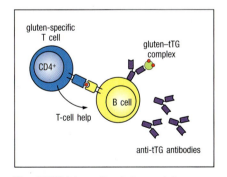

Fig. 14.26 A hypothesis to explain antibody production against tissue transglutaminase (tTG) in the absence of T cells specific for tTG in celiac patients. tTG-reactive B cells endocytose gluten–tTG complexes and present gluten peptides to the gluten-specific T cells. The stimulated T cells can now provide help to these B cells, which produce autoantibodies against tTG.

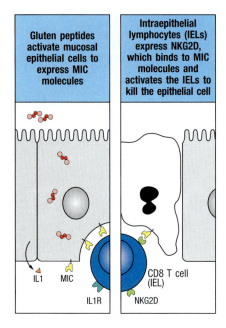

Gluten peptides activate mucosal epithelial cells to express MIC molecules

Intraepithelial lymphocytes (IELs) express NKG2D, which binds to MIC molecules and activates the IELs to kill the epithelial cell

IL1 MIC

IL1R

CD8 T cell (IEL)

NKG2D

Fig. 14.27 The activation of cytotoxic T cells by the innate immune system in celiac disease. Gluten peptides can induce the expression of the MHC class Ib molecules MIC-A and MIC-B on gut epithelial cells and the synthesis and release of IL-1 from these cells. Intraepithelial lymphocytes (IELs), many of which are CD8 cytotoxic T cells, recognize the MIC proteins via the receptor NKG2D, which, together with the co-stimulator IL-1, activates the IELs to kill the MIC-bearing cells, leading to destruction of the gut epithelium.

Chronic T-cell responses against food proteins are normally prevented by the development of oral tolerance (see Section 12-18). Why this breaks down in patients with celiac disease is unknown. The properties of the HLA-DQ2 molecule provide a partial explanation, but there must be additional factors because most HLA-DQ2-positive individuals do not develop celiac disease, and the high concordance rates in monozygotic twins indicate a role for additional genetic factors. The incidence of celiac disease is especially high in individuals with trisomy 21 (Down syndrome), approximately sixfold higher than in the normal population, underscoring the impact of genetic factors on disease prevalence. Polymorphisms in the gene for CTLA-4 or in other immunoregulatory genes have been suggested to be associated with susceptibility. There could also be differences in how individuals digest gliadin in the intestine, so that differing amounts survive for deamidation and presentation to T cells.

The gluten protein also seems to have several properties that contribute to pathogenesis. As well as its relative resistance to digestion, there is mounting evidence that some gliadin-derived peptides stimulate the innate immune system by inducing the release of IL-15 by intestinal epithelial cells. This process is antigen-nonspecific and involves peptides that cannot be bound by HLA-DQ2 molecules or recognized by CD4 T cells. IL-15 release leads to the activation of dendritic cells in the lamina propria, as well as the upregulation of MIC-A expression by epithelial cells. CD8 T cells in the mucosal epithelium can be activated via their NKG2D receptors, which recognize MIC-A, and they can kill MIC-A-expressing epithelial cells via these same NKG2D receptors (**Fig. 14.27**). Triggering of these innate immune responses by α-gliadin may create some intestinal damage on its own and also induce some of the co-stimulatory events necessary for initiating an antigen-specific CD4 T-cell response to other parts of the α-gliadin molecule. The ability of gluten to stimulate both innate and adaptive immune responses may thus explain its unique ability to induce celiac disease.

Summary.

Non-IgE-mediated immunological hypersensitivity reflects normal immune mechanisms that are inappropriately directed against innocuous antigens or inflammatory stimuli. It comprises both immediate-type and delayed-type reactions. Immediate-type reactions are due to the binding of specific IgG antibodies to allergen-modified cell surfaces, as in drug-induced hemolytic anemia, or to the formation of immune complexes of antibodies bound to poorly catabolized antigens, as occurs in serum sickness. Cellular hypersensitivity reactions mediated by T_H1 cells and cytotoxic T cells develop more slowly than immediate-type reactions. The T_H1-mediated hypersensitivity reaction in the skin provoked by mycobacterial tuberculin is used to diagnose previous exposure to *Mycobacterium tuberculosis*. The allergic reaction to poison ivy is due to the recognition and destruction by cytotoxic T cells of skin cells modified by a plant molecule, and to cytotoxic T-cell cytokines. These T-cell-mediated responses require the induced synthesis of effector molecules and develop over 1–10 days.

Summary to Chapter 14.

In susceptible individuals, immune responses to otherwise innocuous antigens can produce allergic reactions upon reexposure to the same antigen. Most allergic reactions involve the production of IgE antibody against common environmental allergens. Some people are intrinsically prone to making IgE antibodies against many allergens, and such people are said to be atopic. IgE production is driven by antigen-specific T_H2 cells; the response is skewed toward T_H2 by an array of chemokines and cytokines that engage specific signaling pathways, including signals that activate ILC2 cells in submucosal tissues at sites of antigen entry. The IgE produced binds to the high-affinity IgE receptor FcεRI on mast cells and basophils. Specific effector T cells, mast cells, and eosinophils, in combination with T_H1 and T_H2 cytokines and chemokines, orchestrate chronic allergic inflammation, which is the major cause of the chronic morbidity of asthma. Failure to regulate these responses can occur at many levels of the immune system, including defects in regulatory T cells. Increasingly successful processes for suppressing allergic responses and reestablishing the ability to tolerate the sensitizing antigen are being developed, raising the hope of reducing the prevalence of allergic disorders. Antibodies of certain isotypes and various antigen-specific effector T cells contribute to allergic hypersensitivity to other antigens.

Questions.

14.1 True or False: Only T_H2 cells can initiate the chain of signals needed to induce B cells to class-switch to IgE.

14.2 Multiple Choice: Which of the following has not been associated with genetic susceptibility to both allergic asthma and atopic eczema?

  A. β subunit of FcεRI

  B. GM-CSF

  C. IL-3

  D. IL-4

  E. IFN-γ

14.3 Different factors affect our susceptibility to allergic diseases. Which of the following is a false statement?

  A. Environmental factors rarely contribute to the development of allergic disease.

  B. The prevalence of atopy has been steadily increasing in the developed world.

  C. Individuals with variant alleles of *GSTP1* and *GSTM1* have higher susceptibility to increased airway hyperactivity.

  D. Children less than 6 months old who are exposed to other children in day care appear to be partially protected against asthma.

14.4 True or False: Like other antibodies, IgE is mainly found in body fluids.

14.5 Matching: Match the following options with the best description.

___ **A.** Prostaglandin and thromboxane	**i.** Produced by the lipoxygenase pathway.
___ **B.** Leukotrienes	**ii.** Inhibit cyclooxygenase activity on arachidonic acid.
___ **C.** TNF-α	**iii.** Produced by the cyclooxygenase pathway.
___ **D.** Histamine	**iv.** Produced in large amounts by mast cells after activation.
___ **E.** Nonsteroidal anti-inflammatory drugs	**v.** Causes dendritic cells to increase their antigen-presenting capacity when it binds to the H1 receptor.

14.6 Which of the following statements is true?

  A. Connective tissue mast cells do not participate in the initiation of an anaphylactic reaction.

  B. Epinephrine should be avoided in patients suffering from anaphylactic shock as it may worsen the patient's condition.

  C. During anaphylactic shock, blood vessels lose permeability, and high blood pressure leads to death.

  D. Penicillin can modify self proteins, causing an immune response with IgE production in some individuals that can lead to anaphylaxis upon re-encountering the drug.

14.7 Multiple Choice: Hypersensitivity reactions can cause pathology through the deposition of immune complexes. Which of the following is a mechanism by which immune complexes can be pathogenic? (More than one may apply.)

  ___ **A.** Immune complexes deposit in blood vessel walls.

  ___ **B.** IgE is cross-linked on the surface of mast cells and basophils, leading to activation.

  ___ **C.** Fc receptor ligation leads to leukocyte activation and tissue injury.

  ___ **D.** The complement system is activated, leading to the production of anaphylatoxin C5a.

  ___ **E.** CD8+ T cells are stimulated to secrete IL-4.

14.8 Fill-in-the-Blanks: There are two phases to a cutaneous allergic response: _____ and _____. The first phase is characterized by activation of T cells by skin antigen-presenting cells called _____, while the second phase invokes release of chemokines and cytokines by _____ upon subsequent antigen exposure.

14.9 Matching: Match each allergic reaction with the corresponding immune process.

___ **A.** Arthus reaction	**i.** Formation of local immune complexes caused by IgG antibodies acting against an antigen in previously sensitized individuals
___ **B.** Poison ivy rash	**ii.** Systemic reaction to injection of large quantities of foreign antigen, primarily IgG-mediated

___ **C.** Serum sickness

iii. Type of allergic contact dermatitis caused by lipid-soluble chemicals that alter intracellular proteins, primarily CD8 T cell-driven

___ **D.** Nickel allergy

iv. Cellular hypersensitivity, primarily T-cell driven; can also invoke inflammatory response by binding to TLR-4

C. Allergic contact dermatitis can be mediated by CD4 or CD8 T cells.

D. Mice deficient in B cells or complement have impaired contact hypersensitivity reactions.

14.11 Short Answer: Describe the perceived need for the endotyping system for asthma.

14.12 True or False: Allergic asthma can be triggered by factors other than the initial specific allergen.

14.10 Multiple Choice: Which of the following is a false statement?

A. The tuberculin test illustrates the prototypic delayed-type hypersensitivity reaction.

B. T_H1 cells are not directly involved in delayed-type hypersensitivity reactions.

General references.

Fahy, J.V.: **Type 2 inflammation in asthma – present in most, absent in many.** *Nat. Rev. Immunol.* 2015, **15**:57–65.

Holgate, S.T.: **Innate and adaptive immune responses in asthma.** *Nat. Med.* 2012, **18**:673–683.

Johansson, S.G., Bieber, T., Dahl, R., Friedmann, P.S., Lanier, B.Q., Lockey, R.F., Motala, C., Ortega Martell, J.A., Platts-Mills, T.A., Ring, J., *et al.*: **Revised nomenclature for allergy for global use: report of the Nomenclature Review Committee of the World Allergy Organization, October 2003.** *J. Allergy Clin. Immunol.* 2004, **113**:832–836.

Kay, A.B.: **Allergy and allergic diseases. First of two parts.** *N. Engl. J. Med.* 2001, **344**:30–37.

Kay, A.B.: **Allergy and allergic diseases. Second of two parts.** *N. Engl. J. Med.* 2001, **344**:109–113.

Valenta, R., Hochwallner, H., Linhart, B., and Pahr, S.: **Food Allergies: the basics.** *Gastroenterology* 2015, **148**:1120-1131.

Section references.

14-1 Sensitization involves class switching to IgE production on first contact with an allergen.

Akuthota, P., Wang, H., and Weller, P.F.: **Eosinophils as antigen-presenting cells in allergic upper airway disease.** *Curr. Opin. Allergy Clin. Immunol.* 2010, **10**:14–19.

Berkowska, M.A., Heeringa, J.J., Hajdarbegovic, E., van der Burg, M., Thio, H.B., van Gahen, P.M., Boon, L., Orfao, A., van Dongen, J.J.M, and van Zelm, M.C.: **Human IgE⁺B cells are derived from T cell-dependent and T cell-independent pathways.** *J. Allergy Clin. Immunol.* 2014, **134**:688–697.

Bieber T: **The pro- and anti-inflammatory properties of human antigen-presenting cells expressing the high affinity receptor for IgE (FcεRI).** *Immunobiology* 2007, **212**:499–503.

Gold, M.J., Antignano, F., Halim, T.Y.F., Hirota, J.A., Blanchet, M.-R., Zaph, C., Takei, F., and McNagny, K.M.: **Group 2 innate lymphoid cells facilitate sensitization to local, but not systemic, T_H2-inducing allergen exposures.** *J. Allergy Clin. Immunol.* 2014, **133**:1142–1148.

He, J.S., Narayanan, S., Subramaniam, S., Ho, W.Q., Lafaille, J.J., and Curotto de Lafaille, M.A.: **Biology of IgE production: IgE cell differentiation and the memory of IgE responses.** *Curr. Opin. Microbiol. Immunol.* 2015, **388**:1–19.

Kumar, V.: **Innate lymphoid cells: new paradigm in immunology of inflammation.** *Immunol. Lett.* 2014, **157**:23–37.

Mikhak, Z., and Luster, A.D.: **The emergence of basophils as antigen-presenting cells in Th2 inflammatory responses.** *J. Mol. Cell Biol.* 2009, **1**:69–71.

Mirchandani, A.S., Salmond, R.J., and Liew, F.Y.: **Interleukin-33 and the function of innate lymphoid cells.** *Trends Immunol.* 2012, **33**:389–396.

Platzer, B., Baker, K., Vera, M.P., Singer, K., Panduro, M., Lexmond, W.S., Turner, D., Vargas, S. O., Kinet, J.-P., Maurer, D., *et al.*: **Dendritic cell-bound IgE functions to restrain allergic inflammation at mucosal sites.** *Mucosal Immunol.* 2015, **8**:516–532.

Spencer, L.A., and Weller, P.F.: **Eosinophils and Th2 immunity: contemporary insights.** *Immunol. Cell Biol.* 2010, **88**:250–256.

Wu, L.C., and Zarrin, A.A.: **The production and regulation of IgE by the immune system.** *Nat. Rev. Immunol.* 2014, **14**:247-259.

14-2 Although many types of antigens can cause allergic sensitization, proteases are common sensitizing agents.

Alvarez, D., Arkinson, J.L., Sun, J., Fattouh, R., Walker, T., and Jordana, M.: **T_H2 differentiation in distinct lymph nodes influences the site of mucosal T_H2 immune-inflammatory responses.** *J. Immunol.* 2007, **179**:3287–3296.

Brown, S.J., and McLean, W.H.I.: **One remarkable molecule: Filaggrin.** *J. Invest. Dermatol.* 2012, **132**:751–762.

Grunstein, M.M., Veler, H., Shan, X., Larson, J., Grunstein, J.S., and Chuang, S.: **Proasthmatic effects and mechanisms of action of the dust mite allergen, Der p 1, in airway smooth muscle.** *J. Allergy Clin. Immunol.* 2005, **116**:94–101.

Lambrecht, B.N., and Hammad, H.: **Allergens and the airway epithelium response: gateway to allergic sensitization.** *J. Allergy Clin. Immunol.* 2014, **134**:499–507.

Nordlee, J.A., Taylor, S.L., Townsend, J.A., Thomas, L.A., and Bush, R.K.: **Identification of a Brazil-nut allergen in transgenic soybeans.** *N. Engl. J. Med.* 1996, **334**:688–692.

Papzian, D., Wagtmann, V.R., Hansen, S., and Wurtzen, P.A.: **Direct contact between dendritic cells and bronchial epithelial cells inhibits T cell recall responses towards mite and pollen allergen extracts *in vitro*.** *Clin. Exp. Immunol.* 2015, **181**:207–218.

Sehgal, N., Custovic, A., and Woodcock, A.: **Potential roles in rhinitis for protease and other enzymatic activities of allergens.** *Curr. Allergy Asthma Rep.* 2005, **5**:221–226.

Sprecher, E., Tesfaye-Kedjela, A., Ratajczak, P., Bergman, R., and Richard, G.: **Deleterious mutations in SPINK5 in a patient with congenital ichthyosiform erythroderma: molecular testing as a helpful diagnostic tool for Netherton syndrome.** *Clin. Exp. Dermatol.* 2004, **29**:513–517.

Wan, H., Winton, H.L., Soeller, C., Tovey, E.R., Gruenert, D.C., Thompson, P.J., Stewart, G.A., Taylor, G.W., Garrod, D.R., Cannell, M.B., *et al.*: **Der p 1 facilitates transepithelial allergen delivery by disruption of tight junctions.** *J. Clin. Invest.* 1999, **104**:123–133.

14-3 Genetic factors contribute to the development of IgE-mediated allergic disease.

Cookson, W.: **The immunogenetics of asthma and eczema: a new focus on the epithelium.** *Nat. Rev. Immunol.* 2004, **4**:978–988.

Illing, P.R., Vivian, J.P., Purcell, A.W., Rossjohn, J., and McCluskey J.: **Human leukocyte antigen-associated drug hypersensitivity.** *Curr. Opin. Immunol.* 2013, **25**:81–89.

Li, Z., Hawkins, G.A., Ampleford, E.J., Moore, W.C., Li, H., Hastie, A.T., Howard, T.D., Boushey H.A., Busse, W.W., Calhoun, W.J., *et al.*: **Genome-wide association study identifies T_H1 pathway genes associated with lung function in asthmatic patients.** *J. Allergy Clin. Immunol.* 2013, **132**:313–320.

Peiser, M.: **Role of T_H17 cells in skin inflammation of allergic contact dermatitis.** *Clin. Dev. Immunol.* 2013, doi:10.1155/2013/261037.

Thyssen, J.P., Carlsen, B.C., Menné, T., Linneberg, A., Nielsen, N.H., Meldgaard, M., Szecsi, P.B., Stender, S., and Johansen, J.D.: **Filaggrin null-mutations increase the risk and persistence of hand eczema in subjects with atopic dermatitis: results from a general population study.** *Br. J. Dermatol.* 2010, **163**:115–120.

Van den Oord, R.A.H.M., and Sheikh, A.: **Filaggrin gene defects and risk of developing allergic sensitisation and allergic disorders: systematic review and meta-analysis.** *BMJ* 2009, **339**:b2433.

Van Eerdewegh, P., Little, R.D., Dupuis, J., Del Mastro, R.G., Falls, K., Simon, J., Torrey, D., Pandit, S., McKenny, J., Braunschweiger, K., *et al.*: **Association of the *ADAM33* gene with asthma and bronchial hyperresponsiveness.** *Nature* 2002, **418**:426–430.

Weiss, S.T., Raby, B.A., and Rogers, A.: **Asthma genetics and genomics.** *Curr. Opin. Genet. Dev.* 2009, **19**:279–282.

14-4 Environmental factors may interact with genetic susceptibility to cause allergic disease.

Culley, F.J., Pollott, J., and Openshaw, P.J.: **Age at first viral infection determines the pattern of T cell-mediated disease during reinfection in adulthood.** *J. Exp. Med.* 2002, **196**:1381–1386.

Deshane, J., Zmijewski, J.W., Luther, R., Gaggar, A., Deshane, R., Lai, J.F., Xu, X., Spell, M., Estell, K., Weaver, C.T., *et al.*: **Free radical-producing myeloid-derived regulatory cells: potent activators and suppressors of lung inflammation and airway hyperresponsiveness.** *Mucosal Immunol.* 2011, **4**:503–518.

Fuchs, C., and von Mutius, E.: **Prenatal and childhood infections: implications for the development and treatment of childhood asthma.** *Lancet Respir. Med.* 2013, **1**:743–754.

Harb, H., and Renz, H.: **Update on epigenetics in allergic disease.** *J. Allergy Clin. Immunol.* 2015, **135**:15–24.

Huang, L., Baban, B., Johnson, B.A. 3rd, and Mellor, A.L.: **Dendritic cells, indoleamine 2,3 dioxygenase and acquired immune privilege.** *Int. Rev. Immunol.* 2010, **29**:133–155.

Meyers, D.A., Bleecker, E.R., Holloway, J.W., and Holgate, S.T.: **Asthma genetics and personalized medicine.** *Lancet Respir. Med.* 2014, **2**:405–415.

Minelli, C., Granell, R., Newson, R., Rose-Zerilli, M.-J., Torrent, M., Ring, S.M., Holloway, J.W., Shaheen, S.O., and Henderson, J.A.: **Glutathione-S-transferase genes and asthma phenotypes: a Human Genome Epidemiology (HuGE) systematic review and meta-analysis including unpublished data.** *Int. J. Epidemiol.* 2010, **39**:539–562.

Morahan, G., Huang, D., Wu, M., Holt, B.J., White, G.P., Kendall, G.E., Sly, P.D., and Holt, P.G.: **Association of IL12B promoter polymorphism with severity of atopic and non-atopic asthma in children.** *Lancet* 2002, **360**:455–459.

Romieu, I., Ramirez-Aguilar, M., Sienra-Monge, J.J., Moreno-Macías, H., del Rio-Navarro, B.E., David, G., Marzec, J., Hernández-Avila, M., and London, S.: **GSTM1 and GSTP1 and respiratory health in asthmatic children exposed to ozone.** *Eur. Respir. J.* 2006, **28**:953–959.

Saxon, A., and Diaz-Sanchez, D.: **Air pollution and allergy: you are what you breathe.** *Nat. Immunol.* 2005, **6**:223–226.

von Mutius, E.: **Allergies, infections and the hygiene hypothesis -- the epidemiologic evidence.** *Immunobiology* 2007, **212**:433-439.

14-5 Regulatory T cells can control allergic responses.

Bohm, L., Meyer-Martin, H., Reuter, S., Finotto, S., Klein, M., Schild, H., Schmitt, E., Bopp, T., and Taube, C.: **IL-10 and regulatory T cells cooperate in allergen-specific immunotherapy to ameliorate allergic asthma.** *J. Immunol.* 2015, **194**:887–897.

Duan, W., and Croft, M.: **Control of regulatory T cells and airway tolerance by lung macrophages and dendritic cells.** *Ann. Am. Thorac. Soc.* 2014, **11** Suppl **5**:S305–S313.

Hawrylowicz, C.M.: **Regulatory T cells and IL-10 in allergic inflammation.** *J. Exp. Med.* 2005, **202**:1459–1463.

Lin, W., Truong, N., Grossman, W.J., Haribhai, D., Williams, C.B., Wang, J., Martin, M.G., and Chatila, T.A.: **Allergic dysregulation and hyperimmunoglobulinemia E in Foxp3 mutant mice.** *J. Allergy Clin. Immunol.* 2005, **116**:1106–1115.

Mellor, A.L., and Munn, D.H.: **IDO expression by dendritic cells: tolerance and tryptophan catabolism.** *Nat. Rev. Immunol.* 2004, **4**:762–774.

14-6 Most IgE is cell-bound and engages effector mechanisms of the immune system by pathways different from those of other antibody isotypes.

Conner, E.R., and Saini, S.S.: **The immunoglobulin E receptor: expression and regulation.** *Curr. Allergy Asthma Rep.* 2005, **5**:191–196.

Gilfillan, A.M., and Tkaczyk, C.: **Integrated signalling pathways for mast-cell activation.** *Nat. Rev. Immunol.* 2006, **6**:218–230.

Mcglashan, D.W., Jr.: **IgE-dependent signalling as a therapeutic target for allergies.** *Trends Pharmacol. Sci.* 2012, **33**:502–509.

Suzuki, R., Scheffel, J., and Rivera, J.: **New insights on the signalling and function of the high-affinity receptor for IgE.** *Curr. Top. Microbiol.Immunol.* 2015, **388**:63–90.

14-7 Mast cells reside in tissues and orchestrate allergic reactions.

Eckman, J.A., Sterba, P.M., Kelly, D., Alexander, V., Liu, M.C., Bochner, B.S., MacGlashan, D.W., and Saini, S.S.: **Effects of omalizumab on basophil and mast cell responses using an intranasal cat allergen challenge.** *J. Allergy Clin. Immunol.* 2010, **125**:889–895.

Galli, S.J., Nakae, S., and Tsai, M.: **Mast cells in the development of adaptive immune responses.** *Nat. Immunol.* 2005, **6**:135–142.

Gonzalez-Espinosa, C., Odom, S., Olivera, A., Hobson, J.P., Martinez, M.E., Oliveira-Dos-Santos, A., Barra, L., Spiegel, S., Penninger, J.M., and Rivera, J.: **Preferential signaling and induction of allergy-promoting lymphokines upon weak stimulation of the high affinity IgE receptor on mast cells.** *J. Exp. Med.* 2003, **197**:1453–1465.

Islam, S.A., and Luster, A.D.: **T cell homing to epithelial barriers in allergic disease.** *Nat. Med.* 2012, **18**:705-715.

Kitamura, Y., Oboki, K., and Ito, A.: **Development of mast cells.** *Proc. Jpn Acad, Ser. B* 2007, **83**:164–174.

Kulka, M., Sheen, C.H., Tancowny, B.P., Grammer, L.C., and Schleimer, R.P.: **Neuropeptides activate human mast cell degranulation and chemokine production.** *Immunology* 2007, **123**:398–410.

Metcalfe, D.: **Mast cells and mastocytosis.** *Blood* 2007, **112**:946–956.

Schwartz, L.B.: **Diagnostic value of tryptase in anaphylaxis and mastocytosis.** *Immunol. Allergy Clin. N. Am.* 2006, **26**:451–463.

Smuda, C., and Bryce, P.J.: **New development in the use of histamine and histamine receptors.** *Curr. Allergy Asthma Rep.* 2011, **11**:94–100.

Taube, C., Miyahara, N., Ott, V., Swanson, B., Takeda, K., Loader, J., Shultz, L.D., Tager, A.M., Luster, A.D., Dakhama, A., *et al.*: **The leukotriene B4 receptor (BLT1) is required for effector CD8+ T cell-mediated, mast cell-dependent airway hyperresponsiveness.** *J. Immunol.* 2006, **176**:3157–3164.

Thurmond, R.L.: **The histamine H4 receptor: from orphan to the clinic.** *Front. Pharmacol.* 2015, **6**:65.

14-8 Eosinophils and basophils cause inflammation and tissue damage in allergic reactions.

Blanchard, C., and Rothenberg, M.E.: **Biology of the eosinophil.** *Adv. Immunol.* 2009, **101**:81–121.

Hogan, S.P., Rosenberg, H.F., Moqbel, R., Phipps, S., Foster, P.S., Lacy, P., Kay, A.B., and Rothenberg, M.E.: **Eosinophils: biological properties and role in health and disease.** *Clin. Exp. Allergy* 2008, **38**:709–750.

Lee, J.J., Jacobsen, E.A., McGarry, M.P., Schleimer, R.P., and Lee, N.A.: **Eosinophils in health and disease: the LIAR hypothesis.** *Clin. Exp. Allergy* 2010, **40**:563–575.

MacGlashan, D., Jr, Gauvreau, G., and Schroeder, J.T.: **Basophils in airway disease.** *Curr. Allergy Asthma Rep.* 2002, **2**:126–132.

Ohnmacht, C., Schwartz, C., Panzer, M., Schiedewitz, I., Naumann, R., and Voehringer, D.: **Basophils orchestrate chronic allergic dermatitis and protective immunity against helminths.** *Immunity* 2010, **33**:364–374.

Schwartz, C., Eberle, J.U., and Voehringer, D.: **Basophils in inflammation.** *Eur. J. Pharmacol.* 2015, doi:10.1016/j.ejphar.2015.04.049.

Tomankova, T., Kriegova, E., and Liu, M.: **Chemokine receptors and their therapeutic opportunities in diseased lung: far beyond leukocyte trafficking.** *Am. J. Physiol. Lung Cell. Mol. Physiol.* 2015, **308**:L603-L618.

14-9 IgE-mediated allergic reactions have a rapid onset but can also lead to chronic responses.

deShazo, R.D., and Kemp, S.F.: **Allergic reactions to drugs and biologic agents.** *JAMA* 1997, **278**:1895–1906.

Nabe, T., Ikedo A., Hosokawa, F., Kishima, M., Fujii, M., Mizutani, N., Yoshino, S., Ishihara, K., Akiba, S., and Chaplin, D.D.: **Regulatory role of antigen-induced interleukin-10, produced by CD4(+) T cells, in airway neutrophilia in a murine model for asthma.** *Eur. J. Pharmacol.* 2012, **677**:154–162.

Pawankar, R., Hayashi, M., Yamanishi, S., and Igarashi, T.: **The paradigm of cytokine networks in allergic airway inflammation.** *Curr. Opin. Allergy Clin. Immunol.* 2015, **15**:27–32.

Taube, C., Duez, C., Cui, Z.H., Takeda, K., Rha, Y.H., Park, J.W., Balhorn, A., Donaldson, D.D., Dakhama, A., and Gelfand, E.W.: **The role of IL-13 in established allergic airway disease.** *J. Immunol.* 2002, **169**:6482–6489.

14-10 Allergen introduced into the bloodstream can cause anaphylaxis.

Fernandez, M., Warbrick, E.V., Blanca, M., and Coleman, J.W.: **Activation and hapten inhibition of mast cells sensitized with monoclonal IgE anti-penicillin antibodies: evidence for two-site recognition of the penicillin derived determinant.** *Eur. J. Immunol.* 1995, **25**:2486–2491.

Finkelman, F.D., Rothenberg, M.E., Brandt, E.B., Morris, S.C., and Strait, R.T.: **Molecular mechanisms of anaphylaxis: lessons from studies with murine models.** *J. Allergy Clin. Immunol.* 2005, **115**:449–457.

Golden, D.B.: **Anaphylaxis to insect stings.** *Immunol. Allergy Clin. North Am.* 2015, **35**:287–302.

Kemp, S.F., Lockey, R.F., Wolf, B.L., and Lieberman, P.: **Anaphylaxis. A review of 266 cases.** *Arch. Intern. Med.* 1995, **155**:1749–1754.

Sicherer, S.H., and Leung, D.Y.: **Advances in allergic skin disease, anaphylaxis, and hypersensitivity reactions to foods, drugs, and insects in 2014.** *J. Allergy Clin. Immunol.* 2015, **35**:357–367.

Weltzien, H.U., and Padovan, E.: **Molecular features of penicillin allergy.** J. *Invest. Dermatol.* 1998, **110**:203–206.

14-11 Allergen inhalation is associated with the development of rhinitis and asthma.

Bousquet, J., Jeffery, P.K., Busse, W.W., Johnson, M., and Vignola, A.M.: **Asthma. From bronchoconstriction to airways inflammation and remodeling.** *Am. J. Respir. Crit. Care Med.* 2000, **161**:1720–1745.

Boxall, C., Holgate, S.T., and Davies, D.E.: **The contribution of transforming growth factor-β and epidermal growth factor signalling to airway remodelling in chronic asthma.** *Eur. Respir. J.* 2006, **27**:208–229.

Dakhama, A., Park, J.W., Taube, C., Joetham, A., Balhorn, A., Miyahara, N., Takeda, K., and Gelfand, E.W.: **The enhancement or prevention of airway hyperresponsiveness during reinfection with respiratory syncytial virus is critically dependent on the age at first infection and IL-13 production.** *J. Immunol.* 2005, **175**:1876–1883.

Finotto, S., Neurath, M.F., Glickman, J.N., Qin, S., Lehr, H.A., Green, F.H., Ackerman, K., Haley, K., Galle, P.R., Szabo, S.J., *et al.*: **Development of spontaneous airway changes consistent with human asthma in mice lacking T-bet.** *Science* 2002, **295**:336–338.

George, B.J., Reif, D.M., Gallagher, J.E., Williams-DeVane, C.R., Heidenfelder, B.L., Hudgens, E.E., Jones, W., Neas, L., Cohen Hubal, E.A., and Edwards, S.W.: **Data-driven asthma endotypes defined from blood biomarker and gene expression data.** *PLoS ONE* 2015, **10**:e0117445.

Gour, N., and Wills-Karp, M.: **IL-4 and IL-13 signaling in allergic airway disease.** *Cytokine* 2015, **75**:68-78.

Haselden, B.M., Kay, A.B., and Larche, M.: **Immunoglobulin E-independent major histocompatibility complex-restricted T cell peptide epitope-induced late asthmatic reactions.** *J. Exp. Med.* 1999, **189**:1885–1894.

Kuperman, D.A., Huang, X., Koth, L.L., Chang, G.H., Dolganov, G.M., Zhu, Z., Elias, J.A., Sheppard, D., and Erle, D.J.: **Direct effects of interleukin-13 on epithelial cells cause airway hyperreactivity and mucus overproduction in asthma.** *Nat. Med.* 2002, **8**:885–889.

Lambrecht, B.N., and Hammad, H.: **The role of dendritic and epithelial cells as master regulators of allergic airway inflammation.** *Lancet* 2010, **376**:835–843.

Lloyd, C.M., and Hawrylowicz, C.M.: **Regulatory T cells in asthma.** *Immunity* 2009, **31**:438–449.

Lotvall, J., Akdis, C.A., Bacharier, L.B., Bjermer, L., Casale, T.B., Custovic, A., Lemanske, R.F., Jr., Wardlaw, A.J., Wenzel, S.E., and Greenberger, P.A.: **Asthma endotypes: a new approach to classification of disease entities within the asthma syndrome.** *J. Allergy Clin. Immunol.* 2011, **127**:355–360.

Meyer, E.H., DeKruyff, R.H., and Umetsu, D.T.: **T cells and NKT cells in the pathogenesis of asthma.** *Annu. Rev. Med.* 2008, **59**:281–292.

Newcomb, D.C., and Peebles, R.S., Jr.: **Th17-mediated inflammation in asthma.** *Curr. Opin. Immunol.* 2013, **25**:755-760.

Peebles, R.S. Jr.: **The emergence of group 2 innate lymphoid cells in human disease.** *J. Leukoc. Biol.* 2015, **97**:469–475.

Robinson, D.S.: **Regulatory T cells and asthma.** *Clin. Exp. Allergy* 2009, **39**:1314–1323.

Yan, X., Chu, J.-H., Gomez, J., Koenigs, M., Holm, C., He, X., Perez, M.F., Zhao, H., Mane, S., Martinez, F.D., *et al.*: **Non-invasive analysis of the sputum transcriptome discriminates clinical phenotypes of asthma.** *Am. J. Respir. Crit. Care Med.* 2015, **191**:1116–1125.

14-12 Allergy to particular foods causes systemic reactions as well as symptoms limited to the gut.

Du Toit, G., Robert, G., Sayre, P.H., Bahnson, H.T., Radulovic, S., Santos, A.D., Brough, H.A., Phippard, D., Basting, M., Feeney, M., *et al.*: **Randomized trial of peanut consumption in infants at risk for peanut allergy.** *N. Engl. J. Med.* 2015, **372**:803–813.

Lee, L.A., and Burks, A.W.: **Food allergies: prevalence, molecular characterization, and treatment/prevention strategies.** *Annu. Rev. Nutr.* 2006, **26**:539–565.

14-13 IgE-mediated allergic disease can be treated by inhibiting the effector pathways that lead to symptoms or by desensitization techniques that aim at restoring biological tolerance to the allergen.

Akdis, C.A., and Akdis, M.: **Mechanisms of allergen-specific immunotherapy and immune tolerance to allergens.** *World Allergy Organ. J.* 2015, **8**:17.

Bryan, S.A., O'Connor, B.J., Matti, S., Leckie, M.J., Kanabar, V., Khan, J., Warrington, S.J., Renzetti, L., Rames, A., Bock, J.A., et al.: **Effects of recombinant human interleukin-12 on eosinophils, airway hyper-responsiveness, and the late asthmatic response.** *Lancet* 2000, **356**:2149–2153.

Dunn, R.M., and Wechsler, M.E.: **Anti-interleukin therapy in asthma.** *Clin. Pharmacol. Ther.* 2015, **97**:55–65.

Haldar, P., Brightling, C.E., Hargadon, B., Gupta, S., Monteiro, W., Sousa, A., Marshall, R.P., Bradding, P., Green, R.H., Wardlaw A.J., et al.: **Mepolizumab and exacerbations of refractory eosinophilic asthma.** *N. Engl. J. Med.* 2009, **360**:973–984.

Lai, T., Wang, S., Xu, Z., Zhang, C., Zhao, Y., Hu, Y., Cao, C., Ying, S., Chen, Z., Li, W., et al: **Long-term efficacy and safety of omalizumab in patients with persistent uncontrolled allergic asthma: a systematic review and meta-analysis.** *Sci. Rep.* 2015, **5**:8191.

Larche, M.: **Mechanisms of peptide immunotherapy in allergic airways disease.** *Ann. Am. Thorac. Soc.* 2014, **11**:S292-S296.

Nair, P., Pizzichini, M.M.M., Kjarsgaard, M., Inman, M.D., Efthimiadis, A., Pizzichini, E., Hargreave, F.E., and O'Byrne, P.M.: **Mepolizumab for prednisone-dependent asthma with sputum eosinophilia.** *N. Engl. J. Med.* 2009, **360**:985–993.

Peters-Golden, M., and Henderson, W.R., Jr: **The role of leukotrienes in allergic rhinitis.** *Ann. Allergy Asthma Immunol.* 2005, **94**:609–618.

Roberts, G., Hurley, C., Turcanu, V., and Lack, G.: **Grass pollen immunotherapy as an effective therapy for childhood seasonal allergic asthma.** *J. Allergy Clin. Immunol.* 2006, **117**:263–268.

Shamji, M.H., and Durham, S.R.: **Mechanisms of immunotherapy to aeroallergens.** *Clin. Exp.. Allergy* 2011, **41**:1235–1246.

Zhu, D., Kepley, C.L., Zhang, K., Terada, T., Yamada, T., and Saxon, A.: **A chimeric human–cat fusion protein blocks cat-induced allergy.** *Nat. Med.* 2005, **11**:446–449.

14-14 Non-IgE dependent drug-induced hypersensitivity reactions in susceptible individuals occur by binding of the drug to the surface of circulating blood cells.

Arndt, P.A.: **Drug-induced immune hemolytic anemia: the last 30 years of changes.** *Immunohematology* 2014, **30**:44–54.

Greinacher, A., Potzsch, B., Amiral, J., Dummel, V., Eichner, A., and Mueller Eckhardt, C.: **Heparin-associated thrombocytopenia: isolation of the antibody and characterization of a multimolecular PF4–heparin complex as the major antigen.** *Thromb. Haemost.* 1994, **71**:247–251.

Semple, J.W., and Freedman, J.: **Autoimmune pathogenesis and autoimmune hemolytic anemia.** *Semin. Hematol.* 2005, **42**:122–130.

14-15 Systemic disease caused by immune-complex formation can follow the administration of large quantities of poorly catabolized antigens.

Bielory, L., Gascon, P., Lawley, T.J., Young, N.S., and Frank, M.M.: **Human serum sickness: a prospective analysis of 35 patients treated with equine anti-thymocyte globulin for bone marrow failure.** *Medicine (Baltimore)* 1988, **67**:40–57.

Davies, K.A., Mathieson, P., Winearls, C.G., Rees, A.J., and Walport, M.J.: **Serum sickness and acute renal failure after streptokinase therapy for myocardial infarction.** *Clin. Exp. Immunol.* 1990, **80**:83–88.

Hansel, T.T., Kropshofer, H., Singer, T., Mitchell, J.A., and George, A.J.: **The safety and side effects of monoclonal antibodies.** *Nat. Rev. Drug Discov.* 2010, **9**:325–338.

Schifferli, J.A., Ng, Y.C., and Peters, D.K.: **The role of complement and its receptor in the elimination of immune complexes.** *N. Engl. J. Med.* 1986, **315**:488–495.

Schmidt, R.E., and Gessner, J.E.: **Fc receptors and their interaction with complement in autoimmunity.** *Immunol. Lett.* 2005, **100**:56–67.

Skokowa, J., Ali, S.R., Felda, O., Kumar, V., Konrad, S., Shushakova, N., Schmidt, R.E., Piekorz, R.P., Nurnberg, B., Spicher, K., et al.: **Macrophages induce the inflammatory response in the pulmonary Arthus reaction through $G\alpha_{i2}$ activation that controls C5aR and Fc receptor cooperation.** *J. Immunol.* 2005, **174**:3041–3050.

14-16 Hypersensitivity reactions can be mediated by T_H1 cells and CD8 cytotoxic T cells.

Fyhrquist, N., Lehto, E., and Lauerma, A.: **New findings in allergic contact dermatitis.** *Curr. Opin. Allergy Clin. Immunol.* 2014, **14**:430–435.

Kalish, R.S., Wood, J.A., and LaPorte, A.: **Processing of urushiol (poison ivy) hapten by both endogenous and exogenous pathways for presentation to T cells *in vitro*.** *J. Clin. Invest.* 1994, **93**:2039–2047.

Mark, B.J., and Slavin, R.G.: **Allergic contact dermatitis.** *Med. Clin. North Am.* 2006, **90**:169–185.

Muller, G., Saloga, J., Germann, T., Schuler, G., Knop, J., and Enk, A.H.: **IL-12 as mediator and adjuvant for the induction of contact sensitivity *in vivo*.** *J. Immunol.* 1995, **155**:4661–4668.

Schmidt, M., Raghavan, B., Müller, V., Vogl, T., Fejer, G., Tchaptchet, S., Keck, S., Kalis, C., Nielsen, P.J., Galanos, C., et al.: **Crucial role for human Toll-like receptor 4 in the development of contact allergy to nickel.** *Nat. Immunol.* 2010, **11**:814–819.

Vollmer, J., Weltzien, H.U., and Moulon, C.: **TCR reactivity in human nickel allergy indicates contacts with complementarity-determining region 3 but excludes superantigen-like recognition.** *J. Immunol.* 1999, **163**:2723–2731.

14-17 Celiac disease has features of both an allergic response and autoimmunity.

Ciccocioppo, R., Di Sabatino, A., and Corazza, G.R.: **The immune recognition of gluten in celiac disease.** *Clin. Exp. Immunol.* 2005, **140**:408–416.

Green, P.H.R., Lebwohl, B., and Greywoode, R.: **Celiac disease.** *J. Allergy Clin. Immunol.* 2015, **135**:1099–1106.

Koning, F.: **Celiac disease: caught between a rock and a hard place.** *Gastroenterology* 2005, **129**:1294–1301.

Shan, L., Molberg, O., Parrot, I., Hausch, F., Filiz, F., Gray, G.M., Sollid, L.M., and Khosla, C.: **Structural basis for gluten intolerance in celiac sprue.** *Science* 2002, **297**:2275–2279.

van Bergen, J., Mulder, C.J., Mearin, M.L., and Koning, F.: **Local communication among mucosal immune cells in patients with celiac disease.** *Gastroenterology* 2015, **148**:1187–1194.

Autoimmunity and Transplantation

15

We have already seen how undesirable adaptive immune responses can be elicited by environmental antigens, and how this can cause serious disease in the form of allergic and atopic reactions (see Chapter 14). In this chapter, we examine responses to other medically important categories of antigens—those expressed by the body's own cells and tissues, by the commensal microbiota, or by transplanted organs. The responses to self antigens or antigens associated with the microbiota that lead to tissue damage and disease are broadly referred to as **autoimmunity**—although, strictly speaking, disease-causing immune responses to the commensal microbiota are a form of **xenoimmunity** because the organisms from which the antigens derive are foreign and are not encoded by the human genome. Nevertheless, here we will consider immune-mediated disease directed against the commensal microbiota to be part of the extended spectrum of **autoimmune diseases**, because the microbiota can be regarded as being part of a 'superorganism' made up of host and commensal microbiota together. The response to nonself antigens on transplanted organs is called **allograft rejection**.

The gene rearrangements that occur during lymphocyte development in the central lymphoid organs inevitably result in the generation of some lymphocytes with affinity for self antigens. Such lymphocytes are normally removed from the repertoire or held in check by a variety of mechanisms. These generate a state of **self-tolerance** in which an individual's immune system does not attack the normal tissues of the body. Autoimmunity represents a breakdown or failure of the mechanisms of self-tolerance. Therefore, we first revisit the mechanisms that keep the lymphocyte repertoire self-tolerant and see how these may fail. We then discuss a selection of autoimmune diseases that illustrate the various pathogenic mechanisms by which autoimmunity can damage the body. How genetic and environmental factors predispose to or trigger autoimmunity are then considered. In the remaining part of the chapter, we discuss the adaptive immune responses to nonself tissue antigens that cause transplant rejection.

IN THIS CHAPTER

The making and breaking of self-tolerance.

Autoimmune diseases and pathogenic mechanisms.

The genetic and environmental basis of autoimmunity.

Responses to alloantigens and transplant rejection.

The making and breaking of self-tolerance.

As we learned in Chapter 8, the immune system takes advantage of surrogate markers of self and nonself to identify and delete potentially self-reactive lymphocytes. Despite this, some self-reactive lymphocytes escape elimination and can subsequently be activated to cause autoimmunity. In addition, many lymphocytes with some degree of self-reactivity can also respond to foreign antigens; therefore, if all weakly self-reactive lymphocytes were eliminated, the function of the immune system would be impaired.

15-1 A critical function of the immune system is to discriminate self from nonself.

The immune system has very powerful effector mechanisms that can eliminate a variety of pathogens. Early in the study of immunity it was realized that these could, if turned against the host, cause severe tissue damage. The concept of autoimmunity was first presented at the beginning of the 20th century by **Paul Ehrlich**, who described it as '*horror autotoxicus.*' Autoimmune responses resemble normal immune responses to pathogens in that they are specifically activated by antigens—in this case self antigens, or **autoantigens**—and give rise to autoreactive effector cells and antibodies, called **autoantibodies**, against the self antigen. When dysregulated reactions to self tissues occur they cause a variety of chronic syndromes called autoimmune diseases. These syndromes are quite varied in their severity, their tissue distribution, and effector mechanisms that are critical in causing tissue damage (**Fig. 15.1**).

Collectively, autoimmune disorders affect approximately 5% of the populations of Western countries, and their incidence is on the rise. Nevertheless, their relative individual rarity indicates that the immune system has evolved multiple mechanisms to prevent self-injury. The most fundamental principle underlying these mechanisms is the discrimination of self from nonself, but this discrimination is not easy to achieve. B cells recognize the three-dimensional shape of an epitope, but an epitope presented by a pathogen can be indistinguishable from one originating in humans. Similarly, short peptides derived from the processing of pathogen antigens can be identical to self peptides. So how does a lymphocyte know what 'self' really is if there are no unique molecular signatures of self?

The first mechanism proposed for distinguishing between self and nonself was that recognition of antigen by an immature lymphocyte leads to a negative signal causing lymphocyte death or inactivation. Thus, 'self' was thought to comprise molecules recognized by a lymphocyte shortly after it began to express its antigen receptor. Indeed, this is an important mechanism of inducing self-tolerance in lymphocytes developing in the thymus and bone marrow. The tolerance induced at this stage is known as **central tolerance** (see Chapter 8). Newly formed lymphocytes are especially sensitive to inactivation by strong signals through their antigen receptor, whereas the same signals activate mature lymphocytes in the periphery.

Fig. 15.1 Some common autoimmune diseases. The diseases listed are among the most common autoimmune diseases and will be used as examples in this part of the chapter. They are listed in order of prevalence.

Disease	Disease mechanism	Consequence	Prevalence
Psoriasis	Autoreactive T cells against skin-associated antigens	Inflammation of skin with formation of scaly patches or plaques	1 in 50
Rheumatoid arthritis	Autoreactive T cells and autoantibodies against antigens localized to joint synovium	Joint inflammation and destruction causing arthritis	1 in 100
Graves' disease	Autoantibodies against the thyroid-stimulating-hormone receptor	Hyperthyroidism: overproduction of thyroid hormones	1 in 100
Hashimoto's thyroiditis	Autoantibodies and autoreactive T cells against thyroid antigens	Destruction of thyroid tissue leading to hypothyroidism: underproduction of thyroid hormones	1 in 200
Systemic lupus erythematosus	Autoantibodies and autoreactive T cells against DNA, chromatin proteins, and ubiquitous ribonucleoprotein antigens	Glomerulonephritis, vasculitis, rash	1 in 200
Sjögren's syndrome	Autoantibodies and autoreactive T cells against ribonucleoprotein antigens	Lymphocyte infiltration of exocrine glands, leading to dry eyes and/or dry mouth; other organs may be involved, leading to systemic disease	1 in 300
Crohn's disease	Autoreactive T cells against intestinal flora antigens	Intestinal inflammation and scarring	1 in 500
Multiple sclerosis	Autoreactive T cells against brain and spinal cord antigens	Formation of sclerotic plaques in brain and spinal cord with destruction of myelin sheaths surrounding nerve cell axons, leading to muscle weakness, ataxia, and other symptoms	1 in 700
Type 1 diabetes (insulin-dependent diabetes mellitus, IDDM)	Autoreactive T cells against pancreatic islet cell antigens	Destruction of pancreatic islet β cells leading to nonproduction of insulin	1 in 800

Tolerance induced to antigens recognized after lymphocytes have left the central or primary lymphoid organs is known as **peripheral tolerance**. An antigenic quality that correlates with self in the periphery is recognition in the absence of 'danger' signals that are produced by the innate immune system as a result of tissue damage or infection. Nearly all cells in the body become senescent and die, and many cells routinely undergo turnover at steady state (for example, hematopoietic cells and epithelial cells of the intestines and skin). Typically, this occurs by programmed cell death, or apoptosis. In contrast to cell death that results from physical or microbial injury, which generate damage- or microbe-associated molecular patterns (DAMPs and MAMPs, respectively), death of senescent cells by apoptosis releases signals to tissue phagocytes that generally promote an anti-inflammatory response and repress presentation of antigens in an activating form. Thus, self antigens recognized in the context of normal, or physiologic, cellular turnover fail to induce pro-inflammatory cytokines (for example, IL-6 or IL-12) and co-stimulatory molecules (for example, B7.1) that would otherwise induce naive T cells to undergo effector differentiation. In these circumstances, the encounter of a naive lymphocyte with a self antigen may lead to no signal at all, or such an encounter can promote the development of regulatory lymphocytes that suppress the development of damaging effector responses. The removal of apoptotic cells by phagocytes is thus important for maintaining tissue homeostasis and activating programs in antigen-presenting cells that promote immunological tolerance. Some of the same mechanisms appear to be involved in the induction of tolerance to antigens of the commensal microbiota in the intestines, where recognition of bacterial antigens typically does not generate inflammation unless there is associated tissue damage.

Thus, several clues are used to distinguish self from nonself ligands: encounter with the ligand when the lymphocyte is immature, recognition of antigen in the context of antigen-presenting cells that have received tolerizing signals from recognition of homeostatic cell turnover signals, and binding of ligand in the absence of inflammatory cytokines or co-stimulatory signals. All of these mechanisms are error-prone because none of them distinguishes a self ligand from a foreign one at the molecular level. The immune system therefore has several additional mechanisms for controlling autoimmune responses should they start.

15-2 Multiple tolerance mechanisms normally prevent autoimmunity.

The mechanisms that normally prevent autoimmunity may be thought of as a succession of checkpoints. Each checkpoint is partially effective in preventing antiself responses, and together the checkpoints act synergistically to provide efficient protection against autoimmunity without inhibiting the ability of the immune system to mount effective responses to pathogens. Central tolerance mechanisms eliminate newly formed, strongly autoreactive lymphocytes. On the other hand, mature self-reactive lymphocytes that do not sense self strongly in the central lymphoid organs—because their cognate self antigens are not expressed there, for example—may be killed or inactivated in the periphery. The principal mechanisms of peripheral tolerance are anergy (functional unresponsiveness), suppression by T_{reg} cells, induction of T_{reg} cell development instead of effector T-cell development (functional deviation), and deletion of lymphocytes from the repertoire due to activation-induced cell death. In addition, some antigens are sequestered in organs that are not normally accessible to the immune system (**Fig. 15.2**).

Each checkpoint strikes a balance between preventing autoimmunity and not impairing immunity too greatly, and in combination, all the checkpoints provide an effective overall defense against autoimmune disease. It is relatively

Fig. 15.2 Self-tolerance depends on the concerted action of a variety of mechanisms that operate at different sites and stages of development. The different ways in which the immune system prevents activation of and damage caused by autoreactive lymphocytes are listed, along with the specific mechanism and where such tolerance predominantly occurs.

Layers of self-tolerance		
Type of tolerance	**Mechanism**	**Site of action**
Central tolerance	Deletion Editing	Thymus (T cells) Bone marrow (B cells)
Antigen segregation	Physical barrier to self-antigen access to lymphoid system	Peripheral organs (e.g., thyroid, pancreas)
Peripheral anergy	Cellular inactivation by weak signaling without co-stimulus	Secondary lymphoid tissue
Regulatory T cells	Suppression by cytokines, intercellular signals	Secondary lymphoid tissue and sites of inflammation; multiple tissues in steady state
Functional deviation	Differentiation of regulatory T cells that limit inflammatory cytokine secretion	Secondary lymphoid tissue and sites of inflammation
Activation-induced cell death	Apoptosis	Secondary lymphoid tissue and sites of inflammation

easy to find isolated breakdowns of one or even more layers of protection, even in healthy individuals. Thus, activation of autoreactive lymphocytes does not necessarily equal autoimmune disease. In fact, a low level of autoreactivity is physiologic and crucial to normal immune function. Autoantigens help to form the repertoire of mature lymphocytes, and the survival of naive T cells and B cells in the periphery requires continuous exposure to autoantigens (see Chapter 8). Autoimmune disease develops only if enough safeguards are overcome to lead to a sustained reaction to self that includes the generation of effector cells and molecules that destroy tissues. Although the mechanisms by which this occurs are incompletely understood, autoimmunity is thought to result from a combination of genetic susceptibility, breakdown in natural tolerance mechanisms, and environmental triggers such as infections (**Fig. 15.3**).

15-3 Central deletion or inactivation of newly formed lymphocytes is the first checkpoint of self-tolerance.

Central tolerance mechanisms, which remove strongly autoreactive lymphocytes, are the first and most important checkpoints in self-tolerance (see Chapter 8). Without them, the immune system would be strongly self-reactive, and lethal autoimmunity would occur early in life. It is unlikely that peripheral tolerance mechanisms would be sufficient to compensate for the failure to remove self-reactive lymphocytes during primary development. Indeed, no known autoimmune diseases are attributable to complete failure of these mechanisms, although some are associated with a partial failure of central tolerance.

For a long time it was thought that many self antigens were not expressed in the thymus or bone marrow, and that peripheral mechanisms must be the only way of generating tolerance to them. It is now clear that many (but not all) tissue-specific antigens, such as insulin, are expressed in the thymus by either thymic epithelial cells in the medulla or a CD8α⁺ subset of dendritic cells, and thus self-tolerance against these antigens can be generated centrally. How these 'peripheral' genes are turned on ectopically in the thymus is not yet completely worked out, but an important clue has been found. A single transcription factor, AIRE (for <u>a</u>uto<u>i</u>mmune <u>re</u>gulator), is thought to be responsible for turning on many peripheral genes in the thymus (see Section 8-23). The *AIRE* gene is defective in patients with a rare inherited form of

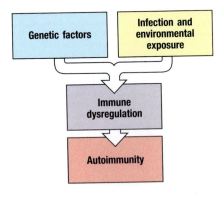

Fig. 15.3 Requirements for the development of autoimmune disease. In genetically predisposed individuals, autoimmunity may be triggered as a result of the failure of intrinsic tolerance mechanisms and/or environmental triggers such as infection.

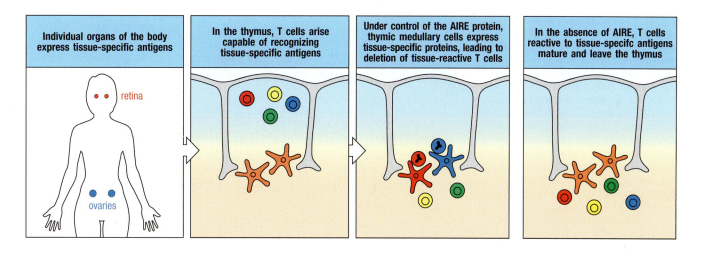

| Individual organs of the body express tissue-specific antigens | In the thymus, T cells arise capable of recognizing tissue-specific antigens | Under control of the AIRE protein, thymic medullary cells express tissue-specific proteins, leading to deletion of tissue-reactive T cells | In the absence of AIRE, T cells reactive to tissue-specifc antigens mature and leave the thymus |

autoimmunity—**APECED (autoimmune polyendocrinopathy–candidiasis–ectodermal dystrophy)**, also known as autoimmune polyglandular syndrome type 1 (APS-1)—that leads to destruction of multiple endocrine tissues, including insulin-producing pancreatic islets, and to fungal infections, particularly candidiasis. Mice engineered to lack the *AIRE* gene fail to express many peripheral genes in the thymus and develop a similar syndrome. This links AIRE to the expression of these genes, and the antigens they encode, indicating that an inability to express these antigens in the thymus leads to autoimmune disease (**Fig. 15.4**). The autoimmunity that accompanies AIRE deficiency takes time to develop and does not always affect all potential organ targets. So as well as emphasizing the importance of central tolerance, this disease shows that other layers of tolerance control have important roles.

15-4   Lymphocytes that bind self antigens with relatively low affinity usually ignore them but in some circumstances become activated.

Most circulating lymphocytes have a low affinity for self antigens but make no effector response to them, and may be considered 'ignorant' of self (see Section 8-6). Such ignorant but latently self-reactive cells can be recruited into autoimmune responses if their threshold for activation is exceeded by co-activating factors. One such stimulus is infection. Naive T cells with low affinity for a ubiquitous self-antigen can become activated if they encounter an activated dendritic cell presenting that antigen and expressing high levels of co-stimulatory signals or pro-inflammatory cytokines as a result of the presence of infection.

A situation in which normally ignorant lymphocytes may be activated is when their autoantigens are also ligands for Toll-like receptors (TLRs). These receptors are usually considered to be specific for microbe-associated molecular patterns (see Section 3-5), but some of these patterns can be found among self molecules. An example is unmethylated CpG sequences in DNA that are recognized by TLR-9. Unmethylated CpG is normally much more common in bacterial than mammalian DNA, but is enriched in apoptotic mammalian cells. In a scenario of extensive cell death coupled with inadequate clearance of apoptotic fragments, B cells specific for chromatin components can internalize CpG sequences via their B-cell receptors. These sequences can be recognized by TLR-9 intracellularly, leading to a co-stimulatory signal that activates the previously ignorant anti-chromatin B cell (**Fig. 15.5**). B cells activated in this way produce anti-chromatin autoantibodies and also can act as antigen-presenting cells for autoreactive T cells. Ribonucleoprotein complexes containing uridine-rich RNA have similarly been shown to activate

Fig. 15.4 The 'autoimmune regulator' gene *AIRE* promotes the expression of some tissue-specific antigens in thymic medullary cells, causing the deletion of immature thymocytes that can react to these antigens. Although the thymus expresses many genes, and thus self proteins, common to all cells, it is not obvious how antigens that are specific to specialized tissues, such as retina or ovary (first panel), gain access to the thymus to promote the negative selection of immature autoreactive thymocytes. It is now known that a gene called *AIRE* promotes the expression of many tissue-specific proteins in thymic medullary cells. Some developing thymocytes will be able to recognize these tissue-specific antigens (second panel). Peptides from these proteins are presented to the developing thymocytes as they undergo negative selection in the thymus (third panel), causing deletion of these cells. In the absence of *AIRE*, this deletion does not occur; instead, the autoreactive thymocytes mature and are exported to the periphery (fourth panel), where they could cause autoimmune disease. Indeed, people and mice that lack expression of *AIRE* develop an autoimmune syndrome called APECED, or autoimmune polyendocrinopathy–candidiasis–ectodermal dystrophy.

Autoimmune Polyendocrinopathy–Candidiasis–Ectodermal Dystrophy (APECED)

Fig. 15.5 Self antigens that are recognized by Toll-like receptors can activate autoreactive B cells by providing co-stimulation. The receptor TLR-9 promotes the activation of B cells that produce antibodies specific for DNA, a common autoantibody in the autoimmune disease systemic lupus erythematosus (SLE) (see Fig. 15.1). Although B cells with strong affinity for DNA are eliminated in the bone marrow, some DNA-specific B cells with lower affinity escape and persist in the periphery but are not normally activated. Under some conditions and in genetically susceptible individuals, the concentration of DNA may increase, leading to the ligation of enough B-cell receptors to initiate activation of these B cells. B cells signal through their receptor (left panel) but also take up the DNA (center panel) and deliver it to an endosomal compartment (right panel). Here the DNA has access to TLR-9, which recognizes DNA that is enriched in unmethylated CpG DNA sequences. Such CpG-enriched sequences are much more common in microbial than eukaryotic DNA and normally this allows TLR-9 to distinguish pathogens from self. DNA in apoptotic mammalian cells is enriched in unmethylated CpG, however, and the DNA-specific B cell will also concentrate this self-DNA in the endosomal compartment. This would provide sufficient ligands to activate TLR-9, potentiating the activation of the DNA-specific B cells and leading to the production of autoantibodies against DNA.

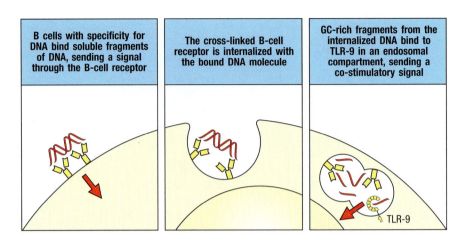

| B cells with specificity for DNA bind soluble fragments of DNA, sending a signal through the B-cell receptor | The cross-linked B-cell receptor is internalized with the bound DNA molecule | GC-rich fragments from the internalized DNA bind to TLR-9 in an endosomal compartment, sending a co-stimulatory signal |

Systemic Lupus Erythematosus

Rheumatoid Arthritis

naive B cells through binding by TLR-7 or TLR-8. Autoantibodies against DNA, chromatin, and ribonucleoproteins are produced in the autoimmune disease **systemic lupus erythematosus** (**SLE**), and this appears to be one mechanism by which self-reactive B cells are stimulated to produce them.

Another mechanism by which ignorant lymphocytes can be drawn into action is by the changing of the availability or form of self antigens. Some antigens are normally intracellular and not encountered by lymphocytes, but they may be released as a result of massive tissue injury or inflammation. These antigens can then activate ignorant T and B cells, leading to autoimmunity. This can occur after myocardial infarction, when an autoimmune response is detectable some days after the release of cardiac antigens. Such reactions are typically transient and cease when the autoantigens have been removed; however, when clearance mechanisms are inadequate, they can continue, causing clinical autoimmune disease.

Additionally, some autoantigens are present in great quantity but are usually in a nonimmunogenic form. IgG is a good example, as there are large quantities of it in blood and extracellular fluids. B cells specific for the IgG constant region are not usually activated, because IgG is monomeric and cannot cross-link the B-cell receptor. However, when immune complexes form following infection or immunization, enough IgG is in multivalent form to evoke a response from otherwise ignorant B cells. The anti-IgG autoantibody they produce is called **rheumatoid factor** because it is often present in rheumatoid arthritis. Again, this response is normally short-lived, as long as the immune complexes are cleared rapidly.

A unique situation can occur when activated B cells undergo somatic hypermutation in germinal centers (see Section 10-7), resulting in some activated B cells increasing their affinity for a self antigen or becoming newly self-reactive (**Fig. 15.6**). There seems, however, to be a mechanism to control germinal center B cells that have acquired affinity for self. In this case, if a hypermutated self-reactive B cell encounters strong cross-linking of its B-cell receptor in the germinal center, it undergoes apoptosis rather than further proliferation.

15-5 Antigens in immunologically privileged sites do not induce immune attack but can serve as targets.

Foreign tissue grafts placed in some body sites do not elicit immune responses. For instance, grafts placed in the brain and anterior chamber of the eye do not induce rejection. Such locations are termed **immunologically privileged sites** (**Fig. 15.7**). It was originally believed that immunological privilege arose from the failure of antigens to leave privileged sites and induce immune responses.

Subsequent studies have shown that antigens do leave these sites and interact with T cells. However, instead of eliciting an effector immune response, they induce a tolerogenic response that does not injure the tissue.

Immunologically privileged sites seem to be unusual in three ways. First, communication between the privileged site and the body is atypical in that extracellular fluid does not pass through conventional lymphatics, although proteins placed in these sites do leave and can have immunological effects. Privileged sites are generally surrounded by tissue barriers that exclude naive lymphocytes. The brain, for example, is guarded by the blood–brain barrier. Second, soluble factors that affect the course of an immune response are produced in privileged sites. The anti-inflammatory transforming growth factor (TGF)-β seems to be particularly important in this regard. Under homeostatic conditions, antigens recognized in concert with TGF-β tend to induce T_{reg} responses, rather than pro-inflammatory T_H17 responses, which are induced by TGF-β in the presence of IL-6 (see Section 9-21). Third, the expression of Fas ligand in immunologically privileged sites may provide a further level of protection by inducing the apoptosis of Fas-bearing effector lymphocytes that enter these sites.

Paradoxically, antigens sequestered in immunologically privileged sites are often targets of autoimmune attack; for example, brain and spinal cord autoantigens such as myelin basic protein are targeted in the autoimmune disease **multiple sclerosis**, a chronic inflammatory demyelinating disease of the central nervous system (see Fig. 15.1). Thus, the tolerance normally shown to these antigens cannot be due to previous deletion of the self-reactive T cells. In **experimental autoimmune encephalomyelitis** (**EAE**), a mouse model for multiple sclerosis, mice become diseased only when they are immunized with myelin antigens and adjuvants, which cause infiltration of the central nervous system with antigen-specific T_H17 and T_H1 cells that induce a local inflammatory response that damages nerve tissue.

Thus, some antigens expressed in immunologically privileged sites induce neither tolerance nor lymphocyte activation in normal circumstances, but if autoreactive lymphocytes are activated elsewhere, these autoantigens can become targets for autoimmune attack. Likely, T cells specific for antigens sequestered in immunologically privileged sites are in a state of immunological ignorance. Further evidence comes from the eye disease **sympathetic ophthalmia** (**Fig. 15.8**). If one eye is ruptured by a blow or other trauma, an autoimmune response to eye proteins can occur, although this happens only rarely. Once the response is induced, it often attacks both eyes. Immunosuppression—and, rarely, removal of the damaged eye, the source of antigen—is required to preserve vision in the undamaged eye.

Unsurprisingly, effector T cells can enter immunologically privileged sites when such sites become infected. Effector T cells can enter most tissues after activation (see Chapter 11), but accumulation of cells is seen only when antigen is recognized in the site, triggering the production of cytokines that alter tissue barriers.

15-6 Autoreactive T cells that express particular cytokines may be nonpathogenic or may suppress pathogenic lymphocytes.

As described in Chapter 9, CD4 T cells can differentiate into various types of effector lineages, namely, T_H1, T_H2, and T_H17 cells. These effector subsets evolved to control different types of infections and orchestrate distinct types of responses, as reflected in their different effects on antigen-presenting cells, B cells, and innate cells such as macrophages, eosinophils, and neutrophils (see Chapters 9–11). A similar paradigm holds true for autoimmunity: certain T-cell-mediated autoimmune diseases such as **type 1 diabetes mellitus** (see Fig. 15.1) depend on T_H1 cells to cause disease, whereas others, such as psoriasis (an autoimmune disease of the skin), depend on T_H17 cells.

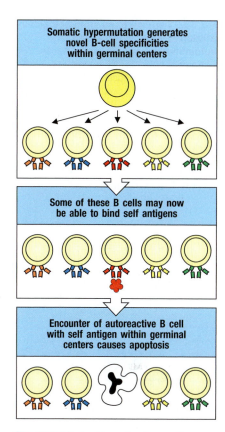

Fig. 15.6 Elimination of autoreactive B cells in germinal centers. During somatic hypermutation in germinal centers (top panel), B cells with autoreactive B-cell receptors can arise. Ligation of these receptors by soluble autoantigen (center panel) induces apoptosis of the autoreactive B cell by signaling through the B-cell antigen receptor in the absence of helper T cells (bottom panel).

Immunologically privileged sites
Brain
Eye
Testis
Uterus (fetus)

Fig. 15.7 Some sites in the body are immunologically privileged. Tissue grafts placed in these sites often last indefinitely, and antigens placed in these sites do not elicit destructive immune responses.

Fig. 15.8 Damage to an immunologically privileged site can induce an autoimmune response. In the disease sympathetic ophthalmia, trauma to one eye releases the sequestered eye antigens into the surrounding tissues, making them accessible to T cells. The effector cells that are elicited attack the traumatized eye, and also infiltrate and attack the healthy eye. Thus, although the sequestered antigens do not induce a response by themselves, if a response is induced elsewhere they can serve as targets for attack.

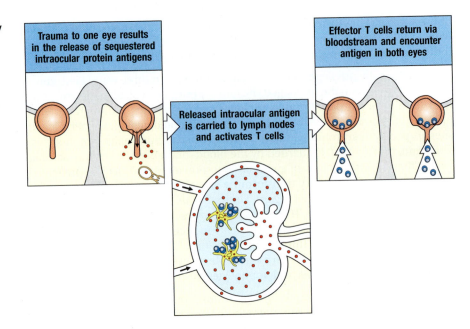

Trauma to one eye results in the release of sequestered intraocular protein antigens

Released intraocular antigen is carried to lymph nodes and activates T cells

Effector T cells return via bloodstream and encounter antigen in both eyes

In murine models of diabetes, when cytokines were infused to influence T-cell differentiation or when knockout mice predisposed to T_H2 differentiation were studied, the development of diabetes was inhibited. In some cases, potentially pathogenic T cells specific for pancreatic islet-cell components, and expressing T_H2 instead of T_H1 cytokines, actually suppressed disease caused by T_H1 cells of the same specificity. So far, attempts to control human autoimmune disease by switching cytokine profiles from one effector cell type to another (for example, T_H1 to T_H2), a procedure termed **immune modulation**, have not proved successful. Another subset of CD4 T cells, T_{reg} cells, might prove to be more important in the prevention of autoimmune disease, and efforts to deviate effector to regulatory T-cell responses are being studied as a novel therapy for autoimmunity.

15-7   Autoimmune responses can be controlled at various stages by regulatory T cells.

Autoreactive cells that escape the tolerance-inducing mechanisms described above can still be regulated so that they do not cause disease. This regulation takes two forms: the first is extrinsic, and is mediated by regulatory T cells that act on activated T cells and antigen-presenting cells; the second is intrinsic, and has its basis in limits on the size and duration of immune responses that are programmed into lymphocytes themselves. We shall first discuss the role of regulatory T cells (introduced in Chapter 9).

Tolerance due to regulatory lymphocytes is distinguished from other forms of self-tolerance by the fact that T_{reg} cells have the potential to suppress self-reactive lymphocytes that recognize antigens different from those recognized by the T_{reg} cell (**Fig. 15.9**). This type of tolerance is known as **regulatory tolerance**. The key feature of regulatory tolerance is that regulatory cells can suppress autoreactive lymphocytes that recognize a variety of different self antigens, as long as the antigens are from the same tissue or are presented by the same antigen-presenting cell. As discussed in Chapter 9, two general types of regulatory T cells have been defined experimentally. 'Natural' T_{reg} (nT_{reg}) cells are programmed in the thymus to express the transcription factor FoxP3 in response to self antigens. When activated by the same antigens in peripheral tissues, nT_{reg} cells inhibit other self-reactive T cells that recognize antigens in the same tissue to prevent their differentiation into effector T cells or prevent their effector function. 'Induced' T_{reg} (iT_{reg}) cells also express FoxP3 but develop in peripheral immune tissues in response to antigens recognized in the presence of TGF-β but in the absence of pro-inflammatory cytokines. Giving animals large amounts of self antigen orally, which induces so-called **oral tolerance** (see Section 12-18), can sometimes lead to unresponsiveness to these antigens when given by other routes, and can prevent autoimmune disease. Oral tolerance is routinely generated to antigens such as food antigens and is accompanied by the generation of iT_{reg} cells in the gut-draining mesenteric lymph nodes. These cells are known to suppress immune responses to the given antigen in the gut itself, but how the suppression in the rest of the peripheral immune system is achieved is unclear. Many investigators have hypothesized that iT_{reg} cells could have therapeutic potential for the treatment of autoimmune disease if they could be isolated or induced to differentiate and then be infused into patients.

The importance of FoxP3—and the T_{reg} cells whose development and function it controls—in the maintenance of immune tolerance is evident from the fact that humans and mice that carry mutations in the gene for FoxP3 rapidly develop severe, systemic autoimmunity (discussed in Section 15-21). A protective role for FoxP3-expressing T_{reg} cells has been demonstrated in several autoimmune syndromes in mice, including diabetes, EAE, SLE, and inflammation of the large intestine, or colon (colitis). Experiments in mouse models of these diseases have established that $FoxP3^+$ T_{reg} cells actively suppress disease in the normal immune system, as depletion of these cells results in multi-organ autoimmune disease. T_{reg} cells have also been shown to prevent or ameliorate other immunopathologic syndromes, such as graft-versus-host disease and graft rejection, which are described later in this chapter.

The importance of regulatory T cells has been demonstrated in several human autoimmune diseases. For example, in some patients with multiple sclerosis or with autoimmune polyglandular syndrome type 2 (a rare syndrome in which two or more autoimmune diseases occur simultaneously), the suppressive activity of $FoxP3^+$ T_{reg} cells is defective, although their numbers are normal. Thus, T_{reg} cells have an important role in preventing autoimmunity, and a variety of functional defects in these cells may lead to autoimmunity.

FoxP3-expressing T_{reg} cells are not the only type of regulatory lymphocyte that has been identified. For example, FoxP3-negative regulatory T cells characterized by their production of IL-10 are enriched in the intestinal tissues, where they may suppress inflammatory bowel disease (IBD) through an IL-10-dependent mechanism. The developmental origins of these cells are not currently understood.

Almost every type of lymphocyte has been shown to display regulatory activity in some circumstance. Even B cells can regulate experimentally induced autoimmune syndromes, including collagen-induced arthritis (CIA) and EAE. This activity is probably mediated in a similar way to that of regulatory CD4 T cells, with the secretion of cytokines that inhibit proliferation and differentiation of effector T cells.

In addition to the extrinsic regulation of autoreactive T and B cells by regulatory cells, lymphocytes have intrinsic proliferation and survival limits that can help restrict autoimmune as well as normal responses (see Section 11-16). This is illustrated by the development of spontaneous autoimmunity that is caused by mutations in pathways that control apoptosis, such as the Bcl-2 pathway or the Fas pathway (see Section 7-23); mutations in these pathways lead to spontaneous autoimmunity. This form of autoimmunity provides evidence that autoreactive cells are normally generated but are then eliminated by apoptosis. This seems to be an important mechanism for both T- and B-cell tolerance.

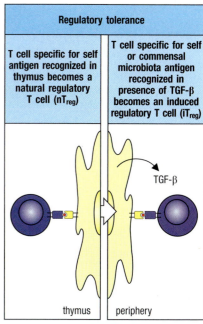

Regulatory tolerance	
T cell specific for self antigen recognized in thymus becomes a natural regulatory T cell (nT_{reg})	T cell specific for self or commensal microbiota antigen recognized in presence of TGF-β becomes an induced regulatory T cell (iT_{reg})

TGF-β

thymus   periphery

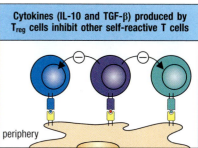

Cytokines (IL-10 and TGF-β) produced by T_{reg} cells inhibit other self-reactive T cells

periphery

Fig. 15.9 Tolerance mediated by regulatory T cells can inhibit multiple autoreactive T cells that all recognize antigens from the same tissue. Specialized autoreactive natural regulatory T (nT_{reg}) cells develop in the thymus in response to stimulation by self antigens that is too weak to cause deletion but is greater than required for simple positive selection (upper left panel). Regulatory T cells can also be induced from naive self-reactive T cells in the periphery if the naive T cell recognizes its antigen and is activated in the presence of the cytokine TGF-β (upper right panel). The lower panel shows how regulatory T cells, both natural and induced, can inhibit other self-reactive T cells. If regulatory T cells encounter their self antigen on an antigen-presenting cell, they secrete inhibitory cytokines such as IL-10 and TGF-β that inhibit all surrounding autoreactive T cells, regardless of their precise autoantigen specificity.

Summary.

Discrimination between self and nonself is imperfect, partly because a proper balance must be struck between preventing autoimmune disease and preserving immune competence. Self-reactive lymphocytes always exist in the natural immune repertoire but are not often activated. In autoimmune diseases, however, these cells become activated by autoantigens. If activation persists, autoreactive effector lymphocytes are generated and cause disease. The immune system has a remarkable set of mechanisms that work together to prevent autoimmune disease (see Fig. 15.2). This collective action means that each mechanism need not work perfectly nor apply to every possible self-reactive cell. Self-tolerance begins during lymphocyte development, when autoreactive T cells in the thymus and B cells in the bone marrow are deleted or, in the case of CD4 T cells, give rise to a subpopulation of self antigen-reactive FoxP3[+] 'natural' or 'thymic' T_{reg} cells that suppress autoimmune responses after exiting the thymus. Mechanisms of peripheral tolerance, such as anergy and deletion, or the extrathymic production of 'induced' or 'peripheral' T_{reg} cells, complement these central tolerance mechanisms for antigens that are not expressed in the thymus or bone marrow. Weakly self-reactive lymphocytes are not removed in the primary lymphoid tissues (thymus and bone marrow), as deletion of weakly autoreactive cells would impose too great a limitation on the immune repertoire, resulting in impaired immune responses to pathogens. Instead, weakly self-reactive cells are suppressed only if they are activated in the periphery, and the mechanisms that suppress them include inhibition by T_{reg} cells, which are themselves autoreactive, although not pathogenic. T_{reg} cells can inhibit self-reactive lymphocytes if the regulatory cells are targeting autoantigens located in the same vicinity as the autoantigens to which the self-reactive lymphocytes respond. This allows regulatory cells to home to and suppress sites of autoimmune inflammation. A final mechanism that controls autoimmunity is the natural tendency of immune responses to be self-limited; intrinsic programs in activated lymphocytes make them prone to apoptosis. Activated lymphocytes also acquire sensitivity to external apoptosis-inducing signals, such as those mediated by Fas.

Autoimmune diseases and pathogenic mechanisms.

Here we describe some common clinical autoimmune syndromes, and ways in which loss of self-tolerance can generate self-reactive lymphocytes that cause tissue damage. The mechanisms of pathogenesis resemble in many ways those that target invading pathogens. Damage by autoantibodies, mediated through the complement and Fc receptor systems, has an important role in some diseases, such as SLE. Similarly, cytotoxic T cells directed at self tissues destroy them much as they would virus-infected cells; this is one way by which pancreatic β cells are destroyed in diabetes. However, unlike most pathogens, self proteins are not typically eliminated, so that, with rare exceptions—such as the islet cells in the pancreas—the response persists chronically. Some pathogenic mechanisms are unique to autoimmunity, such as antibodies against cell-surface receptors that affect their function, as in myasthenia gravis. In this part of the chapter we describe the pathogenic mechanisms of some major autoimmune diseases.

15-8 Specific adaptive immune responses to self antigens can cause autoimmune disease.

In certain genetically susceptible strains of experimental animals, autoimmune disease can be induced by the injection of 'self' tissues that were taken from a genetically identical animal and mixed with strong adjuvants (see Appendix I, Section A-1). This shows directly that autoimmunity can be provoked by inducing a specific adaptive immune response to self antigens. Such experimental systems highlight the importance of the activation of other components of the immune system, primarily dendritic cells, by bacteria contained in the adjuvant. There are drawbacks to the use of such animal models for the study of autoimmunity, however. In humans and genetically autoimmune-prone animals, autoimmunity usually arises spontaneously: that is, we do not know what events initiate the immune response to self that leads to autoimmune disease. By studying the patterns of autoantibodies and particular tissues affected, it has been possible to identify some of the self antigens that are targets of autoimmune disease, although it has yet to be proven that the immune reaction was initiated in response to these same antigens.

Some autoimmune disorders may be triggered by infectious agents that express an epitope resembling a self antigen found in a tissue and that lead to sensitization of the patient against that tissue. There is, however, also evidence from animal models of autoimmunity that some autoimmune disorders are caused by internal dysregulation of the immune system without the apparent participation of infectious agents.

15-9 Autoimmunity can be classified into either organ-specific or systemic disease.

The classification of disease is an uncertain science, especially in the absence of a precise understanding of causative mechanisms. This is well illustrated by the difficulty in classifying autoimmune diseases. From a clinical perspective it is often useful to distinguish between the following two major patterns of autoimmunity: diseases restricted to specific organs of the body, known as 'organ-specific' autoimmune diseases; and those in which many tissues of the body are affected, the 'systemic' autoimmune diseases. In both types, disease has a tendency to become chronic because, with a few notable exceptions (for example, Hashimoto's thyroiditis), autoantigens are rarely cleared from the body. Some autoimmune diseases seem to be dominated by the pathogenic effects of a particular immune effector pathway, either autoantibodies or effector T cells. However, both of these pathways often contribute to the overall pathogenesis.

In organ-specific diseases, autoantigens from one or a few organs are targeted, and disease is limited to those organs. Examples include **Hashimoto's thyroiditis** and **Graves' disease**, which both predominantly affect the thyroid gland; and type 1 diabetes, which is caused by immune attack on insulin-producing pancreatic β cells. Examples of systemic autoimmune disease are SLE and primary **Sjögren's syndrome**, in which tissues as diverse as the skin, kidneys, and brain may all be affected (**Fig. 15.10**).

The autoantigens recognized in these two categories of disease are themselves organ-specific and systemic, respectively. Thus, Graves' disease is characterized by the production of antibodies against the thyroid-stimulating hormone (TSH) receptor, Hashimoto's thyroiditis by antibodies against thyroid peroxidase, and type 1 diabetes by anti-insulin antibodies. By contrast, SLE is characterized by the presence of antibodies against antigens that are ubiquitous and abundant in every cell of the body, such as chromatin and the proteins of the pre-mRNA splicing machinery—the spliceosome complex.

A strict separation of diseases into organ-specific and systemic categories does, however, break down to some extent, because not all autoimmune diseases can be usefully classified in this manner. For example, autoimmune hemolytic anemia, in which red blood cells are destroyed, sometimes occurs

Organ-specific autoimmune diseases
Type 1 diabetes mellitus
Goodpasture's syndrome
Multiple sclerosis Crohn's disease Psoriasis
Graves' disease Hashimoto's thyroiditis Autoimmune hemolytic anemia Autoimmune Addison's disease Vitiligo Myasthenia gravis

Systemic autoimmune diseases
Rheumatoid arthritis
Scleroderma
Systemic lupus erythematosus Primary Sjögren's syndrome Polymyositis

Fig. 15.10 Some common autoimmune diseases classified according to their 'organ-specific' or 'systemic' nature. Diseases that tend to occur in clusters are grouped in single boxes. Clustering is defined as more than one disease affecting a single patient or different members of a family. Not all autoimmune diseases can be classified according to this scheme. For example, autoimmune hemolytic anemia can occur in isolation or in association with systemic lupus erythematosus.

Myasthenia Gravis

as a solitary entity and could be classified as an organ-specific disease. In other circumstances it can occur in conjunction with SLE as part of a systemic auto-immune disease.

A prevalent variant of chronic inflammatory disease is **inflammatory bowel disease** (**IBD**), which includes two main clinical entities—**Crohn's disease** (discussed later in this chapter) and **ulcerative colitis**. We discuss IBD in this chapter because it has many features of an autoimmune disease, even though it is not primarily targeted against self-tissue antigens. Instead, the targets of the dysregulated immune response in IBD are antigens derived from the commensal microbiota resident in the intestines. Strictly speaking, therefore, IBD is an outlier among autoimmune diseases in that the immune response is not directed against 'self' antigens; rather, it is directed against microbial antigens of the resident, or 'self,' microbiota. Nevertheless, features of immune tolerance breakdown are also seen in IBD, and, as with the organ-specific auto-immune diseases, the tissue destruction wrought by the aberrant immune response is primarily localized to a single organ—the intestines.

15-10 Multiple components of the immune system are typically recruited in autoimmune disease.

Immunologists have long been concerned with which parts of the immune system are important in different autoimmune syndromes, because this can be useful in understanding disease etiology and developing therapies. In **myasthenia gravis**, for example, autoantibodies produced against the acetylcholine receptor block receptor function at the neuromuscular junction, resulting in a syndrome of muscle weakness. In other autoimmune conditions, antibodies in the form of immune complexes are deposited in tissues and cause tissue damage as a consequence of the inflammation that results from complement activation and ligation of Fc receptors on inflammatory cells.

Relatively common autoimmune diseases in which effector T cells seem to be the main destructive agents include type 1 diabetes, psoriasis, IBD, and multiple sclerosis. In these diseases, T cells recognize self peptides or peptides derived from the commensal microbiota that are complexed with self MHC molecules. The damage in such diseases is caused by T cells recruiting and activating myeloid cells of the innate immune system to cause local inflammation, or by direct T-cell damage to tissue cells.

When disease can be transferred from a diseased individual to a healthy one by transferring autoantibodies and/or self-reactive T cells, this both confirms that the disease is autoimmune in nature and also proves the involvement of the transferred material in the pathological process. In myasthenia gravis, serum from affected patients can transfer symptoms to animal recipients, thus proving the pathogenic role of the anti-acetylcholine autoantibodies (**Fig. 15.11**). Similarly, in the animal model disease EAE, T cells from affected animals can transfer disease to normal animals (**Fig. 15.12**).

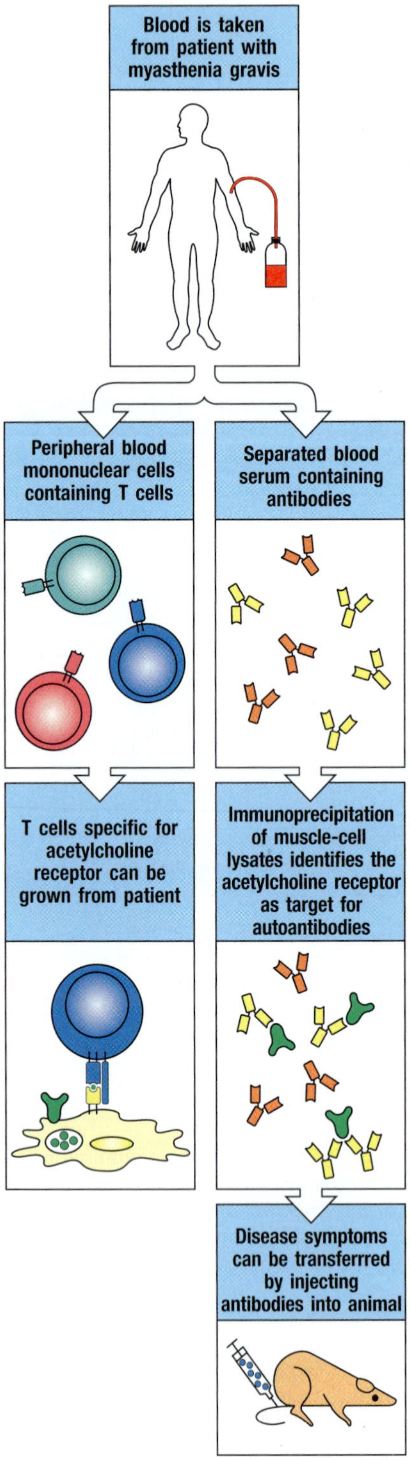

Fig. 15.11 Identification of autoantibodies that can transfer disease in patients with myasthenia gravis. Autoantibodies from the serum of patients with myasthenia gravis immunoprecipitate the acetylcholine receptor from lysates of skeletal muscle cells (right-hand panels). Because the antibodies can bind to both the murine and the human acetylcholine receptor, they can transfer disease when injected into mice (bottom panel). This experiment demonstrates that the antibodies are pathogenic. However, to be able to produce antibodies, the same patients should also have CD4 T cells that respond to a peptide derived from the acetylcholine receptor. To detect them, T cells from patients with myasthenia gravis are isolated and grown in the presence of the acetylcholine receptor plus antigen-presenting cells of the correct MHC type (left-hand panels). T cells specific for epitopes of the acetylcholine receptor are stimulated to proliferate and can thus be detected.

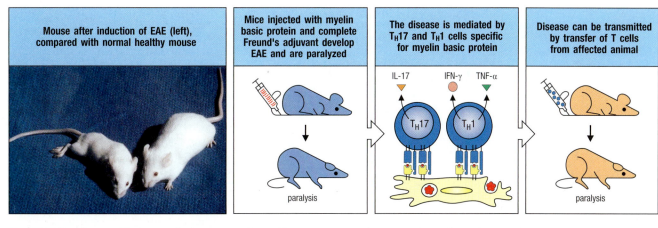

Fig. 15.12 T cells specific for myelin basic protein mediate inflammation of the brain in experimental autoimmune encephalomyelitis (EAE). This disease is produced in experimental animals by injecting them with isolated spinal cord homogenized in complete Freund's adjuvant. EAE is due to an inflammatory reaction in the brain that causes a progressive paralysis affecting first the tail and hind limbs (as shown in the mouse on the left of the photograph, compared with a healthy mouse on the right) before progressing to forelimb paralysis and eventual death. One of the autoantigens identified in the spinal cord homogenate is myelin basic protein (MBP). Immunization with MBP alone in complete Freund's adjuvant can also cause these disease symptoms. Inflammation of the brain and paralysis are mediated by T_H1 and T_H17 cells specific for MBP. Cloned MBP-specific T_H1 cells can transfer symptoms of EAE to naive recipients provided that the recipients carry the correct MHC allele. In this system it has therefore proved possible to identify the peptide:MHC complex recognized by the T_H1 clones that transfer disease. Other purified components of the myelin sheath can also induce the symptoms of EAE, so there is more than one autoantigen in this disease. Photograph from Wraith, D., et al.: Cell 1989, 59:247–255. With permission from Elsevier.

Pregnancy can demonstrate a role for antibodies in disease, as IgG antibodies, but not T cells, can cross the placenta (see Section 10-15). For some autoimmune diseases (Fig. 15.13), transmission of autoantibodies across the placenta leads to disease in the fetus or neonate (Fig. 15.14). This provides proof in humans that autoantibodies cause some of the symptoms of autoimmunity. The symptoms of disease in the newborn infant typically disappear rapidly as

Autoimmune diseases transferred across the placenta to the fetus and newborn infant		
Disease	**Autoantibody**	**Symptom**
Myasthenia gravis	Anti-acetylcholine receptor	Muscle weakness
Graves' disease	Anti-thyroid-stimulating-hormone (TSH) receptor	Hyperthyroidism
Thrombocytopenic purpura	Anti-platelet antibodies	Bruising and hemorrhage
Neonatal lupus rash and/or congenital heart block	Anti-Ro antibodies Anti-La antibodies	Photosensitive rash and/or bradycardia
Pemphigus vulgaris	Anti-desmoglein-3	Blistering rash

Fig. 15.13 Some autoimmune diseases that can be transferred across the placenta by pathogenic IgG autoantibodies. These diseases are caused mostly by autoantibodies against cell-surface or tissue-matrix molecules. This suggests that an important factor determining whether an autoantibody that crosses the placenta causes disease in the fetus or newborn baby is the accessibility of the antigen to the autoantibody. Autoimmune congenital heart block is caused by fibrosis of the developing cardiac conducting tissue, which expresses abundant Ro antigen. Ro protein is a constituent of an intracellular small cytoplasmic ribonucleoprotein. It is not yet known whether it is expressed at the cell surface of cardiac conducting tissue to act as a target for autoimmune tissue injury. Nevertheless, autoantibody binding leads to tissue damage and results in slowing of the heart rate (bradycardia).

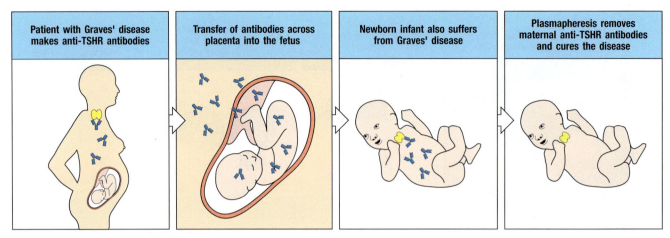

| Patient with Graves' disease makes anti-TSHR antibodies | Transfer of antibodies across placenta into the fetus | Newborn infant also suffers from Graves' disease | Plasmapheresis removes maternal anti-TSHR antibodies and cures the disease |

Fig. 15.14 Antibody-mediated autoimmune diseases can appear in the infants of affected mothers as a consequence of transplacental antibody transfer. In pregnant women, IgG antibodies cross the placenta and accumulate in the fetus before birth (see Fig. 10.30). Babies born to mothers with IgG-mediated autoimmune disease therefore frequently show symptoms similar to those of the mother in the first few weeks of life. Fortunately, there is little lasting damage because the symptoms disappear along with the maternal antibody. In Graves' disease, the symptoms are caused by antibodies against the thyroid-stimulating hormone receptor (TSHR). Children of mothers making thyroid-stimulating antibody are born with hyperthyroidism, but this can be corrected by replacing the plasma with normal plasma (plasmapheresis), thus removing the maternal antibody.

the maternal antibody is catabolized, but in some cases the antibodies cause organ injury before they are removed, such as damage to the conducting tissue of the heart in babies of mothers with SLE or Sjögren's syndrome. Antibody clearance can be accelerated by exchange of the infant's blood or plasma (plasmapheresis), although this is not useful after permanent injury has occurred.

Figure 15.15 lists a selection of autoimmune diseases, along with the parts of the immune response that contribute to their pathogenesis. However, although the diseases noted above are clear examples that a particular effector function can drive disease, most autoimmune diseases are not caused solely by a single effector pathway. It is more useful to consider autoimmune responses, like immune responses to pathogens, as engaging the integrated immune system and typically involving T, B, and innate immune cells. Indeed,

Autoimmune diseases involve all aspects of the immune response			
Disease	**T cells**	**B cells**	**Antibody**
Systemic lupus erythematosus	Pathogenic Help for antibody	Present antigen to T cells	Pathogenic
Type 1 diabetes	Pathogenic	Present antigen to T cells	Present, but role unclear
Myasthenia gravis	Help for antibody	Antibody secretion	Pathogenic
Multiple sclerosis	Pathogenic	Present antigen to T cells	Present, but role unclear

Fig. 15.15 Autoimmune diseases involve all aspects of the immune response. Although some autoimmune diseases have traditionally been thought to be mediated by B cells or T cells, it is useful to consider that, typically, all aspects of the immune system have a role. For four important autoimmune diseases, the figure lists the roles of T cells, B cells, and antibody. In some diseases, such as SLE, T cells can have multiple roles such as helping B cells to make autoantibody and directly promoting tissue damage. B cells can have two roles as well—presenting autoantigens to stimulate T cells and secreting pathogenic autoantibodies.

although autoimmunity research has traditionally focused on identification of the antigen specificity and effector subclass of autoreactive T and B cells, experimental evidence shows that cells of the innate immune system—particularly phagocytic myeloid cells—are critical in mediating tissue damage in most autoimmune diseases. Innate lymphoid cells (ILCs) have also been found in autoimmune lesions, especially those at barrier surfaces. However, the exact role of ILCs, and whether they may be good therapeutic targets in autoimmune disorders, is currently unclear.

15-11 Chronic autoimmune disease develops through positive feedback from inflammation, inability to clear the self antigen, and a broadening of the autoimmune response.

When normal immune responses are engaged to destroy a pathogen, the typical outcome is elimination of the foreign invader, after which the immune response ceases, accompanied by mass extinction of most effector cells and persistence of a small cohort of memory lymphocytes (see Chapter 11). In autoimmunity, however, the self antigen cannot be easily eliminated, because it is in vast excess or is ubiquitous (as is, for example, chromatin). Thus, a very important mechanism for limiting an immune response is abrogated in many autoimmune diseases.

In general, autoimmune diseases are characterized by an early activation phase with the involvement of only a few autoantigens, followed by a chronic stage. The constant presence of autoantigen leads to chronic inflammation. This leads to the release of more autoantigens as a result of tissue damage, and this breaks an important barrier to autoimmunity known as 'sequestration,' by which many self antigens are normally kept apart from the immune system. It also leads to the attraction of nonspecific effector cells such as macrophages and neutrophils that respond to the release of cytokines and chemokines from injured tissues (Fig. 15.16). The result is a continuing and evolving self-destructive process.

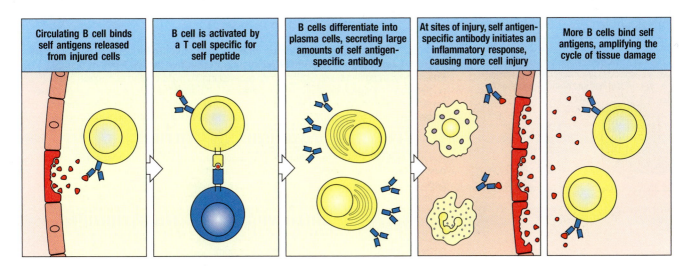

| Circulating B cell binds self antigens released from injured cells | B cell is activated by a T cell specific for self peptide | B cells differentiate into plasma cells, secreting large amounts of self antigen-specific antibody | At sites of injury, self antigen-specific antibody initiates an inflammatory response, causing more cell injury | More B cells bind self antigens, amplifying the cycle of tissue damage |

Fig. 15.16 Autoantibody-mediated inflammation can lead to the release of autoantigens from damaged tissues, which in turn promotes further activation of autoreactive B cells. Autoantigens, particularly intracellular ones that are targets in SLE, stimulate B cells only when released from dying cells (first panel). The result is the activation of autoreactive T and B cells and the eventual secretion of autoantibodies (second and third panels). These autoantibodies can mediate tissue damage through a variety of effector functions (see Chapter 10), resulting in the further death of cells (fourth panel). A positive feedback loop is established because these additional autoantigens recruit and activate additional autoreactive B cells (fifth panel), which in turn can start the cycle over again, as shown in the first panel.

Pemphigus Vulgaris

Progression of the autoimmune response is often accompanied by recruitment of new clones of lymphocytes reactive to new epitopes on the initiating autoantigen, as well as new autoantigens. This phenomenon is known as **epitope spreading**, and is important in perpetuating and amplifying disease. As seen in Chapter 10, activated B cells can internalize cognate antigens by receptor-mediated endocytosis via their antigen receptor, process them, and present the derived peptides to T cells. Epitope spreading can occur in several ways. Because antibody-bound antigens can be more efficiently presented, self antigens that are normally present in concentrations too low to activate naive cell processing of the internalized autoantigen can reveal novel, previously hidden, peptide epitopes called **cryptic epitopes** that the B cell can then present to T cells. Autoreactive T cells responding to these 'new' epitopes will provide help to any B cells presenting these peptides, recruiting additional B-cell clones to the autoimmune reaction, with the consequent production of a greater variety of autoantibodies. In addition, on binding and internalizing specific antigen via their B-cell receptor, B cells will also internalize any other molecules closely associated with that antigen. By these routes, B cells can act as antigen-presenting cells for peptides derived from antigens completely different from the original autoantigen that initiated the autoimmune reaction.

The autoantibody response in SLE initiates these mechanisms of epitope spreading. In this disease, autoantibodies against both the protein and DNA components of chromatin are found. **Figure 15.17** shows how autoreactive B cells specific for DNA can recruit autoreactive T cells specific for histone proteins, another component of chromatin, into the autoimmune response. In turn, these T cells provide help not only to the original DNA-specific B cells but also to histone-specific B cells, resulting in the production of both anti-DNA and anti-histone antibodies.

Another autoimmune disease in which epitope spreading is linked to the progression of disease is **pemphigus vulgaris**, which is characterized by severe blistering of the skin and mucosal membranes. It is caused by autoantibodies against desmogleins, a type of cadherin present in cell junctions (desmosomes) that hold cells of the epidermis together (**Fig. 15.18**). Binding of

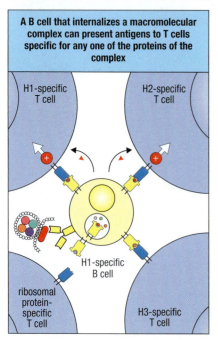

Fig. 15.17 Epitope spreading occurs when B cells specific for various components of a complex antigen are stimulated by an autoreactive helper T cell of a single specificity. In patients with SLE, an ever-broadening immune response is made against nucleoprotein antigens such as nucleosomes, which consist of histones and DNA and are released from dying and disintegrating cells. The upper panel shows how the emergence of a single clone of autoreactive CD4 T cells can lead to a diverse B-cell response to nucleosome components. The T cell in the center is specific for a particular peptide (red) from the linker histone H1, which is present on the surface of the nucleosome. The B cells at the top are specific for epitopes on the surface of a nucleosome, on H1 and DNA, respectively, and thus bind and endocytose intact nucleosomes, process the constituents, and present the H1 peptide to the helper T cell. Such B cells will be activated to make antibodies, which in the case of the DNA-specific B cell will be anti-DNA antibodies. The B cell at the bottom right is specific for an epitope on histone H2, which is hidden inside the intact nucleosome and is thus inaccessible to the B-cell receptor. This B cell does not bind the nucleosome and does not become activated by the H1-specific helper T cell. A B cell specific for another type of nucleoprotein particle, the ribosome (which is composed of RNA and specific ribosomal proteins), will not bind nucleosomes (bottom left) and will not be activated by the T cell. In reality a T cell interacts with one B cell at a time, but different members of the same T-cell clone will interact with B cells of different specificity. The lower panel shows the broadening of the T-cell response to the nucleosome. The H1-specific B cell in the center has processed an intact nucleosome and is presenting a variety of nucleosome-derived peptide antigens on its MHC class II molecules. This B cell can activate a T cell specific for any of these peptide antigens, which will include those from the internal histones H2, H3, and H4 as well as those from H1. This H1-specific B cell will not activate T cells specific for peptide antigens of ribosomes because ribosomes do not contain histones.

autoantibodies to the extracellular domains of these adhesion molecules causes dissociation of the junctions and dissolution of the affected tissue. Pemphigus vulgaris usually starts with lesions in the oral and genital mucosa; only later does the skin become involved. In the mucosal stage, only autoantibodies against certain epitopes on desmoglein Dsg-3 are found, and these antibodies seem unable to cause skin blistering. Progression to the skin disease is associated both with intramolecular epitope spreading within Dsg-3, which gives rise to autoantibodies that can cause deep skin blistering, and with intermolecular epitope spreading to another desmoglein, Dsg-1, which is more abundant in the epidermis. Dsg-1 is also the autoantigen in a less severe variant of the disease, pemphigus foliaceus. In that disease, the autoantibodies first produced against Dsg-1 cause no damage, and disease appears only after autoantibodies are made against epitopes on parts of the protein involved in the adhesion of epidermal cells.

15-12 Both antibody and effector T cells can cause tissue damage in autoimmune disease.

The manifestations of autoimmune disease are caused by effector mechanisms of the immune system being directed at the body's own tissues. As discussed previously, the response can be amplified and maintained by the constant supply of new autoantigen. An exception to this rule is type 1 diabetes, in which the autoimmune response destroys most or all of the target cells. This leads to a failure to produce sufficient insulin to maintain glucose homeostasis, resulting in the symptoms of diabetes.

Historically, the mechanisms of tissue injury in autoimmunity have been classified according to a scheme adopted for 'hypersensitivity' reactions that were defined in the early 1960s, prior to a more modern understanding of immune mechanisms (Fig. 15.19; also see introduction to Chapter 14). We now recognize that the dominant types of immunity that are orchestrated for the clearance of different types of pathogens are the same ones that become dysregulated in autoimmunity, and that both B and T cells, as well as effector cells of the innate immune system, contribute—even in cases where a particular type of response (for example, autoantibody-mediated cellular injury) predominates in causing tissue damage. The antigen, or group of antigens, against which the autoimmune response is directed, and the mechanism by which the antigen-bearing tissue is damaged, together determine the pathology and clinical expression of the disease.

Type 2 immune responses mediated by IgE (previously referred to as type I hypersensitivity) typically cause allergic or atopic inflammatory disease (see Chapter 14) and play no major part in most forms of autoimmunity. By contrast, autoimmunity that damages tissues by autoantibodies—whether by binding of IgG or IgM to autoantigens located on cell surfaces or extracellular matrix (type II hypersensitivity) or by tissue localization of immune complexes composed of soluble autoantigens and their cognate autoantibodies (type III hypersensitivity)—often appears to be linked to dysregulated type 3 (T_H17) or type 1 (T_H1) immunity, or to T-cell-independent generation of IgM-producing B cells. Because antibody-mediated injury can target a specific cell or tissue type (for example, autoimmune thyroiditis), or it can result in immune complexes that are deposited in specific vascular beds (for example, rheumatoid arthritis), disease can be organ-specific or systemic. In some forms of autoimmunity, such as SLE, autoantibodies cause damage by both of these mechanisms. Finally, several organ-specific autoimmune diseases are due to a type 1 response in which T_H1 cells and/or cytotoxic T cells directly cause tissue damage (type IV hypersensitivity; for example, type 1 diabetes); alternatively, some such diseases are due to a type 3 response in which T_H17 cells promote inflammation at barrier tissues (for example, psoriasis or Crohn's disease).

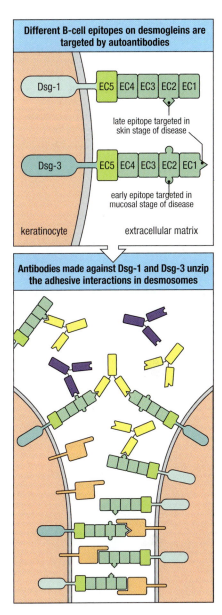

Different B-cell epitopes on desmogleins are targeted by autoantibodies

Dsg-1 — EC5 EC4 EC3 EC2 EC1

late epitope targeted in skin stage of disease

Dsg-3 — EC5 EC4 EC3 EC2 EC1

early epitope targeted in mucosal stage of disease

keratinocyte   extracellular matrix

Antibodies made against Dsg-1 and Dsg-3 unzip the adhesive interactions in desmosomes

Fig. 15.18 Pemphigus vulgaris is a skin-blistering disease caused by autoantibodies specific for desmoglein. An adhesion molecule in the cell junctions that hold keratinocytes together, desmoglein is a cell-surface protein with five extracellular domains (EC1–EC5; upper panel). Early in the autoimmune response, antibodies are made against the EC5 domain of shed desmoglein-3 (Dsg-3), but do not cause disease. However, in time, intra- and intermolecular epitope spreading occurs and IgG antibodies are made against the EC1 and EC2 domains of Dsg-3 and Dsg-1. These autoantibodies can inhibit adhesion of desmoglein in desmosomes (lower panel), and thereby interfere with the physiological adhesive interactions of desmoglein that are necessary for maintaining skin integrity. Consequently, the antibodies cause the outer layers of the skin to separate, producing blisters.

Fig. 15.19 Mechanisms of tissue damage in autoimmune diseases. Autoimmune diseases can be grouped according to the predominant type of immune response and the mechanism by which it damages tissues. In many autoimmune diseases, several immunopathogenic mechanisms operate in parallel. This is illustrated here for rheumatoid arthritis, which appears in more than one category of immunopathogenic mechanism.

Some common autoimmune diseases classified by immunopathogenic mechanism		
Syndrome	**Autoantigen**	**Consequence**
Antibody against cell-surface or matrix antigens		
Autoimmune hemolytic anemia	Rh blood group antigens, I antigen	Destruction of red blood cells by complement and FcR⁺ phagocytes, anemia
Autoimmune thrombocytopenic purpura	Platelet integrin GpIIb:IIIa	Abnormal bleeding
Goodpasture's syndrome	Noncollagenous domain of basement membrane collagen type IV	Glomerulonephritis, pulmonary hemorrhage
Pemphigus vulgaris	Epidermal cadherin	Blistering of skin
Acute rheumatic fever	Streptococcal cell-wall antigens. Antibodies cross-react with cardiac muscle	Arthritis, myocarditis, late scarring of heart valves
Immune-complex disease		
Mixed essential cryoglobulinemia	Rheumatoid factor IgG complexes (with or without hepatitis C antigens)	Systemic vasculitis
Rheumatoid arthritis	Rheumatoid factor IgG complexes	Arthritis
T-cell-mediated disease		
Type 1 diabetes	Pancreatic β-cell antigen	β-cell destruction
Rheumatoid arthritis	Unknown synovial joint antigen	Joint inflammation and destruction
Multiple sclerosis	Myelin basic protein, proteolipid protein, myelin oligodendrocyte glycoprotein	Brain and spinal cord invasion by CD4 T cells, muscle weakness, and other neurological symptoms
Crohn's disease	Antigens of intestinal microbiota	Regional intestinal inflammation and scarring
Psoriasis	Unknown skin antigens	Inflammation of skin with formation of plaques

In most autoimmune diseases, however, several mechanisms of pathogenesis operate. Thus, helper T cells are almost always required for the production of pathogenic autoantibodies. Reciprocally, B cells often have an important role in the maximal activation of T cells that mediate tissue damage or help auto-antibody production. In type 1 diabetes and rheumatoid arthritis, for example, both T-cell- and antibody-mediated pathways cause tissue injury. SLE is an example of autoimmunity that was previously thought to be mediated solely by antibodies and immune complexes but is now known to have a component of T-cell-mediated pathogenesis. Moreover, in virtually all autoimmune diseases, innate immune cells contribute to inflammation and antibody- or T-cell-mediated tissue injury. We will first examine how autoantibodies cause tissue damage, then consider self-reactive T-cell responses and their role in autoimmunity.

15-13 Autoantibodies against blood cells promote their destruction.

IgG or IgM responses to antigens located on the surface of blood cells lead to the rapid destruction of these cells. An example of this is **autoimmune hemolytic anemia**, in which antibodies against self antigens on red blood cells trigger destruction of the cells, leading to anemia. This can occur in two ways (**Fig. 15.20**). Red cells with bound IgG or IgM antibody can be rapidly cleared from the circulation by interaction with Fc or complement receptors, respectively, on cells of the mononuclear–macrophage phagocytic system, particularly in the spleen. Alternatively, the autoantibody-sensitized red blood cells can be lysed by formation of the membrane-attack complex of complement. In **autoimmune thrombocytopenic purpura**, autoantibodies against the GpIIb:IIIa fibrinogen receptor or other platelet-specific surface antigens can cause thrombocytopenia (a depletion of platelets), which can in turn cause hemorrhage.

Lysis of nucleated cells by complement is less common because these cells are better defended by complement-regulatory proteins, which protect cells against immune attack by interfering with the activation of complement components (see Section 2-15). Nevertheless, circulating nucleated cells targeted by autoantibodies are still destroyed by cells of the mononuclear phagocytic system or NK cells via antibody-dependent cell-mediated cytotoxicity (ADCC). Autoantibodies against neutrophils, for example, cause neutropenia, which increases susceptibility to infection with pyogenic bacteria. In all these cases, accelerated clearance of autoantibody-sensitized cells is the cause of their depletion. One therapeutic approach to this type of autoimmunity is removal of the spleen, the organ in which the main clearance of red cells, platelets, and leukocytes occurs. Another is the administration of large quantities of nonspecific IgG (termed IVIG, for intravenous immunoglobulin), which among other mechanisms inhibits the Fc receptor-mediated uptake of antibody-coated cells and activates inhibitory Fc receptors to suppress production of inflammatory mediators by myeloid cells.

15-14 The fixation of sublytic doses of complement to cells in tissues stimulates a powerful inflammatory response.

The binding of IgG and IgM antibodies to cells in tissues causes inflammatory injury by a variety of mechanisms. One of these is the fixation of complement. Although nucleated cells are relatively resistant to lysis by complement, the assembly of sublytic amounts of the membrane-attack complex on their surface provides a powerful activating stimulus. Depending on the cell type, this interaction can cause cytokine release, a respiratory burst, or the mobilization of membrane phospholipids to generate arachidonic acid—the precursor of prostaglandins and leukotrienes, lipid mediators of inflammation.

Most cells in tissues are fixed in place, and innate and adaptive immune cells are attracted to them by chemoattractant molecules. One such molecule is the complement fragment C5a, which is released as a result of complement activation triggered by autoantibody binding. Other chemoattractants, such as leukotriene B4, can be released by the autoantibody-targeted cells. Inflammatory leukocytes are further activated by binding to autoantibody Fc regions and fixed complement C3 fragments on cells. Tissue injury can result from the products of the activated leukocytes and by antibody-dependent cellular cytotoxicity mediated by NK cells (see Section 10-23).

A probable example of this type of autoimmunity is Hashimoto's thyroiditis, in which autoantibodies against tissue-specific antigens are found at extremely high levels for prolonged periods. Direct T-cell-mediated cytotoxicity, which we discuss later, is probably also important in this disease.

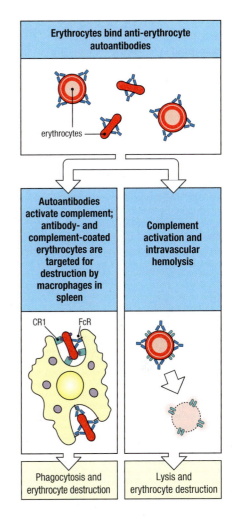

Fig. 15.20 Antibodies specific for cell-surface antigens can destroy cells. In autoimmune hemolytic anemias, red blood cells (erythrocytes) coated with IgG autoantibodies against a cell-surface antigen (upper panel) are rapidly cleared from the circulation by uptake by Fc receptor-bearing macrophages located primarily in the spleen (lower left panel). Erythrocytes coated with IgM autoantibodies fix C3 and are cleared by CR1-bearing macrophages. The binding of certain rare autoantibodies that fix complement extremely efficiently causes the formation of the membrane-attack complex on the red cells, leading to intravascular hemolysis (lower right panel).

Autoimmune Hemolytic Anemia

Fig. 15.21 Feedback regulation of thyroid hormone production is disrupted in Graves' disease. Graves' disease is caused by autoantibodies specific for the receptor for thyroid-stimulating hormone (TSH). Normally, thyroid hormones are produced in response to TSH and limit their own production by inhibiting the production of TSH by the pituitary (left panels). In Graves' disease, the autoantibodies are agonists for the TSH receptor and therefore stimulate the production of thyroid hormones (right panels). The thyroid hormones inhibit TSH production in the normal way but do not affect production of the autoantibody; the excessive thyroid hormone production induced in this way causes hyperthyroidism.

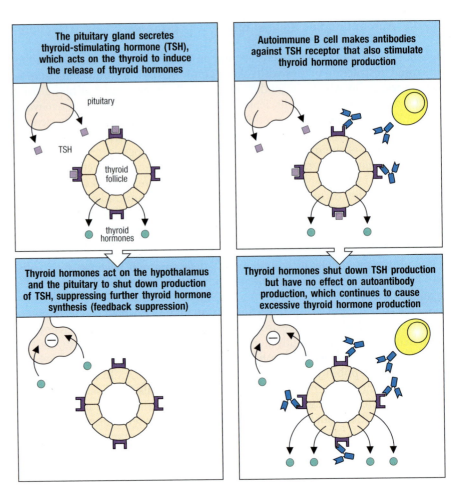

The pituitary gland secretes thyroid-stimulating hormone (TSH), which acts on the thyroid to induce the release of thyroid hormones

Autoimmune B cell makes antibodies against TSH receptor that also stimulate thyroid hormone production

Thyroid hormones act on the hypothalamus and the pituitary to shut down production of TSH, suppressing further thyroid hormone synthesis (feedback suppression)

Thyroid hormones shut down TSH production but have no effect on autoantibody production, which continues to cause excessive thyroid hormone production

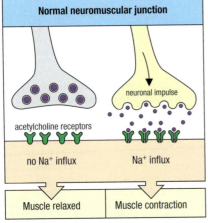

Normal neuromuscular junction

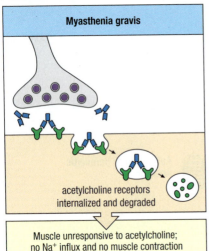

Myasthenia gravis

15-15 Autoantibodies against receptors cause disease by stimulating or blocking receptor function.

Autoimmune disease can occur when autoantibody binds a cell-surface receptor. Antibody binding to a receptor can either stimulate the receptor or block stimulation by the natural ligand. In Graves' disease, autoantibodies against the thyroid-stimulating hormone (TSH) receptor on thyroid cells stimulate excessive production of thyroid hormone. The production of thyroid hormone is normally controlled by feedback regulation; high levels of thyroid hormone inhibit the release of TSH by the pituitary. In Graves' disease, feedback inhibition fails because the autoantibody continues to stimulate the TSH receptor in the absence of TSH, and the patient produces chronically elevated levels of thyroid hormone ('hyperthyroidism'; Fig. 15.21).

In myasthenia gravis, autoantibodies against the α chain of the nicotinic acetylcholine receptor present at neuromuscular junctions in skeletal muscle cells can block stimulation of muscle contraction. The antibodies are believed to drive internalization and degradation of the receptor (Fig. 15.22).

Fig. 15.22 Autoantibodies inhibit receptor function in myasthenia gravis. In normal circumstances, acetylcholine released from stimulated motor neurons at the neuromuscular junction binds to acetylcholine receptors on skeletal muscle cells, triggering muscle contraction (upper panel). Myasthenia gravis is caused by autoantibodies against the α subunit of the receptor for acetylcholine. These autoantibodies bind to the receptor without activating it and also cause receptor internalization and degradation (lower panel). As the number of receptors on the muscle is decreased, the muscle becomes less responsive to acetylcholine.

Diseases mediated by antibodies against cell-surface receptors				
Syndrome	Antigen	Antibody	Consequence	Target cell
Graves' disease	Thyroid-stimulating hormone receptor	Agonist	Hyperthyroidism	Thyroid epithelial cell
Myasthenia gravis	Acetylcholine receptor	Antagonist	Progressive muscle weakness	Muscle
Insulin-resistant diabetes	Insulin receptor	Antagonist	Hyperglycemia, ketoacidosis	All cells
Hypoglycemia	Insulin receptor	Agonist	Hypoglycemia	All cells
Chronic urticaria	Receptor-bound IgE or IgE receptor	Agonist	Persistant itchy rash	Mast cells

Fig. 15.23 Autoimmune diseases caused by autoantibodies against cell-surface receptors. These antibodies produce different effects depending on whether they are agonists (which stimulate the receptor) or antagonists (which inhibit it). Note that different autoantibodies against the insulin receptor can either stimulate or inhibit signaling.

Patients with myasthenia gravis develop potentially fatal progressive weakness as a result of their disease. Diseases caused by autoantibodies that act as agonists or antagonists on cell-surface receptors are listed in Fig. 15.23.

15-16 Autoantibodies against extracellular antigens cause inflammatory injury.

Antibody responses to extracellular matrix molecules are infrequent, but can be very damaging. In **Goodpasture's syndrome**, antibodies are formed against the α_3 chain of basement membrane collagen (type IV collagen). These antibodies bind to the basement membranes of renal glomeruli (Fig. 15.24) and, in some cases, to the basement membranes of pulmonary alveoli, causing a rapidly fatal disease if untreated. The autoantibodies bound to basement membrane ligate Fcγ receptors on innate effector cells, such as monocytes and neutrophils, leading to their activation. These cells, in turn, release chemokines that attract a further influx of monocytes and neutrophils into glomeruli, causing severe tissue injury. The autoantibodies also cause a local activation of complement, which amplifies tissue injury.

Immune complexes are produced when there is an antibody response to a soluble antigen. Normally, such complexes cause little tissue damage because they are cleared efficiently by red blood cells that bear complement receptors and by phagocytes of the mononuclear phagocytic system that have both complement and Fc receptors. This clearance system can, however, fail in circumstances where the production of immune complexes exceeds the capacity

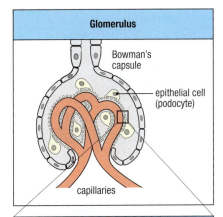
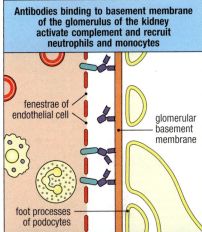
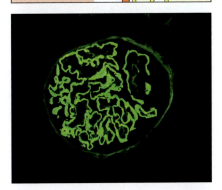
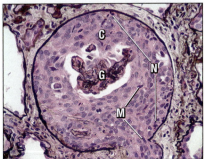

Fig. 15.24 Autoantibodies reacting with glomerular basement membrane cause the inflammatory glomerular disease known as Goodpasture's syndrome. Upper two panels: schematic of antibody-mediated damage to a glomerulus of the kidney. The autoantibody binds to type IV collagen within the basement membrane of the glomerular capillaries, causing complement activation and recruitment and activation of neutrophils and monocytes. Third panel: section of renal glomerulus in biopsy taken from the kidney of a patient with Goodpasture's syndrome. The glomerulus is stained for IgG deposition by immunofluorescence. Anti-glomerular basement membrane antibody (stained green) is deposited in a linear fashion along the glomerular basement membrane. Bottom panel: silver staining of a section through a renal glomerulus shows that the glomerulus capillaries (G) are compressed by the formation of a crescent (C), composed of proliferating epithelial cells and an influx of neutrophils (N) and monocytes (M), that have filled the urinary (Bowman's) space surrounding the glomerular capillaries.

Drug-induced Serum Sickness

of the normal clearance mechanisms, or when there are deficiencies in normal clearance mechanisms. An example of the former is serum sickness (see Section 14-15), which is caused by the injection of large amounts of serum proteins or by small-molecule drugs binding to serum proteins and acting as haptens. Serum sickness is a transient disease, lasting until the immune complexes have been cleared. Similarly, normal clearance mechanisms can be overwhelmed in chronic infections, such as bacterial endocarditis, in which the immune response to bacteria lodged on a cardiac valve is incapable of clearing the infection. The persistent release of bacterial antigens from the valve infection in the presence of a strong antibacterial antibody response causes widespread immune-complex injury to small blood vessels in organs such as the kidney and the skin. Other chronic infections, such as hepatitis C infection, can lead to the production of cryoglobulins and the condition **mixed essential cryoglobulinemia**, in which immune complexes are deposited in joints and tissues. Alternatively, there can be inherited impairment of mechanisms that contribute to normal clearance of immune complexes; such impairment may be caused by reduced expression of or functional defects in specific components of either complement or its receptors, or in Fc receptors, each of which occurs in subsets of patients with SLE.

Indeed, SLE can result from either the overproduction or the defective clearance of immune complexes, or both, at multiple levels (Fig. 15.25). In this disease, there is chronic IgG antibody production directed at ubiquitous self antigens present in nucleated cells, leading to a wide range of autoantibodies against common cellular constituents. The main antigens are three types of intracellular nucleoprotein particles—the nucleosome subunits of chromatin, the spliceosome, and a small cytoplasmic ribonucleoprotein complex containing two proteins known as Ro and La (named after the first two letters of the surnames of the two patients in whom autoantibodies against these proteins were discovered). For these autoantigens to participate in immune-complex formation, they must become extracellular. The autoantigens of SLE are exposed on dead and dying cells released from injured tissues.

In SLE, large quantities of antigen are available, so large amounts of small immune complexes are produced continuously and are deposited in the walls of small blood vessels in the renal glomerular basement membrane, joints, and other organs (Fig. 15.26). This leads to the activation of phagocytic cells through their Fc receptors. A hereditary deficiency of some complement proteins, specifically those for C1q, C2, and C4, is strongly associated with the development of SLE in humans. C1q, C2, and C4 are early components in the classical complement pathway, which is important in antibody-mediated clearance of apoptotic cells and immune complexes (see Chapter 2). If apoptotic cells and immune complexes are not cleared, the chance that their antigens will activate low-affinity self-reactive lymphocytes in the periphery is increased. The consequent tissue damage releases more nucleoprotein complexes, which in turn form more immune complexes. During this process, autoreactive T cells also become activated, although much less is known about their specificity. Animal models for SLE cannot be initiated without the help of T cells, and T cells can also be directly pathogenic, forming part of the cellular infiltrates in the skin and kidney. As discussed in the next section,

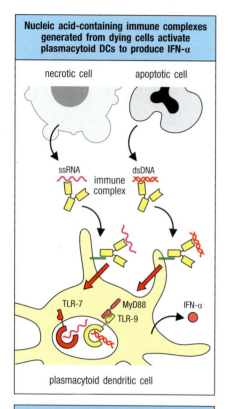

Nucleic acid-containing immune complexes generated from dying cells activate plasmacytoid DCs to produce IFN-α

necrotic cell   apoptotic cell

ssRNA   dsDNA
immune complex

TLR-7   MyD88
TLR-9   IFN-α

plasmacytoid dendritic cell

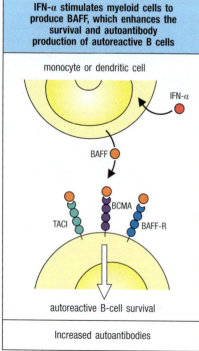

IFN-α stimulates myeloid cells to produce BAFF, which enhances the survival and autoantibody production of autoreactive B cells

monocyte or dendritic cell

IFN-α

BAFF

BCMA
TACI   BAFF-R

autoreactive B-cell survival

Increased autoantibodies

Fig. 15.25 Defective clearance of nucleic acid-containing immune complexes activates overproduction of BAFF and type I interferons that can cause SLE. In SLE, it is believed that antibody:nucleic acid immune complexes containing, for example, ssRNA or dsDNA from dead cells, are bound by FcγRIIa (green rods) on plasmacytoid dendritic cells. The Fc receptor-bound ssRNA and dsRNA are delivered to endosomes, where they activate TLR-7 and TLR-9, respectively, to induce IFN-α production (upper panel). IFN-α increases BAFF production by monocytes and dendritic cells, and BAFF interacts with receptors on B cells. Excess BAFF can increase autoreactive B-cell survival, leading to increased autoantibody production (lower panel).

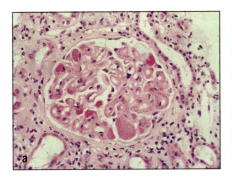
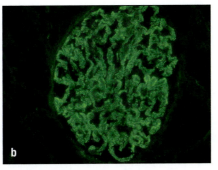
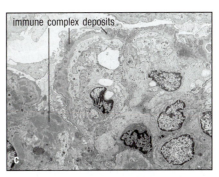

T cells contribute to autoimmune disease in two ways: by helping B cells make antibodies, analogous to a normal T-dependent immune response; and by direct effector functions as they infiltrate and destroy target tissues.

Fig. 15.26 Deposition of immune complexes in the renal glomeruli causes renal failure in systemic lupus erythematosus (SLE). Panel a: a section through a renal glomerulus from a patient with SLE, showing that the deposition of immune complexes has caused thickening of the glomerular basement membrane, seen as the clear 'canals' running through the glomerulus. Panel b: a similar section stained with fluorescent anti-immunoglobulin, revealing immunoglobulin deposits in the basement membrane. In panel c, the immune complexes are seen under the electron microscope as dense protein deposits between the glomerular basement membrane and the renal epithelial cells. Polymorphonuclear neutrophilic leukocytes are also present, attracted by the deposited immune complexes. Photographs courtesy of H.T. Cook and M. Kashgarian.

15-17 T cells specific for self antigens can cause direct tissue injury and sustain autoantibody responses.

Traditionally, it has been more difficult for a number of reasons to demonstrate the existence of autoreactive T cells than the presence of autoantibodies. First, autoreactive human T cells cannot transfer disease to experimental animals because T-cell recognition is MHC-restricted. Second, autoantibodies can be used to stain self tissues to reveal distribution of the autoantigen, whereas T cells cannot be used in the same way. However, the use of fluorophore-labeled peptide–MHC tetramers (see Appendix 1, Section A-24) that can stain antigen-specific T cells for flow cytometry is now providing a means to both identify and track autoreactive T cells *in vivo* in autoimmune diseases. Furthermore, there is already strong evidence for the involvement of autoreactive T cells in many autoimmune diseases. In type 1 diabetes, for example, the insulin-producing β cells of the pancreatic islets are selectively destroyed by cytotoxic T cells. This is borne out by the finding that in the rare cases in which patients with diabetes were transplanted with half a pancreas from an identical twin donor, the β cells in the grafted tissue were rapidly and selectively destroyed by the recipient's T cells. Recurrence of disease can be prevented by the immunosuppressive drug cyclosporin A (see Chapter 16), which inhibits T-cell activation.

Autoantigens recognized by CD4 T cells can be identified by adding cells or tissues containing autoantigens to cultures of blood mononuclear cells, and testing for recognition by CD4 cells derived from an autoimmune patient. If the autoantigen is present, it should be effectively recognized by autoreactive CD4 T cells. The identification of autoantigenic peptides is particularly difficult in autoimmune diseases in which CD8 T cells have a role, because autoantigens recognized by CD8 T cells are not effectively presented in such cultures. Peptides presented by MHC class I molecules must usually be made by the target cells themselves (see Chapter 6); intact cells of target tissue from the patient must therefore be used to study autoreactive CD8 T cells that cause tissue damage. Conversely, the pathogenesis of the disease can itself give clues to the identity of the antigen in some CD8 T-cell-mediated diseases. For example, in type 1 diabetes, the insulin-producing β cells seem to be specifically targeted and destroyed by CD8 T cells (**Fig. 15.27**). This suggests that a protein unique to β cells is the source of the peptide recognized by the pathogenic CD8 T cells. Studies in the NOD (non-obese diabetic) mouse model of type 1 diabetes have shown that peptides from insulin itself are recognized by pathogenic CD8 T cells, confirming insulin as one of the principal autoantigens in this model of diabetes.

Multiple sclerosis (MS) is a T-cell-mediated neurologic disease caused by a destructive immune response against central nervous system myelin antigens,

Multiple Sclerosis

Fig. 15.27 Selective destruction of pancreatic β cells in type 1 diabetes indicates that the autoantigen is produced in β cells and recognized on their surface. In type 1 diabetes there is highly specific destruction of insulin-producing β cells in the pancreatic islets of Langerhans, sparing other islet cell types (α and δ). This is shown schematically in the upper panels. In the lower panels, islets from normal (left) and diabetic (right) mice are stained for insulin (brown), which shows the β cells, and for glucagon (black), which shows the α cells. Note the lymphocytes infiltrating the islet in the diabetic mouse (right) and the selective loss of the β cells (brown), whereas the α cells (black) are spared. The characteristic morphology of the islet is also disrupted with the loss of the β cells. Photographs courtesy of I. Visintin.

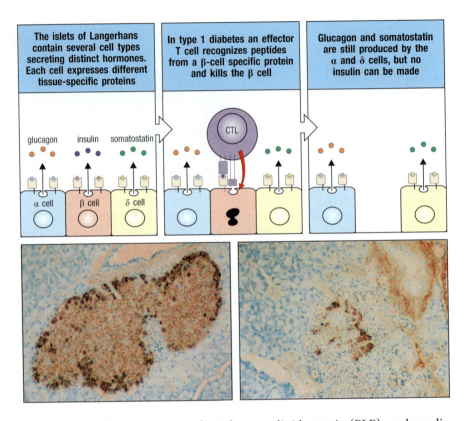

including myelin basic protein (MBP), proteolipid protein (PLP), and myelin oligodendrocyte glycoprotein (MOG) (Fig. 15.28). MS takes its name from the hard (sclerotic) lesions, or plaques, that develop in the white matter of the central nervous system. These lesions show dissolution of the myelin sheath that normally surrounds nerve cell axons, along with inflammatory infiltrates of lymphocytes and macrophages, particularly surrounding blood vessels. Patients with MS may develop a variety of neurological symptoms, including muscle weakness, ataxia, blindness, and paralysis. Normally, few lymphocytes cross the blood–brain barrier, but if this barrier breaks down, activated CD4 T cells specific for myelin antigens and expressing $\alpha_4{:}\beta_1$ integrin can bind vascular cell adhesion molecules (VCAMs) on activated endothelium (see Section 11-3), enabling the T cells to migrate out of the blood vessel. There they reencounter their specific autoantigen presented by MHC class II molecules on infiltrating macrophages or microglial cells (phagocytic macrophage-like cells resident in the central nervous system). Inflammation causes increased vascular permeability, and the site becomes heavily infiltrated by T_H17 and T_H1 effector CD4 T cells, which produce IL-17, IFN-γ, and GM-CSF. Cytokines and chemokines produced by these effector T cells in turn recruit and activate myeloid cells that exacerbate inflammation, resulting in further recruitment of T cells, B cells, and innate immune cells to the lesion. Autoreactive B cells produce autoantibodies against myelin antigens with help from T cells. These combined activities lead to demyelination and interference with neuronal function.

The clinical course of MS both mirrors what is seen in other autoimmune diseases and also displays how the tissue specificity of such conditions affects their progression. Most MS patients experience a disease course characterized by acute attacks (relapse) followed by a reduction in disease activity (remission) that may last for months or years. This relapsing–remitting course is characteristic of many autoimmune diseases (besides MS, Crohn's disease and rheumatoid arthritis, among others), in terms of both the symptoms patients experience and the degree of immune-cell infiltration into the affected organ. Not only are the triggers of relapses not always clear, but the events leading to spontaneous disease remission—even when the autoantigen is still present

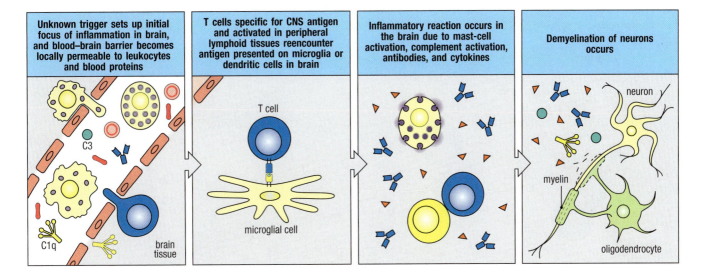

| Unknown trigger sets up initial focus of inflammation in brain, and blood–brain barrier becomes locally permeable to leukocytes and blood proteins | T cells specific for CNS antigen and activated in peripheral lymphoid tissues reencounter antigen presented on microglia or dendritic cells in brain | Inflammatory reaction occurs in the brain due to mast-cell activation, complement activation, antibodies, and cytokines | Demyelination of neurons occurs |

in the organ—remain to be discovered. Furthermore, the relapsing–remitting nature of diseases such as MS makes conducting clinical trials around these disorders especially difficult, as they must be performed over relatively long periods of time to ensure a therapy is effective in preventing relapses and disability.

Ultimately, often after decades, most MS patients change from a relapsing–remitting disease course to 'secondary progressive' MS. In this phase patients begin to undergo a steady neurologic decline without overt periods of remission, and for many patients their disease becomes less responsive to the therapies that effectively target the adaptive immune system in relapsing–remitting MS. The reasons for this are unclear, though it has been suggested that the long-term relapsing–remitting course ultimately exhausts the central nervous system's regenerative capacity, leading to chronic neurodegeneration. Further, prolonged disease may allow immune cells and activated microglia to remain behind the blood–brain barrier, continuing to promote neuronal damage without the need for continuous recruitment of large numbers of inflammatory cells from the periphery.

Rheumatoid arthritis (RA) is a chronic disease characterized by inflammation of the synovium (the thin lining of a joint). As disease progresses, the inflamed synovium invades and damages the cartilage; this is followed by bone erosion (Fig. 15.29), leading to chronic pain, loss of function, and disability. RA was first considered an autoimmune disease driven by B cells producing anti-IgG autoantibodies called rheumatoid factor (see Section 15-4). However, the identification of rheumatoid factor in some healthy individuals, and its absence in some patients with rheumatoid arthritis, suggested that more complex mechanisms orchestrate this pathology. The discovery that RA has an association with particular class II HLA-DR genes of the MHC suggested that T cells are also involved in the pathogenesis of this disease. In RA, as in MS, most data from humans and mouse models indicate that, at least early in disease development, autoreactive T_H17 cells become activated. Autoreactive T cells provide help to B cells to produce arthritogenic antibodies. The activated T_H17 cells also produce cytokines that recruit neutrophils and monocytes/macrophages, which, along with endothelial cells and synovial fibroblasts, are stimulated to produce more pro-inflammatory cytokines such as TNF-α, IL-1, or chemokines (CXCL8, CCL2), and finally matrix metalloproteinases, which are responsible for tissue destruction. IL-17A, which has been found in high concentrations in the synovium and synovial fluid of RA patients, can induce expression of the ligand for receptor activator of NFκB (RANKL), which stimulates the differentiation of osteoclast precursors into mature osteoclasts that resorb bone in affected joints. Although we do not yet know how RA starts, mouse models have shown that both T cells and B cells are needed to initiate disease. Interestingly, interrupting

Fig. 15.28 The pathogenesis of multiple sclerosis. At sites of inflammation, activated T cells autoreactive for brain antigens can cross the blood–brain barrier and enter the brain, where they reencounter their antigens on microglial cells and secrete cytokines such as IFN-γ. The production of T-cell and macrophage cytokines exacerbates the inflammation and induces a further influx of blood cells (including macrophages, dendritic cells, and B cells) and blood proteins (such as complement) into the affected site. Mast cells also become activated. The individual roles of these components in demyelination and loss of neuronal function are still not well understood. CNS, central nervous system.

Rheumatoid Arthritis

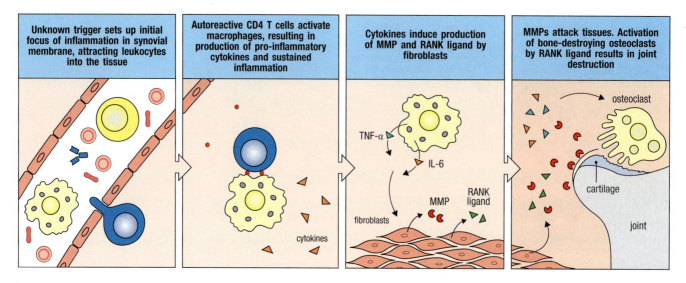

| Unknown trigger sets up initial focus of inflammation in synovial membrane, attracting leukocytes into the tissue | Autoreactive CD4 T cells activate macrophages, resulting in production of pro-inflammatory cytokines and sustained inflammation | Cytokines induce production of MMP and RANK ligand by fibroblasts | MMPs attack tissues. Activation of bone-destroying osteoclasts by RANK ligand results in joint destruction |

Fig. 15.29 The pathogenesis of rheumatoid arthritis. Inflammation of the synovial membrane, initiated by some unknown trigger, attracts autoreactive lymphocytes and macrophages to the inflamed tissue. Autoreactive effector CD4 T cells activate macrophages, and pro-inflammatory cytokines such as IL-1, IL-6, IL-17, and TNF-α are produced. Fibroblasts activated by cytokines produce matrix metalloproteinases (MMPs), which contribute to tissue destruction. The TNF family cytokine RANK ligand, expressed by T cells and fibroblasts in the inflamed joint, is the primary activator of bone-destroying osteoclasts. Antibodies against several joint proteins are also produced (not shown), but their role in pathogenesis is uncertain.

this complex cascade at multiple levels—including therapeutic antibodies against cytokines (TNF-α), B cells, and T-cell activation—have all been successful in treating the symptoms of the disease (discussed in Section 16-8).

Studies of the targets of autoantibodies in RA have yielded insights into how this disease develops, and have also identified a more global mechanism by which self proteins may be seen as foreign in other autoimmune conditions. During inflammation, the amino acid arginine can be converted into citrulline, and this change may result in structural alterations of the self protein that cause the immune system to now view it as nonself (Fig. 15.30). Experimental models have shown that antibodies against these altered proteins can be pathogenic, and diagnostic tests for anti-citrullinated protein antibodies (ACPAs) are highly specific for RA. Interestingly, smoking—long known as the most important environmental risk factor for RA development—has been associated with APCAs in patients with HLA risk alleles, suggesting that this tolerance-breaking mechanism may be an important node in the gene–environment interactions that lead to autoimmunity. Finally, other post-translational modifications (oxidation, glycosylation) of self proteins in the periphery have now been shown to stimulate T- and B-cell responses in other autoimmune diseases.

Summary.

Autoimmune diseases can be broadly classified into those that affect a specific organ and those that affect tissues throughout the body. Organ-specific autoimmune diseases include type 1 diabetes, multiple sclerosis, Graves' disease, and Crohn's disease. In each case the effector functions target autoantigens that are restricted to particular organs—insulin-producing β cells of the pancreas (type 1 diabetes), the myelin sheathing on axons in the central nervous system (multiple sclerosis), and the thyroid-stimulating hormone receptor (Graves' disease)—or, in the case of Crohn's disease, components of the intestinal microbiota. In contrast, systemic diseases such as systemic lupus erythematosus (SLE) cause inflammation in multiple tissues because their autoantigens, which include chromatin and ribonucleoproteins, are found in most cells of the body. In some organ-specific diseases, immune destruction of the target tissue and the unique self antigens it expresses leads to cessation of autoimmune activity, but systemic diseases tend to be chronically active if untreated, because their autoantigens cannot be cleared. Another way of classifying autoimmune diseases is according to the effector functions that are most important in pathogenesis. It is becoming clear, however, that many diseases once thought to be mediated solely by one effector function actually involve several. In this way, autoimmune diseases resemble pathogen-directed immune responses, which typically elicit the activities of multiple effectors—adaptive and innate.

For a disease to be classified as autoimmune, the tissue damage must be shown to be caused by the adaptive immune response to self antigens. Autoinflammatory reactions directed against the commensal microbiota of the intestines, such as those seen in inflammatory bowel diseases (IBDs), are a special case in that the target antigens are not strictly 'self,' but are derived from the 'extended self' of the intestinal microbiota. IBD, nevertheless, shares immunopathogenic features with other autoimmune diseases. The most convincing proof that the immune response is causal in autoimmunity is the transfer of disease by transferring the active component of the immune response to an appropriate recipient. Autoimmune diseases are mediated by autoreactive lymphocytes and their soluble products, pro-inflammatory cytokines, and autoantibodies responsible for inflammation and tissue injury. A few autoimmune diseases are caused by antibodies that bind to cell-surface receptors, causing either excess activity or inhibition of receptor function. In some diseases, transplacental passage of IgG autoantibodies can cause disease in the fetus and neonate. T cells can be involved directly in inflammation or cellular destruction, and they are typically required to initiate and sustain an autoantibody response. Similarly, B cells are important antigen-presenting cells for sustaining autoantigen-specific T-cell responses and causing epitope spreading. In spite of our knowledge of the mechanisms of tissue damage and the therapeutic approaches that this information has engendered, it remains to be determined how autoimmune responses are induced.

The genetic and environmental basis of autoimmunity.

Given the complex mechanisms that exist to prevent autoimmunity, it is not surprising that autoimmune diseases are the result of multiple factors, both genetic and environmental. We first discuss the genetic basis of autoimmunity, attempting to understand how genetic defects perturb various tolerance mechanisms. Genetic defects alone are not, however, always sufficient to cause autoimmune disease. Environmental factors also play a part, although these factors are poorly understood. As we shall see, genetic and environmental factors together can overcome tolerance mechanisms and result in disease.

15-18 Autoimmune diseases have a strong genetic component.

It is increasingly clear that some individuals are genetically predisposed to autoimmunity. Perhaps the clearest demonstration of this is found in inbred mouse strains that are prone to various types of autoimmune diseases. Mice of the NOD strain are very likely to get diabetes, with female mice becoming diabetic faster than males (Fig. 15.31). For reasons that are still unclear, many autoimmune diseases are more common in females than in males (see Fig. 15.37 below), with some disorders (SLE and MS) showing a high degree of sexual dimorphism. Autoimmune diseases in humans also have a genetic component. Some autoimmune diseases, including type 1 diabetes, run in families, suggesting a role for genetic susceptibility. Most convincingly, if one identical (monozygotic) twin is affected, the other twin is quite likely to be affected as well, whereas concordance of disease is much less in nonidentical (dizygotic) twins.

Environmental influences are also clearly involved. For example, although most members of a colony of NOD mice develop diabetes, they do so at different ages. Moreover, disease onset often differs from one animal colony to the next, even though all the mice are genetically identical. Thus, environmental variables must be, in part, determining the rate of diabetes development in genetically susceptible individuals. Particularly striking is the importance of the intestinal microbiota in the development of IBD in mice that

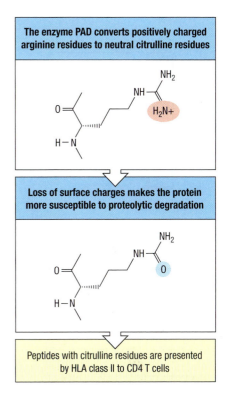

The enzyme PAD converts positively charged arginine residues to neutral citrulline residues

Loss of surface charges makes the protein more susceptible to proteolytic degradation

Peptides with citrulline residues are presented by HLA class II to CD4 T cells

Fig. 15.30 The enzyme peptidyl arginine deiminase converts the arginine residues of tissue proteins to citrulline. In tissues stressed by wounds or infection, peptidyl arginine deiminase (PAD) activity is induced. By converting arginine residues to citrulline, PAD destabilizes proteins and makes them more susceptible to degradation. It also introduces novel B-cell and T-cell epitopes into tissue proteins that can stimulate an autoimmune response.

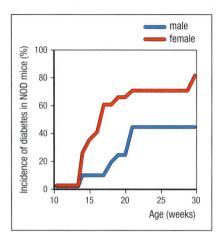

Fig. 15.31 Sex differences in the incidence of autoimmune disease. Many autoimmune diseases are more common in females than males, as illustrated here by the cumulative incidence of diabetes in a population of diabetes-prone NOD mice. Females (red line) get diabetes at a much younger age than do males, indicating their greater predisposition. Data kindly provided by S. Wong.

are genetically predisposed to develop intestinal inflammation. Treatment with broad-spectrum antibiotics that reduce or eliminate many components of the commensal flora can delay or eliminate disease onset, and raising susceptible mice under germ-free conditions (i.e., without a microbiota) eliminates disease. Conversely, certain intestinal microbes—such as segmented filamentous bacteria (SFB)—present in some mouse colonies promote intestinal T_H17 responses that have been linked to intestinal inflammation. Although analogous organisms in humans have not been clearly identified, human studies suggest that components of the microbiota may predispose genetically susceptible individuals to autoimmune disease. For instance, although Crohn's disease incidence in susceptible monozygotic twins is much higher than in dizygotic twins, the concordance rate is not 100%. The explanation for incomplete concordance could lie in variability in the intestinal microbiota, epigenetic differences, or factors yet to be defined.

15-19 Genomics-based approaches are providing new insight into the immunogenetic basis of autoimmunity.

Since the advent of gene knockout technology in mice (see Appendix I, Section A-35), many genes encoding immune system proteins have been experimentally disrupted. Several strains of mice that have been generated show signs of autoimmunity, including autoantibodies and infiltration of organs by T cells. The study of these mice has expanded our knowledge of the pathways that contribute to autoimmunity, and therefore their induced mutations might be candidates for identifying naturally occurring mutations. These mutations likely affect genes that encode cytokines, co-receptors, molecules involved in antigen-signaling cascades, co-stimulatory molecules, proteins involved in apoptosis, and proteins that clear antigen or antigen:antibody complexes. A number of cytokines and signaling proteins implicated in autoimmune disease are listed in **Fig. 15.32**. Other targeted or mutant genes with autoimmune phenotypes in mice are listed in **Fig. 15.33**, as are their corresponding human counterparts, where known.

Fig. 15.32 Defects in cytokine production or signaling that can lead to autoimmunity. Some of the signaling pathways involved in autoimmunity have been identified by genetic analysis, mainly in animal models. The effects of overexpression or underexpression of some of the cytokines and intracellular signaling molecules involved are listed here (see the text for further discussion).

Defects in cytokine production or signaling that can lead to autoimmunity		
Defect	**Cytokine, receptor, or intracellular signal**	**Result**
Overexpression	TNF-α	Inflammatory bowel disease, arthritis, vasculitis
	IL-2, IL-7, IL-2R	Inflammatory bowel disease
	IL-3	Demyelinating syndrome
	IFN-γ	Overexpression in skin leads to SLE
	IL-23R	Inflammatory bowel disease, psoriasis
	STAT4	Inflammatory bowel disease
Underexpression	TNF-α	SLE
	IL-1 receptor agonist	Arthritis
	IL-10, IL-10R, STAT3	Inflammatory bowel disease
	TGF-β	Ubiquitous underexpression leads to inflammatory bowel disease. Underexpression specifically in T cells leads to SLE

Proposed mechanism	Murine models	Disease phenotype	Human gene affected	Disease phenotype
Antigen clearance and presentation	C1q knockout	Lupus-like	*C1QA*	Lupus-like
	C4 knockout		*C2, C4*	
			Mannose-binding lectin	
	AIRE knockout	Multi-organ autoimmunity resembling APECED	*AIRE*	APECED
	Mer knockout	Lupus-like		
Signaling	SHP-1 knockout	Lupus-like		
	Lyn knockout			
	CD22 knockout			
	CD45 E613R point mutation			
	B cells deficient in all Src-family kinases (triple knockout)			
	FcγRIIB knockout (inhibitory signaling molecule)		*FCGR2A*	Lupus
Co-stimulatory molecules	CTLA-4 knockout (blocks inhibitory signal)	Lymphocyte infiltration into organs		
	PD-1 knockout (blocks inhibitory signal)	Lupus-like		
	BAFF overexpression (transgenic mouse)			
Apoptosis	Fas knockout (*lpr*)	Lupus-like with lymphocyte infiltrates	*FAS* and *FASL* mutations (ALPS)	Lupus-like with lymphocyte infiltrates
	FasL knockout (*gld*)			
	Bcl-2 overexpression (transgenic mouse)	Lupus-like		
	Pten heterozygous deficiency			
T_{reg} development/ function	*scurfy* mouse	Multi-organ autoimmunity	*FOXP3*	IPEX
	foxp3 knockout			

Fig. 15.33 Categories of genetic defects that lead to autoimmune syndromes. Many genes have been identified in which mutations predispose to autoimmunity in humans and animal models. These are best understood by the type of process affected by the genetic defect. A list of such genes (or the related protein product) is given here, organized by process (see the text for further discussion). In some cases, the same gene has been identified in mice and humans. In other cases, different genes affecting the same mechanism are implicated in mice and humans. The smaller number of human genes identified so far undoubtedly reflects the difficulty of identifying the genes responsible in outbred human populations.

In humans, genetic susceptibility to autoimmune disorders has been recently assessed by large-scale **genome-wide association studies** (**GWASs**), which look for a correlation between disease frequency and genetic variants (typically **single-nucleotide polymorphisms**, or **SNPs**). Such studies typically involve thousands of patients with a given autoimmune disease diagnosis and

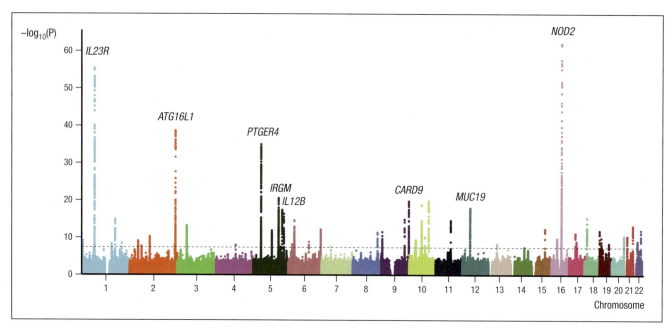

Fig. 15.34 Manhattan plot depicting risk alleles from genome-wide association studies (GWASs) of Crohn's disease. The plot highlights selected gene loci identified by analyses of single nuclear polymorphisms (SNPs) with highly significant disease associations in patients with Crohn's disease compared with healthy controls (see also Section 15-23). The height of the peaks reflects the statistical significance of the association. The dotted line indicates the threshold for significant associations (5×10^{-08}). Figure courtesy of John Rioux and Ben Weaver.

healthy controls in order to identify highly significant associations. Example results from GWASs that identify candidate genes linked to Crohn's disease are shown in the 'Manhattan' plot in **Fig. 15.34**. These plots are so named because they resemble a profile view of skyscrapers in the Manhattan skyline. Here, genomic coordinates are located on the *x*-axis, with the negative logarithm of the *P*-value of the association being given on the *y*-axis, and each assayed SNP is represented by a dot. Thus, the variants with the greatest disease association are the 'tallest skyscrapers' on the plot. Using this approach, hundreds of significant variants have been identified for multiple autoimmune diseases, suggesting that genetic susceptibility to autoimmune disease in humans may be due to a combination of susceptibility alleles at multiple loci.

Analyses of GWASs from multiple autoimmune diseases indicate that certain immune pathways—most notably those involved in T-cell activation and function—are common to multiple different forms of autoimmunity. For example, type 1 diabetes, Graves' disease, Hashimoto's thyroiditis, rheumatoid arthritis, and multiple sclerosis all show genetic association with the *CTLA4* locus on chromosome 2. The cell-surface protein CTLA-4 is produced by activated T cells and is an inhibitory receptor for B7 co-stimulatory molecules (see Section 9-17). Similarly, many of the most common autoimmune disorders have been linked to central factors involved in the development and function of the T_H17 and T_H1 immune pathways (**Fig. 15.35**).

Despite confirming much of our knowledge from experimental immunology, these studies have also revealed our ignorance of gene-regulatory mechanisms that predispose to human disease. For instance, the vast majority of risk alleles identified to date (>80%) are not contained within exons (the protein-coding regions of genes), and many variants reside kilobases away from immunologically relevant genes. Understanding how genetic variation at these noncoding sequences in the genome can contribute to disease is a very active area of research. Recent evidence using computational algorithms, coupled to transcriptional and epigenetic profiling of human immune-cell populations,

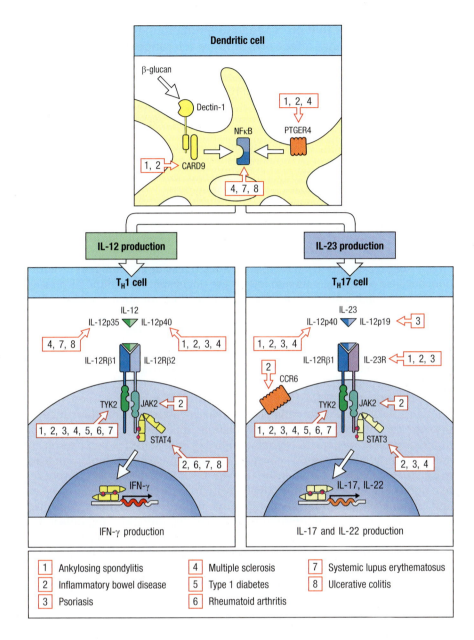

Fig. 15.35 Associations of components of the IL-12R and IL-23R response pathways with autoimmune diseases. Multiple components of the interleukin-12 (IL-12R) and -23 (IL-23R) receptor response pathways show significant genome-wide associations with a broad range of immune-mediated diseases; that is, these components map within genomic intervals that are associated with the respective disease in genome-wide association studies. Although this figure shows these components in the conventional context of T helper 1 (T_H1) and T_H17 lymphocytes, it is now recognized that they are widely expressed in innate lymphoid cells and the specific cell type may vary from phenotype to phenotype. Adapted from Parkes M. *et al.*: *Nat. Rev. Genetics* 2013, **14**:661. With permission from Macmillan Publishers Ltd.

suggests that many of the causal variants are located within critical gene-regulatory elements that control gene expression in immune cells (for example enhancers). Many of these gene-regulatory elements are utilized by effector or regulatory T cells following their activation, further confirming T-cell activation as a key event in the etiology of autoimmune disorders. Ultimately, a deeper understanding of how these variants contribute to disease will require new techniques to experimentally mimic and manipulate risk alleles, either singly or in combination, in order to fully elucidate how they affect the biology of immune-cell populations relevant to disease.

Despite our current ignorance of how most common genetic variants predispose to (or protect from) autoimmune disorders, several other approaches have begun to shed light on the genetic mechanisms of disease. These include the study of mutations that cause overt alterations in molecules regulating tolerance or the innate immune system; the study of patients with rare, monogenic defects of immune tolerance; and investigations into how certain HLA alleles predispose to disease by their ability to present certain self antigens. We will briefly explore each of these in the following sections.

15-20 Many genes that predispose to autoimmunity fall into categories that affect one or more tolerance mechanisms.

Many of the genes identified as predisposing to autoimmunity can be classified as affecting autoantigen availability and clearance; apoptosis; signaling thresholds; cytokine expression or signaling; co-stimulatory molecules or their receptors; or regulatory T cells (see Figs. 15.32 and 15.33).

Genes that control antigen availability and clearance are important both centrally, in the thymus, and in the periphery. In the thymus, genes that control expression of self proteins influence tolerance in developing lymphocytes. In the periphery, hereditary deficiency of some proteins can predispose to autoimmunity—for example, deficiency of early components of the complement cascade is associated with the development of SLE (see Section 15-16). Genes that control apoptosis, such as *FAS*, are important in regulating the duration and vigor of immune responses. Failure to regulate immune responses properly causes excessive destruction of self tissues, releasing autoantigens. In addition, because clonal deletion and anergy are not absolute, immune responses can include some self-reactive cells. As long as their numbers are limited by apoptotic mechanisms, they may not necessarily cause autoimmune disease, but they could cause a problem if apoptosis is not properly regulated.

One of the largest categories of mutations associated with autoimmunity pertains to signals that control lymphocyte activation. These include mutations in co-stimulatory molecules, inhibitory Fc receptors, and inhibitory receptors containing ITIMs, such as PD-1 and CTLA-4 (see Section 15-19). Another subset contains mutations in proteins involved in signal transduction through the antigen receptor itself. Mutations that affect signaling intensity in either direction—making signaling more or less sensitive—can result in autoimmunity. A decrease in sensitivity in the thymus, for example, can lead to a failure of negative selection and thereby to autoreactivity in the periphery. In contrast, increasing receptor sensitivity in the periphery can lead to greater and prolonged activation, resulting in an exaggerated immune response with the side effect of autoimmunity. Additionally, mutations that affect the expression or signaling of cytokines and co-stimulatory molecules have been linked to autoimmunity. A final subset comprises mutations effecting T_{reg}-cell development or function, such as FoxP3 mutations (see Section 15-21).

15-21 Monogenic defects of immune tolerance.

Predisposition to most of the common autoimmune diseases is due to the combined effects of multiple genes, but there are some monogenic autoimmune diseases (Fig. 15.36). Here, the mutant allele confers a very high risk of disease to the individual, but the overall impact on the population is minimal because these variants are rare. The existence of monogenic autoimmune disease was first observed in mutant mice in which the inheritance of an autoimmune syndrome followed a pattern consistent with a single-gene defect. Such alleles are usually recessive or X-linked. For example, the disease APECED is a recessive autoimmune disease caused by a defect in the gene *AIRE* (see Section 15-3).

Immune Dysregulation, Polyendocrinopathy, Enteropathy X-linked Disease

Two monogenic autoimmune syndromes have been linked to defects in regulatory T cells. The X-linked recessive autoimmune syndrome **IPEX** (immune dysregulation, polyendocrinopathy, enteropathy, X-linked) is typically caused by missense mutations in the gene encoding the transcription factor FoxP3, which is key in the differentiation and function of some types of T_{reg} cells (see Section 9-21). This disease is characterized by severe allergic inflammation, autoimmune polyendocrinopathy, secretory diarrhea, hemolytic anemia, and thrombocytopenia, and usually leads to early death. Despite mutation of the *FOXP3* gene, the number of FoxP3+ T_{reg} cells in the blood of individuals with IPEX is comparable to the number in healthy individuals; however, the

Single-gene traits associated with autoimmunity			
Gene	**Human disease**	**Mouse mutant or knockout**	**Mechanism of autoimmunity**
AIRE	APECED (APS-1)	Knockout	Decreased expression of self antigens in the thymus, resulting in defective negative selection of self-reactive T cells
CTLA4	Association with Graves' disease, type 1 diabetes, and others	Knockout	Failure of T-cell anergy and reduced activation threshold of self-reactive T cells
FOXP3	IPEX	Knockout and mutation (*scurfy*)	Decreased function of CD4 CD25 regulatory T cells
FAS	ALPS	*lpr/lpr; gld/gld* mutants	Failure of apoptotic death of self-reactive B and T cells
C1q	SLE	Knockout	Defective clearance of immune complexes and apoptotic cells
ATG16L1	IBD	Hypomorph	Defective autophagy/clearance of bacteria by innate cells in intestines
IL10RA, IL10RB	IBD	Knockout	Defective IL-10 signaling; impaired anti-inflammatory response
INS	Type 1 diabetes	None	Decreased expression of insulin in thymus; impaired negative selection

Fig. 15.36 Single-gene traits associated with autoimmunity. Listed are examples of monogenic disorders that cause autoimmunity in humans. Mice with targeted deletions (knockout) or spontaneous mutations (for example, *lpr/lpr*) in homologous genes have similar disease characteristics and are useful models for the study of the pathogenic basis for these disorders. APECED, autoimmune polyendocrinopathy–candidiasis–ectodermal dystrophy; APS-1, autoimmune polyglandular syndrome 1; IPEX, immune dysregulation, polyendocrinopathy, enteropathy, X-linked syndrome; ALPS, autoimmune lymphoproliferative syndrome. The *lpr* mutation in mice affects the gene for Fas, whereas the *gld* mutation affects the gene for FasL. Adapted from J.D. Rioux and A.K. Abbas: *Nature* 435:584–589. With permission from Macmillan Publishers Ltd.

suppressive function normally displayed by these cells is impaired. A spontaneous frameshift mutation in the mouse *Foxp3* gene (the *scurfy* mutation) that results in loss of the DNA-binding domain of FoxP3 or complete knockout of *Foxp3* leads to an analogous systemic autoimmune disease, in this case associated with the absence of FoxP3$^+$ T$_{reg}$ cells.

Autoimmunity caused by defective development and survival of T$_{reg}$ cells also results from mutation of *CD25*, the high-affinity chain of the IL-2 receptor complex that is constitutively expressed by T$_{reg}$ cells (see Section 9-16). Because deficiency of CD25 affects the development and function of effector T cells as well, in addition to autoimmunity, patients affected by this mutation suffer multiple immunological deficiencies and susceptibility to infections. These findings further confirm the importance of T$_{reg}$ cells in the regulation of the immune system.

An interesting case of a monogenic autoimmune disease is **autoimmune lymphoproliferative syndrome (ALPS)**, a systemic autoimmune syndrome caused by mutations in the gene encoding Fas. Fas is normally present on the surface of activated T and B cells, and when ligated by Fas ligand, it signals the Fas-bearing cell to undergo apoptosis (see Section 11-16). In this way it functions to limit the extent of immune responses. Mutations that eliminate or inactivate Fas lead to a massive accumulation of lymphocytes, especially T cells, and in mice, to the production of large quantities of pathogenic autoantibodies and a disease that resembles SLE. A mutation leading to this autoimmune syndrome was first observed in the MRL mouse strain and named *lpr*, for lymphoproliferation; it was subsequently identified as a mutation in *Fas*. The study of human patients with the rare autoimmune lymphoproliferative syndrome, which is similar to the syndrome in the MRL/*lpr* mice, led to the identification of *FAS* as the mutant gene responsible for most of these cases (see Fig. 15.36).

Autoimmune Lymphoproliferative Syndrome (ALPS)

Fig. 15.37 Associations of HLA and sex with susceptibility to autoimmune disease. The 'relative risk' for an HLA allele in an autoimmune disease is calculated by comparing the observed number of patients carrying the HLA allele with the number that would be expected, given the prevalence of the HLA allele in the general population. For type 1 insulin-dependent diabetes mellitus, the association is in fact with the HLA-DQ gene, which is tightly linked to the DR genes but is not detectable by serotyping. Some diseases show a significant bias in the sex ratio; this is taken to imply that sex hormones are involved in pathogenesis. Consistent with this, the difference in the sex ratio in these diseases is greatest between menarche and menopause, when levels of such hormones are highest.

HLA- and gender-associated risk for autoimmune disease			
Disease	HLA allele	Relative risk	Sex ratio (♀:♂)
Ankylosing spondylitis	B27	87.4	0.3
Type 1 diabetes	DQ2 and DQ8	~25	~1
Goodpasture's syndrome	DR2	15.9	~1
Pemphigus vulgaris	DR4	14.4	~1
Autoimmune uveitis	B27	10	<0.5
Psoriasis vulgaris	CW6	7	~1
Systemic lupus erythematosus	DR3	5.8	10–20
Addison's disease	DR3	5	~13
Multiple sclerosis	DR2	4.8	10
Rheumatoid arthritis	DR4	4.2	3
Graves' disease	DR3	3.7	4–5
Hashimoto's thyroiditis	DR5	3.2	4–5
Myasthenia gravis	DR3	2.5	~1
Type I diabetes	DQ6	0.02	~1

Autoimmune diseases caused by single genes are rare, but are of great interest, as the mutations causing them identify important pathways that normally prevent the development of autoimmune responses.

15-22 MHC genes have an important role in controlling susceptibility to autoimmune disease.

Among genetic loci that contribute to autoimmunity, susceptibility to autoimmune disease has so far been most consistently associated with MHC genotype (Fig. 15.37), particularly MHC class II alleles, thus implicating CD4 T cells in their etiology. The development of experimental diabetes or arthritis in transgenic mice expressing specific human HLA antigens strongly suggests that particular MHC alleles confer disease susceptibility.

As in genome-wide association studies (GWASs), association of MHC with disease is identified by comparing the frequencies of MHC alleles in patients with the disease with the frequencies in the normal population. For type 1 diabetes, this approach demonstrated an association with the HLA-DR3 and HLA-DR4 alleles, identified by serotyping (Fig. 15.38). Such studies also showed that the MHC class II allele HLA-DR2 has a dominant protective

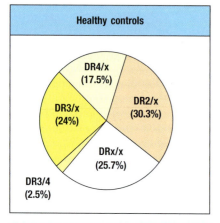

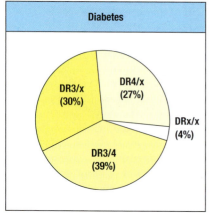

Fig. 15.38 Population studies show association of susceptibility to type 1 diabetes with HLA genotype. The HLA genotypes (determined by serotyping) of patients with diabetes (lower panel) are not representative of those found in the general population (upper panel). Almost all patients with diabetes express HLA-DR3 and/or HLA-DR4, and HLA-DR3/DR4 heterozygosity is greatly overrepresented in diabetics compared with controls. These alleles are linked tightly to HLA-DQ alleles that confer susceptibility to type 1 diabetes. By contrast, HLA-DR2 protects against the development of diabetes and is found only extremely rarely in patients with diabetes. The small letter x represents any allele other than DR2, DR3, or DR4.

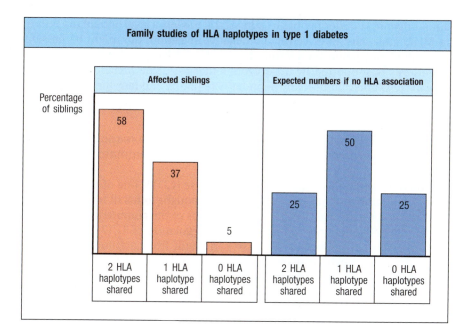

Family studies of HLA haplotypes in type 1 diabetes

	Affected siblings			Expected numbers if no HLA association		
Percentage of siblings	58	37	5	25	50	25
	2 HLA haplotypes shared	1 HLA haplotype shared	0 HLA haplotypes shared	2 HLA haplotypes shared	1 HLA haplotype shared	0 HLA haplotypes shared

Fig. 15.39 Family studies show strong linkage of susceptibility to type 1 diabetes with HLA genotype. In families in which two or more siblings have type 1 diabetes, it is possible to compare the HLA genotypes of affected siblings. Affected siblings share two HLA haplotypes much more frequently than would be expected if the HLA genotype did not influence disease susceptibility.

effect: individuals carrying HLA-DR2, even in association with one of the susceptibility alleles, rarely develop diabetes. It has also been shown that two siblings affected with the same autoimmune disease are far more likely than expected to share the same MHC haplotypes (Fig. 15.39). As HLA genotyping has become more exact through DNA sequencing, disease associations originally discovered by serotyping have been defined more precisely. For example, the association between type 1 diabetes and the DR3 and DR4 alleles is now known to be due to their tight genetic linkage to DQβ alleles that confer susceptibility to the disease. Indeed, susceptibility is most closely associated with polymorphisms at a particular position in the DQβ amino acid sequence that affect the peptide-binding cleft of MHC class II (Fig. 15.40). The diabetes-prone NOD strain of mice also has a serine residue polymorphism at that same position in the homologous mouse MHC class II molecule, known as I-A^{g7}.

The association of MHC genotype with autoimmune disease is not surprising; associations can be explained by a simple model in which susceptibility to an autoimmune disease is determined by differences in the ability of different allelic variants of MHC molecules to present autoantigenic peptides to autoreactive T cells. This would be consistent with what we know of T-cell involvement in particular diseases. In diabetes, for example, there are associations with both MHC class I and MHC class II alleles, consistent with the finding that both CD8 and CD4 T cells mediate the autoimmune response. An alternative hypothesis emphasizes the role of MHC alleles in shaping the T-cell receptor repertoire (see Chapter 8). This hypothesis proposes that self peptides associated with certain MHC molecules may drive the positive selection of developing thymocytes that are specific for particular autoantigens. Such

Position 57 of the DQβ chain affects susceptibility to type 1 diabetes mellitus

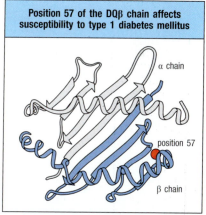

α chain

position 57

β chain

Associated with resistance to T1DM

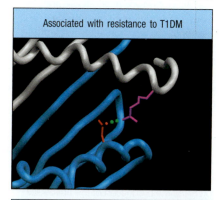

Associated with susceptibility to T1DM

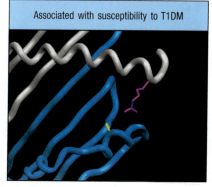

Fig. 15.40 Amino acid changes in the sequence of an MHC class II protein correlate with susceptibility to and protection from diabetes. The HLA-DQβ$_1$ chain contains an aspartic acid (Asp) residue at position 57 in most people; in Caucasoid populations, patients with type 1 diabetes (T1DM) more often have valine, serine, or alanine at this position instead, as well as other differences. Asp 57, shown in red on the backbone structure of the DQβ chain in the top panel, forms a salt bridge (shown in green in the center panel) to an arginine residue (shown in pink) in the adjacent α chain (gray). The change to an uncharged residue (for example, alanine, shown in yellow in the bottom panel) disrupts this salt bridge, altering the stability of the DQ molecule. The non-obese diabetic (NOD) strain of mice, which develops spontaneous diabetes, shows a similar substitution of serine for aspartic acid at position 57 of the homologous I-Aβ chain, and NOD mice transgenic for β chains with Asp 57 have a marked reduction in diabetes incidence. Courtesy of C. Thorpe.

autoantigenic peptides might be expressed at too low a level or bind too poorly to MHC molecules to drive thymic negative selection, but might be present at a sufficient level or bind strongly enough to drive positive selection. This hypothesis is supported by observations that I-A^{g7}, the MHC class II molecule in NOD mice, binds many peptides very poorly and may therefore be less effective in driving thymic negative selection.

15-23   Genetic variants that impair innate immune responses can predispose to T-cell-mediated chronic inflammatory disease.

MOVIE 15.1

Crohn's Disease

As noted earlier in this chapter, Crohn's disease (CD) is one of the two major types of inflammatory bowel disease. CD is thought to result from abnormal hyperresponsiveness of CD4 T cells to antigens of the commensal gut microbiota, rather than to true self antigens. Dysregulation of T_H17 and T_H1 cells is thought to be pathogenic. Disease can result from a failure of mucosal innate immune mechanisms to sequester luminal bacteria from the adaptive immune system, from T-cell-intrinsic defects that cause heightened effector responses, or from failure of T_{reg} cells to suppress microbiota-reactive T_H17 and T_H1 cells (**Fig. 15.41**). Patients with CD have episodes of severe inflammation that commonly affect the terminal ileum, with or without involvement of the colon—hence the alternative name 'regional ileitis' for this disease—but any part of the gastrointestinal tract can be involved. The disease is characterized by chronic inflammation and granulomatous lesions in the mucosa and submucosa of the intestine. Genetic analysis of patients with CD and their families has identified a growing list of disease-susceptibility genes (see Fig. 15.34). One of the earliest to be identified was *NOD2* (also known as *CARD15*), which is expressed predominantly in monocytes, dendritic cells, and the Paneth cells of the small intestine, and is involved in recognition of microbial antigens as part of the innate immune response (see Section 3-8). Mutations and rare polymorphic variants in NOD2 are strongly associated with CD. Mutations in the same gene are also the cause of a dominantly inherited granulomatous disease named **Blau syndrome**, in which granulomas typically develop in the skin, eyes, and joints. Whereas CD results from a loss of function of NOD2, it is thought that Blau syndrome results from a gain of function.

Fig. 15.41 Crohn's disease results from a breakdown of the normal homeostatic mechanisms that limit inflammatory responses to the gut microbiota. The innate and adaptive immune systems normally cooperate to limit inflammatory responses to intestinal bacteria through a combination of mechanisms: a mucus layer produced by goblet cells; tight junctions between the intestinal epithelial cells; antimicrobial peptides released from epithelial cells and Paneth cells; and induction of T_{reg} cells that inhibit effector CD4 T-cell development and promote the production of IgA antibodies that are transported into the intestinal lumen, where they inhibit translocation of intestinal bacteria (not shown). In individuals with impaired homeostatic mechanisms, dysregulated T_H1- and T_H17-cell responses to the intestinal microbiota can result, generating disease-causing chronic inflammation. Crohn's disease susceptibility genes of innate immunity include *NOD2* and the autophagy genes *ATG16L1* and *IRGM*. A major susceptibility gene that affects adaptive immune cells is *IL23R*, which is expressed by T_H17 cells (see also Fig. 15.34).

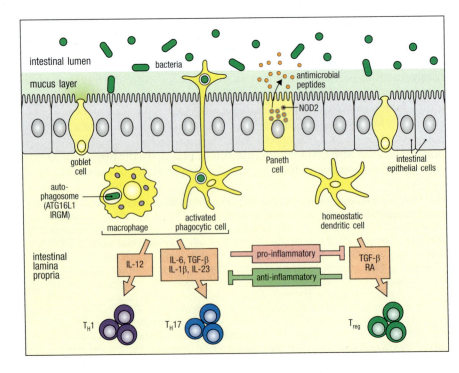

NOD2 is an intracellular receptor for the muramyl dipeptide derived from bacterial peptidoglycan, and its stimulation activates the transcription factor NFκB and the expression of genes encoding pro-inflammatory cytokines and chemokines (see Section 3-8 and Fig. 12.15). In Paneth cells—specialized intestinal epithelial cells in the base of the small intestinal crypts—activation of NOD2 stimulates the release of granules containing antimicrobial peptides that help sequester commensal bacteria to the intestinal lumen, away from the adaptive immune system. Mutant forms of NOD2 that have lost this function limit this innate antibacterial response, thereby predisposing the individual to heightened effector CD4 T-cell responses to the commensal microbiota and consequent chronic intestinal inflammation (see Section 12-22).

In addition to NOD2, other deficiencies in innate immunity have been identified in patients with CD, including defective CXCL8 production and neutrophil accumulation, which can synergize with NOD2 defects to promote intestinal inflammation. Thus, compound defects in innate immunity and the regulation of inflammation may act synergistically to promote immunopathology in CD. GWASs have identified other susceptibility genes for CD that may be linked to impaired innate immune functions (see Fig. 15.34). Defects in two genes (*ATG16L1* and *IRGM*) that contribute to autophagy have been linked to CD, suggesting that other mechanisms that impair clearance of commensal bacteria might predispose to chronic intestinal inflammation. Autophagy, or the digestion of a cell's cytoplasm by its own lysosomes, is important in the turnover of damaged cellular organelles and proteins; autophagy also has a role in antigen processing and presentation (see Section 6-9), and contributes to the clearance of some phagocytosed bacteria.

While defects in important pathways of the innate immune system contribute to CD, genes that regulate the adaptive immune response have also been associated with susceptibility. Most notably, there are variants of the gene for the IL-23 receptor (*IL23R*) that predispose to disease, consistent with heightened T_H17 responses in diseased tissues. Collectively, the growing number of susceptibility genes that confer increased risk for CD point to abnormal regulation of homeostatic innate and adaptive immune responses to the intestinal microbiota as a common factor.

15-24 External events can initiate autoimmunity.

The geographic distribution of autoimmune diseases reveals a heterogeneous distribution among different continents, countries, and ethnic groups. For example, the incidence of disease in the Northern Hemisphere seems to decrease from north to south. This gradient is particularly prominent for diseases such as multiple sclerosis and type 1 diabetes in Europe, where the incidence is greater in the northern countries than in Mediterranean regions. Numerous epidemiologic and genetic associations suggest that this may be partly related to levels of vitamin D. The active form of vitamin D is formed in the skin in response to sunlight—which is less intense and less available in northern latitudes—and has numerous immunoregulatory functions that affect cells of the innate and adaptive immune systems, including suppression of T_H17 cell development. Studies have also shown an increased incidence of autoimmunity in more developed countries, the basis of which is unknown.

Besides vitamin D levels, there are numerous other nongenetic factors contributing to these geographic variations, including socioeconomic status and diet. The contribution of nongenetic factors to disease is exemplified in genetically identical mice, which develop autoimmunity at different rates and severity. There is an emerging appreciation for the diversity of the commensal microbiota having a role in contributing to autoimmune disease—including extraintestinal disease—reflecting the importance of the interplay of the microbiome with the innate and adaptive immune systems in shaping the

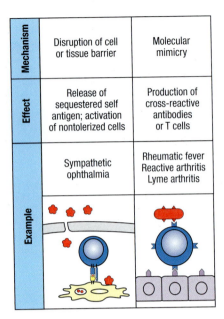

Mechanism	Disruption of cell or tissue barrier	Molecular mimicry
Effect	Release of sequestered self antigen; activation of nontolerized cells	Production of cross-reactive antibodies or T cells
Example	Sympathetic ophthalmia	Rheumatic fever Reactive arthritis Lyme arthritis

Fig. 15.42 Infectious agents could break self-tolerance in several different ways. Left panels: because some antigens are sequestered from the circulation, either behind a tissue barrier or within the cell, an infection that breaks cell and tissue barriers might expose hidden antigens. Right panels: molecular mimicry might result in infectious agents inducing either T- or B-cell responses that can cross-react with self antigens.

systemic immune response. Finally, exposure to infections and environmental toxins may be factors that help trigger autoimmunity. However, it should be noted that epidemiological and clinical studies over the past century have also shown a negative correlation between exposure to some types of infections in early life and the development of allergy and autoimmune diseases. This 'hygiene hypothesis' proposes that a lack of infection during childhood may affect the regulation of the immune system in later life, leading to a greater likelihood of allergic and autoimmune responses (see Section 14-4).

15-25 Infection can lead to autoimmune disease by providing an environment that promotes lymphocyte activation.

How might pathogens contribute to autoimmunity? While an infection is in progress, inflammatory mediators released from activated antigen-presenting cells and lymphocytes and the increased expression of co-stimulatory molecules can affect so-called bystander cells—lymphocytes that are not themselves specific for the antigens of the infectious agent. Self-reactive lymphocytes can become activated in these circumstances, particularly if tissue destruction by the infection leads to an increase in the availability of the self antigen (Fig. 15.42, left panels). Furthermore, pro-inflammatory cytokines, such as IL-1 and IL-6, impair the suppressive activity of regulatory T cells, allowing self-reactive naive T cells to become activated to differentiate into effector T cells that can initiate an autoimmune response.

The perpetuation or exacerbation of autoimmune disease by viral or bacterial infections has been shown in experimental animal models. For example, the severity of type 1 diabetes in NOD mice is exacerbated by Coxsackie virus B4 infection, which leads to inflammation, tissue damage, the release of sequestered islet antigens, and the generation of autoreactive T cells.

We discussed earlier the ability of self ligands such as unmethylated CpG DNA and RNA to directly activate autoreactive B cells via their TLRs and thus break self-tolerance (see Section 15-4 and Fig. 15.25). Microbial ligands for TLRs may also promote autoimmunity by stimulating dendritic cells and macrophages to produce large quantities of cytokines that cause local inflammation and help stimulate already activated autoreactive T and B cells. This mechanism might be relevant to the flare-ups of inflammation that follow infection in patients with autoimmune vasculitis associated with anti-neutrophil cytoplasmic antibodies.

One example of how TLR ligands can induce local inflammation derives from an animal model of arthritis in which injection of bacterial CpG DNA, which is recognized by TLR-9, into the joints of mice induces an arthritis characterized by macrophage infiltration. These macrophages express chemokine receptors on their surface and produce large amounts of CC chemokines, which promote leukocyte recruitment to the site of injection.

15-26 Cross-reactivity between foreign molecules on pathogens and self molecules can lead to antiself responses and autoimmune disease.

Infection with certain pathogens is associated with autoimmune sequelae. Some pathogens express antigens that resemble host molecules, a phenomenon called **molecular mimicry**. In such cases, antibodies produced against a pathogen epitope may cross-react with a self molecule (see Fig. 15.42, right panels). Such structures do not necessarily have to be identical: it is sufficient that they be similar enough to be recognized by the same antibody. Molecular mimicry may also activate autoreactive T cells and result in an attack on self tissues if a processed peptide from a pathogen antigen is similar to a host peptide. A model system to demonstrate molecular mimicry has been generated by using transgenic mice expressing a viral antigen in the pancreas.

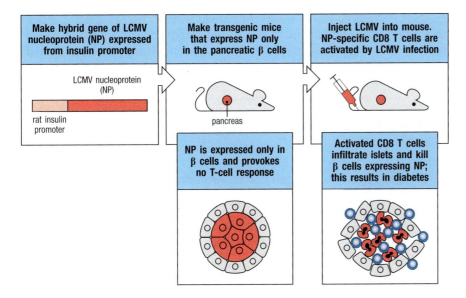

Make hybrid gene of LCMV nucleoprotein (NP) expressed from insulin promoter	Make transgenic mice that express NP only in the pancreatic β cells	Inject LCMV into mouse. NP-specific CD8 T cells are activated by LCMV infection

LCMV nucleoprotein (NP)

rat insulin promoter

pancreas

| | NP is expressed only in β cells and provokes no T-cell response | Activated CD8 T cells infiltrate islets and kill β cells expressing NP; this results in diabetes |

Fig. 15.43 Virus infection can break tolerance to a transgenic viral protein expressed in pancreatic β cells. Mice made transgenic for the lymphocytic choriomeningitis virus (LCMV) nucleoprotein under the control of the rat insulin promoter express the nucleoprotein in their pancreatic β cells but do not respond to this protein and therefore do not develop an autoimmune diabetes. However, if the transgenic mice are infected with LCMV, a potent antiviral cytotoxic T-cell response is elicited, and this kills the β cells, leading to diabetes. It is thought that infectious agents can sometimes elicit T-cell responses that cross-react with self peptides (a process known as molecular mimicry) and that this could cause autoimmune disease in a similar way.

Normally, there is no response to this virus-derived 'self' antigen, but upon infection with the virus that was the source of the transgenic antigen, mice develop diabetes, because the virus activates T cells that are cross-reactive with the 'self' viral antigen (Fig.15.43).

One might wonder why these self-reactive lymphocytes have not been deleted or inactivated by the usual mechanisms of self-tolerance. One reason is that lower-affinity self-reactive B and T cells are not removed efficiently and are present in the naive lymphocyte repertoire as ignorant lymphocytes (see Section 15-4). Pathogens may provide substantially higher local doses of the eliciting antigen in an immunogenic form, whereas normally it would be relatively unavailable to lymphocytes. Some examples of autoimmune syndromes thought to involve molecular mimicry are the **rheumatic fever** that sometimes follows strepto-coccal infection, and the reactive arthritis that can occur after enteric infection.

Once self-reactive lymphocytes have been activated by such a mechanism, their effector functions can destroy tissues. Autoimmunity of this type is some-times transient, and remits when the inciting pathogen is eliminated. This is the case in the autoimmune hemolytic anemia that follows mycoplasma infec-tion. The anemia ensues when antibodies against the pathogen cross-react with an antigen on red blood cells, leading to hemolysis (see Section 15-13). The autoantibodies disappear when the patient recovers from the infection. Sometimes, however, the autoimmunity persists well beyond the initial infec-tion. This is true in some cases of rheumatic fever (Fig. 15.44), which occa-sionally follows a sore throat, scarlet fever, or local skin infections (impetigo)

Autoimmune Hemolytic Anemia

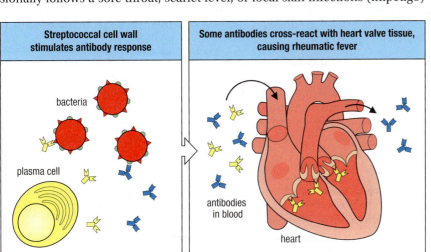

Streptococcal cell wall stimulates antibody response	Some antibodies cross-react with heart valve tissue, causing rheumatic fever

bacteria

plasma cell

antibodies in blood

heart

Fig. 15.44 Antibodies against streptococcal cell-wall antigens cross-react with antigens on heart tissue. The immune response to the bacteria produces antibodies against various epitopes of the bacterial cell surface. Some of these antibodies (yellow) cross-react with the heart valves, whereas others (blue) do not. An epitope in the heart (orange) is structurally similar, but not identical, to a bacterial epitope (red).

caused by *Streptococcus pyogenes.* The similarity of epitopes on streptococcal antigens to self epitopes leads to antibody-mediated, and possibly T-cell-mediated, damage to a variety of tissues, including heart valves and the kidney. Although the tissue injury is typically transient, especially with antibiotic treatment, it can become chronic. Similarly, Lyme disease, an infection with the spirochete *Borrelia burgdorferi,* can be followed by late-developing autoimmunity (Lyme arthritis). In this case, the mechanism is not entirely clear, but it is likely to involve cross-reactivity of pathogen and host components, leading to autoimmunity.

15-27 Drugs and toxins can cause autoimmune syndromes.

Perhaps the clearest evidence of external causative agents in human autoimmunity comes from the effects of certain drugs, which elicit autoimmune reactions in a small proportion of patients. Procainamide, a drug used to treat heart arrhythmias, is notable for inducing autoantibodies similar to those in SLE, although these are rarely pathogenic. Several drugs are associated with the development of autoimmune hemolytic anemia, in which autoantibodies against surface components of red blood cells attack and destroy these cells (see Section 15-13). Toxins in the environment can also cause autoimmunity. When heavy metals, such as gold or mercury, are administered to susceptible strains of mice, a predictable autoimmune syndrome, including the production of autoantibodies, ensues. The extent to which heavy metals promote autoimmunity in humans is debatable, but the animal models show that environmental factors such as toxins could have roles in certain syndromes.

The mechanisms by which drugs and toxins cause autoimmunity are uncertain. For some drugs it is thought that they react chemically with self proteins and form derivatives that the immune system recognizes as foreign. The immune response to these haptenated self proteins can lead to inflammation, complement deposition, destruction of tissue, and finally immune responses to the original self proteins.

15-28 Random events may be required for the initiation of autoimmunity.

Although scientists and physicians would like to attribute the onset of 'spontaneous' diseases to some specific cause, this may not always be possible. There might not be one virus or bacterium, or even any understandable pattern of events that precedes the onset of autoimmune disease. The chance encounter in the peripheral lymphoid tissues of a few autoreactive B and T cells that can interact with each other, at just the moment when an infection is providing pro-inflammatory signals, may be all that is needed. This could be a rare event, but in a susceptible individual such events could be more frequent and/or more difficult to control.

Thus, the onset or incidence of autoimmunity can seem to be random. Genetic predisposition represents, in part, an increased chance of occurrence of this random event. This view, in turn, could explain why many autoimmune diseases appear in early adulthood or later, after enough time has elapsed to permit low-frequency events to occur. It may also explain why, after certain kinds of aggressive therapies, the disease eventually recurs after a long interval of remission.

Summary.

The specific causes of most autoimmune diseases are not known. Genetic risk factors, including particular alleles of MHC class II molecules and polymorphisms or mutations of other genes, have been identified, but many individuals with genetic variants that predispose to a particular autoimmune disease do not get the disease. Epidemiological studies of genetically identical populations of animals have highlighted the role of environmental factors in the initiation of autoimmunity, but although environmental factors have a strong influence on disease, they are not well understood. Some toxins and drugs are known to cause autoimmunity, but their role in the common autoimmune diseases is unclear. Similarly, some autoimmune syndromes can follow viral or bacterial infections. Pathogens can promote autoimmunity by causing nonspecific inflammation and tissue damage, and can sometimes elicit responses to self proteins if they express molecules resembling self, a phenomenon known as molecular mimicry. More research is needed to define specific contributions of environmental factors to autoimmune diseases. It may prove that for most diseases no single environmental trigger that induces disease will be found, but rather a combination of triggers, or even stochastic, or chance, events, will have important roles.

Responses to alloantigens and transplant rejection.

Although transplantation of tissues to replace diseased organs has emerged as an important medical therapy, adaptive immune responses to the grafted tissues are a major impediment. Rejection is caused by immune responses to alloantigens on the graft, which are proteins that vary from individual to individual within a species and are therefore perceived as foreign by the recipient. When tissues containing nucleated cells are transplanted, T-cell responses to the highly polymorphic MHC molecules almost always trigger a response against the grafted organ. Matching the MHC type of the donor and the recipient increases the success rate of grafts, but perfect matching is possible only when donor and recipient are related, and even in these cases, genetic differences at other loci can still trigger rejection, although less severely. Nevertheless, advances in immunosuppression and transplantation medicine now mean that the precise matching of tissues for transplantation is no longer the major factor in graft survival. In blood transfusion, the earliest and most common tissue transplant, MHC matching is not necessary for routine transfusions, because red blood cells and platelets express small amounts of MHC class I molecules and do not express MHC class II molecules; thus, they are not usually T-cell targets. However, antibodies made against platelet MHC class I molecules can be a problem when repeated transfusions of platelets are required. Blood must be matched for ABO and Rh blood group antigens to avoid the rapid destruction of mismatched red blood cells by antibodies in the recipient (see Appendix I, Sections A-5 and A-7), but because there are only four major ABO types and two Rh types, this is a relatively easy form of tissue matching.

In this part of the chapter we examine the immune response to tissue grafts and also ask why such responses do not reject the one foreign tissue graft that is tolerated routinely—the mammalian fetus.

15-29 Graft rejection is an immunological response mediated primarily by T cells.

The basic rules of tissue grafting were first elucidated by skin transplantation between inbred strains of mice. Skin can be grafted with 100% success between different sites on the same animal or person (an **autograft**), or between genetically identical animals or people (a **syngeneic graft**). However, when skin is grafted between unrelated or **allogeneic** individuals (an **allograft**), the graft initially survives but is then rejected about 10–13 days after grafting

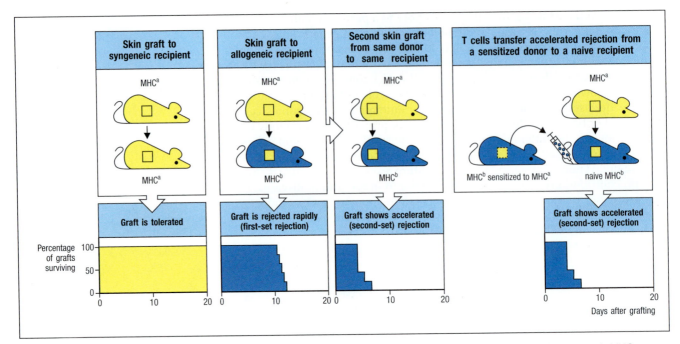

Fig. 15.45 Skin graft rejection is the result of a T-cell-mediated anti-graft response. Grafts that are syngeneic are permanently accepted (first panels), but grafts differing at the MHC are rejected about 10–13 days after grafting (first-set rejection, second panels). When a mouse is grafted for a second time with skin from the same donor, it rejects the second graft faster (third panels). This is called a second-set rejection, and the accelerated response is MHC-specific; skin from a second donor of the same MHC type is rejected equally fast, whereas skin from an MHC-different donor is rejected in a first-set pattern (not shown). Naive mice that are given T cells from a sensitized donor behave as if they had already been grafted (final panels).

(Fig. 15.45). This response is called an **acute rejection**, and it depends on a T-cell response, because skin grafted onto *nude* mice, which lack T cells, is not rejected. The ability to reject skin can be restored to *nude* mice by the adoptive transfer of normal T cells.

When a recipient that has previously rejected a graft is regrafted with skin from the same donor, the second graft is rejected more rapidly (6–8 days) in an **accelerated rejection** (see Fig. 15.45). Skin from a third-party donor grafted onto the same recipient at the same time does not show this faster response but follows a first-set rejection course. The rapid course of second-set rejection can also be transferred to new recipients by T cells from the initial recipient, showing that second-set rejection is caused by a memory-type response (see Chapter 11) from clonally expanded and primed T cells specific for the donor skin.

Immune responses are the major barrier to effective tissue transplantation, destroying grafted tissue by an adaptive immune response to its foreign proteins. These responses can be mediated by either CD8 or CD4 T cells, or both. Antibodies can also contribute to second-set rejection of tissue grafts.

15-30   Transplant rejection is caused primarily by the strong immune response to nonself MHC molecules.

Antigens that differ between members of the same species are known as **alloantigens**, and an immune response against such antigens is known as an **alloreactive** response. When donor and recipient differ at the MHC, an alloreactive immune response is directed at the nonself allogeneic MHC molecule or molecules on the graft. In most tissues these are predominantly MHC class I antigens. Once a recipient has rejected a graft of a particular MHC type, any further graft bearing the same nonself MHC will be rapidly rejected in a second response. The frequency of T cells specific for any nonself MHC molecule is relatively high, making differences at MHC loci the most potent trigger of initial graft rejections (see Section 6-13); indeed, the MHC was originally so named because of its central role in graft rejection.

Once it became clear that recognition of nonself MHC molecules was a major determinant of graft rejection, a considerable amount of effort was put into MHC matching of recipient and donor. Today, with advances in immuno-suppression, MHC matching has become largely irrelevant for most allografts, although it remains important for bone marrow transplantation, for reasons that are discussed in Section 15-36. Even a perfect match at the MHC locus, known as the HLA locus in humans, does not prevent rejection reactions. Grafts between HLA-identical siblings will invariably incite a rejection reaction, albeit more slowly than an unmatched graft, unless donor and recipient are identical twins. This reaction is the result of differences between antigens from non-MHC proteins that also vary between individuals.

Thus, unless donor and recipient are identical twins, all graft recipients must be given immunosuppressive drugs chronically to prevent rejection. Indeed, the current success of clinical transplantation of solid organs is more the result of advances in immunosuppressive therapy (see Chapter 16) than of improved tissue matching. The limited supply of cadaveric organs, coupled with the urgency of identifying a recipient once a donor organ becomes available, means that accurate matching of tissue types is achieved only rarely, with the notable exception of matched-sibling donation of kidneys.

15-31 In MHC-identical grafts, rejection is caused by peptides from other alloantigens bound to graft MHC molecules.

When donor and recipient are identical at the MHC but differ at other genetic loci, graft rejection is not as rapid, but left unchecked it will still destroy the graft (Fig. 15.46). This is the reason that grafts between HLA-identical siblings would be rejected without immunosuppressive treatment. MHC class I and II molecules bind and present a large selection of peptides derived from self proteins made in the cell, and if these proteins are polymorphic, then different peptides will be produced from them in different members of a species. Such proteins can also be recognized as **minor histocompatibility antigens**

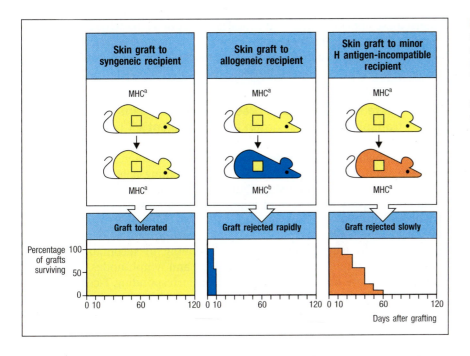

Fig. 15.46 Even complete matching at the MHC does not ensure graft survival. Although syngeneic grafts are not rejected (left panels), MHC-identical grafts from donors that differ at other loci (minor H antigen loci) are rejected (right panels), albeit more slowly than MHC-disparate grafts (center panels).

Fig. 15.47 Minor H antigens are peptides derived from polymorphic cellular proteins bound to MHC class I molecules. Self proteins are routinely digested by proteasomes within the cell's cytosol, and peptides derived from them are delivered to the endoplasmic reticulum, where they can bind to MHC class I molecules and be delivered to the cell surface. If a polymorphic protein differs between the graft donor (shown in red on the left) and the recipient (shown in blue on the right), it can give rise to an antigenic peptide (red on the donor cell) that can be recognized by the recipient's T cells as nonself and elicit an immune response. Such antigens are the minor H antigens.

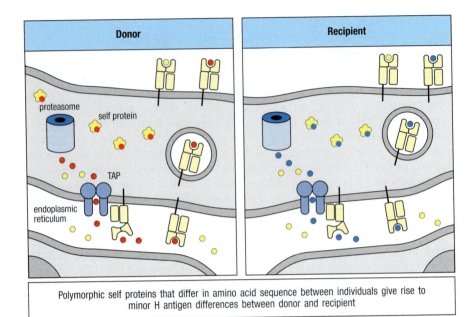

Polymorphic self proteins that differ in amino acid sequence between individuals give rise to minor H antigen differences between donor and recipient

(**Fig. 15.47**). One set of proteins that induce minor histocompatibility responses are encoded on the male Y chromosome. Responses induced by these proteins are known collectively as H-Y. As Y chromosome-specific genes are not expressed in females, female anti-male responses occur; however, male anti-female responses do not occur, because both sexes express X-chromosome genes. One H-Y antigen has been identified in mice and humans as peptides from a protein encoded by the gene *Smcy*. An X-chromosome homolog of *Smcy* (or *Kdm5d*), called *Smcx* (or *Kdm5c*), does not contain these peptide sequences, which are therefore expressed uniquely in males. Most minor histocompatibility antigens are encoded by autosomal genes and their identity is largely unknown, although an increasing number have now been identified at the genetic level.

The response to minor histocompatibility antigens is in many ways analogous to the response to viral infection. However, whereas an antiviral response eliminates only infected cells, a large fraction of cells in the graft express minor histocompatibility antigens, and thus the graft is destroyed in the response against these antigens. Given the virtual certainty of mismatches in minor histocompatibility antigens between two individuals, and the potency of the reactions they incite, it is understandable that successful transplantation requires the use of powerful immunosuppressive drugs.

15-32 There are two ways of presenting alloantigens on the transplanted donor organ to the recipient's T lymphocytes.

Before naive alloreactive T cells can develop into effector T cells that cause rejection, they must be activated by antigen-presenting cells that express both the allogeneic MHC and co-stimulatory molecules. Organ grafts carry with them antigen-presenting cells of donor origin, sometimes called passenger leukocytes, and these are an important stimulus to alloreactivity. This route for sensitization of the recipient to a graft seems to involve donor antigen-presenting cells leaving the graft and migrating to secondary lymphoid tissues of the recipient, including the spleen and lymph nodes, where they can activate those host T cells that bear the corresponding T-cell receptors. Because the lymphatic drainage of solid organ allografts is interrupted by transplantation, migration of donor antigen-presenting cells occurs via the

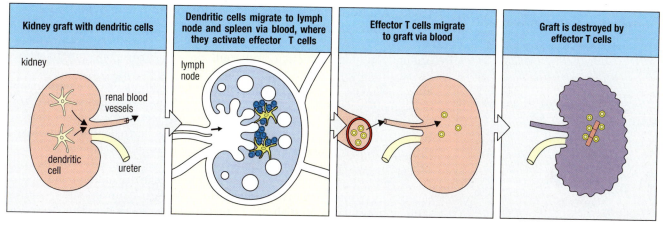

Fig. 15.48 Acute rejection of a kidney graft through the direct pathway of allorecognition. Donor dendritic cells in the graft (in this case a kidney) carry complexes of donor HLA molecules and donor peptides on their surfaces. The dendritic cells are carried via the blood to secondary lymphoid organs (a lymph node is illustrated here), where they move to the T-cell areas. Here, they activate the recipient's T lymphocytes, whose receptors can bind specifically to the complexes of allogeneic donor HLA (both class I and class II) in combination with donor peptides. After activation, the effector T cells travel in the blood to the grafted organ, where they attack cells that display the peptide:HLA molecule complexes for which the T cells are specific.

blood, not lymphatics. The activated alloreactive effector T cells can then circulate to the graft, which they attack directly (Fig. 15.48). This recognition pathway is known as **direct allorecognition** (Fig. 15.49, upper panel). Indeed, if the grafted tissue is depleted of antigen-presenting cells by treatment with antibodies or by prolonged incubation, rejection occurs only after a much longer time.

A second mechanism of allograft recognition leading to graft rejection is the uptake of allogeneic proteins by the recipient's own antigen-presenting cells and their presentation to T cells by self MHC molecules. This is known as **indirect allorecognition** (see Fig. 15.49, lower panel). Peptides derived from both the foreign MHC molecules themselves and minor histocompatibility antigens can be presented by indirect allorecognition.

Direct allorecognition is thought to be largely responsible for acute rejection, especially when MHC mismatches mean that the frequency of directly alloreactive recipient T cells is high. Furthermore, a direct cytotoxic T-cell attack on graft cells can be made only by T cells that recognize the graft MHC molecules directly. Nonetheless, T cells with specificity for alloantigens presented on self MHC can contribute to graft rejection by activating macrophages, which cause tissue injury and fibrosis. T cells with indirect allospecificity are also likely to be important in the development of an antibody response to a graft. Antibodies produced against nonself antigens from the same species are known as **alloantibodies**.

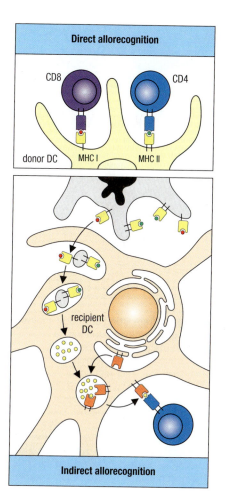

Fig. 15.49 Direct and indirect pathways of allorecognition contribute to graft rejection. Dendritic cells from an organ graft stimulate both the direct and indirect pathways of allorecognition when they travel from the graft to secondary lymphoid tissues. The upper panel shows how the allogeneic HLA class I and II allotypes of donor type on a donor dendritic cell (donor DC) will interact directly with the T-cell receptors of alloreactive CD4 and CD8 T cells of the recipient (direct allorecognition). The lower panel shows how the death of the same antigen-presenting cell produces membrane vesicles containing the allogeneic HLA class I and II allotypes, which are then endocytosed by the recipient's dendritic cells (recipient DC). Peptides derived from the donor's HLA molecules (yellow) can then be presented by the recipient's HLA molecules (orange) to peptide-specific T cells (indirect allorecognition). Presentation by HLA class II molecules to CD4 T cells is shown here. Peptides derived from donor HLA can also be presented by recipient HLA class I molecules to CD8 T cells (not shown).

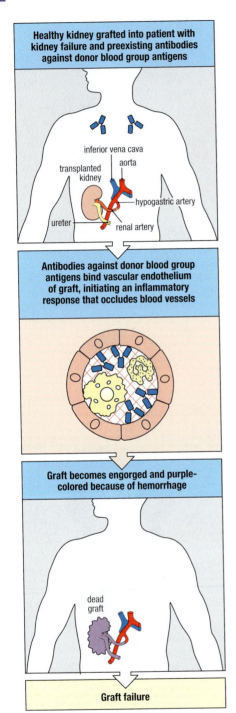

Fig. 15.50 Preexisting antibody against donor graft antigens can cause hyperacute graft rejection. Prior to transplantation, some recipients have made antibodies that react with donor ABO or HLA class I antigens. When the donor organ is grafted into such a recipient, these antibodies bind to the vascular endothelium in the graft, initiating the complement and clotting cascades. Blood vessels in the graft become obstructed by clots and leak, causing hemorrhage of blood into the graft. The graft becomes engorged, turns purple from the presence of deoxygenated blood, and dies.

15-33 Antibodies that react with endothelium cause hyperacute graft rejection.

Antibody responses are an important potential cause of graft rejection. Preexisting alloantibodies against blood group antigens and polymorphic MHC antigens can cause rapid rejection of transplanted organs in a complement-dependent reaction that can occur within minutes of transplantation. This type of reaction is known as **hyperacute graft rejection**. Most grafts that are transplanted routinely in clinical medicine are vascularized organ grafts linked directly to the recipient's circulation. In some cases the recipient may have preexisting antibodies against donor graft antigens. Antibodies of the ABO type can bind to all tissues, not just red blood cells. In addition, antibodies against other antigens can be produced in response to a previous transplant or a blood transfusion. All such preexisting antibodies can cause rapid rejection of vascularized grafts because they react with antigens on the vascular endothelial cells of the graft and initiate the complement and blood clotting cascades. The vessels of the graft become blocked, or thrombosed, causing its rapid destruction. Such grafts become engorged and purple-colored from hemorrhaged blood, which becomes deoxygenated (Fig. 15.50). This problem can be avoided by ABO-matching as well as **cross-matching** donor and recipient. Cross-matching involves determining whether the recipient has antibodies that react with the white blood cells of the donor. Antibodies of this type, when found, have hitherto been considered a serious contraindication to transplantation of most solid organs, because in the absence of any treatment they lead to near-certain hyperacute rejection.

For reasons that are incompletely understood, some transplanted organs, particularly the liver, are less susceptible to this type of injury, and can be transplanted despite ABO incompatibilities. In addition, the presence of donor-specific MHC alloantibodies and a positive cross-match are no longer considered an absolute contraindication for transplantation, as treatment with intravenous immunoglobulin has been successful in a proportion of patients in whom antibodies against the donor tissue were already present.

A similar problem prevents routine use of animal organs—**xenografts**—in transplantation. If xenografts could be used, it would circumvent a limitation in organ replacement therapy: the shortage of donor organs. Pigs have been suggested as a source of organs for xenografting, but most humans have antibodies that react with a ubiquitous cell-surface carbohydrate antigen (α-Gal) of other mammalian species, including pigs. When pig xenografts are placed in humans, these antibodies trigger hyperacute rejection by binding graft endothelial cells and initiating complement and clotting cascades. The problem of hyperacute rejection is exacerbated in xenografts because complement-regulatory proteins such as CD59, DAF (CD55), and MCP (CD46) (see Section 2-16) work less efficiently across a species barrier. A recent step toward xenotransplantation has been the development of transgenic pigs expressing human DAF as well as pigs that lack α-Gal. These approaches might one day reduce or eliminate hyperacute rejection in xenotransplantation.

15-34 Late failure of transplanted organs is caused by chronic injury to the graft.

The success of immunosuppression means that about 90% of cadaveric kidney grafts are still functioning a year after transplantation. There has, however, been little improvement in rates of long-term graft survival: the half-life for functional survival of renal allografts remains about 8 years. Although traditionally the late failure of a transplanted organ has been termed **chronic rejection**, it is typically difficult to determine whether the cause of chronic allograft injury involves specific immune alloreactivity, nonimmune injury, or both.

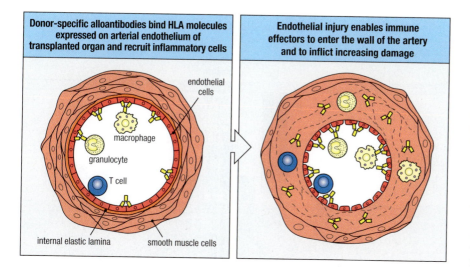

| Donor-specific alloantibodies bind HLA molecules expressed on arterial endothelium of transplanted organ and recruit inflammatory cells | Endothelial injury enables immune effectors to enter the wall of the artery and to inflict increasing damage |

endothelial cells

macrophage

granulocyte

T cell

internal elastic lamina   smooth muscle cells

Fig. 15.51 Chronic rejection in the blood vessels of a transplanted kidney. Left-hand panel: chronic rejection is initiated by the interaction of anti-HLA class I alloantibodies with blood vessels of the transplanted organ. Antibodies bound to endothelial cells recruit Fc receptor-bearing monocytes and neutrophils. Right-hand panel: accumulating damage leads to thickening of the internal elastic lamina and to infiltration of the underlying intima with smooth muscle cells, macrophages, granulocytes, alloreactive T cells, and antibodies. The net effect is to narrow the lumen of the blood vessel and create a chronic inflammation that intensifies tissue remodeling. Eventually the vessel becomes obstructed, ischemic, and fibrotic.

The pattern of chronic injury to transplanted organs is variable, depending on the tissue. A major component of late failure of vascularized transplanted organs is a chronic reaction called **chronic allograft vasculopathy**, which is a prominent cause of injury in heart and kidney allografts. This is characterized by concentric arteriosclerosis of graft blood vessels, which leads to hypoperfusion of the graft and its eventual fibrosis and atrophy (Fig. 15.51). Multiple mechanisms may contribute to this form of vascular injury, although the major cause is thought to be recurring, subclinical acute rejection events, whether due to the development of allospecific antibodies reactive to the vascular endothelium of the graft (so-called donor-specific antibodies), or to allograft-reactive effector T cells, or both. Some forms of immunosuppressive therapy (for example, calcineurin inhibitors such as cyclosporin) also cause vascular injury, although this is typically more limited to very small arteries and causes a different pattern of injury, referred to as arteriolar hyalinosis, that is marked by proteinaceous deposits that narrow the vascular lumen. In transplanted livers, chronic rejection is associated with loss of bile ducts, the so-called 'vanishing bile duct syndrome,' whereas in transplanted lungs, the major cause of late organ failure is accumulation of scar tissue in the bronchioles, termed bronchiolitis obliterans. Alloreactive responses can occur months to years after transplantation, and may be associated with gradual loss of graft function that is hard to detect clinically.

Other important causes of chronic graft dysfunction include: ischemia–reperfusion injury, which can promote sterile inflammatory signals at the time of grafting due to the restoration of blood flow after a period of poor perfusion of the organ to be transplanted; viral infections that emerge as a result of immunosuppression; and recurrence of the same disease in the allograft that destroyed the original organ. Irrespective of etiology, chronic allograft injury is typically irreversible and progressive, ultimately leading to complete failure of allograft function.

15-35 A variety of organs are transplanted routinely in clinical medicine.

Three major advances have made it possible to use organ transplantation routinely in the clinic. First, surgical techniques for performing organ replacement have advanced to the point where such surgeries are now relatively routine in most major medical centers. Second, networks of transplantation centers have been organized to procure healthy organs that become available from cadaveric donors. Third, the use of powerful immunosuppressive drugs

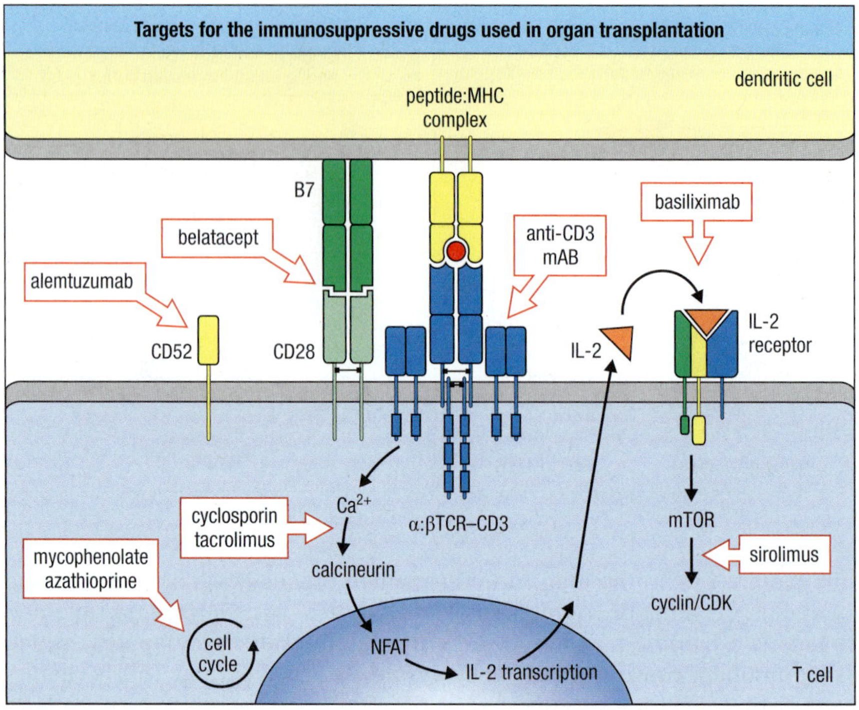

Fig. 15.52 Immunosuppressive drugs act at different stages in the activation of alloreactive T cells. Rabbit anti-thymoglobulin and anti-CD52 monoclonal antibody (alemtuzumab) are used to deplete T cells and other leukocytes before transplantation. Anti-CD3 monoclonal antibody prevents the generation of signaling by the T-cell receptor complex, whereas cyclosporin and tacrolimus interfere with the translocation of nuclear factor of activated T cells (NFAT) to the nucleus by inhibiting calcineurin. The CTLA-4–Fc fusion protein belatacept binds B7 and prevents the generation of co-stimulation via CD28. Basiliximab, an anti-CD25 antibody, binds to the high-affinity IL-2 receptor on partially activated T cells and prevents IL-2 signaling. Sirolimus interferes with activation of the mTOR cascade, which is required for differentiation of effector T cells. Azathioprine and mycophenolate inhibit the replication and proliferation of activated T cells.

Targets for the immunosuppressive drugs used in organ transplantation

that inhibit T-cell activation, thereby limiting the development of anti-allograft effector T cells and antibodies, has markedly increased graft survival rates (Fig. 15.52). The different organs or tissues that are frequently transplanted and allograft survival rates are listed in Fig. 15.53. The most frequently transplanted solid organ is the kidney, the organ first successfully transplanted between identical twins in the 1950s. Transplantation of the cornea is even more frequent; this tissue is a special case because it is not vascularized, and corneal grafts between unrelated people are usually successful without immunosuppression.

Many problems other than graft rejection are associated with organ transplantation. First, donor organs are difficult to obtain. Second, the disease that destroyed the transplant recipient's organ might also destroy the graft, as in the destruction of pancreatic β cells in autoimmune diabetes. Third, the immunosuppression required to prevent graft rejection increases the risk of cancer and infection. The problems most amenable to scientific solution are the development of more effective means of immunosuppression that prevent rejection with minimal impairment of more generalized immunity, the induction of graft-specific tolerance, and the development of xenografts as a practical solution to organ availability.

Tissue transplanted	No. of grafts in USA (2014)*	5-year graft survival
Kidney	17,815	81.4%#
Liver	6729	68.3%
Heart	2679	74.0%
Pancreas	954	53.4%†
Lung	1949	50.6%
Intestine	139	~48.4%
Cornea	~45,000	~70%
HSC transplants	~20,000**	>80%‡

Fig. 15.53 Organs and tissues commonly transplanted in clinical medicine. The numbers of organ and tissue grafts performed in the United States in 2014 are shown. HSC, hematopoietic stem cells (includes bone marrow, peripheral blood HSCs, and cord blood transplants). *Number of grafts includes multiple organ grafts (for example, kidney and pancreas, or heart and lung). For solid organs, 5-year survival of the transplanted graft is based on transplants performed between 2002 and 2007. Data from the United Network for Organ Sharing. #Kidney survival listed (81.4%) is for kidneys from living donors; 5-year survival for cadaveric donor transplants is 69.1%. †Pancreas survival listed (53.4%) is when transplanted alone; 5-year survival when transplanted with a kidney is 73.5%. ** Includes autologous and allogeneic transplants. ‡Successful HSC engraftment is assessed within weeks of transplant, not years. Nearly all solid organ grafts (e.g., kidney, heart) require long-term immunosuppression.

15-36   The converse of graft rejection is graft-versus-host disease.

Transplantation of hematopoietic stem cells (HSCs) from peripheral blood, bone marrow, or fetal cord blood is a successful therapy for some tumors derived from hematopoietic cells, such as certain leukemias and lymphomas. By replacing genetically defective stem cells with normal donor ones, HSC transplantation can also be used to cure some primary immunodeficiencies (see Chapter 13) and other inherited blood cell disorders, such as severe forms of thalassemia. In leukemia therapy, the recipient's bone marrow, the source of the leukemia, must first be destroyed by a combination of irradiation and aggressive cytotoxic chemotherapy.

Graft-Versus-Host Disease

One of the major complications of allogeneic HSC transplantation is **graft-versus-host disease** (**GVHD**), in which mature donor T cells present in preparations of HSCs recognize the tissues of the recipient as foreign, causing a severe inflammatory disease in multiple tissues, but particularly involving the skin, intestines, and liver and characterized by rashes, diarrhea, and liver dysfunction (Fig. 15.54). Because the consequences of GVHD are particularly aggressive when there is mismatch of MHC class I or class II antigens, HLA matching between donor and recipient is more critical than in solid organ transplantation. Most transplants are therefore undertaken only when the donor and recipient are HLA-matched siblings or, less frequently, when there is an HLA-matched unrelated donor. Therefore, GVHD mostly occurs in the context of disparities between minor histocompatibility antigens, so immunosuppression must be used in every HSC transplant.

The presence of alloreactive donor T cells can be demonstrated experimentally by the **mixed lymphocyte reaction** (**MLR**), in which lymphocytes from a potential donor are mixed with irradiated lymphocytes from the recipient. If the donor lymphocytes contain naive T cells that recognize alloantigens on the recipient lymphocytes, they will proliferate or kill the recipient target cells (Fig. 15.55). However, the limitation of the MLR in the selection of HSC donors is that the test does not accurately quantify alloreactive T cells. A more accurate test is a version of the limiting-dilution assay (see Appendix I, Section A-21), which precisely counts the frequency of alloreactive T cells.

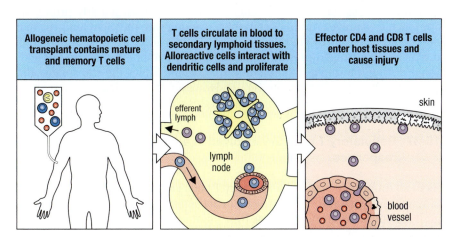

Fig. 15.54 Graft-versus-host disease is due to donor T cells in the graft that attack the recipient's tissues. After bone marrow transplantation, any mature donor CD4 and CD8 T cells present in the graft that are specific for the recipient's HLA allotypes become activated in secondary lymphoid tissues. Effector CD4 and CD8 T cells move into the circulation and preferentially enter and attack tissues of the graft recipient, particularly epithelial cells of the skin, intestines, and liver that have been damaged by the conditioning regimen of chemotherapy and irradiation prior to transplantation.

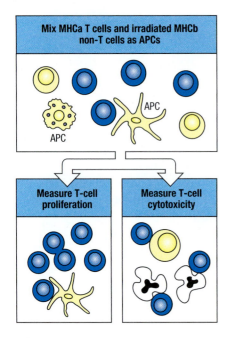

Mix MHCa T cells and irradiated MHCb non-T cells as APCs

APC

APC

Measure T-cell proliferation

Measure T-cell cytotoxicity

Fig. 15.55 The mixed lymphocyte reaction (MLR) can be used to detect histoincompatibility. Peripheral blood mononuclear cells, which include lymphocytes and monocytes, are isolated from the two individuals to be tested. The cells from the person who serves as the stimulator (yellow) are first irradiated to prevent their proliferation. Then they are mixed with the cells from the other person, who serves as the responder (blue), and cultured for 5 days (top panel). In the culture, responder lymphocytes are stimulated by allogeneic HLA class I and II molecules expressed by the stimulator's monocytes and the dendritic cells that differentiate from the monocytes. The stimulated lymphocytes proliferate and differentiate into effector cells. Five days after mixing, the culture is assessed for T-cell proliferation (bottom left panel), which is due to CD4 T cells recognizing HLA class II differences, and for cytotoxic T cells (bottom right panel) produced in response to HLA class I differences. The mixed lymphocyte reaction is instrumental in distinguishing MHC class II from MHC class I.

Although GVHD is harmful to the recipient of an HSC transplant, it can have beneficial effects that are crucial to the success of the therapy. Much of the therapeutic effect of HSC transplantation for leukemia can be due to a **graft-versus-leukemia effect**, in which the donor T cells in the allogeneic preparations of HSCs recognize minor histocompatibility antigens expressed by the leukemic cells and kill the leukemic cells. One of the treatment options for suppressing the development of GVHD is the elimination of mature T cells from the preparations of donor HSCs *in vitro* before transplantation, thereby removing alloreactive T cells. Those T cells that subsequently mature from the donor marrow *in vivo* in the recipient are tolerant to the recipient's antigens. Although the elimination of GVHD has benefits for the patient, there is an increased risk of leukemic relapse, which provides strong evidence in support of the graft-versus-leukemia effect.

Immunodeficiency is another complication of donor T-cell depletion. Because most of the recipient's T cells are destroyed by the combination of chemotherapy and irradiation used to treat the recipient before transplant, donor T cells are the major source for reconstituting a mature T-cell repertoire early after transplant. This is particularly true in adults, who have poor residual thymic function and therefore a limited ability to repopulate their T-cell repertoire from T-cell precursors. Thus, if too many T cells are depleted from the graft, transplant recipients experience, and can die from, opportunistic infections. The need to balance the beneficial effects of the graft-versus-leukemia effect and immunocompetence with the adverse effects of GVHD caused by donor T cells has spawned much research. One particularly promising approach is to prevent donor T cells from reacting with recipient antigens that they could meet shortly after the transplant. This is accomplished by depleting the recipient's antigen-presenting cells. Here, the donor T cells are not activated during the initial inflammation that accompanies the transplant, and thereafter they do not promote GVHD. However, it is unclear whether there would be a graft-versus-leukemia effect in this context.

15-37   Regulatory T cells are involved in alloreactive immune responses.

As in all immune responses, regulatory T cells are thought to have an important immunoregulatory role in the alloreactive immune responses involved in graft rejection. Experiments on the transplantation of allogeneic HSCs in mice have thrown some light on this question. Here, depletion of CD25$^+$ T$_{reg}$ cells in either the recipient or the HSC graft before transplantation accelerated the onset of GVHD and subsequent death. In contrast, supplementing the graft with either fresh or *ex vivo* expanded T$_{reg}$ cells delayed, or even prevented, death from GVHD, with similar results in early human studies. Also, treatment with a low dose of IL-2, which is thought to preferentially expand T$_{reg}$ cells, has shown positive effects in preventing GVHD. Similar observations have been made in experimental mouse models of solid organ transplantation, where the transfer of either naturally occurring or induced T_{reg} cells significantly delayed allograft rejection. These experiments suggest that enriching or generating T_{reg} cells in preparations of donor HSCs might provide a possible therapy for GVHD in the future.

Another class of regulatory T cells, CD8⁺CD28⁻ T cells, have an anergic phenotype and are thought to maintain T-cell tolerance indirectly by inhibiting the capacity of antigen-presenting cells to activate CD4⁺ T cells. These cells have been isolated from transplant patients, and can be distinguished from alloreactive CD8 T cells because they do not display cytotoxic activity against donor cells and express high levels of the inhibitory killer receptor CD94 (see Section 3-25). This suggests that CD8⁺CD28⁻ T cells interfere with the activation of antigen-presenting cells and have a role in the maintenance of transplant tolerance.

15-38 The fetus is an allograft that is tolerated repeatedly.

All of the transplants discussed so far are the result of advances in modern medicine. However, one 'foreign' tissue that is repeatedly grafted and tolerated is the mammalian fetus. The fetus carries paternal MHC and minor histocompatibility antigens that differ from those of the mother (**Fig. 15.56**), and yet a mother can successfully bear many children expressing the same nonself proteins derived from the father. The mysterious lack of fetal rejection has consistently puzzled immunologists, and no comprehensive explanation has yet emerged. One problem is that acceptance of the fetal allograft is so much the norm that it is difficult to study the mechanism that prevents rejection; if the mechanism for rejecting the fetus is rarely activated, how can one analyze the mechanisms that control it?

The mechanisms contributing to 'fetomaternal tolerance' are likely multifactorial and redundant. Although it has been proposed that the fetus is simply not recognized as foreign, women who have borne children often make antibodies directed against the father's MHC and red blood cell antigens. However, the placenta, which is a fetus-derived tissue, seems to sequester the fetus from the mother's T cells. The outer layer of the placenta—the interface between fetal and maternal tissues—is the trophoblast. This does not express MHC class II molecules, and expresses only low levels and a restricted subset of MHC class I molecules, making it resistant to direct alloantigen recognition by maternal T cells. Tissues lacking MHC class I expression are, however, vulnerable to attack by NK cells (see Section 3-25). The trophoblast might be protected from attack by NK cells by the expression of a nonclassical and minimally polymorphic HLA class I molecule, HLA-G, which has been shown to inhibit NK killing.

The placenta may also inhibit the mother's T cells by an active mechanism of nutrient depletion. The enzyme indoleamine 2,3-dioxygenase (IDO) is expressed at a high level by cells at the maternal–fetal interface. This enzyme depletes the essential amino acid tryptophan at this site, and T cells starved of tryptophan show reduced responsiveness. Inhibition of IDO in pregnant mice, using the inhibitor 1-methyltryptophan, causes rapid rejection of allogeneic, but not syngeneic, fetuses.

The cytokine milieu at the maternal–fetal interface also contributes to fetal tolerance. Both the uterine epithelium and the trophoblast secrete TGF-β and IL-10. This combination of cytokines suppresses the development of effector T cells in favor of iT_{reg} cells (see Section 9-23). Regulatory T cells are increased during pregnancy, including iT_{reg} cells in the placenta. These cells are important for suppressing responses to the fetus in mice, as iT_{reg} deficiency promotes fetal resorption—the equivalent of spontaneous abortion in humans—as does induction of T_H1-inducing cytokines (for example, IFN-γ and IL-12). Provocatively, a regulatory element that controls FoxP3 expression in iT_{reg}

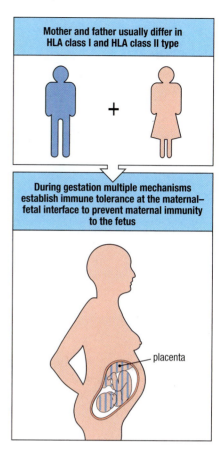

Mother and father usually differ in HLA class I and HLA class II type

+

During gestation multiple mechanisms establish immune tolerance at the maternal–fetal interface to prevent maternal immunity to the fetus

placenta

Fig. 15.56 The fetus is an allograft that is not rejected. With very few exceptions, the mother and father in human families have different HLA types (top panel). When the mother becomes pregnant she carries for 9 months a fetus that expresses one HLA haplotype of maternal origin (pink) and one HLA haplotype of paternal origin (blue) (bottom panel). Although the paternal HLA class I and II molecules expressed by the fetus are alloantigens against which the mother's immune system has the potential to respond, the fetus does not provoke such a response during pregnancy and is protected from preexisting alloreactive antibodies or T cells. Even when the mother bears several children to the same father, no sign of immunological rejection is seen.

cells, but is not required for FoxP3 expression by nT_{reg} cells, has been found only in placental mammals. This suggests that iT_{reg} cells might have evolved to play an important role in maternal–fetal tolerance. Finally, stromal cells of the specialized maternal uterine tissue that directly interfaces with the placenta—the decidua—appear to repress the local expression of key T cell-attracting chemokines. Collectively, then, both maternal and fetal factors contribute to the formation of an immunologically privileged site akin to other sites of local immune suppression that allow prolonged acceptance of tissue grafts, such as the eye (see Section 15-5).

Summary.

Clinical transplantation is now an everyday reality, its success built on MHC matching, immunosuppressive drugs, and advances in surgical techniques. However, even accurate MHC matching does not prevent graft rejection; other genetic differences between host and donor can result in allogeneic proteins whose peptides are presented by MHC molecules on the grafted tissue, and responses to these can lead to rejection. Because we lack the ability to specifically suppress the response to the graft without compromising host defense, most transplants require generalized immunosuppression of the recipient that can increase the risk of cancer and infection. The fetus is a natural allograft that must be accepted for the species to survive. A better understanding of tolerance to the fetus could ultimately provide insights for inducing specific allograft tolerance in transplantation.

Summary to Chapter 15.

Ideally, the effector functions of the immune system would be targeted only to foreign pathogens and never to self tissues. In practice, because foreign and self proteins are chemically similar, strict discrimination between self and nonself is impossible. Yet the immune system maintains tolerance to self. This is accomplished by layers of regulation, all of which use surrogate markers to distinguish self from nonself to properly direct the immune response. When these mechanisms break down, autoimmune disease can result. Minor breaches of single regulatory barriers probably occur every day but are quelled by the effects of other regulatory layers; thus, tolerance operates at the level of the overall immune system. For disease to occur, multiple layers of tolerance have to be overcome and the effect needs to be chronic. These layers begin with central tolerance in the bone marrow and thymus, and include peripheral mechanisms such as anergy, cytokine deviation, and regulatory T cells. Sometimes immune responses do not occur simply because the antigens are not available, as in immune sequestration.

Perhaps because of selective pressure to mount effective immune responses to pathogens, the dampening of immune responses to promote self-tolerance is limited and prone to failure. Genetic predisposition has an important role in determining which individuals will develop an autoimmune disease. Environmental forces also have a significant role, because even identical twins are not always both affected by the same autoimmune disease. Influences from the environment include infections, toxins, and chance events.

When self-tolerance is broken and autoimmune disease ensues, the effector mechanisms are quite similar to those employed in responses to pathogens. Although the details vary from disease to disease, both antibody and T cells can be involved. Much has been learned about immune responses made to tissue antigens by examining the response to nonself transplanted organs and tissues; lessons learned in the study of graft rejection apply to autoimmunity and vice versa. Transplantation has brought on syndromes of rejection that are in many ways similar to autoimmune disease, but the targets are either major or minor histocompatibility antigens. T cells are the main effectors in graft rejection and graft-versus-host disease, although antibodies can also contribute.

For each of the undesirable responses discussed here, the question is how to control the response without adversely affecting protective immunity to infection. The answer may lie in a more complete understanding of the regulation of the immune response, especially the suppressive mechanisms important in tolerance. The deliberate control of the immune response is examined further in Chapter 16.

Questions.

15.1 True or False: Inflammatory bowel disease—Crohn's disease and ulcerative colitis—is a disease in which the adaptive immune system causes tissue damage in response to self antigens.

15.2 Matching: Match the following monogenic autoimmune diseases with the associated defective gene.

A. Autoimmune polyendocrinopathy–candidiasis–ectodermal dystrophy
    i. *Fas*

B. Immune dysregulation, polyendocrinopathy, enteropathy, X-linked
    ii. *FoxP3*

C. Autoimmune lymphoproliferative syndrome
    iii. *AIRE*

15.3 Multiple Choice: Which of the following statements is incorrect?

A. The autoantibodies induced by procainamide, a drug widely used to treat abnormal heart rhythms, are similar to the autoantibodies that characterize systemic lupus erythematosus.

B. Inflammatory mediators released during the course of an infection can lead to activation of self-reactive lymphocytes and thus cause an autoimmune response.

C. Crohn's disease and Blau syndrome are both strongly associated with loss-of-function mutations in *NOD2*, among other causes.

D. *ATG16L1* and *IRGM* are genes that contribute to autophagy under normal circumstances, and defects in these have been linked to Crohn's disease.

15.4 Multiple Choice: Which of the following options correctly describe a transplantation scenario?

A. A syngeneic skin graft from a young mouse is rejected by an adult mouse.

B. An allogeneic skin graft from a male mouse is not rejected by a female mouse.

C. A syngeneic skin graft from a male mouse is rejected by a female mouse.

D. A skin autograft is rejected 3 weeks after transplantation.

15.5 Short Answer: How can graft-versus-host disease (GVHD) be of benefit to patients with leukemia?

15.6 Multiple Choice: Which of the following options incorrectly describes a mechanism used to prevent fetal rejection?

A. High expression of 2,3-dioxygenase (IDO), which starves T cells of tryptophan

B. Absence of MHC class II expression and low levels of MHC class I expression by the trophoblast

C. Downregulation of HLA-G expression by the trophoblast

D. Secretion of TGF-β and IL-10 by the uterine epithelium and the trophoblast.

15.7 Multiple Choice: Which of the following is not a mechanism by which immunologically privileged sites maintain tolerance?

A. Exclusion of effector T cells during infection

B. Tissue barriers that exclude naive lymphocytes (for example, the blood–brain barrier)

C. Anti-inflammatory cytokine production (for example, TGF-β)

D. Expression of Fas ligand to induce apoptosis of Fas-bearing effector lymphocytes

E. Decreased communication via conventional lymphatics

15.8 Multiple Choice: Which of the following is not a mechanism of peripheral tolerance?

A. Anergy
B. Negative selection
C. Induction of T$_{reg}$s
D. Deletion
E. Suppression by T$_{reg}$s

15.9 Short Answer: The phenomenon of epitope spreading occurs in systemic lupus erythematosus (SLE), where anti-DNA autoantibodies are present and may progress to the production of anti-histone antibodies. Describe mechanistically how this occurs.

15.10 Short Answer: Autoimmune polyendocrinopathy–candidiasis–ectodermal dysplasia (APECED) is caused by defects in the transcription factor AIRE, which result in impaired expression of peripheral genes and reduced negative selection (that is, impaired central tolerance). Patients afflicted with APECED suffer destruction of multiple endocrine tissues and exhibit impaired antifungal immunity. However, these autoimmune phenomena take time to develop and do not develop in all potential organ targets in all patients. Explain why this is the case.

15.11 Fill-in-the-Blanks: Autoantibodies that develop in certain autoimmune disorders can act as either antagonists or agonists, depending on whether they inhibit or stimulate a function. In _____, autoantibodies against the _____ receptor block its function in the neuromuscular junction, resulting in a syndrome of muscle weakness. Another example is _____, where autoantibodies against the _____ receptor stimulate excessive production of thyroid hormone.

15.12 Matching: Match the autoimmune disease with its pathophysiology:

A. Rheumatoid arthritis

B. Type I diabetes mellitus

C. Multiple sclerosis

D. Hashimoto's thyroiditis

E. Autoimmune hemolytic anemia

F. Autoimmune thrombocytopenic purpura

G. Goodpasture's syndrome

H. Mixed essential cryoglobulinemia

i. Chronic hepatitis C infection leading to production of immune complexes that deposit in joints and tissues

ii. T-cell-mediated autoimmune attack against central nervous system myelin antigens, leading to demyelinating disease with neurological phenotypes

iii. Autoantibodies against IgG

iv. Autoantibodies against the GpIIb:IIIa fibrinogen receptor on platelets

v. Autoantibodies against red blood cells

vi. T_H1-dependent autoimmune attack of β cells in the pancreas

vii. Autoantibodies against the α_3 chain of basement membrane collagen (type IV collagen)

viii. Cell- and autoantibody-mediated autoimmune attack of the thyroid leading to hypothyroidism

Section references.

15-1 A critical function of the immune system is to discriminate self from nonself.

Ehrlich, P., and Morgenroth, J.: **On haemolysins**, in Himmelweit, F. (ed): *The Collected Papers of Paul Ehrlich*. London, Pergamon, 1957, 246–255.

Janeway, C.A., Jr: **The immune system evolved to discriminate infectious nonself from noninfectious self.** *Immunol. Today* 1992, **13**:11–16.

Matzinger, P.: **The danger model: a renewed sense of self.** *Science* 2002, **296**:301–305.

15-2 Multiple tolerance mechanisms normally prevent autoimmunity.

Goodnow, C.C., Sprent, J., Fazekas de St Groth, B., and Vinuesa, C.G.: **Cellular and genetic mechanisms of self tolerance and autoimmunity.** *Nature* 2005, **435**:590–597.

Shlomchik, M.J.: **Sites and stages of autoreactive B cell activation and regulation.** *Immunity* 2008, **28**:18–28.

15-3 Central deletion or inactivation of newly formed lymphocytes is the first checkpoint of self-tolerance.

Hogquist, K.A., Baldwin, T.A., and Jameson, S.C.: **Central tolerance: learning self-control in the thymus.** *Nat. Rev. Immunol.* 2005, **5**:772–782.

Kappler, J.W., Roehm, N., and Marrack, P.: **T cell tolerance by clonal elimination in the thymus.** *Cell* 1987, **49**:273–280.

Kyewski, B., and Klein, L.: **A central role for central tolerance.** *Annu. Rev. Immunol.* 2006, **24**:571–606.

Nemazee, D.A., and Burki, K.: **Clonal deletion of B lymphocytes in a transgenic mouse bearing anti-MHC class-I antibody genes.** *Nature* 1989, **337**:562–566.

Steinman, R.M., Hawiger, D., and Nussenzweig, M.C.: **Tolerogenic dendritic cells.** *Annu. Rev. Immunol.* 2003, **21**:685–711.

15-4 Lymphocytes that bind self antigens with relatively low affinity usually ignore them but in some circumstances become activated.

Billingham, R.E., Brent, L., and Medawar, P.B.: **Actively acquired tolerance of foreign cells.** *Nature* 1953, **172**:603–606.

Hannum, L.G., Ni, D., Haberman, A.M., Weigert, M.G., and Shlomchik, M.J.: **A disease-related RF autoantibody is not tolerized in a normal mouse: implications for the origins of autoantibodies in autoimmune disease.** *J. Exp. Med.* 1996, **184**:1269–1278.

Kurts, C., Sutherland, R.M., Davey, G., Li, M., Lew, A.M., Blanas, E., Carbone, F.R., Miller, J.F., and Heath, W.R.: **CD8 T cell ignorance or tolerance to islet antigens depends on antigen dose.** *Proc. Natl Acad. Sci. USA* 1999, **96**:12703–12707.

Marshak-Rothstein, A.: **Toll-like receptors in systemic autoimmune disease.** *Nat. Rev. Immunol.* 2006, **6**:823–835.

15-5 Antigens in immunologically privileged sites do not induce immune attack but can serve as targets.

Forrester, J.V., Xu, H., Lambe, T., and Cornall, R.: **Immune privilege or privileged immunity?** *Mucosal Immunol.* 2008, **1**:372–381.

Mellor, A.L., and Munn, D.H.: **Creating immune privilege: active local suppression that benefits friends, but protects foes.** *Nat. Rev. Immunol.* 2008, **8**:74–80.

Simpson, E.: **A historical perspective on immunological privilege.** *Immunol. Rev.* 2006, **213**:12–22.

15-6 Autoreactive T cells that express particular cytokines may be nonpathogenic or may suppress pathogenic lymphocytes.

von Herrath, M.G., and Harrison, L.C.: **Antigen-induced regulatory T cells in autoimmunity.** *Nat. Rev. Immunol.* 2003, **3**:223–232.

15-7 Autoimmune responses can be controlled at various stages by regulatory T cells.

Asano, M., Toda, M., Sakaguchi, N., and Sakaguchi, S.: **Autoimmune disease as a consequence of developmental abnormality of a T cell subpopulation.** *J. Exp. Med.* 1996, **184**:387–396.

Fillatreau, S., Sweenie, C.H., McGeachy, M.J., Gray, D., and Anderton, S.M.: **B cells regulate autoimmunity by provision of IL-10.** *Nat. Immunol.* 2002, **3**:944–950.

Fontenot, J.D., and Rudensky, A.Y.: **A well adapted regulatory contrivance: regulatory T cell development and the forkhead family transcription factor Foxp3.** *Nat. Immunol.* 2005, **6**:331–337.

Izcue, A., Coombes, J.L., and Powrie, F.: **Regulatory lymphocytes and intestinal inflammation.** *Annu. Rev. Immunol.* 2009, **27**:313–338.

Mayer, L., and Shao, L.: **Therapeutic potential of oral tolerance.** *Nat. Rev. Immunol.* 2004, **4**:407–419.

Maynard, C.L., and Weaver, C.T.: **Diversity in the contribution of interleukin-10 to T-cell-mediated immune regulation.** *Immunol. Rev.* 2008, **226**:219–233.

Sakaguchi, S.: **Naturally arising CD4+ regulatory T cells for immunologic self-tolerance and negative control of immune responses.** *Annu. Rev. Immunol.* 2004, **22**:531–562.

Wildin, R.S., Ramsdell, F., Peake, J., Faravelli, F., Casanova, J.L., Buist, N., Levy-Lahad, E., Mazzella, M., Goulet, O., Perroni, L., *et al.*: **X-linked neonatal diabetes mellitus, enteropathy and endocrinopathy syndrome is the human equivalent of mouse scurfy.** *Nat. Genet.* 2001, **27**:18–20.

Yamanouchi, J., Rainbow, D., Serra, P., Howlett, S., Hunter, K., Garner, V.E.S., Gonzalez-Munoz, A., Clark, J., Veijola, R., Cubbon, R., *et al.*: **Interleukin-2 gene variation impairs regulatory T cell function and causes autoimmunity.** *Nat. Genet.* 2007, **39**:329–337.

15-8 Specific adaptive immune responses to self antigens can cause autoimmune disease.

Lotz, P.H.: **The autoantibody repertoire: searching for order.** *Nat. Rev. Immunol.* 2003, **3**:73–78.

Santamaria, P.: **The long and winding road to understanding and conquering type 1 diabetes.** *Immunity* 2010, **32**:437–445.

Steinman, L.: **Multiple sclerosis: a coordinated immunological attack against myelin in the central nervous system.** *Cell* 1996, **85**:299–302.

15-9 Autoimmunity can be classified into either organ-specific or systemic disease.

Davidson, A., and Diamond, B.: **Autoimmune diseases.** *N. Engl. J. Med.* 2001, **345**:340–350.

D'Cruz, D.P., Khamashta, M.A., and Hughes, G.R.V.: **Systemic lupus erythematosus.** *Lancet* 2007, **369**:587–596.

Marrack, P., Kappler, J., and Kotzin, B.L.: **Autoimmune disease: why and where it occurs.** *Nat. Med.* 2001, **7**:899–905.

15-10 Multiple components of the immune system are typically recruited in autoimmune disease.

Drachman, D.B.: **Myasthenia gravis.** *N. Engl. J. Med.* 1994, **330**:1797–1810.

Firestein, G.S.: **Evolving concepts of rheumatoid arthritis.** *Nature* 2003, **423**:356–361.

Lehuen, A., Diana, J., Zaccone, P., and Cooke, A.: **Immune cell crosstalk in type 1 diabetes.** *Nat. Rev. Immunol.* 2010, **10**:501–513.

Shlomchik, M.J., and Madaio, M.P.: **The role of antibodies and B cells in the pathogenesis of lupus nephritis.** *Springer Semin. Immunopathol.* 2003, **24**:363–375.

15-11 Chronic autoimmune disease develops through positive feedback from inflammation, inability to clear the self antigen, and a broadening of the autoimmune response.

Marshak-Rothstein, A.: **Toll-like receptors in systemic autoimmune disease.** *Nat. Rev. Immunol.* 2006, **6**:823–835.

Nagata, S., Hanayama, R., and Kawane, K.: **Autoimmunity and the clearance of dead cells.** *Cell* 2010, **140**:619–630.

Salato, V.K., Hacker-Foegen, M.K., Lazarova, Z., Fairley, J.A., and Lin, M.S.: **Role of intramolecular epitope spreading in pemphigus vulgaris.** *Clin. Immunol.* 2005, **116**:54–64.

Steinman, L.: **A few autoreactive cells in an autoimmune infiltrate control a vast population of nonspecific cells: a tale of smart bombs and the infantry.** *Proc. Natl Acad. Sci. USA* 1996, **93**:2253–2256.

Vanderlugt, C.L., and Miller, S.D.: **Epitope spreading in immune-mediated diseases: implications for immunotherapy.** *Nat. Rev. Immunol.* 2002, **2**:85–95.

15-12 Both antibody and effector T cells can cause tissue damage in autoimmune disease.

Naparstek, Y., and Plotz, P.H.: **The role of autoantibodies in autoimmune disease.** *Annu. Rev. Immunol.* 1993, **11**:79–104.

Vlahakos, D., Foster, M.H., Ucci, A.A., Barrett, K.J., Datta, S.K., and Madaio, M.P.: **Murine monoclonal anti-DNA antibodies penetrate cells, bind to nuclei, and induce glomerular proliferation and proteinuria *in vivo*.** *J. Am. Soc. Nephrol.* 1992, **2**:1345–1354.

15-13 Autoantibodies against blood cells promote their destruction.

Beardsley, D.S., and Ertem, M.: **Platelet autoantibodies in immune thrombocytopenic purpura.** *Transfus. Sci.* 1998, **19**:237–244.

Clynes, R., and Ravetch, J.V.: **Cytotoxic antibodies trigger inflammation through Fc receptors.** *Immunity* 1995, **3**:21–26.

Domen, R.E.: **An overview of immune hemolytic anemias.** *Cleveland Clin. J. Med.* 1998, **65**:89–99.

15-14 The fixation of sublytic doses of complement to cells in tissues stimulates a powerful inflammatory response.

Brandt, J., Pippin, J., Schulze, M., Hansch, G.M., Alpers, C.E., Johnson, R.J., Gordon, K., and Couser, W.G.: **Role of the complement membrane attack complex (C5b-9) in mediating experimental mesangioproliferative glomerulonephritis.** *Kidney Int.* 1996, **49**:335–343.

Hansch, G.M.: **The complement attack phase: control of lysis and non-lethal effects of C5b-9.** *Immunopharmacology* 1992, **24**:107–117.

15-15 Autoantibodies against receptors cause disease by stimulating or blocking receptor function.

Bahn, R.S., and Heufelder, A.E.: **Pathogenesis of Graves' ophthalmopathy.** *N. Engl. J. Med.* 1993, **329**:1468–1475.

Drachman, D.B.: **Myasthenia gravis.** *N. Engl. J. Med.* 1994, **330**:1797–1810.

Vincent, A., Lily, O., and Palace, J.: **Pathogenic autoantibodies to neuronal proteins in neurological disorders.** *J. Neuroimmunol.* 1999, **100**:169–180.

15-16 Autoantibodies against extracellular antigens cause inflammatory injury.

Casciola-Rosen, L.A., Anhalt, G., and Rosen, A.: **Autoantigens targeted in systemic lupus erythematosus are clustered in two populations of surface structures on apoptotic keratinocytes.** *J. Exp. Med.* 1994, **179**:1317–1330.

Clynes, R., Dumitru, C., and Ravetch, J.V.: **Uncoupling of immune complex formation and kidney damage in autoimmune glomerulonephritis.** *Science* 1998, **279**:1052–1054.

Kotzin, B.L.: **Systemic lupus erythematosus.** *Cell* 1996, **85**:303–306.

Lee, R.W., and D'Cruz, D.P.: **Pulmonary renal vasculitis syndromes.** *Autoimmun. Rev.* 2010, **9**:657–660.

Mackay, M., Stanevsky, A., Wang, T., Aranow, C., Li, M., Koenig, S., Ravetch, J.V., and Diamond, B.: **Selective dysregulation of the FcgIIB receptor on memory B cells in SLE.** *J. Exp. Med.* 2006, **203**:2157–2164.

Xiang, Z., Cutler, A.J., Brownlie, R.J., Fairfax, K., Lawlor, K.E., Severinson, E., Walker, E.U., Manz, R.A., Tarlinton, D.M., and Smith, K.G.: **FcgRIIb controls bone marrow plasma cell persistence and apoptosis.** *Nat. Immunol.* 2007, **8**:419–429.

15-17 T cells specific for self antigens can cause direct tissue injury and sustain autoantibody responses.

Feldmann, M., and Steinman, L.: **Design of effective immunotherapy for human autoimmunity.** *Nature* 2005, **435**:612–619.

Firestein, G.S.: **Evolving concepts of rheumatoid arthritis.** *Nature* 2003, **423**:356–361.

Frohman, E.M., Racke, M.K., and Raine, C.S.: **Multiple sclerosis—the plaque and its pathogenesis.** *N. Engl. J. Med.* 2006, **354**:942–955.

Klareskog, L., Stolt, P., Lundberg, K. *et al.*: **A new model for an etiology of rheumatoid arthritis: Smoking may trigger HLA–DR (shared epitope)–restricted immune reactions to autoantigens modified by citrullination.** *Arthritis Rheum.* 2005, **54**:38–46.

Lassmann, H., van Horssen, J., and Mahad, D.: **Progressive multiple sclerosis: pathology and pathogenesis.** *Nat. Rev. Neurol.* 2012, **8**:647–656.

Ransohoff, R.M., and Engelhardt, B.: **The anatomical and cellular basis of immune surveillance in the central nervous system.** *Nat. Rev. Immunol.* 2012, **12**:623–635.

Zamvil, S., Nelson, P., Trotter, J., Mitchell, D., Knobler, R., Fritz, R., and Steinman, L.: **T-cell clones specific for myelin basic protein induce chronic relapsing paralysis and demyelination.** *Nature* 1985, **317**:355–358.

15-18 Autoimmune diseases have a strong genetic component.

Fernando, M.M.A., Stevens, C.R., Walsh, E.C., De Jager, P.L., Goyette, P., Plenge, R.M., Vyse, T.J., and Rioux, J.D.: **Defining the role of the MHC in autoimmunity: a review and pooled analysis.** *PLoS Genet.* 2008, **4**:e1000024.

Parkes, M., Cortes, A., Van Heel, D.A., and Brown, M.A.: **Genetic insights into common pathways and complex relationships among immune-mediated diseases.** *Nat. Rev. Genet.* 2013, **14**:661–673.

Rioux, J.D., and Abbas, A.K.: **Paths to understanding the genetic basis of autoimmune disease.** *Nature* 2005, **435**:584–589.

Wakeland, E.K., Liu, K., Graham, R.R., and Behrens, T.W.: **Delineating the genetic basis of systemic lupus erythematosus.** *Immunity* 2001, **15**:397–408.

15-19 Genomics-based approaches are providing new insight into the immunogenetic basis of autoimmunity.

Botto, M., Kirschfink, M., Macor, P., Pickering, M.C., Wurzner, R., and Tedesco, F.: **Complement in human diseases: lessons from complement deficiencies.** *Mol. Immunol.* 2009, **46**:2774–2783.

Cotsapas, C., and Hafler, D.A.: **Immune-mediated disease genetics: the shared basis of pathogenesis.** *Trends Immunol.* 2013, **34**:22–26.

Duerr, R.H., Taylor, K.D., Brant, S.R., Rioux, J.D., Silverberg, M.S., Daly, M.J., Steinhart, A.H., Abraham, C., Regueiro, M., Griffiths, A., *et al.*: **A genome-wide association study identifies IL23R as an inflammatory bowel disease gene.** *Science* 2006, **314**:1461–1463.

Farh, K.K.-H., Marson, A., Zhu, J., Kleinewietfeld, M., Housley, W.J., Beik, S., Shoresh, N., Whitton, H., Ryan, R.J.H., Shishkin, A.A. *et al.*: **Genetic and epigenetic fine mapping of causal autoimmune disease variants.** *Nature* 2014, **518**:337–343.

Gregersen, P.K.: **Pathways to gene identification in rheumatoid arthritis: PTPN22 and beyond.** *Immunol. Rev.* 2005, **204**:74–86.

Xavier, R.J., and Rioux, J.D.: **Genome-wide association studies: a new window into immune-mediated diseases.** *Nat. Rev. Immunol.* 2008, **8**:631–643.

15-20 Many genes that predispose to autoimmunity fall into categories that affect one or more tolerance mechanisms.

Goodnow, C.C.: **Polygenic autoimmune traits: Lyn, CD22, and SHP-1 are limiting elements of a biochemical pathway regulating BCR signaling and selection.** *Immunity* 1998, **8**:497–508.

Tivol, E.A., Borriello, F., Schweitzer, A.N., Lynch, W.P., Bluestone, J.A., and Sharpe, A.H.: **Loss of CTLA-4 leads to massive lymphoproliferation and fatal multiorgan tissue destruction, revealing a critical negative regulatory role of CTLA-4.** *Immunity* 1995, **3**:541–547.

Wakeland, E.K., Liu, K., Graham, R.R., and Behrens, T.W.: **Delineating the genetic basis of systemic lupus erythematosus.** *Immunity* 2001, **15**:397–408.

Walport, M.J.: **Lupus, DNase and defective disposal of cellular debris.** *Nat. Genet.* 2000, **25**:135–136.

15-21 Monogenic defects of immune tolerance.

Anderson, M.S., Venanzi, E.S., Chen, Z., Berzins, S.P., Benoist, C., and Mathis, D.: **The cellular mechanism of Aire control of T cell tolerance.** *Immunity* 2005, **23**:227–239.

Bacchetta, R., Passerini, L., Gambineri, E., Dai, M., Allan, S.E., Perroni, L., Dagna-Bricarelli, F., Sartirana, C., Matthes-Martin, S., Lawitschka, A., *et al.*: **Defective regulatory and effector T cell functions in patients with FOXP3 mutations.** *J. Clin. Invest.* 2006, **116**:1713–1722.

Ueda, H., Howson, J.M., Esposito, L., Heward, J., Snook, H., Chamberlain, G., Rainbow, D.B., Hunter, K.M., Smith, A.N., DiGenova, G., *et al.*: **Association of the T-cell regulatory gene CTLA4 with susceptibility to autoimmune disease.** *Nature* 2003, **423**:506–511.

Wildin, R.S., Ramsdell, F., Peake, J., Faravelli, F., Casanova, J.L., Buist, N., Levy-Lahad, E., Mazzella, M., Goulet, O., Perroni, L., *et al.*: **X-linked neonatal diabetes mellitus, enteropathy and endocrinopathy syndrome is the human equivalent of mouse scurfy.** *Nat. Genet.* 2001, **27**:18–20.

15-22 MHC genes have an important role in controlling susceptibility to autoimmune disease.

Fernando, M.M.A., Stevens, C.R., Walsh, E.C., De Jager, P.L., Goyette, P., Plenge, R.M., Vyse, T.J., and Rioux, J.D.: **Defining the role of the MHC in autoimmunity: a review and pooled analysis.** *PLoS Genet.* 2008, **4**:e1000024.

McDevitt, H.O.: **Discovering the role of the major histocompatibility complex in the immune response.** *Annu. Rev. Immunol.* 2000, **18**:1–17.

15-23 Genetic variants that impair innate immune responses can predispose to T cell-mediated chronic inflammatory disease.

Cadwell, K., Liu, J.Y., Brown, S.L., Miyoshi, H., Loh, J., Lennerz, J.K., Kishi, C., Kc, W., Carrero, J.A., Hunt, S., *et al.*: **A key role for autophagy and the autophagy gene Atg16l1 in mouse and human intestinal Paneth cells.** *Nature* 2008, **456**:259–263.

Eckmann, L., and Karin, M.: **NOD2 and Crohn's disease: loss or gain of function?** *Immunity* 2005, **22**:661–667.

Xavier, R.J., and Podolsky, D.K.: **Unravelling the pathogenesis of inflammatory bowel disease.** *Nature* 2007, **448**:427–434.

15-24 External events can initiate autoimmunity.

Klareskog, L., Padyukov, L., Ronnelid, J., and Alfredsson, L.: **Genes, environment and immunity in the development of rheumatoid arthritis.** *Curr. Opin. Immunol.* 2006 **18**:650–655.

Munger, K.L., Levin, L.I., Hollis, B.W., Howard, N.S., and Ascherio, A.: **Serum 25-hydroxyvitamin D levels and risk of multiple sclerosis.** *J. Am. Med. Assoc.* 2006, **296**:2832–2838.

15-25 Infection can lead to autoimmune disease by providing an environment that promotes lymphocyte activation.

Bach, J.F.: **Infections and autoimmune diseases.** *J. Autoimmunity* 2005, **25**:74–80.

Moens, U., Seternes, O.M., Hey, A.W., Silsand, Y., Traavik, T., Johansen, B., and Hober, D., and Sauter, P.: **Pathogenesis of type 1 diabetes mellitus: interplay between enterovirus and host.** *Nat. Rev. Endocrinol.* 2010, **6**:279–289.

Sfriso, P., Ghirardello, A., Botsios, C., Tonon, M., Zen, M., Bassi, N., Bassetto, F., and Doria, A.: **Infections and autoimmunity: the multifaceted relationship.** *J. Leukocyte Biol.* 2010, **87**:385–395.

Takeuchi, O., and Akira, S.: **Pattern recognition receptors and inflammation.** *Cell* 2010, **140**:805–820.

15-26 Cross-reactivity between foreign molecules on pathogens and self molecules can lead to anti-self responses and autoimmune disease.

Barnaba, V., and Sinigaglia, F.: **Molecular mimicry and T cell-mediated autoimmune disease.** *J. Exp. Med.* 1997, **185**:1529–1531.

Rose, N.R.: **Infection, mimics, and autoimmune disease.** *J. Clin. Invest.* 2001, **107**:943–944.

Rose, N.R., Herskowitz, A., Neumann, D.A., and Neu, N.: **Autoimmune myocarditis: a paradigm of post-infection autoimmune disease.** *Immunol. Today* 1988, **9**:117–120.

15-27 Drugs and toxins can cause autoimmune syndromes.
&
15-28 Random events may be required for the initiation of autoimmunity.

Eisenberg, R.A., Craven, S.Y., Warren, R.W., and Cohen, P.L.: **Stochastic control of anti-Sm autoantibodies in MRL/Mp-lpr/lpr mice.** *J. Clin. Invest.* 1987, **80**:691–697.

Yoshida, S., and Gershwin, M.E.: **Autoimmunity and selected environmental factors of disease induction.** *Semin. Arthritis Rheum.* 1993, **22**:399–419.

15-29 Graft rejection is an immunological response mediated primarily by T cells.

Cornell, L.D., Smith, R.N., and Colvin, R.B.: **Kidney transplantation: mechanisms of rejection and acceptance.** *Annu. Rev. Pathol.* 2008, **3**:189–220.

Wood, K.J., and Goto, R.: **Mechanisms of rejection: current perspectives.** *Transplantation.* 2012, **93**:1–10.

15-30 Transplant rejection is caused primarily by the strong immune response to nonself MHC molecules.

Macdonald, W.A., Chen, Z., Gras, S., Archbold, J.K., Tynan, F.E., Clements, C.S., Bharadwaj, M., Kjer-Nielsen, L., Saunders, P.M., Wilce, M.C.J., *et al.*: **T cell allorecognition via molecular mimicry.** *Immunity* 2009, **31**:897–908.

Macedo, C., Orkis, E.A., Popescu, I., Elinoff, B.D., Zeevi, A., Shapiro, R., Lakkis, F.G., and Metes, D.: **Contribution of naive and memory T-cell populations to the human alloimmune response.** *Am. J. Transplant.* 2009, **9**:2057–2066.

Opelz, G., and Wujciak, T.: **The influence of HLA compatibility on graft survival after heart transplantation. The Collaborative Transplant Study.** *N. Engl. J. Med.* 1994, **330**:816–819.

15-31 In MHC-identical grafts, rejection is caused by peptides from other alloantigens bound to graft MHC molecules.

Dierselhuis, M., and Goulmy, E.: **The relevance of minor histocompatibility antigens in solid organ transplantation.** *Curr Opin Organ Transplant.* 2009, **14**:419–425.

den Haan, J.M., Meadows, L.M., Wang, W., Pool, J., Blokland, E., Bishop, T.L., Reinhardus, C., Shabanowitz, J., Offringa, R., Hunt, D.F., *et al.*: **The minor histocompatibility antigen HA-1: a diallelic gene with a single amino acid polymorphism.** *Science* 1998, **279**:1054–1057.

Mutis, T., Gillespie, G., Schrama, E., Falkenburg, J.H., Moss, P., and Goulmy, E.: **Tetrameric HLA class I-minor histocompatibility antigen peptide complexes demonstrate minor histocompatibility antigen-specific cytotoxic T lymphocytes in patients with graft-versus-host disease.** *Nat. Med.* 1999, **5**:839–842.

15-32 There are two ways of presenting alloantigens on the transplanted donor organ to the recipient's T lymphocytes.

Jiang, S., Herrera, O., and Lechler, R.I.: **New spectrum of allorecognition pathways: implications for graft rejection and transplantation tolerance.** *Curr. Opin. Immunol.* 2004, **16**:550–557.

Safinia, N., Afzali, B., Atalar, K., Lombardi, G., and Lechler, R.I.: **T-cell alloimmunity and chronic allograft dysfunction.** *Kidney Int.* 2010, **78**(Suppl 119):S2–S12.

Lakkis, F.G., Arakelov, A., Konieczny, B.T., and Inoue, Y.: **Immunologic ignorance of vascularized organ transplants in the absence of secondary lymphoid tissue.** *Nat. Med.* 2000, **6**:686–688.

15-33 Antibodies that react with endothelium cause hyperacute graft rejection.

Griesemer, A., Yamada, K., and Sykes, M.: **Xenotransplantation: immunological hurdles and progress toward tolerance.** *Immunol. Rev.* 2014, **258**:241–258.

Montgomery, R.A., Cozzi, E., West, L.J., and Warren, D.S.: **Humoral immunity and antibody-mediated rejection in solid organ transplantation.** *Semin. Immunol.* 2011, **23**:224–234.

Williams, G.M., Hume, D.M., Hudson, R.P., Jr, Morris, P.J., Kano, K., and Milgrom, F.: **'Hyperacute' renal-homograft rejection in man.** *N. Engl. J. Med.* 1968, **279**:611–618.

15-34 Late failure of transplanted organs is caused by chronic injury to the graft.

Smith, R.N., and Colvin, R.B.: **Chronic alloantibody mediated rejection.** *Semin. Immunol.* 2012, **24**:115–121.

Libby, P., and Pober, J.S.: **Chronic rejection.** *Immunity* 2001, **14**:387–397.

15-35 A variety of organs are transplanted routinely in clinical medicine.

Ekser, B., Ezzelarab, M., Hara, H., van der Windt, D.J., Wijkstrom, M., Bottino, R., Trucco, M., and Cooper, D.K.C.: **Clinical xenotransplantation: the next medical revolution?** *Lancet* 2012, **379**:672–683.

Lechler, R.I., Sykes, M., Thomson, A.W., and Turka, L.A.: **Organ transplantation—how much of the promise has been realized?** *Nat. Med.* 2005, **11**:605–613.

Ricordi, C., and Strom, T.B.: **Clinical islet transplantation: advances and immunological challenges.** *Nat. Rev. Immunol.* 2004, **4**:259–268.

15-36 The converse of graft rejection is graft-versus-host disease.

Blazar, B.R., Murphy, W.J., and Abedi, M.: **Advances in graft-versus-host disease biology and therapy.** *Nat. Rev. Immunol.* 2012, **12**:443–458.

Shlomchik, W.D.: **Graft-versus-host disease.** *Nat. Rev. Immunol.* 2007, **7**:340–352.

15-37 Regulatory T cells are involved in alloreactive immune responses.

Ferrer, I.R., Hester, J., Bushell, A., and Wood, K.J.: **Induction of transplantation tolerance through regulatory cells: from mice to men.** *Immunol. Rev.* 2014, **258**:102–116.

Qin, S., Cobbold, S.P., Pope, H., Elliott, J., Kioussis, D., Davies, J., and Waldmann, H.: **"Infectious" transplantation tolerance.** *Science* 1993, **259**:974–977.

Tang, Q., and Bluestone, J.A.: **Regulatory T-cell therapy in transplantation: moving to the clinic.** *Cold Spring Harb. Perspect. Med.* 2013, **3**:1–15.

15-38 The fetus is an allograft that is tolerated repeatedly.

Erlebacher, A.: **Mechanisms of T cell tolerance towards the allogeneic fetus.** *Nat. Rev. Immunol.* 2013, **13**:23–33.

Samstein, R.M., Josefowicz, S.Z., Arvey, A., Treuting, P.M., and Rudensky A.Y.: **Extrathymic generation of regulatory T cells in placental mammals mitigates maternal-fetal conflict.** *Cell* 2012, **150**:29–38.

Manipulation of the Immune Response

16

In this chapter, we consider the various ways in which the immune system can be manipulated to suppress unwanted immune responses in the form of autoimmunity, allergy, and graft rejection, or to stimulate protective immune responses. Intentional manipulations of the immune system date back over 500 years to the use of variolation as a measure to protect against smallpox. In the late 1800s these measures advanced greatly with the development of numerous vaccines and antisera against other infectious agents. Later progress, against unwanted immune responses, came with the introduction of a number of now-conventional pharmaceutical agents; although these allow only relatively nonspecific control over unwanted immune reactions, they remain an important component of clinical medicine. More recently, these standard therapeutics have been joined by so-called biological therapeutics, or biologics, which are artificially produced versions of natural products, such as hormones, cytokines, and monoclonal antibodies, or derivatives of these such as engineered fusion proteins. These biologics possess extraordinary specificity, and though some have been used for decades, such as the hormone insulin in patients with type 1 diabetes, recent advances in cell biology and engineering have allowed for the introduction of a broad array of new biologics that allow for very precise manipulation of the immune system. Finally, long-standing efforts to deploy the power of the adaptive immune system against tumors have made major advances, and biologics that target negative regulators of immunity to stimulate protective responses against cancers have made a significant impact in clinical medicine. The categories of agents used to manipulate immune responses are listed in Fig. 16.1. This chapter will discuss these approaches, beginning with pharmaceutical agents used in clinical practice. In the first part of the chapter, we focus on efforts to relieve unwanted immune responses, and on advances in cancer treatment based on immune-system therapies. In the last part of the chapter, we discuss current vaccination strategies against infectious diseases, and consider how a more rational approach to the design and development of vaccines promises to increase their efficacy and widen their usefulness and application.

IN THIS CHAPTER

Treatment of unwanted immune responses.

Using the immune response to attack tumors.

Fighting infectious diseases with vaccination.

Agents used to manipulate immune responses	
Type	**Example**
Radiation	
Small molecules	
Drugs	Sirolimus (rapamycin)
Adjuvants	Alum
Macromolecules	
Hormones	Cortisol
Cytokines	Interferon α
Antibodies	Rituximab (anti-CD20 antibody)
Fusion proteins	Abatacept (CTLA-4–Ig)
DNA vaccines	(Experimental)
Subunit vaccines	Hepatitis B vaccine
Conjugate vaccines	Hib (*Haemophilus influenzae* type B) vaccine
Cells and organisms	
Inactivated vaccines	IPV (inactivated poliovirus vaccine)
Live attenuated vaccines	MMR (measles, mumps, rubella) vaccine
Adoptive cell transfer	CAR (chimeric antigen receptor) T cells
Heterologous bone marrow transplant	

Fig. 16.1 Categories of immunomodulating agents.

Treatment of unwanted immune responses.

Unwanted immune responses occur in many settings, such as autoimmune disease, transplant rejection, and allergy, which present different therapeutic challenges. The goal of treatment in all cases is to avoid tissue damage and prevent the disruption of tissue function. Some unwanted immune responses can be anticipated so that preventive measures may be taken, as in the case of allograft rejection. Other unwanted responses may be undetectable until after they become established, as is the case with autoimmune or allergic reactions. The relative difficulty of suppressing established immune responses is seen in animal models of autoimmunity, in which treatments that could have prevented induction of the disease are generally unable to halt it once it is established.

Conventional immunosuppressive drugs—meaning natural or synthetic small-molecule compounds—can be divided into several different

Fig. 16.2 Conventional immunosuppressive drugs in clinical use.

Conventional immunosuppressive drugs in clinical use	
Immunosuppressive drug	**Mechanism of action**
Corticosteroids	Inhibit inflammation; inhibit many targets including cytokine production by macrophages
Azathioprine, cyclophosphamide, mycophenolate	Inhibit proliferation of lymphocytes by interfering with DNA synthesis
Cyclosporin A, tacrolimus (FK506)	Inhibit the calcineurin-dependent activation of NFAT; block IL-2 production by T cells and proliferation by T cells
Rapamycin (sirolimus)	Inhibits proliferation of effector T cells by blocking Rictor-dependent mTOR activation
Fingolimod (FTY270)	Blocks lymphocyte trafficking out of lymphoid tissues by interfering with signaling by the sphingosine 1-phosphate receptor

Fig. 16.2 Conventional immunosuppressive drugs in clinical use.

categories (Fig. 16.2). There are the powerful anti-inflammatory drugs of the corticosteroid family such as prednisone, the cytotoxic drugs such as azathioprine and cyclophosphamide, and the noncytotoxic fungal and bacterial derivatives such as cyclosporin A, tacrolimus (FK506 or fujimycin), and rapamycin (sirolimus), which inhibit intracellular signaling pathways within T lymphocytes. Finally, a recently introduced drug, fingolimod, interferes with signaling by the sphingosine 1-phosphate receptor that controls the egress of B and T cells from lymphoid organs, thus preventing effector lymphocytes from reaching peripheral tissues. Most of these drugs exert broad inhibition of the immune system, and suppress helpful as well as harmful responses. Opportunistic infection is therefore a common complication of immunosuppressive drug therapy.

Newer treatments attempt to target the aspects of the immune response that cause tissue damage, such as cytokine action, while avoiding wholesale immunosuppression, but even these therapeutic agents can affect important components of the response to infectious disease. The most immediate way of inhibiting a particular part of the immune response is via highly specific antibodies, usually directed against specific proteins expressed and/or secreted by immune cells. Approaches of this type that were experimental at the time of previous editions of this book are now part of established medical practice. Anticytokine monoclonal antibodies, such as the drug **infliximab** (anti-TNF-α) used in the treatment of rheumatoid arthritis, can neutralize local excesses of cytokines or chemokines or target natural cellular regulatory mechanisms to inhibit unwanted immune responses. Proteins besides antibodies are also in use to control immune responses, an example being **abatacept**, a fusion protein consisting of the Fc region of an immunoglobulin fused to the extracellular domain of CTLA-4. Abatacept reduces co-stimulation of T cells by binding to B7 molecules and blocking their interactions with CD28, and it is currently used to treat patients with rheumatoid arthritis who fail to respond to anti-TNF-α therapy.

16-1 Corticosteroids are powerful anti-inflammatory drugs that alter the transcription of many genes.

Corticosteroid drugs are powerful anti-inflammatory and immunosuppressive agents that are used widely to attenuate the harmful effects of autoimmune or allergic immune responses (see Chapters 14 and 15), as well as responses against transplanted organs. **Corticosteroids** are derivatives of the glucocorticoid family of steroid hormones that play a crucial role in maintaining the body's homeostasis; one of the most widely used is **prednisone**, a synthetic version of the hormone cortisol. Corticosteroids cross the cell's plasma membrane and bind to intracellular receptors of the nuclear receptor family. Activated glucocorticoid receptors are transported to the nucleus, where they bind directly to DNA and interact with other transcription factors to regulate as many as 20% of the genes expressed in leukocytes. The response to steroid therapy is complex, given the large number of genes regulated in leukocytes and in other tissues. With respect to immunosuppression, corticosteroids exert multiple anti-inflammatory effects, which are briefly summarized in **Fig. 16.3**.

Corticosteroids target the pro-inflammatory functions of monocytes and macrophages and reduce the number of CD4 T cells. They can induce the expression of certain anti-inflammatory genes, such as *AnxaI*, which encodes a protein inhibitor of phospholipase A2 and thereby prevents this enzyme from generating pro-inflammatory **prostaglandins** and **leukotrienes** (see Sections 3-3 and 14-7). Conversely, corticosteroids can also suppress the expression of pro-inflammatory genes, including those encoding the cytokines IL-1β and TNF-α.

The therapeutic effects of corticosteroid drugs are due to their presence at much higher concentrations than the natural concentration of glucocorticoid hormones, causing exaggerated responses with both toxic and beneficial effects. Adverse effects include fluid retention, weight gain, diabetes, bone mineral loss, and thinning of the skin, requiring a careful balance to be maintained between beneficial and harmful effects. These drugs can also decrease in effectiveness over time. Despite these drawbacks, inhaled corticosteroids have proven highly beneficial in the treatment of chronic asthma (see Section 14-13). In the treatment of autoimmunity or allograft rejection, in which high doses of oral corticosteroids are needed to be effective, they are most often administered in combination with other immunosuppressant drugs to keep the corticosteroid dose and side-effects to a minimum. These other drugs include cytotoxic agents that act as immunosuppressants by killing rapidly dividing lymphocytes, and drugs that more specifically target lymphocyte signaling pathways.

16-2 Cytotoxic drugs cause immunosuppression by killing dividing cells and have serious side-effects.

The three cytotoxic drugs most commonly used as immunosuppressants are **azathioprine**, **cyclophosphamide**, and **mycophenolate**. These drugs interfere with DNA synthesis, and their major pharmacological action is on tissues in which cells are continually dividing. Developed originally to treat cancer, these drugs were found to be immunosuppressive after observations that they were cytotoxic to dividing lymphocytes. Azathioprine also interferes with CD28 co-stimulation in T cells, thus promoting T-cell apoptosis (see Section 7-24). The use of these compounds is, however, limited by their toxic effects on all tissues in which cells are dividing, such as the skin, gut lining, and bone marrow. Effects include decreased immune function, as well as anemia, leukopenia, thrombocytopenia, damage to intestinal epithelium, hair loss, and fetal death or injury. As a result of their toxicity, these drugs are used at high doses only when the aim is to eliminate all dividing lymphocytes, as in the treatment of lymphoma and leukemia; in these cases, treated patients require subsequent bone marrow transplantation to restore their hematopoietic function. When used to treat unwanted immune responses such as autoimmune conditions, they are used at lower doses and in combination with other drugs such as corticosteroids.

Azathioprine is converted *in vivo* to the purine analog 6-thioguanine (6-TG), which is metabolized to 6-thioinosinic acid. This competes with inosine monophosphate, blocking the *de novo* synthesis of adenosine monophosphate and guanosine monophosphate, thus inhibiting DNA synthesis. 6-TG is also incorporated into the DNA in place of guanine, and accumulation of

Corticosteroid therapy

Effect on	Physiological effects
↓ IL-1, TNF-α, GM-CSF ↓ IL-3, IL-4, IL-5, CXCL8	↓ Inflammation caused by cytokines
↓ NOS	↓ NO
↓ Phospholipase A₂ ↓ Cyclooxygenase type 2 ↑ Annexin-1	↓ Prostaglandins ↓ Leukotrienes
↓ Adhesion molecules	Reduced emigration of leukocytes from vessels
↑ Endonucleases	Induction of apoptosis in lymphocytes and eosinophils

Fig. 16.3 Anti-inflammatory effects of corticosteroid therapy. Corticosteroids regulate the expression of many genes, with a net anti-inflammatory effect. First, they reduce the production of inflammatory mediators, including cytokines, prostaglandins, and nitric oxide (NO). Second, they inhibit inflammatory cell migration to sites of inflammation by inhibiting the expression of adhesion molecules. Third, corticosteroids promote the death by apoptosis of leukocytes. The layers of complexity are illustrated by the actions of annexin-1 (originally identified as a factor induced by corticosteroids and named lipocortin), which has been shown to participate in all of the effects of corticosteroids listed on the right. NOS, NO synthase.

6-TG increases the DNA's sensitivity to mutations induced by the ultraviolet radiation in sunlight. Thus, patients treated with azathioprine have the long-term side-effect of increased risk of skin cancer. Azathioprine also generates 6-thioguanine triphosphate (6-thio-GTP), which in T cells binds to the small GTPase Rac1 in place of GTP and suppresses its activity. Signaling from CD28 co-stimulation requires Rac1, and T cells therefore do not receive the anti-apoptotic signals from co-stimulation and instead undergo apoptosis. **Mycophenolate mofetil**, the 2-morpholinoethyl ester of mycophenolic acid, is the newest addition to the family of cytotoxic immunosuppressive drugs; it works in a similar fashion to azathioprine. It is metabolized to mycophenolic acid, which inhibits the enzyme inosine monophosphate dehydrogenase, thus blocking the *de novo* synthesis of guanosine monophosphate.

Azathioprine and mycophenolate are less toxic than cyclophosphamide, which is metabolized to phosphoramide mustard, which alkylates DNA. Cyclophosphamide is a member of the nitrogen mustard family of compounds, which were originally developed as chemical weapons. It has a range of highly toxic effects including inflammation of and hemorrhage from the bladder, known as hemorrhagic cystitis, and induction of bladder neoplasia.

16-3 Cyclosporin A, tacrolimus, rapamycin, and JAK inhibitors are effective immunosuppressive agents that interfere with various T-cell signaling pathways.

Three noncytotoxic alternatives to the cytotoxic drugs are available as immunosuppressants and are widely used to treat transplant recipients. These are **cyclosporin A**, **tacrolimus** (previously known as **FK506**), and **rapamycin** (also known as **sirolimus**). Cyclosporin A is a cyclic decapeptide derived from a soil fungus found in Norway, *Tolypocladium inflatum*. Tacrolimus is a macrolide compound from the filamentous bacterium *Streptomyces tsukabaensis*, found in Japan; macrolides are compounds that contain a many-membered lactone ring to which is attached one or more deoxysugars. Rapamycin, another macrolide, is derived from *Streptomyces hygroscopicus*, found on Easter Island ('Rapa Nui' in Polynesian—hence the name of the drug). All three compounds exert their pharmacological effects by binding to members of a family of intracellular proteins known as the **immunophilins**, forming complexes that interfere with signaling pathways important for the clonal expansion of lymphocytes.

MOVIE 16.1 ▶

As explained in Section 7-14, cyclosporin A and tacrolimus block T-cell proliferation by inhibiting the phosphatase activity of the Ca^{2+}-activated protein phosphatase **calcineurin**, which is required for the activation of the transcription factor NFAT. Both drugs reduce the expression of several cytokine genes that are normally induced on T-cell activation (**Fig. 16.4**), including the gene encoding **interleukin (IL)-2**, which is an important growth factor for T cells (see Section 9-16). Cyclosporin A and tacrolimus inhibit T-cell proliferation in response to either specific antigens or allogeneic cells, and are used extensively in medical practice to prevent the rejection of allogeneic organ grafts. Although the major immunosuppressive effects of both drugs are probably the result of inhibition of T-cell proliferation, they also act on other cells and have a large variety of other immunological effects (see Fig. 16.4).

These two drugs inhibit calcineurin by first binding to an immunophilin molecule; cyclosporin A binds to the **cyclophilins**, and tacrolimus to the **FK-binding proteins** (FKBPs). Immunophilins are peptidyl-prolyl *cis–trans* isomerases, but their isomerase activity is not relevant to the immunosuppressive activity of the drugs that bind them. Rather, the immunophilin:drug complexes bind and inhibit the Ca^{2+}-activated serine/threonine phosphatase calcineurin. In a normal immune response, the increase in intracellular calcium ions in response to T-cell receptor signaling activates the calcium-binding protein

Immunological effects of cyclosporin A and tacrolimus	
Cell type	**Effects**
T lymphocyte	Reduced expression of IL-2, IL-3, IL-4, GM-CSF, TNF-α Reduced proliferation following decreased IL-2 production Reduced Ca²⁺-dependent exocytosis of granule-associated serine esterases Inhibition of antigen-driven apoptosis
B lymphocyte	Inhibition of proliferation secondary to reduced cytokine production by T lymphocytes Inhibition of proliferation following ligation of surface immunoglobulin Induction of apoptosis following B-cell activation
Granulocyte	Reduced Ca²⁺-dependent exocytosis of granule-associated serine esterases

Fig. 16.4 Cyclosporin A and tacrolimus inhibit lymphocyte and some granulocyte responses.

calmodulin; calmodulin then activates calcineurin (see Fig. 7.18). Binding of the immunophilin:drug complex to calcineurin prevents the latter's activation by calmodulin; the bound calcineurin is unable to dephosphorylate and activate NFAT (Fig. 16.5). Calcineurin is found in other cells besides T cells, but its levels in T cells are much lower than in other tissues. T cells are therefore particularly susceptible to the inhibitory effects of these drugs.

Cyclosporin A and tacrolimus are effective immunosuppressants but are not problem-free. As with the cytotoxic agents, they affect all immune responses indiscriminately. This can be countered by carefully varying the dose of drug given during the course of a response. During organ transplantation, for example, high doses are required during the time of grafting, but once the graft has become established the dose can be decreased in order to allow useful protective immune responses while maintaining adequate suppression of the residual response to the grafted tissue. This balance is difficult to achieve and requires careful monitoring of the patient. These drugs also have effects on many different tissues and thus can have broad side-effects, such as injury to kidney tubule epithelial cells. Finally, treatment with these drugs is relatively expensive, because they are complex natural products that must be taken for long periods. Nevertheless, at present they are the immunosuppressants of choice in clinical transplantation, and they are also being tested in a variety of autoimmune diseases, especially those that, like graft rejection, are mediated by T cells.

Rapamycin has a different mode of action from either cyclosporin A or tacrolimus. Like tacrolimus, rapamycin binds to the FKBP family of immunophilins, but the rapamycin:immunophilin complex does not inhibit calcineurin activity. Instead, it inhibits a serine/threonine kinase known as **mTOR** (mammalian target of rapamycin), which is involved in regulating cell growth and proliferation (see Section 7-17). The mTOR pathway can be activated by different upstream signaling pathways, including the Ras/MAPK pathway and

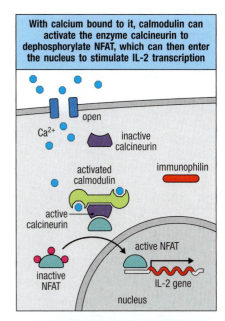

With calcium bound to it, calmodulin can activate the enzyme calcineurin to dephosphorylate NFAT, which can then enter the nucleus to stimulate IL-2 transcription

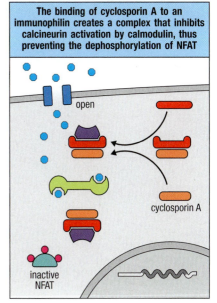

The binding of cyclosporin A to an immunophilin creates a complex that inhibits calcineurin activation by calmodulin, thus preventing the dephosphorylation of NFAT

Fig. 16.5 Cyclosporin A and tacrolimus inhibit T-cell activation by interfering with the serine/threonine-specific phosphatase calcineurin. As shown in the upper panel, signaling via T-cell receptor-associated tyrosine kinases leads to opening of calcium-release-activated calcium (CRAC) channels in the plasma membrane. This increases the concentration of Ca²⁺ in the cytoplasm and promotes calcium binding to the regulatory protein calmodulin (see Fig. 7.18). Calmodulin is activated by binding Ca²⁺ and can then target many downstream effector proteins such as the phosphatase calcineurin. Binding by calmodulin activates calcineurin to dephosphorylate the transcription factor NFAT (see Section 7-14), which then enters the nucleus and transcribes genes that are required for T-cell activation to progress. As shown in the lower panel, when cyclosporin A or tacrolimus or both are present, they form complexes with their immunophilin targets, cyclophilin and FK-binding protein, respectively. These complexes bind to calcineurin, preventing it from becoming activated by calmodulin, and thereby preventing the dephosphorylation of NFAT.

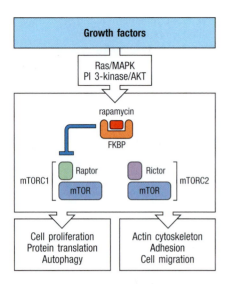

Fig. 16.6 Rapamycin inhibits cell growth and proliferation by selectively blocking activation of the kinase mTOR by Raptor. Rapamycin binds FK-binding protein (FKBP), the same immunophilin protein that binds to tacrolimus (FK506). The rapamycin:FKBP complex does not inhibit calcineurin, but instead blocks one of two complexes that activate mTOR, a large kinase that regulates many metabolic pathways. mTOR is activated downstream of various growth factor signaling pathways, and becomes associated with either of two proteins, Raptor (regulatory associated protein of mTOR) and Rictor (rapamycin-insensitive companion of mTOR). The complex with Raptor, mTORC1, promotes cell proliferation, translation of proteins, and autophagy, and the complex with Rictor, mTORC2, influences cell adhesion and migration by controlling the actin cytoskeleton. The rapamycin:FKBP complex acts to inhibit the Raptor-associated mTORC1, and thereby selectively reduces cell growth and proliferation.

the PI 3-kinase pathway. These pathways activate **AKT**, which phosphorylates and inactivates a regulatory complex called **TSC**. This complex normally acts as an inhibitor of the small GTPase **Rheb**; after TSC is phosphorylated, Rheb is free to activate mTOR (see Fig. 7.22). Two distinct mTOR complexes can be formed, **mTORC1** and **mTORC2**, which are controlled by two regulatory proteins called **Raptor** and **Rictor**, respectively, and which activate different downstream cellular pathways (Fig. 16.6). Rapamycin appears to inhibit only the mTORC1 complex, as the rapamycin:FKBP complex selectively inhibits the Raptor-dependent pathway that regulates this complex (see Fig. 16.6). Blockade of this pathway markedly reduces T-cell proliferation, arresting cells in the G_1 phase of the cell cycle and promoting apoptosis. Rapamycin inhibits lymphocyte proliferation driven by growth factors such as IL-2, IL-4, and IL-6, and increases the number of regulatory T cells, perhaps because these cells use different signaling pathways from those of effector T cells. Rapamycin also selectively reduces the outgrowth of effector T cells while apparently enhancing the formation of memory T cells. Because of this, the use of rapamycin to augment T-cell memory induced by vaccines is being considered.

One recently introduced drug manipulates immune responses by regulating the migration of immune effector cells to the sites of a graft or of autoimmune disease. In Section 9-7, we described how emigration of lymphocytes out of the lymphoid tissues requires recognition of the lipid molecule sphingosine 1-phosphate (S1P) by the G-protein-coupled receptor S1PR1. **Fingolimod** (FTY720), a sphingosine 1-phosphate analog, is a relatively newer drug that causes the retention of effector lymphocytes in lymphoid organs, thus preventing these cells from mediating their effector activities in target tissues. Fingolimod was approved in 2010 for treatment of the autoimmune disease multiple sclerosis, and has shown promise in the treatment of kidney graft rejection and asthma.

Cytokines activate many aspects of the immune response, and many cytokine receptors use **Janus kinases (JAKs)** in signal transduction (see Section 3-16). The four JAK family members, JAK1, JAK2, JAK3, and TYK2, bind and phosphorylate the cytoplasmic regions of cytokine receptors and initiate the activation of different STAT transcription factors. Selective **JAK inhibitors** have been developed over the last decade that can block the kinase activity of one or more members of this family. Since different JAKs binds to different cytokine receptors, JAK inhibitors, or **Jakinibs**, can therefore exert potentially specific effects on the quality of T-cell development. Two Jakinibs are now approved for use in treating inflammatory diseases and are being investigated for their application in cancer. For example, **tofacitinib** inhibits JAK3, interfering with signaling by IL-2 and IL-4, and somewhat more weakly JAK1, interfering with signaling by IL-6. Tofacitinib is approved for treatment of rheumatoid arthritis. **Ruxolitinib** inhibits JAK1 and JAK2 and has been approved for treating myelofibrosis, an abnormal proliferation of bone marrow progenitor cells that causes fibrosis.

16-4   Antibodies against cell-surface molecules can be used to eliminate lymphocyte subsets or to inhibit lymphocyte function.

All the drugs discussed so far exert a general inhibition on immune responses and can have severe side effects, but antibodies can act in a more specific manner and with less direct toxicity. The initial therapeutic use of antibodies extends back to the late 1800s with the development of equine sera for treatment of diphtheria and tetanus. Today, intravenous immunoglobulin (IVIG), a collection of polyvalent IgG antibodies pooled from many blood donors, is still widely used as a treatment for various primary and acquired immune deficiencies. It is also used in some acute infections, where it likely works by providing antibodies that may neutralize certain pathogens or their toxins. Finally, IVIG is also used to treat certain autoimmune and inflammatory diseases, such as immune thrombocytopenia and Kawasaki disease. In these cases, IVIG exerts an immunomodulatory effect that seems to operate through interactions with inhibitory Fc receptors that inhibit immune-cell activation.

Drug-induced Serum Sickness

The relatively recent expansion in the use of antibodies as therapeutic agents has extended their function from targeting pathogens to targeting components of the immune system itself in order to achieve a specific regulatory result. For example, the potential of antibodies to eliminate unwanted lymphocytes is demonstrated by **anti-lymphocyte globulin**, a preparation of polyclonal immunoglobulin from rabbits (and previously horses) immunized with human lymphocytes, which has been used for many years to treat episodes of acute graft rejection. Anti-lymphocyte globulin does not, however, discriminate between useful lymphocytes and those responsible for the unwanted responses and therefore leads to global immunosuppression. Foreign immunoglobulins are also highly antigenic in humans, and the large doses of anti-lymphocyte globulin used in therapy often cause a condition called **serum sickness**, resulting from the formation of immune complexes of the animal immunoglobulin and human antibodies against it (see Section 14-15).

Anti-lymphocyte globulin is nevertheless still used to treat acute rejection, and this has stimulated the quest for monoclonal antibodies (see Appendix I, Section A-7) that would achieve more specifically targeted effects. One such antibody is **alemtuzumab** (marketed as Campath-1H), which is directed at the cell-surface protein CD52 expressed by most lymphocytes. It has similar actions to anti-lymphocyte globulin, causing long-standing lymphopenia. It is also used to eliminate cancer cells in the treatment of chronic lymphocytic leukemia.

Immunosuppressive monoclonal antibodies act by one of two general mechanisms. Some, such as alemtuzumab, trigger the destruction of lymphocytes in vivo and are referred to as **depleting antibodies**, whereas others are **nondepleting** and act by blocking the function of their target protein without killing the cell that bears it. Depleting monoclonal IgG antibodies bind to lymphocytes and target them to macrophages and NK cells, which bear Fc receptors and kill the lymphocytes by phagocytosis or antibody-dependent cell-mediated cytotoxicity (ADCC), respectively. Complement-mediated lysis may also play a part in lymphocyte destruction.

16-5   Antibodies can be engineered to reduce their immunogenicity in humans.

A major impediment to therapy with monoclonal antibodies in humans has been that these antibodies are most readily made by immunizing nonhuman species, such as the mouse, to generate antibodies of the desired specificity (see Appendix I, Section A-7). Humans may develop an antibody response against such nonhuman antibodies, since aggregated forms of foreign antibodies can be immunogenic. Such a reaction not only interferes with the therapeutic actions of the antibodies, but also leads to allergic reactions, and, if treatment is continued, may result in **anaphylaxis** (see Section 14-10). Once a patient has made a response to an antibody, it can no longer be used for future treatment. This problem can, in principle, be avoided by making antibodies that are not recognized as foreign by the human immune system, a process called **humanization**.

Various approaches have been tried to humanize antibodies. The variable regions encoding the antigen-recognition determinants from a murine antibody can be spliced onto the Fc regions of human IgG by gene manipulation. Antibodies of this type are called chimeric antibodies. However, this approach leaves regions within the murine variable regions that could potentially induce

Fig. 16.7 Monoclonal antibodies used to treat human diseases can be engineered to decrease immunogenicity but maintain their antigen specificity.

Fig. 16.7 Monoclonal antibodies used to treat human diseases can be engineered to decrease immunogenicity but maintain their antigen specificity. Antibodies that are derived fully from mice, named with the suffix -omab, are immunogenic in humans. This causes patients to generate antibodies against them, limiting their usefulness over time. This immunogenicity can be reduced by making chimeric antibodies in which the V regions from the mouse are spliced onto human antibody constant regions; such antibodies are named with the suffix -ximab. Humanization is the process of splicing in just the complementarity-determining regions from the mouse antibody, further reducing immunogenicity; humanized antibodies are named with the suffix -zumab. New techniques now allow fully human (-umab) monoclonal antibodies to be derived, which are the least immunogenic type of antibody currently used for treating humans.

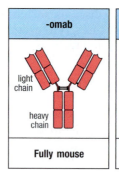
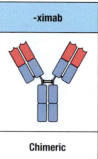
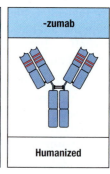
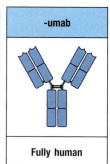

immune responses (Fig. 16.7). Genetically engineered mice that harbor human immunoglobulin genes inserted into their immunoglobulin locus represent one way that human antibodies may be obtained from the immunization of mice. Newer methods are aimed at generating fully human monoclonal antibodies directly from human cells through the use of viral transformation of human primary B-cell lines or antibody-secreting plasmablasts, or by generating human B-cell hybridomas.

Monoclonal antibodies belong to a new class of therapeutic compounds called **biologics**, which includes other natural proteins such as anti-lymphocyte globulin, cytokines, protein fragments, and even whole cells, which are used, for example, in the adoptive transfer of T cells in cancer immunotherapy. Many monoclonal antibodies have been, or are in the process of being, approved for clinical use by the US Food and Drug Administration (Fig. 16.8), and a systematic naming process identifies the type of antibody. Murine monoclonal antibodies are designated by the suffix -**omab**, such as muromomab (originally called OKT3), a murine antibody against CD3. Chimeric antibodies in which the entire variable region is spliced into human constant regions have the suffix -**ximab**, such as basiliximab, an anti-CD25 antibody approved for the treatment of transplantation rejection. Humanized antibodies in which the murine hypervariable regions have been spliced into a human antibody have the suffix -**zumab**, as in alemtuzumab and natalizumab (Tysabri). The latter is directed against the α_4 integrin subunit, and is used to treat multiple sclerosis and Crohn's disease. Antibodies derived entirely from human sequences have the suffix -**umab**, as in adalimumab, an antibody derived from phage display that binds TNF-α; it is used to treat several autoimmune diseases.

16-6 Monoclonal antibodies can be used to prevent allograft rejection.

Antibodies specific for various physiological targets are being used, or are under investigation, to prevent the rejection of transplanted organs by inhibiting the development of harmful inflammatory and cytotoxic responses. For example, alemtuzumab, discussed in Section 16-4, is licensed for the treatment of certain leukemias but is also used in both solid-organ and bone marrow transplantation. In solid-organ transplantation, alemtuzumab may be given to the recipient around the time of transplantation to remove mature T lymphocytes from the circulation. In bone marrow transplantation, alemtuzumab can be used *in vitro* to deplete donor bone marrow of mature T cells before its infusion into a recipient, or used *in vivo* to treat the recipient following infusion. Elimination of mature T cells from donor bone marrow is very effective at reducing the incidence of **graft-versus-host disease** (see Section 15-36). In this disease, the T lymphocytes in the donor bone marrow recognize the recipient as foreign and mount a damaging response, causing rashes, diarrhea, and hepatitis, which can occasionally be fatal. Bone marrow transplantation is also used as a treatment for leukemia, as T cells in the graft can have a so-called graft-versus-leukemia effect where they recognize the leukemic cells as foreign and destroy them. It was originally thought that elimination of mature donor

Monoclonal antibodies developed for immunotherapy			
Generic name	Specificity	Mechanism of action	Approved indication
Rituximab	Anti-CD20	Eliminates B cells	Non-Hodgkin's lymphoma
Alemtuzumab (Campath-1H)	Anti-CD52	Eliminates lymphocytes	Chronic myeloid leukemia
Muromomab (OKT3)	Anti-CD3	Inhibits T-cell activation	Kidney transplantation
Daclizumab	Anti-IL-2R	Reduces T-cell activation	
Basiliximab	Anti-IL-2R	Reduces T-cell activation	
Infliximab	Anti-TNF-α	Inhibit inflammation induced by TNF-α	Crohn's disease
Certolizumab	Anti-TNF-α		Rheumatoid arthritis
Adalimumab	Anti-TNF-α		
Golimumab	Anti-TNF-α		
Tocilizumab	Anti-IL-6R	Blocks inflammation induced by IL-6 signaling	
Canakinumab	Anti-IL-1β	Blocks inflammation caused by IL-1	Muckle–Wells syndrome
Denosumab	Anti-RANK-L	Inhibits activation of osteoclasts by RANK-L	Bone loss
Ustekinumab	Anti-IL-12/23	Inhibits inflammation caused by IL-12 and IL-23	Psoriasis
Efalizumab	Anti-CD11a (α_L integrin subunit)	Block lymphocyte trafficking	Psoriasis (withdrawn from use in United States and European Union)
Natalizumab	Anti-α_4 integrin		Multiple sclerosis
Omalizumab	Anti-IgE	Removes IgE antibody	Chronic asthma
Belimumab	Anti-BLyS	Reduces B-cell responses	Systemic lupus erythematosus (pending approval)
Ipilimumab	Anti-CTLA-4	Increases CD4 T-cell responses	Metastatic melanoma
Raxibacumab	Anti-*Bacillus anthracis* protective antigen (the cell-binding moiety of anthrax toxin)	Prevents action of anthrax toxins	Anthrax infection (pending approval)

Fig. 16.8 Monoclonal antibodies developed for immunotherapy.
A substantial fraction of pharmaceuticals currently under development are antibodies, and additions to this list, current as of this writing, are under development and in clinical trials.

T cells during such procedures would be disadvantageous, as the antileukemic action of the donor T cells could be lost, but this seems to not be the case when alemtuzumab is used as the depleting agent.

Specific antibodies directed against T cells have been used to treat episodes of graft rejection that occur after transplantation. The murine antibody **muromomab** (OKT3) targets the CD3 complex and leads to T-cell immunosuppression by inhibiting signaling through the T-cell receptor. It has been used clinically in solid-organ transplantation but is often associated with a dangerous side-effect, namely, the stimulation of pro-inflammatory cytokine release, and its use is declining. This cytokine release is related to muromomab's intact

Graft-Versus-Host Disease

Fc region, which can activate Fc receptors via cross-linking and activate the cells that bear these receptors. In the antibody called teplizumab, or OKT3γ1 (**Ala-Ala**), amino acids 234 and 235 in the human IgG1 Fc region have been changed to alanines, and this antibody no longer stimulates cytokine release.

Two other antibodies, **daclizumab** and **basiliximab**, approved for treating kidney transplant rejection, are directed against CD25 (a subunit of the IL-2 receptor) and reduce T-cell activation, presumably by blocking the growth-promoting signals delivered by IL-2.

A primate model of kidney transplant rejection showed promising effects for a humanized monoclonal antibody against the **CD40 ligand** expressed by T cells (see Section 9-17). A possible mechanism of protection by this antibody is blockade of the activation of dendritic cells by helper T cells that recognize donor antigens. Only preliminary studies of anti-CD40 ligand antibodies have been performed in humans. One antibody was associated with thromboembolic complications and was withdrawn; a different anti-CD40 ligand antibody was administered to patients with the autoimmune disease **systemic lupus erythematosus** (SLE) without significant complications, but also with little evidence of efficacy.

Systemic Lupus Erythematosus

Rheumatoid Arthritis

In experimental models, monoclonal antibodies against other targets have also had some success in preventing graft rejection, including nondepleting antibodies that bind the CD4 co-receptor or the co-stimulatory receptor CD28 on lymphocytes. Similarly, abatacept, a soluble recombinant CTLA-4–Ig fusion protein which binds to the co-stimulatory B7 molecules on antigen-presenting cells and prevents their interaction with CD28 on T cells, is approved for the treatment of rheumatoid arthritis.

16-7 Depletion of autoreactive lymphocytes can treat autoimmune disease.

In addition to their use in preventing transplantation rejection, monoclonal antibodies can be used to treat certain autoimmune diseases, and the different immune mechanisms targeted are discussed in the next few sections. We start by discussing the use of depleting and nondepleting antibodies to remove lymphocytes nonspecifically. The anti-CD20 monoclonal antibody **rituximab** was originally developed to treat B-cell lymphomas, but has also been tried in treating certain autoimmune diseases. By ligating CD20, rituximab (Rituxan, MabThera) transduces a signal that induces lymphocyte apoptosis and depletes B cells for several months. Certain autoimmune diseases are believed to involve autoantibody-mediated pathogenesis. There is evidence for the efficacy of rituximab in some patients with autoimmune hemolytic anemia, SLE, rheumatoid arthritis, or type II mixed cryoglobulinemia, all of which have autoantibodies as a part of their clinical presentation. Although CD20 is not expressed on antibody-producing plasma cells, their B-cell precursors are targeted by anti-CD20, resulting in a substantial reduction in the short-lived, but not the long-lived, plasma-cell population.

Alemtuzumab, discussed above for its use in treating leukemia and in transplant rejection, has shown some beneficial effect in studies of small numbers of patients with multiple sclerosis. However, immediately after its infusion, most multiple sclerosis patients suffered a frightening, although fortunately brief, flare-up of their illness, illustrating another potential complication of antibody therapy. Alemtuzumab was acting as intended, killing cells by complement- and Fc-dependent mechanisms. However, it also stimulated the release of cytokines, including TNF-α, interferon (IFN)-γ, and IL-6, which transiently block nerve conduction in nerve fibers previously affected by demyelination. This caused the transient but dramatic exacerbation of symptoms. Nevertheless, alemtuzumab may be useful at early stages of the disease, when the inflammatory response is maximal, but this has yet to be determined.

Treating patients suffering from rheumatoid arthritis or multiple sclerosis by using anti-CD4 antibodies has been tried, but with disappointing results. In controlled studies, the antibodies showed only small therapeutic effects but caused depletion of T lymphocytes from peripheral blood for more than 6 years after treatment. The likely explanation for the failure seems to be that these antibodies failed to delete already primed CD4 T cells secreting pro-inflammatory cytokines, and may thus have missed their target. This cautionary tale shows that it is possible to deplete large numbers of lymphocytes and yet completely fail to kill the cells that matter.

16-8 Biologics that block TNF-α, IL-1, or IL-6 can alleviate autoimmune diseases.

Anti-inflammatory therapy either can attempt to eliminate an autoimmune response altogether, as with immunosuppressive drugs or depleting anti-bodies, or it can try to reduce the tissue injury caused by the immune response. This second category of treatment is called **immunomodulatory therapy**, and is illustrated by the use of conventional anti-inflammatory agents such as aspirin, nonsteroidal anti-inflammatory drugs, or low-dose corticosteroids. A newer avenue of immunomodulatory therapy using biologics is illustrated by several FDA-approved antibodies that block the activity of powerful pro-inflammatory cytokines such as TNF-α, IL-1, and IL-6.

Anti-TNF therapy was the first specific biological therapeutic to enter the clinic. Anti-TNF-α antibodies induced striking remissions in rheumatoid arthritis (Fig. 16.9) and reduced inflammation in **Crohn's disease**, an inflammatory bowel disease (see Section 15-23). Two types of established biologics are used to antagonize TNF-α in clinical practice. The first type comprises the anti-TNF-α antibodies, such as infliximab and adalimumab, which bind to TNF-α and block its activity. The second type is a recombinant human TNF receptor (TNFR) subunit p75–Fc fusion protein called **etanercept**, which also binds TNF-α, neutralizing its activity. These are extremely potent anti-inflammatory agents, and the number of diseases in which they have been shown to be effective is growing as further clinical trials are performed. In addition to rheumatoid arthritis, the rheumatic diseases **ankylosing spondylitis**, **psoriatic arthropathy**, and juvenile idiopathic arthritis (other than the systemic-onset subset) respond well to blockade of TNF-α, and this treatment is now routine in many cases.

Rheumatoid Arthritis

Multiple Sclerosis

Crohn's Disease

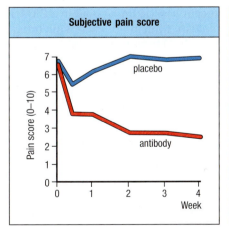

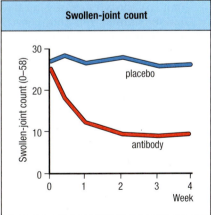

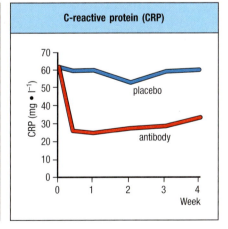

Fig. 16.9 Anti-inflammatory effects of anti-TNF-α therapy in rheumatoid arthritis. The clinical course of 24 patients was followed for 4 weeks after treatment with either a placebo or a monoclonal antibody against TNF-α at a dose of 10 mg • kg⁻¹. The antibody therapy was associated with a reduction in both subjective and objective parameters of disease activity (as measured by pain score and swollen-joint count, respectively) and in the systemic inflammatory acute-phase response, measured as a fall in the concentration of the acute-phase C-reactive protein. Data courtesy of R.N. Maini.

An illustration of the importance of TNF-α in defending against infection is the observation that TNF-α blockade carries a small but increased risk of patients developing serious infections, including tuberculosis (see Section 3-20). Anti-TNF-α therapy has not been successful in all diseases. TNF-α blockade as a treatment for **experimental autoimmune encephalomyelitis** (**EAE**), a mouse model of multiple sclerosis, led to amelioration of the disease, but in human patients with multiple sclerosis treated with anti-TNF-α, relapses became more frequent, possibly because of an increase in T-cell activation.

Antibodies and recombinant proteins against the pro-inflammatory cytokine IL-1 and its receptor have not proved as effective as TNF-α blockade for treating rheumatoid arthritis in humans, despite being equally powerful in animal models of arthritis. An antibody against the cytokine IL-1 has been licensed for clinical use against the hereditary autoinflammatory disease **Muckle–Wells syndrome** (see Section 13-9), and blockade of the IL-1β receptor by the recombinant protein **anakinra** (Kineret) has also proven useful in adults with moderate to severe rheumatoid arthritis.

Another cytokine antagonist in clinical use is the humanized antibody **tocilizumab**; by virtue of being directed against the IL-6 receptor, it blocks the effects of the pro-inflammatory cytokine IL-6. This seems to be as effective as anti-TNF-α in patients with rheumatoid arthritis and also shows promise in treating systemic-onset juvenile idiopathic arthritis, an autoinflammatory condition.

Interferon (IFN)-β (Avonex) is used to treat diseases of viral origin based on its ability to enhance immunity, but is also effective in treating multiple sclerosis, attenuating its course and severity and reducing the occurrence of relapses. Until recently, it was unclear how IFN-β could reduce rather than enhance immunity. In Section 3-9, we described the inflammasome, in which innate sensors of the NLR family activate caspase 1 to cleave the IL-1 pro-protein into the active form of the cytokine (see Fig. 3.19). We now know that IFN-β acts at two levels to reduce IL-1 production. It inhibits the activity of the NALP3 (NLRP3) and NLRP1 inflammasomes and also reduces expression of the IL-1 pro-protein, reducing the substrate available to caspase 1. Thus, IFN-β limits the production of a powerful pro-inflammatory cytokine, which may explain its observed effects on the symptoms of multiple sclerosis.

Hereditary Periodic Fever Syndromes

Systemic Juvenile Idiopathic Arthritis

16-9 Biologic agents can block cell migration to sites of inflammation and reduce immune responses.

Effector lymphocytes expressing the **integrin α₄:β₁** (**VLA-4**) bind to **VCAM-1** on endothelium in the central nervous system, while those expressing **integrin α₄:β₇** (lamina propria-associated molecule 1) bind to **MAdCAM-1** on the endothelium in the gut. The humanized monoclonal antibody **natalizumab** is specific for the α_4 integrin subunit and binds both VLA-4 and $\alpha_4{:}\beta_7$, preventing their interaction with their ligands (Fig. 16.10). This antibody has shown therapeutic benefit in placebo-controlled trials in patients with Crohn's disease or with multiple sclerosis. The early signs that this treatment could be successful illustrate the fact that disease depends on the continuing emigration of lymphocytes, monocytes, and macrophages from the circulation into the tissues of the brain in multiple sclerosis, and into the gut wall in Crohn's disease. However, blockade of $\alpha_4{:}\beta_1$ integrin is not specific and, like anti-TNF therapy, could lead to reduced defense against infection. Rarely patients treated with natalizumab have developed **progressive multifocal leukoencephalopathy** (PML), an opportunistic infection caused by the JC virus. This led to the temporary withdrawal of natalizumab from the market in 2005, but in June 2006 it was again allowed to be prescribed for multiple sclerosis and for Crohn's disease.

A similar problem with multifocal leukoencephalopathy led to the withdrawal from the market in the United States and Europe in 2009 of another

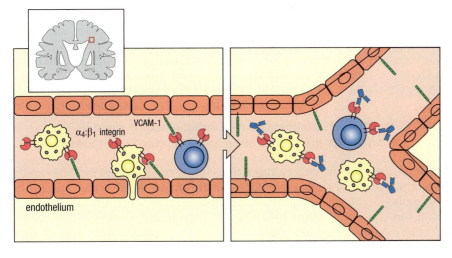

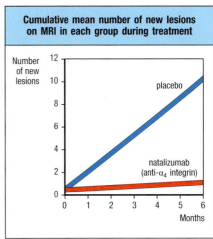

anti-integrin antibody, efalizumab; this drug targets the α_L subunit CD11a and had shown promise in treating **psoriasis**, an inflammatory skin disease driven primarily by T cells that produce pro-inflammatory cytokines.

16-10 Blockade of co-stimulatory pathways that activate lymphocytes can be used to treat autoimmune disease.

Blocking co-stimulatory pathways, as noted above in connection with the prevention of transplantation rejection (see Section 16-6), has also been applied to autoimmune diseases. For example, CTLA-4–Ig (abatacept) blocks the interaction of the B7 expressed by antigen-presenting cells with the CD28 expressed by T cells. This drug is approved for the treatment of rheumatoid arthritis, and also seems to be beneficial in treating psoriasis. When CTLA-4–Ig was given to patients with psoriasis, there was an improvement in the psoriatic rash and histological evidence of loss of activation of keratinocytes, T cells, and dendritic cells within the damaged skin.

Another co-stimulatory pathway that has been targeted in psoriasis is the interaction between the adhesion molecule CD2 on T cells and CD58 (LFA-3) on antigen-presenting cells. A recombinant CD58–IgG1 fusion protein, called **alefacept** (Amevive), inhibits the interaction between CD2 and CD58, and is now a routine and effective treatment for psoriasis. Although memory T cells are targeted by this therapy, responses to vaccination such as antitetanus remain intact.

16-11 Some commonly used drugs have immunomodulatory properties.

Certain existing medications, such as the statins and angiotensin-converting enzyme (ACE) inhibitors widely used in the prevention and treatment of cardiovascular disease, can also modulate the immune response in experimental animals. **Statins** are very widely prescribed drugs that block the enzyme **3-hydroxy-3-methylglutaryl-co-enzyme A (HMG-CoA) reductase**, thereby reducing cholesterol levels. They also reduce the increased level of expression of MHC class II molecules in some autoimmune diseases. These effects may be due to an alteration in the cholesterol content of membranes, thereby influencing lymphocyte signaling. In animal models, these drugs also seem to cause T cells to switch from a more pathogenic T_H1 response to a more protective T_H2 response, although whether this occurs in human patients is not clear.

The hormone vitamin D_3, essential for bone and mineral homeostasis, also exerts immunomodulatory effects. It decreases IL-12 production by dendritic cells and leads to a decrease in IL-2 and IFN-γ production by CD4 T cells, and

Fig. 16.10 Treatment with an anti-α_4 integrin humanized monoclonal antibody reduces relapses in multiple sclerosis. Left panel: interaction between α_4:β_1 integrin (VLA-4) on lymphocytes and macrophages and VCAM-1 expressed on endothelial cells permits the adhesion of these cells to brain endothelium. This facilitates the migration of these cells into the plaques of inflammation in multiple sclerosis. Center panel: the monoclonal antibody natalizumab (blue) binds to the α_4 chain of the integrin and blocks adhesive interactions between lymphocytes and monocytes and VCAM-1 on endothelial cells, thus preventing the cells from entering the tissue and exacerbating the inflammation. The future of this treatment is unclear because of the development of a rare infection as a side-effect (see the text). Right panel: the number of new lesions detected on magnetic resonance imaging (MRI) of the brain is greatly reduced in patients treated with natalizumab compared with a placebo. Data from Miller, D.H., *et al.*: *N. Engl. J. Med.* 2003, **348**:15–23.

protective effects have been demonstrated in a variety of animal models of auto-immunity, such as EAE (see Section 15-5) and diabetes, as well as in transplantation. The major drawback of vitamin D_3 is that its immunomodulatory effects are seen only at dosages that would lead to hypercalcemia and bone resorption in humans. There is a major search under way for structural analogs of vitamin D_3 that retain the immunomodulatory effects but do not cause hypercalcemia.

16-12 Controlled administration of antigen can be used to manipulate the nature of an antigen-specific response.

In some diseases, the target antigen of an unwanted immune response can be identified. It can then be possible to use the antigen itself, rather than drugs or antibodies, to treat the disease, because the manner of antigen presentation can alter the immune response and reduce or eliminate its pathogenic features. As discussed in Section 14-13, this principle has been applied with some success to the treatment of allergies caused by an IgE response to very low doses of antigen. Repeated treatment of allergic individuals with increasing doses of allergen seems to divert the allergic response to one dominated by T cells that favor the production of IgG and IgA antibodies from B cells. These antibodies are thought to desensitize the patient by binding the small amounts of allergen normally encountered and preventing it from binding to IgE.

There has been considerable interest in using peptide antigens to suppress pathogenic responses in T-cell-mediated autoimmune disease. The type of CD4 T-cell response induced by a peptide depends on the way in which it is presented to the immune system (see Section 9-18). For instance, peptides given orally tend to induce **regulatory T cells** through production of transforming growth factor (TGF)-β, but do not induce T_H1 cells or systemic antibody production (see Section 12-14). Indeed, experiments in animals indicate that oral antigens can protect against induced autoimmune disease. Diseases resembling multiple sclerosis or rheumatoid arthritis can be induced in mice by the injection of myelin basic protein (MBP) or collagen type II, respectively, in Freund's complete adjuvant (see Section 16-29). Oral administration of MBP or type II collagen inhibits the development of these diseases in animals, but oral administration of the whole protein in patients with multiple sclerosis or rheumatoid arthritis has had marginal therapeutic effects. Similarly, no protective effect was found in a large study that examined whether giving low-dose parenteral insulin could delay the onset of diabetes in people at high risk of developing the disease.

Multiple Sclerosis

Other approaches using antigen to shift the autoimmune T-cell response to a less damaging T_H2 response have been more effective in humans. The peptide drug glatiramer acetate (Copaxone) is approved for treating multiple sclerosis, in which it may reduce relapse rates by up to 30%. It is a polymer consisting of the four amino acids glutamic acid, alanine, tyrosine, and lysine in ratios that mimic their composition in MBP, and it induces a T_H2-type protective response. A more refined strategy uses **altered peptide ligands** (APLs), in which amino acid substitutions have been made in specific amino acids in an antigenic peptide that are at the T-cell receptor contact positions. APLs can be designed to act as partial agonists or antagonists, or to induce regulatory T cells. But despite the success of APLs in treating EAE in mice, a trial of these peptides for multiple sclerosis in some patients led to exacerbated disease or to allergic reaction associated with a vigorous T_H2 response, and their value in humans remains to be seen.

Summary.

Treatments for unwanted immune responses, such as graft rejection, auto-immunity, or allergic reactions, include conventional drugs—anti-inflammatory, cytotoxic, and immunosuppressive drugs—as well as biologic agents such as monoclonal antibodies and immunomodulatory proteins. Anti-inflammatory drugs, of which the most potent are the corticosteroids, have a broad spectrum of actions and a wide range of toxic side-effects. Their dose must be carefully controlled, and they are normally used in combination with either cytotoxic or immunosuppressive drugs. The cytotoxic drugs work by killing dividing cells and thereby prevent lymphocyte proliferation, but they suppress all immune responses indiscriminately and also kill other types of dividing cells. The immunosuppressive drugs, such as cyclosporin A and rapamycin, interfere in specific signaling pathways and are generally less toxic, but are more expensive and still suppress the immune response somewhat indiscriminately.

Several types of biologic agents are now established in the clinic for treating transplant rejection and autoimmune diseases (Fig. 16.11). Many monoclonal antibodies have been approved for human use that deplete lymphocytes either generally or selectively, or inhibit lymphocyte activation through receptor blockade, or prevent lymphocyte migration into tissues. Immunomodulatory agents also include monoclonal antibodies or fusion proteins that inhibit the inflammatory actions of TNF-α, a triumph of immunotherapy.

Therapeutic agents used to treat human autoimmune diseases				
Target	**Therapeutic agent**	**Disease**	**Disease outcome**	**Disadvantages**
Integrins	α_4:β_1 integrin-specific monoclonal antibody (mAb)	Relapsing/remitting multiple sclerosis (MS) Rheumatoid arthritis (RA) Inflammatory bowel disease	Reduction in relapse rate; delay in disease progression	Increased risk of infection; progressive multifocal encephalopathy
B cells	CD20-specific mAb	RA Systemic lupus erythematosus (SLE) MS	Improvement in arthritis, possibly in SLE	Increased risk of infection
HMG-coenzyme A reductase	Statins	MS	Reduction in disease activity	Hepatotoxicity; rhabdomyolysis
T cells	CD3-specific mAb	Type 1 diabetes mellitus	Reduced insulin use	Increased risk of infection
	CTLA-4-immuno-globulin fusion protein	RA Psoriasis MS	Improvement in arthritis	
Cytokines	TNF-specific mAb and soluble TNFR fusion protein	RA Crohn's disease Psoriatic arthritis Ankylosing spondylitis	Improvement in disability; joint repair in arthritis	Increased risk of tuberculosis and other infections; slight increase in risk of lymphoma
	IL-1 receptor antagonist	RA	Improves disability	Low efficacy
	IL-15-specific mAb	RA	May improve disability	Increased risk of opportunistic infection
	IL-6 receptor-specific mAb	RA	Decreased disease activity	Increased risk of opportunistic infection
	Type I interferons	Relapsing/remitting MS	Reduction in relapse rate	Liver toxicity; influenza-like syndrome is common

Fig. 16.11 New therapeutic agents for human autoimmunity. The immunosuppressive agents listed in Figs. 16.2 and 16.8 can act in one of three general ways. First (red), they can act by depleting cells from inflammatory sites, causing global cell-specific depletion, or blocking integrin interactions, thereby inhibiting lymphocyte trafficking. Second (blue), agents may block specific cellular interactions or inhibit various co-stimulatory pathways. Third (green), agents may target the terminal effector mechanisms, such as the neutralization of various pro-inflammatory cytokines.

Using the immune response to attack tumors.

Cancer is one of the three leading causes of death in industrialized nations, the others being infectious disease and cardiovascular disease. As treatments for infectious diseases and the prevention of cardiovascular disease continue to improve, and the average life expectancy increases, cancer is likely to become the most common fatal disease in these countries. Cancers are caused by the uncontrolled growth of the progeny of transformed cells. A major problem in treating cancer is controlling **metastasis**, or the spread of cancerous cells from one part of the body to unconnected parts. Curing cancer therefore requires that all the malignant cells be removed or destroyed without killing the patient. An attractive way of achieving this would be to induce an immune response against the tumor that would discriminate between the cells of the tumor and their normal-cell counterparts, in the same way that vaccination against a viral or bacterial pathogen induces a specific immune response that provides protection only against that pathogen. Immunological approaches to the treatment of cancer have been attempted for more than a century, but it is only in the past decade that immunotherapy of cancer has shown real promise. An important conceptual advance has been the integration of conventional approaches such as surgery or chemotherapy, which substantially reduce tumor load, with immunotherapy.

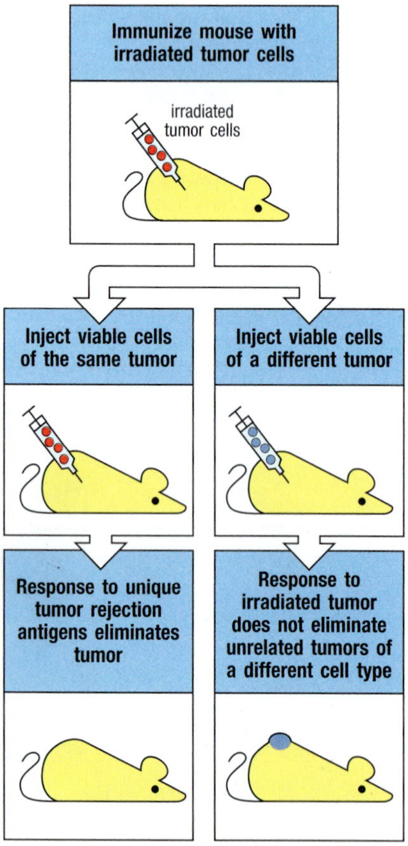

16-13 The development of transplantable tumors in mice led to the discovery of protective immune responses to tumors.

The finding that tumors could be induced in mice after treatment with chemical carcinogens or irradiation, coupled with the development of inbred strains of mice, made it possible to undertake the key experiments that led to the discovery of immune responses to tumors. These tumors could be transplanted between mice, and the experimental study of tumor rejection has generally been based on the use of such tumors. If they bear MHC molecules foreign to the mice into which they are transplanted, the tumor cells are readily recognized and destroyed by the immune system, a fact that was exploited to develop the first MHC-congenic strains of mice. Specific immunity to tumors must therefore be studied within inbred strains, so that host and tumor can be matched for their MHC type.

In mice, transplantable tumors exhibit a variable pattern of growth when injected into syngeneic recipients. Most tumors grow progressively and eventually kill the host. However, if mice are injected with irradiated tumor cells that cannot grow, they are frequently protected against subsequent injection with a normally lethal dose of viable cells of the same tumor (**Fig. 16.12**). There seems to be a spectrum of immunogenicity among transplantable tumors: injections of irradiated tumor cells seem to induce varying degrees of protective immunity against a challenge injection of viable tumor cells at a distant site. These protective effects are not seen in T-cell-deficient mice but can be conferred by adoptive transfer of T cells from immune mice, showing the need for T cells to mediate these effects.

These observations indicate that the tumors express antigens that can become targets of a tumor-specific T-cell response that rejects the tumor. These **tumor rejection antigens** are expressed by experimentally induced murine tumors (in which they are often termed tumor-specific transplantation antigens), and are usually specific for an individual tumor. Thus, immunization with irradiated tumor cells from one tumor usually protects a syngeneic mouse from challenge with live cells from that same tumor, but not from challenge with a different syngeneic tumor (see Fig. 16.12).

Fig. 16.12 Tumor rejection antigens are specific to individual tumors. Mice immunized with irradiated tumor cells and challenged with viable cells of the same tumor can, in some cases, reject a lethal dose of that tumor (left panels). This is the result of an immune response to tumor rejection antigens. If the immunized mice are challenged with viable cells of a different tumor, there is no protection and the mice die (right panels).

16-14 Tumors are 'edited' by the immune system as they evolve and can escape rejection in many ways.

Paul Ehrlich, who received the 1908 Nobel Prize for his work in immunology, was perhaps the first to propose that the immune system could be used to treat established tumors, suggesting that the molecules we call antibodies might be used to deliver toxins to cancer cells. In the 1950s, **Frank MacFarlane Burnet**, recipient of the 1960 Nobel Prize, and **Lewis Thomas** formulated the '**immune surveillance**' hypothesis, according to which cells of the immune system would detect and destroy tumor cells. Since then, it has become clear that the relationship between the immune system and cancer is considerably more complex, and this hypothesis has been modified to consider three phases of tumor growth. The first is the '**elimination phase**,' in which the immune system recognizes and destroys potential tumor cells—the phenomenon previously referred to as immune surveillance (Fig. 16.13). If elimination is not completely successful, what follows is an '**equilibrium phase**,' in which tumor cells undergo changes or mutations that aid their survival as a result of the selection pressure imposed by the immune system. During the equilibrium phase, a process known as **cancer immunoediting** continuously shapes the properties of the tumor cells that survive. In the final '**escape phase**,' tumor cells that have acquired the ability to elude the attentions of the immune system and grow unimpeded become clinically detectable.

Mice with targeted gene deletions or treated with antibodies to remove specific components of innate and adaptive immunity have provided the best evidence that immune surveillance influences the development of certain types of tumors. For example, mice lacking perforin, part of the killing mechanism of NK cells and CD8 cytotoxic T cells (see Section 9-31), show an increased frequency of lymphomas—tumors of the lymphoid system. Strains of mice lacking the RAG and STAT1 proteins, thus being deficient in both adaptive and certain innate immune mechanisms, develop gut epithelial and breast tumors. Mice lacking T lymphocytes expressing γ:δ receptors show markedly increased susceptibility to skin tumors induced by the topical application of carcinogens, illustrating a role for intraepithelial γ:δ T cells (see Section 6-20) in surveying and killing abnormal epithelial cells. Both IFN-γ and IFN-α/β are important in the elimination of tumor cells, either directly or indirectly through their actions on other cells. Studies of the various effector cells of the immune system show that γ:δ T cells are a major source of IFN-γ, which may explain their importance in the removal of cancer cells.

Fig. 16.13 Malignant cells can be controlled by immune surveillance. Some types of tumor cells are recognized by a variety of immune-system cells, which can eliminate them (left panel). If the tumor cells are not completely eliminated, variants occur that eventually escape the immune system and proliferate to form a tumor.

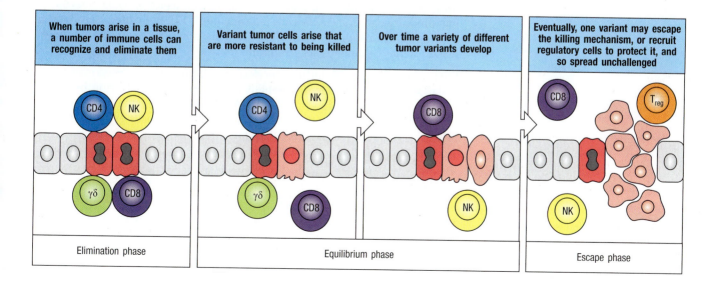

| When tumors arise in a tissue, a number of immune cells can recognize and eliminate them | Variant tumor cells arise that are more resistant to being killed | Over time a variety of different tumor variants develop | Eventually, one variant may escape the killing mechanism, or recruit regulatory cells to protect it, and so spread unchallenged |

Elimination phase | Equilibrium phase | Escape phase

According to the immunoediting hypothesis, those tumor cells that survive the equilibrium phase have acquired additional changes, either due to additional mutations or from selection during the equilibrium phase, that prevent their elimination by the immune system. In an immunocompetent individual, the equilibrium phase of the immune response continually removes tumor cells, delaying tumor growth; if the immune system is compromised, the equilibrium phase quickly turns into escape, as no tumor cells at all are removed. An excellent clinical example to support the presence of the equilibrium phase is the occurrence of cancer in recipients of organ transplants. One study reported the development of melanoma between 1 and 2 years after transplantation in two patients who had received kidneys from the same donor, a patient who had had successfully treated malignant melanoma 16 years before her death. Presumably, melanoma cells, which are known to spread easily to other organs, were present in the donor kidneys at the time of transplantation but were in equilibrium phase with the immune system. If so, this would indicate that the melanoma cells had not been killed off completely by the immunocompetent immune system of the donor, but instead had simply been held in check. Because the recipients' immune systems were immunosuppressed to prevent graft rejection, the melanoma cells were released from equilibrium and began to divide rapidly and spread to other parts of the body.

Acute Infectious Mononucleosis

Another situation in which the suppression of immunity can lead to tumor development is in **post-transplant lymphoproliferative disorder**, which can occur when patients are immunosuppressed after, for example, solid-organ transplantation. It usually takes the form of a B-cell expansion driven by Epstein–Barr virus (EBV) in which the B cells can undergo mutations and become malignant. Here, antiviral immunity functions as a form of cancer immunosurveillance, as it normally eliminates the EBV that leads to B-cell transformation.

Tumors can avoid stimulating an immune response or can evade it when it occurs by means of numerous mechanisms, which are summarized in Fig. 16.14. Spontaneous tumors may initially lack mutations that produce new tumor-specific antigens that elicit T-cell responses (see Fig. 16.14, first panel). And even when a tumor-specific antigen is expressed and is taken up and presented by antigen-presenting cells (APCs), if co-stimulatory signals are absent the antigen-presenting cell will tend to tolerize any antigen-specific naive T cells rather than activating them (see Fig. 16.14, second panel). How long such tumors are treated as 'self' is unclear. Recent sequencing of entire tumor genomes reveals that mutations may generate as many as 10–15 unique antigenic peptides that could be recognized as 'foreign' by T cells. In addition, cellular transformation is frequently associated with induction of MHC class Ib proteins (such as MIC-A and MIC-B) that are ligands for NKG2D, thus allowing tumor recognition by NK cells (see Section 6-17). But cancer cells tend to be genetically unstable, so that clones that are not recognized by an immune response may be able to escape elimination.

Some tumors, such as colon and cervical cancers, lose the expression of a particular MHC class I molecule, perhaps through immune selection by T cells specific for a peptide presented by that MHC class I molecule (see Fig. 16.14, third panel). In experimental studies, when a tumor loses expression of all MHC class I molecules (Fig. 16.15), it can no longer be recognized by cytotoxic T cells, although it might become susceptible to NK cells (Fig. 16.16). Tumors that lose only one MHC class I molecule might be able to avoid recognition by specific CD8 cytotoxic T cells while still remaining resistant to NK cells, conferring a selective advantage *in vivo*.

Tumors also seem to be able to evade immune attack by creating a microenvironment that is generally immunosuppressive (see Fig. 16.14, fourth panel). Many tumors make immunosuppressive cytokines. Transforming growth factor-β (TGF-β) was first identified in the culture supernatant of a

Mechanisms by which tumors avoid immune recognition

Low immunogenicity	Tumor treated as self antigen	Antigenic modulation	Tumor-induced immune suppression	Tumor-induced privileged site
No peptide:MHC ligand No adhesion molecules No co-stimulatory molecules	Tumor antigens taken up and presented by APCs in absence of co-stimulation tolerize T cells	T cells may eliminate tumors expressing immunogenic antigens, but not tumors that have lost such antigens	Factors (e.g.,TGF-β, IL-10, IDO) secreted by tumor cells inhibit T cells directly. Expression of PD-L1 by tumors	Factors secreted by tumor cells create a physical barrier to the immune system

Fig. 16.14 Tumors can avoid immune recognition in a variety of ways. First panel: tumors can have low immunogenicity. Tumors may lack antigens recognized by T cells, may have lost one or more MHC molecules, or may express inhibitory molecules such as PD-L1 that repress T-cell function. Second panel: tumor-specific antigens may be cross-presented by dendritic cells without co-stimulatory signals, inducing a tolerant state in T cells. Third panel: tumors can initially express antigens to which the immune system responds. Such tumors may be eliminated. The genetic instability of tumors allows antigenic change, part of an equilibrium phase, during which tumor cells lacking immunogenic antigens can expand. Fourth panel: tumors often produce molecules, such as TGF-β, IL-10, IDO, or PD-L1, that suppress immune responses directly or recruit regulatory T cells that can secrete immunosuppressive cytokines. Fifth panel: tumor cells can secrete molecules such as collagen that form a physical barrier around the tumor, preventing lymphocyte access.

tumor (hence its name), and, as we have seen, it tends to suppress the inflammatory T-cell responses and cell-mediated immunity that are needed to control tumor growth. Recall that TGF-β induces the development of inducible regulatory T cells (T$_{reg}$ cells; see Section 9-21), which have been found in a variety of cancers and might expand specifically in response to tumor antigens. In mouse models, removal of T$_{reg}$ cells increases resistance to cancer, whereas their transfer into a T$_{reg}$-negative recipient allows cancer cells to more greatly proliferate.

The microenvironments of some tumors also contain populations of myeloid cells, collectively called **myeloid-derived suppressor cells** (MDSCs), which can inhibit T-cell activation within the tumor. MDSCs may be heterogeneous, comprising both monocytic and polymorphonuclear cells, and are incompletely characterized at present. Several tumors of different tissue origins, such as melanoma, ovarian carcinoma, and B-cell lymphoma, have also been shown to produce the immunosuppressive cytokine IL-10, which can reduce dendritic cell activity and inhibit T-cell activation.

Some tumors express cell-surface proteins that directly inhibit immune responses (see Fig. 16.14, fourth panel). For example, some tumors express **programmed death ligand-1** (**PD-L1**), a B7 family member and ligand for the inhibitory receptor **PD-1** expressed by activated T cells (see Section 7-24). Furthermore, tumors can produce enzymes that act to suppress local immune responses. The enzyme **indoleamine 2,3-dioxygenase** (IDO) catabolizes tryptophan, an essential amino acid, in order to produce the immunosuppressive metabolite kynurenine. IDO seems to function in maintaining a balance between immune responses and tolerance during infections, but can be induced during the equilibrium phase of tumor development. Finally, tumor cells can produce materials such as collagen that create a physical barrier to interaction with cells of the immune system (see Fig. 16.14, last panel).

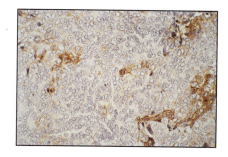

Fig. 16.15 Loss of MHC class I expression in a prostatic carcinoma. Some tumors can evade immune surveillance by a loss of expression of MHC class I molecules, preventing their recognition by CD8 T cells. Shown is a section of a human prostate cancer that has been stained with a peroxidase-conjugated antibody against HLA class I molecules. The brown stain that represents HLA class I expression is restricted to infiltrating lymphocytes and tissue stromal cells. The tumor cells that occupy most of the section show no staining. Photograph courtesy of G. Stamp.

Fig. 16.16 Tumors that lose expression of all MHC class I molecules as a mechanism of escape from immune surveillance are more susceptible to being killed by NK cells. Regression of transplanted tumors is largely due to the actions of cytotoxic T lymphocytes (CTLs), which recognize novel peptides bound to MHC class I antigens on the surface of the cell (left panels). NK cells have inhibitory receptors that bind MHC class I molecules, so variants of the tumor that have low levels of MHC class I, although less sensitive to CD8 cytotoxic T cells, become susceptible to NK cells (center panels). Although *nude* mice lack T cells, they have higher than normal levels of NK cells, and so tumors that are sensitive to NK cells grow less well in *nude* mice than in normal mice. Transfection with MHC class I genes can restore both resistance to NK cells and susceptibility to CD8 cytotoxic T cells (right panels). The bottom panels show scanning electron micrographs of NK cells attacking leukemia cells. The NK cell is the smaller cell on the left in both photographs. Left panel: shortly after binding to the target cell, the NK cell has put out numerous microvillous extensions and established a broad zone of contact with the leukemia cell. Right panel: 60 minutes after mixing, long microvillous processes can be seen extending from the NK cell (bottom left) to the leukemia cell and there is extensive damage to the leukemia cell; the plasma membrane has rolled up and fragmented. Photographs reprinted from Herberman, R., and Callewaert, D: *Mechanisms of Cytotoxicity by Natural Killer Cells*, 1985, with permission from Elsevier.

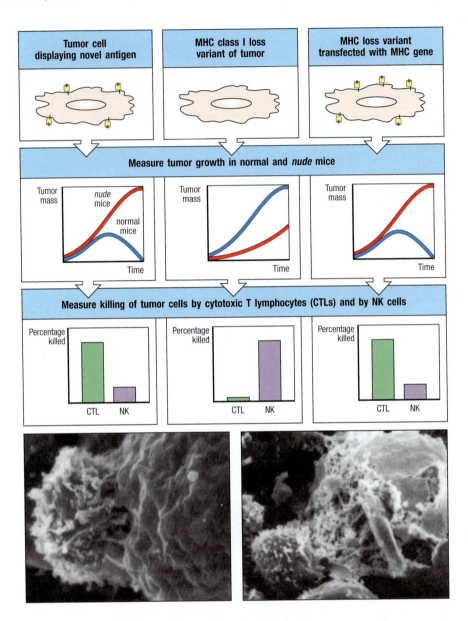

16-15 Tumor rejection antigens can be recognized by T cells and form the basis of immunotherapies.

The tumor rejection antigens recognized by the immune system are peptides of tumor-cell proteins that are presented to T cells on MHC molecules. These peptides become the targets of a tumor-specific T-cell response even though they can also be present on normal tissues. For instance, strategies to induce immunity to the relevant antigens in melanoma patients can induce vitiligo, an autoimmune destruction of pigmented cells in healthy skin. Several categories of tumor rejection antigens can be distinguished (**Fig. 16.17**). One consists strictly of **tumor-specific antigens** that result from point mutations or gene rearrangements that occur during oncogenesis and that affect a particular gene product. Point mutations in the gene for a particular protein may alter the epitope for T cells by altering the specific residues in a peptide that is already able to bind to MHC class I molecules, or they may allow some mutant peptides to bind *de novo* to MHC class I molecules. These peptides are often referred to as **neoepitopes** as they are newly immunogenic versions of normal proteins. Either change may evoke a new T-cell response against the tumor. In B- and T-cell tumors, which are derived from single clones of lymphocytes,

Potential tumor rejection antigens have a variety of origins			
Class of antigen	**Antigen**	**Nature of antigen**	**Tumor type**
Tumor-specific mutated oncogene or tumor suppressor gene	Cyclin-dependent kinase 4	Cell-cycle regulator	Melanoma
	β-Catenin	Relay in signal transduction pathway	Melanoma
	Caspase 8	Regulator of apoptosis	Squamous cell carcinoma
	Surface Ig/idiotype	Specific antibody after gene rearrangements in B-cell clone	Lymphoma
Cancer-testis antigens	MAGE-1 MAGE-3 NY-ESO-1	Normal testicular proteins	Melanoma Breast Glioma
Differentiation	Tyrosinase	Enzyme in pathway of melanin synthesis	Melanoma
Abnormal gene expression	HER-2/neu	Receptor tyrosine kinase	Breast Ovary
	WT1	Transcription factor	Leukemia
Abnormal post-translational modification	MUC-1	Underglycosylated mucin	Breast Pancreas
Abnormal post-transcriptional modification	NA17	Retention of introns in the mRNA	Melanoma
Oncoviral protein	HPV type 16, E6 and E7 proteins	Viral transforming gene products	Cervical carcinoma

Fig. 16.17 Proteins selectively expressed in human tumors are candidate tumor rejection antigens. The molecules listed here have all been shown to be recognized by cytotoxic T lymphocytes raised from patients with the tumor type listed.

a special class of tumor-specific antigen comprises the unique rearranged antigen receptor expressed by the clone. However, not all mutated peptides may be properly processed or be able to associate with MHC molecules and thus ensure that they stimulate an effective response.

The second category of tumor rejection antigens is the **cancer-testis antigens**. These are proteins encoded by genes that are normally expressed only in male germ cells in the testis. Male germ cells do not express MHC molecules, and therefore peptides from these molecules are not normally presented to T lymphocytes. Tumor cells show widespread abnormalities of gene expression, including the activation of genes encoding cancer-testis antigens, such as the **melanoma-associated antigens** (**MAGE**) (see Fig. 16.17). When expressed by tumor cells, peptides derived from these 'germ-cell' proteins can now be presented to T cells by MHC class I molecules. Therefore, these proteins are effectively tumor-specific in their expression as antigens. Perhaps the cancer-testis antigen best characterized immunologically is **NY-ESO-1** (New York esophageal squamous cell carcinoma-1), which is highly immunogenic and is expressed by a variety of human tumors, including melanomas.

The third category is the '**differentiation antigens**' encoded by genes that are expressed only in particular types of tissues. Examples of these are the differentiation antigens expressed in melanocytes and melanoma cells, which include several proteins in the pathways that produce the black pigment melanin, and the CD19 antigen expressed by B cells. The fourth category consists of antigens that are strongly overexpressed in tumor cells compared with their normal counterparts. An example is **HER-2/neu** (also known as c-Erb-2), which is a receptor tyrosine kinase homologous to the epidermal growth factor receptor. HER-2/neu is overexpressed in many adenocarcinomas, including breast and ovarian cancers, where it is associated with a poor clinical prognosis. CD8 T lymphocytes have been found infiltrating solid tumors overexpressing HER-2/neu but are not capable of destroying such tumors *in vivo*. The fifth category of tumor rejection antigens consists of molecules that display abnormal post-translational modifications. An example is underglycosylated mucin, MUC-1, which is expressed by several tumors, including breast and pancreatic cancers. The sixth category consists of novel proteins that are generated when one or more introns are retained in the mRNA transcribed from a gene, which occurs in melanoma. Proteins encoded by viral oncogenes comprise the seventh category of tumor rejection antigens. These oncoviral proteins can have a critical role in the oncogenic process and, because they are foreign, they can evoke a T-cell response. Examples are the human papilloma virus type 16 proteins E6 and E7, which are expressed in cervical carcinoma (see Section 16-18).

In melanoma, tumor-specific antigens were discovered by culturing irradiated tumor cells with autologous lymphocytes, a technique known as the mixed lymphocyte–tumor cell culture. From such cultures, cytotoxic T cells were identified that were reactive against melanoma peptides and would kill tumor cells bearing the relevant tumor-specific antigen. Such studies have revealed that melanomas carry at least five different antigens that can be recognized by cytotoxic T lymphocytes. It seems that cytotoxic T lymphocytes reactive against melanoma antigens are not effective *in vivo*, perhaps due to deficient priming and insufficient effector function, or to downstream resistance mechanisms. However, melanoma-specific T cells can be isolated from peripheral blood, lymph nodes, or directly from lymphocytes infiltrating the tumor and propagated *in vitro*. These T cells do not recognize the products of mutant proto-oncogenes or tumor suppressor genes, but instead recognize antigens derived from the protein products of other mutant genes or from normal proteins that are now displayed on tumor cells at levels detectable by T cells for the first time. Cancer-testis antigens such as the melanoma MAGE antigens discussed above probably represent early developmental antigens reexpressed in the process of tumorigenesis. Only a minority of melanoma patients have T cells reactive to the MAGE antigens, indicating that these antigens either are not expressed or are not immunogenic in most cases.

The most common melanoma antigens are peptides from the enzyme **tyrosinase** or from three other proteins—gp100, MART1, and gp75. These are differentiation antigens specific to the melanocyte lineage. It is likely that overexpression of these antigens in tumor cells leads to an abnormally high density of specific peptide:MHC complexes and it is this that makes them immunogenic. Although tumor rejection antigens are usually presented as peptides complexed with MHC class I molecules, the enzyme tyrosinase has also been shown to stimulate CD4 T-cell responses in some melanoma patients. This is likely because it is ingested and presented by cells expressing MHC class II molecules. Both CD4 and CD8 T cells are likely to be important in achieving immunological control of tumors. CD8 cells can kill the tumor cells directly, while CD4 T cells have a role in the activation of CD8 cytotoxic T cells and the establishment of memory. CD4 T cells may also kill tumor cells by means of the cytokines that they secrete, such as TNF-α.

Other potential tumor rejection antigens include the products of mutated cellular oncogenes or tumor suppressors, such as Ras and p53, and also fusion proteins, such as the **Bcr–Abl tyrosine kinase** that results from the chromosomal translocation (t9;22) found in chronic myeloid leukemia (CML). When present on CML cells, the HLA class I molecule HLA-A*0301 can display a peptide derived from the fusion site between Bcr and Abl. This peptide was detected by a powerful technique known as 'reverse immunogenetics,' in which endogenous peptides were eluted from the MHC binding groove and their sequence was determined by highly sensitive mass spectrometry.

This technique has identified HLA-bound peptides from other tumor antigens, such as the MART1 and gp100 tumor antigens of melanomas, as well as candidate peptide sequences for vaccination against infectious diseases.

T cells specific for the Bcr–Abl fusion peptide can be identified in peripheral blood from patients with CML by using tetramers of HLA-A*0301 carrying the peptide as specific ligands for the antigen-specific T-cell receptor (see Section 7-24). Cytotoxic T lymphocytes specific for this and other tumor antigens can be selected *in vitro* by using peptides derived from the mutated or fused portions of these oncogenic proteins; these cytotoxic T cells are able to recognize and kill tumor cells.

After a bone marrow transplant to treat CML, mature lymphocytes from the bone marrow donor infused into the patient can help to eliminate any residual tumor. This technique is known as **donor lymphocyte infusion** (DLI). At present, it is not clear to what extent the clinical response is due to a graft-versus-host effect, in which the donor lymphocytes are responding to general alloantigens expressed on the leukemia cells (see Section 15-36), or whether a specific antileukemic response is important. It is encouraging that it has been possible to separate T lymphocytes *in vitro* that mediate either a graft-versus-host effect or a graft-versus-leukemia effect. The ability to prime the donor cells against leukemia-specific peptides offers the prospect of enhancing the antileukemic effect while minimizing the risk of graft-versus-host disease.

16-16   T cells expressing chimeric antigen receptors are an effective treatment in some leukemias.

Adoptive T-cell therapy involves *ex vivo* expansion of tumor-specific T cells to large numbers and the infusion of those T cells into patients. Cells are expanded *in vitro* by various methods, such as treatment with IL-2, anti-CD3 antibodies, and allogeneic antigen-presenting cells to provide a co-stimulatory signal. Adoptive T-cell therapy is made more effective when the patient is immunosuppressed before treatment and IL-2 is then administered systemically. Another approach that has excited much interest is the use of retroviral vectors to transfer tumor-specific T-cell receptor (TCR) genes into patients' T cells before reinfusion. This can have long-lasting effects as a result of the ability of T cells to become memory cells, and there is no requirement for histocompatibility because the transfused cells are derived from the patient.

Another form of adoptive immunotherapy also uses retroviruses to introduce genes into a patient's T cells, but involves expressing a novel type of receptor, known as a **chimeric antigen receptor** (**CAR**). CARs are fusion receptors that contain extracellular antigen-specific domains fused to intracellular domains that provide signals for activation and co-stimulation. These receptors are introduced into T cells via retroviral vectors to produce so-called CAR T cells. This approach differs from conventional adoptive T-cell therapies as the use of a CAR allows the T cell's target specificity to be almost any molecule recognizable by an antibody rather than just peptide:MHC complexes. Recently this approach was used to target CD19 as a tumor rejection antigen in treating acute lymphocytic leukemia (ALL), an aggressive cancer of transformed B cells (Fig. 16.18). The CAR used in this case had an extracellular domain of an antibody that recognizes human CD19. The intracellular domain had three ITAMs from the ζ chain of the T-cell receptor CD3 complex (see Chapter 7) fused with a co-stimulation domain from 4-1BB, a member of the TNF receptor superfamily. These CART-19 transduced T cells were expanded *in vitro* and transferred into a patient. The results of this case combined with others have demonstrated that CD8 T cells expressing CART-19 (see Fig. 16.18) are effective at achieving complete clinical remissions in many patients with ALL. This approach is not without its side-effects, however, as it also eliminates normal B cells in patients and they therefore require treatment with IVIG.

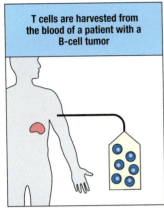

T cells are harvested from the blood of a patient with a B-cell tumor

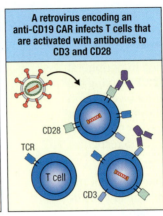

A retrovirus encoding an anti-CD19 CAR infects T cells that are activated with antibodies to CD3 and CD28

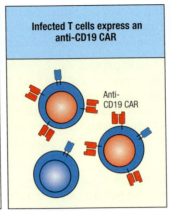

Infected T cells express an anti-CD19 CAR

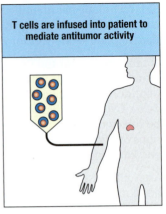

T cells are infused into patient to mediate antitumor activity

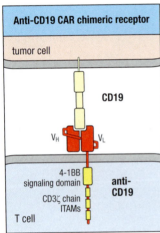

Anti-CD19 CAR chimeric receptor

Fig. 16.18 Chimeric antigen receptors (CARs) expressed in T cells can confer antitumor specificity to a patient's lymphocytes. Bottom panel: a chimeric antigen receptor, CART-19, is composed of an extracellular single-chain antibody that binds to CD19 that is fused to intracellular signaling domains from 4-1BB and the CD3ζ chain. Top row of panels: a lentivirus, a type of retrovirus, is used to express the gene encoding CART-19 in the T cells harvested from a patient diagnosed with ALL. After *in vitro* activation and expansion, the transfected CART-19 cells are infused into the patient and exert cytotoxic actions against CD19-expressing tumor cells as well as nontransformed B cells.

16-17   Monoclonal antibodies against tumor antigens, alone or linked to toxins, can control tumor growth.

Using monoclonal antibodies to destroy tumors requires that a tumor-specific antigen be expressed on the tumor's cell surface, so that antibody can direct the activity of a cytotoxic cell, toxin, or even radioactive nuclide specifically to the tumor (Fig. 16.19). Some of the cell-surface molecules targeted in clinical trials are shown in Fig. 16.20, and some of these treatments have now been licensed. Striking improvements in survival have been reported for breast cancer patients treated with the monoclonal antibody **trastuzumab** (Herceptin), which targets the receptor HER-2/neu. This receptor is overexpressed in about one-quarter of breast cancer patients and is associated with a poor prognosis.

Fig. 16.19 Monoclonal antibodies that recognize tumor-specific antigens have been used to help eliminate tumors. Tumor-specific antibodies of the correct isotypes can lyse tumor cells by recruiting effector cells such as NK cells and activating them via their Fc receptors (left panels). Another strategy has been to couple the antibody to a powerful toxin (center panels). When the antibody binds to the tumor cell and is endocytosed, the toxin is released from the antibody and can kill the tumor cell. If the antibody is coupled to a radioisotope (right panels), binding of the antibody to a tumor cell will deliver a dose of radiation sufficient to kill the tumor cell. In addition, nearby tumor cells could also receive a lethal radiation dose, even though they do not bind the antibody. Antibody fragments have started to replace whole antibodies for coupling to toxins or radioisotopes.

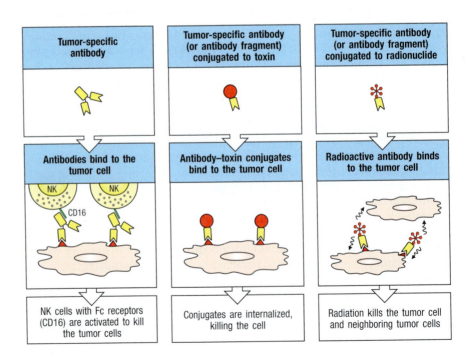

Tumor tissue origin	Type of antigen	Antigen	Tumor type
Lymphoma/ leukemia	Differentiation antigen	CD5 Idiotype CD52 (Campath-1H)	T-cell lymphoma B-cell lymphoma T- and B-cell lymphoma/ leukemia
	B-cell signaling receptor	CD20	Non-Hodgkin's B-cell lymphoma
Solid tumors	Cell-surface antigens Glycoprotein Carbohydrate	CEA, mucin-1 Lewis^y CA-125	Epithelial tumors (breast, colon, lung) Epithelial tumors Ovarian carcinoma
	Growth factor receptors	Epidermal growth factor receptor HER-2/neu IL-2 receptor Vascular endothelial growth factor (VEGF)	Lung, breast, and head and neck tumors Breast, ovarian tumors T- and B-cell tumors Colon cancer Lung, prostate, breast
	Stromal extracellular antigen	FAP-α Tenascin Metalloproteinases	Epithelial tumors Glioblastoma multiforme Epithelial tumors

Fig. 16.20 Examples of tumor antigens that have been targeted by monoclonal antibodies in therapeutic trials. CEA, carcinoembryonic antigen.

Herceptin is thought to act by blocking the binding of the natural ligand (so far unidentified) to this receptor, and by downregulating the level of expression of the receptor. The effects of this antibody can be enhanced when it is combined with conventional chemotherapy. Beyond blocking a growth signal for tumor cells, experiments in mice suggest that some of trastuzumab's antitumor effects also involve innate and adaptive immune responses, such as directing ADCC or inducing antitumor T-cell responses. A monoclonal antibody that has yielded excellent results in the treatment of non-Hodgkin's B-cell lymphoma is the anti-CD20 antibody **rituximab**, which triggers the apoptosis of B cells upon binding to CD20 on their surface (see Section 16-7). ADCC may be another mechanism by which rituximab acts, as its clinical efficacy has been linked to polymorphisms in activating Fc receptors.

Technical problems with monoclonal antibodies as therapeutic agents include inefficient killing of cells after binding the monoclonal antibody, inefficient penetration of the antibody into the tumor mass (which can be improved by using small antibody fragments), and soluble target antigens mopping up the antibody. The efficiency of killing can be enhanced by linking the antibody to a toxin, producing a reagent called an **immunotoxin** (see Fig. 16.19): two favored toxins are ricin A chain and *Pseudomonas* toxin. The antibody must be internalized to allow cleavage of the toxin from the antibody in the endocytic compartment, permitting the freed toxin chain to penetrate and kill the cell. Toxins coupled to native antibodies have had limited success in cancer therapy, but fragments of antibodies such as single-chain Fv molecules (see Section 4-3) show more promise. An example of a successful immunotoxin is a recombinant Fv anti-CD22 antibody fused to a fragment of Pseudomonas toxin, which induced complete remissions in two-thirds of a group of patients with a type of B-cell leukemia known as hairy-cell leukemia in whom the disease was resistant to conventional chemotherapy.

Monoclonal antibodies can also be conjugated to chemotherapeutic drugs or to radioisotopes. In the case of a drug-linked antibody, the binding of the antibody to a cell-surface antigen concentrates the drug to the site of the tumor. After internalization, the drug is released in the endosomes and exerts its cytostatic or cytotoxic effect. For example, the antibody trastuzumab has been linked to the cytotoxic agent mertansine, a drug that inhibits the assembly of microtubules, in a conjugate called **T-DM1**. Since HER-2 is overexpressed only in cancer cells, T-DM1 selectively delivers the toxin specifically to tumor cells.

Another drug–antibody conjugate, brentuximab vedotin, links an anti-CD30 antibody with a different microtubule inhibitor and is approved for certain forms of relapsed lymphomas.

A variation on this approach is to link an antibody to an enzyme that metabolizes a nontoxic pro-drug to the active cytotoxic drug, a technique known as **antibody-directed enzyme/pro-drug therapy** (**ADEPT**). With this technique, a small amount of enzyme localized by the antibody can generate much larger amounts of active cytotoxic drug in the immediate vicinity of tumor cells. Monoclonal antibodies linked to radioisotopes (see Fig. 16.19) have been successfully used to treat refractory B-cell lymphoma, using anti-CD20 antibodies linked to yttrium-90 (ibritumomab tiuxetan). These approaches have the advantage of also killing neighboring tumor cells, because the released drug or radioactive emissions can affect cells adjacent to those that bind the antibody. Monoclonal antibodies coupled to γ-emitting radioisotopes have also been used successfully to image tumors for the purposes of diagnosis and monitoring tumor spread.

16-18 Enhancing the immune response to tumors by vaccination holds promise for cancer prevention and therapy.

In addition to CAR T cells and monoclonal antibody-based therapies, there are two other major approaches to cancer immunotherapy. Cancer vaccines are based on the idea that tumors are intrinsically poorly immunogenic, and the vaccine should act to supply the immunogenicity. A second approach, called checkpoint blockade, which we will discuss in the next section, is based on the idea that the immune system has been primed but is held in check by tolerance mechanisms, which can be blocked by therapeutic interventions.

Many cancers are associated with viral infections, and vaccines that prevent these infections can reduce cancer risk. A major breakthrough in anticancer therapy occurred in 2005 with the completion of a clinical trial involving 12,167 women that tested a vaccine against human papilloma virus (HPV). This trial showed that a recombinant vaccine against HPV was 100% effective in preventing cervical cancer caused by two key HPV strains, HPV-16 and HPV-18, which are associated with 70% of cervical cancers. The vaccine most likely prevents infection of cervical epithelium by HPV through the induction of anti-HPV antibodies (**Fig. 16.21**). Although this trial showed the potential of vaccines to prevent cancer, attempts to use vaccines to treat existing tumors have been less effective. In the case of HPV, certain types of vaccines that have increased immunogenicity for eliciting T-cell responses are beginning to show effectiveness in treating existing intraepithelial neoplasia caused by the virus. Similarly, the majority of liver cancers are associated with chronic hepatitis caused by several viruses. The vaccine against hepatitis B can reduce primary liver cancer due to this virus, although it will not protect against cancers caused by infections by other viruses such as hepatitis C.

Vaccines based on tumor antigens are, in principle, the ideal approach to T-cell-mediated cancer immunotherapy, but they are difficult to develop. For HPV, the relevant antigens are known. For most spontaneous tumors, however, relevant peptides of tumor rejection antigens may not be the same in different patients' tumors and may be presented only by particular MHC alleles. This means that an effective tumor vaccine must include a range of tumor antigens. It is also clear that cancer vaccines for therapy should be used only when the tumor burden is low, such as after adequate surgery and chemotherapy.

The sources of antigens for cell-based cancer vaccines are the individual patients' tumors removed at surgery. These vaccines are prepared by mixing either irradiated tumor cells or tumor extracts with killed bacteria such as Bacille Calmette–Guérin (BCG) or *Corynebacterium parvum*, which act

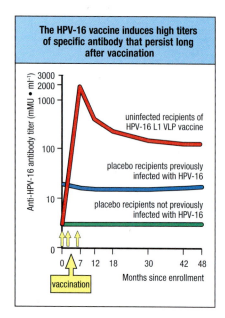

The HPV-16 vaccine induces high titers of specific antibody that persist long after vaccination

Anti-HPV-16 antibody titer (mMU • ml⁻¹) axis: 3000, 2000, 1000, 100, 10, 0

uninfected recipients of HPV-16 L1 VLP vaccine

placebo recipients previously infected with HPV-16

placebo recipients not previously infected with HPV-16

0 7 12 18 30 42 48
Months since enrollment

vaccination

Fig. 16.21 An effective vaccine against human papilloma virus (HPV) induces antibodies that protect against HPV infection. Serotype 16 of HPV (HPV-16) is highly associated with the development of cervical cancer. In a clinical trial, 755 healthy uninfected women were immunized with a vaccine generated from highly purified noninfectious 'viruslike particles' (VLP) consisting of the capsid protein L1 of HPV-16 and formulated with an alum adjuvant (in this case aluminum hydroxyphosphate sulfate). In comparison with the very low titers of antibody in placebo-treated uninfected women (green line), or women previously infected with HPV who received placebo (blue line), the women treated with the viruslike particle vaccine (red line) developed high titers of antibody against the L1 capsid protein. None of these immunized women subsequently became infected by HPV-16. An anti-HPV vaccine marketed as Gardasil is now available and recommended for use in girls and young women as a protection from cervical cancer caused by HPV serotypes 6, 11, 16, and 18. mMU, milli-Merck units.

as adjuvants to enhance the immunogenicity of the tumor antigens (see Appendix I, Section A-41). Vaccination using BCG adjuvants has had variable results in the past, but renewed interest is based on recent appreciation of their interaction with Toll-like receptors (TLRs). Stimulation of TLR-4 by BCG and other ligands has been tested against melanoma and other solid tumors. CpG DNA, which binds to TLR-9, has also been used to increase the immunogenicity of cancer vaccines. In cases where candidate tumor rejection antigens have been identified, for example, in melanomas, experimental vaccination strategies have included the use of whole proteins, peptide vaccines based on sequences recognized by cytotoxic T lymphocytes and helper T lymphocytes (either administered alone or presented by the patient's own dendritic cells), and recombinant viruses encoding these peptide epitopes.

The potency of dendritic cells in activating T-cell responses provides the rationale for yet another antitumor vaccination strategy. The use of antigen-loaded dendritic cells to stimulate therapeutically useful cytotoxic T-cell responses to tumors has now undergone clinical trials in humans with cancer. One such vaccine, **sipuleucel-T** (**Provenge**), was recently approved for treatment of metastatic prostate cancer. In this therapy, a patient's monocytes are extracted from peripheral blood and cultured with a fusion protein containing the antigen **prostatic acid phosphatase** (PAP), which is expressed by most prostate cancers, and the cytokine granulocyte–macrophage colony-stimulating factor (GM-CSF), which induces monocytes to undergo maturation to monocyte-derived dendritic cells. The resulting cells are reinfused into the patient to induce an immune response specific to the PAP antigen. This treatment reduced the risk of death by 22% and improved survival by about 4 months relative to a placebo group. Other methods in clinical trials include loading dendritic cells *ex vivo* with DNA encoding the tumor antigen or with mRNA derived from tumor cells, and the use of apoptotic or necrotic tumor cells as sources of antigens.

16-19 Checkpoint blockade can augment immune responses to existing tumors.

Other approaches to tumor immunotherapy attempt to strengthen the natural immune responses against a tumor by one of two approaches: by making the tumor itself more immunogenic, or by relieving the normal inhibitory mechanisms that regulate these responses. The first kind of approach has been explored by inducing the expression of co-stimulatory molecules, such as B7 molecules, on tumor cells and then using these cells to activate tumor-specific naive T cells. Similarly, tumor cells may be transfected with the gene encoding granulocyte–macrophage colony-stimulating factor (GM-CSF) in order to induce the maturation of tumor-proximal monocytes into monocyte-derived dendritic cells. Once these cells have differentiated and capture antigens from the tumor, they may migrate to the local lymph nodes and activate tumor-specific T cells. No approved therapies have yet emerged from these approaches. In mice, B7-transfected cells seem less potent than the monoctye-derived dendritic cells differentiated by GM-CSF in inducing antitumor responses. This may be because molecules in addition to B7 act in priming naive T cells and these molecules may be expressed only by specific types of cross-presenting dendritic cells.

Another approach to cancer immunotherapy is called **checkpoint blockade**, which attempts to interfere with the normal inhibitory signals that regulate lymphocytes. Immune responses are controlled by several positive and negative immunological checkpoints. A positive checkpoint for T cells is controlled by the B7 co-stimulatory receptors expressed by professional antigen-presenting cells such as dendritic cells, as discussed earlier. Negative immunological checkpoints are provided by inhibitory receptors such as CTLA-4 and PD-1.

CTLA-4 imposes a critical checkpoint for potentially autoreactive T cells by binding to B7 molecules on dendritic cells and delivering a negative signal that must be overcome by other signals before T cells can become activated. Blocking CTLA-4 with antibodies may therefore lower the threshold for T-cell activation. Some evidence also indicates that anti-CTLA-4 antibodies may augment immune responses by eliminating regulatory T cells, which express CTLA-4 on their surface. Whatever the mechanism, the absence of this checkpoint causes self-reactive T cells that are normally held in check to become activated instead and produce a multi-tissue autoimmune reaction, as seen in CTLA-4-deficient mice.

Since checkpoint blockade relies on activation of the patient's own immune system against tumors, its effects are not immediately evident, presenting a challenge in evaluating clinical responses to such therapy. Guidelines for evaluating clinical responses were based on the immediate effects of chemotherapeutic drugs or radiation, whereas checkpoint blockade requires more time for reversing immune inhibition and expanding tumor-specific T cells that then exert their effects within the tumor. Once such issues were considered, it became possible to design clinical trials that could document the effects of checkpoint blockade used in combination with traditional anticancer therapies.

Checkpoint blockade based on the anti-CTLA-4 antibody **ipilimumab** has now been shown to be effective in treating metastatic melanoma and recently received FDA approval for this indication. Patients with metastatic melanoma who were treated with ipilimumab showed an increase in the number and activity of T cells recognizing NY-ESO-1, a cancer-testis antigen expressed by melanoma. Overall only about 15% of patients exhibited a response to ipilimumab, but the treatment appeared to induce long-term remission in responding patients. One side-effect of ipilimumab in these patients seemed to be an increased risk of autoimmune phenomena, in agreement with the role of CTLA-4 in maintaining tolerance of self-reactive T cells.

Another checkpoint involves the inhibitory receptor PD-1 and its ligands PD-L1 and PD-L2. PD-L1 is expressed on a wide variety of human tumors; in renal cell carcinoma, PD-L1 expression is associated with a poor prognosis. In mice, transfection of the gene encoding PD-L1 into tumor cells increased their growth *in vivo* and reduced their susceptibility to lysis by cytotoxic T cells. These effects were reversed by an antibody against PD-L1. In humans, the anti-PD-1 antibody **pembrolizumab** has been shown to be effective in previously treated melanoma patients, giving a nearly 30% response rate. It is FDA-approved for use following treatment with ipilimumab, or in patients with a BRAF mutation after treatment with ipilimumab and a B-raf inhibitor. Another anti-PD-1 antibody, **nivolumab**, is also approved for treatment of metastatic melanoma and is being considered for use in treatment of Hodgkin's lymphoma. Ongoing clinical trials are evaluating checkpoint blockade using antibodies to PD-L1 and PD-L2.

Summary.

Some tumors elicit specific immune responses that suppress or modify their growth. Tumors evade or suppress these responses in several ways, passing through various stages of a process known as immunological editing. Understanding how the immune system promotes and prevents cancer growth has led to new therapies now deployed in the clinic. The possibility of eradicating cervical cancer, for example, has been brought closer by the development of an effective vaccine against specific strains of cancer-causing human papilloma virus. Monoclonal antibodies have also been successfully developed for tumor immunotherapy in several cases, such as an anti-CD20 antibody used to treat B-cell lymphoma. Attempts are also being made to develop vaccines incorporating peptides designed to generate effective cytotoxic and helper T-cell responses. CAR T cells engineered to recognize CD19 expressed on B cells can be an effective treatment for acute lymphocytic leukemia. Checkpoint blockade strategies for CTLA-4 and PD-1 have been approved for treating melanoma, and related strategies are being developed for other biologic targets to stimulate antitumor immune responses or block inhibitory mechanisms that suppress such responses. One vaccine using dendritic cells that present tumor antigens has been approved for treating prostate cancer. A current trend in cancer therapy has been to incorporate immunotherapy with other traditional anticancer treatments to take advantage of the specificity and power of the immune system.

Fighting infectious diseases with vaccination.

The two most important contributions to public health in the past 100 years—sanitation and vaccination—have markedly decreased deaths from infectious disease, and yet infectious diseases remain the leading cause of death worldwide. Modern immunology itself grew from the success of **Edward Jenner's** and **Louis Pasteur's** vaccines against smallpox and chicken cholera, respectively, and its greatest triumph has been the global eradication of smallpox, announced by the World Health Organization in 1979. A global campaign to eradicate polio is now well under way. With the past decade's tremendous progress in basic immunology, particularly in understanding innate immunity, there is now great hope that vaccines for other major infectious diseases, including malaria, tuberculosis, and HIV, are within reach. The vision of the current generation of vaccine scientists is to elevate their art to the level of modern drug design; to move it from an empiric practice to a true 'pharmacology of the immune system.'

The goal of vaccination is the generation of long-lasting and protective immunity. Throughout this book, we have illustrated how the innate and the adaptive immune systems collaborate in the face of infection to eliminate pathogens and generate protective immunity with immunological memory. Indeed, a single infection is often (but not always) sufficient to generate protective immunity to a pathogen. Recognition of this important relationship was recorded more than 2000 years ago in accounts of the Peloponnesian War, during which two successive outbreaks of plague struck Athens. The Greek historian Thucydides noted that people who had survived infection during the first outbreak were not susceptible to infection during the second.

The recognition of this type of relationship perhaps prompted the practice of **variolation** against smallpox, in which an inoculation of a small amount of dried material from a smallpox pustule was used to produce a mild infection that was then followed by long-lasting protection against reinfection. Smallpox itself has been recognized in medical literature for more than 1000 years; variolation seems to have been practiced in India and China many centuries before its introduction into the West (some time in the 1400s–1500s), and it was familiar to Jenner. However, infection after variolation was not always mild: fatal smallpox ensued in about 3% of cases, which would not meet modern criteria of drug safety. It seems there was some recognition that milkmaids exposed to a bovine virus similar to smallpox—cowpox—seemed protected from smallpox infection, and there is even one historical account suggesting that cowpox inoculation had been tried before Jenner. However, Jenner's achievement was not only the realization that infection with cowpox would provide protective immunity against smallpox in humans without the risk of significant disease, but its experimental proof by the intentional variolation of people whom he had previously vaccinated. He named the process **vaccination** (from *vacca*, Latin for cow), and Pasteur, in his honor, extended the term to the stimulation of protection against other infectious agents. Humans are not a natural host of cowpox, which establishes only a brief and limited subcutaneous infection. But the cowpox virus contains antigens that stimulate an immune response that cross-reacts with smallpox antigens and thereby confers protection against the human disease. Since the early 20th century, the virus used to vaccinate against smallpox has been vaccinia virus, which is related to both cowpox and smallpox, but whose origin is obscure.

As we will see, many current vaccines offer protection by inducing the formation of neutralizing antibodies. However, that statement contains a hidden tautology; pathogens for which current vaccines are effective may also be pathogens for which antibodies are sufficient for protection. For several major pathogens—malaria, tuberculosis, and HIV—even a robust antibody response is not fully protective. The elimination of these pathogens requires additional effector activities, such as the generation of strong and durable cell-mediated immunity, which are not efficiently generated by current vaccine technologies. These are the issues that face modern vaccine scientists.

16-20 Vaccines can be based on attenuated pathogens or material from killed organisms.

Vaccine development in the early part of the 20th century followed two empirical approaches. The first was the search for **attenuated** organisms with reduced pathogenicity, which would stimulate protective immunity but not cause disease. This approach continues into the present with the design of genetically attenuated pathogens in which desirable mutations are introduced into the organism by recombinant DNA technologies. This idea is being applied to important pathogens, such as malaria, for which vaccines are currently unavailable, and may be important in the future for designing vaccines for influenza and HIV.

The second approach was the development of vaccines based on killed organisms and, subsequently, on purified components of organisms that would be as effective as live whole organisms. Killed vaccines were desirable because any live vaccine, including vaccinia, can cause lethal systemic infection in immunosuppressed people. Evolving from this approach were vaccines based on the conjugation of purified antigens, as described for *Haemophilus influenzae* (see Section 16-27). This approach continues with the addition of 'reverse immunogenetics' (see Section 16-15) to identify candidate peptide antigens for T cells and with strategies to use ligands that activate TLRs or other innate sensors as adjuvants to enhance responses to simple antigens.

Immunization with such approaches is now considered so safe and so important that most states in the United States require all children to be immunized against several potentially deadly diseases. These include the the viral diseases measles, mumps, and polio, for which live-attenuated vaccines are used, as well as against tetanus (caused by *Clostridium tetani*), diphtheria (caused by *Corynebacterium diphtheriae*), and whooping cough (caused by *Bordetella pertussis*), for which vaccines composed of inactivated toxins or toxoids prepared from the respective bacteria are used. More recently, a vaccine has become available against *H. influenzae* type b (HiB), one of the causative agents of meningitis, as well as two vaccines for childhood diarrhea caused by rotaviruses, and, as described in Section 16-18, a vaccine for preventing HPV infection for protection against cervical cancer. Most vaccines are given to children within the first year of life. The vaccines against measles, mumps, and rubella (MMR), against chickenpox (varicella), and against influenza, when recommended, are usually given between the ages of 1 and 2 years.

Impressive as these accomplishments are, there are still many diseases for which we lack effective vaccines (Fig. 16.22). For many pathogens, natural infection does not seem to generate protective immunity, and infections

Some infections for which effective vaccines are not yet available	
Disease	**Estimated annual mortality**
Malaria	618,248
Schistosomiasis	21,797
Intestinal worm infestation	3304
Tuberculosis	934,879
Diarrheal disease	1,497,724
Respiratory infections	3,060,837
HIV/AIDS	1,533,760
Measles*	130,461

Fig. 16.22 Diseases for which effective vaccines are still needed. *Current measles vaccines are effective but heat-sensitive, which makes their use difficult in tropical countries; heat stability is being improved. Mortality data are the most recent estimated figures available (2014). *Global Health Estimates 2000–2012.* World Health Organization, June 2014.

become chronic or recurrent. In many infections of this type, such as malaria, tuberculosis, and HIV, antibodies are insufficient to prevent reinfection and to eliminate the pathogen, and cell-mediated immunity instead seems to be more important in limiting the pathogen, but alone is still insufficient to provide full immunity. It is not the absence of an immune response to the pathogen that is the problem, but rather that this response does not clear the pathogen, eliminate pathogenesis, or prevent reinfection.

Another obstacle is that even when a vaccine such as that against measles can be used effectively in developed countries, technical and economic problems can prevent its widespread use in developing countries, where mortality from these diseases is still high. For example, simple costs of storage and deployment can be significant barriers to the use of existing vaccines in poorer countries. Therefore, the development of vaccines therefore remains an important goal of immunology, and the latter half of the 20th century saw a shift to a more rational approach based on a detailed molecular understanding of microbial pathogenicity, analysis of the protective host response to pathogenic organisms, and an understanding of the regulation of the immune system to generate effective T- and B-lymphocyte responses.

16-21 Most effective vaccines generate antibodies that prevent the damage caused by toxins or that neutralize the pathogen and stop infection.

Although the requirements for generating protective immunity vary with the nature of the infecting organism, many effective vaccines currently work by inducing antibodies against the pathogen. For many pathogens, including extracellular organisms and viruses, antibodies can provide protective immunity. This is not the case for all pathogens, unfortunately; some may require additional cell-mediated immune responses such as those mediated by CD8 T cells.

Effective protective immunity against some microorganisms requires the presence of preexisting antibody at the time of infection, either to prevent the damage caused by the pathogen or to prevent reinfection by the pathogen altogether. The first case is illustrated by vaccines against tetanus and diphtheria, whose clinical manifestations of infection are due to the effects of extremely powerful exotoxins (see Fig. 10.31). Preexisting antibody against the exotoxin is necessary to provide a defense against these diseases. Indeed, the tetanus exotoxin is so powerful that the tiny amount that can cause disease may be insufficient to lead to a protective immune response; consequently, even survivors of tetanus require vaccination to be protected against the risk of subsequent attack.

The second way in which antibodies can protect is by preventing infection a second time by the same pathogen, as in the case of certain viral infections. While CD8 T cells are able to kill already virally infected cells during an infection, antibodies are able to prevent infection of cells by the virus in the first place. This action is called **neutralization**. The ability of an antibody to neutralize a pathogen may depend on its affinity, its isotype subclass, complement, and the activity of phagocytic cells. For example, preexisting antibodies are required to protect against the polio virus, which infects critical host cells within a short period after entering the body and is not easily controlled by T lymphocytes once intracellular infection has been established. Vaccines against seasonal influenza virus provide protection in this same manner, by inducing antibodies that reduce the chance of a second infection by the same strain of influenza. For many viruses, antibodies produced by an infection or by vaccination can neutralize the virus and prevent further spread of infection, but this is not always the case. In HIV infection, despite the generation of antibodies that can bind to surface viral epitopes, most of these antibodies fail to neutralize the virus. In addition, HIV has many different strains, or clades, and most vaccines based on HIV proteins do not induce antibodies that neutralize all clades, presenting a challenge for effective vaccine design. However, a recent clinical trial suggests that boosting previously vaccinated subjects with protein 5–7 years after immunization may induce some antibodies with cross-clade activity.

Immune responses to infectious agents usually involve antibodies directed at multiple epitopes, and only some of these antibodies, if any, confer protection. The particular T-cell epitopes recognized can also affect the nature of the response. In Section 10-2, we described **linked recognition**, in which antigen-specific B cells and T cells provide mutually activating signals, leading to affinity maturation and isotype switching that may be required for neutralization. This process requires that an appropriate peptide epitope for T cells be presented by the B cells, and typically that the T-cell epitope be contained within the region of protein epitope recognized by the B cell, a fact that must be considered in modern vaccine design. Indeed, the predominant epitope recognized by T cells after vaccination with respiratory syncytial virus induces a vigorous inflammatory response but fails to elicit neutralizing antibodies and thus causes pathology without protection.

16-22 Effective vaccines must induce long-lasting protection while being safe and inexpensive.

A successful vaccine must possess several features in addition to its ability to provoke a protective immune response (**Fig. 16.23**). First, it must be safe. Vaccines must be given to huge numbers of people, relatively few of whom are likely to die of, or sometimes even catch, the disease that the vaccine is designed to prevent. This means that even a low level of toxicity is unacceptable. Second, the vaccine must be able to produce protective immunity in a very high proportion of the people to whom it is given. Third, particularly in poorer countries where it is impracticable to give regular 'booster' vaccinations to dispersed rural populations, a successful vaccine must generate long-lived immunological memory. This means that the vaccine must prime both B and T lymphocytes. Fourth, vaccines must be very cheap if they are to be administered to large populations. Vaccines are one of the most cost-effective measures in health care, but this benefit is eroded as the cost per dose rises.

Another benefit of an effective vaccination program is the '**herd immunity**' that it confers on the general population. By lowering the number of susceptible members of a population, vaccination decreases the natural reservoir of infected individuals in that population and so reduces the probability of transmission of infection. Thus, even unvaccinated members will be protected because their individual chance of encountering the pathogen is decreased. However, the herd immunity effect is seen only at relatively high levels of vaccination within a population; for mumps, the necessary level is estimated to be around 80%, and below this level sporadic epidemics can occur. This is illustrated by a marked increase in mumps in the United Kingdom in 2004-2005 in young adults as a result of the variable use in the mid-1990s of a measles/rubella vaccine rather than the combined MMR, as the combined vaccine was in short supply at that time.

16-23 Live-attenuated viral vaccines are usually more potent than 'killed' vaccines and can be made safer by the use of recombinant DNA technology.

Most antiviral vaccines currently in use consist of either live attenuated or inactivated viruses. Inactivated, or 'killed,' viral vaccines consist of viruses treated so that they are unable to replicate. Inactivated viruses therefore

Features of effective vaccines	
Safe	Vaccine must not itself cause illness or death
Protective	Vaccine must protect against illness resulting from exposure to live pathogen
Gives sustained protection	Protection against illness must last for several years
Induces neutralizing antibody	Some pathogens (such as polio virus) infect cells that cannot be replaced (e.g., neurons). Neutralizing antibody is essential to prevent infection of such cells
Induces protective T cells	Some pathogens, particularly intracellular, are more effectively dealt with by cell-mediated responses
Practical considerations	Low cost per dose Biological stability Ease of administration Few side-effects

Fig. 16.23 There are several criteria for an effective vaccine.

cannot produce proteins in the cytosol of infected cells, so peptides from the viral antigens are not presented by MHC class I molecules. Thus, CD8 T cells are neither efficiently generated nor needed with killed virus vaccines. Live-attenuated viral vaccines are generally far more potent: they elicit a greater number of effector mechanisms, including the activation of CD4 T cells and cytotoxic CD8 T cells. CD4 T cells help in shaping the antibody response, which is important for a vaccine's subsequent protective effect. Cytotoxic CD8 T cells provide protection while infection by the virus itself is under way, and, if maintained, may contribute to protective memory. Attenuated viral vaccines include the routine childhood vaccines in use for polio, measles, mumps, rubella, and varicella. Other attenuated live viral vaccines that are licensed for special circumstances or for use in high-risk populations include influenza, poxvirus (vaccinia), and yellow fever virus.

Traditionally, attenuation is achieved by growing the virus in cultured cells. Viruses are usually selected for preferential growth in nonhuman cells and, in the course of selection, become less able to grow in human cells (Fig. 16.24). Because these attenuated strains replicate poorly in human hosts, they induce immunity but not disease. Although attenuated virus strains contain multiple mutations in genes encoding several of their proteins, it might be possible for a pathogenic virus strain to reemerge by a further series of mutations. For example, the type 3 Sabin polio vaccine strain differs from a wild-type progenitor strain at only 10 of 7429 nucleotides. On extremely rare occasions, reversion of the vaccine to a neurovirulent strain can occur, causing paralytic disease in the unfortunate recipient.

Attenuated viral vaccines can also pose particular risks to immunodeficient recipients, in whom they often behave as virulent opportunistic infections. Immunodeficient infants who are vaccinated with live attenuated polio virus before their inherited immunoglobulin deficiencies have been diagnosed are at risk because they cannot clear the virus from their gut, and there is therefore an increased chance that mutations associated with the continuing uncontrolled replication of the virus in the gut will revert the virus to a virulent form and lead to fatal paralytic disease.

An empirical approach to attenuation is still in use but might be superseded by two new approaches that use recombinant DNA technology. One is the isolation and *in vitro* mutagenesis of specific viral genes. The mutated genes are used to replace the wild-type genes in a reconstituted virus genome, and this

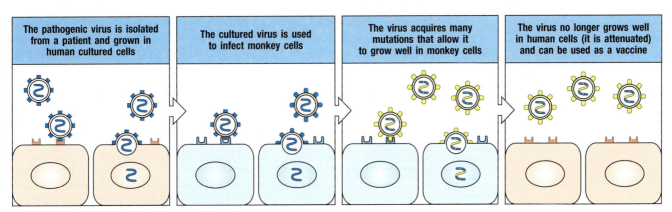

| The pathogenic virus is isolated from a patient and grown in human cultured cells | The cultured virus is used to infect monkey cells | The virus acquires many mutations that allow it to grow well in monkey cells | The virus no longer grows well in human cells (it is attenuated) and can be used as a vaccine |

Fig. 16.24 Viruses are traditionally attenuated by selecting for growth in nonhuman cells. To produce an attenuated virus, the virus must first be isolated by growing it in cultured human cells. The adaptation to growth in cultured human cells can cause some attenuation in itself; the rubella vaccine, for example, was made in this way. In general, however, the virus is then adapted to growth in cells of a different species, until it grows only poorly in human cells. The adaptation is a result of mutation, usually a combination of several point mutations. It is usually difficult to tell which of the mutations in the genome of an attenuated viral stock are critical to attenuation. An attenuated virus will grow poorly in the human host and will therefore produce immunity but not disease.

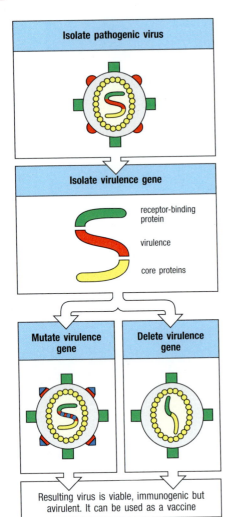

Fig. 16.25 Attenuation can be achieved more rapidly and reliably with recombinant DNA techniques. If a gene in the virus that is required for virulence but not for growth or immunogenicity can be identified, this gene can be either multiply mutated (left lower panel) or deleted from the genome (right lower panel) by using recombinant DNA techniques. This procedure creates an avirulent (nonpathogenic) virus that can be used as a vaccine. The mutations in the virulence gene are usually large, so that it is very difficult for the virus to revert to the wild type.

deliberately attenuated virus can then be used as a vaccine (Fig. 16.25). The advantage of this approach is that mutations can be engineered so that reversion to wild type is virtually impossible.

Such an approach might be useful in developing live influenza vaccines. As described in Chapter 13, the influenza virus can reinfect the same host several times, because it undergoes antigenic shift and thus predominantly escapes the original immune response. A weak protection conferred by previous infections with a different subtype of influenza is observed in adults, but not in children, and is called **heterosubtypic immunity**. The current approach to vaccination against influenza is to use a killed virus vaccine that is reformulated annually on the basis of the prevalent strains of virus. The vaccine is moderately effective, reducing mortality in elderly people and illness in healthy adults. The ideal influenza vaccine would be an attenuated live organism that matched the prevalent virus strain. This could be created by first introducing a series of attenuating mutations into the gene encoding a viral polymerase protein, PB2. The mutated gene segment from the attenuated virus could then be substituted for the wild-type gene in a virus carrying the relevant **hemagglutinin** and **neuraminidase** antigen variants of the current epidemic or pandemic strain. Alternatively, broadly neutralizing antibodies that block the receptor-binding domain of the hemagglutinin can be generated in humans and could be used as a universal vaccine. Public attention has recently been directed toward the possibility of a flu pandemic caused by the **H5N1 avian flu** strain. This strain can be passed between birds and humans and is associated with a high mortality rate; however, a pandemic would occur only if human-to-human transmission could occur. A live-attenuated vaccine would be used only if a pandemic occurred, because to give it beforehand would introduce new influenza virus genes that might recombine with existing influenza viruses.

16-24 Live-attenuated vaccines can be developed by selecting nonpathogenic or disabled bacteria or by creating genetically attenuated parasites (GAPs).

Similar approaches have been used for bacterial vaccine development. The most important example of an attenuated vaccine is that of BCG, which is quite effective at protecting against serious disseminated tuberculosis in children, but is not protective against adult pulmonary tuberculosis. The current BCG vaccine, which remains the most widely used vaccine in the world, was obtained from a pathogenic isolate of *Mycobacterium bovis* and passaged in a laboratory at the beginning of the 20th century. Since then, several genetically diverse strains of BCG have evolved. The level of protection afforded by the BCG vaccine is extremely variable, ranging from none in some countries, such as Malawi, to 50–80% in the UK.

Considering that tuberculosis remains one of the biggest killers worldwide, there is an urgent need for a new vaccine. Two recombinant BCG (rBCG) vaccines intended to prevent infection in unexposed individuals recently passed Phase I clinical trials. One was engineered to overexpress an immunodominant antigen of *M. tuberculosis*, to engender greater specificity toward the human pathogen. The second expresses the pore-forming protein listeriolysin from *L. monocytogenes* to induce the passage of BCG antigens from phagosomes into the cytoplasm and allow cross-presentation (see Section 6-5) on MHC class I molecules, thereby stimulating BCG-specific cytotoxic T cells.

A similar approach is being used to generate new vaccines for malaria. Analysis of different stages of *Plasmodium falciparum*, the major cause of fatal malaria, identified genes that are selectively expressed in sporozoites within the mosquito's salivary gland, where they first become infectious for human

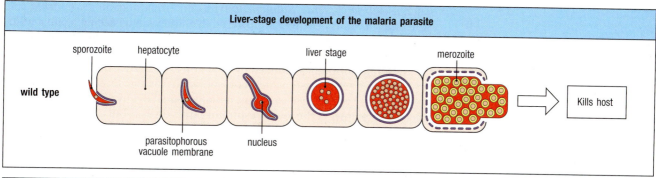

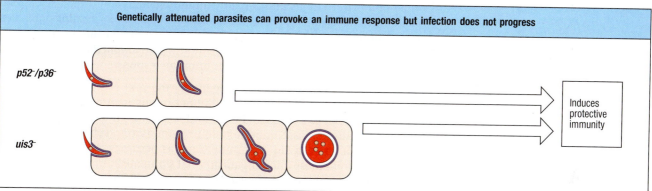

Fig. 16.26 Genetically attenuated parasites can be engineered as live vaccines to provide protective immunity. Top panel: wild-type *Plasmodium* sporozoites transmitted through the bite of an infected mosquito enter the bloodstream and are carried to the liver, where they infect hepatocytes. Each sporozoite multiplies in the liver, killing the infected cell and releasing thousands of merozoites, the next stage in infection. Bottom panels: in mice immunized with sporozoites with targeted disruption of key genes [for example, *p52* and *p36* (*p52⁻/p36⁻*), or *uis3* (*uis3⁻*)], the sporozoites circulate in the bloodstream and mimic an early infection but cannot establish a productive infection in the liver. The mice do, however, produce an immune response against the sporozoites and are protected against a subsequent infection by wild-type sporozoites.

hepatocytes. Deletion of two such genes from the *P. falciparum* genome rendered sporozoites incapable of establishing a blood-stage infection in mice, yet capable of inducing an immune response that protected mice from subsequent infection by wild-type *P. falciparum*. This protection was dependent on CD8 T cells, and to some extent on IFN-γ, indicating that cell-mediated immunity is important for protection against this parasite (Fig. 16.26). This highlights once again the importance of being able to generate vaccines that are capable of inducing strong cell-mediated immunity.

16-25   The route of vaccination is an important determinant of success.

The ideal vaccination induces host defense at the point of entry of the infectious agent. Stimulation of mucosal immunity is therefore an important goal for vaccination against the many organisms that enter through mucosal surfaces. Still, most vaccines are given by injection. This route has several disadvantages. Injections are painful and unpopular, reducing vaccine uptake, and they are expensive, requiring needles, syringes, and a trained injector. Mass vaccination by injection is laborious. There is also the immunological drawback that injection may not be the most effective way of stimulating an appropriate immune response because it does not mimic the usual route of entry of the majority of pathogens against which vaccination is directed.

Many important pathogens infect mucosal surfaces or enter the body through mucosal surfaces. Examples include respiratory microorganisms such as *B. pertussis*, rhinoviruses, and influenza viruses, and enteric microorganisms such as *Vibrio cholerae*, *Salmonella typhi*, enteropathogenic *Escherichia coli*, and *Shigella*. Intranasally administered live-attenuated vaccine against influenza virus induces mucosal antibodies, which are more effective than systemic antibodies in the control of upper respiratory tract infection. However, the systemic antibodies induced by injection are effective in controlling lower respiratory tract disease, which is responsible for the severe morbidity and mortality due to this disease. Thus, a realistic goal of any pandemic influenza vaccine is to prevent the lower respiratory tract disease but accept the fact that mild illness will not be prevented.

The power of the mucosal approach is illustrated by the effectiveness of live-attenuated polio vaccines. The Sabin oral polio vaccine consists of three attenuated polio virus strains and is highly immunogenic. Moreover, just as polio itself can be transmitted by fecal contamination of public swimming pools and other failures of hygiene, the vaccine can be transmitted from one individual to another by the fecal–oral route. Infection with *Salmonella* likewise stimulates a powerful mucosal and systemic immune response.

Presentation of soluble protein antigens by the oral route often results in tolerance, which is important given the enormous load of food-borne and airborne antigens presented to the gut and respiratory tract (see Chapter 12). Nonetheless, the mucosal immune system responds to and eliminates mucosal infections that enter by the oral route, such as pertussis, cholera, and polio. The proteins from these microorganisms that stimulate immune responses are therefore of special interest. One group of powerfully immunogenic proteins at mucosal surfaces is a group of bacterial toxins that have the property of binding to eukaryotic cells and being resistant to proteases. A recent finding of potential practical importance is that certain of these proteins, such as the *E. coli* heat-labile toxin and pertussis toxin, have adjuvant properties that are retained even when the parent molecule has been engineered to eliminate its toxic properties. These molecules can be used as adjuvants for oral or nasal vaccines. In mice, nasal insufflation of either of these mutant toxins together with tetanus toxoid resulted in the development of protection against lethal challenge with tetanus toxin.

16-26 *Bordetella pertussis* vaccination illustrates the importance of the perceived safety of a vaccine.

The history of vaccination against the bacterium that causes whooping cough, *Bordetella pertussis*, illustrates the challenges of developing and disseminating an effective vaccine, as well as the public appeal of acellular conjugate vaccines over attenuated live organisms. At the beginning of the 20th century, whooping cough killed about 0.5% of American children under the age of 5 years. In the early 1930s, a trial of a killed, whole bacterial cell vaccine on the Faroe Islands provided evidence of a protective effect. In the United States, systematic use of a whole-cell vaccine in combination with diphtheria and tetanus toxoids (the DTP vaccine) during the 1940s resulted in a decline in the annual infection rate from 200 to fewer than 2 cases per 100,000 of the population. First vaccination with DTP was typically given at the age of 3 months.

Whole-cell pertussis vaccine causes side-effects, typically redness, pain, and swelling at the site of the injection; less commonly, vaccination is followed by high temperature and persistent crying. Very rarely, fits and a short-lived sleepiness or a floppy unresponsive state ensue. During the 1970s, widespread concern developed after several anecdotal observations that encephalitis leading to irreversible brain damage might very rarely follow pertussis vaccination. In Japan, in 1972, about 85% of children were given the pertussis vaccine, and fewer than 300 cases of whooping cough and no deaths were reported. As a result of two deaths after vaccination in Japan in 1975, the use of DTP was temporarily suspended and then reintroduced with the first vaccination at 2 years of age rather than at 3 months. In 1979, there were about

13,000 cases of whooping cough and 41 deaths. The possibility that pertussis vaccine very rarely causes severe brain damage has been studied extensively, and expert consensus is that pertussis vaccine is not a primary cause of brain injury. There is no doubt that there is greater morbidity from whooping cough than from the vaccine.

The public and medical perception that whole-cell pertussis vaccination might be unsafe provided a powerful incentive to develop safer pertussis vaccines. Study of the natural immune response to *B. pertussis* showed that infection induced antibodies against four components of the bacterium—pertussis toxin, filamentous hemagglutinin, pertactin, and fimbrial antigens. Immunization of mice with these antigens in purified form protected them against challenge with pertussis. This has led to the development of **acellular pertussis vaccines**, all of which contain purified pertussis toxoid—that is, toxin inactivated by chemical treatment, for example with hydrogen peroxide or formaldehyde, or more recently by genetic engineering of the toxin. Some pertussis vaccines also contain filamentous hemagglutinin, pertactin, and/or fimbrial antigens, either alone or in any combination of the three. Current evidence shows that these vaccines are probably as effective as whole-cell pertussis vaccine while lacking its common minor side-effects. The acellular vaccine is more expensive, however, thus restricting its use in poorer countries.

The history of pertussis vaccination illustrates that, first and foremost, vaccines must be extremely safe and free of side-effects; second, the public and medical profession must perceive the vaccine to be safe; and third, careful study of the nature of the protective immune response can lead to acellular vaccines that are safer than whole-cell vaccines but still as effective. Still, public concerns about vaccination remain high. Unwarranted fears of a link between the combined live-attenuated MMR vaccine and autism saw the uptake of MMR vaccine in England fall from a peak of 92% of children in 1995–1996 to 84% in 2001–2002. Small clustered outbreaks of measles and mumps in London since 2002 illustrate the importance of maintaining high uptake of vaccine to maintain herd immunity.

16-27 Conjugate vaccines have been developed as a result of linked recognition between T and B cells.

Many bacteria, including *Neisseria meningitidis* (meningococcus), *Streptococcus pneumoniae* (pneumococcus), and *H. influenzae*, have an outer capsule composed of polysaccharides that are species- and type-specific for particular strains of the bacterium. The most effective defense against these microorganisms is opsonization of the polysaccharide coat with antibody. The aim of vaccination for these organisms is therefore to elicit antibodies against the polysaccharide capsules of the bacteria. However, effective acellular vaccines cannot be made from a single isolated constituent of a microorganism, since generation of an effective antibody response requires the participation of several types of cells, and this fact has led to the development of **conjugate vaccines** (**Fig. 16.27**).

Capsular polysaccharides can be harvested from bacterial growth medium and, because they are T-cell-independent antigens (see Section 10-1), they can be used on their own as vaccines. However, young children under the age

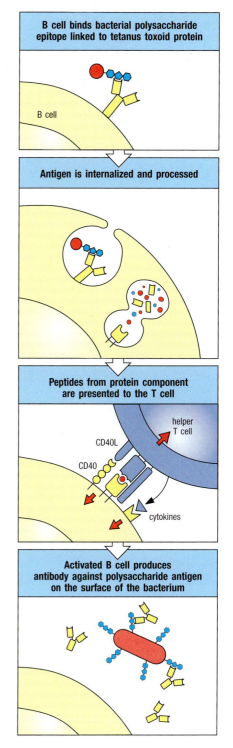

Fig. 16.27 Conjugate vaccines take advantage of linked recognition to boost B-cell responses against polysaccharide antigens. The Hib vaccine against *Haemophilus influenzae* type b is a conjugate of bacterial polysaccharide and the tetanus toxoid protein. The B cell recognizes and binds the polysaccharide, internalizes and degrades the whole conjugate, and then displays toxoid-derived peptides on surface MHC class II molecules. Helper T cells generated in response to earlier vaccination against the toxoid recognize the complex on the B-cell surface and activate the B cell to produce anti-polysaccharide antibody. This antibody can then protect against infection with *H. influenzae* type b.

Fig. 16.28 The effect of vaccination against group C *Neisseria meningitidis* (meningococcus) on the number of cases of group B and group C meningococcal disease in England and Wales. Meningococcal infection affects roughly 5 in 100,000 people a year in the UK, with groups B and C meningococci accounting for almost all the cases. Before the introduction of the meningitis C vaccine, group C disease was the second most common cause of meningococcal disease, accounting for about 40% of cases. Group C disease now accounts for less than 10% of cases, with group B disease accounting for more than 80% of cases. After the introduction of the vaccine, there was a significant decrease in the number of laboratory-confirmed cases of group C disease in all age groups. The impact was greatest in the immunized groups, with reductions of more than 90% in these groups. An impact has also been seen in the unimmunized age groups, with a reduction of about 70%, suggesting that this vaccine has had a herd immunity effect.

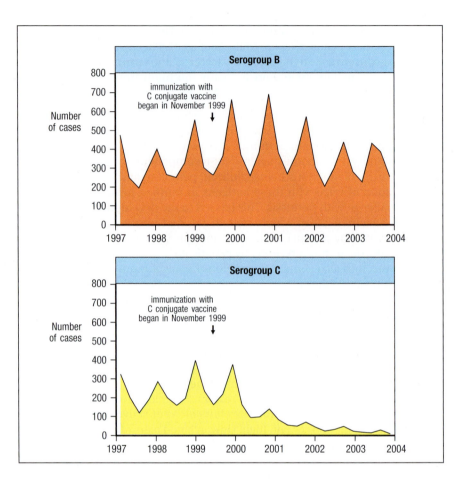

of 2 years cannot make good T-cell-independent antibody responses and cannot be vaccinated effectively with polysaccharide (PS) vaccines. An efficient way of overcoming this problem is to conjugate bacterial polysaccharides chemically to protein carriers (see Fig. 16.27). This carrier protein provides peptides that can be recognized by antigen-specific T cells, thus converting a T-cell-independent response into a T-cell-dependent antipolysaccharide antibody response. Using this approach, various conjugate vaccines have been developed against *H. influenzae* type b, an important cause of serious childhood chest infections and meningitis, and against *N. meningitidis* serogroup C, an important cause of meningitis, and these are now widely applied. The success of the latter vaccine in the United Kingdom is illustrated in Fig. 16.28, which shows that the incidence of meningitis C has been markedly reduced in comparison with meningitis B, against which there is currently no vaccine. Endemic meningitis B is due to diverse serogroup B strains, so an ideal vaccine would target the group B capsular polysaccharide. Unfortunately, group B polysaccharide is identical to some polysialyl polysaccharides on human cells, and is poorly immunogenic due to tolerance of these self antigens. Some strategies to chemically modify the group B polysaccharide for use in a conjugate vaccine have been considered, but a major focus in group B meningococcal vaccine development has been to instead direct immunity against noncapsular antigens, which will be generally effective against endemic disease.

16-28 Peptide-based vaccines can elicit protective immunity, but they require adjuvants and must be targeted to the appropriate cells and cell compartment to be effective.

Another vaccine-development strategy that does not require the whole organism, whether killed or attenuated, identifies the T-cell peptide epitopes that stimulate protective immunity. Candidate peptides can be identified in two ways: in one, overlapping peptides from immunogenic proteins are systematically synthesized and their ability to stimulate protective immunity is tested; alternatively, a reverse immunogenetic approach (see Section 16-15) can be used to predict potential peptide epitopes from a genome sequence. The latter approach has been applied to malaria by using the complete sequence of the *Plasmodium falciparum* genome. The starting point was the association between the human MHC class I molecule HLA-B53 and resistance to cerebral malaria, a relatively infrequent, but usually fatal, complication of infection by *P. falciparum*. It was thought that HLA-B53 might protect against cerebral malaria because it could present peptides that are particularly good at activating naive cytotoxic T lymphocytes. Peptides eluted from HLA-B53 frequently contain a proline as the second of their nine amino acids. On the basis of this information, reverse genetic analysis identified candidate protective peptides from four proteins of *P. falciparum* expressed in the early phase of hepatocyte infection—an important phase of infection to target in an effective immune response. One of the candidate peptides, from liver stage antigen-1, has been shown to be recognized by cytotoxic T cells when bound to HLA-B53 and may be a useful peptide for use in vaccination.

Peptide-based vaccines, although promising, have several drawbacks. First, a particular peptide may not bind to all the MHC molecules present in the population. Because humans are highly polymorphic in the MHC, a large panel of protective peptides would be needed for coverage of most individuals. Second, some direct exchange of short peptides on MHC molecules can occur without physiological antigen processing. If the required antigenic peptides load directly onto MHC molecules on cells other than dendritic cells, this may induce tolerance in T cells rather than stimulating immunity. Third, exogenous proteins and peptides delivered by a synthetic vaccine are efficiently processed for presentation by MHC class II molecules, but require 'cross-presentation' in specific types of dendritic cells to be loaded onto MHC class I molecules (see Section 6-5). Directing peptide-based vaccines to such cells may enhance vaccine efficacy.

Recent peptide-based vaccines have already shown promise in human clinical trials. Patients with established vulvar intraepithelial neoplasia, an early form of vulvar cancer caused by human papilloma virus (HPV), were treated with a vaccine consisting of long peptides covering the entire length of two oncoproteins of HPV-16—E6 and E7—and delivered in an oil-in-water emulsion as adjuvant. By using very long peptides, around 100 amino acids in length, multiple candidate peptide epitopes can be delivered that may also be presented by different MHC alleles. These peptides seem to be too long for direct exchange with peptides on cell surfaces and require processing by dendritic cells in order to be loaded onto MHC class I molecules. This vaccine induced complete clinical remission in one-quarter of the patients, and about half of the treated patients showed significant clinical responses that correlated with *in vitro* evidence of enhanced cell-mediated immunity.

16-29 Adjuvants are important for enhancing the immunogenicity of vaccines, but few are approved for use in humans.

Vaccines based on peptides or purified proteins require additional components to mimic how real infections activate immunity. Such components of a vaccine are known as **adjuvants**, which are defined as substances that enhance the immunogenicity of antigens (see Appendix I, Section A-41). For example, tetanus toxoid is not immunogenic in the absence of adjuvants, and so tetanus toxoid vaccines contain inorganic aluminum salts (**alum**) in the form of noncrystalline gels that bind polyvalently to the toxoid by ionic interactions. Pertussis toxin has adjuvant properties in its own right and, when given mixed as a toxoid with tetanus and diphtheria toxoids, not only protects against whooping cough but also acts as an additional adjuvant for the other two toxoids. This mixture makes up the DTP triple vaccine given to infants in the first year of life.

The antigenic components and adjuvants in a vaccine are not approved for use on their own; they are approved only in the context of the specific vaccine in which they are formulated. At present, alum is the only adjuvant that is approved by the FDA in the United States for use in marketed human vaccines, although some other adjuvant–vaccine combinations are undergoing clinical trials. Alum is the common name for certain inorganic aluminum salts, of which aluminum hydroxide and aluminum phosphate are most frequently used as adjuvants. In Europe, in addition to the alum adjuvants, an oil (squalene)-in-water emulsion called **MF-59** is used as an adjuvant in a formulation of influenza vaccine and is undergoing evaluation in clinical trials. As we described in Section 3-9, alum seems to act as an adjuvant by stimulating one of the innate immune system's bacterial sensor mechanisms, **NLRP3**, thus activating the inflammasome and the inflammatory reactions that are a prerequisite for an effective adaptive immune response.

Several other adjuvants are widely used experimentally in animals but are not approved for use in humans. Many of these are sterile constituents of bacteria, particularly of their cell walls. **Freund's complete adjuvant** is an oil-in-water emulsion containing killed mycobacteria. The peptidoglycan muramyl dipeptide and the glycolipid trehalose dimycolate (TDM) found in mycobacterial cell walls contain much of the adjuvant activity of the whole killed organism. Other bacterial adjuvants include killed *B. pertussis*, bacterial polysaccharides, bacterial heat-shock proteins, and bacterial DNA. Many of these adjuvants cause quite marked inflammation and so are not suitable for use in vaccines for humans.

Many adjuvants seem to work by triggering the innate viral and bacterial sensor pathways in APCs, via TLRs and proteins of the NOD-like receptor family such as NLRP3 (see Chapter 3), and thereby activating them to initiate an adaptive immune response. The TLR-4 agonist **lipopolysaccharide** (LPS) has adjuvant effects, but its use is limited by its toxicity. Small amounts of injected LPS can induce a state of shock and systemic inflammation that mimics Gram-negative sepsis, raising the question whether its adjuvant effect can be separated from its toxic effects. Monophosphoryl lipid A, an LPS derivative and TLR-4 ligand, partly achieves this, retaining adjuvant effects but being associated with much lower toxicity than LPS. Both **unmethylated CpG DNA**, which activates TLR-9, and **imiquimod**, a small-molecule drug that acts as a TLR-7 agonist, can provide adjuvant activity experimentally, but neither is approved as an adjuvant in human vaccines.

16-30 Protective immunity can be induced by DNA-based vaccination.

Surprisingly, when bacterial plasmids were used to express proteins *in vivo* for gene therapy, some were found to stimulate an immune response. Later, it was found that DNA encoding a viral immunogen, when injected intramuscularly in mice, induced antibody responses and cytotoxic T cells that could protect against subsequent infection from the live virus. This response does not seem to damage the muscle tissue, is safe and effective, and, because it uses only a single microbial gene or a stretch of DNA encoding sets of antigenic peptides, does not carry the risk of active infection. This procedure is termed **DNA vaccination**, and can be carried out in various ways. In one, DNA coated onto minute metal particles can be administered by a gene gun, so that particles penetrate the skin and potentially some underlying muscle, but other approaches, such as electroporation, are also possible. Because of DNA's stability, DNA vaccination is suitable for mass immunization. One problem with

DNA-based vaccines, however, is that they are comparatively weak. Mixing in plasmids that encode cytokines such as IL-12, IL-23, or GM-CSF makes immunization with genes encoding protective antigens much more effective. In DNA vaccination, the antigen is produced by cells that are directly transfected, such as skin or muscle, but CD8 T-cell activation requires cross-presentation of the antigen by dendritic cells. Current approaches are identifying how best to transfect DNA into these dendritic cell populations. DNA vaccines are being tested in human trials for malaria, influenza, HIV infection, and breast cancer.

16-31 Vaccination and checkpoint blockade may be useful in controlling existing chronic infections.

There are many chronic diseases in which infection persists because of a failure of the immune system to eliminate disease. Such infections can be divided into two groups: those in which there is an obvious immune response that fails to eliminate the organism, and those that seem to be invisible to the immune system and evoke a barely detectable immune response.

In the first category, the immune response is often partly responsible for the pathogenic effects. Infection by the helminth *Schistosoma mansoni* is associated with a powerful T_H2-type response, characterized by high levels of IgE, circulating and tissue eosinophilia, and a harmful fibrotic response to schistosome ova in the liver, leading to hepatic fibrosis. Other common parasites, such as *Plasmodium* and *Leishmania* species, also cause damage in many patients because they are not eliminated effectively by the immune response. The mycobacterial agents of tuberculosis and leprosy cause a persistent intracellular infection; a T_H1 response helps to contain these infections but also causes granuloma formation and tissue necrosis (see Fig. 11.13).

Among viruses, hepatitis B and hepatitis C infections are commonly followed by a persistent viral burden and hepatic injury, resulting in eventual death from hepatitis or from hepatocellular carcinoma. Infection with HIV, as we have seen in Chapter 13, also persists despite an ongoing immune response. In a preliminary trial involving HIV-infected patients, dendritic cells derived from the patients' own bone marrow were loaded with chemically inactivated HIV. After immunization with the loaded cells, a robust T-cell response to HIV was observed in some patients that was associated with the production of IL-2 and IFN-γ (Fig. 16.29). Viral load in these patients was reduced by 80%,

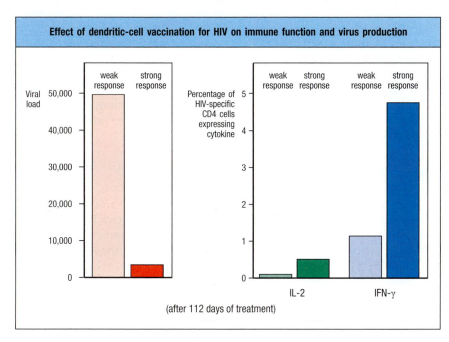

Fig. 16.29 Vaccination with dendritic cells loaded with HIV substantially reduces viral load and generates T-cell immunity. Left panel: viral load is shown for a weak and transient response to treatment (pink); the red bar represents individuals who made a strong and durable response. Right panel: CD4 T-cell IL-2 and interferon-γ production for individuals who made a weak or strong response. The production of both these cytokines, indicating T-cell activity, correlates with the response to treatment.

and in almost half of these patients the suppression of viremia lasted for more than a year. Nonetheless, these responses were not sufficient to eliminate the HIV infection.

In the second category of chronic infection, which is predominantly viral, the immune response fails to clear the infection because of the relative invisibility of the infectious agent to the immune system. A good example is herpes simplex virus type 2, which is transmitted venereally, becomes latent in nerve tissue, and causes genital herpes, which is frequently recurrent. The invisibility of this virus seems to be caused by a viral protein, ICP-47, that binds to the TAP complex (see Section 6-3) and inhibits peptide transport into the endoplasmic reticulum in infected cells. Thus, viral peptides are not presented to the immune system by MHC class I molecules. A similar example in this category of chronic infection is genital warts, caused by certain papilloma viruses that evoke very little immune response, particularly a cell-mediated response. As discussed previously, the results of a recent clinical trial showed that using long-peptide vaccines against HPV-16 was effective in increasing the strength of cell-mediated immune responses to viral antigens, and in reducing or eliminating precancerous lesions associated with the HPV infection (see Section 16-28). These results are a positive indication that vaccines directed at increasing cell-mediated responses to other pathogens may be similarly effective.

Summary.

Vaccination is arguably the greatest success of immunology, having eradicated or virtually eliminated several human diseases. It is the single most successful manipulation of the immune system so far, because it takes advantage of the immune system's natural specificity and inducibility. But important human infectious diseases remain that lack effective vaccines. Most effective vaccines are based on attenuated live microorganisms, but such vaccines carry some risk and are potentially lethal to immunosuppressed or immunodeficient individuals. New techniques are being developed to generate genetically attenuated pathogens for use as vaccines, particularly for malaria and tuberculosis. While most current viral vaccines are based on live attenuated virus, many bacterial vaccines are based on components of the microorganism, including components of the toxins that it produces. Protective responses to carbohydrate antigens, which in very young children do not provoke lasting immunity, can be enhanced by conjugation of the carbohydrate to a protein. Vaccines based on peptides, particularly very long peptides, are just emerging from the experimental stage and are beginning to be tested in humans. A vaccine's immunogenicity often depends on adjuvants that can help, directly or indirectly, to activate antigen-presenting cells that are necessary for the initiation of immune responses. Adjuvants activate these cells by engaging the innate immune system and providing ligands for TLRs and other innate sensors on antigen-presenting cells. The development of oral vaccines is particularly important for stimulating immunity to the many pathogens that enter through the mucosa.

Summary to Chapter 16.

One of the great future challenges in immunology is to be able to control the immune system so that unwanted immune responses can be suppressed and desirable responses elicited. Current methods of suppressing unwanted responses rely, to a great extent, on drugs that suppress adaptive immunity indiscriminately and are thus inherently flawed. We have seen in this book that the immune system can suppress its own responses in an antigen-specific manner and that, by studying these endogenous regulatory events, it has been possible to devise strategies to manipulate specific responses while sparing general immune competence. New treatments, including many monoclonal antibodies, have emerged as clinically important therapies to selectively suppress the responses that lead to allergy, autoimmunity, or the rejection of grafted organs. Similarly, as we understand more about tumors and infectious agents, better strategies to mobilize the immune system against cancer and infection are becoming possible. To achieve all this, we need to learn more about the induction of immunity and the biology of the immune system, and to apply what we have learned to human disease.

Questions.

16.1 Multiple Choice: Which of the following immunomodulators has a similar mechanism to azathioprine?

A. Mycophenolate

B. Cyclophosphamide

C. Abatacept

D. Rapamycin

16.2 Matching: Match the following immunomodulating antibodies with their respective mechanism of action.

A. Natalizumab   i. Prevents allograph rejection by targeting the CD3 complex, which inhibits T-cell receptor signaling.

B. Rituximab   ii. Anti-IL-6 receptor

C. Muromomab   iii. Inhibits cell migration by blocking VLA-4

D. Tocilizumab   iv. Depletion of B cells by targeting CD19

16.3 True or False: Chimeric antigen receptor (CAR) T cells are cells that have been retrovirally transduced with a tumor-specific T-cell receptor in order to treat a leukemia.

16.4 Multiple Choice: Which statement is *false*?

A. The vaccine Provenge is prepared using the patient's own antigen-loaded dendritic cells to induce therapeutic antitumor T-cell responses.

B. Clinical trials of vaccines against HPV-16 and HPV-18 (associated with 70% of cervical cancers) were 100% effective in preventing cervical cancers caused by these viruses.

C. Cell-based cancer vaccines can use the patient's tumor as a source of antigens. In order to enhance immunogenicity these can be mixed with adjuvants such as CpG, which binds to TLR-7.

16.5 Multiple Choice: Which of the following treatments against cancer is a checkpoint blockade therapy? (One or more may apply.)

A. Ipilimumab (anti-CTLA-4 antibody)

B. Trastuzumab (anti-HER-2/neu antibody)

C. Rituximab (anti-CD20 antibody)

D. Pembrolizumab (anti-PD-1 antibody)

E. Sipuleucel-T (patient's dendritic cells cultured with prostatic acid phosphatase tumor antigen and GM-CSF and reinfused into patient)

16.6 True or False: Chimeric antigen receptor (CAR) T cells can recognize other target molecules besides peptide:MHC complexes.

16.7 Matching: Classify the currently used vaccines of the following organisms as live-attenuated (A), toxin-based (T), killed (K), or conjugate polysaccharide (P).

A. ___ *Corynebacterium diphtheriae*

B. ___ *H. influenzae* type B

C. ___ Measles/mumps/rubella (MMR)

D. ___ Bacille Calmette–Guérin (BCG)

E. ___ Influenza A virus

F. ___ Sabin polio vaccine

16.8 Fill-in-the-Blanks: Vaccines have exhibited many phenomena that are beneficial and can be exploited. For example, when an antibody response against a bacterial polysaccharide is desired, it is conjugated to a protein to exploit the phenomenon of _____, thus ensuring T-dependent antibody responses. In addition, vaccines may protect against different subtypes of virus, as in the case of influenza, a phenomenon called _____ immunity. When enough people in a population are vaccinated, _____ immunity is achieved, where even unvaccinated individuals are indirectly protected from infection.

16.9 Short Answer: Explain the three main drawbacks of peptide-based vaccines.

16.10 True or False: All routes of vaccination successfully elicit virtually identical immune responses.

16.11 Matching: Match the adjuvant to the immune receptor it stimulates.

A. Alum   i. TLR-9

B. Freund's complete adjuvant   ii. TLR-4

C. Lipopolysaccharide   iii. NLRP3

D. DNA   iv. NOD2

E. Imiquimod   v. TLR-7/8

General references.

Maus, M.V., Fraietta, J.A., Levine, B.L., Kalos, M., Zhao, Y., and June, C.H.: **Adoptive immunotherapy for cancer or viruses.** *Annu Rev Immunol.* 2014, **32**:189–225.

Feldmann, M.: **Translating molecular insights in autoimmunity into effective therapy.** *Annu. Rev. Immunol.* 2009, **27**:1–27.

Kappe, S.H., Vaughan, A.M., Boddey, J.A., and Cowman, A.F.: **That was then but this is now: malaria research in the time of an eradication agenda.** *Science* 2010, **328**:862–866.

Kaufmann, S.H.: **Future vaccination strategies against tuberculosis: thinking outside the box.** *Immunity* 2010, **33**:567–577.

Korman, A.J., Peggs, K.S., and Allison, J.P.: **Checkpoint blockade in cancer immunotherapy.** *Adv. Immunol.* 2006, **90**:297–339.

Section references.

16-1   Corticosteroids are powerful anti-inflammatory drugs that alter the transcription of many genes.

Kampa, M., and Castanas, E.: **Membrane steroid receptor signaling in normal and neoplastic cells.** *Mol. Cell. Endocrinol.* 2006, **246**:76–82.

Löwenberg, M., Verhaar, A.P., van den Brink, G.R., and Hommes, D.W.: **Glucocorticoid signaling: a nongenomic mechanism for T-cell immunosuppression.** *Trends Mol. Med.* 2007, **13**:158–163.

Rhen, T., and Cidlowski, J.A.: **Antiinflammatory action of glucocorticoids—new mechanisms for old drugs.** *N. Engl. J. Med.* 2005, **353**:1711–1723.

Barnes, P.,J.: **Glucocorticosteroids: current and future directions.** *Br. J. Pharmacol.* 2011, **163**:29–43.

16-2   Cytotoxic drugs cause immunosuppression by killing dividing cells and have serious side-effects.

Aarbakke, J., Janka-Schaub, G., and Elion, G.B.: **Thiopurine biology and pharmacology.** *Trends Pharmacol. Sci.* 1997, **18**:3–7.

Allison, A.C., and Eugui, E.M.: **Mechanisms of action of mycophenolate mofetil in preventing acute and chronic allograft rejection.** *Transplantation* 2005, **802** Suppl.: S181–S190.

O'Donovan, P., Perrett, C.M., Zhang, X., Montaner, B., Xu, Y.Z., Harwood, C.A., McGregor, J.M., Walker, S.L., Hanaoka, F., and Karran, P.: **Azathioprine and UVA light generate mutagenic oxidative DNA damage.** *Science* 2005, **309**:1871–1874.

Taylor, A.L., Watson, C.J., and Bradley, J.A.: **Immunosuppressive agents in solid organ transplantation: mechanisms of action and therapeutic efficacy.** *Crit. Rev. Oncol. Hematol.* 2005, **56**:23–46.

Zhu, L.P., Cupps, T.R., Whalen, G., and Fauci, A.S.: **Selective effects of cyclophosphamide therapy on activation, proliferation, and differentiation of human B cells.** *J. Clin. Invest.* 1987, **79**:1082–1090.

16-3   Cyclosporin A, tacrolimus, rapamycin and JAK inhibitors are effective immunosuppressive agents that interfere with various T-cell signaling pathways.

Araki, K., Turner, A.P., Shaffer, V.O., Gangappa, S., Keller, S.A., Bachmann, M.F., Larsen, C.P., and Ahmed, R.: **mTOR regulates memory CD8 T-cell differentiation.** *Nature* 2009, **460**:108–112.

Battaglia, M., Stabilini, A., and Roncarolo, M.G.: **Rapamycin selectively expands CD4+CD25+FoxP3+ regulatory T cells.** *Blood* 2005, **105**:4743–4748.

Bierer, B.E., Mattila, P.S., Standaert, R.F., Herzenberg, L.A., Burakoff, S.J., Crabtree, G., and Schreiber, S.L.: **Two distinct signal transmission pathways in T lymphocytes are inhibited by complexes formed between an immunophilin and either FK506 or rapamycin.** *Proc. Natl Acad. Sci. USA* 1990, **87**:9231–9235.

Crabtree, G.R.: **Generic signals and specific outcomes: signaling through Ca²⁺, calcineurin, and NF-AT.** *Cell* 1999, **96**:611–614.

Crespo, J.L., and Hall, M.N.: **Elucidating TOR signaling and rapamycin action: lessons from Saccharomyces cerevisiae.** *Microbiol. Mol. Biol. Rev.* 2002, **66**:579–591.

Fleischmann, R., Kremer, J., Cush, J., Schulze-Koops, H., Connell, C.A., Bradley, J.D., Gruben, D., Wallenstein, G.V., Zwillich, S.H., Kanik, K.S. *et al.*: **Placebo-controlled trial of tofacitinib monotherapy in rheumatoid arthritis.** *N. Engl. J. Med.* 2012, **367**:495–507.

Pesu, M., Laurence, A., Kishore, N., Zwillich, S.H., Chan, G, and O'Shea, J.J.: **Therapeutic targeting of Janus kinases.** *Immunol. Rev.* 2008, **223**:132–142.

16-4   Antibodies against cell-surface molecules can be used to eliminate lymphocyte subsets or to inhibit lymphocyte function.

Graca, L., Le Moine, A., Cobbold, S.P., and Waldmann, H.: **Antibody-induced transplantation tolerance: the role of dominant regulation.** *Immunol. Res.* 2003, **28**:181–191.

Nagelkerke, S.Q, and Kuijpers, T.W.: **Immunomodulation by IVIg and the role of Fc-gamma receptors: classic mechanisms of action after all?** *Front. Immunol.* 2015, **5**:674.

Nagelkerke, S.Q., Dekkers, G., Kustiawan, I, van de Bovenkamp, F.S., Geissler, J., Plomp, R., Wuhrer, M., Vidarsson, G., Rispens, T., van den Berg, T.K. *et al*: **Inhibition of FcγR-mediated phagocytosis by IVIg is independent of IgG-Fc sialylation and FcγRIIb in human macrophages.** *Blood* 2014, **124**:3709–3718.

Waldmann, H., and Hale, G.: **CAMPATH: from concept to clinic.** *Phil. Trans. R. Soc. Lond. B* 2005, **360**:1707–1711.

16-5 Antibodies can be engineered to reduce their immunogenicity in humans.

Kim, S.J., Park, Y., and Hong, H.J.: **Antibody engineering for the development of therapeutic antibodies.** *Mol. Cell* 2005, **20**:17–29.

Liu, X.Y., Pop, L.M., and Vitetta, E.S.: **Engineering therapeutic monoclonal antibodies.** *Immunol. Rev.* 2008, **222**:9–27.

Smith, K., Garman, L., Wrammert, J., Zheng, N.Y., Capra, J.D., Ahmed, R., and Wilson, P.C.: **Rapid generation of fully human monoclonal antibodies specific to a vaccinating antigen.** *Nat. Protocols* 2009, **4**:372–384.

Traggiai, E., Becker, S., Subbarao, K., Kolesnikova, L., Uematsu, Y., Gismondo, M.R., Murphy, B.R., Rappuoli, R., and Lanzavecchia, A.: **An efficient method to make human monoclonal antibodies from memory B cells: potent neutralization of SARS coronavirus.** *Nat. Med.* 2004, **10**:871–875.

Winter, G., Griffiths, A.D., Hawkins, R.E., and Hoogenboom, H.R.: **Making antibodies by phage display technology.** *Annu. Rev. Immunol.* 1994, **12**:433–455.

16-6   Monoclonal antibodies can be used to prevent allograft rejection.

Kirk, A.D., Burkly, L.C., Batty, D.S., Baumgartner, R.E., Berning, J.D., Buchanan, K., Fechner, J.H., Jr, Germond, R.L., Kampen, R.L., Patterson, N.B., *et al.*: **Treatment with humanized monoclonal antibody against CD154 prevents acute renal allograft rejection in nonhuman primates.** *Nat. Med.* 1999, **5**:686–693.

Li, X.C., Strom, T.B., Turka, L.A., and Wells, A.D.: **T-cell death and transplantation tolerance.** *Immunity* 2001, **14**:407–416.

Londrigan, S.L., Sutherland, R.M., Brady, J.L., Carrington, E.M., Cowan, P.J., d'Apice, A.J., O'Connell, P.J., Zhan, Y., and Lew, A.M.: **In situ protection against islet allograft rejection by CTLA4Ig transduction.** *Transplantation* 2010, **90**:951–957.

Masharani, U.B., and Becker, J.: **Teplizumab therapy for type 1 diabetes.** *Expert Opin. Biol. Ther.* 2010, **10**:459–465.

Pham, P.T., Lipshutz, G.S., Pham, P.T., Kawahji, J., Singer, J.S., and Pham, P.C.: **The evolving role of alemtuzumab (Campath-1H) in renal transplantation.** *Drug Des. Dev. Ther.* 2009, **3**:41–49.

Sageshima, J., Ciancio, G., Chen, L., and Burke, G.W.: **Anti-interleukin-2 receptor antibodies—basiliximab and daclizumab—for the prevention of acute rejection in renal transplantation.** *Biologics* 2009, **3**:319–336.

16-7   Depletion of autoreactive lymphocytes can treat autoimmune disease.

Coiffier, B., Lepage, E., Briere, J., Herbrecht, R., Tilly, H., Bouabdallah, R., Morel, P., Van Den Neste, E., Salles, G., Gaulard, P., *et al.*: **CHOP chemotherapy plus rituximab compared with CHOP alone in elderly patients with diffuse large-B-cell lymphoma.** *N. Engl. J. Med.* 2002, **346**:235–242.

Coles, A., Deans, J., and Compston, A.: **Campath-1H treatment of multiple sclerosis: lessons from the bedside for the bench.** *Clin. Neurol. Neurosurg.* 2004, **106**:270–274.

Edwards, J.C., Leandro, M.J., and Cambridge, G.: **B lymphocyte depletion in rheumatoid arthritis: targeting of CD20.** *Curr. Dir. Autoimmun.* 2005, **8**:175–192.

Yazawa, N., Hamaguchi, Y., Poe, J.C., and Tedder, T.F.: **Immunotherapy using unconjugated CD19 monoclonal antibodies in animal models for B lymphocyte malignancies and autoimmune disease.** *Proc. Natl Acad. Sci. USA* 2005, **102**:15178–15783.

Zaja, F., De Vita, S., Mazzaro, C., Sacco, S., Damiani, D., De Marchi, G., Michelutti, A., Baccarani, M., Fanin, R., and Ferraccioli, G.: **Efficacy and safety of rituximab in type II mixed cryoglobulinemia.** *Blood* 2003, **101**:3827–3834.

16-8 Biologic agents that block TNF-α, IL-1, or IL-6 can alleviate autoimmune diseases.

Guarda, G., Braun, M., Staehli, F., Tardivel, A., Mattmann, C., Förster, I., Farlik, M., Decker, T., Du Pasquier, R.A., Romero, P., *et al.*: **Type I interferon inhibits interleukin-1 production and inflammasome activation.** *Immunity* 2011, **34**:213–223.

Feldmann, M., and Maini, R.N.: **Lasker Clinical Medical Research Award. TNF defined as a therapeutic target for rheumatoid arthritis and other autoimmune diseases.** *Nat. Med.* 2003, **9**:1245–1250.

Hallegua, D.S., and Weisman, M.H.: **Potential therapeutic uses of interleukin 1 receptor antagonists in human diseases.** *Ann. Rheum. Dis.* 2002, **61**:960–967.

Karanikolas, G., Charalambopoulos, D., Vaiopoulos, G., Andrianakos, A., Rapti, A., Karras, D., Kaskani, E., and Sfikakis, P.P.: **Adjunctive anakinra in patients with active rheumatoid arthritis despite methotrexate, or leflunomide, or cyclosporin-A monotherapy: a 48-week, comparative, prospective study.** *Rheumatology* 2008, **47**:1384–1388.

Mackay, C.R.: **New avenues for anti-inflammatory therapy.** *Nat. Med.* 2002, **8**:117–118.

16-9 Biologic agents can block cell migration to sites of inflammation and reduce immune responses.

Boster, A.L., Nicholas, J.A., Topalli, I., Kisanuki, Y.Y., Pei, W., Morgan-Followell, B., Kirsch, C.F., Racke, M.K., and Pitt, D.: **Lessons learned from fatal progressive multifocal leukoencephalopathy in a patient with multiple sclerosis treated with natalizumab.** *JAMA Neurol.* 2013, **70**:398–402.

Clifford, D.B., De Luca, A., Simpson, D.M., Arendt, G., Giovannoni, G., and Nath, A.: **Natalizumab-associated progressive multifocal leukoencephalopathy in patients with multiple sclerosis: lessons from 28 cases.** *Lancet Neurol.* 2010, **9**:438–446.

Cyster, J.G.: **Chemokines, sphingosine-1-phosphate, and cell migration in secondary lymphoid organs.** *Annu. Rev. Immunol.* 2005, **23**:127–159.

Idzko, M., Hammad, H., van Nimwegen, M., Kool, M., Muller, T., Soullie, T., Willart, M.A., Hijdra, D., Hoogsteden, H.C., and Lambrecht, B.N.: **Local application of FTY720 to the lung abrogates experimental asthma by altering dendritic cell function.** *J. Clin. Invest.* 2006, **116**:2935–2944.

Kappos, L., Radue, E.W., O'Connor, P., Polman, C., Hohlfeld, R., Calabresi, P., Selmaj, K., Agoropoulou, C., Leyk, M., Zhang-Auberson, L., *et al.*: **A placebo-controlled trial of oral fingolimod in relapsing multiple sclerosis.** *N. Engl. J. Med.* 2010, **362**:387–401.

Podolsky, D.K.: **Selective adhesion-molecule therapy and inflammatory bowel disease—a tale of Janus?** *N. Engl. J. Med.* 2005, **353**:1965–1968.

16-10 Blockade of co-stimulatory pathways that activate lymphocytes can be used to treat autoimmune disease.

Fife, B.T., and Bluestone, J.A.: **Control of peripheral T-cell tolerance and autoimmunity via the CTLA-4 and PD-1 pathways.** *Immunol. Rev.* 2008, **224**:166–182.

Ellis, C.N., and Krueger, G.G.: **Treatment of chronic plaque psoriasis by selective targeting of memory effector T lymphocytes.** *N. Engl. J. Med.* 2001, **345**:248–255.

Menter, A.: **The status of biologic therapies in the treatment of moderate to severe psoriasis.** *Cutis* 2009, **84**:14–24.

Kraan, M.C., van Kuijk, A.W., Dinant, H.J., Goedkoop, A.Y., Smeets, T.J., de Rie, M.A., Dijkmans, B.A., Vaishnaw, A.K., Bos, J.D., and Tak, P.P.: **Alefacept treatment in psoriatic arthritis: reduction of the effector T cell population in peripheral blood and synovial tissue is associated with improvement of clinical signs of arthritis.** *Arthritis Rheum.* 2002, **46**:2776–2784.

Lowes, M.A., Chamian, F., Abello, M.V., Fuentes-Duculan, J., Lin, S.L., Nussbaum, R., Novitskaya, I., Carbonaro, H., Cardinale, I., Kikuchi, T., *et al.*: **Increase in TNF-α and inducible nitric oxide synthase-expressing dendritic cells in psoriasis and reduction with efalizumab (anti-CD11a).** *Proc. Natl Acad. Sci. USA* 2005, **102**:19057–19062.

16-11 Some commonly used drugs have immunomodulatory properties.

Baeke, F., Takiishi, T., Korf, H., Gysemans, C., and Mathieu, C.: **Vitamin D: modulator of the immune system.** *Curr. Opin. Pharmacol.* 2010, **10**:482–496.

Okwan-Duodu, D., Datta, V., Shen, X.Z., Goodridge, H.S., Bernstein, E.A., Fuchs, S., Liu, G.Y., and Bernstein, K.E.: **Angiotensin-converting enzyme overexpression in mouse myelomonocytic cells augments resistance to Listeria and methicillin-resistant Staphylococcus aureus.** *J. Biol. Chem.* 2010, **285**:39051–39060.

Ridker, P.M., Cannon, C.P., Morrow, D., Rifai, N., Rose, L.M., McCabe, C.H., Pfeffer, M.A., Braunwald, E: **Pravastatin or Atorvastatin Evaluation and Infection Therapy-Thrombolysis in Myocardial Infarction 22 (PROVE IT-TIMI 22) Investigators: C-reactive protein levels and outcomes after statin therapy.** *N. Engl. J. Med.* 2005, **352**:20–28.

Youssef, S., Stuve, O., Patarroyo, J.C., Ruiz, P.J., Radosevich, J.L., Hur, E.M., Bravo, M., Mitchell, D.J., Sobel, R.A., Steinman, L., *et al.*: **The HMG-CoA reductase inhibitor, atorvastatin, promotes a Th2 bias and reverses paralysis in central nervous system autoimmune disease.** *Nature* 2002, **420**:78–84.

16-12 Controlled administration of antigen can be used to manipulate the nature of an antigen-specific response.

Diabetes Prevention Trial: Type 1 Diabetes Study Group: **Effects of insulin in relatives of patients with type 1 diabetes mellitus.** *N. Engl. J. Med.* 2002, **346**:1685–1691.

Haselden, B.M., Kay, A.B., and Larché, M.: **Peptide-mediated immune responses in specific immunotherapy.** *Int. Arch. Allergy. Immunol.* 2000, **122**:229–237.

Mowat, A.M., Parker, L.A., Beacock-Sharp, H., Millington, O.R., and Chirdo, F.: **Oral tolerance: overview and historical perspectives.** *Ann. N.Y. Acad. Sci.* 2004, **1029**:1–8.

Steinman, L., Utz, P.J., and Robinson, W.H.: **Suppression of autoimmunity via microbial mimics of altered peptide ligands.** *Curr. Top. Microbiol. Immunol.* 2005, **296**:55–63.

Weiner, H.L., da Cunha, A.P., Quintana, F., and Wu, H.: **Oral tolerance.** *Immunol. Rev.* 2011, **241**:241–259.

16-13 The development of transplantable tumors in mice led to the discovery of protective immune responses to tumors.

Jaffee, E.M., and Pardoll, D.M.: **Murine tumor antigens: is it worth the search?** *Curr. Opin. Immunol.* 1996, **8**:622–627.

Klein, G.: **The strange road to the tumor-specific transplantation antigens (TSTAs).** *Cancer Immun.* 2001, **1**:6.

16-14 Tumors are 'edited' by the immune system as they evolve and can escape rejection in many ways.

Dunn, G.P., Old, L.J., and Schreiber, R.D.: **The immunobiology of cancer immunosurveillance and immunoediting.** *Immunity* 2004, **21**:137–148.

Gajewski, T.F., Meng, Y., Blank, C., Brown, I., Kacha, A., Kline, J., and Harlin, H.: **Immune resistance orchestrated by the tumor microenvironment.** *Immunol. Rev.* 2006, **213**:131–145.

Girardi, M., Oppenheim, D.E., Steele, C.R., Lewis, J.M., Glusac, E., Filler, R., Hobby, P., Sutton, B., Tigelaar, R.E., and Hayday, A.C.: **Regulation of cutaneous malignancy by γδ T cells.** *Science* 2001, **294**:605–609.

Kloor, M., Becker, C., Benner, A., Woerner, S.M., Gebert, J., Ferrone, S., and von Knebel Doeberitz, M: **Immunoselective pressure and human leukocyte antigen class I antigen machinery defects in microsatellite unstable colorectal cancers.** *Cancer Res.* 2005, **65**:6418–6424.

Koblish, H.K., Hansbury, M.J., Bowman, K.J., Yang, G., Neilan, C.L., Haley, P.J., Burn, T.C., Waeltz, P., Sparks, R.B., Yue, E.W., *et al.* : **Hydroxyamidine inhibitors of indoleamine-2,3-dioxygenase potently suppress systemic tryptophan catabolism and the growth of IDO-expressing tumors.** *Mol. Cancer. Ther.* 2010, **9**:489–498.

Koebel, C.M., Vermi, W., Swann, J.B., Zerafa, N., Rodig, S.J., Old, L.J., Smyth, M.J., and Schreiber, R.D.: **Adaptive immunity maintains occult cancer in an equilibrium state.** *Nature* 2007, **450**:903–907.

Koopman, L.A., Corver, W.E., van der Slik, A.R., Giphart, M.J., and Fleuren, G.J.: **Multiple genetic alterations cause frequent and heterogeneous human histocompatibility leukocyte antigen class I loss in cervical cancer.** *J. Exp. Med.* 2000, **191**:961–976.

Munn, D.H., and Mellor, A.L.: **Indoleamine 2,3-dioxygenase and tumor-induced tolerance.** *J. Clin. Invest.* 2007, **117**:1147–1154.

Ochsenbein, A.F., Sierro, S., Odermatt, B., Pericin, M., Karrer, U., Hermans, J., Hemmi, S., Hengartner, H., and Zinkernagel, R.M.: **Roles of tumour localization, second signals and cross priming in cytotoxic T-cell induction.** *Nature* 2001, **411**:1058–1064.

Peggs, K.S., Quezada, S.A., and Allison, J.P.: **Cell intrinsic mechanisms of T-cell inhibition and application to cancer therapy.** *Immunol. Rev.* 2008, **224**:141–165.

Peranzoni, E., Zilio, S., Marigo, I., Dolcetti, L., Zanovello, P., Mandruzzato, S., and Bronte, V.: **Myeloid-derived suppressor cell heterogeneity and subset definition.** *Curr. Opin. Immunol.* 2010, **22**:238–244.

Schreiber, R.D., Old. L.J., and Smyth, M.J.: **Cancer immunoediting: integrating immunity's roles in cancer suppression and promotion.** *Science* 2011, **331**:1565–1570.

Shroff, R., and Rees, L.: **The post-transplant lymphoproliferative disorder—a literature review.** *Pediatr. Nephrol.* 2004, **19**:369–377.

Wang, H.Y., Lee, D.A., Peng, G., Guo, Z., Li, Y., Kiniwa, Y., Shevach, E.M., and Wang, R.F.: **Tumor-specific human CD4+ regulatory T cells and their ligands: implications for immunotherapy.** *Immunity* 2004, **20**:107–118.

16-15 Tumor rejection antigens can be recognized by T cells and form the basis of immunotherapies.

Clark, R.E., Dodi, I.A., Hill, S.C., Lill, J.R., Aubert, G., Macintyre, A.R., Rojas, J., Bourdon, A., Bonner, P.L., Wang, L., *et al.*: **Direct evidence that leukemic cells present HLA-associated immunogenic peptides derived from the BCR-ABL b3a2 fusion protein.** *Blood* 2001, **98**:2887–2893.

Comoli, P., Pedrazzoli, P., Maccario, R., Basso, S., Carminati, O., Labirio, M., Schiavo, R., Secondino, S., Frasson, C., Perotti, C., *et al.*: **Cell therapy of Stage IV nasopharyngeal carcinoma with autologous Epstein–Barr virus-targeted cytotoxic T lymphocytes.** *J. Clin. Oncol.* 2005, **23**:8942–8949.

Disis, M.L., Knutson, K.L., McNeel, D.G., Davis, D., and Schiffman, K.: **Clinical translation of peptide-based vaccine trials: the HER-2/neu model.** *Crit. Rev. Immunol.* 2001, **21**:263–273.

Dudley, M.E., Wunderlich, J.R., Yang, J.C., Sherry, R.M., Topalian, S.L., Restifo, N.P., Royal, R.E., Kammula, U., White, D.E., Mavroukakis, S.A., *et al.*: **Adoptive cell transfer therapy following non-myeloablative but lymphodepleting chemotherapy for the treatment of patients with refractory metastatic melanoma.** *J. Clin. Oncol.* 2005, **23**:2346–2357.

Matsushita, H., Vesely, M.D., Koboldt, D.C., Rickert, C.G., Uppaluri, R., Magrini, V.J., Arthur, C.D., White, J.M., Chen, Y.S., Shea, L.K., *et al.*: **Cancer exome analysis reveals a T-cell-dependent mechanism of cancer immunoediting.** *Nature* 2012, **482**:400–404.

Michalek, J., Collins, R.H., Durrani, H.P., Vaclavkova, P., Ruff, L.E., Douek, D.C., and Vitetta, E.S.: **Definitive separation of graft-versus-leukemia- and graft-versus-host-specific CD4+ T cells by virtue of their receptor β loci sequences.** *Proc. Natl Acad. Sci. USA* 2003, **100**:1180–1184.

Morris, E.C., Tsallios, A., Bendle, G.M., Xue, S.A., and Stauss, H.J.: **A critical role of T cell antigen receptor-transduced MHC class I-restricted helper T cells in tumor protection.** *Proc. Natl Acad. Sci. USA* 2005, **102**:7934–7939.

Schultz, E.S., Schuler-Thurner, B., Stroobant, V., Jenne, L., Berger, T.G., Thielemanns, K., van der Bruggen, P., and Schuler, G.: **Functional analysis of tumor-specific Th cell responses detected in melanoma patients after dendritic cell-based immunotherapy.** *J. Immunol.* 2004, **172**:1304–1310.

16-16 T cells expressing chimeric antigen receptors are an effective treatment in some leukemias.

Grupp, S.A., Kalos, M., Barrett, D., Aplenc, R., Porter, D.L., Rheingold, S.R., Teachey, D.T., Chew, A., Hauck, B., Wright, J.F., *et al*: **Chimeric antigen receptor-modified T cells for acute lymphoid leukemia.** *N. Engl. J. Med.* 2013, **368**:1509–1518.

Stromnes, I.M., Schmitt, T.M., Chapuis, A.G., Hingorani, S.R., and Greenberg, P.D.: **Re-adapting T cells for cancer therapy: from mouse models to clinical trials.** *Immunol. Rev.* 2014, **257**:145–164.

16-17 Monoclonal antibodies against tumor antigens, alone or linked to toxins, can control tumor growth.

Bradley, A.M., Devine, M., and DeRemer, D.: **Brentuximab vedotin: an anti-CD30 antibody-drug conjugate.** *Am. J. Health Syst. Pharm.* 2013, **70**:589–597.

Hortobagyi, G.N.: **Trastuzumab in the treatment of breast cancer.** *N. Engl. J. Med.* 2005, **353**:1734–1736.

Kreitman, R.J., Wilson, W.H., Bergeron, K., Raggio, M., Stetler-Stevenson, M., FitzGerald, D.J., and Pastan, I.: **Efficacy of the anti-CD22 recombinant immunotoxin BL22 in chemotherapy-resistant hairy-cell leukemia.** *N. Engl. J. Med.* 2001, **345**:241–247.

Park, S., Jiang, Z., Mortenson, E.D., Deng, L., Radkevich-Brown, O., Yang, X., Sattar, H., Wang, Y., Brown, N.K., Greene, M., *et al.*: **The therapeutic effect of anti-HER2/neu antibody depends on both innate and adaptive immunity.** *Cancer Cell* 2010, **18**:160–170.

Tol, J., and Punt, C.J.: **Monoclonal antibodies in the treatment of metastatic colorectal cancer: a review.** *Clin. Ther.* 2010, **32**:437–453.

Veeramani, S., Wang, S.Y., Dahle, C., Blackwell, S., Jacobus, L., Knutson, T., Button, A., Link, B.K., and Weiner, G.J.: **Rituximab infusion induces NK activation in lymphoma patients with the high-affinity CD16 polymorphism.** *Blood* 2011, **118**:3347–3349.

Verma, S., Miles, D., Gianni, L., Krop, I.E., Welslau, M., Baselga, J., Pegram, M., Oh, D.Y., Diéras, V., Guardino, E., *et al.*: **Trastuzumab emtansine for HER2-positive advanced breast cancer.** *N. Engl. J. Med.* 2012, **367**:1783–1791.

Weiner, L.M., Dhodapkar MV, and Ferrone S.: **Monoclonal antibodies for cancer immunotherapy.** *Lancet* 2009, **373**:1033–1040.

Weng, W.K., and Levy, R.: **Genetic polymorphism of the inhibitory IgG Fc receptor FcgammaRIIb is not associated with clinical outcome in patients with follicular lymphoma treated with rituximab.** *Leuk. Lymphoma* 2009, **50**:723–727.

16-18 Enhancing the immune response to tumors by vaccination holds promise for cancer prevention and therapy.

Kantoff, P.W., Higano, C.S., Shore, N.D., Berger, E.R., Small, E.J., Penson, D.F., Redfern, C.H., Ferrari, A.C., Dreicer, R., Sims, R.B., *et al.*: **Sipuleucel-T immunotherapy for castration-resistant prostate cancer.** *N. Engl. J. Med.* 2010, **363**:411–422.

Kenter, G.G., Welters, M.J., Valentijn, A.R., Lowik, M.J., Berends-van der Meer, D.M., Vloon, A.P., Essahsah, F., Fathers, L.M., Offringa, R., Drijfhout, J.W., *et al.*: **Vaccination against HPV-16 oncoproteins for vulvar intraepithelial neoplasia.** *N. Engl. J. Med.* 2009, **361**:1838–1847.

Mao, C., Koutsky, L.A., Ault, K.A., Wheeler, C.M., Brown, D.R., Wiley, D.J., Alvarez, F.B., Bautista, O.M., Jansen, K.U., and Barr, E.: **Efficacy of human papillomavirus-16 vaccine to prevent cervical intraepithelial neoplasia: a randomized controlled trial.** *Obstet. Gynecol.* 2006, **107**:18–27.

Palucka, K., Ueno, H., Fay, J., and Banchereau, J.: **Dendritic cells and immunity against cancer.** *J. Intern. Med.* 2011, **269**:64–73.

Vambutas, A., DeVoti, J., Nouri, M., Drijfhout, J.W., Lipford, G.B., Bonagura, V.R., van der Burg, S.H., and Melief, C.J.: **Therapeutic vaccination with papillomavirus E6 and E7 long peptides results in the control of both established virus-induced lesions and latently infected sites in a pre-clinical cottontail rabbit papillomavirus model.** *Vaccine* 2005, **23**:5271–5280.

16-19 Checkpoint blockade can augment immune responses to existing tumors.

Ansell, S.M., Lesokhin, A.M., Borrello, I., Halwani, A., Scott, E.C., Gutierrez, M., Schuster, S,J., Millenson, M.M., Cattry, D., Freeman, G.J., *et al.*: **PD-1 blockade with nivolumab in relapsed or refractory Hodgkin's lymphoma.** *N. Engl. J. Med.* 2015, **372**:311–319.

Bendandi, M., Gocke, C.D., Kobrin, C.B., Benko, F.A., Sternas, L.A., Pennington, R., Watson, T.M., Reynolds, C.W., Gause, B.L., Duffey, P.L., *et al.*: **Complete molecular remissions induced by patient-specific vaccination plus granulocyte-monocyte colony-stimulating factor against lymphoma.** *Nat. Med.* 1999, **5**:1171–1177.

Egen, J.G., Kuhns, M.S., and Allison, J.P.: **CTLA-4: new insights into its biological function and use in tumor immunotherapy.** *Nat. Immunol.* 2002, **3**:611–618.

Hamid, O., Robert, C., Daud, A., Hodi, F.S., Hwu, W.J., Kefford, R., Wolchok, J.D., Hersey, P., Joseph, R.W., Weber, J.S., *et al.*: **Safety and tumor responses with lambrolizumab (anti-PD-1) in melanoma.** *N. Engl. J. Med.* 2013, **369**:134–144.

Hodi, F.S., O'Day, S.J., McDermott, D.F., Weber, R.W., Sosman, J.A., Haanen, J.B., Gonzalez, R., Robert, C., Schadendorf, D., Hassel, J.C. *et al.*: **Improved survival with ipilimumab in patients with metastatic melanoma.** *N. Engl. J. Med* 2010; **363**:711–723

Li, Y., Hellstrom, K.E., Newby, S.A., and Chen, L.: **Costimulation by CD48 and B7–1 induces immunity against poorly immunogenic tumors.** *J. Exp. Med.* 1996, **183**:639–644.

Phan, G.Q., Yang, J.C., Sherry, R.M., Hwu, P., Topalian, S.L., Schwartzentruber, D.J., Restifo, N.P., Haworth, L.R., Seipp, C.A., Freezer, L.J., *et al.*: **Cancer regression and autoimmunity induced by cytotoxic T lymphocyte-associated antigen 4 blockade in patients with metastatic melanoma.** *Proc. Natl Acad. Sci. USA* 2003, **100**:8372–8377.

Yuan, J., Gnjatic, S., Li, H., Powel, S., Gallardo, H.F., Ritter, E., Ku, G.Y., Jungbluth, A.A., Segal, N.H., Rasalan, T.S., *et al.*: **CTLA-4 blockade enhances polyfunctional NY-ESO-1 specific T cell responses in metastatic melanoma patients with clinical benefit.** *Proc. Natl Acad. Sci. USA* 2008, **105**:20410–20415.

16-20 Vaccines can be based on attenuated pathogens or material from killed organisms.

Anderson, R.M., Donnelly, C.A., and Gupta, S.: **Vaccine design, evaluation, and community-based use for antigenically variable infectious agents.** *Lancet* 1997, **350**:1466–1470.

Dermer, P., Lee, C., Eggert, J., and Few, B.: **A history of neonatal group B streptococcus with its related morbidity and mortality rates in the United States.** *J. Pediatr. Nurs.* 2004, **19**:357–363.

Rabinovich, N.R., McInnes, P., Klein, D.L., and Hall, B.F.: **Vaccine technologies: view to the future.** *Science* 1994, **265**:1401–1404.

16-21 Most effective vaccines generate antibodies that prevent the damage caused by toxins or that neutralize the pathogen and stop infection.

Levine, M.M., and Levine, O.S.: **Influence of disease burden, public perception, and other factors on new vaccine development, implementation, and continued use.** *Lancet* 1997, **350**:1386–1392.

Mouque, H., Scheid, J.F., Zoller, M.J., Krogsgaard, M., Ott, R.G., Shukair, S., Artyomov, M.N., Pietzsch, J., Connors, M., Pereyra, F., *et al.*: **Polyreactivity**

increases the apparent affinity of anti-HIV antibodies by heteroligation. *Nature* 2010, **467**:591–595.

Nichol, K.L., Lind, A., Margolis, K.L., Murdoch, M., McFadden, R., Hauge, M., Palese, P., and Garcia-Sastre, A.: **Influenza vaccines: present and future.** *J. Clin. Invest.* 2002, **110**:9–13.

16-22 Effective vaccines must induce long-lasting protection while being safe and inexpensive.

Gupta, R.K., Best, J., and MacMahon, E.: **Mumps and the UK epidemic 2005.** *BMJ* 2005, **330**:1132–1135.

Hviid, A., Rubin, S., and Mühlemann, K.: **Mumps.** *Lancet* 2008, **371**:932–944.

Magnan, S., and Drake, M.: **The effectiveness of vaccination against influenza in healthy, working adults.** *N. Engl. J. Med.* 1995, **333**:889–893.

16-23 Live-attenuated viral vaccines are usually more potent than 'killed' vaccines and can be made safer by the use of recombinant DNA technology.

Mueller, S.N., Langley, W.A., Carnero, E., García-Sastre, A., and Ahmed, R.: **Immunization with live attenuated influenza viruses that express altered NS1 proteins results in potent and protective memory CD8+ T-cell responses.** *J. Virol.* 2010, **84**:1847–1855.

Murphy, B.R., and Collins, P.L.: **Live-attenuated virus vaccines for respiratory syncytial and parainfluenza viruses: applications of reverse genetics.** *J. Clin. Invest.* 2002, **110**:21–27.

Pena, L., Vincent, A.L., Ye, J., Ciacci-Zanella, J.R., Angel, M., Lorusso, A., Gauger, P.C., Janke, B.H., Loving, C.L., and Perez, D.R.: **Modifications in the polymerase genes of a swine-like triple-reassortant influenza virus to generate live attenuated vaccines against 2009 pandemic H1N1 viruses.** *J. Virol.* 2011, **85**:456–469.

Subbarao, K., Murphy, B.R., and Fauci, A.S.: **Development of effective vaccines against pandemic influenza.** *Immunity* 2006, **24**:5–9.

Whittle, J.R., Wheatley, A.K., Wu, L., Lingwood, D., Kanekiyo, M., Ma, S.S., Narpala, S.R., Yassine, H.M., Frank, G.M., Yewdell, J.W., *et al.*: **Flow cytometry reveals that H5N1 vaccination elicits cross-reactive stem-directed antibodies from multiple Ig heavy-chain lineages.** *J. Virol.* 2014, **88**:4047–4057.

16-24 Live-attenuated vaccines can be developed by selecting nonpathogenic or disabled bacteria or by creating genetically attenuated parasites (GAPs).

Grode, L., Seiler, P., Baumann, S., Hess, J., Brinkmann, V., Nasser Eddine, A., Mann, P., Goosmann, C., Bandermann, S., Smith, D., *et al.*: **Increased vaccine efficacy against tuberculosis of recombinant *Mycobacterium bovis* bacille Calmette–Guérin mutants that secrete listeriolysin.** *J. Clin. Invest.* 2005, **115**:2472–2479.

Guleria, I., Teitelbaum, R., McAdam, R.A., Kalpana, G., Jacobs, W.R., Jr, and Bloom, B.R.: **Auxotrophic vaccines for tuberculosis.** *Nat. Med.* 1996, **2**:334–337.

Labaied, M., Harupa, A., Dumpit, R.F., Coppens, I., Mikolajczak, S.A., and Kappe, S.H.: ***Plasmodium yoelii* sporozoites with simultaneous deletion of P52 and P36 are completely attenuated and confer sterile immunity against infection.** *Infect. Immun.* 2007, **75**:3758–3768.

Martin, C.: **The dream of a vaccine against tuberculosis; new vaccines improving or replacing BCG?** *Eur. Respir. J.* 2005, **26**:162–167.

Thaiss, C.A., and Kaufmann, S.H.: **Toward novel vaccines against tuberculosis: current hopes and obstacles.** *Yale J. Biol. Med.* 2010, **83**:209–215.

Vaughan, A.M., Wang, R., and Kappe, S.H.: **Genetically engineered, attenuated whole-cell vaccine approaches for malaria.** *Hum. Vaccines* 2010, **6**:1–8.

16-25 The route of vaccination is an important determinant of success.

Amorij, J.P., Hinrichs, W.Lj., Frijlink, H.W., Wilschut, J.C., and Huckriede, A.: **Needle-free influenza vaccination.** *Lancet Infect. Dis.* 2010, **10**:699–711.

Belyakov, I.M., and Ahlers, J.D.: **What role does the route of immunization play in the generation of protective immunity against mucosal pathogens?** *J. Immunol.* 2009, **183**:6883–6892.

Douce, G., Fontana, M., Pizza, M., Rappuoli, R., and Dougan, G.: **Intranasal immunogenicity and adjuvanticity of site-directed mutant derivatives of cholera toxin.** *Infect. Immun.* 1997, **65**:2821–2828.

Dougan, G., Ghaem-Maghami, M., Pickard, D., Frankel, G., Douce, G., Clare, S., Dunstan, S., and Simmons, C.: **The immune responses to bacterial antigens encountered in vivo at mucosal surfaces.** *Phil. Trans. R. Soc. Lond. B* 2000, **355**:705–712.

Eriksson, K., and Holmgren, J.: **Recent advances in mucosal vaccines and adjuvants.** *Curr. Opin. Immunol.* 2002, **14**:666–672.

16-26 *Bordetella pertussis* **vaccination illustrates the importance of the perceived safety of a vaccines.**

Decker, M.D., and Edwards, K.M.: **Acellular pertussis vaccines.** *Pediatr. Clin. North Am.* 2000, **47**:309–335.

Madsen, K.M., Hviid, A., Vestergaard, M., Schendel, D., Wohlfahrt, J., Thorsen, P., Olsen, J., and Melbye, M.: **A population-based study of measles, mumps, and rubella vaccination and autism.** *N. Engl. J. Med.* 2002, **347**:1477–1482.

Mortimer, E.A.: **Pertussis vaccines**, in Plotkin, S.A., and Mortimer, E.A. (eds): *Vaccines*, 2nd ed. Philadelphia, W.B. Saunders Co., 1994.

Poland, G.A.: **Acellular pertussis vaccines: new vaccines for an old disease.** *Lancet* 1996, **347**:209–210.

16-27 **Conjugate vaccines have been developed as a result of linked recognition between T and B cells.**

Berry, D.S., Lynn, F., Lee, C.H., Frasch, C.E., and Bash, M.C.: **Effect of O acetylation of Neisseria meningitidis serogroup A capsular polysaccharide on development of functional immune responses.** *Infect. Immun.* 2002, **70**:3707–3713.

Bröker, M., Dull, P.M., Rappuoli, R., and Costantino, P.: **Chemistry of a new investigational quadrivalent meningococcal conjugate vaccine that is immunogenic at all ages.** *Vaccine* 2009, **27**:5574–5580.

Levine, O.S., Knoll, M.D., Jones, A., Walker, D.G., Risko, N., and Gilani, Z.: **Global status of *Haemophilus influenzae* type b and pneumococcal conjugate vaccines: evidence, policies, and introductions.** *Curr. Opin. Infect. Dis.* 2010, **23**:236–241.

Peltola, H., Kilpi, T., and Anttila, M.: **Rapid disappearance of *Haemophilus influenzae* type b meningitis after routine childhood immunisation with conjugate vaccines.** *Lancet* 1992, **340**:592–594.

Rappuoli, R.: **Conjugates and reverse vaccinology to eliminate bacterial meningitis.** *Vaccine* 2001, **19**:2319–2322.

16-28 **Peptide-based vaccines can elicit protective immunity, but they require adjuvants and must be targeted to the appropriate cells and cell compartment to be effective.**

Alonso, P.L., Sacarlal, J., Aponte, J.J., Leach, A., Macete, E., Aide, P., Sigauque, B., Milman, J., Mandomando, I., Bassat, Q., *et al.*: **Duration of protection with RTS, S/AS02A malaria vaccine in prevention of *Plasmodium falciparum* disease in Mozambican children: single-blind extended follow-up of a randomised controlled trial.** *Lancet* 2005, **366**:2012–2018.

Berzofsky, J.A.: **Epitope selection and design of synthetic vaccines. Molecular approaches to enhancing immunogenicity and cross-reactivity of engineered vaccines.** *Ann. N.Y. Acad. Sci.* 1993, **690**:256–264.

Davenport, M.P., and Hill, A.V.: **Reverse immunogenetics: from HLA-disease associations to vaccine candidates.** *Mol. Med. Today* 1996, **2**:38–45.

Hill, A.V.: **Pre-erythrocytic malaria vaccines: towards greater efficacy.** *Nat. Rev. Immunol.* 2006, **6**:21–32.

Hoffman, S.L., Rogers, W.O., Carucci, D.J., and Venter, J.C.: **From genomics to vaccines: malaria as a model system.** *Nat. Med.* 1998, **4**:1351–1353.

Ottenhoff, T.H., Doherty, T.M., Dissel, J.T., Bang, P., Lingnau, K., Kromann, I., and Andersen, P.: **First in humans: a new molecularly defined vaccine shows excellent safety and strong induction of long-lived *Mycobacterium tuberculosis*-specific Th1-cell like responses.** *Hum. Vaccin.* 2010, **6**:1007–1015.

Zwaveling, S., Ferreira Mota, S.C., Nouta, J., Johnson, M., Lipford, G.B., Offringa, R., van der Burg, S.H., and Melief, C.J.: **Established human papillomavirus type 16-expressing tumors are effectively eradicated following vaccination with long peptides.** *J. Immunol.* 2002, **169**:350–358.

16-29 **Adjuvants are important for enhancing the immunogenicity of vaccines, but few are approved for use in humans.**

Coffman, R.L., Sher, A., and Seder, R.A.: **Vaccine adjuvants: putting innate immunity to work.** *Immunity* 2010, **33**:492–503.

Hartmann, G., Weiner, G.J., and Krieg, A.M.: **CpG DNA: a potent signal for growth, activation, and maturation of human dendritic cells.** *Proc. Natl Acad. Sci. USA* 1999, **96**:9305–9310.

Palucka, K., Banchereau, J., and Mellman, I.: **Designing vaccines based on biology of human dendritic cell subsets.** *Immunity* 2010, **33**:464–478.

Persing, D.H., Coler, R.N., Lacy, M.J., Johnson, D.A., Baldridge, J.R., Hershberg, R.M., and Reed, S.G.: **Taking toll: lipid A mimetics as adjuvants and immunomodulators.** *Trends Microbiol.* 2002, **10**:S32–S37.

Pulendran, B.: **Modulating vaccine responses with dendritic cells and Toll-like receptors.** *Immunol. Rev.* 2004, **199**:227–250.

Takeda, K., Kaisho, T., and Akira, S.: **Toll-like receptors.** *Annu. Rev. Immunol.* 2003, **21**:335–376.

16-30 Protective immunity can be induced by DNA-based vaccination.

Donnelly, J.J., Ulmer, J.B., Shiver, J.W., and Liu, M.A.: **DNA vaccines.** *Annu. Rev. Immunol.* 1997, **15**:617–648.

Gurunathan, S., Klinman, D.M., and Seder, R.A.: **DNA vaccines: immunology, application, and optimization.** *Annu. Rev. Immunol.* 2000, **18**:927–974.

Li, L., Kim, S., Herndon, J.M., Goedegebuure, P., Belt, B.A., Satpathy, A.T., Fleming, T.P., Hansen, T.H., Murphy, K.M., and Gillanders, W.E.: **Cross-dressed CD8α+/CD103+ dendritic cells prime CD8+ T cells following vaccination.** *Proc. Natl Acad. Sci. USA* 2012, **109**:12716–12721.

Nchinda, G., Kuroiwa, J., Oks, M., Trumpfheller, C., Park, C.G., Huang, Y., Hannaman, D., Schlesinger, S.J., Mizenina, O., Nussenzweig, M.C., *et al.*: **The efficacy of DNA vaccination is enhanced in mice by targeting the encoded protein to dendritic cells.** *J. Clin. Invest.* 2008, **118**:1427–1436.

Wolff, J.A., and Budker, V.: **The mechanism of naked DNA uptake and expression.** *Adv. Genet.* 2005, **54**:3–20.

16-31 Vaccination and checkpoint blockade may be useful in controlling existing chronic infections.

Burke, R.L.: **Contemporary approaches to vaccination against herpes simplex virus.** *Curr. Top. Microbiol. Immunol.* 1992, **179**:137–158.

Grange, J.M., and Stanford, J.L.: **Therapeutic vaccines.** *J. Med. Microbiol.* 1996, **45**:81–83.

Hill, A., Jugovic, P., York, I., Russ, G., Bennink, J., Yewdell, J., Ploegh, H., and Johnson, D.: **Herpes simplex virus turns off the TAP to evade host immunity.** *Nature* 1995, **375**:411–415.

Lu, W., Arraes, L.C., Ferreira, W.T., and Andrieu, J.M.: **Therapeutic dendritic-cell vaccine for chronic HIV-1 infection.** *Nat. Med.* 2004, **10**:1359–1365.

Modlin, R.L.: **Th1–Th2 paradigm: insights from leprosy.** *J. Invest. Dermatol.* 1994, **102**:828–832.

Plebanski, M., Proudfoot, O., Pouniotis, D., Coppel, R.L., Apostolopoulos, V., and Flannery, G.: **Immunogenetics and the design of *Plasmodium falciparum* vaccines for use in malaria-endemic populations.** *J. Clin. Invest.* 2002, **110**:295–301.

Reiner, S.L., and Locksley, R.M.: **The regulation of immunity to *Leishmania major*.** *Annu. Rev. Immunol.* 1995, **13**:151–177.

Stanford, J.L.: **The history and future of vaccination and immunotherapy for leprosy.** *Trop. Geogr. Med.* 1994, **46**:93–107.

APPENDICES

The Immunologist's Toolbox

A-1. Immunization.

Natural immune responses are normally directed at antigens borne by pathogenic microorganisms. The immune system can also be induced to respond to simple nonliving antigens, and experimental immunologists have focused on the responses to these simple antigens in developing our understanding of the immune response. The deliberate induction of an immune response is known as **immunization**. Experimental immunizations are routinely carried out by injecting the test antigen into the animal or human subject. The route, dose, and form in which antigen is administered can profoundly affect whether a response occurs and the type of response that is produced. The induction of protective immune responses against common microbial pathogens in humans is often called vaccination, although this term originally referred to the induction of immune responses against smallpox by immunizing with the cross-reactive virus, vaccinia.

To determine whether an immune response has occurred and to follow its course, the immunized individual is monitored for the appearance of immune reactants. Immune responses to most antigens include soluble factors, such as cytokines and specific antibodies, and cellular responses, such as the generation of specific effector T cells. Monitoring the cytokine and antibody responses usually involves the analysis of relatively crude preparations of **antiserum** (plural: **antisera**). **Serum** is the fluid phase of clotted blood, which, if taken from an individual immunized against a particular antigen, is called antiserum. To study immune responses mediated by T cells, blood lymphocytes or cells from lymphoid organs such as the spleen are tested; T-cell responses are more commonly studied in experimental animals than in humans.

Antisera generated by immunization with even the simplest antigen will contain many different antibody molecules that bind to the immunogen in slightly different ways. In addition, antisera contain many antibodies that do not bind at all to the immunizing antigen, because they were present in the individual prior to immunization. These nonspecific antibodies often lead to technical difficulties in using antisera for detecting an immunogen. To circumvent this problem, antibodies that bind to the immunogen can be purified by affinity chromatography using immobilized antigen (see Section A-3). Alternatively, these problems can be avoided by making monoclonal antibodies (see Section A-7).

Any substance that can elicit an immune response is said to be **immunogenic** and is called an **immunogen**. There is a clear operational distinction between an immunogen and an antigen. Immunogens are substances that elicit an adaptive immune response, whereas an antigen is defined as any substance that can bind to a specific antibody. All antigens therefore have the potential to elicit specific antibodies; however, not all antigens are immunogenic. An example of this distinction is evident when considering protein antigens. In spite of the fact that antibodies against proteins are of enormous utility in experimental biology and medicine, purified proteins are not generally immunogenic. This is because purified proteins lack microbial-associated molecular patterns (MAMPs), and therefore do not elicit an innate immune response. To provoke an immune response to a purified protein, the protein must be administered together with an adjuvant (see below).

Certain properties of antigens that favor the initiation of an adaptive immune response have been defined by studying antibody responses to simple natural proteins such as hen egg-white lysozyme, to synthetic polypeptide antigens, and to small organic molecules of simple structure. The study of antibody responses to small organic molecules, such as phenyl arsonates and nitrophenyls, was essential in defining early immunological principles. These molecules do not provoke antibodies when injected by themselves. However, antibodies can be raised against them if the molecule is attached covalently, by simple chemical reactions, to a protein carrier. Such small molecules were termed **haptens** (from the Greek *haptein*, to fasten) by the immunologist **Karl Landsteiner**, who first studied them in the early 20th century. He found that animals immunized with a hapten–carrier conjugate produced three distinct sets of antibodies (**Fig. A.1**). One set comprised hapten-specific antibodies that reacted with the same hapten on any carrier, as well as with free hapten. The second set of antibodies was specific for the carrier protein, as shown by their ability to bind both the hapten-modified and unmodified carrier protein. Finally, some antibodies reacted only with the specific conjugate of hapten and carrier used for immunization. Landsteiner's studies focused primarily on the antibody response to the hapten, as these small molecules could be synthesized in many closely related forms. He observed that antibodies raised against a particular hapten bind that hapten but, in general, fail to bind even very closely related chemical structures. The binding of haptens by anti-hapten antibodies has played an important part in defining the precision of antigen binding by antibody molecules. Anti-hapten antibodies are also important medically because they mediate allergic reactions to penicillin and other compounds that elicit antibody responses when they attach to self proteins (see Section 14-10).

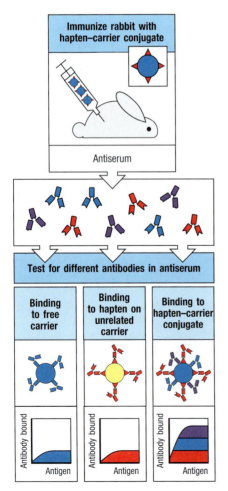

Fig. A.1 Antibodies can be elicited by small chemical groups called haptens only when the hapten is linked to an immunogenic protein carrier. Following immunization with a hapten–carrier conjugate, three types of antibodies are produced. One set (blue) binds the carrier protein alone and is called carrier-specific. One set (red) binds to the hapten on any carrier or to free hapten in solution and is called hapten-specific. One set (purple) binds only the specific conjugate of hapten and carrier used for immunization, apparently binding to sites at which the hapten joins the carrier, and is called conjugate-specific. The amount of antibody of each type in this serum is shown schematically in the graphs at the bottom; note that the original antigen binds more antibody than the sum of anti-hapten and anti-carrier antibodies as a result of the additional binding of conjugate-specific antibody.

The route by which antigen is administered affects both the magnitude and the type of response obtained. The most common routes by which antigen is introduced experimentally or as a vaccine into the body are into tissue by **subcutaneous** (**s.c.**) injection into the fatty layer just below the dermis or by **intradermal** (**i.d.**) or **intramuscular** (**i.m.**) injection; directly into the blood-stream by **intravenous** (**i.v.**) injection or transfusion; into the gastrointestinal tract by oral administration; and into the respiratory tract by **intranasal** (**i.n.**) administration or inhalation.

Antigens injected subcutaneously generally elicit strong responses, most probably because the antigen is taken up by resident dendritic cells in the skin and efficiently presented in local lymph nodes, and so this is the method most commonly used when the object of the experiment is to elicit specific antibodies or T cells against a given antigen. Antigens injected or transfused directly into the bloodstream tend to induce immune unresponsiveness or tolerance unless they bind to host cells or are in the form of aggregates that are readily taken up by antigen-presenting cells.

Antigen administration via the gastrointestinal tract is used mostly in the study of allergy. It has distinctive effects, frequently eliciting a local antibody response in the intestinal lamina propria, while producing a systemic state of tolerance that manifests as a diminished response to the same antigen if subsequently administered in immunogenic form elsewhere in the body (see Chapter 12). This 'split tolerance' may be important in avoiding allergy to antigens in food, because the local response prevents food antigens from entering the body, while the inhibition of systemic immunity helps to prevent the formation of IgE antibodies, which are the cause of such allergies (see Chapter 14).

The immune response to an antigen is also influenced by the dose of immunogen administered. Below a certain threshold dose, most proteins do not elicit any immune response. Above the threshold dose, there is a gradual increase in the response as the dose of antigen is increased, until a broad plateau level is reached, followed by a decline at very high antigen doses (**Fig. A.2**). In general, secondary and subsequent immune responses occur at lower antigen doses and achieve higher plateau values, which is a sign of immunological memory.

Most proteins are poorly immunogenic or nonimmunogenic when administered by themselves. Strong adaptive immune responses to protein antigens almost always require that the antigen be injected in a mixture known as an **adjuvant**. An adjuvant is any substance that enhances the immunogenicity of substances mixed with it. Commonly used adjuvants are listed in **Fig. A.3**.

Adjuvants generally enhance immunogenicity in two different ways. First, adjuvants convert soluble protein antigens into particulate material, which is more readily ingested by phagocytic antigen-presenting cells such as macrophages and dendritic cells. For example, the antigen can be adsorbed on particles of the adjuvant (such as alum), be made particulate by emulsification in mineral oils, or be incorporated into the colloidal particles of immune stimulatory complexes (ISCOMs). Second, and more important, adjuvants contain PAMPs that elicit a strong innate immune response. When taken up by phagocytic cells, the PAMPs in the adjuvant stimulate inflammatory cytokine production and induce the activation of the antigen-presenting cell. The activated antigen-presenting cells upregulate abundant levels of co-stimulatory molecules that are important for activating T cells. Activated antigen-presenting cells also upregulate high levels of MHC class I and class II proteins, plus many of the proteins important for efficient antigen processing and presentation (see Section 3-12). Due to the strong local inflammatory responses induced by adjuvants that contain PAMPs, most of the adjuvants commonly used in experimental animals are not approved for use in humans.

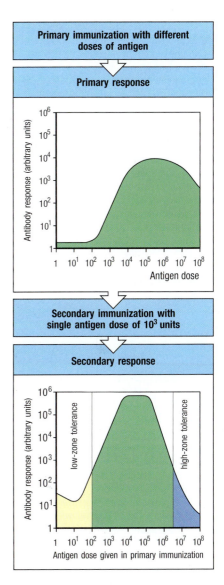

Fig. A.2 The dose of antigen used in an initial immunization affects the primary and the secondary antibody response. The typical antigen dose–response curve shown here illustrates both the influence of dose on a primary antibody response (amounts of antibody produced expressed in arbitrary units) and the effect of the dose used for priming on a secondary antibody response elicited by a dose of antigen of 10^3 arbitrary mass units. Very low doses of antigen do not cause an immune response at all. Slightly higher doses seem to inhibit specific antibody production, an effect known as low-zone tolerance. Above these doses there is a steady increase in the response with antigen dose until an optimum response is reached; this response persists across a broad range of doses. Very high doses of antigen also inhibit immune responsiveness to a subsequent challenge, a phenomenon known as high-zone tolerance.

Fig. A.3 **Common adjuvants and their use.** Adjuvants are mixed with the antigen and usually render it particulate, which helps to retain the antigen in the body and promotes uptake by macrophages. Most adjuvants include bacteria or bacterial components that stimulate macrophages and dendritic cells, aiding in the induction of the immune response. ISCOMs are small micelles of the detergent Quil A; when viral proteins are placed in these micelles, they apparently fuse with the antigen-presenting cell, allowing the antigen to enter the cytosol. Thus, the antigen-presenting cell can stimulate a response to the viral protein, much as a virus infecting these cells would stimulate an antiviral response. Vaccines designed to elicit responses to purified proteins often include compounds that stimulate pattern recognition receptors, such as Toll-like receptors (TLRs), NOD-like receptors (NLRs), or C-type lectin receptors.

Adjuvants that enhance immune responses		
Adjuvant name	**Composition**	**Mechanism of action**
Incomplete Freund's adjuvant	Oil-in-water emulsion	Delayed release of antigen; enhanced uptake by macrophages
Complete Freund's adjuvant	Oil-in-water emulsion with dead mycobacteria that stimulate C-type lectin receptors	Delayed release of antigen; enhanced uptake by macrophages; induction of co-stimulators in macrophages
Freund's adjuvant with MDP	Oil-in-water emulsion with muramyl dipeptide (MDP), a constituent of mycobacteria that stimulates NOD-like receptors	Similar to complete Freund's adjuvant
Alum (aluminum hydroxide)	Aluminum hydroxide gel	Delayed release of antigen; enhanced macrophage uptake
Alum plus *Bordetella pertussis*	Aluminum hydroxide gel with killed *B. pertussis*	Delayed release of antigen; enhanced uptake by macrophages; induction of co-stimulators
Immune stimulatory complexes (ISCOMs)	Matrix of Quil A containing viral proteins	Delivers antigen to cytosol; allows induction of cytotoxic T cells
TLR agonists	Lipopolysaccharide, flagellin, lipopeptides, ds-RNA, unmethylated DNA	Inflammatory cytokine production, induction of co-stimulators, enhanced antigen presentation to T cells
NOD-like receptor (NLR) agonists	Muramyl dipeptide (bacterial cell wall constituent)	Inflammatory cytokine production, induction of co-stimulators, enhanced antigen presentation to T cells
C-type lectin receptor agonists	Mycobacterial cell wall component trehalose-6,6'-dimycolate	Inflammatory cytokine production

Nevertheless, some human vaccines naturally contain microbial antigens that can also act as effective adjuvants. For example, purified constituents of the bacterium *Bordetella pertussis*, which is the causative agent of whooping cough, are used as both antigen and adjuvant in the triplex DPT (diphtheria, pertussis, tetanus) vaccine. In addition, modified TLR ligands, such as monophosphoryl lipid A, a derivative of LPS, or poly(I):poly(C12U), a derivative of polyI:C, are currently included as components of several human vaccines.

A-2   Antibody responses.

B cells contribute to adaptive immunity by secreting antibodies, and the response of B cells to an injected immunogen is usually measured by analyzing the specific antibody produced in a **humoral immune response**. This is most conveniently achieved by assaying the antibody that accumulates in the fluid phase of the blood, or **plasma**; such antibodies are known as circulating antibodies. Circulating antibody is usually measured by collecting blood, allowing it to clot, and then isolating the serum from the clotted blood. The amount and characteristics of the antibody in the resulting antiserum are then determined using the assays described below. Because assays for antibody were originally conducted using antisera from immune individuals, they are commonly referred to as **serological assays**, and the use of antibodies in such testing is often called **serology**.

The most important characteristics of an antibody response are the specificity, amount, isotype or class, and affinity of the antibodies produced. The **specificity** determines the ability of the antibody to distinguish the immunogen from other antigens. The amount of antibody can be determined in many different ways and is a function of the number of responding B cells, their rate of antibody synthesis, and the persistence of the antibody after production. The persistence of an antibody in the plasma and extracellular fluid bathing the tissues is determined mainly by its isotype or class (see Sections 5-12 and 10-14); each isotype has a different half-life *in vivo*. The isotypic composition of an antibody response also determines the biological functions these antibodies can perform and the sites in which antibody will be found. Finally, the strength of binding of the antibody to its antigen in terms of a single antigen-binding site binding to a monovalent antigen is termed its **affinity**; the total binding strength of a molecule with more than one binding site is called its **avidity**. Binding strength is important: the higher the affinity of the antibody for its antigen, the less antibody is required to eliminate the antigen, because antibodies with higher affinity will bind at lower antigen concentrations. All these parameters of the humoral immune response help to determine the capacity of that response to protect the host from infection.

A-3   Affinity chromatography.

The specificity of antigen:antibody binding interactions can be exploited for the purification of a specific antigen from a complex mixture, or alternatively, for the purification of specific antibodies from antiserum containing a mixture of different antibodies. The technique employed is called **affinity chromatography** (Fig. A.4). For purification of an antigen, antigen-specific antibodies are bound, often covalently, to small, chemically reactive beads, which are loaded into a column. The antigen mixture is allowed to pass over the beads. The specific antigen binds; all the other components in the mixture can then be washed away. The specific antigen is then eluted, typically by lowering the pH to 2.5 or raising it to greater than 11. Antibodies bind stably under physiological conditions of salt concentration, temperature, and pH, but the binding is reversible because the bonds are noncovalent. Affinity chromatography can also be used to purify antibodies from complex antisera by using beads coated with specific antigen. The technique is known as affinity chromatography because it separates molecules on the basis of their affinity for one another.

A-4   Radioimmunoassay (RIA), enzyme-linked immunosorbent assay (ELISA), and competitive inhibition assay.

Radioimmunoassay (**RIA**) and **enzyme-linked immunosorbent assay** (**ELISA**) are direct binding assays for antibody (or antigen); both work on the same principle, but the means of detecting specific binding is different. Radioimmunoassays are commonly used to measure the levels of hormones in blood and tissue fluids, while ELISA assays are frequently used in viral diagnostics, for example, in detecting cases of infection with the human immunodeficiency virus (HIV), which is the cause of AIDS. For both of these methods, one needs a pure preparation of a known antigen or antibody, or both, in order to standardize the assay. We will describe the assay that is used to determine the amount of a specific antigen in a sample, for instance, the amount of HIV p24 protein in a patient's serum. For this, a preparation of pure antibody specific for the antigen is required. One can also use RIA or ELISA to determine the amount of specific antibody in a mixture, such as serum; in this case, a preparation of pure antigen is needed as a starting point.

For the determination of antigen concentration using RIA, pure antibody against the antigen is radioactively labeled, usually with ^{125}I; for ELISA, an enzyme is linked chemically to the antibody. The unlabeled component, which in this case would be the solution containing an unknown amount of antigen, is attached to a solid support, such as the wells of a plastic multiwell plate, which will adsorb a certain amount of any protein. Following this, the labeled antibody

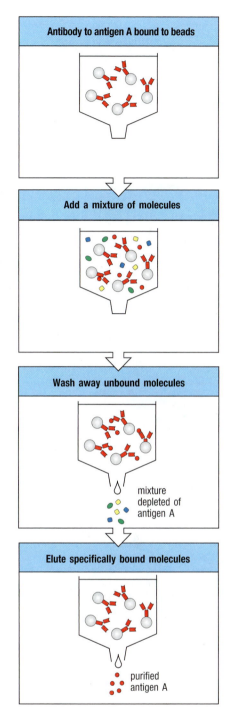

Fig. A.4 Affinity chromatography uses antigen–antibody binding to purify antigens or antibodies. To purify a specific antigen from a complex mixture of molecules, a monoclonal antibody is attached to an insoluble matrix, such as chromatography beads, and the mixture of molecules is passed over the matrix. The specific antibody binds the antigen of interest; other molecules are washed away. Specific antigen is then eluted by altering the pH, which can usually disrupt antibody–antigen bonds. Antibodies can be purified in the same way on beads coupled to antigen (not shown).

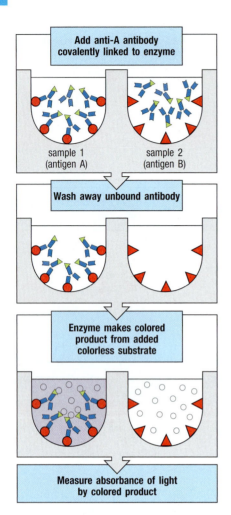

Fig. A.5 The principle of the enzyme-linked immunosorbent assay (ELISA). To detect antigen A, purified antibody specific for antigen A is linked chemically to an enzyme. The samples to be tested are coated onto the surface of plastic wells, to which they bind nonspecifically; residual sticky sites on the plastic are blocked by adding irrelevant proteins (not shown). The labeled antibody is then added to the wells under conditions that prevent nonspecific binding, so that only binding to antigen A causes the labeled antibody to be retained on the surface. Unbound labeled antibody is removed from all wells by washing, and bound antibody is detected by an enzyme-dependent color-change reaction. This assay allows arrays of wells known as microtiter plates to be read in fiberoptic multichannel spectrometers, greatly speeding the assay. Modifications of this basic assay allow antibody or antigen in unknown samples to be measured as shown in Figs. A.6 and A.25.

The figure panels, top to bottom, are labeled:
- **Add anti-A antibody covalently linked to enzyme** — sample 1 (antigen A), sample 2 (antigen B)
- **Wash away unbound antibody**
- **Enzyme makes colored product from added colorless substrate**
- **Measure absorbance of light by colored product**

is added to the well and allowed to bind to the unlabeled antigen under conditions in which nonspecific adsorption is blocked, and any unbound antibody and other proteins are washed away. Antibody binding in RIA is measured directly in terms of the amount of radioactivity retained by the coated wells, whereas in ELISA the binding is detected by a reaction that converts a substrate into a reaction product of a different color (Fig. A.5). The color change can be read directly in the reaction tray, making data collection very easy and providing a quantitative measurement of reaction product concentration; furthermore, ELISA also avoids the hazards of radioactivity. This makes ELISA the preferred method for most direct-binding assays. In a variation of this assay, labeled anti-immunoglobulin antibodies can also be used in RIA or ELISA as a second layer, following the binding of unlabeled antibody to unlabeled antigen-coated plates. The use of such a second layer amplifies the signal, because at least two molecules of the labeled anti-immunoglobulin antibody are able to bind to each unlabeled antibody. As mentioned above, RIA and ELISA can also be carried out in reverse when the goal is to determine the amount of antibody in a solution; in this case, unlabeled antibody is adhered to the plates, labeled antigen is added, and the amount of labeled antigen bound after washing is measured.

A modification of ELISA known as a **capture** or **sandwich ELISA** (or more generally as an **antigen-capture assay**) is commonly used to detect secreted products such as cytokines. Rather than the antigen being directly attached to a plastic plate, antigen-specific antibodies are bound to the plate. These are able to bind antigen with high affinity, and thus concentrate it on the surface of the plate, even with antigens that are present in very low concentrations in the initial mixture. A separate labeled antibody that recognizes a different epitope from that recognized by the immobilized first antibody is then used to detect the bound antigen.

Another variant of the antigen-capture assay, often referred to as a **multiplex** assay, has been developed to allow quantitation of multiple antigens in a single sample. This technique is often utilized to examine the levels of multiple cytokines in clinical serum samples, or in sera from experimental animals, cases in which it is not feasible to assess each cytokine of interest individually. For this type of assay, small microspheres are differentially labeled with fluorescent dyes that can be distinguished based on their distinct emission spectra. Microspheres labeled with a given fluorescent dye are conjugated to antibodies specific for one antigen, for instance, a single cytokine. The microspheres—up to 100 different microspheres with unique identifiers—are added to the sample to capture the antigen. Bound antigen is then detected using a second antibody that binds the antigen at a distinct site. This second antibody is conjugated to a different fluorescent dye, and the magnitude of its fluorescence is a measure of the quantity of bound antigen. The machine that performs this multiplex analysis, the Luminex® analyzer, then measures the amount of fluorescence associated with each differentially labeled microsphere.

These assays illustrate two crucial aspects of all serological assays. First, at least one of the reagents must be available in a pure, detectable form in order to obtain quantitative information. Second, there must be a means of separating the bound fraction of the labeled reagent from the unbound, free fraction so that the percentage of specific binding can be determined. Normally, this separation is achieved by having the unlabeled partner trapped on a solid support. Labeled molecules that do not bind can then be washed away, leaving just the labeled partner that has bound. In Fig. A.5, the unlabeled antigen is attached to the well and the labeled antibody is trapped by binding to it. The separation of bound reagent from the free fraction is an essential step in every assay that uses antibodies.

RIA and ELISA do not allow one to measure directly the amount of antigen or antibody in a sample of unknown composition, because both depend on the binding of a pure labeled antigen or antibody. There are various ways around

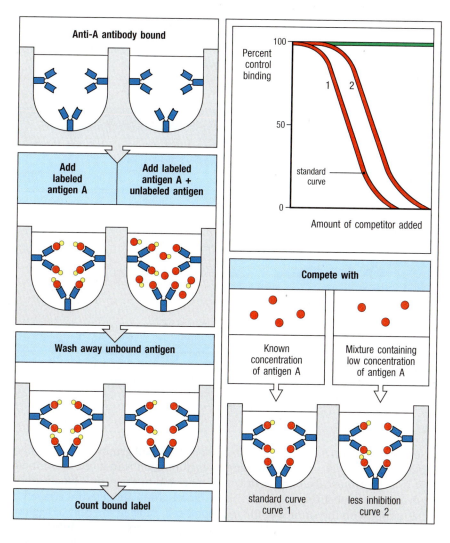

Anti-A antibody bound

Add labeled antigen A

Add labeled antigen A + unlabeled antigen

Wash away unbound antigen

Count bound label

100

Percent control binding

50

0

standard curve

1   2

Amount of competitor added

Compete with

Known concentration of antigen A

Mixture containing low concentration of antigen A

standard curve curve 1

less inhibition curve 2

Fig. A.6 Competitive inhibition assay for antigen in unknown samples.
A fixed amount of unlabeled antibody is attached to a set of wells, and a standard reference preparation of a labeled antigen is bound to it. Unlabeled standard or test samples are then added in various amounts and the displacement of labeled antigen is measured, generating characteristic inhibition curves. A standard curve is obtained by using known amounts of unlabeled antigen identical to that used as the labeled species, and comparison with this curve allows the amount of antigen in unknown samples to be calculated. The green line on the graph represents a sample lacking any substance that reacts with anti-A antibodies.

this problem, one of which is to use a **competitive inhibition assay**, as shown in Fig. A.6. In this type of assay, the presence and amount of a particular antigen in an unknown sample are determined by the antigen's ability to compete with a labeled reference antigen for binding to an antibody attached to a plastic well. A standard curve is first constructed by adding various amounts of a known, unlabeled standard preparation; the assay can then measure the amount of antigen in unknown samples by comparison with the standard. The competitive binding assay can also be used for measuring antibody in a sample of unknown composition by attaching the appropriate antigen to the plate and measuring the ability of the test sample to inhibit the binding of a labeled specific antibody.

A-5   Hemagglutination and blood typing.

The direct measurement of antibody binding to antigen is used in most quantitative serological assays. However, some important assays are based on the ability of antibody binding to alter the physical state of the antigen to which the antibody binds. These secondary interactions can be detected in a variety of ways. For instance, when the antigen is displayed on the surface of a large particle such as a bacterium, antibodies can cause the bacteria to clump, or **agglutinate**. The same principle applies to the reactions used in blood typing, only here the target antigens are on the surface of red blood cells and the clumping reaction caused by antibodies against them is called **hemagglutination** (from the Greek *haima*, blood).

Fig. A.7 Hemagglutination is used to type blood groups and match compatible donors and recipients for blood transfusion. Common gut bacteria bear antigens that are similar or identical to blood-group antigens, and these stimulate the formation of antibodies against these antigens in individuals who do not bear the corresponding antigen on their own red blood cells (left column); thus, type O individuals, who lack A and B, have both anti-A and anti-B antibodies, whereas type AB individuals have neither. The pattern of agglutination of the red blood cells of a transfusion donor or recipient with anti-A and anti-B antibodies reveals the individual's ABO blood group. Before transfusion, the serum of the recipient is also tested for antibodies that agglutinate the red blood cells of the donor, and vice versa, a procedure called a cross-match, which may detect potentially harmful antibodies against other blood groups that are not part of the ABO system.

Serum from individuals of type	Red blood cells from individuals of type			
	O	A	B	AB
	Express the carbohydrate structures			
	R–GlcNAc–Gal | Fuc	R–GlcNAc–Gal–GalNAc | Fuc	R–GlcNAc–Gal–Gal | Fuc	R–GlcNAc–Gal–GalNAc | Fuc + R–GlcNAc–Gal–Gal | Fuc
O Anti-A and anti-B antibodies	no agglutination	agglutination	agglutination	agglutination
A Anti-B antibodies	no agglutination	no agglutination	agglutination	agglutination
B Anti-A antibodies	no agglutination	agglutination	no agglutination	agglutination
AB No antibodies to A or B	no agglutination	no agglutination	no agglutination	no agglutination

Hemagglutination is used to determine the **ABO blood group** of blood donors and transfusion recipients. Clumping or agglutination is induced by antibodies or agglutinins called anti-A or anti-B that bind to the A or B blood-group substances, respectively (Fig. A.7). These blood-group antigens are arrayed in many copies on the surface of the red blood cell, causing the cells to agglutinate when cross-linked by antibodies. Because hemagglutination involves the cross-linking of blood cells by the simultaneous binding of antibody molecules to identical antigens on different cells, this reaction also demonstrates that each antibody molecule must have at least two identical antigen-binding sites.

A-6   Coombs tests and the detection of rhesus incompatibility.

Coombs tests use anti-immunoglobulin antibodies to detect the antibodies that cause **hemolytic disease of the newborn**, or **erythroblastosis fetalis**. Anti-immunoglobulin antibodies were first developed by **Robin Coombs**, and the test for this disease is still called the Coombs test. Hemolytic disease of the newborn occurs when a mother makes IgG antibodies specific for the **rhesus** or **Rh blood-group antigen** expressed on the red blood cells of her fetus. Rh-negative mothers make these antibodies when they are exposed at delivery to Rh-positive fetal red blood cells bearing the paternally inherited Rh antigen. During subsequent pregnancies, these antibodies are transported across the placenta to the fetus. This normal process is generally beneficial, as it protects newborn infants against infection. However, IgG anti-Rh antibodies coat the fetal red blood cells, which are then destroyed by phagocytic cells in the liver, causing a hemolytic anemia in the fetus and newborn infant.

Rh antigens are widely spaced on the red blood cell surface, and so the IgG anti-Rh antibodies do not bind in the correct conformation to fix complement and so do not cause lysis of red blood cells *in vitro*. Furthermore, for reasons that are not fully understood, antibodies against Rh antigens do not agglutinate red blood cells, unlike antibodies against the ABO blood-group antigens. Thus, detecting anti-Rh antibodies was difficult until anti-human immunoglobulin antibodies were developed. With these, maternal IgG antibodies bound to the fetal red blood cells can be detected after washing the cells to remove unbound immunoglobulin that is present in the fetal serum. Adding anti-human immunoglobulin antibodies against the washed fetal red blood cells agglutinates any cells to which maternal antibodies are bound. This is the **direct Coombs test** (**Fig. A.8**), so called because it directly detects antibody bound to the surface of the fetal red blood cells. An **indirect Coombs test** is used to detect nonagglutinating anti-Rh antibody in maternal serum: the serum is first incubated with Rh-positive red blood cells, which bind the anti-Rh antibody, after which the antibody-coated cells are washed to remove unbound immunoglobulin and are then agglutinated with anti-immunoglobulin antibody (see Fig. A.8). The indirect Coombs test allows Rh incompatibilities that might lead to hemolytic disease of the newborn to be detected, and this knowledge allows the disease to be prevented (see Section 15-10). The Coombs test is also commonly used to detect antibodies against drugs that bind to red blood cells and cause hemolytic anemia.

A-7   Monoclonal antibodies.

The antibodies generated in a natural immune response or after immunization in the laboratory are a mixture of molecules of different specificities and affinities. Some of this heterogeneity results from the production of antibodies that bind to different epitopes on the immunizing antigen, but even antibodies directed at a single antigenic determinant such as a hapten can be markedly heterogeneous, as shown by **isoelectric focusing**. In this technique, proteins are separated on the basis of their isoelectric point, the pH at which their net charge is zero. By electrophoresing proteins in a pH gradient for long enough, each molecule migrates along the pH gradient until it reaches the pH at which it is neutral and is thus concentrated (focused) at that point. When antiserum containing anti-hapten antibodies is treated in this way and then transferred to a solid support such as nitrocellulose paper, the anti-hapten antibodies can be detected by their ability to bind labeled hapten. The binding of antibodies of various isoelectric points to the hapten shows that even antibodies that bind the same antigenic determinant can be heterogeneous.

Antisera are valuable for many biological purposes but they have certain inherent disadvantages that relate to the heterogeneity of the antibodies they contain. First, each antiserum is different from all other antisera, even if raised in a genetically identical animal by using the identical preparation of antigen and the same immunization protocol. Second, antisera can be produced in only limited volumes, and thus it is impossible to use the identical serological reagent in a long or complex series of experiments or clinical tests. Finally, even antibodies purified by affinity chromatography (see Section A-3) can include minor populations of antibodies that give unexpected cross-reactions, which confound the analysis of experiments. To avoid these problems, and to harness the full potential of antibodies, it was necessary to develop a way of making an unlimited supply of antibody molecules of homogeneous structure and known specificity. This has been achieved through the production of monoclonal antibodies from cultures of hybrid antibody-forming cells or, more recently, by genetic engineering.

Starting in the 1950s, biochemists in search of a homogeneous preparation of antibody that they could subject to detailed chemical analysis turned to proteins produced by patients with multiple myeloma, a common tumor of plasma cells. It was known that antibodies are normally produced by plasma cells, and because this disease is associated with the presence of large amounts of a homogeneous gamma globulin called a **myeloma protein** in the patient's serum, it seemed likely that myeloma proteins would serve as models for

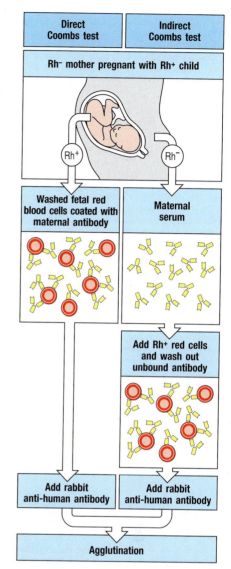

Fig. A.8 The Coombs direct and indirect anti-globulin tests for antibody against red blood cell antigens. An Rh⁻ mother of an Rh⁺ fetus can become immunized to fetal red blood cells that enter the maternal circulation at the time of delivery. In a subsequent pregnancy with an Rh⁺ fetus, IgG anti-Rh antibodies can cross the placenta and damage the fetal red blood cells. In contrast to anti-Rh antibodies, maternal anti-ABO antibodies are of the IgM isotype and cannot cross the placenta, and so do not cause harm. Anti-Rh antibodies do not agglutinate red blood cells, but their presence on the fetal red blood cell surface can be shown by washing away unbound immunoglobulin and then adding antibody against human immunoglobulin, which agglutinates the antibody-coated cells. Anti-Rh antibodies can be detected in the mother's serum in an indirect Coombs test; the serum is incubated with Rh⁺ red blood cells, and once the antibody has bound, the red blood cells are treated as in the direct Coombs test.

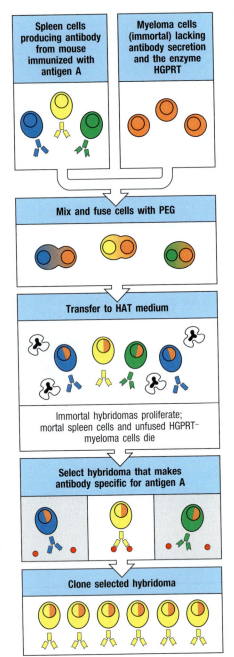

Spleen cells producing antibody from mouse immunized with antigen A	Myeloma cells (immortal) lacking antibody secretion and the enzyme HGPRT

Mix and fuse cells with PEG

Transfer to HAT medium

Immortal hybridomas proliferate; mortal spleen cells and unfused HGPRT− myeloma cells die

Select hybridoma that makes antibody specific for antigen A

Clone selected hybridoma

Fig. A.9 The production of monoclonal antibodies. Mice are immunized with antigen A and given an intravenous booster immunization 3 days before they are killed, in order to produce a large population of spleen cells secreting specific antibody. Spleen cells die after a few days in culture. To produce a continuous source of antibody they are fused with immortal myeloma cells by using polyethylene glycol (PEG) to produce a hybrid cell line called a hybridoma. The myeloma cells are selected beforehand to ensure that they are not secreting antibody themselves and that they lack the enzyme hypoxanthine:guanine phosphoribosyl transferase (HGPRT); without this enzyme, unfused myeloma cells are sensitive to the hypoxanthine–aminopterin–thymidine (HAT) medium, which is used to select hybrid cells. The HGPRT gene contributed by the spleen cell allows hybrid cells to survive in the HAT medium, and only hybrid cells can grow continuously in culture, because of the malignant potential contributed by the myeloma cells combined with the finite life-span of unfused spleen cells. Unfused myeloma cells and unfused spleen cells therefore die in the HAT medium, as shown here by cells with dark, irregular nuclei. Individual hybridomas are obtained by single cell dilution and then screened for antibody production, and single clones that make antibody of the desired specificity can be isolated and grown. The cloned hybridoma cells are grown in bulk culture to produce large amounts of antibody. As each hybridoma is descended from a single cell, all the cells of a hybridoma cell line make the same antibody molecule, which is thus called a monoclonal antibody.

normal antibody molecules. Thus, much of the early knowledge of antibody structure came from studies on myeloma proteins. These studies showed that monoclonal antibodies could be obtained from immortalized plasma cells. However, the antigen specificity of most myeloma proteins was unknown, which limited their usefulness as objects of study or as immunological tools.

This problem was solved by **Georges Köhler** and **César Milstein**, who devised a technique for producing a homogeneous population of antibodies of known antigenic specificity. They did this by fusing spleen cells from an immunized mouse to cells of a mouse myeloma to produce hybrid cells that both proliferated indefinitely and secreted antibody specific for the antigen used to immunize the spleen cell donor. The spleen cell provides the ability to make specific antibody, while the myeloma cell provides the ability to grow indefinitely in culture and secrete immunoglobulin continuously. By using a myeloma cell partner that produces no antibody proteins itself, the antibody produced by the hybrid cells comes only from the immune spleen cell partner. After fusion, the hybrid cells are selected using drugs that kill the myeloma parental cell, while the unfused parental spleen cells have a limited life-span and soon die, so that only hybrid myeloma cell lines, or **hybridomas**, survive. Those hybridomas producing antibody of the desired specificity are then identified and cloned by regrowing the cultures from single cells (**Fig. A.9**). Because each hybridoma is a **clone** derived from fusion with a single B cell, all the antibody molecules it produces are identical in structure, including their antigen-binding site and isotype. Such antibodies are called **monoclonal antibodies**. This technology has revolutionized the use of antibodies by providing a limitless supply of antibody of a single and known specificity. Monoclonal antibodies are now used in most serological assays, as diagnostic probes, and as therapeutic agents. So far, however, only mouse monoclonals are routinely produced in this way, and efforts to use the same approach to make human monoclonal antibodies have met with limited success. 'Fully human' therapeutic monoclonal antibodies are currently made by using phage display technology (described in Section A-8), by using recombinant DNA technology to clone and express antibody genes from human plasma cells (see Section A-9), or by immunizing transgenic mice (see Section A-34) carrying human antibody genes.

A-8   Phage display libraries for antibody V-region production.

In a technique for producing antibody-like molecules, gene segments encoding the antigen-binding variable, or V, domains of antibodies are fused to genes encoding the coat protein of a bacteriophage. Bacteriophages

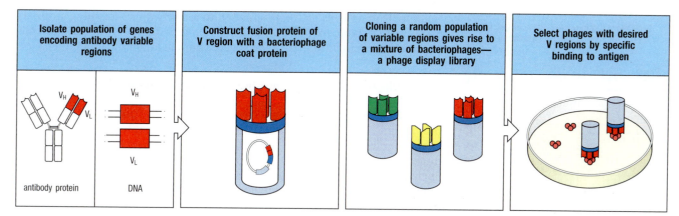

| Isolate population of genes encoding antibody variable regions | Construct fusion protein of V region with a bacteriophage coat protein | Cloning a random population of variable regions gives rise to a mixture of bacteriophages— a phage display library | Select phages with desired V regions by specific binding to antigen |

Fig. A.10 The production of antibodies by genetic engineering. Short primers to consensus sequences in heavy- and light-chain variable (V) regions of immunoglobulin genes are used to generate a library of heavy- and light-chain V-region DNAs by PCR, with spleen DNA as the starting material. These heavy- and light-chain V-region genes are cloned randomly into filamentous phages such that each phage expresses one heavy-chain and one light-chain V region as a surface fusion protein with antibody-like properties. The resulting phage display library is multiplied in bacteria, and the phages are then bound to a surface coated with antigen. The unbound phages are washed away; the bound phages are recovered, multiplied in bacteria, and again bound to antigen. After a few cycles, only specific high-affinity antigen-binding phages are left. These can be used like antibody molecules, or their V genes can be recovered and engineered into antibody genes to produce genetically engineered antibody molecules (not shown). This technology may replace the hybridoma technology for producing monoclonal antibodies, and has the advantage that humans can be used as the source of DNA.

containing such gene fusions are used to infect bacteria, and the resulting phage particles have coats that express the antibody-like fusion protein, with the antigen-binding domain displayed on the outside of the bacteriophages. A collection of recombinant phages, each displaying a different antigen-binding domain on the surface, is known as a **phage display library**. In much the same way that antibodies specific for a particular antigen can be isolated from a complex mixture by affinity chromatography (see Section A-3), phages expressing antigen-binding domains specific for a particular antigen can be isolated by selecting the phages in the library for binding to that antigen. The phage particles that bind are recovered and used to infect fresh bacteria. Each phage isolated in this way will produce a monoclonal antigen-binding particle analogous to a monoclonal antibody (Fig. A.10). The genes encoding the antigen-binding site, which are unique to each phage, can then be recovered from the phage DNA and used to construct genes for a complete antibody molecule by joining them to parts of immunoglobulin genes that encode the invariant parts of an antibody. When these reconstructed antibody genes are introduced into a suitable host-cell line, such as the non-antibody-producing myeloma cells used for hybridomas, the transfected cells can secrete antibodies with all the desirable characteristics of monoclonal antibodies produced from hybridomas.

A-9   Generation of human monoclonal antibodies from vaccinated individuals.

In some cases, human monoclonal antibodies can be made by cloning the rearranged antibody heavy- and light-chain gene sequences from plasma cells isolated from vaccinated individuals. Based on the expression of cell-surface molecules such as CD27 and CD38, human plasma cells can be isolated from the peripheral blood of individuals who were immunized approximately 1 week earlier. Individual plasma cells are sorted into wells of microtiter plates, and the antibody heavy- and light-chain variable-region sequences are cloned from each cell by PCR. These sequences are then inserted into constructs that recreate the full-length antibody heavy- and light-chain genes, and the paired heavy- and light-chain vectors are introduced into an immortalized human

Excitation and emission wavelengths of some commonly used fluorochromes		
Probe	Excitation (nm)	Emission (nm)
R-phycoerythrin (PE)	480; 565	578
Fluorescein	495	519
PerCP	490	675
Texas Red	589	615
Rhodamine	550	573

Fig. A.11 Excitation and emission wavelengths for common fluorochromes.

cell line. Human cells are then screened to identify those secreting antibody proteins that bind to the immunizing antigen. These immortalized cells become a permanent source of the specific human antibody protein.

A-10 Microscopy and imaging using fluorescent dyes.

Because antibodies bind stably and specifically to their corresponding antigen, they are invaluable as probes for identifying a particular molecule in cells, tissues, or biological fluids. Antibody molecules can be used to locate their target molecules accurately in single cells or in tissue sections by a variety of different labeling techniques. When the antibody itself, or the anti-immunoglobulin antibody used to detect it, is labeled with a fluorescent dye (a fluorochrome or fluorophore) and then detected by microscopy, the technique is known as **immunofluorescence microscopy**. As in all serological techniques, the antibody binds stably to its antigen, allowing unbound antibody to be removed by thorough washing. Because antibodies against proteins recognize the surface features of the native, folded protein, the native structure of the protein being sought usually needs to be preserved, either by using only the most gentle chemical fixation techniques or by using frozen tissue sections that are fixed only after the antibody reaction has been performed. Some antibodies, however, bind proteins even if they are denatured, and such antibodies will bind specifically even to protein in fixed tissue sections.

The fluorescent dye can be covalently attached directly to the specific antibody; however, the bound antibody is more commonly detected by fluorescently labeled anti-immunoglobulin, a technique known as **indirect immunofluorescence**. The dyes chosen for immunofluorescence are excited by light of one wavelength, usually blue or green, and emit light of a different wavelength in the visible spectrum. The most commonly used fluorochromes are fluorescein, which emits green light; Texas Red and peridinin chlorophyll protein (PerCP), which emit red light; and rhodamine and phycoerythrin (PE), which emit orange/red light (**Fig. A.11**). By using selective filters, only the light coming from the fluorochrome used is detected in the fluorescence microscope (**Fig. A.12**). Although Albert Coons first devised this technique to

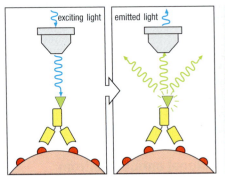
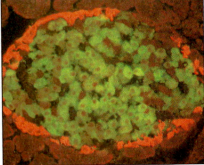

Fig. A.12 Immunofluorescence microscopy. Antibodies labeled with a fluorescent dye such as fluorescein (green triangle) are used to reveal the presence of their corresponding antigens in cells or tissues. The stained cells are examined using a microscope that exposes them to blue or green light to excite the fluorescent dye. The excited dye emits light at a characteristic wavelength, which is captured by viewing the sample through a selective filter. This technique is applied widely in biology to determine the location of molecules in cells and tissues. Different antigens can be detected in tissue sections by labeling antibodies with dyes of distinctive color. Here, antibodies against the protein glutamic acid decarboxylase (GAD) coupled to a green dye are shown to stain the β cells of pancreatic islets of Langerhans. The α cells do not make this enzyme and are labeled with antibodies against the hormone glucagon coupled with an orange fluorescent dye. GAD is an important antigen in type 1 diabetes, a disease in which the insulin-secreting β cells of the islets of Langerhans are destroyed by an immune attack on self tissues (see Chapter 15). Photograph courtesy of M. Solimena and P. De Camilli.

identify the plasma cell as the source of antibody, it can be used to detect the distribution of any protein. By attaching different dyes to different antibodies, the distribution of two or more molecules can be determined in the same cell or tissue section (see Fig. A.12).

The development of the **confocal fluorescent microscope**, which uses computer-aided techniques to produce ultrathin optical sections of a cell or tissue, gives very high resolution (sub-micrometer) fluorescence microscopy without the need for elaborate sample preparation. The light source for excitation (a laser) is focused onto a particular plane in the specimen, and the emitted light is refocused through a 'pinhole' so that only light from the desired plane reaches the detector, thus removing out-of-focus emissions from above or below the plane. This gives a sharper image than conventional fluorescence microscopy, and a three-dimensional picture can be built up from successive optical sections taken along the 'vertical' axis. Confocal microscopy can be used on fixed cells stained with fluorescently tagged antibodies or on living cells expressing proteins tagged with naturally fluorescent proteins. The first of these fluorescent proteins to come into wide use was green fluorescent protein (GFP), isolated from the jellyfish *Aequorea victoria*. The list of fluorescent proteins in routine use now includes those emitting red, blue, cyan, or yellow fluorescence. By using cells transfected with genes encoding different fusion proteins, it has been possible to visualize the redistribution of T-cell receptors, co-receptors, adhesion molecules, and other signaling molecules, such as CD45, that takes place when a T cell makes contact with a target cell (see Fig. 9.37).

Confocal microscopy, however, can penetrate only around 80 μm into a tissue, and at the wavelengths typically used for excitation, the source light will soon bleach the fluorescent label and damage the specimen. This means that the technique is not suitable for imaging a live specimen over a period of time sufficient, for example, to track the movements of cells in a tissue. The more recently developed technique of **two-photon scanning fluorescence microscopy** overcomes some of these limitations. In this technology, ultrashort pulses of laser light of much longer wavelength (and thus with photons of lower energy) are used for excitation, and two of these lower-energy photons arriving nearly simultaneously are required to excite the fluorophore. Excitation will therefore occur in only a very small region at the focus of the microscope, where the beam of light is most intense, and so fluorescence emission will be restricted to the plane of focus, producing a sharp, high-contrast image. The longer-wavelength light (typically in the near infrared) is also less damaging to living tissue than the blue and ultraviolet wavelengths typically used in confocal microscopy, and so imaging can be carried out over a longer period. More of the emitted light is collected than in confocal microscopy, and because single photons scattering within the tissue cannot cause fluorescence and consequent background haze, imaging to greater depths (several hundred micrometers) is possible. Like confocal microscopy, two-photon microscopy produces thin optical sections from which a three-dimensional image can be built up.

To track the movements of molecules or cells over time, confocal or two-photon microscopy is combined with **time-lapse video imaging** using sensitive digital cameras. In immunology, time-lapse two-photon fluorescence imaging has been particularly valuable for tracking the movements of individual T cells and B cells expressing fluorescent proteins in intact lymphoid organs and observing where they interact (see Chapter 10).

A-11 Immunoelectron microscopy.

Antibodies can be used to detect the intracellular location of structures or particular proteins at high resolution by electron microscopy, a technique known as immunoelectron microscopy. Antibodies against the required antigen are labeled with gold particles and then applied to ultrathin sections, which are then examined in the transmission electron microscope. Antibodies labeled with gold particles of different diameters enable two or more proteins to be studied simultaneously (see Fig. 6.12). The difficulty with this technique is in staining the ultrathin section adequately, because few molecules of antigen are present in each section.

A-12 Immunohistochemistry.

An alternative to immunofluorescence (see Section A-10) for detecting a protein in tissue sections is **immunohistochemistry**, in which the specific antibody is chemically coupled to an enzyme that converts a colorless substrate into a colored reaction product *in situ*. The localized deposition of the colored product where antibody has bound can be directly observed under a light microscope. The antibody binds stably to its antigen, allowing unbound antibody to be removed by thorough washing. This method of detecting bound antibody is analogous to ELISA (see Section A-4) and frequently uses the same coupled enzymes, the difference in detection being primarily that in immunohistochemistry the colored products are insoluble and precipitate at the site where they are formed. Horseradish peroxidase and alkaline phosphatase are the two enzymes most commonly used in these applications. Horseradish peroxidase oxidizes the substrate diaminobenzidine to produce a brown precipitate, while alkaline phosphatase can produce red or blue dyes depending on the substrates used; a common substrate is 5-bromo-4-chloro-3-indolyl phosphate plus nitroblue tetrazolium (BCIP/NBT), which gives rise to a dark blue or purple stain. As with immunofluorescence, the native structure of the protein being sought usually needs to be preserved so that it will be recognized by the antibody. Tissues are fixed by the most gentle chemical fixation techniques, or frozen tissue sections are used that are fixed only after the antibody reaction has been performed.

A-13 Immunoprecipitation and co-immunoprecipitation.

To raise antibodies against membrane proteins and other cellular structures that are difficult to purify, mice are often immunized with whole cells or crude cell extracts. Antibodies against the individual molecules are then obtained by using these immunized mice to produce hybridomas making monoclonal antibodies (see Section A-7) that bind to the cell type used for immunization. To characterize the molecules identified by the antibodies, cells of the same type are labeled with radioisotopes and dissolved in nonionic detergents that disrupt cell membranes but do not interfere with antigen–antibody interactions. This allows the labeled protein to be isolated by binding to the antibody in a reaction known as **immunoprecipitation**. The antibody is usually attached to a solid support, such as the beads that are used in affinity chromatography (see Section A-3), or to Protein A, a protein derived from the cell wall of *Staphylococcus aureus* that binds tightly to the Fc region of IgG antibodies. Cells can be labeled in two main ways for immunoprecipitation analysis. All the proteins in a cell can be labeled metabolically by growing the cell in a medium containing radioactive amino acids that are then incorporated into cellular proteins (Fig. A.13). Alternatively, one can label only the cell-surface proteins by radioiodination under conditions that prevent iodine from crossing the plasma membrane and labeling proteins inside the cell, or by a reaction that labels only membrane proteins with biotin, a small molecule that is detected readily by labeled avidin, a protein found in egg whites that binds biotin with very high affinity.

Once the labeled proteins have been isolated by the antibody, they can be characterized in several ways. The most common is polyacrylamide gel electrophoresis (PAGE) of the proteins after they have been dissociated from antibody in the strong ionic detergent sodium dodecyl sulfate (SDS), a technique

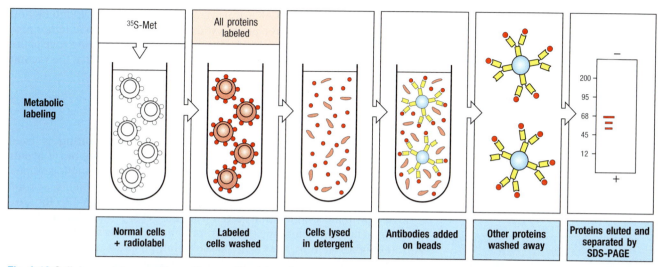

| Metabolic labeling | Normal cells + radiolabel | Labeled cells washed | Cells lysed in detergent | Antibodies added on beads | Other proteins washed away | Proteins eluted and separated by SDS-PAGE |

Fig. A.13 Cellular proteins reacting with an antibody can be characterized by immunoprecipitation of labeled cell lysates. All actively synthesized cellular proteins can be labeled metabolically by incubating cells with radioactive amino acids (shown here for methionine; ^{35}S-Met), or one can label just the cell-surface proteins by using radioactive iodine in a form that cannot cross the cell membrane or by a reaction with the small molecule biotin, detected by its reaction with labeled avidin (not shown). Cells are lysed with detergent and individual labeled cell-associated proteins can be precipitated with a monoclonal antibody attached to beads. After unbound proteins have been washed away, the bound protein is eluted in the detergent sodium dodecyl sulfate (SDS), which dissociates it from the antibody and also coats the protein with a strong negative charge, allowing it to migrate according to its size in polyacrylamide gel electrophoresis (PAGE). The positions of the labeled proteins are determined by autoradiography using X-ray film. This technique of SDS-PAGE can be used to determine the molecular weight and subunit composition of a protein. Patterns of protein bands observed with metabolic labeling are usually more complex than those revealed by radioiodination, owing to the presence of precursor forms of the protein (right panel). The mature form of a surface protein can be identified as being the same size as that detected by surface iodination or biotinylation (not shown).

generally abbreviated as **SDS-PAGE**. SDS binds relatively homogeneously to proteins, conferring a charge that allows the electrophoretic field to drive protein migration through the gel. The rate of migration is controlled mainly by protein size (see Fig. A.13). Proteins of differing charges can be separated using isoelectric focusing (see Section A-7). This technique can be combined with SDS-PAGE in a procedure known as **two-dimensional gel electrophoresis**. For this, the immunoprecipitated protein is eluted in urea, a nonionic solubilizing agent, and run on an isoelectric focusing gel in a narrow tube of polyacrylamide. This first-dimensional isoelectric focusing gel is then placed across the top of an SDS-PAGE slab gel, which is then run vertically to separate the proteins by molecular weight (Fig. A.14). Two-dimensional gel electrophoresis is a powerful technique that allows many hundreds of proteins in a complex mixture to be distinguished from one another.

Immunoprecipitation and the related technique of immunoblotting (see Section A-14) are useful for determining the molecular weight and isoelectric point of a protein as well as its abundance, distribution, and whether, for example, it undergoes changes in molecular weight and isoelectric point as a result of processing within the cell.

Co-immunoprecipitation is an extension of the immunoprecipitation technique and is used to determine whether a given protein interacts physically

Fig. A.14 Two-dimensional gel electrophoresis of MHC class II molecules. Proteins in mouse spleen cells have been labeled metabolically (see Fig. A.13), precipitated with a monoclonal antibody against the mouse MHC class II molecule H2-A, and separated by isoelectric focusing in one direction and SDS-PAGE in a second direction at right angles to the first (hence the term two-dimensional gel electrophoresis). This allows one to distinguish molecules of the same molecular weight on the basis of their charge. The separated proteins are detected by autoradiography. The MHC class II molecules are composed of two chains, α and β, and in the different MHC class II molecules these have different isoelectric points (compare upper and lower panels). The MHC genotype of mice is indicated by lowercase superscripts (k, p). Actin, a common contaminant, is marked a. Photographs courtesy of J.F. Babich.

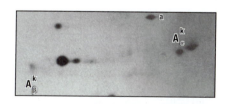

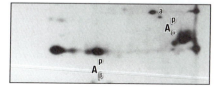

with another given protein. Cell extracts containing the presumed interaction complex are first immunoprecipitated with antibody against one of the proteins. The material isolated by this means is then tested for the presence of the other protein by immunoblotting with a specific antibody against the second protein.

A-14 Immunoblotting (Western blotting).

Like immunoprecipitation (see Section A-13), **immunoblotting** is used for identifying the presence of a given protein in a cell lysate, but it avoids the problem of having to label large quantities of cells with radioisotopes. Unlabeled cells are placed in detergent to solubilize all cell proteins, and the lysate is run on SDS-PAGE to separate the proteins (see Section A-13). The size-separated proteins are then transferred from the gel to a stable support such as a nitrocellulose membrane. Specific proteins are detected by treatment with antibodies able to react with SDS-solubilized proteins (mainly those that react with denatured sequences); the bound antibodies are revealed by anti-immunoglobulin antibodies labeled with an enzyme. The term **Western blotting** as a synonym for immunoblotting arose because the comparable technique for detecting specific DNA sequences is known as Southern blotting, after Edwin Southern, who devised it, which in turn provoked the name 'Northern' for blots of size-separated RNA, and 'Western' for blots of size-separated proteins. Western blots have many applications in basic research and clinical diagnosis. They are often used to test sera for the presence of antibodies against specific proteins, for example, to detect antibodies against different constituents of HIV (**Fig. A.15**).

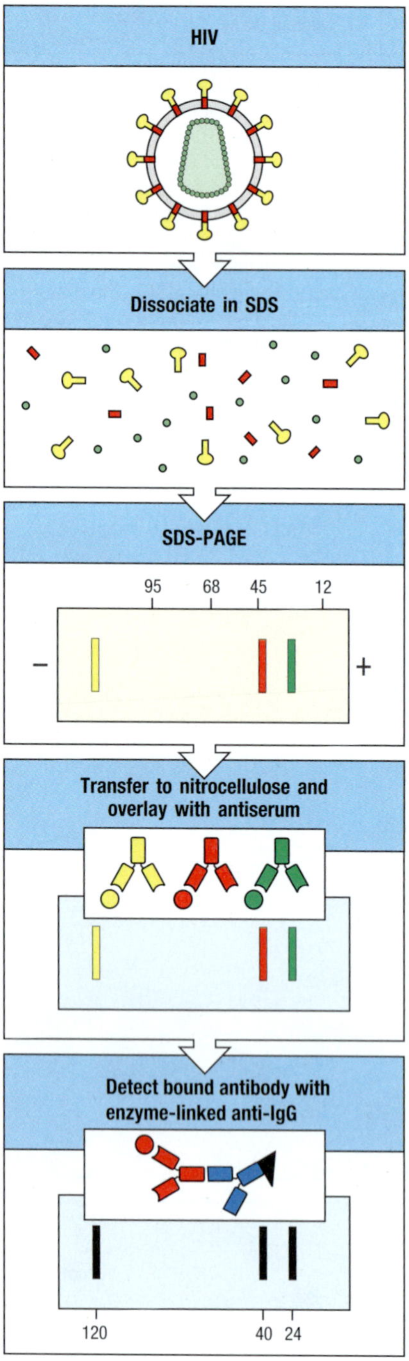

A-15 Use of antibodies in the isolation and characterization of multiprotein complexes by mass spectrometry.

Many of the proteins that function in immune cells are components of multiprotein complexes. This is the case for cell-surface receptors, such as the T-cell and B-cell antigen receptors and most cytokine receptors, as well as intracellular proteins involved in signal transduction, gene expression, and cell death. Antibodies that bind to one member of such a complex can be used to identify the other members of the complex by a process of co-immunoprecipitation followed by Western blotting or **mass spectrometry**.

A mass spectrometer can provide extremely precise measurements of the masses of constituents in a preparation of molecules. To identify unknown proteins in a sample, such as that acquired by co-immunoprecipitation, the sample often is first subjected to one-dimensional SDS-PAGE or two-dimensional gel electrophoresis (see Section A-13) to separate the proteins in the complex for individual analysis. Gel slices are excised and treated with a proteolytic enzyme, such as trypsin, to digest the protein into a series of peptides that can be easily eluted from the gel. The peptide mixture is then introduced into the mass spectrometer, which ionizes the peptides, transfers them to the gas phase, and then separates them under high vacuum by subjecting them to a magnetic field. The separation is based on the mass/charge (m/z)

Fig. A.15 Western blotting is used to identify antibodies against the human immunodeficiency virus (HIV) in serum from infected individuals. The virus is dissociated into its constituent proteins by treatment with the detergent SDS, and its proteins are separated using SDS-PAGE. The separated proteins are transferred to a nitrocellulose sheet and reacted with the test serum. Anti-HIV antibodies in the serum bind to the various HIV proteins and are detected with enzyme-linked anti-human immunoglobulin, which deposits colored material from a colorless substrate. This general methodology will detect any combination of antibody and antigen and is used widely, although the denaturing effect of SDS means that the technique works most reliably with antibodies that recognize the antigen when it is denatured.

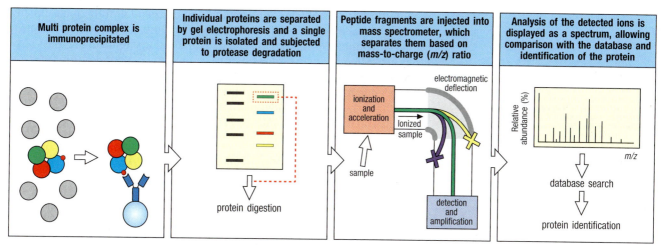

Fig. A.16 Characterization of multiprotein complexes by mass spectrometry. Following immunoprecipitation of a multiprotein complex using antibodies specific for one component of the complex, the individual proteins are separated by gel electrophoresis. An individual band representing one protein is isolated and digested with a protease such as trypsin. The digested protein sample is injected into the mass spectrometer, which ionizes the peptides, transfers them to the gas phase, and then separates them based on differences in their mass to charge (*m/z*) ratio by subjecting them under high vacuum to a magnetic field. A detector collects information on the signal intensity for each peptide ion and displays the information as a histogram. This histogram, usually referred to as a spectrum, is compared with a database containing potential cleavage sites for the proteolytic enzyme used in all known protein sequences, allowing for identification (ID) of the protein in the sample.

ratio of each ion, and a detector collects information on the signal intensity for each ion and displays the data as a histogram (Fig. A.16). This histogram, usually referred to as a spectrum, can be compared with a database containing potential cleavage sites (for the proteolytic enzyme used) in all known protein sequences. Due to the precision of these measurements, and the information derived from multiple peptides comprising the initial protein, the spectrum can often be unambiguously assigned to a unique protein in the database.

Modern multidimensional mass spectrometers (MS/MS) allow peptide ions to be sequenced as well as analyzed by their mass. In these instruments, peptide ions separated in one sector are fragmented in a second sector by collision with other molecules (often an inert gas such as N_2), with the resultant fragments separated in a third section (Fig. A.17). Fragmentation occurs primarily in the

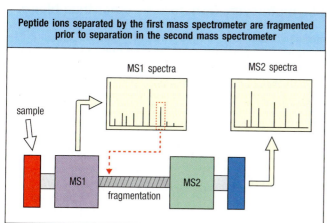

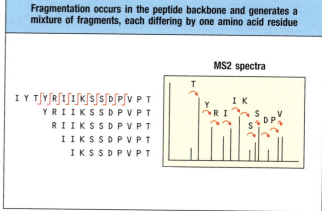

Fig. A.17 Determining the amino acid sequence of a peptide by multidimensional mass spectrometry (MS/MS). Multidimensional mass spectrometers (MS/MS) consist of two mass spectrometers linked in tandem but with an interceding middle sector that fragments ions. In the first sector, the first mass spectrometer separates peptide ions, as shown in Fig. A.16. Each peptide ion from this first separation is then fragmented in the middle sector of the apparatus by collision with other molecules (often an inert gas such as N_2). Since fragmentation occurs primarily in the peptide backbone, a mixture of fragments is generated in which the fragments each differ by one amino acid residue. The resultant fragments are then separated in the second mass spectrometer, the final sector. The sequence of the peptide can be read directly from the second mass spectrum. The order of amino acid residues in the peptide can be deduced because of the exquisite precision of the measurements of each ion together with knowledge of the exact mass of each possible amino acid residue.

peptide backbone, allowing the sequence of the peptide to be read directly from the mass spectrum of the mixture of fragments. In place of gel electrophoresis, a liquid chromatograph can be employed upstream of the mass analyzers (LC-MS/MS), to provide additional separation of peptides prior to mass separation and allow a very complex mixture of thousands of peptides to be sequenced in a single run. It was this latter technique that played an essential role in the early studies that identified the repertoire of peptides bound to MHC molecules on the surface of antigen-presenting cells (see Chapter 6).

A-16 Isolation of peripheral blood lymphocytes by density-gradient fractionation.

The first step in studying lymphocytes is to isolate them so that their behavior can be analyzed *in vitro*. Human lymphocytes can be isolated most readily from peripheral blood by density centrifugation over a step gradient consisting of a mixture of the carbohydrate polymer Ficoll-Hypaque™ and the dense iodine-containing compound metrizamide. A step gradient is made by preparing a solution of Ficoll-Hypaque at a precise density (1.077 g/liter for human cells) and placing a layer of this solution at the bottom of a centrifuge tube. A sample of heparinized blood mixed with saline (heparin prevents clotting) is carefully layered on top of the Ficoll-Hypaque solution. Following centrifugation for about 30 minutes, the components of the blood have separated based on their densities. The upper layer contains the blood plasma and platelets, which remain in the top layer during the short centrifugation. Red blood cells and granulocytes have a higher density than the Ficoll-Hypaque solution and collect at the bottom of the tube. The resulting population, called **peripheral blood mononuclear cells** (**PBMCs**), collects at the interface between the blood and the Ficoll-Hypaque layers and consists mainly of lymphocytes and monocytes (**Fig. A.18**). Although this population is readily accessible, it is not necessarily representative of the lymphoid system, because only recirculating lymphocytes can be isolated from blood.

The 'normal' ranges in the numbers of the different types of white blood cells in blood, along with the normal ranges in concentrations of the various antibody classes, are given in **Fig. A.19**.

A-17 Isolation of lymphocytes from tissues other than blood.

In experimental animals, and occasionally in humans, lymphocytes are isolated from lymphoid organs, such as spleen, thymus, bone marrow, lymph nodes, or mucosal-associated lymphoid tissues—in humans, most commonly the palatine tonsils (see Fig. 12.6). A specialized population of lymphocytes resides in surface epithelia; these cells are isolated by fractionating the

Fig. A.18 Peripheral blood mononuclear cells can be isolated from whole blood by Ficoll-Hypaque™ centrifugation. Diluted anticoagulated blood (left panel) is layered over Ficoll-Hypaque™ and centrifuged. Red blood cells and polymorphonuclear leukocytes or granulocytes are denser and travel through the Ficoll-Hypaque™, while mononuclear cells consisting of lymphocytes together with some monocytes band over it and can be recovered at the interface (right panel).

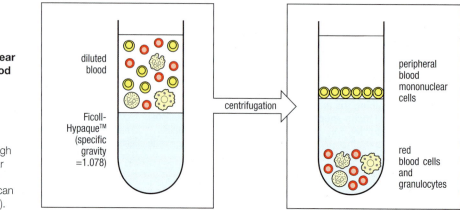

Evaluation of the cellular components of the human immune system			
	B cells	**T cells**	**Phagocytes**
Normal numbers (x10^9 per liter of blood)	Approximately 0.3	Total 1.0–2.5 CD4 0.5–1.6 CD8 0.3–0.9	Monocytes 0.15–0.6 Polymorphonuclear leukocytes: Neutrophils 3.00–5.5 Eosinophils 0.05–0.25 Basophils 0.02
Measurement of function *in vivo*	Serum Ig levels Specific antibody levels	Skin test	—
Measurement of function *in vitro*	Induced antibody production in response to pokeweed mitogen	T-cell proliferation in response to phytohemagglutinin or to tetanus toxoid	Phagocytosis Nitroblue tetrazolium uptake Intracellular killing of bacteria

Evaluation of the humoral components of the human immune system				
	Immunoglobulins			**Complement**
Component	IgG	IgM	IgA	IgE	
Normal levels	600–1400 mg•dl^{-1}	40–345 mg•dl^{-1}	60–380 mg•dl^{-1}	0–200 IU•ml^{-1}	CH$_{50}$ of 125–300 IU•ml^{-1}

Fig. A.19 The major cellular and humoral components of human blood. Human blood contains B cells, T cells, and myeloid cells, as well as high concentrations of antibodies and complement proteins.

epithelial layer after its detachment from the basement membrane. Finally, in situations where local immune responses are prominent, lymphocytes can be isolated from the site of the response itself. For example, in order to study the autoimmune reaction that is thought to be responsible for rheumatoid arthritis, an inflammatory response in joints, lymphocytes are isolated from the fluid aspirated from the inflamed joint space.

Laser-capture microdissection is a technique used to isolate specific populations of cells from an intact tissue sample or histological specimen after visualizing the cells by light microscopy. The cells of interest can be 'captured' by placing a polymer over the sample on the microscope slide, and using an infrared laser to melt the polymer onto the sample in discrete locations. Once completed, the polymer–cell composite can be lifted and DNA, RNA, or proteins can be isolated from the dissected cells (Fig. A.20). A variant of this approach makes use of an ultraviolet (UV) laser, rather than infrared. In this case, the UV laser acts as a molecular cutting tool, and can be used to cut away or even ablate the unwanted portions of the tissue, leaving intact the area of interest.

A-18 Flow cytometry and FACS analysis.

An immensely powerful tool for defining and enumerating populations of immune cells is the flow cytometer, which detects and counts individual cells passing in a stream through a laser beam. A flow cytometer equipped to separate the identified cells is called a **fluorescence-activated cell sorter** (**FACS**). These instruments are used to study the properties of cell subsets that are identified by using monoclonal antibodies against cell-surface or intracellular proteins. Individual cells within a mixed population are first labeled by

Fig. A.20 Laser-capture microdissection. Specific populations of cells from an intact tissue sample or histological specimen can be isolated after visualizing the cells by light microscopy. A polymer called the transfer film is placed over the sample on the microscope slide, and an infrared laser is used to melt the polymer onto the sample in discrete locations. The polymer–cell composite is then lifted and the cells of interest are isolated. DNA, RNA, or proteins can be prepared from the dissected cells.

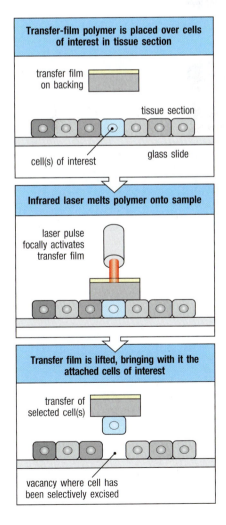

treatment with specific monoclonal antibodies coupled to fluorescent dyes, or by specific antibodies followed by fluorescently tagged anti-immunoglobulin antibodies. The mixture of labeled cells is then suspended in a much larger volume of saline and forced through a small nozzle, creating a fine stream of liquid composed of droplets, each containing a single cell. As each cell passes through a laser beam it scatters the laser light, and any dye molecules bound to the cell are excited and fluoresce. Sensitive photomultiplier tubes detect both the scattered light, which gives information about the size and granularity of the cell, and the fluorescence emissions, which give information about the binding of the labeled monoclonal antibodies and hence about the expression of cell-surface or intracellular proteins by each cell (Fig. A.21).

In the cell sorter, the signals passed back to the computer are used to generate an electric charge, which is passed from the nozzle through the liquid stream at the precise time that the stream breaks up into droplets, each containing no more than a single cell; droplets containing a charge can then be deflected from the main stream of droplets as they pass between plates of opposite charge, so that positively charged droplets are attracted to a negatively charged plate, and vice versa. Once deflected, droplets containing cells are collected in tubes. In this way, specific subpopulations of cells, distinguished by the binding of the labeled antibody, can be purified from a mixed population of cells. Alternatively, to deplete a population of cells, the same fluorochrome can be used to label different antibodies directed at marker proteins expressed by the various undesired cell types. The cell sorter can be used to direct labeled cells to a waste channel, retaining only the unlabeled cells.

When cells are labeled with a single fluorescent antibody, the data from a flow cytometer are usually displayed in the form of a one-dimensional histogram of fluorescence intensity versus cell numbers. If two or more antibodies are used, each coupled to a different fluorescent dye, then the data are usually

Fig. A.21 Flow cytometers allow individual cells to be identified by their cell-surface antigens and to be sorted. Cells to be analyzed by flow cytometry are first labeled with fluorescent dyes (top panel). Direct labeling uses dye-coupled antibodies specific for cell-surface antigens (as shown here), while indirect labeling uses a dye-coupled immunoglobulin to detect unlabeled cell-bound antibody. The cells are forced through a nozzle in a single-cell stream that passes through a laser beam (second panel). Photomultiplier tubes (PMTs) detect the scattering of light, which is a sign of cell size and granularity, as well as emissions from the different fluorescent dyes. This information is analyzed by computer (CPU). By examining a large number of cells, the proportion of cells with a specific set of characteristics can be determined and levels of expression of various molecules on these cells can be measured. The lower part of the figure shows how these data can be represented, the example in this case being the expression of two surface immunoglobulins, IgM and IgD, on a sample of B cells from a mouse spleen. The two immunoglobulins have been labeled with different-colored dyes. When the expression of just one type of molecule is to be analyzed (IgM or IgD), the data are usually displayed as a histogram, as in the left-hand panels. Histograms display the distribution of cells expressing a single measured parameter (for example, size, granularity, fluorescence intensity). When two or more parameters are measured for each cell (IgM and IgD), various types of two-dimensional plots can be used to display the data, as shown in the right-hand panel. All four plots represent the same data, and in each case, the horizontal axis represents intensity of IgM fluorescence, and the vertical axis the intensity of IgD fluorescence. Two-color plots provide more information than histograms; they allow recognition, for example, of cells that are 'bright' for both colors, 'dull' for one and bright for the other, dull for both, negative for both, and so on. For example, the cluster of dots in the extreme lower left portions of the plots represents cells that do not express either immunoglobulin, and are mostly T cells. The standard dot plot (upper left) places a single dot for each cell whose fluorescence is measured. This format works well for identifying cells that lie outside the main groups, but tends to saturate in areas containing a large number of cells of the same type. A second means of presenting these data is the color dot plot (lower left), which uses color density to indicate high-density areas. A contour plot (upper right) draws 5% 'probability' contours, with contour lines drawn to indicate each successive 5% of the population; this format provides the best monochrome visualization of regions of high and low density. The lower right plot is a 5% probability contour map, which also shows outlying cells as dots.

displayed in the form of a two-dimensional scatter diagram or as a contour diagram, where the fluorescence of one dye-labeled antibody is plotted against that of a second, with the result that a population of cells labeled with one antibody can be further subdivided by its labeling with the second antibody (see Fig. A.21). By examining large numbers of cells, flow cytometry can give quantitative data on the percentage of cells bearing different

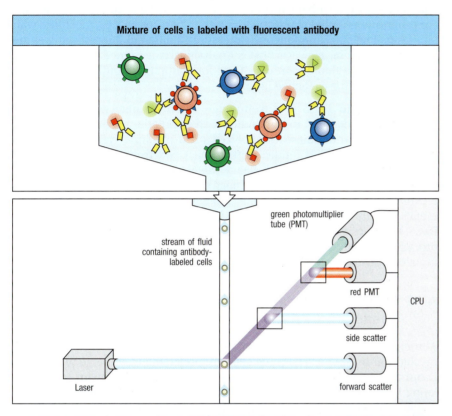

Mixture of cells is labeled with fluorescent antibody

green photomultiplier tube (PMT)

stream of fluid containing antibody-labeled cells

red PMT

CPU

side scatter

forward scatter

Laser

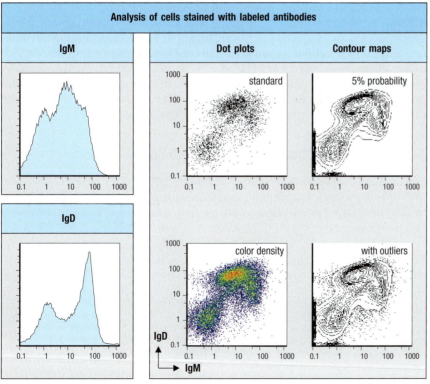

Analysis of cells stained with labeled antibodies

IgM

Dot plots

Contour maps

standard

5% probability

IgD

color density

with outliers

IgD

IgM

proteins, such as surface immunoglobulin, which characterizes B cells; the T-cell receptor-associated molecules known as CD3; and the CD4 and CD8 co-receptor proteins that distinguish the major T-cell subsets. Likewise, FACS analysis has been instrumental in defining stages in the early development of B and T cells. This technology also played a vital role in the early identification of AIDS as a disease in which T cells bearing CD4 are depleted selectively (see Chapter 13). Advances in FACS technology permit progressively more antibodies labeled with distinct fluorescent dyes to be used at the same time. For experiments aimed at cell analysis, rather than cell sorting, machines with four lasers that can simultaneously measure 18 different fluorescent dyes are currently available. However, FACS analysis is constrained by the spectral properties of the fluorescent dyes used for coupling to antibodies, and this technology may have reached its limit.

An alternative to FACS is a technology based on detecting heavy-metal atoms that are coupled to antibodies. Cell populations labeled with heavy metal-coupled antibodies are analyzed on a machine called a CyTOF™, which combines liquid fluidics with a mass spectrometer. As each cell is analyzed, the quantity of each heavy metal associated with that cell, and thus the abundance of the target of each antibody, is measured. In total, this technology is estimated to have the capability to measure 100 distinct heavy metals, greatly expanding the range of analysis that is currently possible by FACS. However, with this technique, the cells are destroyed by the ionization process required for the mass spectrometry analysis, so the CyTOF cannot function as a cell sorter.

A-19 Lymphocyte isolation using antibody-coated magnetic beads.

Although FACS is superb for isolating small numbers of cells in pure form, mechanical means of separating cells are preferable when large numbers of lymphocytes must be prepared quickly. A powerful and efficient way of isolating lymphocyte populations is to couple paramagnetic beads to monoclonal antibodies that recognize distinguishing cell-surface molecules. These antibody-coated beads are mixed with the cells to be separated and are run through a column containing material that attracts the paramagnetic beads when the column is placed in a strong magnetic field. Cells binding the magnetically labeled antibodies are retained; cells lacking the appropriate surface molecule can be washed away (**Fig. A.22**). The retained cells are recovered by removing the column from the magnetic field. In this case, the bound cells are positively selected for expression of the particular cell-surface molecule, while the unbound cells are negatively selected for its absence.

A-20 Isolation of homogeneous T-cell lines.

The analysis of specificity and effector function of T cells depends heavily on the study of monoclonal populations of T lymphocytes. These can be obtained in four distinct ways—T-cell hybrids, cloned T-cell lines, T-cell tumors, and

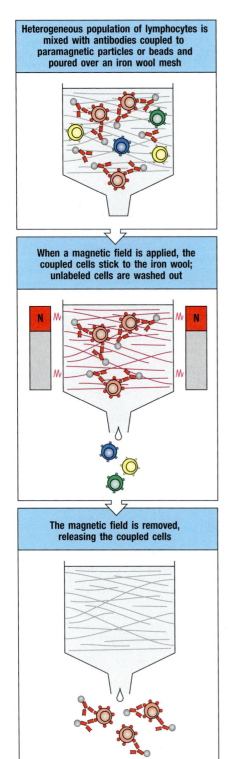

Heterogeneous population of lymphocytes is mixed with antibodies coupled to paramagnetic particles or beads and poured over an iron wool mesh

When a magnetic field is applied, the coupled cells stick to the iron wool; unlabeled cells are washed out

The magnetic field is removed, releasing the coupled cells

Fig. A.22 Lymphocyte subpopulations can be separated physically by using antibodies coupled to paramagnetic particles or beads. A mouse monoclonal antibody specific for a particular cell-surface molecule is coupled to paramagnetic particles or beads. It is mixed with a heterogeneous population of lymphocytes and poured over an iron wool mesh in a column. A magnetic field is applied so that the antibody-bound cells stick to the iron wool while cells that have not bound antibody are washed out; these cells are said to be negatively selected for lack of the molecule in question. The bound cells are released by removing the magnetic field; they are said to be positively selected for presence of the antigen recognized by the antibody.

limiting-dilution culture. By analogy with B-cell hybridomas (see Section A-7), normal T cells proliferating in response to specific antigen can be fused to malignant T-cell lymphoma lines to generate **T-cell hybrids**. The hybrids express the receptor of the normal T cell, but proliferate indefinitely owing to the cancerous state of the lymphoma parent. T-cell hybrids can be cloned to yield a population of cells all having the same T-cell receptor. When stimulated by their specific antigen, these cells release cytokines such as the T-cell growth factor interleukin-2 (IL-2), and the production of cytokines is used as an assay to assess the antigen specificity of the T-cell hybrid.

T-cell hybrids are excellent tools for the analysis of T-cell specificity, because they grow readily in suspension culture. However, they cannot be used to analyze the regulation of specific T-cell proliferation in response to antigen because they are continually dividing. T-cell hybrids also cannot be transferred into an animal to test for function *in vivo* because they would give rise to tumors. Functional analysis of T-cell hybrids is also confounded by the fact that the malignant partner cell affects their behavior in functional assays. Therefore, the regulation of T-cell growth and the effector functions of T cells must be studied using **T-cell clones**. These are clonal cell lines of a single T-cell type and antigen specificity, which are derived from cultures of heterogeneous T cells, called **T-cell lines**, whose growth is dependent on periodic restimulation with specific antigen and, frequently, on the addition of T-cell growth factors (Fig. A.23). T-cell clones also require periodic restimulation with antigen and are more tedious to grow than T-cell hybrids, but because their growth depends on specific antigen recognition, they maintain antigen specificity, which is often lost in T-cell hybrids. Cloned T-cell lines can be used for studies of effector function both *in vitro* and *in vivo*. In addition, the proliferation of T cells, a critical aspect of clonal selection, can be characterized only in cloned T-cell lines, where such growth is dependent on antigen recognition. Thus, both types of monoclonal T-cell lines, T-cell hybrids and antigen-dependent T-cell clones, have valuable applications in experimental studies.

Studies of human T cells have relied largely on T-cell clones because a suitable fusion partner for making T-cell hybrids has not been identified. However, a human T-cell lymphoma line, called Jurkat, has been characterized extensively because it secretes IL-2 when its antigen receptor is cross-linked with anti-receptor monoclonal antibodies. This simple assay system has yielded much information about signal transduction in T cells. One of the Jurkat cell line's most interesting features, shared with T-cell hybrids, is that it stops growing when its antigen receptor is cross-linked. This has allowed mutants lacking the receptor or having defects in signal transduction pathways to be selected simply by culturing the cells with anti-receptor antibody and selecting those that continue to grow. Thus, T-cell tumors, T-cell hybrids, and cloned T-cell lines all have valuable applications in experimental immunology.

Finally, primary T cells from any source can be isolated as single, antigen-specific cells by limiting dilution (see Section A-21) rather than by first establishing a mixed population of T cells in culture as a T-cell line and then deriving clonal subpopulations. During the growth of T-cell lines, particular T-cell clones can come to dominate the cultures and give a false picture of the number and specificities in the original sample. Direct cloning of primary T cells avoids this artifact.

A-21 Limiting-dilution culture.

On many occasions it is important to know the frequency of antigen-specific lymphocytes, especially T cells, in order to measure the efficiency with which an individual responds to a particular antigen, for example, or the degree to which specific immunological memory has been established. There are a

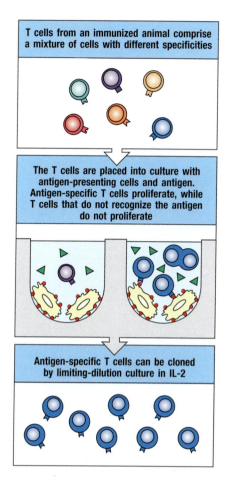

T cells from an immunized animal comprise a mixture of cells with different specificities

The T cells are placed into culture with antigen-presenting cells and antigen. Antigen-specific T cells proliferate, while T cells that do not recognize the antigen do not proliferate

Antigen-specific T cells can be cloned by limiting-dilution culture in IL-2

Fig. A.23 Production of cloned T-cell lines. T cells from an immunized donor, comprising a mixture of cells with different specificities, are activated with antigen and antigen-presenting cells. Single responding cells are cultured by limiting dilution (see Section A-21) in the T-cell growth factor IL-2, which selectively stimulates the responding cells to proliferate. From these single cells, cloned lines specific for antigen are identified and can be propagated by culture with antigen, antigen-presenting cells, and IL-2.

number of methods for doing this, either by detecting the cells directly by the specificity of their receptor, or by detecting activation of the cells to provide some particular function, such as cytokine secretion or cytotoxicity.

The response of a lymphocyte population is a measure of the overall response, but the frequency of lymphocytes able to respond to a given antigen can be determined by **limiting-dilution culture**. This assay makes use of the Poisson distribution, a statistical function that describes how objects are distributed at random. For instance, when a sample of heterogeneous T cells is distributed equally into a series of culture wells, some wells will receive no T cells specific for a given antigen, some will receive one specific T cell, some two, and so on. The T cells in the wells are activated with specific antigen, antigen-presenting cells, and growth factors. After allowing several days for their growth and differentiation, the cells in each well are tested for a response to antigen, such as cytokine release or the ability to kill specific target cells (Fig. A.24). The assay is replicated with different numbers of T cells in the samples. The logarithm of the proportion of wells in which there is no response is plotted against the number of cells initially added to each well. If cells of one type, typically antigen-specific T cells because of their rarity, are the only limiting factor for obtaining a response, then a straight line is obtained. From the Poisson distribution, it is known that there is, on average, one antigen-specific cell per well when the proportion of negative wells is 37%. Thus, the frequency of antigen-specific cells in the population equals the reciprocal of the number of cells added to each well when 37% of the wells are negative. After priming, the frequency of specific cells goes up substantially, reflecting the antigen-driven proliferation of antigen-specific cells. The limiting-dilution assay can also be used to measure the frequency of B cells that can make antibody against a given antigen.

Fig. A.24 The frequency of specific lymphocytes can be determined using limiting-dilution assay. Various numbers of lymphoid cells from normal or immunized mice are added to individual culture wells and stimulated with antigen and antigen-presenting cells (APCs) or polyclonal mitogen and added growth factors. After several days, the wells are tested for a specific response to antigen, such as cytotoxic killing of target cells. Each well that initially contained a specific T cell will make a response to its target, and from the Poisson distribution one can determine that when 37% of the wells are negative, each well contained, on average, one specific T cell at the beginning of the culture. In the example shown, for the unimmunized mouse 37% of the wells are negative when 160,000 T cells have been added to each well; thus the frequency of antigen-specific T cells is 1 in 160,000. When the mouse is immunized, 37% of the wells are negative when only 1100 T cells have been added; hence the frequency of specific T cells after immunization is 1 in 1100, an increase in responsive cells of 150-fold.

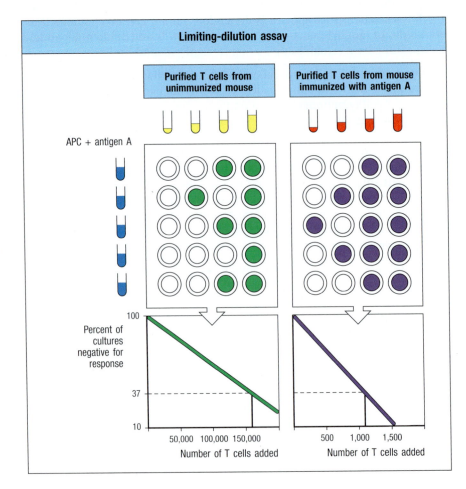

Fig. A.25 The frequency of cytokine-secreting T cells can be determined by the ELISPOT assay. The ELISPOT assay is a variant of the ELISA assay in which antibodies bound to a plastic surface are used to capture cytokines secreted by individual T cells. Usually, cytokine-specific antibodies are bound to the surface of a plastic tissue-culture well and the unbound antibodies are removed (top panel). Activated T cells are then added to the well and settle onto the antibody-coated surface (second panel). If a T cell is secreting the appropriate cytokine, this will then be captured by the antibody molecules on the plate surrounding the T cell (third panel). After a period of time the T cells are removed, and the presence of the specific cytokine is detected using an enzyme-labeled second antibody specific for the same cytokine. Where this antibody binds, a colored reaction product can be formed (fourth panel). Each T cell that originally secreted the cytokine gives rise to a single spot of color, hence the name of the assay. The results of such an ELISPOT assay for T cells secreting IFN-γ in response to different stimuli are shown in the last panel. In this example, T cells from a stem-cell transplant recipient were treated with a control peptide (top two panels) or a peptide from cytomegalovirus (bottom two panels). As can be seen, there are a greater number of spots in the bottom two panels, a clear indication that the patient's T cells are able to respond to the viral peptide and produce IFN-γ. Photographs courtesy of S. Nowack.

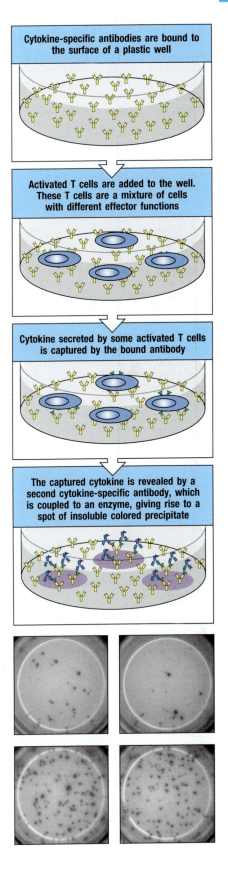

A-22 ELISPOT assay.

A modification of the ELISA antigen-capture assay (see Section A-4), called the **ELISPOT assay**, is a powerful tool for measuring the frequency of T-cell responses and also provides information about the cytokines produced. Populations of T cells are stimulated with the antigen of interest, and are then allowed to settle onto a plastic plate coated with antibodies against the cytokine that is to be assayed (Fig. A.25). If an activated T cell is secreting that cytokine, the cytokine is captured by the antibody on the plastic plate. After a period of time the cells are removed, and a second antibody against the cytokine is added to the plate to reveal a circle ('spot') of bound cytokine surrounding the position of each activated T cell; it is these circles that give the ELISPOT assay its name. By counting each spot and knowing the number of T cells originally added to the plate, one can easily calculate the frequency of T cells secreting that particular cytokine. ELISPOT can also be used to detect specific antibody secretion by B cells, in this case by using antigen-coated surfaces to trap specific antibody and labeled anti-immunoglobulin to detect the bound antibody.

A-23 Identification of functional subsets of T cells based on cytokine production or transcription factor expression.

One problem with the detection of cytokine production on a single-cell level is that the cytokines are secreted by the T cells into the surrounding medium, and any association with the originating cell is lost. Three methods have been devised that allow the cytokine profile produced by individual cells to be determined. The first, that of **intracellular cytokine staining** (Fig. A.26), relies on the use of metabolic poisons that inhibit protein export from the cell. The cytokine thus accumulates within the endoplasmic reticulum and vesicular network of the cell. If the cells are subsequently fixed and rendered permeable by the use of mild detergents, antibodies can gain access to these intracellular compartments and detect the cytokine. The T cells can be stained for other markers simultaneously, and thus the frequency of IL-10-producing CD25$^+$ CD4 T cells, for example, can be easily obtained.

A second method, which has the advantage that the cells being analyzed are not killed in the process, is called cytokine capture. This technique uses hybrid antibodies, in which the two separate heavy- and light-chain pairs from different antibodies are combined to give a mixed antibody molecule in which

Fig. A.26 Cytokine-secreting cells can be identified by intracellular cytokine staining. Fluorochrome-labeled antibodies can be used to detect the cytokines secreted by activated T cells after the cytokine molecules have been allowed to accumulate inside the cell. The accumulation of cytokine molecules to a high enough concentration for efficient detection is achieved by treating the activated T cells with inhibitors of protein export. In such treated cells, proteins destined to be secreted are instead retained within the endoplasmic reticulum (left panel). These treated cells are then fixed, to cross-link the proteins inside the cell and in the cell membranes, so that they are not lost when the cell is permeabilized by dissolving the cell membrane in a mild detergent (center panel). Fluorochrome-labeled antibodies can now enter the permeabilized cell and bind to the cytokines inside the cell (right panel). Cells labeled in this way can also be labeled with antibodies that bind to cell-surface proteins to determine which subsets of T cells are secreting particular cytokines.

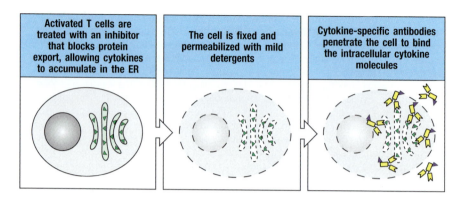

| Activated T cells are treated with an inhibitor that blocks protein export, allowing cytokines to accumulate in the ER | The cell is fixed and permeabilized with mild detergents | Cytokine-specific antibodies penetrate the cell to bind the intracellular cytokine molecules |

the two antigen-binding sites recognize different ligands (Fig. A.27). In the bispecific antibodies used to detect cytokine production, one of the antigen-binding sites is specific for a T-cell surface marker, while the other is specific for the cytokine in question. The bispecific antibody binds to the T cells through the binding site for the cell-surface marker, leaving the cytokine-binding site free. If that T cell is secreting the particular cytokine, it is captured by the bound antibody before it diffuses away from the surface of the cell. It can then be detected by adding to the cells a fluorochrome-labeled second antibody specific for the cytokine.

A third method for identifying which T cells in a population produce a particular cytokine utilizes cytokine gene reporter mice. In these lines of mice, a cDNA clone encoding a readily detectable protein (the 'reporter' protein) is inserted into the 3′ untranslated region of the targeted cytokine gene downstream of a sequence known as an internal ribosome entry site (IRES). The IRES sequence allows translation of the reporter protein from the same mRNA as that encoding the cytokine; thus, the reporter protein is produced only when the cytokine mRNA is expressed (Fig. A.28). Common reporter proteins for this application are fluorescent proteins, such as green fluorescent protein (GFP). In fact, the GFP commonly used for this purpose contains

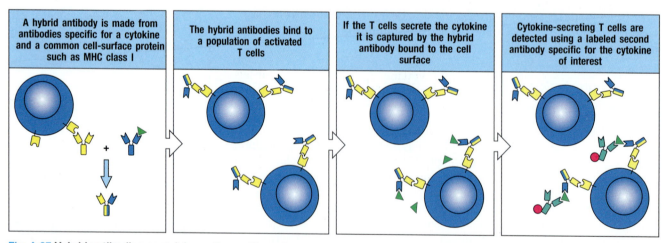

| A hybrid antibody is made from antibodies specific for a cytokine and a common cell-surface protein such as MHC class I | The hybrid antibodies bind to a population of activated T cells | If the T cells secrete the cytokine it is captured by the hybrid antibody bound to the cell surface | Cytokine-secreting T cells are detected using a labeled second antibody specific for the cytokine of interest |

Fig. A.27 Hybrid antibodies containing cell-specific and cytokine-specific binding sites can be used to assay cytokine secretion by living cells and to purify cells secreting particular cytokines. Hybrid antibodies can be made by mixing together heavy- and light-chain pairs from antibodies of different specificities, for example, an antibody against an MHC class I molecule and an antibody specific for a cytokine such as IL-4 (first panel). The hybrid antibodies are then added to a population of activated T cells, and bind to each cell via the MHC class I binding arm (second panel). If some of the cells in the population are secreting the appropriate

cytokine, IL-4, this is captured by the cytokine-specific arm of the hybrid antibody (third panel). The presence of the cytokine can then be revealed, for example, by using a fluorochrome-labeled second antibody specific for the same cytokine but binding to a different site from the one used by the hybrid antibody (last panel). The labeled cells are analyzed by flow cytometry or are isolated using a fluorescence-activated cell sorter (FACS). Alternatively, the second cytokine-specific antibody can be coupled to magnetic beads, and the cytokine-producing cells isolated magnetically.

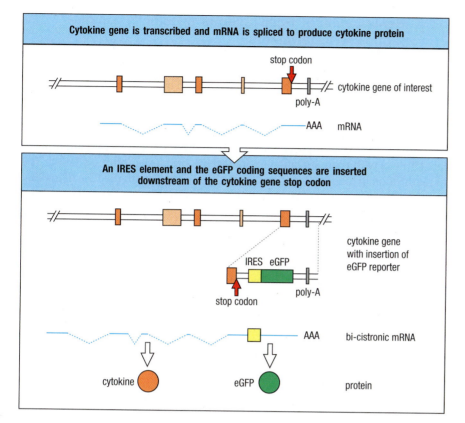

Cytokine gene is transcribed and mRNA is spliced to produce cytokine protein

stop codon

cytokine gene of interest

poly-A

AAA   mRNA

An IRES element and the eGFP coding sequences are inserted downstream of the cytokine gene stop codon

cytokine gene with insertion of eGFP reporter

IRES eGFP

poly-A

stop codon

AAA   bi-cistronic mRNA

cytokine   eGFP   protein

Fig. A.28 Cytokine-expressing cells can be tracked *in vivo* using cytokine gene knock-in reporter mice. To identify cells expressing a specific cytokine in intact animals, the locus encoding the cytokine is modified by homologous recombination (see Fig. A.44 and Section A-35). An internal ribosome entry site (IRES) and the gene for a fluorescent protein such as eGFP are inserted 3′ of the last exon of the cytokine gene, downstream of the cytokine protein stop codon and upstream of the mRNA transcription termination and polyadenylation signal (the poly-A site). The IRES element allows the ribosome to initiate translation of a second protein-coding sequence at an internal site on the mRNA. When the modified locus is transcribed and spliced to form the mature mRNA, both the intact cytokine protein and the fluorescent reporter protein (e.g., eGFP) are produced from the same transcript. This allows the identification and characterization of cytokine-expressing cells, such as by flow cytometry, based on the detection of eGFP.

a point mutation that greatly improves its spectral properties for experimental purposes. This version of GFP is commonly referred to as 'enhanced GFP,' or 'eGFP' for short. eGFP can be detected by FACS or by fluorescent microscopy using the settings designed to detect the commonly used fluorescent dye FITC. Due to broad utility of these fluorescent proteins, a host of GFP derivatives have been developed by genetic engineering of the original GFP protein. Each derivative has distinct fluorescent properties and therefore can be uniquely identified, allowing these proteins to be used in combination with each other to provide information about multiple cytokines at the same time (**Fig. A.29**).

Several of these techniques for measuring cytokine expression by T-cell subsets have been adapted to examine the transcription factors expressed by T cells and other lymphocytes, providing an alternative method for identifying functional lymphocyte subsets. In one approach, antibodies specific for lineage-defining transcription factors are used to label permeabilized cells. As was described above for the intracellular cytokine staining assay, the cells can then be examined by flow cytometry or immunofluorescence microscopy. Lines of reporter mice have also been generated, in which the gene locus encoding a transcription factor is modified to express a fluorescent protein, such as eGFP. For both of these approaches, the advantage of using transcription factors to identify lymphocyte subsets is that there is no need to stimulate the cells prior to antibody staining or to assessing reporter protein expression, as lineage-defining transcription factors are constitutively expressed by the cells. Therefore, this approach is more widely used for the identification of T-cell and other lymphocyte subsets in intact tissues by microscopy.

Eight different fluorescent proteins expressed in bacteria

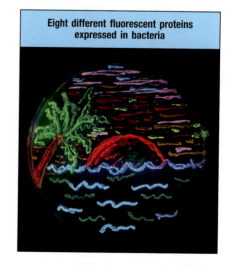

Fig. A.29 Fluorescent proteins are available in a rainbow of colors. Derivatives of GFP and a red-fluorescent coral protein can generate eight different fluorescent colors. A beach scene is drawn with bacterial strains expressing each fluorescent protein. Courtesy of Roger Tsien.

A-24 Identification of T-cell receptor specificity using peptide:MHC tetramers.

For many years, the ability to identify antigen-specific T cells directly through their receptor specificity eluded immunologists. Foreign antigen could not be used directly to identify T cells because, unlike B cells, T cells do not recognize antigen alone but rather the complexes of peptide fragments of antigen bound to self MHC molecules. Moreover, the affinity of interaction between the T-cell receptor and the peptide:MHC complex is in practice so low that attempts to label T cells with their specific peptide:MHC complexes routinely failed. The breakthrough in labeling antigen-specific T cells came with the idea of making multimers of the peptide:MHC complex, so as to increase the avidity of the interaction.

Peptides can be biotinylated using the bacterial enzyme BirA, which recognizes a specific amino acid sequence. Recombinant MHC molecules containing this target sequence are used to make peptide:MHC complexes, which are then biotinylated. Avidin, or its bacterial counterpart streptavidin, contains four sites that bind biotin with extremely high affinity. Mixing the biotinylated peptide:MHC complex with avidin or streptavidin results in the formation of a **peptide:MHC tetramer**—four specific peptide:MHC complexes bound to a single molecule of streptavidin (**Fig. A.30**). Routinely, the streptavidin moiety is labeled with a fluorochrome to allow detection of those T cells capable of binding the peptide:MHC tetramer.

Peptide:MHC tetramers have been used to identify populations of antigen-specific T cells in, for example, patients with acute Epstein–Barr virus infections (infectious mononucleosis), showing that up to 80% of the peripheral T cells in infected individuals can be specific for a single peptide:MHC complex. They have also been used to follow responses over timescales of years in individuals with HIV or, in the example we show, cytomegalovirus infections. These reagents have also been important in identifying the cells responding, for example, to nonclassical MHC class I molecules such as HLA-E or HLA-G, in both cases showing that these nonclassical molecules are recognized by subsets of NK receptors.

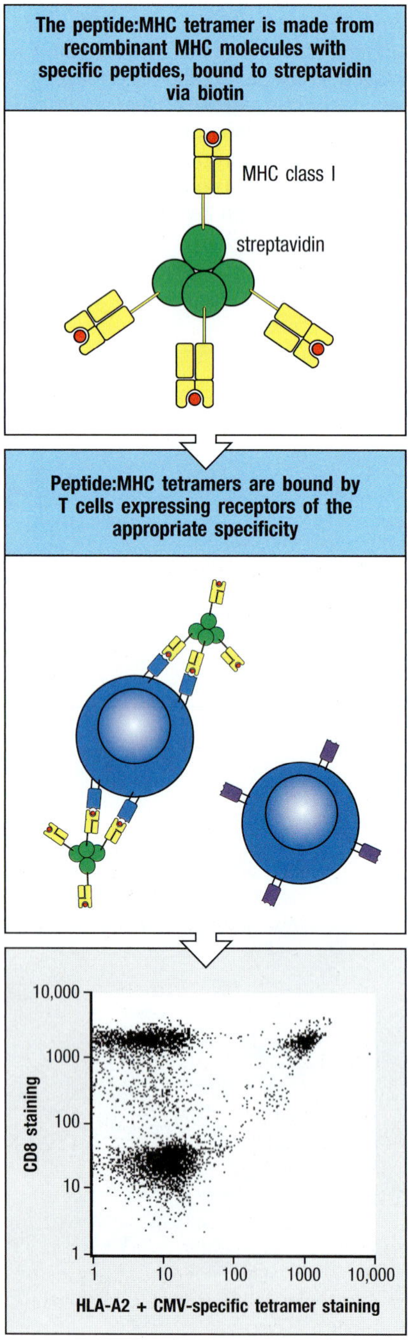

The peptide:MHC tetramer is made from recombinant MHC molecules with specific peptides, bound to streptavidin via biotin

MHC class I

streptavidin

Peptide:MHC tetramers are bound by T cells expressing receptors of the appropriate specificity

CD8 staining

HLA-A2 + CMV-specific tetramer staining

Fig. A.30 Peptide:MHC complexes coupled to streptavidin to form tetramers are able to stain antigen-specific T cells. Peptide:MHC tetramers are formed from recombinant refolded peptide:MHC complexes containing a single defined peptide epitope. MHC molecules that contain biotin can be chemically synthesized, but usually the recombinant MHC heavy chain is linked to a bacterial biotinylation sequence, a target for the *Escherichia coli* enzyme BirA, which is used to add a single biotin group to the MHC molecule. Streptavidin is a tetramer, each subunit having a single binding site for biotin; hence the streptavidin:peptide:MHC complex creates a tetramer of peptide:MHC complexes (top panel). Although the affinity between the T-cell receptor and its peptide:MHC ligand is too low for a single complex to bind stably to a T cell, the tetramer, by being able to make a more avid interaction with multiple peptide:MHC complexes binding simultaneously, is able to bind to T cells whose receptors are specific for the particular peptide:MHC complex (middle panel). Routinely, the streptavidin molecules are coupled to a fluorochrome, so that the binding to T cells can be monitored by flow cytometry. In the example shown in the bottom panel, T cells have been stained simultaneously with antibodies specific for CD3 and CD8, and with a tetramer of HLA-A2 molecules containing a cytomegalovirus peptide. Only the CD3[+] cells are shown, with the staining of CD8 displayed on the vertical axis and the tetramer staining displayed along the horizontal axis. The CD8[−] cells (mostly CD4[+]), on the bottom left of the figure, show no specific tetramer staining, while the bulk of the CD8[+] cells, on the top left, likewise show no tetramer staining. However, a discrete population of tetramer positive CD8[+] cells, at the top right of the panel, comprising some 5% of the total CD8[+] cells, can clearly be seen. Data courtesy of G. Aubert.

A-25 Biosensor assays for measuring the rates of association and dissociation of antigen receptors for their ligands.

Two of the important questions that are always asked of any receptor–ligand interaction are, What is the strength of binding, or affinity, of the interaction, and, What are the rates of association and dissociation? These parameters are generally assessed using purified preparations of proteins. For receptors that are integral membrane proteins in their native state, soluble forms of the proteins are prepared, usually by truncating the proteins to eliminate their membrane-spanning domains. With these purified proteins, binding rates can be measured by following the binding of ligands to receptors immobilized on gold-plated glass slides, using a phenomenon known as **surface plasmon resonance** (**SPR**) to detect the binding (Fig. A.31). A full explanation of surface plasmon resonance is beyond the scope of this textbook, as it is based on advanced physical and quantum mechanical principles. In brief, it relies on the total internal reflection of a beam of light from the surface of a gold-coated glass slide. As the light is reflected, some of its energy excites electrons in the gold coating and these excited electrons are affected by the electric field of any molecules binding to the surface of the glass coating. The more molecules that bind to the surface, the greater is the effect on the excited electrons, and this in turn affects the reflected light beam. The reflected light thus becomes a sensitive measure of the number of atoms bound to the gold surface of the slide.

If a purified receptor is immobilized on the surface of the gold-coated glass slide, to make a biosensor 'chip,' and a solution containing the ligand is streamed over that surface, the binding of ligand to the receptor can be followed until it reaches equilibrium (see Fig. A.31). If the ligand is then washed out, dissociation of ligand from the receptor can easily be followed and the

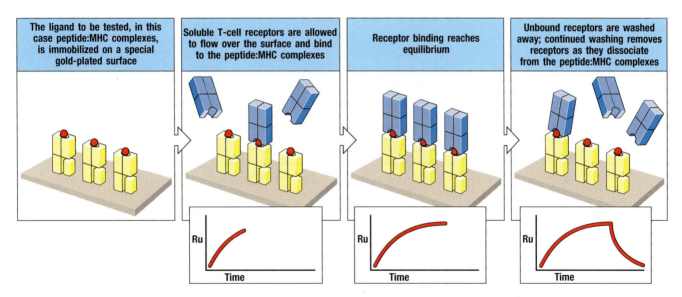

Fig. A.31 Measurement of receptor–ligand interactions can be made in real time using a biosensor. Biosensors are able to measure the binding of molecules on the surface of gold-plated glass chips through the indirect effects of the binding on the total internal reflection of a beam of polarized light at the surface of the chip. Changes in the angle and intensity of the reflected beam are measured in 'resonance units' (Ru) and plotted against time in what is termed a 'sensorgram.' Depending on the exact nature of the receptor–ligand pair to be analyzed, either the receptor or the ligand can be immobilized on the surface of the chip. In the example shown, peptide:MHC complexes are immobilized on such a surface

(first panel). T-cell receptors in solution are now allowed to flow over the surface, and to bind to the immobilized peptide:MHC complexes (second panel). As the T-cell receptors bind, the sensorgram (inset panel below the main panel) reflects the increasing amount of protein bound. As the binding reaches either saturation or equilibrium (third panel), the sensorgram shows a plateau, as no more protein binds. At this point, unbound receptors can be washed away. With continued washing, bound receptors now start to dissociate and are removed in the flow of the washing solution (last panel). The sensorgram now shows a declining curve, reflecting the rate at which receptor and ligand dissociation occurs.

Mitogen	Responding cells
Phytohemagglutinin (PHA) (red kidney bean)	T cells
Concanavalin (ConA) (jack bean)	T cells
Pokeweed mitogen (PWM) (pokeweed)	T and B cells
Lipopolysaccharide (LPS) (*Escherichia coli*)	B cells (mouse)

Fig. A.32 Polyclonal mitogens, many of plant origin, stimulate lymphocyte proliferation in tissue culture. Many of these mitogens are used to test the ability of lymphocytes in human peripheral blood to proliferate.

dissociation rate calculated. A new solution of the ligand at a different concentration can then be streamed over the chip and the binding once again measured. The affinity of binding can be calculated in a number of ways in this type of assay. Most simply, the ratio of the rates of association and dissociation will give an estimate of the affinity, but more accurate estimates can be obtained from the measurements of the binding at different concentrations of ligand. From measurements of binding at equilibrium, a Scatchard plot will give a measurement of the affinity of the receptor–ligand interaction.

A-26 Assays of lymphocyte proliferation.

To function in adaptive immunity, rare antigen-specific lymphocytes must proliferate extensively before they differentiate into functional effector cells in order to generate sufficient numbers of effector cells of a particular specificity. Thus, the analysis of induced lymphocyte proliferation is a central issue in their study. It is, however, difficult to detect the proliferation of normal lymphocytes in response to specific antigen because only a minute proportion of cells will be stimulated to divide. Great impetus was given to the field of lymphocyte culture by the finding that certain substances induce many or all lymphocytes of a given type to proliferate. These substances are referred to collectively as **polyclonal mitogens** because they induce mitosis in lymphocytes of many different specificities or clonal origins. T and B lymphocytes are stimulated by different polyclonal mitogens (Fig. A.32). Polyclonal mitogens seem to trigger essentially the same growth response mechanisms as antigen. Lymphocytes normally exist as resting cells in the G_0 phase of the cell cycle. When stimulated with polyclonal mitogens, they rapidly enter the G_1 phase and progress through the cell cycle. In most studies, lymphocyte proliferation is most simply measured by the incorporation of ^{3}H-thymidine into DNA. This assay is used clinically for assessing the ability of lymphocytes from patients with suspected immunodeficiencies to proliferate in response to a nonspecific stimulus.

An alternative to the use of a radioisotope to measure lymphocyte proliferation is to use a fluorescent assay that can be performed by FACS. For this assay, the lymphocytes are incubated with a fluorescent dye such as carboxyfluorescein succinimidyl ester (CFSE). This dye enters the cell and, once in the cytosol, becomes covalently coupled to lysine residues on cellular proteins. Each time the cell divides, the amount of CFSE is cut in half, as each daughter cell inherits one-half of the CFSE-labeled proteins. When a population of dividing cells is analyzed by FACS, peaks of CFSE fluorescence can be detected, each of which represents cells that have undergone a fixed number of divisions (Fig. A.33). This assay is capable of detecting up to 7–8 cell divisions, after which CFSE fluorescence can no longer be measured.

Fig. A.33 Flow cytometric assay for cell proliferation based on CFSE dilution. Cells are first incubated with a fluorescent dye such as carboxyfluorescein succinimidyl ester (CFSE). This dye enters the cell and, once in the cytosol, becomes covalently coupled to lysine residues on cellular proteins. Each time the cell divides, the amount of CFSE is diluted by one-half, as each daughter cell inherits one-half of the CFSE-labeled proteins. Cell division can then be analyzed by flow cytometry, where a histogram of CFSE fluorescence displays a series of peaks, each of which represents cells that have undergone a fixed number of divisions. Under optimal conditions, this assay is capable of detecting up to 7–8 cell divisions, after which CFSE fluorescence can no longer be measured.

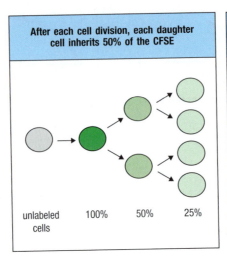

After each cell division, each daughter cell inherits 50% of the CFSE

unlabeled cells   100%   50%   25%

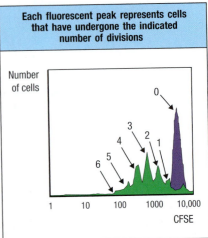

Each fluorescent peak represents cells that have undergone the indicated number of divisions

Number of cells

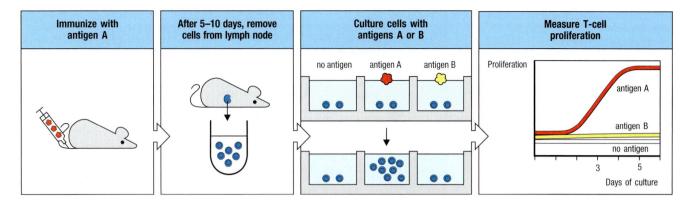

Once lymphocyte culture had been optimized using the proliferative response to polyclonal mitogens as an assay, it became possible to detect antigen-specific T-cell proliferation in culture by measuring ³H-thymidine uptake in response to an antigen to which the T-cell donor had previously been immunized (Fig. A.34). This is the assay most commonly used for assessing T-cell responses after immunization, but it reveals little about the functional capabilities of the responding T cells. These must be ascertained by functional assays, as outlined in Sections A-28 and A-29.

A-27 Measurements of apoptosis.

Apoptotic cells can be detected by a procedure known as **TUNEL** (**T**dT-dependent d**U**TP–biotin **n**ick **e**nd **l**abeling) **staining**. In this technique, the 3′ends of the DNA fragments generated in apoptotic cells are labeled with biotin-coupled uridine by using the enzyme terminal deoxynucleotidyl transferase (TdT). The biotin label is then detected with enzyme-tagged streptavidin, which binds to biotin. When the colorless substrate of the enzyme is added to a tissue section or cell culture, it produces a colored precipitate only in cells that have undergone apoptosis (Fig. A.35).

Additional methods are often used to detect apoptosis of cells in experimental animals. One simple method is to incubate cells with a fluorescently labeled preparation of the protein **Annexin V**. This protein has a high affinity for a specific membrane phospholipid, phosphatidylserine (PS). In healthy cells, PS is found exclusively on the inner leaflet of the plasma membrane, and is therefore inaccessible to extracellular incubation with Annexin V. When cells are undergoing apoptosis, the PS is transported to the outer cell surface, where it can be bound by the fluorescently labeled Annexin V, which is then detected

Fig. A.34 Antigen-specific T-cell proliferation is used frequently as an assay for T-cell responses. T cells from mice or humans that have been immunized with an antigen (A) proliferate when they are exposed to antigen A and antigen-presenting cells but not when cultured with unrelated antigens to which the hosts have not been immunized (antigen B). Proliferation can be measured by incorporation of ³H-thymidine into the DNA of actively dividing cells. Antigen-specific proliferation is a hallmark of specific CD4 T-cell immunity.

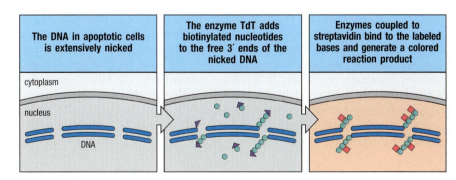

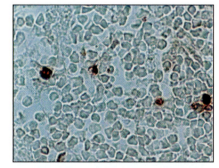

Fig. A.35 In the TUNEL assay, fragmented DNA is labeled by terminal deoxynucleotidyl transferase (TdT) to reveal apoptotic cells. When cells undergo programmed cell death, or apoptosis, their DNA becomes fragmented (first panel). The enzyme TdT is able to add nucleotides to the ends of DNA fragments; most commonly in this assay, biotin-labeled nucleotides (usually dUTP) are added (second panel). The biotinylated DNA can be detected by using streptavidin, which binds to biotin, coupled to enzymes that convert a colorless substrate into a colored insoluble product (third panel). Cells stained in this way can be detected by light microscopy, as shown in the photograph of apoptotic cells (stained red) in the thymic cortex. Photograph courtesy of R. Budd and J. Russell.

Fig. A.36 Detection of apoptotic cells with Annexin V. In healthy cells, the membrane phospholipid phosphatidylserine is oriented with its polar headgroup facing the cytosolic face of the plasma membrane. When cells undergo apoptosis, the enzyme responsible for maintaining phosphatidylserine polarity, called flippase, is no longer active. As a result, phosphatidylserine becomes randomly oriented, with many molecules exposing their polar head groups on the extracellular face of the plasma membrane. The protein Annexin V binds tightly to the exposed phosphatidylserine and, if fluorescently labeled, can be used to detect apoptotic cells by FACS.

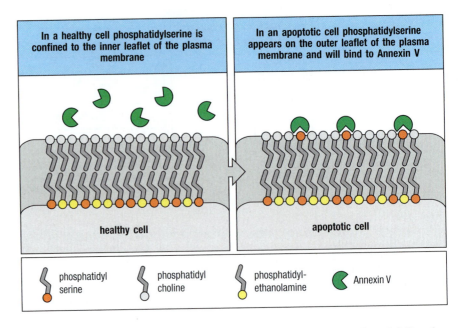

by FACS (**Fig. A.36**). Annexin V staining is often combined with a viability dye such as propidium iodide (PI) or 7-aminoactinomycin D (7-AAD). These two dyes fluoresce when bound to DNA, but are unable to enter viable or apoptotic cells prior to their loss of membrane integrity. Therefore, when combined with Annexin V, cells in the early stages of apoptosis can be identified as Annexin V-positive but PI/7-AAD-negative, whereas cells in the late stages of apoptosis are both Annexin V-positive and PI/7-AAD-positive.

An additional assay that provides a sensitive means of detecting apoptotic cells by FACS analysis is based on the detection of activated caspase 3, a cysteine protease that functions in the execution phase of the apoptotic cell death program. Caspase 3 is initially synthesized by cells in an inactive precursor form called a pro-caspase. When cells are undergoing apoptosis, pro-caspase 3 is cleaved into two subunits that dimerize to form the active enzyme. Antibodies have been generated that detect the active form of caspase 3, but not pro-caspase 3, and fluorescently coupled versions of these antibodies can be used to detect apoptotic cells that have been fixed and permeabilized (**Fig A.37**).

A-28 Assays for cytotoxic T cells.

Activated CD8 T cells generally kill any cells that display the specific pep-tide:MHC class I complex they recognize. Thus, CD8 T-cell function can be determined using the simplest and most rapid T-cell bioassay—the killing of a target cell by a cytotoxic T cell. This is usually detected in a ^{51}Cr-release assay. Live cells will take up, but do not spontaneously release, radioactively labeled sodium chromate, $Na_2{}^{51}CrO_4$. When these labeled cells are killed, the radio-active chromate is released and its presence in the supernatant of mixtures of target cells and cytotoxic T cells can be measured (**Fig. A.38**). In a similar assay, proliferating target cells such as tumor cells can be labeled with ^{3}H-thymidine, which is incorporated into the replicating DNA. On attack by a cytotoxic T cell,

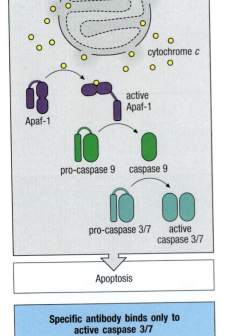

Fig. A.37 Detection of apoptotic cells by intracellular staining for active caspases. An early event in the apoptotic process is the release of cytochrome *c* from mitochondria. Cytochrome *c* binds to Apaf-1, leading to the cleavage of pro-caspase 9 into active caspase 9. Caspase 9 then cleaves pro-caspase 3 and pro-caspase 7 to yield their active forms, 'executioner' caspases that promote cell death. Antibodies that recognize the active caspase 3 or caspase 7, but not the pro-caspase forms of these enzymes, will detect permeabilized cells undergoing apoptosis.

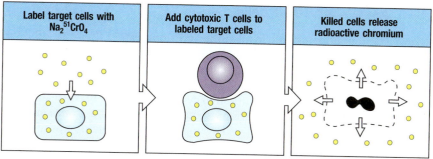

Fig. A.38 Cytotoxic T-cell activity is often assessed by chromium release from labeled target cells. Target cells are labeled with radioactive chromium as $Na_2^{51}CrO_4$, washed to remove excess radioactivity, and exposed to cytotoxic T cells. Cell destruction is measured by the release of radioactive chromium into the medium, detectable within 4 hours of mixing target cells with T cells.

the DNA of the target cells is rapidly fragmented and retained in the filtrate, while large, unfragmented DNA is collected on a filter, and one can measure either the release of these fragments or the retention of ^{3}H-thymidine in chromosomal DNA. These assays provide a rapid, sensitive, and specific measure of the activity of cytotoxic T cells.

An alternative to these *in vitro* cytotoxicity assays is to measure target-cell killing by cytotoxic T cells in intact experimental animals. This assay is generally performed with mice that have been infected with a pathogen known to induce a strong cytotoxic T-cell response, such as a virus. Target cells are incubated with the antigenic peptide, which will bind to MHC class I on the target-cell surface. These cells are then incubated with a low concentration of the fluorescent dye CFSE (see Section A-26). A control population of cells that is not given the antigenic peptide is incubated with a high concentration of CFSE, allowing these cells to be distinguished from the antigen-bearing target cells. The two cell populations are mixed 1:1 and injected into the experimental animals. Four hours later, spleen cells are recovered from the animals and analyzed by FACS, allowing the specific target-cell lysis to be calculated from the ratio of the two CFSE-labeled cell populations (Fig. A.39).

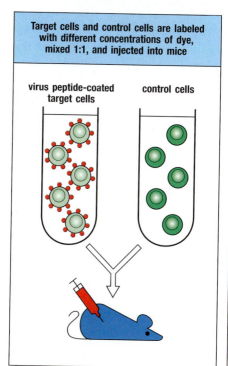

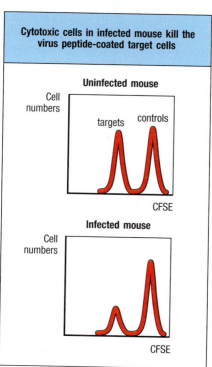

Fig. A.39 Assay for cytotoxic T-cell activity using CFSE-labeled target cells. To detect cytotoxic T-cell activity in intact experimental animals, mice that have been infected with a virus are injected with a mixture of target cells labeled with the fluorescent dye CFSE. One group of target cells is pre-incubated with a viral peptide that binds to MHC class I on the target cells; this group of cells is labeled with a low concentration of CFSE. A second group of cells is incubated with a control (nonviral) peptide and is labeled with a high concentration of CFSE. The two groups of cells are mixed together in a 1:1 ratio, and injected into the infected mice. After 4 hours, the mice are sacrificed and the target cells are recovered and analyzed by flow cytometry. Examination of the ratio of the two target-cell populations provides a measure of specific lysis of viral peptide-coated target cells.

A-29   Assays for CD4 T cells.

CD4 T-cell functions usually involve the activation rather than the killing of cells bearing specific antigen. The activating effects of CD4 T cells on B cells or macrophages are mediated in large part by cytokines, which are released by the T cell when it recognizes antigen. Thus, CD4 T-cell function is usually studied by measuring the type and amount of cytokine released. Because different effector T cells release different amounts and types of cytokines, one can learn about the effector potential of a T cell by measuring the proteins it produces.

Cytokines can be detected by their activity in biological assays of cell growth, where the cytokines serve either as growth factors or as growth inhibitors. A more specific assay is a modification of ELISA known as a capture or sandwich ELISA (see Section A-4). In this assay, the cytokine is characterized by its ability to act as a bridge between two monoclonal antibodies reacting with different epitopes on the cytokine molecule. Cytokine-secreting cells can also be detected by ELISPOT (see Section A-22).

Sandwich ELISA and ELISPOT avoid a major problem of cytokine bioassays, namely, the ability of different cytokines to stimulate the same response in a bioassay. Bioassays must always be confirmed by inhibition of the response with neutralizing monoclonal antibodies specific for the cytokine. Another way of identifying cells actively producing a given cytokine is to stain them with a fluorescently tagged anti-cytokine monoclonal antibody, and then identify and count them by FACS (see Section A-23).

A quite different approach to detecting cytokine production is to determine the presence and amount of the relevant cytokine mRNA in stimulated T cells. This can be done for single cells by *in situ* hybridization and for cell populations by the **reverse transcriptase–polymerase chain reaction** (**RT–PCR**). Reverse transcriptase is an enzyme used by certain RNA viruses, such as HIV, to convert an RNA genome into a DNA copy, or cDNA. In RT–PCR, mRNA is isolated from cells and cDNA copies are made *in vitro* using reverse transcriptase. The desired cDNA is then selectively amplified by PCR by using sequence-specific primers. When the products of the reaction are subjected to electrophoresis on an agarose gel, the amplified DNA can be visualized as a band of a specific size. The amount of amplified cDNA sequence will be proportional to its representation in the mRNA; stimulated T cells actively producing a particular cytokine will produce large amounts of that particular mRNA and will thus give correspondingly large amounts of the selected cDNA on RT–PCR. The level of cytokine mRNA in the original tissue is usually determined by comparison with the outcome of RT–PCR on the mRNA produced by a so-called 'housekeeping gene' expressed by all cells.

A-30   Transfer of protective immunity.

Protective immunity to a pathogen may involve humoral immunity, cell-mediated immunity, or both. For studies in experimental animals such as inbred mice, the nature of protective immunity can be determined by transferring serum or lymphoid cells from an immunized donor animal to an unimmunized syngeneic recipient (that is, a genetically identical animal of the same inbred strain). If protection against infection can be conferred by the transfer of serum, the immunity is provided by circulating antibodies and is called **humoral immunity**. Transfer of immunity by antiserum or purified antibodies provides immediate protection against many pathogens and against toxins such as those of tetanus and snake venom (**Fig. A.40**). However, although protection is immediate, it is temporary, lasting only so long as the transferred antibodies remain active in the recipient's body. This type of transfer is therefore called **passive immunization**, to distinguish it from **active immunization** with antigen, which can provide lasting immunity. A disadvantage of passive

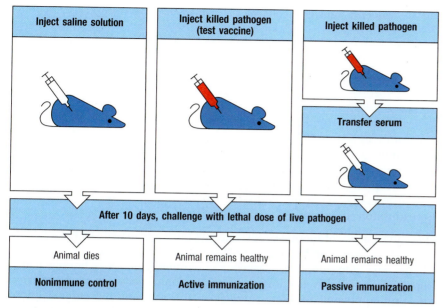

Fig. A.40 *In vivo* assay for the presence of protective immunity after vaccination in animals. Mice are injected with the test vaccine, such as a heat-killed pathogen, or a control such as saline solution. Different groups are then challenged with lethal or pathogenic doses of the test pathogen or with an unrelated pathogen as a specificity control (not shown). Unimmunized animals die or become severely infected (left panel). Successful vaccination is seen as specific protection of immunized mice against infection with the test pathogen. This is called active immunity, and the process is called active immunization (middle panel). If this immune protection can be transferred to a normal syngeneic recipient with serum from an immune donor, then immunity is mediated by antibodies; such immunity is called humoral immunity and the process is called passive immunization (right panel). If immunity can be transferred only by infusing lymphoid cells from the immune donor into a normal syngeneic recipient, then the immunity is called cell-mediated immunity and the transfer process is called adoptive transfer or adoptive immunization (not shown). Passive immunity is short-lived, because antibody is eventually catabolized, but adoptively transferred immunity is mediated by immune cells, which can survive and provide longer-lasting immunity.

immunization is that the recipient may become immunized to the antiserum used to transfer immunity. Horse or sheep sera are the usual sources of anti-snake venoms used in humans, and repeated administration can lead either to serum sickness (see Section 14-5) or, if the recipient becomes allergic to the foreign serum, to anaphylaxis (see Section 14-10).

Protection against many diseases cannot be transferred with serum but can be transferred by lymphoid cells from immunized donors. The transfer of lymphoid cells from an immune donor to a normal syngeneic recipient is called **adoptive transfer** or **adoptive immunization**, and the immunity transferred is called **adoptive immunity**. Immunity that can be transferred only with lymphoid cells is called **cell-mediated immunity**. Such cell transfers must be between genetically identical donors and recipients, such as members of the same inbred strain of mouse, so that the donor lymphocytes are not rejected by the recipient and do not attack the recipient's tissues. Adoptive transfer of immunity is used clinically in humans in experimental approaches to cancer therapy or as an adjunct to bone marrow transplantation; in these cases, the patient's own T cells, or the T cells of the bone marrow donor, are given.

A-31 Adoptive transfer of lymphocytes.

Ionizing radiation from X-ray or gamma-ray sources kills lymphoid and other immune cells at doses that spare the other tissues of the body. This makes it possible to eliminate immune function in a recipient animal before

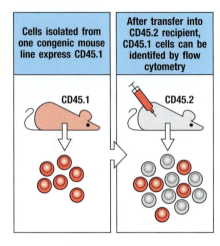

Cells isolated from one congenic mouse line express CD45.1	After transfer into CD45.2 recipient, CD45.1 cells can be identifed by flow cytometry
CD45.1	CD45.2

Fig. A.41 Adoptive transfer of congenically marked cells. Hematopoietic cells can be transferred between genetically identical (or nearly identical) mice. The transferred cells, usually a minority population in the recipient, are identified based on expression of an allelic variant of an abundant cell-surface receptor. One common receptor used for this purpose is CD45, which has two alleles that can be distinguished by allele-specific antibodies. When cells from a CD45.1[+] mouse are transferred into mice of the identical strain (save for their expression of CD45.2), the donor-cell population can easily be identified by antibody staining followed by flow cytometry or immunofluorescence microscopy.

attempting to restore immune function by adoptive transfer, and allows the effect of the adoptively transferred cells to be studied in the absence of other lymphoid cells. **James Gowans** originally used this technique to prove the role of the lymphocyte in immune responses. He showed that all active immune responses could be transferred to irradiated recipients by small lymphocytes from immunized donors.

One common use of adoptive transfer assays takes advantage of the availability of T-cell receptor or B-cell receptor transgenic mice (see Section A-34). In this case, the adoptively transferred lymphocytes are a homogeneous population with a fixed antigen specificity. These cells can be transferred into unmanipulated recipient animals of the same inbred strain without the need to deplete the host immune system, and their ability to respond to immunization or challenge by infection can be monitored. One advantage of this experimental strategy is that relatively small numbers of antigen-specific T cells or B cells can be transferred; after dilution by the recipient's lymphocyte population, the responses of these cells can be examined in the environment of a normal immune response carried out by the host's immune system. Commonly, the transferred cells are 'marked' with an allelic variant of an abundant cell-surface receptor, such as CD45 (**Fig. A.41**). When the donor lymphocytes express one allelic variant of CD45 and the recipient cells express a different variant, the transferred cells can easily be distinguished from the host cells by staining with an antibody that binds to one variant of CD45, but not the other. When two strains of mice are genetically identical with the exception of a single gene, they are said to be **congenic**. In the example above, the donor strain and the recipient strain are referred to as 'CD45 congenics'; it should be noted, however, that in the case where one strain is a T-cell receptor or B-cell receptor transgenic line, this terminology is not completely accurate, as the presence of the transgenic DNA as a genetic difference is conveniently ignored. Such adoptive transfer studies are a cornerstone in the study of the intact immune system. They have provided a rapid and convenient means of determining the effects of many gene deficiencies, such as those in cell-surface receptors, transcription factors, cytokines, and cell survival/cell death genes, on the ability of T cells or B cells to mount protective immune responses.

A-32 Hematopoietic stem-cell transfers.

All cells of hematopoietic origin can be eliminated by treatment with high doses of γ radiation or X rays, allowing replacement of the entire hematopoietic system, including lymphocytes, by transfusion of donor bone marrow or purified hematopoietic stem cells from another animal. The resulting animals are called **radiation bone marrow chimeras**, from the Greek word *chimera*, a mythical animal that had the head of a lion, the tail of a serpent, and the body of a goat. This technique is used experimentally to examine the development of immune-cell lineages, as opposed to their effector functions, and has been particularly important in studying T-cell development. Essentially the same technique is used in humans to replace the hematopoietic system when it fails, as in aplastic anemia or after nuclear accidents, or to eradicate the bone marrow and replace it with normal marrow in the treatment of certain cancers. In humans, bone marrow is the main source of hematopoietic stem cells, but they are increasingly being obtained from peripheral blood after the donor has been treated with hematopoietic growth factors such as GM-CSF, or from umbilical cord blood, which is rich in such stem cells (see Chapter 15).

A-33 *In vivo* administration of antibodies.

Antibodies administered to intact experimental animals, or to humans, provide a potent means of manipulating the immune system. Depending on the target molecule recognized by the antibody, and the intrinsic properties of each antibody, *in vivo* antibody administration can either inhibit the function of the target molecule or, in some cases, lead to the elimination of a cell population that expresses the target molecule.

In animal models, antibodies targeting an individual cytokine have been used to inhibit that cytokine's function during an otherwise intact immune response. Experiments of this type provided some of the first evidence that the cytokine IL-12 provides an important signal in polarizing CD4$^+$ T cells into the T$_H$1 lineage following infection with an intracellular protozoan. This approach has also been used with great success in humans. One of the most common treatments for the inflammatory autoimmune disease rheumatoid arthritis (see Chapter 16) is the administration of an antibody that binds to the cytokine TNF-α; in this case, the inhibition of TNF-α activity provides patients with relief from the symptoms of joint inflammation. The great utility of this antibody therapy has led to the development of related strategies for inhibiting cytokine actions *in vivo*. One successful approach has been to create a hybrid protein that has the ligand-binding domain of the cytokine's receptor fused to the constant-region domains (Fc) of an antibody heavy chain (Fig. A.42). This Fc-fusion protein acquires the stability and long half-life of an antibody, but has the binding properties of the cytokine receptor. When given *in vivo*, the Fc-fusion protein binds to the cytokine, thereby interfering with the cytokine's ability to stimulate its receptor on immune cells. As an example, the Fc-fusion protein containing the TNF-receptor ligand-binding domain has also been used as an effective treatment for patients with rheumatoid arthritis.

Antibody administration can also be used to augment the immune response by interfering with T-cell surface receptors, such as CTLA-4 or PD-1. When engaged by their ligands, these receptors normally function to downregulate the immune response. From experiments in mice, antibodies that bind to and inhibit these receptors have been found to enhance immune responses to tumors, leading in some cases to tumor eradication. Currently, these strategies are being tested in humans for a variety of tumor types, and the initial results have shown great promise.

In vivo administration of antibodies can also be used to deplete specific cell populations. The efficiency with which a given antibody functions for *in vivo* depletion is quite variable, as the mechanism relies on a process known as **antibody-dependent cell-mediated cytotoxicity (ADCC)** (see Section 10-23 and Fig. 10.36). When a cell is coated with antibodies, it becomes a target of natural killer (NK) cells that express the Fc receptor known as CD16 or FcγRIII. The cross-linking of FcγRIII induces the NK cell to kill the antibody-coated target cell. While FcγRIII is a receptor for IgG, it does not bind with equal affinity to all IgG subtypes; thus the efficiency of ADCC after administration of a given antibody is determined by its ability to cross-link FcγRIII and induce NK cell killing. Common uses for this technique include the depletion of CD4$^+$ T cells with an antibody to CD4, or the depletion of CD8$^+$ T cells with an antibody to CD8. In human patients undergoing organ transplantation, T cells are transiently depleted by administering an antibody to the CD3 component of the T-cell receptor complex. This produces a severe, but temporary, state of immunosuppression during the early stages of post-transplantation. As with all *in vivo* antibody depletion regimens, the depleted cell population gradually returns as cells of that subset are replenished by ongoing lymphocyte development.

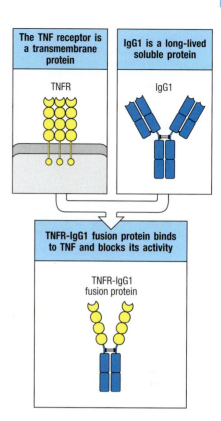

Fig. A.42 *In vivo* administration of antibodies is an effective therapeutic. The cytokine TNF-α contributes to chronic inflammation in a range of conditions, including rheumatoid arthritis, by binding to and triggering signaling of the TNF receptor (TNFR). To treat these conditions, a fusion protein consisting of the constant-region domains of human IgG1 is fused to the extracellular portion of the TNFR, creating a therapeutic known as etanercept. When administered to patients, this fusion protein is effective at binding TNF-α and preventing it from triggering TNFR signaling, thereby reducing inflammation.

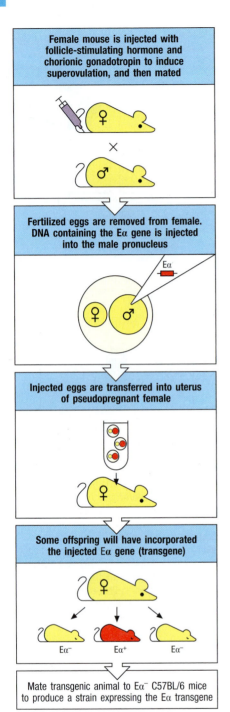

Female mouse is injected with follicle-stimulating hormone and chorionic gonadotropin to induce superovulation, and then mated

Fertilized eggs are removed from female. DNA containing the Eα gene is injected into the male pronucleus

Eα

Injected eggs are transferred into uterus of pseudopregnant female

Some offspring will have incorporated the injected Eα gene (transgene)

Eα⁻   Eα⁺   Eα⁻

Mate transgenic animal to Eα⁻ C57BL/6 mice to produce a strain expressing the Eα transgene

Fig. A.43 The function and expression of genes can be studied *in vivo* by using transgenic mice. DNA encoding a protein of interest, here the mouse MHC class II protein Eα, is purified and microinjected into the male pronuclei of fertilized eggs. The eggs are then implanted into pseudopregnant female mice. The resulting offspring are screened for the presence of the transgene in their cells, and positive mice are used as founders that transmit the transgene to their offspring, establishing a line of transgenic mice that carry one or more extra genes. The function of the Eα gene used here is tested by breeding the transgene into C57BL/6 mice that carry an inactivating mutation in their endogenous Eα gene.

A-34   Transgenic mice.

The function of genes has traditionally been studied by observing the effects of spontaneous mutations in whole organisms and, more recently, by analyzing the effects of targeted mutations in cultured cells. The advent of gene cloning and *in vitro* mutagenesis now makes it possible to produce specific mutations in whole animals. Mice with extra copies or altered copies of a gene in their genome can be generated by **transgenesis**, which is now a well-established procedure. To produce **transgenic mice**, a cloned gene is introduced into the mouse genome by microinjection into the male pronucleus of a fertilized egg, which is then implanted into the uterus of a pseudopregnant female mouse. In some of the eggs, the injected DNA becomes integrated randomly into the genome, giving rise to a mouse that has an extra genetic element of known structure, the transgene (**Fig. A.43**).

This technique allows one to study the impact of a newly discovered gene on development, to identify the regulatory regions of a gene required for its normal tissue-specific expression, to determine the effects of its overexpression or its expression in inappropriate tissues, and to find out the impact of mutations on gene function. Transgenic mice have been particularly useful in studying the role of T-cell and B-cell receptors in lymphocyte development, as described in Chapter 8, and in providing a source of primary T and B lymphocytes of known antigen specificity for adoptive transfer studies (see Section A-31). This utility is largely due to the fact that expression of the transgene-encoded T-cell and B-cell receptors preempts the rearrangement and expression of the endogenous antigen receptor genes during T-cell and B-cell development, respectively, thereby generating homogeneous populations of cells bearing a unique antigen receptor of known specificity.

A-35   Gene knockout by targeted disruption.

In many cases, the functions of a particular gene can be fully understood only if a mutant animal that does not express the gene can be obtained. Whereas genes used to be discovered through the identification of mutant phenotypes, it is now far more common to discover and isolate the normal gene and then determine its function by replacing it *in vivo* with a defective copy. This procedure is known as **gene knockout**, and it has been made possible by two developments: a powerful strategy to select for targeted mutation by homologous recombination, and the development of continuously growing lines of **embryonic stem cells** (**ES cells**). These are embryonic cells that, on implantation into a blastocyst, can give rise to all cell lineages in a chimeric mouse.

The technique of **gene targeting** takes advantage of the phenomenon known as **homologous recombination** (**Fig. A.44**). Cloned copies of the target gene are altered to make them nonfunctional and are then introduced into the ES cell, where they recombine with the homologous gene in the cell's genome, replacing the normal gene with a nonfunctional copy. Homologous recombination is a rare event in mammalian cells, and thus a powerful selection strategy is required to detect those cells in which it has occurred. Most commonly, the introduced gene construct has its sequence disrupted by an inserted antibiotic-resistance gene such as that for neomycin resistance (*neo*ʳ). If this

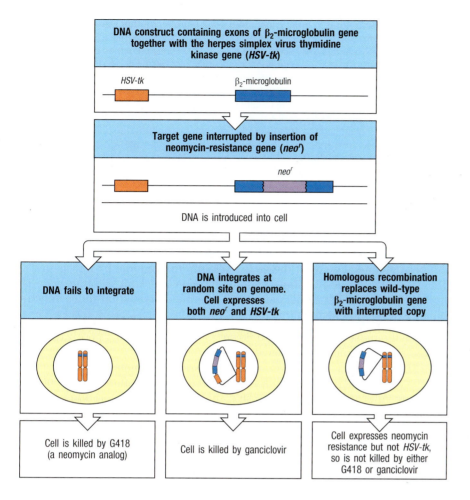

Fig. A.44 The deletion of specific genes can be accomplished by homologous recombination. When pieces of DNA are introduced into cells, they can integrate into cellular DNA in two different ways. If they randomly insert into sites of DNA breaks, the whole piece is usually integrated, often in several copies. However, extrachromosomal DNA can also undergo homologous recombination with the cellular copy of the gene, in which case only the central, homologous region is incorporated into cellular DNA. Inserting a selectable marker gene such as resistance to neomycin (*neo^r*) into the coding region of a gene does not prevent homologous recombination, and it achieves two goals. First, any cell that has integrated the injected DNA is protected from the neomycin-like antibiotic G418. Second, when the gene recombines with homologous cellular DNA, the *neo^r* gene disrupts the coding sequence of the modified cellular gene. Homologous recombinants can be discriminated from random insertions if the gene encoding herpes simplex virus thymidine kinase (*HSV-tk*) is placed at one or both ends of the DNA construct, which is often known as a 'targeting construct' because it targets the cellular gene. In random DNA integrations, *HSV-tk* is retained. *HSV-tk* renders the cell sensitive to the antiviral agent ganciclovir. However, as *HSV-tk* is not homologous to the target DNA, it is lost from homologous recombinants. Thus, cells that have undergone homologous recombination are uniquely resistant to both G418 and ganciclovir, and survive in a mixture of the two antibiotics. The presence of the disrupted gene has to be confirmed by Southern blotting or by PCR using primers in the *neo^r* gene and in cellular DNA lying outside the region used in the targeting construct. By using two different resistance genes, one can disrupt the two cellular copies of a gene, making a deletion mutant (not shown).

construct undergoes homologous recombination with the endogenous copy of the gene, the endogenous gene is disrupted but the antibiotic-resistance gene remains functional, allowing cells that have incorporated the gene to be selected in culture for resistance to the neomycin-like drug G418. However, antibiotic resistance on its own shows only that the cells have taken up and integrated the neomycin-resistance gene. To be able to select for those cells in which homologous recombination has occurred, the ends of the construct usually carry the thymidine kinase gene from the herpes simplex virus (*HSV-tk*). Cells that incorporate DNA randomly usually retain the entire DNA construct including *HSV-tk*, whereas homologous recombination between the construct and cellular DNA, the desired result, involves the exchange of homologous DNA sequences so that the nonhomologous *HSV-tk* genes at the ends of the construct are eliminated. Cells carrying *HSV-tk* are killed by the antiviral drug ganciclovir, and so cells with homologous recombinations have the unique feature of being resistant to both neomycin and ganciclovir, allowing them to be selected efficiently when these drugs are added to the cultures (see Fig. A.44).

To knock out a gene *in vivo*, it is necessary only to disrupt one copy of the cellular gene in an ES cell. These ES cells are then injected into a blastocyst, which is reimplanted into the uterus. The cells carrying the disrupted gene become incorporated into the developing embryo and contribute to all tissues of the resulting chimeric offspring, including those of the germline. The mutated gene can therefore be transmitted to some of the offspring of the original chimera, and further breeding to homozygosity of the mutant gene produces mice that completely lack the expression of that particular gene product (**Fig. A.45**). The effects of the absence of the gene's function can then be

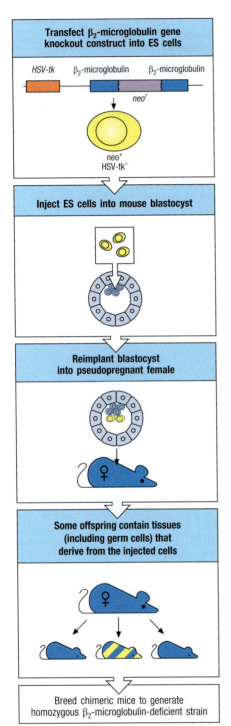

Transfect β₂-microglobulin gene knockout construct into ES cells

HSV-tk   β₂-microglobulin   β₂-microglobulin

neoʳ

neo⁺
HSV-tk⁻

Inject ES cells into mouse blastocyst

Reimplant blastocyst into pseudopregnant female

Some offspring contain tissues (including germ cells) that derive from the injected cells

Breed chimeric mice to generate homozygous β₂-microglobulin-deficient strain

Fig. A.45 Gene knockout in embryonic stem cells enables mutant mice to be produced. Specific genes can be inactivated by homologous recombination in cultures of embryonic stem cells (ES cells). Homologous recombination is performed as described in Fig. A.44. In this example, the gene encoding β₂-microglobulin in ES cells is disrupted by homologous recombination with a targeting construct. Only a single copy of the gene needs to be disrupted. ES cells in which homologous recombination has taken place are injected into mouse blastocysts. If the mutant ES cells give rise to germ cells in the resulting chimeric mice (striped in the figure), the mutant gene can be transferred to their offspring. By breeding the mutant gene to homozygosity, offspring can be tested to determine if a mutant phenotype is generated. In this case, the homozygous mutant mice lack MHC class I molecules on their cells, because MHC class I molecules have to pair with β₂-microglobulin for surface expression. The β₂-microglobulin-deficient mice can then be bred with mice transgenic for subtler mutants of the deleted gene, allowing the effect of such mutants to be tested *in vivo*.

studied. In addition, the parts of the gene that are essential for its function can be identified by determining whether function can be restored by introducing different mutated copies of the gene back into the genome by transgenesis. The manipulation of the mouse genome by gene knockout and transgenesis has revolutionized our understanding of the role of individual genes in lymphocyte development and function.

Because the most commonly used ES cells are derived from a poorly characterized strain of mice known as strain 129, the analysis of the function of a gene knockout often requires extensive back-crossing to another strain. One can track the presence of the mutant copy of the gene by the presence of the *neoʳ* gene. After sufficient back-crossing, the mice are intercrossed to produce mutants on a stable genetic background.

A problem with gene knockouts arises when the function of the gene is essential for the survival of the animal; in such cases the gene is termed a **recessive lethal gene**, and homozygous animals cannot be produced. To study the function of such a gene, tissue-specific or developmentally regulated gene deletion can be employed. This strategy makes use of the DNA sequences and enzymes used by bacteriophage P1 to excise itself from a host cell's genome. Integrated bacteriophage P1 DNA is flanked by recombination signal sequences called *loxP* sites. A recombinase, Cre, recognizes these sites, cuts the DNA, and joins the two ends, thus excising the intervening DNA in the form of a circle. This mechanism can be adapted to allow the deletion of specific genes in a transgenic animal only in certain tissues or at certain times in development. First, *loxP* sites flanking a gene, or perhaps flanking just a single exon, are introduced by homologous recombination (**Fig. A.46**). Usually, the introduction of these sequences into flanking or intronic DNA does not disrupt the normal function of the gene. Next, mice containing such *loxP* mutant genes are mated with mice made transgenic for Cre recombinase that has been placed under the control of a tissue-specific or inducible promoter. When the Cre recombinase is active, either in the appropriate tissue or when induced, it excises the DNA between the inserted *loxP* sites, thus inactivating the gene or exon. Thus, for example, using a T-cell-specific promoter to drive expression of the Cre recombinase, a gene can be deleted only in T cells while remaining functional in all other cells of the animal. This extremely powerful genetic technique was used to demonstrate the importance of B-cell receptors in B-cell survival.

Recently, a new technology has been developed for inducing specific gene disruptions in mice; it is known as the CRISPR/Cas9 system. This technique is adapted from a bacterial system that uses an RNA-based strategy to generate double-stranded DNA breaks in the genomes of invading pathogens or plasmids, a form of bacterial immunity. The Cas9 gene encodes an endonuclease; this has been modified for use in eukaryotic cells by incorporating a nuclear localization signal into the protein-coding sequence of the enzyme. To target mutations to a particular gene, a synthetic guide RNA is produced that incorporates a short sequence (~20 nucleotides) homologous to the gene

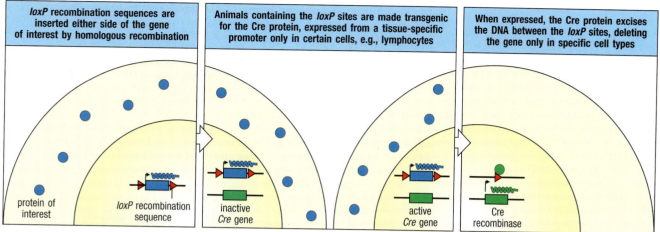

Fig. A.46 The P1 bacteriophage recombination system can be used to eliminate genes in particular cell lineages. The P1 bacteriophage protein Cre excises DNA that is bounded by recombination signal sequences called *loxP* sequences. These sequences can be introduced at either end of a gene by homologous recombination (left panel). Animals carrying genes flanked by *loxP* can also be made transgenic for the gene encoding the Cre protein, which is placed under the control of a tissue-specific promoter so that it is expressed only in certain cells or only at certain times during development (center panel). In the cells in which the Cre protein is expressed, it recognizes the *loxP* sequences and excises the DNA lying between them (right panel). Thus, individual genes can be deleted only in certain cell types or only at certain times. In this way, genes that are essential for the normal development of a mouse can be analyzed for their function in the developed animal and/or in specific cell types. Genes are shown as boxes, RNA as squiggles, and proteins as colored balls.

being targeted along with sequences that bind the Cas9 enzyme. The guide RNA recruits Cas9 to the genomic location, where the endonuclease will produce a double-stranded DNA break (Fig. A.47). When this break is repaired

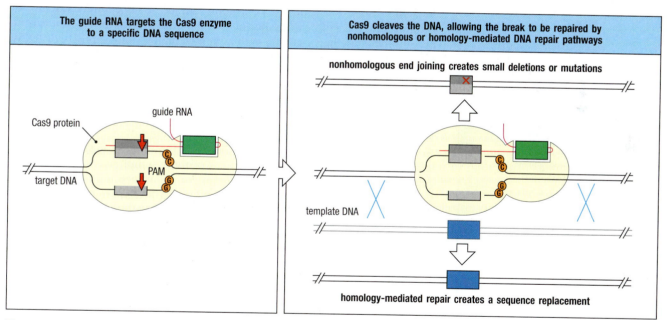

Fig. A.47 Genetic engineering using the bacterial CRISPR/Cas9 system. Genetic engineering can be targeted to a specific gene locus in cells by using two components, the bacterial Cas9 enzyme and a guide RNA (left panel). The guide RNA is a single-stranded RNA that contains two regions of sequence in tandem, the first with homology to the gene being targeted and a second recognized by the Cas9 enzyme. The guide RNA targets the enzyme to the homologous genomic region, promoting double-stranded DNA cleavage by the Cas9 endonuclease 3–4 nucleotides upstream of the protospacer adjacent motif (PAM) sequence (right panel). The PAM sequence required by the Cas9 endonuclease is the dinucleotide GG (CC on the other strand). When the double-stranded DNA break is repaired by the nonhomologous end-joining pathway, small deletions and/or point mutations are introduced into the target gene, often leading to loss of gene function. To induce a specific sequence replacement in the target gene, cells are provided with a template DNA, in addition to Cas9 and the guide RNA. This template is a double-stranded DNA sequence homologous to the target gene, but containing specific nucleotide changes. In the presence of this template, cells will repair the Cas9-mediated double-stranded cleavage using homologous recombination, rather than nonhomologous end-joining, thereby replacing the original sequence with the sequence provided in the template DNA.

by the nonhomologous end-joining DNA repair pathway, small insertions or deletions are commonly introduced, leading to a disruption of the original sequence.

This powerful technique can be used to generate homozygous gene deficiencies in cultured cells and cell lines, but importantly, it can also be used as a single-step means of generating homozygous mutant mice. For this latter purpose, RNA molecules encoding Cas9 are mixed with guide RNAs and injected into single-cell mouse zygotes, using the same technique as that used to generate transgenic mouse lines (see Fig. A.43). Due to the efficiency of the CRISPR/Cas9 system, these embryos frequently harbor mutations on both alleles of the targeted gene. Thus, after transplantation of the embryos into foster mothers, the pups born from these embryos are already homozygous for the targeted gene, without the need for lengthy mouse breeding. A refinement of this technique has been developed that allows specific nucleotide changes to be introduced into the targeted gene, rather than the random changes resulting from nonhomologous end-joining. This is accomplished by introducing a DNA oligonucleotide into the fertilized mouse zygotes along with the Cas9 and guide RNAs. The oligonucleotide contains the desired nucleotide changes flanked by sequences homologous to the targeted gene. When this oligonucleotide is present, the double-strand DNA break introduced by Cas9 is preferentially repaired by a homology-directed process that replaces the damaged DNA with the sequences from the oligonucleotide (see Fig. A.47).

A-36 Knockdown of gene expression by RNA interference (RNAi).

In some cases, the function of a gene can be assessed by reducing, or even eliminating, the expression of that gene in specific cells. This can be accomplished by harnessing a system known as RNA interference, or RNAi, that is present in many eukaryotic cell types. When small double-stranded RNA molecules (referred to as small interfering RNAs, or siRNAs) are introduced into cells, the two RNA strands will be separated and one will bind to an enzyme complex known as RISC (RNA-induced silencing complex). The bound siRNA targets the RISC complex to the mRNA to which it has homology, leading either to translation arrest or to degradation of the mRNA, and thus to silencing of the gene (**Fig. A.48**). For cells that are not easily transfected with siRNA molecules directly, such as primary lymphocytes and myeloid cells, gene silencing can be implemented by using recombinant viruses. In this case, genes encoding small hairpin RNAs (shRNAs) are introduced into viral vectors that can be packaged into infectious viral particles. The shRNAs encode small RNAs that form a double-stranded hairpin structure; these hairpins are processed by enzymes in the cell to generate the siRNAs needed for gene silencing (see Fig. A.48). Since many primary hematopoietic cell types are readily transduced with recombinant viruses, such as retroviruses and lentiviruses, shRNAs can be effectively used to silence genes in these cell types.

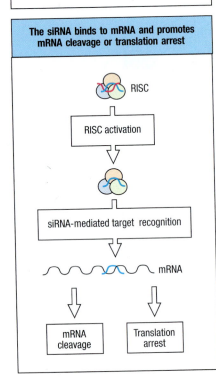

siRNA can be introduced into cells or made from a double-stranded hairpin RNA precursor

Transcribed from expression vector

shRNA
Dicer

siRNA duplex

Transfection of cells

The siRNA binds to mRNA and promotes mRNA cleavage or translation arrest

RISC

RISC activation

siRNA-mediated target recognition

mRNA

mRNA cleavage | Translation arrest

Fig. A.48 Knockdown of gene expression using the RNAi pathway. Small double-stranded RNA molecules with homology to an mRNA transcript will target the mRNA for degradation or translation arrest. This pathway is initiated by expression of a short hairpin RNA (shRNA), which can be produced from an expression vector that is introduced into cells, or by the direct transfection of cells with small double-stranded RNA molecules called siRNA. shRNA molecules are processed by the enzyme Dicer to generate siRNA duplexes. The siRNA duplexes bind to the RISC complex, which separates the two RNA strands, retaining the noncoding strand of the siRNA. This noncoding strand targets the siRNA–RISC complex to the mRNA, leading to mRNA degradation or translation termination.

Appendices II–IV

Appendix II. CD antigens					
CD antigen	Cellular expression	Molecular weight (kDa)	Functions	Other names	Family relationships
CD1a, b, c, d	Cortical thymocytes, Langerhans cells, dendritic cells, B cells (CD1c), intestinal epithelium, smooth muscle, blood vessels (CD1d)	43–49	MHC class I-like molecule, associated with β2-microglobulin. Has specialized role in presentation of lipid antigens		Immunoglobulin
CD2	T cells, thymocytes, NK cells	45–58	Adhesion molecule, binding CD58 (LFA-3). Binds Lck intracellularly and activates T cells	T11, LFA-2	Immunoglobulin
CD3	Thymocytes, T cells	γ: 25–28 δ: 20 ε: 20	Associated with the T-cell antigen receptor (TCR). Required for cell-surface expression of and signal transduction by the TCR	T3	Immunoglobulin
CD4	Thymocyte subsets, helper T cells and T_{reg} cells, some ILC3 cells (LTi cells), some NKT cells, some monocytes and macrophages	55	Co-receptor for MHC class II molecules. Binds Lck on cytoplasmic face of membrane. Receptor for HIV-1 and HIV-2 gp120	T4, L3T4	Immunoglobulin
CD5	Thymocytes, T cells, subset of B cells	67	Attenuates TCR signaling. Enhances Akt signaling in T cells. Required for optimal T_H2 and T_H17 differentiation	T1, Ly1	Scavenger receptor
CD6	Thymocytes, T cells, B cells in chronic lymphatic leukemia	100–130	Binds CD166	T12	Scavenger receptor
CD7	Pluripotential hematopoietic cells, thymocytes, T cells	40	Unknown, cytoplasmic domain binds PI 3-kinase on cross-linking. Marker for T-cell acute lymphatic leukemia and pluripotential stem cell leukemias	GP40, TP41, Tp40, LEU-9	Immunoglobulin
CD8	Thymocyte subsets, cytotoxic T cells (about one third of peripheral T cells), α chain homodimer is expressed on a subset of dendritic cells and intestinal lymphocytes	α: 32–34 β: 32–34	Co-receptor for MHC class I molecules. Binds Lck on cytoplasmic face of membrane	T8, Lyt2,3	Immunoglobulin
CD9	Pre-B cells, monocytes, eosinophils, basophils, platelets, activated T cells, brain and peripheral nerves, vascular smooth muscle	24	Mediates platelet aggregation and activation via FcγRIIa, may play a role in cell migration	MIC3, MRP-1, BTCC-1, DRAP-27, TSPAN29	Tetraspanning membrane protein, also called transmembrane 4 (TM4)
CD10	B- and T-cell precursors, bone marrow stromal cells, and some endothelial cells	100	Zinc metalloproteinase, marker for pre-B acute lymphatic leukemia (ALL)	Neutral endopeptidase, common acute lymphocytic leukemia antigen (CALLA)	
CD11a	Lymphocytes, granulocytes, monocytes, and macrophages	180	αL subunit of integrin LFA-1 (associated with CD18); binds to CD54 (ICAM-1), CD102 (ICAM-2), and CD50 (ICAM-3)	LFA-1	Integrin α
CD11b	Myeloid and NK cells	170	αM subunit of integrin CR3 (associated with CD18): binds CD54, complement component iC3b, and extracellular matrix proteins	Mac-1, Mac-1a, CR3, CR3A, Ly40	Integrin α
CD11c	Myeloid cells	150	αX subunit of integrin CR4 (associated with CD18); binds fibrinogen	CR4, p150, 95	Integrin α

CD antigen	Cellular expression	Molecular weight (kDa)	Functions	Other names	Family relationships
CD11d	Leukocytes	125	αD subunits of integrin; associated with CD18; binds to CD50	ADB2	Integrin α
CDw12	Monocytes, granulocytes, platelets	90–120	Unknown		
CD13	Myelomonocytic cells	150–170	Zinc metalloproteinase	Aminopeptidase N	
CD14	Myelomonocytic cells	53–55	Receptor for complex of lipopoly-saccharide and lipopolysaccharide binding protein (LBP)		
CD15	Neutrophils, eosinophils, monocytes	59	Terminal trisaccharide expressed on glycolipids and many cell-surface glycoproteins	Lewis^x (Le^x)	
CD15s	Leukocytes, endothelium	43	Ligand for CD62E, P	Sialyl-Lewis^x (sLe^x)	poly-*N*-acetyl-lactosamine
CD15u	Subset of memory T cells, NK cells	41	Sulfated CD15		Carbohydrate structures
CD16a	NK cells	50–80	Contributes to phagocytosis and antibody-dependent cell-mediated cytotoxicity as component of low affinity Fc receptor, FcγRIII, expressed by NK cells. Highly similar to CD16b	FcγRIIIa	Immunoglobulin
CD16b	Neutrophils, macrophages	50–80	Contributes to phagocytosis and antibody-dependent cell-mediated cytotoxicity as component of low affinity Fc receptor, FcγRIII, expressed by neutrophils/macrophages. Highly similar to CD16a	FcγRIIIb	Immunoglobulin
CD17	Neutrophils, monocytes, platelets		Lactosyl ceramide, a cell-surface glycosphingolipid		
CD18	Leukocytes	95	Integrin β2 subunit, associates with CD11a, b, c, and d	LAD, MF17, MFI7, LCAMB, LFA-1, Mac-1	Integrin β
CD19	B cells	95	Forms complex with CD21 (CR2) and CD81 (TAPA-1); co-receptor for B cells—cytoplasmic domain binds cytoplasmic tyrosine kinases and PI 3-kinase		Immunoglobulin
CD20	B cells	33–37	Oligomers of CD20 may form a Ca^{2+} channel; possible role in regulating B-cell activation; involved in B-cell development and plasma B-cell differentiation		Contains 4 transmembrane segments
CD21	Mature B cells, follicular dendritic cells	145	Receptor for complement component C3d, Epstein–Barr virus. With CD19 and CD81, CD21 forms co-receptor for B cells	CR2	Complement control protein (CCP)
CD22	Mature B cells	α: 130 β: 140	Binds sialoconjugates	BL-CAM, SIGLEC-2, Lyb8	Immunoglobulin
CD23	Mature B cells, activated macrophages, eosinophils, follicular dendritic cells, platelets	45	Low-affinity receptor for IgE, regulates IgE synthesis; ligand for CD19:CD21:CD81 co-receptor	FcεRII, FCE2, CD23A, CLEC4J, BLAST-2	C-type lectin
CD24	B cells, granulocytes	35–45	Sialoglycoprotein, anchored to cell surface via glycosylphosphatidylinositol (GPI) link	Possible human homolog of mouse heat stable antigen (HSA)	
CD25	Activated T cells, B cells, some ILCs and monocytes	55	IL-2 receptor α chain	Tac, IL2RA	CCP
CD26	Activated B and T cells, macrophages, highly expressed on T$_{reg}$ cells	110	Exopeptidase, cleaves N-terminal X-Pro or X-Ala dipeptides from polypeptides	Dipeptidyl peptidase IV	Type II transmembrane glycoprotein
CD27	Medullary thymocytes, T cells, NK cells, some B cells	55	Binds CD70; can function as a co-stimulator for T and B cells	S152, Tp55, TNFRSF7	TNF receptor
CD28	T-cell subsets, activated B cells	44	Activation of naive T cells, receptor for co-stimulatory signal (signal 2) binds CD80 (B7.1) and CD86 (B7.2)	Tp44	Immunoglobulin and CD86 (B7.2)

CD antigen	Cellular expression	Molecular weight (kDa)	Functions	Other names	Family relationships
CD29	Leukocytes	130	Integrin β1 subunit, associates with CD49a in VLA-1 integrin		Integrin β
CD30	Activated T, B, and NK cells, monocytes	120	Binds CD30L (CD153); cross-linking CD30 enhances proliferation of B and T cells	Ki-1	TNF receptor
CD31	Monocytes, platelets, granulocytes, T-cell subsets, endothelial cells	130–140	Adhesion molecule, mediating both leukocyte–endothelial and endothelial–endothelial interactions	PECAM-1	Immunoglobulin
CD32	Monocytes, granulocytes, B cells, eosinophils	40	Low affinity Fc receptor for aggregated immunoglobulin:immune complexes	FcγRII	Immunoglobulin
CD33	Myeloid progenitor cells, monocytes	67	Binds sialoconjugates	SIGLEC-3	Immunoglobulin
CD34	Hematopoietic precursors, capillary endothelium	105–120	Ligand for CD62L (L-selectin), attaches bone marrow stem cells to stromal cell extracellular matrix		Mucin
CD35	Erythrocytes, B cells, monocytes, neutrophils, eosinophils, follicular dendritic cells	250	Complement receptor 1, binds C3b and C4b, mediates phagocytosis	CR1	CCP
CD36	Platelets, monocytes, endothelial cells	88	Platelet adhesion molecule; involved in recognition and phagocytosis of apoptosed cells	Platelet GPIV, GPIIIb	
CD37	Mature B cells, mature T cells, myeloid cells	40–52	Unknown, may be involved in signal transduction; may play a role in T-cell/B-cell interactions; forms complexes with CD53, CD81, CD82, and MHC class II	TSPAN26	Transmembrane 4
CD38	Early B and T cells, activated T cells, germinal center B cells, plasma cells	45	NAD glycohydrolase, augments B-cell proliferation	T10	
CD39	Activated B cells, activated NK cells, macrophages, dendritic cells	78	Involved in suppressive functions of CD4$^+$ T_{reg} cells; may mediate adhesion of B cells	ENTPD1; ATPDase; NTPDase-1	
CD40	B cells, macrophages, dendritic cells, basal epithelial cells	48	Binds CD154 (CD40L); receptor for co-stimulatory signal for B cells, promotes growth, differentiation, and isotype switching of B cells, and promotes germinal center formation and memory B-cell development; promotes cytokine production by macrophages and dendritic cells	TNFRSF5	TNF receptor
CD41	Platelets, megakaryocytes	Dimer: GPIIba: 125 GPIIbb: 22	αIIb integrin, associates with CD61 to form GPIIb, binds fibrinogen, fibronectin, von Willebrand factor, and thrombospondin	GPIIb	Integrin α
CD42a, b, c, d	Platelets, megakaryocytes	a: 23 b: 135, 23 c: 22 d: 85	Binds von Willebrand factor, thrombin; essential for platelet adhesion at sites of injury	a: GPIX b: GPIbα c: GPIbβ d: GPV	Leucine-rich repeat
CD43	Leukocytes, except resting B cells	115–135 (neutrophils) 95–115 (T cells)	Has extended structure approx. 45 nm long and may be anti-adhesive	Leukosialin, sialophorin	Mucin
CD44	Leukocytes, erythrocytes	80–95	Binds hyaluronic acid, mediates adhesion of leukocytes	Hermes antigen, Pgp-1	Link protein
CD45	All hematopoietic cells	180–240 (multiple isoforms)	Tyrosine phosphatase, augments signaling through antigen receptor of B and T cells, multiple isoforms result from alternative splicing (see below)	Leukocyte common antigen (LCA), T200, B220	Protein tyrosine phosphatase (PTP); fibronectin type III
CD45RO	T-cell subsets (memory T cells), B-cell subsets, monocytes, macrophages	180	Isoform of CD45 containing none of the A, B, and C exons		Protein tyrosine phosphatase (PTP); fibronectin type III
CD45RA	B cells, T-cell subsets (naive T cells), monocytes	205–220	Isoforms of CD45 containing the A exon		Protein tyrosine phosphatase (PTP); fibronectin type III
CD45RB	T-cell subsets, (naive T cells, mouse) B cells, monocytes, macrophages, granulocytes	190–220	Isoforms of CD45 containing the B exon	T200	Protein tyrosine phosphatase (PTP); fibronectin type III

CD antigen	Cellular expression	Molecular weight (kDa)	Functions	Other names	Family relationships
CD46	Hematopoietic and non-hematopoietic nucleated cells	56/66 (splice variants)	Membrane co-factor protein, binds to C3b and C4b to permit their degradation by Factor I	MCP	CCP
CD47	All cells	47–52	Adhesion molecule, thrombospondin receptor	IAP, MER6, OA3	Immunoglobulin
CD48	Leukocytes	40–47	Putative ligand for CD244	Blast-1	Immunoglobulin
CD49a	Activated T cells, monocytes, neuronal cells, smooth muscle	200	α1 integrin, associates with CD29, binds collagen, laminin-1	VLA-1	Integrin α
CD49b	B cells, monocytes, platelets, megakaryocytes, neuronal, epithelial and endothelial cells, osteoclasts	160	α2 integrin, associates with CD29, binds collagen, laminin	VLA-2, platelet GPIa	Integrin α
CD49c	B cells, many adherent cells	125, 30	α3 integrin, associates with CD29, binds laminin-5, fibronectin, collagen, entactin, invasin	VLA-3	Integrin α
CD49d	Broad distribution includes B cells, thymocytes, monocytes, granulocytes, dendritic cells	150	α4 integrin, associates with CD29, binds fibronectin, MAdCAM-1, VCAM-1	VLA-4	Integrin α
CD49e	Broad distribution includes memory T cells, monocytes, platelets	135, 25	α5 integrin, associates with CD29, binds fibronectin, invasin	VLA-5	Integrin α
CD49f	T lymphocytes, monocytes, platelets, megakaryocytes, trophoblasts	125, 25	α6 integrin, associates with CD29, binds laminin, invasin, merosine	VLA-6	Integrin α
CD50	Thymocytes, T cells, B cells, monocytes, granulocytes	130	Binds integrin CD11a/CD18	ICAM-3	Immunoglobulin
CD51	Platelets, megakaryocytes	125, 24	αV integrin, associates with CD61, binds vitronectin, von Willebrand factor, fibrinogen, and thrombospondin; may be receptor for apoptotic cells	Vitronectin receptor	Integrin α
CD52	Thymocytes, T cells, B cells (not plasma cells), monocytes, granulocytes, spermatozoa	25	Unknown, target for antibodies used therapeutically to deplete T cells from bone marrow	CAMPATH-1, HE5	
CD53	Leukocytes	35–42	Contributes to transduction of CD2-generated signals in T cells and NK cells; may play a role in regulation of growth	MRC OX44	Transmembrane 4
CD54	Hematopoietic and non-hematopoietic cells	75–115	Intercellular adhesion molecule (ICAM)-1 binds CD11a/CD18 integrin (LFA-1) and CD11b/CD18 integrin (Mac-1), receptor for rhinovirus	ICAM-1	Immunoglobulin
CD55	Hematopoietic and non-hematopoietic cells	60–70	Decay accelerating factor (DAF), binds C3b, disassembles C3/C5 convertase	DAF	CCP
CD56	NK cells, some activated T cells	135–220	Isoform of neural cell-adhesion molecule (NCAM), adhesion molecule	NKH-1	Immunoglobulin
CD57	NK cells, subsets of T cells, B cells, and monocytes		Oligosaccharide, found on many cell-surface glycoproteins	HNK-1, Leu-7	
CD58	Hematopoietic and non-hematopoietic cells	55–70	Leukocyte function-associated antigen-3 (LFA-3), binds CD2, adhesion molecule	LFA-3	Immunoglobulin
CD59	Hematopoietic and non-hematopoietic cells	19	Binds complement components C8 and C9, blocks assembly of membrane-attack complex	Protectin, Mac inhibitor	Ly-6
CD60a	T cells, platelets, keratinocytes, smooth muscle cells	70	Disialyl ganglioside D3 (GD3)		Carbohydrate structures
CD60b	T cells, platelets, keratinocytes, smooth muscle cells	70	9-O-acetyl-GD3		Carbohydrate structures
CD60c	T cells, platelets, keratinocytes, smooth muscle cells	70	7-O-acetyl-GD3		Carbohydrate structures
CD61	Platelets, megakaryocytes, macrophages	110	Intergrin β3 subunit, associates with CD41 (GPIIb/IIIa) or CD51 (vitronectin receptor), involved in platelet aggregation		Integrin β

CD antigen	Cellular expression	Molecular weight (kDa)	Functions	Other names	Family relationships
CD62E	Endothelium	140	Endothelium leukocyte adhesion molecule (ELAM), binds sialyl-Lewis^x, mediates rolling interaction of neutrophils on endothelium	ELAM-1, E-selectin	C-type lectin, EGF, and CCP
CD62L	B cells, T cells, monocytes, NK cells	150	Leukocyte adhesion molecule (LAM), binds CD34, GlyCAM, mediates rolling interactions with endothelium	LAM-1, L-selectin, LECAM-1	C-type lectin, EGF, and CCP
CD62P	Platelets, megakaryocytes, endothelium	140	Adhesion molecule, binds CD162 (PSGL-1), mediates interaction of platelets with endothelial cells, monocytes and rolling leukocytes on endothelium	P-selectin, PADGEM	C-type lectin, EGF, and CCP
CD63	Activated platelets, monocytes, macrophages	53	Unknown, is lysosomal membrane protein translocated to cell surface after activation	Platelet activation antigen	Transmembrane 4
CD64	Monocytes, macrophages	72	High-affinity receptor for IgG, binds IgG3>IgG1>IgG4>>>IgG2, mediates phagocytosis, antigen capture, ADCC	FcγRI	Immunoglobulin
CD65	Myeloid cells	47	Oligosaccharide component of a ceramide dodecasaccharide		
CD66a	Neutrophils, NK cells	160–180	Inhibits NKG2D-mediated cytolytic function and signaling in activated NK cells	C-CAM, BGP1, CEA-1, CEA-7, MHVR1	Immunoglobulin
CD66b	Granulocytes	95–100	Regulates adhesion and activation of human eosinophils	CEACAM8, CD67, CGM6, NCA-95 (previously called CD67)	Immunoglobulin
CD66c	Neutrophils, colon carcinoma	90	Regulation of CD8$^+$ T-cell responses against multiple myeloma	CEACAM6, NCA	Immunoglobulin
CD66d	Neutrophils	30	Directs phagocytosis of several bacterial species, thought to regulate innate immune response	CEACAM3, CEA, CGM1, W264, W282	Immunoglobulin
CD66e	Adult colon epithelium, colon carcinoma	180–200	Resistance to bacterial and viral infections of the respiratory tract	CEACAM5	Immunoglobulin
CD66f	Macrophages		Upregulates arginase activity and inhibits nitric oxide production in macrophages, induces alternative activation in monocytes, suppresses accessory cell-dependent T-cell proliferation	Pregnancy specific beta-1-glycoprotein 1 (PSG1), SP1, B1G1, DHFRP2	Immunoglobulin
CD68	Monocytes, macrophages, neutrophils, basophils, large lymphocytes	110	Unknown	Macrosialin, GP110, LAMP4, SCARD1	Lysosomal/endosomal-associated membrane glycoprotein (LAMP), scavenger receptor
CD69	Activated T and B cells, activated macrophages and NK cells	28, 32 homodimer	Downregulates S1PR1 to promote retention in secondary lymphoid tissues, may play a role in regulating proliferation, may act to transmit signals in natural killer cells and platelets	Activation inducer molecule (AIM)	C-type lectin
CD70	Activated T and B cells, and macrophages	75, 95, 170	Ligand for CD27, may function in co-stimulation of B and T cells	Ki-24	TNF
CD71	All proliferating cells, hence activated leukocytes	95 homodimer	Transferrin receptor	T9	
CD72	B cells (not plasma cells)	42 homodimer	Ligand for SLAM, NKG2	Lyb-2	C-type lectin
CD73	B-cell subsets, T-cell subsets	69	Ecto-5'-nucleotidase, dephosphorylates nucleotides to allow nucleoside uptake, marker for lymphocyte differentiation	NT5E, NT5, NTE, E5NT, CALJA	
CD74	B cells, macrophages, monocytes, MHC class II positive cells	33, 35, 41, 43 (alternative initiation and splicing)	MHC class II-associated invariant chain	Ii, Iγ	

CD antigen	Cellular expression	Molecular weight (kDa)	Functions	Other names	Family relationships
CD75	Mature B cells, T-cell subsets	47	Lactosamines, ligand for CD22, mediates B-cell–B-cell adhesion	CD76	
CD75s			α-2,6-sialylated lactosamines		Carbohydrate structures
CD77	Germinal center B cells	77	Neutral glycosphingolipid (Galα1→4Galβ1→4Glcβ1→ceramide), binds Shiga toxin, cross-linking induces apoptosis	Globotriaosyl-ceramide (Gb3) Pk blood group	
CD79α, β	B cells	α: 40–45 β: 37	Components of B-cell antigen receptor analogous to CD3, required for cell-surface expression and signal transduction	Igα, Igβ	Immunoglobulin
CD80	B-cell subset	60	Co-stimulator, ligand for CD28 and CTLA-4	B7 (now B7.1), BB1	Immunoglobulin
CD81	Lymphocytes	26	Associates with CD19, CD21 to form B cell co-receptor	Target of anti-proliferative antibody (TAPA-1)	Transmembrane 4
CD82	Leukocytes	50–53	Unknown	R2	Transmembrane 4
CD83	Dendritic cells, B cells, Langerhans cells	43	Regulation of antigen presentation; a soluble form of this protein can bind to dendritic cells and inhibit their maturation	HB15	Immunoglobulin
CD84	Monocytes, platelets, circulating B cells	73	Interacts with SAP (SH2D1A) and FYN, regulates platelet function and LPS-induced cytokine secretion by macrophages	CDw84, SLAMF5, Ly9b	Immunoglobulin
CD85	Dendritic cells, monocytes, macrophages, and lymphocytes		Binds to MHC class I molecules on antigen-presenting cells, inhibits activation	LILR1-9, ILT2, LIR1, MIR7	Immunoglobulin
CD86	Monocytes, activated B cells, dendritic cells	80	Ligand for CD28 and CTLA4	B7.2	Immunoglobulin
CD87	Granulocytes, monocytes, macrophages, T cells, NK cells, wide variety of non-hematopoietic cell types	35–59	Receptor for urokinase plasminogen activator	uPAR	Ly-6
CD88	Polymorphonuclear leukocytes, macrophages, mast cells	43	Receptor for complement component C5a	C5aR	G protein-coupled receptor
CD89	Monocytes, macrophages, granulocytes, neutrophils, B-cell subsets, T-cell subsets	50–70	IgA receptor	FcαR	Immunoglobulin
CD90	CD34+ prothymocytes (human), thymocytes, T cells (mouse), ILCs, some NK cells	18	Adhesion and trafficking of leukocytes at sites of inflammation	Thy-1	Immunoglobulin
CD91	Monocytes, many non-hematopoietic cells	515, 85	α2-macroglobulin receptor		EGF, LDL receptor
CD92	Neutrophils, monocytes, platelets, endothelium	70	Choline transporter		
CD93	Neutrophils, monocytes, endothelium	120	Intercellular adhesion and clearance of apoptotic cells/debris	C1QR1	
CD94	T-cell subsets, NK cells	43	Regulation of NK-cell function	KLRD1	C-type lectin
CD95	Wide variety of cell lines, *in vivo* distribution uncertain	45	Binds TNF-like Fas ligand, induces apoptosis	Apo-1, Fas	TNF receptor
CD96	Activated T cells, NK cells	160	Adhesive interactions of activated T and NK cells, may influence antigen presentation	T-cell activation increased late expression (TACTILE)	Immunoglobulin
CD97	Activated B and T cells, monocytes, granulocytes	75–85	Binds CD55	GR1	EGF, G protein-coupled receptor
CD98	T cells, B cells, natural killer cells, granulocytes, all human cell lines	80, 45 heterodimer	Dibasic and neutral amino acid transporter	SLC3A2, Ly10, 4F2	
CD99	Peripheral blood lymphocytes, thymocytes	32	Leukocyte migration, T-cell adhesion, ganglioside GM1 and transmembrane protein transport, and T-cell death by a caspase-independent pathway	MIC2, E2	

CD antigen	Cellular expression	Molecular weight (kDa)	Functions	Other names	Family relationships
CD100	Hematopoietic cells	150 homodimer	Ligand for Plexin B1, interacts with calmodulin	SEMA4D	Immunoglobulin, semaphorin
CD101	Monocytes, granulocytes, dendritic cells, activated T cells	120 homodimer	Inhibits TCR/CD3-dependent IL-2 production by T cells, induces production of IL-10 by dendritic cells	BPC#4	Immunoglobulin
CD102	Resting lymphocytes, monocytes, vascular endothelium cells (strongest)	55–65	Binds CD11a/CD18 (LFA-1) but not CD11b/CD18 (Mac-1)	ICAM-2	Immunoglobulin
CD103	Intraepithelial lymphocytes, 2–6% peripheral blood lymphocytes	150, 25	αE integrin	HML-1, α6, αE integrin	Integrin α
CD104	CD4$^-$ CD8$^-$ thymocytes, neuronal, epithelial, and some endothelial cells, Schwann cells, trophoblasts	220	Integrin β4 associates with CD49f, binds laminins	β4 integrin	Integrin β
CD105	Endothelial cells, activated monocytes and macrophages, bone marrow cell subsets	90 homodimer	Binds TGF-β	Endoglin	
CD106	Endothelial cells	100–110	Adhesion molecule, ligand for VLA-4 ($\alpha_4\beta_1$ integrin)	VCAM-1	Immunoglobulin
CD107a	Activated platelets, activated T cells, activated neutrophils, activated endothelium, NK cells	110	Influences endosome/vesicle sorting, protects NK cells from degranulation-associated damage	Lysosomal associated membrane protein-1 (LAMP-1)	
CD107b	Activated platelets, activated T cells, NK cells, activated neutrophils, activated endothelium	120	Influences endosome/vesicle sorting, protects NK cells from degranulation-associated damage	LAMP-2	
CD108	Erythrocytes, circulating lymphocytes, lymphoblasts	80	Receptor for PlexinC1, influences both monocyte and CD4 activation/differentiation	GR2, John Milton-Hagen blood group antigen, SEMA7A	Semaphorin
CD109	Activated T cells, activated platelets, vascular endothelium	170	Binds to and negatively regulates signaling of transforming growth factor-β (TGF-β)	Platelet activation factor, GR56	α2-macroglobulin/complement
CD110	Platelets	71	Receptor for thrombopoietin	MPL, TPO R	Hematopoietic receptor
CD111	Myeloid cells	57	Plays a role in organization of adherens and tight junctions in epithelial and endothelial cells	PPR1/Nectin1	Immunoglobulin
CD112	Myeloid cells	58	Component of adherens junctions	PRR2	
CD113	Neurons		May be involved in cell adhesion and neural synapse formation; component of adherens junctions	NECTIN3, PVRL3	Immunoglobulin
CD114	Granulocytes, monocytes	150	Granulocyte colony-stimulating factor (G-CSF) receptor	CSF3R, GCSFR	Immunoglobulin, fibronectin type III
CD115	Monocytes, macrophages	150	Macrophage colony-stimulating factor (M-CSF) receptor	M-CSFR, CSF1R, C-FMS	Immunoglobulin, tyrosine kinase
CD116	Monocytes, neutrophils, eosinophils, endothelium	70–85	Granulocyte-macrophage colony-stimulating factor (GM-CSF) receptor a chain	GM-CSFRα	Cytokine receptor, fibronectin type III
CD117	Hematopoietic progenitors	145	Stem-cell factor (SCF) receptor	c-Kit	Immunoglobulin, tyrosine kinase
CD118	Broad cellular expression		Interferon-α, β receptor	IFN-α, βR	
CD119	Macrophages, monocytes, B cells, endothelium	90–100	Interferon-γ receptor	IFN-γR, IFNGR1	Fibronectin type III
CD120a	Hematopoietic and non-hematopoietic cells, highest on epithelial cells	50–60	TNF receptor, binds both TNF-α and LT	TNFR-I	TNF receptor
CD120b	Hematopoietic and non-hematopoietic cells, highest on myeloid cells	75–85	TNF receptor, binds both TNF-α and LT	TNFR-II	TNF receptor
CD121a	Thymocytes, T cells	80	Type I interleukin-1 receptor, binds IL-1α and IL-1β	IL-1R type I	Immunoglobulin

CD antigen	Cellular expression	Molecular weight (kDa)	Functions	Other names	Family relationships
CD121b	B cells, macrophages, monocytes	60–70	Type II interleukin-1 receptor, binds IL-1α and IL-1β	IL-1R type II	Immunoglobulin
CD122	NK cells, resting T-cell subsets, some B-cell lines	75	IL-2 receptor β chain	IL-2Rβ	Cytokine receptor, fibronectin type III
CD123	Bone marrow stem cells, granulocytes, monocytes, megakaryocytes	70	IL-3 receptor α chain	IL-3Rα	Cytokine receptor, fibronectin type III
CD124	Mature B and T cells, hematopoietic precursor cells	130–150	IL-4 receptor	IL-4R	Cytokine receptor, fibronectin type III
CD125	Eosinophils, basophils, activated B cells	55–60	IL-5 receptor	IL-5R	Cytokine receptor, fibronectin type III
CD126	Activated B cells and plasma cells (strong), most leukocytes (weak)	80	IL-6 receptor α subunit	IL-6Rα	Immunoglobulin, cytokine receptor, fibronectin type III
CD127	Bone marrow lymphoid precursors, pro-B cells, mature T cells, ILCs, monocytes	68–79, possibly forms homodimers	IL-7 receptor	IL-7R	Fibronectin type III
CD128a, b	Neutrophils, basophils, T-cell subsets	58–67	IL-8 receptor	IL-8R, CXCR1	G protein-coupled receptor
CD129	Eosinophils, thymocytes, neutrophils	57	IL-9 receptor	IL-9R	IL2RG
CD130	Most cell types, strong on activated B cells and plasma cells	130	Common subunit of IL-6, IL-11, oncostatin-M (OSM) and leukemia inhibitory factor (LIF) receptors	IL-6Rβ, IL-11Rβ, OSMRβ, LIFRβ, IFRβ	Immunoglobulin, cytokine receptor, fibronectin type III
CD131	Myeloid progenitors, granulocytes	140	Common β subunit of IL-3, IL-5, and GM-CSF receptors	IL-3Rβ, IL-5Rβ, GM-CSFRβ	Cytokine receptor, fibronectin type III
CD132	B cells, T cells, NK cells, mast cells, neutrophils	64	IL-2 receptor γ chain, common subunit of IL-2, IL-4, IL-7, IL-9, and IL-15 receptors	IL-2RG, SCIDX	Cytokine receptor
CD133	Stem/progenitor cells	97	Unknown	Prominin-1, AC133	
CD134	Activated T cells	50	Receptor for OX40L, provides co-stimulation to CD4 T cells	OX40	TNF receptor
CD135	Multipotential precursors, myelomonocytic and B-cell progenitors	130, 155	Receptor for FLT-3L, important for development of hematopoietic stem cells and leukocyte progenitors	FLT3, FLK2, STK-1	Immunoglobulin, tyrosine kinase
CD136	Monocytes, epithelial cells, central and peripheral nervous system	180	Chemotaxis, phagocytosis, cell growth, and differentiation	MSP-R, RON	Tyrosine kinase
CD137	T and B lymphocytes, monocytes, some epithelial cells	28	Co-stimulator of T-cell proliferation	4-1BB, TNFRSF9	TNF receptor
CD138	B cells	32	Heparan sulfate proteoglycan binds collagen type I	Syndecan-1	
CD139	B cells	209, 228	Unknown		
CD140a, b	Stromal cells, some endothelial cells	a: 180 b: 180	Platelet-derived growth factor (PDGF) receptor α and β chains		
CD141	Vascular endothelial cells	105	Anticoagulant, binds thrombin, the complex then activates protein C	Thrombomodulin fetomodulin	C-type lectin, EGF
CD142	Epidermal keratinocytes, various epithelial cells, astrocytes, Schwann cells. Absent from cells in direct contact with plasma unless induced by inflammatory mediators	45–47	Major initiating factor of clotting. Binds Factor VIIa; this complex activates Factors VII, IX, and X	Tissue factor, thromboplastin	Fibronectin type III
CD143	Endothelial cells, except large blood vessels and kidney, epithelial cells of brush borders of kidney and small intestine, neuronal cells, activated macrophages and some T cells. Soluble form in plasma	170–180	Zn^{2+} metallopeptidase dipeptidyl peptidase, cleaves angiotensin I and bradykinin from precursor forms	Angiotensin converting enzyme (ACE)	
CD144	Endothelial cells	130	Organizes adherens junction in endothelial cells	Cadherin-5, VE-cadherin	Cadherin

CD antigen	Cellular expression	Molecular weight (kDa)	Functions	Other names	Family relationships
CD145	Endothelial cells, some stromal cells	25, 90, 110	Unknown		
CD146	Endothelium, T cells, mesenchymal stromal cells (MSCs)	130	Maintenance of hematopoietic stem and progenitor cells, may regulate vasculogenesis	MCAM, MUC18, S-ENDO	Immunoglobulin
CD147	Leukocytes, red blood cells, platelets, endothelial cells	55–65	Activates some MMPs, receptor for CyPA, CypB, and some integrins	M6, neurothelin, EMMPRIN, basigin, OX-47	Immunoglobulin
CD148	Granulocytes, monocytes, dendritic cells, T cells, fibroblasts, nerve cells	240–260	Contact inhibition of cell growth	HPTPη	Fibronectin type III, protein tyrosine phosphatase
CD150	Thymocytes, activated lymphocytes	75–95	Important in signaling in T and B cells, interacts with FYN, PTPN11, SH2D1A (SAP), and SH2D1B	SLAMF1	Immunoglobulin, SLAM
CD151	Platelets, megakaryocytes, epithelial cells, endothelial cells	32	Associates with β1 integrins	PETA-3, SFA-1	Transmembrane 4
CD152	Activated T cells	33	Receptor for B7.1 (CD80), B7.2 (CD86); negative regulator of T-cell activation	CTLA-4	Immunoglobulin
CD153	Activated T cells, activated macrophages, neutrophils, B cells	38–40	Ligand for CD30, inhibits Ig class switching in germinal center B cells	CD30L, TNFSF8L	TNF
CD154	Activated CD4 T cells	30 trimer	Ligand for CD40, inducer of B-cell proliferation and activation	CD40L, TRAP, T-BAM, gp39	TNF receptor
CD155	Monocytes, macrophages, thymocytes, CNS neurons	80–90	Normal function unknown; receptor for poliovirus	Poliovirus receptor	Immunoglobulin
CD156a	Neutrophils, monocytes	69	Metalloprotinease, cleaves TNFαR1	ADAM8, MS2	
CD156b			TNFα converting enzyme (TACE), cleaves pro-TNFα to produce mature TNFα	ADAM17	
CD156c	Neurons		Potential adhesion molecule and known processing amyloid-precursor protein	ADAM10	
CD157	Granulocytes, monocytes, bone marrow stromal cells, vascular endothelial cells, follicular dendritic cells	42–45 (50 on monocytes)	ADP-ribosyl cyclase, cyclic ADP-ribose hydrolase	BST-1	
CD158	NK cells		KIR family		
CD158a	NK-cell subsets	50 or 58	Inhibits NK-cell cytotoxicity on binding MHC class I molecules	p50.1, p58.1	Immunoglobulin
CD158b	NK-cell subsets	50 or 58	Inhibits NK-cell cytotoxicity on binding HLA-Cw3 and related alleles	p50.2, p58.2	Immunoglobulin
CD159a	NK cells	26	Binds CD94 to form NK receptor; inhibits NK-cell cytotoxicity on binding MHC class I molecules	NKG2A	
CD160	T cells, NK cells, intraepithelial lymphochytes	27	Binds classical and non-classical MHC-I molecules, activates phosphoinositide-3 kinase to trigger cytotoxicity and cytokine secretion	NK1	
CD161	NK cells, T cells, ILCs	44	Regulates NK cytotoxicity	NKRP1	C-type lectin
CD162	Neutrophils, lymphocytes, monocytes	120 homodimer	Ligand for CD62P	PSGL-1	Mucin
CD162R	NK cells			PEN5	
CD163	Monocytes, macrophages	130	Clearance of hemoglobin/haptoglobin complexes by macrophages, may function as innate immune sensor for bacteria	M130	Scavenger receptor cysteine-rich (SRCR)
CD164	Epithelial cells, monocytes, bone marrow stromal cells	80	Adhesion receptor	MUC-24 (multi-glycosylated protein 24)	Mucin

CD antigen	Cellular expression	Molecular weight (kDa)	Functions	Other names	Family relationships
CD165	Thymocytes, thymic epithelial cells, CNS neurons, pancreatic islets, Bowman's capsule	37	Adhesion between thymocytes and thymic epithelium	Gp37, AD2	
CD166	Activated T cells, thymic epithelium, fibroblasts, neurons	100–105	Ligand for CD6, involved integrin neurite extension	ALCAM, BEN, DM-GRASP, SC-1	Immunoglobulin
CD167a	Normal and transformed epithelial cells	63, 64 dimer	Binds collagen	DDR1, trkE, cak, eddr1	Receptor tyrosine kinase, discoidin-related
CD168	Breast cancer cells	Five isoforms: 58, 60, 64, 70, 84	Adhesion molecule. Receptor for hyaluronic acid-mediated motility—mediated cell migration	RHAMM	
CD169	Subsets of macrophages	185	Adhesion molecule. Binds sialylated carbohydrates. May mediate macrophage binding to granulocytes and lymphocytes	Sialoadhesin	Immunoglobulin, sialoadhesin
CD170	Neutrophils	67 homodimer	Adhesion molecule. Sialic acid-binding Ig-like lectin (Siglec). Cytoplasmic tail contains ITIM motifs	Siglec-5, OBBP2, CD33L2	Immunoglobulin, sialoadhesin
CD171	Neurons, Schwann cells, lymphoid and myelomonocytic cells, B cells, CD4 T cells (not CD8 T cells)	200–220, exact MW varies with cell type	Adhesion molecule, binds CD9, CD24, CD56, also homophilic binding	L1, NCAM-L1	Immunoglobulin
CD172a		115–120	Adhesion molecule; the transmembrane protein is a substrate of activated receptor tyrosine kinases (RTKs) and binds to SH2 domains	SIRP, SHPS1, MYD-1, SIRP-α-1, protein tyrosine phosphatase, nonreceptor type substrate 1 (PTPNS1)	Immunoglobulin
CD173	All cells	41	Blood group H type 2. Carbohydrate moiety		
CD174	All cells	42	Lewis y blood group. Carbohydrate moiety		
CD175	All cells		Tn blood group. Carbohydrate moiety		
CD175s	All cells		Sialyl-Tn blood group. Carbohydrate moiety		
CD176	All cells		TF blood group. Carbohydrate moiety		
CD177	Myeloid cells	56–64	NB1 is a GPI-linked neutrophil-specific antigen, found on only a subpopulation of neutrophils present in NB1-positive adults (97% of healthy donors) NB1 is first expressed at the myelocyte stage of myeloid differentiation	NB1	
CD178	Activated T cells	38–42	Fas ligand; binds to Fas to induce apoptosis	FasL	TNF
CD179a	Early B cells	16–18	Immunoglobulin iota chain associates noncovalently with CD179b to form a surrogate light chain which is a component of the pre-B-cell receptor that plays a critical role in early B-cell differentiation	VpreB, IGVPB, IGι	Immunoglobulin
CD179b	B cells	22	Immunoglobulin λ-like polypeptide 1 associates noncovalently with CD179a to form a surrogate light chain that is selectively expressed at the early stages of B-cell development. Mutations in the CD179b gene have been shown to result in impairment of B-cell development and agammaglobulinemia in humans	IGLL1, λ5 (IGL5), IGVPB, 14.	Immunoglobulin
CD180	B cells	95–105	Type 1 membrane protein consisting of extracellular leucine-rich repeats (LRR). Is associated with a molecule called MD-1 and forms the cell-surface receptor complex, RP105/MD-1, which by working in concert with TLR4, controls B-cell recognition and signaling of lipopolysaccharide (LPS)	LY64, RP105	Toll-like receptors (TLR)

CD antigen	Cellular expression	Molecular weight (kDa)	Functions	Other names	Family relationships
CD181	Neutrophils, monocytes, NK cells, mast cells, basophils, some T cells		Receptor for CXCL6, CXCL8 (IL-8). Important for neutrophil trafficking	CXCR1, IL8Rα	Chemokine receptor, GPCR class A
CD182	Neutrophils, monocytes, NK cells, mast cells, basophils, some T cells		Receptor for CXCL1, CXCL2, CXCL3, CXCL5, CXCL6, and CXCL8 (IL-8). Neutrophil trafficking and egress from bone marrow	CXCR2, ILRβ	Chemokine receptor, GPCR class A
CD183	Particularly on malignant B cells from chronic lymphoproliferative disorders	46–52	CXC chemokine receptor involved in chemotaxis of malignant B lymphocytes. Binds INP10 and MIG³	CXCR3, G protein-coupled receptor 9 (GPR 9)	Chemokine receptors, G protein coupled receptor
CD184	Preferentially expressed on the more immature CD34⁺ hematopoietic stem cells	46–52	Binding to SDF-1 (LESTR/fusin); acts as a cofactor for fusion and entry of T-cell line; trophic strains of HIV-1	CXCR4, NPY3R, LESTR, fusin, HM89	Chemokine receptors, G protein coupled receptor
CD185	B cells, T_FH cells, and some CD8 T cells		Receptor for CXCL13. B- and T-cell trafficking into B-cell zones in lympoid tissue	CXCR5	Chemokine receptor, GPCR class A
CD186	T_H17 cells, some NK cells, some NKT cells. Some ILC3s		Receptor for CXCL16 and HIV co-receptor	CXCR6	Chemokine receptor, GPCR class A
CD191	Monocytes, macrophages, neutrophils, T_H1 cells, dendritic cells		Receptor for CCL3, CCL5, CCL8, CCL14, and CCL16. Involved in various processes of innate and adaptive immune-cell trafficking	CCR1	Chemokine receptor, GPCR class A
CD192	Monocytes, macrophages, T_H1 cells, basophils, NK cells		Receptor for CCL2, CCL7, CCL8, CCL12, CCL13, and CCL16. Important for monocyte trafficking and T_H1 responses	CCR2	Chemokine receptor, GPCR class A
CD193	Eosinophils, basophils, mast cells		Receptor for CCL5, CCL7, CCL8, CCL11, CCL13, CCL15, CCL24 and CCL28, involved in eosinophil trafficking	CCR3	Chemokine receptor, GPCR class A
CD194	T_H2 cells, T_reg cells, T_H17 cells, CD8 T cells, monocytes, B cells	41	Receptor for CCL17, CCL22, T-cell homing to the skin and T_H2 response	CCR4	Chemokine receptor, GPCR class A
CD195	Promyelocytic cells	40	Receptor for a CC-type chemokine. Binds to MIP-1α, MIP-1β, and RANTES. May play a role in the control of granulocytic lineage proliferation or differentiation. Acts as co-receptor with CD4 for primary macrophage-tropic isolates of HIV-1	CMKBR5, CCR5, CKR-5, CC-CKR-5, CKR5	Chemokine receptors, G protein-coupled receptor
CD196	T_H17 cells, γ:δ T cells, NKT cells, NK cells, T_reg cells, T_FH cells, ILCs		Receptor for CCL20 and CCL21, necessary for gut association lymphoid tissue development and T_H17 responses	CCR6	Chemokine receptor, GPCR class A
CD197	Activated B and T lymphocytes, strongly upregulated in B cells infected with EBV and T cells infected with HHV6 or 7	46–52	Receptor for the MIP-3β chemokine; probable mediator of EBV effects on B lymphocytes or of normal lymphocyte functions	CCR7, EBI1 (Epstein–Barr virus induced gene 1), CMKBR7, BLR2	Chemokine receptors, G protein-coupled receptor
CDw198	Th2 cells, T_reg cells, γ:δ T cells, monocytes, macrophages		Receptor for CCL1, CCL8, and CCL18, necessary for T_H2 immunity and thymopoiesis	CCR8	Chemokine receptor, GPCR class A
CDw199	Intestinal T cells, thymocytes, B cells, dendritic cells		Receptor for CCL25, necessary for gut associated lymphoid tissue development and thymopoiesis	CCR9	Chemokine receptor, GPCR class A
CD200	Normal brain and B-cell lines	41 (rat thymocytes) 47 (rat brain)	Antigen identified by MoAb MRCOX-2. Nonlineage molecules. Function unknown	MOX-2, MOX-1	Immunoglobulin
CD201	Endothelial cells	49	Endothelial cell-surface receptor (EPCR) that is capable of high-affinity binding of protein C and activated protein C. It is downregulated by exposure of endothelium to tumor necrosis factor	EPCR	CD1 major histocompatibility complex
CD202b	Endothelial cells	140	Receptor tyrosine kinase, binds angiopoietin-1; important in angiogenesis, particularly for vascular network formation in endothelial cells. Defects in TEK are associated with inherited venous malformations; the TEK signaling pathway appears to be critical for endothelial cell–smooth muscle cell communication in venous morphogenesis	VMCM, TEK (tyrosine kinase, endothelial), TIE2 (tyrosine kinase with Ig and EGF homology domains), VMCM1	Immunoglobulin, tyrosine kinase

CD antigen	Cellular expression	Molecular weight (kDa)	Functions	Other names	Family relationships
CD203c	Myeloid cells (uterus, basophils, and mast cells)	101	Belongs to a series of ectoenzymes that are involved in hydrolysis of extracellular nucleotides. They catalyze the cleavage of phosphodiester and phosphosulfate bonds of a variety of molecules, including deoxynucleotides, NAD, and nucleotide sugars	NPP3, B10, PDNP3, PD-Iβ, gp130RB13-6	Type II transmembrane proteins, Ecto-nucleotide pyrophosphatase/phosphodiesterase (E-NPP)
CD204	Myeloid cells	220	Mediate the binding, internalization, and processing of a wide range of negatively charged macromolecules. Implicated in the pathologic deposition of cholesterol in arterial walls during atherogenesis	Macrophage scavenger R (MSR1)	Scavenger receptor, collagen-like
CD205	Dendritic cells	205	Lymphocyte antigen 75; putative antigen-uptake receptor on dendritic cells	LY75, DEC-205, GP200-MR6	Type I transmembrane protein
CD206	Macrophages, endothelial cells	175–190	Type I membrane glycoprotein; only known example of a C-type lectin that contains multiple C-type CRDs (carbohydrate-recognition domains); it binds high-mannose structures on the surface of potentially pathogenic viruses, bacteria, and fungi	Macrophage mannose receptor (MMR), MRC1	C-type lectin
CD207	Langerhans cells	40	Type II transmembrane protein; Langerhans cell specific C-type lectin; potent inducer of membrane superimposition and zippering leading to BG (Birbeck granule) formation	Langerin	C-type lectin
CD208	Interdigitating dendritic cells in lymphoid organs	70–90	Homologous to CD68, DC-LAMP is a lysosomal protein involved in remodeling of specialized antigen-processing compartments and in MHC class II-restricted antigen presentation. Up-regulated in mature DCs induced by CD40L, TNF-α and LPS	D lysosome-associated membrane protein, DC-LAMP	Major histocompatibility complex
CD209	Dendritic cells	44	C-type lectin; binds ICAM3 and HIV-1 envelope glycoprotein gp120 enables T-cell receptor engagement by stabilization of the DC/T-cell contact zone, promotes efficient infection in *trans* cells that express CD4 and chemokine receptors; type II transmembrane protein	DC-SIGN (dendritic cell-specific ICAM3-grabbing non-integrin)	C-type lectin
CD210	B cells, T helper cells, and cells of the monocyte/macrophage lineage	90–110	Interleukin 10 receptor α and β	IL-10Rα, IL-10RA, HIL-10R, IL-10Rβ, IL-10RB, CRF2-4, CRFB4	Class II cytokine receptor
CD212	Activated CD4, CD8, and NK cells	130	IL-12 receptor β chain; a type I transmembrane protein involved in IL-12 signal transduction	IL-12R, IL-12RB	Hemopoietin cytokine receptor
CD213a1	B cells, monocytes, fibroblasts, endothelial cells	60–70	Receptor which binds Il-13 with a low affinity; together with IL-4Rα can form a functional receptor for IL-13, also serves as an alternate accessory protein to the common cytokine receptor γ chain for IL-4 signaling	IL-13Rα 1, NR4, IL-13Ra	Hemopoietic cytokine receptor
CD213a2	B cells, monocytes, fibroblasts, endothelial cells		IL-13 receptor which binds as a monomer with high affinity to interleukin-13 (IL-13), but not to IL-4; human cells expressing IL-13RA2 show specific IL-13 binding with high affinity	IL-13Rα 2, IL-13BP	Hemopoietic cytokine receptor
CD215	NK cells, CD8 T cells		Forms complex with IL2RB (CD122) and IL2RG (CD132), enhances cell proliferation and expression of BCL2	IL-15Ra	IL2G
CD217	Activated memory T cells		Interleukin 17 receptor homodimer	IL-17R, CTLA-8	Chemokine/cytokine receptors
CD218a	Macrophages, neutrophils, NK cells, T cells		Signaling induces cytotoxic response	IL-18Ra	Immunoglobulin

CD antigen	Cellular expression	Molecular weight (kDa)	Functions	Other names	Family relationships
CD218b	Macrophages, neutrophils, NK cells, T cells		Signaling induces cytotoxic response	IL-18Rb	Immunoglobulin
CD220	Nonlineage molecules	α:130 β:95	Insulin receptor; integral transmembrane glycoprotein comprising two α and two β subunits; this receptor binds insulin and has a tyrosine-protein kinase activity—autophosphorylation activates the kinase activity	Insulin receptor	Insulin receptor family of tyrosine-protein kinases
CD221	Nonlineage molecules	α:135 β:90	Insulin-like growth factor I receptor binds insulin-like growth factor with a high affinity. It has tyrosine kinase activity and plays a critical role in transformation events. Cleavage of the precursor generates α and β subunits	IGF1R, JTK13	Insulin receptor family of tyrosine-protein kinases
CD222	Nonlineage molecules	250	Cleaves and activates membrane-bound TGFβ. Other functions include internalization of IGF-II, internalization or sorting of lysosomal enzymes and other M6P-containing proteins	IGF2R, CIMPR, CI-MPR, IGF2R, M6P-R (mannose-6-phosphate receptor)	Mammalian lectins
CD223	Activated T and NK cells	70	Involved in lymphocyte activation; binds to HLA class II antigens; role in down-regulating antigen specific response; close relationship of LAG3 to CD4	Lymphocyte-activation gene 3 LAG-3	Immunoglobulin
CD224	Nonlineage molecules	62 (unprocessed precursor)	Predominantly a membrane-bound enzyme; plays a key role in the γ-glutamyl cycle, a pathway for the synthesis and degradation of glutathione. This enzyme consists of two polypeptide chains, which are synthesized in precursor form from a single polypeptide	γ-glutamyl transferase, GGT1, D22S672 D22S732	γ-glutamyl transferase
CD225	Leukocytes and endothelial cells	16–17	Interferon-induced transmembrane protein 1 is implicated in the control of cell growth. It is a component of a multimeric complex involved in the transduction of antiproliferative and homotypic adhesion signals	Leu 13, IFITM1, IFI17	IFN-induced transmembrane proteins
CD226	NK cells, platelets, monocytes, and a subset of T cells	65	Adhesion glycoprotein; mediates cellular adhesion to other cells bearing an un-identified ligand and cross-linking CD226 with antibodies causes cellular activation	DNAM-1 (PTA1), DNAX, TLiSA1	Immunoglobulin
CD227	Human epithelial tumors, such as breast cancer	122 (non-glycosylated)	Epithelial mucin containing a variable number of repeats with a length of 20 amino acids, resulting in many different alleles. Direct or indirect interaction with actin cytoskeleton	PUM (peanut-reactive urinary mucin), MUC.1, mucin 1	Mucin
CD228	Predominantly in human melanomas	97	Tumor-associated antigen (melanoma) identified by monoclonal antibodies 133.2 and 96.5, involved in cellular iron uptake	Melanotransferrin, P97	Transferrin
CD229	Lymphocytes	90–120	May participate in adhesion reactions between T lymphocytes and accessory cells by homophilic interaction	Ly9	Immunoglobulin (CD2 subfamily)
CD230	Expressed in both normal and infected cells	27–30	The function of PRP is not known. It is encoded in the host genome found in high quantity in the brain of humans and animals infected with neurodegenerative diseases known as transmissible spongiform encephalopathies or prion diseases (Creutzfeld–Jakob disease, Gerstmann–Strausler–Scheinker syndrome, fatal familial insomnia)	CJD, PRIP, prion protein (p27-30)	Prion
CD231	T-cell acute lymphoblastic leukemia, neuroblastoma cells, and normal brain neurons	150	May be involved in cell proliferation and motility. Also a cell-surface glycoprotein which is a specific marker for T-cell acute lymphoblastic leukemia. Also found on neuroblastomas	TALLA-1, TM4SF2, A15, MXS1, CCG-B7	Transmembrane 4 (TM4SF also known as tetra-spanins)

CD antigen	Cellular expression	Molecular weight (kDa)	Functions	Other names	Family relationships
CD232	Nonlineage molecules	200	Receptor for an immunologically active semaphorin (virus-encoded semaphorin protein receptor)	VESPR, PLXN, PLXN-C1	Plexin
CD233	Erythroid cells	93	Band 3 is the major integral glycoprotein of the erythrocyte membrane. It has two functional domains. Its integral domain mediates a 1:1 exchange of inorganic anions across the membrane, whereas its cytoplasmic domain provides binding sites for cytoskeletal proteins, glycolytic enzymes, and hemoglobin. Multifunctional transport protein	SLC4A1, Diego blood group, D1, AE1, EPB3	Anion exchanger
CD234	Erythroid cells and nonerythroid cells	35	Fy-glycoprotein; Duffy blood group antigen; nonspecific receptor for many chemokines such as IL-8, GRO, RANTES, MCP-1, and TARC. It is also the receptor for the human malaria parasites *Plasmodium vivax* and *Plasmodium knowlesi* and plays a role in inflammation and in malaria infection	GPD, CCBP1, DARC (duffy antigen/ receptor for chemokines)	Family 1 of G protein-coupled receptors, chemo-kine receptors
CD235a	Erythroid cells	31	Major carbohydrate-rich sialoglycoprotein of human erythrocyte membrane which bears the antigenic determinants for the MN and Ss blood groups. The N-terminal glycosylated segment, which lies outside the erythrocyte membrane, has MN blood group receptors and also binds influenza virus	Glycophorin A, GPA, MNS	Glycophorin A
CD235b	Erythroid cells	GYPD is smaller than GYPC (24 kDa vs 32 kDa)	This protein is a minor sialoglycoprotein in human erythrocyte membranes. Along with GYPA, GYPB is responsible for the MNS blood group system. The Ss blood group antigens are located on glycophorin B	Glycophorin B, MNS, GPB	Glycophorin A
CD236	Erythroid cells	24	Glycophorin C (GPC) and glycophorin D (GPD) are closely related sialoglyco-proteins in the human red blood cell (RBC) membrane. GPD is a ubiquitous shortened isoform of GPC, produced by alternative splicing of the same gene. The Webb and Duch antigens, also known as glycophorin D, result from single point mutations of the glycophorin C gene	Glycophorin D, GPD, GYPD	Type III membrane proteins
CD236R	Erythroid cells	32	Glycophorin C (GPC) is associated with the Gerbich (Ge) blood group deficiency. It is a minor red cell-membrane component, representing about 4% of the membrane sialoglycoproteins, but shows very little homology with the major red cell-membrane glycophorins A and B. It plays an important role in regulating the mechanical stability of red cells and is a putative receptor for the merozoites of *Plasmodium falciparum*	Glycophorin C, GYPC, GPC	Type III membrane proteins
CD238	Erythroid cells	93	KELL blood group antigen; homology to a family of zinc metalloglycoproteins with neutral endopeptidase activity, type II transmembrane glycoprotein	KELL	Peptidase m13 (zinc metallo-proteinase); also known as the neprilysin subfamily
CD239	Erythroid cells	78	A type I membrane protein. The human F8/G253 antigen, B-CAM, is a cell-surface glycoprotein that is expressed with restricted distribution pattern in normal fetal and adult tissues, and is upregulated following malignant transformation in some cell types. Its overall structure is similar to that of the human tumor marker MUC 18 and the chicken neural adhesion molecule SC1	B-CAM (B-cell adhesion molecule), LU, Lutheran blood group	Immunoglobulin

CD antigen	Cellular expression	Molecular weight (kDa)	Functions	Other names	Family relationships
CD240CE	Erythroid cells	45.5	Rhesus blood group, CcEe antigens. May be part of an oligomeric complex which is likely to have a transport or channel function in the erythrocyte membrane. It is highly hydrophobic and deeply buried within the phospholipid bilayer	RHCE, RH30A, RHPI, Rh4	Rh
CD240D	Erythroid cells	45.5 (product—30)	Rhesus blood group, D antigen. May be part of an oligomeric complex which is likely to have a transport or channel function in the erythrocyte membrane. Absent in the Caucasian RHD-negative phenotype	RhD, Rh4, RhPI, RhII, Rh30D	Rh
CD241	Erythroid cells	50	Rhesus blood group-associated glycoprotein RH50, component of the RH antigen multisubunit complex; required for transport and assembly of the Rh membrane complex to the red blood cell surface. Highly homologous to RH, 30 kDa components. Defects in RhAg are a cause of a form of chronic hemolytic anemia associated with stomatocytosis, and spherocytosis, reduced osmotic fragility, and increased cation permeability	RhAg, RH50A	Rh
CD242	Erythroid cells	42	Intercellular adhesion molecule 4, Landsteiner-Wiener blood group. LW molecules may contribute to the vaso-occlusive events associated with episodes of acute pain in sickle cell disease	ICAM-4, LW	Immunoglobulin, intercellular adhesion molecules (ICAMs)
CD243	Stem/progenitor cells	170	Multidrug resistance protein 1 (P-glycoprotein). P-gp has been shown to utilize ATP to pump hydrophobic drugs out of cells, thus increasing their intracellular concentration and hence their toxicity. The MDR 1 gene is amplified in multidrug-resistant cell lines	MDR-1, p-170	ABC superfamily of ATP-binding transport proteins
CD244	NK cells	66	2B4 is a cell-surface glycoprotein related to CD2 and implicated in the regulation of natural killer and T-lymphocyte function. It appears that the primary function of 2B4 is to modulate other receptor–ligand interactions to enhance leukocyte activation	2B4, NK cell activation inducing ligand (NAIL)	Immunoglobulin, SLAM
CD245	T cells	220–240	Cyclin E/Cdk2 interacting protein p220. NPAT is involved in a key S phase event and links cyclical cyclin E/Cdk2 kinase activity to replication-dependent histone gene transcription. NPAT gene may be essential for cell maintenance and may be a member of the housekeeping genes	NPAT	
CD246	Expressed in the small intestine, testis, and brain but not in normal lymphoid cells	177 kDa; after glycosylation, produces a 200 kDa mature glycoprotein	Anaplastic (CD30$^+$ large cell) lymphoma kinase; plays an important role in brain development, involved in anaplastic nodal non-Hodgkin lymphoma or Hodgkin's disease with translocation t(2;5)(p23;q35) or inv2(23;q35). Oncogenesis via the kinase function is activated by oligomerization of NPM1-ALK mediated by the NPM1 part	ALK	Insulin receptor family of tyrosine-protein kinases
CD247	T cells, NK cells	16	T-cell receptor ζ; has a probable role in assembly and expression of the TCR complex as well as signal transduction upon antigen triggering. TCRζ together with TCRα:β and γ: δ heterodimers and CD3-γ, -δ, and -ε, forms the TCR-CD3 complex. The ζ chain plays an important role in coupling antigen recognition to several intracellular signal-transduction pathways. Low expression of the antigen results in impaired immune response	ζ chain, CD3Z	Immunoglobulin

CD antigen	Cellular expression	Molecular weight (kDa)	Functions	Other names	Family relationships
CD248	Adipocytes, smooth muscle	80	Cell adhesion	CD164L1, endosialin	C-type lectin, EGF
CD249	Pericytes and podocytes in the kidney	109	Aminopeptidase	ENPEP, APA, gp160, EAP	Peptidase M1
CD252	Activated B cells, dendritic cells	21	T-cell activation	TNFSF4, GP34, OX40L, TXGP1, CD134L, OX-40L, OX40L	TNF
CD253	B cells, dendritic cells, NK cells, monocytes, macrophages	33	Induction of apoptosis	TNFSF10, TL2, APO2L, TRAIL, Apo-2L	TNF
CD254	Osteoblasts, T cells	35	Osteoclast and dendritic-cell development and function	TNFSF11, RANKL, ODF, OPGL, sOdf, CD254, OPTB2, TRANCE, hRANKL2	TNF
CD256	Dendritic cells, monocytes, CD33+ myeloid cells	27	B-cell activation	TNFSF13, APRIL, TALL2, TRDL-1, UNQ383/PRO715	TNF
CD257	DCs, monocytes, CD33+ myeloid cells	31	B-cell activation	TNFSF13B, BAFF, BLYS, TALL-1, TALL1, THANK, TNFSF20, ZTNF4, ΔBAFF	TNF
CD258	B cells, NK cells	26	Apoptosis, lymphocyte adhesion	TNFSF14, LTg, TR2, HVEML, LIGHT, LTBR	TNF
CD261	B cells, CD8+ T cells	50	TRAIL receptor, induces apoptosis	TNFRSF10A, APO2, DR4, MGC9365, TRAILR-1, TRAILR1	TNF receptor
CD262	B cells, CD33+ myeloid cells	48	TRAIL receptor, induces apoptosis	TNFRSF10B, DR5, KILLER, KILLER/ DR5, TRAIL-R2, TRAILR2, TRICK2, TRICK2A, TRICK2B, TRICKB, ZTNFR9	TNF receptor
CD263	Variety of cell types	27	Inhibits TRAIL-induced apoptosis	TNFRSF10C, DCR1, LIT, TRAILR3, TRID	TNF receptor
CD264	Variety of cell types	42	Inhibits TRAIL-induced apoptosis	TNFRSF10D, DCR2, TRAILR4, TRUNDD	TNF receptor
CD265	Osteoclasts, dendritic cells	66	Receptor for RANKL	TNFRSF11A, EOF, FEO, ODFR, OFE, PDB2, RANK, TRANCER	TNF receptor
CD266	NK cells, CD33+ myeloid cells, monocytes	14	Receptor for TWEAK	TNFRSF12A, FN14, TWEAKR, TWEAK	TNF receptor
CD267	B cells	32	APRIL and BAFF signal through it, B-cell activation	TNFRSF13B, CVID, TACI, CD267, FLJ39942, MGC39952, MGC133214, TNFRSF14B	TNF receptor
CD268	B cells	19	BAFF receptor	TNFRSF13C, BAFFR, CD268, BAFF-R, MGC138235	TNF receptor
CD269	B cells, dendritic cells	20	APRIL and BAFF signal through it, B-cell activation	TNFRSF17, BCM, BCMA	TNF receptor
CD270	B cells, dendritic cells, T cells, NK cells, CD33+ myeloid cells, monocytes	30	Receptor for LIGHT	TNFRSF14, TR2, ATAR, HVEA, HVEM, LIGHTR	TNF receptor
CD271	Mesenchymal stem cells and some cancers	45	Receptor for various neurotrophins	NGFR, TNFRSF16, p75(NTR)	TNF receptor
CD272	B cells, T cells (T_H1, γ:δ T cells)	33	Blunts B and T cell activation	BTLA1, FLJ16065	Immunoglobulin

CD antigen	Cellular expression	Molecular weight (kDa)	Functions	Other names	Family relationships
CD273	Dendritic cells	31	Ligand for PD-1	PDCD1LG2, B7DC, Btdc, PDL2, PD-L2, PDCD1L2, bA574F11.2	Immunoglobulin
CD274	Antigen-presenting cells	33	Binds PD-1	PDL1, B7-H, B7H1, PD-L1, PDCD1L1	Immunoglobulin
CD275	Antigen-presenting cells	33	Binds ICOS, multiple functions in immune system	ICOS-L, B7-H2, B7H2, B7RP-1, B7RP1, GL50, ICOSLG, KIAA0653, LICOS	Immunoglobulin
CD276	Antigen-presenting cells	57	Blunts T-cell activity	B7H3	Immunoglobulin
CD277	T cells, NK cells	58	Blunts T-cell activity	BTN3A1, BTF5, BT3.1	Immunoglobulin
CD278	T cells, B cells, ILC2s, some ILC3s	23	Receptor for ICOSL, multiple functions in immune system	ICOS, AILIM, MGC39850	
CD279	T cells, B cells	32	Inhibitory molecule on multiple immune cells	PD1, PDCD1, SLEB2, hPD-l	Immunoglobulin
CD280	Variety of cell types	166	Mannose receptor, binds extracellular matrix	MRC2, UPARAP, ENDO180, KIAA0709	C-type lectin, fibronectin type II
CD281	Many different immune cells	90	Binds bacterial lipoproteins, dimerizes with TLR2	TLR1, TIL, rsc786, KIAA0012, DKFZp547I0610, DKFZp564I0682	Toll-like receptor
CD282	Dendritic cells, monocytes, CD33+ myeloid cells, B cells	89	Binds numerous microbial molecules	TLR2, TIL4	Toll-like receptor
CD283	Dendritic cells, NK cells, T cells, B cells	104	Binds dsRNA and polyI:C	TLR3	Toll-like receptor
CD284	Macrophages, monocytes, dendritic cells, epithelial cells	96	Binds LPS	TLR4, TOLL, hToll	Toll-like receptor
CD286	B cells, monocytes, NK cells	92	Binds bacterial lipoproteins, dimerizes with TLR2	TLR6	Toll-like receptor
CD288	Monocytes, NK cells, T cells, macrophages	120	Binds ssRNA	TLR8	Toll-like receptor
CD289	Dendritic cells, B cells, macrophages, neutrophils, NK cells, microglia	116	Binds CpG DNA	TLR9	Toll-like receptor
CD290	B cells, dendritic cells	95	Ligand unknown	TLR10	Toll-like receptor
CD292	Variety of cell types, skeletal muscle	60	Receptor for BMPs	BMPR1A, ALK3, ACVRLK3	Type I trans-membrane
CDw293				BMPR1B	
CD294	NK cells	43	Activated by prostaglandin D2	GPR44, CRTH2	GPCR class A receptor
CD295	Mesenchymal stem cells	132	Receptor for leptin	LEPR, OBR	Immunoglobulin, fibronectin type III, IL-6R
CD296	Cardiomyocytes	36	ADP ribosyltransferase activity	ART1, ART2, RT6	
CD297	Erythroid cells	36	ADP ribosyltransferase activity	DO, DOK1, CD297, ART4	
CD298	Variety of cell types	32	Subunit of Na+-K+ ATPase	ATP1B3, ATPB-3, FLJ29027	P-type ATPase
CD299	Endothelium of lymph nodes and liver	45	Receptor for DC-SIGN, DC/T-cell interaction	CLEC4M, DC-SIGN2, DC-SIGNR, DCSIGNR, HP10347, LSIGN, MGC47866	C-type lectin

CD antigen	Cellular expression	Molecular weight (kDa)	Functions	Other names	Family relationships
CD300A	B cells, T cells, NK cells, monocytes, CD33+ myeloid cells	33	Inhibitory receptor on T, B, and NK cells	CMRF-35-H9, CMRF35H, CMRF35H9, IRC1, IRC2, IRp60	Immunoglobulin
CD300C	CD33+ myeloid cells, monocytes	24	Activating receptor on multiple cell types	CMRF-35A, CMRF35A, CMRF35A1, LIR	Immunoglobulin
CD301	Dendritic cells, monocytes, CD33+ myeloid cells	35	Macrophage adhesion and migration	CLEC10A, HML, HML2, CLECSF13, CLECSF14	C-type lectin
CD302	Dendritic cells, monocytes, CD33+ myeloid cells	26	Macrophage adhesion and migration	DCL-1, BIMLEC, KIAA0022	C-type lectin
CD303	Plasmacytoid DC	25	Involved in plasmacytoid dendritic cell function	CLEC4C, BDCA2, CLECSF11, DLEC, HECL, PRO34150, CLECSF7	C-type lectin
CD304	T_{reg} cells, plasmacytoid DCs	103	Cell migration and survival, preferentially expressed on thymic compared with induced T_{reg} cells	Neuropilin-1, NRP1, NRP, VEGF165R	
CD305	Variety of hematopoietic cells	31	Inhibitory receptor on multiple immune cells	LAIR-1	Immunoglobulin
CD306	NK cells	16	Unknown	LAIR2	Immunoglobulin
CD307a	B cells	47	B-cell signaling and function	FCRH1, IFGP1, IRTA5, FCRL1	Immunoglobulin
CD307b	B cells	56	B-cell signaling and function	FCRH2, IFGP4, IRTA4, SPAP1, SPAP1A, SPAP1B, SPAP1C, FCRL2	Immunoglobulin
CD307c	B cells, NK cells	81	B-cell signaling and function	FCRH3, IFGP3, IRTA3, SPAP2, FCRL3	Immunoglobulin
CD307d	Memory B cells	57	B-cell signaling and function	FCRH4, IGFP2, IRTA1, FCRL4	Immunoglobulin
CD307e	B cells, dendritic cells	106	B-cell signaling and function	CD307, FCRH5, IRTA2, BXMAS1, PRO820	Immunoglobulin
CD309	Endothelial cells	151	VEGF signaling, hematopoiesis	KDR, FLK1, VEGFR, VEGFR2	Immunoglobulin, type III tyrosine kinase
CD312	Dendritic cells, NK cells, monocytes, CD33+ myeloid cells	90	GPCR involved in neutrophil activation	EMR2	EGF, GPCR class B
CD314	T cells, NK cells	25	NK- and T-cell activation	KLRK1, KLR, NKG2D, NKG2-D, D12S2489E	C-type lectin
CD315	Smooth muscle	99	Interacts with CD316	PTGFRN, FPRP, EWI-F, CD9P-1, SMAP-6, FLJ11001, KIAA1436	Immunoglobulin
CD316	Keratinocytes	65	Modulates integrin function	IGSF8, EWI2, PGRL, CD81P3	Immunoglobulin
CD317	Variety of hematopoietic cells	20	IFN-induced antiviral protein	BST2	
CD318	Epithelial cells	93	Cell migration and tumor development	CDCP1, FLJ22969, MGC31813	
CD319	B-cells, NK cells, dendritic cells	37	B-cell and NK-cell function and proliferation	SLAMF7, 19A, CRACC, CS1	Immunoglobulin
CD320	B cells	29	Receptor for transcobalamin	8D6A, 8D6	LDL receptor
CD321	Dendritic cells, T cells, NK cells, CD33+ myeloid cells	33	Immune-cell interaction with endothelium, may act as receptor for reovirus	F11R, JAM, KAT, JAM1, JCAM, JAM-1, PAM-1	Immunoglobulin

CD antigen	Cellular expression	Molecular weight (kDa)	Functions	Other names	Family relationships
CD322	Endothelial cells	33	Immune-cell migration across endothelium	JAM2, C21orf43, VE-JAM, VEJAM	Immunoglobulin
CD324	Endothelial cells	97	Cell adhesion, epithelial development	E-Cadherin, CDH1, Arc-1, CDHE, ECAD, LCAM, UVO	Cadherin
CD325	Neurons, smooth muscles, cardiomyocytes	100	Cell adhesion, neural development	N-Cadherin, CDH2, CDHN, NCAD	Cadherin
CD326	Epithelial cells	35	Cell signaling and migration, promotes proliferation	Ep-CAM, TACSTD1, CO17-1A, EGP, EGP40, GA733-2, KSA, M4S1, MIC18, MK-1, TROP1, hEGP-2	
CD327	Neurons	50	Sialic acid binding on multiple immune cells	CD33L, CD33L1, OBBP1, SIGLEC-6	Immunoglobulin, sialic acid binding-type lectin
CD328	NK cells, CD33+ myeloid cells, monocytes	51	Sialic acid binding on multiple immune cells	p75, QA79, AIRM1, CDw328, SIGLEC-7, p75/AIRM1	Immunoglobulin, sialic acid binding-type lectin
CD329	CD33+ myeloid cells, monocytes	50	Sialic acid binding on multiple immune cells	CDw329, OBBP-LIKE, SIGLEC9	Immunoglobulin, sialic acid binding-type lectin
CD331	Variety of cell types	92	Cell proliferation and survival, skeletal development	FGFR1, H2, H3, H4, H5, CEK, FLG, FLT2, KAL2, BFGFR, C-FGR, N-SAM	Immunoglobulin, FGFR, tyrosine kinase
CD332	Variety of cell types	92	Cell proliferation and survival, craniofacial development	FGFR2, BEK, JWS, CEK3, CFD1, ECT1, KGFR, TK14, TK25, BFR-1, K-SAM	Immunoglobulin, FGFR, tyrosine kinase
CD333	Variety of cell types	87	Cell proliferation and survival, skeletal development	FGFR3, ACH, CEK2, JTK4, HSFGFR3EX	Immunoglobulin, FGFR, tyrosine kinase
CD334	Variety of cell types	88	Cell proliferation and survival, bile acid synthesis	FGFR4, TKF, JTK2, MGC20292	Immunoglobulin, FGFR, tyrosine kinase
CD335	NK cells, some ILCs	34	NK-cell function	NKp46, LY94, NKP46, NCR1	Immunoglobulin
CD336	NK cells	30	NK-cell function	NKp44, LY95, NKP44, NCR2	Immunoglobulin
CD337	NK cells	22	NK-cell function	NKp30, 1C7, LY117, NCR3	Immunoglobulin
CD338	Erythroid cells	72	ABC transporter, role in stem cells	ABCG2, MRX, MXR, ABCP, BCRP, BMDP, MXR1, ABC15, BCRP1, CDw338, EST157481, MGC102821	ATP binding cassette transporters
CD339	Variety of cell types	134	Notch receptor ligand	JAG1, AGS, AHD, AWS, HJ1, JAGL1	EGF
CD340	Variety of cell types, certain aggressive breast cancers	134	EGF receptor, promotes proliferation	HER2, ERBB2, NEU, NGL, TKR1, HER-2, c-erb B2, HER-2/neu	ERBB, tyrosine kinase
CD344	Adipocytes	60	Wnt and Norrin signaling	EVR1, FEVR, Fz-4, FzE4, GPCR, FZD4S, MGC34390	GPCR class F
CD349	Variety of cell types	65	Wnt signaling	FZD9, FZD3	GPCR class F
CD350	Variety of cell types	65	Wnt signaling	FZD10, FzE7, FZ-10, hFz10	GPCR class F
CD351	Variety of cell types	57	Fc receptor for IgA and IgM	FCA/MR, FKSG87, FCAMR	Immunoglobulin

CD antigen	Cellular expression	Molecular weight (kDa)	Functions	Other names	Family relationships
CD352	B cells, T cells, NKT cells, NK cells	37	T-, B-, and NKT-cell development and function	SLAMF6, KALI, NTBA, KALIb, Ly108, NTB-A, SF2000	Immunoglobulin
CD353	Variety of cell types	32	B-cell development	SLAMF8, BLAME, SBBI42	Immunoglobulin
CD354	CD33+ myeloid cells, monocytes	26	Amplifies inflammation in myeloid cells	TREM-1	Immunoglobulin
CD355	T cells, NK cells	45	TCR signaling, cytokine production	CRTAM	Immunoglobulin
CD357	Activated T cells	26	Modulates T_{reg} suppressive function	TNFRSF18, AITR, GITR, GITR-D, TNFRSF18	TNF receptor
CD358	Dendritic cells	72	Induces apoptosis	TNFRSF21, DR6, BM-018, TNFRSF21	TNF receptor
CD360	B cells	59	Receptor for IL-21, numerous immune functions	IL21R, NILR	Type I cytokine receptor, fibronectin type III
CD361	Variety of hematopoietic cells	49	Unknown	EVDB, D17S376, EVI2B	
CD362	Endothelial cells, fibroblasts, neurons, and B cells	22	Cell organization, interaction with extracellular matrix	HSPG, HSPG1, SYND2, SDC2	Syndecan proteoglycan
CD363	Variety of cell types, including effector lymphocytes	43	Sphingosine-1-phosphate receptor 1, immune-cell survival, motility, and egression from lymph nodes	EDG1, S1P1, ECGF1, EDG-1, CHEDG1	GPCR class A receptor
CD364	T_{reg} cells		Unknown	MSMBBP, PI16	
CD365	T cells		T-cell activation	HAVCR, TIM-1	Immunoglobulin
CD366	T cells		Induces apoptosis	HAVCR2, TIM-3	Immunoglobulin
CD367	Dendritic cells		HIV receptor, important in cross-priming CD8 T-cell and DC interactions	DCIR, CLEC4A	C-type lectin
CD368	Monocytes, macrophages		Receptor for endocytosis	MCL, CLEC-6, CLEC4D, CLECSF8	C-type lectin
CD369	Neutrophils, dendritic cells, monocytes, macrophages, B cells		Pattern recognition receptor important for antifungal immunity, recognizes glucans and carbohydrates in fungal walls	DECTIN-1, CLECSF12, CLEC7A	C-type lectin
CD370	Dendritic cells, NK cells		Important for cross-priming of CD8 T cells for antiviral immunity	DNGR1, CLEC9A	C-type lectin
CD371	Dendritic cells		Unknown	MICL, CLL-1, CLEC12A	C-type lectin

Compiled by Daniel DiToro, Carson Moseley, and Jeff Singer, University of Alabama at Birmingham. Data based on CD designations made at the 9th Workshop on Human Leukocyte Differentiation Antigens.

Appendix III. Cytokines and their receptors.						
Family	**Cytokine (alternative names)**	**Size (no. of amino acids and form)**	**Receptors (c denotes common subunit)**	**Producer cells**	**Actions**	**Effect of cytokine or receptor knock-out (where known)**
Colony-stimulating factors	G-CSF (CSF-3)	174, monomer*	G-CSFR	Fibroblasts and monocytes	Stimulates neutrophil development and differentiation	G-CSF, G-CSFR: defective neutrophil production and mobilization
	GM-CSF (granulocyte-macrophage colony-stimulating factor) (CSF-2)	127, monomer*	CD116, βc	Macrophages, T cells	Stimulates growth and differentiation of myelomonocytic lineage cells, particularly dendritic cells	GM-CSF, GM-CSFR: pulmonary alveolar proteinosis
	M-CSF (CSF-1)	α: 224 β: 492 γ: 406 active forms are homo- or heterodimeric	CSF-1R (c-fms)	T cells, bone marrow stromal cells, osteoblasts	Stimulates growth of cells of monocytic lineage	Osteopetrosis
Interferons	IFN-α (at least 12 distinct proteins)	166, monomer	CD118, IFNAR2	Leukocytes, dendritic cells, plasmacytoid dendritic cells, conventional dendritic cells	Antiviral, increased MHC class I expression	CD118: impaired antiviral activity
	IFN-β	166, monomer	CD118, IFNAR2	Fibroblasts	Antiviral, increased MHC class I expression	IFN-β: increased susceptibility to certain viruses
	IFN-γ	143, homodimer	CD119, IFNGR2	T cells, natural killer cells, neutrophils, ILC1s, intraepithelial lymphocytes	Macrophage activation, increased expression of MHC molecules and antigen processing components, Ig class switching, supresses T_H17 and T_H2	IFN-γ, CD119: decreased resistance to bacterial infection and tumors
Interleukins	IL-1α	159, monomer	CD121a (IL-1RI) and CD121b (IL-1RII)	Macrophages, epithelial cells	Fever, T-cell activation, macrophage activation	IL-1RI: decreased IL-6 production
	IL-1β	153, monomer	CD121a (IL-1RI) and CD121b (IL-1RII)	Macrophages, epithelial cells	Fever, T-cell activation, macrophage activation	IL-1β: impaired acute-G21 phase response
	IL-1 RA	152, monomer	CD121a	Monocytes, macrophages, neutrophils, hepatocytes	Binds to but doesn't trigger IL-1 receptor, acts as a natural antagonist of IL-1 function	IL-1RA: reduced body mass, increased sensitivity to endotoxins (septic shock)
	IL-2 (T-cell growth factor)	133, monomer	CD25α, CD122β, CD132 (γc)	T cells	T_{reg} maintenance and function, T-cell proliferation and differentiation	IL-2: deregulated T-cell proliferation, colitis IL-2Rα: incomplete T-cell development autoimmunity IL-2Rβ: increased T-cell autoimmunity IL-2Rγc: severe combined immunodeficiency
	IL-3 (multicolony CSF)	133, monomer	CD123, βc	T cells, thymic epithelial cells, and stromal cells	Synergistic action in early hematopoiesis	IL-3: impaired eosinophil development. Bone marrow unresponsive to IL-5, GM-CSF
	IL-4 (BCGF-1, BSF-1)	129, monomer	CD124, CD132 (γc)	T cells, mast cells, ILC2s	B-cell activation, IgE switch, induces differentiation into T_H2 cells	IL-4: decreased IgE synthesis
	IL-5 (BCGF-2)	115, homodimer	CD125, βc	T cells, mast cells, ILC2s	Eosinophil growth, differentiation	IL-5: decreased IgE, IgG1 synthesis (in mice); decreased levels of IL-9, IL-10, and eosinophils
	IL-6 (IFN-B502, BSF-2, BCDF)	184, monomer	CD126, CD130	T cells, B cells, macrophages, endothelial cells	T- and B-cell growth and differentiation, acute phase protein production, fever	IL-6: decreased acute phase reaction, reduced IgA production
	IL-7	152, monomer*	CD127, CD132 (γc)	Non-T cells, stromal cells	Growth of pre-B cells and pre-T cells, and ILCs	IL-7: early thymic and lymphocyte expansion severely impaired
	IL-9	125, monomer*	IL-9R, CD132 (γc)	T cells	Mast-cell enhancing activity, stimulates T_H2 and ILC2 cells	Defects in mast-cell expansion
	IL-10 (cytokine synthesis inhibitory factor)	160, homodimer	IL-10Rα, IL–10Rβc (CRF2-4, IL-10R2)	Macrophages, dendritic cells, T cells, and B cells	Potent suppressant of macrophage functions	IL-10 and IL20Rβc-: reduced growth, anemia, chronic enterocolitis
	IL-11	178, monomer	IL-11R, CD130	Stromal fibroblasts	Synergistic action with IL-3 and IL-4 in hematopoiesis	IL-11R: defective decidualization
	IL-12 (NK-cell stimulatory factor)	197 (p35) and 306 (p40c), heterodimer	IL-12Rβ1c + IL-12Rβ2	Macrophages, dendritic cells	Activates NK cells, induces CD4 T-cell differentiation into T_H1-like cells	IL-12: impaired IFN-γ production and T_H1 responses

Family	Cytokine (alternative names)	Size (no. of amino acids and form)	Receptors (c denotes common subunit)	Producer cells	Actions	Effect of cytokine or receptor knock-out (where known)
Interleukins	IL-13 (p600)	132, monomer	IL-13R, CD132 (γc) (may also include CD24)	T cells, ILC2s	B-cell growth and differentiation, inhibits macrophage inflammatory cytokine production and T_H1 cells, induces allergy/asthma	IL-13: defective regulation of isotype specific responses
	L-15 (T-cell growth factor)	114, monomer	IL-15Rα, CD122 (IL-2Rβ) CD132 (γc)	Many non-T cells	IL-2-like, stimulates growth of intestinal epithelium, T cells, and NK cells, enhances CD8 memory T-cell survival	IL-15: reduced numbers of NK cells and memory phenotype CD8+ T cells IL-15Rα: lymphopenia
	IL-16	130, homotetramer	CD4	T cells, mast cells, eosinophils	Chemoattractant for CD4 T cells, monocytes, and eosinophils, anti-apoptotic for IL-2-stimulated T cells	
	IL-17A (mCTLA-8)	150, homodimer	IL-17AR (CD217)	T_H17, CD8 T cells, NK cells γ:δ T cells, neutrophils, ILC3s	Induces cytokine and antimicrobial peptide production by epithelia, endothelia, and fibroblasts, proinflammatory	IL-17R: reduced neutrophil migration into infected sites
	IL-17F (ML-1)	134, homodimer	IL-17AR (CD217)	T_H17, CD8 T cells, NK cells γ:δ T cells, neutrophils, ILC3s	Induces cytokine production by epithelia, endothelia, and fibroblasts, proinflamatory	
	IL-18 (IGIF, interferon-α inducing factor)	157, monomer	IL-1Rrp (IL-1R related protein)	Activated macrophages and Kupffer cells	Induces IFN-γ production by T cells and NK cells, promotes T_H1 induction	Defective NK activity and T_H1 responses
	IL-19	153, monomer	IL-20Rα + IL-10Rβc	Monocytes	Induces IL-6 and TNF-α expression by monocytes	
	IL-20	152	IL-20Rα + IL–10Rβc; IL-22Rαc + IL-10Rβc	T_H1 cells, monocytes, epithelial cells	Promotes T_H2 cells, stimulates keratinocyte proliferation and TNF-α production	
	IL-21	133	IL-21R, + CD132(γc)	T_H2 cells, T cells, primarily T_{FH} cells	Germinal center maintenance induces proliferation of B, T, and NK cells	Increased IgE production
	IL-22 (IL-TIF)	146	IL-22Rαc + IL-10Rβc	NK cells, T_H17 cells, T_H22 cells, ILC3s, neutrophils, γ:δ T cells	Induces production of antimicrobial peptides; induces liver acute-phase proteins, pro-inflammatory agents; epithelial barrier	Increased susceptibility to mucosal infections
	IL-23	170 (p19) and 306 (p40c), heterodimer	IL-12Rβ1 + IL-23R	Dendritic cells, macrophages	Induces proliferation of T_H17 memory T cells, increased IFN-γ production	Defective inflammation
	IL-24 (MDA-7)	157	IL-22Rαc + IL-10Rβc; IL-20Rα + IL-10Rβc	Monocytes, T cells	Inhibits tumor growth, wound healing	
	IL-25 (IL-17E)	145	IL-17BR (IL-17Rh1)	T_H2 cells, mast cells, epithelial cells	Promotes T_H2 cytokine production	Defective T_H2 response
	IL-26 (AK155)	150	IL-20Rα +IL-10Rβc	T cells (T_H17), NK cells	Pro-inflammatory, stimulates epithelium	
	IL-27	142 (p28) and 229 (EBI3), heterodimer	WSX-1 + CD130c	Monocytes, macrophages, dendritic cells	Induces IL-12R on T cells via T-bet induction, induces IL-10	EBI3: reduced NKT cells. WSX-1: overreaction to *Toxoplasma gondii* infection and death from inflammation
	IL-28A,B (IFN-B502,3)	175	IL-28Rαc + IL-10Rβc	Dendritic cells	Antiviral	
	IL-29 (IFN-λ1)	181	IL-28Rαc + IL-10Rβc	Dendritic cells	Antiviral	
	IL-30 (p28, IL27A, IL-27p28)	243	see IL-27			
	IL-31	164	IL31A + OSMR	T_H2	Pro-inflammatory, skin lesions	IL-31A: elevated OSM responsiveness
	IL-32 (NK4, TAIF)	188	Unknown	Natural killer cells, T cells, epithelial cells, monocytes	Induces TNF-α	
	IL-33 (NF-HEV)	270 heterodimer	ST2 (IL1RL1) + IL1RAP	High endothelial venules, smooth muscle, and epithelial cells	Induces T_H2 cytokines (IL-4, IL-5, IL-13)	IL-33: reduced dextran-induced colitis; reduced LPS-induced systemic inflammatory response
	IL-34 (C16orf77)	242 homodimer	CSF-1R	Many cell types	Promotes growth and development of myeloid cells/osteoclasts	
	IL-35	197 (IL-12α (p35)) + 229 (EB13) heterodimer	IL-12RB2 and gp130 heterodimer	T_{reg} cells, B cells	Immunosuppressive	

Family	Cytokine (alternative names)	Size (no. of amino acids and form)	Receptors (c denotes common subunit)	Producer cells	Actions	Effect of cytokine or receptor knock-out (where known)
Interleukins	IL-36α, β, λ	(20 kDa) 155–169	IL-1Rrp2, Acp	Keratinocytes, monocytes	Pro-inflammatory stimulant of macrophages and dendritic cells	
	IL-36 Ra		IL-1Rp2, Acp		Antagonist of IL-36	
	IL-37	(17–24 kDa) homodimer	IL-18Rα?	Monocytes, dendritic cells, epithelial cells, breast tumor cells	Suppresses dendritic cell/monocyte production of IL-1, -6, -12 etc. cytokines, synergizes with TGFs	siRNA knockdown: increases pro-inflammatory cytokines
	TSLP	140 monomer	IL-7Rα, TSLPR	Epithelial cells, especially lung and skin	Stimulates hematopoietic cells and dendritic cells to induce T_H2 responses	TSLP: resistance to induction of allergies and asthmatic reactions
	LIF (leukemia inhibitory factor)	179, monomer	LIFR, CD130	Bone marrow stroma, fibroblasts	Maintains embryonic stem cells, like IL-6, IL-11, OSM	LIFR: die at or soon after birth; decreased hematopoietic stem cells
	OSM (OM, oncostatin M)	196, monomer	OSMR or LIFR, CD130	T cells, macrophages	Stimulates Kaposi's sarcoma cells, inhibits melanoma growth	OSMR: defective liver regeneration
TNF	TNF-α (cachectin)	157, trimers	p55 (CD120a), p75 (CD120b)	Macrophages, NK cells, T cells	Promotes inflammation, endothelial activation	p55: resistance to septic shock, susceptibility to *Listeria*, STNFαR: periodic febrile attacks
	LT-α (lymphotoxin-α)	171, trimers	p55 (CD120a), p75 (CD120b)	T cells, B cells	Killing, endothelial activation, and lymph node development	LT-α: absent lymph nodes, decreased antibody, increased IgM
	LT-β	Transmembrane, trimerizes with LT-α	LTβR or HVEM	T cells, B cells, ILC3s	Lymph node development	Defective development of peripheral lymph nodes, Peyer's patches, and spleen
	CD40 ligand (CD40L)	Trimers	CD40	T cells, mast cells	B-cell activation, class switching	CD40L: poor antibody response, no class switching, diminished T-cell priming (hyper-IgM syndrome)
	Fas ligand (FasL)	Trimers	CD95 (Fas)	T cells, stroma (?)	Apoptosis, Ca^{2+}-independent cytotoxicity	Fas, FasL: mutant forms lead to lymphoproliferation, and autoimmunity
	CD27 ligand (CD27L)	Trimers (?)	CD27	T cells	Stimulates T-cell proliferation	
	CD30 ligand (CD30L)	Trimers (?)	CD30	T cells	Stimulates T- and B-cell proliferation	CD30: increased thymic size, alloreactivity
	4-1BBL	Trimers (?)	4-1BB	T cells	Co-stimulates T and B cells	
	Trail (APO-2L)	281, trimers	DR4, DR5 DCR1, DCR2 and OPG	T cells, monocytes	Apoptosis of activated T cells and tumor cells, and virally infected cells	Tumor-prone phenotype
	OPG-L (RANK-L)	316, trimers	RANK/OPG	Osteoblasts, T cells	Stimulates osteoclasts and bone resorption	OPG-L: osteopetrotic, runted, toothless OPG: osteoporosis
	APRIL	86	TAC1 or BCMA	Activated T cells	B-cell proliferation	Impaired IgA-class switching
	LIGHT	240	HVEM, LTβR	T cells	Dendritic cell activation	Defective CD8+ T-cell expansion
	TWEAK	102	TWEAKR (Fn14)	Macrophages, EBV transformed cells	Angiogenesis	
	BAFF (CD257, BlyS)	153	TAC1 or BCMA or BR3	B cells	B-cell proliferation	BAFF: B-cell dysfunction
Unassigned	TGF-β1	112, homo- and heterotrimers	TGF-βR	Chondrocytes, monocytes, T cells	Generation of iT_{reg} cells and T_H17 cells, induces switch to IgA production	TGF-β: lethal inflammation
	MIF	115, monomer	MIF-R	T cells, pituitary cells	Inhibits macrophage migration, stimulates macrophage activation, induces steroid resistance	MIF: resistance to septic shock, hyporesponsive to Gram-negative bacteria

* May function as dimers

Compiled by Robert Schreiber, Washington University School of Medicine, St Louis, and Daniel DiToro, Carson Moseley, and Jeff Singer, University of Alabama at Birmingham.

Appendix IV. Chemokines and their receptors.				
Chemokine systematic name	Common names	Chromosome	Target cell	Specific receptor
CXCL (†ELR+)				
1	GROα	4	Neutrophil, fibroblast	CXCR2
2	GROβ	4	Neutrophil, fibroblast	CXCR2
3	GROγ	4	Neutrophil, fibroblast	CXCR2
5	ENA-78	4	Neutrophil, endothelial cell	CXCR2>>1
6	GCP-2	4	Neutrophil, endothelial cell	CXCR2>1
7	NAP-2 (PBP/CTAP-III/ β-B44TG)	4	Fibroblast, neutrophil, endothelial cell	CXCR1, CXCR2
8	IL-8	4	Neutrophil, basophil, CD8 cell subset, endothelial cell	CXCR1, CXCR2
14	BRAK/bolekine	5	T cell, monocyte, B cell	Unknown
15	Lungkine/WECHE	5	Neutrophil, epithelial cell, endothelial cell	Unknown
(†ELR−)				
4	PF4	4	Fibroblast, endothelial cell	CXCR3B (alternative splice)
9	Mig	4	Activated T cell ($T_H1 > T_H2$), natural killer (NK) cell, B cell, endothelial cell, plasmacytoid dendritic cell	CXCR3A and B
10	IP-10	4	Activated T cell ($T_H1 > T_H2$), NK cell, B cell, endothelial cell	CXCR3A and B
11	I-TAC	4	Activated T cell ($T_H1 > T_H2$), NK cell, B cell, endothelial cell	CXCR3A and B, CXCR7
12	SDF-1α/β	10	CD34+ bone marrow cell, thymocytes, monocytes/macrophages, naive activated T cell, B cell, plasma cell, neutrophil, immature dendritic cells, mature dendritic cells, plasmacytoid dendritic cells	CXCR4, CXCR7
13	BLC/BCA-1	4	Naive B cells, activated CD4 T cells, immature dendritic cells, mature dendritic cells	CXCR5>>CXCR3
16	sexckine	17	Activated T cell, natural killer T (NKT) cell, endothelial cells	CXCR6
CCL				
1	I-309	17	Neutrophil (TCA-3 only), T cell ($T_H2 > T_H1$) monocyte	CCR8
2	MCP-1	17	T cell ($T_H2 > T_H1$) monocyte, basophil, immature dendritic cells, NK cells	CCR2
3	MIP-1α/LD78	17	Monocyte/macrophage, T cell ($T_H1 > T_H2$), NK cell, basophil, immature dendritic cell, eosinophil, neutrophil, astrocyte, fibroblast, osteoclast	CCR1, 5
4	MIP-1β	17	Monocyte/macrophage, T cell ($T_H1 > T_H2$), NK cell, basophil, immature dendritic cell, eosinophil, B cell	CCR5>>1
5	RANTES	17	Monocyte/macrophage, T cell (memory T cell > T cell; $T_H1 > T_H2$), NK cell, basophil, eosinophil, immature dendritic cell	CCR1, 3, 5
6	C10/MRP-1	11 (mouse only)	Monocyte, B cell, CD4 T cell, NK cell	CCR 1
7	MCP-3	17	$T_H2 > T_H1$ T cell, monocyte, eosinophil, basophil, immature dendritic cell, NK cell	CCR1, 2, 3, 5
8	MCP-2	17	$T_H2 > T_H1$ T cell, monocyte, eosinophil, basophil, immature dendritic cell, NK cell	CCR1, 2, 5
9	MRP-2/MIP-1γ	11 (mouse only)	T cell, monocyte, adipocyte	CCR1
11	Eotaxin	17	Eosinophil, basophil, mast cell, T_H2 cell	CCR3>>CCR5
12	MCP-5	11 (mouse only)	Eosinophil, monocyte, T cell, B cell	CCR2
13	MCP-4	17	$T_H2 > T_H1$ T cell, monocyte, eosinophil, basophil, dendritic cell	CCR2, 3
14a	HCC-1	17	Monocyte	CCR1, 3, 5
14b	HCC-3	17	Monocyte	Unknown

Chemokine systematic name	Common names	Chromosome	Target cell	Specific receptor
15	MIP-5/HCC-2	17	T cells, monocytes, eosinophils, dendritic cells	CCR1, 3
16	HCC-4/LEC	17	Monocytes, T cells, natural killer cells, immature dendritic cells	CCR1, 2, 5, 8
17	TARC	16	T cells ($T_H2 > T_H1$), immature dendritic cells, thymocytes, regulatory T cells	CCR4>>8
18	DC-CK1/PARC	17	Naive T cells > activated T cells, immature dendritic cells, mantle zone B cells	PITPNM3
19	MIP-3β/ELC	9	Naive T cells, mature dendritic cells, B cells	CCR7
20	MIP-3α/LARC	2	T cells (memory T cells, T_H17 cells), blood mononuclear cells, immature dendritic cells, activated B cells, NKT cells, GALT development	CCR6
21	6Ckine/SLC	9	Naive T cells, B cells, thymocytes, NK cells, mature dendritic cells	CCR7
22	MDC	16	Immature dendritic cells, NK cells, T cells ($T_H2 > T_H1$), thymocytes, endothelial cells, monocytes, regulatory T cells	CCR4
23	MPIF-1/CK-β\8	17	Monocytes, T cells, resting neutrophils	CCR1, FPRL-1
24	Eotaxin-2/MPIF-2	7	Eosinophils, basophils, T cells	CCR3
25	TECK	19	Macrophages, thymocytes, dendritic cells, intraepithelial lymphocytes, IgA plasma cells, mucosal memory T cells	CCR9
26	Eotaxin-3	7	Eosinophils, basophils, fibroblasts	CCR3
27	CTACK	9	Skin homing memory T cells, B cells	CCR10
28	MEC	5	T cells, eosinophils, IgA+ B cells	CCR10>3
C and CX3C				
XCL 1	Lymphotactin	1	T cells, natural killer cells, CD8α + dendritic cells	XCR1
XCL 2	SCM-1β	1	T cells, natural killer cells, CD8α + dendritic cells	XCR1
CX3CL 1	Fractalkine	16	Activated T cells, monocytes, neutrophil, natural killer cells, immature dendritic cells, mast cells, astrocytes, neurons, microglia	CX3CR1

Atypical chemokine receptors

Chemokine ligands	Target cell	Specific receptor
Chemerin and resolvin E1	Macrophages, immature dendritic cells, mast cells, plasmacytoid dendritic cells, adipocytes, fibroblasts, endothelial cells, oral epithelial cells	CMKLR1/chem23
CCL5, CCL19 and chemerin	All hematopoietic cells, microglia, astrocytes, lung epithelial cells	CCRL2/CRAM
Inflammatory CC chemokines	Lymphatic endothelial cells	D6
Various CXC and CC chemokines	Red blood cells, Purkinje cells, blood endothelial cells, kidney epithelial cells	Duffy/DARC
CCL19, CCL21, CCL25	Thymic epithelial cells, lymph node stromal cells, keratinocytes	CCXCKR

Chromosome locations are for humans. Chemokines for which there is no human homolog are listed with the mouse chromosome.

† ELR refers to the three amino acids that precede the first cysteine residue of the CXC motif. If these amino acids are Glu-Leu-Arg (i.e. ELR+), then the chemokine is chemotactic for neutrophils; if they are not (ELR−) then the chemokine is chemotactic for lymphocytes

Compiled by Joost Oppenheim, National Cancer Institute, NIH.

Biographies

Emil von Behring (1854–1917) discovered antitoxin antibodies with Shibasaburo Kitasato.

Baruj Benacerraf (1920–2011) discovered immune response genes and collaborated in the first demonstration of MHC restriction.

Bruce Beutler (1957–) discovered the role of the Toll-like receptor in innate immunity in mice, showing that mice with a spontaneous inactivating mutation in TLR-4 were unresponsive to activating effects of LPS.

Jules Bordet (1870–1961) discovered complement as a heat-labile component in normal serum that would enhance the antimicrobial potency of specific antibodies.

Ogden C. Bruton (1908–2003) documented the first description of an immunodeficiency disease describing the failure of a male child to produce antibody. Because inheritance of this condition is X-linked and is characterized by the absence of immunoglobulin in the serum (agammaglobulinemia), it was called Bruton's X-linked agammaglobulinemia.

Frank MacFarlane Burnet (1899–1985) proposed the first generally accepted clonal selection hypothesis of adaptive immunity.

Robin Coombs (1921–2006) first developed anti-immunoglobulin antibodies to detect the antibodies that cause hemolytic disease of the newborn. The test for this disease is still called the Coombs test.

Jean Dausset (1916–2009) was an early pioneer in the study of the human major histocompatibility complex or HLA.

Peter Doherty (1940–) and Rolf Zinkernagel (1944–) showed that antigen recognition by T cells is MHC-restricted, thereby establishing the biological role of the proteins encoded by the major histocompatibility complex and leading to an understanding of antigen processing and its importance in the recognition of antigen by T cells.

Gerald Edelman (1929–2014) made crucial discoveries about the structure of immunoglobulins, including the first complete sequence of an antibody molecule.

Paul Ehrlich (1854–1915) was an early champion of humoral theories of immunity, and proposed a famous side-chain theory of antibody formation that bears a striking resemblance to current thinking about surface receptors.

James Gowans (1924–) discovered that adaptive immunity is mediated by lymphocytes, focusing the attention of immunologists on these small cells.

Jules Hoffman (1941–) discovered the role of the Toll-like receptor in innate immunity in *Drosophila melanogaster*.

Michael Heidelberger (1888–1991) developed the quantitative precipitin assay, ushering in the era of quantitative immunochemistry.

Charles A. Janeway, Jr. (1945–2003) recognized the importance of co-stimulation for initiating adaptive immune responses. He predicted the existence of receptors of the innate immune system that would recognize pathogen-associated molecular patterns and would signal activation of the adaptive immune system. His laboratory discovered the first mammalian Toll-like receptor that had this function. He was also the principal original author of this textbook.

Edward Jenner (1749–1823) described the successful protection of humans against smallpox infection by vaccination with cowpox or vaccinia virus. This founded the field of immunology.

Niels Jerne (1911–1994) developed the hemolytic plaque assay and several important immunological theories, including an early version of clonal selection, a prediction that lymphocyte receptors would be inherently biased to MHC recognition, and the idiotype network.

Shibasaburo Kitasato (1852–1931) discovered antibodies in collaboration with Emil von Behring.

Robert Koch (1843–1910) defined the criteria needed to characterize an infectious disease, known as Koch's postulates.

Georges Köhler (1946–1995) pioneered monoclonal antibody production from hybrid antibody-forming cells with César Milstein.

Karl Landsteiner (1868–1943) discovered the ABO blood group antigens. He also carried out detailed studies of the specificity of antibody binding using haptens as model antigens.

Peter Medawar (1915–1987) used skin grafts to show that tolerance is an acquired characteristic of lymphoid cells, a key feature of clonal selection theory.

Èlie Metchnikoff (1845–1916) was the first champion of cellular immunology, focusing his studies on the central role of phagocytes in host defense.

César Milstein (1927–2002) pioneered monoclonal antibody production with Georges Köhler.

Ray Owen (1915–2014) discovered that genetically different twin calves with a common placenta, thus sharing placental blood circulation, were immunologically tolerant to one another's tissues.

Louis Pasteur (1822–1895) was a French microbiologist and immunologist who validated the concept of immunization first studied by Jenner. He prepared vaccines against chicken cholera and rabies.

Rodney Porter (1917–1985) worked out the polypeptide structure of the antibody molecule, laying the groundwork for its analysis by protein sequencing.

Ignác Semmelweis (1818–1865) German-Hungarian physician who first determined a connection between hospital hygiene and an infectious disease, puerperal fever, and consequently introduced antisepsis into medical practice.

George Snell (1903–1996) worked out the genetics of the murine major histocompatibility complex and generated the congenic strains needed for its biological analysis, laying the groundwork for our current understanding of the role of the MHC in T-cell biology.

Ralph Steinman (1943–2011) discovered dendritic cells as powerful acitvators of T-cell responses and elaborated the various roles played by these cells in controlling the character and intensity of immune responses to pathogens.

Tomio Tada (1934–2010) first formulated the concept of the regulation of the immune response by 'suppressor T cells' in the 1970s, from indirect experimental evidence. The existence of such cells could not be verified at the time and the concept became discredited, but Tada was vindicated when researchers in the 1980s identified the cells now called 'regulatory T cells.'

Susumu Tonegawa (1939–) discovered the somatic recombination of immunological receptor genes that underlies the generation of diversity in human and murine antibodies and T-cell receptors.

Jürg Tschopp (1951-2011) contributed to the delineation of the complement system and T-cell cytolytic mechanisms, and made seminal contributions to the fields of apoptosis and innate immunity, in particular by discovering the inflammasome.

Don C. Wiley (1944–2001) solved the first crystal structure of an MHC I protein, providing a startling insight into how T cells recognize their antigen in the in the context of MHC molecules.

Photograph Acknowledgments

Chapter 1

Fig. 1.1 reproduced courtesy of Yale University, Harvey Cushing/John Hay Whitney Medical Library. Fig. 1.4 second panel from Tilney, L.G., Portnoy, D.A.: **Actin filaments and the growth, movement, and spread of the intracellular bacterial parasite, *Listeria monocytogenes*.** *J. Cell. Biol.* 1989, **109**:1597–1608. With permission from Rockefeller University Press. Fig. 1.24 photographs from Mowat, A., Viney, J.: **The anatomical basis of intestinal immunity.** *Immunol. Rev.* 1997, **156**:145–166. Fig. 1.34 photographs from Kaplan, G., et al.: **Efficacy of a cell-mediated reaction to the purified protein derivative of tuberculin in the disposal of *Mycobacterium leprae* from human skin.** *PNAS* 1988, **85**:5210–5214.

Chapter 2

Fig. 2.7 top panel from Button, B., et al.: **A periciliary brush promotes the lung health by separating the mucus layer from airway epithelia.** *Science* 2012, **337**:937–941. With permission from AAAS. Fig. 2.12 micrograph adapted from Mukherjee, S., et al.: **Antibacterial membrane attack by a pore-forming intestinal C-type lectin.** *Nature* 2014, **505**:103–107. Fig. 2.35 photographs reproduced with permission from Bhakdi, S., et al.: **Functions and relevance of the terminal complement sequence.** *Blut* 1990, **60**:309–318. © 1990 Springer-Verlag.

Chapter 3

Fig. 3.12 structure reprinted with permission from Jin, M.S., et al.: **Crystal structure of the TLR1-TLR2 heterodimer induced by binding of a tri-acylated lipopeptide.** *Cell* 2007, **130**:1071–82. © 2007 with permission from Elsevier. Fig. 3.13 structure reprinted with permission from Macmillan Publishers Ltd. Park, B.S., et al.: **The structural basis of lipopolysaccharide recognition by the TLR4-MD-2 complex.** *Nature* 2009, **458**:1191–1195. Fig. 3.34 model structure reprinted with permission from Macmillan Publishers Ltd. Emsley, J., et al.: **Structure of pentameric human serum amyloid P component.** *Nature* 1994, **367**: 338–345.

Chapter 4

Fig. 4.5 photograph from Green, N.M.: **Electron microscopy of the immunoglobulins.** *Adv. Immunol.* 1969, **11**:1–30. © 1969 with permission from Elsevier. Fig. 4.15 and Fig. 4.24 model structures from Garcia, K.C., et al.: **An αβ T cell receptor structure at 2.5 Å and its orientation in the TCR-MHC complex.** *Science* 1996, **274**:209–219. Reprinted with permission from AAAS.

Chapter 6

Fig. 6.6 reprinted with permission from Macmillan Publishers Ltd. Whitby, F.G., et al.: **Structural basis for the activation of 20S proteasomes by 11S regulators.** *Nature* 2000, **408**:115–120. Fig. 6.7 bottom panel from Velarde, G., et al.: **Three-dimensional structure of transporter associated with antigen processing (TAP) obtained by single particle image analysis.** *J. Biol. Chem.* 2001 **276**:46054–46063. © 2001 ASBMB. Fig. 6.22 structures from Mitaksov, V.E., & Fremont, D.: **Structural definition of the H-2Kd peptide-binding motif.** *J. Biol. Chem.* 2006, **281**:10618–10625. © 2006 American Society of Biochemistry and Molecular Biology. Fig. 6.25 molecular model reprinted with permission from Macmillan Publishers Ltd. Fields, B.A., et al.: **Crystal structure of a T-cell receptor β-chain complexed with a superantigen.** *Nature* 1996, **384**:188–192.

Chapter 8

Fig. 8.19 photographs reprinted with permission from Macmillan Publishers Ltd. Surh, C. D., Sprent, J.: **T-cell apoptosis detected *in situ* during positive and negative selection in the thymus.** *Nature* 1994, **372**:100–103.

Chapter 9

Fig. 9.12 fluorescent micrographs reprinted with permission from Macmillan Publishers Ltd. Pierre, P., Turley, S.J., et al.: **Development regulation of MHC class II transport in mouse dendritic cells.** *Nature* 1997, **388**:787–792. Fig. 9.38 panel c from Henkart, P.A., & Martz, E. (eds): *Second International Workshop on Cell Mediated Cytotoxicity.* © 1985 Kluwer/Plenum Publishers. With kind permission of Springer Science and Business Media.

Chapter 10

Fig. 10.17 left panel from Szakal, A.K., et al.: **Isolated follicular dendritic cells: cytochemical antigen localization, Nomarski, SEM, and TEM morphology.** *J. Immunol.* 1985, **134**:1349–1359. © 1985 The American Association of Immunologists. Fig. 10.17 center and right panels from Szakal, A.K., et al.: **Microanatomy of lymphoid tissue during humoral immune responses: structure function relationships.** *Ann. Rev. Immunol.* 1989, **7**:91–109. © 1989 Annual Reviews www.annualreviews.org.

Chapter 12

Fig. 12.4 adapted by permission from Macmillan Publishers Ltd. Dethlefsen, L., McFall-Ngai, M., Relman, D.A.: **An ecological and evolutionary perspective on human–microbe mutualism and disease.** *Nature* 2007, **449**:811–818. © 2007. Fig. 12.10 bottom left micrograph from Niess, J.H., et al.: **CX3CR1-mediated dendritic cell access to the intestinal lumen and bacterial clearance.** *Science* 2005, **307**:254–258. Reprinted with permission from AAAS. Fig. 12.10 bottom center micrograph from McDole, J.R., et al.: **Goblet cells deliver luminal antigen to CD103+ DCs in the small intestine.** *Nature* 2012, **483**: 345–9. With permission from Macmillan Publishers Ltd. Fig. 12.10 bottom right micrograph from Farache, J., et al.: **Luminal bacteria recruit CD103+ dendritic cells into the intestinal epithelium to sample bacterial antigens for presentation.** *Immunity* 2013, **38**: 581–95. With permission from Elsevier.

Chapter 13

Fig. 13.20 top left photograph from Kaplan, G., Cohn, Z.A.: **The immunobiology of leprosy.** *Int. Rev. Exp. Pathol.* 1986, **28**:45–78. © 1986 with permission from Elsevier. Fig. 13.37 based on data from Palella, F.J., et al.: **Declining morbidity and mortality among patients with advanced human immunodeficiency virus infection. HIV Outpatient Study Investigators.** *N. Engl. J. Med.* 1998, **338**:853–860. Fig. 13.40 adapted by permission from Macmillan Publishers Ltd. Wei, X., et al.: **Viral dynamics in human immunodeficiency virus type 1 infection.** *Nature* 1995, **373**:117–122.

Chapter 14

Fig. 14.5 top photograph from Sprecher, E., et al.: **Deleterious mutations in SPINK5 in a patient with congenital ichthyosiform erythroderma: molecular testing as a helpful diagnostic tool for Netherton syndrome.** *Clin. Exp. Dermatol.* 2004, **29**:513–517. Fig. 14.14 photographs from Finotto, S., et al: **Development of spontaneous airway changes consistent with human asthma in mice lacking T-bet.** *Science* 2002, **295**:336–338. Reprinted with permission from AAAS. Fig. 14.24 left photograph from Mowat, A.M., Viney, J.L.: **The anatomical basis of intestinal immunity.** *Immunol. Rev.* 1997 **156**:145–166.

Chapter 16

Fig. 16.16 photographs are reprinted from Herberman, R., & Callewaert, D. (eds): *Mechanisms of Cytotoxicity by Natural Killer Cells,* © 1985 with permission from Elsevier.

Glossary

-omab   Suffix applied to fully murine monoclonal antibodies used for human therapies.

-umab   Suffix applied to fully human monoclonal antibodies used for human therapies.

-ximab   Suffix applied to chimeric (i.e., mouse/human) monoclonal antibodies used for human therapies.

-zumab   Suffix applied to humanized monoclonal antibodies used for human therapies.

12/23 rule   Phenomenon wherein two gene segments of an immunoglobulin or T-cell receptor can be joined only if one recognition signal sequence has a 12-base-pair spacer and the other has a 23-base-pair spacer.

α:β heterodimer   The dimer of one α and one β chain that makes up the antigen-recognition portion of an α:β T-cell receptor.

α:β T-cell receptors   *See* T-cell receptor.

α₄:β₁ integrin (VLA-4, CD49d/CD29)   *See* integrins. Properties of individual CD antigens can be found in Appendix II.

α-defensins   A class of antimicrobial peptides produced by neutrophils and the Paneth cells of the intestine.

α-galactoceramide (α-GalCer)   An immunogenic glycolipid originally extracted from marine sponges but actually produced by various bacteria that is a ligand presented by CD1 to invariant NKT (iNKT) cells.

2B4   A receptor belonging to the signaling lymphocyte activation molecule (SLAM) family expressed by NK cells, which binds to CD48, another SLAM receptor. These signal through SAP and Fyn to promote survival and proliferation.

19S regulatory caps   Multisubunit component of the proteasome that functions to capture ubiquitinated proteins for degradation in the catalytic core.

20S catalytic core   Multisubunit component of proteasome responsible for protein degradation.

abatacept   An Fc fusion protein containing the CTLA-4 extracellular domain used in treating rheumatoid arthritis that blocks co-stimulation of T cells by binding B7 molecules.

accelerated rejection   The more rapid rejection of a second graft after rejection of the first graft. It was one of the pieces of evidence that showed that graft rejection was due to an adaptive immune response.

accessory effector cells   Cells that aid in an adaptive immune response but are not involved in specific antigen recognition. They include phagocytes, neutrophils, mast cells, and NK cells.

acellular pertussis vaccines   A formulation of pertussis used for vaccination containing chemically inactivated antigens, including pertussis toxoid.

acquired immune deficiency syndrome (AIDS)   A disease caused by infection with the human immunodeficiency virus (HIV-1). AIDS occurs when an infected patient has lost most of his or her CD4 T cells, so that infections with opportunistic pathogens occur.

activating receptors   On NK cells, a receptor whose stimulation results in activation of the cell's cytotoxic activity.

activation-induced cell death   A process by which autoreactive T cells are induced to die if they complete thymic maturation and migrate to the periphery.

activation-induced cytidine deaminase (AID)   Enzyme that initiates somatic hypermutation and isotype switching by deaminating DNA directly at cytosine in immunoglobulin V regions or switch regions. Loss of AID activity in patients leads to loss of both activities, causing hyper IgM and lack of affinity maturation.

activator protein 1 (AP-1)   A transcription factor formed as one of the outcomes of intracellular signaling by antigen receptors of lymphocytes.

active immunization   Immunization with antigen to provoke adaptive immunity.

acute desensitization   An immunotherapeutic technique for rapidly inducing temporary tolerance to, for example, an essential drug such as insulin or penicillin in a person who is allergic to it. Also called rapid desensitization. When performed properly, can produce symptoms of mild to moderate anaphylaxis.

acute phase   In reference to HIV infection, the period that occurs soon after a person becomes infected. It is characterized by an influenza-like illness, abundant virus in the blood, and a decrease in the number of circulating CD4 T cells.

acute-phase proteins   Proteins with innate immune function whose production is increased in the presence of an infection (the acute-phase response). They circulate in the blood and participate in early phases of host defense against infection. An example is mannose-binding lectin.

acute-phase response   A change in the proteins present in the blood that occurs during the early phases of an infection. It includes the production of acute-phase proteins, many of which are produced in the liver.

acute rejection   The rejection of a tissue or organ graft from a genetically unrelated donor that occurs within 10–13 days of transplantation unless prevented by immunosuppressant treatment.

adaptive immunity   Immunity to infection conferred by an adaptive immune response.

adaptors   Nonenzymatic proteins that form physical links between members of a signaling pathway, particularly between a receptor and other signaling proteins. They recruit members of the signaling pathway into functional protein complexes.

ADCC   *See* antibody-dependent cell-mediated cytotoxicity.

adenoids   Paired mucosa-associated lymphoid tissues located in the nasal cavity.

adenosine deaminase (ADA) deficiency   An inherited defect characterized by nonproduction of the enzyme adenosine deaminase, which leads to the accumulation of toxic purine nucleosides and nucleotides in cells, resulting in the death of most developing lymphocytes within the thymus. It is a cause of severe combined immunodeficiency.

adhesins   Cell-surface proteins on bacteria that enable them to bind to the surfaces of host cells.

adipose differentiation related protein A protein that functions in the maintenance and storage of neutral lipid droplets in many types of cells.

adjuvant Any substance that enhances the immune response to an antigen with which it is mixed.

afferent lymphatic vessels Vessels of the lymphatic system that drain extracellular fluid from the tissues and carry antigen, macrophages, and dendritic cells from sites of infection to lymph nodes or other peripheral lymphoid organs.

affinity The strength of binding of one molecule to another at a single site, such as the binding of a monovalent Fab fragment of antibody to a monovalent antigen. Cf. **avidity**.

affinity hypothesis Hypothesis that proposes how the choice between negative selection and positive selection of T cells in the thymus is made, according to the strength of self-peptide:MHC binding by the T-cell receptor. Low-affinity interactions rescue the cell from death by neglect, leading to positive selection; high-affinity interactions induce apoptosis and thus negative selection.

affinity maturation The increase in affinity for their specific antigen of the antibodies produced as an adaptive immune response progresses. This phenomenon is particularly prominent in secondary and subsequent immunizations.

agammaglobulinemia An absence of antibodies in the blood. *See also* **X-linked agammaglobulinemia (XLA)**.

age-related macular degeneration A leading cause of blindness in the elderly, for which some single-nucleotide polymorphisms (SNPs) in the factor H genes confer an increased risk.

agnathan paired receptors resembling Ag receptors (APARs) Multigene family of genes containing immunoglobulin domains present in hagfish and lamprey, that possibly represent ancestral predecessors of mammalian antigen receptors.

agnathans A class of vertebrate comprising jawless fish lacking adaptive immunity based on the RAG-mediated V(D)J recombination, but possessing a distinct system of adaptive immunity based on somatically assembled VLRs.

agonist selection A process by which T cells are positively selected in the thymus by their interaction with relatively high-affinity ligands.

AID *See* **activation-induced cytidine deaminase**.

AIDS *See* **acquired immune deficiency syndrome**.

AIM2 (absent in melanoma 2) A member of PYHIN subfamily of NLR (NOD-like receptor) family containing an N-terminal HIN domain. It activates caspase 1 in response to viral double-stranded DNA.

AIRE Gene encoding a protein (autoimmune regulator) that is involved in the expression of numerous genes by thymic medullary epithelial cells, enabling developing T cells to be exposed to self proteins characteristic of other tissues, thereby promoting tolerance to these proteins. Deficiency of AIRE leads to an autoimmune disease, APECED.

airway hyperreactivity, hyperresponsiveness The condition in which the airways are pathologically sensitive to both immunological (allergens) and nonimmunological stimuli, such as cold air, smoke, or perfumes. This hyperreactivity usually is present in chronic asthma.

airway tissue remodeling A thickening of the airway walls that occurs in chronic asthma due to hyperplasia and hypertrophy of the smooth muscle layer and mucus glands, with the eventual development of fibrosis. Often results in an irreversible decrease of lung function.

Akt Serine/threonine kinase activated downstream of PI3 kinase with numerous downstream targets involved in cell growth and survival, including activation of the mTOR pathways.

alefacept Recombinant CD58–IgG1 fusion protein that blocks CD2 binding by CD58 used in treatment for psoriasis.

alemtuzumab Antibody to CD52 used for lymphocyte depletion, such as for T-cell depletion during bone marrow allografts used in treating chronic myeloid leukemia.

allele A variant form of a gene; many genes occur in several (or more) different forms within the general population. *See also* **heterozygous**, **homozygous**, **polymorphism**.

allelic exclusion In a heterozygous individual, the expression of only one of the two alternative alleles of a particular gene. In immunology, the term describes the restricted expression of the individual chains of the antigen receptor genes, such that each individual lymphocyte produces immunoglobulin or T-cell receptors of a single antigen specificity.

allergen Any antigen that elicits an allergic reaction.

allergen desensitization An immunotherapeutic technique that aims either to change an allergic immune response to a symptom-free non-allergic response, or to develop immunologic tolerance to an allergen that has been causing unpleasant clinical symptoms. The procedure involves exposing an allergic individual to increasing doses of allergen.

allergic asthma An allergic reaction to inhaled antigen, which causes constriction of the bronchi, increased production of airway mucus, and difficulty in breathing.

allergic conjunctivitis An allergic reaction involving the conjunctiva of the eye that occurs in sensitized individuals exposed to airborne allergens. It is usually manifested together with nasal allergy symptoms as allergic rhinoconjunctivitis or hay fever.

allergic contact dermatitis A largely T-cell-mediated immunological hypersensitivity reaction manifested by a skin rash at the site of contact with the allergen. Often the stimulus is a chemical agent, for example urushiol oil from the leaves of the poison ivy plant, which can haptenate normal host molecules to render them allergenic.

allergic reaction A specific response to an innocuous environmental antigen, or allergen, that is caused by sensitized B or T cells. Allergic reactions can be caused by various mechanisms, but the most common is the binding of allergen to IgE bound to mast cells, which causes the cells to release histamine and other biologically active molecules that cause the signs and symptoms of asthma, hay fever, and other common allergic responses.

allergic rhinitis An allergic reaction in the nasal mucosa that causes excess mucus production, nasal itching, and sneezing.

allergy The state in which a symptomatic immune reaction is made to a normally innocuous environmental antigen. It involves the interaction between the antigen and antibody or primed T cells produced by earlier exposure to the same antigen.

alloantibodies Antibodies produced against antigens from a genetically nonidentical member of the same species.

alloantigens Antigens from another genetically nonidentical member of the same species.

allogeneic Describes two individuals or two mouse strains that differ at genes in the MHC. The term can also be used for allelic differences at other loci.

allograft A transplant of tissue from an allogeneic (genetically nonidentical) donor of the same species. Such grafts are invariably rejected unless the recipient is immunosuppressed.

allograft rejection The immunologically mediated rejection of grafted tissues or organs from a genetically nonidentical donor. It is due chiefly to recognition of nonself MHC molecules on the graft.

alloreactivity   The recognition by T cells of MHC molecules other than self. Such responses are also called alloreactions or alloreactive responses.

altered peptide ligands (APLs)   Peptides in which amino acid substitutions have been made in T-cell receptor contact positions that affect their binding to the receptor.

alternative pathway   A form of complement activation that is initiated by spontaneous hydrolysis of C3 and which uses factor B and factor D to form the unique C3 convertase C3bBb.

alternatively activated macrophages   *See* **M2 macrophages**.

alum   Inorganic aluminum salts (for example aluminum phosphate and aluminum hydroxide); they act as adjuvants when mixed with antigens and are one of the few adjuvants permitted for use in humans.

amphipathic   Describes molecules that have a positively charged (or hydrophilic) region separated from a hydrophobic region.

anakinra   A recombinant IL-1 receptor antagonist (IL-1RA) used to block IL-1 receptor activation and used in treating rheumatoid arthritis.

anaphylactic shock   *See* **anaphylaxis**.

anaphylatoxins   Pro-inflammatory complement fragments C5a and C3a released by cleavage during complement activation. They are recognized by specific receptors, and recruit fluid and inflammatory cells to the site of their release.

anaphylaxis   A rapid-onset and systemic allergic reaction to antigen, for example to insect venom injected directly into the bloodstream, or to foods such as peanuts. Severe systemic reactions can be potentially fatal due to circulatory collapse and suffocation from tracheal swelling. It usually results from antigens binding to IgE bound by Fcε receptors on mast cells, leading to systemic release of inflammatory mediators.

anchor residues   Specific amino acid residues in antigenic peptides that determine peptide binding specificity to MHC class I molecules. Anchor residues for MHC class II molecules exist but are less obvious than for MHC class I.

anergy   A state of nonresponsiveness to antigen. People are said to be anergic when they cannot mount delayed-type hypersensitivity reactions to a test antigen, whereas T cells and B cells are said to be anergic when they cannot respond to their specific antigen under optimal conditions of stimulation.

ankylosing spondylitis   Inflammatory disease of the spine leading to vertebral fusion strongly associated with HLA-B27.

antibody   A protein that binds specifically to a particular substance—called its antigen. Each antibody molecule has a unique structure that enables it to bind specifically to its corresponding antigen, but all antibodies have the same overall structure and are known collectively as immunoglobulins. Antibodies are produced by differentiated B cells (plasma cells) in response to infection or immunization, and bind to and neutralize pathogens or prepare them for uptake and destruction by phagocytes.

antibody combining site   *See* **antigen-binding site**.

antibody-dependent cell-mediated cytotoxicity (ADCC)   The killing of antibody-coated target cells by cells with Fc receptors that recognize the constant region of the bound antibody. Most ADCC is mediated by NK cells that have the Fc receptor FcγRIII on their surface.

antibody-directed enzyme/pro-drug therapy (ADEPT)   Treatment in which an antibody is linked to an enzyme that metabolizes a nontoxic pro-drug to the active cytotoxic drug.

antibody repertoire   The total variety of antibodies in the body of an individual.

antigen   Any molecule that can bind specifically to an antibody or generate peptide fragments that are recognized by a T-cell receptor.

antigen-binding site   The site at the tip of each arm of an antibody that makes physical contact with the antigen and binds it noncovalently. The antigen specificity of the site is determined by its shape and the amino acids present.

antigenic determinant   That portion of an antigenic molecule that is bound by the antigen-binding site of a given antibody or antigen receptor; it is also known as an epitope.

antigenic drift   The process by which influenza virus varies genetically in minor ways from year to year. Point mutations in viral genes cause small differences in the structure of the viral surface antigens.

antigenic shift   A radical change in the surface antigens of influenza virus, caused by reassortment of their segmented genome with that of another influenza virus, often from an animal.

antigenic variation   Alterations in surface antigens that occur in some pathogens (such as African trypanosomes) from one generation to another, which allows them to evade preexisting antibodies.

antigen presentation   The display of antigen on the surface of a cell in the form of peptide fragments bound to MHC molecules. T cells recognize antigen when it is presented in this way.

antigen-presenting cells (APCs)   Highly specialized cells that can process antigens and display their peptide fragments on the cell surface together with other, co-stimulatory, proteins required for activating naive T cells. The main antigen-presenting cells for naive T cells are dendritic cells, macrophages, and B cells.

antigen processing   The intracellular degradation of foreign proteins into peptides that can bind to MHC molecules for presentation to T cells. All protein antigens must be processed into peptides before they can be presented by MHC molecules.

antigen receptor   The cell-surface receptor by which lymphocytes recognize antigen. Each individual lymphocyte bears receptors of a single antigen specificity.

anti-lymphocyte globulin   Antiserum raised in another species against human T cells. It is used in the temporary suppression of immune responses in transplantation.

antimicrobial enzymes   Enzymes that kill microorganisms by their actions. An example is lysozyme, which digests bacterial cell walls.

antimicrobial peptides, antimicrobial proteins   Amphipathic peptides or proteins secreted by epithelial cells and phagocytes that kill a variety of microbes nonspecifically, mainly by disrupting cell membranes. Antimicrobial peptides in humans include the defensins, the cathelicidins, the histatins, and RegIIIγ.

antiserum   The fluid component of clotted blood from an immune individual that contains antibodies against the antigen used for immunization. An antiserum contains a mixture of different antibodies that all bind the antigen, but which each have a different structure, their own epitope on the antigen, and their own set of cross-reactions. This heterogeneity makes each antiserum unique.

antivenin   Antibody raised against the venom of a poisonous snake or other organism and which can be used as an immediate treatment for the bite to neutralize the venom.

aorta-gonad-mesonephros (AGM)   An embryonic region in which hematopoietic cells arise during development.

AP-1   A heterodimeric transcription factor formed as one of the outcomes of intracellular signaling via the antigen receptors of lymphocytes and the TLRs of cells of innate immunity. Most often, contains one Fos-family member and one Jun-family member. AP-1 mainly activates the expression of genes for cytokines and chemokines.

APECED *See* autoimmune polyendocrinopathy-candidiasis-ectodermal dystrophy.

APOBEC1 (apolipoprotein B mRNA editing catalytic polypeptide 1) An RNA editing enzyme that deaminates cytidine to uracil in certain mRNAs, such as apolipoprotein B, and which is related to the enzyme AID involved in somatic hypermutation and isotype switching.

apoptosis A form of cell death common in the immune system, in which the cell activates an internal death program. It is characterized by nuclear DNA degradation, nuclear degeneration and condensation, and the rapid phagocytosis of cell remains. Proliferating lymphocytes experience high rates of apoptosis during their development and during immune responses.

apoptosome A large, multimeric protein structure that forms in the process of apoptosis when cytochrome *c* is released from mitochondria and binds Apaf-1. A heptamer of cytochrome *c*-Apaf-1 heterodimers assembles into wheel-like structure that binds and activates procaspase-9, an initiator caspase, to initiate the caspase cascade.

appendix A gut-associated lymphoid tissue located at the beginning of the colon.

APRIL A TNF family cytokine related to BAFF that binds the receptors TACI and BCMA on B cells to promote survival and regulate differentiation.

apurinic/apyrimidinic endonuclease 1 (APE1) A DNA repair endonuclease involved in class switch recombination.

Artemis An endonuclease involved in the gene rearrangements that generate functional immunoglobulin and T-cell receptor genes.

Arthus reaction A local skin reaction that occurs when a sensitized individual with IgG antibodies against a particular antigen is challenged by injection of the antigen into the dermis. Immune complexes of the antigen with IgG antibodies in the extracellular spaces in the dermis activate complement and phagocytic cells to produce a local inflammatory response.

aryl hydrocarbon receptor (AhR) A basic helix-loop-helix transcription factor that is activated by various aromatic ligands including, famously, dioxin. It functions in the normal activity of several types of immune cells including some ILCs and IELs.

ASC (PYCARD) An adaptor protein containing pyrin and CARD domains involved in activating caspase 1 in the inflammasome.

asymptomatic phase In reference to HIV infection, period in which the infection is being partly held in check and no symptoms occur; it may last for many years.

ataxia telangiectasia (ATM) A disease characterized by a staggering gait and multiple disorganized blood vessels, and often accompanied by clinical immunodeficiency. It is caused by defects in the ATM protein, which is involved in DNA repair pathways that are also used in V(D)J recombination and class-switch recombination.

atopic march The clinical observation that it is common for children with atopic eczema to later develop allergic rhinitis and/or asthma.

atopy A genetically based increased tendency to produce IgE-mediated allergic reactions against innocuous substances.

ATP-binding cassette (ABC) A large family of proteins containing a particular domain for nucleotide-binding that includes many transporters, such as TAP1 and TAP2, but also various NOD members.

attenuation The process by which human or animal pathogens are modified by growth in culture so that they can grow in their host and induce immunity without producing serious clinical disease.

atypical hemolytic uremic syndrome A condition characterized by damage to platelets and red blood cells and inflammation of the kidneys that is caused by uncontrolled complement activation in individuals with inherited deficiencies in complement regulatory proteins.

autoantibodies Antibodies specific for self antigens.

autoantigens A self antigen to which the immune system makes a response.

autocrine Describes a cytokine or other biologically active molecule acting on the cell that produces it.

autograft A graft of tissue from one site to another on the same individual.

autoimmune disease Disease in which the pathology is caused by adaptive immune responses to self antigens.

autoimmune hemolytic anemia A pathological condition with low levels of red blood cells (anemia), which is caused by autoantibodies that bind red blood cell surface antigens and target the red blood cell for destruction.

autoimmune lymphoproliferative syndrome (ALPS) An inherited syndrome in which a defect in the *Fas* gene leads to a failure in normal apoptosis, causing unregulated immune responses, including autoimmune responses.

autoimmune polyendocrinopathy–candidiasis–ectodermal dystrophy (APECED) A disease characterized by a loss of tolerance to self antigens, caused by a breakdown of negative selection in the thymus. It is due to defects in the gene *AIRE*, which encodes a transcriptional regulatory protein that enables many self antigens to be expressed by thymic medullary epithelial cells. Also called autoimmune polyglandular syndrome type I.

autoimmune thrombocytopenic purpura An autoimmune disease in which antibodies against platelets are made. Antibody binding to platelets causes them to be taken up by cells with Fc receptors and complement receptors, resulting in a decrease in platelet count that leads to purpura (bleeding).

autoimmunity Adaptive immunity specific for self antigens.

autoinflammatory diseases Diseases due to unregulated inflammation in the absence of infection; they can have a variety of causes, including inherited genetic defects.

autophagosome A double bilayer membrane structure that functions in macroautophagy by engulfing cytoplasmic contents and fusing with lysosomes.

autophagy The digestion and breakdown by a cell of its own organelles and proteins in lysosomes. It may be one route by which cytosolic proteins can be processed for presentation on MHC class II molecules.

avidity The sum total of the strength of binding of two molecules or cells to one another at multiple sites. It is distinct from affinity, which is the strength of binding of one site on a molecule to its ligand.

avoidance Mechanisms that prevent a host's exposure to microbes, such as anatomic barriers or particular behaviors.

azathioprine A powerful cytotoxic drug that is converted to its active form *in vivo*, which then kills rapidly proliferating cells, including proliferating lymphocytes; it is used as an immunosuppressant to treat autoimmune disease and in transplantation.

B-1 B cells A class of atypical, self-renewing B cells (also known as CD5 B cells) found mainly in the peritoneal and pleural cavities in adults and considered part of the innate rather than the adaptive immune system. They have a much less diverse antigen-receptor repertoire than conventional B cells and are the major source of natural antibody.

B7 molecules, B7.1 and B7.2   Cell-surface proteins on specialized antigen-presenting cells such as dendritic cells, which are the major co-stimulatory molecules for T cells. B7.1 (CD80) and B7.2 (CD86) are closely related members of the immunoglobulin superfamily and both bind to the CD28 and CTLA-4 proteins on T cells.

β1i (LMP2), β2i (MECL-1), β5i (LMP7)   Alternative proteasome subunits that replace the constitutive catalytic subunits β1, β2, and β5 that are induced by interferons and produce the immunoproteasome.

β5t   Alternative proteasome subunit expressed by thymic epithelial cells that substitutes for β5 to produce the thymoproteasome involved in generating peptides encountered by thymocytes during development.

β-defensins   Antimicrobial peptides made by virtually all multicellular organisms. In mammals they are produced by the epithelia of the respiratory and urogenital tracts, skin, and tongue.

β sandwich   A secondary protein structure composed of two β sheets that fold such that one lies over the other, as in an immunoglobulin fold.

β sheets   A secondary protein structure composed of β strands stabilized by noncovalent interactions between backbone amide and carbonyl groups. In 'parallel' β sheets, the adjacent strands run in the same direction; in 'antiparallel' β sheets, adjacent strands run in opposite directions. Immunoglobulin domains are made up of two antiparallel β sheets arranged in the form of a β barrel.

β strands   A secondary protein structure in which the polypeptide backbone of several consecutive amino acids is arranged in a flat, or planar, conformation, and often illustrated as an arrow.

β$_2$-microglobulin   The light chain of the MHC class I proteins, encoded outside the MHC. It binds noncovalently to the heavy or α chain.

B and T lymphocyte attenuator (BTLA)   An inhibitory CD28-related receptor expressed by B and T lymphocytes that interacts with the herpes virus entry molecule (HVEM), a member of the TNF receptor family.

bacteria   A vast kingdom of unicellular prokaryotic microorganisms, some species of which cause infectious diseases in humans and animals, while others make up most of the body's commensal microbiota. Disease-causing bacteria may live in the extracellular spaces, or inside cells in vesicles or in the cytosol.

BAFF   B-cell activating factor belonging to the TNF family that binds the receptors BAFF-R and TACI to promote B cell survival.

BAFF-R   Receptor for BAFF that can activate canonical and non-canonical NF-κB signaling and promote survival of B cells.

bare lymphocyte syndrome   *See* **MHC class I deficiency, MHC class II deficiency**.

base-excision repair   Type of DNA repair that can lead to mutation and that is involved in somatic hypermutation and class switching in B cells.

basiliximab   Antibody to human CD25 used to block IL-2 receptor signaling in T cells for treatment of rejection in renal transplantation.

basophils   Type of white blood cell containing granules that stain with basic dyes. It is thought to have a function similar to mast cells.

BATF3   A transcription factor expressed in dendritic cells belonging to the AP1 family, which includes many other factors such as c-Jun and Fos.

B cells, B lymphocytes   One of the two types of antigen-specific lymphocytes responsible for adaptive immune responses, the other being the T cells. The function of B cells is to produce antibodies. B cells are divided into two classes. Conventional B cells have highly diverse antigen receptors and are generated in the bone marrow throughout life, emerging to populate the blood and lymphoid tissues. B-1 cells have much less diverse antigen receptors and form a population of self-renewing B cells in the peritoneal and pleural cavities.

B-cell antigen receptor, B-cell receptor (BCR)   The cell-surface receptor on B cells for specific antigen. It is composed of a transmembrane immunoglobulin molecule (which recognizes antigen) associated with the invariant Igα and Igβ chains (which have a signaling function). On activation by antigen, B cells differentiate into plasma cells producing antibody molecules of the same antigen specificity as this receptor.

B-cell co-receptor   A transmembrane signaling receptor on the B-cell surface composed of the proteins CD19, CD81, and CD21 (complement receptor 2), which binds complement fragments on bacterial antigens also bound by the B-cell receptor. Co-ligation of this complex with the B-cell receptor increases responsiveness to antigen about 100-fold.

B-cell co-receptor complex   A transmembrane signaling receptor on the B-cell surface composed of the proteins CD19, CD81, and CD21 (complement receptor 2), which binds complement fragments on bacterial antigens also bound by the B-cell receptor. Co-ligation of this complex with the B-cell receptor increases responsiveness to antigen about 100-fold.

B-cell mitogens   Any substance that nonspecifically causes B cells to proliferate.

Bcl-2 family   Family of intracellular proteins that includes members that promote apoptosis (Bax, Bak, and Bok) and members that inhibit apoptosis (Bcl-2, Bcl-W, and Bcl-XL).

Bcl-6   A transcriptional repressor that opposes differentiation of B cells into plasma cells.

BCMA   Receptor of the TNFR superfamily that binds APRIL.

Bcr–Abl tyrosine kinase   Constitutively active tyrosine kinase fusion protein caused by a chromosomal translocation—the Philadelphia chromosome— between *Bcr* with the *Abl* tyrosine kinase genes associated with chronic myeloid leukemia.

BDCA-2 (blood dendritic cell antigen 2)   A C-type lectin expressed selectively as a receptor on the surface of human plasmacytoid dendritic cells.

Berlin patient   A man with HIV who was treated in Berlin with a hematopoietic stem cell (HSC) transplant from a donor deficient in a co-receptor for the virus (CCR5) for an unrelated illness (leukemia). He is thought to be cured of HIV infection, and is one of the only known patients in which the virus is thought to be completely eliminated, a so-called 'sterilizing' cure.

biologics therapy   Medical treatments comprising natural proteins such as antibodies and cytokines, and antisera or whole cells.

Blau syndrome   An inherited granulomatous disease caused by gain-of-function mutations in the *NOD2* gene.

BLIMP-1   A transcriptional repressor that promotes B-cell differentiation into plasma cells and suppresses proliferation, and further class switching and affinity maturation.

BLNK   B-cell linker protein. *See* **SLP-65**.

bone marrow   The tissue where all the cellular elements of the blood—red blood cells, white blood cells, and platelets—are initially generated from hematopoietic stem cells. The bone marrow is also the site of further B-cell development in mammals and the source of stem cells that give rise to T cells on migration to the thymus. Thus, bone marrow transplantation can restore all the cellular elements of the blood, including the cells required for adaptive immune responses.

booster immunization   *See* **secondary immunization**.

bradykinin   A vasoactive peptide that is produced as a result of tissue damage and acts as an inflammatory mediator.

broadly neutralizing antibodies   Antibodies that block viral infection by multiple strains. In reference to HIV, these are antibodies that block binding of the virus to CD4 and/or chemokine co-receptors.

bronchus-associated lymphoid tissue (BALT) Organized lymphoid tissue found in the bronchi in some animals. Adult humans do not normally have such organized lymphoid tissue in the respiratory tract, but it may be present in some infants and children.

Bruton's tyrosine kinase (Btk) A Tec-family tyrosine kinase important in B-cell receptor signaling. Btk is mutated in the human immunodeficiency disease X-linked agammaglobulinemia.

Bruton's X-linked agammaglobulinemia *See* **X-linked agammaglobulinemia**.

bursa of Fabricius Lymphoid organ associated with the gut that is the site of B-cell development in chickens.

butyrate A short chain fatty acid produced abundantly by anaerobic digestion of carbohydrates in the intestine by commensals that can influence host cells in several ways, acting as an energy source for enterocytes and as an inhibitor of histone deacetylases.

C1 complex, C1 Protein complex activated as the first step in the classical pathway of complement activation, composed of C1q bound to two molecules each of the proteases C1r and C1s. Binding of a pathogen or antibody to C1q activates C1r, which cleaves and activates C1s, which cleaves C4 and C2.

C1 inhibitor (C1INH) An inhibitor protein for C1 that binds and inactivates C1r:C1s enzymatic activity. Deficiency in C1INH causes hereditary angioedema through production of vasoactive peptides that cause subcutaneous and laryngeal swelling.

C2 Complement protein of the classical and lectin pathways that is cleaved by the C1 complex to yield C2b and C2a. C2a is an active protease that forms part of the classical C3 convertase C4bC2a.

C3 Complement protein on which all complement activation pathways converge. C3 cleavage forms C3b, which can bind covalently to microbial surfaces, where it promotes destruction by phagocytes.

C3 convertase Enzyme complex that cleaves C3 to C3b and C3a on the surface of a pathogen. The C3 convertase of the classical and lectin pathways is formed from membrane-bound C4b complexed with the protease C2a. The alternative pathway C3 convertase is formed from membrane-bound C3b complexed with the protease Bb.

C3(H$_2$O)Bb *See* **fluid-phase C3 convertase**.

C3a *See* **anaphylatoxins**.

C3b *See* **C3**.

C3b2Bb The C5 convertase of the alternative pathway of complement activation.

C3bBb The C3 convertase of the alternative pathway of complement activation.

C3dg Breakdown product of iC3b that remains attached to the microbial surface, where it can bind complement receptor CR2.

C3f A small fragment of C3b that is removed by factor I and MCP to leave iC3b on the microbial surface.

C4 Complement protein of the classical and lectin pathways. C4 is cleaved by C1s to C4b, which forms part of the classical C3 convertase.

C4b-binding protein (C4BP) A complement-regulatory protein that inactivates the classical pathway C3 convertase formed on host cells by displacing C2a from the C4bC2a complex. C4BP binds C4b attached to host cells, but cannot bind C4b attached to pathogens.

C4b2a C3 convertase of the classical and lectin pathways of complement activation.

C4b2a3b C5 convertase of the classical and lectin pathways of complement activation.

C5 convertase Enzyme complex that cleaves C5 to C5a and C5b.

C5a *See* **anaphylatoxins**.

C5a receptor The cell-surface receptor for the pro-inflammatory C5a fragment of complement, present on macrophages and neutrophils.

C5b Fragment of C5 that initiates the formation of the membrane-attack complex (MAC).

C5L2 (GPR77) Non-signaling decoy receptor for C5a expressed by phagocytes.

C6, C7, C8, C9 Complement proteins that act with C5b to form the membrane-attack complex, producing a pore that leads to lysis of the target cell.

calcineurin A cytosolic serine/threonine phosphatase with a crucial role in signaling via the T-cell receptor. The immunosuppressive drugs cyclosporin A and tacrolimus inactivate calcineurin, suppressing T-cell responses.

calmodulin Calcium-binding protein that is activated by binding Ca^{2+}; it is then able to bind to and regulate the activity of a wide variety of enzymes.

calnexin A chaperone protein in the endoplasmic reticulum (ER) that binds to partly folded members of the immunoglobulin superfamily of proteins and retains them in the ER until folding is complete.

calprotectin A complex of heterodimers of the antimicrobial peptides S100A8 and S100A9, which sequester zinc and manganese from microbes. Produced in abundance by neutrophils, and in lesser amounts by macrophages and epithelial cells.

calreticulin A chaperone protein in the endoplasmic reticulum that, together with ERp57 and tapasin, forms the peptide-loading complex that loads peptides onto newly synthesized MHC class I molecules.

cancer immunoediting A process that occurs during the development of a cancer when it is acquiring mutations that favor its survival and escape from immune responses, such that cancer cells with these mutations are selected for survival and growth.

cancer-testis antigens Proteins expressed by cancer cells that are normally expressed only in male germ cells in the testis.

capping A process occurring in the nucleus in which the modified purine 7-methylguanosine is added to the 5′ phosphate of the first nucleotide of the RNA transcript.

capsular polysaccharides *See* **capsulated bacteria**.

capsulated bacteria Referring to bacteria surrounded by a polysaccharide shell that resists actions of phagocytes, resulting in pus formation at the site of infection. Also called pyogenic (pus-forming) bacteria.

carboxypeptidase N (CPN) A metalloproteinase that inactivates C3a and C5a. CPN deficiency causes a condition of recurrent angioedema.

cardiolipin A lipid found in many bacteria and in the inner mitochondrial membrane that is a ligand recognized by some human γ:δ T cells.

caspase 8 An initiator caspase activated by various receptors that activates the process of apoptosis.

caspase 11 This caspase is homologous to human capsase 4 and 5. Its expression is induced by TLR signaling. Intracellular LPS can directly activate it, leading to pyroptosis.

caspase recruitment domain (CARD) A protein domain present in some receptor tails that can dimerize with other CARD-domain-containing proteins, including caspases, thus recruiting them into signaling pathways.

caspases A family of cysteine proteases that cleave proteins at aspartic acid residues. They have important roles in apoptosis and in the processing of cytokine pro-polypeptides.

cathelicidins Family of antimicrobial peptides that in humans has one member.

cathelin A cathepsin L inhibitor.

cathepsins A family of proteases using cysteine at their active site that frequently function in processing antigens taken into the vesicular pathway.

CC chemokines One of the two main classes of chemokines, distinguished by two adjacent cysteines (C) near the amino terminus. They have names CCL1, CCL2, etc. *See* Appendix IV for a list of individual chemokines.

CCL9 (MIP-1γ) Chemokine made by follicle-associated epithelial cells and binds CCR6, recruiting activated T and B cells, NK cells, and dendritic cells into GALT.

CCL19 Chemokine made by dendritic cells and stromal cells in T-cell zones of lymph nodes that binds CCR7 and functions to attract naive T cells.

CCL20 Chemokine made by follicle-associated epithelial cells and binds CCR6, recruiting activated T and B cells, NK cells, and dendritic cells into GALT.

CCL21 Chemokine made by dendritic cells and stromal cells in T cell zones of lymph nodes that binds CCR7 and functions to attract naive T cells.

CCL25 (TECK) Chemokine made by small-intestinal epithelial cells that binds CCR9 to recruit gut-homing T and B cells.

CCL28 (MEC, mucosal epithelial chemokine) Chemokine made by colonic intestinal cells, salivary gland, and lactating mammary gland cells that binds CCR10 to recruit B lymphocytes producing IgA into these tissues.

CCR1 Chemokine receptor expressed by neutrophils, monocytes, B cells, and dendritic cells, that binds several chemokines, including CCL6 and CCL9.

CCR6 Chemokine receptor expressed by follicular and marginal zone B cells and dendritic cells that binds CCL20.

CCR7 Chemokine receptor expressed by all naive T and B cells, and some memory T and B cells, such as central memory T cells, that binds CCL19 and CCL21 made by dendritic cells and stromal cells in lymphoid tissues.

CCR9 Chemokine receptor expressed by dendritic cells, T cells, and thymocytes, and some γ:δ T cells, that binds CCL25 that mediates recruitment of gut-homing cells.

CCR10 Chemokine receptor expressed by many cells that binds CCL27 and CCL28 that mediates intestinal recruitment of IgA-producing B lymphocytes.

CD1 Small family of MHC class I-like proteins that are not encoded in the MHC and can present glycolipid antigens to CD4 T cells.

CD3 complex The invariant proteins CD3γ, δ, and ε, and the dimeric ζ chains, which form the signaling complex of the T-cell receptor. Each of them contains one or more ITAM signaling motifs in their cytoplasmic tails.

CD4 The co-receptor for T-cell receptors that recognize peptide antigens bound to MHC class II molecules. It binds to the lateral face of the MHC molecule.

CD8 The co-receptor for T-cell receptors that recognize peptide antigens bound to MHC class I molecules. It binds to the lateral face of the MHC molecule.

CD11b (α$_M$ integrin) Integrin expressed by macrophages and some dendritic cells that functions with β2 integrin (CD18) as complement receptor 3 (CR3).

CD19 *See* **B-cell co-receptor**.

CD21 Another name for complement receptor 2 (CR2). *See also* **B-cell co-receptor**.

CD22 An inhibitory receptor on B cells that binds sialic acid-modified glycoproteins commonly found on mammalian cells and contains an ITIM motif in its cytoplasmic tail.

CD23 The low-affinity Fc receptor for IgE.

CD25 Also known as IL-2 receptorα (IL-2Rα), this is the high-affinity component of the IL-2 receptor, which also includes IL-2Rβ and the common γ chain. It is upregulated by activated T cells and is constitutively expressed by T$_{reg}$ cells to confer responsiveness to IL-2.

CD27 A TNF receptor-family protein constitutively expressed on naive T cells that binds CD70 on dendritic cells and delivers a potent co-stimulatory signal to T cells early in the activation process.

CD28 An activating receptor on T cells that binds to the B7 co-stimulatory molecules present on specialized antigen-presenting cells such as dendritic cells. CD28 is the major co-stimulatory receptor on naive T cells.

CD30, CD30 ligand CD30 on B cells and CD30 ligand (CD30L) on helper T cells are co-stimulatory molecules involved in stimulating the proliferation of antigen-activated naive B cells.

CD31 A cell-adhesion molecule found both on lymphocytes and at endothelial cell junctions. CD31–CD31 interactions are thought to enable leukocytes to leave blood vessels and enter tissues.

CD40, CD40 ligand CD40 on B cells and CD40 ligand (CD40L, CD154) on activated helper T cells are co-stimulatory molecules whose interaction is required for the proliferation and class switching of antigen activated naive B cells. CD40 is also expressed by dendritic cells, and here the CD40–CD40L interaction provides co-stimulatory signals to naive T cells.

CD40 ligand deficiency An immunodeficiency disease in which little or no IgG, IgE, or IgA antibody is produced and even IgM responses are deficient, but serum IgM levels are normal to high. It is due to a defect in the gene encoding CD40 ligand (CD154), which prevents class switching from occurring. Also known as X-linked hyper IgM syndrome, reflecting location of gene that encodes CD40L on the X chromosome and phenotype of elevated IgM antibody relative to other immunoglobulins.

CD44 Also known as phagocytic glycoprotein-1 (Pgp1), CD44 is a cell-surface glycoprotein expressed by naive lymphocytes and upregulated on activated T cells. It is a receptor for hyaluronic acid and functions in cell–cell and cell–extracellular matrix adhesion. High expression of CD44 is used as a marker for effector and memory T cells.

CD45 A transmembrane tyrosine phosphatase found on all leukocytes. It is expressed in different isoforms on different cell types, including the different subtypes of T cells. Also called leukocyte common antigen, it is a generic marker for hematopoietically derived cells, with the exception of erythrocytes.

CD45RO An alternatively spliced variant of CD45 that serves as a marker for memory T cells.

CD48 *See* **2B4**.

CD59, protectin Cell-surface protein that protects host cells from complement damage by blocking binding of C9 to the C5b678 complex, thus preventing MAC formation.

CD69 A cell-surface protein that is rapidly expressed by antigen-activated T cells. It acts to down-modulate the expression of the sphingosine 1 phosphate receptor 1 (S1PR1), thereby retaining activated T cells within T-cell zones of secondary lymphoid tissues as they divide and differentiate into effector T cells.

CD70 The ligand for CD27 that is expressed on activated dendritic cells and delivers a potent co-stimulatory signal to T cells early in the activation process.

CD81 *See* **B-cell co-receptor**.

CD84 *See* **SLAM** (signaling lymphocyte activation molecule).

CD86 (B7-2) A transmembrane protein of the immunoglobulin superfamily that is expressed on antigen-presenting cells and binds to CD28 expressed by T cells.

CD94 A C-type lectin that is a subunit of the KLR-type receptors of NK cells.

CD103 Integrin $\alpha_E{:}\beta_7$, a cell-surface marker on a subset of dendritic cells in the gastrointestinal tract that are involved in inducing tolerance to antigens from food and the commensal microbiota.

CD127 Also known as IL-7 receptor α (IL-7Rα), which pairs with the common γ chain of the IL-2 receptor family to form the IL-7 receptor. It is expressed by naive T cells and a subset of memory T cells to support their survival.

celiac disease A chronic condition of the upper small intestine caused by an immune response directed at gluten, a complex of proteins present in wheat, oats, and barley. The gut wall becomes chronically inflamed, the villi are destroyed, and the gut's ability to absorb nutrients is compromised.

cell-adhesion molecules Cell-surface proteins of several different types that mediate the binding of one cell to other cells or to extracellular matrix proteins. Integrins, selectins, and members of the immunoglobulin gene superfamily (such as ICAM-1) are among the cell-adhesion molecules important in the operation of the immune system.

cell-mediated immune responses An adaptive immune response in which antigen-specific effector T cells have the main role. The immunity to infection conferred by such a response is called cell-mediated immunity. A primary cell-mediated immune response is the T-cell response that occurs the first time a particular antigen is encountered.

cellular hypersensitivity reactions A hypersensitivity reaction mediated largely by antigen-specific T lymphocytes.

cellular immunology The study of the cellular basis of immunity.

central lymphoid organs, central lymphoid tissues The sites of lymphocyte development; in humans, these are the bone marrow and thymus. B lymphocytes develop in bone marrow, whereas T lymphocytes develop within the thymus from bone marrow-derived progenitors. Also called the primary lymphoid organs.

central memory T cells (TCM) Memory lymphocytes that express CCR7 and recirculate between blood and secondary lymphoid tissues similarly to naive T cells. They require restimulation in secondary lymphoid tissues to become fully mature effector T cells.

central tolerance Immunological tolerance to self antigens that is established while lymphocytes are developing in central lymphoid organs. Cf. **peripheral tolerance**.

centroblasts Large, rapidly dividing activated B cells present in the dark zone of germinal centers in follicles of peripheral lymphoid organs.

centrocytes Small B cells that derive from centroblasts in the germinal centers of follicles in peripheral lymphoid organs; they populate the light zone of the germinal center.

cGAS (cyclic GAMP synthase) A cytosolic enzyme that is activated by double-stranded DNA to form cyclic guanosine monophosphate-adenosine monophoshate. *See* **cyclic dinucleotides (CDNs)**.

checkpoint blockade Approach to tumor therapy that attempts to interfere with the normal inhibitory signals that regulate lymphocytes.

Chediak–Higashi syndrome A defect in phagocytic cell function caused by a defect in a protein involved in intracellular vesicle fusion. Lysosomes fail to fuse properly with phagosomes, and killing of ingested bacteria is impaired.

chemokines Small chemoattractant protein that stimulates the migration and activation of cells, especially phagocytic cells and lymphocytes. Chemokines have a central role in inflammatory responses. Properties of individual chemokines are listed in Appendix IV.

chemotaxis Cellular movement occurring in response to chemical signals in the environment.

chimeric antigen receptor (CAR) Engineered fusion proteins composed of extracellular antigen-specific receptors (e.g., single-chain antibody) and intracellular signaling domains that activate and co-stimulate, expressed in T cells for use in cancer immunotherapy.

chronic allograft vasculopathy Chronic damage that can lead to late failure of transplanted organs. Arteriosclerosis of graft blood vessels leads to hypoperfusion of the graft and its eventual fibrosis and atrophy.

chronic granulomatous disease (CGD) An immunodeficiency in which multiple granulomas form as a result of defective elimination of bacteria by phagocytic cells. It is caused by defects in the NADPH oxidase system of enzymes that generate the superoxide radical involved in bacterial killing.

chronic infantile neurologic cutaneous and articular syndrome (CINCA) An autoinflammatory disease due to defects in the gene *NLRP3*, one of the components of the inflammasome.

chronic rejection Late failure of a transplanted organ, which can be due to immunological or nonimmunological causes.

CIIV An early endocytic compartment containing MHC class II molecules in dendritic cells.

class I cytokine receptors A group of receptors for the hematopoietin superfamily of cytokines. These include receptors using the common γ chain for IL-2, IL-4, IL-7, IL-15, and IL-21, and a common β chain for GM-CSF, IL-3, and IL-5.

class II-associated invariant chain peptide (CLIP) A peptide of variable length cleaved from the invariant chain (Ii) by proteases. It remains associated with the MHC class II molecule in an unstable form until it is removed by the HLA-DM protein.

class II cytokine receptors A group of heterodimeric receptors for a family of cytokines that includes interferon (IFN)-α, IFN-β, IFN-γ, and IL-10.

class switching, class switch recombination A somatic gene recombination process in activated B cells that replaces one heavy-chain constant region with one of a different isotype, switching the isotype of antibodies from IgM to the production of IgG, IgA, or IgE. This affects the antibody effector functions but not their antigen specificity. Also known as isotype switching. Cf. **somatic hypermutation**.

classes The class of an antibody is defined by the type of heavy chain it contains. There are five main antibody classes: IgA, IgD, IgM, IgG, and IgE, containing heavy chains α, δ, μ, γ, and ε, respectively. The IgG class has several subclasses. *See also* **isotypes**.

classical C3 convertase The complex of activated complement components C4b2a, which cleaves C3 to C3b on pathogen surfaces in the classical pathway of complement activation.

classical MHC class I genes MHC class I genes whose proteins function by presenting peptide antigens for recognition by T cells. Cf. **nonclassical MHC class Ib**.

classical monocyte The major form of monocyte in circulation capable of recruitment to sites of inflammation and differentiation into macrophages.

classical pathway The complement-activation pathway that is initiated by C1 binding either directly to bacterial surfaces or to antibody bound to the bacteria, thus flagging the bacteria as foreign. *See also* **alternative pathway, lectin pathway**.

classically activated macrophage   *See* **M1 macrophages**.

cleavage stimulation factor   A multi-subunit protein complex involved in the modification of the 3′ end of pre-messenger RNA for the addition of the polyadenine (polyA) tail.

clonal deletion   The elimination of immature lymphocytes when they bind to self antigens, which produces tolerance to self as required by the clonal selection theory of adaptive immunity. Clonal deletion is the main mechanism of central tolerance and can also occur in peripheral tolerance.

clonal expansion   The proliferation of antigen-specific lymphocytes in response to antigenic stimulation that precedes their differentiation into effector cells. It is an essential step in adaptive immunity, allowing rare antigen-specific cells to increase in number so that they can effectively combat the pathogen that elicited the response.

clonal selection theory   The central paradigm of adaptive immunity. It states that adaptive immune responses derive from individual antigen-specific lymphocytes that are self-tolerant. These specific lymphocytes proliferate in response to antigen and differentiate into antigen-specific effector cells that eliminate the eliciting pathogen, and into memory cells to sustain immunity. The theory was formulated by Macfarlane Burnet and in earlier forms by Niels Jerne and David Talmage.

clone   A population of cells all derived from the same progenitor cell.

clonotypic   Describes a feature unique to members of a clone. For example, the distribution of antigen receptors in the lymphocyte population is said to be clonotypic, as the cells of a given clone all have identical antigen receptors.

Clostridium difficile   Gram-positive anaerobic toxogenic spore-forming bacterium frequently associated with severe colitis following treatment with certain broad-spectrum antibiotics.

c-Maf   A transcription factor acting in the development of T_{FH} cells.

coagulation system   A collection of proteases and other proteins in the blood that trigger blood clotting when blood vessels are damaged.

coding joint   DNA join formed by the imprecise joining of a V gene segment to a (D)J gene segment during recombination of the immunoglobulin or T-cell receptor genes. It is the joint retained in the rearranged gene. Cf. **signal joint**.

codominant   Describes the situation in which the two alleles of a gene are expressed in roughly equal amounts in the heterozygote. Most genes show this property, including the highly polymorphic MHC genes.

collectins   A family of calcium-dependent sugar-binding proteins (lectins) containing collagen-like sequences. An example is mannose-binding lectin (MBL).

combinatorial diversity   The diversity among antigen receptors generated by combining separate units of genetic information, comprising two types. First, receptor gene segments are joined in many different combinations to generate diverse receptor chains; second, two different receptor chains (heavy and light in immunoglobulins; α and β, or γ and δ, in T-cell receptors) are combined to make the antigen-recognition site.

commensal microbiota, commensal microorganisms   Microorganisms (predominantly bacteria) that normally live harmlessly in symbiosis with their host (for example the gut bacteria in humans and other animals). Many commensals confer a positive benefit on their host in some way.

common β chain   A transmembrane polypeptide (CD131) that is a common subunit for receptor of the cytokines IL-3, IL-5, and GM-CSF.

common γ chain (γc)   A transmembrane polypeptide chain (CD132) that is common to a subgroup of cytokine receptors.

common lymphoid progenitor (CLP)   Stem cell that can give rise to all the types of lymphocytes with the exception of innate lymphoid cells (ILCs).

common mucosal immune system   The mucosal immune system as a whole, the name reflecting the fact that lymphocytes that have been primed in one part of the mucosal system can recirculate as effector cells to other parts of the mucosal system.

common myeloid progenitor (CMP)   Stem cells that can give rise to the myeloid cells of the immune system—macrophages, granulocytes, mast cells, and dendritic cells of the innate immune system. This stem cell also gives rise to megakaryocytes and red blood cells.

common variable immunodeficiencies (CVIDs)   A relatively common deficiency in antibody production in which only one or a few isotypes are affected. It can be due to a variety of genetic defects.

complement   A set of plasma proteins that act together as a defense against pathogens in extracellular spaces. The pathogen becomes coated with complement proteins that facilitate its removal by phagocytes and that can also kill certain pathogens directly. Activation of the complement system can be initiated in several different ways. *See* **classical pathway, alternative pathway, lectin pathway**.

complement activation   The activation of the normally inactive proteins of the complement system that occurs on infection. *See* **classical pathway, alternative pathway, lectin pathway**.

complement proteins   *See* **C1, C2, C3**, etc..

complement receptors (CRs)   Cell-surface proteins of various types that recognize and bind complement proteins that have become bound to an antigen such as a pathogen. Complement receptors on phagocytes enable them to identify and bind pathogens coated with complement proteins, and to ingest and destroy them. *See* **CR1, CR2, CR3, CR4, CRIg**, and the **C1 complex**.

complement regulatory proteins   Proteins that control complement activity and prevent complement from being activated on the surfaces of host cells.

complement system   A set of plasma proteins that act together as a defense against pathogens in extracellular spaces. The pathogen becomes coated with complement proteins that facilitate its removal by phagocytes and that can also kill certain pathogens directly. Activation of the complement system can be initiated in several different ways. *See* **classical pathway, alternative pathway, lectin pathway**.

complementarity-determining regions (CDRs)   Parts of the V domains of immunoglobulins and T-cell receptors that determine their antigen specificity and make contact with the specific ligand. The CDRs are the most variable part of antigen receptor, and contribute to the diversity of these proteins. There are three such regions (CDR1, CDR2, and CDR3) in each V domain.

conformational epitopes, discontinuous epitopes   Antigenic structure (epitope) on a protein antigen that is formed from several separate regions in the sequence of the protein brought together by protein folding. Antibodies that bind conformational epitopes bind only native folded proteins.

conjugate vaccines   Antibacterial vaccines made from bacterial capsular polysaccharides bound to proteins of known immunogenicity, such as tetanus toxoid.

constant Ig domains (C domains)   Type of protein domain that makes up the constant regions of each chain of an immunoglobulin molecule.

constant region, C region   That part of an immunoglobulin or a T-cell receptor that is relatively constant in amino acid sequence between different molecules. Also known as the Fc region in antibodies. The constant region of an antibody determines its particular effector function. Cf. **variable region**.

continuous epitope, linear epitope   Antigenic structure (epitope) in a protein that is formed by a single small region of amino acid sequence. Antibodies that bind continuous epitopes can bind to the denatured protein. The epitopes detected by T cells are continuous. Also called a linear epitope.

conventional (or classical) dendritic cells (cDCs)   The lineage of dendritic cells that mainly participates in antigen presentation to, and activation of, naive T cells. Cf. **plasmacytoid dendritic cells**.

co-receptors   Cell-surface protein that increases the sensitivity of a receptor to its ligand by binding to associated ligands and participating in signaling. The antigen receptors on T cells and B cells act in conjunction with co-receptors, which are either CD4 or CD8 on T cells, and a co-receptor complex of three proteins, one of which is the complement receptor CR2, on B cells.

cortex   The outer part of a tissue or organ; in lymph nodes it refers to the follicles, which are mainly populated by B cells.

corticosteroids   Family of drugs related to natural steroids such as cortisone. Corticosteroids can kill lymphocytes, especially developing thymocytes, inducing apoptotic cell death. They are medically useful anti-inflammatory and immunosuppressive agents.

co-stimulatory molecules   Cell-surface proteins on antigen-presenting cells that deliver co-stimulatory signals to naive T cells. Examples are the B7 molecules on dendritic cells, which are ligands for CD28 on naive T cells.

co-stimulatory receptors   Cell-surface receptors on naive lymphocytes through which the cells receive signals additional to those received through the antigen receptor, and which are necessary for the full activation of the lymphocyte. Examples are CD30 and CD40 on B cells, and CD27 and CD28 on T cells.

CR1 (CD35)   A receptor expressed by phagocytic cells that binds to C3b. It stimulates phagocytosis and inhibits C3 convertase formation on host-cell surfaces.

CR2 (CD21)   Complement receptor that is part of the B-cell co-receptor complex. It binds to antigens coated with breakdown products of C3b, especially C3dg, and, by cross-linking the B-cell receptor, enhances sensitivity to antigen at least 100-fold. It is also the receptor used by the Epstein–Barr virus to infect B cells.

CR3 (CD11b:CD18)   Complement receptor 3. A β2 integrin that acts both as an adhesion molecule and as a complement receptor. CR3 on phagocytes binds iC3b, a breakdown product of C3b on pathogen surfaces, and stimulates phagocytosis.

CR4 (CD11c:CD18)   A β2 integrin that acts both as an adhesion molecule and as a complement receptor. CR4 on phagocytes binds iC3b, a breakdown product of C3b on pathogen surfaces, and stimulates phagocytosis.

CRAC channel   Channels in the lymphocyte plasma membrane that open to let calcium flow into the cell during the response of the cell to antigen. Channel opening is induced by release of calcium from the endoplasmic reticulum.

C-reactive protein   An acute-phase protein that binds to phosphocholine, a constituent of the surface C-polysaccharide of the bacterium *Streptococcus pneumoniae* and of many other bacteria, thus opsonizing them for uptake by phagocytes.

CRIg (complement receptor of the immunoglobulin family)   A complement receptor that binds to inactivated forms of C3b.

Crohn's disease   Chronic inflammatory bowel disease thought to result from an abnormal overresponsiveness to the commensal gut microbiota.

cross-matching   A test used in blood typing and histocompatibility typing to determine whether donor and recipient have antibodies against each other's cells that might interfere with successful transfusion or grafting.

cross-presentation   The process by which extracellular proteins taken up by dendritic cells can give rise to peptides presented by MHC class I molecules. It enables antigens from extracellular sources to be presented by MHC class I molecules and activate CD8 T cells.

cross-priming   Activation of CD8 T cells by dendritic cells in which the antigenic peptide presented by MHC class I molecules is derived from an exogenous protein (i.e., by cross-presentation), rather than produced within the dendritic cells directly. Cf. **direct presentation**.

cryptdins   α-Defensins (antimicrobial peptides) made by the Paneth cells of the small intestine.

cryptic epitopes   Any epitope that cannot be recognized by a lymphocyte receptor until the antigen has been broken down and processed.

cryptopatches   Aggregates of lymphoid tissue in the gut wall that are thought to give rise to isolated lymphoid follicles.

CstF-64   Subunit of cleavage stimulation factor that favors polyadenylation at pAS leading to the secreted form of IgM.

C-terminal Src kinase (Csk)   A kinase that phosphorylates the C-terminal tyrosine of Src-family kinases in lymphocytes, thus inactivating them.

CTLA-4   A high-affinity inhibitory receptor on T cells for B7 molecules; its binding inhibits T-cell activation.

C-type lectins   Large class of carbohydrate-binding proteins that require Ca^{2+} for binding, including many that function in innate immunity.

cutaneous lymphocyte antigen (CLA)   A cell-surface molecule that is involved in lymphocyte homing to the skin in humans.

CVIDs   *See* **common variable immunodeficiencies**.

CX3CR1   Chemokine receptor expressed by monocytes, macrophages, NK cells, and activated T cells that binds CXCL1 (Fractalkine).

CXC chemokines   One of the two main classes of chemokines, distinguished by a Cys-X-Cys (CXC) motif near the amino terminus. They have names CXCL1, CXCL2, etc. *See* Appendix IV for a list of individual chemokines.

CXCL12 (SDF-1)   Chemokine produced by stromal cells in the dark zone of the germinal center that binds CXCR4 expressed by centroblasts.

CXCL13   Chemokine produced in the follicle and the light zone of the germinal center that binds CXCR5 expressed on circulating B cells and centrocytes.

CXCR5   A chemokine receptor expressed by circulating B cells and activated T cells that binds the chemokine CXCL13 and directs cell migration into the follicle.

cyclic dinucleotides (CDNs)   Cyclic dimers of guanylate and/or adenylate monophosphate that are produced by various bacteria as second messengers and detected by STING.

cyclic guanosine monophosphate-adenosine monophosphate (cyclic GMP-AMP or cGAMP)   *See* **cyclic dinucleotides (CDNs)**.

cyclic neutropenia   A dominantly inherited disease in which neutrophil numbers fluctuate from near normal to very low or absent, with an approximate cycle time of 21 days. This is in contrast to severe congenital neutropenia (SCN), in which the inherited defect results in persistently low neutrophil numbers.

cyclic reentry model   An explanation of the behavior of B cells in lymphoid follicles, proposing that activated B cells in germinal centers lose and gain expression of the chemokine receptor CXCR4 and thus move from the light zone to the dark zone and back again under the influence of the chemokine CXCL12.

cyclophilins   A family of prolylisomerases that affect protein folding, that also bind cyclosporin A to produce a complex that associates with calcineurin, preventing its activation by calmodulin.

cyclophosphamide   A DNA alkylating agent that is used as an immunosuppressive drug. It acts by killing rapidly dividing cells, including lymphocytes proliferating in response to antigen.

cyclosporin A (CsA)   A powerful noncytotoxic immunosuppressive drug that inhibits signaling from the T-cell receptor, preventing T-cell activation and effector function. It binds to cyclophilin, and the complex formed binds to and inactivates the phosphatase calcineurin.

cystic fibrosis   Disease caused by defect in *CFTR* gene, leading to abnormally thick mucus and causing serious recurrent infections of the lung.

cytidine deaminase activity (CDA)   An enzymatic activity exhibited by AID-APOBEC family proteins of agnathan species that may mediate rearrangement and assembly of complete VLR genes.

cytokines   Proteins made by a cell that affect the behavior of other cells, particularly immune cells. Cytokines made by lymphocytes are often called interleukins (abbreviated IL). Cytokines and their receptors are listed in Appendix III. Cf. **chemokines**.

cytomegalovirus UL16 protein   A nonessential glycoprotein of cytomegalovirus that is recognized by innate receptors expressed by NK cells.

cytosol   One of several major compartments within cells containing elements such as the cytoskeleton, and mitochondria, and separated by membranes from distinct compartments such as the nucleus and vesicular system.

cytotoxic T cells   T cells that can kill other cells, typically CD8 T cells defending against intracellular pathogens that live or reproduce in the cytosol, but in some cases also CD4 T cells.

daclizumab   Antibody to human CD25 used to block IL-2 receptor signaling in T cells for treatment of rejection in renal transplantation.

DAG   *See* **diacylglycerol**.

damage-associated molecular patterns (DAMPs)   *See* **pathogen-associated molecular patterns (PAMPs)**.

DAP10, DAP12   Signaling chains containing ITAMS that are associated with the tails of some activating receptors on NK cells.

dark zone   *See* **germinal center**.

DC-SIGN   A lectin on the dendritic-cell surface that binds ICAM-3 with high affinity.

DDX41 (DEAD box polypeptide 41)   A candidate DNA sensor of the RLR family that appears to signal through the STING pathway.

death effector domain (DED)   Protein-interaction domain originally discovered in proteins involved in programmed cell death or apoptosis. As part of the intracellular domains of some adaptor proteins, death domains are involved in transmitting pro-inflammatory and/or pro-apoptotic signals.

death-inducing signaling complex (DISC)   A multi-protein complex that is formed by signaling through members of the 'death receptor' family of apoptosis-inducing cellular receptors, such as Fas. It activates the caspase cascade to induce apoptosis.

decay-accelerating factor (DAF or CD55)   A cell-surface protein that protects cells from lysis by complement. Its absence causes the disease paroxysmal nocturnal hemoglobinuria.

Dectin-1   A phagocytic receptor on neutrophils and macrophages that recognizes β-1,3-linked glucans, which are common components of fungal cell walls.

defective ribosomal products (DRiPs)   Peptides translated from introns in improperly spliced mRNAs, translations of frameshifts, or improperly folded proteins, which are recognized and tagged by ubiquitin for degradation by the proteasome.

defensins   *See* **α-defensins**, **β-defensins**.

delayed-type hypersensitivity reactions   A form of cell-mediated immunity elicited by antigen in the skin stimulating sensitized Th1 CD4 lymphocytes and CD8 lymphocytes. It is called delayed-type hypersensitivity because the reaction appears hours to days after antigen is injected. Referred to as type IV hypersensitivity in the historic Gell and Coombs classification.

dendritic cells   Bone marrow-derived cells found in most tissues, including lymphoid tissues. There are two main functional subsets. Conventional dendritic cells take up antigen in peripheral tissues, are activated by contact with pathogens, and travel to the peripheral lymphoid organs, where they are the most potent stimulators of T-cell responses. Plasmacytoid dendritic cells can also take up and present antigen, but their main function in an infection is to produce large amounts of the antiviral interferons as a result of pathogen recognition through receptors such as TLRs. Both these types of dendritic cells are distinct from the follicular dendritic cell that presents antigen to B cells in lymphoid follicles.

dendritic epidermal T cells (dETCs)   A specialized class of γ:δ T cells found in the skin of mice and some other species, but not humans. They express $V_\gamma 5:V_\delta 1$ and may interact with ligands such as Skint-1 expressed by keratinocytes.

dephosphorylation   The removal of a phosphate group from a molecule, usually a protein.

depleting antibodies   Immunosuppressive monoclonal antibodies that trigger the destruction of lymphocytes *in vivo*. They are used for treating episodes of acute graft rejection.

diacyl and triacyl lipoproteins   Ligands for the Toll-like receptors TLR1:TLR2 and TLR2:TLR6.

diacylglycerol   A lipid intracellular signaling molecule formed from membrane inositol phospholipids that are cleaved by the action of phospholipase C-γ after the activation of many different receptors. The diacylglycerol stays in the membrane and activates protein kinase C and RasGRP, which further propagate the signal.

diapedesis   The movement of blood cells, particularly leukocytes, from the blood across blood vessel walls into tissues.

differentiation antigens   Referring to a category of genes with restricted expression patterns that can be targeted as antigens by immunotherapies in treatment of cancers.

DiGeorge syndrome   Recessive genetic immunodeficiency disease in which there is a failure to develop thymic epithelium. Parathyroid glands are also absent and there are anomalies in the large blood vessels.

direct allorecognition   Host recognition of a grafted tissue that involves donor antigen-presenting cells leaving the graft, migrating via the lymph to regional lymph nodes, and activating host T cells bearing the corresponding T-cell receptors.

direct presentation   The process by which proteins produced within a given cell give rise to peptides presented by MHC class I molecules. This may refer to antigen-presenting cells, such as dendritic cells, or to nonimmune cells that will become the targets of CTLs.

dislocation   In reference to viral defense mechanisms, the degradation of newly synthesized MHC class I molecules by viral proteins.

disseminated intravascular coagulation (DIC)   Blood clotting occurring simultaneously in small vessels throughout the body in response to disseminated TNF-α, which leads to the massive consumption of clotting proteins, so that the patient's blood cannot clot appropriately. Seen in septic shock.

diversion colitis　Inflammation and necrosis of intestinal enterocytes following surgical diversion of normal flow of fecal contents due to impaired metabolism resulting from loss of short-chain fatty acids derived from microbiota.

diversity gene segment (D$_H$)　Short DNA sequences that form a join between the V and J gene segments in rearranged immunoglobulin heavy-chain genes and in T-cell receptor β- and δ-chain genes. *See* **gene segments**.

DN1, DN2, DN3, DN4　Substages in the development of CD4+CD8+ double-positive T cells in the thymus. Rearrangement of the TCRβ-chain locus starts at DN2 and is completed by DN4.

DNA-dependent protein kinase (DNA-PK)　Protein kinase in the DNA repair pathway involved in the rearrangement of immunoglobulin and T-cell receptor genes.

DNA ligase IV　Enzyme responsible for joining the DNA ends to produce the coding joint during V(D)J recombination.

DNA transposons　Genetic elements encoding their own transposase that can insert themselves into and excise themselves from the DNA genomes of a host.

DNA vaccination　Vaccination by introduction into skin and muscle of DNA encoding the desired antigen; the expressed protein can then elicit antibody and T-cell responses.

donor lymphocyte infusion (DLI)　Transfer of mature lymphocytes (i.e., T cells) from donor into patients during bone marrow transplantation for cancer treatment to help eliminate residual tumor.

double-negative thymocytes　Immature T cells in the thymus that lack expression of the two co-receptors CD4 and CD8 and represent the progenitors to the remaining T cells developing in the thymus. In a normal thymus, these represent about 5% of thymocytes.

double-positive thymocytes　Immature T cells in the thymus that are characterized by expression of both the CD4 and the CD8 co-receptor proteins. They represent the majority (about 80%) of thymocytes and are the progenitors to the mature CD4 and CD8 T cells.

double-strand break repair (DSBR)　A nonhomologous end joining pathway of DNA repair used in the completion of isotype switching.

double-stranded RNA (dsRNA)　A chemical structure that is a replicative intermediate of many viruses that is recognized by TLR-3.

Down syndrome cell adhesion molecule (Dscam)　*See* **Dscam**.

DR4, DR5　Members of the TNFR superfamily expressed by many cell types that can be activated by the TRAIL to induce apoptosis.

draining lymph nodes　A lymph node downstream of a site of infection that receives antigens and microbes from the site via the lymphatic system. Draining lymph nodes often enlarge enormously during an immune response and can be palpated; they were originally called swollen glands.

Dscam　A member of the immunoglobulin superfamily that in insects is thought to opsonize invading bacteria and aid their engulfment by phagocytes. It can be made in a multiplicity of different forms as a result of alternative splicing.

dysbiosis　Altered balance of microbial species comprising the microbiota resulting from a variety of causes (e.g., antibiotics, genetic disorders) and frequently associated with outgrowth of pathogenic organisms such as *Clostridium difficile*.

dysregulated self　Refers to changes that take place in infected or malignant cells that alter expression of various surface receptors that can be detected by the innate immune system.

E3 ligase　An enzymatic activity that directs the transfer of a ubiquitin molecule from an E2 ubiquitin-conjugating enzyme onto a specific protein target.

early-onset sarcoidosis　Disease associated with activating NOD2 mutations characterized by inflammation in tissues such as liver.

early pro-B cell　*See* **pro-B cells**.

EBI2 (GPR183)　A chemokine receptor that binds oxysterols and regulates B-cell movement to the outer follicular and interfollicular regions during early phases of B-cell activation in lymphoid tissues.

E-cadherin　Integrin expressed by epithelial cells important in forming the adherens junctions between adjacent cells.

edema　Swelling caused by the entry of fluid and cells from the blood into the tissues; it is one of the cardinal features of inflammation.

effector caspases　Intracellular proteases that are activated as a result of an apoptotic signal and mediate the cellular changes associated with apoptosis. To be distinguished from initiator caspases, which act upstream of effector caspases to initiate the caspase cascade.

effector CD4 T cells　The subset of differentiated effector T cells carrying the CD4 co-receptor molecule, which includes the T$_H$1, T$_H$2, T$_H$17, and regulatory T cells.

effector lymphocytes　The cells that differentiate from naive lymphocytes after initial activation by antigen and can then mediate the removal of pathogens from the body without further differentiation. They are distinct from memory lymphocytes, which must undergo further differentiation to become effector lymphocytes.

effector mechanisms　Those processes by which pathogens are destroyed and cleared from the body. Innate and adaptive immune responses use most of the same effector mechanisms to eliminate pathogens.

effector memory T cells (TEM)　Memory lymphocytes that recirculate between blood and peripheral tissues and are specialized for rapid maturation into effector T cells after restimulation with antigen in non-lymphoid tissues.

effector modules　This term refers to a set of immune mechanisms, either cell-mediated and humoral, innate or adaptive, that act together in the elimination of a particular category of pathogen.

effector T lymphocytes　The T cells that perform the functions of an immune response, such as cell killing and cell activation, that clear the infectious agent from the body. There are several different subsets, each with a specific role in an immune response.

electrostatic interactions　Chemical interaction occurring between charged atoms, as in the charged amino acid side chains and an ion in a salt bridge.

elimination phase　Stage of anti-tumor immune response that detects and eliminates cancer cells, also called immune surveillance.

elite controllers　A subset of HIV-infected long-term non-progressors who have clinically undetectable levels of virus without antiretroviral therapy.

ELL2　A transcription elongation factor that favors the polyadenylation at pA$_S$ leading to the secreted form of IgM.

endocrine　Describes the action of a biologically active molecule such as a hormone or cytokine that is secreted by one tissue into the blood and acts on a distant tissue. Cf. **autocrine, paracrine**.

endogenous pyrogens　Cytokines that can induce a rise in body temperature.

endoplasmic reticulum aminopeptidase associated with antigen processing (ERAAP)   Enzyme in the endoplasmic reticulum that trims polypeptides to a size at which they can bind to MHC class I molecules.

endoplasmic reticulum-associated protein degradation (ERAD)   A system of enzymes in the endoplasmic reticulum that recognizes incompletely or misfolded proteins and assures their eventual degradation.

endosteum   The region in bone marrow adjacent to the inner surface of the bone; hematopoietic stem cells are initially located there.

endothelial activation   The changes that occur in the endothelial walls of small blood vessels as a result of inflammation, such as increased permeability and the increased production of cell-adhesion molecules and cytokines.

endothelial cell   Cell type that forms the endothelium, the epithelium of a blood vessel wall.

endothelial protein C receptor (EPCR)   A nonclassical MHC class I protein induced on endothelial cells that can interact with the blood coagulation factor XIV (protein C) and can be recognized by some $\gamma{:}\delta$ T cells.

endothelium   The epithelium that forms the walls of blood capillaries and the lining of larger blood vessels.

endotoxins   Toxins derived from bacterial cell walls released by damaged cells. They can potently induce cytokine synthesis and in large amounts can cause a systemic reaction called septic shock or endotoxic shock.

enteroadherent *Escherichia coli*   Referring to multiple strains of *E. coli* capable of attachment to, and infection and destruction of cells of the intestinal microvilli, causing colitis and diarrheagenic diseases.

eomesodermin   A transcription factor involved in development and function of certain types of NK cells, ILCs, and CD8 T cells.

eosinophilia   An abnormally large number of eosinophils in the blood.

eosinophils   A type of white blood cell containing granules that stain with eosin. It is thought to be important chiefly in defense against parasitic infections, but is also medically important as an effector cell in allergic reactions.

eotaxins   CC chemokines that act predominantly on eosinophils, including CCL11 (eotaxin 1), CCL24 (eotaxin 2), and CCL26 (eotaxin 3).

epitope   A site on an antigen recognized by an antibody or an antigen receptor. T-cell epitopes are short peptide bound to MHC molecules. B-cell epitopes are typically structural motifs on the surface of the antigen. Also called an antigenic determinant.

epitope spreading   Increase in diversity of responses to autoantigens as the response persists, as a result of responses being made to epitopes other than the original one.

equilibrium phase   Stage of anti-tumor immune response when immunoediting allows the immune response to continuously shape the antigenic character of cancer cells.

Erk   Extracellular signal-related kinase, a protein kinase that is the MAPK for one module of the T-cell receptor signaling pathway. Erk also functions in other receptors in other cell types.

ERp57   A chaperone protein involved in loading peptide onto MHC class I molecules in the endoplasmic reticulum.

error-prone 'translesion' DNA polymerases   A DNA polymerase operates during DNA repair, such as Polη which can repair a basic lesion by incorporating untemplated nucleotides into the newly formed DNA strand.

escape mutants   Mutants of pathogens that are changed in such a way that they can evade the immune response against the original pathogen.

escape phase   Final stage of anti-tumor immune response when immunoediting has removed the expression of antigenic targets such that the cancer cells are no loner detected by the immune system.

E-selectin   *See* **selectins**.

etanercept   Fc fusion protein containing the p75 subunit of the TNF receptor that neutralizes TNF-α used for treatment of rheumatoid arthritis and other inflammatory diseases.

eukaryotic initiation factor 2 (eIF2α)   Subunit of eukaryotic initiation factor that helps form the preinitiation complex that begins protein translation from mRNA. When it is phosphorylated by PKR, protein translation is suppressed.

eukaryotic initiation factor 3 (eIF3)   Multisubunit complex that acts in formation of the 43S preinitiation complex. It can bind interferon-induced transmembrane (IFIT) proteins which thereby suppress translation of viral proteins.

exogenous pyrogen   Any substance originating outside the body that can induce fever, such as the bacterial lipopolysaccharide LPS. Cf. **endogenous pyrogens**.

exotoxins   A protein toxin produced and secreted by a bacterium.

experimental autoimmune encephalomyelitis (EAE)   An inflammatory disease of the central nervous system that develops after mice are immunized with neural antigens in a strong adjuvant.

extrachromosomal DNA   DNA not contained within chromosomes, such as the circular DNA produced by V(D)J recombination occurring between RSSs in the same chromosomal orientation and is eventually lost from the cell.

extravasation   The movement of cells or fluid from within blood vessels into the surrounding tissues.

extrinsic pathway of apoptosis   A pathway triggered by extracellular ligands binding to specific cell-surface receptors (death receptors) that signal the cell to undergo programmed cell death (apoptosis).

Fab fragment   Antibody fragment composed of a single antigen-binding arm of an antibody without the Fc region, produced by cleavage of IgG by the enzyme papain. It contains the complete light chain plus the amino-terminal variable region and first constant region of the heavy chain, held together by an interchain disulfide bond.

F(ab')$_2$ fragment   Antibody fragment composed of two linked antigen-binding arms (Fab fragments) without the Fc regions, produced by cleavage of IgG with the enzyme pepsin.

factor B   Protein in the alternative pathway of complement activation, in which it is cleaved to Ba and an active protease, Bb, the latter binding to C3b to form the alternative pathway C3 convertase, C3bBb.

factor D   A serine protease in the alternative pathway of complement activation, which cleaves factor B into Ba and Bb.

factor H   Complement-regulatory protein in plasma that binds C3b and competes with factor B to displace Bb from the convertase.

factor H binding protein (fHbp)   A protein produced by the pathogen *Neisseria meningitidis* that recruits factor H to its membrane, thereby inactivating C3b deposited on its surface, and evading destruction by complement.

factor I   Complement-regulatory protease in plasma that cleaves C3b to the inactive derivative iC3b, thus preventing the formation of a C3 convertase.

factor I deficiency   A genetically determined lack of the complement-regulatory protein factor I. This results in uncontrolled complement activation, so that complement proteins rapidly become depleted. Those with the deficiency suffer repeated bacterial infections, especially with ubiquitous pyogenic bacteria.

factor P Plasma protein released by activated neutrophils that stabilizes the C3 convertase C3bBb of the alternative pathway.

familial cold autoinflammatory syndrome (FCAS) An episodic autoinflammatory disease caused by mutations in the gene *NLRP3*, encoding NLRP3, a member of the NOD-like receptor family and a component of the inflammasome. The symptoms are induced by exposure to cold.

familial hemophagocytic lymphohistiocytosis (FHL) A family of progressive and potentially lethal inflammatory diseases caused by an inherited deficiency of one of several proteins involved in the formation or release of cytolytic granules. Large numbers of polyclonal CD8-positive T cells accumulate in lymphoid and other organs, and this is associated with activated macrophages that phagocytose blood cells, including erythrocytes and leukocytes.

familial Mediterranean fever (FMF) A severe autoinflammatory disease, inherited as an autosomal recessive disorder. It is caused by mutation in the gene (*MEFV*) that encodes the protein pyrin, which is expressed in granulocytes and monocytes. In patients with this disorder, defective pyrin is thought to spontaneously activate inflammasomes.

farmer's lung A hypersensitivity disease caused by the interaction of IgG antibodies with large amounts of an inhaled antigen in the alveolar wall of the lung, causing alveolar wall inflammation and compromising respiratory gas exchange.

Fc fragment, Fc region The carboxy-terminal halves of the two heavy chains of an IgG molecule disulfide-bonded to each other by the residual hinge region. It is produced by cleavage of IgG by papain. In the complete antibody this portion is often called the Fc region.

Fc receptors Family of cell-surface receptors that bind the Fc portions of different immunoglobulins: Fcγ receptors bind IgG, for example, and Fcε receptors bind IgE.

FCAS *See* **familial cold autoinflammatory syndrome**.

FcεRI The high affinity receptor for the Fc region of IgE. Expressed primarily on the surface of mast cells and basophils. When multivalent antigen interacts with IgE that is bound to FcεRI and cross-links nearby receptors, it causes activation of the receptor-bearing cell.

FcγR1 (CD64) Fc receptor highly expressed by monocytes and macrophages that has the highest affinity of the Fc receptors for IgG.

FcγRIIB-1 An inhibitory receptor on B cells that recognizes the Fc portion of IgG antibodies. FcγRIIB-1 contains an ITIM motif in its cytoplasmic tail.

FcγRIII Cell-surface receptors that bind the Fc portion of IgG molecules. Most Fcγ receptors bind only aggregated IgG, allowing them to discriminate bound antibody from free IgG. Expressed variously on phagocytes, B lymphocytes, NK cells, and follicular dendritic cells, the Fcγ receptors have a key role in humoral immunity, linking antibody binding to effector cell functions.

FcRn (neonatal Fc receptor) Neonatal Fc receptor, a receptor that transports IgG from mother to fetus across the placenta, and across other epithelia such as the epithelium of the gut.

FHL *See* **familial hemophagocytic lymphohistiocytosis**.

fibrinogen-related proteins (FREPs) Members of the immunoglobulin superfamily that are thought to have a role in innate immunity in the freshwater snail *Biomphalaria glabrata*.

ficolins Carbohydrate-binding proteins that can initiate the lectin pathway of complement activation. They are members of the collectin family and bind to the *N*-acetylglucosamine present on the surface of some pathogens.

fingolimod Small-molecule immunosuppressive drug that interferes with the actions of sphingosine, leading to retention of effector T cells in lymphoid organs.

FK506 *See* **tacrolimus**.

FK-binding proteins (FKBPs) Group of prolyl isomerases related to the cyclophilins and bind the immunosuppressive drug FK506 (tacrolimus).

flagellin A protein that is the major constituent of the flagellum, the tail-like structure used in bacterial locomotion. TLR-5 recognizes intact flagellin protein that has dissociated from the flagellum.

fluid-phase C3 convertase Short-lived alternative pathway C3 convertase, $C3(H_2O)Bb$, that is continually produced at a low level in the plasma that can initiate activation of the alternative pathway of complement.

fMet-Leu-Phe (fMLF) receptor A pattern recognition receptor for the peptide fMet-Leu-Phe, which is specific to bacteria, on neutrophils and macrophages. fMet-Leu-Phe acts as a chemoattractant.

folic acid A B vitamin, derivatives of folic acid produced by various bacteria can be bound by the nonclassical MHC class Ib protein MR1 for recognition by MAIT cells.

follicle-associated epithelium Specialized epithelium separating the lymphoid tissues of the gut wall from the intestinal lumen. As well as enterocytes it contains microfold cells, through which antigens enter the lymphoid organs from the gut.

follicles An area of predominantly B cells in a peripheral lymphoid organ, such as a lymph node, which also contains follicular dendritic cells.

follicular B cells The majority population of long-lived recirculating conventional B cells found in the blood, the spleen, and the lymph nodes. Also known as B-2 B cells.

follicular dendritic cell (FDC) A cell type of uncertain origin in B-cell follicles of peripheral lymphoid organs that captures antigen:antibody complexes using non-internalized Fc receptors and presents them to B cells for internalization and processing during the germinal center reaction.

follicular helper T cell (T_{FH}) Type of effector CD4 T cell that resides in lymphoid follicles and provides help to B cells for antibody production.

framework regions Relatively invariant regions that provide a protein scaffold for the hypervariable regions in the V domains of immunoglobulins and T-cell receptors.

Freund's complete adjuvant Emulsion of oil and water containing killed mycobacteria used to enhance immune responses to experimental antigens.

fungi A kingdom of single-celled and multicellular eukaryotic organisms, including the yeasts and molds, that can cause a variety of diseases. Immunity to fungi is complex and involves both humoral and cell-mediated responses.

Fyn *See* **Src-family tyrosine kinases**.

γ:δ T cells Subset of T lymphocytes bearing a T-cell receptor composed of the antigen-recognition chains, γ and δ, assembled in a γ:δ heterodimer.

γ:δ T-cell receptors Antigen receptor carried by a subset of T lymphocytes that is distinct from the α:β T-cell receptor. It is composed of a γ and a δ chain, which are produced from genes that undergo gene rearrangement.

γ-glutamyl diaminopimelic acid (iE-DAP) A product of degradation of the peptidoglycan of Gram-negative bacteria. It is sensed by NOD1.

GAP *See* **GTPase-activating proteins**.

GEFs *See* **guanine nucleotide exchange factors**.

gene rearrangement The process of somatic recombination of gene segments in the immunoglobulin and T-cell receptor genetic loci to produce a functional gene. This process generates the diversity found in immunoglobulin and T-cell receptor variable regions.

gene segments   Sets of short DNA sequences at the immunoglobulin and T-cell receptor loci that encode different regions of the variable domains of antigen receptors. Gene segments of each type are joined together by somatic recombination to form a complete variable-domain exon. There are three types of gene segments: V gene segments encode the first 95 amino acids, D gene segments (in heavy-chain and TCRα chain loci only) encode about 5 amino acids, and J gene segments encode the last 10–15 amino acids of the variable domain. There are multiple copies of each type of gene segment in the germline DNA, but only one of each type is joined together to form the variable domain.

genetic locus   The site of a gene on a chromosome. In the case of the genes for the immunoglobulin and T-cell receptor chains, the term locus refers to the complete collection of gene segments and C-region genes for the given chain.

genome-wide association studies (GWASs)   Genetic association studies in the general population that look for a correlation between disease frequency and variant alleles by scanning the genomes of many people for the presence of informative single-nucleotide polymorphisms (SNPs).

germ-free mice   Mice that are raised in the complete absence of intestinal and other microorganisms. Such mice have very depleted immune systems, but they can respond virtually normally to any specific antigen, provided it is mixed with a strong adjuvant.

germinal center   Sites of intense B-cell proliferation and differentiation that develop in lymphoid follicles during an adaptive immune response. Somatic hypermutation and class switching occur in germinal centers.

germline theory   An excluded hypothesis that antibody diversity was encoded by a separate germline gene for each antibody, known not to be true for most vertebrates, although cartilaginous fishes do have some rearranged V regions in the germline.

glycosylphosphatidylinositol (GPI) tail   A glycolipid modification of proteins that can allow attachment to host membranes without the requirement of a transmembrane protein domain.

gnathostomes   The class of jawed vertebrates comprising most fish and all mammals. These possess an adaptive immunity based on the RAG-mediated V(D)J recombination.

gnotobiotic mice   *See* **germ-free mice**.

goblet cells   Specialized epithelial cells located in many sites throughout the body responsible for mucus production; important in protection of the epithelium.

Goodpasture's syndrome   An autoimmune disease in which autoantibodies against type IV collagen (found in basement membranes) are produced, causing extensive inflammation in kidneys and lungs.

gout   Disease caused by monosodium urate crystals deposited in the cartilaginous tissues of joints, causing inflammation. Urate crystals activate the NLRP3 inflammasome, which induces inflammatory cytokines.

G proteins   Intracellular GTPases that act as molecular switches in signaling pathways. They bind GTP to induce their active conformation, which is lost when GTO is hydrolyzed to GDP. There are two kinds of G proteins: the heterotrimeric (α, β, γ subunits) receptor-associated G proteins, and the small G proteins, such as Ras and Raf, which act downstream of many transmembrane signaling events.

G-protein-coupled receptors (GPCRs)   A large class of seven-span transmembrane cell-surface receptors that associate with intracellular heterotrimeric G proteins after ligand binding, and signal by activation of the G protein. Important examples are the chemokine receptors.

G-quadruplex   A structure formed from G-rich regions of DNA in which four guanine bases form a planar hydrogen-bonded network, or guanine tetrad, that can further stack on other guanine tetrads. G-quadruplexes processed from intronic switch region RNA may target AID back to the switch regions during isotype switching.

graft rejection   *See* **allograft rejection**.

graft-versus-host disease (GVHD)   An attack on the tissues of the recipient by mature T cells in a bone marrow graft from a nonidentical donor, which can cause a variety of symptoms; sometimes these are severe.

graft-versus-leukemia effect   A beneficial side-effect of bone marrow grafts given to treat leukemia, in which mature T cells in the graft recognize minor histocompatibility antigens or tumor-specific antigens on the recipient's leukemic cells and attack them.

Gram-negative bacteria   Bacteria that fail to retain crystal violet stain following alcohol wash due to a thin peptidoglycan layer.

Gram-negative binding proteins (GNBPs)   Proteins that act as the pathogen-recognition proteins in the Toll pathway of immune defense in *Drosophila*.

granulocyte-macrophage stimulating factor (GM-CSF)   A cytokine involved in the growth and differentiation of cells of the myeloid lineage, including dendritic cells, monocytes and tissue macrophages, and granulocytes.

granulocytes   White blood cells with multilobed nuclei and cytoplasmic granules. They comprise the neutrophils, eosinophils, and basophils. Also known as polymorphonuclear leukocytes.

granuloma   A site of chronic inflammation usually triggered by persistent infectious agents such as mycobacteria or by a nondegradable foreign body. Granulomas have a central area of macrophages, often fused into multinucleate giant cells, surrounded by T lymphocytes.

Grass   A serine protease of *Drosophila* that functions downstream of peptidoglycan-recognition proteins (PGRPs) and Gram-negative binding proteins (GNBPs) to initiate the proteolytic cascade leading to Toll activation.

Graves' disease   An autoimmune disease in which antibodies against the thyroid-stimulating hormone receptor cause overproduction of thyroid hormone and thus hyperthyroidism.

Griscelli syndrome   An inherited immunodeficiency disease that affects the pathway for secretion of lysosomes. It is caused by mutations in a small GTPase Rab27a, which controls the movement of vesicles within cells.

group 1 ILCs (ILC1s)   The subtype of innate lymphoid cells (ILCs) characterized by IFN-γ production.

GTPase-activating proteins (GAP)   Regulatory proteins that accelerate the intrinsic GTPase activity of G proteins and thus facilitate the conversion of G proteins from the active (GTP-bound) state to the inactive (GDP-bound) state.

guanine nucleotide exchange factors (GEFs)   Proteins that can remove the bound GDP from G proteins, thus allowing GTP to bind and activate the G protein.

gut-associated lymphoid tissues (GALT)   Lymphoid tissues associated with the gastrointestinal tract, comprising Peyer's patches, the appendix, and isolated lymphoid follicles found in the intestinal wall, where adaptive immune responses are initiated, and by lymphatics to mesenteric lymph nodes.

GVHD   *See* **graft-versus-host disease**.

H-2 locus, H-2 genes   The major histocompatibility complex of the mouse. Haplotypes are designated by a lower-case superscript, as in H-2^b.

H-2DM *See* **HLA-DM**.

H-2O *See* **HLA-DO**.

H2-M3 A nonclassical MHC class Ib protein in mice that can bind and present peptides having an *N*-formylated amino terminus for recognition by CD8 T cells.

H5N1 avian flu A highly pathogenic influenza subtype responsible for 'bird flu'.

haploinsufficient Describes the situation in which the presence of only one normal allele of a gene is not sufficient for normal function.

hapten carrier effect Antibody production against a small chemical group, the hapten, following its attachment to a carrier protein for which an immune response has been generated.

haptens Any small molecule that can be recognized by a specific antibody but cannot by itself elicit an immune response. A hapten must be chemically linked to a protein molecule to elicit antibody and T-cell responses.

Hashimoto's thyroiditis An autoimmune disease characterized by persistent high levels of antibody against thyroid-specific antigens. These antibodies recruit NK cells to the thyroid, leading to damage and inflammation.

heavy chain, H chain One of the two types of protein chain in an immunoglobulin molecule, the other being called the light chain. There are several different classes, or isotypes, of heavy chain ($\alpha, \delta, \varepsilon, \gamma$, and μ), each of which confers a distinctive functional activity on the antibody molecule. Each immunoglobulin molecule contains two identical heavy chains.

heavy-chain-only IgGs (hcIgGs) Antibodies produced by some camelid species composed of heavy-chain dimers without an associated light chain that retain antigen binding capacity.

heavy-chain variable region (V_H) Referring to the V region of the heavy chain of an immunoglobulin.

helicard *See* **MDA-5**.

helper CD4 T cells, helper T cells Effector CD4 T cells that stimulate or 'help' B cells to make antibody in response to antigenic challenge. T_H2, T_H1, and the T_{FH} subsets of effector CD4 T cells can perform this function.

hemagglutinin (HA) Substances that can cause hemagglutination, such as human antibodies that recognize the ABO blood group antigens on red blood cells, or the influenza virus hemagglutinin, a glycoprotein that functions in viral fusion with endosome membranes.

hematopoietic stem cells (HSCs) Type of pluripotent cell in the bone marrow that can give rise to all the different blood cell types.

hematopoietin superfamily Large family of structurally related cytokines that includes growth factors and many interleukins with roles in both adaptive and innate immunity.

hemochromatosis protein A protein expressed by intestinal epithelial cells that regulates iron uptake and transport by interacting with the transferrin receptor to decrease its affinity for iron-loaded transferrin.

hemolytic disease of the newborn A severe form of Rh hemolytic disease in which maternal anti-Rh antibody enters the fetus and produces a hemolytic anemia so severe that the fetus has mainly immature erythroblasts in the peripheral blood.

hemophagocytic lymphohistiocytic (HLH) syndrome A dysregulated expansion of CD8-positive lymphocytes that is associated with macrophage activation. The activated macrophages phagocytose blood cells, including erythrocytes and leukocytes.

hepatobiliary route Route whereby mucosally produced dimeric IgA enters the portal veins in the lamina propria, is transported to the liver, and reaches the bile duct by transcytosis. This pathway is not of great significance in humans.

heptamer The conserved seven-nucleotide DNA sequence in the recombination signal sequences (RSSs) flanking gene segments in the immunoglobulin and T-cell receptor loci.

HER-2/neu A receptor tyrosine kinase overexpressed in many cancers, particularly breast cancer, that is the target of trastuzumab (Herceptin) used in its treatment.

herd immunity Protection conferred to unvaccinated individuals in a population produced by vaccination of others and reduction in the natural reservoir for infection.

hereditary angioedema (HAE) A genetic deficiency of the C1 inhibitor of the complement system. In the absence of C1 inhibitor, spontaneous activation of the complement system can cause diffuse fluid leakage from blood vessels, the most serious consequence of which is swelling of the larynx, leading to suffocation.

hereditary hemochromatosis A disease caused by defects in the *HFE* gene characterized by abnormally high retention of iron in the liver and other organs.

herpes virus entry molecule (HVEM) *See* **B and T lymphocyte attenuator**.

heterosubtypic immunity Immune protection against a pathogen conferred by infection with a distinct strain, typically with reference to different influenza A serotypes.

heterotrimeric G proteins *See* **G proteins**.

heterozygous Describes individuals that have two different alleles of a given gene, one inherited from the mother and one from the father.

HFE *See* **hemochromatosis protein**.

high endothelial cells, high endothelial venules (HEV) Specialized small venous blood vessels in lymphoid tissues. Lymphocytes migrate from the blood into lymphoid tissues by attaching to the high endothelial cells in the walls of the venules and squeezing between them.

highly active antiretroviral therapy (HAART) A combination of drugs that is used to control HIV infection. It comprises nucleoside analogs that prevent reverse transcription, and drugs that inhibit the viral protease.

hinge region The flexible domain that joins the Fab arms to the Fc piece in an immunoglobulin. The flexibility of the hinge region in IgG and IgA molecules allows the Fab arms to adopt a wide range of angles, permitting binding to epitopes spaced variable distances apart.

HIP/PAP An antimicrobial C-type lectin secreted by intestinal cells in humans. Also known as RegIIIα.

histamine A vasoactive amine stored in mast-cell granules. Histamine released by antigen binding to IgE antibodies bound to mast cells causes the dilation of local blood vessels and the contraction of smooth muscle, producing some of the symptoms of IgE-mediated allergic reactions. Antihistamines are drugs that counter histamine action.

histatins Antimicrobial peptides constitutively produced by the parotid, sublingual, and submandibular glands in the oral cavity. Active against pathogenic fungi such as *Cryptococcus neoformans* and *Candida albicans*.

HIV *See* **human immunodeficiency virus**.

HLA   The genetic designation for the human MHC. Individual loci are designated by upper-case letters, as in HLA-A, and alleles are designated by numbers, as in HLA-A*0201.

HLA-DM   An invariant MHC protein resembling MHC class II in humans that is involved in loading peptides onto MHC class II molecules. A homologous protein in mice is called H-2M, or sometimes H2-DM.

HLA-DO   An invariant MHC class II molecule that binds HLA-DM, inhibiting the release of CLIP from MHC class II molecules in intracellular vesicles. A homologous protein in mice is called H-2O or H2-DO.

homeostatic chemokines   Chemokines that are produced at steady-state to direct the localization of immune cells to lymphoid tissues.

homing   The direction of a lymphocyte into a particular tissue.

homing receptors   Receptors on lymphocytes for chemokines, cytokines, and adhesion molecules specific to particular tissues, and which enable the lymphocyte to enter that tissue.

homozygous   Describes individuals that have two identical alleles of a given gene, inherited separately from each parent.

host-versus-graft disease (HVGD)   Another name for the allograft rejection reaction. The term is used mainly in relation to bone marrow transplantation when immune cells of the host recognize and destroy transplanted bone marrow or hematopoietic stem cells (HSCs).

human immunodeficiency virus (HIV)   The causative agent of the acquired immune deficiency syndrome (AIDS). HIV is a retrovirus of the lentivirus family that selectively infects macrophages and CD4 T cells, leading to their slow depletion, which eventually results in immunodeficiency. There are two major strains of the virus, HIV-1 and HIV-2, of which HIV-1 causes most disease worldwide. HIV-2 is endemic to West Africa but is spreading.

human leukocyte antigen (HLA)   *See* **HLA**.

humanization   The genetic engineering of mouse hypervariable loops of a desired specificity into otherwise human antibodies for use as therapeutic agents. Such antibodies are less likely to cause an immune response in people treated with them than are wholly mouse antibodies.

humoral   Referring to effector proteins in the blood or body fluids, such as antibodies in adaptive immunity, or complement proteins in innate immunity.

humoral immunity, humoral immune response   Immunity due to proteins circulating in the blood, such as antibodies (in adaptive immunity) or complement (in innate immunity). Adaptive humoral immunity can be transferred to unimmunized recipients by the transfer of serum containing specific antibody.

HVGD   *See* **host-versus-graft disease**.

hydrophobic interaction   Chemical interaction occurring between nearby hydrophobic moieties typically excluding water molecules.

21-hydroxylase   An enzyme of non-immune function but encoded in the MHC locus required for normal cortisol synthesis by the adrenal gland.

3-hydroxy-3-methylglutaryl-co-enzyme A (HMG-CoA) reductase   Rate-limiting enzyme in the production of cholesterol and a target of cholesterol-lowering drugs such as the statins.

hygiene hypothesis   A hypothesis first proposed in 1989 that reduced exposure to ubiquitous environmental microorganisms was a cause of the increased frequency of patients with allergies observed over the course of the mid- to late-20th century.

hyper IgE syndrome (HIES)   Also called Job's syndrome. A disease characterized by recurrent skin and pulmonary infections and high serum concentrations of IgE.

hyper IgM syndrome   A group of genetic diseases in which there is overproduction of IgM antibody, among other symptoms. They are due to defects in various genes for proteins involved in class switching such as CD40 ligand and the enzyme AID. *See* **activation-induced cytidine deaminase, CD40 ligand deficiency**.

hyper IgM type 2 immunodeficiency   *See* **activation-induced cytidine deaminase**.

hyperacute graft rejection   Immediate rejection reaction caused by preformed natural antibodies that react against antigens on the transplanted organ. The antibodies bind to endothelium and trigger the blood-clotting cascade, leading to an engorged, ischemic graft and rapid death of the organ.

hypereosinophilic syndrome   Disease associated with an overproduction of eosinophils.

hypervariable regions   *See* **complementarity-determining regions**.

hypomorphic mutations   Applied to mutations that result in reduced gene function.

IκB   A cytoplasmic protein that constitutively associates with the NFκB homodimer, composed of p50 and p65 subunits. When IκB is phosphorylated by activated IKK (IκB kinase), IκB becomes degraded and allows the NFκB dimer to be released as an active transcription factor.

IκB kinase (IKK)   *See* **IKK**.

iC3b   Inactive complement fragment produced by cleavage of C3b.

ICAMs   ICAM-1, ICAM-2, ICAM-3. Cell-adhesion molecules of the immunoglobulin superfamily that bind to the leukocyte integrin CD11a:CD18 (LFA-1). They are crucial in the binding of lymphocytes and other leukocytes to antigen-presenting cells and endothelial cells.

ICOS (inducible co-stimulatory)   A CD28-related co-stimulatory receptor that is induced on activated T cells and can enhance T-cell responses. It binds a co-stimulatory ligand known as ICOSL (ICOS ligand), which is distinct from the B7 molecules.

ICOSL   *See* **ICOS**.

IFI16 (IFN-γ-inducible protein 16)   A member of the PYHIN subfamily of NLR (NOD-like receptor) family containing an N-terminal HIN domain. It activates the STING pathway in response to double-stranded DNA.

IFIT (IFN-induced protein with tetratricoid repeats)   A small family of host proteins induced by interferons that regulate protein translation during infection in part by interactions with eIF3.

IFITM (interferon-induced transmembrane protein)   A small family of host transmembrane proteins induced by interferons that function in the cell's vesicular compartment to restrain various steps in viral replication.

IFN-α, IFN-β   Antiviral cytokines produced by a wide variety of cells in response to infection by a virus, and which also help healthy cells resist viral infection. They act through the same receptor, which signals through a Janus-family tyrosine kinase. Also known as the type I interferons.

IFN-γ   A cytokine of the interferon structural family produced by effector CD4 T_H1 cells, CD8 T cells, and NK cells. Its primary function is the activation of macrophages, and it acts through a different receptor from that of the type I interferons.

IFN-γ-induced lysosomal thiol reductase (GILT)   An enzyme present in the endosomal compartment of many antigen-presenting cells that denatures disulfide bonds to facilitate the degradation and processing of proteins.

IFN-λ   Also called type III interferons, this family includes IL-28A, IL-28B, and IL-29, which bind a common receptor expressed by a limited set of epithelial tissues.

IFN-λ receptor Receptor composed of a unique IL-28Rα subunit and the β subunit of the IL-10 receptor that recognizes IL-28A, IL-28B, and IL-29.

Igα, Igβ *See* **B-cell receptor**.

IgA Immunoglobulin class composed of α heavy chains that can occur in a monomeric and a polymeric (mainly dimeric) form. Polymeric IgA is the main antibody secreted by mucosal lymphoid tissues.

IgA deficiency The class of immunoglobulin characterized by α heavy chains. It is the most common type of immunodeficiency. It can occur in a monomeric and a polymeric (mainly dimeric) form. Polymeric IgA is the main antibody secreted by mucosal lymphoid tissues.

IgD Immunoglobulin class composed of δ heavy chains that appears as surface immunoglobulin on mature B cells.

IgE Immunoglobulin class composed of ε heavy chains that acts in defense against parasite infections and in allergic reactions.

IgG Immunoglobulin class composed of γ heavy chains that is the most abundant class of immunoglobulin in the plasma.

IgM Immunoglobulin class composed of μ heavy chains that is the first to appear on B cells and the first to be secreted.

IgNAR *See* **immunoglobulin new antigen receptor**.

IgW Type of heavy-chain isotype present in cartilaginous fishes composed of six immunoglobulin domains.

IKK The IκB kinase, IKK, is a multisubunit protein complex composed of IKKα, IKKβ, and IKKγ (or NEMO).

IKKε A kinase that interacts with TBK1 (TANK-binding kinase 1) in the phosphorylation of IRF3 downstream of TLR-3 signaling.

IL-1 family One of four major families of cytokines, this family contains 11 cytokines that are structurally similar to IL-1α, and are largely pro-inflammatory in function.

IL-1β A cytokine produced by active macrophages that has many effects in the immune response, including the activation of vascular endothelium, activation of lymphocytes, and the induction of fever.

IL-6 Interleukin-6, a cytokine produced by activated macrophages and which has many effects, including lymphocyte activation, the stimulation of antibody production, and the induction of fever.

IL-7 receptor (IL-7Rα) *See* **CD127**.

IL-21 A cytokine produced by T cells (e.g., T$_{FH}$ cells) that activates STAT3 and promotes survival and proliferation, particularly germinal center B cells.

ILC1 A subset of innate lymphoid cells characterized by production of IFN-γ.

ILCs (innate lymphoid cells) These are a class of innate immune cells having overlapping characteristics with T cells but lacking an antigen receptor. They arise in several groups, ILC1, ILC2, ILC3, and NK cells, which exhibit properties roughly similar to T$_H$1, T$_H$2, T$_H$17, and CD8 T cells.

Imd (immunodeficiency) signaling pathway A defense against Gram-negative bacteria in insects that results in the production of antimicrobial peptides such as diptericin, attacin, and cecropin.

imiquimod Drug (aldara) approved for treatment of basal cell carcinoma, genital warts, and actinic keratoses known to activate TLR-7, although not approved as an adjuvant for vaccines.

immature B cells B cells that have rearranged a heavy- and a light-chain V-region gene and express surface IgM, but have not yet matured sufficiently to express surface IgD as well.

immediate hypersensitivity reactions Allergic reactions that occur within seconds to minutes of encounter with antigen, caused largely by activation of mast cells or basophils.

immune complexes Complexes formed by the binding of antibody to its cognate antigen. Activated complement proteins, especially C3b, are often bound in immune complexes. Large immune complexes form when sufficient antibody is available to cross-link multivalent antigen; these are cleared by cells of the reticuloendothelial system that bear Fc receptors and complement receptors. Small, soluble immune complexes form when antigen is in excess; these can be deposited in small blood vessels and damage them.

immune evasion Mechanisms used by pathogens to avoid detection and/ or elimination by host immune defenses.

immune modulation The deliberate attempt to change the course of an immune response, for example by altering the bias toward T$_H$1 or T$_H$2 dominance.

immune surveillance The recognition, and in some cases the elimination, of tumor cells by the immune system before they become clinically detectable.

immune system The tissues, cells, and molecules involved in innate immunity and adaptive immunity.

immunodeficiency diseases Any inherited or acquired disorder in which some aspect or aspects of host defense are absent or functionally defective.

immunodominant Describes epitopes in an antigen that are preferentially recognized by T cells, such that T cells specific for those epitopes come to dominate the immune response.

immunoevasins Viral proteins that prevent the appearance of peptide:MHC class I complexes on the infected cell, thus preventing the recognition of virus-infected cells by cytotoxic T cells.

immunogenic Any molecule that, on its own, is able to elicit an adaptive immune response on injection into a person or animal.

immunoglobulin (Ig) The protein family to which antibodies and B-cell receptors belong.

immunoglobulin A (IgA) *See* **IgA**.

immunoglobulin D (IgD) *See* **IgD**.

immunoglobulin domain Protein domain first described in antibody molecules but present in many proteins.

immunoglobulin E (IgE) *See* **IgE**.

immunoglobulin fold The tertiary structure of an immunoglobulin domain, comprising a sandwich of two β sheets held together by a disulfide bond.

immunoglobulin G (IgG) *See* **IgG**.

immunoglobulin-like domain (Ig-like domain) Protein domain structurally related to the immunoglobulin domain.

immunoglobulin-like proteins Proteins containing one or more immunoglobulin-like domains, which are protein domains structurally similar to those of immunoglobulins.

immunoglobulin M (IgM) *See* **IgM**.

immunoglobulin new antigen receptor (IgNAR) A form of heavy-chain-only Ig molecule made by shark species.

immunoglobulin repertoire The variety of antigen-specific immunoglobulins (antibodies and B-cell receptors) present in an individual. Also known as the antibody repertoire.

immunoglobulin superfamily   Large family of proteins with at least one Ig or Ig-like domain, many of which are involved in antigen recognition and cell–cell interaction in the immune system and other biological systems.

immunological ignorance   A form of self-tolerance in which reactive lymphocytes and their target antigen are both detectable within an individual, yet no autoimmune attack occurs.

immunological memory   The ability of the immune system to respond more rapidly and more effectively on a second encounter with an antigen. Immunological memory is specific for a particular antigen and is long-lived.

immunological synapse   The highly organized interface that develops between a T cell and the target cell it is in contact with, formed by T-cell receptors binding to antigen and cell-adhesion molecules binding to their counterparts on the two cells. Also known as the supramolecular adhesion complex.

immunological tolerance   See **tolerance**.

immunologically privileged sites   Certain sites in the body, such as the brain, that do not mount an immune response against tissue allografts. Immunological privilege can be due both to physical barriers to cell and antigen migration and to the presence of immunosuppressive cytokines.

immunology   The study of all aspects of host defense against infection and also of the adverse consequences of immune responses.

immunomodulatory therapy   Treatments that seek to modify an immune response in a beneficial way, for example to reduce or prevent an autoimmune or allergic response.

immunophilins   See **cyclophilins, FK-binding proteins**.

immunoproteasome   A form of proteasome found in cells exposed to interferons. It contains three subunits that are different from the normal proteasome.

immunoreceptor tyrosine-based activation motif (ITAM)   Sequence motifs in the signaling chains of receptors, such as antigen receptors on lymphocytes, that are the site of tyrosine phosphorylation after receptor activation, leading to recruitment of other signaling proteins.

immunoreceptor tyrosine-based inhibition motif (ITIM)   Sequence motifs in the signaling chains of inhibitory receptors that are sites of tyrosine phosphorylation, leading to inhibitory signaling, such as through recruitment of phosphatases that remove phosphate groups added by tyrosine kinases.

immunoreceptor tyrosine-based switch motif (ITSM)   A sequence motif present in the cytoplasmic tails of some inhibitor receptors.

immunotoxin   Antibodies that are chemically coupled to toxic proteins usually derived from plants or microbes. The antibody targets the toxin moiety to the required cells.

indirect allorecognition   Recognition of a grafted tissue that involves the uptake of allogeneic proteins by the recipient's antigen-presenting cells and their presentation to T cells by self MHC molecules.

indoleamine 2,3-dioxygenase (IDO)   Enzyme expressed by immune cells and some tumors that catabolizes tryptophan into kynurenine metabolites that can have immunosuppressive functions.

induced pluripotent stem cells (iPS cells)   Pluripotent stem cells that are derived from adult somatic cells by the introduction of a cocktail of transcription factors.

infectious mononucleosis   The common form of infection with the Epstein–Barr virus. It consists of fever, malaise, and swollen lymph nodes. Also called glandular fever.

inflammasome   A pro-inflammatory protein complex that is formed after stimulation of the intracellular NOD-like receptors. Production of an active caspase in the complex processes cytokine proproteins into active cytokines.

inflammation   General term for the local accumulation of fluid, plasma proteins, and white blood cells that is initiated by physical injury, infection, or a local immune response.

inflammatory bowel disease (IBD)   General name for a set of inflammatory conditions in the gut, such as Crohn's disease and colitis, that have an immunological component.

inflammatory cells   Cells such as macrophages, neutrophils, and effector T_H1 lymphocytes that invade inflamed tissues and contribute to the inflammation.

inflammatory chemokines   Chemokines that are produced in response to infection or injury to direct the localization of immune cells to sites of inflammation.

inflammatory inducers   Chemical structures that indicate the presence of invading microbes or cellular damage, such as bacterial lipopolysaccharides, extracellular ATP, or urate crystals.

inflammatory mediators   Chemicals such as cytokines produced by immune cells that act on target cells to promote defense against microbes.

inflammatory monocytes   An activated form of monocyte producing a variety of pro-inflammatory cytokines.

inflammatory response   See **inflammation**.

infliximab   Chimeric antibody to TNF-α used in the treatment of inflammatory diseases, such as Crohn's disease and rheumatoid arthritis.

inherited immunodeficiency diseases   See **primary immunodeficiencies**.

inhibitory receptors   On NK cells, receptors whose stimulation results in suppression of the cell's cytotoxic activity.

initiator caspases   Proteases that promote apoptosis by cleaving and activating other caspases.

iNKT   See **invariant NKT cells**.

innate immunity   The various innate resistance mechanisms that are encountered first by a pathogen, before adaptive immunity is induced, such as anatomical barriers, antimicrobial peptides, the complement system, and macrophages and neutrophils carrying nonspecific pathogen-recognition receptors. Innate immunity is present in all individuals at all times, does not increase with repeated exposure to a given pathogen, and discriminates between groups of similar pathogens, rather than responding to a particular pathogen. Cf. **adaptive immunity**.

innate lymphoid cells (ILCs)   See **ILCs**.

innate recognition receptors   General term for a large group of proteins that recognize many different inflammatory inducers and that are encoded in the germline and do not need gene rearrangement in somatic cells to be expressed.

inositol 1,4,5-trisphosphate (IP_3)   A soluble second messenger produced by the cleavage of membrane inositol phospholipids by phospholipase C-γ. It acts on receptors in the endoplasmic reticulum membrane, resulting in the release of stored Ca^{2+} into the cytosol.

integrin   Heterodimeric cell-surface proteins involved in cell–cell and cell–matrix interactions. They are important in adhesive interactions between lymphocytes and antigen-presenting cells and in lymphocyte and leukocyte adherence to blood vessel walls and migration into tissues.

integrin α_4:β_7   Integrin binding to VCAM-1, MAdCAM-1, and fibronectin and expressed by various cells, such as IELs, that traffic to intestinal lamina propria.

intercellular adhesion molecules (ICAMs)   See **ICAMs**.

interdigitating dendritic cells   *See* **dendritic cells**.

interferon regulatory factor (IRF)   A family of nine transcription factors that regulate a variety of immune responses. For example, IRF3 and IRF7 are activated as a result of signaling from some TLRs. Several IRFs promote expression of the genes for type I interferons.

interferon stimulated genes (ISGs)   A category of gene induced by interferons, which include many that promote innate defense against pathogens, such as oligoadenylate synthetase, PKR, and the Mx, IFITs, and IFITM proteins.

interferon-α receptor (IFNAR)   This receptor recognizes IFN-α and IFN-β to activate STAT1 and STAT2 and induce expression of many ISGs.

interferon-induced transmembrane protein (IFITM)   *See* **IFITM**.

interferon-producing cells (IPCs)   *See* **plasmacytoid dendritic cells**.

interferons (IFNs)   Several related families of cytokines originally named for their interference of viral replication. IFN-α and IFN-β are antiviral in their effects; IFN-γ has other roles in the immune system.

intergenic control regions   Sites in non-coding regions of genes that control their expression and rearrangement by interactions with transcription factors and chromatin-modifying proteins.

interleukin (IL)   A generic name for cytokines produced by leukocytes. The more general term cytokine is used in this book, but the term interleukin is used in the naming of specific cytokines such as IL-2. Some key interleukins are listed in the glossary under their abbreviated names, for example IL-1β and IL-2. Cytokines are listed in Appendix III.

intraepithelial lymphocytes (IELs)   Lymphocytes present in the epithelium of mucosal surfaces such as the gut. They are predominantly T cells, and in the gut are predominantly CD8 T cells.

intrathymic dendritic cells   *See* **dendritic cells**.

intrinsic pathway of apoptosis   Signaling pathway that mediates apoptosis in response to noxious stimuli including UV irradiation, chemotherapeutic drugs, starvation, or lack of the growth factors required for survival. It is initiated by mitochondrial damage. Also called the mitochondrial pathway of apoptosis.

invariant chain (Ii, CD74)   A polypeptide that binds in the peptide-binding cleft of newly synthesized MHC class II proteins in the endoplasmic reticulum and blocks other peptides from binding there. It is degraded in the endosome, allowing for loading of antigenic peptides there.

invariant NKT cells (iNKT cells)   A type of innate-like lymphocyte that carries a T-cell receptor with an invariant α chain and a β chain of limited diversity that recognizes glycolipid antigens presented by CD1 MHC class Ib molecules. This cell type also carries the surface marker NK1.1, which is usually associated with NK cells.

IPEX (immune dysregulation, polyendocrinopathy, enteropathy, X-linked)   Immune dysregulation, polyendocrinopathy, enteropathy, X-linked syndrome. A very rare inherited condition in which CD4 CD25 regulatory T cells are lacking as a result of a mutation in the gene for the transcription factor FoxP3, leading to the development of autoimmunity.

ipilimumab   Antibody to human CTLA-4 used to treat melanoma, and first checkpoint blockade immunotherapy.

Ir (immune response) genes   An archaic term for genetic polymorphisms controlling the intensity of the immune response to a particular antigen, now known to result from allelic differences in MHC molecules, especially MHC class II molecules, that influence binding of particular peptides.

IRAK1, IRAK4   Protein kinases that are part of the intracellular signaling pathways leading from TLRs.

IRAK4 deficiency   An immunodeficiency characterized by recurrent bacterial infections, caused by inactivating mutations in the *IRAK4* gene that result in a block in TLR signaling.

IRF9   A member of the IRF family of transcription factors that interacts with activated STAT1 and STAT2 to form the complex called ISGF3, which induces transcription of many ISGs.

IRGM3   A protein that functions in the maintenance and storage of neutral lipid droplets in many types of cells in association with adipose differentiation related protein.

irradiation-sensitive SCID (IR-SCID)   A type of severe combined immunodeficiency due to mutations in DNA repair proteins, such as Artemis, that causes abnormal sensitivity to ionizing radiation and defects in V(D)J recombination.

ISGF3   *See* **IRF9**.

isoforms   Different forms of the same protein, for example the different forms encoded by different alleles of the same gene.

isolated lymphoid follicles (ILF)   A type of organized lymphoid tissue in the gut wall that is composed mainly of B cells.

isolation membrane   *See* **phagophore**.

isotype   The designation of an immunoglobulin chain in respect of the type of constant region it has. Light chains can be of either κ or λ isotype. Heavy chains can be of μ, δ, γ, α, or ε isotype. The different heavy-chain isotypes have different effector functions and determine the class and functional properties of antibodies (IgM, IgD, IgG, IgA, and IgE, respectively).

isotype switching   *See* **class switching**.

isotypic exclusion   Describes the use of one or other of the light-chain isotypes, κ or λ, by a given B cell or antibody.

JAK inhibitors (Jakinibs)   Small molecule kinase inhibitors with relative selectivity for one or more of the JAK kinases.

Janus kinase (JAK) family   Enzymes of the JAK–STAT intracellular signaling pathways that link many cytokine receptors with gene transcription in the nucleus. The kinases phosphorylate STAT proteins in the cytosol, which then move to the nucleus and activate a variety of genes.

J chain   Small polypeptide chain made by B cells that attaches to polymeric immunoglobulins IgM and IgA by disulfide bonds, and is essential for formation of the binding site for the polymeric immunoglobulin receptor.

JNK   *See* **Jun kinase**.

Job's syndrome   *See* **hyper IgE syndrome**.

joining gene segment, J gene segment   Short DNA sequences that encode the J regions of immunoglobulin and T-cell receptor variable domains. In a rearranged light-chain, TCRα, or TCRγ genes, the J gene segment is joined to a V gene segment. In a rearranged heavy-chain, TCRβ, or TCRδ locus, a J gene segment is joined to a D gene segment.

Jun kinase   A protein kinase that phosphorylates the transcription factor c-Jun, enabling it to bind to c-Fos to form the AP-1 transcription factor.

junctional diversity   The variability in sequence present in antigen-specific receptors that is created during the process of joining V, D, and J gene segments and which is due to imprecise joining and insertion of nontemplated nucleotides at the joins between gene segments.

κ chain   One of the two classes or isotypes of immunoglobulin light chains.

K63-linkages   In polyubiquitin chains, the covalent ligation of lysine 63 amino group of one ubiquitin protein with the carboxy terminus of a second ubiquitin. This type of linkage is most associated with activation of signaling by formation of a scaffold recognized by signaling adaptors such as TAB1/2.

killer cell immunoglobulin-like receptors (KIRs) Large family of receptors present on NK cells, through which the cells' cytotoxic activity is controlled. The family contains both activating and inhibitory receptors.

killer cell lectin-like receptors (KLRs) Large family of receptors present on NK cells, through which the cells' cytotoxic activity is controlled. The family contains both activating and inhibitory receptors.

kinase suppressor of Ras A scaffold protein in the Raf–MEK1–Erk MAP-kinase cascade that binds to all three members following antigen receptor signaling to facilitate their interactions and to accelerate the signaling cascade.

kinin system An enzymatic cascade of plasma proteins that is triggered by tissue damage to produce several inflammatory mediators, including the vasoactive peptide bradykinin.

Kostmann's disease A form of severe congenital neutropenia, an inherited condition in which the neutrophil count is low. In Kostmann's disease, this is due to a deficiency of the mitochondrial protein HAX1, which leads to apoptosis of developing myeloid cells and persistent neutropenia.

KSR   *See* **kinase suppressor of Ras**.

Ku A DNA repair protein required for immunoglobulin and T-cell receptor gene rearrangement.

Kupffer cells Phagocytes lining the hepatic sinusoids; they remove debris and dying cells from the blood, but are not known to elicit immune responses.

kynurenine metabolites Various compounds derived from tryptophan through the actions of the enzymes indolamine-2,3-dioxygenase (IDO) or tryptophan-2,3-dioxygenase (TDO) expressed in various immune cells or the liver.

λ chain One of the two classes or isotypes of immunoglobulin light chains.

λ5   *See* **surrogate light chain**.

L-selectin Adhesion molecule of the selectin family found on lymphocytes. L-selectin binds to CD34 and GlyCAM-1 on high endothelial venules to initiate the migration of naive lymphocytes into lymphoid tissue.

lamellar bodies Lipid-rich secretory organelles in keratinocytes and lung pneumocytes that release β-defensins into the extracellular space.

lamina propria A layer of connective tissue underlying a mucosal epithelium. It contains lymphocytes and other immune-system cells.

large pre-B cell Stage of B-cell development immediately after the pro-B cell, in which the cell expresses the pre-B-cell receptor and undergoes several rounds of division.

LAT   *See* **linker for activation of T cells**.

late-phase reaction Allergic reactions that occurs several hours after initial encounter with an antigen. Thought to be manifestations of recruitment of multiple leukocyte subsets to the site of allergen exposure.

late pro-B cell Stage in B-cell development in which VH to DJH joining occurs.

latency A state in which a virus infects a cell but does not replicate.

Lck An Src-family tyrosine kinase that associates with the cytoplasmic tails of CD4 and CD8 and phosphorylates the cytoplasmic tails of the T-cell receptor signaling chains, thus helping to activate signaling from the T-cell receptor complex once antigen has bound.

lectin A carbohydrate-binding protein.

lectin pathway Complement activation pathway that is triggered by mannose-binding lectins (MBLs) or ficolins bound to bacteria.

lentiviruses A group of retroviruses that include the human immunodeficiency virus, HIV-1. They cause disease after a long incubation period.

lethal factor An endopeptidase produced by *Bacillus anthracis* that cleaves NLRP1, inducing cell death within the infected cell, typically a macrophage.

leucine-rich repeat (LRR) Protein motifs that are repeated in series to form, for example, the extracellular portions of Toll-like receptors.

leukocyte A white blood cell. Leukocytes include lymphocytes, polymorphonuclear leukocytes, and monocytes.

leukocyte adhesion deficiencies (LADs) A class of immunodeficiency diseases in which the ability of leukocytes to enter sites infected by extracellular pathogens is affected, imparing elimination of infection. There are several different causes, including a deficiency of the common β chain of the leukocyte integrins.

leukocyte adhesion deficiency type 2 Disease causes by defects in the production of sulfated sialyl-Lewis^x that prevent neutrophils from interacting with P- and E-selectin, eliminating their ability to migrate properly to sites of infection.

leukocyte functional antigens (LFAs) Cell-adhesion molecules on leukocytes that were initially defined using monoclonal antibodies. LFA-1 is a β2 integrin; LFA-2 (now usually called CD2) is a member of the immunoglobulin superfamily, as is LFA-3 (now called CD58). LFA-1 is particularly important in T-cell adhesion to endothelial cells and antigen-presenting cells.

leukocyte receptor complex (LRC) A large cluster of immunoglobulin-like receptor genes that includes the killer cell immunoglobulin-like receptor (KIR) genes.

leukocytosis The presence of increased numbers of leukocytes in the blood. It is commonly seen in acute infection.

leukotrienes Lipid mediators of inflammation that are derived from arachidonic acid. They are produced by macrophages and other cells.

LFA-1   *See* **leukocyte functional antigens**.

LGP2 A member of the RLR family, it cooperates with RIG-I and MDA-5 in the recognition of viral RNA.

licensing The activation of a dendritic cell so that it is able to present antigen to naive T cells and activate them.

light chain, L chain The smaller of the two types of polypeptide chains that make up an immunoglobulin molecule. It consists of one V and one C domain, and is disulfide-bonded to the heavy chain. There are two classes, or isotypes, of light chain, known as κ and λ, which are produced from separate genetic loci.

light-chain variable region (V$_L$) Referring to the V region of the light chain of an immunoglobulin.

light zone   *See* **germinal center**.

lingual tonsils Paired masses of organized peripheral lymphoid tissue situated at the base of the tongue, in which adaptive immune responses can be initiated. They are part of the mucosal immune system. *See also* **palatine tonsils**.

linked recognition The rule that for a helper T cell to be able to activate a B cell, the epitopes recognized by the B cell and the helper T cell have to be derived from the same antigen (that is, they must originally have been physically linked).

linker for activation of T cells A cytoplasmic adaptor protein with several tyrosines that become phosphorylated by the tyrosine kinase ZAP-70. It helps to coordinate downstream signaling events in T-cell activation.

LIP10   A cleaved fragment of invariant chain retaining the transmembrane segments that remains bound to MHC class II proteins and helps target the complex to the endosome.

LIP22   The initial cleaved fragment of invariant chain bound to MHC class II molecules.

lipid bodies   Storage organelles rich in neutral lipids within the cytoplasm.

lipocalin-2   An antimicrobial peptide produced in abundance by neutrophils and mucosal epithelial cells that inhibits bacterial and fungal growth by limiting availability of iron.

lipopeptide antigens   A diverse set of antigens derived from microbial lipids typically presented by nonclassical MHC class Ib molecules such as CD1 molecules to invariant T-cell populations, including iNKT cells.

lipopolysaccharide (LPS)   The surface lipopolysaccharide of Gram-negative bacteria, which stimulates TLR-4 on macrophages and dendritic cells.

lipoteichoic acids   Components of bacterial cell walls that are recognized by Toll-like receptors.

long-term non-progressors   HIV-infected individuals who mount an immune response that controls viral loads such that they do not progress to AIDS despite the absence of antiretroviral therapy. *See also* **elite controllers**.

LPS-binding protein   Protein in blood and extracellular fluid that binds bacterial lipopolysaccharide (LPS) shed from bacteria.

Ly49 receptors   A family of C-type lectins expressed by mouse, but not human, NK cells. These can be either activating or inhibitory in function.

Ly49a   *See* **Ly49 receptors**.

Ly49H   *See* **Ly49 receptors**.

Ly108   *See* **SLAM**.

lymph   The extracellular fluid that accumulates in tissues and is drained by lymphatic vessels that carry it through the lymphatic system to the thoracic duct, which returns it to the blood.

lymph nodes   A type of peripheral lymphoid organ present in many locations throughout the body where lymphatic vessels converge.

lymphatic system   The system of lymph-carrying vessels and peripheral lymphoid tissues through which extracellular fluid from tissues passes before it is returned to the blood via the thoracic duct.

lymphatic vessels, lymphatics   Thin-walled vessels that carry lymph.

lymphoblast   A lymphocyte that has enlarged after activation and has increased its rate of RNA and protein synthesis, but is not yet fully differentiated.

lymphocyte   A class of white blood cells that bear variable cell-surface receptors for antigen and are responsible for adaptive immune responses. There are two main types—B lymphocytes (B cells) and T lymphocytes (T cells)—which mediate humoral and cell-mediated immunity, respectively. On antigen recognition, a lymphocyte enlarges to form a lymphoblast and then proliferates and differentiates into an antigen-specific effector cell.

lymphocyte receptor repertoire   All the highly variable antigen receptors carried by B and T lymphocytes.

lymphoid   Describes tissues composed mainly of lymphocytes.

lymphoid organs   Organized tissues characterized by very large numbers of lymphocytes interacting with a nonlymphoid stroma. The central, or primary, lymphoid organs, where lymphocytes are generated, are the thymus and bone marrow. The main peripheral, or secondary, lymphoid organs, in which adaptive immune responses are initiated, are the lymph nodes, spleen, and mucosa-associated lymphoid organs such as tonsils and Peyer's patches.

lymphoid tissue   Tissue composed of large numbers of lymphocytes.

lymphoid tissue inducer (LTi) cells   Cells of the blood lineage, which arise in the fetal liver and are carried in the blood to sites where they will form lymph nodes and other peripheral lymphoid organs.

lymphopenia   Abnormally low levels of lymphocytes in the blood.

lymphopoiesis   The differentiation of lymphoid cells from a common lymphoid progenitor.

lymphotoxins (LTs)   Cytokines of the tumor necrosis factor (TNF) family that are directly cytotoxic for some cells. They occur as trimers of LT-α chains (LT-α3) and heterotrimers of LT-α and LT-β chains (LT-α2:β1).

lysogenic phase   The phase of the viral life cycle in which the virus genome integrates into the host cell genome but remains dormant, employing mechanisms to avoid destroying its cellular host.

lysozyme   Antimicrobial enzyme that degrades bacterial cell walls.

lytic phase, productive phase   The phase of the viral life cycle in which there is active viral replication followed by destruction of the infected host cell as the virus escapes to infect new target cells.

M1 macrophages   The name sometimes given to 'classically' activated macrophages, which develop in the context of type 1 responses and have pro-inflammatory properties.

M2 macrophages   The name sometimes given to 'alternatively' activated macrophages, which develop in in the context of type 2 responses (e.g., parasite infection) and promote tissue remodeling and repair.

macroautophagy   The engulfment by a cell of large quantities of its own cytoplasm, which is then delivered to the lysosomes for degradation.

macrophages   Large mononuclear phagocytic cells present in most tissues that have many functions, such as scavenger cells, pathogen-recognition cells, production of pro-inflammatory cytokines. Macrophages arise both embryonically and from bone marrow precursors throughout life.

macropinocytosis   A process in which large amounts of extracellular fluid are taken up into an intracellular vesicle. This is one way in which dendritic cells can take up a wide variety of antigens from their surroundings.

MAdCAM-1   Mucosal cell-adhesion molecule-1. A mucosal addressin that is recognized by the lymphocyte surface proteins L-selectin and VLA-4, enabling the specific homing of lymphocytes to mucosal tissues.

MAIT cells   *See* **mucosal associated invariant T cells**.

major basic protein   Protein released by activated eosinophils that acts on mast cells and basophils to cause their degranulation.

major histocompatibility complex (MHC)   A cluster of genes on human chromosome 6 that encodes a set of membrane glycoproteins called the MHC molecules. The MHC also encodes proteins involved in antigen processing and other aspects of host defense. The genes for the MHC molecules are the most polymorphic in the human genome, having large numbers of alleles at the various loci.

MAL   An adaptor protein that associates with MyD88 in signaling by TLR-2/1, TLR-2/6, and TLR-4.

mannose-binding lectin (MBL)   Mannose-binding protein present in the blood. It can opsonize pathogens bearing mannose on their surfaces and can activate the complement system via the lectin pathway, an important part of innate immunity.

mannose receptor (MR)   A receptor on macrophages that is specific for mannose-containing carbohydrates that occur on the surfaces of pathogens but not on host cells.

mantle zone   A rim of B lymphocytes that surrounds lymphoid follicles.

Mantoux test A screening test for tuberculosis in which a sterile-filtered glycerol extract of *Mycobacterium tuberculosis* bacilli (Tb) is injected intradermally and the result is read 48–72 hours later. Induration, firm swelling caused by infiltration into the skin of inflammatory cells, can indicate previous exposure to Tb, either prior vaccination or current infection of *M. tuberculosis*. Generally, induration at the site of injection greater than 10 mm in diameter indicates the need for additional tests to assess whether infection with Tb is present.

MAP kinase (MAPK) *See* **mitogen-activated protein kinase**.

MARCO (macrophage receptor with a collagenous structure) *See* **scavenger receptor**).

marginal sinus A blood-filled vascular network that branches from the central arteriole and demarcates each area of white pulp in the spleen.

marginal zone Area of lymphoid tissue lying at the border of the white pulp in the spleen.

marginal zone B cells A unique population of B cells found in the spleen marginal zones; they do not circulate and are distinguished from conventional B cells by a distinct set of surface proteins.

MASP-1, MASP-2, MASP-3 Serine proteases of the classical and lectin pathway of complement activation that bind to C1q, ficolins, and mannose-binding lectin, and function in their activation to cleave C4.

mast cells A large granule-rich cell found in connective tissues throughout the body, most abundantly in the submucosal tissues and the dermis. The granules store bioactive molecules including the vasoactive amine histamine, which are released on mast-cell activation. Mast cells are thought to be involved in defenses against parasites and they have a crucial role in allergic reactions.

mastocytosis The overproduction of mast cells.

mature B cell B cell that expresses IgM and IgD on its surface and has gained the ability to respond to antigen.

MAVS (mitochondrial antiviral signaling protein) A CARD-containing adaptor protein attached to the outer mitochondrial membrane that signals downstream of RIG-I and MDA-5 to activate IRF3 and NFκB in response to viral infection.

MBL-associated serine proteases *See* **MASP-1, MASP-2, MASP-3**.

M cells Specialized epithelial cell type in the intestinal epithelium over Peyer's patches, through which antigens and pathogens enter from the gut.

MD-2 Accessory protein for TLR-4 activity.

MDA-5 (melanoma differentiation-associated 5, also helicard) This protein contains an RNA helicase-like domain similar to RIG-I, and senses double-stranded RNA for detection of intracellular viral infections.

medulla The central or collecting point of an organ. The thymic medulla is the central area of each thymic lobe, rich in bone marrow-derived antigen-presenting cells and the cells of a distinctive medullary epithelium. The medulla of the lymph node is a site of macrophage and plasma cell concentration through which the lymph flows on its way to the efferent lymphatics.

MEK1 A MAPK kinase in the Raf–MEK1–Erk signaling module, which is a part of a signaling pathway in lymphocytes leading to activation of the transcription factor AP-1.

melanoma-associated antigens (MAGE) Heterogeneous group of proteins of diverse or unknown functions characterized by restricted expression limited to tumors (i.e., melanoma) or testis germ cells.

membrane associated ring finger (C3HC4) 1, MARCH-1 An E3 ligase expressed in B cells, dendritic cells, and macrophages that induces the constitutive degradation of MHC class II molecules, regulating their steady-state expression.

membrane attack Effector pathway of complement based on formation of the membrane-attack complex (MAC).

membrane-attack complex (MAC) Protein complex composed of C5b to C9 that assembles a membrane-spanning hydrophilic pore on pathogen surfaces, causing cell lysis.

membrane cofactor of proteolysis (MCP or CD46) A complement regulatory protein, a host-cell membrane protein that acts in conjunction with factor I to cleave C3b to its inactive derivative iC3b and thus prevent convertase formation.

membrane immunoglobulin (mIg) Transmembrane immunoglobulin present on B cells; it is the B-cell receptor for antigen.

memory B cells *See* **memory cells**.

memory cells B and T lymphocytes that mediate immunological memory. They are more sensitive than naive lymphocytes to antigen and respond rapidly on reexposure to the antigen that originally induced them.

mesenteric lymph nodes Lymph nodes located in the connective tissue (mesentery) that tethers the intestine to the rear wall of the abdomen. They drain the GALT.

metastasis Spread of a tumor from its original location to distant organs of the body by traveling through the blood or lymphatics or by direct extension.

2′-O-methyltransferase (MTase) An enzyme that transfers a methyl group to the 2′ hydroxyl of the first and second ribose groups in mRNA. Viruses that acquire MTase can produce cap-1 and cap-2 on their transcripts and thereby evade restriction by IFIT1.

MF-59 A proprietary adjuvant based on squaline and water used in Europe and Canada in conjunction with influenza vaccine.

MHC class I *See* **MHC class I molecules**.

MHC class I deficiency An immunodeficiency disease in which MHC class I molecules are not present on the cell surface, usually as a result of an inherited deficiency of either TAP-1 or TAP-2.

MHC class I molecules Polymorphic cell-surface proteins encoded in the MHC locus and expressed on most cells. They present antigenic peptides generated in the cytosol to CD8 T cells, and also bind the co-receptor CD8.

MHC class II *See* **MHC class II molecules**.

MHC class II compartment (MIIC) The cellular vesicles in which MHC class II molecules accumulate, encounter HLA-DM, and bind antigenic peptides, before migrating to the surface of the cell.

MHC class II deficiency A rare immunodeficiency disease in which MHC class II molecules are not present on cells as a result of various inherited defects. Patients are severely immunodeficient and have few CD4 T cells.

MHC class II molecules Polymorphic cell-surface proteins encoded in the MHC locus are expressed primarily on specialized antigen-presenting cells. They present antigenic peptides derived from internalized extracellular pathogens to CD4 T cells and also bind the co-receptor CD4.

MHC class II transactivator (CIITA) Protein that activates transcription of MHC class II genes. Defects in the *CIITA* gene are one cause of MHC class II deficiency.

MHC haplotype A set of alleles in the MHC that is inherited unchanged (that is, without recombination) from one parent.

MHC molecules Highly polymorphic cell-surface proteins encoded by MHC class I and MHC class II genes involved in presentation of peptide antigens to T cells. They are also known as histocompatibility antigens.

MHC restriction   The fact that a peptide antigen can only be recognized by a given T cell if it is bound to a particular self MHC molecule. MHC restriction is a consequence of events that occur during T-cell development.

MIC-A, MIC-B   MHC class Ib proteins that are induced by stress, infection, or transformation in many cell types and are recognized by NKG2D.

microautophagy   The continuous internalization of the cytosol into the vesicular system.

microbial glycolipids   Diverse class of antigens frequently presented by CD1 molecules to iNKT cells.

microbiome   *See* **commensal microorganisms**.

microbiota   *See* **commensal microorganisms**.

microclusters   Assemblies of small numbers of T-cell receptors that may be involved in the initiation of T-cell receptor activation by antigen in naive T cells.

microfold cells   *See* **M cells**.

microglial cells   An embryonically derived form of tissue macrophage in the central nervous system that is dependent on IL-34 for local self-renewal throughout life.

minor histocompatibility antigens   Peptides of polymorphic cellular proteins bound to MHC molecules that can lead to graft rejection when they are recognized by T cells.

minor lymphocyte stimulating (Mls) antigens   An old term referring to non-MHC antigens responsible for unusually strong T cell responses to cells from different strains of mice, now known to be superantigens encoded by endogenous retroviruses.

mismatch repair   A type of DNA repair that causes mutations and is involved in somatic hypermutation and class switching in B cells.

missing self   Refers to the loss of cell-surface molecules that engage with inhibitory receptors on NK cells, resulting in NK-cell activation.

mitogen-activated protein kinases (MAPKs)   A series of protein kinases that become phosphorylated and activated on cellular stimulation by a variety of ligands, and lead to new gene expression by phosphorylating key transcription factors. The MAPKs are part of many signaling pathways, especially those leading to cell proliferation, and have different names in different organisms.

mixed essential cryoglobulinemia   Disease due to the production of cryoglobulins (cold-precipitable immunoglobulins), sometimes in response to chronic infections such as hepatitis C, which can lead to the deposition of immune complexes in joints and tissues.

mixed lymphocyte reaction (MLR)   A test for histocompatibility in which lymphocytes from donor and recipient are cultured together. If the two people are histoincompatible, the recipient's T cells recognize the allogeneic MHC molecules on the cells of the other donor as 'foreign' and proliferate.

molecular mimicry   The similarity between some pathogen antigens and host antigens, such that antibodies and T cells produced against the former also react against host tissues. This similarity may be the cause of some autoimmunity.

monoclonal antibodies   Antibodies produced by a single clone of B lymphocytes, so that they are all identical.

monocyte   Type of white blood cell with a bean-shaped nucleus; it is a precursor of tissue macrophages.

monomorphic   Describes a gene that occurs in only one form. Cf. **polymorphic**.

motheaten   A mutation in the SHP-1 protein phosphatase that impairs the function of some inhibitory receptors, such as Ly49, resulting in over-activation of various cells, including NK cells. Mice with this mutation have a 'motheaten' appearance due to chronic inflammation.

MR1   A 'non-classical' MHC class Ib molecule that binds certain folic acid metabolites produced by bacteria for recognition by mucosal associated invariant T (MAIT) cells.

MRE11A (meitotic recombination 11 homolog a)   A protein involved in DNA damage and repair mechanisms that also recognizes cytoplasmic dsDNA and can activate the STING pathway.

MSH2, MSH6   Mismatch repair proteins that detect uridine and recruit nucleases to remove the damaged and several adjacent nucleotides.

mTOR (mammalian target of rapamycin)   Serine/threonine kinase that functions in regulating numerous aspects of cell metabolism and function in complex with regulatory proteins Raptor or Rictor. The Raptor/mTOR complex (mTORC1) is inhibited by the immunosuppressive drug rapamycin.

mTORC1, mTORC2   Active complexes of mTOR formed with the regulatory proteins Raptor and Rictor, respectively.

mucins   Highly glycosylated cell-surface proteins. Mucin-like molecules are bound by L-selectin in lymphocyte homing.

Muckle–Wells syndrome   An inherited episodic autoinflammatory disease caused by mutations in the gene encoding NLRP3, a component of the inflammasome.

mucosal associated invariant T cells (MAIT)   Primarily γ:δ T cells with limited diversity present in the mucosal immune system that respond to bacterially derived folate derivates presented by the nonclassical MHC class Ib molecule MR1.

mucosa-associated lymphoid tissue (MALT)   Generic term for all organized lymphoid tissue found at mucosal surfaces, in which an adaptive immune response can be initiated. It comprises GALT, NALT, and BALT (when present).

mucosal epithelia   Mucus-coated epithelia lining the body's internal cavities that connect with the outside (such as the gut, airways, and vaginal tract).

mucosal immune system   The immune system that protects internal mucosal surfaces (such as the linings of the gut, respiratory tract, and urogenital tracts), which are the site of entry for virtually all pathogens and other antigens. *See also* **mucosa-associated lymphoid tissue**.

mucosal mast cells   Specialized mast cells present in mucosa. They produce little histamine but large amounts of prostaglandins and leukotrienes.

mucosal tolerance   The suppression of specific systemic immune responses to an antigen by the previous administration of the same antigen by a mucosal route.

mucus   Sticky solution of proteins (mucins) secreted by goblet cells of internal epithelia, forming a protective layer on the epithelial surface.

multiple sclerosis   A neurological autoimmune disease characterized by focal demyelination in the central nervous system, lymphocytic infiltration in the brain, and a chronic progressive course.

multipotent progenitor cells (MPPs)   Bone marrow cells that can give rise to both lymphoid and myeloid cells but are no longer self-renewing stem cells.

muramyl dipeptide (MDP)   A component of the peptidoglycan of most bacteria that is recognized by the intracellular sensor NOD2.

muromomab   A mouse antibody against human CD3 used to treat transplant rejection; this was the first monoclonal antibody approved as a drug in humans.

mutualism   A symbiotic relationship between two organisms in which both benefit, such as the relationship between a human and its normal resident (commensal) gut microorganisms.

Mx (myxoma resistant) proteins   Interferon-inducible proteins required for cellular resistance to influenza virus replication.

myasthenia gravis   An autoimmune disease in which autoantibodies against the acetylcholine receptor on skeletal muscle cells cause a block in neuromuscular junctions, leading to progressive weakness and eventually death.

mycophenolate   An inhibitor of the synthesis of guanosine monophosphate that acts as a cytotoxic immunosuppressive drug. It acts by killing rapidly dividing cells, including lymphocytes proliferating in response to antigen.

mycophenolate mofetil   Pro-drug used in cancer treatment that is metabolized to mycophenolate, and inhibitor of inosine monophosphate dehydrogenase, thereby impairing guanosine monophosphate, and thus DNA, synthesis.

MyD88   An adaptor protein that functions in signaling by all TLR proteins except TLR3.

myeloid   Refers to the lineage of blood cells that includes all leukocytes except lymphocytes.

myeloid-derived suppressor cells (MDSCs)   Cells in tumors that can inhibit T-cell activation within the tumor.

myelomonocytic series   Innate immune cells derived from myelomonocytic bone marrow precursors, including neutrophils, basophils, eosinophils, monocytes, and dendritic cells.

NADPH oxidase   Multicomponent enzyme complex that is assembled and activated in the phagolysosome membrane in stimulated phagocytes. It generates superoxide in an oxygen-requiring reaction called the respiratory burst.

NAIP2   An NLR protein that, together with NLRC4, recognizes the PrgJ protein of the *Salmonella typhimurium* type III injection system to activate an inflammasome pathway in response to infection.

NAIP5   An NLR protein that, together with NLRC4, recognizes intracellular flagellin to activate an inflammasome pathway in response to infection.

naive lymphocytes   T cells or B cells that have undergone normal development in the thymus or the bone marrow but have not yet been activated by foreign (or self) antigens.

naive T cells   Lymphocytes that have never encountered their specific antigen and thus have never responded to it, as distinct from effector and memory lymphocytes.

nasal-associated lymphoid tissue (NALT)   Organized lymphoid tissues found in the upper respiratory tract. In humans, NALT consists of Waldeyer's ring, which includes the adenoids, palatine, and lingual tonsils, plus other similarly organized lymphoid tissue located around the pharynx. It is part of the mucosal immune system.

natalizumab   Humanized antibody to α4 integrin used to treat Crohn's disease and multiple sclerosis. It blocks lymphocytes' adhesion to endothelium, impairing their migration into tissues.

natural antibodies   Antibodies produced by the immune system in the apparent absence of any infection. They have a broad specificity for self and microbial antigens, can react with many pathogens, and can activate complement.

natural cytotoxicity receptors (NCRs)   Activating receptors on NK cells that recognize infected cells and stimulate cell killing by the NK cell.

natural interferon-producing cells   *See* **plasmacytoid dendritic cells**.

natural killer (NK) cell   A type of ILC that is important in innate immunity to viruses and other intracellular pathogens, and in antibody-dependent cell-mediated cytotoxicity (ADCC). NK cells express activating and inhibitory receptors, but not the antigen-specific receptors of T or B cells.

necrosis   The process of cell death that occurs in response to noxious stimuli, such as nutrient deprivation, physical injury, or infection. To be distinguished from apoptosis, in which the cell activates an internal, or intrinsic, program of death, such as occurs in immune cells as a result of deficiency of cell survival signals.

negative selection   The process by which self-reactive thymocytes are deleted from the repertoire during T-cell development in the thymus. Autoreactive B cells undergo a similar process in bone marrow.

NEMO   *See* **IKK**.

NEMO deficiency   *See* **X-linked hypohidrotic ectodermal dysplasia and immunodeficiency**.

neoepitopes   Type of tumor rejection antigen created by mutations in protein that can be presented by self-MHC molecules to T cells.

neonatal Fc receptor (FcRn)   *See* **FcRn**.

neuraminidase   An influenza virus protein that cleaves sialic acid from host cells to allow viral detachment, a common antigenic determinant, and target of antiviral neuraminidase inhibitors.

neutralization   Inhibition of the infectivity of a virus or the toxicity of a toxin molecule by the binding of antibodies.

neutralizing antibodies   Antibodies that inhibit the infectivity of a virus or the toxicity of a toxin.

neutropenia   Abnormally low levels of neutrophils in the blood.

neutrophil   The most numerous type of white blood cell in human peripheral blood. Neutrophils are phagocytic cells with a multilobed nucleus and granules that stain with neutral stains. They enter infected tissues and engulf and kill extracellular pathogens.

neutrophil elastase   Proteolytic enzyme stored in the granules of neutrophils that is involved in the processing of antimicrobial peptides.

neutrophil extracellular traps (NETs)   A meshwork of nuclear chromatin that is released into the extracellular space by neutrophils undergoing apoptosis at sites of infection, serving as a scaffold that traps extracellular bacteria to enhance their phagocytosis by other phagocytes.

NFκB   A heterodimeric transcription factor activated by the stimulation of Toll-like receptors and also by antigen receptor signaling composed of p50 and p65 subunits.

NFAT   *See* **nuclear factor of activated T cells**.

Nfil3   A transcription factor important during the development of several types of immune cells including certain types of NK cells.

NHEJ   *See* **nonhomologous end joining**.

nitric oxide   A reactive molecular gas species produced by cells—particularly macrophages—during infection, that is toxic to bacteria and intracellular microbes.

nivolumab   Human anti-PD-1 antibody used for checkpoint blockade in treatment of metastatic melanoma.

NK receptor complex (NKC)   A cluster of genes that encode a family of receptors on NK cells.

NKG2   Family of C-type lectins that supply one of the subunits of KLR-family receptors on NK cells.

NKG2D   Activating C-type lectin receptor on NK cells, cytotoxic T cells, and γ:δ T cells that recognizes the stress-response proteins MIC-A and MIC-B.

NLRC4   An NLR family member that cooperates with NAIP2 and NAIP5.

NLRP family   A group of 14 NOD-like receptor (NLR) proteins that contain a pyrin domain and function in the formation of a signaling complex called the inflammasome.

NLRP3   A member of the family of intracellular NOD-like receptor proteins that have pyrin domains. It acts as a sensor of cellular damage and is part of the inflammasome. Sometimes called NALP3.

N-nucleotides   Nontemplated nucleotides inserted by the enzyme terminal deoxynucleotidyl transferase into the junctions between gene segments of T-cell receptor and immunoglobulin heavy-chain V regions during gene segment joining. Translation of these N-regions markedly increases the diversity of these receptor chains.

NOD subfamily   A subgroup of NLR proteins that contain a CARD domain which is used for activation of downstream signaling.

NOD1, NOD2   Intracellular proteins of the NOD subfamily that contain a leucine-rich repeat (LRR) domain that binds components of bacterial cell walls to activate the NFκB pathway and initiate inflammatory responses.

NOD-like receptors (NLRs)   Large family of proteins containing a nucleotide-oligomerization domain (NOD) associated with various other domains, and whose general function is the detection of microbes and of cellular stress.

nonamer   Conserved nine-nucleotide DNA sequence in the recombination signal sequences (RSSs) flanking gene segments in the immunoglobulin and T-cell receptor loci.

non-canonical inflammasome   An alternate form of the inflammasome that is independent of caspase 1, but instead relies on caspase 11 (mice) or caspases 4 or 5 (human).

non-canonical NFκB pathway   A pathway for NFκB activation that is distinct from the one activated by antigen receptor stimulation. This pathway leads to activation of the NFκB-inducing kinase, NIK, which phosphorylates and activates IκB kinase α (IKKα) inducing cleavage of the NFκB precursor protein p100 to form the active p52 subunit.

nonclassical MHC class Ib genes   A class of proteins encoded within the MHC that are related to the MHC class I molecules but are not highly polymorphic and present a restricted set of antigens.

non-depleting antibodies   Immunosuppressive antibodies that block the function of target proteins on cells without causing the cells to be destroyed.

nonhomologous end joining (NHEJ)   DNA repair pathway that directly ligates double-stranded DNA breaks without use of a homologous template.

nonproductive rearrangements   Rearrangements of T-cell receptor or immunoglobulin gene segments that cannot encode a protein because the coding sequences are in the wrong translational reading frame.

nonreceptor kinase   Cytoplasmic protein kinases that associate with the intracellular tails of signaling receptors and help generate the signal but are not an intrinsic part of the receptor itself.

non-structural protein 1 (NS1)   An influenza A virus protein that inhibits TRIM25, an intermediate signaling protein downstream of the viral sensors RIG-I and MDA-5, thereby promoting evasion of innate immunity.

nuclear factor of activated T cells   A family of transcription factors that are activated in response to increased cytoplasmic calcium following antigen receptor signaling in lymphocytes.

nucleotide-binding oligomerization domain (NOD)   A type of conserved domain originally recognized in ATP-binding cassette (ABC) transporters

present present in a large number of proteins, but which also mediates protein homooligomerization.

nude   A mutation in mice that results in hairlessness and defective formation of the thymic stroma, so that mice homozygous for this mutation have no mature T cells.

NY-ESO-1   A particular highly immunogenic cancer-testis antigen expressed by many types of human tumors including melanoma.

occupational allergies   An allergic reaction induced to an allergen to which someone is habitually exposed in their work.

oligoadenylate synthetase   Enzyme produced in response to stimulation of cells by interferon. It synthesizes unusual nucleotide polymers, which in turn activate a ribonuclease that degrades viral RNA.

Omenn syndrome   A severe immunodeficiency disease characterized by defects in either of the *RAG* genes. Affected individuals make small amounts of functional RAG protein, allowing a small amount of V(D)J recombination.

opsonization   The coating of the surface of a pathogen by antibody and/or complement that makes it more easily ingested by phagocytes.

oral tolerance   The suppression of specific systemic immune responses to an antigen by the prior administration of the same antigen by the oral (enteric) route.

original antigenic sin   The tendency of humans to make antibody responses to those epitopes shared between the first strain of a virus they encounter and subsequent related viruses, while ignoring other highly immunogenic epitopes on the second and subsequent viruses.

p50   *See* NFκB.

p65   *See* NFκB.

PA28 proteasome-activator complex   A multisubunit protein complex induced by interferon-γ that takes the place of the 19S regulatory cap of the proteasome and increases the rate of peptides exiting from the proteasome catalytic core.

palatine tonsils   Paired masses of organized peripheral lymphoid tissues located on each side of the throat, and in which an adaptive immune response can be generated. They are part of the mucosal immune system.

Paneth cells   Specialized epithelial cells at the base of the crypts in the small intestine that secrete antimicrobial peptides.

papain   A protease that cleaves the IgG antibody molecule on the amino-terminal side of disulfide linkages, producing two Fab fragments and one Fc fragment.

paracortical areas   The T-cell area of lymph nodes.

paracrine   Describes a cytokine or other biologically active molecule acting on cells near to those that produce it.

parasites   Organisms that obtain sustenance from a live host. In immunology, it refers to worms and protozoa, the subject matter of parasitology.

paroxysmal nocturnal hemoglobinuria   A disease in which complement regulatory proteins are defective, so that activation of complement binding to red blood cells leads to episodes of spontaneous hemolysis.

passive immunization   The injection of antibody or immune serum into a naive recipient to provide specific immunological protection. Cf. **active immunization**.

pathogen   Microorganism that typically causes disease when it infects a host.

pathogen-associated molecular patterns (PAMPs)   Molecules specifically associated with groups of pathogens that are recognized by cells of the innate immune system.

pathogenesis The origin or cause of the pathology of a disease.

pathogenic microorganisms Microorganism that typically causes disease when it infects a host.

patrolling monocyte A form of circulating monocyte that adheres to and surveys the vascular endothelium, distinguished from classical monocytes by its low expression of Ly6C.

pattern recognition receptors (PRRs) Receptors of the innate immune system that recognize common molecular patterns on pathogen surfaces.

PD-1 Programmed death-1, a receptor on T cells that when bound by its ligands, PD-L1 and PD-L2, inhibits signaling from the antigen receptor. PD-1 contains an ITIM motif in its cytoplasmic tail. Target of cancer therapies aimed at stimulating T-cell responses to tumors.

PD-L1 (programmed death ligand-1, B7-H1) Transmembrane receptor that binds to the inhibitory receptor PD-1. PD-L1 is expressed on many cell types and is upregulated by inflammatory cytokines.

PD-L2 (programmed death ligand-2, B7-DC) Transmembrane receptor that binds to the inhibitory receptor PD-1; mainly expressed on dendritic cells.

PECAM *See* **CD31**.

pembrolizumab Human anti-PD-1 antibody used for checkpoint blockade in treatment of metastatic melanoma.

pemphigus vulgaris An autoimmune disease characterized by severe blistering of the skin and mucosal membranes.

pentameric IgM Major form of the IgM antibodies produced by the action of J chain resulting in higher avidity for antigens.

pentraxin A family of acute-phase proteins formed of five identical subunits, to which C-reactive protein and serum amyloid protein belong.

pepsin A protease that cleaves several sites on the carboxy-terminal side of the disulfide linkages, producing the $F(ab')_2$ fragment and several fragments of the Fc region.

peptide-binding cleft The longitudinal cleft in the top surface of an MHC molecule into which the antigenic peptide is bound. Sometimes called the peptide-binding groove.

peptide editing In the context of antigen processing and presentation, the removal of unstably bound peptides from MHC class II molecules by HLA-DM.

peptide-loading complex (PLC) A protein complex in the endoplasmic reticulum that loads peptides onto MHC class I molecules.

peptide:MHC tetramers Four specific peptide:MHC complexes bound to a single molecule of fluorescently labeled streptavidin, which are used to identify populations of antigen-specific T cells.

peptidoglycan A component of bacterial cell walls that is recognized by certain receptors of the innate immune system.

peptidoglycan-recognition proteins (PGRPs) A family of *Drosophila* proteins that bind peptidoglycans from bacterial cell walls that serve to initiate the proteolytic cascade of the TOLL pathway.

periarteriolar lymphoid sheath (PALS) Part of the inner region of the white pulp of the spleen; it contains mainly T cells.

peripheral lymphoid organs, peripheral lymphoid tissues The lymph nodes, spleen, and mucosa-associated lymphoid tissues, in which adaptive immune responses are induced, as opposed to the central lymphoid organs, in which lymphocytes develop. They are also called secondary lymphoid organs and tissues.

peripheral tolerance Tolerance acquired by mature lymphocytes in the peripheral tissues, as opposed to central tolerance, which is acquired by immature lymphocytes during their development.

Peyer's patches Organized peripheral lymphoid organs under the epithelium in the small intestine, especially the ileum, and in which an adaptive immune response can be initiated. They contain lymphoid follicles and T-cell areas. They are part of the gut-associated lymphoid tissues (GALT).

phagocyte oxidase *See* **NADPH oxidase**.

phagocytic glycoprotein-1 (Pgp1) *See* **CD44**.

phagocytosis The internalization of particulate matter by cells by a process of engulfment, in which the cell membrane surrounds the material, eventually forming an intracellular vesicle (phagosome) containing the ingested material.

phagolysosome Intracellular vesicle formed by the fusion of a phagosome (containing ingested material) and a lysosome, and in which the ingested material is broken down.

phagophore A crescent-shaped double-membrane cytoplasmic structure.

phagosome Intracellular vesicle formed when particulate material is ingested by a phagocyte.

phosphatidylinositol 3-kinase (PI 3-kinase) Enzyme involved in intracellular signaling pathways. It phosphorylates the membrane lipid phosphatidylinositol 3,4-bisphosphate (PIP_2) to form phosphatidylinositol 3,4,5-trisphosphate (PIP_3), which can recruit signaling proteins containing pleckstrin homology (PH) domains to the membrane.

phosphatidylinositol kinases Enzymes that phosphorylate the inositol headgroup on membrane lipids to produce phosphorylated derivatives that have a variety of functions in intracellular signaling.

phospholipase C-γ (PLC-γ) Key enzyme in intracellular signaling pathways leading from many different receptors. It is activated by membrane recruitment and tyrosine phosphorylation following receptor ligation, and cleaves membrane inositol phospholipids into inositol trisphosphate and diacylglycerol.

phosphorylation Addition of a phosphate group to a molecule, usually a protein, catalyzed by enzymes called kinases.

phycoerythrin A light harvesting protein pigment made by algae and used in conjunction with flow cytometry, it can also be recognized as a ligand by some γ:δ T-cell receptors.

physiological inflammation The state of the normal healthy intestine, whose wall contains large numbers of effector lymphocytes and other cells. It is thought to be the result of continual stimulation by commensal organisms and food antigens.

pi-cation interactions Chemical interaction between a cation (e.g., Na^+) and the pi-electron system of an aromatic moiety.

pilin An adhesin of *Neisseria gonorrhoeae* allowing attachment to and infection of epithelial cells of urinary and reproductive tracts.

PIP_2 Phosphatidylinositol 3,4-bisphosphate, a membrane-associated phospholipid that is cleaved by phospholipase C-γ to give the signaling molecules diacylglycerol and inositol trisphosphate and is phosphorylated by PI3-kinase to generate PIP_3.

PIP_3 Phosphatidylinositol 3,4,5-trisphosphate, a membrane-associated phospholipid that can recruit intracellular signaling molecules containing pleckstrin homology (PH) domains to the membrane.

PKR Serine/threonine kinase activated by IFN-α and IFN-β. It phosphorylates the eukaryotic protein synthesis initiation factor eIF-2, inhibiting translation and thus contributing to the inhibition of viral replication.

plasma cells   Terminally differentiated activated B lymphocytes. Plasma cells are the main antibody-secreting cells of the body. They are found in the medulla of the lymph nodes, in splenic red pulp, in bone marrow, and in mucosal tissues.

plasmablasts   A B cell in a lymph node that already shows some features of a plasma cell.

plasmacytoid dendritic cells (pDCs)   A distinct lineage of dendritic cells that secrete large amounts of interferon on activation by pathogens and their products via receptors such as Toll-like receptors. Cf. **conventional dendritic cells**.

platelet-activating factor (PAF)   A lipid mediator that activates the blood clotting cascade and several other components of the innate immune system.

pluripotent   Typically referring to the capacity of a progenitor cell to generate all possible lineages of an organ system.

P-nucleotides   Short palindromic nucleotide sequences formed between gene segments of the rearranged V-region gene generated by the asymmetric opening of the hairpin intermediate during RAG-mediated rearrangement.

Polη   An error-prone, 'translesion', DNA polymerase involved in repairing DNA damage caused by UV radiation and in somatic hypermutation.

polyclonal activation   The activation of lymphocytes by a mitogen regardless of antigen specificity, leading to the activation of clones of lymphocytes of multiple antigen specificities.

polygenic   Containing several separate loci encoding proteins of identical function; applied to the MHC. Cf. **polymorphic**.

polymerase stalling   The halting of RNA polymerase during the transcription of a gene at locations within the gene locus, known to be a regulated process, and involved in mechanisms of isotype switching.

polymeric immunoglobulin receptor (pIgR)   The receptor for polymeric immunoglobulins IgA and IgM on basolateral surfaces of mucosal and glandular epithelial cells that transports IgA (or IgM) into secretions.

polymorphic   Existing in a variety of different forms; applied to a gene, occurring in a variety of different alleles.

polymorphism   Applied to genes, variability at a gene locus in which all variants occur at a frequency greater than 1%.

polymorphonuclear leukocytes   *See* **granulocytes**.

polysaccharide capsules   A distinct structure in some bacteria—both Gram-negative and Gram-positive—that lies outside cell membrane and cell wall that can prevent direct phagocytosis by macrophages without the aid of antibody or complement.

polyubiquitin chains   Polymers of ubiquitin covalently linked from lysine residues within one ubiquitin monomer to the carboxy terminus of a second ubiquitin.

PorA   Outer membrane protein of *Neisseria meningitidis* that binds C4BP, thereby inactivating C3b deposited on its surface.

positive selection   A process occurring in the thymus in which only those developing T cells whose receptors can recognize antigens presented by self MHC molecules can mature.

post-transplant lymphoproliferative disorder   B-cell expansion driven by Epstein–Barr virus (EBV) in which the B cells can undergo mutations and become malignant. This can occur when patients are immunosuppressed after, for example, solid organ transplantation.

pre-B-cell receptor   Receptor produced by pre-B cells that includes an immunoglobulin heavy chain, as well as surrogate light-chain proteins, Igα and Igβ signaling subnits. Signaling through this receptor induces the pre-B cell to enter the cell cycle, to turn off the RAG genes, to degrade the RAG proteins, and to expand by several cell divisions.

pre-T-cell receptor   Receptor protein produced by developing T lymphocytes at the pre-T-cell stage. It is composed of TCRβ chains that pair with a surrogate α chain called pTα (pre-T-cell α), and is associated with the CD3 signaling chains. Signaling through this receptor induces pre-T-cell proliferation, expression of CD4 and CD8, and cessation of TCR β chain rearrangement.

prednisone   A synthetic steroid with potent anti-inflammatory and immunosuppressive activity used in treating acute graft rejection, autoimmune disease, and lymphoid tumors.

PREX1   A guanine exchange factor (GEF) activated downstream of small G proteins in response to activation of GPCRs such as the fMLP or C5a receptor.

PrgJ   A protein component of the *Salmonella typhimurium* type III secretion system inner rod used by the bacterium to infect eukaryotic cells. This protein is detected by NLR proteins NAIP2 and NLRC4.

primary focus   Site of early antibody production by plasmablasts in medullary cords of lymph nodes that precedes the germinal center reaction and differentiation of plasma cells.

primary granules   Granules in neutrophils that correspond to lysosomes and contain antimicrobial peptides such as defensins and other antimicrobial agents.

primary immune response   The adaptive immune response that follows the first exposure to a particular antigen.

primary immunization   *See* **priming**.

primary immunodeficiencies   A lack of immune function that is caused by a genetic defect.

primary lymphoid follicles   Aggregates of resting B lymphocytes in peripheral lymphoid organs. Cf. **secondary lymphoid follicle**.

primary lymphoid organs   *See* **central lymphoid organs**.

priming   The first encounter with a given antigen, which generates the primary adaptive immune response.

pro-B cells   A stage in B-lymphocyte development in which cells have displayed B-cell surface marker proteins but have not yet completed heavy-chain gene rearrangement.

pro-caspase 1   The inactive pro-form of caspase 1 that is part of the NLRP3 inflammasome.

pro-inflammatory   Tending to induce inflammation.

profilin   An actin-binding protein that sequesters monomeric actin. Protozoan profilins contain sequences recognized by TLR-11 and TLR-12.

programmed cell death   *See* **apoptosis**.

progressive multifocal leukoencephalopathy (PML)   Disease in immunocompromised patients caused by opportunisitic infection by JC virus, for example as a consequence of immunotherapy.

propeptides   Inactive precursor form of a polypeptide or peptide, which requires proteolytic processing to produce the active peptide.

properdin   *See* **factor P**.

prostaglandins   Lipid products of the metabolism of arachidonic acid that have a variety of effects on tissues, including activities as inflammatory mediators.

prostatic acid phosphatase (PAP) Enzyme expressed by prostate cancer cells used as tumor rejection antigen in the vaccine Sipuleucel-T (Provenge).

proteasome A large intracellular multisubunit protease that degrades proteins, producing peptides.

protein inhibitors of activated STAT (PIAS) A small family of proteins that inhibit STAT family transcription factors.

protein-interaction domains, protein-interaction modules Protein domains, usually with no enzymatic activity themselves, that have binding specificity for particular sites (such as phosphorylated tyrosines, proline-rich regions, or membrane phospholipids) on other proteins or cellular structures.

protein kinase C-θ (PKC-θ) A serine/threonine kinase that is activated by diacylglycerol as part of the signaling pathways from the antigen receptor in lymphocytes.

protein kinases Enzymes that add phosphate groups to proteins at particular amino acid residues: tyrosine, threonine, or serine.

protein phosphatases Enzymes that remove phosphate groups from proteins phosphorylated on tyrosine, threonine, or serine residues by protein kinases.

proteolytic subunits β1, β2, β5 Constitutive components of the proteasome's catalytic chamber.

provirus The DNA form of a retrovirus when it is integrated into the host-cell genome, where it can remain transcriptionally inactive for long periods.

P-selectin *See* **selectins**.

P-selectin glycoprotein ligand-1 (PSGL-1) Protein expressed by activated effector T cells that is a ligand for P-selectin on endothelial cells, and may enable activated T cells to enter all tissues in small numbers.

pseudo-dimeric peptide:MHC complexes Hypothetical complexes containing one antigen peptide:MHC molecule and one self peptide:MHC molecule on the surface of the antigen-presenting cell, which have been proposed to initiate T-cell activation.

pseudogenes Gene elements that have lost the ability to encode a functional protein but that are retained in the genome and may continue to be transcribed normally.

psoriasis Chronic autoimmune disease thought to be driven by T cells manifested in skin, but which can also involve nails and joints (psoriatic arthropathy).

psoriatic arthropathy *See* **psoriasis**.

pTα *See* **pre-T-cell receptor**.

purine nucleotide phosphorylase (PNP) deficiency An enzyme defect that results in severe combined immunodeficiency. The deficiency of PNP causes an intracellular accumulation of purine nucleosides, which are toxic to developing T cells.

purinergic receptor P2X7 An ATP-activated ion channel that allows potassium efflux from cells when activated, which can trigger inflammasome activation in response to excessive extracellular ATP.

pus Thick yellowish-white liquid typically found at sites of infection with some types of extracellular bacteria, which is composed of the remains of dead neutrophils and other cells.

pus-forming bacteria Capsulated bacteria that result in pus formation at the site of infection. Also called pyogenic (pus-forming) bacteria.

PYHIN A family of four intracellular sensor proteins containing an H inversion (HIN) domain in place of the LRR domain found in most other NLR proteins. The HIN domain functions in recognition of cytoplasmic dsDNA. Examples are AIM2 and IFI16.

pyogenic arthritis, pyoderma gangrenosum, and acne (PAPA) Autoinflammatory syndrome that is caused by mutations in a protein that interacts with pyrin.

pyogenic bacteria *See* **pus-forming bacteria**.

pyrin One of several protein interaction domains, structurally related to but distinct from CARD, TIR, DD, and DED domains.

pyroptosis A form of programmed cell death that is associated with abundant pro-inflammatory cytokines such as IL-1β and IL-18 produced through inflammasome activation.

Qa-1 determinant modifiers (Qdm) A class of peptides derived from the leader peptides of various HLA class I molecules that can be bound by the human HLA-E and murine Qa-1 proteins, and then are recognized by the inhibitory NKG2A:CD94 receptor.

quasi-species The different genetic forms of certain RNA viruses that are formed by mutation during the course of an infection.

Rac *See* **Rho family small GTPase proteins**.

radiation-sensitive SCID (RS-SCID) Severe combined immunodeficiency due to a defect in DNA repair pathways, which renders cells unable to perform V(D)J recombination and unable to repair radiation-induced double-strand breaks.

RAE1 (retinoic acid early inducible 1) protein family Several murine MHC class Ib proteins; these are orthologs of human RAET1 family proteins, including H60 and MULT1, and are ligands for murine NKG2D.

RAET1 A family of 10 MHC class Ib proteins that are ligands for NKG2D, and includes several UL16-binding proteins (ULBPs).

Raf A protein kinase in the Raf–MEK1–Erk signaling cascade that is the first protein kinase in the pathway, and is activated by the small GTPase Ras.

RAG-1, RAG-2 Proteins encoded by the recombination-activating genes *RAG-1* and *RAG-2*, which form a dimer that initiates V(D)J recombination.

rapamycin An immunosuppressant drug that blocks intracellular signaling pathways involving the serine/threonine kinase mammalian target of rapamycin (mTOR) required for the inhibition of apoptosis and T-cell expansion. Also called sirolimus.

Raptor *See* **mTORC1**.

Ras A small GTPase with important roles in intracellular signaling pathways, including those from lymphocyte antigen receptors.

reactive oxygen species (ROS) Superoxide anion (O_2^-) and hydrogen peroxide (H_2O_2), produced by phagocytic cells such as neutrophils and macrophages after ingestion of microbes, and which help kill the ingested microbes.

rearrangement by inversion In V(D)J recombination, the rearrangement of gene segments having RSS elements in an opposing orientation, leading to retention.

receptor editing The replacement of a light or heavy chain of a self-reactive antigen receptor on immature B cells with a newly rearranged chain that does not confer autoreactivity.

receptor-mediated endocytosis The internalization into endosomes of molecules bound to cell-surface receptors.

receptor serine/threonine kinases Receptors that have an intrinsic serine/threonine kinase activity in their cytoplasmic tails.

receptor tyrosine kinases Receptors that have an intrinsic tyrosine kinase activity in their cytoplasmic tails.

recombination signal sequences (RSSs) DNA sequences at one or both ends of V, D, and J gene segments that are recognized by the RAG-1:RAG-2 recombinase. They consist of a conserved heptamer and nonamer element separated by 12 or 23 base pairs.

red pulp The nonlymphoid area of the spleen in which red blood cells are broken down.

RegIIIγ An antimicrobial protein of the C-type lectin family, produced by Paneth cells in the gut in mice.

regulatory T cells Effector CD4 T cells that inhibit T-cell responses and are involved in controlling immune reactions and preventing autoimmunity. Several different subsets have been distinguished, notably the natural regulatory T-cell lineage that is produced in the thymus, and the induced regulatory T cells that differentiate from naive CD4 T cells in the periphery in certain cytokine environments.

regulatory tolerance Tolerance due to the actions of regulatory T cells.

Relish A distinct member of the *Drosophila* NFκB transcription factor family that induces the expression of several antimicrobial peptides in response to Gram-negative bacteria.

resistance A general immune strategy aimed at reducing or eliminating pathogens; compare with **avoidance** and **tolerance**.

respiratory burst An oxygen-requiring metabolic change in neutrophils and macrophages that have taken up opsonized particles, such as complement- or antibody-coated bacteria, by phagocytosis. It leads to the production of toxic metabolites that are involved in killing the engulfed microorganisms.

restriction factors Host proteins that act in a cell-autonomous manner to inhibit the replication of retroviruses such as HIV.

retinoic acid Signaling molecule derived from vitamin A with many roles in the body. It is thought to be involved in the induction of immunological tolerance in the gut.

retrotranslocation complex The return of endoplasmic reticulum proteins to the cytosol.

retrovirus A single-stranded RNA virus that uses the viral enzyme reverse transcriptase to transcribe its genome into a DNA intermediate that integrates into the host-cell genome to undergo viral replication.

reverse transcriptase Viral RNA-dependent DNA polymerase that is found in retroviruses and transcribes the viral genomic RNA into DNA during the life cycle of retroviruses (such as HIV).

Rheb A small GTPase that activates mTOR when in its GTP-bound form, and is inactivated by a GTPase-activating protein (GAP) complex TSC1/2.

rheumatic fever Disease caused by antibodies elicited by infection with some *Streptococcus* species. These antibodies cross-react with kidney, joint, and heart antigens.

rheumatoid arthritis (RA) A common inflammatory joint disease that is probably due to an autoimmune response.

rheumatoid factor An anti-IgG antibody of the IgM class first identified in patients with rheumatoid arthritis, but which is also found in healthy individuals.

Rho *See* **Rho family small GTPase proteins**.

Rho family small GTPase proteins Several distinct small GTPase family members that regulate the actin cytoskeleton in response to signaling through various receptors. Examples: Rac, Rho, and Cdc42.

Rictor *See* **mTORC2**.

RIG-I (retinoic acid-inducible gene I) *See* **RIG-I-like receptors (RLRs)**.

RIG-I-like receptors (RLRs) A small family of intracellular viral sensors that use a carboxy terminal RNA helicase-like domain in detection of various forms of viral RNA. These signal through MAVS to activate antiviral immunity. Examples include RIG-I, MDA-5, and LGP2.

RIP2 A CARD domain containing serine-threonine kinase that functions in signaling by NOD proteins to activate the NFκB transcription factor.

Riplet An E3 ubiquitin ligase involved in signaling by RIG-I and MDA-5 for the activation of MAVS.

rituximab A chimeric antibody to CD20 used to eliminate B cells in treatment of non-Hodgkin's lymphoma.

R-loops A structure formed when transcribed RNA displaces the nontemplate strand of the DNA double helix at switch regions in the immunoglobulin constant-region gene cluster. R-loops are thought to promote class switch recombination.

RNA exosome A multisubunit complex involved in processing and editing of RNA.

ruxolitinib An inhibitor of JAK1 and JAK2 approved for treatment of myelofibrosis.

S1PR1 A G protein-coupled receptor expressed on circulating lymphocytes that binds the chemotactic phospholipid, sphingosine 1-phosphate, which forms a chemotactic gradient that promotes the egress of non-activated lymphocytes out of secondary lymphoid tissues into the efferent lymphatics and blood. *See also* **CD69**.

SAP (SLAM-associated protein) An intracellular adaptor protein involved in signaling by SLAM (signaling lymphocyte activation molecule). Inactivating mutations in this gene cause X-linked lymphoproliferative (XLP) syndrome.

scaffolds Adaptor-type proteins with multiple binding sites, which bring together specific proteins into a functional signaling complex.

scavenger receptors Receptors on macrophages and other cells that bind to numerous ligands, such as bacterial cell-wall components, and remove them from the blood. The Kupffer cells in the liver are particularly rich in scavenger receptors. Includes SR-A I, SR-A II, and MARCO.

scid Mutation in mice that causes severe combined immunodeficiency. It was eventually found to be due to mutation of the DNA repair protein DNA-PK.

SCID *See* **severe combined immunodeficiency**.

seasonal allergic rhinoconjunctivitis IgE-mediated allergic rhinitis and conjunctivitis caused by exposure to specific seasonally occurring antigens, for example grass or weed pollens. Commonly called hay fever.

Sec61 A multisubunit transmembrane protein pore complex that resides in the membrane of the endoplasmic reticulum and allows peptides to be translocated from the ER lumen into the cytoplasm.

second messengers Small molecules or ions (such as Ca^{2+}) that are produced in response to a signal; they act to amplify the signal and carry it to the next stage within the cell. Second messengers generally act by binding to and modifying the activities of enzymes.

secondary granules Type of granule in neutrophils that stores certain antimicrobial peptides.

secondary immune response The immune response that occurs in response to a second exposure to an antigen. In comparison with the primary response, it starts sooner after exposure, produces greater levels of antibody, and produces class-switched antibodies. It is generated by the reactivation of memory lymphocytes.

secondary immunization A second or booster injection of an antigen, given some time after the initial immunization. It stimulates a secondary immune response.

secondary immunodeficiencies   Deficiencies in immune function that are a consequence of infection (e.g., HIV infection), other diseases (e.g., leukemia), malnutrition, etc..

secondary lymphoid follicle   A follicle containing a germinal center of proliferating activated B cells during an ongoing adaptive immune response.

secondary lymphoid organs   *See* **peripheral lymphoid organs**.

secondary lymphoid tissues   *See* **peripheral lymphoid organs**.

secretory component (SC)   Fragment of the polymeric immunoglobulin receptor that remains after cleavage and is attached to secreted IgA after transport across epithelial cells.

secretory IgA (SIgA)   Polymeric IgA antibody (mainly dimeric) containing bound J chain and secretory component. It is the predominant form of immunoglobulin in most human secretions.

secretory phospholipase A2   Antimicrobial enzyme present in tears and saliva and also secreted by the Paneth cells of the gut.

segmented filamentous bacteria (SFB)   Referring to commensal Gram-positive Firmicute species and members of the *Clostridiaceae* family that adhere to the intestinal wall of rodents and several other species that induce T_H17 and IgA responses.

selectins   Family of cell-adhesion molecules on leukocytes and endothelial cells that bind to sugar moieties on specific glycoproteins with mucin-like features.

self antigens   The potential antigens on the tissues of an individual, against which an immune response is not usually made except in the case of autoimmunity.

self-tolerance   The failure to make an immune response against the body's own antigens.

sensitization   The acute adaptive immune response made by susceptible individuals on first exposure to an allergen. In some of these individuals, subsequent exposure to the allergen will provoke an allergic reaction.

sensitized   In allergy, describes an individual who has made an IgE response on initial encounter with an environmental antigen and who manifests IgE-producing memory B cells. Subsequent allergen exposure can elicit an allergic response.

sepsis   Bacterial infection of the bloodstream. This is a very serious and frequently fatal condition.

septic shock   Systemic shock reaction that can follow infection of the bloodstream with endotoxin-producing Gram-negative bacteria. It is caused by the systemic release of TNF-α and other cytokines. Also called endotoxic shock.

sequence motif   A pattern of nucleotides or amino acids shared by different genes or proteins that often have related functions.

serine protease inhibitor (serpin)   Class of proteins that inhibit various proteases, originally referring to those specific to serine proteases.

seroconversion   The phase of an infection when antibodies against the infecting agent are first detectable in the blood.

serotypes   Name given to a strain of bacteria, or other pathogen, that can be distinguished from other strains of the same species by specific antibodies.

serum sickness   A usually self-limiting immunological hypersensitivity reaction originally seen in response to the therapeutic injection of large amounts of foreign serum (now most usually evoked by the injection of drugs such as penicillin). It is caused by the formation of immune complexes of the antigen and the antibodies formed against it, which become deposited in the tissues, especially the kidneys.

severe combined immunodeficiency (SCID)   Type of immune deficiency (due to various causes) in which both B-cell (antibody) and T-cell responses are lacking; it is fatal if not treated.

severe congenital neutropenia (SCN)   An inherited condition in which the neutrophil count is persistently extremely low. This is in contrast to cyclic neutropenia, in which neutrophil numbers fluctuate from near normal to very low or absent, with an approximate cycle time of 21 days.

SH2 (Src homology 2) domain   *See* **Src-family protein tyrosine kinases**.

shear-resistant rolling   The capacity of neutrophils to maintain attached to the vascular endothelium under high rates of flow—or shear—enabled by specialized plasma membrane extensions called slings.

shingles   Disease caused when herpes zoster virus (the virus that causes chickenpox) is reactivated later in life in a person who has had chickenpox.

SHIP (SH2-containing inositol phosphatase)   An SH2-containing inositol phosphatase that removes the phosphate from PIP_3 to produce PIP_2.

shock   The potentially fatal circulatory collapse caused by the systemic actions of cytokines such as TNF-α.

SHP (SH2-containing phosphatase)   An SH2-containing protein phosphatase.

signal joint   The noncoding joint formed in DNA by the recombination of RSSs during V(D)J recombination. Cf. **coding joint**.

signal peptide   The short N-terminal peptide sequence responsible for directing newly synthesized proteins into the secretory pathway.

signal transducers and activators of transcription (STATs)   *See* **Janus kinase (JAK) family**.

signaling scaffold   A configuration of proteins and modifications, such as phosphorylation or ubiquitination, that facilitates signaling by binding various enzymes and their substrates.

single-chain antibody   Referring to the heavy-chain-only IgGs produced by camelids or shark species that lack the light chain present in conventional antibodies.

single-nucleotide polymorphisms (SNPs)   Positions in the genome that differ by a single base between individuals.

single-positive thymocytes   A mature T cell that expresses either the CD4 or the CD8 co-receptor, but not both.

single-stranded RNA (ssRNA)   Normally confined to the nucleus and cytoplasm, this normal molecular form serves as a ligand for TLR-7, TLR-8, and TLR-9 when it is present in endosomes, as during parts of a viral life cycle.

sipuleucel-T (Provenge)   Cell-based immunotherapy used to treat prostate cancer that combined prostatic acid phosphatase as a tumor rejection antigen presented by dendritic cells derived from a patient's monocytes.

sirolimus   *See* **rapamycin**.

Sjögren's syndrome   An autoimmune disease in which exocrine glands, particularly the lacrimal glands of the eyes and salivary glands of the mouth, are damaged by the immune system. This results in dry eyes and mouth.

Skint-1   A transmembrane immunoglobulin superfamily member expressed by thymic stromal cells and keratinocytes that is required for the development of dendritic epidermal T cells, which are a type of γ:δ T cell.

SLAM (signaling lymphocyte activation molecule)   A family of related cell-surface receptors that mediate adhesion between lymphocytes, that includes SLAM, 2B4, CD84, Ly106, Ly9, and CRACC.

slings   *See* **shear-resistant rolling**.

SLP-65   A scaffold protein in B cells that recruits proteins involved in the intracellular signaling pathway from the antigen receptor. Also called BLNK.

SLP-76   A scaffold protein involved in the antigen-receptor signaling pathway in lymphocytes.

small G proteins   Single-subunit G proteins, such as Ras, that act as intracellular signaling molecules downstream of many transmembrane signaling events. Also called small GTPases.

small pre-B cell   Stage in B-cell development immediately after the large pre-B cell in which cell proliferation ceases and light-chain gene rearrangement commences.

somatic diversification theories   Gerenal hypotheses proposing that the immunoglobulin repertoire was formed from a small number of V genes that diversified in somatic cells. Cf. **germline theory**.

somatic DNA recombination   DNA recombination that takes place in somatic cells (to distinguish it from the recombination that takes place during meiosis and gamete formation).

somatic gene therapy   The introduction of functional genes into somatic cells to treat disease.

somatic hypermutation   Mutations in V-region DNA of rearranged immunoglobulin genes that produce variant immunoglobulins, some of which bind antigen with a higher affinity. These mutations affect only somatic cells and are not inherited through germline transmission.

spacer   *See* **12/23 rule**.

sphingolipids   A class of membrane lipid containing sphingosine (2-amino-4-octadecene-1,3-diole), an amino alcohol with unsaturated 18-hydrocarbon chain.

sphingosine 1-phosphate (S1P)   A phospholipid with chemotactic activity that controls the egress of T cells from lymph nodes.

sphingosine 1-phosphate receptor (S1P1)   A G-protein-coupled receptor activated by sphingosine 1-phosphate, a lipid mediator in the blood that regulates several physiologic processes, including the trafficking of naive lymphocytes from tissues into the blood.

spleen   An organ in the upper left side of the peritoneal cavity containing a red pulp, involved in removing senescent blood cells, and a white pulp of lymphoid cells that respond to antigens delivered to the spleen by the blood.

S-protein (vitronectin)   Plasma protein that binds incompletely formed MAC complexes, such as C5b67, preventing bystander complement damage to host membranes.

Spt5   A transcription elongation factor required for isotype switching in B cells that functions in associatioin with RNA polymerase to enable recruitment of AID to its targets in the genome.

SR-A I, SR-A II   *See* **scavenger receptors**.

Src-family protein tyrosine kinases   Receptor-associated protein tyrosine kinases characterized by Src-homology protein domains (SH1, SH2, and SH3). The SH1 domain contains the kinase, the SH2 domain can bind phosphotyrosine residues, and the SH3 domain can interact with proline-rich regions in other proteins. In T cells and B cells they are involved in relaying signals from the antigen receptor.

staphylococcal complement inhibitor (SCIN)   Staphylococcal protein that inhibits the activity of the classical and alternative C3 convertases, promoting the evasion of destruction by complement.

staphylococcal enterotoxins (SEs)   Secreted toxins produced by some staphylococci, which cause food poisoning and also stimulate many T cells by binding to MHC class II molecules and the V_β domain of certain T-cell receptors, acting as superantigens.

staphylococcal protein A (Spa)   Staphylococcal protein that blocks the binding of the antibody Fc region with C1, thereby preventing complement activation.

staphylokinase   Staphylococcal protease that cleaves immunoglobulins bound to its surface, thereby preventing complement activation.

STAT (signal transducers and activators of transcription)   A family of seven transcription factors activated by many cytokine and growth factor receptors.

STAT3   *See* **STAT**.

STAT6   *See* **STAT**.

statins   Drug inhibitors of HMG-CoA reductase used to lower cholesterol.

sterile injury   Damage to tissues due to trauma, ischemia, metabolic stress, or autoimmunity, bearing many immune features similar to infection.

sterilizing immunity   An immune response that completely eliminates a pathogen.

STIM1   A transmembrane protein that acts as a Ca^{2+} sensor in the endoplasmic reticulum. When Ca^{2+} is depleted from the endoplasmic reticulum, STIM1 is activated and induces opening of plasma membrane CRAC channels.

STING (stimulator of interferon genes)   A dimeric protein complex in the cytoplasm anchored to the ER membrane that functions in intracellular sensing for infection. It is activated by specific cyclic di-nucleotides to activate TBK1, which phosphorylates IRF3 to induce transcription of type I interferon genes.

stress-induced self   *See* **dysregulated self**.

stromal cells   The nonlymphoid cells in central and peripheral lymphoid organs that provide soluble and cell-bound signals required for lymphocyte development, survival, and migration.

subcapsular sinus (SCS)   The site of lymphatic entry in lymph nodes lined by phagocytes, including subcapsular macrophages which capture particulate and opsonized antigens draining in from tissues.

sulfated sialyl-Lewis^X   A sulfated tetrasaccharide carbohydrate structure attached to many cell surface proteins, it binds the P-selectin and E-selectin molecules on the surface of cells, such as neutrophils, that mediate interactions with the endothelium.

superoxide dismutase (SOD)   An enzyme that converts the superoxide ion produced in the phagolysosome into hydrogen peroxide, a substrate for further reactive antimicrobial metabolites.

suppressor of cytokine signaling (SOCS)   Regulatory protein that interacts with JAK kinases to inhibit signaling by activated receptors.

supramolecular activation complex (SMAC)   Organized structure that forms at the point of contact between a T cell and its target cell, in which the ligand-bound antigen receptors are co-localized with other cell-surface signaling and adhesion molecules. Also known as supramolecular adhesion complex.

surface immunoglobulin (sIg)   The membrane-bound immunoglobulin that acts as the antigen receptor on B cells.

surfactant proteins A and D (SP-A and SP-D)   Acute-phase proteins that help protect the epithelial surfaces of the lung against infection.

surrogate light chain   A protein in pre-B cells, made up of two subunits, VpreB and λ5, that can pair with a full-length immunoglobulin heavy chain and the Igα and Igβ signaling subunits and signals for pre-B-cell differentiation.

switch regions Genomic regions, several kilobases in length each, located between the JH region and the heavy-chain Cµ genes, or in equivalent positions upstream of other C-region genes (except Cδ), containing hundreds of G-rich repeated sequences that function in class switch recombination.

Syk A cytoplasmic tyrosine kinase found in B cells that acts in the signaling pathway from the B-cell antigen receptor.

symbiotic Relationship between two agents, typically diverse species, that confers benefits to both.

sympathetic ophthalmia Autoimmune response that occurs in the other eye after one eye is damaged.

syngeneic graft A graft between two genetically identical individuals. It is accepted as self.

systemic immune system Name sometimes given to the lymph nodes and spleen to distinguish them from the mucosal immune system.

systemic lupus erythematosus (SLE) An autoimmune disease in which autoantibodies against DNA, RNA, and proteins associated with nucleic acids form immune complexes that damage small blood vessels, especially in the kidney.

T10, T22 Murine MHC class Ib genes expressed by activated lymphocytes and recognized by a subset of γ:δ T cells.

T lymphocytes (T cells) *See* **T cell**.

TAB1, TAB2 An adaptor complex that binds K63 linked polyubiquitin chains. TAB1/2 complex with TAK1, targeting TAK1 to signaling scaffolds where it phosphorylates substrates such as IKKα.

TACE (TNF-α-converting enzyme) A protease responsible for cleavage of the membrane-associated form of TNF-α, allowing cytokine release into its soluble form that can enter the systemic circulation.

TACI A receptor for BAFF expressed on B cells that activates the canonical NFκB pathway.

tacrolimus An immunosuppressant polypeptide drug that binds FKBPs and inactivates T cells by inhibiting calcineurin, thus blocking activation of the transcription factor NFAT. Also called FK506.

TAK1 A serine-threonine kinase that is activated by phosphorylation by the IRAK complex, and that activates downstream targets such as IKKβ and MAPKs.

talin An intracellular protein involved in the linkage of activated integrins, such as LFA-1, to the cytoskeleton to allow changes in cellular motility and migration, such as in the diapedesis of neutrophils across the vascular endothelium.

TAP1, TAP2 Transporters associated with antigen processing. ATP-binding cassette proteins that form a heterodimeric TAP-1:TAP-2 complex in the endoplasmic reticulum membrane, through which short peptides are transported from the cytosol into the lumen of the endoplasmic reticulum, where they associate with MHC class I molecules.

tapasin TAP-associated protein. A key molecule in the assembly of MHC class I molecules; a cell deficient in this protein has only unstable MHC class I molecules on the cell surface.

Tbet A transcription factor active in many immune cell types but most typically associated with ILC1 and T_H1 function.

TBK1 (TANK-binding kinase) A serine-threonine kinase activated during signaling by TLR-3 and MAVS and serving to phosphorylate and activate IRF3 for induction of type I interferon gene expression.

T cell, T lymphocyte One of the two types of antigen-specific lymphocytes responsible for adaptive immune responses, the other being the B cells.

T cells are responsible for the cell-mediated adaptive immune reactions. They originate in the bone marrow but undergo most of their development in the thymus. The highly variable antigen receptor on T cells is called the T-cell receptor and recognizes a complex of peptide antigen bound to MHC molecules on cell surfaces. There are two main lineages of T cells: those carrying α:β receptors and those carrying γ:δ receptors. Effector T cells perform a variety of functions in an immune response, acting always by interacting with another cell in an antigen-specific manner. Some T cells activate macrophages, some help B cells produce antibody, and some kill cells infected with viruses and other intracellular pathogens.

T-cell antigen receptor *See* **T-cell receptor**.

T-cell areas Regions of peripheral lymphoid organs that are enriched in naive T cells and are distinct from the follicles. They are the sites at which adaptive immune responses are initiated.

T-cell plasticity Flexibility in the developmental programming of CD4 T cells such that effector T-cell subsets are not irreversibly fixed in their function or the transcriptional networks that underpin those functions.

T-cell receptor (TCR) The cell-surface receptor for antigen on T lymphocytes. It consists of a disulfide-linked heterodimer of the highly variable α and β chains in a complex with the invariant CD3 and ζ proteins, which have a signaling function. T cells carrying this type of receptor are often called α:β T cells. An alternative receptor made up of variable γ and δ chains is expressed with CD3 and ζ on a subset of T cells.

T-cell receptor α (TCRα) and β (TCRβ) The two chains of the α:β T-cell receptor.

T-cell receptor excision circles (TRECs) Circular DNA fragments excised from the chromosome during V(D)J recombination in developing thymoctytes that are transiently retained in T cells that have recently left the thymus.

T-cell zones *See* **T-cell areas**.

T-DM1 An antibody-drug conjugate combining trastuzumab (Herceptin) with mertansine used to treat recurrent metastatic breast cancer previously treated with a different trastuzumab drug conjugate.

TdT *See* **terminal deoxynucleotidyl transferase**.

TEPs *See* **thioester-containing proteins**.

terminal deoxynucleotidyl transferase (TdT) Enzyme that inserts nontemplated N-nucleotides into the junctions between gene segments in T-cell receptor and immunoglobulin V-region genes during their assembly.

tertiary immune responses Adaptive immune response provoked by a third injection of the same antigen. It is more rapid in onset and stronger than the primary response.

T follicular helper (T_{FH}) cell An effector T cell found in lymphoid follicles that provides help to B cells for antibody production and class switching.

T_H1 A subset of effector CD4 T cells characterized by the cytokines they produce. They are mainly involved in activating macrophages but can also help stimulate B cells to produce antibody.

T_H2 A subset of effector CD4 T cells that are characterized by the cytokines they produce. They are involved in stimulating B cells to produce antibody, and are often called helper CD4 T cells.

T_H17 A subset of CD4 T cells that are characterized by production of the cytokine IL-17. They help recruit neutrophils to sites of infection.

thioester-containing proteins Homologs of complement component C3 that are found in insects and are thought to have some function in insect innate immunity.

thioredoxin (TRX) A set of sensor proteins normally bound to thioredoxin-interacting protein (TXNIP). Oxidative stress causes thioredoxin to release TXNIP, which can mediate downstream actions.

thioredoxin-interacting protein (TXNIP) *See* **thioredoxin**.

thymectomy Surgical removal of the thymus.

thymic anlage The tissue from which the thymic stroma develops during embryogenesis.

thymic cortex The outer region of each thymic lobule in which thymic progenitor cells (thymocytes) proliferate, rearrange their T-cell receptor genes, and undergo thymic selection, especially positive selection on thymic cortical epithelial cells.

thymic stroma The epithelial cells and connective tissue of the thymus that form the essential microenvironment for T-cell development.

thymic stromal lymphopoietin (TSLP) Thymic stroma-derived lymphopoietin. A cytokine thought to be involved in promoting B-cell development in the embryonic liver. It is also produced by mucosal epithelial cells in response to helminthic infections, and promotes type 2 immune responses through its actions on macrophages, ILC2s, and T$_H$2 cells.

thymocytes Developing T cells when they are in the thymus. The majority are not functionally mature and are unable to mount protective T-cell responses.

thymoproteasome Specialized form of the proteasome composed of a unique subunit, β 5t, that replaces β 5i (LMP7) and associates with β 1i and β 2i in the catalytic chamber.

thymus A central lymphoid organ, in which T cells develop, situated in the upper part of the middle of the chest, just behind the breastbone.

thymus-dependent antigens (TD) Antigens that elicit responses only in individuals that have T cells.

thymus-independent antigens (TI) Antigens that can elicit antibody production without the involvement of T cells. There are two types of TI antigens: the TI-1 antigens, which have intrinsic B-cell activating activity, and the TI-2 antigens, which activate B cells by having multiple identical epitopes that cross-link the B-cell receptor.

thymus leukemia antigen (TL) Nonclassical MHC class Ib molecule expressed by intestinal epithelial cells and a ligand for CD8α:α.

TI-1 antigens *See* **thymus-independent antigens**.

TI-2 antigens *See* **thymus-independent antigens**.

tickover The low-level generation of C3b continually occurring in the blood in the absence of infection.

tingible body macrophages Phagocytic cells engulfing apoptotic B cells, which are produced in large numbers in germinal centers at the height of an adaptive immune response.

TIR (for Toll–IL-1 receptor) domain Domain in the cytoplasmic tails of the TLRs and the IL-1 receptor, which interacts with similar domains in intracellular signaling proteins.

tissue-resident memory T cells (TRM) Memory lymphocytes that do not migrate after taking up residence in barrier tissues, where they are retained long term. They appear to be specialized for rapid effector function after restimulation with antigen or cytokines at sites of pathogen entry.

TLR-1 Cell-surface Toll-like receptor that acts in a heterodimer with TLR-2 to recognize lipoteichoic acid and bacterial lipoproteins.

TLR-2 Cell-surface Toll-like receptor that acts in a heterodimer with either TLR-1 or TLR-6 to recognize lipoteichoic acid and bacterial lipoproteins.

TLR-3 Endosomal Toll-like receptor that recognizes double-stranded viral RNA.

TLR-4 Cell-surface Toll-like receptor that, in conjunction with the accessory proteins MD-2 and CD14, recognizes bacterial lipopolysaccharide and lipoteichoic acid.

TLR-5 Cell-surface Toll-like receptor that recognizes the flagellin protein of bacterial flagella.

TLR-6 Cell-surface Toll-like receptor that acts in a heterodimer with TLR-2 to recognize lipoteichoic acid and bacterial lipoproteins.

TLR-7 Endosomal Toll-like receptor that recognizes single-stranded viral RNA.

TLR-8 Endosomal Toll-like receptor that recognizes single-stranded viral RNA.

TLR-9 Endosomal Toll-like receptor that recognizes DNA containing unmethylated CpG.

TLR-11, TLR-12 Mouse Toll-like receptor that recognizes profilin and profilin-like proteins.

TNF family Cytokine family, the prototype of which is tumor necrosis factor-α (TNF or TNF-α). It contains both secreted (for example TNF-α and lymphotoxin) and membrane-bound (for example CD40 ligand) members.

TNF-receptor associated periodic syndrome (TRAPS)
An autoinflammatory disease characterized by recurrent, periodic episodes of inflammation and fever caused by mutations in gene that encodes TNF receptor I. The defective TNFR-I proteins fold abnormally and accumulate in cells in such a way that they spontaneously activate production of TNF-α. *See also* **familial Mediterranean fever**.

TNF receptors Family of cytokine receptors which includes some that lead to apoptosis of the cell on which they are expressed (for example Fas and TNFR-I), whereas others lead to activation.

tocilizumab Humanized anti-IL-6 receptor antibody used in treating rheumatoid arthritis.

tofacitinib An inhibitor of JAK3 and JAK1 used to treat rheumatoid arthritis and under investigation in other inflammatory disorders.

tolerance The failure to respond to an antigen. Tolerance to self antigens is an essential feature of the immune system; when tolerance is lost, the immune system can destroy self tissues, as happens in autoimmune disease.

tolerant Describes the state of immunological tolerance, in which the individual does not respond to a particular antigen.

tolerogenic Describes an antigen or type of antigen exposure that induces tolerance.

Toll Receptor protein in *Drosophila* that activates the transcription factor NFκB, leading to the production of antimicrobial peptides.

Toll-like receptors (TLRs) Innate receptors on macrophages, dendritic cells, and some other cells, that recognize pathogens and their products, such as bacterial lipopolysaccharide. Recognition stimulates the receptor-bearing cells to produce cytokines that help initiate immune responses.

tonsils *See* **lingual tonsils, palatine tonsils**.

toxic shock syndrome A systemic toxic reaction caused by the massive production of cytokines by CD4 T cells activated by the bacterial superantigen toxic shock syndrome toxin-1 (TSST-1), which is secreted by *Staphylococcus aureus*.

toxic shock syndrome toxin-1 (TSST-1) *See* **toxic shock syndrome**.

toxoids Inactivated toxins that are no longer toxic but retain their immunogenicity so that they can be used for immunization.

TRAF3 An E3 ligase that produces a K63 polyubiquitin signaling scaffold in TLR-3 signaling to induce type I interferon gene expression.

TRAF6 (tumor necrosis factor receptor-associated factor 6) An E3 ligase that produces a K63 polyubiquitin signaling scaffold in TLR-4 signaling to activate the NFκB pathway.

TRAIL (tumor necrosis factor-related apoptosis-inducing ligand) A member of the TNF cytokine family expressed on the cell surface of some cells, such as NK cells, that induces cell death in target cells by ligation of the 'death' receptors DR4 and DR5.

TRAM An adaptor protein that pairs with TRIF in signaling by TLR-4.

transcytosis The active transport of molecules, such as secreted IgA, through epithelial cells from one face to the other.

Transib A superfamily of transposable elements identified computationally and proposed to date back more than 500 million years and to have given rise to transposons in diverse species.

transitional immunity Referring to the recognition by some adaptive immune system (e.g., MAIT, γ:δ T cells) of non-peptide ligands expressed as a consequence of infection, such as various MHC class Ib molecules.

transitional stages Defined stages in the development of immature B cells into mature B cells in the spleen, after which the B cell expresses B-cell co-receptor component CD21.

transporters associated with antigen processing-1 and -2 (TAP1 and TAP2) *See* TAP1, TAP2.

transposase An enzyme capable of cutting DNA and allowing integration and excision of transposable genetic elements into or from the genome of a host.

trastuzumab Humanized antibody to HER-2/neu used in treatment of breast cancer.

TRECs *See* T-cell receptor excision circles.

TRIF An adaptor protein that alone is involved in signaling by TLR-3, and that when paired with TRAM, functions in signaling by TLR-4.

TRIKA1 A complex of the E2 ubiquitin ligase UBC13 and cofactor Uve1A, that interacts with TRAF6 in forming the K63 polyubiquitin signaling scaffold in TLR signaling downstream of MyD88.

TRIM21 (tripartite motif-containing 21) A cytosolic Fc receptor and E3 ligase that is activated by IgG and can ubiquitinate viral proteins after an antibody-coated virus enters the cytoplasm.

TRIM25 An E3 ubiquitin ligase involved in signaling by RIG-I and MDA-5 for the activation of MAVS.

tropism The characteristic of a pathogen that describes the cell types it will infect.

TSC Protein complex that acts as a GTPase-activating protein (GAP) for Rheb in its non-phosphorylated state. TSC is inactivated when phosphorylated by Akt.

TSLP Thymic stroma-derived lymphopoietin. A cytokine thought to be involved in promoting B-cell development in the embryonic liver.

tumor necrosis factor-α *See* TNF family.

tumor rejection antigens Antigens on the surface of tumor cells that can be recognized by T cells, leading to attack on the tumor cells. TRAs are peptides of mutant or overexpressed cellular proteins bound to MHC class I molecules on the tumor-cell surface.

type 1 diabetes mellitus Disease in which the β cells of the pancreatic islets of Langerhans are destroyed so that no insulin is produced. The disease is believed to result from an autoimmune attack on the β cells. It is also known as insulin-dependent diabetes mellitus (IDDM), because the symptoms can be ameliorated by injections of insulin.

type 1 immunity Class of effector activities aimed at elimination of intracellular pathogens.

type 2 immunity Class of effector activities aimed at elimination of parasites and promoting barrier and mucosal immunity.

type 3 immunity Class of effector activities aimed at elimination of extracellular pathogens such as bacteria and fungi.

type I interferons The antiviral interferons IFN-α and IFN-β.

type II interferon The antiviral interferon IFN-γ.

type III secretion system (T3SS) Specialized appendage of Gram-negative bacteria used to aid infection of eukaryotic cells by direct secretion of effector proteins into their cytoplasm.

tyrosinase Enzyme in melanin synthesis pathway and frequently a tumor rejection antigen in melanoma.

tyrosine phosphatases Enzymes that remove phosphate groups from phosphorylated tyrosine residues on proteins. *See also* CD45.

tyrosine protein kinases Enzymes that specifically phosphorylate tyrosine residues in proteins. They are critical in the signaling pathways that lead to T- and B-cell activation.

UBC13 *See* TRIKA1.

ubiquitin A small protein that can be attached to other proteins and functions as a protein interaction module or to target them for degradation by the proteasome.

ubiquitin ligase Enzyme that attaches ubiquitin covalently to exposed lysine residues on the surfaces of other proteins.

ubiquitin–proteasome system (UPS) A quality control system in the cell that involves K48-linked ubiquitination of target proteins that are then recognized by the proteasome for degradation.

ubiquitination The process of attachment of one or many subunits of ubiquitin to a target protein, which can mediate either degradation by the proteasome, or formation of scaffolds used for signaling, depending on the nature of the linkages.

UL16-binding proteins, or ULBPs *See* RAET1.

ULBP4 *See* RAET1.

ulcerative colitis One of the two major types of inflammatory bowel disease thought to result from an abnormal overresponsiveness to the commensal gut microbiota. *See also* Crohn's disease.

UNC93B1 A mutlipass transmembrane protein that is necessary for the normal transport of TLR-3, TLR-7, and TLR-9 from the ER, where they are assembled, to the endosome, where they function.

unmethylated CpG dinucleotides While mammalian genomes have heavily methylated the cytosine within CpG sequences, unmethylated CpG is more typically characteristic of bacterial genomes, and is recognized by TLR-9 when encountered in the endosomal compartment.

uracil-DNA glycosylase (UNG) Enzyme that removes uracil bases from DNA in a DNA repair pathway that can lead to somatic hypermutation, class switch recombination or gene conversion.

urticaria The technical term for hives, which are red, itchy skin wheals usually brought on by an allergic reaction.

Uve1A See TRIKA1.

V$_\alpha$ Variable region from the TCRα chain.

V$_\beta$ Variable region from the TCRβ chain.

vaccination The deliberate induction of adaptive immunity to a pathogen by injecting a dead or attenuated (nonpathogenic) live form of the pathogen or its antigens (a vaccine).

variability plot A measure of the difference between the amino acid sequences of different variants of a given protein. The most variable proteins known are antibodies and T-cell receptors.

variable Ig domains (V domains) The amino-terminal protein domain of the polypeptide chains of immunoglobulins and T-cell receptor, which is the most variable part of the chain.

variable lymphocyte receptors (VLRs) Nonimmunoglobulin LRR-containing variable receptors and secreted proteins expressed by the lymphocyte-like cells of the lamprey. They are generated by a process of somatic gene rearrangement.

variable region The region of an immunoglobulin or T-cell receptor that is formed of the amino-terminal domains of its component polypeptide chains. These are the most variable parts of the molecule and contain the antigen-binding sites.

variolation The intentional inhalation of or skin infection with material taken from smallpox pustules of an infected person for the purpose of deriving protective immunity.

VCAM-1 An adhesion molecule expressed by vascular endothelium at sites of inflammation; it binds the integrin VLA-4, which allows effector T cells to enter sites of infection.

V(D)J recombinase A multiprotein complex containing RAG-1 and RAG-2, as well as other proteins involved in cellular DNA repair.

V(D)J recombination The process exclusive to developing lymphocytes in vertebrates, that recombines different gene segments into sequences encoding complete protein chains of immunoglobulins and T-cell receptors.

vesicular compartments One of several major compartments within cells, composed of the endoplasmic reticulum, Golgi, endosomes, and lysosomes.

V gene segments Gene segments in immunoglobulin and T-cell receptor loci that encode the first 95 amino acids or so of the protein chain. There are multiple different V gene segments in the germline genome. To produce a complete exon encoding a V domain, one V gene segment must be rearranged to join up with a J or a rearranged DJ gene segment.

viral entry inhibitors Drugs that inhibit the entry of HIV into its host cells.

viral integrase inhibitors Drugs that inhibit the action of the HIV integrase, so that the virus cannot integrate into the host-cell genome.

viral protease Enzyme encoded by the human immunodeficiency virus that cleaves the long polyprotein products of the viral genes into individual proteins.

viral set point In human immunodeficiency virus infection, the level of HIV virions persisting in the blood after the acute phase of infection has passed.

virus Pathogen composed of a nucleic acid genome enclosed in a protein coat. Viruses can replicate only in a living cell, because they do not possess the metabolic machinery for independent life.

virus-neutralizing antibodies Antibodies that block the ability of a virus to establish infection of cells.

vitronectin *See* S-protein.

VLRs *See* variable lymphocyte receptors.

VpreB *See* surrogate light chain.

WAS *See* Wiskott–Aldrich syndrome.

WASp The protein defective in patients with Wiskott–Aldrich syndrome. When activated, WASp promotes actin polymerization.

Weibel–Palade bodies Granules within endothelial cells that contain P-selectin.

wheal-and-flare reaction A skin reaction observed in an allergic individual when an allergen to which the individual has been sensitized is injected into the dermis. It consists of a raised area of skin containing edema fluid, and a spreading, red, itchy inflammatory reaction around it.

white pulp The discrete areas of lymphoid tissue in the spleen.

Wiskott–Aldrich syndrome (WAS) An immunodeficiency disease characterized by defects in the cytoskeleton of cells due to a mutation in the protein WASp, which is involved in interactions with the actin cytoskeleton. Patients with this disease are highly susceptible to infections with pyogenic bacteria due to defects in T-follicular helper cell interactions with B cells.

XBP1 (X-box binding protein 1) A transcription factor that induces genes required for optimal protein secretion by plasma cells, and is part of the unfolded protein response. XBP1 mRNA is spliced from an inactive to an active form by signals produced by ER stress.

XCR1 A chemokine receptor selectively expressed by a subset of dendritic cells that are specialized for cross-presentation whose development requires the transcription factor BATF3.

xenografts Grafted organs taken from a different species than the recipient.

xenoimmunity In the context of immune-mediated disease, refers to immunity directed against foreign antigens of non-human species, such as bacteria-derived antigens of the commensal microbiota that are targets in inflammatory bowel disease (IBD).

xeroderma pigmentosum Several autosomal recessive diseases caused by defects in repair of ultraviolet light-induced DNA damage. Defects in Polη cause type V xeroderma pigmentosum.

xid *See* X-linked immunodeficiency.

X-linked agammaglobulinemia (XLA) A genetic disorder in which B-cell development is arrested at the pre-B-cell stage and no mature B cells or antibodies are formed. The disease is due to a defect in the gene encoding the protein tyrosine kinase Btk, which is encoded on the X chromosome.

X-linked hyper IgM syndrome *See* CD40 ligand deficiency.

X-linked hypohidrotic ectodermal dysplasia and immunodeficiency A syndrome with some features resembling hyper IgM syndrome. It is caused by mutations in the protein NEMO, a component of the NFκB signaling pathway. Also called NEMO deficiency.

X-linked immunodeficiency An immunodeficiency disease in mice due to defects in the protein tyrosine kinase Btk. Shares the gene defect with X-linked agammaglobulinemia in humans, but leads to a milder B cell defect than seen in the human disease.

X-linked lymphoproliferative (XLP) syndrome Rare immunodeficiency diseases that result from mutations in the gene *SH2D1A* (XLP1) or *XIAP* (XLP2). Boys with this deficiency typically develop overwhelming Epstein–Barr virus infection during childhood, and sometimes lymphomas.

X-linked severe combined immunodeficiency (X-linked SCID) An immunodeficiency disease in which T-cell development fails at an early intrathymic stage and no production of mature T cells or T-cell dependent antibody occurs. It is due to a defect in a gene that encodes the γc chain shared by the receptors for several different cytokines.

XLP *See* X-linked lymphoproliferative syndrome.

XRCC4 A protein that functions in NHEJ DNA repair by interacting with DNA ligase IV and Ku70/80 at double-strand breaks.

ζ chain One of the signaling chains associated with the T-cell receptor that has three ITAM motifs in its cytoplasmic tail.

ZAP-70 (ζ-chain-associated protein) A cytoplasmic tyrosine kinase found in T cells that binds to the phosphorylated ζ chain of the T-cell receptor and is a key enzyme in signaling T-cell activation.

ZFP318 A spliceosome protein expressed in mature and activated B cells, but not immature B cells, that favors splicing from the rearranged VDJ exon of immunoglobulin heavy chain to the Cδ exon, thereby promoting expression of surface IgD.

zoonotic Describes a disease of animals that can be transmitted to humans.

zymogens An inactive form of an enzyme, usually a protease, that must be modified in some way, for example by selective cleavage of the protein chain, before it can become active.

Index